Herbert Bucksch

Wörterbuch für Bautechnik und Baumaschinen

Deutsch – Englisch
German – English

1

Dictionary of Civil Engineering and Construction Machinery and Equipment

Achte Auflage

Fourth reprint of the fourth edition

BAUVERLAG GMBH · WIESBADEN UND BERLIN

CIP- Kurztitelaufnahme der Deutschen Bibliothek

Bucksch, Herbert

Wörterbuch für Bautechnik und Baumaschinen = Dictionary of civil engineering and construction machinery and equipment. / Herbert Bucksch. — Wiesbaden; Berlin: Bauverlag.

[Deutsch-englische Ausgabe]

1. Deutsch-englisch. — 8. Aufl. — 1982.
    ISBN 3-7625-2032-1

1. Auflage 1955
2. Auflage 1958
3. Auflage 1965
4. Auflage 1968
5. Auflage 1973
6. Auflage 1976
7. Auflage 1978
8. Auflage 1982

©

Printed in Germany 1982 · Bauverlag GmbH, Wiesbaden

Druck: Hans Meister KG, Kassel

Einband: C. Fikentscher, Darmstadt

ISBN 3-7625-2032-1

VORWORT

Unsere heutige Zivilisation ist ohne den modernen Ingenieurbau nicht mehr denkbar. Durch ihn werden die Bauten zur Ausnützung der Wasserkräfte, wie Talsperren und Wasserkraftwerke, die Verkehrsverbindungen, wie Straßen, Brücken, Kanäle, Eisenbahnlinien und Flugplätze, die Bewässerungs- und Entwässerungsanlagen für die Gewinnung neuer Kulturflächen, die menschlichen Siedlungen in der Vielfalt ihrer Formen usw. geschaffen. Der Mensch verändert die Oberfläche der Erde.

Obwohl bereits im Altertum imposante Bauwerke geschaffen worden sind, wofür die Chinesische Mauer und die Sieben Weltwunder als die bekanntesten Beispiele dienen mögen, brach für den Ingenieurbau jedoch erst mit dem Zeitalter der Erfindungen eine neue Epoche an. Die Lokomotive erforderte Schienenstränge, das Auto verlangte nach Straßen, und für das Flugzeug mußten geeignete Flächen zum Starten und Landen geschaffen werden. Die stetige Verbesserung der Erfindungen bedingte eine Ausweitung der von diesen Erfindungen benötigten Anlagen, und der Ingenieurbau mußte sich zur Erfüllung der ihm in immer rascherem Tempo gestellten Aufgaben eine Reihe von Wissensgebieten erschließen. Systematische Forschung wurde unumgänglich, und bald stellte sich die Notwendigkeit einer Dokumentation der Ergebnisse heraus.

Die Interessen und Ziele der Bauingenieure aller Völker sind die gleichen; sie wollen mit ihren Bauwerken dem Wohl der Menschheit dienen. Daraus ergibt sich die Forderung zu internationaler Zusammenarbeit.

Die heute zur Vertiefung des Wissens auf den einzelnen Teilgebieten des Ingenieurbaues auf internationaler Ebene abgehaltenen Konferenzen sind der Grundstein für diese Zusammenarbeit. Die persönliche Fühlungnahme führt zur Beseitigung des Trennenden und des Mißtrauens. Zwischenstaatliche Bindungen werden jedoch auch auf andere Art und Weise geknüpft. So ist z. B. die „Deutsche Gesellschaft für Erd- und Grundbau" mit der diesbezüglichen Fachorganisation der USA, „ASCE Soil Mechanics and Foundations Division", übereingekommen, einen gegenseitigen Austausch des Wissens auf ihrem Fachgebiet durchzuführen. Als Korrespondenzsprachen wurden Englisch und Deutsch bestimmt.

Die internationale Zusammenarbeit auf dem Gebiete des Ingenieurbaues lenkt nun auch demzufolge das Augenmerk auf die Fachterminologie in bezug auf eine Gegenüberstellung der Begriffe in den verschiedenen Sprachen.

Die heutigen Sprachen erleben täglich die Geburt neuer Wörter. Der Ingenieurbau mit seinen Ausläufern in die verschiedenartigsten Wissensgebiete hat daran, wie überhaupt die Technik, erheblichen Anteil. Hier ergibt sich nun die Forderung nach technischen Wörterbüchern.

Die Zusammenstellung eines technischen Wörterbuches ist wegen der schon in der eigenen Sprache oft unklaren Begriffsbestimmung eines einzelnen technischen Wortes schwierig, und bei der Auswahl des meistens dem gleichen Gesetz verschiedener Definitionsmöglichkeiten unterliegenden fremdsprachlichen Wortes ist daher größte Vorsicht geboten. Selbst wenn technische Wörter frei von jeder Zweideutigkeit in bezug auf ihre Definition sind, ist bei einer Gegenüberstellung mit dem fremdsprachlichen Wort stets zu prüfen, ob sich die Definitionen in beiden Sprachen vom technischen Standpunkt aus decken.

Das vorliegende Wörterbuch wurde mit dem Ziel zusammengestellt, die internationale Zusammenarbeit auf dem Gebiete des Ingenieurbaues zu fördern. Bei der Bearbeitung wurden die neuesten Quellen erschöpft. Viele Begriffe sind hier zum ersten Male lexikographisch erfaßt.

Verlag und Verfasser hoffen mit diesem Wörterbuch einem Bedürfnis aller am Bauwesen interessierten Kreise entsprochen zu haben.

H. Bucksch

PREFACE

Our present civilization is hardly conceivable without the achievements of modern civil engineering. The "use and convenience" to man of all these activities can scarcely be denied. Civil engineers had certainly left their mark on the face of Nature, damming great lakes and constructing hydro-electric plants, highways, bridges, canals, railway lines, airports, towns, etc.

Although remarkable structures have been erected in ancient times, e. g. the Great Wall of China and the Seven Ancient Wonders of the World, it was not until the last century that civil engineering entered upon a new era. The steam locomotive required trackings, the automobile called for high-quality roads and the aircraft for suitable landing, taking off and servicing facilities. Steady improvements of these inventions brought about a further development of these facilities and civil engineering spread itself over a wide range of scientific fields in order to meet the ever increasing demands. Systematic research became imperative with detailed documentation of the results obtained.

As civil engineers of all countries work towards the same aim of "directing the great sources of power in Nature to the use and convenience of man," international co-operation becomes inevitable.

The foundations of this international co-operation are the many conferences convened by the individual civil engineering organizations in which the experience is pooled for the benefit of all concerned. Civil engineers of all nations thus get into touch with each other and this personal contact soon removes any reserve or sense of mistrust. International connections between civil engineers can be initiated in many ways. For instance, the "Deutsche Gesellschaft für Erd- und Grundbau" (German Society for Soil and Foundation Engineering) and the "ASCE Soil Mechanics and Foundations Division" exchange their knowledge in their relevant field by correspondence in English and German.

This international partnership naturally draws attention to the terminology involved and the selection of equivalent terms in the various languages.

To-day's languages frequently experience the introduction of new words and the field of civil engineering with its widespread ramifications into various other related fields contributes to a considerable extent to this phenomenon. Thus, technical dictionaries are required.

The compilation of a technical dictionary is already difficult because the scope of meaning of a single word in one language is not always clearly defined, and when selecting the appropriate term in the foreign language, which might also not be subject to precise definition, great care must be exercised. Even if technical terms are free from any ambiguity as to their meaning, it must always be borne in mind, that the selected relative foreign words must cover the same range of technical definition.

This dictionary has been compiled in an attempt to foster the harmony of international partnership in Civil Engineering, and the latest relevant literature has been searched for its compilation. It contains a multitude of definitions not yet incorporated into any other dictionary available to-day.

It is hoped by the publisher as well as the author, that this dictionary will fill the gap felt in the past.

H. Bucksch

EINFÜHRUNG

Die Wörter sind alphabetisch geordnet. Zum Zwecke einer Darstellung nach technischen Gesichtspunkten sind bei manchen Begriffen weitere außerhalb der alphabetischen Reihenfolge aufgeführt. Beispiel:

Angebot n
Abgabetermin m, Submissionstermin
usw.

bid; proposal (US); tender, offer (Brit.) closing date (for receipt of tenders), deadline for submission of bids, tendering date

Anwendung der Klammern:
a) Eckige Klammern:
  1. *Definition eines Wortes, z. B.*

Aufstieg m, Fisch~ *[stromaufwärts gerichtete Fischwanderung]*

  2. *Bestimmung eines Sachgebiets, z. B.*

(Ab)Dichtungsgewölbe n *[Talsperre]*

b) Runde Klammern:
  3. *Auslassungen, z. B.*

(Ab)Dichten = **Abdichten, Dichten**
oder
Abfluß(mengen)kurve = **Abflußkurve**

  4. *Abkürzungen, z. B*

(Geol.) = Geologie
(Min.) = Mineralogie
(Brit.) = Großbritannien
(US) = USA

Die Tilde ~ *tritt an die Stelle des zu wiederholenden Wortes, z. B.* **Ablaß,** ~**vorrichtung** = **Ablaßvorrichtung**

Die Tilde im Fettdruck ~ *tritt an die Stelle des darüber befindlichen Wortes oder des zu wiederholenden Wortteiles, welcher vor dem Strich steht, z. B.*

  Ablagerung f
  ~ = Ablagerung
  Ablagerungs|boden m
  ~gestein n = Ablagerungsgestein

Das Komma teilt Wörter gleicher Definition.

Das Semikolon teilt Wörter gleicher Definition, die aber einerseits in den USA und andererseits in Großbritannien gebräuchlich sind.

Das Geschlecht der deutschen Wörter ist wie folgt angegeben:

  m = *männlich*
  f = *weiblich*
  n = *sächlich*
  mpl = *das betreffende Wort ist in der Mehrzahl angegeben, wobei die Einzahl männlich ist. Dementsprechend sind fpl und npl zu verstehen.*

INTRODUCTION

The terms are compiled in alphabetical order. Certain terms, however, forming subject titles from a technical point of view are accompanied by a sub-section giving other words relevant to the subject. Example:

Angebot n
   Abgabetermin m, Submissionstermin
       etc.

bid; proposal (US); **tender, offer** (Brit.) closing date (for receipt of tenders), deadline for submission of bids, tendering date

Brackets are used:
a) Square brackets:
  (1) to define a term, e. g.

Aufstieg m, Fisch~ [stromaufwärts gerichtete Fischwanderung]

  (2) to denote the particular field, e. g.

(Ab)Dichtungsgewölbe n [Talsperre]

b) Round brackets:
  (3) for terms or part of terms which can be omitted, e. g.

(Ab)Dichten = **Abdichten, Dichten**
or
Abfluß(mengen)kurve = **Abflußkurve**

  (4) for abbreviations, e. g.

(Geol.) = Geology
(Min.) = Mineralogy
(Brit.) = Great Britain
(US) = USA

The Tilde ~ substitutes the word to be repeated, e. g. **Ablaß,** ~vorrichtung = **Ablaßvorrichtung**

The Tilde ~ substitutes the preceding word or the part of the preceding word printed before the sign | and to be repeated in the following word, e. g.

    Ablagerung f
    ~ = Ablagerung
    Ablagerungs|boden m
    ~gestein n = Ablagerungsgestein

The Comma separates equivalent definitions.

The Semicolon separates equivalent definitions but alternatively used either in the USA or Great Britain.

The gender of the German words is indicated by:
    m = masculine
    f = feminine
    n = neuter
    mpl = the term is given in its plural form, with the singular being masculine; fpl and npl apply accordingly.

The German letter "ß" means "sz".

BAND 1

VOLUME 1

*

DEUTSCH - ENGLISCH

GERMAN - ENGLISH

A

Aal|abstieg *m* = downstream migration of eel, ~ eel migration
~aufstieg *m* = upstream migration of eel, ~ eel migration
~aufstiegrohr *n* [*im Pfeiler eines Wehres eingebaut und mit Faschinen ausgefüllt*] = tube for upstream eel migration
~treppe *f*, Aalleiter *f* [*Vorrichtung für den Aufstieg der Aale vom Unterwasser zum Oberwasser und für den Abstieg vom Oberwasser zum Unterwasser*] = eel ladder
~wanderung *f* = migration of eel(s), eel migration
AASHO-Versuch *m*, AASHO-Probe *f*, AASHO-Prüfung *f*, Proctor-Versuch, Proctor-Probe, Proctor-Prüfung = standard Proctor test, ~ AASHO ~
Abakus *m*, (Säulen)Deckplatte *f*, Kapitellplatte = abacus, raised table [*A slab atop the capital of a column*]
Aba-Lorenz-Bohrpfahl *m*, Aba-Lorenz-Ortpfahl [*Stahlbetonpfahl mit verlorenem Rohr*] = (cast-in-place reinforced-concrete) shell pile Aba-Lorenz, (~ ~) cased ~ ~
abänderungsfähig = capable of being modified

(ab)baggern, ausbaggern, trockenbaggern, ausheben = to dig, to excavate
Abbau *m* → Schlammfaulung *f*
~, Demontage *f*, Abbauen *n*, Zerlegen, Zerlegung *f* = dismantling
~ [*im Bergwerk der Ort, an dem abgebaut wird*] = face
~, Abbauen *n* [*planmäßige Inangriffnahme der Lagerstätte bei der Gewinnung*] = getting, working
~, Abbauen *n* [*Steinbruch*] = quarrying
~, ~ der Querschnittfläche [*Tunnel- und Stollenbau*] = excavation
~ **der Pfeiler,** Abpfeilern *n* (⚒) = robbing pillars
~ **(der Querschnittfläche)** [*Tunnel- und Stollenbau*] = excavation
~ **durch Lösen mit Hochdruckwasserstrahl** = hydraulic mining, hydraulicking
~ **Sprengen,** Abbauen *n* ~ ~ [*Steinbruch*] = quarrying by the use of explosives
~ **eines Lehrgerüstes,** Ausrüsten *n*, Abrüsten, Lehrgerüstabbau = dismantling of falsework, stripping ~ ~
~ **im steilen Flöz,** Abbauen *n* ~ ~ ~ = steep working

Abbau — Abbaumethode

~ mit Unterstützung des Hangenden, Abbauen n ~ ~ ~ ~ = supported stopes

~ unter Tage, Abbauen n ~ ~, untertägiger Abbau, untertägiges Abbauen [*Gewinnung von Bodenschätzen*] = underground getting, ~ working

~arbeit f (⚒) = getting work

abbaubar über Tage = workable by open-cast mining

~ **unter Tage** = workable by underground mining

Abbau|beleuchtung f (⚒) = face lighting

~**dynamik** f (⚒) = strata movement

abbaudynamische Messung f (⚒) = measurement of strata movement

abbauen → demontieren

~, ausbeuten [*Bodenschätze*] = to get, to work

~ [*Steinbruch*] = to strip, to quarry

~ [*Last; Kraft; Spannung*] = to absorb, to minimize

~ **eines Lehrgerüstes**, ausrüsten, abrüsten = to strip the falsework, ~ dismantle ~ ~

~ **im Tagebau**, ~ **über Tage** = to work by the open-mining method, to get ~ ~ ~ ~

~ **über Tage** → ~ **im Tagebau**

Abbauen n, Abbau m, Demontage f, Zerlegen, Zerlegung f = dismantling

~, Abbau m [*planmäßige Inangriffnahme der Lagerstätte bei der Gewinnung*] = working, getting

~, Abbau m [*Steinbruch*] = quarrying

~ **im steilen Flöz**, Abbau m ~ ~ ~ = steep working

~ **mit Unterstützung des Hangenden** Abbau m ~ ~ ~ ~ = supported stopes

~ **unter Tage** → Abbau m ~ ~

(ab)bau|fähig, (ab)bauwürdig, ausbeutbar [*eine Lagerstätte oder ein Lagerstättenteil sind (ab)bauwürdig, wenn sich der Abbau wirtschaftlich lohnt oder zur Bedarfsdeckung notwendig ist*] = paying, workable, profitable

~**fähige Vorräte** mpl, (ab)bauwürdige ~, ausbeutbare ~ = reserves

(Ab)Baufähigkeit f, (Ab)Bauwürdigkeit, Ausbeutbarkeit = profitability, minability

Abbau|förderband n (⚒) = face conveyor, ~ conveyer

~**fortschritt** m, Abbaugeschwindigkeit f [*Maß des täglichen, monatlichen oder schichtweisen Vorrückens des Abbaustoßes*] = rate of face advance

~**front** f (⚒) = (working) face

~**gebiet** n = working zone

Abbauhammer m, Pickhammer, Förderhammer [*ein Preßlufthammer mit Spitzmeißel zum Abbau von Mineralien (DIN 20374 bis 20376)*] = picker, pick hammer

~**gewinnung** f, Pickhammergewinnung, Förderhammergewinnung = picker work, pick hammer ~

~**-Prüfung** f, Abbauhammer-Probe f, Abbauhammer-Versuch m = picker test, pick hammer ~

Abbau|höhe f [*Bagger*] = cutting height

~**kammer** f (⚒) = chamber

~**kante** f (⚒) = rib, fast end [*The face of solid coal along the edge of a panel or of a road(way) driven in the solid. The miner working next to the rib must be an experienced man, as the coal is hard and sometimes a stable must be cut in*]

~**leistung** f **von Belebtschlamm** = sludge activity

~**lok(omotive)** f [*Grubenlok(omotive) die in Abbaustrecken verkehrt*] = underground loco(motive)

~**mächtigkeit** f = pay(ing) thickness, workable ~, profitable ~

(Ab)Baumethode f, (Ab)Bauverfahren n, (Ab)Bauweise f [*Art der (Ab)Bauplanung*] (⚒) = getting method, working ~

Abbaumethode f, Abbauverfahren n, Abbauweise f [*für den Abbau der Querschnittfläche im Tunnel- und Stollenbau*] = excavation method

Abbaureihenfolge *f*, Reihenfolge des Abbaues, (Ab)Baufolge (⚒) = sequence of getting
~ **von oben nach unten** (⚒) = sequence of getting from top to bottom
~ ~ **unten nach oben** (⚒) = sequence of getting from bottom to top
Abbau|richtung *f* [*Richtung des Abbaufortschrittes, entweder schwebend, fallend, streichend, schräg*] (⚒) = direction of working
~**riß** *m* (⚒) = break
~**riß** *m* [*im Hangenden*] = roof break
~**riß** *m* [*im Liegenden*] = floor break
~**schacht** *m* (⚒) = working shaft
~**schwerpunkt** *m* (⚒) = focal point of working
(Ab)Bausohle *f* (⚒) = working level
(Abbau)Stoß *m*, (Abbau)Wand *f* [*Großerdbau und Gewinnung von Bodenschätzen*] = (working) face
Abbau|strecke *f* (⚒) = development road
~**strecke** *f* im Strebbau (⚒) = gate road
~**streckenlok(omotive)** *f* = gathering loco(motive)
~**stufe** *f* [*Abbau der organischen Stoffe in lufthaltigem Wasser*] = decomposition stage
~**- und Aufreißhammer** *m*, Allzweck-Abbau- und Aufreißhammer, Aufreiß- und Abbauhammer = demolition and breaking hammer
~**- und Bohrhammer** *m*, Bohr- und Abbauhammer, Allzweck-Bohr- und Abbauhammer = demolition and drill hammer
~**verfahren** *n* → (Ab)Baumethode *f*
(Abbau)Vorfeld *n* (⚒) = zone in front of the face
Abbauvorgang *m*, Demontagevorgang = dismantling operation
(Abbau)Wand *f* → (Abbau)Stoß *m*
~, (Stein)Bruchwand, Abtragwand = (quarry) face
(Ab)Bauweise *f* → (Ab)Baumethode *f*
(ab)bau|würdig → (ab)baufähig

~**würdige Vorräte** *mpl*, (ab)baufähige ~, ausbeutbare ~ = reserves
(Ab)Bauwürdigkeit *f*, (Ab)Baufähigkeit, Ausbeutbarkeit = profitability
Abbauzone *f* (⚒) = working zone
(ab)beizen [*Anstrich von seinem Untergrund*] = to remove a paint or varnish film with paint remover [*the paint remover softens a paint or varnish film, so that it can be easily scraped or brushed off*]
(Ab)Beiz|fluid *n*, lösendes (Ab)Beizmittel *n*, neutrales (Ab)Beizmittel = solvent-based paint remover
~**mittel** *n*, Abbeizer *m*, Farbentferner, Farbenvertilger = paint remover
Abbiegefahrbahn *f* [*eine getrennt geführte Richtungsfahrbahn an höhengleichen Knotenpunkten, die ausschließlich für den Links- bzw. Rechtsabbiegeverkehr bestimmt ist*] = carriageway for turning traffic
Abbiegen *n* im Gegenverkehr = turning across oncoming traffic
Abbiegespur *f*, Abbiegestreifen *m* [*eine bei Knotenpunkten für den Links- bzw. Rechtsabbiegeverkehr bestimmte und entsprechend gekennzeichnete Fahrspur*] = lane for turning traffic
(Ab)Biegestelle *f* [*Bewehrungsstahl*] = bend(ing) point
Abbiege|streifen *m* → Abbiegespur *f*
~**verbot** *n* = turn-ban
~**verkehr** *m* = turning traffic
Abbiegung(en) *f (pl)* → Aufbiegung(en)
Abbinde|beginn *m*, Erstarrungsbeginn [*hydraulisches Bindemittel*] = initial set(ting)
~**beschleuniger** *m*, Abbindezeitbeschleuniger, Erstarrungs(zeit)beschleuniger, BE *m* [*für hydraulische Bindemittel*] = accelerating agent, ~ admix(ture), accelerator
~**beschleunigung** *f*, Erstarrungs(zeit)beschleunigung, Abbindezeitbeschleunigung [*hydraulisches Bindemittel*] = acceleration of set(ting)
~**ende** *n*, Erstarrungsende [*hydraulisches Bindemittel*] = end of set(ting)

~ende *n* [*Verschnittbitumen*] = end of curing

~festigkeit *f*, Erstarrungsfestigkeit = set(ting) strength, strength of set(ting)

~geschwindigkeit *f*, Erstarrungsgeschwindigkeit [*hydraulisches Bindemittel*] = rate of set(ting), set(ting) rate

~geschwindigkeit *f* [*Verschnittbitumen*] = rate of cure, curing rate

~kraft *f*, Erstarrungskraft [*hydraulisches Bindemittel*] = set(ting) power, power of set(ting)

~kurve *f*, Erstarrungskurve [*hydraulisches Bindemittel*] = curve of set(ting), set(ting) curve

~mechanismus *m*, Erstarrungsmechanismus [*hydraulisches Bindemittel*] = mechanism of set(ting), set(ting) mechanism

~mittel *n* = curing agent [*e. g. as part of an epoxy resin adhesive*]

abbinden, erstarren [*hydraulisches Bindemittel*] = to set

~ [*Verschnittbitumen*] = to cure

~ [*Leim*] = to cure, to set

~, zulegen [*Holzkonstruktion*] = to join, to trim

Abbinden *n*, Abbindung *f*, Erstarren, Erstarrung [*hydraulisches Bindemittel*] = set(ting)

~, Abbindung *f* [*Verschnittbitumen; Bitumenemulsion*] = curing

~, Abbund *m*, Abbindung *f* [*Zimmerei*] = joining, trimming

Abbinde|platz *m*, Abbundplatz = joining yard, trimming ~

~probe *f* → Abbindeprüfung *f*

~prüfgerät *n*, Erstarrungsprüfgerät [*für hydraulische Bindemittel*] = set(ting) tester

~prüfung *f*, Abbindeprobe *f*, Abbindeversuch *m*, Erstarrungsprüfung, Erstarrungsprobe, Erstarrungsversuch [*hydraulisches Bindemittel*] = set(ting) test

~reaktion *f*, Erstarrungsreaktion [*hydraulisches Bindemittel*] = reaction of set(ting), set(ting) reaction

abbinderegelnd, abbindesteuernd = set-controlling

Abbinderegler *m*, Abbindezeitregler, Erstarrungs(zeit)regler [*für hydraulische Bindemittel*] = set(ting)-time controlling agent, set-controlling admix(ture) [*is either a retarder or an accelerator*]

abbindesteuernd, abbinderegelnd = set-controlling

Abbinde|temperatur *f*, Erstarrungstemperatur [*hydraulisches Bindemittel*] = temperature of set(ting), set(ting) temperature

~verhalten *n*, Erstarrungsverhalten [*hydraulisches Bindemittel*] = behavio(u)r of set(ting), set(ting) behavio(u)r

~vermögen *n*, Erstarrungsvermögen [*hydraulisches Bindemittel*] = power of set(ting), set(ting) power

~vermögen *n* [*Verschnittbitumen*] = power of curing, curing power

~versuch *m* → Abbindeprüfung *f*

~verzögerer *m*, Erstarrungsverzögerer, VZ *m*, Abbindezeitverzögerer [*für hydraulische Bindemittel*] = retarding admix(ture), retarder (of set(ting)), cement retarder

~vorgang *m*, Erstarrungsvorgang, Abbindeprozeß, Erstarrungsprozeß, Abbindeverlauf *m*, Erstarrungsverlauf [*hydraulischse Bindemittel*] = process of set(ting), set(ting) process

~vorgang *m*, Abbindeprozeß *m*, Abbindeverlauf *m* [*Verschnittbitumen*] = process of curing, curing process

~wärme *f*, Erstarrungswärme [*hydraulisches Bindemittel*] = heat of set(ting), set(ting) heat

~wasser *n*, Erstarrungswasser [*hydraulisches Bindemittel*] = water of set(ting), set(ting) water

~wert *m*, Erstarrungswert [*hydraulisches Bindemittel*] = set(ting) value

~zeit *f*, Abbindedauer *f* [*Verschnittbitumen*] = curing time, ~ period

~zeit *f*, Erstarrungszeit, Abbindedauer, Erstarrungsdauer *f* [*hydrau-

lisches Bindemittel] = set(ting) time, ~ period
~zeit *f* **des Zements** [*Tiefbohrtechnik*] = waiting-on-cement time
~zeitbeschleuniger *m* → Abbindebeschleuniger
(Ab)zeitverzögerer *m* → Abbindeverzögerer
Abbindung *f* → Abbinden [*hydraulisches Bindemittel*]
~ → Abbinden [*Verschnittbitumen*]
~ → Abbund [*Zimmerei*]
Abbindungsschwindung *f*, Erstarrungsschwindung [*hydraulisches Bindemittel*] = set(ting) shrinkage
Abblasen *n* = blow(ing) down (Brit.); blow off (US) [*Opening a valve in a steam boiler mud drum or other place where boiler sediment collects, so as to eject it*]
Abblasung *f* (Geol.) → Deflation *f*
Abblättern *n*, Abplatzen [*Anstrich*] = flaking [*The detachment of flakes of a coat due to penetration of water behind the coat or greasiness of the surface or, in distempers, by too much size*]
~, Abblätterung *f* [*Feuerfestindustrie*] = shelling, peeling
~ [*Beton*] → Abschuppen
abblenden, abdunkeln = to dim
Abblenden *n*, Abdunkeln = dimming
Abblend|kappe *f* = dim-out cap
~lampe *f* = passlamp
~licht *n* = dimmed light, low beam, passing beam
~schalter *m* = dip switch, dimmer ~
abbohren [*Gelände*] = to test-drill
Abbohren *n* **der (Orts)Brust** [*Tunnel- und Stollenbau*] = drilling shotholes in the face
~ eines Erdölvorkommens mit bekannter Begrenzung = inside pool drilling
~ von Schächten = shaft sinking by drilling, shaft boring
(ab)bölzen, abspreizen, (ab)sprießen, absteifen, (schräg)aussteifen = to brace

~, abspreizen, (ab)sprießen, absteifen, (quer)aussteifen = to strut horizontally
Abbolzen *n* [*Verstärkung des Schacht- und des Streckenausbaues durch Bolzen*] = strata bolting
(Ab)Bölzung *f*, Abspreizung, (Ab-)Sprießung, Absteifung, (Quer)Aussteifung, Verspreizung = horizontal strutting
~, Abspreizung, (Ab)Sprießung, Absteifung, (Schräg)Aussteifung, Verspreizung = bracing
Abbordgerät *n*, Bordkantenschneidgerät, Straßenbordfräs- und -räummaschine *f* = verge cutter, ~ trimmer
(ab)böschen = to slope
(Ab)Böschen *n* = slope work
abbossiert = bush-hammered
Abbrand *m*, (Ab)Brennen *n* [*Zündschnur*] = burning
~ von Legierungsbestandteilen = loss of alloying elements during deposition
~anflug *m*, Aschenanflug [*auf den Ofensteinen*] = kiln scum
~geschwindigkeit *f*, (Ab)Brenngeschwindigkeit [*Zündschnur*] = burning rate, ~ speed, rate of burning, speed of burning
Abbrauseflüssigkeit *f* = rinsing [*The liquid in or with which anything has been rinsed*]
abbrausen = to spray, to rinse (off)
Abbrause|sieb *n* = rinsing screen
~wasser *n* = rinsing water, spraying ~, rinse
Abbrausung *f*, Abbrausen *n* = spraying, rinsing
abbrechen, abreißen, einreißen, abtragen, niederreißen [*Bauwerk*] = to demolish, to pull down, to take down, to tear down, to wreck
Abbrechen *n* → Abbruch *m*
(Ab)Bremsen *n* = braking
abbrennen von Anstrichen, flammstrahlreinigen = to burn off old coats of paint

(ab)brennen — (ab)dämmen

(ab)brennen [*Zündschnur*] = to burn [*fuse*]
(Ab)Brennen *n*, Abbrand *m* [*Zündschnur*] = burning
Abbrennen *n*, Flammstrahlreinigung *f* [*Entfernen von Anstrichen*] = paint burning, burning off paint
~ **von Basaltpflaster**, thermische Aufrauhung *f* = heat treatment of basalt paving setts
(Ab)Brenngeschwindigkeit *f*, Abbrandgeschwindigkeit [*Zündschnur*] = burning speed, ~ rate, rate of burning, speed of burning
abbrennstumpfschweißen = to flash weld
Abbrennstumpf|schweißmaschine *f* = flash-welder, flash-welding machine
~**schweißung** *f*, Abbrennstumpfschweißen *n* = resistance flash welding
abbröckeln = to crumble away
Abbröckeln *n*, Abbröck(e)lung *f* = crumbling away
~, Abbröck(e)lung *f* [*Straßendecke*] = fretting [deprecated:(un-)ravel(ling)] [*break up of the road surface because of binder failure*]
Abbruch *m*, Absturz *m* (Geol.) = fall
~, Abtrag *m*, Abtragung *f*, Abbrucharbeit(en) *f(pl)*, Abbrechen *n*, Abreißen, Abtragen [*Bauwerk*] = demolition work, wrecking, pulling down (work), taking down (work), wreckage, tearing down (work)
~, Ufer~ = eroding of a bank, washing ~ ~ ~
~**arbeiter** *m* = wrecker, demolisher, mattock man, topman, housebreaker
~**firma** *f*, Abbruchgeschäft *n*, Abbruchunternehmung *f*, Abbruchunternehmen *n* = wrecking firm, demolition ~
~**genehmigung** *f* = demolition permit
~**gerät** *n* = wrecker, demolition device
~**geschäft** *n* → Abbruchfirma *f*
~**hammer** *m* = demolition pick (hammer)
~**kolonne** *f*, Abbruchmannschaft *f* = wrecking gang, ~ party, ~ crew, ~ team, demolition ~
~**material** *n*, Abbruchschutt *m* = demolition rubbish (US); ~ waste, ~ spoil (Brit.)
~**meißel** *m* = demolition bit, wrecking ~
abbruchreif = fit for demolition, ~ ~ wrecking
Abbruch|schutt *m* → Abbruchmaterial
~**stelle** *f* = wrecking site, demolition ~
~**ufer** *n*, Prallufer = eroding bank, washing ~
~**unternehmer** *m* = wrecking contractor, demolition ~; housewrecker (US)
~**unternehmung** *f* → Abbruchfirma *f*
~**vertrag** *m* = demolition contract, wrecking ~
~**werkzeug** *n*, Werkzeug für Abbruchhämmer = demolition tool, wrecking ~, wrecker
~**ziegel** *m* = demolition rubbish brick (US); ~ waste ~, ~ spoil ~ (Brit.)
~**zone** *f* (Geol.) = fault(ed) zone
Abbund *m*, Abbinden *n*, Abbindung *f* [*Zimmerei*] = trimming, joining
~**platz** *m*, Abbindeplatz = trimming yard, joining ~
abbürsten = to brush off, to scrub
Abbürsten *n* = brushing off, scrubbing
Abdach *n* → Vordach
Abdachung *f*, Ausbiß *m*, Böschung (Geol.) = scarp, escarpment
~ = coping [*Is the top cap or top course of a wall, designed to shed water, to protect the top of the wall, and to give a finished appearance to the top of the wall*]
Abdachungs|fluß *m* [*Abdachungsflüsse entstammen einseitig abgedachten Gegenden und verlaufen meist parallel zueinander, z.B. Pyrenäenflüsse*] = consequent stream
~**tal** *n* = consequent valley
(ab)dämmen [*Baukonstruktionen gegen Wärme- und Schalldurchgang sichern*] = to insulate

abdämmen, absperren [*durch den Bau einer Talsperre*] = to dam [*to put a dam in*]

Abdämmen *n*, Absperren, Abdämmung *f*, Absperrung [*durch den Bau einer Talsperre*] = damming

~ **von Erschütterungen** = vibration insulation

Abdämmung *f* **durch Sandsäcke**, Absperrung ~ ~, Sandsackabsperrung, Sandsackabdämmung = damming by sand bags

Abdämmungs|becken *n* (Geol.) = basin due to damming, ~ ~ ~ blocking

~**see** *m*, Stausee, Abriegelungssee (Geol.) = obstruction lake, ponded ~

~**stufe** *f* (Geol.) = step due to ponding

(ab)dämpfen, abschwächen [*Geräusch; Erschütterung*] = to dampen

Abdampf *m* = waste steam

~**(aus)nutzung** *f*, Abdampfverwertung = waste steam utilization

~**entöler** *m* = oil separator for waste steam

~**fördermaschine** *f* (⚒) = wastesteam (type) winding engine, ~ (~) winder

~**heizung** *f* [*Sie kann als Dampf- oder Dampfwasserheizung ausgeführt werden. Der Abdampf wird entölt, entwässert und in Abdampfverteiler geleitet*] = waste steam heating

~**injektor** *m* = waste steam injector

~**leitung** *f* = waste steam line

~**probe** *f*, Abdampfprüfung *f*, Abdampfversuch *m* [*für Vergaserkraftstoff und Spezialbenzine*] = gum test

~**rohr** *n*, Dampfauslaßrohr = waste steam pipe

~**rohrnetz** *n* = waste steam pipes network

~**rückstand** *m* [*Für Trink- und Brauchwasser ist ein Abdampfrückstand, also ein nach Eindampfen durch Wägung festzustellender Gehalt an gelöster Substanz von weniger als 500 g/l erwünscht, weil größere Lösungsinhalte durch zu große Härte, Salzgehalt usw. die Verwendungsmöglichkeiten beeinträchtigen*] = evaporation residue

~**stutzen** *m* = waste steam connecting branch

~**turbine** *f* = waste steam turbine

Abdampfung *f*, nasse Destillation *f* = boiling down

(Ab)Deck|blech *n* = cover(ing) plate

~**bohle** *f* = cover(ing) plank

~**brett** *n* = cover(ing) board

abdecken, abräumen, abtragen = to strip [*overburden or thin layers of pay material*]

~, bedecken, zudecken = to cover (over)

~, freilegen, freimachen = to uncover, to (lay) bare, to lay open [*by removing a covering*]

~, auszwicken, verzwicken [*Packe*] = to chink, to choke, to blind, to key

~ [*Mauer*] = to cope [*To cover a wall with stones, bricks, or precast slabs which usually overhang as a protection to the wall from rain*]

~ **mit Plane(n)** = to cover with tarpaulin(s)

~ ~ **Platten** = to cover with slabs

~ ~ **Splitt**, abspritten, verteilen von Splitt = to spread grits, to grit

Abdecken *n*, Abräumen, Abraumabtrag *m*, Abtragung *f*, Abtragen = stripping overburden

~, Abräumen, Abtragung *f*, Abtragen [*Abraum oder ausbeutbare Schicht(en)*] = stripping

~, Abspritten, Splittverteilung *f*, Splittstreuen *n* = gritting, chipping, grit blinding, grit spreading [*deprecated: blinding, dressing*]

~, Freilegen, Freimachen = uncovering [*by removing a covering*]

~ [*Betonnachbehandlung*] = protection of green concrete by curing overlay

~ **einer Böschung** = covering of a slope

~ **mit Platten** = covering with slabs

(Ab)Deck|folie *f* = protecting foil, cover(ing) ~

Abdeckhaube — (Ab)Dichtungsmaterial

~haube *f* = hood
~lage *f*, **(Ab)Deckschicht** *f* = cover(ing) layer
~leiste *f* = cover moulding (Brit.); ~ molding (US) [*A mo(u)lding planted to cover a joint on a flush surface*]
Abdeck|matte *f* [*Nachbehandlung von Beton*] = (concrete-)curing blanket
~papier *n* [*Nachbehandlung von Beton*] = (concrete-)curing paper, overlay ~
(Ab)Deck|platte *f* = cover(ing) slab
~rost *m* = cover(ing) grate, ~ grating
~schicht *f*, **(Ab)Decklage** *f* = cover(ing) layer
~splitt *m*, **(Ab)Streusplitt** = cover aggregate (for seal), cover stone, gritting material; blotter (material) (for bituminous prime coat) (US); surface dressing chipping(s) (Brit.)
~stein *m* = coping stone, cope [*A top stone, generally slightly projecting, to shelter the masonry from the weather, or to distribute the pressure from exterior loading. A projecting covering stone of a wall*]
Abdeckung *f*, Zudeckung, Bedeckung = covering
~ [*Nachbehandlung von Beton*] = curing overlay
~ = coping [*A top course of stone or concrete, generally slightly projecting, to shelter the masonry from the weather, or to distribute the pressure from exterior loading. A projecting covering or cap of a wall*]
~ [*Da Eisenbahnbrücken begehbar sein müssen, erhalten offene Fahrbahnen eine Abdeckung aus Holzbohlen oder Stahlbetonplatten*] = cover(ing)
~ eines Gewölbes, Gewölbeabdeckung = coping of a vault
Abdeckzelt *n* [*Nachbehandlung von Beton(straßen)decken*] = (concrete) protection tent, (~) curing ~
(Ab)Deckziegel *m* = coping brick, cope
(ab)dichten, undurchlässig machen, dicht machen = to seal

~ [*von Grundbauten gegen Grundwasser durch Umhüllung mit einer wasserdichten Haut*] = to lay a waterproof skin
(Ab)Dichten *n* = sealing
(Ab)Dichtung *f* = sealing
~ → (Ab)Dichtungsvorlage *f*
~ des Untergrundes, (Unter)Grund(ab)dichtung = subsoil waterproofing
~ gegen (Boden)Feuchte = dampproofing
~ ~ Grundwasser, Grundwasserisolierung = waterproofing
~ von Wassereinbrüchen = sealing of inrushes of water
(Ab)Dichtungs|anstrich *m*, **(Ab)Dichtungsaufstrich** = sealing (paint) finish, ~ (~) coat, ~ coat of paint
~anstrichmittel *n*, **(Ab)Dichtungsanstrichstoff** *m* = sealing paint
~anstrichstoff *m*, **(Ab)Dichtungsanstrichmittel** *n* = sealing paint
~arbeiten *fpl* = sealing work
~aufstrich *m* → (Ab)Dichtungsanstrich
~bahn *f* = sealing strip
~band *n* = sealing tape
~belag *m* → (Ab)Dichtungsvorlage *f*
~filz *m* = sealing felt
~folie *f* = sealing foil
~fuge *f*, Dichtfuge = sealing joint
~gewölbe *n* [*Talsperre*] = watertight facing arch
~graben *m*, Verherdung *f* = (cut-off) trench [*A (cut-off) trench is carried across the valley under the dam at such a depth as to prevent water from the reservoir percolating underneath the dam and flowing away downstream*]
~gürtel *m* → Einpreß-Dichtwand *f*
~haut *f* → (Ab)Dichtungsvorlage *f*
~lage *f* → Sperrschicht *f*
~latte *f* [*kittlose Verglasung*] = draught fillet, windguard
~masse *f* = sealing compound, sealant
~material *n* → (Ab)Dichtungsmittel

~**mauer** *f*, (Ab)Dichtungssporn *m*, Trennmauer, Sporn, Abschlußwand *f*, Herdmauer, Fußmauer [*Talsperre*] = (toe) cut-off wall, toe ~

~**mittel** *n*, (Ab)Dichtungsstoff *m*, (Ab)Dichtungsmaterial *n* = sealing material, sealant, sealing medium

~**mittel** *n* [*Nachbehandlung von Beton*] = (membrane) curing compound

~**pappe** *f* = sealing felt

~**schicht** *f* → Sperrschicht

~**schirm** *m* → Einpreß-Dichtwand *f*

~**schleier** *m* → Einpreß-Dichtwand *f*

~**schürze** *f* → Einpreß-Dichtwand *f*

~**schürze** *f* → (Ab)Dichtungsvorlage *f*

~**sporn** *m* → (Ab)Dichtungsmauer *f*

~**stoff** *m*, (Ab)Dichtungsmittel *n*, (Ab)Dichtungsmaterial *n* = sealing material, sealant, sealing medium

~**technik** *f* = sealing engineering

~**teppich** *m* → (Ab)Dichtungsvorlage *f*

~**trog** *m*, (Ab)Dichtungswanne *f* = tanking

~**vorlage** *f*, (Ab)Dichtungsteppich *m*, (Ab)Dichtung(sschürze) *f*, Schürze, Außen(haut)dichtung *f*, Oberflächendichtung, (Ab)Dichtungsschicht *f*, (Ab)Dichtungshaut *f*, (Ab)Dichtungsbelag *m* = impervious blanket, (waterproof) ~

~**wanne** *f*, (Ab)Dichtungstrog *m* = tanking

(**ab**)**drehen** [*mit der Schablone. Steinzeugindustrie*] = to jigger

~ = to turn (off)

(**Ab**)**Drehen** *n* [*mit der Schablone. Steinzeugindustrie*] = jiggering

abdrehen der Wellen verbunden mit abflauen = to fall back

(**ab**)**drosseln** = to throttle (down)

Abdruck *m* [*Tiefbohrtechnik*] = impression

~**büchse** *f*, Abdruckstempel *m*, Wachskrone *f* [*Rotarybohren*] = impression block

Abdrückdruck *m* [*Abdrücken von Rohrleitungen und Druckgefäßen mit Wasser*] = water test pressure

abdrücken [*(Beton)Probewürfel; (Beton)Probezylinder*] = to run a compression test

~ **mit kaltem Wasser** [*Rohrleitungen und Druckgefäße*] = to run a hydraulic test, ~ ~ cold water ~

Abdruckverfahren *n* [*Fangarbeiten in der Tiefbohrtechnik*] = impression method

Abdrückversuch *m* für Nahtschweißungen = pillow test; burst ~ (US)

abdunkeln = to dim, to darken

Abdunkeln *n*, Abblenden = dimming

Abend|schule *f* = evening institute, ~ school

~**spitzenverkehr** *m* = evening peak traffic, ~ rush-hour ~

Abessinier(brunnen) *m* → Norton-Brunnen

abfackeln = to flare (to waste)

Abfackeln *n* = flaring (to waste)

abfahren, abkarren = to cart [*to carry or deliver (something) in a cart or other vehicle*]

Abfahrt *f*, Ausfahrt, Autobahn~ = exit point, egress ~

~**rampe** *f* = exit ramp

Abfall *m* (Geol.) = detritus

~, Abfallstoff *m*, Abfallmaterial *n* = waste

~, Rückgang *m* [*z.B. Temperatur; Leistung usw.*] = drop

~, Müll *m*, Abfallstoff *m*, Abfallmaterial *n* = refuse

~**behälter** *m*, Müllbehälter = refuse hopper

~**beseitigung** *f*, Müllbeseitigung = disposal of refuse, refuse disposal

~**boden** *m* → Abfallmauer *f* [*Kammerschleuse*]

~**boden** *m* [*Bei Schuß- und Sturzwehren unterhalb des Staukörpers befindliche, kurze befestigte Flußstrecke*] = downstream floor, ~ apron

~**eisen** *n* = scrap iron

abfallend = descending

abfallende Gesteinsstrecke *f* (⚒) = dipping stone drift, ~ rock ~

abfallende Verwerfung — Abfangen mit Trossen 22

~ **Verwerfung** *f* (Geol.) = fault dipping against the beds

~ **Wetterführung** *f* (⚒) = downcast ventilation

abfallender Ast *m*, absteigender ~ [*Diagramm*] = descending branch

~ **Strahl** *m*, überfallender ~, Überfallstrahl = nappe [*A sheet or curtain of water overflowing a weir, dam, etc. The nappe has an upper and a lower surface*]

~ **Strand** *m* = shelving beach

abfallendes Gelände *n* = sloping ground

~ **Ufer** *n* = shelving shore

~ **Vorland** *n* = piedmont slope

Abfall|erz *n* = waste ore, tailings

~**fett** *n* = used grease

~**gummi** *m* = waste rubber

~**kohle** *f* = waste coal

~**koks** *m* = scrap coke

~**-Lauge** *f*, Ablauge = spent liquor, waste ~

~**material** *n* → Abfall *m*

~**mauer** *f* [*Senkrechter Abfall in der Sohle einer Kammerschleuse, der den Übergang vom Oberhaupt zur Schleusenkammer bildet. Früher auch ,,Abfallboden" genannt*] = forebay

~**mauer** *f*, Abfallwand *f*, Wehrrücken *m* [*Die vom überfallenden Wasser mit oder ohne Berührung verdeckte Luftseite eines Wehres*] = downstream face of a weir

~**öl** *n*, Altöl = used oil

Abfallrohr *n*, Regen(fall)rohr, Fallrohr, Abfallrohr für Regenwasser, Regenablaufrohr = downpipe, downcomer, downspout, fall pipe

~ [*Gebäudeentwässerung*] = soil pipe, ~ stack

~ **für Aborte**, Abortfallrohr = WC soil pipe, ~ ~ stack

~ ~ **Regenwasser** → Abfallrohr

Abfall|rücken *m* [*Kolkschutz einer Stauanlage*] = bucket

~**schlamm** *m* = residual sludge

~**stoff** *m*, Abfall *m* = waste

~**stoff** *m*, Abfall *m*, Müll *m* = refuse

~**(stoff)verwertung** *f* = utilization of waste(s)

~**strom** *m* = off-peak (electrical) energy

~**verbrennung** *f*, Müllverbrennung = refuse incineration, ~ destruction

~**verbrennungsanlage** *f*, Müllverbrennungsanlage, Abfallvernichtungsofen *m*, Müllvernichtungsofen = (refuse) destructor, incinerator (plant)

~**wand** *f*, Abfallmauer *f*, Wehrrücken *m* [*Die vom überfallenden Wasser mit oder ohne Berührung verdeckte Luftseite eines Wehres*] = downstream face of a weir

~**wärme** *f* = waste heat

~**zerkleinerer** *m*, Müllzerkleinerer = rubbish grinder, waste disposer

Abfaltung *f* (Geol.) = downfolding

abfangbares Bohrgestänge *n* = retainable drill pipe

Abfangdrän *m* = intercepting drain

abfangen, eine Schicht ~ [*Tunnel-, Stollen- und Bergbau*] = to hold

~, den Druck ~ [*Tunnel-, Stollen- und Bergbau*] = to resist

~ [*sichern gegen Umfallen, Einsturz, Verschiebung, Knicken und dgl.*] = to hold

~, fassen, fangen [*Quelle*] = to shut off, to intercept

~ **mit Ankerseilen**, abspannen ~ ~, ~ ~ Trossen = to guy [*To guide or steady with a guy or guys*]

Abfangen *n*, Abfangung *f* [*Wasser durch Drän(e)*] = intercepting, interception

~ [*Sichern gegen Einsturz durch drucksichere oder biegefeste Teile*] = holding

~, Abfangung *f* [*Quelle*] = shutting off, interception, intercepting

~ **durch Rechen** = screening [*The removal of relatively coarse floating and suspended solids by straining through racks or screens made of bars, gratings, wires, or perforated plates*]

~ **mit Trossen**, ~ ~ Ankerseilen, Abspannen ~ ~, Trossenabspan-

Abfanggabel — Abfließen

nung *f*, Trossenabfangung, Ankerseilabspannung, Ankerseilabfangung, Seilverspannung, Trossenverspannung = guying

Abfanggabel *f*, Abfangschere *f* = fishing jars

(Ab)Fang|graben *m* → Auffanggraben

~graben *m* [*Rieselfeld*] = pick-up carrier

Abfangkabel *n* → Abfangseil *n*

Abfangkeil *m*, Gestänge~, Rohrklemmkeil, Rohrkeilklemme *f* [*Rotarybohren*] = (rotary) slip

~ auf der Innenseite gezahnt [*Rotarybohren*] = serrated (rotary) slip

Abfang|keile *mpl* **mit auswechselbaren Schalen**, Rohrklemmkeile ~ ~ ~, Rohrklemmen *fpl* ~ ~ ~ [*Rotarybohren*] = (rotary) slips with interchangeable inserts

~leitung *f* = interceptor

(~)Sammler *m* [*Entwässerung*] = outfall drain, collecting ~, intercepting ~, catch(-water) ~, interceptor

~schere *f*, Abfanggabel *f* = fishing jars

~seil *n*, Abfangkabel *n*, Abfangtrosse *f* = guy rope

~straße *f*, Entlastungsstraße *f* = alternative side-street

~träger *m* = holding girder

~trosse *f* → Abfangseil *n*

Abfangung *f*, Abfangen *n* [*Wasser durch Drän(e)*] = interception, intercepting

~, Abfangen *n* [*Quelle*] = interception, intercepting, shutting off

(ab)fasen [*symmetrisches Abschrägen rechteckiger Kanten, d. h. im Winkel von 45°, durch Fräsen, Schleifen oder Brennschneiden*] = to chamfer

(Ab)Fasung *f*, (Ab)Fasen *n* = chamfering

~, Fase *f* = chamfer [*A right-angle corner cut off symmetrically, that is, at 45°. When cut off unsymmetrically, the surface may be called a bevel = Abschrägung*]

(Ab)Fegen *n*, (Ab)Kehren [*von Fahrbahnen, Bürgersteigen, Industriefußböden, Flugplatzdecken und dgl. mit Kehrmaschinen*] = brushing, brooming, (mechanical) sweeping

(ab)feilen = to file (off)

abfertigen [*Fahrgäste; Gepäck*] = to process, to handle

Abfertigung *f* [*Fahrgäste; Gepäck*] = handling, processing

Abfertigungs|anlage *f* [*als Einzelanlage*] [*Flugplatz*] = processing facility, handling ~

~anlage *f* [*als Gesamtheit aller*] → Nahverkehrsbereich [*Flugplatz*]

~bau *m*, Abfertigungsgebäude *n*, Empfangsbau, Empfangsgebäude [*Flughafen*] = (passenger) terminal building

~gebäude *n* **für Auslandverkehr** [*Flughafen*] = international terminal

~gebäude *n* **für Inlandverkehr** [*Flughafen*] = domestic terminal

~vorfeld *n* → Nahverkehrsbereich *m*

abfeuern, abtun, zünden [*Sprengschüsse*] = to fire (the blast)

Abfeuern *n*, Abtun *n*, Zünden *n* [*Sprengschüsse*] = firing (the blast)

abfiltern, (aus)filtern, (ab)filtrieren = to filter (off)

(ab)filtrieren → abfiltern

abflachen = to flatten

Abflachung *f* = flattening

abflanschen, ausklinken, ausflanschen = to notch

Abflanschen *n*, Ausklinken, Ausflanschen [*Örtliches Abtrennen von ganzen Flanschen bei Formstahl*] = notching

Abflanschmaschine *f* → Ausklinkmaschine

abflauen, einschlafen [*Wind; Wellen*] = to die

abfliegender Passagier *m* = departing passenger

abfließen [*Wasser aus einem Einzuggebiet*] = to leave a catchment area

~ [*von einer Fläche*] = to run off

~, ablaufen = to flow off (by gravity)

Abfließen *n* → Abfluß *m*

abfluchten → (ein)fluchten
Abfluchten *n*, Ausfluchten, Abfluchtung *f*, Ausfluchtung = ranging into line, ~ out
Abflug *m*, Start *m* = take-off
Abfluß *m*, Abfluß(wasser)menge *f*, Durchfluß, Durchfluß(wasser)menge, Wassermenge [*Zeichen: Q. Wassermenge, welche in der Sekunde einen Abflußquerschnitt durchfließt. Gemessen in l/s oder m³/s*] = discharge, flow, river ~, stream ~
~, Abfließen *n*, Ablaufen *n*, Ablauf *m* [*Durch die Schwere bedingter Bewegungsvorgang des Niederschlagwassers in einem Wasserlauf, auf dem Boden oder im Boden*] = flow [*Abfluß auf dem Boden = (surface) runoff. Abfluß im Wasserlauf auch discharge*]
~, ~rohr *n*, Ablauf(rohr) *m*, *(n)* [*für Grundstücksentwässerungsanlagen in den Gebäuden und Grundstücken*] = waste-pipe
~, oberirdischer ~ [*jener Anteil des Niederschlages, der oberirdisch abfließt*] = (surface) run-off
~ausgleich *m* = balancing of flow, ~ ~ discharge
~beiwert *m* → Abflußkoeffizient *m*
~beobachtung *f* = flow observation, discharge ~
~diagramm *n* = flow diagram, discharge ~
~fläche *f* = run-off surface
~form *f*, Durchflußform [*Fluß; Strom; Kanal*] = type of discharge, ~ ~ flow
~ganglinie *f*, Wassermengenganglinie, Ablaufganglinie = flow hydrograph
~gebiet *n*, Einzug(s)gebiet [*Zeichen: F*$_E$*. In der Horizontalprojektion gemessenes Gebiet, dem der Abfluß in einem bestimmten Abflußquerschnitt oder eine abflußlose Wasseransammlung entstammt*] = watershed, tributary area (US); catchment area, drainage basin, drainage area, catchment basin, gathering-ground (Brit.)

~gerinne *n*, Abflußrinne *f*, Ablauf(ge)rinne = flume, grip, channel, gutter
~geschwindigkeit *f*, Ablaufgeschwindigkeit = rate of flow, flow rate
~geschwindigkeit *f* [*oberirdischer Abfluß*] = rate of (surface) run-off, (surface) run-off rate
~graben *m*, Ablaufgraben, Abzuggraben = field ditch
~graben *m* → Abflußkanal *m* [*Talsperre*]
~hindernis *n* = flow obstacle, discharge ~
~jahr *n*, hydrographisches Jahr [*Einjähriger, nach hydrologischen Gesichtspunkten festgesetzter Zeitraum. In Deutschland vom 1. November bis zum 31. Oktober des folgenden Kalenderjahres*] = water year, climatic ~ [*A special grouping of the periods of the year to facilitate water supply studies. The United States Geological Survey uses October 1 to September 30*]
~kanal *m*, Ablaufkanal [*Schiffschleuse*] = discharge culvert
~koeffizient *m*, Abflußbeiwert *m*, Durchflußbeiwert, Durchflußkoeffizient [*Ungleichförmigkeit eines Wasserlaufes*] = discharge coefficient, flow ~ (Brit.); coefficient of discharge, coefficient of flow (US)
~koeffizient *m*, Abflußbeiwert *m*, Abflußverhältnis *n*, Abflußzahl *f* [*Verhältniszahl zwischen Abfluß und Niederschlag*] = coefficient of run-off, run-off coefficient [*of a catchment area*]
~kurve *f* → Abflußmengenkurve
~leitung *f*, Entleerungsleitung, Ablaßleitung [*Talsperre*] = discharge conduit
~linie *f* → Abfluß(mengen)kurve *f*
~menge *f* → Abfluß *m*
~menge *f* **bei Mittelwasser**, MQ = average flow, mean ~
~menge *f* **bei niedrigstem Niedrigwasser**, NNQ = low flow record
~menge *f* **bei Niedrigwasser**, NQ = low flow

Abflußmengen|beiwert *m*, Durchflußbeiwert = flow coefficient, discharge ~

~beziehungslinie *f*, Durchflußbeziehungslinie = curve of corresponding discharges, ~ ~ ~ flows

~dauer *f*, Durchflußdauer = flow duration, discharge ~

~dauerkurve *f*, Abflußmengendauerlinie *f* = (flow) duration curve, (~) ~ hydrograph

~häufigkeit *f*, Durchflußhäufigkeit = frequency of flows, ~ ~ discharges

Abfluß(mengen)|kurve *f*, Abfluß(mengen)linie *f*, Schlüsselkurve, Durchflußkurve, Durchflußlinie, Wassermengenkurve, Wassermengenlinie, Ablaufkurve, Ablauflinie [*Bezugskurve zwischen den Wasserständen und den zugehörigen Abflüssen*] = discharge curve, rating ~ [*shows the relation between the discharge and the stage of a water course*]

~messer *m*, Durchfluß(mengen)messer = flow meter, discharge ~

~messung *f* → Abflußmessung

~quotient *m*, Durchfluß(mengen)quotient = flow ratio, discharge ~

~-Registriergerät *n*, Durchfluß(mengen)-Registriergerät = flow recorder

~-Summenlinie *f*, Durchfluß(mengen)-Summenlinie = flow mass curve, discharge ~ ~, summation hydrograph, mass diagram

Abfluß|messung *f*, Abflußmengenmessung, Durchfluß(mengen)messung, Wassermessung [*Fluß; Strom; Kanal*] = discharge measurement, flow ~

~minderung *f* = flow reduction, discharge ~

~öffnung *f*, Ablauföffnung, Durchflußöffnung, Ablaßöffnung, Auslauföffnung [*Wehr*] = discharge opening, sluiceway

~öffnung *f* → Ablauföffnung

~querschnitt *m*, Durchflußquerschnitt, benetzter Querschnitt, durchströmter Querschnitt, Ablaufquerschnitt [*Zeichen: F oder f. Einheit m²*. *Von abfließendem Wasser erfüllter kleinster lotrechter Schnitt*] = discharge section, cross-section of discharge, wetted cross section; area of waterway (US)

~regler *m* = flow regulating device, discharge ~ ~

~rinne *f*, Abflußgerinne *n*, Ablauf(ge)rinne = grip, flume, channel, gutter

~rohr *n*, Ablaufrohr, Abfluß *m*, Ablauf *m* [*für Grundstücksentwässerungsanlagen in den Gebäuden und Grundstücken. Abflußrohre werden aus Blei, Gußeisen, Beton, Steinzeug usw. hergestellt*] = waste-pipe

~rohrbogen *m*, Ablaufrohrbogen = waste-pipe bend

~schwankung *f*, Durchflußschwankung [*Fluß; Strom; Kanal*] = discharge variation, flow ~

~seite *f*, Ablaufseite [*Becken*] = discharge side

~spende *f*, Wasserspende = yield factor (of a catchment), volume of discharge in a second of time for each square mile of tributary area

~spitze *f* [*jener Spitzenanteil des Niederschlages, der oberirdisch abfließt*] = (surface) run-off peak

~statistik *f*, Durchflußstatistik [*Fluß; Strom; Kanal*] = flow records, discharge ~

~stollen *m* → Ablaufstollen

~stopfen *m*, Entleerungsstopfen, Ablaßstopfen = drain plug, ~ stopper, waste ~ [*A bag plug or screw plug*]

~summe *f* = flow mass, discharge ~, cumulative discharge

~-Summenlinie *f* → Abflußmengen-Summenlinie

~tafel *f* [*Tabellarische Zusammenstellung der Koordinaten der Abflußkurve*] = discharge table, rating ~

~verhältnis *n* → Abflußkoeffizient *m*

~verlagerung *f* [*durch eine Talsperre*] = diversion of flow, ~ ~ discharge

~**verminderung** *f*, Durchflußverminderung [*Strom; Fluß; Kanal*] = decrease of discharge, ~ ~ flow

~**vermögen** *n* [*Meeresarm*] = tidal capacity

~**verzögerung** *f*, Durchflußverzögerung [*Strom; Kanal; Fluß*] = retardation of discharge, ~ ~ flow

~**-Vorhersage** *f* = hydrological forecast

~**wasser** *n*, Ablaufwasser = flow water

~**(wasser)menge** *f* → Abfluß *m*

~**zahl** *f* → Abflußkoeffizient *m*

~**zeit** *f* [*Hydraulik*] = discharge time, flow ~

Abfolge *f* (Geol.) = sequence of processes

Abfräsen *n* = milling off

Abfuhr *f*, Müll~ = refuse cartage

~ [*Baggergut*] = cartage of excavated material

~, Abtransport *m* = cartage, hauling away

~ überschüssiger Bodenmassen, ~ auf Kippe, Abtransport *m* ~ ~ = surplus cartage

abführbare Flüssigkeitsmenge *f* [*Rohr*] = capacity

abführen [*z. B. Wasser*] = to evacuate

~ in einer Rohrleitung, ableiten ~ ~ ~ = to pipe away

Abfuhr|fahrzeug *n*, Abtransportfahrzeug = haul-away vehicle

~**gerät** *n*, Abtransportgerät [*Erdbau*] = haul-away equipment

~**gleis** *n*, (Be)Ladegleis [*Erdbau*] = loading track

~**schute** *f*, Baggerschute = (hopper) barge (Brit.); (~) scow (US)

~**wagen** *m*, Abtransportwagen [*Erdbau*] = haul-away wagon

Abführungsvermögen *n* = capacity

Abfüll|maschine *f* für Säcke, Einsackmaschine, Sackfüllmaschine, Absackmaschine = bagging machine

~**pumpe** *f*, Faßpumpe = barrel (exhausting) pump

~**schieber** *m*, Entleer(ungs)schieber [*Silo*] = silo gate

~**waage** *f* für Zement → Zementwaage

Abgabe|seite *f*, Förderseite [*Pumpe*] = delivery side, ~ part

~**termin** *m* → → Angebot *n*

Abgänge *mpl*, Abgang *m*, unhältiges Gut *n* [*Aufbereitungstechnik*] = tailing(s)

~, menschliche ~, menschliche Abgangsstoffe *mpl* = (human) excreta

Abgas *n*, Auspuffgas = exhaust gas

abgasbeheizte Kippmulde *f* = exhaust-heated body

Abgas|deflektor *m* = exhaust deflector

~**gebläse** *n* = exhaust (gas) turboblower

~**nebenstromventil** *n* = exhaust by-pass valve

~**turboaufladung** *f* [*Die Leistung zur Aufladung wird durch eine von den Auspuffgasen des Verbrennungsmotors angetriebene Abgasturbine erzeugt*] = turbocharging

~**turbolader** *m* = exhaust (gas) turbocharger

~**vorwärmer** *m*, Ekonomiser *m* = economizer

abgeblättert, abgeschuppt, abgeplatzt [*Beton*] = scaled

abgeblendeter Scheinwerfer *m* = dimmed headlight

abgebogene (Stahl)Einlagen *fpl*, ~ Bewehrung *f*, ~ Armierung, Schubbewehrung, Schubarmierung, Schub(stahl)einlagen [*in auf Biegung beanspruchten Stahlbeton-Konstruktionsteilen ist eine Schubbewehrung notwendig, welche aus den unteren Trageinlagen durch Abbiegen unter 45-60⁰ erhalten wird*] = bent-up reinforcement, ~ reinforcing bars, ~ re-bars

(ab)geböscht = sloped

abgebraust = rinsed

abgebundenes Gerüst *n*, zugelegtes ~ [*Gerüst für Bauwerk; Kanthölzer zimmermannsmäßig abgebunden und auf/gestellt*] = joined and erected timber scaffold(ing), trimmed ~ ~ ~ ~

abgedichtet gegen = sealed against

abgefangen, abgespannt [*gegen Umfallen durch Ankerseile gesichert*] = guyed
abgefederte Bewegung *f* = cushioned movement, ~ motion
abgeflachtes Rundeisen *n* = flattened round bar
abgehängte Decke *f*, Hängedecke = drop ceiling, false ~, counter ~, hung ~, suspended ~ [*A ceiling which is built with a gap between it and the floor above*]
abgekantetes Drahtgewebe *n* = (woven-)wire cloth with turned up edges, gauze wire ~ ~ ~ ~
~ Geröll(e) *n* = subrounded pebbles
abgeknickte Verteilerpalette *f* [*Palettenwalze eines Betondeckenfertigers*] = double bent blade
abgekürzte Prüfung *f* → Schnellprobe
~ Wetterbeständigkeitsprobe *f*, ~ Wetterbeständigkeitsprüfung *f*, abgekürzter Wetterbeständigkeitsversuch *m*, Bewitterungskurzprüfung, Bewitterungsschnellprüfung, Bewitterungsschnellversuch, Bewitterungskurzversuch, Bewitterungskurzprobe, Bewitterungsschnellprobe = accelerated weathering test
abgekürztes Verfahren *n*, Schnellverfahren, Kurzverfahren, abgekürzte Methode *f*, Schnellmethode, Kurzmethode = accelerated method
abgelagert, angeschwemmt, alluvial = alluvial
abgelagerter Zement *m* = matured cement
abgeleitete Gezeit(enerscheinung) *f*, ~ Tide(erscheinung) *f* = derived tide
abgelenkter Verkehr *m* = deviated traffic
abgenutzt, verschlissen = worn(-out)
abgeplatzt, abgeblättert, abgeschuppt [*Beton*] = scaled
abgeriegeltes Tal *n* = blocked-up valley, obstructed ~
abgerissen, steil abstürzend, schroff, abschüssig, prall [*Küste*] = steep, bold

abgerissener Schuß *m*, totgelaufener ~ = cut-off shot, hangfire
abgerundet = rounded
abgerundete Kante *f* = bullnose [*A rounded edge*]
abgerundetes Ende *n* = bullnose [*A rounded end*]
~ Korn *n*, rundes ~ = round grain, ~ particle
abgesackte Kehlnaht *f* = fillet weld with vertical leg unintentionally shorter than the horizontal leg
abgesackter Zement *m*, eingesackter ~, Sackzement = sacked cement, bagged ~
abgesäuert [*Werkstein*] = etched
abgescherter Erdkeil *m* = slipping soil wedge
Abgeschiedenheit *f* **einer Ortschaft** = remoteness of a community
abgeschliffene Straßendecke *f*, abgeschliffener Straßenbelag *m* = bumpcut pavement, ~ surfacing
abgeschlossene Bohrung *f* = shut in well
abgeschnittener Balken *m* [*beim Balkenwechsel*] = trimmed joist (Brit.); tailpiece, tailbeam (US) [*A beam which frames into a header (joist) (US) / trimmer joist (Brit.) instead of spanning the entire distance between supports*]
abgeschuppt, abgeblättert, abgeplatzt [*Beton*] = scaled
abgeschwemmter Laterit *m* = low-level laterite
abgesenkter Grundwasserspiegel *m* = lowered ground-water table
~ Wasserspiegel *m* [*Talsperre*] = drawn-down water level
abgesetzte Fugen *fpl* [*verschieden gefärbte Stoß- und Lagerfugen*] = masonry joints of different colo(u)rs
abgespannt, abgefangen [*gegen Umfallen durch Ankerseile gesichert*] = guyed
abgesplittet = gritted, grit-blinded
abgestuft, stufenförmig, treppenförmig, abgetreppt = stepped
abgestufter Filter *m*, abgestuftes Filter *n* = graded filter

abgestuftes Material n, gesiebtes ~ = graded material, screened ~, ~ aggregate

abgetopptes Öl n = reduced oil

~ russisches (Roh)Erdöl n, Masut n = mazut, masut, heavy fuel

abgetreppte Bergseite f **des Wehrkörpers** = upstream stepped face, upriver ~ ~

abgetrepptes Fundament n, abgetreppter Gründungskörper m, abgetreppte Gründung f, abgetreppte Fundation f, abgetreppter Fundamentkörper = stepped footing (US); footing [British standard name. B. S. 3589: 1963 A stepped construction to spread the load at the foot of a wall or column]

~ Widerlager n = stepped abutment

abgewalmt = hipped

abgewalmtes Mansardendach n = mansard roof, curb ~ (US); hipped mansard ~ (Brit.) [It slopes in four directions, but there is a break in each slope]

(ab)gewalzt = rolled

(ab)gewalzte (Auf)Schüttung f, Walzdamm m = rolled fill

~ Erd(auf)schüttung f, Walzdamm m = rolled earth fill

(ab)gewalzter Damm m, Walz(stau)damm, (ab)gewalzter Staudamm = rolled earth (fill) dam, ~ earth(work) ~, ~ earthen ~

~ Erdkern m, Walzkern [Felsschüttungs(stau)damm] = rolled earth(en) core

abgewandelt = modified

abgewichene Bohrung f = curved well

abgewickelte Fläche f = developed surface

~ Länge f = developed length

abgewinkelter Ausleger m, Knickausleger, geknickter Ausleger = articulated jib (Brit.); ~ boom (US)

abgewogen = weighed

abgewürgt [Motor] = stalled [engine]

abgiebeln = to gable

(Ab)Glätten n = smoothing

Abgleichbohle f, Abstreichbohle, Abziehbohle [Betondeckenfertiger] = strike-off (screed), screeding beam, level(l)ing beam, level(l)ing screed, front screed

abgleichen [Bodenverfestigung mit Zement usw.] = to shape

~, (ab)schürfen, abziehen [entfernen der obersten Bodenschicht (Humus), 15–30 cm] = to skim

~, abstreichen, abziehen [Beton] = to strike off

Abgleichen n, Abstreichen, Abziehen [Beton] = level(l)ing, striking off, strike-off

Abgleitung f → Bodenfließen n

Abgliederungs|halbinsel f = detached peninsula

~insel f = detached island

abgraben [einen Einschnitt] = to dig, to excavate

Abgrabung f **von Einschnitten**, Abgraben n ~ ~ = digging, excavation

abgraten = to remove the burr

Abgraten n [Vollständiges Trennen überflüssigen Werkstoffes bei Preß- oder Gußteilen] = deburring

Abgrat|maschine f, Abgrater m = burr removing machine, deburring machine, flash trimmer

~meißel m, Putzmeißel = fettling chisel

~presse f = trimming press

~werkzeug n, Abgrater m = deburring tool

Abgrenzung f = delimitation

Abgrund m = precipice

abgrundtief → abyssisch

(Ab)Hämmern n [der Schweiße] = chipping

(Ab)Hang m, Berghang, Berglehne f, Gehänge n = hillside, flank of a hill

Abhängehaken m **am Kehrrad** = disconnecting hook, throw-off ~

abhängen [Tiefbohrtechnik] = to leave the pipe hanging

Abhänger m = hanger

Abhängevorrichtung f [Erdöltiefpumpe] = knock off joint, hook ~ ~

abhängig [*Schaltung*] = interrelated
Abhängigkeitsschaltung *f* = interrelated control
abhaspeln = to unreel, to reel off
Abhauen *n*, **Fallort** *n* (⚒) [*Von oben nach unten in der Lagerstätte zur Vorrichtung hergestellter Verbindungsweg zwischen zwei Sohlen von schwacher bis mittlerer Neigung*] = dip working
Abheben *n* [*Schwimmschicht im Klärbecken*] = skimming
~ **eines Trägers**, **Gleiten** ~ ~ = lifting of a girder
~ **von Rasen und Humus** = clearing, stripping
Abhebung *f* (Geol.) → Deflation *f*
Abhilfe(maßnahme) *f*, **Behebung** *f* = remedial measure, remedy
Abhitze *f* → Abwärme *f*
Abhobelung *f*, Abschleifung, (marine) Abrasion *f* (Geol.) = marine abrasion
Abholort *m* [*Baumaschine; Baustoff*] = point of collection
Abholung *f* [*Baumaschine; Baustoff*] = collection
Abhub *m* → (Mutterboden)Abtrag *m*
abkanten [*Kante herstellen*] = to edge
~ **nach oben**, **umlegen** ~ ~ = to fold up(wards), to tip ~
~ ~ **unten**, **umlegen** ~ ~ = to fold down(wards), to tip ~
Abkanten *n* [*Abbiegen langer Kanten an Blechen mittels Prägestempel und dazu passendem Gesenk in einer Abkantmaschine*] = folding
Abkant|maschine *f* = folding machine
~**presse** *f* = folding press (Brit.); press brake (US)
~**presse** *f* [*Ziegelindustrie*] = edging press
abkarren, **abfahren** = to cart [*to carry or deliver (something) in a cart or other vehicle*]
(Ab)Kehren *n*, **(Ab)Fegen** [*von Fahrbahnen, Bürgersteigen, Industriefußböden, Flugplatzdecken u. dgl. mit Kehrmaschinen*] = brooming, (mechanical) sweeping, brushing
Abkehrfluß *m* = obsequent stream

abkeilen [*Gestein*] = to lift the bed by wedging, to wedge
Abkeilen *n* [*Gestein*] = (bed lifting by) wedging
abkerben, **(ein)kerben** = to notch
abkippen = to tilt down [*the screed of a black top paver*]
~ (Geol.) [*Scholle*] = to tip
~ [*Ladung*] = to tip, to dump by tipping
Abkippen *n* **von Schüttgut mit Kippmulde** = dumping with tilting bucket
Abkipp|höhe *f* [*Schaufellader*] = dumping clearance
~**stellung** *f* [*Schaufellader*] = dumping position
abklappbare Panzerwanne *f* = hinged crankcase guard
abklingen, **ausklingen** [*Hochwasser; Schall; Schwingung usw.*] = to fade
Abklingen *n* **der Setzungen**, **Ausklingen** ~ ~ = fading of the settlements
Abklingzeit *f* [*Schall*] = decay time
Abklopfhammer *m*, **(Ab)Klopfer** *m* = scaling hammer, chipping ~, boiler pick
Abkohlen *n* = coal getting
Abkömmling *m*, **Derivat** *n* = derived product
(ab)köpfen der Packlage = to break off the projecting tops of stones with light hammers
(ab)kratzen, **(ab)schaben** = to scrape (off)
abkreiden, **abschnüren** = to mark by chalk line
Abkreiden *n* **(von Anstrichen)** = chalking
Abkreideprüfer *m* = chalking tester
Abkreuzung *f*, **Kreuzgebälk** *n*, **Kreuzband** *n*, **Kreuzstreben** *fpl*, **Andreaskreuz** *n* = cross stays, saltier cross bars, diagonal struts, St. Andrew's cross
abkröpfen → kröpfen
abkühlen = to cool (off)
Abkühlung *f*, **Abkühlen** *n* = cooling (off)

Abkühl(ungs)beiwert — Ablaßöffnung

Abkühl(ungs)|beiwert *m*, Abkühl(ungs)koeffizient *m* = cooling(-off) coefficient
~fläche *f* = cooling(-off) area
~geschwindigkeit *f* [*Temperaturabnahme in der Zeiteinheit beim Durchlaufen eines bestimmten Temperaturbereiches*] = cooling(-off) rate
~kurve *f*, Abkühl(ungs)linie *f* = cooling(-off) curve, ~ diagram, ~ graph
~oberfläche *f* = cooling(-off) surface
~verlust *m* = cooling(-off) loss
~vorgang *m* = cooling(-off) process
~zone *f* = cooling(-off) zone
Ablade|anlage *f*, Ausladeanlage, Entladeanlage = unloading station
~gleis *n*, Entladegleis, Ausladegleis = unloading track
abladen = to unload
Abladeplatz *m*, Abladestelle *f*, Entladeplatz, Entladestelle, Ausladeplatz, Ausladestelle = unloading point
Ablader *m* = unloader
Abladung *f*, Abladen *n* = unloading
Ablagerung *f*, Inkrustierung = incrustation
~, Akkumulation *f*, Ablagerungsvorgang *m*, Sedimentation *f* [*Der Vorgang des Stoffabsatzes auf dem Festland, im Süßwasser (Seen, Flüsse) und im Meer*] = deposition (process), sedimentation (~)
~, Akkumulation *f*, Ablagerungslandschaft *f* (Geol.) = alluvium, alluvion
~ [*infolge der Schwerkraft abgesetzte Stoffe*] = sediment
~, Sediment *n* (Geol.) = sediment
Ablagerungsbecken *n*, Absetzbecken [*Becken verschiedener Ausbildung hinter einem Triebwassereinlaß und meist auch hinter einem Sandfang zur Ausscheidung der feineren Sinkstoffe*] = sedimentation basin, settling ~
~, Absetzbecken (Geol.) = basin of deposit(ion), ~ ~ sedimentation, depositional basin, depositional trough, sedimentary basin, sedimentary trough
Ablagerungs|boden *m*, alluvialer Boden = alluvial soil
~gestein *n*, Sediment(är)gestein, Absetzgestein, Absatzgestein, Bodensatzgestein, Schichtgestein = bedded rock, sedimentary ~
~landschaft *f*, Akkumulation *f*, Ablagerung *f* = alluvion, alluvium
~vorgang *m* → Ablagerung *f*
ablandiger Wind *m*, Landwind = offshore wind
ablängen = to cut to length
Ablängen *n* [*Bewehrung*] = cutting to length
ablassen, absenken, versenken [*Schwimmkasten*] = to sink
~ [*Spannweg beim Spannbeton*] = to relax, to release
~, absenken [*Talsperre*] = to draw down
Ablassen *n*, Versenken, Absenken [*Schwimmkasten*] = sinking
~, Ablassung *f*, Absenkung, Absenken [*Talsperre*] = drawdown
~ [*Spannbeton*] = release, relaxing (the tension)
~ angesammelter Geschiebemengen, Entkiesung *f* = scouring, flushing, washing out
Ablaß *m*, **~vorrichtung** *f* = discharge system
~, **~vorrichtung** *f* = blow-off [*A waste gate or device for discharging accumulated solids or for emptying a depressed sewer*]
~ auf mittlere Höhe [*Talsperre*] = middle-height discharge tunnel
~hahn *m* → Entleerungshahn
~leitung *f* → Abflußleitung
~öffnung *f* → Ablauföffnung
~öffnung *f*, Auslauföffnung, Ablauföffnung, Abflußöffnung, Durchflußöffnung [*Wehr*] = discharge opening, sluiceway, outlet
~öffnung *f* [*Kraftübertragungssystem*] = drain port

~organ n, Entleerungsorgan [Talsperre] = outlet element
~regulierschütz(e) n, (f) = outlet control gate, ~ regulating ~
~rohr n = discharge pipe
~schieber m, Entleerungsschieber = outlet valve
~schraube f = screw plug, bleeder screw
~stollen m → Ablaufstollen
~stopfen m → Abflußstopfen
~ventil n [Kraftübertragungssystem] dump valve
~ventil n [Druckluft] = unloading valve
~ventil n, Entleerungsventil = drain valve
~verschluß m [Talsperre] = emptying gate, outlet ~
~vorrichtung f, Ablaß m = blow-off [A waste gate or device for discharging accumulated solids or for emptying a depressed sewer]
~vorrichtung f, Ablaß m = discharge system
Ablation f, Abschmelzung f [Gletscher] = wastage, ablation
Ablationsmoräne f = ablation moraine
Ablauf m → Abfluß m
~, Abwick(e)lung f [z. B. eines Versuches] = procedure
~ → Abflußrohr n
~ → Abflußkanal m
~ [Säule] = throat, cavetto
~ [Bestandteil einer Entwässerungsanlage zur Aufnahme des abfließenden Wassers] = discharge
~ = effluent [Sewage, partly or completely treated, flowing out of any sewage-treatment device]
~, Einlauf = gull(e)y
~ [einer Frist] = expiration [of a date]
~anlage f, Ablaufberg m, Eselsrücken m [Teil des Rangierbahnhofes zur Zerlegung der Züge] = hump, double incline
~bahn f [zur Verladung von Fertigpfählen oder Fertigteilen im Brücken- und Hafenbau] = barge-loading runway

~berg m → Ablaufanlage f
~berggleis n, Ablaufgleis = hump track
~breite f [Breite des Geländes und der Wasserfläche vor Helling für Stapellauf eines Schiffes] = launching width
ablaufen, abfließen = to flow off (by gravity)
Ablaufen n → Abfluß m
~ [Farbe; Kleber usw.] = running down
ablaufen lassen [Wasser von einer Fläche] = to shed
Ablauf|ganglinie f → Abflußganglinie
~gerinne n → Abflußkanal m
~gerinne n, Ablaufrinne f, Abfluß(ge)rinne = gutter, channel, grip, flume
~geschwindigkeit f, Abflußgeschwindigkeit = flow rate, rate of flow
~gipfel m [Höchstpunkt des Ablaufberges] = top of hump, ~ ~ double incline
~gleis n, Ablaufberggleis = hump track
~graben m, Abflußgraben, Abzuggraben = field ditch
~grube f = drainage sump
~kanal m, Abflußkanal [Schiffschleuse] = discharge culvert
Ablauf|kasten m [Siebdurchgang] = run-off box
~kurve f → Abfluß(mengen)kurve
~leitung f der Schwenkblöcke [Radschrapper] = swivel joint drain line
~linie f → Abfluß(mengen)kurve f
~loch n = discharge hole
Ablauföffnung f, Abflußöffnung, Durchflußöffnung, Ablaßöffnung, Auslauföffnung [Wehr] = sluiceway, discharge opening, outlet
~, Einlauföffnung [Straßenablauf; Hofablauf] = inlet opening, gull(e)y ~
Ablauf|platte f, Abtropfplatte, Abtropffläche f [am Spülbecken der Küche] = drain(ing) board, drainer

Ablaufprofil — Ablösungserscheinung

~**profil** n [*Längsschnitt durch den Ablaufberg*] = longitudinal hump profile, ~ profile of double incline

~**programm** n [*Maschinensteuerung*] = stop-controlled sequence

~**querschnitt** m → Abflußquerschnitt

~**rinne** f, Ablaufgerinne n, Abfluß(ge)rinne = channel, gutter, grip, flume

Ablauf|(rohr) m, (n) → Abflußrohr

~**rohrbogen** m, Abflußrohrbogen = waste-pipe bend

~**rost** m, Einlaufrost = gull(e)y grid, inlet ~, inlet grating, bar grating

~**seite** f, Abflußseite [*Becken*] = discharge side

~**stollen** m, Entleerungsstollen, Ablaßstollen, Abflußstollen, Auslaufstollen [*Talsperre*] = discharge tunnel

~**tiefe** f [*Die vor einer Helling für den Stapellauf eines Schiffes notwendige Fahrwassertiefe*] = navigable depth for launching

~**wagen** m [*zum Ablaufen von Brückenbauteilen vom Fertigungsplatz auf Pontons für das Einschwimmen zur Einbaustelle*] = launching troll(e)y

~**wasser** n, Abflußwasser = flow water

Ablauge f, Abfall-Lauge = waste liquor, spent ~

Ablaugmittel n, alkalisches (Ab)Beizmittel = soda-based paint remover

Ablegeplatte f = pallet

ableiten [*Wasser in einen Fluß oder Strom*] = to discharge [*water into a stream or river*]

~, umleiten [*Wasser*] = to divert

~, **in einer Rohrleitung**, abführen ~ ~ ~ = to pipe away

ableitfähiger (Fuß)Bodenbelag m = conductive floor(ing), ~ floor finish

Ableitfähigkeit f **für elektrostatische Ladung** = inability to hold static electric charge

Ableitung f [*Formel*] = derivation

~, Umleitung [*Talsperrenbau*] = diversion

Ableitungs|bauwerk n, Umleitungsbauwerk = diversion structure

~**(ge)rinne** f, (n) = diversion channel

Ablenkblech n, Leitblech, Prallblech, Ablenkplatte f, Leitplatte, Prallplatte [*in einer strömungstechnischen Anlage*] = baffle plate, deflecting ~, deflector ~

Ablenken n **von Bohrlöchern** [*um verlorenem Gestänge auszuweichen*] = side tracking

Ablenk|keil m, Richtkeil, Abweichungskeil [*zum Ablenken von Bohrlöchern*] = whip-stock

~**platte** f → Ablenkblech n

~**trommel** f → → Bandförderer

Ablenkung f = deviation

Ablenkungs|kraft f = force of deviation

~**punkt** m = point of deviation

ablesbar = readable

Ablesefehler m = reading error

ablesen = to read (off), to take readings, to take a reading

Ablesen n, Ablesung f = reading (off)

Ablesevorrichtung f, Ablesegerät n = reading (off) device

Ableucht|lampe f = inspection lamp

~**loch** n = lamp hole [*A small vertical pipe or shaft leading from the surface of the ground to a sewer, for admitting a light for purposes of inspection*]

ablösbarer Schuh m [*Simplex-Betonpfahl*] = detachable shoe

Ablösung f, Ablösen n [*flüssiges Bindemittel vom Gestein*] = stripping, displacement

~ **der Schichten**, Ablösen n ~ ~, Aufblätterung ~ ~, Aufblättern ~ ~ (⚒) = bed separation

~ **des (Überfall)Strahles**, Strahlablösung = freeing of the nappe (US); separation of the water layer (Brit.)

Ablösungs|anzeiger m [*Haftfestigkeit zwischen bit. Bindemittel und Mineralmasse*] = displacement indicator, stripping ~

~**erscheinung** f [*Hydraulik*] = freeing phenomenon (US); separation ~ (Brit.)

~feld n, Ablösungsfläche f [*Hydraulik*] = separation area, freeing ~, surface of separation, surface of freeing

~verlust m [*Hydraulik*] = separation loss, eddy ~, freeing ~

~versuch m, Ablösungsprobe f, Ablösungsprüfung f [*bituminöses Bindemittel*] = stripping test, displacement ~

~widerstand m [*bituminöses Bindemittel*] = stripping resistance, displacement ~

abloten, absenkeln = to plumb

Abloten n → Absenkeln n

Abluft f = exit air

~entstaubung f = dust extraction from exit air

~kanal m, Entlüftungskanal = exit air duct

~lüfter m = exit air fan

~öffnung f, Entlüftungsöffnung = exit air opening

~rohr n, Entlüftungsrohr = exit air pipe

~rost m, Entlüftungsrost = outlet grille, exit air ~

~schacht m, Entlüftungsschacht = exit air shaft

~schornstein m, Entlüftungsschornstein, Abluftschlot m, Entlüftungsschlot = exit air chimney

Abmarkung f **der Uferlinie** f = marking of the bank line, staking out ~ ~ ~

Abmaß n, Maßabweichung f, Genauigkeitsgrad m = deviation, off-size, margin

Abmeißeln n [*Entfernen von Werkstoff mit Meißel(n)*] = chiselling off

abmessen, zuteilen, zumessen, dosieren = to batch, to proportion, to measure

Abmessung f = dimension

~ → Dosierung

Abmessungen fpl **über alles** = overall dimensions

Abmessungs|norm f, Maßnorm [*betrifft Abmessungen von Erzeugnissen*] = dimension standard

~überwachung f = dimension control

Abmeßanlage f, Zuteilanlage, Dosieranlage, Zumeßanlage = batch(ing) plant, proportioning ~, measuring ~, ~ installation

~ **für (Beton)Zuschlagstoffe** → Dosier(ungs)anlage ~ ~

~ ~ **Zement** → Dosier(ungs)anlage ~ ~

~ **in Turmanordnung**, Zuteilanlage ~ ~, Zumeßanlage ~ ~, Dosieranlage ~ ~, Abmeßturm m, Dosierturm, Zuteilturm, Zumeßturm = batch(ing) tower, measuring ~, proportioning ~

~ **mit dreimaligem Halten der Fahrzeuge**, Dosieranlage ~ ~ ~ ~ ~, Zumeßanlage ~ ~ ~ ~ ~, Zuteilanlage ~ ~ ~ ~ ~ = three-stop batch(ing) plant, ~ proportioning ~, ~ measuring ~, ~ ~ installation

~ ~ **einmaligem Halten der Fahrzeuge**, Dosieranlage ~ ~ ~ ~ ~, Zumeßanlage ~ ~ ~ ~ ~, Zuteilanlage ~ ~ ~ ~ ~ = one-stop batch(ing) plant, ~ proportioning ~, ~ measuring ~, ~ ~ installation

~ ~ **zweimaligem Halten der Fahrzeuge**, Dosieranlage ~ ~ ~ ~ ~, Zuteilanlage ~ ~ ~ ~ ~, Zumeßanlage ~ ~ ~ ~ ~ = two-stop batch(ing) plant, ~ proportioning ~, ~ measuring ~, ~ ~ installation

Abmeß|apparat m → Dosier(ungs)gerät n

~automat m → Dosier(ungs)automat

~automatik f, Dosierautomatik, Zuteilautomatik, Zumeßautomatik = auto(matic) batch(ing) system, ~ measuring ~, ~ proportioning ~

~band n, Dosierband, Zuteilband, Zumeßband = batch(ing) conveyor belt, proportioning ~ ~, measuring ~ ~, ~ conveyer ~

~bandwaage f → Dosierbandwaage

~bunker m → Dosier(ungs)bunker

~**förderschnecke** *f*, Zuteilförderschnecke, Zumeßförderschnecke, Dosierförderschnecke = batch(ing) screw, proportioning ~, measuring ~

~**gefäß** *n*, Dosiergefäß, Zumeßgefäß, Zuteilgefäß = batch(ing) container, proportioning ~, measuring ~

~**gerät** *n* → Dosier(ungs)gerät

~-**Hygrometer** *n*, Dosier-Hygrometer, Zuteil-Hygrometer, Zumeß-Hygrometer = batch(ing) hygrometer, proportioning ~, measuring ~

~**karre(n)** *f*, (*m*) **für (Beton)Zuschläge**, Dosierkarre(n) ~ ~, Zuteilkarre(n) ~ ~, Zumeßkarre(n) ~ ~ = batch(ing) cart, measuring ~, proportioning ~

~**kiste** *f*, Zumeßkiste, Zuteilkiste, Dosierkiste [*Betonherstellung*] = batch(ing) box, gauge ~, measuring ~, ~ frame

~**programmwähler** *m* → Dosier(ungs)programmwähler

~**pumpe** *f*, Dosierpumpe, Zuteilpumpe, Zumeßpumpe, Bindemittel~ = binder metering pump, proportioning ~, ~ measuring ~

~**rinne** *f* → Dosier(ungs)rinne

~**schleuse** *f*, Zuteilschleuse, Zumeßschleuse, Dosierschleuse = batch(ing) lock, measuring ~, proportioning ~

~**schnecke** *f* → Dosier(ungs)schnecke

~**schnecke** *f* **für Zement**, Dosierschnecke ~ ~, Zumeßschnecke ~ ~, Zuteilschnecke ~ ~, Zementabmeßschnecke, Zementzuteilschnecke, Zementdosierschnecke, Zementzumeßschnecke = cement batch(ing) screw, ~ proportioning ~, ~ measuring ~

~**silo** *m* → Dosier(ungs)silo

~**spiel** *n* → Dosierspiel

~**stern** *m* → Dosier(ungs)stern

~**teil** *m, n*, Zumeßteil, Dosierteil [*bituminöse Mischanlage*] = batch(ing) section, measuring ~, proportioning ~

~**trommel** *f* → Dosier(ungs)trommel

~**turm** *m* → Abmeßanlage *f* in Turmanordnung

~**vorrichtung** *f* → Dosier(ungs)gerät *n*

~**waage** *f* → Dosier(ungs)waage

Abminderungs|beiwert *m*, Abminderungszahl *f*, Abminderungsfaktor *m* = reduction factor, ~ coefficient

~**kurve** *f* = reduction curve

~**zahl** *f* = reduction factor

abmontieren → demontieren

Abnahme *f*, Minderung *f* = decrease, diminution

~ [*Die Übernahme des Bauwerkes oder Gebäudes nach Fertigstellung. Schweiz: Kollaudierung f*] = acceptance

~ **durch Augenschein** = acceptance inspection

~**band** *n*, Abnahmeförderband = take-off belt

~**bedingungen** *fpl* → Abnahmevorschriften *fpl*

~**bescheinigung** *f*, Abnahmebescheid *m*, Abnahmeschein *m*, Abnahmeniederschrift *f*, Abnahmeprotokoll *n* = certificate of acceptance, acceptance certificate

~**dose** *f* [*zur Stromabnahme bei Stromaggregaten*] = socket outlet

~**(förder)band** *n* = take-off belt

~**kommission** *f* = acceptance committee

~**(prüf)lehre** *f* = inspection gauge (Brit.); ~ gage (US)

~**unterlagen** *fpl* = acceptance documents

~**versuch** *m*, Abnahmeprobe *f*, Abnahmeprüfung *f* = acceptance trial, ~ test

~**verweigerung** *f* = rejection

~**vorschriften** *fpl*, Normalien *fpl*, Abnahmebedingungen *fpl* = (contract) specifications, (~) specs

~-**Zeichnung** *f*, Revisions-Zeichnung [*Zeichnung mit Kennzeichnung der für die Revision (Abnahme) wichtigen Maße*] = acceptance drawing

abnehmbar, herausnehmbar = detachable, withdrawable, removable

abnehmen — abräumen

abnehmen, wegnehmen, entfernen = to detach, to withdraw, to remove
Abnutzung f **durch mechanische Einwirkung**, Verschleiß m = wear (and tear)
Abnutz(ungs)|festigkeit f = wear(ing) strength
~**fläche** f = wear(ing) area
~**geschwindigkeit** f = rate of wear
~**grad** m = degree of wear
~**prüfung** f, Abnutz(ungs)probe f, Abnutz(ungs)versuch m = wear(ing) test
~**schaden** m = damage of wear
~**widerstand** m = wear(ing) resistance
Аböl n = waste oil, used ~
Abort m, Klosett n, Abtritt m = closet [*A privy, water closet, etc.*]
~**anlage** f, Abtrittanlage, Klosettanlage = toilet
~**becken** n, Abtrittbecken, Klosettbecken, Spülbecken = (water) closet bowl, wc ~, WC ~
~**druckspüler** m = flushing valve (Brit.); flushometer (US)
~**fallrohr** n, Abfallrohr für Aborte = WC soil stack, ~ ~ pipe
~**grube** f, Fäkaliengrube, Latrinengrube, Senkgrube, Abtrittgrube = f(a)eces pit, privy ~
~**spülkasten** m → Spülkasten
~**spülung** f = flushing
abpfählen, abstecken, abpflocken, verpflocken, verpfählen [*von Bauprofilen*] = to peg out, to set ~, to stake ~ [*of building sites*]
Abpfählen n, Abstecken, Abpflocken, Verpflocken, Verpfählen, Abstekkung(sarbeiten) f (pl), Abpfählung f, Verpfählung, Abpflockung = pegging out, setting ~, site-marking
Abpfeilern n, Abbau m der Pfeiler (✕) = robbing pillars
abpflocken → abpfählen
abplatzen, zerspringen [*Keramikindustrie*] = to spall, to chip
Abplatzen n, Abplatzung f, Zerspringen [*Keramikindustrie*] = spalling, chipping
~ [*Anstrich*] → Abblättern

~ [*Beton*] → Abschuppen
abpressen → verpressen
Abpressen n → Injektion f
abpumpen [*Brunnen*] = to develop
Abputzhammer m = waller's hammer, walling ~
abrammen = to ram
Abrammen n = ramming
Abrams'scher (Feinheits)Modul m, ~ Feinmodul = Abrams fineness modulus, F. M.
Abrasion f, marine ~, Abschleifung f, Abhobelung (Geol.) = marine abrasion
Abrasions|bucht f, Abschleifungsbucht = abrasion embayment, corrasion ~
~**fläche** f, Abrasionsebene f = abrasion plain, corrasion ~, marine (pene)plain, plain of marine denudation (or erosion)
Abraum m, Überlagerung f, Deckschutt m, Deckgebirge n, Abraumdecke f = overburden, shelf, overlay, uncallow, topspit; tir [*in Scotland*]
~**abtrag** m, Abdecken n, Abräumen, Abtragung f, Abtragen, Abtrag = stripping overburden
~**arbeit** f → Abraumbaggerung f
~**bagger** m = overburden excavator
~**baggerung** f, Abraumarbeit f, Abraumbetrieb m = overburden stripping
~**betrieb** m → Abraumbaggerung f
~**boden** m = overburden soil
~**böschung** f = overburden slope
~**brücke** f, Abraumförderbrücke, Förderbrücke für Abraum = overburden conveying bridge
~**-Dampflok(omotive)** f, Dampf-Abraumlok(omotive) = overburden steam loco(motive)
~**decke** f → Abraum m
~**-Elektrolok(omotive)** f, Elektro-Abraumlok(omotive) = electric overburden loco(motive)
abräumen, abtragen, abdecken = to strip [*overburden or thin layers of pay material*]

Abräumen *n*, Abdecken, Abraumabtrag *m*, Abtragung *f*, Abtragen, Abtrag = stripping overburden

~, Bereißen, Abtreiben, Beräumen, Nachbruch *m*, Nachbrechen [*Entfernen loser Brocken und Schalen von Firste und Stößen mit Brechstange, Hacke oder Abbruchhammer, vor allem bei Schichtbeginn und nach dem Schießen*] = ripping, brushing

~, Abdecken, Abtragung *f*, Abtragen [*Abraum oder ausbeutbare Schicht(en)*] = stripping

Abräumer *m*, Abraumgerät *n* = overburden equipment

Abraum|(förder)brücke *f*, Förderbrücke für Abraum = overburden conveying bridge

~**förderer** *m*, Abraumfördermittel *n* = overburden transporter, ~ conveyer, ~ conveyor

~**förderung** *f*, Abraumtransport *m* = overburden transport

~**(förder)wagen** *m* [*DIN 22618*] = overburden wagon

~**gerät** *n*, Abräumer *m* = overburden equipment

~**halde** *f* → Aussatzhalde

~**-Hochlöffel(bagger)** *m*, Großlöffel-(-Abraum)bagger, Abraum-Löffelhochbagger = stripper, (long-boom) stripping shovel

~**kippe** *f* [*Die Ablagerungsstelle und -vorrichtung zum Verkippen des in Tagebauen abgeräumten Deckgebirges*] = overburden tip and disposal installation

(~)**Kippenpflug** *m* → Planierpflug

~**lage** *f* → Abraumschicht *f*

~**lok(omotive)** *f* **für den Tagebau** = overburden loco(motive)

~**luftseilbahn** *f*, Abraum-Seilschwebebahn, Abraumseilhängebahn = overburden (aerial) ropeway (Brit.); ~ (~) tramway (US)

~**pflug** *m* → Planierpflug

~**salz** *n*, Kalisalz = potassic salt, potassium ~

~**schicht** *f*, Abraumlage *f*, Überlagerungsschicht, Deckgebirgslage, Deckgebirgsschicht = overburden layer

~**schrapper** *m*, Abraum-Schleppschrapper, gewöhnlicher Abraumschrapper, (gewöhnliche) Abraumschrapperanlage *f* = overburden (power) drag scraper (machine), ~ power scraper excavator, ~ (drag) scraper installation

~**-Schürfkübel-Schreitbagger** *m* = walking dragline stripper

~**-Seilschwebebahn** *f*, Abraumluftseilbahn, Abraumseilhängebahn = overburden (aerial) ropeway (Brit.); ~ (~) tramway (US)

~**sprengung** *f*, Abraumsprengen *n* = overburden blasting

~**standseilbahn** *f* = overburden funicular railway (Brit.); ~ ~ railroad (US)

~**transport** *m*, Abraumförderung *f* = overburden transport

~**wagen** *m*, Abraumförderwagen [*DIN 22618*] = overburden wagon

abrechnen = to settle the accounts for construction work

Abrechnungs|form *f*, Abrechnungsweise *f* = method of payment for construction work

~**menge** *f*, kostenvergütbare Menge, kostenvergütete Menge = pay(able) quantity

Abreiben *n* [*eine Betonfläche um gutes Aussehen zu erzielen*] = rubbing down

~, Abrieb *m*, Abreibung *f* = abrasion, attrition

Abreibungsversuch *m*, Abreibungsprobe *f*, Abreibungsprüfung *f*, Abriebversuch, Abriebprobe, Abriebprüfung = abrasion test, attrition ~

Abreißbewehrung *f*, Abreißarmierung, Abreiß(stahl)einlagen *fpl* [*Wenn sich eine Stahlbetonplatte parallel zu einem Unterzug spannt, so muß als Übergang zur Decke senkrecht zum Unterzug eine obere Bewehrung eingelegt werden, um das Abreißen der Decke vom Unterzug zu verhindern. DIN 1045*] = top reinforcement, upper ~

abreißen — Absackung eines Bauwerkes

abreißen (✵) [*Firste der Strecke*] = to break [*roof*]

~ → abbrechen [*Bauwerk*]

Abreißen n [*Saugüberlauf*] = un-priming (Brit.); stop siphoning (US)

~ → Abbruch m [*Bauwerk*]

Abreiß|hebel m [*Motor*] = contact point bumper block

~widerstand m, Haftung f, Haftfestigkeit f [*Putzmörtel*] = adhesion

(ab)richten [*auf genaue Form bringen*] = to dress

Abrieb m, Abreibung f, Abreiben n = attrition, abrasion

~ [*ungewollt zerkleinertes Gut*] = breakage

abriebfeste Spitze f & **Stiefel** m [*Aufreißer*] = abrasion tip & boot

Abriebfestigkeit f = abrasion resistance

abriebfreie Siebung f, abriebfreies Sieben n = screening without breakage

Abrieb|härte f = abrasion hardness, attrition ~

~prüfmaschine f = abrasion tester, attrition ~, ~ testing machine

~-Trommelwäscher m = super-scrubber

~verlust m = abrasion loss, attrition ~

Abriegelungssee m, Abdämmungssee, Stausee (Geol.) = obstruction lake, ponded ~

Abriß m (✵) [*Firste der Strecke*] = roof break

~punkt m, Festpunkt, Fixpunkt = bench mark

Abroll|bahn f, Abrollweg m [*Flughafen*] = turn-off, exit taxiway

~brücke f, Rollklappbrücke, Schaukelbrücke = roller bascule bridge

Abrollen n, Umrollen [*Gleiskette*] = roll-over

~ [*Flugzeug*] = exit taxiing

~ **der Mahlkörper übereinander**, Kaskadenwirkung f = cascading (action)

~ **des Siebgutes auf dem Siebboden** = cascading of material along the deck

~ **von Glasuren** = crawling of glazes

Abroll|gerät n für Bitumen-Fertigbahnen = P.B.S. laying machine, lick roller, stamp licker

~halbmesser m, Abrollradius m [*eines Abrollweges auf einem Flughafen*] = radius of curvature for turn-off, ~ ~ ~, ~ exit taxiway

~weg m, Abrollbahn f [*Flughafen*] = exit taxiway, turn-off

~wiege f [*Ziegelindustrie*] = cradle iron, rocking ~

Abrostbürste f, Entrostbürste = rust removing brush

Abrosten m, Entrosten = rust removal

abrunden = to round (off), to radius

Abrunden n, Abrundung f = rounding (off), radiusing

Abrundungskoeffizient m [*Kies*] = roundness index

abrüsten, ausrüsten = to remove a scaffold(ing), ~ ~ scaffold(ing)s

~, ausrüsten, abbauen eines Lehrgerüstes = to dismantle the falsework, ~ strip ~ ~

Abrüsten n → Abbau m eines Lehrgerüstes

~, Ausrüsten = removal of scaffold(ing)s

Abrutsch m → Bergrutsch

~gefahr f [*z. B. einer Wand im Steinbruch*] = cave-in risk

Abrutschung f → Bodenfließen n

Absackanlage f, Einsackanlage = bag packing plant, bagging ~

absacken, einsacken = to bag

Absacken n, Einsacken, Absackung f, Einsackung = bagging, bag packing

~ **der Betonkanten** [*Betondeckenbau mit Gleitschalungsfertiger*] = edge slump(ing)

~ **eines Bauwerkes** → Bauwerksetzung

Absack|maschine f, Einsackmaschine, Abfüllmaschine für Säcke, Sackfüllmaschine = bagging machine, bag packing ~, bag packer

~schnecke f, Einsackschnecke = bagging screw, bag packing ~

Absackung f **eines Bauwerkes** → Bauwerksetzung f

Absackwaage *f* → Einsackwaage
(ab)sanden, mit Sand bestreuen = to (cover with) sand
absanden, sandstrahlen = to sandblast
Absanden *n*, Absandung *f* = covering with sand, sand dressing
~, Sandstrahlen = sand blasting
~ [*Eine Fläche „sandet ab", wenn sich beim Abfegen oder Abreiben aus ihrer Oberfläche immer wieder Bestandteile in Sand- oder Staubform ablösen*] = dust formation
Absatz *m*, Bank *f* [*Tunnel- und Stollenbau*] = bench
~ [*Gezeitensaum*] = nip
~, Podest *n*, Treppen~ = landing [*Resting space usually arranged at the top of any flight of stairs*]
absatzartiges Fließen *n* = slug flow
Absatz|band *n* → Abwurfband
~**boden** *m*, Kolluvialboden, Derivatboden, umgelagerter Boden = transported soil, colluvial ~ [*soil consisting of alluvium in part and also containing angular fragments or the original rocks*]
~**förderung** *f* = market development
~**gestein** *n* → Absetzgestein
~**schwelle** *f* [*Thenard'sche Hubschütze und Chanoineklappe*] = toothed bar, tripping ~
~**tank** *m* [*Tiefbohrtechnik*] = settling tank

absatzweise arbeitende Aushubmaschine *f*, ~ arbeitendes Aushubgerät *n* = intermittently working excavator
~ ~ **Mischanlage** *f*, Chargen-Mischanlage, Perioden-Mischanlage, Stoß-Mischanlage = intermittent weighbatch (mixing) plant, batch(-mix) type plant
~ **arbeitendes Gerät** *n* = intermittently working equipment
~ ~ **Gleisrückgerät** *n*, Gleisrückmaschine *f* für absatzweisen Betrieb = intermittent (type) track shifting machine
~ **Zwangsmischung** *f*, absatzweises Zwangsmischen *n*, Chargen-Zwangsmischen, Chargen-Zwangsmischung = batch pugmill mixing
absatzweiser Transport *m* = discontinuous transport
absatzweises Mehrlagenschweißen *n* = block welding (Brit.); ~ sequence (US)
~ **Schweißen** *n* = skip welding, intermittent ~
~ **(Ver)Wiegen** *n* = intermittent weighing, batch ~
Absäuern *n* = acid(-etch)ing, etching
Absaugebewetterung *f*, Saugbewetterung, saugende Bewetterung, Bewetterung durch Unterdruck = exhaust ventilation

---

Absaugegerät *n* nach ASTM C 91–51
  Absaugeprüfung *f*
  Anrühren *n*
  Ausbreitmaß *n* nach Absaugen
  Ausbreittisch *m*
  Bestimmung *f* des Ausbreitmaßes
  Dreiwegehahn *m*
  Druckregler *m*
  Filterpapier *n*
  Gummidicht(ungs)ring *m*
  Lineal *n*
  Mauerzement *m*
  Nutsche *f*, Lochplatte *f*
  Quecksilbermanometer *n*
  Rührschüssel *f*

apparatus for the water retention test
  water retention test
  mixing
  flow after suction
  flow table
  flow determination
  three-way stop-cock
  pressure control device
  filter paper
  rubber gasket
  straightedge
  masonry cement
  perforated dish
  mercury manometer
  mixing bowl

(Setz)Trichter *m* — funnel
Spachtel *m*, *f* — spatula
Wasserhaltewert *m*, Wasserhaltevermögen *n* — water retention value
Wasserstrahlpumpe *f* — water aspirator

(ab)saugen [*Späne und Staub*] = to extract, to exhaust, to suck
Absaugen *n*, Vakuumbehandlung *f*, Saugbehandlung [*Beton*] = vacuum treatment
Absauger *m*, Exhaustor *m* = exhaust(ing) fan, extract ~, induced-draught ~
(Ab)Saug|haube *f* = suction hood
~stutzen *m* = suction tube
abschaben, abkratzen = to scrape off
Abschabspachtel *m*, *f* = flexible knife
Abschälen *n* [*Anstrich*] = peeling [*The dislodgement of paint from its backing*]
~, Abschälung *f* [*Keramikindustrie*] = scaling
~ der Obermörtelschicht [*Betondecke*] = scaling
~ ~ (Straßen)Bankette auf die ursprüngliche Höhe = level(l)ing of soft shoulders
~ des Rasens, Rasenabtrag *m* = turf stripping
abschalten [*Maschine*] = to disengage
~ [*Strom*] = to cut off the power
Abschalten *n* [*Maschine*] = disengagement
~, Ausschalten [*allgemein*] = disconnecting
~ des Hinterradantriebes bei Straßenfahrt = rear axle disconnecting
Abschaltstellung *f* → Ausschaltstellung
Abschaltung *f*, Ausschaltung [*allgemein*] = disconnection
~, Stillsetzung *f* = shutdown [*of a power plant*]
Abschälung *f*, Abschälen *n* [*Keramikindustrie*] = scaling
abscheidbar [*Staub*] = precipitable
(Ab)Scheider *m* [*Grundstückentwässerung*] = trap (Brit.); interceptor (US)

Abscheider *m*, Separator *m* = separator
Abscheidung *f* **von Wasser** → Bluten *n* [*Beton*]
abscheren = to shear (off)
Abscherung *f*, Abscheren *n* = shearing (off)
Abschiebevorrichtung *f* [*Transportvorrichtung*] = kick-off
Abschiebung *f*, Verwerfung, Bruch *m*, Sprung *m*, Paraklase *f*, Absenkung [*Grabenrand*] (Geol.) = fault (plane), geological fault, faulting
Abschirm|beton *m*, Strahlenschutzbeton = radiation shield concrete, (biological) shield(ing) ~
~block *m*, Strahlenschutzblock, Abschirmstein *m*, Strahlenschutzstein = radiation shield block, (biological) shield(ing) ~
abschirmen = to shield
Abschirm|mauer *f*, Abschirmwand *f* = radiation shield wall, (biological) shield(ing) ~
~tür *f*, Strahlenschutztür = radiation shield door, (biological) shield(ing) ~
Abschirmung *f*, Strahlenschutz *m* = (radiation) shielding, biological ~
~ [*allgemein*] = shielding
Abschirmungsanlage *f*, Strahlen~, (Reaktor)Strahlenschutzanlage, Schutzanlage = (reactor) shield, structure for shielding atomic plants
Abschlag *m*, Sprengabschnitt *m*, Sprengstrecke *f* [*Vortrieb je Angriff*] = round
Abschlagen *n* [*Steine*] = napping
Abschlag|länge *f* [*Sprengen*] = length of round

~zahlung *f* = payment on account, instalment payment, progress payment

(Ab)Schlämmanlage *f* [*Ziegelindustrie*] = settling basin

abschlämmbar, absetzbar, ausschlämmbar = settleable

abschlämmbare Bestandteile *mpl*, Abschlämmbares *n* [*Betonzuschläge*] = settleable solids

abschlämmen, (sich) absetzen, ausschlämmen = to settle (out)

Abschlämmen *n*, Absetzen, Ausschlämmen = settling (out), sedimentation

Abschlämmgeschwindigkeit *f*, Sinkgeschwindigkeit von Teilchen in einer Flüssigkeit = settling velocity, ~ rate

(Ab)Schlämmung *f* [*Korngrößenbestimmung*] = wet analysis [*The mechanical analysis of soil particles smaller than 0.06 mm (the smallest convenient sieve, BS 200 mesh)*]

Abschleifen *n*, Abschleifung *f* = grinding (off)

~ hoher Stellen [*auf einer Betondecke*] = bump-cutting

Abschleifung *f*, (marine) Abrasion *f*, Abhobelung (Geol.) = marine abrasion

Abschleifungsbucht *f*, Abrasionsbucht = abrasion embayment, corrasion ~

Abschlepp-Drahtseil *n* = towing wire cable, ~ ~ rope

abschleppen [*eine Fläche zur Beseitigung von Unebenheiten*] = to drag

Abschlepp|-Fahrzeug *n*, Abschleppkran *m*, Abschleppwagen *m*, Fahrzeugrettungskran = recovery vehicle, breakdown van, crash truck, tow truck

~seil *n*, Schlepptau *n* = towing rope

Abschließen *n* [*Meer(es)arm; Flußarm*] = closing

~, Aufrichten (der Böcke) [*Nadelwehr*] = raising, closing

~ des Durchflusses, Absperrung *f* ~ ~ [*in einer Rohrleitung*] = closing a flow

Abschließungsküste *f* = concordant coast

Abschluß *m*, Randeinfassung *f* [*Fahrbahn*] = (carriageway) edging

~bauwerk *n* = closure structure

~block *m* [*bei Balken aus Blöcken*] = end-block

~damm *m*, Abschlußdeich *m* = closure embankment

~damm *m* **der Spülfläche** *f*, **~ des Spülfeldes,** **~ der Ablagerungsstelle** [*Naßbaggerung*] = surrounding embankment (of a spoil ground) (Brit.); retaining dike, retaining level (US)

~fehler *m* [*Vermessung*] = closing error, error of closure

~flansch *m* → Deckelflansch

~hahn *m*, Abstellhahn, Absperrhahn = shut off cock, (stop) ~

~körper *m* [*bei einem Schieber*] = (valve-)closure member, ~ element

~mauer *f* [*Trockendock*] = head wall

~mörtel *m* = seal(ing) mortar

~nadel *f* [*Ringschieber*] = needle

~nadelsitz *m* [*Ringschieber*] = needle seat, ~ support

~organ *n* → Absperrorgan

~schieber *m*, Absperrschieber = stop valve, shut-off ~ [*A device to open or close a flow*]

~schieber *m* [*Tellermischer*] = discharge door

~schütz(e) *n*, (*f*), Grundablaßverschluß *m* [*Talsperre*] = lower gate, ~ sluice, sluice (gate), sluiceway gate

~spundwand *f* = cut-off sheet pile wall, ~ ~ piling

~vorrichtung *f* → Absperrorgan *n*

~wand *f*, Trennmauer *f* = diaphragm

~wand *f*, Herdmauer *f*, Fußmauer, (Ab)Dichtungsmauer, (Ab)Dichtungssporn *m*, Trennmauer, Sporn [*Talsperre*] = (toe) cut-off wall

Abschmelzung *f*, Ablation *f* [*Gletscher*] = ablation, wastage

(Ab)Schmierdienst *m* = lubrication (and greasing) service

(ab)schmieren, einschmieren = to lubricate

~, einschmieren, (ein)fetten = to grease
(Ab)Schmieren n, (Ab)Schmierung f, Einschmieren, Einschmierung = lubrication, lubricating
~, (Ab)Schmierung f, (Ein)Fetten, (Ein)Fettung, Einschmieren, Einschmierung = greasing
(Ab)Schmierer m = greaseboy, oiler, lube man
Abschmier|fahrzeug n = greasing and lubricating service vehicle, ~ ~ ~ rig; mobile lube rig (US)
~fett n = car grease
~-LKW m [fahrbarer Abschmierdienst auf der Baustelle] = greasing and lubricating truck (US); ~ ~ ~ lorry (Brit.)
Abschneideautomat m [Ziegelindustrie] = auto(matic) cutting(-off) table
abschneiden = to cut off
Abschneiden n = cutting off
~ einer Krümmung = cutting out of a bend, ~ ~ ~ ~ curve
~ von Pfählen und Spundwänden unter Wasser = cutting of piles and sheet piling under water
Abschneider m, Ziegel~ = (brick) cutter
~ mit Messern, Messerabschneider [Ziegelindustrie] = knife-cutter
~draht m, Abschneidedraht, Ziegel~ = (brick) cutting wire
Abschnitt m [in einer Norm, einem Vertrag und dgl.] = clause
~, (Teil)Strecke f, Teilstück n = section, stretch
~fuge f, Baufuge [Brückenbau] = construction joint
abschnittsweises Kernen n = discontinuous coring
abschnüren → abkreiden
abschotten = to seal off (walls to prevent transmission of airborne sound)
abschrägen = to bevel
Abschrägung f = bevel [A surface meeting another surface at an angle which is not a right angle]

~, Druckschlag m [Gewölbe] = splaying, bevel
abschrauben = to screw off, to unscrew
~ [Rotarybohren] = to break the joint
abschreiben = to depreciate (Brit.); to deplete (US)
Abschreibung f = depreciation (Brit.); depletion (US)
Abschreibungssatz m = depreciation rate (Brit.); depletion ~ (US)
Abschrot(er) n, (m) [keilförmiges Aufsteckstück aus Stahl. Dient als Unterteil beim Abhauen von Rund- oder Flacheisen] = blacksmith's hardy
Abschuppen n, Abblättern, Abplatzen, Abschuppung f, Abblätterung, Abplatzung = (surface) scaling [(Surface) Scaling of concrete may be of several types representing many causes. The following types can be recognized: (a) Manipulation scaling, (b) frost scaling, (c) progressive scaling]
(ab)schürfen, abgleichen, abziehen [entfernen der obersten Bodenschicht (Humus), 15–30 cm] = to skim
(Ab)Schürfen n → Abziehen
abschüssig, prall, abgerissen, steil abstürzend, schroff [Küste] = bold, steep
Abschußstelle f, Raketen~ = launching site
abschwächen, (ab)dämpfen [Geräusch; Erschütterung] = to dampen
abschwellen, fallen, sinken [Wasserspiegel] = to fall
abschwimmen → einschwimmen
Abschwimmpunkt m [eines Schwimmkörpers] = floating point
Absendeort m, Versandort = point of shipping, ~ ~ shipment
Absenkbrücke f [zum Absenken von Tunnelstücken] = trestle for lowering tunnel sections
absenkeln, abloten = to plumb
Absenkeln n, Abloten = plumbing [Transferring a point at one level to a point vertically below or above it, usually with a plumb bob or plumb rule]

(ab)senken [*Grundwasserspiegel*] = to lower [*ground-water table*]

absenken, herunterlassen [*1. eine Last mit einem Hebezeug; 2. den Kranausleger*] = to lower

~, ablassen [*Talsperre*] = to draw down

~, ablassen, versenken [*Schwimmkasten*] = to sink

Absenken *n*, Herunterlassen [*1. eine Last mit einem Hebezeug; 2. den Kranausleger*] = lowering

~, Ablassen, Versenken [*Schwimmkasten*] = sinking

~ → Ablassen [*Talsperre*]

(Ab)Senken *n* [*Grundwasserspiegel*] = lowering [*groundwater table*]

Absenken *n* **eines geometrischen Ortes**, Absinken ~ ~ ~ [*z.B. Tnw durch Flußbauwerke*] = subsidence of a locus

~ vorgefertigter Tunnelstücke im Wasser = lowering precast tunnel sections

(Absenk)Insel *f* [*Caissongründung*] = artificial island, sand ~

Absenk|kegel *m* → Absenkungskegel

~methode *f*, Absenkverfahren *n* [*Nach dem Ausschachten oder dem Ausbaggern des Bodens bis zum Grundwasser wird der Schachtbrunnen aufgemauert und unter gleichzeitigem Aushub des Bodens bis zur erforderlichen Brunnentiefe abgesenkt*] = dug well method

(Ab)Senkung *f*, Grundwasser~ = ground-water lowering

Absenkung *f* → Ablassen *n* [*Talsperre*]

~, Abschiebung, Verwerfung, Bruch *m*, Sprung *m*, Paraklase *f* [*Grabenrand*] (Geol.) = fault (plane), geological fault, faulting

~ des Hangenden, Absenken *n* ~ ~ (✵) = squeeze, crush, pressure, weight

(Ab)Senkung *f* **des Wasserspiegels durch Wind**, ~ der Wasseroberfläche ~ ~ = lowering of the water level by wind

Absenkungs|anlage *f*, Grundwasser-(ab)senkungsanlage = ground water lowering installation, dewatering ~

~bereich *m* [*Grundwasserhaltung und Grundwassergewinnung*] = pumping depression area

~faktor *m* (✵) = subsidence factor

~geschwindigkeit *f* (✵) = rate of subsidence, subsidence rate

Absenk(ungs)kegel *m* [*Grundwasser-(ab)senkung*] = cone of depression, draw-down cone

Absenkungs|kurve *f*, Absenkungslinie *f* [*Hydraulik*] = drawdown curve

~theorie *f* (✵) = theory of subsidence

Absenk(ungs)ziel *n* [*Grundwasser(ab)senkung*] = draw-down level

~, Mindeststau(höhe) *m*, (*f*) [*Talsperre. Niedrigster zulässiger Oberwasserstand*] = minimum water storage elevation, drawdown level

Absenk|verfahren *n* → Absenkmethode

~walze *f*, Versenkwalze [*absenkbarer Verschlußkörper eines Walzenwehres*] = submersible gate

Absetzanlage *f* → Bandabsetzer *m*

~ [*für die Klärung von Abwasser durch Absetzen der Schmutzstoffe*] = settling plant, sedimentation ~

Absetz|apparat *m* → Bandabsetzer *m*

~band *n*, (Seiten)Absetz(förder)band, Auslegerförderband = stacker belt, ~ conveyer, ~ conveyor

absetzbar, abschlämmbar, ausschlämmbar = settleable

absetzbarer Stoff *m*, Sinkstoff = settleable solid, settling ~ [*Suspended solid which will subside in quiescent sewage in a reasonable period. Two hours is a common arbitrary period*]

Absetzbecken *n*, Ablagerungsbecken [*Becken verschiedener Ausbildung hinter einem Triebwassereinlaß und meist auch hinter einem Sandfang zur Ausscheidung der feineren Sinkstoffe*] = settling basin, sedimentation ~

Absetzbecken — (ab)sieben

∼, Klärbecken = sedimentation tank, settling ∼, ∼ basin, subsiding ∼ [*A tank or basin in which sewage, partly treated sewage, or other liquid containing settleable solids is retained long enough, and in which the velocity is low enough, to bring about sedimentation of a part of the suspended matter, but without a sufficient detention period to produce anaerobic decomposition*]
∼ (Geol.) → Ablagerungsbecken

Absetzbehälter *m*, Wechselbehälter = demountable tank, detachable ∼
∼ → Absetzkippbehälter

Absetzbrunnen *m*, Klärbrunnen [*Behälter mit trichterförmiger Sohle, dessen Tiefe im Verhältnis zur Oberfläche groß ist und den das Abwasser in aufsteigender Richtung durchfließt*] = vertical-flow sedimentation tank with hopper bottom

absetzen [*Verrohrung in der Tiefbohrtechnik*] = to set casing, ∼ land ∼
∼ → abschlämmen
∼ [*Last*] = to place load
∼, sich ∼ (⚒) = to settle

Absetzen *n*, Hinsetzen [*z.B. Fertigteile beim Montagebau*] = placing
∼, Abschlämmen, Ausschlämmen = settling (out), sedimentation
∼ [*Aushuberde längs der Ausschachtung*] = sidecasting [*Piling spoil alongside the excavation from which it is taken*]
∼ **von Paraffin** = deposition of paraffin

Absetzer *m* → Band∼
∼**ausleger** *m*, Bandausleger = boom
∼**kippe** *f* = stacker spoil dump

Absetz|(förder)band *n*, Auslegerförderband, Seiten-Absetz(förder)band = stacker belt, ∼ conveyor, ∼ conveyer
∼**gerät** *n* → Bandabsetzer *m*
∼**geschwindigkeit** *f*, Klärgeschwindigkeit = rate of settling, ∼ ∼ sedimentation, settling velocity, sedimentation velocity
∼**gestein** *n*, Sedimentgestein, Ablagerungsgestein, Absatzgestein, Bodensatzgestein, Schichtgestein, neptunisches Gestein = sedimentary rock, bedded ∼
∼**glas** *n* [*dient zur Bestimmung der Raummenge der während gewisser Zeit absetzbaren Schwebestoffe in cm^3/l*] = settling cone, ∼ glass
∼**grad** *m* **von Markierungsfarbe** = degree of settling of traffic paint
∼**(kipp)behälter** *m*, Wechsel(kipp)behälter, Absetz(kipp)mulde *f*, Wechsel(kipp)mulde = lift-on lift-off (bulk) tipping container
∼**kipper** *m* = tipping container vehicle
∼**klärung** *f* [*Abwasser*] = sedimentation, settling
∼**methode** *f*, Absetzverfahren *n* = settling method, sedimentation ∼
∼**raum** *m*, Klärraum = settling chamber, sedimentation compartment [*of a two-storey tank, as in the Imhoff tank*]
∼**stelle** *f* [*Spülbaggern*] = disposal area, ∼ point
∼**teich** *m* → Schlammteich

Absetzung *f*, (Erd)Rutschung = (land)slide [*Earth slip upon a large scale*]

Absetzungsgeschwindigkeit *f* (⚒) = settling rate, rate of settling

Absetz|verfahren *n*, Absetzmethode *f* = sedimentation method, settling ∼
∼**versuch** *m*, Absetzprobe *f*, Absetzprüfung *f*, Sedimentationsprobe, Sedimentationsversuch, Sedimentationsprüfung, Sedimentierprobe, Sedimentierversuch, Sedimentierprüfung, Schlämmanalyse *f* = decantation test (US); sedimentation ∼
∼**vorgang** *m* = sedimentation process, settling ∼
∼**zeit** *f* = settling time, ∼ period, sedimentation ∼

Absickern *n* [*von unterirdischem Wasser durch eine mächtige lufthaltige Zone*] = inward seepage, influent ∼

(ab)sieben = to screen

Absieben — abspannen

Absieben *n* von feuchtem Gut, Naß-(ab)siebung *f* = wet screening
Absiebung *f*, Absieben *n*, Klassierungssiebung = screening, size separation by screening
absiegeln, versiegeln [*Straßenbau*] = to seal
Absiegeln *n*, Versiegeln [*Straßenbau*] = seal coating
Absieg(e)lung *f*, Versieg(e)lung, Porenschluß *m* [*Straßenbau*] = seal
Absieg(e)lungs|mittel *n*, Versieg(e)-lungsmittel, Porenschlußmittel [*Straßenbau*] = surface sealant
~schicht *f*, Versieg(e)lungsschicht, Porenschlußschicht [*Straßenbau*] = seal coat
~schicht *f* [*Bodenverfestigung*] = protective coat
~schicht *f* als Tränkung [*Bodenverfestigung*] = protective coat by penetration
Absinkzeit *f* [*zwischen der zweiten und dritten Marke beim Blaine-Gerät*] = time of flow
absolut trocken, atro = absolutely dry
absolute Festigkeit *f* = absolute strength
~ Feuchtigkeit *f* der Luft, Luftfeuchtigkeit = absolute humidity, humidity of air
~ Höhe *f* über dem Meeresspiegel = altitude
~ Lichtweite *f* [*Siebtechnik*] = nominal width of aperture, ~ internal width, ~ aperture width
~ (Wand)Rauhigkeit *f* [*Rohr*] = absolute roughness
absoluter statischer Luftdruck *m* → barometrischer Druck
Absolut|bestimmung *f* = absolute determination
~druck *m* = absolute pressure
~geschwindigkeit *f* = absolute velocity
~rauhigkeit *f*, Oberflächenrauhigkeit [*Hydraulik*] = surface roughness
~verformung *f* = absolute deformation

~verschiebung *f* = absolute displacement
~wert *m* = absolute value
Absonderung *f* (Geol.) = structure, parting, jointing
Absonderungs|fläche *f* [*Gestein*] = division(al) plane, plane of division, joint
~kluft *f* = diaclase
absorbieren = to absorb
Absorption *f*, Aufzehrung *f* [*Einsaugen eines Stoffes in das Innere eines anderen ohne chemische Vereinigung*] = absorption
~ von Gammastrahlen, ~ der Gammastrahlung = gamma absorption
Absorptions|anlage *f* = absorption installation
~beiwert *m* = absorption coefficient
~färbung *f*, Einsaugfärbung = absorption colo(u)ring
~gefäß *n* = absorption vessel, ~ tube
~grad *m*, Schluckgrad [*Schallschutz im Hochbau*] = absorption coefficient
~(kälte)maschine *f* = absorption refrigerator
~mittel *n*, Absorptionsstoff *m* = absorbent
~-Naturgasolin *n* = absorption gasoline
~öl *n* = scrubbing oil
~spektrum *n* = asborption spectrum
~stoff *m*, Absorptionsmittel *n* = absorbent
~turm *m* = absorption tower
~verlust *m* = absorption loss
~vermögen *n*, Absorptionsfähigkeit *f* = absorptive capacity, absorption ~, ~ property
~wasser *n* = absorbed water
absorptiv = absorptive
abspachteln = to screed
Abspaltung *f* [*Magma*] = differentiation
~ [*Gebirgsbildung*] = rift building, rifting
Abspanndraht *m* = stay wire
abspannen mit Ankerseilen → abfangen ~

Abspannen *n* mit Trossen → Abfangen ~ ~

Abspanner *m*, Reduziertransformator *m*, Abspanntransformator = step-down transformer

Abspann|gerüst *n* [*Brückenbau*] → Montagegerüst

~**seil** *n*, Trosse *f*, Ankerseil = standing rope, guy-rope, guy cable, (back-)stay cable

Absperr|bauwerk *n* = damming structure

~**bewegung** *f*, Schließbewegung [*Rohrleitung*] = shutting off motion, ~ ~ movement

absperren [*Förderbohrung*] = to close in a producing well

~, abdämmen [*durch den Bau einer Talsperre*] = to dam [*to put a dam in*]

Absperren *n*, Abdämmen, Absperrung *f*, Abdämmung [*durch den Bau einer Talsperre*] = damming

~, Schließen [*Rohrleitung*] = shutting off

Absperr|gerät *n*, Absperrvorrichtung *f* [*Eine senkrechte Leitvorrichtung, die eine Sperrstelle kenntlich macht und den Verkehr am Befahren oder Betreten der Sperrstelle hindert*] = barrier

~**hahn** *m* → Abstellhahn

~**kegel** *m*, Leitkegel, Gummihut *m* = rubber cone

~**klappe** *f* = flap

~**mittel** *n* [*Mittel, um Einwirkungen von Stoffen aus dem Untergrund auf den Anstrich oder umgekehrt oder innerhalb des Anstrichaufbaues zwischen den einzelnen Anstrichschichten zu verhindern. Die hierfür noch gebrauchte Bezeichnung ,,Isoliermittel'' ist zu vermeiden, um Verwechslungen mit Wärme- und Schalldämmstoffen und elektrischen Isolierstoffen zu vermeiden*] = sealer, sealant

~**organ** *n*, Absperrvorrichtung *f*, Abschlußorgan, Abschlußvorrichtung, (Betriebs)Verschluß *m* [*Talsperre*] = shut-off unit

~**organ** *n* am Bohrloch, Preventer *m*, Schieber *m* = (blowout) preventer, cellar control

~**schicht** *f* [*in der Anstrichtechnik*] = sealing coat

~**schieber** *m*, Abschlußschieber = stop valve, shut-off ~ [*A device to open or close a flow*]

Absperrung *f*, Abdämmung, Absperren *n*, Abdämmen [*durch den Bau einer Talsperre*] = damming

~ [*Straße*] = barricading, barrier

~ **des Durchflusses**, Abschließen *n* ~ ~ [*in einer Rohrleitung*] = closing a flow

~ **durch Sandsäcke** → Abdämmung ~

Absperr|ventil *n* = shut-off valve

~**vorrichtung** *f* → Absperrgerät

~**vorrichtung** *f* → Absperrorgan *n*

Abspiegeln *n*, Spiegelprobe *f* [*Rohr*] = inspection with the aid of mirrors

absplitten, abdecken mit Splitt, verteilen von Splitt = to grit, to spread grits

Absplitten *n*, Abdecken, Splittverteilung *f*, Splittstreuen = gritting, chipping, grit blinding, grit spreading [*deprecated: blinding, dressing*]

~ **von Hand**, Handsplittverteilung *f* = hand chipping, ~ gritting, ~ grit blinding, ~ grit spreading

absplittern = to splinter, to spall, to chip, to break up

Absplittern *n*, Absplitterung *f* = splintering, spalling, chipping, breaking up

Absplitt|fahrzeug *n*, Splittstreufahrzeug = grit spreading vehicle

~**-LKW** *m*, Splittstreu-LKW = gritting lorry (Brit.); ~ truck (US)

Absprache *f* unter Bietern → → Angebot *n*

abspreizen → (ab)bölzen

Abspreizung *f* → (Ab)Bölzung

Absprengen *n* = chipping off, splintering ~ [*Alteration of the surface of a construction material*]

(ab)sprießen → (ab)bölzen

Absprießung *f* → (Ab)Bölzung

Absprießwinde *f*, bewegliche Quersteife *f*, bewegliche (Graben)Steife, beweglicher Stempel *m* = trench brace [*extensible metal support*]

Abspritzen *n* **mit Hydromonitoren**, Gewinnung *f* ~ ~, hydraulische Gewinnung, Abspritzgewinnung, Druckstrahlbaggerung = hydraulic mining, (~) sluicing, hydraulicking, jetting

abspritzen mit Schlauch = to hose down

Abspülungsfaden *m* (Geol.) = rain rill, wet-weather ~

Abspulwagen *m* = spool-cart [*for laying out prestressing wire*]

Abspunden *n* = cutting off by sheet pile wall(s)

abstampfen = to tamp

Abstand *m* = spacing, distance

~, **Spiel** *n* = clearance

~ **der Auflagermitten** = centre-to-centre distance (Brit.); center-to-center ~ (US)

~ ~ **Strebepfeilerachsen** [*Vielfachbogensperre*] = buttress spacing; ~ centers (US); ~ centres (Brit.)

~ **von Oberkante Schar bis Unterkante Drehkranz** [*Motor-Straßenhobel*] = circle throat clearance

~ **zwischen Laufrollenflansch und Bolzenauge** [*Gleiskette*] = pin boss-roller flange clearance

~**anzeige** *f* [*Kübelhub beim Radschrapper*] = clearance indicator

~**block** *m* = spacer block

~**halter** *m* = distance piece, spacer, separator

~**halter** *m* [*für Bewehrung*] = bar spacer

Abstechdrehbank *f* = slicing lathe, trimming ~

Abstechen *n* [*Abtrennen von Metallteilen mittels Werkzeugstahles auf einer Drehbank*] = slicing, parting off

Abstechstahl *m* = parting off tool

abstecken → abpfählen

(Absteck)Pflock *m*, **Absteckpfahl** *m* = (setting-out) peg, ~ stake [*A short pointed rod driven into the ground to mark a line or a level. A nail driven into the top of the peg usually shows the position of the point*]

Absteckung *f* → Abpfählen *n*

~ **von Geraden, Richtungsübertragungen und**|**oder Kurven** = large-scale setting out

Absteck(ungs)|**arbeiten** *fpl* → Abpfählen *n*

~**element** *n* = setting out element

~**größe** *f* = setting out value

abstehender Schenkel *m* [*Winkelstahl*] = outstanding flange

absteifen → (ab)bölzen

Absteifung *f* → (Ab)Bölzung

absteigender Ast *m*, **abfallender** ~ [*Diagramm*] = descending branch

Abstell|**anordnung** *f*, **Abstellsystem** *n* [*für Flugzeuge*] = aircraft parking system

~**druckknopf** *m* = stop button

abstellen [*eine arbeitende Maschine*] = to stop, to switch off

Abstellen *n*, **Parken** [*Fahrzeug*] = parking

Abstell|**fläche** *f*, **Abstellplatz** *m*, **Standplatz** = (hard-)standing, hardstand, parking apron [*airfield*]

~**fläche** *f*, **Parkfläche** [*für Kraftfahrzeuge*] = parking area

~**gleis** *n* = storage track

~**hahn** *m*, **Absperrhahn**, **Abschlußhahn** = shut off cock, stop (~)

~**platz** *m* → Abstellfläche *f*

~**raum** *m*, **Abstellkammer** *f* [*gehört zu einer Wohnung oder dgl. und dient zum Abstellen nicht ständig benötigter Gegenstände*] = store chamber

~**-Solenoid** *n* = shut-off solenoid

~**ventil** *n* = cut-off valve

Abstieg *m*, **Fisch**~ [*stromabwärts gerichtete Fischwanderung*] = downstream fish migration, ~ migration of fish

~ [*Taucher*] = going down

~**stollen** *m*, **Zugangstollen**, **Fensterstollen**, **Seitenstollen**, **Tagesstollen**, **Hilfsstollen für den Bau**, **Baustollen** = (construction) adit, access Aunnel, (side) drift, approach adit

Abstoßvorrichtung *f*, Förderwagen-⁓ (⚒) = shunt back
Abstreichbohle *f* → Abgleichbohle
abstreichen, abgleichen, abziehen [*Beton*] = to strike off
Abstreichen *n* → Abgleichen
Abstreicher *m* → Bandreiniger
Abstreich|gut *n* [*Schwimmschicht*] = skimmings
⁓**vorrichtung** *f* [*Abheben der Schwimmschicht eines Klärbeckens*] = skimmer, scum-collector
Abstreifdichtung *f* = wiping seal
Abstreifen *n* [*Tiefbohrtechnik*] = joint failure
Abstreifer *m* → Bandreiniger
⁓ = strike-off (screed)
Abstreif|ring *m* [*Kolbenstange*] = scraper ring
⁓**- und Glättbohle** *f* → Glättbohle
⁓**- und Glättelement** *n* → Glättbohle *f*
abstreuen mit Salz = to salt
(Ab)Streusplitt *m* → (Ab)Decksplitt
Abstreuung *f*, Bestreuung, Mineral⁓ [*Dachpappe*] = mineral surface, ⁓ granules
Abstufung *f*, Korn⁓ = gradation, grading
Abstufung *f* [*Mineralmasse*] = gradation, grading
Abstufungsschichtung *f* (Geol.) = graded bedding
abstumpfen [*Werkzeug*] = to dull, to blunt
⁓, anrauhen, aufrauhen [*Straßendecke*] = to skidproof
⁓ **der freien Säuren** = to attenuate the effects of free acids
abstumpfender Stoff *m*, Streugut *n* [*wird auf Fahrbahnen aufgebracht, um ein Rutschen von Fahrzeugen zu verringern bzw. zu verhindern*] = abrasive (material), gritting ⁓, abrading ⁓, grit(s)
Abstumpfung *f*, Anrauhung, Aufrauhung [*Straßendecke*] = skidproofing
Absturz *m* [*eine Gefällestufe in offener oder geschlossener Freispiegelleitung zur Energieverzehrung des Wassers*] = drop

⁓, Absturzschacht *m* [*Bestandteil eines Abwassernetzes*] = drop manhole [*A shaft in which sewage falls from a sewer to a lower level*]
⁓, Abbruch *m* (Geol.) = fall
⁓, festes Wehr *n*, Absturzbauwerk *n* [*Aus Mauerwerk oder Beton gebaut, mit Tosbecken und Sohlsicherung gegen Kolk*] = fixed weir
⁓, Bergrutsch *m*, Bergsturz *m* = debris-slide, rock avalanche, fall of rock, rock slide, rock-fall, rock slip
⁓**becken** *n*, Sturzbett *n* = (downstream) apron
⁓**schacht** *m* → Absturz *m*
⁓**-Sicherungsapparat** *m* = safety belt, ⁓ harness
Abstütz|arm *m* → Stützarm
⁓**bock** *m*, A-Rahmen *m*, A-Bock = A-frame
abstützen [*zur Verbesserung der Standsicherheit von Baggern, Kränen oder Ladern*] = to stabilize
Abstützen *n* → Abstützung *f*
⁓ **mit Hydraulikstützen** [*Kran; Bagger; Lader*] = stabilizing by hydraulic jacks
Abstütz|fuß *m* → Stützfuß
⁓**gerät** *n* = shoring equipment
⁓**platte** *f* → Stützfuß *m*
Abstützung *f* [*Schaufellader; Bagger; Kran*] = stabilizers, outriggers
⁓, Abstützen *n* = shoring [*Giving temporary support with shores to a building being repaired, altered, or underpinned, or a trench*]
Abstützwinde *f* = stabilizing jack
Abszisse *f*, Auftraglinie *f* = abscissa
Abszissen|achse *f*, X-Achse = axis of abscissa, horizontal scale
⁓**element** *n* = element measured along axis of x
Abteil *n* = compartment
Abteilung *f* → Stau⁓
⁓ [*in einer Werkstatt*] = bay, department [*in a machine shop*]
⁓ **Wasserbau des Ministeriums** = Water Department
Abteilungsleiter *m* = department head

Abteufanlage — Abtragmaterial

Abteuf|anlage *f* = sinking installation, ~ plant
~arbeit *f* = sinking work
~-Baustelle *f* = sinking site
~bohrhammer *m* → Abteufhammer
~bühne *f* [*Ist eine den Schachtquerschnitt ausfüllende Stahlplatte, an starken Seilen hängend, von der aus der Schacht beim Abteufen ausgemauert und mit Aus- und Einbau versehen wird. Sie hängt etwa 50 m über der Schachtsohle*] = sinking platform
~einrichtung *f*, Abteufvorrichtung = sinking equipment
(ab)teufen, herunterbringen, niederbringen, schlagen = to drill, to sink, to put down, to bore
(Ab)Teufen *n*, (Ab)Teufung *f*, Niederbringen, Herunterbringen, Schlagen = sinking, drilling, boring, putting down
abteufende Bohrung *f*, im (Ab)Teufen befindliche Bohrung = drilling well
Abteuf|förderung *f*, Haufwerkabförderung = hoisting muck, ~ spoil
~gerüst *n*, Abteufördergerüst, Fördergerüst für Schachtaushubmassen, (Übertage-)Arbeitsgerüst = sinking head gear
~hammer *m*, Abteufbohrhammer [*schwerer Bohrhammer mit Selbstumsatz für Bohrungen nach unten*] = sinker (drill)
~kübel *m*, Ladekübel, Förderkübel für Schachtaushubmassen, Aufzugkübel = kibble, sinking bucket, bowk, skip, hoppit
~lüfter *m*, Grubenlüfter, Schachtventilator *m* = pit ventilator
~pumpe *f*, Senkpumpe, Bohrlochpumpe [*mehrstufige Vertikalpumpe für eine nutzbare Förderhöhe bis zu 60 m, die mit einem Elektromotor auf einem Gleitrahmen zusammengebaut ist*] = sinking pump
~rohr *n* = sinking tube
~schacht *m* = shaft during sinking
~turm *m* = sinking derrick

~verfahren *n*, Abteufmethode *f* = sinking method, drilling ~, boring ~, putting down ~
~vorrichtung *f*, Abteufeinrichtung = sinking equipment
~winde *f* = sinking winch
abtorfen = to dig peat
Abtrag *m*, Abbruch *m*, Abtragung *f*, Abbrucharbeit(en) *f(pl)*, Abtragarbeit(en) [*Bauwerk*] = wrecking, demolition work, pulling down (work), taking down (work)
~ → Abraum~
~, Abtragen *n*, Einschnitt *m*, Abtragung *f* [*Bodenentnahme über zukünftigem Planum*] = cut(ting)
~, Mutterboden~ = stripping (of top soil), top soil stripping
~arbeit(en) *f(pl)* → Abtrag *m* [*Bauwerk*]
~böschung *f*, Aushubböschung, Einschnittböschung, Böschung im Einschnitt, Böschung im Abtrag, Böschung im Aushub = slope of cutting, cutting slope, excavation slope
Abtragegerät *n* [*Betonsteinindustrie*] = block take-off
abtragen = to carry [*as of an idler a conveyor belt*]
~, abräumen, abdecken = to strip [*overburden or thin layers of pay material*]
~ → abbrechen [*Bauwerk*]
~ von Fels unter Wasser = to cut rock under water
Abtragen *n* [*Bauwerk*] → Abbruch *m*
~, Abtragung *f*, Abraumabtrag *m*, Abräumen, Abdecken = stripping overburden
~, Abtragung *f*, Abräumen, Abdecken [*Abraum oder ausbeutbare Schichten*] = stripping
~, Abtrag *m*, Einschnitt *m*, Abtragung *f* [*Bodenentnahme über zukünftigem Planum*] = cut(ting)
Abtrag|kanal *m* → Kanal im Einschnitt
~kubatur *f*, Einschnittkubatur = volume of cut(ting)
~material *n* [*Erdbau*] = cut(ting)

Abtragstation — (Ab)Walztemperatur

~station *f*, (Trag)Rollenstation [*Förderband*] = idler set

Abtrag *m* **und Auftrag** *m* = cut and fill

Abtragung *f*, **Abtrag** *m*, **Abbruch** *m*, **Abbrucharbeit(en)** *f(pl)* [*Bauwerk*] = pulling down (work), taking ~ (~), wrecking, demolition work

~, Abtragen *n*, Abraumabtrag *m*, Abräumen, Abdecken = stripping overburden

~, Abtragen *n*, Abdecken, Abräumen [*Abraum oder ausbeutbare Schichten*] = stripping

~, Einschnitt *m*, Abtragen *n*, Abtrag *m* [*Bodenentnahme über zukünftigem Planum*] = cut(ting)

~ → → Denudation *f*

~ durch Wind → Deflation *f*

Abtragungs|fläche *f*, Einebnungsfläche, Abtragungsebene *f* = degradation level, ~ plain, ~ surface, denudation ~, destruction ~, erosion(al) ~, gradational ~, peneplain

~**gebirge** *n* = mountain of erosion

~**gelände** *n*, Einschnitt *m* = cut(ting)

~**küste** *f* = abrasion coast

~**terrasse** *f*, Denudationsterrasse (Geol.) = denudation terrace, structural rock bench

(Abtrag)Wand *f*, Abbauwand, (Stein-)Bruchwand = (quarry) face

~ [*die mit einem Hochlöffelbagger abgebaggert wird*] = face excavated by shovel

Abtransport|fahrzeug *n*, Abfuhrfahrzeug = haul-away vehicle

~**gerät** *n*, Abfuhrgerät [*Erdbau*] = haul-away equipment

~**wagen** *m*, Abfuhrwagen [*Erdbau*] = haul-away wagon

Abtreiben *n* → Abräumen

Abtrennen *n* der n-Paraffine durch Adduktbildung mit Harnstoff = dewaxing with urea

abtreppen [*terrassieren des zu überschüttenden Geländes bei großen Neigungen*] = to bench, to terrace

Abtreppung *f*, Abtreppen *n*, Terrassieren, Terrassierung = benching, terracing

Abtreppungsriß *m* [*Bauwerksenkung*] = angular crack

Abtrieb *m* [*Schaufelradbagger*] = drift

~**flansch** *m* = output flange

~**leistung** *f* = engine output

~**sonnenrad** *n* = output sun gear

~**untersetzung** *f* = final drive transmission

~**welle** *f*, Ausgangswelle = output shaft

~**welle** *f*, Getriebe~ [*Doppelkettenförderer*] = head shaft

~**(zahn)rad** *n*, Ausgangs(zahn)rad = output gear

Abtrift *f* [*Motor-Straßenhobel*] = side pull

Abtritt *m*, Klosett *n*, Abort *m* = closet [*A privy, water closet, etc.*]

~**anlage** *f*, Abortanlage, Klosettanlage = toilet

~**becken** *n* → Abortbecken

~**grube** *f*, Abortgrube, Fäkaliengrube, Latrinengrube, Senkgrube = f(a)eces pit, privy ~

~**rost** *m* = door scraper

Abtropf|kegel *m*, Abtropfkonus *m* = dewatering cone

~**platte** *f* → Ablaufplatte

~**sieb** *n*, Entwässerungssieb = draining screen

~**stein** *m*, Stalaktit *m* = stalactite

~**wasser** *n*, Tropfwasser = dripping water

abtun, abfeuern, zünden [*Sprengschüsse*] = to fire (the blast)

Abtun *n*, Abfeuern *n* Zünden *n* [*Sprengschüsse*] = firing (the blast)

abwalmen = to hip

(ab)walzen = to roll

Abwalzen *n*, Druckverdichtung *f*, Walzverdichtung, Walzkompression *f*, Walzung, Walzen = (compaction by) rolling

~ mit Gummiwalzen → → Walze *f*

~ mit Stahlmantelwalzen → → Walze *f*

(Ab)Walz|prüfung *f* [*zum Nachprüfen der Walzverdichtung f*] = proof rolling

~**temperatur** *f* [*Straßenbau*] = rolling temperature

abwandern — Abwasserfachmann 50

abwandern, auswandern (⚒) = to move
Abwärme *f*, Abhitze *f* = waste heat
~**heizung** *f*, Abhitzeheizung [*Durch Zwischenschaltung von Wärmespeichern wird der fast wertlose Abdampf von Fördermaschinen, Dampfhämmern usw. für Heizzwecke genutzt. Auch Auspuffgase sind verwendbar*] = heating by waste heat
~**kessel** *m*, Abhitzekessel = waste heat boiler
~**rückgewinnung** *f*, Abhitzrückgewinnung = waste heat recovery, ~ ~ reclamation
~**verwertung** *f*, Abhitzeverwertung = waste heat utilization
Abwärts|bewetterung *f* = downcast ventilation
~**bohren** *n* (⚒) = down hole drilling
~**bohrer** *m* = down hole drill
~**fahrgeschwindigkeit** *f* **des Hakens** [*Tiefbohrtechnik*] = lowering speed of the hook
~**fluß** *m* = down-flow
~**förderbandanlage** *f* = downhill conveyor, ~ conveyer, downward ~
~**förderung** *f* = downward conveying, downhill ~
~**schrappen** *n* [*Radschrapper*] = downhill loading, downward ~
~**schürfen** *n* **mit dem Brustschild** = downhill (bull)dozing, downward ~
~**schweißen** *n*, Abwärtsschweißung *f* = downhand welding, downhill ~
~**siebung** *f* = downhill screening
abwaschbar = washable
Abwaschbarkeit *f* = washability
Abwaschen *n* = washing down
Abwasser *n*, Abwässer *npl* [*Oberbegriff für Niederschlagwasser und Schmutzwasser*] = sewage
~ **nach der Behandlung**, behandeltes Abwasser = effluent sewage
~**ableitung** *f*, Abwasserabführung *f* = sewage disposal, disposal of sewage [*The act of disposing of sewage by any method*]
~**abzweig** *m*, Abwasserendstrang *m* = dead end sewer
~**anfall** *m* = sewage flow
~**anlagen** *fpl* = sewerage works [*Comprehensive term, including all constructions for collection, transportation, pumping, treatment and final disposal of sewage*]
~**art** *f* = type of sewage
~**auslauf** *m* **ins Meer** = ocean outfall
~**bandrechen** *m*, (Abwasser)Siebband *n* = belt sewage screen
~**behandlung** *f* = sewage treatment [*Any artificial process to which sewage is subjected in order to remove or so alter its objectionable constituents as to render it less offensive or dangerous*]
~**behandlungs- und -beseitigungsanlage** *f* = sewage-treatment works, sewerage-treatment ~ [*Treatment plant and means of disposal*]
~**belüftung** *f*, Abwasserdurchlüftung = sewage aeration
~**beschaffenheit** *f* = state of sewage
~**-Betonkanal** *m* → Betonkanal
~**-Betonrohr** *n*, Beton-Abwasserrohr = concrete sewer pipe
~**bewässerung** *f* → Bewässerung mit Abwasser
~**biologie** *f* = sewage biology
~**chemie** *f* = chemistry of sewage
~**chlorung** *f* = sewage chlorination
~**desinfektion** *f*, Abwasserentkeimung *f* = sewage disinfection, ~ sterilization
~**-Doppeldüker** *m*, Doppel-Abwasserdüker = depressed sewer with two pipes, inverted siphon ~ ~ ~ ~
~**düker** *m* = inverted siphon sewer, depressed ~ [*A sewer, often crossing beneath a valley or a watercourse, which runs full or under greater than atmospheric pressure because its profile is depressed below the hydraulic grade line*]
~**durchlüftung** *f* → Abwasserbelüftung
~**einleitung** *f* **in den Vorfluter** = sewage flow into the receiving water, ~ ~ ~ ~ recipient
~**fachmann** *m* = sewage engineer

Abwasserfaulraum — Abweichung

~faulraum m → Faulraum
~feinrechen m = fine sewage screen
~fett n = sewage grease
~filter m, n = sewage filter
~fischteich m = sewage stock pond
~gas n = sewage gas
~grobrechen m = coarse sewage screen
~heben n, Abwasserhebung f = raising of sewage
~hebewerk n, Abwasser-Überpumpwerk = sewage raising plant
~installation f [eines Gebäudes] = soil stack installation
(~)Kanal m, Leitungskanal = conduit-type sewer
(~)Kanal m aus Ziegelmauerwerk = brick (conduit-type) sewer
~klärung f = sewage clarification
~last f [ist die Zahl der an Ortsentwässerungen angeschlossenen Einwohner und der dazu gehörenden Einwohnergleichwerte der Industrie, die auf 1 l/s der mittleren Niederwasserführung des als Vorfluter dienenden Flusses kommen] = sewage pollution value
~leitung f = sewer [A pipe or conduit, generally closed, but normally not flowing full, for carrying sewage and other waste liquids]
Abwasserleitungs|bemessung f = sewer design
~gefälle n = sewer grade, ~ slope
~größe f = sewer size
~hydraulik f = hydraulics of sewers
~leistung f = sewer capacity
~querschnitt m = sewer section
~sohle f = sewer invert
Abwasser|menge f = sewage quantity
~netz n, Kanal(isations)netz, Schwemmkanalisation f = sewer network
~organismus m = pollutional organism
~pilz m = sewage fungus
~probe f = sewage specimen
~pumpe f = sewage(-handling) pump
~pumpwerk n = sewage pumping station, ~ ~ plant
~rechen m = sewage screen

~recht n = sewage legislation
~reinigung f = sewage purification [1. The removal or mineralization of all putrescible organic matter and the removal of all infectious and offensive matter; 2. Loosely used for sewage treatment]
~reinigungsanlage f, Abwasserreinigungswerk n, Kläranlage, Klärwerk = sewage purification plant
~rohr n, Kanal(isations)rohr = sewer pipe, sewage ~
~rohrleitung f = pipe sewer
~rücknahme f = recirculation of sewage
~sammler m = main sewer
~sammlung f = sewage collection
~scheibenrechen m = disc sewage screen (Brit.); disk ~ ~ (US)
~schlauch m = sewage hose
~schrägrechen m = inclined sewage screen
(~)Siebband n, Abwasserbandrechen m = belt sewage screen
~siebrechen m = mesh sewage screen
~stabrechen m = bar sewage screen
~stollen m = sewer tunnel
~stoß m, Beschickungsmenge f = sewage batch, ~ dose
~technik f = sewage engineering
~teich m = oxidation pond
~trommelrechen m = drum sewage screen
~-Überpumpwerk n, Abwasserhebewerk = sewage raising plant
~untersuchung f = sewage investigation
~verregnung f, Beregnung mit Abwasser = sewage sprinkling
~verwertung f = sewage utilization
~wesen n = sewerage and sewage treatment and disposal
~wirtschaft f = sewage practice
abwechseln, wechsellagern (Geol.) = to alternate
Abweichung f = deviation
~ des Bohrlochs gegen das Einfallen der Schichten fpl = upstructure deflection of the (bore) hole

Abweichungs|keil m, Ablenkkeil, Richtkeil [*zum Ablenken von Bohrlöchern*] = whip-stock
~messer m, Neigungsmesser [*Tiefbohrtechnik*] = driftmeter, drift recorder, inclinometer
~messung f, Neigungsmessung [*Tiefbohrtechnik*] = verticality survey, deviation measurement
~richtung f, Azimuth m der Abweichung [*Tiefbohrtechnik*] = direction of the angle of deviation, azimuth ~ ~ ~ ~, ~ ~ ~ ~ ~ deflection [Brit. deflexion]
~winkel m, Ausschlagwinkel = angle of deflection, ~ ~ deviation
Abweis(e)blech n = flashing [*A strip of flexible metal which excludes water from the junction between a roof covering and another surface (usually vertical)*]
abweisen [*Wasser durch Dichtungsmittel*] = to repel
Abweis(e)streifen m = flashing [*Usually made of flexible metal, but sometimes made of roofing felt*]
Abweis|mittel n = repellent
~stein m, Prellstein, Leitstein = spur post
Abweisungsblech n, Prallblech [*im Fangkessel beim Druckluftbetonförderer*] = (central) baffle
abwerben [*Arbeitskräfte*] = to „pirate"
abwerfen, ausschütten = to discharge by dumping, to dump
Abwetter n (⚒) = exit mine air
Abwickel|anlage f, Abwickelvorrichtung f, Abwickler m = unwinding installation, winding off ~
~maschine f [*Draht*] = swift
abwickeln = to unwind, to wind off
~ [*z.B. eine Wölbungsleibung*] = to develop
Abwick(e)lung f, Ablauf m [*z.B. eines Versuches*] = procedure
~, Abwickeln n = unwinding, winding off
~ der Wölbungsleibung f, ~ ~ Wölbungslaibung = development of soffit

Abwick(e)lungslänge f = developed length
~ = unwinding length, winding-off ~
Abwiegen n, Abwiegung f, (Ver)Wiegen, (Ver)Wiegung = weighing
Abwind m = katabatic wind
Abwinkelung f des (Förder)Bandes = superelevation of the belt
Abwitterung f, Verwitterung = weathering, process of rock wastage
Abwitterungs|hang m = erosion slope
~produkte npl → Verwitterungskruste f
Abwölbung f (Geol.) = downward bowing
Abwurf m, Ausschütten n = dumping
~band n, Absatzband = discharge belt
~halbmesser m, (Aus)Schütthalbmesser = dumping radius
~höhe f, Kipphöhe, Ausleerhöhe, (Aus)Schütthöhe [*Schaufellader*] = dumping height
~reichweite f [*Schaufellader*] → (Aus-)Schüttreichweite
~rolle f → Ausschüttrolle
~vorrichtung f → Ausschüttvorrichtung
~wagen m → → Bandförderer
~winkel m [*Schaufellader*] → (Aus-)Schüttwinkel
~zeit f [*Schaufellader*] → (Aus)Schüttzeit
Abwürgen n [*Motor*] = stalling [*engine*]
abyssisch, plutonisch, grundlos, abgrundtief (Geol.) [*In der Tiefe, besonders in der Tiefsee gebildet*] = abyssal, abysmal, plutonic
abyssisches Gestein n, Tiefengestein, plutonisches Gestein, Plutonit m, subkrustales Gestein, Intrusivgestein, älteres Eruptivgestein = abysmal rock, intrusive ~, (igneous) plutonic ~, abyssal ~
Abzapfbrunnen m, Entlastungsbrunnen [*in artesisch gespanntem Grundwasser*] = bleeder well
Abzäunung f = fencing

Abzieh|band *n*, Abzugband = drawing belt conveyor, ~ ~ conveyer
~band *n* **im Haldentunnel** = tunnel conveyor (belt), ~ conveyer (~)
~bild *n* = transfer, decal
~bohle *f* → Abgleichbohle [*Betondeckenfertiger*]
~bohle *f*, Glättbohle [*Schwarzbelageinbaumaschine*] = screed plate
abziehen, glätten [*mit Kelle*] = to smooth by trowe(l)ling
~, abstreichen, abgleichen [*Beton*] = to strike off
~ [*von einem Silo oder Bunker*] = to draw
~ → (ab)schürfen
~ [*Schweißtechnik*] = to drag [*the electrode along the joint line*]
Abziehen *n* → Abgleichen
~, (Ab)Schürfen, (Ab)Schürfung *f*, Flachbaggerung, Flachbaggern = skimming [*the removal of the top layer or of the irregularities in the ground surface*]
~ des Flotationsschaumes unter Teilsortenbildung = multiple-increment test
~ (~ Gutes), Abzug *m*, Gutabzug [*von einem Silo oder Bunker*] = drawing
Abzieh|lager *n* = detachable bearing, removable ~, withdrawable ~
~latte *f*, Kardätsche *f* = levelling board (Brit.); leveling ~ (US); screed(ing) ~
~schleppe *f*, (Planier)Schleppe [*zur Beseitigung von Bodenunebenheiten*] = drag
~schnecke *f*, Abzugschnecke = drawing screw (conveyor), ~ ~ conveyer
~silo *m*, Abzugsilo = live storage silo
~stein *m*, Polierstein [*zum Feinnacharbeiten (= Polieren) von Werkzeugschneiden*] = oilstone
~verteilgerät *n* → zweiachsiger Motor-Straßenhobel *m*
~vorrichtung *f*, Abzugvorrichtung = drawing device
~walze *f* = roller screed

Abzug *m*, Abziehen *n* (des Gutes), Gutabzug [*von einem Silo oder Bunker*] = drawing
~band *n* → Abziehband
~drehteller *m* [*Gießereimaschine*] = rotating discharge plate
~graben *m*, Ablaufgraben, Abflußgraben = field ditch
~kanal *m*, Durchlaß *m*, Drumme *f* = culvert
~kanalrohr *n*, Durchlaßrohr, Drummenrohr = culvert pipe
abzugrabende Sohle *f* [*Druckluftgründung*] = bottom to be excavated
Abzug|rolle *f* (⚒) = finger opening
~schieber *m* → Schieberschütz(e)
~schnecke *f* → Abziehschnecke
~silo *m* → Abziehsilo
~vorrichtung *f*, Abziehvorrichtung = drawing device
~vorrichtung *f* (⚒) = pusher
Abzweig *m* = branch
~, Abzweigung *f*, Streckenast *m* [*Bahn*] = branch line
~, Abzweigung *f*, Gab(e)lung *f*, Abzweigstelle *f* = bifurcation
Abzweiger *m* [*Kanalisationssteinzeug*] = junction
Abzweig|kasten *m* = junction box
~leitung *f* = branch line
~mast *m* = branch mast
~punkt *m*, Abzweigstelle *f* = branch point
~rohr *n* [*In Rohrleitung eingebautes Formstück, um Durchflußmenge selbsttätig auf zwei Stränge unter verschiedenen Winkeln zu verteilen oder Seitenzufluß aufzunehmen*] = branch pipe
~stelle *f* → Abzweig *m*
~stelle *f*, Abzweigpunkt *m* = branch point
~stück *n*, Verteilerblock *m* = junction block [*of a power train*]
Abzweigung *f* → Abzweig *m*
~, Abzweig *m*, Streckenast *m* [*Bahn*] = branch line
~ = junction
Abzweig|ventil *n* = branch valve
~winkel *m* = bifurcation angle

Achat|mörser m, Achatreibschale f = agate mortar
~schwüle f = thunder egg
~textur f = agate texture
Achs|abstände mpl [*Förderband*] = ... ft centres
~anordnung f = axle arrangement
Achse f, Mittellinie f = axis; centre line (Brit.); center line (US)
~, Rad~ = axle
~ mit Planetenantrieb, Planetenantriebachse = planetary drive axle
Achsen|drehung f = axis rotation
~kreuz n, Achsensystem n, Koordinatensystem, Koordinatenkreuz = coordinate axis, ~ system (of axes)
~öl n = axle oil
~richtung f = axis direction
~system n → Achsenkreuz n
~zahl f, Anzahl der Achsen = number of axles
Achsgehäuse n = axle housing
Achsialdruck m = axial pressure
achsiale Vorspannung f = axial pretensioning
Achsial|kolbenpumpe f = axial plunger pump
~kraft f = axial force
~schub m, Seitenschub, achsrechter Schub = axial thrust
~zug m = axial tension
Achslager n = axle bearing
Achslast f, Achsdruck m, Last pro Achse, Druck pro Achse = axle load
~äquivalenzfaktor m = traffic equivalence factor
~gruppe f = axle load weight group
~verteiler m = weight transfer unit, ~ ~ attachment, traction control
~waage f = axle load scale
Achs|maß n [*bei der Maßordnung*] = unit spacing
~pendelzapfen m = axle housing trunnion
~pfahl m, Achspflock m = centre line peg (Brit.); center ~ ~ (US)
~querträger m = axle beam
achsrechter Schub m, Seitenschub, Achsialschub = axial thrust

Achsrohr n = axle shaft
Achsschenkel m = steering knuckle
~bolzen m = axle king pin, steering knuckle ~
Achs|stand m = axle base, ~ spacing
~stand m [*Fehlname*] → Radstand
~strebe f = axle strut
achsversetzter Kegeltrieb m, Hypoidantrieb = hypoid axle
Achs|welle f = axle shaft
~wellen-Sonnenrad n = axle sun gear
~zählung f = axle count
~zapfen m, Achsstumpf m, Achsstummel m, Achshals m [*Teil der Achse, der im Lager ruht*] = journal
Achteckfliese f = octagonal tile
achteckiger Querschnitt m, Achteckquerschnitt = octagonal section
Achteckschornstein m = octagonal chimney
achtfacher Hammerseilzug m = eight-part hammer line
achtfaches Einscheren n [*Seile*] = eight line reeving [*ropes*]
achtflügeliger Lüfter m = eight blade fan
Acht-Gang-Getriebe n = eight-speed gearbox
Achtkant|querschnitt m = octagonal section
~stahl m = octagonal steel
Acht|lochziegel m = eight-hole brick
~-Stunden-Schicht f = eight-hour shift
~-Stunden-Tag m = eight-hour day
~-Tage-Tachograph m = eight-day tachograph
achttourig = eight-speed
Achtwertigkeit f = octovalence
Achtzyklon(staub)abscheider m = eight-cyclone dust collector
ackerfähig, kultivierbar, anbaufähig, urbar, kulturfähig, anbauwürdig [*Land*] = arable [*land*]
Acker|gleiboden m = tilled field gley soil
~(gleis)kettenschlepper m, Ackerbulldog m auf (Gleis)Ketten = agricat

(tractor) (US); crawler-mounted farm tractor

Ackermann-Decke *f* [*Rippendecke mit Tonhohlsteinen. Steinmaße $10 \times 30 \times 25$ cm*] = Ackermann ribbed floor

Acker|schlepper *m*, Ackerbulldog *m*, Ackertrecker *m*, Ackertraktor *m* = farm tractor, agricultural ~

~walze *f* = agricultural roller, farm ~

Acrylglas *n* = acrylic glass

~tafel *f* = acrylic sheet

Acrylharz *n* = acrylic resin

~emulsion *f* = acrylic resin emulsion

addierte Gewichtsprozente *npl*, Gesamtdurchgang *m* in % [*Siebtechnik*] = % wt. cumulative, cumulative weight percentage

~ Prozente *npl*, ~ Hundertteile *mpl*, *npl* = cumulative percentage

Addition *f* **von Kräften**, Zusammenzählung *f* ~ ~ = summation of forces

Additionsregel *f*, Zusammenzählregel, Additionsgesetz *n* = rule of addition

Ader|gneis *m*, Arterit *m* = vein gneiss

~marmor *m* = veined marble

Adhäsion *f*, scheinbare Haftfestigkeit *f*, Anhaftung *f*, Flächenanziehung, Grenzflächenkraft *f*, Haftfähigkeit = adhesion, adhesiveness, bond, interface strength

Adhäsionsbeiwert *m*, Adhäsionszahl *f* = adhesion coefficient

adhäsionsfördernder Zusatzstoff *m* → Netzhaftmittel *n*

Adhäsions|grenze *f* = adhesion limit

~strahl *m*, vollangesaugter Strahl [*Hydraulik*] = adherent nappe, weir ~

~verbesserer *m* → Netzhaftmittel *n*

adiabatische Linie *f*, Druck-Volumenkurve *f*, Adiabate *f*, adiabatische Kurve = adiabatic line

adiabatischer Versuch *m*, adiabatische Probe *f*, adiabatische Prüfung *f* = adiabatic test

~ Vorgang *m* = adiabatic process

Adinol *m*, Bergkiesel *m* = adinole

Adjustagemaschine *f* [*Rohrherstellung*] = finishing machine

adsorbieren = to adsorb [*To condense and hold a gas on the surface of a solid, particularly metals. Also to hold a mineral particle within a liquid interface*]

Adsorption *f*; Wasseranlagerungsvermögen *n*; Gasanlagerungsvermögen = adsorption

Adsorptions|druck *m* = adsorption pressure

~fähigkeit *f* = adsorption power

~geschwindigkeit *f* = adsorption rate

~gleichgewicht *n* = adsorption equilibrium

~hülle *f* = adsorptive hull

~wärme *f* = heat of adsorption

~wasser *n* → Benetzungswasser

adsorptiv gebundenes Wasser *n* → Benetzungswasser

Adular *m* = adularia, moonstone [*A colourless and translucent variety of orthoclase usually in pseudo-orthorhombic crystals*]

Advektionsnebel *m*, Mischungsnebel = advection fog

Adventivkegel *m* = adventive cone

AEA-Beton *m*, belüfteter Beton, Luftporenbeton, Bläschenbeton = air-entraining concrete, air-entrained ~

AEA-Mörtel *m*, belüfteter Mörtel, Luftporenmörtel, Bläschenmörtel = air-entraining mortar, air-entrained ~

aërisch, äolisch = aerial, eolian, windblown, wind-deposited, wind-drift, wind-laid

~ Verwitterungsschutt *m*, äolischer ~ = eolian mantle rock

aerober Abbau *m* [*der Abwasserstoffe*] = aerobic decomposition

Aerobier *m* = aerobe

Aerophotogrammetrie *f* → Luftbildmessung *f*

Aerosol|filtration *f* = aerosol filtration

~teilchen *n* = aerosol particle

Aerotriangulation *f* = aero-triangulation

Aethan n, C_2H_6 oder $CH_3 \cdot CH_3$ = ethane

Aethanol n = ethanol

Affe m, Erdwinde f, Hand(seil)winde, mechanische Hand(trommel)winde, handbetriebene (Trommel)Winde = windlass

Affinität f, Verbindungsstreben n = affinity, liking

Affinitätsachse f = axis of affinity

Agens n = agent

Agglomerat n (Geol.) = agglomerate

Agglomerationsgebiet n → Ballungsgebiet

Agglomerieranlage f = agglomerating plant, ~ installation

Aggregat n, Maschinen~ [*Bezeichnung für einen gekuppelten Maschinensatz, z.B. Pumpen-Aggregat = Elektromotor mit Kreiselpumpe, Druckluft-Aggregat = Dieselmotor mit Kompressor, usw.*] = unit, set-up, assembly

~, Strom(erzeugungs)~, Stromerzeuger~, Elektro~, Elektrozentrale f [*besteht aus Antriebsmaschine mit gekuppeltem Generator*] = generating set, electric generator (set), electric set

aggressiv, angreifend = aggressive

aggressive Lufteinflüsse mpl = aggressive atmospheric conditions

Aggressivbeständigkeit f = resistance to aggressive agencies

Aggressivität f, Aggressiveinfluß m = aggressive action

Aggressivwasser n → Schadwasser

ägyptisches Kreuz n → Taukreuz

AH m → A-Horizont m

Ähnlichkeits|bedingung f = condition of similarity, ~ ~ similitude

~gesetz n = law of similarity, ~ ~ similitude

~versuch m, Ähnlichkeitsprobe f, Ähnlichkeitsprüfung f [*wasserbauliches Versuchswesen*] = similarity test, similitude ~

A-Horizont m, Oberboden m, AH, Auslaugungshorizont, Auslaugungsschicht f [*humoser Oberboden (im humiden Klima meist ein eluvialer, d.h. ausgewaschener Horizont)*] = A-horizon, zone of removal

Airdox-Verfahren n = Airdox system [*A coal bursting system in which the pressure is applied to the steel tube by a pipeline from a surface compressor*]

Airy'sche Spannungsfunktion f = Airy's stress function

Akanthus m, Säulenlaubwerk n = acanthus

~blatt n = acanthus leaf

Akkordarbeit f = piecework; gyppo (US)

Akkordeontür f = accordion door

Akkordlohn m, Stücklohn = piece wage

~satz m, Stücklohnsatz = piece rate

Akku|blättchen n = battery separator

~ladegerät n = battery charging equipment

~lok(omotive) f, Akkumulatorenlok(omotive) = (storage-)battery loco(motive), accumulator ~

Akkumulation f, Ablagerung f, Ablagerungsvorgang m [*Der Vorgang des Stoffabsatzes auf dem Festland, im Süßwasser (Seen, Flüsse) und im Meer*] = sedimentation (process), deposition (~)

~, Ablagerung f, Ablagerungslandschaft f = alluvion, alluvium

Akku(mulatoren)|batterie f, Speicherbatterie, Sammler m, Akku(mulator) m = accumulator battery, storage ~

~karre(n) f, (m) = battery operated industrial truck

~raum m = battery room

~-Triebwagen m = battery railcar

a-Kohle f → aktive Kohle

Aktenaufzug m = office dumbwaiter

Akt m **Höherer Gewalt** = Act of God

Aktinolith m, Strahlstein m = actinolite

Aktionshalbmesser m, Fahrbereich m [*Fahrzeug*] = (cruising) range, operational ~, operating ~, travel(l)ing ~

Aktionshalbmesser — Alaungips

~ einer **Erdölförderbohrung**, Entölungshalbmesser = drainage radius of an oil well, radius of oil drainage
Aktions|radius *m* → Bereich *m*
~**turbine** *f* → Druckturbine
Aktivator *m*, Beschleuniger *m*, Anreger *m* = activator
aktive Kohle *f*, a-Kohle, Aktivkohle, A-Kohle = activated carbon, active ~, ~ (char)coal
~ **Rankine'sche Zone** *f* = active Rankine zone
aktiver Bruch *m* = active rupture
(~) **Erddruck** *m* = active soil pressure, ~ earth ~
~ **Rankine'scher Druck** *m* = active Rankine pressure
~ ~ **Zustand** *m* = active Rankine state
aktivieren, anregen = to activate
Aktivieren *n*, Anregen = activating
aktivierter (Kohlen)Hobel *m* = activated (coal) plough, Huwood slicer, oscillating plough
Aktivitätsindex *m* [*Zement*] = index of activity
Aktivstreifen *m* [*Bei der Dehnungsmessung werden jeweils zwei elektrische Widerstandsmeßstreifen verwendet, von welchem der eine so angebracht ist, daß er die gesamten Verformungen der betreffenden Meßstrecke des zu messenden Körpers mitmacht (Aktivstreifen), während der zweite Meßstreifen (Kompensationsstreifen) im Prüfkörper so befestigt ist, daß er zwar alle Temperaturänderungen mitmacht, aber sich frei bewegen kann*] = active ga(u)ge
Akustik *f* = acoustics
~**bauweise** *f*, Schallschluckbauweise = acoustic construction (method)
~**decke** *f*, Schallschluckdecke = acoustic ceiling
~**deckenplatte** *f*, Schallschluckdeckenplatte = acoustic ceiling board
~**folie** *f*, Schallschluckfolie = acoustic foil, sound absorbing ~
~**ingenieur** *m* = acoustics engineer
~**matte** *f*, Schallschluckmatte = acoustic blanket
~**platte** *f*, Schallschluckplatte = acoustic board, ~ tile
~**putz** *m*, Schallschluckputz = acoustic plaster
~**tafel** *f*, Schallschlucktafel = acoustic panel
~**wandplatte** *f*, Schallschluckwandplatte = acoustic wallboard
akustische Eigenschaft *f* = acoustic property
~ **Energie** *f*, Schallenergie = sound energy
~ **Fernmessung** *f* = acoustic remote measurement
~ **Warnung** *f* = acoustic warning
akustischer Anzeiger *m* = acoustic indicator
~ **Baustoff** *m*, Schallschluckbaustoff = acoustic material
~ **Dehnungsmesser** *m* = acoustic strain gauge
akustisches Lot *n*, Echolot, Schall-Peilgerät *n* = echo sounder, sonic ~, sound-ranging altimeter, sonic log
~ **Signal** *n*, Hörsignal = audible signal, sound ~, acoustic ~
akzessorisches Mineral *n*, zufälliges ~, Übergemengteil *m*, *n* = accessory mineral
Alabaster *m* = alabaster
~**bruch** *m* = alabaster quarry
Alarm|anlage *f*, Alarmvorrichtung *f* = alarm system
~**klingel** *f* = call bell, alarm ~
Alaun *m*, Kalium-Aluminium-Sulfat *n*, schwefelsaure Tonerde *f* [*Fällungsmittel in der Wasseraufbereitungstechnik*] = potassium aluminum sulfate (US); ~ aluminium sulphate (Brit.)
~ [*schwefelsaures Doppelsalz des Aluminiums mit einem anderen Metall*] = alum
~**gips** *m*, Keene-Zement *m*, Keene'scher Zement, Hartalabaster *m*, alaunisierter Gips, Marmorzement, Marmorgips = Keene's cement

Alaunschiefer — allgemeiner Geschiebetrieb 58

~schiefer m = alum shale
Albedo n **des Schnees** = albedo of snow
Albertschlag m, Gleichschlag, Parallelschlag, Längsschlag = Lang lay
Albino-Bitumen n, helles Bitumen = albino asphalt (US); ~ bitumen (Brit.)
Albit m, $NaAlSi_3O_8$ = albite, white schorl
~**-Chloritschiefer** m = albitite
Albuminoidstickstoff m, Eiweißstickstoff = albuminoid nitrogen
Aldehydharz n = aldehyde resin
Alge f = alga
Algen|anwuchs m [*Schiff*] = growth of algae
~**bekämpfung** f = algae control
~**bekämpfungsmittel** n, Algengift n = algaecide
~**pilz** m = phycomycete
~**wachstum** n, Algenwuchs m = algae growth
Alhidade f = alidade, sighting rule
A-Linie f, Einflußlinie der Auflagerkraft = influence line of reactions, A-line
Alit n = alite [*A term given by Tornebohm in 1897 to the principal constituent of portland cement, later identified as C_3S*]
Alkali n = alkali
~**alterung** f = alkali ag(e)ing
~**augit** m = soda pyroxene
~**basalt** m = alkali basalt
~**beseitigung** f = alkali elimination
alkalibeständig = alkali resistant
Alkali|beständigkeit f [*Anstrichfilm*] = alkali resistance
~**blau** n = alkali blue
alkaliempfindlicher Betonzuschlag(stoff) m = (alkali) reactive material in concrete
Alkali|feldspat m = alkali fel(d)spar
~**fluorid** n = alkali fluoride
~**gehalt** m, Alkalianteil m = alkali content
~**gestein** n = alkali rock
~**granit** m = alkali-granite
~**hornblende** f = soda amphibole
~**kalkbasalt** m = calc-alkali basalt
~**kalkgestein** n, Kalkalkaligestein = calc-alkali rock
~**lösung** f = alkali solution
~**metall** n [*die Leichtmetalle Kalium und Natrium*] = alkali metal
~**phase** f = alkali phase
alkalische Reaktion f **der Zuschläge mit dem Zement**, Alkali-Zuschlagstoff-Reaktion [*Umsetzung zwischen reaktionsfähiger Kieselsäure der Zuschlagstoffe und dem Alkali des Zements*] = alkali-aggregate reaction (or expansion)
~ **Salmiakfumarole** f = alkaline ammoniacal fumarole
alkalischer Boden m = alkaline soil
alkalisches (Ab)Beizmittel n, Ablaugmittel = soda-based paint remover
~ **Lösungsmittel** n = alkaline solvent
~ **Mineralwasser** n = alkaline mineral water
Alkali|syenit m = alkali syenite
~**umwälzung** f = alkali circulation
~**-Zuschlagstoff-Reaktion** f → alkalische Reaktion der Zuschläge mit dem Zement
Alkoholverfahren n, Alkoholmethode f = alcohol slug process [*for increasing oil recovery*]
Alkoven m = alcove [*Recess in a room*]
Alkydharz n = alkyd resin
~**lack** m, Alkydkunstharzlack = alkyd (resin) based varnish
Allee f = alley
allein eingebauter Packer m = single packer
Allein|druck m, reiner Druck = direct compression
~**vertretung** f = sole agency
Allerfeinstes n [*Siebtechnik*] = finest sizes
Alles-Rot-Zeit f = all-red period
allgemeine Formel f = general formula
allgemeiner Geschiebebetrieb m, vollkörniger ~ = general bed load transport

allgemeines Abscheren n, allgemeine Abscherung f = general shear failure

Allgemeinversorgung f = general supply

Alligator|spitze f [*Ogivale aufklappbare Spitze des Simplexpfahlrohres, die sich unter dem Druck des Stampfens öffnet*] = alligator point

~-(Zahn)Ringdübel m, (Zahn)Ringdübel „Alligator", Alligator-Dübel, Alligatorring = toothed-ring

allmähliche Erweiterung f **eines Flusses** = progressive widening of a stream

~ **Lastaufbringung** f, ~ **Belastung** = progressive loading

~ **Verengung** f **eines Flusses** = progressive narrowing of a stream, ~ contracting ~ ~ ~

allochthon → bodenfremd

allotriomorph = allotriomorphic, anhedral

~-**körnig** = allotriomorphic-granular, granitic, granitoid, xenomorphic-granular

Allrad|antrieb m = all-wheel drive

~**antrieb-Vierradschlepper** m, Vierrad-Allradantrieb-Schlepper = four-wheel drive tractor

~-**Autoschütter** m = all-wheel dumper

~-**Bremse** f = all-wheel brake

~-**Druckluftbremse** f, Allrad-Preßluftbremse = all-wheel (compressed-)air brake, ~ pneumatic ~

~-**Fahrzeug** n = all-wheel drive vehicle

~**kipper** m = all-wheel tipping lorry (Brit.); ~ ~ truck (US)

~-**Lenkung** f = all-wheel steer(ing)

~-**Lenkung** f, Vierrad-Lenkung = four-wheel steer(ing)

~-**Lenkung** f, Sechsrad-Lenkung = six-wheel steer(ing)

~-**Planeten-Radantrieb** m = all-wheel planetary wheel drive

~**räumer** m = all-wheel dozer

~**schlepper** m, Allradtrecker m, Allradtraktor m = all-wheel tractor

allseitiger Druck m = confining pressure

Allsichtkanzel f → Rundblickkanzel

Allstrommotor m, Universalmotor = a.c./d.c. motor, universal (electric) ~, universal AC/DC ~

alluvial, abgelagert, angeschwemmt = alluvial

alluviale Seife f (Geol.) = alluvial placer

alluvialer Boden m, Ablagerungsboden, Alluvialboden = alluvial soil

~ **Laterit** m = alluvial laterite

~ **Sand** m = alluvial sand

~ **Verwitterungsschutt** m = alluvial mantle rock

Alluvial|-Einpressung f = alluvium grouting

~**gold** n, Seifengold, Waschgold = alluvial gold, placer

~**kegel** m, Alluvialfächerschutt m = alluvial cone, ~ fan (accumulations) [*Water-deposited material at the place where a mountain stream debouches on to a plain. An alluvial cone carries the suggestion of finer material than a debris cone, which is a mixture of all sizes and kinds*]

~**ton** m = alluvial clay

Alluvion n → Alluvium

Alluvionsschleier m, injizierter ~ = grouted alluvium

Alluvium n, Anlandung f, Anschütte f, Stromablagerung, Verlandung f, angeschwemmtes Land, Alluvion n (Geol.) = alluvium, alluvial deposit [*Deposits of mud and silt commonly found on the flat lands along the lower courses of rivers*]

Allvibration f [*Doppelvibrationswalze*] = all-vibration

Allwetter|flugplatz m = all-weather airfield

~-**Piste** f, Allwetter-Start- und -Landebahn f = all-weather runway

~**sicht** f = all-weather visibility

~-**Start- und -Landebahn** f → Allwetter-Piste f

~-**Straße** f = all-weather road, ~ highway

Allzweck|-Abbau- und Aufreißhammer *m* → Abbau- und Aufreißhammer
~-Baumaschine *f* = jack-of-all-trades rig
~-Bohr- und Abbauhammer *m*, Bohr- und Abbauhammer, Abbau- und Bohrhammer = demolition and drill hammer
~-Hebezeug *n* = universal lifting gear
~-Kipper *m* [*LKW*] = general-purpose tipper
~-Maschine *f* = universal machine
~straße *f* = all-purpose road, ~ highway
~-Trockner *m* = multi-purpose dryer; ~ drier (US)
Alpen|durchstich *m*, Alpentunnel *m* = Alpine tunnel
~gneis *m* = protogene gneiss, protogenic ~, protogin(e ~)
~granit *m* = Alpine granite
~kalk(stein) *m* = Alpine limestone
~see *m* = Alpine lake
~straße *f* = Alpine road, ~ highway
~tunnel *m*, Alpendurchstich *m* = Alpine tunnel
~vorland *n* = Alpine piedmont
Alpha-Tonerde *f*, Korund *m* = corundum
Alstonit *m*, Bariumaragonit *m* (Min.) = alstonite
Altarm *m* → Altwasser *n*
Alter *n* der Tide, Nacheilen *n* ~ ~, ~ ~ Gezeit, Gezeitenalter, Tide(n)alter = age of the tide, lag ~ ~ ~
~ Mann *m* (⚒) = goaf, waste (area); gob [*Wales*]; condie, cundy [*Scotland*]
älteres Eruptivgestein *n* → plutonisches Gestein
altern = to age
~, verwittern, „wettern" [*Steinzeugindustrie*] = to age, to store
Altern *n*, Alterung *f* = aging (US); ageing (Brit.)
Alternativ|angebot *n* → → Angebot
~bauweise *f*, Alternativbauverfahren *n* = alternative construction method, ~ building ~

~drehzahl *f* = alternative speed, ~ RPM
~projekt *n* = alternative project
~trasse *f* = alternative route, ~ location line
Altersheim *n* = old people's home, aged persons' ~
Alterung *f* → Altern
~ → Rekristallisation *f* [*Schraube; Niet*]
alterungs|beständig → haltbar
~beständiger Straßenteer *m*, Wetterteer = weather-resistant tar
Alterungs|beständigkeit *f*, Haltbarkeit = durability
~schwindung *f* = shrinkage due to ag(e)ing
~versuch *m* durch Freibewitterung, Alterungsprobe *f* ~ ~, Alterungsprüfung *f* ~ ~ = outdoor exposure test
Alt|kupfer *n* = scrap copper
~material *n* = salvage, arisings
~messing *n* = scrap brass
~öl *n*, Abfallöl *n* = used oil
~reifengummi *m* = scrap tyre rubber (Brit.); ~ tire ~ (US)
~sand *m*, gebrauchter Formsand = used sand
~stadt *f* = old quarters
~stadtsanierung *f*, Erneuerung = rehabilitation, clearance
~stahl *m* [*Der aus zerstörten Bauteilen und Bauwerken geborgene Baustahl*] = scrap structural steel
alttertiär = old tertiary
Alt|wasser *n*, toter Arm *m*, Totarm *m*, Altarm = oxbow (lake), dead stream branch; bayou (US)
~wert *m* = salvage value
Aluminat *n* = aluminate
aluminatisch = aluminous
aluminatische Schmelze *f* → Trikalziumaluminat *n*
Alumina(t)zement *m* → Tonerdezement
Alu(minium) *n* = aluminium (Brit.); aluminum (US); alum.
~-Abstützplatte *f*, Alu(minium)-Stützfuß *m* [*eines (Ab)Stützarmes*]

zur Erhöhung der Standsicherheit von Baggern, Kränen und Schaufelladern] = aluminium stabilizer base, ~ outrigger ~, ~ ~ plate (Brit.); aluminum ~ ~ (US)
~abweis(e)blech *n* = alumin(i)um flashing
~anstrich *m* = alumin(i)um paint coat
~arsenat *n* = alumin(i)um arsenate
~auskleidung *f* → Alu(minium)bekleidung
~-Aussichtswagen *m* = alumin(i)um dome car
~balken(träger) *m* = aluminium beam (Brit.); aluminum ~ (US)
~-Bauelement *n* = alumin(i)um unit
~bekleidung *f*, Alu(minium)verkleidung, Alu(minium)auskleidung = alumin(i)um lining
~blättchen *n* = aluminium flake (Brit.); aluminum ~ (US)
~blech *n* = alumin(i)um sheet
~bronze *f* = alumin(i)um bronze
~brücke *f* = alumin(i)um bridge
~chlorid *n*, $AlCl_3$ = alumin(i)um chloride
~dach *n* = aluminium deck (roof) (Brit.); aluminum ~ (~) (US)
~-Dach(ein)deckung *f* = aluminium roofing (Brit.); aluminum ~ (US)
~dachplatte *f* = aluminium deck unit (Brit.); aluminum ~ ~ (US)
~decke *f* = alumin(i)um ceiling
~draht *m* = alumin(i)um wire
~-Druckguß-Legierung *f* = alumin(i)um die-casting alloy (Brit.); aluminum ~ ~ (US)
~-Druckgußteil *m, n* = alumin(i)um die casting
~farbe *f* = alumin(i)um paint
~fassade *f* = alumin(i)um façade, ~ front, ~ face
~fenster *n* = alumin(i)um window
~folie *f* = alumin(i)um foil
~form *f* = aluminium mould (Brit.); aluminum mold (US)
~geländer *n* = aluminum (hand-)railing (US); aluminium ~ (Brit.)
~grieß *m* = alumin(i)um powder

~-Gußlegierung *f* = alumin(i)um casting alloy
~-Gußteil *m, n* = alumin(i)um casting
~hydroxyd *n*, $Al(OH)_3$ = aluminium hydroxide (Brit.); aluminum ~ (US)
~kabelmantel *m* = aluminium cable sheath (Brit.); aluminum ~ ~ (US)
~-Knetlegierung *f* = wrought alloy of alumin(i)um, ~ alumin(i)um alloy
~-Kokillenguß-Legierung *f* = alumin(i)um permanent mo(u)lding alloy, ~ ~ mo(u)ld casting alloy
~legierung *f* = alumin(i)um alloy
~legierungsgußteil *m, n* = aluminium alloy casting (Brit.); aluminum ~ ~ (US)
~legierungsprofil *n* = alumin(i)um alloy (structural) section, ~ ~ (~) shape
~lochdecke *f* = aluminium perforated ceiling (Brit.); aluminum ~ ~ (US)
~lot *n* = alumin(i)um solder
~mantel *m* = alumin(i)um jacket
~-Mehrstoffbronze *f* = alumin(i)um multi-compound bronze
~mulde *f* [*Erdtransportfahrzeug*] = alumin(i)um body
~oxyd *n*, Tonerde *f*, Al_2O_3 = alumina, alumin(i)um oxide
~oxydhydrat *n* → Bauxit *m*
~platte *f* = alumin(i)um plate
~profil *n* = alumin(i)um (structural) section, ~ (~) shape
~pulver *n* [*zerstampfte Aluminiumblättchen; als Pigment für Innen- und Außenanstrich; Treibmittel für Leichtbeton*] = aluminium powder (Brit.); aluminum ~ (US)
~rohr *n* = alumin(i)um pipe
~-Sandguß-Legierung *f* = aluminium sand-casting alloy (Brit.); aluminum ~ ~ (US)
~schalung *f* = alumin(i)um formwork
~schmelzelektrolyseanlage *f* = plant for the electrolytic production of alumin(i)um
~schmelzwerk *n*, Alu(minium)schmelze *f* = alumin(i)um smelting plant

~**seifenfett** n von glänzendem Aussehen und hoher Transparenz = lucid compound

~**silikat** n = alumin(i)um silicate

~**sprengkapsel** f = lead azide alumin(i)um detonator, tetryl-azide ~

~**-Stützfuß** m → Alu(minium)-Abstützplatte f

~**sulfat** n, schwefelsaure Tonerde f = aluminium sulphate (Brit.); aluminum sulfate (US)

~**tragwerk** n [Teil eines Bauwerkes] = aluminium load-bearing member, ~ ~ part (Brit.); aluminum ~ ~ (US)

~**tragwerk** n [Teil eines Gebäudes] = aluminium structure (Brit.); aluminum ~ (US)

~**verkleidung** f → Alu(minium)bekleidung

~**walzprofil** n = rolled alumin(i)um alloy (structural) section, ~ ~ (~) shape

~**walzwerk** n = alumin(i)um rolling plant

~**wand** f = aluminium wall (Brit.); aluminum ~ (US)

~**wellblech** n = corrugated aluminium sheet (Brit.); ~ aluminum ~ (US)

~**-Well(dach)eindeckung** f = corrugated aluminium roofing (Brit.); ~ aluminum ~ (US)

~**wolkenkratzer** m = aluminium skyscraper (Brit.); aluminum ~ (US)

~**wolle** f = alumin(i)um wool

Aluminoszement m → Tonerdezement

aluminothermisches Schweißen n → Thermitschweißen

Alumosilikat n = aluminosilicate, aluminous silicate

Alunit m [tonerdereicher Alaunschiefer] = alunite, alumstone

Alunitisation f = alunitization

Alveolinenkalkstein m = alveolina limestone

Amalgamation f = amalgamation (process)

Amalgamationstisch m = amalgamating table

A-Mast m, **A-Rahmen** m, **A-Bock** m = A-frame

Ambursen|(stau)mauer f, Ambursen-Pfeiler-Plattensperre f, Ambursen-Pfeilerplattensperrmauer, Pfeiler-Plattensperre Bauweise Ambursen, Ambursensperre = Ambursen (type) dam

~**wehr** n [nach dem schwedischen Ingenieur Ambursen genannt. Die offene Bauart besteht darin, daß die Stahlbetonwand kurz unterhalb der Wehrkrone nach Unterwasser aufhört] = Ambursen-type weir

~**zellkerndamm** m [Der Kern besteht aus Stahlbeton mit von oben bis zur Grundfläche führenden senkrechten Zellen, in denen Sicker- und etwaiges Überlaufwasser unschädlich nach Luftseite abgeleitet wird] = cellular-wedge core wall Ambursen dam

Ameisensäure f, Formylsäure = formic acid

amerikanischer Brunnen m → Norton-Brunnen

amerikanisches Pfeilergerät n [Brückenbau] = standard unit steel trestles

Ammongelatine f = ammonia gelatin(e)

Ammoniak n, NH_3 = ammonia

~**wasser** n = crude ammonia liquor

Ammonit m, Ammonsalpetersprengstoff m ohne Nitroglyzerin = ammonite

Ammoniumnitrat n mit großem Schwadenvolumen, Ammonsalpeter m = ammonium nitrate

~**-Heizölgemisch** n als Sprengstoff, AN-Heizölgemisch ~ ~ = AN/FO blasting agent

Ammonsalpeter m → Ammoniumnitrat mit großem Schwadenvolumen

Ammonsalpetersprengstoff m = ammonium nitrate (blasting) explosive

~ **mit Nitroglyzerin**, Donarit m = donarite

~ **ohne Nitroglyzerin**, Ammonit m = ammonite

Ammonsalz n = ammonium salt

amorpher Quarz m, Kieselsäure-Glas n = vitreous silica
amorphes Bleioxyd n → Bleigelb n
Amosit(asbest) m, gelber Hornblendeasbest (Min.) = Amosite [*this name embodies the initials of the company exploiting this material in the Transvaal, viz. the «Asbestos Mines of South Africa»*]
Amphibol m = amphibole
Amphibolit m = amphibolite
Amphitheater n = amphitheatre, amphitheater
Amplitude f, Schwing(ungs)weite f, Ausschlag(weite) m, (f), Vibrationsamplitude = amplitude
anaerober Abbau m → Schlammfaulung f
Anaerobier m = anaerobe
Anaglyphen|bild n = anaglyph
~karte f = anaglyph map
Analog(ie)|-Rechenmaschine f, Analog(ie)-Rechengerät n = analogy computer
~-Verfahren n = analogy method
Analyse f = analysis
Analysenwaage f = analytical scale
Anatexis f, Palingenese f, Einschmelzungsmetamorphose f = anatexis, refusion
Anbau m, Gebäude~, Erweiterungsbau = annex
~-Arbeitsgerät n [*Fehlname*] → Anbaugerät
~-Aufreißer m = scarifying attachment
~bagger m, (Anbau)Baggergerät n = digging attachment, excavating ~
~-Besenwalze f, Anbau-Bürstenwalze, Anbau-Fegewalze, Anbau-Kehrwalze = rotary brush attachment, rotating ~ ~
~-Bohrgerät n, Bohr(-Anbau)gerät, Bohr-Zusatzgerät = drill attachment
anbauen (✗) [*eine Schicht, besonders Nachfallpacken*] = to hold
anbaufähig → ackerfähig
~ [*Straße*] = subject to further development

Anbau|gerät n → Austauschgerät
~-Grabenbagger m, Anbau-Grabenaushubmaschine f = trench digging attachment
~kompressor m = compressor attachment
~-Planiervorrichtung f = level(l)ing attachment
~platte f [*für hydraulische Frontgeräte, z. B. an einem Schlepper oder an einer Zugmaschine*] = plate for attaching hydraulic front attachments
~schürfkübel m [*beim Motor-Straßenhobel*] = graderscraper
~-Seilwinde f = cable winch attachment
~-Streugutverteiler m, Anbaustreuer m = tailboard gritter, ~ abrasive (material) spreader
~-Tiefaufreißer m = ripper attachment
~vorrichtung f → Austauschgerät n
~werkzeug n → Austauschgerät n
~-Winde f = winch attachment
anbauwürdig → ackerfähig
(an)bieten = to tender
(an)bietende Firma f → Angebot n
Anbinde|kette f, Anschlagkette, Anhängekette [*Kranarbeit*] = slinging chain
~mittel n, Anschlagmittel, Anhängemittel [*Kranarbeit*] = slinging means
Anbinden n [*von frischem Zementmörtel oder Beton an erhärteten Zementmörtel oder Beton*] = placing fresh material on hardened material [*cement mortar or concrete*]
~, Anschlagen, Anhängen [*Verbindung von Last und Lasthaken an Kranen durch den Anbinder mittels Anbindemittel*] = slinging
Anbinder m, Anschläger [*Kranarbeit*] = (crane) slinger, ~ rigger, craneman
Anbindeseil n, Anschlagseil, Anhängeseil [*Kranarbeit*] = slinging rope
Anblatten n, Einblatten [*Verbindung zweier Kanthölzer durch ein Blatt*] = halving

Anbohren n, Vorbohren [*Der Anfang einer Bohrung*] = collaring
~ **einer Lagerstätte**, ~ ~ **Fundstätte**, ~ **eines Vorkommens**, ~ **eines Vorkommnisses**, **Anritzen** n ~ ~ = drilling in
Anbohr|meißel m = starting bit, starter ~
~stahl m = starter steel, starting ~
anböschen = to bank against, to fill and grade
Anbringen n **einer Kraft**, Kraftanbringung f = application of a force
Anbruchsgebiet n = region where an avalanche starts
andecken [*Böschung mit Mutterboden*] = to cover a slope with topsoil
~ **mit (Rasen)Soden** → besoden
Anderton-Schrämlader m, Anderton-Schrämlademaschine f = Anderton shearer-loader
Änderung f **der Stromrichtung** f = change of direction of the current
Andesin n (Min.) = andesine
Andesit m = andesite [*A very small-grained, dark-grey, often porphyritic rock of volcanic origin, containing phenocrystals of fel(d)spar and darker minerals, but never quartz*]
~asche f = andesite ash
Andrangwasser n, Einbruchwasser = inrush water
Andreaskreuz n, Kreuzstreben fpl, Kreuzband n, Abkreuzung f, Kreuzgebälk n = diagonal struts, St. Andrew's cross, saltier cross bars, cross stays
andrehen → anwerfen
Andrehkurbel f, Anwerfkurbel, Startkurbel, Anwurfkurbel, Anlaßkurbel = (manual) starting crank
Andruck m [*Radschrapper*] = down pressure
~ [*Hammerbohrmaschine*] = feed pressure
Andrücken n [*Eimerleiter eines Baggers durch die Winde*] = downcrowd
Aneinanderkleben n [*Von Leimen in Filmform wird gefordert, daß benachbarte Lagen gleichen oder verschiedenen Stoffs während der Lagerung kein unerwünschtes Aneinanderkleben unter dem Einfluß von Wärmeplastizität oder von Hygroskopizität erfahren*] = blocking
Anemometer n, Windmesser m = anemometer
Aneroidbarometer n, Dosenbarometer, Federbarometer, Aneroid n, Metallbarometer = aneroid barometer
Anfahr|beschleunigung f = starting acceleration
~(dreh)moment n, Anlauf(dreh)moment, Anlaß(dreh)moment, Anzugsmoment = starting torque
anfahren, **antreffen** [*eine Formation beim Bohren*] = to encounter
~, **antransportieren**, **anliefern** = to haul to (the site), to transport ~ (~ ~)
~ **mit LKW**, **antransportieren** ~ ~ = to transport to with lorries (Brit.); to truck to (US)
Anfahren n, Anlauf m, Anlaufen n = (machine) start-up, (~) start(ing), run-up
~ **auf maximale Leistung** = run-up to maximum power
~ **einer Pumpe** = priming
Anfahr|geschwindigkeit f = starting speed
~kessel m = starting up boiler
~punkt m, Drehmomentwandler-Abwürgemoment n = torque converter stall point
~punkt m = stall point
~straße f, Anfahrweg m, Anfuhrstraße, Anfuhrweg [*zur Baustelle*] = access road, haul ~
~strom m, Anlaufstrom, Anlaßstrom = starting current
~wandlung f, Anlaufwandlung = stall torque ratio
~zeit f, Anlaufzeit, Anlaßzeit = starting time, ~ period
Anfallgespärre n, Anfallgebinde [*Walmdach*] = hip rafters, angle ridges, angle rafters [*The rafters forming hips. The jack-rafters meet on them*]

anfällig — Anfressung

anfällig = vulnerable, liable to attack
Anfall|punkt *m* [*Walmdach*] = hip ridge
~sparren *m* = hip rafter, angle ~, angle ridge
Anfänger *m*, Kämpfer(stein) *m* = impost
anfängliche Vorspannung *f* = initial prestress
Anfangs|aufwölbung *f* = initial camber, ~ hog(ging)
~bedingung *f* = initial condition
~belastung *f* = initial loading, early ~
~beton *m* [*Mit Flurfördermitteln eingebrachter erster Beton bei Wasserbauwerken*] = initial concrete
~dehnung *f* = initial strain, early ~
~druck *m* = initial pressure, early ~
~druckfestigkeit *f* = initial compressive strength, early ~ ~
~(er)härtung *f* = initial hardening, early ~
~festigkeit *f* = initial strength, early ~
~feuchtigkeit *f*, Anfangsfeuchte *f* = initial moisture
~förderung *f* [*Erdölgewinnung*] = initial production
~geschwindigkeit *f* = initial speed
~gewicht *n*, Ursprungsgewicht = original weight
~haftung *f*, Anfangsverband *m* = initial bond, early ~
~härtung *f*, Anfangserhärtung = initial hardening, early ~
~isochrone *f* = initial isochrone
~lage *f*, Anfangsstellung = initial position
~last *f* = initial load
~luftgehalt *m* = initial air content
~schwindung *f*, Anfangsschwinden *n* = initial shrinkage, early ~
~spannung *f* = initial stress, early ~
~spannungszustand *m* = initial state of stress, early ~ ~ ~
~spitzenförderung *f* [*Erdölgewinnung*] = flush production
~stadium *n* = initial stage, early ~
~tagesförderung *f* [*Erdölgewinnung*] = initial daily production
~tragkraft *f*, Klemmlast *f* (⚒) = setting load, initial load-bearing capacity
~verschiebung *f* = initial displacement, early ~
~volumen *n* = initial volume
~vorspannung *f* = initial prestress, early ~
~wassergehalt *m* = initial moisture content
~wert *m* = initial value, early ~
~winkel *m* = initial angle
~zusammendrückung *f* = initial compression, early ~
~zustand *m* = initial state, early ~
Anfeuchtanlage *f* für Flugasche = humidifying plant for fly ash
anfeuchten, befeuchten = to moisten, to humidify
Anfeuchten *n*, Anfeuchtung *f*, Befeuchten, Befeuchtung = moistening, humidification, humidifying
~ → Anfeuchtung *f* [*Betonnachbehandlung*]
Anfeuchter *m*, Anfeucht(ungs)maschine *f*, Befeuchter, Befeucht(ungs)maschine = humidifier, humidifying machine, moistening machine
Anfeuchtstoff *m* = moistening agent, humidifying ~
Anfeuchtung *f*, Feuchthalten *n*, Anfeuchten, Feuchthaltung, Naßhaltung, Befeuchtung, Naßhalten [*Betonnachbehandlung*] = wet job-site after-treatment, ~ ~ curing, moist ~, water ~
Anfeucht(ungs)anlage *f*, Befeucht(ungs)anlage = humidifying installation, ~ plant, moistening ~
anflanschen = to flange-mount
Anflanschen *n* = flange-mounting
Anflug *m*, Lande~ = approach
~, Ofen~, Verschmauchung *f* = kiln scum
~befeuerung *f* = approach lighting
anfressen, korrodieren, angreifen = to corrode
Anfressung *f*, Angriff *m*, Korrosion *f*, Angreifen *n*, Anfressen = corrosion

Anfressungsbeständigkeit — Angebot

Anfressungsbeständigkeit *f*, Korrosionsbeständigkeit = corrosion resistance
Anfuhr *f*, Antransport *m*, Anlieferung *f* = haulage, hauling (operations)
~-LKW *m*, Anfuhr-Last(kraft)wagen *m* = delivery truck (US); ~ lorry (Brit.)
~straße *f* → Anfahrstraße
~weg *m* → Anfahrstraße *f*
Angaben *fpl* = data
angebaut = attached

Angebot *n* — bid; proposal (US); tender, offer (Brit.)
  Abgabetermin *m*, Submissionstermin — closing date (for receipt of tenders), deadline for submission of bids, tendering date
  Absprache *f* unter Bietern, Preisabsprache — combining to eliminate competition, price fixing, collusion among bidders or (tenderers)
  Alternativ~ — alternative ~, alternate ~
  ~ für Entwurf und Bau — "turn-key" bid, all-in tender
  (an)bietende Firma *f* → Bieter
  Angebotsauswertung *f* — evaluation of bids
  Angebotsblankett *n* — form
  Angebotsgegenüberstellung *f* — summary (or schedule) of ~s
  Angebotseröffnungstermin *m*, Submissionstermin, Einreichungstermin — opening date, submission date, tendering date, date set for the opening of tenders
  Angebotseinholung *f*, Ausschreibung — competitive tender(ing) action, bidding procedure; adjudication (US)
  Angebotsposition *f* — bid item
  Angebotssumme *f*, Angebotspreis *m* — ~ sum
  Angebotsunterlagen → Ausschreibungsunterlagen
  Auftragerteilung *f*, Vergabe *f* — contract award, award (or letting) of contract, contract letting
  Auftraggeber *m*, Bauherr *m*, Bauherrschaft *f* — promoter, purchaser, owner, sponsor, employer, client
  Auftragnehmer *m* — successful bidder, ~ tenderer
  Auftragschreiben *n* — letter (or notice) of award
  Ausschreibung *f* → Angebotseinholung
  Ausschreibungsunterlagen *fpl*, Verdingungsunterlagen, Angebotsunterlagen — ~ documents
  Bauherr *m* → Auftraggeber
  beschränkte Ausschreibung *f*, begrenzte Ausschreibung — limited submission, limited public tender action (or tendering)
  Bieter *m*, Bewerber *m*, (an)bietende Firma *f* — bidder, tenderer
  Bietungsgarantie *f* — initial guarantee; bid bond (US)
  Gegen~ — competitive ~
  Haupt~ — base ~
  Leistungsbeschreibung *f*, Vorschreibung — spec(ification)

Leistungsposition *f*, kostenvergütete Position	pay item
Leistungsverzeichnis *n*	schedule of prices, list of bid items and quantities
Massenverzeichnis *n*, Mengenverzeichnis	bill of quantities
Mindestfordernder *m*	lowest bidder (or tenderer)
öffentliche Ausschreibung *f*	open tendering
Preisabsprache *f* → Absprache unter Bietern	
Submission *f*	submission (of competitive tenders)
Submissionstermin *m* → Abgabetermin	
unterbieten	to undercut
Vergabe *f* → Auftragerteilung	
Vertragsblankett *n*	form of contract
Vertragsunterlagen *fpl*	contract documents
Zuschlag ~	winning bid (US); ~ tender (Brit.)
zusätzlich	"extra over"

angeflanscht = flange mounted
angefügte Ebene *f* = tied plain
angegriffenes Fossil *n*, angelöstes ~ = corroded fossil, partly dissolved ~
angehauenes Widerlager *n* = reduced abutment
angehobenes Kabel *n* [*Spannbeton*] = deflected strand, raised ~, ~ cable
angelagertes Wasser *n* → Benetzungswasser
angelenkt = hinged, articulated
~ = pinned, pin jointed [*A description of a beam or a column which has a hinge at the end*]
(angelenkte) Blechplatte *f*, (~) Stahlplatte, (angelenktes) Blechglied *n*, (angelenktes) Stahlglied [*Gliederband*] = hinged steel plate, articulated ~ ~
(~) Holzplatte *f*, (angelenktes) Holzglied *n* [*Gliederband*] = hinged wood(en) slat, articulated ~ ~
(~) Platte *f*, (angelenktes) Glied *n* [*Gliederband*] = hinged plate, articulated ~ [*called "slat" when of wood*]
angelernte Arbeitskräfte *fpl* = semiskilled labour (Brit.); ~ labor (US)
angelernte Arbeiter *m* = semi-skilled worker, ~ workman, improver
angelöstes Fossil *n* → angegriffenes ~

angelsächsisches Gewölbe *n*, normannisches ~, Fächergewölbe, Trichtergewölbe, Palmengewölbe, Strahlengewölbe = fan vaulting
angenommene Last *f*, Lastannahme *f*, rechnerisch vorgesehene Last = assumed load, design ~
angereichert = enriched
angesäte Böschung *f*, eingesäte ~ = seeded slope
angesaugte Luftmenge *f*, (An)Saugleistung *f*, Ansaug(e)menge [*Kompressor*] = actual volume of the cylinder
angesaugter freier Strahl *m*, gesenkter ~ ~ [*Hydraulik*] = apparent free nappe
(an)geschärft [*Grubenstempel*] = sharpened
angeschlossene Bevölkerung *f* = contributary population, population served
angeschütteter Boden *m*, aufgeschütteter ~, aufgefüllter ~, aufgetragener ~ = filled soil
~ Untergrund *m* → aufgeschütteter ~
angeschüttetes Gelände *n* → aufgefülltes ~
angeschweißte Kreisringmembran(e) *f* = welded torus
~ Zahnspitze *f* = weld-on tip

angeschweißter Zahnhalter m = weld-on tip adapter, ~ ~ adaptor
angeschwemmt, abgelagert, alluvial = alluvial
angeschwemmter Seetang m = cast-up seaweed
angeschwemmtes Land n → Alluvium n
(an)gespannter Grundwasserspiegel m = perched (or false) (ground-)water table
angespitzte Stange f = pointed rod
angespülter Flußschutt m = stream wash
angestrahlt [*durch Scheinwerfer*] = floodlit
angetrieben [*z. B. Welle; Zahnrad; Förderbandrolle u. dgl.*] = live, driven
angetriebener Spülkopf m [*Rotarybohren*] = power swivel
angewandte Geologie f = applied geology
angewittert (Geol.) = weathered
angewitterter Geschiebelehm m = mesotil
angezogen [*Schraube*] = tight
Anglesit m, Vitriolbleierz n, $PbSO_4$ = anglesite [*Orthorhombic. Specific gravity 6.2–6.4. Hardness 3. Lustre adamantine and colour white when pure and crystallized. Otherwise may be dull and gray*]
angreifen, beanspruchen = to act
~, korrodieren, anfressen = to corrode
angreifend, aggressiv = aggressive
~, korrosiv, korrodierend, anfressend = corrosive
angreifende Kraft f = acting force
angreifender Stoff m, Schadstoff = aggressive matter, ~ substance
angreifendes Wasser n → Schadwasser
Angrenzer m = adjoining owner
Angriff m [*Tunnel- und Stollenbau. Herstellen der Bohrlöcher, Laden, Schießen, Lüften und Schuttern*] = attack, heading
~, Korrosion f = corrosion
~linie f = line of application
~moment n = applied moment
~punkt m [*einer Kraft*] = point of application
~punkt m → Angriffstelle f

~punkt m **des Erddruckes**, Erddruckangriffpunkt = point of application of active soil pressure, ~ ~ ~ earth ~
~punkt m **des Erdwiderstandes**, Erdwiderstandangriffpunkt = point of application of passive earth pressure, ~ ~ ~ ~ ~ soil ~
~schacht m [*Tunnel- und Stollenbau*] = working shaft
~stelle f, Angriffpunkt m [*Tunnel- und Stollenbau*] = point of attack, ~ ~ heading
Anguß m, Auge n, Nabe f = boss
(an)haften = to bond, to adhere
Anhaftewasser n → Benetzungswasser
Anhaftung f → Adhäsion f
anhaken, einhaken = to hook on
anhalten [*den Lauf einer Maschine oder eines Motors*] = to stop
Anhänge|aufreißer m → Aufreißer
~ausführung f, Schleppausführung = mobile model (trailer type), towing model
~-Bagger m, Schlepp-Bagger = trailer excavator
~-Besenwalze f → Anhänge-Bürstenwalze
~-Betonmischer m, Schlepp-Betonmischer = trailer-type concrete mixer
~-Bodenentleerer m, bodenentleerender Erdtransportwagen m, Bodenentleerer, Schlepp-Bodenentleerer, Bodenentleerer-Anhänger m = bottom-dump wagon, ~ tractor-truck, ~ tractor-trailer
~-Bürstenwalze f, Anhänge-Fegewalze, Anhänge-Kehrwalze, Schlepp-Bürstenwalze, Schlepp-Fegewalze, Schlepp-Kehrwalze, Anhänge-Besenwalze, Schlepp-Besenwalze [*mit 2 hinteren Laufrädern*] = trailer brush, traction-driven ~
~chassis n, Anhängerchassis, Schleppfahrgestell n, Anhängefahrgestell, Anhänger-Fahrgestell, Anhänge(r)-Untergestell, Anhänge(r)-(Grund-)Rahmen m, Schlepp-Untergestell, Schlepp-(Grund)Rahmen = trailer chassis

(Anhänge)Deichsel — Anhänge-Rahmen

(~)Deichsel *f* = tow-bar, tongue
~-Doppelseitenkipper *m*, Anhänge-Zweiseitenkipper, Schlepp-Zweiseitenkipper, Schlepp-Doppelseitenkipper [*Erdbaufahrzeug*] = two-way dump trailer wagon
~-Dreiseitenkipper *m*, Schlepp-Dreiseitenkipper, Dreiseitenkippanhänger [*Erdbaufahrzeug*] = three-way dump trailer wagon
~-Erdförderwagen *m*, Schlepp-Erdförderwagen = trailer wagon
~-Erdhobel *m* → Anhänge-Straßenhobel
~fahrgestell *n* → Anhängechassis *n*
~fahrzeug *n*, Anhänger *m*, Schleppfahrzeug [*Fahrzeuge, die nach ihrer Bauart dazu bestimmt sind von Kraftfahrzeugen gezogen zu werden, werden in 3 Hauptgruppen unterteilt: Vollanhänger, Sattelanhänger und Arbeitsanhänger*] = trailer, towed vehicle
~-Fegemaschine *f* → Anhänge-Kehrmaschine
~-Förderwagen *m*, Schlepp-Förderwagen [*Erdbau*] = wagon
~gerät *n*, Schleppgerät = trailer-type unit, towed ~
~grabenbagger *m*, Schleppgrabenbagger = detachable ditcher
~-(Grund)Rahmen *m* → Anhänge-Chassis *n*
~gummiradwalze *f*, Schlepp-Gummiradwalze, Anhänge-Gummi(reifenvielfach)walze, Schlepp-Gummi-(reifenvielfach)walze = tractor-drawn multi-rubber-tire roller (US); ~ multi-tyred ~ (Brit.); towed-type ~ ~
~-Hinterkipper *m* → Hinterkipp(er)anhänger
~-Kehrmaschine *f*, Anhänge-Fegemaschine, Schlepp-Kehrmaschine, Schlepp-Fegemaschine, Anhänge-Straßenbesen *m*, Anhänge-Straßenkehrmaschine, Schlepp-Straßenbesen, Schlepp-Straßenkehrmaschine = traction-driven rotary sweeper, trailer(-type) ~ ~

~-Kehrwalze *f* → Anhänge-Bürstenwalze
~kette *f* → Anbindekette
~kran *m*, Schleppkran = trailer crane
~-Luftreifenwalze *f*, Schlepp-Luftreifenwalze, Luftreifen-Schleppwalze, Luftreifen-Anhängewalze = pneumatic-tyred tractor-drawn roller (Brit.); pneumatic-tired ~ ~ (US)
~maschine *f*, Schleppmaschine = trailer-type machine, towed ~
~mischer *m*, Schleppmischer, Anhänge-Mischmaschine *f*, Schlepp-Mischmaschine = trailer mixer, fast-towing ~
~mittel *n* → Anbindemittel
~-(Motor)Bodenfräse *f*, Schlepp-(Motor)Bodenfräse = tractor-drawn rotavator, ~ rotary hoe, ~ rotary tiller, ~ soil pulverizer
Anhängen *n*, Anschlagen, Anbinden [*Verbindung von Last und Lasthaken an Kranen durch den Anbinder mittels Anbindemittel*] = slinging
Anhänger *m* → Anhängefahrzeug *n*
~ [*Einschienen-Transportbahn*] = trailer wagon
~ als Prüfwagen für Straßenbrücken = deadweight trailer for testing road bridges
~-Bau *m* = trailer construction
~-Betonpumpe *f* = trailer concrete pump
~-Chassis *n* → Anhängechassis
~-Fahrgestell *n* → Anhängechassis
~geschwindigkeit *f*, Schleppgeschwindigkeit = towing speed
Anhänge|-Radschrapper *m*, Schlepp-Radschrapper, Schrapperwagen *m*, (Anhänge-)Schürf(kübel)wagen, Schlepp-Schürf(kübel)wagen, Schürfwagenanhänger *m*, gezogener Radschrapper = four-wheel(ed) scraper, crawler-tractor-drawn ~, rubber-mounted tractor-drawn ~, hauling ~, pull(-type) ~, towed (four wheel) ~
~-Rahmen *m* → Anhängechassis *n*

∼-Rüttelwalze *f*, Anhänge-Vibrationswalze, Anhänge-Schwing(ungs)walze, Schlepp-Vibrationswalze, Schlepp-Rüttelwalze, Schlepp-Schwing(ungs)walze, Anhänge-Vibrierwalze, Schlepp-Vibrierwalze = trailer (type) vibration roller, ∼ (∼) vibratory ∼

∼-Sandstreuer *m*, Schlepp-Sandstreuer = trailer (type) gritter

(∼)Schürf(kübel)wagen *m* → Schrapperwagen

∼seil *n* → Anbindeseil

∼-Seitenentleerer *m*, Schlepp-Seitenentleerer, Anhänge-Seitenkipper, Schlepp-Seitenkipper, seitenentleerender Erdtransportwagen *m*, Seitenkippanhänger = side-dump waggon

∼-Straßenbesen *m* → Anhänge-Kehrmaschine

∼-Straßenhobel *m*, geschleppter Erdhobel, Anhänge-Wegehobel, Anhänge-Erdhobel, Schlepp-Straßenhobel, Schlepp-Wegehobel, Schlepp-Erdhobel, Anhänge-Straßenplanierer *m*, Schlepp-Straßenplanierer, geschleppter Wegehobel, geschleppter Straßenhobel, geschleppter Straßenplanierer = towed-type grader, pull ∼

∼-Streugutverteiler *m*, Schlepp-Streugutverteiler, Anhänge-Streuer *m*, Schleppstreuer = towed-type gritter

∼verdichter *m*, Schleppverdichter = trailer compactor

∼-Vibrations-Schaffußwalze *f* = sheepsfoot trailer vibrating roller

∼walze *f* → → Walze

∼-Wasserwagen *m*, Schlepp-Wasserwagen = water trailer

∼-Wegehobel *m* → Anhänge-Straßenhobel

Anhäufung *f*, (Gang)Stock *m* (Geol.) = stocklike dike (US); ∼ dyke (Brit.)

Anhebekettenbahn *f* → Schrägkettenbahn

Anheben *n* **der Kabel** = bending up cables, deflecting of strands, deflected-strand technique [*prestressed concrete*]

∼ von Wannen [*Straßenbau*] = raising of sags

Anhebevorrichtung *f*, Kabel∼ [*Spannbeton*] = deflecting device, strand ∼ ∼

AN-Heizölgemisch *n* **als Sprengstoff**, Ammoniumnitrat-Heizölgemisch ∼ ∼ = AN/FO blasting agent

anhydrisch, wasserfrei = anhydrous, anhydric

Anhydrisierungsmittel *n* = dehydrating agent

Anhydrit *m*, wasserfreier Gipsstein *m*, $CaSO_4$ = anhydrite

∼ I *m* → Estrichgips *m*

∼ II *m*, totgebrannter Gips *m*, unlöslicher Anhydrit = deadburnt gypsum

∼binder *m* = anhydrite plaster

∼gips *m* = anhydrous gypsum plaster

∼gruppe *f* = Middle Muschelkalk

∼knolle *f* = anhydritic concretion, concretion of anhydrite

∼region *f* = anhydrite zone

∼schnur *f* = anhydrite band

Anilin|farbe *f* = aniline dye

∼punkt *m* [*DIN 51775*] = aniline point

∼rot *n* = aniline red, anileine, tyraline

∼schwarz *n* = aniline black

∼sulfat *n*, schwefelsaures Anilin *n* = aniline sulphate, anilic ∼ (Brit.); ∼ sulfate (US)

∼violett *n* = aniline violett, regina purple

anisotrope Platte *f* = anisotropic plate

Anisotropie *f* = anisotropy, anisotropism [*The property or state of being anisotropic*]

ankeilen = to wedge two pieces together

Anker *m*, ∼bolzen *m*, Bolzen-, Gesteinsanker, Gebirgsanker [*Tunnel- und Stollenbau*] = roof (suspension) bolt

∼ = anchor [*Any means whereby a building part is held by a masonry mass*]

~ → Ankerstab m
~ = dead man [*A timber, log, or beam buried in the ground for anchorage*]
~ausbau m [*Tunnel- und Stollenbau*] = strata bolting, roof ~, bolted supports
~balken m = anchor beam
~band n = anchor strap
~block m, Verankerungsblock = stay block, anchor ~
~bock m [*Kettenhängebrücke*] = chain chair, ~ standard
~bohrloch n → Ankerloch
~boje f = mooring buoy
~(bolzen) m, Bolzen, Gesteinsanker, Gebirgsanker [*Tunnel- und Stollenbau*] = roof (suspension) bolt
~eisen n, Fugenanker m = tie bar, anchor ~, steel ~ ~
~ende n [*Spannbeton*] = anchored end
~grund m, Ankerplatz m = anchorage
~gurt m = waling (Brit.); wale (US)
~höhenverhältnis n → → Spundwand f
~kammer f, Verankerungskammer [*Hängebrücke*] = anchorage chamber
~kegel m, Verankerungskegel, Ankerkonus m, Verankerungskonus = anchoring cone
~kette f, Spannkette [*Hängebrücke*] = tension chain, tightening ~, anchorage ~
~klotz m, Ankerstein m [*Hafen*] = fixed mooring, deadman
~körper m [*Spannbetonverfahren Freyssinet*] = cylindrical block
~kraft f = anchoring force
~linie f → → Spundwand f
~loch n, Verankerungsloch = anchor hole
~loch n, Ankerbohrloch [*Tunnel- und Stollenbau*] = roofhole
~loch-Bohrgerät n = roof-pinning rock drill
~mutter f, Verankerungsmutter = tie nut
~pfahl m, Verankerungspfahl = stay pile [*A pile driven or cast in the ground as an anchorage for a land tie holding back a sheet-pile wall, etc.*]
~pfeiler m, Verankerungspfeiler = anchorage pier
~platte f [*Wenn die Haftung zwischen Beton und Bewehrung zur Übertragung der Haftspannungen nicht ausreicht, werden die Enden der Stähle mit Ankerplatten versehen. Ankerplatten werden häufig als Endverankerung beim Spannbeton verwendet*] = anchor plate; fixing device [*when used in prestressed concrete*]
~platte f [*Industrie(fuß)boden*] = anchor plate
~riegel m, Ankersplint m = tie cotter
~ring m, Verankerungsring = anchorloop (Brit.); U-bolt (US)
~rohr n = anchoring pipe, ~ tube
~rohr n [*Tiefbohrtechnik*] = surface casing [*serves as an anchor string for blowout preventer equipment*]
~rohrfahrt f = anchor string
~rührer m = anchor agitator
~schacht m [*Hängebrücke*] = anchorage shaft
~scheibe f, Verankerungsscheibe = anchor ring (Brit.); form anchor (US)
~schelle f = clamp, guy ~
~schiene f, Verankerungsschiene = anchor(ing) rail
~schlitz m, Verankerungsschlitz = anchor slot
~seil n, Trosse f, Abspannseil = guy-rope, standing rope, guy cable, (back-)stay cable
~seilabspannung f → Abfangen n mit Trossen
~spannschraube f [*Leitungsmast*] = stay tightener, swivel; slack puller (US)
~splint m, Ankerriegel m = tie cotter
~(spund)wand f → → Spundwand
~stab m, (Zug)Anker m, (Anker)Zugstange f, Ankerstange, Verankerungsstab, Verankerungsstange = anchor bar, ~ tie, holding down rod
~stein m, Ankerklotz m [*Hafen*] = deadman, fixed mooring

Ankerstein — Anlaßschalter

~stein m, Durchbinder m [Hohlmauer] = wall tie closer, through-stone
~wand f → → Spundwand
~wand f → Verankerungswand
~winde f = anchor winch
~zug m → → Spundwand f
(~)Zugstange f → Ankerstab m
~zugverhältnis n → → Spundwand f
ankippen = to tilt up [the screed of a black top paver]
Ankippwinkel m [Schürflader] = bucket angle for raising and carrying a load
ankommender Passagier m = arriving passenger
Ankopplungsmittel n → → zerstörungsfreie Betonprüfung
ankörnen = to centre punch (Brit.); ~ center ~ (US)
Ankörnen n = centre punching (Brit.); center ~ (US)
Ankörner m = centre punch (Brit.); center ~ (US)
Ankörnung f = punch mark; center ~ (US); centre ~ (Brit.)
Ankündigungsschild n = advance direction sign
Ankupplung f, Sattel m = hitch
ankurbeln → anwerfen
Anlage f → Betriebs~
~ [als Maschine oder Maschinengruppe] = machine, set-up, equipment
~ [im Sinne einer elektrischen, pneumatischen oder hydraulischen Anlage einer Maschine oder eines Fahrzeuges und dgl.] = system
~ zur Abschwächung der Zersetzung durch Elektrolyse = electrolysis mitigation system, ~ ~ plant, ~ ~ installation
~kapital n, Investitionskapital = invested capital
~kosten f = initial cost
~-Planung f = plant layout
Anlagerungswasser n → hygroskopisches Wasser
Anlageteil m, n = part of an installation, ~ ~ a plant, ~ ~ a facility
anlandendes Ufer n → Anwachsufer

Anlandevorrichtung f [Kai, wenn genügend Wassertiefe am ausgebauten Ufer vorhanden ist, oder Landungsbrücke, wenn seichtes Wasser überbrückt werden muß] = berthing structure
Anlandung f → Alluvium n
anlassen [Wärmebehandlung von Metallen] = to temper
~, starten = to start
Anlassen n, Starten [Motor] = starting
~ vom Fahrersitz aus = in-seat starting
Anlasser m, Anlaßvorrichtung f = starting gear, ~ apparatus
~bedienung f = starter control
~kupplung f = starter clutch
~ritzel n, Starterritzel = starter pinion
~schlüssel m = starter key
Anlaß|batterie f = starting battery
~(dreh)moment n, Anzugsmoment, Anfahr(dreh)moment, Anlauf(dreh)moment [Motor] = starting torque
~-Elektromotor m = cranking motor, starting ~
~energie f [Anlasser] = starting energy
~flüssigkeit f, Startflüssigkeit = starting fluid
~hilfe f, Starthilfe = starting aid
~kompressor m, Anlaßverdichter m = starting compressor, auxiliary ~
~kraft f = power required for starting
~kraftstoff m = priming fuel
~kurbel f → Andrehkurbel
~(luft)behälter m, Startluftbehälter = starting air vessel
~motor m, Anlaßhilfsmotor, Anwurfmotor, Anwerfmotor, Startmotor = starting engine, cranking ~
~motorgetriebe n = starting engine transmission
~öl n = tempering oil
~patrone f, Startpatrone [Verbrennungsmotor] = starting cartridge
~schalter m, Startschalter = starter switch

~strom m, Anlaufstrom, Anfahrstrom = starting current
~temperatur f, Starttemperatur = starting temperature
~verdichter m, Anlaßkompressor m = starting compressor, auxiliary ~
~vorrichtung f → Anlasser m
~wert m [*Suspension*] = yield value
~zeit f, Anfahrzeit, Anlaufzeit = starting period, ~ time
Anlauf m → Anfahren n
~ [*Säule*] = inverted cavetto
~, Anzug m, Dossierung f = taper, batter
~(dreh)moment n, Anfahr(dreh)moment, Anlaß(dreh)moment, Anzugsmoment [*Motor*] = starting torque
Anlaufen n [*Korrosionsbegriff; wenn in unreiner Luft bei normaler Temperatur erzeugt*] = tarnishing
~ [*Korrosionsbegriff; wenn bei erhöhter Temperatur erzeugt*] = heat-tinting
~ → Anfahren
Anlauf|hafen m = port of call
~strom m, Anfahrstrom, Anlaßstrom = starting current
~wandlung f, Anfahrwandlung = stall torque ratio
~zeit f, Anfahrzeit, Anlaßzeit = starting period, ~ time
~zeit f [*für ein Bauvorhaben*] = lead time
Anlege|aufzug m = mobile self-supporting hoist
~brücke f, Landungssteg m, Ladebrücke, Ladesteg, Landungsbrücke = jetty, wharf, staging [*Usually of timber. Sometimes also called "pier", particularly when of solid construction, and commonly so called in North American ports*]
~brücke f für Schwimmdock = floating dock-berth
anlegen, festmachen [*Schiff*] = to berth
Anlegen n, Festmachen [*Schiff*] = berthing, landing, docking, mooring
Anlege|stelle f, Anlegeplatz m, Festmacheplatz, Schiff~ [*Hafen*] = berth [*Place for mooring*]

~vorrichtung f → Schiffhaltevorrichtung
Anleitungs-Fachmann m, Vorführer m = demonstrator
Anlernling m = trainee
anliefern → antransportieren
Anlieferung f → Anfuhr f
Anlieferungszustand m = state of delivery
anliegender Grundbesitz m = neighbo(u)ring property, adjoining ~
Anlieger m; Anstößer m [*Schweiz*] = riparian owner, ~ proprietor
~beitrag m, Perimeterbeitrag, Erschließungsbeitrag, Anliegerkosten f [*Planungs- und vor allem Erschließungsmaßnahmen können auch Wertsteigerungen von Grundstücken bewirken. In diesen Fällen werden die Grundstückseigentümer zur Leistung von Kostenbeiträgen verpflichtet*] = riparian owner's contribution
~fahrbahn f, Ortsfahrbahn [*Eine Fahrbahn, die neben einer Fahrbahn für den durchgehenden Verkehr liegt und in erster Linie dem Anliegerverkehr dient*] = frontage carriageway
~funktion f [*Straße*] = land service function
~grundstück n = riparian property
~kosten f → Anliegerbeitrag
~rechtwasser n → garantiertes Brauchwasser
~straße f = service street
~verkehr m = no through traffic
~zone f = riparian zone
Anmachholz n = kindling wood
anmachen, mischen = to mix
Anmachen n, Mischen = mixing
Anmach(e)wasser n, Mischwasser = mix(ing) water
~erwärmung f, Mischwassererwärmung = mix(ing) water heating
~prüfung f, Anmach(e)wasserversuch m, Anmach(e)wasserprobe f = mix(ing) water test
~waage f, Mischwasserwaage = water weighing device [*Central concrete mixing plant*]
anmontiert = attached

(An)Näherung f = approximation
(An)Näherungs|annahme f = approximation hypothesis
~formel f, (An)Näherungsgleichung f = approximation formula, ~ equation
~gleichung f, (An)Näherungsformel f = approximation equation, ~ formula
~lösung f = approximate solution
~probe(stück) f, (n) = approximate sample
~rechnung f, Berechnung der (An-)Näherungswerte, (an)näherungsweise Berechnung = approximation
~verfahren n = approximate method
~wert m = approximate value
Annahme f, Hypothese f = hypothesis
~bunker m, Aufnahmebunker, Einfüll(ungs)bunker, Füllbunker = receiving bin
~trichter m → Aufnahmetrichter
annietbar = rivet-on
Anoden|gießmaschine f = anode casting machine
~kupfer n = anode copper
~platten-Chargiermaschine f = anode plate charging machine
~schlamm m = anodic slime
anodische Behandlung f = anodising
~ Oxydation f, Eloxierung f, Eloxieren n, Anodisieren n = anodising, anodic oxidation
~ Schicht f = anodic coating
Anodisieren n → anodische Oxydation f
anordnen [Maschinenteile] = to arrange
Anordnung f [von Maschinenteilen] = arrangement
~ der (Stahl)Einlagen, Armierungsanordnung, Bewehrungsanordnung = reinforcement system
~ im offenen Schacht f = (arrangement for an) open wheel pit setting [reaction turbine]
~ von Grubenstempeln = arrangement of props, ~ ~ posts
~ ~ (Straßen)Leuchten = siting of lanterns, arrangement ~ ~

An-Ort-Bringen n (der Schleppschaufel) = spotting (the dragline bucket)
Anorthit m, Kalkfeldspat m, $CaAl_2Si_2O_5$ = anorthite, lime fel(d)spar
~diorit m = anorthite diorite
An-Ort|-Mischverfahren n → Bodenmischverfahren
~-Schäumen n, An-Ort-Schäumung f = in-situ foaming
~-Verbrennung f = in-situ combustion
anpassungsfähig = flexible
Anpassungsfähigkeit f = flexibility
~ einer Mahlanlage durch automatische Wichtekontrolle = mill circuit flexibility with automatic density control
Anpassungsstück n = adapter, adaptor
anpasten, anteigen = to make a paste, to form into a paste
anpflocken [z. B. Flachrasen] = to peg down
Anplaggen n, Ansoden = sodding
Anprall m, Aufprall [Prallzerkleinerung] = impact
~energie f, Aufprallenergie [Prallzerkleinerung] = impact force
~winkel m, Aufprallwinkel [Prallzerkleinerung] = impact angle
Anpreß|druck m [Motor-Straßenhobel] = blade pressure
~stempel m der Nietmaschine, Gegenhalter m ~ ~, Nietstock m = head cup, stationary rivet die
anrampen = to ramp
anrauhen, abstumpfen, aufrauhen [Straßendecke] = to skidproof
Anrauhung f → Abstumpfung
anregen, aktivieren = to activate
Anregen n, Aktivieren = activating
Anreger m, Beschleuniger m, Aktivator m = activator
Anreicherung f = concentration
~ sehr kleiner Stoffmengen nach dem spezifischen Gewicht = gravity concentration of small quantities of materials
~ von Grundwasser, Grundwasseranreicherung f = ground water recharge

Anreicherungs|horizont *m* → B-Horizont
~schicht *f* → B-Horizont *m*
Anreißstarter *m* = recoil starter
Anrisse *mpl* **zeigen** [*Gestein*] = to show slight breaks
Anrißspannung *f* = initial cracking stress
Anritzen *m* **einer Lagerstätte**, ~ ~ Fundstätte, ~ eines Vorkommens, ~ eines Vorkommnisses, Anbohren *n* ~ ~ = drilling in
anrühren [*Farbe*] = to stir
(an)säen = to seed, to sow
(An)Säen *n*, Besamung *f*, Ansaat *f*, Besämung = seeding, sowing
Ansanden *n* → Versandung
Ansatz *m* = scale, incrustant
~, Kessel~, Kesselstein *m* = boiler scale, ~ incrustant
~, Schutzwulst *f* = protective lug
~ [*Radreifen*] = lip [*of a wheel tyre*]
~bildung *f* = incrustation
~bohrung *f* = lip bore
~punkt *m* = starting point
~ring *m* [*Zementbrennen*] = clinker ring
~rohr *n*, Verlängerungsrohr = extension pipe
Ansaug(e)menge *f*, (An)Saugleistung *f*, angesaugte Luftmenge [*Kompressor*] = actual volume of the cylinder
ansaugen = to suck
Ansaugen *n*, Ansaugung *f* = suction
~, Ansaugung *f* [*Pumpe*] = priming
(An)Saug|höhe *f* = suction lift, ~head
~hub *m*, (An)Saugtakt *m* = suction stroke
~krümmer *m* [*Motor*] = intake manifold, induction ~, inlet ~
~leistung *f*, angesaugte Luftmenge *f*, Ansaug(e)menge [*Kompressor*] = actual volume of the cylinder
~luft *f* = suction air
~öffnung *f* **mit Gliederkappe** [*Lüftung*] = louver
~rohr *n* = suction tube, ~ pipe
~seite *f* [*Pumpe*] = suction part, ~ side

~takt *m*, (An)Saughub *m* = suction stroke
Ansaugung *f*, Ansaugen *n* = suction
~, Ansaugen *n* [*Pumpe*] = priming
anscheren, spleißen = to splice
anschießen [*das Hangende*] = to blow down
Anschlag *m*, Falz *m* = rabbet, rebate
~, ~vorrichtung *f* = stop, dog
~, Nase *f* [*(Gleis)Kettenlaufwerk*] = lug
~ (⚒) = landing [*A point in a shaft at which the cage can be loaded or unloaded with materials or men*]
~, Fenster~ = (window) rabbet, (~) rebate; (~) check [*Scotland*]
~, Tür~ = (door) rabbet, (~) rebate; (~) check [*Scotland*]
~ [*Dammbalkenwehr*] = stop
~block *m* = stop block
~bolzen *m* = trip dog, stop ~
~brett *n*, Anschlagtafel *f*, „schwarzes Brett" = notice board, bulletin ~
~büchse *f* = stop bush
anschlagen [*Stempel*] (⚒) = to tighten
Anschlagen *n*, Anbinden, Anhängen [*Verbindung von Last und Lasthaken an Kranen durch den Anbinder mittels Anbindemittel*] = slinging
Anschläger *m* (⚒) = banksman [*The man at the pit top in charge of banking and shaft signalling*]
~, Anbinder [*Kranarbeit*] = (crane) slinger, ~ rigger, craneman
anschlaggesteuert = dog-controlled
Anschlag|kette *f* → Anbindekette
~laden *m*, Schlagladen [*Fenster*] = hinged (window) shutter
~mittel *n*, Anbindemittel, Anhängemittel [*Kranarbeit*] = slinging means
~mutter *f* = stop nut
~nocken *m* = trip dog
~platte *f* = stop plate
~schalter *m*, Begrenzungsschalter, Endschalter, Grenzschalter = limit switch, stop ~
~scheibe *f* = stop collar
~schiene *f* = stop bar
~schiene *f* → Backenschiene

Anschlagschraube — Anschrauben

~**schraube** *f* = stop screw
~**seil** *n* → Anbindeseil
~**stift** *m* = trip pin, stop ~
~**ventil** *n* = cushion valve
~**vorrichtung** *f*, Anschlag *m* = dog, stop
~**winkel** *m* [*Winkelmaß aus Holz, Eisen, Metall oder Kunststoff für Schlosser, Zimmerleute usw. zum Messen von rechten Winkeln an Arbeitsstücken*] = (try) square
anschließen [*Stab; Träger*] = to connect, to join
Anschließen *n* = connecting
Anschliff *m* (Min.) = polished section
Anschluß *m* = connection [*e. g. connecting beams to girders, beams to columns, girders to columns, etc.*]
~ [*bei Fachwerkstäben*] = connection, joint
~, ~**stück** *n* = connecting piece
~, Strom~ = connection to electrical supply
~ [*Autobahn*] → ~punkt
~ **für Gußeisenrohr an Steinzeugrohr** = increaser
~**armierung** *f* → Anschlußbewehrung
~**bahn** *f*, Anschlußstreifen *m*, Nachbarbahn, Nachbarstreifen [*Schwarz- und Betondeckenbau*] = adjacent strip
~**bewehrung** *f*, Anschlußarmierung, Anschluß(stahl)einlagen *f pl*, Steckeisen, Anschlußeisen *npl* = connection rebars (US); projecting reinforcement
~**blech** *n*, Anschlußeisen *n*, Knotenblech, Knoteneisen = connecting plate, gusset ~, binding ~, ~ iron
~**bohrung** *f*, Bohrung zur Feststellung der Ausdehnung eines Ölfeldes = field extension well
~**dose** *f*, Steckdose = connection box, power socket, power outlet
~**einlagen** *f pl* → Anschlußbewehrung *f*
~**eisen** *npl* → Anschlußbewehrung *f*
~**eisen** *n* → Anschlußblech *n*
~**flügel** *m* = buttress, wing, side ~
~**gleis** *n*, Gleisanschluß *m*, Stichgleis, privater Bahnanschluß *m* = spur track; rail(way) spur (Brit.); rail(road) ~ (US); (private) siding, connecting track
~**kanal** *m*, Verbindungskanal = connection canal, connecting ~
~**klemme** *f* = terminal
~**leitung** *f* = connecting line
~**maß** *n*, Anschlußabmessung *f* [*Maß eines Bauteiles zum Anschluß an einen anderen*] = connection dimension
~**niet** *m* = connecting rivet
~**pfahl** *m*, Verbindungspfahl = connection pile
~**punkt** *m*, Anschlußstelle *f*, Zufahrtstelle, Anschluß *m*, Auffahrt *f* [*Autobahn*] = access point, interconnecting roadway, interchange
~**rampe** *f* [*Autobahn*] = access ramp, interchange ~
~**rohr** *n* = connecting tube, ~ pipe
~**schweißung** *f* = tie-in welding
~**stab** *m* = projecting bar, connecting ~
~**strang** *m* = service pipe [*The whole of the pipe (communication pipe + supply pipe = Endstrang + Grundstückstrang) between the main and consumers' premises*]
~**stück** *n*, Anschluß *m* = connecting piece
~**stutzen** *m* = connection socket
~**teil** *m, n* Verbindungsteil, Zwischenstück *n* [*Gleiskettenlaufwerk*] = adapter, mounting, adaptor
~**wert** *m* [*Elektrotechnik*] = connected load, connection ~
~**winkel** *m*, Befestigungswinkel = clip angle, cangle) cleat, angle bracket, (end) connection angle
Anschnitt *m* [*Bergseitig ist die Straße im Gelände eingeschnitten und talseitig über dem Gelände aufgeschüttet*] = hillside cut and fill
anschraubbar = bolt-on
anschrauben = to screw on, to bolt to
~ [*eine Stange an das Bohrgestänge*] = to add a new joint to the drill pipe
Anschrauben *n* = screwing on, bolting to

Anschreiben — (an)streichen

Anschreiben n, Begleitbrief m = covering letter
Anschriftendruckerei f = adrema
(An)Schub|fahrzeug n, Schub(hilfe)gerät n, (Schürfgang-)Schubmaschine f, (Schürfgang-)Stoßmaschine [*unterstützt einen Radschrapper beim Schürfgang*] = pusher, pushloader, scraper ~
~-Radschlepper m, Stoß-Radschlepper [*unterstützt einen Radschrapper beim Schürfgang*] = wheel(ed) pusher, ~ pushloader, ~ scraper ~
~raupe f, (An)Schub-(Gleis)Kettenschlepper [*unterstützt einen Radschrapper beim Schürfgang*] = crawler pusher, ~ pushloader, ~ scraper ~
Anschütte f (Geol.) → Alluvium n
anschütten → (auf)schütten
(An)Schütten n, Auftragen, Auffüllen, Aufschütten = placing the fill(ing), depositing ~ ~, ~ ~ fill material
Anschüttung f → Schüttung
Anschütt(ungs)höhe f → Auftraghöhe
anschweißen = to weld to
anschwellen = to swell
~, steigen [*Wasserspiegel*] = to rise
Anschwellung f [*bei Gewässern infolge Vorbeiganges einer Tidewelle*] = rise (or raise) in the water level due to the passage of a tidal wave
~, Schwelle f, Bodenwelle f [*Geomorphologie*] = hillock
anschwemmen → auflanden
Anschwemmschicht f [*Filter*] = precoat
Anschwemmung f → Auflandung
Ansenkung f, Senkbohrung = counterbore
ansetzen = to locate [*a well*]
Ansetzen n [*Bohrung*] = collaring
Ansicht f, Aufriß m, Vertikalprojektion f = elevation
Ansiedescherben m = scorifier
Ansoden n, Anplaggen = sodding
(An)Spannen n = tensioning, stretching
Anspitzmaschine f [*Rohrherstellung*] = pipe pointer

Anspringen n [*Saugüberlauf*] = priming
Anspring|nase f [*Saugüberlauf*] = priming nose (Brit.); sealing bucket (US)
~-Wasserspiegelstand [*Saugüberlauf*] m = priming level
(an)stauen [*Wasser*] = to impound, to pond, to back up
(An)Stauung f, (An)Stau m, (Wasser-)Spiegelhebung f = pondage
Ansteck|achse f = detachable axle
~deichsel f = detachable tow-bar
anstehender Boden m = in-situ soil, site ~, in-place ~
~ Fels m, gewachsener ~, Kernfels, Sprengfels, anstehendes Gestein n, fester Fels, gewachsenes Gestein = bed rock, solid ~ [*Bed rock may be exposed on the surface or it may be several hundred feet deep*]
~ Kies m = in-situ gravel, in-place ~
~ Sand m = in-situ sand, in-place ~
anstehendes Erz n = visible mineral
(an)steigen [*der HW-Zeit gegenüber dem mittleren HW-Intervall*] = to rise
Ansteigen n der Betonfestigkeit = rise of the concrete strength
ansteigende Gesteinsstrecke f (⚒) = rising stone drift, ~ rock ~
~ Straße f = road with rising gradient, highway ~ ~ ~; street ~ ~ ~
ansteigender Bandförderer m = upgrade belt conveyor, ~ ~ conveyer, uphill ~ ~, ~ band ~
(an)steigendes Wasser n = rising water
Anstellen n [*Annähern von Werkstück und Werkzeug zwecks Spanabnahme*] = positioning, setting
anstoßen = to foul
Anstößer m [*Schweiz*]; Anlieger m = riparian owner, ~ proprietor
anstrahlen = to floodlight
Anstrahlen n, Anstrahlung f = floodlighting
anstreichbar = paintable
Anstreichbarkeit f = paintability
(an)streichen, malen = to paint

(An)Streichen — Antimonnickelglanz

(An)Streichen n, Anstrich m, Malen n = painting [*The use of paints, varnishes, and stains for protection or decoration of materials*]
Anstreicher m, Maler = (house) painter
~**arbeiten** fpl, Malerarbeiten, Anstricharbeiten = painting work
~**gerüst** n → Malergerüst
~**werkzeug** n, Malerwerkzeug = (house) painter's tool
Anstrich m, Farbschicht f, Farbanstrich, Aufstrich, Farbaufstrich [*Ein gleichmäßig verteilter Auftrag von Anstrichstoff(en) auf einem Untergrund, auf dem er nach dem Trocknen haftet*] = paint coat, (~) finish, coat of paint
~ → (An)Streichen n
~**farbe** f [*ein pigmenthaltiger Anstrichstoff*] = paint [*It consists of pigment and medium*]
anstrichfertig = ready-to-paint, prepared for use, P.F.U.
Anstrich|film m, Farbfilm, Anstrichschicht f, Farbschicht = paint film
~**film** m, Lackfilm = varnish film
~**fläche** f = surface to be painted
~**gerät** n = painting device
~**grund** m = base for painting
~**-Markierungsstreifen** m = painted road stripe, ~ highway ~; ~ street ~
~**schaden** m, Anstrichmangel m = paint coat failure
~**stoff** m, Anstrichmittel n = No equivalent term in English. "Paint" (Anstrichfarbe) is the nearest term, but paint always is a pigmented material.
~**system** n, mehrschichtiger Anstrich m, mehrlagiger Anstrich = paint system
~**technik** f = painting practice
anteigen → anpasten
Anteil m, Gehalt m = content
~, Prozentsatz m = percentage
Anteil m **der Vorfertigung** = prefabrication content, factory ~
Antennengang m [*Rundfunkturm*] = aerial gallery

~**mast** m, Antennenturm m = radio tower, ~ mast, wireless ~
Anthrazen n, $C_{14}H_{10}$ = anthracene
~**öl** n [*Die bei 300° bis 350° C übergehenden Anteile des Steinkohlenteeres, die für Ölfeuerungen verwendet werden*] = anthracene oil, green ~
Anthrazit m = anthracite coal, hard ~
Anti-Abschuppungsmittel n, Anti-Abblätterungsmittel [*Beton*] = antispalling agent [*e. g. linseed oil*]
antieben = antiplane
Anti|-Epizentrum m = anticentre (Brit.); anticenter (US)
~**eruptions-Verrohrungskopf** m, Sicherheits-Verrohrungskopf = control casing head
~**friktionsrolle** f, Gleitrolle = antifriction roller
Antikglas n = antique glass
Anti|-Kleb(e)mittel n = parting agent
~**klinale** f → Sattel m der Faltung f
~**klinalkamm** m = anticlinal ridge
~**klinallager** n, ~**stätte** f, Antiklinalvorkommen n, Antilinalvorkommnis n, Antiklinalfundstätte = anticlinal deposit
~**kline** f → Sattel m der Faltung f
~**klopfmittel** n, Klopfbremse f = antiknock agent
~**kondensationsputz** m = anti-condensation plaster [*See remark under "Putz"*]
Antimon n, Sb = antimony
~**blei** n, Hartblei = antimonial lead, hard ~
~**blende** f, Kermesit m (Min.) = kermesite, pyrostibnite
~**blüte** f, Valentinit m (Min.) = valentinite, Sb_2O_3
~**erz** n = antimony ore
~**glanz** m, Antimonit m = antimonite; antimony glance [*obsolete term*]
~**grube** f, Antimonbergwerk n = (native) antimony mine
~**metall** n = antimony metal
~**nickelglanz** m, Ullmanit m (Min.) = ullmanite, nickel antimony glance, NiSbS

~salz n = antimonate

~silberblende f, dunkles Rotgültigerz n, Pyrargyrit m, Ag_3SbS_3 (Min.) = pyrargyrite, ruby silver ore, dark-red silver ore

Anti|oxydationsmittel n, Antioxydans n = antioxidant, oxidation inhibitor

~parallelkurbeln fpl, Gegendrehungskurbeln, gegenläufige Kurbeln = cranks moving in opposite directions, crossed parallelogram

~passat m, Gegenpassat = antitrade (wind)

~pol m, Gegenpol = antipole, reciprocal pole

~polare f, Gegenpolare = antipolar, reciprocal polar

~-Schall-Tür f = sound-proof door

~schaummittel n → Schaumverhütungsmittel

~schmiermittel n = antilubricating agent

Antizyklon m, Hoch(druckgebiet) n = anticyclone, high(-pressure area)

Antoniuskreuz → Taukreuz

antragen [*Putz*] = to lay on, to apply [*plaster*]

Antransport m, Anfuhr f, Anlieferung f = hauling (operations), haulage

antransportieren, anfahren, anliefern = to haul to (the site), to transport ~ (~ ~)

~ mit LKW, anfahren ~ ~ = to truck to (US); to transport to with lorries (Brit.)

antreffen, anfahren [*eine Formation beim Bohren*] = to encounter

antreiben = to drive

antreibender Teil m, antreibendes ~ n = driving part

Antrieb m = drive

~, Antriebsgruppe f, Triebsatz m = power unit, drive (~), driving gear, power pack

~ mit Ritzel und Zahnstange = rack and pinion drive

Antriebs|achse f, Triebachse, Treibachse = drive axle, driving ~, live ~, power ~

~art f = type of drive

~aufteilung f = drive lines

~element n, Antriebsteil m, n = driving part

~ende n = drive end

~flansch m = drive flange

~gehäuse n = gear case

~gehäuse n des Drehkranzes [*Motor-Straßenhobel*] = circle reverse control housing

~gruppe f, Antrieb m, Triebsatz m = power unit, drive (~), driving gear, power pack

~kette f, Treibkette, Triebkette = drive chain, driving ~

~ketten- und -riemenkennzeichnung f = drive chain and belt identification, driving ~ ~ ~ ~

~ketten(zahn)rad n, Antriebsturas m = driving (chain) sprocket, drive (~) ~, ~ sprocket wheel, bull wheel

~ketten(zahn)rad n, Antriebsturas m [*für eine (Gleis)Kette*] = crawler drive sprocket

~kraft f = driving power, drive ~

~kupplung f = drive clutch, driving ~

~kupplung f der Lichtmaschine = dynamo drive coupling, generator ~ ~

~leistung f = driving output

~maschine f für mehrere Tiefpumpenböcke = band wheel power unit

~mechanismus m = driving mechanism, drive ~

~mittel n = means of driving

Antriebsrad n, Antriebszahnrad = input gear

~, Triebrad, Treibrad = driving wheel, driving ~, traction ~

~, Antriebszahnrad, Triebling m, Ritzel n = pinion, drive gear, driving gear

~ mit wechselndem Zahneingriff, Antriebsturas m ~ ~ ~ = hunting tooth type sprocket

~-Flankenspiel n [*Gleiskettenlaufwerk*] = sprocket backlash

~-Hohlwelle f [*Gleiskettenlaufwerk*] = sprocket hub

Antriebsradlager — anwenden

~lager n [Gleiskettenlaufwerk] = sprocket bearing
~-Zahnkranz m [Gleiskettenlaufwerk] = sprocket rim
Antriebs|riemen m, Treibriemen = drive belt, driving ~
~ritzel n = drive pinion, driving ~
~ritzel n [Seitenantrieb eines Gleiskettenlaufwerks] = input pinion
~rolle f → Antriebsscheibe f
~rollenkette f = roller drive chain
Antriebsscheibe f, Treibscheibe, Treibrolle f, Antriebsrolle = drive pulley, driving ~
~ [pennsylvanisches Seilbohren] = band wheel
~ der Verrohrungsseiltrommel [pennsylvanisches Seilbohren] = calf wheel tug rim
~ des Flaschenzughaspels [pennsylvanisches Seilbohren] = calf wheel tug rim
Antriebsseil n [Tiefbohrtechnik] = bull rope
~ der Bohrseiltrommel [pennsylvanisches Seilbohren] = bull rope, traction ~, travel(l)ing ~
~scheibe f [Tiefbohrtechnik] = tug pulley
Antriebsseite f = drive side, driving ~
Antriebsstation f [Bandförderer] = drive head
~ = driving station
Antriebs|technik f = transmission engineering
~teil m, n, Antriebselement n = driving part, drive ~
~trommel f → → Bandförderer
Antriebsturas m, Antriebsketten(zahn)rad n = driving sprocket wheel, drive ~ ~, ~ (chain) sprocket, bull wheel
~, Antriebsketten(zahn)rad n [für eine (Gleis)Kette] = crawler drive sprocket, ~ driving ~
~ aus Borstahl = boralloy sprocket
~ mit wechselndem Zahneingriff, Antriebsrad n ~ ~ ~ = hunting tooth type sprocket

Antriebs|turm m → (Kabelkran-) Maschinenturm
~übersicht f = shafting and drive chain chart
~ und Hebewerk n → (Rotary-) Hebewerk
~vorgelege n = drive back gear
~wagen m [Einschienen-Transportbahn] = power wagon
Antriebswelle f, Triebwelle [die Eingangswelle bei Getrieben] = input shaft
~, Triebwelle [eine Welle, die ein Fahrzeug in Bewegung setzt] = drive shaft, driving ~
~ [pennsylvanisches Seilbohren] = band-wheel shaft
~ der Nockenwelle, Triebwelle ~ ~ = drive shaft of the camshaft
~ des Scharsteuerantriebes [Motor-Straßenhobel] = power control drive shaft
Antriebs(zahn)rad n, Triebling m, Ritzel n = pinion, drive gear, driving gear
~ = input gear
~ der Nockenwelle = camshaft gear
Antriebszylinder m, Arbeitszylinder = working cylinder
Antritt m, ~stufe f = starting step
~pfosten m = starting newel (post)
~stufe f, Antritt m = starting step
An- und Abschleppen n [Hubinsel] = towing to and from the site
anvisieren = to take the bearings of, to align the sights on
Anwachsen n des Ufers = accretion (of earth on the bank)
Anwachsufer n, anlandendes Ufer, verlandendes Ufer = accreting bank, deposition bluff
Anwärmen n [Bindemittel; Zuschlagstoffe; Anmach(e)wasser] = heating
Anwärmofen m = heating furnace
Anweisung f = instruction
Anweisungen fpl für Maßnahmen vor Inbetriebnahme = pre-starting instructions
Anwendbarkeit f = applicability
anwenden = to use, to apply

Anwendung *f* = use, application
Anwendungs|beispiel *n* = example of use, ~ ~ application
~gebiet *n*, **Anwendungsbereich** *m*, **Anwendungsfeld** *n* = field of use, ~ ~ application, scope
~sicherheit *f* = safety of use, ~ ~ application
~temperatur *f* = application temperature, temperature of use
anwerfen, ankurbeln, andrehen = to crank [*To start by a crank*]
Anwerf|kurbel *f* → Andrehkurbel
~motor *m* → Anlaßmotor
anwinkeln = to angle
anwuchsverhindernde Schiffbodenfarbe *f* = anti-fouling (bottom) paint, ~ (composition), ~ compound
anwuchsverhindernder Schiffbodenanstrich *m* → vergifteter Anstrich
Anwurf|kurbel *f*, **Andrehkurbel**, **Startkurbel** = starting crank
~motor *m* → Anlaßmotor
~turbine *f* = pony turbine
Anwürgen *n* → → 1. Schießen; 2. Sprengen
Anwürgzange *f* → → 1. Schießen; 2. Sprengen
Anzahl *f* **der Achsen**, **Achsenzahl** = number of axles
~ pro Gerät, ~ ~ **Maschine** [*in einer Ersatzteilliste*] = quantity, qty.
Anzapf|dampfmaschine *f*, **Anzapfdampfmotor** *m* = bleeder (type) steam engine
~hahn *m* = bleeder
~kondensationsanlage *f* = bleeder type condensing plant
Anzapfung *f* [*Fluß*] = tapping, abstraction
Anzeige|fehler *m* = error of indication
~genauigkeit *f* = indicating accuracy
~gerät *n* **für explosive Dünste** = explosive vapo(u)r indicator
~nadel *f* = indicating needle, ga(u)ge ~
Anzeiger *m*, **Messer** *m*, **Anzeigevorrichtung** *f*, **Anzeigegerät** *n*, **Anzeigeinstrument** *n* = indicator, meter, indicating instrument
Anzeigeskala *f* = graduated scale
~ für Querneigung = crossfall indicator plate [*asphalt paver*]
Anzeige|stab *m*, **Meßstab**, **Peilstab** [*Flüssigkeitsstandmessung*] = dipstick, dip rod
~tafel *f* = telltale board, indicator ~
~vorrichtung *f* **für Bunkerfüllungen** → Bunkerstandanzeiger *m*
~vorrichtung *f* **für Silofüllungen** → Silostandanzeiger *m*
anziehen [*Schraube; Mutter*] = to snug, to tighten
Anziehen *n* [*Mörtel*] = initial hardening
Anzug *m*, **Anlauf** *m*, **Dossierung** *f*, **Neigung** = batter, taper
Anzugs|folge *f* = tightening sequence
~moment *n*, **Anlauf(dreh)moment**, **Anfahr(dreh)moment**, **Anlaß(dreh)moment** [*Motor*] = starting torque
~moment *n* **einer Mutter** = nut torque
~moment *n* **einer Schraube** = bolt torque
anzünden = to light up, to ignite [*a gas-fired furnace*]
Anzündlitze *f* = ignitor cord
äolisch, aërisch = eolian, aerial, wind-blown, wind-deposited, wind-drift, wind-laid
äolische Abtragung *f* → Deflation *f*
Apatit *m*, $Ca_5(F,Cl)(PO_4)_3$ = apatite [*Hexagonal. Specific gravity 3.15–3.20. Hardness 5*]
Apfelsinenschalengreif(er)korb *m*, **Apfelsinenschalengreifer** *m* = orange-peel grab, ~ bucket, four-bladed circular bucket, four-bladed circular grab, four-segment circular grab, four-segment circular bucket [*Is used for excavating inside cylinders (except in clay) and for large boulders*]
Apiezon|fett *n* = Apiezon grease
~öl *n* = Apiezon oil
Aplit *m* = aplite [*A light-coloured, finely grained granite made up largely of quartz and fel(d)spar*]

Apophyse — Arbeiterschaft

Apophyse *f* (Geol.) [*ein von einer größeren subkrustalen Masse abzweigender Gang*] = apophysis [*a veinlike offshoot from an igneous intrusion*]
Apparat *m* = apparatus
Apparatebau *m* = apparatus engineering
Apparate(n)|haus *n* [*Schweiz*]; Schieberhaus [*Haus bei Wasserkraftanlagen, das die Verschlüsse einer oder mehrerer Druckleitungen nebst ihren Hilfseinrichtungen enthält*] = valve house
~kammer *f* [*Schweiz*]; Schieberkammer = valve chamber
Aprolith *m*, Spratzlava *f*, Blocklava, Schollenlava, Zackenlava = block lava, aa (~) [*It consists of a rough tumultuous assemblage of clinkerlike scoriaceous masses*]
Apsis *f*, Apside *f* = apsis, apse
~-Halbkuppel *f*, Apside-Halbkuppel = apse roof
A-Pylon *m*, A-Turm *m*, A-Pylone *f* [*Hängebrücke*] = A-shaped pylon, ~ tower
Aquädukt *m*, Wasserleitungsbrücke *f* = aqueduct, water conduit bridge, water-carrying bridge
Aquarell *n* [*farbige Darstellung mit Wasserfarben*] = aquarelle
~farbe *f*, Wasserfarbe = water colo(u)r
aquatisches Sedimentgestein *n*, neptunisches ~ = waterborne sediment
Äquatorialströmung *f* = equatorial current
Äquinoktialtide *f*, Äquinoktialgezeit *f* = equinoctial tide
Äquipotential|fläche *f* = equipotential surface
~linie *f* = equipotential line
~linie *f* [*geophysikalisches Aufschlußverfahren*] = equipotential contour
~linienverfahren *n* [*geophysikalisches Aufschlußverfahren*] = equipotential method, ~ prospecting, ~ exploration
äquivalente Ablagerung *f*, homotaxe ~ (Geol.) = homotactic bed, equiva-

lent ~, contemporary ~, homotaxeous ~, homotaxial ~, homotaxic ~, synchronous ~
~ Korngröße *f*, gleichwertige ~ = equivalent grain size, ~ particle ~
~ Rundlochweite *f* = equivalent round hole diameter
äquivalenter Verdichtungsdruck *m*, gleichwertiger ~ = equivalent compaction pressure
Äquivalentkorndurchmesser *m* = equivalent grain diameter
äquiviskose Temperatur *f*, Äquiviskositätstemperatur [*Vergleichstemperatur bei der Teer eine Viskosität von 50 Sekunden hat*] = equi-viscous temperature [*abbrev. E. V. T.*]
arabischer Bogen *m* → Hufeisenbogen
Aräometer *n*, Senkwaage *f*, Spindel *f* = hydrometer
~meßtechnik *f* = hydrometry [*The use of hydrometers*]
A-Rahmen *m*, Abstützbock *m*, A-Bock = A-frame
Arbeiten *n* [*Die unangenehme Eigenschaft des Betons unter klimatischen Einwirkungen die Abmessungen und Formen zu verändern*] = working, movement
~ *fpl* **der öffentlichen Hand** = public works
~ *fpl* **für Brückenzufahrten** = bridge access work
~ *n* **mit versetzter Spur** [*Motor-Straßenhobel*] = grading with frame offset
arbeitende Siebfläche *f* → Siebboden *m*
Arbeiter *m* = worker, workman
~dorf *n* [*bei einem Bergwerk*] = miner's village
~-Eingang *m* = workers' entrance
Arbeiterin *f* = female worker, ~ hand
Arbeiter|kolonne *f* → Kolonne
~lager *n*, Baustellenlager, Unterkunftslager, Lagerunterkünfte *fpl* = contractor's construction camp, workers' ~, construction village
~schaft *f* [*einer Fabrik*] = labo(u)r, workers [*on the payroll of a factory*]

Arbeitersiedlung — Arbeitsfugenband

~siedlung *f* = workers' estate
~stunde *f* = man-hour
~unterkunft *f* = barracks (US); workers' accomodation (Brit.) [*in civil engineering, a temporary building for quartering workmen*]
~wohnung *f* = workman's dwelling, working-class ~
Arbeit|geber *m* = employer
~nehmer *m* = employee
~nehmerwechsel *m*, Personalwechsel = employee turnover
Arbeits|ablauf *m*, Arbeitsfolge *f*, Arbeitsgang(folge) *m*, (*f*), Arbeitsgänge *mpl*, Verfahrensgang(folge), Verfahrensgänge, Verfahrensablauf, Verfahrensfolge = sequence of operations, operational sequence
~abmessungen *fpl*, Arbeitsmaße *npl* [*Maschine*] = working dimensions
~amt *n* = labo(u)r exchange
~angriff *m* [*Tunnel- und Stollenbau*] = attack, heading
~arm *m* [*Schürflader*] = lifting beam
~aufwand *m* = expenditure of work
~ausrüstung *f* → Austauschgerät *n*
~ausschuß *m*, Arbeitsgruppe *f* = working committee
~bank *f*, Werkbank, Arbeitstisch *m*, Werktisch [*z. B. für Schlosser, Schreiner, Dreher usw.*] = bench
~bereich *m*, Arbeitshalbmesser *m*, Arbeitsradius *m*, Schwenkbereich, Schwenkhalbmesser, Drehbereich, Drehradius, Drehhalbmesser [*Auslegerkran; Auslegerbagger*] = working range, operational ~
~beschaffung *f* [*um die Arbeitslosigkeit zu verringern*] = provision of employment
~betonfuge *f* → Arbeitsfuge
~bewegung *f* [*Maschine*] = working movement, ~ motion
~breite *f* = working width, operating ~
~breite *f*, Einbaubreite [*Straßenfertiger*] = operating width, working ~, laying ~
~brücke *f* = staging

~bühne *f*, Arbeitsplattform *f* [*allgemein*] = work(ing) deck, ~ platform
~bühne *f*, Arbeitsplattform *f* [*bei einem Lastträger eine Plattform mit Geländer zum Ausführen von Arbeiten in großer Höhe*] = work(ing) platform
~bühne *f*, Montagebühne, Arbeitsplattform *f*, Montageplattform = erecting deck, ~ platform
~bühne *f* → → Fertigbehandlung
(~)**Bühne** *f*, Arbeitsplattform *f* [*im Bergbau Gerüst oder Schachtabsatz als Arbeitsplatz*] = work(ing) platform
(~)**Bühne** *f*, Arbeitsplattform *f* [*Arbeitsplatz als Gerüst beim Abteufen von Schächten*] = shaft sinking stage
~bunker *m* → Aufgabebunker
~dampf *m* = working steam
~dauer *f* [*Leim*] → Topfzeit *f*
~druck *m*, Betriebsdruck = service pressure, operating ~, working ~
~einsatz *m* → Geräteeinsatz
~einsparung *f*, Arbeitsersparnis *f* = saving in labo(u)r, labo(u)r saving
~einstellung *f* = stoppage
~element *n*, Betriebsteil *m*, *n*, Arbeitsorgan *n* = working part
arbeitsfähig [*Arbeiter*] = employable, able to work
Arbeits|fähigkeit *f*, Energie *f* [*Hydraulik*] = energy
~festigkeit *f* → Ermüdungsfestigkeit
~fläche *f* = work(ing) area
~fläche *f* [*Siebtechnik*] = nominal working area
~flüssigkeit *f* = working fluid
~fluß *m*, Arbeitsstetigkeit *f* = continuity of work
~folge *f* → Arbeitsablauf *m*
~folge *f* → Baufolge
~fortgang *m* → Baufortschritt *m*
~fortschritt *m* → Baufortschritt
~fuge *f*, Betonier(ungs)fuge, Arbeitsbetonfuge, Betonarbeitsfuge = (concrete) construction joint, (~) stopend ~
~fugenband *n*, Betonier(ungs)fugenband, Arbeitsbetonfugenband, Be-

Arbeitsgang — Arbeitsrichtung 84

tonarbeitsfugenband = construction joint tape, stop-end ~ ~
~**gang** *m*, Arbeitsphase *f* = operation, process, pass, step
~**gang** *m*, Walzgang, Übergang [*Walze*] = pass
~**gang** *m* [*Lader*] = work range, low ~
~**gänge** *mpl* → Arbeitsablauf *m*
~**gang(folge)** *m*, (*f*) → Arbeitsablauf *m*
~**gemeinschaft** *f*, Arge *f*, Baufirmengruppe *f* = contracting combine, bidding combination, contractor combination, amalgamation of contractors, partnership, joint venture (firm)
 federführende Firma *f* = sponsor (US); pilot firm, lead firm
~**gerät** *n* → Austauschgerät
~**gerüst** *n* → Abteufgerüst
~**gerüst** *n* → (Bau)Gerüst
~**geschwindigkeit** *f* [*im Gegensatz zur Transportgeschwindigkeit einer Baumaschine*] = operating speed, working ~, on-the-job travel ~, speed of working [*construction machinery*]
~**gewicht** *n*, Dienstgewicht, betriebsfertiges Gewicht, Betriebsgewicht = service weight, operating ~
~**grube** *f* → Reparaturgrube
~**gruppe** *f*, Arbeitsausschuß *m* = working committee
~**güte** *f* = quality of (the) work
~**halbmesser** *m* → Arbeitsbereich *m*
~**halde** *f* = surge pile
~**höhe** *f* [*Kran*] = working height
~**hub** *m*, Expansionshub, Ausdehnungshub = working stroke, expansion ~, firing ~, power ~
~**hygiene** *f* = factory hygiene, industrial ~
Arbeitskammer *f*, Druck(luft)kammer [*pneumatische Gründung*] = (caisson) working chamber
~**decke** *f* = caisson ceiling
~**druck** *m* = (caisson)working chamber pressure
Arbeitskräfte *fpl* = labo(u)r (force), manpower

~**mangel** *m* = labo(u)r (force) shortage, manpower ~
Arbeits|lärm *m* = industrial noise
(~)**Leistung** *f* = performance, output, capacity
~**leistung** *f*, Zughakenleistung = drawbar performance, ~ horsepower, ~ pull
~**linie** *f* = stress-deformation diagram
arbeitslos = unemployed
Arbeits|losigkeit *f* = unemployment
~**maschine** *f* = machine
~**maße** *npl*, Arbeitsabmessungen *fpl* [*Maschine*] = working dimensions
~**methode** *f*, Arbeitsverfahren *n* = working method
~**pfahl** *m*, rechter Pfahl [*Naßbagger*] = working spud
~**phase** *f*, Arbeitsgang *m* = process, pass, operation, step
~**plan** *m* → Bau(zeit)plan
~**plan** *m* → Ausführungsplan
Arbeitsplattform *f*, Bohrbühne *f* = drilling platform
~, Arbeitsbühne *f*, Montageplattform, Montagebühne = erecting deck, ~ platform
~, Arbeitsbühne *f* [*allgemein*] = work(ing) deck, ~ platform
~, Arbeitsbühne *f* [*bei einem Lastträger eine Plattform mit Geländer zum Ausführen von Arbeiten in großer Höhe*] = work(ing) platform
~, (Arbeits)Bühne *f* [*im Bergbau Gerüst oder Schachtabsatz als Arbeitsplatz*] = work(ing) platform
~, (Arbeits)Bühne *f* [*Arbeitsplatz als Gerüst beim Abteufen von Schächten*] = shaft sinking stage
Arbeits|prinzip *n*, Funktionsprinzip [*Maschine*] = operating principle, work(ing) ~
~**puffer** *m*, Speicherpuffer [*Sieb*] = rubber buffer, working ~ ~
~**raum** *m* = workroom
~**rhythmus** *m* [*Maschine*] = production cycle, work(ing) ~
~**richtung** *f* [*Bodenverdichter*] = direction of compaction

~runde *f*, Umlauf *m*, Rundfahrt *f* [*Radschrapper*] = round trip
~rüstung *f* → (Bau)Gerüst *n*
~schacht *m* = working shaft
(~)Schicht *f* = shift, work(ing) ~, (working) tour
~schutz(be)kleidung *f* = protective clothing
~schutzdach *n* → → Fertigbehandlung *f*
~seite *f*, Siebseite = screening side
~sitzung *f* = session
~spalt *m* [*Magnetscheider*] = air gap
~spiel *n* [*Maschine*] = work(ing) cycle, operating ~
~spielzeit *f* [*Maschine*] = cycling time
~stellung *f* [*Baumaschine*] = working position
~stellung *f* für Böschungsherstellung [*Straßenhobel*] = pitch
~stumpffuge *f*, Betonier(ungs)stumpffuge = construction butt joint, stop-end ~ ~
~stunde *f* = working hour
~tag *m* = working day
~tag *m*, Einsatztag [*Maschine*] = working day of a machine
~takt *m* = (operating) cycle
~technik *f* = working practice
~temperatur *f* = working temperature
~tempo *n*, Bautempo = rate of performance, ~ ~ progress (of the construction work)
~tiefe *f*, (Auf)Reißtiefe [*Aufreißer*] = digging depth of tines below ground level, working depth, depth of penetration
~tiefe *f* = working depth, depth of penetration
Arbeitstisch *m* [*für Präzisionsarbeiten, z. B. Uhrmacherarbeiten*] = work table
~, Werktisch, Arbeitsbank *f*, Werkbank [*z. B. für Schlosser, Schreiner, Dreher usw.*] = (work) bench
(Arbeits)Trupp *m* → Kolonne *f*
Arbeits|- und Schlafzimmer *n* **kombiniert** = study-bedroom

~unfall *m*, Betriebsunfall = on-the-job accident, occupational ~
~untersuchung *f*, Arbeitsstudie *f* [*Maschine*] = equipment production study
~ventil *n* einer Tiefpumpe = travel(l)ing valve, working(-barrel) ~
~vereinfachung *f* = work simplification
~verfahren *n*, Arbeitsmethode *f* = working method
~versäumnis *n* = absenteeism
~vertrag *m* = employment contract
~volumen *n* [*z. B. Umfang der Bauarbeiten*] = volume of work
~vorbereiter *m* = setter out, marker ~
~vorgang *m* [*Maschine*] = action
~vorrichtung *f* → Austauschgerät *n*
~weise *f* [*Maschine*] = mode of operation
~werkzeug *n* → Austauschgerät *n*
~wissenschaft *f* = work science
~woche *f* = working-week
~zeichnung *f* → Ausführungsplan *m*
~zeit *f* = working hours
~zeitkarte *f* [*für ein in der Werkstatt in Reparatur befindliches Gerät*] = time card
~zelt *n* → → Fertigbehandlung *f*
~zimmer *n* = study [*A room designated for study, writing, reading, etc., usually with books, a desk and similar furnishings*]
~zug *m* (**von Großgerät**) = equipment train
~zylinder *m*, Antriebszylinder = working cylinder
Archimedes'sches Prinzip *n* = principle of Archimedes
Architekten|gebühr *f* = architect's fee
~gemeinschaft *f* = architect partnership
~gruppe *f* = architectural team
~kammer *f* = association of architects
~leistung *f* = architectural work
~liste *f* = list of architects
architektonisch = architecturally

archtitektonischer Fertigteil — Armierungsquerschnitt 86

architektonischer Fertigteil *m*, architektonisches ~ *n* = precast architectural (concrete) member
architektonisches Bauelement *n* = architectural unit
~ Betonbauelement *n* = architectural concrete unit
Architektur *f*, Baukunst *f* = architecture
~beton *m*, Sichtbeton = exposed concrete, fair-faced ~
~platz *m* = place of outstanding architectural merit, architecturally beautiful square
~wettbewerb *m* = architectural competition
~zeichner *m* = architectural draughtsman
~zeichnung *f* = architectural drawing
Architrav *m* = architrave
Archiv *n* = archives
Archivolte *f* = archivolt
Arge *f* → Arbeitsgemeinschaft *f*
Argentit *m* → Silberglanz *m*
Argillit *m* [*natürlich entwässerter Tonschiefer*] = argillite
Argonarc-Verfahren *n* → WIG-Verfahren
(arithmetische) Reihe *f* = arithmetical progression
(arithmetisches) Mittel *n*, Mittelwert *m* = arithmetical mean
Arkade *f*, Bogengang *m*, Bogenlaube *f* = archway, arcade [*Is a series of arches and includes the supporting members between the arches, also as piers, columns, etc.*]
Arkose *f*, feldspatreicher Sandstein *m* = arkose
arm [*Erz*] = low-grade, poor
Arm *m* = arm
~, Eingab(e)lung *f*, Tragstück *n*, Halter *m* = bracket
~ → (Treppen)Lauf *m*
~ der Lüfterriemenspannscheibe = fan belt adjusting pulley bracket
Armaturen|beleuchtung *f* = dash lamp, ~ light, instrument ~

~brett *n*, Schalttafel *f*, Steuer(ungs)tafel, Instrumentenbrett, Schaltbrett, Instrumententafel = instrument cluster, control panel, dash(board), instrument panel
Armgas *n* [*von Gasolin befreites Erdgas*] = stripped gas, residue ~
armieren, bewehren, mit (Stahl)Einlagen *fpl* versehen = to reinforce
Armieren *n*, Bewehren = reinforcing
armiert, bewehrt, mit (Stahl)Einlagen = reinforced
armiertes Kabel *n*, bewehrtes ~, Panzerkabel = armo(u)red cable
Armierung *f* → Bewehrung
Armierungs|anordnung *f*, Bewehrungsanordnung, Anordnung der (Stahl-)Einlagen = reinforcement system
~anteil *m*, Bewehrungsanteil, Armierungsgehalt *m*, Bewehrungsgehalt, Armierungsprozentsatz *m*, Bewehrungsprozentsatz = reinforcement percentage
~arbeiten *fpl*, Bewehrungsarbeiten = reinforcement work
~block *m* → Bewehrungskorb *m*
~bündel *n*, Bewehrungsbündel = bundle of reinforcement
~gehalt *m* → Armierungsanteil *m*
~gerippe *n* → Bewehrungskorb *m*
~gerüst *n* → Bewehrungskorb *m*
~kolonne *f*, Bewehrungskolonne; Eisenlegergruppe *f* [*Schweiz*] = crew of steel fixers, gang ~ ~ ~, team ~ ~ ~, party ~ ~ ~
~korb *m* → Bewehrungskorb
~lage *f*, Armierungsschicht *f*, Bewehrungslage, Bewehrungsschicht = reinforcement layer
~matte *f* → Baustahlmatte
~menge *f*, Bewehrungsmenge, Stahlmenge, Menge der (Stahl)Einlagen = amount of reinforcement
~netz *n*, Bewehrungsnetz, Netzarmierung *f*, Netzbewehrung = mat reinforcement
~plan *m* → Bewehrungsplan
~prozentsatz *m* → Armierungsanteil
~querschnitt *m*, Bewehrungsquerschnitt = reinforcement cross-section

~ring *m*, Bewehrungsring = reinforcement ring
~stab *m* → Bewehrungsstab
~stahl *m* → Betonstahl
~stange *f* → Bewehrungsstab *m*
~stumpfschweißen *n*, Bewehrungsstumpfschweißen = reinforcement butt welding
~verhältnis *n*, Bewehrungsverhältnis = ratio of reinforcement
~zeichnung *f* → Bewehrungsplan *m*
Armstütze *f* = arm rest
ärodynamische Labilität *f* = aerodynamic instability
~ **Standsicherheit** *f* = aerodynamic stability
Aromaten *mpl*, aromatische Kohlenwasserstoffe *mpl*, Benzolkohlenwasserstoffe = aromatic hydrocarbons, aromatics
~bestimmung *f* = determination of aromatics
aromatisches Lösungsmittel *n* = aromatic solvent
Arretierschraube *f* = lock bolt
Arretierung *f* [*Laufgewichtswaage*] = beam lifters
Arsensulfid *n* = arsenic sulphide (Brit.); ~ sulfide (US)
Art *f* **der Luftporenbildung** = air-void characteristics
Arterit *m*, Adergneis *m* = vein gneiss
artesische Bedingung *f* = artesian condition
~ **Druckhöhe** *f* = artesian (pressure) head
~ **Quelle** *f*, artesischer Brunnen *m*, Überlaufbrunnen, frei ausfließende Wasserbohrung *f* [*Bohrbrunnen, dessen Wasser durch eigenen Überdruck zur Oberfläche gelangt*] = artesian spring, ~ well, pressure well, "blow-well"
artesischer Druck *m* = artesian pressure
artesisches Becken *n* = artesian basin
~ **(Grund)Wasser** *n*, Druckwasser, artesisch gespanntes (Grund)Wasser [*Grundwasser, das mit artesischem Brunnen gewonnen werden kann. Das artesische Wasser ist eine Sonderform des gespannten Grundwassers; aufsteigendes Wasser wird nicht zum artesischen Wasser gerechnet*] = artesian (ground-)water
Arzneischleuse *f*, Krankenschleuse = medical lock
ärztliche Bescheinigung *f* = medical certificate
~ **Fürsorge** *f*, ~ **Betreuung** *f* = medical attention, ~ attendance
~ **Hilfe** *f* = medical assistance
~ **Überwachung** *f* = medical control, ~ supervision
Asbest *m* [*Asbest ist ein Mineral, ein in feinsten Fasern kristallisierter Naturstein. Die einzelne Faser ist dünner als jede tierische, pflanzliche oder künstliche Faser, weil es sich bei Asbest um ein Fadenmolekül handelt. Der Asbest kommt in der Natur im Gestein wie die Kohle im Flözen vor, nur daß die Asbestflöze gegenüber den Kohlenflözen sehr viel weniger mächtig sind und nur eine Höhe von wenigen Millimetern bis zu einigen Zentimetern besitzen. Die Faserflöze werden in Brechanlagen vom umgebenden Gestein getrennt, in Mühlen in Faserbündel aufgelöst und dann in Sichteranlagen entsprechend der Faserlänge sortiert. Chemisch besteht Asbest aus kompliziert aufgebauten Silikathydraten. Technisch von Bedeutung sind die reinen Magnesiumsilikathydrate (der Serpentin- oder Chrysotil-Asbest) und die Natriummagnesiumeisensilikathydrate (Blau-Asbest). Alle Asbeste sind feuerfest, von hoher chemischer Beständigkeit, beständig gegenüber Angriffen von Bakterien und Pilzen und praktisch absolut witterungsbeständig, weil sie nur im Verlauf von Jahrtausenden einer geringen Erosion und Korrosion unterliegen. Die Reißfestigkeit ist abhängig von der Asbestsorte und kann zwischen wenigen kg pro mm^2 und der Festigkeit von Stahl liegen. Für die Eternit-*

Herstellung kommen nur Asbeste höchster Festigkeit zur Verwendung, insbesondere hochwertige Chrysotil- und Blau-Asbestsorten] = asbestos
~aufbereitung *f* = asbestos dressing
~bauplatte *f* = asbestos building board
~bekleidung *f* = asbestos clothing
~bergbau *m* = asbestos mining
~bergwerk *n*, Asbestgrube *f* = asbestos mine
~bruch *m* = asbestos quarry
~(-Dach)schindel *f* = asbestos roof shingle
~dämmplatte *f* = insulating asbestos board
~dämmschicht *f* = asbestos blanket
~dämmung *f* = asbestos lagging (Brit.); ~ insulation
~einlage *f* = asbestos insert
~erzeugnis *n* = asbestos product
~faser *f* = asbestos fibre (Brit.); ~ fiber (US)
~-Fausthandschuh *m* = asbestos mitter
~filz *m* = asbestos felt
~garn *n* = asbestos yarn
~gestein *n* = asbestos rock
~gewebe *n* = asbestos cloth
~grobgespinst *n*, Asbestgrobgarn *n* = asbestos roving, ~ rove
~grube *f*, Asbestbergwerk *n* = asbestos mine
~lage *f*, Asbestschicht *f* = asbestos layer
~mehl *n* = asbestos powder
~mörtel *m* = asbestos mortar
Asbestosis *f*, Asbestose *f* = asbestosis [*disease of the lungs due to inhalation of asbestos particles*]
Asbest|pappe *f* = asbestos felt
~platte *f* = asbestos sheet
~prüfmaschine *f* = asbestos tester, ~ testing machine
~ring *m* = asbestos ring
~schaumbeton *m* = asbestos foamed concrete
~schiefer *m* = asbestos slate
~schnur *f* = asbestos cord
~schürze *f* = asbestos apron
~staub *m* = asbestos dust
~strick *m*, Asbestzopf *m* = asbestos rope
~tafel *f* = asbestos panel
~wolle *f* = asbestos wool
Asbestzement *m* [*Eine Mischung aus fein aufgeschlossenen Asbestfasern und Zement, die die für das Abbinden des Zementes nötige Menge Wasser enthält. Manchmal auch ,,Kunstschiefer'' genannt*] = asbestos cement
~-Abfluß(rohr) *m*, (*n*), Asbestzement-Ablauf(rohr) = asbestos-cement waste pipe
~-Abwasserrohr *n* = asbestos-cement pipe sewer, ~ sewer pipe
~(dach)eindeckung *f* = asbestos-cement roofing
~-Dachplatte *f* = asbestos-cement roof(ing) sheet
~-Dachrinne *f* = asbestos-cement eave(s) gutter
~-Druckrohr *n* = asbestos-cement pressure pipe
~erzeugnis *n* = asbestos-cement product
~-Faulgrube *f* = asbestos-cement septic tank
~(fuß)boden *m* = asbestos-cement floor(ing), ~ floor finish, ~ floor covering
~-Gummi-Fliese *f*, Asbestzement-Gummi-Platte *f* = asbestos rubber tile [*An asbestos-cement tile surfaced with rubber*]
~lochrohr *n* = perforated asbestos-cement pipe
~mörtel *m* = asbestos-cement mortar
~platte *f* = asbestos-cement sheet
~platten *fpl* = asbestos-cement sheeting [*Corrugated, reeded, or otherwise patterned or plain sheets for roofing and wall cladding, made of asbestos cement*]
~-Rauchkanal *m* = asbestos-cement flue
~rauchrohr *n* = asbestos cement flue pipe [*B. S. 567*]

~**-Regen(fall)rohr** n, Asbestzement-Fallrohr, Asbestzement-Abfallrohr (für Regenwasser) = asbestos-cement downcomer, ~ downpipe, ~ downspout, ~ fall pipe
~**rohr** n = asbestos-cement pipe
~**schindel** f = asbestos shingle [*Asbestos shingles are made of asbestos fiber and portland cement under pressure*]
~**tafel** f = asbestos-cement panel
~**ware** f = mass-produced asbestos-cement product
~**wellplatte** f = corrugated asbestos-cement sheet

Aschen|anflug m, Abbrandanflug [*auf den Ofensteinen*] = kiln scum
~**bahn** f = cinder track
~**fall** m, Aschenregen m (Geol.) = ash shower, ~ fall, ~ rain
~**fall** m → Aschenraum m
~**gehalt** m = ash content
~**grube** f → Aschenraum m
~**kegel** m = ash cone, cinder ~ [*The conical hill or mountain built up with the ejected material from a volcano*]
~**lage** f, Aschenschicht f = layer of ashes
~**pumpe** f = ash-handling pump
~**raum** m, Aschenfall m, Aschengrube f [*Kessel*] = ash pit
~**regen** m, Aschenfall m (Geol.) = ash shower, ~ fall, ~ rain
~**rinne** f = ash channel
~**rohr** n = ash pipe
~**schaumbeton** m = foam-ash concrete
~**schieferton** m = ashy shale
~**sinter** m = sintered flyash
~**tonschiefer** m = ash slate
~**trichter** m = ash hopper
~**tuff** m = vitric tuff
~**tür** f, Entaschungstür = ash door

Asphalt m, Natur~, natürliches Bitumen-Mineral-Gemisch n, natürliche Bitumen-Mineral-Mischung f [*das natürlich vorkommende Bitumen mit seinem Begleitgestein*] = (natural) asphalt (Brit.); rock asphalt (US)
~, künstliches Bitumen-Mineral-Gemisch n, künstliche Bitumen-Mineral-Mischung f = (artificial) asphalt (Brit.); (~) asphalt-aggregate mix(ture) (US)
~**anlage** f → Walzasphalt(misch)anlage
~**anstrich** m = asphaltic paint coat
~**aufbereitungsanlage** f → Walzasphaltmischanlage
~**auskleidung** f, Asphaltbekleidung, Asphaltverkleidung, Bitumenauskleidung, Bitumenbekleidung, Bitumenverkleidung = asphaltic lining, ~ facing (US); (~) bitumen ~ (Brit.)
~**axt** f = asphalt cutter (and marker)
asphaltbasisches Erdöl n → bitumenbasisches ~
Asphalt|becken n [*Becken mit Asphaltdichtung*] = asphalt reservoir
Asphaltbelag m → Asphaltdecke f
~**einbaumasse** f, Asphaltdeckenmischgut n, Asphaltdeckenmischung f, Asphaltbelagmischgut, Asphaltdeckeneinbaumasse, Asphaltdeckengemisch n, Asphaltbelaggemisch, Asphaltbelagmischung, Bitumenbelageinbaumasse, Bitumendeckenmischung, Bitumendeckenmischgut, Bitumenbelagmischgut, Bitumendeckeneinbaumasse, Bitumendeckengemisch, Bitumenbelaggemisch, Bitumenbelagmischung = asphaltic paving mix(ture) (US); (~) bitumen pavement ~ (Brit.)
Asphalt|bergwerk n → Asphaltgrube f
~**bestimmung** f = hard asphalt determination
Asphaltbeton m, Bitumenbeton [*Gemisch aus Splitt, Natursand (oder Brechsand) und Füller mit Bitumen als Bindemittel*] = asphaltic concrete (US); (~) bitumen ~ (Brit.)
~**belag** m, Asphaltbetondecke f, Bitumenbetonbelag, Bitumenbetondecke = asphalt(ic) concrete pavement, ~ ~ paving (US); (asphaltic) bitumen concrete surfacing (Brit.)

~dichtung f, Bitumenbetondichtung, Asphaltbetonabdichtung, Bitumenbetonabdichtung, Asphaltbeton-(Ab)Dichtungsschicht f, Asphaltbeton-(Ab)Dichtungslage f, Bitumenbeton-(Ab)Dichtungsschicht, Bitumenbeton(-Ab)Dichtungslage, Asphaltbeton-(Ab)Dichtungsvorlage, Bitumenbeton-(Ab)Dichtungsvorlage, Asphaltbeton-(Ab)Dichtungsschürze f, Bitumenbeton-(Ab)Dichtungsschürze, Asphaltbetonschürze, Bitumenbetonschürze, Asphaltbeton-(Ab)Dichtungsteppich m, Bitumenbeton-(Ab)Dichtungsteppich, Asphaltbeton-Außen(haut)dichtung, Bitumenbeton-Außen(haut)-dichtung, Asphaltbetonoberflächendichtung, Bitumenbetonoberflächendichtung, Asphaltbetondichtungsbelag m, Bitumenbetondichtungsbelag, Asphaltbeton(ab)dichtungshaut f, Bitumenbeton(ab)dichtungshaut = impervious asphaltic concrete layer (US); ~ (~) bitumen ~ ~ (Brit.)

~gemisch n, Asphaltbetonmischung f = asphaltic concrete mix(ture)

~mineralgerüst n = asphaltic concrete (mineral) skeleton

~mischanlage f, Bitumenbetonmischanlage = asphaltic concrete mixing plant

~mischer m, Bitumenbetonmischer = asphaltic concrete mixer

~schürze f → Asphaltbetondichtung

~teppich m, Bitumenbetonteppich = asphaltic concrete carpet (US); (~) bitumen ~ ~ (Brit.)

~tragschicht f, obere ~ = asphaltic concrete base (US); (~) bitumen ~ ~ (Brit.)

~verschleißschicht f, Bitumenbetonverschleißschicht = asphaltic plank wearing course, ~ ~ ~ surface (US); (~) bitumen ~ ~ (Brit.)

Asphalt|binder(schicht) m, (f), Bitumenbinder(schicht) = asphaltic binder (course) (US); (~) bitumen ~ (~) (Brit.)

~bodenplatte f, Asphaltfußbodenplatte, Asphalt(fuß)bodenfliese f = asphalt floor finish tile, ~ flooring ~

~-Bordstein m, Asphalt-Hochbordstein, Asphalt-Bordschwelle f = asphalt curb; ~ kerb (Brit.); ~ ~ above ground

~brot n, Naturasphalt-Mastixbrot = mastic block

~bügeleisen n = asphalt smoother, ~ smoothing iron

~dachpappe f → Bitumendachpappe

~-Dachschindel f, Bitumen-Dachschindel = asphaltic roof(ing) shingle (US); (~) bitumen ~ ~ (Brit.)

~decke f, Asphaltstraßendecke, Bitumen(straßen)decke, Asphalt(straßen)belag m, Bitumen(straßen)belag = asphaltic (road) surfacing (US); (~) bitumen (~) ~ (Brit.)

~deckenerhitzer m, Bitumendeckenerhitzer = asphaltic pavement heater, asphalt surface heater (US); road heater, road burner (Brit.)

~dichtung f = asphaltic sealing

~eimer m = asphalt bucket

~einfärbung f = colo(u)ring of asphalt

~eingußdecke f → Mastix-Vergußdecke

~emulsion f → Bitumenemulsion

Asphalten n, Hartasphalt m [*dieser Ausdruck für den unlöslichen Rückstand im Bitumen stammt aus der Zeit, in der Bitumen mit Asphalt bezeichnet wurde*] = asphaltene

Asphalt|erzeugnis n = asphaltic product

~estrich m = asphalt (jointless) floor(ing), ~ composition ~, ~ screed

Asphalteur m, Gußasphaltstreicher m = spreader (Brit.); asphalter (US)

Asphaltfeinbeton m = fine(-graded) asphaltic concrete (US); fine asphalt (Brit.)

~belag m, Asphaltfeinbetondecke f = fine(-graded) asphaltic concrete pavement (US); fine asphalt surfacing (Brit.)

~platte *f*, Asphaltfeinbetonfliese *f* = fine(-graded) asphaltic concrete tile (US); fine asphalt ~ (Brit.)

~teppich(belag) *m* = fine asphalt carpet, ~ ~ mat (Brit.); fine(-graded) asphaltic concrete carpet (US)

Asphalt|feinbinder *m*, Bitumenfeinbinder = fine-grained asphaltic binder course

~fertiger *m*, Guß~ [*auf Schienen fahrbar mit Handkurbelantrieb, Arbeitsbreite durch Einsatzstücke verstellbar, Propangasheizung für Bohle*] = mastic asphalt finisher

~filzplatte *f* = asphaltic felt panel

~fliese *f*, Asphaltplatte *f* = asphalt tile

Asphalt(fuß)boden *m* = asphalt flooring, ~ floor finish [*Asphalt floorings are of three types, i. e. asphalt tile, heavy asphalt mastic, and light asphalt mastic*]

~platte *f*, Asphalt(fuß)bodenfliese *f* = asphalt flooring tile, ~ floor (finish) ~

Asphalt|gabel *f* = asphalt fork

~gestein *n* → Natur~

~gesteinmehl *n* = (natural) rock asphalt powder

~(gestein)vorkommen *n*, Asphalt(gestein)lager(stätte) *n*, (*f*), Asphalt(gestein)fundstätte, Asphalt(gestein)vorkommnis *n* = (natural) rock asphalt deposit

~grobbeton *m* = coarse(-graded) asphaltic concrete (US); coarse asphalt (Brit.)

~grobbinder *m*, Bitumengrobbinder = coarse-grained asphaltic binder course

~grube *f*, Asphaltbergwerk *n* = (natural) rock asphalt mine

Asphaltieren *n*, Asphaltierung *f*, Asphaltierungsarbeiten *fpl* = asphalt work

asphaltisches Säureanhydrid *n* = asphaltous acid anhydride

Asphaltisolierplatte *f* = asphaltic insulating slab

Asphaltkalkstein *m*, bituminöser Kalkstein, Stinkkalkstein = bituminous limestone, asphaltic ~

~grube *f* = bituminous limestone quarry, asphaltic ~ ~

Asphalt|kaltbeton *m* = Damman cold asphalt, fine ~ ~, cold fine asphalt (Brit.); Damman cold asphaltic concrete (US)

~kies *m* → Bitu(men)kies

~kitt *m*, Bitumenkitt = asphalt putty (US); (asphaltic-)bitumen ~ (Brit.)

~klebemasse *f* = asphaltic adhesive compound

~kocher *m*, Guß~, Mastixkocher, Asphalt(-Schmelz)kessel *m* = mastic cooker, ~ asphalt mixer

~kocher *m* = potman [*A skilled man who prepares and heats solid asphalt in a cauldron with a suitable amount of grit*]

~lack *m*, Bitumenlack = black Japan, bituminous varnish (Brit.); asphalt enamel, asphalt varnish (US)

~lage *f*, Asphaltschicht *f* = asphalt(ic) layer

Asphaltmakadam *m*, Bitumenmakadam, bituminöser Makadam = (asphaltic) bitumen macadam (Brit.); asphalt ~ (US)

~decke *f*, Asphaltmakadambelag *m*, Bitumenmakadamdecke, Bitumenmakadambelag = (asphaltic) bitumen macadam surfacing (Brit.); asphalt ~ ~ (US)

(Asphalt)Mastix *m* = asphalt mastic (US); mastic asphalt (Brit.) [*B. S. 1446*]

~brot *n* = mastic block

~prisma *n* = mastic asphalt beam (Brit.); asphalt mastic ~ (US)

Asphalt|matte *f* = asphaltic mattress

~mattenverlegeschute *f* = asphaltic mattress laying vessel

~mauerwerk *n* [*Mauerwerk mit Asphaltmörtel*] = asphaltic-mortar walling

Asphaltmehl — Atlant

~mehl n [*gemahlenes bitumenhaltiges Gestein, das zusammen mit Bitumen für Stampfasphalt verwendet wird*] = powdered asphalt (US)
~meißel m = asphalt cutter
~(misch)anlage f → Walz~
~mischer m = asphalt mixer
~mischmakadam m, Bitumenmischmakadam = (asphaltic) bitumen macadam (Brit.); asphaltic macadam (US)
~mörtel m, Bitumenmörtel [*Mörtel mit Bitumen als Bindemittel*] = asphaltic mortar (US); (asphaltic) bitumen ~ (Brit.)
~öl n = asphaltic residual oil
~papier n → Bitumenpapier
~pappdach n → Bitumenpappdach
~pappe f → Bitumen(dach)pappe
Asphaltplatte f, Asphaltfliese f = asphalt(ic) tile
~ [*Wasserbau*] = asphalt(ic) slab
Asphalt|plattenrinnstein m = asphalt(ic) tile gutter
~pulver n [*gemahlener Asphaltkalkstein*] = bituminous limestone powder, asphaltic ~ ~
~rechen m = asphalt rake
~sand m → Bitumensand
~sandstein m, bituminöser Sandstein, bitumenhaltiger Sandstein = bituminous sandstone, asphaltic ~
~schicht f, Asphaltlage f = asphalt(ic) layer
~schiff n = asphalt-mattress laying vessel
~schindel f → Bitumenschindel
~see m, Pechsee = asphalt lake, pitch ~
~splittdecke f → Bitumensplitt-Teppich(belag) m
~stadtstraße f = asphalt street
~stampfer m = asphalt tamper
~straße f = asphalt road, ~ highway
~straßenbau m, Bitumenstraßenbau = asphalt road construction, ~ highway ~
~(straßen)belag m → Asphaltdecke f
~straßendecke f → Asphaltdecke

~streumakadam m, Asphalteinstreudecke f = asphalt dry penetration surfacing, ~ ~ process penetration macadam
~-Torkret m = asphalt gunite
~tragschicht f → obere ~
~tränkmakadam m, Bitumentränkmakadam = (asphaltic) bitumen grouted macadam (Brit.); ~ penetration ~
~überzug m = asphalt(ic) overlay
~vergußmasse f = asphalt filler (US) [*An asphaltic product used for filling cracks and joints in pavements*]
~verschleißschicht f, Asphaltverschleißlage f = asphalt(ic) wearing course
~wandplatte f, Asphaltwandfliese f = asphalt(ic) wall tile
~wasserbau m, bituminöser Wasserbau = bituminous blankets and structures for hydraulic engineering structures (Brit.); asphalt ~ ~ ~ ~ ~ ~ (US)

Ästelzaun m, Spalierzaun = lattice fence
Ästhetikfrage f = aesthetic aspect
Ast-Molindecke f = Ast-Molin ribbed floor
Ästuarium n, Ästuar m = estuary
Asynchronmotor m für Drehstrom, Induktionsmotor, Drehstrommotor = asynchronous motor, polyphase induction ~
Atem|luft f = breathing air
~schutzgerät n, Atmungsgerät = respiratory device, ~ apparatus, breathing ~
Äther|kapsel f = ether capsule
~starthilfe f, Ätheranlaßhilfe = ether starting aid
~zerstäuber m = ether discharger
Äthyläther m, $C_2H_5OC_2H_5$ = ethylether
Äthylen n, CH_2CH_2 = ethylene
Atlant m, Simsträger m, Telamon m = atlante [*Carved male figure used instead of columns to support an entablature in classic architecture*]

Atlas-Copco-Bohreinheit *f*, Gesteinsbohreinheit, Doppel-Bohraggregat *n* = twin-drill

Atmosphärilien *fpl* [*Die physikalisch und chemisch wirksamen, in der Atmosphäre vorkommenden Stoffe, wie z. B. Kohlensäure, Ozon, Sauerstoff, Salpetersäure, Ammoniak, Wasser und Wasserdampf*] = climatic agencies

atmosphärische Korrosion *f* = atmospheric corrosion

atmosphärischer Druck *m*, barometrischer ~ = atmospheric pressure

~ **Niederschlag** *m* [*aus der Lufthülle in flüssiger oder fester Form ausgeschiedenes Wasser; Hauptformen: Regen und Schnee*] = atmospheric precipitation

Atmungs|gerät *n*, Atemschutzgerät = breathing apparatus, respiratory ~, ~ device

~schlauch *m* = breathing tube

Atoll *n*, Ringriff *n*, Lagunenriff = atoll

Atomantrieb *m*, Kernenergie-Antrieb = nuclear propulsion

atomare Wasserstoffschweißung *f*, atomares Lichtbogenschweißen *n* = atomic-hydrogen welding

Atombunker *m*, Atomschutzbunker = atomic shelter

atomgetrieben, mit Atomantrieb = with nuclear propulsion

Atomgewicht *n* = atomic weight

atomistische Natur *f* **der Bruchvorgänge** [*Zerkleinerungstechnik*] = atomic mechanics of fracture

Atom|kraftwerk *n*, Atomenergieanlage *f*, Kernkraftwerk, Atom-Elektrizitätswerk, Atom-Eltwerk = atomic power station, nuclear power plant, atomic electric plant

~schiff *n*, Kernenergieschiff = nuclear ship

~(schutz)bunker *m* = atomic shelter

~sprengungsaushub *m* = atomic blast excavation

Atrium *n* = atrium

atro, absolut trocken = absolutely dry

Atterberg'sche Konsistenzgrenzen *fpl* → Konsistenzgrenzen nach Atterberg

Attika *f* = attic [*In architecture, a low wall or stor(e)y above the cornice of a classical façade*]

~-Riegel *m* [*Süddeutschland*]; Attika-Element *n* = attic cladding element

attischer Säulenfuß *m* = Attic base, moulded ~

Ätz-Alkalität *f* = caustic alkalinity

Ätzkali *n*, Kaliumhydroxyd *n*, Kaliumhydrat *n*, KOH = potassium hydroxide, ~ hydrate, caustic potash

Ätzkalk *m* → Kalziumhydroxyd

Ätznatron *n*, kaustische Soda *f*, kaustisches Soda *n*, Natriumhydroxyd *n*, NaOH = caustic soda

~natronprobe *f*, Ätznatronprüfung *f*, Ätznatronversuch *m* [*Betonzuschlag*] = organic test for fine aggregate, test for organic matter, Abrams' test, extraction with caustic soda, colorimetric test (for organic impurities)

Audiometrie *f* [*die Auswirkung von Geräuschen auf die Hörsamkeit*] = audiometry

Aueboden *m* [*Gleiboden in Überschwemmungsgebieten*] = riverside soil

Auelehm *m* → Silt *m*

Auenebene *f*, Flußaue *f*, Hochflutebene *f* = flood plain

auf der Baustelle geschweißt, montagegeschweißt = field-welded, site-welded

Aufbau *m*, bauliche Anordnung *f*, Konstruktion *f*, Ausführung, Bau [*Maschine*] = design, version

~, Zusammensetzung *f* [*Beton; Mörtel*] = composition

~, Montage *f*, Aufstellen *n*, Aufstellung *f*, Aufbauen = erection

~ [*Teil eines Bauwerkes oder Gebäudes über Geländeoberfläche. Gegenteil: Fundament = foundation, substructure*] = superstructure [*The part of a structure or building above the surface of the ground*]

Aufbau — aufblasen 94

~ [*Teer; Bitumen*] = chemical constitution

~-**Bagger** *m*, Schlepper-~ = tractor-mounted excavator

~**bagger** *m*, Last(kraft)wagenbagger, LKW-Bagger = truck(-mounted) excavator (US); lorry(-mounted) ~ (Brit.); fast-travel ~

(~-)**Baggergerät** *n*, Aufbaubagger *m* = truck-mounted digging attachment, ~ excavating ~ (US); lorry-mounted ~ ~ (Brit.)

~**damm** *m* = earth(en) dam of composite (cross-)section, composite earth(en) dam

aufbauen [*eine Bohranlage*] = to rig up

Aufbau|gerät *n* = truck(-mounted) attachement (US); lorry(-mounted) ~ (Brit.)

~-**Grabenbagger** *m* → Last(kraft)wagen-Grabenbagger

~**kran** *m*, Last(kraft)wagenkran, LKW-Kran = truck(-mounted) crane (US); lorry(-mounted) ~ (Brit.); fast-travel ~

~**lader** *m* → Last(kraft)wagenlader

Aufbäumen *n* **der Welle**, Hochklettern ~ ~ = uprush, swash

~ ~ ~ **im flachen Wasser** = increase in wave height in shallow water

Aufbau|plan *m*, Aufbauprogramm *n* = development plan, ~ program(me)

~**querschnitt** *m* [*Staudamm*] = composite (cross-)section

~**zeit** *f* [*Bohranlage*] = rigging up time

aufbereiten [*Erze*] = to dress

~ [*Kohle*] = to prepare coal

~ [*Wasser*] = to purify, to treat, to condition

~ [*Zuschlagstoffe*] = to prepare aggregate(s)

~, herstellen, erzeugen [*Beton; Mörtel; Schwarzdeckenmischgut*] = to fabricate, to prepare [*sometimes 'incorrectly called "to mix"*]

Aufbereitung *f*, Herstellung, Erzeugung [*Beton; Mörtel; Schwarzdeckenmischgut*] = preparation, fabrication [*sometimes incorrectly called "mixing"*]

~ **von Erzen**, Erzaufbereitung = mineral dressing

~ ~ **Kohle**, Kohle(n)aufbereitung = coal preparation

~ ~ **Wasser**, Wasseraufbereitung = water purification, ~ treatment, ~ conditioning

Aufbereitungsanlage *f* **für Erze**, Erzaufbereitungsanlage = mineral dressing plant, ~ ~ installation

~ ~ **Kohle**, Kohle(n)aufbereitungsanlage = coal preparation plant

~ ~ **saure Abwässer** = sour-water stripper

~ ~ **Wasser**, Wasseraufbereitungsanlage = water purification plant, ~ treatment ~, ~ conditioning ~

~ ~ **Zuschlagstoffe**, Zuschlagstoff-Aufbereitungsanlage, Aufbereitungsanlage für Betonzuschläge = aggregate(s) preparation plant, ~ installation

Aufbereitungsmaschine *f* **für Erze**, Erzaufbereitungsmaschine = mineral dressing machine

~ ~ **Kohle**, Kohle(n)aufbereitungsmaschine = coal preparation machine

~ ~ **Zuschlagstoffe**, Zuschlagstoff-Aufbereitungsmaschine = aggregate(s) preparation machine

Aufbereitungstechnik *f* = dressing and preparation engineering

Aufbeton *m* → Beton

Aufbeulung *f* (Geol.) = boss, stock

aufbiegen [*Bewehrung*] = to bend up

Aufbiegung(en) *f(pl)*, aufgebogene Stäbe *mpl*, Abbiegung(en) [*Schubbewehrungsart*] = bent-up bars

(**Auf)Blähen** *n*, (Auf)Blähung *f* [*Herstellung von Leichtzuschlägen*] = expansion, expanding; bloating (Brit.)

aufblasbare Füllwand *f* = pneu-bin panel [*Trade name*]

aufblasbarer Gummischlauch *m* **für Kanäle im Beton** = deflatable rubber tube, inflatable ~ ~, ~ ~ core, ductube

aufblasen [*Reifen*] = to inflate

Aufblätterung *f* **(der Schichten)** (Geol.) = bed separation

Aufbohren *n* **des Zementstopfens** = drilling out of the cement plug

Aufbohrrollenmeißel *m*, Erweiterungsrollenmeißel = enlarging roller bit

aufbördeln, (auf)weiten [*Rohr*] = to expand

aufbrechen, aufreißen = to break up, to disintegrate, to loosen [*to disturb superficially hard formations ranging from rock, sandstone, shale, tarmac, heavy clay, laterite and other compacted materials, preparatory to excavation by scraper or dozer*]

~ **der Straßendecke durch Frost** = to heave

Aufbrechen *n*, Aufreißen, Aufbruch *m* = disintegrating, breaking up, disintegration, loosening, cutting

~, Aufbruch *m*, Aufreißen [*mit Aufreiß(anbau)vorrichtung*] = scarifying

~ **der Straßendecke durch Frost**, Aufbruch *m* ~ ~ ~ ~ = frost heave

aufbringen, auftragen [*Anstrichstoff*] = to spread, to apply

~ [*Spannkraft beim Spannbeton*] = to apply, to establish

~ [*Bindemittel; Splitt; Mischgut*] = to apply, to spread, to distribute

Aufbringen *n* → Belasten

~, Verteilung *f* [*Schotter; Splitt; Bindemittel; Mischgut usw. im Straßenbau*] = application, spreading, distribution

~ **der Vorspannung**, ~ ~ (Vor-)Spannkraft = stretching, tensioning, stressing

Aufbringung *f* → Belasten *n*

Aufbringungsrate *f* = application rate

Aufbruch *m* → Aufbruchmassen *fpl*

~ → Aufbrechen *n*

~, Aufreißen *n*, Aufbrechen [*mit Aufreißvorrichtung*] = scarifying

~, Hochbruch, Überbruch, überbrochener Schacht *m* [*Blindschacht von unten nach oben hergestellt*] = raise

~ **der Straßendecke durch Frost**, Aufbrechen *n* ~ ~ ~ ~ = frost heave

~ **durch Eislinsenbildung im Boden** → Frosthebung *f*

~**arbeiten** *fpl* = loosening work

~**bauweise** *f*, Dreizonenbauweise, österreichische Bauweise, österreichisches Verfahren *n*, österreichische Methode *f*, Sohlenort *n* und Firste *f* = (English-)Austrian method

~**hammer** *m* → Straßen~

~**massen** *fpl*, Aufbruch(material) *m*, (*n*) [*Straßenbau*] = broken up material

Aufdach *n* → Dachkappe *f*

aufdrehen, aufschwenken [*Drehbrücke*] = to open [*swing bridge*]

Aufdrehen *n*, Aufschwenken [*Drehbrücke*] = opening [*swing bridge*]

aufdüsen, aufsprühen = to spray on by atomizing

Aufeinanderfolge *f* = sequence

~ **der Schichten** (Geol.) = sequence of strata

Aufenthalt *m*, Durchfluß *m* [*Wasser in einem Becken*] = retention, detention

Aufenthaltsraum *m* = rest room

~ **für Besatzungen** [*Flughafen*] = crew rest room, ~ ~ facility

Aufenthaltszeit *f*, Durchflußdauer *f* [*Wasser in einem Becken*] = retention period, detention ~

auffahren [*Tunnel-, Stollen- und Bergbau*] = to drive

~ **im vollen Querschnitt** [*Tunnel-, Berg- und Stollenbau*] = to drive in full section

Auffahren *n* → Vortrieb *m*

Auffahr|geschwindigkeit *f* [*Verkehrsunfall*] = contact speed

~**richtung** *f* [*Tunnel-, Berg- und Stollenbau*] = direction of driving

Auffahrt *f* [*zur Autobahn*] → Anschlußpunkt *m*

~**rampe** *f* = access ramp

Auffahrunfall *m* = running-up accident

Auffahrung *f*, Auffahren *n* [*Tunnel-, Berg- und Stollenbau*] = driving

Auffaltung *f* (Geol.) = upthrust

Auffang|becken *n* [*Oberflächenwasserabführung*] = catch basin (structure)
~graben *m*, Sammelgraben, (Ab-)Fanggraben, Hintergraben, Saumgraben [*zur Aufnahme des talwärts fließenden Oberflächenwassers*] = intercepting ditch, catch-water ~, diversion ~
~kessel *m*, Fangkessel, Rohrfänger *m* [*Druckluftbetonförderer*] = discharge box
~rinne *f*, Hangrinne = intercepting channel, catch-water ~
auffinden [*eine Lagerstätte*] = to discover [*a deposit*]
Aufflanschung *f* [*Tiefbohrtechnik*] = well-head
Auffrischung *f* = bright pickling
(Auf)Füllboden *m*, (Auf)Schütt(ungs)boden, Auftragboden = fill earth, ~ soil, fill(ing)
Auffüllböschung *f* → Dammböschung
auffüllen = to fill
~ → (auf)schütten
Auffüllen *n* → (An)Schütten
Auffüll|höhe *f* → Auftraghöhe
~-Lage *f* → Auftragschicht *f*
~masse *f* → Auftragmaterial *n*
~material *n* → Auftragmaterial
~schicht *f* → Auftragschicht
Auffüllung *f* [*Strohlehm, Bimskies usw. zur Wärme- und Schalldämmung beim Einschub von Holzbalkendecken*] = pug(ging), deafening, deadening
~ → Schüttung
Auffüllzone *f*, Auftragzone, (Auf-)Schütt(ungs)zone = fill zone
Aufgabe *f*, Beschickung *f*, Material~, Beschicken *n*, Aufgeben *n* = feed(ing), material(s) ~ [*The process of supplying material to a conveying or processing unit*]
~ = problem
~apparat *m* → Aufgeber *m*
~aufzug *m*, Beschickungsaufzug = feeder skip hoist
~aufzug *m* für Silos, Beschickungsaufzug ~ ~ = feeder skip hoist for silos

~band *n* → Bandaufgeber *m*
~becherwerk *n*, Beschickungsbecherwerk, Aufgabeelevator *m*, Beschickungselevator = feed(ing) (bucket) elevator
~bunker *m*, Arbeitsbunker, Beschickungsbunker = surge bin, live storage ~
~dosierung *f*, Beschickungsdosierung = feed batching, ~ proportioning, ~ measuring
~ende *n*, Beschickungsende = feed(ing) end
~feines *n*, Beschickungsfeines [*Siebtechnik*] = fine feed
~gehäuse *n* = feed-end housing
~grobes *n*, Beschickungsgrobes [*Siebtechnik*] = coarse feed
Aufgabegut *n*, Aufgabematerial *n*, Beschickungsgut, Beschickungsmaterial = feed (material), ~ materials
~ [*(Hart)Zerkleinerung*] = crude, feed (material)
~, Haufwerk *n*, Siebgut, Einlaufgut [*Siebtechnik*] = material to be screened, (screen) feed, (screen) head
Aufgabe|kübel *m*, Beschickungskübel = feed(ing) bucket
~menge *f*, Beschickungsmenge = feed(ing) quantity
~öffnung *f*, Beschickungsöffnung = feed opening
~plattenband *n*, Plattenbandspeiser *m*, Plattenbandaufgeber, Plattenbandbeschicker, Lamellen-Bandaufgeber, Lamellen-Bandspeiser, Lamellen-Bandbeschicker, Aufgabegliederband, Gliederbandspeiser, Gliederbandaufgeber, Gliederbandbeschicker = apron(-conveyor) feeder, apron-conveyer ~, apron-type mechanical ~
~rinne *f*, Beschickungsrinne, Speiserinne = feeder channel
~rohr *n*, Beschickungsrohr, Speiserohr = feed pipe [*The pipe through which the feed material to a mill, kiln, tank, etc. is conveyed*]
~rost *m* → Beschickungsrost
~schnecke *f* → Beschickungsschnecke

Aufgabeschurre — aufgesattelter Kippanhänger

~schurre *f*, Beschickungsschurre = feed(ing) chute (US); ~ shoot (Brit.); ~ spout
~silo *m*, Beschickungssilo = feed(ing) silo
~teil *m, n*, Beschickungsteil = feed(ing) unit
~trichter *m*, Beschickungstrichter = feed(ing) hopper
~trog *m*, Trogaufgeber *m* = trough feeder, feed(ing) trough
~vorrichtung *f* → Aufgeber *m*
~vorrichtung *f* für Zuschlagstoffe, ~ ~ Zuschläge, Beschickungsvorrichtung ~ ~ = aggregate feeder
~walze *f* → Beschickungswalze
aufgeben [*eine Bohrung*] = to abandon [*a well*]
~, beschicken, speisen [*Das Material wird „aufgegeben". Die Maschine usw. wird „beschickt"*] = to feed [*The material is fed into the machine. The machine is fed with material*]
Aufgeben *n* → Aufgabe *f*
Aufgeber *m*, Aufgabeapparat *m*, Beschickungsapparat, Beschicker, Aufgabevorrichtung *f*, Beschickungsvorrichtung, Material~, Speiser *m*, Beschickervorrichtung *f* = (material(s)) feeder, (~) feeding unit, (~) feeding device
aufgeblähte Hochofenschlacke *f*, schaumige ~, geschäumte ~, Hochofenschaumschlacke [*Die gebrochene geschäumte Hochofenschlacke wird Hüttenbims oder Kunstbims genannt*] = foamed (blast-furnace) slag (Brit.); expanded (~) ~ (US)
aufgeblähter Beton *m* → Blähbeton
~ **Schieferton** *m* → Blähschieferton
~ **Ton** *m*, Blähton *m* = bloating clay (Brit.); expanded ~
~ **Tonschiefer** *m* → Blähtonschiefer
aufgebogene Stäbe *mpl*, Aufbiegung(en) *f(pl)*, Abbiegung(en) [*Schubbewehrungsart*] = bent-up bars
aufgebrachte Last *f* = applied load
aufgebrochen, aufgerissen = broken up, disintegrated, loosened, cut

aufgefüllter Boden *m* → angeschütteter ~
~ **Untergrund** *m* → aufgeschütteter ~
aufgefülltes Gelände *n*, aufgeschüttetes ~, angeschüttetes ~, Anschüttung *f*, (Auf)Schüttung *f*, Auftrag *m*, Auffüllung *f*, aufgetragenes Gelände = filled ground, filled-up ~, made-up ~, fill
aufgehängte Taucherglocke *f* = suspended diving bell
aufgehende Zahl *f*, rationale ~ = rational number, ~ quantity
aufgehender Teil *m*, aufgehendes ~ *n* = rising part
aufgehendes Mauerwerk *n* über Geländeoberkante = above-grade masonry
aufgekohlt = carburized
aufgelagert = supported
aufgelöste Bauweise *f* [*Tunnel- und Stollenbau*] = method (of attack) with several drifts
~ **Gewichtsperre** *f* → Pfeilersperre
~ **(Stau)Mauer** *f* → Pfeilersperre
aufgelöstes Widerlager *n*, Sparwiderlager, Hohlwiderlager = hollow abutment
aufgenommene Last *f* = accepted load
~ **Leistung** *f* [*El-Motor*] = input
aufgepumpt = inflated
aufgerissen, aufgebrochen = disintegrated, broken up, loosened, cut
aufgesattelt, Aufl(i)eger-... = semitrailer type, gooseneck ~, semimounted ~
aufgesattelte Treppe *f* [*Stufen auf die zahnartig ausgeschnittenen Wangen aufgelegt*] = stair(s) with treads fitted on strings, saddled stair(s)
aufgesattelter Anhänger *m* → Sattelanhänger
~ **Bodenentleerer** *m* → Halbanhänger mit Bodenentleerung
~ **Kippanhänger** *m*, Kipp-Halbanhänger, Kipp-Aufl(i)eger *m*, Kipp-Sattel(schlepp)anhänger, Kipp-Aufsattel-Anhänger = dump semitrailer

aufgesattelter Schrapperwagen — (Auf)Ladegleis 98

~ **Schrapperwagen** *m* → Aufsattel-Schrapperwagen

~ **Schürfkübel(wagen)** *m* → Aufsattel-Schrapperwagen

aufgeschlossen (Geol.) = disclosed

aufgeschütteter Boden *m* → angeschütteter ~

~ **Untergrund** *m*, aufgefüllter ~, aufgetragener ~, angeschütteter ~ = artificial subgrade

aufgeschüttetes Gelände → aufgefülltes ~

aufgeschweißt = welded-on

aufgesetzte (Wehr)Schwelle *f* = renewable sill

aufgetragener Boden *m* → angeschütteter ~

~ **Untergrund** *m* → aufgeschütteter ~

aufgetragenes Gelände *n* → aufgefülltes ~

aufgewalztes Gewinde *n* = rolled(-on) thread

aufgewickelte Länge *f* [*Stahldraht*] = coiling length

aufgezwungene Schwingung *f*, erzwungene ~, Kraftschwingung ~ = forced vibration

aufgleisen = to rerail

Aufgleisen *n* = rerailing

Aufgleiskran *m* = rerailing crane

Aufgleitregen *m* = warm-front rain

aufgraben, schürfen [*Baugrundaufschluß*] = to investigate by digging, to test ~ ~

Aufgraben *n*, Schürfen, Beschürfung *f*, Schurf *m*, Schürfarbeit(en) *f(pl)* [*Baugrundaufschluß*] = subsurface exploration by test pits, foundation ~ ~ ~ ~, ~ reconnaissance ~ ~ ~, ~ investigation ~ ~ ~

aufhacken = to pick

Aufhänge|brücke *f* [*Wehr*] = supporting bridge, upper support for frames

~-**Fender** *m* = suspension(-type) fender

~**haken** *m* = suspension hook

~**kabel** *n*, Aufhängeseil *n* = suspension rope

~**kette** *f* = suspension chain

(~)**Nase** *f* [*Dachziegel*] = nib

~**rolle** *f* der Rollenbahn [*Wehr mit Hubschützen*] = suspension pulley of the roller train

~**stange** *f* = suspension rod

~**stange** *f*, Pleuelstange [*Klappenwehr mit unterer Achse*] = suspension link

~**vorrichtung** *f* = suspension device

aufhaspeln = to reel up

Aufhauen *n*, Überhauen, Überbruch *m* [*von unten nach oben in der Lagerstätte zur Vorrichtung hergestellter Verbindungsweg zwischen zwei Abbaustrecken*] = rise working

aufheizen, erwärmen [*Bindemittel; Wasser; Zuschläge und dgl.*] = to heat

Aufheiz- und Mischtrog *m* [*für Teer und Bitumen*] = heating and mixing kettle for tar and (asphaltic) bitumen (Brit.) / asphalt (US)

Aufheizung *f*, Erwärmung = heating

aufhellen [*Straßendecke durch Zusatz heller Zuschlagstoffe*] = to relieve the dark appearance, to lighten, to brighten

Aufhellungsvermögen *n* = capacity for relieving the dark appearance

Aufhöhung *f* [*Deich; Damm; Talsperre*] = heightening

aufholen [*aus dem Bohrloch*] = to withdraw

Aufkadung *f*, Aufkastung = temporary heightening of a dyke (Brit.); ~ ~ ~ ~ dike (US)

aufkanten, hochkanten = to fold up

Aufkastung *f*, Aufkadung = temporary heightening of a dyke (Brit.); ~ ~ ~ ~ dike (US)

Aufkipp-Bauweise *f*, Richtaufbauweise = tilt-up method, ~ construction

aufklaffen, (auseinander)klaffen = to gape

Aufklaffen *n*, (Auseinander)Klaffen [*Fuge*] = opening

(**Auf**)**Lade|band** *n* → Bandlader *m*

~**gebläse** *n* → Auflader *m*

~**gerät** *n* → Auflader *m*

~**gleis** *n*, Beladegleis = loading siding

~grad *m* = supercharge pressure, boost [*supercharging engines*]
~kompressor *m* → Auflader *m*
~maschine *f* → Auflader *m*
aufladen = to load
Aufladen *n* = loading
Auflader *m*, (Auf)Ladegerät *n*, (Auf-)Lademaschine *f*, Lader, Belademaschine, Beladegerät, (Ver)Ladegerät = (material) loader, (~) loading machine
~, Aufladekompressor *m*, Aufladegebläse *n*, Vorverdichter *m* = supercharger, blower
~ für Schüttgut → Becher(werk)auflader
Aufladerampe *f* = loading ramp
Aufladung *f*, Vorverdichtung [*Dieselmotor*] = supercharging
Auflage|fläche *f* [*(Gleis)Kette*] = area of ground contact
~fläche *f* **des (Gleis)Kettengliedes** = link shoe strap
~länge *f* [*(Gleis)Kette*] = length of track on ground
Auflager *n* [*die Stelle, an der ein Träger auf der Unterkonstruktion aufliegt und die von ihm belastet wird*] = support
~ [*Nadelwehrbock*] = stop
~bedingung *f* = condition of support, support condition
~druck *m* → Auflagerkraft
~fläche *f* = surface of support
(~)Knagge *f* = (angle) cleat [*A small bracket of angle section fixed in a horizontal position, normally to a wall or stanchion, to support or to locate a structural member*]
~kraft *f*, Stützenwiderstand *m*, Stützendruck *m*, Gegendruck, Auflagerdruck, Stützkraft, Auflagerreaktion *f*, Auflagerwiderstand = bearing pressure, ~ reaction, reaction of the support
~länge *f* = length of support, supporting length
~mauerwerk *n* = bearing masonry
~mitte *f* = centre of support (Brit.); center ~ ~ (US)
~nachgiebigkeit *f* = resilience of the support
~platte *f* = plate [*A plate which is over 2 in. thick is called bearing plate*]
~punkt *m* = point of support
~quader *m* [*Auflagerkörper aus Beton, Stahlbeton oder Naturstein. Er nimmt den Lagerdruck einer Brücke auf und verteilt ihn auf das Widerlager*] = bearing block
~quadrat *n* = square of support
~reaktion *f* → Auflagerkraft *f*
~stein *m* = bearing stone
~widerstand *m* → Auflagerkraft
~winkel *m* [*Stahlbau*] = angle
Auflagerung *f* = support
Auflage|stufe *f* → Trittstufe
~zeit *f* [*Siebbelag*] = service time, ~ period
auflanden, kolmatieren, anschwemmen, aufschwemmen, aufschlicken = to silt up
auflandiger Wind *m*, Seewind = onshore wind
Auflandung *f*, Anschwemmung, Aufschwemmung, Aufschlickung, (natürliche) Kolmation *f*, Kolmatierung, Verschlickung, Verlandung (Geol.) = filling up, silting up, accretion through alluvium, aggradation, sedimentation, deposition of sediments
Auflandungsteich *m* → Schlammteich
Auf-Länge-Schneiden *n*, Ablängen *n* = cutting-to-length
Auflast *f* = surcharge
~höhe *f* = depth of surcharge, surcharge depth
~verhältnis *n* → → Spundwand *f*
Auflauf|bremse *f* = over-run braking gear, ~ (controlled) brake, brake operating on over-ride principle
~klappe *f* [*bewegliche Brücke*] = approach flap
~zunge *f* = tilting rail
Aufleger *m* → Sattelanhänger
~sprengung *f* = plaster shooting (Brit.); mudcapping, bulldozing (US)

Auflichtmikroskop n = reflecting microscope
aufliegen [*Siebbelag*] = to be in service
Aufl(i)eger m → Sattelanhänger
~-..., aufgesattelt = gooseneck type, semi-trailer ~, semi-mounted ~
Aufl(i)eger m **für Baggertransport**, Baggertiefladehalbanhänger m = semi-trailer for transport of excavating machines
~-Förderwagen m, Halbanhänger-Förderwagen [*Erdbau*] = gooseneck (-type) wagon
~-Hinterkipper m → Halbanhänger-Rückwärtskipper
~-Rückwärtskipper m → Halbanhänger-Rückwärtskipper
~-Schrapperwagen m → Aufsattel-Schrapperwagen
~-Schürfkübel(wagen) m → Aufsattel-Schrapperwagen
~-Seitenkipper m → Halbanhänger-Seitenkipper
auflockern = to bulk [*Excavated material increases its volume above the volume of the excavation from which it came, often about 50%*]
Auflockern [*einer Masse durch Rühren*] = agitating
Auflockerung f [*durch maschinelle Bearbeitung des Bodens*] = loosening
~ [*erfolgt beim Lösen von gewachsenem Boden oder Fels*] = bulking
~ [*Tunnel- und Stollenbau*] = breaking up
Auflockerungs|beiwert m [*Aushubmaterial*] = bulking coefficient, coefficient of bulking
~druck m [*Tunnel-, Stollen- und Bergbau*] = breaking-up pressure
~schnecke f, Verteilerschnecke [*Schwarzbelageinbaumaschine*] = spreading screw, spreader ~, feed ~, (cross) auger, agitator
~sprengung f = bursting
~tiefe f [*maschinelle Bearbeitung des Bodens*] = loosening depth
Auflösbarkeit f, Löslichkeit = solubility
(auf)löslich = soluble

Auflösung f, Zerfall m, Zerrüttung = disintegration
(Auf)Lösungsmittel n, (Auf)Lösemittel = solvent
Aufmaß n [*Schweiz: Ausmaß*] = measurement (of quantities), site measuring, quantity survey
gemeinsames ~ = joint ~
~ und Abrechnung f = measurement and payment
(auf)mauern = to brick up
Aufmauerung f [*Caissongründung*] = rising masonry
Aufnahme f, Bau~ = survey (of a structure)
~, geodätische ~ = (geodetic) survey
~ der Querprofile = cross-sectioning, taking cross-sections
~ von natürlichen Baustoffvorkommen für den Straßenbau = highway materials survey, road ~ ~
~becherwerk n → Becherwerk
~behälter m = receiving container
~behälter m, Mischgutbehälter [*Schwarzbelageinbaumaschine*] = (receiving) hopper
~bunker m → Annahmebunker
~elevator m → Becherwerk n
~fähigkeit f, Aufnahmevermögen n [*z. B. die Fähigkeit eines Bodens Wasser aufzunehmen*] = absorptive capacity
~(förder)band n = receiving conveyor, ~ conveyor
~gerät n, Aufnahmemaschine f, Aufnehmer m, Rückverladegerät, Rückverlademaschine, Rückverlader = rehandling unit, reclaiming ~
~kratzer m, Rückverladekratzer = reclaiming scraper, rehandling ~
~mischen n = travel plant mixing, travel(l)ing ~
~mischer m → → Bodenvermörtelungsmaschine
~mischverfahren n [*Bodenverfestigung*] = travel plant method, travel(l)ing ~ ~
~trichter m, Annahmetrichter, Einfüll(ungs)trichter, Fülltrichter = receiving hopper

aufnehmen [*Last*] = to accept, to sustain, to carry
~ von Profilen = to profile
Aufnehmen *n*, Rückverladen = reclaiming, rehandling
Aufnehmer *m* → Becher(werk)auflader
~ = recorder
~ → Aufnahmegerät *n*
~mischer *m* → → Bodenvermörtelungsmaschine *f*
Aufprall *m* → Anprall
Aufpressen *n* [*zur Befestigung von gebohrten Teilen auf Wellen, Achsen und Zapfen*] = forcing on
aufpumpen = to inflate
Aufpumpen *n*, Aufspülen = filling ground by pumping dredged materials
(auf)quellen, schwellen = to swell
(Auf)Quellen *n*, Quellung *f*, Schwellbewegung, Schwellung, Schwellen, Aufquellung = swelling
~ des Liegenden, Sohlendruck *m* (⚒) = pucking, creep, pooking [*Rising of the floor of a coal mine when the floor is soft and there is much squeeze. The literary word is "heave"*]
Aufragung *f* [*Geomorphologie*] = eminence
aufrahmbarer Stoff *m* [*Abwasserwesen*] = matter forming a scum
aufrauhen = to roughen
~ → anrauhen
Aufrauh|gerät *n*, Aufrauhmaschine *f*, Anrauhgerät, Anrauhmaschine, Abstumpfgerät, Abstumpfmaschine = roughening machine
~splitt *m* = chip(ping)s for roughening treatment
Aufrauhung *f* → Abstumpfung
(Auf)Räumen *n*, Nachnehmen *n* = reaming
(Auf)Räumer *m* → Nachnehmer
aufrechte Falte *f* (Geol.) = upright fold
Aufrechterhaltung *f* **der Fahrwassertiefe** = maintaining the navigable depth
aufreiben [*ein Loch*] = to ream

Aufreiber *m*, Reibahle *f*, Räumer *m* = reamer
Aufreiß(anbau)vorrichtung *f*, Aufreißer *m* = scarifier [*An accessory on a grader, roller, or other machine, used chiefly for shallow loosening of road surfaces*]
aufreißen → aufbrechen
~ [*mit Anhängeaufreißer*] = to rip
~ [*mit Aufreiß(anbau)vorrichtung*] = to scarify
Aufreißen *n*, Aufbruch *m*, Aufbrechen [*mit Aufreiß(anbau)vorrichtung*] = scarifying
~, Aufbrechen, Aufbruch *m* = breaking up, disintegrating, disintegration, loosening, cutting
~ = drawing of elevations and sections
Aufreißer *m*, Aufreißgerät *n* [*allgemein*] = loosening equipment
~, Anhänge**~**, Schlepp**~**, Tief**~** = rooter, ripper [*A towed machine equipped with teeth, used primarily for loosening hard soil and soft rock*]
~, Aufreiß(anbau)vorrichtung *f* = scarifier [*An accessory on a grader, roller, or other machine, used chiefly for shallow loosening of road surfaces*]
~aufhängung *f* = ripper bracket, rooter **~**
~ausschaltung *f* = ripper kick-out, rooter **~**
~gestänge *n* = ripper linkage, rooter **~**
~leitung *f* = ripper line, rooter **~**
~querbalken *m* = shank beam
~träger *m* = ripper carriage bar, rooter **~ ~**
~zahn *m*, Aufreiß(er)zinke(n) *f*, (*m*), Aufreißzahn = tooth, tine
~zahnspitze *f*, Aufreißzahnspitze = tine tip, tooth **~**
Aufreiß|gerät *n*, Aufreißer *m* [*allgemein*] = loosening equipment
~gut *n* [*mit Anhängeaufreißer aufgerissenes Material*] = ripped material
~gut *n* [*mit Aufreiß(anbau)vorrichtung aufgerissenes Material*] = scarified material

Aufreißhammer — aufschließen

~**hammer** *m* → Straßenaufbruchhammer

~**tiefe** *f*, Arbeitstiefe, Reißtiefe [*Aufreißer*] = digging depth of tines below ground level, working depth, depth of penetration

~**- und Abbauhammer** *m* → Abbau- und Aufreißhammer

~**widerstand** *m* = resistance to breaking up, ~ ~ disintegrating, ~ ~ disintegration, ~ ~ cutting, ~ ~ loosening

~**winkel** *m* = angle of cutting, ~ ~ loosening, ~ ~ breaking up, ~ ~ disintegrating, ~ ~ disintegration

Aufreiter *m* → Dachkappe *f*

Aufrichten *n* **(der Böcke)**, Abschließen [*Nadelwehr*] = closing, raising

~ **des Mastes** = tilting up of the mast

Aufrichter *m* [*(Pfahl)Ramme*] = frame erecting gear, ~ raising ~

Aufrichtungsboot *n*, Boot zur Aufrichtung des Wehrs = working barge (Brit.); maneuver boat (US)

Aufriß *m*, Ansicht *f*, Vertikalprojektion *f* = elevation

aufrollen (⚒) = to work out a whole district in one direction

Aufsattelanhänger *m* → Sattelanhänger

aufsattelbarer Bodenentleerer *m* → Halbanhänger mit Bodenentleerung

Aufsattel|-Bodenentleerer *m* → Halbanhänger mit Bodenentleerung

~**-Schrapperwagen** *m*, Aufsattel-Schürfkübel(wagen) *m*, Aufl(i)eger-Schrapperwagen, Aufl(i)eger-Schürfkübel(wagen), aufgesattelter Schürfkübel(wagen), aufgesattelter Schrapperwagen = two-wheel(ed) scraper

~**-Schürfkübel(wagen)** *m* → Aufsattel-Schrapperwagen

Aufsatz *m* **für die Ladegabel** = fork standards

~**klappe** *f*, Eisklappe [*Wehr*] = ice shutter

~**schloß** *n* = rim lock

~**schlüssel** *m* = socket wrench

~**silo** *m* [*landwirtschaftlicher Silo*] = pit silo

~**teil** *m, n*, drehbarer Oberteil *m*, (drehbares) Oberteil *n*, Schwenkteil, schwenkbarer Oberteil, schwenkbares Oberteil, Oberwagen *m*, Drehteil, drehbarer Oberwagen, schwenkbarer Oberwagen [*Bagger; Kran*] = revolving superstructure, upper machinery

aufsaugen = to imbibe

Aufsaugung *f*, Aufsaugen *n* = imbibition

aufschäumbares Polystyrol *n* = expandable polystyrene

aufschäumen [*STYROPOR*] = to expand

Aufschieber *m* **für Förderkorbbeschikkung** = decking ram

~**motor** *m* (⚒) = decking engine

Aufschiebling *m* = tilting fillet, tilting piece, eaves board, doubling piece, arris fillet; cant strip (USA) [*A board of triangular cross-section nailed to the rafters or roof boarding under the double eaves course to tilt it slightly less steeply than the rest of the roof and to ensure that the tails of the lowest tiles bed tightly on each other*]

aufschlagen = to impinge [*To strike directly upon (as a flame may impinge on the load in a kiln)*]

Aufschlämmbares *n* → Feinstkorn *n*

aufschlämmen, aufschwemmen = to suspend

Aufschlämmung *f*, Aufschwemmung, Suspension *f* = suspension

Aufschleppe *f*, Schlipp *m*, Slip, (Schiffauf)Schleppe [*geneigte Ebene mit Schleppwagen auf einer Schienenbahn zum Herausziehen der Schiffe aus dem Wasser. Herstellung als Quer- oder Längsaufschleppe*] = slip(way)

Aufschleppen *n* **von Schiffen** = hauling up the slip(way)

aufschlicken → auflanden

aufschließen [*eine Lagerstätte auf (Ab-) Bauwürdigkeit und Umfang durch Schürfbaue und/oder Tiefbohrungen untersuchen*] = to explore

~, verflüssigen [*Steinzeugindustrie*] = to deflocculate, to dissolve, to disintegrate
Aufschließung *f* → Aufschluß *m*
Aufschließungs|arbeit(en) *f(pl)*, Aufschlußarbeit(en) = exploration work, exploratory ~
~**bohrgerät** *n*, Aufschlußbohrgerät = exploration drill, exploratory ~
~**bohrloch** *n*, Aufschlußbohrloch = exploratory drill hole, exploration ~ ~
~**bohrung** *f*, (Aufschluß)Bohrung, Suchbohrung, Prospektionsbohrung Pionierbohrung, Untersuchungsbohrung, Erkundungsbohrung = exploration drilling, ~ boring, exploratory ~
~**bohrung** *f* **mit Erdbohrer**, Aufschlußbohrung ~ ~ = auger drilling, ~ boring
~**schacht** *m*, Aufschlußschacht, Untersuchungsschacht, Suchschacht, Pionierschacht, Prospektionsschacht, Erkundungsschacht = exploration shaft, exploratory ~
~**stollen** *m*, Aufschlußstollen, Suchstollen, Pionierstollen, Untersuchungsstollen, Prospektionsstollen, Untersuchungsstollen = exploration gallery, exploratory ~
~**vorhaben** *n*, Aufschlußvorhaben, Untersuchungsvorhaben, Prospektionsvorhaben, Erkundungsvorhaben = exploration project, exploratory ~
Aufschlüsselung *f* [*Kosten, Arbeiten*] = breakdown
Aufschluß *m*, Verflüssigung *f* [*Steinzeugindustrie*] = defloccuation
~, Aufschließung *f* [*Untersuchung einer Lagerstätte auf (Ab)Bauwürdigkeit und Umfang durch Schürfbaue und/oder Tiefbohrungen*] = exploration
~ → Ausgehende *n*
~ [*jener Ort in der Geologie oder Hydrologie, der einen Einblick in Lagerung der Gesteine und Schichten gewährt*] = exploration point

Aufschluß *m*, Erschließung *f* [*Bodenschätze*] = winning [*The act of building shafts, levels and passages to attain seams*]
~ **von Ton** = dispersion of clay
~**bohrung** *f* = preliminary hole
~**mittel** *n*, Verflüssigungsmittel [*Steinzeugindustrie*] = deflocculant
(Auf)Schottern *n* [*Schweiz*]; Schuttern *n*, Schutterung *f*, Fortschaffen *n* der Ausbruchmassen [*Tunnel- und Stollenbau*] = mucking
aufschraubbarer Gestängeverbinder *m* [*Rotarybohren*] = screwed-on tool joint
Aufschraubrohr *n* = screwed-on pipe
Aufschrumpfen *n* = shrinking on
aufschütten, auftragen [*das Gelände*] = to fill up
(auf)schütten, auftragen, anschütten, auffüllen [*den Boden*] = to place fill material, to deposit ~ ~, ~ ~ fill(ing)
Aufschütten *n* → (An)Schütten
(Auf)Schüttung *f* → Schüttung
(Auf)Schütt(ungs)|arbeiten *fpl*, Auftragarbeiten, Einbau *m*; (Auf-)Schütt(ungs)- und Hinterfüll(ungs)arbeiten = fill work
~**boden** *m* → (Auf)Füllboden
~**gelände** *n*, (Auf)Schütt(ungs)fläche *f*; (Auf)Schütt(ungs)- und Hinterfüll(ungs)fläche, (Auf)Schütt(ungs)- und Hinterfüll(ungs)gelände = ground to be filled
~**höhe** *f* → Auftraghöhe
~**lage** *f* → Auftragschicht *f*
~**masse** *f* → Auftragmaterial *n*
~**material** *n* → Auftragmaterial
~**schicht** *f* → Auftragschicht
~**zone** *f* → Auffüllzone
aufschwemmen, aufschlämmen = to suspend
~ → auflanden
Aufschwemmung *f* → Auflandung
~ → Aufschlämmung
aufschwenken, aufdrehen [*Drehbrücke*] = to open [*swing bridge*]
Aufschwenken *n*, Aufdrehen [*Drehbrücke*] = opening [*swing bridge*]

Aufschwimmklassierer *m*, Naßzyklon *m*, Zyklonnaßklassierer = cyclone classifier

Aufsetzen *n* **des Kopfstückes** [*Kletter-Turmdrehkran*] = mounting of the crane head

Aufsetzrost *m* [*Rost auf dem sich Schiffe bei fallender Gezeit (oder Tide) aufsetzen*] = careening grid, gridiron

Aufsicht *f* = supervision

Aufsichts|beamter *m* = inspector

~**behörde** *f* = supervising authority, supervisory ~

~**personal** *n* = supervising staff, ~ personnel, supervisory ~

aufspachteln, verstreichen = to float

Aufspalten *n* [*Holzrammpfahl*] → „Perücke" *f*

Aufspanner *m*, Aufspann-Transformator *m* = step-up transformer

Aufspannstation *f* = step-up substation

(auf)speichern, einlagern, bevorraten = to store up

~ [*Talsperre*] = to store (water)

(Auf)Speicherung *f*, Bevorratung, (Material)Lagerung, Lagerhaltung, Einlagerung = (materials) storage, storing, storekeeping

~ → → Talsperre *f*

~ **während einer Tide**, ~ ~ ~ Gezeit = tidal storage

Aufsperrpistole *f* **für verschlossene YALE-Schlösser** = picking pistol

Aufspitzen *n*, Stocken = stone granulating, bush hammering

aufspritzen [*Beton; Mörtel*] = to gun into position

Aufspritzen *n*, Bindemittelverteilung *f*, Bindemittelaufbringung, Bindemittelaufspritzung = application of binder(s), spreading ~ ~, binder application, binder spreading

~ [*Meereswelle*] = spouting, bursting

aufsprühen, aufdüsen = to spray on by atomizing

Aufspülen *n*, Aufpumpen = filling ground by pumping dredged materials

(Auf)Spülung *f*, Spülverfahren *n* [*Herstellung von Dämmen*] = hydraulicking

Aufstandfläche *f*, Reifen~ = contact area; tire ~ ~ (US); tyre ~ ~ (Brit.)

(auf)stapeln = to pile (up), to stack

Aufstau *m* → → Talsperre *f*

(auf)stauen = to retain, to impound

Aufstauung *f* → → Talsperre *f*

aufsteckbar = slip-on type

aufstecken = to slip on, to mount, to insert, to place on to, to slip over

Aufsteck|flansch *m* = slip-on flange

~**getriebe** *n* = slip-on transmission

(~)**Hülse** *f* **aus Blech**, Blechhülse [*Betondeckenfuge*] = light metal cap

~**mäkler** *m*, Aufstechläuferrute *f*, Aufstecklaufrute = false leader

aufsteigende Feuchtigkeit *f*, ~ Feuchte *f* = capillary moisture

~ **Tide** *f* → Flut *f*

aufsteigender Ast *m* [*Diagramm*] = ascending branch

Aufstellen *n*, Aufstellung *f*, Montage *f*, Aufbau *m* = erection

Aufstell|geräte *npl* **für Turmdrehkrane** = tower crane erection equipment

~**gerüst** *n* [*Brückenbau*] → Montagegerüst

~**punkt** *m* [*Theodolit*] = set-up

Aufstellrichtung *f* [*Flugzeug*] = aircraft-parking configuration, airplane-parking ~

bugauswärts = nose-out
bugauswärts im Winkel = angled nose-out
bugeinwärts = nose-in
bugeinwärts im Winkel = angled nose-in
parallel = parallel

Aufstellung *f* = listing

~ [*Maschine*] = erection

~ **unter Wasser** [*Überdruckturbine*] = open flume setting

Aufstellungsarten *fpl* **der Fahrzeuge** [*Parken*] = parking configurations

Aufstell(ungs)gerüst *n* [*Brückenbau*] → Montagegerüst
Aufstellungsplan *m*, Montageplan = erection drawing
Aufstemmen *n*, Übersichbrechen *n* = overhead stoping
Aufstieg *m*, Fisch~ [*stromaufwärts gerichtete Fischwanderung*] = upstream migration of fish, ~ fish migration
~ [*Taucher*] = going up
~**geschwindigkeit** *f* = ascending velocity
Aufstocken *n* = heightening of a building
aufstreichbar = brushable
Aufstreichen *n*, Aufstrich *m* [z. B. *einer Dichtungsmasse*] = brush application, brushwork
Aufstrich *m* → Anstrich
~, Aufstreichen *n* [z. B. *einer Dichtungsmasse*] = brush application, brushwork
~**masse** *f* → Klebemasse
Aufsuchen *n* → Prospektion *f*
~ **im Küstenvorland**, Suche *f* ~ ~, Prospektion *f* ~ ~, Lagerstättenforschung *f* ~ ~ = off-shore prospecting
Auftanken *n* = fuel(l)ing
auftauchen [*Taucher*] = to (rise to the) surface
auftauen = to thaw (out)
Auftauen *n*, Frostaufgang *m* = thaw(ing)
(Auf)Tausalz *n* [*früher: Streusalz*] = de-icing salt, ice-control ~, ice-removal agent, highway salt, road salt
Auftauschaden *m* = damage due to thaw(ing)
Aufteilung *f* **der Schrägstähle** [*Einige untere Zugstähle werden zur Schubspannungsdeckung aufgebogen. In Balkenlängsrichtung werden sie so aufgestellt, daß sie dem Schubspannungsdiagramm entsprechen*] = spacing of bent-up bars
Auftorkretieren *n* → Torkretverfahren *n*
Auftrag *m* = order

~, Aufbringen *n* [*Tätigkeit des Auftragens, z. B. von Anstrichstoff*] = application, spreading
~ → Schüttung *f*
~ [*bereits vollendet, z. B. von Farbe*] = coat(ing)
~, Schicht *f* [*Putz*] = coat
~**arbeiten** *fpl* → (Auf)Schütt(ungs)arbeiten
~**boden** *m* → (Auf)Füllboden
~**böschung** *f* → Dammböschung
auftragen = to plot
~, aufschütten [*das Gelände*] = to fill up
~, aufschütten [*den Boden*] = to deposit fill material, to place ~ ~, ~ ~ fill(ing)
~, aufbringen [*Anstrichstoff*] = to apply, to spread
Auftragen *n* → (An)Schütten
~ **der Ergebnisse** = plotting of the results
~ **einer Kurve** = plotting a curve
Auftrag|erteilung *f*, Vergabe *f* = contract award, award of contract, letting of contract, contract-letting
~**geber** *m*, Bauherr *m*, Bauherrschaft *f* = promoter, purchaser, owner, sponsor, employer, client
~**höhe** *f*, (Auf)Schütt(ungs)höhe, Auffüllhöhe, Anschütt(ungs)höhe = depth of fill(ing)
~**linie** *f*, Abszisse *f* = abscissa
~**maschine** *f*, Beschichter *m* [*zur Kunststoffbeschichtung*] = coater, coating machine
~**material** *n*, (Auf)Schütt(ungs)masse *f*, (Auf)Schütt(ungs)material, Auftragmasse, (Auf)Füllmasse, (Auf-)Füllmaterial = fill material, fill(ing)
~**nehmer** *m* = successful bidder, ~ tenderer
~**schicht** *f*, Auftraglage *f*, Auffüllschicht, Auffüll-Lage, (Auf)Schütt(ungs)lage, (Auf)Schütt(ungs)schicht = (fill) lift
~**schreiben** *n* = letter of award, notice ~ ~
~**schweißung** *f*, Auftragschweißen *n* = building up by welding (Brit.); surfacing (US)

Auftragvergabe — aufzeichnen

~vergabe *f* = contract letting
~wert *m*, Auftragsumme *f* = contract price
~zone *f* → Auffüllzone
Auftreffwinkel *m* = angle of impact
Auftreten *n* **von Schrägrissen**, Schrägrissebildung *f* = oblique cracking
Auftrieb *m* [*A t, kg, g*] = buoyancy [*The reduction in weight of a body immersed in a fluid. It is equal to the weight of the fluid displaced by the body*]
~, Sohlenwasserdruck *m* = uplift, foundation water pressure
~ **durch Gaskappe** [*Erdölförderung*] = gas-cap drive
~ ~ **gelöstes Gas** [*Erdölförderung*] = dissolved-gas drive
~**sicherheit** *f*, Sicherheit gegen Sohlenwasserdruck = safety against uplift, ~ ~ foundation water pressure
Auftritt *m*, Auftrittmaß *n* [*ist die Breite der Stufe, auf der Ganglinie waagerecht gemessen von Vorderkante Setzstufe zur Vorderkante Setzstufe der nächsthöheren Stufe (senkrecht projiziert)*] = run
~ ~ → Trittstufe *f*
(Auf)Trittfläche *f*, Auftritt *m* = (stair) tread [*The horizontal top surface of a step*]
Auftropfstein *m*, Stalagmit *m* = stalagmite
aufvulkanisieren eines neuen Laufstreifens, besohlen = to topcap, to retread
aufwältigen [✂, *eine Strecke*] = to reopen
aufwalzen [*z. B. Rippen auf einen Stahldraht*] = to roll on
Aufwalzen *n* **einer neuen Decklage** [*Straße*] = resurfacing
aufwärmen [*Bindemittel; Zuschlagstoffe*] = to heat
Aufwärmen *n*, Aufwärmung *f* [*Bindemittel; Zuschlagstoffe*] = heating
Aufwärts|(bohr)hammer *m* = stoper [*A rock drill for upward drilling*]

~**bohrloch** *n* (✂) = upper (US) [*A hole drilled upwards*]
~**bohrung** *f*, Aufwärtsbohren *n* = upward drilling
~**schweißen** *n* = upward welding, uphand ~, uphill ~
~**siebung** *f* = uphill screening
~**strom** *m* = upward current
~**transport** *m* = upward transport
(auf)weichen, durchweichen = to soften
Aufweichung *f*, Erweichung, Durchweichung = softening
Aufweitbank *f* = expanding bench
(auf)weiten, aufbördeln = to expand
Aufweiter *m* = expanding device
Aufweitewalzwerk *n* [*Rohre*] = becking mill, expanding ~
Aufweitpresse *f* = expanding press
Aufwellen *n* [*Straßenoberfläche*] = washboarding
aufwickeln = to wind up
Aufwickeln *n*, Aufwick(e)lung *f* = winding up
Aufwickeltrommel *f* = winding up drum
Aufwinden *n*, Hochwinden [*mit Hebebock*] = jacking (up)
aufwölben, sich ~ = to warp upward
Aufwölbung *f* (Geol.) = uplift
~ = hog(ging), camber [*Upward bending, that is a shape which is curved on the upper side, the opposite of sag. A beam may be built with hog to counteract its sag*]
Aufwölbungs|moment *n* = hog(ging) moment, negative ~ [*A bending moment which tends to cause hog such as occurs at the supports of a beam*]
~**unterschied** *m* = differential hog(ging), ~ camber [*hog or camber is an upward bending, that is a shape which is curved on the upper side, the opposite of sag. A beam may be built with hog to counteract its sag, like many prestressed concrete beams*]
Aufzehrung *f* → Absorption *f*
aufzeichnen, registrieren = to record

Aufzeichnen *n*, Registrieren, Aufzeichnung *f*, Registrierung = recording
Aufzeichnung *f* = record
(auf)ziehen des Bohrzeuges, herausziehen ~ ~, ~ der Bohrgarnitur *f* [*pennsylvanisches Seilbohren*] = to hoist the drilling tools out, to pull ~ ~ ~
Aufziehen *n* **des Seiles** [*Radschrapper*] = cable threading
Aufzug *m* = lift (Brit.); elevator (US) [*An enclosed platform for carrying goods or passengers from one level to another in a tall building*]
~, Lasten~, Material~ = goods lift (Brit.); ~ elevator (US)
~, Personen~, Fahrstuhl *m* = passenger lift (Brit.); ~ elevator (US)
~, Bau~ = hoist, contractors' ~, builders' ~
~ **von Hand**, Handaufzug [*für Maurerarbeiten*] = hand hoist
~**-Abspannseil** *n* = hoist guy (rope)
~**bahn** *f* → → Betonmischer *m*
~**beschickungskübel** *m* → → Betonmischer *m*
~**bügel** *m* [*dreiteiliger Steinwolf*] = shackle
~**führer** *m* = lift attendant (Brit.); elevator ~ (US)
~**kasten** *m* → → Betonmischer *m*
~**-Klein(beton)mischer** *m* → → Betonmischer *m*
~**kübel** *m* → Abteufkübel
~**kübel** *m* → → Betonmischer *m*
~**kübelrahmen** *m* → → Betonmischer *m*
~**kübelwaage** *f* → → Betonmischer *m*
~**mast** *m*, Lasten~ = hoisting mast
~**motorraum** *m* = lift machine room (Brit.); elevator ~ ~ (US)
~**mulde** *f* → → Betonmischer *m*
~**öffnung** *f* [*durch eine Tür verschlossen*] = landing opening, ~ entrance
~**podest** *n* = lift landing (Brit.); elevator ~ (US)
~**schacht** *m* → Fahr(stuhl)schacht
~**schienen** *fpl* → → Betonmischer *m*
~**seil** *n*, Förderseil, Schacht~ = hoist rope, winding ~
~**seil** *n* = hoist rope
~**steuerung** *f* = lift control (Brit.); elevator ~ (US)
~**treppe** *f* [*Schweiz*]; Schiebetreppe = disappearing stairway (US)
~**tür** *f* = lift-car door (Brit.); elevator cage ~ (US)
Aufzugturm *m*, Bau~ = hoist tower
~ [*z. B. in einem Klärwerk*] = isolated lift tower (Brit.); ~ elevator ~ (US)
~ **aus Holz**, Bau~ ~ ~ = timber hoist tower
~ ~ **Stahl**, Bau~ ~ ~ = steel hoist tower
Aufzugwinde *f*, Bau~ = hoist winch
~ = hoist winch
Augbolzen *m* → Augenbolzen
Auge *n*, Anguß *m*, Nabe *f* = boss
~, Öffnung *f* = eye
~ **der Kuppel**, Kuppelauge, Laternenöffnung *f* = lantern opening
Augenblick *m* **des Bruches** = instant of failure
augenblickliche Korrosionsgeschwindigkeit *f* = instantaneous corrosion rate
Augen|bolzen *m*, Augenbolzen, Schraubenbolzen mit Ring, Ringbolzen, Ösenbolzen = eye bolt
~**gneis** *m* = eye gneiss, augen-gneiss
~**höhe** *f*, Sichthöhe = eye level
~**kohle** *f* = eye coal, birdseye ~, circular ~, curley cannel
~**ring** *m*, Seilring = grommet, rope eyelet
~**scheinnahme** *f*, Beurteilung *f* nach Augenschein = visual inspection, ~ examination, ~ check
~**stab** *m* [*U-Stahl, Flachstahl oder am Ende flach ausgeschmiedeter Rundstahl mit einem Bolzenloch in Stabendnähe, Auge genannt, für einen gelenkigen Anschluß mit Gelenkbolzen*] = eye bar
~**waschanlage** *f* = eyewash shower
Augit *m* = augite
~**diorit** *m* = augite diorite, augitic ~

Augitfels — Ausbaumethode

~**fels** m = augite rock, augitic ~
~**granit** m = augite granite, augitic ~
~**granophyr** m = augite granophyre, augitic ~
Augitit m = augitite
Augit|melaphyr m = augite melaphyre, augitic ~
~**porphyr** m, schwarzer Porphyr = augite porphyry, augitic ~
~**syenit** m = augite syenite, augitic ~
aus zwei Elementen bestehend → binär
(aus)baggern, abbaggern, trockenbaggern = to dig, to excavate
~ [*Naßbaggerung*] = to dredge
Ausbau m [*nicht im Sinne von Erweiterung, z.B. Hafenausbau, d. h. einen Hafen erweitern, aber „Ausbau der Wasserkräfte" heißt die Wasserkräfte zur Stromerzeugung „ausbauen"*] = development
~ [*alle technischen Mittel zum Offenhalten der unter Tage geschaffenen Hohlräume*] = permanent lining, ~ support
~, Erweiterung f = development
~, ~**arbeiten** fpl [*im Gegensatz zum Rohbau*] = interior work
~ **der Wasserkräfte** = development of hydro-electric resources
~ **einer Strommündung in See** = prolongation of a river mouth seawards
~**(arbeiten)** m, (fpl) [*im Gegensatz zum Rohbau*] = interior work
~**arbeiten** fpl, Ausbauen n [*Ausbau erstellen, d. h. die unter Tage geschaffenen Hohlräume mit Ausbau versehen*] = permanent lining work, ~ support ~
~**arbeiten** fpl, Erweiterungsarbeiten development work
~**bedarf** m **bei Straßen** = highway needs, road ~
ausbauchen = to bulge
Ausbauchen n = bulging
Ausbauchung f, Bauch m = bulge
~ **in cm³** [*nach Trauzl*] = expansion of the axial hole in a large lead cylinder

Ausbauelement n [*Fehlname: Ausbaueinheit. Tunnel-, Stollen- und Schachtbau*] = permanent support unit, ~ lining ~
ausbauen, erweitern = to develop
~, nutzen [*einen Fluß durch Wasserkraftanlage(n)*] = to develop
~ [*Bohrgestänge*] = to come out of the (bore) hole
~ [*Verrohrung in der Tiefbohrtechnik*] = to pull the casing
~ [*einen unter Tage geschaffenen Hohlraum*] = to line permanently
Ausbauen n, Ausbauarbeiten fpl [*Ausbau erstellen, d. h. die unter Tage geschaffenen Hohlräume mit Ausbau versehen*] = permanent lining work, ~ support ~
ausbaufähige Druckhöhe f [*Wasserkraft*] = developable head, ~ fall
~ **Wasserkräfte** fpl = developable hydropower
Ausbau|fallhöhe f [*diejenige Fallhöhe bei Wasserkraftwerken, bei deren Auftreten gleichzeitig gerade die höchste hydraulische Leistung aller Turbinen der Wasserkraftanlage erreicht wird*] = developed head, ~ fall
~**form** f **einer Wasserkraftanlage** = type of (hydroelectric) development
~**geschwindigkeit** f, Berechnungsgeschwindigkeit, Entwurfsgeschwindigkeit [*Straße*] = design speed
~**leistung** f [*Elektrizitätswerk*] = installed capacity
~**maß** n [*für den fertigen Bau*] = actual size, ~ dimension
Ausbaumaterial n, Ausbaustoff m [*Holz, Stahl, Mauerwerk (Ziegel- und Naturstein) oder Beton (Gußbeton, Stampfbeton, Stahlbeton, Betonformsteine)*] = permanent lining material, ~ support ~
~, Ausbaustoff m [*im Gegensatz zum Rohbaustoff*] = material for interior work
Ausbau|methode f, Ausbauverfahren n [*zum Offenhalten eines unter Tage geschaffenen Hohlraumes*] = permanent lining method, ~ support ~

Ausbauprogramm — Ausbrechen

~**programm** n, Ausbauplan m = development program(me)
~**projekt** n, Ausbauvorhaben n = development project
~**schema** n [*Tunnel-, Stollen- und Schachtbau*] = permanent support pattern, ~ lining ~
Ausbaustoff m, Ausbaumaterial n [*Holz, Stahl, Mauerwerk (Ziegel- und Naturstein) oder Beton (Gußbeton, Stampfbeton, Stahlbeton, Betonformsteine)*] = permanent support material, ~ lining ~
~, Ausbaumaterial n [*im Gegensatz zum Rohbaustoff*] = material for interior work
Ausbau|strecke f [*Wasserkraftanlage*] = developed section
~**stück** n = sample cut from a surfacing
~**stufe** f → Bauabschnitt m [*zeitlich gesehen*]
~**wasser** n [*Brückenbau*] = construction water level
~**zeit** f, Bohrgestänge-Ziehzeit [*Rotarybohren*] = time of "coming out"
ausbessern, instandsetzen = to reinstate
Ausbessern n, Flicken n [*Beton*] = patching [*concrete*]
Ausbesserung f, Instandsetzung = reinstatement
~ [*Straße*] → Ausflicken n
Ausbesserungs|arbeiten fpl, Instandsetzungsarbeiten = reinstatement work
~**haken** m, Dachhaken, Dachknappe m, Reparaturhaken = roof hook
~**kai** m, Reparaturkai = repair quay
~**mörtel** m, Flickmörtel = repair mortar, patching ~
~**werft** f, Reparaturwerft = repair shipyard
ausbetonieren = to fill with concrete
Ausbetonieren n = filling with concrete
ausbeutbar → (ab)baufähig
ausbeutbare Vorräte mpl, (ab)bauwürdige ~, (ab)baufähige ~ = reserves

Ausbeutbarkeit f → (Ab)Baufähigkeit
Ausbeutefaktor m [*Der aus Erdöllagerstätten gewonnene Prozentsatz des ursprünglichen Ölgehalts der Lagerstätte bei völliger Erschöpfung des betreffenden Ölfeldes*] = ultimate recovery
ausbeuten = to exploit
Ausbeutung f [*Bodenschätze*] = exploitation
Ausbiegen n, Knickung f, Ausbiegung, Knicken = buckling
Ausbildung f, Berufs~ = training
~ **auf dem Arbeitsplatz** = training on the job, on-the-job training
~ **der Frostmauer** [*Gefrierverfahren*] = drilling and forming the ice wall round the shaft (or excavation)
~ **fern vom Arbeitsplatz** = off-the-job training, training off the job
Ausbildungs|lehrgang m = training course
~**stätte** f = training centre (Brit.); ~ center (US)
Ausbiß m → Ausgehende n
~, Abdachung f, Böschung (Geol.) = scarp, escarpment
Ausbläser m [*Sprengen*] = blown-out shot, blow
~ [*Tunnel- und Stollenbau mit Druckluft*] = air piping
Ausblasevorrichtung f [*Bindemittelspritzmaschine*] = air scavenging gear, pneumatic ~ ~
ausbleichen, verbleichen [*Farbton*] = to fade [*tone*]
ausblühen = to effloresce
Ausblühung(en) f(pl) [*manche Säurelösungen laugen aus dem Zement den Kalk aus, was an weißen Ausblühungen zu erkennen ist*] = efflorescence
Ausbohlung f, Ausbohlen n [*Bohlwand*] = insertion of horizontal timber sheeting, ~ ~ ~ ~ lagging
ausbrechen [*Fels*] = to excavate [*rock*]
~ [*Bohrung*] = to blow-out, to rupture
Ausbrechen n [*des Hangenden*] = flaking [*of the roof*]

~ **untertägiger Räume**, Ausbruch *m* = underground excavation
ausbreiten, verteilen [*eine Schicht*] = to spread
Ausbreit|maß *n*, Sackmaß, Setzmaß, Fließmaß [*Betonprüfung. DIN 1048*] = slump
~**tisch** *m* mit Trichter [*Betonprüfung. DIN 1048*] = flow table (with cone)
Ausbreit(ungs)geschwindigkeit *f* [*m/s. Kennziffer der dynamischen Eigenschaften eines Baugrundes bei dynamischer Baugrunduntersuchung durch künstlich erregte sinusförmige Schwingungen mit veränderlicher Frequenz*] = wave velocity
~**versuch** *m*, Ausbreit(ungs)probe *f*, Ausbreit(ungs)prüfung *f*, Setzversuch, Setzprobe, Setzprüfung [*DIN 1048*] = slump test (for consistency)
Ausbrennen *n* **der Brennstoffhülsen** = burn-out
Ausbringen *n*, Mengen~ [*Siebtechnik*] = yield, recovery, output
Ausbruch *m*, Eruption *f*, gewaltsames Austreten *n* [*von Erdgas oder Erdöl in Bohrlöchern aus großen Tiefen*] = blow-out
~, Ausbrechen *n* untertägiger Räume = underground excavation
~, Eruption *f* [*Vulkan*] = emission of lava, eruption
~**gestein** *n* → Ergußgestein
~, Haufwerk *n*, Schutter *m* = muck (pile), broken rock
~, (Strecken)Vortrieb *m* [*Tunnel-, Berg- und Stollenbau*] = heading, progress, advance
~**profil** *n* [*Tunnel- und Stollenbau*] = minimum excavation line
~**querschnitt** *m*, gesamtes Querprofil *n* [*Tunnel- und Stollenbau*] = full section
~**stelle** *f*, Vortriebstelle, (Orts)Brust *f*, Ort *n* [*Tunnel- und Stollenbau*] = face, working ~
~**verhüter** *m*, Ausbruchschieber *m*, Eruptionsstopfbüchse *f* = blow-out preventer

ausbuchtendes Ufer *n*, konvexes ~, inneres ~, Ausbuchtung *f*, ausbuchtende Seite *f* = convex bank
ausdehnbar = expansi(b)le
ausdehnen = to expand
Ausdehnung *f*, Ausdehnen *n* = expansion
Ausdehnungsbeiwert *m*, Ausdehnungskoeffizient *m*, Ausdehnungszahl *f* = expansion coefficient, coefficient of expansion
(Aus)Dehn(ungs)fuge *f* → Raumfuge [*Betonstraßenbau*]
~ → Trennfuge
Ausdehnungs|gefäß *n*, Expansionsgefäß = expansion vessel
~**hub** *m* → Arbeitshub
~**verhalten** *n* = expansivity
Ausdornen *n* = broaching
ausdrücken, auspressen, ausquetschen = to squeeze out
Ausdrückvorrichtung *f* = coke discharging ram
auseinander|bauen [*ein Kernrohr*] = to dismantle [*a core barrel*]
~**klaffen**, aufklaffen = to gape
Auseinanderklaffen *n*, Aufklaffen [*Fuge*] = opening
auseinander|nehmen → demontieren
~**schrauben** = to unscrew
~**schrauben** [*Gestänge in der Tiefbohrtechnik*] = to break the joint
Auseinanderschrauben *n* = unscrewing
auseinanderziehen = to telescope
ausfachen, ausfüllen [*Süddeutschland: ausriegeln*] = to fill in
Ausfachung *f*, Ausfüllung [*Süddeutschland: Ausriegelung*] = panel wall, infill walling
~ [*Träger*] = bracing
Ausfachungs|mauerwerk *n*, Ausfüllungsmauerwerk = infill masonry
~**stab** *m* = stay rod
~**stein** *m*, Ausfüllungsstein = infiller block, infilling ~
~**tafel** *f*, Ausfüllungstafel, raumbildende Tafel = infilling panel, ~ slab, infiller ~

~ziegel *m*, Ausfüllungsziegel = infiller brick, infilling ~
(ausfahrbare) (Seiten)Stütze *f* → Hilfsstütze
Ausfahrgleis *n* = departure track
Ausfahrt *f*, Abfahrt, Autobahn~ = exit point, egress ~
~verbot *n* [*Planzeichen*] = exit-ban sign
Ausfall *m*, Geräte~, Panne *f* = (equipment) breakdown, (~) failure
(aus)fällen [*Chemie*] = to precipitate
(Aus)Fällen *n*, (Aus)Fällung *f* [*Chemie*] = precipitation
(Aus)Fällbecken *n*, (Aus)Fällungsbecken [*in einem Ausfällbecken hält sich das Wasser mehrere Stunden auf und wird grob vorgereinigt*] = precipitation tank, ~ basin
Ausfall|kornbeton *m*, Auslaßkornbeton = omitted-size (type) grain concrete
~körnung *f*, Auslaßkörnung [*Eine Mischung bei der aus der normalen Sieblinie Zuschläge gewisser Korngrößen ausgeschaltet sind*] = omitted-size (type) grain mix(ture)
~straße *f* [*eine Hauptverkehrsstraße, die das Zentrum einer Stadt oder Ortschaft mit ihren Außenbezirken bzw. mit dem Fernverkehrsstraßennetz verbindet. Schweiz: Radialstraße*] = outward-bound road, exit ~
(Aus)Fällung *f*, (Aus)Fällen *n* [*Chemie*] = precipitation
Ausfällung *f*, Sinkstoff *m* = sediment, deposit
(Aus)Fällungsverfahren *n* = precipitation method
(Aus)Faulung *f* → Schlammfaulung
ausfiltern → abfiltern
ausflanschen, ausklinken, abflanschen = to notch
Ausflanschen *n* → Ausklinken
Ausflicken *n*, Flickverfahren *n*, Ausbesserung *f*, Flickarbeit(en) *f(pl)* [*Straße*] = patch(ing) work, mending, patching
Ausflocken *n*, Ausflockung *f* = flocculation

ausfluchten → (ein)fluchten
Ausfluchten *n*, Abfluchten = ranging out, ~ into line
Ausfluchtung *f*, Bauflucht *f*, Abfluchtung, Einfluchtung = alignment; alinement (US)
Ausfluß|beiwert *m*, Ausflußziffer *f*, Ausflußkoeffizient *m* [*Dient zur Berechnung der Ausfluß- oder Überfallmenge des Wassers bei Behältern oder Wehren, weil dieses nicht den ganzen Querschnitt der Öffnung oder des Überfalles ausfüllt (Strahleinschnürung) und weil infolge ungleichmäßiger Geschwindigkeitsverteilung die wirkliche Geschwindigkeit die theoretische nicht erreicht*] = coefficient of discharge, discharge coefficient
~düse *f* = discharge nozzle
~kanal *m* = discharge canal
~rohr *n* = discharge pipe
~strahl *m* = jet
~-Viskosimeter *n*, Auslaufviskosimeter = orifice visco(si)meter
~ziffer *f* → Ausflußbeiwert *m*
Ausformen *n* = demo(u)lding
ausfugen, verfugen, ausstreichen = to point
Ausfugen *n*, (Ver)Fugen, Ausstreichen, (Aus)Fugung *f*, Verfugung Ausstreichung = pointing (operation)
Ausfugkelle *f* → Fugenkelle
(Aus)Fugung *f* → Ausfugen *n*
Ausfugung *f*, Verfugung, Ausstrich *m* [*als Ergebnis des Ausfugens*] = pointing
Ausführbarkeit *f*, Durchführbarkeit = feasibility
ausführender Ingenieur *m* = field engineer
Ausführung *f*, Durchführung = execution
~ [*als Bauweise*] = method
~, Bau *m*, Aufbau, bauliche Anordnung, Konstruktion *f* [*Maschine*] = design, version
Ausführungs|art *f*, Durchführungsart = method of execution

Ausführungsentwurf — ausgefugt

~entwurf m = winning design, working ~
~frist f = specified construction time
~garantie f = completion bond, performance ~
~phase f, Ausführungsstadium n, Bauphase, Baustadium = stage of construction, construction stage
~plan m, Ausführungszeichnung f, Arbeitsplan, Arbeitszeichnung = working drawing
~unterlagen fpl = construction documents
ausfüllen = to fill in
~, ausfachen [*Süddeutschland: ausriegeln*] = to fill in
Ausfüllen n, Pflasterfugenverfüllung f, Pflasterfugenausfüllung = feeding, jointing [*sett paving*]
Ausfüllung f → Ausfachung
Ausfüllungs|mauerwerk n, Ausfachungsmauerwerk = infill masonry
~stein m, Ausfachungsstein = infilling block, infiller ~
~system n [*Träger*] = system of bracing
~tafel f, Ausfachungstafel, raumbildende Tafel = infilling panel, cladding ~, ~ slab
~ziegel m → Ausfachungsziegel
ausfüttern, ausfuttern [*Ofen*] = to line
~, verrohren, ausfuttern [*Bohrloch*] = to case
Ausfuttern n [*Ausstampfen eines Kanalbettes mit Ton zur Dichtung des Kanals*] = lining a canal bottom with clay
Ausfütterung f, Verrohrung, Ausfutterung [*Bohrloch*] = casing
Ausfütterungsschuß m, Ausfutterungsschuß, Verrohrungsschuß [*Bohrloch*] = casing element
Ausgabebeleg m = issue voucher
Ausgabeln n [*Schotter*] = sorting by fork(s)
Ausgang m = exit
Ausgangs|beleuchtung f = exit lighting
~bitumen n, Grundbitumen = base asphalt (US); ~ asphaltic bitumen (Brit.)

~brennstoff m = basic fuel
~erz n = basic ore
~erzeugnis n = basic product
~gestein n = parent rock
~gleichung f = initial equation
~gut n, Rohhaufwerk n, Rohgut [*Siebtechnik*] = raw material
~hafen m = port of origin
~kohle f [*noch aufzubereitende Kohle*] = feed coal
~lage f, Ausgangsstellung f, Ausgangsposition f [*eines beweglichen Maschinenteiles*] = initial position, original ~
~material n, Ausgangswerkstoff m = base material
~mischung f, Ausgangsgemisch n = design mix(ture)
~position f → Ausgangslage
~rad n → Ausgangszahnrad
~stellung f → Ausgangslage
~temperatur f = original temperature
~tür f = exit door
~versuch m, Ausgangsprobe f, Ausgangsprüfung f = original test
~welle f, Abtriebwelle = output shaft
~werkstoff m → Ausgangsmaterial n
~werte mpl = basic data
~(zahn)rad n, Abtrieb(zahn)rad = output gear
~zeit f = original time
~zustand m = original state
Ausgasung f (🜍) = gas emission
ausgebaggerte Sohle f = dredged bottom
ausgebaute Wasserkraft f = developed water power, ~ hydro ~
ausgefahren [*Erdstraße*] = rutted, worn into ruts, worn down
ausgefahrene Spuren fpl [*in einer Straßendecke*] = rutting (in the wheel paths), ~ of the pavement
ausgefallener Arbeitstag m, Wartetag = shut down day
ausgefaulter Schlamm m, Faulschlamm = digested sludge
ausgefugt, verfugt, verstrichen = pointed

ausgehen — Ausgleichkolben

ausgehen, ausstreichen, zutage treten (Geol.) = to come out to the day, ~ ~ up to grass, ~ crop out, ~ outcrop

Ausgehende *n*, Ausstreichende, Aufschluß *m*, Ausstrich *m*, Ausbiß *m* [*das Zutagetreten einer Lagerstätte (Ausstreichen)*] = basset, course of outcrop, line of outcrop, outcrop(ping)

ausgehender Fels *m*, ausstreichender ~ = outcropping rock

ausgehobenes Material *n*, Aushub(material) *m*, (*n*) = excavated material

ausgekragt, vorgekragt = cantilevered

ausgekuppelte Raupe *f*, ~ (Gleis)Kette *f* = disconnected (crawler) track

ausgelastet sein [*Maschine; Fahrzeug*] = to be given a full work-out

ausgelaufenes Lager *n* = burnt bearing

ausgelaugte Braunerde *f* → degradierte ~

ausgepreßte Schüttung *f*, injizierte ~, verpreßte ~, eingepreßte ~ = injected fill

ausgepreßter Alluvionsschleier *m* → (injizierter) ~

ausgerückt [*Kupplung*] = disengaged

ausgeschnittene Seitenwand *f* [*Lader*] = cut-back side plate

ausgesteift, versteift, verstärkt = stiffened, reinforced

ausgewalzter Stahlstab *m* = rolled steel bar

ausgewaschener Horizont *m*, Eluvialhorizont = eluvial horizon

ausgewogener Drehofen *m* = balanced rotary kiln

ausgezogene Linie *f* [*Fahrbahnmarkierung*] = continuous line

Ausgiebigkeit *f* [*Ein Maß für die Fläche, die mit einer Mengeneinheit des Anstrichstoffes und mit einem Anstrich in vereinbarter Arbeitsweise versehen werden kann, wird z. B. in m²/kg oder m²/Liter angegeben*] = spreading rate [*This term is now preferred to covering power*]

ausgießen, vergießen [*Fuge*] = to pour

Ausgießen *n*, Vergießen [*Fuge*] = pouring

Ausgleichbalken *m*, Doppelhebel *m* [*Gleichgewichtklappe*] = balance beam

Ausgleichbecken *n*, Ausgleichbehälter *m* [*zum Ausgleich des Unterschiedes zwischen Zufluß und Abfluß beim Trink- und Brauchwasser*] = equalizing reservoir, make-up ~, ~ tank

~, Ausgleichweiher *m* [*Speicher zum Umformen eines stark schwankenden Zuflusses in einen wenig schwankenden oder gleichmäßigen Abfluß*] = balancing reservoir, compensating ~, equalizing ~, regulating ~

Ausgleichbehälter *m*, Ausgleichbecken *n* [*zum Ausgleich des Unterschiedes zwischen Zufluß und Abfluß beim Trink- und Brauchwasser*] = equalizing reservoir, make-up ~, ~ tank

~ [*Hydraulikanlage*] = expansion tank

Ausgleich|beton *m* = level(l)ing concrete

~decke *f* → Ausgleichschicht *f*

~düse *f* [*Vergaser*] = compensating jet

ausgleichen [*Unebenheiten in einer Oberfläche*] = to average out, to correct

Ausgleichen *n* [*Unebenheiten in einer Oberfläche*] = averaging out, correcting

Ausgleich|feder *f*, Spannfeder = compensator spring, equalizer ~

~feuchtigkeit *f*, Ausgleichfeuchte = equilibrium moisture content

~flansch *m* [*Tiefbohrtechnik*] = spacer spool [*for well-head assembly*]

~gewicht *n*, Gegengewicht = counterweight

~kolben *m*, Entlastungskolben = dummy piston, balance ~, equalizer ~ [*piston to balance out-of-line moments*]

~kolben *m* **und Haltering,** Entlastungskolben ~ ~ = balance piston and retainer, dummy ~ ~ ~

~korngröße *f* nach Heidenreich-Paul → Trennkorngröße HP
~masse *f* → Nivelliermasse
~-Querbalken *m* [*(Gleis)Kettenschlepper*] = equalizer bar
~rahmen *m* = equalizer frame
~rohr(leitung) *n*, (*f*) [*Ringschieber*] = equalizing pipe (Brit.); ~ port for needle (US)
~schicht *f*, Ausgleichlage *f*, Ausgleichdecke *f*, Vorprofildecke [*Straßenbau*] = leve(l)ing course, underlay ~, regulating ~, level regulating course
~schlangenrohr *n* = coiled expansion pipe
~schleife *f* = loop expansion pipe
~traverse *f* mit Wiegebett [*Gleiskette*] = rolling contact-type equalizer bar
~-Übersetzungsgetriebe *n* = unequal torque-splitting differential
~ventil *n* = equalizer valve, make-up ~, differential ~
~welle *f* = balancer
~wellrohr *n* = corrugated expansion pipe
~zement *m* [*Fehlname*] → Nivelliermasse *f*
ausglühbare Bestandteile *mpl*, organische Verunreinigungen *fpl* = organic matter (present), ~ impurities (~)
ausgraben = to unearth
Ausgrabungsfund *m* = article of antiquity
Ausguß *m*, Küchen~, Ausgußbecken *n* [*ein an die Wasserableitung angeschlossenes Becken für die Aufnahme der Haushaltabwässer*] = (kitchen) sink, cleaners' ~, slop ~
~hahn *m* = sink bib; bib nozzle (US) [*The tap at a kitchen sink*]
~masse *f* → Vergußmasse
~wasser *n* → Küchenabwasser
Aushärtezeit *f* [*Betonkleber*] = polymerization time, setting ~
Aushebemagnet *m* = pick-up magnet
~scheider *m* = pick-up separator, ~ machine

Aushebeverfahren *n* [*Magnetscheidung*] = pick-up method
ausheben, graben, ausschachten [*von Hand oder maschinell. Wenn mit Bagger = (aus)baggern*] = to excavate, to dig
aushöhlen = to hollow out
Aushöhlen *n*, Aushöhlung *f* = hollowing out
Aushub *m* → Einschnitt *m*
~, ausgehobenes Material *n*, Aushubmaterial = excavated material
~, Aussschachtung *f*, Aushubarbeit(en) *f*(*pl*), Ausschachtungsarbeit(en) = excavation, digging
~, Boden~ = soil digging, ~ excavation
~ in gewachsenem Boden = primary excavation [*Digging in undisturbed soil, as distinguished from rehandling stockpiles*]
Aushubarbeit(en) *f*(*pl*) → Ausschachtung *f*
~ [*Baugrubenaushub*] = pit excavation, ~ digging
~ [*Grabenaushub*] = 1.) trenching, trench digging; 2.) ditching, ditch digging
Aushub|bereich *m* [*Bagger*] = digging radius, radius of excavation
~(boden) *m*, Aushuberde *f*, Bodenaushub [*durch Trockenbaggerung gewonnener Baustoff oder Aussatzmaterial*] = excavated earth, muck; spoil [*Aussatzmaterial*]
~böschung *f*, → Abtragböschung
~fläche *f* = excavated area
~gerät *n*, Aushubmaschine *f* = (excavator) digger
~grenze *f* = limit of excavation, excavation line
~höhe *f* = excavation level
~masse *f*, Aushubmenge *f* = quantity of excavation
~(material) *m*, (*n*), ausgehobenes Material *n* = excavated material
~querschnitt *m*, Ausschachtungsquerschnitt = section of excavation
~schacht *m* [*beim Baggersenkkasten*] = digging well

~schleuse *f* [*dient zum Herausschaffen des Aushubes aus der Arbeitskammer bei Druckluftgründungsarbeiten*] = spoil air-lock

~stellung *f*, Grabstellung = digging position, excavating ~

~tiefe *f*, Grabtiefe = digging depth, excavation ~

~verfahren *n* = excavating method, digging ~

~wand *f*, Ausschachtungswand = digging face, excavating ~

auskeilen (Geol.) = to die away, ~ ~ out, to end off, to peter out, to pinch (out), to taper out, to thin out, to wedge out, to thin away

Auskeilen *n*, seitliches Aufhören, Dünnerwerden, Sichausspitzen (Geol.) = pinching (out), thinning out

auskeilende Wechsellagerung *f* (Geol.) = dovetailing, interdigitation, interfingering

~ **Zwischenlage** *f* (Geol.) = interfingering member

Auskernung *f* = demolition of backcourt housing

Auskesselung *f* → **Kessel** *m*

auskippen = to dump by tilting

Auskipp|-Hydraulikzylinder *m* [*Schürflader*] = dump(ing) ram

~stellung *f*, Ausschüttstellung = dump(ing) position

~werk *n* [*Schürflader*] = dump(ing) mechanism

~winkel *m* [*Ladeschaufel*] = dump(ing) angle

auskleiden, bekleiden, verkleiden = to face, to line

Auskleidung *f*, Verkleidung, Bekleidung = lining, facing

~, Verkleidung, Bekleidung [*Baugrube*] = sheeting, lining

~, Verkleidung, Bekleidung, nicht tragende Mauerung geringer Dicke [*Tunnel- und Stollenbau*] = lining, facing

~ **mit Gußeisentübbingen**, Bekleidung ~ ~, Verkleidung ~ ~, Gußeisenauskleidung, Gußeisenverkleidung, Gußeisenbekleidung = cast-iron lining

Auskleidungs|arbeit(en) *f(pl)*, Verkleidungsarbeit(en), Bekleidungsarbeit(en) = lining work

~beton *m*, Verkleidungsbeton, Bekleidungsbeton = lining concrete

~-Betonfertigteil *m, n*, Bekleidungs-Betonfertigteil, Verkleidungs-Betonfertigteil = precast (concrete) lining unit, ~ (~) ~ member, cast-concrete ~ ~

~blech *n*, Verkleidungsblech, Bekleidungsblech = lining plate

~bohle *f*, Verkleidungsbohle, Bekleidungsbohle = lining plank

~platte *f*, Bekleidungsplatte, Verkleidungsplatte = lining slab

~maschine *f* = lining machine, liner

~mauer *f* → **Futtermauer**

ausklingen, abklingen [*Hochwasser; Schall; Schwingung usw.*] = to fade

Ausklingen *n* **der Setzungen**, Abklingen ~ ~ = fading of the settlements

ausklingende Falte *f* (Geol.) = lessening fold

~ **vulkanische Tätigkeit** *f* = declining volcanic activity

ausklinken = to release, to disengage

~, abflanschen, ausflanschen = to notch

Ausklinken *n*, Abflanschen, Ausflanschen [*örtliches Abtrennen von ganzen Flanschen bei Formstahl*] = notching

Ausklink|maschine *f*, Abflanschmaschine [*Stanzmaschine, um Flansche an den Enden von Formstählen oder Flanschenteile an beliebigen Punkten eines Trägers herauszuschneiden oder Ausklinkungen aus dem Profilstahlsteg durchzuführen*] = notching machine

~vorrichtung *f* = disengaging device, releasing ~

ausknicken, (ein)knicken = to buckle, to yield to axial compression

Ausknicken *n*, (Ein)Knicken, Ausknickung *f*, (Ein)Knickung = buckling

Auskoffern n, Auskofferung f = road bed excavation
auskohlen = to work out, to extract, to get, to win
auskolkbar = erodible
(aus)kolken, auswaschen, unterspülen, unterwaschen, wegspülen = to scour, to underwash, to undermine
(Aus)Kolkgrenze f = limit of scour(ing), ~ ~ underwashing, ~ ~ undermining, ~ ~ wash, ~ ~ subsurface erosion
(Aus)Kolkung f, Unterspülung, Unterwaschung, Wegspülung, Kolk m, Auswaschung [*Einwirkung von strömendem Wasser auf Erdmassen*] = scour(ing), undermining, underwashing, subsurface erosion, wash(-out)
~ **durch Eis** = exaration, glacial scour
Auskolkungsgefahr f = risk of scour(ing), ~ ~ undermining, ~ ~ underwashing, ~ ~ wash, ~ ~ subsurface erosion
auskragen, vorkragen = to corbel outwards, to cantilever (out), to project
Auskragen n, Vorkragen, Ausladen = cantilevering
Auskragung f, Ausladung, Vorkragung = cantilever
Auskratzeisen n = scraper
Auskratzen n [*Fuge*] = raking out
~ **von Ziegelmauerwerkfugen** = brick raking [*to remove old mortar*]
auskreuzen [*Nietkopf*] = to cut out the rivet head with a cross-cut chisel
Auslade|anlage f, Abladeanlage, Entladeanlage = unloading station
~**arbeit** f = unloading work
~**band** n, Bandauslader m, Entladeband, Bandentlader = unloading belt conveyor, ~ ~ conveyer
~**fläche** f = unloading area
~**gerät** n, Auslademittel n [*Gerät, welches ein Transportfahrzeug entlädt*] = unloading unit
~**gleis** n, Entladegleis, Abladegleis = unloading track
~**kran** m, Entladekran = unloading crane
ausladen = to unload

Ausladen n, Ausladung f = unloading
~, **Auskragen, Vorkragen** = cantilevering
ausladender Balken m → Kragbalken
Auslade|stelle f, Entladestelle = unloading point
~**vorrichtung** f, Entladevorrichtung = unloading device
Ausladung f, Ausladen n = unloading
~, **Reichweite** f, **Ausleger**~ = (working) radius, (out)reach, rad.
~, **Reichweite** f [*Lader*] = dumping reach (at 45° discharge angle)
~, **Auskragung, Vorkragung** = cantilever
ausländische Arbeitskräfte [*pl*] = imported labo(u)r, ~ manpower
Auslands|erz n, Einfuhrerz, Importerz = foreign ore
~**patent** n = foreign patent
~**tätigkeit** f = out-of-the-country job
~**warteraum** m [*Flughafen*] = international waiting room
Auslastung f [*Maschine*] = work-out
Auslaß m → Auslauf m
~**körnung** f → Ausfallkörnung
~**öffnung** f [*Hydraulikanlage*] = outlet port
~**schieber** m = outlet valve
~**ventil** n = poppet valve
Auslauf m [*für einen Aufzug*] = over-run [*for a lift*]
~, **freier** ~ [*Erdöl*] = natural flow
~ = over-run, stopway [*airfield*]
~, **Auslaß** m [*ohne Bauwerk*] = outlet [*e.g. of a river into the sea*]
~, ~**bauwerk** n, Auslaß(bauwerk) = sewer outfall [*The structure at the lower end of an outfall sewer*]
~**bauwerk** n, Betriebsauslaß m, Auslaß(bauwerk) [*Talsperre*] = (river) outlet, (reservoir) outlet (works), outlet structure, outfall structure
~**bauwerk** n, Ausmündungsbauwerk, Auslauf m, Auslaß(bauwerk) m, (n) [*Seitenöffnung an Wasserläufen mit Verschlußvorrichtung*] = outlet structure

~**(bauwerk)** *m*, (*n*), Auslaß(bauwerk) = sewer outfall [*The structure at the lower end of an outfall sewer*]

Ausläufer *m*, Bergsporn *m* = spur

~ [*Wetterkarte*] = extension

Auslauf|(ge)rinne *f*, (*n*), Auslaufkanal *m* = outlet channel

~**hahn** *m* = orifice tap

~**kanal** *m* → Auslaufrinne

~**kasten** *m* = discharge box

Auslaufkonus *m* [*Betonmischer*] = discharge cone

~ [*Grundablaß-Drosselklappe*] = outlet bellmouth

Auslauföffnung *f* → Ablauföffnung

~, Ablauföffnung, Abflußöffnung, Durchflußöffnung, Ablaßöffnung [*Wehr*] = discharge opening, sluiceway

Auslauf|quelle *f* = descending spring

~**regulierungsbauwerk** *n* = escape regulator

~**rinne** *f*, Auslaufgerinne *n*, Auslaufkanal *m* = outlet channel

~**schurre** *f*, Austrag(ungs)schurre, Entleerungsschurre, Auslaufrutsche *f*, Austragung(ungs)rutsche, Entleerungsrutsche = discharge chute

~**stelle** *f* [*Abwasser ins Meer*] = ocean outfall

~**stollen** *m* → Abflußstollen

~**stollen** *m* [*von Turbine zum Ableitungstunnel*] = draft tube tunnel, ~ ~ gallery (US); draught ~ ~ (Brit.); discharge branch

~**tisch** *m* = discharge table, out-end ~

~**versuch** *m*, Bremsversuch, Auslaufprüfung *f*, Auslaufprobe *f*, Bremsprüfung, Bremsprobe [*Bestimmung des Reibungsbeiwertes auf der Straße*] = braking test, stopping distance ~

~**-Vertiefung** *f*, Fußbecken *n* [*Saugüberlauf*] = outlet bucket (Brit.); lower lip (US)

~**viskosimeter** *n* → Ausfluß-Viskosimeter

Auslaugbarkeit *f* = leachability

auslaugen = to leach (out)

Auslaugung *f*, Auslaugen *n* = leaching (out) [*The removal of soluble substances, e.g. from soil or timber, by percolation and/or diffusion*]

Auslaugungs|horizont *m* → A-Horizont

~**see** *m* = sink (hole) lake

Ausleer|höhe *f*, (Aus)Schütthöhe, Kipphöhe, Abwurfhöhe [*Schaufellader*] = dump(ing) height

~**reichweite** *f* [*Schaufellader*] → (Aus-)Schüttreichweite

~**winkel** *m* [*Schaufellader*] → (Aus-)Schüttwinkel

~**zeit** *f* [*Schaufellader*] → (Aus)Schüttzeit

Ausleger *m* [*Kran; Bagger*] = boom (US); jib (Brit.)

~, Baggerleiter *f*, Eimerleiter(-Ausleger), Becherleiter = digging ladder, ditcher ~, bucket ~, bucket flight

~, Baggerleiterrahmen *m*, Eimerleiterrahmen = ladder frame

~ [*Straßenbetoniermaschine*] = (delivery) boom, distributing ~, elevating ~

~ = cantilever [*A beam or slab fixed at one extremity and unsupported at the other*]

~, Oberleiter *f* [*Eimerkettenbagger*] = top ladder, jib

~ **der Grabenfräse**, Fräsbalken *m* = trench cutting boom

~**anlenkpunkt** *m* = fulcrum of the jib (Brit.); ~ ~ ~ boom (US)

~**ansatz** *m* = jib boom (US); ~ extension

(~-)**Ausladung** *f*, (Ausleger-)Reichweite *f* = working radius, (out-)reach

~**ausziehzylinder** *m* = boom extension ram (US); jib ~ ~ (Brit.)

~**bagger** *m* = mechanical shovel

~**balken(träger)** *m*, Kragbalken(träger) = cantilever beam

~**bogen** *m*, Kragbogen = cantilever arch

~**bogenbrücke** *f*, Kragbogenbrücke = cantilever arch(ed) bridge

Auslegerbogen(träger) — Auslegerstiel

~bogen(träger) m, Kragbogen(träger) = cantilever arch(ed girder)

~-Bohrwagen m, Bohrjumbo m, Groß-Bohrwagen, Ausleger-Bohrjumbo = (tunnel) jumbo, drilling ~

~brücke f, Kragbrücke = cantilever bridge

~-Drehkran m → Schwenkkran

~drehwinkel m, Auslegerschwenkwinkel = boom swing angle (US); jib ~ ~ (Brit.)

~endschalter m = jib limit switch (Brit.); boom ~ ~ (US)

~förderband n, Absetz(förder)band = stacker belt, ~ conveyer, ~ conveyor

~fuß m = lower end of the jib (Brit.); ~ ~ ~ boom (US)

~fußbolzen m = boom pin (US); jib ~ (Brit.)

~-Fußgängerbrücke f, Krag-Fußgängerbrücke = cantilever pedestrian bridge, ~ footbridge

~fußtrommel f = boom foot spool (US); jib ~ ~ (Brit.)

~gerüst n, Auslegerrüstung f, Kraggerüst, Kragrüstung, Ausleger-Baugerüst, Krag-Baugerüst [*Für Ausbesserungsarbeiten an Dachrinnen, Gesimsen usw. Konstruktion: Ausleger durch Fenster gesteckt, durch Stiele auf Fußboden gestützt, Kragenden eventuell durch Streben gestützt*] = bracket scaffold(ing)

~-Gießturm m → (Beton)Gießturm

~höchstausladung f = boom-out (US)

~höhe f = boom height (US); jib ~ (Brit.)

~hubendschalter m = boom hoist limit switch (US); jib ~ ~ ~ (Brit.)

~-Hubseil n = boom lift cable (US); jib ~ ~ (Brit.)

~hubtrommel f = boom lift drum (US); jib ~ ~ (Brit.)

~-Hubventil n = boom lift valve (US); jib ~ ~ (Brit.)

~-Hubzylinder m = boom lift ram (US); jib ~ ~ (Brit.)

~kopf m = boom point (US); jib ~ (Brit.)

~kopf m mit Festpunkt = boom point with fixed point (US); jib ~ ~ ~ ~ (Brit.)

~kran m = boom crane (US); jib ~ (Brit.)

~kran m, Kran mit nicht drehbarem Ausleger = crane with non-slewing jib (Brit.); ~ ~ ~ boom (US)

~kran m auf Chassis mit Eigenantrieb → Auto(mobil)-Auslegerkran

~kratzer m = boom scraper

~kübel m [*Straßenbetoniermaschine*] = (self-spreading) bucket, boom ~, operating ~

~länge f = jib length (Brit.); boom ~ (US)

~lastkatze f, Auslegerlaufkatze, Auslegerkatze = boom crab (US); jib ~ (Brit.)

~(licht)mast m, Auslegerbeleuchtungsmast = (lighting) column with bracket (Brit.); mast with arm (US)

~neigung f = jib inclination (Brit.); boom ~ (US)

~pfostenaufzug m = crane hoist

(~-)Reichweite f, (Ausleger-)Ausladung f = (out)reach, working radius

~-Reißzahn m, Ausleger-Aufreißhaken = jib tooth, ~ tine (Brit.); boom ~ (US)

~rohr n [*Hydraulik-Mobilkran*] = shipper

~schlitz m = saddle block

~schrapper m mit Ladewagen für den Streckenvortrieb = scraper and loader

~seil n = boom cable (US); jib ~ (Brit.)

~seiltrommel f = boom drum (US); jib ~ (Brit.)

~seitenseilscheibe f = boom side sheave (US); jib ~ ~ (Brit.)

~spitze f = tip of the jib (Brit.); ~ ~ ~ boom (US)

~stellung f = jib position (Brit.); boom ~ (US)

~stiel m → Löffelstiel

Auslegerteil — Ausnutzen

~teil *m, n* = jib component (Brit.); boom ~ (US)

~träger *m*, Kragträger = cantilever girder

~trommel *f* = boom drum (US); jib ~ (Brit.)

~verschiebewelle *f* = boom shipper (shaft) (US); jib ~ (~) (Brit.)

~-Verstellgeschwindigkeit *f* = jib raising and lowering speed (Brit.); boom ~ ~ ~ (US)

~(verstell)motor *m* = jib raising and lowering motor (Brit.); boom ~ ~ ~ ~ (US)

~verstellseil *n* = jib raising and lowering cable (Brit.); boom ~ ~ ~ (US)

~verstellung *f*, Auslegerverstellen *n* = jib raising and lowering (Brit.); boom ~ ~ ~ (US)

~(verstell)windwerk *n* = jib raising and lowering gear (Brit.); boom ~ ~ ~ ~ (US)

~winde *f* = boom winch (US); jib ~ (Brit.)

~winde *f*, Baggerwindwerk *n* = shovel boom winch (US); ~ jib ~ (Brit.)

~winde *f* → Eimerleiterwinde

~winkel *m* = jib inclination angle (Brit.); boom ~ ~ (US)

(~)Wippen *n* = derrick(ing) motion, ~ movement

(~-)Zwischenstück *n* = (jib) section (Brit.); (boom) ~ (US)

~zylinder *m* = boom ram (US); jib ~ (Brit.)

Auslegung *f* = interpretation

Auslegungsirrtum *m*, Auslegungsfehler *m* = interpretation error

Ausleihe *f* = lending library

Auslenkung *f* = deflection

(Aus)Lesen *n* → Klauben

Auslitern *n*, Eichung *f* [*Flüssigkeitsbehälter*] = dipping, calibrating

Auslöse|balken *m* (✷) = chock release

~haken *m* = detaching hook

~haken *m* [(*Pfahl*)*Ramme*] = trip hook

~hebel *m* = disengaging lever

~mechanismus *m* = trip mechanism

Auslöser *m*, Sperrklinke *f*, Sperrzahn *m* = detent, catch, (locking) pawl

Auslösung *f* [*Ein Arbeitnehmer, der auf einer Baustelle außerhalb des Betriebssitzes arbeitet, erhält eine Auslösung, wenn die tägliche Rückkehr zum Wohnort nicht zumutbar ist*] = subsistence allowance, cost of living allowance (Brit.); subsistence, living allowance (US); separation allowance

Auslösungsbeben *n* = simultaneous earthquake

Ausmaß *n* [*Schweiz*] → Aufmaß

~ der Eisbedeckung = ice cover ratio

Ausmauerung *f*, (Tunnel)Mauerung [*Endgültige tragende Abstützung des ausgebrochenen Hohlraumes. Nur im günstigsten Fall eines völlig standfesten Gebirges bleibt der Tunnel unverkleidet. In allen anderen Fällen ist eine Ausmauerung notwendig. Die leichteste Form besteht in der Abdeckung des Gesteins mit Spritzbeton. Bei den Ausmauerungen mit Mauerwerk oder Beton kommen, je nach den Verhältnissen, Wanddicken von 20 cm bis rund 1 m, schließlich auch Stahlbeton zur Anwendung*] = permanent supporting system

Ausmitte *f*, Ausmittigkeit *f*, Exzentrizität *f* = eccentricity

ausmittig, exzentrisch = eccentric; off-center (US); off-centre (Brit.)

ausmittiges Baggern *n* [*ein an einem Schlepper angebrachter Baggerausleger wird aus Fahrzeugmittellinie nach der einen oder anderen Seite hin verschoben*] = offset digging

ausmörteln, bemörteln = to fill with mortar

Ausmündungs|bauwerk *n* → Auslaufbauwerk

~stelle *f* [*Entwässerung*] = outlet

ausnagen, erodieren = to erode

Ausnagung *f* → → Denudation *f*

Ausnutzen *n* der Wärme = heat utilization

Ausnutzungsfaktor m = utilization factor

auspressen, injizieren, einpressen, verpressen = to inject

~, **ausdrücken, ausquetschen** = to squeeze out

Auspressen n **der Spannglieder** = tendon grouting, ~ bonding

Auspressung f → Injektion f

Auspreß|anlage f → Injektionsanlage

~**-Diaphragma** n → Einpreß-Dichtwand f

~**-Dicht(ungs)wand** f → Einpreß-Dichtwand

~**druck** m → Injektionsdruck

~**-Entlüftungsbohrung** f → Injektions-Entlüftungsbohrung

~**gang** m, Injektionsgang, Verpreßgang, Einpreßgang, Injiziergang = injection gallery

~**gerät** n → Verpreßkessel m

~**gründung** f, Einpreßgründung, Verpreßgründung, Injektionsgründung, Injiziergründung = injected foundation

~**gürtel** m → Einpreß-Dichtwand f

~**gut** n → Injektionsgut

~**kammer** f → Injektionskammer

~**lanze** f → Injektionslanze

~**loch** n → Injektionsloch

~**mörtel** m → Injektionsmörtel

~**pumpe** f → Injektionspumpe

~**ring** m → Injektionsring

~**rohr** n → Injektionsrohr

~**schirm** m → Einpreß-Dichtwand f

~**schlauch** m → Einpreßschlauch

~**-Schleier** m → Einpreß-Dichtwand f

~**schürze** f → Einpreß-Dichtwand f

~**-Sperrwand** f → Einpreß-Dichtwand

~**stutzen** m → Injektionsstutzen

~**verfahren** n → Injektionsverfahren

~**zement** m → Injektionszement

Auspuff|dämpfer m, **Auspufftopf** m = muffler, silencer

~**deflektor** m = exhaust deflector

~**gas** n, Abgas = exhaust gas

~**hub** m, Auspufftakt m = exhaust stroke, scavenging ~

~**-Klappenbremse** f, Motorbremse = exhaust brake

~**krümmer** m = exhaust manifold

~**leitung** f = exhaust line

~**rauch** m = exhaust smoke

~**rohr** n = exhaust pipe ~ stack

~**rohrdeckel** m = flapper-type rain cap

~**rohrverlängerung** f = exhaust extension

~**schlauch** m = flexible exhaust pipe

~**schlitz** m = exhaust port

~**takt** m, Auspuffhub m = exhaust stroke, scavenging ~

~**topf** m, Auspuffdämpfer m = muffler, silencer

auspumpen = to pump out

Auspumpen n = pumping out

ausquetschen, ausdrücken auspressen = to squeeze out

ausrammen, umspunden [*Baugrube mit Spundbohlen*] = to close a building pit by means of sheetpiling

Ausrammung f, Einspundung, Spundwandumschließung = closed sheeting

ausräumende Tätigkeit f **der Wellen** fpl = quarrying action of the waves

Ausräumer m = (Schlamm)Kratzer m

ausrichten → (ein)fluchten

Ausrichten n [*Kabelspinnen*] = adjusting

~ [*einen Bauteil in die Soll-Lage bringen*] = adjusting

~ [*Pfahl*] = (pile) pitching

~, Visieren = boning [*Operation of levelling or regulating the straightness of trenches by sighting along the tops of a series of tee-pieces, known as boning rods*]

Ausrichte|seil n [*zum Ausrichten von Pfählen*] = pile pitching rope

~**tafel** f, Visiertafel = boning rod

Ausrichtung f (⚒) = development work in stone

~, Nacheichung, Spur f = alignement (Brit.); alinement (US)

Ausriegelung f → Ausfachung

Ausrodung f → Rodung

(Aus)Rollgrenze f → Konsistenzgrenzen nach Atterberg

Ausrücken n, Lösen = disengagement

Ausrück|gabel *f* = yoke assy, clutch fingers

~hebel *m*, **Stößel** *m* = lifter

~lager *n* = throw-out bearing, release ~

Ausrundung *f*, **Ausrundungsbogen** *m*, Ausrundung des Neigungswinkels [*Straßengradiente*] = vertical curve, transition between gradients

~ des Neigungswinkels → Ausrundung

Ausrundungshalbmesser *m* [*Straßengradiente*] = vertical curve radius

ausrüsten, ausstatten, bestücken = to fit out, to equip

~, abrüsten = to remove a scaffold(ing), ~ ~ scaffold(ing)s

~, abrüsten, abbauen eines Lehrgerüstes = to strip the falsework, ~ dismantle ~ ~

Ausrüsten *n* → Abbau *m* eines Lehrgerüstes

~, Abrüsten = removal of scaffold(ing)s

Ausrüstung *f*, Ausstattung, Bestückung = equipment, outfit

Ausrüstungs|becken *n* = fitting-out basin

~hafen *m* = fitting-out port

~kai *m* [*Nach dem Stapellauf des Schiffes werden am Ausrüstungskai die Maschinen- und Kesselanlagen mit Hilfe der Ausrüstungskrane eingebaut*] = fitting-out quay

~kran *m* = fitting-out crane

Aussalzen *n* = soap/salt displacement

Aussatz|halde *f*, Abraumhalde = waste pile

(~)Kippe *f* [*an der Aussatzkippe werden unbrauchbare Bodenmassen abgekippt. Als Aussatzkippen eignen sich sehr gut verlandete Tümpel, ausgebeutete Kiesgruben, ausgebeutete Sandgruben, ausgebeutete Braunkohlengruben usw.*] = spoil dump, spoil deposit, waste bank waste site, waste area, spoil area, spoil bank, (spoil) tip, shoot [*deprecated*: chute]

~material *n* = spoil, waste

ausschachten, ausheben, graben [*von Hand oder maschinell. Wenn mit Bagger* = (*aus*)*baggern*] = to dig, to excavate

Ausschachtung *f*, Aushub *m*, Aushubarbeit(en) *f* (*pl*), Ausschachtungsarbeit(en) = digging, excavation

Ausschachtungs|querschnitt *m*, Aushubquerschnitt = section of excavation

~wand *f*, Aushubwand = digging face

ausschalen, entschalen = to strip

Ausschalen *n*, Entschalen, Ausschalung *f*, Entschalung = stripping, formwork removal

Ausschaler *m*, Entschaler = (formwork) stripper

Ausschal|festigkeit *f* [*Beton*] = release strength

~frist *f* → Ausschal(ungs)zeit

~kolonne *f*, Ausschaltrupp *m* = (formwork) stripping crew, (~) ~ gang, (~) ~ party, (~) ~ team

ausschalten, abschalten [*Elektrotechnik*] = to switch off, to disconnect, to cut out, to put out of circuit, to break the circuit

Ausschalten *n*, Abschalten [*allgemein*] = disconnecting

Ausschalt|-Hauptzylinder *m* = kick-out master cylinder

~hebel *m* = kick-out lever

Ausschaltrupp *m* → Ausschalkolonne *f*

Ausschaltstellung *f*, Nullstellung, Abschaltstellung = open position, neutral ~, (switch-)off ~

Ausschaltung *f*, Abschaltung [*allgemein*] = disconnection

~ [*Hydraulikanlage*] = kick-out

~ der Haftwirkung [*beim Spannbeton durch Hüllrohre erzielt*] = bond-breaking

Ausschalung *f* → Ausschalen *n*

Ausschal(ungs)zeit *f*, Ausschal(ungs)frist *f* = stripping time

Ausscheidungs|lager(stätte) *n*, (*f*), Ausscheidungsvorkommen *n*, Ausschei-

Ausscheidungssediment(gestein) — Aussetzgerät

dungsfundstätte, Ausscheidungsvorkommnis n = precipitated deposit
~sediment(gestein) n → chemisches Sediment(gestein)
ausscheren [Seil] = to unreeve
Ausschlachten n [Maschine] = cannibalization
Ausschlag m, Schwingungsweite f, Amplitude f, Ausschlagweite, Vibrationsamplitude = amplitude of vibration
~rost m = knocking-out grid
~winkel m, Abweichungswinkel = angle of deflection, ~ ~ deviation
ausschlämmbar, abschlämmbar, absetzbar = settleable
ausschlämmen → abschlämmen
Ausschlämmen n, Abschlämmen, Absetzen = settling (out), sedimentation
Ausschleudern n = centrifuging, spinning out
Ausschleusen n [Druckluftgründung] = locking out
~ des Bodenaushubs [Druckluftgründung] = locking out of the spoil
Ausschleushahn m [Druckluftschleuse] = locking out cock
Ausschneiden n [Herstellen von Löchern usw. in Stahlbauteilen] = cutting out
Ausschneider m, Eimerreiniger m [Eimerkettenbagger] = (scraper) bucket cleaner
Ausschnitt m, Kreissektor m = sector
~ [rechteckig oder dreieckig in einem Meßwehr zur Messung der überlaufenden Wassermengen] = notch
ausschreiben, Angebot(e) einholen = to tender out, to put out to tender
Ausschreibung f → → Angebot n
Ausschreibungs|bedingungen fpl = tender conditions
~entwurf m = tender design, ~ project, official ~
~unterlagen fpl → → Angebot n
Ausschuß m = committee
~ = reject, throw-out

ausschütten, abwerfen = to dump, to discharge by dumping
Ausschütten n, Abwurf m = dumping
(Aus)Schütt|halbmesser m, Abwurfhalbmesser = dumping radius
~höhe f, Ausleerhöhe, Kipphöhe, Abwurfhöhe [Schaufellader] = dumping height
~höhe f bei größter (Aus)Schütt(reich)weite = dumping height at maximum radius
~(reich)weite, Ausleerreichweite, Kippreichweite, Abwurfreichweite [Schaufellader] = dumping radius
~(reich)weite f bei größter (Aus-)Schütthöhe = dumping radius at maximum height
Ausschütt|rolle f, Abwurfrolle [Schleppschaufelbagger] = dump sheave
~stellung f, Auskippstellung = dump(ing) position
~vorrichtung f, Abwurfvorrichtung = dumping device
~vorrichtung f für Torladewagen, Abwurfvorrichtung ~ ~ = straddle dump device, dumping device for straddle carrier
(Aus)Schütt|weite f → (Aus)Schüttreichweite
~winkel m, Ausleerwinkel, Kippwinkel, Abwurfwinkel [Schaufellader] = dump(ing) angle, angle of dump
~zeit f, Ausleerzeit, Kippzeit, Abwurfzeit [Schaufellader] = dumping time
(Aus)Schwitzen n → Bluten
Aussehen n = appearance
Aussetzbagger m → Aussetzgerät n
aussetzende Betonförderung f, aussetzender Betontransport m [Ein Fördermittel bringt eine bestimmte Menge Beton in einem Gefäß zur Schalung, schüttet den Beton in die Schalung und kehrt mit leerem Gefäß zum Mischer zurück] = discontinuous distribution of concrete, ~ concrete transport
Aussetz|gerät n, Eingefäßgerät, Aussetzbagger m, Eingefäßbagger = intermittent equipment

~**-Naßbagger** *m*, Aussetz-Schwimmbagger, Eingefäß-Naßbagger, Eingefäß-Schwimmbagger [*Schwimmbagger, der im Aussetzbetrieb mit einem einzigen Grabwerkzeug arbeitet, das in regelmäßigem Arbeitsspiel gräbt, fördert und entleert*] = intermittent dredger (Brit.); ~ dredge (US)

~**-Trockenbagger** *m*, Eingefäß-Trockenbagger = intermittent excavator

Aussetzung *f* **gegenüber Schlagregen** = exposure to driving rain

Aussetzungsgrad *m* = degree of exposure

Aussichts|gang *m* [*bei einem Turm*] = viewing gallery, observation ~

~**plattform** *f* = observation platform, viewing ~

~**raum** *m* = look-out room

~**straße** *f* = scenic boulevard

~**turm** *m* = observation tower

aussieben = to screen out

Aussieben *n*, **Aussiebung** *f* = screening out

Aussortieren *n*, **Aussortierung** *f* = sorting out

Ausspachteln *n* [*von Rissen*] = screeding of cracks

Ausspalten *n* [*Steine*] = cleaving

aussparen (✂) [*z. B. eine Fluchtstrecke*] = to carry, to maintain [*e.g. an escape road*]

Aussparung *f* = recess, blockout, embrasure, cavity

Aussparungskasten *m* [*Schornsteinbau im Gleitschalungsverfahren*] = box-out

Ausspülbecken *n* = washout closet, unsanitary ~

Ausspülen *n* **der Bodenteilchen durch die Sohlfläche eines Belastungsfilters** [*Bodenmechanik*] = escape of soil particles through the voids of a loaded inverted filter

Ausspülung *f*, innere Abscheuerung = internal erosion

ausstatten, ausrüsten, bestücken = to equip, to fit out

Ausstattung *f*, Satz *m* = kit

~ → Ausrüstung

aussteifen → (ab)bölzen

aussteifende Pfette *f* = bracing purlin(e)

Aussteifung *f* → (Ab)Bölzung

~, Verstifung, Aussteifen *n*, Versteifen = stiffening

Aussteifungs|rost *m* = stiffening grillage

~**träger** *m*, Versteifungsträger, Verstärkungsträger = stiffening girder, reinforcing ~

~**wand** *f*, Versteifungswand = stiffening wall

~**winkel** *m*, Versteifungswinkel = stiffening angle

Aussteigluke *f* = hatch

Ausstellung *f* = exposition, exhibition

Ausstellungs|architektur *f* = exhibition architecture, exposition ~

~**bau** *m*, Ausstellungsgebäude *n* = exhibition building, exposition ~

~**gelände** *n* = exhibition ground, exposition ~

~**halle** *f* = exhibition block, ~ hall, exposition ~

~**palast** *m* = exposition palace, exhibition ~

~**raum** *m* = showroom

~**stand** *m* = exhibition stand, exposition ~

Ausstockung *f* → Rodung

Ausstoßbremsventil *n* = ejector brake valve

ausstoßen, auswerfen = to eject

Ausstoßen *n*, Auswerfen = ejecting

~ [*Radschrapper*] = (load) ejection

Ausstoßer *m*, Auswerfer = ejector

~**anschlag** *m* = ejector stop

~**folgeventil** *n* = ejector sequence valve

~**leitungen** *fpl* = ejector lines

~**platte** *f* = ejector plate

~**seil** *n* [*Radschrapper*] = ejector cable

Ausstoß|gas *n* [*Düsenflugzeug*] = blast

~**geschwindigkeitswechselventil** *n* = ejector speed change valve

~**hebel** *m* = ejector (control) lever

~**kupplungsventil** *n* = ejector clutch valve

Ausstoßmaschine — Austragspalt 124

~maschine *f* [*für Blöcke oder Knüppel*] = ejector
~öffnung *f* [*Radschrapper*] = ejection opening
~steuerventil *n* = ejector control valve
Ausstoßung *f*, Auswerfung, Ausstoßen *n*, Auswerfen = ejection
Ausstoß|ventil *n* = ejector valve
~zylinder *m* [*Radschrapper*] = ejector jack
ausstrahlender Verkehr *m* = outbound traffic
ausstreichen (Geol.) → ausgehen
~, verfugen, ausfugen = to point
Ausstreichen *n* → Ausfugen
Ausstreichende *n* → Ausgehende
ausstreichender Fels *m*, ausgehender ~ = outcropping rock, rock outcrop
Ausstrich *m* → Ausgehende *n*
~, Ausfugung *f*, Verfugung [*als Ergebnis des Ausfugens*] = pointing
ausströmendes Wasser *n* [*bei Flut*] = outgoing water
Ausströmerelektrode *f* = discharge electrode
Ausströmung *f*, Entleerungsströmung [*Meer*] = emptying current, draining
austauchen [*Taucher*] = to go up
Austausch *m*, Austauschung *f*, Auswechs(e)lung = interchange
austauschbar, auswechselbar = interchangeable
austauschbare Bohrkrone *f*, auswechselbare ~, lösbare ~ = detachable bit
Austausch|barkeit *f*, Auswechselbarkeit = interchangeability
~bau *m* [*Maschine*] = interchangeable design, ~ version
~bild *n* = exchange pattern [*in the case of densimetric exchange between parallel flows*]
Austauschen *n*, Auswechseln = interchanging
Austausch|energie *f* = displacement energy [*power supply*]
~fähigkeit *f*, Austauschvermögen *n* [*Die Menge der austauschbaren Kationen in einem Boden*] = exchange capacity [*the quantity of exchangeable cations in a soil*]
~fläche *f* [*Wärme(aus)tauscher*] = heat transfer area
~gerät *n*, Austauschwerkzeug *n*, Austauschvorrichtung *f*, Umbaugerät, Umbauwerkzeug, Umbauvorrichtung, Arbeitsgerät, Arbeitswerkzeug, Arbeitsvorrichtung, Anbaugerät, Zusatzvorrichtung, Zusatzwerkzeug, Anbauvorrichtung, Zusatzgerät, Anbauwerkzeug, Arbeitsausrüstung *f* = interchangeable equipment item, attachment, rig, conversion equipment item
~satz *m* [*Alle Verschleißteile werden in einem vormontierten Austauschsatz geliefert*] = replacement set
~stoff *m* = interchangeable material
Austenit-Ferrit-Verbinder *m* = austenite-ferrite joint
austiefen, vertiefen = to deepen
Austiefen *n*, Vertiefen = deepening
Austiefung *f*, Vertiefung = deepening
Austiefungssee *m* = erosion lake, lake due to erosion
Austrag *m*, Austragung *f*, Austragen *n* [*aus einem Brecher, einer Mühle, einem Bunker, einem Silo und dgl.*] = discharge
~ende *n* = delivery end, tail ~
Austragevorrichtung *f* für Koks = coke extractor
Austrag|fähigkeit *f* [*Spülung beim Rotarybohren*] = carrying capacity, load ~
~höhe *f* = height of discharge
~mischer *m*, (Freifall-)Durchlaufmischer, Mischer mit Schurrenaustrag(ung) = closed drum (concrete) mixer with tipping chute discharge
~rutsche *f* → Auslaufrutsche *f*
~schnecke *f* = discharge screw (conveyer), ~ ~ conveyor
~schurre *f* → Auslaufschurre
~schurre *f* → → Betonmischer *m*
~spalt *m* → → Backenbrecher

~stirnwand *f* = discharge front end
~vorrichtung *f* = discharge device
~wand *f* [*Kugelmühle; Rohrmühle*] = end liner
Austritt *m*, ~ende *n* = discharge end, out-end
~ des Betons = discharge of concrete
~(ende) *m*, (*n*) = out-end, discharge end
~gefälle *n* = exit gradient
~geschwindigkeit *f* = discharge velocity
~öffnung *f* = discharge opening, orifice
~öffnung *f* → → Backenbrecher
~produkt *n* = discharge product
~punkt *m* von Grundwasser, Quelle *f* = spring
~spalt *m* → → Backenbrecher
~stelle *f* = discharge point
~stufe *f* = last step, top ~
austrocknen = to dry out
Austrocknung *f*, Austrocknen *n* = drying out
Austrocknungsvorgang *m* = drying out process
Aus- und Ankleideraum *m*, Umkleideraum = changing room, dressing ~
Aus- und Einbauen *n* von Bohrgestänge = ,,round trip", pulling and running drill pipe
Aus- und Rückkippen *n* der Mulde mit Hilfe der Schwerpunktlage = gravity discharge and return of the skip
Auswahl *f* = selection
auswandern, abwandern (⚒) = to move
~ [*Gebirge im Tunnel-, Stollen- und Schachtbau*] = to creep
Auswasch|delta *n* = outwash delta
~ebene *f* = outwash plain, (over-)wash ~
auswaschen, (aus)kolken, unterspülen, unterwaschen, wegspülen = to underwash, to scour, to undermine
~ (aus) = to wash out (of)
Auswasch|fächer *m* = outwash fan
~maschine *f* für Waschbetonplatten, Plattenwaschmaschine, Waschbetonmaschine [*mit Einzel- oder Ringbürsten*] = scrubbing machine for concrete tiles
~probe *f*, Auswaschprüfung *f*, Auswaschversuch *m* = washing test
~schotter *m* = (glacial) outwash gravel
Auswaschung *f*, Kolk *m*, Wegspülung, Unterwaschung, Unterspülung, (Aus)Kolkung, Kolkbildung = scour(ing), undermining, underwashing, subsurface erosion, wash(-out)
~ des Bodens, Podsolierung = podsolization
Auswaschungen *fpl* [*Tiefbohrtechnik*] = eroded formations
Auswaschungs|boden *m* → Eluvialboden
~tasche *f*, Auswaschungstrichter *m* = sink
auswechselbar, austauschbar = interchangeable
auswechselbare Bohrkrone *f*, austauschbare ~, lösbare ~ = detachable bit
Auswechselbarkeit *f*, Austauschbarkeit = interchangeability
Auswechseln *n*, Austauschen = interchanging
Auswechs(e)lung *f*, Austausch(ung) *m*, (*f*) = interchange
~, Balkenwechsel *m* = framing (around an opening)
auswehen (Geol.) = to deflate
Auswehung *f* (Geol.) = deflation
Ausweiche *f*, Ausweichstelle *f*, Kreuzungsstelle [*für ein Schiff*] = passing-place, layby
~ = lay-by; turn-out (US) [*A wide place or side bay on a road enabling vehicles to pass or lie up*]
ausweichen (⚒) = to flow
~ [*Bodenmechanik*] = to stretch
Ausweichen *n* [*eines Sandkörpers in waag(e)rechter Richtung. Bodenmechanik*] = stretching
Ausweich|fahrbahn *f* längs der bestehenden Trasse = deviation alongside the existing alignment [*road engineering*]
~flughafen *m* = alternate airport

Ausweichflugplatz — Außendock

~flugplatz m, Ausweichlandeplatz = alternate airfield, ~ landing ground
~gleis n, Wartegleis = passing track, lay-by [*A part of a railway out of the traffic tracks, where trains may wait*]
~stelle f, Ausweiche f, Kreuzungsstelle [*für ein Schiff*] = layby, passing place
~stelle f [*Fläche, die von einem oder mehreren Fahrzeugen benutzt werden kann, um einem oder mehreren anderen Fahrzeugen die Fortsetzung seiner oder ihrer Fahrt zu ermöglichen*] = passing area
Ausweichung f [*Horizontalkomponente des Weges eines Wassertropfens im Meer infolge der Gezeiten*] = elongation (of molecules)
Ausweitung f [*Tunnelbau*] = enlarging
auswerfen, Ausstoßen = to eject
Auswerfen n, Ausstoßen = ejecting
~ [*Schürfkübelbagger*] = spotting [*dragline bucket*]
Auswerfer m, Ausstoßer = ejector
Auswerfkranz m = throw-out collar
Auswerfung f, Ausstoßung = ejection
Auswertegerät n [*zur Auswertung von Luftbildern*] = interpreting device
auswerten [*Luftbilder*] = to interpret
Auswerteverfahren n [*für Luftbildauswertung*] = interpretation method
Auswertung f [*von Luftbildern*] = interpretation
~, Bewertung = evaluation
~ **von Meißelschäden** = dull bit evaluation
~ **von Straßenverkehrsunfällen** = evaluating road accident data
Auswirkung f = effect
Auswitterung f → → Denudation f
auswuchten = to balance the weights
Auswuchten n, Auswuchtung f = balancing
Auswuchtmaschine f = balancing machine
außen verstärktes Bohrgestänge n = external upset drill pipe
Außen|abmessungen fpl, Außenmaße npl [*Länge x Höhe x Breite*] = overall dimensions

~anlage f, Freiluftanlage = outdoor installation
~ansicht f = exterior view, external ~
~anstrich m, Außen(an)streichen n = exterior painting, external ~, outdoor ~ [*The external use of paints, varnishes, and stains for protection or decoration of materials*]
~anstrich m, Außenfarbschicht f, Außenfarbanstrich = exterior paint coat, external ~ ~, outdoor ~ ~, ~ (paint) finish, ~ coat of paint
~(anstrich)farbe f = exterior paint, external ~, outdoor ~
~antenne f = outdoor aerial
~arbeiten fpl [*beim Bau eines Hauses*] = exterior work, external ~
~architektur f = outdoor architecture, external ~, exterior ~
~aufzug m, Außenpersonenaufzug, Außenfahrstuhl m = exterior passenger lift (Brit.); ~ ~ elevator (US)
~aufzug m → Außenbauaufzug
~backenbremse f = external cheek brake, exterior ~ ~
~(bau)aufzug m = external (contractors') hoist
~(bau)platte f = external building panel, exterior ~ ~
~bekleidung f, Außenverkleidung, Außenauskleidung = exterior lining, ~ facing
~beleuchtung f = outside lighting, external ~
~-Betonkübelaufzug m, Außen-Mulden(bau)aufzug, Außen-Kübel(bau)aufzug = external skip-hoist
~bohrung f [*Bohrung außerhalb der nachgewiesenen Ölführung eines Feldes*] = outpost well
~böschung f, Außenseite f [*bei Deichen, die dem Gewässer zugekehrte Böschung*] = outside slope
~dämmung f = external insulation, exterior ~
~dichtung f → (Ab)Dichtungsvorlage f
~dock n = open dock

~druck m [auf Rohrleitungen durch Verfüllmaterial und/oder Verkehrslasten] = surcharge pressure
~durchmesser m = outside diameter
~(ein)rütt(e)lung f, Schalungs(ein)rütt(e)lung [Beton] = external vibration, consolidation (of concrete) by external vibration
~fahrstuhl m, Außen(personen)aufzug m = exterior passenger elevator (US); ~ ~ lift (Brit.)
~fenster n = external window
~-Fensterbank = external window sill; ~ ~ cill (Brit.)
~fläche f, Stirn f, Haupt n [Quader] = face
~flansch m der Felge = rim outer flange
~ganghaus n, Laubenganghaus = gallery apartment building
~gewinde n = external thread
~gewindekegel m = male cone
~(gleis)kette f, Außenraupe f = outside crawler
~haupt n [Seeschleuse] = outer gates, lower ~
~haut f von Gebäuden = exterior walls of buildings, external ~ ~ ~
~(haut)dichtung f → (Ab)Dichtungsvorlage f
~hydrant m = outside hydrant
~(kern)rohr n = outer cylinder, ~ tube
~klima n = external climate, exterior ~
~konus m, Außenkegel m = taper shank
~kräfte fpl [Tiefbohrtechnik] = compression forces on a casing
~-Kübel(bau)aufzug m, Außen-Mulden(bau)aufzug, Außen-Betonkübelaufzug = external skip-hoist
~lager n = outboard (ball) bearing
~lärmdämmung f = external noise insulation, exterior ~ ~
~lastenaufzug m = external materials lift
außenliegendes Zugband n, freiliegendes ~ = exterior tieback

Außenluft f [Die aus dem Freien angesaugte Luft. Der Ausdruck „Frischluft" sollte zur Vermeidung von Verwechslungen nicht gebraucht werden] = atmospheric air, outside ~
~verunreinigung f, Außenluftverschmutzung = atmospheric pollution
Außenmantel m = outer jacket
~ [Erdkruste] = litosphere
Außen|maß n = overall dimension
~mauer f, Umfassungsmauer [Mauer, die ein Gebäude nach außen abschließt. Sie kann eine Decke tragen oder deckenunbelastet sein] = exterior wall, external ~
~-Mulden(bau)aufzug m, Außen-Kübel(bau)aufzug, Außen-Betonkübelaufzug = external skip-hoist
~öffnung f [Brücke] = end span
~(personen)aufzug m, Außenfahrstuhl m = exterior passenger elevator (US); ~ ~ lift (Brit.)
~putz m [siehe englische Anmerkung unter „Putz"] = external finish, exterior ~, rendering (~), external rendering, exterior rendering
~putzmörtel m = stuff for exterior rendering
~reede f, offene Reede = open roadstead
~riß m = external crack
~rohr n, Außenkernrohr = outer tube, ~ cylinder
~rüstung f, Außen(bau)gerüst n = outside scaffold(ing), external ~, exterior ~
~rüttler m, Außenvibrator m, Schalungsrüttler, Schalungsvibrator = external vibrator, exterior ~
~schale f [Hohlmauer] = outer shell, external leaf
~schalung f = external formwork, exterior ~ [deprecated: shuttering]
~schal(ungs)platte f = external formwork panel, exterior ~ ~
~schaufel f [Turbomischer] = external mixing blade
~schiene f = outside rail
~schutz m = outside protection

Außenseite — Ausziehverrohrung

~seite f = outside
~seite f → Außenböschung f
~spannglied n = external tendon, exterior ~
~stauchung f [Rohr] = external upset
~strand m = outer shore
~streifen m → unbefestigter Seitenstreifen
~stuck m = exterior stucco, external ~
~stütze f = outside column
~temperatur f = outside temperature, external ~, outdoor ~, exterior ~, ~ ambient ~
~temperaturfühler m = outdoor temperature sensing device, outside ~ ~ ~, external ~ ~ ~, exterior ~ ~ ~
~treppe f = outdoor staircase (Brit.); ~ stair (US)
~tür f = outside door
~türrahmen m = outside door frame
~verglasung f = external glazing
~verschleiß m = external wear
~verwendung f = external use, exterior ~
außenverzahntes Stirnrad n, Außenrad = external-teeth gear
Außen|vibration f, Außenrüttlung f = exterior vibration, external ~
~vorspannung f = outer pre-stress
Außenwand f = external wall
~putz m = external wall finish, exterior ~ ~, ~ ~ rendering
~stütze f = external wall column
~verkleidung f, Außenwandbekleidung, Außenwandauskleidung = external wall lining
Außen|wange f = outer string
~werbung f an Straßen = road-side advertisement, ~ advertising, ~ publicity
~zone f = outer zone
außer Betrieb = out of service
äußere Bodenplatte f [Unterkellerung] = external floor slab
~ Bogenfläche f, Bogenrücken m = extrados, back
~ Bremsnabe f = outer brake hub

~ Dossierung f [Gleitschalungsbau] = diameter taper
~ Fensterleibung f, ~ Fensterlaibung = window reveal
~ Fliesenarbeiten fpl = flagging work
~ Gewölbefläche f, Gewölberücken m = back of vault, extrados ~ ~
~ Kraft f = external force, exterior ~
~ Türleibung f, ~ Türlaibung = door reveal
~ Wärmeübergangszahl f = outside film coefficient [BTU/h sqft. F]
äußeres Planetenrad n = outer planetary gear
~ Produkt n, vektorielles ~ = vector product [of two vectors]
~ Steuerventil n = exterior valve
~ Ufer n → einbuchtendes ~
außermittige Stegplatte f [Gleiskette] = offset grouser shoe
außermittiges Einrücken n = over-center engagement (US); over-centre ~ (Brit.)
Auszieharm m, Teleskoparm = telescoping arm
ausziehbar, teleskopierbar = telescopic
ausziehbarer Bohrturm m = telescoping (drilling) derrick, ~ ~ tower
Ausziehdeichsel f, Ausziehzugstange f, Teleskopdeichsel f, Teleskopzugstange = telescopic drawbar
ausziehen, teleskopieren = to telescope
(aus)ziehen, herausziehen [Pfahl; Spundbohle] = to pull out, to extract
Auszieher m = puller
Auszieh|feder f = drawing pen
~gleis n → Sortiergleis
~leiter f = extension ladder
~rohr n = telescoping tube
~schacht m, ausziehender Schacht [Schacht, der nur zur Ableitung der Abwetter zur Tagesoberfläche dient] = upcast shaft, discharge air ~
~schalung f, Teleskopschalung = telescoping formwork
~strecke f (⚒) = return gate road
~verrohrung f, Teleskopverrohrung = telescoping casing

Ausziehversuch — Autobusbahnhof

~versuch m, Ausziehprobe f, Ausziehprüfung f [zur Feststellung der Haftung zwischen Bewehrung und Beton] = pull-out test

~widerstand m = pull-out resistance

Auszugrohr n ~→ Talsperre f

auszwicken, abdecken, verzwicken [Packe] = to choke, to chink, to blind, to key

(Aus)Zwickstein m [für Packe] = (rock) spall

autallotriomorph (Geol.) → panidiomorphkörnig

Autarkie f = self-sufficiency

Auto n, ~mobil n = automobile

Autoanhänger-Einachs-Mischer m → (luftbereifter) Einachs(-Schnell)mischer

(Auto-)Aufbau-Grabenbagger m → Last(kraft)wagen-Grabenbagger

(Auto-)Aufbaulader m → Last(kraft)wagenlader

Auto-Auslegerkran m → Automobil-Auslegerkran

Autobagger m → Automobilbagger

Autobahn f = motorway

~ mit Mittelstreifen = dual (or double) carriageway motorway

~ mit vier Spuren ohne Mittelstreifen = four-lane single-carriageway motorway

~ ohne Mittelstreifen = single-carriageway motorway

~abfahrt f → Ausfahrt

~abschnitt m, Autobahnteilstück n, Autobahnstrecke f = motorway section

~anschluß m → Anschlußpunkt m

~ausfahrt f → Ausfahrt

~bau m = motorway construction

~baustelle f = motorway construction site

~beleuchtung f = motorway lighting

~bepflanzung f = motorway plantation

~-Betonbaustelle f = motorway concrete construction site

~-Betonfahrbahn f = motorway concrete carriageway

~(-Beton)mischer m → Brückenmischer

~brücke f = motorway bridge

~(-Brücken)mischer m → Brückenmischer

~decke f, Autobahnbelag m = motorway surfacing (Brit.); turnpike pavement, turnpike paving (US)

~-Fegemaschine f, Autobahn-Kehrmaschine = motorway sweeper

~hängebrücke f = motorway suspension bridge

~-Hochbrücke f, Autobahn-Talbrücke, Autobahn-Viadukt m = motorway viaduct

~kirche f = motorway church

~meisterei f; Autobahnwerkhof m [Schweiz] = motorway breakdown and repair depot

~mischer m → Brückenmischer

~netz n = motorway network

~planung f = motorway planning

~querschnitt m = motorway cross-sectional profile

~randstreifen m = hardened outside verge

~ring m = motorway ring

~talbrücke f = motorway viaduct

~trassierung f = motorway location

~tunnel m = motorway tunnel

(~)Unterlagpapier n, Straßenbaupapier = waterproof paper, concreting ~

~verkehr m = motorway traffic

~(verkehrs)spur f, Autobahn(verkehrs)streifen m = motorway traffic lane

~verkehrszeichen n = motorway traffic sign

~werkhof m [Schweiz]; Autobahnmeisterei f = motorway breakdown and repair depot

Auto-Betonpumpe f, Transport-Betonpumpe = self-propel(l)ed concrete pump

Autobrecher m → Auto(mobil)(stein)brecher

Autobus m, (Omni)Bus = bus, passenger service vehicle

~bahnhof m → Busbahnhof

Autobushaltestelle — automatische Entaschungsvorrichtung 130

~haltestelle *f*, (Omni)Bushaltestelle = bus stop, ~ halt
~linie *f*, (Omni)Buslinie = bus line
autochthon → bodeneigen
Autoderrick(-Kran) *m* = rubber-mounted derricking jib crane (Brit.); ~ ~ boom ~ (US); ~ derrick (~)
Autodiesel(motor) *m* = automobile diesel engine
Autodrehkran *m*, Automobildrehkran = two-engine rubber-mounted slewing crane
Autodrehscheibe *f*, Automobildrehscheibe = driveway turntable
Autofahrbahn *f* = motor vehicle carriageway
Autofähre *f*, Automobilfähre = car-ferry, drive-on drive-off ship
Autofährenstation *f* = car ferry terminal
(Auto)Fahrer *m*, Kraftfahrer = (motor vehicle) driver; chauffeur (US)
Auto-Fegemaschine *f* → Automobil-Fegemaschine
autogenes Mahlen *n* → Autogenmahlung *f*
~ **Schneiden** *n* → Brennschneiden
~ **Schweißen** *n* → Gasschweißen
Autogen|härtemaschine *f*, Brennhärtemaschine, Flammhärtemaschine = autogenous hardening machine, flame-hardening ~
~**härten** *n*, Flammhärten, Brennhärten = autogenous hardening, flame-hardening
~**mahlung** *f*, Autogenmahlen *n*, autogenes Mahlen, autogene Mahlung = autogenous grinding
~**schweißen** *n* → Gasschweißen
Autoheber *m* → Autowinde *f*
Autohof *m* → Busbahnhof
Auto-Kino *n* = drive-in (motion picture) theatre
Auto-Kipper *m* → Kipper
Autoklav *m*, Druckkessel *m*, (Dampf-)Härtekessel, Dampfdruckerhitzer *m* = pressure vessel, autoclave
~**gips** *m* = autoclaved gypsum

Autoklavisierung *f*, Dampfdruckhärtung = autoclaving
Autoklavprüfung *f*, Autoklavprobe *f*, Autoklavversuch *m* = autoclave test
Autokran *m* → Automobilkran
~ **auf Chassis mit Eigenantrieb**, schnellfahrender Autokran = truck crane
Automatik|heizung *f* = auto(matic) heating
~**kessel** *m* = auto(matic) boiler
~**spill** *n* [*Tiefbohrtechnik*] = auto(matic) cathead
~**steuerung** *f*, automatische Steuerung = auto(matic) control system, ~ controls
automatisch, selbsttätig = auto(matic)
~ **eingetragenes Schweißgut** *n* = auto(matic) weld deposit
automatische Abmessung *f*, ~ Dosierung, ~ Zumessung, ~ Zuteilung = auto(matic) measuring, ~ batching, ~ proportioning
~ **Abmeßanlage** *f*, ~ Dosieranlage, ~ Zumeßanlage, ~ Zuteilanlage = auto(matic) batching plant, ~ proportioning ~, ~ measuring ~
~ **Aufreißerausschaltung** *f* = auto(-matic) ripper kickout, ~ rooter ~
~ **Ausschüttwaage** *f* = auto(matic) dumping batcher scale
~ **Beschickung** *f*, ~ (Material)Aufgabe *f* = auto-feed, automatic feed
~ **Betondachsteinmaschine** *f*, Betondachsteinautomat *m* = auto(matic) concrete roof-tile-making machine, ~ roofing tile ~
~ **Betonmischanlage** *f* = auto(matic) concrete mixing plant
~ **Brettzufuhr** *f* [*Steinformmaschine*] = auto(matic) pallet feed
~ **Datenuntersuchung** *f* = auto(matic) data logging
~ **Drehzahlreg(e)lung** *f* = auto(matic) speed control
~ **Entaschungsvorrichtung** *f*, Entaschungsautomat *m* = auto(matic) ash remover

automatische Entleerung — automatisches Vorschubgerät

~ **Entleerung** *f*, Selbstentleerung = auto(matic) emptying
~ **Fernsteuerung** *f* = auto(matic) remote control
~ **Feuerung** *f* = auto(matic) furnace
~ **Formanlage** *f* [*Gießereimaschine*] = auto(matic) mo(u)lding plant
~ **Gestängezange** *f*, Motorzange [*Rotarybohren*] = power tongs
~ **Gewichtsdosieranlage** *f*, ~ Gewichtszuteilanlage, ~ Gewichtszumeßanlage, ~ Gewichtsabmeßanlage = auto(matic) weighbatching plant
~ **Höhensteuerung** *f* [*Gleitschalungsfertiger*] = auto(matic) grade control system
~ **Hubabstellung** *f* → Hubausschaltautomat *m*
~ **Kippmulde** *f* [*Erdbaulast(kraft)wagen*] = auto(matic) dump body
~ **Krankatze** *f*, ~ Lastkatze, ~ Laufkatze = auto(matic) crab
~ **Kübelarretierung** *f* [*Radschrapper*] = auto(matic) bowl latch
~ **Raumteildosieranlage** *f* = auto(matic) volume batching plant, ~ ~ measuring ~, ~ ~ proportioning ~
~ **Schaltung** *f* [*Kraftübertragung*] = auto(matic) shift
~ **Schweißmaschine** *f*, Schweißautomat *m* = auto(matic) welder, ~ welding machine
~ **Schwimmschaltung** *f* [*Schaufellader*] = auto(matic) bucket level(l)ing device
~ **Spachtelmaschine** *f*, Spachtelautomat *m* [*zum Spachteln von Terrazzoplatten*] = auto(matic) grouting machine
~ **Sperren** *f pl* und Zuteiler *m pl* [*Förderanlage*] = pre-selective sorters and traffic controls
~ **Steuerung** *f* = auto(matic) control
~ **Stützenbetoniermaschine** *f*, Stützenbetoniermaschine *m* = auto(matic) concrete column pourer

~ **Transportbetonherstellung** *f* = auto(matic) ready-mix(ed) concrete fabrication
~ **(Ver)Wiegeanlage** *f* = auto(matic) weighing plant
~ **Waage** *f*, (Ver)Wiegeautomat *m* = auto(matic) scale
~ **Wasserhebemaschine** *f* → Widder *m*
~ **Zählung** *f* = auto(matic) counting
automatischer Abschneider *m*, Abschneideautomat *m* [*Ziegelindustrie*] = auto(matic) cutting(-off) table
~ **Befeuchter** *m*, Befeucht(ungs)automat *m* = auto(matic) humidifier
~ **Betrieb** *m* = auto(matic) operation
~ **Bohrhammervorschub** *m* = auto(matic) drill (hammer) feed
~ **Bremseinsteller** *m* = auto(matic) brake adjuster
~ **Dampferzeuger** *m*, Dampfautomat *m* = auto(matic) steam generator
~ **Förderer** *m*, Förderautomat *m* = auto(matic) conveyor, ~ conveyer
~ **Regler** *m*, Regelautomat *m* = auto(matic) governor
~ **Rückspeiser** *m* [*Dampfheizung*] = auto(matic) return trap
~ **Schaufeleinsteller** *m* [*Lader*] = auto(matic) bucket positioner
~ **Schwimmer** *m* = auto(matic) float
~ **Vorschub** *m* = auto(matic) feed
automatisches (An)Saugen *n* = auto(matic) priming
~ **Ausstoßgeschwindigkeitssteuerventil** *n* [*Radschrapper*] = ejector auto(matic) speed selector valve
~ **Ausstoßgeschwindigkeitswechselventil** *n* = auto(matic) speed change valve
~ **Nivellier(instrument)** *n* = auto(matic) level
~ **Schweißen** *n* = auto(matic) welding
~ **(Straßen)Verkehrszählgerät** *n* = auto(matic) traffic counter
~ **Transportbetonwerk** *n*, ~ Lieferbetonwerk, ~ Betonlieferwerk = auto(matic) ready-mix works; ~ depot (Brit.)
~ **Vorschubgerät** *n* [*Bohrhammer*] = auto(matic) feed unit

automatisches Wehr — Autoschütter

~ **Wehr** *n* [*bewegliches Wehr, das den Oberwasserstand selbsttätig reguliert*] = self-regulating barrage, ~ movable weir

automatisierte Magazinierung *f* = automated storage

Auto(mobil) *n* = automobile

~**-Auslegerkran** *m*, Auslegerkran auf Chassis mit Eigenantrieb = two-engine rubber-mounted crane with non-rotating jib (Brit.)/boom (US)

~**bagger** *m*, Umbau-~, Universal-~, Vielzweck-~, Mehrzweck-~, Bagger auf Chassis mit Eigenantrieb [*Der Unterteil (Unterwagen) dieses Baggers ist ein teilweise abgeändertes schweres LKW-Fahrgestell mit Lenkachse und Führerhaus. Der Bagger hat zwei Motoren, von denen der kleinere Baggermotor im Oberteil nur für die Bewegungen der Arbeitsvorrichtungen dient, während der größere Fahrmotor im Fahrgestell das Fahrwerk antreibt. Die Fahrgeschwindigkeit beträgt bis zu 60 km/Std. Für die Bedienung sind zwei Mann erforderlich*] = two-engine rubber-mounted excavator

~**drehkran** *m* = two-engine rubber-mounted slewing crane

~**drehscheibe** *f* = driveway turntable

~**fähre** *f* = car-ferry, drive-on drive-off ship

~**-Fegemaschine** *f*, Auto(mobil)-Kehrmaschine = engine-driven rotary sweeper, ~ scavenging machine, ~ road sweeper, ~ mechanical sweeper, ~ road sweeping machine, ~ street sweeper, power(-driven) ~ ~

~**kipper** *m* → Kipper

~**kran** *m*, Kran auf Chassis mit Eigenantrieb [*es gibt Auto-Auslegerkrane (mit nicht drehbarem Ausleger) und Autodrehkrane (mit drehbarem Ausleger)*] = two-engine rubber-mounted crane

~**motor** *m*, Kraftfahrzeugmotor, Kfz.-Motor = automobile engine

~**rennbahn** *f* = automobile race course

~**-Sandstreuer** *m* = self-propelled sand spreader

~**scheinwerfer** *m* = automobile headlight

~**-Splittstreuer** *m* = self-propelled gritter

~**-Sprengwagen** *m* → Bitumen-Sprengwagen

~**-Spritzwagen** *m* → Bitumen-Sprengwagen

~**(stein)brecher** *m*, Fahrbrecher = self-propelled crushing plant (US); rubber-mounted crusher, rubber-mounted crushing plant

~**-Straßensprengwagen** *m*, Auto(mobil)-Wassersprengwagen = self-propelled street sprinkler

~**-Teerspritzmaschine** *f* = self-propelled tar distributor

~**verkehr** *m* → Autoverkehr

~**versuchsstrecke** *f* = automobile test ground

~**-Wassersprengwagen** *m*, Auto(mobil)-Straßensprengwagen = self-propelled street sprinkler

~**werkstatt** *f* = automobile workshop

Automotor *m* → Automobilmotor

(Auto)Parkplatz *m*, (Auto)Parkfläche *f* = parking place, car park, parking lot

Autorennbahn *f*, Automobilrennbahn, Auto(mobil)rennstrecke *f* = car racing track

Auto-Sandstreuer *m*, Automobil-Sandstreuer = self-propelled sand spreader

Autoscheinwerfer *m*, Automobilscheinwerfer = automobile headlight

Autoschrapper *m* mit 2-Rad-Schlepper → Einachs-Motorschrapper

~ **4-Rad-Schlepper**, Motor-Schürfzug *m* = tractor scraper, motor(ized) ~ (with 4-wheel traction)

Autoschütter *m*, Motorkübelwagen *m*, Kopfschütter, Vorkopf-Schütter, Frontkipper, Motorkipper, (Auto-) Vorderkipper, Klein-Förderwagen *m* für Baubetrieb = dumper, shuttle ~, front tipper

Autoschütter-Anbaugerät — Azidität

~-Anbaugerät *n* = attachment for dumpers
Autosilo *m*, Parkturm *m* = parking tower, pidgeonhole
Auto-Splittstreuer *m*, Automobil-Splittstreuer = self-propelled gritter
Autosteinbrecher *m* → Automobilsteinbrecher
Autostraße *f* [*Eine Straße, die in Konstruktion und Gestaltung den Bedürfnissen einer vorwiegenden Benutzung durch Kraftfahrzeuge gerecht wird*] = motor road
Auto-Straßensprengwagen *m* → Auto(mobil)Wassersprengwagen
Autostraßentunnel *m* = motor road tunnel
Auto-Teerspritzmaschine *f*, Automobil-Teerspritzmaschine = self-propelled tar distributor
Auto-Transportbeton-Mischer *m* → Transportmischer
Autotunnel *m*, Kraftfahrzeugtunnel, Motorfahrzeugtunnel = tunnel for automobile traffic
Auto-Turm(dreh)kran *m* = rubber-mounted (mono)tower crane, ~ rotating ~ ~, ~ revolving ~ ~
Autoverkehr *m*, Automobilverkehr, Kraftverkehr, Motorfahrzeugverkehr = automobile traffic
Autoversuchsstrecke *f*, Automobilversuchsstrecke = automobile test ground
Auto-Vorderkipper *m* → Autoschütter *m*
Autowaschplatz *m* = washdown yard for cars, car wash, (vehicle) washdown
Auto-Wassersprengwagen *m* → Automobil-Wassersprengwagen
Autowerkstatt *f*, Automobilwerkstatt = automobile workshop
Autowinde *f*, Wagenwinde, Autoheber *m*, Wagenheber, Autohebebock *m*, Wagenhebebock = car jack
axial beaufschlagte Turbine *f* = axial-flow (type) wheel, ~ (~) turbine
axialer Druck *m* = axial pressure

axiales (Erd)Beben *n* = linear earthquake
Axialgebläse *n* = axial-flow blower
~ **mit Spaltflügelprofilen** = slotted-blade axial-flow blower
~**rad** *n* = axial-flow impeller
Axial|-Gleitlager *n* **mit abgestuften parallelen ebenen Tragflächen** = stepped thrust-bearing
~**-Hochdruck-Ventilator** *m*, Propeller-Hochdruck-Ventilator, Schrauben-Hochdruckventilator, = axial-flow high-pressure fan, disc ~ ~, propeller ~ ~; disk ~ ~ (US)
~**kolbenmaschine** *f* = axial piston machine
~**kolbenmotor** *m* = axial piston engine
~**-Kolbenpumpe** *f* = axial piston pump
~**kreiselverdichter** *m* = axial-flow turbo compressor
~**ölpumpe** *f* = axial-flow oil pump
~**pumpe** *f* = axial (flow) pump, propeller ~
~**-Radialrad** *n* = mixed-flow-centrifugal compressor
~**turbine** *f* = axial-flow turbine
~**ventilator** *m*, Propellerventilator, Schraubenventilator = axial-flow fan, disc ~, propeller ~; disk ~ (US)
~**verdichter** *m* = axial-flow compressor
~**zugfestigkeit** *f* = axial tensile strength
Azetylen *n* = acetylene (gas)
~**erzeugungsanlage** *f* = acetylene producing plant
~**-Flammstrahlbrenner** *m* **zum Aufrauhen von Basaltpflaster** = acetylene heat sett treatment unit
~**flasche** *f*, Dissousgasflasche = acetylene cylinder, ~ bottle
~**(-Leucht)boje** *f* = acetylene gas lighted buoy
~**sauerstoff(schneid)brenner** *m* = oxyacetylene cutting torch
Azidität *f*, Säuregehalt *m*, Neutralisationszahl *f*, Säuregrad *m* = acidity, acidic content

Azimut n, m, Scheitelkreis m = azimuth
~ der Abweichung, Abweichungsrichtung f [*Tiefbohrtecnihk*] = direction of the angle of deviation, azimuth
~ ~ ~ ~ ~, ~ ~ ~ ~ ~ deflection
~**winkel** m = azimuth bearing angle
Azohumussäure f = azo-humic acid

B

Baaderprüfung f, Baaderprobe f, Baaderversuch m = Baader copper test

Backenbrecher m [*früher: Backenquetsche f*]
  Arbeitsorgan n, Arbeitselement n
  Austragsspaltweite f → (untere Brech)Spaltweite
  Austrittsöffnung f, Austragsspalt m, Austrittspalt, Entleerungsspalt, Entleerungsöffnung, Austragsöffnung
  ~ mit gradlinigen Backenplatten
  ~ mit Rollenlagerung
  ~ mit unten befestigter Schwingenbrechbacke
  Backenplatte f
  bewegliche Brechbacke f → Schwingen(brech)backe(n)
  (Blake'scher) (Doppel-)Kniehebel(backen)brecher m → Doppelkniehebelbrecher
  (Brech)Backe f, (Brech)Backen m
  Brechelement n, Brechorgan n
  Brech(er)rahmen m, Brech(er)körper m, Brech(er)gehäuse n, Gehäuse
  (Brecher)Schwinge f, Brechschwinge [*Einschwingenbrecher*]
  Brechfläche f
  Brechmaul n, Maul, Eintrittsöffnung f

  Brechraum m, Brecherraum, Brech(er)kammer f
  Brechspalteinstellung f → (untere Brech)Spaltweite f

BAB, Bundesautobahn f = federal motorway
BAB-Talbrücke f = Federal motorway viaduct
Bach m = brook
~**brücke** f = brook bridge
~**durchlaß** m, Bachabzugkanal m = brook culvert
~**geröll(e)** n = rolled pebbles
~**kläranlage** f = brook retention basin
Bächlein n, kleiner Bach m = brooklet
Bachspeisung f [*Kanal*] = canal feeding by brook(s)
Backe f, Backen m = jaw

jaw (type) crusher, ~ (~) breaker

working element

discharge opening, ~ point, outlet ~

straight plate ~ ~
roller bearing ~ ~
Dodge ~ ~

jaw (crushing) plate

(crushing) jaw
crushing member
base, (main) frame

pitman

crushing surface [*crushing jaw*]
feed opening, receiving opening, crusher opening, feed mouth, receiving mouth, crusher mouth
crushing chamber, breaking chamber

Backenbrecher

Doppel∼, Doppelschwingenbrecher, Zweischwingenbrecher
twin ∼ ∼, two movable ∼ ∼, double movable ∼ ∼

Doppelkniehebelbrecher *m*, Zweipendelbackenbrecher, (Blake'scher) (Doppel-)Kniehebel(backen)brecher [*erfunden von dem Amerikaner Blake im Jahre 1858*]
Blake (type jaw) crusher (or breaker); double toggle crusher, compound toggle lever stone breaker (Brit.); swing jaw crusher

Drei∼
three-jaw crusher

dreiteilige Druckplatte *f*, ∼ Kniehebelplatte
three-piece toggle plate

Druckplatte *f* → Kniehebelplatte

Einschwingen(backen)brecher *m*
(Dalton) overhead eccentric (type jaw) crusher

Einschwingen-Granulator *m*, Granulator-Splittbrecher *m*
single-toggle granulator, single-toggle type jaw granulator, chipping(s) breaker

Einstellzahn *m*
adjusting sprocket

Eintrittsöffnung *f*, (Brech)Maul *n*
feed opening, receiving opening, crusher opening, (feed) mouth, receiving mouth, crusher mouth

Einzug *m*
nip

Einzugswinkel *m*
angle of nip

Entleerungsspalt *m* → Austrittsöffnung *f*

Exzenterlager *n* der Zugstange *f*, Zugstangenlager
pitman eccentric bearing

Exzenter(welle) *m*, (*f*), Antriebswelle
(overhead) eccentric (shaft)

Exzenterwellenlager *n*
eccentric shaft bearing

Fein∼, Nach∼
secondary ∼ ∼, reduction ∼ ∼, fine ∼ ∼

Festbacken *m* → Stirnwand(brech)backe *f*

Festbackenplatte *f*
fixed jaw (crushing) plate, front crushing plate

Flachriemen-Antriebsscheibe *f*, Schwungscheibe für Flachriemen(an)trieb
flywheel crowned for flat belt

Gehäuse *n* → Brech(er)rahmen *m*

geradlinige Backenplatte *f* [*Grobbackenbrecher*]
straight jaw (crushing) plate

gewölbte Backenplatte *f* [*Feinbackenbrecher*]
curved jaw (crushing) plate

Gleitklotz *m* [*Blake-Brecher*]
sliding wedge block

Granulator-Splittbrecher *m* → Einschwingen-Granulator *m*

Groß∼, Grob∼, Vor∼
(large) primary ∼ ∼

Herausspringen *n* des Brechgutes aus dem Brechmaul
belching

Backenbrecher

Hub *m*	stroke
Keilriemen-Antriebsscheibe *f*, Schwungscheibe für Keilriemenantrieb	flywheel grooved for V-belt
Kniehebel~ → Doppelkniehebelbrecher	
Kniehebelplatte *f*, Druckplatte	toggle plate
Kurbelschwingen(backen)brecher	single toggle jaw crusher
Maulbreite *f*	opening across the width of the jaws
Maulweite *f* [*Spaltweite der Eintrittsöffnung f*]	jaw opening [*distance between the jaws at the top of the crusher opening*]
Nach~ → Fein~	
(Nach)Stellkeil *m*, verstellbarer Keil	adjusting wedge, push wedge, pull wedge
oberer Seitenkeil *m*	upper cheek plate
reibender Brechvorgang *m*	rubbing action
Schwingachse *f*, Schwingenwelle *f* [*Blake-Brecher*]	swing jaw shaft
Schwingbackenplatte *f*	swing jaw (crushing) plate
Schwinge *f*, Brecherschwinge [*Einschwingenbrecher*]	pitman
Schwingenbrechbacke *f*, Schwingbacken *m*, bewegliche Brechbacke	swing(ing) jaw, movable jaw, moving jaw
Schwingenlager *n* [*Einschwingenbrecher*]	pitman eccentric bearing
Schwungrad *n*, Schwungscheibe *f*	flywheel
Seitenkeil *m*	cheek plate, side cheek
Spaltweite *f* → untere Brechspaltweite	
Stauung *f*	packing, choking
Stellkeil *m* → Nach~	
Stirnwand(brech)backe *f*, Festbacken *m*, fest(stehend)e Brechbacke, Körperbrechbacke(n)	stationary jaw
Tiefrahmen-Doppelkniehebelbrecher *m*	deep frame swing ~ ~
(untere Brech)Spaltweite *f*, Spaltweite der Austrittsöffnung *f*, Austragsspaltweite, Brechspalteinstellung *f*	jaw setting, clearance [*distance between the jaws at the discharge point at the bottom of the crushing chamber*]
unterer Seitenkeil *m*	lower cheek plate
verstellbarer Keil *m* → (Nach)Stellkeil	
Wälzlagerung *f*	anti-friction bearings
Zugstange *f* [*Blake-Brecher*]	pitman
Zugstangenlager *n* → Exzenterlager *n* der Zugstange *f*	
Zugstangenwelle *f* [*Blake-Brecher*]	pitman shaft

Zweipendel~ → Doppelkniehebelbrecher
Zweischwingenbrecher *m* → Doppel~

Backenbremse *f*, Klotzbremse = shoe brake
backende Kohle *f*, Backkohle, weiche Kohle [*Fettkohle, die gut backenden, festen Hüttenkoks liefert*] = caking coal
Backen|feinbrecher *m*, Backennachbrecher, Feinbackenbrecher, Nachbackenbrecher = secondary jaw crusher, reduction ~ ~, fine ~ ~
~**schiene** *f*, Anschlagschiene, Stockschiene, Mutterschiene, Stammschiene [*Fahrschiene der Weiche, an die sich die Zunge anlegt*] = stock rail
~**schmiege** *f*, Wangenschmiege, Kleb(e)schmiege = bevelling cut of jack rafter
~**steinbrecher** *m* = stone jaw crusher, rock ~ ~, ~ ~ breaker
Back|kohle *f* → backende Kohle
~**kork** *m* = baked cork
~**korkplatte** *f* = baked corkboard, ~ cork slab
Backstein *m*, Ziegel *m* [*früher: Mauerziegel*] = (building) brick
~**auskleidung** *f*, Backsteinbekleidung, Backsteinverkleidung, Ziegelauskleidung, Ziegelverkleidung, Ziegelbekleidung = brick lining
~**bogen** *m*, Ziegelbogen = brick arch
~**brecher** *m*, Ziegelbrecher = brick crusher, ~ breaker
~**gewölbe** *n*, Ziegelgewölbe = brick vault
~**mauerwerk** *n*, Ziegelmauerwerk = brick masonry (US); brickwork (Brit.)
~**schornstein** *m*, Ziegelschornstein = brick chimney
~**ton** *m*, Töpferton, Ballton = ball clay, potters' ~
(~)**Verband** *m*, Ziegelverband = (brick) bond

Bad *n* = bath
~ → Baderaum *m*
~, Badewanne *f* = bath (tub)
~**ablauf** *m* = bathroom gull(e)y
(~)**Brause** *f*, Dusche *f* = shower
Bade|anstalt *f* = public baths
~**armatur** *f* = bath accessory
~**einrichtung** *f* = bathroom equipment
~**ofen** *m* = geyser
~**raum** *m*, Badezimmer *n*, Bad *n* = bathroom
~**raumtafel** *f*, Badezimmertafel = bathroom panel
~**speicher** *m*, Vorratswasserheizer *m* = storage-type geyser
~**strand** *m* = bathing beach
~**wanne** *f*, Bad *n* = bath (tub)
Badewasser *n* = bath water
~**pflege** *f* = bath water treatment
Badezimmer *n*, Baderaum *m*, Bad *n* = bathroom
~**leuchte** *f* = bathroom lighting fixture
Badzubehör *m* = bathroom accessories
Bagger *m*, Trocken~ = excavator
~, Naß~, Schwimm~ = dredger (Brit.); dredge (US)
~, Löffel~, Stand~ = (mechanical) shovel
~, Stetig~, Dauer~ = bucket excavator
~ **auf Chassis mit Eigenantrieb** → Auto(mobil)bagger
~**achse** *f*, Löffel~ = shovel axle
~**-Anbau-Bohrgerät** *n* = drill attachment for excavators
~**-Anbaugerät** *n*, Bagger-Zusatzgerät = attachment for excavators
~**antrieb** *m* = excavator drive
~**arbeit** *f* = excavator work
~**ausleger** *m*, Löffel~ = shovel boom (US); ~ jib (Brit.)
~**becher** *m*, Baggereimer *m* = bucket

Baggerbedienungsstand — Baggerleiter

Baggerbedienungsstand m, Löffel ~
  Ausleger-Windwerk n
  Auslegerwindwerk-Sicherheitssperrklinke f
  Auslösung f für Verriegelung und Löffelklappe
  Bremspedalfeststellung f
  Grabblockierung f
  Hubkupplung f
  Hubwerkbremse f
  Kupplungen fpl für die Reservetrommel
  Lenkkupplungen fpl
  Motorbedienungen fpl
  Motorkupplung f
  Oberwagensperre f
  Reibungskupplungen fpl für Schwenk- und Fahrwerk
  Reservetrommelbremse f
  Schwenk- und Fahrwerkklauenkupplungen fpl

shovel control station
  boom hoist (US); jib ~ (Brit.)
  boom hoist safety pawl (US); jib ~ ~ ~ (Brit.)
  dipper trip

  brake pedal lock
  digging lock
  hoist clutch
  hoist brake
  second drum clutches

  steering clutches
  engine controls
  engine clutch
  swing lock
  swing and propel friction clutches

  second drum brake
  swing and propel jaw clutches

Bagger|brunnen m, Baggerschacht m = dredging well
~deck n, Baggerplattform f, Bagger-Grundplatte f = excavator deck
~eimer m, Baggerbecher m = bucket
~einsatz m, Löffel~ = shovel use
~-El(ektro)motor m, Bagger-Elt-Motor = excavator motor
~-El(ektro)motor m, Löffel~ = shovel motor
baggerfähiges Material n = ground suitable for mechanical excavation
Bagger|fett n = excavator grease
~fibel f = excavator primer
~-Flachlöffel m → Flachlöffel
(~-)Flachlöffelwinde f = skimmer hoist
~führer m = excavator driver
~führer m, Löffel~ = shovel(l)er
~gerät n, Anbau~, Anbaubagger m = excavating attachment, digging ~
~gerät n, Aufbau-~, Aufbaubagger m = truck-mounted digging attachment, ~ excavating ~ (US); lorry-mounted ~ ~ (Brit.)

~gleis n, Stetig~ = bucket excavator track
~greifer m = grab for excavating [*For excavating fairly loose ground below water, within cylinders, or other deep excavation well below the level of the machine, a grab suspended from a crane is used*]
~gut n = mechanically-excavated material
~gut-Hinterfüllung f [*Hafen*] = fill(ing) of dredged material
~hafen m = dredger port (Brit.); dredge ~ (US)
(~-)Hochlöffel m = shovel bucket
~kabine f = excavator cab(in)
~kette f, Löffel~ = shovel chain
~kette f, Stetig~ = bucket excavator chain
~kran m → Greif(bagg)er m
~lader m, Bagger und Lader, Lader und Bagger, Ladebagger, Schürflader-Heckbagger-Kombination f = excavator-loader
Baggerleiter f, Ausleger m, Eimerleiter(-Ausleger), Becherleiter [*Grabenbagger mit Eimern*] = digging

ladder, ditcher ~, bucket ~, bucket flight

~, Eimerleiter [*allgemein*] = digging ladder, bucket ~, bucket flight

~, Eimerleiter, Becherleiter [*Naßbagger*] = bucket ladder, digging ~, dredging ~, bucket flight

~rahmen *m*, Eimerleiterrahmen, Ausleger *m* = ladder frame

~winde *f*, Becherleiterwinde, Auslegerwinde, Eimerleiterwinde [*z. B. bei einem Grabenbagger mit Eimern*] = ladder hoist, boom ~

Baggerlöffel *m*, Grablöffel, Grabgefäß *n* = shovel, (shovel) dipper, (excavator) bucket

~inhalt *m* → Grablöffelinhalt

Bagger|löffelstiel *m* → Löffelstiel

~maschinen *fpl* = excavating and dredging machinery

(~)**Matratze** *f* → Baggerrost *m*

~motor *m* = excavator engine

~motor *m*, Löffel~ = shovel engine

baggern, aus~, ab~, trocken~, ausheben = to excavate, to dig

~, aus~ [*Naßbaggerung*] = to dredge

Bagger|-Oberteil *m,n*, Aufsatzteil, Bagger-Oberwagen *m*, Bagger-Schwenkteil = upper machinery, revolving superstructure

~-Planierkübel *m* → Flachlöffel *m*

~-Planierlöffel *m* → Flachlöffel

~-Planierschaufel *f* → Flachlöffel *m*

~planum *n*, Baggersohle *f* = excavator base

~pumpe *f*, Förderpumpe = dredging pump, dredge(r) ~

~(reiß)zahn *m* = excavator tooth; shovel ~

~rost *m*, (Bagger)Matratze *f*, Fahrbahnmatte *f*, Fahrbahnplatte *f* = excavator (supporting) mat, shovel (~) ~, platform

~sand *m* = dredged sand

~schacht *m*, Baggerbrunnen *m* = dredging well

~schmierer *m*, Löffel~ = shovel oiler

~schute *f*, Abfuhrschute = barge (Brit.), scow (US); hopper ~

~schute *f* ohne Antrieb = dumb barge, non-propelled ~ (Brit.); dumb scow, non-propeled scow (US)

~schutenwinde *f* = scow winding gear (US); hopper barge ~ ~ (Brit.)

~schwelle *f* = sleeper for crane navvy (Brit.); tie for power shovel (US)

(~-)**Schwung** [*Naßbagger*] = digging swing

~seil *n*, Löffel~ = shovel cable

~-Senkkasten *m*, Gründungsbrunnen *m* mit Scheidewänden, Senkbrunnen mit Scheidewänden, offener Senkkasten = open caisson (with cross walls), drop shaft (~ ~ ~), caisson with digging wells

~sohle *f* [*Naßbaggerung*] = dredging bottom

~sohle *f*, Baggerplanum *n* = excavator base

~stampfer *m* → (Freifall-)Kranstampfer

~stampfplatte *f*, Fallplatte = tamping plate, falling ~

~stelle *f*, Grabstelle = digging site, ~ point

~tiefladehalbanhänger *m*, Aufl(i)eger *m* für Baggertransport = semitrailer for tranport of excavating machines

~tieflöffel *m* → Tieflöffel

~torf *m* = dredged peat, lard ~

~transportwagen *m*, Baggertief(bett)-lader *m*, Baggertiefladeanhänger, Baggertiefbettanhänger, Baggertiefladewagen = trailer for transport of excavating machines, excavator low-loader

~trommel *f* = excavator drum

~ und Lader *m* → Baggerlader

Baggerung *f* = mechanical excavation

(**Bagger-**)**Unterwagen** *m*, (Bagger-)Unterteil *m, n* = mounting unit, travel ~

Bagger|verschleißteil *m,n* = excavator wearing part

~windwerk *n*, Auslegerwinde *f* = shovel boom winch (US); ~ jib ~ (Brit.)

Baggerzahn — Bajonettsicherung

~zahn m, Baggerreißzahn = shovel tooth; excavator ~
~-Zusatzgerät n, Bagger-Anbaugerät = attachment for excavators
Bahn f, Eisen~ = railway (Brit.); railroad (US)
~, Streifen m, Einbau~ [Schwarz- und Betondeckenbau] = strip
~achse f, Bahnmittellinie f, Eisen~ = railway centre line (Brit.); railroad center ~ (US)
~anlieferung f, Eisen~ = rail delivery
~anschluß m, Eisen~ [Hafen] = railway connection (Brit.); railroad ~ (US)
~betrieb m mit dritter Schiene = third-rail traction
~brücke f, Eisen~ = railway bridge (Brit.); railroad ~ (US)
~damm m, Eisen~, (Eisen)Bahnaufschüttung f = railway embankment (Brit.); railroad ~ (US)
~dreieck n = triangular area occupied by railway installations
~einschnitt m, Eisen~ = railroad cutting (US); railway ~ (Brit.)
~fähre f, Eisen~ = train ferry
~fahrzeug n → Eisen~
~gelände n = railway territory, ~ property (Brit.); railroad ~ (US)
Bahnhof m, ~anlage f = railway station (Brit.); railroad ~ (US)
~gebäude n = station building
~gleis n = station track
~halle f = railway station hall (Brit.); railroad ~ ~, concourse (US)
~markt m = terminal market
~streckenverbreiterung f [U-Bahn] = widening at a station, ~ ~ stations
~tunnel m [U-Bahn] = station tube
~(vor)platz m = station forecourt
Bahn|körper m = railway right-of-way (Brit.); railroad ~ (US)
~krone f, Eisen~ [Die von den Böschungen begrenzte Fläche in Höhe der Schwellenoberkante] = railway embankment crest (Brit.); railroad ~ ~ (US)

~linie f, Bahnstrecke f, Eisen~ = trainway
~meisterwagen m → Draisine f
~netz n, Eisen~ = railway network (Brit.); railroad ~ (US)
~oberbau m, Eisen~ = permanent way
~profil n → Ladeprofil
~schiene f, Eisen~ = railway rail (Brit.); railroad ~ (US)
~schotter m → Bettungsschotter
~schwelle f, Eisen~, (Gleis)Schwelle f, Schienenschwelle = sleeper (Brit.); tie (US)
Bahnsteig m = station platform
~aufzug m = station platform lift (Brit.); ~ ~ elevator (US)
~dach n, Bahnsteigüberdachung f = (station) platform roof
~gleis n = station platform track
~karre(n) f, (m) = baggage truck
~rampe f [zur Verbindung der Bahnsteige mit tiefer oder höher gelegenen Punkten] = station platform ramp
~treppe f = station platform staircase (Brit.); ~ ~ stair (US)
~tunnel m = subway leading to platforms, inter-platform subway
Bahn|strecke f, Bahnlinie f, Eisen~ = trainway
~strom m = traction current
~tangente f, Bahnberührende f = tangent of motion, ~ ~ movement, tangential path
~transport m, Eisen~ = rail transport
~tunnel m, Eisen~ = railway tunnel (Brit.); railroad ~ (US)
~überführung f, Eisen~ = overpass
~übergang m → niveaugleicher ~
~unterführung f, Eisen~ = underpass
bahnverladbar = suitable for rail transport
Bahnversand m, Eisen~ = rail shipment
(Bahn)Waggon m, Güterwagen m = wagon (Brit.); freight car (US)
Bajonettsicherung f [Druckluftwerkzeug] = lug shank retainer

Bake *f* [*festes Seezeichen*] = beacon
~ [*ILS-Anflugschema*] = ILS marker, low-power fan ~
~ [*Bei Landstraßen Zeichen als Warnung zur Vorankündigung planngleicher Bahnübergänge*] = warning sign
~, Fluchtstab *m* = range pole, ~ rod, ranging~, banderolle
Bakelitleim *m* = phenolic-resin glue
bakteriologische Reinigung *f* = bacteriological purification
Balanzier *m*, Schwengel *m* = beam
Balatariemen *m* = balata belt
Baldachin *m* = baldacchino
Balgdichtung *f* = bellow-type seal
Balgengaszähler *m* = positive total-flow gas meter
Balken *m*, ~träger [*Jeder Balken ist ein Träger, aber jeder Träger ist nicht notwendigerweise ein Balken; denn es gibt Balken(träger) und Bogen-(träger)*] = beam
~ [*Einer der beiden benachbarten Balken beim Balkenwechsel, in welche der Wechsel eingezapft ist*] = trimmer (US); trimming joist (Brit.)
~, Fertiger~, (Fertiger)Bohle *f* = (finisher) beam
~ **aus Blöcken**, Blockbalken = beam made of precast hollow blocks
~ **mit gleichbleibendem Querschnitt**, prismatischer Balken = prismatic beam
~ ~ **großer Konstruktionshöhe**, ~ ~ ~ Bauhöhe = deep beam
~ ~ **halber Vorspannung** = half-prestressed beam
~ ~ **T-Querschnitt** = T-beam
~**abstand** *m* = beam distance
~**achse** *f*, Balkenmittellinie *f* = beam centre line (Brit.); ~ center ~ (US)
~**anker** *m*, Kopfanker, Schlauder *f* = beam tie
~**art** *f* = beam type, type of beam
~**auflager** *n* = beam support
~**bemessung** *f* = beam design, design of beams
~**bemessungsformel** *f* = beam design formula
~**betoniergerät** *n* = beam pourer
~**bewehrung** *f*, Balkenarmierung, Balken(stahl)einlagen *fpl* = beam reinforcement
~**biegepresse** *f* = beam bending press
~**biegung** *f* = beam bending
~**binder** *m* [*gitterförmiges Tragwerk aus Vollholz*] = wood(en) truss, timber ~
~**brause** *f* [*Schweiz*] → Sprengrampe *f*
~**breite** *f* = beam width
~**bruchprüfung** *f*, Balkenbruchprobe *f*, Balkenbruchversuch *m* = beam test to failure, ~ ~ ~ rupture
~**brücke** *f* = beam bridge
~**bügel** *m* = beam stirrup
~**dach** *n* = beam roof
~**decke** *f* = beam floor
~**decke** *f* mit dicht verlegten Balken = joist floor
~**diagramm** *n*, Balkendarstellung *f* = bar graph (diagram), ~ chart
~**durchbiegung** *f* = beam deflection; ~ deflexion (Brit.)
~**einschub** *m* → Einschub
~**endbereich** *m* = extreme part of beam
~**fach** *n*, Balkenfeld *n* [*Holzdecke*] = space between beams
~**feld** *n*, Balkenfach *n* [*Holzdecke*] = space between beams
~**festigkeit** *f* = beam strength
~**form** *f* = beam mould (Brit.); ~ mold (US)
~**gesims** *n* = open cornice (US); ~ eaves (Brit.)
~**hälfte** *f*, Halbbalken *m* = half-beam
~**höhe** *f* = depth of beam
~**holz** *n* = beam timber, ~ wood
~**kammer** *f* = beam aperture, wall pocket, beam pocket
~**kopf** *m* [*Balkenendquerschnitt am Auflager*] = beam head
~**kreuzung** *f* = beam crossing
~**lage** *f*, Gebälk *n* = system of binders and joists
~**maschine** *f*, Balkenfertigungsmaschine, Balkenformmaschine = beam-making machine; beam-moulding ~ (Brit.); beam-molding ~ (US)

Balkenmitte — Bandabmeßwaage

~mitte *f* = centre of beam (Brit.); center ~ ~ (US)
~moment *n* = beam moment
~platte *f* = beam slab
~profil *n* = beam profile
~prüfung *f*, Balkenprobe *f*, Balkenversuch *m* = beam-breaking test
~querschnitt *m* = beam cross section
~rippe *f* = beam rib
~rüttler *m*, Balkenvibrator *m* = beam vibrator
~schalung *f* = beam formwork
~schwingung *f* = beam vibration
~-Spannbett *n* = beam bed [*For the manufacture of prestressed concrete beams*]
~steg *m* = beam web
~stoß *m* = beam butt joint
~-Straßenbrücke *f*, Straßen-Balkenbrücke = highway beam bridge, road ~ ~
~theorie *f* = beam theory
~träger *m* → Balken *m*
~untersicht *f*, Balkenunterseite *f* = beam soffit, ~ bottom
~vibrator *m*, Balkenrüttler *m* = beam vibrator
~waage *f* = (weigh) beam scale
~wechsel *m*, Auswechs(e)lung *f* = framing (around an opening)
~wirkung *f* = beam action [*The modulus of elasticity of a concrete slab is much greater than that of the foundation material, so that a major portion of the load-carrying capacity is derived from the slab itself. This is often referred to as beam action*]
Balkon *m* = balcony
~brüstung *f* = balcony parapet
~fenster *n* = balcony window
~platte *f* = balcony slab
~tür *f* = balcony door
~verkleidung *f* = balcony lining
Ballast|arm *m*, Hinterarm, Sackbalken *m* [*Klappbrücke*] = back lever, balance ~, back arm
~kohle *f* = low-grade coal
~seite *f*, Gegengewichtsseite = counterweight side

Ballen|drücker *m*, obenliegender Drücker [*liegt auf der Oberseite des Handgriffs*] = outside lever, top ~ [*(compressed-)air throttle control*]
~pflanze *f* = balled plant
Ballon|methode *f* → Gummiblasenmethode
~reifen *m* = balloon tyre (Brit.); ~ tire, doughnut tire (US)
Ballsaal *m* = ball room
~foyer *n* = ball room foyer
Ballton *m* → Backsteinton
Ballungs|gebiet *n*, Agglomerationsgebiet [*Gebiet mit überdurchschnittlich hoher Wohn-, Verkehrs- und Betriebsdichte*] = pressure area
~kern *m*, Agglomerationskern [*Zentrum eines Ballungsgebietes*] = pressure core
Balme *f* (Geol.) = overhang
Balneologie *f* = balneology
Balsam-Terpentinöl *n* = balsam terpentine
Baluster *m*, Docke *f* = baluster [*In architecture, a dwarfed column or pillar supporting a railing or coping. A series of balusters form a balustrade*]
Balustrade *f*, Brüstungsgeländer *n* = balustrade, balustrading
Bananen|schiff *n* = banana vessel, ~ ship
~schuppen *m* = banana shed
~umschlag *m* = banana handling
Band *n* = band
~ → Bandförderer *m*
~, Stahlnippel *m* [*Schweißtechnik*] = backing strip, ~ ring
~, Gabelholz *n* = forked wood
~ mit Nylongarn im Schuß → Bandförderer
~abmeßanlage *f*, Banddosieranlage, Bandzuteilanlage, Bandzumeßanlage [*für Zement oder Zuschlagstoffe*] = belt type proportioning unit, ~ ~ measuring ~, ~ ~ batching ~
~abmeßwaage *f*, Banddosierwaage, Bandzuteilwaage, Bandzumeßwaage = conveyor type batcher scale, conveyer ~ ~ ~, ~ ~ batching ~

Bandabsetzer — Bandförderer

~absetzer m, Absetzanlage f, (Schwenk)Absetzer, Bandabwurfgerät n, Absetzer für Hoch-, Tief- und Dammschüttung, Absetzapparat m, Absetzgerät [*Maschine mit schwenkbaren oder festem Oberteil zum Verteilen von Erde oder Anschütten derselben. DIN 22266*] = stacker

~abzug m [*aus einem Silo oder Bunker*] = drawing by belt

~abzug m [*Fertigbehandlung von Beton*] = finishing by belt, belting

Bandage f, Mantel m, Stahl~ [*Straßenwalze*] = (steel) facing

Bandagierung f = covering of pipe(s) with protective material

Band|aufgeber m, Bandbeschicker, Aufgabeband n, Beschickungsband, Bandspeiser m = feed(ing) belt, ~ conveyor, ~ conveyer, feeder ~, belt feeder [*A short conveyor belt that supplies material to a long belt*]

~auslader m → Ausladeband n

~ausleger m, Absetzerausleger = boom

~bebauung f = ribbon building, ~ development

~(becher)elevator m → Gurtbecherwerk n

~becherwerk n → Gurtbecherwerk

~belag m → Bandförderer

~bergförderung f = conveyor hoisting, conveyer ~

~betonierturm m = belt concreting tower

~bremse f = band brake

~bruch m = belt failure [*belt conveyor*]

~brücke f [*Teil eines Bandförderers*] = conveyor belt frame, conveyer ~ ~

~brücke f = belt bridge, ~ conveyor ~, ~ conveyer ~

~decke f → Bandförderer

~dichtung f = strip sealing

~dosiergerät n, Dosierapparat m = belt batcher

~durchschlagfestigkeit f → → Bandförderer

~eisen n (Min.) = taenite

~eisen n → Bandstahl m

~eisenprofil n → Bandstahl m

~elevator m → Gurbecherwerk n

~entlader m → Ausladeband n

Bänder|gneis m, Lagengneis = banded gneiss, ribbon ~

~ton m, dünngeschichteter Ton = varve(d) (glacial) clay, pellodite, banded clay, leaf clay, book clay, ribbon clay

~ton m mit Jahresringen = seasonally banded clay, ~ stratified ~

Bandfilter m, n = band filter

Bandförderer m, Gurtförderer	belt conveyor, band ~, ~ conveyer
Abfallverluste mpl	spillage off a belt
abgestrichenes Material n	scrapings
Ablenktrommel f	snub(bing) pulley, snubber ~
Abstreicher m, Abstreifer, Band~	(belt) wiper, (belt) scraper, (belt) cleaner
Abtragstation f, Rollenstation, Tragrollenstation	idler set
Abwurfende n	discharge end
Abwurftrommel f, Kopftrommel	head pulley
Abwurfwagen m	tripper
Antriebstrommel f	drive pulley, motorized ~
aufvulkanisieren	to vulcanize in place
ausziehbar	extendible
Band n, Gurt m	belt(ing), band

Bandförderer 144

Bandabwanderung f, Gurtabwanderung	wandering
Bandbelag m, Banddecke f	belt cover, band ∼
Bandbrücke f	conveyor belt frame, conveyer ∼∼
Bandbrücke f	belt bridge, ∼ conveyor ∼, ∼ conveyer ∼
Band n mit Querrippen	belt with a cleated surface
Band n aus Baumwollkettfäden und Nylongarn im Schuß	nylon-filled belt, cotton-warp nylon-weft(ed) belt
Band n mit Nylongarn im Schuß	nylon-weft(ed) belt
Banddecke f → Bandbelag m	
Banddurchschlagfestigkeit f	resistance to impact
Bandförderung f, Bandtransport m	belt conveying, ∼ conveyance
Bandlauf m	belt run
Bandmulde f	belt trough
Bandrahmen m	boom of a conveyor, ∼ ∼ ∼ conveyer, conveyor boom, conveyer boom
Bandrücklaufrolle f, unbelastete Tragrolle	return idler, belt return roller
Bandspannung f, Straffung f, Gurtspannung	tension, belt ∼
Bandstraße f, Transport-∼	sectional ground conveyor frame, ∼ ∼ conveyer ∼
Bandtragrahmen m	carryable belt conveyor, ∼ ∼ conveyer
Bandtrommel f → Trommel	
(Band)Umlenkrolle f → Trommel	
Baumwolleband n, Baumwollegurt m	cotton (fabric duck) belt (or band), ∼ reinforced belt (or band)
Baumwolle(gewebe)zwischenlage f, Baumwolle(gewebe)einlage	cotton ply
Baumwolle-Nylon-Band n	cotton/nylon belt
Baumwollegarn n	cotton yarn
Baumwollegewebe n	cotton duck, ∼ fabric
Baumwollekettgarn n	cotton-warp yarn
Baumwolleschußfestigkeit f	cotton weft(ed) strength
Baumwolleschußgarn n	cotton weft yarn
Beaufschlagung f	impact
(belastete) Tragrolle f, Gurt(ab)tragrolle	carrier (idler), supporting roller
Berührungsfläche f	area of contact, contact area
Berührungsbogen m, Umspannungsbogen	arc of contact, contact arc
Beschädigung f des Gurtes durch den Materialaufschlag	impact damage to the belt
Bürste f mit Borsten	bristle brush
Cordband n, Cordgurt m	cord (or fabric) belt(ing)
Deckplattenmasse f	cover compound
diagonaler Reinigungsriegel m	diagonal cleaner bar

Bandförderer

Drehzapfen *m* im Mittelpunkt	center swivel (US), centre ~ (Brit.)
Drei-Lagen-Band *n*	three-ply belt
dreiteilige Rollenstation *f*	troughed idler, three-pulley ~, 3-spindle troughing ~, three-roller idler set
dünne Gummischicht *f*	skin
durchdrehen, schlüpfen [*Trommel*]	to spin, to slip
Durchhangseite *f* → unbelasteter Strang	
Einlage *f*, Zwischenlage	ply
einteilige Tragrolle *f*	flat idler
Eintrommelantrieb *m* mit Ablenktrommel	snub(bing) pulley drive, snubber ~ ~
Eintrommel-Kopfantrieb *m*	head pulley drive
(End)Trommel *f* → Trommel	
fahrbarer Gurtförderer *m*, Fahrband *n*, fahrbares Förderband	portable belt (or band) conveyor (or conveyer)
Schwenkradsatz *m*	swivel wheels
Ferntransportband *n*, Langstreckenförderbandanlage *f* zum Transport von Gesteinen, Streckenband	"rock road"
Festigkeits/Querschnitt-Wert *m*	strength/bulk characteristic
Flachband *n*, Flachgurt *m*	flat belt, ~ band
flache Tragrolle *f*, glatte ~	flat idler, ~ roller
Förderbandrolle *f*, Fördergurtrolle	(conveyor) idler, (idler) roller, conveyer idler
Förderband *n* zur Bodenentladung von Güterwagen	belt type car unloader
Förderband-Tragwerk *n*	conveyor gantry, conveyer ~
Förderseite *f* → oberer Strang *m*	
Führungs- und Tragrolle *f*, selbstausrichtende Rolle	guidler, training idler, self-aligning (guide) idler
Gewebe *n*	fabric, (belting-)duck
Gegenseite *f* → unbelasteter Strang	
Ganz-Baumwollgewebe *n*	all-cotton duck, ~ fabric
Ganz-Baumwollgurt *m*	all-cotton carcase belt, ~ carcass ~
Gewichtsspannvorrichtung *f*	gravity takeup
gezackte Gummirolle *f*	serrated rubber roll
Glasrollenförderband *n*	belt conveyor with glass idlers, ~ conveyer ~ ~ ~
Glättung *f* [*des Bandes*]	smoothing
Gummibelag *m*, Gummidecke *f*, Gummischicht *f*	rubber cover
Gummibelag *m* [*Antriebstrommel*]	rubber lagging
Gummi(förder)gurt *m*, Gummi(förder)band *n*, Gummi-Transportband	rubber (conveyor) (or conveyer) belt(ing) (or band)
gummigelagerte Tragrolle *f*	rubber-cushioned idler
Gurt *m*, Band *n*	belt(ing), band
Gurtgewebe *n*	belt(ing) duck, ~ fabric
Gurtspannung *f* → Bandspannung	

Bandförderer 146

Gurtspannung *f* der Durchhangseite *f*	slack side tension
Gurtzugkraft *f*	drive traction
Haftfestigkeit *f*	bond stength
Haftung *f* zwischen den einzelnen Gewebelagen, Bindung ~ ~ ~ ~	interply adhesion
Haltevermögen *n* für die Verbinder	fastener-holding ability
Kettdehnung *f*	warp(way) elongation, belt stretch
Kette *f*	warp
Kettfestigkeit *f*	warp(way) strength
Kettgarn *n*	warp yarn
Klaubeband *n* → Leseband	
Kopfantriebstrommel *f*	motorized (or drive) head pulley, head drive ~
Kopftrommel *f*, Abwurftrommel	head pulley
Kräuselung *f*	crimp
Kunstseidenband *n*	rayon carcase belt, ~ carcass ~
Kunstseideneinlage *f*	rayon fabric ply, ~ duck ~
Kunstseidengarn *n*	rayon yarn
Langstreckenbandförderanlage *f* zum Transport von Gesteinen, Ferntransportband *n*, Streckenband	"rock road"
Leitrolle *f* [*Führungsrollensatz*]	spool
Leittrommel *f* → Trommel	
Leseband *n*, Klaubeband	picking belt (or band) conveyor (or conveyer)
mechanischer Reiniger *m*	mechanical cleaner
Muldenband *n* → Trogband	
Mulden(band)rolle *f*	belt troughing roller, troughed idler
Muldengurtförderband *n*	trough belt (or band) conveyor (or conveyer)
Muldungsfähigkeit *f*	transverse stability, lateral ~
oberer Strang *m*, Tragseite *f*, Obergurt *m*, Förderseite	carrier side, upper strand, carrying strand
Rahmenteil *m, n*	frame member
Rolle *f* → Förderbandrolle	
Rollenstation *f*, Abtragstation, Rollenbock *m*	idler set
rotierende Bürste *f*	brush outfit for cleaning carrier side
Rücklaufsperre *f*	holdback ratchet
scharfkantiges Fördergut *n*	abrasive material
Schlupf *m*	slippage
schlüpfen → durchdrehen	
schmiegsame Tragrolle *f*	flexible (belt) conveyor (or conveyer) idler, limberoller
Schurre *f*	chute
Schuß *m*	weft (of the cloth)
Schußfestigkeit *f*	weft(way) strength
Schüttbeanspruchung *f*	loading impact

Schwanztrommel f	tail pulley, return ~
Seitenbord n	skirt, chute skirt board
selbstausrichtende Rolle → Führungs- und Tragrolle	
Spannrolle f, Spanntrommel f	takeup pulley
Spannschraube f	threaded adjustment to move the tail pulley in or out
Spannstation f, Spannvorrichtung f	takeup (set), take-up gear
Straffung f, Bandspannung f Gurtspannung	tension
Strang m	strand
Stückgut-Bandförderer m	package (belt) conveyor (or conveyer)
stückiges Schüttgut n	lumpy material
technische Kunstseide f	rayon
Textilband n, Textilgurt m	fabric carcass belt, ~carcase ~
Tragrolle f → belastete ~	
Tragrollenstation f, Tragrollensatz m	carrier idler set
Trageseite f → oberer Strang m	
Transportbandanlage f	fixed conveyor, ~ conveyer
Transport-Bandstraße f → Bandstraße	
Trogband n, Muldenband, Troggurt m, Muldengurt	trough belt, ~ band
Trommel f, Endtrommel, Bandtrommel, (Band)Umlenkrolle f, Endumlenktrommel, Leittrommel	pulley
Trommel(elektro)motor m	pulley (electric) motor
Übertragungsband n	transition belt
Umlenkblech n	baffle
Umlenkrolle f → Trommel f	
Umschlingungswinkel m	degree of wrap
Umspannungsbogen m, Berührungsbogen	arc of contact
unbelastete Tragrolle f, Bandrücklaufrolle	return idler, belt return roller
unbelasteter Strang m, unterer ~, Untergurt m, Gegenseite f, Durchhangseite	lower strand, return ~, low tension side, slack (tension) side
Verformungsspannung f des Gurtes	flexing strain on the belt
Zug m der Schwerkraft f	pull of gravity
Zurücklaufen n [Band]	backsliding
zweiteiliges Seitenbord n	dual skirt
Zweitrommelantrieb m	two-pulley drive, tandem ~
Zwischenlage f, Einlage	(carcase) ply, carcass ~
Zwischenlagen(an)zahl f, Einlagen(an)zahl	textile bulk, number of plies in the carcase (or carcass)

Band|(förder)schnecke *f* = open-spiral worm conveyor, ~ ~ conveyer, antifriction ~ ~
~förderung *f* → Bandtransport
~gesims *n* = platband
~gießmaschine *f* **für Kupferlegierungen** = strand casting machine for copper alloys
~kanal *m* → Bandstollen *m*
~kanal *m* → Haldentunnel *m*
~kühlsystem *n*, Kühlbandanlage *f* = cooling belt system
~kupplung *f* = rim clutch (coupling)
~lader *n*, Bandaufspader, Aufladeband *n*, (Be)Ladeband = conveyor loader, conveyer ~
~magnetscheider *m* = belt magnetic separator
~maß *n*, Meßband *n* = measuring tape
~mulde *f* → → Bandförderer
~porphyr *m* = banded porphyry
~reiniger *m*, (Band)Abstreifer, (Band)Abstreifer = (belt) wiper, (~) scraper, (~) cleaner
~rücklaufrolle *f* → → Bandförderer
~schaufellader *m* [*Die Schaufel kippt ihren Inhalt in einen Bunker, an dessen Boden sich ein Doppelketten-Kratzerförderer befindet, dessen Abwurfende hinten über das Fahrwerk hinausragt*] = overhead loader with armoured chain conveyor (or conveyer)
~scheider *m* = belt separator
~schelle *f* = ribbon clip
~schleifenwagen *m* = tripper car
~schnecke *f* → Bandförderschnecke
~schrapper *m* [*ist ein Handschrapper, der in Verbindung mit einem leichten Förderband eingesetzt ist*] = hand scraper with belt
~-Selbstauflader *m* = belt type bucket elevator loader, self-loading tractor bucket elevator
~sieb *n* = band screen [*An endless belt or band of wire mesh, bars, plates, or other screening medium, which passes around upper and lower rollers or guides*]

~sinteranlage *f* = strand sintering plant
~spannung *f* → → Bandförderer
~stadt *f*, lineare Stadt [*das Rückgrat dieser Stadtform ist ein Verkehrsband an das sich hintereinander gestaffelt die Zonen der menschlichen Betätigung reihen. Die Theorie stammt von dem Spanier Soria y Mala aus dem Jahre 1882*] = linear city, ~ town
Bandstahl *m*, Bandeisen *n*, Bandeisenprofil *n*, Bandstahlprofil [*9,5 bis 450 mm breit und 0,75 bis 8 mm dick. Bandeisenprofile werden verwendet zur Versteifung und Verbindung der Knotenpunkte im Stahlskelettbau*] = strip steel (section)
~abschneidemaschine *f* = strip-steel cutting-off machine
~dach *n* = (steel) roof deck
~haspel *f*, *m* **für Langbunde**, Langbundhaspel = strip steel coiler for long coils
~haspel *f*, *m* **für Rundbunde**, Rundbundhaspel = strip steel coiler for round coils
~profil *n* → Bandstahl *m*
~richtmaschine *f* = strip steel level(l)er, ~ ~ straightening machine
~walzwerk *n* = hoop mill, strip ~
~wickelmaschine *f* = strip steel coiler
Band|stollen *m*, Bandstraßenstollen, Band(straßen)kanal *m*, Band(straßen)tunnel *m* = belt conveyor gallery, ~ conveyer ~
~straße *f* → → Bandförderer
~(straßen)kanal *m* → Bandstollen
~strecke *f* (⚒) = bottom road [*with a belt conveyor*]
~strömung *f*, laminare Strömung = streamline flow, laminar ~
~strömungsgeschwindigkeit *f* = streamline velocity
~tragrahmen *m* → → Bandförderer
~transport *m*, Bandförderung *f* = belt conveyance, ~ conveying
~transportanlage *f* = belt conveying system, ~ conveyance ~
~trieb *m*, Riementrieb = belt drive

~**trockner** *m* = belt dryer; ~ drier (US)
~**trommel** *f* → → Bandförderer
~**tunnel** *m* → Bandstollen *m*
~**tunnel** *m* → Haldentunnel
(~)**Umlenkrolle** *f* → → Bandförderer
~**waage** *f*, Wiegeband *n* = conveyor type scale, conveyer ~ ~
~**zerteilanlage** *f* [*für Bandstahl*] = strip sizer
~**zuteilung** *f*, Bandabmessung, Banddosierung, Bandzumessung = proportioning by conveyor belt, ~ ~ conveyer ~, batching ~ ~ ~, measuring ~ ~ ~
Bank *f* (Geol.) = bed
~, Barre *f* = bar; batture (US) [*Silt, sand, or gravel dropped at the mouth of a river or stream*]
~, Flach *n* [*über Meeresgrund aufragende, aber die Meeresoberfläche nicht erreichende Erhebung*] = bank, bar
~, Absatz *m* [*Tunnel- und Stollenbau*] = bench
~ **in der Mitte** [*Meer*] = middle ground
~**eisen** *n* = cramp iron
Bankett *n*, Straßen~ = (side) verge
~ → Streifenfundament *n*
~ → Grundbank *f* [*Gründung*]
~, Berme *f* [*Horizontaler Absatz in Damm- oder Einschnittböschung*] = berm(e), banquette, bench
~ → (unbefestigter) Seitenstreifen
~**fräse** *f*, Rasenfräse für Bankettunterhaltung = verge cutter, siding machine
~**(kanten)schneiden** *n* = verge cutting
~**rasen** *m* = verge turf
~**rasen-Beseitigung** *f* = mowing of verge turf
Bank|garantie *f* = bank bond
~**gebäude** *n*, Bankbau *m* = bank building
bankig, gebankt (Geol.) = bedded
bankrecht (Geol.) = normal to the stratification
bankschräg (Geol.) = at an oblique angle to the stratification

Bankungsspalte *f* (Geol.) = bedding joint
Banndeich *m* → Hauptdeich
Bär *m*, Hammer *m*, Ramm~ = (pile) hammer
Baracke *f* = hut
Barackenlager *n* = hut camp
Barchan → Bogendüne *f*
Bärführung *f*, Hammerführung = (pile) hammer guide
Bärgewicht *n*, Hammergewicht = (pile) hammer weight
Barium|fett *n* = barium base grease
~**hydroxyd** *n* = barium hydroxide
~**sulfat** *n* = barium sulphate, $BaSO_4$
~**sulfaterz** *n*, Naturbaryt *m* = barytes ore
~**zement** *m*, Edelzement = baritic cement
Bärkatze *f*, Hammerkatze = traveller
Bärkopf *m*, Hammerkopf = (pile) hammer head
Barograph *m*, Registrier-Barometer *n*, Schreibbarometer, selbstaufzeichnender Luftdruckmesser *m* = barograph
Barometer *n*, Wetterglas *n* = barometer
~**druck** *m* → barometrischer Druck
barometrische Höhenbestimmung *f*, ~ Höhenaufnahme *f*, ~ Höhenmessung, physikalische ~ = barometric level(l)ing
barometrischer Druck *m*, atmosphärischer ~, Barometerdruck, absoluter statischer Luftdruck = atmospheric pressure
Barre *f*, Flutbrandung *f* = bore
~, Bank *f* = bar [*Silt, sand or gravel dropped at the mouth of a river or stream*]
Barren *m* [*durch Gießen hergestellter Block aus Metall*] = billet
~**hafen** *m* = bar port
Barriereriff *n*, Wallriff = barrier reef
Bärschiene *f*, Hammerschiene = (pile) hammer rail
Bärschlag *m*, Hammerschlag = (pile) hammer blow

Bärseil — Batteriezündung 150

Bärseil *n*, Hammerseil = (pile) hammer rope
Bärtrommel *f*, Hammertrommel = (pile) hammer drum
Baryt *m* (Min.) → Schwerspat *m*
~feldspat *m*, Hyalophan *m* (Min.) = hyalophane
~(o)salpeter *m* = nitrobarite
Basal|konglomerat *n* = base conglomerate, basal ~
~sandstein *m* = basal sandstone
~scholle *f* (Geol.) = overridden mass
Basalt *m* [*grobkörnige Basalte heißen Dolerit. Basalt ist oft nur in 1 cm bis mehrere Meter dicken Säulen abgesondert. Fälschlich auch ,,schwarzer schwedischer Granit" genannt*] = basalt
~bruch *m* = basalt quarry
~eisenerz *n* = basaltic iron ore
~lava *f*, Lungstein *m*, Lavaschlacke *f*, Lavakrotze *f*, Lavakies *m*, Schaumlava [*basaltische feinporige bis blasige vulkanische Auswurfmasse. ,,Lavalit" ist ein geschützter Handelsname für gebrochene Lavaschlacke, die in verschiedenen Körnungen aufbereitet ist*] = volcanic cinders
~mehl *n* = basalt meal, ~ powder, powdered basalt
~pflaster(decke) *n*, (*f*) = basalt sett paving
~pflasterstein *m* = basalt (paving) sett
~platte *f* = basalt slab
~schotter *m* = crushed basaltic stone
~schutt *m* = basaltic débris, ~ scree
~splitt *m* = basalt chip(ping)s
~splittbeton *m* = basalt chip(ping)s concrete
~tuff *m*, Trapptuff = basaltic tuff, trapp ~
~wacke *f* = basaltic wacke
~wand *f* = basalt face
~wolle *f* = basalt wool
Basalzement *m* (Geol.) = basal matrix, ~ cement(ing material)
Basen|austausch *m* = base exchange, ion ~

~austauschverfahren *n* = base exchange method, ion-exchange ~
~austauschvermögen *n* = base exchange capacity
~grad *m*, Schlackenzahl *f*, Schlackenziffer *f* $\dfrac{CaO + MgO}{SiO_2}$ = basicity factor
Basilika *f* = basilica
~kirche *f* = basilican church
Basis *f* [*Grundplatte, auf der eine Säule, ein Pilaster, eine Stütze, oder ein Pfeiler steht*] = base (plate)
~ (Geol.) = base of a formation
~ der natürlichen Logarithmen [*Eytelwein'sche Beziehung für Seilrutsch bei der Treibscheibenförderung*] = base of hyperbolic or Naperian logarithms
~bruch *m* = toe failure
basischer Diorit *m* = basic diorite
~ feuerfester Stein *m*, ~ ff. ~ = basic brick
basisches Gestein *n* = basic rock
~ Kupferazetat *n* → Grünspan *m*
~ Magma *n* = basic magma [*contains less than 50 per cent silica*]
Basis-Viereck *n* = base quadrilateral
Bastardasbest *m*, Pikrolith *m* = picrolite
Bastion *f* (Geol.) = ledge
Bastit *m* → Schillerspat *m*
Bast|kohle *f* = bast coal
~matte *f* = bast mat
Batholith *m*, Riesenstock *m*, Liegendkörper *m* = batholith, bathylith
Batterie|beleuchtung *f* = battery lighting
~deckel *m* = battery cover
batterie-elektrisch = battery-electric
Batterie|kasten *m* = battery box
~ladeaggregat *n* = battery charging set
~ladung *f* = battery charging
~schalungen *fpl* [*Sie ermöglichen die gleichzeitige Fertigung von Betonelementen auf kleinstem Raum*] = battery of formwork
~zündung *f* = battery ignition

Batzenpresse *f* = wad-box
Bau *m*, Errichtung *f* [*z.B. einer Talsperre*] = construction, building
~ → Gebäude *n*
~ → Bauwerk *n*
~, Aufbau, bauliche Anordnung *f*, Konstruktion *f*, Ausführung [*Maschine*] = design, version
Bauabnahme *f* = acceptance of the work
Bauabschnitt *n*, Ausbaustufe *f*, Bauetappe *f* [*zeitlich gesehen*] = stage (of construction)
~ [*räumlich gesehen*] =
1) section under construction;
2) section to be executed;
3) completed section
Bauabsteckung *f* = setting out
Bauabteilung *f* = building department, construction ~
Bauachse *f* → Bau(werk)achse
Bauakustik *f*, Raumakustik = architectural acoustics
Bauarbeiten *fpl* = construction operations, ~ work
~ der öffentlichen Hand, öffentliche Bauarbeiten = public works
~ von privater Seite = non-public works
Bauarbeiter *m* = construction worker
Bau-Arbeitskräfte *fpl* = construction labo(u)r
Bauart *f*, Type *f* [*Maschine*] = type
(Bau)Aufnahme *f* = survey (of a structure)
Bauaufseher *m* = construction supervisor
Bauaufsicht *f*, Bauüberwachung *f* = construction supervision
Bauaufsichts|beamter *m* = building official (US); ~ inspector (Brit.)
~**behörde** *f*, Baupolizei *f*, Bauaufsichtsorgan *n* = construction supervising authority
Bauauftrag *m* = construction order
(Bau)Auftraggeber *m* → Bauherr *m*
(Bau)Aufzug *m* = (contractor's) hoist, builders' ~
~**turm** *m* **aus Holz** = timber hoist tower
~**turm** *m* **aus Stahl** = steel hoist tower
~**winde** *f* = hoist winch
bauausführende Firma *f* = winning contractor
Bauausführung *f*, Baudurchführung *f* = execution of (construction) work
Bau(ausführungs)|plan *m*, Bauzeichnung *f* = structural drawing
~**zeitraum** *m*, Baudurchführungszeitraum = construction time, ~ period
Bauausstellung *f* = construction show, ~ exhibit(ion), building ~
Bau(aus)trocknung *f* = drying(-out), dehumidifying
Bau(aus)trocknungsofen *m* → Kokskorb *m*
Baubahn *f*, Baustellenbahn, Industriebahn = site railway (Brit.); ~ railroad (US)
Baubaracke *f*, Baustellenbaracke = site hut
Baubaracken|heizung *f* = site hut heating
~**lampe** *f* = site hut lamp
Baubeamter *m* = building official, ~ officer, construction ~
Baubeendigung *f*, Beendigung der Bauarbeiten = termination of (construction) work, ~ ~ building ~
Baubeginn *m*, Beginn der Bauarbeiten = commencement of (construction) work, ~ ~ building ~
Baubehörde *f* = building authority, construction ~
Bauberatung *f* = construction consultation service
Baubeschläge *mpl* = builders' hardware [*This is a collective noun having no singular*]
Baubeschlagfabrik *f*, Baubeschlägefabrik = builders' hardware factory
Baubeschränkung *f* = bulk zoning
Baubeschreibung *f*, Baubeschrieb *m* = technical spec(ification)s
(Bau)Bestandsplan *m*, (Bau)Bestandszeichnung *f* = "as completed" drawing

Baubestimmung — Bauelement

Baubestimmung *f*, Bauvorschrift *f* = construction regulation, building ~

Baubetrieb *m* = construction management

Baubetriebsforschung *f* = construction management research

Bau(betriebs)-LKW *m* = contractor's lorry (Brit.); ~ truck (US)

Baubetriebsplan *m* → Bau(zeit)plan

Baubetriebsunfall *m* mit tödlichem Ausgang = fatal construction mishap (or accident)

Baubewilligung *f*, Bauerlaubnis *f*, Baugenehmigung = construction permit, building ~

Baublech *n*, Konstruktionsblech [*zum Unterschied von Kesselblech, Schiffblech usw.*] = construction sheet metal, ~ metal sheet

Baublei *n*, Konstruktionsblei = construction lead

Baubohle *f* = construction plank

Baubranntkalk *m*, Baubrennkalk = construction quicklime

Baubrücke *f*, Dienstbrücke, Bausteg *m* = service gangway (Brit.); distributing bridge, overhead track way, construction trestle (US)

Baubuch *n*, Bautagebuch, Baujournal *n* = job record

Baubude *f*, Leutebude = workmen's shelter

Baubüro *n* = construction office
~, Baustellenbüro = site office, job ~

Bauch *m*, Ausbauchung *f* [*das Vorstehen über eine Fläche oder Linie*] = bulge

Bauchemie *f* = construction chemistry

Baudämmstoff *m* = construction insulating material

Baudecke *f* [*Straßenbau*] = temporary pavement

Baudenkmal *n*, historisches Bauwerk *n* = monument

~pflege *f* = preservation of monuments

Baudichte *f*, Bebauungsdichte [*Verhältnis der Fläche der Baugrundstücke zur unbebauten Fläche einer Ortschaft*] = building density

Bau(dichtungs)folie *f* = construction foil

Baudicke *f* [*z.B. Straßendecke*] = thickness of construction

Bau-Diesellok(omotive) *f* → Baustellen-Diesellok(omotive)

Bau-Dokumentation *f* = documentation of building topics

Baudränage *f* = construction drainage system

Bau-Drehkran *m*, Bau-Schwenkkran, Dreh-Baukran, Schwenk-Baukran = slewing construction crane, rotary ~ ~, rotating ~ ~

Bau-Druckluterzeuger *m*, Bau-Luftverdichter *m*, Bau-Kompressor *m*, Baustellen-Druckluterzeuger, Baustellen-Luftverdichter, Baustellen-Kompressor = contractor's (air) compressor (set), builder's (~) ~ (~)

Baudübel *m*, Konstruktionsdübel = construction dowel

Baudurchführung *f*, Bauausführung *f* = execution of (construction) work, ~ ~ building ~

~ durch Einzelunternehmer [*Im Gegensatz zur Baudurchführung durch einen Generalunternehmer*] = separate trade contractor system

Baudurchführungszeitraum *m*, Bau(ausführungs)zeitraum = construction period, ~ time

Baueinheit *f* [*Fehlname*] → Bauelement *n*

Baueinrichtung *f* → Baustelleneinrichtung

Baueinrichtungsplan *m* → Baustelleneinrichtungsplan

Baueinstellung *f*, Bauunterbrechung = suspension of the work

Baueisen *n* = construction iron
~waren */pl* = ironmongery for building trades

Bauelement *n* [*Fehlname: Baueinheit f*] = unit [*Building material which is formed as a single article complete in itself but which is intended to be*

part of a compound unit or building or structure. Examples are brick, block, tile, lintel]
~, Krannormteil *m*, *n* = standard crane piece
bauen = to build, to construct
~ [*eine Maschine*] = to fabricate
Bauen *n* = construction, constructing
~ **mit Fertigteilen**, Montagebau *m*, Fertigteilbau = precast construction
Bauerlaubnis *f* → Baubewilligung
Bauetappe *f* → Bauabschnitt *m*
Baufach|ausdruck *m* = construction term
~**buch** *n* = construction book
~**zeitschrift** *f* = construction journal
baufähig → ab~ (✖)
baufähige Schicht *f* → (ab)bauwürdige ~
Baufähigkeit *f* → (Ab)Bauwürdigkeit
Baufahrzeug *n*, Baustellenfahrzeug = construction vehicle
~**-Waage** *f* = contractor's weigh bridge
baufällig = delapidated, ramshackle
Baufehler *m* = construction fault, building ~
Baufertigstellung *f* = completion of the work
Baufeuchtigkeit *f*, Baufeuchte *f* = construction material moisture, building ~ ~
Bau-Feuchtigkeitsmesser *m* [*zur Messung des Feuchtigkeitsgehaltes von Baustoffen*] = construction material moisture meter, building ~ ~ ~
Baufilz *m* = construction felt
Baufinanzierung *f* = construction financing
Baufirma *f*, bauindustrielle Unternehmung *f*, bauindustrielles Unternehmen *n*, Bauunternehmung, Bauunternehmen = construction company, ~ firm, building ~
Baufirmengruppe *f* → Arbeitsgemeinschaft *f*
Baufläche *f* = construction area
Baufucht *f*, Abfluchtung *f*, Einfluchtung, Ausfluchtung = alignment; alinement (US)

Bau(flucht)linie *f*, Fluchtlinie [*die von Amts wegen festgelegte Linie, bis an die heran Gebäude errichtet werden dürfen*] = building line
Baufolge *f*, Arbeitsfolge „Aufeinanderfolge der (Bau)Arbeiten = building sequence, construction ~
~ (✖) → Abbaureihenfolge
Baufolie *f*, Baudichtungsfolie = construction foil
Bauforschung *f* = construction research
Bauforschungs|anstalt *f*, Institut *n* für Bauforschung, Bauforschungszentrum *n*, Bauforschungsinstitut = building research station, ~ ~ institute
~**institut** *n* → Bauforschungsanstalt *f*
~**zentrum** *n* → Bauforschungsanstalt *f*
Baufortschritt *m*, Arbeitsfortschritt, Fortschritt der (Bau)Arbeiten, Baufortgang *m*, Fortgang der (Bau)Arbeiten, Arbeitsfortgang = progress of (construction) work, (construction) work progress
Baufortschrittplan *m* → Bau(zeit)plan
Baufrist *f*, Bauzeit *f* = completion period (or time), construction ~ (~ ~)
Baufristenplan *m* → Bau(zeit)plan
(Bau)Fristverlängerung *f* = extension of time for completion
Baufuge *f*, Abschnittfuge [*Brückenbau*] = construction joint
Bauführer *m* [*besorgt die Bauarbeiten unternehmerseitig*] = contractor's supervisor
Baugelände *n*, Bauland *n* = building land
Baugenossenschaft *f* = cooperative (or benefit) building society
Baugeologie *f* → Baugrundgeologie
Baugerät *n* [*als Einzelgerät*] = construction equipment item
Baugeräteliste *f*, Geräteliste für die Bauwirtschaft = equipment rental compilation (US); list of hiring charges

Baugeräteliste-Formular — Baugrundbodenprobe 154

~-**Formular** n = form for hiring charges

(**Bau**)**Gerüst** n, (Bau)Rüstung f, Arbeitsgerüst, Arbeitsrüstung = scaffold(ing)

Baugesetzgebung f, Baurecht n = building legislation

Baugestein n → Baustein

Baugewerbe n = construction trade(s), building ~ [*One of or all the occupations of tradesmen in building*]

Baugips m [*DIN 1168*]; gebrannter Gips für Bauzwecke [*Schweiz. SIA Nr. 115*] = calcined gypsum plaster [*ASTM C 28-50*]; building plaster [*B. S. 1191*]

~- **und Gipsplattenwerk** n = integrated building plaster mill and plaster board plant

~**werk** n = (building) plaster mill

Bauglas n = construction glass

Baugleis n, Baustellengleis, Industriebahngleis = contractor's track

Bauglied n = construction(al) member

Baugrenze f = limit of building area
~ (⚒) = boundary of the mine take, ~ ~ ~ ~ taking

Baugröße f → Baumaße npl

Baugrube f, Tunnel~ = (tunnel) trench
~ = building pit, excavation, cut
~ **mit Böschungen**, (ab)geböschte Baugrube = sloped building pit, ~ excavation, ~ cut

Bauguben|aufzug m, Grubenaufzug = building pit materials hoist

~**auskleidung** f, Baugrubenverkleidung, Baugrubenbekleidung = cut lining, excavation ~, building pit~

~**aussteifung** f → (Ab)Bölzung

~**böschung** f = building pit slope

~**entwässerung** f, Baugrubenwasserhaltung, Trockenlegung der Baugrube = building pit drainage

~**herstellung** f = building pit construction

~**pumpe** f = building pit pump

~**schrägaussteifung** f, Baugrubenverstrebung = raking shoring of building pit, ~ ~ ~ excavation, ~ ~ ~ cut

~**sohle** f, Bausohle = bottom of building pit, ~ excavation, ~ cut

~**sprengung** f = excavation blasting, building pit ~, cut ~

~-**Tauchpumpe** f = contractor's submersible pump, contractors' ~

(~)**Umspundung** f = sheeting an excavation with sheet piles, ~ a building pit ~ ~ ~, ~ a cut ~ ~ ~

~**verstrebung** f → Baugrubenschrägaussteifung

~**wand** f = building pit side, excavation ~, cut ~

~**wand** f, Tunnel~ = (tunnel) trench side

~**wasser** n = building pit water

Baugrund m, Untergrund [*Schicht auf der gegründet wird*] = ground, supporting medium, foundation (material) [*The soil or rock on which a structure or building rests. The word "foundation" is also used as an equivalent to the word "substructure" and then means "Unterbau" in German*]

~(**ab**)**dichtung** f = waterproofing of foundation (material), ~ ~ supporting medium, ~ ~ ground

~**aufschlußbohrung** f, Baugrunduntersuchungsbohrung = foundation exploration drilling, subsurface ~

~**beanspruchung** f = stressing of ground, ~ ~ supporting medium, ~ ~ foundation (material)

~**belastung** f = loading of ground, ~ ~ supporting medium, ~ ~ foundation (material)

~**boden** m, Gründungsboden = supporting soil, foundation ~, (sub)soil

~**bodenprobe** f, Gründungsbodenprobe = supporting soil sample, foundation ~ ~, (sub)soil ~

Baugrundbodenschicht — Bauhilfsgeräte

~**bodenschicht** *f*, Gründungsbodenschicht = foundation soil stratum, supporting ~ ~, (sub)soil ~

~**bohrung** *f*, Untergrundbohrung = ground drilling, foundation ~

~**entwässerung** *f* = drainage of the ground, ~ ~ ~ foundation (material), ~ ~ ~ supporting medium

~**fels** *m*, Gründungsfels, felsiger Baugrund *m* = supporting rock, foundation ~

~**felsprobe** *f*, Gründungsfelsprobe = supporting rock sample, foundation ~ ~

~**forschung** *f* = ground research

~**geologie** *f*, Ingenieurgeologie, Baugeologie = engineering geology

~**karte** *f* = ground map, foundation ~

~**klasse** *f* = ground class

~**probe** *f* = foundation (material) sample, ground ~, supporting medium ~

~**sondierung** *f* **mit indirektem Verfahren**, geophysikalische Baugrunduntersuchung, indirekte Baugrundsondierung = geophysical subsurface investigation, ~ ~ exploration, ~ ~ reconnaissance, ~ foundation ~

~**spannung** *f*, Untergrundspannung = ground stress, foundation (material) ~

~**stück** *n*, Bauplatz *m* = site

~**stückerschließung** *f* = site development

~**untersuchung** *f*, Baugrundaufschluß *m* = foundation exploration, ~ reconnaissance, ~ investigation, subsurface ~

~**untersuchungsbohrung** *f*, Baugrundaufschlußbohrung = subsurface exploration drilling, foundation ~ ~, reconnaissance ~, ~ investigation ~

~**verbesserung** *f* = improvement of the mechanical properties of foundations, ~ ~ ~ ~ ~ ~ foundation materials

~**verdichtung** *f*, künstliche ~ = ground compaction

~**verdichtung** *f*, natürliche ~ = ground consolidation

~**vereisung** *f*, Gefriergründung *f*; Gefrierverfahren *n* [*Bergbau*] = freezing (process), ~ method

~**verfestigung** *f*, (Boden)Verfestigung [*Die Festigkeitseigenschaften der gewachsenen oder gestörten Böden werden durch Injektionen, elektrochemische Behandlungen oder Gefrieren erhöht*] = ground stabilization

~**verfestigung** *f* **durch Einpressen von Zementmilch** = ground stabilization by cement grouting

~**verfestigung** *f* **durch Elektro-Osmose** = electro-osmotic ground stabilization

~**verhältnisse** *f*, Untergrundverhältnisse = underground conditions

Bau-Grundwasserstand *m* = ground water level during the construction

Baugruppe *f* [*Teil einer Maschine*] = unit

Bau-(Gummi)Reifenschlepper *m* → Bau-Radschlepper

Bauhammer *m*, Installationshammer [*kraftgetrieben*] = contractors' hammer, builders' ~

Bau(hand)karren *m*, Bau(hand)karre *f* = builders' hand cart

Bauhauptstoff *m*, Hauptbaustoff = direct material [*A material which is physically incorporated in the work(s)*]

Bauhebezeug *n*, Baustellen-Hebezeug = contractor's lifting gear

Bauheizung *f* [*Hochbau*] = heating for building operations

~ [*Ingenieurbau*] = heating for construction operations

Bauherr *m*, Bauherrschaft *f*, (Bau-)Auftraggeber *m* = purchaser, promoter, (building) owner, sponsor, employer, client

Bauhilfs|geräte *npl* = auxiliary building (or construction) equipment

Bauhilfsstoff — Baulandbeschaffung

~stoff m, Hilfsbaustoff = indirect material [*A material which is not physically incorporated into the work(s)*]

Bauhof m, Gerätepark m = plant depot, contractor's yard, equipment (repair) yard, equipment pool, builder's yard; dump (US)

~**reparatur** f = depot repair, plant ~

Bauhöhe f, Konstruktionshöhe = height of construction

~, Konstruktionshöhe [*Träger*] = construction depth

Bau-Hubschrauber m = construction helicopter, skycrane

Bauhygiene f = architectural hygiene, sanitary science as applied to buildings

Bauindex m = construction index

Bauinstandsetzungsmaterial n = structural repair material

Bauindustrie f = construction industry

bauindustrielle Unternehmung f → Baufirma f

Bauingenieur m = construction engineer

~**wesen** n = civil engineering and building construction, construction engineering

Baujournal n, Bau(tage)buch n = job record

Baukalk m = building lime, construction ~

~**hydrat** n = trowel trades hydrated lime

Baukantine f → Bau(stellen)kantine

Baukapazität f = construction capacity

Baukarre(n) f, (m) → Bau(hand)karre(n)

Baukasten|förderer m, Baukastensystemförderer = unit constructed conveyor, ~ ~ conveyer

~**system** n = component system

~**(system)anlage** f, halbstationäre bituminöse Mischanlage = sectional type bituminous mixing plant, ~ ~ ~ ~ installation

Baukaufmann m = purchasing clerk

Baukeramik f = structural ceramics, heavy ~

~**(fuß)bodenplatte** f = structural clay flooring tile, heavy ~ ~ ~

baukeramische Fertigteile mpl, npl = precast structural clay units, heavy ~ ~ ~

~ **Industrie** f = structural clay industry, heavy ~ ~

baukeramisches Erzeugnis n = structural clay product, heavy ~ ~

Bauklammer f = dog

Bauklasse f = construction class

Bauklebstoff m, Baukleber m = building adhesive, ~ bonding agent

Baukleinkran m, Baustellenkleinkran = midget construction crane

Baukompressor m → Bau-Drucklufterzeuger m

Baukonsortium n = consortium of contractors

Baukonstruktions|forschung f = structural research

~**lehre** f = structural design theory

~**verfahren** n, Baukonstruktionsmethode f, Baukonstruktionsweise f = building system [*An arrangement of building units to form a connected whole*]

Baukörper m = part of a structure

Baukosten f = construction cost

~**index** m = construction cost index

Baukraftwerk n = site power plant, ~ ~ station, ~ generating ~

Baukran m = construction crane

~, Hoch~ = building crane

Baukreiselpumpe f = contractors' centrifugal (water) pump

Baukreissäge(maschine) f = contractors' circular saw(ing machine), ~ ~ saw machine

Baukunststoff m = building plastic

Baukupfer n = construction copper

Bauland n, Baugelände n = building land

~**beschaffung** f, Grunderwerb m [*in Österreich: Grundeinlösung f*] = acquisition of building land, building land purchase

Baulandenteignung — Baumischbelag

~enteignung *f* = expropriation of building land
~erschließung *f*, Baureifmachung = building land development
Baulänge *f* = length of structure
~ [*Rohr*] = effective length (of a pipe), laying ~ (~ ~ ~)
Baulärm *m* = construction noise
Baulastträger *m* **der öffentlichen Hand** = public construction agency
Bauleistung *f* = construction work
Bauleistungs|buch *n*, Lastenheft *n*, Bauleistungsverzeichnis *n* [*Schweiz: Übernahmebedingungen fpl*] = contract spec(ification)s
~umfang *m* = scope of (construction) work, ~ building ~
Bauleiter *f* = bricklayers' ladder, builder's ~, pole ~
~ *m*, Firmen~ = contractor's agent
Bauleitung *f* (**der Bauherrschaft**) = resident engineer, Engineer
Bauleuchte *f* = site lantern (Brit.); ~ luminaire (US); ~ lighting fixture
baulich = structurally
bauliche Anordnung *f*, Aufbau *m*, Konstruktion *f*, Ausführung, Bau [*Maschine*] = design, version
~ **Maßnahme** *f* = structural measure
~ **Schutzmaßnahme(n)** *f(pl)* = structural protection
baulicher Luftschutz *m* → bautechnischer ~
~ **Schallschutz** *m* = acoustic conc-struction
Baulinie *f* → Baufluchtlinie
Baulok(omotive) *f* → Feldbahnlok(omotive)
Baulos *n*, Los = contract section
Baulücke *f* [*An einer anbaufähigen Straße zwischen bebauten Grundstücken unbebautes Grundstück*] = gap site
Bau-Luftverdichter *m* → Bau-Druckluft erzeuger

Baumarkt *m* = construction market, building ~
Baumaschine *f* = construction machine
Baumaschinen|-Achse *f* = construction machine axle
~ausstellung *f* = construction machinery and equipment show
~fibel *f* = construction machinery and equipment primer
~-**Handbuch** *n* = construction equipment manual
~messe *f* = construction machinery and equipment fair
~park *m*, Gerätebestand *m*, Gerätepark, Baumaschinenbestand = (contractor's) spread, (equipment) fleet
~sitz *m* = construction machine seat
~-**Tieflader** *m*, Baumaschinen-Tiefladeanhänger = low-loader for construction machinery and equipment
Baumaschinen *fpl* **und Baugeräte** *npl* = construction machinery, ~ equipment, contractor's plant and machinery, constructional plant, building and civil engineering plant, public works equipment, contracting plant, contractor equipment, builder's equipment, builder's plant and machinery
Baumasse *f* [*für den Feuerungsbau*] = refractory compound
Baumaße *npl*, Baugröße *f* = size [*size means the measurements of a body expressed in terms of dimension, area, or volume*]
Baumaterial *n* → Baustoff *m*
Baumbepflanzung *f* = tree planting
Baumeister *m* = master builder
Baumerkmale *npl* [*Maschine*] = construction details [*machine*]
Baumetall *n* = construction(al) metal
Baumethode *f* → Bauverfahren *n*
~ → Ab~ (⚒)

Baumischbelag *m*
~ mit dicht abgestuftem Korngemenge, dichter ~

road mix surface
graded aggregate type ~ ~ ~, dense-graded ~ ~ ~

∼ von der Art des Makadam, offener ∼, hohlraumreicher ∼ = macadam aggregate type ∼ ∼ ∼, open-graded ∼ ∼ ∼

Baumischer *m* = construction mixer
Baumlandschaft *f* = tree-scape
Baumonat *m* = construction month
Bau-Montageverfahren *n* = building erection system
Baumörtel *m* = building mortar, construction ∼
Baumpflanzung *f* = tree planting
Baumreihe *f*, Reihenbepflanzung *f* = avenue planting
Baumsäger *m*, Raupenschlepper *m* mit Sägenase = saw nose dozer, tree saw
Baum|scheibe *f* = root feeding area
∼schieber *m* = treedozer
Baumstein *m*, Dendrit *m* = dendrite
Baumstumpf *m* → Stumpf
Baumuster *n* [*Maschine*] = type, model
Baumwolle|band *n* → → Bandförderer
∼cord *n* = cotton cord
∼gurt *m* → Bandförderer
∼-Nylon-Band *n* → → Bandförderer
Baunagel *m* = construction nail
Baunickel *n* = construction nickel
Baunivellier(instrument) *m*, (*n*), Baustellennivellier(instrument) = builders' level
Baunorm *f* = construction standard (specification)
Bauobjekt *n* → (Bau)Projekt *n*
Bauordnung *f*, Ortsbausatzung [*dient zur Regelung zukünftiger Ausnutzung des Grund und Bodens und zur Einschränkung der willkürlichen Benutzung*] = building code (US); bylaws (Brit.)
Baupapier *n* = building paper, waterproof ∼
∼ **am beweglichen Auflager eines Sturzes** = building paper bond barrier over lintel bearing area
∼ **im Bereich der Wandauflagerzone** = building paper bond barrier over wall bearing area
Baupflege *f*, Baudenkmalpflege = preservation of monuments

∼amt *n*, Baudenkmalpflegeamt = Office for the Preservation of Monuments
Bauphase *f* → Ausführungsphase
Bauplan *m*, Bauzeichnung *f*, Bauausführungsplan = structural drawing
∼ → Bau(zeit)plan
Bauplanung *f* = project planning
Bauplanungsrecht *n* = construction planning legislation
(Bau)Platte *f* = building board
∼ **für den Industriebau**, Industrie-Bauplatte = building board for industrial construction
Bauplatz *m*, Baugrundstück *n* = site
Baupolitik *f* = construction policy
Baupolizei *f* → Bauaufsichtsbehörde *f*
Baupraxis *f* = construction practice, building ∼
Baupreis *m* = construction price
∼buch *n* = schedule of prices
∼index *m* = construction price index
∼recht *n* = construction prices legislation
Bauprofil *n* → Profil
(Bau)Profilhöhe *f* = depth of (structural) section
Bauprogramm *n* → Bau(zeit)plan *m*
(Bau)Projekt *n*, (Bau)Vorhaben *n*, Bauobjekt *n* = construction(al) project, (building) ∼
Baupumpe *f*, Pumpe für die Bauwirtschaft, Baustellenpumpe = contractors' pump, contractor's ∼
Bau-Radschlepper *m*, Bau(stellen)-(Gummi)Reifenschlepper, Bau(stellen)-(Gummi)Reifentrecker *m*, Bau(stellen)-(Gummi)Reifentraktor *m*, Bau-Radtraktor, Bau-Radtrecker, Baustellen-Radschlepper, Baustellen-Radtraktor, Baustellen-Radtrecker = contractor's rubber-mounted tractor, contractor-type wheel ∼

Baurampe *f* [*provisorische Baustellenzufahrt*] = temporary site access ramp

Bauraupe *f* → → Bulldozer

Baurecht *n*, Baugesetzgebung *f* = building legislation

Bau-Reifenschlepper *m* → Bau-Radschlepper

baureifer Plan *m*, baureife Zeichnung *f* = drawing "issued for construction", working drawing

Baureifmachung *f*, Baulanderschließung = building land development

(Bau)Reihe *f*, Typenreihe [*Maschine*] = series

(Bau)Rüstung *f* → (Bau)Gerüst *n*

Bausachverständiger *m* = construction expert, building ∼

Bausäge *f* = contractors' saw

Bausaison *f* = construction season, building ∼

Bausand *m* = building sand

Bauschaden *m* = building failure

Bauschaum *m* [*für die Herstellung von Schaumbeton*] = concrete foam

Bauschiene *f*, Baustellenschiene, Feldbahnschiene = contractor's rail

Bauschild *n* = construction-site sign

Bauschlepper *m*, Bautrecker, Baustellenschlepper, Baustellentrecker, Baustellentraktor = contractor's tractor

Bau-Schnellmontageverfahren *n* = speedy (building) erection system

Bauschraube(nwinde) *f* → Gerüstspindel

Bauschreiner *m*, Bautischler *m* = joiner

Bauschutt *m* = building rubbish, rubble (Brit.); waste (US)

Bau-Schwenkkran *m* → Bau-Drehkran

Bau-Seilbahn *f* = construction ropeway

bauseitig geliefert, bauseits ∼ = supplied free issue

bauseitige Leistungen *fpl* = services and facilities provided by the owner

Bausicherheitsvorschriften *fpl* = constructional safety regulations, building ∼ ∼

Bausohle *f*, Ab∼ (⚒) = working level ∼ → Baugrubensohle

Bausparkasse *f*, Bausparverein *m*, Bau- und Darlehnsverein = building and loan association, ∼ society

Bausperre *f* = construction site barrier
∼ [*Bauerlaubnis wird nicht erteilt*] = refusal of building licence
∼ [*erteilte Bauerlaubnis wird zurückgezogen*] = cancellation of building licence

Bauspindel *f* → Gerüstspindel

Bauspundwand *f* [*dient zur Herstellung des Bauwerkes und wird gezogen*] = temporary sheetpiling

Baustadium *n* → Ausführungsphase *f*

Baustahl *m* = structural steel, construction(al) ∼

∼**gewebe** *n* → Baustahlmatte

∼**gitterträger** *m* = steel lattice web girder

∼**matte** *f*, (geschweißte) Bewehrungsmatte, (geschweißte) Armierungsmatte, Stahlbeton-Armierungsmatte, Stahlbeton-Bewehrungsmatte, Betonstahlmatte [*Baustahlgewebe ist der Markenname für Bewehrungsmatten mit unverschieblichen Knotenpunkten aus kaltgezogenem Sonderbetonstahl IV nach DIN 1045*] = steel fabric (mat), welded ∼ (∼), steel wire mesh, reinforcing mat, reinforcement mat, steel wire reinforcement, reinforcing steel mesh

Baustatik *f* = (statical) analysis, design ∼, structural ∼, statics for structural engineering [*The early part of a structural design, which consists of determining what forces are carried by all the parts of a structure or building and what proportions they bear to the loads on them*]

baustatische Theorie *f* = theory of structures

Bausteg *m* → Baubrücke *f*

Baustein *m* [*bearbeiteter Naturstein*] = building block

~, **Baugestein** *n*, **bautechnisches Gestein** [*Naturstein unbearbeitet*] = (building) stone, structural ~

~maschine *f*, **Steinformmaschine** = block making machine

~verwitterung *f* = weathering of building stones, ~ ~ structural ~

Baustelle *f* = site (of work(s)), jobsite, project site, construction site, work site, work location (US)

~ des Hochbaues = building site

~ ~ Ingenieurbaues = civil engineering site

Baustellen|-Aggregat *n* contractors' generating set, builders' ~ ~

~anstrich *m* = field (paint) coat, ~ (~) finish, ~ coat of paint

~arbeit *f* = site work

~arbeiter *m* = field hand

~(auf)räumung *f* = job cleanup, cleanup and move out, clearance of site on completion

~bahn *f* → Baubahn

~baracke *f*, **Baubaracke** = site hut

~begehung *f*, **Begehung** [*vor Angebotabgabe*] = site inspection

~-Beheizung *f* = site heating

~-Belastungsversuch *m* → Baustellen-Lastversuch

~belegschaft *f*, **Baustellenpersonal** *n* = site personnel

~beleuchtung *f* = site lighting

~beton *m* → → Beton

~-Betonbereitungsanlage *f* [*Anlage zur (gewichtsdosierten) Beschickung des Mischers mit Zuschlagstoffen und Zement*] = central concrete mixing plant, ~ ~ installation

~bunker *m* = site bin

~büro *n*, **Baubüro** = job office, site ~

~dichte *f* [*Boden*] = field density

~dichteprüfung *f*, **Baustellendichteprobe** *f*, **Baustellendichteversuch** *m* = field density test

~-Diesellok(omotive) *f* **Dieselfeldbahnlok(omotive)**, **Bau-Diesellok(omotive)** = Diesel field loco(motive)

~-Drucklufterzeuger *m* → Bau-Drucklufterzeuger

~-Druckluftwerkzeug *n*, **Baustellen-Preßluftwerkzeug**, **pneumatisches Baustellenwerkzeug** = contractor's pneumatic tool, ~ (compressed-)air ~, builders' ~ ~

~-Druckprüfmaschine *f* = contractor's compression tester, builders' ~ ~, ~ ~ testing machine

~einfahrt *f* = site approach

~einfriedung *f* = site enclosure

~einrichtung *f*, **Baueinrichtung** = site installations, ~ facilities, ~ plant

~einrichtungsplan *m*, **Baueinrichtungsplan** = site installations plan, ~ facilities ~, ~ plant ~

~einsatz *m* = construction-site service

~fahrzeug *n*, **Baufahrzeug** = construction vehicle

~fertigungsplatz *m* [*für Pfähle, Träger usw.*] = on-site casting yard

~-Flachbodenselbstentlader *m* = narrow ga(u)ge flat dump car

~gleis *n* → Baugleis

~-(Gummi)Reifenschlepper *m* → Bau-Radschlepper

~-Hebezeug *n* → Bau-Hebezeug

~-Instandsetzungs-Werkstatt *f* = field (work)shop

~kantine *f*, **Baukantine** = site canteen

~kipper *m* = site tipper

~kleinkran *m*, **Baukleinkran** = midget construction crane

~-Kompressor *m* → Bau-Drucklufterzeuger *m*

~kosten *f* = site cost

~labor(atorium) *n* = field lab(oratory), on-job ~

~lager *n* → Arbeiterlager

~-Last(kraft)wagen *m*, **Baustellen-LKW** *m* = contractor's lorry (Brit.); ~ truck (US); builders' ~

~-Lastversuch *m*, **Baustellen-Belastungsversuch**, **Baustellen-Belastungsprüfung** *f*, **Baustellen-Lastprobe** *f* = in-situ loading test

~**-LKW** *m* → Baustellen-Last(kraft)wagen *m*
~**lok(omotive)** *f*, Feldbahnlok(omotive), Baulok(omotive) = field loco(motive), contractor's ~
~**lore** *f* → Feldbahnlore
~**-Luftverdichter** *m* → Bau-Drucklufterzeuger
~**-Metallbaracke** *f* = metal site hut
~**mikroskop** *n* = field microscope
~**mischanlage** *f* = (on-)site mixing plant
~**mischer** *m* = (on-)site mixer
~**-Mischungsformel** *f* = site mix design, ~ ~ formulation
~**montage** *f* = erection on the site, field erection
~**niet** *m* = field rivet, site ~
~**nietung** *f* = site riveting, field ~
~**nivellier(instrument)** *m*, (*n*), Baunivellier(instrument) = builders' level, contractors' ~
~**organisation** *f* = job "housekeeping" (US); site organization
~**personal** *n*, Baustellenbelegschaft *f* *f* = site personnel
~**-Pfahlfabrik** *f* = on-site pile casting yard
~**praxis** *f* = site practice
~**-Preßluftwerkzeug** *n* → Baustellen-Druckluftwerkzeug
~**probe** *f* → Baustellenprüfung *f*
~**prüfung** *f*, Baustellenprobe *f*, Baustellenversuch *m*, Feldversuch, Freifeldprüfung = site test
~**pumpe** *f* → Baupumpe
~**-Radschlepper** *m* → Bau-Radschlepper
~**-Radtraktor** *m* → Bau-Radschlepper
~**-Radtrecker** *m* → Bau-Radschlepper
~**räumung** *f* → Baustellen(auf)räumung
~**-Reifenschlepper** *m* → Bau-Radschlepper
~**schiene** *f*, Bauschiene, Feldbahnschiene = contractors' rail
~**-Schienenwagen** *m* → Feldbahnlore *f*
~**schlepper** *m* → Bauschlepper
~**schweißung** *f*, Feldschweißung = field welding

~**-Selbstentlader** *m* = narrow ga(u)ge dump car
~**silo** *m* = site silo
~**stoß** *m*, Baustoß [*Stahlbau*] = field connection, ~ joint, site ~
~**straße** *f*, Baustellenweg *m*, Fahrweg auf einer Großbaustelle, Baustraße = builders' road, site ~
~**-Stromerzeuger** *m* = contractor's generating set, builders' ~ ~
~**traktor** *m* → Bauschlepper
~**trecker** *m* → Bauschlepper
~**überflutung** *f*, Baustellenüberschwemmung = site flooding
~**-Unterhaltung** *f* = field maintenance
~**unterkunftswagen** *m* = mobile site cabin
~**verbolzung** *f* = site bolting, field ~
~**verfahren** *n* = site method, field ~
~**verkehr** *m*, Bauverkehr = building traffic, construction ~
~**versuch** *m* → Baustellenprüfung *f*
~**vorfertigung** *f* = on-site prefabrication
~**waage** *f*, Bauwaage = contractor's scale, builders' ~
~**wagen** *m* = site trailer
~**wechsel** *m*, Ortswechsel, Transport *m* von Baustelle zu Baustelle, Standortwechsel, Standortveränderung *f* = job-to-job hauling, ~ transport
~**weg** *m*, Baustellenstraße *f*, Fahrweg auf einer Großbaustelle = builders' road, site ~
~**weiche** *f*, Bauweiche = contractor's switch, builders' ~
~**werkstatt** *f*, Bauwerkstatt = site workshop, builder's ~, contractor's ~
~**werkzeug** *n* → Bauwerkzeug
~**zwischentransport** *m* = transportation on the site
Baustil *m* = architectural style
~ (Geol.) = type of mountain structure
Baustoff *m*; Baumaterial *n* [*Schweiz*] = building material, construction(al) ~, structural ~

Baustoffabrikant *m* → Baustoffhersteller *m*

Baustoff|anfuhr *f*, Baustoffantransport *m*, Baustoffanlieferung *f* = construction materials delivery, building ~ ~, structural ~ ~

~aufbereitung *f* = building materials processing, construction ~ ~, structural ~ ~

~ausstellung *f* = building materials show, construction ~ ~, structural ~ ~

~bedarf *m* = building materials requirement, construction ~ ~, structural ~ ~

~-Bindemittel *n* = binder for construction materials, ~ ~ building ~, ~ ~ structural ~

~einlagerung *f*, Baustoffbevorratung = construction materials storage, building ~ ~, structural ~ ~

~einzelhändler *m* = building material dealer, construction ~ ~, structural ~ ~

~ersparnis *f* = building materials saving, construction ~ ~, structural ~ ~

~erzeugung *f*, Baustoffherstellung = building material production, construction ~ ~, structural ~ ~

~großhändler *m* = building material distributor, construction ~ ~, structural ~ ~

~güte *f* = building materials quality, construction ~ ~, structural ~ ~

~(güte)überwachung *f* = construction material quality control, building ~ ~ ~, structural ~ ~ ~

~hersteller *m*, Baustofferzeuger *m*, Baustoffabrikant *m* = building material producer, construction ~ ~, structural ~ ~, ~ ~ manufacturer

~herstellung *f* → Baustofferzeugung

~(-Herstellungs)maschine *f* = construction material machine, building ~ ~, structural ~ ~

~industrie *f* = construction materials industry, building ~ ~, structural ~ ~

~ingenieur *m* = construction material engineer, building ~ ~, structural ~ ~

~lager *n* = construction materials store, building ~ ~, structural ~

~markt *m* = building materials market, construction ~ ~, structural ~ ~

~maschine *f*, Baustoff-Herstellungsmaschine = building material machine, construction ~ ~, structural ~ ~

~presse *f* → Baustoffprüfpresse

~probe *f* = construction materials sample, building ~ ~, structural ~ ~

~probe *f*, Baustoffprüfung *f*, Baustoffversuch *m* = building materials test, construction ~ ~, structural ~ ~

~prüfanstalt *f*, Baustoffprüfungsanstalt = building materials testing institute, construction ~ ~ ~, structural ~ ~ ~

~prüfgerät *n* = building material testing device, construction ~ ~, structural ~ ~ ~

~prüfmaschine *f* = building material testing machine, construction ~ ~, structural ~ ~ ~

~(prüf)presse *f* = compression machine, ~ tester

~prüfung *f* → Baustoffprobe *f*

baustoffschädlich = aggressive

Baustoff|technik *f* = building materials practice, construction ~ ~, structural ~ ~

~überwachung *f* → Baustoffgüteüberwachung

~versuch *m* → Baustoffprobe *f*

~vorkommen *n* = building materials deposit, construction ~ ~, structural ~ ~

~waage *f* = building materials scale, construction ~ ~, structural ~ ~

Baustollen *m* → Abstiegstollen

Baustoß *m* → Baustellenstoß

Baustraße *f* → Baustellenstraße

Baustrecke *f* = construction section

Bausumme *f*, Vertragssumme = contract sum
Bausystem *n* = construction system
Bautafel *f* = building panel, ~ sheet
Bau(tage)buch *n*, Baujournal *n* = job record
Bautätigkeit *f* = construction activity
Bautechnik *f* [*als betriebstechnische Anwendung*] = construction(al) technique
~ [*als Wissenschaft*] = construction(al) engineering
bautechnische Auskunftsstelle *f*, Bauzentrum *n* = building information centre (Brit.); ~ ~ center (US)
~ **Bodenkunde** *f* → Bodenmechanik *f*
(~) **Bodenuntersuchung** *f* = (sub)soil investigation, ~ study
~ **Bodenverhältnisse** *npl* = engineering soil conditions
~ **Eigenschaft** *f* [*Naturstein*] = structural property
~ **Weisheit** *f* = structural philosophy
bautechnischer Luftschutz *m*, baulicher ~ = civil defense construction (US); ~ defence ~ (Brit.)
bautechnisches Gestein *n* → Baustein
Bauteil *m*, *n* [*Eine aus einem oder mehreren Baustoffen geformte Sache, die geeignet ist, Bestandteil einer baulichen Anlage zu sein*] = component [*A section, unit, or compound unit*]
~, Kranteil = crane piece
Bautemperatur *f*, Ein~ [*Straßenbau*] = laying temperature
Bautempo *n*, Arbeitstempo = rate of progress (of the construction work), ~ ~ performance
Bauten *f* = structures
~**schutz** *n* = preservation of structures and buildings
(~)**Schutzfolie** *f* = protective foil
(~)**Schutzmittel** *n* = preservative for structures and buildings
Bautermin *m* = deadline, target date
Bauterrakotta *f* = architectural terra cotta
Bautischler *m*, Bauschreiner *m* = joiner

Bautoleranz *f* = permissible deviation of construction(al) elements
Bautraktor *m* → Bauschlepper
Bautrecker *m* → Bauschlepper
Bautrockner *m*, Bautrocknungsgerät *n* = dehumidifier, contractors' air heater, builders' air heater [*for drying(-out) new buildings*]
Bautrocknung *f*, Bauaustrocknung = dehumidifying, drying(-out)
Bautrocknungsgerät *n* → Bautrockner *m*
Bautrupp *m* = railway maintenance gang, ~ ~ crew, ~ ~ team, ~ ~ party (Brit.); railroad ~ ~ (US)
Bauüberwachung *f*, Bauaufsicht *f* = construction supervision
Bauumfang *m*, Bauvolumen *n* = volume of work
Bau- und Darlehnsverein *m* → Bausparkasse *f*
Bau- und Montageverfahren *n* = construction and erection method
Bauunfall *m* = construction accident
Bau-Unterabschnitt *m*, Unterabschnitt für den Abbau, Feldesteil *m*, *n* (⚒) = panel [*The area of coal extracted by a pair of development roads in longwall or pillar methods of mining. It is usually limited at the extreme end farthest from the shaft by the boundary of the property*]
Bauunterbrechung *f*, Baueinstellung = suspension of the work
Bauunterhaltung *f* = maintenance of structures
Bauunternehmer *m* = (public works) contractor, construction ~, building ~
~**verband** *m* = general contractors association
Bauunternehmung *f* → Baufirma *f*
Bauventilation *f* [*Schweiz*]; Bewetterung *f* [*Tunnel- und Stollenbau*] = ventilation during construction work
Bauverfahren *n* → (Ab)Baumethode *f* (⚒)
~, Baumethode *f*, Bauweise *f* [*Das Verfahren der Herstellung eines Bauwerkes*] = construction method, building ~

Bauverfahren — Bauwerkstatt 164

~ mit maßeinheitlichen Bauteilen = modular method
Bauverkehr *m*, Baustellenverkehr = construction traffic, building ~
Bauvertrag *m* = construction contract, building ~, memorandum of agreement
Bauverwaltung *f* [*im Falle als Auftraggeber*] = promoting authority
Bauvolumen *n*, Bauumfang *m* = volume of work
Bauvorbereitung *f* = preliminary work for actual construction work
Bauvorgang *m* = construction operation
Bauvorhaben *n* → (Bau)Projekt *n*
~ der öffentlichen Hand = public works project
Bauvorlagen *fpl* [*Zeichnungen und Festigkeitsberechnung(en)*] = construction documents
Bauvorschrift *f*, Baubestimmung *f* = building regulation, construction ~
Bauwaage *f*, Baustellenwaage = contractors' scale, builders' ~
Bauwagen *m* = construction-site trailer
Bauwasser *n* = water occurring during the construction
~**stand** *m* = level of water occurring during the construction, water level during construction
Bauweg *m* = construction road
Bauweiche *f*, Baustellenweiche = contractor's switch
Bauweise *f*, Baumethode *f*, Bauverfahren *n* [*Das Verfahren der Herstellung eines Bauwerkes*] = construction method
~ → (Ab)Baumethode *f* (⚒)
~, Bebauungsweise [*Bebauungsform mit Bezug auf den Abstand von den seitlichen Nachbargrenzen. In der geschlossenen Bauweise sind die Gebäude ohne Abstand, in der offenen mit beiderseitigem Abstand und in der halboffenen mit einseitigem Abstand von den seitlichen Nachbargrenzen zu errichten*] = type of development as regards the building limits

~ **Hoyer**, (Stahl)Saitenbeton *m* = Hoyer method, prestressed concrete with thin wires
Bauwerk *n*, Bau *m* = structure [*An organized combination of connected units constructed to perform a function or functions requiring some measure of rigidity. A structure need not necessarily provide shelter from the weather*]
~ [*Fehlname*] → Gebäude *n*
~ mit großer Stützweite = long-span structure
~**abdichtung** *f* gegen Bodenfeuchte = dampproofing
~**abdichtung** *f* gegen Grundwasser = waterproofing
~**achse** *f*, Bauachse, Symmetrieachse = centre line (Brit.); center ~ (US); axis
~**art** *f* = type of structure
~**auspressung** *f*, Bauwerkinjektion *f*, Bauwerkeinpressung, Bauwerkverpressung = grouting of structures
~**beton** *m*, Konstruktionsbeton = structural concrete
~**bezeichnung** *f* = designation of a structure
~**dehnung** *f* = elongation of a structure
~**fuge** *f* = joint of a structure
~**hinterfüllung** *f* = backfilling of a structure
~**instandsetzung** *f* = reinstatement of a structure
~**last** *f* = structure load
~**pfahl** *m*, Tragpfahl [*im Gegensatz zum Verdichtungs- und Dalbenpfahl*] = structural pile
~**schwingungen** *fpl* = vibrations of a structure
~**setzung** *f*, Bauwerksenkung, Setzen *n* eines Bauwerkes, Absackung eines Bauwerkes, Absacken eines Bauwerkes = settlement of a structure
~**spannungen** *fpl* = civil engineering stresses
Bauwerkstatt *f*, Baustellenwerkstatt = builder's workshop, site ~, contractor's ~

Bauwerkteil — Bebauungsweise

Bauwerk|teil *m*, *n* = element of construction [*Part of a structure having its own functional identity. Examples are foundation, floor, roof, wall*]
~trennwand *f* = structure partition [*A wall whose primary function is to divide space within a structure*]
~unterseite *f*, Bauwerksohle *f* = base of a structure, underside ~ ~ ~
~verzeichnis *n* = list of structures
Bauwerkzeug *n*, Baustellenwerkzeug [*z.B. Spaten, Hacke, Schaufel usw.*] = contractor's tool, construction ~, builders' ~
Bauwesen *n* = civil engineering and building construction, construction
~versicherung *f* = all risks erection insurance
Bauwinde *f* = contractors' winch, builders' ~
Bauwissenschaft *f* = building science
Bau-Wochenzeitschrift *f* = building weekly
Bau-Wohn(schlaf)wagen(anhänger) *m* = accomodation trailer for construction sites (US); sleeping ~ ~ ~ ~ (Brit.); house ~ ~ ~ ~
Bauwolle *f* = construction wool
bauwürdig → (ab)baufähig
bauwürdige Vorräte *mpl*, ab~ ~, (ab)baufähige ~, ausbeutbare ~ = reserves
Bauwürdigkeit *f* → Ab~ (✗)
Bauxit *m*, Alu(minium)oxydhydrat *n*, $Al_2O_3 \cdot 2H_2O$ = b(e)auxite
~bergwerk *n*, Bauxitgrube *f* = b(e)auxite mine
~erz *n* = b(e)auxite ore
~-Erzeugnis *n* = b(e)auxite refractory
~landzement *m* → Kühlzement
~werk *n* = b(e)auxite works
Bauzaun *m* = site fence
Bauzeichen *n* = structural drawing
Bauzeichner *m* = structural draughtsman (Brit.); ~ draffsman (US)
Bauzeichnung *f*, Bau(ausführungs)plan *m* = structural drawing

Bauzeit *f* → Baufrist *f*
~hochwasser *n*, bauzeitliches Hochwasser = construction flood
~plan *m*, Arbeitsplan *m*, Baufristenplan *m*, Bau(betriebs)plan *m*, Bauprogramm *n*, Baufortschrittplan = construction (or work, or time) schedule, schedule of construction operations. job plan, job schedule, phasing, (phased) program(me) of works, (programme &) progress chart, progress schedule
Bauzentrum *n*, bautechnische Auskunftsstelle *f* = building information centre (Brit.); ~ ~ center (US)
Bauzug *m* [*Eisenbahn*] = work train
Bazin|-Geschwindigkeitsformel *f* = Bazin (weir) formula
~-Wehr *n* = Bazin-type sharp-crested weir
BE → Abbindebeschleuniger *m*
beanspruchen, angreifen = to act
Beanspruchung *f* = stress and strain
Bearbeiten *n* **von Kalkstein** [*z.B. Sägen, Drehen, Hobeln*] = milling of limestone [*e.g. sawing, turning, planing*]
bearbeiteter Granit *m* = milled granite
Bearbeitung *f* **mit Werkzeug** [*z.B. Naturstein*] = tooling
Beaufschlagung *f* [*Turbine; Pumpe*] = admission
~ = inlet loading [*of a filter or dust collector*]
Bebakung *f*, Besetzen *n* mit Baken = beaconing
bebautes Gelände *n*, ~ Gebiet *n* = built-up area
Bebauung *f* = development
Bebauungs|dichte *f* → Baudichte
~plan *m*, Fluchtlinienplan, Ortsbauplan, Durchführungsplan [*zeichnerisch dargestellter Gesamtplan und schriftliche Erläuterung zur Regelung der Bebauungsmöglichkeiten und Anlage der Straßen und Plätze*] = development plan
~weise *f* → Bauweise

Bebout-Klappen(stau)wehr — Becherwerk(auf)lader

Bebout-Klappen(stau)wehr n = Bebout wicket (dam) (US); ~ shutter (~) (Brit.)
Becher m, Eimer m, Grab~ = (digging) bucket
~, Eimer m, Elevator~ = (elevator) bucket
~**auflader** m → Becherwerk(auf)lader
~**befestigungsschiene** f, Eimerbefestigungsschiene = rail for fixing elevator buckets
~**(be)lademaschine** f → Becherwerk(auf)lader
~**(be)lademaschine** f **mit Schneckenbeschickung** = screw feed type bucket elevator loader
(~)**Elevator** m → Becherwerk n
~**förderwerk** n → Becherwerk
~**glas** n = beaker
~**-Grabenbagger** m, Eimer-Grabenbagger = bucket ditcher, ~ trencher
~**-Gummiband** n, Eimer-Gummiband = bucket elevator rubber belt
~**gurt** m, Becherwerkgurt = elevator belt(ing), ~ conveyor ~, ~ conveyer ~
~**kette** f, Eimerkette, Elevatorkette, Becherwerkkette = ladder chain carrying the buckets, bucket elevator chain, bucket line chaine
~**koralle** f = corallite, cup coral
~**lademaschine** f → Becherwerk(auf)lader
~**leiter** f, Baggerleiter, Ausleger m, Eimerleiter(-Ausleger) [*Grabenbagger mit Eimern*] = digging ladder, ditcher ~, bucket ~, bucket flight
~**leiter** f, Eimerleiter, Baggerleiter [*Naßbagger*] = digging ladder, bucket ~, dredging ~, bucket flight
~**leiter** f, Eimerleiter [*Gleisschotter-Reinigungsmaschine*] = bucket flight, ~ ladder
~**leiterwinde** f → Baggerleiterwinde
~**rad** n, Schaufelrad, Eimerrad, Kreiseimerleiter f [*Bagger*] = bucket wheel, digging ~
~**rad** n, Schöpfrad, Eimerrad [*Waschmaschine*] = dewatering wheel
~**radbagger** m, Schaufelradbagger = bucket wheel excavator
~**rad-Lademaschine** f, Becherrad-Lader m, Lader mit Becherrad = bucket wheel loader, ~ ~ loading machine
~**reihe** f → Becherstrang m
~**strang** m, Eimerstrang, Becherreihe f, Eimerreihe = line of buckets, bucket line
~**turbine** f = bucket (type) wheel, ~ (~) turbine
Becherwerk n, Eimerketten-Aufzug m, Becherförderwerk, (Becher-)Elevator m, Aufnahmebecherwerk, Aufnahmeelevator, Becherwerkförderer m = (bucket) (type) elevator, scoop flight conveyor, scoop flight conveyer
~ **in Becher-an-Becher-Ausführung** → Reihenbecherwerk
Becherwerk(auf)lader m, (Eimerketten-)Fahrlader, Selbstauflader, (Auf-)Lader für Schüttgut, Aufnehmer, selbstaufnehmendes Förderband n, Becherwerk(fahr)lader, Becher(be)lademaschine f, Becherwerk-Ladeschaufler m, Fahrlade-Gerät n, Motor-Selbstlader für Schüttgüter, Eimerkettenlader, Eimerkettenlademaschine, Fahrlade-Maschine, Becherauflader = bucket elevator loader, portable ~, multi-bucket ~, conveyor (or conveyer) type bucket ~, multiple bucket ~
~ **auf LKW montiert** = lorry-mounted bucket (elevator) loader (Brit.); truck-mounted ~ (~) (US)
~ ~ **Luftreifen** = pneumatic-tyred bucket (elevator) loader
~ ~ **(Gleis)Ketten**, ~ ~ Raupen, (Gleis)Ketten-Becherwerk(auf)lader, Raupen-Becherwerk(auf)lader = crawler bucket (elevator) loader, track-laying (type) ~(~)~, caterpillar (type) ~ (~) ~, creeper-mounted (type) ~ (~) ~
~ **mit Band** = multi-bucket loader with belt conveyor (or conveyer) discharge arrangement

Becherwerk|-Ausleger m, Elevatorausleger = bucket elevator boom
~**förderer** m → Becherwerk
~**fuß** m, Elevatorfuß = boot [*The lower end of a bucket elevator from which the buckets lift material*]
~**fußtrommel** f, Elevatorfußtrommel = boot pulley
~**grube** f, Elevatorgrube = bucket elevator pit
~**gurt** m, Bechergurt = elevator belt(ing), ~ conveyor ~, ~ conveyer ~
~**kopf** m, Elevatorkopf = bucket elevator head
~**lader** m → Becherwerk(auf)lader
~**-Ladeschaufler** m → Becherwerk(auf)lader
~**silo** m = silo fed by bucket elevator
~**teil** m, n = bucket elevator part
Becherzahn m, Eimerzahn = bucket tooth
Becken n [*allgemein*] = basin
~ → → Talsperre
~ [*Obstbewässerung*] = basin
~, Pumpspeicher m = pool, reservoir, (storage) basin
~ [*Abwasserwesen*] = basin [*A shallow tank or natural or artificial depression through which liquids pass or in which they are detained for treatment*]
~ (Geol.) [*Großer Ablagerungsraum, der meist von schüsselartig gelagerten, oft nachträglich dislozierten Schichten, erfüllt ist; z.B. Wiener Becken, Mainzer Becken, Pariser Becken*] = basin
~, Hafen~ [*zwischen zwei Ladezungen*] = dock, basin
~**bewässerung** f = basin flooding, ~ method of irrigation
~**bildung** f (Geol.) = basining
~**damm** m = reservoir embankment
~**fundament** n = reservoir foundation
~**landschaft** f (Geol.) = basin topography
~**sohle** f = reservoir bottom
~**urinal** n = pedestal urinal

~**wand** f = reservoir wall
~**wüste** f = basin desert
bedachen, (ein)decken = to roof
Bedachen n → Bedachung f
Bedachung f, Dacheindeckung, Eindeckung, Bedachen n, Eindecken = placing the roofing
Bedachungs|material n = roofing material
~**platte** f → Dachelement n
~**zubehör** m, n, (Dach)Eindeckungszubehör = roofing accessories
Bedampfung f = steaming
Bedarf m = requirement
Bedarfs|dränung f, Bedarfsdränage f, Teildränung, Teildränage = partial drainage
~**haltestelle** f = request stop
~**spitze** f, Belastungsspitze = load peak, peak load
bedecken, abdecken, zudecken = to cover (over)
bedienen [*Maschine*] = to operate
Bedienen n = operation
Bediener m → Bedienungsmann m
~**kabine** f, Bedienungskabine, Führerkabine = operator's cab(in)
~**sitz** m, Führersitz = operator's seat
Bedienung f = operation
Bedienungs|anweisung f, Bedienungsanleitung = operating instructions, operator's handbook
~**bequemlichkeit** f = operating comfort, ease of operation
~**brücke** f, Bedienungssteg m [*(Stau-)Wehr*] = gangway (Brit.); walkway (US); operating bridge
~**bühne** f = (control) platform, service ~
~**element** n, Bedienungsorgan n [*Maschine*] = operating element
~**fehler** m = operating error, ~ mistake
~**gestänge** n = operating linkage
~**gestänge** n **des Anlaßmotors** = starting engine operating linkage

Bedienungshebel — befestigt

~hebel *m* = operating lever
~hebel *m* **für Starterkupplung und Ritzel** = clutch and pinion operating lever
~hilfe *f* = operator aid
~kabine *f* → Bedienerkabine
~kammer *f* [*Talsperre*] = operating chamber
~kette *f* = operating chain
~knopf *m*, Druckknopf = push-button
~knopfanlasser *m*, Druckknopfanlasser = push-button starter
~knopfkasten *m*, Druckknopfkasten = push-button box
~knopfschalter *m*, Fingerdruckschalter, Druckknopfschalter = push-button switch
~mann *m*, Maschinenführer *m*, (Geräte-)Bediener *m* = (plant) operator, driver, equipment operator, plant hand; operative (Brit.)
~mannschaft *f* = operating gang, ~ team, ~ crew, ~ party
~pedal *n* = operating pedal
~personal *n* = operating labour (Brit.); ~ labor (US)
~plattform *f*, Bedienungspodest *n*, Bedienungsbühne *f* = operator's platform, operating ~, service ~
~podest *n* → Bedienungsplattform *f*
~pult *n* = operating desk
~schalttafel *f* = operating panel
~schild *n* = instruction plate
~schlüssel *m* = operating key
~schwierigkeit *f* = operating difficulty
~stand *m* operator's station, operating ~
~stange *f* = operating rod
~steg *m* → Bedienungsbrücke *f*
~steg *m* [*in einem Verkehrstunnel*] = service walkway
~tür *f* [*gewährt Zugang zur Wartung von Maschinenteilen*] = access door
~vorrichtung *f* = operating device
~vorschrift *f* = operating instructions

~wagen *m* [*Wehr*] = working carriage (Brit.); traveling crane (US)
~welle *f* = operating shaft
~winde *f* [*Wehr*] = working winch (Brit.); maneuvering ~ (US)
Bedingung *f* = condition
Bedrucken *n* [*Linoleum*] = surface printing
Bedürfnisanstalt *f* = public convenience
Beendigung *f* **der Bauarbeiten**, Baubeendigung = termination of the (construction) work, ~ ~ ~ building ~
Beet|bewässerung *f*, Furchenbewässerung = furrow irrigation [*A method of irrigating by small ditches or furrows leading from a header or supply ditch*]
~einfassung *f* = garden surround
~schlacke *f* = bank slag
~winkel *m* = corner garden edging
Befähigungs|nachweis *m* = experience record
~schreiben *n* = letter of capacity
befahrbar = runnable
~, schiffbar = navigable; boatable (US)
befahrbare Schachtabdeckung *f* = carriageway cover
Befahrbarkeit *f* = trafficability, rideability
Befahrbarkeitswert *m* [*Straße*] = present serviceability index, PSI
Befahren *n*, Befahrung *f* [*Bergmannsausdrücke für Besichtigen*] = inspection
Befahrungsabgabe *f* [*künstliche Wasserstraße*] = toll
Befensterung *f* = fenestration
befestigen [*Ufer im Wasserbau*] = to protect, to stabilize
~ [*Straßen- und Flugplatzbau*] = to pave
Befestigen *n*, Festmachen = fixing
befestigen mit Bolzen, verbolzen = to bolt
~ ~ **Nieten**, vernieten = to rivet
befestigt [*Verkehrsfläche*] = paved, hard

befestigte Düne f, ruhende ~ = stabilized dune
~ Standspur f [Straße] = hard shoulder, paved ~
befestigter Radspurstreifen m, Spurstreifenstraße f = strip(e) road, creteway, trackways
(~) Seitenstreifen m, (~) Seitenraum m = shoulder
Befestigung f = fixing
~, Fahrbahn~ [Wenn nur eine Schicht vorhanden ist] = pavement, surfacing; paving (US)
~, Fahrbahn~ [Die Gesamtheit aller Schichten] = (layered) system
~, Fahrbahn~ [allgemeiner Ausdruck ohne Rücksicht auf die Anzahl der Schichten und ohne Einschränkung auf nur die „Decke"] = pavement, surfacing; paving (US)
~ von Böschungen = slope stabilization
Befestigungs|arbeiten fpl = fixing work
~bolzen m = fixing bolt
~bügel m = fixing stirrup
~flansch m = fixing flange
~fuß m = fixing foot
~haken m = fixing hook
~keil m = fixing wedge
~klemme f = fixing clip
~leiste f = fixing strip
~loch n = fixing hole
(~)Pfropfen m [Verfahren Freyssinet] = plug
~platte f = fixing plate
~punkt m = fixing point
~schraube f [dient zur festen, aber lösbaren Verbindung von Teilen] = fixing screw, retaining bolt
~steg m = fixing web
~stift m = fixing pin
~stück n = fixing piece
~verfahren n, Befestigungsweise f = fixing technique, ~ method
~wand f = fixing wall
~winkel m → Anschlußwinkel
befeuchten → anfeuchten
Befeuchtung f → Anfeuchtung [Betonnachbehandlung]

Befeucht(ungs)|anlage f, Anfeucht(ungs)anlage = humidifying installation, ~ plant
~automat m, automatischer Befeuchter m = auto(matic) humidifieir
~-Elektroabscheider m = wet type electric precipitator
~maschine f, Anfeuchter m, Anfeucht(ungs)maschine, Befeuchter = humidifying machine, humidifier
Befeuerung f = lighting
Befeuerungskabel n = lighting cable
beflechten [mit frischen Weidenruten oder Erlenzweigen und Pfählen. Wasserbau] = to line with wickerwork, ~ ~ ~ wattle-work, ~ ~ ~ basket-work
Befördern n von Beton, Beförderung f ~ ~ = concrete handling, ~ transport
Beförderung f, Transport m = handling, transport
~ mittels Dampf = transport by steam power, steam traction
~ von Menschenmassen, öffentlicher Massenverkehr m = mass transit, public ~, transport(ation)
Beförderungs|armierung f → Transportbewehrung
~mittel n, Transportmittel = means of transport
Befragungsmethode f auf der Straße f, Verkehrsbefragung f ~ ~ ~ = external (type of O & D) survey, roadside interview method
~ durch Postkarten, Postkartenzählung f, schriftliche Befragungsmethode, Postkartenmethode = motor driver postal card method, post-card survey
~ von Haus zu Haus, Verkehrsbefragung f ~ ~ ~ ~ = dwelling unit interview method, internal (type of O & D) survey, home-interview method
(Be)Füllen n, (Be)Füllung f = filling
begehbares Dach n = roof subjected to foot traffic
Begehung f, Baustellen~ [vor Angebotabgabe], = site inspection

Begehungsgang *m* [*bei einer Maschinenanlage*] = walkway

Begichtungs|anlage *f* = charging installation

~aufzug *m* = charging hoist

~kübel *m* = charging bucket

beginnen [*Bohrung beim pennsylvanischen Seilbohren*] = to spud in

beginnender Bruch *m* = incipient failure, ~ rupture

begleitbeheiztes Erzeugnis [*Erdölwerk*] = steam traced product

Begleit|brief *m*, Anschreiben *n* = covering letter

~bruch *m* (Geol.) = auxiliary fracture

Begleiterstraße *f* = frontage road

Begleit|heizung *f* [*Erdölwerk*] = steam tracing

~verwerfer *m*, Nebenverwerfung *f* (Geol.) = auxiliary fault, branch ~, minor ~, companion ~

begrabener Boden *m*, fossiler ~ = fossil soil

begrenzte Ausschreibung *f* → → Angebot

Begrenzungs|fläche *f* = limiting surface

~linie *f* = limiting line

~linie *f*, Begrenzungsstrich *m* [*Fahrbahnmarkierung*] = carriageway line

~schalter *m*, Anschlagschalter, Endschalter, Grenzschalter = limit switch, stop ~

~stein *m*, Grenzstein = border stone

~strich *m* → Begrenzungslinie

Begriffsbestimmung *f* = definition

begrünen, berasen [*durch Ansaat*] = to sow down to grass

~, berasen [*durch Andecken mit Rasensoden*] = to sod

Begrünen *n* → Berasen

Begrünung *f* **durch Ansaat**, Berasung ~ ~ = sowing down to grass

~ ~ Sodenandeckung, Berasung ~ ~ = sodding

begutachten = to give an expert opinion

Begutachtung *f*, Sachverständigenuntersuchung = expert opinion, study, expertise

Behaglichkeitsgefühl *n* = feeling of comfort, comfortable feeling

Behälter *m*, Gefäß *n* [*Aufnahmegefäß für Dämpfe, Gase und Flüssigkeiten*] = tank, vessel

~, Transportgefäß *n*, Transportbehälter = container

~, Tank *m* = tank

~ für 3 Körnungen = three-aggregate bin

~ drucklose Lagerung = atmospheric storage tank

~ ~ „Haus-zu-Haus-Verkehr" = traffic container for door-to-door transport

~ mit Schwimmdecke = floating roof tank

~-Anhänger *m*, Tankanhänger = tank trailer

~bau *m* = tank construction

~boden *m* = tank bottom, ~ floor

(~)Decke *f*, Tankdecke, (Behälter-)Dach *n*, Tankdach = roof, tank ~

~(ein)füllstutzen *m* = tank filler, ~ filling spout, ~ filling tube

~fahrzeug *n* = tank vehicle

~fahrzeug *n* → Silofahrzeug

~fassungsvermögen *n* = capacity of tank

~-Fegemaschine *f*, Behälter-Kehrmaschine = collecting sweeper

~fuß *m*, Fußring *m* = (tank) base joint

~fuß *m* **mit elastischem Band**, Fußring *m* ~ ~ ~ = elastomeric (tank) (base) joint

~fußnut *f*, Fußringnut = foundation-wall keyway [*to receive a tank wall*]

~geometrie *f* = tank geometry

~gips *m* = bulk gypsum

~-Kehrmaschine *f*, Behälter-Fegemaschine = collecting sweeper

~kraftwagen *m* → Silokraftwagen

~kuppel *f* = domed tank roof

~magazin *n*, Behälterlager *n* = container store

~rand *m* = tank edge

~rauminhalt *m* = volume of tank

~schale f = tank shell
~schalung f = formwork for tank construction
~schiff n = container ship [*The ship carries dry cargo in sealed containers of standard size and shape*]
~-Straßenfahrzeug n → Silokraftwagen m
~transport m, Behälterverkehr m = container traffic
~verkehr m von Haus zu Haus, Behältertransport m ~ ~ ~ ~ = door-to-door container traffic
~waage f, Gefäßwaage, Bunkerwaage, Wiegetrichter m, Dosierwaage, Zuteilwaage, Zumeßwaage, Dosierapparatwaage = bucket scale, hopper ~
~wagen m → Silofahrzeug n
~wand f = tank wall
~-Wechselsystem n = multi-bucket system
(~)Wickelmaschine f = (tank) winding machine
~zement m, Silozement, loser Zement, unverpackter Zement, ungesackter Zement = bulk cement
behandeltes Abwasser n, Abwasser nach der Behandlung = effluent
Behandlung f = treatment
Behandlungs|bau m, Behandlungstrakt m [*Krankenhaus*] = treatment block, theatre ~
~dauer f = treatment time
Beharrung f, Aufrechterhaltung f des Widerstandes [*Gummi-Bitumen-Mischung*] = tenacity
Beharrungsvermögen n, Trägheitsvermögen, Trägheit f, Trägheitskraft f, Beharrungskraft = inertia
Behausungsziffer f, Behausungszahl f, Behausungsdichte f = average number of inhabitants per building
Behebung f, Abhilfe(maßnahme) f = remedial measure
beheizbarer Draht m = heated wire
beheizte Schalung f = heated formwork

~ Straße f, geheizte ~ = heated highway, ~ road
beheizter (Fuß)Boden m = heated flooring
~ Tiefofen m = live pit
beheiztes Sieb n = heated screen
Beheizung f [*Industrieofen*] = firing
Behelfs|ausbesserung f [*Kraftfahrzeug auf der Straße*] = roadside repair
~bau m = temporary structure
~bau m, Behelfsgebäude n = temporary building
~brücke f, Notbrücke = emergency bridge
~garage f = carport
~gebäude n, Behelfsbau m = temporary building
~gerüst n, Behelfsrüstung f = auxiliary scaffold(ing)
behelfsmäßiger Flachbagger m → (Front-)Ladeschaufel f
Behelfs|methode f, Behelfsverfahren n = makeshift method
~rüstung f, Behelfsgerüst n = auxiliary scaffold(ing)
~schutz m = auxiliary civil defense (US); ~ ~ defence (Brit.)
~silo m, Leichtbausilo [*landwirtschaftlicher Silo*] = fence silo
beherrschen [*Spannungen usw. bei einem Bauwerk*] = to control
Beherrschung f der Spannungen = control of stresses
~ des Hangenden, ~ ~ Gebirges [*Tunnel- und Stollenbau*] = roof control, strata ~
beiderseitige Kante f [*Sieb*] = double selvedge
Beifahrer m = co-driver
Beigeschoß n, Halbgeschoß, Zwischengeschoß, Mezzanin n = mezzanine
Beil n, Hand~ = hatchet
Beilage f, Richtplatte f [*Gleiskettenlaufwerk*] = adjustment plate
Beilagscheibe f, Paßblech n = shim
Beileitung f, Überleitung [*von Zuschußwasser aus einem benachbarten Einzugsgebiet*] = diversion (of water by a conduit (or conduits) from one catchment to another)

Beimischung — Belagbemessung

Beimischung *f*, Zumischung, Zusatz(stoff) *m* = admix(ture), addition, additive
Bein *n*, Türstock~ (⚒) = leg
~, Kohlen~ = pillar (of coal)
~ [*Hubinsel*] = caisson, leg
~**schwarz** *n* = ivory black
Beitrag *m*, Referat *n* = paper
Beiwagen *m*, Seitenwagen = side-car
Beiwert *m* der Auflockerung, Auflockerungsbeiwert [*Aushubmaterial*] = bulking coefficient
~ der wirksamen Kohäsion *f* → Kohäsionsfaktor *m*
Beiz|anlage *f* [*zur chemischen Entrostung*] = pickling plant
~**bad** *n* = pickling bath
~**bottich** *m* = pickling vat
Beize *f* [*zur chemischen Entrostung*] = pickle
beizen → ab~
Beizereienabwasser *n* = pickling acid waste, spent pickling solution, waste pickle liquor, spent pickle liquor
Beiz|fluid *n* → Ab~
~**mittel** *n*, Ab~ = paint remover
Bekämpfung *f* = suppression
bekiest = covered with gravel
bekleiden, auskleiden, verkleiden = to line, to face
Bekleidung *f*, Auskleidung, Verkleidung [*Baugrube*] = sheeting, lining
~, Auskleidung, Verkleidung = lining, facing
~, Auskleidung, Verkleidung, nicht tragende Mauerung geringer Dicke [*Tunnel- und Stollenbau*] = facing, lining
Bekleidungs|arbeit(en) *f(pl)* → Auskleidungsarbeit(en)
~**beton** *m*, Auskleidungsbeton, Verkleidungsbeton = lining concrete
~**-Betonfertigteil** *n* → Auskleidungs-Betonfertigteil
~**blech** *n* → Auskleidungsblech
~**bohle** *f* → Auskleidungsbohle
~**mauer** *f* → Futtermauer
~**platte** *f* → Auskleidungsplatte
bekohlen = to coal

Bekohlung *f*, Bekohlen *n* = coaling
Bekohlungs|anlage *f*, Bekohlungsstation *f* = coaling installation, ~ plant
~**hafen** *m* = coaling port
~**kran** *m*, Kohlenkran = coaling crane
~**schiff** *n* = coaling vessel
~**station** *f*, Bekohlungsanlage *f* = coaling installation, ~ plant
Bekriechungsraum *m* → Kriechraum
(Be)Ladeanlage *f* = loading installation, ~ plant
(Be)Ladeband *n* → Bandlader *m*
~**ausleger** *m* = loading belt jib (Brit.); ~ ~ boom (US)
(Be)Lade|brücke *f* = loading bridge
~**bunker** *m* = loading hopper
~**gerät** *n* → Auflader *m*
~**gleis** *n*, Auflagegleis = loading siding, ~ track
~**gleis** *n*, Abfuhrgleis [*Erdbau*] = loading track, ~ siding
~**maschine** *f* → Auflader *m*
(be)laden = to load
(Be)Laden *n* = loading
(Be)Lade|silo *m* = loading silo
~**stelle** *f*, (Be)Ladeplatz *m* = point of loading, loading point
~**vorgang** *m* = loading operation
~**vorrichtung** *f* = loading device
Belag *m*, Decke *f*, Straßen~ = (road) surfacing, (~) pavement, highway ~; street ~
~, Besatz *m* [*Aufnahmefähigkeit einer Freiluft-Trocknungsanlage einer Ziegelei*] = capacity
~ auf einem Betonüberbau, Decke *f* ~ ~ ~ [*Brücke*] = concrete deck pavement, ~ ~ surfacing
~ ~ ~ Stahlüberbau, Decke *f* ~ ~ ~ [*Brücke*] = steel deck pavement, ~ ~ surfacing
~**arbeiten** *fpl* → Deckenarbeiten
~**art** *f*, Deckenart *f* = type of pavement, ~ ~ surfacing
~**bau** *m*, Deckenbau, Straßen~ = (road) surfacing construction, (~) pavement ~
~**bemessung** *f* → Deckenbemessung

Belagbestandteil — Belastungs-Setzungsdiagramm

~bestandteil m → Deckenbestandteil
~blech n = revetment plate
~bohle f, (Holz)Bohle = plank
~dicke f, Deckendicke = surfacing thickness, pavement ~
~einbau m, Deckeneinbau [*Straßenbau*] = pavement laying
~eisen n → Belagstahl
~ermüdungsbruch m → Deckenermüdungsbruch
~fuge f → Deckenfuge
~gut n → Belagmaterial n
~güte f, Deckengüte = pavement quality, surfacing ~
~konstruktion f, Deckenkonstruktion [*Straßenbau*] = paving system (US); pavement ~
~material n, Deckenmaterial, Straßenbelageinbaumasse f, Straßendeckeneinbaumasse, Belaggut n, Deckengut = surfacing material, pavement ~
~oberfläche f → Deckenoberfläche
(~)Platte f [*früher: Fliese f*] = tile
~querschnitt m → Deckenquerschnitt
~stahl m, Zoreseisen n, Trogplatte f, Belageisen = trough plate
~stoff m = revetment material
~verhalten n → Deckenverhalten
Belanger'sche Grenzgeschwindigkeit f = Belanger's critical velocity [*Is that condition in open channels for which the velocity head equals one-half the mean depth*]
Belastbarkeit f = loadability
belasten = to load; to apply a load (to); to apply a load train (to)
Belasten n, (Last)Aufbringung f, Lastangriff m, (Last)Aufbringen n, Belastung, Lasteintragung, Lasteintragen = loading, load application
belastete Fläche f = loaded area
~ Mauer f, lasttragende ~ = load-bearing wall
(~) Tragrolle f → → Bandförderer
~ Wand f, lasttragende ~ = load-bearing partition
Belästigung f = nuisance [*A condition which is offensive to the sense of sight or smell; e. g. a polluted stream*]

Belastung f → Belasten n
~ [*Kraftwerk*] = load
~ am Tage [*Kraftwerk*] = daytime load
~ in der Nacht [*Kraftwerk*] = nighttime load
Belastungs|alter n = age of loading
~änderung f = change of loading
~annahme f, Lastannahme, Entwurflast f = design load
~anzeiger m, Belastungsmesser [*Tiefbohrtechnik*] = weight indicator, load ~
~bedingung f = loading condition
~beiwert m = loading coefficient
~diagramm n = load diagram
~eigenschaft f = loading property
~fahrzeug n [*zur Prüfung der Tragfähigkeit von Brücken*] = load vehicle
~faktor m, Lastfaktor = load factor
~fall m, Lastfall = condition of loading
~filter m, n, umgekehrter Filter m, umgekehrtes Filter n [*Bodenmechanik*] = loaded filter, inverted ~, loaded inverted ~
~filter m, n = heavy filter
~fläche f, Lastfläche = load surface, ~ area
~geschwindigkeit f = rate of loading, loading rate
~glied n = load term
~grenze f = loading limit
~größe f = magnitude of loading
~länge f = loaded length
~messer m, Belastungsanzeiger [*Tiefbohrtechnik*] = load indicator, weight ~
~meßdose f = load cell
~metamorphose f (Geol.) = load metamorphism
~platte f, Last(versuchs)platte = load plate
~probe f, Belastungsprüfung f, Belastungsversuch m = load(ing) test
~ring m = loading ring
~-Setzungsdiagramm n, Last-Setzungskurve f = load-settlement diagram

Belastungsspannung — Belüftungsleitung

~spannung *f* = loading stress
~spitze *f*, Bedarfsspitze = load peak, peak load
~spitze *f*, Höchstbelastung *f* = peak of loading, maximum ~
~stempel *m* = load piston, loading ~, ~ column
~stufe *f*, Laststufe = load increment
~stuhl *m* = loading platform [*pile testing*]
~versuch *m* → Belastungsprobe *f*
~vorgang *m* = process of loading
~vorrichtung *f* [*Probebelastung von Pfählen*] = loading system
~zustand *m* = loading state
belebter Boden *m* = organic soil
~ **Schlamm** *m* = activated sludge [*Sludge settled out of sewage previously agitated in the presence of abundant atmospheric oxygen*]
Belebungs|anlage *f* = activated-sludge plant, ~ installation
~**verfahren** *n* = activated-sludge method, ~ process [*Sewage treatment in which sewage standing in or flowing through a tank is brought into intimate contact with air and with biologically active sludge, previously produced by the same process. The effluent is subsequently clarified by sedimentation*]
Belegschaft *f*, Personal *n* = staff, personnel
belegt [*Bitumenpappe*] = self finished
Belegungs|dauer *f* [*Flughafen*] = gate occupancy time
~**dichte** *f* = accomodation density, amount of housing, occupancy rate [*The number of persons per habitable room in a dwelling or group of dwellings*]
beleuchten = to light
Beleuchtung *f* = lighting
Beleuchtungs|art *f* = type of lighting
~**ingenieur** *m* = lighting engineer
~**körper** *m* = light fitting
~**mast** *m*, (Straßen)Leuchtenmast, (Licht)Mast *m* = (lighting) column (Brit.); mast (US)
~**stärke E** *f* = illumination (E)

~**-Versuchsstraße** *f* = lighting test road, ~ ~ highway
~**zweck** *m* = purpose of lighting
belgische Bauweise *f* → Unterfangungsbauweise mit First- und Sohlstollen
belgischer Zinkdestillations(-Doppel)-Ofen *m* = Belgian zinc distilling furnace
Belichtung *f* [*von Wohn- und Arbeitsräumen*] = natural lighting, daylight illumination
Belüften *n* → Belüftung *f*
Belüfter *m* [*Abwasserwesen*] = aerator
belüfteter Beton *m*, AEA-Beton, Luftporenbeton, Bläschenbeton = air-entrained concrete, air-entraining ~
~ **Mörtel** *m* → AEA-Mörtel
(~) **Tauchkörper** *m* = contact aerator [*A device consisting of a crate holding broken stone, coke, brushwood, or other media, which is placed in a single or two-story sedimentation tank and through which sewage is made to flow upward and return on the outside and become activated by the admission of compressed air below*]
~ **Überfallstrahl** *m* = aerated nappe (US); ~ sheet of water (Brit.)
Belüftung *f*, Durchlüftung, Belüften *n*, Luftzufuhr *f* = aeration [*The process or method of bringing about intimate contact between air and a liquid by allowing finely divided air to pass through the liquid or the finely divided liquid to pass through air*]
Belüftungs|becken *n*, Belüftungsraum *m*, Durchlüftungsbecken, Durchlüftungsraum *m* = aeration tank, ~ basin, aerator ~
~**dauer** *f*, Durchlüftungsdauer = aeration time, ~ period
~**element** *n* [*Korrosion unter Sauerstoffverbrauch*] = differential aeration cell [*oxygen-consumption type of corrosion*]
~**leitung** *f*, Durchlüftungsleitung [*Abwasserwesen*] = aeration line, aerator ~

Belüftungsleitung — Bemessungswärmeverlust

~leitung *f*, Luftventil *n* = air intake (Brit.); aeration conduit [*auto-(matic) siphon*]

~messer *m*, Luftmengenmesser, Luftmeßgerät *n*, Luftporenmesser, Luftporenprüfer, Beton-Belüftungsmesser, Beton-Luftmengenmesser [*Beton*] = air (entrainment) meter, entrained air indicator

~mittel *n*, Betonbelüfter *m*, Lufteinschlußmittel, LP *m*, luftporenbildender Zusatzstoff *m*, Luftporenerzeuger *m*, luftporenbildendes Zusatzmittel, Luftporenbildner *m* = air-entraining agent, air entrainment agent, air-entraining admix(ture) for concrete, air-entraining compound, AEA

~stollen *m* = ventilation gallery

~verfahren *n* [*zur Belüftung des Abwassers in Belebungsbecken*] = aeration method

bemannen = to man

bemessen, dimensionieren = to design

bemessene Lebensdauer *f* = design life

Bemessung *f*, Bemessen *n*, Dimensionierung, Dimensionieren [*Der Vorgang der Bestimmung der Abmessungen eines Trägers oder einer Platte auf der Grundlage der Schnittkräfte (Moment, Querkraft und Längskraft)*] = (structural) design

~ die die verminderte Tragfähigkeit des anstehenden Bodens während der Tauperiode berücksichtigt = reduction in subgrade strength design, reduced-strength ~, frost condition reduced subgrade strength ~, design for loss of strength during the frost-melt period

~ ~ nur eine begrenzte Frosteindringung in den anstehenden Boden zuläßt = limited subgrade frost penetration design

~ gegen Auftrieb = (structural) design for submergence

~ im Stahlbetonbau, Stahlbetonbemessung = design in reinforced concrete, reinforced-concrete design

~ von Schwarzdecken = flexible pavement design, ~ surfacing ~

~ ~ Stahlbauten, Stahlbemessung = steel design

~ ~ Straßendecken = pavement design, surfacing ~

~ ~ Verbundbauwerken Stahl/Beton = composite design (in steel and concrete)

Bemessungs|annahme *f* = (structural) design hypothesis, (~) ~ assumption

~aufgabe *f* = design problem

~beispiel *n* = design example

~biegemoment *n* = design bending moment

~blatt *n* = design data sheet

~daten *f* = design data

~diagramm *n*, Bemessungstafel *f* = design chart

~dicke *f* = design thickness

~erdbebenkraft *f* = design seismic force

~faktor *m* = design factor

~formel *f* = (structural) design formula

~frostindex *m* = design (air) freezing index

~gleichung *f* = (structural) design equation

~grundlagen *fpl* = basic design data

~grundsatz *m* = design principle

~kurve *f* = (structural) design curve

~last *f* = design load

~liegezeit *f* [*Straßendecke; Flugplatzdecke*] = design life

~moment *n* = design moment

~radlast *f* = design wheel load

~richtlinien *fpl* = design specification(s)

~tabelle *f* = design table

~tafel *f*, Bemessungsdiagramm *n* = design chart

~theorie *f* = design theory

~verfahren *n* = method of design, design method

~verfahren *n* für Schwarzdecken = flexible pavement design method, ~ surfacing ~ ~

~wärmeverlust *m* = design heat loss

bemörteln, ausmörteln = to fill with mortar

Benennung *f* [*in einer Ersatzteilliste*] = description

Benetzbarkeit *f*, Umhüllbarkeit [*eines Gesteinskorns mit Bindemittel*] = wettability

benetzen = to wet

~, umhüllen [*mit Bindemittel*] = to coat, to wet

Benetzen *n* = wetting

~, Umhüllen, Bindemittelumhüllung *f* = coating, wetting; wet mixing (US)

benetzter Querschnitt *m* → Abflußquerschnitt

~ Umfang *m* **(des Querschnittes)** [*Zeichen: U. Einheit: m. Der von der Flüssigkeit eingenommene Teil des Umfangs eines offenen oder geschlossenen Gerinnes*] = wetted perimeter

Benetzungs|eigenschaft *f* [*Bindemittel*] = wetting property, coating ~

~fähigkeit *f*, Benetzungsvermögen *n* = wetting power, coating ~

~mittel *n* = coating agent, wetting ~

~wärme *f* = wetting heat, coating ~

~wasser *n*, Anhaftewasser, angelagertes Wasser, Haftwasser, Adsorptionswasser, adsorptiv gebundenes Wasser [*hygroskopisches Wasser = Kapillarwasser*] = solidified water, adsorbed ~, adsorption ~, thin-film ~

~winkel α *m* [*Wasseranstieg in Kapillarrohren und Spalten*] = contact angle α

Benne *f* [*in der Schweiz*] → Schiebkarre(n) *f*, (*m*)

Bentonit *m* [*ist ein an feinsten Teilchen < 0,001 mm reicher Kolloid-Ton*] = bentonite (clay), bentonitic ~

~aufschwemmung *f*, Bentonitsuspension *f* = bentonite suspension, bentonitic ~

~(bohr)spülung *f*, Bentonit(bohr)schlamm *m*, Bentonitbohrflüssigkeit *f*, Bentonitspülflüssigkeit, Bentonitspülschlamm = bentonite drilling mud, ~ ~ liquid, ~ rotary ~, bentonitic ~ ~, bentonite-laden ~

~flockung *f* = bentonite flocculation, betonitic ~

~suspension *f*, Bentonitaufschwemmung *f* = bentonite suspension, bentonitic ~

~vermahlung *f* = intergrinding of bentonite

Benutzer *m* = user

Benzin *n*, Motoren~, Otto-Kraftstoff *m* = gas(oline) (or gasolene) (naphtha), naphtha (US); petrol, motor-spirit (Brit.); Otto type fuel

~(ab)scheider *m*, Benzinfang *m*, Benzintraps *m* [*Grundstücksentwässerung*] = petrol trap, ~ intercepting chamber (Brit.); gas(oline) interceptor, gasolene interceptor (US)

~-Abwasser *n* = petrol sewage (Brit.); gas(olene) ~, gasoline ~ (US)

~behälter *m*, Benzintank *m* = petrol tank (Brit.); gas(oline) ~, gasolene ~ (US)

bezinbetrieben = gas(oline)-driven, gasolene-driven (US); petrol-driven (Brit.)

„Benzin-Brikett" *n* = gas(oline) brick, gasolene ~ (US)

benzinelektrischer Antrieb *m* = petrol-electric drive (Brit.); gas(oline)-electric ~, gasolene-electric ~ (US)

Benzin|fallzuleitung *f* = gravity petrol feed (Brit.); ~ gas(oline) ~, ~ gasolene ~ (US)

~filter *m, n* = petrol filter (Brit.); gas(olene) ~, gasoline ~ (US)

~hahn *m* = petrol cock (Brit.); gas(olene) ~, gasoline ~ (US)

~kanister *m* = jerrycan, petrol can (Brit.); gas(oline) (storage) can, gasolene (~) ~ (US)

~lager *n* = petrol dump (Brit.); gas(olene) ~, gasoline ~ (US)

~motor *m*, Vergasermotor, Otto-(Vergaser)motor = gas(oline) engine, gasolene ~ (US); petrol ~ (Brit.); Otto cycle engine

Benzin(motor)|-(Bau)Aufzug *m* = petrol hoist (Brit.); gas(olene) ~, gasoline ~ (US)

~-**Betonmischer** m = petrol concrete mixer (Brit.); gas(olene) ~ ~, gasoline ~ ~ (US)

~-**(Gesteins)Bohrhammer** m, Benzin(motor)-Felsbohrhammer = petrol (hand) hammer (rock) drill, ~ (rock) drill (hammer) [US = gas(olene) or gasoline]

~-**Hochfrequenzinnenrüttler** m, Benzin(motor)-Hochfrequenzinnenvibrator m = petrol-driven high-frequency immersion (or needle, or poker) vibrator, petrol-driven high-frequency pervibrator [US = gas(olene)-driven or gasoline-driven]

~-**Hubkarre(n)** f, (m), Benzin(motor)-Hubwagen m = gas(olene)-powered lift truck, gasoline-powered ~ ~ (US); petrol-powered ~ ~ (Brit.)

~-**Innenrüttler** m, Benzin(motor)-Tauchrüttler, Benzin(motor)-Innenvibrator m, Benzin(motor)-Tauchvibrator, Benzin(motor)-Tiefenrüttler, Benzin(motor)-Tiefenvibrator, Benzin(motor)-Nadelrüttler, Benzin(motor)-Nadelvibrator = gasolene internal vibrator, gas(oline) ~ ~ (US); petrol ~, petrol immersion ~, petrol poker ~, petrol needle ~, petrol pervibrator, petrol spud vibrator for concrete, petrol concrete vibrator for mass work (Brit.)

~**karre(n)** f, (m), Benzinflurfördermittel n, Benzinförderkarre(n), Benzintransportkarre(n) = petrol-driven truck (Brit.); gas(oline-)driven ~, gasolene-driven ~ (US)

~-**Rüttler** m, Benzin(motor)-Vibrator m = gasolene vibrator, gas(oline) ~ (US); petrol ~ (Brit.)

~**winde** f = gas(oline)-powered winch, gasolene-powered ~ (US); petrol-powered ~ (Brit.)

Benzin|pumpe f = petrol pump (Brit.); gas(oline) ~, gasolene ~ (US)

(~)**Rüttelmotor** m = vibrating petrol engine (Brit.); ~ gas(olene) ~, ~ gasoline ~ (US)

~**standanzeiger** m, Benzin(tank)uhr f = petrol level indicator (Brit.); gas(olene) ~ ~, gasoline ~ ~ (US)

~**startanlage** f, Benzinanlasser m = gasoline starting system, gas(olene) ~ ~ (US); petrol ~ ~ (Brit.)

~**startmotor** m, Benzinanlaßmotor = gas(oline) starting engine, gasolene ~ ~ (US); petrol ~ ~ (Brit.)

~-**Stromerzeuger** m = petrol generating set (Brit.); gas(olene) ~ ~, gasolene ~ ~ (US)

~**tank** m → Benzinbehälter m

~**tankstelle** f = petrol (filling) station (Brit.); gas(olene) ~, gasoline ~ (US)

~**verschmutzungen** fpl [auf einer Fahrbahn] = petrol drippings (Brit.); gas(olene) ~, gasoline ~ (US)

~**walze** f = petrol-driven roller (Brit.); gas(olene)-driven ~, gasoline-driven ~ (US)

~**zusatz** m = gas(olene) additive, gasolene ~ (US); petrol ~ (Brit.)

Benzol n, C_6H_6 [Klopffester Vergaserbrennstoff; durch Destillation des Steinkohlenteeres oder durch Auswaschen von Kokereigas gewonnen] = benzene, benzole

~**kohlenwasserstoffe** mpl → Aromaten mpl

~-**Unlösliches** n = insoluble in benzene, benzene insolubles, extrinsic insolubles

Beobachtungs|brunnen m, Beobachtungsrohr n, Standrohr [zur Beobachtung des Grundwasserspiegels in der Umgebung von Brunnen in den Boden eingetriebenes gelochtes Rohr] = piezometer tube, piezometric ~

~**bühne** f = observation platform

~**kammer** f = inspection chamber

~**rohr** n = Beobachtungsbrunnen m

~**rohr** n = inspection pipe

~**rohrspiegelhöhe** f, Standrohrspiegelhöhe = piezometric level

~**schacht** m, Kontrollschacht [Talsperre] = inspection shaft

~spanne f → Beobachtungszeitraum
~stollen m, Kontrollgang m [*Talsperre*] = inspection gallery, ~ tunnel; footway (Brit.)
~strecke f → Versuchsstrecke
~turm m = observation tower
~tür f = observation door
~zeitraum m, Beobachtungsspanne f = period of record
Bepflanzung f **mit Schilf(rohr)** = reed planting
beplaggen → besoden
berasen, begrünen [*durch Andecken mit Rasensoden*] = to sod
~, **begrünen** [*durch Ansaat*] = to sow down to grass
Berasen n, **Berasung** f, **Begrünen, Begrünung** f = turfing
Berasung f **durch Ansaat, Begrünung** ~ ~ = sowing down to grass
~ ~ **Sodenandeckung, Begrünung** ~ ~ = sodding
beratender Geologe m = consulting geologist
~ **Ingenieur** m, **Beratungsingenieur** [*Österreich: Ingenieurkonsulent m*] = consulting engineer
Beratung f = consultation, advice
Beratungsbüro n = consultation office
beräumen, bereißen (⚒) = to brush, to rip
Beräumen n → Abräumen
Beräumstelle f, **Bereißstelle** (⚒) = caunch, ripping lip
berechnen = to compute
berechnetes Gefälle n, **berechnete Fallhöhe** f = theoretical height of fall
Berechnung f = computation
~ **der (An)Näherungswerte** → (An-)Näherungsrechnung
~ ~ **Fließgeschwindigkeit** = flow computation
~ **von Rohrsträngen** [*Tiefbohrtechnik*] = casing string design
~ ~ **selektiven Netzwerken nach der synthetischen Theorie** = network synthesis
Berechnungs|beispiel n, Rechenbeispiel = example of calculation

~**gang** m, Rechengang [*Baustatik*] = design operation
~**geschwindigkeit** f → Ausbaugeschwindigkeit
~**grundlage** f = design principle
~**-Hochwasser** n, maßgebendes Hochwasser = design flood, project ~ [*Design-flood hydrographs are needed in the design of reservoirs, spillways, flood channels, and other water control structures*]
~**-Niederschlag** m, maßgebender Niederschlag = design precipitation, project ~
~**regen** m, maßgebender Regen [*Regen bestimmter Häufigkeit, Stärke und Dauer, welcher der Berechnung einer Entwässerungsanlage zugrunde gelegt werden soll (DIN 4045)*] = design storm, ~ rain, project ~
~**tafel** f, Rechentafel = chart
Beregnung f, **Feld**~ = (agricultural) sprinkler irrigation
~ **mit Abwasser**, Abwasserverregnung = sewage sprinkling
Beregnungs|anlage f, **Feld**~ = (agricultural) sprinkler (irrigation) system
~**gerät** n, **Regner** m = sprinkler
Bereich m, **Aktionsradius** m, **Reichweite** f = range, reach
~ **der möglichen Hebung** [*Bodenmechanik*] = zone of potential heave
Bereifung f = formation of hoar frost
~ = tire(s) (US); tyre(s) (Brit.)
bereißen, beräumen (⚒) = to rip, to brush
Bereißen n → Abräumen
Bereißstelle f, **Beräumstelle** (⚒) = ripping lip, caunch
Berg m, **taubes Gestein** n, erzfreies Gestein, **Gangart** f, **Ganggestein** = gangue
bergab = downgrade, downhill
Bergab-Bandförderanlage f = downhill conveyor, ~ conveyer
Bergakademie f = mining academy, ~ college, school of mines
Bergarbeiter m, Grubenarbeiter, Bergmann m = miner

Bergarbeitersiedlung — Bergrutsch

~siedlung *f* = miners' housing estate
~wohnungen *fpl*, Grubenarbeiterwohnungen = miners' dwellings
bergauf = upgrade, uphill
Bergauf-Bandförderanlage *f* = uphill conveyor, ~ conveyer
Bergbau *m* = mining
~arbeiten *fpl* = mining work
~ausrüstungen *fpl* = mining equipment
~betrieb *m*, Grube *f*, Bergwerk *n* = mine
bergbaufremd = not due to mining
Bergbau|gebiet *n* = mining region
~geologe *m* = mining geologist
~geologie *f* = mining geology
~getriebe *n* = gearing for miningduties
~industrie *f* = mining industry
~ingenieur *m* = mining engineer
~maschinen *fpl* und -geräte *npl*, Bergwerkmaschinen und -geräte = mining machinery and equipment
~schacht *m*, Grubenschacht, Bergwerkschacht = mining shaft
~senkung *f* → Bergsenkung
~sprengstoff *m*, gewerblicher Sprengstoff für den Bergbau = mining (blasting) explosive
~tätigkeit *f* = mining activity
~technik *f* = mining practice
Bergbelag *m*, Bergdecke *f* = mountain road surfacing (Brit.); ~ ~ pavement (US)
Bergdorf *n* = mountain village
Berge *f* (⚒) = waste
~fortbringen *n* [*bei der Flotation*] = gangue rejection
~klein *n* (⚒) = dirt
~kübel *m* [*Schachtbau*] = muck (sinking) bucket, ~ (~) skip; ~ kibble, ~ bowk, ~ hoppit (Brit.)
~mauer *f*, Trockenmauer, Bergedamm *m* [*eine aus Gesteinsstücken ohne Mörtel aufgesetzter Mauer*] (⚒) = pack (wall)
~mittel *n*, Zwischenmittel (⚒) = dirt (band), parting [*Stone in a coal mine, gangue in a metal mine, usually called muck in American and British metal mines*]

~rippe *f* (⚒) = roadside stone packing
~rolle *f*, Sturzrolle = ore pass
~wirtschaft *f* [*Erzaufbereitung*] = tailings disposal
Bergfahrer *m* [*Flußschiff, das stromaufwärts fährt*] = ascending barge
Bergfeste *f*, Feste (⚒) = barrier pillar
bergfeuchtes Gestein *n*, bergfeuchter Fels *m* = fresh rock
Bergfeuchtigkeit *f* = connate water, quarry sap
Bergfried *m* → Donjon *m*
Berggegend *f* = mountainous region
Berggelb *n* = iron ochre
Berggesetz *n* = mining law
Berggold *n* = free gold
Berggrün *n* [*Malachit*] = mineral green, green basic copper carbonate
~, Chrysokoll *n* = chrysocolla
Berggut *n*, Fossil *n* = fossil
Berghang *m*, Berglehne *f*, (Ab)Hang = flank of a hill, hillside
Bergholz *n* = timber fender
bergig, gebirgig = mountainous
Bergingenieur *m* = mining engineer
Bergkiesel *m*, Adinol *m* = adinole
Bergkristall *m* = mountain crystal, (quartz) rock ~
Bergland *n* [*Entwurfklasse für deutsche Autobahnen*] = mountainous region
Bergmann *m*, Bergarbeiter *m*, Grubenarbeiter *m* = miner
bergmännisch gewinnen = to mine
~ **vortreiben** [*Tunnel oder Stollen*] = to construct without ground disturbance, ~ drive by underground means, to bore
bergmännisches Verfahren *n* [*z. B. Schildvortrieb beim Tunnel- und Stollenbau*] = mining technique
Bergmannspfahl *m* [*Tunnel- und Stollenbau*] = forepoling board
Bergmehl *n*, Kieselgur *f* fossil meal, kieselguhr
Bergmilch *f* = rock milk
Bergordnung *f* = mining regulations
Bergrecht *n* = Mining Code
Bergrutsch *m*, Bergsturz *m*, Absturz, Rutschung *f*, Abrutsch = debris-

Bergschaden — Beruhigung

slide, rock avalanche, fall of rock, rock slide, rock-fall, rock slip
Bergschaden m, Senkungsschaden = mining damage, subsidence ∼
Bergschlag m, Steinschlag, Felsabbruch m, Felssturz m = rock fall, popping rock
Bergschraffen fpl [Karte] = hill hachures
Bergsee m = mountain lake, (∼) tarn
bergseitig, oberwasserseitig = upstream
Bergsenkung f, Bergbausenkung, Setzung in Bergwerkgebieten = ground subsidence due to coal extraction, mining subsidence
Bergsenkungs|gebiet n, bergbauliches Senkungsgebiet = mining subsidence region, ∼ area
∼schaden m = effect on buildings of ground subsidence due to coal extraction
Bergsporn m, Ausläufer m = spur
Bergstation f [Seilbahn] = mountain station
Bergsteigfähigkeit f [Fahrzeug] = climbing ability
Bergstraße f, Gebirgsstraße = mountain road, ∼ highway
Bergsturz m → Bergrutsch
Bergteer m = mineral tar, pissasphalt(um), maltha
Bergtobel m (Geol.) = mountain gulch
Bergtunnel m, Felstunnel, Gebirgstunnel = rock tunnel
Bergung f = salvage
Bergungs|arbeiten fpl = salvage work
∼geräte npl = salvage appliances
∼Kolonne f → Bergungstrupp
∼kran m → Unfallkran
∼pumpe f = salvage pump
∼schiff n = salvage ship
∼trupp m, Bergungskolonne f, Bergungsmannschaft f = salvage crew, ∼ gang, ∼ team, ∼ party
∼winde f = salvage winch
Bergwerk n, Grube f, Bergbaubetrieb m = mine
∼maschinen fpl und -geräte npl, Bergbaumaschinen und -geräte = mining machinery and equipment
∼pumpe f, Grubenpumpe = mine pump
∼schacht m, Bergbauschacht, Grubenschacht = mining shaft
Bergzinn n = vein-tin
Berichter m = reporter
Berichtigung f, Berichtigen n, Justierung, Justieren, Feineinstellung, Feineinstellen [*Berichtigen von Meßgeräten vor ihrer Anwendung, um Meßfehler durch falsche Einstellung der Geräte zu vermeiden*] = (fine) adjusting, precise ∼, ∼ setting
Berichtswesen n = recordkeeping
Beries(e)lung f = sprinkling
(Be)Ries(e)lung f, Rieselverfahren = flooding (method) [*irrigation*]
Beries(e)lungsanlage f, Wasser∼ [*Walze*] = sprinkler system
beripptes Rohr n → Rippenrohr
Berme f, Bankett n [*Horizontaler Absatz in Damm- oder Einschnittböschungen*] = banquette, berm(e), bench
Bermudas-Asphalt m = Bermudez asphalt
Bernoulli'sche Gleichung f, Bernoulli-Gleichung, Bernoulli-Satz m = Bernoulli('s) theorem [*A proposition advanced by Daniel Bernoulli that the energy head at any section in a flowing stream is equal to the energy head at any other downstream section plus the intervening losses*]
Bernstein m [*fossiles Harz*] = amber
Berstdruck m = bursting pressure
∼versuch m, Berstdruckprobe f, Berstdruckprüfung f = bursting pressure test
(Berufs)Ausbildung f = training
Berufs|krankheit f = occupational disease
∼pendler m = commuter
∼unfall m = occupational accident
beruhigter Stahl m = stabilized steel
Beruhigung f, Energievernichtung, Energieverzehrung [*die Bewegungsenergie des stürzenden oder schießen-*

Beruhigungsbecken — Beschickungsmaterial

den Wassers wird durch Walzenbildung (Tosen) verzehrt] = stilling, absorption, energy dissipation
Beruhigungsbecken *n*, Tosbecken = stilling basin, absorption ~ (Brit.); stilling pool (US)
~, wellenbrechendes Becken [*Hafen*] = stilling basin
Beruhigungsrechen *m* = stilling screen
Berührungs|druck *m* = contact pressure
~**fläche** *f* = contact area, ~ surface
~**korrosion** *f* = crevice corrosion at a contact with non-metallic material
~**spannung** *f* = contact stress
~**überhitzer** *m* = contact superheater
~**winkel** *m* = contact angle
Besamung *f* → Ansäen *n*
besanden = to (cover with) sand
Besatz *m*, Belag *m* [*Aufnahmefähigkeit einer Freiluft-Trocknungsanlage einer Ziegelei*] = capacity
~, Verdämmung *f* [*Schußloch*] = stemming, tamping
~**maschine** *f*, Verdämmaschine, Verdämm-Maschine [*gewerbliche Sprengtechnik*] = stemming machine, tamping ~
~**stoff** *m*, Verdämmstoff = stemming material, tamping ~
Beschädigung *f* = distress
Beschaffenheit *f*, Zustand *m* = condition
~ [*im Text einer Norm*] = absence of defects
Beschaffung *f* = provision, procurement
Bescheinigung *f* für Abschlagzahlung = (building) certificate [*A statement signed by the architect (or engineer) that the builder is entitled to an instalment on work done. Certificates are usually completed and paid monthly during the progress of building*]
Beschichter *m*, Auftragmaschine *f* [*zur Kunststoffbeschichtung*] = coating machine, coater
Beschichtung *f* **mit Kunststoff** = coating with plastic

beschicken [*Brennofen*] = to charge, to feed
~ → aufgeben
Beschicken *n* → Aufgabe *f*
Beschicker *m* → Aufgeber *m*
~**kasten** *m* → → Betonmischer
~**kübel** *m* → Betonmischer
~**vorrichtung** *f* → Aufgeber *m*
Beschickung *f* → Aufgabe *f*
~ [*Brennofen*] = feeding, charging
Beschickungs|apparat *m* → Aufgeber *m*
~**aufzug** *m*, Aufgabeaufzug = feeder skip hoist
~**aufzug** *m* **für Silos**, Aufgabeaufzug ~ ~ = feeder skip hoist for silos
~**bahn** *f* → → Betonmischer *m*
~**band** *n* → Bandaufgeber *m*
~**becherwerk** *n* → Aufgabebecherwerk
~**behälter** *m*, Dosierbehälter = dosing tank, ~ chamber [*A tank into which raw or partly treated sewage is introduced and held until the desired quantity has been accumulated, after which it is discharged at such a rate as may be necessary for distribution essential to the subsequent treatment*]
~**bühne** *f*, Ladeflur *m* [*Ofenhaus*] = charging platform
~**bunker** *m*, Arbeitsbunker, Aufgabebunker = live storage bin, surge ~
~**dosierung** *f*, Aufgabedosierung = feed measuring, ~ proportioning, ~ batching
~**elevator** *m* → Aufgabebecherwerk *n*
~**ende** *n*, Aufgabeende = feed(ing) end
~**feines** *n*, Aufgabefeines [*Siebtechnik*] = fine feed
~**(förder)band** *n* → Bandaufgeber *m*
~**grobes** *n*, Aufgabegrobes [*Siebtechnik*] = coarse feed
~**gut** *n* → Aufgabegut
~**kasten** *m* → → Betonmischer *m*
~**kran** *m*, Chargierkran, Muldenbeschickungskran = charging crane
~**kübel** *m* → → Betonmischer *m*
~**kübel** *m*, Aufgabekübel = feed(ing) bucket
~**material** *n* → Aufgabegut *n*

~menge *f*, Aufgabemenge = feed(ing) quantity

~menge *f*, Abwasserstoß *m* = sewage dose, ~ batch

~öffnung *f*, Aufgabeöffnung = feed opening

~rinne *f*, Aufgaberinne, Speiserinne = feeder channel

~rohr *n* → Aufgaberohr

~rost *m*, Aufgaberost, Speiserost = charging grate

~schnecke *f*, Schneckenaufgeber *m*, Aufgabeschnecke, Speiseschnecke, Zubringerschnecke, Schneckenzubringer, Schneckenbeschicker, Schneckenspeiser = feed(ing) screw, screw feeder

~schurre *f*, Aufgabeschurre = feed(ing) shoot (Brit.); ~ chute (US); ~ spout

~silo *m*, Aufgabesilo = feed(ing) silo

~teil *m, n*, Aufgabeteil = feed(ing) unit

~trichter *m*, Aufgabetrichter = feed(ing) hopper

~vorrichtung *f* → Aufgeber *m*

~vorrichtung *f* für Zuschläge, ~ ~ Zuschlagstoffe, Aufgabevorrichtung ~ ~ = aggregate feeder

~wagen *m*, Konverter-~ = (converter) charging car

~walze *f*, Aufgabewalze, Speisewalze, Zubringerwalze = feed(ing) roll

~werk *n* [*Chargen-Betonmischer*] = skip-hoist, batch elevator

beschildern = to sign-post

Beschilderung *f* = sign-posting, signing

Beschlag-Einlaßmaschine *f* = machine for letting in mounts, hardware mounting machine

beschleunigen = to accelerate

Beschleuniger *m*, Aktivator *m*, Anreger *m* = activator

~ **für Ablaufberge** = accelerating device for humps

beschleunigte Heißwasserheizung *f* = accelerated hot-water heating

Beschleunigung *f*, Geschwindigkeitszunahme *f* [*in der Zeiteinheit*] = acceleration

Beschleunigungs|bogen *m* = acceleration curve

~**(förder)band** *n* = accelerating conveyor, ~ conveyer

~**kraft** *f* = force of acceleration

~**spur** *f*, Beschleunigungsstreifen *m* [*Straße*] = acceleration lane

~**vergütung** *f* = bonus for completing work before the target date

beschränkt vorgespannt, teilweise ~ partially prestressed

beschränkte Ausschreibung *f* = limited submission, ~ public tender action, ~ public tendering

~ **Platzverhältnisse** *f* = confined quarters

Beschriftung *f* = lettering

Beschürfung *f* → Aufgraben *n* [*Baugrundaufschluß*]

Beschweren *n* [*Steine auf ein Sinkstück werfen, um es zu beschweren*] = ballasting

Beseitigung *f* **alter Anstriche** = removing old paints

~ **des überschüssigen Wassers auf einer Betondecke mit einem Rohr** = pipe floating the surface

~ **von Elendsvierteln** = slum clearance

Besen|abzug *m*, Besenstrich *m* [*Fertigbehandlung von Betondecken*] = broom finishing, brooming

~**putz** *m*, Stippputz = regrating skin

~**schleppe** *f* = broom drag

~**strich** *m* → Besenabzug

~**walze** *f* → Bürstenwalze

besetzen mit Diamantsplittern = to surface set diamonds

Besetzen *n* **mit Baken**, Bebakung *f* = beaconing

Besichtigungswagen *m* **mit Antriebsmotor** [*zur Nachprüfung des Zustandes von Stahlbauten*] = motorized inspection troll(e)y, self-propelled ~ ~

besiedeln = to populate

besoden, beplaggen, andecken mit (Rasen)Soden, andecken mit (Rasen)Plaggen = to sod

besohlen, aufvulkanisieren eines neuen Laufstreifens = to topcap, to retread
Besondere Bedingungen /pl eines Bauvertrages = Special Conditions
Besonnung f = incoming radiation
Bessemerstahl m = Bessemer steel
Bestand m, Lager~ = stock
Bestandes|aufnahme f = stock checking
~**plan** m, Bestandeszeichnung f = as-built drawing
~**reg(e)lung** f = stock control
~**umfang** m = stock level
Beständigkeit f, Haltbarkeit, Dauerhaftigkeit = durability, resistance
~, Stabilität f [*Emulsion*] = stability
~ [*Hydraulik*] = continuity
Beständigkeits|grad m = degree of resistance, ~ ~ durability
~**probe** f, Beständigkeitsversuch m, Beständigkeitsprüfung f = resistance test, durability ~
Bestands|plan m → Bau~
~**zeichnung** f → (Bau)Bestandsplan
Bestandteil m, Komponente f = component
~ **einer Mischung**, Komponente f ~ ~, Mischungsbestandteil, Mischungskomponente = ingredient
~ **eines Straßenkörpers** = pavement component
bestehende Trasse f = existing route
besteigbarer Schornstein m = man-sized chimney
Bestellung f = order

Bestgeschwindigkeit f, Freihandgeschwindigkeit = hands off speed
bestimmen, ermitteln = to determine
Bestimmung f, Ermitt(e)lung = determination
~ **der Gleitfläche mittels logarithmischer Spirale** [*Bodenmechanik*] = logarithmic-spiral method for calculating a sliding surface
Bestimmungs|gleichung f = conditional equation, equation of condition
~**hafen** m = port of destination
~**ort** m = place of destination
bestmögliche Verdichtung f = optimum compaction
Bestrahlung f = irradiation
Bestreuung f → Abstreuung
bestrichene Windbahn f, Seeraum m = fetch
bestücken, ausstatten, ausrüsten = to fit out, to equip
Bestückung f, Ausrüstung, Ausstattung = equipment, outfit
Besucher|strom m = flow of visitors
~**Zimmer** n = visitors' room
~**zustrom** m = influx of visitors
Beta|eisen n = beta iron
~**-Messing** n = beta brass
Betanken n = fuel(l)ing
Betankungsfeld n [*auf einem Luftfahrtgelände*] = refuel(l)ing apron
Betatonerde f = beta alumina
Beting|knie n = bitt standard, Samson knee
~**stange** f = bitt pin
~**stützplatte** f = bitt bracket

Beton m, Zement~ concrete, cement ~
   abgebundener ~ set ~
   AEA-~ → belüfteter ~
   AEROKRET-Gas~ AEROCRETE [*Trademark*]
   antiseptischer ~ antiseptic ~
   Architektur ~ → Sicht~
   Auf~, Ober~ top (course) ~
   aufgeblähter ~ → Zellen~
   Baryt~ barytes aggregate ~
   Baustellen~ field ~, job-mix ~
   Bauwerk~ → Konstruktions~

Beton

belüfteter ~, AEA-~, Luftporen~, Bläschen~	air-entrained ~, air-entraining ~
~ mit Epoxyharzbindemittel	epoxy ~
~ mit leichten Zuschlagstoffen	lightweight aggregate ~
~ niederer Festigkeit	low-strength ~
~ niederer Güte	low-grade ~
~ ohne (Stahl)Einlagen	plain ~
~ ohne Wirkstoff	plain ~
~ unzureichender Güte	rejected ~
Bims~, Naturbims~	pumice ~
Blähschiefer~	expanded shale ~
Blähton~	expanded clay ~
Bläschen~ → belüfteter ~	
Bruchstein~	rubble ~
Dämm~, Isolier~	insulating ~
Dampfgas~	steam-cured gas-formed ~
Decken~ → Straßendecken~	
Dekorativ~, Zier~	ornamental ~, decorative ~
Einkorn~	short range aggregate ~, single-size material ~
Eis~	ice ~
entfeinter ~, Schütt~, ~ ohne Feinkorn	no-fines ~ (Brit.); popcorn ~ (US)
erdfeuchter ~	no-slump ~, (earth-)dry ~
erhärteter ~, Fest~	hardened ~
Farb~	colo(u)red ~
Fertig~ → Transport~	
Fertigteil~	precast ~
Fest~, erhärteter ~	hardened ~
flüssiger ~	fluid ~, wet ~, sloppy ~, earth-damp ~
Frisch~, unabgebundener ~	fresh(ly-mixed) ~, unset ~
frühstandfester ~	~ with high early stability
frühtragfester ~	~ with high early strength, high-early-strength ~
Füll~ → Mager~	
Füll~ [*der ein Rohr umschließt*]	haunching ~
Fundament~, Gründungs~	foundation ~
Gas~	gas(-formed) ~
Gefälle~	sloping ~
gering beanspruchter ~	non-stressed ~
gerissener ~	cracked ~
gewöhnlicher ~ → Schwer~	
Gießharz~	resin ~
Glas~, Glasstahl~	glazed reinforced ~
Grob~	coarse ~
Großkorn~	~ with large aggregate
Gründungs~, Fundament ~	foundation ~
grüner ~ → junger ~	
Guß~, Rinnen~, gießfähiger ~	chuted ~

Beton

handgemischter ~	hand-mixed ~
Hart~	~ made with hard aggregates
Hartgußeisenschrot~	chilled (cast) iron (shot) ~
hitzebeständiger ~	heat-resistant ~
Hochbau~	building ~
(Hochofen)Schlacken~	slag ~
Holzfaser~	wood fibre ~ (Brit.); ~ fiber ~ (US)
Holzmehl~, Sägemehl~	sawdust ~
Holzspäne~	wood-chip ~, wood-particle ~
Holzwolle~	wood-wool ~
Hüttenbims~	foamed (blast-furnace) slag ~ (Brit.); expanded (~) ~ ~ (US)
Isolier~, Dämm~	insulating ~
junger ~, grüner ~	green ~
Kern~, Unter~ [*Betonwerksteinplatte*]	backing ~
Kesselschlacken~, Lösch(e)~	(boilerhouse) cinder ~, ashes ~; (boiler) clinker ~ (Brit.)
Kies~	~ made from natural aggregates, gravel ~
Kokslösch(e)~	breeze ~
Kolloid~, kolloidaler ~	colloidal ~
Konstruktions~, Bauwerk~, nichtmassiver ~	structural ~
Konstruktions-Leicht~	structural lightweight ~
Kontraktor~	tremie ~, fixed ~
Kunstharz~	~ with artifical resin admix(ture)
Leicht~ [*in Deutschland Raumgewichte von 250 bis 1800 kg/m³*]	light(-weight) ~
Liefer~ → Transport~	
Luftporen~ → belüfteter ~	
Mager~, Füll~, Spar~, zementarmer ~, magerer ~	lean(-mixed) ~
Magerwalz~	(dry) lean (rolled) ~
Magnetit~	magnetite ~
Markierungs~	~ for carriageway markings
maschinell gemischter ~	machine-mixed ~
Massen~, Massiv~	mass ~, bulk ~, concrete-in-mass
nagelbarer ~, Nagel~	nailing ~
(Natur)Bims~	pumice ~
nichtmassiver Beton → Konstruktions~	
Normal~ → Schwer~	
Ober~, Auf~	top (course) ~
Ort~	cast-in-place ~, ~ cast in position, ~ cast in situ, poured-in-place ~, cast-in situ ~
plastischer ~ → Weich~	
Poren~ → Zellen~	

Beton

Portlandzement~ | Portland cement ~, P. C. C.
Prepakt~, vorgepackter ~, Schlämm~, Skelett~ | Prepakt ~, prepacked ~, (grout-) intruded ~
Preß~ | ~ for pressure grouting
Pump~ | pumping ~, pumped ~, pumpcrete
Quell~, Expansiv~, Schwell~ | self-stressed ~
Rinnen~ → Guß~ |
rißbewehrter ~, rißarmierter ~ | crack-reinforced ~
Rüttel~, Vibrations~ | vibrated ~
Rüttelgrob~, Vibrationsgrob~ | vibrated coarse ~
Sack~ unter Wasser | underwater ~ filled in bags
Sägemehl~ → Holzmehl~ |
Sand~ [*1. im Zuschlagstoff kein oder nur wenig grobes Korn> 7 mm; 2. Zement-Sand-Gemisch mit einem Mischungsverhältnis von 1:4 bis 1:20*] | sand ~ [*1. concrete with no or little coarse aggregate > 7 mm; 2. cement-sand mixture from 1:4 up 1:20*]
Saug~ → Vakuum~ |
Schalstocher~ | rodded ~, puddled ~
Schamotte~ | fire-clay ~
Schaum~ [*Poren mit Luft gefüllt*] | foam(ed) ~, ~ aerated with foam
Schill~ [*Seemuschelschalen als Zuschlagstoff*] | oyster-shell ~
Schlacken~, Hochofenschlacken~ | slag ~
Schlämm~ → Prepakt~ |
Schleuder~ | centrifugally cast ~, spun ~
Schock~, Stoß~ [*250 Schläge/min.*] | ~ compacted by jolting
Schotter~ | ballast ~, ~ made with crushed stone aggregates
Schraub~ | screwcrete
Schrot~ | shot ~
Schütt~ → entfeinter ~ |
Schwer~, ~ mit geschlossenem Gefüge, gewöhnlicher ~, Normal~ [*in Deutschland Raumgewichte von 1800 bis 2750 kg/m³*] | heavy ~, high-density ~, heavyweight ~, normal ~
Schwerst~ [*in Deutschland Raumgewichte von 2750 bis 5000 kg/m³*] | super-heavy ~
Sicht~, Architektur~ | exposed ~, fair-faced ~
Skelett~ → Prepakt~ |
Spann~, ~ vorgespannter ~ |
Spar~ → Mager~ |
Sperr~ | waterproof ~
Splitt~ | ~ made from natural and broken aggregates, chip(ping)s ~
Spritz~, Torkret~ | jetcrete, gun-applied concrete, airplaced concrete, shotcrete, gunite, pneumatically placed ~, gunned ~, cement-gun ~, pneumatically-applied mortar

Stahl~ [*früher: Eisen~*]	(steel) reinforced ~, R.C. [*deprecated: ferroconcrete*]
Stampf~	tamped ~
statisch bewehrter ~, statisch armierter ~	statically reinforced ~
steifer ~	stiff ~
Stoß~ → Schock~	
(Straßen)Decken~	pavement ~; paving ~ (US)
Torkret~ → Spritz~	
Transport~, Fertig~, Liefer~	ready-mixed ~, pre-mixed ~
Nachmischen *n*	agitating
Trocken- und Naßmischen *n* während der Fahrt *f*	truck mixing
Naßmischen *n* während der Fahrt *f*	shrink-mixing, partial mixing
Transportmischer~	transit-mix(ed) ~, truck-mixed ~
Trichter~	tremie(d) ~, tremie(d) grout
unabgebundener ~ → Frisch~	
unbewehrter ~, nicht armierter ~	non-reinforced ~, unreinforced ~, plain ~
Unter~	rough ~, sub-~, seal-coat ~
Unter~ → Kern~	
Unterwasser~	underwater ~
Vakuum~, Saug~	VACUUM CONCRETE [*Trademark*]
Vibrations~, Rüttel~	vibrated ~
Vibrationsgrob~, Rüttelgrob~	vibrated coarse ~
vorgepackter ~ → Prepakt~	
vorgespannter ~, Spann~	pre-stressed (reinforced) ~
Vorsatz~	face ~
Walz~	rolled ~
Wärmedämm~	heat-insulating ~
Weich~ [*früher: plastischer ~*]	high-slump ~, plastic(ised) ~, buttery ~
Zellen~, Poren~, aufgeblähter ~, Bläh~	cellular(-expanded) ~, aerated ~
zementarmer ~ → Mager~	
Zier~, Dekorativ~	ornamental ~, decorative ~
Zyklopen~	cyclopean ~

Beton *m* **mit nur leichten Zuschlägen** = all-lightweight-aggregate concrete

~ von geringer Güte = low-grade concrete

~abbruch *m*, Betonabtrag *m* = concrete demolition work

~abdeckung *f* = concrete cover(ing)

~abflußrohr *n*, Betonablaufrohr, Betonabfluß *m*, Betonablauf *m* = concrete waste-pipe

~ablauf(ge)rinne *f*, (*n*) = concrete dish, ~ catch-gutter, ~ surface channel

Betonablaufrohr — Betonbalkenformmaschine

~ablaufrohr *n* → Betonabflußrohr
~abmessung *f* → Betondosierung
~abmessungen *fpl* = dimensions of the concrete
~abmeßanlage *f* → Betondosieranlage
~abmeßwaage *f* → Betondosierwaage
~abschirmung *f* = concrete radiation shield
~abschnitt *m*, Betonstrecke *f* [*einer Straße*] = concrete section
~abstandhalter *m* = distance piece, separator [*A part used for maintaining the position and spacing of formwork during concreting*]
~abtrag *m*, Betonabbruch *m* = concrete demolition work
~-Abwasserkanal *m* → Betonkanal
~-Abwasserrohr *n*, Abwasser-Betonrohr = concrete sewer pipe
betonaggressiv = aggressive to concrete
Beton|alter *n* = age of concrete
~anfuhr *f* = concrete delivery
~anhäufung *f* im Mischer = "build up" in the concrete mixer
~anker *m* = anchor embedded in concrete
~ansatz *m* = concrete droppings
~anstrich *m* [*Schutzanstrich der Oberfläche des Betons gegen chemische Angriffe*] = protective coat of concrete
~arbeiten *fpl*, Betonbetrieb *m* = concrete work
~arbeiter *m*, Betonfacharbeiter, Betonbauer *m* = concrete labo(u)rer
~arbeitsfuge *f* → Arbeitsfuge
~arbeitsfugenband *n* → Arbeitsfugenband
~architektur *f* = concrete architecture
~armierung *f* → Bewehrung *f*
~art *f*, Betonsorte *f* = type of concrete
~aufbau *m*, Betonzusammensetzung *f* = concrete composition
~(auf)bereitung *f*, Betonherstellung, Betonerzeugung = concrete fabrication, ~ production

~(auf)bereitungsanlage *f*, Betonmischanlage = concrete mixing plant, ~ fabrication ~, ~ production ~
~(auf)bereitungstemperatur *f*, Betonmischtemperatur = concrete mixing temperature
~(auf)bruch *m* = concrete breaking
~aufbruchhammer *m*, Betonaufbrechhammer, Betonbrecher *m*, Betonaufreißhammer = concrete breaker
~aufbruchstahl *m* = concrete breaker point, ~ ~ steel
~auflast *f* = concrete surcharge
~aufrauhgerät *n* = concrete roughening unit
~aufreißhammer *m* → Betonaufbruchhammer
~aufzug *m* = concrete hoist, hoist for concrete
~ausbesserung *f*, Betonreparatur *f* = concrete reintegration, ~ patching, ~ repair
~ausbesserungsstoff *m* = concrete repair material, ~ reintegration ~
~ausfachungselement *n*, Betonausfüll(ungs)element *f* = concrete cladding unit, ~ infilling ~
~ausgleich(schicht) *m*, (*f*), Betonausgleichlage *f* = concrete level(l)ing course, ~ ~ layer
~auskleidung *f*, Betonverkleidung, Betonbekleidung = concrete lining, ~ facing
~auskleidung *f*, Betonverkleidung, Betonmantel *m*, nicht tragende Betonmauerung *f* geringer Dicke, Betonbekleidung [*Felshohlraumbau*] = concrete lining, ~ facing
~auswahl *f* = concrete selection
~autobahn *f* = concrete motorway
~automat *m* → → Betonmischer *m*
~(bahn)automat *m* **ABG** = concrete vibrator and finisher with rotating (or rotary) grading screed, revolving paddle finisher
~bahnfertiger *m* → Betondeckenfertiger
~balkenformmaschine *f* = concrete beam making machine

Betonbalken(träger) — (Beton)Blockwerkplatz

~balken(träger) *m* = concrete beam

~balken(träger)brücke *f* = concrete beam bridge

~bau *m* = concrete construction(al work)

~bauer *m* → Betonarbeiter *m*

~bauindustrie *f* = concrete constructional industry

~baumaschinen *fpl* = concrete machinery

~baustelle *f* = concrete (construction) site

~bauwerk *n* = concrete structure

~-Beckenwand *f* = reservoir wall of concrete

~behälter *m* = concrete reservoir, ~ tank

~behälter *m* für Flüssigkeiten, Beton-Flüssigkeitsbehälter = concrete tank for liquids, ~ reservoir ~ ~

~bekleidung *f* → Betonauskleidung

~bekleidung *f* einer Böschung = concrete cover of a slope

~belag *m* → Betondecke *f*

~belagfuge *f* → Deckenfuge

~belagplatte *f*, Betonplatte, Betonfliese *f* = concrete tile

~belagspannung *f*, Betondeckenspannung = rigid pavement stress, concrete ~ ~, ~ surfacing ~; ~ paving ~ (US)

~-Beleuchtungsmast *m* → Beton-Lichtmast

~belüfter *m* → Belüftungsmittel *n*

~belüftungsmesser *m* → Belüftungsmesser

~belüftungsmittel *n* → Belüftungsmittel

~beratungsdienst *m* = concrete advisory service

~bereitung *f* → Betonaufbereitung

~bereitungsanlage *f*, Betonaufbereitungsanlage, Betonmischanlage = concrete fabrication plant, ~ mixing ~, ~ production ~

~bereitungstemperatur *f*, Betonaufbereitungstemperatur = concrete mixing temperature

~(be)schädigung *f* = concrete distress

~beschicker *m*, Betontransporter *m* [*Betonsteinindustrie*] = concrete loader

~bestandteil *m*, Betonkomponente *f* = concrete ingredient

~bestimmungen *fpl* = concrete code

~betrieb *m*, Betonarbeiten *fpl* = concrete work

~bett(ung) *n*, (*f*) = concrete bed

~-Bewässerungsrohr *n*, Bewässerungsbetonrohr = concrete irrigation pipe

~bewehrung *f* → Bewehrung

~binder *m*, Betondachbinder, Betonfachwerkbinder = concrete roof truss

~-Blasrohrleitung *f*, Blasrohrleitung für Beton [*für Druckluft-Betonförderung*] = pneumatic line for concrete distribution

(Beton)Block *m* = (concrete) block, (~) building ~, cement ~

~automat *m* → (Beton)Stein(form)automat

~form *f* → (Beton)Steinform

~fertigung *f* → (Beton)Steinfertigung

~hersteller *m* → (Beton)Steinhersteller

~herstellung *f* → (Beton)Steinfertigung

~maschine *f* → (Beton)Stein(form)maschine

~mauerwerk *n* → (Beton)Steinmauerwerk

~pflaster *n* = concrete block paving

~prüfpresse *f*, (Beton)Steinprüfpresse = (concrete) block crushing testing machine

~schneider *m* → (Beton)Steintrennmaschine *f*

~stampfmaschine *f*, (Beton)Steinstampfmaschine, Stampfmaschine = machine tamper

~trennmaschine *f* → (Beton)Steintrennmaschine

~werk *n*, (Beton)Steinwerk = (concrete) block factory

~werkplatz *m*, (Beton)Steinwerkplatz = (concrete) blockyard

Betonboden — Betondecke 190

Betonboden *m*, Betonfußboden = concrete floor(ing), ~ floor finish, ~ floor covering
Betonbogen *m* = concrete arch
~, Betonrohrbogen, Betonkrümmer *m*, Betonbogenrohr *n* = concrete (pipe) bend
~**bordsteinform** *f* = radius kerb mould (Brit.); ~ curb mold (US)
~**brücke** *f* = concrete arch(ed) bridge
~**gewichts(stau)mauer** *f* = concrete arch(ed) gravity dam
~**rippe** *f* = concrete arch rib
~**rohr** *n*, Beton(rohr)bogen *m*, Betonkrümmer *m* = concrete (pipe) bend
~**trägerbrücke** *f* **mit Zugband** = concrete bowstring arch(ed) bridge
Betonbohren *n* = drilling of concrete
~ **mit Sauerstofflanze** = oxygen lance concrete drilling
Beton|bohrer *m* = concrete drill
~**bohrkern** *m*, Beton(probe)kern *m* = concrete test core
~**bohrpfahl** *m* → Bohrpfahl
~**bordschwelle** *f*, Beton(hoch)bordstein *m*, Betonschrammbord *n* = concrete curb; ~ kerb (Brit.)
Betonbordstein *m* → Betonbordschwelle *f*
~ → Betontiefbordstein
~**form** *f* = kerb mould (Brit.); curb mold (US)
Beton|-Bord- und -Pflasterstein-Automat *m* = auto(matic) concrete curb and (paving) sett making machine
~**brecher** *m* → Betonaufbruchhammer *m*
~**brechsand** *m*, zerkleinerter Feinzuschlag(stoff) *m* = crushed fine aggregate, ~ (concreting) sand, ~ concrete sand
~**bruch** *m* = concrete failure
~**bruch** *m* → Betonaufbruch
~**brücke** *f* = concrete bridge
~**brücke** *f* [*in einer Betonfuge*] = concrete droppings
~**brückenbau** *m* = concrete bridge construction, ~ bridgebuilding

~**-Brückenpfeiler** *m* = concrete bridge pier, ~ ~ support
~**-Brunnengründung** *f* = concrete well foundation
~**-Brunnenring** *m*, Brunnenring aus Beton = concrete ring
~**-Brustmauer** *f*, Beton-Brustwehr *f* = concrete parapet
~**-Brüstungselement** *n* = concrete parapet element
~**büchse** *f*, Betonlöffel *m*, Zementierbüchse [*Tiefbohrtechnik*] = cement dump, dump bailer
~**buhne** *f* = concrete groyne (Brit.); ~ groin (US)
~**bunker** *m* = concrete bin, ~ hopper
~**bunker** *m*, Betonschutzbunker = concrete shelter
~**-Bürgersteigplatte** *f* → Bürgersteig-Betonplatte
(~)**Charge** *f* = (concrete) batch
~**dach** *n* = concrete roof
~**(dach)binder** *m*, Betonfachwerkbinder = concrete roof truss
~**dachdecke** *f*, Betonunterkonstruktion *f* = concrete (roof) deck
~**dachplatte** *f* = concrete roof slab
Betondachstein *m*, (Zement)Dachstein, Zement(dach)ziegel *m*, Zement(dach)stein = concrete (roof(ing)) tile, cement (~) ~, (precast) cement roof(ing) shingle, (precast) concrete roof(ing) shingle
~**automat** *n*, automatische Betondachsteinmaschine *f* = auto(matic) concrete roofing tile machine, ~ roof-tile-making ~
~**farbe** *f*, (Zement)Dachsteinfarbe, Zement(dach)ziegelfarbe, Zementsteinfarbe = concrete (roof(ing)) tile paint, cement (~) ~ ~
~**maschine** *f* → Dachsteinmaschine
~**prüfgerät** *n* = concrete roofing tile tester
Betondecke *f*, Betonbelag *m*, starre Decke, starrer Belag = rigid pavement, concrete ~, ~ surfacing [*Is made up of cement concrete and may*

Betondecke — Betondosierung

or may not have a base course between the pavement and the subgrade]

~, Betonhochbaudecke, Betonstockwerkdecke, Betonetagendecke, Betongeschoßdecke, Beton-Gebäudedecke = concrete floor

Betondecken|bemessungskurve *f* = rigid-pavement design curve, rigid-surfacing ~ ~

~**fertiger** *m*, Beton(fahr)bahnfertiger, Betonstraßen(bohlen)fertiger, Brückenfertiger = concrete finisher

~**fertiger** *m* **mit Diagonal-Glättbohle**, Diagonalbohlenfertiger = diagonal screed finisher

~**fertigteil** *m, n* = precast concrete floor unit

~**fuge** *f* → Deckenfuge

~**fugenvergußmasse** *f* = concrete pavement joint sealing compound

~**geräte** *npl* = concrete paving equipment

~**hebegeräte** *npl* = concrete slab raising equipment

~**hebeverfahren** *n* **durch Einpressen von einem aus Feinsand, Mo, Schluff und Kolloidmaterial als Zuschlag, Zement als Bindemittel und Wasser als Verflüssiger bestehendem Gemisch bei Setzung von Betonfahrbahnplatten**, Mud-Jack-Verfahren = mud-jack(ing) (of pavements slabs), mud-jack treatment

~**hohlkörper** *m*, Beton-Deckenhohlblock *m*, Beton-Deckenhohlstein *m* = concrete hollow filler tile, ~ ~ ~ block

~**innenrüttler(-Fertiger)** *m*, Betondeckeninnenvibrator(-Fertiger), Tauchrüttler-Betonstraßen-Fertiger = full depth internal concrete pavement (or paving/US) vibrator, ~ ~ ~ slab vibrator; paving vibrator (US); internal vibrating machine (Brit.)

~**nachbehandlung** *f* = concrete pavement curing

~**platte** *f*, Betonfeld *n*, Betonplatte = concrete pavement slab; ~ paving ~ (US)

~**platte** *f*, Deckenplatte [*für Beton-(geschoß)decken*] = concrete floor slab

~**schalung** *f*, Straßenbauschalung, Seitenschalung = road formwork, highway ~, ~ forms

~**schleifen** *n* = bumpcutting

~**schleifgerät** *n*, Betonschleifgerät = concrete planer, bumpcutter

~**spannung** *f* → Betonbelagspannung

~**überzug** *m* = concrete resurfacing

~**verteiler** *m* → Betonverteiler

Beton|deckung *f*, Betonüberdeckung [*Maß der Überdeckung der Bewehrungsstähle des Stahlbetons mit Beton aus Gründen der Rostsicherheit*] = concrete cover, cover to reinforcement, concrete protection for reinforcement, concrete protective covering

~**dehnung** *f* = concrete strain

~**-Dehn(ungs)fuge** *f*, Beton-Raumfuge = concrete expansion joint

~**diagonale** *f* → Betonschräge *f*

~**diaphragma** *n* → Betonschirm *m*

~**dichte** *f* = concrete density

~**dichtung** *f* = concrete waterproofing

~**(dichtungs)kern** *m*, Betonkernmauer *f*, massive Kernmauer [*Talsperre*] = concrete core wall

(~)**Dichtungsmittel** *n*, Sperrzusatz *m*, wasserabweisendes Mittel, DM, wasserabweisender Zusatz = waterproofer, concrete waterproofing compound, water-repellent, water-repelling agent, densifier, densifying agent, dampproofing and permeability reducing agent

~**dichtungsschürze** *f* **im Inneren**, Zement~ ~ ~ = concrete curtain

~**dicke** *f* = concrete thickness

~**dosieranlage** *f*, Betonabmeßanlage, Betonzumeßanlage, Betonzuteilanlage = concrete batching plant, ~ measuring ~, ~ proportioning ~

~**dosierung** *f*, Betonabmessung, Betonzumessung, Betonzuteilung = concrete batching, ~ measuring, ~ proportioning

~**dosierwaage** *f*, Betonabmeßwaage, Betonzumeßwaage, Betonzuteilwaage = concrete batching scale, ~ measuring ~, ~ proportioning ~

~**dränrohr** *n*, Dränbetonrohr = concrete drain pipe; ~ ~ tile (US)

~**druckbereich** *m*, Betondruckzone *f* = concrete compressive zone

~**druckfestigkeit** *f* = concrete compressive strength

~**druckkraft** *f* = compressive force of concrete

~**druck(leitungs)rohr** *n* = concrete pressure pipe

~**druckluftförderer** *m*, pneumatischer Betonförderer, Druckluftbetonförderer = pneumatic concrete placer, ~ ~ placing machine

~**-(Druck)Luftschleuse** *f*, Betonschleuse [*Caisson*] = concreting airlock

~**druckprobe** *f*, Betondruckprüfung *f*, Betondruckversuch *m* = concrete compressive test

~**druckprüfung** *f* → Betondruckprobe *f*

~**druckrohr** *n*, Betondruckleitungsrohr = concrete pressure pipe

~**-Druckspannung** *f* = compressive stress in the concrete

~**druckversuch** *m* → Betondruckprobe *f*

~**druckzone** *f* = concrete compressive zone

~**dübel** *m* = concrete dowel

~**durchlässigkeit** *f* = permeability of concrete

~**durchlaufträgerbrücke** *f* = concrete through girder bridge

~**ebner** *m* = straight edge

~**eigenschaft** *f* = concrete property

Betoneinbau *m*, Betoneinbringung *f*, Betonierung, Betonieren *n*, Betonverarbeitung, Einbringen *n* des Betons, Betonierungsarbeiten *fpl* = placement of concrete, pouring ~ ~, concrete placement, concrete placing, concrete pour(ing), concreting (operations)

~**kurve** *f* → Betonierkurve

~**maschine** *f* = concreting machine, concrete placer

~**plan** *m*, Betonier(ungs)plan = progress chart for concrete work, concrete progress chart

~**temperatur** *f* = concreting temperature

Beton|einbringer *m* → Betoniermaschine *f*

~**einbringung** *f* → Betoneinbau *m*

~**(einfüll)trichter** *m* [*Betonpumpe*] → Trichter

~**einpreßmaschine** *f*, Betoninjektor *m*, Betoneinspritzapparat *m* = concrete grouter, ~ pressure grouting machine

~**(ein)rütt(e)lung** *f*, Beton(ein)rütteln *n* = concrete vibration, consolidation of concrete by vibration

~**einspritzapparat** *m* → Betoneinpreßmaschine *f*

~**-Einzelfundament** *n* = concrete single base, ~ individual ~

~**eisen** *n*, Betonstahl *m* = reinforcing steel

~**-Eisenbahnschwelle** *f* → Betongleisschwelle

Betoneisen|biegemaschine *f* → Biegemaschine

~**bieger** *m*, Betonstahlbieger = reinforcement bar-bender, rod bender

~**(hand)schneider** *m*, Betonstahlschere *f*, Betoneisenschere, Betonstahl-Schneider = hand operated bar cutter, bar cutter, bar cutting shears, reinforcement bar shear cutter, rod shears

~**eisenschere** *f* → Betoneisen(hand)schneider *m*

~**schneid(e)maschine** *f*, Betonstahlschneid(e)maschine = (reinforcement) bar cutting machine

~**schneiden** *n*, Betonstahlschneiden = cutting of reinforcement

~**stab** *m*, Betonstahlstab = reinforcing rod, (~) bar, re-bar

Beton|elastizitätsmodul *m* = concrete modulus of elasticity

Betonemulsion — Beton-Fertiggarage

~emulsion *f* = concrete emulsion
(~)**Entschalungsemulsion** *f*, (Beton-)Schalemulsion = release emulsion
(~)**Entschalungsmittel** *n*, (Beton-)Schalmittel = release agent
~**-Entschalungsöl** *n* → Entschalungsöl
(~)**Entschalungspaste** *f*, (Beton-)Schalpaste = release paste
~**(er)härtung** *f* = concrete hardening
~**ermüdung** *f* = fatigue of the concrete
~**erwärmung** *f* **mit Elektroden**, elektrische Widerstandserwärmung des Betons = electrical resistivity heating of concrete
~**erzeugung** *f*, Beton(auf)bereitung, Betonherstellung = concrete fabrication, ~ production
~**estrich** *m* = concrete screed
~**(etagen)decke** *f*, Betonhochbaudecke, Betonstockwerkdecke, Betongeschoßdecke = concrete floor
~**fabrik** *f*, Großbetonanlage *f* = concrete mixing plant, concrete central-mix plant, central batching and mixing station, central weigh batching and mixing plant, wet and dry batch concrete plant, automatic batch plant
~**fabrik** *f* **als Reihenanlage** = concrete mixing plant with belt conveying
~**fabrik** *f* **als Turmanlage** → Betonturm *m*
~**fabrik** *f* **System JOHNSON** → JOHNSON-Turm *m*
~**facharbeiter** *m* → Betonarbeiter
~**fachwerk** *n*, Betonskelett *n* = concrete framework, ~ skeleton
~**fachwerkbau** *m*, Betonskelettbau = concrete framework construction, ~ skeleton ~, framed ~
~**fachwerkbinder** *m*, Beton(dach)binder = concrete roof truss
~**fachwerkträger** *m* = concrete truss
Betonfahrbahn *f*, Zement~ = concrete carriageway
~**decke** *f*, Zement~ = concrete carriageway pavement, ~ ~ surfacing

Beton(fahr)bahnfertiger *m* → Betondeckenfertiger
~ **mit Diagonal-Glättbohle** → Diagonalbohlenfertiger
Betonfahrbahnplatte *f* [*Brücke*] = concrete deck slab
~ = concrete carriageway slab
Beton|-Fahrbahntafel *f* [*Brücke*] = concrete deck
~**-Fahrradstand** *m* = concrete (cycle) stand [*B. S. 1716*]
~**-Fahrsilo** *m* = concrete horizontal silo [*in agriculture*]
~**faltwerkkuppel** *f* = folded plate concrete dome
~**falzziegel** *m* = concrete interlocking tile
~**fang(e)damm** *m* [*Die Seitenwände aus Holz – nur beim Bau als Schalung dienend – werden nach Erhärtung des Betons wiedergewonnen*] = double-wall cofferdam with concrete fill
~**farbe** *f* = concrete paint
~**farbstoff** *m* = colo(u)red pigment for concrete
~**fassade** *f* = concrete façade
~**feinkies** = fine gravel for concrete
~**feinsand** *m* [< *1 mm*] = crushed stone sand
~**feld** *n* [*in einem Rahmen*] = concrete bay
~**feld** *n*, Beton(decken)platte *f* = concrete pavement slab, ~ surfacing ~; ~ paving ~ (US)
~**fenster** *n*, Stahl~ [*Fenster mit Stahlbetonrahmen*] = concrete window
~**fensterrahmen** *m*, Stahl~ = concrete window frame
~**fertigbalken(träger)** *m* = precast (concrete) beam
(~)**Fertigbehandlung** *f* = finishing operations
~**fertiger** *m* **mit Diagonal-Glättbohle**, Beton(fahrbahn)fertiger ~ ~, Diagonalbohlenfertiger = diagonal screed finisher
~**-Fertiggarage** *f* = precast (concrete) garage

Beton-Fertiglüftungskanal — (Beton)Formenöl 194

~-**Fertiglüftungskanal** m = precast (concrete) vent(ilation) duct
~**fertigpfahl** m, Stahl~ = precast (concrete) pile
(~)**Fertigrahmen** m = structural precast concrete building frame, precast (concrete) frame
~**fertigrohr** n, Fertig-Betonrohr = precast (concrete) pipe
(~)**Fertigteil** m, n, Fertigbetonteil, (Beton)Montageteil, Montagebetonteil = precast (concrete) member, cast-concrete ~, (concrete) casting
(~)**Fertigteilbau** m → (Beton)Montagebau
~**fertigteil-Daubensilo** m = precast (concrete) stave silo [*B. S. 2810*]
~**fertigteildecke** f = precast concrete floor
(~)**Fertigteil-Fachwerk** n = precast (concrete) framework, ~ (~) skeleton
(~)**Fertigteil-Fachwerkträger** m = precast (concrete) truss
~**fertigteilgebäude** n, Betonfertigteilhaus n = precast (concrete) building
~**fertigteilhaus** n, Betonfertigteilgebäude n = precast (concrete) building
~**fertigteilindustrie** f → Fertigteilindustrie
(~)**Fertigteil(wohn)haus** n = precast (concrete) home
~-**Fertigträger** m, Fertig-Betonträger = precast (concrete) girder
~**fertigwand** f → Betonmontagewand
~**festigkeit** f = concrete strength
~**fibel** f = concrete primer
~**firststein** m, Betonfirstziegel m = concrete ridge tile
~**fläche** f = concrete area
~**flachkübel** m = flat-bottomed concrete skip
~-**Flaschenrüttler** m → Innenrüttler
~-**Flaschenvibrator** m → Innenrüttler
~**fliese** f, Betonplatte f, Betonbelagplatte = concrete tile
~**fliesenbekleidung** f → Betonplattenbekleidung

~**fliesen(fuß)boden** m → = Betonplattenboden
~**fliesenpresse** f, Betonplattenpresse = concrete tile press
~**fliesenwerk** n → Betonplattenwerk
~**fluder** m = concrete flume
~**flugplatzdecke** f, Betonflugplatzbelag m = rigid airport pavement, concrete ~ ~, ~ airfield ~, ~ surfacing
~-**Flüssigkeitsbehälter** m, Betonbehälter für Flüssigkeiten = concrete tank for liquids, ~ reservoir ~ ~
~**förderanlage** f = concrete handling installation
~**förderapparat** m, Betonförderer m, Betontransporteur m, Betonfördermaschine f, Betonfördergerät n = concrete placing machine, ~ handling ~, ~ placer
~**förderer** m → Betonförderapparat m
~**fördergerät** n → Betonförderapparat m
~**förderleitung** f → Betonförderrohrleitung
~**fördermaschine** f → Betonförderapparat m
~**(förder)pumpe** f, Förderpumpe = concrete pump
~**förderrohr** n, Betontransportrohr = concrete discharge pipe, ~ delivery ~
~**förder(rohr)leitung** f, Betontransport(rohr)leitung f = concrete discharge pipework, ~ delivery ~
Betonförderung f → Betontransport m
~ **durch Druckluft**, Betontransport m ~ ~ = pneumatic concrete conveying
~ **in Rohren**, Betontransport m ~ ~ = concrete conveying through pipes
Beton|form f = concrete mould (Brit.); (~) mold (US)
~-**Formanker** m = deformed tie-bar
~**formenfabrik** f, Betonformenwerk n = concrete mo(u)lding equipment factory
(~)**Formenöl** n = (concrete) mould oil (Brit.); (~) mold ~ (US)

~**formentischler** *m* = joiner for concrete moulds (Brit.); ~ ~ ~ molds (US)
~**formling** *m* = concrete cast(ing); ~ moulding (Brit.); ~ molding (US)
~**formstahl** *m* = deformed (concrete) (reinforcing) bar(s)
Betonformstein *m* → (Beton)Stein
~**maschine** *f* → Blockfertiger *m*
~**silo** *m* = hollow concrete block silo
Beton|fräsmaschine *f* = concrete miller, ~ milling machine
~**frostschutzmittel** *n* = concrete antifreezer
~**fuge** *f* = concrete joint
~**fugendübel** *m* → Fugendübel
(~)**Fugensäge** *f* → Fugenschleifgerät
~**fugenschleifgerät** *n* → Fugenschleifgerät
(~)**Fugenschneider** *m* → Fugenschleifgerät *n*
(~)**Fugenschneidemaschine** *f* → Fugenschleifgerät *n*
~**fugenvergußmasse** *f* = concrete joint sealing compound
~**füllung** *f* = concrete fill
Betonfundament *n*, Betongründung *f*, Betonfundation *f*, Betongründungskörper, Betonfundamentkörper, Gründungsbetonkörper *m* = concrete foundation (structure), ~ engineering foundation
~**körper** *m* → Betonfundament *n*
~**pfeiler** *m* → Betonpfeiler
~**platte** *f* → Plattenfundament *n*
~**sockel** *m* → Betonsockelgründung *f*
Beton|fundation *f* → Betonfundament
(~)**Fundationsplatte** *f* → Plattenfundament *n*
~**fundationssockel** *m* → Betonsockelgründung *f*
~**fundierung** *f*, Betongründung, Betonfundation *f* = concrete foundation
~**(fuß)boden** *m* = concrete floor(ing), ~ floor finish, ~ floor covering
~**-Fußwegplatte** *f* → Bürgersteig-Betonplatte

~**gammaschild** *m*, Gammaabschirmung *f* aus Beton = concrete gamma shield
~**-Gasbehälter** *m* = concrete gasholder
Betongebäude *n*, Betonhaus *n* = concrete building
~**decke** *f* → Betondecke
~**fabrik** *f* → Häuserfabrik
Beton|gefüge *n* = concrete texture
~**gegengewicht** *n* = concrete counterweight
~**-Gehbahnplatte** *f* → Bürgersteig-Betonplatte
~**-Gehsteigplatte** *f* → Bürgersteig-Betonplatte
~**-Gehwegplatte** *f* → Bürgersteig-Betonplatte
~**gelenk** *n* = concrete hinge
~**gemenge** *n* → Betongemisch *n*
~**gemisch** *n*, Betonmischung *f*, Betongemenge *n* = concrete mix(ture)
~**geräte** *npl* und Betonmaschinen *fpl* = concrete equipment
(~)**(geschoß)decke** *f*, Betonhochbaudecke, Betonstockwerkdecke, Betonetagendecke, Beton-Gebäudedecke = concrete floor
~**gewände** *n* = concrete jamb
~**gewicht** *n* = concrete weight
~**-Gewichts(stau)mauer** *f* = concrete gravity dam
~**gewölbe** *n* = concrete vault
(~)**Gießmast** *m* = concrete chuting (or placing) mast, gin pole type concrete spouting plant
(~)**Gießrinne** *f*, Gußbeton-Rinne = concrete chute
(~)**Gießrinnenanlage** *f*, Gußbeton-Rinnenanlage = (concrete) spouting plant, (~) chuting ~, (~) ~ installation [*Distribution of concrete by a chuting installation is a convenient method when fairly large areas have to be covered, whether at ground level, as in reservoir floors, or at various elevations as in large multi-story buildings. The chutes swivel about a central hoist-tower*]

(Beton)Gießrohr — Betonhohlblock 196

(~)Gießrohr n, Gußbeton-Rohr = pipe for concrete placing by gravity

(~)Gießturm m, Ausleger-Gießturm = concrete chuting (or placing) tower, elevator ~, cage type concrete spouting plant, tower concrete spouting plant

~glas n, Glasbetonstein m, Glaskörper m für Glasstahlbeton [gepreßter Glaskörper für den Einbau in Tragwerke aus Glasstahlbeton oder verglaste Stahlbetongerippe] = glassconcrete block

~glätter m = float

~gleisschwelle f, Beton(schienen)schwelle, Beton-Eisenbahnschwelle = concrete tie (US); ~ sleeper, ~ cross-tie (Brit.); ~ cross-tie

~(gleis)schwellenwerk n = concrete tie works (US); ~ sleeper ~, ~ cross-sill ~ (Brit.); ~ cross-tie ~

~-Glockenturm m = concrete bell tower, ~ belfry

~granit m = granitic finish [A face mix, resembling granite, on precast concrete]

~granitplatte f = precast concrete slab with granitic finish

betongrau = concrete gray, ~ grey

Betongrob|kies m = coarse gravel for concrete

~sand m [3/7 mm] = crushed gravel sand

(Beton)Grobzuschlag(stoff) m = coarse (concrete) aggregate

Betongrund|pfahl m = concrete pile completely in the ground

~pfeiler m → Betonpfeiler

~platte f → Plattenfundament n

~sockel m → Betonsockelgründung f

Betongründung f → Betonfundament n

Betongründungs|körper m → Betonfundament n

~platte f → Plattenfundament n

~sockel m → Betonsockelgründung f

Betongüte f, Betonqualität f = concrete quality

~klasse f = concrete grade

~überwachung f = concrete quality control

Beton|härter m, Betonhärtemittel n, Härtungsmittel, (Beton)Hartstoff m = concrete surface hardener, ~ floor ~

~hartstoff m [Fehlname] → Hartbetonzuschlag(stoff)

(~)Hartstoff m → Betonhärter

~härtung f, Betonerhärtung = concrete hardening

~härtungsgerät n = concrete hardening device

~haus n, Betongebäude n = concrete building

~häuserfabrik f → Häuserfabrik

~hebeanlage f, Betonhebewerk n = concrete-elevating plant (or gear)

~herstellung f, Beton(auf)bereitung, Betonerzeugung = concrete fabrication, ~ production

~herstellungsgeräte npl = concrete fabrication equipment, ~ production ~

~hinterfüllung f = concrete back filling

~-Hochbau m = concrete building construction

~(hochbau)decke f, Betonstockwerkdecke, Betonetagendecke, Betongeschoßdecke, Beton-Gebäudedecke = concrete floor

~(hoch)bordstein m → Betonbordschwelle.f

~hochförderung f, Hochförderung von Beton = upward transport of concrete

~höhe f [unter der Bohle eines Betondeckenfertigers] = level of (concrete) mix

~hohlbalken(träger) m = hollow concrete beam

~hohlbauweise f = hollow concrete method

Betonhohlblock m, Betonhohlkörper m, Betonhohlstein m = concrete hollow block, ~ ~ tile, ~ pot, hollow concrete block, hollow concrete tile

Betonhohl|diele *f* = hollow concrete slab
~körper *m* → Betonhohlblock *m*
~mast *m* = concrete hollow mast
(Beton)Hohlstein *m* → Betonhohlblock
Betonierabschnitt *m*, Betonierblock *m* [*bei Massenbeton*] = concreting section, pouring ~
~ [*zeitlich*] = concreting stage, pour(ing) ~, casting ~
Betonier|aggregat *n* [*Fehlname*] → Betoniermaschine *f*
~anlage *f*, Betoniereinrichtung *f*, Betonierungsanlage, Betonierungseinrichtung = concreting plant
~art *f* = concreting method
~bandturm *m* = concrete belt (type) placing tower, ~ ~ (~) pouring ~
~bett *n* = casting bed
~block *m*, Betonierabschnitt *m* [*bei Massenbeton*] = concreting section
~brücke *f*, Betoniersteg *m*, Betoniergerüst *n* = concreting scaffold(ing)
~bühne *f* = concrete placing platform, ~ pouring ~
~bunker *m* = concreting hopper
~einrichtung *f* → Betonieranlage *f*
betonieren *v* = to concrete, to pour
Betonieren *n* → Betoneinbau
~ **bei Frost** = cold weather concreting, ~ pouring
~ ~ **warmem Wetter** = hot weather concreting, ~ ~ pouring
~ **im Freivorbau** = pouring in cantilever work, concreting ~ ~ ~
~ **in Schutzbuden** = concreting under shelter, pouring ~ ~
~ **massiger Baukörper** = mass concreting, ~ pouring
~ **mit Pumpe(n)**, Einbringen von Pumpbeton = concrete pumping
~ **unter Vibration** = vibratory concreting, ~ pouring
~ ~ **Wasser**, Unterwasserbetonieren = underwater concreting, ~ pouring
Betonier|folge *f* = placing sequence, concreting ~, pouring ~
~folie *f* = concreting foil
~fuge *f* → Arbeitsfuge

~gerät *n* → Betoniermaschine *f*
~gerät *n* [*nicht notwendigerweise eine Maschine*] = concreting device
~gerüst *n* → Betonierbrücke *f*
~kolonne *f*, Betoniermannschaft *f*, Betoniertrupp *m*, Betonkolonne = concreting gang, ~ team, ~ party, ~ crew, pouring ~
~kran *m* = concreting crane, (concrete) pouring ~
~kübel *m* → Betonkübel
~kurve *f*, Betoneinbaukurve, Betonleistungskurve [*gibt die eingebauten Betonmengen im Verhältnis zur Zeit an*] = concreting curve, casting ~, pouring ~
~lage *f*, Betonierschicht *f* = (concreting) lift, pouring ~
~leistung *f* = concreting output
~maschine *f*, Betoniergerät *n*, Betoneinbringer *m*; Betonieraggregat *n* [*Fehlname*] = concreting machine
~plan *m* → Betonierungsplan
~schacht *m* = concreting shaft
~schalung *f* [*Schweiz*]; (Beton)Schalung = formwork
~schicht *f*, Betonierlage *f* = (concreting) lift, pouring ~
~schleuse *f*, Betonschleuse = concreting air-lock, pouring ~
~steg *m* → Betonierbrücke *f*
~stelle *f*, Betonierungsstelle = concreting site, pouring ~
~stumpffuge *f* → Arbeitsstumpffuge
~technik *f* = concreting technique
~turm *m* [*dieser Ausdruck wird fälschlich für eine Turmbetonzentrale verwendet*]
Betonierung *f* → Betoneinbau *m*
Betonierungs|anlage *f* → Betonieranlage
~arbeiten *fpl* → Betoneinbau *m*
~einrichtung *f* → Betonieranlage *f*
Betonier(ungs)|fläche *f* = pouring area, concreting ~, casting ~
~fuge *f* → Arbeitsfuge
~fugenband *n* → Arbeitsfugenband
~kübel → Betonkübel

Betonier(ungs)plan — Beton-Klärgrube 198

~plan m, Betoneinbauplan = concrete progress chart, progress chart for concrete work
~platz m, Herstellungsplatz, Werkplatz [*für Fertigteile*] = casting yard
~stahl m → Betonstahl
~stelle f = concreting site, pouring ~
~verfahren n = concreting method
~vorgang m = concreting operation
~zyklus m = casting cycle, pouring ~, concreting ~

Beton(ier)zug m, fahrbare Stollenbetonieranlage f, ortsbewegliche Stollenbetonieranlage = tunnel concreting train, ~ pouring ~
~ [*Straßen- und Flugplatzbau*] = concreting train; paving ~ (US)

Beton|-Imprägnier(ungs)mittel n = concrete impregnation agent
~industrie f = concrete industry
~ingenieur m = concrete engineer
~injektor m → Betoneinpreßmaschine
~innen(ein)rütt(e)lung f = pervibration (of concrete), internal vibration (~ ~)
~innenrüttler m → Innenrüttler
~innenvibrator m → Innenrüttler m
~-Kabel(hochbahn)kran m, Kabelkran mit Beton-Gießvorrichtung = concrete-chuting cable crane
~kai m = concrete quay
~kamintür f = concrete chimney door
~kämpfer m = concrete rein, ~ impost, ~ springer
~kanal m, Beton-Abwasserkanal, Betonleitungskanal, Abwasser-Betonkanal = conduit-type concrete sewer
~kanone f → Betonspritzmaschine f
~karre(n) f, (m), Beton(rund)kipper m, Japaner(karren) m, Japaner-(Kipp)Karre, Kipp-Betonkarre(n), Betonvorderkipper, Kipp(er)karre(n), Betonkippkarre(n) = concrete buggy, (hand) concrete-cart, rocker-dump hand cart, circular-tipping concrete skip, two-wheeled concrete cart

~karrenaufzug n = (cart and) barrow hoist
~-Kassettenplatte f = concrete cored slab, ~ waffle ~
~kasten m [*für Gründungen unter Wasser*] = concrete caisson
~kern m, Betonbohrkern, Betonprobekern = concrete test core
~kern m, Betondichtungskern, Betonkernmauer f, massive Kernmauer [*Talsperre*] = concrete core wall
~kern m, Stahl~, Betonkernbauwerk n, (Gebäude)Kern = reinforced-concrete core
~kernbauwerk n → Betonkern m
(~)Kernbohrmaschine f, (Beton-)Kernbohrgerät n, Bohrgerät n zur Kernentnahme aus Straßendecken, Bohrmaschine zur Kernentnahme aus Straßendecken = (concrete) core (cutting) machine, coring machine, core drill(ing machine), (concrete) coring drill, core-cutting rig
~kerndamm m, Kernmauerdamm = concrete core (wall) (type) dam
~kernmauer f, Beton(dichtungs)kern m, massive Kernmauer [*Talsperre*] = concrete core wall

Betonkies m, natürlicher Grobzuschlag(stoff) m [*Rückstand auf dem Rundlochsieb 7 nach DIN 1170*] = natural coarse aggregate, ~ gravel [*This term means an aggregate mainly retained on a 3/16 in. B. S. test sieve*]
~sand m = all-in aggregate

Betonkipper m → Betonrundkipper
~ [*Kipper auf Feldbahngleis, 600-mm-Spur*] = rail-mounted tipping concrete skip

Betonkipp|karre(n) f, (m) → Betonkarre(n)
~kübel m [*beim Muldenaufzug*] → Kippmulde f
~mulde f [*beim Muldenaufzug*] → Kippmulde

Beton|kirche f = concrete church
~-Klärgrube f = concrete septic tank

Betonklebstoff — Betonmauer

~**klebstoff** *m*, Betonkleber *m* = concrete bonding agent, ~ adhesive
~**kolben** *m* [*die in die Rohrleitung eingeschobene Kesselfüllung eines Druckluftbetonförderers*] = batch
~**kolbenpumpe** *f* = concreting piston pump
~**kolonne** *f* → Betonierkolonne
~**komponente** *f*, Betonbestandteil *m* = concrete ingredient
(~)**Konsistenz** *f* → (Beton)Steife *f*
~**konsole** *f* = concrete bracket, ~ corbel
~**körper** *m* = body of concrete
~**korrosion** *f*, Zerstörung *f* des Betons = concrete corrosion
~**korrosionsschutz** *m* = protecting concrete against corrosion
~**kragarm** *m*, Betonkragstück *n* = concrete cantilever arm
~**kran(bahn)träger** *m* [*Betonträger für die Fahrschiene eines Laufkranes*] = concrete girder for crane runway
~**kreisplatte** *f*, Betonrundplatte = concrete circular slab
~**kreisrohr** *n* → Betonrundrohr
~**(kreis)säge** *f* → Fugenschleifgerät *n*
~**(kreis)säge** *f* [*zum Durchsägen von Beton*] = concrete saw
~**kriechen** *n* = creep of concrete
~**kriechtheorie** *f* = theory of concrete creep
~**krümmer** *m*, Beton(rohr)bogen *m*, Betonbogenrohr *n* = concrete (pipe) bend
~**kubatur** *f* = concrete cubage, ~ cubic yardage

Betonkübel *m*, Betonschüttkübel, (Kabel)Krankübel, Betontransportkübel, Betonier(ungs)kübel [*manchmal fälschlich auch „Betonsilo" genannt*] = concrete (placing) bucket, ~ (~) skip, ~ pouring ~
~ [*Anbauvorrichtung an einem Schaufellader*] = concreting skip attachment, ~ bucket ~
~**aufzug** *m* → Mulden(bau)aufzug
Beton|kuchen *m* = concrete pat
~**kühlanlage** *f* = concrete cooling installation, ~ ~ plant
~**kühlung** *f* = concrete cooling
~**kühlverfahren** *n*, Betonkühlmethode *f* = concrete cooling method
~**kuppel** *f* = concrete dome
~**labor(atorium)** *n* = concrete lab(oratory)
~**lage** *f*, Betonschicht *f* = concrete course, ~ layer
~**landebahn** *f* = concrete landing runway
~**längskraft** *f* = longitudinally acting force of concrete
(~)**Leichtzuschlag(stoff)** *m* = lightweight (concrete) aggregate
~**leistungskurve** *f* → Betonierkurve
~**leitplanke** *f* → Beton-(Sicherheits-)Leitplanke
~**leitstreifen** *m* → Leitstreifen
~**leitung** *f* = concrete line
~**leitungskanal** *m* → Betonkanal
~**leuchtturm** *m* = concrete lighthouse
~**-Lichtmast** *m*, Beton-Beleuchtungsmast = concrete (lighting) column (Brit.); ~ mast (US) [*B. S. 1308*]
~**lieferwerk** *n* → Transportbetonwerk
~**löffel** *m*, Betonbüchse *f*, Zementierbüchse [*Tiefbohrtechnik*] = cement dump, dump bailer
~**lösungsmittel** *n*, Betonlöser *m*, Betonlösemittel = concrete solvent, ~ remover [*used to soften and remove built-up concrete on mixers, moulds, barrows, etc.*]
(~)**Luftmengenmesser** *m*, (Beton)Belüftungsmesser = (concrete) air entrainment meter
~**-Luftschleuse** *f* → Beton-Druckluftschleuse
~**mantel** *m*, Betonauskleidung *f*, Betonverkleidung, nicht tragende Betonmauerung *f* geringer Dicke, Betonbekleidung [*Felshohlraumbau*] = concrete lining, ~ facing
~**masse** *f* = mass of concrete, concrete mass
~**mast** *m* = concrete column (Brit.); ~ mast (US)

Betonmauer *f* = concrete wall
~, Betonstaumauer, Massiv~, Betonsperre *f*, Staumauer, Talsperren-

(stau)mauer, (Beton)Sperrmauer, Betontalsperre = (massive) concrete dam
~**stein** *m*, Mauerstein aus Beton = concrete wall block
Beton|mauerung *f* [*Tunnel- und Stollenbau*] = permanent concrete supporting
~**mauerwerk** *n* = concrete blockwork
~**meißel** *m* [*Abbruchhammer*] = concrete breaking chisel
Betonmischanlage *f*, Beton(auf)bereitungsanlage = concrete mixing plant, ~ fabrication ~, ~ production ~

~ **in Punktanordnung** → Betonturm *m*
~ ~ **senkrechtem Aufbau** → Betonturm *m*
Beton|mischautomat *m* → → Betonmischer *m*
(~)**Mischblech** *n* = concrete mixing plate
Betonmischen *n*, Betonmischung *f* = concrete mixing
~ **von Hand**, Betonmischung *f* ~ ~, Hand-Betonmischen, Hand-Betonmischung = hand concrete mixing, manual ~ ~
~ **während des Transports**, Betonmischung *f* ~ ~ ~ = truck mixing

---

Betonmischer *m*, Betonmischmaschine *f*

absatzweise arbeitender Mischer → Periodenmischer
Anhängedeichsel *f*
Aufzugkasten *m*, (Schräg)Aufzugkübel *m*, Kippkübel, Vorfüllkasten, Mischer(beschickungs)kübel, Beschickerkübel, Beschikkerkasten, (Beschickungs)Kübel, Beschicker *m*, Aufzugmulde *f*, Aufzugbeschickungskübel
Aufzug-Klein(beton)mischer

Aufzugkübelrahmen *m*, (Schräg-) Aufzugschienen *fpl*, Beschickungsbahn *f*, (Kübel)Aufzugbahn
Aufzugkübelwaage *f*, Mischer(beschickungs)kübelwaage
Auslaufseite *f*
Austragmischer → Durchlaufmischer
Austrag(ung) *m*, (*f*)
Austrag(ungs)schurre *f*, (Auslauf-) Schurre
Auto-Transportbeton-Mischer → Transportmischer
Beschickungskübel *m* → Aufzugkasten *m*
Beschickungsbahn *f* → Aufzugkübelrahmen *m*

concrete mixer, ~ mixing machine

drawbar, towbar, towing pole
(loading) skip, mixer skip, hopper open-end skip, folding weigher, power-operated skip, side-loader, (loading) hopper, (batch) loader (itself), loader skip

small-capacity (concrete) mixer with winch
inclined guides, trackway for skip rollers, loader runway, guide rails

hopper scale, skip scale, loader scale

discharge side

discharge
(discharge) chute, (~) spout

Beschickungsstellung *f*	load position
Beschickungswerk *n*	(batch) loader (unit), loader raising gear
Beschickungswinde *f*, Beschickerwinde	loading winch
Betonautomat *m* → Durchlauf-(Beton)Zwangsmischer	
Betonmischautomat *m* → Durchlauf-(Beton)Zwangsmischer	
Betonmischer mit Handschrapperbeschickung des Aufzugkübels	concrete mixer fitted with mechanical skip, loader attachment, scraper-fed concrete mixer
Betontransportwagen *m* mit aufgebautem Rührwerk(mischer) → Nachmischer	
Betonzwangsmischer → Zwangsmischer	
Charge *f*, Mischung *f*, Gemisch *n*	batch
Chargen-Betonmischer mit Wendetrommel → Umkehr(trommel)(beton)mischer	
Chargen-Beton-Zwangsmischer	batch-type concrete pug-mill (mixer)
Chargenmischer → Periodenmischer	
Chargenzähler *m*, Mischungszähler	batch counter, batchmeter
Doppel-Konus-Kipptrommelmischer	duo-cone tilting (type) mixer
Doppeltrommel(beton)mischer	double-drum (concrete) mixer
Doppelwellenmischer, Zweitrog-Zwangsbetonmischer, Doppelrührwellen-Zwangsmischer	two-shaft pug(-)mill concrete mixer
Durchlaufmischer, Mischer mit Schurrenaustrag(ung), Austragmischer, Freifall-Durchlaufmischer	closed drum (concrete) mixer with tipping chute discharge
Durchlauf-(Beton)Zwangsmischer, Stetig(beton)zwangsmischer, kontinuierlicher (Beton-)Zwangsmischer, Betonautomat *m*, Konti-Mischer, Beton-Mischautomat, Dauer-(Beton)Zwangsmischer	constant-flow (or continuous) (concrete) pug mill mixer
Einachsfahrwerk *n*	2-wheeled axle
Einlaufseite *f*	feeding side, loading side
Einseilaufzug *m*	single rope (or cable) hoist
Eintrog-Zwangsmischer, Einwellenmischer	single-shaft pug(-)mill concrete mixer
Entleerstellung *f*	discharge position
Entleerung *f*	discharge
Fahrmischer → Transportmischer	

Betonmischer

Freifall-Chargen-Kipptrommelmischer [*ortsfest*], Fabrikat THE T. L. SMITH COMPANY, MILWAUKEE 45, WISCONSIN, USA	SMITH TILTER [*Trademark*]
Freifallchargenmischer	free fall (type) batch mixer, gravity batch mixer, rotary batch mixer
Freifall-Durchlaufmischer → Durchlaufmischer	
Freifallmischen *n*	gravity mixing
Freifallmischer → Trommelmischer	
Freifallmischer mit Umkehraustragung → Umkehr(trommel)(beton)mischer	
Gegenstrom-Schnellmischer, Gegenstrom-Betonzwangsmischer	counter-current revolving-pan mixer, contra-flow paddle type mixer
Gemisch *n*, Charge *f*, Mischung *f*	batch
Großbetonmischer, Groß-Betonmischmaschine *f*	volume-production concrete mixer, large mixer
Gummireibradantrieb *m*	rubber friction drive
Handpumpe *f*	hand-pump
Handrad *n*	hand wheel
Kippkübel *m* → Aufzugkasten *m*	
Kipptrommel *f*	tilting drum
Kipptrommelmischer	tilting (drum) mixer, tilt(-drum) mixer
Knetmischer → Zwangsbetonmischer mit vertikalem Rührwerk	
kombinierter Nach- und Liefermischer, Fabrikat THE T. L. SMITH COMPANY, MILWAUKEE 45, WISCONSIN, USA [*als „Liefermischer" mit der Anlage für die Wasserzugabe*]	SMITH MOBILE [*Trademark*]
Konti-Mischer → Durchlauf-(Beton-)Zwangsmischer	
kontinuierlicher (Beton)Zwangsmischer → Durchlauf-(Beton-)Zwangsmischer	
Kratzschaufel *f*	scraper blade
Kübel *m* → Aufzugkasten *m*	
Kübelaufzugbahn *f* → Aufzugkübelrahmen	
Kübelaufzugmischer	elevating hopper mixer
Laufkranz *m*	roller path
Leistung *f* in fertiger Mischung	mixed batch cap., $\sim$ $\sim$ capacity
Liefermischer → Transportmischer	
Mischer mit 280 Liter Fassungsvermögen für ungemischtes Material und 195 Liter Leistung in fertiger Mischung	mixer 10/7 [*capacity is ten cubic feet unmixed and seven cubic feet mixed*]

Betonmischer

Mischerführer m, Mischerbediener	mixer driver, mixer operator
Mischerkübel m → Aufzugkasten m	
Mischer m mit Schurrenaustragung → Durchlaufmischer	
Mischermotor m	mixer engine
Mischerrahmen m	mixer frame, ~ framing
Mischschaufel f, Mischerschaufel	mixing (or helical, or baffle) blade
Mischschnecke f	mixing screw
Mischstellung f	mix position
Mischstern m	mixing star
Mischtabelle f	table of mixes
Mischteil m, n	mixing unit
(Misch)Trog m	pan
Mischtrommel f	mixing drum
Mischung f, Charge f, Gemisch n	batch
Mischungszähler m, Chargenzähler	batchmeter, batch counter
Mischwelle f	mixing shaft
Nachmischer, Betontransportwagen m mit aufgebautem Rührwerk(-mischer) [*zentrale Betonaufbereitung*]	agitating (or agitator) conveyor (or truck), truck agitator (US); agitating lorry; agitator [*deprecated*] (Brit.)
Nachmischer [*ohne Fahrzeug*]	agitator, remixer
Periodenmischer, Chargenmischer, Stoßmischer, absatzweise arbeitender Mischer	batch (type concrete) mixer
Reibrad n	friction wheel
Reibradbetonmischer	friction wheel-drive concrete mixer
Rücklaufentleerung f, Umkehrentleerung	reverse discharge, reversing drum ~
Rührwerkmischer → Zwangsbetonmischer mit vertikalem Rührwerk	
Rührwerkmischer → Zwangsbetonmischer mit horizontalem Rührwerk	
Rüttler m, Vibrator m	(skip) shaker (gear)
Sackaufschneider m	bag cutter
(Schräg)Aufzugkübel m → Aufzugkasten m	
Seelemann-Regulus-Mischer, REGULUS-Beton-Mischautomat m [*Bindemittel und Zuschlagstoffe laufend im Mischungsverhältnis zugeteilt*]	Seelemann-Regulus mixer [*aggregate and cement are measured by feed screws, giving a continuous output of concrete*]
Seitenentleerung f	side discharge
Stetig(beton)zwangsmischer → Durchlauf-(Beton-)Zwangsmischer	
Stoßmischer → Periodenmischer	
Straßen(beton)mischer, Straßenbetoniermaschine f	rotating drum paver mixer, travel(l)ing (concrete) mixer (plant), combined mixing and paving ma-

Betonmischer

Stützrad n	front stand wheel
Stützradvorrichtung f	front stand
Tellerzwangsmischer	turbomixer
Transportmischer, Liefermischer, Fahrmischer, Auto-Transportbeton-Mischer [*zentrale Abmessung, Mischen während der Fahrt*]	truck mixer, transit-mixer (truck) (US); mixer lorry (Brit.)
Trog m, Mischtrog	pan
Trogmischer m	open-pan mixer
Trommelantrieb m	drum drive
Trommeldrehzahl f	drum speed, speed of drum
Trommelfüllung f	drum cap. per batch, unmixed (batch) capacity
(Trommel)Kippvorrichtung f	(drum) tilt(ing) mechanism (or gear, or unit, or device)
Trommelmischer, Freifallmischer	rotary (drum) mixer, rotary type mixer, (revolving) drum (type) mixer, gravity mixer
Trommelsteuerung f	drum control
Trommeltragachse f	drum shaft
Umkehrentleerung f, Rücklaufentleerung	reversing drum discharge, reverse $\sim$
Umkehrtrommel f	non-tilting drum
Umkehr(trommel)(beton)mischer, Freifallmischer mit Umkehraustrag(ung), Mischer mit Reversier-Mischtrommel, Chargen-Betonmischer mit Wendetrommel, Umkehrtrommel-Freifallmischer, Freifall-Betonmischer mit Rücklaufentleerung, Reversiermischer	closed drum (concrete) mixer with reverse discharge, (free fall type) non-tilt(ing) (drum) mixer, N. T. mixer
Verriegelung f	lock mechanism
Vibrator m → Rüttler	
Vorfüllkasten m → Aufzugkasten	
Wasserbehälter m	water (measuring) tank
Wasserdosiergerät n, Wasserdosator m	batch water meter
Wasserzähler m, Wassermesser m	water meter (system), $\sim$ flowmeter
Werk n	unit, mechanism, gear, device
Zwangsbetonmischer mit vertikalem Rührwerk, Kneter m, Knetmischer, Rührwerkmischer	pug(-)mill (mixer)
Zwangsbetonmischer mit horizontalem Rührwerk, Rührwerkmischer	pan type concrete mixer
zweiräd(e)rige Anhängerausführung f	2-wheel(ed) portable model

(continued from previous page: ...chine, (combined) paver, paving mixer, (concrete) paver)

Betonmischer — Betonpfostenform

Zweitrog-Zwangsbetonmischer, Doppelwellenmischer = two-shaft pug(-)mill concrete mixer

Betonmischer *m* **auf Brücke** → Brückenmischer
~bau *m* = concrete mixer construction
~fahrzeug *n* → Mischerfahrzeug
~führer *m* = concrete mixer operator
Betonmisch|temperatur *f*, **Beton(auf)-bereitungstemperatur** = concrete mixing temperature
~turm *m* → Betonturm
~- und -dosierturm *m* → Betonturm
Betonmischung *f*, **Betongemisch** *n*, **Betongemenge** *n* = concrete mix(ture)
~, Betonmischen *n* = concrete mixing
Beton|misch(ungs)verhältnis *n*, **Betonmisch(ungs)formel** *f* = concrete mix proportions
~mischzentrale *f* → Betonzentrale
~modell *n* = concrete model
(Beton)Montage|bau *m*, **(Beton)Fertigteilbau** = precast (concrete) construction
~teil *m, n* → (Beton)Fertigteil
~teilindustrie *f* → Fertigteilindustrie
~wand *f*, **Betonfertigwand, Montagebetonwand** = prefab(ricated) concrete wall
Beton|mulde *f* [*für Transportbeton*] = concrete bowl
~muldenaufzug *m* → Mulden(bau)aufzug
~nachbehandlung *f* → Nachbehandlung
~nachbehandlungsmaschine *f* [*Betondeckenbau*] = concrete curing machine
(~)Nachbehandlungsmittel *n* = (concrete) curing agent
(~)Nachbehandlungszeitraum *m* = (concrete) curing period
(~)Nachmischer *m* → (Beton)Rührfahrzeug
~nadelrüttler *m* → Innenrüttler
~nadelvibrator *m* → Innenrüttler *m*

~nest *n* [*Hohlraum im Beton infolge ungenügender Verdichtung oder falscher Kornzusammensetzung*] = cavity in the concrete
Betonnung *f* = buoyage
Beton|oberfläche *f* = concrete surface
~ortpfahl *m* → Ortpfahl
~packlage *f*, **Betonschotter** *m*, **Kleinschlag** *m* **aus Betonbrocken** = (aggregate of) broken concrete, concrete hardcore, crushed concrete
Betonpfahl *m* = concrete pile
~ mit konischen teleskopartigen Rohrschüssen = step taper concrete pile
~fabrik *f* → Betonpfahlwerk *n*
~fundament *n* → Betonpfahlgründung *f*
~fundation *f* → Betonpfahlgründung
~gründung *f*, **Betonpfahlfundament** *n*, **Betonpfahlfundation** *f*, **Betonpfahlgründungskörper** *m*, **Betonpfahlfundamentkörper** *m* = concrete pile foundation (structure)
~-Ramme *f* = concrete pile driver
~rost *m* [*Fehlname: Betonrost*] = concrete piled pier
~werk *n*, **Betonpfahlfabrik** *f* = concrete pile making plant
Betonpfeiler *m*, **Betonfundamentpfeiler, Betongrundpfeiler** = concrete (foundation) pier
~ ohne Bewehrung, ~ ~ Armierung, ~ ~ (Stahl)Einlagen = plain concrete (foundation) pier
Betonpfette *f* = concrete purlin(e)
Betonpflaster|(decke) *n*, (*f*) = concrete sett paving
~platte *f* = concrete paving slab
~stein *m* = concrete paving sett
~steinmaschine *f* = concrete paving sett machine
Beton|pfosten *m* = concrete post
~pfostenform *f* = concrete post mo(u)ld

Betonpfropfen — (Beton)Probewürfelpresse 206

~**pfropfen** m, Betonplombe f = concrete plug
Betonplatte f = concrete slab
~, Betondeckenplatte, Betonfeld n = concrete pavement slab, ~ surfacing ~; ~ paving ~ (US)
~ [*Hochbau*] = concrete structural slab
~, Betonfliese f, Betonbelagplatte = concrete tile
Betonplatten|(ab)deckung f, Betonplattenbelag m [*Wasserseite eines Erdstaudammes*] = concrete slab facing (US); ~ ~ ~ membrane (Brit.)
~**auskleidung** f → Betonplattenbekleidung
~**bau** m, Plattenbau = (concrete) slab construction
~**bekleidung** f, Betonplattenauskleidung, Betonplattenverkleidung, Betonfliesenbekleidung, Betonfliesenverkleidung, Betonfliesenauskleidung = concrete tile lining
~**bekleidung** f, Betonplattenauskleidung, Betonplattenverkleidung = concrete slab lining
~**belag** m, Betonplatten(ab)deckung f [*Wasserseite eines Erdstaudammes*] = concrete slab facing membrane (Brit.); ~ ~ facing (US)
~**boden** m, Betonplattenfußboden, Betonfliesen(fuß)boden = concrete tile flooring, ~ ~ floor finish
~**brücke** f = concrete slab bridge
~**deckung** f, Betonplattenabdeckung, Betonplattenbelag m [*Wasserseite eines Erdstaudammes*] = concrete slab facing (US); ~ ~ ~ membrane (Brit.)
~**ende** n = concrete slab end
~**fundament** n → Plattenfundament
~**fundamentkörper** m → Plattenfundament n
~**fundation** f → Plattenfundament m
~**fußboden** m → Betonplattenboden
~**gründung** f → Plattenfundament n
~**gründungskörper** m → Plattenfundament n

~**heben** n, Regulieren von abgesunkenen Betonfahrbahnplatten, Plattenheben = raising of sunken (concrete) slabs
~**presse** f, Betonfliesenpresse = concrete tile press
~**presse** f = concrete slab press
~**pumpen** n → Plattenpumpen
~**rand** m = concrete slab edge
~**reibung** f, Bodenreibung, Plattenreibung = (frictional) restraint of (the) subgrade, subgrade restraint
~**überzug** m = overslabbing
~**verkleidung** f → Betonplattenbekleidung
~**werk** n, Betonplattenfabrik f, Betonfliesenwerk, Betonfliesenfabrik = concrete tile works, ~ ~ factory, ~ ~ plant
Beton|plattform f = concrete platform
~**plombe** f, Betonpfropfen m = concrete plug
~**polier** m = concrete foreman
~**portalrahmen** m = concrete portal frame
~**pressung** f [*Spannung eines auf Druck beanspruchten Betonteiles in kg/cm²*] = compressive stress of concrete
~**prinzip** n, Hohlraumminimumprinzip [*Asphaltbeton und Teerbeton sind nach diesem Prinzip hergestellt. Das Gegenteil ist „Makadamprinzip"*] = principle of minimum void(s) content
~**prisma** n = concrete prism
Betonprobe f → Betonprobekörper m
~, Betonprüfung f, Betonversuch m = concrete test
~**balken** m, Betonprüfbalken = concrete test beam
~**kern** m, Beton(bohr)kern, Deckenkern = concrete test core
~**probekörper** m, Betonprobe f, Betonprobestück n = concrete sample
~**nahme** f = sampling of concrete
~**stück** n → Betonprobekörper m
(Beton)Probewürfel m → (Beton-)Würfel
~**presse** f → Druckpresse

(Beton)Probezylinder *m*, Betonzylinder, (Beton)Versuchszylinder = concrete test cylinder, compression ~

Betonprofil *n* = concrete section

Betonprüf|gerät *n* = concrete tester, ~ test apparatus

~hammer *m*, Betonschlaghammer, Rückprall-Härteprüfer *m*, Rückprallhammer = concrete test hammer, rebound tester, scleroscope [*A tester indexing the compressive strength of concrete by the height of the elastic rebound*]

~maschine *f* = concrete testing machine

~prüfpresse *f* → Druckpresse

Betonprüfung *f*, Betonprobe *f*, Betonversuch *m* = concrete test

(Beton)Prüfwürfel *m* → (Beton)Würfel

Betonpumpe *f*, Betonförderpumpe, Förderpumpe = concrete pump

~ mit zwei Pumpaggregaten = two-valve concrete pump

~ System TORKRET, Torkretpumpe *f* [*Trademark*] = TORKRET concrete pump

Betonpumpen *n*, Einbringen von Pumpbeton = pumping of concrete

~rohr *n* = concrete pumping pipe

~(rohr)leitung *f* = concrete pump pipeline

Beton|qualität *f*, Betongüte *f* = concrete quality

~querschnitt *m* = concrete section

(~)Querverteiler *m* → Betonverteilungswagen *m*

~rahmenkonstruktion *f* = concrete-frame construction

~rammpfahl *m* = (impact-)driven concrete pile

~rammpfahl *m* **System Mast**, Mast-Betonrammpfahl = Mast (impact-)driven concrete pile

~randspannung *f* = concrete edge stress

(~)Randstein *m* = (concrete) edging

(~)Randstreifen *m* → Leitstreifen

~-Raumfachwerk *n* = concrete space truss

~-Raumfuge *f*, (Beton-)Dehn(ungs)fuge = concrete expansion joint

~raumtragwerk *n* = concrete space structure

~-(Reaktor)Strahlenschutzanlage *f*, Beton-Schutzanlage, Beton-(Strahlen)Abschirmungsanlage = concrete (reactor) shield, ~ structure for shielding atomic plants

~reparatur *f*, Betonausbesserung *f* = concrete patching, ~ reintegration, ~ repair

~rezept *n* = concrete formulation

~ring *m* = concrete ring

~rinne *f* = concrete channel

~-Rinnenstein *m* = concrete gutter (paving) sett

~rinnstein *m* = concrete gutter

~rippe *f* = concrete rib

~rippendecke *f* = concrete ribbed floor

~rippenplatte *f* = concrete ribbed slab

(~)Rippenstahl *m*, quergerippter Betonformstahl [*Die gleichmäßig auf die ganze Stablänge verteilten Querrippen stehen rechtwink(e)lig zur Stabachse und erstrecken sich über den ganzen Stabumfang*] = ribbed (re-)bars, ~ reinforcing bars, ~ (reinforcing) rods

Betonrohr *n*, Zementrohr = concrete pipe

~ mit eiförmigem Abflußquerschnitt, eiförmiges Betonrohr = egg-shaped concrete pipe

~ ~ kreisförmigem Abflußquerschnitt → Betonrundrohr

~automat *m* = auto(matic) concrete pipe machine

~bogen *m*, Betonbogenrohr *n*, Betonkrümmer *m*, Betonbogen = concrete (pipe) bend

~drän *m* = concrete pipe drain; ~ tile ~ (US)

~düker *m* = concrete pipe siphon

~durchlaß *m*, Betonrohrabzugkanal *m* = concrete pipe culvert

Betonrohr(-Einsteig)schacht — Betonrundrohr

~**(-Einsteig)schacht** *m*, Betonrohr-Mannloch *n* = concrete pipe manhole

~**fertigung** *f*, Betonrohrherstellung = = concrete pipe making

~**fertigungsmaschine** *f* → Betonrohrmaschine

Beton-Rohrfestpunkt *m* = concrete thrust block

Betonrohr|form *f*, Zementrohrform = concrete pipe mould (Brit.); ~ ~ mold (US)

~**formmaschine** *f* → Betonrohrmaschine

Beton(rohr)formstück *n*, Beton(rohr)-fitting = concrete fitting

Betonrohr|hersteller *m*, Zementrohrhersteller = concrete pipe maker

~**herstellung** *f*, Betonrohrfertigung = concrete pipe making

~**herstellung** *f* **in Rohrgräben** = pipecasting in trenches

~**herstellungsmaschine** *f* → Betonrohrmaschine

~**leitung** *f*, Betonrohrstrang *m* = concrete pipe line

~**-Mannloch** *n* → Betonrohr-(Einsteig)schacht *m*

~**maschine** *f*, Betonrohrfertigungsmaschine, Betonrohrformmaschine, Betonrohrherstellungsmaschine, Rohr(fertigungs)maschine, Rohrformmaschine, Rohrherstellungsmaschine = (concrete) pipe making machine

~**presse** *f*, Zementrohrpresse = concrete pipe press

~**prüfgerät** *n* = concrete pipe tester

~**prüfpresse** *f* = concrete pipe compression tester

~**schacht** *m* → Betonrohr-Einsteigschacht

~**schale** *f*, Betonschale = concrete split duct

~**schleudermaschine** *f* = concrete pipe spinning machine

~**stampfmaschine** *f* → Zementrohrstampfmaschine

~**strang** *m*, Betonrohrleitung *f* = concrete pipe line

~**-Verlegeanlage** *f*, Betonrohr-Verlegegerät *n* = concrete pipe laying unit

~**-Vibriermaschine** *f*, Betonrohr-Vibrator *m*, Vibriermaschine für Betonrohre, Vibrationsrohrmaschine, Betonrohrvibrationsmaschine = concrete pipe vibrating machine

Beton|rost *m* [*Fehlname*] → Betonpfahlrost

(~)**Rührfahrzeug** *n*, (Beton)Nachmischer *m*, (Beton)Rührwagen *m*, Fertigbeton-Nachmischer, Transportbeton-Nachmischer, Lieferbeton-Nachmischer, Fahrmischer [*ein Fahrzeug, das zur Aufnahme des in einer Mischmaschine fertig hergestellten Betons eingerichtet ist und den Beton bis zum Augenblick der Ablieferung in einwandfreiem Zustand erhält*] = agitator, remixer

(~)**Rührwagen** *m* → (Beton)Rührfahrzeug *n*

(Beton)**Rundeisen** *n*, (Beton)Rundstahl *m* = round re-bars, ~ reinforcing bars, ~ reinforcing rods

~**bündel** *n*, (Beton)Rundstahlbündel = bundle of round reinforcing bars

~**stab** *m*, (Beton)Rund(stahl)stab, Betonstab = round re-bar, ~ reinforcing bar, ~ reinforcing rod

Beton(rund)kipper *m*, Betonkarre(n) *f*, (*m*), Japaner(karren), Japaner-(Kipp)Karre, Kipp-Betonkarre(n) = (hand) concrete-cart, concrete buggy, rocker-dump hand cart, circular-tipping concrete skip, two-wheeled concrete cart

~, (Beton)Schüsselwagen *m* [*als Schienenfahrzeug*] = concrete narrow-gage railcar (US); (~ tip) wagon (Brit.)

Betonrund|platte *f*, Betonkreisplatte = concrete circular slab

~**rohr** *n*, Betonkreisrohr, Betonrohr mit kreisförmigem Abflußquerschnitt, kreisförmiges Betonrohr, rundes Betonrohr = concrete cylindrical pipe [*B. S. 556*]

(Beton)Rundstahl — Betonschleuderung

(Beton)Rundstahl *m* → (Beton)Rundeisen *n*

Beton|rüttelstampfer *m* [*Straßenbeton*] = (road) vibrating tamper, concrete tamping and screed board vibrator

~**rütt(e)lung** *f* → Beton(ein)rütteln

~**rüttelverfahren** *n*, Betonrüttelmethode *f* = concrete vibration method

(~)**Rüttler** *m*, (Beton)Vibrator *m* = (concrete) vibrator

~**sack** *m* = concrete filled bag

Betonsäge *f* → Fugenschleifgerät *n*

~, Betonkreissäge [*zum Durchsägen von Beton*] = concrete saw

Beton|sand *m*, natürlicher Feinzuschlag(stoff) *m* [*Durchgang durch das Rundlochsieb 7 nach DIN 1170*] = natural fine aggregate, ~ (concreting) sand, ~ concrete sand

~**säule** *f* = concrete column

~**schacht** *m* = concrete shaft

~**schachtring** *m* = concrete shaft ring

~**schaden** *m* → Betonschädigung *f*

~**schädigung** *f*, Betonbeschädigung, Betonschaden, Betonschäden *mpl* = concrete distress

betonschädliches Wasser *n* = water aggressive to concrete

Beton|schale *f*, Betonrohrschale = concrete split duct

~**schale** *f* = concrete shell

(~)**Schalemulsion** *f*, (Beton)Entschalungsemulsion = release emulsion

~**schalendach** *n* = concrete shell roof

(~)**Schalmittel** *n*, (Beton)Entschalungsmittel = release agent

(~)**Schalpaste** *f*, (Beton)Entschalungspaste = release paste

~**schalplattenbauweise** *f* = construction with concrete slabs used as formwork

(Beton)Schalung *f* = (concrete) formwork, (~) shuttering, (~) forms

~ → Seitenschalung

(Beton)Schal(ungs)|geräte *npl* = concrete forming equipment, ~ shoring ~

~**platte** *f* = shuttering panel, formwork ~

~**-Reiniger** *m*, (Beton)Schal(ungs)-Reinigungsmaschine = formwork cleaner, ~ cleaning machine, shuttering cleaner, shuttering cleaning machine

~**rüttler** *m*, (Beton)Schal(ungs)vibrator *m* = (concrete) formwork vibrator (~) shuttering ~

~**schiene** *f* → Seitenschalung *f*

~**steinsilo** *m* = silo of concrete blocks acting as lost formwork

Betonscheibe *f* → Betonschirm *m*

~ **als Stützenelement** [*Brücke*] = flat concrete column

Beton|schicht *f*, Betonlage *f* = concrete layer, ~ course

~**(schienen)schwelle** *f* → Betongleisschwelle

~**schiff** *n* = concrete vessel

~**schirm** *m*, Betondiaphragma *n*, Betonscheibe *f*, Betonschürze *f* = concrete diaphragm, ~ shield, ~ apron

~**schlaghammer** *m* → Betonprüfhammer

(~)**Schlämmschicht** *f*, Zementmilch *f*, Feinschlämme *f* = laitance

~**schleifgerät** *n*, Betondeckenschleifgerät = concrete planer, bumpcutter

(~)**Schleifmaschine** *f* **mit Biegewelle** = flexible grinder, ~ shaft concrete grinding machine

Betonschleuder|anlage *f*, Schleuderbetonanlage = installation for centrifugally cast concrete

~**mast** *m* → Schleuderbetonmast

~**pfahl** *m*, Schleuderbetonpfahl = centrifugally-cast concrete pile, (centrifugally-)spun ~ ~

~**rammpfahl** *m* → Schleuderbetonrammpfahl

~**rohr** *n*, Schleuderbetonrohr = centrifugally-cast concrete pipe, (centrifugally-)spun ~ ~

Betonschleuderung *f* = centrifugal casting of concrete

Betonschleudervorspannrohr — Betonsockelgründung

Betonschleuder|vorspannrohr n, Schleuderbetonvorspannrohr = prestressed centrifugally-cast concrete pipe, ~ (centrifugally-)spun ~ ~
~**werk** n, Betonschleuderfabrik f, Schleuderbetonwerk, Schleuderbetonfabrik = works for centrifugally cast concrete
Betonschleuse f → Beton-(Druck)Luftschleuse
~ [*Schiffahrtschleuse*] = concrete lock
(**Beton**)**Schneckenverteiler** m = screw distributor, ~ spreader, ~ spreading machine
~ **mit Glättelement** = combination screw-screed spreader, ~ ~ distributor, ~ ~ spreading machine
Beton|schotter m, Betonpacklage f, Kleinschlag m aus Betonbrocken, Betonsteinschlag = (aggregate of) broken concrete, concrete hardcore
~**schräge** f, Betondiagonale f = (precast) concrete diagonal
~**schrammbord** n → Betonbordschwelle f
~**schürze** f → Betonschirm m
(~)**Schüsselwagen** m, Beton(rund)-kipper m [*als Schienenfahrzeug*] = (concrete tip) wagon (Brit.); narrow-gage railcar (US)
~**schüsselwagen** m, LKW m für fertigen Beton = concrete bowl truck (US); ~ ~ lorry (Brit.)
~**schüttkübel** m → Betonkübel
~**schüttung** f **unter Wasser** [*im Schutz von Spundwänden oder Fang(e)-dämmen*] = underwater concreting, ~ pour(ing) (of concrete)
~**schutz** m = concrete protection
~-**Schutzanlage** f → Beton-(Strahlen-)Abschirm(ungs)anlage
Betonschutz|anstrich m, Betonschutzfarbschicht f, Betonschutzfarbanstrich = concrete protective paint coat, ~ ~ coat of paint, ~ ~ (paint) finish
~**bunker** m, Betonbunker = concrete shelter

~**farbanstrich** m → Betonschutzanstrich
~**farbschicht** f → Betonschutzanstrich m
~**mittel** n = concrete protective agent
~**schicht** f = protective concrete layer
Beton|schwelle f → Betongleisschwelle
~**schwellenwerk** n → Betongleisschwellenwerk
~-**Schwergewichts-Stützmauer** f = concrete gravity retaining wall
(~-)**Schwerzuschlag(stoff)** m = heavy-weight (concrete) aggregate
~**schwimmdock** n = floating concrete (dry) dock
~-**Schwimmkasten** m = concrete box caisson
~-**Schwimmstoffabweiser** m, Beton-Tauchwand f = concrete scum slab
~**schwindung** f = concrete shrinkage
~**schwitzen** n → Wasserabstoßen
~**seitenstreifen** m, Betonseitenraum m = concrete shoulder
~-(**Sicherheits-**)**Leitplanke** f = concrete safety fence, ~ guard ~, ~ protection ~, ~ guard rail
~**sichtfläche** f = visible concrete surface
~**silo** m = concrete silo
~**silo** m [*Fehlname*] → Betonkübel m
~**skelett** n, Betonfachwerk n = concrete skeleton, ~ framework
~**skelettbau** m, Betonfachwerkbau = concrete framed construction, ~ skeleton ~, ~ framework ~
Betonsockel m → Betonsockelgründung f
~**fundament** n → Betonsockelgründung
~(**fundamentkörper**) m → Betonsockelgründung f
~**fundation** f → Betonsockelgründung f
~**gründung** f, Betonsockelfundament n, Betonsockelfundation f, Betonsockel(fundamentkörper) m, Betonsockelgründungskörper, Betongründungssockel, Betongrundsockel, Betonfundamentsockel, Betonfunda-

tionssockel = plain concrete (single) base, ~~ individual ~, ~~ block
~**gründungskörper** m → Betonsockelgründung f
Beton|sorte f, Betonart f = type of concrete
~**sortenwahl** f = preselection of concrete
~**spannung** f = concrete stress
~**sperre** f, (Massiv)Beton(stau)mauer f, Staumauer, Talsperren(stau)mauer, (Beton)Sperrmauer = (massive) concrete dam
~**sperrenbaustelle** f = (massive) concrete dam (construction) site
(~)**Sperrmauer** f, Betonsperre f, (Massiv)Beton(stau)mauer, Staumauer, Talsperren(stau)mauer, (Beton)Sperrmauer = (massive) concrete dam
~**spirale** f [*Turbine*] = concrete spiral casing, ~ volute ~
~**splitt** m [*7/30 mm*] = crushed gravel
~**splitterschutzwand** f = splinterproof concrete wall, anti-splinter ~
~**sprengung** f = concrete blasting
~-**Spritzarbeit** f, Torkretieren n, Torkretierung f = guniting
~**spritzgerät** n → Betonspritzmaschine f
~**spritzmaschine** f, (Torkret-)Zement(mörtel)kanone f, Torkretbeton-Spritzmaschine f, Druckluft-Spritzgerät n, Zementmörtel-Spritzapparat m nach dem Torkretverfahren, Tektor m, Torkretkanone, Betonkanone, Betonspritzgerät = (pneumatic) concrete gun, cement gunite machine, air placing machine, concrete placing gun, jetcrete gun, Boulder Pneumatic Concretor, compressed-air ejector
~**spritzverfahren** n → Torkretverfahren
~**spund(wand)bohle** f = precast (concrete) sheet pile
~**stab** m → (Beton)Rundeisenstab

Betonstahl m, Bewehrungsstahl, Armierungsstahl, Betonierungsstahl, Bewehrungseisen n [*früher: Moniereisen n*] = reinforcing steel
~-**Abstandhalter** m = bar spacer, reinforcing rod ~
~**bearbeitungsanlage** f = reinforcement cutting and bending plant, steel ~ ~ ~ ~
~**biegemaschine** f → Biegemaschine
~**bieger** m, Betoneisenbieger = rod bender, reinforcement bar-bender
~**(hand)schneider** m → Betoneisen(hand)schneider
~**lager** n = reinforcement storage yard, steel ~ ~
~**matte** f → Baustahlmatte
~**schere** f → Betoneisen(hand)schneider
~**schneid(e)maschine** f, Betoneisenschneid(e)maschine = (reinforcement) bar cutting machine
~**schneiden** n, Betoneisenschneiden = cutting of reinforcement
~**schneider** m → Betoneisen(hand)schneider
~**stab** m, Betoneisenstab = reinforcing rod, (~)bar, re-bar
~**verarbeitungsanlage** f, zentraler Betonstahlverarbeitungsplatz m = reinforcement yard
~-**Verbinder** m = bar binder, reinforcing rod ~
Beton|stampfer m [*aus Stahl mit Holzstiel. Dient zum Verdichten des erdfeuchten Betons von Hand*] = concrete tamper
~**startbahn** f = concrete take-off runway
~**(stau)mauer** f, Massiv~, Betonsperre f, Staumauer, Talsperren(stau)mauer, (Beton)Sperrmauer = (massive) concrete dam
~**steg** m [*Zug- und Druckflansch eines Betonbalkens sind durch den Betonsteg miteinander verbunden, der die Schubspannungen aufnimmt*] = concrete web

(Beton)Steife — Betonstollen

(∼)Steife f, (Beton)Konsistenz f = (concrete) consistency, (∼) consistence, (∼) fluidity

Betonstein m [Sammelbegriff für das fabrik- oder werkstattmäßig hergestellte Erzeugnis aus Beton] = precast concrete

∼, Betonblock(stein) m, Betonformstein, Stein, Block = concrete (building) block, cement (∼) ∼, block

∼arbeiter m = precast concrete worker

∼automat m → (Beton)Stein(form)automat

∼betrieb m = precast (concrete) plant

∼betrieb m = (concrete) block plant

∼-Bodenfertiger m → Bodenfertiger

∼dach n, (Zement)Ziegeldach = concrete tile roof

∼drehtischpresse f = revolving-table (concrete) block press

∼fertiger m → Blockfertiger

∼fertigung f, (Beton)Blockfertigung, (Beton)Steinherstellung, (Beton-)Blockherstellung = (concrete) blockmaking

∼fertigungsmaschine f → Blockfertiger m

∼form f, (Beton)Blockform = (concrete) block mould (Brit.); (∼) ∼ mold (US)

∼(form)automat m, (Beton)Blockautomat, (Beton)Steinfertigungsautomat = auto(matic) (concrete) blockmaking machine

∼formmaschine f → Blockfertiger m

∼hersteller m = precast (concrete) = (concrete) block producer, (∼) ∼ manufacturer, (∼) ∼ maker

∼hersteller m = precast (concrete) manufacturer

∼herstellung f → (Beton)Steinfertigung

∼industrie f = precast (concrete) industry

∼kongreß m = precast concrete symposium, congress on precast concrete

∼maschine f → Blockfertiger m

∼mauern n = blocklaying

∼mauerwerk n, (Beton)Blockmauerwerk = blockwork, concrete-block masonry

∼maurer m = block-layer

∼pflaster n = concrete sett paving

∼presse f = (concrete) block press

∼prüfpresse f, (Beton)Blockprüfpresse, Steinprüfpresse = (concrete) block crushing testing machine

∼-Rauchkanal m = precast concrete block flue

Betonsteinschlag m → Betonschotter m

∼ [als Betonzuschlag(stoff). 30/70 mm] = crushed stone

(Beton)Stein|schneider m → (Beton-)Steintrennmaschine f

∼silo m = concrete block silo

∼stampfmaschine f, (Beton)Blockstampfmaschine, Stampfmaschine = machine tamper

∼trennmaschine f, (Beton)Blocktrennmaschine, (Beton)Steinschneider m, (Beton)Blockschneider, (Beton)Blockschneidemaschine, (Beton)Steinschneidemaschine = block splitter, ∼ cutter, ∼ cutting machine, ∼ splitting machine

∼vibrationsmaschine f = concrete block vibrating machine

∼werk n, (Beton)Blockwerk = (concrete) block factory, precast ∼

∼werkplatz m, (Beton)Blockwerkplatz = (concrete) blockyard

Beton|stirnmauer f [Durchlaß] = concrete head wall

∼stochereisen n = punner, hand rammer [A steel bar plunged up and down in wet concrete to compact it]

∼stocherverdichtung f = punning (Brit.); rodding (US)

∼(stockwerk)decke f, Betonhochbaudecke, Betonetagendecke, Betongeschoßdecke = concrete floor

∼stoff m = concrete material

∼stollen m = concrete gallery

213 Beton-(Strahlen)Abschirmungsanlage — Betontrennwandstein

~-(Strahlen)Abschirmungsanlage *f*, Beton-(Reaktor)Strahlenschutzanlage, Beton-Schutzanlage = concrete (reactor) shield, ~ structure for shielding atomic plants
~straße *f*, Zement~ = concrete road, ~ highway
~-Straßenablauf *m*, Beton-Straßeneinlauf [*DIN 4053*] = concrete gull(e)y
Betonstraßenbau *m*, Zement~ = concrete road construction, ~ highway ~
~maschinen *fpl* und -geräte *npl*, Zement~ = concrete road construction machinery and equipment, ~ highway ~ ~ ~
Betonstraßen|belag. *m* → Betonstraßendecke *f*
~(bohlen)fertiger *m* → Betondeckenfertiger
~decke *f*, Betonstraßenbelag *m* = rigid road pavement, ~ highway ~, concrete ~ ~, ~ surfacing
~fertiger *m* → Betondeckenfertiger
~platte *f* = concrete road slab, ~ highway ~
~verteiler *m* → Betonverteiler
Beton|strecke *f*, Betonabschnitt *m* [*einer Straße*] = concrete section
~streifen *m* = concrete strip, ~ lane Pilot~, Richt~ = pilot ~ ~ Zwischen~ = intermediate ~
~sturz *m* = concret lintel, ~ lintol
~sturzbett *n* = downstream concrete apron
~stütze *f* = concrete column
~tabelle *f* = concrete table
Betontafel *f* = concrete panel
~ mit bloßgelegten Zuschlägen = (exposed) aggregate panel
Beton|talsperre *f* → Betonmauer *f*
~tauchrüttler *m* → Innenrüttler
~tauchvibrator *m* → Innenrüttler *m*
~-Tauchwand *f*, Beton-Schwimmstoffabweiser *m* = concrete scum slab

~technik *f* = concrete engineering
~technologe *f* = concrete technologist
~technologie *f* = concrete technology
~(tief)bordstein *m* = flush concrete curb; ~ ~ kerb (Brit.)
~tiefenrüttler *m* → Innenrüttler
~tiefenvibrator *m* → Innenrüttler *m*
Betonträger *m* = concrete girder
~brücke *f* = concrete girder bridge
~- und -plattenbrücke *f* = concrete girder and slab bridge
Beton|tragschicht *f* → obere ~
Betontransport *m*, Betonförderung *f* = distribution of concrete, concrete transport [*Transporting concrete from the mixer to the formwork*]
~ durch Druckluft, Betonförderung *f* ~ ~ = pneumatic concrete conveying
~ in Rohren, Betonförderung *f* ~ ~ = concrete conveying through pipes
Beton|transporter *m*, Betontransportfahrzeug *n* = concrete hauling unit, ~ hauler
~transporter *m*, Betonbeschicker [*Betonsteinindustrie*] = concrete loader
~transporteur *m* → Betonförderapparat *m*
Betontransport|fahrzeug *n*, Betontransporter *m* = concrete hauler, ~ hauling unit
~geräte *npl* = (concrete) distributing plant
~kübel *m* → Betonkübel
~leitung *f* → Betonförder(rohr)leitung
Beton-Transportmischer *m* → Transportmischer
Betontransportrohr *n*, Betonförderrohr = concrete delivery pipe, ~ discharge ~
~leitung *f*, Betonförderrohrleitung = concrete discharge pipework, ~ delivery ~
Beton|trennmaschine *f* = concrete cutter, ~ cutting machine
~trennsäge *f* = concrete saw(ing) machine, ~ saw
~trennwandstein *m* = concrete block for partitions

~treppe *f* = concrete stair (US); ~ staircase (Brit.)
~-Treppenweg *m* = stepped concrete path
~trichter *m* [*Betonpumpe*] → Trichter
~trichter *m* = concrete placement funnel
~trümmer *mpl* = demolished concrete
~tübbing *m* = concrete ring, ~ tubbing
~tunnel *m* [*im Zuge von Verkehrswegen*] = concrete (traffic) tunnel
~turm *m* = concrete tower
~turm *m*, Betonmischturm, Turmbetonzentrale *f*, Betonfabrik *f* als Turmanlage, Betonmisch- und -dosierturm, Betonmischanlage in Punktanordnung, Betonmischanlage in senkrechtem Aufbau, Betonzentrale in Punktanordnung, Betonzentrale in senkrechtem Aufbau = concrete mixing tower
~überdeckung *f* → Betondeckung
~überzug *m* = concrete overlay
~überzugeinbau *m* [*Einbau einer Betondecke auf einer Straßenbefestigung*] = overslabbing
Betonumhüllung *f*, Betonummantelung = haunching
betonummantelt, betonumhüllt = concrete-encased
Beton|ummantelung *f*, Betonumhüllung = haunching
~umschlag *m* = concrete re-handling
~undurchlässigkeit *f* = impermeability of concrete
~unterbau *m* [*Straßenbau*] = concrete foundation
Betonunterboden *m* = concrete (slab) sub-floor
~ mit Blindboden = suspended concrete sub-floor
~ ohne Blindboden = concrete (slab) sub-floor laid at ground level
Betonunter|konstruktion *f*, Betondachdecke *f* = concrete (roof) deck
~lage *f* [*Unterlage aus Beton*] = concrete supporting medium

~lage *f* [*Unterlage für Beton*] = supporting medium for concrete
(Beton)Verarbeitbarkeit *f* = workability (of concrete)
Beton|verarbeitung *f* → Betoneinbau *m*
~verdichtung *f* = concrete compaction
~verdichtungsgerät *n*, Betonverdichter *m*, Betonverdichtungsmaschine *f* = concrete compactor
~verflüssiger *m*, Plastifizierungsmittel *n*, plastifizierendes Betonzusatzmittel, BV *m*, Weichmacher *m* = plasticizer, workability agent; wetting agent (Brit.)
~verflüssigung *f* = plasticizing of concrete
~verformung *f* = deformation of concrete
~vergütung *f* = improvement of concrete quality
~verkleidung *f*, Betonauskleidung, Betonbekleidung = concrete facing, ~ lining
~verkleidung *f*, Betonauskleidung, Betonbekleidung, Betonmantel *m*, nicht tragende Betonmauerung *f* geringer Dicke [*Felshohlraumbau*] = concrete lining, ~ facing
~verkrustung(en) *f(pl)* = built-up concrete
~verladesilo *m* = concrete loading hopper
~verschleißschicht *f* = concrete wearing layer, ~ ~ course
~-Verschlußschwelle *f* [*Wehr*] = concrete gate sill
(~)Versenkkasten *m* = underwater concreting box
(~)Versenkrohr *n* = trémie pipe
~-Versetzblock *m* = concrete block
~versuch *m*, Betonprobe *f*, Betonprüfung *f* = concrete test
~versuchsabschnitt *m* = concrete test section
~versuchsfeld *n* = concrete farm
(~)Versuchswürfel *m* → (Beton)Würfel
(~)Versuchszylinder *m* = (Beton-) Probezylinder

Betonverteiler m, Betondeckenverteiler, Betonstraßenverteiler, Betonverteilerwagen m = concrete (pavement) spreader, ~ (~) distributor, ~ (~) spreading machine

~ **mit hin- und hergehender (Verteiler)Schaufel** → Wendeschaufelverteiler

~ **mit Rand-Tauchvibratoren** = concrete spreader with side form vibrators, ~ distributor ~ ~ ~, ~ spreading machine ~ ~ ~

~**wagen** m → Betonverteiler m

Betonverteilung f = concrete spreading

Betonverteilungswagen m, (Beton)-Querverteiler m, Kübelverteiler, Kübel-Betonverteiler = trough-type concrete distributor, ~ ~ spreader, ~ ~ spreading machine, hopper spreader, hopper distributor, hopper spreading machine, box spreader, box distributor, box spreading machine

~ **mit Bodenplatte** = controlled-discharge door concrete spreader, ~ ~ ~ distributor, ~ ~ ~ spreading machine, spreading machine fitted with bottom doors

~ **ohne Bodenplatte** = open-bottomed concrete spreader, ~ ~ distributor, ~ ~ spreading machine, spreading machine without bottom doors

Beton|-Vibrationsmischer m, Vibrationsmischer für Beton = vibratory concrete mixer, vibrating ~ ~

(~)**Vibrator** m, (Beton)Rüttler m = (concrete) vibrator

(~-)**Vierfußkörper** m = (concrete) tetrapod

~**vollbauweise** f = solid concrete method

(~-)**Voll(block)stein** m = solid concrete block

~**vorderkipper** m → Betonkarre(n)

~-**Vorfabrikation** f [*Schweiz*] = precasting of concrete units, ~ ~ ~ members

~-**Vorfertigungsstelle** f, bewegliches Montagebetonwerk n = precast concrete manufacturing yard

~**wälzgelenk** n = concrete rolling contact joint

~**wand** f = concrete wall

~**ware** f [*Erzeugnis aus Zementbeton in Massenfertigung hergestellt, z. B. Hohlblockstein, Betonpflasterstein usw.*] = mass-produced precast (concrete) product, concrete product, concrete article

Betonwaren fpl = concrete ware

~**hersteller** m = concrete ware manufacturer

~**werk** n; Zementwarenwerk [*Schweiz*] = concrete ware plant

Beton|-Wasserbehälter m = concrete water tank

~**weg** m = concrete path

~**wehr** n = concrete weir

~-**Wellenbrecher** m = concrete breakwater

~**wendeltreppe** f = concrete spiral stair (US); ~ ~ staircase (Brit.)

Betonwerk n [*Produktionsstätte für Zementbetonerzeugnisse mit industriellem Charakter*] = (precast) concrete factory, (~) ~ works, (~) ~ plant, precasting works, precasting factory, precasting plant

~**ausrüstungen** fpl = precast concrete equipment

Beton(werk)stein m = cast stone; reconstructed ~, patent ~ (Brit.) [*Cast stone consists of mo(u)lded blocks of concrete with special surface treatment. They may be formed in any of the shapes obtained by cutting the natural stone and may have surfac finishes which resemble the rubbed finish commonly used on limestone and other stones, or any of the tooled finishes*]

~**arbeiten** fpl = cast stone work

~**treppe** f = cast stone stair(case); reconstructed ~ ~, patent ~ ~ (Brit.)

Beton|-Windscheibe f = concrete shear wall

(Beton)Wirkstoff — (Beton)Zuschlagstoffe

(~)Wirkstoff *m*, (Beton)Zusatzmittel *n*, (Beton)Zusatz *m* [*wird dem Beton beigegeben*] = concrete additive; ~ admix(ture) (US)
~wirtschaftsweg *m* = concrete farm track
~wulst *f*, *m* [*Der Teil des Betons, der aus einem Vergußspalt entfernt wird*] = fillet of concrete
(Beton)Würfel *m*, (Beton)Probewürfel, (Beton)Prüfwürfel, (Beton-)Versuchswürfel = concrete (test) cube
~druckpresse *f* → Druckpresse
~prüfmaschine *f* → Druckpresse *f*
~(prüf)presse *f* → Druckpresse
Betonzentrale *f*, Betonmischzentrale, zentrale Betonmischanlage *f* = central concrete mixing plant, concrete central-mix ~
~ **in Punktanordnung** → Betonturm *m*

~ ~ **senkrechtem Aufbau** → Betonturm *m*
Beton|zersetzung *f* = concrete disintegration
~**zerstörung** *f* = concrete destruction
~**-Zertrümmerungsmaschine** *f* = concrete pavement shattering machine
Betonzug *m* = concrete tension
~**riß** *m* = concrete tension crack
~**spannung** *f* = concrete tensile stress
~**zone** *f* = concrete tensile zone
Beton|zulauf *m* [*Rohrteil*] = concrete junction
~**zumessung** *f* → Betondosierung
~**zumeßanlage** *f* → Betondosieranlage
~**zumeßwaage** *f* → Betondosierwaage
~**zusammensetzung** *f*, Betonaufbau *m* = concrete composition
(~)**Zusatz** *m* → (Beton)Wirkstoff *m*
(~)**Zusatzmittel** *n* → (Beton)Wirkstoff *m*

(Beton)Zuschlagstoffe *mpl*, (Beton-)Zuschläge *mpl*, **Beton-Zuschlagmaterial** *n*
  natürliche Stoffe *mpl*, Naturmaterial *n*
    natürlicher Feinzuschlag *m*, Betonsand *m* [*Durchgang durch das Rundlochsieb 7 nach DIN 1170*]

    natürlicher Grobzuschlag *m*, Betonkies *m* [*Rückstand auf dem Rundlochsieb 7 nach DIN 1170*]

  zerkleinerte Stoffe *mpl*, gebrochenes Material *n*
    zerkleinerter Feinzuschlag *m*, Betonbrechsand *m*
      Betonfeinsand *m* [< *1 mm*]
      Betongrobsand *m* [*3/7 mm*]
    zerkleinerter Grobzuschlag *m*
      Betonsplitt *m* [*7/30 mm*]
      Betonsteinschlag *m* [*30/70 mm*]
  Betonzuschlag(stoff) *m* nicht größer als 37 mm

(cement) (concrete) aggregate
  natural aggregate
    natural fine aggregate, ~ sand, ~ concreting sand, ~ concrete sand [*This term shall mean an aggregate mainly passing a 3/16 in. B.S. test sieve*]
    natural coarse aggregate, natural gravel [*this term shall mean an aggregate mainly retained on a 3/16 in. B.S. test sieve*]
  crushed aggregates
    crushed fine aggregate, ~ sand, ~ concreting sand, ~ concrete sand
      crushed stone sand
      crushed gravel sand
    crushed coarse aggregate
      crushed gravel
      crushed stone
  concrete ballast (Brit.) [*containing nothing larger than $1^1/_2$ in.*]

einkörniges Material *n*, Einkornzuschlag(stoff) — single-size material, short range aggregate [*material a major proportion of whose particles are of sizes lying between narrow limits*]

künstliche Stoffe *mpl* — artifical aggregates

Beton|zuteilanlage *f* → Betondosieranlage
~zuteilung *f* → Betondosierung
~zuteilwaage *f* → Betondosierwaage
~zwangsmischer *m* → → Betonmischer
~zwillingspumpe *f* = dual concrete pump
~zylinder *m* → (Beton)Probezylinder
(~)Zylinder-Druck(prüf)presse *f* = (concrete) cylinder tester, (~) ~ testing machine
Betrag *m* **der Kronenüberhöhung, Größe** *f* ~ ~ = amount of crown
Betrieb *m* [*Maschine*] = operation, service
Betriebs|ablauf *m* = operational sequence
(~)Anlage *f*, (Betriebs)Einrichtung *f* = installation
~anleitung *f*, Betriebsanweisung = operating instructions
~anweisung *f*, Betriebsanleitung = operating instructions
~auslaß *m*, Auslaufbauwerk *n* [*Talsperre*] = (reservoir) outlet (works), (river) outlet, outlet structure, outfall structure
~bedingungen *fpl*, Betriebsverhältnisse *npl* = operating conditions
~bereich *m* = operating range
betriebsbereiter Zustand *m* → Betriebsfähigkeit *f*
Betriebs|bereitschaft *f* → Betriebsfähigkeit *f*
~boden *m*, Betriebsfußboden, Industrie(fuß)boden, Werk(fuß)boden, Fabrik(fuß)boden = industrial floor finish, ~ flooring
~brauchbarkeit *f* = serviceability
~bremse *f* = service brake, power ~
~dampf *m* = operating steam
~dauer *f*, Betriebszeit *f* = operating time, ~ period
~drehzahl *f* = operating speed
~drehzahlbereich *m* = operating speed range
~druck *m*, Arbeitsdruck = operating pressure, service ~, working ~
(~)Einrichtung *f*, (Betriebs)Anlage *f* = installation
~erfahrung *f* [*mit einem Baustoff*] = service experience
~erfordernisse *npl* [*z. B. eines Flugplatzes*] = operational requirements
~ergebnisse *npl* = operating results, practical ~
betriebsfähiger Zustand *m* → Betriebsfähigkeit *f*
Betriebsfähigkeit *f*, Betriebsbereitschaft *f*, betriebsfähiger Zustand *m*, betriebsbereiter Zustand = operating order, running ~, ~ condition
betriebsfertiges Gewicht *n* → Arbeitsgewicht
Betriebsfläche *f* = operational area
~ für den Verkehr = service area [*road*]
Betriebs|flüssigkeit *f* = operating liquid
~führung *f* = management
~fußboden *m* → Betriebsboden
~gebäude *n* = operational building
~geschoß *n* [*Fernmeldeturm*] = apparatus floor
~geschwindigkeit *f*, Fahrgeschwindigkeit im Betrieb [*Bagger; Lader usw.*] = travel(l)ing speed during working
~gewicht *n* → Arbeitsgewicht
~hof *m* = machinery and equipment yard
~ingenieur *m* = operating engineer

Betriebsingenieur — Bettsohle

~ingenieur m [*Schweiz*]; betriebswissenschaftlich ausgebildeter Ingenieur = industrial engineer, production ~; efficiency expert (US)
~jahr n = year of operation
~kapital n = working capital
~kern m = service core [*Service cores contain lift towers and stairwells or either of them*]
~kompressor m = service compressor
~kontrolle f, Betriebsüberwachung f = checking of operations
~kosten f = operating cost(s)
~küche f = kitchen of works canteen
~labor(atorium) n = factory lab(oratory)
~leiter f, Dienstleiter = service ladder
~mangel m = operating failure
~pumpe f [*im Gegensatz zur Reservepumpe*] = operating pump
~ruhe f = shutdown
~schema n = operating scheme
betriebssicher = safe to operate
Betriebs|sicherheit f = operating safety
~soziologie f = industrial sociology
~stellung f = operating position
~störung f = breakdown
~streuung f [*Streuung der Druckfestigkeiten von Betonwürfeln aus der laufenden Betriebsproduktion*] = scatter of routine results
~studie f = management study
~stunde f = operating hour, hour of operation
~stundenzähler m = hourmeter
~stundenzählerstand m = hourmeter reading
~teil m, n → Arbeitselement n
~temperatur f [*Kompressor*] = compressed-air temperature
~überwachung f, Betriebskontrolle f = checking of operations
~umstellung f = operational change over
~- und Unterhaltungsanweisung f = service and maintenance manual
~unfall m → Arbeitsunfall
~verhältnisse npl, Betriebsbedingungen fpl = operating conditions
~verlust m = operation waste [*Water lost from an irrigation system through spillways or otherwise, after being diverted into it*]
(~)Verschluß m → Absperrorgan n
Betriebswasser n, Brauchwasser, Rohwasser = raw water, non-potable ~
~behälter m → Brauchwasserbehälter
~netz n → Brauchwassernetz
~pumpwerk n → Brauchwasserpumpstation f
~stollen m → Triebwasserstollen
~versorgung f, Brauchwasserversorgung, Rohwasserversorgung = raw water supply, non-potable ~ ~
Betriebs|wirtschaftlichkeit f = operating economy
~wissenschaft f = business and management
betriebswissenschaftlich ausgebildeter Ingenieur m [*in der Schweiz: Betriebsingenieur*] = industrial engineer, production ~; efficiency expert (US)
Betriebs|woche f = service week
~zeit f, Betriebsdauer f = operating time, ~ period
~zentrale f = central control station
~zustand m = operating state
Bett n [*allgemein*] = bed
~ [*Fluß; Kanal; Strom*] = bed
~beständigkeit f [*Fluß; Kanal; Strom*] = bed stability
~beständigkeitsziffer f = bed stability factor
Bettenhaus n [*Teil eines Krankenhauses*] = ward block
Bett|erosion f = bed erosion
~füllungszahl f = stage fluctuation in percent of the total stage range
~hang m [*Fluß*] = slope of bed
~material n, Sohlenmaterial = bed material
~schicht f, (Unter)Bettungsschicht, Bettung f = bed(ding course), underlay, cushion (course); subcrust (Brit.)
~sohle f [*Fluß; Kanal; Strom*] = bottom of bed

Bettung f → Bettschicht f
Bettungs|dicke f [*Fehlname: Bettungsstärke*] = bed(ding course) thickness
~**koeffizient** m → Bettungszahl f
~**material** n, Bettungsstoff m = bed(ding course) material
~**methode** f → Bettungs(zahl)verfahren n
~**mörtel** m, Verlegemörtel = bed(ding course) mortar
~**schicht** f → Bettschicht
~**räumer** m → Bettungsreinigungsmaschine f
~**reinigungsmaschine** f, Gleisschotter-Reinigungsmaschine, Bettungsreiniger m, Bettungsräumer, Bettungsräumungsmaschine = ballast cleaner, ~ cleaning machine
Bettungsschotter m, Schienenschotter, Gleisschotter, (Eisen)Bahnschotter = (track) ballast
~**reinigung** f = (track) ballast cleaning
Bettungs|stoff m, Bettungsmaterial n = bed(ding course) material
~**verfahren** n → Bettungszahlverfahren
~**zahl** f, Bettungsziffer f, Bodenziffer f, Planumsmodul m, Druck-Setzungs-Quotient m, Bettungskoeffizient m, Bodenkonstante k f, Unterlageziffer = coefficient of subgrade reaction, modulus of foundation support
~**(zahl)verfahren** n, Bettungs(zahl)methode f = coefficient of subgrade reaction method
~**ziffer** f → Bettungszahl f
Bettwiderstand m = bed resistance
Beulbelastung f = bulging loading
beulen, aus~, ausbauchen = to bulge
Beulen n, Aus~, Ausbauchen = bulging
Beul|last f = bulging load
~**probe** f → Beulversuch m
~**prüfung** f → Beulversuch m
~**sicherheit** f = safety against bulging
~**spannung** f = bulging stress
Beulung f, Ausbauchung, Bauch m, Ausbeulung = bulge

Beul|versuch m, Beulprobe f, Beulprüfung f = bulging test
~**wert** m = bulging coefficient
Be- und Entlüftung f, Lüftung = ventilation, ventilating
Be- und Entlüftungs|abschnitt m, Lüftungsabschnitt [*Tunnel; Stollen*] = ventilating section, ventilation ~
~**anlage** f, Lüftungsanlage = ventilating installation, ventilation ~
~**bauwerk** n, Lüftungsbauwerk = ventilating structure, ventilation ~
~**betrieb** m, Lüftungsbetrieb = ventilating service, ventilation ~
~**kanal** m, Lüftungskanal = ventilating duct, ventilation ~
~**methode** f, Be- und Entlüftungsverfahren n, Lüftungsmethode, Lüftungsverfahren = ventilating method, ventilation ~
~**schacht** m, Lüftungsschacht = ventilating shaft, ventilation ~
~**station** f, Lüftungsstation = ventilating station, ventilation ~
~**stein** m, Lüftungsstein = ventilating block, ventilation ~
~**stollen** m, Lüftungsstollen = ventilation gallery, ventilating ~
~**verfahren** n → Be- und Entlüftungsmethode f
Beurteilung f **nach Augenschein**, Augenscheinnahme f = visual examination, ~ inspection
Bevölkerungs|dichte f = density of population, population density
~**schutz** m = civil defense (US); ~ defence (Brit.)
~**zunahme** f = growth of population, population growth
bevorraten, (ein)lagern, (auf)speichern = to store (up)
Bevorratung f → (Auf)Speicherung
bewachsene Bank f [*im Meer*] = grassy ridge
Bewachsung f, Bewuchs m = vegetation
bewaldet, waldbestanden = forested, forest-clad
bewässern = to irrigate

Bewässerung *f*, Bewäßerung, Boden~ [*Ober- oder unterirdische Zuführung von Wasser oder Abwasser zum Zwecke der Ertragssteigerung. Auch Grabeneinstau und Grabenanstau werden zur Bewässerung gerechnet*] = irrigation

~ **im Schwergewichtfluß**, Boden~ ~ ~ = irrigation by gravity, gravity irrigation

~ **mit Abwasser**, Abwasserbewässerung, Oberflächenberieselung mit Abwasser = land treatment, sewage irrigation; broad irrigation (US) [*The disposal of sewage by application to farm land, involving the incidental benefit to crops growing out of the irrigation and fertilization resulting from the application of the sewage. It differs from sewage farming in that the primary purpose is the disposal of sewage*]

~ **während der Wachstumsruhe** = dormant-season irrigation

Bewässerungs|anlage *f* = irrigation installation

~**-Anpflanzung** *f* = irrigated plantation

~**art** *f*, Bewässerungsverfahren *n*, Bewässerungsmethode *f* = irrigation method

~**betonrohr** *n*, Beton-Bewässerungsrohr *n* = concrete irrigation pipe

~**brunnen** *m* = irrigation well

~**einrichtung** *f* = irrigational facility

~**fläche** *f* = irrigated area

~**furche** *f*, Rieselfurche = irrigation furrow

~**gebiet** *n* = irrigated region

~**graben** *m* = irrigation ditch

~**kanal** *m*, Wässerwasserkanal = irrigation canal

~**land** *n* = irrigation land

~**leitung** *f* = irrigation line

~**methode** *f* → Bewässerungsart *f*

~**monat** *m* = irrigation month

~**nebenkanal** *m*, Bewässerungsseitenkanal = irrigation lateral (canal)

~**netz** *n* = irrigation network

~**periode** *f* = irrigation period

~**plan** *m* = irrigation scheme

~**praxis** *f* = irrigation practice

~**probe** *f*, Bewässerungsprüfung *f*, Bewässerungsversuch *m* = irrigation test

~**projekt** *n* = irrigation project

~**prüfung** *f* → Bewässerungsprobe *f*

~**pumpe** *f* = irrigation pump

~**pumpen** *n* = irrigation pumping

~**rohr** *n* = irrigation pipe

~**seitenkanal** *m*, Bewässerungsnebenkanal = irrigation lateral (canal)

~**speicher(becken)** *m*, (*n*) = irrigation pool

~**speicherung** *f* = irrigation storage

~**sperre** *f*, Bewässerungstalsperre = irrigation dam

~**system** *n* = irrigation system

~**(tal)sperre** *f* = irrigation dam

~**technik** *f* = irrigation engineering

~**teilnehmer** *m* = irrigator [*One who applies water to land for growing crops*]

~**umfang** *m* = perimeter of irrigation

~**verband** *m* = irrigation district [*An organization operating under legal regulations for financing, constructing, and operating an irrigation system*]

~**verfahren** *n* → Bewässerungsart *f*

~**versuch** *m* → Bewässerungsprobe *f*

~**wasser** *n*, Wässerwasser = irrigation water [*The quantity of water artificially applied in the process of irrigation. It does not include precipitation*]

~**wasserbedarf** *m*, Wässerwasserbedarf = irrigation requirement [*The quantity of water, exclusive of precipitation, that is required for crop production. It includes economically unavoidable wastes*]

~**wassertransport** *m*, Wässerwassertransport = conveyance of irrigation water

~**wehr** *n*, Wässerwehr [*Wehr in einem Wasserlauf, das der Entnahme von Wässerwasser dient*] = irrigation weir

~**wissenschaft** *f* = irrigation science

bewässerungswürdig — Bewegungswiderstand

bewässerungswürdig = irrigable
Bewässerungszeit *f* = time of irrigation
Bewäßrung *f* → Bewässerung
bewegen, fördern [*Erdmassen*] = to move, to shift
bewegliche Arbeitskammer *f*, wiedergewonnene ~, Taucherglocke *f* = diving-bell
~ **(Brech)Backe** *f* → → Backenbrecher *m*
~ **Brücke** *f* = opening bridge, movable ~, drawbridge
~ **Front** *f* [*Hydraulik*] = non-stationary front
~ **(Graben)Steife** *f* → Absprießwinde *f*
~ **Quersteife** *f* → Abspießwinde *f*
~ **Schalung** *f* = movable formwork (Brit.); ~ shuttering
~ **Schaufeln** *fpl*, Rotor *m* [*Bohrturbine*] = rotor
~ **Sohle** *f* [*Strom; Fluß*] = shifting bottom
~ **Steife** *f* → Abspießwinde *f*
beweglicher Stempel *m* → Abspießwinde *f*
~ **Teil** *m*, bewegliches ~ *n* = moving part
~ **Wehrverschluß** *m* = movable sluice
~ **Zughaken** *m* = swinging drawbar
bewegliches Bolzenkipplager *n* [*für Hoch- und Tiefbauten*] = movable rocker bearing
~ **Fahrwasser** *n* = shifting channel
~ **Gerüst** *n* [*Brückenbau*] = movable scaffold(ing)
bewegliches Lager *n* [*für Hoch- und Tiefbauten*] = movable bearing
~ ~ **der Rollenbahn** [*Wehr mit Hubschützen*] = moving roller path (Brit.); ~ ~ track (US)
~ **Montagebetonwerk** *n*, Beton-Vorfertigungsstelle *f* = precast concrete manufacturing yard
~ **Montagespannbetonwerk** *n*, Spannbeton-Vorfertigungsstelle *f* = precast prestressed concrete manufacturing yard

~ **Montagestahlbetonwerk** *n*, Stahlbeton-Vorfertigungsstelle *f* = precast reinforced concrete manufacturing yard
~ **Netz** *n*, labiles ~ [*ebenes Fachwerk*] = unstable frame
~ **(Stau)Wehr** *n* = movable weir, barrage
~ **Tangentiallager** *n* [*für Hoch- und Tiefbauten*] = movable tangential bearing
~ **Tangentialkipplager** *n* [*für Hoch- und Tiefbauten*] = movable tangential rocker bearing
~ **Teil** *n*, beweglicher ~ *m* = moving part
~ **Wehr** *n*, ~ Stauwehr = barrage, movable weir
Beweglichkeit *f* [*Fahrzeug*] = mobility
bewegt, uneben, unruhig [*Gelände*] = undulated, undulating
Bewegung *f* zusammendrückbarer Medien = compressible flow
Bewegungs|energie *f*, kinetische Energie = energy of motion, ~ ~ movement
~**freiheit** *f* = freedom of motion, ~ ~ movement
~**fuge** *f* [*Breite je nach Größe der zu erwartenden Setzungen, mindestens aber 15 mm*] = settlement joint
~**größe** *f*, Impuls *m*, Kraftstoß *m* [*Hydraulik*] = impulse
~**größe** *f* = magnitude of motion, ~ ~ movement
~**lehre** *f*, Zwanglauflehre, Kinematik *f* = kinematics
~**richtung** *f* = direction of movement, ~ ~ motion
~**übertragung** *f* = transmission of motion, ~ ~ movement
~**umkehr** *f* = reversal, return of motion, return of movement
~**verdichtung** *f* → Vibrationsverdichtung
~**vorrichtung** *f* [*Wehr mit oberer Bedienungsbrücke*] = releasing gear
~**widerstand** *m*, Fahrwiderstand = rolling resistance

Bewegungs-Zeit-Analyse — Bewetterung durch Überdruck

~-Zeit-Analyse *f* = motion-time analysis
~zustand *m* = state of motion, ~ ~ movement
bewehren, armieren = to reinforce
Bewehren *n*, **Armieren** = reinforcing
bewehrt, armiert, mit (Stahl)Einlagen = reinforced
bewehrte Backsteinwand *f* → armierte
~ **Fundamentplatte** *f* → Plattenfundament
~ **Platte** *f*, armierte ~ = reinforced slab
~ **Torkretschicht** *f*, armierte ~, Torkretschicht mit Stahl(ein)lagen = reinforced gunite
bewehrtes Kabel *n*, armiertes ~, Panzerkabel = armo(u)red cable
~ **Mauerwerk** *n* = reinforced masonry
~ **Ziegelmauerwerk** *n*, armiertes ~, bewehrte Ziegelkonstruktion *f*, armierte Ziegelkonstruktion = reinforced brickwork
Bewehrung *f*, Armierung, Beton~, Stahleinlagen *fpl*, (Eisen)Einlagen = (concrete) reinforcement
Bewehrungs|anordnung *f*, Armierungsanordnung, Anordnung der (Stahl-)Einlagen = reinforcement system
~**anteil** *m* → Armierungsanteil
~**arbeiten** *fpl*, Armierungsarbeiten = reinforcement work
~**block** *m* → Bewehrungskorb *m*
~**bündel** *n*, Armierungsbündel = bundle of reinforcement
~**eisen** *n* → Betonstahl *m*
~**gehalt** *m* → Armierungsanteil *m*
~**gerippe** *n* → Bewehrungskorb *m*
~**gerüst** *n* → Bewehrungskorb *m*
~**kolonne** *f* → Armierungskolonne
~**korb** *m*, Armierungskorb, vorgefertigte Bewehrung *f*, vorgefertigte Armierung, vorgefertigte (Stahl)Einlagen *fpl*, Bewehrungsgerüst *n*, Armierungsgerüst, vorgefertigte Eiseneinlagen, Bewehrungsgerippe *n*, Armierungsgerippe, Bewehrungsblock, Armierungsblock = cage of reinforcing steel, reinforcing cage

~**lage** *f* → Armierungslage
~**matte** *f* → Baustahlmatte
~**matten-Ausleger** *m*, Mattenverleger = mesh laying jumbo, fabric ~ ~
~**menge** *f*, Armierungsmenge = amount of reinforcement
~**netz** *n* → Armierungsnetz
~**plan** *m*, Bewehrungszeichnung *f*, Armierungsplan, Armierungszeichnung = reinforcement drawing
~**prozentsatz** *m* → Armierungsanteil *m*
~**querschnitt** *m*, Armierungsquerschnitt = reinforcement cross-section
~**ring** *m*, Armierungsring = reinforcement ring
~**schicht** *f* → Armierungslage *f*
~**stab** *m*, Armierungsstab, Bewehrungsstange *f*, Armierungsstange = (steel) reinforcing bar, re-bar, (concrete) reinforcing rod, reinforcing steel bar
~**stab** *m* **aus einem Knüppel hergestellt** = billet bar, natural ~, ~ rod
~**stahl** *m* → Betonstahl
~**stange** *f* → Bewehrungsstab
~**stumpfschweißen** *n*, Armierungsstumpfschweißen = reinforcement butt welding
~**verhältnis** *n*, Armierungsverhältnis = ratio of reinforcement
~**zeichnung** *f* → Bewehrungsplan *m*
Bewerber *m* → Angebot *n*
Bewertungs|maßstab *m* = rating
~**verfahren** *n*, Auswertungsverfahren = evaluating method
bewettern = to ventilate
Bewettern *n* → Bewetterung *f* (der Grubenbaue)
Bewetterung *f*; Bauventilation *f* [*Schweiz*] [*Tunnel- und Stollenbau*] = ventilation during construction work
~ **(der Grubenbaue)**, Wetterführung, Wetterung, Wetterversorgung, Bewettern *n*, Grubenbewetterung = mine ventilation
~ **durch Überdruck** → Druckbewetterung

~ ~ Unterdruck → Absaugebewetterung
Bewetterungs|anlage *f*, Gruben~ = mine ventilation plant
~station *f*, Gruben~ = mine ventilation station
~strom *m* = ventilation current, ventilating ~
bewirtschaftete Böschung *f* = cultivated slope
Bewirtschaftung *f*, Zwangs~ = rationing
~ der Einzuggebiete = watershed management (US)
Bewitterung *f* = weathering
Bewitterungskurz|probe *f* → abgekürzte Wetterbeständigkeitsprobe *f*
~prüfung *f* → abgekürzte Wetterbeständigkeitsprobe *f*
~versuch *m* → abgekürzte Wetterbeständigkeitsprobe *f*
Bewitterungsprüfung *f*, Bewitterungsprobe *f*, Bewitterungsversuch *m* = weathering test
Bewitterungsschnell|probe *f* → abgekürzte Wetterbeständigkeitsprobe
~prüfung *f* → abgekürzte Wetterbeständigkeitsprobe *f*
~versuch *m* → abgekürzte Wetterbeständigkeitsprobe *f*
Bewitterungs|schrank *m* = weatherometer
~spiel *n* = weathering cycle
bewohnbar = habitable
bewohnt, bezogen [*Haus*] = lived in
Bewölkung *f* = cloudiness, cloud cover
Bewuchs *m*, Bewachsung *f* = vegatation
bezahlter Urlaub *m* = reimbursed leave
Bezeichnung *f* = designation
Bezeichnungsschild *n*, Typenschild = name-plate
Bezettelungsmethode *f* = windshield card method, two-cordon ~
Beziehungen *fpl* **zwischen Moment und Krümmung** = moment-curvature law
Beziehungswert *m* = correlation value
bezogen, bewohnt [*Haus*] = lived in

bezogene Verformung *f* = unit deformation
Bezugs|ablesung *f* = reference reading
~achse *f* = reference axis
~ebene *f*, Leitebene = datum surface
~größe *f* = reference value
~-Halbzelle *f*, Bezugselektrode *f* = reference half-cell
~höhe *f* = datum
~horizont *m*, Leithorizont (Geol.) = datum horizon, key ~, "marker"
~länge *f* = reference length
~linie *f* = reference line
~probe *f* = reference sample
~punkt *m* = referring object, reference mark, reference point
~quellennachweis *m* = directory
~temperatur *f* = reference temperature
~zeichnung *f* = reference drawing
B-Horizont *m*, Anreicherungshorizont, Einwaschungshorizont, Anreicherungsschicht *f*, Illuvialhorizont, Einschwemmungshorizont, Unterboden *m* = B-horizon, zone of concentration, zone of illuviation
Biberschwanz *m*, Flachwerk *n*, Flachziegel(stein) *m*, Flachstein, Zungenstein, Biber(schwanzziegel) *m* [*auf der Strangpresse hergestellter ebener Dachziegel. Segmentschnitt ist üblich, Sonderformate sind Gotisch, Gradschnitt, Halbkreis, Rautenspitz und Sechseck*] = plain (roof) tile
~-Betondachstein *m* = plain concrete (roof) tile
~dach *n* → Flachziegeldach
Bibliothek *f*, Bücherei *f* = library
Bicke *f*, Picke, (Erd)Hacke *f* = pick
Bickford-Zündschnur *f*, gewöhnliche Zündschnur, Pulverzündschnur = safety fuse, blasting ~ [*This device is a train of enclosed black powder which is used as a medium to convey a flame to an explosive. The flame will travel at a uniform rate*]
biegbar, biegungsfähig, biegefähig = capable of being bent, pliant, pliable

Biegbarkeit *f*, Biegefähigkeit *f*, Biegevermögen *n* = bendability, pliability, bending capacity

Biege|achse *f* [*Balken*] = neutral axis

~**arbeitsfestigkeit** *f* → Biege-Dauerfestigkeit

~**arm** *m* = bending arm

~**balken** *m*, Probebalken, Prüfbalken, Versuchsbalken, Biegeversuchsbalken = bending test beam, flexure ~ ~

biegebeansprucht, biegungsbeansprucht = subjected to bending (stress)

Biege|beanspruchung *f*, Biegungsbeanspruchung = bending stress and strain

~**beiwert** *m*, Biegezahl *f* = bending coefficient

~**belastung** *f* → Biegungsbelastung

~**breite** *f* = bending width

Biegebruch *m* = bending failure

~**kurve** *f* = bending failure curve

~**spannung** *f* = bending failure stress

~**theorie** *f* = theory on failure by bending

Biege|dauer *f*, Biegezeit *f* = bending time

~**-Dauerfestigkeit** *f*, Biege-Ermüdungsfestigkeit, Biege-Arbeitsfestigkeit = flexural fatigue strength, ~ ~ resistance

~**dorn** *m* = mandrel for bending

~**drillknicken** *n* = torsional-flexural buckling

Biegedruck *m* = bending compression, ~ pressure

~**festigkeit** *f* = bending compression strength, ~ pressure ~

~**spannung** *f* = compressive stress due to bending

~**zone** *f* = bending compression zone, ~ pressure ~

Biege|elastizität *f* = bending elasticity

~**ermüdung** *f* = fatigue in flexure, flexural fatigue endurance

~**ermüdungsgrenze** *f* = flexural fatigue endurance limit

biegefähig → biegbar

biegefest → biegesteif

Biege|festigkeit *f*, Biegungsfestigkeit, Querfestigkeit, Horizontalfestigkeit [*gewöhnlich „Tragkraft" genannt. Die bis zum Bruch des Probekörpers auftretende höchste Biegespannung*] = bending strength, cross-breaking ~, flexural ~, ~ resistance

~**fläche** *f* [*Verformung der Mittelfläche einer Platte nach der Belastung*] = bent surface

~**formel** *f* = bending formula, flexure ~

~**gelenk** *n* = snaker

~**gleichung** *f* = flexural equation

~**grenze** *f* = bending limit

~**größe** *f*, Biegebetrag *m* = amount of bend

~**-Haftfestigkeits-Versuch** *m*, Biege-Haftfestigkeits-Prüfung *f*, Biege-Haftfestigkeits-Probe *f* = flexural bond test

~**halbmesser** *m*, Biegungshalbmesser = bending radius

~**hebel** *m* [*Arbeitsgerät zum Biegen von Rundstählen*] = bending lever

~**knickung** *f* = flexural bending

~**körper** *m*, Prisma *n* zur Ermitt(e)lung der Biegefestigkeit = bending test specimen

~**kraft** *f*, Biegungskraft = bending force, ~ power

~**last** *f* = bending load

~**linie** *f*, Biegungslinie = bending line

~**linienverfahren** *n* → → Spundwand *f*

~**liste** *f*, Eisen~, Stahl~ [*für Bewehrung*] = bending schedule [*A list of reinforcement which accompanies a reinforcement detail drawing, prepared by a reinforced concrete designer*]

~**maschine** *f*, Profileisen~, Betoneisen~, Betonstahl~, Profilstahl~ = reinforcement (bar-)bending machine, power-operated bar-bender, powered bender, powered rod bender

Biegemaschine — Biegetisch

~maschine *f* = bending machine; bulldozer (US) [*In fabricating steel, a machine in which angles are bent in small circular arcs by pressure between two supports*]

~maschine *f* für Bewehrungsmatten = mat reinforcement bender, ~ ~ bending machine

~modul *m* = modulus of bending

~moment *n*, Biegungsmoment [*Die Summe der Wirkungen aller angreifenden Kräfte mal ihrem Hebelarm bezogen auf einen bestimmten Querschnitt*] = bending moment

~momente *npl* bei Hängebrücken nach der Theorie 2. Ordnung = bending interaction in suspension bridges

~momentenlinie *f* = bending-moment diagram [*A diagram which shows for one loading the amount of the bending moment at any point along a beam. From this diagram, the position and amount of the maximum bending moment can be immediately seen*]

~momententheorie *f* = bending moment theory

~momentverhältnis *n* = bending-moment ratio

biegen [*Bewehrungsstahl*] = to bend

Biegen *n*, Biegung *f* = flection, flexion, flexing, bending

~ der Bewehrung = reinforcement bending

~ von Krümmern auf der Baustelle = field bending

Biege|plan *m*, Eisen~, Stahl~ [*für Bewehrung*] = reinforcement detail drawing

~platz *m*, Eisen~, Stahl~ = reinforcement-bending yard, steel-bending ~

~presse *f* = bending press

~probe *f* → Biegeversuch *m*

~prüfmaschine *f*, Biegeprüfgerät *n*, Biegeprüfer *m* = bending tester, ~ testing machine

~prüfung *f* → Biegeversuch *m*

Bieger *m*, Eisen~, Stahl~ [*Stahlbetonbau*] = bar bender [*He bends his reinforcement according to the schedule and places it according to the drawing*]

Biegerei *f* [*Betonsteinwerk*] = reinforcement shop

Biege|riß *m* = flexural crack

~schwingung *f*, Biegungsschwingung = flexural oscillation, bending ~, ~ vibration

Biegeschwingungs|bruch *m* = flexural oscillation failure, bending ~, ~ vibration ~

~festigkeit *f* = flexural oscillation strength, bending ~ ~, ~ vibration ~

~versuch *m*, Biegeschwingungsprobe *f*, Biegeschwingungsprüfung *f* = flexural oscillation test, bending ~, ~ vibration ~

Biege|spannung *f*, Biegungsspannung = bending stress, flexural ~

~spannungsformel *f* = bending stess formula, flexural ~ ~

~stab *m* = flexural member

biegesteif, starr, biegefest, biegungssteif, biegungsfest = resistant to bending, bending-resistant, rigid, stiff, bendproof, flexurally rigid

biegesteife Form *f*, spannsteife ~ [*elektrothermisches Spannen von Spannstahl*] = power form

Biege|steifigkeit *f*, Biegungswiderstand *m*, Biegewiderstand, Biegungssteifigkeit *f* = flexural rigidity, bending resistance, flexural stiffness, resistance to bending

~stelle *f*, Ab~ [*Bewehrungsstahl*] = bend(ing) point

~theorie *f* [*Schalen und Faltwerke erhalten außer Schub- und Normalspannungen auch Biegespannungen, die nach der Biegetheorie berechnet werden*] = bending theory, flexure ~

~theorie *f* des Balkens [*Bodenmechanik*] = theory of elastic beams on a continuous elastic support

~tisch *m* = bending table

Biegetorsion — Biegsamkeitszahl

~torsion *f*, Biegungstorsion = flexural torsion
~träger *m* = girder subjected to bending
~verformung *f* = deformation due to bending
~vermögen *n* → Biegbarkeit *f*

Biegeversuch *m*, Biegeprobe *f*, Biegeprüfung *f*
  Hin- und Her-~

  Schlag~
  Kerb~
  Einkerbhin- und -her~
  Kerbschlag~

  Abschreck~, Härtungs~

  Kalt~
  Warm~

bending test, flexure ~
  alternating ~ ~, test by bending in opposite directions, to-and-fro ~ ~
  shock ~ ~, blow ~ ~
  notch ~ ~
  alternating notch ~ ~
  notch shock test, shock test with notched test piece; nick-break test (US)
  ~ ~ in tempered (or quenched) state
  cold~ ~
  warm ~ ~

Biege|(versuch)balken *m* → Biegebalken
~wechselfestigkeit *f* = alternate bending strength
„biegeweicher" Stahlbogen *m* = flexible steel arch
Biegewelle *f*, Schlauchwelle, biegsame Welle = flexible shaft
Biegewellen|antrieb *m*, Schlauchwellenantrieb = flexible drive
~-(Beton)Innenrüttler *m*, Biegewellen-(Beton)Innenvibrator *m*, Biegewellen-(Beton)Tauchrüttler, Biegewellen-(Beton)Tauchvibrator, Biegewellen-(Beton)Nadelrüttler, Biegewellen-(Beton)Nadelvibrator, Biegewellen-(Beton)Tiefenrüttler, (Biegewellen-(Beton)Tiefenvibrator = flexible shaft (drive) internal vibrator ~ ~ (~) immersion ~, ~ ~ (~) concrete mass ~, ~ ~ (~) (concrete) poker ~, ~ ~ (~) needle ~, ~ ~ (~) pervibrator, ~ ~ (~) concrete vibrator for mass work, ~ ~ (~) spud vibrator for concrete

Biege|wert *m* = bending value
~widerstand *m* → Biegesteifigkeit *f*
~winkel *m* = bending angle
~zahl *f*, Biegebeiwert *m* = bending coefficient
~zeit *f*, Biegedauer *f* = bending time
Biegezug *m* = bending tension
~bruch *m* = bending tension failure
~festigkeit *f* = bending tension strength
~-Prüfgerät *n*, Biegezugfestigkeitsprüfer *m* = machine for flexural tensile tests
~spannung *f* = flexural tensile stress
~versuch *m*, Biegezugprobe *f*, Biegezugprüfung *f* = flexural tensile test
~zone *f* = zone of bending tension
biegsam *a* = flexible
biegsame Verbindung *f*, Gleitfuge *f* [*Motor*] = slip joint
~ Welle *f* → Biegewelle
Biegsamkeit *f* = flexibility
Biegsamkeitszahl *f*, Biegsamkeitsziffer *f* [*Maß für die Biegung eines Stabes; abhängig vom Elastizitätsmodul, dem Trägheitsmoment, der*

Stablänge und den Einspannverhältnissen des Stabes] = flexibility number

Biegung *f*, Biegen *n* = flexion, flection, flexing, bending

~, Krümmung [*im Sinne von einer Kurve einer Straße, eines Wasserlaufes, eines Schienenstranges usw.*] = bend, turn, curvature

~, Falte *f* = fold, flexure

~ **mit Achsdruck** = combined bending and axial loading

biegungsbeansprucht, biegebeansprucht = subjected to bending (stress)

Biegungs|beulung *f* = bulging due to bending

~**ebene** *f*, Biegeebene [*Bezugsebene für die Biegemomente*] = bending plane, plane of bending

~**elastizität** *f* = flexional elasticity

~**ermüdung** *f* = fatigue in flexure

biegungs|fähig → biegbar

~**fest** → biegesteif

biegungsfrei = not subjected to bending

biegungsfreier gezogener Draht *m* = dead-drawn wire

Biegungs|momentdiagramm *n* = bending moment diagram, B. M. D. [*The B. M. D. is a diagram with the centre line of the member as base line and ordinates representing the bending moment at successive right sections, positive bending moment being plotted upwards and negative downwards*]

~**spannung** *f*, Biegespannung = bending stress, flexural ~

~**theorie** *f* = theory of bending

~**winkel** *m* = angle of bending

Bienenwabenverband *m*, Pflasterverband, gabbroider Verband, granoblastischer Verband, Mosaikverband (Geol.) = granoblastic texture

bieten, an~ = to tender

bietende Firma *f* → → Angebot *n*

Bieter *m* → → Angebot *n*

Bietungsgarantie *f* → → Angebot *n*

bifunktioneller Katalysator *m* = duofunctional catalyst

Bikalziumsilikat *f* → Dikalziumsilikat

Bild|auswerter *m* = photo interpreter

~**auswertung** *f* = photo interpretation (technique)

~**ebene** *f* [*Perspektivzeichnen*] = picture plane

~**flug** *m*, Vermessungs~ = mapping flight

~**flugzeug** *n*, Vermessungs~ = mapping (air)plane

~**güte** *f* = image quality

~**güteprüfsteg** *m* = penetrameter, radiographic image quality indicator

~**gütezahl** *f* = radiographic quality index

~**hauerhammer** *m* = stone hammer

~**hauermarmor** *m*, Statuenmarmor = statuary marble

bildliche Analyse *f*, graphische ~ = graphical analysis

~ **Berechnung** *f*, graphische ~, bildliches Rechnen *n*, graphisches Rechnen = graphical calculation

~ **Darstellung** *f*, graphische ~ = graphical representation

~ **Gleichgewichtslehre** *f*, graphische ~, ~ Statik *f*, Graphostatik = graphical structural analysis, ~ statics, graphostatics

~ **Statik** *f* → → Gleichgewichtslehre *f*

~ **Untersuchung** *f*, graphische ~ = graphical investigation

Bildmessung *f*, Photogrammetrie *f* = photogrammetry

bildsam = plastic

bildsame Verformung *f* → plastische Formänderung

~ **Zone** *f*, plastische ~, bildsamer Bereich *m*, plastischer Bereich = plastic zone

bildsamer Bereich *m* → bildsame Zone *f*

~ **Zustand** *m*, plastischer ~, Plastizitätszustand, Bildsamkeitszustand = state of plasticity

Bildsamkeit *f* → Bildsamkeitszahl *f*

~, Plastizität *f* = plasticity

Bildsamkeits|theorie *f*, Plastizitätstheorie = theory of plasticity
~zahl *f*, Plastizitätszahl, Bildsamkeit *f*, Plastizitätsindex *m*, Bildsamkeitsindex = plasticity index
Bildstein *m*, Agalmatolith *m* (Geol.) = pagodite, agalmatolite
Bildungs|wärme *f* = heat of formation
~weise *f* [*z. B. eines Fachwerkes*] = arrangement
Billner-Verfahren *n*, Vakuum-Behandlung *f* nach Billner [*Vakuumbeton*] = Billner method
Bims *m*, **~stein** *m* = pumice (stone)
Bimsbaustoff *m* = pumice building material, ~ construction ~
~werk *n*, Bimsbaustoff-Fabrik *f* = pumice building material factory, ~ construction ~ ~
Bimsbeton *m* = pumice concrete
~dachplatte *f* = pumic concrete roof slab
~-Deckenhohlkörper *m* = pumice concrete hollow block for floors, ~ ~ ~ tile ~ ~
~diele *f* = pumice concrete sheathing board
~Hohl(block)stein *m*, Bimsbeton-Hohlblock *m*, Bimsbeton-Hohlkörper *m* = pumice concrete hollow tile, ~ ~ ~ block, ~ ~ pot
~-Hohldiele *f*, Stegdiele = pumice concrete hollow slab
~-Kassettenplatte *f* [*mit architektonisch ausgestatteter kassettierter Untersicht*] = pumice concrete waffle panel
~platte *f* = pumice concrete slab
~stein *m* = pumice concrete block
~trägerdecke *f* = pumice concrete girder floor
~-Vollstein *m* = pumice concrete solid block
~wand *f* = pumice concrete wall
Bimsbrecher *m* = pumice stone breaker, ~ ~ crusher
Bimsindustrie *f* = pumice industry
Bimskies *m* [*Bims mit mehr als 7 mm Korndurchmesser*] = pumice gravel

Bimsmehl *n* = pumice meal, ~ dust, ~ powder, powdered pumice (stone)
Bimssand *m* [*Bims mit 0-7 mm Korndurchmesser*] = pumice sand
Bimssplitt *m* = pumice chip(ping)s
Bimsstein *m*, Bims *m* = pumice (stone)
~tuff *m* = pumiceous tuff, pumice (stone) ~
Bimswalzenbrecher *m* = pumice roll crusher, ~ ~ breaker
binär, aus zwei Elementen bestehend, zweiglied(e)rig, Zweistoff .... = binary
Binärstelle *f*, Dualstelle = bit, binary digit
Binde *f* [*z. B. für Korrosionsschutz*] = jacketing, blanket
~blech *n* [*Stahlbau*] = batten (plate), stay ~, tie ~
~draht *m*, Rödeldraht = binding wire, lashing ~, annealed ~, iron ~ [*Soft steel wire used for binding reinforcement*]
~drahtanker *m*, Rödeldrahtanker = twisted wire tie
~draht-Vorratstrommel *f*, Rödeldraht-Vorratstrommel [*der Arbeiter trägt sie am Gürtel und kann somit Bewehrungen leichter montieren*] = binding wire coil, lashing ~ ~, annealed ~ ~, iron ~ ~
~eigenschaft *f* = binding property
~erde *f*, Bodenmörtel *m*, haftendes Lockergestein *n*, Bodenbinder *m*, Binderton *m* = soil matrix, ~ binder, ~ mortar, (clay) binder, binder soil
~holz *n*, Brustholz, Bundholz, zugeschnittenes Bauholz [*Zimmerei*] = scantlings
~kraft *f* → Kohäsion *f*
Bindemittel *n*, (Straßenbau)Binder *m*, Straßenbauhilfsstoff *m*, Straßenbaubindemittel = road binder
~, Binder *m* [*bei einer wassergebundenen Schotterdecke*] = stone dust and water; slag ~ ~ ~

Bindemittel — Bindemittelfaß

~, verkittende Zwischensubstanz *f* [*Gestein*] = matrix, cement, cementing material, cementitious material

~, Binder *m* = cementitious material, binder, cementing material, binding medium

~ [*einer Anstrichfarbe*] = medium, binding agent

~ auf Harzbasis, Binder *m* ~ ~ = resin-based binder

~ ~ Kohlenwasserstoffbasis, Binder *m* ~ ~, Kohlenwasserstoffbindemittel, Kohlenwasserstoffbinder, Schwarzbindemittel, Schwarzbinder = hydrocarbon binder

~ aus Eisen (Geol.) = ferruginous cement(ing material)

~ aus Kalk (Geol.) = calcareous cementing material (or cement)

~ mit Füller, Binder *m* ~ ~, gefülltes Bindemittel, gefüllter Binder = fillerized binder

~ ~ Haftanreger, Binder *m* ~ ~ = doped binder

~ ~ hydraulischem Zuschlag, Binder *m* ~ ~ ~, ~ ~ Puzzolane = puzzolan-filled binder

~ zur Vorumhüllung, Binder *m* ~ ~ = binder for pre-coating

~abmessung *f*, Bindemitteldosierung, Bindemittelzuteilung, Bindemittelzumessung = binder batching, ~ proportioning, ~ measuring

~abmeßapparat *m* → Bindemitteldosierapparat

~-Abmeßpumpe *f* → Abmeßpumpe

~-Abmeßschnecke *f* → Bindemittel-Zuteilschnecke

~anlage *f* = binder bulk storage and heating installation

~anteil *m*, Bindemittelgehalt *m* = binder content

bindemittelarme Decke *f*, bindemittelarmer Belag *m* = lean pavement; ~ paving (US); ~ surfacing (Brit.)

Bindemittel|aufbringung *f* → Aufspritzen *n*

~aufheizer *m* als Anhänge(r)fahrzeug → Bindemittelerhitzer ~ ~

~aufheizung *f*, Bindemittelerwärmung, Bindemittelerhitzung = binder heating

~aufheizungsanlage *f* → Bindemittelerhitzungsanlage

~aufspritzung *f* → Aufspritzen *n*

~behälter *m*, Bindemittellagertank *m* = binder storage tank

~beigabe *f*, Bindemittelzugabe = addition of binder

~bestimmung *f* = binder content determination

~-Bevorratungsanlage *f*, Bindemittellagerungsanlage = binder storage installation

~dosierapparat *m*, Bindemitteldosierer *m*, Bindemittelzuteilapparat, Bindemittelzumeßapparat, Bindemittelabmeßapparat = binder-batcher

~dosierpumpe *f* → Abmeßpumpe

~-Dosierschnecke *f* → Bindemittel-Zuteilschnecke

~dosierung *f* → Bindemittelabmessung

~-Eindüsapparatur *f*, Zerstäuber *m* = binder spray(ing) device

~einspritzung *f* = binder injection

~einspritzvorrichtung *f* = binder injecting device, ~ injection ~

~emulsion *f* = binder emulsion

~erhitzer *m* als Anhänge(r)fahrzeug, Bindemittelwärmer ~ ~, Bindemittelaufheizer ~ ~ = binder heating trailer

~erhitzung *f* → Bindemittelerwärmung

~erhitzungsanlage *f*, Bindemittelerwärmungsanlage, Bindemittelaufheizungsanlage = binder heating installation

~erwärmer *m* als Anhänge(r)fahrzeug → Bindemittelerhitzer ~ ~

~erwärmung *f*, Bindemittelaufheizung, Bindemittelerhitzung = binder heating

~erwärmungsanlage *f* → Bindemittelerhitzungsanlage

~faß *n* = binder barrel

Bindemittelfilm — Binderabstand

~film *m*, Bindemittelhaut *f*, Bindemittelhäutchen *n* = binder film
~gehalt *m*, Bindemittelanteil *m* = binder content
~gemisch *n*, Bindemittelmischung *f* = binder mix(ture)
~haftung *f* = bond between binder and aggregate
~haut *f* → Bindemittelfilm *m*
~kocher *m* → Kocher
~lagertank *m*, Bindemittelbehälter *m* = binder storage tank
~lagerungsanlage *f*, Bindemittel-Bevorratungsanlage *f* = binder storage installation, ~ ~ plant
~leitungen *fpl* [*Tank-Spritzmaschine*] = binder tubes, ~ pipes
~-Mineral-Masse *f*, Bindemittel-Mineral-Gemisch *n* = binder-aggregate mix(ture)
~mischung *f* → Bindemittelgemisch *n*
~pumpe *f* = binder (metering) pump (Brit.); bitumen (metering) pump (US)
(~-)Schmelzkessel *m* → Kocher *m*
~-Schnellbestimmung *f* = accelerated binder content determination
~silo *m*, Zementsilo = cement silo
~Sprengrohr *n* = distributor bar
~-Spritzmaschine *f*, Bindemittelverteiler *m*, Bindemittel-Spritzgerät *n* = binder spraying machine, ~ distributor, bitumen ~, bituminous ~, tar/bitumen (Brit.)/asphalt (US) ~, maintenance distributor
~teilchen *n*, Binderteilchen = binder particle
~thermometer *n* = binder thermometer
~überfluß *m* = excess of binder
~umhüllung *f*, Umhüllen *n*, Benetzen = wetting, coating; wet mixing (US)
~verteilung *f* → Aufspritzen *n*
(~)Vorumhüllung *f* = precoating
~waage *f*, Zement(silo)waage, Zementabfüllwaage, Abfüllwaage für Zement = cement weigh(ing) batcher, ~ batcher scale

~waage *f* [*für Teer und Bitumen*] = binder (weighing) batcher, binder weigh batcher (Brit.); bitumen (weighing) batcher, bitumen weigh batcher (US)
~zugabe *f*, Bindemittelbeigabe = addition of binder
~zumessung *f* → Bindemittelabmessung
~zumeßapparat *m* → Bindemitteldosierapparat
~zumeßpumpe *f* → Abmeßpumpe
~-Zumeßschnecke *f* → Bindemittel-Zuteilschnecke
~zuteilapparat *m* → Bindemitteldosierapparat
~zuteilpumpe *f* → Abmeßpumpe
~-Zuteilschnecke *f*, Bindemittel-Zumeßschnecke, Bindemittel-Abmeßschnecke, Bindemittel-Dosierschnecke, Zementabmeßschnecke, Zementdosierschnecke, Zementzuteilschnecke, Zementzumeßschnecke = cement proportioning screw
~zuteilung *f* → Bindemittelabmessung

Binder *m*, Strecker *m*, Binderstein *m* = header, binder, bondstone
~ [*für Putz und Mauerwerk*] = cementing material
~, Bindemittel *n* [*bei einer wassergebundenen Schotterdecke*] = slag dust and water; stone ~ ~ ~
~, Dach~ [*Die Ausbildung der Dachbinder ist als Vollwandträger oder Fachwerkträger. Der englische Begriff „roof truss" ist allerdings nur für den Binder als Fachwerkträger anzuwenden. Ein Binder als Vollwandträger heißt „plate roof girder"*] = roof truss
~ → ~schicht *f*
~, Bindemittel *n*, Straßenbau~, Straßenbauhilfsstoff *m* = road binder
~, Bindemittel *n* = cementitious material, cementing ~, binder, binding medium
~abstand *m*, Dach~, Fachwerk~ = distance between roof trusses

~balken *m*, Bundbalken [*Balken, der mit zwei Sparren und den Stuhlsäulen den hölzernen Dachstuhl bildet*] = tie beam
~form *f* → Dach~
~gemisch *n* → Bindermaterial *n*
~gespärre *n*, Bundgespärre = principals, principal rafters
~kopf *m*, Streckerkopf = head
~lage *f* → Binderschicht *f*
~material *n*, Bindergemisch *n*, Bindermischung *f* [*Straßenbau*] = binder course mix(ture)
~mischung *f* → Bindermaterial *n*
~schicht *f*, Binder *m*, Binderlage *f* [*Straßen- und Flugplatzbau*] = (surface) base course, base-course (of the surfacing) (Brit.); binder course, base coat, binding course, bottom coat, bottom course; [*deprecated: black base*] [*The binder course is a transitional layer between the base course and the surface course. A tack coat is applied at the interface of the surface and binder courses*]
~schicht *f*, Bindersteinschicht, Streckerschicht = header course, heading ~, bonder ~, bondstone ~
~sparren *m* = common rafter, intermediate ~, rafter spar
Binderstein *m* → Binder *m*
~schicht *f* → Binderschicht
Binder|stiel *m*, Stuhlsäule *f*, Ständer *m*, Bundpfosten *m* = post
~teilchen *n*, Bindemittelteilchen = binder particle
~ton *m*, Bodenbinder *m*, haftendes Lockergestein *n*, Bindeerde *f*, Bodenmörtel *m* = soil matrix, ~ binder, ~ mortar, binder soil, (clay) binder
~verband *m*, Streckerverband [*Die Steine liegen mit Längsseite senkrecht zur Mauerfläche; die Schichten sind zueinander um $^1/_4$ Stein verschoben*] = heading bond
~verteilung *f*, Dach~, Fachwerk~ = spacing of roof trusses
Bindeschicht *f* [*Straße*] = binder coat, tack ~

~anspritzung *f* = tack coating
~-Spritzapparat *m*, Bindeschicht-Spritzgerät *n*, Bindeschicht-Spritzmaschine *f* = tack-coater
Binde|ton *m* = ball clay
~wert *m* **unter Wassereinwirkung**, BWW = binding value under the action of water [*bituminous binder*]
~zone *f*, Übergangszone = weld junction
bindig, kohäsiv, kohärent [*Boden*] = cohesive, plastic
bindige Beimengung *f*, kohärente ~, kohäsive ~ = cohesive matter
bindiger Boden *m* **mit Eigengewicht** = cohesive soil with weight
Bindigkeit *f* → Kohäsion *f*
Bindung *f* [*Siebtechnik*] = crossing of wires, interlacing ~ ~, joints
Bindungskraft *f* = binding force
Binnen|bankett *n*, Binnenberme *f* [*Deich*] = inner berm, ~ bench
~barre *f* = inner bar
~berme *f*, Binnenbankett *n* [*Deich*] = inner bench, ~ berm
~böschung *f* [*bei Deichen, die dem Polder zugekehrte Böschung*] = inner slope
~düne *f* → Winddüne
~eisdecke *f*, Eismantel *m*, Eisschild *n*, Inlandeis(decke), Binneneis *n* = continental ice sheet
~flotte *f* = inland fleet
~gebiet *n* = inland region
~gewässer *n* = inland water
~gewässerfahrzeug *n* = inland craft
~gewässerkunde *f* → Limnologie *f*
~hafen *m* = inland harbour (Brit.); ~ harbor (US)
~hafen *m* [*innerer Teil eines Seehafens*] = inner harbo(u)r
~haupt *n* [*Seeschleuse*] = inner gates, upper ~
~schiff *n* = inlad craft
~schiffahrt *f* = inland navigation
~schiffahrtkanal *m*, künstliche Wasserstraße *f* = inland (navigation) canal, artifical navigation ~, barge ~

Binnenschiffahrtschleuse — Bitumenaufstrich

~schiffahrtschleuse f = navigation lock, (inland) ~
~schiffahrtstraße f, Binnenwasserstraße = inland waterway
~schiffbestand m = number of inland vessels available
(~)See m = lake
(~)Seewasser n = lake water
~seite f [*Deich*] = inner side
~tor n, Ebbetor = ebb tide gate
~wasserkraft f → Wasserkraft
~wasserstraße f, Binnenschiffahrtstraße f = inland waterway
~wasserstraßennetz n, Binnenschiffahrtstraßennetz = inland waterway network
Binsenmatte f = rush mat
Bioaktivität f = bioactivity
biochemische Wirkung f = biochemical action [*Chemical action resulting from the growth or metabolism of living organisms*]
biochemischer Sauerstoffbedarf m, B.S.B. = biochemical oxygen demand, BOD [*The quantity of oxygen required for biochemical oxidation in a given time at a given temperature, the determinations usually being for 5 days at 20° C*]
Bio-Filter m, n, biologischer Körper m = bacteria filter, trickling ~, percolating ~, sprinkling ~, continuous ~, biological ~, bio-filter, bacteria bed, contact bed
Biofilterung f, Biofiltern n = biofiltration [*The recirculation of sewage from a trickling filter to a settling tank preceding the filter. This process is covered by the Jenks patent*]
biologisch abbaubar = biodegradable
biologische Abwasserbehandlung f → ~ Reinigung
~ Betonabschirmung f = concrete biological shield
~ Nachbehandlung f → ~ Reinigung
~ Reinigung f, ~ Abwasserbehandlung, ~ Nachbehandlung, biologische Abwasserreinigung, biologisches Verfahren n = bacteria treatment, biological ~, ~ purification

~ Schutzanlage f, ~ Abschirmungsanlage = biological (reactor) shield, structure for biological shielding of atomic plants
~ Wasseruntersuchung f = biological water analysis
biologischer Körper m → Bio-Filter m, n
~ Rasen m [*Tropfkörper*] = biological slime
biologisches Verfahren n → biologische Reinigung
Biotit m, dunkler Glimmer m (Min.) = biotite
~gneis m = biotite gneiss
~granit m [*früher: Granitit m*] = biotitegranite, granitite
~schiefer m, Dunkelglimmerschiefer = biotite schist
biquadratische Gleichung f, Gleichung vierten Grades = biquadratic equation
Birken|besen m = birch broom
~faschine f = birch faggot, ~ fascine
Bischofsmütze f → Gehrung f
Biskuit m, Schrühbrandscherben m = biscuit
Bismarckbraun n = Bismarck brown
Bismit m, Wismutocker m, Bi_2O_3 (Min.) = bismuth ochre, bismite
Bismuthin n, Wismutglanz m = bismuthinite
Bitterwasser n = bitter water, aperient ~
Bitukies m → Bitumenkies
Bitumen n [*DIN 55946*] = (asphaltic) bitumen (Brit.); asphalt (US)
~-Anhänger m = asphalt trailer (US); (asphaltic) bitumen ~ (Brit.)
~anstrich m = (asphaltic) bitumen paint coat (Brit.); asphalt ~ ~ (US)
~anstrichmittel n = asphalt paint (US); (asphaltic) bitumen ~ (Brit.) [*An asphalt product sometimes containing small amounts of other materials such as lampblack, and mineral pigments*]
~anteil m → Bitumengehalt m
~aufstrich m, Bitumendeckschicht f = asphalt coat(ing) (US); (asphaltic) bitumen ~ (Brit.)

Bitumenauskleidung — Bitumenemulsion

~auskleidung *f* → Asphaltauskleidung

~band *n* [*als Rohrdichtung*] = asphalt tape (US); (asphaltic) bitumen ~ (Brit.)

bitumenbasisches Erdöl *n*, asphaltbasisches ~ = asphaltic(-base) (crude) petroleum, ~ crude

Bitumen|behälter *m* → Bitumentank *m*

~bekleidung *f* → = Asphaltauskleidung

Bitumenbelag *m* → Asphaltdecke *f*

~einbaumasse *f* → Asphaltbelageinbaumasse

~gemisch *n* → Asphaltbelageinbaumasse *f*

~mischgut *n* → Asphaltbelageinbaumasse *f*

~mischung *f* → Asphaltbelageinbaumasse *f*

Bitumenbeton *m* → Asphaltbeton

~abdichtung *f* → Asphaltbetondichtung

Bitumenbeton(ab)dichtungs|haut *f* → Asphaltbetondichtung *f*

~lage *f* → Asphaltbetondichtung *f*

~schicht *f* → Asphaltbetondichtung *f*

~schürze *f* → Asphaltbetondichtung *f*

~teppich *m* → Asphaltbetondichtung *f*

~vorlage *f* → Asphaltbetondichtung *f*

Bitumenbeton|-Außen(haut)dichtung *f* → Asphaltbetondichtung

~belag *m* → Asphaltbetonbelag

~decke *f* → Asphaltbetonbelag *m*

~dichtung *f* → Asphaltbetondichtung *f*

Bitumenbetondichtungs|belag *m* → Asphaltbetondichtung *f*

~haut *f* → Asphaltbetondichtung *f*

~lage *f* → Asphaltbetondichtung *f*

~schicht *f* → Asphaltbetondichtung *f*

~schürze *f* → Asphaltbetondichtung *f*

~teppich *m* → Asphaltbetondichtung *f*

~vorlage *f* → Asphaltbetondichtung *f*

Bitumenbeton|mischanlage *f*, Asphaltbetonmischanlage = asphaltic concrete mixing plant, ~ ~ ~ installation

~mischer *m*, Asphaltbetonmischer = asphaltic concrete mixer

~oberflächendichtung *f* → Asphaltbetondichtung

~schürze *f* → Asphaltbetondichtung *f*

~teppich *m* → Asphaltbetonteppich

~verschleißschicht *f* → Asphaltbetonverschleißschicht

Bitumen|binder(schicht) *m*, (*f*) → Asphaltbinder(schicht)

~blasanlage *f* = air-rectification installation, (air-)blowing ~, oxidizing ~

~(dach)pappe *f*, Asphalt(dach)pappe = (asphaltic-)bitumen (roofing) felt (Brit.); asphalt (~)~ , asphaltsaturated (~)~ (US)

~-Dachschindel *f* → Asphalt-Dachschindel

Bitumendecke *f* → Asphaltdecke

Bitumendecken|einbaumasse *f* → Asphaltbelageinbaumasse

~erhitzer *m* → Asphaltdeckenerhitzer

~gemisch *n* → Asphaltbelageinbaumasse *f*

~mischung *f* → Asphaltbelageinbaumasse *f*

Bitumen|deckschicht *f*, Bitumenaufstrich *m* = (asphaltic) bitumen coat(ing) (Brit.); asphalt ~ (US)

~-Dichtungshaut *f* = waterproofing membrane; asphalt ~ (US); (asphaltic) bitumen ~ (Brit.)

~-Einspritzregelventil *n* = asphalt injection control valve (US); (asphaltic) bitumen ~ ~ ~ (Brit.)

Bitumenemulsion *f*, Asphaltemulsion, emulgiertes Bitumen *n*, Kaltasphalt *m*, Kaltbindemittel *n* = asphaltic emulsion, emulsified asphalt (US); (asphaltic) bitumen (road) emulsion, cold (bitumen) emulsion, cold asphaltic bitumen emulsion (Brit.)

~ **mit unlöslichem Emulgator**, ~ ~ festem ~ [*vielfach unrichtig als "Bitumensuspension" bezeichnet*] = (asphaltic) bitumen emulsion with colloidal emulsifier (Brit.); asphalt ~ ~ ~ ~ (US)

Bitumen|farbe *f* = (asphaltic-)bitumen paint, bituminous ~ (Brit.); asphalt ~ (US)
~feinbinder *m*, Asphaltfeinbinder = fine-grained asphaltic binder course
~fertigbahn *f* → Bitumengewebebahn
~film *m* = (asphaltic) bitumen film (Brit.); asphalt ~ (US)
~filz *m* = (asphaltic) bitumen-impregnated felt (Brit.); asphalt-impregnated ~]US)
~formling *m* = (asphaltic) bitumen briquette (Brit.); asphalt ~ (US)
~fuge *f* [*Spannbeton*] = asphalt joint (US); (asphaltic) bitumen ~ (Brit.)
~fundstätte *f* → Bitumenvorkommen *n*
~gehalt *m*, Bitumenanteil *m* = (asphaltic) bitumen content (Brit.); asphalt ~ (US)
bitumen|getränkt = asphalt-impregnated (US); (asphaltic) bitumen-impregnated (Brit.)
~getränkter Fugenstreifen *m* = bituminous filler type expansion joint material (Brit.); asphalt-impregnated ~ ~ ~ ~ ~ (US)
Bitumen|gewebebahn *f*, Bitumenfertigbahn, Bitumengewebe *n* = prefabricated (asphaltic) bitumen surfacing (Brit.); ~ asphalt ~ (US)
~grobbinder *m*, Asphaltgrobbinder = coarse-grained asphaltic binder course
~-Gummi-Gemisch *n*, Bitumen-Gummi-Mischung *f*, Gummi-Bitumen-Gemisch, Gummi-Bitumen-Mischung = rubberized asphalt (US); ~ (asphaltic) bitumen (Brit.)
bitumenhaltig = asphaltic (US); containing (asphaltic) bitumen (Brit.)
Bitumen|-Kalksandstein *m* = asphalt-impregnated sand lime brick (US); (asphaltic) bitumen-impregnated ~ ~ (Brit.)
~kanone *f* [*Bitumenunterpressung*] = asphalt gun (US); (asphaltic) bitumen ~ (Brit.)
~-Kesselwagen-Erhitzer *m* = asphalt tank car heater (US); (asphaltic) bitumen ~ ~ ~ (Brit.)

~kies *m*, Asphaltkies, Bitukies = asphalt-coated gravel (US); asphaltic bitumen-coated ~ (Brit.)
~-Kiespreßpappe *f* = asphaltic bitumen-gravel roofing (Brit.); asphalt-gravel ~ (US)
~kitt *m* → Asphaltkitt
~klebemasse *f* = bonding compound
~kocher *m* → Bitumenschmelzkessel *m*
~kügelchen *n* = (asphaltic) bitumen globule (Brit.); asphalt ~ (US)
~lack *m* → Asphaltlack
~lager *n* → Bitumenvorkommen *n*
~lagerstätte *f* → Bitumenvorkommen *n*
~lösung *f* = bituminous solution, (asphaltic-)bitumen ~ (Brit.); asphalt ~ (US)
Bitumenmakadam *m* → Asphaltmakadam
~belag *m* → Asphaltmakadamdecke *f*
~decke *f* → Asphaltmakadamdecke
Bitumen|masse *f* = (asphaltic-)bitumen composition (Brit.); asphalt ~ (US)
~Mineralgemisch *n*, Bitumen-Mineralmischung *f* = (asphaltic) bitumen-aggregate mix(ture) (Brit.); asphalt-aggregate ~ (US)
~mischmakadam *m*, Asphaltmischmakadam = asphalt macadam (US); (~) bitumen ~ (Brit.)
~mörtel *m*, Asphaltmörtel [*Mörtel mit Bitumen als Bindemittel*] = (asphaltic) bitumen mortar (Brit.); asphaltic ~ (US)
~-Mörtelschmiere *f* [*klebt an den Teilen eines Schwarzdeckenfertigers und ist zu beseitigen*] = fat
~papier *n*, Asphaltpapier = (asphaltic) bitumen-impregnated paper (Brit.); asphalt-impregnated ~ (US)
~pappdach *n*, Asphaltpappdach = (asphaltic-)bitumen felt roof (Brit.); asphalt ~ (US)
~pappe *f* → Bitumendachpappe
~probe *f* → Bitumenprüfung *f*

~prüfung *f*, Bitumenprobe *f*, Bitumenversuch *m* = (asphaltic) bitumen test (Brit.); asphalt ~ (US)

~pumpe *f* = asphalt pump (US); (asphaltic) bitumen ~ (Brit.)

~rostschutzfarbe *f* = asphalt-based rust protective paint (US); (asphaltic) bitumen-based ~ ~ ~ (Brit.)

~sand *m*, Bitusand, Asphaltsand = asphaltic bitumen-coated sand (Brit.); asphalt-coated ~ (US)

~schicht *f* = (asphaltic) bitumen layer (Brit.); asphalt ~ (US)

~schindel *f*, Asphaltschindel = asphalt shingle (US); (asphaltic) bitumen ~ (Brit.); [*Asphalt shingles are made of heavy asbestos felt and rag felt saturated or coated with asphalt, and with crushed slate or other material embedded in an asphalt coating to form the exposed surface*]

~schlämme *f* = (asphaltic) bitumen slurry (Brit.); asphalt ~ (US)

~schlämmeabsieg(e)lung *f*, Bitumenschlämmeversieg(e)lung, Bitumenschlämme-Porenschluß *m* = (asphaltic) bitumen slurry seal (Brit.); asphalt ~ ~ (US)

~schmelzkessel *m*, Bitumenkocher *m* = asphalt heater, ~ heating kettle, ~ melting tank (US); (asphaltic) bitumen heater, (asphaltic) bitumen heating kettle, (asphaltic) bitumen melting tank, (asphaltic) bitumen melting boiler (Brit.)

~schotter *m* = asphalt-coated crushed (road) metal (Brit.)

~-Schutzanstrich *m* = asphaltic bitumen protective coat (Brit.); asphalt ~ ~ (US)

~sorte *f* = (asphaltic) bitumen grade (Brit.); asphalt ~ ~ (US)

~-Spachtel(masse) *m*, (*f*) = asphaltic bitumen screed(ing compound) (Brit.); asphalt ~ (~) (US)

Bitumensplitt *m* = bituminised chippings (Brit.); asphalt-coated ~ (US)

~decke *f* → Bitumensplitt-Teppich(belag) *m*

~-Teppich(belag) *m*, Bitumensplittdecke *f*, Asphaltsplittdecken = (asphaltic) bitumen-coated chip(ping)s carpet (Brit.); asphalt-coated ~ ~ (US)

Bitumen|-Sprengwagen *m*, Druckverteiler *m*, Tankspritzmaschine *f*, Großspritzgerät *n*, Automobil-Sprengwagen, Drucksprengwagen *m*, Drucktankwagen, Motor-Spritzwagen, Rampen-Tankspritzmaschine *f*, Automobil-Spritzwagen = bulk (asphaltic) bitumen distributor, pressure tank lorry, pressure spray tanker (Brit.); pressure distributor truck, asphalt truck distributor (US); pressure tanker, truck mounted distributor, road oil distributor tanker

~-Spritzkanone *f* = asphalt gunite equipment (US); (asphaltic) bitumen ~ ~ (Brit.)

~-Spritzmaschine *f* → Bitumenverteiler *m*

~-Straßenbau *m*, Asphaltstraßenbau = asphalt highway construction, ~ road ~

~(straßen)belag *m* → Asphaltdecke *f*

~(straßen)decke *f* → Asphaltdecke

~streifen *m* = asphalt-impregnated strip (US); (asphaltic) bitumen-impregnated ~ (Brit.)

~suspension *f* [*Fehlname*] → Bitumenemulsion *f* mit unlöslichem Emulgator

~tank *m*, Bitumenbehälter *m* = (asphaltic) bitumen tank (Brit.); asphalt tank (US)

~teer *m* → Teer-Bitumen-Gemisch *n*

~teppich *m* = asphalt carpet (US); (asphaltic) bitumen ~ (Brit.)

~tragschicht *f* → (obere) Asphalttragschicht

~tränkmakadam *m*, Asphalttränkmakadam = (asphaltic) bitumen grouted mcadam (Brit.); ~ penetration ~ (US)

Bitumentränkung — bituminöse Mischanlage

~tränkung f = asphalt impregnation (US); (asphaltic) bitumen ~ (Brit.)
~trommel f = drum for (asphaltic) bitumen (Brit.); ~ ~ asphalt (US)
~-Umschlaganlage f = bulk (asphaltic) bitumen installation, ~ (~) ~ reception facility (Brit.); ~ asphalt ~ ~ (US)
~-Umwälzerhitzer m = circulating heater for (asphaltic) bitumen (Brit.); ~ ~ ~ asphalt (US)
~- und Teerkocher m → Kocher
~- und Teerschmelzkessel m → Kocher m
~-Unterpressung f = asphalt underseal(ing), ~ subseal(ing) (US); (asphaltic) bitumen ~ (Brit.)
~verfestigung f = (asphaltic) bitumen stabilisation (Brit.); asphalt stabilization, soil asphalt (US)
~verkleidung f → Asphaltauskleidung
~vermörtelung f → → Baugrundverbesserung
~versuch m → Bitumenprüfung f
~verteiler m, Bitumen-Spritzmaschine f = asphalt distributor (US); (asphaltic) bitumen ~ (Brit.)
~vorkommen n, Bitumenlager n, Bitumenlagerstätte f, Bitumenfundstätte, Bitumenvorkommnis n = deposit of (asphaltic) bitumen (Brit.); ~ ~ asphalt (US)
~vorkommnis n → Bitumenvorkommen n
~wellpappe f = corrugated bituminous board (Brit.); ~ asphalt ~ (US)
~zementestrich m = cement-bitumen (jointless) floor(ing), ~ composition ~ (Brit.); cement-asphalt ~ (US)
bituminieren = to bituminize
Bituminieren n, Bituminierung f = bituminizing
bituminierter Zement m = bituminized cement
bituminös = bituminous [*containing bitumen or tar, or a mix(ture) of bitumen and tar*]
~ gebunden = bituminous bound [*bonded with the aid of bituminous material*]

bituminöse Arbeiten fpl = bituminous work
~ Bauwerk(ab)dichtung f = bituminous damp-proofing and waterproofing
~ Binderschicht f, bituminöser Binder m [*Straßenbau*] = bituminous binder (course), ~ level(l)ing ~
~ Bindeschicht f [*Straßenbau*] = bituminous tack coat
~ Bodenvermörtelung f, ~ Bodenverfestigung, Bodenvermörtelung mit Schwarzbindemittel, Bodenverfestigung mit Schwarzbindemittel = bituminous soil stabilization
~ Brücken(fahrbahn)decke f → bituminöser Brücken(fahrbahn)belag m
~ Dachpappe f = bituminous felt
~ Decke f, bituminöser Belag m, Schwarzdecke, Schwarzbelag = bituminous pavement, ~ surfacing, black-top ~
~ Deckenbautechnik f, ~ Belagbautechnik = asphalt paving practice, ~ pavement ~ (US); asphaltic-bitumen surfacing ~ (Brit.)
~ Emulsion f = bituminous emulsion [*A liquid mix(ture) in which minute globules of bitumen (US) are held in suspension in water or a watery solution*]
~ Fertigbahn f, ~ Gewebebahn = prefabricated bituminous surfacing, P. B. S., prefabricated bituminized hessian surfacing, bituminized jute hessian cloth
~ Fertigfuge f = prefab(ricated) bituminous joint
~ Gewebebahn f → Fertigbahn
~ Heißmischdecke f, bituminöser Heißmischbelag m = hot-mix(ed) bituminous pavement, ~ ~ surfacing
~ Kaltmischdecke f, bituminöser Kaltmischbelag m = cold-mix(ed) bituminous pavement, ~ ~ surfacing
~ Mischanlage f → Schwarzdeckenmischanlage

bituminöse Mischdecke — bituminöser (Tränk)Makadam

~ **Mischdecke** *f*, bituminöser Mischbelag *m* = plant-mix(ed) bituminous pavement, ~ ~ surfacing, ~ black-top ~

~ **OB**, ~ Oberflächenbehandlung *f* = bituminous surface treatment

~ **Straßenbauemulsion** *f* = bituminous road emulsion

~ **Straßenbaumischung** *f*, bituminöses Straßenbaugemisch *n* = bituminous road mix(ture), ~ highway ~

~ **Straßendecke** *f*, bituminöser Straßenbelag *m* = bituminous highway pavement, ~ ~ surfacing

~ **Tragschicht** *f*, Schwarz-Tragschicht = flexible base (course), hydrocarbon ~ (~), bituminous ~ (~)

~ **(Tränk)Makadamtragschicht** *f* = bituminous macadam base (course) *f*

~ **Verschleißschicht** *f* = bituminous wearing course

bituminöser (Bau)Stoff *m*, schwarzer ~ [*Solche Stoffe auf der Basis von Bitumen oder Teer werden im Bauwesen als Bindemittel, Klebstoffe und Sperrstoffe verwendet. Chemische Ausdrücke sind „Kohlenwasserstoffgemisch" und „Kohlenwasserstoffmischung"*] = bituminous material; bitumen (US)

~ **Belag** *m* → bituminöse Decke *f*

~ ~ [*auf einer Tonbetonstraße*] = wearing carpet

~ **Beton** *m*, Kohlenwasserstoffbindemittelbeton, Schwarzbeton = bituminous concrete

~ **Binder** *m*, bituminöse Binderschicht *f* [*Straßenbau*] = bituminous binder (course), ~ level(l)ing ~

~ **Brücken(fahrbahn)belag** *m*, bituminöse Brücken(fahrbahn)decke *f* = asphalt bridge carriageway pavement, ~ ~ paving, ~ deck ~ (U); asphaltic-bitumen bridge carriageway surfacing, asphaltic-bitumen deck surfacing (Brit.)

~ **Grund(ier)anstrich** *m* → ~ Voranstrich

~ **Heißmischbelag** *m*, bituminöse Heißmischdecke *f* = hot-mix(ed) bituminous pavement, ~ ~ surfacing

~ **Kaltmischbelag** *m*, bituminöse Kaltmischdecke *f* = cold-mix(ed) bituminous pavement, ~ ~ surfacing

~ **Kalkstein** *m* → Asphaltkalkstein

~ **Makadam** *m* → Asphaltmakadam

~ **Mischbelag** *m* → bituminöse Mischdecke *f*

~ **Mischmakadam** *m* = mixed bituminous macadam

~ **Mörtel** *m* = bituminous mortar

(~) **Ölschiefer** *m* = bituminous oil shale, (asphaltic) pyrobituminous ~

~ **Porenschlußbelag** *m* = bituminous seal coat [*A thin bituminous carpet coat seldom over 1/2-inch thick*]

~ **Sandstein** *m*, Asphaltsandstein = asphaltic sandstone, bituminous ~

~ **Schutzanstrich** *m* = bituminous protective coat(ing)

~ **Stoff** *m* → ~ Baustoff

~ **Straßenbau** *m*, Schwarzstraßenbau = bituminous road construction, flexible ~ ~, ~ highway ~

~ **(Straßen)Belag** *m*, Schwarzbelag = bituminous surface course, ~ surfacing [*The top course of a bituminous pavement. It may be fine aggregate type, graded aggregate type, or coarse aggregate type*]

~ **Teppich(belag)** *m* = bituminous carpet coat

~ **(Tränk)Makadam** *m* = bituminous macadam [*A type of highway construction in which a broken stone aggregate of relatively coarse and uniform size fragments is first spread and interlocked by compaction after which the individual stones are coated and bound together with hot bituminous cement which is applied at the surface but penetrates the layer of stone before it cools. A bituminous macadam surface course is finished* *off with a surface treatment of bituminous cement and cover of stone chips to seal the surface. In the construction of a bituminous macadam base the seal coat is omitted]*

~ **Überzug** m **auf Pflaster** = bituminous topping on sett paving

~ **Unterbau** m *[Straßenbau]* = bituminous foundation

~ **Voranstrich** m, ~ Grund(ier)anstrich, ~ Grundierüberzug m = bituminous prime coat(ing)

~ **Wasserbau** m → Asphaltwasserbau

bituminöses Anstrichmittel n = bituminous paint

~ **Bindemittel** n, bituminöser Binder m, Schwarzbindemittel, Schwarzbinder = bituminous (road) binder

~ **Grundiermittel** n = bituminous primer *[A liquid bituminous road material of low viscosity which upon application to a nonbituminous surface is completely absorbed. Its purpose is to waterproof the existing surface and prepare it to serve as a base for the construction of a bituminous carpet or surface course]*

~ **Kontrollsystem** n *[Talsperre]* = bituminous filter layers

~ **Mischgut** n, Schwarzmaterial n, Schwarz(decken)(misch)gut, Schwarz(decken)mischung f, Schwarz(decken)gemisch n, Schwarz(belag)einbaumasse f = bituminous mix(ture)

(~) **Mischwerk** n = central bituminous mixing plant

~ **Straßenbaugemisch** n, bituminöse Straßenbaumischung f = bituminous road mix(ture)

Bitusand m → Bitumensand

Blähbeton m, Zellenbeton, Porenbeton, aufgeblähter Beton = cellular(-expanded) concrete, aerated ~

Blähdruck m *[Tunnel- und Stollenbau. Bei bestimmten Gesteinen wie Gipsstein oder Mergel durch Wasseraufnahme entstehender allseitiger Druck]* = swelling pressure, squeeze

Blähen n, Blähung f, Auf~ *[Herstellung von Leichtzuschlägen]* = expanding, expansion; bloating (Brit.)

Blähgrad m *[Kohle]* = swelling index

Blähkork m = expanded cork

Blähmittel n → gaserzeugendes Mittel

Blähschieferton m, aufgeblähter Schieferton = expanded shale, burned ~; bloating ~ (Brit.)

~**-Grobzuschlag(stoff)** m = expanded-shale coarse aggregate

Blähschlamm m *[Abwasserwesen]* = bulking sludge

Blähton m, aufgeblähter Ton = expanded clay; bloating ~ (Brit.)

Blähtonbeton m *[Im Schacht- oder Drehofen hergestellter Blähton erlangt bimsähnliche Struktur. Der Drehofenblähton erlangt durch die Art des Herstellungsverfahrens rundliche Kornformen in den bei der Betonbereitung erforderlichen Korngrößen. Die Blähtonpatzen aus dem Schacht- oder Ringofen werden im Steinbrecher zerkleinert und durch Sieben in die erforderlichen Körnungen zerlegt. Das Maß der Aufblähung und damit die Eigenfestigkeit des Blähtons ist bis zu einem gewissen Grad regelbar]* = expanded clay concrete; bloating ~ ~ (Brit.)

~**platte** f = expanded clay concrete slab

~**-Schiffskörper** m = expanded clay concrete hull

Blähton|fabrik f, Blähtonwerk n = expanded clay factory

~**schiefer** m, aufgeblähter Tonschiefer = expanded slate, burned ~; bloating ~ (Brit.)

~**werk** n, Blähtonfabrik f = expanded clay factory

Bläh-Zuschlag(stoff) m = expanded concrete aggregate; bloating ~ ~ (Brit.)

Blaine|-Feinheit f = Blaine fineness

~**-Gerät** n = Blaine (air permeability) apparatus *[Air permeability apparatus for measuring the surface area of a finely ground cement, raw material, or other product]*

(Blake'scher)(Doppel)Kniehebel(backen)brecher m → → Backenbrecher
blank, nackt, nichtumhüllt = uncoated, bare
blanke Schraube f = turned bolt, finished ~
Blank|glühen n = bright annealing
~glühofen m = bright-annealing furnace
~lack m, Klarlack = (clear) varnish
~lackieren n, Klarlackieren = varnishing
~stahl m = bright steel, cold drawn ~
Bläschen|beton m, belüfteter Beton, AEA-Beton, Luftporenbeton = air-entraining concrete, air-entrained ~
~mörtel m → AEA-Mörtel
Blase f = bubble
~ [*Gußasphalt*] = blister
blasen = to blow
~ [*den Zement aus den Behälterwagen in Silos*] = to convey by air
Blasen n [*Erdölbitumen*] = air-rectification, (air-)blowing, oxidizing
~abscheider m = air bubble eliminator
~bildung f = blistering
blasend [*Bohrhammer*] = blowing
blasende Bewetterung f → Druckbewetterung
~ Hammerbohrmaschine f = blowing rock drill drifter
~ Wetterführung f → Druckbewetterung
blasender Lüfter m = blower fan
Blasen|kupfer n = blister copper
~schiefer m = amygdaloidal Zechstein dolomite, blasenschiefer
~tang m, Fukus m = bladder wrack, fucus vesiculosus
Blasevorrichtung f, Luftspülvorrichtung [*Bohrhammer*] = puff blowing device
Blasgut n [*z. B. Zement*] = material conveyed by air
Blasigwerden n **der Glasur** = blebbing, blistering, bubbling (of the glaze)
~ des Scherbens = bloating (of the stone)

Blaskopf m, Spülluftkopf [*Bohrhammer*] = puff blowing head
Blasrohrleitung f **für Beton**, Beton-Blasrohrleitung [*für Druckluft-Betonförderung*] = pneumatic line for concrete distribution
Blasschatten m (⚒) = cavity in the stowing
Blaspistole f [*zum Säubern verschmutzter Maschinenteile mit Druckluft*] = cleaning gun
Blasversatz m, Blasversetzen n, Verblasen (⚒) = pneumatic stowing, (compressed-)air ~
~anlage f (⚒) = pneumatic stowing plant, ~ ~ installation, (compressed-)air ~ ~
~maschine f (⚒) = pneumatic stowing machine, (compressed-)air ~ ~
~rohr n (⚒) = pneumatic stowing pipe, (compressed-)air ~ ~
Blasversetzen n → Blasversatz m (⚒)
Blatt n, Masse~ [*zum Überdrehen. Steinzeugindustrie*] = bat [*for jiggering*]
~ [*Holzverbindung*] = halved joint
~aluminium n = alumin(i)um-leaf
~erde f = leaf mould
Blätter|kohle f, Laubkohle = foliated coal, laminated ~, leaf ~
~ton m, blätt(e)riger Ton = foliated clay
~torf m, Laubtorf = leaf peat
Blätterung f (Geol.) = foliation
Blatt|feder f = laminated (or flat, or leaf, or blade, or plate) spring
~federaufhängung f = laminated spring mounting
~gold n = gold-leaf
~meißel m, Schneidenmeißel = bit with wings, ~ ~ blades, drag bit
~metall n = metal-leaf
~silber n = silver-leaf
~verschiebung f (Geol.) = lateral fault
Blau|asbest m, Kapasbest = blue asbestos, Cape ~
~brenner m, Bunsenbrenner = Bunsen (gas) burner
~eisenerde f, Blaueisenerz n, Vivianit m (Min.) = (earthy) vivianite

blaue Mineralfarbe *f* = blue pigment
blauer Ton *m*, Blauton = blue clay
blaues Öl *n* = blue oil, pressed distillate
blaufarbiges gebrochenes Material *n* [*Straßenbau*] = blue metal [*Australia*]
Blau|pause *f* = blue print
~**salz** *n*, K₃Fe(CN)₆ = potassium ferrocyanide
~**ton** *m*, blauer Ton = blue clay
Blech *n* erster Wahl = prime sheet
~ zweiter Wahl = mender
~**abdeckung** *f*, Blechbelag *m* = plate cover(ing)
~**abfälle** *mpl* = plate clippings
~**anreißmaschine** *f* = plate marking machine
~**arbeiten** *fpl* = plate work
~**auskleidung** *f*, Blechbekleidung, Blechverkleidung, Blechhaut *f*, Blechbelag *m* = plate lining, ~ skin
~**bearbeitung** *f* = plate working
~**bearbeitungsmaschine** *f* = plate working machine
~**behälter** *m* = plate container
~**behälter** *m* **für Flüssigkeiten** = plate tank, ~ reservoir
~**bekleidung** *f* → Blechauskleidung
~**belag** *m* → Blechauskleidung *f*
~**belag** *m*, Blechabdeckung *f* = plate cover(ing)
~**(beton)schalung** *f* = plate shuttering; ~ formwork (Brit.)
~**biegemaschine** *f* = plate bending machine
~**bogen(träger)** *m* = plate arch(ed girder)
~**brückenträger** *m*, Brückenblechträger = bridge plate girder, plate girder span
~**bunker** *m* = plate hopper, ~ bin
~**bürstmaschine** *f* = plate brushing machine
~**dach** *n* = plate roof
~**dichtungsstreifen** *m* = sheet-metal water stop
~**dicke** *f* = plate thickness
~**doppler** *m* = plate doubler

~**-Dreigelenkbogen(träger)** *m* = plate-webbed arch(ed girder) with three hinges
~**folie** *f*, Metallfolie = metal foil
~**futter** *n* = armour plating (Brit.); armor ~ (US)
(~)**Gleitkanal** *m* → Blechröhre *f*
~**glied** *n* → (angelenkte) Blechplatte *f* [*Gliederband*]
~**gliederband** *n*, Plattenband(förderer) *n*, (*m*), Gliederband(förderer) = (plate) apron conveyor, (~) ~ conveyer
~**handschere** *f* = tin snips
~**hängedach** *n*, hängendes Blechdach = plate suspension roof
~**haube** *f* = plate hood, ~ cowl
~**haut** *f* → Blechauskleidung *f*
~**hebezeug** *n*, Blechheber *m* = lifting gear for plates
~**hülse** *f*, (Aufsteck)Hülse aus Blech [*Betondeckenfuge*] = light metal cap
~**hülse** *f* → Pfahlrohr *n*
~**hülsenpfahl** *m*, Mantelrohrpfahl = shell pile, cased ~
~**hülsen-Preßpfahl** *m*, Mantelrohr-Preßpfahl = cased pressure pile, shell ~ ~
~**kabine** *f* = plate cab(in)
~**kanister** *m* = jerrycan
~**kastenträger** *m*, Kastenblechträger = box plate girder
~**klammer** *f* = plate cramp, ~ staple
~**konstruktion** *f* = platework
~**konstruktionsteile** *npl*, *mpl* = plate metal work
~**lehre** *f* = plate ga(u)ge
~**lochsieb** *n* = perforated plate sieve, ~ metal screen
~**lutte** *f*, Blechwetterlutte, Blechluttenrohr = plate air conduct, ~ ~ pipe, ~ ~ tube, ~ ~ channel
~**luttenrohr** *n* → Blechlutte *f*
~**mantel** *m* → Blechröhre *f*
~**pakettrennmaschine** *f* = sheet pack separator
~**platte** *f* = metal plate
~**platte** *f* → angelenkte ~ [*Gliederband*]

~prüfmaschine *f* = plate testing machine
~richtmaschine *f*, Grob~ = plate level(l)ing machine
~rinne *f* = plate gutter
~rohr *n* = plate pipe
~röhre *f*, (Blech)Gleitkanal *m*, Rohr *n* aus dünnem Blech, Hülle *f* aus Stahlblech, Blechmantel *m*, Hüllrohr, Umhüllungsrohr = (sheet-)metal sheath [*prestressed concrete*]
~rohrleitung *f* = plate pipeline
~schalung *f*, Blechbetonschalung = plate formwork (Brit.); ~shuttering
~schere *f*, Tafelschere = plate shears
~schneidmeißel *m* = ripping chisel
~schornstein *m* = plate chimney, metal ~
~schraube *f* = plate screw
~schrott *m* = plate scrap
~-Schwimmstoffabweiser *m*, Blech-Tauchwand *f* [*Abwasserwesen*] = plate scum collector
~sieb *n* = plate sieve
~silo *m* = plate silo
~spirale *f* [*Turbine*] = steel volute casing, ~ spiral ~
~stapler *m* = plate and sheet piler
~stegträger *m* = plate-web girder
~steifigkeit *f* = plate stiffness
~straße *f*, Blechstrecke *f* = plate train
~strecke *f*, Blechstraße *f* = plate train
~streifen *m* = plate strip
~-Tauchwand *f*, Blech-Schwimmstoffabweiser *m* [*Abwasserwesen*] = plate scum collector
Blechträger *m* = (built-up) plate girder
~brücke *f* = plate girder bridge
~-Deckbrücke *f* = plate girder deck bridge
~gurt *m* = plate girder flange, ~ ~ chord, ~ ~ boom
~höhe *f* = plate girder depth
~steg *m* = plate girder web
~trogbrücke *f* = trough plate girder bridge
Blech|trichter *m* = plate funnel
~trockentisch *m* = plate drying table
~trommel *f* = plate drum

~verbindung *f* = plate connection
~verkleidung *f* → Blechauskleidung
~walze *f*, Rundbiegemaschine *f* für Bleche = plate roll
~walzwerk *n*, Fein~ = sheet (rolling) mill
~walzwerk *n*, Grob~ = plate (rolling) mill
~wetterlutte *f* → Blechlutte
~zange *f* = plate gripping tongs
~zylinder *m* [*Walzenverschluß*] = plate (roller) gate cylinder
Blei *n* = lead
~abflußrohr *n*, Bleiablaufrohr = lead waste-pipe
Bleiauskleidung *f*, Bleiverkleidung, Bleibekleidung, Bleihaut *f*, Bleibelag *m* = lead lining, ~ facing
Bleiazid *n* = lead azide
Bleibekleidung *f*, Bleiauskleidung, Bleiverkleidung, Bleihaut *f*, Bleibelag *m* = lead lining, ~ lining
Bleibelag *m* → Bleiauskleidung *f*
bleibende Auflockerung *f* = final bulking
~ Belastung *f*, Dauerbelastung = permanent loading
~ Dehnung *f* = permanent strain
~ Formänderung *f* → plastische ~
~ Härte *f*, permanente ~, Dauerhärte = permanent hardness
~ Last *f* → Dauerlast
~ Verformung *f* → plastische Formänderung
Bleibenzin *n* = leaded petrol (Brit.)
Bleibergwerk *n*, Bleigrube *f* = lead mine
Bleiblech *n*, Tafelblei *n*, Walzblei = sheet lead
Bleiblockausbauchung *f* nach Trauzl = Trauzl method
Bleibronze *f* [*DIN 1716*] = lead bronze
~büchse *f* = lead bronze wrapped bush
bleichen, entfärben = to bleach
Bleichen *n*, Entfärben = bleaching
Bleicherde *f*, Podsolboden *m* = podzol
Bleicherei *f* = bleach works

Bleichkalk — Bleiweiß(anstrich)farbe

Bleichkalk m, Chlorkalk, Bleichpulver n = bleaching powder, chloride of lime, chlorinated lime
Bleichlorid n, Chlorblei n = phosgenite
Bleichmittel n, Entfärbungsmittel = bleaching agent
Bleichpulver n → Bleichkalk m
Bleichromat n → Chromgelb n
Bleidämmpappe f = insulating lead felt
Bleidichtung f, Bleiverstemmung = lead wool ca(u)lking
Bleidraht m = lead wire
Bleieinlage f, Bleizwischenlage f = lead filler, ⁓ insert
Bleierz n = lead ore
Bleifarbe f = lead (base) paint
Bleifolie f = lead foil
bleiführend = lead-bearing, plumbiferous
Bleigang m = lead vein
Bleigelb n, Königsgelb, amorphes Bleioxyd n = massicot, yellow lead
Bleigelenk n = lead hinge
Bleigeruchverschluß m, Bleisiphon m, Bleisyphon = lead (stench) trap
Bleigewicht n → Lot n
Bleigießerei f, Bleihütte f = lead foundry
Bleiglanz m, Galenit m, PbS = galena, lead glance
⁓trübe f = lead-glance pulp
Bleiglas n = lead glass
bleiglasiert = lead-glazed
Bleiglätte f, Bleioxyd n, PbO = lead oxide
Bleigrube f, Bleibergwerk n = lead mine
Bleihaut f → Bleiauskleidung f
Blei(hochofen)schlacke f = lead slag
Bleihütte f, Bleigießerei f = lead foundry
Bleikabel n = lead-covered cable
⁓mantel m, Bleimantel = lead (cable) sheath
Bleikrankheit f = plumbism
Bleilot n, Schnurlot, Senkblei n = plumb-bob, plummet
⁓, Lötblei n = plumber's solder, coarse ⁓ [*An alloy of lead and tin varying from 1 : 1 to 3 : 1*]

Bleilöten n = lead soldering
Bleimantel m → Bleikabelmantel
(Blei)Mennige f, Bleisuperoxyd n = red lead
Bleioxyd n, Bleiglätte f, PbO = lead oxide
Bleipflaster n = lead paving
Bleipigment n = lead pigment
Bleiplatte f = lead plate
Bleiplombe f = lead seal
Bleiplombengießmaschine f = lead seal casting machine
Bleiprofilpresse f = lead section press
Bleirohr n = lead pipe
⁓presse f = lead pipe press
Bleisäurebatterie f = lead-acid battery
Bleischeibe f, Bleiunterlegscheibe = lead washer
Bleischlacke f, Bleihochofenschlacke = lead slag
Bleischmelzgerät n = lead melting equipment
Bleischrot m = lead shot
Bleisiphon m, Bleigeruchverschluß m, Bleisyphon = lead (stench) trap
Bleistein m [*Metallurgie*] = lead matte
⁓ = lead regulus, reguline of lead
Blei|tafel f = lead sheet
⁓tetraäthyl n, BTÄ, Pb(C$_2$H$_5$)$_4$ = tetraethyl lead, TEL
⁓(unterleg)scheibe f = lead washer
⁓vergiftung f = lead poisoning
Bleiverglasung f = lead glazing, leaded light [*A light in which small diamond-shaped panes of glass are held in lead cames*]
bleivergossen [*Fuge*] = lead-sealed
Bleiverhüttung f = lead smelting
Bleiverkleidung f, Bleiauskleidung, Bleibekleidung, Bleibelag m, Bleihaut f = lead facing, ⁓ lining
Bleiverstemmung f, Bleidichtung = lead wool ca(u)lking
Bleiweiß n = white lead
Bleiweiß(anstrich)farbe f, Bleiweißölfarbe = white lead paint

Bleiwolle f = lead wool
Bleiziegel m [für den Atomreaktorbau] = lead block
Blei-Zinn-Lot n, Zinn-Blei-Lot = tin-lead solder, lead-tin ~
Bleizwischenlage f, Bleieinlage f = lead insert, ~ filler
Bleizylinder m = lead cylinder
Blendboden m, Blindboden [Bretterlage über der Deckenkonstruktion als Unterlage für Riemen- oder Parkett-(fuß)böden] = wood(en) sub-floor; rough floor (US)
Blendbogen m, Bogenblende f, Nischenbogen, Schildbogen [Mauerbogen, der flache Nischen, die nicht durch die ganze Mauerdicke hindurchdringen, überdeckt] = blind arch
Blende f [Keilschieber] = follower-ring
~, Stauscheibe f = blind
~ [flache Mauervertiefung zur Aufnahme von Skulpturen, Bildern usw.] = flat niche
blendehaltig [Mineral] = blendous
blenden = to glare [to shine with a strong, steady, dazzling light]
Blenden n → Blendung f
blendend = glaring
Blendfassade f = lined façade, (stone-)faced ~
blendfrei = glare-resistant
Blend|freiheit f = absence of glare
~**gefahr** f = risk of glare
~**mauer** f = facing wall
~**rahmen** m → Fensterrahmen
Blendschutz m = protection from glare
~**hecke** f = anti-dazzle hedge
~**vorrichtung** f = anti-glare device
~**zaun** m = anti-glare fence
Blendstein m → Verblendstein
Blendung f, Blenden n = glare [A strong, steady, dazzling light or brillant reflection, as from sunlight on ice]
Blendungsart f = type of glare
Blendziegel m → Verblendstein m
Blickfeld n, Gesichtsfeld = field of vision

blinder Arm m = false channel
~ **Pfeiler** m [Schornsteinbau im Gleitschalungsverfahren] = local column of weak concrete around each jacking rod
~ **Stab** m → Blindstab
blindes Widerlager n = blind abutment
Blind|boden m → Blendboden
~**decke** f = false ceiling
~**flansch** m, Deckelflansch [Deckel, der ein Rohr abschließt] = blind flange
~**ort** n (⚒) = dummy road
~**ortversatz** m, Bruchortversatz (⚒) = dummy-road packing
~**schacht** m = blind shaft
~**schloß** n, eingelassenes Schloß = dummy lock, mortise ~, mortice ~, rabbeted ~, flush encased ~
~**schraube** f = expansion plug
~**stab** m, spannungsloser Stab, blinder Stab [Gitterträger] = unstrained member
~**stopfen** m = welch plug
~**zarge** f = blind casing, ~ trim
Blinkdauer f, Blinkzeit f [Warnblinker] = flashing time, ~ period
Blinkfeuer n, Blinklicht n, Blinker m [allgemein] = intermittent light (or beacon)
~, Blinklicht n [Lichtdauer größer als Dunkeldauer] = occulting light (or beacon) [A light regularly eclipsed in which the duration of the light is greater than the duration of darkness]
Blinklaterne f, Warnblinker m = flash lamp
Blitz|ableiter m = lightning conductor, ~ rod
~**feuer** n = flashing light [A light regularly eclipsed in which the duration of light is shorter than the duration of darkness]
~**schutz** m = lightning protection
~**schutzanlage** f = lightning protective system
~**trocknungsverfahren** n [für Klärschlamm] = flash drying system

Block — Blockgründung

Block *m*, Quader *m* = cut-stone, ashlar, ashler
~ = block (US) [*An area of land bounded by streets*]
~, Klotz *m* = block
~ → Stein
~, (Seil)Flasche *f*, (Seil)Kloben *m*, Unterflasche = block
~ [*nach Fischer und Udluft (Jahr 1936), nach Niggli (Jahr 1938), nach Gallwitz (Jahr 1939) und nach Grengg (Jahr 1942; heißt hier „grobsteiniger Boden")* 2000 bis 200 mm] = boulder [*2000 to 300 mm*]
~**ausstoßmaschine** *f* = ingot ejector
~**ausziehkran** *m* = ingot drawing crane
~-**Ausziehmaschine** *f* = ingot drawing machine
~**automat** *m* → (Beton)Stein(form)automat
~**balken** *m*, Balken aus Blöcken = beam made of precast hollow blocks
~**brammenstraße** *f* = blooming slabbing train
~**brecher** *m* = ingot crusher, ~ breaker
~-**Chargierkran** *m* → Blockeinsetzkran
~**drehbank** *f* = ingot lathe
~**drücker** *m* = ingot pusher
~**druckfestigkeit** *f* → Steindruckfestigkeit
~**einsetzkran** *m*, Block-Chargierkran = ingot charging crane
~**entstapler** *m* = bloom and slab unpiler
~**feld** *n*, Felsenmeer *n*, Blockmeer = block field, stone ~
~**fertiger** *m*, (Beton)Steinfertiger, Blockfertigungsmaschine *f*, (Beton-)Steinfertigungsmaschine, Blockherstellungsmaschine, Block(form)maschine, Stein(form)maschine, Betonsteinmaschine, Betonblock(stein)maschine, Betonformsteinmaschine, Betonsteinformmaschine = (concrete) block making machine
~**fertigung** *f* → (Beton)Steinfertigung
~**fertigungsmaschine** *f* → Blockfertiger *m*
~**festigkeit** *f* → Steinfestigkeit
~**form** *f* → (Beton)Steinform
~**formmaschine** *f* → Blockfertiger *m*
~**gletscher** *m* = rock glacier, ~ train, talus train
~**greifer** *m* [*Holzindustrie*] = log grapple, ~ grab

Blockgründung *f* [*für Hafendämme, Wellenbrecher und Kaimauern an der See. Die Blöcke werden auf einem Werkplatz hergestellt und an Ort und Stelle mit und ohne Verzahnung versetzt*]	block foundation
Ankerblock *m*, Verankerung *f*	dead-man, anchor block
Aufhängevorrichtung *f*	T bar (for lifting blocks); lifting gear, suspension gear
Blockaufhängung *f*	block lifting
Blöcke *mpl* zu Wasser lassen	underwater block setting
Blockförderung *f*, Blocktransport *m*	transport of blocks (Brit.); transportation ~ ~ (US)
Blockform *f*, (Gieß)Form	block making form, ~ ~ mo(u)ld
Blockreihe *f*	tier of blocks, row ~ ~, course ~ ~
Blockverlegung *f*	block setting
Blockverlegung *f* vor Kopf	setting in advance
Blockverschiebung *f*	block handling

Blockhaus — Blockhersteller

Blockwagen *m*, Rollwagen für den Blocktransport	block-carrying bogie, ~ truck
Bruchsteinunterlage *f*, Steinschüttung *f*	rubble bed
Eintauchen *n*, Versenken mit schwimmendem Mastkran	setting by pontoon shears, ~ ~ floating ~
Form *f*, Gießform, Blockform	block making mo(u)ld, ~ ~ form
geneigte Blocklagen *fpl*	inclined courses, sliced blockwork, sloping blockwork
Gerüst *n*, Gerüstbrücke *f*	gantry, staging; construction trestle (US)
(Gieß)Form *f*, Blockform	block making mo(u)ld, ~ ~ form
Hängekette *f*	lifting chain
(Hebe)Schacht *m*	lifting hole, slot
Herstellungsplatz *m*, Werkplatz	block yard
Portalkran *m*	gantry crane, **tra**vel(l)ing ~
Rolle *f*, Walze *f*	travel(l)ing wheel
Rollwagen *m* für den Blocktransport, Blockwagen	block-carrying truck, ~ bogie
Schacht *m*, Hebeschacht	slot, lifting hole
schwimmender Mastkran *m*	pontoon shears, floating ~
Steinschüttung *f*, Bruchsteinunterlage *f*	rubble bed
Stützwerk *n* für den Ausleger-Verlegekran	supporting frame
Stützwerkverankerung *f*	frame anchorage
Titan-Gegenausleger *m*	counterbalance arm of the Titan
Titankran *m*	Titan crane
Titan-Laufgerüst *n*	Titan truck, ~ travel(l)ing pedestal
Verankerung *f*, Ankerblock *m*	anchor block, dead-man
Verlegung *f* von einem festen Gerüst aus	setting from temporary gantry, ~ ~ ~ staging; ~ ~ ~ ~ construction trestle (US)
Versenken *n* mit schwimmendem Mastkran, Eintauchen	setting by floating shears, ~ ~ pontoon ~
Versenken *n* von Prahmen aus	setting by barge
Vollblock *m*	solid block
Waagebalken *m*	spreader, yoke, swingle-tree
waag(e)rechte Blocklagen *fpl*	horizontal block courses
Walze *f*, Rolle *f*	travel(l)ing wheel
Werkplatz *m*, Herstellungsplatz	blockyard
Zellenblock *m*	hollow block

Block|haus *n*, Blockhütte *f* = log cabin, ~ house, ~ hut

~heizung *f*, Fernheizung = district heating

~heizwerk *n*, Fernheizwerk = district heating plant, ~ ~ installation, central ~ ~

~hersteller *m* → (Beton)Steinhersteller

Blockherstellung — Bockstütze

~herstellung *f* → (Beton)Steinfertigung

~herstellungsmaschine *f* → Blockfertiger *m*

~-Herstellungsplatz *m*, Block-Werkplatz = blockyard

~-Hohlstufe *f* = hollow block step

blockieren = to lock, to block

Blockieren *n* = locking, blocking

Blockierkeil *m* = (b)locking wedge

blockiertes Bremsen *n*, blockierte Bremsung *f* = locked-wheel braking

~ **Rad** *n* = locked wheel

Blockierverhinderer *m* = anti-(wheel-)locking device, ~ unit

Block|kaimauer *f* = block quay wall

~kipper *m* = ingot tilter, ~ tipping device

~lava *f*, Schollenlava, Zackenlava, Aprolith *m*, Spratzlava = block lava, aa (~)

~maschine *f* → Blockfertiger *m*

~mauerwerk *n* → (Beton)Steinmauerwerk

~mauerwerk *n*, Quadermauerwerk = cut-stone masonry, ashlar ~, ashler ~

~meer *n* → Blockfeld *n*

~presse *f*, Steinpresse [*Betonsteinindustrie*] = block press

~prüfpresse *f* → (Beton)Steinprüfpresse

~schere *f* = bloom shears, bar-cutting machine

~schneider *m* → (Beton)Steintrennmaschine *f*

~stampfmaschine *f* → (Beton)Steinstampfmaschine

~stapler *m* = ingot piler

~stapler *m* [*Betonsteinindustrie*] = block-lifting machine

Block(stein) *m* → Stein

~druckfestigkeit *f*, Steindruckfestigkeit = tile compression strength, block ~ ~

~festigkeit *f*, Steinfestigkeit = tile strength, block ~

Block|straße *f* = blooming train

~strom *m*, Steinstrom (Geol.) = rock flow, ~ stream

~stufe *f* = block step

~trennmaschine *f* → (Beton)Steintrennmaschine

~verband *m* = English bond

~vollstufe *f*, Massivstufe = solid block step

~wagen *m* = ingot (transfer) car, ~ buggy, ~ chariot

~wagen *m*, Rollwagen [*Betonsteinindustrie*] = block bogie

~walzwerk *n* = blooming mill, cogging ~

~wender *m* = ingot manipulator

Blockwerk *n*, Betonwerk~ = (concrete) block factory

~platz *m* → (Beton)Steinwerkplatz

Blockwinde *f* → Flaschenzug *m*

bloßgelegte Antiklinale *f* = exposed anticline

Blume *f*, (Gesteins)Staub *m*, Steinmehl *n*, Gesteinsmehl = flour, stone dust

Blumen|becken *n* = flower basin

~guirlande *f* = flower festoon

~kasten *m* = flower box

Bluten *n*, (Aus)Schwitzen [*Überfluß an Bindemittel bei Schwarzdecken*] = bleeding, ponding, fatting(-up)

~, Wasserabstoßen *n*, Wasserabsonderung *f*, Abscheidung von Wasser, ungenügende Anmach(e)wasserhaltung [*Beton*] = bleeding, sweating, water gain

Blutregen *m* = blood rain

Bo(a)rt *m* → Boort

Bock *m* [*Gleitschalung*] = (formwork) yoke, steel ~, jack ~

~, Bockgerüst *n* = trestle

~dalbe *m* = raker dolphin

~-Derrick(kran) *m* = stiff-leg derrick (crane) (US); Scotch ~ (~) (Brit.)

~kran *m*, Gerüstkran [*Portalkran zum Aufstellen von Stahlkonstruktionen, insbesondere Hallen oder Brücken, zur Montage von Großbaggern usw.*] = gantry crane for erection purposes, portal (jib) ~ ~ ~ ~

~pfette *f* = lean-to roof purlin(e)

~rolle *f* = rigid-type castor

~säule *f* [*Dach*] = lean-to roof strut

~stütze *f* = lean-to trussed strut

Bocksystem — bodeneigen

~**system** n = jack system
(~)**Trommel(seil)winde** f → Kanalwinde
~**winde** f → Kanalwinde
~**winde** f, Schrauben(hebe)bock m, Schraubenwinde, (Schrauben)Spindel f = screw jack, jack(ing) screw
~**winde** f → Zahnstangenwinde
Boden m, Lockergestein n, Erdstoff m, Erde f = soil, earth [*terms used by engineers*]; mantle rock [*term used by geologists*]
~, Fuß~ = floor finish, floor(ing), floor covering
~ → Dachgeschoß n
~**abdichtung** f = soil (or earth) waterproofing
~**ablaßventil** n, Bodenentleerungsventil = bottom drain valve
~**abstand** m → Bodenfreiheit f
~**analyse** f = soil analysis
~**andeckung** f = soil covering, earth ~, soiling
~**anlagen** fpl [*Flugplatz*] = ground installations, ~ facilities
~**(an)schwellung** f = swelling of soil
~**anteil** m = soil fraction
~**art** f = soil type, type of soil
~**aufschluß** m = soil exploration
~**aufschüttung** f → Erdschüttung
~**auftrag** m → Erdschüttung f
~**auftrieb** m = (ground) heaving
(~)**Aushub** m, Erdaushub [*als Tätigkeit*] = soil excavation, ~ digging
~**aushub** m → Aushub(boden)
~**aushubarbeiten** fpl, Erdaushubarbeiten = soil excavation work, ~ digging ~
~**bakterie** f = soil bacterium
~**beanspruchung** f = stress and strain of the ground
~**bedeckung** f, Geländebedeckung = ground cover, ~ vegetation
~**belag** m, Fuß~ = floor cover(ing), flooring, floor finish
~**belüftung** f = soil aeration
~-**Bentonit-Gemisch** n, Boden-Bentonit-Mischung f = soil-bentonite mix(ture)
~**beschaffenheit** f, Bodenzustand m = nature of soil
~**bestandteil** m = soil constituent
~**beton** m → Zementverfestigung f
~**bewachsung** f, Pflanzendecke f, Pflanzenbewuchs m = vegetable cover, vegetation ~, vegetative ~, plant ~, mantle of vegetation
~**bewässerung** f → Bewässerung
~**bewässerung** f im Schwergewichtfluß → Bewässerung ~ ~
~**bewegung** f, Erdbewegung, Bodenverlagerung, Erdverlagerung, Bodenförderung, Erdförderung = earth moving
~**binder** m, Bindeton m, haftendes Lockergestein n, Bodenmörtel m, Bindeerde f = soil matrix, ~ binder, ~ mortar, (clay) binder, binder soil
~-**Bitumen-Gemisch** n = soil-asphalt mix(ture) (US); soil-(asphaltic) bitumen ~ (Brit.)
~**blech** n = floor plate
~**böschung** f, Erdböschung = bank [*An earth slope formed or trimmed to shape*]
~**brett** n, Fuß~ = flooring board
~**chemie** f = soil chemistry
~**dichte** f = soil density
~**dichtung** f [*Unterkellerung*] = floor skin
~**druck** m, Bodenpressung f, Flächenpressung, Flächendruck [*z.B. einer (Gleis)Kette auf den Untergrund*] = ground pressure, bearing load, ground-bearing ~
~**druck(meß)dose** f, Erddruck(meß)dose = soil pressure cell, earth-pressure (measuring) ~
~**düse** f [*Tiefbohrtechnik*] = bottom-hole choke
~**ebene** f, Vorfeldebene [*Abfertigungsgebäude*] = ground level
bodeneigen, autochthon, bodenständig [*Kohlenflöze, deren Ursprungpflanzen am Orte der Inkohlung gewachsen waren*] = autochthonous, grown in situ

Bodeneigenschaft — Bodenfördergerät

Boden|eigenschaft *f* = soil property
~einbau *m*, Erdeinbau = soil placement
~einlauf *m* = (rainwater) gull(e)y
~-Einschnitt *m* = soil cutting
~einsetzwagen *m*, Konverter-~ = converter bottom jacking car
~einteilung *f*, Bodenklassifikation *f*, Bodenklassifizierung, Bodenunterteilung = soil classification
bodenentleerende Schute *f* = drop-bottom barge
bodenentleerender Betonierkübel *m* = bottom opening concreting skip
~ Erdtransportwagen *m* → Halbanhänger *m* mit Bodenentleerung
~ Erdtransportwagen *m*, Anhänge-Bodenentleerer *m*, Bodenentleerer, Schlepp-Bodenentleerer, Bodenentleerer-Anhänger *m* = bottom-dump wagon, ~ tractor-truck, ~ tractor-trailer, bottom-discharge ~
Bodenentleerer *m* → bodenentleerender Erdtransportwagen *m*
~ [*Schachtfördergefäß*] = skip, skep, dumping shaft ~
~-Anhänger *m* → bodenentleerender Erdtransportwagen *m*
~-Halbanhänger *m* → Halbanhänger mit Bodenentleerung
Bodenentleerung *f*, Untenentleerung = bottom dump discharge, bottom-dumping
Bodenentleer(ungs)|klappe *f*, Bodenklappe = bucket gate, ~ bottom ~
~kübel *m* = bottom-dump skip, botton-discharge ~
~tür *f* → Bodenklappe *f*
~ventil *n*, Bodenablaßventil = bottom drain valve
(Boden)Entnahme *f* = (soil) borrow
~einschnitt *m* = (soil) borrow cut(ting)
~kubatur *f* = (soil) borrow yardage
~stelle *f* = (soil) borrow source
Boden|entwässerung *f* → Untergrundentwässerung
~entwicklungsprozeß *m* = soil-building process, soil-forming ~
~erhaltung *f* = soil conservation
~erosion *f* = soil erosion
~erschütterung *f* = ground shock
~erwärmung *f* = soil heating
~farbe *f*, Fuß~ = floor(ing) paint, floor finish ~
~fauna *f* = soil fauna
~fertiger *m*, Boden-Vibrations-Betonsteinmaschine *f*, Boden-Schwingverdichter, Betonstein-Bodenfertiger = egg layer, egg-laying type (block) machine, on floor placing machine, travel(l)ing block machine, ground laying machine
~festigkeit *f* = soil strength
~feuchte *f*, Bodenfeuchtigkeit *f*, Erdfeuchte, Erdfeuchtigkeit [*Zeichen: f. Einheit:* %. *Anteil der im Boden vorhandenen Wassermenge als Trockengewicht- oder Porenanteil*] = soil moisture, ground ~
~feuchtemesser *m*, Bodenfeuchtigkeitsmesser = soil moisture meter
~feuchtespannung *f* = soil moisture tension
~feuchtigkeitsgehalt *m*, Bodenfeuchtegehalt = soil moisture content
~feuchtigkeitsstufe *f*, Bodenfeuchtestufe = soil moisture grade, ground ~ ~
~filtration *f* [*Abwasserwesen*] = intermittent sand filtration, land ~
~flansch *m* [*Tiefbohrtechnik*] = casing spool
~fliese *f* → Bodenplatte *f*
~fließen *n*, Erdfließen, (Fließ)Rutschung *f*, Abgleitung, Abrutschung, Solifluktion *f*, Bodenkriechen, Gekriech *n* = soil-flow, solifluction, earth flow, solifluxion, soil-creep, creepwash [*Soil creep takes place mainly at the surface. It is frequently known as terminal creep and may give an erroneous reading for true dip of the strata. Soil creep is common in clay*]
~fördergerät *n*, Erdbewegungs(groß)gerät, Erdbewegungsmaschine *f*, Erdfördergerät = dirtmover (US); earthmoving gear (Brit.); earthmover

~förderung *f* → Bodenbewegung
~förderung *f* mit **Pferd und Wagen** = leading, teaming
~förderung *f* mit **Schubkarren** = wheeling
~form *f*, Geländeform = land form
~forschung *f* = soil research
~fräse *f*, Motor-Bodenfräse = rotavator, rotary hoe, rotary tiller, soil pulverizer, rotary cultivator
Bodenfreiheit *f*, Bodenabstand *m* = ground clearance (of a vehicle) [*The vertical distance between the lowest point of a vehicle and a horizontal ground surface on which it stands*]
~ des **Drehkranzes** = circle clearance
bodenfremd, allochthon = allochthonous
Boden|frost *m* = soil freezing, ground ~, ~ frost
~fruchtbarkeit *f* = soil fertility
~fuge *f*, Fuß~ = flooring joint, floor (finish) ~
~fundament *n* = soil foundation
~gare *f* = tilth
~gefüge *n* → Bodenstruktur *f*
~gemenge *n* → Bodengemisch *n*
~gemisch *n*, Bodenmischung *f*, Bodengemenge *n* = soil mix(ture), ~ aggregate
~geologie *f* = agrogeology
~gerüst *n*, Bodenskelett *n* = soil skeleton
~gewinnung *f*, Lösen *n* und Laden des Bodens [*im Hand-, Trockenbagger- oder Schwimmbaggerbetrieb*] = excavation and loading of the soil, digging ~ ~ ~ ~
~güte *f* = soil quality
~haftung *f*, Kraftschluß *m*; Bodenhaftigkeit *f* [*Schweiz*] [*Fahrzeug*] = traction, adhesion
~haftungsbeiwert *m*, Kraftschlußbeiwert; Bodenhaftigkeitsbeiwert [*Schweiz*] = traction coefficient, adhesion ~
bodenhaltiges Wasser *n* = soil-laden water
Boden|härtemittel *n*, Fuß~ = flooring hardener

~-Härteprüfgerät *n*, Fuß~ = flooring hardness tester
~heizung *f* → Fuß~
~hydraulik *f* = hydraulics of soils
~ingenieur *m* → Bodenmechaniker *m*
~kabel *n*, Erdkabel, erdverlegtes Kabel, bodenverlegtes Kabel = buried cable
~karte *f* = soil map
~kennzeichnung *f* = identification of soils
~kern *m*, Bodenprobekern = soil core, core of soil
~klappe *f*, Bodenentleer(ungs)klappe = bucket gate, ~ bottom ~
~klappe *f*, Bodentür *f*, Bodenentleer(ungs)klappe, Bodenentleer(ungs)tür [*Halbanhänger mit Bodenentleerung*] = trailer door
~klappenkübel *m* = bottom-gate skip
~klasse *f* = soil class
~klassifikation *f* → Bodeneinteilung *f*
~klassifizierung *f* → Bodeneinteilung
~klassifizierung *f* nach **A. Casagrande** = airfield soil classification, AC system
~kleber *m*, Fuß~ = flooring adhesive
~klumpen *m* → Bodenscholle *f*
~kolloid *n*, Bodenquellstoff *n* [*0,002 bis 0,00002 mm*] = (soil) colloid, soil colloidal particle
~konstante k *f* → Bettungszahl *f*
~korn *n*, Erdkorn, Bodenteilchen *n*, Erdteilchen, Bodenkörnchen *n*, Erdkörnchen = soil particle, ~ grain
~körper *m*, Erdkörper [*Erdbau*] = soil body
~korrosion *f* = soil corrosion, underground ~
~korrosionsversuch *m*, Bodenkorrosionsprobe *f*, Bodenkorrosionsprüfung *f* = soil corrosion test
~(kratz)schaufel *f* [*Betonmischer*] = floor scraper blade
~kriechen *n* → Bodenfließen
~krume *f*, Mutterboden *m* [*die oberste, Humus und Kleinlebewesen enthaltende Bodenschicht*] = top soil, surface ~

Bodenkrümel — Bodenplatte 250

~krümel m, Erdkrümel = soil crumb
(~)**Kultivator** m = (agricultural) cultivator, field ~
~**kunde** f → Bodenmechanik f
~**kunde** f, Pedologie f = pedology
bodenkundliche Untersuchung f = pedological study
~ **Vorarbeiten** fpl = preliminary pedological work
Bodenleitung f → Erdleitung
bodenlos, ohne Boden = bottomless
Boden|lösung f, Lösung des Bodens [*der Begriff „Bodenlösung" ist von dem Begriff „Bodengewinnung" nicht scharf zu trennen, da mit dem Vorgang des Lösens vielfach auch das Laden auf Transportmittel (z.B. beim Bagger) verbunden ist*] = soil digging, ~ excavation
~**luft** f = soil air
~**lysimeter** m = soil lysimeter
~**markierung** f [*Schweiz*]; Straßenmarkierung, Bodenzeichen n = traffic-line marking, traffic-zoning, horizontal marking
~**markierungsmaschine** f [*Schweiz*]; Fahrbahnmarkierungsmaschine = traffic-line marking machine
~**masse** f, Erdmasse — soil mass, earth ~
(~)**Massenermitt(e)lung** f, Erdmassenermitt(e)lung = soil quantity determination
(~)**Massenverteilung** f → Erdmassenverteilung
~**material** n, Fuß~ = floor finish material, flooring ~
~**mechanik** f, Erdbaumechanik, Grundbaumechanik, (bautechnische) Bodenkunde f = soil mechanics
~**mechaniker** m, Bodeningenieur m = soil(s) engineer, ~ mechanician
bodenmechanische Untersuchung f = soil mechanics investigation
bodenmechanischer Gutachter m = soil mechanics expert
Boden|merkmal n = soil characteristic, ~ feature

~**mischer** m, Mehrwellen-Boden-Zwangsmischer, Bodenmischmaschine f = nonelevating roadmixer, pulverizing mixer, rotary (speed) mixer, soil-mixer, soil mixing machine, mix-in-travel plant, mix-in-place machine, mix-in-place travel(l)ing plant, rotary soil mixer
~**mischung** f, Bodengemisch n, Bodengemenge n = soil aggregate, ~ mix(ture)
~**mischverfahren** n, An-Ort-Mischverfahren = mix(ed)-in-place (method), site mixing in place
~**monolith-Lysimeter** m, Bodenmonolith-Versickerungsmesser m, Lysimeter mit gewachsenem Boden, Versickerungsmesser mit gewachsenem Boden = monolith lysimeter
~**mörtel** m, Bindeerde f, haftendes Lockergestein n, Bodenbinder m, Binderton m = soil mortar, ~ matrix, ~ binder, (clay) binder, binder soil
~**müdigkeit** f = soil sickness
~**nebel** m = ground fog
~**neigung** f [*Bunker; Silo*] = bottom slope
~**nutzbelag** m → Nutzschicht f
~**nutzschicht** f → Nutzschicht
~**oberfläche** f = soil surface
~**öffnung** f = bottom opening
~**physik** f = soil physics
bodenphysikalische Untersuchung f = soil physics investigation
Bodenplatte f [*Holzgerüst*] = base plate
~ [*Hohlkastenträger*] = soffit slab
~, Bodenfliese f, Fuß~ = floor (finish) tile, flooring ~
~ [*Unterkellerung*] = floor slab
~ [*Dock- und Kaimauerbau*] = platform, deck
~, (Gleis)Kettenplatte, Raupenplatte = (track) shoe, crawler ~
~ **mit Doppel(winkel)greifern,** (Gleis-)Kettenplatte ~, Raupenplatte ~~ = double grouser (track) shoe, ~~ crawler ~

Bodenplatte — Bodenschwellung

~ ~ **Gummipolster**, (Gleis)Kettenplatte ~ ~, Raupenplatte ~ ~ = rubber track shoe, ~ (crawler) ~

Bodenplatten|aufsatz m [*Gleiskettenlaufwerk*] = shoe attachment

~**presse** f, Fuß~ = floor-tile press

Bodenpressung f → Bodendruck m

~ [*Grundbau*] = subgrade reaction

Bodenprobe f = soil specimen

~, Gesamt~ = soil sample

~ → Bodenprüfung

~ [*Fließgrenzenermitt(e)lung*] = soil pat

Bodenprobe(ent)nahme f = soil sampling

~**probe(ent)nahmegerät** n, Bodenprobenehmer m = soil sampler

(Bodenproben)Entnahmerohr n = tube sampler, soil pencil

Boden|(probe)kern m = core of soil, soil core

~**probenahme** f **mit Flügelbohrer** = rotating-auger test, (auger-)vane ~

~**probenehmer** m, Bodenprobe(ent)nahmegerät n = soil sampler

~**profil** n [*Bauwesen*] = soil profile

~**profil** n [*Landwirtschaft*] = pedological profile

~**prüfgerät** n, Erdprüfgerät, Bodenprüfer m, Erdprüfer = soil tester, ~ test(ing) apparatus

Bodenprüfung f, Bodenprobe f, Bodenversuch m = soil test

~ **an Ort und Stelle** = in situ soil test

~ **durch Strahlungsmessung** = nuclear soil test

Boden|quelle f = spring of intermediate depth

~**quellstoff** m → Bodenkolloid n

~**reibung** f, (Beton)Plattenreibung = subgrade restraint, (frictional) restraint of (the) subgrade

~**reinigungsmittel** n, Fuß~ = floor(ing) cleaner, floor finish ~

~**rückstrahler** m, Fahrbahnrückstrahler, Katzenauge n = cat's eye, reflective traffic marker in a pavement, reflective traffic button, reflector stud

~**rüttelplatte** f = vibratory soil compacting plate

~**rüttler** m → Bodenschwingungsrüttler

~**satz** m = dregs, grounds, sediment, lees, deposit [*The particles that settle to the bottom of a liquid*]

~**satzbildung** f = dregs formation

~**satzgestein** n, Ablagerungsgestein, Sedimentgestein, Absetzgestein, Absatzgestein = sedimentary rock, bedded ~

~**satzprobe** f, Bodensatzprüfung f, Bodensatzversuch m [*Leuchtpetroleum*] = flock test

bodensatzverhindernde Eigenschaft f [*z.B. von Talk*] = antisetting property

Boden|sau f, Ofensau [*Hochofen*] = furnace sow, salamander

~**saugvermögen** n = absorptive capacity of the soil

~**schalung** f [*Träger*] = bottom formwork

~**schatz** m = mineral deposit

~**schaufel** f, Bodenkratzschaufel [*Betonmischer*] = floor scraper blade

~**schicht** f (Geol.) = soil stratum

~**schicht** f = soil layer

~**schichtenverzeichnis** n = soil section sheet, ~ ~ diagram

~**schichtung** f = soil stratification

~**schleifmaschine** f, Fuß~ = floor(ing) grinder, ~ grinding machine

~**schnellvermörtelung** f **für militärische Zwecke** = expeditious military soil stabilization

~**scholle** f, Erdscholle, Bodenklumpen m, Erdklumpen = soil clod

~**schürftransport** m, Erdschürftransport = (bull)dozing

~**schütter** m = bottom-dump wagon

~**schüttung** f → Erdschüttung

~**schutz** m = soil protection

~**schwelle** f → Drempel m

~**schwellung** f, Bodenanschwellung = swelling of soil

Bodenschwingungsrüttler — (Boden)Verfestigung

~schwingungsrüttler *m*, Bodenschwingverdichter *m*, Bodenrüttler = soil vibrator, vibratory (soil) compactor
~schwingverdichter *m* → Bodenschwingungsrüttler *m*
~-Schwingverdichter *m* → Bodenfertiger
~seilbahn *f* → Seil-Standbahn
~senke *f* → (Gelände)Senke
~sicht *f* = ground visibility
~skelett *n*, Bodengerüst *n* = soil skeleton
~sonde *f*, Sondenbohrer *m* = probing staff, pricker ~, piercing ~, ~ rod, soil test probe
~sondierung *f* = probing, piercing [*Pushing or driving a pointed steel rod up to 20 ft long into the ground for determining the position of bedrock or hard lumps*]
~spannung *f* = soil stress
~spekulation *f* = land speculation
~stabilisator *m* = soil stabilizer
~stabilisierung *f* → (Boden)Verfestigung
~standfestigkeit *f*, Bodenstandsicherheit *f* = soil stability
bodenständig → bodeneigen
Boden|stauwasser *n* = internal stagnant water
~stein *m* [*Hochofen*] = block, bottom
~stein *m* [*Kollergang*] = bedstone, base stone, bedder
~strahlungsheizung *f* → (Fuß)Bodenheizung
~struktur *f*, Bodengefüge *n* [*Anordnung der Bodenteilchen zueinander*] = soil structure
~stück *n* [*Spannbetonverfahren Coff-Roebeling, Wayss & Freytag und Dischinger*] = end pot
~teilchen *n* → Bodenkorn *n*
~temperatur *f* = ground heat
~textur *f* [*Korngestaltung des Bodens*] = soil texture
~tragwert *m* = soil support value, SSV
~tränkung *f* = impregnation [*Impregnation of a soil from its surface may be successfully resorted to when it is desired to improve the properties of a relatively thin upper layer*]
~transport *m*, Erdtransport = soil transport; ~ transportation (US)
~treppe *f* = attic stair(case)
~tür *f* → Bodenklappe *f*
~überdeckung *f*, Erdüberdeckung [*Rohrgraben*] = soil cover, earth ~
~unebenheiten *fpl* → Geländeunebenheiten
~untersuchung *f*, bautechnische ~ = (sub)soil investigation, ~ study
~untersuchung *f* an der Oberfläche *f* = surface reconnaissance
~untersuchung *f* mit Spülstangengerät = jet probing, rod sounding
~untersuchungsbohrung *f* = boring for soil investigation
~untersuchungsgerät *n* = (sub)soil investigation device, ~ study ~
~unterteilung *f* → Bodeneinteilung
bodenverankert, erdverankert = anchored in the ground, ground-anchored
Boden|verbesserung *f* = improvement of the mechanical properties of soils
~verdichter *m*, Bodenverdichtungsgerät *n*, Bodenverdichtungsmaschine *f* = soil (or earth) compacting machine
~verdichtung *f*, künstliche ~ = soil (or earth) compaction (or densification)
~verdichtung *f*, natürliche ~, Eigensetzung, Eigenverfestigung, Konsolidation *f*, Konsolidierung = (soil) (or earth) consolidation
~verdichtung *f* durch Einschlämmen, Einschlämmen *n*, Einspülverfahren *n*, Spülspritzverfahren, Spülkippverfahren, Wasserstrahlverfahren = sluicing
(~)Verdichtungspfahl *m* = compaction pile
~verdrängung *f*, Erdverdrängung = soil displacement, earth ~
~verdunstung *f* = soil evaporation
(~)Verfestigung *f* → Baugrundverfestigung

(Boden)Verfestigung — Bodenvermörtelungsmaschine

(~)Verfestigung *f* [*Schweiz: Bodenstabilisierung*] = soil stabilization
(~)Verfestigungsverfahren *n* mit zentralem ortsfesten Mischer = plant-mix stabilization method
~verfestigungswalze *f* → Knetverdichter *m*
~verhalten *n* = soil behavio(u)r

~verlagerung *f* → Bodenbewegung
bodenverlegte Leitung *f* → Erdleitung
bodenverlegtes Kabel *n* → Bodenkabel
Bodenverluste *mpl* [*Elektrolyse*] = losses in soil
~vermörteler *m* → Bodenvermörtelungsmaschine *f*
~vermörtelung *f* = soil stabilization

Bodenvermörtelungsmaschine *f*, Bodenvermörtelungsgerät *n*, Bodenvermörteler *m*	soil stabilizing machine, soil stabilizer, road mixer, soil stabilization machine
Bodenmischmaschine *f*, Bodenmischer *m*, Mehrwellen-Bodenzwangsmischer	non-elevating road mixer, pulverizing mixer, rotary (speed) mixer, soil mixing machine, mix-in-place travel(l)ing plant, (rotary) soil mixer
Bodenmischmaschine SAWOE, Raupen-Schneckenfräse *f* [*Hersteller: LINNHOFF, BERLIN*]	SAWOE single-pass machine
Raupenschlepper *m*	caterpillar tracks for propulsion
Fräse *f*	tined rotor
Mischschnecke *f*	gathering rotor
Verteilerschnecke *f*	spreading rotor
Vibratorschleppe *f*	vibrating screed board
Bodenmischmaschine VOEGELE	VOEGELE three-tined rotor machine
Bodenmischmaschine STRABAG	STRABAG single-pass mix-in-place machine
Verteiler *m*	tamping beam
Dreiwellenmischer *m*	three mixing rotors
Bodenmischmaschine SONTHOFEN	SONTHOFEN mix-in-place machine
Vertikal-Zwangsmischer *m*	four vertical shaft rotors
Ein-Gang-Mischer (oder Ein-Gang-Bodenvermört(e)ler) HARNISCHFEGER	SINGLE PASS SOIL STABILIZER
Bodenmischmaschine, Fabrikat ROTARY HOES LTD.	HOWARD SINGLE-PASS EQUIPMENT
Mehrgang(boden)mischer SEAMAN	SEAMAN MIXER
Aufnahmemischer, Aufnehmermischer, Bodenvermörtelungsmaschine mit hochliegendem Zwangsmischer	windrow-type travel(l)ing (or travel) (asphalt) (or road) plant, elevating roadmixer, travel(l)ing (mixing) plant, travel(l)ing mixer for soil stabilization, road pug travel-mix plant
Aufnahmemischer REISER	REISER travel-mix soil-stabilization machine
Aufnahmemischer WOOD	WOOD road-mixer

Aufnahmemischer BARBER-GREENE

BARBER-GREENE travel mixing plant

Boden|verölung *f* = oil pollution of the ground
~verschluß *m* = bottom gate
~verstein(er)ung *f* [*Der Baugrund wird durch Einpressen von Zement oder Kieselsäuregel in ein Netz von Einpreßlöchern zu einem Gesteinsblock umgewandelt*] = (artificial) cementation, ~ solidification, ~ cementing, ~ stabilization, soil ~, ground ~, earth ~ [*sometimes popularly and incorrectly referred to as soil consolidation*]
~versuch *m* → Bodenprüfung *f*
~-Vibrations-Betonsteinmaschine *f* → Bodenfertiger *m*
~volumen *n* = soil volume
~wachs *n*, Fuß~ = floor wax
~walze *f*, Fuß~ = floor(ing) roller
~wanne *f* [*zur Ansammlung von Schmiersinkstoffen*] = bottom sump
~wasser *n* = gravitational water [*Is that part in excess of hygroscopic and capillary water which will move out of the soil if favourable drainage is provided*]
~welle *f*, Schwelle *f*, Anschwellung *f* [*Geomorphologie*] = hillock
~welligkeit *f* → Geländeunebenheiten *fpl*
~widerstand *m* = soil resistance
~wind *m* = ground wind, surface ~
~zeichen *n* → Bodenmarkierung *f* [*Schweiz*]
~zement *m* → Zementverfestigung *f*
~-Zement-Gemisch *n*, Boden-Zement-Mischung *f*, Boden-Zement-Gemenge *n* = soil-cement mix(ture)
~ziegel *m*, Fuß~ = flooring brick
~ziffer *f* → Bettungszahl *f*
~zustand *m*, Bodenbeschaffenheit *f* = nature of soil
Böenwindkanal *m* = gust tunnel
Boetonasphalt *m* = buton asphalt

Bogen *m* [*In der Baukonstruktion: Gewölbtes Tragwerk, das eine Öffnung überspannt*] = arch
~, ~träger *m*, Bogentragwerk *n* [*Träger, der bei senkrechten Lasten nach außen gerichtete Stützkräfte (inclined reactions) auf die Auflager überträgt*] = arch(ed girder)
~, Rohr~, Bogenrohr *n*, (Rohr-) Krümmer *m* = (pipe) bend
~ mit ausgesteiften Zwickeln → Bogenträger *m* ~ ~ ~
~ ~ durch Zugband aufgehobenem Horizontalschub → Bogenträger *m* ~ ~ ~ ~ ~
~ ~ Durchzug → Bogenträger *m* ~ ~
~ ~ Einzellast → Bogenträger *m* ~ ~
~ ~ gebrochenen Linien → Bogenträger *m* ~ ~ ~
~ ~ Kämpfergelenken → Bogenträger *m* ~ ~
~ ~ konstantem Horizontalschub → Bogenträger *m* ~ ~ ~
~ ~ Parabelgurtung → Bogenträger *m* ~ ~
~ ~ Rohrquerschnitt → Bogenträger *m* ~ ~
~ ~ vermindertem Horizontalschub → Bogenträger *m* ~ ~ ~
~ ~ Zugband → Bogenträger *m* ~ ~
~abschnitt *m* = arch section
~-Abzugkanal *m* → Bogen-Durchlaß *m*
~achse *f* = arch centre line (Brit.); ~ center ~ (US)
~anfang *m* der Innenleibung [*Vielfachbogen(stau)mauer*] = springing of intrados, ~ ~ soffit, intrados springing line (US); springing of intrados (or soffit) (Brit.)
~anfang *m* des Gewölberückens [*Vielfachbogen(stau)mauer*] = springing of extrados (Brit.); extrados springing line (US)

Bogenanfänger — Bogenrohr

~anfänger m → Gewölbekämpfer m
~art f = type of arch
~aussteifung f, Bogenversteifung = arch stiffening
~aussteifungselement n, Bogenversteifungselement = arch stiffener
~bau m = arcuated construction, arched ~
~betonieren n = arch pouring, ~ casting, ~ concreting
~binder m = Belfast (roof) truss (Brit.); bowstring (~) ~ (US)
~blende f → Blendbogen m
~brücke f = arch(ed) bridge
~brückenbau m = arch(ed) bridge construction
~dach n = arch(ed) roof
~decke f = arch(ed) ceiling
~differential n = differential of arc
~düne f, Sicheldüne [*In Turkesten „Barchan" genannt*] = barkhan, barchan(e)
~-Durchlaß m, Bogen-Abzugkanal m, überwölbter Durchlaß, überwölbter Abzugkanal = arch(ed) culvert
~durchlaßrohr n = arch(ed) culvert pipe
~ebene f = arch plane
~einheit f = radian
~element n = arch element
~fachwerk(träger) n, (m) mit parallelen Gurt(ung)en, Fachwerkbogen(träger) m ~ ~ ~ = braced arch(ed girder) with parallel booms, trussed ~ ~ ~ ~ ~, ~ ~ ~ ~ ~ chords, ~ ~ ~ ~ ~ ~ flanges
~fachwerk(träger)brücke f, Fachwerk-Bogen(träger)brücke = braced arch(ed girder) brigde, trussed ~ ~ ~
~fenster n = arch(ed) window
~form f = form of arch
~gang m → Arkade f
~gewichtssperre f, Bogen(schwer)gewichts(stau)mauer f, Gewölbe-Gewichtssperre = arch-gravity dam
~gewichts(stau)mauer f → Bogengewichtssperre f
~gleis n = curved track

~gurt(ung) m, (f), gekrümmter Gurt, gekrümmte Gurtung = arched boom, boom of an arch(ed) girder)
~gurt(ungs)winkeleisen n = arched boom angle iron
~halbmesser m = arch radius
~höhe f → (Bogen)Pfeil m
(~)Kämpfer(stein) m = (arch) impost
~krümmung f, Schweifung = bowing
~küste f = curving beach
~lager n [*für Hoch- und Tiefbauten*] = arch bearing
~lampe f = arc lamp
~länge f = arch length
~laube f → Arkade f
~lehre f = arch template
~lehrgerüst n, Bogenrüstung f = arch falsework
~leibung f, innere Bogenfläche f = intrados, soffit
~linie f, Bogenprofil n, Wölbung f = outline of arch
~linie f = curved line
~maß n = radian measure, circular measure (by radians)
~mauer f → Bogensperrmauer
~moräne f = crescentic moraine
~öffnung f = included angle of arch, central ~ ~ ~
(~)Pfeil m, (Bogen)Stich m, Bogenhöhe f = rise (of arch), upward camber, versed sine
~pfeiler m, Strebebogen m, Strebepfeiler = arched buttress, flying ~
~pfeiler(stau)mauer f, Gewölbereihen(stau)mauer, Vielfachbogensperre f, Pfeilergewölbe(stau)mauer = multiple arch(ed) dam
~profil n, Bogenlinie f, Wölbung f = outline of arch
~rahmen m = arch(ed) frame
~reihe f = arcade, continuous arches [*A line of arcade carried on a colonnade*]
~rippe f = arch rib
~rippenträger m = ribbed arch(ed girder)
~rohr n, (Rohr)Bogen m, (Rohr-)Krümmer m = (pipe) bend

Bogenrücken — Bogenzwickel

~rücken m, äußere Bogenfläche f = extrados, back of arch

~rüstung f, Bogenlehrgerüst n = arch falsework

~schäkel m = bow shackle

~schalung f = arch formwork

~scheitel m, Scheitel(punkt) m = crown, vertex, apex, key, top

~schenkel m = haunch

~schluß m [als Bauvorgang] = completion of the arch

~schluß(stein) m = keystone

~schub m, Horizontalschub = horizontal thrust

~(schwer)gewichts(stau)mauer f → Bogengewichtssperre

~sehne f = chord of an arc, arc chord

~sehnenträger m, Segmentträger = polygonal bowstring girder, segmental ~

~sieb n, DSM-~ = DSM screen, Dutch Staatsmijnen ~

~spannweite f = arch span

~sperre f → Bogensperrmauer f

~sperrmauer f, Bogen(stau)mauer, Bogensperre f, Gewölbe(stau)mauer, Einfach-Gewölbesperre, Gewölbesperrmauer = arch(ed) dam [*A curved dam, convex upstream, that depends on arch action for its stability. The load is transferred by the arch to the canyon walls, or other abutments*]

~stab m = bar of arch

~(stau)mauer f → Bogensperrmauer

~stein m = arch-stone

~stich m → (Bogen)Pfeil m

~strömung f, Wirbel m = vortex, whirl, eddy

Bogen(träger) m, Bogentragwerk n [*Träger, der bei senkrechten Lasten nach außen gerichtete Stützkräfte (inclined reactions) auf die Auflager überträgt*] = arch(ed girder)

~ mit aufgeständerter Fahrbahn = barrel arch(ed girder)

~ ~ ausgesteiften Zwickeln = arch(ed girder) with braced spandrels, ~ ~ ~ reinforced ~, ~ ~ ~ stiffened ~

~ ~ durch Zugband aufgehobenem Horizontalschub = arch(ed girder) without horizontal thrust

~ ~ Durchzug = arch(ed girder) with intermediate tie

~ ~ Einzellast = single-load arch(ed girder)

~ ~ gebrochenen Linien = arch(ed girder) with polygonal outlines

~ ~ Kämpfergelenken, Zweigelenkbogen(träger) = double-hinged arch(ed girder), two-hinged ~ ~, arch(ed girder) hinged at the abutments

~ ~ konstantem Horizontalschub = arch(ed girder) with invariable horizontal thrust

~ ~ Parabelgurtung = arch(ed girder) with parabolic chord

~ ~ Rohrquerschnitt = tubular arch(ed girder)

~ ~ vermindertem Horizontalschub = arch(ed girder) with diminished horizontal thrust

~ ~ Zugband = arch(ed girder) with tieback

Bogen|tragwerk n → Bogen(träger) m

~treppe f = arch(ed) stair (US); ~ staircase (Brit.)

~tür f = arch(ed) door

~überlauf m = arch spillway

~verankerung f = grappling of arch

~verband m = arch bond

~versteifung f → Bogenaussteifung

~versteifungselement n, Bogenaussteifungselement = arch stiffener

~viadukt n = arch-type viaduct

~wechsel m [*Talsperre*] = change of curvature

~widerlager n = arch abutment

~wirkung f, Gewölbewirkung = arch action, arching (effect)

~zwickel m = arch spandrel

~zwickel m mit Füll(ungs)stäben = braced arch spandrel

Bohle *f*, Belag~, Holz~ = plank
~, Spund(wand)~ = sheet pile
~, Abgleich~, Abstreich~, Abzieh~ [*Betondeckenfertiger*] = strike-off (screed), screeding beam, level(l)ing beam, level(l)ing screed, front screed
~, Stampf~, Schlag~, Verdichtungs-~ [*Betondeckenfertiger*] = tamping beam, tamper ~
~, Balken *m*, Fertiger~ = (finisher) beam
Bohlen|bahn *f* → Bohlensteg
~**belag** *m*, Holz~ = plank covering
~**boden** *m*, Bohlenfußboden = plank flooring
~**dach** *n*, Bohlensparrendach, geschweiftes Dach = curved plank roof
~**decke** *f* = plank floor
~**dicke** *f*, Spund~ = sheetpile thickness
~**(fuß)boden** *m* = plank flooring
~**klappe** *f* [*Klappbrücke*] = trap door
~**kopf** *m*, Spund~ = sheetpile head
~**rost** *m* = plank foundation platform
~**rüttler** *m*, Bohlenvibrator *m* = beam vibrator
~**sparrendach** *n* → Bohlendach
~**steg** *m*, Bohlenbahn *f* = plank road, ~ run(way), ~ track [*Used to permit distribution of concrete by carts and barrows*]
~**vibrator** *m*, Bohlenrüttler *m* = beam vibrator
~**weg** *m* = plank road, ~ track
~**zarge** *f* = plank trim, ~ casing
Bohl|wand *f*, Bohlwerk *n* = horizontal timber sheeting, ~ ~ lagging [*Wide-flange steel H piles are driven from the surface into the ground, 5 to 6 ft. on centres, before the excavation is begun. Horizontal timber sheeting or lagging is then inserted, board by board, behind the flanges of the H piles as excavation progresses*]
~**werk** *n* → Bohlwand *f*
böhmisches Gewölbe *n* = Bohemian vault

~ **Kappengewölbe** *n*, böhmische Kappe *f* = flat vaulted ceiling
Bohr|abteilung *f* = drilling department
~**achse** *f* = drilling centre line (Brit.); ~ center ~ (US)
~**akte** *f* = drill-hole record
~**(-Anbau)gerät** *n* → Anbau-Bohrgerät
~**anlage** *f*, Bohreinrichtung *f* = drilling installation, ~ rig
~**arbeit(en)** *f(pl)* = drilling operation(s), ~ work
~**arbeiter** *m* [*ein im Bohrturm arbeitender Mann*] = rough neck
~**ausrüstung** *f* = drilling equipment, ~ outfit
bohrbar = drillable
Bohr|barkeit *f* = drillability
~**bedingung** *f* = drilling condition
~**belegschaft** *f*, Bohrmannschaft Bohrtrupp *m* = drilling gang, ~ crew, ~ team, ~ party
~**bereich** *m* = drilling range, boring ~
~**bericht** *m*, Bohrliste *f*, Bohrjournal *n*, Bohrregister *n*, Tagesrapport *m* des Bohrmeisters, Tagesbericht des Bohrmeisters, täglicher Bericht des Bohrmeisters, Bohrprotokoll *n* = daily drilling report, driller's tour ~, drilling record sheet
~**betonpfahl** *m* = Bohrpfahl
~**bock** *m* [*Bei untiefen Bohrungen über dem Bohrloch befindlich, zur Aufnahme der Seilrolle für das Gestängefördern*] = breast derrick
~**boot** *n*, Bohrschiff *n* = boring vessel, drilling ~, blue water ~
~**brunnen** *m*; Rohrbrunnen [*veralteter Begriff*] = drilled well, bore-well, tube well
~**bühne** *f*, Arbeitsplattform *f* = drilling platform
~**dauer** *f*, Bohrzeit *f* = drilling time
~**diagramm** *n* → Bohrleistungsdiagramm
~**diamant** *m*, schwarzer Diamant = carbon(ado), black diamond, drilling diamond

Bohrdreibein — Bohrfutter

~dreibein n, Bohrgerüst n = drilling tripod

Bohrdruck m, Bohrgestängelast f, Werkzeugauflast [*auf dem Meißel lastendes Gestängegewicht*] = weight on the bit, drill pipe load

~diagramm n = weight indicator chart

~luft f = compressed air for drilling

~messer m, Bohrdruckmeßgerät n, Drillometer n = weight indicator, drilling ~, drillometer

~regulierung f, Bohrdruckreg(e)lung = drilling control

Bohr|einrichtung f, Bohranlage f = drilling rig, ~ installation

~-Elektromotor m = drilling motor

bohren = to drill, to bore

~ [*eines neuen Loches bei starker Abweichung*] = to redrill, to drill a new well

Bohren n, Bohrung f = drilling, boring

~ mit belüftetem Wasser als Spülflüssigkeit = drilling with aerated water

~ ~ Bohrgreifer(n), Greiferbohren = hammer grab work

~ ~ Gegenstrom-Umlauf (der Bohrspülung) → ~ ~ Verkehrspülung

~ ~ Luftspülung = air drilling

~ ~ parallelen Bohrlöchern = drilling with parallel holes

~ ~ Sauerstofflanze, Sauerstoffbohren = oxygen lancing

~ ~ Verkehrspülung, ~ ~ Gegenstrom-Umlauf (der Bohrspülung), Gegenstrombohrverfahren n = counterflush drilling, drilling with reversed circulation

~ ~ Wasserspülung = wet drilling

~ ohne geologische Vorarbeiten = blind drilling

~ vom verankerten Schiff = drilling from floating vessels, ship-side drilling

Bohrer m = drill

~, Bohrstahl m [*für Bohrhammer*] = (drill) steel

~ zum Gesteinsbohren durch elektrische Entladungen = electric arc drill

(~)**Einsteckende** n, Bohrstahleinsteckende = shank [*That part of the drill steel which is inserted in the chuck of the drill*]

~futter n → Bohrfutter

Bohrergebnis n = drilling result

Bohrer|halter m = drill holder

~längenstufe f = steel change

Bohrerlaubnis f = well permit

Bohrer|satz m = drill steel set

~schärfmaschine f = drill sharpening machine

~schärf- und -stauchmaschine f = drill sharpening and shanking machine

~schleifmaschine f = drill grinding machine

~schneide f → (Bohr)Schneide

~stütze f → Bohrstütze

Bohr(er)vorschub m = feeding

Bohrer|wechsel m, Bohrstahlwechsel = changing (drill) steels

~werkstatt f, Bohrstahlwerkstatt [*Tunnel- und Stollenbau*] = (drill) steel shop

~-Ziehring m = steel puller

Bohrfeld n, Bohrgelände n = drilling field

~ausrüstungen fpl = drilling field equipment

Bohr|firma f, Bohrunternehmen n, Bohrunternehmung f = drilling firm

~flur m, (Bohr)Turmflur = (drilling) derrick floor

~flüssigkeit f → Bohrschlamm m

~fortschritt m, Bohrgeschwindigkeit f [*Beim Bohren gemessene Eindringtiefe je Zeiteinheit (cm/min.) mit Angabe des Bohrhammers, Bohrerschneidendurchmessers, Luftdruckes usw.*] = drilling progress, ~ rate, rate of drilling progress

Bohrfortschritts|diagramm n, Bohrfortschrittsschaubild n = drilling progress chart

~wechsel m = drilling (time) break

Bohrfutter n, Bohrerfutter [*für Gewindebohrer*] = tap holder

~, Bohrerfutter [*für Spiralbohrer*] = drill chuck

Bohrgarnitur — Bohrinsel

Bohr|garnitur *f* → Bohrzeug *n*
~gelände *n*, Bohrfeld *n* = drilling field
Bohrgerät *n* → Anbau-Bohrgerät
~ = drilling equipment
~ zur Kernentnahme aus Straßendecken → (Beton)Kernbohrmaschine *f*
Bohrgerüst *n* [*Tunnel- und Stollenbau*] = drilling platform
~, Bohrdreibein *n* = drilling tripod
Bohrgeschwindigkeit *f* → Bohrfortschritt *m*
Bohrgestänge *n* = (string of) drill pipe
~ einlassen = to lower the drill stem
~ (heraus)ziehen, **~ aufholen**, **~ ausbauen** = to pull the drill pipes, to withdraw ~ ~ ~, ~ ~ ~ stem
~ im Turm stehend = racked drill pipe
~ zusammenschrauben = to couple drill pipes [*Muffen*]; to make the joint [*Verbinder*]
~abfangkeil *m* = drill pipe slip
~anschlußstück *n* = substitute
~-Drehbank *f* = drill-pipe lathe
~-Drehschlüssel *m*, Gestängehaken *m*, Hakenschlüssel, Drehschlüssel für Bohrgestänge = pipe hook
~fahrstuhl *m*, Bohrgestängeelevator *m* = drill pipe elevator
~fett *n* = drill pipe thread dope
~-Förderversuch *m* [*Erdölförderung*] = drill-stem test
~gewinde *n* = drill pipe thread
~last *f* → Bohrdruck *m*
~-Rückschlagventil *n* = drill pipe float valve
~strang *m* = string of drill pipes, drill pipe string, drill stem, drilling stem, string (of rods), drilling string
~übergang *m* = drill pipe sub
~-Verbindungsmuffe *f* = drill pipe coupling
~zapfen *m* = drill pipe pin
~-Ziehzeit *f*, Ausbauzeit [*Rotarybohren*] = time of coming out, coming-out time

~zug *m*, Gestängeschuß *m* [*Rotarybohren*] = stand of drill pipe
~(zug)verbinder *m* [*Rotarybohren*] = (drilling) tool joint
Bohr|greifer *m* = hammer grab
~grube *f* [*Unter(flur)bohrung*] = horizontal drilling pit
~gut *n* → Bohrklein *n*
~haken *m* = drilling hook, rotary ~
Bohrhammer *m*, Gesteins~, Fels~ [*schlagend wirkende Freihandbohrmaschine*] = (hand) hammer (rock) drill, (rock) drill (hammer)
~ mit Trockenabsaugung durch Saugkopf = rock drill with dry suction head
~ Wasserspülung = wet type rock drill
~ ~ ~ durch den Spülkopf = rock drill with water flushing head
~ ~ zentraler Absaugung = rock drill with centre dry suction
~ ~ ~ kombinierter Wasser-Luftspülung = rock drill with combined centre water and air flushing
~ ~ ~ Wasserspülung = rock drill with centre water flushing
~probe *f* → Bohrhammerprüfung *f*
~prüfung *f*, Bohrhammerprobe *f*, Bohrhammerversuch *m* = rock drill test
~stütze *f* → Bohrstütze
~versuch *m* → Bohrhammerprüfung *f*
Bohr|haube *f* mit Staubbeutel = drill guard with dust bag
~hebewerk *n* → (Rotary-)Hebewerk
~hilfsvorrichtung *f* = auxiliary drilling device
~hindernis *n* = drilling obstacle
„Bohrhitze", reine Bohrzeit *f* = actual boring time, ~ drilling ~
Bohrhülsenpfahl *m*, Hülsenbohrpfahl = bored cased pile, ~ shell ~
Bohrinsel *f* = (mobile) drilling platform, (~) boring ~, off-shore (~) ~, drilling island, sea-going drilling platform
~ für Einzelbohrung, **~ mit unverschieblichem Bohrturm** = single-well platform

Bohrinsel — Bohrlochabsperr(ungs)vorrichtung

∼ mit unverschieblichem Bohrturm, ∼ für Einzelbohrung = single-well platform

Bohr|inspektor *m* = drilling superintendent

∼**journal** *n* → Bohrbericht *m*

∼**jumbo** *m*, Ausleger-Bohrwagen *m*, Groß-Bohrwagen, Jumbo-Bohrwagen = drill(ing) jumbo, (tunnel) ∼

∼**kampagne** *f*, Bohrperiode *f* = drilling campaign

Bohrkern *m* → (Probe)Kern

∼, (Bohr)Kernprobe *f* [*Das beim Abteufen von Bohrungen als Probe gewonnene Gestein*] = well core, drill ∼

∼**analyse** *f*, Bohrkernuntersuchung *f* = core investigation

∼**-Druckfestigkeit** *f* = drilled core crushing strength

∼**entnahme** *f*, (Bohr)Kerngewinnung *f* [*der Vorgang nach dem Bohren*] = core lifting

∼**gewinnung** *f*, (Bohr)Kernentnahme *f* [*der Vorgang nach dem Bohren*] = core lifting

∼**probe** *f* → Bohrkernprüfung *f*

∼**probe** *f* → Bohrkern *m*

∼**prüfung** *f*, Bohrkernprobe *f*, Bohrkernversuch *m* = core test

∼**untersuchung** *f*, Bohrkernanalyse *f* = core investigation

∼**versuch** *m* → Bohrkernprüfung *f*

Bohrklappmast *m* = jack knife mast

Bohrklein *n*, Bohrgut *n*, Bohrschmant *m* [*Tiefbohrung*] = wet rock cuttings, ∼ drill ∼, ∼ well ∼, ∼ drillings

∼, Bohrgut *n*, Bohrmehl *n* [*Trockenbohrung*] = dry rock cuttings, ∼ drill ∼, ∼ well ∼, ∼ drillings

Bohrkleinabführung *f* = removal of rock cuttings

∼ **durch Bewegung der Bohrstange** [*Die Bohrstange ist als rückwärts fördernde Schnecke ausgebildet*] = removal of rock cuttings by screw-shaped (drill) steel extension

∼ ∼ **Schwerkraft** [*ist nur bei steil ansteigenden Bohrlöchern möglich*] = removal of rock cuttings by gravity

∼ ∼ **Spülung** = removal of rock cuttings by drilling mud

(Bohr)Knarre *f*, (Bohr)Ratsche *f* = ratchet

Bohrknecht *m* → Bohrstütze *f*

∼ **für Bohren im Aufbruch** = raising (pusher) leg

Bohr|kontrakt *m*, Bohrvertrag *m* = boring contract, drilling ∼

∼**kopf** *m* = drill point

∼**kragen** *m*, Schwerstange *f* [*Rotarybohren*] = drill collar

∼**kragenführung** *f*, Schwerstangenführung = (drill collar) stabilizer

∼**kran** *m* → (Bohr)Turmkran

∼**kreuz** *n* = Christmas tree

∼**krone** *f* = (drill) bit

Bohrkronen|durchmesser *m* = (drill) bit diameter

∼**-Schleiflehre** *f* = (drill) bit grinding ga(u)ge

∼**-Schleifmaschine** *f* = (drill) bit grinder, (∼) ∼ grinding machine

Bohr|lafette *f*, (Vorschub)Schlitten *m*, (Vorschub)Lafette [*Bohrwagen*] = cradle, slide [*wagon drill*]

∼**länge** *f* [*Unterführungsbohrer*] = drilling length

Bohrleistung *f* pro Mineur und Schicht = drilling performance per driller and shift

∼ ∼ **Schicht** = drilling performance per shift

Bohr|(leistungs)diagramm *n* = drilling performance diagram

∼**liste** *f* → Bohrbericht *m*

Bohrloch *n* [*zur Gewinnung von Erdgas oder Erdöl*] = well

∼, Schußloch, Spreng(bohr)loch, Schießloch, Ladeloch = drill hole, bore ∼

∼ **mit schiefer Richtung** = angling well, slanting ∼

∼**ablenkung** *f* um verlorenem Gestänge auszuweichen = side tracking

∼**absperr(ungs)vorrichtung** *f* = blow out preventer

Bohrlochabstand — Bohrmannschaft

~abstand *m* = well spacing
~abstand *m*, Schußlochabstand, Spreng(bohr)lochabstand, Schießlochabstand, Ladelochabstand = drill hole spacing, bore ~ ~
~abweichung *f* gegen das Schichtenfallen = up structure deflection of the well
~abweichung *f* mit dem Schichtenfallen = down structure deflection of the well
Bohr(lochansatz)punkt *m*, Bohrstelle *f* = location of the well, well-location, site of the well, drilling site [*place where a well is to be drilled*]
(Bohrloch)Ausfütterung *f*, (Bohrloch-)Verrohrung *f* = casing string, (well) casing, (well) csg.
Bohrloch|auskesselung *f* = well cavity
~behandlung *f* = well servicing
~besatzstoff *m*, Schußlochbesatzstoff, Spreng(bohr)lochbesatzstoff, Schießlochbesatzstoff, Ladelochbesatzstoff = stemming material, tamping ~
~(-Bildgüteprüf)steg *m* = plaque-type penetrameter
~druck *m* = well pressure
~durchmesser *m* = well diameter
~durchmesser *m*, Schußlochdurchmesser, Spreng(bohr)lochdurchmesser, Schießlochdurchmesser, Ladelochdurchmesser = drill hole diameter, bore ~ ~
~fertigstellung *f* = well completion
~flansch *m* = casing flange
~fußpunkt-Projektion *f* auf die Erdoberfläche = bottom well location
~geschoßperforator *m* = gun-perforator
~kamera *f* = bore-hole camera
~kartenzeichen *n*, Bohrlochsymbol *n* = well symbol
~kopf *m* = well head
~kopfdruck *m* = well head pressure
~kopfgas *n* = well head gas
~krater *m* = well crater
~lage *f*, Lage der Bohrung = well location
~längsgeber *m* = borehole axial strain indicator

~-Log *n* = well log
~neigung *f* = deviation of a borehole, drift ~ ~
~neigungsmessung *f* = borehole surveying, surveying of borehole
~packer *m* = well packer, ~ packing
~plan *m* → Bohrplan
~probe *f* = borehole sample
~quergeber *m* = borehole diametral strain indicator
~richtung *f* = well direction
~schieber *m* = master gate
~sohle *f* = well bottom
~sprengung *f* = shooting, torpedoing [*Exploding a high explosive in the oil bearing formation to increase the flow of oil*]
~steg *m*, Bohrloch-Bildgüteprüfsteg = plaque-type penetrameter
~symbol *n*, Bohrlochkartenzeichen *n* = well symbol
~temperatur *f* = well temperature
~tiefe *f* = well depth
~tiefe *f* mit bezug auf die Tagesoberfläche = well depth below the surface (of the ground)
~überprüfer *m* [*fahrbares Gerät zum Messen der Erdölförderung von Förderbohrungen*] = well checker
~überwachung *f* = well surveying, ~ logging
(~)Verrohrung *f*, (Bohrloch)Ausfütterung = casing string
~wand-Kernentnehmer *m* = (side) wall sampler, (~) ~ tester, (~) ~ sampling gun, verifier
~wand(ung) *f* = (side) wall
~wand(ung) *f* mit Schlamm verkleiden = to plaster the wall of the hole with mud
~wasser *n* = well water
~wassersäule *f* = water column in the well
~zementierung *f* = well cementing
Bohr|löffel *m* [*aus Stahl mit winkelrechter Löffelfläche zum Entfernen des Bohrschmants aus Bohrlöchern*] = sand bucket
~mannschaft *f*, Bohrbelegschaft *f*, Bohrtrupp *m* = drilling crew, ~ gang

Bohrmaschine — Bohrrohr

Bohrmaschine *f* = drilling machine
~ für Pfahlgründungen = boring machine for pile foundations
~ ~ Steinbruchbetrieb = quarry drilling machine
~ mit pneumatischem Vorschub = stoper (drill)
~ ~ Spindelvorschub = drifter
~ zur Kernentnahme aus Straßendecken → (Beton)Kernbohrmaschine
Bohr|mast *m* = drilling mast
~mehl *n*, Bohrklein *n* [*Trockenbohrung*] = dry rock cuttings, ~ well ~, ~ drill ~, ~ drillings
Bohrmeister *m* = driller (US); boring master, head driller, foreman-driller
~stand *m* = driller's station, ~ position (US); boring master's ~
(Bohr)Meißel *m* = bit, drill(ing) ~
~blatt *n* = bit blade
~fanghaken *m* = bit hook
~lehre *f* → (Bohr)Meißelschablone *f*
~probe *f* [*am (Bohr)Meißel haftende Gebirgsprobe*] = bit sample
~schablone *f*, (Bohr)Meißellehre *f* = bit ga(u)ge
~schaft *m* = bit shank, shank of bit
~schneide *f* → (Bohr)Schneide
~wechsel *m* = changing of the bit
~zapfen *m* = bit pin
Bohr|meter *n* = drilled metre (Brit.); ~ meter (US)
~methode *f*, Bohrverfahren *n* = boring system, drilling ~, method of boring, method of drilling
Bohrmotor *m* = drilling engine
~leistung *f* = footage
Bohr|ort *m* → Bohrstelle *f*
~periode *f*, Bohrkampagne *f* = drilling campaign
Bohrpfahl *m*, Betonbohrpfahl, gebohrter Ortpfahl, Bohrbetonpfahl [*Es wird ein Rohr im Bohrverfahren niedergebracht. Beton und erforderliche Stahleinlagen werden unter Ziehen des Rohres eingebracht*] = bore(d) pile, drilled ~, shelless (cast-in-place) (concrete) ~, uncased (cast-in-place) (concrete) ~

(Bohr)Pfahlwand *f* = bore(d) diaphragm, ~ (~) wall, ~ cut-off
Bohrpfeiler *m* = drilled-in caisson [*It is made by driving a heavy steel pipe with a cutting shoe down to bedrock and as far into the rock as it will go. In this respect the caisson is a pile. After refusal is met, the soil encased in the pipe is removed, a hole is drilled through the decomposed top layer into sound rock by means of a churn drill, and the hole and the pipe are filled with concrete. These procedures are characteristic for piers*]
~-Stützwand *f* = drilled-in caisson retaining wall
Bohr|plan *m*, Bohrlochplan, Schablone *f* = drilling pattern [*A plan showing the location, direction, length and firing sequence of the drill holes in a round*]
~plattform *f* = drilling platform
~probe *f*, Bohrprüfung *f*, Bohrversuch *m* = drill(ing) test
~probe *f* [*Dem Baugrund entnommene Probe, aus der die Beschaffenheit des Baugrundes ermittelt werden kann*] = drilled specimen
~probe *f* [*Das beim Abteufen von Bohrungen als Spül- oder Kernprobe gewonnene Gestein*] = well sample
~probenanzeiger *m* = sample logger
~profil *n* [*Darstellung der von einer Bohrung durchteuften Gebirgsformationen in Schnitten*] = well-log, drill log, profile of a well
~programm *n* = drilling program(me)
~protokoll *n* → Bohrbericht *m*
~prüfung *f* → Bohrprobe *f*
~punkt *m* → Bohrstelle *f*
(~)Ratsche *f*, (Bohr)Knarre *f* = ratchet
~register *n* → Bohrbericht *m*
~richtung *f* = drilling direction
~rohr *n*, Futterrohr = casing tube
~rohr *n*, Vortreibrohr [*gebohrter Ortpfahl*] = bore(d) casing (tube), ~ lining (~), drilled ~ (~)

~rohre *npl* für Probebohrungen = casing for lining exploratory drill holes

~rohr-Fahrstuhl *m*, Bohrrohr-Elevator *m* = casing elevator

(~)Rohrschuh *m* = casing shoe

~rohr-Zange *f* = casing tongs

~schacht *m* [*durch Bohren abgeteufter Schacht*] = trepan-sunk shaft

~schappe *f*, Schappe, Schappenbohrer *m* = (shell) auger, mud ~

~schappe *f*, einschneidige ~ = soil sampler with one cutting edge

~schappe *f*, zweischneidige aufklappbare ~ = split spoon sampler

~schere *f* → Rutschschere

~schicht *f* [*Arbeitsschicht, in der gebohrt wird*] = drilling shift

~schiff *n*, Bohrboot *n* = boring vessel, drilling ~, blue water ~, under water drilling ~

Bohrschlamm *m*, Bohrflüssigkeit *f*, Spülflüssigkeit, Spülschlamm = drilling mud, rotary ~

~ mit Erdgas durchsetzt = gas-cut mud

~ ~ Wasser verdünnen = to dilute the mud with water

~aufbereitung *f* [*Rotarybohren*] = mud conditioning

~dichte *f*, Bohrschlammgewicht *n* [*Rotarybohren*] = density of the drilling mud, weight ~ ~ ~

~gewicht *n*, Bohrschlammdichte *f* [*Rotarybohren*] = weight of the drilling mud, density ~ ~ ~

Bohr|schlammischer *m* = mud mixer

~schlauch *m*, Spülschlauch, Rotaryschlauch [*bei Rotary-Bohranlagen die Verbindung zwischen Spülkopf und Steigleitung*] = drilling hose, rotary ~

~schmant *m*, Bohrklein *n*, Bohrgut *n* [*Tiefbohrung*] = wet rock cuttings, ~drill ~, ~ well ~, ~ drillings

~schnecke *f*, Schnecken-Erdbohrer *m* = earth auger, soil ~, auger drill

~schneckenverlängerung *f* = earth auger extension, soil ~ ~

(~)Schneide *f*, (Bohr)Meißelschneide, Bohrerschneide = (drill) bit (cutting) edge, drill point, chisel-type bit cutting edge

Bohrschwengel *m*, Schwengel = drilling beam, walking ~, working ~

~bock *m* [*pennsylvanisches Bohrverfahren*] = sam(p)son post

~bockstütze *f*, Bohrschwengelbockstrebebalken *m* = sam(p)son post brace

Bohrseil *n*, Ölfelddrahtseil = rotary drilling line

~ = drilling cable, ~ rope, ~ line

~rolle *f*, Bohrseilscheibe [*pennsylvanisches Seilbohren*] = crown pulley, rope ~, cable ~, line ~

Bohrseiltrommel *f* [*pennsylvanisches Seilbohren*] = bull wheel spool, ~ ~ drum, cable drum, rope drum, line drum

~bockstütze *f* [*pennsylvanisches Seilbohren*] = bull wheel post brace

~bremsband *n* [*pennsylvanisches Seilbohren*] = bull wheel brake band

~bremshebel *m* [*pennsylvanisches Seilbohren*] = bull wheel brake lever

Bohr|sondierung *f*, Sondierbohrung, Sondierung mit direktem Verfahren, Direktsondierung = drilling for ground investigation

~spindel *f* [*Teil einer Säulendrehbohrmaschine, sie bewirkt den Differentialvorschub*] = drill spindle

~spitze *f* = tip

~spülprobe *f* = wet rock cuttings sample

(~)Spülung *f* [*für den Spülungsumlauf beim Spülbohren verwendete Flüssigkeiten oder Gase*] = oil well fluid

Bohrstahl *m*, Bohrer *m* [*für Bohrhammer*] = (drill) steel

~einsteckende *n* → (Bohrer)Einsteckende

~wechsel *m*, Bohrerwechsel = changing (drill) steels

~werkstatt *f* → Bohrerwerkstatt

Bohrstange *f* [*Rotarybohren*] = auger stem, one length of drill pipe

Bohrstange — Bohrung

~, Verlängerungsbohrstahl *m* = (drill) steel extension

Bohrstangen|-Ablagebock *m* = cage for drill steels

~**durchmesser** *m* = (drill) steel extension diameter

~**führung** *f* = (drill) steel retainer, (~) ~ guide

Bohr|staub *m* [*tritt bei der Luftspülung aus dem Bohrloch aus*] = drilling dust

~**stelle** *f*, Bohr(lochansatz)punkt *m*, Bohrort *m* = well-location, site of the well, drilling site, location of the well, drilling point

~**stellung** *f* = drilling position

~**stütze** *f*, Bohrerstütze, (Bohr-)Hammerstütze, Bohrknecht, (Vorschub-)Stütze = (pusher) leg

~**technik** *f* = drilling engineering

~**tiefe** *f*, Bohrteufe = drilling depth

Bohrtisch *m*, Drehtisch, Rotary(bohr)tisch = rotary table, turn ~, rotary machine

~ **mit Kettenantrieb** = chain-driven rotary table

~ ~ **Wellenantrieb** = shaft-driven rotary table

Bohr|trupp *m*, Bohrmannschaft *f*, Bohrbelegschaft *f* = drilling gang, ~ crew, ~ party, ~ team

~**tunnel** *m* = drilled tunnel

~**turbine** *f*, Turbinenbohrer *m*, Turbo-Bohrmaschine *f* = turbodrill

(Bohr)Turm *m* = (drilling) derrick, (~) tower

~**bühne** *f* = (drilling) derrick platform, (~) tower ~, working ~

~**eckpfosten** *m*, (Bohr)Turmmast *m* = (drilling) derrick leg, (~) tower ~

~**flur** *m*, Bohrflur = (drilling) derrick floor, (~) tower ~

~**fundament** *n*, (Bohr)Turmsockel *m* = (drilling) derrick foundation, (~) tower ~

~**galgen** *m* = gin-pole of a (drilling) derrick, ~ ~ ~ (~) tower

~**helfer** *m* = helper, roustabout

~**keller** *m*, (Bohr)Turmschacht *m* = (drilling) derrick cellar, (~) tower ~

~**kran** *m*, Bohrkran [*pennsylvanisches Bohrverfahren*] = (drilling) derrick crane, (~) tower ~

~**kranz** *m* = (drilling) derrick cornice, (~) tower ~

~**krone** *f* = (drilling) derrick crown, (~) tower ~

~**mast** *m*, (Bohr)Turmeckpfosten *m* = (drilling) derrick leg, (~) tower ~

~**querträger** *m*, (Bohr)Turmtragbalken *m*, (Bohr)Turmbindebalken = (drilling) derrick girder, (~) ~ girt, (~) tower ~

~**rolle** *f*, Kronenrolle = crown sheave

~**rollenblock** *m*, Kronenblock = crown block

~**rost** *m* = (drilling) derrick grillage, (~) tower ~

~**schacht** *m*, (Bohr)Turmkeller *m* = (drilling) derrick cellar, (~) tower ~

~**sockel** *m*, (Bohr)Turmfundament *n* = (drilling) derrick foundation, (~) tower ~

~**steiger** *m* = (drilling) derrick man, (~) tower ~

~**strebe** *f*, ~**balken** *m* = (drilling) derrick brace, (~) tower ~

~**tragbalken** *m*, (Bohr)Turmquerträger *m*, (Bohr)Turmbindebalken = (drilling) derrick girt, (~) ~ girder, (~) tower ~

~**tragpfosten** *m* = (drilling) derrick foundation post, (~) tower ~ ~

~**unterbau** *m* = (drilling) derrick substructure, (~) tower ~

Bohr|- und Abbauhammer *m*, Abbau- und Bohrhammer, Allzweck-Bohr- und Abbauhammer = demolition and drill hammer

~**- und Schießdaten** *f* = drilling and blasting data

~**unfall** *m* = drilling accident

Bohrung *f*, **Bohren** *n* = boring, drilling

~, Zylinder~ = bore

~, Sonde *f*; Schacht *m* [*in Galizien*] = well
~ = Aufschließungs~
~ im Küstenvorland, Bohren *n* ~ ~ = off-shore drilling
~ ~ Stadium der Anfangsspitzenförderung, Sonde *f* ~ ~ ~; Schacht *m* ~ ~ ~ ~ [*in Galizien*] = flush producer
~ mit Hilfe von Druckgas fördernd, Sonde *f* ~ ~ ~ ~ ~; Schacht *m* ~ ~ ~ ~ ~ [*in Galizien*] = producing well put on gas lift, ~ boring ~ ~ ~, ~ bore ~ ~ ~ ~
~ ~ ~ ~ Druckluft fördernd, Sonde *f* ~ ~ ~ ~ ~ [*in Galizien*] = producing well put on air lift, ~ bore ~ ~ ~ ~, ~ boring ~ ~ ~ ~
~ ~ sehr kleiner Förderung = stripper well, marginal ~
~ ohne Filterrohre, Sonde *f* ~ ~; Schacht *m* ~ ~ [*in Galizien*] = barefooted well, ~ bore, ~ boring
~ zum Pumpen vorrichten, Sonde *f* ~ ~ ~; Schacht *m* ~ ~ ~ [*in Galizien*] = to place the well "on the beam" for pumping, to put ~ ~ "~ ~ ~"
~ zur Untersuchung der Ausdehnung eines Ölfeldes, Anschlußbohrung = field extension well
Bohrungen *fpl* mit Luftspülung bei Prospektierungen = prospect sampling by airflush drill
Bohr|unternehmen *n*, Bohrunternehmung *f*, Bohrfirma *f* = drilling firm
~unternehmer *m* = drilling contractor
~unterwagen *m* [*nur der Wagen an sich, also ohne Aufbauten*] = drill carriage
~verfahren *n*, Bohrmethode *f* = drilling system, boring ~, method of drilling, method of boring
~verpflichtung *f* = drilling obligation, ~ requirement, boring ~
~versuch *m* → Bohrprobe *f*

~vertrag *m*, Bohrkontrakt *m* = drilling contract, boring ~
~vorgang *m* = drilling operation
~vorrichtung *f* = drill rig [*Drill rigs are usually mounted on self-propelled wheels or caterpillar treads, but occasionally they must be pulled from one location to another by truck or tractor*]
~vorschub *m*, Bohrervorschub = feeding
~wagen *m*, fahrbares Bohrgerät *n*, fahrbare Bohrmaschine *f* [*Im Gegensatz zu den Drehbohrmaschinen für Großbohrlöcher arbeiten die Bohrvorrichtungen der Bohrwagen nach dem Bohrhammerprinzip, d. h. schlagend und drehend (umsetzend). Jedoch kann bei einigen Typen neben den Bohrhämmern auch mit Drehbohrmaschinen gebohrt werden*] = wagon drill
~wagenfahrgestell *n* = (wagon) drill frame, main ~
~wasser *n*, Spülwasser = drilling water
~werk *n* → (Rotary-)Hebewerk
~werktrommel *f*, Hebwerktrommel, Windwerktrommel, Rotary-~ = (rotary) drawworks drum
~werkzeug *n* = drilling tool
~widerstand *m* = drilling resistance
~wirbel *m*, Drehkopf *m*, Rotationskopf, Spülkopf [*Rotarybohren*] = rotary swivel
~zeit *f*, Bohrdauer *f* = drilling time
~zeug *n*, Bohrgarnitur *f* [*beim pennsylvanischen Seilbohren bestehend aus Meißel, Schwerstange, Schlagschere und Seilhülse*] = drilling string, string of tools [*consisting of bit, drill stem, drilling jars and rope socket*]
~zubehör *m*, *n* = drilling accessories
~-Zusatzgerät *n* → Anbau-Bohrgerät
Boiler *m*, Heißwassergerät *n*, Heißwassererhitzer *m* = calorifier
Boje *f* = buoy
Boltonkreisel *m* = Simplex turbine aerator

Bol(us) *m*, Boluserde *f* = bolus, bole
bölzen → ab~
Bolzen *m* = bolt
~, Gelenk~ = pin
~, Anker(bolzen), Gesteinsanker *m*, Gebirgsanker [*Tunnel- und Stollenbau*] = roof bolt
~ **mit Anzug**, konischer Bolzen = conical bolt
~**gelenk** *n* = pin hinge
~**schießgerät** *n*, Bolzenschießer *m*, Schußgerät, Einschießgerät = cartridge-powered tool, powder-actuated ~, bolt driving gun, cartridge hammer
~**verbindung** *f*, Gelenk~ = pin joint [*A hinge in a structure*]
~**verlaschung** *f* = fishplating
Bölzung *f* → Ab~
Bombenkrater *m*, Bombentrichter *m* = bomb crater
bombensicher = bomb-resistant, bomb-proof
Bombentrichter *m*, Bombenkrater *m* = bomb crater
Bommerband *n* → Federband
bonder(isiere)n = to bonderize
Bondern *n* = bonderizing
Bonder-Verfahren *n*, Bonderiteverfahren = bonderizing method
Bonnellfeder *f* = Bonnell type spring
Boort *m*, Bo(a)rt, Bortz, Bowr [*Bohrdiamant*] = bo(a)rt
Boot *n* **zur Aufrichtung des Wehrs**, Aufrichtungsboot = maneuver boat (US); working barge (Brit.)
~ **zur Stützung des Wehres**, Stützboot = barge acting as support
~**aufschleppe** *f* = boat incline
~**wendegetriebe** *n* = marine reverse gear
Borax|kalk *m* = calcium borate
~**see** *m* = borax lake
Bördel *m*, Krempe *f* = flange
~**blech** *n*, Krempblech, Kümpelblech = flanged plate
~**maschine** *f*, Kümpelpresse *f* = flanging machine
bördeln, um~, (um)krempen = to flange

Bördelprobe *f*, Bördelprüfung *f*, Bördelversuch *m* = flanging test
Bord|kantenschneidgerät *n*, Abbordgerät, Straßenbordfräs- und -räummaschine *f* = verge cutter, ~ trimmer
~**kran** *m* = shipboard cargo crane
~**rinne** *f*, Rinnstein *m* = channel, gutter
~**rinnenform** *f* → Rinnsteinform
~**schwelle** *f*, (Hoch)Bordstein *m*, Schrammbord *n* = curb; kerb (Brit.); ~ above ground
~**schwelle** *f* **monolithisch mit Betondecke**, (Hoch)Bordstein *m* ~ ~ ~, Schrammbord *n* ~ ~ ~ = integral curb (above ground); ~ kerb (~ ~) (Brit.)
~**schwellenbeleuchtung** *f* = curb lighting; kerb ~ (Brit.)
Bordstein *m*, Hoch~, Bordschwelle *f*, Schrammbord *n* = curb; kerb (Brit.); ~ above ground
~, Tief~ = flush curb; ~ kerb (Brit.); inverted ~; curb below ground; kerb below ground (Brit.)
~ → Bordziegel *m*
~**-Bogenform** *f* = radius kerb mould (Brit.); ~ curb mold (US)
~**fertiger** *m* = power curber; ~ kerber (Brit.) [*A machine for the rapid forming and laying of concrete curbs and gutters*]
~**form** *f* = kerb mould, curb ~ (Brit.); curb mold (US)
~**führung** *f* = kerb alignment (Brit.); curb ~
~**greifer** *m* = kerb grab(bing) equipment (Brit.); curb ~ ~
~**markierung** *f* = curb marking; kerb ~ (Brit.)
~**maschine** *f* = kerb machine (Brit.); curb ~
~**parken** *n* = curb parking; kerb ~ (Brit.)
~**presse** *f* = kerb press (Brit.); curb ~
~**- und Plattenpresse** *f*, Platten- und Bordsteinpresse = slab and kerb press (Brit.); ~ ~ curb ~

Bordziegel m, Bordstein m, Saumziegel, Saumstein, Kanten(dach)ziegel, Kanten(dach)stein = marginal tile (for gables)
Bore f, Mascaret f, Gezeitenbrandung f, Sprungwelle f, Sturzwelle = bore, eager
borsaurer Kalk m = borate of lime
Borstahl m = boron steel
Borstenbürste f = bristle brush
Bort(z) m → Boort
böschen → ab~
Böschen n, Ab~ = slope work
Böschung f, Abdachung, Ausbiß m (Geol.) = escarpment, scarp
~ = slope [*The inclined face of a cutting or embankment*]
~ **1 : 1** f = pitched slope 1 : 1
~ **im Abtrag**, Einschnittböschung, Böschung im Einschnitt, Abtragböschung, Aushubböschung, Böschung im Aushub = slope of cutting, cutting slope, excavation slope
~ ~ **Auftrag** → Dammböschung
~ **mit handgesetzter Stein(ab)deckung** ~ ~ ~ Steinvorlage, ~ ~ handgesetztem Steindeckwerk = pitched slope
Böschungs|abflachung f = slope flattening
~**abschnitt** m = slope portion
~**abtrag** m = slope removal
~**abtreppung** f, Böschungsterrassierung = berm construction
~**auskleidung** f, Böschungsverkleidung, Böschungsbekleidung, Böschungsbelag m = slope revetment, ~ lining
~**auskleidungsmaschine** f = slope lining machine, ~ liner
~**bau** m = slope construction, ~ work
~**befestigung** f = slope stabilization
~**befestigung** f **durch Berasung**, ~ ~ Begrünung = slope stabilization by seeding
~**begradigung** f = slope rectification
~**bekleidung** f → Böschungsauskleidung
~**belag** m → Böschungsauskleidung f

~**bepflanzung** f = slope planting
~**berasung** f **durch Sodenandeckung**, Böschungsbegrünung ~ ~ = slope sodding
~**berechnung** f = slope design, ~ calculation, ~ computation
~**besamung** f = sowing a slope down to grass
~**betoniermaschine** f = (single) slope (canal) concrete paver
~**bruch** m = slope failure [*Special case of toe failure in which hard strata limit the extent of the failure surface*]
~**einsturz** m, Böschungsnachfall m = slope caving, ~ cave(-in)
~**entwässerung** f = slope drainage
~**erosion** f = slope erosion
~**fertiger** m [*Asphaltwasserbau*] = slope finisher, ~ finishing machine
~**fläche** f = slope area
~**flügel** m [*Widerlager*] = retaining wing
~**fuß** m = toe of slope
~**fußgraben** m = toe ditch
~**fußkreis** m = toe circle
~**geräte** npl = slope construction equipment
~**herstellung** f = slope trimming
~**hobel** m, Hang-Kippschar f = (back-)sloper, slope grader
~**kegel** m = cone of slope
~**kopf** m [*Stein*] = embankment crown retainer
~**kreis** m = slope circle
~**krone** f = slope crown
~**leiter** f [*Aus zwei Brettern mit Lattensprossen zum Verlegen und Anpatschen der Rasensoden an Böschungen*] = slope ladder
~**linie** f [*Geländedarstellung*] = slope line
~**-Mäher** m = slope mower
~**mähwerk** n [*Aufsetzgerät zum UNIMOG, zur Pflege des Rasens an Straßen-, Damm-, Deich- und Kanalböschungen*] = slope mowing attachment
~**mauer** f [*Eine Stützmauer vor aufgeschüttetem Boden*] = retaining wall

Böschungsmittelpunkt — Brahms-Chézy'sche Formel 268

~mittelpunkt m = mid-point of a slope
~nachfall m, Böschungseinsturz m = slope caving, ~ cave(-in)
~neigung f, Böschungsverhältnis n = inclination of slope, slope inclination
~pflaster n = slope sett paving
~pflasterstein m = slope sett
~planiermaschine f = (single) slope (canal) trimmer (or trimming machine)
~planierschild n, Böschungsprofilierschild = slope trimming blade
~planier- und verdichtungsmaschine f = slope shaping and compaction machine
~profil n = profile of slope
~profiliermaschine f = slope trimming machine, (single) slope (canal) trimmer (~)
~profilierschild n, Böschungsplanierschild = slope trimming blade
(~)Rutschung f, Gleitflächenbruch m = slide
~schrapper m = slope scraper
~schulter f = slope edge
~schutz m, Böschungssicherung f = slope protection
~sicherung f, Böschungsschutz m = slope protection
~standfestigkeit f, Böschungsstandsicherheit = slope stability
~terrassierung f, Böschungsabtreppung = berm construction
~torkretieren n = slope guniting
~unterhaltung f = slope maintenance
~verdichtung f = slope compaction
~verhältnis n, Böschungsneigung f = inclination of slope, slope inclination
~verkleidung f → Böschungsauskleidung
~wagen m [*Asphaltwasserbau*] = finisher feeding truck
~walze f = slope roller
~winkel m [*ist der Neigungswinkel, unter dem sich der Boden normalerweise abböscht*] = angle of slope
~ziehen n mit Straßenhobel = sloping with grader

Bossage-Mauerwerkbildung f, Rustika f = rustication, rusticated ashlar, rusticated ashler
Bossen m = roughly dressed ashlar, ~ ashler
~werk n = bossage
bossieren = to bush-hammer
Bossieren n = bush-hammering
Bossierhammer m = bush hammer
Bossierung f = bush-hammered finish, ~ face
Boulé'sche Gleittafel f = sliding shutter Boulé
Boulèwehr n = Boulé dam, ~ weir
Bourdonmanometer n, Bourdondruckmesser m = Bourdon (pressure) gauge
Boussinesq'sche Gleichung f = Boussinesq's equation
Bouteillenstein m = moldavite
Bowdenzug m = Bowden cable, ~ wire
Bowr m → Boort
Bow'scher Kräfteplan m, reziproker ~ [*ebenes Fachwerk*] = reciprocal force polygon, Bow's ~
Boxer|motor m = engine with opposing cylinders
~zylinder m mit zwei Kolben = double ended cylinder
BPA f [*Beginnende Paraffinausscheidung. Kriterium für die Kältebeständigkeit von Dieselkraftstoffen nach DIN 51772*] = cloud point
Brach|land n, Brachacker m = fallow land
~tag m, Stillstandtag [*Maschine*] = stoppage day, downtime ~
~zeit f, Stillstandzeit [*Zeitspanne, während der Maschinen nicht produktiv sind*] = stoppage time, no-productive time, downtime
brackig = brackish, saltish
Brackwasser n [*Mischung von Süßwasser und Salzwasser im Tidegebiet eines Stromes*] = brackish water
~kalk m = brackish water limestone
~zone f = brackish water zone
Brahms-Chézy'sche Formel f = Chezy formula

Brammen|entstapler m = slab unpiler
~schere f = slab shears
~stapler m = slab piler
~walzwerk n = slab cogging mill, slabbing ~
Brand m, Feuer n = fire
~, Brennen n [*Keramik*] = burning
~abschnitt m [*DIN 14011*] = fire lobby
~ausbreitung f = fire spread, spread of fire
~ausbruch m = fire outbreak, outbreak of fire
~bekämpfung f, Feuerlöschen n = fire fighting
~blende f = fire barrier
~damm m (✗) = fire wall, ~ dam
~dauer f = fire duration
branden, brechen [*einer Welle*] = to break
Brand|forschungsinstitut n = fire research station
~gefahr f, Feuergefahr = fire risk
~giebel m = fire gable
~herd m = seat of a fire
~hohlmauer f, Hohl-Brandmauer = cavity party wall
~klasseneinteilung f = fire(-resistance) grading
~klassenwert m = fire rating
~leiter f, Feuerleiter = escape ladder
~mauer f, gemeinschaftliche Giebelmauer, Kommunmauer, Feuermauer = party wall
~meister m [*Feuerwehr*] = fire chief
~probe f → Brandversuch m
~prüfung f → Brandversuch m
~riß m, Brennriß [*Keramik*] = burning crack
~ruine f = fire-gutted structure
~schiefer m, Brennschiefer = combustible shale
Brandschutz m, Feuerschutz [*Maßnahmen, um Bauteile feuerbeständig zu machen*] = fire proofing
~, Feuerschutz [*Maßnahmen zur Verhütung von Bränden*] = fire prevention, ~ protection
~-Anstrich m, Feuerschutz-Anstrich = fire-coat

~tür f = fire door; armour ~ (Brit.); armor ~ (US)
Brand|sicherheit f [*Baustoff*] = fire-safety
~tür f → Brandschutztür
Brandung f der Wellen = breaking of waves
~ durch Herabstürzen = plunging, breaking by falling
~ in Volutenform = plunging, breaking by curling
Brandungs|boot n = surf boat
~breccie f [*diese Schreibweise ist zu vermeiden*], Brandungsbrekzie = surf breccia
~ebene f = abrasion plain
~erosion f, marine Erosion = wave erosion, coast(al) ~, shore ~, marine ~, wave quarrying
~gebilde n = wave-cut topography
~gürtel m, Brandungszone f = surf zone
~kehle f = notch
~kluft f = wave-cut chasm
~linie f = surf line
~platte f → Strandterrasse f
~schutt m = wave-worn material
~strom m, Wellenströmung f = wave current
~stromversetzung f, Stromversetzung bewirkt durch Brandungsbänke = drift due to waves breaking along the length of a coast
~welle f = breaking wave, breaker
~zone f, Brandungsgürtel m = surf zone
Brand|ursache f = cause of a fire
~verhalten n [*Baustoff*] = behavio(u)r in fire
~versuch m, Brandprüfung f, Brandprobe f = fire test
~versuchsabteilung f [*einer Materialprüfanstalt*] = fire test department
~verzug m, Verziehen n im Feuer [*Keramik*] = deformation during burning
~vorschrift f, Feuervorschrift, feuerpolizeiliche Vorschrift = fire (prevention and fire proofing) regulation
~wache f, Feuerwache = fire station

Brandwand — Brausekabine

~wand *f* = fire wall
Brannt|gips *m*, gebrannter Gips = calcined gypsum
~**kalk** *m*, gebrannter Kalk, Brennkalk = (quick)lime, anhydrous lime, common lime, calcined calcium carbonate
Brauchbarkeit *f* = serviceability
Brauchbarkeitsziffer *f* [*Straßendecke*] = serviceability index, ~ level
Brauchwasser *n*, Betriebswasser, Rohwasser = non-potable water, raw ~
~**behälter** *m*, Betriebswasserbehälter, Rohwasserbehälter = non-potable water tank, raw ~ ~
~**netz** *n*, Betriebswassernetz, Rohwassernetz = non-potable water network, raw ~ ~
~**pumpstation** *f*, Brauchwasserpumpwerk *n*, Betriebswasserpumpstation, Betriebswasserpumpwerk, Rohwasserpumpstation, Rohwasserpumpwerk = non-potable water pumping station, raw ~ ~ ~
~**versorgung** *f*, Betriebswasserversorgung, Rohwasserversorgung = non-potable water supply, raw ~ ~
Brauerei *f* = brewery
~**abwasser** *n* = brewery waste
Braunerde *f* = brown earth
~**boden** *m* = brown earth soil
Braunfleckigkeit *f* [*Mörtelfuge*] = brown staining [*mortar joint*]
Braunkohle *f* = brown coal [*In the USA the term "brown coal" also includes the German term "Pechkohle"*]
Braunkohlen|abbau *m* = brown coal getting, ~ ~ working
~**asche** *f* = brown coal ash
~**bagger** *m*, Braunkohlen-Löffelbagger = brown coal shovel
~**bagger** *m*, Braunkohlen-Stetigbagger, Braunkohlendauerbagger = continuous brown coal excavator
~**brikett** *n*, Preßbraunkohle *f* = brown coal briquette
~**brikettfabrik** *f*, Braunkohlenbrikettwerk *n* = brown coal briquetting plant, ~ ~ ~ installation
~**(dauer)bagger** *m* → Braunkohlen-Stetigbagger
~**feuerung** *f* = brown coal furnace
~**flöz** *n* = brown coal seam
~**-Förderanlage** *f*, Braunkohlen-Transportanlage = brown coal conveyor, ~ ~ conveyer
~**fundstätte** *f*, Braunkohlenlager(stätte) *n*, (*f*), Braunkohlenvorkommen *n*, Braunkohlenvorkommnis *n* = brown coal deposit
~**grube** *f* = brown coal mine
~**-Hochdruck-Kraftwerk** *n* = brown coal-fired high pressure power plant, ~ ~ ~ ~ ~ station, ~ ~ ~ ~ generating ~
~**kabelbagger** *m* = brown coal cableway excavator
~**lager(stätte)** *n*, (*f*) → Braunkohlenfundstätte
~**(-Löffel)bagger** *m* = brown coal shovel
~**mühle** *f* = brown coal (grinding) mill
~**sandstein** *m* = brown coal grit
~**schwelkoks** *m* = brown coal low temperature coke
~**-Stetigbagger** *m*, Braunkohlen(dauer)bagger = continuous brown coal excavator
~**tagebau** *m* = brown coal open-cast mining
~**tagebau(grube)** *m*, (*f*) = brown coal open-cast mine
~**teer** *m* → → Teer
~**tiefgrube** *f* = brown coal underground mine
~**-Transportanlage** *f*, Braunkohlen-Förderanlage = brown coal conveyor, ~ ~ conveyer
~**vorkommen** *n* → Braunkohlenfundstätte *f*
Braun-Kurve *f* [*Hydraulik*] = Braun curve
Braunquarzsandstein *m* = brownstone
Braun'sche Röhre *f* = cathode ray tube
Brause *f*, Bad~, Dusche *f* = shower
~**bad** *n* = shower bath
~**kabine** *f* → Brausenische *f*

Brausenische — Brechgut

~nische *f*, Duschnische, Brausekabine *f*, Duschkabine, Brausezelle *f*, Duschzelle = shower cubicle
~raum *m*, Massendusche *f* = shower room without cubicles
~rosette *f*, Duschrosette = shower rose, spray ~
~säule *f*, Duschsäule = shower column
~sieb *n* = spraying screen, rinsing ~
~tülle *f*, Duschtülle = shower rose
~vorrichtung *f* [*Vibrationssieb*] = rinsing device
~wanne *f* = shower stall
~wasserpumpe *f* = rinsing water pump
~zelle *f* → Brausenische *f*
Brech|anlage *f*, Grobzerkleinerungsanlage, Vorzerkleinerungsanlage, Brecheranlage = breaking plant, crushing ~, installation
~barkeitsprüfung *f* mit Splitt = breaking time test with chippings [*asphalt emulsion* (US)]
~barkeitsprüfung *f* nach Caroselli = breaking time test of Caroselli [*asphalt emulsion* (US)]
~bewegung *f* = crushing movement, ~ motion, breaking ~
~druck *m* = crushing pressure, breaking ~
~eisen *n* → Brechstange
~element *n* → → Backenbrecher
brechen, zerfallen [*Emulsion*] = to break
~ [(*Hart*)*Zerkleinerung*] = to break, to crush
~, branden [*einer Welle*] = to break
~ [*Gestänge in der Tiefbohrtechnik*] = to unscrew
Brechen *n*, Zerfall *m* [*Emulsion*] = breakdown, breaking
~ [(*Hart*)*Zerkleinerung in Brechern*] = crushing, breaking
Brecher *m* [*heftiger Zusammenbruch der Welle mit Schaumbildung*] = breaker
~, sich brechende See *f* = breaking sea

~, Grobzerkleinerungsmaschine *f*, Brechmaschine, Vorzerkleinerungsmaschine = crushing machine, breaking ~, crusher, breaker
~ an der Küste = coastal breaker
~ in kabbeliger See = breaker due to clapotis
~anlage *f* → Brechanlage
~aufgabebehälter *m* = breaker feed hopper, crusher ~
(~)Aufgabegut *n*, Brechgut = uncrushed material, unbroken ~
~aufgeber *m*, Brecherspeiser, Brecherbeschicker = crusher feeder, breaker ~
~band *n* [*Trenn- und Waschanlage*] = conveyor from the breaker, ~ ~ ~ crusher, conveyer ~ ~ ~
~beschicker *m*, Brecherspeiser, Brecheraufgeber = crusher feeder, breaker ~
~-Fertiggut *n* = crushed product, ~ material
Brech(er)|gehäuse *n* → → Backenbrecher
~kammer *f* → Brechraum
~körper *m* → → Backenbrecher
~rahmen *m* → → Backenbrecher
~raum *m* → Brechraum
Brecher|schlacke *f* = crushed slag, broken ~
~speiser *m*, Brecheraufgeber, Brecherbeschicker = breaker feeder, crusher ~
~staub *m* = crusher dust, breaker ~
~teil *m*, *n* [*Brech- und Siebanlage*] = breaker unit, crusher ~
~wand *f* = breaker wall, crusher ~
Brechfein|kies *m* = fine crushed gravel
~sand *m* = fine crushed sand
Brech|fläche *f* → → Backenbrecher
~gehäuse *n* → → Backenbrecher
~geschwindigkeit *f* = crushing rate, ~ speed, breaking ~
~grad *m* = degree of crushing, ~ ~ breaking
~grobsand *m* = coarse crushed sand
Brechgut *n*, ungesiebtes gebrochenes Material *n* = crushed material, broken ~

Brechgut — Breieis

~, (Brecher)Aufgabegut = unbroken material, uncrushed ~
Brech|kammer *f* → Brechraum
~kegel *m* = crushing cone, breaking ~
~kies *m*, Quetschkies = crushed gravel, broken ~
~kopf *m* = breaking head, crushing ~
~körper *m* → → Backenbrecher
~lava *f* = crushed lava, broken ~
~leistung *f* = crushing output, breaking ~
~maschine *f* → Brecher
(~)Maul *n* → → Backenbrecher
~organ *n* → → Backenbrecher
~probe *f*, Brechprüfung *f*, Brechversuch *m* = breaking test, crushing ~
~prüfung *f* → Brechprobe
Brechpunkt *m* [*Punkt, an dem sich die in längeren Strecken, Straßen usw. gleichmäßige Neigung ändert*] = break
~ [*Temperatur bituminöser Stoffe, bei der die Plastizität schwindet und die Probe bei Biegebeanspruchung bricht. DIN 1995 - U 6*] = brittle point, breaking ~
~ nach Fraas → Brech-Temperatur *f*
Brech|rahmen *m* → → Backenbrecher
~raum *m*, Brech(er)kammer *f*, Brecherraum = crushing chamber, breaking ~
~ring *m* [*Brechwalzwerk*] = (crushing) roll segment, breaking ~ ~
~sand *m*, Steinsand [*umfaßt in den USA den Kornstufenbereich 0 = 4,76 mm*] = screening(s), stone ~, crusher ~, artificial sand, manufactured sand, (crushed) stone sand
(~)Schotter *m* → Steinschlag
~spalteinstellung *f* → → Backenbrecher
~spill *n* [*Tiefbohranlage*] = break-out cathead
~splitt *m* = broken chip(ping)s, crushed ~

~stange *f*, Brecheisen *n*, Hebebaum *m*, Hebeeisen = dwang (Brit.); crowbar
~station *f*, Brecherei *f* = crushing station, breaking ~
~stufe *f* = crushing stage, breaking ~
Brechtel-Preßbeton(bohr)pfahl *m*, Preßbeton(bohr)pfahl System Brechtel = Brechtel pressure pile
Brech-Temperatur *f* nach Fraas, Brechpunkt *m* ~ ~ = Fraas brittle temperature, ~ breaking point, ~ brittle point
Brech- und Mahlanlage *f* = crushing and grinding plant, ~ ~ ~ installation, breaking ~ ~ ~
Brech- und Siebanlage *f* = crushing and screening plant, breaking ~ ~, ~ ~ ~ installation
Brechung *f* = refraction
Brechungs|seismik *f* → Brechungsverfahren
~verfahren *f*, Refraktionsverfahren, elastisches Verfahren, Brechungsseismik *f*, Refraktionsseismik, seismische Bodenerforschung *f* nach dem Refraktionsverfahren = seismic refraction survey
Brech|versuch *m* → Brechprobe
~vorgang *m*, Bruchvorgang = crushing operation, breaking ~
~walze *f* = crushing roll, breaking ~
(~)Walzenmantel *m* = (crushing) roll shell, breaking ~ ~
~walzwerk *n* → Walzwerk
~weinstein *m* = tartar emetic
~werkzeug *n* [*Backen; Kegel; Hammer*] = crushing tool, breaking ~, ~ element
~zone *f* = crushing zone, breaking ~
Brei *m* = slip, paste
~eis *n*, Eisschweb *m* = frazil ice, slush ~ [*Granular or spicular ice which forms in agitated water. Riffles and rapids are prolific sources of such ice during protracted freezing temperatures*]

Breitband|straße *f* = wide strip train, broad ~ ~

~walzwerk *n* = wide strip mill, broad ~ ~

breitbeiniger Streckenbogen *m* (⚒) = splay-legged (roadway) arch

Breiteisen *n* = broad-tool, bolster

Breite *f* über alles, Gesamtbreite [*Maschine*] = over-all width

Breit|felgenreifen *m* = wide-base tire (US); wide-base tyre (Brit.)

~flachstahl *m* = universal plate

breitflanschig = wide-flanged

Breit|flanschprofil *n*, Breitflanschträger *m* = wide-flange section, WF ~ [*or H sections for some sizes*]

~folie *f* = wide foil

~fuge *f* = wide joint

~fußschiene *f*, breitfüßige Schiene = flange(d) rail, flat-bottomed ~

~hacke *f*, Breithaue *f*, Platthacke = mattock

~kammermühle *f* = wide chamber mill

breitkantiges Wehr *n* = broad-crested weir [*An overflow structure on which the nappe is supported for an appreciable length; a weir with a significant dimension in the direction of the stream*]

Breit|masche *f* = wide mesh

~reifen *m* = super single tyre (Brit.); ~ ~ tire (US)

~spur *f* [*größer als 1,435 m*] = broad gauge

~spurbahn *f* = broad gauge railway

~spurnetz *n* = broad gauge network

~trommelwinde *f* = wide drum winch

Brekzie *f*, „Naturbeton" *m*, Bretschie, Breccie [*die Trümmerbestandteile sind eckig und kantig*] = breccia

Brems|ansprechzeit *f* = brake lag

~arm *m* = brake arm

~arretierhebel *m* = brake lock control lever

~arretierung *f* = brake lock

~aufhängung *f* = brake support

~backe(n) *f*, (*m*) = brake shoe

Bremsband *n* = brake band

~ mit Selbstverstärkungswirkung, Servobremsband = self energizing brake band, ~ activating ~ ~

~spanner *m* = brake band tightener

Brems|bedienungswelle *f* = brake control shaft

~beiwert *m* = braking force coefficient

Bremsbelag *m*, Bremsfutter *n* = (brake) lining, (~) facing, (~) covering

~ aus Asbestpreßgewebe = woven asbestos (brake) lining, ~ ~ (~) facing, ~ ~ (~) covering

~fläche *f* → Bremsennutzfläche

~größe *f* = (brake) lining size, (~) facing ~, (~) covering ~

~satz *m* = lining service group, facing kit

Bremsbereich *m* = braking range

Bremsberg *m* = gravity plane, brake incline, self-acting incline, go-devil plane, jinny

~winde *f*, Bremsberg-Haspel *m*, *f* = gravity plane winch

Brems|betätigung *f* = brake application, ~ control

~block *m* = brake block

~dauer *f* [*Zeitdauer vom Beginn bis zum Ende des Bremsvorganges*] = brake application time, ~ control ~

~drehmoment *n* = brake torque

~druckleitung *f* = brake pressure line

~druckluft *f* = compressed air for brake control

~drucksammler *m*, Bremsdruckspeicher *m* = brake accumulator

~druckventil *n* = brake pressure valve

Bremse *f* auf der Getriebewelle = transmission-mounted drive line brake

~ mit Reserve = oversize brake

Brems|eindruck *m* → Bremsspur *f*

~einsteller *m* = brake adjuster

~einstellmutter *f* = brake adjusting nut

~einstellschraube *f* = brake adjusting bolt

~einstellung *f* = brake adjustment

~einstellwinde *f* = brake adjusting jack

Bremsen *n*, Ab~ = braking
~ **mit Auspuff-Klappenbremse** = exhaust braking
~**nutzfläche** *f*, wirksame Bremsenfläche, Bremsbelagfläche = braking surface
Bremsentlüftung *f* = bleeding of the brake
Bremser *m* [*Feldbahnwagen mit eingebauter Bremse*] = brake-equipped wagon
~ = brakeman, car rider
~**häuschen** *n* = brakeman's box, car rider's ~; caboose (US)
Brems|fahrzeug *n* [*es übt bei Versuchen eine bremsende Wirkung auf das Zugfahrzeug aus*] = brake vehicle
~**feder** *f* = brake spring
~**fläche** *f* [*einer Bandbremse*] = braking surface
~**flüssigkeit** *f* = brake liquid
~**flüssigkeitsbehälter** *m*, Bremsflüssigkeitstank *m* = brake (liquid) supply tank
~**folgeventil** *n* = relay valve
~**fußhebel** *m* → Bremspedal *n*
~**fußhebelventil** *n*, Bremspedalventil = brake pedal valve
~**futter** *n* → Bremsbelag *m*
~**gehäuse** *n* = brake housing
~**gestänge** *n* = brake rods, ~ linkage, ~ rigging
~**gewicht** *n* = brake counterweight
~**handgriff** *m* → Bremshebelgriff
~**handhebel** *m* = brake hand lever
~**hauptzylinder** *m* = brake master cylinder
Bremshebel *m* = brake (selector) lever, ~ operating ~
~**griff** *m*, Bremshandgriff = brake handle
~**kette** *f* [*Rotary-Bohrverfahren*] = brake lever chain
Brems|hilfe *f*, Bremsservovorrichtung *f* = brake booster
~**kabel** *n*, Bremsseil *n* = brake cable
~**kammer** *f* = brake chamber
~**kammer** *f* [*Schleuse*] = stilling chamber
~**kegel** *m* = brake cone

~**keil** *m* → (Hemm)Schuh *m*
~**klotz** *m* [*Stahl- oder Gußgraußklotz, der an einem Schienenfahrzeug zum Abbremsen der Fahrgeschwindigkeit durch Reibung zwischen ihm und dem Rad dient*] = brake block
~**klotz** *m* [*Tiefbohrtechnik*] = scotch
Bremskraft *f* = braking force
~ [*Brücke*] = force due to braking
~**ableitung** *f* [*Brücke*] = transfer of the force due to braking
~**beiwert** *m* = braking force coefficient
~**messer** *m*, Kraftmesser *m* = dynamometer
Brems|kreis *m* = brake circuit
~**lamelle** *f* = brake plate
~**leitung** *f* = brake piping, ~ pipe
~**lenkung** *f* [(*Gleis*)*Kettenfahrwerk*] = steering by brake(s)
~**licht** *n*, Bremsleuchte *f* = stop light
~**luftbehälter** *m* = brake air supply tank
~**luftventil** *n* = brake shuttle valve, ~ air ~
~**magnet** *m* = braking magnet
~**manschette** *f* = brake sealing cup
~**mechanismus** *m* = brake mechanism
~**nachstellschraube** *f* = brake adjustment screw
~**nachstellventil** *n* = adjuster valve
~**nocken** *m* = brake cam
~**ölkühler** *m* = brake cooler
~**pedal** *n*, Bremsfußhebel *m* = brake pedal
~**pedalventil** *n*, Bremsfußhebelventil = brake pedal valve
~**pfeiler** *m*, Schikane *f* [*Kolkschutz*] = energy dissipator
~**probe** *f* → Auslaufversuch *m*
~**prüfung** *f* → Auslaufversuch *m*
~**-PS** *f* = BHP, brake horsepower, FHP, friction horsepower
~**reibung** *f* = braking friction
~**ring** *m* = brake ring
~**rolle** *f* = brake roll
~**sand** *m* = track sand
~**schacht** *m* (⚒) = brake shaft

Bühne — Bulldozer

~, Arbeits~, Arbeitsplattform *f* [*Arbeitsplatz als Gerüst beim Abteufen von Schächten*] = shaft sinking stage

~, Arbeits~, Arbeitsplattform *f* [*im Bergbau Gerüst oder Schachtabsatz als Arbeitsplatz*] = work(ing) platform

~, Brücke *f*, Rammwagen *m*, Rammbrücke, Rammbühne, (Ramm-)Schlitten *m* = additional transverse carriage on rails

Buhne *f* = groin (US); groyne (Brit.)

~ **senkrecht zum Strom**, ~ ~ ~ **Fluß** = groin at right angles to the bank (US); groyne ~ ~ ~ ~ ~ ~ (Brit.)

Buhnen|abstand *m* = distance between groins (US); ~ ~ groynes (Brit.)

~**bau** *m* = groyning (Brit.); groining (US)

Bühnen|beleuchtung *f* → → Theaterbau

~**haus** *n* [*Theater*] = stage and backstage (of a theatre)

Buhnenkopf *m* = groyne head (Brit.); groin ~ (US)

Bühnen|turm *m* [*Theater*] = stage shaft, ~ well

~**werkstatt** *f* [*Theater*] = theatre workshop

Buhnenwurzel *f*, **Landanschluß** *m* der Buhne = groyne root (Brit.); groin ~ (US)

Bühnloch *n* (⚒) = hole made in the floor to put props in

Bulldog(-Holz)verbinder *m*, **Bulldogdübel** *m* = claw plate, bulldog ~

Bulldozer *m*, Fronträumer *m*, Schürfschlepper *m*, Planierschlepper

  Bulldozer auf Luftreifen, Radschlepper-Bulldozer, Reifentrecker *m* mit Planierschild, Radschlepper *m* mit Planierschild, gummibereifter Bulldozer, Planierreifenschlepper, Reifenplaniergerät *n*

  Bulldozer auf Raupen, Planierraupe *f*, Bulldozer auf Gleisketten, Schürfraupe, Planier-(Gleis)kettengerät *n*, Bauraupe [*Die Ausdrücke „Raupen" sowie „Planierraupe" sollten nicht mehr verwendet werden, dafür „Gleisketten" und „Planier-Gleiskettengerät"*]

    Planier-Gleiskettengerät *n* [*früher: Planierraupe f*] für Stubbenrodearbeiten

      Stubbenrodeschild *n*

    Planier-Gleiskettengerät *n* [*früher: Planierraupe*] für Baumfällarbeiten, Baumschieber *m*

      Baumfällschild *n*, dreiseitiges Stoßschild, dreiseitiges Streichschild

(bull)dozer(-equipped tractor), tractor dozer

  pneumatic-tired (bull)dozer (US); pneumatic-tyred (bull)dozer (Brit.); (bull)dozer fitted to wheel tractor, wheeled (bull)dozer

  (bull)dozer fitted to track-laying (type) tractor, caterpillar (bull)dozer, crawler tractor with (bull)dozer, (bull)dozer-equipped track-type tractor

    rootdozer

      dozer blade for rooting work

    treedozer

      dozer blade for tree felling work, stumper, three-side(d) (bull)dozer; three-side(d) mold-board (US)

Brustriegel — Bühne

~riegel m, Sohlbankriegel [Fach(werk)wand] = sill rail, breast ~
~schild n → Bulldozer m
~schwelle f → Sattelschwelle
Brüstung f [Niedrige Mauer zum Schutz für Fuhrwerke und Fußgänger an Ufern und auf Brücken. Im Hochbau bei Balkons, Dachaltanen und Terrassen] = parapet (wall)
Brüstungs|geländer n, Balustrade f = balustrading, balustrade
~gitter n = parapet grille
~höhe f = parapet (wall) height
~mauer f. = breast [The wall under the sill of a window, down to floor level]
~mauerhöhe f = breast height
~stein m = parapet (wall) block
~ziegel m = parapet (wall) brick
Brustwehr f → Brustmauer f
Brutto|fahrzeuggewicht n, Gesamtgewicht = gross vehicle weight, GVW, vehicle gross weight
~gefälle n, Rohfallhöhe f = total head
~geschoßfläche f = gross floor area
~gewicht n, Rohgewicht = gross weight
~last f = gross load
~lohn m = total wage
B. S. B. m → biochemischer Sauerstoffbedarf
Buchbesprechung f = book review
Buchenschindel f = beech shingle
Bücherei f, Bibliothek f = library
Bücher|leseraum m = reading room
~speicher m [Teil einer Bibliothek] = stack block
~turm m [Bauteil einer Bibliothek] = book tower
Büchsbohren n, Erdbohren mittels Büchse = bailer-boring
Buchse f, Lager~ = bush(ing)
~ [Spannpatrone] = quill
~ [Bohrwinde; Dorn] = quill, sleeve
~, Ketten~ = barrel
~, Zylinder~ = liner
Bucht f (Geol.) = bight, inlet
Buchungszeitraum m = accounting period

Buckel|blech n, geknickte Platte f = dished steel plate, buckle(d) ~
~(blech)spundwand f = buckle(d) plate sheet piling
~schweißen n [Schweiz: Dellenschweißen] = projection welding
~spundwand f → Buckelblechspundwand
Buckmann-Herd m, Kippherd = tilting frame
Bude f = hut
Bug m → Kopfband
Bügel m [allgemein] = stirrup
~ = stirrup, binder [A small diameter steel rod usually about $1/4$ or $3/8$ in. dia. used for holding together the main steel in a reinforced-concrete beam or column]
~ [Eimerhenkel] = bail
~, Umfang~ [Säule] = tie
~ [Exzenter] = strap
~ [Kettentrommel] = cleat, stirrup
~, Schaltgabel f = (shifting) fork
~abstand m = distance between stirrups
~anordnung f = spacing of stirrups
~armierung f → Bügelbewehrung
~bewehrung f, Bügelarmierung, Bügel(stahl)einlagen fpl = circumferential reinforcement in the form of ties, lateral ~ ~ ~ ~ ~, lateral ties
~bieger m, Bügelbiegemaschine f = stirrup bending machine
~draht m = stirrup wire
~einlagen fpl → Bügelbewehrung f
~eisen n = smoothing iron
~querschnittfläche f = cross-sectional area of stirrup
~raum m [Hotel] = ironing room
~spannung f = stirrup stress
~(stahl)einlagen fpl → Bügelbewehrung f
~zugkraft f = stirrup tensile force
Bugrad n [Flugzeug] = nose wheel
Bugsier-Schlepper m = berthing tug(boat), ~ towboat
Bühne f [Theater] = stage

Brunnen|becken n, Spring~ = fountain basin
~**bekleidung** f → Brunnenauskleidung
~**bohren** n → Brunnenbohrung f
~**bohrgerät** n = well drill(ing rig)
~**bohrindustrie** f = well-drilling industry
~**bohrung** f, Brunnenbohren n = well drilling
~**bohrunternehmer** m = well-drilling contractor
~**bohrwerkzeug** n = well-drilling tool
~**charakteristik** f = well characteristics
~**durchmesser** m = well diameter
~**ergiebigkeit** f = (water) yield of wells, (~) ~ ~ a well
~**filter** m, n = well screen
~**filterrohr** n = (well) screen pipe
~**formel** f = well formula
~**fundament** n → Brunnengründung f
~**fundamentkörper** m → Brunnengründung f
~**fundation** f → Brunnengründung f
~**futter** n = well casing [of a drainage well]
~**galerie** f, Brunnenreihe f = row of wells, line ~ ~, gang ~ ~, well field
~**gleichung** f = well equation
~**gräber** m, Brunnenbauer [als Techniker] = well builder
~**gräber** m [als Arbeiter] = well digger, ~ sinker, excavator
~**greifer** m = well grab
~**grund** m, Brunnensohle f = well bottom
~**gründung** f, Brunnenfundament n, Brunnenfundation f, Brunnengründungskörper m, Brunnenfundamentkörper m, Senk~, Senkfundation mit Brunnen, Senkfundament mit Brunnen, Senkgründung mit Brunnen, Senkgründungskörper mit Brunnen, Senkfundamentkörper mit Brunnen = open caisson foundation (structure), drop shaft ~ (~)
~**gründungskörper** m → Brunnengründung f
~**halbmesser** m = well radius

~**hydraulik** f = hydraulics of wells
~**industrie** f = well industry
~**klinker** m = clinker brick for well construction
~**kopf** m = well head, ~ top
~**kranz** m, Schlenge f, Brunnenschlinge f = cutting curb, drum ~, shoe [A curb on which a drop shaft is built]
~**loch** n = well hole
~**mantel** m = well lining
~**pumpe** f = well pump
~**rand** m = well edge
~**reihe** f, Brunnengalerie f = line of wells, row ~ ~, gang ~ ~, well field
~**reinigung** f = well cleaning
~**ring** m aus Beton, Beton-Brunnenring = concrete ring
~**rohr** n [*Elektroosmose*] = well electrode
~**schacht** m [*bei der Brunnengründung*] = sunk well, caisson
~**schacht** m = well shaft
~**schalung** f = well formwork
~**schlinge** f → Brunnenkranz m
~**sohle** f, Brunnengrund m = well bottom
~**staffel** f [*Grundwasserabsenkung*] = well points in series
~**stein** m, Radialstein, radialer Formstein, Schachtstein, Brunnenziegel m = radial brick
~**tiefe** f = well depth
~**verkleidung** f → Brunnenauskleidung
~**wand(ung)** f = shell of well
~**wasser** n = well water
~**widerstand** m = well resistance, ~ loss
~**ziegel** m → Brunnenstein m

Brust f, Orts~, Vortriebstelle f, Ausbruchstelle [*Tunnel- und Stollenbau*] = (working) face
~**holz** n [*Grabenverbau; Baugrubenverbau*] = soldier beam
~**holz** n → Bindeholz
~**mauer** f, Brustwehr f [*Am Rande von Molen und Wellenbrechern gegen das Überschlagen der Wellen gebaute Mauer*] = parapet (wall)

Brückenschalung — Brunnenbauwerkzeug

~schalung *f* = bridge formwork

~schwingung *f* = bridge oscillation, ~ vibration

~seil *n* → Brückenkabel *n*

~stahl *m*, Brückenbaustahl = bridge steel

~stau *m*, Pfeilerstau = backwater effect of bridge pier

~steg *m*, Stegbrücke *f* = catwalk

~stein *m* [*besonders niedriger Großpflasterstein*] = bridge (paving) sett

~steinpflaster *n* = bridge sett paving

~tafel *f* → Brückenüberbau *m*

~trasse *f* = line taken by a bridge

~treppe *f* = bridge staircase (Brit.); ~ stair (US)

~-Tunnel-Straße *f* [*Überquerung einer Bucht*] = bridge-tunnel highway, ~ road

~überbau *m* [*als Gesamtheit der Konstruktion über den Pfeilern, also einschließlich Pylonen bei Hängebrücken*] = bridge superstructure

~überbau *m*, Brückentafel *f*, Brückenplatte *f* [*als Gesamtheit der Konstruktion zur Aufnahme der Beläge für Fahrbahn(en), Bürgersteig(e) und Radweg(e)*] = (bridge) deck(ing)

~überbau *m* (**oder Brückentafel** *f*, **oder Brückenplatte** *f*) **mit ebenen Blechen die durch darunter geschweißte parallele Träger unterstützt sind**, ~ (~ ~, ~ ~) **mit ebenen Deckblechen auf Walzträgern** = battle deck (bridge) floor, battle(ship) (bridge) deck(ing)

~vermessung *f* = bridge survey

~waage *f*, Fahrzeugwaage = weighbridge, platform weighing machine

~wiederaufbauprogramm *n* = bridge rebuilding program(me)

~zeichnung *f* = bridge drawing

~zufahrt *f*, Brückenrampe *f* = bridge approach

Brüden *m*, Wrasen *m*, Schwaden *m* = water vapo(u)r

Brunnen *m* → Gründungs~

~ [*als Schachtbrunnen zum Aufsuchen oder zur Gewinnung von Grundwasser*] = well

~ [*Grundwasser(ab)senkung*] = drainage well [*There are two types of drainage well, viz. the wellpoint and the filter well. Drainage wells are commonly lined with steel tubes called well casings, which are perforated where they are in contact with water-bearing strata. If the casing has a diameter less than about $2^1/_2$ in., it is called a well point. If the diameter of a drainage well is 12 in. or more, the water is commonly pumped out through a suction tube with a very much smaller diameter, and the space between the tube and the wall of the hole is filled with coarse sand or gravel. Such wells are known as filter wells*]

~ **in artesisch gespanntem Grundwasser** = artesian well, ~ spring [*A well bored down to a point, usually at great depth, where the water pressure, owing to the conformation of the strata, is so great as to force the water to the surface*]

~ **mit gleichmäßigem Durchmesser** = well of one diameter

~absenkung *f* → Brunnenabteufung

~abteufung *f*, Brunnenabsenkung = well sinking, ~ lowering

~anlage *f* = well system

~anordnung *f* = grouping of wells

~ausbau *m* = development of a well [*The restoration or increase of its capacity*]

~auskleidung *f*, Brunnenbekleidung, Brunnenverkleidung = well lining

~ausmauerung *f* = well masonry

~batterie *f* = battery of wells

Brunnenbau *m* = construction of wells, well construction

~anlage *f* = well construction plant

Brunnenbauer *m*, Brunnengräber [*als Techniker*] = well builder

Brunnenbau|gerät *n* = well construction rig

~werkzeug *n* = well construction tool

Brückenbaustelle — Brückenrost

~baustelle *f* = bridge (construction) site, ~ building ~
~baustoff *m* = bridge building material, ~ construction ~
~bautechnik *f* = bridge construction practice, ~ building ~
~belag *m* → Brückenfahrbahnbelag
~belagbaustoff *m*, Brückendeckenbaustoff = (bridge) deck(ing) paving material (US); (~) ~ pavement ~, (~) ~ surfacing ~
~betonmischer *m* → Brückenmischer
~(beton)platte *f* [*im Sinne einer Betonplatte als Teil einer Brückentafel*] = concrete deck slab
~bildung *f* von Material = build-up of material, bridging (~ ~)
~blechträger *m*, Blechbrückenträger = plate girder span, bridge plate girder
~bogen *m* = bridge arch (span), arch span
~decke *f* → Brücken(fahrbahn)belag *m*
~deckenbaustoff *m* → Brückenbelagbaustoff
~dichtung *f*, Brückenabdichtung = bridge seal(ing)
~draht *m* = wire for bridge construction, ~ ~ ~ building
~einweihung *f* = bridge dedication
~endträger *m* = anchor span, bridge anchor girder
~entwässerung *f* = bridge drainage
~fachwerkträger *m* = bridge truss, truss-span
~fahrbahn *f* = bridge carriageway
~(fahrbahn)belag *m*, Brücken(fahrbahn)decke *f* = bridge carriageway surfacing, deck ~, bridge carriageway pavement, deck pavement, deck covering; bridge carriageway paving, deck paving (US)
~(fahrbahn)decke *f* → Brücken(fahrbahn)belag *m*
(~)Feld *n*, (Brücken)Öffnung *f* = (bridge) span, (~) opening

~fertiger *m* → Betondeckenfertiger
~geländer *n* = bridge railing
~gleis *n* = bridge track
~gradiente *f* = bridge gradient
~hals *m*, Schwanenhals [*Eine Tiefladebrücke ruht während des Transportes über anschraubbaren Brückenhälsen auf zwei Fahrwerken*] = gooseneck
~isolierung *f* = bridge insulation
~joch *n* = bridge bent
~kabel *n*, Brückenseil *n* = (suspension) bridge cable, (~) ~ rope, (~) ~ strand
~konstrukteur *m* → Brückenbauer *m*
~kran *m* = gantry crane
~lager *n* = bridge bearing
~leichtbelag *m*, Brückenleichtdecke *f* = lightweight bridge paving (US); ~ ~ pavement, ~ ~ surfacing
~mischer *m*, Autobahn-Betonmischer, RAB-Brückenmischer, Autobahn-(brücken)mischer, Brückenbetonmischer = bridge(-type) travel(l)ing concrete) mixer, bridge type travel(l)ing mixer plant, superhighway bridge-mixer
~modell *n* = bridge model
~montage *f* = bridge erection
~montagegerät *n* = erection rig, bridge assembly equipment
(~)Öffnung *f*, (Brücken)Feld *n* = (bridge) span, (~) opening
~pfahl *m* = bridge pile
~pfeiler *m* = bridge pier, ~ support
~pfeilerplatte *f* = bridge (pier) cap, ~ support ~
~pfropfen *m* [*Tiefbohrtechnik*] = bridge
~platte *f*, Brückenbetonplatte [*im Sinne einer Betonplatte als Teil einer Brückentafel*] = concrete deck slab
~platte *f* → Brückenüberbau *m*
~pylon *m* = bridge pylon
~querschnitt *m* = bridge cross-section
~rampe *f*, Brückenzufahrt *f* = bridge approach
~rost *m* = bridge grating

~scherspannung *f* = ultimate shear stress
~schleppung *f* (Geol.) = fault drag
~see *m* = fault basin lake
~sicherheit *f* = safety against rupture, ~ ~ failure
~sicherheitsgrad *m* = degree of safety against failure, ~ ~ ~ ~ rupture
(~)Sohle *f*, Steinbruchsohle = (quarry) floor
~spaltenbildung *f* (Geol.) = rifting
~spannung *f* = rupture stress, failure ~
~spannungsbedingung *f* = rupture stress condition, failure ~ ~
Bruchstein *m* = quarry stone
~ [*wird vom Maurer zurechtgerichtet wenn für Mauerwerk verwendet*] = rubble stone
~beton *m* = rubble concrete
Bruchsteine *mpl* = rubble
Bruchstein|gewölbe *n* = rubble vault
~mauer *f* = rubble wall
~mauerwerk *n* [*Aus natürlichen, lagerhaften Steinen mit Mörtel hergestellt*] = rubble masonry
~schüttung *f* = rubble fill
~(stau)mauer *f* in Bogenform = masonry arch gravity dam
~unterlage *f* = rubble bed
Bruch|stelle *f* = point of failure, ~ ~ rupture
~stempel *m* (⚒) = breaker prop, ~ post
~stempelreihe *f* (⚒) = breaker prop row, ~ post ~
~stück *n* [*Rohr*] = fragment (of pipe)
~tal *n* = fault(-zone) valley
~torf *m* = fen peat [*richer in mineral content than moor peat*]
~umwandlungstemperatur *f* für elastische und plastische Beanspruchung = fracture transition temperature for elastic and plastic loading
~versuch *m* → Bruchprüfung *f*
~vorgang *m*, Brechvorgang = breaking operation, crushing ~

~wand *f*, Stein~, (Abbau)Wand = (quarry) face
~werfen *n* → Hereinbrechen
~wert *m* = ultimate value
~widerstand *m* → Bruchlast *f*
~winkel *m* (⚒) = angle of break
~zeitpunkt *m* = moment of rupture, ~ ~ failure
~zone *f* = rupture zone, failure ~
~zone *f*, Sprödigkeitszone, Zone der Öffnungen, Ruschelzone, Zerrüttungszone (Geol.) = zone of fracture, fracture(d) zone
~-Zugkraft *f* [*Spannbetonkabel*] = ultimate strength
~zustand *m* = state of failure, ~ ~ rupture
Brücke *f* = bridge
~, Rammwagen *m*, Bühne *f*, Rammbrücke, Rammbühne, Schlitten *m*, Rammschlitten = additional transverse carriage on rails
~ auf Eis gebaut = ice bridge
~ mit einer Öffnung = single-span bridge
~ ~ großer Spannweite = long-span bridge
~ ~ mehreren Öffnungen = multi(ple)-span bridge
~ ~ oben liegender Fahrbahn, Deckbrücke = deck bridge
~ ~ Stahlüberbau = steel deck bridge
~ ~ untenliegender Fahrbahn → Trogbrücke
Brücken|(ab)dichtung *f* = bridge seal(ing)
~bau *m* = bridge construction, ~ building
~bauarbeiter *m* = bridgeman
~bauer *m*, Brückenkonstrukteur *m* = bridge constructor, ~ designer, ~ builder
~baufirma *f* = bridge building firm, ~ construction ~
~baugerät *n* = bridge building equipment, ~ construction ~
~bauingenieur *m* = bridge construction engineer, ~ building ~
~(bau)stahl *m* = bridge steel

Bruch — Bruchschermoment

~ = fracture [*The texture of the surface of a mineral broken across the normal line of cleavage*]

~, Bresche *f* [*Deich; Damm*] = breach, rupture

~, Fenn *n* = fen, marshy ground

~ → Hereinbrechen *n* (⚒)

~ = failure, rupture

~, Stein~ = (rock) quarry, stone ~

~beanspruchung *f* = ultimate stress and strain, failure ~ ~ ~, rupture ~ ~ ~

~bedingung *f* = ultimate condition, rupture ~, failure ~

~belastung *f* = ultimate loading, failure ~, rupture ~

~bemessung *f* = ultimate strength design

~biegemoment *n* = ultimate bending moment

~bild *n* = fracture pattern

~dehnung *f* = ultimate strain, failure ~, strain at failure, elongation at break

~ebene *f*, Bruchfläche *f* = plane of rupture, ~ ~ failure

~erscheinung *f* = phenomenon of rupture, ~ ~ failure

~faltengebirge *n* = mountain formed of disrupted folds

~feld *n* (⚒) = caved goaf, ~ gob [*Wales*]; ~ condie, ~ cundy [*Scotland*]; ~ waste (area)

~festigkeit *f* = ultimate strength

~festigkeit *f* [*Rohr*] = bursting strength

~festigkeit *f* **in sprödem Zustand**, Sprödbruch *m* = brittle fracture, ~ failure

~fläche *f*, Bruchebene *f* = plane of failure, ~ ~ rupture

~fläche *f*, Verwerfungsfläche (Geol.) = fault plane

~form *f* = type of failure, ~ ~ rupture

~gebirge *n* = faulted mountains

~gefahr *f* = risk of failure, ~ ~ rupture

~gestein *n* = quarry rock

~grenze *f* [*Druckspannung im Beton bzw. Zugspannung im Stahl, die den Bruch verursacht*] = limit of the ultimate strength, failure limit

~gut *n* [*das ungebrochene und ungesiebte Material aus einem Steinbruch*] = quarry run

~kante *f* [*des Hangenden*] = breaking edge [*of the roof*]

~kreis *m* [*Bodenmechanik*] = circle of rupture, ~ ~ failure

~kriterium *n*, Bruchmerkmal *n* = failure criterion

Bruchlast *f*, Grenzlast, Tragfähigkeit *f* beim Bruch, Bruchwiderstand *m* = breaking load, failure ~, ultimate ~, collapse ~

~ [*Pfahl*] = ultimate resistance

~ [*Drahtseil*] = nominal breaking strength

~theorie *f* = ultimate strength theory

Bruch|linie *f* [*Mohr'scher (Spannungs-)Kreis*] = envelope of failure, line of rupture, Mohr's envelope, strength envelope envelope of rupture

~linie *f*, Verwerfungslinie = fault line

~liniental *n* = fault line valley

~linientheorie *f* [*Berechnungsverfahren für Platten*] = fracture line theory

~massen *fpl* (⚒) = caved material, debris, débris

~merkmal *n*, Bruchkriterium *n* = failure criterion

~modul *m* = modulus of rupture

~moment *n* = ultimate moment

~ortversatz *m*, Blindortversatz (⚒) = dummy-road packing

~probe *f* → Bruchprüfung *f*

~prüfung *f*, Bruchprobe *f*, Bruchversuch *m* = failure test, rupture ~

~querschnitt *m* = failure section

~rechnung *f* [*Mathematik*] = fractions

~sattel *m* (Geol.) = fault saddle

~scherfestigkeit *f* = ultimate shear strength

~schermoment *n* = ultimate shear moment

~tür *f* = ledged door, batten ~
~tür *f* mit **Bug** = ledged and braced door [*A batten door which is diagonally braced between the ledges*]
~verkleidung *f*, Bretterauskleidung, Bretterbekleidung = board lining, ~ facing
~wand *f* = board partition
~zaun *m* = board fence
Brett|greifer *m* → Brettergreifer
~heber *m*, Unterlag~ [*Betonsteinindustrie*] = pallet lifter
~silo *m*, Unterlag~ [*Betonsteinindustrie*] = pallet feeder
~ste(i)g *m*, Bretterste(i)g = board runway
Briefeinwurf *m* = letter plate
Brikett *n*, Preßkohle *f*, Kohlenbrikett = (coal) briquet(te)
~fabrik *f*, Brikettwerk *n*, Preßkohlenfabrik, Preßkohlenwerk, Kohlenbrikettfabrik, Kohlenbrikettwerk = (coal) briquetting plant
Brikettier|kohle *f* = briquetting coal
~presse *f* → Brikettpresse
~probe *f* → Brikettierversuch
~verfahren *n* = briquetting method
~versuch *m*, Brikettierprobe *f*, Brikettierprüfung *f* = briquetting test
Brikett|pech *n* = briquetting pitch
~presse *f*, Brikettierpresse, Kohlen~, Kohlenpresse = briquetting press
~werk *n* → Brikettfabrik *f*
Brillenflansch *m* = tongued flange
Brinell|-Härte *f*, HB = Brinell hardness
~-Härtezahl *f* = B. H. N. [*Brinell Hardness Number*]
~presse *f*, Brinell'sche Kugeldruckpresse = Brinell hardness testing machine
~-Probe *f*, Brinell-Prüfung *f*, Brinell-Versuch *m*, Kugeldruckversuch nach Brinell, Kugeldruckprobe nach Brinell, Kugeldruckprüfung nach Brinell = Brinell hardness test
Bringungsweg *m* [*in Österreich*] → Förderweite *f*
brisanter Sprengstoff *m*, detonierender ~, hochbrisanter ~ = high (strength) blasting explosive

Brisanz *f*, Brisanzwert *m* [= *Ladedichte* × *spez. Druck* × *Detonationsgeschwindigkeit*] = brisance
britisches Kartennull *n* = Ordnance Datum
Brock *m* → → Grobkies *m*
Bröckelfels *m* = crumbly rock
bröckelig = crumbly
Bröckel|torf *m* = crumble peat
~tuff *m*, Puzzolanerde *f* = pozzuolana, pozzolana, puzzolano, puzzolana
Brocken|material *n*, Tropfkörpermaterial = filtering material, contact ~
~mergel *m* = clastic marl
Brodel-Luftanschluß *m* [*Druckluftbetonförderer*] = booster air pipe
~vorrichtung *f* [*Druckluftbetonförderer*] = booster
Bromgold *n* = gold bromide
Bromzahl *f* = bromine number
Bronze|draht *m* = bronze wire
~dübel *m* = bronze dowel
~farbe *f* = bronze paint
~fenster *n* = bronze window
~folie *f* = bronze foil
~form *f* = bronze mo(u)ld
~profil *n* = bronze section
~pulver *n* = bronze powder
~ring *m* = bronze ring
~schraube *f* = bronze screw
~tür *f* = bronze door
~(unterleg)scheibe *f* = bronze washer
~-Wolkenkratzer *m* = bronze skyscraper
bronzieren = to bronze
Bronzieren *n*, Bronzierung *f* = bronzing
Brownmillerit *m*, Tetrakalziumaluminatferrit *n*, 4 CaO x Al_2O_3 x Fe_2O_3 [*abgekürzt C_4FA oder $CeAF$*] = tetracalcium aluminoferrite
~zement *m*, Ferrarizement = tetracalcium aluminoferrite cement
Brown'sche Bewegung *f* = Brownian movement, ~ motion
Bruch *m*, Verwerfung *f*, Abschiebung, Sprung *m*, Paraklase *f*, Absenkung [*Grabenrand*] (Geol.) = fault (plane), geological fault, faulting

Brenngut — Bretterste(i)g 276

~gut n, gebranntes Gut = burned product

~härtemaschine f, Flammhärtemaschine, Autogenhärtemaschine = flame-hardening machine, autogenous hardening ~

~härten n, Flammhärten, Autogenhärten = autogenous hardening, flame ~

~hilfsmittel npl [Keramik] = kiln furniture

~kalk m → Branntkalk

~kammer f = combustion chamber

~kapsel f [Keramik] = saggar, sagger

~kurven fpl = burning diagram

Brennofen m [Keramikindustrie] = burning kiln

~, Röstofen, Kalzinierofen [Verhüttung] = calcining kiln

Brenn|öl n [Rüböl, das früher vorwiegend als Leuchtöl benutzt wurde. Heute ist es durch Leuchtpetroleum völlig verdrängt] = burning oil

~prozeß m, Brennvorgang m = burning process

~punkt m = fire point [deprecated: ignition point]

~raum m, Kalzinierraum = calcining compartment

~riß m, Brandriß [Keramik] = burning crack

~schiefer m, Brandschiefer = combustible shale

brennschneiden, schneidbrennen = to flame-cut, to oxygen-cut

Brenn|schneiden n [früher: Autogenes Schneiden] = autogenous cutting

~schneider m, Schneidbrenner m = autogenous cutter, ~ cutting blowlamp

~schneider m, Brenner m [als Arbeiter] = cutter

~schneidemaschine f = autogenous cutting machine

~schneid- und -schweißgerät n = autogenous cutting and welding apparatus

~schwinden n = burning shrinkage

~spiritus m = methylated spirit

~spirituskocher m = primus stove (Brit.)

Brennstoff m = fuel

~chemie f = fuel chemistry

~element n, Brennstoffzelle f = fuel element, ~ cell

~-Geologie f = fuel geology

~nebel m = atomized fuel spray

~pumpenhebel m [Dieselbär] = injection pump lever

~raum m [bei einem Kesselhaus] = fuel room

~technologie f = fuel technology

~versorgung f = fuel supply

~zelle f, Brennstoffelement n = fuel cell, ~ element

Brenn|temperatur f = burning temperature

~torf m = combustible peat

~vorgang m, Brennprozeß m = burning process

~wachsen n = burning expansion

Brennzone f = burning zone

~, Kalzinierzone, Röstzone, (Ver-)Glühzone [Verhüttung] = calcining zone, calcination ~

Bresche f, Bruch m [Deich; Damm] = rupture, breach

~ = breach

Breschmauer f, entlastete Futtermauer = counter-arched revetment

Bretschie f → Brekzie

Brett n, Unterlag~ [Betonsteinindustrie] = pallet

~ zur Aufnahme der Füllung [Holzbalkendecke] = pugging board [wood-joist floor]

Bretter|auskleidung f → Bretterverkleidung

~bekleidung f → Bretterverkleidung

~bude f, Bretterhütte f, Bretterschuppen m = wood(en) hut, timber ~

~greifer m, Brettgreifer = board grapple

~hütte f → Bretterbude f

~schuppen m → Bretterbude f

~steg m, Schubkarrenbahn f, Schubkarrensteg m = (wheel) barrow run

~ste(i)g m, Brettste(i)g = board runway

Bremsschaltventil — Brenngrad

~schaltventil *n* = brake selector (valve)
~scheibe *f* mit Belag = brake disc with lining; ~ disk ~ ~ (US)
~schlauch *m* = expander tube
~schlupf *m* = braking slip
~schlüssel *m* = brake spanner
~schuh *m*, (Hemm)Schuh, Bremskeil *m* = brake shoe, skid-pan, wheel block, chock
~seil *n*, Bremskabel *n* = brake cable
~sektor *m*, Bremszahnbogen *m* = brake pinion, ~ sector
~servovorrichtung *f*, Bremshilfe *f* = brake booster
~sicherungshebel *m* = brake lock
~spur *f*, Bremseindruck *m* = (braking) skid mark
~stahllamelle *f* = brake steel plate
~stand *m*, Prüfstand = test stand, ~ bench, ~ bed; torque stand (US)
~stange *f* = brake rod
~steuerung *f* = brake control
~träger *m* = girder to resist braking
~trägerverband *m* = group of girders to resist braking
~trommel *f* = brake drum
~ventil *n* = brake valve
~verband *m* = braking bracing
~verlustzeit *f* = brake lag
~versuch *m* → Auslaufversuch
~verzögerung *f* = braking inertia
~vorgang *m* = braking action
~weg *m* = braking distance
~welle *f* = brake shaft
~wirkung *f* = braking efficiency
~zone *f* [*Staudamm*] = transition zone
~zylinder *m* = brake cylinder
Brennanlage *f* = burning plant, ~ installation
brennbar = combustible
Brenn|bereich *m* [*Keramikindustrie*] = burning range, maturing ~
~bohren *n* → Feuerstrahlbohren
brennen [*Keramikindustrie*] = to burn ~, ab~ [*Zündschnur*] = to burn [*fuse*]
Brennen *n*, Ab~, Abbrand *m* [*Zündschnur*] = burning

~, Kalzinieren, Rösten, (Ver)Glühen [*Verhüttung*] = calcining, calcination
~ **des Kalksteins**, Kalzinieren ~ ~ = burning of limestone, calcination ~ ~, calcining ~ ~
brennende Bohrung *f*, ~ Sonde *f*; brennender Schacht *m* [*Galizien*] = burning well, well on fire
~ **Sonde** *f* → brennende Bohrung *f*
brennender Schacht *m* [*Galizien*] → brennende Bohrung *f*
Brenner *m* = burner
~ = blowlamp [*for paint burning, plumbing, silver soldering, etc.*]
~, Brennschneider *m* [*als Arbeiter*] = cutter
~ **mit Gebläse** [*Schwarzbelageinbaumaschine*] = heater and blower unit
~bohrer *m* = combination burner tool
~einbruch *m*, Michiganeinbruch = burn cut, shatter ~, Michigan ~ [*Cut holes for tunnel blasting which are heavily charged, close together, and parallel. About four cut holes are used. They pull a cylindrical hole of completely shattered rock which provides an excellent free face for the remaining holes*]
~einbruch-Bohrloch *n*, Michiganeinbruch-Bohrloch = cut hole (US); easer (Brit.)
~flamme *f* = burner flame
~gebläse *n* = burner blower
~muffelstein *m* = burner muffle block
~mündung *f*, Brennermaul *n* = burner opening
~rampe *f* = bank of burners
~schlauch *m* = burner hose
~spitze *f*, Brennermundstück *n* = tip
Brenn|farbe *f* = burning colo(u)r
~gas *n* = fuel gas
brenngeschnitten = flame-cut, oxygen-cut
Brenn|geschwindigkeit *f*, Ab~, Abbrandgeschwindigkeit [*Zündschnur*] = burning speed, ~ rate, rate of burning, speed of burning
~grad *m* = degree of burning

Schubstange *f*; Stoßbarren *m* [*Österreich*]	pusher bar
Planier-Gleiskettengerät *n* [*früher: Planierraupe*] mit (Unterholz-) Roderechen	caterpillar (bull)dozer with brush rake (or clearing rake)
Planier-Gleiskettengerät *n* [*früher: Planierraupe*] mit schneepflugartiger Anordnung des Planierschildes, Trassenschäler *m*, Planier-Gleiskettengerät mit Schwenkschild, Seitenräumer *m*	roadbuilder, trailbuilder, gradebuilder, angledozer, (bull)dozer with angling blade, sidedozer, crawler tractor fitted with hydraulic angledozer, angle-blade dozer, bullgrader [*the word "angledozer" is also used for the equipment which, when mounted on a tractor, constitutes a complete machine*]
Erdschürftransport *m*	(bull)dozing
Seitenräumer-Planierschild *n*, Schwenkschild, winkelbares Schild, Schrägschild, Winkel-Planierschild, Schwenk-Vorbauschild, winkelbare Planiervorrichtung *f*, winkelbarer Schild *m*	angledozer, angling blade, angle dozer blade
Planierschild *n*, Brustschild, Fronträumer-Planierschild, Vorbau-Brustschild, Frontschild, Bulldozer-Räumschild, Schürf-Brustschild, Schürfschlepper-Brustschild, Bulldozer-Brustschild	dozer blade, (straight) pusher-blade (Brit.); (dozer) apron, (bull)dozer (blade)
Bulldozer(-Lade(schaufel *f*, Bulldozer-Schaufellader *m*	(bull)dozer-shovel, dozer-loader
Querschild-Abstützung *f*	(bull)dozer stabilizer
Felsrechen *m*, Steinharke *f*, Felsharke, Steinrechen	rock rake
Bulldozer(-Lade)schaufeleinrichtung *f*	shovel dozer attachment
Heckaufreißer *m*	rear scarifier (attachment)
Querschild *n*, festes Schild, Gerad-Schild	(bull)dozer [*The word "(bull)dozer" is also used for the equipment which, when mounted on a tractor, constitutes a complete machine*]
Seilsteuerwinde *f*	cable control unit

Bund *m*, Wellen~ = (shaft) collar

~**balken** *m* → Binderbalken

~**bolzen** *m* = collar stud, flange bolt, shoulder stud

Bündel *n* **aus Drähten**, Drahtbündel, Spannkabel *n*, Spann(draht)bündel, Vorspannbündel, Stahldrahtbündel, Spannbetonkabel, Vorspannkabel = prestressing (strand) cable

Bündeldraht — buntknochiger Bernstein

~draht *m*, (Vor)Spann(bündel)draht, Spann(beton)(kabel)draht = prestressing wire
~ende *n* [*Spannbeton*] = cable end
~leiter *m* = bundle(d) conductor
~pfeiler *m* = multiple rib pillar
Bundes|autobahn *f*, BAB = Federal motorway
~(fern)straße *f* = Federal highway, ~ road
~straße *f* → Bundesfernstraße
~wasserstraße *f* = Federal waterway
Bund|flansch *m* = union flange, coupling ~
~förderer *m* = coil conveyor, ~ conveyer
~gespärre *n*, Bindergespärre = principal rafters, principals
~greifer *m* = coil grab
~holz *n* → Bindeholz
bündig mit, niveauebven = flush to, ~ with, dead level
Bund|lager *n* = collar end bearing
~pfosten *m* → Binderstiel *m*
~ring *m* = end ring, ~ collar
~säule *f*, Bundständer *m* [*Fach(werk)wand*] = stud, principal post
~schraube *f* = collar screw
~stahl *m* = fag(g)ot steel
~ständer *m* → Bundsäule *f*
Bungalow *m* = bungalow
Bunker *m*, Material~, Trichter~ = bin, hopper
~ = bunker [*shallow container*]
~abzug *m* = drawing from bin(s), ~ ~ hopper(s)
~abzugsorgan *n* = bin drawing device, hopper ~ ~
~abzugsrinne *f* = bin drawing channel, hopper ~ ~
~aktivität *f* = free-running bin capacity, live ~ ~, ~ hopper ~
~anlage *f* = bin plant, ~ installation, hopper ~
~auslauf *m* = bin outlet, hopper ~
~batterie *f*, Bunkergruppe *f* = group of bins, ~ ~ hoppers
~beschicker *m*, Bunkeraufgeber, Bunkerspeiser = bin feeder, hopper ~
~boden *m* = bin bottom, hopper ~

~entleerung *f* = bin discharge, hopper ~
~form *f* = shape of bin, ~ ~ hopper, bin shape, hopper shape
~füllung *f*, Bunkergut *n* = bin filling, hopper ~
~grube *f*, Erdbunker *m*, Tiefbunker, versenkter Bunker [*Materiallagerung*] = storage pit, bunker in the ground
~gruppe *f* → Bunkerbatterie *f*
~gut *n*, Bunkerfüllung *f* = bin filling, hopper ~
~lagerung *f*, Bunkerung *f* = bin storage, hopper ~
~öl *n* = bunker oil
~rüttler *m*, Bunkervibrator *m*, Klopfhammer *m*, Haftrüttler, Haftvibrator = hopper vibrator, bin ~
~stand *m*, Füll(ungs)grad *m*, Füllstand = material level (in bin), ~ (~ hopper), bin level, hopper level
~standanzeiger *m*, Bunkerstandwächter *m*, Anzeigevorrichtung *f* für Bunkerfüllungen = (bin) level detector, storage (~) ~ ~
~standanzeiger *m* mit umlaufendem Flügel = rotating-paddle type of material-level indicator
~tasche *f*, Silotasche, Bunkerwabe *f*, Silowabe *f* = bin compartment, hopper ~
Bunkerung *f* → Bunkerlagerung
Bunker|verschluß *m* = bin gate, hopper ~
~vibrator *m* → Bunkerrüttler *m*
~waage *f* → Behälterwaage
~wabe *f* → Bunkertasche *f*
~wagen *m* = travel(l)ing hopper, ~ bin
~wand *f* = bin wall, hopper ~
Bunsenbrenner *m* → Blaubrenner
Bunt|asphalt *m* → Farbasphalt
~beton *m*, Farbbeton, farbiger Beton, bunter Beton = colo(u)red concrete
bunter Verband *m* (Geol.) = poikiloblastic texture
buntknochiger Bernstein *m* = mottled osseous amber

Buntmetall *n* = brass and bronze
~**erzeugnis** *n* = non-ferrous casting
Bunt|pigment *n* = tinting pigment, staining ~ [*is used to colour the paint*]
~**sandstein** *m* = variegated sandstone (Brit.); mottled ~, bunter ~, Lower Triassic
~**wacke** *f* = sparagmite
Burette *f* = buret(te) [*A graduated glass tube with a stopcock at the bottom, used by chemists, etc. for measuring small quantities of liquid or gas*]
Bürgersteig *m*, Gehsteig, Gehbahn *f*, Fuß(gänger)weg *m*, Gehweg = footway (Brit.); sidewalk, banquette (US) [*That part of the street reserved for pedestrians*]
~-**Betonplatte** *f*, Beton-Bürgersteigplatte, Gehsteig-Betonplatte, Beton-Gehsteigplatte, Gehweg-Betonplatte, Beton-Gehwegplatte, Gehbahn-Betonplatte, Beton-Gehbahnplatte, Fußweg-Betonplatte, Beton-Fußwegplatte = concrete flag(stone)
~**konsole** *f* → Gehwegkonsole
~**platte** *f* → Gehbahnplatte
~**schalter** *m* [*Bank*] = sidewalk teller
Burgunderpech *n* = Burgundy pitch
Büro *n* = office
~**baracke** *f* = office hut
~**block** *m* = office block
~**etage** *f* → Bürogeschoß *n*
~**gebäude** *n*, Bürohaus *n* = office building
~**geschoß** *n*, Büroetage *f*, Bürostockwerk *n* = office stor(e)y, ~ floor
~**haus** *n*, Bürogebäude *n* = office building
~**hochhaus** *n* = office tower
~**ingenieur** *m* = offsite engineer
~**personal** *n* = office staff
~**schalldämmung** *f* = office quieting
~**stockwerk** *n* → Bürogeschoß *n*
~- **und Geschäftsviertel** *n* = shopping and business area
~- **und Wohnhaus** *n*, Wohn- und Bürohaus = flats-and-offices building, office-and-flat block
~**wagen** *m* = office trailer, mobile site office

~**wolkenkratzer** *m* = office skyscraper
Bürste *f* = brush
Bürsten|belüftung *f* = brush aeration
~**motor** *m* [*einer Spachtelmaschine*] = brush motor
~**sieb** *n* = brush screen
~**walze** *f* [*Belebungsverfahren*] = brush-aerator
~**walze** *f*, Kehrwalze, Fegewalze, Besenwalze = rotating brush, rotary ~
~**waschbeton** *m* = scrubbed concrete
Bus *m* → Auto~
Busbahnhof *m*, Omni~, Auto~, Autohof = bus terminal, ~ terminus
Busch|land *n* = bushland
~**packwerk** *n*, (Pack)Faschinat *n*, (Faschinen)Packwerk, Faschinenlage *f* = fascine work
~**pflüger** *m* → Buschschneider *m*
~**räumer** *m* → Buschschneider *m*
~**räumung** *f* = bush clearing
~**schneider** *m*, Buschpflüger *m*, Buschräumer *m*, Gestrüpp-Pflug *m* = brush cutter
Bushaltestelle *f*, Omni~, Auto~ = bus stop, ~ halt
Buslinie *f*, Auto~, Omni~ = bus route
Butan|brenner *m* = butane blowlamp [*for paint burning, plumbing, silver soldering, etc.*]
~**gaslager** *n* = butane storage
Butylkautschuk *m* = butyl rubber
BV *m* → Betonverflüssiger *m*

C

Caisson *m*, Druckluftsenkkasten *m*, Preßluftsenkkasten, pneumatischer Senkkasten = (pneumatic) caisson, (compressed-)air ~
~**arbeiter** *m*, Senkkastenarbeiter = sand hog
~**decke** *f*, Senkkastendecke = caisson ceiling
~**gründung** *f* → pneumatische Gründung
~**krankheit** *f*, Preßluftkrankheit, Drucklufterkrankung *f*, Sekkastenkrankheit = compressed-air sick-

ness, caisson disease, compressed air illness, bends, compressed-air disease

~schneide *f*, (Senk)Kastenschneide = caisson edge

Calgon *n* = calgon, sodium hexametaphosphate

Caliche *f* → erdiges Rohsalz *n*

Calyx-Bohrer *m* = calyx core drill

Candela *f*, cd. [*Einheit der Lichtstärke*] = candela, cd.

Carbonado *m*, schwarzer Diamant *m*, Karbon *m* = carbonado diamond

Carcasse *f*, Karkasse [*Gewebeunterbau des Reifens als Festigkeitsträger*] = casing

Cardox|rohr *n* = Cardox pipe

~-**Schießen** *n*, Cardox-Verfahren *n* = Cardox method

Carnaubawachs *n*, Karnaubawachs = Carnauba (wax)

Castigliano-Prinzip *n* [*Von allen möglichen Gleichgewichtszuständen tritt derjenige wirklich ein, für den die Formänderungsarbeit ein Minimum wird*] = minimizing principle of Castigliano

CBR-Form *f* = CBR mo(u)ld

CBR-Gerät *n* mit Dreifuß und Meßuhr zum Messen der Probenschwellung unter Wasser = CBR expansion test equipment

CBR-Verfahren *n* = California Bearing Ratio (and Expansion) Test

CBR-Wert *m*, kalifornisches Tragfähigkeitsverhältnis *n*, kalifornischer Index *m* [*rein empirischer Tragfähigkeitsbeiwert für den Unterbau und Untergrund nichtstarrer Decken*] = CBR value

Centipoise *f* = centipoise

Cermet *n* auf Carbidbasis = carbide-base(d) cermet

~**überzug** *m* = cermet coating

Chalkopyrit *m* → Kupferkies *m*

Chance|-Sinkscheider *m*, Chance-Tieftrogscheider, Chance-Kegel *m* = Chance cone separator, ~ washer [*cone-shaped coal washer*]

~-**Verfahren** *n* = Chance process, sand flotation

Chanoine'sches Klappen(stau)wehr *n* = Chanoine wicket (dam) (US); ~ shutter (~) (Brit.) [*It consists of a curtain formed by timber leaves or wickets about 3 ft. 8 in. in width and inclined somewhat downstream*]

Chanoine-Schütz(en)tafel *f*, Chanoine-Stautafel, Chanoine-Schütz(e) *n*, (*f*) = Chanoine wicket (US); ~ shutter (Brit.)

Charakteristik *f* eines Stempels, Kennlinie *f* ~ ~, Lastwegkurve *f* ~ ~ (⚒) = resistance-yield curve, load-yield ~

charakteristischer Reststrom *m* [*Meer*] = characteristic residual current

Charge *f* = batch

~, **Beton~** = (concrete) batch

Chargen|beschickerwerk *n* [*Betonmischer*] = batch elevator, skip-hoist

~-**Betonmischer** *m* mit Wendetrommel *m* → ~ Betonmischer

~**bunker** *m* = batch holding bin, ~ ~ hopper

~**förderung** *f*, Chargentransport *m* = batch handling

~-**Last(kraft)wagen** *m*, LKW *m* mit Zuschlagstoffabteilen, Last(kraft)-wagen zur Aufnahme mehrerer Chargen = batch truck (US); ~ lorry (Brit.)

~**mahlung** *f* → satzweise Mahlung

~**messer** *m* → Chargen-Zählvorrichtung *f*

~-**Mischanlage** *f*, absatzweise arbeitende Mischanlage, diskontiunierliche Mischanlage = batch(-mix) type plant, intermittent weigh-batch (mixing) ~

chargenmischen = to batch-mix

Chargen|mischer *m* → Periodenmischer

~**mischzeit** *f* = batch mixing time

~**mischzeitmesser** *m* = batchmeter

~**registriergerät** *n* → Chargenzählvorrichtung

~-**Schaufelwellenmischer** *m* = batch (type) paddle mixer

~transport *m*, Chargenförderung *f* = batch handling
~(ver)wiegen *n* = batch weighing
~(ver)wiegevorrichtung *f* = batch weighing equipment
~volumen *n* = batch volume
chargenweise = batch-wise
Chargen|-Zählvorrichtung *f*, Chargenzähler *m*, Chargenmesser *m*, Chargenregistriergerät *n* = batch counter, ~ recorder
~zusammensetzung *f* = composition of batch, batch composition
~zwangsmischen *n*, absatzweise Zwangsmischung *f*, diskontinuierliche Zwangsmischung = batch pugmill mixing
~zwangsmischer *m* = batch pugmill
Chargierkran *m*, (Mulden)Beschickungskran = charging crane
Chassis *n* → Fahrgestell *n*
Chaussierung *f* → Steinschlagdecke
Chefingenieur *m* = chief engineer
Chefmonteur *m* [*Schweiz*]; Montagemeister *m* [*Stahlbaumontage*] = chief steel erector
Chemie|abfall *m* → Chemiemüll *m*
~bau *m* = chemical construction
~kalk *m* = chemical lime
~kombinat *n* = chemical combine
~müll *m*, chemischer Müll, Chemieabfall *m*, chemischer Abfall = chemical refuse
~werk *n* → chemische Fabrik *f*
Chemikal *n* = chemical
Chemikalienauspressung *f* → Chemikalienverpressung
chemikalienbeständig = resistant to chemical(s)
Chemikalien|dosierung *f* = chemical dosing
~einpressung *f* → Chemikalienverpressung
~gebäude *n* = chemical building [*water filtration plant*]
~injektion *f* → Chemikalienverpressung *f*
~schlämme *f* = chemical grout, ~ slurry

~verpressung *f*, Chemikalienauspressung, Chemikalieneinpressung, Chemikalieninjektion *f* = chemical injection, ~ grouting
chemisch beständiger Stahl *m*, nichtrostender ~ = stainless steel
~ gebundener Wasserstoff *m* = chemically combined hydrogen, fixed ~
chemische Analyse *f* = chemical analysis
~ Ausfällung *f*, ~ Klärung [*Abwasser*] = chemical precipitation [*Sedimentation accelerated by the coagulation of suspended or colloidal matter through the addition of chemicals*]
~ Betontechnologie *f* = chemical concrete technology
~ Fabrik *f*, chemisches Werk *n*, Chemiewerk = chemical plant
~ Gesteinsauflösung *f* → ~ Verwitterung
~ Kinetik *f* = chemical kinetics
~ Klärung *f* → ~ Ausfällung
(~) Umsetzung *f* [*Schwarzpulver*] = reaction
~ Verbindung *f* = chemical compound
~ Verfestigung *f* = chemical stabilization
~ Verwitterung *f*, Gesteinszersetzung, chemische Gesteinsauflösung = chemical weathering
~ Vorspannung *f* = chemical prestressing
(~) Zersetzung *f* [*Sprengstoff*] = decomposition
~ Zusammensetzung *f* = chemical composition
chemischer Angriff *m* = chemical attack
chemisches Schüttgut *n* = chemical bulk product
~ Sediment(gestein) *n*, (Aus)Fällungsgestein, Präzipitatgestein, Ausscheidungssediment(gestein) *n*, chemisches Absatzgestein = chemically deposited sedimentary rock, ~ formed ~
~ Steinzeug *n* = chemical stoneware
~ Werk *n* → chemische Fabrik *f*

Chezy|-Beiwert m = Chézy('s) coefficient

~**-Formel** f = Chézy('s) formula

chinesisches Wasserrad n **aus Bambus** = Chinese bamboo water wheel, ~ noria, ~ float wheel

Chlor n = chlorine

Chloratit m, **Chlorat-Sprengstoff** m = chlorate blasting explosive

Chlorator m, **Chlorzusatzgerät** n = chlorinator

Chlor|bindungsvermögen n = chlorine absorptive property, ~ demand, I.C.D., initial chlorine demand

~**blei** n, Bleichlorid n = phosgenite

chloriertes Lösungsmittel n = chlorinated solvent

Chlor(ier)ung f = chlorination [*Treatment with chlorine or bleaching powder for the purpose of disinfection, the retardation of decomposition, or the oxidation of organic matter*]

Chlorit m (Min.) = chlorite, green earth

~**bildung** f = chloritization

chloritisches Tonmineral n = chloritic mineral

Chloritschiefer m = chlorite-schist

Chloritoidschiefer m = chloritoid schist

Chlor|kalk m → Bleichkalk

~**kautschuk** m = chlorinated rubber

~**kautschuk(anstrich)farbe** f = chlorinated rubber paint

~**magnesiumlauge** f, Magnesiumchloridlösung f = liquid magnesium chloride

~**methyl** n → Methylchlorid n

~**silber** n, AgCl = silver chloride

Chorgang m = ambulatory aisle [*basilica*]

C-Horizont m, **Rohboden** m [*Er wird von der Verwitterung nicht oder nur wenig angegriffen, weil er von Luft Wasser, Wind, Wärme und Frost kaum erreicht wird*] = C horizon

Chromeisen|erz n → Chromit m

~**stein** m → Chromit m

Chrom|(erz)stein m = ferrochrome brick

~**farbe** f = chrome pigment

~**gelb** n, Parisergelb, chromsaures Blei n, Bleichromat n = chrome yellow, Leipzig ~, lead chromate, PbCrO$_4$

~**grün** n, Milorigrün = chrome green

Chromit m, Chromeisenerz n, Chromeisenstein n = chromite

~**-Dolomit-Erzeugnis** n = chromite-dolomite refractory

~**-Erzeugnis** n = chromite refractory

~**-Korund-Erzeugnis** n = chromite-corundum refractory

~**-Magnesia-Erzeugnis** n = chromite-magnesia refractory

~**-Silika-Erzeugnis** n = chromite-silica refractory

~**-Sillimanit-Erzeugnis** n = chromite-sillimanite refractory

~**stein** m = chromite brick

Chrommagnesit m = chrome magnesite

~**stein** m = chrome-magnesite brick

Chrom|mörtel m = chrome mortar

~**oxyd** n = oxide of chromium, chromium oxide

~**rot** n = chrome red

~**stein** m, Chromerzstein = ferrochrome brick

Chrysokoll n, Berggrün n = chrysocolla

Chrysotilasbest m, Kanadaasbest, H$_4$Mg$_3$Si$_2$Og (Min.) = Canadian asbestos, chrysotile ~

CH$_4$-Gehalt m **der Wetter** = percentage of gas in the air, amount ~ ~ ~ ~ ~

CH$_4$-Meßgerät n = firedamp indicating detector [*It shows or attempts to show the percentage of gas in the air*]

CH$_4$-Schreiber m = recording firedamp indicating detector

Cif-Wert m **Maschinen** = landed cost of plant and equipment

Cipollettiwehr n = Cipolletti (trapezoidal) weir

Clapeyron'sche Gleichung f → Dreimomentengleichung

cm^3-Teilung f = cubic centimetre graduation (Brit.); ~ entimeter ~ (US)

Colcrete|-Beton m = Colcrete

~-Ein-Trommelmischer m = single-drum Colcrete mixer
~-Pfahl m = Colcrete pile
~-Rollermischer m = Colcrete roller mixer
~-Verfahren n = Colcrete process
~-Zweitrommelmischer m = double-drum Colcrete mixer
COLGROUT-Mörtel m, Einpreßmörtel [*Prepaktbeton*] = COLGROUT
Colgrunite-Verfahren n, Naßspritzverfahren mit kolloidalem Mörtel und Beton hergestellt nach dem Colcrete-Verfahren = Colgunit process
Container m einer Reaktoranlage, Einschlußbauwerk n ~ ~ = (atomic) reactor containment structure, container, reactor housing
Containerschale f = reactor shell
~ mit eingebettetem Boden = embedded reactor shell
~ ~ elastischer Zwischeneinspannung = elastically supported reactor shell
Contractor-Verfahren n, Kontraktorverfahren = tremie method
Conferventorf m = conferva peat
Cordband n = cord belt
Cordgewebe n = cord fabric
Cordierit m, $Mg_2Al_3[AlSi_5O_{15}]$ = cordierite
Coslettieren n = Coslettizing
coslettiert = Coslettized
Cottrellfilter m, n, Industrieelektrofilter = Cottrell (pipe) precipitator, single-stage ~
Coulomb'sche (Erddruck)Theorie f = Coulomb's theory, wedge ~, Coulomb's sliding wedge analysis
~ Gleichung f = Coulomb's equation
Coulomb'scher Modul m = Coulomb's modulus
Coulomb'sches Bruchprisma n = Coulomb's soil-failure prism
Cowper m, Winderhitzer m = Cowper
Craeger-Schwelle f = Craeger's sill
Cremona'scher Kräfteplan m, Kräfteplan nach Cremona [*zeichnerisches Verfahren zur Ermitt(e)lung der Stabkräfte eines Fachwerkträgers*] = Cremona's polygon of forces

Cross-Verfahren n, Cross-Methode f [*Rechnungsverfahren zur Ermitt(e)lung der Momente von Rahmen und Durchlaufträgern. Ein Iterationsverfahren mit schrittweiser Verbesserung der Ergebnisse*] = Cross method
CRYPTO-System n [*schwedisches Ölbrennverfahren für Ring- und Zickzacköfen* [*Trademark*]] = Swedish impulse oilfiring system [*brick manufacture*]
Cu/CuSO$_4$-Halbzelle f = copper-copper sulfate half-cell
Culmann'sche E-Linie f = Culmann line
Culmann-Verfahren n = Culmann's graphical construction, ~ ~ method, ~ ~ procedure
Cutter(saug)bagger m, Cuttersauger, (stationärer) Schneidkopf(saug)bagger, (stationärer) Saugbagger mit Schneidkopf = suction-cutter dredger, cutter suction ~, cutterhead (pipeline) (hydraulic) ~, clay cutter (suction) ~, hydraulic pipe line ~ [*US = dredge*]
Cyanidlaugung f = cyaniding

D

D, Deklination f, Mißweisung f [*Die Abweichung gegenüber dem geodätischen Meridian*] = declination
Dach n = roof
~, Gebäude~ = (building) roof
~ mit Wiederkehr, Wiederkehrdach, Kehlendach = intersecting roofs, roof with valley
~(ab)dichtung f, Dachisolierung = roof insulation
~(ab)dichtungsbahn f = roof insulation strip, ~ sealing ~
~ablauf m, Dachentwässerung f = roof drainage
~anschluß m = roof connection
~antenne f = overhouse aerial, roof ~
~aufsatzfenster n → Dachkappe
~ausmitt(e)lung f → Dachzerlegung
~ausstieg m, Dachausstiegluke f = exit opening, trap door on roof

Dachaufsatzfenster — Dachhöhe

~aufsatzfenster n → Dachkappe f
~ausbesserung f = roof repair
~bahn f, (rollbare) (bituminöse) ~ = prepared roofing, roll ~, ready ~, composition ~; strip of asphalt-saturated felt (US) [*Prepared roofings consist of asbestos felt or rag felt saturated with asphalt (US)/bitumen (Brit.) and assembled with asphalt (US)/bitumen (Brit.) at the factory to form strips of about 1 yd. wide and 12 yd. long*]
~balken(träger) m = roof-beam
~bau m = roof construction
~baustoff m = roof material
~belag m → Dachdeckung f
~(belag)folie f = roof foil
~beleuchtung f = daylighting by the roof
Dachbinder m → Binder
~abstand m → Binderabstand
~form f = roof truss mould (Brit.); ~ ~ mold (US)
~verteilung f → Binderverteilung
Dach|bitumen n = roofing asphalt (US); ~ asphaltic bitumen (Brit.)
~blech n = roof plate
~boden m → Dachgeschoß n
~decke f, Unterkonstruktion f = roof deck
Dachdecker m = roofer
~hammer m = pick hammer
~kran m = roofer's crane
~nagel m = roofer's nail
Dachdeckung f, Dacheindeckung, Dachhaut f, (Ein)Deckung, Dachbelag m = roof cladding, ~ covering, roofing
Dachdeckungs|arbeiten fpl = roofing work
~material n → Dach(ein)deckungsstoff m
~stoff m → Dacheindeckungsstoff
Dach|dichtung f, Dachabdichtung, Dachisolierung = roof insulation
~dichtungsbahn f, Dachabdichtungsbahn = roof insulation strip, ~ sealing ~
~eindeckung f → Dachdeckung
(~)Eindeckung f, Bedachung, Bedachen n, Eindecken = placing the roofing
~(ein)deckungsstoff m, Dach(ein)deckungsmaterial n = roofing product, ~ material
(~)Eindeckungszubehör m, n, Bedachungszubehör = roofing accessories
~element n = roof member
~element n, Dachplatte f, Dachtafel f, Bedachungsplatte = roof(ing) slab, ~ panel
~entwässerung f, Dachablauf m = roof drainage
~entwurf m → Dachzerlegung f
~estrich m = roof (insulation) screed
~etage f → Dachgeschoß n
~falzziegel m → Falzziegel
~farbe f = roof paint
~fenster n = roof window
~filz m = roofing felt
~filznagel m = roofing felt nail
~first m → First
~fläche f, Dachseite f [*Österreich: Dachresche f. Beim geneigten Dach*] = pane of a roof, roof pane
~fläche f [*obere Begrenzung einer geologischen Schicht*] = superface
~folie f, Dachbelagfolie = roof foil
~form f = roof shape
dachförmiger Verbrennungsraum m = penthouse combustion chamber
dachförmiges Gefälle n → Dachprofil
~ Querprofil n → Dachprofil
Dach|formquerschnitt m → Dachprofil
~garten m = roof garden
(~)Gaupe f, (Dach)Gaube f = (roof) dormer (window)
~gebinde n = roof course
~gefälle n → Dachneigung f
~geschoß n, Dachstockwerk n, Dachetage f, (Dach)Boden m, Speicher m = roof floor, ~ stor(e)y, uppermost ~, topmost ~
~gesims n = string course, cornice of a roof, eaves mouldings
~grundriß m = roof plan
~haken m → Ausbesserungshaken
~haube f → Dachkappe f
~haut f → Dachdeckung f
~höhe f = roof level

~(-Hubschrauber)flughafen m = elevated heliport, roof-top ~
~isoliermaterial n = roof insulation material
~isolierung f, Dach(ab)dichtung = roof insulation
~kabel n [*Hängedach*] = suspension cable
~kandel m, f → Dachrinne
~kappe f, Dachreiter m, Dachhaube f, Aufreiter, Dachaufsatzfenster n, Dachlaterne f, Aufdach n, Überdach = lantern light, ridge turret
(~)Kehle f, (Dach)Kehlung f, Einkehle = (roof) valley
~knappe m → Ausbesserungshaken m
~konstruktion f = roof structure, ~ system
~kran m = roof crane
~kühler m, Scheitel-Kühler [*Motor*] = cabane radiator; saddle ~ (US)
~laden m = rooftop shop
~last f = roof load
~laterne f → Dachkappe f
~latte f = roof batten
(~)Lattung f = roof battens, ~ lathing
~leitungen fpl [*einer Blitzschutzanlage*] = air termination network
~licht n, Dachoberlicht = rooflight
~neigung f, Dachgefälle n [*Österreich: Dachresche f. Gefälle der Dachfläche, meist durch den Neigungswinkel bezeichnet*] = roof pitch, ~ slope, pitch of roof, slope of roof
~neigungswinkel m = angle of roof pitch, ~ ~ ~ slope
~(ober)licht n = rooflight
~pappe f = black roofing, (roofing) felt
~pappen-Klebemasse f = (roofing) felt adhesive, black roofing ~
~parkplatz m = roof-top parking deck, ~ car park
(~)Pfette f = purlin(e), binding rafter
~platte f → Dachelement n
~platte f [*für Metalldächer*] = deck unit
~platte f in Form von Doppel-T-Trägern = double T roof slab

~platte f in Form von T-Trägern = single T roof slab
~profil n, dachförmiges Gefälle n, dachförmiges Querprofil, Dachformquerschnitt m [*Straßendecke*] = straight finish from the shoulder to the centre line, ~ crossfalls
~profil n mit Firstausrundung [*Straßenquerschnitt*] = section with two straight lines joined by an easy curve, sloped camber, straight side slopes joined by a central parabolic curve
~raum m = roof-space
~raum m [*Fehlname*] → Dachzimmer n
~reiter m → Dachkappe f
~resche f → Dachneigung f
~resche f → Dachfläche f
~rinne f, Dachkandel m, f = eave(s) gutter
~rippe f = roof rib
(~-)Rohrpfette f = tubular purlin
~schale f = roof shell
~schalung f = roof boards, ~ boarding, ~ sheathing
~schicht f, unmittelbares Hangendes n (⚒) = immediate roof
Dachschiefer m [*ist die bautechnische, aber auch geologische Bezeichnung für sich als Dachdeckstoff eignenden Tonschiefer*] = roof(ing) slate, slate
~bergwerk n, Dachschiefergrube f = slate mine
~bruch m = slate quarry
~schneider m = slate cutter
Dach|schifter m = hip rafter, angle ridge, angle rafter
~schindel f = roof shingle
~schornstein m = chimney stack, roof chimney [*That part of a chimney that is external to a building*]
~seite f → Dachfläche f
(~)Sparren m [*Schweiz: Rofen m*] = rafter
~sparrenfalte f (Geol.) = chevron fold
Dachstein m → Dachziegel m
~ → Beton~
~aufzug m → Dachziegelaufzug

Dachstein(aus)formung — Dämmasse

~(aus)formung *f* → Dachziegel(aus)formung

~farbe *f* → Beton~

~form *f* → Dachziegelform

~formung *f* → Dachziegel(aus)formung

~karre(n) *f*, (*m*), Dachziegelkarre(n) = roof(ing) tile truck

~maschine *f*, Zement~, Beton~ = concrete roof(ing) tile machine

~prüfer *m* = roof tile tester, ~ ~ testing machine

Dach|stockwerk *n* → Dachgeschoß *n*

~stube *f* → Dachzimmer *n*

~stuhl *m* = roof framing

~tafel *f* → Dachelement *n*

~tennisplatz *m* = tennis court on roof

~terrasse *f*, Terrassendach *n* = terrace roof, roof terrace

~träger *m* = roof girder

~traufe *f* → Traufe

~tribüne *f* = covered spectator's stand

~überstand *m* = roof overhang

~verankerung *f* [*Tunnel- und Stollenbau*] = roof bolting

~verband *m* = roof system

~verblechung *f* = plate roofing

~verfallung *f* → Dachzerlegung

~verglasen *n* = roof glazing

~verglasung *f* = roof glazing

~wasser *n* = roof water [*Storm water from roofs*]

~wehr *n* → Doppelklappenwehr

~wehrklappe *f* = leaf

~zerfallung *f* → Dachzerlegung

~zerlegung *f*, Dachausmitt(e)lung, Dachentwurf *m*, Dachverfallung, Dachzerfallung [*die waag(e)rechte Projektion der Dachformen, wie sie sich durch das Festlegen der Begrenzungslinien für die Dachflächen ergibt*] = roof design

Dachziegel *m*, Dachstein *m* = roof(ing) tile

~(-Ab)schneider *m* = roof(ing) tile cutter

~aufzug *m*, Dachsteinaufzug = roof(ing) tile hoist

~(aus)formung *f*, Dachstein(aus)formung = roof(ing) tile moulding

Dachziegelei *f*, Dachziegelfabrik *f* = roof tile factory

Dachstein|form *f*, Dachsteinform = roof(ing) tile mo(u)ld

~formung *f* → Dachziegelausformung

~karre(n) *f*, (*m*), Dachsteinkarre(n) = roof(ing) tile truck

~schneider *m*, Dachziegel(-Ab)schneider = roof(ing) tile cutter

Dachzimmer *n*, Dachstube *f* [*fälschlicherweise auch „Dachraum" genannt*] = attic; garret [*deprecated*] [*A habitable room entirely within the roof-space of a building*]

Dalbe *m*, Duck~, Dück~, Pfahlbündel *n*, Pfahlgruppe *f* [*im Wasser stehender Bock aus eingerammten Pfählen zum Festmachen oder Führen von Schiffen*] = dolphin

Damm *m*, Dammschüttung *f* [*allgemein*] = fill

~ (⚒) = pack (wall) [*a dry-stone wall built in a colliery or metal mine to support the roof*]

~ [*zur Abgrenzung einer Stauabteilung oder eines Beckens bei der Obstbewässerung*] = controlling ridge, border, levee [*An earth ridge built to hold irrigation water within prescribed limits in a field or orchard*]

~, Verkehrs~ = embankment, bank [*A ridge of rock or earth thrown up to carry a road, railway, or canal*]

~, Stau~, Talsperren~, Sperr~ = fill dam, embankment (type) ~

~, Deich *m*, Fluß~ = stream dike, ~ levee (US); ~ dyke (Brit.)

~, Deich *m*, See~, Meer(es)~ = sea dike, ~ levee (US); ~ dyke (Brit.)

~ **aus Erdschüttung** → → Talsperre

~ ~ **Felsschüttung** → → Talsperre

~ ~ **Kiesschüttung** → → Talsperre

~achse *f*, Dammittellinie *f*, Damm-Mittellinie = fill centre line (Brit.); ~ center ~ (US)

Dämmasse *f*, Dämm-Masse = insulating compound

Dämmatte *f*, Dämm-Matte = insulating blanket
Damm|auflager *n* → Dammbasis *f*
~aufschüttung *f* bis zukünftiges Planum → Schüttung
~balken *m*, Staubalken = stop log
~balkenwehr *n*, Staubalkenwehr [*Verschluß durch waag(e)recht liegende Balken aus Holz, Stahl, oder Stahlbeton, die in Falze der Pfeiler und Wehrwangen eingelassen sind und herausgenommen werden können*] = stop log weir
~basis *f*, Dammsohle *f*, Dammauflager *n* = fill base, base of fill
Dammbau *m*, Verkehrs~ = bank construction, embankment ~, embanking
~, Deichbau, Fluß~ = stream dike construction, ~ levee ~ (US); ~ dyke ~ (Brit.)
~, Deichbau, See~, Meer(es)~ = sea dike construction, ~ levee ~ (US); ~ dyke ~ (Brit.)
~, Sperr~, Stau~, Talsperren~ = fill dam construction, embankment (type) ~ ~
~stelle *f* → → Talsperre
Dämmbeton *m* = insulating concrete
~-Spritzauskleidung *f* = sprayed insulating concrete lining
Dämmblock *m* = insulating block
Dammböschung *f*, Auffüllböschung, Auftragböschung, Böschung im Auftrag = fill slope
Dammböschungs|winkel *m* = angle of fill slope
~ziehen *n* mit Straßenhobel = fill sloping with grader
Dammbreite *f* = fill width
~ (✘) = pack (wall) width
~, Verkehrs~ = embankment width, bank ~
~, Stau~, Talsperren~, Sperr~ = fill dam width, embankment (type) ~ ~
~, Deichbreite = dike width, levee ~ (US); dyke ~ (Brit.)

Damm|bruch *m*, Deichbruch, Dammeinbruch, Deicheinbruch = dyke breach, ~ blow (Brit.); dike ~, levee ~ (US)
(~)Durchstoßgerät *n* → Rohrdurchstoßgerät
~einbruch *m* → Dammbruch
dämmen, ab~ [*Baukonstruktionen gegen Wärme- und Schalldurchgang sichern*] = to insulate
Dämmen *n* [*Die Sicherung der Baukonstruktionen gegen Wärme- und Schalldurchgang. Den Schutz gegen das Eindringen von (Boden)Feuchtigkeit und Grundwasser nennt man Isolieren*] = insulating
Dämm|estrich *m* = insulating screed
~fähigkeit *f*, Dämmvermögen *n* = insulating property
Dammfalz *m*, Mauerfalz [*Dammbalkenwehr*] = stop log groove
Dämmfilz *m* = insulating felt
Damm|fuß *m* = fill toe, ~ foot
~herstellung *f*, Dammschüttung *f* = fill (construction)
~höhe *f* = fill height
Dammittellinie *f* → Dammachse *f*
Dammkanal *m*, Kanal im Auftrag = canal on embankment
Dammkern *m* [*bei der Seitenschüttung*] = central fill
~, (Dichtungs)Kern, Innendichtung *f*, Kerndichtung [*Erdstaudamm*] = (impervious) core, ~ diaphragm
(Damm)Kernmauer *f* = core wall [*A wall of masonry, sheet piling, or puddled clay built inside an embankment-type dam to reduce percolation*]
Dämmkork *m* = insulating cork
~platte *f* = insulating cork slab, ~ corkboard
Dammkörper *m*, Deichkörper = dike fill, levee ~ (US); dyke ~ (Brit.)
~, Stützkörper, Mitte *f*, Kern *m* [*nicht zu verwechseln mit Dichtungskern beim Staudamm = core*] = body
Dammkrone *f* = fill crest
Dämmlage *f*, Dämmschicht *f* = insulating layer, insulation ~

Damm|lage *f*, Dammschicht *f* = fill layer
~länge *f* = fill length
Dämm|-Masse *f*, Dämmasse = insulating compound
~-Matte *f*, Dämmatte = insulating blanket
Damm-Mittellinie *f* → Dammachse *f*
Dammodell *n*, Damm-Modell, Modelldamm *m* = fill model, model fill
Dämm|papier *n* = insulating paper
~platte *f*, Dämmtafel *f* = insulating board, insulation ~
~plattenverkleidung *f*, Dämmtafelverkleidung, Dämmplattenauskleidung, Dämmtafelauskleidung, Dämmplattenbekleidung, Dämmtafelbekleidung = insulating board lining
Dammprofil *n* = fill section
Dammriff *n* → Wallriff
Dämmputz *m* = insulating plaster
Damm|rutschung *f*, Dammrutsch *m* = fill slide
~sackung *f* = fill settlement
Dämm-Schalung *f* = insulated formwork [*deprecated: shuttering*] [*for cold weather concreting*]
Dammschicht *f*, Dammlage *f* = fill layer
Dämmschicht *f*, Dämmlage *f* = insulating layer, insulation ~
Damm|schulter *f* [*Übergang der waag-(e)rechten Dammfläche zur Dammböschung*] = ogee curve
~schütter-Bandabsetzer *m*, Hochabsetzer = stacker for building up fills
Dammschüttung *f*, Damm *m* [*allgemein*] = fill
~, Dammherstellung = fill (construction)
Dammschüttungs|boden *m* = fill soil
~fläche *f* = fill (construction) area
Dammsohle *f*, Dammbasis *f*, Dammauflager *n* = base of fill, fill base
Dämmstein *m* = insulating brick
Dämmstoff *m* = insulation material, insulant, insulation ~
~ gegen Erschütterungen = vibration insulation material

Dammstraße *f* = causeway
~ aus Felsschüttung = rock causeway
Dammstrecke *f* [*einer Straße*] = causeway section
Dämmtafel *f*, Dämmplatte *f* = insulating board, insulation ~
~verkleidung *f* → Dämmplattenverkleidung
Dämmtür *f* = insulating door
Dammüberhöhung *f* = heightening of fill
Dämmung *f* = insulation
Damm|verbreiterung *f* = widening of fill
~verdichtung *f* = fill compaction
Dämm|vermögen *n*, Dämmfähigkeit *f* = insulating property
~wandplatte *f* = insulating wallboard
~wert *m* = insulating value
Dampf *m*, Wasser~ = steam
~ [*im weiteren Sinne*] = vapo(u)r
~abführung *f*, Dampfableitung, Dampfaustritt *m*, Dampfausströmung, Dampfabgang *m* = steam discharge
~abgang *m* → Dampfabführung *f*
~ableitung *f* → Dampfabführung
~-Abraumlok(omotive) *f*, Abraum-Dampflok(omotive) = overburden steam loco(motive)
~absperrhahn *m* = steam stop cock
~(absperr)schieber *m* = steam parallel slide stop valve; ~ gate ~ (US)
~absperrventil *n* = steam stop valve
~anlage *f* = steam plant
~anschluß *m* = steam connection
~arbeit *f* = work done by steam
~aufsteigrohr *n* = steam riser
~auslaßrohr *n*, Abdampfrohr = exhaust steam pipe
~auslegerkran *m* = steam jib crane (Brit.); ~ boom ~ (US)
~ausnützung *f* = utilization of steam
~ausströmung *f* → Dampfabführung
~ausström(ungs)periode *f*, Dampfaustrittperiode = steam exhaust period
~austritt *m* → Dampfabführung *f*
~automat *m*, automatischer Dampferzeuger *m* = auto(matic) steam generator

Dampfbagger *m*, Dampflöffelbagger, Löffel-Dampfbagger, Dampfschaufel *f* = steam shovel; ∼ navvy (Brit.)
∼, Dampf-Dauerbagger, Dampf-Stetigbagger = continuous (type) steam excavator
Dampfbär *m* [*der Zylinder schlägt*] = steam (pile) hammer
∼ **mit halbautomatischer Steuerung** = semi-automatic steam (pile) hammer, single-acting ∼ (∼) ∼
Dampf|-Baulok(omotive) *f* = contractor's steam loco(motive)
∼**behälter** *m* = steam receiver
dampfbehandelt, wärmebehandelt [*Beton*] = low-pressure steam-cured, cured by atmospheric steam
Dampfbehandlung *f*, Wärmebehandlung [*es wird Dampf mit Temperaturen unter 100° C als Wärmeträger benutzt*] = low-pressure steam curing, steam curing at atmospheric pressure [*concrete*]
∼**bereiter** *m*, Dampferzeuger = steam generator
dampfberührte Heizfläche *f* Dampfheizfläche = heating surface in contact with the steam
Dampf|bildung *f*, Dampfentwick(e)lung = steam formation
∼**blase** *f* = steam bubble
∼**bohranlage** *f* = steam drilling plant
∼**bohrung** *f*, Dampfsonde *f*, Dampfquelle *f* = steam well
∼**bremse** *f* → Dampfsperre *f*
∼**-Dauerbagger** *m* → Dampfbagger
∼**diagramm** *n*, Dampfdruckdiagramm = steam diagram
∼**diffusionsbremse** *f* → Dampfsperre *f*
∼**dom** *m*, Kesseldom = steam dome
∼**-Doppeltrommelwinde** *f*, Doppeltrommel-Dampfwinde = double-drum steam winch
∼**-Drehramme** *f*, Dampf-Schwenkramme, Schwenk-Dampframme, Dreh-Dampframme = steam slewing and raking pile driving plant

∼**-Dreitrommelwinde** *f*, Dreitrommel-Dampfwinde = three-drum steam winch
∼**(druck)diagramm** *n* = steam diagram
Dampfdruck|erhitzer *m* → Autoklav *m*
∼**härtung** *f*, Autoklavisierung = autoclaving
∼**minderer** *m* = steam pressure reducer
∼**pumpe** *f*, Pulsometerpumpe = (steam) pulsometer pump
Dampf|düse *f* = steam jet
∼**dynamo(maschine)** *m*, (*f*) = steam dynamo
∼**einlaßbüchse** *f* = steam chest
∼**einlaßseite** *f* = steam admission side
dämpfen, ab∼, abschwächen [*Geräusch; Erschütterung*] = to dampen
Dämpfen *n*, Dämpfung *f* [*Allmähliches Abnehmen und Verschwinden der Amplitude einer Schwingung*] = damping
Dampf|entnahme *f* = drawing off of steam, bleeding ∼
∼**entöler** *m* = oil separator for steam
∼**entwick(e)lung** *f*, Dampfbildung = steam formation
Dämpfer *m*, Kurbel∼ = damper
∼**verkleidung** *f*, Kurbel∼ = damper guard
Dampferzeuger *m*, Dampfbereiter = steam generator
∼, Kesselwagenerhitzer *m* = tank car heater, booster ∼
Dampf|erzeugung *f* = steam raising, ∼ generation
∼**erzeugungsanlage** *f* = steam raising plant, ∼ generation ∼
∼**erzeugungsturm** *m* = steam raising tower, ∼ generation ∼
∼**-Feldbahnlok(omotive)** *f* = narrow-ga(u)ge steam loco(motive)
∼**gasbeton** *m* → → Beton
dampfgehärtet [*Beton*] = cured by autoclaving, autoclaved
Dampf|-Greif(er)-Naßbagger *m*, Dampf-Schwimmbagger mit Zweischalen-Greifkorb = clamshell

Dampfgrenzkurve — Dampfsammler

(bucket) steam dredge (US); ~ (~) ~ dredger (Brit.); ~ (~) ~ dredging craft
~**grenzkurve** *f* = steam-limit curve
~**-(Grund)Sauger** *m*, Dampf-Saugbagger, Dampf-Pumpen(naß)bagger = steam suction dredger (Brit.); ~ ~ dredge (US)
~**hahn** *m* = steam cock
~**härtung** *f*, Härten *n* im Druckhärtekessel, Behandlung mit Hochdruckdampf = autoclaving, steam curing at high pressure
~**haube** *f* [*wird über ein Betonrohr zur Erhärtung desselben gestülpt*] = steam curing hood
~**hauptleitung** *f* = steam main
~**heizfläche** *f*, dampfberührte Heizfläche = heating surface in contact with the steam
~**heizmantel** *m* = steam jacket
~**heizung** *f* = steam heating
~**heizungsanlage** *f* = steam heating system
~**-Holzaufbringungswinde** *f* = steam logging winch
~**kammer** *f* [*Betonsteinfertigung*] = curing chamber, ~ room
Dampfkessel *m* = (steam) boiler
~**gasbrenner** *m* = boiler gas burner
~**kohle** *f* = steaming coal
~**speisung** *f* = boiler feeding
~**überwachungsverein** *m* = Steam Boiler Supervising Association
~**wirkungsgrad** *m* = boiler efficiency
Dampf|kolben *m* = steam piston
~**kraftwerk** *n* = steam (generating) plant, ~ power ~, ~ electric generating ~, ~ electric station, ~ (power) station
~**kran** *m* = steam crane
~**kreis** *m* = steam circuit
~**leitung** *f* = steam line
~**leitungswiderstand** *m*, Widerstand in der Dampfleitung = resistance to steam flow
~**löffelbagger** *m* → Dampfbagger
~**lok(omotive)** *f* = steam loco(motive)
~**mantel** *m* = steam jacket

~**mantelrohr** *n* = steam-jacketed pipe
~**mantelzylinder** *m* = jacketed cylinder
~**maschine** *f* = steam engine
~**mehrzweckbagger** *m* → Dampfuniversalbagger
~**minderventil** *n*, Dampfreduzierventil = steam reducing valve
~**-Naßbagger** *m*, Dampf-Schwimmbagger = steam dredge (US); ~ dredger (Brit.); ~ dredging craft
~**nietmaschine** *f* = steam riveting machine
~**öler** *m* = steam lubricator
~**(pfahl)ramme** *f* = steam (pile) driving plant, ~ piling ~
~**pfeife** *f* = steam whistle
~**phasen-Schmierung** *f* = vapo(u)r phase lubrication
~**plungerpumpe** *f* = steam plunger pump
~**pochwerk** *n* = steam stamp battery, ~ gravity stamp
~**pumpe** *f* = steam pump
~**-Pumpen(naß)bagger** *m* → Dampf-(Grund)Sauger
~**quelle** *f* (Geol.) = steam vent
~**quelle** *f* → Dampfbohrung *f*
dampfraffiniert = steam-refined
Dampf|ramme *f* → Dampfpfahlramme
~**rammhammer** *m* = steam piling hammer
~**-Rammwinde** *f*, Ramm-Dampfwinde = steam piling winch
~**raum** *m* [*Kessel*] = steam space
~**reduzierventil** *n*, Dampfminderventil = steam reducing valve
~**regulierventil** *n* = steam regulating valve
~**reinigung** *f* = steam purification
~**reinigungssieb** *n* = steam strainer
~**reserve** *f*, Dampfvorrat *m* = steam reserve
~**rohr** *n* = steam pipe
(~)**Röhrentrockner** *m*, rotierender dampfbeheizter Röhrentrockner = steam-tube rotary dryer; ~ ~ drier (US)
~**sack** *m* = steam pocket
~**sammler** *m* = steam collector

~**-Saugbagger** *m* → Dampf-(Grund)-Sauger
~**-Sauger** *m* → Dampf-Grundsauger
~**-schaufel** *f* → Dampfbagger *m*
~**schieber** *m* → Dampfabsperrschieber
~**-Schienengreifer(kran)** *m*, Greifer(-kran) auf Schienen mit Dampfantrieb = rail-mounted steam grab(bing crane), ~ ~ (grab) bucket crane
~**schiffahrt** *f* = steam navigation
~**schlange** *f* = steam coil
~**schlauch** *m* = steam hose
~**schleierfeuerung** *f* = steam-fan furnace, furnace with steam jets above grate
~**schleife** *f*, Wasserkreislauf *m* = automatic return of water
~**schleppboot** *n* [*Schweiz*] → Dampfschlepper
~**schlepper** *m*, Schleppdampfer *m*; Dampfschleppboot *n* [*Schweiz*] = steam tug (boat), ~ towboat
~**schmierapparat** *m* = steam lubrication apparatus, duplex plunger lubricator
~**schmierung** *f* = steam lubrication
~**-Schwenkramme** *f* → Dampf-Drehramme
~**-Schwimmbagger** *m* → Dampf-Naßbagger
~**schwimmkran** *m* = steam floating crane
~**sirene** *f* = steam siren
~**sonde** *f* → Dampfbohrung *f*
~**speisepumpe** *f* = steam feed pump
~**sperre** *f*, Dampfsperrschicht *f*, Dampfbremse *f*, Dampfdiffusionsbremse [*Eine Schicht mit hohem Durchlaßwiderstand gegen Wasserdampf. Sie wird zwischen Dachdecke und Wärmedämmbelag angeordnet und hat die Aufgabe, eine Durchfeuchtung des Wärmedämmbelages durch Bau- und Nutzungsfeuchte zu verhindern*] = vapour barrier (Brit.); vapor ~ (US); ~ seal
~**-Stetigbagger** *m* → Dampfbagger

Dämpfstoff *m* [*gegen Erschütterungen*] = damping material
Dampfstrahl *m* = steam jet
~**elevator** *m*, saugender Injektor *m* [*zum Wasserfassen bei Lokomotiven und zum Lenzen in Wasserfahrzeugen*] = lifting injector
~**gebläse** *n* = steam jet blower
~**pumpe** *f*, Injektor *m* [*Speisung von Dampfkesseln*] = (steam) injector [*boiler feeding*]
~**reiniger** *m*, Dampfstrahlreinigungsgerät *n*, Dampfwäscher *m* = steam (jet) cleaner
~**reinigung** *f* = steam jet cleaning
~**rührer** *m* = steam mixer
~**zerstäuber** *m* = steam jet sprayer, ~ ~ atomizer
Dampf|strom *m* = steam stream
~**tellertrockner** *m* = steam-heated plate dryer (Brit.); ~ ~ drier (US)
~**temperaturanstieg** *m* = steam temperature rise, rise in steam temperature
~**(trommel)winde** *f* = steam drum winch
~**turbine** *f* = steam turbine
Dampfturbinen|kraftwerk *n*, Dampfturbinenanlage *f* = steam turbine power station, ~ ~ ~ plant, ~ ~ generating ~
~**satz** *m* = steam turbine generating set
Dampf|überdruck *m* = effective steam pressure, steam pressure above atmospheric pressure
~**überhitzer** *m* = steam superheater
~**überhitzung** *f* = steam superheating
~**umbaubagger** *m* → Dampfuniversalbagger
Dämpfung *f*, Dämpfen *n* [*Allmähliches Abnehmen und Verschwinden der Amplitude einer Schwingung*] = damping, attenuation
Dämpfungs|beiwert *m* [*einer Schwingung*] = damping coefficient
~**faktor** *m* = damping factor
~**feder** *f* = check spring, cushioning ~
~**lage** *f*, Dämpfungsschicht *f* = cushioning layer

Dämpfungsrille — Dauerbiegebruch

~rille *f*, Ventilnut *f* = throttle slot
~schicht *f*, Dämpfungslage *f* = cushioning layer
~vermögen *n* = damping capacity
~wirkung *f* = damping effect
~zylinder *m* = dash pot
Dampf|universalbagger *m*, Dampfmehrzweckbagger, Dampfvielzweckbagger, Dampfumbaubagger = steam universal excavator
~versuch *m* [*Korrosionsprüfung*] = steam test
~verteil(er)anlage *f* = steam distributing system
~verteilung *f* = steam distribution
~vielzweckbagger *m* → Dampfuniversalbagger
~vorrat *m*, Dampfreserve *f* = steam reserve
~vulkanisator *m* = steam vulcanizer
~walze *f* = steam roller
~wärme *f* = steam heat
~wäscher *m* → Dampfstrahlreiniger *m*
~winde *f*, Dampftrommelwinde = steam drum winch
~wirtschaft *f* = steam engineering
~zerstäubungsbrenner *m* = steam atomizing oil burner
~zuleitungsrohr *n* = steam supply pipe
~zweigleitung *f* = branch steam line
~zylinder *m* = steam cylinder
Darcy'sches Filtergesetz *n*, Filtergesetz von Darcy = d'Arcy's law, Darcy' ~ [*1856*]
Darrboden *m*, Plandarre *f* = hot floor dryer; ~ ~ drier (US)
Darre *f*, Röstplatte *f* = hot plate
Darrprobe *f*, Darrprüfung *f*, Darrversuch *m* [*Wenn bei Verwendung feuchter Zuschlagstoffe nach dem Wassergehalt der Mischung und damit nach dem wirklichen WZF gefragt wird, dann muß neben der Wasserzugabe auch die im Zuschlagstoff enthaltene Wassermenge bekannt sein. Hierfür wird eine Durchschnittsprobe von 5 kg aus den verschiedenen Körnungen im festgelegten Verhältnis zunächst genau gewogen, dann völlig getrocknet und wieder gewogen. Der Gewichtsverlust ist gleich dem Wassergehalt und kann in Prozent oder gleich in Liter pro Mischung bzw. pro Kubikmeter Festbeton ausgedrückt werden.
Beispiel:*
  Feuchtgewicht   5,000 kg
  Trockengewicht   4,810 kg
  *Wassergehalt   0,190 kg = 3,9%
  vom Feuchtgewicht oder 4,0%
  vom Trockengewicht.
Für eine Mischung mit 640 kg feuchtem Zuschlagstoff sind das*
  $\frac{0{,}19 \times 640}{5{,}0} = 24$ *Liter*
*und für 1 cbm Festbeton
bei 1900 kg* $\frac{0{,}19 \times 1900}{5{,}0} = 72$ *Liter*] = weighing and drying test
Darstellungsverfahren *n* [*für ein Bauwerk oder eine Maschine auf einer Zeichnung*] = method of representation
Daten *f* = data
Dauben|rohr *n* = stave pipe
~silo *m* = stave silo
Dauer *f* = duration
~ der Eigenschwingung = period of natural vibration
~abfluß *m* = perennial flow
~antrieb *m* = life power
~aufgeber *m* → Dauerspeiser
~bagger *m*, (Stetig)Bagger = bucket excavator
~behelfsbrücke *f* = semi-permanent bridge
~belag *m* → Dauerdecke *f*
~belastung *f*, bleibende Belastung, stetige Belastung = fatigue loading
~beschicker *m* → Dauerspeiser
~betrieb *m* = continuous operation
~bewegung *f* = continuous movement, ~ motion
~biegebruch *m* = fatigue bending failure

Dauerbiegeversuch — Dauerspeiser

~biegeversuch *m*, Dauerbiegeprobe *f*, Dauerbiegeprüfung *f* = fatigue bending test
~bremse *f* = retarder brake
~bruch *m* = fatigue failure
~brücke *f* = permanent bridge
~decke *f*, schwere (Straßen)Decke, Dauerbelag *m*, schwerer (Straßen-)Belag = high-type pavement, ~ surfacing
~eingriff *m* = constant mesh
~eingriffgetriebe *n* = constant mesh gearing
~elastizität *f* = permanent elasticity
~energie *f* = permanent energy
~festigkeit *f*, Ermüdungsfestigkeit, Arbeitsfestigkeit = endurance, fatigue strength, fatigue resistance [*Is defined as the number of cycles of loading imposed before fracture terminates the test*]
~festigkeitsprüfung *f*, Dauerfestigkeitsprobe *f*, Dauerfestigkeitsversuch *m* = endurance test
dauerfettgeschmiert = greased-for-life
Dauer|fettschmierung *f* = life grease lubrication, single-shot ~
~feuchtigkeit *f*, Dauerfeuchte *f* = permanent humidity, ~ moisture
~förderer *m* → Stetigförderer
~förderung *f* → Fließförderung
~formgießerei *f* = gravity die foundry
~frost *m* = permafrost
~frostboden *m* → dauernd gefrorener Boden
dauergeschmiert = sealed-for-life
Dauergeschwindigkeit *f* = permanent speed
dauerhaft, haltbar, beständig = durable
Dauer|haftigkeit *f* → Beständigkeit
~härte *f*, permanente Härte, bleibende Härte = permanent hardness
~kurve *f* → Dauerlinie *f*
~lader *m*, Stetiglader, kontinuierlicher Lader, Massenlader = continuous loader, ~ loading machine
~last *f*, bleibende Last, ständige Last, Stetiglast = permanent load

~linie *f*, Dauerkurve *f* = duration curve, ~ hydrograph [*A graphical representation of the number of times given quantities are equalled or exceeded during a certain period of record. For example, if, in a 10-yr. record of daily stream flow, the percentage of time the flow was above certain values (100, 200, 300 cu.ft. per sec., etc.) was plotted against flow, the graph would constitute a duration curve for that stream and period. From it could be read the percentage of time the flow was greater or less than any given value within the range that occurred during the period cited. The duration curve is the integral of the frequency curve*]
~magnet *m* = permanent magnet
~mischer *m* → Fließmischer
dauernd gefrorener Boden *m*, ewiggefrorener ~, Dauerfrostboden [*in der Schweiz auch „Permafrostzone" f genannt*] = permafrost soil
dauernde Strömung *f* → stationäre ~
dauernder Direktantrieb *m* = live power drive
~ Schichtausgleich *m* → → Schwarzbelageinbaumaschine
Dauer|parken *n* = day parking
~parker *m* = day parker
~probe *f*, Dauerprüfung *f*, Dauerversuch *m* = continuous test
~prüfung *f* → Dauerprobe *f*
~pumpversuch *m* [*Probebrunnen*] = continuous pumping test
~schalung *f*, verlorene Schalung = permanent formwork
~schlagbiegeversuch *m*, Dauerschlagbiegeprobe *f*, Dauerschlagbiegeprüfung *f* = repeated impact bending test
~schlagprobe *f*, Dauerschlagversuch *m*, Dauerschlagprüfung *f* = continuous impact test
~schmierung *f* = life-lubrication
~speiser *m*, Daueraufgeber, Dauerbeschicker, Stetigspeiser, Stetigaufgeber, Stetigbeschicker = continuous feeder

Dauerströmung — Deckenbeleuchtung 302

~strömung *f* [*Meer*] = permanent current

~tauchversuch *m*, Dauertauchprobe *f*, Dauertauchprüfung *f* [*Korrosionsprüfung*] = total immersion test

~verformung *f* = permanent deformation

~versuch *m* → Dauerprobe *f*

~vorspannung *f* = permanent prestress

Dauerwühler *m* (✲) = continuous miner

~ **Bauart Dosco** = Dosco continuous miner

~ **Bauart Joy** = Joy continuous miner

Dauer|zugversuch *m*, Dauerzugprobe *f*, Dauerzugprüfung *f* = repeated tensile test

~zustand *m* = permanent state

~-**Zwangsmischer** *m*, Durchlauf-Zwangsmischer, Stetig-Zwangsmischer, kontinuierlicher Zwangsmischer, stetig arbeitender Zwangsmischer = continuous pugmill

Daumen|drücker *m* = thumb lever [(*compressed-*)*air throttle control*]

~schraube *f* = thumbscrew

~sprung *m* = thumb jump

~welle *f*, Nockenwelle = camshaft

Dazit *m* = dacite

DE *f* → Drahterkennbarkeit *f*

DECAUVILLE-Feldbahnlok(omotive) *f* = DECAUVILLE motor tractor

Dechet *n* [*Periode der Tide, während der sich der Tidehub verringert, von Springwasser bis Nippwasser*] = decrease

Deck *n*, Plattform *f*, Grundplatte *f* [*Bagger; Kran*] = deck

~anstrich *m* = finish(ing) coat

~blech *n*, Ab~ = cover(ing) plate

~blech *n* [*Brückenbau*] = deck plate

~bohle *f*, Ab~ = cover(ing) plank

~brett *n*, Ab~ = cover(ing) board

~brett *n*, Fensterbrett, Latteibrett, Simsbrett = window board

~brücke *f*, Brücke mit oben liegender Fahrbahn = deck bridge

Decke *f*, Geschoß~, Stockwerk~, Etagen~, Hochbau~, Gebäude~ = floor

~, Belag *m*, Straßen~ = (road) surfacing, (~) pavement

~, Behälter~, Tank~ = (tank) roof

~, Verschleißschicht *f* [*Tonbetonstraße, mechanisch verfestigt*] = surface course

~ **auf einem Betonüberbau**, Belag *m* ~ ~ ~ [*Brücke*] = concrete deck pavement; ~ ~ paving (US)

~ ~ ~ **Stahlüberbau**, Belag *m* ~ ~ ~ [*Brücke*] = steel deck pavement; ~ ~ paving (US)

Deckel *m* = lid, cover

~flansch *m*, Blindflansch, Abschlußflansch [*Deckel, der ein Rohr abschließt*] = blind flange

decken [*positive Biegemomente*] = to resist

~, ein~, bedachen = to roof

Deckenanker *m*, Felsanker, Gebirgsanker [*Tunnel- und Stollenbau*] = roof bolt

~, Maueranker, Mauerhaken *m* [*Verbindungsglied der Deckenbalken mit der Außenmauer*] = wall tie

Decken|arbeiten *fpl*, Belagarbeiten = (road) pavement work, (~) surfacing, ~ highway

~art *f*, Belagart = type of pavement, ~ ~ surfacing

~auskleidung *f* → Deckenbekleidung

~ausschalen *n* = striking of the floor formwork

~balken *m* [*für Balkendecken mit dicht verlegten Balken*] = joist

~balken(träger) *m* = floor beam

Deckenbau *m* → Belagbau

~geräte *npl* → Deckenbaumaschinen *fpl*, Straßen~ = road pavement construction machinery, highway ~ ~ ~

Decken|bekleidung *f*, Deckenverkleidung, Deckenauskleidung = ceiling lining

~belastung *f* [*Hochbau*] = floor loading

~beleuchtung *f* = ceiling lighting

~bemessung *f*, Belagbemessung = pavement design, surfacing ~

~bestandteil *m*, Belagbestandteil = pavement component, surfacing ~

~betonierung *f* [*Geschoßdecke*] = floor pouring, ~ concreting

~betonierung *f* [*Straßenbau*] = pavement pouring, ~ concreting, surfacing ~

~bogen *m* = floor arch

~bügel *m* = floor stirrup

~dicke *f* [*Hochbaudecke*] = floor thickness

~dicke *f*, Belagdicke = pavement thickness, surfacing ~

~einbau *m*, Belageinbau [*Straßenbau*] = pavement laying, surfacing ~

~einbauzug *m* = paving train (US); surfacing ~

~einbruch *m*, Straßen~ = pavement sinkage, surfacing ~

~einschub *m* → Einschub

~ermüdungsbruch *m*, Belagermüdungsbruch = pavement failure by fatigue, surfacing ~ ~

~erneuerung *f*, Straßen~ = resurfacing (operations)

~fach *n* → Deckenfeld *n*

~feld *n*, Kassette *f*, Deckenfach *n* = waffle, core

~fertigbauelement *n*, Deckenfertigteil *m*, *n* = precast floor unit

~fertiger *m*, (Straßen)Fertiger, Straßendeckenfertiger = (road) finisher, (~) finishing machine, (~) laying and finishing machine

~fertigteil *m*, *n*, Deckenfertigbauelement *n* = precast floor unit

~fläche *f* = floor area

~fliese *f*, Deckenplatte *f* = ceiling tile

~fuge *f* = ceiling joint

~fuge *f* = floor joint

~fuge *f*, Belagfuge, Beton~ = concrete surfacing joint, ~ pavement ~

(~)Füllkörper *m*, Deckenkörper, Deckenstein *m* = floor filler, filling member, filler tile, filler block, infilling block, floor block, floor tile

~füllstoff *m* = floor filling material

Deck(en)gebirge *n* = overthrust mountain

Decken|gewicht *n* = floor weight

~granit *m* = mantle granite

~gut *n* → Belagmaterial *n*

~güte *f*, Belaggüte, Straßen~ = pavement quality, surfacing ~

~heizung *f*, Deckenstrahlungsheizung = ceiling heating

~hohlkörper *m*, Deckenhohlstein *m* [*Fehlname: Deckenhohlziegel m*] = hollow filler tile, ~ ~ block

~hohlplatte *f* = hollow (core) floor slab

~hohlstein *m* → Deckenhohlkörper *m*

~hohlziegel *m* [*Fehlname*] → Deckenhohlkörper

~isolierung *f* = floor insulation

~kern *m* → Betonprobekern

~konstruktion *f*, Belagkonstruktion [*Straßenbau*] = pavement system; paving ~ (US)

~kran *m*, Hängekran = suspension crane

~last *f* = floor load

~-Leimfarbe *f* = ceiling distemper

~leuchte *f* = ceiling fitting

(~)Liegedauer *f* [*Straßenbau*] = service life

~los *n* [*Straßenbau*] = contract section

~markierungsfarbe *f*, Straßen~ = striping paint

~material *n* → Belagmaterial

~mörtelbelag *m*, Deckenputz *m* = ceiling plaster

~oberfläche *f*, Belagoberfläche [*Straßen- und Flugplatzbau*] = pavement surface, surfacing ~

~öffnung *f* = floor opening

~platte *f* [*Hochbau*] = structural slab, floor ~, ~ panel

~platte *f*, Deckenfliese *f* = ceiling tile

~platte *f*, Beton~ = concrete pavement slab, ~ surfacing ~

~prüfgerät *n*, Straßen~ = pavement tester

Deckenputz — Deckung

~putz m, Deckenmörtelbelag m = ceiling plaster
~putzmörtel m = ceiling stuff
~querschnitt m, Belagquerschnitt = geometric section of the pavement, ~ ~ ~ ~ surfacing
~querschnitt m [*Hochbau*] = floor cross-section
~rippe f = floor rib
~rüstung f = floor center(s), ~ cent(e)ring (US); ~ centre(s) (Brit.)
~rüttler m, Deckenvibrator m [*Betonstraßenbau*] = pavement vibrator
~schalung f = floor formwork
~schalungsträger m = floor centre
~schalungs-Transportgerät n = floor formwork handling rig
~schleifmaschine f = ceiling grinder, ~ grinding machine [*it is designed to facilitate dry-grinding concrete overheads*]
~schluß m [*Fertigbehandlung von Beton*] = (surface) finish
~schlußbohle f → Glättbohle
~schlußübergang m = finishing pass
~stein m → (Decken)Füllkörper
~(strahlungs)heizung f = ceiling heating
~stütze f = (slab) prop
~stütze f aus Stahl, Stahlrohr-Deckenstütze f = tubular steel (slab) prop
~system n [*Hochbau*] = floor system
~system n [*Straßenbau*] = pavement system
~tafel f [*zur Bekleidung von Deckenuntersichten*] = ceiling panel
~tafel f = precast floor panel
~träger m = floor girder
~unterbau m = (road) pavement foundation, (~) surfacing ~, highway ~ ~
~untersicht f, sichtbare Decke f = floor soffit
(~)Unterzug m = ceiling joist, floor ~, bearer
~verhalten n, Belagverhalten = pavement performance, surfacing ~

~verkleidung f → Deckenbekleidung
~versieg(e)lung f → Porenschluß m
(~)Verteiler m, (Decken)Verteilergerät n [*Straßenbau*] = pavement spreader
~vibrator m → Deckenrüttler m
~weite f = floor span
~ziegel m = floor brick
Deck|folie f, Ab~ = cover(ing) foil, protecting ~
~gebirge n → Abraum m
~gebirge n [*Tunnel- und Stollenbau*] = surrounding material
~gebirgslage f → Abraumschicht f
~gebirgsschicht f → Abraumschicht
~haube f, Ab~ = hood
~haus n [*Naßbagger*] = deckhouse
~kraft f, Deckvermögen n = hiding power [*The power of a paint or paint material as used to obscure a surface painted with it. In this definition the word obscure means to render invisible or to cover up a surface so that it cannot be seen. The term "covering" power should not be used because of its ambiguity*]
~kran m = deck crane
~lage f [*Bitumendachpappe*] = cap sheet
~lage f → Verschleißschicht f
~lage f → Ab~
~lage f [*Schweißtechnik*] = surface layer
~lasche f = butt-strap (Brit.); cover-plate (US); splice plate
Deckleiste f → Ab~
~, Fugen~ = joint cover(ing) strip
Deck|platte f, Ab~ = cover(ing) slab
~platte f → Pilzkopfplatte
~rost m, Ab~ = cover(ing) grating, ~ grate
Deckschicht f → Ab~
~, Laufschicht, Gehschicht [*Fußboden*] = wearing course
Deck|schute f = flat-top barge
~schutt m → Abraum m
~splitt m → Ab~
~stein m → Ab~
Deckung f → Dach~

Deck|vermögen *n* → Deckkraft *f*
~ziegel *m*, Ab~ = cope, coping brick
Deflagration *f* [*Eine langsame Umsetzung eines undetonierten Ladungsrestes*] = deflagration, smouldering
deflagrieren = to deflagrate, to smoulder
Deflation *f*, Abblasung *f*, Abtragung durch Wind, Abhebung, Winderosion *f*, Windabtragung, äolische Abtragung (Geol.) = (wind) deflation
Deformation *f*, Verformung *f* [*Änderung einer Dimension eines Körpers unter dem Einfluß physikalischer Beanspruchung(en)*] = deformation
Deformations|geschwindigkeit *f* → Formänderungsgeschwindigkeit
~messer *m*, Verformungsmesser, Deformator *m* = deformeter, deformability meter
~methode *f* [*Baustatik*] = equilibrium method
Deformator-Klinograph *m* Galileo = Galileo slide-ga(u)ge
degradierte Braunerde *f*, ausgelaugte ~, stark ausgelaugter Boden *m* = strongly leached soil, (gray brown) podzolic ~, leached brown earth
~ Schwarzerde *f*, degradierter Tschernosem *m*, tschernosemartiger grauer Waldboden *m* = prairie-forest soil, degraded black-earth ~
degradierter Tschernosem *m* → degradierte Schwarzerde *f*
degressive Abschreibung *f* = declining balance method (of depreciation)
dehnbarer Boden *m* = dilative soil, dilatable ~
Dehnbarkeit *f* → Streckbarkeit
dehnen = to elongate
Dehnfähigkeit *f* = extensibility
Dehnfuge *f* → Trennfuge
Dehngeschwindigkeit *f* = rate of strain
Dehnung *f*, Längung [*positiv*] = extension
Dehnungsberechnung *f* = strain calculation
Dehn(ungs)|fuge *f* → Trennfuge
~fugenband *n* = expansion joint tape

Dehnungs|messer *m* = strain ga(u)ge
~messung *f* = strain measurement
~übertragung *f* = strain transfer
~verteilung *f* = strain distribution
Dehnungs/Vorspannkraft-Diagramm *n* = elongation/prestressing force diagram
Dehnweg *m* = trajectory of strain
Dehnzement *m* → Quellzement
Deich *m*, Damm *m*, Fluß~ = stream dike, ~ levee (US); ~ dyke (Brit.)
~, Damm *m*, See~, Meer(es)~ = sea dike, ~ levee (US); ~ dyke (Brit.)
Deichbau *m*, Dammbau, Fluß~ = stream dike construction, ~ levee ~ (US); ~ dyke ~ (Brit.)
~, Dammbau, See~, Meer(es)~ = sea dike construction, ~ levee ~ (US); ~ dyke ~ (Brit.)
Deich|breite *f*, Dammbreite = dyke width (Brit.); dike ~, levee ~ (US)
~bruch *m*, Damm(ein)bruch, Deicheinbruch = dyke blow, ~ breach (Brit.); dike ~, levee ~ (US)
~einbruch *m* → Deichbruch
~hauptmann *m* = dike-master (US); dyke-master (Brit.)
~körper *m*, Dammkörper = levee fill, dike ~ (US); dyke ~ (Brit.)
~land *n* = dykeland (Brit.); innings
Deichscharte *f* = dike opening (US); dyke ~ (Brit.)
~ für eine Straße = street opening
Deichsel *f*, Anhänge~, Fahr~, Zug~ = tow-bar, tongue, pole
deichselgelenkt [*Handwalze*] = hand-steered
deklinante Buhne *f* = groyne (Brit.)/groin (US) pointing slightly downstream
Deklination *f*, Mißweisung *f*, D [*Die Abweichung gegenüber dem geodätischen Meridian*] = declination
Dekompressions|hebel *m* = compression release lever
~welle *f* = compression release shaft
Dekor *n*, Muster *n* [*auf einer Platte*] = pattern

Dekor(ations)oberfläche — Derrick(kran)

Dekor(ations)oberfläche *f*, Zieroberfläche = decorative surface
dekorativ = of attractive appearance
Dekorativbeton *m*, Zierbeton = ornamental concrete, decorative ∼
Dekor-Platte *f* = ornamental panel, ∼ sheet
Dekrement *n* = decrement
Delle *f* = dell, birdbath
Dellenschweißen *n* [*Schweiz*]; Buckelschweißen = projection welding
Delta *n*, Mündungskegel *m* = delta
∼aufschüttung *f*, Deltabildung = delta building, ∼ fill
∼damm *m* = delta(ic) embankment
∼niederung *f* = delta flats
Demontage *f*, Abbau *m*, Abbauen *n*, Zerlegen, Zerlegung *f* = dismantling
∼vorgang *m*, Abbauvorgang = dismantling operation
demontieren, abbauen, auseinandernehmen, abmontieren, zerlegen = to disassemble, to strip down, to dismantle
denaturiertes Salz *n* = denatured salt
Denkmalschutz *m* = preservation of monuments

Denudation *f*

Verwitterung *f*, Abwitterung
  chemische Verwitterung *f*, Gesteinszersetzung, chemische Gesteinsauflösung
    Einwitterung *f*
    Auswitterung *f*
  mechanische Verwitterung *f*, physikalische Verwitterung, Gesteinszerfall *m*
Erosion *f*, (flächenhafte)Ausnagung *f*, Abtragung
Transport *m*, Verfrachtung *f*

denudation (Brit.); **abrasion** [*The making bare of the surface of the earth*]
weathering, process of rock wastage
chemical weathering

weathering from top to bottom
weathering from bottom to top
mechanical weathering, physical weathering, rock disintegration

erosion, geological erosion

transport

Denudations|ebene *f*, Fastebene, Rumpffläche *f* = denudation plain, peneplain, peneplane
∼terrasse *f* (Geol.) → Abtragungsterrasse
Denver-Flotationszelle *f* = Denver sub-A machine, Fahrenwald ∼
Deponie *f*, Haldenlagerung *f*, Freilagerung im (oder in) Schütthaufen = stockpiling (in the open air), ground storage
∼förderband *n* → Hochabsetzer
Depotanlage *f* = storage depot
Depressionstrichter *m* = depression cone
Derbyshire-Basalt *m* = toadstone
Derivat *n* → Abkömmling *m*
∼boden *m*, Absatzboden, Kolluvialboden, umlagerter Boden = transported soil, colluvial ∼ [*Soil consisting of alluvium in part and also containing angular fragments or the original rocks*]

Derrick(kran) *m*, Mastenkran, Ladebaum *m*

A-∼, A-Mastschwenker *m*

derricking jib crane (Brit.); ∼ **boom** ∼ (US); **derrick (crane)** [*Invented by Henderson in 1845*]
A-∼ ∼ ∼, jinniwink

Bock-~ [*Schwenkwinkel auf etwa 280° beschränkt*]
  Bockstrebe *f*
Derrickwinde *f*
Eisenbahn-~

Gitter-~
Holz-~
hydraulischer ~
Montage-~; Vorbau-~ [*im Brückenbau*]
nahtloser ~
Scheren-~ ohne Ausleger
Schwergut-~
Schwimm-~, Derrick-Ponton *m*
Trossen-~, Seil-~ [*Ausleger mit 360° Schwenkwinkel*]
Trossen-~ ohne Ausleger, Hebezeug-Mast *m*

Unfall-~

stiff-leg ~ ~ (US); Scotch ~ ~ ~ (Brit.)
  back-stay, stiff-leg
derrick winch
railway car ~ ~ ~ (Brit.); railroad car ~ ~ ~ (US)
lattice(d) ~ ~ ~
wood ~ ~ ~, timber ~ ~ ~
hydraulic ~ ~ ~
erecting ~ ~ ~

weldless ~ ~ ~
shears ~ ~ ~
heavy ~ ~ ~
floating ~ ~ ~, barge ~ ~ ~
guy ~ ~ ~

gin pole (derrick) (crane) [*guyed derrick pole (or mast) used in conjunction with either pulley-blocks or a winch*]
auto wrecking ~ ~ ~

Desinfektionsmittel *n* = sterilizing agent, disinfectant
Desintegrator *m*, Stiftenschleudermaschine *f* = disintegrator, disintegrating mill
Dessinblech *n* = fancy sheet metal, show ~ ~
Destillations|benzin *n* = straight-run gasolene (or gasoline) (US); ~ petrol (Brit.)
~kolben *m* = distillation flask
destillierter Teer *m* = distilled tar
Detektor *m* = detector
~kopf *m* [*Ultraschallzählgerät*] = transducer, detector head
Detergentie *f* = detergent
Detonation *f* = detonation
Detonations|geschwindigkeit *f* = velocity of detonation
~übertragungsprüfung *f* = gap test
~welle *f* = wave of detonation
Detonator *m* [*Tiefbohrtechnik*] = godevil
detonierende Zündschnur *f* → → 1. Schießen *n*; 2. Sprengen *n*
deutsche Asphalteingußdecke *f* → Asphalteingußdecke

~ Bauweise *f*, Kernbauweise = core method of tunnel construction
deutscher Tübbingausbau *m* = German tubbing
~ Türstock *m* **mit Kniestempelverstärkung am Oberstempel** → K-Bau
Deval|-Probe *f*, Deval-Prüfung *f*, Deval-Versuch *m* = Deval test
~-Trommelmühle *f*, Deval-Abnutzungstrommel *f* = Deval (attrition) machine, ~ testing ~
Devon *n* = Devonian Period
Dewargefäß *n*, Weinholdgefäß = Dewar flask
Dextrin *n* = dextrine, starch gum
Dezentrationszone *f* [*Boden*] = zone of removal
Dezimal|klassifikation *f*, DK = decimal index system, UDC
~stelle *f* = decimal place
Diabas *m* = diabase
~-Mandelstein *m* = amygdaloidal diabase
~tuff *m*, Grünsteintuff = diabasic tuff, greenstone ~

Diagenese — dichten

Diagenese *f* = diagenesis [*The consolidation of clastic rocks by static metamorphism involving not much, if any, recrystallization. Examples may be muds to shales and clays to slates*]

Diagonal|abstreifer *m* = diagonal screed

~aussteifung *f*, Diagonalversteifung = diagonal bracing

~bohlenfertiger *m*, Betondeckenfertiger mit Diagonal-Glättbohle, Beton-(fahrbahn)fertiger mit Diagonal-Glättbohle = diagonal screed finisher, ~ ~ finishing machine

~masche *f* = square diamond mesh

~rippe *f* = diagonal rib

~schnittpunkt *m* = intersection point of diagonals

~stab *m* → Schräge *f*

~strebe *f* = diagonal brace

~strebenauflager *n* [*Gleiskettenlaufwerk*] = diagonal brace bearing

~versteifung *f*, Diagonalaussteifung, Schrägversteifung, Schrägaussteifung, Diagonalverband *m*, Schrägverband = diagonal bracing

Diagramm *n*, Schaubild *n* = diagram, graphic representation, diagrammatic representation, graph, chart

~ der waag(e)rechten Wasserdrücke = diagram of horizontal water pressures

~bogen *m* → Pegelbogen

~papier *n* = plotting paper

Diamant|bergbau *m* = diamond mining

~bohrausrüstung *f* = diamond drilling outfit

~bohren *n* = diamond drilling

~bohrer *m* = diamond drill

~bohrgerät *n*, Diamantbohrmaschine *f* = diamond drill(ing) machine

~bohrkern *m* = diamond (drill) core

~(bohr)krone *f* = diamond(-drill) bit

~bohrloch *n* = diamond drill hole

~bohrung *f* = diamond drilled well

~bohrverfahren *n*, Diamantbohrmethode *f* = diamond (system of) drilling

diamanthaltiger Kies *m* = diamond-bearing gravel

Diamantkern *m* [*mit Diamantkrone gebohrter Kern*] = diamond core

~bohren *n*, Diamantkernbohrung *f* = diamond coring, ~ core drilling

~bohrer *m* = diamond core drill

~rohr *n* = diamond core barrel

Diamant|(kreis)sägeblatt *n*, Diamanttrennscheibe *f* = diamond (circular) saw blade

~säge *f* [*Säge mit diamantbesetztem Blatt*] = diamond saw

~trennscheibe *f*, Diamant(kreis)sägeblatt *n* = diamond (circular) saw blade

~-Vollmeißel *m* = non-coring type diamond bit, solid ~ ~ ~

~werkzeug *n* = diamond tool [*saw; drill etc.*]

Diaphragma *n* → Dichtwand *f*

Dia(phragma)|-Baupumpe *f*, Membran(e)-Baupumpe = diaphragm contractor's pump, membrane ~ ~

~-Pumpe *f*, Membran(e)pumpe = membrane pump, diaphragm ~

Diaphthorese *f* (Geol.) → rückläufige Metamorphose *f*

Diapositiv *n* = slide

Diapumpe *f* → Diaphragma-Pumpe

Diatomee *f*, Kieselalge *f*, Stabalge = diatom

Diatomeen|erde *f*, Diatomit *n*, Kieselalgenerde *f*, Kieselgur *f* = diatom(aceous) earth, diatomite, Kieselgur

~schlamm *m*, Kiesel(algen)schlamm, Diatomeenschlick *m* = diatom ooze

dicht, hohlraumarm, geschlossen (abgestuft) = dense(-graded)

~, undurchlässig = impervious

~ machen, undurchlässig **~**, (ab-) dichten = to seal

dichte Verbindung *f* **von Glas mit Metall** = glass-to-metal seal

dichten, ab**~** [*von Grundbauten gegen Grundwasser durch Umhüllung mit einer wasserdichten Haut*] = to lay a waterproof skin

dichten, ab~ undurchlässig machen, dicht machen = to seal
dichter Beton m = dense concrete
~ **Gips** m → kompakter ~
~ **Kalkstein** m = compact limestone, mountain ~
~ **polierbarer Kalkstein** m → technischer Marmor m
~ **Sandasphalt** m, heißer ~ = sheet asphalt
~ **(Straßen)Verkehr** m = bumper-to-bumper traffic, close-packed ~, heavy ~
dichtes Gestein n → kompaktes ~
Dichte f = density
~**anzeiger** m = density indicator
~**bestimmung** f = density determination
~**messer** m → hydrostatische Senkwaage f
Dichten n, Ab~ = sealing
dichtes Wohngebiet n = heavily developed populated area
Dicht|fläche f = sealing face
~**fuge** f → (Ab)Dicht(ungs)fuge
dicht|gebrannte feinkeramische Masse f = vitreous whiteware body
~**gelagert** = tight, compact
Dichtigkeit f, Undurchlässigkeit = imperviousness
Dichtigkeits|-Prüfgerät n = density control unit [*for checking asphaltic concrete during construction*]
~**prüfung** f, Dichtigkeitsversuch m, Dichtigkeitsprobe f = permeability test
Dicht|ring m → Dichtungsirng
~**schweißung** f = ca(u)lk welding, seal ~
~**sitz-Gestängeverbinder** m [*Tiefbohrtechnik*] = seal grip tool joint
~**teil** m, n [*Tiefbohrtechnik*] = seal
Dichtung f, Ab~ = seal(ing)
~ → (Ab)Dichtungsvorlage f
~ **des Untergrundes**, Ab~ ~ ~, (Unter)Grund(ab)dichtung = subsoil waterproofing
~ **gegen (Boden)Feuchte**, Ab~ ~ ~ = dampproofing

~ ~ **Grundwasser**, Ab~ ~ ~, Grundwasserisolierung = waterproofing
~ **von Wassereinbrüchen**, Ab~ ~ ~ = sealing of inrushes of water
Dichtungsanstrich m → Ab~
~**mittel** n, Dichtungsanstrichstoff m, Ab~ = sealing paint
~**stoff** m, Dichtungsanstrichmittel n, Ab~ = sealing paint
Dichtungs|arbeiten fpl, Ab~ = sealing work
~**bahn** f, Ab~ = sealing strip
~**balken** m, Sohlen~, hölzerne Dichtungsleiste f [*Walzen(stau)wehr*] = timber stop
~**band** n, Ab~ = sealing tape
~**belag** m → (Ab)Dichtungsvorlage f
~**binde**, Dichtbinde, Ab~ = sealing jacket
~**blech** n [*Walzen(stau)wehr*] = staunching plate (Brit.); skin ~ (US)
~**boden** m [*Erdstaudamm*] = soil for (impervious) core
~**filz** m, Ab~ = sealing felt
~**folie** f, Ab~ = sealing foil
~**fuge** f → Ab~
~**gewölbe** n, Ab~ [*Talsperre*] = watertight facing arch
~**graben** m, Ab~, Verherdung f = (cut-off) trench [*A (cut-off) trench is carried across the valley under the dam at such a depth as to prevent water from the reservoir percolating underneath the dam and flowing away downstream*]
~**gürtel** m → Einpreß-Dichtwand f
~**halter** m = seal retainer
~**haut** f → (Ab)Dichtungsvorlage f
~**haut** f, Emulsionsanstrich m, Filmüberzug m, Nachbehandlungsfilm m [*Beton*] = concrete curing membrane, emulsion coating, curing seal
(~)**Kern** m, Dammkern, Innendichtung f, Kerndichtung [*Erdstaudamm*] = (impervious) core, ~ diaphragm
~**kitt** m = sealing putty

Dichtungslage — Dienstgewicht

~lage *f* → Sperrschicht *f*
~latte *f*, Ab~ [*kittlose Verglasung*] = windguard, draught fillet
dichtungslose Pumpe *f* = packingless pump
Dichtungs|manschette *f* = gasket
~masse *f*, Ab~ = sealing compound, sealant, sealer
~material *n*, Ab~ = sealing material, sealant
~mauer *f*, Ab~, Abschlußwand *f*, Herdmauer, Fußmauer, (Ab)Dichtungssporn *m*, Trennmauer, Sporn [*Talsperre*] = (toe) cut-off wall
~mittel *n*, Beton~, Sperrzusatz *m*, wasserabweisendes Mittel, Dichtungszusatz = concrete waterproofing compound, waterproofer, water-repellent, water-repelling agent, densifier, densifying agent, dampproofing and permeability reducing agent
~mittel *n*, Ab~ [*Nachbehandlung von Beton*] = (membrane) curing compound
~mittel *n*, Dichtungsstoff *m*, Ab~ = sealing material
~nut *f* = seal groove
~packung *f* = gland packing
~pappe *f*, Ab~ = sealing felt
~pulver *n* [*für Mörtel und Beton*] = waterproofing powder
~ring *m*, Dichtring = seal(ing) ring, packing ~, washer
~schicht *f*→ Sperrschicht
~schirm *m*→ Einpreß-Dichtwand *f*
~schleier *m* → Einpreß-Dichtwand *f*
Dichtungsschürze *f* → Dichtungsvorlage *f*
~ → Einpreß-Dichtwand *f*
Dichtungssporn *m* → Dichtungsmauer
Dicht(ungs)spundwand *f* = sheet pile cut-off, ~ ~ diaphragm, ~ ~ (~) wall
Dichtungs|stoff *m*, Dichtungsmittel *n*, Ab~ = sealing material
~strick *m*, Dichtstrick = sealing rope
~technik *f*, Ab~ = sealing engineering

~teppich *m* → Dichtungsvorlage *f*
~trog *m*, Dichtungswanne *f*, Ab~ = tanking
~vorlage *f*, Ab~, (Ab)Dichtungsteppich *m*, (Ab)Dichtungsschürze *f*, Schürze, Außen(haut)dichtung *f*, Oberflächendichtung = (waterproof) blanket, impervious ~
Dichtungswand *f* → Dichtwand
~ aus Pfählen → Pfahlwand
~ ICOS → ICOS-Wand
Dichtungs|wanne *f*, Dichtungstrog *m*, Ab~ = tanking
~zusatz *m* → Dichtungsmittel *n*
Dichtwand *f*, Dichtungswand, Sperrwand, Diaphragma *n* = impervious diaphragm, impermeable ~, underground ~, ~ (~) wall, foundation cut-off
~ aus Pfählen → Pfahlwand
dicke Metallplatte *f* = heavy metal plate
Dicken|änderung *f* = thickness change
~bemessung *f* = thickness design
~faktor *m* [*Straßenbau*] = structural number, thickness index
~herabsetzung *f* = decrease in thickness
Dickglas *n* = heavy crystal sheet glass
Dickschlammverfahren *n* [*Zementherstellung*] = slurry process
Dickspülung *f* [*Rotarybohren*] = drilling mud, rotary ~, mud slush
Dickstoffpumpe *f*, Schlammpumpe *f* = solids-handling pump, slush ~, mud ~
dickwandiges Entnahmegerät *n* [*für ungestörte Bodenproben*] = thick-walled sampler
~ Rohr *n* = heavy wall pipe
diebessicher = thief-proof, pilfer-proof
diebstahlgeeignet = theft-prone
Dielektrizitätskonstante *f* = specific inductive capacity, S.I.C., dielectric constant
Dielen(fuß)boden *m* = (wood) strip flooring, (~) ~ floor finish
Dienst|brücke *f* → Baubrücke
~fahrzeug *n* = service vehicle
~gewicht *n* → Arbeitsgewicht

Dienstleiter — Diesel(ramm)bär

~leiter *f*, Betriebsleiter = service ladder
~vertrag *m* = service-contract
Diesel|aggregat *n*, Diesel-Strom(erzeugungs)aggregat, Diesel-Elektroaggregat = diesel generating set
~-Ankerwinde *f* = diesel anchor winch
~-Autoschütter *m* = diesel (shuttle) dumper
~bär *m*, Dieselhammer *m*, Diesel-Rammbär, Diesel-Rammhammer = diesel (pile) hammer
~betrieb *m* = diesel traction
~-Doppeltrommelwinde *f*, Doppeltrommel-Dieselwinde = double-drum diesel winch
~-Einachs-Räumer *m* = diesel-driven hand-guided two-wheel dozer
(~-)Einspritzpumpe *f* = (diesel) injection pump
diesel-elektrisch angetriebenes Rad *n* = electric wheel
diesel-elektrische Lok(omotive) *f* = diesel-electric loco(motive)
diesel-elektrischer Antrieb *m*, Verbundantrieb = diesel-electric drive
Diesel-Elektro|aggregat *n* → Dieselaggregat
~mobilkran *m* = diesel-electric mobile crane
Diesel|fahr(antrieb)motor *m* = diesel propulsion engine
~fahrzeug *n* = diesel engine vehicle
~feldbahnlok(omotive) *f*, Bau(stellen)-Diesellok(omotive) = diesel field loco(motive)
~-Gleisbaukran *m* = diesel-driven track laying crane
~-(Gleis)Kettenschlepper *m* → Diesel-Raupenschlepper
~grubenlok(omotive) *f* = diesel mine loco(motive)
~hammer *m* → Dieselbär *m*
~karre(n) *f*, (*m*) = diesel truck
~-Kettenschlepper *m* → Diesel-Raupenschlepper
~kompressor *m* = diesel (air) compressor
~kompressorschlepper *m* = diesel universal compressor tractor, diesel tractor-compressor
~-Kraftstoff *m* = diesel fuel, (automotive) gasoil
~löffelbagger *m* = diesel shovel
~-Lok(omotive) *f* = diesel loco(motive)
~-Lok(omotive) *f* für den Streckendienst = main line diesel loco(motive)
dieselmechanischer Antrieb *m* = diesel-mechanical drive
Dieselmotor *m* = diesel (engine), compression-ignition ~; Diesel motor (US)
~ mit Auflading = supercharged diesel (engine), turbodiesel (~), turbocharged diesel ~
~ mit luftloser Einspritzung = direct-injection diesel unit
~-Straßenfahrzeug *n* = derv, diesel-engine road vehicle
~(straßen)walze *f* → → Walze
Dieselöl|-Feinfilter *m, n* = fuel oil secondary filter
~-Grobfilter *m, n* = fuel oil primary filter
Diesel|ölzement *m*, DOC [*Ist eine Mischung von Zement, Dieselöl und oberflächenaktiven Reagenzien. Dieselölzement verfestigt sich durch Aufnahme von Wasser und wird in Erdölbohrungen vorwiegend für Druckzementationen verwendet, um Wasserzuflüsse in Erdöltragern abzusperren*] = diesel oil cement
~(pfahl)ramme *f* = diesel (pile) driving plant, ~ piling ~
~pfahlzieher *m* = diesel pile-puller, ~ pile extractor
~-Radschlepper *m* = diesel wheel-type tractor, ~ wheeled ~
~-Rammbär *m* → Dieselbär *m*
~(ramm)bär *m* mit Schlagzerstäubung, Diesel(ramm)hammer *m* ~ ~ = diesel (pile) hammer operating on the impact atomizing principle

Dieselramme — dimensionslose Zahl

~ramme f, Dieselpfahlramme = diesel (pile) driving plant, ~ piling ~

~-Rammhammer m → Dieselbär m

~-Raupengreifer(kran) m, Greifer(kran) mit Gleisketten mit Dieselantrieb = caterpillar (or tracklaying) diesel-driven grab(bing crane), ~ (~ ~) ~ (grab) bucket crane

~-Raupenschlepper m, Diesel-Raupentraktor m, Diesel-Raupentrecker m, Diesel-(Gleis)Kettenschlepper, Diesel-(Gleis)Kettentraktor m = diesel crawler tractor

~-Rollmaterial n = diesel stock

~-rüttler m, Dieselvibrator m = diesel-operated vibrator

~-Schienengreifer(kran) m, Greifer(kran) auf Schienen mit Dieselantrieb = rail-mounted diesel-driven grab(bing crane), ~ ~ (grab) bucket crane

~-Schlepper-Kran m = diesel tractor crane

~schlosser m = diesel mechanic

~-Schmieröl n = diesel lube (oil), ~ lubricating oil

~schwimmsaugbagger m, Dieselnaßsaugbagger = diesel suction dredge(r)

~-Straßenbahnwagen m = diesel tramcar (US); ~ streetcar (Brit.)

~-Strom(erzeugungs)aggregat n → Dieselaggregat

~-Triebwagen m = (self-propelled) diesel railcar

~-Triebwagenbetrieb m = diesel multiple-unit operation

~universalbagger m = diesel universal excavator

~vibrator m, Dieselrüttler m = diesel-operated vibrator

~walze f → ~ Walze

~winde f = diesel winch

~zug m = diesel train

~-Zylinderlaufbuchse f = diesel liner

Differential n = differential

~-Eingangswelle f = differential input shaft

~-Flaschenzug m = differential pulley block

~gehäuse n = differential compartment, ~ case, ~ housing

~gleichung f = differential equation

~lagerkasten m = differential carrier

~-Rechnung f = differential calculus

~sperre f = differential lock

~-Teilgerät n = differential indexing head

~-Thermo-Analyse f, DTA = differential thermal analysis

~trieb m = differential drive

~-Vakuumheizung f = sub-atmosphere heating system

~wandler m = torque divider

~wandlergetriebe n = torque divider transmission

~wandler-Steuerventil n = torque divider transmission control valve

~wasserschloß n, Steigrohrwasserschloß = differential surge tank

Differenzmethode f [*Berechnung von Tragwerken mit veränderlichen Trägheitsmomenten*] = constant segment method, analysis by finite differences [*Analysis of non-uniform structural members*]

Diffusions|konstante f = diffusion constant

~luftpumpe f, Quecksilber-Diffusionspumpe = mercury-vapo(u)r pump

digitale Abschreibung f = years-digits method

~ **Anzeige** f = display in digital form

Digitalrechner m = digital computer

Dihydrat n, Rohgips m, $CaSO_4 \times 2H_2O$ [*Ausgangsmaterial für gebrannten Gips*] = raw gypsum

Dikalziumsilikat n, Bikalziumsilikat n, $2\,CaO \times SiO_2$ [abgek. C_2S] = dicalcium silicate

dimensionieren, bemessen = to design

Dimensionieren n → Bemessung f

Dimensionierung f → Bemessung

dimensions|frei, dimensionslos = dimensionless

~lose Zahl f = pure number

Dinasstein *m* [*veraltete Bezeichnung*]; Silikastein, Quarzitstein, Quarzkalkziegel *m* = silica brick, acid firebrick; gannister brick (Brit.); ganister brick (US)

Diopsid *m* = diopside

Diorit *m* = diorite

Dipyrschiefer *m*, Schmelzsteinschiefer = dipyre (or dipyrite, or mizzonite) slate

direkt gekuppelte Pumpe *f* = close-coupled pump

direkte Wärmeschichtung *f* → Tropenschichtung

direkter Auslaufstollen *m* [*von Schieberkammer zum Ableitungstunnel unter Umgehung des Kraftbauses*] = penstock drain, drain(age) tunnel

direktgesteuerte Bauart *f* [*Ventil*] = direct acting design

Direkt|analyse *f* = direct analysis

~emaillierung *f* = direct-on enameling

~sondierung *f* → Bohrsondierung

~stufe *f* = direct drive speed

~stufenkupp(e)lung *f* = direct drive clutch

~stufenschaltventil *n* = direct drive shift valve

Dischinger-Vorspannsystem *n* [*Vorspannung durch hydraulische Pressen*] = Dischinger pre-stressing method

Disken(bohr)meißel *m*, Scheiben(bohr)meißel = disc bit; disk ~ (US)

diskontinuierlich abgestufte Mineralmasse *f*, unstetig ~ ~ = gap-graded aggregate

diskontinuierliche Kornabstufung *f*, unstetige ~ = gap grading, discontinuous ~, discontinuous granulometry

~ Mischanlage *f* → Chargen-Mischanlage

~ Zwangsmischung *f* → Chargen-Zwangsmischen *n*

Diskusegge *f*, Scheibenegge = disc harrow; disk ~ (US)

Dislokation *f* → Krustenbewegung

Dislokationsbeben *n*, tektonisches Beben = tectonic earthquake

dispergieren, feinverteilen = to disperse

Dispersion *f*, Feinverteilung *f* = dispersion

~, Verteilung *f* = dispersion [*A method of disposal of the suspended solids in sewage or effluent by scattering them widely in a stream or other body of water*]

Dispersionsmittel *n* = dispersion agent

Dissousgasflasche *f*, Azetylenflasche = acetylene cylinder, ~ bottle

Distanz|ring *m* = circular spacer

~stück *n* = spacer

divergentstrahlig-körnige Textur *f*, ophitische ~, Intersertaltextur = ophitic texture

DM → (Beton)Dichtungsmittel *n*

DOC → Dieselölzement *m*

Docht|schmierung *f* = wick oiling

~wirkung *f* **der Kapillaren** = capillary effect

Dock|anlage *f* = dock installation

~ausbau *m*, Dockerweiterung *f* = dock development

Docke *f*, Baluster *m* = baluster [*In architecture, a dwarfed column or pillar supporting a railing or coping. A series of balusters form a balustrade*]

Dock|erweiterung *f*, Dockausbau *m* = dock development

~kran *m* = shipbuilding crane, building slip(way) ~, docksider

~mauer *f* = dock wall

~pumpe *f* = dock pump

~schlepper *m* = dock tug

~schleuse *f* = entrance lock

Döglingtran *m*, Entenwaltran, arktisches Spermacetiöl *n* = bottlenose oil

Dolerit *m* = dolerite [*A basic hypabyssal rock of medium-grained texture, consisting of plagioclase feldspar, augite, iron ores, and frequently olivine*]

Doline *f*, Karsttrichter *m* = doline, dolina

Dolinensee *m* = sink (hole) lake

Döll-Klappen(stau)wehr — Doppeleisen 314

Döll-Klappen(stau)wehr n = Döll wicket (dam) (US); ~ shutter (~) (Brit.)
Dolomit m → Dolomitgestein n
~asche f = earthy dolomite
~gestein n, Dolomit(fels) m, Dolomitstein m = dolomite (rock) [*A limestone containing in excess of 40 per cent of magnesium carbonate as the dolomite molecule*]
Dolomitisierung f = dolomitization
Dolomit|kalk m → Graukalk
~marmor m = dolomite marble [*A crystalline variety of limestone, containing in excess of 40 per cent of magnesium carbonate as the dolomite molecule*]
~marmor m **mit zwischen 5–40% Magnesiumkarbonat** = dolomitic marble, magnesian ~
~mergel m = dolomitic marl
~sand m = dolomitic sand
~stein m = dolomite brick
~stein m → Dolomitgestein
~steinpresse f = dolomite brick press
~werk n = dolomite (processing) plant
~-Zerfall m **bei thermischem Gleichgewicht** = equilibrium thermal decomposition of dolomite
Dom m [*Tellerradgehäuse*] = oil guard [*bevel gear case*]
~gebirge n = dome(d) mountain
Donarit m, Ammonsalpetersprengstoff m mit Nitroglyzerin = donarite
donlägiger Schacht m, tonnlägiger ~ = sloping shaft
Doppel|-Abwasserdüker m, Abwasser-Doppeldüker = inverted siphon sewer with two pipes, depressed ~ ~ ~ ~ ~
~abzweig m = double branch pipe, ~ Y
~achse f = twin axle
~antrieb m = dual drive
~aufzug m → Doppelbauaufzug
~ausleger-(Licht)Mast m = (lighting) column with two brackets (Brit.); mast with two arms (US)

~backenbrecher m → → Backenbrecher
~backenbremse f = double-shoe brake
~(bau)aufzug m = double (contractors') (or builders') hoist
~baugrubenaufzug m = double building pit hoist, ~ foundation ~
~bereifung f = dual tyres (Brit.); ~ tires (US); double ~
~bindung f [*Bitumenchemie*] = double linking
~bogen(stau)mauer f = double arch(ed) dam
~bohlenfertiger m = twin-screed finisher, two-screed ~, double-screed ~, ~ finishing machine
~-Bohraggregat n, Gesteinsbohreinheit f, Atlas-Copco-Bohreinheit = twin-drill
~brechung f [*Doppelspat*] = double refraction
~brücke f = twin bridge
~dach n = close-boarded battened roof, double ~
~decker m, Doppelstufensieb n, Sieb mit zwei Siebfeldern, Zweidecker, zweistufiges Sieb, Doppeldeckersieb, Doppeldecksiebmaschine f = double deck screen, DD ~
~decker-O(mni)bus m, Doppelstockbus = double-decker bus
~deckung f [*Dach*] = close-boarded battened roofing
~drahtgewebe n = double (woven-) wire cloth, ~ gauze wire ~
~drehbrücke f = double swing(-span) bridge
~drehkran m, Duplexkran, Doppelschwenkkran = duplex crane
~düker m = inverted siphon with two pipes, depressed ~ ~ ~ ~
~ebenenbetrieb m, Zweiebenenbetrieb [*Flughafen*] = two-level operation
~ebenensystem n → Zweiebenensystem
~eisen n, Doppelhobeleisen = double plane iron

~endklassierung *f* [*mechanische Trennung festflüssiger Mischsysteme*] = tayloring

~entsandungsbecken *n* = double sand catching basin, dual ~ ~ ~, twin ~ ~ ~

~falzziegel *m*, Pfannenziegel = single Roman tile

~fenster *n* = double(-glazed) window, storm ~

~flachlitze *f* = double flattened strand

~flanschrolle *f* [*Gleiskette*] = double flange roller

~förderrinne *f* → Doppelschwingförderrinne

~gefäß(förder)pumpe *f* = dual-tank type pump, ~ pneumatic conveying system

~generator *m* = double current generator

doppelgleisige Bahnlinie *f*, zweigleisige ~ = double-track line

doppelgleisiger Privatanschluß *m* = two-line siding, double-track ~

Doppel|hammerbrecher *m*, Doppelrotorenhammerbrecher = double hammer breaker, ~ ~ crusher

~hammermühle *f* = two-shaft hammer mill

~handgriff *m* = double hand grip

~haus *n* = pair of semi-detached houses

~hebel *m* = dual lever

~hebel *m*, Ausgleichbalken *m* [*Gleichgewichtsklappe*] = balance beam

~-H-Stahlstütze *f* = double H steel column

~hydrat *n* → Rohgips *m*

~käfigmotor *m* = double-squirrel cage motor

~keilanker *m* = sliding wedge bolt

~kernrohr *n* = double tube (core) barrel

~ketten(kratzer)förderer *m* (⚒) [*Der Ausdruck „Panzerförderer" ist eine geschützte Firmenbezeichnung*] = armo(u)red face conveyor, ~ ~ conveyer, snaking ~

~klappenwehr *n*, Dachwehr = beartrap (gate), ~ weir, roof weir [*It consists essentially of two leaves, an upstream leaf hinged and sealed along its upstream edge and a downstream leaf hinged and sealed along its downstream edge*]

~kniehebelbrecher *m* → → Backenbrecher

~konus-Kipptrommelmischer *m* → → Betonmischer

~konustrommel *f* = double-cone drum, dual-cone ~

~kopf-Ellira-Schweißmaschine *f*, Zweikopf-Ellira-Schweißmaschine = twin-head submerged-arc welding machine

~kopfschiene *f*, Stuhlschiene = double headed rail, chair ~

~kragbinder *m* = double-cantilever truss

~kran *m* = combined tower crane and derrick

~kurbelsieb *n* = screen with twin crank-shaft drive

doppellagige Dach(pappen)eindeckung *f*, zweilagige ~ = 2-ply roofing

Doppel|laschenkettenbecherwerk *n* = bucket elevator with double roller type chains

~laschennietung *f* = double butt strap joint

doppellippiger Simmerring *m* = double lip seal

Doppel(material)umschlag *m* = double-handling

doppelmäuliger Schraubenschlüssel *m* → Doppelschlüssel

Doppelmeißel|bohrkrone *f* = double chisel (drill) bit

~schneide *f* = two-point chisel type bit cutting edge

Doppel|membran(e)pumpe *f* = dual diaphragm pump

~motorenschürfzug *m* = tandem powered scraper, twin-powered ~

doppelmotorig, zweimotorig = twin-powered, tandem powered

doppelmotoriger Radschrapper — Doppel(schwing)förderrinne 316

doppelmotoriger Radschrapper *m*, Motorschürf(kübel)wagen *m* mit zusätzlichem Heckantrieb, Schürfwagen(zug) *m* mit besonderem Schürfkübelmotor = twin-powered scraper, tandem powered ~

Doppel|muffe *f* [*Verbindungsstück für zwei Bohrgestängezapfen*] = double box

~**niedrigwasser** *n* = double low waters

~**pendellager** *n*, Zweipendellager = double tumbler bearing, ~ pendulum ~

~**pendellager** *n*, Zweipendellager [*für Hoch- und Tiefbauten*] = double rocker bearing

~**pentagon-Winkelprisma** *n* = double prism optical square

~-**Prallbrecher** *m* → Doppel-Rotoren-Prallbrecher

~**preventer** *m* = double unit type (blow-out) preventer

~**pumpe** *f* = dual pump

~-**Querbohlen(beton)fertiger** *m* = double-screed transverse finisher

~**rad** *n*, Zwillingsrad = dual wheel

~**radlenkbock** *m* = dual wheel steering gear

~**reifenrad** *n*, Zwillingsreifenrad = dual-tyre wheel (Brit.); dual-tire ~ (US)

~**rinnenförderer** *m* → Doppelschwingförderrinne

~**rohrabhitzekessel** *m* = double-tube waste heat boiler

~**rohr-Durchlaß** *m* → Zweirohr-Durchlaß

~**rohrentnahmegerät** *n*, Rotations-Kern-Gerät [*für ungestörte Bodenproben*] = double tube sampler

~**rohrförderer** *m* = twin-tube conveyor, ~ conveyer

~(**rohr**)**tunnel** *m* → Doppeltunnel

~**rollenlager** *n*, Zweirollenlager [*für Hoch- und Tiefbauten*] = two-roller bearing

~**rotorenhammerbrecher** *m* → Doppelhammerbrecher

~(-**Rotoren**)-**Prallbrecher** *m*, Zweiwalzen-Prallbrecher, Prallbrecher mit zwei Rotoren = double impeller impact breaker

~-**Rotoren-Prallmühle** *f*, Zweiwalzen-Prallmühle = double impeller impact mill

~**rührwerk** *n* = double agitator

~**rüttelwalze** *f* → Doppelvibrationswalze

~**rüttler** *m* → Doppelvibrator *m*

~**sackbohrer** *m* = two-pocket bore

~**sägeblattanordnung** *f* (✗) = double saw-tooth system

~**schacht** *m* = twin shaft

~**schale** *f* = double shell

~**schaufellader** *m* = twin rocker shovel

~**scheibenkupplung** *f* = twin-plate clutch

~**scheibenrad** *n* = double-disc wheel; double-disk ~ (US)

~**schichtenbetrieb** *m* = double-shifts operation

~**schiebetischpresse** *f* = double sliding table press

~-**Schiebetür** *f*, Schiebetür mit zwei Flügeln = double sliding door

~**schlagbohlenfertiger** *m*, Doppelstampfbohlenfertiger = double tamping beam finisher

~**schleuse** *f*, Zwillingsschleuse = twin locks

~**schlüssel** *m*, doppelmäuliger Schraubenschlüssel = double-ended spanner (Brit.); double head wrench (US)

~-**Schmelzkessel** *m* [*Fugenvergußmasse*] = double-boiler kettle

~**schütz(e)** *n*, (*f*) = double-leaf gate

~**schwenkkran** *m*, Duplexkran, Doppeldrehkran = duplex crane

~-**Schwertwäsche** *f*, Doppel-Schwertauflöser *m* = double log washer

~**schwingenbrecher** *m* → → Backenbrecher

~(**schwing**)**förderrinne** *f*, Doppel(schwing)rinnenförderer *m*, Doppelvibrationsrinne, Doppelvibrierrinne, Doppelschwingrinne = two-

317 Doppelschwingrinne — Doppeltr.(beton)straßenmischer

trough vibrating conveyor, ~ ~ conveyer

~schwingrinne f, Dosiervibrationsrinne = vibrating trough batcher

~schwing(ungs)walze f → Doppelvibrationswalze

~seilgreifer(krankorb) m → Zweiseilgreifer(krankorb)

~seilschwebebahn f, Zweiseilschwebebahn [*sogenannte deutsche Bauart*] = double ropeway, ~ cableway

~seiltrommel f = twin rope drum

doppelseitiger Ofen m = double-sided (gas-fired) furnace

Doppel|sitzventil n = dual-seat(ed) valve

~-skisprung-Überlauf m über die Krafthausdecke → → Talsperre f

~-S-Kurve f = two-chain curve

~spat m, Kalkspat von Island, isländischer Kalkspat [*Ein klar durchsichtiges Spaltstück von Kalkspat läßt von einer darunterliegenden Schrift zwei Bilder erscheinen*] = double refraction calcspar

~spitzhacke f, Doppelspitzpickel m, Zweispitz m = (clay) pick [*pointed at both ends*]

~-Sprungschanze f → → Talsperre

~spurkranzrad n = double-flanged wheel

~stampfbohlenfertiger m, Doppelschlagbohlenfertiger = double tamping beam finisher

~stegblechträger m = double-webbed girder

~stiefelhandpumpe f = double barrel hand pump

~stockbus m, Doppeldecker-O(mni)bus = double-decker bus

~stufensieb n → Doppeldecker m

doppelstumpfgeschweißt = double butt welded

Doppel|stütze f = twin column

~tandemfahrgestell n, Doppeltandemradpaar n [*Flugzeug*] = dual-in-tandem (landing gear) assembly

~taschensilo m, Doppelwabensilo = two-compartment bin

~-Tauchkolben m = 2 plungers

~-T-Balken m, I-Balken = I-beam

doppelt beaufschlagt = double-acting

~ gekrümmte Bogen(stau)mauer f = double-curvature arch(ed) dam

~ gekrümmte Schale f = doubly curved (shell) dome, ~ ~ shell, double curvature shell

doppelt|-logarithmischer Maßstab m = log-log scale

~-logarithmisches Körnungsnetz n = log-log coordinate for particle curves

doppelt symmetrischer Stützenquerschnitt m = column section symmetrical about two axes

~ untersetzter Seitenantrieb m [*Gleiskettenlaufwerk*] = double reduction final drive

doppelte Einschubdecke f = double sound-boarded floor

~ Oberflächenbehandlung f, ~ OB [*Straße*] = double (bituminous) surface treatment, armo(u)r coat, two-pass surface treatment, inverted penetration surface treatment, multiple-lift treatment

~ Spundwand f → Kastenwand

~ Stegverlaschung f = double-strap web joint

doppelter Hängebock m, doppeltes Hängewerk n = queen (post) truss [*Is used for short spans in connection with wood construction*]

~ Luftreiniger m = twin air cleaner

~ Muldenboden m [*Erdbaulastfahrzeug*] = double skin floor

doppeltes Hängewerk n → doppelter Hängebock m

Doppel-T-|Deckenelement n = double T floor unit

~-Montageträger m, Montageträger I = double T beam without cast in situ concrete

~-Profil n = I-beam section

~-Stahl m = I-steel

Doppel|treppe f = double stair(s)

~trommelbetoniermaschine f → → Straßen(beton)mischer m

~trommel(beton)straßenmischer m → → Straßen(beton)mischer

Doppeltrommel-Dampfwinde — Dosieranlage

~trommel-Dampfwinde *f*, Dampf-Doppeltrommelwinde = double-drum steam winch

~trommel-Dieselwinde *f*, Diesel-Doppeltrommelwinde = double-drum diesel winch

~trommelmischer *m* = dual-drum mixer, double-drum ~

~trommelrammwinde *f* = double-drum pile driving winch

~trommelseilwinde *f* = double-drum cable winch, dual-drum ~ ~

~tunnel *m*, Zwillingstunnel, Doppelrohrtunnel = twin(-bore) tunnel

~tür *f* = double door

doppeltwirkend, dpw = double-acting

doppeltwirkende Kolbenmembran(e)-pumpe *f* = double-acting combination piston and diaphragm pump

~ Pumpe *f* = double-acting pump

Doppel|umschlag *m*, Doppelmaterialumschlag = double-handling

~-Unterdükerung *f* = double-sag crossing

~untersetzung *f* = double reduction

~ventil-Spülspitze *f* = dual-valve jetting tip

~verglasen *n*, Doppelverglasung *f* = double-glazing

~-Vibrations(ramm)bär *m*, Doppel-Vibrations(ramm)hammer *m* = dual vibratory (pile) hammer, ~ vibrating (~) ~

~vibrationswalze *f*, Doppelrüttelwalze, Doppelschwing(ungs)walze = double vibrating roller, ~ vibratory ~, vibration ~

~vibrator *m*, Doppelrüttler *m* = dual vibrator, double ~, twin ~

~vibrierrinne *f* → Doppel(schwing)-förderrinne

~wabensilo *m*, Doppeltaschensilo = two-compartment bin

~walzwerk *n*, Doppelwalzenbrecher *m* = four-roll crusher, ~ breaker

doppelwandiger Spundwandfang(e)-damm *m*, doppeltes Spundwandbauwerk *n* = double-wall(ed) sheetpile cofferdam

Doppelwellen|-Knetmischer *m* → Zweiwellen-Zwangsmischer

~mischer *m* = double shaft mixer, twin ~ ~

~-Schwertwäsche *f* = two-log washer

~sieb *n* = twin shaft screen

~-Zwangsmischer *m* → Zweiwellen-Zwangsmischer

Doppel|winde *f* = donkey

~winkelsteg *m* = double angle web

~zapfen *m* [*Verbindungsstück für zwei Bohrgestängemuffen*] = double pin

~zimmer *n* = double bedroom

~zug *m* [*aus zwei Stangen bestehender Gestängezug beim Rotarybohren*] = double

~zugbrücke *f* = double draw bridge

~zündung *f* = dual ignition

~zweckhydraulikpumpe *f* = double unit hydraulic pump

Döpper *m* = rivet snap

~schelle *f* = rivet snap retaining spring

dorisches Kyma *n*, dorische Kymation *f* = Doric cyma(tium), cyma(tium) recta

Dor NII-Verankerung *f* mit innerem Konusring = Dor NII anchorage system, ~ ~ anchoring ~

DORR-Klärer *m* = DORR thickener, ~ clarifier [*A proprietary device placed in a tank by which a system of scrapers, driven by a central shaft, revolves slowly, pushing the deposited sludge to a central outlet, from which it is removed by gravity or pumping*]

Dortmund-Brunnen *m* = Dortmund tank [*A vertical-flow sedimentation tank with hopper bottom. The sewage, introduced near the bottom, rises and overflows at the surface, and the sludge may be removed from the bottom before it becomes septic*]

Dosen|barometer *n* → Aneroidbarometer

~libelle *f* = circular spirit level

~sextant *m* = box sextant

Dosieranlage *f* → Abmeßanlage

Dosieranlage in Turmanordnung — Dosier(ungs)silo

~ in Turmanordnung → Abmeßanlage ~ ~

~ mit dreimaligem Halten der Fahrzeuge → Abmeßanlage ~ ~ ~ ~

~ ~ einmaligem Halten der Fahrzeuge → Abmeßanlage ~ ~ ~ ~

~ ~ zweimaligem Halten der Fahrzeuge → Abmeßanlage ~ ~ ~ ~

Dosier|apparat m, Banddosiergerät n = belt batcher

~automatik f → Abmeßautomatik

~band n → Abmeßband

~bandwaage f, Abmeßbandwaage, Zuteilbandwaage, Zumeßbandwaage = batch(ing) conveyor belt scale, proportioning ~ ~ ~, measuring ~ ~ ~, ~ conveyer ~ ~

~behälter m → Beschickungsbehälter [Abwasserwesen]

dosieren, zuteilen, zumessen, abmessen = to proportion, to batch, to measure

Dosier|förderschnecke f → Abmeßförderschnecke

~gefäß n → Abmeßgefäß

~-Hygrometer n → Abmeß-Hygrometer

~karre(n) f, (m) für (Beton)Zuschläge → Abmeßkarre(n) ~ ~

~kiste f → Abmeßkiste

~pumpe f → Abmeßpumpe

~schleuse f → Abmeßschleuse

~schnecke f für Zement → Abmeßschnecke ~ ~

~spiel n, Zumeßspiel, Zuteilspiel, Abmeßspiel = batch(ing) cycle, proportioning ~, measuring ~

~teil m, n → Abmeßteil

~turm m → Turmdosieranlage

Dosierung f, Zuteilung, Zumessung, Abmessung = proportioning, batching, measuring

Dosier(ungs)anlage f für Zement, Abmeßanlage ~ ~, Zumeßanlage ~ ~, Zuteilanlage ~ ~, Zement-Dosier(ungs)anlage, Zement-Abmeßanlage, Zement-Zumeßanlage, Zement-Zuteilanlage = cement batch(ing) plant, ~ proportioning ~, ~ measuring ~, ~ installation

~ ~ **Zuschlagstoffe**, Abmeßanlage ~ ~, Zuteilanlage ~ ~, Zumeßanlage ~ ~, Zuschlagstoff-Dosier(ungs)anlage, Zuschlagstoff-Abmeßanlage, Zuschlagstoff-Zuteilanlage, Zuschlagstoff-Zumeßanlage = aggregate batch(ing) plant, ~ proportioning ~, ~ measuring ~

Dosier(ungs)|apparat m → Dosier(ungs)gerät n

~automat m, Zuteilautomat, Zumeßautomat, Abmeßautomat = auto(matic) batch(ing) system, ~ proportioning ~, ~ measuring ~

~bunker m, Zuteilbunker, Zumeßbunker, Abmeßbunker = batch(ing) bin, proportioning ~, measuring ~

~gerät n, Zuteilgerät, Zumeßgerät, Dosier(ungs)apparat m, Zuteilapparat, Abmeßgerät, Abmeßapparat, Zumeßapparat, Dosier(ungs)vorrichtung f, Zuteilvorrichtung, Zumeßvorrichtung, Abmeßvorrichtung = batch(ing) unit, proportioning ~, measuring ~, batcher

~programmwähler m, Zuteilprogrammwähler, Zumeßprogrammwähler, Abmeßprogrammwähler = batch(ing) selector, proportioning ~, measuring ~

~rinne f, Abmeßrinne, Zumeßrinne, Zuteilrinne = batch(ing) trough conveyor, ~ ~ conveyer, measuring ~ ~, proportioning ~ ~, conveyor trough, conveyer trough

~schnecke f, Zuteilschnecke, Zumeßschnecke, Abmeßschnecke, Schneckenzuteiler m, Schneckendosierer = proportioning worm conveyor, ~ ~ conveyer, batch(ing) ~ ~, measuring ~ ~, screw ~, batch(ing) screw, proportioning screw, measuring screw

~silo m, Zuteilsilo, Zumeßsilo, Abmeßsilo = batch(ing) silo, proportioning ~, measuring ~

Dosier(ungs)stern — Drahtstift

~**stern** *m*, Abmeßstern, Zumeßstern, Zuteilstern = batch(ing) star, proportioning ~, measuring ~

~**trommel** *f*, Abmeßtrommel, Zumeßtrommel, Zuteiltrommel, Trommeldosierer *m*, Trommelzuteiler = batch(ing) drum, proportioning ~, measuring ~

~**vorrichtung** *f* → Dosier(ungs)gerät *n*

~**waage** *f*, Abmeßwaage, Zuteilwaage, Zumeßwaage = batch(ing) weigh gear, measuring ~ ~, proportioning ~ ~, ~ scale, batcher scale, scale batcher

Dossierung *f*, Anlauf *m*, Anzug *m*, Neigung = batter, taper

Dowsongas *n*, Mischgas, Stadtgas = Dowson gas

dpw, doppeltwirkend = double-acting

Draht|-Bildgüteprüfsteg *m*, Drahtsteg = wire (image) penetrameter

~**bindemaschine** *f* = wire tying machine

~**bindung** *f* = wire tie

~**bügel** *m* = wire stirrup

~**bündel** *n*, Bündel aus Drähten, Spannkabel *n*, Spann(draht)bündel = prestressing cable, group of wires, strand

~**bürste** *f* = wire brush

~**durchmesser** *m* = wire diameter

~**erkennbarkeit** *f*, DE *f* [*bei einem Drahtsteg*] = radiographic sensitivity, penetrameter ~ (US); wire ~ (Brit.)

~**folge** *f* [*bei einem Drahtsteg*] = seven metal wires of graduated thickness

~**geflecht** *n* = wire mesh, ~ netting

Drahtgewebe *n* = (woven-)wire cloth, gauze wire ~

~ **mit geschnittenen Kanten** = (woven-)wire cloth with cropped edges, gauze wire ~ ~ ~

~ ~ **umgelegtem Blechfalz** = (woven-)wire cloth with folded edges and sheet metal, gauze wire ~ ~ ~

~ ~ **umgelegten Kanten** = (woven-) wire cloth with folded edges, gauze wire ~ ~ ~ ~

~**normen** *fpl* = (woven-)wire cloth standards (Brit.); wire fabric ~ (US)

Draht|glas *n* = wired glass

~**gurtförderer** *m* = wire mesh belt conveyor, ~ ~ ~ conveyer

~**gurtvollelevator** *m* = continuous wire-belt elevator

~**haspel** *m*, *f* = wire reel

~**hose** *f*, Schutzgitter *n*, Baumrost *m* = tree-guard

~**kabel** *n* → Drahtseil *n*

~**klemme** *f* = wire grip

~**korb** *m* = wire basket

~**kreuzung** *f* [*Siebboden*] = crossing of wires, interlacing ~ ~

~**lage** *f* = layer of wire

~**lage** *f* [*Spannbeton*] = layer of wires, group ~ ~

~**lehre** *f* = wire ga(u)ge

(~)**Litze** *f* = strand

~**nagel** *m* = wire nail

~**querschnitt** *m* = wire section

~**richt- und -schneidmaschine** *f* = wire straightening and cutting machine

~**rolle** *f*, Drahtring *m* = coil of wire, wire coil

~**rolle** *f* = wire reel

~**schere** *f* = wire cutter

~**schrott-Wickler** *m* = cobble-baller

~**seele** *f* = core

Drahtseil *n*, Drahtkabel *n*, Stahl(draht)seil = (steel) wire rope, (~) cable

~**bahn** *f* = ropeway

~**besen** *m* [*beim Verguß*] = "brush" end

~**machart** *f* = wire rope construction

~**messer** *n* = wire rope knife

Draht|siebgewebe *n* = (woven-)wire screen cloth, gauze wire ~ ~

~**sorte** *f* = wire grade

~**speichenrad** *n* = spoked wire wheel

~**spritzpistole** *f* = wire-spray pistol, ~ gun

~**steg** *m*, Draht-Bildgüteprüfsteg = wire (image) penetrameter

~**stift** *m* = wire brad [*A small headless nail made from wire*]

Draht- und Funksprechzentrale — drehen

~- und Funksprechzentrale *f* = WT/RT radio station
~verhau *m* = wire entanglement
~wort *n*, Katalogbezeichnung *f*, Modellbezeichnung = code (word)
~ziegelgewebe *n*, Stauß-Ziegelgewebe = clay lath(ing) [*B.S. 2705*]
Draisine *f*, Eisenbahndraisine, Bahnmeisterwagen *m* = trackmotor car (US); platelayers' troll(e)y (Brit.)
Drall *m* = twist
~ [*Hydraulik*] = moment of momentum, ~ ~ velocity
~drossel *f* [*Verdichter*] = intake guide unit, guide vanes
drallfrei = non-spinning
Drän *m* = drain [*A conduit or pipe, usually underground, for carrying off, by gravity, liquids other than sewage and industrial wastes, and including ground or subsoil water, surface water and storm water*]
~abstand *m* = drain spacing
Dränage *f*, ~system *n* = drainage system [*A system of drains*]
~, Entwässerung *f*, Wasserentzug *m* [*Der Vorgang, der im Boden zum Endzustand kapillaren Gleichgewichtes führt*] = drainage
~arbeiten *fpl* [*Schweiz: Dränarbeit*] = drainage work
~löffel *m* = (back)hoe dipper for drainage work (US); backacter ~ ~ ~ (Brit.)
~rohr *n* = drain tile (US); ~ pipe [*Pipe of burned clay, concrete, etc., in short lengths, usually laid with open joints to collect and remove drainage water*]
~system *n*, Dränage *f* = drainage system [*A system of drains*]
~wasser *n* = drainage water
Drän|arbeiten *fpl* = drainage work
~betonrohr *n*, Betondränrohr = concrete drain tile (US); ~ ~ pipe
~graben *m* = drain trench
~grabenaushub *m* = excavating drain trenches
(~)Grabenfräse *f* → Grabkettenbagger

~grabenmaschine *f* = mole plough (Brit.); ~ plow (US)
drainieren = to drain
Drän|maschine *f* = Kjellmann-Franki machine
~rohr *n*, Entwässerungsrohr = drain pipe; ~ tile, sewer tile (US)
~rohrverlegemaschine *f* = tile-laying machine (US); drain pipe-laying ~
~schicht *f* → Filterschicht
~spaten *m* = drain-trench spade
~tonrohr *n*, Tondränrohr = clay drain pipe; ~ ~ tile (US)
~werkzeug *n* = drainage tool
Draufsicht *f* = plan view, top side ~
Dreharm *m* [*Nichols-(Etagen)Ofen*] = mechanical rake
~ [*Segmentwehr*] = arm
Dreh|ausleger *m* = swivelling jib (Brit.); swiveling boom (US)
~bagger *m*, Schwenkbagger = slewing excavator
~bagger *m*, Schwenkbagger [*Eimerkettenbagger mit um 360° schwenkbarem Oberteil*] = slewing bucket-ladder excavator
drehbare Laufrolle *f*, Schwenkrolle = caster (wheel), castor (~)
drehbarer Oberteil *m* → Aufsatzteil
~ Oberwagen *m* → Aufsatzteil
~ Radschrapper *m* = rotary scraper
Dreh|-Baukran *m* = Bau-Drehkran
~bereich *m* → Arbeitsbereich
~bohren *n* = rotating drilling
~bohrschneide *f* = rotary bit edge, ~ drill point
~bohrverfahren *n* mit Wasserspülung, Drehspülbohrverfahren [*Brunnenbau*] = hydraulic rotary method
~brücke *f* = swing bridge
~bühne *f* = rotating deck
~bühne *f* [*Theater*] = revolving stage, rotary ~, rotating ~
~-Dampframme *f* → Dampf-Drehramme
~-Elektromotor *m* → Elektro-Schwenkmotor
drehen, ab~, über~ [*mit der Schablone. Steinzeugindustrie*] = to jigger

Drehen — Drehrichtung

Drehen *n*, Ab~, Über~ [*mit der Schablone. Steinzeugindustrie*] = jiggering

Dreh|feuer *n* = rotating lens beacon
~**filter** *m, n* = revolving filter
~**flügelfenster** *n* **nach außen** = outswinging casement window
~**flügelfenster** *n* **nach innen** = inswinging casement window
~**flügelpumpe** *f* = rotary vane pump
~**flügelsonde** *f* = rotating-auger tester
~**gasgriff** *m* = twist grip throttle control
~**gelenk** *n* = swivel joint
~**gestell** *n* = bogie (truck)
~**gestellfederung** *f* = bogie bolster
~**gestellwagen** *m* = bogie vehicle
~**greifer** *m* = rotary grab
~**griff** *m* = turning handle
~**halbmesser** *m* → Arbeitsbereich *m*
~**kappe** *f* = rotocap
~**kolbenmotor** *m* = rotary piston motor
~**kolbenpumpe** *f*, Kapselpumpe, Rotationspumpe = rotary pump
~**kolbenverdichter** *m* = rotary piston compressor

Drehkopf *m*, Bohrwirbel *m*, Rotationskopf, Spülkopf [*Rotarybohren*] = rotary swivel
~**bolzen** *m* [*Rotarybohren*] = swivel pin
~**kugel** *f* [*Rotarybohren*] = swivel ball

Drehkran *m* → Schwenkkran

Drehkranz *m*, Schwenkkranz = slewing ring, circular pathway
~**antriebswelle** *f* [*Motorstraßenhobel*] = circle reverse drive shaft
~**aufhängung** *f* [*Motorstraßenhobel*] = circle support
~**halter** *m*, Zugstange *f* [*Motorstraßenhobel*] = (circle) drawbar

Dreh|kübel *m*, Schwenkkübel = rotary bucket, ~ skip, rotating ~, revolving ~
~**kühler** *m* = rotary cooler
~**kuppel** *f* = revolving dome
~**meißel** *m* = rotary bit

Drehmoment *n* = torque

~**anstieg** *m* = torque rise
~**belastung** *f* = torque load
~**feder** *f* = torque spring
~**leistung** *f* = torque performance
~**messer** *m* = torquemeter
~**schraubenschlüssel** *m* = torque wrench
~**stufe** *f* = torque rate
~**verzweigung** *f*, Leistungsverzweigung = torque division, ~ split
~**-Wahlschalter** *m* = torque selector
~**wandler** *m* = torque converter
~**wandler-Abwürgemoment** *n*, Anfahrpunkt *m* = torque converter stall point
~**wandlerantrieb** *m* = torque converter drive
~**wandlerbehälter** *m* = torque converter tank
~**wandlergetriebe** *m* = torque-converter transmission
~**wandlung** *f* = torque conversion

Drehofen *m*, Rotierofen = rotary kiln
~**-Blähton** *m* = rotary kiln expanded clay; ~ ~ bloating ~ (Brit.)
~**futter** *n*, Rotierofenfutter = rotary kiln lining
~**kühler** *m*, Rotierofenkühler = rotary kiln cooler
~**mantel** *m* = rotary kiln shell
~**material** *n* = rotary kiln stone [*lime industry*]

Drehofenrohr *n* = rotary kiln tube, ~ ~ drum

Drehpunkt *m* = pivoting point
~, Momentenpunkt, Momentenpol *m* = centre of rotation (Brit.); center ~ ~ (US); point about which a moment is taken, moment pole
~ → → Schwarzbelageinbaumaschine
~**-Hebelwirkung** *f* [*Radschrapper*] = rear-draft fulcrum leverage

Dreh|radius *m* → Arbeitsbereich *m*
~**rahmen** *m* = revolving frame
~**ramme** *f*, Schwenkramme = slewing and raking pile driving plant
~**restaurant** *n* = revolving(-floor) restaurant
~**richtung** *f* [*volle Drehungen*] = rotational direction

Drehriegel — Drehwäschetrockner

~**riegel** m = turning bolt, spagnolet, sash fastener [*window*]
~**rohrofen-Verfahren** n = rotary kiln method
~**säule** f → Kranmast m
~**säule** f, Schwenksäule [*Anbaubagger und -lader*] = slew post
~**schale** f → drehsymmetrische Schale
~**scheibe** f = turntable
~**scheibenkipper** m = turntable tip
~**schemel** m = pivoted bogie
~**schemelkurvenfahrwerk** n = trave(l)ling gear with privoted bogie for running on curved rails
~**schemelwagen** m = bogie wagon
~**schieber** m = rotary valve
~**schlagbohren** n, Schlagdrehbohren = rotary impact drilling, percussive rotary ~
~**schlagbohrmaschine** f, Schlagdrehbohrmaschine [*arbeitet kombiniert schlagend und drehend, also umsetzend*] = rotary impact drilling machine, percussive rotary ~ ~
~**schlagwerkzeug** n = rotary percussive tool
~**schlüssel** m **für Bohrgestänge**, Bohrgestänge-Drehschlüssel, Hakenschlüssel, Gestängehaken m = pipe "hook", ~ wrench
~**schrauber** m = nutrunner
~**schurre** f = swivel chute
~**sitz** m, Schwenksitz = swivel seat
~**spannung** f → Torsionsspannung
~**sprenger** m [*Abwasserwesen*] = rotary distributor, rotating ~, revolving ~, ~ sprinkler
~**spülbohrverfahren** n, Drehbohrverfahren mit Wasserspülung [*Brunnenbau*] = hydraulic rotary method
~**spulenmesser** m = moving-coil meter
~**spülkopf** m = power-swivel
~**strahlregner** m, Weitstrahlregner = rotating sprinkler, rotary ~, revolving ~
Drehstrom m, Dreiphasenwechselstrom = three-phase current, polyphase ~, rotary ~

~**aggregat** n = A.C. (generating) set, A.C. (generating) plant
~**-Frequenz- und Spannungsumformer** m = three-phase frequency and voltage converter
~**generator** m = A.C. generator
~**lichtmaschine** f = (charging) alternator
~**nebenschlußmotor** m = polyphase shunt motor
Drehstuhl m [*Motor-Straßenhobel*] = circle
drehsymmetrische Biegung f = axisymmetric bending
~ **Membran(e)** f = rotary symmetrical diaphragm
~ **Schale** f, Drehschale = shell of rotational symmetry, ~ ~ revolution (subjected to loads of rotational symmetry)
Dreh|teil m, n → Aufsatzteil
~**-Tiefaufreißer** m = revolving ripper, ~ rooter, rotary ~
Drehtisch m, Bohrtisch, Rotarytisch = turntable, rotary table, rotary machine
~**einsatz** m [*Rotarybohren*] = table bushing, split (master) ~
~**presse** f = revolving (table) press, rotating (~), rotary (~) ~
~**vorgelegewelle** f = rotary table transmission countershaft
Drehtrieb m, Schwenktrieb [*Kran; Bagger*] = rotating drive, slewing ~
Drehtür f = revolving door, swing ~
Drehung f → Torsion
Drehventil n **mit selbsttätiger Rückstellung** = twist throttle with automatic return [*(compressed-)air throttle control*]
~ **ohne selbsttätige Rückstellung** = twist throttle without automatic return [*(compressed-)air throttle control*]
Dreh|verwerfung f (Geol.) = pivotal (or rotary, or tension) fault
~**waage** f, Eötvös'sche ~ = Eötvös torsion balance
~**wäschetrockner** m = rotary clothes dryer, revolving ~ ~

Drehwerk *n* [*Bagger; Kran*] = slewing gear, swing ~
~bremse *f* [*Bagger; Kran*] = swing brake, slewing ~
Drehwerkzeug *n* = rotary tool [*grinder; sander; drill etc.*]
Drehwinkel *m* = angle of rotation
~, Schwenkwinkel [*Auslegerbagger; Auslegerkran*] = angle of boom swing (US); ~ ~ jib ~ (Brit.)
~ausgleich *m* = equilibrium of angles of rotation
~-Iterationsverfahren *n* = angle-balancing method
Drehzahl *f*, (minutliche) Umlaufzahl, Umdrehungszahl, Tourenzahl = revolutions per minute ,(rotational) speed, r. p. m.
~abfall *m* = speed drop
drehzahl|abhängig = speed sensing
~abhängige Abstellvorrichtung *f* = overspeed shut-off
~abhängiger Spritzversteller *m* = speed sensing (variable) timing unit
Drehzahl|begrenzer *m* = speed limiter
~erhöhung *f*, Drehzahlzunahme *f* = speed pick-up
~fühler *m* = speed sensing device
~messer *m*, Tourenzähler *m*, Tachometer *n* = tachometer, revolution counter, speedometer
~messerantrieb *m* = tachometer drive
~reg(e)lung *f* = speed control
~regler *m* = speed controller
~regulierhebel *m* = speed lever
~verminderung *f*, Verzögerung *f* = deceleration
~zunahme *f*, Drehzahlerhöhung *f* = speed pick-up
Drehzapfenlager *n* [*für Hoch- und Tiefbauten*] = slewing journal, king pin bearing
drehzapfenmontiert = trunnion-mounted
Drehzylinder *m*, Schwenkzylinder [*Kran; Bagger*] = rotating ram, slewing ~, swing ~
Dreiachs|auto(mobil)kran *m* = three-axle rubber-mounted crane

~fahrgestell *n*, Dreiachsunterwagen *m*, Dreiachsfahrwerk *n* = three-axle carrier
~-Fahrzeug *n* = three-axle vehicle
~-Hinterkipper *m* = three-axle rear-dump truck; ~ ~ lorry (Brit.)
dreiachsige Type *f* = three-axle model
Dreiachs|-Lkw *m* = three-axle truck; ~ lorry (Brit.)
~(-Tandem)walze *f* → Walze
dreiaxiale Druckzelle *f* → Triaxialgerät *n*
Drei|axialversuch *m*, Triaxialversuch, Dreiachsenversuch = triaxial test
~backenbrecher *m* → → Backenbrecher
~baum *m*, Montage-~ = shear-legs, tripod, sheer-legs, sheers
~bogenbrücke *f* = three-arch bridge
~decker(sieb) *m*, (*n*), Dreistufensieb = triple-deck screen
dreidimensional, räumlich = three-dimensional
dreidimensionale Rohrleitung *f* → räumlich verlegte ~
dreidimensionales Anspannen *n* [*Spannbeton*] = three-dimensional stressing
~ Element *n*, räumliches ~ = three-dimensional element
Dreiebenen|betrieb *m* [*Flughäfen*] = three-level operation
~system *n* [*Passagier- und Gepäckabfertigung auf Flughäfen*] = three-level (build**ing** operational) system
Dreieck|anordnung *f* → Dreiecksystem
~anordnung *f* (✕) = triangular system, diagonal ~
~binder *m* = triangular roof truss
~diagramm *n*, Dreieckdarstellung *f* = triangular diagram
~fachwerkträger *m* = triangular truss
~gitterkonstruktion *f* = triangular lattice(d) construction
dreieckiger Ausschnitt *m* [*In einem Meßwehr zur Messung der überlaufenden Wassermengen*] = V notch
Dreieck|kastenkonstruktion *f* = triangular construction

Dreiecklenker — Dreimomentengleichung

~lenker m = triangular steering control arm
~masche f = triangular mesh
~meßwehr n = triangular measuring weir
~platte f = triangular slab
~schaltung f = delta connection
~sprengwerk n = triangulated truss, triangular ~
~stütze f = triangular column
~system n, Dreieckanordnung f = triangular system, triangulated ~
~träger m = triangulated girder, triangular ~
~verband m = triangulated bracing, triangular ~
dreietagig, dreistöckig, dreigeschossig = three-floored
dreietagige Kreuzung f, dreistöckige ~, dreigeschossige ~ = triple-deck grade separation structure, three-level ~ ~ ~
dreifach eingeschertes Seil n = three-part line
~ sichere Auslegerwinde f [Bagger] = triple-safe boom hoist (US); ~ jib ~ (Brit.)
Dreifach|antrieb m = triple drive
~kastenprofil n = triple box section
~-Rollenkette f = three strand roller chain
~stollen m [(Gleis)Kette] = triple grouser
~ventil n = triple valve
~verglasen n = triple glazing
~wellblech n, Tripelwellblech = triple corrugated (sheet) iron
Drei|farbenschreiber m = three-colo(u)r recorder
~feldbrücke f = three-span bridge
~feldträger m = three-span girder
~flächengleitlager n = three-lobe bearing
~flügelbohrkrone f = three wing (drill) bit
~-Füllungstür f = three-panel door
~fuß m = tripod
~fuß-Grabenramme f = three-legged (or triplex) (backfill) tamper (or rammer)

Dreigang|getriebe n = three-speed gear
~zwischengetriebe n = three-speed transfer case
Dreigelenk|bogen m, Korbbogen = three-centred arch (Brit.); three-centered ~ (US)
~fachwerkrahmen m = trussed frame with three hinges
~rahmen m = frame with three hinges
~-Rippenkuppel f = three-pinned arch-ribbed dome
dreigeschossig, dreistöckig, dreietagig = three-floored, three-stor(e)y
Drei|gutscheidung f = three-product separation
~kammer-Hohlkörper m, Dreikammer-Hohlblock m, Dreikammer-Hohl(block)stein m = three-cell hollow block, ~ tile, ~ pot
~kanter m (Geol.) = dreikanter, ventifact, glyptolith, gibber
~kantleiste f = triangular fillet
~kantlitze f → Flachlitze
~kegel-Rollenmeißel m = tri-cone rock bit
~kolbenpumpe f = triple piston (type) pump
dreilagige Dach(pappen)eindeckung f = 3-ply roofing
dreilagiger Mörtelbelag m, ~ Putz m = three-coat plaster
~ Putz m auf Putzträgergewebe, ~ Mörtelbelag m ~ ~ = render, float, and set; [*The first or render-coat is applied to produce a level surface; when this coat is dry, the surface is scored, and divided into 8-ft. squares by screeds of coarse stuff, 8 in. wide, carefully levelled. These are then filled in with coarse stuff, floated carefully level. The finishing coat is then put on as usual and trowelled to a smooth surface*]
Dreimomentengleichung f, Clapeyron'sche Gleichung, Dreimomentensatz m [*Berechnungsverfahren zur Ermitt(e)lung der Stützmomente von*

Durchlaufträgern] = three-moment equation, theory of three moments
Dreimotoren-Elektrobagger *m* = three-motor all-electric shovel
~-Laufkran *m* [*für elektrischen Betrieb*] = three-motor travel(l)ing crane
dreimotoriger Elektroantrieb *m* = three-motor drive
Dreiphasen|motor *m* = three-phase motor
~wechselstrom *m*, Drehstrom = three-phase current, polyphase ~, rotary ~
Drei|plungerpumpe *f* = triple throw plunger pump
~punkt-Aufhängung *f*, Dreipunktabstützung *f* = three-point suspension (mounting)
~punktgelenkabstützung *f* = three-point linkage
~quartier *n*, Dreiviertelstein *m* = three quarters
Dreirad *n* = tricycle
~-Bohrwagen *m* mit Lafette = three-wheel(ed) wagon drill
~fahrwerk *n* [*Flugzeug*] = tricycle (landing) gear
~-Walze *f* → → Walze
Drei|rohrleitung *f* = three-pipe line
~rollenmeißel *m* = three-roller bit
dreischiffig = triple-aisle
Drei|seilgreifer(krankorb) *m*, Dreiseilgreif(er)korb = three-rope suspension grab(bing) (crane) bucket, ~ ~ grab
~-Seil-Schrapper *m*, Schleppschrapperanlage *f* mit 3-Seil-Anordnung = rapid shifting drag scraper machine, rapid-shifter
Dreiseiten|entleerung *f* = three-way dump discharge
~kippanhänger *m*, Anhänge-Dreiseitenkipper, Schlepp-Dreiseitenkipper [*Erdbaufahrzeug*] = three-way dump trailer wagon
~-Kipper *m* = three-way tipper
Dreispurabschnitt *m* = three-lane section
dreispurig [*Fahrbahn*] = three-lane

dreispurige Straße *f* = undivided two-way road with three lanes altogether
~ Verkehrsführung *f* = three-lane traffic handling
dreistegig = triple web, three-webbed
Drei|stegplatte *f* [(*Gleis*)*Kette*] = triple-grouser shoe
~stellungsventil *n* = three-position valve
dreistöckig, dreietagig, dreigeschossig = three-floored
Dreistufen|-Brechen *n* = three-stage crushing, ~ breaking
~sieb *n*, Dreidecker(sieb) *m*, (*n*) = triple-deck screen
dreistufig = triple-stage
dreiteilig ausgebildeter Rahmen *m* = three-section framing bent
dreiteilige Rollenstation *f* → → Bandförderer
~ Schnittkante *f* = three-piece cutting edge
dreiteiliger Bogen *m* (⚒) = three-element arch
~ Steinwolf *m* = three-leg(ged) lewis
Dreitrommel|-Dampfwinde *f*, Dampf-Dreitrommelwinde = three-drum steam winch
~-Straßen(beton)mischer *m*, Dreitrommel-Straßenbetoniermaschine *f* = three-batch paver
~winde *f* = three-drum winch
Dreiviertel|säule *f* = three-quarter column
~stein *m*, Dreiquartier *n* = three-quarters
Dreiwalzenbrecher *m* = triple roll crusher, ~ ~ breaker, ~ ~ crushing machine, ~ ~ breaking machine
Dreiwege|ausgleichventil *n* = three-way compensating valve
~hahn *m* = three-way tap
~schalter *m* = three-way switch
~ventil *n* = three-way valve
Dreizahnaufreißer *m* = 3-tine unit
Dreizonen|bauweise *f* → Aufbruchbauweise
~förderung *f* [*Es wird gleichzeitig aus drei Erdölhorizonten gefördert*] = triple (well) completion

Drei|zug-Flammrohr-Rauchrohrkessel *m* = three-pass economic boiler
~zwecklöffel *m* = 3 in 1 (multi-purpose) bucket
~zylinder(dampf)maschine *f*, Drillings(dampf)maschine = triple (cylinder) steam engine
~zylindermotor *m* = three-cylinder engine
Drempel *m*, Grundschwelle *f*, Bodenschwelle, Sohlschwelle, Unterschwelle [*niederdeutsch: Süll m, n*] = (ground) sill
~, Kniestock *m* [*Schwelle im Torboden einer Schleuse, gegen die sich das geschlossene Schleusentor stützt*] = pointing sill (Brit.); miter ~ (US)
~kopf *m* = sill head
~wand *f*, Kniestockwand, Versenkungswand = jamb wall
Dressierwalzwerk *n* = temper pass mill, skin ~ ~
Driftholztorf *m* = driftwood peat
Driftströmung *f* = drift current
Drillbewehrung *f*, Drillarmierung, Drill(stahl)einlagen *fpl*, Torsionsbewehrung, Torsionsarmierung, Torsions(stahl)einlagen = torsion reinforcement
drilliertes Gewebe *n* = stranded wire cloth, twisted ~ ~
Drillingsseilpumpe *f* = three throw high speed pump
Drillknickung *f*, Torsionsknickung, Drillknicken *n*, Torsionsknicken = torsional buckling
Drillometer *m*, Bohrdruckmesser *m*, Bohrdruckmeßgerät *n*, Gewichts-Kontrollmesser = drilling indicator, weight ~, drillometer
~-Meßdose *f* = drillometer capsule
Drill(stahl)einlagen *fpl* → Drillbewehrung
Drillung *f* → Torsion *f*
Drill(ungs)|moment *n*, Torsionsmoment = twisting moment, moment due to torsion, torsional moment
~winkel *m*, Torsionswinkel = angle of twist, twist angle
Drittbrechen *n*, Nachbrechen = tertiary crushing, ~ breaking [*The tertiary crushing stage*]
dritte Zerkleinerungstheorie *f* = third theory of comminution
Drittel|dach *n* = roof with pitch of 1 : 3
~punkt *m* = third point
~punktbelastung *f* = third-point loading
Dritter *m* [*im Gegensatz zu zwei Vertragspartnern*] = disinterested party
Dritt|mahlen *n*, Drittmahlung *f*, Nachmahlen, Nachmahlung = tertiary grinding [*The tertiary grinding operation*]
~zerkleinerung *f*, Nachzerkleinerung = tertiary reduction, ~ comminution
Drossel *f* [*Durch eine solche Drossel fließt nur eine ganz bestimmte Menge Erdöl in langsamem Strome*] = flow bean
~, Luftklappe *f* [*Vergaser*] = strangler (Brit.); choke (US)
~druckreduzierventil *n* = throttle pressure (reducing) valve
~gestänge *n* = throttle linkage
~handhebel *m* = hand throttle (lever)
~klappe *f* = butterfly valve
~klappe *f* **mit Fallgewichtantrieb** → Fallgewichtdrosselklappe
~klappengehäuse *n* = butterfly valve body
~knopf *m* = choke button
drosseln, ab~ = to throttle
Drossel|organ *n* = choke control
~turbine *f* = throttling turbine
~ventil *n* = throttle valve
~welle *f* = butterfly spindle
Druck *m* **pro Achse** → Achslast *f*
Druckabfall *m*, Druckabnahme *f*, Druckverminderung *f* = pressure drop
~ [*Hydraulik*] = reduction of pressure head, drop in ~ ~
Druck|anschluß *m* [*Pumpe*] = discharge connection delivery
~anstieg *m*, Druckzunahme *f*, Druckerhöhung *f* = pressure rise
druckarme Lagerstätte *f* [*Erdöl; Erdgas*] = low-pressure formation

Druck|armierung *f* → Druckbewehrung
- **~aufbau** *m* = pressure build-up
- **~aufbaukurve** *f* = pressure build-up curve
- **~aufbauverfahren** *n* [*Erdöllagerstätte*] = pressure restoration method
- **~ausgleich** *m* = equalization of pressure
- **~ausgleich** *m* [*Durch die Wasserverbindung der Druckzylinder beim Preßkolben-Hebewerk*] = hydraulic balance
- **~auswaschung** *f* = pressure washing operation
- **~beanspruchung** *f*, Druckkraft *f* = compressive stress
- **~begrenzungsventil** *n* = relief valve
- **~behälter** *m*, Druckgefäß *n* = pressure vessel, ~ tank
- **~bereich** *m* = pressure range
- **~beton** *m* = compression concrete
- **~bewehrung** *f*, Druckarmierung, Druck(stahl)einlagen *fpl* = compressive reinforcement, compression ~
- **~bewetterung** *f*, Bewetterung durch Überdruck, blasende Bewetterung, Einblasebewetterung, blasende Wetterführung *f*, Überdruckbewetterung = blowing ventilation
- **~bolzen** *m* = thrust bolt
- **~bolzen** *m* [*Kupplung*] = clutch fork ball support
- **~bruch** *m* = compression failure
- **~dislokation** *f* (Geol.) = pressure fault
- **~diagonale** *f* → Druckschräge *f*
- **~diagramm** *n* = pressure diagram
- **~dose** *f* → (Boden)Druck(meß)dose
- **~dose** *f* = load cell
- **~einlagen** *fpl* → Druckbewehrung
- **drücken** → spülen [*Naßbaggergut*]
- **drückende Gebirgsschichten** *fpl* [*Tiefbohrtechnik*] = abnormal pressure formations

Druck|energie *f* [*Hydraulik*] = pressure energy
- **~entlastung** *f* = decompression
- **~entlastungshebel** *m* = compression release lever
- **~entlastungsventil** *n* = pressure relief valve
- **~entleerung** *f* = pressure discharge
- **~erhaltungsverfahren** *n* [*Erdöllagerstätte*] = pressure maintenance method
- **~erhöhung** *f* → Druckanstieg *m*
- **~ermüdung** *f* = fatigue in compression
- **~erzeugungsanlage** *f* [*Pumpen, Leitungen und Druckpressen als Teil einer Belastungsvorrichtung für die Probebelastung von Pfählen*] = load-generating system
- **~faltung** *f* (Geol.) = compressional folding
- **~faser** *f*
- **~faser** *f* fibre in compression (Brit.); fiber ~ ~ (US)
- **~feder** *f* = compression spring
- **~festigkeit** *f* [*Für den üblichen Konstruktions-Schwerbeton gibt es in den USA folgende Güteklassen (bezogen auf die Druckfestigkeit von Zylindern $h = 30$ cm, $d = 15$ cm, im Alter von 28 Tagen oder in einem früheren Alter, falls dann bereits die volle Beanspruchung auftritt):*
  2500 3000 4000 5000 psi
  Auf den 20 cm-Würfel umgerechnet sind dies rd.
  200 250 330 410 kp/cm²
  Bei dieser Umrechnung (Faktor 0,07/ 0,85 = 0,082) wäre für Schwerbeton nicht berücksichtigt, daß Zylinder hieraus nach den amerikanischen Normen bis zum Alter von 28 Tagen dauernd feucht gelagert werden. Der Schwerbeton der Zylinder ist also bei der Prüfung in den USA feuchter als der der Würfel nach DIN 1048, der nur 7 Tage feucht gehalten wird. Die Zylinderdruckfestigkeit von Schwerbeton dürfte daher – als Würfeldruckfestigkeit nach DIN 1048 ausgedrückt – noch etwas größer sein als oben angegeben ist. Das trifft jedoch nicht für Zylinder aus Konstruktions-Leichtbeton zu, da diese wie die Würfel nach DIN 1048 7 Tage feucht und dann an der Luft gelagert werden. Der Umrechnungsfaktor von 0,082 dürfte daher für den Leichtbeton zutreffen] = compressive strength, crushing ~

Druckfilter — Druckknopfsteuerung

~filter *m, n*, geschlossener Schnellfilter *m*, geschlossenes Schnellfilter *n* = pressure filter
~filtration *f* = pressure filtration
~fläche *f* = surface under pressure, ~ taking up pressure, area of pressure, pressure area
~flaschenzement *m* = bottled cement [*Steel bottles contain cement under pressure. The cement is shot from the bottles through an air line to a cement bin*]
~flüssigkeit *f*, Preßflüssigkeit, Hydraulikflüssigkeit = pressure liquid, hydraulic ~
~flüssigkeitsbehälter *m*, Preßflüssigkeitsbehälter, Hydraulikflüssigkeitsbehälter = hydraulic liquid tank, pressure ~ ~
Druckgas *n*, verdichtetes Gas = compressed gas
~behälter *m* = pressure gas vessel, ~ ~ tank
~flasche *f*, Stahlflasche [*für Propan oder Butan*] = steel cylinder
~förderverfahren *n*, Preßgasförderverfahren, Gastrieb-Verfahren = gas-lift
~kabel *n* = gas-pressure cable
Druck|gebläse *n* = pressure blower
~gefälle *n*, piezometrisches Gefälle = hydraulic gradient, gradient of piezometric head [*The slope of the hydraulic grade line. The slope of the surface of water flowing in an open conduit*]
~gefälle *n* → Fallhöhe
~gefäß *n*, Druckbehälter *m* = pressure tank, ~ vessel
~gefäßversuch *m* [*Korrosionsprüfung*] = test at pressure above atmospheric
~gerät *n* → Oedometer *n*
druckgeschmiert = pressure-lubricated, force-lubricated
Druck|gießmaschine *f* = die casting machine
~griff-Einlaß *m* = pressure grip throttle [*(compressed-)air throttle control*]

~gurt *m* = compressive flange, ~ boom, ~ chord, compression ~
Druckguß *m* [*früher: Spritzguß*] = pressure die casting process
~stück *n* = pressure die casting
~stück-Gießerei *f* = pressure die foundry
~teil *n, m* [*früher: Spritzgußteil*] = pressure die casting
druckhaftes Gebirge *n* [*Tunnel-, Stollen- und Schachtbau*] = swelling ground
Druck|haltung *f* = maintenance of pressure, pressure maintenance
~härtekessel *m* = autoclave
Druckhöhe *f* [*Bodenmechanik*] = hydraulic gradient
~ → Fallhöhe
~ = pressure level
Druckhöhen|bereich *m*, Fallhöhenbereich = range of head
~unterschied *m*, hydraulischer ~ = (hydraulic) head
Druck|hülse *f* = pressure sleeve
~kammer *f*, Druckluftkammer, Arbeitskammer [*pneumatische Gründung*] = (caisson) working chamber
~kessel *m* → Treibkessel
~kessel *m* → Autoklav *m*
~kissen *n* → (Preß)Kissen
~kluft *f* (Geol.) = pressure joint
Druckknopf *m*, Schaltknopf = push-button
~anlasser *m*, Bedienungsknopfanlasser = push-button starter
~-Innensteuerung *f* für Selbstfahrer-(aufzug) = automatic push-button control [*A method of control by means of buttons at the landings and in the lift-car, the momentary pressure of which will cause the lift-car to start and automatically stop at the landing corresponding to the button pressed*]
~kasten *m*, Bedienungsknopfkasten = push-button box
~schalter *m*, Fingerdruckschalter, Bedienungsknopfschalter = push-button (or press-button) switch
~steuerung *f*, Fingerdrucksteuerung, Druckknopfbedienung *f*, Druck-

Druckkolben — Druckluftbremse

knopfbetätigung *f*, Fingerdruckbedienung, Fingerdruckbetätigung = push-button (or press-button) control, finger tip ~

Druck|kolben *m* = pressure piston, load ~

~**kraft** *f*, Druckbeanspruchung *f* = compressive stress

~**kraft** *f* **des Wassers**, Gesamtwasserdruck *m*, hydrostatische Druckkraft = total normal pressure, hydrostatic force

~**kristallisation** *f* = piezocrystallization

~**kugellager** *n* = thrust ball bearing

~-**Kugelventil** *n*, Kugel-Druckventil [*Pumpe*] = ball discharge valve, ~ delivery ~

~**kupp(e)lung** *f* [*Rohr*] = compression union

~**lagen** *f pl* = pressure partings [*in coal*]

~**lager** *n* = thrust bearing

~**last** *f* = compressive load, pressure ~

~**leitung** *f*, Triebwasserleitung = penstock

~**leitung** *f* = pressure line, ~ conduit

~**linie** *f*, piezometrische Linie = piezometric line, hydraulic grade ~ [*In a closed conduit a line joining the elevations to which water could stand in rivers. In an open conduit, the hydraulic grade line is the water surface*]

~**linie** *f*, Mittelkraftlinie, Stützlinie = pressure line

~**lok(omotive)** *f* = train-pushing engine

drucklose Leitung *f*, Freispiegelleitung = free-surface flow line

Druckluft *f*, Preßluft, komprimierte Luft = compressed air

~-**Abbauhammer** *m*, Druckluft-(Kohlen-)Pickhammer, Druckluft-Förderhammer = pneumatic (or air) (coal) picker (or pick hammer)

~**abteufung** *f*, Druckluftabsenkung *f* = pneumatic sinking

~-**Ankerwinde** *f* = (compressed-)air anchor winch, pneumatic ~ ~

~**anlage** *f*, Preßluftanlage = (compressed-)air installation, ~ system

~**anlasser** *m* = (compressed-)air starter, pneumatic ~

~**antrieb** *m* = (compressed-) air drive, pneumatic ~

~**arbeit** *f pl* = compressed air work

~**arbeiter** *m* = sand hog

~-**Aufreißhammer** *m*, Druckluft-Aufbruchhammer, Druckluft hammer für Aufreißarbeiten, Straßen-~ = pneumatic (or air) (pavement) breaker, air hammer, pneumatic pick

~**aufzug** *m* = (compressed-)air hoist, pneumatic ~

~**ausgleich** *m* = air pressure compensation

~-**Außenrüttler** *m*, Druckluft-Außenvibrator *m* = pneumatic external vibrator, (compressed-)air ~ ~

Druckluftbär *m* → Druckluft(-Ramm)hammer *m*

~ **mit halbautomatischer Steuerung**, Druckluft(-Ramm)hammer *m* ~ ~ ~ = single-acting (compressed-)air (pile) hammer, ~ pneumatic (~) ~

Druckluft|becken *n* = diffused air tank

~**behälter** *m*, Druckluftkessel *m*, Wind(druck)kessel, Druckluftspeicher *m*, Speicherkessel = (compressed-)air vessel, air receiver

~**belüftung** *f* = diffused air aeration [*activated sludge process*]

druckluftbetätigt = air actuated

Druckluftbeton|förderer *m* → Betondruckluftförderer

~**förderung** *f*, pneumatischer Betontransport *m*, pneumatische Betonförderung, Preßluft-Betonförderung = pneumatic concrete placing, air ~

druckluftbetriebene Fettpresse *f* = (compressed-)air-operated grease unit

Druckluft|-Bodenramme *f* = pneumatic (or air) soil (or earth) rammer

~-**Bohrhammer** *m* → Druckluft-Gesteinsbohrhammer

~-**Bohrknecht** *m* → pneumatische (Bohr)Hammerstütze

~**bremse** *f*, Luftdruckbremse = (compressed-)air brake, pneumatic ~

Druckluft einpreßgerät — Druckluftlok(omotive)

~einpreßgerät *n* → Verpreßkessel

~einspritzung *f* = air injection

~erkrankung *f*, Preßluftkrankheit *f*, Caissonkrankheit = compressed-air sickness, caisson disease, compressed-air illness, bends

~erzeuger *m*, Luftverdichter *m*, Kompressor *m* = (air) compressor

~-Fernsteuerung *f*, Druckluft-Fernbetätigung *f*, pneumatische Fernbedienung *f*, pneumatische Fernsteuerung, pneumatische Fernbetätigung, Preßluft-Fernsteuerung, Preßluft-Fernbedienung, Preßluft-Fernbetätigung, Druckluft-Fernbedienung = pneumatic remote control

~-Feststellbremse *f* = pneumatic parking brake, (compressed-)air ~ ~

~feuerung *f* = furnace with forced draught (Brit.); ~ ~ ~ draft (US)

~förderanlage *f*, pneumatische Druckförderanlage = (compressed-)air conveying system, pneumatic ~ ~

~-Förderhammer *m* → Druckluft-Abbauhammer

~-Förderrinne *f*, (Gebläse)Luftförderrinne, pneumatische Förderrinne, Luft-Rutsche *f* = airslide

~förderung *f*, Preßluftförderung = conveying by compressed air

~förderverfahren *n* [*Erdöl*] = air lift method, ~ pumping ~

~-Gegenhalter *m*, Druckluftvorhalter = pneumatic dolly, ~ holder-up

~gerät *n*, Druckluftwerkzeug *n* = (compressed-)air tool, pneumatic ~

~geräte *npl* = (compressed-)air equipment, pneumatic ~

~gerätelärm *m* = pneumatic equipment exhaust noise

~-(Gesteins)Bohrhammer *m*, Druckluft-Felsbohrhammer = pneumatic (hand) hammer (rock) drill, ~ (rock) drill (hammer), (compressed-)air (hand)hammer (rock)drill, (compressed-)air (rock) drill (hammer), hammer-type pneumatic hand-held drill

druckluftgesteuert = pneumatically controlled, (compressed-)air ~

Druckluft|-Glättbohle *f* = pneumatic vibrated (finishing) screed (or smoother)

~gründung *f*, pneumatische Gründung = pneumatic foundation work, foundation work under compressed air, pneumatic sinking

~hammer *m* → Druckluft-Rammhammer

~hammer *m* für Aufreißarbeiten → Druckluft-Aufreißhammer

~hammer *m* mit halbautomatischer Steuerung → Druckluftbär ~ ~ ~

~-Handstampfer *m*, Hand-Druckluftstampfer = (compressed-)air (mechanical) tamper, pneumatic ~

~heber *m* = (compressed-)air jack, pneumatic ~

~heber *m*, Preßluftheber [*Abwasserwesen*] = (compressed-)air ejector, air-jet lift, (pneumatic) ejector

~hebezeug *n*, Preßlufthebezeug = pneumatic lifting device

~injektor *m* → Verpreßkessel

~-Innenrüttler *m*, Druckluft-Innenvibrator *m* = pneumatic internal vibrator, (compressed-)air ~ ~

~insel *f* = compressor platform [*for re-injecting natural gas into the oil reservoir to increase oil recovery*]

~kammer *f*, Arbeitskammer, Druckkammer [*pneumatische Gründung*] = (caisson) working chamber

~kessel *m* → Druckluftbehälter *m*

~-Kesselsteinabklopfer *m*, pneumatischer Kesselsteinabklopfer = pneumatic scaling hammer, ~ ~ chipper

~kipper *m* = pneumatic tip(ping) wagon

~-Kohlenstaubförderung *f* durch Rohre = pulverized coal transport through pipes

~kühlung *f* = forced-draught cooling (Brit.); forced-draft ~ (US)

~leitung *f* = (compressed-)air line

~lok(omotive) *f*, Preßluftlok(omotive) = (compressed-)air loco(motive), pneumatic ~

Druckluftmaschine — Druckluftwurfschaufellader 332

~**maschine** *f* = (compressed-)air machine, pneumatic ~
~**messer** *m* = air pressure gauge (Brit.); ~ ~ ga(u)ge (US)
~**motor** *m* = (compressed-)air motor, pneumatic ~
~**-Niethammer** *m*, Preßluft-Niethammer = pneumatic riveting hammer
~**nietung** *f*, Preßluftnietung = pneumatic riveting
~**öler** *m*, Preßluftöler = air lubricator
~**pegel** *m* [*Zeichen: Sd*] = pressure (type of recording) ga(u)ge
~**(pfahl)ramme** *f* = pneumatic pile driver, (compressed-)air ~ ~
~**pflasterramme** *f*, Preßluftpflasterramme = pneumatic (sett) paving rammen, (compressed-)air (~) ~ ~
~**pistole** *f*, Preßluftpistole = (compressed-)air gun
~**pumpe** *f*, Preßluftpumpe = (compressed-)air pump, air lift ~
~**ramme** *f*, Preßluftramme [*für Verdichtungsarbeiten*] = pneumatic rammer
~**(-Ramm)hammer** *m*, Druckluft-(-Ramm)bär *m* = (compressed-)air (pile) hammer, pneumatic (~) ~
~**rüttler** *m*, Preßluftrüttler, Druckluftvibrator *m*, Preßluftvibrator, pneumatischer Rüttler, pneumatischer Vibrator = (compressed-)air vibrator, pneumatic ~
~**schaltung** *f*, Drucklufsteuerung *f*, Preßluftschaltung, Preßluftsteuerung, pneumatische Schaltung, pneumatische Betätigung *f*, pneumatische Bedienung *f* = pneumatic control, (compressed-)air ~
~**schießen** *n*, nicht detonates Sprengen = bursting by compressed air
~**-Schildvortrieb** *m*, pneumatischer Schildvortrieb = pneumatic shield driving, (compressed-)air ~ ~
~**-Schlaghammer** *m* = (compressed-)air impact hammer
~**schlauch** *m* = (compressed-)air hose
~**schlauchleitung** *f* = flexible (compressed-)air line

~**schleuse** *f*, Luftschleuse = (compressed-)air lock, pneumatic ~
~**senkkasten** *m* → Caisson
(~-)**Senkkastengründung** *f*, pneumatische Gründung = (compressed-) air caisson foundation, (pneumatic) ~ ~
~**servoanlage** *f* = (compressed-)air booster, pneumatic ~
~**spatenhammer** *m*, Preßluftspatenhammer = pneumatic (clay) digger, ~ (~) spade, ~ spader, ~ spade hammer, air(-operated clay) spade
~**speicher** *m* → Druckluftbehälter
~**spritzgerät** *n* → Betonspritzmaschine
~**stampfer** *m*, Preßluftstampfer = pneumatic tamper, (compressed-)air ~
~**-Startanlage** *f* = air starting system
~**steuerung** *f* → Druckluftschaltung
~**strahl** *m* = (compressed-)air jet
~**stütze** *f*, Druckluftbohrstütze = pneumatic leg, airleg
~**-Sumpfpumpe** *f* = (compressed-)air sump pump, pneumatic ~ ~
~**-Treibkessel** *m* → Treibkessel
~**-Überkopf-(Schaufel)Lader** *m* → Druckluftwurfschaufellader
(~**- und**) **Fallbremse** *f* = trailer-type overrun brake
druckluftunterstützt = (compressed-) air-assisted, (compressed-)air-boosted
Druckluft|verfahren *n* **zur Abwasserbelüftung** = air-diffusion method, ~ aeration, diffused-air aeration
(~)**Verpreßgerät** *n* → Verpreßkessel
~**-Verputzgerät** *n* = pneumatic plaster-throwing machine, ~ plastering ~
~**verteiler** *m* = air (junction) manifold
~**vibrator** *m* → Druckluftrüttler
~**-Vorschubmotor** *m* [*Hammerbohrmaschine*] = air feed motor
~**werkzeug** *n* → Druckluftgerät
~**winde** *f* = pneumatic winch, (compressed-)air ~
~**wurfschaufellader** *m*, Druckluft-Überkopf-(Schaufel)Lader = pneu-

matic rocker (type) shovel (loader), ~ overshot loader, ~ overhead loader, ~ flipover bucket loader, ~ overloader
~-Zahnrad-Drehbohrmaschine *f* = pneumatic gear rotary (or rotating) drilling machine
~zementtransport *m*, Druckluftzementförderung *f* = pneumatic cement handling
~zug *m* = (compressed-)air hoist, pneumatic ~
~zylinder *m* = pneumatic ram
Druck|messer *m* = pressure ga(u)ge
~messer *m* → Piezometerrohr
~(meß)dose *f* = pressure capsule, ~ cell
~(meß)dose *f* zur Bestimmung der ungefähren Lage der Sickerlinie = hydrostatic pressure cell, ~ ~ capsule
~minderventil *n*, Druckminderer *m* = pressure reducing valve, ~ relief ~
~mutter *f* = thrust nut
(druck)nachgiebig = yielding (under pressure)
Druck|öl *n*, Preßöl, Hydrauliköl = pressure oil, hydraulic ~, pressurized ~
~öler *m* = force feed oiler
~ölmotor *m*, Preßölmotor, Hydraulikölmotor = hydraulic oil engine, pressure ~ ~, pressurized ~ ~
~ölpumpe *f*, Preßölpumpe, Hydraulikölpumpe = hydraulic oil pump, pressure ~ ~, pressurized ~ ~
~ölpresse *f*, Hydraulikölpresse. Preßölpresse = hydraulic oil jack, pressure ~ ~, pressurized ~ ~
~ölung *f* = pressure feed lubrication, forced ~
~ölzufluß *m*, Preßölzufluß, Hydraulikölzufluß = pump supply oil flow
~ölzylinder *m*, Preßölzylinder, Hydraulikölzylinder = hydraulic oil ram, pressure ~ ~, pressurized ~ ~
~pfahl *m* = pressure pile, non-uplift ~
~pistole *f* = pressure gun

~platte *f*, Druckschicht *f* [*Teil einer Massivdecke, der die Druckspannungen aufnimmt*] = slab
~platte *f* → → Backenbrecher
~platte *f* [*einer Druckprüfmaschine*] = platen
~platte *f* = pressure plate, thrust ~
~platte *f*, Preßplatte [*Gleitschalungsfertiger*] = (con)forming plate, pressure ~
druckpneumatisch = compressed-pneumatic
Druck|polster *m, n* [*Absenken eines Tunnelelementes*] = crushing pad
~-Porenziffer-Diagramm *n* = pressure-voids ratio diagram
~-Porenziffer-Verhältnis *n* = pressure-voids ratio ratio
~presse *f*, (Druck)Prüfpresse, Druckprüfmaschine *f*, (Beton)Würfel(prüf)presse, Betonprüfpresse, (Beton)Würfelprüfmaschine, (Beton-) Probewürfelpresse, Betonwürfeldruckpresse = (concrete) compression machine, (~) ~ tester, cube testing machine, cube tester [*In the USA cylinders are used instead of cubes and the machine is called a "cylinder tester"*]
~probe *f* → Druckversuch
~prüfmaschine *f* → Druckpresse
(~)Prüfpresse *f* → Druckpresse
~prüfung *f* → Druckversuch
~prüfung *f* [*Gestein*] = crushing (or compressive) (strength) test (on rock specimens)
~prüfungswert *m* [*Straßenbaugestein*] = aggregate crushing value
~pumpe *f* = (lift and) force pump, pressure ~
~querschnitt *m*, gedrückter Querschnitt = cross section under compression
~reg(e)lung *f* = pressure control
~regler *m* = pressure control valve
~riegel *m* = compression member
~ring *m*, Halterung = retainer ring
Druckrohr *n*, Spülrohr [*Naßbaggerei*] = dredging pipe
~ = pressure pipe

Druckrohrleitung — Druckstrahlbaggerung 334

~leitung *f* = pressure pipe line, ~ conduit
~leitung *f*, Triebwasserrohrleitung = pipe penstock
Druckrüttler *m*, Schwinggerät *n* [*Rütteldruckverfahren*] = vibroflot (machine)
Drucksammler *m* → Druckspeicher
~ des Direktstufenschaltventils = direct drive shift valve accumulator
~ ~ Schnellstufenschaltventils = overdrive shift valve accumulator
~ladeventil *n*, Druckspeicherladeventil = accumulator charging valve
Druck|säule *f* [*Dammbalkenwehr*] = holding down beam
~schacht *m*, Druckleitungsschacht, Fallschacht = penstock shaft, pressure ~
~scheibe *f* = thrust washer
~schicht *f*, Druckplatte *f* [*Teil einer Massivdecke, der die Druckspannungen aufnimmt*] = slab
~schieferung *f*, Vorzerklüftung *f* = induced cleavage
~schlag *m*, Abschrägung *f* [*Gewölbe*] = bevel, splaying
Druckschlauch *m* = pressure hose
~ [*Saug- und Druckpumpe*] = delivery hose, discharge ~
~, Meßschlauch [*Zählgerät mit einem über die Fahrbahn gelegten Luftschlauch*] = pneumatic tube laid across the road surface
~filter *m*, *n* = pressure (type) cloth filter dust collector
Druck|schmierung *f*, Preßölschmierung = forced lubrication, pressure ~
~schräge *f*, Druckdiagonale *f*, gedrückte Diagonale, gedrückte Schräge = diagonal in compression, compression diagonal
~schreiber *m* = pressure recorder
~schütz(e) *n*, (*f*) = sluice valve with bottom release
~schwankung *f* = change in pressure
~schweißung *f* = pressure welding
~schwellbelastung *f* = pulsating compressive loading
~schwimmer *m* = pressure float

~-Setzungskurve *f* = load-settlement curve
~-Setzungs-Quotient *m* → Bettungszahl
~-Setzungsversuch *m* → Kompressionsversuch
~sieben *n*, Drucksiebung *f*, Sieben unter Druck, Siebung unter Druck = forced screening
~sonde *f* = static sounding rod
~sondierung *f*, Drucksondieren *n* = static (subsurface) sounding, ~ penetration testing
~spannung *f* = compressive stress
~spannungsfeld *n* = compressive stress field
~spannungstextur *f* (Geol.) = shear structure
Druckspeicher *m*, Drucksammler = accumulator
~ladeventil *n*, Drucksammlerladeventil = accumulator charging valve
Druck|spirale *f* [*Vorspannung*] = spiral duct
~sprengwagen *m* → Bitumen-Sprengwagen
~spüler *m* [*WC*] = flush valve
~stab *m* = bar in compression
~(stahl)einlagen *fpl* → Druckbewehrung *f*
~steigerungspumpwerk *n* = booster station
~steuergruppe *f* = pressure control group
~steuerventil *n* = pressure control valve
~steuerventil *n* [*Wandler*] = inlet relief valve
~stollen *m* → → Talsperre
~störung *f* = pressure disturbance
Druckstoß *m*, Druckwelle *f* [*im Bohrloch*] = pressure surge
~, Wasserschlag *m*, Wasserstoß = water hammer
Druckstrahl *m* = jet of water
~bagger *m*, Hydromonitor *m* = monitor, giant, jetting gear
~baggerdüse *f* = jetting nozzle
~baggerung *f* → Abspritzen *n* mit Hydromonitoren

Druck|strebe *f* = compression strut
~**stufe** *f* = pressure stage
~**stütze** *f* = compression column
~**tankwagen** *m* → Bitumen-Sprengwagen
~**teller** *m* [*unterer Teil einer Seitenstütze eines Autokranes, Autobaggers oder Schaufelladers*] = jacking pad
~**turbine** *f*, Gleich~, Aktionsturbine = action turbine, impulse ~
~**übersetzer** *m* [*Hydraulikanlage*] = pressure transmitter
~**umformer** *m* = pressure transformer
~**umlagerung** *f* = pressure redistribution
~**umlaufschmierung** *f* = full pressure lubrication
~**unterschied** *m* = pressure difference
~**ventil** *n* = pressure valve
~**ventil** *n* [*Pumpe*] = discharge valve, delivery ~
~**verdichtung** *f*, Walzverdichtung, Walzkompression *f*, Walzung, Walzen *n* = (compaction by) rolling
~**verdichtung** *f* = compaction by compression
~**verlagerung** *f* = pressure transfer
~**verlust** *m*, Strömungsenergieverlust [*technische Hydromechanik*] = head loss
~**verstärker** *m* = booster [*A compressor or pump inserted into a pipeline for compressed air or a liquid, so as to increase the pressure*]
~**verstärkerpumpe** *f*, Verstärkungspumpe *f* = booster pump
~**verstärker(pumpen)anlage** *f*, Druckverstärkerstation *f*, Verstärkungspumpenanlage, Druckerhöhungs(pumpen)anlage *f* = booster station, ~ plant, boosting ~
Druckversuch *m*, Druckprobe *f*, Druckprüfung *f* [*an Betonwürfeln oder Betonzylindern*] = compression test
~ **bei seitlich unbehinderter Ausdehnung** *f* = unconfined compression test
Druck|verteiler *m* → Bitumen-Sprengwagen *m*

~**verteilung** *f* = distribution of pressure, pressure distribution
~**verteilungsplatte** *f* → Verteilungsplatte
~**-Volumenkurve** *f*, adiabatische Linie *f* = adiabatic line
Druckwasser *n*, Preßwasser, Hydraulikwasser = hydraulic water, pressure ~, pressurized ~
~ → ~ artesisches (Grund)Wasser
~ → Körwasser
druckwasserhaltende Isolierung *f* = water-proofing course
Druckwasser|reaktor-Einschlußbauwerk *n*, Druckwasser-Reaktor-Container *m* = PWR plant container, pressurized water reactor ~
~**stollen** *m* → → Talsperre
Druckwegnahme *f* [*Hydraulikanlage*] = bleed-off
Druckwelle *f* = pressure wave
~, Druckstoß *m* [*im Bohrloch*] = pressure surge
~ **im Boden** = compressional wave in the soil
Druckwiederaufbau *m* = pressure restoration, ~ replacement
~**verfahren** *n*, Wiederaufbau *m* des Druckes in einem Erdölvorkommen = repressuring, Marietta process
Druck|zerkleinerung *f* = compression reduction
~**zerstäubungsbrenner** *m* = mechanical atomizing oil burner
~**zone** *f* = compressed zone, compression ~
~**zone** *f*, Versorgungszone [*Wasserversorgung*] = service district
~**zug** *m* = forced draught (Brit.); ~ draft (US)
~**zunahme** *f*, Druckanstieg *m*, Druckerhöhung *f* = pressure rise
~**zwiebel** *f*, Spannungsdiagramm *n* nach Boussinesq = bulb of pressure, pressure bulb
Drum(lin) *m*, Drümmel *m*, Rückenberg *m* (Geol.) = drumlin
Drumme *f* → Abzugkanal *m*
Drusentextur *f* = drusy structure

(DSM-)Bogensieb n = Dutch Staatsmijnen screen, DSM ~

DTA, Differential-Thermo-Analyse f = differential thermal analysis

Dualstelle f, Binärstelle = binary digit, bit

Dübel m = dowel
~**loch** n = dowel hole
~**stein** m = dowel brick
~**treibgerät** n, Dübelschießgerät = dowel driver
~**wirkung** f = dowel action

Duckdalbe f → Dalbe
Dückdalbe f → Dalbe
Düker m → Rohr~
~**bogen** m = sag bend, over-bend
~**rohr** n = crossing pipe, sag ~

Duktilität f, Streckbarkeit f [*Bitumen*] = ductility

Duktilometer m, Streckbarkeitsmesser m [*Bitumenmessung*] = ductilometer

Düllspaten m = socket spade

Düne f = dune

Dünge|kalk m = aglime, agricultural lime, agstone, liming material
~**mittelfabrik** f, Düngemittelwerk n = fertilizer plant, ~ factory, ~ works

düngende Beregnung f = fertilizer application by sprinkling
~ **Bewässerung** f = organic irrigation
~ **Wirkung** f = fertilizing action

Dungkran m = dung crane

Dunit m (Geol.) = dunite, olivine-rock

Dunkelglimmerschiefer m, Biotitschiefer = biotite schist

dunkler Glimmer m, Biotit m = biotite

dünnbankig = thinly stratified

Dünne f = thinness

dünne Schale f = thin shell

dünner Balken(träger) m = slender beam

Dünnerwerden n → Auskeilen

dünnes Abwasser n = dilute sewage, weak ~ [*Sewage containing a relatively small quantity of organic matter*]
~ **Schalendach** n = thin-shell roof

dünn|glied(e)rige Betonkonstruktion f = thin walled concrete structure

~**plattige Spaltung** f (Geol.) = slaty cleavage

Dünnschliff m, Gesteinsplättchen n von etwa 0,02–0,03 mm = rock slice

dünn|viskoses Schmieröl n = spindle oil
~**wandiges Entnahmegerät** n [*für ungestörte Bodenproben*] = thin-walled sampler
~**wandiges Profil** n [*Hohlträger*] = thin-walled profile

Dunstabzughaube f = cooker hood

Duo|-Kaltwalzwerk n = two-high cold reduction mill
~**-Sieb** n = Duo screen
~**-Trocknungstrommel** f = double drying drum

Duplex|kette f = duplex chain
~**kran** m, Doppeldrehkran, Doppelschwenkkran = duplex crane
~**pumpe** f = duplex pump
~**-Rippenrohr** n = finned duplex tube
~**-Rollenkette** f = double strand roller chain

Durchbiegung f = bending, flexure [*The curvature of a beam about its axis or central plane*]

Durchbiegungs|kurve f = bending curve, flexure ~
~**messer** m = deflectometer
~**moment** n [*Balken*] = sagging (bending) moment, positive ~ [*A bending moment which causes a beam to sink in the middle. Usually described as a positive moment*]
~**theorie** f = yield line theory

Durch|binder m, Ankerstein m [*Hohlmauer*] = wall tie closer, throughstone
~**blasen** n [*Erdöl*] = blowing through [*gas lift*]; by-passing [*gas drive*]
~**bohren** n **von Überlagerungen** = overburden drilling
~**bruch** m = break-through
~**bruchgestein** n → Ergußgestein
~**bruchgesteintuff** m → vulkanischer Tuff
~**drehen** n, Radschlupf m = (wheel) spin, (~) slip

durchdrehsicheres Differential n = non-slip differential
durchfädeln = to thread through [*wires of a prestressing cable*]
Durchfahrt|höhe f = clear height, headroom
~portal n = portal type gantry arranged to allow passage of rolling stock
~profil n, lichter Raum m = clearance
~zeit f [*Straßenverkehrstechnik*] = through journey time
Durchfeuchtung f = moisture penetration, soaking
Durchfluß m → Abfluß
~, Aufenthalt m [*Wasser in einem Becken*] = retention
~begrenzer m = flow limiter
~beiwert m, Abflußmengenbeiwert = discharge coefficient, flow ~
~beziehungslinie f, Abflußmengenbeziehungslinie = curve of corresponding discharges, ~ ~ ~ flows
~dauer f, Aufenthaltzeit f [*Wasser in einem Becken*] = retention period, detention ~
~dauer f, Abflußmengendauer = discharge duration, flow ~
~erhitzer m, Durchlauferhitzer = flow type calorifier
~form f, Abflußform [*Kanal; Strom; Fluß*] = type of flow, ~ ~ discharge
~geschwindigkeit f = flowing-through velocity
~häufigkeit f, Abflußmengenhäufigkeit = frequency of discharges, ~ ~ flows
~kammer f = flowing-through chamber [*The upper compartment of a two stor(e)y sedimentation tank*]
~körper m [*Hydraulik*] = velocity block
~kurve f → Abfluß(mengen)kurve
~linie f → Abfluß(mengen)kurve f
~menge f → Abfluß
Durchfluß(mengen)|ausgleicher m = flow compensator
~messer m, Abfluß(mengen)messer = flow meter
~messung f → Abflußmessung
~quotient m, Abfluß(mengen)quotient = discharge ratio, flow ~
~-Registriergerät n, Abfluß(mengen)-Registriergerät = flow recorder
~regler m [*Zum Gleichhalten einer bestimmten, einstellbaren Durchflußmenge in Rohrleitungen*] = liquid level float cage
~-Summenlinie f, Abfluß(mengen)-Summenlinie = discharge mass curve, flow ~ ~
Durchflußmesser m → Durchflußmengenmesser
~ nach dem Aufheizverfahren = heat-balance flow meter
Durchfluß|messung f → Abflußmessung
~öffnung f, Ablaßöffnung, Auslauföffnung, Ablauföffnung, Abflußöffnung [*Wehr*] = discharge opening, sluiceway
~querschnitt m → Abflußquerschnitt
~quotient m → Durchflußmengenquotient
~-Registriergerät n → Durchflußmengen-Registriergerät
~richtung f = flow direction
~schieber m, Steuerschieber, Regulierschieber = control valve, discharge ~ [*A valve for reducing or increasing the flow in a pipe, as opposed to a stop valve*]
~schwankung f, Abflußschwankung [*Strom; Fluß; Kanal*] = flow variation, discharge ~
~statistik f → Abflußstatistik
~-Summenlinie f → Durchflußmengensummenlinie
~verminderung f, Abflußverminderung [*Strom; Fluß; Kanal*] = decrease of flow, ~ ~ discharge
~versuch m [*Korrosionsprüfung*] = continuous flow test
~verzögerung f → Abflußverzögerung
~(wasser)menge f → Abfluß m
Durchführung f, Ausführung = execution
Durchführungs|art f, Ausführungsart = method of execution
~plan m → Bebauungsplan

Durchgang|bahnhof *m* = through station
~fahrt *f* [*Straßenverkehrstechnik*] = through trip
~hafen *m*, Transithafen = transit harbo(u)r
~höhe *f* → Kopfhöhe
~kennlinie *f*, Durchgangkennkurve *f* = screening characteristic, cumulative size distribution curve of screen underflow
~lager *n* = throwout bearing
~-Lieferwagen *m* = walk-through delivery van
~rohr *n* = barrel [*That portion of a pipe throughout which the internal diameter and thickness of wall remain uniform*]
~sieb *n* = limiting screen
~verkehr *m* = through traffic
durchgehende Fuge *f* = continuous joint
~ Schweißnaht *f* = continuous weld
durchgehendes Gefälle *n* = unchanging grade
Durch|gehgeschwindigkeit *f* [*Turbine*] = runaway speed
~halten *n* **einer Schicht** (Geol.) = continuity of a bed, ~ ~ ~ stratum, persistence ~ ~ ~
~hang *m* = sag(ging), dip
~hieb *m* (✻) = cut-through
~körner *m* = pin punch
durchlässige Unterlage *f* = permeable base
Durchlässigkeit *f* = permeability
Durchlässigkeitbeiwert *m*, Durchlässigkeitszahl *f* = permeability coefficient, coefficient of permeability
Durchlässigkeitsgerät *n* **mit abnehmender Wasserhöhe (oder abnehmendem Wasserdruck)** = falling-head permeameter, variable-head ~
~ ~ gleichbleibender Wasserhöhe (oder gleichbleibendem Wasserdruck) = constant-head permeameter
Durchlässigkeits|gleichung *f* = permeability equation, equation of permeability
~messer *m* = permeameter

Durchlässigkeitsversuch *m* **mit abnehmendem Wasserdruck** = falling-head permeability test
~ ~ gleichbleibendem Wasserdruck = constant-head permeability test
Durchlässigkeits|zahl *f* → Durchlässigkeitsbeiwert *m*
~zelle *f* = permeability cell
Durchlaß *m*, Abzugkanal *m* = culvert
~rohr *n*, Abzugkanalrohr = culvert pipe
Durchlauf *m* [*bei Prüfsiebung*] = screen underflow
~auflagerung = continuous support
~balken(träger) *m* = continuous beam [*Is supported at three or more points*]
~balken(träger) *m* **über drei Öffnungen** = three-span continuous beam
~-(Beton-)Zwangsmischer *m* → → Betonmischer
~-Blankglühofen *m* = continuous bright-annealing furnace
~-Bogen(träger) *m* = continuous arch(ed girder)
durchlaufend = continuous
~ bewehrt, ~ armiert = continuously-reinforced
durchlaufende Bewehrung *f*, ~ Armierung = continuity reinforcement
~ Flachgründung *f*, Streifenfundament *n* geringer Gründungstiefe = shallow continuous footing
durchlaufenes Kolbenvolumen *n*, Hubvolumen [*Kompressor*] = swept volume, piston displacement
Durchlauf|erhitzer *m*, Durchflußerhitzer = flow type calorifier
~gitterträgerbrücke *f* = continuous lattice girder bridge, ~ truss ~
~-Hohlkastenträger *m* = continuous box girder
~kanal *m* [*Ölhydraulik*] = gallery
~-Konstruktion *f* = continuity construction
~kühlung *f* = flow cooling
~mahlung *f*, Durchlaufmahlen *n* = open-circuit grinding
~mischer *m*, Freifall-~, Austragmischer, Mischer mit Schurrenaus-

trag(ung) = closed drum (concrete) mixer with tipping chute discharge
~mischer *m* → Fließmischer
~pfette *f* = continuous purlin(e)
~platte *f* = continuous slab
~rahmen *m* = continuous frame
~tank *m* an der Bohrung zur Trennung von Öl, Wasser und Gas = gun barrel
~träger *m* = continuous girder [*Is supported at three or more points*]
~träger *m* über drei Öffnungen = three-span continuous girder, continuous girder with three spans
~waschmaschine *f*, Durchlaufwäsche *f* = flow washer (or washing machine)
~wirkung *f* [*z. B. bei Stahlbetonhohldielen*] = continuous effect
~-Zwangsmischer *m* → → Betonmischer
~-Zwangsmischer *m* → Dauer-Zwangsmischer
Durchlicht *n* = transmitted light
durchlochen [*Die Verrohrung im Bohrloch mit Geschossen durchlochen*] = to gun-perforate
Durchlöcherung *f* [*Korrosionserscheinung*] = perforation
Durchlüftung *f* → Belüftung
Durchlüftungs|becken *n* → Belüftungsbecken
~dauer *f*, Belüftungsdauer = aeration period, ~ time
~leitung *f*, Belüftungsleitung [*Abwasserwesen*] = aerator line, aeration ~
~raum *m* → Belüftungsbecken *n*
durchörtern (⚒) = to work through
~ [*Tunnel- und Stollenbau*] = to drive through
Durch|querung *f* = crossing
~querungsbauwerk *n* = crossing installation, ~ structure
~satz *m* [*Material*] = throughput
~satzleistung *f*, Durchsatzmenge *f* = throughput per hour
~schlagen *n* der Federn = "bottom" of springs

~schlagen *n* von Erdgas aus dem Ringraum in die Steigrohre = surging
~schlagfestigkeit *f* des Gases [*Elektrofilter*] = (electrical) breakdown-strength of gas
~schlagröhre *f*, vulkanischer Schlot *m*, Eruptionskanal *m*, vulkanischer Neck *m*, Esse *f* = (volcanic) neck, pipe(-like conduit)
~schlagspannung *f* = breakdown voltage, sparking potential
~schlagtür *f*, Pendeltür = double acting door
durchschleusen = to lock
Durchschleusen *n*, (Schiff)Schleusung *f* = lockage, locking
durchschnittlicher Korndurchmesser *m*, mittlerer ~ = average grain diameter
~ **Wasserstand** *m*, mittlerer ~, gewöhnlicher ~ = average stage
Durchschnittverkehr *m* = average traffic
durchsenken (⚒) = to dint
durchsichtig = transparent
Durch|sickerung *f*, Versickerung = percolation
~steckschraube *f*, Bolzen *m* mit Kopf und Mutter = bolt and nut
~stich *m* [*zur Beseitigung von Windungen in einem Wasserlauf*] = cut
~stoßgerät *n* → Rohr~
~strahlungsprüfung *f*, Durchstrahlungsprobe *f*, Durchstrahlungsversuch *m* = radiographic inspection, ~ examination
durchströmter Querschnitt *m* → Abflußquerschnitt
Durch|teufung *f* = sinking through
~trittsöffnung *f* [*Steigrohrwasserschloß*] = port
~trümmerungszone *f* (Geol.) = detritus zone
~tunnelung *f* = tunnel(l)ing through
~wärmung *f*, Vorwärmung, Schmauchen *n* [*Ziegelbrennen*] = initial dehydration, water-smoking
durchweichen, (auf)weichen = to soften

Durch|weichung *f* → Aufweichung
~wurf(sieb) *m*, (*n*) = riddle
~zugvermögen *n* [*Motor*] = lugging ability
Durit *m* = durain
dürr, trocken = arid [*A term applied to lands or climates that lack sufficient water for agriculture without irrigation*]
Dusche *f*, (Bad)Brause *f* = shower
Dusch|kabine *f* → Brausenische *f*
~nische *f* → Brausenische
~raum *m* = shower room
~rosette *f*, Brauserosette = shower rose, spray ~
~säule *f* → Brausesäule
~zelle *f* → Brausenische *f*
Düse *f* = jet
~ am Eruptionskopf = top-hole choke, ~ flow bean
~ ~ Steigleitungsfuß = bottom-hole choke, ~ flow bean
Düsen|bohren *n* → Feuerstrahlbohren
~bomber-Piste *f*, Düsenbomber-Start- und Landebahn *f* = jet-bomber runway
~flugplatz *m* = jet airfield
~flugzeug *n* = jet (air) plane, ~ aircraft
~flugzeugpiste *f*, Düsenflugzeug-Start- und Landebahn *f* = jet runway
~flugzeugverkehr *m* = jet plane traffic
~halter *m* [*Motor*] = fuel injection valve body
~kreuz *n* [*Tiefbohrtechnik*] = choke manifold, flow bean ~
~meißel *m* [*Rollenmeißel mit als Düsen ausgebildeten Spülkanälen. Dadurch erhält der Spülstrom eine Geschwindigkeit von 60–90 m/s*] = jet bit
~motor *m* = jet engine
~mundstück *n* = nozzle
~öffnung *f* = nozzle opening
~schieber *m* = jet valve
~strang *m* → Sprengrampe *f*
düsentreibstoffeste Fugenvergußmasse *f* = jet and fuel resisting joint sealing compound, JFR ~

Düsenwerkstatt *f* [*Flughafen*] = workshop hangar for jets
Dutzendfeile *f* = dozen file
Dy *m*, Torfschlamm *m* [*vollständig zersetzter Torf*] = dy
dynamische Ähnlichkeit *f* = dynamic similarity, ~ similitude
~ Baugrunduntersuchung *f* = dynamic foundation exploration, ~ ~ reconnaissance, ~ ~ investigation, ~ subsurface ~
~ Belastung *f* = dynamic loading
~ Geologie *f* [*nicht verwechseln mit „Geodynamik"*] = physical geology
~ Rammformel *f* = dynamic (pile) formula
~ Viskosität *f*, ~ Zähflüssigkeit *f* = dynamic viscosity
~ Viskositätszahl *f*, dynamischer Viskositätsbeiwert *m* = dynamic viscosity coefficient
dynamischer Elastizitätsmodul *m*, E-Modul, dynamische Elastizitätszahl *f* = dynamic modulus
(dynamischer Pfahl)Rammwiderstand *m* = dynamic (pile-driving) resistance
dynamisches elastisches Verhalten *n* = dynamic elastic behavio(u)r
Dynamit *n* = dynamite
Dynamometamorphose *f*, Pressungsumwandlung *f*, Pressungsumprägung, Stauungsmetamorphose = dynamic metamorphism
Dynamometer-Stempel *m* (⚒) = dynamometer prop, ~ post

E

Ebbe *f*, Tidefall *m* [*das Fallen des Wassers vom Tidehochwasser zum folgenden Tideniedrigwasser*] = falling tide, ebb ~
~intervall *n* = ebb interval
~marke *f*, Ebbelinie *f* = low water mark
~rinne *f* = scour produced by the ebb(-current)
~(strom)rinne *f* [*Rinne im Mündungsgebiet eines Stromes oder im Watt,*

die vorwiegend vom Ebbestrom durchflossen wird] = ebb(-tide) channel
Ebb(e)strömung *f* = ebb(-tide) current, outgoing tide
Ebbetor *n*, Binnentor = ebb tide gate
Ebbe- und Flut-Kraftwerk *n* → Flutkraftwerk
Ebbe- und Flut-Zone *f*, Litoral *n* = littoral zone
Ebbewassermenge *f* = volume of water discharging on the ebb-tide, volume of the ebb
Ebbstromstärke *f* = ebb strength
Ebene *f*, Flachland *n* = plain
~ [*Baustatik*] = plane
ebene Biegung *f* = plane bending
~ **Elastizität** *f* = two-dimensional elasticity
~ **Formänderung** *f*, ~ Verformung = plane deformation
~ **Geometrie** *f*, Geometrie der Ebene, Planimetrie = planimetry, plane geometry
~ **Gleitfläche** *f* = plane surface of sliding
~ **Spannung** *f* = plane stress
~ **Strömung** *f* = two-dimensional flow
~ **Verformung** *f*, ~ Formänderung = plane deformation
~ **zusammengesetzte Biegung** *f* = combined plane bending and compression
ebener Aluminiumprofilteil *m*, ebenes Aluminiumprofilteil *n* = alumin(i)um plate element
~ **Rahmen** *m* **mit starren Ecken** = rigid-jointed plane framework
~ **Spannungszustand** *m* = uniform stress
ebenerdige Kreuzung *f* → plangleiche ~
ebenerdiger Hubschrauberflughafen *m* = ground-level heliport
ebenes elastisches Tragwerk *n* = plane elastic system
Ebenflächigkeit *f*, (Plan)Ebenheit *f* = accuracy of levels, evenness; surface smoothness (US)
Ebenheits|abmaß *n* = (surface) planeness tolerance

~**messer** *m* = roughness indicator [*for highway surface roughness measurements*]
~**messung** *f* **mit dem Richtscheit** = straightedging
ebenmäßiger Angriff *m* [*Korrosion*] = uniform attack
ebnen → ein~
Echo|impuls *m* = reflected pulse
~**lot** *n*, akustisches Lot, Schall-Peilgerät *n* = sonic altimeter, sound-ranging ~, echo sounder, sonic log, sonic sounder
~**lotung** *f*, akustische Lotung = acoustic sounding, echo ~, sonic logging
echowidrig = anechoic
echter kristallinischer Marmor *m*, Urkalk(stein) *m* = true marble, recrystallized limestone
~ **Löß** *m* → Urlöß
~ **Obsidian** *m* = true obsidian, Iceland agate
Eck|(absperr)ventil *n* = angle (shut-off) valve
~**anschluß** *m* = corner connection
~**armierung** *f* → Eckbewehrung
~**aussteifung** *f* [*Kastenträger*] = corner truss
~**(bade)wanne** *f* = corner bath
~**bewehrung** *f*, Eckarmierung, Eck(stahl)einlagen *fpl* = corner reinforcement
~**blech** *n* = gusset, knee bracket, corner plate
~**bohle** *f*, Eckspundbohle = corner sheetpile
~**einlagen** *fpl* → Eckbewehrung *f*
Eckenbohrmaschine *f* = corner drill
Eck|fenster *n* = corner window
~**förmigkeit** *f*, Winkeligkeit, Kantigkeit [*z.B. Brekzie*] = angularity [*e. g. breccia*]
eckig, kantig = angular
eckiges Bruchstück *n*, kantiges ~ (Geol.) = angular fragment
Eck|leiste *f* **für Decken** = cove
~**naht** *f* = corner weld, ~ joint
~**niethammer** *m* = corner riveting hammer

Eckpfahl — Eigenlast

~pfahl m = corner pile
~schrauber m = corner nutrunner
~schutzleiste f = angle bead
~(spund)bohle f = corner sheetpile
~(stahl)einlagen fpl → Eckbewehrung f
~stein m = quoin
~stütze f = corner column
~verband m [Mauerwerk] = corner bond
~verbindung f = corner connection
~versteifung f = corner bracing
~wanne f, Eckbadewanne = corner bath
~winkel m, Saumwinkel = angle bracket, corner angle

Edel|gas n = noble gas, rare ~
~metall n = precious metal
~rost m, Patina f [Bildet sich infolge des Einflusses der in der atmosphärischen Luft enthaltenen Säuren auf Metallen] = patina
~splitt m = double broken and double screened chip(ping)s, twice crushed and screened ~
~zement m, Bariumzement = baritic cement
~zement m, Strontiumzement = strontium cement

effektive Pferdestärke f, nutzbare ~, Nutzpferdestärke = brake horsepower
effektiver Querschnitt m, wirksamer ~, Strömungsquerschnitt = effective cross section

Effusivgestein n → Ergußgestein
EH, Einpreßhilfe f [Verminderung des Absetzens des Zementmörtels im Spannkanal bzw. mäßiges Quellen. Verbesserung des Fließens des Mörtels beim Einpressen und Verminderung des Wasseranspruchs des Mörtels] = grouting aid, intrusion ~
Eichenschindel f = oak shingle
Eich|gerät n = calibrating device
~tabelle f, Eichprotokoll n = calibration chart, ~ table
Eichung f, Eichen n = calibration
~, Auslitern n [Flüssigkeitsbehälter] = calibrating, dipping

Eierschalenmatt|glanz m = egg-shell gloss
~lack m = egg-shell flat varnish
Eierschalentextur f [Fehler im Steinzeug] = egg shell
eiförmiger Querschnitt m = egg-shaped section
eiförmiges Betonrohr n, Betonrohr mit eiförmigem Abflußquerschnitt, Ei-profil-Betonrohr = egg-shaped concrete pipe
~ Durchgangrohr n mit Fuß = egg-shaped barrel with base
~ Profil n, Eiprofil = egg-shaped profile
~ Rohr n → Eiprofilrohr
Eiformrohr n, Eiprofilrohr = egg-shaped pipe
Eigenantrieb-Autofahrwerk n, Eigenantrieb-Unterwagen m [Autokran] = rubber-tire (or rubber-tyre) carrier (or chassis) with its own engine and automotive-type drive
eigener Gleiskörper m, Eigenfahrbahnkörper = own right-of-way
Eigen|festigkeit f = natural strength
~feuchtigkeit f der Zuschläge, ~ ~ Zuschlagstoffe = water contained in aggregate
~frequenz f = natural frequency
~füller m = return filler
~gewicht n, Totlast f, Totgewicht, Eigenlast = dead (or fixed) weight (or load)
~gewicht n [Nutzfahrzeug] = unladen weight
~gewichtsmoment n = dead weight moment, ~ load ~, fixed ~ ~
~gewichtsspannung f = dead weight stress, ~ load ~, fixed ~ ~
~gewichtswalze f → → Walze
~heimaufzug m = home lift
~heimbauindustrie f, Eigenhausbau-industrie = home building industry
~heimhandwerker m = (home)handyman [Somebody who does jobs about his home and garden]
~konsolidation f → Konsolidierung f
~last f → Eigengewicht n

Eigenlastdurchbiegung — Eimer(ketten)-Naßbagger

~lastdurchbiegung *f* = dead-load deflexion (Brit.); ~ deflection (US)
~potential *n*, S. P = spontaneous potential, natural ~, self-potential
~potentialkurve *f* = self-potential curve, natural potential ~, spontaneous potential ~
~potentialverfahren *n*, Eigenpotentialmethode *f* [*Geoelektrik*] = self-potential method, ~ exploration, ~ prospecting, natural potential ~, spontaneous potential ~
~schaft *f* = property
~schwingung *f* = natural vibration, ~ oscillation, characteristic ~
~setzung *f* → Konsolidierung
eigensicher = inherently safe
Eigen|standfestigkeit *f* = natural stability
~steifigkeit *f* = natural stiffness
eigentliche Raumfuge *f* → Raumfuge
Eigentumswohnung *f*, Stockwerkeigentum *n* = horizontal property, freehold flat
Eigen|überwachung *f* [*Güteüberwachung im Werk*] = factory control
~verfestigung *f* → Konsolidierung
~widerstand *m* = natural resistance
Eignung *f* = suitability
Eignungsprüfung *f*, Eignungsprobe *f*, Eignungsversuch *m* = qualification test
Eikanal *m* = egg oval sewer, ~ ~ drain
Eilpumpe *f* = high-speed pump, express ~
Eimer *m* = bucket
~, Hand~ = pail
~, Becher *m*, Elevator~ = (elevator) bucket
~, Becher *m*, Grab~ = (digging) bucket
~, Schlepplöffel *m*, Schleppschaufel *f* = dragline bucket
~aufhängung *f* [*Schlepplöffelbagger*] = dragline bucket suspension
~-Ausleger *m*, Schlepplöffel-Ausleger, Schleppschaufel-Ausleger = dragline bucket boom (US); ~ ~ jib (Brit.)
~-Ausrüstung *f*, Eimer-Vorrichtung, Schleppschaufel-Ausrüstung, Schleppschaufel-Vorrichtung, Schlepplöffel-Ausrüstung, Schlepplöffel-Vorrichtung, Eimerseilausrüstung = dragline (bucket) attachment
~bagger *m* → Eimerketten(trocken)bagger
~-Gleisbagger *m* → Eimerketten-Gleisbagger
~-Gleiskettenbagger *m*, Eimer(ketten)bagger auf (Gleis)Ketten, Eimer(ketten)raupenbagger, Eimer(ketten)-Gleiskettenbagger = crawler bucket ladder excavator
~-Grabenbagger *m*, Becher-Grabenbagger, Grabenbagger mit (Grab)Eimern, Grabenbagger mit (Grab)Bechern = bucket trencher, ~ ditcher
~kette *f*, Becher(werk)kette, Elevatorkette = ladder chain carrying the buckets, bucket (elevator) chain, bucket line chain
~kette *f* → → Grabenbagger mit Eimern
Eimer(ketten)|aufzug *m* → Becherwerk
~bagger *m* → Eimerkettentrockenbagger
~bagger *m* auf (Gleis)Ketten → Eimer-Gleiskettenbagger
~fahrlader *m* → Becher(werk)auflader
~fahrlader *m* mit Drehabsatzband = swivel conveyor bucket loader, ~ conveyer ~ ~
~-Gleisbagger *m*, Eimer(ketten)-Schienenbagger = rail-mounted bucket ladder excavator
~-Gleiskettenbagger *m* → Eimer-Gleiskettenbagger
~grabenbagger *m* → → Grabenbagger mit Eimern
~-Hochbagger *m* = face bucket ladder excavator, bucket ladder excavator working above ground level
~lademaschine *f* → Becherwerk(auf)lader
~lader *m* → Becherwerk(auf)lader

Eimer(ketten)raupenbagger — einachsige Spannung

~-**Naßbagger** *m* → Eimer(ketten)-Schwimmbagger
~**raupenbagger** *m* → Eimer-Gleiskettenbagger
~-**Schienenbagger** *m*, Eimer(ketten)-Gleisbagger = rail-mounted bucket ladder excavator
~-**Schwimmbagger** *m*, Eimer(ketten)-Naßbagger = multi-bucket dredger, bucket-ladder ~, ladder (bucket) ~, elevator ~, endless chain ~, chain-bucket ~ [*US = dredge*]
~-**Schwimmbagger** *m* **mit Förderband**, Eimer(ketten)-Naßbagger ~ ~ = stacker dredge (US); ~ dredger (Brit.)
~-**Schwimmbagger** *m* **mit Spülrinne**, Eimer(ketten)-Naßbagger ~ ~ = sluice dredger, flume ~ (Brit.); ~ dredge (US)
~-**Tiefbagger** *m*, Tief-Eimerketten-(trocken)bagger = bucket ladder excavator working below ground level
~**(trocken)bagger** *m*, Eimerbagger = bucket ladder excavator, dredger ~, chain bucket ~, multibucket ~, continuous land bucket dredger [*US = dredge*]
Eimerlatrine *f* = bucket latrine
Eimerleiter *f*, Unterleiter [*Eimerkettenbagger*] = (main) ladder
~, Baggerleiter, Becherleiter [*Naßbagger*] = digging ladder, bucket ~, dredging ~, bucket flight, boom
~, Baggerleiter [*allgemein*] = bucket ladder, ~ flight, digging ladder
~, ~-Ausleger *m*, Ausleger, Baggerleiter, Becherleiter [*Grabenbagger mit Eimern*] = digging ladder, ditcher ~, bucket ~, bucket flight, boom
~, Becherleiter = bucket ladder, boom
~, Becherleiter [*Gleisschotter-Reinigungsmaschine*] = bucket flight, ~ ladder
~(-Ausleger), Ausleger *m*, Baggerleiter *f*, Becherleiter [*Grabenbagger mit Eimern*] = digging ladder, ditcher ~, bucket ~, bucket flight

~**grabenbagger** *m* → → Grabenbagger mit Eimern
~**rahmen** *m*, Baggerleiterrahmen, Ausleger *m* = ladder frame, boom
~**winde** *f*, Baggerleiterwinde, Becherleiterwinde, Auslegerwinde = digging ladder hoist, boom ~
Eimer-Naßbagger *m* → Eimer(ketten)-Schwimmbagger
Eimerrad *n*, Becherrad, Schaufelrad, Kreiseimerleiter *f* [*Bagger*] = bucket wheel, digging ~
~, Schöpfrad, Becherrad [*Waschmaschine*] = dewatering wheel
Eimer|reihe *f* → Eimerstrang *m*
~**reiniger** *m*, Ausschneider *m* [*Eimerkettenbagger*] = (scraper) bucket cleaner
~-**Schienenbagger** *m* → Eimer(ketten)-Gleisbagger
~-**Schwimmbagger** *m* → Eimerketten-Schwimmbagger
~**seilausrüstung** *f* → Eimer-Ausrüstung
~**seilbagger** *m* → Schürfkübelbagger
~**seilkübel** *m* → Schleppschaufel
~**seil-Schreitbagger** *m* → (Schleppschaufel-)Schreitbagger
~**strang** *m*, Becherstrang, Eimerreihe *f*, Becherreihe = line of buckets, bucket line
~**trockenbagger** *m* → Eimerkettentrockenbagger
~-**Vorrichtung** *f* → Eimer-Ausrüstung
~**zahn** *m*, Becherzahn = bucket tooth
Einachs|-Anhänge-Bagger *m*, Einachs-Schlepp-Bagger = single-axle trailer excavator, two-wheel ~ ~
~-**Anhänger** *m* = single-axle trailer, two-wheel ~
Einachser *m*, Einachsfahrzeug *n* = single-axle unit, two-wheel ~
Einachs|fahrwerk *n* = single-axle travel(l)ing gear, ~ undercarriage
~-**Gußasphaltkocher** *m*, Einachs-Mastixkocher = semi-trailer type mastic asphalt boiler (or cooker)
einachsige Spannung *f* = uniaxial stress

einachsiger Erdtransportwagen *m* mit **Einachs-Radschlepper** = motor wagon
Einachs|lenkung *f* = one-axle steering
~**-Luftverdichter** *m* = trailer compressor
~**-Mastixkocher** *m*, Einachs-Gußasphaltkocher = semi-trailer type mastic asphalt boiler (or cooker)
~**mischer** *m* → (luftbereifter) Einachs(-Schnell)mischer
~**-Motor-Schrapper** *m*, Autoschrapper mit 2-Rad-Schlepper, Schrapper mit gummibereiftem Einachs-Traktor, Motor-Schürf(kübel)wagen *m* mit Einachs-Schlepper, motorisierter Schürfkübel *m* mit Einachs-Traktor, Motor-Schürfzug *m*, selbstfahrender Schürfwagen = tractor-scraper, motor(ized) scraper, motor scraper (with two-wheel traction)
~**radschlepper** *m* → Einachsschlepper
~**räumer** *m* = hand-guided two-wheel dozer, ~ single-axle ~
~**-Reifen-Sattelschlepper** *m* → (Zweirad-)Vorspänner *m*
~**-Schlepp-Bagger** *m*, Einachs-Anhänge-Bagger = single-axle trailer excavator, two-wheel(ed) ~ ~
~**schlepper** *m*, Einachsradschlepper, Zweirad-Schlepper, Einachszugwagen *m*, Zweirad-Trecker *m*, Zweirad-Traktor *m* = two-wheel(ed) tractor, single-axle ~

Einachs-Schlepperkran *m*, **Einachs-Krananhänger** *m*
Einachs-Schlepperkran *m*, Fabrikat Le TOURNEAU-WESTINGHOUSE COMPANY, PEORIA, ILLINOIS, USA

tractor-drawn crane, lift-and-carry ~

TOURNACRANE, Le TOURNEAU portable crane [*Trademark*]

Einachs-Zugwagen *m* → Einachsschlepper *m*
Ein-Arbeitsgang-Verfahren *n* [*Bodenvermörtelung*] = single-pass mixing
Einbahn|straße *f* = one-way road
~**straßenpaar** *n* = one-way pair
Ein-Bahnsystem *n* [*Flugplatz*] = single runway (system)
Einbahnverkehr *m* = one-way traffic, uni-directional ~
Einbau *m* → (Auf)Schütt(ungs)arbeiten *fpl*
~**arbeiten** *fpl* = placing operations
~**art** *f*, Einbauverfahren *n* [*für Massenbaustoffe, z.B. Beton, Schwarzdeckenmischgut, Schüttmaterial usw.*] = type of placing, method ~ ~
~**(bade)wanne** *f* = built-in bath
(~)**Bahn** *f*, (Einbau)Streifen *m* [*Schwarz- und Betondeckenbau*] = strip, lane
~**breite** *f*, Arbeitsbreite [*Straßenfertiger*] = operating width
~**-Dieselmotor** *m* = built-in diesel (engine), ~ ~ motor
einbauen, einbringen, verarbeiten [*Baustoff*] = to place
Einbauen *n* des Bohrgestänges, Einlassen ~ ~ = running in of the drill pipe
Einbau|fähigkeit *f* [*Beton*] = placeability
~**folge** *f* = placing sequence
~**gerät** *n* [*Straßenbau*] = laying machine
~**gerät** *n* für Gerinneauskleidungen → → Kanalbaumaschinen *fpl*
~**geschwindigkeit** *f* [*Beton*] = rate of placement, ~ ~ pouring, ~ ~ concreting
~**geschwindigkeit** *f*, Arbeitsgeschwindigkeit [*Straßenfertiger*] = laying rate
~**-Kleiderschrank** *m*, Wand-Kleiderschrank = built-in wardrobe

Einbaukolonne — Ein-Blattausgleichfeder 346

~kolonne f, Einbaumannschaft f, Einbautrupp m [Straßendeckeneinbau von Hand] = hand-laying gang, ~ crew, ~ team, ~ party

~kosten f [Straßendecke] = laying cost, cost of laying

~kosten f [Bodenvermörtelung] = processing cost

~kosten f = placement cost

~kupplung f = built-in clutch, inbuilt ~

(~)Lage f, (Einbau)Schicht f = lift, layer, course, pour

~leistung f [Straßenfertiger] = laydown rate [road finisher]

~mannschaft f → Einbaukolonne f

~maschine f für Gußasphalt, Gußasphaltfertiger m = mastic asphalt finisher, ~ ~ finishing machine

~möbel npl = fitment, fitting, built-in furniture, fitted furniture, inbuilt furniture [*Furniture fixed, often by the builder, as opposed to the loose furniture bought by the occupier*]

(~)Schicht f, (Einbau)Lage f = lift, layer, course, pour

~schrank m, Wandschrank = built-in cupboard, inbuilt ~

~stelle f, Vortriebstelle f [Straßenbau] = spreading site, placing point, laying ~, placement ~; working face (Brit.)

~stelle f, Verarbeitungsstelle [Baustoff] = point of placement

(~)Streifen m, (Einbau)Bahn f [Schwarz- und Betondeckenbau] = strip

~streuer m, Einbau-Streugutverteiler m = built-in bottom (type) gritter, ~ ~ (~) abrasive (material) spreader

~-Streugutverteiler m → Einbaustreuer m

~technik f [Straßenbau] = laying technique

~teil m, n → → Schwarzbelageinbaumaschine f

~temperatur f, Bautemperatur [Straßenbelag] = laying temperature, placing ~

Einbauten f [*im Abwassernetz*] = sewer appurtenances [*The more common sewer appurtenances include manholes, junction chambers, street inlets, catch basins, and flush tanks*]

~ [*allgemein*] = internal fittings, ~ equipment

Einbau|teufe f der Ankerrohre [*Tiefbohrtechnik*] = setting depth of surface casing

~trupp m → Einbaukolonne f

~verfahren n → Einbauart f

~vermögen n → → Schwarzbelageinbaumaschine f

~wanne f, Einbaubadewanne = built-in bath, inbuilt ~

~wassergehalt m = placement moisture (or water) content

~zeichnung f = installation drawing

~zeit f = placing period

~zug m [Straßenbau] = paving (plant) train, train of paving plant

einbetonieren, einbetten in Beton = to concrete in, to encase in concrete

Einbettungsmasse f [*Imprägnations-(bohr)krone*] = matrix

Einbettzimmer n = single-bed room

Einbeulen n [*Tiefbohrtechnik*] = collapse

Einbindelänge f, Pfahltiefe f, Eindring-(ungs)tiefe f = depth of penetration, embedded length

~ → Haftlänge

einbinden [*Straße in die Landschaft*] = to fit [*the road into the landscape*]

~ = to feather [*to blend the edge of new material smoothly into the old surface*]

Einbinden n, Anschließen = connecting to existing work

~, Einbindung f = stub-in [*welding of branch lines to headers*]

Einbindung f [*Pfahl*] = embedment

Einblasebewetterung f → Druckbewetterung

Einblasung f = insufflation [*Practice of adding dust to the coal in a burner pipe*]

Ein-Blattausgleichfeder f = single leaf equalizer spring

Einblatten *n* → Anblatten
Einblockausführung *f* = monobloc design
Ein-Bogenbrücke *f* = single arch bridge
Ein-Bohlensteg *m*, Ein-Bohlenbahn *f* = single-plank track, ~ run(way), ~ road
Einbrennlack *m* = stoving paint, ~ enamel
einbringen, einbauen, verarbeiten [*Baustoff*] = to place
Einbringen *n* [*Rohre in einen Graben*] = lowering in
~ **der Zuschläge** = prepacking
~ **des Betons** → Betoneinbau *m*
~ **von Pumpbeton**, Betonieren mit Pumpe(n) = concrete pumping
Einbruch *m* [*in die Kohle*] = holing [*The British term for undercutting or overcutting coal. In Britain coal may not be blasted down if it has not been holed*]
~**alarmanlage** *f* = burglar alarm system
~**caldera** *m* = engulfment caldera
~**wasser** *n*, Andrangwasser = inrush water
einbuchtendes Ufer *n*, konkaves ~, äußeres ~, Konkave *f*, Einbuchtung *f* = concave bank, outer ~
Einbuchtung *f* → einbuchtendes Ufer *n*
~ = embayment
~ [*Schweißtechnik*] = surface cavity
einbühnen [*Stempel*] (⚒) = to sink the foot of props into the floor, ~ ~ ~ ~ ~ posts ~ ~ ~
einbürsten = to brush in place, to broom ~ ~
(ein)decken, bedachen = to roof
Eindecken *n* → Bedachung *f*
Eindecker *m*, Einstufensieb *n*, Eindeckersieb, Sieb mit einem Siebfeld, Eindecksiebmaschine *f* = single deck screen
~**-Schwingsieb** *n*, Einstufen-Schwingsieb = single-deck vibratory screen
(Ein)Deckung *f* → Dachdeckung
Eindeckungszubehör *m*, *n*, Dach~, Bedachungszubehör = roofing accessories
eindeichen = to bank in
Eindicker *m* = thickener
~ **mit Kratzer-Klassierer** = spiral rake thickener, ~ scraper ~, classifier for fine separations
Eindickkegel *m* = thickening cone
Eindickung *f* = thickening
eindimensional = one-dimensional, unidimensional
eindrahtige Verankerung *f* = single-wire anchorage
eindringen = to penetrate
Eindringen *n* = penetrating
Eindringsfestigkeit *f*, Eindringungsfestigkeit *f* = penetration strength
~**gerät** *n* → Eindring(ungs)messer *m*
~**härte** *f* = penetration hardness
~**mörtel** *m* → Injektionsmörtel
~**öl** *n* = penetrating oil
~**querschnitt** *m* [*Die drehmomentübertragende Kraft eines Greifmesserzahnes wirkt auf eine Fläche, die man als Eindringquerschnitt bezeichnet. Bei der Handhabung von Steigrohren*] = penetration cross section
~**stempel** *m* = penetration piston
~**tiefe** *f*, Eindringungstiefe = depth of penetration, penetration depth
~**tiefe** *f* **pro Schlag** [*Pfahlrammung*] = set per blow
Eindringung *f*, Penetration *f* = penetration
~ **des Reißzahns** = tooth penetration
Eindring(ungs)|festigkeit *f* = penetration strength
~**kurve** *f* = penetration curve
~**messer** *m*, Penetrometer *n*, Eindringgerät *n* = penetrometer
~**spitze** *f* = penetration tip
~**tiefe** *f* = depth of penetration, penetration depth
~**tiefe** *f*, Pfahltiefe, Einbindelänge *f* = embedded length, depth of penetration
~**vermögen** *n*, Eindring(ungs)kraft *f* = force of penetration, penetration force
~**wert** *m* = penetration value

Eindring(ungs)widerstand — Einfach-Trommel(seil)winde 348

~widerstand *m*, Verdrängungswiderstand [*Rammen*] = penetration resistance, driving ~

~widerstand *m* = penetration resistance

~zone *f* der Bohrlochflüssigkeit = zone of invasion of bore hole liquid

Eindrücken *n* [*des Bogens in die Firste*] = spiring [*of an arch*]

Eindruck *m* = impression

~-**Scherfestigkeits-Prüfung** *f* = punching shear stability test

~**tiefe** *f* = depth of impression

Eindüsrohr *n* [*Mischer für Schwarzdeckenmischgut*] = spray tube

Einebenen|betrieb *m* [*Flughafen*] = one-level operation

~**finger** *m* [*Flughafen*] = one-stor(e)y finger

~**system** *n* [*Passagier- und Gepäckabfertigung auf Flughäfen*] = one-level (building operational) system

einebnen, (ein)planieren, (ver)ebnen = to spread and level

Einebnungs|fläche *f* → Abtragungsfläche

~**pflug** *m* → Planierpflug

Eineinhalbebenensystem *n* [*Passagier- und Gepäckabfertigung auf Flughäfen*] = one and a half level (building operational) system

Einengung *f* = narrowing

einetagig, eingeschossig, einstöckig = single-stor(e)y, single-storied

einfach eingeschertes Seil *n* = one-part line

~ **gekrümmte Schale** *f* = single curvature shell

einfache Bohl(en)wand *f*, Schurzholzwand = single plank wall

~ **Gewichts(stau)mauer** *f* → → Talsperre *f*

~ **Klärung** *f*, ~ mechanische ~ [*Abwasser*] = plain settlement, ~ sedimentation, ~ settling

~ (**mechanische**) **Klärung** *f* [*Abwasser*] = plain settlement, ~ sedimentation, ~ settling

~ **Proctordichte** *f* = Standard Proctor dry density

~ **Vergußfuge** *f* = poured joint

~ **Wand** *f* = solid wall

einfacher Hängebock *m* → einsäuliges Hängewerk *n*

~ **Heizröhrenkessel** *m* = single cylinder multitubular boiler

~ **Lasthaken** *m*, ~ **Hubhaken** = single hook

~ **Strebenderrick(kran)** *m*, Strebenderrick(kran) üblicher Bauart = three-legged derrick

~ **symmetrischer Rahmen** *m* = single symmetrical frame

~ **Träger** *m* = single girder

einfaches Hängewerk *n* → einsäuliges ~

Einfach-Abmeßvorrichtung *f* = single-material batcher

~ **mit Waage** = single-material scale batcher

Einfach|-Bereifung *f* = single tyre (Brit.); ~ tire (US)

~**bohrturbine** *f* = standard turbodrill

~**dampfraumkessel** *m*, Kessel mit einem Dampfraum = boiler with single steam space

~**dünger** *m* = single-nutrient fertilizer

~**federring** *m* = single coil washer

~**fenster** *n* = single-glazed window

~-**Gewölbesperre** *f* → Bogensperrmauer

~**kernrohr** *n*, einfaches Kernrohr = single tube (core) barrel

~**kernrohr** *n* mit „Spinne", ~ ~ grobgezahnter Stirn = basket-type core barrel

~**kernrohrloch** *n* [*zum Entweichen der Spülung*] = weep hole

einfach-logarithmischer Maßstab *m* = log scale

Einfach|meißelbohrkrone *f*, einfache Meißelschneide *f* = single chisel (drill) bit

~-**Rollenkette** *f* = single strand roller chain

~-**Schiebetür** *f*, Schiebetür mit einem Flügel = single sliding door

~-**Spurkranzrad** *n* = single-flanged wheel

~-**Trommel(seil)winde** *f* = single-drum winch

Einfachverglasen — Einfuhrkohle

~verglasen n = single glazing
~verglasung f = single glazing
~-Vibrationswalze f, Einfach-Rüttelwalze, Einfach-Schwing(ungs)walze = single-vibration roller
einfachwirkend = single-acting
Einfädeln n = threading
~ [*Straßenverkehrstechnik*] = merging, weaving
Einfädelungs|länge f, Verflechtungslänge [*Straßenverkehrstechnik*] = weaving distance, ~ length, merging ~
~raum m, Verflechtungsraum [*Straßenverkehrstechnik*] = weaving space, ~ area, ~ section, merging ~
Einfahr|gleis n [*Verschiebebahnhof*] = receiving track
~pulver n = bon ami powder, break-in ~
Einfahrt f = entrance
~, Autobahn-~ = motorway entrance
~schleuse f = (dock) entrance lock
~tor n zur gebührenpflichtigen Autobahn = toll gate
Einfahrzeit f = breaking-in period
Einfallen n von Störungen (⚒) = hade
(Ein)Fallwinkel m [*wird zwischen einer Fallinie und ihrer Horizontalprojektion gemessen*] (Geol.) = angle of dip, ~ ~ inclination
Einfamilienhaus n = house [*British standard name. B.S. 3589:1963. A building forming one self-contained dwelling. It would not be correct in Great Britain to describe a building containing more than one dwelling as a house. Such a building might be a pair of (semi-detached) houses, a row of (terrace) houses or a block of flats or of maisonettes*]
~ aus Betonfertigteilen = precast concrete house
einfärben [*Beton*] = to dye [*concrete during casting*]
Einfärben n [*Beton*] = dyeing
Einfärbevermögen n = coloring power (US); colouring ~ (Brit.)

Einfassung f = surround
Einfassungsplatte f = surround slab
Einfederung f [*Gummireifen*] = cushioning
Einfeld|balken(träger) m = single-span beam
~brücke f = single-span bridge
~platte f = single-span slab
~rahmen m = frame of one bay
~träger m = single-span girder
(ein)fetten, (ab)schmieren, einschmieren = to grease
(Ein)Fetten n → (Ab)Schmieren
Ein-Flanschrolle f [*Gleiskettenlaufwerk*] = single flange roller
(ein)fluchten, ausrichten, ausfluchten, abfluchten = to align, to range into line, to range out; to aline (US)
Einfluchtung f, Bauflucht f, Abfluchtung f = alignment; alinement (US)
einflügelige Tür f = single wing door
einflügeliges Fenster n = single-sashed window
Einflugschneisenfeuer n, Anflugfeuer, Einflugschneisenleuchte f, Anflugleuchte = approach light
Einfluß|beiwert m, Einflußfunktion f = influence coefficient
~dreieck n = triangle of influence
~faktor m auf die Wirtschaftlichkeit, Wirtschaftlichkeitsfaktor = economic factor
~feld n = influence surface
~fläche f = influence area
~line f = influence line
~linie f der Auflagerkraft, A-Linie = A-line, influence line of reactions
~linie f des Bogenschubs, H-Linie = line of horizontal stresses
~linientafel f, Einflußtabelle f = influence chart
~linienverfahren n = influence line analysis
~zone f = zone of influence
Einfrieren n des Bodens, Gefrieren ~ ~ = soil freezing
Einfuhr|erz n, Importerz, Auslandserz = foreign ore
~kohle f = foreign coal

Einführungslehrgang m = introduction course
Einfuhrzoll m = tariff duty
Einfüllbunker m → Annahmebunker
einfüllen, verfüllen [z. B. *Graben*] = to backfill, to fill in, to re-fill
Einfüll|öffnung f, Eingabeöffnung = filler opening
~**schraube** f = filler plug
~**stelle** f, Eingabestelle = filling point
~**stutzen** m = filler, filling spout, filling tube
~**trichter** m → Aufnahmetrichter
Einfüll(ungs)|bunker m → Annahmebunker
~**trichter** m → Aufnahmetrichter
Ein-Füllungstür f = one-panel door
Einfüllverschluß m = filler cap
Eingabelung f, Arm m, Tragstück n, Halter m = bracket
Eingabe|öffnung f, Einfüllöffnung = filler opening
~**stelle** f, Einfüllstelle = filling point
Eingang m = entrance
Ein-Gang-(Boden)Mischer m, Ein-Gang-(Boden)Mischmaschine f, Ein-Gang-Bodenverfestigungsmaschine, Ein-Gang-Bodenvermörteler = single-pass soil mixer, one-pass ~ ~, ~ travel ~, ~ ~ mixing machine, ~ rotary tiller and spreader, ~ stabilizer, ~ stabilizing machine
(Eingangs)Halle f, Empfangshalle = (entrance) hall, reception ~ [*An entrance space, often containing a staircase*]
Eingangswelle f = input shaft
eingebaut = built-in, inbuilt
eingebautes Deckenmischgut n = laid pavement mix(ture)
eingebundene Säule f = engaged column, attached ~ [*Column attached to a wall, with at least half of its diameter projecting from the wall face*]
eingebundener Pfeiler m = attached pier, engaged ~ [*A projection bonded with a brick wall to provide stability, or support for a beam, roof truss, etc.*]

eingedeicht = diked (US); dyked (Brit.)
eingeeist, vereist [*Hafen*] = ice-bound
Eingefäß|bagger m → Aussetzgerät n
~**(förder)pumpe** f = single-tank type pump, ~ pneumatic conveying system
~**gerät** n → Aussetzgerät
~**-Naßbagger** m → Aussetz-Naßbagger
~**-Schwimmbagger** m → Aussetz-Naßbagger
~**-Trockenbagger** m, Aussetz-Trockenbagger = intermittent excavator
eingeführte Luftpore f = entrained air void
eingefurchter Kegel (Geol.) = trenched cone
eingegossener Ringträger m [*Motor*] = cast-in iron band
eingelassenes Schloß n → Blindschloß
eingepreßte Schüttung f → ausgepreßte ~
eingepreßter Alluvionsschleier m → (injizierter) ~
eingerammt = driven
eingerückt [*Kupplung*] = engaged
eingerüttelt = vibrated
eingesackter Zement m, Sackzement, (ab)gesackter Zement = bagged cement, sacked ~
eingeschalte Betonhöhe f = depth of concrete supported in formwork
eingeschaltete Sandsteinbänke fpl = intercalated beds of sandstone
eingeschliffene Betondeckenfuge f = sawn joint, cut ~
eingeschlossene Feuchtigkeit f, ~ Feuchte f = entrapped moisture
~ **Luftpore** f = natural air void, entrapped ~ ~
eingeschlossenes Öl n = blocked oil
eingeschneit, verschneit = snowed-up, snow-bound
eingeschnürter Querschnitt m = contracted cross-section
eingeschobene Treppe f, Leitertreppe [*Die Stufen werden in Nuten der Wangen eingeschoben*] = staircase with treads between strings (Brit.); stair ~ ~ ~ ~ (US)

eingeschobenes Gewölbe n = interposed vault

eingeschossener Stiftbolzen m = power-driven stud

eingeschossig → einetagig

eingesetzt = inserted

eingesetzte Lagerbüchse f = slip-in liner

eingespannt, gelenklos [*Träger; Bogen*] = fixed, encastré, with fixed ends, rigid, without articulation

eingespanntes Lager n → festes ~

eingespült [*Filterbrunnen; Pfahl*] = jetted

eingespülte Auffüllung f, ~ Aufschüttung = hydraulic fill

~ **Steinfüllung** f = hydraulic rock fill

eingespülter Pfahl m, Spülpfahl = jetted pile

~ **wasserdichter Kern** m, Sumpf m [*Erddamm*] = core pool

eingestängig (✵) = single-track, single-line

eingestemmte Treppe f = staircase mortised into strings, ~ morticed ~ ~ (Brit.); stair ~ ~ ~ (US)

eingesumpfter Kalk m, Sumpfkalk, Grubenkalk = pit lime

eingetauchter Körper m = body immersed in a liquid

eingewalzter Splitt m = rolled-in stone chip(ping)s

eingezogene Bohrrohre npl, muffenlose Verrohrung f = inserted joint casing

eingleisig = single-line, single-track

Eingreifen n, Einrücken = engagement

~ → Eingriff m

Eingriff m, Eingreifen n, Kämmen n [*Zähne eines Getriebes*] = mesh

Einguß m, Tränken n, Tränkung f, [*Straßenbau*] = grouting (Brit.); penetration

einhaken, anhaken = to hook on

Einhand|griff m = single hand grip

~**stein** m = one-hand block

Einhänge|binder m = drop-in truss

~**träger** m = drop-in girder

einhängiges Dach n, Schleppdach, Halbdach, Flugdach, Pultdach = shed roof

Einhebelbedienung f, Einhebelsteuerung, Einhebelbetätigung = single lever control (or operation)

~ **für Kübel und Schürze** [*Radschrapper*] = single bowl-apron lever

Einheit f **im Bogenmaß** = radian unit

Einheits|beiwert m **der Tidehöhe eines Hafens** = semi-range of the tide at a port on a day of mean equinoctial springs

~**last** f = unit load

~**preis** m = unit rate, ~ price

~**preisangaben** fpl = unit pricing

~**vektor** m = unit vector

~**verschiebung** f = unit displacement

einhüftiger Bogen m, steigender ~ = inclined arch, rising ~, rampant ~

Einhüllende f = envelope

Ein-Joch-Giebelrahmen m = single-bay gable frame

Einkammer-Hohlkörper m, Einkammer-Hohlblock m, Einkammer-Hohl(block)stein m = single-cell hollow block, ~ ~ tile, ~ pot

einkammeriges Druckgefäß n → Treibkessel m

Einkammer|kessel m **mit Doppelröhren** = single header boiler with circulating water tubes, ~ ~ ~ ~ ~ Field ~

~**rohrmühle** f = single compartment mill

Ein-Karrenaufzug m = single-barrow hoist (concrete elevator)

Einkauf|fahrt f = shopping trip

~**straße** f = shopping street

~**zentrum** n, Einkaufsviertel n = shopping center (US); ~ centre (Brit.)

Einkehle f → (Dach)Kehle

(ein)kerben, abkerben = to notch

Einketten|-Greifer(krankorb) m, Einketten-Greif(er)korb = single-chain suspension grab(bing) (crane) bucket, ~ ~ grab

~**-Raupe(nfahrwerk)** f, (n) = single crawler unit

Einkieselung — Einlauf(bauwerk)

Einkieselung *f*, Verkieselung = silification, silicating

Einkipphöhe *f* [*Baustellensilo*] = bin feeding height

einknicken, ausknicken = to buckle, to yield to axial compression

(Ein)Knicken *n* → Ausknicken

Einkornbeton *m* = like-grained concrete

einkörnig, gleichkörnig = uniform, like-grained

Einkornmörtel *m* = like-grained mortar

Einkurbelpumpe *f* = single throw pump

Einladebrücke *f*, Schiff~, Landungsbrücke = place of embarkation, ~ embarcation

Einlage *f*, Zwischenlegscheibe *f* = shim

~, Quetschholz *n* (✗) = lid, crusher block [*A short horizontal wood or steel piece placed over a single post to support the roof*]

~, Kern *m*, Seele *f* [*Drahtseil*] = core

~**blech** *n* [*Blechbogen(träger)*] = stiffening plate

Einlagen *fpl* → Bewehrung *f*

einlagern, bevorraten, (auf)speichern = to store up

Einlagerung *f* → (Auf)Speicherung

~ (✗) = inclusion

~ (Geol.) → Einschluß *m*

einlagig, einschichtig = single-layer, one-layer

einlagige Dach(pappen)eindeckung *f* = single-layer roofing, one-layer ~

einlagiger Betoneinbau *m*, einschichtiger ~ = full-course (construction) work

~ **Putz** *m*, ~ Mörtelbelag *m* = single-coat plaster

~ ~ **auf Putzträgergewebe**, ~ Mörtelbelag *m* ~ ~ = lath-and-plaster, one-coat-work [*A coat of coarse stuff, $^1/_4$ in. to $^3/_8$ in. thick laid on the lathing and smoothed off with the trowel*]

Einlassen *n* **des Bohrgestänges**, Einbauen ~ ~ = running in of the drill pipe

Einlaß *m* → Einlauf(bauwerk) *m*, (*n*)

~ **mit geradem Flansch** [*Kanalisationsteinzeug*] = square saddle junction piece

~ ~ **schrägem Flansch** [*Kanalisationsteinzeug*] = oblique saddle junction piece

~ **ohne Flansch** [*Kanalisationsteinzeug*] = junction piece without saddle

~**arretierung** *f* = throttle lock [*(compressed-)air throttle control*]

~**bauwerk** *n* → Einlauf(bauwerk) *m*, (*n*)

~**bauwerk** *n* **zum Kraftwerk**, Einlaufbauwerk ~ ~ [*als Sperre ausgebildet*] = power plant intake structure, penstock dam

~**deckel** *m* = inlet head

~**hebel** *m*, Einlaßventil-Steuerhebel = inlet cam roller lever

~**leitung** *f*, Einlaufleitung = intake line

~**nockenwelle** *f* = inlet camshaft

~**öffnung** *f* [*Motor*] = inlet port

~**regler** *m* = throttle regulator [*(compressed-)air throttle control*]

~**rohr** *n*, Einlaufrohr = inlet pipe, intake ~

~**schacht** *m*, Einlaufschacht = intake shaft, inlet ~

~**schieber** *m* = inlet valve

~**steuerung** *f* = throttle control [*for compressed air*]

~**turm** *m* → → Talsperre *f*

~**ventil-Steuerhebel** *m*, Einlaßhebel = inlet cam roller lever

~**verschluß** *m* → → Talsperre *f*

Einlauf *m* = intake, inlet

~ [*In einem Schieberschacht zur Trinkwasserversorgung sind mehrere Einläufe mit Schiebern vorhanden*] = sevice main inlet

~ = inlet [*A connection between the surface of the ground and a sewer for the admission of surface or storm water*]

~, Ablauf = gully

~**(bauwerk)** *m*, (*n*), Einlaß(bauwerk),

Entnahmebauwerk = intake (structure), inlet, ~ works, head ~
~(bauwerk) m, (n) zum **Kraftwerk**, Einlaß(bauwerk) ~ ~ [*als Sperre ausgebildet*] = penstock dam, power plant intake structure
~becken n = inlet reservoir
~becken n [*Oberflächenwasserversorgung*] = intake basin
Einlaufen n [*Motor*] = break-in
~ des Stranges = running in [*rotary drilling*]
Einlauf|(ge)rinne f, (n) → Einlaufkanal m
~gut n → Aufgabegut [*Siebtechnik*]
~kanal m, Einlauf(ge)rinne f, (n), Zulauf(ge)rinne = inlet channel
~kasten m = feed box
~leitung f, Einlaßleitung = intake line
~öffnung f, Ablauföffnung [*Straßenablauf; Hofablauf*] = inlet opening, gully ~
~rechenreiniger m = trashrack rake
~regulierungsbauwerk n = inlet regulator
~rinne f → Einlaufkanal
~rohr n, Einlaßrohr = inlet pipe, intake ~
~rost m, Ablaufrost = gully grid, inlet ~, inlet grating, bar grating
~schacht m, Einlaßschacht = intake shaft, inlet ~
~schacht m = inlet well [*A well or opening at the surface of the ground to receive surface water which is thence conducted to a sewer*]
~schacht m, Fallkessel m [*Düker*] = gully, sump
~schieber m → Einlaßschieber
~schmieröl n = running-in compound
~schütz(e) m, (f) = inlet sluice
~schwelle f = inlet sill
~trichter m [*Betonpumpe*] → Trichter
~trompete f → Talsperre f
~turm m → Talsperre f
Einlegen n = placing
Einlegescheibe f → Unterlegscheibe
Einlochdüse f, Einspritzdüse mit einer Öffnung = single orifice fuel injection valve

Ein-Mann-Abfangkeil m für Bohrgestänge = drill pipe retaining slip for one-man handling
~ ~ **Schwerstangen** [*Rotarybohren*] = drill collar retaining slip for one-man handling
Einmann|-Bedienung f = one-man operation
~-**Handschrapper** m = one-man scraper
Einmassenrüttler m, Einmassenvibrator m = single-mass vibrator
Einmotor|bagger m = one-power unit excavator
~greiferhubwerk n = grab(bing) hoisting gear with one power unit
Einmündung f [*Straße*] = junction
Einölen n = oiling
Einpfahldalbe f = single-pile dolphin
Einphasenmotor m = single-phase motor
einplanieren, einebnen, verebnen = to spread and level
Einplanieren n → Planieren
einpressen → injizieren
Einpressen n [*z.B. eine horizontal geführte Spundbohle mit einer Presse*] = jacking
Einpressung f → Injektion f
Einpreß|analge f → Injektionsanlage
~arbeiten fpl, Injektionsarbeiten = grouting work, ~ operations
~bohrung f = input well, intake ~, inlet ~, index ~, key ~
~-**Diaphragma** n → Einpreß-Dichtwand f
~-**Dichtwand** f, Einpreß-Dichtungswand, Auspreß-Dicht(ungs)wand, Verpreß-Dicht(ungs)wand, Injektions-Dicht(ungs)wand, Einpreß-Sperrwand, Auspreß-Sperrwand, Verpreß-Sperrwand, Injektions-Sperrwand, Einpreß-Diaphragma n, Auspreß-Diaphragma, Verpreß-Diaphragma, Injektions-Diaphragma, Einpreß-Schleier m, Auspreß-Schleier, Verpreß-Schleier, Injektions-Schleier, Einpreßschürze f, Auspreßschürze, Verpreßschürze, Injektionsschürze, Einpreßschirm m,

Einpreßdruck — einrüsten

Auspreßschirm, Injektionsschirm, Verpreßschirm, (Ab)Dichtungsschleier, (Ab)Dichtungsschirm, (Ab-)Dichtungsgürtel m, Einpreßgürtel, Verpreßgürtel Auspreßgürtel, Injektionsgürtel, (Ab)Dichtungsschürze = grout curtain, grouted cut-off wall

~druck m [*Erdölbohrung*] = input pressure

~druck m → Injektionsdruck

~-Entlüftungsbohrung f → Injektions-Entlüftungsbohrung

~gang m → Auspreßgang

~gas n [*Tiefbohrtechnik*] = input gas

~gerät n → Verpreßkessel m

~gerät n, Auspreßgerät, Verpreßgerät, Injektionsgerät = grouting machine, grouter

~gründung f → Auspreßgründung

~gürtel m → Einpreß-Dichtwand f

~gut n → Injektionsgut

~hilfe f, EH = grouting aid, intrusion ~

~kammer f → Injektionskammer

~lanze f → Injektionslanze

~loch n → Injektionsloch

~mörtel m → Injektionsmörtel

~mörtel m, COLGROUT-Mörtel [*Prepaktbeton*] = COLGROUT

~öffnung f → Injektionsloch n

~prüfung f [*Prüfmethode für Hochdruck-Schmiermittel*] = press-fit test

~pumpe f, Injektionspumpe, Verpreßpumpe, Auspreßpumpe = injection pump

~ring m → Injektionsring

~rohr n → Injektionsrohr

~schirm m → Einpreß-Dichtwand f

~schlauch m, Verpreßschlauch, Auspreßschlauch, Injektionsschlauch = grouting hose, injection ~

~-Schleier m → Einpreß-Dichtwand f

~schürze f → Einpreß-Dichtwand f

~-Sperrwand f → Einpreß-Dichtwand

~-Spülverfahren n = jacking-jetting process

~stutzen m → Injektionsstutzen

~verfahren n → Injektionsverfahren

~zement m → Injektionszement

Einpudern n **mit Zement** = dusting with cement

einräd(e)riger Unebenheitsmeßanhänger m = single-wheel(ed) bump integrator

einräd(e)riges Fahrgestell n, Einzelrad n [*Flugzeug*] = single-wheel(ed) assembly, one-wheel ~

Einrad|-Fahrvorrichtung f = single-wheel(ed) troll(e)y

~-Nachläufer m = single-wheel(ed) trailer

~-Rüttelwalze f, Einrad-Vibrationswalze, Einrad-Schwing(ungs)walze, Einrad-Vibrierwalze = single-wheel(ed) vibrating roller, ~ vibratory ~

~-Wagenschieber m = single-wheel(ed) railway-wagon shifter

~walze f = single-wheel(ed) roller

(Ein)Rammen n, Eintreiben, Einschlagen [*Pfahl; Spundbohle*] = driving

Einreichungstermin m → → Angebot n

einreihige Nietung f = single (row) rivet(ed) joint

Einrichten n **der Baustelle** = set-up

Einrichtung f → Betriebs~

Einrichtungen fpl **des Fernmeldeverkehrs** = telecommunication facilities

Einrichtungsplaner m = production layout engineer

Einrohrkessel m = monotube boiler

einrollen = to roll into [*e.g. The sheets are rolled into a hot bituminous cement*]

Ein-Rollenlager n [*für Hoch- und Tiefbauten*] = single-roller bearing

Einrotorenprallbrecher m → Einwalzen-Prallbrecher

Einrücken n, Eingreifen = engagement

Einrück|hebel m = clamp lever, engaging ~

~schraube f = engagement screw

einrüsten [*Arbeitsgerüst aufstellen*] = to scaffold

einrüsten — einschlafen

~ [*Schalungsgerüst oder Lehrgerüst aufstellen*] = to erect centring, ~ ~ centers (US); ~ ~ centering (Brit.)
Einrütteln *n*, Einrütt(e)lung *f* = vibrating
Einrütt(e)lungsverdichtung *f* → Vibrationsverdichtung
einsacken, absacken = to bag
Einsacken *n*, Absacken, Absackung *f*, Einsackung = bagging
Einsack|maschine *f*, Absackmaschine, Abfüllmaschine für Säcke = bagging machine, bag-packing ~, bag packer
~schurre *f* = bagging chute
~trichter *m* = bagging hopper
~waage *f*, Absackwaage [*Waage zur Bestimmung des Sackfüllgewichtes eines pulverförmigen oder körnigen Massenschüttgutes, z. B. Zement, Mehl, Kaffee usw.*] = bagging scale
Einsanden *n* [*Pflasterfuge*] = feeding, rejointing of sett paving
Einsatz *m* [*Gießereitechnik*] = charge
~, Arbeits~ = employment
~ = insert
~ für Hydraulikpumpe bzw. Aufladegebläse = cartridge
einsatzbereit = ready to operate
Einsatz|halter *m* [*Rotarytisch*] = retaining bar
~härte *f* = case hardness
~stück *n*, Distanzstück = insert
~tag *m*, Arbeitstag [*Maschine*] = working day of a machine
~versuch *m* [*Trennungsflotationsversuch in Laborzelle*] = batch test
~werkzeug *n* = inserting tool
~zeit *f* [*Maschine*] = period of use
Einsaugfärbung *f*, Absorptionsfärbung = absorption colo(u)ring
einsäuliger Hängebock *m* → einsäuliges Hängewerk *n*
einsäuliges Hängewerk *n*, einfaches ~, einfacher Hängebock *m*, einsäuliger Hängebock = king (post) truss [*Is used for short spans in connection with wood construction*]
einschalen, verschalen = to form

Einschalen *n* = formwork erection
Einschaler *m*, Schalungsarbeiter, Schalungssetzer = formwork setter
einschalige Tafelwand *f* = one-leaf precast concrete slab wall
Einschalt|mineral *n* = released mineral
~spannung *f* = cut-in voltage
(Ein)Schalung *f*, Betonschalung = formwork
Einschalungsarbeit *f* = formwork assembly work
Einscharungsmoräne *f*, Scheidemoräne = subglacial moraine
Ein-Schaufelwellenmischer *m*, Einwellen-Knetmischer, Einwellen-Zwangsmischer = single shaft pug mill, long ~ ~
einschäumen = to foam into place
Einscheibentrockenkupp(e)lung *f* = single plate dry clutch
einscheren [*Seile*] = to reeve [*ropes*]
Einscheren *n* [*Seile*] = reeving [*ropes*]
Einschichtbelag *m* = single-layer course, one-layer ~
einschichtig, einlagig = single-layer, one-layer
einschichtige Betondecke *f*, einlagige ~ = full-course concrete pavement, ~ ~ surfacing
einschichtiger Betrieb *m* = single-shift work, ~ operation, one-shift ~
Einschienenbahn *f* = monorail system, ~ transporter
~station *f* = monorail station
Einschienen|-Baubahn *f* = monorail transporter equipment for civil engineering work
~-Hängebahn *f*, Einschienen-Schwebebahn = overhead monorail
~laufkatze *f*, Einschienenlastkatze = single-rail troll(e)y
~-Laufkran *m* = overhead monorail crane
Einschieß|gerät *n* → Bolzenschießgerät
~stand *m* [*Militärflugplatz*] = firing-in butt
einschlafen, abflauen [*Wind; Wellen*] = to die

Einschlagen *n*, Eintreiben, (Ein)Rammen [*Spundbohle; Pfahl*] = driving
Ein-Schlagniethammer *m* = one shot riveting hammer
Einschlämmen *n* → Bodenverdichtung *f* durch ~
~ [*Straßenbau*] = flushing
einschlämmen mit Wasser = to flush with water
~ ~ ~ [*wassergebundene Schotterdecke*] = to wash in the fines by sluicing [*waterbound macadam*]
Einschleifen *n* **von Scheinfugen** = joint sawing
Einschleifpaste *f* = grinding compound
Einschleusen *n* [*Druckluftgründung*] = locking in
Einschluß *m*, Einlagerung *f* (Geol.) [*Sonderbildung in Mineralien oder Gesteinen*] = inclusion, inlier
~**bauwerk** *n* **einer Reaktoranlage**, Container *m* ~ ~ = (atomic) reactor containment structure, container, reactor housing
Einschmelzen *n* = meltdown
Einschmelzungsmetamorphose *f* → Anatexis *f*
(ein)schmieren, abschmieren = to lubricate
einschmieren, (ab)schmieren, (ein)fetten = to grease
Einschmieren *n* → (Ab)Schmieren
Einschmierung *f* → (Ab)Schmieren *n*
einschneiden [*Braunkohlenabbau*] = to break into a face
(einschneidige) Bohrschappe *f* = soil sampler with one cutting edge
Einschnitt *m*, Kerbe *f* = notch
~, Abtrag *m*, Abtragen *n*, Abtragung *f*, Aushub *m* [*Bodenentnahme über zukünftigem Planum*] = cut(ting)
~, Abtragungsgelände *n* = cut(ting)
~ **und Damm** *m*, Abtrag *m* und Auftrag *m* (oder Aufschüttung *f*) = cut(ting) and fill
~**aushub** *m* = excavation of cutting
~**bahn** *f* = railway in a cut (Brit.); railroad ~ ~ ~ (US)
~**böschung** *f* → Böschung im Abtrag

einschnittige Verbindung *f*, einschnittiges Gelenk *n* [*Stahlbau*] = single-lap joint, single-shear ~
Einschnitt|kanal *m* → Kanal im Einschnitt
~**kubatur** *f*, Abtragkubatur = volume of cut(ting)
~**straße** *f* = sub-surface road, ~ highway
Einschrauben *n* = screwing
Einschraubende *n* = tap end; pipe-tap (US)
Einschub *m*, Balken~, Decken~, Fehlboden *m*, Streifboden = sound boarding [*Horizontal boards fitted closely between joists and resting on them in the thickness of the floor. They carry pugging, which increases the sound and heat insulation of the floor*]
Einschubdecke *f* = sound-boarded floor
~ **mit geraden Fugen** = sound-boarded floor with straight joints
~ ~ **schrägen Fugen** = sound-boarded floor with slanting joints
Einschub|gerüst *n* [*Tiefbohrtechnik*] = removable substructure
~**seite** *f* [*Ofen*] = feed end
~**widerstand** *m* (⚒) = resistance to yield
Einschwemmungshorizont *m*, B-Horizont, Anreicherungshorizont, Anreicherungsschicht *f*, Illuvialhorizont, Illuviationszone *f* = B-horizon, zone of concentration, zone of illuviation
einschwimmen, abschwimmen = to barge to (location), to scow ~ (~), to float into position, to float out to location, to float out to the site, to float to
Einschwimmen *n* [*Brückenteil*] = floating in (on barges), ~ ~ on a barge, ~ into position, ~ into place
~ [*Rotarybohren*] = floating in
Einschwimmverfahren *n* [*Tunnelbau*]. *Die Bauteile werden im Trockendock erstellt, schwimmen an die Baustelle gebracht, dort auf ein vorbereitetes*

Bett abgesenkt und zusammengefügt] = floating method

Einschwingen|brecher *m* → → Bakkenbrecher

~-Granulator *m*, Granulator-Splittbrecher *m* = single-toggle (type jaw) granulator, chip(ping)s breaker, chip(ping)s crusher

Einseil|aufzug *m* → → Betonmischer

~förderung *f* (✂) = single-rope winding

~greifer(krankorb) *m*, Einseilgreif(er)korb = one-rope suspension grab(bing) (crane) bucket, ~ ~ grab

~haken *m* = single-line hook

~-Schrapper *m* = single-rope scraper

~schwebebahn *f* [*sogenannte englische Bauart*] = single ropeway, ~ cableway

Einseiten|pflug *m* → Schild(schnee)pflug

~räumer *m* → Schild(schnee)pflug *m*

einseitig einfallender Bergkamm *m*, ~ ~ Felsrücken *m* = hogback

~ eingespannter und einseitig freiaufliegender Balken(träger) *m* = propped (or propt) cantilever beam (US); beam simply supported at one end and fixed at the other

~ ~ ~ ~ ~ Träger *m* = propped (or propt) cantilever girder (US); girder simply supported at one end and fixed at the other

~ zusammengesetztes Faltengebirge *n* = unilaterally compound folded mountain

einseitige Besonnung *f* = unilateral sunshine

~ Kante *f* [*Sieb*] = single selvedge

einseitiger Korbbogen *m* = single-centred compound curve

~ Pflug *m* → Schild(schnee)pflug

~ Querschnitt *m* → Überhöhung *f*

einseitiges Schweißen *n* = one-side welding

~ Stumpfschweißen *n* = one-side butt welding

Einsetzen *n* **der Flut** = entering of flood

Einsinken *n* **eines Rades** = wheel penetration

Einsink|weg *m*, Zusammenschub *m* (✂) = yield [*of props, arches, etc.*]

~widerstand *m*, Zusammenschubwiderstand *m* (✂) = resistance to yield

Einspannbedingung *f* = end condition

einspannen [*festhalten eines Bauteils an einem anderen*] = to fix, to restrain

Einspann|-Endmoment *n* = fixed-end moment

~höhe *f* [*Kletterkran*] = height between stories

~lager *n* (✂) = point of clamping

Einspannung *f*, **Einspannen** *n* [*Festhalten eines Bauteiles an einem anderen*] = restraint, fixity

Einspann(ungs)moment *n* = end moment, terminal ~, fixing ~

Einsprengling *m* [*beim Porphyr*] = ange, inset, metacryst, phenocryst, porphyritic crystal, show crystal, exnocryst, xenolith

einspringender Winkel *m* = re-entering angle

einspringendes Widerlager *n* = re-entering abutment

Einspritz|anlage *f* [*Motor*] = fuel injection system

~düse *f*, Ölspritzdüse *f* [*Dieselbär*] = fuel jet in piston head

~düse *f* **mit einer Öffnung**, Einlochdüse = single orifice fuel injection valve

~düsenmundstück *n* = fuel injection valve nozzle

~pumpe *f* = fuel injection pump

~pumpengehäuse *n* = fuel injection pump barrel

~pumpenstößel *m* = fuel injection pump plunger

~pumpenverteiler *m* = fuel pump manifold

~ventil *n* [*Motor*] = fuel injection valve

~versteller *m* [*Motor*] = fuel injection timing mechanism

~vorrichtung *f*, Kaltstartanlage *f* = primer

~zeitpunktversteller *m*, Spritzversteller = variable timing unit

~zeitpunktverstellung *f* [*Motor*] = injection timing advance

(Ein)Spülen *n* [*Pfahl*] = pile-sinking with the water jet, pile jetting

~ [*Pfahl; Spundbohle*] = water jetting, water-jet driving

Einspülöffnung *f*, Spül(spitzen)öffnung = jetting orifice

(Ein)Spülrohr *n* [*Pfahl*] = jetting pipe

Einspül|verfahren *n*, (Bodenverdichtung *f* durch) Einschlämmen *n*, Spülspritzverfahren, Spülkippverfahren, Wasserstrahlverfahren = sluicing

~wasser *n* = jetting water

Einspundung *f*, Ausrammung, Spundwandumschließung = closed sheeting

Einständerfahrgestell *n* = single-pole travel(l)ing gear

Einstau *m*, Stau = pondage

~filter *m, n*, Füllkörper *m* = contact bed, ~ filter, coarse-grained filter [*An artificial bed of coarse material, such as broken stone or clinkers, in a water-tight basin provided with controlled inlet and outlet. It is operated in cycles of filling with sewage, standing in full contact, emptying, and resting empty, in order to remove some of the suspended matter and oxidize organic matter by biochemical agencies*]

Einstechmesser *n* [*Radschrapper*] = stringer bit

Einsteck|ende *n* → Bohrer~

~schloß *n* = mortise lock, mortice ~

~werkzeug *n* = inserted tool

einstegig = single-webbed

Ein-Steg-Spezialplatte *f* [*Gleiskette*] = special application single grouser

Einsteig|luke *f* = hatch(way)

~schacht *m* [*Tunnelbau mit vorgefertigten Elementen*] = stub riser for the access hatch

~schacht *m*, Zugangschacht = access shaft

(Einsteig)Schacht *m*, Reinigungsschacht = manhole, inspection chamber [*A local enlargement to a sewer or duct arranged so that a man can enter it from the surface*]

~ aus Stahlbetonfertigteilen, Reinigungsschacht ~ ~ = precast reinforced concrete manhole

~abdeckung *f*, Mannlochabdeckung = manhole cover, inspection chamber ~

~abdeckungsrahmen *m*, Mannlochabdeckungsrahmen = manhole frame

~deckel *m*, Mannlochdeckel = manhole cover, inspection chamber ~

~kammer *f*, Mannlochkammer = manhole chamber, chamber of manhole

~körper *m*, Mannlochkörper = manhole walls and invert

~sohle *f*, Mannlochsohle = manhole invert, invert of a manhole

~steigeisen *npl*, Mannlochsteigeisen = manhole step(-iron)s

~wand *f*, Mannlochwand = manhole wall, inspection chamber ~

ein-Stein-starke Wand *f* = one-brick wall

Einstell|arm *m*, Einstellstange *f* [*dient zum Einstellen des Anstellwinkels eines Löffels zum Stiel, eines Schildes oder einer Schar*] = pitch arm, ~ brace, ~ rod

~barkeit *f*, Verstellbarkeit = adjustability

~bereich *m* = range of adjustment

Einstellehre *f* für Zahnstange = rack setting ga(u)ge

Einstellen *n*, Einstellung *f* = adjustment

~, Einstellung *f* [*eine Arbeit*] = abandonment

Einstellgewicht *n*, Laufgewicht = poise weight, sliding ~, jockey ~, moving poise

einstellige Zahl *f* = number with one digit

Einstell|kompaß *m* → Regelkompaß

~marke *f*, (Einsteuer)Marke [*Motor*] = timing mark

~mutter *f* = adjusting nut

~platte *f*, Verstellplatte = adjusting plate

Einstellschraube — Einteilung

~schraube *f* = adjusting screw
~schraube *f* [*CBR-Gerät*] = adjustable stem
~stange *f* → Einstellarm *m*
Einstellung *f* der Arbeiten = suspension of work
~ mit Hilfe von Beilagscheiben = shim adjustment
Einsteuermarke *f*, Einstellmarke [*Motor*] = timing mark
Einstichstraße *f*, Zick-Zack-Duostraße = staggered mill
Einstieg *m*, ~öffnung *f* = access opening
einstöckig → einetagig
einstrahlender Verkehr *m* = city-bound traffic
Ein-Strang-Trassierung *f* = single-line piping layout
Einstreichfeile *f*, Schraubenkopffeile = screw-head file, slitting ~, feather-edge(d) ~
Ein-Streifen-Leistungsfähigkeit *f* [*Straße*] = single lane capacity
Einstreudecke *f*, Streumakadam *m* = dry penetration surfacing, ~ process penetration macadam, ~-bound macadam
Einstrich *m* (⚒) = bunton, divider [*Horizontal timbers in a rectangular or circular shaft which may carry the cage guides, water and air pipes, and cables, and separate the shaft into compartments*]
Einströmung *f* (der Tidewelle) = introduction of the tidal wave, entering ~ ~ ~ ~
Einstrom-Wäscher *m*, Gleichstrom-Wäscher = concurrent scrubber
Einstufen|abscheider *m* = single-stage separator
~destillation *f* [*Erdölaufbereitung*] = single-flash distillation
~-Schwingsieb *n*, Eindecker-Schwingsieb = single deck vibratory screen
~sieb *n*, Eindecker *m*, Eindeckersieb, Sieb mit einem Siebfeld = single deck screen

einstufige Druckturbine *f* = single-stage impulse turbine
~ (Hart)Zerkleinerung *f* = single-stage comminution, ~ reduction
~ Pumpturbine *f* = single-stage pump turbine
~ Verdichtung *f* [*Kompressor*] = single-stage compression
Einstufung *f*, Klassifizierung = classification, indexing
Einsturz *m*, Nachfall *m* = cave(-in), caving
~ [*unter einer Last*] = collapse, fall
~beben *n* = subsidence earthquake
~becken *n* = subsidence basin
~doline *f* = collapse dolina, ~ doline
einstürzen [*unter einer Last*] = to collapse, to fall
Einsturzsee *m* = sink lake
Einsumpfen *n*, Teichverfahren *n*, Teichmethode *f* [*Betonnachbehandlung*] = ponding
~ → Naßlöschverfahren *n*
Eintafelschütz(e) *n*, (*f*) = single leaf (sluice) gate
eintägige Welle *f* = diurnal wave
Eintauch|fläche *f*, Inundationsfläche (Geol.) = flood plain
~lötung *f*, Eintauchlöten *n* = dip soldering
~rohr *n* = immersion pipe
~schmierung *f*, Tauchschmierung = flood lubrication
~thermometer *n* = immersion thermometer
~ung *f*, Eintauchen *n* = immersion
~versuch *m*, Eintauchprobe *f*, Eintauchprüfung *f* = immersion test
Eintauschreifen *m* = trade-in tyre (Brit.); ~ tire (US)
einteilige Bauweise *f* [*Maschine*] = one-piece design
~ Trocken- und Stetigmischanlage *f* = continuous volumetric-type one-piece drying and mixing plant
einteiliger Zahn *m* = one-piece tooth
Einteilung *f* der Gesteine *npl* in Handelsgruppen, Gesteinseinteilung ~ ~ = trade grouping of rocks

eintouriger Elektromotor — Einzelhändler 360

eintouriger Elektromotor m = constant speed motor
eintouriges Schloß n = single turn lock
Ein-Träger-Handlaufkran m = hand-operated troll(e)y hoist
Eintreiben n, Einschlagen, (Ein)Rammen [*Pfahl; Spundbohle*] = driving
Eintrittverlust m [*Hydraulik*] = entrance loss
eintrocknender See m = evanescent lake
Ein-Trommel|straßen(beton)mischer m → Straßen(beton)mischer
~winde f = single drum winch
eintrümiger Schacht m = undivided shaft, single ~
(Ein)Visieren n = boning(-in), sighting [*Setting out intermediate levels on the straight line joining two given level pegs*]
Einwaage f = original sample weight
Einwägen n → geometrische Höhenmessung
einwalzen [*Splitt*] = to roll in
Einwalzen|brecher m = single roll crusher, ~ ~ breaker
~-Prallbrecher m, Prallbrecher mit einer Schlagwalze, Einrotorenprallbrecher = single impeller impact breaker, ~ ~ ~ crusher
einwandfrei [*in bezug auf Güte*] = sound
einwandiger Blechbogen(träger) m = single-webbed plate arch(ed girder)
~ Obergurt m = single-webbed top chord
Einwaschungshorizont m → B-Horizont
Einwegkupp(e)lung f = one-way clutch
Einweihungsfeier f = dedication ceremony
Einweisen n [*LKW*] = spotting
Einweiser m [*beim Abkippen*] = spotter, dumpman
Einweiser m [*Erdbau*] = turnboy
Ein-Wellen|mischer m = single shaft mixer
~-Zwangsmischer m, Einwellen-Knetmischer, Ein-Schaufelwellenmischer = single shaft pug mill, long ~ ~

einwirken [*Last*] = to act
Einwirkungs|bereich m (⚒) = zone of affected overburden
~fläche f (⚒) = area of extraction, ~ ~ working, ~ ~ winning, ~ ~ getting
~schwerpunkt m (⚒) = focal point of subsidence
Einwitterung f → → Denudation f
Einwölbung f auf Kuf = barrel vaulting
~ ~ Schwalbenschwanz = dovetail vaulting
Einzahn|-Anbau-Aufreißer m = single tooth ripper, ~ tine ~, ~ shank ~
~aufreißer m = single tooth unit, ~ tine ~, ~ shank ~, one-tyne ripper
~-Reißbalken m = single shank beam
Einzäunung f = fencing
Einzel|abstellfläche f [*Flugplatz*] = dispersed (hard)standing, dispersal
~achsantrieb m = single-axle drive
~achse f [*Achse mit Einzelantrieb*] = single axle
~achslast f [*Last, die nur von einer Achse ausgeübt wird*] = single-axle load
~antrieb m = single drive
~aufbereitung f = individual processing, ~ preparation
~ausbau m = individual development
~brenner m [*Schweißen*] = torch with non-variable head, non-variable (head) torch
~drahtspannpresse f = single wire prestressing jack
~draht-Spannverfahren n = single wire prestressing system
~druckknopfsteuerung f = single push-button control
~fahrbahn f = single carriageway
~fahrer m [*Binnenschiffahrt*] = single barge
~fall m, Sonderfall = special problem
~feld n = single span
~gänger-Schürfer m = lone-wolf prospector
~gerät n → Geräteeinheit f
~händler m = dealer

Einzel-Hinterachse — Eisbahn

~-Hinterachse *f* = single rear axle
~kamin *m*, Einzelschornstein *m* = isolated chimney
~ketten-Entladegreifer *m* = single chain unloading (grab) bucket
~korn *n*, Gesteinskorn, Partikelchen *n*, Einzelteilchen, Körnchen = particle, grain
~kreuzung *f*, Kreuzung von zwei Straßen = crossing
~kristall *n* = unit crystal
~last *f*, einzeln wirkende Last, einzeln konzentrierte Last = concentrated load, single ~, load concentrated at a point; point load (US)
~leiter *m* = single conductor
~pfahl *m* = single pile, individual ~
~-Pfahlfundament *n* = pile footing, footing on piles
~prämiensystem *n* = single incentive bonus scheme
~preventer *m* = single unit type (blow-out) preventer
~punktsteuerung *f* [*Werkzeugmaschinensteuerung*] = point-to-point positioning system
~rad *n* → einrädriges Fahrgestell *n*
~radbremse *f* = single-wheel brake
~radlast *f* = single-wheel load
~schleuse *f* = single (navigation) lock, ~ inland ~
~stabspannpresse *f* = single rod prestressing jack, ~ bar ~ ~
~stütze *f* = single support, ~ column
~tauchkolbenhebel *m* = single plunger lever
~-Trinkwasserversorgung *f* = individual drinking water supply
~tunnel *m* [*im Gegensatz zum Doppeltunnel*] = single tunnel
~ventil *n* = single valve
~vulkan *m* = central volcano, solitary ~
~welle *f* = solitary wave
~zerkleinerung *f* = single particle comminution, ~ grain ~
~zimmer *n* [*Hotel*] = single bedroom
(ein)zementieren [*Tiefbohrtechnik*] = to cement

einziehbarer Pfosten *m* = tubular column, (supporting) caisson [*mobile barge*]
Einziehen *n* von Deckenankern, Ankerausbau *m* zur Sicherung der Firste = roof-bolting
einziehender (Wetter)Schacht *m*, (Wetter)Einziehschacht (⚒) = downcast shaft
Einzieh|kupp(e)lung *f* [*Bagger*] = retract clutch
~strecke *f* (⚒) = intake gate road
Einzonenbauweise *f*, englische Bauweise, Longarinenbauweise, englisches Verfahren *n*, englische Methode *f* = English method, ~ system of timbering
Einzug|kessel *m* = single-pass boiler
~gebiet *n* → Abflußgebiet
Einzweck|-Baumaschine *f* = single-purpose construction machine
~straße *f* = single-purpose road, ~ highway
Ein-Zylinder-|Dieselmotor *m* = single-cylinder diesel (engine), ~ ~ motor
~-Motor *m* = single-cylinder engine
~-Zweitaktmotor *m* = single-cylinder two-cycle engine
Eiprofil *n*, eiförmiges Profil = egg(-shaped) profile
~-Betonrohr *n* → eiförmiges Betonrohr
~rohr *n*, Eiformrohr, eiförmiges Rohr = egg-shaped pipe
Eis|abfuhr *f* → Eisgang *m*
~abführung *f* → Eisgang *m*
~abgang *m* → Eisgang
~abwehr *f* = ice control
~abweiser *m* in Form eines Schiffbugs, Eisbrechpflug *m* = ice-breaking ram of a vessel
~(auf)bruch *m* = debacle, breaking-up of the ice, iceboom [*A breaking up of ice in a river, stream, canal, lake, etc.*]
~aufreißgerät *n* = ice breaking machine
~aufreißwerkzeug *n* = ice breaking tool
~bahn *f* = ice-skating rink

Eisbarre — Eisenerzmahlen

~barre *f* = ice-barrage, ice-dam
~berg *m* = iceberg
~beton *m* = ice concrete [*Is a type of lightweight cellular concrete in which the cells are formed by ice*]
~bildung *f* = ice formation, ~ segregation, segregation of ice, formation of ice
~blink *m* = iceblink
~blumenglas *n* = floral pattern glass
~brecher *m* = ice-breaker [*ship*]
~brecher *m* = ice guard
~brechpflug *m*, Eisabweiser *m* in Form eines Schiffbugs = ice-breaking ram of a vessel
~bruch *m* → Eisaufbruch
~decke *f* = ice cover
~druck *m* = ice pressure
(Eisen)Bahn|achse *f* → Bahnachse
~anlage *f* = railway facility (Brit.); railroad ~, RR ~ (US)
~lieferung *f* = rail delivery
~anschluß *m* [*Hafen*] = railroad connection (US); railway ~ (Brit.)
~aufschüttung *f* → Bahndamm *m*
~brücke *f* = railroad bridge (US); railway ~ (Brit.)
~damm *m* → Bahndamm
~draisine *f* → Draisine
~durchgangsprofil *n* → Ladeprofil
~einschnitt *m* = railway cutting (Brit.); railroad ~ (US)
(Eisen)Bahnersiedlung *f* [*bei einem Hafen*] = railway colony (Brit.); railroad ~ (US)
(Eisen)Bahn|fähre *f* = train ferry
~fahrzeug *n* = railway vehicle (Brit.); railroad ~ (US)
~hochbrücke *f* = high-level railway bridge (Brit.); ~ railroad ~ (US)
~-Kesselwagen *m* = rail tanker, tanker wagon (Brit.); railroad tank car (US)
~knotenpunkt *m* = rail center (US); ~ centre (Brit.)
~krone *f* → Bahnkrone
~linie *f*, (Eisen)Bahnstrecke *f* = trainway
~mittellinie *f* → Bahnachse *f*

~netz *n* = railroad network (US); railway ~ (Brit.)
~oberbau *m* = permanent way, p. w.
(~-)Rollmaterial *n* = rolling stock
~schiene *f* = railroad rail (US); railway ~ (Brit.)
~schotter *m* → Bettungsschotter
~schwelle *f* → Bahnschwelle
~strecke *f* → (Eisen)Bahnlinie
~technik *f* = railway engineering (Brit.); railroad ~ (US)
~transport *m* = rail transport
~tunnel *m* = railroad tunnel (US); railway ~ (Brit.)
~überführung *f* = overpass
~- und Straßentunnel *m*, Straßen- und (Eisen)Bahntunnel = road-railway tunnel, railway-road ~ (Brit.); road-railroad ~, railroad-road ~ (US)
~-Unfallkran *m* = breakdown crane, accident ~, permanent way crane (Brit.); wrecking ~ (US)
~unterführung *f* = underpass
Eisenbeizerei *f* = pickling department
Eisenbiegeliste *f* → Biegeliste
Eisenbiegen *n* = bending of reinforcement
Eisenbiege|plan *m* → Biegeplan
~platz *m* → Biegeplatz
Eisen|bieger *m* → Bieger
~blech *n* [*Fehlname*]: Schwarzblech = black sheet [*Eisenfeinblech*]; ~ plate [*Eisengrobblech*]
~bogen *m* = iron arch
~bohle *f*, Eisenspundbohle = iron sheetpile
~chlorür *n*, Ferrochlorid *n*, FeCl$_2$ + 4H$_2$O = ferrous chloride
~-III-Sulfat *n* = ferric sulphate
~einlagen *fpl* → Bewehrung *f*
Eisenerz|aufbereitung *f* = iron ore processing, ~ ~ dressing, ~ ~ preparation
~flotation *f* = iron ore flotation
~grube *f*, Eisenerzbergwerk *n* = iron ore mine
~mahlen *n*, Eisenerzmahlung *f* = iron ore grinding

Eisenerz-Tagebaugrube — Eisglätte

~-**Tagebaugrube** *f* = iron ore strip mine
~**zement** *m* = iron-ore cement
Eisen|feilspäne *mpl* = iron filings, ~ turnings
~**fenster** *n* = iron window
~**flechter** *m* [*Stahlbetonbau*] = (reinforcing) ironworker, steel fixer
eisenführend = iron-bearing
Eisen|glas *n*, Fayalit *m*, Fe_2SiO_4 (Min.) = fayalite, iron-olivine
~**glimmer** *m* [*erhöht die Wetterfestigkeit von Rostschutzfarben*] = micaceous iron ore
~**glimmerschiefer** *n* (Geol.) = itabaryte, itabarite
~**graupen** *fpl* (Min.) = granular bog iron ore
eisen|haltiger Sand *m* = iron-bearing sand
~**haltiges Wasser** *n* = iron-bearing water
Eisen|hochofen-Stückschlacke *f* = lump slag
~**hüttenschlacke** *f* → Metall(hütten)schlacke
~**karbid** *n*, Zementit *m*, Fe_3C [*sehr harte Eisenkohlenstofflegierung*] = cementite
~**kies** *m* → Schwefelkies
~**kitt** *m*, Graphitzement *m* = iron cement
~**klammer** *f*, Eisenkrampe *f* = iron dog
~**krampe** *f*, Eisenklammer *f* = iron dog
~**lebererz** *n* (Min.) = hepatic iron ore
~**legergruppe** *f* [*Schweiz*] → Armierungskolonne *f*
~**legierung** *f* = iron alloy
~**modul** *m* → Tonerdemodul
~**nickelkies** *m* (Min.) = nicopyrite, pentlandite
~**ölseife** *f* = iron oleate
Eisenoxyd *n*, Ferrioxyd = iron oxide
~**farbe** *f* = iron oxide paint
~**gelb** *n* = yellow (iron) oxide
~**rot** *n*, Eisenrot *n* = red (iron) oxide
~**überzug** *m* = iron oxide coating

Eisen|pfahl *m* = iron pile
~**portlandzement** *m*, EPZ [*maximal 30% Schlacke*] = Eisen-Portland cement, iron Portland ~, German portland slag ~
~**profil** *n*, Formeisen *n*, Fassoneisen, Profileisen = section iron, iron section
~**pulver** *n* = iron powder
~**rad** *n*, eisenbereiftes Rad = metal wheel, steel-rimmed ~
~**rohr** *n* = iron pipe
~**sandstein** *m* → Sandstein
~**schalung** *f* = iron formwork
~**schmelze** *f* = iron-smelting factory
~**schrotbeton** *m* = iron shot concrete
~**schwamm-Herstellung** *f* **im Tunnelofen** = tunnel kiln sponge iron process
~**(spund)bohle** *f* = iron sheetpile
~**teile** *mpl*, *npl* **einer pennsylvanischen Seilbohranlage** = rig irons
~**ton** *m*, Toneisenstein *m* = clay ironstone, iron clay
~**treppe** *f* = iron staircase (Brit.); ~ stair (US)
~**trübung** *f* = red water trouble
~**turm** *m* = iron tower
~**verbindung** *f* = iron compound
~**wichte** *f* = iron density
~**zarge** *f* = iron trim, ~ casing
~**zuschlag(stoff)** *m* = iron aggregate
~-**II-Sulfat** *n* = ferrous sulphate (Brit.); ~ sulfate (US)
Eis|fabrik *f*, Eiswerk *n* = ice-plant
~**feld** *n* = ice field, ice-floe, ice-pack, ice-jam
~**film** *m* = ice film
~**führen** *n* → Eisgang *m*
~**gang** *m*, Eistreiben *n*, Eisführen *n*, Eistrieb *m*, Eisabgang *m*, Eisabfuhr *f*, Eisabführung *f* [*massenhaftes Abschwimmen von Eis, das vorher in Ruhe war*] = ice-drift(ing), embacle
~**gischt** *f* = freezing sprays
~**glätte** *f* [*Bildet sich durch gefrorenes Wasser und setzt keinen unmittelbaren Niederschlag voraus. Sie entsteht durch gefrorenes Schmelzwasser, gefrorene Regenwasserpfützen oder ver-*

schüttetes und dann gefrorenes Wasser] = ice caused by the freezing of a wet road surface
~haut *f* = ice film
~kern *m* = freezing nucleus
~klappe *f*, Aufsatzklappe [*Wehr*] = ice shutter
~klüftigkeit *f* = cracking by frost, splitting ~ ~
~kristall *m* = ice crystal
~lawine *f*, Gletscherlawine = ice-avalanche
~linse *f*, Frostlinse = ice lens, frost ~
~mantel *m* → Binneneisdecke *f*
~mauer *f* → Frostmauer
~nadel *f* = ice needle
~nebel *m* = ice fog
~palast *m* → Halleneisbahn
~pfropfen *m* = downstream margin of unbroken ice cover
~pickel *m* = ice pick
~räumung *f* = ice removal
~riß *m* = split of ice cover
~schild *n* → Binneneisdecke *f*
~-Schmelzvorgang *m* = melting of ice
~scholle *f* = ice flow, block of ice
~schranke *f* = ice barrier
~schrumpfung *f* = glacial shrinkage
~schürze *f* → Frostmauer *f*
~schweb *m* → Breieis *n*
~sprengung *f*, Eissprengen *n* = ice blasting
~stollen *m* [*Gleiskette*] = ice grouser
~stopfung *f*, Eisstau(ung) *m*, (*f*), Eisversetzung *f* = ice jam
~tisch *m*, Gletschertisch = glacier table
Eistollen *m* = egg-shaped gallery
Eis|tor *n* = ice cave
~treiben *n* → Eisgang *m*
~überzug *m* = ice layer
~verbreitungszentrum *n* = center of ice dispersal (US); centre ~ ~ ~ (Brit.)
~verhältnis *n* [*Hydraulik*] = nucleation ratio
~versetzung *f*, Eisstopfung, Eisstau(ung) *m*, (*f*) = ice jam
~wall *m* = ice rampart
~wand *f* → Frostmauer *f*
~wasser *n* = ice water
~wasseranlage *f* = ice-water installation, ~ plant
~werk *n*, Eisfabrik *f* = ice-plant
~wüste *f* = ice desert
Eiseinkalk *m*, oolithischer Kalkstein *m*, Oolithkalk, Kalkoolith *m* = oölitic limestone, oölite
eiszapfenförmig = icicle shaped
Eiweißstickstoff *m*, Albuminoidstickstoff = albuminoid nitrogen
Ekonomiser *m*, Abgasvorwärmer *m* = economizer
Elaeolith *m*, Ölstein *m* (Min.) = elaeolite
~syenit *m* = elaeolite-syenite
El-(Antriebs)Motor *m*, Elektromotor = (electric) motor
elastikbereift = solid-tyred (Brit.); solid-tired (US)
Elastikreifen *m*, Vollgummireifen = solid tire (US); ~ tyre (Brit.); hard-rubber ~
elastisch = flexible, elastic
~ eingespannt [*Baustatik*] = elastically fixed, ~ restrained
~ gebetteter Balken *m*, elastischer Gründungsbalken = elastic beam
elastische Bettung *f* [*Bodenmechanik*] = continuous elastic support
~ Dehnung *f* = elastic strain
~ Durchbiegung *f* = deflection; deflexion (Brit.)
~ Einspannung *f* = elastic end-restraint
~ Formänderung *f*, federnde ~, ~ Verformung *f* = elastic deformation
~ Kupp(e)lung *f* = flexible coupling
~ Pfahlzusammendrückung *f* = temporary compression in the pile
~ Platte *f* [*Grundbau*] = flexible slab
~ ~, ~ Scheibe *f* [*Baustatik*] = elastic plate
~ Rißdehnung *f* = elastic cracking strain
~ Scheibe *f*, ~ Platte *f* [*Baustatik*] = elastic plate
~ Sohldruckverteilung *f* = elastic distribution of bearing pressure, ~ bearing pressure distribution

elastische Unterlage — elektrische Verriegelung

~ **Unterlage** *f* = elastic support
~ **Zugdehnung** *f* = elastic tensile strain
elastischer Balken(träger) *m* = elastic beam
~ **Gleichgewichtzustand** *m* = state of elastic equilibrium
~ **Gründungsbalken** *m*, elastisch gebetteter Balken = elastic beam
~ **Körper** *m* = elastic body
~ **Untergrund** *m* = dense liquid subgrade
~ **Zustand** *m* = elastic state
elastisches Erdharz *n* → Elaterit *m*
~ **Erdpech** *n* = elastic mineral pitch
~ **Gleichgewicht** *n* = elastic equilibrium
~ **System** *n*, ~ Tragwerk *n* = elastic system
~ **Verfahren** *n* → Brechungsverfahren
~ ~ = elastic method
elastisch|-isotroper Halbraum *m* = semi-infinite elastic solid
~**plastisch** = elastoplastic
~**-plastischer Bereich** *m* = elastoplastic range
Elastizität *f* = resiliency, elasticity
Elastizitäts|aufgabe *f* = elasticity problem
~**bedingung** *f* = elasticity condition
~**grenze** *f* = elasticity limit
~**grenze** *f* [*Stahl*] = yield stress
~**modul** *m*, Elastizitätsmaß *n*, Elastizitätszahl *f*, Modul für Beton nach Young, E-Modul = modulus of elasticity, Young's modulus (Brit.); module of elasticity; Young's module (US)
~**theorie** *f* = theory of elasticity, elastic theory
Elaterit *m*, elastisches Erdharz *n* = elastic bitumen, mineral caoutchouc, elaterite [*Old name: elastic mineral pitch. Variety of bitumen which, when fresh, is elastic, but which on exposure becomes hard and brittle. First described "Fungus subterraneus" by M. Lister*]
elektrifiziert = electrified
Elektrifizierung *f* = electrification

~ **mit Industriefrequenz** = industrial-frequency electrification
elektrisch beheiztes Sieb *n* = electrically heated screen
~ **geschweißt** = electrically welded
~ **gesteuert** = electrically controlled
~ **leitende (Fuß)Bodenfliese** *f* = conductive tile
elektrische (Be)Heizung *f* **von Straßen** = electrical heating of roads, ~ ~ ~ highways
~ **Bodenerforschung** *f* → Geoelektrik *f*
~ **Bodenverfestigung** *f* = electric (curtain) stabilization
~ **Bohrlochmessung** *f*, elektrisches Kernen *n* = electrical logging, ~ coring
~ **Bohrlochmessungen** *fpl* **ohne Ziehen des Gestänges** = drillstem logging
~ **Direktstartanlage** *f*, ~ Direktanlasseranlage = direct electric starting system
~ **Durchschlagsfestigkeit** *f* = (di)electric strength
~ **Gleichstromlok(omotive)** *f* = electric D.C. locomotive
~ **Härtung** *f* **des Betons** = electrothermal curing
~ **Klingel** *f* = electric bell
~ **Kraftlenkung** *f* = electric positive power steer(ing)
~ **Leitfähigkeit** *f* = electrical conductivity
~ ~ **des Bodens** = electrical resistivity of the soil, ~ resistance ~ ~ ~
~ **Methode** *f* → Geoelektrik *f*
~ **Prospektion** *f* → Geoelektrik *f*
~ **Sicherung** *f* = fuse
~ **Stumpfschweißmaschine** *f* = electric butt welding apparatus
~ **Tiefbrunnenpumpe** *f*, Elektro-Tiefbrunnenpumpe = electric deep well turbine pump
~ **Untersuchung** *f* → Geoelektrik *f*
~ **Verdrahtung** *f* = electrical wiring
~ **Verrieg(e)lung** *f* = electrical interlocking

elektrische Widerstandserwärmung — Elektro-Betonmischer

~ **Widerstandserwärmung** *f* **des Betons**, Betonerwärmung mit Elektroden = electrical resistivity heating of concrete
~ **Widerstandsmessung** *f* → Widerstandsmessungsverfahren
~ **Zusatzlenkung** *f* = supplemental electric steer(ing)
elektrischer Aufschluß *m* → Geoelektrik *f*
~ **Dehnungsmesser** *m* = electrical (strain) ga(u)ge, ~ resistance ~
~ **Fahrmotor** *m* = traction motor
~ **Fernpegel** *m*, ~ **Schreibpegel** [*Zeichen: Se*] = electrical ga(u)ge [*registers electrically the stage of a stream or river at any distance from the ga(u)ging station*]
~ **Leiter** *m* = electrical conductor
~ **Mehrmotorenantrieb** *m*, mehrmotoriger Elektroantrieb = multiple (electric) motor drive
~ **Mehrmotorenkran** *m* = multiple motor crane
~ **Oberleitungstriebwagen** *m* = electric railcar
~ **Porenwasserdruckmesser** *m* → elektrisches Porenwasserdruck-Meßgerät *n*
~ **Schreibpegel** *m* → ~ Fernpegel
~ **Strom** *m* = electric current
~ **Thermometer** *m*, Thermoelement *n*, elektrisches Thermometer *n* = electrical thermometer, electrothermic ga(u)ge
~ **Widerstand** *m* = electrical resistivity, ~ resistance
~ **Zünder** *m* = electric detonator
~ **Zündschnurzünder** *m* = electric powder fuse
elektrisches Analogieverfahren *n* = electrical analogy method
~ **Aufschlußverfahren** *n* → Geoelektrik *f*
~ **Kernen** *n*, elektrische Bohrlochmessung *f* = electrical coring, ~ logging
~ **Netzwerk** *n* [*Planungsrechnung*] = electrical network
~ **Porenwasserdruck-Meßgerät** *n*, elektrischer Porenwasserdruckmesser *m* = electrical(ly operating) pore water pressure cell (or ga(u)ge)
~ **Profil** *n* = electric log
~ **Schweißen** *n*, Elektroschweißen = electric welding
~ **Verfahren** *n* → Geoelektrik *f*
Elektrizitäts|gesellschaft *f* = electricity supply company
~**werk** → Kraftwerk
Elektro|-Abraumlok(omotive) *f*, Abraum-Elektrolok(omotive) = electric overburden loco(motive)
~**aggregat** *n* → Aggregat
~**-Ankerwinde** *f* = electric anchor winch
~**anlage** *f* [*einer Maschine*] = electrical equipment, ~ system
~**anlasser** *m* = electric starter
~**-Aufbruchhammer** *m*, Elektro-Aufreißhammer = electric paving breaker
~**aufzug** *m* = electric lift (Brit.); ~ elevator (US)
~**-Aufzugtreppe** *f* [*Schweiz*]; Elektro-Schiebetreppe = electric (disappearing) stairway
~**-Außenrüttler** *m*, Elektro-Außenvibrator *m* = electric external vibrator
~**(bau)aufzug** *m* = electric hoist
~**-Baugerät** *n*, Elektro-Bauwerkzeug *n* = electrical construction tool
~**-Bauhammer** *m*, Elektro-Installationshammer = electric builders' hammer, ~ contractors' ~
~**-Baupumpe** *f* = electric contractors' pump, ~ builders' ~
~**-Bauwerkzeug** *n*, Elektro-Baugerät *n* = electrical construction tool
~**bedampfung** *f*, Elektro-Dampfhärtung, Elektro-Dampfbehandlung [*Dampfbehandlung von Beton durch elektrisch beheizte Dampferzeuger*] = electric steam curing
~**-Beton(förder)pumpe** *f*, Elektro-Förderpumpe = electric concrete pump
~**-Betonmischer** *m* = electric concrete mixer

~bodenheizung *f* → Elektrofußbodenheizung

~bohrer *m* = electrodrill, electric drill

~-Bohrhammer *m* → Elektro-Gesteins-Bohrhammer

~-Bohrmaschine *f* = electric drilling machine

~boiler *m*, Elektro-Heißwassergerät *n*, Elektro-Heißwassererhitzer *m* = electric cylinder, ~ calorifier, ~ domestic water heater

~-Bunkerrüttler *m*, Elektro-Bunkervibrator *m* = electric hopper vibrator

elektrochemische Korrosion *f* = electrochemical corrosion

~ **Verfestigung** *f* **von Ton** = electrochemical solidification (or hardening) of clay

Elektro|-Dampfbehandlung *f* → Elektrobedampfung

~-Dampfhärtung *f* → Elektrobedampfung

Elektrode *f* **mit starkem Einbrand** = deep-penetration electrode

Elektroden|-Dampfkessel *m* = electrode boiler

~fräser *m* = electrode dresser

Elektro|-Drehbohrmaschine *f*, Elektro-Drehbohrgerät *n* = electric rotary drilling machine, ~ rotating ~ ~

~-Drehmotor *m* → Elektro-Schwenkmotor

elektrodynamisches Verfahren *n* = electro-dynamic method

Elektro|-(Einschienen)hängebahn *f* = telpher

~(end)osmose *f* = electro-osmosis, electrical endosmose

~entwässerung *f* = electrical drainage, electro-osmotic ~

~fahrsteiger *m* (⚒) = chief electrician

~fahrzeug *n* = electric vehicle

~fahrzeug *n* → Elektrokarre(n)

~-Farbspritzpistole *f* = electric paint spray gun

~-Felsbohrhammer *m* → Elektro-(Gesteins)Bohrhammer

~-Fettpresse *f*, Elektro-Schmierpresse = power-operated grease gun

~filter *m*, *n* [*Unter einem Elektrofilter versteht man ganz allgemein eine Einrichtung, mit deren Hilfe es möglich ist, feste und flüssige Schwebeteilchen aus dem Trägergas mit Hilfe einer künstlichen elektrischen Aufladung abzuscheiden. Die aufgeladenen Schwebeteilchen wandern zur Niederschlagselektrode und werden dann mechanisch aus dem Filter entfernt. Für diesen Vorgang der elektrischen Staubabscheidung findet man häufig die Wortbildung „elektrostatischer" Staubaustrag. Diese Bezeichnungsform ist unzutreffend und sollte möglichst vermieden werden. Der Ausdruck „elektrostatisch" ist wahrscheinlich dadurch entstanden, daß man zunächst gar nicht den hierbei auftretenden Stromfluß beachtete. Im Verlaufe der technischen Entwicklung haben sich zwei Systeme herausgebildet: a) das Raumluftelektrofilter (Trionfilter) b) das Industrieelektrofilter (Cottrellfilter)*] = electric precipitator, electrostatic ~

~flurfördermittel *n* → Elektrokarre(n)

~-Förderpumpe *f*, Elektro-Beton(förder)pumpe = electric concrete pump

~(fuß)bodenheizung *f* = solid-floor heating by electricity, floor-warming ~ ~, (under)floor heating ~ ~

~-(Gesteins)Bohrhammer *m*, Elektro-Felsbohrhammer *m* = electric (hand) hammer (rock) drill, ~ (rock) drill (hammer)

~-Getriebemotor *m* = geared electric motor

~glas *n* = electro-copper glazing (Brit.); copper(lite) ~ (US)

~-Glättmaschine *f*, Elektro-Glättkelle *f* = electric trowel, ~ float

~gleisstopfer *m*, Elektroschotterstopfer = electric tie tamper (US); ~ track ~ (Brit.); ~ ballast ~

~hammer *m* = electric hammer

~-Handschrapper *m*, Elektro-Räumschaufel *f*, Elektro-Schrapperschaufel, Elektro-Schrapperwinde *f* mit

Elektro(-Hand)stampfer — Elektronenstrahlschmelzofen 368

Handschaufel = electrically-operated hand scraper, ~ manually guided drag skip
~(-Hand)stampfer *m*, Hand-Elektrostampfer = electric (mechanical) tamper
~-Hängeleuchte *f* = electrolier
~heber *m* = electric jack
~-Heißwassergerät *n* → Elektroboiler
~heizgerät *n* → Elektrowärmegerät
~heizung *f* = electric heating
~-Hochlöffel(bagger) *m* = electric shovel, electric (operated) excavator
elektrohydraulisch = hydraulic electric, hydrauleltric
Elektro|-Industrieofen *m* = industrial electric furnace
~-Innenrüttler *m*, Elektro-Innenvibrator *m*, Elektro-Tauchrüttler, Elektro-Tauchvibrator, Elektro-Nadelrüttler, Elektro-Nadelvibrator = electric (or electro-magnet) internal vibrator, ~ (~ ~) immersion ~, ~ (~ ~) concrete mass ~, ~ (~ ~) (concrete) poker ~, ~ (~ ~) needle ~, ~ (~ ~) pervibrator, ~ (~ ~) concrete vibrator for mass work, ~ (~ ~) spud vibrator for concrete
~-Innenvibrator *m* → Elektro-Innenrüttler *m*
~-Installationshammer *m* → Elektro-Bauhammer
~kabel *n* = (electricity) cable, electric (power) ~
~karre(n) *f*, (*m*), Elektroflurfördermittel *n*, Elektrowagen *m*, Elektrofahrzeug *n* = electrical industrial truck, electric freight truck
~karrenanhänger *m* = electric industrial truck (or freight) trailer
~karrenkran *m* = electric (industrial) truck crane
~kernbohrer *m* = electric core drill
~-Kleinwerkzeug *n* = electric hand tool
~Klingel *f* = electric bell
~-Kompressor *m* = electric (air) compressor
~kran *m* = electric crane

~-Lasthebemagnet *m* = electric lifting magnet
~(leer)rohr *n* = electrical conduit, conduit for electrical wiring(s)
~-Lenkmotor *m* = electric steering motor
~-Leuchtboje *f* = electrically lighted buoy
~-Linde-Rapidverfahren *n* → Unterpulverschweißen
~-Lok(omotive) *f*, El-Lok(omotive) = electric locomotive
elektrolytisch erzeugte Oxydschicht *f* = electrolytical oxide layer
Elektro|magnetbremse *f* = electromagnetic brake
~-Magnethammer *m* = electro-magnetic hammer
elektromagnetische Fernsteuerung *f*, ~ Fernbedienung, ~ Fernbetätigung = electro-magnetic remote control
~ Strahlung *f* = electromagnetic radiation
elektromagnetisches Verfahren *n* = electromagnetic method, ~ exploration, ~ prospecting
Elektro|mischer *m* = electrically-driven mixer
~-Mobilkran *m* = mobile electric crane
~motor *m*, El-(Antriebs)Motor = motor, electric ~
~-Nadelrüttler *m* → Elektro-Innenrüttler
~-Nadelvibrator *m* → Elektro-Innenrüttler *m*
Elektronen|beweglichkeit *f* = electron mobility
~-Mikroskop *n* = electron microscope
elektronenmikroskopische Untersuchung *f* = electron-microscopical examination
Elektronen|schaltung *f* = electronic tube control
~strahloszillograph *m*, Kathodenstrahloszillograph = electron (or cathode) beam (or ray) oscillograph
~strahlschmelzofen *m* = electron beam melting furnace

Elektronenstrahlschweißen — Elektroschere

~strahlschweißen n = electron beam welding

elektronisch gesteuertes Verkehrssignal n = computer-controlled traffic signal

elektronische Berechnung f = electronic computation

~ Datenverarbeitungsmaschine f = electronic data processing machine

~ Wiegeeinrichtung f = electronic weighing system

elektronisches Bandwiegesystem n = electronic belt weighing system

Elektro|ofen m = electric furnace

~-Osmose f, Dränung f des Bodens durch Anwendung eines elektrischen Potentials, Elektro(-Osmose-)Entwässerung f, elektro-osmotische stationäre Entwässerung = electro-osmosis, electro-osmotic drainage, subsoil drainage by the electro-osmotic method, drainage by electro-osmosis

    elektrisches Potential n = electric(al) potential

    Filterbrunnen m mit Kathodenspannung = wellpoint cathode, brass ~

    Ladungsteilchen n = charge

    Minuspol m = negative pole

    Rohr n mit Anodenspannung = pipe anode, iron ~

    Zwei-Leiter-System n [Bei kapillar gehaltenem Wasser haftet eine dünne Wasserschicht an der Wandung des Kapillarröhrchens, während sich innerhalb des dadurch gebildeten Wasserröhrchens eine Restmenge Wasser bewegen kann] = (electric) double layer [In a cylindrical capillary tube filled with water, we must distinguish between the free water and a boundary film of water adjacent to the capillary wall. This explanation was given by Helmholtz in 1879]

elektro-osmotische Baugrundverbesserung f → → Baugrundverbesserung

Elektro|(pfahl)ramme f = electric (pile) driving plant, ~ piling ~

~photographie f = xerography

~-Plattformkarre(n) f, (m) = electric platform truck

elektro-pneumatisch gesteuert = electro-pneumatically controlled

Elektro|pumpe f = electric(ally) driven pump

~punktschweißmaschine f = electric spot welding machine

~-Radnabenmotor-Antrieb m = motorized wheel drive

~ramme f → Elektropfahlramme

~-Rammwinde f, Ramm-Elektrowinde = electric piling winch

~-Räumschaufel f → Elektro-Handschrapper m

~-Rohrgerüstkran m, Rohrgerüst-Elektrokran = electric tubular crane

~rührer m = electric stirrer

~-Rüttelaufgeber m, Elektro-Rüttel-Beschicker, Elektro-Rüttelspeiser = electro-vibrating feeder

~-Rüttel-Beschicker m → Elektro-Rüttelaufgeber

~rüttelbohle f, Elektro-Vibratorbohle-Elektro-Vibrationsbohle = electric vibrating beam, ~ vibratory ~

(~-)Rüttelmotor m = vibrating motor

~rüttelplatte f, Elektro-Vibratorplatte, Elektro-Vibrationsplatte = electric vibrating plate, ~ vibratory ~

~rüttelstampfbohle f, Elektro-Vibrationsstampfbohle = vibrating tamper with electrically operated units, electric vibrating tamper, electric vibratory tamper

~rüttelstampfer m = electric vibrating tamper, ~ vibratory ~

~rüttler m, Elektrovibrator m, Magnetrüttler, Magnetvibrator = electric vibrator, electro-magnet ~

~-Saugbagger m = electric suction dredger (Brit.); ~ ~ dredge (US)

~-Schalungsrüttler m, Elektro-Schalungsvibrator m = electric formwork vibrator

~schere f = electric shears

Elektro-Schiebetreppe — Elektrozugkarre(n)

~-Schiebetreppe *f*; Elektro-Aufzugtreppe [*Schweiz*] = electric (disappearing) stairway

~-Schienenbohrmaschine *f* = electric rail drilling machine

~-Schienengreifer(kran) *m*, Greifer-(kran) auf Schienen *fpl* mit Elektroantrieb *m* = rail-mounted electric grab(bing crane), ~ ~ (grab) bucket crane

~-Schienensägemaschine *f* = electric rail sawing machine

~-Schienenstoßhobelmaschine *f* = electric rail joint planing machine

~-Schlagbohrgerät *n*, Elektro-Schlagbohrmaschine *f* = electric percussion drill

~-Schlagbohrmaschine *f*, Elektro-Schlagbohrgerät *n* = electric percussion drill

~-Schlaghammer *m* = electric impact hammer

~-Schlepper *m* = electric tractor

~-Schmierpresse *f*, Elektro-Fettpresse = power-operated grease gun

~-schotterstopfer *m* → Elektrogleisstopfer

~-Schrämmaschine *f* [*Steinbruch*] = electrical channel(l)ing machine

~-Schrapperschaufel *f* → Elektro-Handschrapper

~-Schrapperwinde *f* mit Handschaufel → Elektro-Handschrapper

~schweißaggregat *n* = electric welding set

~schweißen *n*, elektrisches Schweißen = electric welding

~schweißer *m* = electric welder

~schweißmaschine *f* = electric welding machine

~-Schweißtransformator *m* = electric welding transformer

~schweißung *f* = electric welding

~-Schwenkmotor *m*, Elektro-Drehmotor, Schwenk-Elektromotor, Dreh-Elektromotor [*Bagger; Kran*] = electric slewing motor

~-Speisenaufzug *m* = electric service lift

~-Spundwandramme *f* = electric sheet-pile driver

~stampfer *m* → Elektro-Handstampfer

~start *m* = electric starting

elektrostatische Scheidung *f*, elektrostatisches Trennen *n* = electrostatic separation

elektrostatischer Scheider *m* = electrostatic separator

Elektro-Tauch|pumpe *f* = electric submersible pump

~rüttler *m* → Elektro-Innenrüttler

~vibrator *m* → Elektro-Innenrüttler

elektrothermisches Spannen *n* von Spannstahl, Vorspannung *f* auf elektrischem Wege = electrical heating of high-tensile bars, electro-thermal pretensioning, electrical prestressing

Elektro|-Tiefbrunnenpumpe *f*, elektrische Tiefbrunnenpumpe = electric deep well turbine pump

~-Vibrationsstampfbohle *f* → Elektrorüttelstampfbohle

~vibrator *m* → Elektrorüttler *m*

~-Vibratorbohle *f*, Elektrorüttelbohle = electric vibrating beam, ~ vibratory ~

~vibrierschurre *f*, Vibrier-Elektroschurre = electrically vibrated chute

~-Vospannung *f* = electro-thermal stressing

~wagen *m* → Elektrokarre(n) *f*, (*m*)

~wärmegerät *n*, Elektroheizgerät = electric heating appliance, ~ ~ device, ~ heater

~wärmetauscher *m* = electrical heat exchanger

~werkzeug *n* = electric tool

~winde *f* = electric winch

~zählernische *f* = electric meter enclosure

~zement *m* = electric cement

~zentrale *f* → Aggregat *n*

~zentralheizung *f* = electric central heating

~zugkarre(n) *f*, (*m*) = electric towing truck

~-Zusatzlenkung *f* = supplemental electric steering
~zwangsmischer *m* = electric pugmill
elementarer Kohlenstoff *m* = uncombined carbon
Elementar-Tragwerk *n* = elementary structure
Elendsviertel *n* = slum
~beseitigung *f* = slum clearance
Elevator *m* → Becherwerk *n*
~ → Gestänge~
~, Umlaufaufzug *m*, Paternoster (-Aufzug) *m* = paternoster
~ausleger *m*, Becherwerk-Ausleger = bucket elevator boom
(~)Becher *m*, (Elevator)Eimer *m* = (elevator) bucket
~fuß *m* → Becherwerkfuß
~fußtrommel *f*, Becherwerkfußtrommel = boot pulley
~grube *f*, Becherwerkgrube = bucket elevator pit
~gurtband *n*, Becherwerk-Gurtband = bucket elevator belt
~kette *f* → Becherkette
~kopf *m*, Becherwerkkopf = bucket elevator head
~teil *m, n* = elevator component
Ellipsen|-Aufgeber *m*, Ellipsen-Beschicker, Ellipsen-Speiser = feeder with elliptic(al) movement, ~ ~ ~ motion
~bogen *m* = elliptic(al) arch
~gewölbe *n* = elliptic(al) vault
~platte *f*, Ellipsenscheibe *f* = elliptic(al) plate
~querschnitt *m* = elliptic(al) cross-section
~ring *m* = elliptic(al) ring
~schacht *m* = elliptic(al) shaft
~scheibe *f*, Ellipsenplatte *f* = elliptic(al) plate
~-Schwingsieb *n* = elliptic(al)-motion screen, screen having elliptic(al) vibratory action, elliptically vibrating screen, vibratory screen with elliptic(al) movement
~stollen *m* = elliptical gallery
elliptische Paraboloidschale *f* = elliptic(al) paraboloid shell

elliptischer Bogen *m*, Ellipsenbogen = elliptic(al) arch
elliptisches Integral *n* = elliptic(al) integral
Ellira|-Schweißmaschine *f* = submerged-arc welding machine
~-Verfahren *n* → Unterpulverschweißen
Ellis-Verfahren *n* [*technisches Spaltverfahren*] = tube and tank process
El-Motor *m* → Elektromotor
Elmsfeuer *n* = St. Elmo's fire, Saint ~ ~
Eloxalschicht *f* = anodised finish
Eloxier|anlage *f* = anodising installation
~dauer *f*, Eloxierzeit *f* = anodising time
eloxieren = to anodise
Eloxierhaut *f*, Oxydhaut = anodic film
Eloxierung *f*, Eloxieren *n*, anodische Oxydation *f* = anodising, anodic oxidation
~ (oder Eloxierverfahren *n***) mit direkt im Eloxalbad erzeugten lichtbeständigen Farbtönen** = integral colo(u)r anodising (process)
Eloxier|verfahren *n* = anodising process
~zeit *f*, Eloxierdauer *f* = anodising time
Eltwerk *n* → Kraftwerk
Eluvial|boden *m*, Verwitterungsboden, Verwitterungslockergestein *n*, Auswaschungsboden = residual soil (or earth), residuary ~ (~ ~), eluvial ~ (~ ~), derived ~ (~ ~), sedentary ~ (~ ~)
~horizont *m*, ausgewaschener Horizont = eluvial horizon
Eluviationszone *f* = zone of eluviation
Emaillier|ofen *m* = enamel stove
~ton *m* = enamel(l)ing clay
Emissionsspektrographie *f* = emission spectrographic method
E-Modul *m* → Elastizitätsmodul
Empfangs|gebäude *n* → Abfertigungsbau *m*
~halle *f* → (Eingangs)Halle

Empfangsort — endloser Förderer

~ort *m* [*Baumaschine; Baustoff*] = receiving point
~raum *m* [*Messestand*] = promotional lounge
~schallkopf *m* = receiving transducer
Empfindlichkeit *f*, Genauigkeit [*Waage*] = weighing sensibility
empirisch = empirical
Emplektit *m*, Kupferwismutglanz *m*, $CuBiS_2$ (Min.) = emplectite
Emscher|brunnen *m*, Imhoffbrunnen = Emscher tank, Imhoff ~
~filter *m, n* = Emscher filter
Emulgation *f*, Emulgieren *n*, Emulgierung *f* = emulsification
Emulgator *m*, Emulgiermittel *n* = emulsifying agent, emulsifier
~molekül *n* = emulsifier molecule
~schicht *f* = emulsifier layer
emulgierbar = emulsifiable
Emulgierbarkeit *f* = emulsifiability
emulgieren = to emulsify
Emulgiermittel *n*, Emulgator *m* = emulsifier, emulsifying agent
emulgiertes Bitumen *n* → Bitumenemulsion *f*
Emulsion *f* = emulsion
~ mit Haftmittelzusatz = doped emulsion
Emulsionsanstrich *m* = emulsion coating
~, Dichtungshaut *f*, Filmüberzug *m*, Nachbehandlungsfilm *m* [*Beton*] = emulsion coating, concrete curing membrane, curing seal
Emulsions|farbe *f* = emulsion paint
~wasser *n* = emulsification water
End|abschalter *m* [*Radschrapper*] = cable saver
~absenkung *f* (⚒) = final subsidence
~abtrieb *m* = transmission extension drive
~antrieb *m*, letzte Getriebestufe *f* = final drive
~auflager *n* [*frei drehbar*] = end support [*pin-jointed*]
~ausschalter *m* für Hubhöhe [*Turmdrehkran*] = height cut-out
~bahnhof *m* = railway terminal (Brit.); railroad ~ (US)

~bearbeitung *f*, Fertigbehandlung [*Betondecke*] = (surface) finishing, final ~
~bindeblech *n* [*Stütze*] = end tie-plate
~block *m* = end-block of a beam
~bogen *m*, Widerlagerbogen = abutment arch
~diagonale *f*, Endschräge *f* = end diagonal
~dichtung *f* [*Wehr*] = end seal
~drehung *f* = end rotation
~druck *m* [*Baustatik*] = end thrust
~druck *m* [*Verdichter*] = delivery pressure, working ~
~druck *m* [*Hochdruckverdichter*] = discharge pressure
~einspannung *f* = end fixity, ~ restraint
~ergebnis *n* = final result
~erzeugnis *n* [*Siebtechnik*] = (end) product, finished ~
~feld *n*, Endöffnung *f* [*Brücke*] = end span, ~ opening, anchor ~
~festigkeit *f* = final strength
~getriebekasten *m* = final drive gearbox
~gewinde *n* = end thread
endgültige Fertigstellung *f* **einer Bohrung** = permanent well completion
~ Vorspannung *f* = final prestress(ing)
endgültiger Ausbau *m* (⚒) = permanent support
endgültiges Anspannen *n* = final prestressing
End|haken *m* [*Das um 180° herumgebogene Ende eines Betonstahles*] = end hook
~isochrone *f* = final isochrone
~lagerflansch *m* = end bearing flange
endlich = finite
endliche Biegung *f* = finite bending
~ Dehnung *f* = finite strain
~ Länge *f* = finite length
~ Verdrehung *f* = finite twisting
~ Verformung *f* = finite deformation
~ Verschiebung *f* = finite displacement
endloser Förderer *m* = endless conveyor, ~ conveyer

End|moment n = terminal moment, final ~, end ~
~moräne f = terminal moraine, end ~
~mühle f = finish mill [*1. Usually a tube mill in which the final stages of clinker grinding are accomplished. 2. The entire finish grinding department*]
~öffnung f, Endfeld n [*Brücke*] = end opening, ~ span, anchor ~
~pfeiler m = end pier
~planeten-Getriebe n = final planetary gearing, secondary epicyclic train [*of a double reduction gearing*]
~platte f [*Bremse*] = end plate
~produkt-Förderband n [*Brech- und Siebanlage*] = finished product (or material, or grade) conveyor (or conveyer)
~punkt m, Endstation f [*mitunter ist damit ein Bahnhof verbunden*] = terminal, terminus
~querträger m = extreme cross girder
~rahmen m → Portal(verband) n, (m)
~reaktion f = terminal reaction, final ~
~schalter m, Begrenzungsschalter, Anschlagschalter = limit switch, stop ~
~schräge f → Enddiagonale f
~schub m **der Kurbelwelle**, Kurbelwellenendschub = crankshaft end thrust
~schwelle f = end sill
~sieb n, Fertigsieb = finishing screen
~station f → Endpunkt m
~steife f [*Blechträger*] = end stiffener, ~ stiffening angle
~strang m = house sewer, ~ connection [*A pipe conveying the sewage from a single building to a common sewer or point of immediate disposal*]
~strang m = communication pipe [*That portion of the service pipe from the main to the boundary of consumers' premises*]
~strang m [*allgemein*] = dead ended piping, ~ ~ section, dead-end ~
~stück n **für Planiermesser** = end bit

~stufe f [*(Hart)Zerkleinerung*] = final stage
~stütze f = end column
~temperatur f = final temperature
~termin m = final date
~übersetzung f [*in der Radnabe*] = final drive
endverankert = end-anchored
(End)Verankerung f [*Spannbeton*] = (end) anchoring system, (~) anchorage ~
End|wert m, Schlußwert = terminal value, final ~, end ~
~zustand m = final condition, ~ state
Energie f, Arbeitsfähigkeit f [*Hydraulik*] = energy
~, Kraft f = power
~ der Lage, potentielle Energie, latente Energie = energy of position, potential energy, position energy, geodetic head
~bedarf m, Kraftbedarf [*Maschine*] = (horse)power requirement
~einfuhr f = energy import
~erzeugung f, Stromerzeugung = power production, electricity ~, ~ generation
~höhe f, spezifische Energie f [*Hydraulik*] = energy head [*The elevation of the hydraulic grade line at any section plus the velocity head of the mean velocity of the water in that section. The energy of a unit height of the stream. The energy head may refer to any datum, or to an inclined plane, such as the bed of a conduit*]
~kapazität f = generating capacity
~linie f = energy line, line of total head [*A line joining the elevations of the energy heads of a stream. The energy line is above the hydraulic grade line a distance equivalent to the velocity heads at all sections along the stream*]
~minister m = Minister of Power
~netz n = national grid
~politik f = power policy
~projekt n = power scheme
~quelle f = power source

~speicherung *f* = power storage
~verbrauch *m* = power consumption
~vernichter *m* ↠ Talsperre *f*
~vernichtung *f* ↠ Beruhigung
~vernichtungspfeiler *m*, Energieverzehrungspfeiler, Störpfeiler = baffle-pier [*A pier on the apron of an overflow dam, to dissipate energy and prevent scour*]
~versorgung *f*, Kraftversorgung, Stromversorgung = power supply
~verteilung *f* = power distribution
~verzehrung *f* ↠ Beruhigung
~wirtschaft *f* = economics of energy, power economy

Engbohrloch *n* [*Ein Rotary-Bohrloch mit kleinem Durchmesser*] = slim hole, rat ~

enge Grubenbaue *mpl* (⚒) = narrow workings

englische Methode *f* ↠ Einzonenbauweise *f*

englischer Quarzit *m* = gannister (Brit.); ganister (US)
~ **Schraubenschlüssel** *m*, Franzose *m*, Engländer *m* = agricultural wrench
~ **Tripel** *m* = rottenstone
~ **Tübbingausbau** *m* = English tubbing

englisches Verfahren *n* ↠ Einzonenbauweise *f*

Englischrot *n* = English vermilion

Englochbohren *n* = slim hole drilling, rat ~ ~

Engobe *f* = engobe, clay-coated finish
~**ton** *m* = coating clay

Engobieren *n* = slip coating, clay ~

Engobiermaschine *f* = slip coating machine, clay ~ ~

Engstelle *f* [*Hüllrohr*] = narrowed portion of duct

Entaktivierung *f* = desensitisation

Entaschung *f* = ash removal

Entaschungs|automat *m*, automatische Entaschungsvorrichtung *f* = auto(matic) ash remover
~**tür** *f*, Aschentür = ash door

Entbindungsheim *n* [*Teil eines Krankenhauses*] = maternity unit

Entdeckungsbohrung *f*, Fundbohrung, Erschließungsbohrung = discovery well

Entdröhnung *f* = anti-drumming treatment, sound-deadening

Enteisenung *f* = iron removal, removal of iron

Enteisung *f* = defrosting, de-icing

Entemulgierbarkeit *f* = demulsibility

Entenschnabellader *m* (⚒) = duckbill loader

entfärben, bleichen = to bleach

Entfärben *n*, Bleichen = bleaching

Entfärbungsmittel *n*, Bleichmittel = bleaching agent

entfernen, wegnehmen, abnehmen = to detach, to withdraw, to remove

Entfernungsfeuer *n*, Abstandfeuer [*Flugplatz*] = distance marking light

Entfetten *n* **durch Glühen** = grease burning

Entfettung *f*, Entfetten *n* = degreasing [*The process of removing fats and greases from sewage, waste, or sludge*]

Entfettungs|anlage *f* = degreasing plant, ~ installation
~**mittel** *n* = degreasing agent

entfeuchten = to dehumidify

Entfeuchten *n* = dehumidifying

Entfeuchter *m* = dehumidifier

Entfeuchtung *f*, Entfeuchten *n* = dehumidification, de-humidifying

Entfeuchtungs|leistung *f* = dehumidifying capacity, dehumidification ~
~**schlange** *f* = dehumidifier coil

entflammbar, entzündlich = inflammable

Entformen *n* = demo(u)lding

Entfroster|düse *f* = de-mister [*In a driver's cab(in)*]
~**gebläse** *f* = defroster

Entgaser *m* = degasser

Entgasung *f*, Entgasen *n* = degassing
~, **trockene Destillation** *f*, Zersetzungsdestillation = destructive distillation, dry ~ [*sometimes referred to as "pyrolysis"*]

entgegen dem Uhrzeigersinn, linksläufig = counterclockwise
Entglasung *f* = devitrification
entgleisen = to derail
Entgleisungsbeiwert *m* = derailment coefficient
Enthärter *m* = softening material
Enthärtung *f* = softening
Enthärtungsanlage *f* [*für Wasser*] = (water) softening plant
Entkiesung *f*, Ablassen *n* angesammelter Geschiebemengen = washing out, scouring, flushing
Entkupplungshammer *m* = uncoupling hammer
Entlade|anlage *f*, Abladeanlage, Ausladeanlage = unloading station
~band *n* → Ausladeband
~gleis *n*, Abladegleis, Ausladegleis = unloading track
~kran *m*, Abladekran, Ausladekran = unloading crane
Entladen *n* → Ausladen
Entlade|pumpe *f* [*für pulverförmiges Schüttgut, z. B. Zement*] = unloader pump
~schaufel *f*, Kraftschaufel, Waggonschrapper *m*, Waggonschaufel für Schüttgüter, Entladekraftschaufel = hand scraper (for unloading railway cars), manually-guided drag skip (~ ~ ~ ~), hand scraper unloader
~schleuder *m* für Schüttgüter, (fahrbare) Schüttgut-Schleuder = lorry dump piler (Brit.); truck ~ ~ (US)
~stelle *f*, Ausladestelle, Abladestelle = unloading point
~vorrichtung *f*, Ausladevorrichtung = unloading device
~weite *f*, (Funken)Schlagweite, Funkenstrecke *f* = sparking distance, striking ~, spark gap
Entladungslampe *f* = discharge lamp
Entlastungs|bogen *m* = relieving arch
~bohrung *f* = relief pocket
~brunnen *m*, Abzapfbrunnen [*in artesisch gespanntem Grundwasser*] = bleeder well

~-Dränageloch *n* → Entwässerungsschlitz *m*
~leitung *f* = relief sewer [*A sewer built to relief an existing sewer of inadequate capacity*]
~methode *f* [*Sprengen weicher Baugrundmassen*] = relief method, ~ blasting
~platte *f* = relieving platform
~pumpe *f* = relief pump
~stollen *m* → → Talsperre *f*
~verfahren *n* [*Sprengverfahren, durch welches weiche Baugrundmassen mit großer Wucht auseinandergetrieben und gestört werden, so daß die aufgeschütteten tragfähigen Massen auf die tragenden Lockergesteine absacken können*] = relief method
entleeren, absenken, ablassen [*Talsperre*] = to draw down, to empty
Entleeren *n*, Ablassen, Ablassung *f*, Entleerung, Absenkung, Absenken [*Talsperre*] = drawdown, emptying
Entleerung *f* → Entleeren *n* [*Talsperre*]
Entleer(ungs)|hahn *m*, Ablaßhahn = drain cock, "bleeder" [*A cock placed at the lowest point of a water system, through which the system can be drained when required*]
~hebel *m* = discharge lever
~organ *n*, Ablaßorgan [*Talsperre*] = outlet element
~rutsche *f* → Auslaufschurre *f*
~schieber *m*, Abfüllschieber [*Silo*] = silo gate
~schurre *f* → Auslaufschurre
~stollen *m* → Ablaufstollen
~stopfen *m* → Abflußstopfen
~strömung *f*, Ausströmung [*Meer*] = draining current, emptying ~
~ventil *n*, Ablaßventil = drain valve
Entlösungsdruck *m* [*z. B. von Gas aus Erdöl*] = expulsive force
Entlüfter *m* = breather
~kappe *f* = breather cap
~sieb *n* = breather screen
entlüftetes Wasser *n* = air-free water
Entlüftungs|öffnung *f* = bleed hole, vent ~, ~ opening

Entlüftungsrost — Entschalungspaste 376

~rost m, Abluftrost = exit air grille, outlet ~
~schacht m, Luftschacht, Wetterschacht = ventilation shaft, air ~, ventilating ~
~schornstein m, Abluftschornstein, Entlüftungsschlot m, Abluftschlot = exit air chimney, ~ ~ stack
~schraube f = bleeder
~ventil n = bleed valve, vent ~
Entminung f = mine removal
entmischter Grobzuschlag(stoff) m = loose core [*Large aggregate which has become segregated from the concrete by mishandling between mixing and placing*]
Entmischung f, Entmischen n = segregation, de-mixing [*The mix must not suffer by separation of its components*]
Entnahme f, Boden~ = borrow, soil ~
~ **ungestörter Bodenproben** = undisturbed (soil) sampling
~bauwerk n → → Talsperre f
~boden m = borrow soil
~büchse f [*Bodenprobe*] = sampling tin
~einschnitt m → Boden~
~gerät n für Bodenproben = soil sampler, sampling tool
~grube f = borrow pit; reclaim area [*Australia*]
~kubatur f → Boden~
~material n = borrow(ed) material
~rohr n, Bodenproben~ = tube sampler
~sand m = sandy borrow
~schacht m, Entnahmeturm m [*Trinkwasserbecken*] = reservoir water tower
~stelle f → Boden~
~stollen m → Haldentunnel m
~stutzen m [*Bodenprobenahme*] = sampling tube, tube sampler, sample(r) tube
~tunnel m → Haldentunnel
~turm m, Entnahmeschacht m [*Trinkwasserbecken*] = reservoir water tower

Entnietungs|dorn m = rivet punch
~hammer m = rivet buster, ~ cutter
Entölungs|halbmesser m, Aktionshalbmesser einer Erdölförderbohrung = radius of oil drainage, drainage radius of an oil well
~kegel m [*um ein Bohrloch herum*] = oil drainage cone
Entparaffinierungsanlage f = dewaxing plant, ~ installation
Entrattung f = deratization
entrohren, Rohre ziehen [*Tiefbohrtechnik*] = to withdraw casing, to pull ~
Entrohren n der Bohrung nach erfolgter Einstellung = "pulling the well"
Entrostbürste f, Abrostbürste = rust removing brush
Entrosten n, Abrosten = rust removal
entrostet, abgerostet = derusted
Entrostungs|meißel m, Abrostungsmeißel = scaling chisel
~mittel n, Abrostungsmittel = rust remover
entsalztes Wasser n, entsalzenes ~ = demineralized water
Entsalzung f = demineralization
entsanden [*Brunnen*] = to develop
Entsander m, Sandfang m = sand trap (Brit.); desilter (US); sand catcher, desander
Entsandung f, Entsanden n = sand removal
Entsandungsbecken n = sand catching basin
Entsäuerungs|zeitformel f = decarbonating time formula
~zone f = decarbonating zone
entschalen, ausschalen = to strip
Entschalen n, Ausschalen, Entschalung f, Ausschalung = (formwork) stripping
Entschaler m → Ausschaler
Entschalungs|emulsion f → Beton~
~mittel n → Beton~
~öl n, Schalöl = formwork oil; mould ~ (Brit.); mold ~ (US)
~öl-Sprühgerät n, Schalöl-Sprühgerät = formwork oil atomizing device
~paste f → Beton~

~stellung *f* = stripping position
Entschlammen *n* [*Abwasserwesen*] = desludging
entschlammtes Abwasser *n*, geklärtes ~ = clarified sewage
Entschlämmungs|apparat *m*, Schlämmapparat [*Materialaufbereitung*] = desiltor
~sieb *n* = desliming screen
Entschwef(e)lung *f* = desulfurization
Entschwef(e)lungsanlage *f* = distillate desulfurizer unit
entseuchtes Abwasser *n* = disinfected sewage [*Crude sewage, or a sewage plant effluent which has been treated with a disinfecting agent, commonly chlorine or "bleach", resulting in the destruction of bacteria sufficiently to reduce materially the danger of infection*]
Entseuchung *f* = disinfection [*The partial destruction, ordinarily by the use of some chemical, of the microorganismus likely to cause infection and disease*]
entspannen [*Spannbeton*] = to detension
Entspannen *n* = stress relieving
Entspannungs|ort *n* (⚒) = relief roadway
~-Spaltanlage *f* = flashing plant
~turm *m* = flash tower
~-Verfahren *n* [*technisches Spaltverfahren*] = flashing process
entspringen [*Gewässer*] = to originate in, to spring from
Entstapler *m* = de-stacker
Entstaubung *f* = dust collection, dedusting
Entstaubungs|anlage *f*, Staubabsaugungsanlage, Staubsammleranlage = dust collection plant, ~ exhaust ~, ~ arrestor ~
~siebung *f* = dirt screening
Entstehungs|material *n* [*Boden*] = parent material
~temperatur *f* = original temperature
Entteerungsanlage *f* = detarring plant
Enttrümmerung(sarbeiten) *f (pl)*, Schutträumung(sarbeiten), Trümmerräumung(sarbeiten) = blitzed site clearance
entwässern, trockenlegen = to drain, to dewater, to unwater
Entwässern *n*, Trockenlegen = draining, dewatering, unwatering
entwässerter Scherversuch *m*, entwässerte Scherprüfung *f*, entwässerte Scherprobe *f* = drained shear(ing) test
Entwässerung *f* [*in der Aufbereitungstechnik*] = dewatering
~, Dränage *f*, Wasserentzug *m* [*Der Vorgang, der im Boden zum Endzustand kapillaren Gleichgewichtes führt*] = drainage
Entwässerungs|anlage *f* = drainage system
~apparat *m* [*Materialaufbereitung*] = dewatering unit, dewaterer, dehydrator
~arbeiten *fpl*, Trockenlegungsarbeiten = drainage work
~bauwerk *n* = drainage structure
~bunker *m* = dewatering bin
~fall *m* = condition of drainage
~(ge)rinne *f*, (*n*), Entwässerungsgraben *m* = drainage channel, open canal, cut drain [*for surface water*]
~graben *m* → Entwässerungs(ge)rinne *f*, (*n*)
~kammer *f* [*Waschmaschine*] = drain bin
~leitung *f* = drainage line
~netz *n* = drainage network
~probe *f*, Entwässerungsprüfung *f*, Entwässerungsversuch *m* [*Bitumenemulsion*] = dehydration test
~prüfung *f*, Entwässerungsprobe *f*, Entwässerungsversuch *m* [*Bitumenemulsion*] = dehydration test
~pumpe *f* = drainage pump
~pumpwerk *n* = drainage pumping station
~rechenklassierer *m* = rake classifier for dewatering
~rinne *f* → Entwässerungsgerinne *n*
~rohr *n* → Dränrohr
~rohrnetz *n* = drainage pipe system, ~ ~ network

Entwässerungsrohrstollen — Epoxy(d)harzkleber

~rohrstollen m = drainage pipe gallery
~schacht m = drainage shaft
~schicht f → Filterschicht
~schlitz m, Sickerschlitz m, Entlastungs-Dränageloch n = weep hole, weeper, drainage opening, drainage hole
~schürze f → → Talsperre f
~sieb n, Abtropfsieb = draining screen
~sieb n für Schwertrübe = heavy medium draining screen
~silo m = dewatering silo
~stollen m = drainage gallery
~stutzen m = drain connection
~technik f = drainage engineering
~teppich m = drainage blanket
~verband m = drainage district [*An organization operating under legal regulations for financing, constructing, and operating a drainage system*]
~versuch m, Entwässerungsprüfung f, Entwässerungsprobe f [*Bitumenemulsion*] = dehydration test
~vorrichtung f [*in der Aufbereitungstechnik*] = dewaterer
Entwerfer m, Entwurfverfasser m, Konstrukteur m = designer
entwickeln, erzeugen [*Azetylen*] = to produce, to generate
Entwick(e)lungs|kurve f [*Bodensenkung*] = development curve
~maß n des Flusses = degree of bed development
Entwicklungsland n = developing country, emergent ~, emerging ~, pre-industrial ~, young industrial ~ [*The term "underdeveloped country" is employed only in contexts where there can be no risk of offending a friendly government*]
Entwurf m, zeichnerische Ausarbeitung f, zeichnerische Konstruktion f = design
~ über die Bestimmung von Bauwerken = functional design of structures
~aufgabe f = design problem

~geschwindigkeit f → Ausbaugeschwindigkeit [*Straße*]
~grundlagen fpl = design principles
~merkmal n = design standard
~raster m = planning grid [*not to be confused with a grid plan = Rasterplan*]
~richtlinien fpl = design specifications
~spannung f = design stress
~stadium n = design stage
~- und Bauauftrag m = "turn key" type of contract
~- und Konstruktionsausstellung f = design engineering show
~verfasser m, Entwerfer m, Konstrukteur m = designer
Entzinkung f = dezincification
Entzunderung f = descaling
Entzunderungs|anlage f = descaling plant, ~ installation
~ofen m = wash heating furnace
entzündlich, entflammbar = inflammable
(Eötvös'sche) Drehwaage f = Eötvös torsion balance
Eozänton m = Eocene clay
Epikote-Kunstharzkleber m = Epikote resin-based adhesive compound
Epizentrum n = epicentre
Epoxy(d)harz n bei der Erdöldestillation als Nebenprodukt gewonnen = epoxy resin of petrochemical origin
~ mit Bitumen = epoxy-asphalt material (US)
~ ~ Kohlenteer = epoxy-coal tar material
~ ~ Schwarzbindemittel = epoxy-bitumen material (US)
~bindemittel n = epoxy (resin) binder [*This term is used to describe a formulation used as a cementing agent to bind particles of aggregate into a mass that is originally plastic but later becomes rigid*]
~kleber m = epoxy (resin) adhesive [*This term is used to describe a formulation for bonding discrete portions of concrete*]

Epoxy(d)harzklebnaht — erdbebensicher

~klebnaht f = epoxy joint
~mörtel m = epoxy-based mortar
Epoxy(d)überzugharz n [*Oberbegriff für 1. Epoxyharzbindemittel und 2. Epoxyharzkleber*] = epoxy coating resin
Epoxyzahl f = epoxide number
Epoxy|gemisch n = epoxy mix(ture), ~ combination, ~ composition, ~ system
~zusammensetzung f = epoxy formulation
Epurée-Asphalt m → Trinidad-Epuré
EPZ m → Eisenportlandzement
Erbsenstein m, Pisolith m = pisolite, pisolitic limestone
Erbs|kies m → Perlkies
~kohle f → Grießkohle
Erd|abdeckung f [*Nachbehandlung einer Beton(straßen)decke*] = earth cover(ing)
~achse f = axis of the earth
~anziehung f = earth's attraction
~arbeiten fpl → Erdbau m
~arbeiter m = navvy [*A labourer who specializes in digging sewers, drains, roads, railway foundations and trenches with pick, shovel and graft*]
~auflast f = soil surcharge
~aufschüttung f → Erdschüttung
~auftrag m → Erdschüttung f
~aushub m → (Boden)Aushub
~aushubarbeiten fpl, Bodenaushubarbeiten = soil digging work, ~ excavation ~
~balsam m = earth balsam [*A variety of asphalt from Pechelbronn, Alsace*]
Erdbau m, Erdarbeiten fpl = earthwork(s) [*The construction of large open cuttings, or excavation involving both cutting and filling, in material other than rock, is called earthwork(s)*]
~fahrzeug n, Fahrzeug für den Bodentransport, Erdbaulastfahrzeug, Erdtransportfahrzeug, Erdtransportwagen m = earthwork(s) vehicle
~firma f = earthmoving contracting firm

~geräte npl, Erdbaumaschinen fpl, Erdbaugroßgeräte = earthwork(s) plant, ~ equipment, ~ machinery, earthworking ~
~labor(atorium) n = soils lab(oratory), earth materials ~, soil mechanics ~, soil testing ~
~lastfahrzeug n → Erdbaufahrzeug
~lastfahrzeug n → Muldenkipper m
~last(kraft)wagen m → Muldenkipper m
~los n, Erdlos = earthwork(s) contract section
~luftreifen m = pneumatic tire for earthwork(s) (US); ~ tyre ~ ~ (Brit.)
~maschinen fpl, Erdbau(groß)geräte npl = earthworking machinery, ~ equipment, ~ plant, earthwork(s) ~
~mechanik f → Bodenmechanik
~stelle f = earthwork(s) site
~technik f, Erdbauwesen n = earthwork(s) engineering
~walze f → → Walze
~werk n = earth structure, soil ~
~wesen n, Erdbautechnik f = earthwork(s) engineering
Erdbeben n = earthquake
~bemessung f = seismic design, earthquake ~
~erschütterung f = earthquake vibration
~flutwellen fpl, Tsunanis f = seismic sea waves, tsunanis
~fortpflanzung f = earthquake propagation
~herd m, Hypozentrum n = origin, focus
~kunde f, Seismik f, Seismologie f = seismology
~reaktion f eines Bauwerkes oder Gebäudes = earthquake response
erdbebensicher [*Gestaltung der Bauwerke gegen plötzliche und rasch vorübergehende Erschütterungen durch tektonische und vulkanische Beben*] earthquake-resistant, quakeproof, aseismic

Erdbeben|technik *f* = earthquake engineering
~welle *f* = earthquake wave
Erd|behälter *m*, Erdtank *m* = buried tank
~beschleunigung g *f* → Fallbeschleunigung g
~beton *m* → Zementverfestigung *f*
~bett *n*, Erdlager *n*, Erdpolster *n* [*Österreich: m*] [*Rohrverlegung im Erdreich*] = bedding layer of soil
~bettung *f* [*Rohr*] = bedding in soil
~bewegung *f* → Bodenbewegung
Erdbewegungs|gerät *n* → Bodenfördergerät
~(groß)gerät *n*, Erdbewegungsmaschine *f*, Bodenfördergerät, Erdfördergerät = earthmover; dirtmover (US); earthmoving gear (Brit.)
~ingenieur *m* = earthmoving engineer
~maschine *f* → Erdbewegungsgroßgerät *n*
Erd|bildmessung *f*, terrestrische Photogrammetrie *f*, Erd-Photogrammetrie = terrestrial photogrammetry
~bildungskunde *f* → Geologie *f*
~boden *m* [*als Fläche*] = ground
~bogen *m* → Gegenbogen
~bohren *n* mittels Büchse, Büchsbohren *n* = bailer-boring
~bohrer *m* = (soil) auger, earth ~
~bohrgerät *n* [*tragbar*] = carryable (soil) auger, ~ earth ~
~böschung *f*, Bodenböschung = bank [*An earth slope formed or trimmed to shape*]
~bunker *m* → Bunkergrube *f*
Erddamm *m* = earth embankment, ~ bank, soil ~ [*A ridge of earth thrown up to carry a road, railway, or canal*]
~, Erddeich *m* [*Fluß; Strom*] = earth levee; ~ dyke (Brit.); ~ dike (US)
~, Erd(-See)deich *m*, Erd-Meer(es)deich, Erd-Meer(es)damm = earth sea dyke (Brit.); ~ ~ dike (US); ~ ~ levee
~, (Erdschüttungs)Staudamm, Erdstaudamm, Erdschüttungsdamm = earth dam, ~ fill ~, earth-work ~

Erd|dampf *m* = natural steam
~dampfkraftwerk *n*, geothermisches Kraftwerk, Vulkankraftwerk = natural steam power plant, ~ ~ ~ station
~deich *m* → Erddamm *m*
~drehung *f* = rotation of the earth
Erddruck *m*, aktiver ~ = active earth pressure, ~ soil ~
~angriffpunkt *m*, Angriffpunkt des Erddruckes = point of application of active earth pressure, ~ ~ ~ ~ ~ soil ~
~beiwert *m* = coefficient of active earth pressure, ~ ~ ~ soil ~
~berechnung *f*, Erddruckermitt(e)lung = earth-pressure computation
~meßdose *f* → Bodendruck(meß)dose
~theorie *f* = earth-pressure theory
~umlagerung *f* = active earth-pressure redistribution, ~ soil-pressure ~

Erde *f* → Boden *m*
Erdeinbau *m*, Bodeneinbau = soil placement, earth ~
erden [*Elektrotechnik*] = to earth
Erden *n*, Erdung *f*, Erdverbindung [*Elektrotechnik*] = earthing
Erd|falte *f* = nappe, earth fold
~fang(e)damm *m*, Lockergesteinsfang(e)damm = earth(work) cofferdam; dike-type ~ (US)
~farbe *f* = natural earth, ~ pigment
~feuchte *f* → Bodenfeuchte
erdfeuchter Beton *m* = no-slump concrete, (earth-)dry ~, earth-damp ~
Erd|feuchtigkeit *f*, Erdfeuchte *f* = ground moisture
~fließen *n* → Bodenfließen
~fördergerät *n* → Bodenfördergerät
~fördergerät *n* → Erdbewegungs(groß)gerät
~förderung *f* → Bodenbewegung
Erdgas *n*, Naturgas = rock gas, natural ~

~ das an seiner Fundstelle nicht in Berührung mit Öl ist = non-associated natural gas

~ das in Erdöl gelöst ist = natural gas in solution

Erdgas — Erdölanzeichen

~ das zwar in Berührung mit Erdöl, aber nicht in ihm gelöst ist = associated natural gas
~ mit Schwefelverbindungen = sour natural gas
~ ohne Schwefelverbindungen = sweet natural gas
~benzin n, Naturgasolin n, Naturgasbenzin = casing-head gasoline, natural ~
~brenner m, Naturgasbrenner = natural gas burner
~-Durchschlag m aus dem Ringraum in die Steigrohre, Naturgas-Durchschlag ~ ~ ~ ~ ~ ~ = surging
~gasfernleitung, Naturgasfernleitung = natural gas pipeline
~fundstätte f, Erdgaslager(stätte) n, (f), Erdgasvorkommen n, Erdgasvorkommnis n = natural gas deposit
~kraftwerk n, Naturgaskraftwerk = natural gas power plant, ~ ~ ~ station
Erdgasometer m, Erd-Gassammler m, Erd-Gasbehälter m, unterirdischer Gasspeicher m = underground gasholder, ~ gas-tank, ~ gasometer
Erdgas|schwefel m, Naturgasschwefel = sulfur from natural gas (US); sulphur ~ ~ ~ (Brit.)
~sonde f → produzierende ~
~-Tanker m, Naturgas-Tanker, Erdgas-Tankschiff n, Naturgas-Tankschiff = methane ship
~vorkommen n → Erdgasfundstätte
Erdgeschoß n = ground floor
~balken(träger) m = ground floor beam
~bau m = building with ground floor only
~decke f = ground-floor floor
~stütze f = ground floor column
erdgestütztes Bauwerk n = soil-supported structure, earth-supported ~
(Erd)Hacke f, Bicke f, Picke = pick
Erd|hobel m → Straßenhobel
~-Hochwasserdamm m = earth(en) flood bank
~hügel m = earth(en) mound

erdiger Gips m, erdiges Gipsgestein n, feinporiges Gipsgestein = gypsite, earthy gypsum, friable gypsum
erdiges Rohsalz n, Caliche f = caliche
Erd|kabel n → Bodenkabel
~kappe f [*Klostergewölbe als umgekehrtes Gewölbe bei Gründungen*] = cloister vault in the ground
~keil m = soil wedge
~kern m [*Felsschüttungs(stau)damm*] = earth(en) core, soil ~
~kippe f = soil tip
~klumpen m = Bodenscholle f
~kobalt m (Min.) → Kobaltschwärze
~kohle f = earth coal
~korn n, Bodenkorn, Bodenteilchen, Erdteilchen n = soil grain, ~ particle
~körper m, Bodenkörper [*Erdbau*] = soil body, earth ~
~kratzer m, Schrapper m = scraper
~kruste f → Erdrinde f
~lager n → Erdbett n
~leitung f, Bodenleitung, bodenverlegte Leitung, erdverlegte Leitung, eingeerdete Leitung = buried line
~los n, Erdbaulos = earthwork(s) contract section
~mantel m = earth mantle
~masse f, Bodenmasse = earth mass, soil ~
~massenermitt(e)lung f, (Boden)Massenermitt(e)lung = soil quantity determination, earth ~ ~
~massenverteilung f, (Boden)Massenverteilung [*Erdbau*] = spreading of soil, ~ ~ earth
~-Meer(es)damm m → Erddamm
~-Meer(es)deich m → Erddamm m
Erdöl n, Roh(erd)öl, Öl m, (Roh)Petroleum n = (crude) oil, rock ~, petroleum
~abkömmling m, Erdölderivat n = petroleum derivative, oil ~
~anreicherung f = accumulation of petroleum, ~ ~ oil, oil accumulation, petroleum accumulation
~anzeichen npl an der Tagesoberfläche = surface oil indications

Erdölbergbau — (Erdöl)Lagerstättendruck

~bergbau m [Der Bergbau ermöglicht die größte Erdölausbeute. Sie liegt beim Sickerbetrieb bei etwa 60% und beim Abbau bei fast 100 %. Beim Sikkerbetrieb durchziehen Bergbaustrekken die Ölsandschichten in ganzer Breite und Höhe. Durch diese Strekken sickert das Erdöl langsam zum Schacht, sammelt sich dort und wird mit Pumpen nach oben gefördert. Beim Abbau wird der Ölsand nach oben gefördert und gewaschen. Zum Erdölbergbau gehört auch die Gewinnung des Ölschiefers] = oil mining
~bergwerk n, Erdölgrube f = oil mine
~bitumen n = refinery (asphaltic) bitumen (Brit.); petroleum asphalt (US) [Such bitumen is sometimes called "residual bitumen" (i.e. the residue on distillation of petroleum) in order to distinguish it from the bitumens which occur naturally. The undesirable overtones attached to the word "residual" have led the bitumen industry to prefer the use of the term "straight run", which, however, is nowadays misleading, as few modern petroleum bitumens are produced by simple distillation without some further treatment]
~bohranlage f = oil drilling installation
~bohren n, (Erd)ölbohrung f, Ölbohren n = oil drilling
~bohrschieber m = valve for oil drilling
~brunnen m → Erdölbohrung f
~bohrung f, Ölbohrung, (Erd)Ölsonde f, (erd)ölfündige Bohrung, (Erd)Ölbrunnen m = oil well
~derivat n, Erdölabkömmling m = petroleum derivative, oil ~
~destillat n = petroleum distillate
~destillationsanlage f = oil-refining still
~-Entsalzungsanlage f = desalter
~erzeugnis n = oil product, petroleum ~
~feld n, Ölfeld, Petroleumfeld = oil field

~fernleitung f, Ölfernleitung = petroleum pipeline, oil ~
~förderkolben m = oil well swab
~förderpumpe f = oil well (plunger) pump
~förderschieber m = valve for oil production
~fördertechnik f = petroleum production engineering
~förderung f, Erdölproduktion f = oil production, ~ recovery, petroleum ~
~förderung f mittels unterirdischer Verbrennung, Erdölproduktion f ~ ~ ~ = oil production by underground combustion, petroleum ~ ~ ~
~fraktion f = petroleum fraction, oil ~
(erd)öl|führende Schicht f → Erdölträger m
~führender geologischer Horizont m, Ölhorizont, Erdölhorizont = oil horizon
~führendes Gebiet n [in diesem Gebiet ist die Ölführung nachgewiesen] = proved acreage, ~ oil land, ~ area
~fündige Bohrung f → Erdölbohrung
Erdöl|geologe m = rock hound
~geologie f = petroleum geology, oil ~
~grube f, Erdölbergwerk n = oil mine
~hafen m, Ölhafen = oil port, petroleum ~
erdölhöffiges Gebiet n, ölhöffiges ~ [ein Gebiet in dem Ölführung möglich ist oder ein Gebiet, das eine möglicherweise ölführende Struktur überlagert] = prospective acreage, ~ oil land, area overlying a possibly oil-bearing structure
Erdöl|horizont m, Ölhorizont, erdölführender geologischer Horizont = oil horizon
~lagerstätte f, Erdölvorkommen n, Erdölvorkommnis n, Erdölfundstätte, Erdöllager n = oil reservoir
(~)Lagerstättendruck m = (oil) reservoir pressure

(~)Lagerstättentechnik *f* = (oil) reservoir engineering
~lagerstättenwasser *n* = oil pool waters
~muttergestein *n*, Muttergestein = source rock, mother ~, source bed
~produktion *f* → Erdölförderung *f*
~raffinerie *f*, Erdölwerk *n* = petroleum refinery, oil ~
~sand *m*, Ölsand = "pay" (sand), oil ~
~sonde *f* → Erdölbohrung *f*
~suche *f*, Ölsuche, (Erd)Öl-Lagerstätten-Forschung *f* = petroleum prospecting
~träger *m*, Ölträger, (erd)ölführende Schicht *f* = petroliferous bed, oil-bearing stratum
~verladebrücke *f*, Ölverladebrücke = oil jetty, petroleum ~
~wanderung *f*, (Erd)Ölmigration *f*, Ölwanderung = oil migration, petroleum ~
~werk *n*, Erdölraffinerie *f* = oil refinery, petroleum ~
~zement *m*, Ölbohrzement, Tiefbohrzement = oil well cement
Erd|pfeiler *m*, Erdpyramide *f* = hoodoo, earth pillar
~-Photogrammetrie *f* → Erdbildmessung *f*
(Erd)Planum *n* → Planum
~automat *m* = automatically controlled (subgrade) trimmer, ~ ~ (~) spreader
~beschaffenheit *f* = subgrade condition (US); formation ~ (Brit.)
~breite *f* → Planumbreite
~fertiger *m* = concrete bay subgrader (Brit.); (sub)grade planer, (sub-)grading machine (US); (power) finegrader, (precision) subgrader, mechanical fine-grader, formgrader, subgrade trimmer
 Schneidmesser *n*, Hobelmesser = cutter bar
 Kratzerkette *f* = transverse conveyor, ~ conveyer
 Abgleichbohle *f* = strike-off (screed)

~genauigkeit *f* = subgrade accuracy (US); formation ~ (Brit.)
~herstellung *f* = subgrading, subgrade preparation (US); formation work (Brit.)
~höhe *f* = subgrade elevation
~unebenheit *f* = subgrade irregularity (US); formation ~ (Brit.)
Erd|polster *n* [*Österreich: m*] → Erdbett *n*
~prüfgerät *n* → Bodenprüfgerät
~pyramide *f*, Erdpfeiler *m* = earth pillar, hoodoo
~rampe *f* = soil ramp, earth ~
~reich *n*, Untergrund *m* = (natural) ground, earth formation(s)
~rinde *f*, Erdkruste *f*, Lithosphäre *f* = lithosphere, crust of the earth
 äußerer Gesteinsmantel *m* = outer rocky earth shell
 Sima *n* = sima
 S(i)al *n* = sial
~rutsch *m* [*in kleinem Umfang*] = slip, earth ~
(~)Rutschung *f*, Absetzung, Erdrutsch *m* = (land)slide [*earth slip upon a large scale*]
~sauganlage *f* zu Lande, landgängige Erdsauganlage = dry-land suction dredging plant
~schieber *m*, Verfüllschild *n* [*Anbaugerät*] = backfilling blade, planing ~, backfiller ~
~scholle *f* → Bodenscholle
~schürftransport *m*, Bodenschürftransport = (bull)dozing
~schüttung *f*, Erdaufschüttung, Erdauftrag *m*, Boden(auf)schüttung, Bodenauftrag, Erdanschüttung, Erdauffüllung, Bodenaufschüttung, Bodenauffüllung = earth fill(ing)
~schüttungsdamm *m* → Erddamm
~schutzwall *m* = earth traverse
~(-See)deich *m* → Erddamm *m*
~staudamm *m* → Erddamm
~stoff *m* → Boden *m*
~stollen *m* [*Gleiskette*] = dirt grouser
~stoß *m* = earth shock
~straße *f* = earth (surface) road (Brit.); dirt ~ (US); soil ~

~strom *m*, Streustrom, vagabundierender Strom = current from external source, stray current, leakage current
~stütz(bau)werk *n* = soil-supporting structure, earth-supporting ~, soil-retaining ~, earth-retaining ~
~tank *m*, Erdbehälter *m* = underground storage tank
~teilchen *n* → Erdkorn *n*
~transport *m*, Bodentransport = soil transport; ~ transportation (US)
~transportfahrzeug *n* → Erdbaufahrzeug
~transportfahrzeug *n* für Fremdbeladung → Muldenkipper *m*
~transportwagen *m* → Erdbaufahrzeug *n*
~überdeckung *f*, Bodenüberdeckung [*Rohrgraben*] = soil cover, earth ~
~- und Grundbau *m* = soil (and foundation) engineering
Erdung *f* → Erden *n*
Erdungsmaterial *n* = earthing material
erdverankert, bodenverankert = anchored in the ground

Erd|verbindung *f* → Erden *n*
~verdrängung *f*, Bodenverdrängung = earth displacement, soil ~
~verlagerung *f* → Bodenbewegung
erdverlegte Leitung *f* → Erdleitung
erdverlegtes Kabel *n* → Bodenkabel
~ Rohr *n*, eingeerdetes ~, Erdrohr, Bodenrohr, bodenverlegtes Rohr = buried pipe
Erd|wachs *n* → Ozokerit *m*
~wärme *f* = terrestrial heat, ground ~
~weg *m* = earth path (Brit.); dirt ~ (US); soil ~
~widerstand *m*, passiver Erddruck *m* = passive earth (or soil) pressure
~widerstandangriffpunkt *m*, Angriffpunkt des Erdwiderstandes = point of application of passive earth pressure, ~ ~ ~ ~ soil ~
~widerstandbeiwert *m* = coefficient of passive earth pressure, ~ ~ ~ soil ~
~widerstandspannung *f* = passive earth pressure stress, ~ soil ~ ~
~winde *f* → Affe *m*
~zement *m* → Zementverfestigung *f*

Erfahrung *f*
praktische ~
Erfahrungsschatz *m*
Erfahrungsformel *f*, empirische Formel
Erfahrungsziffer *f*, Erfahrungswert *m*

experience
actual ~, practical ~
accumulated and recorded ~
empirical formula

empirical value

---

erforderliche Festigkeit *f* → geforderte ~
ergänzte Einflußlinie *f* = complementary influence line
Ergänzungswinkel *m*, Komplementwinkel = complementary angle
ergebnislose Bohrung *f* → Fehlbohrung
Ergiebigkeit *f* = yield
Ergußgestein *n*, vulkanisches Gestein, Durchbruchgestein, Effusivgestein, Ausbruchgestein, Extrusivgestein, Oberflächen(erguß)gestein, suprakrustales Erstarrungsgestein, Vulkanit *m* [*frühere Bezeichnung: jüngeres Eruptivgestein*] = lava flow, extrusive rock, (igneous) volcanic rock [*cooled on the Earth's surface*]
erhärten [*Putz(mörtel)*] = to set hard
~ = to harden
erhärteter Beton *m*, Festbeton = hardened concrete
~ Mörtel *m*, Festmörtel = hardened mortar
~ Zementleim *m*, Zementstein *m* = hardened cement paste

Erhärtung — Erosionsneigung

Erhärtung *f*, **Erhärten** *n* = hardening
~ **des Betons** *m* **ohne Nachbehandlung durch Feuchthalten** = self-curing
Erhärtungs|alter *n* = age of hardening
~**beschleuniger** *m* = accelerator for hardening, hardening accelerator
~**beschleunigung** *f* **durch Hitze** = acceleration of hardening by heat
~**energie** *f* = energy of hardening
~**fortschritt** *m* = progress of hardening
~**geschwindigkeit** *f* = hardening rate
~**kurve** *f* = rate of hardening curve
~**mechanismus** *m* = mechanism of hardening
~**probe** *f* → Erhärtungsprüfung *f*
~**prüfung** *f*, Erhärtungsprobe *f*, Erhärtungsversuch *m* = hardening test
~**reaktion** *f* = reaction of hardening
~**schwindung** *f*, Trocknungs-Schrumpfung *f* = drying shrinkage
~**spannung** *f* = hardening stress
~**temperatur** *f* = hardening temperature
~**theorie** *f* = theory of hardening
~**vermögen** *n* = hardening characteristic
~**versuch** *m* → Erhärtungsprüfung *f*
~**zeit** *f* = hardening time, ~ period
erhellen = to brighten
Erhitzer *m*, **Vorwärmer** *m* = (pre-)heater
~**schlange** *f*, Heizschlange, Wärmeschlange = heating coil
Erhitzungsgeschwindigkeit *f* [*Kalkbrennen*] = heating rate
Erhöhen *n* = heightening
erhöhte Ausweichzahlen *fpl* **von Jäger** = Jäger's increased deflection values
~ **Kübelseitenwand** *f* [*Radschrapper*] = bowl side wall extension
Erholungsgebiet *n* = recreation area
erkaltete Hochofenschlacke *f* = air-cooled blast-furnace slag
erkennbar = indicated [*resources*]
Erker(fenster) *m*, (*f*) = bay window
Erkundung *f* = reconnaissance

Erkundungs|bohrer *m* = hand auger
~**bohrung** *f* → Aufschließungsbohrung
~**schacht** *m* → Aufschließungsschacht
~**vorhaben** *n* → Aufschließungsvorhaben
Erläuterungsbericht *m* = explanatory report
Erlenmeyer-Kolben *m* = Erlenmeyer flask
Ermitt(e)lung *f* **eines Stromliniennetzes, Bestimmung** ~ ~ = construction of flow net
Ermüdung *f* = fatigue
Ermüdungs|beiwert *m* = coefficient of fatigue
~**beständigkeit** *f* = fatigue resistance
~**bruch** *m* = fatigue fracture
~**festigkeit** *f*, Dauerfestigkeit, Arbeitsfestigkeit *f* = fatigue strength
~**grad** *m* = fatigue degree
~**grenze** *f* = fatigue (endurance) limit, endurance ~
~**probe** *f* → Ermüdungsversuch *m*
~**prüfung** *f* → Ermüdungsversuch *m*
~**riß** *m* = fatigue crack
~**verhalten** *n* = fatigue behaviour (Brit.); ~ behavior (US)
~**versuch** *m*, Ermüdungsprüfung *f*, Ermüdungsprobe *f* = fatigue test
Erneuerung *f*, (Altstadt)Sanierung = clearance, rehabilitation
Erneuerungsgebiet *n*, Sanierungsgebiet [*Städtebau*] = rehabilitation zone, clearance ~
erodieren, ausnagen = to erode
erodierende Tätigkeit *f* → Erosionstätigkeit
Eröffnungssitzung *f* = opening session
Erosion *f* → → Denudation *f*
~ **im engeren Sinne, fluviatile Erosion, Flußerosion, Stromerosion** = fluviatile erosion
Erosions|anfälligkeit *f* → Erosionsneigung *f*
~**grenzgeschwindigkeit** *f* = critical velocity of sediment movement
~**kraft** *f* = erosive power
~**neigung** *f*, Erosionsanfälligkeit *f* = erodibility [*Is not an inherent prop-*

erty of any mature and fertile soil; it is a property induced, most commonly, by human interference]
~**tätigkeit** *f*, erodierende Tätigkeit = erosional action, ~ activity, ~ work, erosive ~, work of erosion
~**tiefe** *f* = depth of erosion
~**topf** *m* → Gletschermühle *f*
erratischer Block *m* → Findling(stein)
~ **Sandsteinblock** *m* = graywether, sarsen stone, Saracen's stone
errechnete Aufgabe *f*, ~ Beschickung *f* = calculated feed(ing)
erregen [*z.B. Vibrationswalze*] = to energize
Erreger|maschine *f* = exciter
~**stoff** *m* [*z.B. Kalkhydrat*] = activator
erreichte Festigkeit *f*, vorhandene ~ = actual strength
errichten, bauen = to build, to construct
Errichtung *f*, Bau *m* [*z.B. einer Talsperre*] = building, construction
Ersatz|kraft *f* → Mittelkraft *f*
~**mittel** *n* = substitute
~**rad** *n* = spare wheel
~**stab** *m* [*Baustatik*] = spare member
Ersatzteil *m*, *n* = replacement part, spare (~), plant spare
~**bestand** *m* = (spare) parts inventory
~**dienst** *m* = (spare) parts service
~**haltung** *f* = keeping (spare) parts
~**lieferant** *m* = (spare) parts supplier
~**liste** *f* = (spare) parts manual
~**sortiment** *n* = (spare) parts collection
erschließbar [*Gebiet; Baugelände*] = developable, capable of being opened up
erschließen [*Gebiet; Baugelände*] = to open up, to develop
Erschließung *f* [*Gebiet; Baugelände*] = development
~, Aufschluß *m* [*Bodenschätze*] = winning [*The act of building shafts, levels and passages to attain seams*]
Erschließungs|beitrag *m* → Anliegerbeitrag
~**kosten** *f* = development cost

~**sonde** *f*, Entdeckungsbohrung *f*, Fundbohrung = discovery well
~**stollen** *m* = exploratory drift
~**straße** *f* = development road
erschlossenes Bauland *n*, ~ Baugelände *n* = developed ground
erschmelzen [*Erz*] = to (s)melt
erschöpfte Bohrung *f*, ~ Sonde *f*; erschöpfter Schacht *m* [*in Galizien*] = exhausted well, depleted ~, ~ bore, ~ boring
~ Sonde *f* → ~ Bohrung *f*
erschöpfter Schacht *m* [*in Galizien*] → → erschöpfte Bohrung *f*
Erschütterungen *fpl* durch Verkehr = vibrations due to traffic
Erschütterungs|messung *f* = vibration measurement
~**schutz** *m* = protection against vibrations
~**stelle** *f* [*geophysikalisches (Aufschluß) Verfahren*] = shot point
~**welle** *f* [*geophysikalisches (Aufschluß) Verfahren*] = seismic wave
Erschwernis *f* = hardship
ersoffenes Bohrloch *n*, ersoffene Bohrung *f* = drowned well
erstarren, abbinden [*hydraulisches Bindemittel*] = to set
Erstarren *n* → Abbindung *f*
Erstarrung *f* → Abbindung
~, Erstarren *n* [*Magma*] = solidification
Erstarrungs|beginn *m* → Abbindebeginn
~**beschleuniger** *m* → Abbindebeschleuniger
~**beschleunigung** *f* → Abbindebeschleunigung
~**dauer** *f* → Abbindezeit *f*
~**ende** *n* → Abbindeende
~**festigkeit** *f*, Abbindefestigkeit = setting strength
~**geschwindigkeit** *f* → Abbindegeschwindigkeit
~**gestein** *n*, Massengestein, Glutflußgestein, Magmagestein = igneous rock, primary ~
~**kraft** *f* → Abbindekraft
~**kurve** *f* → Abbindekurve

Erstarrungsmechanismus — Erweichungspunkt

~mechanismus *m* → Abbindemechanismus
~niveau *n* [*Magma*] = level of solidification
~probe *f* → Abbindeprüfung *f*
~prozeß *m* → Abbindevorgang *m*
~prüfgerät *n* → Abbindeprüfgerät
~prüfung *f* → Abbindeprüfung
~reaktion *f* → Abbindereaktion
~regler *m* → Abbinderegler
~schwindung *f* → Abbindungsschwindung
~temperatur *f* → Abbindetemperatur
~verhalten *n* → Abbindeverhalten
~verlauf *m* → Abbindevorgang
~vermögen *n* → Abbindevermögen
~versuch *m* → Abbindeprüfung *f*
~verzögerer *m* → Abbindeverzögerer
~vorgang *m* (Geol.) = magmatic differentiation
~vorgang *m* → Abbindevorgang
~wärme *f* → Abbindewärme
~wasser *n*, Abbindewasser [*hydraulisches Bindemittel*] = water of set(ting)
~wert *m* → Abbindewert
~zeit *f* → Abbindezeit
~(zeit)beschleuniger *m* → Abbindebeschleuniger
~(zeit)beschleunigung *f* → Abbindebeschleunigung
~(zeit)regler *m* → Abbinderegler
Erst|belastung *f* = initial loading
~dampf *m*, Primärdampf = primary steam
erste fündige Bohrung *f*, ~ ~ Sonde *f*; erster fündiger Schacht *m* [*in Galizien*]; Fundbohrung = discovery well, ~ bore, ~ boring
~ Hilfe *f* = first aid to the injured
~ Rohrfahrt *f* → Leitrohrtour *f*
erster Bauabschnitt *m*, erste Ausbaustufe *f* = first-stage development, ~ construction
~ fündiger Schacht *m* [*in Galizien*] → erste fündige Bohrung
~ Gang *m* = first gear
~ Setzdruck *m*, ~ Hauptdruck, ~ Periodendruck (⚒) = first weight, ~ squeeze, ~ crush, ~ pressure

„Erster Spatenstich" *m* = ground breaking ceremony
Erstreckung *f* des Ölfilms = oil film extent
Erst|sprengung *f* = primary (drilling and) blasting
~wanderung *f*, Erstmigration *f* [*Erdöl*] = primary migration
ertrunkener Fluß *m* = drowned stream
Eruption *f* [*Bohrloch*] → Ausbruch *m*
~, Ausbruch *m* [*Vulkan*] = eruption, emission of lava
Eruptions|düse *f* [*Tiefbohrtechnik*] = flow bean, ~ nipple, ~ plug
~kanal *m* → Durchschlagröhre *f*
~kopfdüse *f* [*Tiefbohrtechnik*] = top-hole choke
~krater *m* einer Bohrung, Ausbruchkrater ~ ~ = well crater
~kreuz *n*, Produktionskreuz = christmas tree
~stopfbüchse *f*, Ausbruchverhüter *m*, Ausbruchschieber *m* = blow-out preventer
Eruptiv|gestein *n* [*umfaßt die Tiefengesteine (früher: ältere Eruptivgesteine) und die Ergußgesteine (früher: jüngere Eruptivgesteine)*] = eruptive rock
~gesteinstafel *f* = eruptive sheet
~sonde *f* → freiausfließende Bohrung *f*
~stock *m* = eruptive stock
erwärmen, aufheizen, erhitzen [*Bindemittel; Wasser; Zuschläge und dgl.*] = to heat
Erwärmung *f*, Aufheizung, Erhitzung = heating
Erweichung *f* → Aufweichung
~ der Kristallmitte *f* [*Natronkalkfeldspat*] = kernel-decomposition [*plagioclase*]
Erweichungspunkt *m* [*diejenige Temperatur, bei der das Glas etwa eine Viskosität von 10^7 bis 10^8 Poisen hat*] = Littleton-point, softening point
~ [*Segerkegel*] = squatting temperature
~ = softening point
~ R. und K., ~ Ring und Kugel = ring and ball softening point

erweitern, ausbauen = to develop
~ [*Bohrloch*] = to enlarge
Erweiterung *f*, Ausbau *m* = development
~ [*Bohrloch*] = enlarging
Erweiterungs|arbeiten *fpl*, Ausbauarbeiten = development work
~**bau** *m*, (Gebäude)Anbau = annex
~**bohrer** *m* [*Schachtbohrverfahren nach Honigmann*] = trepan
~**meißel** *m*, Erweiterungskrone *f*, Erweiterungsbohrer, Nachschneidemeißel, Aufbohrmeißel = enlarging bit
~**rollenmeißel** *m*, Aufbohrrollenmeißel = enlarging roller bit
~**verlust** *m* [*Hydraulik*] = diffuser loss
Erz *n* = ore [*Is a metal-bearing mineral, or a mix(ture) of such minerals with other substances, so situated that it is profitable to work*]
~**abbau** *m* = ore mining
~**abbau-Tagebetrieb** *n* = open-cast ore mining
~**aufbereitung** *f*, Aufbereitung von Erzen = mineral dressing
~**aufbereitungsanlage** *f*, Aufbereitungsanlage für Erze = mineral dressing plant, ~ ~ installation
~**aufbereitungsmaschine** *f*, Aufbereitungsmaschine für Erze = mineral dressing machine
~**behälter** *m*, Erzsilo *m*, Erzbunker *m* = ore bin
~**bergwerk** *n*, Erzgrube *f* = ore mine
~**brecher** *m* = ore crusher, ~ breaker
~**bunker** *m*, Erzsilo *m*, Erzbehälter *m* = ore bunker
~**entlader** *m* = ore unloader
erzeugen, herstellen, aufbereiten [*Beton; Mörtel; Schwarzdeckenmischgut*] = to fabricate, to prepare [*sometimes shortly, but incorrectly, called "to mix"*]
~, entwickeln [*Azetylen*] = to generate, to produce
Erzeugende *f*, erzeugende Linie *f* = generatrice

Erzeuger *m*, Fabrikant *m*, Hersteller *m* = manufacturer, fabricator
~**werk** *n* → Lieferwerk
Erzeugung *f*, Herstellung, Aufbereitung [*Schwarzdeckenmischgut; Beton; Mörtel*] = fabrication, preparation [*sometimes shortly, but incorrectly, called "mixing"*]
Erz|fall *m* = ore shoot
~**feste** *f* = ore barrier pillar
~**formation** *f* = ore formation
~**frachter** *m*, Erzschiff *n* = ore carrier
erzfreies Gestein *n* → Berg *m*
Erz|fundstätte *f* → Erzvorkommen *n*
~**gewinnung** *f* = ore getting
~**hafen** *m* = ore harbour (Brit.); ~ harbor (US)
~**halde** *f* = mass(es) of gangue, ~ ~ dead ores
~**klauber** *m* = ore picker
~**korn** *n* = ore grain, ~ particle
~**körper** *m*, Erzstock *m* = ore-body
~**lagerbehälter** *m* = ore-storage reservoir
~**lager(platz)** *n*, (*m*) = ore (storage) yard
~**lagerstätte** *f*, Erzvorkommen *n*, Erzlager *n*, Erzfundstätte, Erzvorkommnis *n*, Erzfeld *n* = ore deposit, mineral ~
~**lineal** *n* = elongated lens, pod
~**magma** *n* = ore magma
~~**Naßmühle** *f* = ore wet (grinding) mill
~**pellet** *n* = ore pellet
~**pfeiler** *m* = ore pillar
~**schlacke** *f* = ore slag
~**schlauch** *m* = ore-pipe, ore-chimney
~**schnur** *f* = stringer, belt of ore
~**schurre** *f* = ore chute
~**silo** *m* → Erzbehälter *m*
~**suche** *f* = ore-search
~**tagebau** *m* = ore open pit mining (US); ~ open-cast ~, ~ surface ~, ~ open-cut ~, ~ adit ~ (Brit.)
~**trübe** *f* = ore pulp, mineral ~
~**trum** *n* = stringer-head
~**umschlag** *m*, Erzverladung *f* = ore handling

~umschlaganlage *f*, Erzverladeanlage = ore handling plant, ~ ~ installation

~verhüttung *f* = metallurgical working of ores

~verladebrücke *f*, Erzumschlagbrücke, Erztransportbrücke, Erzlagerplatzbrücke = ore bridge

(~)Wäsche *f*, (Erz)Waschwerk *n* = mineral washer

~weg *m* = ore channel

erzwungene Schwingung *f*, aufgezwungene ~, Kraftschwingung = forced vibration

Eselsrücken *m* → Ablaufanlage *f*

Eskaladierfalte *f* (Geol.) = overlapping fold

Espagnoletteverschluß *m* = espagnolette bolt, cremorne ~

Esse *f* → Durchschlagröhre *f*

~ → Schlot

Esskohle *f* = steam coal

Estobitumen *n* = esto-bitumen, esto-asphalt

estländischer Brennschiefer *m*, Kukersit *m* = kukersit, Esthonian (oil) shale

Estrich *m* = floor topping, (~) screed, topping finish

~arbeiter *m* = composition floor layer

~gemisch *n*, Estrichmischung *f* = topping mix(ture)

~gips *m*, Anhydrit I *m* [*Hochgebrannter Gips (950° C); langsamabbindend, zementartig erhärtend. DIN 1168*] = anhydrite (gypsum) plaster [*B.S. 1191*]

~-Glättmaschine *f* = trowel(l)ing machine, mechanical trowel, power trowel, rotary finisher

~mörtel *m* = (floor) screed mortar

~zusatz *m* **für zementgebundene schwimmende Estriche** = additive for cement-bound floating screeds

Eßfisch *m* = food fish

Eßküche *f* = dining kitchen, kitchen dining room

Etage *f*, Stockwerk *n*, Geschoß *n*, Stock *m* = stor(e)y [*The space between two floors or between a floor and a roof*]

~ [*in einem Nichols-(Etagen)Ofen*] = hearth

Etagen|bau *m* → Etagenhaus *n*

~brücke *f*, zweigeschossige Brücke = double-deck bridge

~deckenteil *m, n* → Hochbaudeckenteil

~haus *n*, Geschoßhaus, Stockwerkhaus, Etagenbau *m*, Geschoßbau, Stockwerkbau = multi-stor(e)y building

~heizung *f*, Stockwerkheizung, Geschoßheizung = single-stor(e)y heating system

~höhe *f* → Geschoßhöhe

~trockenofen *m* [*führt aber auch die Verbrennung des Schlammes zu Asche durch*] = multiple-hearth sludge incinerator

~wohnung *f*, Geschoßwohnung, Stockwerkwohnung = flat [*A self-contained dwelling on one storey in a larger building*]

Etikettierung *f* = label(l)ing

EUCLID-Lademaschine *f*, Förderband-Anhängeschürfwagen *m*, Schürfwagen mit Förderbandausleger, EUCLID-Lader *m*, Fabrikat THE EUCLID ROAD MACHINERY CO., CLEVELAND 17, OHIO, USA

Böschungsarbeiten *fpf*
Förderbahnleisten *fpl*
Förderband *n*

EUCLID LOADER [*Trademark*]

slope work
skirts
belt

gelenkige Brückenkonstruktion *f*, Pivotverbindung *f*	pivot connection
Kopfrolle *f*, obere Umlenktrommel *f*	head pulley
Lader *m*	conveyor, conveyer
Pflugschar *f*	mold board plow
Schnittiefe *f*, Schnitt-Tiefe	depth of cut
Schnittwinkel *m*	angle of cut
Schürfschneide *f*	cutting edge, ~ blade
Schwanzrolle *f*, untere Umlenktrommel *f*	tail pulley
Traktorführer *m*, Schlepperfahrer *m*	tractor operator
Zugraupe *f*, (Gleis)Kettenschlepper *m*	puller (crawler) tractor, towing caterpillar tractor

---

Eu-durit *m* = dull clarain
Euler'sche Knickspannung *f* = Euler crippling stress
eupelagische Meeresablagerung *f*, Tiefseeschlamm *m* = pelagic deposit, deep-sea ~, deep-sea ooze, ozeanic deposit, thalassic deposit
eutropher See *m* = eutroph lake
Evolute *f* = evolute
Evolutionstheorie *f* = theory of evolution
Evolvente *f* = involute
Evolventen|pumpe *f* = involute pump
~**rad** *n* = involute gear
~**verzahnung** *f* = involute toothing
Evorsionsbecken *n* = evorsion basin, basin due to evorsion
ewiggefrorener Boden *m* → dauernd gefrorener ~
Exakt-Bagger *m* [*hat einen zweifach geknickten Ausleger und dadurch eine exakte Löffelführung*] = precision-type excavator
Exaration *f* = glacial denudation, ~ erosion
Exerzierplatz *m* = parade ground
Exhaustor *m*, Absauger *m* = exhaust(ing) fan, extract ~, induced-draught ~
exotherme Reaktion *f* = exothermic reaction [*Chemical reaction in which heat is given off after the action commences. Examples: hydration of ce-*
ment, and clinkering in the burning zone in kilns]
expandierter Korkschrot *m* = expanded granulated cork
Expansions|bandbremse *f* = expanding band brake
~**gefäß** *n*, Ausdehnungsgefäß = expansion vessel
~**hub** *m* → Arbeitshub
Expansiv|beton *m*, Schwellbeton, Quellbeton = self-stressed concrete
~**zement** *m*, Quellzement, Schwellzement = self-stressing cement, high-expansion ~, expanding ~, expansive ~
Experimentalverfahren *n* → Versuchsmethode *f*
experimenteller Fehler *m* = experimental error
Exploration *f* = exploration
Explorationsseismik *f* = exploration seismology
explosibles Gemisch *n*, explosible Mischung *f* = explosive mix(ture)
Explosions|caldera *m* = caldera of subsidence
~**druckbelastung** *f* = (internal) blast loading
~**fähigkeit** *f* = explosibility, explosiveness
~**graben** *m* (Geol.) = explosion fissure, ~ trench
~**ramme** *f* → Brennkraft-Handramme

~raum *m* = explosion chamber
~röhre *f*, Explosionsöffnung *f* (Geol.) = (explosion) pipe, (~) vent
~stampfer *m*, Hand-~ = heavy-duty internal combustion tamper
Explosiv|stoff *m* = explosive
~verbrennung *f* = combustion by explosion
Exponential|formel *f* = exponential formula
~gesetz *n* **der Kornverteilung nach Rosin-Rammler-Sperling**, Kornverteilungsgesetz = law of particle size distribution according to R. R. S.
Expositionsbedingung *f* = exposure condition
Expreßstraße *f* → innerstädtische Autobahn
Exsikkator *m*, Trockenapparat *m* = desiccator
Extraktions|apparat *m*, Extraktor *m* = extractor
~kolben *m* = extraction flask
~methode *f* **nach Abson** = Abson recovery method
Extruder-Schnecke *f* = extruder-screw, screw-extruder
Extrusiv|gestein *n* → Ergußgestein
~masse *f* = extrusive (mass)
Exzenter *m* = eccentric
~anlauf *m* = eccentric catch
~antrieb *m* = eccentric drive
~bohrkrone *f* = eccentric (drill) bit
~hebel *m* = eccentric lever
~meißel *m* = eccentric bit
~presse *f* = eccentric press
~ring *m* = eccentric strap
~stange *f* = eccentric arm, ~ rod
~welle *f* = eccentric shaft
exzentrisch → ausmittig
Exzentrizität *f*, Ausmitte *f*, Ausmittigkeit *f* = eccentricity

F

Fabrikat *n* = make
Fabrikations|anlagen *fpl*, Fabrikationsbetrieb *m* = manufacturing facilities, ~ plant, ~ works, ~ installation(s)
~forschung *f* = manufacturing research
Fabrik|ausrüstung *f*, Werkausrüstung = factory equipment, plant ~, works ~
~boden *m* → Betriebsboden
~esse *f* → Fabrikkamin *m*
~fenster *n* = factory window
~(fuß)boden *m* → Betriebsboden
~gebäude *n*, Werkgebäude = factory building
~gelände *n*, Werkgelände = factory ground
~halle *f*, Industriehalle, Werkhalle = factory hangar, ~ shed
~kamin *m*, Fabrikschornstein *m*, Fabrikesse *f*, Fabrikschlot *m* = factory chimney, ~ stack
fabrikmäßig hergestellter Fertigteil *m*, ~ hergestelltes ~ *n* = factory-produced precast element
fabrikmäßige Vorfertigung *f* = precast factory manufacture
Facettengeschiebe *n* → Pyramidalgeröll(e) *n*
Fach *n* = (infilling) panel
~arbeiter *m*, Spezialarbeiter = skilled worker
~dolmetscher *m*, technischer ~ = technical interpreter
Fächer|-Ankerzone *f* [*Spannbeton*] = anchorage zone of fanned-out wires
~brücke *f* = radiating bridge
~einbruch *m*, italienischer Einbruch = fan cut
fächerförmig spleißen = to fan out
Fächer|gewölbe *n*, Trichtergewölbe, angelsächsisches Gewölbe, normannisches Gewölbe, Palmengewölbe, Strahlengewölbe = fan vaulting
~-Lichtstrahl *m* = fan-shaped beam
~pflaster(decke) *n*, (*f*), Kleinpflaster(decke) in Fächerform = fanwise paving, circular ~, DURAX ~, radial sett ~, random ~
~verankerung *f* = fan anchorage
fachgerechte Ausführung *f* = expert workmanship
Fach|literatur *f* = professional literature
~mann *m* = expert

Fachsimpelei — Fahrbahnüberflutung durch Regen

~simpelei *f* = shop talk
fachsimpeln = to talk shop
Fach|übersetzer *m*, **technischer ~** = technical translator
Fachwerk *n*, **Skelett** *n* = (structural) framework, skeleton, framed structure
~**balken(träger)** *m* = trussed beam
~**bildungsweise** *f* = arrangement of framework
~**binder** *m* [*siehe Anmerkung unter „Binder"*] = roof truss
~**bogen(träger)** *m* **mit parallelen Gurt(ung)en**, Bogenfachwerk(träger) *n*, (*m*) ~ ~ ~ = trussed arch(ed) girder) with parallel booms, braced ~~~~~, ~~~~~ chords, ~ ~ ~ ~ ~ flanges
~**-Bogen(träger)brücke** *f*, Bogenfachwerk(träger)brücke *f* = trussed arch(ed girder) bridge, braced ~ ~ ~
~**gurt** *m* = chord of a truss, boom ~ ~, flange ~ ~ ~
~**hängebrücke** *f* = truss-stiffened (suspension) bridge
~**kragarm** *m* = truss(ed) cantilever arm
~**pfette** *f* = trussed purlin
~**rahmen** *m* = trussed frame
~**rostsystem** *n* = two-way truss system
~**schalung** *f* = box-truss formwork
~**sichelbogen(träger)** *m* = sickle-shaped trussed arch(ed girder)
~**träger** *m* = truss
~**trägerbrücke** *f* = truss bridge
~**trägerstab** *m* = truss member
~**träger-Trogbrücke** *f* = trough truss bridge, open ~ ~
~**versteifungsträger** *m* = stiffening truss
~**wand** *f* = framed partition
~**wand** *f*, Riegelwand = framed wall
Fachzeitschrift *f* = trade magazine, ~ journal
Fackel|gas *n* = flare gas
~**turm** *m* = flare stack
Fagergren-Zelle *f* = Fagergren flotation machine
Fahlerz *n* → Graugültigerz

Fahrantrieb *m* = travel drive
~**getriebekasten** *m* = drive gearbox
Fahr-Arbeitsplattform *f*, **Fahrbühne** *f* = travelling stage
Fahrbahn *f* [*Straße*] = carriageway [*deprecated: road, road-way, road surface*]
(~)**Befestigung** *f* [*Die Gesamtheit aller Schichten*] = (layered) system
(~)**Befestigung** *f* [*Wenn nur eine Schicht vorhanden ist*] = surfacing (Brit.); pavement, paving (US)
(~)**Befestigung** *f* [*allgemeiner Ausdruck ohne Rücksicht auf die Anzahl der Schichten und ohne Einschränkung auf nur die „Decke"*] = pavement, surfacing; paving (US)
~(**be)heizung** *f* **auf Brücken** = bridge heating
~**belag** *m* → Fahrbahndecke *f*
~**breite** *f* = carriageway width
~**decke** *f*, Fahrbahnbelag *m* = carriageway surfacing, ~ pavement; ~ paving (US)
~**deckenzement** *m* = carriageway cement
~**markierung** *f* = carriageway marking
~**markierungen** *fpl* **aus plastischer weißer Masse** = plastic white line markings
~**markierungsmaschine** *f* [*in der Schweiz: Bodenmarkierungsmaschine*] = carriageway marking machine, pavement ~ ~
~**matte** *f* → Baggerrost *m*
~**platte** *f* [*allgemein*] = carriageway slab
~**platte** *f* → Baggerrost *m*
~**randmarkierung** *f* = carriageway edge marking
~**reibung** *f* = road friction
~**rost** *m* [*Brücke*] = floor system; ~ grid (US)
~**rückstrahler** *m* → Bodenrückstrahler
~**überflutung** *f* **durch Regen** = carriageway flooding by rain

Fahrbahnübergang — Fahrer

~übergang m [*Brücke*] = movable joint, expansion ~
~unterbau m = carriageway foundation
~unterhaltung f = carriageway maintenance
Fahrband n, fahrbarer Gurtförderer m, fahrbares Förderband = elevating belt conveyer, ~ ~ conveyer, belt elevator
fahrbar, ortsbeweglich = in portable form, mobile, portable
fahrbare Abfüllwaage f → ~ (Hänge-)Waage
~ **Anlage** f = portable plant, ~ installation
~ **Anlage** f **für Beton(auf)bereitung**, Mobilanlage = low-profile (central) plant, ~ batching and mixing ~
~ **Behälterwaage** f → (Hänge)Waage
~ **Bohrmaschine** f → Bohrwagen m
~ **Bohr- und Produktionswinde** f, fahrbares Bohrgerät n mit separatem Mast auf getrennten Fahrzeugen = (dual-type) rambler rig
~ **Chargen-Mischanlage** f **mit Doppeldecker-Horizontal-Vibratorsieb** = portable batch plant with gradation control, mixer-gradation unit
~ **Dosieranlage** f, Mobil-Dosieranlage = low-profile batching plant, ~ measuring ~, ~ apportioning ~, ~ gauging ~
~ **(Hänge)Waage** f, ~ Abfüllwaage, ~ Behälterwaage = travelling type suspended weighbatcher (Brit.); traveling ~ (US)
~ **Ölfeldwinde** f = well pulling machine
~ **Pumpe** f = portable pump
~ **Rohrfabrik** f → Rohrfabrik auf (Gleis)Kettenfahrzeugen
~ **Schüttgut-Schleuder** f → Schüttgut-Schleuder
~ **Stollenbetonieranlage** f, ~ Tunnelbetonieranlage, Betonierzug m = tunnel concreting train
~ **Tunnelbetonieranlage** f → ~ Stollenbetonieranlage
~ **Waage** f → ~ Hängewaage
~ **Wagenunterkunft** f = caravan

fahrbarer Bunker m → Fahrsilo m
~ **Bürgersteig** m, Rollbürgersteig = passenger conveyor, ~ conveyer, travolator
~ **Gurtförderer** m → Fahrband n
~ **Kabelkranturm** m, Kabelkran-Gegenturm = tail tower
~ **Roheisenmischer** m = mixer type hot metal car
~ **Silo** m → Fahrsilo
fahrbares Bohrgerät n → Bohrwagen
~ **Förderband** n → Fahrband
~ **Rohrwerk** n → Rohrfabrik auf (Gleis)Kettenfahrzeugen
Fahr|bequemlichkeit f [*Straße*] = riding comfort
~bereich m → Aktionshalbmesser
~brecher m, fahrbare Brechanlage f = portable crusher, ~ breaker
~bremse f [*im Gegensatz zur Lenkbremse bei einem Raupenschlepper*] = travel brake, traction ~, travelling~(Brit.); traveling~(US)
Fährbrücke f, Schwebefähre = aerial ferry, ferry bridge
Fahr|bühne f, Fahr-Arbeitsplattform f = travelling stage
~bunker m → Fahrsilo m
Fährdampfer m, Dampffähre f = steam ferry
(Fahr)Deichsel f, Zugdeichsel = towbar
Fährdienst m = ferry boat service
Fahr|draht m, Oberleitungsdraht, Fahrleitung f = contact wire, troll(e)y ~, (overhead) contact line
~-Drehmoment n = propelling torque
~drosseldruck m = throttle pressure
~drosselgruppe f = throttle pressure control group
~ebene f → Fahrfläche f
~eigenschaft f [*Straßendecke*] = riding characteristic, ~ quality
Fahren n, Ver~ [*Baumaschine; Laufkatze*] = travel(l)ing
Fährenstation f = ferry terminal
Fahrer m, Auto~, Kraft~ = (motor vehicle) driver; chauffeur (US)
~, Führer m [*Bediener eines fahrenden Gerätes*] = driver, operator

Fahrerbeanspruchung — Fahrlade-Gerät

~beanspruchung f, Fahrerermüdung, Führerbeanspruchung, Führerermüdung = driver fatigue, operator ~

~-Drehsitz m, Fahrer-Schwenksitz, Führer-Drehsitz, Führer-Schwenksitz = driver's swivel seat, operator's ~ ~

~ermüdung f → Fahrerbeanspruchung

~flucht f = "hit and run" offence

~haus n, Fahrerkabine f, Führerhaus, Führerkabine = cab(in), compartment, driver's ~, operator's ~

~haus-Überdachung f, Schutzkappe f für das Fahrerhaus, Fahrerhaus-Schutzdach n [*Erdbauwagen*] = full-width cab-protection plate

~haus-Warmluftheizer m = cab(in) heater, compartment ~

~kabine f → Fahrerhaus n

fahrerlos = driverless, operatorless

Fahrer|-Schwenksitz m → Fahrer-Drehsitz

~sicht f = driver vision, operator ~

~sitz m, Führersitz = driver's seat, operator's ~

~sitz m mit Schaumgummipolsterung = sponge rubber bucket-type seat

~stand m, Führerstand = control platform

Fahr|feld n (⚒) = travelling track, ~ way

~fläche f, Fahrebene f [*Fläche auf der sich ein Fahrzeug bewegt*] = riding surface

~fußhebel m, Gaspedal n, Fußgashebel = accelerator (pedal), gas ~, foot throttle, foot accelerator

Fährgast m = ferry passenger

Fahrgast m, Reisender m, Passagier m = passenger

~abfertigung f = handling of passengers, processing ~ ~

~anlegestelle f, Fahrgastanlegeplatz m, Fahrgastfestmacheplatz = passenger berth

~bahnhof m → Personenbahnhof

~fluß m = passenger flow

~hafen m → Personenhafen

~schiff n, Personenschiff, Passagierschiff = passenger ship

~transport m, Personentransport = transportation of passengers

~verkehr m, Personenverkehr, Passagierverkehr = passenger traffic

Fahr|gelderhebung f = fare collection

Fahrgeschwindigkeit f = drive speed, travel(l)ing ~

~ [*Straßenverkehrstechnik*] = journey speed, trip ~

~ **auf der Straße**, Straßengeschwindigkeit = road speed, highway ~

~ **im Betrieb**, Betriebsgeschwindigkeit [*Bagger; Lader usw.*] = travel(l)ing speed during working

Fahrgestell n, Untergestell, Chassis n, (Grund)Rahmen m = carrier, chassis

~, Fahrwerk n [*Flugzeug*] = (landing) gear, (~) assembly

~ [*für ein Strom(erzeugungs)aggregat*] = wheeled trailer, two-wheel rubber-tyred troll(e)y

~, Fahrwerk n [*doppelachsig für einen Tieflader*] = bogie

~, Fahrwerk n [*einachsig für einen Tieflader*] = dolly

~ **für Dreh(gerüst)ramme** = turntable undercarriage

Fahr|-Getriebegruppe f, Fahrmechanismus m = travel gear

~gleis n = rail track

Fährhafen m = ferry harbour (Brit.); ~ harbor (US)

Fahrkarten|halle f [*Bahnhof*] = ticket hall

~schalter m = ticket counter

Fahr|katze f, Laufkatze = crab

~katzenausleger m, Laufkatzenausleger = boom with crab (US); jib ~ ~ (Brit.)

~korb m [*Aufzug*] = lift-car (Brit.); elevator cage (US)

~kran m = portable crane [*A crane which is not self-propelling but can be moved about on wheels*]

~kupplung f = traction clutch

Fahrlade-Gerät n → Becher(werk)auflader m

~ → Hublader *m*
Fahrlade-Maschine *f* → Becherwerk-(auf)lader
~ → Hublader *m*
Fahrlader *m* → Becher(werk)auflader
~ → Hublader
Fahr|lehrer *m* = driving instructor
~**leitung** *f* → Fahrdraht *m*
~**mechanismus** *m*, Fahr-Getriebegruppe *f* = travel gear
~**mischer** *m* → (Beton)Rührfahrzeug *n*
~**mischer** *m* → Transportmischer
~**motor** *m* = travelling engine (Brit.); traveling ~ (US)
~**motor-Drehzahl** *f* = travel(l)ing engine speed
~**oberfläche** *f* = running surface, riding ~
~**planenergie** *f* = scheduled energy
Fahrrad *n* = pedal cycle, (bi)cycle
~**stand** *m* = cycle stand [*B.S. 1716*]
~**radweg** *m*, Rad(fahr)weg = (bi-)cycle track, pedal cycle ~
Fahr|richtung *f*, Fahrtrichtung = direction of travel
~**richtungsstabilität** *f* = stability of travel direction
Fahrrinne *f*, Durchfahrt *f* = channel
~, Fahrwasser *n*, Schiffahrtweg *m* [*von Natur gegebener oder künstlich hergestellter ziemlich langer Wasserweg, eingefaßt von Hindernissen. Er dient als genügend tiefer und sicherer Zugang zu einem Hafen oder zu einer Reede*] = channel
~, Fahrwasser *n*, Stromrinne, Flußrinne [*Fläche im Fluß oder Strom, in der die Schiffahrt die erforderliche Fahrtiefe vorfindet*] = channel
~, Spurstreifen *m*, Radeindruck *m*, (Rad)Spur *f* = wheelers [*in an earth road*]
Fahr|schacht *m* → Fahrstuhlschacht
~**schiene** *f* = running rail
Fährschiff *n* = ferry-boat
Fahr|schrapper *m* → Schrapplader
~**schüler** *m* = driving pupil
Fahrsilo *m*, Fahrbunker *m*, fahrbarer Silo, fahrbarer Bunker [*für Baustellen*] = portable bin

~, Flachsilo, Grünfutter-~, oberirdischer Grabensilo = horizontal silo
~ **mit aufgerichteten Wandplatten**, Flachsilo ~ ~ ~ = horizontal silo with erected slabs
~ **über Geländeoberkante**, Flachsilo ~ ~ = surface silo
Fahr|sprenger *m*, Wandersprenger [*Abwasserwesen*] = movable distributor, travel(l)ing ~, reciprocating ~
~**(Spur)** *f*, (Fahr)Streifen *m*, Verkehrsspur, Verkehrsstreifen = (traffic) lane
~**spurlinie** *f*, Verkehrsspurlinie = (traffic) lane line
~**spurmarkierung** *f*, Fahrstreifenmarkierung = lane marking
~**steiger** *m* (⚒) = foreman
~**stellung** *f* = travel position
~**steuerung** *f* [*Eisenbahn; U-Bahn; Schnellbahn*] = driving control
(~)**Streifen** *m* → (Fahr)Spur *f*
~**streifenmarkierung** *f*, Fahrspurmarkierung = lane marking
~**stufe** *f* = travel range
Fahrstuhl *m* → (Personen)Aufzug *m*
~**aufzug** *m* → Turmgerüstaufzug
Fahr(stuhl)schacht *m*, Aufzugschacht = passenger lift shaft, ~ ~ well, ~ ~ pit, ~ ~ tower (Brit.); ~ elevator ~ (US)
Fahrstuhltür *f* = passenger lift door (Brit.); ~ elevator ~ (US)
Fahrt *f* [*Straßenverkehrstechnik. Die Bewegung in einer Richtung von der Fahrtquelle zum Fahrtziel*] = trip, journey
~, Übergang *m* [*z. B. mit einer Walze*] = pass
~**beziehung** *f* [*Straßenverkehrstechnik*] = travel pattern
Fahrte *f* (⚒) = mining ladder
Fahrten|baum *m* (⚒) = mining ladder standard
~**buch** *n* = log book
(Fahrt)End(aus)schalter *m* = limit switch
Fahrtenschreiber *m* = recording tachograph

Fahrtreppe — Fallbeschleunigung g

Fahrtreppe *f* → Rolltreppe
Fahr(t)richtung *f* = direction of travel
Fahrtsprosse *f* (⚒) = mining ladder rung, ~ ~ rundle, ~ ~ tread
Fahrtüchtigkeit *f* [*Straßenverkehrsmittel*] = road worthiness
Fahrtunnel *m* [*im Gegensatz zum Bahnsteigtunnel bei der U-Bahn*] = non-station tunnel
Fahrtweite *f* = trip length, journey ~
Fährverkehr *m* = ferry traffic
Fahr|versuch *m* [*Prüfung eines Straßenkörpers*] = traffic test
~**versuchszeitraum** *m* = traffic testing period
~**wasser** *n* → Fahrrinne *f*
~**weg** *m* **auf einer Großbaustelle**, Baustellenweg, Baustellenstraße *f* = builders' road, site ~
Fahrwerk *n*, Fahrgestell *n* [*einachsig für einen Tieflader*] = dolly
~, Fahrgestell *n* [*Flugzeug*] = gear, assembly, landing ~
~, Fahrgestell *n* [*doppelachsig für einen Tieflader*] = bogie
~**klaue** *f* [*Bagger*] = walking jaw
Fahr|widerstand *m*, Bewegungswiderstand = rolling resistance
~**zeit** *f* = journey time, trip ~
Fahrzeug *n* = vehicle
~ **für den Bodentransport** → Erdbaufahrzeug
~**abmessungen** *fpl* = vehicle dimensions
~**achse** *f* = vehicle axle
~**art** *f* = type of vehicle
~**bagger** *m*, gummibereifter selbstfahrbarer Universalbagger [*kann Luftbereifung oder Vollgummibereifung haben*] = rubber-mounted mobile excavator
~**bau** *m* = vehicle construction
~**bauer** *m* = vehicle builder
~**beleuchtung** *f* = vehicle lighting
~**bestand** *m*, Fuhrpark *m* = vehicle fleet
~**brücke** *f* = vehicular bridge
~-**Diesel(motor)** *m* = vehicle diesel engine
~**feder** *f* = vehicle spring
~**federung** *f* = vehicle suspension
~**(front)scheinwerfer** *m* = vehicle headlamp
~**gewicht** *n* = vehicle weight
(~)**Insasse** *m* = occupant
~**kolonne** *f* = caravan of vehicles
~**konstante** *f* = vehicle constant
~**kran** *m*, gummibereifter selbstfahrbarer Kran [*kann Luftbereifung oder Vollgummibereifung haben*] = rubber-mounted mobile crane
~**last** *f* = vehicular load
~**motor** *m* = vehicle engine
~**neigung** *f* = vehicle tilt
~**panne** *f* = vehicular breakdown
~**rad** *n* = vehicle wheel
~**rettungskran** *m* → Abschlepp-Fahrzeug *n*
~-**Scheinwerferbeleuchtung** *f* = vehicle headlighting
~**statistik** *f* = vehicular statistics
~**strom** *m* = vehicle traffic stream
~**(trommel)winde** *f* = drum winch mounted on a vehicle
~**tunnel** *m* = vehicle tunnel
~**umlauf** *m* = vehicle circulation
~**unfall** *m* = vehicular accident
~**verkehr** *m* = vehicular traffic
~**waage** *f* → Brückenwaage
Fahrzielanzeiger *m* [*beim Autobus*] = destination indicator
Fäkaldünger *m* = rotted manure
Fäkalien *f*, Fäkalstoffe *mpl* = feces, faeces
~**abfuhr** *f* = scavenging service
~**grube** *f*, Abortgrube, Latrinengrube, Abtrittgrube = f(a)eces pit, privy ~
~**versickerung** *f* = f(a)eces seepage
Fakultät *f* **für Architektur und Bauwesen** = Department of Architecture, Civil Engineering and Building Construction
Fallbär *m* → Ramm~
~, Stampfbär, Fallgewicht *n* [*Rüttelstampfmaschine*] = tamper
Fällbecken *n* → Aus~
Fall|beschleunigung g *f*, Erdbeschleunigung g, Schwere-Beschleunigung g [m/s^2] = acceleration g

Fallbremse — Faltenverwerfung

~bremse *f*, Druckluft- und ~ = trailer-type overrun brake
~bügelschreiber *m* = pen recorder
Falle *f* → (Schütz)Tafel *f*
~ = trap
~, Schürze *f* [*Radschrapper*] = apron
fallen, ein~ (Geol.) = to dip, to fall
~, abschwellen, sinken [*Wasserspiegel*] = to fall
Fällen *n*, Fällung *f*, Aus~ [*Chemie*] = precipitation
Fallenwehr *n* → Schützenwehr
(Fall)fangschere *f* = fishing jars
Fall|gewicht *n*, Stampfbär *m*, Fallbär [*Rüttelstampfmaschine*] = tamper
~gewichtdrosselklappe *f*, Drosselklappe mit Fallgewichtantrieb = falling-weight (type) butterfly valve
~hammer *m* → Rammfallbär *m*
Fallhöhe *f*, Druckgefälle *n*, Druckhöhe, Gefällhöhe, Gefälle *n*, Niveaudifferenz *f*, Niveauunterschied *m* [*Zeichen: H. Talsperre. Höhenunterschied zweier Punkte*] = head
~, Gefälle *n*, Schleusen~ = lift (of a lock), fall (~ ~ ~)
Fallhöhen|bereich *m*, Druckhöhenbereich = range of head, head range
~schwankung *f*, Druckhöhenschwankung = head variation
Fall|kessel *m*, Einlaufschacht *m*, Absturzschacht [*bei einem Durchlaß*] = gully, sump
~klappe *f* = drop gate
~meißel *m*, Gewichtsmeißel [*Felsbaggerung unter Wasser*] = drop chisel
Fällmittel *n* = precipitant
Fall|ort *n* → Abhauen *n* (⚒)
~platte *f*, Baggerstampfplatte, Kranstampfplatte = falling plate, tamping ~, crane ~ ~
~plattenkran *m* → Freifallkranstampfer *m*
~ramme *f* → Frei~
~rohr *n*, Ab~ (für Regenwasser), Regen(fall)rohr = fall pipe, downpipe, downspout, downcomer
~rohr *n*, Lauge~ [*Gefrierschachtbau*] = brine pipe

~schacht *m* = drop shaft
~schacht *m* [*Tunnelbau*] = downhole
~schirm *m* = parachute
~-Stampfverdichtungsmaschine *f* = dropping-weight compaction machine, ~ compactor
~stromvergaser *m* = down-draught carbure(t)tor
~tür *f* = trap door
Fällung *f*, Fällen *n*, Aus~ [*Chemie*] = precipitation
Fäll(ungs)becken *n* → (Aus)Fällbecken
Fällungs|gestein *n* → chemisches Sediment(gestein) *n*
~verfahren *n*, Aus~ = precipitation method
Fall|-Viskosimeter *n* = sinker visco(si)meter
~wind *m* = down wind, katabatic ~, fall ~
~winkel *m* → Ein~
~zuleitung *f* = gravity feed (line)
falsch geschnittene Kante *f* [*Sieb*] = false selvedge
falsches Abbinden *n*, vorzeitiges ~, falsche Abbindung *f*, vorzeitige Abbindung = false set; ~ setting (Brit.) [*A manifestation of an abnormal early hydration reaction in which rigidity or partial setting of the paste occurs in a few minutes. When the stiffened paste is remixed even without the further addition of water it resumes its plasticity and no loss of strength occurs*]
Falschluft *f* = false air
Falt|balg *m*, Membran(e)balg *m* = membrane bellows
~bohrmast *m*, Klappbohrmast = collapsible drilling mast, folding ~, jackknife ~ ~
~brücke *f* = folding bridge
~dach *n* = folded-plate roof, folded-slab ~
Falte *f*, Biegung *f* = flexure, fold
Falten|gebirge *n* = folded mountain
~rohrbogen *m* = creased pipe bend
~verwerfung *f* = folded fault

Falt|kuppel *f*, Faltwerkkuppel = folded-plate dome, folded-slab ~, folded-shell ~

~-Tank *m*, Falt-Behälter *m* = collapsible tank

~tür *f* = folding door

~türbeschläge *mpl* = folding door furniture

Faltung *f* (Geol.) = folding, bending

Faltungs|gebirge *n* = mountain formed by folding

~überschiebung (Geol.) = overthrust

Faltversuch *m*, Normalversuch mit der Raupe im Zug, Biegeprobe *f* = face bend test, normal ~ ~

~ mit der Wurzel im Zug = root bend test

Faltwerk *n*, prismatische Schale *f* [*Aus ebenen Flächen zusammengesetztes Tragwerk*] = folded slab (roof), plate (structure), prismatic shell, hipped plate structure, prismatic structure, folded plate (roof), prismatic slab, prismatic roof

~beton *m* = folded concrete

Falz *m*, Anschlag *m* = rabbet, rebate

~fuge *f* [*Betondeckenfuge*] = Walker interlocking joint

~(hüll)rohr *n* [*dient zur Aufnahme der Spannbetonstähle*] = rebated (sheet-)metal sheath

~leiste *f*, Schiene *f* [*kittlose Verglasung*] = glazing bar

~leiste *f* mit Drahtstiften befestigt = wire brad bead

~leiste *f* mit Rundkopfschrauben befestigt = screw bead

~ziegel *m*, Falzstein *m*, Dach~ = interlocking (roofing) tile

Familie *f* (Geol.) = clan

fangen, instrumentieren [*Tiefbohrtechnik*] = to fish

~, ab~, fassen [*Quelle*] = to intercept, to shut off

Fang *m* = trap

~arbeit *f*, Instrumentation *f* [*Tiefbohrtechnik*] = fishing job

~blech *n* → Schutzblech

~bohle *f*, Vorpfändbohle = spile, forepole, forepoling board

~dorn *m*, Spitzfänger *m* [*Stahldorn mit gehärtetem Schneidgewinde zur Ausführung von Fangarbeiten*] = fishing tap(er)

~dornführungsrohr *n*, Spitzfängerführungsrohr = bowl for fishing tap(er)

Fang(e)damm *m* = cofferdam

~ aus Holzstapeln = timber-crib type cofferdam

~abschließung *f* = cofferdamming

~-Spundwand *f* = cofferdam sheeting, ~ sheet pile wall, ~ bulkhead, ~ piling

Fänger *m* für Schöpfbüchsenbügel = latch jack

Fang|gerät *n*, Fangwerkzeug *n* [*Tiefbohrtechnik*] = fishing tool, ~ instrument

~glocke *f*, Schraubentute *f*, Fangmuffe *f* = overshot

~graben *m* → Auffanggraben

~haken *m* = extractor, grapple

~muffe *f*, Schraubentute *f*, Fangglocke *f* = obershot

~schere *f*, Fallfangschere = fishing jars

~schiene *f*, Vorpfändschiene = forepoling girder

~seil *n*, Fangtau *n* = guy rope

~speer *m*, Löffelhaken *m* [*Eine mit Widerhaken versehene Eisenstange für verschiedene Fangarbeiten*] = fishing spear

~tau *n*, Fangseil *n* = guy rope

~werkzeug *n* → Fanggerät *n*

Farb|anstrich *m* → Anstrich

~asphalt *m*, Buntasphalt = colo(u)red asphalt

~aufstrich *m* → Anstrich

~behälter *m* = paint tank

~beton *m* → Buntbeton

~bindemittel *n* = vehicle, cold water paint cement, paint vehicle

~druck *m* [*Farbspritzen*] = spraying pressure

Farbe *f* [*Anstrichfarbe = paint*] = colour (Brit.); color (US)

Farben|fabrik *f*, Farbenwerk *n* = paint plant, ~ factory

Farbenindustrie — Fasersiedestein

~industrie *f* = paint industry
Farbentferner *n*, Farbenvertilger, Lacklöser, Lackbeize *f*, Abbeizer, (Ab)Beizmittel *n* [*vom Anstreicher kurz „Lauge" genannt*] = paint remover
Farbenwerk *n*, Farbenfabrik *f* = paint plant, ~ factory
Farb|fernsehstudio *n* = colo(u)r television studio
~film *m*, Anstrichfilm = paint film
~fliese *f* → Farbplatte
~haftung *f* [*eines Untergrundes*] = paintability
~haut *f* → Farbüberzug *m*
farbiger Beton *m* → Buntbeton
Farb|lack *m* = lacquer
~lichtsignal *n* = colo(u)r-light signal
~linie *f* → Verkehrslinie
~musterblatt *n* = shade card
~pigment *n* = colour pigment (Brit.); color ~ (US)
~platte *f*, Farbfliese *f* = colour tile (Brit.); color ~ (US)
~pumpe *f* = paint pump
~schicht *f* → Anstrich *m*
~schlauch *m* = paint hose
~spritzen *n* = paint spraying
~spritzpistole *f* = (paint) spray gun, (~) air brush, (~) spraying pistol, paint spray, aerograph
~stoff *m* = colo(u)ring material, ~ matter
~straßenbelag *m*, Farbstraßendecke *f* = colo(u)red road surfacing, ~ highway ~, ~ ~ pavement
~streifen *m* → Verkehrslinie *f*
~strich *m* → Verkehrslinie *f*
~strichziehmaschine *f* → Gerät *n* zum Anzeichnen der Verkehrslinien
~ton *m* = paint clay
(~)Tönung *f* = shade
~überzug *m*, Farbhaut *f* = paint coat
Färbversuch *m*, Färbprobe *f*, Färbprüfung *f* = disco(u)loration test
Farb|zement *m* = colo(u)red cement
~zerreibung *f* [*in einer Kugelmühle*] = pigment dispersion and grinding
~zerstäubungsmaschine *f* = paint spraying machine

Farm-zum-Markt-Straße *f* = (farm-to-)market road
Faschinat *n*, Pack~, Buschpackwerk *n*, (Faschinen)Packwerk, Faschinenlage *f* = fascine work
Faschine *f* = (brushwood) fascine, (~) faggot
Faschinen|damm *m* = fascine dike (US); ~ dyke (Brit.)
~floß *n* → Sinkstück
~gründung *f* = fascine foundation
~lage *f* → Faschinat *n*
~matte *f* = fascine mattress (weighted with rubble)
(~)Packwerk *n* → Faschinat *n*
~wehr *n* = weir of fascines
(~)Wippe *f*, Faschinenwurst *f* = wipped fascine, saucisse, saucisson
(~)Wippenrost *m*, Rost aus (Faschinen)Würsten = grillage of fascine poles
Fase *f* → Fasung *f*
fasen → ab~
Faser|dämmstoff *m* = fibre insulating material, ~ lagging ~ (Brit.); fiber insulating ~ (US)
~fett *n* = fibre grease (Brit.); fiber ~ (US)
~filter *m*, *n* = fibrous filter
~filter *m*, *n* zum Abscheiden von Nebelteilchen = fiber mist eliminator (US); fibre ~ ~ (Brit.)
~füllstoff *m* [*Fugenvergußmasse*] = fibrous filler
~gips *m*, spätiger Gips = fibrous gypsum, satin spar
~glas *n* = fiber glass (US); fibre ~ (Brit.) [*Glass drawn into thin threads for spinning and weaving*]
faserige (Fugen)Einlage *f* = fibrous jointing material
Faser|kalk *m* = fibrous calcite
~kiesel *m*, Sillimanit *m*, Fibrolith *m* (Min.) = fibrolite, sillimanite
~kohle *f* = fusain
~metallurgie *f* = fiber metallurgy (US); fibre ~ (Brit.)
~putz *m* = fibered plaster
~siedestein *m*, Spreustein (Min.) = fibrous zeolite

Faserstoff — Faulgrube

~stoff m = fibrous material

~stoffänger m [Zellstoffindustrie] = catch-all, save-all

~stoffdicht(ungs)ring m = fibrous composition seal(ing) ring, ~ ~ packing ~, ~ ~ washer

~torf m, filziger Torf, Wurzeltorf = fibrous peat

~verstärkung f [Werkstoff] = fibre reinforcement (Brit.); fiber ~ (US)

Fassade f = façade [*The face of a building*]

Fassaden|auskleidung f → Fassadenbekleidung

~bekleidung f, Fassadenauskleidung, Fassadenverkleidung = façade lining, ~ facing

~element n = façade unit

~klinker m → Fassadenstein m

~platte f = façade slab, ~ panel

~-Reinigungsmittel n = façade cleaning agent

~stein m, Fassadenziegel m, Fassadenklinker m = hard-burnt façade brick

~verkleidung f → Fassadenbekleidung

~ziegel m → Fassadenstein m

fassen, (ab)fangen [Quelle] = to shut off, to intercept

Fasson|stahl m, Formstahl, Profilstahl, Stahlprofil n = steel section, section steel

~stahlwalzwerk n, Formstahlwalzwerk, Profilstahlwalzwerk, Stahlprofilwalzwerk = structural steel rolling mill

Fassungsvermögen n, Stauraum m, Speicherfähigkeit f = (storage) capacity [reservoir]

~, Inhalt m = capacity

Fastebene f → Denudationsebene

Fasung f, Fasen n, Ab~ = chamfering

~, Ab~, Fase f = chamfer [*A right-angle corner cut off symmetrically, that is, at 45°. When cut off unsymmetrically, the surface may be called a bevel ~ Abschrägung*]

Faß|aufzug m, Faßkran m [Teer- und Bitumenspritzmaschine] = barrel hoist

~bereich m [Rotary-Zange] = size range

~bitumen n = barreled asphalt (US); barrelled asphaltic bitumen (Brit.)

~erhitzer m, Faßwärmer m = barrel heater

~kran m, Faßaufzug m [Teer- und Bitumenspritzmaschine] = barrel hoist

~pumpe f, Abfüllpumpe = barrel pump

~reifenprinzip n = wooden barrel principle

~reinigungsanlage f, Faßwaschanlage = barrel washing device

~sprenger m = hand-operated sprayer (or spraying machine) for drawing direct from drums or barrels, direct from drum type sprayer

~spritzer m für Kaltasphalt = hand-operated cold emulsion spraying machine (or sprayer) for drawing direct from drums or barrels, direct from drum type cold emulsion sprayer (or spraying machine)

~vorwärmehaube f → Faßwärmeschrank m

~wärmeschrank m, Faßvorwärmehaube f [Teer- und Bitumenspritzmaschine] = barrel heating pocket

Faul|becken n → Schlamm~

~behälter m, Faulraum m, Schlamm~ [einer zweistöckigen mechanischen Kläranlage] = digestion chamber, ~ tank

~behälter m, Faulraum m, Schlamm~ = septic tank, hydrolising ~

faulendes Abwasser n → fauliges ~

faules Gestein n, fauler Fels m, faules Felsgestein n = soft rock

Faulfähigkeit f [Abwasser] = digestibility

Faulgas n = digester gas, sludge ~, sewage ~

~, Kanalgas = sewer gas

~verwertung f = utilization of digester gas, ~ ~ sludge ~, ~ ~ sewage ~

Faulgrube f = septic tank

fauliges Abwasser n, faulendes ~ = septic sewage [*Sewage undergoing putrefaction in the absence of oxygen*]

Faulkammer f, Faulraum m = sludge-digestion chamber [*1.) Any chamber used for the digestion of sludge 2.) The lower story of an Imhoff tank or Travis tank*]

Fäulnis f → Schlammfaulung f

~fähigkeit f [*Faulschlamm*] = digestibility, putrescibility

fäulnis|fest, fäulnisbeständig = rotproof

~unfähiges Abwasser n → haltbares ~

Faulraum m → Faulkammer f

~, Faulbehälter m, **Schlamm~** [*einer zweistöckigen mechanischen Kläranlage*] = digestion tank, ~ chamber

~, Faulbehälter m, **Schlamm~** = septic tank, hydrolysing ~

~heizung f, Faulbehälterheizung = digestion tank heating, ~ chamber

Faulschlamm m (Geol.) = sapropel, putrid ooze, rotten slime

~, ausgefaulter Schlamm = digested sludge

Faul|turm m = digestion tower

~verfahren n [*Abwasserwesen*] = septic tank method, anaerobic ~

~zeit f [*Die vom frischen Abwasserschlamm benötigte Zeit um auszufaulen*] = digestion period, ~ time

Faust|achse f = stub axle

~lager n [*Gleiskettenlaufwerkrahmen*] = frame inner bearing, inner track roller frame bearing

~regel f, Faustformel f, empirische Regel = rule of thumb

Fayalit m → Eisenglas n

Fayence f = faïence

~-Platte f, Fayence-Fliese f = faïence tile

Fazies f (Geol.) = facies

~verzahnung f = intergrowth along the facies

Feder|anschlag m = recoil pad

~band n, Bommerband, Pendeltürband = double-acting hinge [*It allows a door to swing through 180°, and usually makes the door self-closing by a spring contained in it*]

~barometer n → Aneroidbarometer

~belastung f = spring loading

federbetätigt = spring actuated

Feder|bolzen m = spring(-loaded) bolt

~dichtung f [*Absenkwalze*] = spring water seal

~druck m = spring load, ~ pressure, ~ compression

~drucksonde f = spring-pressure sounding apparatus

~erz n, Jamesonit m (Min.) = jamesonite, feather ore

federführende Firma f = sponsor(ing firm) (US); pilot firm, "lead" firm, management sponsor

Feder|führung f = sponsoring, management

~gehäuse n = spring housing

federgespannt = spring loaded

Feder|haltestift m, Raste f = detent

~hülse f [*Kernfänger*] = core catcher gland

~keil m = captive key

~kissen n = spring cushion

~kraft f, Federkennung f = spring rate

~lasche f der Motorhaube = hood latch

federlose Aufhängung f = springless suspension

federnde Formänderung f, ~ Verformung, elastische ~ = elastic deformation

Feder|puffer m, Federpufferung f = spring buffer

~ring m, Federscheibe f, federnde Unterlegscheibe = lock washer, spring ~, washholder

~sieb n = spring screen

~spannung f = spring tension

~sperre f = spring catch device

~stahldrahtgewebesieb n, Stahlsieb = spring steel wire screen

~teller m = spring plate, collar ~

Federung f, Ab~ = springing, spring suspension

Feder|waage *f* = spring balance
~weg *m* = spring travel
Federzahl *f* **bei Drehschwingung** = spring constant in rotation
~ in einfacher Verschiebung = spring constant in translation
Federzentrierkappe *f* = spring centering cap
federzentriert = spring centered
Fege|breite *f*, Kehrbreite = sweeping width
~maschine *f* → (Straßen)Kehrmaschine
Fegen *n* → Ab~
Fegewalze *f*, Kehrwalze, Bürstenwalze = rotary brush, rotating ~
Fehl|aushub *m* = excess excavation
~ausrichtung *f* = misalignment
~austrag *m* [*Siebtechnik*] = faulty separation, ~ extraction
~baggerung *f*, Fehlnaßbaggerung = excess dredging
~boden *m* → Einschub *m*
~bohrung *f*, ergebnislose Bohrung, trockene Bohrung, sterile Bohrung, sterile Sonde *f*; steriler Schacht *m* [*in Galizien*] = duster, dry hole, dry well, unproductive boring
~brand *m* → Kalkkern *m*
Fehler|dreieck *n* = triangle of error
~fortpflanzung *f* = propagation of error(s)
~grenze *f* = limit of error
fehlerhafte Naht *f* [*Schwarzdecke*] = unsound joint, "cold" ~
Fehlersuche *f* = trouble-shooting
Fehlgut *n* [*Ist der Fehlaustrag bei der Sortierung*] = misplaced material
Fehlkorn *n* [*Ist der Fehlaustrag bei der Klassierung. Unter Fehlkorn sind Überkorn und Unterkorn zu verstehen. Überkorn ist zu Grobes im Siebdurchgang, Unterkorn zu Feines im Siebübergang*] = misplaced size, outsize
~anteil *m*, Fehlkorngehalt *m* = proportion of outsize, ~ ~ misplaced size
~bestimmung *f* [*Der prozentuale Gehalt an Fehlkorn wird durch Prüfsiebung bestimmt*] = outsize determination, misplaced size ~
~gehalt *m*, Fehlkornanteil *m* = proportion of outsize, ~ ~ misplaced size
Fehl|(naß)baggerung *f* = excess dredging
~schuß *m*, Fehlzündung *f*, Versager *m* = misfire(d shot), misfiring
~stelle *f* = crack, defect
~überkorn *n* = oversize in the undersize
~unterkorn *n* = undersize in the oversize
~zündung *f*, Fehlschuß *m*, Versager *m* = misfiring, misfire(d shot)
Feiertagsschicht *f* = holiday shift
feilen, ab~ = to file (off)
Feilenhärte *f* = file hardness
Feilspäne *mpl* = filings
Fein|(ab)siebung *f* = fine screening
~(abwasser)rechen *m* → → Abwasserwesen
~ausbringen *n* → Feinkornausbringen
~backenbrecher *m* → Backenfeinbrecher
~bestandteil *m* [*Boden*] = fine constituent
~beton *m* = fine-grained concrete
~betonieren *n* = fine casting, ~ pouring, ~ concreting
~bewegungsschraube *f* [*Theodolit*] = slow motion screw
Feinblech *n* = sheet
~ [*bei der Weiterverarbeitung geglüht*] = works annealed sheet
~ [*mehrmals gebeizt*] = full pickled sheet
~ [*sehr dünn*] = tagger
~ [*über A_3 hinaus geglüht*] = full annealed sheet, true ~ ~
~reckmaschine *f* = sheet stretcher
~walzwerk *n* = sheet rolling mill
Fein|boden *m* = fine-grained soil
~brechen *n* = fine crushing, ~ breaking
~brecher *m* = fine crusher, ~ breaker, ~ crushing machine, ~ breaking machine

~druckmesser m = micro-pressure ga(u)ge
~druckregler m = precision type pressure regulator
feine Bürstenwaschbetonoberfläche f = fine scrubbed (concrete) surface, ~ ~ (~) finish
~ Rißbildung f, Maronage f, Netzrißbildung = crazing, craze [deprecated: crocodiling (Brit.)]; map cracking (US)
feiner Emailbrei m, Schlicker m = slip
~ Mo m → ~ Mehlsand m
~ Staub m [Staub nicht im Sinne von „Schmutz"] = fine powder
feines Siebkorn IVa → Mittelsand m
Fein|einstellen n → Berichtigung f
~einstellung f → Berichtigung
fein|fühlig = sensitive
~gemahlen = finely ground
~gemahlener vulkanischer Tuffstein m → Traß m
Feingips m, Feingut n = finely-ground gypsum, land plaster
feingriffig = fine-gripping [surfacing]
Fein|gut n [Siebtechnik] = fine fraction, fines
~gut n → Feingips
~(gut)korn n = fine grain
~hammerbrecher m = split hammer rotary granulator, swing ~ ~ ~
~heitsmodul m, Körnungsmodul = fineness modulus
~kalk m, Kalkpulver n [Branntkalk in feingemahlener Form] = ground quicklime, powdered ~, quicklime in powder form
~kegelbrecher m mit segmentförmigem Brechkopf = fine gyrasphere crusher, ~ ~ breaker
~keramik f, feinkeramische Industrie f = pottery industry
feinkeramischer Ofen m = pottery kiln
Fein|kies m [7—30 mm Korngröße] = fine gravel from 7 to 30 mm grain size
~klärzone f = fine clarification zone
~kohle f → Formkohle
~kohlenentwässerung f = fine-coal dewatering

~kohlenfilterkuchen m = fine-coal filter cake
~kohlenfiltern n = fine-coal filtering
~korn n, Feingutkorn = fine grain
~korn n, Feingut n = fine particles, ~ grains, ~ material
~(korn)ausbringen n, Unterkornausbringen [Siebdurchgang, Fehlkorn abgerechnet] = fines output, true undersize recovery
~kornbereich m = fine fraction range
feinkörnig = fine-grain(ed)
~-dicht, felsitisch = felsitic
feinkörniger Geschiebetrieb m, partieller ~ = partial bed load movement
Feinkorn|linie f, Feinkornkurve f = fine fraction curve, ~ ~ line
~sieb n = fine grain screen
Feinkreiselbrecher m, Feinkegelbrecher, Fein-Rundbrecher = fine reduction gyratory (crusher), fine reduction cone crusher
feinkristallin, kryptokristallin = cryptocrystalline
Fein|mahlen n, Feinmahlung f = fine grinding
~mahlkammer f = fine grinding compartment
feinmaschig = fine-mesh
feinmaschiges Platinnetz n mit Rhodiumzusatz = platinum-rhodium gauze (pad)
Fein|material n, Feinstoffe mpl = fines
~mörtel m = fine-grained mortar
~mühle f = finishing mill
~nivellement n, Präzisionsnivellement = precision levelling, precise ~
feinplanieren = to fine grade
Feinplanieren n mit Straßenhobel f = refinishing a flat section, final grading, fine-grading, finish grading
feinporiges Gipsgestein n → erdiger Gips
Feinputz m → Oberputz
~mörtel m = fine stuff
~-Spritzgerät n, Feinputzwerfer m = fine plaster throwing machine
Feinrechen m = fine rack [A relative term, but generally used when the

Feinsand — Feld 404

clear space between the bars is 1 in. or less]

Feinsand *m* [*nach Niggli (Jahr 1938) und International Society of Soil Science (here called "fine sand") 0,2–0,02 mm*] = fine sand and coarse silt (Brit.)

~, **Staubsand** *m* [*nach DIN 4022 0,2 bis 0,1 mm; nach Atterberg 0,2–0,02 mm; nach Fischer und Udluft (Jahr 1936) 0,2–0,02 mm (heißt hier „Silt")*] = fine sand [*Division of Soil Survey (US) 0.25–0.05 mm; A.S.E.E. fraction 0.25 bis 0.074 mm; British Standard 0.2 bis 0.06 mm*]; fine sand and coarse silt (Brit.) [*0.2–0.02 mm*]

~**klassierung** *f* = classification of fine sand

Feinschlämme *f*, **Zementmilch** *f*, (Beton-)**Schlämmschicht** *f* = laitance

Feinschluff *m* [*nach Deutsche Geologische Landesanstalt 0,005–0,002 mm, nach Degebo und Terzaghi 0,006–0,002 mm, nach Fischer und Udluft (Jahr 1936) 0,005–0,002 mm*] = fine silt [*British Standard 0.006 to 0.002 mm*]

~ [*nach Niggli (Jahr 1938) 0,002 bis 0,0002 mm; dieselbe Korngröße heißt nach Grengg (Jahr 1942) „Schlamm", nach Fischer und Udluft „Sink", nach Terzaghi „Kolloidschlamm", im übrigen meistens „(Roh)Ton"*] = clay

Fein|schmiedemaschine *f* = precision forging machine

~**schotter** *m* = $1^1/_2$ inch size broken stone

~**schotterboden** *m* → → **Kies** *m*

~**schüttgut** *n*, **Fein-Massengut** = fine bulk material

~**senkbremse** *f* = lowering brake

~**sieb** *n* [*Abwasserwesen*] = fine screen [*A relative term, but generally used for a screen with openings $^1/_4$ in., or less in least dimension*]

~**sieben** *n*, **Feinsiebung** *f* = fine screening

~**spaltrost** *m*, **Spaltsieb** *n* = slotted (-hole) screen

~**splitt** *m* = fine chip(ping)s

Feinstes *n* → **Feinstkorn** *n*

Feinsteinbrecher *m*, **Nachsteinbrecher** = reduction stone crusher, fine ~~, secondary~~, ~ breaker

Feinst|gut *n* → **Feinstkorn** *n*

~**korn** *n*, **Aufschlämmbare** *n*, **Feinstgut** *n*, **Feinstes** *n* = ultra-fine particles, ~ grains, ~ material

~**korn** *n* = finest grain

~**mahlung** *f*, **Feinstmahlen** *n* = pulverizing, pulverization

feinstratigraphisch = micro-stratigraphical

Feinst|riß *m*, **Mikroriß** = micro-crack

~**sand** *m* = ultra-fine sand

~**sichtung** *f* = classification of ultra-fine particles by air, separation ~~ ~ ~

~**zerkleinerung** *f* → → **Feinzerkleinerung**

feinstzerteilter Feststoff *m* = finely divided solid

Feintrennung *f* [*Siebung*] = fine separation

feinverteilen, **dispergieren** = to disperse

Feinverteilung *f*, **Dispersion** *f* = dispersion

feinverwachsen [*Erz*] = disseminated

Fein|waage *f* = sensitive balance

~**zerkleinerung** *f* [*Korngröße 10 bis 0,5 mm. Die Herstellung von feinkörnigem Gut unter 0,5 mm Korngröße wird in Deutschland „Feinstzerkleinerung" genannt. Die englischen Begriffe gelten für Feinzerkleinerung und Feinstzerkleinerung*] = fine reduction, ~ comminution [*reduction to $^1/_4$-in. or finer*]

~**ziehschliff** *m* = superfinish

Feld *n* [*Fachwerk*] = panel

~, **Beton~**, **Beton(decken)platte** *f* = (road) panel, (road) bay, concrete bay, concrete (pavement) slab

~ [*Bodenmechanik*] = field [*Every section of a flow channel located between two equipotential lines*]

~, Öffnung f, Brücken~ = (bridge) span, (~) opening
~arbeit f [*Vermessungswesen*] = field work
~aufnahme f [*Photogrammetrie von der Erde aus*] = terrestrial photograph
Feldbahn|(anlage) f = light railroad system, narrow gage railroad (US); jubilee track system, light railway system, field railway system (Brit.); narrow-ga(u)ge track system
~gleis n = field railway track (Brit.); ~ railroad ~ (US)
~lok(omotive) f, Bau(stellen)lok(omotive) = field loco(motive)
~lore f, Muldenkipper m, (Mulden)-Kipplore, Muldenwagen m, Baustellenlore, Baustellen-Schienenwagen, Förderwagen, Feldbahnwagen = tip(ping) wagon, (rocker) dump car, skip, side tip(ping) wagon; jubilee wagon, jubilee skip (Brit.); industrial rail car (US)
~schiene f, Bau(stellen)schiene = field rail, contractor's ~
~wagen m → Feldbahnlore f
(Feld)Beregnung f = (agricultural) sprinkler irrigation
(Feld)Beregnungsanlage f = (agricultural) sprinkler (irrigation) system
Feld|brandofen m → Feldofen
~buch n [*Vermessung*] = field book
~elektrode f → Stromelektrode
~erweiterungsbohrung f = extension well, stepout ~
Feldesteil m, n → Bau-Unterabschnitt m (⚒)
Feld|fabrik f = portable factory, site ~, field ~
~fernsprecher m = field telephone
~-Feuchtigkeits-Äquivalent n [*Ist derjenige Wassergehalt, ausgedrückt geteilt durch Gewicht der Festmasse, bei dem ein Tropfen Wasser auf der ebenen Oberfläche des Bodens nicht sogleich absorbiert wird, sondern sich ausbreitet und der Oberfläche ein glänzendes Aussehen verleiht*] = field moisture equivalent [*is defined as the minimum moisture content, expressed as a percentage of the weight of the ovendried soil, at which a drop of water placed on a smooth surface of the soil will not immediately be absorbed by the soil, but will spread out over the surface and give it a shiny appearance, abbrev. FME, symbol W/me*]
~flugplatz m, Frontflugplatz = advanced landing ground, A.L.G., forward airfield
~hütte f, Roll~ [*Militärflugplatz*] = crew hut, dispersal ~
~leitung f [*Diese Leitung einer (Feld-)Beregnungsanlage ist das Bindeglied zwischen der Zuleitung vom Pumpwerk und der Regnerleitung*] = secondary (pipe) line
feldmäßiges Erkennungsverfahren n **zum Benennen der natürlichen mineralischen Bodenarten** = field method for the identification of naturally occuring mineral soils
Feld|mitte f [*Brücke; Träger*] = midspan
~moment n = moment of span
~ofen m, Feldbrandofen, Meiler m [*Ziegelherstellung*] = clamp (Brit.); scove kiln (US)
~schmiede f = forge
~schweißung f, Baustellenschweißung = field welding
~spat m = fel(d)spar
~spatbasalt m = fel(d)spathoidal basalt
feldspatreicher Sandstein m, **Arkose** f = arkose
Feld|spatvertreter m = fel(d)spathoid mineral
~stecher m → Fernglas n
~steinmauerwerk n = rough rubble masonry, ordinary ~ ~ [*Masonry composed of unsquared or field stones laid without regularity of coursing*]
~trupp m = field team, ~ gang, ~ crew, ~ party
~- und Industriebahngerät n = light railway material

Feldverdichtung (von Boden) — Felsoberfläche

~verdichtung *f* (von Boden) = field compaction (of soil)
~versuch *m* → Freifeldprüfung *f*
~weg *m* = field path
~wegbrücke *f* = field path bridge
Felge *f* = rim
Felgenbreite *f* = rim width
Fels *m*, Festgestein *n* = rock
~abbruch *m* = Bergschlag *m*
~abtrag *m* → Felseinschnitt *m*
~anker *m*, Deckenanker = roof bolt, rock ~
~arbeiten *fpl* [*Lösen und Laden von Fels*] = rock work
~art *f* = type of rock
~aufreißspitze *f* [*Auf reiß(er)zahn*] = rock (point) tip
~aushub *m* für das Planum = rock cutting in formation
~auspressung *f* → Kluftinjektion *f*
~bagger *m*, Steinbruch(hochlöffel)-bagger = rock shovel, quarry ~, heavy-duty ~
~baggerung *f*, Felsaushub *m*, Felsausbruch *m* = (solid) rock excavation
Felsbau *m* = construction in rock
~ über Tage = surface construction in rock
~ unter Tage, Felshohlraumbau = underground construction in rock
Fels|beseitigung *f* = rock excavation work
~block *m*, Steinblock = block of rock, ~ ~ stone
~boden *m* = rocky ground
~bohrbarkeit *f*, Gesteinsbohrbarkeit = rock drillability
~bohren *n*, Felsbohrung *f*, Gesteinsbohren, Gesteinsbohrung = rock drilling
~bohrhammer *m* → (Gesteins)Bohrhammer
~böschung *f* = rock slope
~brecher *m* = rock cutter, ~ breaker
~brecherschiff *n* = rock cutter vessel, ~ breaker ~, chisel breaker ~
~damm *m* = rock embankment, ~ bank [*A ridge of rock thrown up to carry a road, railway, or canal*]
~einpressung *f* → Kluftinjektion *f*

~einschnitt *m*, Felsabtrag *m* = rock cut
Felsen|alaun *m* = rock alum
~fenster *n*, Fels(en)tor *n*, Felsfenster = natural arch
~meer *n* → Blockfeld *n*
~riff *n* = rocky reef, ~ ledge
~tempel *m* = rock temple
~tor *n*, Fels(en)fenster *n*, Felstor = natural arch
Fels|(förder)kübel *m* = stone skip
~forke *f*, Gesteinsforke = rock fork
~formation *f* = rock formation
~garten *m*, Steingarten = rock garden
(~)Geröll(e) *n* → Gesteinstrümmer *m*
~grat *m* = high rocky ridge
~-Groß-Förderwagen *m* → Fels-Muldenkipper *m*
~gründung *f* = foundation in rock
~harke *f* → Felsrechen *m*
~(hoch)löffel *m* = rock dipper, ~ shovel bucket
~hohlraumbau *m*, Felsbau unter Tage = underground construction in rock
felsiger Baugrund *m* → Gründungsfels
Felsinjektion *f* → Kluftinjektion
Felsit *m* = felsite [*obsolete term: felstone*]
felsitisch, feinkörnig-dicht = felsitic
Felsitporphyr *m* = felsitic porphyry
Fels|kanal *m* = canal tunnelled in rock
~kübel *m* → Felsförderkübel
~löffel *m*, Felshochlöffel = rock dipper, ~ shovel bucket
~masse *f* = rock mass
~mechanik *f*, Geomechanik = rock mechanics
~meißel *m* [*Schnellschlagbär mit Meißelspitze*] = rock chisel
~mulde *f*, Steinbruchmulde = rock body
~muldenkipper *m*, Fels-Groß-Förderwagen *m*, Felstransportfahrzeug *n* für Fremdbeladung, Steinbruch-Muldenkipper, Steinbruch-Groß-Förderwagen = off-road hauler for quarries, off-highway ~ ~ ~
~oberfläche *f* = rock surface

Felsquerschnitt — Fensterriegel

~querschnitt m = section through rock
~rechen m, Steinharke f, Felsharke, Steinrechen = rock rake
~ritze f = crevice
~schaufel f, Felskübel m, Fels-Ladeschaufel, Fels-Ladekübel [*Schürflader*] = rock bucket
~schicht f, Gesteinsschicht = rock formation
~schrägwand f = inclined rock face
~schüttungs-Fang(e)damm m = rock cofferdam
~schüttungsstaudamm m → → Talsperre f
~sprengung f = rock blasting
~sturz m → Bergschlag m
~transportfahrzeug n für Fremdbeladung → Felsmuldenkipper m
~transport-Halbanhänger m = rock hauling semi-trailer
~transport-Kippmulde f [*Erdbau-Lastfahrzeug*] = rock-type dump body
~trümmer mpl → Gesteinstrümmer
~tunnel m, Gebirgstunnel, Bergtunnel = rock tunnel
~verpressung f → Kluftinjektion f
~wand f = rock face
~widerlager n = rock abutment
~zahn m = rock-cutting tooth
~-Zuganker m [*Zuganker im Fels eingebaut*] = rock anchor
Feltbase n = felt-base floor-covering
Fender m = bumper, fender
~pfahl m = fender pile
Fenn n, Bruch m = marshy ground, fen
Fenster n = window
~ [*Seitenöffnung bei einem Stollen*] = side opening
~achse f = window centre line (Brit.); ~ center ~ (US)
(~)Anschlag m = (window) rabbet, (~) rebate; (~) check [*Scotland*]
~bank f, (Fenster)Sohlbank = window sill; ~ cill (Brit.)
~bankziegel m, Fensterbankstein m = window sill brick; ~ cill ~ (Brit.)
~beschlag m = window fitting
~beschläge mpl = window hardware, ~ fittings
~blei n = window lead
~blumenkasten m = window box
~bogen m = window arch
~brett n, Deckbrett, Latteibrett, Simsbrett = window board
~brüstung f = window parapet
~chen n = fenestella
~dichtung f = window sealing
~dichtungsstrick m = window sealing rope
~feststeller m = peg stay [*A casement stay which can be held in place by a peg through one if its holes*]
~fläche f = window area
~flügel m = light
~flügel m mit Scharnieren auf nur einer Seite = casement
(~)Flügelrahmen m = sash
~futter n → Fensterrahmen m
~gitter n = window grate
~glas n = window-glass
~kitt m → Glaserkitt
~kreuz n = window cross
(~)Laden m = (window) shutter, ~ blind
~leibung f, Fensterlaibung = window reveal
~loch n → Fensteröffnung
fensterlos, fensterfrei = windowless
Fenster|nische f = window niche, ~ recess
~nischenbogen m = window niche arch, ~ recess ~
~öffnung f, Fensterloch n = window opening
~pfeiler m [*tragender Mauerteil zwischen zwei Fenstern*] = window pier
~pfosten m = mullion, munnion
~putzwagen m = window cradles machine [*Motorized carriage running vertically on rails let into the window frames*]
~rahmen m, Blendrahmen, Fensterfutter n, Futterrahmen = window frame [*The outer part of the window which is solidly fixed to the wall and supports the sash*]
~riegel m = window head

(Fenster)Scheibe — Ferrioxyd

(~)Scheibe f = (window) pane
~sohlbank f → Fensterbank
~sprosse f = glazing bar, sahs ~, astragal [*A rebated wood or metal bar which holds the panes of glass in a window. The particular term "glazing bar" is often kept for roof lights or for patent glazing. Metal windows usually have no glazing bars and thus let in more light*]
~stift m = glazing sprig, glazier's point, brad
~stollen m → Abstiegstollen
~sturz m = window lintel, ~ lintol
~sturz-Maschine f = window lintel machine, ~ lintol ~
~verglasung f = window glazing
~wand f = window wall
~zarge f = window trim, ~ casing
Fern|anzeige = remote indication
~anzeiger m = remote reading indicator
~beben n = distant earthquake
Ferner m, **Gletscher** m = glacier
Fernfahrer m = long-distance driver
Ferngas n = grid gas
~leitung f, Gasfernleitung = gas transmission line
ferngesteuertes Kraftwerk n = telecontrolled power station, ~ generating ~, ~ ~ plant
Fern|glas n, Feldstecher m = binoculars, field glasses
~heizung f, Blockheizung = district heating
~heizwerk n, Blockheizwerk = district heating installation, ~ ~ plant, central ~ ~
~meldeanlage f = communications facility
~meldeturm m = post office tower, telecommunication ~
~messung f = telemetering, remote metering [*The remote indication, registration or integration of meter readings*]
~meßanlage f = telemetry system
~pegel m → elektrischer ~
~reg(e)lung f, Fernsteuerung = remote control

~registriergerät n = remote recorder
~schreiber m = teleprinter
~schutz m **gegen Korrosion**, Kathodenschutz = cathodic protection
~schwimmer m = remotely controlled float
Fernseh|-Bohrloch-Kamera f = television bore-hole camera
~studio n = television studio
~turm m = TV tower, television ~, ~ torch
~-Überwachung f = television control, TV ~
~verkehrslenkung f = television monitoring, TV ~
~zentrale f = television control centre (Brit.); ~ ~ center (US); TV ~ ~
Fernsprech|kabel n = communication cable, telephone ~
~zentrale f = telephone exchange
Fern|start m = remote starting
~steuerung f, Fernreg(e)lung = remote control
~straße f, Fernverkehrsstraße = interregional highway, ~ road
~thermometer n = heat indicator
~transport m = long-distance haulage work
~transportband n → Langstreckenförderbandanlage f zum Transport von Gesteinen
~überwachung f = remote monitoring
~verkehr m = long-distance traffic
~verkehrsbeschilderung f = long-distance signing
~(verkehrs)straße f = interregional highway, ~ road
~versorgung f = long-distance supply
~wasser(rohr)leitung f = long-distance water pipeline
~wasserversorgung f = long-distance water supply
~wirktechnik f (⚒) = control and communication engineering
Ferrarizement m, Brownmilleritzement = tetracalcium aluminoferrite cement
Ferrioxyd n, Eisenoxyd = iron oxide

Ferrochlorid — Fertigteil-Fachwerk

Ferrochlorid n, Eisenchlorür n, $FeCl_2 + 4H_2O$ = ferrous chloride
fertig eingebaut = complete in place
Fertig|balkendecke f, Montagebalkendecke = precast beam floor
~**balken(träger)** m = precast beam
~**bauteil** m, n → Fertigteil
~**bauweise**, Montagebauweise = prefab(ricated) construction method
~**bearbeitung** f = finishing
~**behandlung** f, Endbearbeitung [*Betondecke*] = (final) finishing, surface ~
Fertigbeton m → Transportbeton
~**-Nachmischer** m → (Beton)Rührfahrzeug n
Fertig-Beton|platte f, Plattenelement n = precast slab, ~ panel
~**rohr** n, Betonfertigrohr = precast (concrete) pipe
Fertigbetonteil m, n → (Beton)Fertigteil
~**industrie** f → Fertigteilindustrie
Fertig|-Betonträger m, Beton-Fertigträger = precast girder
~**decke** f, Montagedecke = precast floor
~**-Endblock** m = precast end-block
Fertiger m → Decken~
~**für Gußasphalt** → Gußasphaltfertiger
~ **mit fester Arbeitsbreite** = fixed width finisher
~ ~ **selbstverstellbarer Arbeitsbreite** = self-widening finisher
(~)**Balken** m, (Fertiger)Bohle f = (finisher) beam, paver ~
(~)**Bohle** f, (Fertiger)Balken m = (finisher) beam, paver ~
~**rahmen** m = paver frame, finisher ~
Fertig|erzeugnis n = finished product
~**-Fliesentrennwand** f, Fertig-Fliesenzwischenwand = ready-made tiled partition, pre-tiled ~ [*It eliminates on-site tiling*]
~**gewölbe** n = precast concrete vault
~**gut** n [(*Hart*)*Zerkleinerung*] = finished product
~**gut** n [*Siebtechnik*] = final product, screened ~, finished ~, finishes
~**gut** n [*Beton; bituminöses Gemisch usw.*] = mixed product

~**haus** n = prefab(ricated) home
~**hotel** n = prefab(ricated) hotel
~**kies** m = prepared gravel
~**kuppel** f = precast (concrete) dome
~**pfahl** m = prefab(ricated) pile
~**platte** f = precast slab, ~ panel
~**platte** f, Fertigtafel f = precast panel
~**putz** m = ready-mixed plaster [*It requires only the addition of water to be ready for use*]
~**rahmen** m = precast frame
~**rippe** f = precast rib
~**schale** f = precast shell
~**sieb** n, Endsieb = finishing screen
~**spitzbogen** m = precast concrete pointed arch
Fertigstellung f **einer Bohrung** = well completion
~ ~ **offenen Bohrung** = open hole completion
~ ~ **verrohrten Bohrung** = cased hole completion
~ **eines Bauwerkes** = structural completion
Fertigstellungs|technik f **für Bohrungen** = well completion practice
~**termin** m = completion date, target ~
~**verfahren** n **für Bohrungen** = well completion method
Fertig|sturz m = precast lintol, ~ lintel
~**stütze** f = precast column
~**tafel** f, Fertigplatte f = precast panel
Fertigteil m, n → Beton~
~, Montageteil, Fertigbauteil = prefab(ricated) member
~**bau** m, Montagebau, Bauen n mit Fertigteilen = precast construction
~**-Beleuchtungsmast** m → Fertigteil(-Licht)mast
~**-Bogenbrücke** f = precast arch bridge
~**-Brückenbau** m = precast bridge construction
~**decke** f, Montagedecke = precast floor
~**-Fachwerk** n, Beton~ = precast concrete skeleton, ~ ~ framework

Fertigteil-Fachwerkträger — festgewordene Verrohrung

~-**Fachwerkträger** m, Beton~ = precast concrete truss

~-**Faltwerk** n, prismatische Schale f aus Fertigteilen = precast folded plate

~**form** f [*Betonsteinindustrie*] = mould for precast work (Brit.); mold ~ ~ ~ (US)

~**gebäude** n, Fertigteilhaus n = precast building

~**haus** n, Fertigteilgebäude n = precast building

~**industrie** f, Beton~, Fertigbetonteilindustrie, (Beton-)Montageteilindustrie = precast (concrete) member industry, ~ structural ~ ~, cast-concrete ~ ~

~-**Konstruktionsbeton** m = precast structural concrete

~-**Leuchtenmast** m → Fertigteil-(-Licht)mast

~-(**Licht**)**mast** m, Fertigteil-Beleuchtungsmast, Fertigteil-(Straßen)Leuchtenmast = precast (lighting) column (Brit.); ~ mast (US)

~**mast** m → Fertigteil-Lichtmast

~**schalung** f = formwork for precast work

~-**Silo** m = precast silo

~-(**Straßen**)**Leuchtenmast** m → Fertigteil(-Licht)mast

~**stütze** f = precast column

~**werk** n = precast(ing) plant, ~ ~ factory

~**werkplatz** m, Fertigungsplatz = casting yard, precasting site

~**wohnungsbau** m, Montagewohnungsbau = prefab(ricated) housing

Fertigträger m → Montageträger

Fertigung f **kleiner Serien** = small batch production

Fertigungs|**gemeinkosten** f = manufacturing overhead (US); ~ overheads (Brit.)

~**platz** m, Fertigteilwerkplatz = casting yard

Fertigwand f, Montagewand = prefab(ricated) wall

Fesselballon m = captive balloon

Fest|**beton** m, erhärteter Beton = hardened concrete

~**bettverfahren** n [*technisches Spaltverfahren*] = fixed bed process

~**brennstoffheizung** f = solid fuel (central heating) system

Feste f, Berg~ (⚒) = barrier pillar

feste Brücke f = non-opening bridge

~ **Einlage** f → Fugeneinlage

~ **Energiemenge** f, ~ Strommenge, ~ Leistungsvorhaltung f = firm power, prime ~

~ **Kohle** f = solid coal

~ **Kupp(e)lung** f = (solid) coupling

~ **Leistungsvorhaltung** f → ~ Energiemenge

~ **Strommenge** f → ~ Energiemenge

~ **Verbindung** f [*Brücke; Tunnel*] = fixed link

fester Emulgator m, unlöslicher ~ = colloidal emulsifier

~ **Fels** m → anstehender ~

~ **Grund** m = firm bottom

~ (**Kabelkran**)**Turm** m → (Kabelkran)Maschinenturm

~ **Körper** m = solid body

~ **Schmierstoff** m = solid lubricant

~ **Ton** m = stiff clay

~ **Turm** m → (Kabelkran)Maschinenturm

festes Lager n, eingespanntes ~ [*für Hoch- und Tiefbauten*] = fixed bearing

~ **Schild** n → Querschild

~ **Tangentialkipplager** n [*für Hoch- und Tiefbauten*] = fixed tangential rocker bearing

~ **Wehr** n, Absturz m [*Aus Mauerwerk oder Beton gebaut, mit Tosbecken und Sohlsicherung gegen Kolk*] = fixed weir

Fest|**feuer** n = fixed light [*aerodrome*]

~**form** f = solid mould (Brit.); ~ mold (US)

festgelagert, kompakt [*z.B. Sand*] = compact, firm

Festgestein n → Fels m

festgewordene Verrohrung f, ~ Rohrfahrt f [*Tiefbohrtechnik*] = stuck (string of) casing, "frozen" (~ ~) ~

festgewordener Meißel m, ~ Bohrmeißel = stuck bit, "frozen" ~
festgewordenes Bohrgestänge n = stuck drill pipe, "frozen" ~ ~
Festhalle f = civic auditorium
Festigkeit f, Material~ = (material) strength
~ **der Karkasse** = casing strength
~ **gegen Beschädigungen** = damage resistance
~ **gekerbter Bauteile** = notch-rupture strength
Festigkeitsberechnung f = strength analysis
festigkeitentwickelnde Eigenschaft f = strength-developing characteristic
Festigkeitsentwick(e)lung f = development of strength
festigkeitserhöhend = strength-increasing
Festigkeits|gewinn m = gain in strength, strength gain
~**klasse** f, Güteklasse [Betonbaustein] = strength class
~**lehre** f = strength theory
~**parameter** m = strength parameter
~**prüfgerät** n = strength tester
~**prüfung** f, Festigkeitsversuch m, Festigkeitsprobe f = strength test
~**schweißung** f = full-strength welding
~**verlust** m = loss of strength
~**wert** m = strength value
~**zuwachs** m = strength increase
Fest|kornmischung f = solid-solid mix(ture)
~**körper** m = solid (body)
~**körperreibung** f, Trockenreibung = solid friction
~**landsdüne** f, Winddüne = inland dune [deposition by wind]
~**machen** n, Befestigen = fixing
~**machen** n, Anlegen [Schiff] = berthing, docking, landing, mooring
~**macheplatz** m → Anlegestelle f
~**mörtel** m, erhärteter Mörtel = hardened mortar
~**-Paraffin** n = paraffin
~**preis** m = fixed price
~**preisvertrag** m = lump-sum contract, firm price ~, ficed price ~
Festpunkt m → Abrißpunkt
~ → Rohr~
~ [Durchlaufträger] = fixed point
~ **des Greiferhubseils** = bucket closing cable socket
~ ~ **Greiferseils** = bucket holding cable socket
~**beton** m → Rohr~
~**mauerwerk** n → Rohr~
~**netz** n = observation grid
Fest|saugung f [Der durch den Luftdruck verursachte Widerstand von Gegenständen, die aus dem Schlamm herausgehoben werden] = suction
~**sieb** n = fixed screen
~**spielhaus** n = festival theatre
~**sprenger** m [Abwasserwesen] = fixed distributor
~**stellbolzen** m = locking bolt
~**stellbremse** f = parking brake
~**stellvorrichtung** f = locking device
Feststoff m = solid matter
~**brennkammer** f = solid-fuel combustion chamber
Feststoffe mpl [Mechanische Beimengungen des oberirdischen Wassers. Eis gehört nicht dazu] = debris, detritus [Any material, including floating trash, suspended sediment or bed load, moved by a flowing stream. Sometimes erroneously called "silt"]
Fest|stütze f = fixed column
~**teufenbestimmung** f [Tiefbohrtechnik] = locating stuck point
Festungsverband m → Stromverband
Fest|volumen n = volume of solids
~**walze** f [Walzenbrecher] = immovable roll
~**werden** n [Rohr oder Bohrzeug im Bohrloch] = freezing
~**wert** m = fixed value, constant
~**zeitsteuerung** f [Verkehrssignalreg(e)lung] = switching on a fixed-time-cycle basis
~**zurren** n [Ladung] = lashing down
fett [Mischung] = rich

Fett|abscheider m [*Grundstückentwässerung*] = grease trap (Brit.); ~ interceptor (US)
~behälter m = grease tank
~büchse f = screw down cap
fette Mischung f, fettes Gemisch n = rich mix(ture)
fettes Gemisch n, fette Mischung f = rich mix(ture)
Fetten n → (Ab)Schmieren
Fett|fang m = grease trap [*A device by means of which the grease content of sewage is cooled and congealed so that it may be skimmed from the surface*]
~gas n → Ölgas
~kalk m → Weißkalk
~kammer f = grease chamber
~kohle f = fat coal
~nippel m, Fettschmiernippel = grease nipple, ~ fitting
~presse f, Fettspritze f = (grease) pressure gun, grease ~
~prüfmaschine f = grease testing machine
~(schmier)nippel m = grease nipple, ~ fitting
~(schmier)pumpe f = grease (lubricating) pump
~spritze f, Fettpresse f = (grease) pressure gun
~stoß m = grease shot, shot of grease
Fettung f → (Ab)Schmieren
feucht nachbehandelt [*Beton*] = moist cured
feuchtes Erdgas n, ~ Naturgas, nasses ~, Naßgas, Reichnaturgas = combination gas, wet natural ~, casinghead ~, natural gas rich in oil vapo(u)rs
~ Naturgas n → ~ Erdgas
Feucht(ab)siebung f [*nicht verwechseln mit „Naß(ab)siebung"*] = moist screening
Feuchte f, Feuchtigkeit f = moisture, humidity
~erscheinung f, Feuchtigkeitserscheinung = dampness
~isolierung f → Feuchtigkeitsisolierung

~meßdose f, Feuchtigkeitsmeßdose = moisture cell
~sperre f, Feuchtigkeitssperre = water seal, ~ stop
Feucht|halten n → Feuchthaltung f
~haltung f, Naßhaltung, Anfeuchtung, Feuchthalten n, Anfeuchten [*Betonnachbehandlung*] = wet jobsite after-treatment, ~ ~ curing, moist ~, water ~
Feuchtigkeit f, Feuchte f = humidity, moisture
Feuchtigkeits|bestimmung f, Feuchtebestimmung = moisture determination
~erscheinung f, Feuchterscheinung = dampness
~gehalt m → Wassergehalt
~isolierung f, Feuchteisolierung = dampproofing [*The prevention of moisture penetration by capillary action, in contrast to "waterproofing" which prevents the actual flow of water through a wall, etc.*]
~kammer f, Feuchtraum m = humidity chamber
~korrosion f = aqueous corrosion
~meßdose f, Feuchtemeßdose = moisture cell
~-Meßgerät n, Feuchte-Meßgerät, Feuchtigkeitsmesser m, Feuchtemesser = moisture meter
~schaden m, Feuchteschaden = moisture damage
~sperre f, Feuchtesperre = water seal, ~ stop, ~ barrier, moisture ~, humidity ~
~wanderung f = moisture migration
~zufuhr f = ingress of moisture
Feucht|lagerung f = damp storage
~lagerversuch m, Feuchtlagerprobe f, Feuchtlagerprüfung f [*Korrosionsprüfung*] = high humidity and condensation test
~siebung f → Feuchtabsiebung
Feuer n, Brand m = fire
~ [*als Zeichen*] = light
Feuerberg m, Vulkan m = volcano
~asche f, vulkanische Asche = volcanic ash

~sand m → vulkanischer Sand
~tuff m → vulkanischer Tuff
feuerbeständiger Bauteil m, feuerbeständiges ~ n = fire-resisting component

feuerfest [*in bezug auf Festgesteine und Tone zur Herstellung feuerfester Produkte, Schmelztemperatur oberhalb 1600 Grad Celsius*] refractory
    halbfeuerfest [*Schmelztemp. zwischen 1500 und 1600 Grad Celsius*] semi-refractory
    hochfeuerfest [*Schmelztemp. über 1800 Grad Celsius*] high-refractory

feuerfeste Gießmasse f = castable refractory
~ Stampf-, Flick- und Spritzmassen fpl [*auf der Basis von Schamotte (fireclay), Sillimanit, Korund, Magnesit, Chrommagnesit, Chromerz, Siliziumkarbid*] = plastic refractories, castable ~
feuerfester Baustoff m = refractory construction material, ~ building ~
~ Beton m = refractory concrete
~ Isolierstein m, **~ Ziegel** = insulating fire brick, refractory ~
~ Mörtel m = refractory mortar
~ Ton m → Schamotte f
~ Tonerde-Kieselsäure-Stein m = alumina-silica refractory
~ Ziegel m, **~ Isolierstein** m = refractory brick, insulating fire ~
feuerfestes Erzeugnis n, Feuerfestmaterial n = refractory
Feuer|festmaterial n, feuerfestes Erzeugnis n = refractory
~fortschritt m [*Ziegelofen*] = fire travel
~gas n, Rauchgas, Verbrennungsgas = flue gas
~gefahr f → Brandgefahr
~gefährlichkeit f → Entzündbarkeit
~hahn m, Zapfhahn, Wandhydrant m = wall hydrant
feuerhemmend = fire-retarding
feuerhemmender Bauteil m, feuerhemmendes ~ n = fire-retarding component

Feuer|beständigkeit f = fire resistance
~braunkohle f = fuel brown coal [*incorrectly termed "non-bituminous brown coal"*]
~büchse f → Heizkammer

Feuer|leiter f. Brandleiter = escape ladder
~löschschaum m = fire (extinguishing) foam
~löschwasser n = firewater
~löschwasserständer m [*DIN 14244*] = fire hydrant
feuerlose Dampflok(omotive) f = fireless steam loco(motive)
Feuer|mauer f → Brandmauer
~meldeanlage f = fire alarm system
~opal m (Min.) = fire opal
feuerpolizeiliche Vorschrift f → Brandvorschrift
Feuer|raum m, Feuerungsraum, Heizkammer f = fire box
~rost m, (Kessel)Rost = grate, boiler~
~roststab m, (Kessel)Roststab = grate bar, fire ~
Feuersbrunst f = actual fire of severe intensity
Feuerschiff n = lightship, light-vessel
Feuerschutz m, Brandschutz [*Maßnahmen, um Bauteile feuerbeständig zu machen*] = fire proofing
~, Brandschutz [*Maßnahmen zur Verhütung von Bränden*] = fire prevention
~-Anstrich m, Brandschutz-Anstrich = fire-coat
~farbe f, Flammschutzfarbe, Brandschutzfarbe = fire protection paint
~platte f = fire protection slab
~tür f = fire protection door

Feuersetzen im Bergbau — filtrieren

Feuersetzen n **im Bergbau** = fire setting
Feuerstein m → Flint m
~knollen m, Flintknollen = flint pebble
Feuersteine mpl **in der Kreide** = chalk flints
Feuerstrahlbohren n, Strahlbohrung f, Düsenbohren, Flammstrahlbohren, thermisches Bohren, Brennbohren [*früher: Schmelzbohren*] = jet-piercing, heat drilling
Feuerung f = furnace
~, Feuern n = firing
Feuer(ungs)raum m, Heizkammer f = fire box
Feuerungsrost m = fire grate
~schlacke f = cinders
~schlackenbeton m = cinders concrete
feuerverzinkter Draht m = hot-dip galvanized steel wire
Feuer|vorschrift f → Brandvorschrift
~wache f, Brandwache = fire station
~wehrschlauch m = firemen's hose
Fibel f = primer
Fibrolith m (Min.) → Faserkiesel m
Fichtenschindel f = spruce shingle
Filigran|decke f = filigree floor
~träger m = filigree (reinforcement) girder
filmbildend = film-forming
Film|überzug m → Emulsionsanstrich
~verdampfer m = film evaporator
Filter|ablauf m, Filtrat n = filtrate, filter effluent
~becken n = filtering basin
~beton m = pervious concrete
~betriebszeit f = filter run
~boden m = soil for filter layer
~brunnen m = filter well
~drän m → Sickerdrän
~einsatz m = filter element, kit ~
~fähigkeit f = filterability
~fläche f = filtering surface
~füllung f, Sickerfüllung = porous backfill(ing)
~gehäuse n = filter housing
~geschwindigkeit f [*Bodenmechanik*] = discharge velocity
~gesetz n **von Darcy**, Darcy'sches Filtergesetz = Darcy's law, d'Arcy's ~ [*1856*]

~gewebe n = filter(ing) cloth
~hilfe f = filter aid [*Chemical added to slurry to facilitate filtration*]
~kasten m [*zur Luftreinigung*] = cabinet air purifier
~kies m = filter gravel
(Filter)Kuchen m, Kruste f = filter cake
~-Waschkreislauf m = filter-cake washing circuit
Filter|kühler m = filter cooler
~material n = filtering medium
filtern → ab~
Filter|packung f = filter bed
~patrone f = filter cartridge
~platte f [*Abwasserwesen*] = diffuser plate, air diffuser
~presse f, Schlammpresse [*Abwasserwesen*] = sludge press
~rohr n = screen pipe
~sand m = filter sand
~schicht f, Dränschicht, Entwässerungsschicht, Sickerschicht = filter layer, drainage ~
~schlauch m = filter hose
~sieb n, Saugkorb m, Seiher m = filtering screen, strainer, filter screen
~sockel m = filter base
~stein m = porous stone disc [*triaxial compression test chamber*]
~stein m [*Luftrutsche*] = porous block
~strang m = filter line
Filterung f, Filtern n, Filtration f = filtration
Filter|vlies n = woven filter medium
~wand f = filter wall, ~ diaphragm
~wasser n = filtered water
~werk n = filtration plant, ~ station, filter ~
~widerstand m [*Hydraulik*] = screen loss
~zone f = filter zone
Filtrat n, Filterablauf m = filtrate, filter effluent
Filtration f, Filterung f, Filtern n = filtration
Filtrationsbeschleuniger m [*Paraffin*] = wax modifier
filtrieren → abfiltern

Filtrier|geschwindigkeit f = rate of filtration, filtering rate
~**papier** n = filter paper
~**papierscheibe** f = filter paper disk (US); ~ ~ disc
filtriertes Rückstandsöl n = filtered stock
Filtrosplatte f = filtros [*The trade name applied to an artificial porous stone made of carefully graded siliceous sand by molding, pressing, firing, annealing, and grinding. It is used as a filtering medium and for diffusing air in the activated-sludge process*]
Filz m = felt
~, Hochmoor n = high moor
~**einlage** f = felt insert
filziger Torf m, Fasertorf, Wurzeltorf = fibrous peat
Filz|pappe f = felt
~**streifen** m = felt strip
~**unterlage** f = felt base, ~ backing
~**unterlegscheibe** f = felt washer
Findling(stein) m, erratischer Block m, Erratiker m [*in den Alpen auch „Geißberger" genannt*] = erratic block
Finger m [*Flughafen*] = finger
~**abzug** m = trigger
~**abzug** m → Innendrücker m
Fingerdruck|schalter m → Druckknopfschalter
~**steuerung** f → Druckknopfsteuerung
Finger|form f [*Flughafen*] = finger layout
~**kopf** m [*Flughafen*] = finger head
~**nagelprobe** f = finger-nail indentation test
~**system** n [*Flughafen*] = finger system
Fink-Träger m = Fink truss, French ~
Finsterwalder-Vorspannsystem n [*Vorspannung durch das Eigengewicht der Brücke selbst*] = Finsterwalder prestressing method [*the dead load produces pretension in the reinforcement*]
(Firmen)Bauleiter m = contractor's agent
Firmenzeitschrift f = house journal, ~ magazin

Firn m = firn [*Swiss name for the granular, loose or consolidated snow of the high altitudes before it forms glacial ice*]
First m, Firstlinie f, Dach~, Förste f = ridge [*The apex of a roof, usually a horizontal line*]
~**abdeckung** f = ridge covering, ~ capping
~**anfänger** m, Gratanfänger = starting tile
~**anschlußziegel** m = under-ridge tile
~**balken** m = ridge beam
~**ecke** f, Gratecke = ridge corner tile
Firstendruck m [*Felshohlraumbau*] = roof pressure
~**gewölbe** n (✸) = roof (pressure) arch
First|linie f → First m
~**lüfter** m = ventilating ridge tile
~**pfette** f, Scheitelpfette = ridge purlin(e)
~**platte** f [*Stahldach*] = ridge plate
~**punkt** m = ridge point
~**stein** m, Firstziegel m = ridge tile
~**winkel** m [*Stahldach*] = ridge channel
~**ziegel** m, Firststein m = ridge tile
Fisch m [*im Bohrloch verlorener Eisenteil*] = fish
(~)**Abstieg** m [*stromabwärts gerichtete Fischwanderung*] = downstream migration of fish, ~ fish migration
(~)**Aufstieg** m [*stromaufwärts gerichtete Fischwanderung*] = upstream migration of fish, ~ fish migration
~**bauch** m = fish-belly [*The form taken by some girders or trusses where the bottom flange or chord is convex downward*]
fischbauchförmig = fish-bellied
Fischbauch|klappe f = fish-belly gate, bascule ~
~**träger** m, Linsenträger = fish-bellied girder, fish-belly ~
~**untergurt** m = fish-belly bottom flange, ~ ~ chord, ~ ~ boom
Fischerei|hafen m = fishing harbour (Brit.); ~ harbor (US)
~**kai** m = fishing quay

Fischfräsen — flaches Korn

Fisch|fräsen n [*Tief bohrtechnik*] = milling up the fish
~**grätendränage** f = herringbone drainage, lateral ~; mitre ~ (Brit.); miter ~ (US)
~**grätenverband** m, Kornährenverband = herringbone bond
~**leiter** f → Fischtreppe f
~**rechen** m = fish screen [*A barrier to prevent fish entering a channel*]
~**rost** m = fish screen [*A device intended to prevent the entrance of fish into a conduit*]
~**schleuse** f = fish lock
~**schwanz(bohr)meißel** m = fish-tail bit
~**sterben** n = kill-out of fish
~**sterblichkeit** f = fish mortality
~**teich** m = stock pond
~**treppe** f, Fischleiter f = fish ladder, overfall type fish pass
~**wanderung** f = migration of fish, fish migration
~**weg** m = fish pass, fishway
Fitting m, Formstück n = (pipe) fitting [*Bends, couplings, crosses, elbows, unions, etc.*]
Fixpunkt m → Abrißpunkt
Flach n, Bank f [*über Meeresgrund aufragende, aber die Meeresoberfläche nicht erreichende Erhebung*] = bar, bank
~**asbestplatte** f = asbestos flat building sheet
flachaufbrechen, flachaufreißen = to scarify
Flachaufbrechen n → oberflächennahes Aufreißen
flachaufreißen, flachaufbrechen = to scarify
Flach|aufreißen n → oberflächennahes Aufreißen
~**bagger** m, Planiergerät n = surface digging machine; planer (Brit.); leveler (US)
~**bagger-Lader** m → Pflugbagger m
~**baggern** n → Abziehen
~**baggerung** f → Abziehen n
~**band** n → → Bandförderer m
~**bau** m = flat building

~**baugrube** f = shallow excavation, ~ building pit, ~ cut
Flachboden m = flat bottom, ~ floor
~**bunker** m = flat-bottom bin
~**-Kippritsche** f [*LKW*] = combination dump and platform body
~**selbstentlader** m = track wheel flat dump car
~**silo** m = flat-bottom silo
Flach|bohle f, Flachspundbohle = straight-web sheetpile, flat-web ~
~**bohren** n = shallow drilling
~**bohrung** f, Flachsonde f; Flachschacht m [*in Galizien*] = shallow well
~**brunnen** m = shallow well
~**brunnenhydraulik** f = hydraulics of shallow wells
~**brunnenpumpe** f = shallow well pump
Flachdach n = flat roof [*Scotland: platform roof. USA: barrack roof. A roof the pitch of which is 10° or less to the horizontal*]
~ **ohne Neigung zur Entwässerung** = dead-level roof
~**dämmasse** f, Flachdachdämm-Masse f = flat roof(ing) insulating compound
~**platte** f = flat slab for roofs
~**ziegel** m [*Tonziegel für flachgeneigte Dächer*] = flat interlocking tile
Flach|decke f = horizontal floor
~**dichtung** f = gasket
~**draht** m = flat wire
~**drahtlitze** f → Flachlitze
flache Bauhöhe f (✂) = projection of face length in the direction of full dip
~ **Bodenplatte** f [*Gleiskette*] = flat shoe
~ **Einbautiefe** f [*Rohr*] = placement near the surface of the ground
~ **Hochebene** f = tableland
~ **Tragrolle** f → → Bandförderer m
flacher Bogen m → Stichbogen
~ **gestalten** [*ein Gefälle*] = to flatten
flaches Gelände n = flat ground
~ **Korn** n → plattes ~

Flächen|anziehung f → Adhäsion f
~aufteilungsplan m → Flächennutzungsplan
~druck m [*in einem Stempelschloß* ⚒] = contact pressure, surface ~
~einheit f = unit of area
flächenhafte Ausnagung f → → Denudation f
Flächenheizkörper m = panel heating unit
Flächeninhalt m = area, superficial content, surface content
~ einer Fläche = surface area
~ eines Kreises = area of a circle
Flächen|lager n [*für Hoch- und Tiefbauten*] = surface bearing
~last f = area load
~leistung f = yardage
~messer m, Planimeter n = planimeter
~moment n = area moment
~nutzungsplan m, Leitplan, Wirtschaftsplan, Flächenaufteilungsplan [*Dient zur Darstellung der räumlichen und zeitlichen Entwicklung eines Gemeindegebietes einschließlich Außengebiet. Maßstab 1:10000 bis 1:5000*] = plan of economic and use zones
~porosität f = superficial porosity
~pressung f → Bodendruck m
~raum m = area
~rütt(e)lung f, Ober~ = surface vibration
~rüttler m, Ober~, (Ober)Flächenvibrator m = surface vibrator, ~ vibrating machine
~schleifmaschine f = surface grinder
~tragwerk n [*Plattenförmiger Bauteil, welcher sich nicht nur nach einer Achse abspannt, sondern nach allen Richtungen flächenartig abstützt*] = plane (load-)bearing structure
~verhältnis n = surface ratio
~vibrator m → Flächenrüttler m
~winkel m = dihedral angle, interfacial ~
Flach|feile f, gewöhnliche Feile = flat file
~fundament n → Flachgründung f

~fundamentkörper m → Flachgründung f
~fundation f → Flachgründung f
~furche f [*Bewässerung*] = shallow furrow, corrugation
~furchengerät n [*zur Herstellung von Flachfurchen für Bewässerungszwecke*] = corrugator
~furchenwalze f = roller type corrugator [*It compresses and compacts the soil as a means of making shallow furrows*]
flach(gängig)es Gewinde n, Flachgewinde = square thread
Flachgelenk n = flat hinge
flach|geneigt = with a slight slope
~geneigtes Dach n = low-pitch roof
Flach|glas n [*Sammelbegriff für alle in flacher Form hergestellten Glasarten*] = flat glass
~graben m = shallow excavation trench
~gründung f, Flachfundation f, Flachfundament n, Flachgründungskörper m, Flachfundamentkörper m = shallow foundation (structure)
~gründungskörper m → Flachgründung f
Flächhammer m = flat lump hammer
Flach|hänger m, Flachhängeeisen n [*Hängedecke*] = flat hanger
~kolben m = flat-topped piston
~kopf m = flat head
~kopfniet m = flat-head rivet
~kübel m [*Radschrapper*] = low bowl
~kübel m mit Zwangsschürfung [*Radschrapper*] = positive-action low bowl
~kuppel f, Kugelkappe f = shallow dome, flat ~
~küste f = flat low-lying coast
~land n, Ebene f = plain
~lasche f = flat fishplate
~litze f, Flachdrahtlitze, Litze mit dreieckigem Querschnitt, Dreikantlitze = flattened strand
~litzen(-Draht)seil n, Dreikantlitzen(-Draht)seil = flattened strand cable

Flachlöffel — Flammhärtemaschine

~**löffel** *m*, Planierkübel *m*, Planierschaufel *f*, Planierlöffel, Bagger-~ = skimmer

~**löffelausrüstung** *f*, Planierkübelausrüstung, Planierschaufelausrüstung, Planierlöffelausrüstung, Bagger-~ = skimmer attachment

~**löffelbagger** *m* → Planierbagger

~**löffelwinde** *f*, Bagger-~ = skimmer hoist

~**meißel** *m* = flat chisel

Flachmoor *n*, Nieder(ungs)moor = shallow moor, lowland ~

~**torf** *m* = bottom peat

Flach|platte *f* [*im Gegensatz zur Wellplatte*] = flat panel

~**pol** *m*, flacher Magnetpol [*Kreuzbandscheider*] = lower flat pole

~**-Probe(ent)nahmeapparat** *m* = shallow sampler, ~ sample taker

~**rampe** *f* = ramp with a slight pitch

~**relief** *n* = bas relief

~**riemen** *m* = flat belt

~**riemenscheibe** *f* = flat belt pulley

~**riemen(an)trieb** *m* = flat-belt drive

~**rippen-Streckmetall** *n* = flat rib lath(ing, ~ ~ expanded metal

~**rohr** *n* = rectangular pipe

~**rollenlager** *n* [*für Hoch- und Tiefbauten*] = cylindrical roller bearing

~**rost** *m* = flat grate

~**schacht** *m* [*in Galizien*]; Flachbohrung *f*, Flachsonde *f* = shallow well

~**(schleif)scheibe** *f* = flat abrasive wheel

~**schüttung** *f* = dumping in thin layers, thin-layer fill

flachseitiges Ziegel(stein)pflaster *n* = flat brick paving

Flach|sieb *n* → Plansieb

~**silo** *m* → Fahrsilo

~**sonde** *f*, Flachbohrung *f*; Flachschacht *m* [*in Galizien*] = shallow well

~**spaten** *m* = flat spade

~**(spund)bohle** *f* = straight-web sheetpile, flat-web ~

~**spundwand** *f* = straight-web sheetpiling, flat-web ~

~**stab** *m* = flat bar

~**stahl** *m* = flat rolled steel, ~ (steel) bars, flats

~**-Stangen(transport)rost** *m*, Flachstangenaufgeber *m* = flat (or level) bar screen, ~ (~ ~) screen of bars, ~ (~ ~) grizzly

~**stein** *m* → Biberschwanz *m*

~**steindach** *n* → Flachziegeldach

~**stollen** *m* [*(Gleis)Kette*] = flat (track) shoe

~**wasser** *n* = shallow water, transitional ~

~**wasserzone** *f* [*vor dem Strand*] = nearshore

~**werk** *n* → Biberschwanz *m*

~**werkdach** *n* → Flachziegeldach

~**wulststahl** *m* = beaded flat steel

~**zange** *f* = flat pliers

~**ziegel** *m* → Biberschwanz

~**ziegeldach** *n*, Flachwerkdach, Flachsteindach, Biberschwanzdach, Zungensteindach = plain tile roof

~**ziegel(stein)** *m* → Biberschwanz *m*

Fladenlava *f* → Stricklava

Flakturm *m* = anti-aircraft tower, A. A. ~

flämischer Verband *m* = Flemish bond

Flammenbildung *f* = flame formation

flammengespritzt = flame-sprayed

Flammen|photometer *n* = (hydrogen) flame photometer [*An instrument used to determine elements (sodium and potassium in portland cement) by the color intensity of their unique flame spectra resulting from introducing a solution of a compound of the element into a flame*]

~**photometrie** *f* = flame photometry

~**überwachung** *f* = flame failure detection

~**weg** *m* = flame travel

Flamm|entrostung *f* = rost removal by flame

~**gerät** *n* für den Winterbau = flame heater

~**härtemaschine** *f*, Brennhärtemaschine, Autogenhärtemaschine = autogenous hardening machine, flame-hardening ~

Flammhärten — Flickverfahren

~härten *n*, Brennhärten, Autogenhärten = autogenous hardening
~kohle *f* = flame coal, free-burning ~
~ofenschlacke *f* = reverb slag
Flammpunkt *m* = flash point
~prüfer *m* = flash-point apparatus
~prüfer *m* **Abel geschl. Tiegel** = Abel c. c. testing apparatus
Flamm|ruß *m* = carbon block
~schutzfarbe *f* ≙ Feuerschutzfarbe
~strahlbohren *n* → Feuerstrahlbohren
~strahlbohrgerät *n*, thermisches Bohrgerät = jet piercer
flammstrahlreinigen, abbrennen von Anstrichen = to burn off old coats of paint
Flammstrahlreinigung *f*, Abbrennen *n* [*Entfernen von Anstrichen*] = paint burning
flammwidrig = flame-proof
Flanken|(schweiß)naht *f* = groove weld
~spiel *n* = back lash
Flansch *m* = flange
~, Fuß *m*, Gurt *m*, Träger~ = flange, chord, boom
~blech *n* = flange plate
~-Elektromotor *m*, Elektro-Flanschmotor = flanged motor
~dichtung *f*, Flanschdichtung = flange gasket
~neigung *f* = flange slope
~rohr *n*, Flanschenrohr = flanged pipe
~schraube *f* = flange bolt
~verbindung *f* = flange connection
~wulsteisen *n* = bulb rail iron
Flasche *f*, Unter~, Block *m*, Seilflasche, (Seil)Kloben *m* = block
~ [*für Druckluft; Gas usw.*] = storage bottle, ~ cylinder
Flaschen|abfüllung *f* = bottling
~förderer *m* = bottle conveyor, ~ conveyer
~fülldruck *m* = filling pressure [*of a gas bottle*]
~gas *n* = bottled gas
Flaschen|(hebe)bock *m*, Flaschenwinde *f* = bottle jack
~karre(n) *f*, (*m*) = bottle truck

~post *f* = drift bottle
~rüttler *m* → Innenrüttler
~spülanlage *f*, Flaschenwaschanlage = bottle-washing plant
~vibrator *m* → Innenrüttler *m*
~winde *f*, Flaschen(hebe)bock *m* = bottle jack
~zug *m*, Blockwinde *f*, Seilzug *m*, Rollenzug; Talje *f* [*seemännischer Ausdruck*] = (block and) tackle, (rope) pulley block, lifting block, hoist block, rigging with blocks
~zughaspel *m*, Verrohrungs(seil)trommel *f* = calf wheel
Flattern *n* **von Getrieben** = chattering of gears, gear chatter
flatternde Betonplatte *f* = flexing concrete slab
Flecht|strömung *f*, turbulente Strömung, turbulentes Fließen *n* = spiral flow, turbulent ~, sinous ~, tortuous ~
~werkmantel *m*, Mantelfläche *f* [*Kuppel*] = surface shell, trellis casing
~winkel *m*, Seil-Steigungswinkel = angle of twist
~zaun *m*, Flechtwerk *n*, Schlickzaun = wicker (work) fence
Fleckschiefer *m* = fleckschiefer, maculose rock, mottled schist, spotted schist
Fledermausluke *f*, Schwalbenschwanzfenster *n* = oval luthern
Fleurometer *m* = flourometer
Flick|anlage *f*, Flick-Mischanlage [*Straßenbau*] = patch(ing) plant, ~ machine
~arbeiten *fpl* → Ausflicken *n*
~gerät *n*, Flickmaschine *f*, Schwarzdecken-Instandsetzungsmaschine = patcher, patching unit
~maschine *f* → Flickgerät *n*
~(-Misch)anlage *f* [*Straßenbau*] = patch(ing) plant, ~ machine
~mörtel *m*, Ausbesserungsmörtel = patch(ing) mortar, repair ~
~stelle *f* = patch
~technik *f* = patching technique
~verfahren *n* → Ausflicken *n*

Flickwagen — fließender Verkehr

~wagen m [*Straßenbau*] = spot-mix (asphalt and bituminous mixing) plant
fliegend gelagert = taper bore mounted
Flieger m, Schrägbandförderer-Verlängerung f = extension flight conveyor, swivel (~) ~, thrower belt unit
~ → Gegengewichtrinne f
~ **mit Einlauftrichter** = swivel-piler [*thrower belt unit with hopper*]
~**benzin** n, Flugbenzin = aviation petrol, ~ spirit (Brit.); ~ gasoline, avgas (US)
~**horst** m, Luftstützpunkt m, Militärflugplatz m = base, station
Fliehgewicht n = flyweight, spider
~ = centrifugal weight
Fliehkraft f = centrifugal force
~**abscheider** m = centrifugal separator
~**abscheider** m → Zyklon m
~**antrieb** m = centrifugal drive
~**bremse** f = centrifugal brake
~**entstauber** m → Zyklon m
~**klassierer** m = centrifugal classifier
~**komponente** f = component of the centrifugal force
~**lüfter** m, Schleuderlüfter, Zentrifugallüfter, Radiallüfter = centrifugal fan
~**mühle** f, Schleudermühle = centrifugal (force) mill
~**mühle** f **mit Walzen**, Schleudermühle ~ ~, Pendelmühle = suspended roller mill
~**(naß)klassierer** m = centrifugal classifier
~**regler** m = centrifugal governor
~**reglergewicht** n = governor weight
~**-Rollenmühle** f → Pendelmühle
~**-Staubabscheider** m → Zentrifugal-Staubabscheider
Fliehmoment n = centrifugal moment
Fliese f [*veralteter Ausdruck*]; (Belag-)Platte f = tile
Fliesen|fabrik f → Fliesenwerk n
~**fertiger** m → Plattenpresse
~**legen** n = (on-site) tiling
~**leger** m = floor-and-wall tiler

~**presse** f, Plattenpresse = tile press
~**schleifautomat** m, Plattenschleifautomat = auto(matic) tile grinding machine, ~ ~ grinder
~**schleifmaschine** f, Plattenschleifmaschine = tile grinding machine, ~ grinder
~**schneider** m → Fliesentrennmaschine
~**tafel** f = tile-faced panel
~**-Trennfertigwand** f, Fliesen-Zwischenfertigwand = prefab(ricated) tiled partition (wall)
~**trennmaschine** f, Plattentrennmaschine, Fliesenschneider m, Plattenschneider, Plattenschneidemaschine = tile cutter, ~ cutting machine
~**- und Plattenpresse** f, Platten- und Fliesenpresse = tile and slab press, slab and tile ~
~**vorschub** m, Plattenvorschub [*Spachtelmaschine*] = tile feed
~**werk**, Plattenwerk, Fliesenfabrik f, Plattenfabrik = tile making works, ~ ~ factory, ~ ~ plant
~**-Zwischenfertigwand** f, Fliesen-Trennfertigwand = prefab(ricated) tiled partition (wall)
Fließ|arbeit f = line work
~**barkeit** f → Fließfähigkeit
~**bedingung** f = flow condition
~**bett** n = fluidized bed
~**bettanlage** f = fluidization installation
~**bett-Trommeltrockner** m **mit waag(e)rechter Achse** = rotary trommel fluidized dryer/cooler
~**bild** n = flow chart, ~ sheet
~**decke** f [*als Ergebnis des Bodenfließens*] = soil-flow cover
~**dehnung** f [*Stahl*] = yield strain
fließen [*Stahl*] = to yield
Fließen n [*das Überwinden des Gleichgewichtszustandes in allen Punkten eines Körpers zu gleicher Zeit*] = flow
~ (Geol.) = rock flowage
fließender Perlkies m = running pea ballast
~ **Verkehr** m = moving traffic

~ Zustand *m* [*Boden*] = liquid state
Fließ|erscheinung *f* [*hauptsächlich bei Feinsand oder Schluff*] = boil, blow, quick condition
~fähigkeit *f*, Fließbarkeit, Fließvermögen *n* = flowability, ability to flow
~fähigkeitsmesser *m* **für kolloidalen Mörtel** = Colgrout visco(si)meter
~förderer *m* → Stetigförderer
~förderung *f*, kontinuierliche Förderung, Dauerförderung, Stetigförderung, stetige Förderung = continuous conveying
~formel *f* = formula of flow, flow formula
~gefüge *n* → Fluidalgefüge
~gelenk *n* = plastic hinge
~gelenk-Methode *f* = plastic-hinge method
~grenze *f* [*Stahl*] = yield point
~grenze *f*, unterer Plastizitätszustand *m*, *wf* [*Boden*] = liquid limit [*symbol* w_L]
~grenzenermitt(e)lung *f* = liquid-limit determination
~grenzengerät *n* → Gerät zur Bestimmung der Fließgrenze *f*
~kegel *m* = flow cone [*for measuring the fluidity of the grout for prestressed concrete*]
~maß *n*, Sackmaß, Setzmaß, Ausbreitmaß [*Betonprüfung. DIN 1048*] = slump
~mischer *m*, kontinuierlicher Mischer, stetiger Mischer, Kontimischer, Stetigmischer, Durchlaufmischer, Dauermischer = continuous mixer
~pressen *n* **von Stahl** = cold extrusion of steel
~probe *f*, Fließprüfung *f*, Fließversuch *m* [*Beton*] = flow test
~prüfung *f* → Fließprobe *f*
~punkt *m* [*Kälteverhalten von Dieselkraftstoff*] = pour point
~punkt *m* [*Anstrichstoff*] = yield value
~richtung *f* = flow direction
(~)Rutschung *f* → Bodenfließen *n*

~sand *m*, Quicksand, Schwimmsand = quick sand
~sicherheit *f* **von Vollwand-Verbundkonstruktionen** = yield safety of solid-web composite structures
~spannung *f* [*Metall*] = yield stress
~ton *m* → Quickton
~tracht *f* → Fluidalgefüge *n*
~vermögen *n* → Fließfähigkeit *f*
~versatz *m* (⚒) = controlled-gravity stowing, flow ~
~versuch *m* → Fließprobe *f*
~wert *m* **nach Marshall**, Marshallfließwert = Marshall flow value
~widerstand *m* = resistance to flow
~zeit *f* [*Hydraulik*] = time of concentration
Flint *m*, Feuerstein *m* = flint
~steinrohrmühle *f* = flint (stones) tube mill, pebble ~ ~
~-Ton *m* = flint clay
Flinzgraphit *m* = flake graphite
Flocken|bildung *f* = floc formation
~eis *n* = flake ice
~eismaschine *f* = flake-ice machine
~gefüge *n* → Wabentextur *f* 2. Ordnung
~graphit *m* = flake graphite
~schlamm *m* [*Abwasserwesen*] = flocculated sludge
~schnee *m* = flaky snow
Flockungsbecken *n* = flocculation tank
flohmiger Bernstein *m* = oily-looking dim amber
Floßbrücke *f* = raft bridge
flößen = to raft
Flößen *n* = rafting
Floßkanal *m* = canal for rafting wood
Flotation *f*, Schwimmaufbereitung *f* = flotation
~ [*Abwasserwesen*] = flotation [*A method of collecting suspended matter in a tank as a scum at the surface by the evolution of gas by chemicals, electrolysis, heat, or bacterial decomposition*]
Flotations|anlage *f* = flotation plant
~gerät *n* = flotation machine
~kinetik *f* = kinetics of flotation

Flotationsmaschine — Flügelsonden-Scherfestigkeit

~maschine *f* = flotation machine
~mittel *n* = flotation (re)agent
~sammler *m* = flotation collector
~trübe *f* = pulp
~zelle *f* = flotation cell
Flöz *n* = seam
~faltung *f* = roll
flözleerer Sandstein *m* = farewell-rock
Flöz|mächtigkeit *f* = thickness of seam
~schlag *m* = coal burst
Flucht *f* [*eine gerade Linie oder Ebene*] = line
fluchten → ein~
Flucht|linie *f* → Bau(flucht)linie
~linienplan *m* → Bebauungsplan
~punkt *m* [*Perspektivzeichnen*] = vanishing point
~schnur *f*, Maurerschnur = bricklayer's line
~stab *m*, Bake *f* = range pole, ~ rod, ranging ~, banderolle
~weg *m* = escape route
Fluder *n*, Holzrinne *f* = wood(en) flume
Flug|asche *f* = pulverized fuel ash, fly ~
~aschensinter *m* = sintered (pulverized) fly ash
~aschezement *m* = fly-ash cement
~benzintankanlage *f*, Fliegerbenzintankanlage = multi-tank aviation spirit installation (Brit.); ~ ~ gas(oline) ~, ~ ~ gasolene ~ (US)
~bewegung *f* = aircraft movement, ~ operation [*A landing or take-off*]
~dach *n*, Pultdach, Halbdach, Schleppdach, einhängiges Dach = shed roof
Flügel *m* **nach Unterwasser** → Gegenklappe *f*
~anfänger *m*, Fußstein *m*, Vorsetzstein [*Brückenwiderlager*] = wing toe, ~ footing
~anzeigedämpfer *m*, Flügelzeigerdämpfer = paddle dampening device
~bohrer *m* [*Durch drehende Bewegung schraubt sich die spiralförmig ausgebildete Bohrerspitze in den Boden, und die angeschärften Flügel schneiden von der Bohrlochwandung den Boden ab, wobei der Raum zwischen den Flügeln gefüllt wird*] = rotating auger
~damm *m* [*Damm bei Stauanlagen, der das Staubauwerk seitlich fortsetzt*] = wing embankment
~deckstein *m* [*Brückenwiderlager*] = wing coping
~deich *m* = spur (dyke) (Brit.); wing dam (US)
~fenster *n* = casement window
~füßlerschlamm *m*, Pteropodenschlamm = pteropod ooze
~krone *f* = soft formation cutter head
~mauer *f* [*schließt eine Böschung gegenüber einem Durchlaß ab*] = wing wall; turnout ~ (US)
~mauer *f* [*Brückenwiderlager*] = abutment wall, wing ~
~meißel *m* = bit with wings
~mutter *f*, Flügelschraubenmutter [*DIN 813*] = (butter)fly (screw) nut, wing (~) ~
~pumpe *f*, Flügelradpumpe = semirotary pump, vane ~, wing ~
~(rad) *m, (n)* [*Bunkerstandanzeiger*] = rotating paddle
~rad *n* = vane, propeller
~radausräumer *m* = reclaiming paddle
~radausräumwagen *m* = paddle type bunker discharge carriage
~radwasserzähler *m* = rotary meter [*This type of meter is used to measure the amount of water supplied to bulk consumers. The flow through the meter rotates a vane or propeller which drives the mechanism and registers on a dial*]
~rahmen *m*, Fenster~ = sash
~schiene *f* = wing rail
~sonde *f* = vane borer, ~ apparatus, ~ penetrometer, ~ shear tester [*An apparatus for determining the shear strength of clay soils directly in the ground*]
~sonden-Scherfestigkeit *f* = vane shearing strength

~sondenversuch *m*, Flügelsondenprüfung *f*, Flügelsondenprobe *f* = vane shear test

~tür *f*, zweiflügelige Tür = double wing door

~zellenpumpe *f* = multi-stage impeller pump

Fluggast *m*, Flugreisender *m* = air passenger

~aufkommen *n* = volume of air passengers

Fluggesellschaft *f* = airline

Flughafen *m* = airport

~bau *m*, Flughafengebäude *n* = airport building

~bau *m* = airport construction

~bauer *m* = airport designer

~betrieb *m* = airport activities, ~ operation

~fläche *f* = airport area

~gebäude *n*, Flughafenbau *m* = airport building

~gelände *n* = airport property

~leitung *f* = airport management

~planung *f* = airport planning

~randgebiet *n* = airport surrounding area

~verwaltung *f* = airport administration

Flug|kolbenluftverdichter *m* → Freikolbenkompressor

~kraftstoffanlage *f* → Flugzeugtreibstofftankanlage

~lärm *m* = aircraft noise

~leitungsturm *m*, (FS-)Kontrollturm = control tower

~navigation *f* = aerial navigation

Flugplatz *m*, Landeplatz = landing ground, airfield

~befeuerung *f* = airfield lighting

Flug|reisender *m*, Fluggast *m* = flight passenger

~sicherung *f* = air traffic control

~staub *m* = airborne dust

~steig *m* = gate

~verkehr *m*, Luftverkehr = air traffic

~wetterwarte *f* = airfield meteorological station

Flugzeug|führung *f* im Hafen = routing of aircrafts

~-Gravimeter *m* = airborne gravity meter

~halle *f* = aircraft hangar

~hallentor *n* = aircraft hangar door

~schleppwinde *f* = airplane handling winch

~treibstofftankanlage *f*, Flugkraftstoffanlage *f* = (bulk) aviation fuel installation, multi-tank ~ ~ ~

Fluidalgefüge *n*, Fließgefüge, fluidale Textur *f*, Fluidaltextur, Fließtracht *f*, Flußgefüge, Fluktationsgefüge = fluidal arrangement, fluxion structure, flow structure

Fluor *n* [*chemisches Element F, Atomgewicht 19*] = fluorine

Fluoreszenz-Indikator *m* = fluorescent indicator

fluoreszierendes Pigment *n* = fluorescent pigment

Fluorit *m*, Flußspat *m*, CaF$_2$ = fluor(-spar), fluorite, calcium fluoride, fluoride of calcium

Fluor|kalzium *n* → Kalziumfluorid *n*

~kohlenwasserstoff *m* = fluorcarbon

~natrium *n* = sodium fluoride

~salz *n* = fluoride

~wasserstoffsäure *f*, Flußsäure *f* = hydrofluoric acid, fluorhydric ~

Flur *m* → Korridor *m*

~förderer *m*, Industrie-Lastkarre(n) *f*, (*m*), Flurfördermittel *n* = mobile industrial handling equipment

~ofen *m* [*Ziegelherstellung*] = horizontal brick kiln

~säule *f* [*Flachplatten-Schieber*] = extended stem

~schaden *m* = surface and trespass damage, damage to crops, injury done to the fields

Flüssigballast *m* = liquid ballast

flüssige Kohlensäure *f* = compressed carbon dioxide

~ Phase *f* = liquid phase

~ Vaseline *f* = liquid petrolatum

flüssiger Beton *m* [*DIN 1048*] = wet concrete, sloppy ~

~ Binder *m* → flüssiges Bindemittel

flüssiger Brennstoff — Flußdeichbau

~ **Brennstoff** m = liquid fuel
~ **Rauhreif** m = liquid rime
~ **Sauerstoff** m = liquid oxygen
~ **Schlamm** m [*Abwasserwesen*] = liquid sludge [*Sludge containing sufficient water to permit it to flow by gravity (ordinarily above 80%)*]
~ **Treibstoff** m [*Heizöle und Treibstoffe*] = liquid fuel
~ **Zement** m → Zementleim
~ **Zustand** m = liquid state, ~ condition

flüssiges Bindemittel n, flüssiger Binder m = liquid binder, ~ binding agent
~ **Erdgas** n, ~ Naturgas = LPG, liquefied petroleum gas
~ **Fett** n = liquid grease
~ **Magnesium** n = liquid magnesium
~ **Naturgas** n → ~ Erdgas

Flüssiggas n, verflüssigtes Gas = L-P gas, liquefied petroleum ~, LPG
~**anlage** f = L-P gas system, LPG ~
~**behälter** m = L-P gas tank, LPG ~
~**leitung** f = LPG line, L-P gas ~

Flüssigkeits|behälter m, (Sammel)Behälter, (Lager)Tank m = storage tank
~**bremse** f → Hydraulikbremse
~**dämpfer** m = viscous-type damper
~**dämpfung** f = liquid damping
flüssigkeitsdicht = liquid-tight
Flüssigkeits|dichtung = liquid gasket
~**druck** m = liquid pressure
~**durchlässigkeit** f = permeability to liquids
~**grenzkurve** f = limit curve of a liquid
~**heber** m, Syphon m, Heberrohr n = siphon
~**innenkühlung** f = internal liquid cooling
~**kreislauf** m = liquid cycle
~**kupp(e)lung** f = liquid clutch
~**-(Kurbel)Dämpfer** m = liquid-type damper
~**meniskus** m = meniscus of liquid
~**motor** m → Hydraulikmotor
~**pumpe** f = liquid handling pump

~**reibung** f = liquid friction
~**säule** f = liquid column
~**spiegel** m = surface of liquid
~**stand** m = level of liquid
~**strömung** f = liquid flow
~**-Transportfahrzeug** n, Flüssigkeits-Transporter m, Straßentanker m, Straßentankwagen m, Straßentransporter m für flüssiges Gut, Straßen-Flüssigkeitstransporter [*selbstfahrend*] = liquid hauler, ~ hauling truck
~**-Wärmeaustauscher** m = liquid heat exchanger
~**zylinder** m [*Kolbenpumpe*] = liquid cylinder

Flüssigmist m = liquid manure
Flüstergewölbe n = whispering dome
Fluß m [*ein kleiner Strom*] = stream [*A small river*]
~**abschnitt** m, Flußstrecke f = stream stretch
flußabwärts = downstream
Fluß|anzapfung f, Flußenthauptung = stream beheading, ~ betrunking
~**arm** m = stream arm, ~ branch
~**aue** f → Auenebene f
flußaufwärts = upstream
Fluß|ausbau m = stream development
~**auslauf** m [*ohne Bauwerk*] = stream outlet
~**bagger** m, Fluß-Naßbagger, Fluß-Schwimmbagger = stream dredger (Brit.); ~ dredge (US)
~**bank** f = stream bar
Flußbau m = stream engineering
~**arbeiten** fpl = stream work
~**ingenieur** m = stream conservancy engineer
~**stein** m = stone for stream work
~**werk** n = stream structure
Fluß|bett n, Flußschlauch m = stream bed
~**brücke** f = stream bridge
~**damm** m → Deich m
~**dammbau** m → Dammbau
~**deich** m → Deich
~**deichbau** m → Deichbau

424

Flußecke — Flutfeld

~ecke *f* = stream corner
~(einzug)gebiet *n* = stream basin [*The total area drained by a stream and its tributaries*]
~enthauptung *f*, Flußanzapfung = stream betrunking, ~ beheading
~erosion *f* → Erosion im engeren Sinne
~feinsand *m* = fine stream sand
~fisch *m* = stream fish
~gab(e)lung *f*, Flußspaltung = stream bifurcation
~gebiet *n* → Flußeinzuggebiet
~gefüge *n* → Fluidalgefüge
~gold *n* = alluvial gold, placer ~
~hafen *m* = stream port
~haltung *f* = stream reach
~hydraulik *f* = stream hydraulics
~kabel *n* = stream cable
(~)Kanalisierung *f*, Staureg(e)lung = stream canalization
~kies *m* = stream gravel
~kraftwerk *n* → Lauf(kraft)werk
~krümmung *f* = stream bend
~lauf *m* = stream course
(~)Mäander *m* → Flußschlinge *f*
~mitte *f* = mid-stream
~mittel *n* = flux
~modell *n*, flußbauliches Modell = stream model
~morphologie *f* = stream morphology
~mündung *f* = stream mouth
(~)Oberlauf *m* = upper (stream) course
~pfeiler *m* = stream pier
~regulierung *f* → Reg(e)lung
~rinne *f* → Fahrrinne
~sand *m* = stream sand
~säure *f*, Fluorwasserstoffsäure = hydrofluoric acid, fluorhydric ~
~säureflasche *f* [*Bohrlochneigungsmessung*] = acid bottle
~schlauch *m*, Flußbett *n* = stream bed
~schlinge *f*, (Fluß)Mäander *m*, starke Flußwindung *f* = meander
~schwerspat *m*, Barytflußspat = fluorspar of baryte
~seite *f* [*Brückenbau*] = channel side
~sohle *f* = stream floor, ~ bottom

~sohlenerhöhung *f* = raising of the stream floor, ~ ~ ~ ~ bottom
~sohlenvertiefung *f* = deepening of the stream floor, ~ ~ ~ ~ bottom
~spat *m*, Fluorit *m*, CaF_2 = fluorspar, fluorite
~stadium *n* = streamhood
~stahl *m* = mild steel, soft ~
~stahlrohr *n* = mild steel pipe, soft ~ ~
~strecke *f*, Flußabschnitt *m* = stream stretch
~strömung *f* = stream current
~tal *n* = stream valley
~trasse *f* = lie of a stream
~tunnel *m* = stream tunnel
~überbau *m* [*Brücke*] = stream superstructure
~ufer *n* = stream bank
~uferschutz *m* = stream-bank protection
~umleitung *f* = stream diversion
(~)Unterlauf *m* = lower (stream) course
~unterquerung *f* = crossing below a stream
~verkehr *m* = stream traffic
~wasser *n* = stream water
~wasserfassung *f* = stream intake
~wasserstand *m* = stream level

Flut *f* [*das Steigen des Wassers vom Tideniedrigwasser zum folgenden Tidehochwasser*] = high tide of water, flood tide, flowing tide, rising tide
~, Hochwasser *n* = flood, high water, spate
~becken *n*, Dockhafen *m* = wet dock, closed basin
~berg *n* = tidal elevation, ~ swelling
~brandung *f*, Barre *f* = bore
~brücke *f* = flood bridge
Fluten *n*, Wassertreibverfahren *n* [*Erdölförderung*] = water drive, ~ flooding
~ = flooding, filling with water
fluterzeugendes Gestirn *n* = generating celestial body
Flut|feld *n* → Uferfeld

Flutgröße — Fördererlaubnis

~größe *f*, Flutintervall *n*, Tidehub *m*, Gezeitenhub, Thb = range of tide

~hafen *m*, Tidehafen, Gezeitenhafen [*offener Hafen im Tidegebiet, der infolge geringer Tiefe nur bei höheren Wasserständen benutzt werden kann*] = tide harbo(u)r, tidal ~

~intervall *n* → Flutgröße *f*

~kraftwerk *n* Gezeitenkraftwerk, Tidekraftwerk, Ebbe- und Flut-Kraftwerk, ozeanisches Flutkraftwerk = tidal power plant, ~ ~ station, ~ generating ~

~lackierung *f* = flow coating

~lehm *m* = flood loam

~lichtanlage *f* = flood-lighting system

~lichtbeleuchtung *f* = floodlighting

~lichtstrahler *m* = flood-light

~mündung *f* = tidal estuary

~öffnung *f* → Uferfeld *n*

~öffnung *f* = flooding sluice

~raum *m* = tidal capacity

~rinne *f* [*Rinne im Mündungsgebiet eines Flusses oder im Watt, die vorwiegend vom Flutstrom durchflossen wird*] = flood(-tide) channel

~schleuse *f* = tide lock

~schreiber *m*, Gezeitenschreiber = registering tide ga(u)ge, marigraph

~(straßen)verkehr *m*, Verkehr auf Wechselstreifen, Wechselspurverkehr = tidal traffic

~strömung *f*, Flutstrom *m* = flood tide, raising ~

~-Stundenlinienkarte *f* = cotidal chart

~verkehr *m* → Flut(straßen)verkehr

~vorhersage *f* = flood forecasting

~wasser *n* = tide water

~wassermenge *f* = volume of water entering on the flood-tide, ~ ~ the flood

~wechsel *m* = turn of the tides

~weg *m* = floodway

~wulst *f*, *m* = tidal bulge

~zerreißung *f* = tidal disruption

fluvial = fluvial

fluviatile Erosion *f*, Erosion im engeren Sinne, Flußerosion = fluviatile erosion, stream ~

fluxen [*Durch Beigabe von Fluxmitteln wird der Erweichungspunkt bituminöser Stoffe erniedrigt*] = to flux

Fluxöl *n* = flux-oil

Folge|bohrung *f*, Nach~ = follow-up well

~steuerung *f* = sequence control

~ventil *n* = sequence valve

~zylinder *m* [*Bremse*] = slave cylinder

Folie *f* = foil

Folien|bahn *f* = foil sheet

~walzwerk *n* = foil (rolling) mill

Förder|abfallkurve *f* [*Erdöl; Erdgas*] = production decline curve

~anlage *f*, Transportanlage = conveying plant, ~ installation

~aufseher *m* (⚒) = rope runner

~automat *m*, automatischer Förderer *m* = auto(matic) conveyor, ~ conveyer

~band *n* → Fördergurt *m*

~band-(Anhänge)Schürfwagen *m* → Pflugbagger *m*

~bandbeton *m* = belt-conveyed concrete

~bandbrücke *f* = belt conveyor bridge, ~ conveyer ~

~bandrolle *f* → → Bandförderer *m*

~bandstollen *m* = belt conveyor tunnel, ~ conveyer ~, ~ ~ gallery

~beginn *m* der Einspritzpumpe *f* = FPI, fuel pump injection

~bohrung *f*, Fördersonde *f*, produzierende Bohrung = output well, production ~, paying ~

~brücke *f* = transporter bridge, conveyor ~, conveying ~, conveyer ~

~brücke *f* für Abraum, Abraum(förder)brücke = overburden conveying bridge, ~ transporter ~

~druck *m* [*Die am Druckmesser der Pumpe angezeigte Höhe*] = head

~düse *f* [*Erdöl*] = flow nipple, ~ bean

~einrichtungen *fpl* für Baustellen = material handling devices for construction sites

Förderer *m*, Fördergerät *n*, Material~ = conveyor, conveyer

Förder|erlaubnis *f* [*Erdöl; Erdgas*] = production licence, ~ license

Fördergefäß — Fördersonde

~gefäß n, Schürfgefäß, Kübel m [*Radschrapper*] = (scraper) bowl, (skimmer) scoop
~gerät n → Förderer m
~gerüst n (⚒) = headgear
~gerüst n für Schachtaushubmassen → Abteufgerüst
~gestell n, Förderkorb m, Förderschale f (⚒) = cage
~gurt m, Förderband n = conveyor belt(ing), ~ band, conveyer ~
~gut n = material to be conveyed
~gut n [*Pumpe*] = material pumped, substance ~
~haken m für Pumpengestänge = sucker rod hook
~hammer m → Abbauhammer
~höhe f, Hubhöhe = lift(ing) height
~horizont m = underground formation [*oil or gas production from a well*] conveyor, conveyer, conveyor trough, conveyer trough
~kanal m, Förderrinne f = trough
~karre(n) f, (m), (Transport-)Karre(n) = cart
~kessel m → Treibkessel
~kette f (⚒) = mine-hoist chain
~kohle f, Rohkohle = rough coal, run-of-the-mine ~
~korb m → Fördergestell n
~korbzwischengeschirr n, Kübelzwischengeschirr = communicator for sinking-kibble
~kübel m = skip
~kübel m für Schachtaushubmassen → Abteufkübel
~leitung f = conveying line
~maschine f (⚒) = winding engine, winder [*The engine at the head of a shaft which rolls up the hoisting rope on its drum or sheave, and thus pulls the cage or skip up the shaft. A small winder is called a hoist*]
~menge f [*Pumpe*] = capacity
fördern, gewinnen [*Bodenschätze*] = to extract, to win
~, bewegen [*Erdmassen*] = to shift, to move
~ [*Erdöl; Erdgas*] = to produce

~, umschlagen = to handle
fördernde Gasbohrung f → gasproduzierende Sonde f
~ Gassonde f → gasproduzierende Sonde
Förder|packer m [*Tiefbohrtechnik*] = production packer
~periode f [*Erdöl; Erdgas*] = producing period
~pumpe f, Beton(förder)pumpe = concrete pump
~pumpe f, Baggerpumpe = dredging pump
~rad n [*pennsylvanisches Seilbohren*] = bull wheel
~radbock m = bull wheel post
~radbockstütze f = bull wheel post brace
~radbremsband n = bull wheel brake band
~radbremshebel m = bull wheel brake lever
~rad-Hauptlagerstütze = jack post
~radsegment n = cant, segment of the bull wheel, bull wheel segment
~rinne f, Schwingförderrinne, (Schwing)Rinnenförderer m, Vibrationsrinne = vibrating trough conveyor, ~ ~ conveyer
~rohr n, Beton~ = concrete discharge pipe
~rohr n → Gebläseluft-Förderrohr
~rohr n, Schwing~, (Schwing-)Rohrförderer m, Vibrationsrohr = vibrating circular pipeline
~rutsche f, Schurre f = chute
~schacht m, Treibschacht = main pit, ~ shaft, hoisting ~
~schachtseil n, Treibschachtseil = shaft cable
~schale f → Fördergestell n
~schnecke f → Schneckenförderer m
~seil n [*Rotarybohren*] = drilling line
~seil n [*Einseilbahn*] = haulage rope
~seil n, Aufzugseil, Schacht~ = hoist rope, winding ~
~sonde f, Förderbohrung f, produzierende Bohrung = output well, production ~, paying ~

Förderstollen — Form(en)teil

~stollen *m* [*unter einer Materialhalde*] → Haldentunnel *m*

~stollen *m*, Sohlstollen = bottom drift

~strang *m* [*Erdöl*] = production string of casing, ~ casing string

~stuhl *m* für Steigrohre, Steigrohrelevator *m* = tubing elevator

~technik *f* = mechanical handling

~technik *f* [*Erdöl; Erdgas*] = producing practice

~tour *f*, letzte Rohrfahrt *f* außer den Filterrohren = oil string

~trommel *f* (⚒) = winding (engine) drum, hoisting ~

~tunnel *m* [*unter einer Materialhalde*] → Haldentunnel

~turm *m* für Erdölgewinnung = production derrick

~- und Hebeindustrie *f* = mechanical handling industry

Förderung *f*, Transport *m* [*Erdbau*] = hauling, haulage

~ = production [*of an oil or gas well*]

Förderversuch *m* [*Erdölförderung*] = production test

~ durch das Bohrgestänge = drill-stem test

Förderwagen *m* → Schutterwagen

~ → Feldbahnlore

~ [*Erdbau*] = truck

~ mit Hinterkippwanne → Muldenkipper *m*

(~-)Abstoßvorrichtung *f* (⚒) = shunt back

~-Behälter *m* → Mulde

~bremse *f* (⚒) = tub retarder

Förder|weite *f*, Förderstrecke *f*, Transportweite [*in Österreich: Bringungsweite*] [*Bodenförderung*] = lead, run (Brit.); haul(age distance)

~winkel *m* = haulage angle [*belt conveyor*]

~würdigkeit *f* [*Erdöl; Erdgas*] = productivity

Forellenstein *m* = troctolite

Form *f*, Gußform = mo(u)ld

~ [*Steinformmaschine*] = mould box (Brit.); mold ~ (US)

~, Messing~ [*Streckbarkeitsmessung von Bitumen*] = ductility-test mo(u)ld

~, Körperform, Gestalt *f* = shape

~ für Sichtbauelement = face (unit) mould (Brit.); ~ (~) mold (US)

~änderung *f*, Verformung, Deformation *f* = deformation

~änderungsbedingung *j* = deformation condition

~änderungsgeschwindigkeit *f*, Verformungsgeschwindigkeit, Deformationsgeschwindigkeit = deformation rate, rate of deformation

~änderungsmechanismus *m*, Verformungsmechanismus = deformation mechanism

Format *n* = size

~änderung *f* = size change

Formations|brechen *n* [*Erdölgewinnung*] = fracturing

~grenze *f* (Geol.) = contact between two formations

~grenze *f* [*Karte*] = formation(al) boundary, systemic ~

~karte *f* = formation map

~name *m* (Geol.) = formation name

~tabelle *f* (Geol.) = geologic calendar, table of strata

~-und Schichtenkunde *f*, Stratigraphie *f*, historische Geologie *f* = stratigraphy, historical geology, stratigraphical geology

Formeisen *n* → Fassoneisen

Formel *f* = expression, formula

~ zur Filtration = filtration formula

~zeichen *n* = symbol

Form(en)|bau *m* = mold construction, ~ manufacture (US); mould ~ (Brit.)

~blech *n* = plate for mo(u)ld construction, ~ ~ ~ manufacture

~farbe *f* = mo(u)ld paint

~öl *n* = mould oil (Brit.); mold ~ (US)

~rüttler *m*, Form(en)vibrator *m* = mould vibrator (Brit.); mold ~ (US)

~teil *m, n* = mold portion (US); mould ~ (Brit.)

Form|faktor *m* = shape factor [*wind pressure on buildings*]
~gebung *f* = shaping
~gips *m*, Keramikgips, Formengips = pottery plaster
~kasten *m*, Formenkasten = mould box (Brit.); mold ~ (US)
~kasteneinsatz *m*, Formenkasteneinsatz = mould insert (Brit.); mold ~ (US)
~kohle *f*, Feinkohle, Klarkohle, Rieselkohle = crumble coal, fine ~
~lineal *n* → Profillehre *f*
Formling *m*, frischgeformtes Erzeugnis *n* = freshly-made product
~ = green brick
(Form)Mundstück *n*, Profilmundstück [*Strangpresse*] = (pre-formed) die
Form|oberfläche *f*, Formenoberfläche = mould surface (Brit.); mold ~ (US)
~öl *n* → Formenöl
formschön = aesthetically pleasing

Formstahlwalzwerk *n*, Fassonstahlwalzwerk, Profilstahlwalzwerk, Stahlprofilwalzwerk = structural steel rolling mill
Formstein *m* → (Beton)Stein
~, Formziegel *m*, Profilstein, Profilziegel = purpose-made brick
Formstück *n*, Rohr~, Fitting *m* = fitting
formtreu = non-deformable
formtreues prismatisches Faltwerk *n* = undistorted prismatic structure of folded units
Formung *f* von Tonkügelchen, Tonkügelchenherstellung = pellet formation
Formylsäure *f*, Ameisensäure = formic acid
Formziegel *m*, profilierter Ziegel, Profilziegel, Formstein *m*, Profilstein = purpose-made brick
Forscher *m* = researcher, research worker

Forschung *f*
~ im eigenen Land
Grundlagen~
Fabrikations~
angewandte ~
~ im Außendienst
Zweck~
Forschungsauftrag *m*
Entscheidungs~

research
internal ~
basic ~
manufacturing ~
applied ~
field ~
directed ~
~ contract
operations ~

Forschungs|aufgabe *f* = research problem
~auftrag *m* = research task
~gebiet *n* = research topic
~gruppe *f* = research team
~institut *n* = research institute
~labor(atorium) *n* = research lab(oratory)
~mittel *f* = research funds
Förste *f* → First *m*
Forsterit *m*, Magnesiumolivin *n* (Min.) = forsterite
Förster-Sonde *f*, Induktionsfluß-Magnetometer *n* = airborne magnetometer
Fort *n* = fort

~bewegung *f* = locomotion
~gang *m* der (Bau)Arbeiten → Baufortschritt *m*
fortlaufendes Kernen *n* = continuous coring
Fortschaffen *n* der Ausbruchmassen → (Auf)Schottern *n* [*Schweiz*]
fort|schreitende Welle *f*, progressive ~ = translation wave, propagating ~
~schreitender Bruch *m* = progressive failure
Fort|schritt *m* der (Bau)Arbeiten → Baufortschritt
~schrittbericht *m* = progress report
fossiler Boden *m*, begrabener ~ = fossil soil

fossiler Regentropfen — freier Auslauf

~ **Regentropfen** m = rain-drop impression
Fossilienmehl n = fossil meal (or farina, or flour)
Fotozelle f [*als Verkehrszählgerät*] = photoelectric device
Fourierreihe f = Fourier series
Foyer n = foyer
Frachtabfertigung f = cargo handling, ~ processing
Frachter m, **Frachtschiff** n = freighter cargo vessel
Frachtgebäude n, **Frachtbau** m [*Flugplatz*] = cargo (storage) building
Fragment n, **Trümmerstück** n (Geol.) = fragment
Fraktion f → **Kornklasse** f
Fraktionierkolonne f = fractionating column
fraktionierte Destillation f, stufenweise ~ = fractional distillation
Fraktions|gewicht n [*Siebtechnik*] = % fraction weight
~**grenze** f [*Siebtechnik*] = size limit
~**größe** f → **Korngröße**
Francis|-Formel f = Francis (weir) formula
~**-Pumpturbine** f = Francis (reversible) pump turbine, ~ ~ unit, reversed Francis turbine
~**-Turbine** f = Francis water turbine
Frankfurter Schwarz n = Frankfort black
Franki|pfahl m = Franki pile
~**(-Pfahl)ramme** f = Franki pile driver, ~ piling plant
Franzose m → **englischer Schraubenschlüssel** m
französisches Blatt n, schräges Hakenblatt [*Holzverbindung*] = French scarf (joint), oblique ~ (~)
Fräsbalken m, Ausleger m der Grabenfräse = trench cutting boom
Fräser m, Fräskrone f [*Tiefbohrtechnik*] = milling tool, ~ shoe
~, **Kratzer** m [*beim Kratzbagger*] = cutter
~**bagger** m → **Schrämbagger**
~**kette** f, Kratzerkette = cutter chain

~**kopf** m → mechanisch angetriebener Schneidkopf [*Naßbagger*]
Fräs|krone f, Fräser m [*Tiefbohrtechnik*] = milling shoe, ~ tool
~**maschine** f = milling machine
~**rillendrän** m, Schlitzdrän = mole drain
~**ring** m = milling ring
~**werkzeug** n = milling tool
frei abfallender Strahl m, freier überfallender ~ = free nappe
~ **ausfließende Bohrung** f, ~ ~ **Sonde** f, Eruptivsonde; frei ausfließender **Schacht** m [*in Galizien*] = flowing well, ~ bore, ~ boring, well producing by flow, bore producing by flow, boring producing by flow
~ **ausfließender Schacht** m [*in Galizien*]; siehe: ~ **ausfließende Bohrung** f
~ **Baustelle** = delivered site
~**aufliegend** [*Träger*] = simply supported, supported at both ends
Frei|bad n = open-air pool
~**balken** m → **Kragbalken**
freibewegliches Lager n [*für Hoch- und Tiefbauten*] = freely movable bearing
Frei|beweglichkeit f, Manövrierfähigkeit = manoeuverability (Brit.); maneuverability (US)
~**bord** n = free board [*The distance between the normal operating level and the top of the sides of an open conduit, the crest of a dam, etc., left to allow for wave action, floating debris, or emergency, without overtopping the structure*]
(frei)drehen [*Steinzeugindustrie*] = to throw, to turn freely
(Frei)Drehen n [*Steinzeugindustrie*] = throwing
freie Sicht f = unobstructed vision
~ **Siebung** f, freies Sieben n = free screening
~ **Verpflegung** f **und Unterkunft** f = free board and room
freier Auslauf m [*Erdöl*] = natural flow

freier Kalk — Freilaufrohr

~ **Kalk** m, freies CaO = free lime
~ **Kohlenstoff** m [*früher: Unlösliches n*] = free carbon [*deprecated*]; toluene insolubles
~ **(Überfall)Strahl** m = free nappe (US); ~ sheet of water (Brit.)
~ **Überlauf** m, Freistrahlüberlauf = free waste weir, ~ spillway; open spillway (US)
~ **Wasserspiegel** m = free-water level
(~) ~, Grundwasserhorizont m, Grundwasserspiegel = (ground-)water table, phreatic surface
~ **Wettbewerb** m, Ideenwettbewerb = design competition
~ **Wurf** m [*Mühle*] = cataracting
freies Sieben n, freie Siebung f = free screening
Freifahrung f (⚒) = freeing
(Frei)Fallbär m → Rammfallbär
Freifall|bohrer m = drop boring tool
~-**Chargenmischer** m → → Betonmischer
(~)**Durchlaufmischer** m, Austragmischer, Mischer mit Schurrenaustrag(ung) = closed drum (concrete) mixer with tipping chute discharge
~(-**Kran)stampfer** m, Rammplattenbagger m, Baggerstampfer, Stampfbagger, Fallplattenkran m, Stampfeinrichtung f für Bagger, Baggerstampfgerät n = excavator-operated stamper, dropping weight machine, tamping-crane rammer
~**mischer** m → → Betonmischer
~**mischer** m auf Gleisketten für Betondeckenbau → → Straßen(beton)mischer
~**mischer** m mit Umkehraustragung → → Betonmischer
~**pochwerk** n = gravity stamp
(Frei)Fallramme f, indirekt wirkende Ramme [*Die Antriebsenergie wirkt auf eine Winde, die den Bären hochzieht*] = drop hammer (pile) driving plant, ~ ~ piling ~
Freifallstampfer m → (Freifall-)Kranstampfer
Freigabe f **für den Verkehr**, Verkehrsübergabe = opening to traffic

freigeben für den Verkehr = to open to traffic
Frei-Gefällrohrleitung f = overhead gravity pipeline
Freigelände n [*Messe*] = open-air ground, ~ exhibition space
~-**Aufstellung** f **für Busse** = open-air parking of busses
freigelegte Zuschläge mpl, ~**Zuschlagstoffe** mpl = exposed-aggregate finish
Freihafen m, (zoll)freier Hafen [*im Zollausland gelegener Teil eines Hafens, insbesondere bei Seehäfen*] = free port
~**niederlage** f = bonded warehouse
Freihand|geschwindigkeit f, Bestgeschwindigkeit [*die Geschwindigkeit, bei der die zur Fahrbahnquerneigung parallelen Komponenten von Fliehkraft und Schwerkraft im Gleichgewicht sind*] = hands off speed
~**zeichnen** n = freehand drawing
freihändige Vergabe f = negotiated contract
Frei|heitsgrad m = degree of freedom
~**kolbenkompressor** m, Freikolbendrucklufterzeuger m, Freikolbenluftverdichter m, Flugkolbenluftverdichter, Flugkolbendruckluflterzeuger, Flugkolbenkompressor = free-piston air compressor
~**kolbenmotor** m = free-piston engine
~**kolben-Turboanlage** f = free-piston gas generator
~**lagerplatz** m = open storage ground, ~ ~ area
~**lagerung** f = outside storage
~**lagerung** f im (oder in) **Schütthaufen**, Haldenlagerung, Deponie f = ground storage, stockpiling (in the open air)
~**länge** f [*Kragträger*] = unsupported length
~**laßventil** n, Freilaufventil, Leerlaufventil = no-load valve
Freilauf m = free-wheel(ing system)
~**gruppe** f = free-wheel assembly
~**kupplung** f = overrun clutch
~**rohr** n, Freispiegelrohr = free-surface flow pipe

Freilaufschütz(e) — Freivorbau 432

~schütz(e) *n*, (*f*), Leerlaufschütz(e) = waste sluice (gate)
~stall *m* = loose-housing shed
~stollen *m*, Freispiegelstollen = free-surface flow tunnel, ~ ~ gallery
~trommel *f*, Trommel mit Freilauf [*Tiefbohrtechnik*] = free-wheeling drum
~ventil *n* → Freilaßventil
freilegen, abdecken = to (lay) bare, to lay open, to uncover [*by removing a covering*]
Freilegen *n*, Abdecken = uncovering [*by removing a covering*]
Frei|legung *f* von Baugelände *n* → Räumung(sarbeiten) *f*pl
~leitung *f* [*Stromversorgung*] = (electrical) transmission line, overhead ~ ~
~leitungskabel *n* = overhead transmission line cable
~leitungsmast *m*, Überlandmast = electric power pylon, transmission line tower
~lichtbühne *f* = open-air theatre
~lichtkino *n* = open-air cinema
frei|liegende Ladung *f* = blasting hole without stemming (material)
~liegendes Zugband *n*, außenliegendes ~ = exterior tieback
Freiluft|anlage *f*, Außenanlage = outdoor installation
~kessel *m* = outdoor boiler
~-Krafthaus *n*, Oberflächen-Krafthaus = surface power house
~lagerung *f* [*Materialprobe*] = outdoor exposure
~schaltanlage *f*, Umspannwerk *n*, Schalthof *m* = switchyard, outdoor switchgear
~schuppen *m* [*Ziegelindustrie*] = permanent hack, drying shed, open air corridors
~-Schwimmbad *n* = open-air swimming pool
~-Tanzfläche *f* = open-air dance floor
~-Trafostation *f* = outdoor substation
~trocknung *f*, Freilufttrocknerei *f* = open-air drying

freimachen → abdecken
Freimachen *n* → Abdecken
freischaffender Architekt *m* = independent architect
Freischwing(er)sieb(maschine) *f* [*Kreisschwingsieb mit Umwuchtantrieb*] = flexible-drive screen
Freispiegel *m* = free surface
~abfluß *m* = free-surface flow
~brunnen *m* = well in aquifer with free water surface
~kanal *m*, Freispiegelgerinne *n*, Freispiegelrinne *f* = free-surface flow channel
~leitung *f*, drucklose Leitung = free-surface flow line
~rohr *n*, Freilaufrohr = free-surface flow pipe
~stollen *m*, Freilaufstollen [*Stollen, der im planmäßigen Betrieb nicht bis zum Scheitel des Querschnittes vom Wasser gefüllt wird*] = free-surface flow gallery, ~ ~ tunnel
freistehend = free-standing
freistehender Pfahl *m*, Langpfahl [*Pfahl, der über den Baugrund emporragt und nur mit seinem unteren Ende im Baugrund steht. Er wird vorwiegend bei Pfahlgründungen im Wasser verwendet*] = pile partly in the ground
freistehendes Gerüst *n*, freistehende Rüstung *f* = independent scaffold(ing)
~ **Kraftwerk** *n*, Oberflächenkraftwerk, Übertagekraftwerk = overground (power) station, ~ (~) plant, surface (~)
Freistrahl|turbine *f* = free jet turbine, impulse ~
~überlauf *m*, freier Überlauf = open spillway (US); free waste weir, free spillway
Freiterrasse *f* = open-air terrace
freitragende Spannweite *f* = clear span [*The horizontal distance, or clear unobstructed opening, between two supports of a beam. It is always less than the effective span*]
Freivorbau *m* = cantilevering

~schalung *f*, Kragschalung = cantilever formwork
Freiwange *f*, Lichtwange, Öffnungswange, Treppenlochwange = outer string; face ~ (US)
Fremden|verkehr *m* = tourist trade, tourism
~verkehrsort *m* = tourist resort
fremdgestaltig → xenomorph
Fremd|leitung *f* = foreign line
~rost *m* = rust from external sources
~stoff *m* = extraneous matter, foreign material
~strom *m* = outside power
~stromzufuhr *f* = outside power supply
~wasser *n* [*Schädliche Bodennässe entsteht bei Eintritt von Fremdwasser*] = percolating water
Fressen *n* = scuffing
Freyssinet|-Presse *f*, Preßkissen *n*, Plattenpresse nach Freyssinet, (Freyssinet-)Kapselpresse = Freyssinet-type jack, flat ~
~-Verankerung *f* [*Spannbeton*] = Freyssinet anchoring system, ~ anchorage ~
Fries *m* [*Wandstreifen oder -feld mit gemalten oder plastischen Ornamenten oder Figuren*] = frieze
Friktions|kupplung *f*, Reib(ungs)kupplung, Rutschkupplung = friction clutch (coupling)
~spill *n* [*auf dem Ende einer Schlämm-(seil)trommel*] = friction cathead
~winde *f*, Reibradwinde, Reib(ungs)winde = friction hoist
(frischer) Zementbrei *m* [*Zement plus Wasser*] = (wet) cement paste, wet paste
frisches Abwasser *n* = fresh sewage [*Sewage of recent origin still containing free dissolved oxygen*]
Frisch|beton *m* → Beton
~betondruck *m*, Schalungsdruck = pressure developed by concrete on formwork, concrete pressure
~betongewicht *n* = wet-mix weight
~betonpfahl *m* = pile of freshly placed concrete
~betonprofilometer *m* = wet surface profilometer
~betonsteife *f*, Frischbetonkonsistenz *f* = wet consistency
~beton-Verladetrichter *m* = wet hopper
~dampf *m* = live steam
frisch|gefallener Schnee *m* = fresh snow
~geformtes Erzeugnis *n*, Formling *m* = freshly-made product
Frischluft|-Heizgerät *n* = fresh-air heater
~kanal *m* = fresh-air duct
~schornstein *m* = fresh-air inlet stack
~zufuhr *f* = admission of fresh air
Frisch|mörtel *m* = wet mortar, fresh(ly-mixed) ~
~schlamm *m* [*unzersetzter Schlamm, wie er z.B. in Absetzbecken anfällt*] = green sludge, fresh ~, primary ~
~wasserkühlung *f* = raw water cooling
Friseursalon *m* = hairdressing saloon
Frontachse *f*, Vorderachse = front axle
frontaler Niederschlag *m* = frontal precipitation
Frontalzusammenstoß *m* = head-on collision
Front|-Anbaugerät *n*, Front-Zusatzvorrichtung *f*, Front-Arbeitsgerät, Front-Austauschwerkzeug *n*, Front-Zusatzwerkzeug, Front-Anbauvorrichtung, Front-Austauschgerät, Front-Zusatzgerät = front (attachment), ~ rig
~-Anbauvorrichtung *f* → Front-Anbaugerät *n*
~antrieb *m* = front drive
~antriebsgruppe *f*, Fronttriebsatz *m*, Frontantrieb *m* = front power unit, ~ drive (~), ~ driving gear
~-Arbeitsgerät *n* → Front-Anbaugerät
~aufhängung *f* = front suspension
~-Austauschgerät *n* → Front-Anbaugerät
~-Austauschwerkzeug *n* → Front-Anbaugerät *n*

Frontbremse — Frosthebung

~bremse *f* = front brake
~-Eigengewicht *n* = front-end dead weight
~-Fahrlader *m* mit (Gleis)Kettenfahrwerk → (Gleis)Kettenschlepper-Frontlader
~flugplatz *m* → Feldflugplatz
~gewicht *n* = front-end weight
~-Hochlader *m* → Front-Ladeschaufel *f*

Frontkipper *m* → Autoschütter
~ [*allgemeiner Ausdruck*] = front tipper [*universal term*]

Front-Ladeschaufel *f*, Front-Schaufellader *m*, Frontlader, Front-Fahrlader, Front(schaufel)(-Fahr)lader, Front-Hochlader, Vorderlader, Frontladegerät = front-end (tractor) loader, head-end type tractor-loader, forward shovel
~ mit Kufen, Gleitkufen-Frontlader = skid shovel
~einrichtung *f* = dozer-shovel unit

Front|motor *m* = front engine
~platte *f* = facing slab
~querträger *m* des Kübels [*Radschrapper*] = spreader bar
~räumer *m* → Bulldozer
~räumer *m* mit Schwenkschild → Trassenschäler
~(räumer)-Planierschild *n* → → Bulldozer *m*
~-Schaufellader *m* → Front-Ladeschaufel *f*
~scheinwerfer *m* = head light, headlamp
~schild *n* → → Bulldozer *m*
~seilwinde *f* = front cable winch
~sitz *m* = front seat
~stoßstange *f* = front bumper
~triebsatz *m* → Frontantriebsgruppe *f*
~walze *f*, Vorderwalze, Frontwalzrad *n*, Vorderwalzrad = front roller wheel
~zapfwellen-Anschluß *m*, vorderer Kraftanschlußstutzen *m* für Anbaugeräte [*Schlepper*] = front power take-off
~zughaken *m* = front pull hook

~-Zusatzgerät *n* → Front-Anbaugerät
~-Zusatzvorrichtung *f* → Front-Anbaugerät *n*
~-Zusatzwerkzeug *n* → Front-Anbaugerät *n*

Frosch *m*, Klammer *f* = clip
~ [*Gleisbau*] = frog
~, DELMAG-Stampfer *m*, DELMAG-Explosionsramme *f*, Frosch-Ramme = jumping frog, leaping ~, frog ramming machine, frog (type jumping) rammer; DELMAG frog tamper [*Trademark*]
~maul *n* [*stehendes Dachfenster*] = semicircular dormer window

Frost|anzeiger *m*, Frostindikator *m* = ground frost indicator
~aufbruch *m* → Frosthebung *f*
~aufgang *m*, Auftauen *n*, Tau *m* = thaw
~auftreibung *f* → Frosthebung
~auftritt *m* = incidence of frost
~bau *m* = construction work in freezing weather
~beginn *m* → Frostschwelle *f*
~bemessung *f* = design for protection against frost action
frostbeständig, frostunempfindlich = frost-resistant, non-frost-active
Frost|beständigkeit *f*, Frostunempfindlichkeit *f* = frost resistance
~beule *f* = frost boil
~dauer *f* = duration of freeze
~eindringtiefe *f* = frost penetration depth
~eindringung *f* = frost penetration
~eindringzone *f* = frost penetration zone
~einwirkung *f*, Frosteinfluß *m* = frost action
frostempfindlich = frost-susceptible
Frostempfindlichkeit *f* = susceptibility to frost action
frostfrei [*ist die Tiefe, in die der Frost nicht eindringt*] = frost-free, non-freezing
Frost|grenze *f* [*Baugrund*] = frost line
~hebung *f*, Frostaufbruch *m*, Hebung durch Eislinsenbildung im Boden,

Frostindex — Fuge

Aufbruch durch Eislinsenbildung im Boden, Frostauftreibung = frost heaving, ~ heave, break-up
~**index** m = (air) freezing index
~**indikator** m, Frostanzeiger m = ground frost indicator
~**körper** m [*Gefrierverfahren*] = ice mass
~**kriterium** n = frost criterion
~**linse** f, Eislinse = frost lens, ice ~
~**mauer** f, Eismauer, Frostschürze f, Eiswand f, Frostwand, Eisschürze [*Schachtabteufung im Gefrierverfahren*] = ice wall
~**probe** f, Frostprüfung f, Frostversuch m = freezing test
~**probe** f → Frost-Tau-Prüfung f
~**prüfung** f, Frostprobe f, Frostversuch m = freezing test
~**-Prüfung** f → Frost-Tau-Prüfung f
~**rissebildung** f = frost cracking
~**riß** m = frost crack
~**schaden** m = damage by frost
~**schub** m = frost thrust
~**schürze** f → Frostmauer f
Frostschutz m = protection from frost
~**beregnung** f = sprinkling for frost protection
~**flüssigkeit** f = antifreeze liquid
~**kies** m = frost blanket gravel
~**lösung** f = antifreeze solution
~**maßnahme** f = frost precaution
~**mittel** n, Gefrierschutzmittel = antifreeze (mixture), frost protective, antifreezer, frost-protection agent
~**pulver** n = antifreeze powder
~**schicht** f [*Sandschicht unter der unteren Tragschicht, die eine Eislinsenbildung verhindert*] = antifrost layer, antifreeze ~, frost blanket
Frostschwelle f, Frostbeginn m = initiation of freezing
frostsicher [*Jedes Bauwerk muß frostsicher gegründet werden, d.h., seine Sohle muß tiefer liegen als die 0°-Isotherme im Boden*] = frost-proof
Frost|-Tau-Prüfung f, (Tau-)Frost-Prüfung, Gefrier-Auftau-Prüfung,

Frostprobe f, Frostversuch m = freezing and thawing test
~**-Tau-Wechsel** m, Gefrier-Auftau-Folge f = freezing and thawing cycle
~**tiefe** f = depth of frost penetration, frost-penetration depth
~**versuch** m, Frostprobe f, Frostprüfung f = freezing test
~**versuch** m → Frost-Tau-Prüfung f
~**wand** f → Frostmauer f
~**wechsel** m = intermittent freezing
~**wetter** n = frosty weather, freezing ~
~**widerstand** m = resistance to freezing, freezing resistance
~**zone** f = frost zone
Froude'sches Ähnlichkeitsgesetz n = Froude's law
Froude-Zahl f = Froude's number
Fruchttransportschiff n = fruit carrier
frühe Rißbildung f = early cracking
Frühfestigkeit f = early strength
frühhochfest = high-early-strength
frühhochfester Portlandzement m → hochwertiger ~
Frühjahrs|bewässerung f = early spring irrigation
~**hochwasser** n = spring flood
~**müll** m = spring refuse
~**schnee** m = spring snow
~**-Tagundnachtgleiche** f = vernal equinox
Frühschicht f = fore shift
frühtragend (⚒) = early-bearing
Frühzündung f = advance ignition, pre-ignition
(FS-)Kontrollturm m, Flugleitungsturm = control tower
Fuchsit m, Chromglimmer m (Min.) = fuchsite
Fuchs(kanal) m [*das gemauerte Verbindungsstück zwischen großen Kesselanlagen und dem Schornstein*] = flue
Fuchsöffnung f [*Schornstein*] = flue opening
Fuge f = joint
~ **zur Verringerung der Aufwölbungsspannungen** = hinge joint, warping ~

Fugeisen — (Fugen)Stützkorb

Fugeisen n → Fugenkelle f
Fugen n → Aus~
~**abstand** m = distance between joints
~**anker** m, Ankereisen n = anchor bar, tie ~, steel ~ ~
~**anordnung** f = jointing arrangement
~**ausbildung** f = joint design
~**band** n = joint tape, tape to seal joints
~**bankett** n, Fugenschwelle f [*Betondeckenfuge*] = sleeper
~**bläser** m mit Voranstrichvorrichtung = combined air jet and priming unit
~**brett** n, Holz(fugen)einlage f, Holzbrett = wood(en) (joint) filler, filler board
(~)**Deckleiste** f, (Fugen)Deckstreifen m = joint (covering) strip
fugendicht = joint-tight
Fugen|dübel m, Beton~, Rundstahldübel = dowel (bar)
~**einbaugerät** n = joint installing machine, ~ placing ~
~**einlage** f, feste Einlage, Fugenstreifen m = pre-formed (joint) filler, joint sealing strip, (expansion) joint filler, (expansion) joint(ing) strip, pre-mo(u)lded (strip joint) filler, strip of pre-formed filling material
~**eisen** n, Hohl~ = expansion joint cap strip (US); metal sleeve, metal cap(ping), (metal) capping (strip) (Brit.); joint-forming metal strip
~**füllstoff** m = joint filler
~**halter** m [*Betondeckenfuge*] = supporting device, guard
~**herstellung** f = joint construction
~**hobeln** n = gouging
~**hobler** m, Autogen-~ = (autogenous) gouging attachment
~**kelle** f [*Pflasterfuge*] = sett jointer, ~ teeder
~**kelle** f, Streicheisen n, Fugeisen, Ausfugkelle [*Maurerwerkzeug*] = pointing trowel
~**kitt** m → Verguẞmasse f
fugen|los miteinander verbinden = to join monolithically

~**loser (Fuß)Boden** m, Spachtel(fuß)boden, Estrich m = jointless floor(ing), composition ~, ~ floor finish, flooring coat, compofloor, screed [*In spite of the name, many jointless floors are best laid with joints at about 8 ft. apart, particularly those containing cement, to allow for shrinkage*]
~**loses Betonbauwerk** n, Monolithbetonbauwerk = monolithic concrete structure
Fugen|mastix m = joint sealing mastic, building ~ [*other than mastic asphalt = Gußasphalt*]
~**maurer** m [*Herstellung von Betondeckenfugen*] = finisher
~**mörtel** m, Mauermörtel = joint mortar, masonry ~, pointing ~
~**mörtelzement** m = mortar cement, masonry ~
~**pflug** m [*Fugenräumgerät mit Pflugstahl, Schaber und Drahtbürste*] = plough-type raking machine (Brit.); plow-type ~ ~ (US)
~**plan** m = joint plan
~**profil** n = joint profile
~**reiniger** m, Fugenreinigungsgerät n = joint cleaner
~**reißer** m = joint raker
~**säge** f [*Fehlname*] → Fugenschleifgerät n
~**schleifgerät** n [*Diese Bezeichnung ist technisch richtiger als die meist üblichen „Fugenschneidmaschine", „Fugensäge", „Beton(kreis)säge" und „Fugenschneider"*] = (concrete) joint cutter, joint-cutting machine, (contraction) joint sawing machine, concrete cutter; concrete saw (US)
~**schneider** m [*Fehlname*] → Fugenschleifgerät n
~**schneidmaschine** f [*Fehlname*] → Fugenschleifgerät n
~**schwelle** f, Fugenbankett n [*Betondeckenfuge*] = sleeper
~**spalt** m = groove, slot
~**streifen** m → Fugeneinlage f
(~)**Stützkorb** m, Stahldrahtkorb = wire chair, metal ~, ~ cradle

~verdübelungsgerät n = mechanical dowel and tie-bar installer
~vergußarbeiten fpl = joint grouting work, ~ pouring; topsealing (US)
~vergußgerät n = (joint) sealing machine, (~) pouring machine, (~) applicator
(~)Vergußgerät n mit druckhafter Füllung = pressure applicator
~vergußkanne f = pouring can, ~ pot
~vergußmasse f → Vergußmasse
~verwerfung f = joint faulting
~wand(ung) f = joint face
fugenzeigender Fußboden m = joint flooring, ~ floor finish
Fugmörtelbrett n [Maurerwerkzeug] = hawk, handboard, mortar board
Fugung f → Ausfugen n
Fühler m = sensing element, ~ device, sensor
~, Taster m, Höhen~ [Gleitschalungsfertiger] = (grade) sensing device, (~) sensor
Fühl|lehre f = feeler gauge
Fühlnadel f [Einspritzdüse] = feeling pin
führen [Straße; Eisenbahn; Kanal] = to route
Führer m, Fahrer [Bediener eines fahrbaren Gerätes] = operator, driver
~aufzug m = attendant-controlled lift (Brit.); attendant-controled elevator (US)
~beanspruchung f → Fahrerbeanspruchung
~bedienung f [Fahrstuhl] = attendant-operated control
~-Drehsitz m → Fahrer-Drehsitz
~ermüdung f → Fahrerbeanspruchung
~haus n → Fahrerhaus
führerhausbedient = cab-operated
Führerkabine f → Fahrerhaus n
~, Bedienungskabine, Bedienerkabine = operator's cab(in)
Führer|schein m = driver's licence (Brit.); ~ license (US)
~scheinentzug m = suspension of driver's license (US); ~ ~ ~ licence (Brit.)

~-Schwenksitz m → Fahrer-Drehsitz
~sitz m, Fahrersitz = driver's seat
~sitz m, Bedienersitz = operator's seat
~stand m, Fahrerstand = control platform
~standlaufkatze f Führerstandlastkatze = man troll(e)y, driver ~, driver-seat crab
Führungs|bohle f = guide plank
~draht m = guide wire
~handgriff m = guiding handle
~keil m = adaptor, adapter
~konsole f [Absenken eines Tunnelelementes] = coupler arm
~korb m [Tiefbohrtechnik] = centralizer
~lager n = clutch release bearing
~lager n = pilot bearing
~leiste f [Bandförderer] = skirt board
~mast m = guide mast
~nut f = guiding groove
~nut f [Absenken eines Tunnelelementes] = coupler seat
~pfahl m = guide pile
~platte f = guide plate
~rahmen m = guide frame
~ring m = guide ring
~rippe f = guide rib
~rohr n = guide tube
~rolle f = guide roller
~rollenbolzen m = guide roller bolt
~säule f [TOURNACRANE] = tilting track, elevator ~
~schiene f = guide rail
~schlitten m, Führungsblock m = guide block
~schuh m = guide shoe
~stahl m [Bei einem Bohrwagen eine Führung, die den anschneidenden Bohrer richtig ausgerichtet hält] = steel centralizer
~ständer m [Bohrwagen] = guide shell, mast
~stift m = dowel (pin)
~stück n, Führung f = guide
~system n = guide system
~- und Tragrolle f [Bandförderer] = guidler
~winkel m = guide angle

Fuhr|unternehmer m → Transportunternehmer
~weg m = cart-path
~werkverkehr m, Zugtierverkehr, Gespannverkehr = animal-drawn traffic
Fukoidensandstein m = fucoid(al) sandstone
Fukus m → → Tang m
Füllage f [*Schweißtechnik*] = filler bead
Füll|beton m [*der ein Rohr umschließt*] = haunching concrete
~beton m → Hinterfüllungsbeton
~boden m → Auf~
~bunker m → Annahmebunker
~draht-Machart f = filler wire construction
~drahtmachart f 6×25 mit Drahtseilseele = 6×25 filler wire with independent wire rope core
~drahtseil n = filler wire rope
~eisen n, Plessit m = plessite
~element n [*Betonstein-Fassadenverkleidung*] = infilling panel, spandrel ~
Füllen n, Füllung f, Be~ = filling
Füller m, Mineral~, Füllstoff m, porenfüllender Zusatz m [*Straßenbau*] [*Schweiz: Filler*] = (mineral) filler, ~ dust
~anteil m, Füllergehalt m = filler content
~aufgabeelevator m, Fülleraufgabebecherwerk n = filler (bucket) elevator
Füllererde f, Walkerde = fuller's earth
Füllergehalt m, Flleranteil m = filler content
Fuller|-(Kinyon)Pumpe f, Zement(staub)pumpe, Fuller-Zementpumpe, Fuller-Entladepumpe = Fuller-Kinyon unloader (pump), F-K (dry) pump
~kurve f [*Bindemittel eingeschlossen*] = Fuller's best mix curve
~-Luftrutsche f, Fuller-(Gebläse)Luftförderrinne f = Fuller airslide
~parabel f [*Bindemittel ausgenommen*] = Fuller's parabola

Füller|-Rückgewinnung f = filler reclamation
~schnecke f = filler screw
Fuller-Schrägrostkühler m = Fuller grate cooler
Füll|höhe f, Füllstand m = material level (elevation)
~hohlkörper m aus gebranntem Ton, Decken~ ~ ~ ~ = hollow clay block
~inhalt m, Nutzinhalt, Nenninhalt [*Betonmischer*] = unmixed capacity, dry capacity per batch
~kasten m [*Steinformmaschine*] = feed box
~körper m → Deckenkörper
~körper m → Einstaufilter m, n
~körpersäule f [*Destillier- und Rektifiziertechnik*] = packed column
~masse f → Auftragmaterial n
~masse f → Nivelliermasse
~masse f, Vergußmasse [*zur Vermuffung von Rohren*] = pipe joint(ing) compound, sewer ~ ~, joint cement
~material n [*Abwasserwesen*] = contact material
~material n → Auftragmaterial
~mauerwerk n = filling-in work
~mörtel m → Magermörtel
~rutsche f = charging chute, charge ~
~schacht m, Mülleinwurfschacht [*zum Füllen einer Müllverbrennungsanlage*] = charge well, charging ~
~splitt m → Keilsplitt
~stab m, Füllungsstab, Gitterstab, (Träger)Stab = member, bar, rod
~stand m, Füllhöhe f = material level (elevation)
(~)Standanzeiger m, (Füll)StandMeßgerät n, Füllungsmelder m = (material) level indicator
~stand-Fernanzeiger m = remote level indicator
~standmessung f = measurement of level
~stellung f = filling position

Füllstrom — Fundationstiefe

~**strom** m [*Meer*] = filling current
~**stutzen** m = filling spout, ~ tube, filler
~**trichter** m → Aufnahmetrichter
~**tür** f [*zum Füllen einer Anlage, z.B. einer Müllverbrennungsanlage*] = charge door, charging ~
~**umlauf** m [*Schleuse*] = filling culvert
Füllung f, Tür~ = (door) panel
~[*Kastenfang(e)damm; Zellenfang(e)damm*] = fill
~, **Füllen** n, Be~ = filling
Füllungs|melder m → (Füll)Standanzeiger m
~**stab** m, Gitterstab, Füllstab, (Träger)Stab = member, bar, rod
Füll|verfahren n [*Abwasserreinigung*] = contact method of sewage treatment
~**zement** m [*Fehlname*] → Nivelliermasse f
Fundament n, Gründung f, Fundation f, Gründungskörper m, Fundamentkörper m = foundation (structure), engineering foundation [*The part of a structure or building below the surface of the ground. It bears directly on, and transmits the structure load or building load to, the supporting soil or rock*]
~**absatz** m → Grundbank f
~**absteckung** f = setting-out a foundation
Fundamentalzeit f = standard time
Fundament|armierung f → Fundamentbewehrung
~**art** f, Gründungsart, Fundationsart = type of foundation
~**bau** m, Gründungsbau = foundation construction
~**baugrube** f = foundation excavation, ~ cut, ~ building pit
~**belastung** f, Gründungsbelastung = foundation loading
~**bemessung** f, Gründungsbemessung = foundation design
~**beton** m, Gründungsbeton = foundation concrete

~**bewehrung** f, Fundamentarmierung, Fundament(stahl)einlagen fpl = foundation reinforcement
~**einlagen** fpl → Fundamentbewehrung f
~**graben** m = foundation trench
~**höhe** f = foundation level
~**hohlpfeiler** m → Hohlfundamentpfeiler
Fundamentierung f, Gründung, Fundierung, Fundation f = foundation
Fundament|körper m → Fundament n
~**last** f, Gründungslast = foundation load
~**mauer** f, Grundmauer = foundation wall
~**mauerwerk** n, Grundmauerwerk = foundation masonry
~**oberkante** f = top of foundation
~**pfeiler** m, Grundpfeiler = foundation pier
~**plan** m, Gründungsplan = foundation layout plan
~**platte** f → Plattenfundament n
~**schalung** f = foundation formwork
~**sockel** m → Sockelgründung f
~**stabilisierung** f = foundation stabilization
~(**stahl**)**einlagen** fpl → Fundamentbewehrung f
~**verbreiterung** f = extension of the foundation
~**zeichnung** f = foundation plan
Fundation f → Fundament n
~, **Fundierung** f, Gründung, Fundamentierung = foundation
Fundations|arbeiten fpl → Gründungsarbeiten
~**art** f, Gründungsart, Fundamentart = type of foundation
~**kote** f, Fundationshöhe f = level of foundation
~**platte** f → Plattenfundament n
~**sockel** m → Sockelgründung f
~**technik** f, Gründungswesen n, Grundbau m, Grundbautechnik = foundation engineering
~**tiefe** f → Gründungstiefe

Fundbohrung *f* → erste fündige Bohrung
fundieren, gründen = to found
Fundieren *n*, Gründen = founding
Fundierung *f*, Gründung, Fundation *f*, Fundamentierung = foundation
Fundierungsarbeiten *fpl* → Gründungsarbeiten
Fündigkeit *f* auf Erdöl = discovery of petroleum
Fündigwerden *n* = strike
Fundstätte *f* → Vorkommen *n*
Fünf|-Füllungstür *f* = five-panel door
~punktanordnung *f* [*bei sekundärer Förderung von Bohrlöchern*] = five-spot pattern
Fünfteldach *n* = roof with pitch of 1 : 5
Fünf|trommelwinde *f*, Fünftrommelwindwerk *n* = five-drum winch
~zahnaufreißer *m* = 5-tine unit
funikulares Wasser *n* = funicular water
Funken|bildung *f* = sparking
~erosion *f* = spark erosion
~fänger *m* = spark arrester (US); ~ arrestor (Brit.)
~strecke *f*, (Funken)Schlagweite *f*, Entladeweite *f* = sparking distance, striking ~, spark gap
Funk|feuer *n* = beacon light, radio beacon
~meßstation *f* im Meer = radar island, offshore radar platform
~senderempfänger *m* = transceiver
Funktionsprinzip *n*, Arbeitsprinzip [*Maschine*] = work(ing) principle, operating ~
Furche *f* [*Fließgrenzenermitt(e)lung eines Bodens*] = groove
~ = furrow
~ [*Bewässerung*] = furrow, small ditch
Furchen|bewässerung *f*, Beetbewässerung, Furchen(be)rieselung = furrow irrigation [*A method of irrigating by small ditches or furrows leading from a header or supply ditch*]
~bewässerung *f* mit kleinen Furchen = corrugation method (US)

~spachtel *m, f* [*Fließgrenzenermitt(e)lung eines Bodens*] = grooving tool
~ziehung *f* [*Fließgrenzenermitt(e)lung eines Bodens*] = cutting of grooves
Fürsorgeerziehungsheim *n* = remand home
Furt *f* = ford
Furtstrecke *f*, Übergangsstrecke = inflexion stretch
Fusel-Öl *n* Nr. 2, Solaröl = solar oil
Fuß *m*, Rohr~ = base, pipe ~
~ [*Pfahl*] = base
~ [*Damm; Deich*] = toe
~, Flansch *m*, Gurt *m*, Träger~ = flange, chord, boom
~ausbildung *f*, Fußform *f* [*Pfahl*] = base shape [*pile*]
~becken *n*, Auslauf-Vertiefung *f* [*Saugüberlauf*] = lower lip (US); outlet bucket (Brit.)
~bedienung *f* = foot control
(Fuß)Boden *m*, (Fuß)bodenbelag *m* = flooring, floor (finish), floor cover(ing)
~brett *n* = flooring board
~farbe *f* = floor(ing) paint, floor finish ~
~fliese *f*, (Fuß)Bodenplatte *f* = floor (finish) tile, flooring ~
~fuge *f* = floor(ing) joint
~härtemittel *n* = floor(ing) hardener
~-Härteprüfgerät *n* = floor(ing) hardness tester
~heizung *f*, (Fuß)Bodenstrahlungsheizung = solid-floor heating, floor-warming, (under)floor heating
~kleber *m* = flooring adhesive
~material *n* = flooring material, floor (finish) ~
~nutzbelag *m* → Nutzschicht
~nutzschicht *f* → Nutzschicht
~platte *f*, (Fuß)Bodenfliese *f* = flooring tile, floor (finish) ~
~plattenpresse *f* = floor-tile press
~reinigungsmittel *n* = floor(ing) cleaner
~schleifmaschine *f* = floor(ing) grinder, ~ grinding machine

~(strahlungs)heizung *f* = solid-floor heating, floor-warming, (under)floor heating
~wachs *n* = floor wax
~walze *f* = floor(ing) roller
~ziegel *m* = flooring brick
Fuß|breite *f* [*Rohrfuß*] = base width
~bremse *f* = footbrake
~druckknopf *m* = foot button
~form *f*, Fußausbildung *f* [*Pfahl*] = base shape [*pile*]
Fußgänger *m* = pedestrian
~bezirk *m* = pedestrian precinct
~brücke *f* = foot bridge, pedestrian ~, walkover
~-Einkaufstraße *f* = shopping street for pedestrians only
~-Fährgast *m* = pedestrian ferry passenger
~-Schutzgeländer *n* = pedestrian guard rail, protection fence, guard fence, safety fence
~-Schutzinsel *f*, Stützinsel für Fußgängerüberweg, Fußgänger-Stützinsel = refuge (Brit.); [*deprecated: island*]; ~ island, pedestrian island (US)
~sicherheit *f* = pedestrian safety
~steg *m*, Fußgängerüberführung *f*, Fußwegüberführung, Fußwegübergang *m* = overpass for pedestrians
~straße *f*, Gehstraße = pedestrian way
~tunnel *m*, Fußgängerunterführung *f* = pedestrian subway, ~ (under-)pass
~tunnel *n* unter Wasser = underwater walkway for pedestrians, ~ pedestrian tunnel
~überführung *f* → Fußgängersteg *m*
~überweg *m* = pedestrian crossing, ~ crosswalk
~unfall *m* = pedestrian accident
~unterführung *f* → Fußgängertunnel *m*
~verkehr *m* = pedestrian traffic, ~ circulation
~weg *m* → Bürgersteig *m*
~zentrum *n* = pedestrian city centre (Brit.); ~ ~ center (US)

Fußgas|hebel *m* → Fahrfußhebel
~regler *m* = accelerator-decelerator
Fuß|gelenk *n* [*Bogenträger; Rahmen*] = hinge
~gesims *n* = base moulding [*The moulding immediately above the plinth of a wall, column, etc.*]
~hebel *m* → Fußpedal *n*
~kupplung *f* = foot clutch
~leiste *f*, Scheuerleiste = skirting (board) (Brit.); base plate [*Scotland*]; mopboard, washboard, scrub board, base board (US)
~leistenheizung *f*, Scheuerleistenheizung = domestic skirting heating (Brit.); ~ base plate ~ [*Scotland*]; ~ base board ~ (US)
~mauer *f*, Herdmauer, Abschlußwand *f*, (Ab)Dichtungsmauer, (Ab-)Dichtungssporn *m*, Trennmauer, Sporn [*Talsperre*] = (toe) cut-off wall
~pedal *n*, Fußhebel *m*, Pedal = foot pedal
~pfette *f*, Sparrenschwelle *f* [*Pfettendach*] = inferior purlin(e)
~platte *f* [*Stütze*] = base plate
~platte *f* → Rüttelplatte
~pumpe *f* = foot-pump
Fußpunkt *m* = base
~ **der Kraft** *f* = tracing point of the force
~ **des Horizonts** = nadir of the horizon
~ **einer Senkrechten**, ~ ~ Normalen = foot of a perpendicular
Fuß|raste *f* = foot rest
~ringschale *f* [*Brückenpfeiler*] = bottom shell
~schaltung *f* für Umkehrgetriebe plus Drehmomentwandler = instant reverse
~schmiege *f* = horizontal cut (of jack rafter)
~schraube *f* [*Theodolit*] = foot screw, levelling ~, plate ~
~sieb *n* → Saugkorb *m*
~stein *m* → Flügelanfänger
~teil *m, n* [*eines Streckenbogens*] = stilt

Fußventil — Galerie

~ventil *n* [*beim Schmandlöffel*] = check valve

~ventil *n* [*Das Rückschlagventil am Ansaugende eines Pumpensaugschlauches*] = foot valve

~wärme *f* [*Fußboden*] = treading warmth

~waschbecken *n* = foot-bath

Fußweg *m*, Gehbahn *f*, Gehsteig *m*, Gehweg, Bürgersteig = footway (Brit.); sidewalk (US)

~, Gehweg = footpath

~-Betonplatte *f* → Bürgersteig-Betonplatte

~konsole *f* → Gehwegkonsole

~platte *f* → Gehbahnplatte

~überführung *f* → Fußgängersteg *m*

~übergang *m* → Fußgängersteg *m*

Futter *n* → Bohr~

~blech *n* = lining plate

~brett *n* → → Setzstufe

~-Langtrog *m* = manger

~mauer *f*, Verkleidungsmauer, Bekleidungsmauer, Auskleidungsmauer [*Eine Stützmauer vor gewachsenem Erdreich*] = revetment wall [*The revetment wall differs from the retaining wall in that its function is to protect the earth, not to withstand its thrust*]

~rahmen *m* → Fensterrahmen

~rohr *n*, Bohrrohr = casing tube

~rohrkrone *f* [*Bei Schürfbohrungen zum Einbohren von Futterrohrfahrten*] = casing shoe

~rohrkrone *f* [*Zum Nachbohren eines Bohrloches vor dem Einbau der Futterrohre bei zum Nachfall neigenden Formationen oder zum Überbohren eines Fisches*] = casing bit

~rohrspülkopf *m* [*Tiefbohrtechnik*] = casing water swivel

~rohrstrang *m* [*Tiefbohrtechnik*] = casing string

~rohrzentrierung *f* = casing centralizer

~stein *m*, Futterziegel *m* = lining brick

~stufe *f* → Setzstufe

~vorratsraum *m* = feeding store

~ziegel *m* → Futterstein

G

Gabbro *m* = gabbro

~aplit *m* = gabbro aplite, beerbachite

~diorit *m* = gabbro-diorite

~gang *m* = gabbroitic dike (US); ~ dyke

gabbroider Verband *m* (Geol.) → Bienenwabenverband

Gabbro|lagergang *m* = gabbro sill

~linse *f* = gabbroic lens

~magma *n* = gabbro magma

~nelsonit *m* = gabbro nelsonite

Gabel *f* = fork

~abzweiger *m* [*Kanalisationssteinzeug*] = double junction, breech

~anker *m* = forked tie

~antrieb *m* [*Seilbahn*] = rope fork drive

~bolzen *m* = clevis pin

~hebel *m* = forked lever

~holz *n*, Band *n* = forked wood

~kreuzung *f*, Straßeneinmündung *f*, Straßengab(e)lung = fork junction

~probe *f* [*Korrosionsprüfung*] = fork-test bar

~schurre *f* = two-way chute, flap ~

~stange *f* = fork rod

~stapler *m* = fork-lift, fork truck, high-lift fork stacking truck, fork lift truck type conveyancer

~stapler *m* für Rohrtransport = pipe (lift-)truck

~stapler *m* mit leichter Lenkbarkeit = stand-up fork truck

Gab(e)lung *f*, Abzweig *m*, Abzweigung = bifurcation

~ = bifurcation gate [*In hydraulics, a structure that divides the flow between two conduits*]

Gahnit *m*, Zinkspinell *m*, $ZnAl_2O_4$ (Min.) = zinc spinel, gahnite

Gaize *f*, Pseudokieselgur *f* = gaize

~zement *m*, Pseudokieselgurzement = gaize cement

Galenit *m*, Bleiglanz *m*, PbS = galena, lead glance

Galerie *f* [*z. B. Kunstgalerie*] = gallery

~ [*Theater*] = gallery

Galileo-Mikrotelemeter n mit der Methode f des Bleisenkels = Galileo microtelemeter with the method of the plumb rule
Gallkette f = Gall's chain, bush roller ∼
galvanische Verzinkung f = cold galvanizing
Galvanisieren n = electroplating
Gamann-Klappen(stau)wehr n = Gamann wicket (dam) (US); ∼ shutter (∼) (Brit.)
Gamma|-Gamma-Sonde f zur Dichtesondierung von Böden = gamma-scattering soil density gauge
∼**strahl** m = gamma ray
∼**strahlenkurve** f = gamma ray curve
Gang m, Korridor m, Flur m = corridor [*A comparatively narrow enclosed thoroughfare within a building*]
∼, Lauf m [*einer Maschine*] = operation
∼ [*Kraftübertragung*] = gear, speed
∼**art** f → Berg m
∼**bereich** m = speed range
∼**gestein** n → Berg m
∼**hebel** m = Gangschalthebel
∼**höhe** f, Gewindesteigung f [*Schraube*] = pitch
∼**kupp(e)lung** f = range clutch
∼**linie** f = hydrograph [*A graph showing the stage, flow, velocity, or other property of water, with respect to time*]
∼**linie** f, Gehlinie, Lauflinie [*Treppe*] = walking line
∼**planetenträger** m = range carrier
∼**schaltgetriebe** n = primary transmission
∼(**schalt**)**hebel** m, Gangwechselhebel = gear (change) lever, (∼) shift(ing) ∼, transmission shift(ing) ∼
∼**schaltmechanismus** m = gear shift(ing) mechanism
∼**schalttabelle** f = speed chart
∼**schaltung** f, Gangwechsel m = gear change, speed ∼, gear shifting
∼**schaltventil** n = speed change valve
∼**schaltzeit** f, Gangwechselzeit = gear change time, ∼ shift(ing) ∼

(∼)**Spill** n = capstan Zahnrad-∼ = geared ∼
∼**steuerventil** n = range selector valve
(∼)**Stock** m → Anhäufung f
∼**wahl** f = gear selection
∼**wahlhebel** m = (gear) selector lever
∼**wahlventil** n = speed selector valve
Gangwechsel m, **Gangschaltung** f = speed change, gear ∼
∼**hebel** m → Gang(schalt)hebel
∼**zeit** f, Gangschaltzeit = gear change time
Gänsehals m, Spülkopfkrümmer m, Spülkopfknie n [*Rotarybohren*] = gooseneck (bend)
Ganzdecke f, Seitenmoräne f, Wallmoräne, Randmoräne = lateral moraine
ganzer Walm m = whole hip
Ganz|glastür f = tempered plate glass door
∼**kunststoff-Straßentanker** m = all-plastic road tanker (vehicle)
∼**metallbehälter** m = all-metall container
∼**schweiß-Konstruktion** f = all-welded construction
∼**stahlbauweise** f, Vollstahl-Bauweise = all-steel construction
∼**stahl-Drahtseil** n = wire rope with wire core
∼**stahlkette** f = all-steel chain
∼**stahlkipper** m = all-steel tipper [*lorry*]
∼**stahlmäkler** m, Ganzstahl-Läuferrute f, Ganzstahllaufrute = all-steel leader
∼**tagsarbeit(en)** f(pl) [*24 Stunden durchlaufend*] = round-the-clock work
Garage f = garage
(**Garagen**)**Kipptor** n = overhead garage door, up and over ∼ ∼
Garagentor n = garage door
∼**beschläge** mpl = garage door furniture
garantiertes Brauchwasser n, Anliegerrechtwasser, Pflichtwasser = compensation water
Garantiezeit f = period of warranty

Garderobenvorraum m [*Hotelzimmer*] = antechamber with built-in wardrobe
gären = to ferment
Gärfaulverfahren n = fermentation-septization process
Garn n = yarn
Garnieren n [*Steinzeugindustrie*] = sticking on
Garnierholz n, Stauholz, Garnierung f = dunnage [*loose wood or other material used in a ship's hold for the protection of cargo*]
Garten|figur f = garden figure
~gestalter m = garden architect
~hof m = garden court
~mauer f = garden wall
~platte f → Gartenwegplatte
~stadt f = garden city
~weg m = garden path
~(weg)platte f = garden flag(stone)
gärtnerische Arbeiten fpl = garden work
Gärung f = fermentation
Gärungs|behälter m = fermentation tank
~erreger m = ferment
gärungshemmend = anti-fermentative
Gas|absaugungsanlage f = gas suction plant
~abscheider m → Gasseparator m
~anker m [*in einer Erdölsonde*] = gas anchor
~anlagerungsvermögen n → Adsorption f
~-Aschen-Silikatbeton m = gas-ash-silicate concrete
~ausbruch m = gas outburst
~auslaß m = gas vent, ~ slot [*1.) A passage to permit the escape of gases of decomposition. 2.) An opening which allows gas, liberated in an Imhoff tank sludge-digestion chamber, to reach the atmosphere without passing up through the sewage in the settling chamber*]
~austritt m = gas show [*Escape of natural gas at the ground surface, sometimes an indication of oil beneath*]

~ballastpumpe f = gas ballast pump
~behälter m, Gassammler m, Gasometer m = gas-holder, gas-tank, gasometer, gas vessel
~beleuchtung f = gas lighting
Gasbeton m [*Gasbeton wird aus feinkörnigem Sand und Zement oder Zement und Baukalk hergestellt. Wird der Gasbeton im Sinn des Kalksandsteinverfahrens dampfgehärtet, so kann das Bindemittel bei Verwendung von Quarzsand auch lediglich aus Baukalk bestehen. Dem weich bis zähflüssig angemachten Beton wird ein Zusatzstoff beigegeben, der ein Gas entwickelt, das den frischen Beton aufbläht und so mit Poren durchsetzt. Als gaserzeugende Mittel sind bis jetzt in Anwendung: Aluminiumpulver, Kalzium-Karbid-Pulver oder Schmelzen aus Aluminium und Kalzium-Karbid, ferner Wasserstoffsuperoxyd plus Chlorkalk*] = gas(-formed) concrete
~-Fertig(bau)teil m, n, Gasbeton-Montageteil = gas concrete precast unit, ~ ~ ~ element
~-Montageteil m, n → Gasbeton-Fertig(bau)teil
~platte f = gas concrete slab
~probe f = gas concrete specimen
~stein m = gas concrete block
gasbildend, gasentwickelnd [*Gasbeton*] = generating bubbles of gas, gas-forming
Gas|bildner m → gaserzeugendes Mittel n
~bildung f, Gasentwick(e)lung [*Gasbeton*] = gas formation, formation of gas bubbles
~blase f = gas bubble
~bohren n, Gasbohrung f = gas drilling
~bohrung f, Gassonde f, Gasbrunnen m = gas well
~bohrungsgerät n = gas drilling rig
~brenner m = gas burner
~brunnen m, Gassonde f, Gasbohrung f = gas well

Gas-Chromatographie — Gaskohle

~-Chromatographie *f* = gas chromatography
~decke *f* → Abwasserwesen *n*
~detektor *m* (⚒) = gas detector, firedamp ~ [*It shows that gas is present*]
gasdicht = gas-tight
Gas|dichtigkeit *f*, Gasundurchlässigkeit = imperviousness to gas
~druckregler *m* = gas pressure regulating governor
gasdurchlässig = pervious to gas
Gas|durchlässigkeit *f* = permeability to gas
~-Durchlauf-Wasserheizer *m* = gas circulator
~durchsetzung *f* der Spülung [*Tiefbohrtechnik*] = cutting of mud by gas
~einpreßbohrung *f* = gas input well, ~ intake ~
~einpreßleitung *f* = gas input line, ~ intake ~
~einpreßverfahren *n*, Rückdruckverfahren [*Erdöllagerstätte*] = repressuring (method) [*returning gas under pressure to the sands*]
~einschlüsse *mpl* = gas occlusions
~entladungslampe *f* = gas discharge lamp
~entlösungspunkt *m*, Kochpunkt = bubble point
~entschwefelung *f* = desulfurization of gas (US); desulphurisation ~ ~ (Brit.)
gasentwickelnd, gasbildend [*Gasbeton*] gas-forming, generating bubbles of gas
Gas|entwick(e)lung *f* → Gasbildung
~entwickler *m* → gaserzeugendes Mittel *n*
~entwickler *m* [*Schweißen*] = gas generator
gaserzeugendes Mittel *n*, (Metall-)Treibmittel, Blähmittel, Gasentwickler *m*, Gasbildner [*Gasbeton*] = gas-forming agent, metallic additive [*powdered alumin(i)um or zinc*]
Gas|erzeuger *m*, Gasgenerator *m* = (gas) producer

~fänger *m* = gas trap
~feld *n* [*Eine aufgeschlossene Erdgaslagerstätte*] = gas field
~fernleitung *f*, Ferngasleitung = gas transmission line
~filter *m, n* = gas filter
~filtration *f* = gas filtration
~flammkohle *f* = free-burning gas coal, gas-flame ~
~flasche *f*, Stahlflasche = gas cylinder, ~ bottle
gasfördernde Bohrung *f* → gasproduzierende Sonde *f*
Gasförderung *f*, Gasproduktion *f* = gas production
gas|förmiges Produkt *n* = gaseous product
~führender Horizont *m*, Gashorizont = gas horizon
Gasgehalt *m* = gas content
gasgekühlt = gas-cooled
Gas|gemisch *n* mit wenig kondensierbaren Anteilen = dry gas, lean ~
~generator *m*, Gaserzeuger *m* = producer, gas ~
~gerät *n* = gas appliance
~gesellschaft *f* = gas company
~gestänge *n* [*Motor*] = governor control
~gewindebohrer *m*, Rohrgewindebohrer, Handgewindebohrer mit Rohrgewinde = pipe tap
~gewinderohr *n* = screwed gas pipe
~hebel *m* = throttle, governor lever
~herd *m* = gas range
~horizont *m*, gasführender Horizont = gas horizon
~installation *f* = (internal) installation (Brit.) [*The gas pipes and appliances on the consumer's side of the control cock at the board's gas meter*]
~installation *f* = gas installation
~kappe *f* [*einer Struktur*] = gas cap [*of a structure*]
~kegelbildung *f* in Öllagerstätten = gas-coning in oil reservoirs
~kocher *m* = gas cooker
~kohle *f* = gas coal

Gaskreislaufverfahren — Gasversorgung

~kreislaufverfahren *n* [*Sekundärverfahren zur Förderung von Erdöl und Erdgas*] = recycling
~kühlung *f* = cooling of gas
~laterne *f* = gas lantern
~leitungsbaugerät *n* = gas pipeline construction rig
~leitungshahn *m* = plug cock, ~ tap
~liftförderung *f* = gas-lift recovery
~lift-Leitungsanschluß *m* = gas-lift hook-up
~liftsonde *f* = gas-lift well
~luftheizer *m* = gas-fired unit heater
~meßstation *f* = gas metering station, ~ measuring ~
~motor *m* = gas engine
~ofen *m* = gas furnace [*for testing refractory bricks under load at high temperatures*]
Gasöl *n* = gas oil
~ für Absorptionszwecke = straw oil
gasolinfreies Naturgas *n* → trockenes ~

Gas-Öl|-Kontakt *m*, Gas-Öl-Grenze *f* = gas/oil contact line
~-Separator *m* → Gasseparator
~-Verhältnis *n*, GÖV = gas/oil ratio, GOR
Gas|pedal *n* → Fahrfußhebel *m*
~polster *m, n* [*Tiefpumpenförderung von Erdöl*] = gas lock
~produktion *f*, Gasförderung *f* = gas production
gasproduzierende Sonde *f*, ~ Bohrung, gasfördernde ~, fördernde Gasbohrung, fördernde Gassonde = producing gas well, gas producing ~
Gas|prospektion *f*, Gassuche *f* = gas prospection
~radiator *m* = gas radiator
~reduktionsgebäude *n* = gas reduction building
~reduzierpedal *n* = decelerator
~reinigung *f* = gas purification
~rohr *n* [*Elektroosmose*] = rod electrode
~rohr *n* = gas pipe
~rohr-Handlauf *m* = gas barrel handrail, ~ tubing ~, ~ tube ~
~ruß *m* = gas black

~sammler *m* → Gasbehälter *m*
~sand *m* = gas sand
~säule *f* = gas column
~schiefer *m* = gas shale
~schlauch *m* = gas hose, ~ flexible conduit
~schlupf *m* = gas slippage, by-passing of gas
~schmelzschweißen *n* → Gasschweißen
~schmierung *f* = gas lubrication
~schweißen *n*, Gasschmelzschweißen [*die Begriffe „autogenes Schweißen" und „Autogenschweißen" sind veraltet*] = gas welding, autogenous ~
~separator *m*, Gas/Öl-Separator, Gasabscheider *m* = gas trap, ~ separator
~sonde *f*, Gasbohrung *f*, Gasbrunnen *m* = gas well
~spürhammer *m* = gas detection hammer
~station *f* = gas station
~strömung *f* = migration of gas
~suche *f*, Gasprospektion *f* = gas prospection
~technik *f* = gas engineering
Gästezimmer *n* = guest room
~ [*Hotel*] = bedroom
Gas|trennung *f* = separation of gases
Gastriebverfahren *n*, Gastreibverfahren *n* = gas drive
~ mit Expansion der Gaskappe = gas cap drive
~ ~ im Erdöl gelösten Erdgas = dissolved gas drive
Gastrocknungsanlage *f* = gas drying plant
Gasturbinen|kraftwerk *n* = gas turbine power station, ~ ~ ~ plant
~läufer *m* = gas turbine rotor
~motor *m* = gas turbine engine
Gasse *f*, Gäßchen *n* = lane
Gasundurchlässigkeit *f*, Gasdichtigkeit = imperviousness to gas
gasuntersättigt = undersaturated with gas
Gas|versorgung *f* = gas supply, provision of gas

~waschanlage *f* = gas washing plant
Gaswasser|heizer *m* = gas water-heater, geyser, gas-fired water heater
~heizung *f* = gas water heating
Gas|weg *m* = gas travel
~werk *n*, Gasanstalt *f* = gas undertaking
Gaszähler *m* = gas meter
~nische *f* = gas meter enclosure
Gaszerlegung *f* = separation of gases
Gattierungswaage *f*, Waage für alle Gemengeteile, Gattierungsgefäßwaage, Gemischwaage, Mischwaage = multiple(-material) (scale) batcher, cumulative (weigh) ~ [*one common hopper for all materials*]
Gaube *f* → (Dach)Gaupe
Gaupe *f* → Dach~
Gauss'sche Normalverteilungskurve *f* = Gauss normal distribution curve
Gebälk *n*, Balkenlage *f* = system of binders and joists
gebändert (Geol.) = banded [*A term applied to rocks having thin and nearly parallel bands of varying colo(u)rs, minerals or textures*]
gebankt, bankig (Geol.) = bedded
Gebäude *n*, Haus *n*, Bau *m* [*manchmal fälschlicherweise ,,Bauwerk" genannt*] = building [*A structure or combination of structures erected where it is to stand and having as one of its main purposes the provision of shelter from the weather*]
~abbruch *m*, Hausabbruch = demolition of building(s) [*also erroneously called "housebreaking"*]
~absteckung *f* = setting-out a building
~abwasser *n* → Hausabwasser
(~)Anbau *m*, Erweiterungsbau, Hausanbau = annex
~bemessung *f* = design of buildings, ~ of a building, building design
~brand *m*, Hausbrand = building fire
(~)Dach *n* = (building) roof
~decke *f* → Decke

~deckenteil *m*, *n* → Hochbaudeckenteil
~eingang *m*, Hauseingang = building entrance
~fabrik *f* → Häuserfabrik
~fundament *n*, Gebäude-Gründungskörper *m*, Hausfundament, Haus-Gründungskörper = building foundation (structure), ~ substructure [*The part of a building below the surface of the ground. It bears directly on, and transmits the building load to, the supporting soil or rock*]
~-Gründungskörper *m* → Gebäudefundament *n*
~isolierung *f*, Hausisolierung = building insulation
(~)Kern *m*, Betonkernbauwerk *n*, (Stahl)Betonkern, Hauskern = reinforced-concrete core, building ~
~last *f*, Hauslast = building load
~plan *m* = plan of building
~-Rohrleitung *f*, Haus-Rohrleitung = pipe run
~schornstein *m*, Hausschornstein = building chimney
~schutz *m*, Hausschutz = preservation of buildings
~schutzmittel *n*, Hausschutzmittel = building preservative
~schwingung *f*, Hausschwingung = building vibration
~setzung *f*, Haussetzung = settlement of building(s), building settlement
~teil *m*, *n*, Hausteil = element of construction [*Part of a building having its own functional identity. Examples are foundation, floor, roof, wall*]
~trennwand *f*, Haustrennwand = building partition [*A wall whose primary function is to divide space within a building*]
~umbau *m* = building alteration
Geber *m* = sender, transmitter
~ **zur Messung von Schwingungsbeanspruchungen** = vibration strain pickup

Gebietsverdunstung *f* [*Die gesamte Wasserabgabe eines Gebietes an die Lufthülle*] = evapo-transpiration [*Combined loss of water from soils by evaporation and plant transpiration*]

Gebirge *n* [*Tunnel-, Stollen-, Schacht- und Bergbau*] = ground, strata, formation

gebirgig, bergig = mountainous

Gebirgs|anker *m* → Gesteinsanker

~**beschreibung** *f*, Orographie *f* = orography

~**bildung** *f*, Orogenese *f*, Tektogenese *f* = mountain building, orogeny

~**druck** *m* [*Tunnel-, Stollen-, Schacht- und Bergbau*] = ground pressure, pressure of the strata

~**druck** *m*, orogener Druck (Geol.) = pressure of mountain mass

~**druckingenieur** *m* (⚒) = strata control engineer

~**(eisen)bahn** *f* = mountain railway (Brit.); ~ railroad (US)

~**faltung** *f* = folding

~**festigkeit** *f* = formation strength, ground ~, strata ~

~**fluß** *m* = mountain stream

~**kette** *f*, Gebirgszug *m* = chain of mountains

~**landschaft** *f* = mountain landscape

~**massiv** *n* = massif

~**niederschlag** *m* = mountain precipitation

~**schlag** *m* [*Die schlagartig und plötzlich auftretende Bewegung und Erschütterung im Gebirge um bergmännisch geschaffene Hohlräume als Folge von Entspannungsvorgängen im Gebirgskörper*] = rock burst

~**straße** *f*, Bergstraße = mountain highway, ~ road

~**tunnel** *m*, Felstunnel, Bergtunnel = rock tunnel

~**zug** *m*, Gebirgskette *f* = chain of mountains

Gebläse *n* = blower

Gebläseluft-Förderrohr *n*, Luft-Rutsche *f*, (pneumatisches) Förderrohr, Luftförderrohr = airslide

~ **mit Filtersteinen** = porous block conveyor, ~ ~ conveyer

~ ~ **Segeltuch** = canvas fabric airslide

Gebläsemaschine *f* = mechanical blower

geblasenes Bitumen *n*, oxydiertes ~ = air-rectified (asphaltic) bitumen, (air-)blown (~) ~, oxidized (~) ~ (Brit.); (air-)blown asphalt, oxidized ~ (US)

~ **Öl** *n* = blown oil

Gebläse|sand *m* = (sand) blast sand

~**wind** *m* = blast (air)

gebleiter Kraftstoff *m* = leaded fuel

gebleites Benzin *n* = ethyl gasoline (US); ethylized fuel (Brit.)

gebogener Träger *m* = bow girder

~ ~ **mit Zugband** = bowstring girder

~ **(Träger)Flansch** *m*, ~ (Träger)Fuß *m* = saddle flange, ~ chord, ~ boom

gebogenes Drahtgewebe *n* = curved woven wire cloth

gebohrter Kabeltunnel *m*, ~ Kabelstollen *m* = bored cable tunnel

~ **Ortpfahl** *m* → Bohrpfahl

gebäscht, ab~ = sloped

geböschte Baugrube *f* → ab~ ~

gebräch = friable

gebrannter Gips *m*, Branntgips = calcined gypsum

~ ~ **für Bauzwecke** [*Schweiz*] → Baugips

~ **Kalk** *m* → Branntkalk

~ **Scherben** *m* = fire-clay body

~ **Ton** *m* = fired clay

gebranntes Gut *n*, Brenngut = burned product

~ **Schieferton-Erzeugnis** *n* = fired shale product

gebrauchsfertig = ready-to-use

Gebrauchssieb *n* = service screen

~**gewebe** *n* = service screen cloth

Gebrauchsspannung *f* = working stress

gebrauchter Formsand *m*, Altsand = used sand

Gebrauchtmaschine — gedrungen

Gebrauchtmaschine *f* = second-hand machine
gebrochenes Gestein *n* = crushed rock
~ Material *n* → → (Beton)Zuschlagstoffe *mpl*

~ ~ = metal [*Australia. All broken material excepting dust*]
Gebühren|autobahn *f* → gebührenpflichtige Autobahn
~brücke *f* = toll bridge

Gebührenerhebung *f*, **Wegegelderhebung** [*Autobahn*] — toll collection
  Abfertigung *f* — transaction, handling
  Ausfahrtstand *m* — exit booth
  Einfahrtstand *m* — entry booth, entrance booth
  Einnahmen *fpl* — revenue
  Fahrt *f* — vehicle-trip
  Fernregistriergerät *n* — remote recording machine, ~ recorder
  Gebührenberechnungsmaschine *f* — validating machine, validator
  Gebührenerhebungsanlage *f* [*Gesamtkomplex der Gebäude*] — toll station, toll plaza
  Gebührenerhebungs- und -prüfanlage *f* — toll-collection and audit system, interchange equipment
  Kartenausgabemaschine *f* — ticket-issuing machine
  Pauschaljahresausweis *m* — flat-rate annual permit
  Plakatausweis *m* — sticker (on the vehicle)
  Photoapparat *m* — camera
  Registrierkasse *f* — cash totalizer
  ständiger Benutzer *m* — commuter
  Wegeabgabe *f*, (Wege)Gebühr *f*, Wegegeld *n* — toll, fare
  Zählschwelle *f* — treadle

---

gebühren|frei, **gebührenlos** [*Straße; Brücke; Tunnel*] = toll-free
~pflichtige Autobahn *f*, Gebührenautobahn, Wegegeldautobahn = toll road, ~ (super)highway, ~ (turn)pike
Gebühren(Verkehrs)tunnel *m*, Wegegeld(-Verkehrs)tunnel = toll tunnel
gechlort = chlorinated
gedachter Punkt *m* → ideeller ~
gedeckter Abzugkanal *m* → Platten-Abzugkanal
~ Durchlaß *m* → Platten-Abzugkanal
~ Güterwagen *m* = railway box wagon (Brit.); box freight car (US)
~ Säulengang *m* → Hypostylos
Gedenktafel *f* = memorial tablet

gedoptes Verschnittbitumen *n*, Verschnittbitumen mit Haftanreger = doped cutback
gedrehte Schraube *f* = turned bolt
gedrückt = compressed
gedrückte Diagonale *f* → Druckschräge *f*
~ Schräge *f* → Druckschräge
gedrückter Bauteil *m*, gedrücktes ~ *n* [*unter Druck stehend*] = component under compression
~ Querschnitt *m* = cross section under compression
~ Spitzbogen *m*, stumpfer ~ = drop arch, blunt ~
gedrungen [*Bauart einer Maschine*] = compact

gedrungener Balken(träger) m = short deep beam
Gefahren|feuer n = hazard beacon
~**stelle** f, Gefahrenpunkt m = danger point, hazard ~
gefährlicher Querschnitt m, kritischer ~ = critical section
Gefälle n → Steigung f
~ [*Straße*] = downhill grade, favo(u)rable ~
~, Fallhöhe f, Schleusen~ = fall (of a lock), lift (~ ~ ~)
~ → Fallhöhe f
~ [*Darcy'sches Filtergesetz*] = hydraulic gradient
~, Relativ~ [*Hydraulik*] = gradient [*Change of elevation, velocity, pressure, or other characteristic per unit length; slope*]
~**beton** m = sloping concrete
~**dach** n = sloping roof
~**keil** m [*beim Gefälledach*] = sloping wedge
~**-Leichtdach** n = sloping lightweight roof
~**leitung** f = chute [*A high-velocity conduit for conveying to a lower level*]
~**verzerrung** f [*Hydraulik*] = distortion of slope
~**wechsel** m = change of slope
Gefäll|höhe f → Fallhöhe
~**kurve** f = gradient curve, curve of water erosion
~**messer** m **zur Fernmessung** = distant-reading (in)clinometer
~**strecke** f, Strecke in Gefälle = downhill section, downgrade ~
gefalzte Dichtung f, halbgespundete ~ = rebated wood(en) flooring
gefärbtes Schutzglas n [*Schweißen*] = welding glass, filter ~
Gefäß n, Behälter m [*Aufnahmegefäß für Dämpfe, Gase, Flüssigkeiten*] = vessel, tank
~**fördereinrichtung** f (⚒) = skip winder
~**(förder)pumpe** f = tank type pump, ~ ~ pneumatic conveying system
~**förderung** f (⚒) = skip hoisting

~**rauminhalt** m, Behälterrauminhalt volume of vessel ~ ~ tank
~**waage** f → Behälterwaage
gefettetes Metallbearbeitungsöl n = metal-working compound
gefiltertes Abwasser n = filtered sewage [*The effluent of a sewage filter*]
Geflügelhof m = poultry farm
gefluxtes Bitumen n = fluxed asphalt (US); ~ = asphaltic bitumen (Brit.) [*the flux used is a residual product*]
~ **Trinidadbitumen** n = fluxed Lake Asphalt
geforderte Festigkeit f, erforderliche ~, rechnerische ~ = specified strength
geforderter Siebschnitt m, ~ Trennschnitt = calculated screen cut, theoretical ~ ~
Gefrier-Auftau|-Folge f, Frost-Tauwechsel m = freezing and thawing cycle
~**-Prüfung** f Frost-Tau-Prüfung, Tau-Frost-Prüfung = freezing and thawing test
Gefrieren n **des Bodens**, Einfrieren ~ ~ = soil freezing
Gefrier|gründung f, Baugrundvereisung f, Gefrierverfahren n = freezing (process), ~ method, artificial freezing of ground
~**punkt** m = freezing point
~**punkterniedrigung** f [*Durch Zusatz von Frostschutzmitteln wird der Gefrierpunkt des Betonanmachewassers etwas herabgesetzt*] = lowering of the freezing point
~**rohr** n, Steigrohr, Vereisungsrohr [*Gefrierverfahren*] = freeze pipe
~**schacht** m = freezing shaft
~**schachtbauverfahren** n, Gefrierschachtbaumethode f = freezing process for shaft sinking
~**schutzmittel** n → Frostschutzmittel
~**temperatur** f = freezing temperature
~**trocknung** f = freeze-drying
~**- und Auftauzyklus** m, Tau-Frost-Wechsel m, Frost-Tau-Wechsel = freezing and thawing cycle

Gefrierverfahren — Gegenpol

~verfahren n → Gefriergründung f
~vitrine f = freezer case
gefrorener Boden m, ~ Untergrund m = frozen ground, ~ subsoil
Gefügespannung f [*Beton*] = structure stress
gefüllertes Bindemittel n → Bindemittel mit Füller
gegen den Uhrzeigersinn, linksläufig = counterclockwise
Gegenangebot n → → Angebot
Gegenbogen m, Gegenkurve f, Gegenkrümmung f, S-Kurve = reverse curve
~, Grundbogen, Erdbogen, Konterbogen = inverted arch
Gegendampf m = counter steam
~bremse f = counter steam brake
Gegen|diagonale f, Gegenschräge f, Wechselstab m = counter brace, ~ diagonal
~drehungskurbeln fpl, Antiparallelkurbeln, gegenläufige Kurbeln = crossed parallelogram cranks working in opposite directions
Gegendruck m Vorlagedruck = back pressure
~ → Auflagerkraft f
~kolben m, Entlastungskolben Ausgleichkolben = dummy piston
~regelventil n = back-pressure regulating valve
~ring m = back up ring
~turbine f = back-pressure turbine
gegeneinander betonieren [*Fertigteile*] = to cast in contact with each other
Gegenfeder f = opposing spring, reacting ~
~blatt n = counter leaf
Gegen|feuer n, Rückbrennung f = passing of the flames to the front of the furnace
~flansch m = companion flange
~gefälle n [*in Längsneigung einer Straße*] = reverse gradient
Gegengewicht n, Ausgleichgewicht ~ counterweight
~ausleger m, Gegengewichtarm m = counterweight jib (Brit.); ~ boom (US)

~kasten m = balance box
~rinne f, Flieger m [*Gußbetoneinrichtung*] = counter-balanced chute (or spout)
~segment n = segment of counterweight
~seil n = counterweight rope
~seite f, Ballastseite = counterweight side
~trommel f = counterweight drum
~turm m = counterweight tower
Gegen|halter m der Nietmaschine → Anpreßstempel f ~ ~
~hebel m Winkelhebel mit Wälzflächen = angular lever with rolling surfaces
~induktivität f = mutual inductance
~kette f, Rückhaltekette [*Brücke*] = back chain
~klappe f, Flügel m nach Unterwasser, untere Tafel f, Unterklappe [*Dachwehr*] = downstream leaf, lower ~
~klappe f, Gegenschütz(e) n, (f), Stellklappe [*Trommelwehr*] = tailgate
~kraft f = counter force
~krümmung f → Gegenbogen m
~kurbel f = crank with drag link
~kurve f → Gegenbogen m
gegenläufig = contra-rotating
~ gewickelt = wound in both directions [*right lay and left lay*]
gegenläufige Kurbeln fpl → Gegendrehungskurbeln
Gegen|laufturbine f = turbine with nozzles and blades rotating in opposite directions
~laufdoppelturbine f = double turbine with blade wheels running in opposite directions
~mutter f, Klemmmutter, Doppelmutter, Kontermutter = lok nut, check ~, jam ~
~ort n (⚒) = road driven in a reciprocal direction to an existing one
~passat m, Antipassat = antitrade (wind)
~pol m, Antipol = reciprocal pole, antipole

Gegenpolare — Gehwegplatte

~**polare** *f*, Antipolare = reciprocal polar, antipolar
~**schlaghammer** *m* = counterblow hammer
~**schwingrahmen** *m* = counterbalanced (screen) frame
~**schwingsieb** *n*, Kontrasieb = screen with two counterbalanced frames
gegenseitige Einspannungen *fpl* = reciprocal restraints
Gegen|spülverfahren *n*, Gegenspülmethode *f* [*Tiefbohrtechnik*] = reverse circulation method
~**spur** *f* [*Straße*] = opposing lane
Gegenstrom|bohrverfahren *n* → Bohren *n* mit Verkehrtspülung
~**-Kessel** *m* = countercurrent boiler
~**klassierer** *m* = countercurrent classifier
~**-Mischanlage** *f* = countercurrent mixing plant
~**-Schnellmischer** *m* → → Betonmischer
~**-Sonderschaltung** *f* [*Turmdrehkran*] = reverse current lowering connection
~**verfahren** *n* = counterflow process
~**-Wäscher** *m* = counterflow scrubber, countercurrent ~
~**-Waschfiltern** *n* = countercurrent filtration washing
~**waschmaschine** *f*, Gegenstrom-(Trommel)Wäsche *f*, Gegenstrom-Waschtrommel *f* = contraflow washer ~ washing machine
Gegen|taktschaltung *f* = push-pull connection, countercontact ~, push-pull arrangement, push-pull valve operation
~**turm** *m* [*Kabelkran*] = tail tower
~**ufer** *n* [*Fluß*] = opposite bank
~**verkehr** *m* = opposing traffic, oncoming ~
~**welle** *f*, Vorgelegewelle = counter shaft
~**wind** *m*, Stirnwind = headwind
~**winkel** *m*, Wechselwinkel = opposite angle, alternate ~
Gehalt *m*, Anteil *m* = content

~ **an freiem Wasser in Schnee** = liquid water content of snow
Gehänge *n* → (Ab)Hang *m*
~, Schwebekörper *m* [*Flußbau*] = tree retard
~**lehm** *m* = slope wash
~**schutt** *m* Schutthang *m* = detrital slope, scree
Gehäuse *n* = housing casing
~ **des Radsturzantriebs** [*Motor-Straßenhobel*] = front wheel lean control housing, ~ ~ ~ ~ casing
~ ~ **Scharhubgetriebes** = blade lift control housing, ~ ~ ~ ~ casing
Gehbahn *f* → Bürgersteig *m*
~**-Betonplatte** *f* → Bürgersteig-Betonplatte
~**konsole** *f* → Gehwegkonsole
~**platte** *f*, Gehwegplatte, Fußwegplatte, Bürgersteigplatte, Gehsteigplatte = flag(stone)
geheizte Handtuchschiene *f*, beheizte ~ = heated towel rail
~ **Straße** *f*, beheizte ~ = heated road, ~ highway
Gehlinie *f*, Ganglinie, Lauflinie [*Treppe*] = walking line
gehobener Strand *m* = raised beach (Brit.); elevated shore line (US)
gehont = honed
Gehrungsstoß *m* = mitred joint (Brit.); mitered ~ (US)
Gehschicht *f* → Deckschicht [*Fußboden*]
Gehsteig *m* → Bürgersteig
~**-Betonplatte** *f* → Bürgersteig-Betonplatte
~**konsole** *f* → Gehwegkonsole
~**platte** *f* → Gehbahnplatte
Gehstreifen *m* = walking strip
Gehweg *m* → Bürgersteig *m*
~, Fußweg = footpath
~**-Betonplatte** *f* → Bürgersteig-Betonplatte
~**konsole** *f*, Bürgersteigkonsole, Gehsteigkonsole, Gehbahnkonsole, Fußwegkonsole [*Brücke*] = footway cantilever (Brit.); sidewalk ~ (US)
~**platte** *f* → Gehbahnplatte

Geiger-Müller-Zähler m = Geiger-Mueller counter
Geiß|berger m → Findling(stein) m
~fuß m → Nageleisen
gekapselt = encapsulated, clad
gekernte Strecke f = Kernstrecke [*Tiefbohrtechnik*] = cored interval
geklärtes Abwasser n, entschlammtes ~ = clarified sewage [*Loosely used for sewage from which suspended matter has been partly or completely removed*]
geknickte Platte f, Buckelblech n = buckle(d) plate, dished steel ~
geknickter (Polygon)Zug m, ~ Standlinienzug, geknicktes Polygon n = chain traverse
gekochtes Leinöl n = boiled linseed oil
Gekörn n = grain
gekörnte Hochofenschlacke f → granulierte ~
~ **Masse** f, Schweißpulver n [*Ellira-Schweißen*] = granulated material, ~ welding composition
gekrackt = cracked
gekräuselte Schichten fpl (Geol.) = contorted beds
Gekriech n → Bodenfließen n
gekröpfter Draht m = crimped wire
Gekrösestein m = tripestone
gekrümmte Gurtung f → Bogengurt(ung) m, (f)
~ **Schale** f = curved shell
~ **Schichtungen** fpl (Geol.) = contorted strata
~ **Staumauer** f → → Talsperre f
gekrümmter Gurt m → Bogengurt(ung)
~ **Saugschlauch** m [*Wasserkraftwerk*] = elbow (draft) tube (US); ~ (draught) ~ (Brit.)
gekühltes Wasser n = cooled water
Gel n, Gerüst n = gel(-type structure)
Gelände n = ground, terrain, land
geländeaktiv = full ground contact
Gelände|auffüllung f, Geländeaufhöhung, Geländeaufschüttung = land fill
~**beschaffenheit** f, Topographie f = topography

~**bewuchs** m, Bodenbewuchs, Geländebedeckung f, Bodenbedeckung = ground cover, ~ vegetation
~**bohrwagen** m = pneumatic-tyred wagon drill (Brit.); pneumatic-tired ~ ~ (US)
~**einsatz** m = cross-country operation
~**fahrzeug** n, geländegängiges Fahrzeug = off-(the-)highway vehicle, cross-country ~
~**form** f, Geländegestaltung f, Bodenform = land form, configuration (or conformation) of the ground
~**gang** m = lo gear (US); low ~
geländegängige Erdbewegung f, gleislose ~ = railless earthmoving, trackless ~
geländegängiger Förderwagen m **in Kraftwagen-Bauart** → Muldenkipper m
~ **Hinterkipper** m = off-(the-)highway rear dump truck
~ **luftbereifter Erdtransportwagen** m → Muldenkipper m
geländegängiges Transportfahrzeug n = off-(the-)highway hauler
Gelände|gängigkeit f = off-(the-)highway manoeuvrability (Brit.); ~ maneuverability (US); cross-country performance
~**gefälle** n = slope of the ground
~**gestaltung** f → Geländeform f
~**-Grundbruch** m → Grundbruch
~**höhe** f = ground level (elevation)
~**karte** f = land map
~**kette** f = tyre chain for cross-country operation (Brit.); tire ~ ~ ~ ~ (US)
~**oberfläche** f = ground surface
~**oberkante** f = ground level
~**-PKW** m = landrover
Geländer n = railing [*An open fence made of posts and rails*]
Gelände|regen m = orographic rain
~**reifen** m, Reifen mit Geländeprofil = off-(the-)highway (or off-(the-)road) ground tire (US); earthmover tyre (Brit.); grip ~; off-road equipment ~

(Gelände)Senke — gemeinnütziges Wohnungsunternehmen

(~)Senke *f*, Vertiefung *f*, Bodensenke, Geländemulde *f* = (ground) depression, natural ~

~silhouette *f* = natural skyline

~sprung *m* = land wave, undulation (in the topography)

~stollenreifen *m* = directional tire (US); ~ tyre (Brit.)

~streifen *m* = strip of ground

~unebenheiten *fpl*, Bodenunebenheiten, Geländewelligkeit *f*, Bodenwelligkeit = surface roughness of the terrain

~welligkeit *f* → Geländeunebenheiten

Geländer|baustoff *m* = railing material

~docke *f* = baluster supporting a railing

~rohr *n* = railing tube, ~ barrel, ~ tubing

gelatinös-plastische Sprengstoffgemenge *npl* → → 1. Schießen; 2. Sprengen

Gelbbleierz *n*, Wulfenit *m*, $PbMoO_4$ (Min.) = wulfenite

gelbe Arsenblende *f* (Min.) = orpiment

gelber Hornblendeasbest *m* → Amosit(asbest)

Gelbildung *f* = gelling

Gelböl *n* = straw oil

Gelenk *n* = hinge

~blatt *n*, Pendelblech *n* [*bei einem Pendelgelenk*] = pendulum leaf, ~ plate

~bogen *m* (⚒) = articulated (roadway) arch

(Gelenk)Bolzen *m* = pin

~ der Auslegerspitze = boom point pin (US); jib ~ ~ (Brit.)

~verbindung *f* = pin joint [*A hinge in a structure*]

Gelenk|dichtung *f* [*Dachwehr*] = sliding seal, ~ hinge

~drehscheibe *f* = articulated turntable

~fahrzeug *n* [*Erdbau*] = prime-mover and semi-trailer

gelenkiger Kippwagen *m* = articulated dump truck

Gelenk|kappe *f* (⚒) [*Stahlkappen die sich durch eine entsprechende Vorrichtung an andere von Stempeln*

unterstützte Stahlkappen ankuppeln lassen] = link bar, hinged ~, articulated ~

~kettenspanner *m* = rotary chain tightener

~lader *m*, Gelenklademaschine *f* = loader with center pivot steer(ing) (US); ~ ~ centre ~ ~ (Brit.)

gelenklos, eingespannt [*Bogen; Träger*] encastré, fixed, with fixed ends, without articulation, rigid, hingeless

Gelenk|pfette *f* = articulated purlin

~rahmen *m* = hinged frame

~rinne *f* = articulated vibrating trough conveyor, ~ ~ ~ conveyer

~stab *m* = hinged bar, linked ~

~träger *m* → Gerberträger

~wagen-Bodenentleerer *m* → Halbanhänger *m* mit Bodenentleerung

~welle *f* = universal drive shaft

Gelfestigkeit *f* = gel strength

gelochte Dachunterlagsbahn *f* = perforated under layer, vented ~ ~

gelochtes Stahlblechsieb *n* = punched steel plate screen, perforated ~ ~ ~

~ Startbahnblech *n* = P S P, pierced steel planking

gelöschter Kalk *m* → Kalziumhydroxyd *n*

gelöster Stoff *m* [*Abwasserwesen*] = dissolved solid, ~ matter, matter in solution, solid in solution

Gelpore *f* = gel pore

Gelteilchen *n* = gel particle

Geltungsbereich *m* [*Norm*] = scope

gelüftet [*Bremse*] = released

Gelwasser *n* = gel water

gemahlene Schamotte *f* = grog

gemahlener Branntkalk *m* → ungelöschter gemahlener Kalk

~ massiger Bimsstuff *m* → Traß *m*

gemauerter Reinigungsschacht *m*, ~ Einsteig(e)schacht, gemauertes Mannloch *n* = brickwork manhole

gemeinsames Aufmaß *n* = joint measurement

Gemeinkosten *f* = overhead charges

gemeinnütziges Wohnungsunternehmen *n* = mutual benefit building society

gemeinschaftliche Giebelmauer f → Brandmauer

~ Omnibuslinien fpl = co-ordinated bus services

Gemeinschafts|antenne f = communal television aerial

~raum m = common room

~stand m, Gruppenstand [*auf einer Messe oder Ausstellung*] = group stand

Gemengewaage f **in Laufgewichtbauart** = beam and jockey-weight type weigh-batcher (Brit.); beam scale (US)

Gemengteil m, n [*Gestein*] = constituent mineral [*rock*]

gemessen = measured [*resources*]

Gemisch n, **Mischung** f = mix(ture)

gemischtbasisches Erdöl n = mixed-base (crude) petroleum, ~ crude

gemischtes Sedimentgestein n, gemischt mechanisch-chemisches ~ = mixed sedimentary rock

gemischtkörnig = mixed-grained

gemuldet = troughed

gemustertes Glas n, **Musselinglas** = muslin glass

Genauigkeit f, **Empfindlichkeit** [*Waage*] = weighing sensibility

geneigt gerammt, schräg ~ = driven on the rake

geneigte Blocklagen fpl = sloping bond, sliced work [*breakwater construction*]

~ Ebene f, **schiefe ~** [*Schiffhebewerk, bei dem die Schiffe auf Wagen gesetzt und auf einer geneigten Ebene vom Unter- zum Oberwasser und umgekehrt befördert werden*] = inclined plane

~ Laufbahn f, **Schrägbahn** [*Walzenwehr*] = sloping rack

~ Oberseite f **des Wehrkörpers** = battered upstream face

~ Treppenuntersicht f = sloping soffit of a stair

geneigter Tragstiel m [*Pfahljoch*] = raker, batter post

geneigtes Sturzbett n = sloping apron

Generalbebauungsplan m = master plan for development of a city

generalüberholt = completely overhauled

Generalunternehmer m, **Hauptunternehmer, Gesamtunternehmer** = prime contractor, chief ~, general ~

~verkehrsplan m = traffic master plan

Generator m = generator

~aggregat n = generating set

~gas n = producer gas

Genesungsheim n = convalescent home

genietete Verbindung f, **genieteter Anschluß** m = riveted joint, ~ connection

~ Verrohrung f, **Nietverrohrung** [*Bohrloch*] = riveted casing, "stove pipe"

genormt = standardized

genormter U-Stahl m = standard channels

genormtes Profil n = standard section

~ Sieb n, **Normensieb** = standard screen

genutete (Beton)Fuge f → gespundete ~

~ Deckenschalung f = grooved match ceiling boarding

Geochemie f, **geochemisches Prospektieren** n, **geochemischer Aufschluß** m, **geochemische Lagerstättenforschung** f, **geochemische Suche** f, **geochemisches Aufsuchen** n, **geochemische Methode** f = geochemical prospecting, ~ exploration, ~ method, ~ survey, geochemistry

Geochemiker m [*zuweilen auch Paläontologe*] = mud smeller

geochemisches Verfahren n, **geochemische Methode** f = geochemical method

(geodätische) Aufnahme f = (geodetic) survey

~ Druckhöhe f = geodetic pressure head

~ Förderhöhe f [*lotrechter Höhenunterschied zwischen dem Druck- und Saugwasserspiegel (DIN 4044)*] = geodetic head

~ Saughöhe f = geodetic suction head

geodätisches Meßinstrument n = surveyor's instrument, surveying ~
Geodimeter m = geodimeter
geoelektrische Methode f → Widerstandsmessungsverfahren
geoelektrisches Verfahren n → Widerstandsmessungsverfahren
geographische Barre f = geographic bar
Geohydrologie f, Grundwasserkunde f = hydrogeology
Geoid n = geoid [*A hypothetical figure of the earth with the entire surface represented as at mean sea level*]
Geologe m = geologist
~ [*Erforscher von Ölvorkommen*] = rock hound
Geologenhacke f = geologist's pick, prospector's ~
Geologie f, Erdbildungskunde f = geology
geologische Aufeinanderfolge f = geological succession
~ **Aufnahme** f = geological survey
~ **Formation** f → → Geologie f
~ **Karte** f = geological map
~ **Kartierung** f = geological mapmaking, ~ mapping (work)
~ **Schürfbohrung** f, ~ Strukturbohrung = structural test hole
~ **Strukturbohrung** f, ~ Schürfbohrung = structural test hole
~ **Untergrunduntersuchung** f = geological ground investigation
~**Vergangenheit** f = geological history

geologische Zeitskala f
  Neozoikum n
    Quartär n
      Alluvium n
      Pleistozän n, Diluvium n
    Tertiär n
      Pliozän n
      Miozän n
      Oligozän n
      Eozän n
      Paläozän n
  Mesozoikum n
    Kreide f
      Ober-Kreide f
      Unter-Kreide f
    Jura m
      Malm m
      Dogger m
      Lias m
    Trias m
      Keuper m
      Muschelkalk m
      Buntsandstein m
  Paläozoikum n
    Perm n
      Zechstein m
      Rotliegendes n
    Karbon n
      Ober-Karbon n
      Unter-Karbon n, Kulm m
    Devon n
      Ober-Devon n

geologic scale of time
  Cenozoic Era
    Quarternary Period
      Recent Epoch, Holocene Epoch
      Pleistocene Epoch
    Tertiary Period
      Pliocene Epoch
      Miocene Epoch
      Oligocene Epoch
      Eocene Epoch
      Paleocene Epoch
  Mesocoic Era
    Cretaceous Period
      Upper Cretaceous
      Lower Cretaceous
    Jurassic Period
      Upper Jurassic
      Middle Jurassic
      Lower Jurassic
    Triassic Period
      Upper Triassic
      Middle Triassic
      Lower Triassic
  Paleozoic Era
    Permian Period
      Upper Permian
      Lower Permian
    Carboniferous Period
      Upper Carboniferous
      Lower Carboniferous
    Devonian Period
      Upper Devonian

Mittel-Devon *n*	Middle Devonian
Unter-Devon *n*	Lower Devonian
Silur *n*	Silurian Period
Gotlandium *n*	Gotlandian
Ordovizium *n*	Ordovician
Kambrium *n*	Cambrian Period
Ober-Kambrium *n*	Upper Cambrian
Mittel-Kambrium *n*	Middle Cambrian
Unter-Kambrium *n*	Lower Cambrian
Jung-Präkambrium *n*, Algonkium *n*	Early Precambrian Era, Archeozoic Era
Alt-Präkambrium *n*, Archaikum *n*	Late Precambrian Era, Proterozoic Era

geologischer Schnitt *m* = geological (cross-)section
geologisches Profil *n*, Schichtenprofil = geological profile, strata ~
Geomagnetik *f*, (geo)magnetisches Aufschlußverfahren *n* = (geo)magnetic method, ~ exploration, ~ prospecting
(geo)magnetisches Aufschlußverfahren *n* → Geomagnetik *f*
Geomechanik *f*, Felsmechanik = rock mechanics
Geometrie *f* **der Ebene**, ebene Geometrie, Planimetrie = plane geometry, planimetry
geometrisch gestufte (Sieb)Reihe *f*, geometrische Abstufung *f* der Maschenweiten = geometrically progressing series
geometrische Ähnlichkeit *f* = geometric similarity, ~ similitude
~ Höhenmessung *f*, Nivellieren *n*, Einwägen, Höhenmessung mit horizontalen Geraden = geometrical levelling
(~) Reihe *f* = geometrical progression
geometrischer Ort *m* **der Hochwasser** = locus of high water
~ ~ ~ Niedrigwasser = locus of low water
geometrisches Mittel *n* = geometrical mean
geomikrobiologische Prospektion *f* = petroleum prospecting by microbiology, oil ~ ~ ~
Geomorphologie *f* = geomorphology

Geophon *n*, Vertikal-Seismograph *m* = pick-up
Geophysik *f* = geophysics
geophysikalische Aufschlußarbeiten *fpl* → ~ Prospektion *f*
~ Baugrunduntersuchung *f* → Baugrundsondierung mit indirektem Verfahren
~ Methode *f*, geophysikalisches Verfahren *n* = geophysical method
~ Prospektion *f*, geophysikalischer Aufschluß *m*, geophysikalische Aufschlußarbeiten *fpl* = geophysical prospecting, ~ exploration, ~ survey
~ Untersuchung *f* = geophysical investigation
geophysikalischer Aufschluß *m*, geophysikalische Prospektion *f* = geophysical exploration, ~ prospecting, ~ survey
geophysikalisches Aufschlußinstrument *n* = geophysical prospecting instrument
~ Aufschlußverfahren *n*, ~ Aufschließungsverfahren *n* = geophysical exploration method, ~ prospecting ~, ~ survey ~
~ Verfahren *n*, geophysikalische Methode *f* = geophysical method
Geophysiker *m* = geophysicist
Geosynklinale *f*, Senkungstrog *m*, Senkungswanne *f* = geosyncline
geotechnische Untersuchung *f* = geotechnical investigation
geotechnisches Verfahren *n* = geo-

technical method, ~ process [*It means processes which change the properties of soils, for instance compaction, injection, groundwater lowering, freezing, electro-osmosis etc.*]

Geothermik *f* = geothermal investigations [*Geophysical surveys of particular interest to oil prospectors. They are of two types, first the surveying of the temperatures in oil-wells where thermal equilibrium has not yet been reached. The second type of survey is along the ground surface*]

geometrische Tiefenstufe *f* [*Die Tiefe, die in der Erde zu durchteufen ist, um eine Temperaturzunahme von 1° C festzustellen. Sie ist örtlich sehr verschieden und beträgt im Durchschnitt etwa 30 m*] = geothermal gradient, geothermic degree

Geoseismik *f* → seismisches Verfahren *n*

geothermisches Kraftwerk *n* → Erddampfkraftwerk

Gepäck|abfertigung *f* = baggage handling
~**abfertigungsanlage** *f* = baggage-handling facility
~**aufzug** *m* = baggage lift; ~ elevator (US)
~**ausgaberaum** *m* = baggage claim area
~**bahnsteig** *m* = baggage platform
~**durchlauf** *m* = baggage flow
~**förderanlage** *f* = baggage-conveyance system
~**karre(n)** *f*, *(m)*, Gepäckwagenanhänger *m* = baggage cart, ~ truck
~**keller** *m* = baggage basement
~**schalter** *m* = baggage-handling counter
~**stück** *n* = piece of baggage
~**tunnel** *m* = baggage way
~**wagen** *m* [*Eisenbahn*] = baggage car, ~ van
~**wagenanhänger** *m*, Gepäckkarre(n) *f*, *(m)* = baggage cart, ~ truck
~**weg** *m*, Gepäckdurchlaufroute *f* = baggage flow route

gepackter Steindamm *m* → → Talsperre *f*

gepanzertes Wälzgelenk *n* [*Im Bereich der Abwälzung ist der Beton mit Stahlblech gepanzert*] = armo(u)red rolling contact joint

gepflastert = sett-paved

gepulvert, pulverförmig = powdered

geradbeiniger Streckenbogen *m* (✕) = straight-legged (roadway) arch

Gerade *f*, gerade Strecke *f* = straight (line)

gerade Gewichts(stau)mauer *f* = straight gravity dam
~ **Plattenbrücke** *f* = right slab bridge

gerader Bogen *m* → scheitrechter ~

gerades (Stau)Wehr *n*, normales ~ = rectangular weir
~ **Tonnengewölbe** *n*, Kufengewölbe = straight barrel vault

geradlinig = straight

geradlinige Abschreibung *f*, lineare ~ = straight line method (of depreciation)
~ **Anordnung** *f* [*von Stempeln*] (✕) = in-line system
~ **Beziehung** *f* = straight-line relation
~ **Treppe** *f*, geradläufige ~, gerade ~ = straight staircase (Brit.); ~ stair (US)

geradliniger Spannungsverlauf *m* über die Höhe = straight-line no tension

gerammter (Beton)Ortpfahl *m*, Ramm-Ortpfahl, Ort-Rammpfahl = (impact-)driven (cast-)in-situ pile, cast-in-place ~, ~ cast-in-the-ground ~

Gerät *n* für allgemeine Wärmeleitfähigkeitsbestimmungen an feuerfesten Werkstoffen = thermal conductivity apparatus

Gerät *n* zum Anzeichnen der Verkehrslinien */pl*, Straßenmarkierungsmaschine *f*, (Farb)Strichziehmaschine, Strichziehgerät = pavement-marking machine, traffic-line marking machine, road-marking machine, safety-line (street) marker, stripe painter, line marker, highway

Gerät — Geräteträger

	marker, white line machine, carriageway (or highway, or street) marking machine, striping machine, striper
gestrichelte Linie *f*	intermittent line
ausgezogene Linie *f*	continuous line
eine alte gestrichelte Linie unter Synchronisierung nachziehen	to synchronize with the old intermittent line when repainting
Verkehrs(-Markierungs)farbe *f*, Straßenmarkierungsfarbe; Signierfarbe [*Schweiz*]	traffic(-zoning) paint, zone marking ∼, road-marking ∼, line-marking ∼, road line ∼

Gerät *n* **zur Bestimmung der Fließgrenze** *f*, **Fließgrenzengerät** = (mechanical) liquid limit device, apparatus for plasticity test, Casagrande liquid limit machine

Messingschale *f*	brass dish
Nockenwelle *f*	cam shaft
Hartgummiunterlage *f*	hard rubber block
Bodenprobe *f*	soil cake
Furche *f*	groove
Fließkurve *f*	flow curve
semilogarithmisches Netz *n*	semilog plot
Schlag *m*	shock

Gerät *n* **zur Bestimmung des Verdichtungsquotienten** [*Beton*] = compacting factor test apparatus
∼ ∼ ∼ ∼ **W/Z-Faktors** = concrete mix electric testing apparatus, ratiometer
∼ ∼ **Messung der Durchbiegung von nichtstarren Straßendecken** = leverarm deflection indicator, Benkelman beam (apparatus)
Gerät *n* **zur Messung von Querschnittsgeschwindigkeiten** [*Straßenverkehrstechnik*] = spot speed survey device
Geräte *npl* **für die Material(be)förderung, (Material)Umschlaggeräte** = materials handling machinery (or equipment)
  absatzweise arbeitende Fördermittel *npl* = intermittent movement equipment, ∼ ∼ machinery
  stetig arbeitende Fördermittel *npl*, Dauerförderer *mpl*, Stetigförderer = constant flow equipment, ∼ ∼ machinery, intermittent movement ∼
Geräte|abteilung *f* = equipment division, ∼ department
(∼)**Ausfall** *m*, **Panne** *f* = breakdown, equipment ∼
(∼)**Bediener** *m* = equipment operator, plant ∼, driver; operative (Brit.)
∼**bestand** *m*, **Baumaschinenpark** *m*, **Gerätepark** = (equipment) fleet
∼**bude** *f* = equipment storage hut
∼**einheit** *f*, **Einzelgerät** *n*, **Maschineneinheit, Einzelmaschine** *f* = equipment unit (or item), piece of equipment, plant item
∼**einsatz** *m*, **(Arbeits)Einsatz** = employment of plant, disposition of equipment, equipment utilization
∼**einsatztafel** *f* = disposition board
∼**liste** *f* = list of equipment, plant register
∼**liste** *f* **für die Bauwirtschaft, Baugeräteliste** = list of hiring charges; equipment rental compilation (US)
∼**miete** *f* = hire charge
∼**mietsatz** *m* = plant-hire rate
∼**park** *m* → **Gerätebestand** *m*
Geräteträger *m* = tool bar

Geräteträger — Gerüstkran

~ mit Unterteil und Oberteil, Grundtype f [*früher: Grundbagger m*] = basic excavator, ~ machine
~klammer f = tool bar clamp
~-Planierschild n = tool bar bulldozer
Geräte- und Maschinenunterhaltung f = plant maintenance work
~vermietung f = equipment renting
~vorhaltung f = commissioning of plant [*Keeping construction machinery and equipment in fit condition for use on a construction contract*]
~zug m = train of plant items, equipment train
geräumte Fläche f = cleared area
Geräusch n = diffuse sound
geräuscharme Rammanlage f = silent pile driving system
Geräusch|bekämpfung f = diffuse sound control
~fortleitung f = diffuse sound transmission
~minderung f = diffuse sound reduction
Gerber|brücke f = Gerber girder bridge
~gelenk n, Gerber'sches Gelenk = Gerber hinge
~pfette f = Gerber purlin(e)
Gerber'sche Momentfläche f = Gerber's diagram of moments
Gerberträger m, Gelenkträger = Gerber girder; slung span continuous beam (Brit.)
geregelte Drehzahl f [*Motor*] = engine governed speed
~ Kreuzung f = controlled crossing
gereinigter Trinidad-Asphalt m, Trinidad-Epuré n = Trinidad refined asphalt, ~ épuré, parianite
gerichtete Bohrung f, ~ Sonde f; gerichteter Schacht m [*in Galizien*] = directional well
~ Schwingung f = directional vibration
~ Sonde f → ~ Bohrung f
gerichteter Schacht m [*in Galizien*] → gerichtete Bohrung f

gerichtetes Ablenken n von Bohrungen = directional deviation of borings
~ Bohren n [*Steuern des Gestänges durch Richtkeile*] = controlled well drilling, directional ~ ~, directed ~ ~
Gerichtsgebäude n = courthouse
geringe Tiefe, seicht, untief = shallow
Gerinne n, (offener) Kanal m, Rinne f [*nicht im Sinne eines Schiffahrtkanals = canal*] = channel [*An elongated open depression in which water may, or does, flow*]
~auskleidung f, Kanalauskleidung = channel lining
~auskleidungsmaschine f, Kanalauskleidungsmaschine = channel lining machine, ~ paver
~hydraulik f = channel hydraulics
gerissene Zugzone f = cracked tension zone
Germaniumgleichrichter m = germanium rectifier
Geröll(e) n → Gesteinstrümmer f
Geröllhalde f (Geol.) = talus, scree
Gerstner-Welle f, trochoidale Dünung f = trochoidal wave
Geruch|belästigung f = odour nuisance (Brit.); odor ~ (US)
~beseitigung f = odour suppression (Brit.); odor ~, odor neutralization (US)
~verschluß m, Traps m, Syphon m = (stench) trap [*A flap in a frame which opens to admit cellar drainage to a sewer and then closes to prevent sewer air from entering the building*]
Gerüst n → Bau~
~, Ramm~ = (pile) frame, piling ~
~, Schacht m [*Bauaufzug*] = scaffold tower
~, Gel n = gel(-type structure)
~bauer m = scaffolder
~baupolier m = foreman scaffolder
~belag m = working platform of a scaffold(ing)
~brett n = scaffold(ing) board
~kette f = scaffold(ing) chain
~kran m → Bockkran

(~)Pfosten m = (scaffold) pole, standard [*An upright of a scaffold whether of wood or metal*]
~rohr n = scaffold(ing) tube
~schraube f → Gerüstspindel f
~schüttung f [*Dammbau*] = dumping embankment(s) from trestle(s)
~spindel f, Bauschraube(nwinde) f, Gerüstschraube, Spindel, Schraube(nspindel) = (conctractors') jack screw, jack, screw jack
~stange f = scaffold(ing) pole
~turm m, Stahlrohr~ = tubular steel tower
~verbinder m = scaffold(ing) coupler
gerüttelter Steindamm m → → Talsperre f
gesalzene Straße f [*Bodenverfestigung mit hygroskopischen Salzen (Magnesiumchlorid, Kalziumchlorid) als Bindemittel*] = salt-stabilized road
gesammelte Wassermengen fpl [*Talsperre*] = stored water, accumulated ~
Gesamt|abfluß m = total flow
~abmessungen fpl = over-all dimensions, total ~
~ansicht f = total view
~auftrag m [*umfaßt alle Arbeiten bis zur schlüsselfertigen Übergabe*] = turn-key contract
~aushubbodenmenge f = total volume of excavated soil
~bauwerk n = over-all structure
~belastung f = total loading
~betonspannung f = total concrete stress
(~)Bodenprobe f = soil sample
~breite f, Breite über alles = over-all width, total ~
~dicke f = combined thickness, total ~
~druck m = total pressure
~durchbiegung f = total deflection; ~ deflexion (Brit.)
~durchgang m in %, addierte Gewichtsprozente npl [*Siebtechnik*] = cumulative weight percentage, % wt. cumulative
gesamte Formänderung f, ~ Verformung = total deformation
gesamtes Querprofil n, Ausbruchquerschnitt m [*Tunnel- und Stollenbau*] = full section
Gesamt|entstaubungsgrad m = overall collection efficiency
~festigkeit f = overall strength
~gefälle n [*Talsperre*] = total head
~gewicht n = total weight
~gewicht n, Bruttofahrzeuggewicht = vehicle gross weight, gross vehicle ~, GVW
~härte f = total hardness
~herstellungskosten f = total production cost
~höhe f, Höhe über alles = total height, over-all ~
~integral n = total integral
~kerngewinn m = over-all core recovery
~kriechen n = total creep
~kürzung f, Gesamtschwindung = total shrinkage
~länge f, Länge über alles = over-all length, total ~
~last f → Lastensumme
~lösung f = total solution
~moment n = total moment
~normalspannung f = total normal stress
~oxydationsverfahren n = total oxidation process
~porenraum m = total pore space
~porosität f, wahre Porosität, Undichtigkeitsgrad m = total porosity
~säureverbrauch m eines Wassers [*früher: Gesamt-Alkalität eines Wassers, Methylorange-Alkalität eines Wassers*] = total alkalinity
~schneelast f = total snow load
~schub m = total shear
~schubspannung f = total shear(ing) stress
~schubwiderstand m = total shear(ing) resistance
~schüttungshöhe f [*die einzelnen Schüttlagenhöhen zusammen*] = depth of fill
~schwindung f, Gesamtkürzung = total shrinkage

~spannung *f* = total stress

~spannungsverlust *m* = loss of total stress

~starrheit *f* [*Straßenbefestigung*] = overall rigidity

~untersetzung *f* = total reduction

~vorspann(ungs)verlust *m* = total loss of prestress, ~ prestress loss

~wärmefluß *m* = total heat flux, ~ ~ flow

~wasserdruck *m*, Druckkraft *f* des Wassers, hydrostatische Druckkraft = hydrostatic force, total normal pressure

~wassergehalt *m* = total humidity

~wasserhaushalt *m* [*ober- und unterirdisches Wasser*] = surface flow and underground storage water

~zuschlag(stoff) *m*, Zuschlagstoffgemenge *n* = total aggregate, combined ~

Geschäfts|gebäude *n*, Geschäftshaus *n* = commercial building [*A building used principally for business or professional practice*]

~gebiet *n*, Kerngebiet = central area, business ~, ~ district

~haus *n* → Geschäftsgebäude *n*

~straße *f* = business (local) street

geschärft, an~ [*Grubenstempel*] = sharpened

geschäumte Hochofenschlacke *f* → aufgeblähte ~

geschichtete Blattfeder *f* = laminated leaf spring

geschichteter See *m* = stratified lake

geschichtetes Bruchsteinmauerwerk *n* = coursed rubble masonry [*Masonry composed of roughly shaped stones fitting approximately on level beds and well bonded*]

~ **Sedimentgestein** *n* = stratified rock

Geschiebe *n* → ~fracht

~anfall *m* = bed load production

~durchfluß *m* = rate of bed load transport

~falle *f*, Geschiebefang *m*, Schotterfalle, Schotterfang = bed load trap

~fracht *f*, Geschiebe *n* [*Die in einem längeren Zeitabschnitt, meist einem Jahre, durch einen Flußquerschnitt hindurchgehende Geschiebemenge, gemessen in m3/Zeit oder kg/Zeit*] = bed load [*The quantity of silt, sand, gravel, or other detritus rolled along the bed of a stream or river, often expressed as weight or volume per time*]

~führung *f* → Geschiebetrieb *m*

~kraft *f*, Schleppkraft = tractive force

~lehm *m* = glacial loam

~mergel *m* = boulder clay

~rückstand *m* = nappe outlier

~(stau)sperre *f* = debris dam [*A barrier built across a stream channel to store sand, gravel etc.*]

~trieb *m*, Geschiebeführung *f*, Geschiebetransport *m* = movement of bed material, bed load transport, bed load movement

geschlagener Wechsel *m* [*Dränage*] = punched drain pipe bend

geschleifter Spitzbogen *m*, Tudorbogen, Kielbogen = four-centred arch (Brit.); four-centered ~ (US); Tudor ~

geschleppter Erdhobel *m* → Anhänge-Straßenhobel

~ **Straßenhobel** *m* → Anhänge-Straßenhobel

~ **Straßenplanierer** *m* → Anhänge-Straßenhobel

~ **Wegehobel** *m* → Anhänge-Straßenhobel

geschleuderter säurefester Mantelpfahl *m* = acid-resisting spun friction(al) pile

geschlitzte Unterlagscheibe *f* = split washer

geschlitztes Überfallsturzbecken *n* = slotted spillway bucket

geschlossen [*Verschleißschicht*] = close-textured [*wearing course*]

geschlossen [*Maschine*] = self contained

~ **abgestufter Asphaltbeton** *m* = close-graded aggregate type asphaltic concrete

~ **abgestuftes Mineralgemisch** *n*, ~ abgestufte Mineralmasse *f* = dense(-graded) (mineral) aggregate, close-graded (~), ~, DGA

geschlossene Decke *f*, geschlossener Belag *m* [*Straßenbau*] = dense(-graded) pavement, close-graded ~, ~ surfacing

~ **Tafel** *f* [*Eisenbahnbrücke*] = solid deck

geschlossener Außenring *m* [*Tunnel- und Stollenbau*] = continuous external ring

~ **Gammastrahl** *m* = shielded gamma-ray

(~) **Kanal** *m* = duct

~ **Kreislauf** *m* = closed circuit

~ **Schnellfilter** *m*, geschlossenes Schnellfilter *n*, Druckfilter = pressure filter

geschlossenes Becherwerk *n* **mit senkrechter Leiter** = totally enclosed vertical elevator

~ **Kapillarwasser** *n* = capillary water rising directly from the ground water table

Geschmackbeseitigung *f* = taste suppression

geschmolzenes Schweißgut *n* = weld metal

geschnittener Nagel *m*, Schnittnagel = cut steel nail [*B.S. 1202*]

geschnittenes Gewinde *n* = cut thread

Geschoß *n*, Etage *f*, Stock(werk) *m*, (*n*) = stor(e)y [*The space between two floors or between a floor and a roof*]

~**bau** *m* → Etagenhaus *n*

~**decke** *f*, Hochbaudecke, Stockwerkdecke, Gebäudedecke, Etagendecke = floor

~**deckenteil** *m, n* → Hochbaudeckenteil

~**einschalung** *f* [*als Ergebnis der Einschalungsarbeit*] = formwork for an individual stor(e)y

~**einschalung** *f*, Geschoßeinschalen *n* [*als Arbeitsvorgang*] = setting the forms for an individual stor(e)y

~**grundriß** *m* = floor plan

~**haus** *n* → Etagenhaus

~**heizung** *f* → Etagenheizung

geschoßhoch, stockwerkhoch = of stor(e)y height

Geschoß|höhe *f*, Etagenhöhe, Stockwerkhöhe = floor headroom

~**perforator** *m* → Geschoßrohrlocher *m*

~**rohrlocher** *m*, Geschoßperforator *m* [*Tiefbohrtechnik*] = gun-perforator, casing gun

~**wohnung** *f* → Etagenwohnung

geschränktes Köpergewebe *n* = twilled double warp

geschütteter Steindamm *m* → → Talsperre

geschwächter Querschnitt *m* = weakened cross section

geschweiftes Dach *n* → Bohlendach

(geschweißte) Armierungsmatte *f* → Baustahlmatte

(~) **Bewehrungsmatte** *f* → Baustahlmatte

geschweißter Behälter *m* → Schweißbehälter

geschweißtes Gefäß *n* → Schweißbehälter *m*

~ **Profil** *n*, Schweißprofil = welded section

~ **Rohr** *n* = welded pipe

Geschwindigkeit *f* = speed

~ **der Hochwasserwelle** = flood-wave velocity

~ **des Temperaturanstiegs** = rate of temperature rise

Geschwindigkeits|abfall *m* = speed drop

~**beiwert** *m*, Schnelligkeitszahl *f* = (Chezy's) velocity factor

~**bereich** *m* = speed range

~**erhöhung** *f*, Geschwindigkeitszunahme *f* [*Kraftübertragung*] = speed pick-up

~**erhöhung** *f* [*Hydraulik*] = velocity increase

Geschwindigkeitsfläche — (Gestänge)Elevator

~**fläche** *f* [*Hydraulik*] = velocity surface
~**formel** *f* = velocity formula
~**gefälle** *n* [*Hydraulik*] = velocity gradient
~**höhe** *f* [*Hydraulik*] = velocity head [*The distance a body must fall freely under the force of gravity to acquire the velocity it possesses*]
~**messer** *m* = speedmeter
~**messung** *f* [*Abflußmessung in der Gewässerkunde*] = velocity measurement
~**plan** *m* [*Alle Punkte einer kinematischen Kette erfahren bei Verschiebung eines Punktes eine zwangsläufige Richtung und Geschwindigkeit. Diese werden zeichnerisch mit dem Geschwindigkeitsplan bestimmt*] = diagram of velocities
~**potential** *n* [*Hydraulik*] = velocity potential
~**potential-Funktion** *f* [*Hydraulik*] = velocity potential function
~**vektor** *m* = velocity vector
~**versuchsstrecke** *f* = high-speed test track
~**verteilung** *f* [*Hydraulik*] = velocity distribution
~**wechsel** *m* = speed change
~**zahl** *f* [*Hydraulik*] = velocity factor
~**zunahme** *f*, Beschleunigung *f* [*in der Zeiteinheit*] = acceleration
~**zunahme** *f*, Geschwindigkeitserhöhung *f* [*Kraftübertragung*] = speed pick-up
Gesellschaftsraum *m* [*Hotel*] = general public room
gesenkgeschmiedet = drop-forged
gesenkter freier Strahl *m*, angesaugter ~ ~ [*Hydraulik*] = apparent free nappe
Gesetz *n* **zur Reinhaltung der Luft,** Luftreinhaltegesetz *n* = air pollution prevention code
gesetzlich zulässige Geschwindigkeit *f* = legal speed
Gesichts|feld *n*, Blickfeld *n* = field of vision
~**schutzschild** *n* = face shield

gesiebtes Material *n* → abgestuftes ~
Gesims *n*, Sims *m*, *n* = corbel
gesinterter Zusatzstab *m* = sintered rod
gesintertes Nylon *n* = sintered nylon
gespannter Grundwasserspiegel *m* → angespannter ~
~ **Sattdampf** *m* = high pressure live steam
gespanntes Kabel *n* = tensioned strand
Gespannverkehr *m*, Fuhrwerkverkehr, Zugtierverkehr = animal-drawn traffic
Gesperre *n* = pawl and ratchet mechanism
gespülter (Erd)Damm *m*, ~ Staudamm, (Voll)Spül(stau)damm = hydraulic-fill dam [*This dam is essentially an earth(en) dam in the construction of which the materials are transported on to the site by water and distributed to their final position in the dam by water; a pond of mud and water being maintained on the top of the dam during construction*]
gespundete (Beton)Fuge *f*, genutete ~ = tongue-and-groove joint, keyed ~
Gestade *n*, Strand *m* = shore, beach
Gestalt|änderung *f* → Verformung
~**änderungsenergie** *f* = deformation energy
~**festigkeit** *f* = structural strength
Gestaltung *f* **der Brücken** = aesthetic design of bridges
Gestänge *n* = linkage
~ [*Die in den Grubenräumen verlegten Gleise für die Förderwagen*] = tracks
~**abfangkeil** *m* → Abfangkeil
~**abfangschere** *f* [*Rotarybohren*] = rotary slide tongs
~**bruch** *m* [*Rotarybohren*] = (drill pipe) "twist-off", breaking of the drill stem
~**bühnen-Arbeiter** *m* [*Tiefbohrtechnik*] = derrick man
(~)**Elevator** *m* [*Ein mit Drehbolzen versehenes Schellenpaar, das beim Ein-*

Gestängefett — Gesteinsgefüge

und Ausbau der Bohrgestänge und Futterrohre diese unterhalb ihrer Gewindeverbindung erfaßt] = (drill pipe) elevator
~**fett** *n* = drill pipe thread dope
~**haken** *m*, Hakenschlüssel *m*, Drehschlüssel für Bohrgestänge, Bohrgestänge-Drehschlüssel = pipe hook
~-**Konusfahrstuhl** *m*, Gestänge-Konuselevator *m* [*Rotarybohren*] = drill pipe cone elevator
~**kulisse** *f*, Quergelenk *n* = cross link assy
gestängeloses elektrisches Bohrgerät *n*, gestängeloser am Seil arbeitender Elektrobohrer *m* = wireline electrodrill, pipeless electric drill
Gestängeprüfung *f*, Gestängeprobe *f*, Gestängeversuch *m* = drill-stem test
~ **im offenen Bohrloch** = open-hole test
~ ~ **verrohrten Bohrloch** = cased-hole-test
Gestänge|rohr *n* [*Rotarybohren*] = drill pipe, hollow (drill) rod
~**schlagbohrung** *f*, Gestängeschlagbohren *n* = percussive rod boring
~**schuß** *m*, Bohrgestängezug *m* [*Rotarybohren*] = stand of drill pipe
(~-)**Sicherheitsverbinder** *m* = safety joint
~**übergang** *m* [*Rotarybohren*] = drill pipe sub
~**verbinder** *m* [*zwischen zwei Zügen*] = (drilling) tool joint [*rotary drilling*]; boring rod joint
~**verlegung** *f* (⚒) = track laying
~**zange** *f* [*Rotarybohren*] = pipe tongs, rotary ~
Gestängezug *m*, Gestängeschuß *m* [*Rotarybohren*] = stand of drill pipe
~ **aus drei Stangen bestehend** = thrible
~ ~ **vier Stangen bestehend** = fourble
~ ~ **zwei Stangen bestehend** = double
Gesteins|anker *m*, Gebirgsanker, Bolzen *m*, Anker(bolzen) [*Tunnel- und Stollenbau*] = roof bolt
~**aufbereitungsanlage** *f* = aggregate production plant, ~ producing ~

Gesteinsaufbereitungsmaschinen *fpl*	**aggregate production (or producing) equipment**
(Hart)Zerkleinerungsmaschinen	crushing and grinding equipment, comminution equipment
Siebmaschinen	screening equipment
Waschmaschinen	washing equipment

Gesteins|auspressung *f* → Kluftinjektion *f*
~**balken** *m* (⚒) = rock beam
gesteinsbildend = rock-forming
Gesteins|bindung *f* [*auf bituminösen Verschlußdecken*] = aggregate retention [*on bituminous seal coats*]
~**bohrbarkeit** *f*, Felsbohrbarkeit = rock drillability
~**bohreinheit** *f*, Atlas-Copco-Bohreinheit, Doppel-Bohraggregat *n* = twin drill
~**bohren** *n*, Gesteinsbohrung *f*, Felsbohren, Felsbohrung = rock drilling
(~)**Bohrhammer** *m*, Felsbohrhammer = (hand) hammer (rock) drill, (rock) drill (hammer)
~**bohrstahl** *m*, Felsbohrstahl = rock drill steel
~**drehbohren** *n*, Felsdrehbohren, Rotations-Gesteinsbohrung *f* = rotary (or rotating) rock drilling
~**einpressung** *f* → Kluftinjektion *f*
~**entstehungslehre** *f*, Petrologie *f* = petrology
~**forke** *f*, Felsforke = rock fork
~**gefüge** *n* = rock fabric

Gesteinsgemenge — Getreideumschlag

~gemenge *n* → Mineralgemisch *n*
~gemengteil *m* → Gemengteil
~gemisch *n* → Mineralgemisch
~gerüst *n* → Mineralgemisch *n*
~hauer *m* → Mineur
~injektion *f* → Kluftinjektion
~kern *m* = borehole core, drill ~
~korn *n* → Einzelkorn
(~)Korngemisch *n* → Mineralgemisch
(~)Kornmasse *f* → Mineralgemisch
(~)Kornmischung *f* → Mineralgemisch
~mantel *m* = surrounding rock
~masse *f* → Mineralgemisch *n* [*künstlich*]
~mehl *n* → Blume *f*
~mikroskopie *f* = petrographie microscopy
~probe *f* → Gesteinsprüfung *f*
~prüfung *f*, Gesteinsprobe *f*, Gesteinsversuch *m* = rock test, aggregate ~
~schicht *f* = rock stratum, ~ layer
(~)Schutt *m* → Gesteinstrümmer *mpl*
~sprengstoff *m* = rock explosive
(~)Staub *m* → Blume *f*
~strecke *f* (⚒) = stone drift, rock ~
~trockner *m* → Trocken-Trommel *f*
~trümmer *f*, (Gesteins)Schutt *m*, Felstrümmer, Trümmermaterial *n*, (Fels)Geröll(e) *n* = rock debris, rubble
~verpressung *f* → Kluftinjektion *f*
~versuch *m* → Gesteinsprüfung *f*
~vorkommen *n* = rock deposit
~waage *f* (mit Wiegesilo), Zuschlagstoffwaage, Gemengewaage = aggregate weigh-batcher (Brit.); aggregate weighing batcher (with scale) (US)
~waschmaschine *f*, Gesteinswäsche *f* = rock washing machine
~zement *m* [*amerikanischer Naturzement*] = rock cement
~zerfall *m* → physikalische Verwitterung *f*
~zersetzung *f* → chemische Verwitterung
Gestell *n* [*zur Lagerung*] = rack

gesteuerter Überlauf *m* = controlled weir (Brit.); ~ spillway
gestoßener Unterzug *m* = butted bridging joist
gestreckter Griff *m* = body grip
gestreute Gammastrahlung *f* in Bohrlöchern = gamma ray scattering in boreholes
gestrichener Inhalt *m*, gestrichenes Fassungsvermögen *n* [*gestrichen voll*] = struck capacity, flush ~
Gestrüpp|harke *f*, (Unterholz)Roderechen = brush rake, clearing ~
~-Pflug *m* → Buschschneider *m*
Gestück *n* → Setzpacklageschicht *f*
gestundeter Streb *m* (⚒) = standing face, stationary ~
gesund [*Gebirge im Tunnel-, Stollen-, Schacht- und Bergbau*] = firm
Gesundheits|ingenieur *m* = sanitary engineer
~kontrolle *f* [*Flughafen*] = health control
~polizei *f* = public health service
~schutz *m* in der Industrie = industrial health
~technik *f* = sanitary engineering
~techniker *m* = sanitary engineer
Getäfel *n* → Täfelung
Getäfer *n* → Täfelung
geteertes Segeltuch *n* → Persenning *f*
geteilte Bodenentleerung *f* [*Fahrzeug*] = split-bottom dump
~ Kupp(e)lung *f* = split-second clutch (coupling)
getöntes Glas *n* = tinted glass
Getreide|elevator *m*, Schiffelevator, Getreideheber *m*, Getreidebecherwerk *n*, Schiffbecherwerk = grain elevator
~mühle *f* = flour mill
~silo *m* = grain silo
~silozelle *f* = grain silo bin
~speicher *m*, Getreideschuppen *m* = grain warehouse, granary
~speicherung *f* = warehousing of grain
~trocknungsanlage *f* = grain-drying plant
~umschlag *m* = grain handling

getrennte Lagerung f von Zuschlagstoffen nach Korngrößen = segregated aggregate stockpiling (or storage)
~ **Schlammfaulung** f = separate sludge digestion [*The digestion of sludge in basins or tanks to which it is removed from the basins or tanks in which it originally settled*]
~ **Turbinen** fpl **und Pumpen** fpl **auf einer Welle** = separate pumps and turbines on a shaft, ~ pump turbine unit
getrenntes Abfertigungsgebäude n [*Flughafen*] = unit terminal building
~ **Grundablaßbauwerk** n → → Talsperre f
Getriebe|abtriebswelle f = transmission output shaft
(~)**Abtriebswelle** f [*Doppelkettenförderer*] = head shaft
~**armaturenbrett** n, Getriebeinstrumentenbrett = transmission panel
~**ausschaltventil** n = transmission neutralizer valve
~**bedienung** f = transmission control
~**gehäuse** n = transmission case
~**hydrauliksteuerung** f = transmission hydraulic control
~**instrumentenbrett** n, Getriebearmaturenbrett = transmission panel
~**kühlung** f = transmission cooling
~**kupplung** f = transmission clutch
~**motor** m = geared motor
~**ölsumpf** m = transmission oil sump
~**pumpe** f = gear-type pump
~**pumpenüberdruckventil** n = transmission pump relief valve
~**schalthebel** m = transmission range selector lever
~**schmierpumpe** f = transmission lube pump, ~ lubricating ~
~**schmiersystem** n = transmission lube system, ~ lubricating ~
~**schmierventil** n = transmission lubricating valve, ~ lube ~
~**schutz** m = transmission guard
~**sonderuntersetzung** f = optional transmission gear

~**steuerventil** n = transmission control valve
~**tunnel** m = transmission tunnel
~**zimmerung** f = forepoling
gewachsener Boden m, natürlicher ~ = natural soil, unmade ground, undisturbed ~
~ **Fels** m → anstehender ~
~ **Zustand** m, ungestörte natürliche Lagerung [*Boden*] = natural state
gewachsenes Gestein n → anstehender Fels m
Gewährleistung f = guarantee
gewaltsames Austreten n → Ausbruch
gewalzt, ab~ = rolled
gewalzte (Auf)Schüttung f → ab~ ~
~ **Erd(auf)schüttung** f, ab~ ~, Walzdamm m = rolled earth fill
gewalzter Damm m → ab~ ~
~ **Erdkern** m → ab~ ~
~ **Rundstahl** m für Stahlbeton [*DIN 488*] = rolled steel bars, ~ re-bars
Gewände n = jamb
~**aufstand** m = jamb footing
Gewässer n = water(s), body of water
~**bett** n = bed
Gewebe n, Sieb~, gewebter Siebboden m = (screen) cloth; (~) fabric (US)
~ **mit Rahmen** → Sieb~ ~ ~
~ ~ **rechteckigen Maschen** → Sieb~ ~ ~ ~
~**ausführung** f, Sieb~ = finish of (screen) cloth; ~ ~ (~) fabric (US)
~**filter** m, n = cloth filter
~**güte** f, Gewebequalität f, Sieb~ = quality of (screen) cloth; ~ ~ (~) fabric (US)
~**lage** f = fabric ply
~**liste** f, Sieb~ = catalog of (screen) cloth; ~ ~ (~) fabric (US)
~**nummer** f, Sieb~ = mesh number
~**qualität** f → Gewebegüte
gewebter Siebboden m (Sieb)Gewebe n = (screen) cloth; (~) fabric (US)
gewellt [*Platte*] = corrugated
gewerblicher Sprengstoff m = commercial explosive
~ ~ **für den Bergbau**, Bergbausprengstoff = mining explosive

Gewichts|abmessung f, Gewichtsdosierung f, Gewichtszuteilung f, Gewichtsteilmischung f, Gewichtszumessung f, Gewichtsgattierung f, Gewichtszugabe f = gravimetric batching, proportioning by weight, weigh-batching, batching by weight
~**abmeßapparat** m → Gewichtszumeßapparat
~**anzeiger** m = weight indicator
~**ausbringen** n [*Siebtechnik*] = yield (by weight), recovery ~ ~, output ~ ~
~**dosierapparat** m → Gewichtszumeßapparat
~**kontrollmesser** m → Drillometer m
gewichtsloser bindiger Boden m = weightless cohesive soil
Gewichts|mauer f → → Talsperre f
~**meißel** m, Fallmeißel [*Felsbaggerung unter Wasser*] = drop chisel
~**prozente** npl, Gewichtsteile mpl = percentage by weight
Summe f der ~ = cumulative ~ ~ ~
~**rückstand** m = residue by weight
~**sperre** f → → Talsperre
~**staumauer** f → → Talsperre f
~**stützmauer** f [*Siehe Anmerkung unter „Stützmauer"*] = gravity retaining wall
~**teilmischung** f → Gewichtsabmessung
~**teilstrich** m = weight graduation (mark)
~**verhältnis** n = weight ratio
~**verlagerung** f = weight transference
~**verminderung** f, Gewichtsabnahme f, Gewichtsverringerung f = weight reduction
~**verteilung** f = weight distribution
~**zumessung** f → Gewichtsabmessung
~**zumeßapparat** m, Gewichtsabmeßapparat, Gewichtsdosierapparat, Gewichtszuteilapparat = weighbatcher
~**zuteilapparat** m → Gewichtszumeßapparat
~**zuteilung** f → Gewichtsabmessung

Gewinde|-Anschlußstab m = threaded projecting bar
~**bolzen** m = screwed bolt
~**ende** n [*Steinschraube*] = threaded part
~**flansch** m = thread(ed) flange
~**kern** m, Gewindefuß m = root of thread
~**muffe** f → Schraubmuffe
~**platte** f = threaded plate
~**rohr** n = screwed pipe
~**schneider** m, Gewindeschneidapparat m = tapper
~**stange** f, Gewindestab m = screw-threaded rod
~**verbinder** m [*Tiefbohrtechnik*] = tool joint
(~)**Verschraubung** f des Bohrgestänges = joint
gewinnbare Ölvorräte mpl = recoverable oil reserves
gewinnen [*Wasser*] = to draw water from an aquifer
~ → herein~
Gewinnung f → Herein~
~ **aus Seifen** = placer mining
~ **mit Hydromonitoren** → Abspritzen n ~ ~
Gewinnungs|gerät n, Herein~ = winning equipment
~**ingenieur** m [*Erdöl; Erdgas*] = exploitation engineer
~**sonde** f → Produktionsbohrung
Gewitterfront f = thunderstorm front
gewobener Gurt m, ~ **Riemen** m = texrope belt
(gewöhnliche) Abraumschrapperanlage f = Abraumschrapper m
~ **Feile** f, Flachfeile f = flat file
~ **Springtide** f = ordinary spring tide
~ **Zündschnur** f → Bickford-Zündschnur
gewöhnlicher Abraumschrapper m → Abraumschrapper
~ **Portlandzement** m [*in Deutschland Z 225*] = Portland cement for general use, normal Portland cement, ordinary Portland cement; Portland cement type I (US)

(gewöhnlicher) Schrapper — Gezeitenströmung

(~) **Schrapper** m, Schleppschrapper = (power) drag scraper (machine), ~ scraper excavator
~ **Syenit** m, Kalkalkalisyenit = calc-alkali syenite
~ **Wasserstand** m, durchschnittlicher ~, mittlerer ~ = average stage
Gewölbe n = vault
~ → Sattel der Faltung
~**abdeckung** f → Gewölbedeckung
~**anfänger** m → Gewölbekämpfer m
~**art** f = type of vault
~**auskleidung** f, Gewölbebekleidung, Gewölbeverkleidung = vault lining
~**bau** m = vault construction
~**deckung** f, Gewölbeabdeckung, (Ab-)Deckung des Gewölbes = vault covering
~**-Gewichtssperre** f, Bogengewichtssperre, Bogen(schwer)gewichts(stau)mauer f = arch-gravity dam
~**grat** m = groin
~**kämpfer** m, Bogenkämpfer, Kämpfer(stein) m, Gewölbeanfänger m, Bogenanfänger = rein, impost, springer
~**kern** m (⚒) = arch-core [*of pressure arch*]
~**leibung** f, innere Gewölbefläche f, Gewölbelaibung = intrados of vault, soffit ~ ~, vault soffit
~**mauer** f → Bogensperrmauer
~**mauerwerk** n = vaulting masonry, stone arching
~**profil** n = outline of vault
~**reihen(stau)mauer** f, Bogenpfeiler(stau)mauer, Vielfachbogensperre f, Pfeilergewölbe(stau)mauer, Gewölbereihensperre = multiple arch(ed) dam
~**rippe** f = vault rib
~**rücken** m, äußere Gewölbefläche f = extrados of vault, back ~ ~
~**schalung** f = vault formwork
~**schenkel** m = haunch of vault
~**schub** m = vault thrust
~**spannweite** f = vault span
~**sperrmauer** f → Bogensperrmauer
~**stein** m → Wölbstein
~**wirkung** f, Bogenwirkung = arching (effect), arch action
~**ziegel** m → Wölbziegel
~**zwickel** m = vault spandrel
gewölbte Kappe f (⚒) = cambered girder
~ **Schale** f = arched shell
Geysir m, Geyser, pulsierende Springquelle f, intermittierende Springquelle, isländischer Quellsprudel m = geyser, gusher
Gezeit f, Tide(erscheinung) f, Gezeitenerscheinung = tide, tidal phenomenon
Gezeiten|alter n → Alter der Tide
~**becken** n, Tidebecken = tidal basin
~**beiwert** m, Tide(n)beiwert = tidal coefficient
~**berechnung** f = tidal calculation
~**bewegung** f, Tide(n)bewegung = tidal movement, ~ motion
~**brandung** f → Bore f
~**diagramm** n, Gezeitenkurve f, Tidenkurve f = tidal curve, ~ diagram (Brit.); marigram (US)
~**energie** f, Tide(n)energie = tidal power
~**erosion** f = tidal erosion
~**erscheinung** f → Gezeit f
~**erzeuger** m = tide-generator
~**feuer** n = tidal light
~**hafen** m → Fluthafen
~**hub** m → Flutgröße f
~**kraft** f = tidal power
~**kraftwerk** n → Flutkraftwerk
gezeitenlos, tidelos = tide-free
Gezeiten|merkmal n = tidal feature
~**modell** n = tidal model
~**signal** n, Tide(n)signal = tide-ball
~**spiel** n = tide cycle
~**strand** m = tide beach
~**strecke** f [*Strom*] = tidal portion, ~ reach
~**strom** m = tidal river
~**stromfigur** f → Strömungskreislauf
~**strömung** f, Tidestromung, Gezeitenstrom m, Tidestrom = tide current, tidal ~
~**strömung** f, starke ~ = race

~strömung *f* in Flußmündungen = estuarine flow
~tabelle *f* = tidal table
~überschwemmung *f* = tidal inundation
~verfrühung *f* = acceleration, priming [*the shortening of the interval between the time of two high tides*]
~verzögerung *f* = tide retardation
~wechsel *m*, Tidewechsel = tidal range
~welle *f*, Tidewelle = tide wave, tidal ~

gezogener Bauteil *m*, gezogenes ~ *n* [*unter Zug stehend*] = component under tension

~ Radschrapper *m* → Anhänge-Radschrapper

GGL *n* → Gußeisen *n* mit Lamellengraphit

Giebel *m* = gable [*The triangular-shaped piece of wall, closing the end under a gable roof*]
~chen *n* = gablet
~dach *n* = gabled roof
~fenster *n* = gable window
~gaupe *f* = gable dormer
~mauer *f*, Giebelwand *f* = gable wall
~rahmen *m* = gable(d) frame
~turm *m* = gable tower
~wand *f*, Giebelmauer *f* = gable wall

Gießbühne *f* = pouring floor [*foundry*]
Gießen *n* von Beton = concrete placing by gravity, chuting of concrete
Gießerei *f* mit Modellformerei in Sand und Masse = sand foundry
~-Kunstharz *n* = foundry resin
~sand *m* = foundry sand
~schwärze *f*, Gießerschwärze = blacking, blackwash, blackening, founder's black, facing

gießfähiger Beton *m* → → Beton
Gießharz *n* = casting resin
~beton *m* = resin concrete
~mörtel *m* = resin mortar
Gieß|katze *f* = ladle crane troll(e)y
~maschine *f* für Nichteisenmetalle = casting machine for non-ferrous metals
~mast *m* → (Beton)Gießmast

~pfanne *f* = pouring ladle [*foundry*]
~pfannenziegel *m*, Pfannenstein *m* = ladle brick
~rinne *f* = concrete placing chute
~rinnenanlage *f* → (Beton)Gießrinnenanlage
~rohr *n*, Beton~, Gußbeton-Rohr = pipe for concrete placing by gravity
~schlicker *m* = casting-slip
~ton *m* = foundry clay, ladle ~
~turm *m* → (Beton)Gießturm
~turmkübel *m*, Beton-~ = tower hoist bucket

giftfrei = nontoxic
Giftgehalt *m*, Giftigkeit *f* = toxicity
Gilsonitasphalt *m*, Uintait *m* = uintaite, gilsonite
Gips|abbindeverzögerer *m* = gypsum retarder
~aufbereitungsanlage *f*, Gipswerk *n* = gypsum plant
~bauplatte *f* → Gipsplatte
~baustein *m* = gypsum partition tile (US); ~ ~ block
~baustoff *m* = gypsum building material
~beton-Fertigmörtel *m* = gypsum concrete
~bett *n* [*Prüfung der Scheiteldrucklast von Rohren*] = fillet of plaster of Paris
~brecher *m* = gypsum crusher, ~ breaker
~brei *m* = gypsum putty, ~ paste
~brennen *n* = gypsum calcination
~bruch *m* = gypsum quarry
~deckenplatte *f* = gypsum ceiling board
~diele *f* [*großflächige Gipsplatte*] = gypsum plank
~diele *f* mit Pflanzenfasereinlage, ~ ~ Naturfasereinlage = gypsum plank with fiber (US); fibrous plaster, stick and rag work (Brit.)
~diele *f* mit Schilfrohreinlage = gypsum plank with reed
~(dreh)ofen *m* = (rotary) calciner
~erde *f* = gypsum earth, gypsite
~erzeugnis *n* = gypsum product

Gipsestrich — Glanzkobalt

~estrich m = anhydrite (jointless) floor(ing), ~ composition ~
~fels m → kompakter Gips
~form f = plaster mould (Brit.); ~ mold (US)
~gehalt m = gypsum content
~kalkmörtel m = gypsum-lime mortar
~kartonplatte f = sandwich type (gypsum) plasterboard, ~ ~ ~ board [*It consists of a core of processed gypsum rock sandwiched between two sheets of heavy tough paper*]
~kern m = gypsum core
~kocher m, Gipskessel m = calcining kettle, gypsum ~ ~
~lager n, Gipsvorratslager = gypsum stor(ag)e
~mergel m = gypseous marl
~mörtel m = gypsum mortar
~mörtelbelag m, Gipsputz m = gypsum plaster
~ofen m → Gipsdrehofen
~pfanne f = (gypsum) calcining pan
~platte f, Gipsbauplatte = (gypsum) plasterboard, ~ board
~punkt m = plaster dab [*A small lump of gypsum plaster which sticks to brickwork or lathing and is used as a fixing for wall tiles, marble facing, or joinery*]
~putz m, Gipsmörtelbelag m = gypsum plaster
~putzmörtel m = gypsum stuff
~sand m = gypsum sand
~sandmörtel m = gypsum-sand mortar
~sinter m = stalactitical gypsum
~spat m, kristalliner Gips m = crystalline gypsum
~staub m = raw gypsum dust
~stein m = gypsum rock
~tonerdezement m = sulpho-aluminate cement
~(vorrats)lager n = gypsum stor(ag)e
~wand(bau)platte f, Wand(bau)platte aus Gips [*DIN 18163*] = gypsum wallboard

~wand(bau)platte f **als Oberputzträger** = gypsum lath
~wand(bau)platte f **für Sichtflächen** = gypsum wallboard designed to be used without the addition of plaster
~werk n → Gipsaufbereitungsanlage f
~zwischenwandplatte f, Zwischenwandplatte aus Gips = gypsum partition board
Gitter n [*Siebboden*] = weave
~, **~werk** n = lattice(work)
~ = grille
~ausleger m = (steel-)lattice(d) (crane) boom (US)/jib (Brit.)
~fenster n = lattice(d) window
~gerüst n [*Kran*] = lattice(work) undercarriage [*crane*]
~mast m = lattice(d) mast
~netzkarte f = gridded map
~pfette f = latticed purlin(e)
~rohrausleger m = tubular lattice boom (US); ~ ~ jib (Brit.)
~rost m = grating
~rost(fuß)boden m = grid paving
~sieb n = weave screen
~stab m = latticed bar, ~ rod
~stab m, Füll(ungs)stab, (Träger)Stab = member, bar, rod
~träger m [*Fachwerkträger mit einer Vielzahl sich kreuzender Diagonalstabzüge, die ein Netzwerk zwischen den beiden Gurten bilden*] = lattice truss, ~ girder
~trägerbrücke f = lattice truss bridge, ~ girder ~
~tür f = lattice door
~(werk) n = lattice(work)
Glanz|anstrichfarbe f = gloss paint
~blende f, Alabandin n, Manganblende f (Min.) = alabandite, mangan-blende
~braunstein m, Schwarzmanganerz n, Hausmannit m (Min.) = hausmannite
glänzende Schubfläche f → (Rutsch-) Harnisch m
Glanz|kobalt m, Kobaltglanz m, Cobaltin n, CoAsS (Min.) = cobalt glance, cobaltite

Glanzkohle — Glasstahlbetondiele

~kohle *f* = glance coal, lustrous ~; peacock ~ (US)
~pech *n*, Maniak *n* = glance pitch
~ruß *m* = shining soot
~verlust *m* [*Anstrich*] = loss of gloss
Glas|(bau)stein *m*, Glasziegel *m* = glass brick, ~ block
~bausteindecke *f* → Glassteindecke
~bedachung *f* → Glas(dach)eindeckung
~betonstein *m* → Betonglas *n*
~bodenplatte *f*, Glasfußbodenplatte = glass floor(ing) tile
~bruchstück *n* = glass fragment
~(dach)eindeckung *f*, Glasbedachung = glass roofing
~dachstein *m*, Glas(dach)ziegel *m* = glass tile, ~ slate
~eindeckung *f*, Glas(dachein)deckung, Glasbedachung = glass roofing
Glaser *m* = glazier
~kitt *m*, Ölkitt, Fensterkitt, Verglasungskitt = glazing mastic, ~ putty, back ~, glazier's ~
Glasfabrik *f*, Glaswerk *n* = glass plant
Glasfaser *f* = glass fibre (Brit.); ~ fiber (US)
~-Balkenform *f* = glass fibre beam mould (Brit.); ~ fiber beam mold (US)
~form *f* = (reinforced) glass fibre mould (Brit.); (~) ~ fiber mold (US)
~gewebe *n* = glass fabric
~matte *f* = glass fibre mattress (Brit.); ~ fiber ~ (US)
~putz *m* = glass fibered plaster
~schalung *f* = glass fibre formwork (Brit.); ~ fiber ~ (US)
~stab *m* = glass fibre rod (Brit.); glass fiber ~ (US)
~technik *f* = glass fibre technique (Brit.); ~ fiber ~ (US)
glasfaserverstärktes Kunstharz *n* = glass fibre-reinforced (synthetic) resin (Brit.); ~ fiber-reinforced (~) ~ (US)
Glas|faservlies *n* → Glasvlies
~fassadengebäude *n* = glass-fronted building, glass-façade ~

~fliese *f* = glass tile
~füllung *f* [*Tür*] = glass panel
~(fuß)boden *m* = glass floor(ing)
~(fuß)bodenplatte *f* = glass floor(ing) tile
~gang *m* = glazed corridor
~gebäude *n*, Glashaus *n* = glass clad building
~gespinst *n* = spun glass
~gestein *n* = glassy rock
~gewebe-RUBEROID *n* = fibre glass cored bituminous felt RUBEROID
~haus *n*, Glasgebäude *n* = glass clad building
Glasieren *n* = glazing
glasierte Wandplatte *f* = ceramic glazed wall facing tile
glasig [*Grundmasse*] = glassy, vitreous, hyaline [*ground-mass*]
glasiger Scherben *m*, glasige Scherbe *f* = vitreous body
Glas|kassette *f* = glass core, ~ waffle
~kolben *m* = glass flask
~körper *m* für Glasstahlbeton → Betonglas *n*
~lichtkuppel *f* = glass domed roof light, ~ saucer dome
~mosaik *n* = glass mosaic
glaspapierrauhe Straßenoberfläche *f* = sandpaper surface [*A road surface from which sharp pieces of aggregate protrude not more than 3/16 in.*]
Glas|perle *f* = glass bead
~perlen-Markierungsstreifen *m* = beaded stripe
~rohr *n* = glass tube
~-Rolltreppe *f* = crystalator, glass-balustrade escalator
~sand *m* = glass sand
~scheibe *f*, Glastafel *f* = pane (of glass), glass pane
~schiebetür *f* = sliding glass door
~schiebewand *f* = sliding glass wall
~schneider *m* = glas cutter
~seide *f* = glass silk
~splitter *m* = glass splinter
Glasstahlbeton *m* = glass-concrete
~bau *m* = glass-concrete construction
~diele *f* = glass-concrete plank

Glasstahlbeton-Raumtragwerk — Glaukonitsandstein

~-**Raumtragwerk** n = space structure in glass-concrete
Glas|stahlbimsbetondiele f = glass-pumice concrete plank
~**stein** m, Glasbaustein, Glasziegel m = glass brick, ~ block
~**steindecke** f, Glasbausteindecke [*Decke aus Glasprismen, zwischen denen Stahlbetonrippen angeordnet sind*] = glass brick floor, ~ block ~
~**steinwand** f [*DIN 4103*] = glass brick wall, ~ block ~
~**tafel** f, Glasscheibe f = pane (of glass), glass pane
~**terrasse** f = glazed terrace
~**trennwand** f, Glaszwischenwand = glass partition (wall)
~**tür** f = glass door
glasumgeben = glass enclosed
Glasur f = glaze(d finish), glaze coat
Glas|veranda f = glazed veranda(h); sun parlor (US)
~**vlies** n, Glasfaservlies, Glasvlieseinlage f = glass fiber wrapping material (US); ~ fibre ~ ~ (Brit.)
~-**Vorhangwand** f = glass curtain wall
~**wand** f = glass wall
~**wandtafel** f = glass wall panel
~**wanne** f = glass tank crown
~**wannenofen** m = glass tank furnace
~**wannenregenerator** m = glass tank checker
~**werk** n, Glasfabrik f = glass plant
~**wolle** f [*düsengeblasen*] = glass wool
~**ziegel** m, Glasdachstein m, Glasdachziegel = glass tile, ~ slate
~**ziegel** m → Glasstein m
~**zwischenwand** f, Glastrennwand = glass partition (wall)
glatte Bindung f [*Siebtechnik*] = smooth joints
~ **Schalung** f = wrought formwork
glatter Bewehrungsstahl m, ~ Armierungsstahl = plain (steel) bars
glattes Bohrgestänge n [*Rotarybohren*] = flush joint drill pipe

Glättband n, Glättriemen m = smoothing belt, finishing ~
Glättbohle f, Glättelement n, Deckenschlußbohle, Abstreif- und Glättelement, Glätter m [*Beton(decken)fertiger*] = smoothing screed, finishing ~, float pan, float plate, floating pan, (finishing) float ~, Abziehbohle [*Schwarzbelageinbaumaschine*] = screed plate
Glättbohlenfertiger m, Putzbohlenfertiger = smoothing beam (or screed) finisher (or finishing machine)
Glätte f, Schlüpfrigkeit f = slipperiness
Glatteis n [*Ein sehr glatter, meist durchsichtiger Eisüberzug auf einer Fahrbahn oder einem Gehweg, der dadurch entsteht, daß unterkühlte Wassertropfen beim Auftreffen auf den Boden sofort gefrieren oder daß gewöhnliche Wassertropfen auf die unterkühlte Fläche fallen und dort festfrieren*] = glazed frost
~**bildung** f = glazed frost formation
~**schicht** f = glazed frost layer
Glättelement n [*Beton(decken)fertiger*] → Glättbohle f
glätten, abziehen [*mit Kelle*] = to smooth by trowelling
Glätter m → Glättbohle f
Glättkelle f = float, smoothing trowel, finishing ~
Glattmantel|-Anhängewalze f → → Walze
~-**Schleppwalze** f → → Walze
Glätt|maschine f [*für Beton*] = mechanical trowel, ~ float
~- **und Verdichtungsbohle** f [*Kanalauskleidungsmaschine*] = ironing-screed
Glatt|walze f → → Walze
~**walzwerk** n, Glattwalzenbrecher m = smooth-shell crushing rolls
Glaucherz n = poor ore
Glaukonit|kalkstein m = glauconitic limestone
~**sandstein** m = glauconitic sandstone

Glazialerosion — Gleisbaumaschinen

Glazialerosion *f*, Gletschererosion = glaciation
glaziale Flußablagerung *f* = glacial outwash
glaziales und fluvoglaziales Sedimentgestein *n* = ice-borne sediment
Glazialton *m* = glacial clay
Gleiboden *m* = gley soil
Gleichdruck|-Axialturbine *f* = axial flow impulse turbine
~-Kondensationsturbine *f* = impulse condensing turbine
~turbine *f* → Druckturbine
gleichförmig belastet = uniformly loaded
gleichförmiger Abfluß *m* = uniform flow
~ Sand *m* = uniform sand, uniformly graded ~, closely graded ~
Gleichförmigkeitsbeiwert *m* = homogenizing coefficient [*road traffic*]
~ von Kramer [*Hydraulik*] = Kramer's uniformity factor
Gleichgewicht *n* = equilibrium, balance
Gleichgewicht|anzeiger *m*, Gleichgewichtanzeigeapparat *m* = balance indicator
~bedingung *f* = equilibrium condition, balance ~
~hebewerk *n*, lotrechtes (Schiff)Hebewerk mit Gewichtausgleich = balanced vertical lift
~kammer *f* [*Sektor(stau)wehr*] = balancing chamber (Brit.); gate recess
~polygon *n* = equilibrium polygon
~zustand *m* = state of equilibrium, ~ ~ balance
gleichkörniges Gut *n*, monodisperses ~ = grains of equal size, uniformly sized grains
Gleich|last *f* = uniformly-distributed load
~linienlast *f* → Gleichstreckenlast
gleichmäßige Strömung *f*, Strömung gleichmäßiger Geschwindigkeit = uniform flow
Gleich|mäßigkeit *f* g_1 **und** g_2 [*Lichttechnik*] = uniformity of illumination

~richtung *f* = rectification
gleichschenk(e)liger Auflagerwinkel *m* [*Stahlbau*] = equal-leg angle
~ Winkelstahl *m* = angles with equal legs
Gleich|schlag *m* → Albertschlag
~streckenlast *f*, Gleichlinienlast = uniformly distributed line(ar) load, ~ ~ knife-edge ~
Gleichstrom|aggregat *n* = D.C. (generating) set, D.C. (~) plant
~anlage *f* [*einer Maschine*] = D.C. equipment, direct current ~
~(-Aufschluß)verfahren *n* = direct current method, ~ ~ exploration, ~ ~ prospecting
~-Rohrschieber *m* [*beim Druckluftwerkzeug*] = direct flow hollow spool valve
~verfahren *n*, Gleichstrom-Aufschlußverfahren = direct current method, ~ ~ exploration, ~ ~ prospecting
~-Wäscher *m*, Einstrom-Wäscher = concurrent scrubber, uniflow ~
Gleichung *f* **vierten Grades**, biquadratische Gleichung = biquadratic equation
gleichwertige Korngröße *f*, äquivalente ~ = equivalent grain size, ~ particle ~
gleichwertiger Verdichtungsdruck *m*, äquivalenter ~ = equivalent compaction pressure
Gleichwinkel(stau)mauer *f* → → Talsperre *f*
Gleis *n*, Schienenstrang *m*, Schienenweg *m* = rail(way) track (Brit.); rail(road) ~ (US); tracking, line of rails, two-rail surface track
~anlage *f* = track system
~anschluß *m* → Anschlußgleis *n*
~arbeiten *fpl* = railway track-work (Brit.); railroad ~ (US)
~bagger *m*, Schienenbagger = rail-mounted excavator
~baukran *m* = track laying crane
~baumaschinen *fpl* = track laying machinery, permanent way construction ~

Gleisbildstellwerk — (Gleis)Kettenschlepper-Frontlader

~bildstellwerk n = power signalling installation

~bohrwagen m, Schienenbohrwagen = rail-mounted wagon drill, track-mounted ~ ~

~bremse f = car retarder

~erneuerung f = track renewal

~fahrzeug n, Schienenfahrzeug = rail-mounted (or rail-guided) vehicle

gleisgebunden, schienengeführt, schienengebunden = rail-mounted, rail-guided

Gleis|hebewinde f = track (lifting) jack

~kette f → Raupenkette

~kette f für LKW = truck track

~kette f mit Platten = track group

~kette f ohne Platten = track link assembly

(Gleis)Ketten|anhängerwagen m → Raupenwagen

~antriebskette f, Raupenantriebskette = crawler drive chain

~antriebskupplung f, Raupenantriebskupplung = crawler drive clutch

~antriebsturas m, Raupenantriebsturas = crawler drive sprocket

~antriebsturaswelle f, Raupenantriebsturaswelle = crawler pivot shaft

~-Aufstandfläche f = crawler bearing area

~bagger m, Raupenbagger = crawler (type) excavator

~band n = track assembly

~-Becherwerk(auf)lader m → Becherwerk(auf)lader auf (Gleis)Ketten

~-Bohrwagen m, Raupen-Bohrwagen = crawler wagon drill

~bolzen m, Raupenbolzen = track pin

~büchse f = track bushing

~dichtung f = track seal

~durchhang m = track slack

(gleis)kettenfahrbar, raupenfahrbar = crawler-mounted

(Gleis)Ketten|-Fahrlader m → Gleisketten-Ladeschaufel f

~fahrstreifen m, Raupenfahrstreifen, (Gleis)Kettenfahrspur f, Raupenfahrspur [*Gleitschalungsverfahren beim Betondeckenbau*] = track path

~fahrwerk n → Raupenfahrwerk

~fahrzeug n, Raupenfahrzeug, Raupe f = track laying vehicle, crawler-type ~; tracked ~ [*deprecated*]

~fertiger m [*früher: Raupenfertiger m*] = caterpillar (or track-laying type) (asphalt) finisher (or paver, or bituminous paving machine, or bituminous spreading-and-finishing machine, or black-top spreader, or asphaltic concrete paver)

~-Frontlader m = front end crawler shovel

~führungsschutz m = track guiding guard

~gerät n [*früher: Raupengerät n, Raupe f*] = caterpillar, tracklayer

~gerätfahrer m, Raupenfahrer = cat-jockey

~glied n → Raupenschuhplatte f

~gliedlauffläche f = track link rail

~-Hoch(löffel)bagger m → Hochlöffel-Raupenbagger

~kraftschluß m, Raupenkraftschluß = traction of tracks [*gripping action between tracks and the surface*]

~kran m, Raupenkran, Kranraupe f = crawler(-mounted) crane

~kupplung f, Raupenkupplung = crawler clutch

~lademaschine f → Raupenlader m

~lader m → Raupenlader

~laufrolle f = track roller

~laufwerk n → Raupenfahrwerk

~laufwerkrahmen m = track roller frame

~platte f → Bodenplatte

~rad n → Kettenrad

~-Rahmen m = track frame

~rolle f = track roller, tractor ~

~-Schaufelradgrabenbagger m, Raupen-Schaufelradgrabenbagger = crawler-wheel type machine, ~ trencher

~-Schlepper m → Kettenschlepper m

~schlepper-Frontlader m, Gleiskettenschlepper-Frontladeschaufel f, (Gleis-)Ketten-Fahrlader m mit

(Gleis)Ketten-Schlepperwalze — Gleitflächenkrümmung

Frontladeschaufel, Front-Fahrlader m mit (Gleis)Kettenfahrwerk = crawler tractor(-mounted front end loader)
~-Schlepperwalze f = tracked roller
~spanner m, Raupenspanner = crawler take-up
~spannfeder f = track adjusting spring
~spannung f = track tension
~spannzylinder m = track adjusting cylinder
~teilung f = track pitch
~traktor(en)kran m, Gleiskettenschlepperkran m, Raupenschlepperkran = crawler tractor crane
~turas m → Kettenturas
~-Turmdrehkran m, Raupen-Turmdrehkran = crawler-mounted (mobile) tower crane
~verbindung f = track joint
~wagen m → Raupenwagen
~-Wurfschaufellader m, Gleisketten-Über-Kopf-(Schaufel-)Lader m = crawler (or caterpillar, or tracklaying type) overhead loader (or overshot loader, or rocker shovel, or flip-over bucket loader, or overloader)
~zahnrad n → Kettenrad
~-Zugmaschine f → Kettenschlepper m

Gleiskran m, Schienenkran = rail-mounted crane
gleislos = railless, trackless
gleislose Erdbewegung f, geländegängige = railless earthmoving, trackless ~, ~ soil shifting operations
gleisloser Förderwagen m → Muldenkipper m
gleisloses Fahrzeug n = free-wheeled vehicle
Gleis|oberbaugerät n = permanent way construction equipment
~rad n = track wheel
Gleisrückmaschine f **für absatzweisen Betrieb,** absatzweise arbeitendes Gleisrückgerät n = intermittent type track shifting machine
~ ~ **kontinuierlichen Betrieb** m = continuous type track shifting machine
Gleisschotter m → Bettungsschotter
~-Reinigungsmaschine f → Bettungsreinigungsmaschine
~-Reinigungs- und Lademaschine f = (track) ballast cleaning and loading machine
(Gleis)Schwelle f → Bahnschwelle
Gleis|spur f, Spur(weite) f = (track) ga(u)ge, rail ~
~stopfer m, Schotterstopfer = tie tamper (US); track ~ (Brit.); ballast ~
~stopfkolonne f = boxing up gang, ~ ~ team, ~ ~ crew, ~ ~ party
~stopfmaschine f, Schotterstopfmaschine = ballast tamping machine, track ~ ~
~verlegung f = track laying
~waage f, Waggonwaage = trackscale, railway-track scale
~zubehör m, n = track accessories

Gleit|aufreißer m [*an einer Walze*] = sliding scarifier
~bahn f [*Lager für Hoch- und Tiefbauten*] = sliding track, slide ~
~bereich m [*Bodenmechanik*] = sliding zone
~betonschalung f → Gleitschalung
~bewegung f = sliding movement
~bogen m (✵) = yielding roadway arch, sliding ~ ~
~(bohr)meißel m = sliding drill bit
~ebene f = plane of sliding
Gleiten n [*Bodenmechanik*] = sliding
~ → Laminarbewegung
~ **eines Trägers,** Abheben ~ ~ = lifting of a girder
gleitend, verschiebbar = sliding
gleitende Reibung f = sliding friction
Gleitfähigkeit f **des Betongemisches** = placeability
Gleitfläche f = sliding surface, surface of sliding
~ = slip plane
~ → (Rutsch)Harnisch m
Gleitflächen|krümmung f = curve of sliding

Gleitmethode — Gleitwegstrahl

~methode f = method of slices
~rutschung f = landslip due to tectonic movement
~schar f [Boden] = network of slip lines
Gleit|form f = slip mould (Brit.); ~ mold (US)
~fuge f, biegsame Verbindung f [Motor] = slip joint
~fuß m (⚒) = stilt
~geschwindigkeit f [Gleitschalung] = rate of slide
~geschwindigkeit f, laminare Fließgeschwindigkeit = laminar velocity
~kanal m, Blech-~, Rohr n aus dünnem Blech, Blechröhre f, Hülle f aus Stahlblech = (sheet-)metal sheath
~kappenausbau m (⚒) = slide-bar system
~keil m [Bodenmechanik] = wedge of failure, sliding wedge
~klausel f = variation clause, fluctuation ~
~kolben m = floating sealed piston
~kreisanalyse f = friction circle analysis, slip ~ ~
Gleitkufe f = skid
~ über die ganze Schildbreite = full-width skid
Gleitkufenrahmen m = skid type base (frame)
Gleitlager n [für Hoch- und Tiefbauten] = slide bearing, sliding ~
~ = friction bearing, plain (~) ~, sliding ~, journal ~
~ von endlicher Breite = finite journal bearing
~ ~ unendlicher Breite = infinite journal bearing
Gleit|linie f = line of sliding, sliding line
~masse f, Rutschmasse = sliding mass
~meißel m, Gleitbohrmeißel = sliding drill bit
Gleitorfboden m = peat-gley soil
Gleit|platten-Mikroviskosimeter n = sliding plate micro-viscometer
~punkt m [Rohrleitung] = skid
~reibung f = sliding friction
Gleitreibungsbeiwert m = coefficient of sliding friction
~ [Straße] = kinetic coefficient of friction
Gleitring|dichtung f = duo-cone seal, floating ring ~
~-Stopfbuchse f = slip-ring gland
Gleit|riß m = slip crack
~rolle f, Antifriktionsrolle = anti-friction roller
~schalung f = slipform, sliding form
Gleitschalungsbau m = slipform construction, sliding form ~
~werk n = slip-formed structure
Gleitschalungs|beton m = slip-formed concrete
~fertiger m = slip form paver, sliding ~ ~, formless paving machine
~montage f = slipform erection
~verfahren n = slipform construction method, sliding form ~ ~
Gleit|schicht f (Geol.) = flow sheet (Brit.); sheet of drift (US)
~schiene f = sliding rail
~schutz m, Gleitschutzbehandlung f [Straße] = non-skid treatment
~schutzkette f, Schneekette, Schutzkette, Reifenkette, Schneeschutzkette = non-skid chain, snow ~
~schutzteppich m = non-skid carpet, ~ mat, skid-proof ~
gleitsicher, rutschfest, griffig = non-skid, non-slip(pery), skid-proof
Gleit|sicherheit f → Griffigkeit
~stange f [Gleitschalung] = jack(ing) rod
~theorie f [Stützwand] = theory of rupture
~unterbau m [Motor] = skid base
~verbindung f der Antriebswelle = drive shaft spline
Gleitweg m [ILS-Anflugschema] = glide path
~bake f = middle marker, LMM
~sender m = glide slope (beam) radio transmitter
~senderantenne f = glide slope antenna
~strahl m = glide slope radio beam

Gleit|wert *m* [*Ein Wert bei der Übertragbarkeit von Kräften zwischen Reifen und Straße*] = slipping coefficient
~widerstand *m* = resistance to sliding
~zahnhalter *m* = runner tooth adapter, ~ ~ adaptor
Gletscher *m*, **Ferner** *m* = glacier
~eis *n* = glacial ice, glacier ~
~ende *n* = front, snout
~erosion *f*, Glazialerosion = glaciation
~kunde *f* = glaciology
~lawine *f*, Eislawine = ice-avalanche
~mühle *f*, Gletschertopf *m*, Gletschertrichter *m*, Erosionstopf = pot-hole, glacier mill
~quelle *f* = glacier spring
~schnee *m* = glacier snow
~sohle *f* = sole of the glacier
~tisch *m*, Eistisch = glacier table
~topf *m* → Gletschermühle *f*
~trichter *m* → Gletschermühle *f*
~vorstoß *m* = glacier forward movement
~zunge *f* = glacier tongue
Glied *n* = member
~ → (angelenkte) Platte *f* [*Gliederband*]
Gliederband|aufgeber *m* → Aufgabeplattenband *n*
~beschicker *m* → Aufgabeplattenband *n*
~(förderer) *n*, (*m*), Blechgliederband, Plattenband(förderer), Gliederförderer = (plate) apron conveyor, (~) ~ conveyer
~speiser *m* → Aufgabeplattenband *n*
Glieder|förderer *m* → Gliederband(förderer) *n*, (*m*)
~heizkörper *m*, Radiator *m* = radiator
~kessel *m* = sectional boiler
~kette *f* = link (type) chain
~-Maßstab *m* → Zollstock *m*
Gliedseitenfläche *f* [*Gleiskettenlaufwerk*] = link face
Glimmentladung *f*, Koronaentladung = corona emission
Glimmer *m* = mica

glimmer(halt)ig, glimmerführend = micaceous
Glimmer|plättchen *n* = mica flake
~porphyr *m* = mica(ceous) porphyry
~sandstein *m* = mica(ceous) sandstone
~ton *m* = shiny clay, illite, hydrous mica
Globigerinenschlamm *m* = globigerina ooze
Glocke *f* [*See-Schallzeichen*] = aerial bell
Glocken|boje *f*, Glockentonne *f* = bell buoy
~kuppel *f* = bell-shaped dome
~muffe *f* = spigot and socket joint
~ofen *m* = bell kiln
~rad *n*, Ringrad = ring gear
~stube *f* = belfry [*Upper room in a tower or steeple of a church containing the bells and framing*]
~stuhl *m* = bell frame
~turm *m* = bell tower, belfry
Glühen *n* [*Verhüttung*] → Kalzinieren
glühfadenfreie Armaturenbeleuchtung *f* = luminous dial lighting
Glühfadenpyrometer *n* = disappearing filament pyrometer
Glühkerze *f* = glow plug, heater ~
Glühkopfmotor *m*, Semi-Diesel *m* = mixed cycle engine, semi-diesel (~)
~, Semi-Dieselmotor [*ein Teil des Verbrennungsraumes nicht gekühlt*] = hot-bulb engine, surface-ignition ~
Glüh|kopfzündung *f* = hot-bulb ignition
~lampe *f* = filament lamp
~verlust *m* = ignition loss [*The percentage loss in weight when an as-received sample is ignited to constant weight at 900-1000 deg. C. for short periods of time*]
~zone *f* → Brennzone
~zündung *f* = pre-ignition
Glutflußgestein *n* → Erstarrungsgestein
Glyzerinkitt *m* = glycerine mastic
Gneis *m* = gneiss

~textur *f* = gneissose texture, gneissic

~

Goldbeckdose *f* → Luft(-Erddruck)-Meßdose nach Goldbeck

Gold|bronze *f* = (powdered) gold bronze

~**erz** *n* = gold ore

goldführender Kies *m* = gold-bearing gravel

Goldseifensand *m* = gold-placer sand

Gooch-Tiegel *m* = Gooch filter, ~ crucible

Goslarit *m*, Zinkvitriol *n*, *m* (Min.) = goslarite, white vitriol, white copperas

gotischer Bogen *m*, Spitzbogen = equilateral arch, Gothic ~ [*The radius of the intrados equals the span, and the centres are therefore on the springing lines*]

~ **Verband** *m*, polnischer ~ = double Flemish bond

GÖV *n*, Gas-Öl-Verhältnis *n* = GOR, gas/oil ratio

(Grab)Becher *m*, (Grab)Eimer *m* = (digging) bucket

Grabbreite *f* → → Grabenbagger mit Eimern

(Grab)Eimer *m*, (Grab)Becher *m* = (digging) bucket

Grabeinfassung *f* = grave surround

graben, ausheben, ausschachten [*von Hand oder maschinell. Wenn mit Bagger = (aus)baggern*] = to excavate, to dig

Graben *m*, Tiefscholle *f*, Grabenbruch *m*, Grabensenke *f* (Geol.) = graben, trough fault, rift valley

~ [*mit abgeböschten Wänden*] = ditch

~ [*mit senkrechten Wänden*] = trench

~ für Abwasserleitung = sewer trench

~**abdeckung** *f* = trench cover

~**arbeiten** *fpl* = trench work

Grabenaushub *m* [*angeböschter breiter Graben*] = ditching, ditch excavation

~ [*schmaler Graben mit senkrechten Wänden*] = trenching, trench excavation

~**maschine** *f*, Grabenbagger *m* = pipe line excavator (Brit.); trenching plant, trench excavating plant, ditch digger, ditcher, trencher, trench digger, trench excavator, ditching and trenching machine

Grabenaussteifung *f* = temporary (system of) bracing of a trench

---

Grabenbagger *m* **mit Eimern, Grabenziehmaschine** *f*	bucket trencher, trenching machine of the bucket elevator type, trenchliner, endless-bucket trencher, ditcher
Eimerkettengrabenbagger, Eimerleitergrabenbagger	boom type trenching machine, boom type ditcher, boom type trencher, ladder ditcher, ladder-type trenching machine
Grabenbagger mit vertikaler Eimerleiter	vertical boom type trenching machine, vertical boom ditcher, vertical ladder ditcher
Grabenbagger mit schräger Eimerleiter	slanting boom type trenching machine, slanting boom ditcher, inclined boom type trench excavator, slanting ladder ditcher
Grabenbagger mit Eimerrad, Grabenbagger mit Kreiseimerleiter, Schaufel-Rad-Grabenbagger, Kreiseimerleiter-Grabenbagger	wheel (type) trenching machine, wheel (type) ditcher, wheel (type) trencher, rotary scoop trencher, rotary scoop ditcher

Grabenböschung — Grabensenke

Grabenbagger mit Eimerrad, Fabrikat THE CLEVELAND TRENCHER CO., CLEVELAND 17, OHIO, USA	CLEVELAND trencher
Grabenbagger mit seitlich herauskragendem Eimerrad zur Verbreiterung von Straßen	road widener [wheel type ditcher with the wheel overhanging the chassis so that the machine runs on the carriageway while excavating]
Grabbecher m, Eimer(becher), Grabenbaggereimer m	digging bucket, trencher bucket, ditcher bucket
Grabenbagger auf (Gleis)Ketten, ~ auf Raupen	track-type trenching machine, track-type ditcher
Eimerrad n, Kreiseimerleiter f, Becherrad, Schaufelrad	digging wheel
Grabbreite f	digging width
Baggerleiterantrieb m	bucket-line drive
Baggerleiterwinde f, Auslegerwinde, Becherleiterwinde	boom hoist, ladder hoist
Baggerleiter f, Ausleger m, Eimerleiter, Eimerleiter-Ausleger, Becherleiter	digging ladder, ditcher ladder, bucket flight, trencher ladder
Anhänge-Grabenbagger	detachable ditcher, detachable trencher
Aufbau-Grabenbagger	truck-mounted ditcher (US); lorry-mounted ditcher (Brit.)
Abwurfband n	spoil conveyor, spoil conveyer

Graben|böschung f = bank of ditch
~**bruch** m (Geol.) → Graben m
~**durchquerung** f = ditch crossing
~**einfüllung** f, Grabenverfüllung = backfilling (of trench(es))
~**fräse** f → Grabkettenbagger m
~**füller** m = trench filler
~**füllung** f = trench backfill
~**greifer** m = ditching grab
~-**Holzeinbau** m = trench timbering
~-**Holzverkleidung** f = trench timber sheeting, ~ wood(en) ~
~**leitung** f = trench conduit
~-**Methode** f [Sprengmethode, durch welche weiche Baugrundmassen mit großer Wucht auseinandergetrieben und gestört werden so daß die aufgeschütteten tragfähigen Massen auf die tragenden Lockergesteine absacken können] = trench method
~**pflaster** n = ditch sett paving

~**pflug** m, Rigolpflug; Rajolpflug [dieser Ausdruck sollte nicht mehr angewendet werden] = plough type ditcher, trench-plough, trenching plough, deep-plough (Brit.) [US = plow]
~-**Profilschaufel** f = formed ditch digging bucket
~**ramme** f = backfill (trench) rammer
~**rand** m (Geol.) → Verwerfung
~**räumung** f → Grabenreinigung
~**reiniger** m, Grabenreinigungsmaschine f, Grabenräumer m = ditch cleaner, ~ cleaning machine
~**reinigung** f, Grabenräumung, Grabensäuberung = ditch cleaning
~**säuberung** f → Grabenreinigung
(~)**Schrägsteife** f, Strebe f = diagonal brace (US); ~ strut (Brit.)
~**senke** f (Geol.) → Graben m

Grabensilo — Granit

~silo *m* [*Landwirtschaft*] = trench silo
~silo *m* in Putztechnik = trench silo with concrete plastered walls
~stampfer *m* = backfill tamper
~stauvorrichtung *f* = ditch check
(~)Steife *f* → Quersteife
~tieflöffel *m* = ditch digging bucket
~verbau *m* = trench sheeting and bracing, support to the trench sides
~verdichter *m* = backfill compactor
~verfüllung *f*, Grabeneinfüllung = backfilling
~vibrationsverdichter *m* = vibratory backfill compactor, ~ trench ~
~walze *f* = trench roller
~wand *f* = trench side
~wasserstand *m* = ditch water level
~ziehen *n* = trench cutting
~ziehen *n* mit Straßenhobel = ditching by grader
~ziehmaschine *f* → Grabenbagger *m* mit Eimern
Grab|gabel *f* = digging fork
~gefäßinhalt *m* → Grablöffelinhalt
~geschwindigkeit *f* = digging speed
~hebel *m* [*Bagger*] = digging lever
~kettenbagger *m*, (Drän)Grabenfräse *f* = trench cutting machine
~kraft *f* [*Bagger*] = digging power, ~ force
~löffelinhalt *m*, Grabgefäßinhalt, (Bagger)Löffelinhalt = shovel bucket capacity, (~) dipper ~, (excavator) bucket ~
~naßbagger *m*, Grab-Schwimmbagger = digging dredge (US); ~ dredger (Brit.)
~scheit *n*, Spaten *m* = spade
~seil *n* = digging line
~seilscheibe *f* am Auslegerkopf = digging line boom point sheave (US); ~ ~ jib ~ ~ (Brit.)
~stelle *f* → Baggerstelle
~stellung *f*, Aushubstellung = digging position
~tiefe *f*, Aushubtiefe = excavation depth, digging ~
~trommel *f* = digging drum
~werkzeug *n* = digging tool
~winkel *m* = digging angle
Grabzahn *m* [*Grabenbagger*] = cutting tooth
~halter *m* = digger tooth adapter, ~ ~ adaptor
Grabzylinder *m* = digging ram
Grad *m* der Störung *f* = degree of remo(u)lding [*soil mechanics*]
Gradiente *f* → Längsgefälle *n*

Gradierwerk *n*, Rieselwerk	cooler, cooling stack, graduation works
offenes ~	open ~ ~
geschlossenes ~	enclosed ~ ~
Reisig~	brushwood ~ ~
Latten~	lattice ~ ~
Ventilator~	~ ~ with fan
Gradierfall *m*	graduation, trickling of the brine
Gradierwand *f*	graduating wall, thorn-wall

Grad-Tag *m* = degree day
Grammatit *m* (Min.) = grammatite [*this term is also used as a synonym for tremolite, but this use is undesirable*]
Granat *m* (Min.) = garnet
~fels *m*, Granatgestein *n* = garnet rock, garnetyte
Granatit *m* (Min.) = granatite
Granit *m* = granite
feinkörniger ~ = fine-grained ~, granitel(le)
zersetzer ~ = gowan, disintegrated granite

Granitaplit — Gratecke

~aplit m = granite-(h)aplite
~familie f = granite clan
~geröllhalde f = granite boulder slope
~gneis m → körniger Gneis
~grobzuschlag(stoff) m = granite coarse aggregate
Granitit m, Biotitgranit m = biotite-granite, granitite
Granit|mauerwerk n = granite walling
~pflaster n = pitcher paving, granite sett ~
~pflasterstein m = granite (paving) sett, pitcher
~platte f = granite slab
~porphyr m = porphyroid granite, granite-porphyry
~sand m = granite sand
~schotter m = broken granite
~splitt m = granite chip(ping)s
~stein m = granite block
granoblastischer Verband m (Geol.) → Bienenwabenverband
Granulator-Splittbrecher m, Einschwingen-Granulator m = single-toggle (type jaw) granulator, chip(ping)s breaker, chip(ping)s crusher
Granulieren n = pelleting
granulierte Hochofenschlacke f, (künstlicher) Schlackensand m, gekörnte Hochofenschlacke f = slag sand, granulated (blast-furnace) slag
Granulierturm m = nodulizing tower
Granulierung f, Körnungsverfahren n = granulation (process)
Granulit m (Geol.) = granulite
Granulometrie f, Kornabstufung f = (granulometric) grading
graphisch auftragen, ~ darstellen, ~ aufzeichnen = to represent graphically
graphische Analyse f, bildliche ~ = graphical analysis
~ Berechnung f → bildliche ~
~ Darstellung f, bildliche ~ = graphical representation
~ Gleichgewichtslehre f → bildliche ~

~ Lösung f = graphical solution
~ Methode f → bildliche ~
~ Statik f → bildliche Gleichgewichtslehre f
~ Untersuchung f, bildliche ~ = graphical investigation
graphisches Rechnen n → bildliche Berechnung f
~ Verfahren n, zeichnerisches ~ = graphical method, ~ procedure, ~ construction
Graphit m (Min.) = graphite, plumbago, black lead
Graphitierung f, Spongiose f = graphitic corrosion, graphitization
Graphit|schmierstab m = graphite lubricating rod
~zement m, Eisenkitt m = iron cement
Gras|mittelstreifen m [Autobahn] = grassed central reserve, ~ ~ reservation, ~ ~ strip, ~ median (strip), ~ medial strip
~narbe f = grass cover
~narbenrollfeld n, Rasenrollfeld = grass landing area, ~ ~ ground
~plagge f → (Gras)Sode f
~randstreifen m [Straße] = grass margin
~saatgut n, Grassame(n) m = grass seed
(~)Sode f, (Rasen)Plagge f, Rasensode, Grasplagge = (grass) sod, turf ~
(Gras)Soden|fläche f = sodded area
~schneider m, Rasensodenschneider, (Rasen)Plaggenschneider, Rasenziegelschneider = (grass) sod cutter, turf ~ ~
Grat m, Gratlinie f [Dach] = arris
~ [Die Schnittlinie zweier Gewölbeflächen] = groin [The curved line at which the soffits of two vaults are seen to intersect]
~ = burr
~anfänger m, Firstanfänger = starting tile
~ecke f, Firstecke = ridge corner tile

Grätenfeld — Greifkorb

Gräten|feld *n* [*Streckmetall*] = herringbone mesh opening
~struktur *f* [*Streckmetall*] = herringbone (pattern) mesh
Grat|(linie) *m*, (*f*) [*Dach*] = arris
~schifter *m* = hip jack (rafter)
~sparren *m* = hip rafter
~ziegel *m* = hip tile
  kantig = angular
  rund = round
  gewölbt = bonnet
Grau|braunstein *m* (Min.) = gray manganese(-ore), native manganic hydrate
~gültigerz *n*, (Kupfer)Fahlerz = fahl-erz, fahl-ore, grey copper ore, tetrahedrite
Grauguß *m* → Gußeisen *n* mit Lamellengraphit
~stück *n* = gray iron casting
Grau|kalk *m*, Dolomitkalk, Magerkalk = gray (quick-)lime, dolomitic ~
~kobalterz *n* (Min.) = jaipurite, grey cobalt ore
Graup [*nach Fischer und Udluft (Jahr 1936) Boden von 20–2 mm, groß 20 bis 10 mm, mittel 10–5 mm, klein 5–2 mm*] = medium gravel and fine gravel (US) [*25.4 mm to 2.00 mm, "medium" 25.4 to 9.52 mm, "fine" 9.52–2.00*]
Graupelschauer *m* = graupel shower
Grau|silber *n* (Min.) = grey silver, carbonate of silver, selbite
~stein *m* = grey-stone
~wacke *f* = greywacke, graywacke
~wackenkalk *m*, Übergangskalk = greywacke limestone, transition-lime
~wackensandstein *m* = trap sandstone
~wackenschiefer *m* = greywacke slate
gravidosieren = to weigh-batch
Gravimeter *n* = gravity meter
gravimetrische Messung *f*, Schweremessung [*geophysikalisches (Aufschluß) Verfahren*] = gravimetric survey

Gravitationswasser *n*, spannungsfreies Überschußwasser = gravitational water
Greif|(bagg)er *m*, Kranbagger, Baggerkran *m* = grab(bing) excavator [*This is actually a crane from which a grab for excavating is suspended*]
~baggern *n* = grabbing
Greifen *n* = grabbing
Greifer *m*, ~krankorb *m*, Greif(er)korb = grab(bing) (crane) bucket, grab
~, Greifbagger *m* = grab(bing) excavator [*This is actually a crane from which a grab for excavating is suspended*]
~ → Greiferkran *m*
~, **~stollen** *m* = grouser
~arbeit *f* = grabbing (work)
~bohren, Bohren mit Bohrgreifer(n) = hammer grab work
~drehkran *m*, Greiferschwenkkran = grab slewing crane
(~-)Halteseil *n* = hold(ing) line, ~ rope
~katze *f*, Greiferlaufkatze = grab troll(e)y
~korb *m*, Greifer(krankorb), Greifkorb = grab(bing) (crane) bucket, grab
~(kran) *m* = grab(bing) crane
~(krankorb) *m*, Greif(er)korb = grab(bing) (crane) bucket, grab
~-Naßbagger *m* = Schwimmgreifer mit Zweischalen-Greif(er)korb
~schale *f* [*beim Apfelsinenschalengreifer*] = segment, blade
~schale *f* [*beim Zweischalengreifer*] = halfscoop
~schwenkkran *m* → Greiferdrehkran
~-Schwimmbagger *m* → Schwimmgreifer mit Zweischalen-Greif(er)korb
~(stollen) *m* = grouser
~vorrichtung *f* = grab(bing) equipment
~zahn *m* = grab tooth
Greif|korb *m*, Greiferkorb, Greifer(krankorb) = grab(bing) (crane) bucket, grab

Greiflast — Grobblecheinsetzmaschine

~last *f* = grabbing load
~messer *n* [*Rotary-Zange*] = die
~-Naßbagger *m* → Schwimmgreifer mit Zweischalen-Greif(er)korb
~naßbaggern *n* = grab-dredging
~profil *n* = traction tread
~raupe *f*, Raupenschlepper *m* mit Greifschild = bullclam shovel
~schneidkante *f* = clamp cutting edge
~-Schwimmbagger *m* → Schwimmgreifer mit Zweischalen-Greif(er)korb
~zug *m* = comealong
Greisser-Walzenwehr *n* = Greisser (rolling) gate
Grenze *f* (Geol.) = recurrence horizon
Grenz|fläche *f* = interface
~flächenkraft *f* → Adhäsion *f*
~geschwindigkeit *f* [*Hydromechanik*] = critical velocity
~korn *n* = near-mesh grain
~korngröße *f* = limit (screen) size
~last *f* → Bruchlast
~linie *f* = limit line
~schalter *m* → Begrenzungsschalter
~scherspannung *f* = boundary-shear stress
~schleppkraft *f* → kritische Schleppkraft
~schmierung *f* = boundary lubrication
~sieblinie *f* = gradation limit, limiting grading curve, particle size (distribution) limit
~stein *m*, Begrenzungsstein, Kandelstein = boundary stone
~tiefe *f* [*Hydromechanik*] = critical depth
~tragfähigkeit *f*, Grenztragvermögen *n*, Grenztragleistung *f*, Grenztragkraft *f*, Grenztragwiderstand *m* [*Baugrund*] = ultimate bearing power, ~ ~ capacity, ~ supporting ~ [*The minimum load which will cause failure of a foundation by actual rupture of the foundation material*]
~tragfähigkeitswert *m* [*Baugrund*] = ultimate bearing value, ~ supporting ~

~tragkraft *f* → Grenztragfähigkeit *f*
~tragleistung *f* → Grenztragfähigkeit *f*
~tragvermögen *n* → Grenztragfähigkeit *f*
~tragwiderstand *m* → Grenztragfähigkeit *f*
~wassertiefe *f*, kritische Wassertiefe = critical depth
~wert *m* = limit value, ultimate ~
~winkel *m* (✵) = angle of draw, limit angle
~zustand *m* = boundary condition
~zustand *m*, kritischer Zustand [*Hydraulik*] = competent condition, competence
Grieß|kohle *f*, Perlkohle, Erbskohle = pea coal
~mühle *f*, Rohrmühle = tube mill
griffig *a* → gleitsicher
Griffigkeit *f*, Gleitsicherheit, Straßen~ = skid-resisting property, non-skid ~, road-skid ~, anti-skid ~, pavement grip, skid(ding) resistance, resistance to skid(ding)
Griffigkeits|-Erneuerung *f* = restoration of skid resistance
~messung *f* → Griffigkeitsuntersuchung
~meßgerät *n*, Griffigkeitsmesser *m* = skid-resistance tester
~untersuchung *f*, Griffigkeitsmessung = skidding investigation, measurement of skidding resistance, investigation of the friction between tyre and road
Griff|stange *f* = handlebar
~zeit *f*, Nebenzeit [*bei der Zeitnahme*] = downtime
Gritt *m* [*nach Fischer und Udluft (Jahr 1936) 2,0–0,2 mm, nach Niggli (Jahr 1938) 2,0–0,2 mm (heißt hier ,,Grobsand''), nach Gallwitz 2,0–0,2 mm (heißt hier ,,Sand'')*] = coarse sand and medium sand (Brit.) [*2.0–0.2 mm*]; coarse sand and medium sand (US) [*2.0–0.25 mm*]
Grobblech *n* = (heavy) plate
~einsetzmaschine *f* = (heavy) plate charging machine

~kanten-Fräsmaschine f = (heavy) plate edge milling machine
~kanten-Hobelmaschine f = (heavy) plate edge planing machine
~magnetkran m = (heavy) plate-handling magneto crane
~schreckmaschine f = (heavy) plate stretcher
~richtmaschine f, Blechrichtmaschine = (heavy) plate level(l)ing machine
~richtpresse f = (heavy) plate level(l)ing press
~schere f = (heavy) plate shears
~verformungspresse f = (heavy) plate forming press
~walzwerk n = (heavy) plate (rolling) mill
Grob|brechen n = coarse crushing, ~ breaking
~brecher m = coarse crusher, ~ crushing machine, ~ breaker, ~ breaking machine
~brechzone f = coarse-crushing zone, coarse-breaking ~
grobe Schätzung f = guesstimate
grobes Schlämmkorn Va n [nach Dücker (Jahr 1948) 0,06–0,02 mm] = coarse silt [A.S.E.E. fraction 0.074–0.02 mm; British Standard 0.06–0.02 mm]
~ **Siebkorn** n → Kiesel(stein) m
Grob|ausbringen n → Überkornausbringen
~brechen n → → Hartzerkleinerung
~filter m, n, Vorfilter = prefilter
Grobgut n = oversized material, coarse fraction
~korn n, Grobkorn = coarse grain
Grob|kalk m → Pariser Kalkstein m
~kegelbrecher m mit segmentförmigem Brechkopf = coarse gyrasphere crusher
~kies m [nach DIN 1179 und DIN 4022 (Jahr 1936) 70–30 mm, nach DIN 4220 70–20 mm, nach DEGEBO 20–ff0 mm, nach Niggli (Jahr 1938) 200–20 mm, nach Dienemann 20–10 mm (heißt hier „Grobschotterboden"), nach Fischer und Udluft (Jahr 1936) 200–20 mm (heißt hier „Brock"), nach Gallwitz (Jahr 1939) 200–20 mm (heißt hier „Schotter")] = coarse gravel [A.S.E.E. fraction 76.2 to 25.4 mm]; ballast (Brit.)
~klärzone f = coarse clarification zone
~korn n, Grobgutkorn = coarse grain
~körner npl die sich einander berühren [Prepakt-Beton] = point-to-point contact of preplaced coarse aggregate
grob|körnig = coarse-grained
~körniges Eisen n = open-grained iron
~körniges Gestein n für Makadamdecken = macadam aggregate
Grob|kornzwischengut n = coarse middlings
~mahlen n, Grobmahlung f = coarse grinding
~mörtel m = coarse-grained mortar, coarse sand ~
~putz m, Grobputzmörtelbelag m = coarse plaster, plaster of coarse stuff
~putzmörtel m = coarse stuff
~rechen m = coarse rack [A relative term, but generally used when the clear space between bars is 2 in. or more]
~rechen m = coarse screen [A relative term, but generally used when openings are greater than 1 in. in least dimension, except in the case of racks]
~rechengut n = trash [The material removed from combined and stormwater sewers by coarse racks]
~sand m, Schottersand [nach Atterberg 2,0–0,2 mm, nach Fischer und Udluft (Jahr 1936) 2,0–0,2 mm (heißt hier „Gritt"), nach Niggli (Jahr 1938) 2,0–0,2 mm, nach Gallwitz (Jahr 1939) 2,0–0,2 mm (heißt hier „Sand"), nach Dücker (Jahr 1948) 2,0–0,2 mm (heißt hier „mittleres Siebkorn IIIb" 2,0 bis 0,6 mm und „feines Siebkorn IVa" 0,6–0,2 mm, nach DIN 4022 2,0 bis 1,0 mm)] = coarse sand [International Society of Soil Science

Grobschluff — Groß-Förderwagen

2.0–0.2 mm, A.S.E.E. fraction 2.00–0.59 mm, British Standard 2.0–0.6 mm, Division of Soil Survey (US) 2.0–0.25 mm]; [according to Atterberg, Fischer and Udluft, Niggli, Gallwitz, the size from 2.0 to 0.2 mm as classed by them and called by various names as indicated corresponds to the British Standard "coarse sand and medium sand" 2.0–0.2 mm]

~schluff m [nach Niggli (Jahr 1938) 0,02–0,002 mm, dieselbe Korngröße heißt nach Atterberg, Terzaghi, DIN 4022, Fischer und Udluft, Gallwitz und Grengg „Schluff", bei Dücker (Jahr 1948) umfaßt sie „grobes Schlämmkorn Vb" 0,02–0,006 mm und „mittleres Schlämmkorn VIa" 0,006–0,002 mm] = medium silt and fine silt [British Standard "medium silt" 0.02 to 0.006 mm and "fine silt" 0.006 to 0.002 mm, "fine silt" A.S.E.E. fraction 0.02 mm to no size limit]

~schotter m = coarse crushed stone
~schotterboden m → → Grobkies m
~sieb n, Stückgutscheider m = scalper, scalping screen
~siebung f, Stückgutscheidung = scalping
~spalten n [Steine mit Keilen] = blocking, sledging
~splitt m = coarse chip(ping)s
grobsteiniger Boden m → → Block m
Grob|walzwerk n = blooming-mill
~zerkleinerung f [Korngröße über 100 mm] = coarse comminution, ~ reduction [minimum size of product discharged 4 in.]
~zerkleinerungsanlage f → Brechanlage
~zerkleinerungsmaschine f → Brecher
Groß|-Abmeßanlage f → Groß-Dosieranlage
~anlage f = high-capacity plant, ~ facility, ~ installation
~bandanlage f = large-capacity belt conveying plant

~baustelle f = large-scale (project) site
~bauvorhaben n, Großprojekt n = large-scale project
~betonanlage f → Betonfabrik f
~betonmischer m, Großbetonmischmaschine f = volume-production concrete mixer
~bohrlochverfahren n = auger-mining
~bohrwagen m, Ausleger-Bohrwagen, Bohrjumbo m = drilling jumbo, (tunnel) ~
~boot n = long boat
~brückenbau m = construction of large bridges
~bunker m = large bin
~-Dosieranlage f, Groß-Abmeßanlage, Groß-Zuteilanlage, Groß-Zumeßanlage = high-capacity batching plant, ~ proportioning ~, ~ measuring ~, ~ ~ installation, ~ ~ facility

große Druckhöhe f, ~ Fallhöhe = high-head
~ Fallhöhe f, ~ Druckhöhe = high-head
~ Hauptspannung f = major principal stress
~ Reifenverschleißwirkung f = tire-killing (US); tyre-killing (Brit.)

Größe f der Kronenüberhöhung, Betrag m ~ ~ = amount of crown
Größen|bereich m = size range
~faktor m = size factor
größenverstellbar = multisize
Großfertiger m, Groß-Straßenfertiger = large (road) finisher, ~ (~) finishing machine, ~ (~) laying and finishing machine
großflächiges (Stahl)Betonfundament n → Plattenfundament
Groß|flughafen m für interkontinentalen Verkehr = intercontinental airport [An airport to serve the longest range nonstop flights in the transcontinental, transoceanic and intercontinental categories]
~-Förderwagen m → Muldenkipper m

Großform — Grubenglühen

~form *f* = large mould (Brit.); ~ mold (US)
~geräte *npl* = high-powered equipment, ~ plant, heavy construction ~, major ~, capital ~
~hafen *m* = major port
~händler *m* = distributor
~lochbohrmaschine *f* = large diameter drilling machine
~lochdrehbohren *n* = sinking large diameter mine shafts by rotary drilling
~löffel(-Abraum)bagger *m* → Abraum-Hochlöffel(bagger)
~luftschutzbunker *m* = large-capacity air-raid shelter
~markthalle *f* = municipal market, wholesale ~
~motorkahn *m* = large-engined vessel, ~ barge
~mühle *f* = large-capacity (grinding) mill
~muldenkipper *m* → Muldenkipper
~pfahl *m* = large pile
~pfeiler *m* = large-diameter pier
~pflaster(decke) *n*, (*f*) = large sett paving
~pflasterstein *m* = large (paving) sett
~projekt *n* → Großbauvorhaben *n*
Großraum *m* = metropolitan area
~-(Gleis)Kettenwagen *m* → Raupenwagen
~kartierung *f* = large-scale mapping
Großraumlore *f* auf (Gleis)Ketten → Raupenwagen
~ ~ Raupen → Raupenwagen
Großraum|mischer *m*, Großraummischmaschine *f* = large-capacity mixer, ~ mixing machine
~-Raupenwagen *m* → Raupenwagen
~schotterverteiler *m* = large-capacity stone spreader box
Groß|schiffahrtweg *m* → Großwasserweg
~silo *m* = large silo
~speicher *m* → Talsperre *f*
~sprengung *f* = big blast
~-Sprengwagen *m* → Groß-Wassersprengwagen

~spritzgerät *n* → Bitumen-Sprengwagen *m*
~-Straßenfertiger *m* → Großfertiger
~tafelbauweise *f*, Plattenbauart *f* = slab method
Größt|korn *n* = top size of the aggregate
~moment *n* = maximum moment
Groß|turmdrehkran *m* [*ein Portal läßt den Raum zwischen den Schienen frei*] = portal-type (mono)tower crane, ~ revolving tower crane, ~ rotating tower crane
~verbraucher *m* = bulk consumer, large-scale ~
(~)Verkehrsader *f*, Hauptfernverkehrsstraße *f* = traffic artery
großvolumiger Luftreifen *m* → Riesenluftreifen
Groß|wasserweg *m*, Großwasserstraße *f*, Großschiffahrtweg, Großschiffahrtstraße = international waterway
~wasserzähler *m* = water meter for bulk quantities
~-Zumeßanlage *f* → Groß-Dosieranlage
~-Zuteilanlage *f* → Groß-Dosieranlage
Grübchen *n* = pitting
Grube *f* [*über Tage*] = pit
~, Bergwerk *n*, Bergbaubetrieb *m* = mine
Gruben|arbeiter *m*, Bergarbeiter, Bergmann *m* = miner
~aufzug *m*, Bau~ = building pit materials hoist
~baue *mpl* (⚒) = workings
~bewetterung *f* → Bewetterung (der Grubenbaue)
(~)Bewetterungsanlage *f* = mine ventilation plant
(~)Bewetterungsstation *f* = mine ventilation station
~bogen *m* (⚒) = colliery arch
~elektriker *m* (⚒) = mine-electrician
~feld *n* = claim
~(förder)wagen *m* → Grubenwagen
~gasabsaugung *f* = methane drainage
~glühen *n* = pit annealing

Grubenkalk — Grundeigentümer

~kalk m → Sumpfkalk
~kies m, Wandkies = pit(-run) gravel
~lampe f = mining lamp
~lok(omotive) f = mine loco(motive), mining ~
~luft f (⚒) = mine air
~mechaniker m (⚒) = mine-mechanician
~pumpe f, Bergwerkpumpe = mine pump
~rettungswesen n = mine rescue work
~sand m = pit sand
~schacht m, Bergwerkschacht, Bergbauschacht = mine shaft
~schachtauskleidungsplatte f = mineshaft liner
~schiene f (⚒) = colliery rail
~schlacke f = pit slag
~steiger m (⚒) = section foreman
~(stein)bruch m = pit quarry
~wagen m, Grubenförderwagen = mine car

Gruftgewölbe n = burial vault
Grünbleierz n (Min.) → Pyromorphyt m

Grund m für Putzarbeiten → Putz(unter)grund
~(ab)dichtung f, Unter~, (Ab)Dichtung des Untergrundes = subsoil waterproofing
~ablaß m = deep sluice, undersluice
~ablaßstrecke f = deep sluice section, undersluice ~
~ablaßverschluß m, Abschlußschütz(e) n, (f) [Talsperre] = lower sluice, ~ gate, sluice (gate), sluiceway gate
~anfeuchtung f, Untergrundbewässerung = (natural) sub-irrigation, sub-surface ~
~armierung f → Grundbewehrung
~art f = basic type
~auslaß m → Talsperre f
~ausleger m = basic jib (Brit.); ~ boom (US)
~bagger m → Geräteträger m mit Unterteil und Oberteil
~balken m = strap (beam)
~balkengründung f, Grundbalkenfundament n, Grundbalkenfundation f, Grundbalkenfundamentkörper m, Grundbalkengründungskörper m = connected footing, strap ~
~bank f, Bankett n, Fundamentabsatz m [*Massive Mauern erhalten eine absatzweise Verbreiterung. Die obere Fläche eines Absatzes wird Bankett usw. genannt*] = footing of the foundation
~bau m → Grundbautechnik f
~bauer m = foundation engineer
~baumechanik f → Bodenmechanik
~bautechnik f, Grundbau m, Gründungswesen n, Fundationstechnik = foundation engineering
~belastung f = basic loading
~bewehrung f, Grundarmierung, Grund(stahl)einlagen fpl, Hauptbewehrung, Hauptarmierung, Haupt(stahl)einlagen = main steel, ~ reinforcement
~bitumen n → Ausgangsbitumen
~bogen m → Gegenbogen
~breite f = standard width

Grundbruch m, Gelände-~ [*DIN 4017. Ein Grundbruch tritt ein, wenn ein Gründungskörper so stark belastet wird, daß sich unter ihm im Untergrund mehr oder weniger ausgeprägte Gleitbereiche bilden, in denen der Scherwiderstand des Bodens überwunden wird*] = shear failure, base ~
~bedingung f in wassergesättigtem Boden = failure condition in saturated soil
~berechnung f = shear failure calculation, base ~ ~
~formel f = shear failure formula, base ~ ~
~last f = shear failure load, base ~ ~
~sicherheit f = safety against shear failure, ~ ~ base ~

Grund|buch n = land register
~buhne f, Tauchbuhne [*eine vom Ufer ausgehende Grundschwelle*] = ground sill rooted in the bank
~dichtung f → Grundabdichtung
~eigentümer m = property owner

Grundeinlagen — Grundschwelle

~einlagen /pl → Grundbewehrung /
~eis n = anchor ice [*Ice that forms on the bed of a water-course*]
gründen, fundieren = to found
~ auf Boden, ~ ~ Lockergestein, fundieren ~ ~ = to lay a foundation on soil
~ ~ gewachsenem Fels, fundieren ~ ~ ~ = to carry a foundation to bed rock
Gründen n, Fundieren = founding
Grund|erwerb m → Baulandbeschaffung /
~fahrwerk n = basic travel(l)ing gear
~fahrzeug n [*beim Wechselbehältersystem*] = parent vehicle
~fläche / = ground area
~fläche / [*zum Aufstellen einer Maschine*] = floor space
~gerät n, Grundmaschine / = basic machine, ~ unit
~gewicht n = basic weight
~gewölbe n, verkehrtes Gewölbe, umgekehrtes Gewölbe = invert arch [*Used to spread the loads of piers between openings evenly on the foundation*]
~gewölbe n mit ebener Unterseite = invert arch with flat underside
~gewölbe n mit gewölbter Unterseite = invert arch with arched underside
~gleichung / der Hydrodynamik, hydrodynamischer Euler-Satz m = basic equation of hydrodynamics, Eulers' hydrodynamical law
~gleichung / der Hydrostatik, hydrostatischer Euler-Satz m = basic equation of hydrostatics, Euler's hydrostatical law
~höhe / = standard height
~hohlpfeiler m → Hohlfundamentpfeiler
Grundier|mittel n, Grundanstrichmittel = primer
~(ungs)auftrag m, Grundieren n = prime coat application
Grund|länge /, Standardlänge = basic length
~lastenergie / = base-load electricity

~lastkraftwerk n = base-load station, ~ plant, ~ installation
~lauf m [*Schleuse*] = floor culvert
~lawine /, Schlaglawine = ground avalanche
~linie / = base line
~lohn m = basic wage
~luftzone / = zone of aeration
~maschine /, Grundgerät n = basic machine, ~ unit
~masse / (Geol.) = groundmass
~mauer /, Fundamentmauer = foundation wall
~mauerwerk n, Fundamentmauerwerk = foundation masonry
~metall n, Mutterwerkstoff m, unedles Metall = base metal
~moräne /, Untermoräne = ground moraine
~modul M m, Modul von 10 cm, Modul von 4 Zoll = standard module [*It means a module with the dimension of 4 inches*]
~pfahl m [*Pfahl, der in ganzer Länge im Baugrund steht*] = pile completely in the ground
~pfeiler m, Fundamentpfeiler, Gründungspfeiler = foundation pier
Grundplatte / = base plate, bed ~
~ [*Rammgerüst*] = frame base
~ → Pritsche /
~, Maschinen~ = machine base plate, ~ bed ~
~ → Plattenfundament n
~, Plattform /, Deck n [*Kran; Bagger*] = deck
Grund|rahmen m [*Sieb*] = base frame
(~)Rahmen m → Fahrgestell n
~rechtsbreite / [*Straße*] = right-of-way width, width of the road reservation
~reparatur / = basic repair
~riß m = (ground-)plan
~rißanordnung / = layout
~rißfläche / = plan area
~rißsichtweite / [*Straße*] = sight distance in plan
~schwelle /, Sohlschwelle, Stauschwelle [*in der Wasserlaufsohle*

schräg oder quer zur Stromrichtung gelegene Schwelle] = ground sill

~schwelle *f* [*bei einem Holzdrempel*] = cap

~sockel *m* → Sockelgründung *f*

~spannung *f* [*Baustatik*] = basic stress

~spannungszustand *m* = basic stress state

~(stahl)einlagen *fpl* → Grundbewehrung *f*

~stein *m* = foundation stone

~strecke *f*, untere Abbaustrecke (⚒) = bottom development road

~strom *m* = bottom current

~strömung *f* = bottom flood, bed current

~stück *n* → Parzelle *f*

~stückstrang *m* = supply pipe [*That portion of the service pipe lying within the consumers' premises*]

~type *f* → Geräteträger *m* mit Unterteil und Oberteil

~überholung *f* = major overhaul

Gründung *f*, Fundierung, Fundation *f*, Fundamentierung = foundation

~ → Fundament *n*

Gründungs|arbeiten *fpl*, Fundierungsarbeiten, Fundationsarbeiten = foundation work

~art *f*, Fundamentart, Fundationsart = type of foundation

~aufgabe *f* = foundation problem

~aushub *m* = excavation for foundation

~bau *m*, Fundamentbau = foundation construction

~belastung *f*, Fundamentbelastung = foundation loading

~bemessung *f*, Fundamentbemessung = foundation design

~beton *m*, Fundamentbeton = foundation concrete

~betonkörper *m* → Betonfundament *n*

~boden *m*, Baugrundboden = foundation soil, supporting ~, subsoil

~bodenprobe *f* → Baugrundbodenprobe

~bodenschicht *f* → Baugrundbodenschicht

~bohrung *f* = foundation drilling

~breite *f* = foundation width, ~ breadth

~brunnen *m*, (Senk)Brunnen = open caisson, drop shaft [*open both at the top and bottom*]

~brunnen *m* **mit Scheidewänden,** Senkbrunnen ~ ~, Bagger-Senkkasten *m* = open caisson with cross walls, drop shaft ~ ~ ~

~fels *m*, Baugrundfels, felsiger Baugrund *m* = foundation rock, supporting ~

~felsprobe *f*, Baugrundfelsprobe = foundation rock sample, supporting ~ ~

~fläche *f* = bearing area of a foundation

~höhe *f* = founding level, foundation ~

~körper *m* → Fundament *n*

~last *f*, Fundamentlast = foundation load

~methode *f* = foundation method

(~)Pfahl *m* = (foundation) pile

~pfeiler *m*, Fundamentpfeiler, Grundpfeiler = foundation pier

~plan *m*, Fundamentplan = foundation layout plan

~platte *f* → Plattenfundament

(~)Rost *m* = (foundation) grillage

~schicht *f*, Fundamentschicht = foundation stratum

~sockel *m* → Sockelgründung *f*

~sohle *f* = foundation bed, bearing surface

~streifen *m*, Streifenfundament *n* = continuous footing, strip ~

~tiefe *f*, Fundationstiefe = depth of foundation, foundation depth, founding depth

~wesen *n*, Fundationstechnik *f*, Grundbau *m*, Grundbautechnik *f* = foundation engineering

Grundverbindung *f* **der organischen Chemie** = organic compound

Grundwasser *n* = ground water, phreatic ~

Grundwasserabdichtung — Grünsteintuff

~abdichtung f → Abdichtung
~(ab)senkung f = ground-water lowering
~(ab)senkung f im Vakuumverfahren = vacuum method, ground-water lowering by the ~ ~
~(ab)senkungsanlage f = ground-water lowering installation, ~ ~ plant
~andrang m, Grundwassereinbruch m, Grundwasserzutritt m = ingress (or inflow, or inrush) of ground (or phreatic) water
~anreicherung f, Anreicherung von Grundwasser = ground-water recharge, ~ replenishment
~becken n = ground-water basin
~bewegung f = ground-water flow
~chemie f = ground-water chemistry
~dichtungsschicht f, Grundwasserisolierschicht, Wannenisolierung f, Grundwasser(schutz)wanne f = tanking, basement waterproofing [*refers to all basement construction which is made completely waterproof below ground level*]
~drän m = ground-water drain [*A drain which carries away ground water*]
~erschließung f, Grundwassersuche f = ground-water exploration
~gefälle n = ground-water gradient
~gleiboden m = ground-water gley soil
~horizont m → Grundwasserspiegel m
~hygiene f = ground-water hygiene
~isolierschicht f → Grundwasserdichtungsschicht
~isolierung f, (Ab)Dichtung gegen Grundwasser = waterproofing
~kunde f, Geohydrologie f = hydrogeology
~lauf m, Grundwasserstrom m = underground stream, ground-water ~
~physik f = ground-water physics
~reserve f = ground-water resource
~(schutz)wanne f → Grundwasserdichtungsschicht f

~senkung f, Grundwasserabsenkung = ground-water lowering
~speicher m, Grundwasserstockwerk n = ground-water reservoir
~speicherung f = ground-water storage
~spiegel m, (freier) Wasserspiegel, Grundwasserhorizont m = phreatic surface, (ground-)water table
~spiegelhöhe f = (ground-)water level [*The elevation of the (ground-) water table at a given point*]
~stockwerk n, Grundwasserspeicher m = ground-water reservoir
~strom m, Grundwasserlauf m = underground stream, ground-water ~
~strömung f = ground-water flow, phreatic water ~
~suche f, Grundwassererschließung f = ground-water exploration
~träger m = aquifer [*Water-bearing formations that create a ground-water reservoir*]
~verunreinigung f = ground-water pollution
~vorkommen n = ground-water supply
~wanne f → Grundwasserdichtungsschicht f
~zutritt m → Grundwasserandrang m
Grund|wehr n, Stauschwelle f = drowned weir, submerged ~
~werkstoff m [*Schweißen*] = base material, parent metal
~zustand m [*Baustatik*] = basic condition
grüne Ausblühung f, Vanadiumausblühung = green efflorescence
„grüne Welle" f = traffic pacer
Grün|fläche f = green area
~futter-Fahrsilo m → Fahrsilo
~gürtel m = green belt
~ling m = green product
~moostorf m = green moss peat
~schiefer m = green schist
~span m, basisches Kupferazetat n = copper rust, verdigris
~stein m = greenstone
~steintuff m → Diabastuff

Grünstreifen — Gummilager

~streifen *m*, Rasentrennstreifen [*Straße*] = landscaped strip, grassy median ~, central grass reserve

Grün & Bilfinger-Preßbeton(bohr)-pfahl *m*, Preßbeton(bohr)pfahl System Grün & Bilfinger = Grün & Bilfinger pressure pile

Gruppen|bepflanzung *f* = group planting

~**indexmethode** *f* (**für die Dimensionierung von Straßendecken**) = group index method (of pavement design)

~**stand** *m*, Gemeinschaftsstand [*auf einer Messe oder Ausstellung*] = group stand

Grus *m* = breeze

Gufferlinie *f*, Mittelmoräne *f* = medial moraine, median ~

Guillochierung *f* = guilloche [*A decorative border design in which two or more lines or bands are interwoven so as to make circular spaces between them*]

Gullybogen *m* [*Formstück*] = street el

Gummi|anschlag *m* = rubber stop

~**auskleidung** *f*, Gummibekleidung, Gummiverkleidung = rubber lining, ~ facing

~**band** *n* → Gummi(förder)band

~**becher** *m* → Gummi-Elevatorbecher

~**bekleidung** *f* → Gummiauskleidung

~**belag** *m* = rubber covering

~**belag** *m*, Gummisiebbelag = rubber screen plate

gummibereifte Spurbahn *f* = pneumatic-tyred trackway system (Brit.); pneumatic-tired ~ (US)

gummibereifter Grabenbagger *m* = rubber tired ditcher (US); rubber-tyred ~ (Brit.)

~ **selbstfahrbarer Kran** *m*, Fahrzeugkran [*kann Luftbereifung oder Vollgummibereifung haben*] = rubber-mounted mobile crane

~ ~ **Universalbagger** *m*, Fahrzeugbagger [*kann Luftbereifung oder Vollgummibereifung haben*] = rubber-mounted mobile excavator

gummibereiftes Erdbau-Lastfahrzeug *n* → Muldenkipper *m*

Gummi-Bitumen|-Gemisch *n* → Bitumen-Gummi-Gemisch

~**-Vergußmasse** *f* = rubber-asphaltic bitumen sealing compound (Brit.); rubber-asphalt ~ ~ (US)

Gummi|blasenmethode *f*, Ballonmethode [*Nachprüfung der Verdichtung f*] = (waterfilled) rubber membrane method, rubber balloon ~

~**boden** *m*, Gummifußboden = rubber floor (finish), ~ flooring; rubberplate (US)

~**-(Brücken)lager** *n* = rubber (bridge) bearing

~**dichtung** *f* = rubber seal, ~ stop

~**dicht(ungs)ring** *m* = rubber seal(ing) ring, ~ packing ~, ~ washer, ~ gasket

~**dicht(ungs)streifen** *m* = rubber water stop, ~ ~ seal

~**-(Elevator)becher** *m* = rubber elevator bucket

~**fender** *m*, Reifenfender = (rubber) tyre fender (Brit.); (~) tire ~ (US)

~**fliese** *f*, Gummiplatte *f* = rubber tile

~**-(förder)band** *n*, Gummi(förder)gurt *m* = rubber (conveyor) belt

~**(fuß)boden** *m* = rubber flooring, ~ floor (finish); rubberplate (US)

~**(gleis)kette** *f*, Laufband *n* = rubber crawler

~**gurtbecherwerk** *n* = belt(-type bucket) elevator, band(-type bucket) ~

~**handgriff** *m* = rubber hand grip

~**haut** *f* = rubber skin

~**hut** *m*, Absperrkegel *m*, Leitkegel = rubber cone

gummiisoliert = rubber-insulated

Gummi|kabel *n*, Gummischlauchleitung *f* = rubber cable

~**kissen** *n* = rubber cushion

~**-(Kurbel)Dämpfer** *m* = rubber-type damper

~**lager** *n* = rubber bearing

~**lager** *n* [*Erdbauwagen*] = rubber pad, ~ cushion [*for supporting body on chassis*]

~lagerplatte *f* = rubber bearing pad
~masse *f* = rubber composition, ~ compound
~matte *f* = rubber mat
~mehl *n* = rubber powder
~membran(e) *f* = rubber diaphragm
~-Metall-Element *n* = rubber-metal product
~-Misch(er)schaufel *f* = rubber blade, ~ paddle
~-Mörtelwanne *f*, Gummi-Mörteltrog *m* = rubber hod
~pflaster *n* = rubber sett paving
~platte *f*, Gummifliese *f* = rubber tile
~platte *f* = rubber pad
~puffer *m* = rubber buffer
~puffer *m* [*Gleiskette*] = rubber block support
~rad-Schaufellader *m* = rubber-tyred shovel loader (Brit.); rubber-tired ~ ~ (US)
~radwalze *f* → → Walze
~reifen *m* = rubber tire (US); ~ tyre (Brit.)
~reifen *m* mit hoher Hysterese = high-hysteresis tyre (Brit.); ~ tire (US)
gummireifenfahrbar = rubber-tyre mounted (Brit.); rubber-tire ~ (US)
(Gummi)Reifenlader *m*, (Gummi-)Reifenlademaschine *f* = wheel-mounted loader, ~ loading machine
Gummi|ring *m* = rubber grommet
~-Schalhülse *f* [*Leoba-Spannglied*] = rubber sleeve
~schicht *f* = rubber layer
~schieber *m*, Schwabber *m* = squeegee
~schlauch *m* = rubber hose
~schlauch *m* (✖) = bagging [*rubber hose for water, compressed air, or steam*]
~schürze *f* = rubber apron
~(sieb)belag *m* = rubber screen plate
~spülschlauch *m* [*Rotarybohren*] = rubber rotary hose
~-Steinkohlenteerpech-Emulsion *f* = rubberized coal tar pitch emulsion
~torsionslager *n* = torsional rubber mount

~-Transportband *n* → → Bandförderer
~trog *m* = rubber trough
~tür *f* = rubber door
~unterlegscheibe *f* = rubber washer
~verkleidung *f* → Gummiauskleidung
Gurt *m*, Band *n* [*Bandförderer*] = belt ~, Fuß *m*, Flansch *m*, Träger~ = flange, chord, boom
~-Abtragrolle *f* → Gurttragrolle
~aussteifung *f* → Gurtversteifung
~becherwerk *n*, Gurtelevator *m*, Bandbecherwerk, Band(becher)elevator = belt(-type bucket) elevator
~bestandteil *m* = flange component, chord ~, boom ~
~blech *n* → Gurtlamelle *f*
~bogen *m*, Wandbogen = wall arch
~eisen *n*, Gurtwinkel *m* = flange angle, chord ~, boom ~
~elevator *m* → Gurtbecherwerk *n*
Gürtel|radbodenverdichter *m* → → Walze *f*
~reifen *m* = belted tyre (Brit.); radial tire (US)
Gurt|förderer *m* → Bandförderer
~förderer *m* auf Raupen, Bandförderer ~ ~, Gurtförderband *n* ~ ~, ~ auf (Gleis)Ketten = crawler-mounted belt conveyor, ~ ~ conveyer
~förderer *m* mit Seilschrapperkasten = belt conveyor with cable-hauled bucket, ~ conveyer ~ ~ ~
~gesims *n*, Gurtsims *m*, *n* = string course
~gewebe *n* = belting fabric
~gewölbe *n*, versetztes Gewölbe, Zonengewölbe = vault with dentated springing lines
~lamelle *f*, Lamelle, Kopfplatte *f*, Gurtplatte, Verstärkungsplatte, Gurtblech *n* = flange plate, cover ~, boom ~, chord ~
~platte *f* → Gurtlamelle *f*
~reihenbecherwerk *n* → Gurtvollelevator *m*
~scheibe *f* = tie disc; ~ disk (US)
~sims *m*, *n*, Gurtgesims *n* = string course

Gurtstab — Güteprobe

~**stab** m = boom member, chord ~, flange ~, ~ rod, ~ bar
~**tragrolle** f = carrier idler
~**versteifung** f, Gurtaussteifung = boom bracing, chord ~, flange ~, ~stiffening
~**vollelevator** m, Gurtreihenbecherwerk n, Gurtvollbecherwerk = continous belt(-type bucket) elevator
~**winkel** m, Gurteisen n = chord angle, boom ~, flange ~
Guß m [*Betonsteinindustrie*] = cast
Gußasphalt m, Streichasphalt, Mastixasphalt = mastic asphalt [*deprecated: floated ~*]; bituminous mastic concrete (US)
~**decke** f, Gußasphaltbelag m = mastic asphalt surfacing (Brit.); bituminous mastic concrete paving (or pavement) (US)
~**-Einbaumaschine** f → Gußasphaltfertiger m
~**fertiger** m, Gußasphalt-Einbaumaschine f, Einbaumaschine für Gußasphalt, Fertiger für Gußasphalt, Asphaltfertiger = mastic asphalt finisher, ~ ~ finishing machine
~**kocher** m, Mastixkocher, Asphaltkocher = mastic cooker, ~ asphalt mixer, ~ melting boiler, ~ asphalt cauldron
~**kocherei** f, stationäre Asphaltkochanlage f = stationary mastic cooking plant
(~)**Motorkocher** m = engine-driven mastic asphalt mixer
~**streicher** m, Asphalteur m = spreader (Brit.)
~**transportmaschine** f, Gußasphalt-Transportwagen m = mastic asphalt mixer (and) transporter
Gußbeton m → → Beton
~**einrichtungen** fpl = concrete spouting equipment
~**-Rinne** f, (Beton)Gießrinne = concrete chute
~**-Rinnenanlage** f → (Beton)Gießrinnenanlage

~**-Rohr** n, (Beton)gießrohr = pipe for concrete placing by gravity
Gußeisen n mit **Lamellengraphit**, Grauguß m, GGL n = gray cast-iron, grey ~
~**auskleidung** f, Gußeisenverkleidung, Gußeisenbekleidung = cast-iron lining, ~ facing
~**brücke** f = cast-iron bridge
~**deckel** m = cast-iron cover
~**-Fischbauchträger** m = cast-iron fish-bellied girder
~**platte** f = cast-iron plate
~**radiator** m = cast-iron radiator
~**rohr** n = cast-iron pipe
~**-Rundstütze** f = cast-iron round column
~**schuh** m = cast-iron shoe
~**spirale** f [*Turbine*] = cast-iron volute casing, ~ spiral ~
~**strecke** f [*eine Strecke mit Gußeisentübbings im Tunnel- und Stollenbau*] = cast-iron (tunnel-lining) segment section
~**stütze** f = cast-iron column
~**träger** m = cast-iron girder
~**tübbing** m = cast-iron (tunnel-lining) segment
~**verkleidung** f → Gußeisenauskleidung
~**zylinder** m = cast-iron cylinder
Guß|legierung f = casting alloy
~**ringsäule** f [*Schachtabteufen*] = cast-iron tubbing
~**stück** n = casting
(**Gut**)**Abzug** m, Abziehen n (des Gutes) [*von einem Silo oder Bunker*] = drawing
Gutachten n = report
Gutbewegung f [*Siebtechnik*] = travel of material over the deck
Güte|grad m, Trennungsgrad [*Siebtechnik*] = screening efficiency, efficiency of separation, separation sharpness
~**klasse** f, Sorte f = grade
~**klasse** f, Festigkeitsklasse [*Betonbaustein*] = strength class
~**klausel** f = quality clause
~**probe** f → Güteprüfung f

Güteprüfung — Hafenbahnhof

~prüfung *f*, Güteprobe *f*, Güteversuch *m* = quality test, ~ control check
Güterbahn|hof *m* = goods yard
~**steig** *m* = goods platform
Güter|fernverkehr *m* = long-haul traffic
~**hafen** *m*, Warenhafen = cargo harbour (Brit.); ~ harbor (US)
~**halle** *f* → Güterschuppen *m*
~**kraftverkehr** *m* = road haulage
~**nahverkehr** *m* = short-haul traffic
~**schuppen** *m*, Warenspeicher *m*, Güterhalle *f* = freight house, warehouse
~**transport** *m* = commercial transport, transportation of freight
~**umschlag** *m*, Warenverkehr *m*, Umschlag(verkehr) [*Hafen*] = cargo traffic, goods ~, harbo(u)r ~
~**wagen** *m*, (Bahn)Waggon *m* = freight car (US); wagon (Brit.)
~**zug** *m* = freight train
Güte|sicherung *f* **und Abnahme** *f* [*in einer Norm*] = quality control and basis of acceptance
~**überwachung** *f*, Qualitätsüberwachung *f* = quality control
~**überwachung** *f* **bei der Herstellung**, Qualitätsüberwachung ~ ~ ~ = procedural control
~**versuch** *m* → Güteprüfung *f*
~**vorschrift** *f* = quality specification
Guttaperchazündschnur *f*, wasserdichte Zündschnur = white centered guttapercha waterproof fuse

H

Haar|filz *m* = hair-felt
~**kalkmörtel** *m*, Kalkhaarmörtel = hair(ed) lime mortar
~**kies** *m*, Millerit *m*, NiS (Min.) = millerite, capillary pyrite
~**längsriß** *m* = longitudinal hair crack
~**nadelkurve** *f* = hairpin curve
~**nadelschleife** *f* [*Fluß*] = hairpin bend, horseshoe ~
~**putz** *m*, Haarmörtelbelag *m* = hair plaster
~**riß** *m* = hair(line) crack
~**röhrchen** *n* → Kapillarrohr *n*
~**röhrchenkraft** *f* → Kapillarkraft
~**röhrchenspannung** *f* → Kapillarspannung
~**röhrchenwasser** *n* → Kapillarwasser
~**sieb** *n* = (horse) hair sieve
Hackboden *m* [*Erdbau*]; mildes Gebirge *n* [*Tunnelbau*] = hacking (Brit.)
Hacke *f* → Erd~
Hackprobe *f* → Schlitzprobe
Häckseltorf *m* = chaff peat
Hafen *m* = harbour (Brit.); harbor (US); port [*An area of sheltered water where ships can lie and, possibly, load or unload. It may be natural or artificially sheltered (by breakwaters)*]
~ **mit Paralleldämmen**, ~ ~ parallen Hafendämmen = harbo(u)r with parallel jetties
~ **zur Betankung mit Rohöl** = oil-fuel bunkering port, fuel-oil ~ ~
~**abgaben** *fpl*, Hafengebühren *fpl*, Hafengeld *n* = harbo(u)r dues
~**anlagen** *fpl*, Hafenbetriebseinrichtungen *fpl*, Hafenbetriebsmittel *npl* = harbo(u)r installations, ~ facilities
~(**anschluß**)**bahn** *f*, Hafeneisenbahn = port railway (Brit.); ~ railroad (US)
~**ausbau** *m*, Hafenerweiterung *f* = harbo(u)r improvement, ~ development
~**ausrüstung** *f* = harbo(u)r equipment
~**außenwerk** *n*, Mole *f*, Hafendamm *m* [*gibt der Schiffahrt eine gegen Wind, Seegang und Strömung gesicherte Zufahrt zum Hafen und schützt die Hafeneinfahrt vor Verlandung*] = breakwater; mole [*this word is used almost exclusively in connection with British Mediterranean ports, e. g. moles at Gibraltar and Malta are equivalent to breakwaters*]
~**bahn** *f* → Hafenanschlußbahn
~**bahnhof** *m*, Seebahnhof = marine terminal

Hafenbarkasse — Haftlänge

~barkasse *f* = harbo(u)r launch
~bau *m* = harbo(u)r construction
~bauarbeiten *fpl* = harbo(u)r work
~bauingenieur *m* = harbo(u)r construction engineer
~baustelle *f* = harbo(u)r construction site
~bauwerk *n* = harbo(u)r structure
(~)Becken *n* [*zwischen zwei Ladezungen*] = basin, dock
~befeuerung *f* = harbo(u)r lighting
~behörde *f* = harbo(u)r authority, port ~
~betrieb *m* = harbo(u)r working
~betriebseinrichtungen *fpl* → Hafenanlagen *fpl*
~betriebsmittel *npl* → Hafenanlagen *fpl*
~damm *m* → Hafenaußenwerk *n*
~einfahrt *f*, Hafenmündung *f* = harbo(u)r entrance
~eisenbahn *f* → Hafen(anschluß)bahn
~erweiterung *f* → Hafenausbau *m*
~gebäude *n* = harbo(u)r building
~gebiet *n* = harbo(u)r area
~gebühren *fpl* → Hafenabgaben *fpl*
~geld *n* → Hafenabgaben *fpl*
~(güter)umschlagmittel *n*, Hafen(güter)fördermittel *m* = cargo handling appliance, ~ ~ gear
~innenteil *m, n* = inner harbo(u)r
~kran *m* = harbo(u)r crane
~mauer *f* = harbo(u)r wall
~mündung *f* → Hafeneinfahrt *f*
~polizei *f* = port police (force)
(~)Poller *m*, Haltepfahl *m* = bollard
~portalkran *m*, Uferportalkran = harbo(u)r portal crane
~schlepper *m* = harbo(u)r tug
~schuppen *m*, Kaischuppen = shed
(~)Schutzwerk *n* = protective (harbo(u)r) structure
~seite *f* = harbo(u)r side
~siedlung *f* = port colony
~silo *m* = harbo(u)r silo
~sohle *f* = harbo(u)r bottom
~speicher *m* = harbo(u)r warehouse
~stadt *f* = harbo(u)r town
~technik *f* = harbo(u)r engineering
~umschlagbetrieb *m* = cargo handling

~umschlagmittel *n* → Hafengüterumschlagmittel
(~)Umschlagverkehr *m*, Güterumschlag *m* = harbo(u)r traffic
~vertiefung *f* = harbo(u)r deepening
~verwaltungsbüro *n* = harbo(u)r administrative office, ~ administration ~
~vorfeld *n* = harbo(u)r approaches
~zeit *f* = establishment of a port [*The high-water lunitidal interval at full and change of moon (Interval, H. W. F. & C.)*]
Haft|ablösung *f*, Verbundablösung = bond failure
~anreger *m* → Netzhaftmittel *n*
~anregung *f*, Haftfähigkeitsverbesserung *f* [*von plastischen Bindemitteln am Gestein*] = doping of binders, promotion of binder adhesion
~bruch *m*, Verbundbruch = bond failure [*between steel and concrete*]
~brücke *f* → Klebnaht *f*
~emulsion *f* = bonding emulsion
haften, an~ = to adhere, to bond
Haften *n*, Haftung *f* [*Anstrichstoff*] = adhesion
~ → Haftung *f*
haftendes Lockergestein *n*, Bodenmörtel *m*, Bindeerde *f*, Bodenbinder *m*, Binderton *m* = soil matrix, ~ binder, ~ mortar, binder soil, (clay) binder
Haft|ermüdung *f*, Verbundermüdung [*Beton*] = fatigue of bond
~fähigkeit *f* → Adhäsion *f*
~fähigkeitsverbesserung *f* → Haftanregung
~festigkeit *f*, Verbundfestigkeit = bond strength
~festigkeit *f*, Haften *n*, Haftung *f*, Abreißwiderstand *m* [*Putzmörtel*] = adhesion
~festigkeitsverbesserer *m* → Netzhaftmittel *n*
~fläche *f* = bonding area
~grund *m* = bonding adhesive
~länge *f*, Verbundlänge, Einbindelänge = bond length, grip ~, transfer ~ [*reinforcing bar*]

~**masse** *f* [*zur Imprägnierung der Papierisolation von Hochspannungskabeln*] = non-draining compound
~**mittel** *n* → Netz~
~**probe** *f* → Haftversuch *m*
~**prüfung** *f* → Haftversuch *m*
~**putz** *m* = bond plaster [*Is a plaster with high adhesive properties which is made especially for use as a first coat on interior concrete surfaces. It is a ready-mixed plaster which requires only the addition of water to be ready for use*]
~**reibung** *f* = static friction
~**reibungsbeiwert** *m* = static coefficient of friction
~**rüttler** *m* → Bunkerrüttler
~**schlupf** *m*, Verbundschlupf = bond slip
~**spannung** *f*, Verbundspannung = bond stress [*The force of adhesion per unit area of contact between two bonded surfaces such as concrete and reinforcing steel or any other material such as foundation rock*]
~**- und Planierzement** *m* (schnell erhärtend) zum Spachteln von Unterböden sowie zum Füllen von Löchern und Rissen = sub-floor surfacing & levelling underlayment (rapid-hardening)
Haftung *f*, Verbund *m*, Haften *n* = bond
~, Haftfestigkeit *f*, Abreißwiderstand *m*, Haften *n* [*Putzmörtel*] = adhesion
Haft|untersuchung *f*, Verbunduntersuchung = bond study
~**verankerung** *f* [*Spannbeton*] = self-anchoring
~**verbund** *m* → Haftvermögen *n*
~**verhalten** *n*, Verbundverhalten = bond performance
~**verlust** *m* = loss of bond
~**vermögen** *n* [*Anstrichstoff*] = adhesion property
~**vermögen** *n*, Haftung *f*, Haftfestigkeit *f*, Haftverbund *m* [*Bewehrung an Beton*] = bonding, gripping

~**versuch** *m*, Haftprobe *f*, Haftprüfung *f* = bond test
~**vibrator** *m* → Bunkerrüttler *m*
~**wasser** *n* → Benetzungswasser
~**wert** *m* [*Fehlname*] → Höchstwert
~**wirkung** *f*, Verbundwirkung = bond action
~**zerstörung** *f*, Verbundzerstörung = destruction of bond
Hahn *m* = cock [*A valve for controlling a pipeline of water, gas, or other fluid*]
~, Niederschraubventil *n* = screw-down valve
~**armaturen** *fpl* = plug cocks, fullway valves and screw-down valves
Hahnenbalken *m*, Katzenbalken, Spitzbalken, Hainbalken = top beam
(Hahn)Küken *n*, drehbarer Kegel *m* = (tapered) plug
Hainbalken *m* → Hahnenbalken
Haken *m* = hook
~**barre** *f* = baymouth bar
~**blatt** *n* → hakenförmige Überblattung *f*
~**bolzen** *m* = hook bolt
~**bügel** *m* [*Dammbalkenwehr*] = stirrup (Brit.); U-bolt (US)
~**flasche** *f* [*Flaschenzug*] = hook type bottom block
~**flügel** *m* = return wing
hakenförmige Überblattung *f*, Hakenblatt *n* = hooklike halving
Haken|geschirr *n* = hook accessories
~**geschwindigkeit** *f*, Last~ = lift(ing)
~**gewicht** *n*, Belastungsgewicht des Hakens = hook weight
~**greifer** *m* → Motorgreifer(korb) *m*
~**höhe** *f* = height under hook
~**hubgeschwindigkeit** *f* = hook lifting rate
~**kabel** *n* → Hakenseil *n*
~**kette** *f* = hook chain
~**(-Last)katze** *f*, Haken-Laufkatze = crab with hook
~**maul** *n*, Hakenweite *f* = hook opening width
~**nadel** *f*, Nadel mit Haken [*Nadelwehr*] = needle with hook
~**reichweite** *f* = hook reach

~rolle f = hook roller

~schlüssel m, Drehschlüssel für Bohrgestänge, Bohrgestänge-Drehschlüssel, Gestängehaken m = pipe hook, ~ wrench

~schlüssel m [für Ringmuttern mit Nuten] = hook(ed) spanner (wrench)

~schraube f = T-headed bolt

~schütz(e) n, (f) = lifting hook-type gate

~seil n, Hakenkabel n [Kabelkran] = hook rope, ~ cable

~stein m = toed voussoir

~stellung f, Hakenlage f = hook position

~tragkraft f, Nutzlast f = hook capacity

~verankerung f = hook anchorage

~weg m = hook travel

~weite f, Hakenmaul n = hook opening width

~winde f = hook winch

halb in Verband verlegte Blöcke [Hafenbau] = roughly set blocks; query-random blockwork, random blockwork, random blocks (US)

halbe Tide(erscheinung) f, ~ Gezeit(enerscheinung) f = half-tide

halber Abschlag m [Sprengen] = half round

~ **Walm** m, Krüppelwalm = half hip(ped end), partial ~ ~

Halbachse f, Steckachse = half-axle

~ **der Zentralellipse** = semi-axis of the central ellipse

Halbanhänger m, Aufl(i)eger m, aufgesattelter Anhänger, Sattelanhänger, Aufsattel-Anhänger, Sattelfahrzeug n = semi-trailer, articulated trailer

~ **mit Bodenentleerung**, Aufsattel-Bodenentleerer m, aufsatteliter Bodenentleerer, aufgesattelter Bodenentleerer, Bodenentleerer-Halbanhänger, bodenentleerender Erdtransportwagen, Gelenkwagen-Bodenentleerer, aufgesattelter Erdtransportwagen m mit Bodenentleerung, Sattel(fahrzeug)-Bodenentleerer = bottom-dump wagon, ~ semi-trailer

~-**Achse** f, Aufl(i)eger-Achse = semi-trailer axle

~-**Drehschemel** m, Aufl(i)eger-Drehschemel = semi-trailer bogie

~-**Förderwagen** m, Aufl(i)eger-Förderwagen [Erdbau] = gooseneck(-type) wagon

~-**Hinterkipper** m → Halbanhänger-Rückwärtskipper

~-**Rad** n, Aufl(i)eger-Rad = semi-trailer wheel

~-**Rückwärtskipper** m, Rückwärtskipper-Halbanhänger, Rückwärtskipper-Aufl(i)eger, Aufl(i)eger-Rückwärtskipper, Aufl(i)eger-Hinterkipper, Hinterkipper-Aufl(i)eger, Halbanhänger-Hinterkipper, Hinterkipper-Halbanhänger = rear-dump semi-trailer

~-**Seitenkipper** m, Seitenkipper-Halbanhänger, Aufl(i)eger-Seitenkipper, Seitenkipper-Aufl(i)eger = side-dump semi-trailer

~-**Tieflader** m für Baumaschinen = semi-trailer for construction equipment

halbautomatisch, halbselbsttätig = semi-auto(matic)

halbautomatische Schweißmaschine f = semi-auto(matic) welding machine

~ **Schweißung** f, halbautomatisches Schweißen n = touch welding

Halb|balken(träger) m, Balkenhälfte f = half-beam

halbblanke Schraube f = semi-finished screw, half-bright ~

Halb|dach n, Pultdach, Flugdach, Schleppdach, einhängiges Dach = shed roof

halbdurchlässig = semi-permeable

Halb|ebene f = half-plane

~**edelstein** m = semi-precious stone

~**eisen** n = half rounds

~**elliptikfeder** f = semi-elliptic(al) (leaf-)spring

~**erdbehälter** m, Halberdtank m = half-buried tank

~**erzeugnis** n → Halbfabrikat

~**etage** f → Halbgeschoß n

~fabrikat n, Halberzeugnis n = semi-finished product
halb|fahrbar, halbortsbeweglich = semi-portable
~fest = semisolid
~flüssig = semi-liquid
Halb|freiluft-Kraftwerk n → → Talsperre f
~fusit m = semi-fusain, vitri-fusain
~gasfeuerung f = half-gas fired furnace, semi-producer type ~
halb|gebrochen [Zuschläge] = half-broken, half-crushed
~gelatinöser Sprengstoff m = semi-gelatine (type of) explosive
~geschlossene Schicht f [Bodenmechanik] = half-closed layer [The water can escape through only one surface]
Halb|geschoß n, Beigeschoß, Zwischengeschoß, Mezzanin n, Halbstockwerk n, Halbetage f = mezzanine, entresol
~geschoßhaus n [z. B. an Berghängen] = split-level house
halbgespundete Dielung f, gefalzte ~ = rebated wood flooring
Halbglanz m = semigloss
Halb(gleis)|kette f, Halbraupe f = half-track, half-crawler
~kettenfahrzeug n, Halbraupenfahrzeug = semi-tracked vehicle
halb|graphisch = semi-graphical
~hart = semi-hard
Halb|holz n = half-timber
~hydrat n, ~gips m, verzögerter Halbhydratgips, $CaSO_4 \cdot 1/2\ H_2O$ = retarded hemihydrate gypsum(-plaster)
~hydraulik-Bagger m = semi-hydraulic excavator
halbhydraulisch = semi-hydraulic
halbiertes Roheisen n, meliertes ~ = mottled pig iron
halbiertes Rohr n, Rinne f [Kanalisationssteinzeug] = channel
Halbierung f = bisecting
Halbierungslinie f eines Winkels, Winkelhalbierende f = bisecting line of an angle

Halb|kellergarage f, Halbtiefgarage = semi-underground garage
~kette f → Halbgleiskette
Halbketten|fahrzeug n, Halbgleiskettenfahrzeug, Halbraupenfahrzeug = semi-tracked vehicle
~-Traktor m, Halbketten-Trecker, Halbketten-Schlepper = semi-tracked tractor
halbkontinuierliche Förderanlage f = power and free conveyor, ~ ~ ~ conveyer
Halbkreis|bogen m → Rundbogen
~fang(e)damm m = semicircular cofferdam
halbkreisförmig = semicircular
Halbkreis-Rippenkalotte f = semi-circular ribbed roof
halbkristallin = hypocrystalline
Halbkugel f = hemisphere
~boden m = hemispherical bottom
Halbkugeltiegel m = hemispherical crucible, basin (shape ~)
~-Kippofen m = basin tilting furnace, hemispherical crucible ~ ~, basin tilter
Halb|last f = half-load
~leiter m [für einen Transistor] = semiconductor
halblichtdurchlässig = semi-transparent
Halbmassivdecke f = semi-solid floor
halb|massives Wehr n = semi-solid weir
~mondartiges Schrapp(er)gefäß n, halbmondartiger Schrapp(er)kübel m = crescent (scraper) bucket
~offen [Straßendecke] = medium-textured semi-open
Halbopal m = semiopal
halborts|beweglich, halbfahrbar = semi-portable
~fest, halbstationär = semi-stationary
Halb|parabelträger m = hog backed girder, semi-parabolic ~
~parabelträger m mit abgeschrägten Enden = semi-parabolic girder with sloping end posts, hog backed ~ ~ ~ ~ ~

halbparabolische Vibrierbohle *f* = semi-parabolical vibrating screed
Halb|parameter *m* = half parameter
~**podest** *n*, *m* = half-landing; half-space landing [*deprecated*]
~**portalkran** *m*, Halbtorkran = semi-portal (type of pedestal) crane
~**querentlüftung** *f* [*Tunnel*] = semi-transversal ventilation
~**rahmen** *m* = half-frame
halbrauh = semi-rough, half-rough
Halbraum *m* = semi-infinite mass, ~ solid, halfspace
~**modell** *n* = halfspace model
~**oberfläche** *f* = semi-infinite mass surface
Halb|raupe *f* → Halb(gleis)kette *f*
~**raupenfahrzeug** *n*, Halb(gleis)kettenfahrzeug = semi-tracked vehicle
~**rohrschale** *f* → Halbschale
Halbrund|eisen *n* = half-round iron
~**feile** *f* = half-round file
~**niet** *m* = cup head rivet
~**schraube** *f* = half-round screw, button-head(ed) ~
~**stahl** *m* = half-round steel
Halb|säule *f* = half column [*Half a circular column, having a capital and a base and projecting from a wall to which it is bonded. It forms a part of the structure*]
~**schale** *f*, Halbrohrschale = centre split pipe (Brit.); center ~ ~ (US); channel, half-round section
~**schale** *f* [*Greiferkorb*] = half-scoop (Brit.); clamshell jaw (US)
halbschwer = semi-heavy
Halbschwingung *f* = semi-vibration
halb|seitiger Einbau *m* = constructing in half-road width
~**selbsttätig**, halbautomatisch = semi-auto(matic)
~**sphärisches Gewölbe** *n* → → Talsperre *f*
Halbspüldamm *m* = semihydraulic fill dam
halb|stabil, mittelschnellbrechend [*Bitumenemulsion*] = normal-setting, medium-setting, semistable, medium-breaking

~**stationär**, halbortsfest = semi-stationary
~**stationäre bituminöse Mischanlage** *f*, Baukasten(system)anlage = sectional type bituminous mixing plant, ~ ~ ~ ~ installation
~**steil** = semi-steep
Halb|stein *m*, Halbziegel *m*, Zweiquartier *m* = half bat, two quarters
~**stein** *m* [*Beton(bau)stein*] = half-block
halbsteinstark [*Ziegelmauer*] = half-brick thick
Halb|stockwerk *n* → Halbgeschoß *n*
~**-Stoper** *m* → Hammerbohrmaschine *f*
halbtägige Gezeit *f*, ~ Tide *f* = semi-diurnal tide
Halb|tags-Arbeitskraft *f* = part-timer
~**tauchbrücke** *f* = semi-high level bridge
halbtechnischer Betriebsversuch *m* = pilot test
Halb|tidebecken *n* = half-tide basin
~**tidefang(e)damm** *m* = half-tide cofferdam
~**tidehafen** *m* [*Sonderform des Dockhafens, bei dem das Binnenwasser um ein gewisses Maß schwankt, um die Zeit der Zugänglichkeit zu verlängern*] = half-tide harbo(u)r
~**torkran** *m* → Halbportalkran
~**träger** *m* = half span of the girder, half-girder
~**tränkmakadam** *m*, Halbtränkdecke *f* = semi-penetration macadam
~**tränkung** *f* [*Straßendecke*] = semi-grouting (Brit.); semi-penetration treatment
halbversenkt [*Niet*] = half sunk
Halbvitriolblei *n* = lanarkite
halbweich = semi-soft
Halb|zeugwalzwerk *n* = rolling mill for semi-finished products
~**ziegel** *m*, Halbstein *m*, Zweiquartier *m* = two quarters, half bat
~**zylinder** *m* = half-cylinder
Halde *f*, Vorrats~, Material~ = stockpile, (storage) pile, stock-heap

~, Aussatz~, Abraum~ = waste pile, ~ heap
Halden|abzugband *n* = reclaiming conveyer, ~ conveyor
~bahn *f* = stockpile ropeway
~beschickungsbandanlage *f* = belt conveying plant for storage piles
~lagerung *f*, Deponie *f*, Freilagerung im (oder in) Schütthaufen = stockpiling (in the open air), ground storage
~material *n* = (stock-)pile(d) material
~rückverlader *m* = stockpile loader
~rückverladung *f* = handling from storage, stockpile handling
~sand *m* [*gekörnte auf die Halde gestürzte Hochofenschlacke*] = stockpiled granulated blast-furnace slag
~schlacke *f* = discarded slag
~schütter-Bandabsetzer *m* → Hochabsetzer
~schüttung *f* = stockpiling
~schüttungs-Förderband *n* → Hochabsetzer *m*
~seilbahn *f* = storage pile ropeway
~sturz *m* [*das Ablagern von Massengütern auf Halden*] = stockpiling
~tunnel *m*, Vorratstunnel, Entnahmetunnel, Bandkanal *m*, Entnahmestollen *m*, Fördertunnel, Förderstollen *m* = (stock) reclaiming tunnel, stockpile ~, conveyor ~, recovery ~, conveyer ~
~tunnel-Abzugband *n* = (reclaiming) tunnel conveyor, (~) ~ conveyer
Halemaumau-Lavasee *m* = Halemaumau fire-pit
Halitkainit *m*, Hartsalzkainit = thanite
Halle *f* = hall
~, Eingangs~ = hall [*An entrance space, often containing a staircase*]
~, Flugzeug~ = (aircraft) hangar
~, Markt~ = covered market
hallen = to resound
Hallen|bad *n*, Hallenschwimmbad, Hallenbadeanstalt *f* = indoor swimming pool, ~ ~ bath, covered ~ ~, swimming baths

~bahnhof *m* = covered railway station (Brit.); ~ railroad ~ (US)
~bau *m* = hall construction
~binder *m* [*siehe Anmerkung unter "Binder"*] = hall roof truss
~dach *n* = hall roof
~eisbahn *f*, Eispalast *m* = ice palace, covered skating rink, indoor ice rink
~fläche *f* = hall ground area
~kirche *f* [*Eine mehrschiffige Kirche, bei der die Seitenschiffe ebenso oder annähernd so hoch sind wie das Mittelschiff, im Gegensatz zur Basilika*] = church with aisles of equal height
~kran *m* = hall crane
(~)Schiff *n* = nave
~(schwimm)badgebäude *n* = swimming baths building
~stahltor *n* = hall steel door
~tor *n* = hall door
~vorfeld *n* → Vorfeld
Hallinger-Schild *m* = Hallinger shield
Halochalzit *m* = atacamite
Hals *m* = neck(ing) [*The connecting moulding between the capital and the shaft of a column, or the plain part between the mouldings and the shaft*]
~ → Schlot *m* [*Vulkan*]
~band *n* [*Schleusentor*] = collar (strap)
~mutter *f* = round neck nut
~ring *m* = collar
~zapfen *m* [*hölzernes Schleusenstemmtor*] = top pivot, ~ trunnion, gudgeon pin
halt|bar, alterungsbeständig, dauerhaft = durable
~bares Abwasser *n*, fäulnisunfähiges ~ = stable effluent [*A treated sewage that contains enough oxygen to satisfy its oxygen demand*]
Haltbarkeit *f*, Beständigkeit, Dauerhaftigkeit = resistance, durability
Haltbarmachung *f*, Konservierung = conservation
Halte|bucht *f* [*Autobus*] = draw-in
~bügel *m* **der Ladegabel** = fork clamp
~klinke *f* = latch retainer
~konsole *f* = supporting bracket

~kraft *f* = holding strength, ~ power
~kraft *f* → Kohäsion *f*
~linie *f* = stop line
~mutter *f* = retaining nut
~pfahl *m*, (Hafen)Poller *m* = bollard
(~)Pfahl *m* [*Naßbagger*] = spud
(~)Pfahlseil *n* [*Naßbagger*] = spud rope, ~ cable, ~ line
Halter *m* [*Holzgerüst*] = lashing, whip, bond [*A short length of fibre or steel rope for tying scaffold timbers*]
~, Halterung *f* = retainer
~, Tragstück *n*, Arm *m*, Eingab(e)lung *f* = bracket
Halte|raste *f* = lock-out
~ring *m*, Druckring = retainer ring
Halterung *f*, Halter *m* = retainer, fastener
Halte|seil *n* → Greifer~
~sichtweite *f* = stopping distance
~stelle *f* = halt, stop(ping-place)
~stelleninsel *f* = stop island
~stift *m* = retaining pin
~verbot *n* [*als Zeichen*] = "no waiting"
~zeichen *n* = stop sign, halt ~
Haltigkeit *f* **einer Lagerstätte** = yield of a deposit
Haltung *f*, Streckenabschnitt *m*, Staustrecke *f*, Stauhaltung *f* [*horizontale oder nahezu horizontale Wasserlaufstrecke im Schiffahrtkanal oder kanalisiertem Strom zwischen zwei Stauanlagen*] = (level) reach
Hammer *m*, Bär *m*, Ramm~ = (pile) hammer
~ **aus Kunstharzpreßstoff** = composition hammer
~bohren *n* = hammer drilling
Hammerbohrmaschine *f*, überschwerer Bohrhammer *m*, Schwerstbohrhammer, Halb-Stoper *m* = (rock drill) drifter [*Is an air-operated percussion-type drill, similar to a jackhammer, but so large that it requires mechanical mounting*]
~ **mit Schlagwerk unmittelbar hinter dem Bohrkopf** [*zum schlagenden Großlochbohren*] = hammer-down-the-hole machine

~ ~ **Trockenabsaugung durch Saugkopf** = (rock drill) drifter with dry suction head
~ ~ **Wasserspülung durch Spülkopf** = (rock drill) drifter with water flushing head
~ ~ **zentraler Absaugung** = (rock drill) drifter with centre dry suction
~ ~ ~ **kombinierter Wasser-Luft-Spülung** = (rock drill) drifter with combined centre water and air flushing
~ ~ ~ **Wasserspülung** = (rock drill) drifter with centre water flushing
Hammer|brecher *m* = (swing-)hammer crusher, ~ breaker, rotary hammer(-type) breaker, rotary hammer(-type) crusher
~fertiger *m* → Stampfhammerfertiger
~führung *f*, Bärführung *f* = (pile) hammer guide
~fundament *n*, Hammergründung *f*, Hammerfundation *f*, Hammergründungskörper *m*, Hammerfundamentkörper = hammer foundation (structure), base for drop-hammer
~fundamentkörper *m* → Hammerfundament *n*
~fundation *f* → Hammerfundament *n*
~gewicht *n*, Bärgewicht = (pile) hammer weight
~gründung *f* → Hammerfundament *n*
~gründungskörper *m* → Hammerfundament *n*
~katze *f*, Bärkatze = traveller
~kopf *m*, Bärkopf = (pile) hammer head
~kopf *m* [*Am unteren Ende des Rundstahls eines Ankers für Stahlbaustützen angestauchter oder angeschweißter verbreiterter Teil in Form des Kopfes eines Hammers. Der Anker wird durch den Ankerkanal und der H. mit seiner Schmalseite durch den Ankerbarren gesteckt u. der Anker durch Drehen mit dem dann querliegenden H. am Barren festgelegt*] = hammerhead
~(kopf)kran *m* = hammerhead crane, giant cantilever ~

~**kopfpfeiler** *m* = hammerhead pier

~**mühle** *f*, Rotor-~ = hammer mill [*Secondary mill in which rapidly rotating bars or hammers pass between grates to crush material by impact. The tolerance between hammers and grate usually is 1 inch or less*]

~**mühlenschläger** *m* = swing hammer

~**nietmaschine** *f* = hammer riveting machine

~**nietung** *f*, Hammernieten *n* = hammer riveting

~**rammung** *f*, Hammerrammen *n* = hammer-driving

hammerrechtes Schicht(en)mauerwerk *n* = block-in-course masonry, hammer-dressed ashlar ~, hammer-dressed ashler ~ [*Squared-stone masonry laid in regular courses with the stones roughly squared with a hammer*]

Hammer|schiene *f*, Bärschiene = (pile) hammer rail

~**schlag** *m*, Bärschlag = (pile) hammer blow

~**schlag** *m*, Abbrand *m*, Zunder *m*, Walzhaut *f* = scale

~**schweißnaht** *f* = forge welding seam

~**schweißung** *f* [*Schweiz: Feuerschweißung*] = forge welding

~**seil** *n*, Bärseil = (pile) hammer rope

~**stütze** *f* → Bohrstütze

~**trommel** *f*, Bärtrommel, Ramm~ = (pile) hammer drum

Hampelmann *m* → Nageleisen *n*

hand|abgebauter Stoß *m* (⚒) = hand-mined face

~**abgezogener Beton** *m* = hand-finish concrete

Hand|anlasser *m* = hand starter

~**antrieb** *m* = hand drive

~**arbeit** *f* = hand work, manual ~

~**aufgabe** *f*, Handbeschickung *f*, Handbeschicken *n*, Handaufgeben = hand feed(ing), manual ~

~**aufzug** *m*, Aufzug von Hand [*für Maurerarbeiten*] = hand hoist

~**aufzug** *m* → Handlastenaufzug

~**aushub** *m* = manual excavation, hand ~

~**auspreßrohr** *n* → Handverpreßrohr

~**baugerüst** *n* → Handrüstung *f*

handbedient → handbetätigt

Hand|bedienung *f* → Handbetrieb *m*

(~)**Beil** *n* = hatchet

handbeschickt = hand-fe(e)d

Handbeschickung *f* → Handaufgabe *f*

handbetätigt, handbedient, handbetrieben = hand-operated, manually operated, man-handled

Hand|betätigung *f* → Handbetrieb *m*

~**-Betonmischen** *n* → Betonmischen von Hand

~**betrieb** *m*, Handbetätigung *f*, Handbedienung *f* = hand operation, manual ~

handbetrieben → handbetätigt

Hand|biegemaschine *f* = hand bending machine

~**biegen** *n* = hand bending

~**blechschere** *f* = hand plate shears

~**-Blockwinde** *f* → Handflaschenzug *m*

~**bodenfräse** *f* = hand rotary tiller (or rotary hoe, or rotovator, or soil pulverizer)

~**bohle** *f* [*Verdichtungsgerät*] = hand-operated compacting beam

~**bohren** *n* = hand drilling, ~ boring

~**bohrer** *m* [*in Holz*] = hand (boring) bit

~**bohrer** *m*, Handbohrmaschine *f* = hand drill(ing machine), ~ hammer drill

~**bohrmaschine** *f* [*Holzbau*] = hand borer, portable ~, ~ boring machine

~**brause** *f*, Handdusche *f*, Schlauchbrause, Schlauchdusche = hand spray, ~ shower, movable ~

~**bremse** *f* mit Zahnrastenarretierung = hand brake of ratchet and rod construction

~**bremslüftgerät** *n* = manual brake release device

~**buch** *n* = manual, handbook

~**drehbohrmaschine** *f* = hand rotary drilling machine

Handdrehbohrung — handgestampftes Rohr

~drehbohrung *f* = hand rotary drilling

~-Druckluftrüttelbohle *f*, Hand-Preßluftrüttelbohle = manually operated pneumatic tamper (or tamping beam, or compacting beam)

~-Druckluftstampfer *m*, Druckluft-Handstampfer = (compressed-)air (mechanical) tamper

~dusche *f*, Handbrause *f*, Schlauchdusche *f*, Schlauchbrause = hand spray, ~ shower, movable ~

~einbau *m* [*Straßenbaustoff*] = hand spreading, ~ placement, ~ placing, ~ laying, manual ~

~einpreßrohr *n* → Handverpreßrohr

~(-Einschienen)-Hängebahn *f* = hand-operated suspended monorail system

~einspritzpumpe *f* für Kaltwetterstart = coldweather engine primer pump

~einstellung *f* = manual adjustment, hand ~, ~ set(ting)

~-Elektrostampfer *m*, Elektro(-Hand)stampfer = electric (mechanical) tamper

~elektrowerkzeug *n* = hand electric tool

Handelsbaustahl *m* [„*St H W*"; *DIN 1050 für Stahlhochbau. Nicht geschweißt. Zugfestigkeit 34 bis 50 kp/mm²*] = commercial structural steel

Handelsguß *m*, Kundenguß = jobbing work, ~ casting

~gießerei *f*, Kundengußgießerei = jobbing foundry

~-Sandgießerei *f*, Kundenguß-Sandgießerei = jobbing sand foundry

~teil *m, n* = casting of commerce

Handels|güte *f* = commercial quality, ~ grade

~hafen *m* = commercial port, ~ harbo(u)r

~länge *f* = commercial length

~stahl *m* = commercial grade steel

handelsübliches Maß *n* = geometrical contents

Hand|-Explosionsramme *f* → Brennkraft-Handramme

(~)**Explosionsstampfer** *m* = heavy-duty internal combustion tamper

~fahrwerk *n* [*Hängebahnwaage*] = hand travel(l)ing gear

~fahrzeug *n* = hand cart

~fäustel *m* = club hammer, lump ~; mash ~ [*Scotland*]

~felsbohrer *m*, Handgesteinsbohrer = hand hammer rock drill

~fertiger *m* = hand finisher

~fettpresse *f* = hand (type) grease gun

~feuerlöscher *m* = fire drencher

~(flaschen)zug *m*, Hand-Blockwinde *f*, Handrollenzug = hand-operated hoist block, ~ pulley ~, ~ lifting ~, ~ block (and tackle)

~form *f* [*Ziegelherstellung*] = hand mould (Brit.); ~ mold (US)

~form(geb)ung *f* = hand moulding (Brit.); ~ molding (US)

~fuge *f* = hand-formed joint

~gashebel *m* = hand throttle

handgeformter Grauguß *m* = hand-mo(u)lded grey iron castings

handgeführt [*z. B. Rüttelwalze, Rüttelfertiger, Stampfer u. dgl., also sich selbst auf einer Unterlage fortbewegende Geräte*] = hand-guided, manually-guided

~ [*z. B. Bohrhammer, Niethammer u. dgl., also Geräte, welche der Bediener in der Hand hält und dabei ihr Gewicht ganz oder teilweise trägt*] = hand-held, hand-operated, manually-held, manually-operated

Hand|gepäckaufbewahrung *f* = left-luggage office, cloakroom; checkroom (US)

~gerät *n* für Betonoberflächenbehandlung = concrete surfacer

~gerüst *n*, Handrüstung *f*, Handbaugerüst = hand-erected scaffold(ing)

hand|gesetzte Stein(ab)deckung *f*, ~ Steinvorlage, handversetzte ~, handversetzes Steindeckwerk = pitched stonework, hand-packed rock facing

~gestampftes Rohr *n* = hand-tamped pipe

Handgesteinsbohrer — Handramme

Handgesteinsbohrer *m*, Handfelsbohrer = hand hammer rock drill
handgesteuert = hand-controlled, manually controlled
Hand|gewindebohrer *m* **mit Rohrgewinde**, Rohrgewindebohrer, Gasgewindebohrer = pipe tap
~griff *m* = hand grip
~hammer *m* = hammer
~-Hängebahn *f*, Hand-Einschienen-Hängebahn = hand-operated suspended monorail system
~hebebock *m* → Handwinde *f*
~hebel *m* = hand lever
~hebelbremse *f* = hand lever brake
~hebelfettpresse *f* = hand-lever grease gun
~hubwagen *m*, Handhubkarre(n) *f*, (*m*) = hand lift-truck
~hubwerk *n* = hand-operated lifting machine, manual hoist
~injektionsrohr *n* → Handverpreßrohr
~injizierrohr *n* → Handverpreßrohr
~-Kaltasphaltspritzapparat *m* = cold emulsion hand sprayer
~karre(n) *f*, (*m*), Handwagen *m* = hand-truck
~kettenzug *m* = hand-chain block
~kippkarre(n) *f*, (*m*), Handkippwagen *m* = buggy
~kippvorrichtung *f* = hand tilting device, manual ~ ~
~kippwagen *m*, Handkippkarre(n) *f*, (*m*) = buggy
~kolbenpumpe *f* = hand operated piston pump
~kompressor-Spritzmaschine *f* = spraying machine (or sprayer) with hand-operated (air) compressor and hand lance
~körner *m*, Punktiereisen *n* = prick punch
~-Kraftstampfer *m*, Kraft-Hand-stampfer = mechanical tamper
~kraftwerkzeug *n* = powered hand tool
~kran *m*, handbetriebener Kran = hand-operated crane
~kurbel *f* = manual starting crank

~laden *n* (⚒) = hand filling [*shovelling coal or ore by shovel rather than with loaders*]
~lampe *f*, Handleuchte *f* = hand lamp
~langerstunde *f*, Hilfsarbeiterstunde = common labo(u)rer hour
~(lasten)aufzug *m*, Hand-Materialaufzug = hand-operated goods elevator (US); ~ ~ lift (Brit.)
Handlauf *m* = handrail, guard rail [*A rail forming the top of a balustrade on a balcony, bridge, stair, etc.*]
~, Treppen-~ = stair-rail, hand-rail
~kran *m* = hand-operated travel(l)ing bridge crane
~spirale *f* = handrail scroll [*A spiral ending to a handrail*]
Hand|-Leichtbohrhammer *m* = lightweight hand rock drill
~leiste *f*, Handlauf *m* [*Geländer*] = hand rail
~lenkung *f* = hand steer(ing)
~leuchte *f*, Handlampe *f* = hand lamp
~linse *f* = hand lens
~-Materialaufzug *m* → Hand(lasten)aufzug
~mischen *n*, Handmischung *f* = hand-mixing
~nietmaschine *f* = hand riveting machine
~nietung *f*, Handnieten *n* [*von Hand geschlagene Nietung mit Handdöpper und Hammer*] = riveting by hand, hand riveting
~packe *f*, Setzpacke = hand-packed rubble
~pfahlramme *f* → Handramme
~presse *f*, Handschmierpresse = hand gun, ~ grease ~
~(prüf)siebung *f* = hand test-sieving
~pumpe *f* = manual pump, hand ~
~rad *n* = hand wheel
~ramme *f* = punner (Brit.); hand rammer
~ramme *f*, Katze *f*, Handpfahlramme [*aus Holz mit vier Eisengriffen als 2-Mann-Ramme oder mit vier Holzstielen als 4-Mann-Ramme zum Eintreiben von Pfählen und Pfosten*] = hand-operated driver

Handratsche — Handstein

~ratsche f = hand ratchet
~reibahle f = hand reamer
~-Reifenpumpe f = hand pump for tyre inflation (Brit.); ~ ~ ~ tire ~ (US)
(~)Rohrkarre(n) f, (m) = pipe (hand) truck [*for the transport of pipes in precasting factories*]
~rollenzug m → Hand(flaschen)zug
~rührwerk n, Handrührer m = hand(-operated) stirring gear, ~ stirrer
~rüstung f, Hand(bau)gerüst n = hand-erected scaffold(ing)
~rüttelbohle f → Handrüttelfertiger m
~rüttelfertiger m, Handrüttelbohle f, Handvibratorfertiger, Handvibratorbohle, Handvibrationsfertiger, Handvibrationsbohle, Handschwing(ungs)bohle, Handvibrierbohle, Handvibrierfertiger = hand(-guided) vibrating (concrete) finisher, manually-guided ~ (~) ~, ~ ~ (~) screed, hand-wheel propelled ~ (~) ~, hand-propelled vibro-finisher
~rüttelwalze f, Handvibrationswalze, Handschwing(ungs)walze = vibrating hand-roller
~rüttler m, Handvibrator m = hand vibrator
~schacht f = hand excavation, manual ~
~schaltung f [*Kraftübertragung*] = manual shift
~schaltungsgetriebe n = straight shift transmission
~scheinwerfer m = hand-held flood light
~schild n, Schweiß(er)schild = face shield, ~ screen, hand ~
~schleifmaschine f = hand grinder
~(schmier)presse f = hand (grease) gun
~schneidmaschine f = hand(-operated) cutter, ~ cutting machine
~schotter m = hand-broken metal (Brit.); ~ stone (US)
~schrämmaschine f = hand coal cutting machine
~schrapper m, Räumschaufel f, Schrapperschaufel, (Schrapperwinde f mit) Handschaufel, Kraftschaufel = hand scraper, manually guided drag skip, drag loader shovel; mixer-drive shovel
~schrapperwinde f = handscraper winch
~schuhfach n = glove locker, ~ compartment
~schutterung f, Handschuttern n = hand mucking, manual ~
~schweißen n, Handschweißung f = hand welding
~schweißanlage f = hand welding installation
~schweißer m = hand welder
Handschwing(ungs)|bohle f → Handrüttelfertiger m
~fertiger m → Handrüttelfertiger
~walze f → Handrüttelwalze
Hand|seil n = hand rope, ~ cable
~(seil)winde f → Affe m
~sieb n = hand sieve
~siebung f, Handprüfsiebung = hand test sieving
~spachtelung f = screeding by hand
~spindelbremse f = hand spindle brake
~splittstreuer m, Splittstreukarre(n) f, (m), Handsplittverteiler, Handsplittstreugerät n = hand-operated chipping(s) spreader, barrow-type ~ ~, ~ gritter
~splittverteilung f, Absplitten n von Hand = hand gritting, ~ chipping, ~ grit blinding, ~ grit spreading
~spritzgerät n, Handspritzmaschine f = small pressure (or hand) sprayer (or distributor)
~spritzpumpe f = hand-operated spray(ing) pump
~stampfbohle f = hand-operated tamping beam
~stampfer m, Handstampfgerät n = hand tamper
~stampfung f = hand tamping
~stapler m = hand stacker
~start m = manual starting, hand ~
~stein m → Hand(strich)ziegel m

~stein *m* → Spüle *f*
~steuerung *f*, Handsteuergerät *n* = manual controls, hand ~, ~ control unit, ~ control system
~steuerung *f* = manual control, hand ~
~steuerung *f*, Handreg(e)lung = manual control, hand ~
~(strich)ziegel *m*, Hand(strich)stein *m* = hand-formed brick, handmade ~, struck ~
~stück *n* [*Gestein*] = hand specimen
~teerspritzgerät *n*, Handteerspritzmaschine *f* = hand tar spraying machine
~tuchstange *f*, Handtuchstange *f* = towel rail
~-Verkehrszählung *f* = hand traffic count
~verpreßrohr *n*, Handinjektionsrohr, Handeinpreßrohr, Handauspreßrohr, Handinjizierrohr = grouting lance
(~)Versatz *m* (⚒) = (hand) packing
handversetzte Stein(ab)deckung *f* → handgesetzte ~
Handvibrations|bohle *f* → Handrüttelfertiger *m*
~fertiger *m* → Handrüttelfertiger
~walze *f*, Handrüttelwalze, Handschwing(ungs)walze = vibrating hand-roller, vibratory ~
Handvibrator *m*, Handrüttler *m* = hand(-manipulated) vibrator
~bohle *f* → Handrüttelfertiger *m*
~fertiger *m* → Handrüttelfertiger
Handvibrier|bohle *f* → Handrüttelfertiger *m*
~fertiger *m* → Handrüttelfertiger
(Hand)Vollversatz *m* (⚒) = solid (hand) packing
Hand|vorschub *m* = hand feed(ing)
~wagen *m* = hand cart
~wagen *m*, Handkarre(n) *f*, *(m)* = hand-truck
~walze *f* = hand-drawn roller
~werkzeug *n* = hand tool
Handwinde *f* = hand winch
~, Handhebebock *m* = hand jack
~ → Affe *m*

Hand|ziegel *m* → Handstrichziegel
~zug *m* → Handflaschenzug
Hanf|gewebe *n* = hessian (Brit.); burlap (US) [*Strong coarse material woven from hemp*]
~seele *f* = hemp core
~seil *n*, Hanfstrick *m* = hemp rope
Hang *m*, Ab~, Berghang, Berglehne *f* = flank of a hill, hillside
~arbeiten *fpl* = side-hill work
~(be)ries(e)lung *f*, wilde (Be)Ries(e)lung = uncontrolled flooding, "wild" ~
~böschung *f* = hillside slope
~bunker *m* = hillside bin
Hängebahn *f*, Schienen~, Schwebe~ = suspension rail conveying system
~waage *f* = suspension rail type travel(l)ing weighbatcher
Hänge|balken(träger) *m* = suspension beam
~band *n* = flat suspension rod
~baugerüst *n* → Hängerüstung *f*
~brücke *f* = suspension bridge
~brücke *f* mit Versteifungsträger = stiffened suspension bridge
~bühne *f* [*Gleitschalungsbau*] = suspended platform
~bühne *f* [*Theater*] = suspension stage
~bunker *m* = suspension bunker, ~ bin, suspended ~
~dach *n* = suspended roof, hung ~
~decke *f*, abgehängte Decke = false ceiling, drop ~, counter ~, hung ~, suspended ~ [*A ceiling which is built with a gap between it and the floor above*]
~eisen *n*, Hänger *m* [*Hängedecke*] = hanger
~feld *n*, Hängeöffnung *f* [*Brücke*] = suspended span
~gefäßwaage *f*, Hängebehälterwaage = suspension hopper scale, ~ bucket ~
~gerüst *n* → Hängerüstung *f*
~gletscher *m*, Gehängegletscher = hanging glacier, tongue-shaped ~
~gurt(ung) *m*, *(f)* = suspension chord, ~ boom, ~ flange
Hang|einschnitt *m* = side-hill cut

Hang-(Eisen)Bahndamm — Hangstraße 508

~-(Eisen)Bahndamm *m* = railway embankment on sloping ground (Brit.); railroad ~ ~ ~ ~ (US)
Hänge|kette *f* = hanging chain
~**kettenaufgeber** *m*, Hängekettenbeschicker, Hängekettenspeiser = chain curtain feeder
~**klaue** *f* = suspension claw
~**konsole** *f* = hanging bracket
~**konstruktion** *f* = suspension system
~**kran** *m*, Deckenkran = suspension crane
~**kuppel** *f* → Kugelgewölbe *n*
~**last** *f* = hanging load
~**-Läuferrute** *f*, Hänge-Mäkler *m*, Hänge-Laufrute = hanging leader
~**-Laufgang** *m* = catwalk [*A suspended gangway*]
~**leiter** *f* = suspension ladder
~**leuchte** *f* = pendant light fitting
~**mäkler** *m*, Hänge-Läuferrute *f*, Hänge-Laufrute = hanging leader
Hangendes *n*, Hangendschicht *f* (⚒) = roof
hängendes Blechdach *n*, Blechhängedach = plate suspension roof
hängendes Flöz *n* = overlying seam
hängendes Ventil *n* = overhead valve, valve in head
Hangend|geber *m* (⚒) = roof strain indicator
~**schicht** *f*, Hangendes *n* (⚒) = roof
Hangentnahme *f* = hillside borrow
Hänge|öffnung *f*, Hängefeld *n* [*Brücke*] = suspended span
~**pfosten** *m*, Hängesäule *f* [*Bogenbrücke*] = suspension post, suspender
Hänger *m*, Hängeeisen *n* [*Hängedecke*] = hanger
Hänge|rinne *f* [*DIN 18460*] = hanging gutter
~**rüstung** *f*, Hänge(bau)gerüst *n* = boat scaffold(ing), (travel(l)ing) cradle
~**säule** *f* = 1. king post, broach ~; joggle ~ (US) [*of a king post truss*]; 2. queen post [*of a queen post truss*]; 3. truss post [*is either of the two*]

~**säule** *f*, Hängepfosten *m* [*Bogenbrücke*] = suspender, suspension post
~**schale** *f* = suspended shell
~**schalung** *f* = suspended formwork, ~ shuttering
~**schiene** *f* = suspension rail, overhead ~
~**seil** *n*, Hängekabel *n* = suspender cable, ~ rope, suspension ~
~**sieb** *n* = suspended screen, hung type ~
~**sprengwerk** *n*, zusammengesetztes Hängewerk, vereinigtes Hänge- und Sprengwerk = composite truss frame
~**stange** *f* = suspension rod, suspender
~**strebe** *f* = suspension stay, ~ strut
~**tal** *n* = hanging valley
~**überbau** *m* [*Brücke*] = suspended superstructure
~**waage *f* mit fester Aufhängung unter den Bunkertaschen** = weigh-batcher permanently fixed under the bin discharge gates
~**werk** *n* = hanging truss
~**werkbrücke** *f* = hanging truss bridge
~**zange** *f* = hanging brace
Hang|fels *m*, Tal~, (Tal)Flankenfels = valley slope rock, ~ side ~
~**graben** *m* → Hangrinne *f*
~**graben** *m* = hillside ditch
~**kanal** *m* = canal cut on sloping ground
~**-Kippschar** *f* → Böschungshobel *m*
~**kriechen** *n* = hillside creep
~**rampe** *f* = hillside ramp
~**ries(e)lung** *f* → Hangberies(e)lung
~**rinne** *f*, Auffangrinne, Hanggraben *m*, Auffanggraben [*Straße*] = catchwater channel, intercepting ~
~**schutt** *m*, Gehängeschutt (Geol.) = talus material
~**sickerung** *f* = side-hill seepage
~**silo** *m* = hillside silo
~**stadt** *f* = hill town
~**steinbruch** *m* → → Steinbruch
~**straße** *f* = hillside road

Hangtag(es)wasser — Hartlöt(fluß)mittel

~tag(es)wasser n = hillside surface water
~terrasse f = graded terrace
Harfen|masche f, Schlitzmasche = harp mesh
~sieb n = harp(-type) screen
~(sieb)gewebe n = harp mesh cloth; ~ ~ fabric (US)
Harke f, Rechen m = rake
Harnisch m ~ Rutsch~
Harpune f zum Fassen von Seilenden = devil's pitch fork
Hartalabaster m → Alaungips
Hartasphalt m → Asphalten n
~ [*Mastix für Fahrflächen und Gehbahnen*] = mastic asphalt
Hartbelag m, Hartdecke f = hard pavement, ~ surfacing
Hartbeton m, ~mörtel m [*DIN 1100*] = hard-aggregate mortar
~(belag) m, Hartbetondecke f [*früher: Hartestrich*] [*DIN 1100*] = hard-aggregate concrete pavement, ~ ~ surfacing
~(fuß)boden m = hard-aggregate concrete floor(ing)
~mörtel m → Hartbeton
~platte f, Hartsteinplatte aus Beton, Hartgesteinplatte [*zusätzlich Hartbetonstoffe in der Deckschicht (1:1), je qm etwa 2,5 kp Hartstoff*] = hard-aggregate concrete slab
~stoff m → Hartbetonzuschlag(stoff)
~zuschlag(stoff) m, Hartbetonstoff, Härtezuschlag(stoff), Härtemittel n [*Fehlname: Betonhartstoff. DIN 1100*] = hard aggregate
Hart|blei n → Antimonblei
~braunstein m = braunite
~decke f → Hartbelag m
Härte f = hardness
~bestimmung f = determination of hardness
~grad m = degree of hardness, hardness degree
~kessel m, Autoklav m = autoclave
~klasse f = class of hardness, hardness class

~mittel n → Hartbetonzuschlag(stoff)
Härten n im Druckhärtekessel, Dampfhärtung f, Behandlung mit Hochdruckdampf = steam curing at high pressure, autoclaving
Härte|öl n, Vergüteöl = hardening oil, annealing ~
~probe f → Härteprüfung
~prüfmaschine f, Härteprüfer m, Härteprüfgerät n = hardness testing machine, ~ tester
~prüfung f, Härteprobe f, Härteversuch m = hardness test
~skala f nach Mohs = Mohs hardness scale
~versuch m → Härteprüfung
~zuschlag(stoff) m → Hartbetonzuschlag(stoff)
hartes Festgestein n, Hartgestein, Härtling m = hard rock
Hart|estrich m → Hartbeton(belag)
~formation f = hard formation
~formationskopf m = hard formation cutter head
hartgebrannt = hard-burnt
Hart|gestein n, hartes Festgestein, Härtling m = hard rock
~gesteinsplitt m = hard rock chip(ping)s
~glas n, vorgespanntes Glas, gehärtetes Glas = toughened glass, tempered ~, heat-strengthened ~
Hartguß m = chilled (cast) iron
~asphalt m = hand-laid stone-filled asphalt, hard mastic ~
~eisenschrot m = chilled cast iron shot
Hart|holzstift m = wood(en) brad
~kobalterz n, Hartkobaltkies m = skutterudite
Härtling m → Hartgestein
Hart|lot n, Schlaglot = brazing solder, hard ~, brazing alloy
~löten n = brazing
~löten n mit Silberlot = silver brazing, ~ soldering
~löt(fluß)mittel n = brazing flux

Hartlötlampe — Häufigkeitskurve

~lötlampe *f* = brazing lamp
~manganerz *n* = psilomelane
Hartmetall *n* = hard metal
~besatz *m*, Hartmetallauflage *f*, Hartmetallbestückung *f* = hard facing
~bohr(er)schneide *f* → Hartmetallschneide
~(bohr)krone *f* → Hartmetallmeißel
~-Gesteinsbohrer *m* [*besteht aus Hartmetallkrone und Bohrstange*] = (tungsten) carbide (tipped) drill
~-Kreuzschneide *f* = carbide cross bit cutting edge
~meißel *m*, Hartmetall(bohr)krone *f* = (tungsten) carbide (tipped drill) bit, drill bit with tungsten carbide insert (or tip)
~schneide *f*, Hartmetallbohr(er)schneide *f* = carbide (drill) bit cutting edge
Hart|mörtel *m* = hard mortar
~papier *n*, Kraftpapier = Kraft paper
~papierfolie *f*, Kraftpapierfolie = Kraft paper foil
~paraffin *n* = solid melting wax, high ~ ~
~putzgips *m* [*DIN 1168*] = hard plaster
~putzmörtel *m* = hard stuff
~salzkainit *m*, Halitkainit = thanite
Hartschaum|platte *f* **aus Polystyrol**, Polystyrol(-Hartschaum)platte = expanded polystyrene board
~stoffplatte *f* **aus Polystyrol**, Polystyrol-Hartschaumstoffplatte = expanded polystyrene felt covered panel
~stoffplatte *f* **aus Styropor**, Styropor-Hartschaumstoffplatte = Styropor expanded polystyrene felt covered panel
Hart|schicht *f* (Geol.) = hardpan
~schnee *m* = hard(-packed) snow
~schotter *m* = crushed hard rock, broken ~ ~
~spat *m* = andalusite
~splitt *m* = hard rock chip(ping)s
~stein *m* → Hartziegel *m*
~steinplatte *f* **aus Beton** → Hartbetonplatte

~tantalerz *n* = ildefonsite
Härtungsmittel *n* → Betonhärter *m*
(Hart) Zerkleinerung *f* = reduction, comminution
~ **im Durchlauf** [*Das aus der Zerkleinerungsmaschine ausgetragene Gut ist Fertiggut*] = open-circuit comminution, ~ reduction
~ **mit Umlauf** [*Das aus der Zerkleinerungsmaschine ausgetragene Gut wird in Umlauf- und Fertiggut klassiert*] = closed-circuit reduction, ~ comminution
(Hart) Zerkleinerungs|maschine *f* = comminuter
~stufe *f* = reduction step, comminution ~
Hartziegel *m*, Hartbrandziegel, Hart(brand)stein *m* [*Für Fundamente und im Wasserbau. Mindestwürfelfestigkeit nach DIN 105 = 250 kp/cm². Wasseraufnahme nicht über 8%*] = hard brick, well-burned ~
Harz *n*, Natur~ [*Aus Pflanzensäften stammend; Kohlenstoffverbindungen; erhärten durch Polymerisation; Grundstoff für Lacke und Firnisse*] = natural resin
~, Kunst~ [*Grundstoff für Preßmassen, Baubeschläge, Fenster, Türen usw.; Bindemittel für Bauplatten, Bodenbeläge usw.*] = synthetic resin
~ → Kolophonium *n*
~firnis *m* = spirit varnish
~lack *m* = oil varnish
Haspel *m*, *f* → Kanalwinde *f*
~ (⚒) = mining winch
~pfosten *m*, Lagerstütze *f* der Schlämmseiltrommel [*pennsylvanisches Seilbohren*] = knuckle post
Haspen und Krampe *f*, Überwurf *m* = hasp and staple
Hauben|dach *n* = capped roof
~futter *n* [*Ramme*] = dolly
Häufigkeit *f* = frequency
Häufigkeits|diagramm *n* = frequency diagram
~kurve *f*, Häufigkeitslinie *f* [*allgemein*] = frequency curve
~kurve *f* → Häufigkeitslinie *f*

~linie *f*, Häufigkeitskurve *f* = frequency curve [*A graphical representation of the frequency of occurrence of specific events; for example, if the number of rainstorms of certain designed magnitudes (1, 2, 3 in., etc.) that occurred in a ten-year period are plotted against those magnitudes, the resulting graph would constitute a frequency curve of rainfall for that locality and period. The event that that occurs most frequently is termed the "mode". When this coincides with the mean value the curve is symmetrical, but when it differs from the mean, the curve lacks symmetry and is termed a "skew frequency curve"*]

~linie *f*, Häufigkeitskurve *f* [*allgemein*] = frequency curve

~verteilung *f* = frequency distribution

Haufwerk *n*, Aufgabegut *n*, Siebgut, Einlaufgut [*Siebtechnik*] = material to be screened, (screen) feed, (screen) head

~ [*Steinbruch*] = rock pile

~, Ausbruch *m*, Schutter *m* = muck (pile), broken rock

~, gesprengtes Gestein *n*, geschossenes Gestein = blasted rock, muck

~ (Geol.) = heap of debris

~abförderung *f*, Abteufförderung = hoisting spoil, ~ muck

~lader *m*, Haufwerklademaschine *f* = muck loader

~porigkeit *f* = internal porosity of the aggregate particles

~probe *f* (⚒) = muck sample

Haupt *n*, Außenfläche *f*, Kopffläche, Kopfseite *f*, Stirn(fläche) *f* [*Die Sichtfläche der Steine beim Naturwerkstein-Mauerwerk*] = face

~ [*bei Schleusen und Docks*] = entrance

~ablaß *m* [*Talsperre*] = main outlet

~abwasserleitung *f*, (Haupt)Sammler *m*, Sammelleitung *f* = main sewer, trunk ~, intercepting ~, interceptor [*A sewer which receives one or more branch sewers as tributaries*]

~abzweig *m* = main branch

~achse *f* = main centre line (Brit.); ~ center ~ (US)

~angebot *n* → → Angebot

~anlage *f* = main installation, principal ~, ~ facility

~anschluß *m* = main connection

~ansicht *f* = main view

~arm *m* [*Fluß*] = main branch

~armierung *f* → Grundbewehrung

~armierungsstab *m*, Hauptbewehrungsstab = main reinforcement bar

~auslaufkanal *m* = main outlet channel

~bahn *f* = main line

~bahnhof *m* = central station, main ~

~bahn-Lok(omotive) *f* = main-line locomotive

~balken *m* [*eines doppelten Hängewerkes*] = main beam

~balken *m*, Zugbalken, Trambalken, Spannbalken [*eines einsäuligen Hängewerkes*] = tie beam, main ~

~band *n* [*Kreuzbandscheider*] = feed belt

~baustoff *m* → Bauhauptstoff

~bauwerk *n* = main structure, principal ~

~bauzeit *f* = principal construction time

~beton *m*, Kernbeton [*Hauptmasse des Talsperrenbetons im Gegensatz zum Vorsatzbeton*] = mass concrete

~bett *n*, Mutterbett = main bed [*of a river or stream*]

~-Bewässerungskanal *m* = main irrigation canal

~bewehrung *f* → Grundbewehrung

~bewehrungsstab *m*, Hauptarmierungsstab = main reinforcement bar

~binder *m* = principal truss

~bolzen *m* = master pin

~büchse *f* = master bushing

~büro *n* = main office

~deich *m*, Winterdeich, Banndeich = main dyke, ~ dike

~diagonale *f*, Hauptschräge *f* = principal diagonal

Hauptdrän — Hauptschild

~**drän** m = leader drain, main ~
~**druck** m, Setzdruck Periodendruck (⚒) = periodic weight
~**druckspannung** f = principal compression stress
~**eingang** n = main entrance
~**einlagen** fpl → Grundbewehrung f
~**einsatz** m [*Rotarytisch*] = master bushing
~**entwässerungsgraben** m = main drainage ditch
~**etage** f → Hauptgeschoß n
~**fahrgestell** n [*Flugzeug*] = main (landing) gear
~**feld** n, Hauptöffnung f [*Brücke*] = main span
~**fernverkehrsstraße** f → (Groß-)Verkehrsader f
~**fluß** m = main stream
~**gebäude** n = main building, principal ~
~**gebläse** n = main blower
~**gerinne** n, Hauptkanal m = main channel
~**geschäftsviertel** n = central business district
~**geschoß** n, Hauptetage f, Hauptstockwerk n = principal story (US); ~ storey
~**gesims** n = entablature [*The upper portion borne by the columns or pilasters, consisting of architrave, frieze and cornice*]
~**gleis** n = main track
~**graben** m [*Grabeneinstau*] = permanent supply ditch
~**hafen** m = main harbo(u)r, principal ~
~**hahn** m = main cock
~**halle** f = main hall
~**hangendes** n (⚒) = main roof
~**hydraulikpumpe** f = main hydraulic pump
~**kanal** m = main canal
~**kanal** m, Hauptgerinne n = main channel
~**kanal** m = main duct
~**kehle** f [*Dach*] = main valley
~**kettenglied** n [*Gleiskette*] = master track section, ~ ~ link

~**klappe** f, Flügel m nach Oberwasser, obere Tafel f, Oberklappe [*Dachwehr*] = upstream leaf, upper ~
~**kraft** f = main force
~**kühlwasserpumpe** f = main cooling water pump
~**kupplung** f = master clutch, flywheel ~
~**lager** n = main bearing
~**lagerplatz** m = main storage yard
~**landebahn** f = main landing runway
~**last** f = principal load
~**leistung** f [*Bauvertrag*] = main work
~**leitung** f = main line
~**leitungsrohr** n = mainline pipe
~**linie** f = main line
~**lüfter** m = main fan
~**magazin** n = general store
~**maß** n = main dimension
hauptmaßgebende Start- und Landebahn f = basic runway
Haupt|meßort n (⚒) = main measuring point
~**moment** n = principal moment
~**montage** f = main assembly
~**montagestraße** f = main assembly line
~**öffnung** f, Hauptfeld n [*Brücke*] = main span
~**ölpumpe** f = main oil pump
~**pfeiler** m = main pier
~**pumpwerk** n = main pumping station
~**rahmen** m = main frame
~**richtung** f = main direction
~**rohr** n [*unter der Straße*] = street main
~**rohr** n, Wasser~, Hauptwasserrohr = main [*A pipe for the general conveyance of water as distinct from the conveyance to individual premises*]
~**(rund)stab** m, Längs(rund)stab [*Betonstahlmatte*] = main (round) bar
(~)**Sammler** m → Hauptabwasserleitung f
~**schiffahrtrinne** f = main navigation channel
~**schild** n [*Autobahnausfahrt*] = advance direction sign

Hauptschräge — Hausbrand

~**schräge** *f*, Hauptdiagonale *f* = principal diagonal
~**schwelle** *f* [*bei einem Holzdrempel*] = principal beam, tie ~
~**spannung** *f* = principal stress
~**spannungsunterschied** *m* = difference of principal stress
~**sparren** *m* = principal (rafter)
~**speisegraben** *m*, Hauptzubringer *m* [*Schiffahrtkanal*] = main feeding ditch
~**speisesaal** *m* [*Hotel*] = main restaurant
~**stab** *m* → Hauptrundstab
~**stadtstraße** *f* = major street, ~ highway, arterial (street)
~**(stahl)einlagen** *fpl* → Grundbewehrung *f*
~**stockwerk** *n* → Hauptgeschoß *n*
~**stollen** *m* = main tunnel, ~ gallery
~**strang** *m* = main, header
~**straße** *f* = main road
~**straßenpaar** *n* = arterial pair
~**strecke** *f* = trunk route
~**strecken-Bahnhof** *m* = main-line railway station (Brit.); ~ railroad ~ (US)
Hauptstrom *m* [*Kraftstoff*] = full flow
~**filter** *m, n*, Vollstromfilter *m* = full flow filter
~**filterung** *f* = full flow filtering
~**leitung** *f* = common line
Haupt|stütze *f* = main column, ~ support
~**system** *n* = main system, principal ~
~**träger** *m* = main girder
~**trägheitsachse** *f* = principal axis of inertia
~**trägheitsmoment** *n* = principal moment of inertia
~**tragwerk** *n* = main load-bearing system
~**treppe** *f* = main staircase (Brit.); ~ stair (US)
~**turm** *m* → Donjon *m*
~**uhr** *f* = master clock
~**verkehrsstraße** *f* = major road, ~ highway

~**verkehrsstromlinie** *f* = major desire line
~**verkehrsstunden** *fpl*, Hauptverkehrszeit *f* = rush hours
~**verkehrsweg** *m* = itinerary
~**verteilungsleitung** *f* = main supply line
~**verwaltung** *f* [*einer Baufirma*] = executive offices, headquarters
~**wasserrohr** *n* → Hauptrohr
~**wasserstraße** *f* = main navigable water course, ~ waterway
~**welle** *f* = main shaft
~**werkstatt** *f* = central (work)shop
~**winde** *f* = main hoist
~**windrichtung** *f* = main wind direction
~**wohnlager** *n* = workers' main camp
~**zeichen** *n* [*Baustatik*] = main symbol
~**zubringer** *m*, Hauptspeisegraben *m* [*Schiffahrtkanal*] = main feeding ditch
~**zugbewehrung** *f*, Hauptzugarmierung, Hauptzug(stahl)einlagen *fpl* = main tensile reinforcement
~**zugspannung** *f* = principal tensile stress
~**zumeßdüse** *f* [*Vergaser*] = main jet
~**zylinder** *m* = master cylinder
Haus *n* → Gebäude *n*
~**abbruch** *m* → Gebäudeabbruch
~**abwasser** *n*, Gebäudeabwasser, Haushaltabwasser = soil-sewage, domestic sewage, house sewage, sanitary sewage [*The sewage from watercloset, slop sinks and urinals*]
~**anbau** *m* → (Gebäude)Anbau
Hausanschluß *m*, Gebäudeanschluß = branch to a building
~**kasten** *m*, Gebäudeanschlußkasten = branch box to a building
~**keller** *m*, Gebäudeanschlußkeller [*DIN 18012*] = cellar for branches
~**leitung** *f*, Gebäudeanschlußleitung = branch line to a building
~**raum** *m*, Gebäudeanschlußraum = room for branches to a building
Haus|brand *m* = fuel for household use

~brand *m*, Gebäudebrand = building fire
~briefkasten *m* = private letter-box
~dach *n* = building roof
~decke *f* → Gebäudedecke
~eigentümer *m* = property owner, landlord
~eingang *m*, Gebäudeeingang = building entrance
~entwässerung *f*, Gebäudeentwässerung = drainage of a building
Häuserfabrik *f*, Gebäudefabrik, Beton~ = factory for precast concrete buildings
Haus|fernsprechanlage *f* im Hausflur = porter system [*With such a system in the hall, tenants can speak to callers, order from tradesmen, deal with casual callers, etc.*]
~fernsprecher *m* = internal telephone
~filter *m, n*, Kleinfilter = domestic filter
~flur *m* = hall
~fundament *n* → Gebäudefundament
~gang *m* [*zwischen Hauseingangstür und Treppenhaus*] = hallway
~garage *f* = garage of a building
~garten *m* = back garden
~-Gründungskörper *m* → Gebäudefundament *n*
Haushalt|abwasser *n* → Hausabwasser
~chemikalie *f* = household chemical
~-Heizöl *n* = home heating oil
~herd *m* = domestic range
~müll *m* = house refuse, domestic ~
~warmwasser *n* = domestic hot water, household ~ ~
~wasser *n* = household water, domestic ~
Haus|installation *f* = system of sanitation
~isolierung *f*, Gebäudeisolierung = building insulation
~kern *m* → (Gebäude)Kern
~lärm *m*, Gebäudelärm = internal noise
~last *f*, Gebäudelast = building load
~laube *f*, Loggia *f* = loggia
~meisterwohnung *f* = porter's flat
~müll *m* = garbage, household refuse

~müllanfall *m* = volume of garbage, ~ ~ household refuse
~nummer *f*, Gebäudenummer = building number, street ~
~-Rohrleitung *f*, Gebäude-Rohrleitung = pipe run
~schornstein *m*, Gebäudeschornstein = building chimney
~schutz *m*, Gebäudeschutz = preservation of buildings
~schutzmittel *n*, Gebäudeschutzmittel = building preservative
~schwingung *f*, Gebäudeschwingung = building vibration
~setzung *f*, Gebäudesetzung = building settlement, settlement of building(s)
~technik *f* = sanitation of buildings
~teil *m, n* → Gebäudeteil
Haustein *m*, Werkstein [*ein vom Steinmetz bearbeiteter natürlicher Stein*] = ashlar, ashler, cut stone
~mauerwerk *n*, Werksteinmauerwerk = ashlar masonry, ashler ~, cut stone ~ [*Masonry of sawed, dressed, tooled, or quarry-face stone with proper bond*]
~mauerwerk *n* mit Bruchsteinhintermauerung, Werksteinmauerwerk ~ ~ = rubble ashlar (masonry), ~ ashler (~), ~ cut stone (~) [*An ashlar-faced wall, backed with rubble*]
~setzen *n*, Werksteinsetzen = ashlaring, ashlering [*Setting ashlars*]
~vorsatz *m*, Werksteinvorsatz = ashlar facing, ashler ~, cut stone ~ [*Sawed or dressed squared stones used in facing masonry walls*]
Haus|trennwand *f* → Gebäudetrennwand
~tür *f*, Gebäudetür = building door
~türtelefon *n* = front door telephone, intercom(munication system) [*The installation in houses to facilitate answering the door*]
~- und Grundstückentwässerung *f* = sanitation of buildings (and premises), sanitary engineering
~waschküche *f* = wash-house

~wasserversorgung *f* = water supply to buildings, ~ distribution ~ ~
~wirtschaftswasser *n* = water for domestic use
~zeile *f*, Gebäudezeile = row of buildings
Haworth-Verfahren *n*, Haworth-Methode *f* = bio-aeration, mechanical agitation [*A modification of the activated-sludge process in which the sewage and sludge are agitated and aerated by mechanical means, such as paddle wheels or turbines*]
HB, Brinell-Härte *f* = Brinell hardness
HC-Koks *m* = high carbon coke
HD-Öl *n* = heavy-duty oil
Hebe|arm *m* → Hubarm
~arme *mpl* → Hubarme
~baum *m* → Brechstange
~-Becherrad *n*, Hub-Becherrad = lifting bucket wheel
~bock *m*, Schraubenspindel *f*, (Schrauben)Winde *f*, Hubstempel *m* = screw jack
~bock *m*, (hydraulische) Winde *f*, Hydraulikwinde, Hydrowinde, Hydraulikheber *m*, (hydraulische) Presse *f*, Hydraulikpresse, Hydropresse, Hubstempel *m* = hydraulic jack
~bügel *m* für Pump(en)gestänge, Pump(en)gestängefahrstuhl *m* [*Tiefbohrtechnik*] = sucker rod elevator
~bühne *f* = lifting platform
~eisen *n* → Brechstange
~gabel *f*, Hubgabel, Lastgabel [*Gabelstapler*] = lift(ing) fork
hebegerecht [*Last zum Heben*] = machine-sized [*load for lifting*]
Hebe|geschwindigkeit *f*, Hubgeschwindigkeit = lifting speed
~gestell *n* → Hubarme *mpl*
~haken *m*, Hubhaken = lift(ing) hook
Hebel *m* für die Oberwagensperre [*Bagger*] = swing lock lever
~arm *m* = lever arm [*The arm of a bending moment, that is the bending moment divided by the force producing the moment*]
~bremse *f* = lever brake
Hebelei *f* = arrangement of levers
Hebel|konsole *f* = lever bracket
~probe *f* [*Korrosionsprüfung*] = lever-test bar
~säge *f* = lever saw
~steuerung *f* = lever control
~steuerung *f* für Aufzug mit Führer(begleitung) = car-switch control (Brit.); cage-switch ~ (US)
~wirkung *f* = leverage action
Heber|leitung *f* für Abwasser = true sewage siphon [*It carries the sewage flow above the hydraulic gradient*]
~rohr *n*, Flüssigkeitsheber *m*, Syphon *m* = siphon
~überfall *m*, Saugüberfall, Wasseregel *m*, Heberüberlauf *m* [*Als Heber ausgebildeter Überlauf zur Begrenzung der Schwankungen eines Wasserspiegels mit Einrichtungen zum selbsttätigen Anspringen und Abreißen. DIN 4048*] = siphon(ic) spillway
~überlauf *m* → Heberüberlauf *m*
~wehr *n* = siphon(ic) weir
Hebe|transportstapler *m* → Hubstapler
~tür *f* = lifting door [*moved by lifting appliance*]
~-U-Eisen *n* [*z.B. an einem Wechselbehälter*] = lifting channel
~vorrichtung *f* = lifting device
~werk *n* → Rotary-~
~werk *n* → Schiff-~
~werkspillrolle *f* [*Rotarybohren*] = draw work cat head
~werktrommel *f*, Windwerktrommel, Bohrwerktrommel, Rotary-~ = (rotary) draw works drum, hoisting ~
~zeug *n* = elevating plant, lifting appliance, hoisting machine, lifting tackle
~zeuggetriebe *n* = transmission for lifting gear
heb- und senkbarer Kastenausleger *m* [*TOURNACRANE*] = sliding boom on tilting track
Hebung *f* [*Bodenmechanik*] = heave, lift, creep

Hebung der Sohle — Heißbad

~ **der Sohle,** Quellen n ~ ~, ~ **des Liegenden (⚒)** = floor-lift, floor-heave, floor-creep

~ **des Liegenden** → ~ der Sohle

~ **durch Eislinsenbildung im Boden** → Frosthebung

Hebungsbruch m [*Bodenmechanik*] = failure by heave

Heckachse f, Hinterachse = rear axle

Heck-Anbau|gerät n, Heck-Arbeitsgerät, Heck-Zusatzvorrichtung f, Heck-Austauschgerät, Heck-Austauschwerkzeug n, Heck-Austauschvorrichtung, Heck-Umbaugerät, Heck-Umbauwerkzeug, Heck-Umbauvorrichtung, Heck-Arbeitswerkzeug, Heck-Arbeitsvorrichtung, Heck-Zusatzwerkzeug, Heck-Anbauvorrichtung, Heck-Zusatzgerät, Heck-Anbauwerkzeug = rear interchangeable equipment item, ~ attachment, ~ rig, ~ conversion equipment item

~**vorrichtung** f → Heck-Anbaugerät n

~**werkzeug** n → Heck-Anbaugerät

Heckantrieb m = rear drive

~**rad** n = rear-drive wheel

~**fahrzeug** n = rear-drive vehicle

~**gruppe** f, Hecktriebsatz m, Heckantrieb m = rear power unit, ~ drive (~), ~ driving gear

~**maschine** f = rear-drive machine

~-**(Motor)Straßenhobel** m = rear-drive grader

Heck-Arbeits|gerät n → Heck-Anbaugerät

~**vorrichtung** f → Heck-Anbaugerät n

~**werkzeug** n → Heck-Anbaugerät n

Heck-Austausch|gerät n → Heck-Anbaugerät

~**vorrichtung** f → Heck-Anbaugerät n

~**werkzeug** n → Heck-Anbaugerät n

Heck|bagger m = rear digging attachment

~**bremse** f = rear brake

Hecke f = hedge

Heck|greifer m = rear grab

~**kran** m = rear-mounted crane

~**lenkung** f = rear steer(ing)

~**löffel** m = rear (excavator) bucket

heckmontiert = rear mounted

Heck|motor m = rear engine

~**motor** [*Autokran*] = base engine, rear ~

~**rad** n [*Flugzeug*] = tail wheel

~**schild** n = rear blade

~**schubblock** m = rear push block

heckseitig montierter Hydraulikbagger m = rear-mounted hydraulic digger attachment

Heck|triebsatz m, Heckantriebgruppe f, Heckantrieb m = rear power unit, ~ drive (~), ~ driving gear

~-**Umbaugerät** n → Heck-Anbaugerät

~-**Umbauvorrichtung** f → Heck-Anbaugerät n

~-**Umbauwerkzeug** n → Heck-Anbaugerät n

~**winde** f = rear-end winch

~**zapfwelle** f = rear power take-off

~-**Zusatzgerät** n → Heck-Anbaugerät

~-**Zusatzvorrichtung** f → Heck-Anbaugerät n

~-**Zusatzwerkzeug** n → Heck-Anbaugerät n

Heften n [*provisorisches Verbinden von Teilen beim Zusammenbau von Konstruktionen durch Verdornen, Verschrauben, Nieten oder Verschweißen mit kurzen Raupen*] = tacking

Heft|niet m = tack rivet

~**nieten** n, Heftnietung f = tack riveting

~**schraube** f = tacking screw

~**schweißen** n, Heftschweißung f = tack welding

Heidetorf m = moor peat

Heil|bad n [*als Ort*] = watering place

~**kraft** f [*Heilquellenwasser*] = curative power

~- **und Pflegeanstalt** f = mental institution

Heißasphalt|beton m = hot asphaltic concrete

~**kies** m → Heißbitu(men)kies

~**sand** m → Heißbitu(men)sand

Heißbad n [*Zum Reinigen von Motorenguß- und -stahlteilen*] = hot solvent bath

Heißbadreinigung — Heizelement

~reinigung f [*von Motorenguß- und -stahlteilen*] = hot solvent tank cleaning
Heiß|bitumen n = "penetration-grade" (asphaltic) bitumen, refinery (asphaltic) bitumen of penetration-grade (Brit.); penetration-grade asphalt (US)
~bitu(men)kies m, Heißasphaltkies = hot asphalt-coated gravel (US); ~ asphalt bitumen-coated ~ (Brit.)
~bitumenlage f [*zur Dichtung*] = mopping of hot asphalt (US)
~bitu(men)sand m, Heißasphaltsand = hot asphalt-coated sand (US); ~ asphaltic bitumen-coated ~ (Brit.)
~bügeln n [*Schwarzdeckennaht*] = hot ironing
~dampfzylinderöl n = super-heated steam cylinder oil
~einbau m [*Straßenbau*] = hot laying
~einbaubelag m, Heißeinbaudecke f = hot-laid surfacing, ~ pavement; ~ paving (US)
~elevator m, Heißbecherwerk n = hot elevator
heiße Schweißraupe f = hot pass
heißer Sandasphalt m → dichter ~
Heiß|flurtrocknung f [*Ziegel*] = hot-floor drying [*brick*]
~gasanlage f = hot gas plant
~gasleitung f = hot gas main
~gasschamotterost m = refractory hot-gas grate
~gemisch n, Heißmischung f, Heißmischgut n = hot plant mix(ture)
~gutförderer m = conveyor for hot materials, conveyer ~ ~ ~
~gut(förder)gurt m, Heißgut(förder)band n = (conveyor) belt for hot materials, conveyer ~ ~ ~ ~
~lagerfett n = hot bearing grease
~lagerung f = hot storage
~laufen n = running hot
~leim m = hot glue, ~ adhesive
~luftbehandlung f [*Beton*] = hot air treatment
~luftkanal m = hot-air flue

~makadam m = hot-mix(ed) macadam
~mischanlage f → Walzasphaltmischanlage
~mischen n = hot(-plant) mixing
~mischung f, Heißmischung n, Heißgemisch n = hot plant mix(ture)
~mischverfahren n = hot-(-plant) mixing method
~-Planiermaschine f [*zur Erneuerung abgenutzter bituminöser Decken unter Verwendung des alten Deckenmaterials*] = heater planer
~schornstein m = high-temperature chimney, ~ stack
~spritzmaschine f = hot spraying machine
~-Spritzpistole f = hot spraying pistol, ~ ~ gun
~teer m = hot tar, TH
~teerung f = hot tarring
~walzenfett n = hot neck grease
Heißwasser n = hot water
~bad n [*Motorenreinigung*] = hot water rinsing bath
~behälter m = hot water tank
~erhitzer m → Heißwassergerät n
~gerät n, Heißwassererhitzer m, Boiler m, Heißwassererzeuger m = calorifier
~heizung f = hot water heating
~rücklauf m = hot water return
~speicher m = cylinder, storage calorifier, storage-type domestic water heater
~vorlauf m = hot water flow
Heißwind|leitung f = hot-blast main
~schieber m = hot blast slide valve
Heitholz n → Kappe (✥)
Heizanlage f = heating system
heizbare Mulde f [*Erdbauwagen*] = heated body
Heiz|batterie f = heating battery [*The heating surface of a calorifier*]
~bohle f = hot screed
~element n = heating element [*That part of an electric heater which consists of a wire which is heated by an electric current*]

Heizer *m*, Kesselbediener *m*, Kesselwärter *m* = fireman, stoker
~stand *m*, Kesselwärterstand, Kesselbedienerstand = fireman's platform, stoker's ~
Heiz|fläche *f* = heating surface
~gas *n* = heating gas
~gehäuse *n* = heater housing, ~ casing
~gerät *n* = heating appliance, ~ device, heater
~gitter *n* = heating grid
~kabel *n* = heating cable
~kabel *n* **einer Elektro(fuß)bodenheizung** = floor-warming cable
~kammer *f*, Feuer(ungs)raum *m*, Feuerkammer, Feuerbüchse *f* [*Dampfkessel*] = fire box
~kanal *m* = heating duct
~keller *m*, Heizungskeller = heating basement, ~ cellar
~kessel *m* = boiler
~kessel *m* → Kocher *m*
Heizkörper *m* [*Das elektrische Bauelement, z.B. eines Bügeleisens, in dem durch den elektrischen Strom Wärme erzeugt wird*] = heating element
~ [*zur Raumerwärmung dienende Heizfläche einer Heizungsanlage, z.B. Radiator, Konvektor usw.*] = heating unit
~, Radiator *m* = radiator [*The heating industry has accepted the definition of the term "radiator" as a heating unit exposed to view within a room or space heated, although such units transfer heat by convection as well as radiation*]
~-Absperrventil *n*, Radiator-Absperrventil = radiator shut-off valve
~nische *f* = heating unit niche
~-Regulierventil *n*, Heizkörper-Regelventil, Radiator-Regulierventil, Radiator-Regelventil = radiator control valve
~ventil *n*, Radiatorventil = radiator valve
~verkleidung *f* = heating unit guard
Heiz|kreis *m* = heating circuit
~leistung *f* = heating output
~leitung *f* = heating line
Heizmantel *m* = heating mantle, ~ jacket
~rohr *n* = jacket(ed) pipe, mantle(d) ~
Heiz|maschine *f* = heating machine
~öl *n* = fuel oil, heating ~
(Heiz)Öl|behälter *m* → (Heiz)Öl-(lager)tank *m*
(~)lagerraum *m* = (fuel) oil store, heating ~
~(lager)tank *m*, (Heiz)Ölbehälter *m* = (fuel) oil (storage) tank, heating ~ (~)
~pumpstation *f* = fuel oil pumping station, heating ~ ~
~-Seitenturm *m* = fuel oil stripper, heating ~ ~
~versorgungsanlage *f* = (fuel) oil supply system, heating ~ ~ ~
~-Zentralheizung *f* = (fuel) oil-fired central heating
Heiz|platte *f* = heating plate
~röhrenkessel *m*, Heizrohrkessel, Feuerröhrenkessel = fire (or smoke) tube boiler, multitubular ~, multitube ~
~röhrenplatte *f*, Röhrenheizplatte = radiant panel
~saison *f* = heating season
~schalter *m* = heat switch
~schalung *f* = heated formwork
~schlange *f*, Erhitzerschlange, Wärmeschlange = heating coil
~startschalter *m* = heat-start switch
~system *n*, Heizungssystem = heating system
Heizung *f* = heating
Heizungs|anlage *f* = heating system
~dampf *m* = heating steam
~firma *f* = heating firm
~industrie *f* = heating industry
~ingenieur *m* = heating engineer
~keller *m*, Heizkeller = heating basement, ~ cellar
~pumpe *f* = heating pump
~system *n*, Heizsystem = heating system
~techniker *m* = heating technician

Heiz|wasser n = heating water
~wert m, unterer ~, Hu = lower calorific value, L.C.V.
~zentrale f = central heating installation
Helfer m = helper
Helgen m, **Helling** m, f, **Helge** f = building slip(way)
Heliotrop m [*Lichtsignal für Triangulationsmessungen. Es leitet mit seinem Spiegel aufgefangene Sonnenstrahlen durch ein Fadenkreuz zum Theodolitstandpunkt*] = heliograph
hellbrauner Glimmer m → Phlogopit
helle (Farb)Tönung f = light shade
helles Aussehen n = bright appearance
hell|farbiges Bitumen n → Albino-Bitumen
~gestrichen = light-painted
~graues Gußeisen n = mottled cast iron
~hörig = not sound proof
Helligkeit f = brightness
Helligkeitskontrast m = brightness contrast
Helling f, m → Helgen m
~winde f, Helgenwinde = building slip(way) winch
Helm m, Sand~, Sandhafer m, Strandhafer = sand-sedge
~stange f, Kaiserstiel m [*Dachkonstruktion*] = broach post
hemipelagische Meeresablagerung f = continental deposit, hemipelagic ~
(Hemm)Schuh m, Bremsschuh, Bremskeil m = brake shoe, skidpan, wheel block, chock
Hemmstoff m = inhibitor
Heraushebung f **eines Flußgebietes** = uplift of a drainage area
herauspräparieren (Geol.) = to etch into relief
Heraussickern n = outward seepage, effluent ~ [*The water emerges from the ground along some rather extensive line or surface*]
(her)ausziehen, ziehen [*Spundbohle; Pfahl*] = to extract, to pull out

~ des Bohrzeuges n, aufziehen ~ ~, **~ der Bohrgarnitur** f [*pennsylvanisches Seilbohren*] = to hoist the drilling tools out, to pull ~ ~ ~
Herberge f = hostel
Herbst|bewässerung f = fall irrigation
~müll m = autumn refuse
~niederschlag m = fall precipitation
Herd|mauer f, Abschlußwand f, Fußmauer, (Ab)Dichtungsmauer, (Ab-)Dichtungssporn m, Trennmauer, Sporn, Verherdung f [*Talsperre*] = (toe) cut-off wall
~schmelzofen m = hearth melting furnace
~wagenofen m = bogie hearth furnace
Hereinbrechen n, Bruch m, Zusammenbrechen, (Zu)Bruchwerfen [*des Hangenden im Bruchbau*] (⚒) = caving in
(herein)gewinnen [*nutzbares Mineral oder Gestein aus seinem natürlichen Verband lösen*] = to get, to win
(Herein)Gewinnung f [*die Lösung des nutzbaren Gesteins oder Minerals aus seinem natürlichen Verband*] = getting, winning
~ der Kohle, Kohle(n)gewinnung = coal getting, ~ winning
~ des Erzes, Erzgewinnung = ore getting, ~ winning
~ über Tage, übertägige (Herein-)Gewinnung = open cast getting, ~ ~ winning
~ und Förderung = getting and extraction, winning ~ ~
~ unter Tage, untertägige (Herein-)Gewinnung = underground getting, ~ winning
hereinschieben, sich ~ [*Gestein*] (⚒) = to flow (into)
herkömmlicher Bagger m → Mehrzweckbagger
~ Werkstoff m = traditional material
herstellen = to fabricate, to make
~, erzeugen, aufbereiten [*Beton; Mörtel; Schwarzdeckenmischgut*] = to prepare, to fabricate [*sometimes shortly, but incorrectly, called "to mix"*]

herstellen einer Gebäudewanne — Hilfsseil

~ **einer Gebäudewanne** = to tank

~ **in offener Baugrube** = to build by the open cut method, ~ ~ ~ ~ cut-and-cover ~

Hersteller m, Erzeuger m, Fabrikant m = fabricator, manufacturer, maker

~**zeichen** n = maker's mark

Herstellung f, Erzeugung, Aufbereitung [*Beton; Mörtel; Schwarzdeckenmischgut*] = fabrication, preparation [*sometimes shortly, but incorrectly, called "mixing"*]

~ **benachbarter Plattenfelder in zeitlichem Abstand**, Wechselfeldeinbau m [*Betondeckenbau*] = alternate bay method, ~ (concrete) bay construction

~ **von architektonischen Fertigteilen** = architectural modelling

Herstellungs|firma f = manufacturing firm

~**platz** m, Betonier(ungs)platz, Werkplatz [*für Fertigteile*] = casting yard

~**preis** m = cost price, actual cost, product cost

~**werk** n = manufacturing plant

herumlegen, umschlagen [*Drahtseil*] = to twist

~ [*eine Schlinge um eine Ladung*] = to lap [*a sling around a load*]

herunterbringen, (ab)teufen, niederbringen [*eine Bohrung*] = to sink a well, ~ bore ~ ~, to drill ~ ~, ~ put down ~ ~

~ → (ab)teufen

Herunter|bringen n → (Ab)Teufen

~**lassen** n, Absenken [*Ausleger*] = lowering

herunterschalten, zurückschalten = to shift down

hervorstehend = proud [*of the surface*]

Herz|brett n, Kernbrett = heart plank

~**schnitt** m [*Destillatfraktion eines Erdölerzeugnisses mit engen Siedegrenzen*] = heart cut

heterogenes Gebirge n [*Tunnel-, Stollen- und Schachtbau*] = heterogenous ground, ~ formation, ~ strata

Heultonne f, Heulboje f = whistling buoy

Hexakalziumaluminat n = hexacalciumaluminate

HHQ [*größte überhaupt bekannte Abflußmenge*] = flood intensity

HHW → höchster Hochwasserstand m

hiesiges Ufer n, diesseitiges ~ = home bank

Hilfs|antrieb m = auxiliary drive

~**arbeiten** fpl = auxiliary work

~**arbeiter** m = unskilled worker, labo(u)rer

~**armierung** f → Hilfsbewehrung

~**ausleger** m = jib boom, boom extension (US); jib extension (Brit.)

~**baustoff** m → Bauhilfsstoff

~**bauwerk** n [*z. B. Fang(e)damm*] = temporary structure

~**bewehrung** f, Hilfsarmierung, Hilfs(stahl)einlagen fpl, Hilfseisen npl = subsidiary reinforcement

~**einlagen** fpl → Hilfsbewehrung f

~**eisen** npl → Hilfsbewehrung f

~**flugplatz** m → Hilfslandeplatz

~**gerüst** n [*Brückenbau*] → Montagegerüst

~**gerüst** n [*zum Rammen*] = temporary staging

~**gewölbe** n [*Tunnel- und Stollenbau*] = auxiliary vault

~**kesselhaus** n = auxiliary boiler house

~**kolben** m [*Verfahren Freyssinet*] = auxiliary ram

~**landeplatz** m, Hilfsflugplatz, Nebenlandeplatz, Nebenflugplatz = auxiliary airfield, satellite ~, ~ landing ground

~**polier** m = charge hand

~**presse** f [*Druckpresse bei der Spannpresse mit Doppelwirkung*] = auxiliary prestressing jack

~**schacht** m = auxiliary shaft

~**schachtmeister** m = sub-ganger

~**schiff** n **mit Sonderwerkstätten und Vorrichtungen zum Auswechseln von Brennstoffelementen** [*dieses Hilfsschiff gehört zu einem Atomschiff*] = barge for "hot" cargoes

~**seil** n, Reserveseil [*Seilschwebebahn*] = emergency rope, ~ cable

Hilfsspill — Hinterkippvorrichtung

~spill n [*Tiefbohranlage*] = make-up cathead
~(stahl)einlagen *fpl* → Hilfsbewehrung *f*
~stempel *m* (✻) = catch prop
~stollen *m* für den Bau → Abstiegstollen
~stützbock *m* [*Tieflöffelbagger*] = jack boom
~stütze *f*, (ausfahrbare) (Seiten-) Stütze [*Kran; Bagger; Schaufellader*] = stabiliser, outrigger
Hindernis|befeuerung *f* = obstruction lighting, warning ~
~feuer *n* = (aviation) obstruction light, aircraft warning ~
~freiheit *f* = obstruction clearance
Hinsetzen *n*, Absetzen [*z. B. Fertigteile beim Montagebau*] = placing
Hinterachse *f*, Heckachse *f* = rear axle
~ mit Planetengetriebe in den Hinterradnaben = planetary rear axle
Hinter|achsantrieb *m* = final drive, rear axle ~
~achswelle *f* = rear axle shaft
~arm *m* → Ballastarm
~ausgang *m* = rear exit
~balkon *m* = rear balcony
~böschung *f* = back slope
hintere Kurbeldichtung *f* = crankshaft rear seal
hinterer Kämpferdruck *m* rückwärtiger ~ (✻) = rear abutment pressure back ~ ~
~ Kraftanschluß-Stutzen *m* für Anbaugeräte, Zapfwellen-Anschluß *m* ~ ~ [*Schlepper*] = rear power take-off
~ Lenkwinkelantrieb *m* = rear bellcrank steering
~ Planetenträger *m* = rear carrier
~ Tragseilmast *m* Gegenturm *m* [*Schlaffseil-Kabelbagger*] = tail anchor ~ tower
hinteres Kettengehänge *n* hinterer Kettenzaun *m* [*Schrapp(er)gefäß*] = rear bridle chains
Hinterfront *f* → Hoffront
hinterfüllen = to backfill around
Hinter|füllung *f* = structural fill, backfill
(~)Füll(ungs)beton *m*, Kernbeton, Unterbeton [*einer zweischichtigen Platte in der Betonsteinindustrie*] = core concrete
(~)Füll(ungs)beton *m* = backfill concrete
~füllungsdruck *m* = backfill pressure
~füllungskeil *m* = backfill wedge
~füllungssand *m* = backfill sand
~gebäude *n*, Hinterhaus *n* = rear building
~graben *m* → Auffanggraben
~grund *m* = background
~hof *m* = backyard, rear yard
~kante *f* = trailling edge [*of the finishing screed*]
~kippentleerung *f* = end-dumping
Hinterkipper *m*, Rückwärtskipper = rear-dump truck (US); ~ lorry (Brit.)
~ → Hinterkipp(er)anhänger
~anhänger *m*, Anhänge-Hinterkipper, (Schlepp-)Hinterkipper, Rückwärtskipper(-Anhänger), Hinterkippanhänger [*Erdbaufahrzeug*] = rear-dump wagon, end-dump ~
~-Aufl(i)eger *m* → Halbanhänger-Rückwärtskipper
~-Halbanhänger *m* → Halbanhänger-Rückwärtskipper
~-Last(kraft)wagen *m*, Hinter-Kipper(-LKW) = end-tip(ping) (motor) lorry (Brit.); end-dump truck, rear-dump truck (US)
Hinter|kippfahrzeug *n* = end-tipping vehicle
~kippmulde *f*, Hinterkippwanne *f*, Rückwärtskippmulde, Rückwärtskippwanne = rear dump body
~kippung *f*, Hinterkippen *n*, Rückwärtskippung, Rückwärtskippen = rear dumping, end ~
~kippvorrichtung *f* mit Doppelpressenaggregat [*Erdbauwagen*] = twin two-stage telescopic double-acting rams, ~ hydraulic two-stage telescopic rams allowing power return of the body

Hintermauerung — Hochbau-Baustelleneinrichtung

~mauerung *f*, Hintermauerwerk *n* = backing (masonry), back-up ~

~mauer(ungs)block *m* = backing block, back-up ~

~mauer(ungs)material *n* = backing [*Is the material between the facing and the back*]

~mauer(ungs)ziegel *m*, Hintermauer(ungs)stein *m* = backing brick, back-up ~, common stock ~

hinterpacken (⚒) = to backfill [*behind arches*]

Hinterrad *n* = rear wheel

~antrieb *m* = rear wheel drive

~-Drehmoment *n* = rear wheel torque

~federung *f* = rear wheel suspension

~lenkung *f* = rear wheel steer(ing)

Hinter|rahmen *m* [*Baumaschine*] = rear frame

~seite *f* → Hoffront *f*

~spritzkessel *m* = bougie (Brit.); (pressure) grouting pan, pressure pot

~tür *f* = rear door

~walze *f* = rear roll

hin- und hergehende Bewegung *f*, Rüttelbewegung = reciprocating motion

~ Deckenschlußbohle *f* [*Betondeckenfertiger*] = oscillating plate

Hinweiszeichen *n* → → (Straßen)Verkehrszeichen

historische Geologie *f* → Formations- und Schichtenkunde *f*

historisches Bauwerk *n*, Baudenkmal *n* = monument

Hitzebeständigkeit *f* = heat resistance

hitzefest, hitzebeständig = heat-resistant

Hitzegrad *m* = degree of heat

H-Linie *f*, Einflußlinie des Bogenschubs = line of horizontal stresses

Hobelblatt *n* → Hobelmesser *n*

Hobelmesser *n*, (Hobel)Schar *f*, Hobelblatt *n*, Planierschar, Planierschaufel *f*, (Schäl)Messer, Schälblatt *n* = (planing) blade, grader ~

~ [*Kohlenhobel*] = plough blade

~ an der Zugstange eines Schleppers angebracht = drawbar grader, rear mounted grader attachment for tractors

~ mit Evolventenprofil = roll-away blade

~einstellung *f* [*Straßenhobel*] = blade setting

(Hobel)Schar *f* → Hobelmesser *n*

Hobel|schnitt *m* (⚒) = web

~späneteer *m* = wood shavings tar

Hoch|absetzer *m*, Dammschütter-Bandabsetzer, Hochschütter, Haldenschütter-Bandabsetzer, Haldenschüttungs-Förderband *n*, Deponieförderband = stacker for building up fills

~altar *m* = high altar

~ausleger *m* → Hochbauausleger

~autobahn *f* = elevated motorway, high-level ~

~bagger *m* → Löffelhochbagger

~bagger *m* [*baggert oberhalb seiner Standfläche*] = exavator digging above grade

~baggerung *f*, Hochschnitt *m* [*Löffelbagger*] = excavation of material which has a face

~baggerung *f*, Hochschnitt *m* [*Eimerkettentrockenbagger*] = digging down, ~ in height, excavating above rail or ground level

~bahn *f* [*Bezeichnung für Viaduktstrecken der Stadtschnellbahnen im Gegensatz zur Untergrundbahn*] = overhead railway (Brit.); ~ railroad, El, L (US); elevated ~

Hochbau *m* = building (design and) construction

~arbeiten *fpl* = building operations, ~ work

~arbeiter *m* = building worker

~-Arbeitskräfte *fpl* = building labo(u)r

~ausleger *m*, Hochausleger = boom for building construction (US); jib ~ ~ ~ (Brit.)

~-Baustelle *f* = building construction site

~-Baustelleneinrichtung *f* = building construction site installation

522

Hochbaudeckenteil — Hochdruckkessel

~deckenteil *m, n*, Gebäudedeckenteil, Geschoßdeckenteil, Etagendeckenteil, Stockwerkdeckenteil = floor unit

~element *n* → Hochbau-Fertigteil

~-Fertigteil *m, n*, Hochbauelement *n* = precast component for building construction, building component

~-Fördermittel *n* = transporting means for building construction

~industrie *f* = building industry

~ingenieur *m* = building engineer

~-Isolierungen *fpl* = waterproofing and dampproofing of buildings

~kletterkran *m* → Kletter(-Turmdreh)kran

~klinker *m* = clinker brick for building construction

~kosten *f* = building cost

~kran *m*, Baukran = (rotating) building crane

~raupenkran *m* [*dieser Ausdruck ist besser durch „Hochbaugleiskettenkran" zu ersetzen*] = track(-laying) type long boom crane (US); ~ ~ ~ jib ~ (Brit.)

~stoff *m* = building construction material

~techniker *m* = building technician

Hochbauten *f* = rising structures

Hochbau|vertrag *m* = building contract

~vorgang *m* = building operation

~winde *f* = building winch

~zeichnen *n* = building drawing

Hoch|becken *n*, Oberbecken [*Pumpspeicherwerk*] = upper pool, ~ reservoir, ~ (storage) basin

~behälter *m*, Wasser-~, Wasserturm *m* = elevated reservoir, water tower [*Is a service reservoir elevated above ground level in order to perform its function*]

~behälter *m* = elevated tank

~bevorratung *f* [*Wasserversorgung*] = elevated storage

hochbildsam, hochplastisch = highly plastic

Hoch|bocken *n* = jacking up

~bohrung *f* (⚒) = rising borehole

(~)Bordstein *m*, Bordschwelle *f*, Schrammbord *n*, Hochbord = kerb (Brit.); curb

~brause *f*, Hochdusche *f* = overhead shower

~bruch *m*, Aufbruch, Überbruch, überbrochener Schacht *m* [*Blindschacht von unten nach oben hergestellt*] = raise

~brücke *f* = high-level bridge

~bunker *m* = high-level storage bin

Hochdruck *m* = high pressure

~anlage *f*, Hochdruck(wasser)kraftwerk *n* = high-head development

~-Asphaltplatte *f*, Hochdruck-Asphaltfliese *f* = asphalt tile made under high pressure

~-Axialkolbenpumpe *f* = high-pressure axial piston pump

~chemie *f* = high-pressure chemistry

~dampf *m* = high-pressure steam

~dampfheizung *f* = high-pressure steam heating

~dampfleitung *f* = high-pressure steam line

~drillingspumpe *f* = high-pressure three-throw pump, ~ triplex ~

~-Einspritzpumpe *f* = high pressure injection pump

~fett *n* = high-pressure lubricating grease

~fettschmierung *f* = high-pressure grease lubrication

~-Gas(rohr)leitung *f* = high-pressure gas pipe line

~(gas)verfahren *n* zur Förderung von Erdöl durch Fluten mit mischbaren Phasen = high-pressure gas injection method, miscible displacement ~, miscible drive ~

~-Heißdampf *m* = high-pressure superheated steam

~-Heißwasserheizung *f* = H.P.H.W. system, high-pressure hot water (heating) (~ ~)

~-Hydraulik-Handpumpe *f* = high-pressure hydraulic hand pump

~-Kaltwasser *n* = high-pressure cold water

~kessel *m* = high-pressure boiler

Hochdruckklimaanlage — Hochfrequenz-Vibrationsverdichter

~klimaanlage *f* = high-pressure air-conditioning system
~kompressor *m*, Hochdruck(luft)-verdichter *m* = high-pressure (air) compressor
~-Kraftwerk *n* = high-head (power) plant, ~ (~) station
~lüfter *m*, Hochdruckventilator *m* = high-pressure fan
~(luft)reifen *m* = high-pressure tire (US); ~ tyre (Brit.)
~öl *n* → Hypoidöl
~ölbrenner *m* = high-pressure oil burner
~pumpe *f* = high-pressure pump
~-Quecksilberdampflampe *f*, Quecksilberdampf-Hochdrucklampe *f* = high-pressure mercury (discharge) lamp
~-Reaktionskammer *f* [*Erdölwerk*] = soaker
~regner *m* = high-pressure sprinkler
~reifen *m* → Hochdruckluftreifen
~schlauch *m* = high-pressure hose
~-Schlauchleitung *f* = high-pressure hose line
~schmiermittel *n* = extreme pressure lubricant, EPL
~schmierung *f* = extreme pressure lubrication
~-Schneckenpumpe *f* = high-pressure screw pump
~-Spritzausrüstung *f* = high-pressure spraying set-up
~turbine *f* = high-pressure turbine
~verdichter *m* → Hochdruckkompressor *m*
~verfahren *n* → Hochdruckgasverfahren
~(wasser)kraftanlage *f*, = high-head development
~wasserleitung *f* = high-pressure main
~wasserstrahl *m* = high pressure water jet
~-Zahnradpumpe *f* = high-pressure gear (type) pump
Hoch|dusche *f*, Hochbrause *f* = overhead shower
~ebene *f*, Hochplateau *n* = high plain, plateau

~einbau *m* [*Straßenbau*] = construction without removal of existing subgrade
hoch|empfindlich = highly sensitive
~fahren [*Pumpe; Turbine*] = to run up
Hochfahren *n* eines Siebes = starting up a screen
hochfest = high-strength
hochfeuerbeständiger Bauteil *m*, hochfeuerbeständiges ~ *n* = highly fire-resisting component
hochfeuerfest → feuerfest
Hoch|flutebene *f* → Auenebene
~fördergerät *n* = elevating unit
~förderung *f* von Beton, Betonhochförderung = upward transport of concrete
Hochfrequenz|fertiger *m*, Hochfrequenzschwingverdichter *m* = high-frequency finisher, ~ finishing machine
~induktionsofen *m* = high-frequency induction furnace
~-Innen(ein)rütt(e)lung *f* = high-frequency internal vibration
~-Innenrüttler *m*, Hochfrequenz-Innenvibrator *m* = high-frequency internal vibrator
~motor *m* = high-frequency motor
~-Rüttelverdichter *m* → Hochfrequenz-Schwing(ungs)verdichter
~rüttler *m*, Hochfrequenzvibrator *m* = high-frequency vibrator
~schweißen *n* = radio-frequency welding (Brit.); heatronic ~ (US)
~-Schwing(ungs)verdichter *m*, Hochfrequenz-Vibrationsverdichter, Hochfrequenz-Rüttelverdichter, Hochfrequenz-Vibratorverdichter, Hochfrequenz-Vibrierverdichter = high-frequency vibratory compactor, ~ vibrating ~
~schwingverdichter *m* → Hochfrequenzfertiger
~-Vibrationstechnik *f* = high-frequency vibration technique, ~ ~ technic
~-Vibrationsverdichter *m* → Hochfrequenz-Schwing(ungs)verdichter

~vibrator m, Hochfrequenzrüttler m = high-frequency vibrator
~werkzeug n = high-frequency tool
Hochgenauigkeits-Zahnrad n = high precision gear
Hochglanz m = high gloss
Hochhaus n, Turmhaus = tall building, tower ~, ~ block, high rise ~
~-Baustelle f = tall building site, tower ~ ~
~kletterkran m → Kletter(-Turmdreh)kran
~kran m = rotary tower crane
hoch|hydraulischer Kalk m = eminently hydraulic lime
~kant = on edge
~kanten, aufkanten = to fold up
Hoch|kipper m → Hubkipper
~klettern n der Welle, Aufbäumen ~ ~ = swash, uprush
hochkompliziert = highly complicated
Hoch|lader m → Schaufellader
~ladeschaufel f → Schaufellader m
~lauf m = roller conveyor (or conveyer) and troughs for installation above floor level
Hochleistungs|bohrhammer m = high-capacity hammer drill
~brunnen m = high-capacity well
~feuer n = high-intensity light [aerodrome]
~filterung f = high-rate filtration
~-Intraflügelpumpe f = high-capacity intra vane pump
~-Laufwerk n = heavy duty undercarriage
~maschine f = high output machine
~mischanlage f = high-production central-mixing plant
~ofen m = high-production kiln
~-Schall-Siebmaschine f = high-capacity sonic screening machine
~-Spannrolle f [Gleiskette] = heavy duty idler
~-Tropfkörper m, hochbelasteter Tropfkörper = high-rate trickling filter, high-capacity ~ ~
~-Zahnradpumpe f = high-capacity gear pump

~-Zwangsmischer m = high-capacity pugmill (mixer)
hochliegender Pfahlrost m, Stelzenfundament n = elevated pile foundation grillage
Hochloch n = vertical hole, ~ perforation
~(ingenieurbau)klinker m = perforated clay engineering brick
~leichtziegel m = perforated lightweight brick
~ziegel m = perforated brick
Hochlöffel m, Bagger~, Hochräumer m = shovel bucket
~-Ausleger m = shovel bucket boom (US); ~ ~ jib (Brit.)
~ausrüstung f, Löffelhochausrüstung = shovel bucket attachment
~(bagger) m → Löffelhochbagger
~einrichtung f, Löffelhocheinrichtung = face shovel attachment, crowd shovel fitting
~-(Gleis)Kettenbagger m → Hochlöffel-Raupenbagger
~-Kettenbagger m → Hochlöffel-Raupenbagger
~-Raupenbagger m, Hochlöffel-(Gleis)Kettenbagger, Raupen-Hoch(löffel)bagger, (Gleis)Ketten-Hoch(löffel)bagger = crawler(-mounted) (face) shovel
~stellung f = face shovelling position
Hochmischer m = elevated mixer
Hochmoor n, Filz n = high moor
~torf m = highmoor peat
Hochofen m = blast furnace
~gas n, Gichtgas = top-gases, blast-furnace gas
~schaumschlacke f → aufgeblähte Hochofenschlacke
~schlacke f = (iron-ore) blast-furnace slag
(~)Schlackenbeton m = slag concrete
~schlacken-Pflasterstein m = blast furnace slag sett
~schlackensand m → granulierte Hochofenschlacke
~schlackenstein m → Hüttenstein
~zement m [abgekürzt HOZ; max. 85% Schlacke] = Hochofen cement

hochplastisch — Hochtemperatur-Technologie 526

hochplastisch, hochbildsam = highly plastic
Hoch|plateau n → Hochebene f
~puffen n **der Sohle** (✕) = boiling up of the floor
~punkt m [*Rohrleitung*] = summit, high point
hochquellfähig = high-swell
Hoch|räumer m → Hochlöffel m
~relief n = high relief
~rohrleitung f = overhead pipeline
~schaltbereich m = upshift range
hochschalten = to upshift
Hochschnitt m → Hochbaggerung f
~, Hochbaggerung f [*Eimerkettentrockenbagger*] = digging in height, ~ down, excavating above rail or ground level
Hochschütter m → Hochabsetzer
hochschwefelhaltige Ammonsulfoseife f **aus Schieferöl** = ichthyol
Hochsee|schiff n = seagoing vessel
~-Transport m **verflüssigten Erdgases** = ocean transportation of liquefied natural gas
hochsiedendes Leuchtöl n, hochraffiniertes aromatenarmes Solaröl = mineral seal oil
~ Lösungsbenzin n = heavy petroleum spirit
hochspannbarer Bolzen m, Hochspannungsbolzen = high strength steel bolt
Hochspannungs|bolzen m, hochspannbarer Bolzen = high strength steel bolt
~(frei)leitung f = high-voltage transmission line
~kabel n = high-voltage cable
~(-Überland)mast m = high-voltage transmission mast
hochstabil [*Bitumenemulsion*] = fully-stable
Höchst|achslast f = maximum axle load
~belastung f = safe loading
~belastung f, Belastungsspitze f, Maximalbelastung = maximum loading, peak of ~

~biegungsmoment n = maximum bending moment, max. b. m.
~drehmoment n = maximum torque
~drehzahl f = maximum governed speed
~druckverdichter m = hypercompressor
hochstegig [*Träger*] = deep webbed
höchster Hochwasserstand m, HHW, oberster Grenzwert der Wasserstände = highest recorded level
höchstes Drehmoment n = maximum torque
~ Hochwasser n, Hochwasserspitze f = flood peak
Höchst|gefälle n, Höchstlängsneigung f = maximum gradient
~geschwindigkeit f, Spitzengeschwindigkeit = maximum speed
~glanz m = super gloss, full ~
~längsneigung f, Höchstgefälle n = maximum gradient
Hochstraße f = overhead roadway, ~ highway, road on stilts, highway on stilts, elevated highway, elevated roadway; skyway (US)
Höchstschub m = maximum shear
~zone f = region of maximum shear
Höchst|spannung f = maximum stress
~verkehr m = peak traffic
~wert m [*Ein Wert bei der Übertragbarkeit von Kräften zwischen Reifen und Straße. Der Höchstwert wird oft „Haftwert" genannt, aber dies ist im Wortsinn nicht zutreffend; denn aus dem Abweichen der Kraft vom proportionalen Verlauf mit zunehmendem Schlupf muß man schon vorher auf Gleitungen „im ganzen" schließen*] = peak coefficient
~zugspannung f = maximum tensile stress
höchstzulässige Gradiente f = limiting (or ruling) gradient
Hochtemperatur|element n → Hochtemperaturzelle
~|-Technologie f = high temperature technology

~zelle f, Hochtemperaturelement n [*Brennstoffzelle*] = high temperature fuel cell
Hochufer n = high bank
~linie f = high bank line
Hoch- und Tiefbau m = building construction and civil engineering
hoch|verdrillte Schraube f = high-torqued bolt
~verschleißfest = highly wear-resistant
~viskos = high-viscosity
Hochwasser n = flood
~ der Springtide, Spring(tide)hochwasser = high water spring tide, ~ ~ (level) of ~ ~, H.W.S.T.
~abfluß m, ~menge f = flood flow, ~ discharge, maximum flow of a river, maximum flow of a stream, maximum discharge of a river, maximum discharge of a stream
~ablagerung f = flood plain deposit
~alarmsystem n, Hochwasserwarn(ungs)dienst m = flood warning system
~anstieg m = flood rise
~(auffang)becken n, Hochwasserspeicher m = flood basin, ~ storage ~, ~ pool, ~ storage reservoir, flood control ~
~(auffang)sperre f, Hochwasserschutztalsperre = flood (control) dam
~bett n Hochwasserschlauch m = flood bed, major ~
~bogen m [*Brücke*] = flood arch
~deich m, Leitdeich = levee
~entlastungsanlage f → → Talsperre f
~entlastungskanal m → Überlaufkanal
~entlastungsstollen m = spillway gallery
hochwasserfrei = flood-free
Hochwasser|gebiet n = flood region, ~ area
~grenze f = maximum water level
~häufigkeit f = flood frequency
~kanal m = flood canal
~katastrophe f = flood disaster, ~ calamity

~linie f = flood line
~schieber m = flood valve
~pegel m = flood gauge (Brit.); ~ gage (US)
~schaden m = flood damage
~schätzung f = flood estimation
~schlauch m, Hochwasserbett n = major bed, flood ~
~schutz m, Hochwasserabwehr f = flood protection
~schutzbauwerk n = flood-protection works, ~ structure
~schutztalsperre f, Hochwasser(auffang)sperre = flood (control) dam
~speicher m → Hochwasser(auffang)becken n
~speicherung f → → Talsperre f
~sperre f → Hochwasserauffangsperre
~spitze f, höchstes Hochwasser n = flood peak
~spitzenabfluß m, ~menge f = peak-flood discharge, ~ flow
~-Sprengdeich m = "fuse plug" levee
~stand m, HW = high water level
~statistik f = records of actual floods
~stollen m Überlaufstollen = tunnel-(type) discharge carrier
~überlauf m → → Talsperre f
~verhütung f = flood prevention
~versicherung f = flood insurance
~vorhersage f = flood prediction
~welle f = flood wave
~zeit f = flood season
hochwertiger Portlandzement m, früh-hochfester ~, Schnellerhärter m [*in Deutschland Z 325 und Z 425*] = high early (or initial) strength Portland cement, rapid hardening ~ ~; Portland cement type III (US)
~ Stahl m = high-grade steel
hochwertiges Leuchtöl n, ~ Brennöl = long-time burning oil
Hochwinden n, Aufwinden [*mit Hebebock*] = jacking (up)
Hochziehen n von Gleitschalung = slide
hochzugfest = high-tensile
Hoesch-Profil n = Hoesch section

Hof — Hohlblock-Deckenbogen

Hof m = court(yard) [*An unroofed space with access wholly or mainly enclosed by a building or buildings*]
~**ablauf** m, Hofeinlauf = yard gull(e)y
Hoffmann'scher Ringofen m = Hoffmann kiln
Hof|front f, Hofseite f, Hinterfront, Hinterseite = back elevation, elevation facing yard
~**seite** f → Hoffront f
Höft n → Pier
Höhe f, Ordinate f, Kote f = level, elevation
~ **des mittleren Hochwassers bei Nipptide**, mittleres Nipphochwasser n = height of mean high water of neap tide
~ ~ ~ ~ **Springtide**, mittleres Springhochwasser n = height of mean high water of spring tide
~ ~ ~ ~ **Niedrigwassers bei Nipptide**, mittleres Nippniedrigwasser n = height of mean low water of neap tide
~ ~ ~ ~ **Springtide**, mittleres Springniedrigwasser n = height of mean low water of spring tide
~ **über alles**, Gesamthöhe = over-all height, total ~
~ ~ ~ **Normalnull** → Meer(es)höhe
Höhen|ablesung f = level reading [*e.g. taken on driven piles and existing substructures during pile-driving operations*]
~**absteckung** f = setting out levels
~**bestimmung** f **durch Messung von Vertikalwinkeln**, trigonometrische Höhenmessung = trigonometrical level(l)ing
~**bezug** m [*Straßenbau*] = grade reference
~**differenz** f → Höhenunterschied m
(~)**Fühler** m → (Höhen)Taster m
~**genauigkeit** f = accuracy of level(s)
höhen|gerecht verdichten = to compact to final elevation

~**gleiche (oder niveauebene) Straßenkreuzung** f **mit Lichtzeichenregelung**, signalgesteuerter Knoten m = signalized street intersection, signal-controlled intersection
Höhen|kurort m = high altitude resort
~**lage** f = elevation
~**linie** f, (Höhen)Schichtlinie = contour
~**linienkarte** f, Schichtenplan m = contour(ed) plan, ~ map, layered ~
~**messung** f = level(l)ing
~**messung** f **mit horizontalen Geraden** → geometrische Höhenmessung
~**pfahl** m, Höhenpflock m = finishing (or grade, or gradient, or level, or fill) peg (or stake)
~**plan** m → Längenschnitt m
(~)**Schichtlinie** f, Höhenlinie = contour
~**steuerung** f = control of level(s) [*asphalt paver*]
(~)**Taster** m, (Höhen)Fühler m [*Gleitschalungsfertiger*] = sensing device, grade sensor
~**unterschied** m, Höhendifferenz f = difference in level (or elevation)
höhenverstellbar = provided with rise and fall adjustment
hohe Stelle f [*in einer Fläche*] = bump
Höhere Gewalt f = force majeure
höheres Hochwasser n, HThw = higher high water
hohes Bauwerk n = high-rise structure
~ **Gebäude** n = high-rise building
~ **Tideniedrigwasser** n, HTnw = higher low water
Hohlbalken|decke f = hollow beam floor
~**maschine** f = hollow beam machine
~**(träger)** m → Kastenträger
Hohlblock m, Hohlkörper m, Hohl(block)stein m = hollow tile, ~ block, ~ body , pot
~**decke** f → Hohlkörperdecke
~**-Deckenbogen** m → Hohlkörper-Deckenbogen

Hohlblockfertiger — Hohlkugel

~fertiger *m*, Hohlsteinfertiger = hollow tile making machine, ~ block ~ ~

~form *f* = hollow-block mould (Brit.); ~ mold (US)

~stein *m* → Hohlblock *m*

~stein *m* aus Leichtbeton → Leichtbeton-Hohlblock(stein)

~(stein)-Mauerwerk *n*, Hohlkörper-Mauerwerk, Hohlstein-Mauerwerk = hollow-tile masonry, hollow-block ~, pot ~

~(stein)prüfmaschine *f*, Hohlkörperprüfmaschine, Hohlsteinprüfmaschine = hollow tile tester, ~ block ~, ~ ~ testing machine, pot tester, pot testing machine

~steinwand *f* → Hohlkörperwand

~trennwand *f*, Hohlkörpertrennwand, Hohl(block)steintrennwand = hollow partition (wall), ~ division [*A partition built of hollow blocks*]

~wand *f* → Hohlkörperwand

Hohl|bohrer *m* = hollow drill

~(bohr)gestänge *n*, Gestängerohre *npl* = hollow (drill) rods

~(bohr)stange *f* = hollow (drill) rod

~brandmauer *f*, Brandhohlmauer = cavity party wall, hollow ~ ~

~decke *f* = hollow floor

~deckenkörper *m* = hollow floor(ing) block, ~ ~ tile, ~ ~ body, ~ pot

~deckenplatte *f* = hollow floor slab

~element *n* = hollow building component

Höhlenwasser *n*, Kavernenwasser = cavern water

Hohl|falzziegel *m* = Spanish tile

~felge *f* = hollow rim

~fuge *f* = keyed joint

~fugeneisen *n* → Fugeneisen

~fundmantpfeiler *m*, Hohlgrundpfeiler, Fundamenthohlpfeiler, Grundhohlpfeiler = hollow foundation pier

~gestänge *n* → Hohlbohrgestänge

~glas(bau)stein *m* = hollow glass block

~glaserzeugung *f* = glass container manufacture

~(grund)pfeiler *m* → Hohlfundamentpfeiler

~kant *m* = hollow chamfer

~kastengründungskörper *m*, Hohlkastenfundament *n*, Hohlkastengründung *f*, Hohlkastenfundation *f*, Hohlkastenfundamentkörper [*besteht aus einem biegesteifen Trägerrost, welcher mit einer Boden- und Deckenplatte monolithisch zu einem verwindungssteifen Hohlkasten verbunden ist*] = box section foundation (structure), ~ ~ footing, ~ ~ engineering foundation

~kastenplatte *f* = hollow (box) slab

~kasten(träger) *m* → Kastenträger

~kasten(träger)verbundbrücke *f* = hollow (box) girder composite bridge, ~ (~) beam ~ ~

Hohl|kehle *f* = cavetto [*A concave mo(u)lding used in architecture*]

~kehlnaht *f* = concave fillet weld

~kern *m* = hollow core

~kolbenpresse *f* = hollow piston jack

Hohlkörper *m*, Hohlblock *m*, Hohl(block)stein *m* = hollow block, ~ tile, ~ body, pot

~decke *f*, Hohlblockdecke, Hohl(block)steindecke = hollow-block floor, hollow-tile ~, pot ~, hollow-body ~

~-Deckenbogen *m*, Hohlblock-Deckenbogen, Hohl(block)stein-Deckenbogen = hollow-tile floor arch, hollow-block ~ ~, pot ~ ~, hollow-body ~ ~

~presse *f* = hollow body press

~prüfmaschine *f* → Hohlblock(stein)-prüfmaschine

~trennwand *f* → Hohlblocktrennwand

~wand *f*, Hohlblockwand, Hohl(block)steinwand = hollow-block wall, hollow-tile ~, pot ~, hollow-body ~, hollow masonry ~ [*it consists wholly or in part of hollow masonry units*]

Hohlkugel *f* = hollow ball

Hohlmauer — Holzbadewanne

Hohlmauer *f*, Hohlwand *f*, (Zwei-)Schalenmauer, (Zwei)Schalenwand = cavity wall, hollow ~
~ → → Talsperre *f*
(~)**Schale** *f*, Mauerschale = leaf
~**werk** *n* = cavity masonry
Hohl|pfahl *m* = hollow pile
~**pfeiler** *m* = hollow pier
~**platte** *f*, Stahlbeton-~ = hollow slab, reinforced-concrete ~ ~
~**plattenrahmen** *m*, Stahlbeton-~ = hollow slab frame, reinforced-concrete ~ ~ ~
~**profil** *n* = hollow section
~**profilträger** *m* = hollow section girder
~**querschnitt** *m* = hollow section
Hohlraum *m* [*beim Hohlziegel; ist größer als das Loch beim Lochziegel*] = cell (US)
~, Pore *f* = pore, void, interstice
~ [*zwischen zwei Mauerschalen*] = cavity
~**anteil** *m*, Hohlraumprozentsatz *m* = percentage of voids
hohlraum|arm → dicht
~**ausbildender Vibrationsdorn** *m* = vibrating void-forming mandrel
Hohlraum|bildung *f* → Kavitation *f*
~**druck** *m*, Porenwasserdruck = pore water pressure
hohlraumfrei = void-free, voidless
Hohlraum|gehalt *m* = voids content
~**minimum** *n* = maximum density
~**minimumprinzip** *n* → Betonprinzip
~**prozentsatz** *m*, Hohlraumanteil *m* = percentage of voids
~**volumen** *n* = cavity volume
Hohl|säule *f* = hollow column
~**schaft** *m* = hollow shaft
~**schraube** *f* = banjo bolt
~**schraub(en)pfahl** *m*, Schraub(en)hohlpfahl = hollow screw pile
~**schwimmer** *m* = hollow float
~**sog** *m* → Kavitation *f*
~**spindelvorstoß** *m* = shear ga(u)ge with hollow spindle
~**stange** *f*, Hohlbohrstange = hollow (drill) rod
Hohlstein *m* → Hohlkörper *m*
~**decke** *f* → Hohlkörperdecke

~**fertiger** *m*, Hohlblockfertiger, Hohlkörperfertiger = hollow block making machine, ~ tile ~ ~, ~ body ~ ~, pot ~ ~
~**prüfmaschine** *f* → Hohlblock(stein)prüfmaschine
~**trennwand** *f* → Hohlblocktrennwand
~**wand** *f* → Hohlkörperwand
Hohlstrahl *m* = hollow jet
~**schieber** *m* = hollow-jet (needle) valve
Hohl|stufe *f* = hollow step
~**stütze** *f* = hollow stanchion
~**tauchkolben** *m* = hollow plunger
~**träger** *m* = hollow girder
~**trennwand** *f*, Trennhohlwand = hollow partition, double ~ [*for a sliding door*]
~**trennwand** *f*, Trennhohlwand = hollow partition [*Hollow partitions are those formed of clay, terracotta, or breeze, hollow blocks for lightness or for sound insulation. They may also be formed by staggering the studs*]
~**ufer** *n* = concave bank
~**wand** *f* → Hohlmauer *f*
~**wandblock** *m*, Hohlwandkörper *m* = hollow block for walls, ~ tile ~ ~, ~ body ~ ~, pot ~ ~
~**wehr** *n* = hollow weir
~**widerlager** *n*, Sparwiderlager, aufgelöstes Widerlager = hollow abutment
~**ziegel** *m*, Lochziegel = perforated brick (Brit.); multicored tile (US)
holländischer Verband *m* = Dutch (cross) bond, English ~ ~
Höllenstein *m*, Silbernitrat *n* = lunar caustic
Holm *m* = capping, cross beam
Holz|architrav *m* = wood(en) architrave, timber ~
~**asbest** *m* = mountain wood
~**ausbau** *m* (⚒) = wood(en) supports, timber ~
~**ausbau** *m*, Schachtausbau in Holz = shaft timbering
~**badewanne** *f* = wood(en) bath (tub), timber ~ (~)

Holzbahn — Holzkastenträger

~bahn f, Holzsteg m = timber road, ~ run(way), ~ track [Used to permit distribution of concrete by carts and barrows]
~balkendecke f = wood(en) (beam) floor, timber (~)
~balken-Hourdidecke f, Holzbalken-Tonhohlplattendecke, Holzbalken-Tonhohlkörperdecke = burnt-clay hollow-tile floor with wood(en) beams, ~ hollow-block ~ ~ ~, ~ pot ~ ~ ~ ~
~balken(träger) m = wood(en) beam, timber ~
~baluster m, Holzdocke f = wooden turned baluster [in Scotland: wooden turned banister]
~betonestrich m = cement-wood (jointless) floor(ing), ~ composition
~-Beton-Pfahl m [Besteht aus einem unteren Holzpfahl und einem oberen Betonpfahl] = wood composite pile
~binder m, Holzdachbinder, Holzfachwerkbinder = wood(en) roof truss, ~ timber ~
(~)Bohle f, Belagbohle = plank
(~)Bohlenbelag m = plank covering
~-Bohrturm m, Holzturm = wooden (drilling) derrick, ~ drill tower
~brett n → Fugenbrett
~brücke f = wood(en) bridge, timber ~
~-Brückenpfeiler m = timber bridge pier, ~ ~ support, wood(en) ~ ~
~-Brunnenkranz m, Holz-Schlenge f = timber cutting curb, ~ drum ~, wood(en) ~ ~, ~ shoe
~bunker m = wood(en) bin, timber ~
~dachbinder m → Holzbinder
~daubenrohr n = machine-banded pipe, continuous stave-pipe, wood-stave pipe [A pipe made of wooden staves. The assembly is held together in a machine and tightly wrapped with wire. The pipe is made in definite lengths and joined in the field by couplings. Such pipe is rarely made larger than 24 in. in diameter]
~docke f → Holzbaluster m

~einbau m [Baugrube] = timbering
~einbauten fpl [Kühlturm] = wood(en) bars, timber ~
~einlage f = timber insert, wood(en) ~
~einlage f → Fugenbrett n
~einlagen fpl (⚒) = wood(en) inserts, layers of wood(en) blocks
hölzerne Dichtungsleiste f, (Sohlen-) Dichtungsbalken m [Walzen(stau)wehr] = timber stop, wood(en) ~
~ Pflasterramme f → Holzstößel m
hölzerner Pfahlrost m → Holzpfahlrost
Holzfachwerk|binder m → Holzbinder
~träger m = timer truss (girder), wood(en) ~ (~)
~trägerbrücke f = timber truss bridge, wood(en) ~ ~
Holz|falle f → Holzgleitschütz(e) n, (f)
~fasergips m = wood-fibered plaster
~fender m = wood(en) fender
~fenster n = wood(en) window
~-Fenstersprosse f = wood(en) glazing bar, ~ sash ~, ~ astragal
~fertigteil m, n = prefabricated timber section
~form f = wood(en) mould, timber ~ (Brit.); ~ mold (US)
~(fugen)einlage f → Fugenbrett n
~gebäude n, Holzhaus n = timber building, wood(en) ~
~gerüst n, Holzrüstung f = timber scaffold(ing), wood(en) ~
~gerüstbauer m = timber scaffolder
~-Gleisschwelle f → (Holz)Schwelle
~gleitschütz(e) n, (f), Holzfalle f = timber sliding gate, ~ slide ~, wood(en) ~ ~
~glied n → (angelenkte) Holzplatte f [Gliederband]
~griff m = wood(en) handle
~haus n → Holzgebäude n
~kappe f (⚒) = wood(en) (roof) bar, timber (~) ~
~kastenkipper m = wood tip(ping) wagon, ~ skip, track wheel wood dump wagon
~kastenträger m = timber box girder ~ ~ beam, wood(en) ~ ~

Holzkeilschneider — Holzträger

~keilschneider *m* = wood wedge cutter

~kupfererz *n* = olivenite

~ladegabel *f*, Ladegabel für Rundholz = logger, log fork

~läufer *m* (⚒) = wood(en) runner, timber ~

~lehrgerüst *n*, Holzrüstung *f* = timber falsework, wood(en) ~

~leiste *f* [*für Betondeckenfugen*] = timber fillet

~nadel *f* [*Nadelwehr*] = timber needle, wood(en) ~

~opal *m* = wood opal

~pfahl *m* = wood(en) pile, timber ~

~pfahlrost *m*, hölzerner Pfahlrost [*Fehlname: Holzrost*] = wood(en) pile foundation grillage

~pfeiler *m* [*Ausbauelement aus kreuzweise übereinandergelegten dichten Kantholzlagen*] (⚒) = cog, chock (Brit.); crib, pigsty (US)

~pflock *m*, Holzpfahl *m* [*für Absteckungsarbeiten*] = wood(en) peg, timber ~

~platte *f* → angelenkte ~ [*Gliederband*]

~plattenbelag *m* = wood(en) board covering

~positiv *n* [*Formenbau*] = timber positive, wood(en) ~

~rammpfahl *m* = (impact-)driven wood(en) pile, ~ timber ~

~rinne *f*, Fluder *n* = wood(en) flume

~rolladen *m* = wood(en) rolling shutter, ~ roller ~

~rost *m* [*Fehlname*] → Holzpfahlrost

~rost *m* = grating [*A wood(en) grillage (foundation)*]

~rüstung *f*, Holzgerüst *n* = timber scaffold(ing), wood(en) ~

~rüstung *f*, Holzlehrgerüst *n* = wood(en) falsework, timber ~

~schale *f* = timber shell, wood(en) ~

~schalendach *n* = timber shell roof, wood(en) ~ ~

~schalung *f* = wood(en) formwork

~schalung *f* [*Dach*] = wood(en) sheathing

~-Schienenschwelle *f* → (Holz-)Schwelle

~schindel *f* = wood(en) shingle

~schindeldach *n* = wood(en)-shingle roof

~-Schlenge *f* → Holz-Brunnenkranz

~schober *m* = stack of wood

~-Schraub(en)pfahl *m*, Schraub(en)-Holzpfahl = wood(en) screw pile

~schutzanstrich *m* = timber protection coat, wood ~ ~

(~)Schwelle *f*, Holz-Gleisschwelle, Holz-Schienenschwelle, (Eisen-)Bahn(holz)schwelle = wood(en) sleeper (Brit.); ~ (railroad) tie (US)

(~)Schwelle *f* [*Beim Holzfachwerk der untere waag(e)rechte Balken der Riegelwand*] = sill, sole plate (Brit.); abutment piece (US); cill [*deprecated*] (Brit.). [*The lowest horizontal member of a framed partition, or frame construction*]

~-Schwimmkasten *m* = timber box caisson, wood(en) ~ ~

~spachtel *m, f*, Reibebrett *n*, Streichbrett [*Gußasphalt*] = wood(en) float

~spundwand *f*, Holz(spund)bohlenwand = timber sheet-piling, wood(en) ~, ~ sheet pile wall, ~ sheeting, ~ bulkhead

~stäbchen *n* [*beim Rolladen*] = wood(en) slat

~stabgewebe *n* als Putz(mörtel)träger = wood(en) lath(ing)

~steg *m*, Holzbahn *f* = timber run(way), ~ road, ~ track [*Used to permit distribution of concrete by carts and barrows*]

~stempel *m* (⚒) = wooden prop

~stößel *m*, Stampfe *f*, hölzerne Pflasterramme *f* = wood(en) rammer, ~ tamper, ~ punner

~stütze *f* = timber column, wood(en) ~, ~ post

~torf *m* = wood peat

~träger *m* = wood(en) girder, timber ~

~trägerbrücke *f* = timber girder bridge, wood(en) ~ ~
~trennwand *f*, Holzzwischenwand = wood(en) partition (wall)
~treppe *f* = wood(en) staircase (Brit.); ~ stair (US); timber ~
~tür *f* = wood(en) door
~turm *m* → Holz-Bohrturm
~unterboden *m* = wood(en) sub-floor, timber ~
~-Unterkonstruktion *f* [*Putztechnik*] = wood(en) furring, ~ firring
~unterlage *f* = wooden(en) matting, timber ~ [*placed under vehicles, cranes, or excavators, to prevent sinking in soft surface soil*]
~verbau *m* = timbering
~verbinder *m* = (metal) connector, timber ~
~verkleidung *f* [*Baugrube*] = timber sheeting, wood(en) ~
~-Wasserturm *m* = wood(en) water tower, timber ~ ~
~wehr *n* = timber weir, wood(en) ~
~wollefilter *m, n* = excelsior filter (US); wood wool ~ (Brit.)
~zarge *f* = timber trim, wood(en) ~
~zerstörung *f* = wood destruction
(~)Zimmerung *f* = timbering
~zinn *n* = wood tin
~zwischenwand *f*, Holztrennwand = wood(en) partition (wall)
Homogen-Asphaltplatte *f*, Homogen-Asphaltfliese *f* = homogeneous asphalt tile, dense ~ ~
homogener Erd(stau)damm *m* = homogeneous (earth) dam
Homogenisier|maschine *f* [*Emulsionsherstellung*] = colloid mill
~silo *m* = homogenizer, blending bin [*A bin in which fluids or powders are thoroughly mixed and blended by compressed air, paddles or rakes*]
homotaxe Ablagerung *f* → äquivalente ~
Honigmann-Schacht|ausbau *m* = Honigmann method of shaft lining
~bohrverfahren *n* = Honigmann method of shaft sinking
Honigstein *m* → Melilith *m*

Hooke'sches Gesetz *n* [*Spannung ist proportional zur Dehnung*] = Hooke's law
Hopperbagger *m* → Schacht(pumpen)bagger
Höppler-Kugelfall-Viskosimeter *n* = rolling sphere instrument, Höppler visco(si)meter, falling ball ~
Hör|barkeit *f* = audibility
~bereich *m* = audible range
Hordengestell *n* [*Betonsteinherstellung*] = curing rack, rack for curing concrete block
Horizont *m*, Förder~ = underground formation [*Oil or gas production from a well*]
horizontal durchflossenes Becken *n* = = horizontal-flow tank, ~ basin [*A tank or basin, with or without baffles, in which the direction of flow is generally horizontal*]
Horizontal|achse *f* = horizontal axis
~ausleger *m* = cantilever jib (Brit.); ~ boom (US)
~beleuchtung *f* = horizontal lighting
~bohrmaschine *f* = horizontal drilling machine
~bohrung *f*, Söhligbohrung = horizontal drilling
horizontale Aussteifung *f* = cross-bracing
horizontaler Querverband *m* → Längsverband [*Träger*]
Horizontal|festigkeit *f* → Biegefestigkeit
~filterbrunnen *m* = horizontal screen well
~gerinne *n* → Horizontal(strom)sichter *m*
~gesamtwasserdruck *m* = total horizontal water pressure
~komponente *f* = horizontal component
~-Preßanlage *f* → Rohrdurchstoßgerät
~rahmen *m* = horizontal frame
~schiebefenster *n*, Waag(e)recht-Schiebefenster = horizontal sliding window

~schnitt *m* = sectional plan, horizontal section

~schub *m*, Bogenschub = horizontal thrust

~schub *m* [*Kletterkran*] = horizontal load reaction

~schubkomponente *f* = thrust

~sichter *m* → Horizontalstromsichter

~sieb *n*, Horizontal-Vibrations-Plansieb = horizontal vibrating screen, level ~ ~

~-Strömungs-Trockenelektroabscheider *m* = horizontal gas flow dry precipitator

~verband *m* [*Träger*] → Längsverband

~-Vibratorsieb *n* = horizontal vibrating screen

~winkel *m* = horizontal angle

horizontbeständig (Geol.) = stratigraphically consistent

hornartiges Gefüge *n* [*Boden*] = puddled structure

Horn|barre *f* = cuspate bar

~blei *n* = phosgenite

Hornblende *f* (Min.) = hornblende

~asbest *m*, Amphibolasbest (Min.) = amphibole asbestos

~gneis *m* = amphibolic gneiss, hornblende-gneis

~porphyr *m* = hornblende porphyry

~schiefer *m* = horn(blende)-schist, schistous amphibolite

~silber *n*, Silberhornerz *n*, AgCl (Min.) = cerargyrite, clorargyrite, hornsilver

Hornfels(gneis) *m* = hornfels

Hornstein *m* = chert

~kies *m* = chert-gravel

Hör|saal *m* [*Universität*] = auditorium, lecture room

~schall *m* = audible sound

~signal *n*, akustisches Signal = audible signal, acoustic ~, sound ~

Horst *m* (Geol.) = horst

~verwerfung *f* (Geol.) = horst fault

Hosen|rohr *n*, Hosentrichter *m*, zweiarmiger Auslauf *m* = breeches piece

~trichter *m* → Hosenrohr *n*

Hoteleingang *m* = hotel entrance

Hourdi *m* → Tonhohlkörper *m*

~decke *f*, Tonhohlplattendecke, Tonhohlkörperdecke = burnt-clay hollow-tile floor, ~ hollow-block ~, ~ pot ~

~wand *f*, Tonhohlplattenwand, Tonhohlkörperwand = burnt-clay hollow-tile wall, ~ hollow-block ~, ~ pot ~

Howe'scher Träger *m*, Howe-Träger [*Ein Fachwerkträger, bei dem durch Vorspannen der Pfosten erreicht wird, daß die gekreuzten Streben stets nur auf Druck beansprucht sind und damit auf einfache Weise angeschlossen werden können*] = Howe truss

HOZ *m* → Hochofenzement

HP *n*, hyperbolisches Paraboloid = hyperbolic paraboloid

H-Pfahl *m*, (Stahl)Trägerpfahl = structural-steel pile, H-pile

H-Profil *n* = H section

HP-Schale *f*, hyperbolische Paraboloidschale = hyperbolic paraboloid(al) shell

HP-Schalendach *n* = hyperbolic paraboloid (concrete shell) roof

H-Stütze *f* = H section column

HThw, höheres Hochwasser *n* = higher high water

HTnw *n*, hohes Tideniedrigwasser = higher low water

H-Träger *m* = H-girder

Hub *m*, Zylinder~ = stroke

~ [*Schiffshebewerk*] = lift

~arm *m*, Hebearm = lift arm

~arme *mpl*, Hubgestell *n*, Hebearme, Hebegestell [*Schaufellader*] = lift arms

~armverlängerung *f*, Hebearmverlängerung = lift arm extension

~ausschaltautomat *m*, automatische Hubabstellung *f* [*Kübellader*] = auto(matic) kickout at full lifting height

~ausschalter *m* = (bucket) lift kickout

~-Becherrad *n*, Hebe-Becherrad = lifting bucket wheel

~-Betondecke *f* = lift-slab concrete floor

Hubbremse — Hubtrommel

~bremse *f* [*Bagger*] = hoist brake, hoisting gear ~
~brücke *f* = (vertical) lift bridge, hoist ~
~drehtor *n* = turning-lifting gate
~gabel *f*, Hebegabel, Lastgabel [*Gabelstapler*] = lift(ing) fork
~geschwindigkeit *f*, Hebegeschwindigkeit = lifting speed
~gestell *n* → Hubarme *mpl*
~haken *m*, Hebehaken = lift(ing) hook
~hebel *m*, Hubwerkhebel = hoist lever
~hebel *m* = (bucket) lift lever
~höhe *f*, Förderhöhe = lift(ing) height
~(-Hydraulik)zylinder *m* [*Hublader*] = hoist(ing) ram, lift ~
~insel *f* = mobile barge, oceangoing platform, mobile work platform, floating and self-erecting platform, civil engineering construction barge, jack-up rig
~kabel *n* → Hubseil *n*
~karre(n) *f*, *(m)*, Hubwagen *m* = (industrial) lift truck
~kipper *m*, Hochkipper = (tip and) hoist [*tip which unloads above the level of railway lines*]
~kolben-Bügelsäge *f* = pneumatic hack saw
~kolben-Nietpresse *f* = pneumatic compression riveter
~kolben-Stichsäge = pneumatic sabre saw
~kolbenverdichter *m* [*Kältemaschine*] = reciprocating compressor
~kolbenwerkzeug *n* = reciprocating piston tool without percussion
~kraft *f*, Hubvermögen *n* [*Kran*] = lifting capacity
~lader *m*, Ladeschaufel *f*, Schaufellader, Kübelauflader, Hochlader, Hochladeschaufel, Fahrlader, Fahrladegerät *n*, Fahrlademaschine *f* [*als Oberbegriffe für Nur-Hublader, Schürflader und Wurfschaufellader*] = loading shovel, bucket loader, tractor shovel

~laster *m* → Torladewagen *m*
~magnet *m* → Lastmagnet
~öse *f* = lifting eye, ~ lug
~platte *f* = lift slab
~plattenbauweise *f*, Hubplattenverfahren *n* = lift-slab method, ~ construction [*A method of concrete construction where floor and roof slabs are cast on or at ground level and hoisted into position by jacking*]
~pumpe *f* mit Rohrkolben = lift(ing) pump with hollow plunger
~pumpe *f* mit Ventilkolben = lift(ing) pump with bucket valve piston
~raum 1.720 cm³ *m* [*Der vom Kolben auf dem Weg von einem zum anderen Totpunkt verdrängte Raum*] = cap. 1,720 c.c., displacement ~ ~
~schaufel *f* → Schaufel
Hubschrauber *m* = helicopter
~-Dachflugplatz *m* = roof (top) heliport, roof-top helicopter airport
~flughafen *m* = heliport
~-Kran *m*, Kran-Hubschrauber *m* = helicopter for crane work
~kufe *f* = helicopter skid
~verkehr *m* = helicopter traffic
Hubschütz(e) *n*, *(f)* = lifting gate (Brit.); vertical lift ~ (US)
Hubseil *n*, Hubkabel *n* = hoist rope, ~ cable, ~ line
~muffe *f* [*Schleppschaufelbagger*] = hoist line socket, ~ rope ~, ~ cable ~
~scheibe *f* am Auslegerkopf = hoist line boom point sheave (US); ~ ~ jib ~ (Brit.)
Hub|senktor *n* = vertical lift(ing) gate
~-Spannbetonplatte *f* = prestressed lift slab
~spindel *f* → Schrauben(hebe)bock *m*
~-Stahlbetonplatte *f* = reinforced-concrete lift slab
~stapler *m*, Hebetransportstapler = stacker truck, lift ~
~steg *m* = lifting walkway
~stempel *m* → Hebebock *m*
~stoß *m* = upward stroke
~tor *n* [*Schleuse*] = (vertical) lift gate
Hubtrommel *f* = lifting drum

~bremse *f* = lifting drum brake
~welle *f* = lifting drum shaft
Hub|vermögen *n*, Hubkraft *f* [*Kran*] = lifting capacity
~wagen *m* → Hubkarre(n) *f*, *(m)*
~weg *m* [*Hydraulikpresse*] = range
~werk *n* = hoist mechanism
~(werk)hebel *m* = hoist lever
~winde *f* = hoist winch
~windenmotor *m* = hoist winch engine
~zylinder *m* → Hub-Hydraulikzylinder
~zylinder *m* zum Heben des Frontraumer-Planierschildes = dozer ram
~zylinderdrehzapfen *m* = lift cylinder trunnion
Huckepackverkehr *m* = transport on railway wagons of road lorries and trailers loaded with goods; piggyback (transport) (US)
Hufeisen|bogen *m*; maurischer Bogen, arabischer Bogen = Moorish arch; horseshoe ~ [*Characteristic of Saracenic architecture. The soffit consists of two segments struck from centres some distance above the springing. When one centre only is used the soffit is a little more than a semicircle and is called a horseshoe arch*]
~form *f* = horseshoe shape
~gletscher *m* = corrie glacier, horseshoe-shaped ~
~magnet *m* = horseshoe magnet
~querschnitt *m*, Hufeisenprofil *n* = horseshoe section
~tal *n* → Kahntal
Hügelland *n*, hüg(e)liges Gelände *n* = hilly ground, ~ terrain, ~ country, rough, ~ undulating ~
Hülle *f* aus Stahlblech, Rohr *n* aus dünnem Blech, (Blech)Gleitkanal *m*, Blechröhre *f*, Hüllrohr = (sheet-) metal sheath [*prestressed concrete*]
Hüll|kurve *f* = enveloping curve, envelope (~)
~rohr *n* [*Gleitschalungsbau*] = sliding sleeve [*The jacking rods remain in the cast concrete but are sheathed in larger separate sliding sleeves in the green concrete within the depth of the slipform, to facilitate adjustment of the thickness and taper*]
~rohr *n* → Hülle *f* aus Stahlblech [*Spannbeton*]
Hülse *f* aus Blech, Aufsteck~ ~ ~, Blechhülse [*Betondeckenfuge*] = light metal cap
Hülsen|bohrpfahl *m*, Bohrhülsenpfahl = bored shell pile, ~ cased ~
~pfahl *m*, Mantelpfahl = cased pile, shell ~
~rammpfahl *m*, Rammhülsenpfahl = (impact-)driven shell pile, ~ cased ~
Hummersieb *n* [*Die senkrecht schwingende Siebfläche wird durch elektromagnetische Schwingungen erregt*] = Hummer screen
Humphreys-Spirale *f*, Humphreys-Wendelscheider *m* = Humphrey's spiral concentrator
Humus *m* = humus
~karbonatboden *m* → Rendzinaboden
~säure *f*, Huminsäure = humic acid
Hund *m* → Schutterwagen *m*
H. ü. N. N. *f* → Meer(es)höhe *f*
Hunt *m* → Schutterwagen *m*
Hupen|knopf *m* = horn button
~ventil *n* = horn valve
Hütten|aluminium *n*, Reinaluminium = pure aluminium (Brit.); ~ aluminum (US)
~bims *m*, Kunstbims = crushed foamed (blast-furnace) slag (Brit.); ~ expanded (~) ~ (US)
~bimsbeton *m*, Kunstbimsbeton = crushed foamed (blast-furnace) slag concrete (Brit.); ~ expanded (~) ~ ~ (US)
~bimsmehl *n* = foamed (blast-furnace) slag powder
~faser *f*, Schlackenfaser = slag fiber (US); ~ fibre (Brit.)
~hartstein *m* → Hüttenstein
~koks *m* = metallurgical coke
~mauerstein *m* → Hüttenstein
~platte *f* [*Gleiskette*] = steel mill shoe
~sand *m* [*granulierte oder gekörnte Hochofenschlacke*] = slag sand

~schaufel *f* = steel mill bucket
~schlacke *f* → Metall(hütten)schlacke
~schwemmstein *m* [*früher: Hochofenschwemmstein*] = foamed slag aggregate concrete block
~stein *m*, Hüttenmauerstein, Hüttenhartstein [*früher: Hochofenschlakkenstein. DIN 398*] = slag aggregate block
~wolle *f* [*früher: Schlackenwolle*] = slag wool, silicate cotton
~zement *m*, Schlackenzement = slag cement, artifical pozzolana ~
Hvorslev'sches Flächenverhältnis *n* = Hvorslev area ratio C_a
HW, Hochwasserstand *m* = high water level
Hyalophan *m* → Barytfeldspat *m*
hybrides Gestein *n*, Mischgestein = hybrid rock
Hydrant *m* [*DIN 3221 und 3222*] = hydrant
Hydrantenschacht *m* = hydrant pit
Hydratation *f* = hydration [*Chemical combination of cement with water*]
Hydratations|geschwindigkeit *f* = rate of hydration
~wärme *f* = heat of hydration [*The heat given off by cement paste during the chemical combination of cement with water. An exothermic process*]
Hydraulefaktor *m* → Silikatbildner *m*
Hydraulik *f* = hydraulics
~-Anbauaufreißvorrichtung *f*, Hydro-Anbauaufreißvorrichtung = hydraulic scarifier
~-Anbaubagger *m*, Hydro-Anbaubagger, Hydraulik-(Anbau-)Baggergerät *n*, Hydro-(Anbau-)Baggergerät = hydraulic digging attachment
~-Anhängeaufreißer *m* → Hydraulik-Tiefaufreißer
~anlage *f*, hydraulisches System *n*, Hydroanlage, Hydrauliksystem [*Baumaschine*] = hydraulic equipment, ~ system
~-Anlaßanlage *f*, Hydraulik-Startanlage = hydraulic starting system

~-Anschlußstück *n* = hydraulic fitting
~antrieb *m* → Hydroantrieb
~-Arbeitsbühne *f* → Hydraulik-Hebebühne
~-Aufbaubagger *m* → Hydro-Aufbaubagger
~-Aufbaukran *m*, Hydro-Aufbaukran, Hydraulik-Last(kraft)wagenkran, Hydro-Last(kraft)wagenkran, LKW-Hydrokran, Hydraulik-LKW-Kran, Hydro-LKW-Kran = fast-travel hydraulic crane
~-Aufbaulader *m*, Hydro-Aufbaulader, Hydraulik-Last(kraft)wagenlader, Hydro-Last(kraft)wagenlader, LKW-Hydrolader = fast-travel hydraulic loader
~-Aufreißer *m* → Hydraulik-Tiefaufreißer
~ausleger *m*, Hydroausleger = hydraulic boom (US); ~ jib (Brit.)
~bagger *m*, Hydrobagger = hydraulic excavator
~-Baggergerät *n* → Hydraulik-Anbaubagger *m*
~-Baggerlader *m* → Hydraulik-Bagger und -Lader
~-Bagger und -Lader *m*, Hydraulik-Lader und -Bagger, Hydraulik-Baggerlader, Hydraulik-Ladebagger = hydraulic excavator-loader
~behälter *m* = hydraulic liquid tank
~-Betonstahlschneider *m*, Hydro-Betonstahlschneider = hydraulic reinforcement bar cutter
~bremse *f*, Hydrobremse, hydraulische Bremse, Flüssigkeitsbremse = hydraulic brake
~-Fahrbagger *m* (mit Schürfmulde) → Hydro-Fahrbagger (~ ~)
~-Fahrladegerät *n* → Hydraulik-Hublader *m*
~-Fahrlademaschine *f* → Hydraulik-Hublader *m*
~-Fahrlader *m* → Hydraulik-Hublader
~flüssigkeit *f* → Druckflüssigkeit
~-Gerüst *n*, Hydro-Gerüst, Hydraulik-Rüstung *f*, Hydro-Rüstung = hydraulic scaffold(ing)

Hydraulik-Getriebe — Hydraulik-Presse 538

~-Getriebe n, Hydrogetriebe = hydraulic transmission

~getriebe-Steuerorgan n = hydraulic transmission control

~-(Gleis)Kettenbagger m → Hydro-Raupenbagger

~greifer m, Hydrogreifer = hydraulic grab

~gruppe f = hydraulic component

~-Hebebühne f, Hydraulik-Arbeitsbühne, Hydro-Hebebühne, Hydro-Arbeitsbühne = hydraulic (working) platform

~heber m → Hebebock m

~-Heckbagger m, Hydro-Heckbagger = hydraulic rear digging attachment

~-Hocharbeitsbühne f, Hydro-Hocharbeitsbühne = hydraulic lifting work platform

~-Hochlader m → Hydraulik-Hublader

~-Hochladeschaufel f → Hydraulik-Hublader m

~-Hochlöffelbagger m, Hydro-Hochlöffelbagger = hydraulic shovel

~-Hublader m, Hydraulik-Ladeschaufel f, Hydraulik-Schaufellader, Hydraulik-Kübelauflader, Hydraulik-Hochlader, Hydraulik-Hochladeschaufel, Hydraulik-Fahrlader, Hydraulik-Fahrladegerät n, Hydraulik-Fahrlademaschine f, Hydro-Hublader, Hydro-Ladeschaufel, Hydro-Schaufellader, Hydro-Kübelauflader, Hydro-Hochlader, Hydro-Hochladeschaufel, Hydro-Fahrlader, Hydro-Fahrladegerät, Hydro-Fahrlademaschine [*siehe Anmerkung unter "Hublader"*] = hydraulic loading shovel, ~ bucket loader, ~ tractor shovel

~-Hubplattform f, Hydro-Hubplattform = hydraulic lift(ing) platform

~-Hubtisch m, Hydraulik-Hebetisch = hydraulic elevating table

~-Kettenbagger m → Hydro-Raupenbagger

~kran m, Hydrokran = hydraulic crane, hydrocrane

~kreis m = hydraulic circuit

~-Kübelauflader m → Hydraulik-Hublader

~-Ladebagger m → Hydraulik-Bagger und -Lader

~-Lader und -Bagger m → Hydraulik-Bagger und -Lader

~-Ladeschaufel f → Hydraulik-Hublader m

~-Last(kraft)wagenbagger m → Hydro-Aufbaubagger

~-Last(kraft)wagenkran m → Hydraulik-Aufbaukran

~-Last(kraft)wagenlader m → Hydraulik-Aufbaulader

~leitung f, Hydroleitung = hydraulic line

~-Lenkung f, Hydrolenkung, hydraulische Lenkung = hydraulic steering

~lenkung f mit konstanter Lenkgeschwindigkeit = constant speed steering

~-LKW-Kran m → Hydraulik-Aufbaukran

~löffeltiefbagger n → Hydrauliktieflöffel(bagger)

~-Magnetventil n = hydraulic solenoid valve

~maschine f, Hydromaschine = hydraulic machine

~-Mobilbagger m, Hydro-Mobilbagger, Mobil-Hydraulikbagger, Mobil-Hydrobagger = hydraulic mobile excavator

~-Mobilkran m, Hydro-Mobilkran, Mobil-Hydraulikkran, Mobil-Hydrokran = hydraulic mobile crane

~motor m, Hydromotor, Flüssigkeitsmotor = hydraulic motor

~öl n, Drucköl, Preßöl = hydraulic oil, pressure ~

~ölmotor m → Drucköl motor

~ölpresse f → Drucköl presse

~ölpumpe f → Drucköl pumpe

~ölzufluß m → Drucköl zufluß

~ölzylinder m → Drucköl zylinder

~-Presse f, hydraulische Presse, Hydropresse = hydraulic press

Hydraulikpresse — hydraulische Energiequelle

~presse *f* → Hebebock *m*
~-Prüfgerät *n* = hydraulic test box
~pumpe *f*, Hydropumpe = hydraulic pump
~Raupenbagger *m* → Hydro-Raupenbagger
~rohr *n*, Hydrorohr = hydraulic tube, ~ pipe
~rüstung *f* → Hydraulikgerüst *n*
~-Schaufellader *m* → Hydraulik-Hublader
~schlauch *m*, Hydroschlauch = hydraulic hose
~-Schleppaufreißer *m* → Hydraulik-Tiefaufreißer
~-Schürflader *m* → Hydro-Fahrbagger (mit Schürfmulde)
~-Schwenkmotor *m*, Hydro-Schwenkmotor = hydraulic slewing motor
~-Servomotor *m*, Hydro-Servomotor = hydraulic servomotor
~speicher *m*, Hydrospeicher = hydraulic accumulator
~-Startanlage *f*, Hydraulik-Anlaßanlage = hydraulic starting system
~-Stempel *m*, hydraulischer Stempel (⚒) = hydraulic prop, ~ post
~steuerung *f*, Hydrosteuerung = hydraulic control
~system *n* → Hydraulikanlage *f*
~tank *m* = hydraulic tank, ~ reservoir
~-Tiefaufreißer *m*, Hydraulik-(Schlepp)aufreißer, Hydraulik-Anhängeaufreißer, Hydro-Tiefaufreißer, Hydro(-Schlepp)aufreißer, Hydro-Anhängeaufreißer = hydraulic ripper, ~ rooter [*Designed to operate behind tractors fitted with hydraulic equipment*]
~tiefbagger *m* → Hydrauliktieflöffel(bagger)
~-Tieflöffel *m*, Hydro-Tieflöffel = hydraulic back-hoe
~tieflöffel(bagger) *m*, Hydrotieflöffel(bagger), Hydrauliklöffeltiefbagger, Hydrolöffeltiefbagger, Hydrauliktiefbagger, Hydrotiefbagger = hydraulic trench-forming shovel, ~ backacter, ~ backacting shovel, ~ backacting excavator, ~ dragshovel, ~ trencher, ~ ditcher, ~ trenching hoe, ~ drag bucket (Brit.); ~ (back-)hoe, ~ trench-hoe, ~ pullshovel (US); ~ pullstroke trenching machine, ~ ditching shovel, ~ back digger
~-Universal-Mobilbagger *m*, Hydro-Universal-Mobilbagger = hydraulic universal mobile excavator
~-Universal-Raupenbagger *m*, Hydraulik-Universal-(Gleis)Kettenbagger, Hydro-Universal-Raupenbagger, Hydro-Universal-(Gleis)-Kettenbagger = hydraulic universal crawler (type) excavator
~-Waage *f*, Hydrowaage = hydraulic scale
~wasser *n*, Druckwasser, Preßwasser = hydraulic water
~winde *f* → (hydraulische) Winde
~zylinder *m* = (hydraulic) ram
~zylinderdichtung *f* = (hydraulic) ram seal
hydraulisch abbindende Masse *f* [*Schamotteindustrie*] = castable
~ **angetriebenes Rad** *n* = hydraulic wheel
~ **betätigt** = hydraulically operated
~ **günstigster Querschnitt** *m* = optimum hydraulic cross section
hydraulische Arbeitsbühne *f* = hydraulic working platform
~ **Belastungsvorrichtung** *f* [*Baugrundprüfung*] = hydraulic ground-testing machine
~ **Beton(förder)pumpe** *f*, ~ Förderpumpe = hydraulic concrete pump
~ **Bördelmaschine** *f*, ~ Kümpelpresse *f* = hydraulic flanging machine
~ **Bremse** *f* → Hydraulikbremse
~ **Drehbohrmaschine** *f* = hydraulic rotary drilling machine
~ **Druckhöhe** *f* = hydraulic pressure head
~ **Eigenschaft** *f* = hydraulic property
~ **Energiequelle** *f* = hydropower source, source of hydropower

hydr. Energiespeicherung — hydr. Druckhöhenunterschied

~ **Energiespeicherung** *f*, Pumpspeicherung = pumped-storage
~ **Erhärtung** *f* = hydraulic hardening
~ **Flüssigkeit** *f* = hydraulic liquid
~ **Folgesteuerung** *f* = hydraulic sequence control
~ **Förderpumpe** *f*, ~ Beton(förder)pumpe = hydraulic concrete pump
~ **Förderung** *f* = hydraulic transport, ~ conveying
~ **Formsteinmaschine** *f* = hydraulic block-making machine
~ **Gewinnung** *f* → Abspritzen *n* mit (Hydro)Monitor(en)
~ **(Gleis)Kettenspannung** *f*, ~ Raupenspannung = hydraulic track tensioning
~ **Gradiente** *f*, hydraulisches (Druck-)Gefälle *n* = hydraulic gradient
~ **Hand-Kippvorrichtung** *f* = hydraulic hand tipping gear
~ **Hubklappe** *f* [*LKW*] = hydraulic lift tailgate, ~ end loader
~ **Kettenspannung** *f* → ~ Gleiskettenspannung
~ **Korntrennung** *f*, Korntrennung im Naßverfahren = hydraulic classification
~ **Kraftlenkung** *f* = hydraulically-assisted steering
~ **Lenkung** *f*, Hydraulik-Lenkung, Hydrolenkung = hydraulic steering
~ **Mehrzylinder-Spundwandramme** *f* = multi-ram hydraulic sheet pile driver
~ **Nietmaschine** *f* = hydraulic riveting machine
~ **Presse** *f*, Hydraulik-Presse, Hydropresse = hydraulic press
(~) ~ → Hebebock *m*
~ **Pressung** *f* **von Tonpulver mit 5-6% Feuchtigkeit** [*Herstellung von Fußbodenplatten*] = dust-pressed process
~ **Randbedingung** *f* = hydraulic boundary condition
~ **Raupenspannung** *f*, ~ (Gleis)Kettenspannung = hydraulic track tensioning

~**Reibung** *f* = hydraulic friction
~ **Rohrbiegemaschine** *f* = hydraulic pipe bender
~ **Scharseitenverstellung** *f* = hydraulic moldboard side shift
~ **Scheibenbremse** *f* = hydraulic disc-type brake; ~ disk-type ~ (US)
~ **Schürfwinkelverstellung** *f* **der Schar** = hydraulic moldboard tip
~ **Schwerstange** *f*, hydraulischer Bohrkragen *m* = hydraulic drill collar
~ **Steuerung** *f* = hydraulic control
~ **Unterstützung** *f* [*Lenkung*] = hydraulic power assistance
~ **Verlegung** *f* **der nicht verwendeten Erdmassen in den Abraum** = open (hydraulic) fill
~ **Verrohrungsmaschine** *f* = hydraulic casing machine
(~) **Verzögerungsbremse** *f* = (hydraulic) retarder
(~) **Winde** *f*, (hydraulischer) Hebebock *m*, Hydraulikwinde, Hydrowinde = hydraulic jack
hydraulischer Anlasser *m* = hydraulic starter
~ **Antrieb** *m* → Hydroantrieb
~ **Bauteil** *m*, hydraulisches Bauteil *n* [*einer Maschine*] = hydraulic component
~ **Beton** *m* = hydraulic concrete
~ **Betonförderer** *m*, ~ Betonförderapparat *m* = hydraulic concrete placer, ~ ~ placing machine, ~ ~ handling machine
~ **Binder** *m*, hydraulisches Bindemittel *n* = hydraulic binder
~ **Bohrkragen** *m*, hydraulische Schwerstange *f* = hydraulic drill collar
~ **Bolzenzieher** *m* [*Aufreißer*] = hydraulic pin puller [*ripper*]
~ **Drehmomentwandler** *m* = hydraulic torque converter
(~) **Druckhöhenunterschied** *m* = (hydraulic) head

hydraulischer Grundbruch — Hydro-Baggergerät

~ **Grundbruch** m [*Wenn der Strömungsdruck so groß wird, daß es zum Aufwirbeln des Bodens kommt, verliert dieser jeglichen Widerstand und die Teilchen werden weggeschwemmt*] = (failure by) piping
~ **Halbmesser** m → ~ Radius
(~) **Hebebock** m → (hydraulische) Winde f
~ **Jahresspeicher** m → Jahresspeicher
~ **Kalk** m [*früher: Zementkalk*] = semi-hydraulic lime
~ **Modellversuch** m = hydraulic model test
~ **Mörtel** m, Wassermörtel = hydraulic mortar
~ **Radius** m, ~ Halbmesser, Profilradius, Profilhalbmesser [*R m. Abflußquerschnitt geteilt durch den benetzten Umfang:* $R = \frac{F}{U}$] = hydraulic radius
~ **Regler** m = hydraulic governor
~ **Stoßdämpfer** m = hydraulic shock absorber
~ **Verlust** m, ~ Widerstand m = hydraulic loss, ~ resistance
~ **Vorschub** m, ~ Vorstoß m, hydraulisches Vorstoßen n [*Bagger*] = hydraulic crowd(ing)
~ ~ [*Bohrmaschine*] = hydraulic feed
(~) **Widder** m = hydraulic ram, (water) ~ [*A machine for raising water by utilizing the momentum of water flowing by gravity through a pipe to lift a portion of the water to an elevation greater than the source of supply*]
~ **Widerstand** m, ~ Verlust m = hydraulic resistance, ~ loss
~ **Zahnstangenbegrenzer** m = hydraulic rack limiter
~ **Zusatz** m [*zum Portlandzement*] = hydraulic additive
~ **Zyklon** m, Hydrozyklon = hydraulic cyclone
hydraulisches Aufbrechen n = hydraulic fracturing

~ **Bauteil** n, hydraulischer Bauteil m [*einer Maschine*] = hydraulic component
~ **Bindemittel** n, hydraulischer Binder m = hydraulic binder
~ **(Druck)Gefälle** n, hydraulische Gradiente f = hydraulic gradient
~ **Gefälle** n, ~ Druckgefälle, hydraulische Gradiente f = hydraulic gradient
~ **Perforieren** n [*der Verrohrung und unter Umständen der Zementierung*] = hydraulic perforating
~ **Pumpenaggregat** n = liquid pressure pump set
~ **Strangpressen** n = hydraulic extrusion
~ **System** n, Hydraulikanlage f [*Baumaschine*] = hydraulic system, ~ equipment
Hydro[-Anbaubagger m → Hydraulik-Anbaubagger
~**Anhängeaufreißer** m → Hydraulik-Tiefaufreißer
~**anlage** f → Hydraulikanlage
~**antrieb** m, Hydraulikantrieb, hydraulischer Antrieb = hydraulic drive
~**-Arbeitsbühne** f → Hydraulik-Hebebühne
~**-Aufbaubagger** m, Hydro-Last(kraft)wagenbagger, Hydraulik-Aufbaubagger, Hydraulik-Last(kraft)wagenbagger, LKW-Hydrobagger = hydraulic truck(-mounted) excavator (US); ~ lorry (-mounted) ~ (Brit.); ~ fast-travel ~
~**-Aufbaukran** m → Hydraulik-Aufbaukran
~**-Aufbaulader** m → Hydraulik-Aufbaulader
~**-Aufreißer** m → Hydraulik-Tiefaufreißer
~**ausleger** m → Hydraulikausleger
~**bagger** m, Hydraulikbagger = hydraulic excavator
~**-Baggergerät** n → Hydraulik-Anbaubagger

Hydro-Betonstahlschneider — Hydrologie

∼**-Betonstahlschneider** m, Hydraulik-Betonstahlschneider = hydraulic reinforcement bar cutter

∼**-Blitz 907 SV Betonsteinautomat** m = hydro automation 907 SV super concrete blockmaking machine

∼**bremse** f, Hydraulikbremse = hydraulic brake

∼**dynamik** f = hydrodynamics

hydrodynamische Druckkraft f = hydrodynamic force

∼ **Schmierung** f = hydrodynamic lubrication

hydrodynamischer Druck m = hydrodynamic pressure

∼ **Euler-Satz** m, Grundgleichung f der Hydrodynamik = Euler's hydrodynamical law, basic equation of hydrodynamics

hydrodynamisches Netz n, Strömungsbild n = flow net, ∼ pattern [of water]

Hydro|-Fahrbagger m (mit Schürfmulde), Hydraulik-Fahrbagger, Hydro-Schürflader m, Hydraulik-Schürflader = hydraulic loading shovel, ∼ shovel loader, ∼ bucket loader, ∼ tractor-shovel

∼**-Fahrladegerät** n → Hydraulik-Hublader m

∼**-Fahrlademaschine** f → Hydraulik-Hublader m

∼**-Fahrlader** m → Hydraulik-Hublader

∼**geologie** f = hydrogeology

∼**gerät** n = hydraulic equipment

∼**getriebe** n → Hydraulik-Getriebe

∼**-(Gleis)Kettenbagger** m → Hydro-Raupenbagger

∼**granat** m = hydrogarnet

∼**graphie** f = hydrography [Water surveys. The art of measuring, recording, and analyzing the flow of water, and of measuring and mapping water courses, shore lines, and navigable waters]

hydrographisches Jahr n → Abflußjahr

Hydro|graphischnull n = tidal datum plane

∼**greifer** m, Hydraulikgreifer = hydraulic grab

∼**-Hebebühne** f → Hydraulik-Hebebühne

∼**heber** m → hydraulischer Hebebock m

∼**-Heckbagger** m → Hydraulik-Heckbagger

∼**-Hocharbeitsbühne** f → Hydraulik-Hocharbeitsbühne

∼**-Hochlader** m → Hydraulik-Hublader

∼**-Hochladeschaufel** f → Hydraulik-Hublader m

∼**-Hochlöffelbagger** m, Hydraulik-Hochlöffelbagger = hydraulic shovel

∼**-Hublader** m → Hydraulik-Hublader

∼**-Hubplattform** f → Hydraulik-Hubplattform

∼**kalkstein** m → Kunstkalkstein

∼**-Kettenbagger** m → Hydro-Raupenbagger

∼**kran** m, Hydraulikkran = hydrocrane, hydraulic crane

∼**-Kübelauflader** m → Hydraulik-Hublader

∼**-Ladeschaufel** f → Hydraulik-Hublader m

∼**-Last(kraft)wagenbagger** m → Hydro-Aufbaubagger

∼**-Last(kraft)wagenkran** m → Hydraulik-Aufbaukran

∼**-Last(kraft)wagenlader** m → Hydraulik-Aufbaulader

∼**leitung** f, Hydraulikleitung = hydraulic line

∼**lenkung** f, Hydrauliklenkung, hydraulische Lenkung = hydraulic steering

∼**-LKW-Kran** m → Hydraulik-Aufbaukran

∼**löffeltiefbagger** m → Hydrauliktieflöffel(bagger)

Hydrologie f = hydrology [The science treating of waters of the earth in their various forms. Precipitation, evaporation, run-off, and ground water]

hydro|logische Vorarbeiten /pl = preliminary hydrologic work

~logischer Kreislauf m = hydrologic circle

~logisches Jahr n [vom 1. November bis 31. Oktober] = hydrologic year

Hydro|lyse f = hydrolysis

~maschine f, Hydraulikmaschine = hydraulic machine

~mechanik f = hydromechanics

~meteorologie f = hydrometeorology

~metrie f, Wassermeßwesen n = hydrometry [*The measurement and analysis of the flow of water*]

hydrometrischer Flügel m, (Wasser-)Meßflügel [*Anzeigegerät für Anströmgeschwindigkeit*] = screw current meter

Hydro|-Mobilbagger m, Hydraulik-Mobilbagger = hydraulic mobile excavator

~-Mobilkran m → Hydraulik-Mobilkran

~monitor m, Druckstrahlbagger m = giant, monitor, jetting gear

~motor m, Hydraulikmotor = hydraulic motor

~oxygengas n → Knallgas

~phon n, piezoelektrischer Seismograph m = hydrophone

~ponik f, Wasserkultur f = hydroponics [*The science of growing plants without soil in water and chemicals*]

~presse f, hydraulische Presse, Hydraulik-Presse = hydraulic press

~presse f → Hebebock m

~pumpe f, Hydraulikpumpe = hydraulic pump

~-Raupenbagger m, Hydro-(Gleis)Kettenbagger, Hydraulik-Raupenbagger, Hydraulik-(Gleis)Kettenbagger = hydraulic crawler excavator

~rohr n, Hydraulikrohr = hydraulic pipe, ~ tube

~-Schaufellader m → Hydraulik-Hublader

~schlauch m, Hydraulikschlauch = hydraulic hose

~-(Schlepp)Aufreißer m → Hydraulik-Tiefaufreißer

~-Schürflader m → Hydro-Fahrbagger (mit Schürfmulde)

~schwenkmotor m, Hydraulik-Schwenkmotor = hydraulic slewing motor

~speicher m, Hydraulikspeicher = hydraulic accumulator

~statik f = hydrostatics

hydrostatische Druckhöhe f = hydrostatic head

~ Druckkraft f, Gesamtwasserdruck m, Druckkraft des Wassers = hydrostatic force, total normal pressure

hydrostatischer Antrieb m = hydrostatic drive

~ Druck m = hydrostatic pressure

~ Euler-Satz m, Grundgleichung f der Hydrostatik = Euler's hydrostatical law, basic equation of hydrostatics

~ Überdruck m = excess hydrostatic pressure

hydrostatisches Lager n, Schwimmlager = hydrostatic bearing, floating ~

~ Paradoxon n = hydrostatical paradoxon

Hydrosteuerung f, Hydrauliksteuerung = hydraulic control

hydrothermal = hydrothermal

Hydro|-Tiefaufreißer m → Hydraulik-Tiefaufreißer

~tiefbagger m → Hydrauliktieflöffel-(bagger)

~-Tieflöffel m, Hydraulik-Tieflöffel = hydraulic back-hoe

~tieflöffel(bagger) m → Hydrauliktieflöffel(bagger)

~-Universal-Mobilbagger m, Hydraulik-Universal-Mobilbagger = hydraulic universal mobile excavator

~-Universal-Raupenbagger m → Hydraulik-Universal-Raupenbagger

~waage f → Hydraulik-Waage

~winde f → (hydraulische) Winde

~zinkit n, Zinkblüte f, $Zn_5[(OH)_3/CO_3]_2$ = hydrozincite, zinc bloom

~-Zyklon m = hydraulic cyclone

Hygrometer *m*, (Luft)Feuchtemesser *m* = hygrometer

hygroskopisches Wasser *n*, Anlagerungswasser, angelagertes Wasser = hygroscopic water, ~ moisture [*Is on the surface of the soil grains and is not capable of movement by the action of gravity or capillary forces*]

Hygroskopizität *f* = hygroscopicity

hyperbolische Paraboloidschale *f*, Hp-Schale = hyperbolic paraboloid(al) shell, hypar ~

hyperbolisches Paraboloid *n*, HP = hyperbolic paraboloid

Hyperboloidschale *f* = hyperboloidal shell

Hyperschallkanal *m* = hypersonic wind tunnel

Hypoid|achsantrieb *m*, achsversetzter Kegeltrieb = hypoid axle

~öl *n*, Hochdrucköl = hypoid lubricant, ~ oil

Hypo|kaustum *n* = hypocaust

~stylos *m*, gedeckter Säulengang *m*, Säulenhalle *f*, Tempel *m* mit Säulengang = hypostyle [*A hall of which the roof rests on columns characteristic of Egyptian temples*]

~these *f*, Annahme *f* = hypothesis

~trachelion *n*, Unterhals *m* = hypotrachelium, neck(ing) [*The junction of the shaft with the capital of a column*]

~zentrum *n*, Erdbebenherd *m* = focus, origin

Hysteresis *f*, Nachwirkung *f* = hysteresis

~eigenschaft *f* = hysteresis property

~schleife *f*, Hystereseschleife, Verlustschleife = hysteresis loop, ~ curve

I

I-Balken *m*, Doppel-T-Balken = I-beam

ICOS-Wand *f*, Dicht(ungs)wand ICOS, Sperrwand ICOS = ICOS diaphragm, ~ (~) wall, ~ cut-off

ideale Flüssigkeit *f*, vollkommene ~, reibungslose ~, perfekte ~ = ideal liquid, perfect ~. nonviscous ~

~ Kornscheide *f* [*Siebtechnik*] = cut-point

idealer Ton *m* = ideal clay

ideales Mittelkorn *n* = ideal middle-sized grain, ~ ~ particle

idealplastisch = perfectly plastic

Idealsieb *n* = idealized screen

~linie *f*, Sollsieblinie = ideal grading curve

Ideal|spannung *f* = ideal stress

~zustand *m* = ideal state

ideeller Punkt *m*, gedachter ~, imaginärer ~ = mathematical point

Ideenwettbewerb *m*, freier Wettbewerb = design competition

Igelwalze *f* → → Walze

Ikonographie *f* = iconography [*The study of ancient mosaic work, frescoes, statues, etc.*]

Ikosaeder *n*, Zwanzigflach *n*, Zwanzigflächner *m* = icosahedron

Illuvialhorizont *m* → B-Horizont

Ilmenit *m* = ilmenite [*A mineral, iron titanate ($FeTiO_3$), the ore of which is commonly used as aggregate in high density concrete*]

im Freien = outdoor

im Uhrzeigersinn, rechtsläufig = clockwise

imaginäre Zahl *f* = imaginary number, ~ quantity

im Gegenverkehr arbeiten [*z. B. Seilbahn*] = to ply (between two points)

imaginärer Punkt *m* → ideeller ~

imaginäres Gelenk *n*, gedachtes ~ = imaginary hinge

Imbißküche *f* [*Messestand*] = kitchen for the preparation of snacks

Imhoffbrunnen *m*, Emscherbrunnen = Imhoff tank, Emscher ~

imitieren = to simulate

Impact-Verfahren *n* [*Herstellung bituminöser Gemische*] = atomization method

Impfen *n* **von Abwasserschlamm** [*Verhütet die saure Gärung des Frisch-*

schlammes und leitet unmittelbar seine Methangärung ein] = mixing the fresh and ripe (sewage) sludge, seeding green (sewage) sludge with ripe (sewage) sludge

Impfschlamm m [*in Methangärung befindlicher Faulschlamm*] = ripe (sewage) sludge

Importerz n, Auslandserz, Einfuhrerz = foreign ore

Imprägnations|(bohr)krone f, diamantimprägnierte (Bohr)krone [*Diamantsplitter in Grundmasse eingebettet*] = diamond impregnated cemented carbide bit

Imprägnieren n, Tränken, Imprägnierung f, Tränkung = impregnation, impregnating

Imprägniermittel n = impregnation agent

imprägnierter Filz m = impregnated felt

Imprägnier(ungs)anlage f = impregnation plant

Impuls m, Bewegungsgröße f, Kraftstoß m [*Hydraulik*] = impulse

~**echoverfahren** n = pulse transmission-reflection method

~**folge** f = train of pulses

~**strom** m = pulsating electric current

~**verfahren** n = pulse technique

in gegenseitige Beziehung setzen = to interrelate

inaktiv, träge = inert

Inbetriebnahme f = putting into operation

Inbusschraube f = cap screw, sockethead ~

Indexbolzen m = index pin, ~ finger

Indikatorgas n = tracer gas

indirekt wirkende Ramme f → (Frei-)Fallramme

~ **wirkender Bär** m → Rammfallbär

indirekte (Baugrund)Sondierung f → (Baugrund)Sondierung mit indirektem Verfahren

~ **Wärmeschichtung** f, Polarschichtung = indirect temperature lamination, polar ~ ~

indische Schaufel f = Indian shovel

Indizierung f, Messen n der Leistung durch Kreislaufschreiben = indicating

Induktions|(-Aufschluß)verfahren n = induction method, ~ exploration, ~ prospecting

~**fluß-Magnetometer** n, Förster-Sonde f = airborne magnetometer

~**härtung** f = induction hardening

~**motor** m → Asynchronmotor für Drehstrom

~**rinnenofen** m = channel-type induction furnace

~**schwelle** f [*Verkehrszählgerät*] = inductive loop detector, induction vehicle ~

~**tiegelofen** m = induction crucible furnace

~**verfahren** n, Induktions-Aufschlußverfahren = induction exploration, ~ method, ~ prospecting

induktive Beeinflussung f [*Sieb*] = inductive effect [*screen*]

~ **Erwärmung** f = induction heating

industrialisierte Bauindustrie f = industrialized construction industry

industrialisierter Wohnungsbau m = industrial housing

industrialisiertes Bauen n = industrialized building, ~ construction

Industrialisierung f **des Bauwesens** = industrialization of building and civil engineering, ~ ~ construction industry

Industrie|abfall m = industrial refuse

~**abwasser** n, Industrieabwässer, industrielles Abwasser, industrielle Abwässer npl = industrial waste(s)

~**anlage** f = industrial plant, ~ installation, ~ facility

~**anstrich** m = industrial painting [*The use of paints, varnishes, and stains for protection of industrial equipment*]

~**architektur** f = industrial architecture

~**aufzug** m = industrial goods elevator (US); ~ ~ lift (Brit.)

Industriebahn — Industrie-Wasserturm

~bahn *f* → Baubahn
~bahngleis *n* → Baugleis
Industriebau *m* = industrial construction
~, Industriegebäude *n* = industrial building
~platte *f*, Bauplatte für den Industriebau = building board for industrial construction
~stelle *f* = industrial construction site
~stoff *m*, Kunstbaustoff = man-made construction material, ~ building ~
Industrie|bauten *f* = industrial buildings and structures
~bauvorhaben *n*, Industriebauprojekt *n* = industrial construction project
~behälter *m*, Industrietank *m* = industrial tank, ~ reservoir
~boden *m* → Betriebsboden
~dach *n* = industrial roof
~-Dach(ein)deckung *f* = industrial roofing
~decke *f* = industrial floor
~diamant *m* = industrial diamond
~dunst *m*, Stadtdunst, Rußnebel *m* = smog
~elektrofilter *m, n*, Cottrellfilter = Cottrell precipitator
~fahrzeug *n* = industrial vehicle
~fernsehen *n* = industrial television
~fett *n* = industrial grease
~(fuß)boden *m* → Betriebsboden
~(fuß)bodenplatte *f* = industrial floor(ing) tile
~gebäude *n*, Industriebau *m* = industrial building
~gebiet *n* = industrial zone
~geschoßbau *m*, Industriegeschoßgebäude *n* = industrial storied building
~gleis *n*, Anschlußgleis [*Fabrik*] = private siding
~-(Gleis)Kettenschlepper *m*, Industrie-Raupenschlepper *m* = industrial crawler tractor, ~ caterpillar ~
~-(Gummi)Reifenschlepper *m*, Industrie-Radschlepper = industrial wheel tractor

~hafen *m* [*Hafenteil zur Ansiedlung von industriellen Unternehmungen, meist im Zollausland*] = industrial port
~hafen *m* [*Fluß- oder Seehafen, der ausreichende Industriefläche und gute Verkehrsanschlüsse braucht*] = industrial harbo(u)r
~halle *f* = industrial hall
~heizung *f* = industrial heating
~hygiene *f* = industrial hygiene
~kalk *m* = industrial lime
~konsortium *n* = industrial consortium
~lader *m*, Industrielademaschine *f* = industrial loader, ~ loading machine
~lärm *m* = industrial noise
~-Lastkarre(n) *f, (m)*, Flurförderer *m*, Flurfördermittel *n* = mobile industrial handling equipment
industrielle Bebauung *f* = non-residential development
Industrie|luftreifen *m* = pneumatic tire for industrial use (US); ~ tyre ~ ~ ~ (Brit.)
~müll *m* = industrial rubbish, ~ refuse
~ofen *m* = industrial furnace
~pumpe *f* = industrial pump
~-Radschlepper *m*, Industrie-(Gummi)Reifenschlepper = industrial wheel tractor
~-Raupenschlepper *m*, Industrie-(Gleis)Kettenschlepper = industrial crawler tractor, ~ caterpillar ~
~rohr *n* = industrial pipe
~schornstein *m* = big chimney, industrial ~, stack
~sieb *n* = industrial screen
~standort *m* = industrial site
~staub *m*, technischer Staub = industrial dust
~unfall *m* = industrial accident
~-Wandfliese *f* = industrial wall tile
~wasser *n* = non-potable water for industrial purposes, industrial water
~-Wasserturm *m*, Industrie(Wasser)-Hochbehälter *m* = industrial water tower

induzierte Polarisation *f* = induced polarisation

ineinandergreifen, eingreifen, kämmen [*Zahnräder*] = to mesh

Ineinandergreifen *n*, Verspannung *f* [*von Schüttsteinen*] = interlocking

Infiltration *f*, Versickerung *f* = infiltration [*Rainfall reaching the ground moves through the soil surface, a process which is called infiltration*]

~ = infiltration [*The leaching of water from the ground into a sewer*]

Infiltrations|gebiet *n*, Versickerungsgebiet [*Dieses Gebiet unterscheidet sich vom Niederschlaggebiet durch Abströmbereich des versickernden Wassers*] = infiltration basin

~**wert** *m* = infiltration value

~**zahl** *f* [*Verhältnis Sickerwassermenge zu auf bestimmter Bodenoberfläche gelangender Niederschlagmenge*] = infiltration index

Informationsstand *m* = information counter

Infrarot|heizung *f* = infra-red heating

~**-Nahterwärmer** *m* [*Schwarzdeckenbau*] = infra red joint heater

~**-Spektroskopie** *f* = infrared spectroscopy [*Use of spectrophotometer for determination of infrared absorption spectra (2.5 to 18 micron wave lengths) of materials. Used for detection, determination, and identification of organic materials and the reaction between organic admixture and concrete components*]

Infraschall *m* = infrasonics

~**bohrwerkzeug** *n*, Schallbohrwerkzeug = sonic drill

Infusorienerde *f* = infusorial earth

Ingenieur *m* für Außendienst = field engineer

~ ~ **Innendienst** = desk engineer

Ingenieurbau *m* = civil engineering

~**arbeiten** *fpl* = civil engineering work

~**beratungsfirma** *f* = consulting civil engineers

~**firma** *f* = civil engineering firm, engineer-construction ~

~**klinker** *m* = clay engineering brick

~**stelle** *f* = (civil) engineering site

~**stoff** *m* = civil engineering material

~**werk** *n* = (civil) engineering structure

Ingenieur|biologie *f* = engineering biology

~**-Geographie** *f* = engineering geography

~**geologe** *f* = engineering geologist

~**geologie** *f*, Baugrundgeologie = engineering geology

~**-Gewässerkunde** *f* = engineering hydrology

~**-Hilfskraft** *f* = technician (US)

~**konsulent** *m* [*Österreich*] → beratender Ingenieur *m*

~**mathematik** *f* = engineering mathematics

~**-Meteorologie** *f* = engineering meteorology

~**personal** *n* = professional grades (US)

~**wesen** *n*, Technik *f*, Ingenieurwissenschaft *f* = engineering (science)

Inhaltwertangabe *f* = declaration of value

Injektion *f*, Auspressung *f*, Einpressung, Verpressung, Abpressen *n*, Injizieren = injection, grouting

Injektions|anlage *f*, Auspreßanlage, Verpreßanlage, Einpreßanlage, Injizieranlage = injection installation, ~ plant, grouting ~

~**arbeiten** *fpl*, Einpreßarbeiten, Verpreßarbeiten, Auspreßarbeiten, Injizierarbeiten = grouting work

~**-Diaphragma** *n* → Einpreß-Dichtwand *f*

~**-Dicht(ungs)wand** *f* → Einpreß-Dichtwand

~**druck** *m*, Verpreßdruck, Einpreßdruck, Auspreßdruck, Injizierdruck = injection pressure, grouting ~

~**-Entlüftungsbohrung** *f*, Auspreß-Entlüftungsbohrung, Einpreß-Entlüftungsbohrung, Verpreß-Entlüftungsbohrung, Injizier-Entlüftungsbohrung = injection air vent, grouting ~ ~

Injektionsgang — Innenanstrich

~**gang** *m* → Auspreßgang
~**gerät** *n* → Einpreßgerät
~**gründung** *f* → Auspreßgründung
~**gürtel** *m* → Einpreß-Dichtwand *f*
~**gut** *n*, Auspreßgut, Einpreßgut, Verpreßgut, Injiziergut = injection material, grouting ~
~**hilfsmittel** *n*, Auspreßhilfsmittel, Einpreßhilfsmittel, Verpreßhilfsmittel, Injizierhilfsmittel = grouting aid
~**kammer** *f*, Verpreßkammer, Auspreßkammer, Einpreßkammer, Injizierkammer = injection chamber, grouting ~
~**lanze** *f*, Auspreßlanze, Verpreßlanze, Einpreßlanze, Injizierlanze = injection rod, grouting ~
~**loch** *n*, Einpreßloch, Auspreßloch, Verpreßloch, Injektionsöffnung *f*, Einpreßöffnung, Auspreßöffnung, Verpreßöffnung = injection hole, grouting ~
~**mörtel** *m*, Einpreßmörtel, Auspreßmörtel, Verpreßmörtel, Eindringmörtel, Injiziermörtel = intrusion mortar, grouting ~, injection ~
~**öffnung** *f* → Injektionsloch *n*
~**pumpe** *f*, Verpreßpumpe, Einpreßpumpe, Auspreßpumpe, Injizierpumpe = injection pump, grouting ~
~**ring** *m*, Verpreßring, Auspreßring, Einpreßring Injizierring [*Tunnelund Stollenbau*] = injected annulus, ~ ring, treated ~, grouted ~
~**rohr** *n*, Einpreßrohr, Auspreßrohr, Verpreßrohr, Injizierrohr = injection pipe, grouting ~
~**schirm** *m* → Einpreß-Dichtwand *f*
~**schlauch** *m* → Einpreßschlauch
~**-Schleier** *m* → Einpreß-Dichtwand *f*
~**schürze** *f* → Einpreß-Dichtwand *f*
~**-Sperrwand** *f* → Einpreß-Dichtwand
~**stutzen** *m*, Auspreßstutzen, Verpreßstutzen, Einpreßstutzen, Injizierstutzen = injection socket, grouting ~

~**verfahren** *n*, Einpreßverfahren, Verpreßverfahren, Auspreßverfahren, Injizierverfahren = injection method, grouting ~
~**zement** *m*, Verpreßzement, Einpreßzement, Auspreßzement, Injizierzement = grouting cement, injection ~
Injektor *m*, Dampfstrahlpumpe *f* [*Speisung von Dampfkesseln*] = (steam) injector [*boiler feeding*]
~ → Verpreßkessel
~**brenner** *m* = low pressure torch
injizieren, auspressen, einpressen, verpressen = to inject
Injizier|gang *m* → Auspreßgang
~**gründung** *f* → Auspreßgründung
injizierte Schüttung *f* → ausgepreßte ~
(injizierter) Alluvionsschleier *m*, ausgepreßter ~, eingepreßter ~, verpreßter ~ = grouted alluvium
inklinante Buhne *f*, stromauf(wärts) gerichtete ~ = groyne (Brit.)/groin (US) pointing slightly upstream
inkohärent → nichtbindig
Inkohlung *f* = coalification, incoalation
Inkrustierung *f*, Ablagerung, Inkrustation *f* = incrustation
Inland|düne *f* → Windüne
~**eis(decke)** *n*, (*f*) → Binneneisdecke
~**erz** *n* = home ore
~**flughafen** *m* = domestic airport
~**-Niederlassung** *f* = domestic office
~**passagier** *m* = domestic passenger
~**tätigkeit** *f* [*einer Baufirma*] = domestic activities
~**warteraum** *m* [*Flughafen*] = domestic waiting room
innen glattes Bohrgestänge *n* [*Rotarybohren*] = internal flush joint drill pipe
~ **verstärktes Bohrgestänge** *n* = internal upset drill pipe
Innen|ansicht *f* = interior view
~**(an)streichen** *n* → Innenanstrich *m*
~**anstrich** *m*, Innen(an)streichen *n* = interior painting [*The interior use*

548

Innenanstrich — Innenputz

of paints, varnishes and stains for protection or decoration of materials]
~**anstrich** *m*, Innenfarbschicht *f*, Innenfarbanstrich = interior paint coat, ~ (~) finish, ~ coat of paint
~**(anstrich)farbe** *f* = interior paint
~**anwendung** *f* → Innenverwendung
~**architekt** *m* = interior decorator
~**aufzug** *m* → Innenbauaufzug
~**ausbau** *m* = interior finish, ~ trim
~**backenbremse** *f* = expanding brake, internal (expanding) shoe-brake
~**(bau)aufzug** *m* = internal (contractors') hoist
~**(bau)platte** *f* = building panel for interior use
~**bekleidung** *f*, Innenauskleidung, Innenverkleidung = internal lining
~**-Betonkübelaufzug** *m*, Innen-Kübel(bau)aufzug, Innen-Mulden(bau)aufzug = internal skip-hoist
~**betonstütze** *f* = interior concrete column
~**dichtung** *f* → (Dichtungs)Kern *m*
~**druck** *m* [*Tiefbohrtechnik*] = bursting
~**druck** *m* [*Rohrleitung*] = internal pressure
~**drücker** *m*, untenliegender Drücker, Fingerabzug *m* [*liegt auf der Unterseite des Handgriffs*] = inside lever, bottom ~ [*(compressed-)air throttle control*]
~**druck-Rohrprüfgerät** *n* = hydraulic test machine (Brit.); permeability ~ ~, hydrostatic ~ ~ (US)
~**druckversuch** *m*, Innendruckprobe *f*, Innendruckprüfung *f* [*Rohr*] = hydraulic test (Brit.); permeability ~, hydrostatic ~ (US)
~**durchmesser** *m* **des zusammengebauten Nockenlagers** = cam bearing assembled ID
~**(ein)rüttelung** *f* [*Beton*] = internal vibration, consolidation (of concrete) by internal vibration
~**farbanstrich** *m* → Innenanstrich
~**farbe** *f*, Innenanstrichfarbe = interior paint
~**farbschicht** *f* → Innenanstrich *m*

~**fenster** *n* = borrowed light [*A window in an internal wall or partition*]
~**fensterbank** *f* = internal window sill
~**fläche** *f* [*Rohr*] = internal surface
~**flansch** *m* **der Felge** = rim inner flange
~**(fuß)boden** *m* = interior flooring
~**gestaltung** *f* = interior decorating
~**getriebe** *n* [*Paarung eines außen verzahnten Stirnrades mit einem innen verzahnten Stirnrad*] = internal-teeth spur gearing
~**gewinde** *n* = female (thread)
innengezackter Abfangkeil *m*, ~ Rohrklemmkeil, innengezackte Rohrkeilklemme *f* [*Rotarybohren*] = serrated (rotary) slip
Innen|-Holzeinbauten *fpl* [*bei einem Gebäude*] = internal joinery
~**kanal** *m* [*Hydraulikanlage*] = cored passage
~**kegel** *m*, Innenkonus *m* = internal taper, taper socket
~**(kern)rohr** *n* = inner cylinder, ~ tube
~**kessel** *m* [*Gußasphaltmotorkocher*] = pan
~**konus** *m*, Innenkegel *m* = taper socket, internal taper
~**korrosion** *f* = internal corrosion
~**-Kübel(bau)aufzug** *m*, Innen-Mulden(bau)aufzug, Innen-Betonkübelaufzug = internal skip-hoist
~**-Laufring** *m* = ball race
~**leitung** *f* = internal line
~**leitung** *f* = supply piping [*The portion lying within the consumers' premises*]
~**moräne** *f* = englacial moraine
~**motor** *m* = internal motor
~**-Mulden(bau)aufzug** *m*, Innen-Kübel(bau)aufzug, Innen-Betonkübelaufzug = internal skip-hoist
~**platte** *f*, Innenbauplatte = building panel for interior use
~**putz** *m* [*siehe englische Anmerkung unter „Putz"*] = internal finish, plastering (~), internal plastering, interior plaster

Innenrad — (innerer) Reibungswinkel

~rad *n*, innenverzahntes Stirnrad = internal teeth gear

~ring *m*, Kernring [*Tunnel- und Stollenbau*] = internal ring

~rohr *n*, Innenkernrohr = inner tube, ~ cylinder

~rüstung *f*, Innengerüst *n* = internal scaffold(ing)

~-Rüttelbeton *m* = internally vibrated concrete

(~)Rüttelflasche *f*, Vibrationsflasche = vibrating cylinder, ~ head

Innenrüttler *m*, Innenvibrator *m*, Flaschenrüttler, Flaschenvibrator, Tauchrüttler, Tauchvibrator, Tiefenrüttler, Tiefenvibrator, Nadelrüttler, Nadelvibrator, Beton~ [*Die Ausdrücke "Tiefenrüttler" und "Tiefenvibrator" sollten nicht mehr verwendet werden, um eine Verwechslung mit dem Tiefenrüttler für Bodenverdichtungen zu vermeiden*] = internal vibrator, immersion ~, concrete mass ~, (concrete) poker ~, needle ~, pervibrator, concrete vibrator for mass work, spud vibrator for concrete

~ mit starrer Welle, Innenvibrator *m* ~ ~ ~ = stiff-shaft vibrator

Innen|schalung *f* = internal formwork

~schale *f* [*Hohlmauer*] = internal leaf

~schaufel *f* [*Turbomischer*] = internal mixing blade

~scheibe *f* [*Kupplung*] = internal disc; ~ disk (US)

~scheitel *m* = crown [*The inside top of a sewer*]

~schiebefenster *n* = sliding inner window

~schiene *f* = internal rail

~schweißen *n* = internal welding

~seite *f* = inner face

~spannung *f* = internal stress

~(-Stahl)stütze *f* = interior stanchion, ~ stauncheon

~stempel *m*, Oberstempel (⚒) = upper prop, ~ post

~streichen *n* → Innenanstrich *m*

~stütze *f* = interior column

~stütze *f* → Innen-Stahlstütze

~taster *m*, Lochtaster = inside calipers

~temperatur *f*, Raumtemperatur = indoor temperature

~treppe *f* = interior staircase (Brit.); ~ stair (US)

~tür *f* = inside door, internal ~

~verglasung *f* = internal glazing

~verkehr *m* = internal traffic

~verwendung *f*, Innenanwendung = internal use, inside ~, ~ application, interior ~

~vibrator *m* → Innenrüttler *m*

~vibrator *m* mit starrer Welle, Innenrüttler *m* ~ ~ ~ = stiff-shaft vibrator

~vorspannung *f* = inner prestress

~wand *f* = internal wall, interior ~, inside ~

~wandputz *m* [*siehe Anmerkung unter "Putz"*] = internal wall plaster(ing)

~wandtafel *f* = internal wall panel

~zylinder *m* [*Stoßdämpferzylinder*] = rod assembly [*suspension cylinder*]

innerbetrieblicher Transport *m* = intraplant transportation

innere Abscheuerung *f*, Ausspülung = internal erosion

~ **Bodenplatte** *f* [*Unterkellerung*] = loading coat, ~ slab [*A concrete slab laid over asphalt tanking, to ensure that it is not pushed upwards by water pressure below it*]

~ **Bogenfläche** *f*, Bogenleibung *f* = soffit, intrados

~ **Bremsnabe** *f* = inner brake hub

~ **Gewölbefläche** *f*, Gewölbeleibung *f* = soffit of vault, intrados ~ ~

~ **Kraft** *f* = internal force

~ **Molekularspannung** *f*, Molekularanziehung = molecular attraction

~ **Wärmeübergangszahl** *f* = inside film coefficient [*BTU/h sqft. F*]

innerer Lagerring *m*, innerer Laufring = inner race

(~) **Reibungswinkel** *m*, Winkel der inneren Reibung = angle of internal friction

innerer Widerstand — Intersertaltextur

~ **Widerstand** m [*hervorgerufen durch die Zapfen- und Lagerreibung*] = internal friction, influence on rolling resistance
inneres Planetenrad n, Planeteninnenrad = inner planetary gear
~ **Steuerventil** n = interior valve
~ **Ufer** n → ausbuchtendes ~
inner|städtische Autobahn f, Expreßstraße f [*Schweiz*]; Stadtautobahn = urban motorway
~**städtischer Schnellstraßenring** m = network of urban motorways
Inproduktionssetzungsversuch m [*Erdöl; Erdgas*] = production starting test
Insasse m, Fahrzeug~ = occupant, passenger
Insekten|befall m = infestation (US); insect outbreak, insect attack
~**vertilgungsmittel** n = insecticide
Insel f, Absenk~ [*Caissongründung*] = sand island, artificial ~
~**bahnhof** m = island station
~**bahnsteig** m = island platform
~**schutz** m = protection of islands
instabil = instable, unstable
Instabilitätsschauer m = instability shower
Installation f = building equipment, installation [*Services and other plant used in the completed building*]
Installationshammer m → Bauhammer
installierte Leistung f, Leistungssoll n = installed (nameplate) capacity
~ **Turbinenleistung** f = installed wheel capacity, ~ turbine ~
Instand|haltung f → Unterhaltung
~**setzung** f, Ausbesserung = reinstatement, putting in working order
~**setzungsarbeiten** fpl, Ausbesserungsarbeiten = reinstatement work
~**setzungskosten** f, Ausbesserungskosten = reinstatement cost
instationäre Strömung f = unsteady flow
Instellungbringen n [*Pfahl*] = pitching
~ = positioning
Institut n **für Bauforschung** → Bauforschungsanstalt f

Instrumentation f, Fangarbeit f, Fang m, Instrumentierung f [*Tiefbohrtechnik*] = fishing job
Instrumenten|anflug m = ILS approach
~**brett** n → Armaturenbrett
~**-Landebahn** f = instrument runway
~**landesystem** n = instrument(al) landing system, ILS
~**netz** n = instrument network, network of instruments [*e. g. to measure rainfall in a catchment area*]
~**tafel** f → Armaturenbrett n
instrumentieren, fangen [*Tiefbohrtechnik*] = to fish
Instrumentierung f → Instrumentation f
Integrale f = intergral
Interkontinentalflughafen m = intercontinental airport
inter|kristalline Korrosion f, Korngrenzenkorrosion = intercrystalline corrosion, intergranular ~
~**kristalliner Riß** m, Korngrenzriß [*Korrosion*] = intercrystalline crack
~**metallische Verbindung** f = intermetallic compound
~**mittierende Gasliftförderung** f = slug-lifting
~**mittierende Springquelle** f → Geysir m
~**mittierender Filter** m, intermittierendes Filter n = intermittent filter [*A natural or artifical bed of sand or other fine-grained material to which sewage is intermittently applied in doses and through which it flows, opportunity being given for filtration and also oxidation of the organic matter by biochemical agencies*]
internationale Ausschreibung f = global tendering
Internationaler Bausparkassenverband m = International Union of Building Societies and Savings and Loan Associations
interne Vorspannung f = internal prestress
Intersertaltextur f → divergentstrahlig-körnige Textur

Intraflügelpumpe — Isolierstein 552

Intraflügelpumpe *f* = intra vane pump
intra|kristalline Korrosion *f*, transkristalline ~ = intracrystalline corrosion, transcrystalline ~
~kristalliner Riß *m*, transkristalliner ~ [*Korrosion*] = transcrystalline crack, intracrystalline ~
~zonaler Boden *m* = intrazonal soil
Intrusion *f* (Geol.) = intrusion
Intrusionsbeben *n*, kryptovulkanisches Beben = cryptovulcanic earthquake
Intrusiv|gestein *n* → plutonisches Gestein
~lager *n* (Geol.) = intrusive sheet
Inundationsfläche *f*, Eintauchfläche (Geol.) = flood plain
Invar|band *n* = invar tape
~draht *m* = invar wire
~stahl *m* = invar steel
inventarisieren = to inventory
Inventur *f* = stocktaking
Investitionsfreibetrag *m* = investment allowance
Ion *n* = ion [*An electrically charged particle, atom, or group of atoms*]
Ionenaustausch *m* = ion exchange
ionenaustauschender Stoff *m* = ion-exchange material [*Is a substance so loosely bound chemically that, when it is placed in a solution of greater ionic concentration, cations will be exchanged by cation exchangers and anions will be exchanged by anion exchangers*]
ionische Ordnung *f* = Ionic order
~ Volute *f* = Ionic volute
ionisches Kyma *n*, ionische Kymation *f* = cyma(tium) reversa
ionisieren = to ionize
Ionisierungskammer *f* [*Reaktoranlage*] = ion chamber
Irdengut *n* [*Scherben porös, wassersaugend, nicht durchscheinend*] = earthenware
irländisches Moos *n*, Perlmoos, Knorpeltang *m* = carragheen, Irish moss
irrationale Zahl *f*, nicht aufgehende ~ = irrational number, surd ~, ~ quantity

Irrstrom *m*, vagabundierender (Erd-)Strom = stray current
isländischer Kalkspat *m* → Doppelspat
~ Quellsprudel *m* → Geysir *m*
isländisches Moos *n* → Lungenmoos
Isobare *f* = isobar, line of equal pressure
Isobathe *f* = isobath
Isochromate *f*, Farbgleiche *f* = isochromatic line
Isohyete *f* = isohyetal line
Isoklinaltal *n*, Scheidetal = isoclinal valley
Isokline *f*, Neigungsgleiche *f* = isoclinic line
Isokonzentrationslinie *f* = line of equal sediment concentration
Isokracken *n* = isocracking
Isolator *m* = insulator
Isolier|anstrich *m* = insulating paint coat, ~ coat of paint
~band *n* = rubber tape
Isolieren *n* [*Die Sicherung der Baukonstruktionen gegen das Eindringen von (Boden)Feuchtigkeit und Grundwasser. Den Schutz gegen Wärme- und Schalldurchgang nennt man Dämmung*] = dampproofing (against (soil) moisture); waterproofing (against ground water)
Isolier|estrich *m* = insulation screed
~fähigkeit *f* = insulating property
~folie *f* = insulating foil
~glas *n* = insulating glass
~kitt *m* = electrical insulating putty
~masse *f* **zum Isolieren von Restfeuchtigkeit in Estrichen und Wandflächen sowie kapillarer Feuchtigkeit in nicht unterkellerten Räumen** = damp-resistant compound
~öl *n* = (electrical) insulating oil
~papier *n* = insulating paper
~pappe *f* = insulating felt
~platte *f* = insulating slab
~-Profil *n* = insulating section
~pulver *n* = insulating powder
~stein *m* = insulating refractory [*A good grade of refractory fireclay brick with a large percentage of open pore space. This open pore space may be*

about 70–75 percent as compared to approximately 20 percent for high-duty fireclay brick]
~streifen m = insulating strip
isolierte Fuge f = insulated joint
isolierter Heizleiter m aus Widerstandsdraht [*Elektrofußbodenheizung*] = insulated heater wire
Isolierung n gegen Wasser = waterproofing [*It prevents the actual flow of water through a wall, etc.*]
Isolier|ungsarbeiten fpl = insulation work
~(ungs)stoff m = insulating material
~wert m = insulation value
isometrische Darstellung f, maßgleiche ~ = isometric representation
Isostate f, Spannungsgleiche f = isostatic line
Isotache f = equivelocity contour
Isotherme f = isotherm, line of equal temperature, isothermal line
Isotopenindikator m, Isotopengerät n = radioactive tracer
isotropes kreiszylindrisches Schalendach n = isotropic cylindrical shell roof
Ist-Abmessung f, Ist-Maß n = actual dimension
ISTEG-Stahl m = ISTEG (reinforcing) steel
Ist-Größe f = actual size
Istlinie f, Sieblinie des Rohmaterials = actual grading curve
Ist-Maß n, Ist-Abmessung f = actual dimension
Ist-Querschnitt m = actual section
Ist-Spiel n = actual play
Ist-Wert m = actual value
Iterationsfolge f = iteration sequence
Iterationsverfahren n = iteration method

J

Jahres|niederschlag m = annual precipitation
~speicher m, hydraulischer ~ [*Dient zum Ausgleich von Wasserdargebot und Wasserbedarf über ein Jahr*] = annual storage reservoir
~umschlag m [*Hafen*] = annual traffic
jährliche Unterhaltung f, Jahresunterhaltung = annual upkeep, ~ maintenance
jährliches Hochwasser n = annual flood
Jalousie f [*Die feststehenden schrägen Brettchen in hölzernen Klappladen, die Luft durchlassen, aber den Einblick verwehren*] = louver, louvre, ~ board
~ [*Aus verstellbaren waag(e)rechten Blättchen, mit Schnürzug bedient*] = lever board, adjustable louvre, adjustable louver
Jamesonit m → Federerz n
Japaner(karren) m, Japaner-(Kipp)-Karre f, (Kipp-)Betonkarre(n), Beton(rund)kipper = concrete buggy, hand concrete-cart, rocker-dump hand cart
japanischer Lack m = Japanese lacquer [*A glossy lacquer obtained by tapping the sap from the Japanese varnish tree or sumach*]
Japanlack m [*fetter Holzöl-Kombinations-Harzlack*] = (black) Japan
Jaspé n [*Linoleumart*] = jaspe
~bahn f = length of jaspe sheet
jenseitiges Ufer n, gegenüberliegendes ~ = far bank
J-Kante f = J-edge
Joch n = bent
~holz n [*Geviertausbau*] = wall plate
Jodsilber n = iodargyrite
JOHNSON-Turm m [*vollautomatische Betonmischanlage*] = JOHNSON mixing tower
Joosten-Verfahren n = Joosten process
Jörgensen-Mauer f → → Talsperre f
Judenpech n = bitumen judaicum, Jew's pitch
Jumbo-Bohrwagen m, Bohrjumbo m, Ausleger-Bohrwagen, Groß-Bohrwagen [*schwerer Bohrwagen auf Gleis mit schweren Bohrhämmern*] = (tunnel) jumbo, drilling ~

Jungfer — Kabelschuh

Jungfer f → Ramm~
jungfräulicher Boden m = virgin soil
Junggesellenwohnung f = bachelor flat
Jura|kalkstein m = Jurassic limestone
~riff n = Jurassic reef
~sandstein m = Jurrassic sandstone
Justieren n → Berichtigung f
Justierung f → Berichtigung
Jute|bahn f = hessian mat, burlap ~
~fabrik f, Jutewerk n = jute mill, ~ works, ~ plant, ~ factory
~(faser) f [*Bastfaser, gewonnen aus in Indien und Amerika wachsenden krautigen Gewächsen. Die Jute(faser) dient u.a. als Träger für bituminöse Dichtungsstoffe und Linoleummasse*] = jute (fiber) (US); ~ fibre (Brit.)
~filz m = felted jute
~gewebe n = hessian (Brit.); burlap (US) [*Strong coarse material woven from jute*]
juveniles Wasser n = juvenile water, magmatic ~ [*It is given off by igneous rocks during cooling and consolidation, occasionnally met in tunnelling*]

K

Kabel n, Elektro~ = (electric) cable ~, Seil n = cable, rope
~ aus 48 Drähten von 7 mm ⌀ = cable of 48-7 mm wires
~abdeckplatte f, Kabelabdeckstein m = cable cover
~abzug m = cable draw-off gear
~anhebeverfahren n [*Spannbeton*] = deflected-strand technique
(~)Anhebevorrichtung f [*Spannbeton*] = (strand) deflecting device
~aussparung f [*Spannbeton*] = opening formed for the passage of a cable
~bagger m = cable excavator
~baggerturm m = cable excavator tower
~brücke f → Kabelhängebrücke
~bündel n = bunched cables
~durchhang m, Seildurchhang = cable sag
~formstück n, Kabelformstein m = conduit tile (US); (multiple way) cable duct (or conduit, or subway), multitubular slab for cable
~graben m = cable trench
~(hänge)brücke f, Seil(hänge)brücke = cable suspension bridge, rope ~ ~
~hängedach n, Seilhängedach = = rope(-suspended) roof, cable (-suspended)
~hochführungsschacht m = cable chute
~höhe f, Drahtseilhöhe = cable level, rope ~
~kanal m = cable duct
~kette f = cable chain
Kabelkran m = blondin (Brit.); tautline cableway, (aerial) cable-way, cable-crane, overhead cableway
~ mit Beton-Gießvorrichtung, Beton-Kabel(hochbahn)kran = concrete-chuting cable crane
~-Bedienungshaus n = cableway control house
~gegenturm m, fahrbarer Kabelkranturm = tail tower
~kübel m → Betonkübel
(~)Maschinenturm m, fester (Kabelkran)Turm, Antriebsturm = head tower
~winde f = cable-crane winch
Kabel|lage f, Seillage = strand layer, layer of strands
~mantelpresse f = cable sheathing press
~merkstein m = cable marker
~netz n = cable network, ~ grid
~schacht m = cable draw pit
~schelle f = cable clamp, ~ band
~schmiermittel n = cable lubricant
~schranke f = cable barrier
~schrapper m, Seilschrapper = cable excavator
~schrapperkasten m, Seilschrapperkasten = scraper, cable-hauled bucket
~schuh m = strand shoe

Kabelschutzhaube — Kaitreppe

~schutzhaube *f* [*DIN 279*] = cable tile
~schutzrohr *n* = electric cable pipe
~spinnen *n*, Seilspinnen [*Brückenbau*] = (cable) spinning (operation)
~steg *m*, Seilsteg [*Kabelspinnen beim Brückenbau*] = footbridge on which the cables are strung
~stollen *m*, Kabeltunnel *m* = cable tunnel
~strebe *f*, Seilstrebe [*Hängebrücke*] = cable stay
~suchgerät *n* = cable tracing set, cable detector
~trommel *f*, Seiltrommel = rope drum, cable ~
~trommeltransportanhänger *m* = cable drum trailer, rope ~ ~
~trommelwagen *m* = cable reel truck
~tunnel *m*, Kabelstollen *m* = cable tunnel
~turm *m* → Pylone
~verankerung *f*, Seilverankerung [*Brückenbau*] = cable anchorage
~verankerungspfeiler *m* [*Hängebrücke*] = cable anchorage pier
~vergußmasse *f* = electrical filling compound, ~ compound for cable isolations
~verlegeanlage *f* = cable laying plant
~verlegemaschine *f* = cable laying machine
~verlegewinde *f* = cable laying winch
~verlegung *f* = cable laying
~verteilerkasten *m* = cable distribution box
~verzweigung *f* = cable branch(ing)
~winde *f* → Kanalwinde
Kabine *f* = cab(in)
Kachel *f* [*veralteter Begriff*] → keramische Wandfliese *f*
~ofen *m* = tiled stove, Dutch ~
Kadmium *n* = cadmium
~gelb *n* = cadmium yellow
Kahnentladebecherwerk *n* = barge elevator
Kähner *m*, Kandel *m*, Kändel *m* [*Die Dachrinne zur Aufnahme des Regen- und Schneewassers*] = (roof) gutter

~, Kandel *m*, Kändel *m* [*Rechteckiges Gerinne beliebiges Materials zum Überqueren eines Wasserlaufes oder einer Geländemulde für Zwecke der Bewässerung*] = rectangular flume
Kahntal *n*, Hufeisental = canoe(-shaped) valley, cigar-shaped ~
Kai *m* [*in Ostfriesland „Kajung" und an der Unterweser „Kaje" genannt*] = quay
~anlagen *fpl* = quay(side) appliances, ~ accomodations
~fläche *f*, Kaiplanum *n* = quay surface (Brit.); surface of esplanade (US)
~gleis *n* = quay track
~kante *f* = arris of quay
~kran *m* = quay crane
Kaimauer *f* = quay wall
~ **auf Gewölben** = quay wall on arches
~ ~ **Senkkästen** = quay wall on caissons
~ **aus geneigten Blocklagen** = quay wall of sliced blockwork, ~ ~ sloping ~
~ ~ **im Verband verlegten Blöcken** = quay wall of set blocks
~ ~ **Stahlbeton-Fertigteilen** = quay wall of precast reinforced concrete units
~ ~ **Stahlspundwänden** = quay wall of steel sheet piling
~ **mit Steinverkleidung** = quay wall with stone facing
Kai|pfeiler *m* = quay pier
~planum *n* → Kaifläche *f*
~platz *m*, Liegeplatz = berth
~platzsignal *n*, Liegeplatzsignal = berthing signal
~schuppen *m* → Hafenspeicher *m*
Kaiserdach *n* → Zwiebeldach
KAISER-Decke *f* = KAISER floor
Kaiser|stiel *m*, Helmstange *f* [*Dachkonstruktion*] = broach post
~-**Wilhelm-Kanal** *m*, Nord-Ostsee-Kanal = North Sea and Baltic Canal, Kiel ~
Kaitreppe *f* = quay staircase (Brit.); ~ stair (US)

Kaizunge — Kalkglas

Kaizunge *f* → Pier
Kaje *f* → Kai *m*
Kajung *f* → Kai *m*
Kalben *n* [*Gletscher; Eisberg*] = calving
Kalfaterer *m* = ca(u)lker [*A person who ca(u)lks boats, ships, etc.*]
Kalfaterhammer *m* = ca(u)lking mallet
kalfatern = to caulk; to calk (US) [*to stop up cracks of windows, pipes, etc. with a filler*]
Kalfatern *n* = ca(u)lking
Kalfaterwerkzeug *n* = ca(u)lker
Kali|alaun *m* = potash alum
~**bergbau** *m* = potash mining
~**bergwerk** *n*, Kaligrube *f* = potash mine
kalifornischer Index *m* → CBR-Wert *m*
kalifornisches Tragfähigkeitsverhältnis *n* → CBR-Wert *m* [*The ratio of (1) the force per unit area required to penetrate a soil mass with a 3 sq. in. circular piston at the rate of 0.05 in. per min to (2) the force required for corresponding penetration of a standard crushed rock base material; the ratio is usually determined at 0.1 in. penetration*]
Kali|grube *f*, Kalibergwerk *n* = potash mine
~**salpeter** *m* → Kaliumnitrat *n*
~**salz** *n* [*früher: Abraumsalz*] = potash salt, rubbish ~, abraum ~
Kalium|-Aluminium-Sulfat *n* → Alaun *m*
~**-Feldspat** *m*, Orthoklas *m*, $K_2Al_2Si_6O_{16}$ = orthoclase
~**karbonat** *n*, Pottasche *f*, kohlensaures Kalium *n*, K_2CO_3 = potassium carbonate
~**nitrat** *n*, Kalisalpeter *m*, KNO_3 = nitre, (India) salpetre, potassium nitrate, potassic nitrate, saltpeter
~**permanganat** *n*, KMn_nO = potassium permanganate
Kalk *m* = lime
kalkablagernd, kalkabscheidend, kalkausfällend, kalkausscheidend = lime-depositing, lime-precipitating, lime-secreting
Kalk|alabaster *m* = calcareous alabaster
~**alkaligestein** *n*, Alkalikalkgestein = calc-alkali rock
~**alkaligranit** *m* = calc-alkali granite
~**alkalisyenit** *m*, gewöhnlicher Syenit = calc-alkali syenite
~**beton** *m* = lime concrete
kalkbildend = calcigenous
Kalk|bindung *f* = binding of lime
~**blau** *n* = lime blue
~**brei** *m* → Sumpfkalk *m*
~**brennen** *n* = lime burning, calcining
~**brennerei** *f*, Kalkfabrik *f*, Kalkwerk *n* = quicklime manufacturing plant, lime ~
Kalk(brenn)ofen *m* = lime kiln
~ **für das Fließverfahren** = fluidizing calciner
~ ~ ~ ~ = Dorr Fluo-Solide Reactor [*Trademark*]
Kalk|drehofen *m*, Kalkrotierofen = rotary lime kiln
~**eisenstein** *m* = ferruginous limestone, calcareous iron-stone
kalken, tünchen, weißen = to limewash, to whitewash, to whiten
Kalk|erde *f* = calcareous earth
~**erzeugnis** *n* = lime product
~**estrich** *m*, russischer Mörtelestrich = lime mortar flooring
~**fabrik** *f* → Kalkbrennerei *f*
~**faktor** *m* = lime factor
~**fazies** *f* = calcareous facies, limestone ~
~**feldspat** *m* → Anorthit *m*
~**fett** *n*, Kalkseifenfett, kalkverseiftes Fett = lime base grease, calcium ~
kalkgebundener Dinasstein *m* = English dinas
Kalkgips|mörtel *m* = lime-gypsum mortar
~**putz** *m*, Kalkgipsmörtelbelag *m* = lime-gypsum plaster
~**putzmörtel** *m* = lime-gypsum stuff
Kalk|glas *n* = lime glass

Kalkgrube — Kalksandmörtel

~grube f, Kalklöschgrube, Löschgrube = lime pit, (lime) slaking ~
~grün n, Grünerde f = lime green
~haarmörtel m, Haarkalkmörtel = hair(ed) lime mortar
kalk|haltig, kalkig = calcareous, limey
~(halt)iger Ton m = calcareous clay, limey ~
Kalk|harmotom m = lime-harmotome, phillipsite, christianite
~hornfels m = lime silicate rock, calcariferous petrosilex
~hydratisiermaschine f = lime hydrating machine
~kern m, (Kalk)krebs m, Fehlbrand m = cove, unburned lime
~kies m = calcareous gravel, limey ~
~kiesel m (Min.) = calcareous silex
~kitt m = lime cement
~krebs m → Kalkkern m
~krücke f, Kalkrührer m = lime raker, ~ stirrer
~leichtbeton m = lightweight lime concrete
~leichtbetonstein m = lightweight lime concrete block
~licht n → Knallgaslicht
(~)Löschbank f = (lime) slaking trough
~löschen n = lime slaking
~(lösch)grube f, Löschgrube = (lime) slaking pit, lime ~
~löschmaschine f = lime-slaking machine
(~)Löschpfanne f = slaking pan, ~ vessel
(~)Löschtrommel f, (Kalk-)Trommel-Löscher m = (lime) slaking drum
kalklösende Kohlensäure f = agressive carbonic acid
Kalk|manganspat m = mangano-calcite
~marmor m = calcitic marble, limestone ~, crystalline limestone
~mehl n, Kalksteinmehl = flour lime
~mehlfüller m, Kalksteinfüller = limestone filler
~mehl-Trocknungstrommel f = limestone dust drying mill

~mergel m [75-90% $CaCO_3$] = lime marl, calcareous ~
~milch f, Tünche f, Weiße f = limewash, whitewash, whitening
Kalkmörtel m = lime mortar, L. M.
~belag m → Kalkputz m
~putz m → Kalkputz
Kalk|mühle f = quicklime mill
~niere f → Kalkschwüle f
~ofen m, Kalkbrennofen = lime kiln
~oligoklas m = calcic oligoclase
~onyx m → Onyx(marmor) m
~oolith m, Oolithkalk m, oolithischer Kalkstein m, Eisteinkalk m = öolite, öolitic limestone
~phyllit m = limestone phyllite
~pulver n, Feinkalk m [*Branntkalk in feingemahlener Form*] = powdered quicklime, ground ~, quicklime in powder form
~pumpe f = lime paste handling pump
Kalkputz m, Kalkmörtelbelag m, Kalkverputz, Kalkmörtelputz = lime plaster
~mörtel m = lime stuff
Kalk|rot n = lime red
~rotierofen m, Kalkdrehofen = rotary lime kiln
~rührer m, Kalkkrücke f = lime stirrer, ~ raker
~rührwerk n = lime agitator
~salpeter m, salpetersaurer Kalk m, Kalziumnitrat n, $Ca(NO_3)_2$ = nitrate of lime
Kalksand m = calcareous sand, lime ~
~beton m, Silikatbeton = silicate concrete
~hartstein m, KSH m = highly resistant sand-lime brick
~hohlblockstein m, KSHbI m = hollow sand-lime brick
~leichtstein m = lightweight sand-lime brick, ~ lime-sand ~
~lochstein m, KSL m = perforated sand-lime brick, ~ lime-sand ~
~mörtel m = mild mortar, non-hydraulic ~ [*composed of lime and sand*]

Kalksandstein — Kalksteinvorkommen

Kalksandstein *m*, weißer Mauerstein, Kalksandziegel *m* = sand-lime brick, lime-sand ~ [*Sand-lime bricks are made of a mixture of sand and hydrated lime pressed into shape in mo(u)lds and cured in an atmosphere of steam which causes chemical action to take place between the sand and lime thereby cementing the materials together*]

~, kalkiger Sandstein (Geol.) = sandy limestone, arenaceous ~
~maschine *f* = sand-lime brick machine, lime-sand ~ ~
~prüfmaschine *f* = sand-lime brick tester, ~ ~ testing machine
~splitt *m* = sandy limestone chip(ping)s, arenaceous ~ ~
Kalksandvollstein *m*, KSV = solid sand-lime brick

Kalkschachtofen *m*
obere Vorwärmzone *f* und Vorratssilo *m*
untere Vorwärmzone *f*
Pufferzone *f* [*neutrale Schicht*]
Brennzone *f*
Kühlzone *f*
Abzugbühne *f*

vertical lime kiln
upper preheat and storage zone

lower preheat zone
buffer zone
burning zone
cooling zone
draw floor

Kalkschiefer *m* → Plattenkalkstein *m*
~ton *m* = calcareous shale
Kalk|schlamm *m* (Geol.) = calcareous mud, shell-sand
~schwüle *f*, Kalkniere *f* = calcareous concretion, lime(stone) ~, limy ~, ~ nodule, concretion of lime
~(seifen)fett *n*, kalkverseiftes Fett = calcium grease, lime base ~
~silikat *n* = lime silicate, calcium ~
~silikatfels *m* = lime silicate rock
~silikathydrat *n* = hydrated calcium silicate
~sinter *m* = calcareous sinter, calcsinter
~silo *m* = lime silo
~-Soda-Verfahren *n* → → Wasserreinigung *f*
Kalkspat *m* = calcspar
~ von Island → Doppelspat
~-Kalkstein *m* = calcitic limestone
Kalkspessartit *m* = calc-spessartite
Kalkstabilisierung *f*, Kalkverfestigung = lime stabilization
Kalkstein *m* = limestone
~ mit dolomitischen Bestandteilen, Dolomitkalkstein = dolomitic limestone

~aufbereitungsanlage *f* = limestone preparation plant
~braunerde *f* = limestone brown loam
~brecher *m* = limestone crusher
~bruch *m* = limestone quarry
~bruchbesitzer *m* = limestone quarry operator
~füller *m*, Kalkmehlfüller = limestone filler
~kies *m* = limestone gravel
~korngröße *f* = limestone particle size
~lager *n*, Kalksteinvorratslager = limestone stor(ag)e
~mastix *m* = limestone mastic
~mauerwerk *n* = limestone masonry
~pulver *n* = limestone powder, pulverized limestone
~roterde *f* = limestone red earth
~schotter *m* = crushed limestone
~splitt *m* = limestone chip(ping)s
~teermakadam *m* = limestone tarmacadam
~vorkommen *n*, Kalksteinvorkommnis *n*, Kalksteinlager(stätte) *n*, (*f*), Kalksteinfundstätte = limestone deposit

Kalkstein(vorrats)lager — Kältemittel

~(vorrats)lager n = limestone stor(ag)e
Kalk|streuer m = bulk lime spreader
~suspension f = lime suspension
~teig m → Kalziumhydroxyd n
Kalkton|erdegestein n = cement rock
~erdesulfat n, Zementbazillus m = Candlot's salt, Michaelis' ~
~granat m = grossular
~schiefer m = calcareous slate, limestone ~
~stein m = calcilutyte, lime mud rock
Kalk|traßmörtel m = lime-trass mortar
(~-)Trommel-Löscher m, (Kalk-)Löschtrommel f = (lime) slaking drum
~tuff m = tufaceous limestone, (calcareous) tufa, calc tufa
Kalkulationsvordruck m = form for pricing construction works
Kalkulator m = estimator, calculator, quantity surveyor, cost estimator, estimating engineer
Kalk- und Zement-Verteilgerät n, Kalk- und Zement-Verteiler m, Zement- und Kalk-Verteilgrät, Zement- und Kalk-Verteiler = lime and cement spreader, cement and lime ~, ~ ~ ~ spreading machine
Kalkung f = liming, lime spreading
Kalk|uranglimmer m, Kalkuranit m = autunite
~verfahren n → Naß(sand)verfahren
~verfahren n [*Wasserenthärtung*] = lime process
~verfestigung f, Kalkstabilisierung = lime stabilization
~verputz m → Kalkputz
~verteiler m = lime spreader
~wasser n = lime water
~werk n → Kalkbrennerei f
~zementmörtel m, verlängerter Zementmörtel = cement-lime mortar
~zerfall m = lime disintegration
Kalotte f, Kugelhaube f, Kugelkappe f [*ein kreisförmig begrenzter Teil der Kugeloberfläche*] = calotte
kalottenförmiges geriffeltes Blech n = dimpled plate

kalottenförmige Riffelung f = dimpling
Kaltasphalt m → Bitumenemulsion f
~spritzmaschine f, Kaltbindemittelspritzmaschine = (cold) emulsion spraying machine, (~) ~ sprayer
Kalt|aufweiten n, Kaltausweiten [*Rohr*] = cold expanding
~bad n [*Motorenreinigung*] = cold bath
~band n [*kaltgewalzter Bandstahl, bis 5 mm dick und 630 mm breit*] = cold-rolled strip
~becherwerk n, Kaltelevator m = cold elevator
~beton m [*Bei Temperaturen unter $0°$ C abbindender Beton*] = concrete setting below $0°$ Centigrade
kaltbiegen = to cold bend
Kalt|biegeversuch m, Kaltbiegeprobe f, Kaltbiegeprüfung f = cold-bend test
~bindemittel n → Bitumenemulsion f
~bindemittelspritzmaschine f, Kaltasphaltspritzmaschine = (cold) emulsion sprayer, (~) ~ spraying machine
~brüchigkeit f = cold shortness
kalte Druckprobe f, Wasserdruckprobe = water test, hydraulic ~
~ Lawine f → Staublawine
Kalteinbau m [*Straßenbau*] = cold laying
kalteinbaufähiger Asphaltbeton m = cold-laid asphaltic concrete
~ Asphaltfeinbeton m = cold-laid fine asphaltic concrete
~ Asphaltgrobbeton m = cold-laid coarse asphaltic concrete
~ Teerfeinbeton m = cold-laid fine tar concrete
~ Teergrobbeton m = cold-laid coarse tar concrete
Kälteleistung f = refrigeration capacity
Kaltelevator m, Kaltbecherwerk n = cold elevator
Kälte|maschine f = refrigeration machine
~mittel n = refrigerant

Kältemittelleitung — Kalziumkarbidpulver

~mittelleitung *f* = refrigerant line
Kaltfrontregen *m* = cold front rain
kalt|gereckt = cold-worked
~**gewalzt** = cold-rolled
~**gezogen** = cold-drawn
Kaltklebemasse *f*, Kaltleim *m* = cold glu(e)ing compound
Kaltluft *f* = cold air
~**einbruch** *m*, Kaltlufteinfall *m* = influx of cold air
~**kanal** *m* = cold-air duct
Kaltmisch|anlage *f* = cold mix (asphalt) plant (US); bituminous macadam and tarmacadam mixing plant (Brit.)
~**verfahren** *n* = cold plant-mixing
Kalt|nieten *n* = cold riveting
~**preßschweißen** *n* = cold-pressure welding
~**schlamm** *m* [*Bildet sich beim unterkühlten Betrieb des Motors als steife Wasser-Öl-Emulsion*] = low temperature sludge, ~ ~ engine deposit
~**schornstein** *m* = cold chimney
kaltschweißen = to weld without preheating
Kaltschweißung *f* = cold welding
~, Kaltschweißstelle *f* [*Schweißfehler*] = lack of fusion [*Due to metal being deposited on a layer of oxide that has fused but not dispersed*]
Kaltstart *m* = cold weather starting
~**anlage** *f*, Einspritzvorrichtung *f* = primer
Kaltteer *m* [*DIN 1995*] [*ein mit leichten Ölen verschnittener Straßenteer*] = cold tar, TC
kalt|verfestigt = strain-hardened, strengthened by cold working
~**verformbar** = cold-workable
Kaltverformung *f* [*Metall*] = cold working, ~ reduction
Kaltwasser *n* = cold water
~**behälter** *m* = cold water storage tank
~**farbe** *f* = cold water paint
~**leitung** *f* = cold water line
kaltwasserlöslich = soluble in cold water

Kaltwasser|versorgung *f* = cold water supply
~**versuch** *m* [*Prüfung von Zement auf Raumbeständigkeit. DIN 1164*] = soundness test by immersion in cold water
Kaltwindschieber *m* = cold blast slide valve
Kalzinator *m* = calcinator [*Machine for drying and preheating slurry through intimate contact with hot kiln exit gases, passing in counterflow through a vessel charged with heat-exchanging elements. Normally little if any calcination (liberation of CO_2) takes place*]
Kalzinieren *n*, Rösten, Brennen, (Ver-)Glühen [*Verhüttung*] = calcining, calcination
Kalzinierraum *m*, Brennraum = calcine compartment
(kalzinierte) Soda *f*, wasserfreies Natriumkarbonat *n*, kohlensaures Natrium *n*, Na_2CO_3 = soda-ash
Kalzinierzone *f* → Brennzone
Kalzit *m* = calcite
Kalzium *n* = calcium
~**aluminat** *n* = calcium aluminate
~**aluminatsulfathydrat** *n* = calcium sulphoaluminate hydrate
~**bisulfit** *n*, doppelschwefligsaures Kalzium *n*, Kalziumhydrogensulfit *n*, saurer schwefligsaurer Kalk *m* = bisulphite of lime (Brit.); bisulfite ~ ~ (US)
~**chlorid** *n*, Chlorcalcium *n*, $CaCl_2$ = calcium chloride
~**-Harzseife** *f* = calcium rosinate
~**hydroxyd** *n*, gelöschter Kalk *m*, Ätzkalk, Löschkalk, Kalkteig *m*, Kalkbrei *m* [*Branntkalk + Wasser; $CaO + H_2O = Ca(OH)_2$*] = hydrated lime, calcium hydroxide, lime hydrate, slaked (quick) lime, calcic hydrate, hydrate of lime, lime paste
~**karbid** *n*, CaC_2 = carbide of calcium, calcium carbide
~**karbidpulver** *n* = powdered calcium carbide

~**karbonat** *n*, kohlensaurer Kalk *m*, Kreide *f*, CaCO₃ [*DIN 1280*] = calcium carbonate, chalk
~**nitrat** *n* → Kalksalpeter *m*
~**ölseife** *f* = calcium oleate
~**oxyd** *n*, CaO = calcium oxide
Kalziumsulfat *n* = calcium sulfate (US); ~ sulphate (Brit.)
~**ansatz** *m* = calcium sulphate incrustant, ~ ~ scale (Brit.); ~ sulfate ~ (US)
Kalzium|sulfoaluminat *n* [*mineralogisch: Ettringit*] = calcium sulfoaluminate
~**verbindung** *f* = calcium compound
Kamin *m* [*Offene, mit dem Schornstein verbundene Feuerstelle in einem Raum, in der Regel in einer Wandnische*] = fireplace (opening) [*A recess in a chimney breast or wall to receive a fire*]
~, Schornstein *m* = chimney [*A construction containing one or more flues*]
~**aufsatz** *m*, Schornsteinaufsatz, Kaminhaube *f*, Schornsteinhaube = (chimney) cowl
~**-Betonstein** *m* = chimney block
~**(ge)sims** *n* = mantelshelf, mantelslab
~**haube** *f* → Kaminaufsatz *m*
~**kühlturm** *m*, Steigschachtkühlturm = chimney-type cooling tower
~**schieber** *m* = soot door
~**sims** *n*, Kamingesims = mantelshelf, mantelslab
~**stein** *m*, Kaminziegel *m* = chimney brick
~**tür** *f* → Reinigungstür
~**verkleidung** *f* = mantel (piece) [*The facing of stone, marble, etc. about a fireplace, including a projecting shelf or slab above it*]
~**ziegel** *m*, Kaminstein *m* = chimney brick
Kamm *m* [*Querverbindung ungleich hoch liegender Hölzer*] = cocking, cogging, corking
~ [*für Wandputz*] = comb, scratcher
Kämmen *n* → Eingriff *m*

~, Einritzen [*Wandputz*] = scratching [*wall plaster*]
Kammer *f* [*Hohlblockstein*] = cell
~ [*Abwasserwesen*] = chamber [*A general term for a space enclosed by walls or a compartment, often prefixed by a descriptive word as "grit chamber" or "screen chamber" indicating its contents, or "discharge chamber" or "flushing chamber" indicating its office*]
~ **für niedergelegte Wehrböcke** [*Nadelwehr*] = recess for housing end frames
~**boden** *m* = chamber floor, ~ bottom
~**boden** *m* → (Schleusen)Kammersohle
~**feuerung** *f* = furnace with combustion chamber, locomotive type fire box
~**-Pfeiler-Bau** *m* (⚒) [*Es werden zuerst die Kammern hergestellt und dann die Pfeiler (herein)gewonnen*] = room and pillar system
~**schleuse** *f*, Kesselschleuse = chamber (navigation) lock, ~ inland ~, lift (~), ~
~**schleuse** *f* [*Siebtechnik*] = fixed screen with side and bottom apertures
~**sohle** *f* → Schleusen~
~**sprengen** *n* → → 1. Schießen *n*; 2. Sprengen *n*
~**waschmaschine** *f*, Kammerwäsche *f* = multi-compartment washer, ~ washing machine
Kamm|gebirge *n* = ridge mountain(s)
~**putz** *m* = scratched plaster
~**walzengerüst** *n* = pinion stand
~**walzengetriebe** *n* = pinion stand gear
Kampanile *m*, freistehender Glockenturm *m* = campanile
Kämpfer *m*, ~**stein** *m*, Anfänger *m* = impost
~, Losholz *n*, Kämpferholz [*Fensterrahmen*] = transom
~ (⚒) = abutment
~**druck** *m* = impost pressure
~**druck** *m* (⚒) = abutment pressure

Kämpferdruckzone — Kanalbogen

~druckzone f (🗲) = abutment zone
~fuge f = impost joint
~gelenk n = impost hinge
~höhe f → Kämpferpunkthöhe
~holz n → Kämpfer m
~linie f = springline, springing line
~punkt m = point of springing, springing (point)
~(punkt)höhe f = springing level, ~ height
~randstörung f = disturbance at the springing
~(stein) m, Anfänger m = impost
Kanadaasbest m, Chrysotilasbest = Canadian asbestos, chrysotile
kanadisches Bohren n, Vollgestängebohrung f = percussive rod drilling, percussion ~ ~
Kanal m, künstlicher ~ [*Künstlicher ins Gelände eingeschnittener oder auf Dämmen geführter Wasserlauf, in dem Wasserbewegung durch natürliches Gefälle bei freiem Wasserspiegel erfolgt. Man unterscheidet je nach Verwendung Schiffahrtkanal, Werkkanal, Entlastungskanal, Bewässerungskanal und Entwässerungskanal. Ein Kanal dient oft gleichzeitig mehreren Zwecken*] = canal
~ [*einer Pumpe*] = passage
~, geschlossener ~ = duct [*A tube or other provision for the passage of air, gas or services*]
~ → Gerinne n
~, Abwasser~, Leitungs~ = conduit-type sewer
~ aus Ziegelmauerwerk, Abwasserkanal ~ ~ = brick (conduit-type) sewer
~ für Lastkähne = canal for goods-carrying barges
~ ~ Postleitungen [*Telefon und Telegraf*] = G.P.O. duct (Brit.)
~ im Auftrag, Dammkanal = canal on embankment
~ ~ Einschnitt, ~ ~ Abtrag, Einschnittkanal, Abtragkanal = canal in cutting (Brit.); ~ ~ a cut (US)
~ mit Schleusen, Schleusenkanal = canal with locks

~ ~ totem Wasserspiegel = sleeping canal, dead ~
~abdeckung f → Schachtabdeckung
~abzweigung f [*Schiffahrtkanal*] = canal branch
~arbeiter m [*Abwasserwesen*] = sewer man
~auskleidung f, Gerinneauskleidung, Kanalbekleidung, Gerinnebekleidung, Kanalverkleidung, Gerinneverkleidung = channel lining
~auskleidung f, Kanalbekleidung, Kanalverkleidung = canal lining
~auskleidungsmaschine f, Gerinneauskleidungsmaschine = channel lining machine
~bagger m = canal dredger (Brit.); ~ dredge (US)
Kanalbau m = canal construction
~, Gerinnebau = channel construction
~ [*Abwasserwesen*] = conduit-type sewer construction
~ [*für öffentliche Versorgungsleitungen*] = duct construction
Kanalbaumaschinen fpl
Kanalplaniermaschine f
Kanal-Betoniermaschine f, Einbaugerät n für Gerinneauskleidungen in Zementbeton
Einbaugerät n für Gerinne-Auskleidungen, Kanalauskleidungsmaschine f
Kanal-Fugenschleifgerät n
Arbeitsgerät n, Hilfsgerät
canal trimming and lining equipment, canal building machinery
canal trimmer
canal concrete paver; traveling template form for concrete (US)
canal paver; traveling template form, canal paving rig (US); slip form canal lining machine
canal joint cutter
finishing rig
Kanal|bekleidung f, Kanalauskleidung, Kanalverkleidung = canal lining
~bogen m, Kanalkurve f, Kanalkrümmung f = canal curve

Kanalbogen — Kanalsystem

~bogen *m*, Gerinnebogen, Kanalkurve *f*, Gerinnekurve, Kanalkrümmung *f*, Gerinnekrümmung = channel curve
~böschung *f*, Kanalufer *n* = canal bank, ~ slope
~böschung *f*, Gerinneböschung, Kanalufer *n*, Gerinneufer = channel slope, ~ bank
~böschungsschutz *m*, Kanaluferschutz = canal slope protection, ~ bank ~
~brücke *f*, Kanalüberführung *f* = canal aqueduct, ~ bridge
~damm *m* = canal embankment
~deckel *m* → Schachtabdeckung
~deckel *m* = duct cover
~diele *f* = trench sheetpile
~dock *n* = canal dock
~einlaßbauwerk *n* = canal headworks
~gas *n*, Faulgas = sewer gas
~gefälle *n* = canal gradient
~-Gleisbagger *m* → Kanal-Schienenbagger
~graben *m* = duct trench
~hafen *m* = canal port
Kanalisation *f*, Ortsentwässerung *f* = sewerage system, system of sewerage, system of sewers [*A collecting system of sewers and appurtenances*]
Kanalisations|graben *m* = sewer trench
~netz *n* → Kanalnetz
~rohr *n*, Abwasserrohr = sewer pipe, sewage ~
~steinzeug *n* = sewer stoneware
~zone *f* = sewer zone
kanalisieren [*Fluß; Strom*] = to canalize, to control a river (or stream) by locks
kanalisierte Mündung *f* = canalized estuary
~ **Strecke** *f* = canalisation section
kanalisierter Fluß *m* = canalized stream
~ **Strom** *m* = canalized river
Kanalisierung *f*, Staureg(e)lung [*Bei der Kanalisierung wird das Gefälle treppenartig aufgelöst und damit bei jeder Wasserführung die erforderliche Mindestwassertiefe gewährleistet. Bei der Regulierung bleibt das vorhandene Gefälle mit Ausnahme der Durchstiche bestehen*] = canalization
Kanal|kraftwerk *n* = diversion type river power plant, ~ ~ ~ ~ station, ~ ~ ~ generating ~
~krümmung *f* → Kanalbogen *m*
~kurve *f* → Kanalbogen *m*
~messer *m* = measuring flume [*e.g. Venturi flume and Parshall flume*]
~mündung *f* = canal mouth
~-Naßbagger *m*, Kanal-Schwimmbagger = canal dredger (Brit.); ~ dredge (US)
~netz *n*, Kanalisationsnetz, Abwassernetz = sewer network, water-carriage system
~-Profileur *m* = canal trimmer
~pumpwerk *n* = canal pumping station
~querschnitt *m* = canal cross-section
~radpumpe *f* = open impeller (sludge) pump
~reinigung *f* = sewer cleaning
~reinigungsgerät *n* = sewer cleaning device
~schalung *f* = sewer formwork
~-Schienenbagger *m*, Kanal-Gleisbagger = canal-bank rail-mounted chain bucket excavator
~schiffahrt *f* = canal navigation
~schleuse *f* = canal (navigation) lock, ~ inland ~
~schleusenkammer *f* = coffer, canal lock chamber
~-Schwimmbagger *m*, Kanal-Naßbagger = canal dredger (Brit.); ~ dredge (US)
~sohle *f* = canal bottom, ~ bed
~sohle *f*, Gerinnesohle = channel bottom
~speisung *f* **durch Aufpumpen** = canal feeding by pumping
~spüler *m* = sewer flusher
~spülung *f* = sewer flushing
~stauvorrichtung *f* = canal check
~stein *m*, Kanalziegel *m* = sewer brick
~system *n* [*Abwasserwesen*] = network of conduit-type sewers

Kanalsystem — Kapillargeschwindigkeit

~system *n* [*Wasserbau*] = network of canals
~trockner *m* = tunnel dryer; ~ drier (US)
~tunnel *m* = canal tunnel
~überführung *f*, Kanalbrücke *f* = canal aqueduct, ~ bridge
~ufer *n* → Kanalböschung *f*
~verbreiterung *f* = canal widening
~verkleidung *f*, Kanalauskleidung, Kanalbekleidung = canal lining
~wasser *n* = canal water
~winde *f*, Wellbaum *m*, (Bock)Trommel(seil)winde, Haspel *m*, *f*, Seilwinde, Kabelwinde, Bockwinde [*Zum Hochziehen oder Niederlassen von Aushubmaterial oder Baustoffen im Brunnen- und Kanalbau*] = (hand) crab
~ziegel *m*, Kanalstein *m* = sewer brick
Kandel *m* → Kähner *m*
Kändel *m* → Kähner *m*
Kandelaber *m* = candelabra
Kanister *m* = can
Kannel-durit *m* = durain of the cannel coal
kannelieren = to flute [*to make long, rounded grooves in a column, etc.*]
Kannelierung *f*, Riefelung = fluting
Kannelkohle *f* = cannel coal, lantern ~
Kannelur *f*, Riefe *f*, Kannelure *f* [*lotrechte Vertiefung am Säulenschaft*] = flute
Kanonen|ofen *m*, eiserner Ofen = iron stove
~spat *m* = calcite
Kante *f* [*Berührungsgrade von zwei nicht in einer Ebene liegenden Flächen*] = arris
Kanten|abrundung *f* = arris rounding
~brecher *m* = arris trowel, arrising tool, bullnose trowel
~(dach)stein *m* → Bordziegel *m*
~(dach)ziegel *m* → Bordziegel *m*
~druck *m* = arris pressure
~festigkeit *f* = edge-holding power
~geröll *n* → Pyramidalgeröll(e)
~geschiebe *n* → Pyramidalgeröll(e) *n*

~kiesel *m* → Pyramidalgeröll(e) *n*
~länge *f* = arris length
~pressung *f* (⚒) = abutment pressure due to a pillar edge or coal rib
~pressung *f* = arris compression
~profil *n* = arris section
~riegel *m* = flush bolt
~riß *m* = arris crack
~rundstab *m* = staff bead
~schutz *m* = arris protection
~schutzleiste *f* = arris cover strip
~schutzwinkel *m* = arris cover angle
~spannung *f* = arris stress
~stein *m* → Bordziegel *m*
~viertelstab *m* = ovolo
~wirkung *f* [*Bodenmechanik*] = edge action
~ziegel *m* → Bordziegel
kantig → → Kornform *f*
Kantigkeit *f*, Winkeligkeit, Eckförmigkeit = angularity
Kantvorrichtung *f*, Kippvorrichtung = tilting device
Kaolin *m* → Porzellanton *m*
Kaolinisierung *f* = kaolinisation
Kaolinit *m* (Min.) = kaolinite
kaolinitischer Ton *m* = kaolin clay
Kaolinsand *m* = kaolin sand
Kapillar|attraktion *f* → Kapillarität *f*
~bewegung *f* = capillary movement, ~ motion
kapillarbrechend = destroying capillary action, anti-capillary
Kapillar|bremse *f*, Kapillarsperre *f* = capillary break, ~ groove
~druck *m* = capillary pressure, seepage force
Kapillare *f* = capillary
kapillare Heberwirkung *f* = capillary siphoning
~ Sättigung *f* = capillary saturation
~ Steighöhe *f* = height of capillary rise
kapillarer Aufstieg *m*, ~ Wasseranstieg = capillary rise, ~ lift
~ Wasseranstieg *m*, ~ Aufstieg = capillary rise, ~ lift
kapillares Gleichgewicht *n* = capillary equilibrium
Kapillargeschwindigkeit *f* = capillary (time) rate

Kapillarimeter *m* = capillary apparatus, capillarimeter
Kapillarität *f*, Kapillarattraktion *f*, Porensaugwirkung *f* = capillarity, capillary attraction
Kapillar|konstante *f* = capillary constant
~kraft *f*, Haarröhrchenkraft = capillary force
~potential *n* = capillary potential
~raum *m* = capillary space
~rohr *n* = capillary tube
~röhrchenhohlraum *m* = capillary bore
~röhrchenwandung *f* = capillary wall
~saugversuch *m* = capillary water absorption test
~saum *m*, Saugsaum, Kapillarzone *f*, Saugraum *m* = capillary fringe
~spannung *f*, Haarröhrchenspannung = capillary tension
~sperre *f*, Kapillarbremse *f* = capillary groove, ~ break
~-Viskosimeter *n* = capillary visco(si)meter
~wasser *n*, Haarröhrchenwasser, Porensaugwasser = capillary water, ~ moisture, water of capillarity
~zone *f* → Kapillarsaum *m*
Kapitäl *n*, Kapitell = capital
Kapitellplatte *f*, Säulendeckplatte, Abakus *m* = raised table, abacus [*A slab atop the capital of a column*]
Kaplan|-Laufrad *n* = feathering propeller (runner)
~turbine *f* [*Flügelradturbine mit verstellbaren Schaufeln*] = Kaplan (water) turbine, feathering propeller ~
Kappe *f* (✶) [*Ausbauelement aus Holz oder Stahl zum Abfangen des Daches bzw. der Firste*] = (roof) bar, cross ~
~ = arched roof
~, kurzes Joch(holz) *n*, Heitholz [*Geviertausbau*] = end plate
Kappen|führung *f* [*Druckluftwerkzeug*] = composite tool holder and retainer
~gewölbe *n* = sectroid

~kette *f*, Kappenreihe *f* (✶) = row of (roof) bars, ~ ~ cross ~
~quarz *m* = capped quartz
Kapp|schuh *m*, Kappwinkel *m* (✶) [*Ausbauelement zur Verbindung der Stahlkappen mit den Holz- oder Stahlstempeln*] = jointing shoe
~winkel *m* → Kappschuh *m*
Kapsel *f* [*Brennen von Wandkeramikplatten*] = fireclay container, sagger, saggar, seggar
~düse *f* [*Motor*] = capsule-type injector
~gebläse *n* = rotary blower
~presse *f* → (Preß)Kissen *n*
~pumpe *f* → Drehkolbenpumpe
~ton *m* = sagger clay, saggar ~, seggar ~
Kapselung *f* = enclosure
Kar *n*, Zirkustal *n* = corrie, cirque
Karabinerhaken *m* = snap hook, spring ~
Karbid *n* = carbide
~-Gießanlage *f* = carbide casting plant
~kohle *f* = carbide carbon
Karbokohlenteer *m* = carbocoal tar
Karbolineum *n* = carbolineum, peterlineum, coal tar creosote
Karbol|öl *n* = carbolic oil
~pech *n*, Kresolharz *n*, Phenolpech = cresol pitch
~säure *f* = Phenol *n*
Karbon *m* → Carbonado *m*
Karbonat *n*, kohlensaures Salz *n* = carbonate
~härte *f* = temporary hardness, bicarbonate ~
karbonatisches Gestein *n*, Karbonatgestein *n* = carbonate rock
karbonisches Gestein *n* = carbon rock
Karbonisieren *n* = carbonating, carbonation
Karborund *n*, Siliziumkarbid *n*, SiC = silicon carbide, carbide of silicon, carborundum, carbon silicide
Kardan *n* → Kreuzgelenk *n*
~drehzapfen *m* = knuckle
~gelenk *n* → Kreuzgelenk
~tisch *m* = photogrammetric plotter

Kardätsche — Kastenfenster

Kardätsche *f* → Abziehlatte *f*
Karfunkel *m* = pyrope
Karkasse *f*, Carcasse [*Gewebeunterbau des Reifens als Festigkeitsträger*] = casing
Karnaubawachs *n*, Carnaubawachs = Carnauba (wax)
Karnies *n* = ogee
Karosserie *f* = car body
Karrbohle *f* = runway plank
Karre(n) *f*, *(m)* Förder~, Transport~ = cart
Karren|aufzug *m* = barrow-hoist (concrete elevator)
~**mischer** *m* → Schub~
Karst|trichter *m*, Doline *f* = doline, dolina
~**wasser** *n* = karstic water
Kartätsche *f* [*schmales Brett mit Griff zum Abziehen des Putzes*] = darby (US); ~ float, derby float (Brit.)
Karte *f* mit Höhenlinien = layered map, contour ~
Karteiarchivierung *f* auf Mikrofilm = microfilming of card indexes
Karten|(gitter)netz *n* = map gird
~**herstellung** *f*, Kartierung = map-making, mapping (work)
~**herstellung** *f* mit Hilfe von Luftaufnahmen = aerial mapping, mapping by use of aerial photography
~**null** *n* = chart datum
~**schalter** *m* = ticket counter
~**winkelmesser** *m* = map protractor
Kartondrän *m* = card-board wick
Kartusche *f* = cartouch(e) [*A scroll-like ornament or tablet*]
~ = cartouch(e) [*On Egyptian monuments, an oval figure containing the name or title of a ruler or deity*]
Karyatide *f* = caryatide [*A column in the form of a female figure*]
Kaseinfarbe *f* = casein-bound distemper (paint)
Kasematte *f* = basement-area, casemate
Kasernen|bau *m* = construction of barracks
~**gebäude** *n* = barrack block
Kaskade *f* = cascade

Kaskaden|-Abzugkanal *m* → Treppen-Durchlaß *m*
~**-Durchlaß** *m* → Treppen-Durchlaß
~**falte** *f* (Geol.) = zigzag fold
~**mühle** *f* = cascade (grinding) mill
~**wirkung** *f*, Abrollen *n* der Mahlkörper übereinander = cascading (action)
Kasselerbraun *n*, Van-Dyk-Braun = Cassel brown, Vandyk ~
Kassensaal *m* [*Bank*] = banking hall
Kassette *f*, Deckenfeld *n*, Deckenfach *n* = core, waffle
Kassetten|balken(träger) *m* = cored beam, waffle ~
~**bauweise** *f* = cored-slab construction, waffle slab ~
~**decke** *f* = cored floor, waffle ~
~**platte** *f* = cored slab, waffle ~
kassettieren [*eine Betonplatte*] = to core
Kassettierung *f* [*Betonplatte*] = coring
Kästel|betonmauer *f* = pierced concrete wall
~**mauer** *f* = pierced wall, ~ screen
~**mauerwerk** *n* = pierced masonry
~**verband** *m* = pierced bond
Kasten *m* [*für Gründungen unter Wasser*] = caisson
~**-Abzugkanal** *m*, Kasten-Durchlaß *m* = box culvert
kastenartiges Schrapp(er)gefäß *n*, kastenartiger Schrapp(er)kübel *m*, Schürfkasten *m* = box-type scraper bucket
Kasten|ausleger *m* = box type boom (US); ~ ~ jib (Brit.)
~**balken(träger)** *m* → Kastenträger
~**bauweise** *f*, Kastenbauart *f* [*Brückenbau*] = box-type construction
~**blechträger** *m*, Blechkastenträger = box plate girder
~**drän** *m* = box drain
~**-Durchlaß** *m*, Kasten-Abzugkanal *m* = box culvert
~**fang(e)damm** *m* → Kastenwand *f*
~**fenster** *n*, Doppelfenster, Winterfenster, Vorfenster = double window, winter ~

~form *f* = box mould (Brit.); ~ mold (US)
~formmäkler *m* = box-section leader
~fundament *n* → Kastengründungskörper *m*
~fundation *f* → Kastengründung *f*
~-Gasreiniger *m* = box gas purifier
~gitterträger *m* = box lattice girder
~gründung *f*, Kastenfundation *f* = caisson foundation [*Foundation with either box caisson, or open caisson, or pneumatic caisson*]
~gründungskörper *m*, Kastenfundament *n* = caisson foundation (structure), ~ substructure, ~ engineering foundation
~kipper *m* = box type skip, ~ ~ tip wagon, ~ (~ track wheel) dump wagon
~konstruktion *f* = box design
~pfahl *m* = box pile
~platte *f* [*Brückenbau*] = box slab
~profil *n* = box section
~profilrahmen *m* = box section frame
~querschnitt *m* = box section
~rinne *f* = box gutter, trough ~, parallel ~, concealed (box) ~, hidden ~
~rippe *f* [*einer Hinterkippmulde*] = channel rib
~rohr *n* = box-shaped pipe
~schalung *f* = box type formwork
~scherversuch *m*, Kastenscherprobe *f*, Kastenscherprüfung *f* [*Bodenmechanik*] = box shear test
~schloß *n* = rim lock
~schneide *f* → Senk~
~spüler *m* = flush tank
~spundbohle *f* → → Spundwand *f*
~spundwand *f* → Kastenwand
~stahlstütze *f* = box-form (built) steel column
~stütze *f* = box-shaped column
~träger *m*, Hohlkasten(träger) *m*, Hohlbalken(träger) *m*, Kastenbalken(träger) = hollow (box) girder, ~ (~) beam

~trägerbrücke *f* = box-girder bridge, box-beam ~
~trägerkragbrücke *f* = box-girder cantilever bridge, box-beam ~ ~
~verteiler *m* → Verteilerkasten *m*
~wand *f* = caisson wall
~wand *f*, Kastenspundwand, doppelte Spundwand, Kastenfang(e)damm *m* = double-walled (sheet-pile) cofferdam
Kataklasit *m* (Geol.) = cataclastic rock, kataklastic ~
kataklastische Gesteinsform *f*, Kataklase *f* = kataklastic structure, cataclastic ~
Katakombe *f*, unterirdischer Friedhof *m* = catacomb, underground cemetery
Katalogbezeichnung *f*, Modellbezeichnung, Drahtwort *n* = code (word)
Katalysator *m* = catalyst
~-Zwischenschicht *f* = intermediate catalyst film
Katalyse *f* = catalysis
katalytische Entschwef(e)lung *f* = catalytic desulfurization
katalytisches Krackbett *n* = catalytic bed
~ Spaltverfahren *n* = catalytic cracking, catcracking
~ ~ mit Natur-Bleicherde als Kontakt = THERMOPHOR catalytic cracking
~ ~ ~ natürlichem Bauxit als Kontakt und Wasserdampf zur Verminderung der Koksabscheidung = CYCLOVERSION catalytic cracking
~ ~ ~ ~ Reaktionsraum mit Mischsystem = fluid catalytic cracking
Kataster *n* = land-register, register of real estates
~gebühr *f* = land-register fee
~karte *f* = cadastral map
~parzelle *f* = registered plot (Brit.); ~ lot (US)
~vermessung *f* = cadastral survey
Katastrophen|-Hochwasser *n* = catastrophic flood (Brit.); super flood, record flood (US)

Katastrophenhochwasser — Kegelbrecher

~hochwasser *n* in der Zeit vor Aufnahme regelmäßiger Pegelmessungen = historic flood

~pumpe *f* = emergency pump

Katathermometer *n* = kata-thermometer [*An instrument for measuring the cooling power of air on the human body*]

Kathedrale *f*, Bischofskirche *f* = cathedral

Kathedralglas *n* = cathedral sheet

Kathoden|schutz *m*, Fernschutz(wirkung) *m*, *(f)* = cathodic protection

~strahloszillograph *m*, Elektronenstrahloszillograph = cathode ray oscillograph, ~ beam ~, electron ~ ~

~-Tragschiene *f* = cathode headbar

Kation *n* = cation

Kationen|austauscher *m*, Kationenaustauschfilter *m*, *n* = cation exchanger

~-Koordination *f* = cation coordination

kationisch = cationic

kationische Emulsion *f*, saure ~ = cationic emulsion, acid ~

Katze *f* → Handramme *f*

~ → Lauf~

Katzen|auge *n* → Bodenrückstrahler *m*

~balken *m* → Hahnenbalken

~kopf *m* [*Pflasterstein*] = nigger head [*rounded cobble stone*]

Katz|laufbahn *f*, Laufkatzenbahn, Lastkatzenbahn = crab runway

~seil *n* = crab rope, ~ cable, ~ line

~träger *m*, Laufkatzenträger, Lastkatzenträger = crab girder

Kaufhaus *n*, Warenhaus = department store

Kauf-Mietvertrag *m* = (equipment) rental contract with purchase option

Kausche *f*, Kausse [*in Seilenden eingesetzter Metallring*] = thimble

Kauschen|einband *m* = capping

~-Wiedereinband *m* = re-capping

Kaustifizierung *f* = caustification

kaustisch = caustic

Kautschuk|milch-Zement-Vergußmasse *f* = (rubber) latex-cement sealing compound

~pulver *n* = flocculated latex crumb, rubber powder

Kaverne *f* (Geol.) = cavern

~ = underground chamber

Kavernen|aushub *m* = excavation of the underground (machine) hall

~krafthaus *n*, Kavernenturbinenhaus, Untergrundkrafthaus, Untergrundturbinenhaus = underground power house

~kraftwerk *n*, Untergrundkraftwerk = underground power station, ~ ~ plant, ~ (electric) generating ~

~wasser *n*, Höhlenwasser = cavern water

Kavitation *f*, Hohlsog *m*, Hohlraumbildung *f* = cavitation

Kavitations|-Charakteristik *f* = cavitation characteristics

~erosion *f* = cavitation-erosion, cavitational pitting

K-Bau *m*, deutscher Türstock *m* mit Kniestempelverstärkung am Oberstempel [*Ist ein ganzer oder halber Türstock mit einem k-förmigen Einbau als Kniestempelverstärkung*] = K-support

Keene'scher Zement *m* → Alaungips

Kegel *m* = cone

~bahn *f* = bowling alley

~bahnhalle *f* = bowling centre (Brit.); ~ center (US)

~bildung *f* in Öllagerstätten = coning in oil reservoirs

~boden *m*, Trichterboden [*Behälter*] = hoppered bottom, ~ floor, conical ~

~(bohr)meißel *m*, Konus(bohr)meißel = cone bit

Kegelbrecher *m*, Kreiselbrecher, Rundbrecher = cone(-type) crusher, gyratory ~ ~ [*It sizes at the closed side setting*]

~, Kreiselbrecher, Rundbrecher = gyratory crusher [*It sizes at the open side setting*]

Kegelbrecher — Kehren

~ **mit segmentförmigem Brechkopf** = gyrasphere crusher
Kegel|dach n = conical (broach) roof
~**drucksonde** f = conical penetrometer
~**eindringungsapparat** m, **Kegelgerät** n, **Kegeldruckapparat** [*stabilisierte Mischung mit bituminösem Binder*] = cone penetrometer, conical ~
~**gewölbe** n = conical vault
Kegeligkeit f → **Kegelneigung** f
Kegel|kupp(e)lung f = cone clutch
~**meißel** m → **Kegelbohrmeißel**
~**mischer** m, **Konusmischer** = cone-type mixer
~**muffe** f = conical socket
~**neigung** f, **Verjüngungsverhältnis** n **beim Kegel**, **Konizität** f, **Kegeligkeit** f, **Kegelsteigung** f, **Kegelverhältnis** n = taper
~**rad** n, **Winkelrad**, **konisches Rad** [*Dient zur Kraftübertragung zwischen zwei Wellen, deren Achsen sich schneiden*] = bevel gear
~**ritzel** n, **Kegelzahnritzel**, **Konus-(zahn)ritzel** = bevel (gear) pinion
~**rollenlager** n, **Konusrollenlager** = taper roller bearing (Brit.); TIMKEN bearing (US) [*Trademark*]
~**rollenlagernabe** f = taper-roller-bearing hub
(~)**Rollenmeißel** m, **Rollkrone** f = roller bit, cone ~, rotary ~, toothed ~
~**schale** f = conical shell
~**schmelzpunktbestimmung** f = cone deformation study
~**sonde** f = conical penetrometer
~**steigung** f → **Kegelneigung**
~**stumpf** m = frustum of a cone, conical frustum
~**trommel** f, **Konustrommel** = conical-shaped drum
~**umlenktriebgehäuse** n = bevel pinion housing
~**verankerung** f, **Konusverankerung** = cone anchorage
~**verfahren** n → **Viertelung** f
~**verhältnis** n → **Kegelneigung** f
~**viskosimeter** m = cone viscosimeter

~**walm** m = conical hip(ped end)
Kehlbalken m = collar beam, top ~, span piece [*A horizontal tie-beam of a roof*]
~**dach** n = collar-beam roof, trussed rafter ~ [*Common rafters, joined half-way up their length by a horizontal tie beam. This roof gives more headroom in the centre of the room than a close-couple roof*]
Kehl|blech n = flexible-metal for valleys
~**bohle** f = valley board
Kehle f, **Kehlung** f, **Dach**~ = (roof) valley
Kehlendach n → **Dach mit Wiederkehr**
Kehl|linie f = valley line
~**nahtschweißung** f = fillet welding
~**neigung** f = valley slope
~**rinne** f = valley gutter
~**schifter** m = valley jack (rafter)
~**schindel** f = valley shingle
~(**schweiß**)**naht** f = fillet weld
~(**schweiß**)**naht** f **am Überlappungsstoß** [*Schweiz: Überlappnaht*] = lap-joint fillet weld
~(**schweiß**)**naht** f **mit Fugenvorbereitung** [*Österreich: versenkte Kehl(schweiß)naht; Schweiz: Kehl(schweiß)naht mit Kantenvorbereitung*] = bevel weld in a Tee joint, J butt weld in a corner joint
~**sparren** m = valley rafter
~**stein** m, **Kehlziegel** m = valley tile
~**stoß** m [*Zimmerei*] = bolection moulding, balection ~ (Brit.); bilection molding (US) [*A mo(u)lding fixed round the edge of a panel and projecting beyond the surface of the framing in which the panel is held*]
Kehlung f, **Kehle** f, **Dach**~ = (roof) valley
Kehlziegel m, **Kehlstein** m = valley tile
Kehrbreite f, **Fegebreite** = sweeping width
Kehren n → (**Ab**)**Fegen**
~ [*Abhalten des Überflutens von Niederungen durch Anlage von Deichen*] = protection by dykes

Kehrmaschine — Kellergitter

Kehrmaschine *f* → Straßen~
Kehrrad|-Abhängehaken *m*, Kehrrad-Aushängehaken [*Erdölförderung*] = disconnecting hook, throw-off ~
~pumpantrieb *m* [*Erdölförderung*] = push and pull pumping power, central ~ ~
Kehr|saugmaschine *f*, Saugkehrmaschine = suction scavenger, ~ scavenging machine
~walze *f*, Fegewalze, Bürstenwalze, Besenwalze = rotary brush, rotating ~
Keil *m* = wedge
~anker *m* = cone anchor
~befestigung *f* = wedge method of attachment
~dicke *f* = wedge thickness
~einbruch *m* = wedge cut
~fänger *m* [*Tiefbohrtechnik*] = slip socket
~klaue *f*, (Stein)Wolf *m* = (lifting) lewis, ~ pins
~klauenloch *n*, (Stein)Wolfloch = lewis hole
~-Naturstein *m*, Wölb-Naturstein [*(Natur)Steinbogen*] = arch-stone
~pflug *m*, Schnee~, V-Pflug = V-type snow plough
~profil *n*, Wellenkeil, Verkeilung *f* = (shaft) spline
~profildraht *m* = wedge-wire
~profildrahtgewebe *n* = wedge-wire cloth
~riemen(an)trieb *m* = texrope drive (US); V(ee)-belt ~ (Brit.)
~schlitz *m* = wedged slot
~schloß *n*, Keil *m* und Lösekeil = gib and cotter
~sitz *m* = key seat
~splitt *m*, Füllsplitt, Dichtungssplitt = choke (or intermediate) aggregate (or stone), keystone, filler stone, blinding stone
~stein *m*, Wölbstein = voussoir [*An arch-stone in a stone arch or an arch-brick in a brick arch or a well lining*]
~stufe *f* = wedge-shaped step
~topf-Abfangkeil *m* [*Tiefbohrtechnik*] = spider slip

~ und Federn *fpl* (✗) = plug and feathers, multiple wedge
~ ~ Lösekeil, Keilschloß *n* = gib and cotter
~stück *n* [*dreiteiliger Steinwolf*] = (dovetailed) side piece
~verankerung *f* [*Spannbeton*] = wedge anchoring system, ~ anchorage ~
~welle *f* = splined shaft
~wirkung *f* = wedge action
~ziegel *m*, Wölbziegel = arch-brick [*A wedge-shaped brick for building an arch or lining a well*]
Keim *m* = germ
~tötung *f* = sterilization
~zählung *f* = enumeration of bacteria, bacterial count
Kelle *f* = trowel
Kellen|glattstrich *m* = trowel finish(ing)
~putz *m* = trowel plaster
Keller *m* = cellar, basement [*A space within a building or structure and below ground level, designed for storage, boiler room or purposes other than habitation*]
~ablauf *m*, Kellersinkkasten *m* = cellar gull(e)y, basement ~
~boden *m*, Kellerfußboden = cellar flooring, basement ~
~decke *f* = cellar floor, basement ~
~eingang *m* = entrance to a cellar, ~ ~ ~ basement, cellar entrance, basement entrance
~entwässerung *f* = cellar drainage, basement ~
~fenster *n* = cellar window, basement ~
~(fuß)boden *m* = basement flooring, cellar ~
~garage *f* = basement garage
~geschoß *n* = basement [*A storey wholly or mainly below ground level*]
~geschoßhalle *f* [*beim Satellitsystem, Flughafen*] = basement concourse
~gewölbe *n* = cellar vault, basement ~
~gitter *n* = cellar grating, basement ~

Kellerhals — Kern

~hals m [*eine außenliegende Kellertreppe*] = outdoor basement staircase
~lichtschacht m = basement air shaft, ~ light well, cellar ~ ~
~mauer f, Kellerwand f = cellar wall, basement ~
~mauerwerk n = cellar masonry, basement ~
~nische f = cellar niche, ~ recess, basement ~
~raum m = cellar room, basement ~
~siel n = waste drainage pipe in basement
~sinkkasten m, Kellerablauf m = basement gull(e)y, cellar ~
~treppe f = cellar staircase, basement ~ (Brit.); ~ stair (US)
~tür f = cellar door, basement ~
~wand f, Kellermauer f = basement wall, cellar ~
~wohnung f = basement flat
Kennedy'sche Grenzgeschwindigkeit f = Kennedy's critical velocity [*Is that in open channels which will neither deposit nor pick up silt*]
Kenn|farbe f, Signierfarbe = marking paint
~feuer n = identification beacon
~größe f [*Siebtechnik*] = characteristic index
~linie f [*Siebtechnik*] = characteristic curve
~linie f eines Stempels, Charakteristik f ~ ~, Lastwegkurve f (⚒) = resistance-yield curve, load-yield ~
~linienverfahren n [*Energieversorgung*] = tie-line-bias
~wert m = characteristic value
~zahl f nach Terra [*Siebtechnik*] = probable deviation according to Terra
~zeichnung f = marking
Kenntlichmachung f des Fahrwassers, Fahrwasserbezeichnung = channel marking
Kenotaph m = cenotaph
kentern = to capsize
Keramik f = ceramics
~docht n zur Bodenbewässerung = water wick

Keramikergips m → Formgips
Keramikfliesenkleber m = ceramic tile adhesive
keramische Dichtung f = ceramic seal
~ Erzeugnisse npl mit Zellenbildung, keramische Leichterzeugnisse = cellulated ceramics
~ Fliese f, ~ Platte f = clay tile
~ (Fuß)Bodenfliese f, ~ (Fuß-) Bodenplatte f = clay flooring tile
~ Glasur f = ceramic glaze
~ Platte f, ~ Fliese f = clay tile
~ Strangpresse f = ceramic extrusion machine
~ Wandfliese f, ~ Wandplatte f [*veralteter Begriff: Kachel f*] = clay wall tile
keramischer Baustoff m = ceramic construction(al) material, ~ building ~
~ Piezoteil m, keramisches Piezoteil n = piezoelectric ceramic material
~ Scherben m = white-ware
keramisches Futter n in Sulfitzellstoffkochern = ceramic digester linings in sulfide pulp mills
~ Leichtbauerzeugnis n = lightweight structural clay product
Kerbe f = notch; nick (US)
Kerbempfindlichkeit f = notch sensitivity
kerben, ein~, ab~ = to notch
Kerb|marke f auf einem Schraubenkopf = dash on the head of a bolt
~schlagbiegeversuch m, Kerbschlagbiegeprobe f, Kerbschlagbiegeprüfung f = nick-break test (US); notch shock ~, shock test with notched test piece
~schlagzähigkeit f [*Stahl*] = notch toughness
~spannung f = notch stress
~stab m = notched bar
~tal n = V-shaped valley
Kern m → Damm~
~, Mitte f, Stützkörper m, Dammkörper [*nicht zu verwechseln mit Dichtungskern beim Staudamm = core*] = body
~ → → Talsperre f

~, Gebäude~, (Stahl)Beton~, Betonkernbauwerk n, Hauskern = reinforced-concrete core, building~
~ [von der Mitte der Querbewehrung begrenzter Teil des Querschnitts einer Betonstütze] = core, kern
~ → Probe~
~, Einlage f, Seele f [Drahtseil] = core
~ **des Querschnittes** = core of the section
~**analyse** f = well-core analysis
~**apparat** m [setzt sich aus Einfach- oder Doppelkernrohr und Bohrkrone zusammen] = barrel and bit
~**ausstoßhammer** m = core pickhammer
~**barkeit** f = corability
~**baugrube** f [Staudammbau] = core pit
~**bauweise** f, deutsche Bauweise = core method of tunnel construction
~**(bauwerk)** m, (n) → Stahlbeton~
Kernbeton m, Hauptbeton [Hauptmasse des Talsperrenbetons im Gegensatz zum Vorsatzbeton] = mass concrete
~ → (Hinter)Füll(ungs)beton
Kernbohren n, Kernbohrung f [Zur Gewinnung von Bohrproben in Form von Gesteinskernen aus den durchteuften Formationen] = core drilling
~ **zur Bestimmung der Lagerungsbedingungen** = structural core drilling
Kernbohr|gerät n → Kernbohrmaschine f
~**krone** f, Kernbohrwerkzeug n = core(-barrel) bit, rock core ~, core drill ~
~**loch** n, Kernbohrung f, Kernloch n = core(d) hole
~**maschine** f, Kernbohrgerät n, Beton~ = (concrete) core (cutting) machine, coring machine, core drill(ing machine), (concrete) coring drill
Kernbohrung f, Kernbohren n [Zur Gewinnung von Bohrproben in Form von Gesteinskernen aus den durchteuften Formationen] = core drilling

~, Kern(bohr)loch n = core(d) hole
Kern|bohrwerkzeug n → Kernbohrkrone f
~**brett** n, Herzbrett = heart plank
~**büchse** f = core barrel
~**damm** m → Talsperre f
~**dichtung** f → (Dichtungs)Kern m
~**durchmesser** m = diameter of the core
kernen, einen Kern bohren = to (cut a) core
Kernen n = coring
Kernenergie|-Antrieb m, Atomantrieb = nuclear propulsion
~**schiff** n, Atomschiff = nuclear ship
Kern|entnahme f → (Bohr)Kerngewinnung f
~**fangring** m = core trap-ring
~**fels** m → anstehender Fels
~**folge** f, Kernsatz m, Kernsuite f [Tiefbohrtechnik] = suite of cores
~**gebiet** n, Geschäftsgebiet = business area, central ~, ~ district
~**gewinnung** f → Bohr~
~**halbmesser** m = radius of the core
~**härte** f = core hardness
~**kraftwerk** n → Atom-Elektrizitätswerk
~**kraftwerkschieber** m = atomic energy valve
~**krone** f [Erdstaudamm] = crest of core
~**linie** f, Kernumriß m, Kernumfang m = core line, ~ circumference
~**loch** n, Kernbohrloch, Kernbohrung f = core(d) hole
kernmagnetische Bohrlochüberwachung f = nuclear magnetic logging, ~ magnetism ~
Kernmauer f, Damm~ = core wall
~**damm** m, Betonkerndamm = concrete (wall) (type) dam
Kern|probe f → (Probe)Kern
~**punkt** m [Grenzpunkt der Kernfläche in einer Hauptachse des Querschnitts] = core point
~**querschnitt** m [z. B. bei umschnürten Stahlbetonstützen] = core cross-section, kern ~

~resonanz-Durchflußmesser m = flowmeter utilizing nuclear magnetic resonance
~ring m, Innenring [*Tunnel- und Stollenbau*] = internal ring
Kernrohr n = core barrel [*Is a cylindrical chamber for receiving and retaining the core as drilling progresses*]
~, Rohrkern m [*Teil eines Betonrohres*] = core pipe, pipe core
~kopf m = drill barrel head
~krone f = drill barrel shoe
Kern|sand m = core sand
~satz m, Kernfolge f, Kernsuite f [*Tiefbohrtechnik*] = suite of cores
~schale f [*zum Aufbewahren der Bohrkerne*] = core tray
~spaltvorrichtung f = core splitter
~spirale f [*Verfahren Freyssinet*] = helical spring
~strecke f, gekernte Strecke [*Tiefbohrtechnik*] = cored interval
~suite f, Kernsatz m, Kernfolge f [*Tiefbohrtechnik*] = suite of cores
~umfang m → Kernlinie f
~umriß m → Kernlinie f
~verlust m [*elektrisches Kernen*] = core loss
~verlust m [*Rotarybohren*] = loss of (drill) core
~weite f = core dimension
kerosingefluxt = kerosine-fluxed
Kerzen|teer m = candle tar
~zündung f = sparking plug ignition
Kessel m = boiler
~ (Geol.) = ca(u)ldron, fault, pit
~, Auskesselung f (⚒) [*Die Einlagerungen in das Hangende oder das Liegende, die oft von anderer Beschaffenheit und Zusammensetzung als das sie umgebende Gestein sind und sich mit diesem nur in losem Zusammenhang befinden*] = dome, pot-hole
~ **mit einem Dampfraum**, Einfachdampfraumkessel = boiler with single steam space
~ ~ **eingebauter Feuerung** = independent type boiler

(~)**Ansatz** m, Kesselstein m = boiler incrustant, ~ scale
~batterie f = battery of boilers
~bediener m → Heizer m
~bedienerstand m → Heizerstand
~bekohlungsanlage f = boiler coaling installation
~bereich m = boiler area
~blech n = boiler plate
~brunnen m, Schachtbrunnen [*Durch Handschacht oder Baggern hergestellter besteigbarer Brunnen mit gemauerter oder betonierter Wandung*] = dug well
~dampfmaschine f, Lokomobile f = locomobile
~feuerung f = boiler furnace
~gerüst n = boiler supporting structure
~gips m = kettle-calcined gypsum
~haus n = boiler house
~heizfläche f = boiler heating surface
~mantel m = boiler shell
~prüfer m = boiler inspector
~raum m = boiler room
~reg(e)lung f, Kesselsteuerung = boiler control
~rost m, Feuerrost m = (boiler) grate
~schießen n, Vorkesseln n = sprung borehole method, springing, enlarging the hole with explosives, chambering, squibbing
~schlacke f → Schlacke
~schlackenbetonplatte f = clinker concrete slab
~schlackenbildung f = clinkering
~schleuse f → Kammerschleuse
~speisewasser n = boiler feed water
~speise(wasser)pumpe f = boiler feed-water pump, ~ feeding ~, donkey (~)
~stahl m = boiler steel
~stein m, (Kessel)Ansatz m = boiler incrustant, ~ scale
~steingegenmittel n = boiler fluid, ~ composition
~stein- und Rostabklopfer m = boiler scale and rust chipper
~steuerung f, Kesselreg(e)lung = boiler control

Kesselturm — Kettenkran

~turm m, Kesselhaus in Turmform = boiler tower
~wagenerhitzer m, Dampferzeuger m = tank car heater, booster ~
~wärter m → Heizer m
~wärterstand m → Heizerstand
~wasserspeisepumpe f = feedpump
~wirkungsgrad m = boiler efficiency
~wirkungsgradprüfung f = boiler house efficiency test
Kessener'sche Walzenbürste f = Kessener revolving brush
Kettdraht m [*Siebtechnik*] = warp wire, longitudinal wire of the cloth
Kette f = chain
~ → Raupen~
Ketten|anhängerwagen m → Raupenwagen
~(an)trieb m = chain drive
~antriebskette f, Gleis~, Raupenantriebskette = crawler drive chain
~antriebskupplung f, Gleis~, Raupenantriebskupplung = crawler drive clutch
~antriebsturas m, Gleis~, Raupenantriebsturas = crawler drive sprocket
~antriebsturaswelle f, Gleis~, Raupenantriebsturaswelle = crawler pivot shaft
~aufgabe f, Kettenbeschickung f = chain feed(ing)
~aufgeber m, Kettenbeschicker, Kettenspeiser [*Vor dem Bunkerraum freihängende schwere Ketten, die über eine Kettenwalze langsam umlaufen und dadurch das Gut gleichmäßig aufgeben*] = chain feeder
~auflager n [*Hängebrücke*] = chain truck, ~ saddle
~bagger m, Gleis~, Raupenbagger = crawler (type) excavator
~bahn f = chain creeper
~becher m, Ketteneimer m = chain bucket
~becherwerk n, Ketten(becher)elevator m = chain (type) (bucket) elevator
~-Becherwerk(auf)lader m → Becherwerk(auf)lader auf (Gleis)Ketten

~beschicker m → Kettenaufgeber
~beschickung f, Kettenaufgabe f = chain feed(ing)
~bogen m [*Hängebrücke*] = catenary
~-Bohrwagen m, Gleis~, Raupen-Bohrwagen = crawler wagon drill
~bohrwerk n → Kettenhebewerk
~bolzen m → chain pin
~bolzen m, Gleis~, Raupenbolzen = track roller shaft
~bruch m [*Mathematik*] = fractional progression
~brücke f, Kettenhängebrücke = chain suspension bridge
~egge f = chain harrow
~eimer m, Kettenbecher m = chain bucket
~elevator m → Kettenbecherwerk n
kettenfahrbar, gleis~, raupenfahrbar = crawler-mounted
Ketten|fahrzeug n, Gleis~, Raupe f, Raupenfahrzeug = tracked vehicle, crawler-type ~
~fett n, Kettenschmiere f = chain grease
~förderer m = chain conveyor, ~ conveyer
~gehänge n = chain curtain
~glied n → Raupenschuhplatte f
~glied n = chain link
~gliedlaufschiene f [*Gleiskette*] = rail
~greif(er)korb m = chain grab (bucket)
~(hänge)brücke f = chain suspension bridge
~-Hängesieb n = chain hung type screen
~hebewerk n, Kettenwindwerk, Kettenbohrwerk, Rotary-~ = (rotary) chain draw works
~hebezeug n = chain-hoisting device
~-Hoch(löffel)bagger m → Hochlöffel-Raupenbagger
~kraftschluß m, Gleis~, Raupenkraftschluß = traction of tracks [*gripping action between tracks and the surface*]
~kran m, Gleis~, Kranraupe f = crawler-mounted crane

Kettenkupp(e)lung — Kienöl

~kupp(e)lung *f*, Gleis~, Raupenkupp(e)lung = crawler clutch
~lademaschine *f* → Raupenlader *m*
~lader *m* → Raupenlader
~laufwerk *n* → Raupenfahrwerk
~linie *f* = catenary
~messung *f*, Kettenvermessung = chain surveying, chaining
~platte *f* → Bodenplatte
~prüfmaschine *f* = chain testing machine, ~ tester
~rad *n* = chain sprocket
~reihenelevator *m* → Kettenvollbecherwerk *n*
~rost *m* = chain grate
~-Schaufelradgrabenbagger *m* → Gleis~
~schlepper *m*, Gleisketten-Zugmaschine *f*, Gleisketten-Schlepper [*Die Ausdrücke "Raupenschlepper", "Kettenraupe", "Raupentraktor" und "Raupentrecker" sind nicht mehr anzuwenden*] = crawler(-tread) tractor, (self-laying) track-type (industrial) tractor, caterpillar tractor, tracked tractor, track laying tractor
~schlußbolzen *m* = master pin
~schmiere *f*, Kettenfett *n* = chain grease
~schrämmaschine *f* = chain coal-cutting machine, ~ coalcutter
~schuh *m* → Raupenschuhplatte *f*
~spanner *m* = chain adjuster, ~ tensioner
~spanner *m*, Gleis~, Raupenspanner = crawler take-up
~spannfeder *f* [*Gleiskette*] = recoil spring
~spannfederanschlag *m* [*Gleiskette*] = recoil spring stop
~spannfederführung *f* [*Gleiskette*] = recoil spring pilot
~spannung *f* = chain tension
~speiser *m* → Kettenaufgeber
~stab *m* [*Hängebrücke*] = flat link of chain
~-Standbahn *f* = endless-chain haulage system
~steinwurf *m*, Steinkette *f* [*Kettensteinwürfe bestehen aus möglichst gleich großen, durch 1 m lange Ketten verbundenen Steinen. Sie werden in 2 sich kreuzenden Lagen über die lose Steinschüttung ausgebreitet*] = chained blocks
~stern *m*, Turas *m* = (chain) sprocket
~stollen *m*, Gleis~ = (dirt) grouser
~strang *m* [*Becherwerk*] = endless chain
~system *n* in einem Zementofen = chain system installation in a cement kiln
~teilung *f* = chain pitch
~trieb *m* → Kettenantrieb
~trommel *f* = chain drum
~-Turmdrehkran *m* → Gleis~
~verankerung *f* = chain anchorage, ~ anchoring
~vollbecherwerk *n*, Kettenreihenelevator *m* = continuous chain (type bucket) elevator
~-Vorgelege *n* [*Tiefbohranlage*] = chain compound [*deep drilling rig*]
~vorschub *m*, Kettenvorstoß(en) *m*, (*n*) [*Hochlöffel*] = chain crowd(ing)
~-Vorschub *m* [*Hammerbohrmaschine*] = chain feed
~wagen *m* → Raupenwagen
~walze *f* [*beim Kettenaufgeber*] = chain roll
~windwerk *n* → Kettenhebewerk
~wirbel *m* = chain swivel
~zahnrad *n* → Kettenturas *m*
~zange *f* = chain tongs
~zug *m* = chain hoist, ~ block (and tackle), ~ lifting block, ~ pulley block
Kfz. *n* → Kraftfahrzeug
~-Fährschiff *n* = car ferry
~-Motor *m* → Auto(mobil)motor
~-Parkplatz *m* = vehicular parking area
Kielbogen *m* → geschleifter Spitzbogen
kielholen = to careen
Kiel|kühler *m* = keel cooler
~pumpe *f* = bilge pump
~stapel *m* = keel block
Kienöl *n* = pine oil

Kies — Kieslager(stätte)

Kies *m* [*nach Atterberg, Gallwitz 20 bis 2 mm; nach DIN 4022 70–2 mm; nach Niggli 200–2 mm; nach I. Kopecki größer als 2 mm*] = gravel [*International Society of Soil Science, United States Public Roads Administration, Massachussetts Institute of Technology, British Standard Institution 2–60 mm; A.S.E.E. 76.2–2 mm; Bureau of Reclamation 76.2–4.76 mm;* Casagrande *(1947) 2000 to 2 mm*]
~**ablagerung** *f* (Geol.) = gravel fill
~**abstreuung** *f* = gravel blotter (US)
~**asphalt** *m*, Kieswalzasphalt = gravel asphalt
~**auffüllung** *f* → Kiesschüttung
~**aufschüttung** *f* → Kiesschüttung
~**auftrag** *m* → Kiesschüttung *f*
~**-Ausleger** *m*, Kies-Verteiler = gravel distributor, ~ spreader
~**bank** *f* = gravel bar
~**betrieb** *m*, Kieswerk *n* = gravel works
~**ballast** *m* = gravel ballast
~**ballastkasten** *m* = gravel ballast container
~**beton** *m* = gravel concrete
~**boden** *m* = gravelly soil
~**brechanlage** *f*, Kiesbrechstation *f* = gravel-breaking plant, gravel-crushing ~, ~ installation
~**bunker** *m* = gravel bin
~**damm** *m* → Talsperre *f*
~**decke** *f*, Kiesbelag *m* = gravel surfacing, ~ pavement; ~ paving (US)
~**deponie** *f* = gravel stockpiling
Kiesel|alge *f* → Diatomee *f*
~**algenerde** *f* → Kieselgur *f*
~**breccie** *f* [*dieser Ausdruck sollte nicht mehr verwendet werden*]; Kieselbrekzie *f* = silica breccia
~**gallerte** *f*, Kieselsäuregel *n* = silica gel
~**galmei** *m*, Zinkglas(erz) *n*, Kieselzink(erz) *n* (Min.) = siliceous calamine, willemite
~**gur** *f*, Bergmehl *n* = kieselgur, fossil meal

~**gurplatte** *f* = kieselgur panel, ~ slab
kieseliges Bindemittel *n*, kieselige verkittende Zwischensubstanz *f* (Geol.) = siliceous matrix, ~ cement, ~ cementing material
Kiesel|kalk(stein) *m*, kieseliger Kalkstein = siliceous limestone
~**kreide** *f* = chalk flint
~**lage** *f* = pebble bond
~**mangan** *n*, Rotbraunstein *m*, Rotspat *m* (Min.) = rhodonite
~**salz** *n* = silicate
~**sand** *m* = siliceous sand
kieselsauer = silicic
Kieselsäure *f* = silicic acid
~**gel** *n* → Kieselgallerte *f*
~**-Glas** *n*, amorpher Quarz *m* = vitreous silica
~**pulver** *n* = silica powder
kieselsaure Tonerde *f* = hydrous (or hydrated) silicate of aluminum, siliceous earth
Kiesel|schiefer *m* → Lydit *m*
~**schotter** *m* (Geol.) = siliceous oolite gravels
~**schwamm** *m* = siliceous sponge
~**sinter** *m* = siliceous sinter, geyserite
~**(stein)** *m* [*nach Dücker (Jahr 1948) „grobes Siebkorn IIa"* 60–20 mm *und „grobes Siebkorn IIb"* 20–6 mm] = pebble [60–6 mm]
~**substanz** *f* [*Gestein*] = silica cement
~**zinkerz** *n* = zinc silicate
Kies|feld *n*, Sandfeld = outwash plain
~**filter** *m*, *n*, ~**lage** *f*, ~**schicht** *f* = gravel filter layer
~**(füll)damm** *m* → → Talsperre *f*
~**füllung** *f* [*Kastenfang(e)damm; Zellenfang(e)damm*] = gravel fill
~**fundstätte** *f* → Kiesvorkommen *n*
~**greifer(korb)** *m*, Kiesgreifkorb = gravel grab(bing) bucket, ~ grab
~**grube** *f* = gravel pit
~**kern** *m* = gravel core
~**klebedach** *n* = bonded gravel roof
~**lage** *f*, Kiesschicht *f* = gravel layer
~**lager(stätte)** *n*, *(f)* → Kiesvorkommen *n*

~-**Naßbagger** m, Kies-Schwimmbagger = gravel dredger (Brit.); ~ dredge (US)
~**nest** n, Steinnest [*Beton*] = rock pocket [*A porous, mortar-deficient portion of hardened concrete consisting primarily of coarse aggregate and open voids; caused by leakage from formwork, or separation during placement or insufficient consolidation, or both*]
~**plombe** f, Kiespfropfen m = gravel plug
~**preßdach** n = felt-and-gravel roof
~**pumpe** f → Stauchbohrer m
~**sand** m = gravelly sand
~**schicht** f, Kieslage f = gravel layer
~**schiff** n = gravel barge
~**schotter** m, Steinschotter = ballast
~**schutt** m (Geol.) = gravel(l)y detritus, ~ wash
~**schüttdamm** m → → Talsperre f
~**schüttung** f, Kiesaufschüttung, Kiesauffüllung, Kiesauftrag m = gravel fill(ing)
~**schüttung** f [*Brunnen*] = gravel wall, ~ packing, ~ envelope
~**schüttungsbrunnen** m = gravel-wall well, grave-packed ~ [*A gravel envelope around the outside of the casing is frequently used where excessive quantities of fine sand exist in the aquifer*]
~**schüttungsdicke** f [*Brunnen*] = gravel-pack thickness
~**schüttungs(stau)damm** m → → Talsperre f
~-**Schwimmbagger** m, Kies-Naßbagger = gravel dredger (Brit.); ~ dredge (US)
~**silo** m = gravel silo
~**splitt** m = gravel chip(ping)s
~**splittwerk** n, Kiessplittbetrieb m = gravel chip(ping)s works
~**stock** m = pyritic rock
~**straße** f [*Decke aus Kies*] = gravel road
~**streugerät** n mit Schleuderverteilung = spinner type distributor for stone, ~ ~ spreader ~ ~, spinner

~**terrasse** f = gravel terrace
~**tragschicht** f = gravel base (course)
~**trockenbagger** m = gravel excavator
~**trockner** m = gravel dryer; ~ drier (US)
~- und **Sandmischanlage** f = plant for producing graded gravel and sand mix(tur)es
~- und **Splittwerk** n = gravel and chip(ping)s plant
~**unterbau** m [*Straßenbau*] = gravel substructure
~-**Verteiler** m, Kies-Ausleger = gravel spreader, ~ distributor
~**vorkommen** n, Kieslager(stätte) n, (f), Kiesfundstätte, Kiesvorkommnis n = gravel deposit
~(**walz**)**asphalt** m = gravel asphalt [*dense hot-rolled asphalt with gravel as aggregate*]
~**wäsche** f, Kieswaschmaschine f = gravel washing machine, ~ washer
~**weg** m = gravel path
~**werk** n, Kiesbetrieb m = gravel works
~**wüste** f = stony desert
Kilometer|**stein** m = kilometre post
~**zähler** m = odometer
Kilometrierung f = kilometrage, stationing in kilometres
Kilo|**schlüssel** m = torqometer
~**wattstunde** f = Kwhr.
Kimme f, sichtbarer Horizont m = sensible (or apparent, or visible) horizon
Kimm|**schlitten** m = bilge block
~**tiefe** f, Kimmung f = dip of the horizon, apparent depression ~ ~

Kind-Chaudron-Schachtbohrverfahren n = Kind Chaudron method of shaft sinking
Kinder|**genesungsheim** n = convalescent home for children
~**spielplatz** m → Spielplatz
Kinematik f → Bewegungslehre f
kinematische Ähnlichkeit f = kinematic similarity, ~ similitude

~ **Viskosität** *f* = kinematic viscosity
Kinetik *f* = kinetics
kinetische Energie *f*, Bewegungsenergie = energy of motion, ~ ~ movement
Kinkenbildung *f* = kinking
Kino *n*, Lichtspielhaus *n*, Lichtspieltheater *n* = cinema
Kipp|achse *f* [*Theodolit*] = trunnion axis
~**anhänger** *m* = dump(-body) trailer, tipping ~
~**aufbau** *m* = lorry-mounted tipping unit (Brit.); truck-mounted ~ ~ (US)
~**-Aufl(i)eger** *m* → Kipp-Halbanhänger *m*
~**-Aufsattel-Anhänger** *m* → Kipp-Halbanhänger
kippbare Ladeschaufel *f*, Kippwanne *f*, Kippmulde *f* [*Frontlader*] = tilting front end bucket
Kipp|behälter *m* = (bulk) tipping container
(~-)**Betonkarre(n)** *f*, *(m)*, Beton(rund)kipper *m*, Japaner(karren) *m*, Japaner(kipp)karre(n), Kipp-Japaner = concrete buggy, hand concrete cart, rocker-dump hand cart
~**bettanhänger** *m* = tilting platform trailer
Kippe *f* [*Erdbau*] = dump (area)
~ → Aussatz~
Kippen *n* [*Schwimmdock*] = careening
~, Über~ [*Bauwerk*] = overturning
Kipper *m*, Kipplast(kraft)wagen *m*, Kipp-LKW *m*, Autokipper = tipper, dump lorry, tip (motor) lorry, tipping lorry (Brit.); dump truck, tip(ping) truck (US)
~ [*Wagenkipper*] = tip
~ → Mulden~
~**karre(n)** *f*, *(m)* → Betonkarre(n)
Kippflügel *m* = bottom-hinged sash
~**fenster** *n* = bottom-hinged sash window
Kipp|form *f* = tilting mould (Brit.); ~ mold (US)

~**gestänge** *n* [*Schaufellader*] = tipping link
~**-Halbanhänger** *m*, aufgesattelter Kippanhänger, Kipp-Aufl(i)eger *m*, Kipp-Sattel(schlepp)anhänger, Kipp-Aufsattel-Anhänger, Sattelkipper *m* = dump semi-trailer, ~ articulated trailer
~**halde** *f* = dump pile
Kipphebel *m* = rocker
~ [*Schaufellader*] = bucket tilt lever
~**deckel** *m* = rocker cover
~**welle** *f* = rocker shaft
Kipp|höhe *f*, Abwurfhöhe, Ausleerhöhe, (Aus)Schütthöhe [*Schaufellader*] = dumping height
~**kabine** *f* = tilt(ing) cab(in)
~**karre(n)** *f*, *(m)* → Betonkarre(n)
~**kübel** *m* → → Betonmischer *m*
~**kübel** *m* [*beim Muldenaufzug*] → Kippmulde *f*
~**kübelaufzug** *m* → Mulden(bau)aufzug
~**lager** *n* [*für Hoch- und Tiefbauten*] = rocker bearing, tilting ~, pivoting ~
~**last** *f* = overturning load
~**-LKW** *m*, Kipper *m* = dump truck; ~ lorry (Brit.)
~**löffel** *m* = tilting (dipper) bucket
~**lore** *f* → Muldenkipper *m*
~**moment** *n* = overturning moment
Kippmulde *f* [*Erdbau-Lastfahrzeug*] = (dump) body
~, Kippkübel *m*, Beton~ [*beim Muldenaufzug*] = (concrete) tipping skip, (~) ~ hopper, tip-over (concrete) bucket [*There are one-way and two-way side tipping skips*]
~ [*Rollofen*] = tipping cradle
Kipp|muldenaufzug *m* [*am Betonmischer*] = concrete skip hoist
~**ofen** *m*, Schaukelofen = (crucible) tilting furnace, tilter
~**-Pflug** *m* → Planierpflug
~**räumer** *m* → Planierpflug *m*
~**reichweite** *f* [*Schaufellader*] → (Aus)Schüttreichweite

Kipp-Sattel(schlepp)anhänger — Klappenstütze

~-Sattel(schlepp)anhänger m → Kipp-Halbanhänger
~schalter m = key switch, switching key (Brit.); toggle switch (US)
~schrapper m = tilting-type drag scraper
~schraube f [Nivellierinstrument] = tilting level screw
~sicherheit f = safety against overturning
~sicherungskette f [Autoschütter] = check chain
~tieflader m → Tieflader mit Kipp(tief)ladebrücke
~tisch m = tilting table
~tor n, Garagen~ = up and over garage door, overhead ~ ~
~trommelmischer m → → Betonmischer
~überdruckventil n = tilt relief valve
~vorrichtung f, Kippwerk n [LKW] = tipping gear
~weiche f = tilting point
~winkel m [Kipp-LKW] = tipping angle
~winkel m [Schaufellader] → (Aus-)Schüttwinkel
~zapfen m [Brückenlager] = rocker pin
~zeit f [Schaufellader] → (Aus-)Schüttzeit
~zylinder m = tilt ram
Kirchen|architektur f = church architecture, religious ~
~bau(kunst) m, (f) = ecclesiology, church building
kirchenbaulich = ecclesiologic(al)
Kirchen|fenster n = church window
~mittelschiff n = church nave
~portal n = church portal
~raum m = church hall
Kirchturm m mit Spitze = (church) steeple
~ ohne Spitze = church tower
~spitze f = spire
Kissen n → Preß~
Kisten|öffner m = pinch bar, claw ~, wrecking ~, jenny, case opener
~verschluß m, Spannverschluß f = drawbolt (US); toggle bolt (Brit.); box closure [generic term]

Kitt m = putty
~dichtung f = putty seal(ing)
~falz m, Glasfalz = rebate for putty
kittloses Verglasen n = patent glazing, dry ~, puttyless ~
Kitt|messer n = putty knife
~verglasung f = putty glazing
klaffender Riß m = gaping crack
Klammer f = clamp
~ → Frosch m
~ (⚒) [Unterhängezimmerung] = hanging bolt, hanger
~lasche f = fishplate
klammern [zwei Mauerschalen] = to bond
Klangfehlerhaftigkeit f [Ziegel] = dunting
Klapp|(bohr)mast m, Faltbohrmast = jackknife (drilling) mast, folding (~) ~, collapsible (~) ~
~brücke f, Wippbrücke = balance bridge, bascule ~
~bügel m = folding stirrup
Klappe f [Dachwehr] = leaf
~, Tafel f, Stau~ [Klappen(stau)-wehr] = wicket (US); shutter (Brit.)
Klappen|büchse f, Klappenschöpfbüchse = disc valve bailer; disk ~ ~ (US)
~pfeiler m [Klappbrücke] = bascule pier
~(schöpf)büchse f = disc valve bailer; disk ~ ~ (US)
~seil n [Bagger] = discharge gate rope, ~ ~ line
~(stau)wehr n = shutter weir, ~ dam (Brit.); hinged-leaf gate (US)
~flügel m = top-hinged sash
~flügelfenster n = top-hinged sash window
~form f = folding mould (Brit.); ~ mold (US)
~laden m = boxing shutter, folding ~
~mast m = Klappbohrmast
~schute f = bottom-dump scow (US); ~ barge (Brit.); drop-bottom ~, hopper ~
~stütze f [Zur Erhöhung der Standsicherheit von Baggern, Kränen und Schaufelladern] = folding stabilizer

Klappenstütze — Klebmasse

~stütze *f* [*Strom(erzeugungs)aggregat*] = leg stand
~tor *n* [*Schleuse*] = trap gate
~tür *f* = trap door
Klär|anlage *f*, Klärwerk *n*, Abwasserreinigungsanlage, Abwasserreinigungswerk *n* = sewage purification plant
~becken *n* → Absetzbecken
~becken *n* [*Tiefbohrtechnik*] = settling pit
Klarbernstein *m* = clear amber
Klärbrunnen *m* → Absetzbrunnen
Klärer *m* = clarifier
Klärgeschwindigkeit *f*, Absetzgeschwindigkeit = sedimentation velocity, settling ~, rate of settling, rate of sedimentation
Klar|glas *n* = clear glass
~kohle *f* → Formkohle
~lack *m*, Luftlack, Blanklack = (clear) varnish
Klär|raum *m* → Absetzraum
~teich *m* → Schlammteich
~turm *m*, Stoffangtrichter *m* [*Papierfabrik*] = clarifying tower
Klärung *f*, mechanische ~ [*Abwasserwesen*] = settlement, sedimentation, settling, clarification
Klarwasser *n* = clear water
~spülung *f* [*als Verfahren beim Rotarybohren*] = clear water circulation
Klärwerk *n* → Kläranlage *f*
Klassenzimmer *n*, Klassenraum *m* = classroom
Klassierer *m* = classifier
~-Luftrutsche *f* = air-slide classifier
Klassierung *f*, Klassieren *n* [*Das Zerlegen eines Gutes in Kornklassen*] = sizing
Klassifizierung *f*, Einstufung = indexing, classification
Klassifizierungs|eigenschaft *f* [*Bodenmechanik*] = index property, classification ~
~prüfung *f*, Klassifizierungsprobe *f*, Klassifizierungsversuch *m* [*Bodenmechanik*] = classification test, index ~

klastischer Kalktuff *m* = detrital lime tuff, clastic ~
klastisches Eruptivgestein *n* = clastic eruptive rock
~ **Gestein** *n* = clastic rock
~ **Sedimentgestein** *n* = clastic sedimentary rock
Klatsche *f*, Pritschbläuel *m*, Praker *m*, Patsche *f*, Tatsche *f*, Schlagbrett *n* = muller and plate (Brit.)
Klaubeband *n*, Leseband, Sortierband = picking belt conveyor, ~ ~ conveyer
Klauben *n*, (Aus)Lesen [*Bestimmte Bestandteile werden von Hand aus dem Aufgabegut ausgelesen*] = picking, hand ~, sorting
Klaubetisch *m*, Lesetisch, Sortiertisch = picking table
Klauen|bremse *f* = jaw brake
~kupp(e)lung *f* = jaw clutch, denture ~, positive ~, dog ~
Kleb(e)|anstrich *m* = adhesive coat
~band *n* = adhesive tape
~eigenschaft *f* = adhesive property
~kitt *m* → Klebezement *m*
~masse *f* → Klebmasse
~masse *f* für Dachpappen = roofing-felt cement
~mittel *n* → Klebmasse *f*
~mittelverbindung *f* → Klebnaht *f*
Kleben *n* = glueing
Kleb(e)|papier *n* = adhesive paper
~prüfung *f*, Kleb(e)probe *f*, Kleb(e)versuch *m* = bond test
Kleber *m* → Klebmasse *f*
Klebe|schicht *f* [*Straßenbau. In der Schweiz*] = prime coat, priming ~, ~ membrane
~schmiege *f* → Backenschmiede
~zement *m*, Kunstkautschuk-Kontakt-Kleber *m*, Klebekitt *m* = adhesive putty, special ~, bonding ~, ~ cement
~zettel *m* = gummed label
Kleb|fuge *f* → Klebnaht *f*
~masse *f*, Klebemasse, Aufstrichmasse, Klebstoff *m*, Kleb(e)mittel *n*, Kleber *m* = adhesive (compound), bonding agent

Klebmittel — Klein-Planierschlepper

~mittel n → Klebmasse f
~naht f, Haftbrücke f, Klebfuge f, Kleb(e)mittelverbindung f = adhesive-bonded joint
~stoff m → Klebmasse f
Klee|blatt n, ~kreuzung f, Renaissancekreuzung = clover-leaf (flyover) junction, ~ intersection
~blattbogen m = trefoil arch
~säure f → Oxalsäure
Klei m [*Der aus dem Schlick im besonderen durch Wasserverlust entstandene feste Tonboden der Marsch*] = clay containing sea silt
Kleiderspind m = clothes locker
kleine Brechstange f = pinch bar
~ **Furche** f [*Bewässerung*] = corsett
~ **Hauptspannung** f = minor principal stress
kleiner Bach m, Bächlein n = brooklet
Klein|-Aggregat n, Klein-Elektroaggregat = small generating set
~anhänger m = midget trailer
~auflader m = small-size bucket (elevator) loader
~aufzug m, Klein-Lastenaufzug, Klein-Materialaufzug = small-type goods lift (Brit.); ~ ~ elevator (US)
~bagger m = midget excavator
~bahn f, Feldbahn = narrow-gauge railway (Brit.); narrow-gage railroad (US)
~baracke f = small-sized hut
~baustelle f = small-scale (building, or project, or job) site
~-Betonmischer m = midget concrete mixer
~brecher m = small crusher, ~ breaker, ~ crushing machine, ~ breaking machine
~-Bulldozer m → Klein-Planierschlepper
~bus m = minibus
~eisenzeug n = builders' iron supplies, builders' ironmongery [*also loosely termed "hardware"*]
~(-Elektro)-Aggregat n = small generating set

~elektromotor m = small type motor
Kleine'sche (Stahlstein)Decke f = Kleine (hollow-brick) floor
Klein|filter m, n, Hausfilter = domestic filter
~-Förderwagen m für Baubetrieb → Autoschütter
~format n = small size
~-Frontträumer m → Klein-Planierschlepper
~garten m, Schrebergarten = allotment
~-(Gleis)Kettenschlepper m → Kleinraupe f
~-Kettenschlepper m → Kleinraupe f
~kompressor m, Kleindrucklufterzeuger m, Klein(luft)verdichter m [*unter 1 PS*] = fractional horsepower (air) compressor
~küche f, Kochnische f = kitchenet(te)
~-Lastenaufzug m → Kleinaufzug
~lieferwagen m = light van
~lochbohren n = small-hole drilling
~maschine f = small size machine
~maß n = small dimension
~-Materialaufzug m → Kleinaufzug
~mietwohnung f = small-sized flat
~mischer m = small(-capacity) mixer
~mischer m ohne Beschickungsaufzug, (Schub)Karrenmischer = baby concrete mixer
~modell n = reduced model
~motorwalze f = midget self-propel(l)ed roller
Kleinpflaster(decke) n, (f) = peg-top paving [*Paving with very small sets*]
~ **in Fächerform** = radial (small stone) sett paving, random ~, durax ~, fanwise ~, circular ~, fanshaped ~
Klein|pflasterstein m = small (paving) sett
~-Planierschlepper m, Klein-Bulldozer m, Klein-Frontträumer, Klein-Schürfschlepper, Klein-Schürftrekker, Klein-Schürftraktor, Klein-Planiertraktor, Klein-Planiertrekker, Klein-Räumer = baby bulldozer, skipdozer

Klein-Planiertrecker — Kletter(-Turmdreh)kran 582

~-**Planiertrecker** *m* → Klein-Planierschlepper
~-**platte** *f* = small slab
~-**ramme** *f* = midget (pile) driver
~-**Räumer** *m* → Klein-Planierschlepper *m*
~**raupe** *f*, Klein-(Gleis)Kettenschlepper *m* = crawler-mounted baby bulldozer, ~ calfdozer, ~ skipdozer
~-**reparatur** *f* = small repair
~-**Rüttelplatte** *f* = midget vibrating plate, ~ ~ pan, ~ ~ slab, ~ vibration ~, ~ vibratory ~
~-**Schaffußwalze** *f* = midget sheepsfoot roller
~**schlag** *m* = hardcore
~**schlag** *m* aus Betonbrocken, Betonpacklage *f*, Betonschotter *m* = (aggregate of) broken concrete, concrete hardcore
~**schlepper** *m*, Kleintraktor *m*, Kleintrecker *m* = midget tractor
~-**Schürfschlepper** *m* → Klein-Planierschlepper
~-**Schürftraktor** *m* → Klein-Planierschlepper
~-**Schürftrecker** *m* → Klein-Planierschlepper
~-**Sender/Empfänger** *m* = miniature transmitter/receiver, ~ transceiver
Kleinst|abstand *f* = least distance
~**(elektro)motor** *m* = pilot motor, fractional H. P. ~
kleinster Trägheitshalbmesser *m* = least radius of gyration
Kleinst|hydrozyklon *m* = miniature hydrocyclone
~**maß** *n* = least dimension
~**mietwohnung** *f* = minimum-sized flat
~**spiel** *n* = least play
Klein|teile *npl*, *mpl* [*Gleiskettenlaufwerk*] = hardware
~**tektonik** *f* = micro-tectonics
~**tunnelofen** *m*, Platten-Tunnelofen = pusher-type kiln
~**verdichter** *m* → Kleinkompressor *m*
~**verkehrsstraße** *f* → Nebenstraße
~**walze** *f* = midget roller
~**wasserkraftwerk** *n* = small-type hydroelectric power plant
~**werkzeug** *n* = small tool
~**wohnungsbau** *m* = small scale housing
~-**zellenfang(e)damm** *m* = gabion, small cellular cofferdam
~**zylinderverfahren** *n* = small-scale cylinder test method
Kleister *m* = paste
Klemmbacke *f* [*Klemmgriff-Lastenträger*] = side shoe
Klemmkasten *m* = terminal box
Klemm|griff-Lastenträger *m* → Torladewagen *m*
~**kraft** *f* [*von Nieten auf Blechen*] = contact pressure, clamping force
~**last** *f*, Anfangstragkraft *f* [*Grubenstempel*] = initial load-bearing capacity, setting load
~**leim** *m*, WHK-Leim [*nach dem Erfinder Klemm benannter Kaltleim*] = Klemm glue, ~ adhesive, ~ cement
~**platte** *f* [*Feldbahn*] = base plate, sleeper clamping ~, tie clamping ~
~**ring** *m* = clamping ring
~**schieber** *m* [*Anbaugerät für palettenlosen Transport*] = push-pull equipment
~**schraube** *f* = clamping screw
Klemmutter *f* → Kontermutter
Klemmverankerung *f* = grip anchorage system
Klempner *m* = plumber
Kletter|-Derrickkran *m* = creeper derrick
~**eisen** *n* = pole climber, climbing iron
~**eisen** *n*, Steigeisen = access hook, step iron
~**(film)verdampfer** *m* = climbing film evaporator
~**gerüst** *n* = climbing scaffold(ing)
~**moor** *n* = climbing bog
~**schalung** *f* = climbing formwork, jumping ~, leaping ~
~**stange** *f* [*Kletterschalung*] = climbing rod
~**(-Turmdreh)kran** *m*, Hochhauskletterkran, Hochbaukletterkran = climbing (tower) crane

Klettervermögen — Klopfholz

~vermögen n = climbing power
~weiche f = climbing switch
Kliff|hang m (Geol.) = chine
~(küste) n, (f) → Steilküste
~schutt m = undercliff
Klima|anlage f = air conditioning system
~anlagenraum m = (air) conditioning equipment room
~daten f = climatic data
klima|geregelt = air-conditioned
~geregelte Luft f = conditioned air
Klima|kanal m = air-conditioning duct
~karte f [*zur Bemessung von Heizungs- und Klimaanlagen*] = design temperature map
~reg(e)lung f, Klimatisierung f = air conditioning
~schwankung f = climatic variation
~spiel n = climate cycle
~station f = air-conditioning substation
klimatisiert = air-conditioned
Klimatisierung f, Klimareg(e)lung = air conditioning
Klima(tisierungs)anlage f = air conditioning installation (or plant)
Zuluftanlage = induction system
Abluftanlage = extraction system
Flatterjalousie f = automatic ventilating louvre
klimatologische Verhältnisse npl = climatological conditions
Klingstein m (Geol.) = clinkstone
Klinik f = clinic
Klinkenvorschub m = pawl feed(ing)
Klinker m, ~ziegel m, ~stein m = arch brick, clinker ~ [*The bricks which immediately surround the fire are usually overburned and are badly warped and discoloured. These are known as arch or clinker bricks and are suitable for use in foundations or similar places, for they are very durable but unattractive*]
~, Zement~ = (cement) clinker
~brecher m = clinker crusher, ~ breaker

~brennen n = burning to clinker
~freilager n = outdoor clinker storage area
~korn n, Klinkerteilchen n, Zement~ = clinker grain
~kühler m, Klinkerkühlvorrichtung f = clinker cooler
~lager n, Zement~ = clinker store, ~ storage (area)
~mahlanlage f → Klinkermühle f
~mahlung f, Klinkermahlen n = clinker grinding
~mauerwerk n = clinker masonry
~mühle f, Klinkermahlanlage f, Zement~ = clinker(-grinding) mill
~pflaster(decke) n, (f) = clinker paving
~platte f = clinker slab
~schnellkühler m, Klinkerschnellkühlvorrichtung f = rapid clinker cooler
~stein m → Klinker m
~straße f = clinker road
~teilchen n → Klinkerkorn n
~ung f, Teilverglasung = partial vitrification
~verblendung f [*Die Ummantelung eines Bauwerks mit Klinkern bietet einen guten Schutz im Wasser, auch bei Schadwasser. Voraussetzung ist allerdings, daß der Mörtel dicht und bei Schadwasser widerstandsfähig ist*] = arch brick lining, clinker ~ ~
~(vorrats)lager n = clinker stor(ag)e
~ziegel m → Klinker m
Klinograph m = clinograph
Klippenbai f = rocky bay
Kloben m, Unterflasche f, Block m, (Seil)Flasche, Seilkloben = block
Klopfbremse f, Antiklopfmittel n = anti-knock agent
Klopfen n [*Otto-Kraftstoff*] = knocking, pinking
Klopf|festigkeit f, Klopffreiheit f = anti-knock performance, ~property
~hammer m → Bunkerrüttler m
Klopfholz n, Stemmknüppel m [*Dient als Schlagwerkzeug für das Stemm- und Locheisen*] = beechwood mallet

Klosett *n*, Abtritt *m*, Abort *m* = closet [*A privy, water closet, etc.*]
~anlage *f*, Abortanlage, Abtrittanlage = toilet
~becken *n* → Abortbecken
~spülkasten *m* → Spülkasten
Klostergewölbe *n* = cloister vault
Klothoide *f* = spiral (transition) curve, transition spiral, highway spiral
Klothoidentafel *f* = spiral table for highway design
Klotz *m*, Block *m* = block
~bremse *f*, Backenbremse = shoe brake
Klubhaus *n* = club building
Kluft|auspressung *f* → Kluftinjektion *f*
~einpressung *f* → Kluftinjektion *f*
~fallwinkel *m* (Geol.) = fault dip
klüftiger Sandstein *m* = sandstone containing open seams
Kluft|injektion *f*, Kluftauspressung *f*, Kluftverpressung, Klufteinpressung, Felsinjektion, Felsauspressung, Felsverpressung, Felseinpressung, Gesteinsinjektion, Gesteinsauspressung, Gesteinsverpressung, Gesteinseinpressung [*Injektionen von Klüften, Spalten und größeren Hohlräumen im Fels, wobei meist mit Zementmilch, Zement-Sand-Gemischen, Tonsuspensionen oder eventuell chemischen Lösungen die Hohlräume ausgefüllt werden*] = rock sealing
~quelle *f*, (Verwerfungs)Spaltenquelle = fracture spring, fissure ~
~verpressung *f* → Kluftinjektion *f*
~wasser *n*, Spaltenwasser = fissure water, crack ~
Klumpfuß *m* [*Pfahl*] = bulb, pedestal
~pfahl *m* = bulb pile, pedestal ~
klumpig, großstückig = lumpy
Knabbenkohle *f*, Knorpelkohle, Stückkohle = brown coal occurring in the form of lumps
Knagge *f* [*Ein kurzes Holzstück, das in Form einer Konsole im Holzfachwerkbau ein darüberliegendes Konstruktionsglied unterstützt*] = wood bracket, wooden ~

~, Auflager~ = (angle) cleat
Knallblei *n* = fulminating lead
„knallen" [*von Grubenstempeln*] = to crack
Knallgas *n*, Hydrooxygengas = oxyhydrogen (gas)
~licht *n*, Kalklicht = limelight
Knall|quecksilber *n*, Quecksilbercyanat *n*, Merkuricyanat, Hg(CNO)$_2$ = fulminate (of mercury), mercuric isocyanate
~zündschnur *f* → → 1. Schießen *n*; 2. Sprengen *n*
Knäpper *m* = block of stone drilled for blasting
~(bohr)hammer *m* = block holer
knäppern = to pop-shoot (Brit.); to block-hole (US)
Knäpper|schießen *n*, Nachknäppern *n*, Sprengen *n* mit Knäpperschuß = boulder blasting
~schuß *m* = pop shot, plaster ~
Knappheit *f*, Verknappung *f* = shortage
Knarre *f*, (Bohr)Ratsche *f* = ratchet
knarren = to creak
Knetaufbereitung *f* [*Feuerfestindustrie*] = kneading
Kneten *n* → Zwangsmischen
Knet|mischer *m*, Kneter *m*, Zwangsmischer mit vertikalem Rührwerk, Vertikalrührwerk = pug(-)mill (mixer)
~verdichter *m*, Knetwalze *f*, Erdbauwalze, Verdichtungswalze, Bodenverfestigungswalze = compaction roller, soil ~ ~
~walze *f* → Knetverdichter *m*
~wirkung *f* → Verkehrskomprimierung *f*
~wirkung *f* [*Walze*] = kneading action
Knick|aufgabe *f* = buckling problem
~ausleger *m*, abgewinkelter Ausleger = articulated boom (US); ~ jib (Brit.)
~bedingung *f* = buckling condition
~beiwert *m*, Knickzahl *f* = buckling coefficient
~belastung *f* = buckling loading

Knickberechnung — Knotenschiefer

~berechnung *f* = calculation of the buckling strength
knicken → aus~
Knicken *n* → Aus~
Knick|festigkeit *f*, Knickwiderstand *m* = buckling strength, ~ resistance, ultimate column ~
~form *f* = type of buckling
~formel *f* = buckling formula
~gefahr *f* = buckling risk
~kraft *f* = buckling force
~kriterium *n* = buckling criterion
~länge *f* = effective length
~last *f* = buckling load
~lenkung *f*, Zentrallenkung = center pivot steer(ing) (US); centre ~ ~ (Brit.)
~modul *m* = buckling modulus
~probe *f* → Knickversuch *m*
~prüfung *f* → Knickversuch *m*
~punkt *m* [*geophysikalisches (Aufschluß) Verfahren*] = break point
~punkt *m* [*Baustatik*] = buckling point
~punktchlorung *f* = break-point chlorination
~sicherheit *f* = safety against buckling
~spannung *f* = buckling stress
~stabilität *f* = buckling stability
~steifigkeit *f* = buckling stiffness
Knickung *f* → Ausknicken *n*
Knick|verhalten *n* = buckling behavio(u)r
~versuch *m*, Knickprobe *f*, Knickprüfung *f* = buckling test
~widerstand *m* → Knickfestigkeit *f*
~zahl *f* → Knickbeiwert
Knie *n* = elbow
~gelenk *n* = knuckle joint
Kniehebel *m* = toggle
~kupp(e)lung *f* = overcenter-type clutch (US); overcentre-type ~ (Brit.)
~nietmaschine *f* = toggle joint riveting machine, ~ ~ riveter
~platte *f* = toggle plate
~presse *f* = toggle plate press
Knie|rohr *n* → Rohrkrümmer *m*
~schützer *m* = knee pad

~stock *m*, Drempel *m* [*Schwelle im Torboden einer Schleuse gegen die sich das geschlossene Schleusentor stützt*] = pointing sill (Brit.); miter ~ (US)
~stockwand *f* → Drempelwand
~stück *n* = elbow
knirsch stoßen [*Mauerwerk*] = to be laid touching
Knirschfuge *f*, Trockenfuge = nonbonded joint, dry ~
Knisterkohle *f* = coal altered by an igneous intrusion, which on heating explodes with violence
knistern, krebsen [*Gebirge im Tunnel-, Stollen- und Schachtbau*] = to creak
Knochen|bett *n*, Knochenlager *n* (Geol.) = bonebed
~leim *m* [*genormtes Kurzzeichen: KN*] = bone glue
~öl *n* = Dippels oil, bone ~
~säge *f* = butcher's saw
~schwarz *n* = bone black, ivory ~
~teer *m* = bone tar
Knollen|bildung *f* [*Zement*] = lumpiness
~brecher *m* = lump crusher, ~ breaker
~kalk(stein) *m* = nodular lime(stone), calcareous nodule
Knorpel|kohle *f* → Knabbenkohle
~tang *m*, Perlmoos *n*, irländisches Moos = carraghen, Irish moss
Knoten *m* [*mehrere in einem Punkt zusammengeschlossene Stäbe eines Stockwerkrahmens bilden einen Knoten*] = point of junction, ~ ~ intersection of members, assemblage point, joint
Knotenblech *n* → Anschlußblech
~verbindung *f* = gusseted connection
Knoten|dolomit *m* = knotty dolomite
~eisen *n* → Anschlußblech *n*
knotenförmig = nodular
Knoten|gelenk *n* = multiple joint, tiebar ~
~punkt *m*, Systempunkt = system point, knot; system centre (Brit.); system center (US)
~schiefer *m* = nodular shale

Knotenverbindung — Kohle(n)aufbereitung

~verbindung *f* = assemblage point connection, joint ~

knüpfen [*Bewehrung*] = to bind in position

Knüppel|ausstoßmaschine *f* = billet ejector

~damm *m* → Knüppelweg *m*

~schere *f* = billet shears

~walzwerk *n* = billet mill

~weg *m*, Prügelweg, Knüppeldamm *m* = corduroy road

Koagulation *f* = coagulation

Kobalt|arsenkies *m* = danaite

~blau *n* = cobalt blue

~blüte *f* (Min.) = cobalt bloom, erythrite

~glanz *m*. Glanzkobalt *m* (Min.) = cobalt glance, cobaltite

~kies *m* = linnaeite

~manganerz *n* = asbolite

~schwärze *f*, Asbolan *m*, Erdkobalt *m* (Min.) = asbolane, asbolite

~-60-Speicher- und Bestrahlungsanlage *f* = cobalt-60 storage garden and irradiation facility

~seife *f* = cobalt soap

Kocher *m*, Bindemittel~, (Bindemittel-)Schmelzkessel *m*, Teer- und Bitumenkocher, Teer- und Bitumen-Schmelzkessel, Bitumen- und Teerkocher, Bitumen- und Teerschmelzkessel = binder cooker

~ mit Rührwerk = heater-mixer

Kochfest verleimt = boilproof-glued

Koch|geruch *m* = cooking smell

~kessel *m* = (heating) kettle

~punkt *m*, Gasentlösungspunkt = bubble point

~salz *n* → Natriumchlorid *n*

~temperatur *f* [*Mastix*] = cooking temperature

~versuch *m* = corrosion test in boiling liquids

~versuch *m* [*Prüfung von Zement auf Raumbeständigkeit*] = soundness test by immersion in boiling water, boiling test

Kodierung *f* = coding

Koepe|-Förderanlage *f* = Koepe winder, ~ winding installation

~-Förderung *f*, Treibscheibenförderung = Koepe (system of) winding, ~ system

~scheibe *f*, Treibscheibe = Koepe pulley

KOH → Ätzkali *n*

kohärent → bindig

kohärente Tiefen *fpl* = conjugate depths

Kohäsion *f*, Bindigkeit *f*, Bindekraft *f*, Haltekraft [*vom Druck unabhängige Scherfestigkeitskomponente*] = cohesion, cohesiveness, no load shear strength

Kohäsions|faktor *m*, Beiwert der wirksamen Kohäsion = cohesional coefficient

~festigkeit *f* = cohesive resistance

~kraft *f* = cohesive force

kohäsions|los = nichtbindig

~loser Boden *m* mit Eigengewicht = cohesionless soil with weight

kohäsiv → bindig

Kohle|anode *f* = carbon anode

~aufbereitung *f* = coal preparation

~elektrode *f*, Kohlenelektrode = carbon electrode

~-Entwässerung *f* = coal dewatering

~flotation *f* = coal flotation

kohle|führend = coal-bearing

~haltige Auflandung *f* = coaly fill

Kohle|heizung *f* = coal heating

~herd *m* = coal-fired range

~hüllenabdruck *m* = carbon replica

~kraftwerk *n* = coal-fired power plant, ~ ~ station, ~ (electric) generating ~

~lichtbogen-Schweißung *f*, Lichtbogenschweißen *n* mit Kohleelektrode = (electric-)arc welding with a carbon electrode

Kohlenasche *f* = coal ash

~schlacke *f* = coal-ash slag

Kohle(n)auf|bereitung *f*, Aufbereitung von Kohle = coal preparation

Kohle(n)aufbereitungsanlage — Kohlenschiff

~bereitungsanlage *f*, Aufbereitungsanlage für Kohle = coal preparation plant
~bereitungsmaschine *f*, Aufbereitungsmaschine für Kohle = coal preparation machine
~zug *m* = coal lift (Brit.); ~ elevator (US)
Kohlen|becken *n* = coal basin
(~)**Bein** *n* = pillar (of coal), coal pillar
~**bergbau** *m* = coal mining
~**bergbaugebiet** *n* = coal mining region
~**bergwerk** *n*, Zeche *f* = coal mine, colliery
(~)**Bleispat** *m* → Bleispat (Min.)
~**bleivitriolspat** *m* (Min.) = lanarkite
~**brecher** *m* = coal breaker, ~ crusher, ~ breaking machine, ~ crushing machine
~**brikett** *n* → Brikett
~**brikettfabrik** *f* → Brikettfabrik
~**brikett(ier)presse** *f* → Brikettpresse
~**brikettierung** *f* = coal briquetting
~**brikettwerk** *n* → Brikettfabrik *f*
~**bunker** *m* = coal bunker, ~ bin
~**bürste** *f*, Kohlebürste = carbon brush
~**dioxyd** *n*, Kohlensäure *f*, CO_2 = carbon dioxide
~**eisenstein** *m*, Schwarzstreif *m* = blackband (ironstone), carbonaceous ironstone
~**elektrode** *f*, Kohleelektrode = carbon electrode
~**feste** *f* = coal barrier pillar
~**flöz** *n* = coal seam
~**förderbandstollen** *m* = coal conveyor tunnel, ~ conveyer ~
~**förderung** *f* durch Rohrleitungen = coal piping, ~ pumping
Kohle(n)gewinnung *f* über Tage = coal stripping
Kohle(n)gewinnungsmaschine *f* = coal miner
~**greifer(korb)** *m*, Kohlengreifkorb = coal grab(bing bucket), ~ grab bucket

Kohle(n)grießwiderstandsofen *m* = carbon resistance furnace
Kohlenhafen *m* = coal port
kohlenhaltiges Gebirge *n* [*Tunnel-, Stollen- und Schachtbau*] = carboniferous ground, ~ stata, ~ formation
Kohlen|händler *m* = coal merchant
~**hobel** *m* = coal plough (Brit.); ~ plow (US)
~**kalkstein** *m* = carboniferous limestone
~**keller** *m* = coal cellar
~**kessel** *m* = coal-fired boiler
~**kipper** *m* = coal tip
~**kraftwerk** *n* = coal burning power station, ~ ~ generating ~, ~ ~ ~ plant
~**kran** *m* → Bekohlungskran
~**lagerplatz** *m* = coal storage yard
~**lager- und -einsackanlage** *f* = coal storage and bagging plant
~**lagerung** *f* = coal storage
~**mehl** *n* = coal powder
~**mischanlage** *f* = coal mixing plant
~**mühle** *f* = coal (grinding) mill, ~ grinder
~**oxyd** *n* = carbon oxide
~**oxydgas** *n*, CO = carbon monoxide
~**pfeiler** *m* = coal pillar
~**presse** *f* → Brikettpresse
~**rohrleitung** *f* = coal pipeline
~**rutsche** *f*, Kohlenschütte *f* = coal chute
~**säure** *f*, Kohlendioxyd *n*, CO_2 = carbon dioxide
~**säureschnee** *m*, Trockeneis *n* = dry ice, solid carbon dioxide
kohlensaurer Kalk *m*, Kreide *f*, Kalziumkarbonat *n*, $CaCO_3$ [*DIN 1280*] = calcium carbonate, chalk
kohlensaures Kalium *n* → Kaliumkarbonat *n*
~ **Natron** *n* (oder **Natrium** *n*), Na_2CO_3 = carbonate of soda
~ **Salz** *n* → Karbonat *n*
Kohlen|schiff *n* = collier [*Vessel designed to carry coal, also used in carrying ore and other bulk commodities*]

~schlamm *m* = coal slurry
~schrapper *m* = coal scraper
~schupper *m* = coal-shed
~schütte *f*, Kohlenrutsche *f* = coal chute
~sieberei *f* = coal screening plant, ~ ~ installation
~skipanlage *f* = coal skip-winding plant
~sorte *f* = grade of coal
~sprengen *n* = coal blasting
~staub *m* = coal dust
~staub *m* = powdered coal
~staubanlage *f* = coal pulverizing installation, pulverized fuel plant
~staubbekämpfung *f* = coal dust suppression
~staubexplosion *f* = coal dust explosion
~staub/Luft-Suspension *f* = coal-in-air suspension
~staub-Mahlanlage *f* = coal pulverizer
~stoff *m* = carbon
~stoffgehalt *m* = carbon content
~stoff-Oleosol *m* = carbon-oleosole
~stoffrückstand *m*, Verbrennungsrückstand [*Motor*] = carbon deposit
~stoffstahl *m*, unlegierter Stahl = (plain) carbon steel
~stoffstein *m*, Kohlenstoffziegel *m* = carbon brick
~stoff-Zustellung *f* = carbon lining
(~)Stoß *m* = (coal) face
~tagebau *m* = coal open pit mining (US); ~ open-cast ~, ~ surface ~, ~ open-cut ~, ~ adit ~ (Brit.)
~trockentrommel *f* = coal rotary dryer, ~ cylindrical ~; ~ ~ drier (US)
~turm *m* = coal storage tower
~untertagegrube *f* = coal mine
~vergasung *f* unter Tage = underground gasification of coal
~verladebrücke *f* = coal(ing) bridge
~wäsche *f* = coal washing plant
~wassergas-Anlage *f* = carburetted water gas installation
~wassergasteer *m* = dehydrated water-gas tar

~wasserstoff *m* = hydrocarbon
~wasserstoffbindemittel *n* → Bindemittel auf Kohlenwasserstoffbasis
~wasserstoffbindemittelbeton *m*, bituminöser Beton, Schwarzbeton = bituminous concrete
~wasserstoff-Lösungsmittel *n* = hydrocarbon solvent
~wertstoff *m* = coal by-product
~zertrümmerungsverfahren *n* = coal bursting system
Kohle|ofen *m* = coal-fired stove
~ringdichtung *f* = carbon ring seal
~schlamm *m* = coal slurry
~trocknung *f* = coal drying
Kokardenerz *n* = crust ore
Kokerei *f* = coking plant
~gas *n* = coke oven gas
~teer *m* → → Teer
Kokillen|gießautomat *m*, automatische Kokillengießmaschine *f* = auto(matic) (gravity) die casting machine, ~ chill ~ ~
~gußteil *m, n* = permanent mo(u)ld casting
~schmiere *f* = mo(u)lding grease
Kokosfasermatte *f* = coconut mat(ting)
Koks|gabel *f* = coke fork
~gasverdichter *m* = coke-oven gas compressor
~kastenband *n* = char bucket conveyor, ~ ~ conveyer
~korb *m*, Bauaustrocknungsofen *m* = (fire) devil, brazier
kolben [*eine Bohrung*] = to swab a well
Kolben *m* [*Laborgerät*] = flask
~ = piston
~ [*beim Rammhammer*] = ram
~ [*für das Kolben als Erdölfördeverfahren*] = (rubber) swab
~ *n* → Pistonieren *n*
~ *m* mit gußeisernem Ringträger = piston with cast-in ring band
~anzeigedämpfer *m* → Kolbenzeigerdämpfer
~boden *m* = piston head
~bohrer *m*, Kolbenprobenehmer *m*, Kolbenentnahmegerät *n* [*Bodenprobenahme*] = piston sampler

~bolzen m = piston pin, wrist ~
~entnahmegerät n → Kolbenbohrer m
~feder f = piston spring
~flugzeug n = piston aircraft
~gebläse n = piston blowing engine
~geschwindigkeit f = piston speed
~hemd n = piston skirt
~hub m = piston stroke
~kompressor m = piston compressor, reciprocating ~
~kühldüse f = piston cooling jet
~kühlung f = piston cooling
~lagerdämpfer m = piston travel limiter
~manschette f = piston packing, ~ sleeve
~maschine f = piston machine
~membran(e)motorpumpe f = combination piston and diaphragm power pump
~membran(e)pumpe f = (combination) diaphragem and piston pump, (~) piston and diaphragm ~
~motor m = piston engine
~nut f = piston ring groove
~packung f = piston packing
~pochwerk n = piston stamp
~probenehmer m → Kolbenbohrer m
~pumpe f = piston pump
~ring m = piston ring
~ring m mit verchromter Lauffläche = chrome faced piston ring
kolbenringartige Dichtung f = piston ring type seal
Kolben|ringlücke f = piston ring gap
~sitzfläche f = ring seat
~spielraum m = piston clearance
~stange f = piston stem, ~ rod
Kolbenstangen|-Abstreifring m = piston-rod scraper ring
~kreuzkopf m = cross-head of a piston rod
~packung f = rod packing
Kolben|verdichter m = piston compressor
~wirkung f [Tiefbohrtechnik] = swabbing effect [Thick filter cake on the well has a swabbing effect upon removal of tools]

~zeigerdämpfer m, Kolbenanzeigedämpfer = dash pot
Kolk m → Auswaschung f
~abwehr f = scour prevention
~bildung f → Auswaschung
kolken, aus~, auswaschen, unterspülen, unterwaschen, wegspülen = to underwash, to scour
Kolk|geschwindigkeit f = scouring velocity, ~ rate
~grenze f, Aus~ = limit of scour
~höhe f = scour level
~schutz m = scour protection
~schutzsperre f = check dam
~strömung f = scour-provoking current
~tiefe f = depth of scour
Kolkung f → Auswaschung
Kolk|versuch m = scouring test
~vertiefung f = eroded hole (Brit.); scoured ~
~wirkung f = scouring effect
Kollaudierung f [Schweiz]; Abnahme f [Die Übernahme des Bauwerkes oder Gebäudes nach Fertigstellung] = acceptance
Kollektor m = commutator
Koller m, Walze f, Läufer m [Kollergang] = runner
~gang m, Kollermühle f, Läufermühle f = edge-runner, pan grinder, grinding mill of the edge runner type, pan (grinding) mill
Kolliliste f = packing list
Kollimation f = collimation [A collimating or being collimated]
Kollimations|achse f, Zielachse [Theodolit] = line of collimation
~fehler m = error of collimation
~fernrohr n = collimator field glass
~instrument n mit fester Basis und mit 80facher Vergrößerung = fixed-base 80-magnification collimator
Kollimator m = collimator
Kollisionsdiagramm n = collision graph
Kollodiumwolle f, lösliche Schießbaumwolle = collodion cotton
Kolloid n = colloid [1. The finely divided suspended matter which will

not settle and the apparently dissolved matter which may be transformed into suspended matter by contact with solid surfaces or precipitated by chemical treatment. 2. Substance which is soluble as judged by ordinary physical tests but will not pass throug a parchment membrane]

Kolloidalbeton m = colloidal concrete
kolloidale Emulsion f = colloidal emulsion
kolloidaler Mörtel m, **Kolloidmörtel** m = colloidal mortar, Colgrout
~ Stoff m = colloidal matter [*Colloids or matter colloidal in nature and action*]
kolloidales System n = colloidal system
Kolloid|chemie f, **Kolloidik** f, **Kolloidlehre** f = colloidal chemistry
~mühle f = colloid(al) mill
~schlamm m → → Feinschluff m
Kolluvialboden m → Absatzboden
kolluvialer Verwitterungsschutt m = colluvial mantle rock
kolmatieren → auflanden
Kolmatierung f → Auflandung f
Kolmation f → Auflandung f
Kolonnade f, **Säulenhalle** f = colonnade [*In architecture, a series of columns*]
Kolonne f, **Arbeiter~**, **Mannschaft** f, **(Arbeits)Trupp** m = party, gang, team, crew
~, Turm m [*Erdölwerk*] = column
Kolophonium n, **Geigenharz** n [*wird in der Praxis vielfach nur als „Harz" bezeichnet*] = resin, rosin, colophony
Kombinations|gerät n = combination equipment item
~herd m = combination range
~schloß n = combination lock
~strang m, **kombiniertes Bohrgestänge** n = tapered string of drill pipe
~wagen m → Kombiwagen
~zange f = footprints, pipe tongs (Brit.); combination pliers
Kombinatorik f = combinatorial analysis

kombinierte Axial- und Radialpumpe f = combined axial and radial flow pump
(~) Gleich-Gegenstrom-Trockentrommel f = combined contra-flow uniflow drying drum
~ Lage- und Höhenmessung, f, **Tachymetrie** f = tacheometry
~ Rohrfahrt f, **zusammengesetzte ~** [*Tiefbohrtechnik*] = combinated string of casing
~ Saugluft-Druckluft-Förderanlage f, **kombinierter pneumatischer Förderer** m = combined suction and pressure conveying system, suction-jet ~ ~
~ Sieb- und Schlämmanalyse f = mechanical analysis, combined sieve and sedimentation test
~ Zuschlagstoff-Zement-Dosierwaage f = combination weighing unit for aggregate and cement, concentric aggregate-cement batcher
kombinierter Brenner m = combined burner
~ Förderer m **mit Mitnehmern** = conveyor (or conveyer) and elevator with (scraper) flights
~ Stampfbohlen- und Hammerfertiger m → (Stampf)Hammerfertiger
kombiniertes Bohrgestänge n, **Kombinationsstrang** m = tapered string of drill pipe
~ Seil- und Drehbohrgerät n = combination rig
Kombi|schaufel f = multi-purpose bucket
~wagen m, **Kombinationswagen** = station wagon
Kommunalgebäude n = communal building
Kommunmauer f → Brandmauer
kompakt, festgelagert = firm
kompakter Gips m, **dichter ~**, **Gipsfels** m = gypseous solid rock, compact gypsum
kompaktes Gestein n, **dichtes ~**, **Urgestein, kompakter Fels** m = compact rock, solid ~
Kompaktmotor m = compact engine

Kompensationstreifen *m* [*siehe Anmerkung unter „Aktivstreifen"*] = compensating ga(u)ge
kompensieren [*Kompaß*] = to box
Komplementwinkel *m* → Ergänzungswinkel
komplettes Hüttenwerk *n* = integrated iron and steel plant
komplizierter Querschnitt *m* = intricate section, ~ shape, complicated
Komponente *f*, Bestandteil *m* = component
~, Seitenkraft *f*, Zweigkraft [*Baustatik*] = component
Komponenten|gleichung *f*, Seitenkraftgleichung = equation connecting the components
~**zerlegung** *f* = decomposition
Kompositkapitell *n* = composite capital
Kompost *m* = compost
Kompostieren *n*, Kompostbereitung *f* = composting
Kompressions|belag *m* → Makadamdecke *f*
~**decke** *f* → Makadamdecke
~**kältemaschine** *f* = refrigeration compressor
~**-Kälteverfahren** *n* = vapo(u)r compression method
~**nocken** *m* [*Motor*] = compression release
~**ring** *m* = compression ring
~**strömung** *f* = compressible fluid flow
~**verlust** *m* [*Motor*] = blow-by gases
~**versuch** *m*, Zusammendrück(ungs)versuch, Kompressionsprobe, Zusammendrückprobe, Kompressionsprüfung *f*, Zusammendrück(ungs)prüfung, Druck-Setzungsversuch = compression test
~**zündung** *f* = compression ignition, C.I.
Kompressor *m*, (Luft)Verdichter *m*, Druckluterzeuger *m*, Luftkompressor = (air) compressor
~, Verdichter = compressor
~ **mit 6 Druckluftwerkzeugen** = six-tool compressor

Kompressor(en)|öl *n* = compressor luboil
~**station** *f*, Verdichterstation = compressor station
kompressorloser Dieselmotor *m* = compressor-less injection Diesel (engine)
~ **Motor** *m* = solid-injection engine
Kompressor|-Regler *m* = compressor governor
~**schlauch** *m* → Preßluftschlauch
~**schlepper** *m* = universal compressor tractor, tractor-compressor
~**träger** *m* = air compressor mounting
~**zylinder** *m*, Verdichterzylinder = compressor cylinder
Koncha *f* = concha [*The half dome covering an apse*]
Kondensat *n* = condensate
~**austritt** *m* = condensate outlet
~**entölung** *f* = oil removal from condensation water
Kondensations|kern-Größenspektrometer *n* = condensation nucleus size spectrometer
~**wärme** *f*, Verdampfungswärme [*des in Verbrennungsprodukten enthaltenen Wassers*] = latent heat (of the water vapo(u)r)
Kondensator *m* = condenser
~**zündmaschine** *f*, Kondensationszündmaschine, Kondensatorzündapparat *m*, Kondensationszündapparat = condenser discharge (type) blasting machine, CD ~ ~, battery-condenser exploder
Kondensat|pumpe *f* = condensate pump
~**rohr** *n* = condensate pipe
Kondensieren *n* = condensing
Kondens|leitung *f* = condensate line
~**milchfabrik** *f* = condensery
~**topf** *m* = steam trap, condensate ~
~**wasser** *n*, Schwitzwasser, Tauwasser = condensating water, condensation ~, dripping moisture
Konditionierung *f* = conditioning
Konditorei *f* = confectionary store

Konglomerat — Konsolidierungszone 592

Konglomerat n, Naturbeton m (Geol.) = conglomerate

konglomeratischer Sandstein m = conglomeratic sandstone

Kongreßhalle f = conference hall

Königsbolzen m = king pin
 ~**lager** n = king pin bearing

Königs|gelb n → Bleigelb
 ~**pfahl** m [senkrechter Mittelpfahl einer Dückdalbe] = vertical pile of a dolphin
 ~**stück** n [bei einem Holzdrempel] = post
 ~**stuhl** m = pivot post group
 ~**stuhllager** n = pivot post bearing
 ~**zapfen** m = king pin; center journal (US)

konisch = tapered

konische Dichtungsscheibe f = saucer-type disc
 ~ **Fördertrommel** f (⚒) = conical hoisting drum
 ~ **Keilwelle** f = tapered spline
 ~ **Tellerfeder** f = coned disc spring; ~ disk ~ (US)

konischer Bolzen m, Bolzen mit Anzug = conical bolt
 ~ **Schaft** m = tapering shaft
 ~ **Siebzylinder** m → Sortierkonus m

konisches Lager n = cone assy
 ~ **Rad** n → Kegelrad

Konizität f → Kegelneigung f

konjugierte Gerade f, zugeordnete ~ = conjugate line

Konkave f → einbuchtendes Ufer

konkaves Ufer n → einbuchtendes ~

konkordante Schicht f = conformable stratum

Konoid|gewölbe n, konoidisches Gewölbe n = conoidal vault
 ~**schale** f = conoidal shell

Konservierung f, Haltbarmachung = conservation

Konservierungsmittel n, Schutzmittel = preservative

Konsistenz f, Steife f [Beton; Mörtel] = consistency

 ~**beiwert** m, Steifebeiwert = coefficient of consistency
 ~**grenzen** fpl nach Atterberg, Atterberg'sche Konsistenzgrenzen, Atterberg-Grenzen = consistency limits, Atterberg limits of soil
  Fließgrenze f, W_f = liquid limit, LL [symbol w_L]
  (Aus)Rollgrenze, obere Plastizitätsgrenze, W_r = plastic limit, PL [symbol w_P]
  Schrumpfgrenze f, W_s = shrinkage limit, SL [symbol w_S]
 ~**messer** m = consistometer
 ~**probe** f, Konsistenzprüfung f, Konsistenzversuch m = consistency test

Konsole f = bracket [A horizontal projection from a vertical or near vertical surface to support a load]
 ~ [Brückenbau] = cantilever(ed) slab

Konsolidation f → Konsolidierung f

Konsolidationstheorie f = consolidation theory

konsolidierende Schicht f, ~ **Lage** f = consolidating layer

Konsolidierung f, Eigenverfestigung, Eigensetzung, Konsolidation f, natürliche Bodenverdichtung, Eigenkonsolidation = consolidation, soil ~, earth ~

Konsolidierungs|aufgabe f = consolidation problem
 ~**druck** m = consolidation pressure
 ~**geschwindigkeit** f = rate of consolidation
 ~**grad** m, Verfestigungsgrad [Bodenmechanik] = degree of consolidation
 ~**kurve** f = consolidation curve
 ~**spannung** f = consolidation stress
 ~**theorie** f = theory of consolidation
 ~**verlauf** m = progress of consolidation
 ~**versuch** m, Konsolidierungsprobe f, Konsolidierungsprüfung f = consolidation test
 ~**vorgang** m = process of consolidation
 ~**zone** f = zone of consolidation

Konsol|kran *m*, Wandkran = wall crane
~träger *m* → Kragträger
Konstante *f* = constant
Konstantenwert *m* = constant value
konstantes Anpreßmoment *n* **durch eine Feder** = constant-force output
konstitutionelles Molekularwasser *n* → Konstitutionswasser
Konstitutionswasser *n*, konstitutionelles Molekularwasser = chemically bound water, water of constitution, connate water
Konstrukteur *m*, Entwerfer *m*, Entwurfverfasser = designer
Konstruktion *f* = system
~, Aufbau *m*, bauliche Anordnung *f*, Ausführung, Bau = version, design
Konstruktions|abteilung *f* = engineering department
~aufgabe *f* = structural problem
~beton *m*, Bauwerkbeton = structural concrete
~betonarbeiten *fpl* = structural concrete work
~blech *n* → Baublech
~blei *n* → Baublei
~büro *n* = design office
~dübel *m*, Baudübel = construction dowel
~einzelheit *f* = design detail
~element *n* = structural element
~faktor *m* = design factor
~fehler *m* = design error
~gewicht *n* = design weight
~höhe *f*, Bauhöhe [*Träger*] = (construction) depth
~höhe *f*, Bauhöhe = height of construction
~ingenieur *m* = structural engineer
~leichtbeton *m* = structural lightweight concrete, lightweight structural ~
~merkmal *n* = design feature
~-Natursteinmauerwerk *n* = constructional masonry
~norm *f* = design standard spec(ification)

~system *n* = design system
konstruktionstechnisch = constructional
Konstruktions|teil *m*, *n* = structural part, member
~unterkante *f*, Überbauunterkante [*Brücke*] = deck soffit
~zeichnung *f* = design drawing
konstruktive Ausbildung *f* = structural design
~ Durchbildung *f* = structural system
konstruktiver Bauteil *m*, konstruktives Bauteil *n* = (structural) member [*A component of a structural assembly*]
~ Fertigteil *m*, konstruktives ~ *n* = precast structural (concrete) member
~ Ingenieurbau *m* = structural engineering
konstruktives Bauelement *n* = structural unit
~ Bauteil *n*, konstruktiver Bauteil *m* = (structural) member [*A component of a structural assembly*]
~ Betonbauelement *n* = structural concrete unit
Kontakt *m* (*Geol.*) = contact influence
~ablagerung *f* (*Geol.*) = contact deposit
Kontaktdraht *m* [*Gleitschalungsfertiger*] = (pre-erected) grade wire, control ~
~ nur auf einer Seite [*Gleitschalungsfertiger*] = single (pre-erected) grade wire, ~ control ~
Kontakt|erosion *f* = erosion of electrical contacts
~fläche *f* = mating surface, contact ~
~gestein *n* = contact(-altered) rock
~hof *m* (*Geol.*) = metamorphic aureole
~kleber *m* = contact adhesive
~korrosion *f* = corrosion at a contact with a second metal
~linie *f* [*Gleitschalungsfertiger*] = (pre-erected) grade line

Kontaktmetamorphose — Kontrollstück

~metamorphose *f*, Berührungsumprägung *f*, Berührungsumwandlung *f* (Geol.) = contact metamorphism
~schwelle *f* = vehicle detector pad
~schwelle *f* mit ortsfestem Einbau = permanent type detector
~trocknung *f* = direct-heat drying
~zone *f* (Geol.) = contact
Konter|bogen *m*, Gegenbogen, umgekehrter Gurtbogen, Erdbogen = inverted arch
~mutter *f*, Klemmutter, Gegenmutter = lock nut, check ~
Kontimischer *m* → Fließmischer
Kontinental|kante *f* = continental fringe
~schelf *n*, *m* = continental shelf
kontinuierlich bewehrte Betonfahrbahn *f*, durchlaufend armierte ~ = continuously reinforced concrete carriageway, ~ ~ ~ pavement [*A pavement without transverse joints, except tied construction joints placed between successive day's concreting, with sufficient longitudinal reinforcement, adequately lapped to develop tensile continuity, so that transverse cracks will be held tightly closed, resulting in the central portion of the pavement, exclusive of the end 400 to 500 ft, being in substantially complete restraint*]
kontinuierliche Band(ver)wiegung *f* = continuous belt weighing
~ Dicht(ungs)wand *f* → ~ Sperrwand
~ Dosieranlage *f* = continuous batching plant
~ Förderung *f* → Fließförderung
~ Kornabstufung *f*, stetige ~ = continuous grading
~ Raumteilzumessung *f*, stetige ~ = continuous batching by volume
~ Sperrwand *f*, ~ Dicht(ungs)wand, kontinuierliches Diaphragma *n* = impervious diaphragm of the continuous wall type, impermeable ~ ~ ~ ~ ~, underground ~ ~ ~ ~ ~

kontinuierlicher Balken(träger) *m*, durchlaufender ~, Durchlaufbalken(träger) = continuous beam
~ Lader *m* → Dauerlader
~ Mischer *m* → Fließmischer
~ Ofen *m*, Durchlaufofen, Stetigofen = continuous kiln
~ Zwangsmischer *m* → Dauer-Zwangsmischer
kontinuierliches Betonieren *n* = continuous pour, ~ concreting
~ Diaphragma *n* → kontinuierliche Sperrwand
~ Stollenbohrgerät *n* = continuous miner, ~ gallery machine
Kontinuitätsbedingung *f* = continuity condition
Kontraktionsfuge *f* [*Schweiz*]; Schwindfuge; ~ = shrinkage joint
Kontraktorverfahren *n*, Contractor-Verfahren = tremie method
Kontrasieb *n* → Gegenschwingsieb
Kontrollampe = pilot lamp
~ für Fernlicht = high beam indicator light
Kontrolle *f* = check
Kontroll|gang *m*, Beobachtungsstollen *m* [*Talsperre*] = footway (Brit.); inspection gallery, inspection tunnel
~kabel *n* = check cable
~messung *f* = confirmatory measurement
~prüfung *f*, Kontrollprobe *f*, Kontrollversuch *m* = control test
~rechen *m*, Profiltaster *m* = subgrade tester, (~) scratch template, point template
~schacht *m*, Beobachtungsschacht [*Talsperre*] = inspection shaft
~schacht *m*, (Einsteig)Schacht, Mannloch *n*, Prüfschacht = (inspection) manhole
~schieber *m* = sensing piston
~schraube *f* [*Hydraulikölbehälter*] = level plug
~stück *n* = check piece

Kontrollturm — Kopfhaube

~turm *m*, FS-~, Flugleitungsturm = control tower
Kontur *f* = contour line
Konus|anker *m*, Verankerungskegel *m* = anchoring cone
~(bohr)meißel *m*, Kegel(bohr)meißel = cone bit
~meißel *m* → Konusbohrmeißel
~mischer *m*, Kegelmischer = cone-type mixer
~ritzel *n* → Kegelritzel
~rollenlager *n* → Kegelrollenlager
~senkung *f* [*Beton*] = slump
~trommel *f*, Kegeltrommel = conical-shaped drum
~überlauf *m* = bell-mouthed spillway, swallow-hole
~verankerung *f*, Kegelverankerung = cone anchorage
Konvektion *f*, Wärmeströmung *f* = convection
Konvektions|heizung *f* = convection heating
~strom *m* = convection current
~wärme *f* = convected heat
~wärmeverlust *m* = convective heat loss
konvektiver Regen *m* = convection rain
Konvektor *m* = convector radiator
Konventionalstrafe *f* = contract delay penalty, ~ overrun ~
Konvergenz|-Magnetfeld *n* = convergent magnetic field
~messer *m*, Konvergenzgeber = convergence indicator
~schreiber *m* = convergence recorder
(Konverter-)|Beschickungswagen *m* = (converter) charging car
(~)Bodeneinsetzwagen *m* = converter bottom jacking car
Konverterstein *m* = Bessemer matte
konvexes Ufer *n* → ausbuchtendes ~
Konzentrat *n* = concentrate
Konzentration *f* [*Abwasser*] = concentration, strength
Konzentrationsfaktor *m* nach Fröhlich, Ordnungszahl *f* der Spannungsverteilung *f* = Fröhlich's (stress-) concentration factor

Konzentriermaschine *f* = concentrator
konzentriertes Abwasser *n* = strong sewage [*Sewage containing above the normal quantity of organic matter*]
Konzerthalle *f*, Konzerthaus *n*, Tonhalle = concert hall
Koog *m* → Polder *m*
Koordinaten|achse *f* = coordinate axis
~papier *n* = coordinate paper, quadrille ~
~system *n*, Achsenkreuz *n*, Koordinatenkreuz, Achsensystem = sheaf of coordinate axes
koordiniertes Signalisierungssystem *n* [*Straßenverkehr*] = coordinated control system, linked ~
Kopal *m* = copal
Köper *m* mit Wechsel = broken twill weave
~gewebe *n* = twilled weave
Kopf *m* (⚒) = buttock [*A short face in longwall mining as long as the depth of the web of coal to be removed, cut into the coalface at right angles to its general lines as a working face for a cutter loader*]
~abstand *m* [*von zwei Fahrzeugen*] = headway
~anker *m*, Balkenanker, Schlauder *f* = beam tie
~ausbildung *f* [*Säule; Stütze*] = head attachment, ~ construction
~bahnhof *m*, Sackbahnhof = dead-end station, reversing ~; stub-end ~ (US)
~band *n*, Kopfbiege *f*, Bug *m*, Winkelband, Strebeband = strut, (angle) brace
~biege *f* → Kopfband *n*
köpfen der Packlage → ab~ ~ ~
Kopffläche *f*, Haupt *n*, Kopfseite *f*, Stirnfläche [*Die Stirnfläche der Steine beim Naturwerkstein-Mauerwerk*] = face
kopfgesteuerter Motor *m* = valve-in-head engine
Kopf|haube *f* aus Gußeisen [*Holzbau*] cast-iron head

Kopfhöhe — Kornanalyse

~höhe *f*, Durchganghöhe [*Lichter Abstand von Vorderkante Trittstufe bis Unterseite einer darüberliegenden Treppe*] = headroom, headway

~holz *n* (⚒) = lid, wooden crusher block

~klammer *f* = forked clamp

~leitung *f* [*Grundwasser(ab)senkung*] = header (pipe), ~ main [*It interconnects the upper ends of all well points*]

Kopfloch *n* [*zur Schweißung bei strangweiser Absenkung von Rohrleitungen*] = bell hole

~schweißen *n* = bell hole welding

Kopfmauerwerk *n* = stone-faced rubble masonry

Kopfplatte *f* → Gurtlamelle *f*
~ → Pfahl~

Kopf|platten-Form *f* [*zur Herstellung von Kopfplatten*] = cap form

~raum *m* [*Der freie Raum für die Köpfe der Insassen in einem Fahrzeug, oder der Arbeiter beim Vortreiben eines Tunnels usw.*] = head clearance

~rolle *f* [*Rammgerüst*] = head sheave [*pile frame*]

~schraube *f* = set screw

~schütter *m* → Autoschütter

~schutz *m* = head protection

~seite *f* → Kopffläche

Kopfstein *m* = cobble stone

~pflaster(decke) *n*, (*f*) = cobble stone paving

Kopf|strecke *f*, obere Abbaustrecke (⚒) = upper development road, top ~, tail gate

~verband *m* [*Mauerwerk*] = head bond

~waschbrause *f*, Kopfwaschdusche *f* = shampoo shower, ~ spray

~wipper *m* = rotary end tip

Korallen|erz *n* = hepatic cinnabar

~fels *m* [*Jura-Kalkstein als riffartiges Vorkommen im oberen weißen Jura*] = coral limestone

~riff *n* = coral reef

~sand *m* = coral sand

~schlamm *m* = coral mud

~zuschlag(stoff) *m* = coral aggregate

Korb *m* [*Greifbagger*] = bucket

Korbbogen *m* → Dreigelenkbogen
~ = compound curve, more-centre ~; three-centre ~

Korbseiher *m* [*beim Regenrohranschluß*] = wire gutter top

Kord|festigkeit *f* = cord strength

~gehalt *m* = cord content

korinthische Säulenordnung *f* = Corinthian order

korinthisches Kapitell *n*, ~ Kapitäl = Corinthian capital

Kork|belag *m* = cork cover(ing)

~beton *m* = concrete with cork aggregate

~boden *m* → Korkfußboden

~dämmung *f* = cork insulation

~farbe *f* = cork paint

~fliese *f*, Korkplatte *f* = cork tile

~fliesen(fuß)boden *m*, Korkplatten(fuß)boden = cork tile flooring

~(fuß)boden *m* = cork flooring, ~ floor finish

~linoleum *n* = cork carpet

~mehl *n* = cork powder

~pflasterstein *m* = cork (paving) sett

~platte *f* = cork slab, corkboard

~platte *f*, Korkfliese *f* = cork tile

~platten(fuß)boden *m*, Korkfliesen(fuß)boden = cork tile flooring

~schrot *m*, geschroteter Kork *m* = granulated cork

~schüttung *f* [*unter Fußböden*] = cork fill

~stein *m* = cork brick

~tritt(stufe) *m*, (*f*) = cork (stair) tread

Korn *n* = particle, grain

~, Getreide *n* = cereal

~, Körner *npl* = particles, grains, material

~abstufung *f*, Granulometrie *f* = (granulometric) grading

~ährenverband *m*, Fischgrätenverband = herringbone bond

~analyse *f* = grading analysis, grain-size ~, particle-size ~, test for grading

Kornaufbau — körnig-kristalliner Gips

~aufbau m, Kornzusammensetzung f, Kornanordnung = granulometric composition, granulometry
~aufbau m der Zuschläge nach unstetigen Sieblinien, diskontinuierliche Kornabstufung f = discontinuous grading, gap ~
~bereich m → Kornklasse f
~beschaffenheit f = nature of grain(s)
~durchmesser m = grain diameter
~eis n = granular ice

Körner npl, Korn n = grains, particles, material
~kollektiv n = collective (body of) grains
~leim m = grain glue, ~ adhesive, ~ cement
~sumpf m, Nußkleinsumpf = collecting sump for undersize of the nut sizing screens
Korn|feinheit f = fineness of grains, ~ ~ particles
~festigkeit f = grain strength

Kornform f, Korngestalt f
  plattig, flach, scherbig
  kubisch
  rund, kugelig, abgerundet
  nadelförmig
  kantig
  unregelmäßig,
  Kornformfaktor m

particle shape, grain ~
  laminated, flaky
  cubical shaped, cube-shaped
  rounded
  elongated
  angular
  irregular
  particle shape factor, grain shape factor

Korn|formprüfung f, Kornformprobe f, Kornformversuch m = particle shape test, grain ~ ~
~gemisch n → Mineralgemisch
korngerecht = of standard particle size distribution, exactly conform to the nominal size, correctly sized
Korn|gerüst n [z. B. Sand] = granular skeleton
~gestalt f → Kornform f
~grenze f = grain-size limit
~grenzenkorrosion f, interkristalline Korrosion = intergranular corrosion, intercrystalline ~
~grenzriß m, interkristalliner Riß = intercrystalline crack
~größe f [*Größenbezeichnung für ein Korn in mm oder μm. Das Meßverfahren ist anzugeben. Ist kein Meßverfahren angegeben, dann gilt die Loch- bzw. Maschenweite nach DIN 1170 bzw. 4188*] = particle size, grain ~
Korngrößen|analyse f im Bereich unter 1 Mikron = size analysis of submicron particles

~bereich m → Kornklasse f
~bestimmung f = grain-size determination, particle-size ~
~intervall n = sizes interval, fraction
~kennziffer d' [*RRB-Diagramm*] = particle size distribution d', average grain diameter d' [*R.R.B. diagram*]
~verteilung f, Kornverteilung = particle-size distribution, grain-size ~, grading
Korngruppe f → Kornklasse f
körniger Boden m = granular soil
~ Dämmstoff m = granular insulating material
~ Gneis m, Granitgneis = granite gneiss, granitic ~, granitized ~, granitoid
~ Kalkstein m, Massenkalkstein = granular limestone
körniges Gut n = granular product, ~ material
körnig-kristalliner Gips m = granular crystalline gypsum

Korn|klasse *f*, Korngruppe *f*, Fraktion *f*, Kornspanne *f*, Korn(größen)bereich *m* [*Ein aus einer Körnung herausgegriffener Korngrößenbereich mit Angabe seiner gröbsten und feinsten Korngröße in mm oder μm*] = size range, ~ bracket, (~) fraction, screening portion

~**klassenbild** *n* → Körnungskennlinie *f*

~**masse** *f* → Mineralgemisch *n*
~**mischung** *f* → Mineralgemisch *n*
~**oberfläche** *f* = grain surface
~**schüttung** *f* = grains in bulk, heap of granular material
~**spanne** *f* → Kornklasse *f*

kornstabiler (Straßen)Belag *m*, kornstabile (Straßen)Decke *f* = dense (bituminous) surfacing, ~ (~) pavement

Körnung *f*, Fraktion *f* [*Korngemisch mit Angabe der gröbsten und feinsten Korngrößen in mm oder μm*] = grain mix(ture)

Körnungs|analyse *f*, Kornverteilungsanalyse *f* [*Bestimmung der Kornzusammensetzung einer Körnung in Gew.-% unter Angabe des Verfahrens (Siebung mit Sieben nach DIN 1170 bzw. 4188, Sichtung oder Sedimentation)*] = size analysis

~**kennlinie** *f*, Kornklassenbild *n* [*Schaubild der Kornzusammensetzung einer Körnung, bei dem auf der Abszisse die Korngröße in mm oder μm und auf der Ordinate derjenige Anteil der Körnung in Gew.-% aufgetragen wird, der gröber oder / und feiner als die betreffende Korngröße ist*] = screening characteristic, sizing ~, plot of screen test

~**komponente** *f* [*Boden*] = soil separate

~**kurve** *f*, Siebkurve, Körnungslinie *f*, Sieblinie = grading curve, sieve ~

~**modul** *m*, Feinheitsmodul = fineness modulus

~**netz** *n*, RRB-Netz = size distribution diagram, system of coordinates for representation of size distribution curves

~**verfahren** *n* → Granulierung *f*

Korn|verbesserungsboden *m*, Zusatzboden = complementary soil

~**verteilung** *f* = grain-size distribution, particle-size ~

Kornverteilungs|analyse *f* → Körnungsanalyse

~**gesetz** *n*, Exponentialgesetz der Kornverteilung nach Rosin-Rammler-Sperling = law of particle size distribution according to R.R.S.

~**kennzahl** *n* [*R.R.B.*] = size distribution factor n [*R.R.B. diagram*]

~**kurve** *f*, Kornverteilungslinie *f* = grain-size distribution curve, particle-size ~ ~

Korn-zu-Korn-Druck *m*, wirksame Spannung *f*, wirksamer Druck, Kontaktdruck, Berührungsdruck = effective stress

Kornzusammensetzung *f* → Kornaufbau *m*

Koronaentladung *f* → Glimmentladung

Korowkin-Becherverankerung *f* [*Spannbeton*] = Korowkin anchorage system, ~ anchoring ~

Körper *m* → Stein
~**farbe** *f* = body pigment

körperlicher Kraftaufwand *m* = physical exertion, ~ effort

Körperschall *m* = structure-borne sound

~**dämmstoff** *m* = insulation material against structure-borne sounds

~**dämmung** *f* = sound-insulation against structure-borne sounds, structural sound insulation, impact (sound) insulation

Korrasion *f*, Sandschliff *m*, Windkorrasion (Geol.) = corrasion

Korrelation *f*, geologische ~ [*Tiefbohrtechnik*] = geologic correlation

~ **von Bohrprofilen** = correlation of well logs

Korridor *m*, Gang *m*, Flur *m* = corridor [*A comparatively narrow enclosed thoroughfare within a building*]

korrodieren, anfressen = to corrode

Korrosion *f*, Angriff *m*, Anfressung *f* = corrosion
~ **unter gleichzeitiger mechanischer Beanspruchung** = corrosion under mechanical stress
~ ~ **Sauerstoffverbrauch** = oxygen-consumption type of corrosion
~ ~ **Wasserstoffentwicklung** = hydrogen-evolution type of corrosion
Korrosionsabmaß *n* = corrosion allowance
korrosionsbeständige Schutzschicht *f* [*Aluminiumveredlung*] = corrosion-resistant anodic coating
Korrosions|element *n* = corrosion cell
~**ermüdung** *f* = corrosion fatigue
~**festigkeit** *f*, Korrosionsbeständigkeit = corrosion resistance
~**geschwindigkeit** *f* = corrosion rate
~**probe** *f* → Korrosionsprüfung *f*
~**prüfung** *f*, Korrosionsprobe *f*, Korrosionsversuch *m* = corrosion test
~**schutz** *m* = protection against corrosion
~**schutzanstrich** *m* = corrosion protection coat(ing), anti-corrosive ~
~**schutzbinde** *f* = corrosion protection jacketing, ~ ~ blanket, anti-corrosive ~, anti-corrosion ~
~**schutzfarbe** *f* = corrosion protection paint, anti-corrosive ~
~**schutzmörtel** *m* = corrosion-inhibiting mortar
~**schutzschlämme** *f* = corrosion-inhibiting slurry
~**verhalten** *n* = corrosion behavio(u)r
~**verhütung** *f* = anti-corrosion
~**versuch** *m* → Korrosionsprüfung *f*
Korund *m*, Alpha-Tonerde *f* = corundum
~-**Erzeugnis** *n* = corundum refractory product
korundführendes Gestein *n* = boulder corundum
Korundstein *m* = corundum refractory brick
Körwasser *n*, Druckwasser, Kuverwasser, Qualmwasser, Truhwasser [*Dieses Wasser dringt bei hohen Außenwasserständen durch den Untergrund in eingedeichtes Binnenland*] = rushing out water
Kosten|aufgliederung *f* = cost breakdown
~**denken** *n* = cost-mindedness
~**ersparnis** *f*, Kosteneinsparung *f* = savings of cost
~**erstattungsvertrag** *m* = cost-reimbursement contract
~**gefüge** *n* = cost structure
~**rechnung** *f* (**im Baubetrieb**) = costing (for construction work)
~**stelle** *f* = cost item
~**überwachung** *f* = cost control
~**vergleich** *m* = cost comparison
kostenvergütbare Menge *f*, kostenvergütete Menge, Abrechnungsmenge = pay(able) quantity
kostenvergütete Bodenmassen *fpl* = pay dirt
~ **Ladung** *f* = payload
~ **Position** *f*, Leistungsposition = pay item
Kotangente *f* = cotangent
Kote *f*, Höhe *f*, Ordinate *f* = level
Kotflügel *m*, Schutzblech *n*, Fangblech = mudguard
KP-Gerät *n* = Casagrande type of consolidometer for combined compressibility and permeability tests
Krabbe *f*, Steinblume *f* = crocket [*A carved ornament, usually in the form of curved leaves or flowers, decorating the angles of roofs, gables, cornices, etc., especially in Gothic architecture*]
Krack|anlage *f* = cracking plant
~**benzin** *n* = cracked gasoline
~**destillat** *n*, Druckdestillat = pressure distillate, P. D.
~**gas** *n* = crackgas, cracking gas, cracker gas
Kraemer-Sarnow-Erweichungspunkt *m* = K.S. softening point
Kraft *f*, Energie *f* = power
~ [*Statik*] = force
~**abnahme** *f* = (power) take-off
~**anbringung** *f*, Anbringen *n* einer Kraft = application of a force
kraftangetrieben = power-driven
Kraft|antrieb *m* = power drive

Kraftarm — Kraftschaufel

~arm m [*Der Teil eines Hebels zwischen Drehpunkt und Kraftangriffpunkt*] = power arm
~aufzug m = power-operated hoist
~bedarf m, Energiebedarf [*Maschine*] = power requirement, horse~ ~
kraftbetrieben, kraftgetrieben = power-operated
Krafteck n → Kräftepolygon n
Kräfte|diagramm n = force diagram
~dreieck n, Krafteck = triangle of forces, force triangle
~gleichgewicht n = equilibrium of forces
~moment n = moment of a force
~paar n, Drehpaar = couple [*Is a force system consisting of two equal co-planar, parallel forces acting in opposite directions*]
~parallelepiped n → Kräfterechtkant
~parallelogramm n = parallelogram of forces
~plan m nach Cremona, Cremona'scher Kräfteplan = Cremona's polygon of forces, ~ force polygon
~polygon n, Kräftevieleck n, Krafteck = force polygon, polygon of forces
~rechtkant n, Kräfteparallelepiped n = parallelepiped of forces
~system n = system of forces
~vieleck n → Kräftepolygon n
Kraft|erzeugung f → Energieerzeugung
~fahrer m, (Auto)Fahrer = chauffeur (US); (motor vehicle) driver
~fahrerhotel n, Hotel für Motorisierte = motel
~verstoß m = motoring offence
Kraftfahrzeug n, Kfz., Motorfahrzeug = motor-car, motor-vehicle
~betriebsstoffsteuer f, Betriebsstoffsteuer = motor fuel tax
~halle f, Kfz-Halle = motor vehicle hangar
~motor m → Auto(mobil)motor
~tunnel m → Autotunnel
Kraft|feld n = field of force, force field
~fluß m = power flow
~gas n = power gas

~-Gleisstopfer m, Kraft-Schotterstopfer = power tie tamper (US); ~ track ~ (Brit.); ~ ballast ~
~hammer m = power hammer
~(-Hand)ramme f = power rammer, mechanical ~
~-Handstampfer m, Hand-Kraftstampfer = mechanical tamper
Krafthaus n, Maschinenhaus, Turbinenhaus = powerhouse
~aushub m = power house (structure) excavation
~kaverne f, Maschinenhauskaverne = powerhouse cavern
~überbau m = power house superstructure
Kraft|hubwerk n = power-operated lifting gear
~kabelwinde f → Krafttrommelwinde
~karre(n) f, (m) = power(-driven) truck
~komponente f = force component
~lenkung f → Lenkhilfe f
~linie f = line of force
~luftheizung f [*Die Warmluft wird mit Ventilatoren in die zu beheizenden Räume gedrückt*] = fan-assisted warm air heating
~maschine f = prime mover
~messer m, Brems~ = dynamometer
~methode f [*Baustatik*] = compatibility method
~-Momenten-Diagramm n = force-moment diagram
~nagler m = power nailer, ~ nailing machine
~niet m = power rivet
~nietung f = power riveting
~papier n, Hartpapier = Kraft paper
~papierfolie f, Hartpapierfolie = Kraft paper foil
~quelle f = source of power, ~ ~ energy
~ramme f → Krafthandramme
~reserve f = power reserve
~richtung f = direction of force
~rührwerk n = power-operated stirring gear
~schaufel f → Handschrapper m

~schaufel *f*, Entlade(kraft)schaufel = hand scraper unloader
kraftschlüssiger Antrieb *m* = non-positive drive
Kraftschluß *m*, Bodenhaftung *f* = adhesion, traction [*Gripping action between tracks or wheels and the surface*]
~ **zwischen Fahrbahn und Reifen**, Reifenkraftschluß = grip between tyre (Brit.)/ tire (US) and road, rimpull
~**beiwert** *m* → Bodenhaftungsbeiwert
Kraft|-Schotterstopfer *m* → Kraft-Gleisstopfer
~**schwingung** *f*, aufgezwungene Schwingung, erzwungene Schwingung = forced vibration
~**seilwinde** *f* → Krafttrommelwinde
Kraftstoff *m*, Treibstoff = fuel
~ **für Otto-Motor**, (Motoren-)Benzin *n*, Otto-Kraftstoff [*A.S.T.M. D 288–39*] = gas(oline) (or gasolene) (naphtha) (US); petrol (Brit.); Otto type fuel
~**abstellhahn** *m* = fuel shut-off
~**behälter** *m*, Kraftstofftank *m* = fuel tank
~**druck** *m* = fuel pressure
~**druckmesser** *m* = fuel pressure ga(u)ge
~**druckventil** *n* = fuel pressure valve
~**einfüllverschluß** *m* = fuel tank filler cap
~**einspritzpumpe** *f* = fuel injection pump
~**einspritzventil** *n* = fuel injection valve
~**entlüfterpumpe** *f* = fuel priming pump
~**-Filter** *m*, *n*, Kraftstoffilter = fuel filter
~**-Förderpumpe** *f*, Kraftstofförderpumpe = fuel transfer pump
~**gehalt** *m* **in gebrauchten Motorenölen** = fuel dilution of used-engine oils
~**hahn** *m* = fuel cock
~**kerosin** *n*, Traktorenkraftstoff *m*, Motorenpetroleum *n* = power kerosene, tractor fuel, tractor oil
~**leitung** *f* = fuel line
~**messer** *m* = fuel level indicator
~**meßstab** *m* = fuel level plunger
~**regler** *m* = fuel ratio control
~**-Schwefelgehalt** *m* = fuel sulphur content (Brit.); ~ sulfur ~ (US)
~**tank** *m*, Kraftstoffbehälter *m* = fuel tank
~**zusatz** *m* = fuel additive
Kraft|stollen *m* → Triebwasserstollen
~**stoß** *m*, Impuls *m*, Bewegungsgröße *f* [*Hydraulik*] = impulse
~**strom** *m* = power current
~**stromkabel** *n* = power cable
~**trommelwinde** *f*, Kraft(trommel)-windwerk *n*, Kraftseilwinde, Kraftkabelwinde, (mechanische) Kraftwinde = power (drum) winch
~**übertragung** *f* [*als Tätigkeit*] = power transmission, mechanical transmission of power
~**übertragung** *f* [*als Vorrichtung*] = power train
~**verkehr** *m* → Autoverkehr
kraftverschraubt = power-tight
Kraft|versorgung *f*, Stromversorgung, Energieversorgung = power supply
~**verteilung** *f* = force distribution
Kraftwerk *n*, E-Werk, Elektrizitätswerk, Eltwerk = power plant, ~ station, (electric) generating ~
~ **mit liegender (Turbinen)Welle** → Rohrturbinenkraftwerk
~ ~ **stehender (Turbinen)Welle**, ~ ~ **senkrechter** ~, ~ ~ **vertikaler** ~ = vertical-shaft type power plant
~**baustelle** *f* = power site
~**-Bekohlungsanlage** *f* = power station coaling plant
~**kaverne** *f* = power station chamber
~**sperre** *f* = hydro dam
~**-Steuersaal** *m* = power station control room
Kraft|werkzeug *n* = power-actuated tool
~**winde** *f* → Krafttrommelwinde
~**windwerk** *n* → Krafttrommelwinde *f*
Krag|arm *m* = cantilever arm

Kragbalken(träger) — Kran-Hubschrauber

~balken(träger) m, Auslegerbalken(träger) = cantilever beam [Is supported at one end only but is rigidly held in position at that end]
~-Baugerüst n → Auslegergerüst
Kragbinder m = cantilever truss
~ mit Druckstrebe = cantilever strutted roof truss
~ ~ Zugband = cantilever truss with tension rod
Kragbogen m, Auslegerbogen = cantilever arch
~brücke f, Auslegerbogenbrücke = cantilever arch(ed) bridge
~(träger) m = cantilever arch(ed girder)
Krag|brücke f, Auslegerbrücke = cantilever bridge
~dach n = cantilever roof
~ende n = cantilevered end
~faltwerk n = cantilevered folded plate (roof)
~-Fußgängerbrücke f, Ausleger-Fußgängerbrücke = cantilever footbridge, ~ pedestrian bridge
~gerüst n → Auslegergerüst
~-Gitterträgerbrücke f = cantilever lattice girder bridge, ~ ~ truss ~
~länge f = cantilevered length
~moment n = cantilever moment
~platte f = cantilever slab
~rüstung f → Auslegergerüst n
~schalung f, Freivorbauschalung = cantilever formwork
~-Stangen(transport)rost m, Krag-Stangenaufgeber m = cantilever screen of bars, ~ bar screen, ~ grizzly [Is fixed at one end only, the discharge end being overhung and free to vibrate by the impact of the material on the grizzly]
~stein m = cantilever block
~träger m, Auslegerträger, einseitig eingespannter Träger, vorkragender Träger, frei ausladender Träger, Freiträger, Konsolträger = cantilever grider [Is supported at one end only but is rigidly held in position at that end]

~tritt m, Kragstufe f = cantilever step
Kran m auf Chassis mit Eigenantrieb → Auto(mobil)kran
~ mit festem Ausleger → Festauslegerkran
~ ~ Laufkatzenausleger = saddle jib monotower crane (Brit.); ~ boom ~ ~ (US)
~ ~ nicht drehbarem Ausleger, Auslegerkran = crane with non-slewing jib (Brit.); ~ ~ boom (US)
~ ~ veränderlicher Ausladung → Festauslegerkran
~-Anbaugerät n, Kran-Zusatzgerät = attachment for cranes
~arbeiten fpl = crane work
~aufbau m, Kran-Oberwagen m = crane superstructure
~ausleger m = crane boom (US); ~ jib (Brit.)
~bagger m → Greif(bagg)er m
~bahn f = crane runway
~bahnschiene f = crane rail
~bahnträger m = crane (runway) girder
~bediener m, Kranführer m = crane operator
~belastung f → Kranlast f
~bewegung f = crane movement, ~ motion
~deck n, Kranplattform f, Kran-Grundplatte f = crane deck
~führer m, Kranbediener m = crane operator
~führerkabine f → Kranhaus n
~geld n, Krangebühr(en) f = cranage
krängen [Schiff] = to heel
Kran|gleis n = crane track
~grundausleger m = basic crane boom (US); ~ ~ jib (Brit.)
~-Grundplatte f → Krandeck n
Krängung f [Schiff] = heel
Kran|haken m = crane hook
~haus n, Kranführerkabine f, Kranführerhaus n = crane cab(in)
~haus n = gantry house
~-Hubschrauber m, Hubschrauber-Kran m = helicopter for crane work

Kranhydraulikmotorpumpe — Kratzputz

~hydraulikmotorpumpe *f* = crane pump
~katze *f* = (crane) crab
Kranken|aufzug *m* = hospital lift (Brit.); ~ elevator (US)
~auto *n*, Krankenwagen *m* = motor ambulance
~hausbau *m*; Spitalbau [*Schweiz*] = hospital construction
~schleuse *f*, Arzneischleuse = medical lock
~urlaub *m* = sick leave
~wagen *m*, Krankenauto *n* = motor ambulance
~zimmer *n* = patients' room, ward
krankheitserregend = disease-producing, pathogenic
Kran|kübel *m* → Betonkübel
~kübel *m* = crane bucket, ~ skip
~längsträger *m* = longitudinal crane girder
~last *f*, Kranbelastung *f* [*genormte Belastung durch Kran*] = crane load(ing)
~leistung *f* = crane lifting capacity
~mast *m*, Turm *m*, Drehsäule *f* = (mono)tower
~mittellinie *f* = crane centre (Brit.); ~ center (US)
~monteur *m* = crane fitter
~normteil *m*, *n*, Bauelement *n* = standard crane piece
~-Oberwagen *m*, Kranaufbau *m* = crane superstructure, ~ upper structure
~(pfahl)ramme *f* = crane pile driver
~plattform *f* → Krandeck *n*
~ponton *m* = crane barge
(~)Portal *n* = (crane) portal, gantry
~ramme *f*, Kranpfahlramme = crane pile driver
~raupe *f*, (Gleis)Kettenkran *m* = crawler-mounted crane
~rüttler *m*, Kranvibrator *m* = crane-operated vibrator
~schiene *f*, Kranbahnschiene = crane rail
~schmiere *f* = crane grease
~schmierer *m* = crane oiler
~schwerpunktlage *f* auf abschüssigem

Gelände = crane balance on slope
~seil *n* = crane cable, ~ rope
~stampfer *m* → Freifall-~
~stampfplatte *f*, Fallplatte = crane tamping plate
~standfestigkeit *f* = crane stability
~teil *m*, *n*, Bauteil = crane piece
~tragkraft *f* = crane capacity
~unterbau *m* [*Portalkran*] = crane pedestal
~unterwagen *m* = crane carrier
~vibrator *m*, Kranrüttler *m* = crane-operated vibrator
~waage *f* = crane weigher
~winde *f* = crane winch
Kranz *m* [*Schachtbau*] = walling crib, ~ curb, wedging ~ [*A curb on which the lining of a circular shaft is built*]
~bogen *m* = annular arch
~bremse *f* = rim brake
~fundament *n*, Kranzgründung *f*, Kranzfundation *f*, Kranzgründungskörper *m*, Kranzfundamentkörper = ringwall foundation (structure)
~fundamentkörper *m* → Kranzfundament *n*
~gründung *f* → Kranzfundament *n*
~gründungskörper *m* → Kranzfundament *n*
~tau *n* = strap
Kran|zubehör *m* = crane accessories
~-Zusatzgerät *n*, Kran-Anbaugerät = attachment for cranes, crane attachment
Krater *m* = crater
~sand *m* → vulkanischer Sand
Kratz|bagger *m* → Schrämbagger
~becherwerk *n* = elevator with scraping buckets
Kratzer *m* → Schlamm~
~ [*Gütefehler*] = scratch
~ [*Fördermittel*] = drag conveyor, ~ conveyer, scraper ~
~ [*zum Auskratzen*] = scraper
~, Fräser [*beim Kratzbagger*] = cutter
~kette *f*, Fräserkette = cutter chain
Kratz|grund *m*, Sgraffito *m* = sgraffito
~putz *m* = rough cast; harling [*Scotland*]

Kratzschaufel — kreisförmiges Rohr

~schaufel *f* [*Betonmischer*] = scraper blade [*The floor and wall scraper blades turn materials from the edge of the pan into the path of the mixing blades so that the batch is systematically and thoroughly mixed throughout*]
Krautung *f* = water-weed removal
Krebs *m* → Kalkkern *m*
krebsen → knistern
Kreide *f*, Kalziumkarbonat *n*, kohlensaurer Kalk *m*, $CaCO_3$ [*DIN 1280*] = chalk, calcium carbonate
~**abbau** *m* = chalk quarrying
~**boden** *m* = chalky soil
~**fels** *m* = solid chalk
~**mergel** *m* = chalk marl
kreiden [*Anstrich*] = to chalk
Kreiden *n* [*Anstrich*] = chalking [*The break-up of pigmented films on exposure. The binder is so much decomposed by the weather that the pigment can be removed by lightly rubbing it. The term is used for all colours, not only for near-white colours, although it originates from those. Chalking looks like fading, but the colour can be restored by a coat of varnish*]
Kreideschnur *f* = chalk line
Kreis|-Abwasserleitung *f*, Rund-Abwasserleitung = circular sewer
~**bahn-Siebschwingung** *f*, Kreiswurf *m* = circle(-)throw (gyratory movement)
~**behälter** *m*, Rundbehälter = circular tank
~**bogen** *m*, Kreiskurve *f* [*Straße*] = circular arc, ~ curve
~**deckel** *m*, Runddeckel = circular cover
~**diagramm** *n* = circle diagram
~**eimerleiter** *f* → Eimerrad *n*
Kreisel|brecher *m* → Kegelbrecher
~**läufer** *m* = gyroscope rotor
~**lotgerät** *n* [*Tiefbohrtechnik*] = gyroscopic well surveying device
~**platz** *m* → Kreisplatz
Kreiselpumpe *f*, Schleuderpumpe, Zentrifugalpumpe = centrifugal pump

~ **mit Flügelrad** = centrifugal screw (or propeller) pump
~ ~ **senkrechter Schaufelwelle** = vertical-shaft centrifugal pump
~ ~ **spiralförmigen Gehäuse** = centrifugal volute pump
~ ~ **waag(e)rechter Schaufelwelle** = horizontal-shaft centrifugal pump
~ **ohne Leitvorrichtung** = centrifugal pump without guide passage
Kreisel|-Salzstreuer *m* = centrifugal salt spreader
~**-Sandstreuer** *m* = centrifugal sand spreader
~**-Splittstreuer** *m* = centrifugal chip(-ping)s spreader
~**theodolit** *m* = gyrotheodolite
kreisender (Schlamm)Ausräumer *m* = revolving sludge scraper
kreisendtaumelnde Bewegung *f* [*Kegelbrecher*] = gyratory movement
Kreis|etage *f* → Rundgeschoß *n*
~**fahrbahn** *f* = circular carriageway
kreisfahrbarer Kabelkran *m* **mit einem ortsfesten und einem radial verfahrbaren Turm** → Schwenkkabelkran
Kreisfang(e)damm *m* = circular cofferdam
Kreisförderer *m*, Kreistransporteur *m* = overhead chain conveyor, ~ ~ conveyer
~ **mit zwei übereinander angeordneten in gleicher Richtung laufenden Fördersträngen** = power and free system
kreisförmige Krone *f*, Wölbung *f* [*Straße*] = (barrel) camber
~ **Lage** *f* [*Spanndraht*] = circular layer
~ **Mahlbahn** *f*, Mahlring *m* = grinding ring
~ **Platte** *f*, Kreisplatte, kreisförmige Scheibe *f*, Kreisscheibe = circular plate
kreisförmiger Gründunsbrunnen *m* **kleiner Abmessung** → runder ~ ~ ~
kreisförmiges Betonrohr *n* → Betonrundrohr
~ **Rohr** *n* → Kreisrohr

Kreis|frequenz f = rotational frequency
~**fundamentplatte** f, Kreisplattenfundament n, Kreisgründungsplatte = circular footing, ~ foundation slab, spread footing with circular base
~**funktion** f = circular function
~**geschoß** n → Rundgeschoß
~**gewölbe** n = circular vault
~**gründungsplatte** f → Kreisfundamentplatte
~**kanal** m für Abwasser, Rundkanal ~ ~ = circular conduit-type sewer
~**-Kanaltunnel** m, Rund-Kanaltunnel = circular tunnel
~**kipper** m → Rundkipper
~**kurve** f, Kreisbogen m [Straße] = circular curve, ~ arc
~**last** f = circular load
Kreislauf m = cycle
~**mahlung** f → Umlaufmahlung
~**schmierung** f = centralized lubricating system
~**vorgang** m, Zyklus m = cycle
Kreis|messerschere f = circle-cutting shears, circular ~
~**platte** f = circular slab
~**plattenfundament** n → Kreisfundamentplatte f
~**platz** m, Kreisverkehrsplatz, Kreiselplatz, Verkehrskreisel m = traffic roundabout [deprecated: roundabout, gyratory junction, traffic circus]
~**querschnitt** m, Rundquerschnitt = circular cross section
~**ringscherapparat** m → Ringschergerät n
~**rohr** n, Rundrohr, kreisförmiges Rohr, rundes Rohr = cylindrical pipe, circular ~
~**schale** f, Rundschale = circular shell
~**schwinger** m, rotierender Exzenter m [Rüttelwalze] = eccentrically loaded rotating shaft
~**schwingsieb** n, Kreisschwinger m = vibrating screen with circular movement, vibratory ~ ~ ~ ~
~**seiltrieb** m = continuous rope drive
~**sieb** n = cylindrical screen
~**silo** m, Rundsilo = cylindrical silo
~**stockwerk** n → Rundgeschoß n
~**stollen** m = circular tunnel
~**straße** f = district road
~**transporteur** m → Kreisförderer m
~**verkehr** m = gyratory traffic
~**verkehrsplatz** m → Kreisplatz
~**wurf** m, Kreisbahn-Siebschwingung f = circle(-)throw (gyratory movement)
~**zeigerwaage** f, Zifferblattwaage = circular dial-type scale
~**zelle** f = circular cell
~**zellenfang(e)damm** m = circular type cellular cofferdam
~**zylinderschale** f = circular cylindrical shell
~**zylinder-Shedschale** f = circular cylindrical northlight roof shell
kreiszylindrisch = circular cylindrical
kreiszylindrische Sperre f → → Talsperre
Krempblech n → Bördelblech
Krempe f, Bördel m = flange
krempen → bördeln
Krempziegel m = flap tile
~**deckung** f = flap tile roofing
Kreosotierung f = impregnation with creosote
Kreosotöl n = creosote oil [deprecated: creosote]
Kreuzband n → Kreuzstreben fpl
~ [Beschlag] = T-hinge strap
~ [Magnetscheider] = cross belt
~**scheider** m = cross belt separator
Kreuz(bohr)meißel m = cross bit
kreuzförmige Stütze f = cross-shaped column
Kreuzgebälk n → Kreuzstreben fpl
Kreuzgelenk n, Kardangelenk, Verbindung f komplett = spider trunnion, universal joint, U-joint, universal coupling
~**satz** m = spider kit
kreuz|gerippt = cross-ribbed
~**geschichtet** [Geol.] = current-bedded
Kreuz|gewölbe n [Entsteht durch Durchdringung zweier Zylinderflächen] = groin(ed) vault, cross ~

Kreuzhacke — Kriechversuch

~hacke f, Kreuzpicke(l) f, (m) = pickaxe [has one point and one flat end]
~holz n = quarter timber
~kamm m [Holzbau] = cross cogging, ~ corking, ~ cocking
~libelle f = cross-bubble
~meißel m, Kreuzbohrmeißel = cross bit
~picke(l) f, (m) → Kreuzhacke f
~poller m = cruciform bollard
~rollendrehkranz m = cross-roll slewing ring, ~ circular pathway
~schichtung f (Geol.) = cross-bedding, false-bedding
~schlag m [Seil] = regular lay, standard ~, right ~, cross ~
~schneide f = cross bit cutting edge
~schneiden-Bohrkrone f = cross (drill) bit
~stake f = herringbone strut
~stein m (Min.) → Staurolith m
~strebe f = diagonal strut
~streben fpl, Andreaskreuz n, Kreuzband n, Abkreuzung f, Kreuzgebälk n = St. Andrew's cross, diagonal struts, saltier cross bars, cross stays
~stück n [Fitting] = cross
~stück n für Kardangelenk, Spinne f = spider
Kreuzung f von zwei Straßen, Einzelkreuzung f = crossing
Kreuzungs|anlage f = interchange
~bahnhof m = interchange station
~bauwerk n → plankreuzungsfreie Kreuzung f
~rampe f = interchange ramp
~stelle f, Ausweiche f, Ausweichstelle [für ein Schiff] = passing place, layby
~verkehr m = traffic at intersections
~zufahrt f = intersection approach
Kreuzverband m = cross bond
~, Pfostenfachwerk n [mit gekreuzten Streben] = cross bracing
kreuzweise Bewehrung f, ~ Armierung, ~ (Stahl)Einlagen fpl = crosswise reinforcement
Kriech|beiwert m, Kriechzahl f = coefficient of creep

~dehnung f = creep strain, inelastic ~
~durchbiegung f = creep deflection (Brit.); ~ deflection
~eigenschaft f = inelastic property, creep ~, ~ characteristic
kriechen, wandern [Schiene] = to creep
Kriechen n [fester Werkstoff] = creep
~ [Bitumen] = volume flow
~ von Glasuren, Abrollen ~ ~ = crawling of glazes
Kriechfaser f = creep fibre (Brit.); ~ fiber (US)
kriechfest = creep-resisting
Kriech|gang-Selbstfahrwerk n = creeper speed self-propelled carrier
~geschwindigkeit f [Beton; Mörtel] = rate of creep, velocity ~ ~
~geschwindigkeit f, Steiggeschwindigkeit, Kriechgang-Geschwindigkeit [Fahrzeug] = creep speed
~grenze f = creep limit
~größe f = magnitude of creep
~mechanismus m = mechanism of creep, creep mechanism
~modul m = modulus of creep
~probe f, Kriechversuch m, Kriechprüfung f = creep test
~prüfmaschine f = creep test machine
~prüfung f → Kriechprobe f
~raum m, Bekriechungsraum = crawl space, crawlway [An underfloor space providing access to ducts, pipes and other services hung or laid therein and of a height sufficient for crawling]
~schwankung f = creep variation
~streifen m [Der Ausdruck ,,Kriechspur" sollte nicht mehr verwendet werden] = climbing lane
~theorie f = theory of creep
~umlagerung f = redistribution of creep
~verformung f = creep deformation
~verhalten n = creep behavio(u)r
~vermögen n = creep capacity
~versuch m, Kriechprobe f, Kriechprüfung f = creep test

Kriechvorgang — Kronenschneide

~vorgang m = creeping
~wert m = creep value
~zahl f, Kriechbeiwert m = coefficient of creep
~zunahme f = creep growth
Krippmaschine f [für die Vorwellung von Schußdrähten für Siebböden] = crimping machine
kristalliner Gips m, Gipsspat m = crystalline gypsum
~ Schiefer m = (crystalline) schist
kristallines Aggregat n = aggregate of crystals
~ Gestein n = crystalline rock
Kristallisations|druck m = pressure of crystallization
~versuch m, Kristallisationsprüfung f, Kristallisationsprobe f = crystallization test
kristallisierbar = crystallizable
Kristall|klangzelle f = piezoelectric phonometer
~korn n = crystal(line) grain
~-Kronleuchter m = crystal chandelier
~quarzsand m = silica sand, grit
~wachstum n = crystal growth
~wasser n = water of crystallization
kritische Belastung f = critical loading
~ Dichte f [Straßenverkehr] = critical density
~ Druckhöhe f, kritisches Gefälle n [Talsperre] = critical head
~ ~ [Bodenmechanik] = critical hydraulic gradient
~ Schleppkraft f, Grenzschleppkraft [Fluß; Strom] = critical tractive force [The force at which movement of sand particles begins]
~ Sohlengeschwindigkeit f [Strom; Fluß] = critical bottom velocity
~ Wassertiefe f, Grenzwassertiefe = critical depth
kritischer Außendruck m [Tiefbohrtechnik] = collapse resistance (of a casing)
~ Druck m, Grenzdruck = limit(ing) pressure
~ (Gleit)Kreis m = critical circle

~ Innendruck m [Tiefbohrtechnik] = internal yield pressure
~ Querschnitt m, gefährlicher ~ = critical section
~ Regen m = critical rain
~ Wert m = critical value
~ Zustand m, Grenzzustand [Hydraulik] = competence, competent condition
kritisches Gefälle n, kritische Druckhöhe f [Talsperre] = critical head
Krone f [Damm; Talsperre; Wehr] = crest
~ [Straße] = crown
Krönel m, Krönle n, Kröneleisen n = patent pick, ~ axe, roughing hammer
kröneln = to tool with the roughing hammer, ~ ~ ~ ~ patent pick, ~ ~ ~ ~ patent axe
Kronen|abwick(e)lung f [Talsperre; Wehr] = crest development
~block m, (Bohr)Turmrollenblock, Oberblock = crown block [In rotary drilling, a sheave set arranged at the top of a derrick]
~block m [Talsperre; Wehr; Damm] = crest block
~bohrer m = crown drill
~breite f [Straße] = crown width
~breite f [Damm; Talsperre; Wehr] = crest width
~dach n, Ritterdach = high-pitched roof
~erhöhung f [Talsperre; Wehr; Damm] = heightening of the crest
~fräser m = milling tool, ~ shoe
~gelenk n = crown hinge
~gewölbe n = crown vault
~höhe f [Damm; Talsperre; Wehr] = crest level
~länge f [Damm; Talsperre; Wehr] = crest length
~mutter f = horned (screw) nut, castellated (~) ~
~nippel m = core bit connection
~rolle f, (Bohr)Turmrolle [Rotarybohren] = crown sheave
~schneide f = crowned (drill) bit cutting edge, ~ ~ point

Kronenverschluß — Kübel(-Beton)verteiler

~verschluß m = crest gate [*A gate on the crest of a dam*]
Kronglas n = crown-glass
kröpfen, ab~ [*Einen Profil- oder Stabstahl aus seiner ursprünglichen Ebene in eine innerhalb der Konstruktion dazu versetzte Ebene örtlich biegen*] = to offset
~, ab~ [*Kurbeln biegen*] = to crank
Kropfstück n, Krümmling m [*am Treppengeländer*] = string wreath
Kröpfung f, Ab~ [*Profil- oder Stabstahl*] = off-set
Krücke f [*Trommel(stau)wehr*] = back prop
Krückeisen n [*hölzernes Schleusenstemmtor*] = T strap
Krümelgefüge n → Wabentextur zweiter Ordnung
Krümme f, Rundung f = rounding
Krümmer m [*Motor*] = manifold
~, Bogen m, Rohr~, Bogenrohr n = (pipe) bend
~ [*Wasserkraftwerk*] = elbow bend, penstock elbow
~druck m [*Motor*] = manifold pressure
~verlust m, Krümmerwiderstand m = elbow loss
Krümmling m, Kropfstück n [*am Treppengeländer*] = string wreath
Krümmung f, Biegung [*Im Sinne von einer Kurve einer Straße, eines Wasserlaufes, eines Schienenstranges usw.*] = turn, bend, curvature
Krümmungs|anfangspunkt m, Krümmungsendpunkt [*Straße*] = tangent point
~grad m [*Straße*] = degree of curvature
~halbmesser m, Krümmungsradius m [*Straße; Bahnlinie*] = radius at bend, ~ of curvature, ~ at turn
~halbmesser m, Krümmungsradius m [*Talsperre*] = locus of centres (Brit.); line of centers (US)
~-Scheitelpunkt m = apex of bend
Krüppelwalm m, halber Walm = partial hip(ped end), half-hip(ped) ~

~dach n = gambrel roof, half-hip(ped)
~ [*A roof having a gablet near the ridge and the lower part hipped*]
~dach n = jerkin-head roof, hipped-gable ~, shread-head ~; clipped gable ~ (US) [*A roof which is hipped from the ridge halfway to the eaves and gabled from there down, the contrary of a gambrel roof*]
Krupp-Profil n = Krupp section
Kruste f, (Filter)Kuchen m [*An der Bohrlochwand(ung) durch die Bohrspülung abgesetzt*] = cake
Krusten|bewegung f, Dislokation f, Lagerungsstörung f (Geol.) = dislocation
~brechhammer m = crust breaker
kryptokristallin, feinkristallin = cryptocrystalline
kryptovulkanisches Beben n, Intrusionsbeben = cryptovulcanic earthquake
KSH m, Kalksandhartstein m = highly resistant sand-lime brick
KSHbl m, Kalksandhohlblockstein m = hollow sand-lime brick
KSL m, Kalksandlochstein m = perforated sand-lime brick
KSV m, Kalksandvollstein m = solid sand-lime brick
Kubatur f, Rauminhalt m = cubic yardage, cubage
Kübel m = bucket
~, Schürf~, Fördergefäß n, Schürfgefäß [*Radschrapper*] = (scraper) bowl, (skimmer) scoop
~ [*(Beton)Querverteiler*] = trough
~, Schaufel f, Lade~ [*Schürflader*] = bucket, loading ~
~ [*Schachtbau*] = sinking bucket, skip; kibble, bowk, hoppit (Brit.)
~ → → Betonmischer m
~anschlag m = bowl stop
~arm m = bucket arm
~aufzug m → Mulden(bau)aufzug
~aufzugbahn f → → Betonmischer m
~(bau)aufzug m → Mulden(bau)aufzug
~(-Beton)verteiler m → Betonverteilungswagen m

Kübelbremsventil — Kugelfallprobe

~bremsventil *n* = bowl brake valve
~fettpresse *f* = volume compressor
~gehänge *n* = bucket suspension tackle
~höhenverstellung *f* = bowl level adjustment
~hub *m* = bowl lift
~hubhebel *m* = bowl lift lever, ~ control ~
~hubseil *n* = bowl lift cable, ~ control ~, ~ ~ line
~kupp(e)lungsventil *n* = bowl clutch valve
~leitung *f* [*Radschrapper*] = bowl line
~reinigungsanlage *f* [*zum Reinigen von Betonierkübeln*] = skip cleaning station
~seitenwand *f* = bowl side wall
~steuerventil *n* = bowl control valve
Kübeltransport|einrichtung *f* = bucket transfer equipment
~katze *f* = bucket handling crab
~sperre *f* = bowl travel(l)ing lock
~wagen *m* = bucket transfer car
Kübel|überlaufschutz *m* = ejector overflow guard
~ventil *n* [*Radschrapper*] = bowl valve
~verteiler *m* → Betonverteilungswagen *m*
~zwischengeschirr *n*, Förderkorbzwischengeschirr = communicator for sinking bucket
~zylinder *m* [*Radschrapper*] = bowl ram, ~ jack
~zylindersperrventil *n* [*Radschrapper*] = (bowl) carry-check valve
~zylinderventilgruppe *f* [*Radschrapper*] = bowl valve group, ~ ~ assembly
Kubik|inhalt *m* → Rauminhalt
~yards in ungestörter Lagerung [*Kubikyards (zu je 0,76 cbm) in ungestörter, natürlicher Lagerung gemessen*] = bank yards
kubische Abmessung *f* = cubic dimension
~ Parabel *f* = cubic parabola
kubischer Ausdehnungsbeiwert *m*, räumlicher ~ = coefficient of cubic expansion

~ Splitt *m* = cubical chip(ping)s
kubisches Endkorn *n*, ~ Endprodukt [*Hartzerkleinerung*] = cubical product
~ System *n* = cubical system
Kuchen *m* → Kruste *f*
Küchenabfall *m* = kitchen waste, domestic food ~
~zerkleinerer *m* = waste-disposal unit
Küchen|abwasser *n*, Ausgußwasser = slop water, sink ~, kitchen waste [*Liquid culinary waste*]
~ausguß *m* → Spüle *f*
~be- und -entlüftung *f* = kitchen ventilation
~geschoß *n*, Küchenstockwerk *n*, Küchenetage *f* = kitchen floor
Kuchenprobe *f* [*Zement*] = pat test
Küchen- und Barwagen *m* [*Eisenbahnwagen*] = kitchen buffet car
Kufengewölbe *n*, gerades Tonnengewölbe = straight barrel vault
kufen|montiert = skid-mounted
~montierter Kastenverteiler *m* = skid-mounted box(-like) spreader
Kufenrahmen *m* = skid frame
Kugel|achsgelenk *n* = constant velocity universal joint
~bakentonne *f* = globe buoy
~bodenbunker *m* = spherical-bottom bin
~bolzen *m* = pivot pin
~damm *m* (⚒) = spherical dam
~druckhärte *f* = hardness given by the dynamic indentation test
~druckhärteprüfung *f* = dynamic indentation ball test
~drucklager *n* = ball-thrust bearing
~-Druckventil *n*, Druck-Kugelventil [*Pumpe*] = ball delivery valve, ~ discharge ~
~druckversuch *m* nach Brinell, Kugeldruckprobe *f* ~ ~, Kugeldruckprüfung *f* ~ ~, Brinell-Probe, Brinell-Versuch, Brinell-Prüfung = Brinell hardness test
~fallprobe *f* = dropping-ball (penetration) test

Kugelfallviskosimeter — Kugelspeiseventil

~fallviskosimeter n = falling ball viscosimeter, drop ~~, ~ sphere ~
~fangglas n, kugelsicheres Glas = bullet-resisting glass
~flechtwerk n = trellis dome work
~formtrommel f = drum pelletizer [*clay*]
~füllung f [*Mühle*] = ball load
~gasbehälter m, Kugelgasometer m, Kugelgassammler m = spherical gasometer
kugelgelagertes Rad n = ball-bearing wheel
Kugelgelenk n, Kugelpfanne f = ball socket, ~ joint
~ des Schartragrahmens, Kugelpfanne f ~ ~ =circle draft frame ball joint, ~ ~ ~ ~ socket
~kappe f (⚒) = ball-joint (roof) bar
~stempel m (⚒) = ball-joint prop, ~ post
Kugel|gewölbe n, Hängekuppel f, Stutzkuppel, Stichkuppel = spherical vault, ~ dome
~gießmaschine f = ball-casting machine
~gleitlager n = spherical bearing
~hahn m = ball cock
~härteprüfer m = indentation machine
~haube f, Kalotte f, Kugelkappe f [*ein kreisförmig begrenzter Teil der Kugeloberfläche*] = calotte
~hebel m = ball lever
kugelig → → Kornform f
Kugel|kalotte f = spherical dome
~kappe f, Flachkuppel f = flat dome, shallow ~
~kappe f → Kugelhaube f
~keil m = spherical cone
~kipplager n [*für Hoch- und Tiefbauten*] = rocking ball bearing
~kohle f, Mugelkohle = coal pebbles, pebble coal
~kopf m für Drehkranzhalter [*Motor-Straßenhobel*] = ball and socket
~lager n = ball bearing
~lagerfett n = ball bearing grease
~lageröl n = ball bearing luboil
~mühle f = tumbling mill, ball ~

~mühle f mit abrollenden Mahlkörpern, ~ ohne freien Fall der Mahlkörper = non-cataracting ball mill, ~ tumbling ~
~mühlenmahlung f = ball milling, ~ grinding
~mühlenpanzerung f = ball mill lining, ~ ~ liners
~nische f = spherical-headed niche
~pech n, Perlpech = pellet(ed) pitch
~perforator m [*Tiefbohrtechnik*] = bullet perforator
~pfanne f, Kugelgelenk n = ball socket, ~ joint
~ring m → Laufring
~rippe f = rib of a dome
~rohrmühle f, Stahl~ = ball tube mill, steel ~ ~
~rückschlagventil n = ball check valve
~rückschlagventil n des hydraulischen Kettenspanners = track adjuster ball check valve
~-Saugventil n, Saug-Kugelventil [*Pumpe*] = ball suction valve
~schale f, sphärische Schale = spherical shell
~schaufler m, (Be)Lademaschine f mit kugelförmigem Kopf = rotary (head) excavator
~schauflerkopf m = rotary head, digging ~
~schicht f, Kugelzone f = zone of a sphere, spherical segment between two parallel circles
~schieber m = globe valve
~schlagbohren n = jet pump pellet drilling
~schlaghärteprüfer m, Kugelschlaggerät n = dynamic ball-impact tester
~schlagprüfung f [*Beton*] = dynamic ball-impact test method
kugelsicheres Glas n, schußsicheres ~, Kugelfangglas = bullet-resisting glass, bullet proof ~
Kugel|sicherheitsventil n = spherical safety valve
~speiseventil n = spherical feed valve

Kugelstrahlen — Kühltank

~strahlen n = shot peening
~tisch m = ball table [*Table mounted on a number of balls*]
~traglager n für den Königsbolzen = king bolt ball and socket
~ventil n = globe valve, ball ~
~ventilkammer f = ball-valve chamber, globe-valve ~
~ventilpumpe f = ball-valve pump, globe-valve ~
~verformtechnik f = pelletizing technique, ~ technic [*clay*]
~walm m = spherical hip(ped end)
~zapfen m = ball stud
~zapfenkipplager n [*für Hoch- und Tiefbauten*] = ball jointed rocker bearing
~zone f → Kugelschicht f
Kühl|anlage f = cooling system
~bandanlage f, Bandkühlsystem n = cooling belt system
~becken n, Kühlteich m = cooling pond, spray ~
~behälter m, Kühltank m = cooling tank
~bettschere f = cooling bed shears
~einrichtung f = cooling equipment
Kühler m = cooler
~ [*Auto*] = radiator
~ mit schrägstehenden Kühlrohren = cated design radiator
~abdeckung f = radiator shroud
~block m = radiator core
~gitter n = radiator grille
~kopf m = cooler inlet [*evaporation cooler*]
~netz n [*Motor*] = radiator block
~rippe f = cooling fin, ~ rib
Kühlerschutz m = radiator guard, ~ protection, front ~
~ mit Scharnieren = hinged radiator guard
~decke f [*Grobleinen oder Segeltuch*] = bonnet rug
Kühlerventilator m = radiator fan
~ einschließlich Lichtmaschine = fan-generator
Kühlerverkleidung f = radiator cowling
Kühl|fläche f = cooling area
~flüssigkeit f = cooling liquid
~gebläse n = cooling fan
~haus n, Kühllagerhaus = refrigerated warehouse, cold store, refrigeration storage house
~kanal m = cooling duct
~kasten m [*Hochofen*] = cooling box
~lamelle f, Kühlstreifen m, Kühlrippe f = cooling fin
~leistung f = cooling capacity
~luft f = cooling air
~lüfter m = cooling fan
~luftführung f = cooling air circulation
~luftstrom m = cooling air current
Kühlmantel m = cooling jacket
~wasser n = cooling jacket water
Kühlmittel n, Kühlöl n [*Werkzeugkühlung*] = cutting oil, cooling ~
~, Kühlstoff m = coolant
~pumpe f, Kühlstoffpumpe = coolant pump
~rohr n, Kühlstoffrohr = coolant tube
~system n, Kühlstoffsystem = coolant compound
Kühl|nische f [*im Motorblock*] = cooling shelf
~öl n, Kühlmittel n [*Werkzeugkühlung*] = cooling oil, cutting ~
~raum m = cooling chamber
~rippe f → Kühllamelle f
~rohr n = cooling pipe
~schacht m [*Abkühlen von Trink- und Brauchwasser*] = cooling shaft
~schild n [*Drehstromlichtmaschine*] = heat sink [*alternator*]
~schirm m [*Turbolader*] = heat shield [*turbocharger*]
~schlange f = cooling coil, serpentine cooler
~schrank m = refrigerator
~silo m = cooling silo
~stoff m, Kühlmittel n = coolant
~stoffsystem n, Kühlmittelsystem f = coolant compound
~streifen m → Kühllamelle f
~system n = cooling system
~tank m, Kühlbehälter m = cooling tank

Kühlthermostat — Kunstgalerie

~thermostat *m*, Wassertemperaturregler *m* = water temperature regulator

Kühltrommel *f*, Trommelkühler *m* = rotary cooler, cooling drum

~schleuder *f* = cooling drum centrifuge

Kühlturm *m*, Kühlwerk *n* = cooling tower

~gebläse *n* = cooling tower fan

(~)Steigschacht *m*, Zuflußsteigschacht = (cooling tower) chimney

~tasse *f* = cooling tower saucer

Kühlung *f* = cooling

Kühl|ventilator *m* = cooling fan

~vorrichtung *f* = cooling device

Kühlwasser *n* = cooling water

~einlauf(bauwerk) *m*, *(n)* = cooling water intake (structure)

~einlaufkanal *m* = cooling water inlet duct

~kanal *m* = cooling water canal

~kreislauf *m* = cooling water cycle, ~ ~ circuit

~mantel *m* = cooling water jacket

~pumpe *f* = cooling water pump

~pumpenhaus *n* = cooling water pump(ing) station

~pumpenkammer *f* = cooling water pump chamber

~rohr *n* = cooling water pipe

~rohrleitung *f* = cooling-water piping

~rücklaufpumpwerk *n* = cooling water return pumping plant

~schlauch *m* = cooling water hose

~stollen *m* [*Wasserkraftwerk*] = condensation water tunnel, condensing ~ ~

~thermometer *n* = cooling system temperature ga(u)ge

~umlauf *m* = cooling water circulation

Kühl|werk *n*, Kühlturm *m* = cooling tower

~zement *m*, Bauxitlandzement = bauxite cement

~zone *f* = cooling zone

~zyklon *m* = cooling cyclone

kühne Konstruktion *f* [*Brückenbau*] = bold design

Küken *n*, Hahn~, drehbarer Kegel *m* = (tapered) plug

Kukersit *m*, estländischer Brennschiefer *m* = kukersit

Kulm *m* (Geol.) = culm

Kultivator *m*, Boden~ = (agricultural) cultivator, field ~

kultivierbar → ackerfähig

Kulturbau *m* = agricultural engineering

~arbeiten *fpl* = agricultural engineering work

~ingenieur *m* = agricultural engineer

Kultur|boden *m* = agricultural soil, cultivated ~

~denkmal *n* = ancient monument, historic landmark

~fläche *f* = cultivated area, ~ land, farmland

Kumaronharz *n* = c(o)umarone(-indene) resin

Kümmerfluß *m* = underfit stream

Kümpel|blech *n* → Bördelblech

~presse *f*, Bördelmaschine *f* = flanging machine

Kunstbernstein *m* = bastard amber, impure ~, imperfect ~

Kunden|dienst *m* = after-sales service

~guß *m*, Handelsguß = jobbing casting, ~ work

~gußgießerei *f*, Handelsgußgießerei = jobbing foundry

~guß-Sandgießerei *f*, Handelsguß-Sandgießerei = jobbing sand foundry

Kunst|baustoff *m*, Industriebaustoff = man-made construction material, ~ building ~

~bauwerk *n* [*Kunstbauten einer Straße sind alle Bauwerke zur Aufnahme kreuzender Verkehrswege, Wasserläufe, Leitungen usw., auch Stützmauern, Hangverbauungen, Tunnel usw.*] = structure

~bims *m* → Hüttenbims

~feinkorn *n* [*Siebtechnik*] = finely ground grain, artificial fine ~

~galerie *f* = art gallery

Kunstgranitbordstein — künstlicher Verkehr

~granitbordstein *m* mit weißem Quarzitkorn = precast kerb (Brit.) (or curb) of white granite aggregate
Kunstharz *n* → Harz
~auftrag *m* = synthetic resin deposition
~belag *m* = synthetic resin covering
~beton *m* = (synthetic-)resin concrete, resin-based ~
~binder *m* = synthetic resin binder
~-Dicht(ungs)ring *m* = synthetic-resin seal(ing) ring, ~ packing ~, ~ washer
~emulsion *f* = synthetic resin emulsion
~filter *m, n* = synthetic resin filter
~leim *m* = synthetic resin glue, ~ adhesive
~mörtel *m* = resin-based mortar
kunstharzverleimtes Sperrholz *n* = resin-bonded plywood
Kunst|kalkstein *m*, Hydrokalkstein = artifical stone, reconstructed ~
~kautschuk-Kontakt-Kleber *m* → Klebezement *m*
~körper *m*, Bauwerk *n*, Baukörper *m* = structure
Künstlergarderobe *f* [*Theater*] = dressing room
künstlich (aus)gefällter Kalzit *m* = artificially precipitated calcite
~ feuerfester Baustoff *m* = artificially refractory construction material, ~ ~ building ~
künstliche Anlandung *f* [*Strand-Wiederherstellung*] = nourishment [*beach rehabilitation*]
~ Atmung *f* = artificial respiration
~ Austrocknung *f*, künstliches Austrocknen *n* = artificial drying out
(~) Baugrundverdichtung *f* = ground compaction
~ Beleuchtung *f* = artificial lighting
~ Bewetterung *f* = artificial ventilation
~ Bitumen-Mineral-Mischung *f*, künstliches Bitumen-Mineral-Gemisch *n*, Asphalt *m* = artificial asphalt (Brit.)

~ Bodenverdichtung *f* = (soil) (or earth) compaction (or densification), artificial consolidation
~ Grundwassererzeugung *f*, ~ Grundwasseranreicherung = water spreading, artificial recharge of ground water [*The artificial application of water to lands for the purpose of storing it in the ground for subsequent withdrawel*]
~ Kälte *f* = artificial cold
~ Lunge *f* = iron lung
~ Mineralfarbe *f* = artificial (mineral) pigment
~ Puzzolane *f* = pozzolanic material
~ Sandinselschüttung *f* für Senkkasten = sand-island method
~ Schiffahrtstraße *f* → ~ Wasserstraße
~ Schlammtrocknung *f* = artificial sludge drying
(~) Verdichtung *f*, (künstliches) Verdichten *n* = compaction, compacting
~ Wasserstraße *f*, ~ Schiffahrtstraße, künstlicher Schiffahrtweg *m*, künstlicher Wasserweg = artificial navigable waterway, ~ ~ water course
künstlicher Anhydrit *m*, löslicher ~ = artificial anhydrite
~ (Beton)Zuschlag(stoff) *m* = artificial aggregate
~ Diamant *m*, synthetischer ~ = synthetic diamond, man-made ~
~ Feinzuschlag(stoff) *m* = artificial fine aggregate
~ Grobzuschlag(stoff) *m* = artificial coarse aggregate
~ Hafen *m* = man-made harbo(u)r
~ Kanal *m* → Kanal
~ Mauerstein *m*, ~ Wandbaustein [*Betonbaustein; Mauerziegel; Kalksandstein*] = artificial masonry unit
~ Quarz *m*, synthetischer ~ = man-made quartz, synthetic ~
(~) Schlackensand *m* → granulierte Hochofenschlacke *f*
~ Stein *m*, Kunststein = cast stone
~ Verkehr *m*, Versuchsverkehr = test traffic

künstlicher Wandbaustein — Kupfernickel

~ **Wandbaustein** *m* → Mauerstein
~ **Wasserlauf** *m* = artificial waterway, man-made ~, ~ water course
~ **Wasserweg** *m* → künstliche Wasserstraße *f*
~ **Winddruck** *m* = simulated wind pressure
künstliches Austrocknen *n*, künstliche Austrocknung *f* = artificial drying out, ~ desiccation
~ **Bitumen-Mineral-Gemisch** *n*, künstliche Bitumen-Mineral-Mischung *f*, Asphalt *m* = artificial asphalt (Brit.)
~ **Hindernis** *n* = man-made obstacle
~ **Meer(es)wasser** *n*, ~ Seewasser = artificial seawater
Kunst|marmor *m*, künstlicher Marmor = artificial marble
~**schiefer** *m* → Asbestzement *m*
~**schmiedearbeit(en)** *f(pl)* = ornamental ironwork, ~ metalwork
Kunststein *m*, künstlicher Stein = cast stone
~**fliese** *f*, Kunststeinplatte *f* = cast stone tile
~**maschine** *f* = cast stone machine
~**-Pflasterplatte** *f* = cast-stone (paving) flag
~**platte** *f*, Kunststeinfliese *f* = cast stone tile
~**platte** *f* = cast stone slab, ~ ~ panel
Kunststoff *m* = plastic
~**bahn** *f* = plastic sheet
~**band** *n* = plastic tape
~**belag** *m* = plastic cover
kunststoff|bereift = plastic-tyred (Brit.); plastic-tired (US)
~**bezogen**, kunststoffbeschichtet = plastic-faced
Kunststoff|dekor(ations)platte *f* = ornamental plastic slab
~**dicht(ungs)ring** *m* = plastic seal(ing) ring, ~ packing ~, ~ washer, ~ gasket
~**dränrohr** *n* = plastic drainage pipe
~**dübel** *m* = plastic dowel
~**-Einfamilienhaus** *n* = plastic house
~**emulsion** *f* = plastic emulsion

Kunststoffenster *n* = plastic window
Kunststoff|-Form *f* = plastic mould (Brit.); ~ mold (US)
~**gras** *n* = plastic grass
~**-Handlauf** *m* = plastic handrail
~**hülle** *f* [*Spannbeton*] = plastic sheath
~**kabel** *n* = plastic cable
~**kanal** *m* [*für Spannbetonkabel*] = plastic sheath(ing)
~**-Kleber** *m* = plastic adhesive
~**lichtkuppel** *f* = plastic saucer dome, ~ domelight
~**-Markierungsstreifen** *m* → plastischer Markierungsstreifen
~**prüfmaschine** *f* = plastic testing machine
~**rohr** *n* = plastic pipe
~**rohr-Verlegeanlage** *f* = plastic pipe laying plant
~**schalung** *f* = plastic formwork
~**schaum** *m* = foamed plastic, plastic foam
~**schaumdämmplatte** *f* = insulating plastic foam board
~**zahnrad** *n* = plastic gear
~**zusatz(mittel)** *m*, (*n*) = plastic additive
Kunst|vaseline *f* = petrolatum (made by mixture)
~**wachs** *n* → Zeresin(wachs) *n*
Künzelstab *m* = Künzel sounding rod
Kupfer|dach(ein)deckung *f*, Kupferdeckung = copper roofing
~**dichtungsstreifen** *m* = copper sealing strip, ~ water stop
~**drahtwalzwerk** *n* = copper wire mill
~**erz** *n* = copper ore
(~)**Fahlerz** *n* → Graugültigerz
kupferführend = copper-bearing
Kupfer|glanz *m*, Cu_2S = chalcocite, redruthite, copper glance
~**guß** *m* = copper casting
~**indig** *n* = indigo copper
~**kies** *m*, Chalkopyrit *m*, $CuFeS_2$ = copper pyrite, chalcopyrite (Brit.); copper pyrites (US)
~**legierung** *f* = copper(-base) alloy
~**leiter** *m* = copper conductor
~**löslichkeit** *f* = copper solubility
~**nickel** *n* → Rotnickelkies *m*

Kupferplatte — Kuppenausrundung

~platte f = copper plate
~plattierung f, Verkupferung = coppering, copper plating
~profil n = copper section
~rohr n = copper tube
~schalenprüfung f, Kupferschalenprobe f, Kupferschalenversuch m = copper dish test
~schindel f = copper shingle
~schlacke f = copper slag, cast ~
~schlackenblock m = copper slag block, cast ~ ~
~schlackenstein m = cast slag sett, copper ~ ~
~silberglanz m (Min.) = stromeyerite
~sprengkapsel f → → 1. Schießen n; 2. Sprengen n
~stein m = copper matte
~streifenversuch m, Kupferstreifenprüfung f, Kupferstreifenprobe f = copper strip test
~uranit m, Uranglimmer m = copper uranite, cupro-uranite, tobernite
~vitriol n, m, Chalkantit n, $CuSO_4 \times 5H_2O$ (Min.) = blue vitriol, chalcanthite, blue stone
~wellblech n = corrugated copper sheet
Kupolofen m = cupola furnace
~-Begichtungsanlage f = cupola charger
~beschickungskran m = cupola furnace charging crane
~schlacke f = cupola slag
Kuppe f [Längenschnitt einer Straße] = summit
Kuppel f, halbsphärisches Gewölbe n = dome, cupola
~auge n, Auge der Kuppel, Laternenöffnung f = lantern opening
~bau n = domed building, domic(al) ~, domy ~
~boden m [Behälter] = domed floor, ~ bottom
~dach n = dome(-shape)d roof
~gewölbe n = domic(al) vault, domed ~, domy ~
~mauer f, Kuppelstaumauer, Kuppelsperre f, Kuppelsperrmauer = dome(-shape)d dam, cupola ~

~schale f = dome shell
~sperre f → Kuppelmauer f
Kupp(e)lung f = clutch
~ mit Reserve = oversize clutch
Kupp(e)lungsausrück|büchse f = clutch throwout socket
~kolben m = clutch release piston
~muffe f = clutch release sleeve
Kupp(e)lungs|belag m = clutch lining
~betätigung f = clutch control
~bock m = clutch bracket
~bremse f = clutch brake
~druckkolben m = clutch piston
~druckscheibe f = clutch pressure plate
~einrückschraube f = clutch engagement screw
~einstellmutter f = clutch adjusting nut
~feder f = clutch spring
~führungslager n = clutch pilot bearing
~fußhebel m, Kupplungspedal n = clutch pedal
~gehäuse n = clutch housing
~gestänge n = clutch operating device
~hebel m = clutch lever
~hebelrolle f = clutch lever roller
~joch n = clutch yoke
~lamelle f = clutch plate
~muffe f = clutch collar
~nachstellung f = clutch adjustment
~packung f = clutch pack
~pedal n, Kupplungsfußhebel m = clutch pedal
~ring m = clutch carrier
~schaltventil n = transmission declutch valve
~scheibe f (mit Belag) = clutch disc; ~ disk (US)
~schlupf m = clutch slippage
~servovorrichtung f = clutch control booster
~spiel n = clutch pedal clearance
~trommel f = clutch drum
Kuppen|ausrundung f = convex transition between gradients, summit curve

Kuppen- und Wannenausrundungen — kurzflammig 616

∼- **und Wannenausrundungen** /pl = transition curves between gradients, ∼ **radii on** ∼, vertical curves
kuppen- und wannenreiche Straße /, wannen- und kuppenreiche ∼ = undulating road, ∼ highway
Kurbel / [*einarmiger Hebel zum Drehen einer Welle*] = crank
(Kurbel)Dämpfer m = damper
∼**verkleidung** / = damper guard
Kurbelgehäuse n = crankcase
∼**-Entlüftung** / = crankcase breather
∼**-Schaudeckel** m = crankcase inspection cover
∼**schmieranlage** / = crankcase lube system
Kurbel|lager n = main bearing
∼**presse** / = crank press
∼**schere** / = crank shears
∼**sieb** n = crank-shaft drive screen
∼**trieb** m = crank drive
∼**triebschmierdruck** m = crankcase lube pressure
∼**- und Kreuzkopf-Schreitausrüstung** / [*Schreitbagger*] = crosshead type walking mechanism
∼**wanne** /, Kurbelgehäuse n = crankcase
∼**welle** / = crankshaft
∼**welle** / **mit Verbundstirnrad** = crankshaft with integral timing gear
∼**wellenantrieb** m = crankshaft drive
∼**wellenendschub** m, Endschub der Kurbelwelle = crankshaft end thrust
∼**(wellen)lager** n = crankshaft bearing
∼**wellenriemenscheibe** / = crankshaft pulley
∼**wellenschwunggewicht** n = integral counterweight of crankshaft
∼**zapfen** m = crank pin
∼**zapfenlager** n = crank-pin bearing
Kuron'sche Hygroskopizität / = Kuron's hygroscopicity
Kurort m = health resort
Kursbake / [*ILS-Anflugschema*] = outer marker, LOM
Kurve / = curve
∼ **gleicher Selektivität** = isoselective line

∼ **mit kleinstem Halbmesser** [*Straße*] = curve of minimum radius
Kurven /pl **gleicher Setzung** = curves of equal settlement
Kurven|abflachung / = curve easement
∼**absteckung** / = setting out curves
∼**band** n, Kurvenförderband, Falten-(förder)band = curve-negotiating belt conveyor, ∼ ∼ conveyer
∼**bild** n = curve diagram
∼**brücke** / = curved bridge
kurvenfahrbar = curve-negotiating, able to negotiate a curve
Kurven|fahrt / = curvilinear travel
∼**fahrwerk** n [*Turmdrehkran*] = curve-negotiating bogies, curve-going ∼, ∼ gear
∼**familie** /, Kurvenschar / = set of curves, family ∼ ∼
∼**förderband** n → Kurvenband
kurvengängig, kurvenläufig [*Förderband*] = snaking
Kurven|gleis n = curved track
∼**konsistenzmesser** m = curved-trough flow-device
kurvenläufig, kurvengängig [*Förderband*] = snaking
Kurven|lineal n = drawing curve, French ∼
∼**maximum** n, Kurvenspitze / = peak of a curve
∼**schar** /, Kurvenfamilie / = set of curves, family ∼ ∼
∼**schiene** / = curved rail
∼**sicht(barkeit)** / = visibility on curves
∼**spitze** /, Kurvenmaximum n = peak of a curve
∼**stein** m = curve block
∼**überhöhung** / = superelevation, cant(ing), banking
∼**verbreiterung** / = curve enlargement, ∼ widening
Kurz|arbeiter m = part-timer
∼**ausleger** m = short boom (US); ∼ jib (Brit.)
∼**fahrt** / = short trip
kurzes Joch(holz) n (⚒) → Kappe
kurz|flammig = short-flame

~fristig lieferbar = available for prompt delivery
~fristige Hochwasservorhersage f = short-term flood prediction
Kurz|intervallzünder m = short-delay (electric) blasting cap, ~ (~) detonating ~, ~ (~) explosive ~, ~ (~) detonator, short delay-action detonator
~intervallzündung f = short-delay firing
~keilriemenantrieb m = short-centre vee-rope drive
~lichtbogenschweißen n = short arc welding
~methode f → abgekürztes Verfahren
~parken n = short-term parking
~probe f → Schnellprobe
~prüfung f → Schnellprobe f
~riemenantrieb m = short belt drive
kurzschließen = to short-circuit
Kurzschluß m = short (circuit)
~knopf m = ignition cut-out
Kurz|speicherung f [*Talsperre*] = short-term storage
~verfahren n, Schnellverfahren, abgekürztes Verfahren = accelerated method
~versuch m → Schnellprobe f
~zeichen n = symbol
~zeitlast f = short-term load
~zeitversuch m, Kurzzeitprobe f, Kurzzeitprüfung f = test of short duration
Küste f = seaside, sea-coast, seabord
Küsten|badeort m = coastal resort
~bauwerk n, Küstenanlage f = shore structure
~bauwesen n = coastal engineering
~befeuerung f = coastal lighting
~drift f → Küstenstromversetzung f
~düne f, Stranddüne = coastal dune, shore ~
~ebene f = coastal plain
~eis n, Landeis = shore ice
~erhaltung f → Küstenschutz m
~erosion f = coast erosion
~fahrerkai m = coasting vessel quay
~fahrzeug n, Küstenschiff n, Küstenfahrer m = coasting vessel

küstenferne Barre f = offshore bar
Küsten|feuer n = coasting light
~fluß m = coastal stream
~funkmeßanlage f = shore radar installation
~funkmeßgerät n = shore-based radar unit
~gebiet n = coastal region, ~ territory
~gebirge n = coastal range, ~ mountain
~gewässer n, Küstenvorland n = coastal waters, off-shore ~, ~ areas
~gewässeranlage f [*z.B. eine Bohrinsel*] = offshore installation
~kanal m = coastal canal
~kies m = seashore-gravel, (bench) shingle, gravel without fines
~kliff n, Seekliff = coastal cliff
~land n = coastal land, coastland
~linie f = coastline, shore line
~profil n = shore profile
~sand m = coast sand
~schiff n → Küstenfahrzeug n
~schiffahrt f = coastwise shipping
~schutz m, Küstensicherung f, Küstenerhaltung, Küstenverteidigung = coast protection, sea defense (work), shore protection, coast defense work, coastal protection
~schutzbauwerk n = coast protection structure, sea defense ~
~schutzmauer f = seawall
~schutzwesen n = coastal engineering
~sicherung f → Küstenschutz
~stadt f = coastal city (or town)
~straße f = coastal road, ~ highway
~strich m = coastline
~strom m = coastal river
~strömung f, Küstenstrom m = littoral current, coastal ~, shore ~
~stromversetzung f, Stromversetzung an der Küste, Küstendrift f = littoral drift
~tanker m, Küstentankschiff n = coasting tanker
~terrasse f → Strandterrasse
~überschwemmung f = coastal flooding

~- und Stromuferschutzbauten f = coast and river-bank protection works
~verteidigung f → Küstenschutz m
~vorgebirge n, Kap n = cape
~vorland n, Küstengewässer n = off-shore areas, coastal ~, ~ waters
~zone f = coastal zone
Kutter-Formel f, Kutter'sche Formel = Kutter('s) formula
Kuverwasser n → Körwasser
Kyanisierung f, Kyanisieren n = kyanizing
Kyanverfahren n = Kyan's process
Kyma(tion) n, (f) = cyma(tium) [*A moulding of a cornice, whose profile is a line partly convex and partly concave*]

L

labil, unstabil, schnellbrechend [*Bitumenemulsion*] = quick-breaking, labile, rapid-setting
labile Schwimm-Gleichgewichtslage f = instable buoyancy equilibrium
labiles Netz n, bewegliches ~ [*ebenes Fachwerk*] = unstable frame
Labilität f = instability
Labilitätszustand m = state of instability
Laborant m = laboratory technician
Labor(atorium) n für Spannungsoptik = lab(oratory) for photoelasticity
Labor(atoriums)|-Anhänger m = lab(oratory) trailer
~ausrüstung f = lab(oratory) equipment
~-Bewitterungsmethode f = lab(oratory) exposure
~mischer m = lab(oratory) mixer
~rührer m = lab(oratory) stirring device, ~ stirrer
~-Siebgeräte npl = lab(oratory) sieving equipment
~wagen m = lab(oratory) vehicle
Labyrinthdichtung f [*als Dichtungsart*] = labyrinth seal
~ [*als Material*] = labyrinth packing
Lachenbildung f auf Tropfkörpern = ponding

Lachsspaß m, Lachstreppe f = salmon-ladder, salmon-stair
Lack m = varnish
~beize f → Farbentferner
~benzin n = white spirit
~film m, Anstrichfilm = varnish film
~löser m → Farbentferner
~muspapier n = litmus paper
~schicht f = varnish coat
Lade|anlage f, Be~ = loading installation, ~ plant, ~ facility
~bagger m → Baggerlader
~band n → Bandlader m
~bandausleger m → Be~
~baum m → Derrick-Kran m
~beanspruchung f [*Erdbauwagen*] = loading stress
~breite f = loading width
~brücke f, Tief~ = deck, platform
~brücke f → Anlegebrücke
~brücke f, Be~ = loading bridge
~bühne f → Laderampe f
~bühne f = charging platform
~bunker m, Be~ = loading hopper
~drucksteuerung f [*Motor*] = pressure ratio control
~flur m, Beschickungsbühne f [*Ofenhaus*] = charging platform
~flüssigkeit f [*Motor*] = charging liquid
Ladegabel f = loading fork
~ für Papierholz = pulp wood fork
~ ~ Rundholz, Holzladegabel = log fork, logger
~ ~ Schnittholz = lumber fork
Lade|gebläse n, Vorverdichter m = supercharger
~gerät n → Auflader m
~geschirr n = ships' gear
~gestänge n = loader linkage
~gewicht n = weight loaded
Ladegleis n, Be~, Abfuhrgleis [*Erdbau*] = loading track
~ → Auf~
Lade|gut n = material to be loaded
~hafen m, Versandhafen, Verschiffungshafen = port of shipping, ~ ~ shipment
~höhe f = loading height
~hydraulik f = loading hydraulic system

Ladekipper — Lagenbauweise

~kipper m = self-loading motor(ized) buggy
~kran m = loading and unloading crane
Ladekübel m, Ladeschaufel f, Schaufel, Kübel [*Schürflader*] = (loading) bucket
~ → Abteufkübel
Ladeloch n → Bohrloch
~abstand m → Bohrlochabstand
~besatzstoff m → Bohrlochbesatzstoff
~durchmesser m → Bohrlochdurchmesser
Ladeluftkühler m = aftercooler
~ im Zylinderkopf = in-head aftercooler
~netz n, Ladeluftkühlereinsatz m [*Motor*] = aftercooler core
Lademaschine f → Auflader m
~ mit Becherrad → Becherrad-Lademaschine
Lade|maschinenbediener m (⚒) = loader end man
~maß n → Ladeprofil n
laden, be~ = to load
Laden n, Be~ = loading
~ m = store, shop
~ n [*Sprengtechnik*] = charging
~ m, Fenster~ = (window) shutter, ~ blind
~ n aus der Böschung oder Wand = bank loading
~bau m = shopfitting
~bauer m = shopfitter
~etage f → Ladengeschoß n
~front f, Schaufensterfront = shop front, store ~
~gang m = shopping arcade
~geschoß n, Ladenetage f, Ladenstockwerk n = shop stor(e)y
~stockwerk n → Ladengeschoß n
~straße f = retail business street, shopping ~
Lade|panzer m = armo(u)red loader
~platz m → Ladestelle f
~profil n, Eisenbahndurchgangprofil, Bahnprofil, Lademaß n, Lichtraumprofil, Lichtraumumgrenzung f [*Umgrenzung lichten Raumes auf Haupt- und Nebenbahnen, der zur* Durchfahrt der Eisenbahnfahrzeuge vorhanden sein muß. Höhe 4,80 bis 5,50 m und Breite 4,40 bis 5,00 m] = (railway) clearance (Brit.); (railroad) ~ (US)
Lader m = Auf~
~ für Schüttgut → Becher(werk)auflader
~ mit Becherrad → Becherrad-Lademaschine
~ und Bagger → Baggerlader
Lade|rampe f, Ladebühne f, Ver~ = loading and unloading ramp
~raum m [*Hopperbagger; Schute*] = hopper (compartment)
~raupe f = crawler loader
Laderrahmen m = loader frame
Lade|schaufel f, Schaufel, Kübel m, Kippschaufel [*Schürflader*] = (loading) bucket
~silo m, Be~ = loading silo
~station f für Batterien = charging station
~steg m → Anlegebrücke f
~stelle f, Ladeplatz m = point of loading and unloading, loading and unloading point
~stellung f = loading position
~stoß m = loading shock
~straße f = loading and unloading road
~strecke f (⚒) = (loader) gate (road)
~tür f = charging door
~ventil n = charging valve
~vorgang m, Be~ = loading operation
~vorrichtung f, Be~ = loading device
~zunge f → Pier
~zylinder m = load cylinder
Ladung f [*Fahrzeug*] = load, cargo
Lafette f → Bohr~
Lage f [*Schweißlage*] = layer, pass
~ der Bohrung, Bohrlochlage = well location
lagegenau = trued for position, alignment and level [*bridge construction*]
Lagen|bauweise f, Schichtenbauweise, Lagenschüttung f [*Erdbau*] = layer(ed) construction

Lagenenergie — Lagerplatzkran

~energie f, potentielle Energie = potential energy
~gneis m, Bändergneis = banded gneiss, ribbon ~
~schüttung f → Lagenbauweise f
~verdichtung f = compaction in layers
lagenweise = in layers
Lageplan m = location plan [*A plan which shows the dimensions and position of a construction site, usually also the structure(s) proposed*]
~-Massenermitt(e)lung f, Umrißverfahren n [*Die Erdmassen werden aus dem Lageplan ermittelt*] = contour method of grade design and earthwork(s) calculation
Lager n, Vorkommen n, Lagerstätte f, Vorkommnis n, Fundstätte (Geol.) = deposit
~, Magazin n = store
~ [*für Hoch- und Tiefbauten*] = bearing
~ [*die untere oder obere Fläche eines Werksteines*] = bed
~ [*bei Maschinen und Fahrzeugen*] = bearing
~aufnahme f, Lagerkäfig m = cage, bearing ~
~außenring m = bearing shell
(~)Bestand m = stock
~beständigkeit f der Ottokraftstoffe = gas(oline) storage stability, gasolene ~ ~ (US); petrol ~ ~ (Brit.)
   Punkt m der Druckzeitkurve f = break point
   Induktionszeit f = induction period
   Bombenharzwert m = potential gum
~beständigkeitsprüfung f durch Siebung [*Emulsion*] = residue-on-sieving test
~bock m → Lagerstuhl m
~bock m [*Gleiskettenrolle*] = bearing collar
~bock m der Kübelachse [*Radschrapper*] = bowl axle support block
~bohrung f = bearing bore
(~)Buchse f = (bearing) bush(ing)

~bunker m = storage bin
~deckel m = bearing cap
~deckel m [*Motor*] = end frame
~einrichtung f = store equipment
~endspiel n = bearing end play
lagerfähiges Mischgut n [*Straßenbau*] = pre-mix(ed) material
Lager|fähigkeit f [*Leim*] = shelf life, storage ~, storage property
~fixierung f = bearing adjustment
~fläche f [*Lager für Hoch- und Tiefbauten*] = bearing surface
~fläche f = storage area
~flansch m = bearing flange
~fries m [*Fußboden*] = border
~fuge f [*Mauerwerk*] = bed joint, course ~, horizontal ~
~gang m (oder Sill m, oder Intrusivlager n) zwischen diskordant aufgelagertem Sediment und seinem Untergrund (Geol.) = interformational sill
~gut n = product stored
~halbschale f = bearing half
~halter m → Magazinhalter
~haltung f → (Auf)Speicherung
~haus n, Speicher m = warehouse [*A building used principally for storage of goods or materials*]
~hof m, Lagerplatz m = storage yard
~hofkran m, Lagerplatzkran = storage yard crane
~hülse f = bearing jacket
~käfig m, Lageraufnahme f = (bearing) cage
~laufring m = bearing race
~laufsitz m des Nadellagers = roller pattern
~luft f = bearing slackness
~metall n = bearing metal
lagern, ein~, speichern, bevorraten = to store
Lagerplatte f = bearing pad
~, Sohlplatte = bed plate, sole ~
Lagerplatz m [*Hafen*] = open stacking ground
~, Lagerhof m = storage yard
~kran m, Lagerhofkran = storage yard crane

Lager|quader *m*, Auf~ = bearing block
~**ring** *m*, Laufring = bearing race
~**schale** *f* = split bearing
~**schmiermittel** *n* = bearing lubricant
~**schmieröl** *n* = bearing luboil
~**schraube** *f* = bearing screw
~**schuppen** *m* = storage shed
(~)**Silo** *m* = (storage) silo
~**sitz** *m* = bearing point
~**spiel** *n* = bearing clearance
~(**stätte**) *n*, (*f*), Vorkommen *n*, Vorkommnis *n*, Fundstätte = deposit
~**stätte** *f*, Erdöl~ = (oil) reservoir
Lagerstätten|druck *m* [*Erdgas; Erdöl*] = reservoir pressure
~**druckgefälle** *n* = reservoir pressure gradient
~**forschung** *f* → Prospektion *f*
~**forschung** *f* **im Küstenvorland** → Aufsuchen *n* ~ ~
~**kunde** *f* [*Sie betrachtet die Minerallagerstätten unabhängig von ihrer technischen Verwertbarkeit*] = science of mineral deposits
~**lehre** *f* [*Sie umfaßt die Minerallagerstätten, die zur Zeit oder in naher Zukunft technisch verwertbar sind*] = economic geology
~**technik** *f*, Erdöl~ = (oil) reservoir engineering
Lager|stuhl *m*, Lagerbock *m* [*Lager*] = bearing chair, ~ block, ~ stool
~**stütze** *f* **der Schlämmseiltrommel**, Haspelpfosten *m* [*pennsylvanisches Seilbohren*] = knuckle post
~**tank** *m*, Lagerbehälter *m* = storage tank
~**turm** *m* = storage tower
Lagerung *f* → (Auf)Speicherung
~ (Geol.) = bedding
Lagerungs|dichte *f* = compactness
~**störung** *f* → Krustenbewegung
~**übersicht** *f* = locator system [*for yard(s) and/or shed(s)*]
~**verhältnisse** *npl* (⚒) = structural conditions
Lager|unterkünfte *fpl* → Arbeiterlager
~**verankerung** *f* = bearing saddle
~**vorspannung** *f* = bearing preloading

~**weißmetall** *n* = anti-friction metal, white ~, babbitt ~, Babbitt's ~
~**welle** *f* = bearing shaft
~**wölbung** *f* = camber of bearing
~**zapfen** *m* = bearing journal
Lagune *f* = lagoon
Lagunen|riff *n*, Ringriff, Atoll *n* = atoll
~**sand** *m* = lagoon sand
Lahnung *f* [*ins Meer hineingebauter Damm*] = sea embankment
Laibung *f* → Leibung
Lakunar *n* = lacunar [*A ceiling made up of sunken panels*]
Lamelle *f*, Vorbauabschnitt *m* [*Brückenbau*] = cantilever(ed) segment
~ → Gurt~
Lamellen|-Bandaufgeber *m* → Aufgabeplattenband *n*
~**-Bandbeschicker** *m* → Aufgabeplattenband *n*
~**-Bandspeiser** *m* → Aufgabeplattenband *n*
~**bremse** *f* = disc brake; disk ~ (US)
~**-Fenster** *n* = louvred window
~**filter** *m*, *n* = laminated filter
~**kühler** *m* = tube and fin radiator
~**kupp(e)lung** *f*, Mehrscheibenkupp(e)lung = multiple disc clutch; ~ disk ~ (US)
~**rohr** *n* → Rippenrohr
~**stempel** *m* (⚒) = lamellar (pit) prop, ~ (~) post, ~ mine ~
Laminar|bewegung *f*, laminares Fließen *n*, Gleiten *n* [*Flüssigkeitsbewegung, bei der sich alle Teilchen in nebeneinander liegenden Schichten bewegen, die sich weder durchsetzen noch vermischen*] = laminar flow, streamline ~, viscous ~
~**-Grenzschicht** *f* = laminar boundary layer
Laminarie *f*, Riementang *m* = laminaria
Lampe *f* = lamp
Lampen|marke *f* (⚒) = (lamp) check
~**öl** *n* → Petroleum *n*
~**schacht** *m* = lamphole
~**stube** *f* (⚒) = lamproom

~stubenarbeiter m (✖) = lampman, lamproom man
Land|anschluß m der Buhne, Buhnenwurzel f = groin root (US); groyne ~ (Brit.)
~asphalt m [*Trinidad*] = land asphalt, ~ pitch, shore ~
~auffüllungsvorhaben n, Landauffüllungsprojekt n = land fill project, ~ ~ scheme
~aufhöhung f = raising of an area
~behandlung f = land treatment
~beschaffung f = acquisition of land
~drän m = land drain, field ~, agricultural (pipe) ~
(Lande)Anflug m = approach
Landebahn f = landing runway
~befeuerung f = landing runway lighting
~grundlänge f = basic landing runway length
Lande|brücke f → Anlegebrücke
~hilfe f = landing aid
landeinwärts = upstream from the sea, inland
Landeis n, Küsteneis = shore ice
Lande|kreuz n = landing cross
~kurssender m [*ILS-Anflugschema*] = localizer (beam) radio transmitter
~kurssenderantenne f = localizer antenna
Landen n, Landung f [*Flugzeug*] = landing
Landenge f, Isthmus m = isthmus
Landenteignung f = expropriation of land
Landeplatz m, Flugplatz = airfield, landing ground
Landesplanung f = state planning, provincial ~
Landestreifen m = landing strip
Land|fläche f der Erde = land surface of the earth
~flughafen m = terrestrial airport
landgängige Erdsauganlage f → Erdsauganlage zu Lande
Land|gemeinde f = rural community
~gewinnung f, Neu~ = land reclamation, reclamation of land
~ durch Eindeichung = ~ by enclosure
~ durch Aufschüttung = ~ by filling
~herrichtung f [*Landwirtschaft*] = land forming and smoothing
landkartenmäßig erfassen = to map
Land|kies m = bank gravel
~klima n = land-controlled climate
~krankenhaus n = rural hospital
ländlicher Weg m, Wirtschaftsweg, landwirtschaftlicher Weg = farm track
~ Wegebau m, Wirtschaftswegebau = construction of farm tracks
Land|marke f, festes Seezeichen n = land mark
~pfeiler m = land pier
~poller m, Verholpoller = checking bollard
landschaftlicher Aussichtspunkt m = beauty spot
~ Erdbau m = earthmoving for landscape purposes
Landschaftsbild n = scenery
landschaftsgärtnerisch gestalten = to landscape
landschaftsgärtnerische Arbeiten /pl = landscaping work
Landschafts|gestalter m = landscape architect
~gestaltung f = landscaping, landscape treatment (or development)
~gestaltung f an Straßen = roadside improvement
~pflege f = landscape preservation
~schutz m = landscape protection
landseitige Brückenöffnung f, landseitiges Brückenfeld n = land-side(d) bridge span, landward ~ ~, ~ ~ opening
Landstraße f = (public) highway, (~) road, rural ~
~, Provinzstraße = provincial road, ~ highway
(Land)Straßen|beleuchtung f = highway lighting, road ~
~leuchte f = road(way) lantern, highway ~ (Brit.); ~ luminaire (US); ~ lighting fixture
Landungs|brücke f → Anlegebrücke

Landungsplatz — Längsbewegung

~**platz** m = unloading wharf, discharge ~
~**ponton** m = embarkation quay, embarcation ~
~**steg** m → Anlegebrücke f
Land|verdunstung f = evaporation from land surfaces
~**verkehr** m = surface transport
~**vermessung** f = land surveying
~**wind** m, ablandiger Wind = offshore wind
landwirtschaftlich genutzte Fläche f, landwirtschaftliche Anbaufläche = agricultural land, ~ area
landwirtschaftlicher Wasserbau m = agricultural hydraulic engineering ~ Weg m = ländlicher ~
landwirtschaftliches Betriebsgebäude n = agricultural building
Landwirtschaftswasser n = non-potable water for agricultural purposes
Landzunge f, Nehrung f = spit (of land)
Lang|band n [*Bautischlerei*] = crossgarnet
~**bundhaspel** f, m, Bandstahlhaspel für Langbunde = strip steel coiler for long coils
Länge f **über alles**, Gesamtlänge = total length, over-all ~
Längen|änderung f = length variation
~**einheit** f, Längenmaß n = unit of length
~**fehler** m [*bei Messungen*] = linear error
~**maß** n, Längeneinheit f = unit of length
~**maßstab** m = length scale
~**messung** f = length measurement
~**profil** n → Längenschnitt m
~**schnitt** m, Längsschnitt, Längenprofil n, Längsprofil, Höhenplan m = longitudinal section, ~ profile
~**verstellbarkeit** f [*Stempel*] (✼) = extensibility
Langfahrt f = long trip
langfristige Hochwasservorhersage f = long-term flood prediction
langfristiger Plan m = long-range plan

Lang|hobel m, Rauhbank f [*DIN 7218*] = adjustable iron fore plane
~**hubniethammer** m = long stroke riveting hammer
~**hubpumpe** f = long stroke pump
~**lochziegel** m, Langlochstein m [*Löcher parallel der Ziegellänge, in der Mauer gleichlaufend mit der Lagerfuge. DIN 105*] = brick with horizontal perforations
~**mahd** f → Längsreihe f
~**mahdplanierungsschablone** f → Schwadenglätter m
~**masche** f = long mesh
~**maschengewebesieb** n = long mesh woven wire screen
~**maschen(sieb)gewebe** n = long mesh wire cloth
~**pfahl** m → freistehender Pfahl
längs geneigte Ebene f → Längsschrägaufzug
Längs|abstreifer m [*Betondeckenbau*] = longitudinal strike-off (blade)
~**achse** f = longitudinal axis
langsamabbindend [*Verschnittbitumen*] = slow-curing
Langsambinder m = slow-curing liquid asphaltic material, SC
langsambrechend, stabil [*Bitumenemulsion*] = slow-setting, stable, slow-breaking
Langsam|filter m, n = slow filter
~**sandfilter** m, n = slow sand filter
~**schleppen** n [*einer Baumaschine als Anhängerausführung*] = slow-(speed) towing, ~ trailing
~**-Versuch** m, Langsam-Prüfung f, Langsam-Probe f [*zur Bestimmung der Reibungsfestigkeit bindiger Böden*] = drained (shear) test, slow ~
Längs|arbeitsfuge f = longitudinal construction joint
~**armierung** f → Längsbewehrung
~**bauwerk** n → Längswerk
~**belüftung** f → → Tunnelbau m
~**bewegung** f **des Stützpunktes** [*Auflager*] = longitudinal displacement of the bearing point

Längsbewehrung — Längsströmung

~bewehrung f, Längsarmierung, Längs(stahl)einlagen fpl = longitudinal reinforcement, ~ steel
~biegemoment n, Längsbiegungsmoment = longitudinal bending moment
~binder m = longitudinal truss
~bohlenfertiger m = longitudinal (or bullfloat) finishing machine, ~ floating ~
Langschiff n [Kirche] = nave
Längs|dehnung f = longitudinal strain
~dichtung f [Wehr] = longitudinal seal
~druck m = longitudinal pressure
~düne f, Strichdüne = linear dune
~einlagen fpl = Längsbewehrung f
Langseitenschifter m = longitudinal jack rafter
Längs|entlüftung f → → Tunnelbau m
~fuge f = longitudinal joint
~fuge f [Betonstraßenbau] = longitudinal joint, lane ~
~gefälle n, Gradiente f, Längsneigung f = longitudinal slope, (~) gradient
~kette f [Teil einer Reifenkette] = side-chain
~kraft f = longitudinal force
~mole f = lee breakwater
~moment n = longitudinal moment
~naht f [Schwarzdecke] = longitudinal joint
~naht f = longitudinal seam
~neigung f = Längsgefälle n
~neigungsfühler m [Schwarzdeckenfertiger] = grade sensor
~profil n → Längenschnitt m
~profil n durch die Flutwelle = instantaneous profile
~profil n eines Wasserlaufes mit halbtägiger Gezeit = longitudinal profile of a water course having semi-diurnal tides
~profilometer m = slopemeter
~profilzeichner m, Längsprofilograph m = longitudinal profilometer, ~ profilograph
~reihe f, Schwaden m, Längsmahd f, Langmahd, Streifhaufen m = windrow

~reihenverteilung f = windrowing
~richtung f = longitudinal direction
~rippe f = longitudinal rib
~riß m = longitudinal crack
~rißbildung f, Längsrissebildung = longitudinal cracking
~(rund)stab m, Haupt(rund)stab [Betonstahlmatte] = main (round) bar
~sammler m = longitudinal collector
~-(Schiffauf)Schleppe f, Längss(ch)lipp f = longitudinal slipway
~schlag m → Albertschlag
~schnitt m → Längenschnitt
~schnitt m 10-fach überhöht = longitudinal section with vertical dimensions drawn to 10 times larger scale
~-Schrägaufzug m, längs geneigte Ebene f [Schiffhebewerk] = inclined plane with longitudinally travelling caissons
~schrumpfriß m = longitudinal shrinkage crack
~schweißmaschine f = longitudinal welding machine
~schweißnaht f = longitudinal weld
~schweißung f = longitudinal welding
~-Schwingarm m [mit dem (Gleis)Ketten-Rahmen verschweißt] = diagonal brace [welded to the track frame]
~seite f eines Läufer(steins) = stretcher face, long ~
~spannglied n = longitudinal tendon
~spannung f = longitudinal stress
~stab m = longitudinal bar
~stab m → Längsrundstab
~stähle mpl → Längsbewehrung f
~(stahl)einlagen fpl → Längsbewehrung f
~stange f [Holzgerüst] = ledger [A horizontal pole, parallel to the wall in wooden scaffolding, lashed to the standards and carrying the putlogs. In tubular scaffolding similar terms are used, but different fixings]
~steife f [Kastenträger] = longitudinal (web) stiffener
~steifigkeit f = longitudinal stiffness
~strömung f [Meer] = longitudinal current

~träger m = longitudinal girder
Langstreckenförderbandanlage f zum Transport von Gesteinen, Ferntransportband n = "rock road"
Längs|verband m, Horizontalverband, horizontaler Querverband [*Träger*] = longitudinal bracing, horizontal ~
~**verformung** f = longitudinal deformation
längsverschiebbar = slideable in longitudinal direction, longitudinally traversable
Längs|verwerfung f, streichende Verwerfung (Geol.) = strike fault
~**vorspannung** f = longitudinal prestress
~**werk** n, (Leit-)Parallelwerk, Richtwerk, Streichwerk, Längsbau(werk) m, (n) = longitudinal structure
Längung f, Dehnung [*positiv*] = extension
Langzeit|probe f → Langzeitprüfung f
~**prüfung** f, Langzeitprobe f, Langzeitversuch m = test of long duration
~**verhalten** n = long-term behavior (US); ~ behaviour (Brit.)
~**versuch** m → Langzeitprüfung f
Lanzettbogen m → lanzettförmiger Spitzbogen
~**fenster** n = lancet window [*A narrow, sharply pointed window without tracery, set in a lancet arch*]
lanzettförmiger Spitzbogen, überhöhter ~, Lanzettbogen = lancet arch, acute ~
Laplace'sche Gleichung f = Laplacian equation
Läppen n = lapping
Lappen m [*T-Scharnier*] = long member
~ [*Ösenteil eines Scharniers*] = loop
Läppöl n = lapping oil
Lärchenschindel f = larch shingle
Lärm|bekämpfung f, Lärmabwehr f = noise suppression
~**belästigung** f, Lärmstörung = noise nuisance

~**belästigungsfrage** f, Lärmbelästigungsproblem n [*Einer Ortschaft durch Flugzeuge*] = community noise aspect
~**dämmung** f = noise insulation
~**dämpfer** m = noise abatement device
~**forschung** f = noise research
~**minderung** f = noise abatement
~**pegel** m = noise level
~**physik** f = noise physics
~**quelle** f = noise source
~**schutz** m = protection against noise
~**sperre** f = acoustical barrier [*aerodrome*]
~**störung** f, Lärmbelästigung = noise nuisance
Larssen|-Profil n, Larssenbohle f = Larssen section
~**(-Stahl)spundwand** f = Larssen steel sheet pile wall, ~ ~ ~ piling
Lasche f = butt strap
Laschen|breite f = width of butt strap
~**dicke** f = thickness of butt strap
~**nietung** f, Laschennietverbindung = riveted butt joint
~**nietverbindung** f, Laschennietung = riveted butt joint
~**stoß** m, Laschenverbindung f = butt strap joint
~**verbindung** f, Laschenstoß m = butt strap joint
Last f **je Längeneinheit** = surcharge per unit of length
~ **pro Achse** → Achslast
~**abfall** m = load drop
~**angriff** m → Belasten n
~**angriffpunkt** m = point of load application
~**anhänger** m, LKW-Anhänger = truck-trailer (US); lorry-trailer (Brit.)
~**annahme** f, rechnerisch vorgesehene Last f, angenommene Last = design load, assumed ~
~**anzeiger** m = load indicator
~**arm** m [*Der Teil eines Hebels zwischen Drehpunkt und Lastangriffpunkt*] = work arm

Lastart — Last(kraft)wagenbagger

~art *f* = type of load

(~)Aufbringen *n* → Belasten

(~)Aufbringung *f* → Belasten *n*

~aufnahme *f* = load-bearing capacity

~aufteilungsverfahren *n* des Bureau of Reclamation = trial load method of analyzing arch dams

~aufzug *m* → (Lasten)Aufzug

~bedingung *f* = load condition

~beiwert *m*, Lastzahl *f* = load coefficient

~bügel *m* mit zwei Haken = double lifting hooks

~-Dehnungskurve *f* = load-strain curve

~eintragen *n* → Belasten

~eintragung *f* → Belasten *n*

Lastenanordnung *f*, Lastenschema *n* = loading diagram

(Lasten)Aufzug *m*, Materialaufzug, Lastaufzug = goods elevator (US); ~ lift (Brit.)

~ mit Personenbegleitung, Materialaufzug ~ ~ = attendant-operated goods elevator (US); ~ ~ lift (Brit.)

~mast *m*, Materialaufzugmast = hoisting mast

~raum *m*, Materialaufzugraum = goods elevator room (US); ~ lift ~ (Brit.)

Lasten|bewegung *f*, Lastenumschlag *m* = load handling, handling of load(s)

~greifer *m*, Steinklammer *f* [*Betonsteinindustrie*] = loading clamp

~gruppe *f* = group of loads

~heft *n* → Bauleistungsbuch *n*

~schema *n*, Lastenanordnung *f* = loading diagram

~summe *f*, Summe aller Lasten, Gesamtlast *f* = total (of all the) loads, ~ load

~umschlag *m* → Lastenbewegung *f*

~zug *m* [*Baustatik*] = train of loads, load train

~zug *m* [*Brücke*] = train of loading

Lastfahrzeug *n* [*als allgemeiner Ausdruck und nicht auf den LKW beschränkt*] = load vehicle

Lastfahrzeug *n*, LKW *m*, Last(kraft)wagen *m* = lorry (Brit.); (auto) truck (US)

~waage *f* = scale for load vehicles

~waage *f*, LKW-Waage, Last(kraft)wagenwaage = lorry scale (Brit.); (auto) truck ~ (US)

Last|faktor *m* = load factor [*i. e. the ratio of the ultimate strength of the beam or slab to its working load*]

~fall *m* [*Baustatik*] = loading case, case of loading

~feld *n* = area of loading, loaded area, field of load

~fortleitung *f* = transmission of load

~freiheit *f* = absence of load

~gabel *f* → Hubgabel

~gesetz *n* = rule for determining loads

~grenze *f* = safety limit

~größe *f* = magnitude of load

~haken *m* = lifting hook, load ~

(~)Hakengeschwindigkeit *f* = lift(ing) speed

~hebedampfmaschine *f* = hoisting steam engine

~hebemagnet *m* → Lastmagnet

~hebetrommel *f*, Lasthubtrommel = load drum

~kahn *m* = goods-carrying barge

~kahnhöhe *f* über Wasserspiegel = overall height of a goods-carrying barge above water level

~kahnkai *m* = quay for goods-carrying barges

~katze *f*, Laufkatze = crab

~katzenausleger *m* → Fahrkatzenausleger

~katzenkran *m*, Laufkatzenkran = crab crane

~komponente *f* = load component

Last(kraft)wagen *m*, LKW *m*, Lastfahrzeug *n* = (auto) truck (US); lorry (Brit.)

~ zur Aufnahme mehrerer Chargen → Chargen-Last(kraft)wagen

~anhänger *m*, LKW-Anhänger = truck trailer (US); lorry ~ (Brit.)

~bagger *m*, Aufbaubagger = lorry(-mounted) excavator (Brit.); truck(-mounted) ~ (US); fast-travel ~

Last(kraft)wagenfahrer — Lastwagenanhänger

~fahrer *m* = trucker
~-Grabenbagger *m*, (Auto-)Aufbau-Grabenbagger = truck(-mounted) ditcher (US); lorry(-mounted) ~ (Brit.); fast-travel ~
~-Hydraulikkran *m* → Hydraulik-Aufbaukran
~-Hydrokran *m* → Hydraulik-Aufbaukran
~kran *m*, Aufbaukran = lorry(-mounted) crane (Brit.); truck(-mounted) ~ (US); fast-travel ~
~lader *m*, (Auto-)Aufbaulader = truck(-mounted) loader (US); lorry(-mounted) ~ (Brit.); fast-travel ~
~waage *f* → Lastfahrzeugwaage
Last|linie *f* = load line
~magnet *m*, Lasthebemagnet, Hubmagnet = (electric) lifting magnet(o), crane ~
~meßring *m* = proving ring [*Load measuring device which relies for its operation on the elastic deformation, due to loading along a diameter, being indicated by a dial gauge usually situated on the same diameter*]
~meßvorrichtung *f* = load measuring device
~moment *n* = load moment
~mulde *f* → Mulde
~platte *f* = circular plate, bearing ~, circular bearing ~, loading ~
~plattendurchmesser *m* = plate size, ~ diameter
~plattenversuch *m*, Lastplattenprobe *f*, Lastplattenprüfung *f*, Platten(belastungs)versuch, Plattenbelastungsprobe, Plattenbelastungsprüfung, Plattendruckversuch, Plattendruckprobe, Plattendruckprüfung = plate(-loading) test, „k" ~, plate (load) bearing ~; Navy method (US)
~riß *m* = crack resulting from load
~-Ruhekurve *f*, Lastsetzungslinie *f* [*Pfahl*] = load/settlement graph, ~ curve, ~ diagram
~schaltgetriebe *n* = power shift
~scheide *f* = load separation point
~schwankung *f* = load variation

~senkungsschreiber *m* (✿) = load-yield recorder
~setzungskurve *f*, Lastsenkungskurve = load/settlement graph, ~ curve, ~ diagram
~setzungslinie *f*, Last-Ruhekurve *f* [*Pfahl*] = load/settlement graph, ~ curve, ~ diagram
~spannung *f* [*durch äußere Last entstehende Spannung*] = load stress
~spiel *n*, Lastwechsel *m* = load cycle, ~ repetition
~spielanzahl *f* = number of repetitions of load
~stärke *f* = intensity of load
~stellung *f* = load position
~streifen *m* = load strip
~stufe *f* = load stage, ~ increment
~system *n* = system of loads
~tabelle *f* = load table
(last)tragende Decke *f*, belastete ~ [*Flugplatz- und Straßenbau*] = load-bearing pavement, ~ surfacing
~ Mauer *f*, belastete ~ = load-bearing wall
~ Wand *f*, belastete ~ = load-bearing partition
Last|trommel *f*, Lastwindentrommel = winch drum
~übertragung *f* = load transfer(ence), transfer(ence) of load
~umstellung *f* = varying load
~-Verformungskurve *f* = load deformation curve
~verteiler *m* [*Energieversorgung*] = load dispatcher
~verteilung *f* = load distribution
Lastverteilungs|fundament *n* = load-distributing foundation, load distribution ~
~kurve *f* = load distribution curve, load-distributing ~
~schema *n* = [*Straßenentwurf*] = weight distribution pattern
~vermögen *n* = load spreading ability
Lastwagen *m*, Lastkraftwagen, LKW *m*, Lastfahrzeug *n* = (auto) truck (US); lorry (Brit.)
~anhänger *m* → Lastkraftwagenanhänger

40*

Lastwagen-Hydraulikkran — Laufbahn 628

~-Hydraulikkran *m* → Hydraulik-Aufbaukran
~-Hydrokran *m* → Hydraulik-Aufbaukran
~kipper *m*, LKW-Kipper = tipping truck (US); ~ lorry (Brit.)
~kran *m* → Autobagger *m*
~waage *f* → Lastfahrzeugwaage
~zug *m* → Lastzug
Last|wechsel *m* → Lastspiel *n*
~wechselanzahl *f* = number of load cycles, ~ ~ ~ repetitions
~wegkurve *f*, Charakteristik *f* eines Stempels, Kennlinie *f* eines Stempels (⚒) = load-yield curve, resistance-yield ~
~wegnahme *f* = removal of load, load removal
~winde *f* = winch
~(winden)trommel *f* = winch drum
~zahl *f*, Lastbeiwert *m* = load coefficient
~zug *m*, LKW-Zug, Lastwagenzug = full trailer combination
Lasur *f* = glaze [*Semi-transparent film of coloured varnish applied over a non-absorbent surface so that it does not conceal the colour of the undercoating completely*]
lateinisches Kreuz *n* [*mit längerem Unterarm*] = Latin cross
latente Energie *f* → Energie der Lage
~ Wärme *f* = latent heat
Lateral|erosion *f* → Seitenschurf *m*
~kanal *m*, Seitenkanal = lateral canal
Laterit *m* = laterite
~gel *n* = laterite gel
lateritischer Ton *m* = lateritic clay
Laterit|kies *m* = lateritic gravel
~lager *n*, Lateritvorkommen *n*, Lateritfundstätte *f*, Lateritvorkommnis *n* = laterite deposit
~splitt *m* = laterite chip(ping)s
Laterne *f* [*auf einem Dach*] = lantern (light)
Laternen|dach *n* = lantern (light) roof
~öffnung *f*, Auge *n* der Kuppel, Kuppelauge = lantern opening
~öl *n* = burning oil, signal ~

~pfahl *m* mit Brunnen → Brunnenkandelaber *m*
Laterolog *n* = guard electrode log [*For the resistivity logging method with electrodes using an automatic focusing system*]
Latexkleber *m* = latex mastic
Latrine *f*, Torfstreuabort *m* = peat-litter privy, ~ earth closet, ~ latrine
~, Trockenabort *m* = privy, earth closet, latrine
Latrinen|eimer *m* = latrine bucket
~grube *f*, Senkgrube, Fäkaliengrube, Abtrittgrube, Abortgrube = f(a)eces pit, privy ~
Latte *f* = lath
Latteibrett *n* → Fensterbrett
Latten|gitter *n*, Rungen *fpl* [*am LKW*] = stakes
~kiste *f*, (Latten)Verschlag *m* = crate
~pegel *m*, Skalenpegel = (staff) gage [*A staff graduated to indicate the elevation of a water surface*]
~tür *f* = batten door, ledged ~
(~)Verschlag *m*, Lattenkiste *f* = crate
~zaun *m*, Staket *n* = pale fencing, ~ fence
Latthammer *m*, Spitzhammer = scabb(l)ing pick, ~ hammer
Lattung *f*, Dach~ = roof lathing, ~ battens
Laubengang *m*, offener Gang = access balcony [*A balcony intended to give access to a number of separate dwellings above the first stor(e)y*]
~haus *n*, Außenganghaus = gallery apartment building
Laub|kohle *f*, Blätterkohle = laminated coal, foliated ~, leaf ~
~torf *m*, Blättertorf = leaf peat
Lauf *m*, Gang *m* [*einer Maschine*] = operation, running
~ → Treppen~
~achse *f* = free axle
~bahn *f* [*Schütz(e)*] = track
~bahn *f* [*Lager*] = race

Laufbahnträger — Laufrollenmantel

~bahnträger *m* [*Kran*] = runway girder
~band *n*, Gummi(gleis)kette *f* = rubber crawler
~bohle *f* = walk plank, run(way) ~
~breite *f* [*Treppe*] = flight width
~brett *n* = walk board, run(way) ~
~brunnen *m* = flowing well, running ~
~büchse *f*, Zylinder~ = cylinder liner
laufende Produktion *f*, Massenfertigung *f* = mass production
~ Unterhaltung *f*, Wartung = routine maintenance
laufender Draht *m* [*Kabelspinnen beim Brückenbau*] = live wire
laufendes Meter *n*, Laufmeter, ml = linear metre, Lin. M. (Brit.); linear meter (US)
Läufer *m* (⚒) = runner
~, ~stein *m* = stretcher
~, Rotor *m*, Verteilerfinger *mpl* [*Motor*] = rotor
~ = pump rotor
~, Walze *f*, Koller *m* [*Kollergang*] = runner
~mühle *f*, Kollergang *m*, Kollermühle = pan grinder, grinding mill of the edge runner type, edge-runner
~rute *f*, Mäkler *m*, Laufrute *f* = leader
~rutentrommel *f* → Mäklertrommel
~schicht *f* = stretching course
~(stein) *m* = stretcher
~verband *m*, Schornsteinverband = stretching bond, chimney ~
~welle *f* = rotor shaft
Lauffläche *f* [*Motor*] = journal
~ [*Reifen; (Gleis) Kette; Rad*] = tread
Laufflächen|gummi *m* = tread rubber
~material *n* = tread material
~profil *n* → Reifen~
Lauf|freiheit *f* = freeness [*of a drive shaft*]
~gang *m* = gangway, walkway [*A narrow walking way designed to facilitate access, operation or maintenance, e.g. of a roof or machinery*]
~gewicht *n*, Einstellgewicht = poise weight, sliding ~, jockey ~, moving poise
~gewichtsbalken *m* = poise beam [*for each material to be weighed*]
~katze *f*, Fahrkatze, (Last)Katze, Krankatze = crab
~katzenausleger *m* → Fahrkatzenausleger
~katzenkran *m*, Lastkatzenkran, Fahrkatzenkran = crab crane
~(kraft)werk *n*, Laufwasserkraftanlage *f*, Niederdruck(-Wasserkraft)anlage, Stromkraftwerk, Flußkraftwerk = run-of-river scheme, low-head power plant, river power plant, stream power plant
Laufkran *m* = overhead crane; (~) travelling ~ (Brit.); (~) traveling ~ (US)
~träger *m* = overhead crane girder; travelling ~ ~ (Brit.); traveling ~ ~ (US)
~winde *f* = overhead crane winch
Lauf|kranz *m*, Rollenkranz = circular roller path
~kranz *m* [*Drehbrücke*] = live ring
~länge *f* [*Treppe*] = flight length
~linie *f*, Gehlinie, Ganglinie, Teilungslinie [*Treppe*] = walking line
~meter *n* → laufendes Meter
~rad *n* = non-driven wheel
~rad *n*, Laufrolle *f* [*Drehbrücke*] = (travel(l)ing) roller
~rad *n* [*(Gleis) Kette*] = truck wheel (US); tread ~
~rad *n* [*Turbine*] = (wheel) runner, rotor [*sometimes also called "wheel"*]
~radschaufel *f* [*Turbine*] = runner blade, (wheel) vane, rotor blade
~ring *m*, Lagerring, Kugel(lager)ring = bearing race
Laufrolle *f* [*(Gleis) Kette*] = tread roller; truck ~ (US)
~, Laufrad *n* [*Drehbrücke*] = (travel(l)ing) roller
~ = castor
Laufrollen|dichtung *f* [*Gleiskette*] = roller seal
~körper *m* [*Gleiskette*] = roller rim
~lagerbock *m* [*Gleiskette*] = end collar
~mantel *m* [*Gleiskette*] = roller shell

Laufrollenwelle — Leckwasserverlust

~welle *f* [*Gleiskette*] = roller shaft
Lauf|rute *f*, Läuferrute, Mäkler *m* = leader
~rutentrommel *f* → Mäklertrommel
~schaufel *f* [*Überdruckturbine*] = runner vane, stay ~
~schicht *f* → Deckschicht [*Fußboden*]
~schiene *f* = running rail
(~)Steg *m* = walkway, duckboard
~stellung *f* [*Motor*] = run position
~streifen *m*, Protektor *m* = sidewall protection [*pneumatic tyre*]
~wasserkraftanlage *f* → Lauf(kraft)werk *n*
~werk *n* → Raupenfahrwerk
~werk *n* → Laufkraftwerk
~widerstand *m* = rolling resistance
~zeit *f* = travel time
~zeitkurve *f*, Wegzeitkurve *f* = travel time curve
~zeitmessung *f* = transit-time measurement
(Lauge)Fallrohr *n* [*Gefrierschachtbau*] = brine pipe
Laugen *n* = neutralizing
Laughlinfilter *m*, *n*, Magnetitfilter = Laughlin filter
Läuterung *f* = refining
Laut|stärke *f*, Stärke eines Schalles = loudness (of sound), ~ ~ tone
~stärkemesser *m* = noise meter [*measurement of loudness*]
Lava *f* = lava
~bombe *f*, vulkanische Bombe = volcanic bomb
~erstarrung *f* = solidification of lava
~fluß *m* = lava flow
~kies *m*, poröse Lava *f* = foamed lava gravel
~krotze *f* → Lavaschlacke *f*
Lavalverfahren *n* = Laval acid treatment
Lava|schlacke *f*, Lungstein *m*, Lavakrotze *f*, Schaumlava *f*, vulkanische Schlacke, poröse Lava [*Basaltische feinporige bis blasige vulkanische Auswurfmasse. „Lavalit" ist ein geschützter Handelsname für gebrochene Lavaschlacke, die in verschiedenen Körnungen aufbereitet ist*] = scoria(ceous lava), foamed lava
~stromsee *m* = lake ponded up by lava
Lawine *f* = avalanche
Lawinen|brecher *m* = avanlanche brake, ~ structure
~schutt *m* = avalanche débris, ~ cone [*The mass of material deposited where an avalanche has fallen, including snow, ice, rock, and all other objects which have been carried away by it*]
~schutz *m*, Lawinenverbau(ung) *m*, (*f*) = protection against avalanches
~verbau(ung) *m*, (*f*) → Lawinenschutz *m*
~wind *m* = avalanche wind
Lazarett *n* = military hospital
LBV-Stoff *m* → luftporenbildender Beton-Verflüssiger *m*
lebendige Düne *f*, Wanderdüne = blowing dune, inland-moving ~, marching ~, migratory ~, shifting ~, travel(l)ing ~, wandering ~, dune on the march
Lebensdauer *f*, Nutzungsdauer = working life, service(able) ~, useful ~
~ des freien Ausflusses einer Bohrung = flowing lift of a well
~schmierung *f* = life-time lubrication
Lebertorf *m* = liver peat
LECA-Block *m*, LECA-Stein *m* = LECA block [*LECA = lightweight expanded clay aggregate*]
leck = leaky
Leckanzeiger *m* = leakage indicator
leckdicht, lecksicher = leakproof
lecken = to leak
Leck|ölleitung *f* = bleed line
~prüfung *f*, Leckprobe *f*, Leckversuch *m* = leakage test
lecksicher, leckdicht = leakproof
Leck|stelle *f*, undichte Stelle = leak(age)
~suchgerät *n*, Lecksucher *m* = leakage detector, ~ locator
~wasserverlust *m* = waste of water by leakage

Lederdicht(ungs)|ring m = leather seal(ing) ring, ~ packing ~, ~ washer
~streifen m = leather seal(ing) strip, ~ packing ~
Leder|fett n = leather dubbing
~handschutz m = leather hand pad
~hartverputzen n [*Steinzeugindustrie*] = fettling
~leim m = leather glue, ~ cement, ~ adhesive
~membran(e) f = leather diaphragm
~schürze f = leather apron
~teer m = leather tar
~tür f = leather door
Ledigenheim n = bachelor quarter
Leer|-Fahrgeschwindigkeit f [*Radschrapper*] = empty return speed
~gespärre n, Leergebinde n, Leersparren mpl, Zwischensparren mpl = intermediate rafters
Leergewicht n, Taragewicht n = tare ~ [*Fahrzeug*] = unladen weight
Leerlauf m [*Maschine*] = idle run, idling ~; coasting (Brit.)
~drehzahl f = idle speed, idling ~
~düse f = idler jet
~reg(e)lung f [*Verdichter*] = no-load control
~turas m, Spannrad n = idler
leer|pumpen = to pump dry
~schöpfen [*eine Bohrung*] = to bale a hole dry, ~ ~ ~ well ~, ~ ~ ~ bore ~, ~ ~ ~ boring ~
Leerseil n → Rück(hol)seil
~umlenkrolle f [*Tiefen- und Langstreckenförderer*] = backhaul-line guide block
Leersparren mpl → Leergespärre n
Legehaken m [*für Dränrohre*] = land-drain hook, ~ layer
Legen n **der Naht** [*Schweißtechnik*] = making the weld
~ von Grassoden, ~ ~ (Rasen)Soden = planting sod
Legende f, Zeichenschlüssel m, Zeichenerklärung f = key, legend
legiertes Motorenöl n, HD-Öl, HD-Motorenöl = heavy-duty oil
Legierung f = alloy

Legierungs|baustahl m = alloy structural steel
~stahl m = alloy steel, special ~
~zusatz m = alloying addition
Lehm m = loam
~bau m → Lehmstampfbau
~baustein m; Luftziegel m [*Fehlname*] = adobe
~boden m = loamy soil
~dichtung f = loam seal(ing)
~heide f = loam heath
~hütte f = mud hut
~keil m = loam wedge
~kern m [*Staudamm*] = loam core
~mergel m = loamy marl
~raspler m = loam cutter
~schindel f = loam shingle
~schlag m, Tonschlag, Puddle m, Lettenschlag = puddle clay, pug, clay puddle [*Plastic clay used for waterproofing. It is used for lining ponds or ditches, in coffering or in cut-off walls to dams*]
~(schlag)ummantelung f, Ton-(schlag)ummantelung, Puddleummantelung, Umstampfung mit Lehm und Ton, Letten(schlag)ummantelung = puddle clay lining, clay puddle ~, pug ~
~(stampf)bau m = rammed loam construction
~stein m, Lehmziegel m = loam brick
~straße f = loam road
~wand f = loam wall
~wüste f = loam desert
Lehranstalt|bau m, Bau von Lehranstalten = educational building
~gebäude n = educational building
Lehr|bau m → Lehrgebäude n
~baustelle f = training site
~becken n [*Hallen(schwimm)bad*] = teaching pool
~bogen m [*Abgestützte Schalung auf der ein Bogen oder Gewölbe gebaut werden*] = arch formwork
~buch n = text book, instruction ~
Lehre f = ga(u)ge
~ [*Lehrling*] = apprenticeship
~ **der Festgesteine**, Petrographie f = petrography

Lehr|film *m* = training film
~flügel *m* = teaching wing
~gang *m* = course
Lehr|gebäude *n*, Lehrbau *m* [*Schulbau*] = teaching block, classroom ~, lecture room ~
~gerüst *n*, Rüstung *f* = falsework [*Falsework may be defined as the temporary structure necessary to support portions of a permanent structure during its erection; hence falsework actually includes concrete formwork, but in practice it is taken to mean only that part which is the supporting structure on which the formwork rests*]
~gerüstabbau *m* → Abbau eines Lehrgerüstes
~gespärre *n* = guiding rafters
~lingsausbildung *f* = apprenticeship training
~lingswerkstatt *f*, Lehrlingswerkstätte *f* = apprentice (work)shop
~schwimmbecken *n* = training pool
~schwimmhalle *f* = training pool hall
~sparren *m* = guiding rafter
~werkstatt *f*, Lehrwerkstätte *f* = training (work)shop
Leibung *f*, Laibung = reveal [*That part of the side of an opening for a window or door which is between the outer edge of the opening and the frame of the window or door*]
Leibungsdruck *m*, Lochleibungsdruckfestigkeit *f*, Lochaufweitung *f* = bolt-bearing property
leichte Feinstoffe *mpl* [*Beton*] = light-weight fines
leichter Aufreißer *m* **mit Reißschenkeln und Zähnen** = ripper-scarifier
~ Gebirgsschlag *m* (⚒) = bump
~ lockerer Boden *m* [*mit Schaufel oder Spaten lösbar*] = loose ground
~ (Stahl)Gitterträger *m* = open web steel joist
leichtes Heizöl *n* = light fuel oil
leichteste Spaltrichtung *f* [*Gestein*] = rift of rock
Leicht|bär *m* = light pile hammer
~bau *m* = lightweight construction
~(bau)aufzug *m* = light hoist
~baudach *n* = light roof
~bau(dreh)kran *m*, Leicht-Turmdrehkran = light (mono)tower crane, ~ rotating tower crane, ~ revolving tower crane
~(bau)platte *f* = light-weight building board
~bausilo *m*, Behelfssilo = fence silo
~baustoff *m* = light construction material
~bauweise *f* = light construction method
~bauwinde *f* = light construction winch
~benzin *n* = low boiling naphtha
Leichtbeton *m* = light-weight concrete
~-Baustein *m* → Leichtbetonstein
~fertigteil *m,n* = precast light-weight (concrete) unit
~-Hängedach *n* = suspended roof of light-weight concrete
~-Hohlblock(stein) *m*, Leichtbeton-Hohlkörper *m*, Leichtbeton-Hohlstein, Hohl(block)stein aus Leichtbeton = light-weight concrete hollow block
~kern *m* = light-weight concrete core
~körper *m* → Leichtbetonstein *m*
~-Mischer *m*, Leichtbeton-Mischmaschine *f* = light-weight concrete mixer
~stein *m*, Leichtbetonblock(stein) *m*, Leichtbetonkörper *m*, Leichtbeton-Baustein = light-weight (precast) concrete block
~träger *m* = light-weight concrete girder
~-Vollstein *m*, Leichtbeton-Vollblock(stein), Leichtbeton-Vollkörper *m* = light-weight (precast) concrete solid block
~wand *f* = light-weight concrete wall
~wandbauplatte *f* = light-weight concrete wall slab
~zuschlag(stoff) *m* = light-weight aggregate
~zuschlag(stoff)werk *n* = light-weight (concrete) aggregate works

~-Zweikammer-Hohlblock(stein) *m*, Leichtbeton-Zweikammer-Hohlkörper *m*, Leichtbeton-Zweikammer-Hohlstein = light-weight-concrete two-cell hollow block, ~ ~ ~ tile, ~ ~ pot

Leichtblock(stein) *m*, Leichtstein = light block

Leichtbohr|hammer *m* = light-weight rock drill

~wagen *m* = light wagon drill

Leicht|dach *n* = light-weight roof

~dämmbeton *m* = light-weight insulating concrete

Leichterhafen *m* = bunder [*India*]

Leichtfahrbahn *f*, Stahlleichtbau *m*, orthotrope Platte *f*, Stahlblechfahrbahn, mitwirkende Stahlfahrbahn, orthotroper (Brücken)Überbau *m*, orthotrope (Brücken)Tafel *f*, orthotrope (Brücken)Platte *f* [*„orthotrop" ist eine Abkürzung für „orthogonalanisotrop" und besagt, daß die Steifigkeit der Platte quer und längs zur Brückenachse verschieden ist*] = light gauge carriageway, orthotropic(al) plate, fine meshed carriageway grid, light gauge decking, orthotropic (bridge) deck(ing)

leichtflüssiger Teer *m* = high-viscosity tar

Leicht|gerüst *n*, Leichtrüstung *f* = light-weight scaffold(ing)

~(gerüst)ramme *f* = light type frame pile driving plant, ~ ~ ~ piling ~

~gestänge *n* [*Tiefbohrtechnik*] = light-weight drill tubing

~gewölbe *n* = light vault

~grobzuschlag(stoff) *m* = light-weight coarse aggregate

~-Hohlblock(stein) *m*, Leicht-Hohlkörper *m* = light-weight hollow tile, ~ ~ block, ~ pot

~-Hohlkörper *m*, Leicht-Hohlblock(stein) *m* = light-weight hollow tile, ~ ~ block, ~ pot

~kalkbeton *m* [*z. B. Turrit, Mikroporit, Ytong*] = light-weight lime concrete

~kalksandstein *m* = light sand-lime brick

~legierungs-Fachwerkträger *m* = light-alloy truss

~legierungs-Kastenträger *m* = light-alloy box girder, ~ ~ ~ beam

~materialschaufel *f* = light-material bucket

~metallbau *m* = light-metal construction

~metall-Bauprofil *n* = light-metal structural section

~metall-Fahrband *n* = light-metal elevating belt conveyor (or conveyer)

~metallfenster *n* = light-metal window

~metall-Legierung *f* = light-metal alloy

~metallprofil *n* = light-weight metal section

~metallrohrgerüst *n*, Leichtmetallrohrrüstung *f* = light-metal tubular scaffold(ing)

~öl *n* = light oil

~perlit *m* = popped perlite

~perlit-Aufbereitung *f* = perlite popping

~platte *f* → Leichtbauplatte

~profil *n* = light-weight section

~ramme *f* → Leichtgerüstramme

~rammgerüst *n*, Rammleichtgerüst = light type (pile) frame, ~ ~ piling ~

~rüttelbohle *f*, Leichtvibrierbohle, Leichtvibrationsbohle = light vibrating screed, ~ vibratory ~

~rüttelplatte *f*, Leichtvibrierplatte, Leichtvibrationsplatte = light vibrating plate, ~ vibratory ~

~spannbeton *m* = prestressed lightweight concrete

~spannbeton-Dachplatte *f* = prestressed lightweight concrete roof slab

~-Stahltafel *f* → Leicht-Stahlüberbau *m*

~-Stahlüberbau *m*, Leicht-Stahltafel *f*, Leicht-Stahlplatte *f* [*Brücke*] = lightweight steel (bridge) deck(ing)

~stein *m*, Leichtblock(stein) = light block

~stein m, Leichtziegel m = light brick
~strangpreßteil m, n = lightweight extrusion
~turmdrehkran m = light tower crane
~vibrationsbohle f → Leichtrüttelbohle
~vibrationsplatte f → Leichtrüttelplatte
~wand f, Trennwand = partition (wall)
~wandbauplatte f → Trennwandbauplatte
~winde f = light-type winch
~ziegel m, Leichtstein m = light brick
~zuschlag m für Konstruktionsbeton = structural light-weight aggregate
~zuschlagbeton m = light-weight aggregate concrete
~zuschlag-Konstruktionsbeton m = structural light-weight aggregate concrete
~zuschlag(stoff) m → Beton~
~zuschlag(stoff)-Werk n = light-weight-aggregate works
Leier f [zum Einwölben] = rotating templet
Leihbibliothek f = lending library
Leim|farbe f = (glue-)size distemper (paint), size-bound ~ (~)
~farbenanstrich m = distemper coating
Leinöl n = linseed oil
~anstrich m = linseed oil coating
~basis f = linseed oil base
~emulsion f = linseed oil emulsion
~firnis m [schnelltrocknendes Leinöl, hergestellt durch Zusatz von Sikkativlösungen in der Kälte] = bunghole oil
~firnis m [schnelltrocknendes Leinöl, hergestellt durch Zusatz von Trockenstoffen zum Leinöl durch Verkochen] = boiled (linseed) oil
~kitt m = linseed oil putty
~lösung f = linseed oil solution
~standöl n = linseed stand oil
Leinpfad m, Treidelweg m = towpath, towing path

Leistung f, Arbeits~ = performance, output, capacity
~, Motor~ = rated horsepower
~ in t/h m² nutzbare Siebfläche, spezifischer Siebdurchsatz m = rated throughput in t/h m² of effective screen area
Leistungs|bericht m über den Fortschritt der Bauarbeiten = progress report
~beschreibung f → → Angebot n
~fähigkeit f bei voller Sättigung [Straße] = saturation capacity
~fertiger m [im Gegensatz zum Versuchsmodell] = production (model) paver, ~ (~) finisher
~modell n = production model
~position f, kostenvergütete Position = pay item
~preistarif m [Energieversorgung] = demand charge rate (US)
~reserve f = undeveloped potential of production power, reserve power
~schild n = data plate
~soll n, installierte Leistung f = installed (nameplate) capacity
leistungsstark = powerful
Leistungs|tabelle f = performance table
~tafel f = output table
~vergleich m = performance comparison
~verzeichnis n → → Angebot n
~verzweigung f, Drehmomentverzweigung = torque division, ~ split
Leit|achse f, Lenkachse = steering axle
~baum m, Spurlatte f, (Schacht)Führung f (✂) = guide
~blech n → Ablenkblech
~bohle f = wale (US); waling (Brit.)
~damm [Wasserbau] = longitudinal embankment
~deich m, Hochwasserdeich = levee
~ebene f, Bezugsebene = datum surface
leiten [Bauarbeiten durch die Bauleitung der Bauherrschaft] = to supervise

Leiter *f* **mit Rückenschutz** = caged ladder
~**bohrmethode** *f* → schwedische ~
~**gerüst** *n* [*DIN 4411*] = ladder scaffold(ing)
~**haken** *m* = ladder hook
~**schacht** *m* (⚒) = footway
~**sprosse** *f* → Sprosse
~**treppe** *f* → eingeschobene Treppe
~**winde** *f* [*Winde zum Heben und Senken der Eimerleiter bei Eimerkettenbaggern*] = (bucket) ladder winch
Leitfähigkeit *f* = conductivity
leitfähigkeitsmodulierter Silizium-Leistungsgleichrichter *m* = high-voltage conductivity modulated silicon rectifier
Leit|flöz *n* (⚒) = leading seam
~**horizont** *m* (Geol.) → Bezugshorizont
~**insel** *f* = directional island
~**kegel** *m*, Absperrkegel, Gummihut *m* = rubber cone
~**kranz** *m* = stator
~**kranz** *m* [*Motor*] = nozzle ring
~**linie** *f* [*Fahrbahnmarkierung*] = dotted line
~**linie** *f* **am Fahrbahnrand** = side-of-pavement line
~**mauer** *f* [*Wasserbau*] = training wall, longitudinal ~
(~)**Parallelwerk** *n* → Längswerk
~**pflock** *m*, Leitpfosten *m* [*Straße*] = delineator
~**pfosten** *m*, Leitpflock *m* [*Straße*] = delineator
~**plan** *m* → Flächennutzungsplan
~**planke** *f* → Sicherheits~
~**planken-Waschmaschine** *f* = guard rail washing machine (or washer)
~**platte** *f* → Ablenkplatte
Leitrad *n* [(*Gleis*)*Kettenlaufwerk*] = front idler
~ **nach außen verkantet** [*Gleiskette*] = idler toe-out
~ ~ **innen verkantet** [*Gleiskette*] = idler toe-in
~**federung** *f* [*Gleiskette*] = idler travel
~**flansch** *m* [*Gleiskette*] = idler flange
~**lauffläche** *f* [*Gleiskette*] = idler tread
~**mittelflansch** *m* [*Gleiskette*] = idler center flange (US); ~ centre ~ (Brit.).
~**nabe** *f* [*Gleiskette*] = idler hub
~**pendelung** *f* [*Gleiskette*] = idler oscillation
~**seitenplatte** *f* [*Gleiskette*] = idler side plate
~**sturz** *m* [*Gleiskette*] = idler tilt
~**träger** *m* [*Gleiskette*] = idler support beam
Leit|rohrtour *f*, **erste Rohrfahrt** *f*, **Standrohr** *n* [*Tiefbohrtechnik*] = conductor string, surface ~
~**rolle** *f* **des Ausstoßers** = ejector guide roller
~**rollen-Drehzapfen** *m* = swivel sheaves pivot shaft
~**rollenträger** *m* = swivel sheaves support bracket
~**schaufel** *f* [*Kaplanturbine*] = guide vane
~**schaufelkranz** *m* = gate apparatus [*ring of gate vanes*]
~**schiene** *f*, Zwangsschiene = check rail, safety ~, side ~, guard ~
~**schiene** *f* = guide rail
~**schienengleis** *n* = track with check rail, ~ ~ safety ~, ~ ~ side ~, ~ ~ guard ~
~**schwelle** *f* = guard kerb (Brit.); ~ curb
~**seil** *n* [*Bagger*] = tagline
~**stein** *m*, Abweisstein, Prellstein = spur post
~**streifen** *m*, Beton~, Randeinfassung *f* aus Beton, (Beton) Randstreifen [*Straße*] = (concrete) marginal strip
Leitung *f* = line, conduit
Leitungs|brücke *f* = conduit bridge, line ~
~**draht** *m* = line wire
~**graben** *m* = service trench
~**kanal** *m* = duct for services, mains subway, service duct
~**kanal** *m*, (Abwasser)Kanal = conduit-type sewer
(~)**Netz** *n* = line network
~**prüfer** *m* [*Sprengtechnik*] = circuit tester

Leitungsquerschnitte — Lenktrieb

~querschnitte *mpl* des Wasserbaues = hydraulic conduit sections
~rohr *n* = conduit pipe, line ~
~verlust *m* [*Bewässerung*] = conveyance loss [*Loss of irrigation water from a conduit due to seepage, evaporation, or evapotranspiration*]
~wasser *n* = tap water
Leit|vorrichtung *f* [*Kreiselpumpe*] = guide passage
~wand *f* [*einfache Kammerschleuse*] = upper round head, guiding wall
~wand *f* → Prellwand
~werk *n* = piled fendering, timber staging (Brit.); fenders (US)
~werk *n* = training structure
~zeichen *n* = guidance sign
lemniskatenförmiger Läufer *m* [*Rootsgebläse*] = lobed impeller
Lenk|achse *f*, Leitachse = steering axle
~achsenlagerzapfen *m* = steering axle pivot pin
~anlage *f* = steering system, ~ mechanism
~anschlag *m* = anti-jackknife stop
~arm *m* = steering arm
lenkbar = steerable
Lenk|bremse *f* = steering brake
~bügel *m* = steering link
~bügellager *n* = link bolster
lenken = to steer
~, überwachen = to monitor
Lenken *n* = steering
Lenker|konstruktion *f* = wheel control linkage
~schubstange *f* = drag link
Lenk|gestänge *n* = steering linkage
~getriebe *n* = steering gear
~(gleis)kette *f* → Lenkraupe *f*
~griff *m* = steering handle
~hebel *m* = steering lever
~hilfe *f*, Kraftlenkung *f*, Servolenkung *f*, = powered steering, power-(-assisted) ~, steering booster
~hilfspumpe *f*, Servolenkpumpe = steering booster pump
~hilfszylinder *m* = booster steering cylinder

~hilfszylinderarm *m* = steering cylinder pivot pin
~hydraulik *f* = steering hydraulic system
~kette *f* → Lenkraupe *f*
~konsole *f* = steering console
~kupp(e)lung *f* = steering clutch
Lenkkupp(e)lungs|-Ausrücklager *n* = steering clutch release bearing
~betätigung *f* = steering clutch control
~gestänge *n* = steering clutch linkage
~steuerventil *n* = steering clutch control valve
~-Überdruckventil *n* = steering clutch relief valve
Lenk|mechanismus *m* = steering mechanism
~muffe *f* = steering sleeve
~mutter *f* = steering nut
~ölhydraulikpumpe *f* = steering (hydraulic) oil pump
~pedal *n* = steering pedal
~pumpe *f* = steering pump
Lenkrad *n* = steering wheel
~, Tastrad = guide wheel
~ mit versenkter Nabe = deep-dish steering wheel
~knopf *m* = wheel spinner
Lenk|raupe *f*, Lenk(gleis)kette *f* = steerable crawler
~rolle *f* = swivel-type castor
~säule *f* = steering column
~säulenkeil *m* = steering column shaft spline
~schenkel *m* = steering arm
~schnecke *f* = steering worm
~schneckenmuffe *f* = steering nut gear
~schubstange *f* = drag link
~segment *n* = steering sector
~seil *n* = steering cable
~stange *f* [*z.B. bei einem Bodenverdichter*] = steering bar
~steuerventil *n* = steering control valve
~stock *m* = steering arm block
~stoßdämpfer *m* = steering anti-kickback snubber
~trieb *m* = steering gear

Lenktriebwelle — Lichtbogenhobler

~triebwelle *f* = steering gear shaft
Lenkung *f* = steering system, steer
~, Lenken *n* = steering
~ mit 2 Untersetzungen = dual-ratio steering
Lenk|untersetzung *f* = steering ratio
~ventil *n* = steering (control) valve
~vorrichtung *f* = steering device
~walze *f*, Lenkwalzrad *n*, Lenkwalzenzylinder *m* = steerable roll(er wheel)
~welle *f* = steering (control) shaft
~winkelhebel *m* = steering bellcrank, intermediate ~
~zylinder *m* = steering cylinder
Lenzpumpe *f* = bilge pump
LEOBA-Spannglied *n* = LEOBA stressing apparatus [*Trademark*]
Leonardschaltung *f*, Ward-Leonard-Schaltung = Ward-Leonard system of variable voltage control
Lepidokrokit *m*, Rubinglimmer *m* [*hieß ursprünglich ,,Goethit''*] = lepidocrocite (Min.)
Lesbarkeit *f* = legibility
Lesbarkeitsweite *f* = legibility distance
Leseband *n* → Klaubeband
Lesen *n*, Klauben *n* = hand picking
Lese|raum *m* = reading room
~tisch *m*, Klaubetisch = picking table
Letten *m* = potters' clay, plastic ~
~kohle *f* = laminated coal
~schlag *m* → Lehmschlag
Lettner *m* [*Kirchenbau*] = jube
letzte Getriebestufe *f*, Endantrieb *m* = final drive
~ Rohrfahrt *f* außer den Filterrohren, Fördertour *f* = oil string
~ Schweißraupe *f* = finish cover pass
Leucht|beton *m* = reflectorizing concrete
~boje *f* = lighted buoy
~bordstein *m* = reflectorizing curbstone; ~ kerb(stone) (Brit.)
~decke *f* = illumated ceiling, luminous ~
~dichte B *f* = luminance (B)
~dichtefaktor *m* = luminance factor
~dichtepyrometer *n* = photometric pyrometer
Leuchte *f*, Straßen~ = lantern (Brit.); luminaire (US); lighting fixture [*A complete lighting device consisting of a light source together with its direct appurtenances, such as globe, reflector, refractor, housing, and such support as is integral with the housing. The pole, post, or bracket is not considered a part of the luminaire*]
Leuchtenmast *m* → Beleuchtungsmast
Leucht|farbe *f*, Leuchtmasse *f*, nachtleuchtender Stoff *m* = brilliant paint
~feuer *n* = marine light, navigation ~
~folie *f* = reflectorizing foil
~Inselpfosten *m* → Leuchtsäule *f*
~masse *f* → Leuchtfarbe *f*
~mast *m* → Beleuchtungsmast
~petroleum *n* = lamp kerosine, burning oil
~pfad *m*, Landepfad = flare path
~poller *m* → Leuchtsäule *f*
~säule *f*, Leuchtpoller *m*, Leucht-Inselpfosten *m* = illuminated bollard
~schild *n* = illuminated sign
~stofflampe *f* = fluorescent lamp
~stoffröhre *f* [*allgemein auch ,,Neonröhre'' genannt*] = tubular fluorescent lamp
~torf *m* = torch peat
~turm *m* = lighthouse
Leutebude *f* → Baubude
Leuzit *m* [*früher: weißer Granat m*] = leucite
~basalt *m* = leucite-basalt
~basanit *m* = leucite-basanite
~trachit *m* (Geol.) = amphegenyte
~trephit *m* = leucite-trephite
Libelle *f* = level
Libellenblase *f* = bubble
Licht|anlage *f* = light plant
~band *n* = strip lighting element
~blitz *m* = light flash
Lichtbogen|hobeln *n* = flame gouging
~hobler *m* = arc gouging torch

Lichtbogenschweißen *n*, Lichtbogen-Schweißung *f* = arc welding, electric fusion ~
~ **mit Kohleelektrode**, Kohlelichtbogen-Schweißung *f* = (electric-)arc welding with a carbon electrode
~ ~ **Metallelektrode**, Metall-Lichtbogenschweißung *f* = metal-arc welding
Licht|brechung *f* = optical refraction
~**durchlässigkeit** *f* = light permeability
lichtecht = lightfast
Lichtechtheit *f* = lightfastness
lichte Durchfahrthöhe *f* [*Schleuse*] = craft clearance height
~ **Höhe** *f*, Lichthöhe = headroom, headway
~ **Maschenweite** *f*, Lichtweite [*Siebtechnik*] = aperture width, internal ~ ~, clear mesh, clear opening
~ **Weite** *f*, Lichtweite, Innendurchmesser *m* = internal dia(meter)
~ ~, Lichtweite, Spannweite = clear span
lichter Abstand *m* = clear distance
~ **Ausbruchquerschnitt** *m* [*Strecke*] (⚒) = excavated section
~ **Raum** *m*, Durchfahrtprofil *n* = clearance
lichtes Graumanganerz *n* (Min.) = polianite
~ **Rotgültigerz** *n*, Proustit *m*, Ag$_3$AsS$_3$ (Min.) = proustite
Licht|einfallwinkel *m* = angle of incidence, incidence angle
~**gitter** *n* = light-admitting grille
~**hof** *m* = patio [*A courtyard or inner area open to the sky*]
~**höhe** *f* → lichte Höhe
~**installation** *f* = lighting system
~**kuppel** *f* = saucer dome, domelight
~**leistung** *f* = light output
~**maschine** *f* = dynamo, generator
(~)**Mast** *m* → Beleuchtungsmast
~**mastschleuderform** *f* = spun lighting column mould (Brit.); ~ ~ mast mold (US)
~**messung** *f*, Lichtstärkemessung = photometry

~**niveau** *n* [*wird bestimmt von der mittleren Leuchtdichte, dem Farbton der Straßendecke und der Art des Hintergrundes*] = brightness level, level of illumination
~**pausapparat** *m* = printing machine, copier, copying machine
~**pause** *f* = print
~**quelle** *f* = light source
~**raumprofil** *n* → Ladeprofil
~**raumumgrenzung** *f* → Ladeprofil *n*
~**schacht** *m* = air shaft, light well [*An unroofed space within a large building which admits some light and a little bad air to windows facing on it. These light wells should be avoided by open planning*]
~**schalter** *m* = light switch
~**signal** *n* = light signal
~**signalanlage** *f* = traffic signal system
~**signalsteuerung** *f*, (Licht)Signalreg(e)lung, Signalisierung, (Verkehrs-)Signal-Steuerung = signal control
~**spielhaus** *n*, Lichtspieltheater *n*, Kino *n* = cinema
~**stärke I** *f* = luminous intensity (I) [*Sometimes loosely called the candlepower*]
~**(stärke)messung** *f* = photometry
~**strahl** *m* = light beam
lichtstreuende Rückstrahlfläche *f* [*Reflektor*] = diffuse finish
Lichtstrom *m* = luminous flux (F), light ~ ~
~ ~ = lighting current
Licht|verteilung *f* = light distribution
~**wange** *f* → Freiwange
~**weite** *f*, lichte Weite = clear span
~**weite** *f*, lichte Weite = internal dia(meter)
~**weite** *f* [*Siebtechnik*] → lichte Maschenweite
Liebermanngewebe *n* [*Siebtechnik*] = Liebermann cloth
Lieferbeton *m* → Transportbeton
~**-Nachmischer** *m* → (Beton)Rührfahrzeug
~**werk** *n* → Transportbetonwerk
Liefer|menge *f* [*Pumpe*] = capacity

Liefermischer — linksläufig

~mischer *m* → Transportmischer
liefern und ein(zu)bauen = furnish and install, F & I
Liefer|wagen *m* [*Automobil*] = delivery van
~**werk** *n*, Erzeugerwerk = producer works
Liegedauer *f* → Liegezeit
liegend [*Flöz*] = underlying [*seam*]
liegender Behälter *m*, liegendes Gefäß *n* [*Aufnahmegefäß für Dämpfe, Gase, Flüssigkeiten*] = horizontal tank, ~ vessel
Liegendes *n* (⚒) = floor [*of a seam etc.*]
liegendes Gefäß *n* → liegender Behälter *m*
Liegendkörper *m* (Geol.) → Batolith *m*
Liegeplatz *m* [*Kanal*] = lay by
~, Kaiplatz = berth
~**signal** *n*, Kaiplatzsignal = berthing signal
Liege|stelle *f* = basin, tying-up place (Brit.); anchorage basin (US)
~**zeit** *f*, Liegedauer *f* [*Straßendecke*] = life, service ~, service (time)
Lignit *m*, bituminöses Holz *n*, Xylit *m* = lignite
lignitische Braunkohle *f* = lignite coal
~ **Weichbraunkohle** *f* = low-grade lignite coal
Limbus *m* = limb, lower plate [*theodolite*]
Limnologie *f*, Seenkunde *f*, Binnengewässerkunde = limnology
Limonit *m* = limonite [*An iron ore composed of a mixture of hydrated ferric oxides occasionally used in high density concrete because of its water of crystallization which contributes to its effectiveness in radiation shielding*]
lineare Abschreibung *f*, g(e)radlinige ~ = straight line method (of depreciation)
~ **Ausdehnung** *f* = linear expansion
~ **Schwindung** *f* = linear shrinkage
~ **Stadt** *f* → Bandstadt
~ **Strömung** *f* = linear flow
~ **Verschiebung** *f* = linear displacement
~ **Vorspannung** *f* = linear prestressing [*Prestressing as applied to linear members, such as beams, columns, etc.*]
~ **Wärme(aus)dehn(ungs)zahl** *f* = thermal coefficient of linear expansion
~ **Wärmeschichtung** *f* → Tropenschichtung
linearer Ausdehnungsbeiwert *m* = coefficient of linear expansion
~ **Überfall** *m* → Proportional-Überfall
lineares Meßwehr *n* = measuring weir with linear discharge relation
Liner *m* = liner
Linie *f* **der lotrechten Wasserpressungen**, lotrechte Wasserdrucklinie = diagram of vertical unit pressures
~ ~ **Maximalmomente infolge Verkehr** = diagram of maximum moments due to live load
~ ~ **resultierenden Drücke** [*Hydraulik*] = diagram of normal unit pressures
Linien|blitz *m* = forked lightning
~**führung** *f* **im Aufriß** = vertical alignment (Brit.)/alinement (US) of highways (or roads)
~**führung** *f* **im Grundriß** = horizontal alignment (Brit.)/alinement (US) of highways (or roads)
~**last** *f*, Streckenlast = line(ar) load, knife-edge ~
~**pendellager** *n* [*für Hoch- und Tiefbauten*] = linear rocker bearing, straight ~ ~
linker Pfahl *m*, Schreitpfahl [*Naßbagger*] = walking spud
linkes Seitenwindenseil *n*, ~ Schwenkseil [*Naßbagger*] = left swing line, port ~ ~
links einschlagende Tür *f* = left-hand door
Links|abbiegen *n* = left-hand turning
~**abbiegeverbot** *n* = left-turn ban
linksläufig, gegen den Uhrzeigersinn = counterclockwise

Linoleum(fuß)boden — Lochplattensieb

Linoleum|(fuß)boden m = lino(leum) flooring
~teer m = lino(leum) tar
~zement m, Linoleumkitt m = lino(leum) cement
Linoxyn n, oxydiertes Leinöl n = linoxyn, oxidized linseed oil
linsenförmig = lens-shaped, lenticular
Linsenträger m, Fischbauchträger = fish-belly girder, fish-bellied ~
Lisene f [*pfeilerartiger, wenig vortretender Mauerstreifen*] = pilaster strip
Listen|bauholz n, Bauholz nach Liste = cutting-list structural timber, ~ building ~
~preis m = list price
Literleistung f [*Kraftfahrzeug*] = engine capacity
lithiumverseiftes Schmierfett n = lithium-type (pressure) gun grease
Lithographenkalk(stein) m = lithographe stone
Lithopone f = lithopone
Litoral n, Ebbe- und Flutzone f = littoral zone
litorale Meeresablagerung f, landnahe ~ = littoral deposit
Litze f, Draht~ = strand
~ mit dreieckigem Querschnitt → Flachlitze
Litzendraht m, Seildraht = strand(ing) wire
LKW m, Last(kraft)wagen m, Lastfahrzeug n = lorry (Brit.); (auto) truck (US)
~ für fertigen Beton, Betonschüsselwagen m = concrete bowl lorry (Brit.); ~ ~ truck (US)
~ mit Zuschlagstoffabteilen → Chargen-Last(kraft)wagen m
~-Anhänger m, Last(kraft)wagenanhänger m = truck trailer (US); lorry ~ (Brit.)
~-Bagger m → Aufbaubagger
~-Bahnhof m = truck terminal (US); lorry ~ (Brit.)
~-Chassisbagger m → → Autobagger
~-Diesel(motor) m = truck diesel engine (US); lorry ~ ~ (Brit.)
~-Fahrgestell n = truck chassis (US); lorry ~ (Brit.)
~-Grabenbagger m → Aufbau-Grabenbagger
~-Güterfernverkehr m = line-haul trucking
~-Hydrobagger m → Hydro-Aufbaubagger
~-Hydrokran m → Hydraulik-Aufbaukran
~-Hydrolader m → Hydraulik-Aufbaulader
~-Kipper m, Last(kraft)wagenkipper = tipping truck (US); ~ lorry (Brit.)
~-Kran m → Aufbaukran
~-Lader m → Aufbaulader
~-Ladung f = lorry-load (Brit.); truck-load (US)
~-Motor m = truck engine (US); lorry ~ (Brit.)
~-Streifen m = heavy vehicle lane, slow ~; lorry ~ (Brit.); truck ~ (US)
~-Transport m, Straßenanlieferung f, Last(kraft)wagentransport m = haulage by truck, truck haul(ing) (US); haulage by lorry, lorry haul(ing) (Brit.)
~-Waage f → Lastfahrzeugwaage
~-Zufahrt f = lorry entrance (Brit.); truck ~ (US)
~-Zug m → Lastzug
Loch n, Lochung f [*Lochziegel*] = core (US)
~aufweitung f → Leibungsdruck m
~(blech)sieb n = perforated screen, hole ~
~durchmesser m, Lochweite f [*Siebtechnik*] = diameter of opening, opening of hole, hole aperture, hole size, width of opening
~fraß m [*Korrosion*] = pit
~leibungsdruckfestigkeit f → Leibungsdruck m
~-Liner m = perforated liner
~platte f [*zur Bekleidung von Deckenuntersichten*] = perforated tile
~plattensieb n [*Abwasserwesen*] = (perforated) plate screen

Lochreihe — Löffel(seil)haspel

~reihe f [*Siebtechnik*] = series of holes
~rohr n = perforated pipe
~säge f → Spitzsäge
~sieb n, Lochblechsieb = hole screen, perforated ~
~stein m = hole block, perforated ~
~tafel f = perforated panel
~taster m → Innentaster
~teilung f [*Siebtechnik*] = hole pitch
Lochung f, Loch n [*Lochziegel*] = core (US)
Loch|walzwerk n = piercing mill, piercer
~weite f, Lochdurchmesser m [*Siebtechnik*] = opening of hole, diameter of opening, hole aperture, hole size, width of opening
~weite f, Seitenlänge f der quadratischen Öffnung [*Sieb*] = mesh size
~ziegel m, Hohlziegel = perforated brick (Brit.); (multi)cored tile, (multi)cored brick (US)
lockere Ablagerung f, unverfestigte ~ (Geol.) = unconsolidated deposit
lockeres Gebirge n (⚒) = loose ground, ~ strata
Lockergestein n, Boden m = earth, soil [*terms used by engineers*]; mantle rock [*term used by geologists*]
Löffel m [*Bagger*] = (shovel) dipper, shovel, (dipper) bucket
(Löffel)Bagger m = (mechanical) shovel
~achse f = shovel axle
~ausleger m = shovel jib (Brit.); ~ boom (US)
~bedienungsstand m → Baggerbedienungsstand
~einsatz m = shovel use
~-Elektromotor m = shovel motor
~führer m = shovel(l)er
~kette f = shovel chain
~motor m = shovel engine
~schmierer m = shovel oiler
~seil n = shovel cable
Löffel|baggerung f = excavation by shovel(s)
~(boden)probe f = spoon-auger sample
~bohrer m, (offene) Schappe f, Probelöffel m = spoon-auger, sampling spoon
~-Dampfbagger m → Dampfbagger
~flasche f = bucket sheave
~grabbreite f = shovel bucket digging width
~grabwinkel m = shovel bucket digging angle
~haken m, Fangspeer m [*Eine mit Widerhaken versehene Eisenstange für verschiedene Fangarbeiten*] = fishing spear
~haspel m, f → Löffelseilhaspel
~hochausrüstung f, Hochlöffelausrüstung = shovel bucket attachment
~hochbagger m, Hochlöffel(bagger) m, Hochbagger = power navvy, crane navvy, crowd shovel (Brit.); (face) shovel, (swing) dipper shovel; luffing-boom shovel (US)
~inhalt m → Grab~
~klappe f [*Hochlöffel*] = (shovel) dipper door, shovel ~, (dipper) bucket ~
~klappenauslöser m, Löffelklappensperre f = (shovel) dipper trip, shovel ~, (dipper) bucket ~ [*A device which unlatches the door of a shovel bucket to dump the load*]
löffeln, schlämmen, schöpfen = to bail
Löffeln n, Schlämmen n, Schöpfen n = bailing
Löffel-Naßbagger m → Naßlöffelbagger
~probe f, Löffelbodenprobe = spoon-auger sample
~rolle f → Löffelseilrolle
~-Schwimmbagger m → Naßlöffelbagger
~seil n, Schöpfseil, Schlämmseil [*pennsylvanisches Seilbohren*] = bailing rope, ~ line, sand ~ [*Rope used for removing the sand and liquid from deep wells while being drilled with cable tools*]
~(seil)haspel m, f, Schlämm(seil)haspel [*pennsylvanisches Seilbohren*] = sand reel

Löffel(seil)haspelbremse — Lokomotivschuppen

~(seil)haspelbremse *f*, Schlämm(seil)haspelbremse [*pennsylvanisches Seilbohren*] = post brake, back ~

~(seil)haspelhebel *m*, Schlämm(seil)haspelhebel [*pennsylvanisches Seilbohren*] = sand reel lever

~(seil)rolle *f*, Schlämm(seil)rolle, Schlämm(seil)scheibe *f*, Schöpf(seil)scheibe [*Tiefbohrtechnik*] = sand-line pulley, sand pump ~, sand sheave, bailing pulley, bailing sheave

~(seil)trommel *f*, Schlämm(seil)trommel, Schöpf(seil)trommel [*Tiefbohrtechnik*] = sand-line spool, sand reel drum, bailing drum

~stiel *m*, Bagger~, Auslegerstiel = arm, handle, stick, (shovel) dipper ~, shovel ~, (dipper) bucket ~

~stieltasche *f* = (shovel) dipper stick sleeve, (~) ~ arm ~, (~) ~ handle ~, shovel ~ ~, (dipper) bucket ~ ~

(~)Stielzylinder *m* = (dipper) stick ram

~tiefbagger *m*, Tieflöffel(bagger) *m*, Tiefbagger = trench-forming shovel, backacter, backacting shovel (or excavator), dragshovel, trencher, ditcher, trenching hoe, drag bucket (Brit.); (back-)hoe, trenchhoe, pullshovel (US); pullstroke trenching machine, ditching shovel, back digger

~trockenbagger *m* → Trockenlöffelbagger

~trommel *f* → Löffelseiltrommel

~zahn *m* = (shovel) dipper tooth, shovel ~, (dipper) bucket ~

~zylinder *m* = (shovel) dipper ram, shovel ~, (dipper) bucket ~

logarithmische Spirale *f* = log(arithmic) spiral

~ Teilung *f* = log(arithmic) scale

logarithmischer Maßstab *m* = log(arithmic) scale

logarithmisches Dekrement *n* = log(arithmic) decrement

logarithmisches Wahrscheinlichkeitsgesetz *n* = log-probability law

log-Diagramm *n* = log-diagram

Loggia *f*, Hauslaube *f* = loggia

Lohn|abbau *m* = wage cutting, cutting of wages

~anteil *m* = wage fraction

~arbeiter *m* = wage earner, hourly paid worker

~aufwand *m* → Lohnkosten *f*

~büro *n* = wage(s) office

~empfänger *m* = wage earner; ~ worker (US)

~ermitt(e)lung *f* = wage(s) determination

~faktor *m* = wage(s) determinant, ~ factor

~fuhrgewerbe *n* = carrier's trade

lohngebundene Kosten *f* = cost incidental to wages

Lohn|gleitklausel *f* = wages variation clause, ~ fluctuation ~

~klasse *f* = wage category

~kosten *f*, Lohnaufwand *m* = wage(s) cost, ~ expenses

~liste *f* = wage(s) payroll

~satz *m* = wage(s) rate

~steigerung *f* = wage(s) increase

~tag *m* = pay day

~tüte *f* = pay envelope

~- und Materialkosten *f*, Lohn- und Stoffkosten *f* = flat cost [*The cost of labo(u)r and material only*]

Lokalelement *n* [*elektrochemische Korrosion*] = local cell

Lokförderung *f* = locomotive haulage

Lokomobile *f*, Kesseldampfmaschine *f* = locomobile

Lokomotiv|drehkran *m*, Drehkran auf Normalspur = locomotive (jib) crane (Brit.); ~ (boom) ~ (US); railway jib ~ (Brit.); railroad boom ~ (US)

~kessel *m* = locomotive (type) boiler

Lokomotivschuppen *m*
Rechteckschuppen

locomotive shed
straight shed

Rechteckschuppen mit einem Zugang	one-ended shed
Rechteckschuppen mit zwei Zugängen	trough shed
Ringschuppen	round shed
Aufstellgleis *n*	stabling track
Einzeltoreinfahrtgleis *n*	radiating track
Drehscheibe *f*	turntable
Untersuchungsgrube *f*, Arbeitsgrube	inspection pit
Rauchabführung *f*	smoke disposal

Lomaxverfahren *n* = light oil maximizing method, Lomax ∼

Londoner Ton *m* = London clay

∼ **Verkehrsbetriebe** *f* = London Transport

Longarinenbauweise *f*, englische Bauweise, Einzonenbauweise = English system (of timbering), ∼ method (∼ ∼)

Lore *f* = wagon

Lorenz-(Beton)Bohrpfahl *m* = Lorenz bored pile

lösbare Bohrkrone *f*, austauschbare ∼, auswechselbare ∼ = detachable bit

lösbarer Rohrkrebs *m* [*Tiefbohrtechnik*] = releasable casing spear

Löschbank *f*, Kalk∼ = slaking trough, lime ∼ ∼

Lösch(e) *m*, (*f*) → Kesselschlacke *f*

löschen [*Schiff*] = to unload

∼ [*Kalk*] = to slake

∼ [*Feuer*] = to extinguish

Löschen *n* [*Kalk*] = slaking

∼ [*Schiff*] = unloading [*ship*]

∼ [*Feuer*] = extinguishing

∼ **auf der Baustelle** [*Kalk*] = site slaking

Lösch|grube *f*, Kalk(lösch)grube = (lime) slaking pit, lime ∼

∼**hafen** *m* = port of discharge, landing port

∼**kalk** *m* → Kalziumhydroxyd *n*

∼**kopf** *m* → Löschplatz *m*

∼**maschine** *f* = slaking machine, hydrator

∼**pfanne** *f*, Kalk∼ = slaking vessel, ∼ pan

∼**platz** *m*, Löschkopf *m* [*Ölumschlaganlage*] = unloading assembly

∼**schnecke** *f*, kalk∼ = slaking screw

∼**trommel** *f* → Kalk∼

∼**verhalten** *n* [*Kalk*] = slaking behavio(u)r

∼**vorgang** *m* [*Kalk*] = slaking process

∼**wasser** *n* = water for fire-fighting purposes

Losdrempel *m* → Losständer *m*

lose Sohle *f* [*Fluß; Strom*] = loose bottom

Lösemittel *n* → (Auf)Lösungsmittel

lösen [*Baggergrund beim Naßbaggern*] = to dislodge

∼ [*Bremse*] = to release

∼, losschrauben = to unscrew

Lösen *n* [*Erdbau*] = loosening

∼, Ausrücken = disengagement, release

lösendes Abbeizmittel *n*, neutrales ∼, Abbeizfluid *n* = solvent-based paint remover

Lösen *n* **und Laden des Bodens** → Bodengewinnung *f*

loser Boden *m* = loose soil

∼ **Zement** *m* → Behälterzement

∼ **Zustand** *m* = loose state

loses Gebirge *n* (⚒) = unconsolidated strata, ∼ ground, ∼ formation

Löse- und Ladefähigkeit *f* [*Boden*] = loadability

Löseventil *n*, Raubventil (⚒) = (chock) release valve

losgesprengter Fels *m* = blasted rock

Losholz *n*, Kämpfer *m* [*Fensterrahmen*] = transom(e)

loskuppeln = to disengage

löslich — Lösungswärmeverfahren

löslich, auf~ = soluble
lösliche Schießbaumwolle *f*, Kollodiumwolle = collodion cotton
löslicher Anhydrit *m*, künstlicher ~ = artificial anhydrite
Löslichkeit *f*, Auflösbarkeit = solubility
los|lösen, sich ~ (✄) = to break loose
~schrauben, lösen = to unscrew
Losständer *m*, Losdrempel *m*, Setzpfosten *m* = removable sluice pillar

~wehr *n*, Ständerwehr = sluice weir with removable sluice pillars
Lösung *f* = solution
~ des Bodens → Bodenlösung
Lösungs|festigkeit *f* [*Boden*] = loosening strength
~mittel *n* → Auf~
lösungsmittel|fliehend = lyophobe
~frei = solvent-less, solvent-free
Lösungsmittel-Raffinationsanlage *f* = refining plant for solvents

---

Lösungswärmeverfahren *n*
[*Bestimmung der Hydratationswärme von Zement. Die Hydratationswärme wird aus dem Unterschied der Lösungswärme des angelieferten (nichthydratisierten) Zements und der hydratisierten Proben im Alter von 7 oder 28 Tagen errechnet. BSS 1370 und ASTM C 186*]

Ablesung *f*, Ablesen *n*
Ausrührperiode *f*
Beckmann'sches Thermometer *n*

Eingabe *f*, Zugabe [*von Zinkoxyd*]
Endtemperatur *f*
Fluorwasserstoffsäure *f*
5min-Intervall *n*
Gesamtgewicht *n*
Geschwindigkeitsregler *m*
Glasflasche *f*
Glasrührer *m*
glühen [*Zinkoxyd*]
Hydratationswärme *f*
hydratisierter Zement *m*
Kalorimeter *n*
Korkstopfen *m*
korrigierter Temperaturanstieg *m*
Lösungsperiode *f*

Lösungswärme *f*
Lupe *f*
nichthydratisierter Zement *m*, angelieferter Zement

heat of solution method

reading
final rating period
Beckmann type differential thermometer
introduction
final temperature
hydrofluoric acid
5-min. interval
total weight
geared speed reducer
glass vial
glass stirrer
to heat
heat of hydration
hydrated cement
calorimeter
cork stopper
corrected temperature rise
solution period [*The period from the time the introduction of the ZnO is started to the time the rate of temperature change becomes constant is termed the solution period and should last not longer than 30 min.*]
heat of solution
reading lens
dry cement

Löß — Lotrechte

Paraffinwachs n	paraffin wax
Raumtemperatur f	room temperature
Rührer m	stirrer
Rührmotor m	stirring motor
Rührschaufel f	stirrer blade
Rührwärme f	heat of stirring
säurebeständiger Überzug m	acid-resistant coating
Säuremischung f	acid mix(ture)
Temperatur f auf 0,001° genau ablesen	to observe and record the temperature to the nearest $0.001°$ C
Temperaturänderung f	temperature change
Trichter m	funnel
Wärmehaltungsvermögen n	thermal leakage into or out of the calorimeter
Wärmekapazität f	heat capacity
Zementbrei m	cement paste
Zinkoxyd n	zinc oxide, ZnO
Zinkoxyd in gleichmäßigen Gaben quantitativ durch den Trichter in die Säuremischung geben	to introduce the prepared ZnO through the funnel at a uniform rate
Zugabe f, Eingabe [von Zinkoxyd]	introduction
2n-Salpetersäure f	nitric acid (2.00 N), $2.00\ N\ HNO_3$

Löß m = loess, löss [*diluvial deposit of fine loam*]
~**bildung** f = loess formation
~**boden** m, lößartiger Boden = loess soil
~**haus** n = loess dwelling
~**lehm** m = loess loam
~**mergel** m = marl loess
~**region** f = loessical region
~**steilufer** n = loess bluff
~**wind** m = loess depositing wind
Lot n, Bleigewicht n [*an einer mit durch Knoten unterteilten Leine zur lotrechten Wassertiefenmessung*] = hand lead, sounding ~
~ [*zum Löten*] = solder
~ → Senkblei
~**abweichung** f = deviation from the vertical
~**blei** n = hand lead, sounding ~
Lotblei n → Bleilot
Lötdraht m = solder wire
Loten n, Lotung f = sounding
Überschall~ = supersonic ~
Löten n, Hart~ = brazing
~, Weich~ = (soft) soldering

Löt|fitting m, Lötformstück n [*aus Rotguß oder Kupfer*] = braze-on fitting [*Fitting for light-gauge copper tube, many of them not made of copper, but of brass or gunmetal*]
~**flußmittel** n = soldering flux
~**kolben** m = soldering iron, copper bit, soldering bolt, heated bit, heated iron
~**lampe** f = soldering lamp, blowlamp, blowpipe lamp (Brit.); blowtorch (US)
Lotleine f = lead line, sounding ~
Lötofen m = sweat-soldering oven
Lötpaste f = solder paste, ~ cream
Lotpfahl m = vertical pile, plumb ~
lotrecht, senkrecht = vertical, perpendicular, plumb
Lotrechte f [*Abflußmengenmessung mit Meßflügel*] = vertical, perpendicular (boundary line)
~, Pfosten m, Vertikale f, Vertikalstab m [*Fachwerkbinder*] = vertical web member, ~ strut ~, ~ (rod)

lotrechte Fliehkraftkomponente *f*, vertikale Komponente der Fliehkraft = vertical component of the centrifugal force
~ **Holzbohle** *f* [*Baugrubenauskleidung*] = vertical wooden plank
~ **Normalspannung** *f* [*Baugrund*] = vertical normal stress
~ **Wasserdrucklinie** *f*, Linie der lotrechten Wasserpressungen = diagram of vertical unit pressures
lotrechter Abstandhalter *m* = vertical spacer
~ **Aufzug** *m*, (Schiff)Hebewerk *n* mit Gegengewichten, Seilhebewerk = vertical lift with suspension gear
~ **Bügel** *m* = stirrup placed perpendicular to the longitudinal reinforcement
lotrechtes Geschwindigkeitsverteilungsdiagramm *n* [*Hydraulik*] = vertical velocity distribution diagram
~ **(Schiff)Hebewerk** *n* = nonlock type vertical barge lift
~ ~ **mit Gewichtsausgleich,** ~ **Gleichgewichtshebewerk** = balanced vertical lift
Lotse *m* = pilot
Lotsen|flagge *f* = pilot flag
~**gebühren** *fpl* = pilotage charges
Löt|wasser *n* = zinc chloride, killed spirits
~**zange** *f* = brazing tongs, soldering ~, soldering-tweezers, hawk-bill (pliers)
~**zinn** *n*, Zinnlot *n* = fine solder [*An alloy of* $^2/_3$ *tin,* $^1/_3$ *lead*]
LP *m* → Belüftungsmittel *n*
LP-Stoff *m* → Belüftungsmittel *n*
L-Stufe *f* = L-shaped step
Lückenbau *m* [*Ein Gebäude, welches in eine Baulücke zwischen Häusern gebaut wird*] = lock-up
Luft|ansaugstutzen *m*, Lufteintritt *m* = air intake
~**aufnahme** *f* → Luftbild *n*
~**aufnahmetechnik** *f*, Luftbildtechnik *f* = aerial photography, ~ engineering

~**auftrieb** *m* = air buoyancy
~**austritt** *m* = air escape
~**austrittstutzen** *m* = air pipe, vent ~
~**bedarf** *m* = air demand, ~ requirement
~**befeuchter** *m* = air humidifier
~**befeuchtung** *f* = air humidification
~**behälter** *m* = air vessel, ~ tank
luftbereift = pneumatic-tired (US); pneumatic-tyred (Brit.)
(luftbereifter) Einachs(-Schnell)mischer *m*, Autoanhänger-Einachs-Mischer = single-axle trailer mixer
Luftbewegung *f* = air movement, ~ motion
Luftbild *n*, Luftaufnahme *f* = flight photo(graph), aerial ~, air ~
~**auswerter** *m* für Vermessungszwecke = photogrammetrist
~**auswertungsingenieur** *m* = photogrammetric engineer
~**karte** *f* = aerial picture map
~**messung** *f*, Luftbildtechnik *f*, Aerophotogrammetrie *f*, Luftphotogrammetrie = aerial photography, ~ engineering, ~ surveying
~**technik** *f* → Luftbildmessung *f*
~**vermessung** *f* [*als Vorgang, nicht als Wissenschaft*] = aerial survey
~**vermessungsflugzeug** *n* = mapping plane
~**vermessungsingenieur** *m* = aerial engineer
~**vermessungskamera** *f* = mapping camera
~**vermessungsverfahren** *n* = aerial method
~**bindermörtel** *m* = air-setting mortar
~**blase** *f* = air bubble
~**bohren** *n*, Bohren mit Luft = air drilling
~**brücke** *f* = air lift
luftdicht = airtight
Luft|druck *m* [*beim Sprengen*] = air blast
~**druckimpuls** *m* = pulse of air
~**durchdringungsmesser** *m* = permeameter

~durchlässigkeitsverfahren n = air permeability method [*measurement of specific surface*]
~einlaß m = air inlet
~einlaßventil n = air inlet valve
~einschluß m = air entrainment
~einschlußmittel m → Belüftungsmittel
~eintritt m, Luftansaugstutzen m = air intake
~eintrittrohr n = air inlet pipe
~embolie f = air embolism
~(-Erddruck-)Meßdose f nach Goldbeck, Goldbeckdose = Goldbeck (pressure) cell
~(-Erddruck-)Meßdose f nach Ritter = Ritter (pressure) cell
~erhitzer m, Warmluftgerät n, Luftheizapparat m = (air) heater, space ~, unit ~
Lüfter m = fan
~ auf dem Kühler montiert = radiator mounted fan
~ zum Absaugen von Rauchgasen und Dämpfen = fire ventilator [*It opens automatically at a pre-determined temperature, releasing smoke, heat and fumes, keeping the fire localized*]
~antrieb m = fan drive
~haube f, Schirmblech n = fan shroud
~-Kühlturm m, Ventilator-Kühlturm = forced draught cooling tower
~kühlung f = fan cooling
~riemenscheibe f = fan pulley
~riemenspanner m = fan belt tightener
~riemen-Spannrad n = fan adjusting pulley
~rost m = ventilator grate
~schutzkorb m = fan guard
~spannrad n = fan idler
~spannradträger m = fan idler bellcrank
~welle f = fan (drive) shaft
Luft|fährdienst m = air ferry service
~feder f = air spring
~federbalg m = air bellows
~federkissen n = air spring cushion

~federung f [*Auto*] = air suspension
~federventil n [*Luftfederung*] = air spring levelling valve
~feuchtigkeit f, Luftfeuchte f = air moisture
(~)Feuchtemesser m, Hygrometer n, m = hygrometer
~feuchtigkeit f = air humidity
~filter m, n = air filter, ~ cleaner
~filter(an)saugrohr n = air cleaner inlet, ~ ~ intake, ~ filter ~
~filterschale f = air cleaner tray, ~ filter ~
~filterung f = air filtration
~förderrohr n → Gebläseluft-Förderrohr
luftförmig = aeriform
Luft|frachtbrief m = airway bill
~führungskanal m [*Luftrutsche*] = air chamber
~gehalt m = air content
~gehalt-Prüfgerät n → Belüftungsmesser m
luft|gekühlte Schlacke f = air-cooled slag
~gekühlter Kondensator m = air-cooled condenser
~geschmiertes Lager n, Luftlager = air bearing
~getragene Feinststoffe mpl = air float fines
Luft|gitter n → Luftrost
~hebe-Bohranlage f = air lift drilling rig
~hebe-Bohrverfahren n = air lift drilling
~heber m [*Scheidekonus*] = air lift [*separatory cone*]
~heizapparat m → Lufterhitzer m
~heizung f = warm air heating
lufthydraulisch = hydropneumatic
Luft|insel f = still air space
~isolierung f = air insulation
~kabel n = overhead electric cable
~kalk m, nichthydraulisches Bindemittel n, Luftmörtelbildner m = non-hydraulic lime, ~ binder, air-slaked lime
~kalkmörtel m = ordinary lime mortar

Luftkanal — luftreifenbahrbarer Betontransportwagen

~kanal m [*Klimaanlage*] = airway, air(-handling) duct, air flue
~kissen n → Luftpolster
~kissenzelt n = air tent
~klappe f, Drossel f [*Vergaser*] = choke (US); strangler (Brit.)
~klappenzug m = strangler cable (Brit.); choke ~ (US)
~kohlensäure f = air carbon dioxide
~kühler m = air cooler
~kühlung f = air cooling
~lager n, luftgeschmiertes Lager = air bearing
~lagerung f = air storage
~lärm m = air-borne noise
~leistung f = air(-handling) capacity
~loch n, Zugloch = air hole, ~ vent, port
~löschen n [*Kalk*] = air slaking
luftloses Farbspritzen n = airless paint spraying
Luft|menge f = air volume, ~ quantity
~mengenmesser m → Beton~
~meßgerät n = Belüftungsmesser m
~nest n = air pocket [*An air space which accidentally occurs in concrete work*]
~photogrammetrie f → Luftbildmessung f
~photograph m = aerial photographer
~polster n [*Österreich: m*], Luftkissen n, Luftschicht f = cushion of air, air cushion
~pore f = air void

~porenanteil m, scheinbare Porosität f = air space ratio
~porenbeton m, Bläschenbeton, AEA-Beton, belüfteter Beton = air-entraining concrete, air-entrained ~
luftporenbildender Beton-Verflüssiger m, LBV-Stoff m = combined plasticizer and air-entraining agent
~ Zusatzstoff m → Belüftungsmittel n
Luftporen|bildner m → Belüftungsmittel n
~erzeuger m → Belüftungsmittel n
~größe f = air void size
~messer m → Belüftungsmesser
~mörtel m → AEA-Mörtel
~mörtel m aus Brechsand = entrained stone sand mortar
~prüfer m → Belüftungsmesser
~system n = air void system, system of air voids
~verteilung f = air void spacing
~zement m = air-entraining cement
Luft|probensammelgerät n = air sampler
~pumpe f = air pump
~raum m, Luftschicht f = air space [*A cavity or space in walls or between various structural elements*]
~reifen-Anhängewalze f → Anhänge-Luftreifenwalze
~reifen-Doppelrad n = pneumatic dual tired wheel (US); ~ ~ tyred ~ (Brit.)
~reifen-Drehkran m → Pneu-Drehkran
luftreifenfahrbar = pneumatic-mounted

---

luftreifenfahrbarer Betontransportwagen m mit Mischer

Betontransportwagen m mit (aufgebautem) Rührwerk(smischer), Nachmischer m, Rührwagen [*zentrale Betonmischung*]
Transportmischer m, Liefermischer [*zentrale Abmessung, Mischen während der Fahrt*]

lorry-mounted mixer (Brit.); motomixer, mixer-type truck, ready-mix (cement) truck (US); transit (concrete) mixer
  agitator (or agitating) conveyor (or truck), truck agitator (US); agitating lorry; agitator [*deprecated*] (Brit.)
  truck mixer, transit-mixer (truck) (US); mixer lorry (Brit.)

Luft|reifenfahrzeug — Lufttrennung

Transportmischer mit querliegender Mischtrommel = side discharge truck paver

Luft|reifenfahrzeug n → Pneufahrzeug
~**reifengerät** n → Pneugerät
~**reifenjumbo** m → Pneujumbo
~**reifenkran** m → Pneukran
~**reifenlader** m → Pneu-Lader
~**reifen-Löffelbagger** m → Pneu-Löffelbagger
~**-Mobilkran** m → Pneu-Mobilkran
~**reifenrad** n = pneumatic-tired wheel (US); pneumatic-tyred ~ (Brit.)
~**reifenschlauch** m, (Reifen)Schlauch = air chamber
(~)**Reifenschlepper** m → Pneuschlepper
~**reifen-Schleppwalze** f → Anhänge-Luftreifenwalze
~**reifenschwenkkran** m → Pneu-Drehkran
~**reinhaltegesetz** n, Gesetz zur Reinhaltung der Luft = air pollution prevention code
~**reinhaltung** f = air pollution prevention
~**reinhaltungsgesetzgebung** f = air pollution legislation
~**reinhaltungsverordnung** f = air pollution code
~**reinheit** f = air cleanliness, ~ purity
~**reiniger** m, Luftfilter n, m = air cleaner, ~ filter
~**reinigung** f = air purification
~**rost** m, Luftgitter n = grille
~**-Rutsche** f → Gebläseluft-Förderrohr n
~**schacht** m → Entlüftungsschacht
Luftschall m = airborne noise, ~ sound
~**dämmung** f, Luftschallschutz m = sound proofing to prevent sound transmission from air to air through walls etc., airborne sound insulation
Luft|schicht f → Luftraum m
~**schicht** f → Luftpolster
~**schiffhalle** f = airship hangar
~**schleuse** f, Druck~ = pneumatic lock, air ~
~**schmierung** f = air lubrication
~**schraubenschlitten** m, Propeller-Schlitten = propeller sled
~**schutz** m = civil defense against air raids (US); ~ defence ~ ~ ~ (Brit.)
~**schutzraum** m = air raid shelter
~**seilbahn** f = aerial rope-way
luftseitige Sohle f = downstream bottom
Luft|sichtung f → Sichtung
~**spalt** m = air gap
~**spinnen** n [*Brückenbau*] = aerial cable spinning
~**spül(bohr)hammer** m = puff blowing piston rock drill
~**spülkolben** m = puff blowing piston
~**spülung** f [*Bohrhammer*] = puff blowing
~**spülvorrichtung** f, Blasevorrichtung [*Bohrhammer*] = puff blowing device
~**stein** m = ventilating block
~**sterilisation** f = air sterilization
~**staublech** n = air baffle
~**stoßschutzbauten** f = blast-resistant civil defense structures (US); ~ ~ defence ~ (Brit.)
~**strahl** m [*pneumatische Förderanlage*] = conveying air
~**strahl-Prallzerkleinerer** m → pneumatische Prallmühle f
~**strahlpumpe** f = air jet pump
~**straße** f = airway
~**strom** m = air stream, ~ current, stream of air, current of air
~**strom-Gebläse** n im Zementsilo = cement blower
~**stromsichter** m = air classifier to which the product removed from the mill is fed by air
~**stützpunkt** m, Militärflugplatz m, Fliegerhorst m = station, base
~**transport** m = air transport
~**trennung** f = air separation

Lufttriebverfahren — Luxuszug

~triebverfahren n [*Erdöllagerstätte*] = air drive
~umwälzung f = air circulation
Lüftung f, Be- und Entlüftung = ventilation, ventilating
Lüftungs|abschnitt m, Be- und Entlüftungsabschnitt [*Tunnel; Stollen*] = ventilating section, ventilation ~
~anlage f, Be- und Entlüftungsanlage = ventilating installation, ventilation ~
~bauwerk n, Be- und Entlüftungsbauwerk = ventilating structure, ventilation ~
~betrieb m, Be- und Entlüftungsbetrieb = ventilating service, ventilation ~
~kanal m, Be- und Entlüftungskanal = ventilating duct, ventilation ~
~methode f → Lüftungsverfahren n
~rohr n = vent pipe
~schacht m, Be- und Entlüftungsschacht = ventilating shaft, ventilation ~
~station f, Be- und Entlüftungsstation = ventilating station, ventilation ~
~stein m, Be- und Entlüftungsstein = ventilation block, ventilating ~
~stollen m, Be- und Entlüftungsstollen = ventilation gallery, ventilating ~
~verfahren n, Lüftungsmethode f, Be- und Entlüftungsverfahren, Be- und Entlüftungsmethode = ventilating method, ventilation ~
~verdichter m, Kompressor m, Drucklufterzeuger m = (air) compressor
~verflüssigung f = liquefaction of air
Luft|verkehr m, Flugverkehr = air traffic
~verkehrsnetz n = air traffic network
~vermessung f = aerial surveying, air ~
~verschlechterung f = air vitiation
~verschmutzung f, Luftverseuchung, Luftverunreinigung = air pollution
~verseuchung f, Luftverschmutzung, Luftverunreinigung = air pollution

~verteiler m = diffuser [*A porous plate or other device through which air is forced and enters the sewage in the form of minute bubbles*]
~verunreinigung f, Luftverseuchung, Luftverschmutzung = air pollution
~vorfilter m, n = air pre-cleaner
~vorwärmer m, Luvo m = air preheater
~waffenbauverwaltung f = Works Directorate of the Air Force
~wascher m = air washer
~wechsel m = air change
~widerstand m = air resistance
~wirbelkammer f = energy cell
~zerlegung f = air fractionation
~ziegel m = ventilating brick, ventilation ~ [*A brick which has been cored to provide an air passage for ventilating purposes*]
~ziegel m [*Fehlname*]; Lehmbaustein m = adobe
~zufuhr f → Belüftung f
~zylinder m = rotochamber
Luke f = hatch(way)
Lukenübermaß n, Übermaß in Luken = excess of hatch(way)
Lungenmoos n, isländisches Moos = Iceland moss
Lungstein m → Lavaschlacke f
Lunker m = shrinkage cavity, shrink hole, blow hole, sink hole
~ im erstarrten Stahlblock = pipe
LURGI-Gebläse n [*einfaches Strahlgebläse für kleine Bewetterungsaufgaben*] = LURGI blower
Luttenrohr n = underground ventilation pipe, ~ ventilating ~
Luvo m, Luftvorwärmer m = air preheater
Luvseite f = exposed side
Luxus|ausführung f = de luxe model
~(bade)wanne f [*in Feuerton oder Marmor*] = luxury bath(tub)
~-Etagenwohnung f = luxury flat
~(fuß)boden m = luxury flooring, ~ floor finish
~hotel n = luxury hotel
~zug m = Pullmann train

L. V.-Menge *f*, Leistungsverzeichnis-Menge = bill quantity
Lydit *m*, Kieselschiefer *m* = Lydian stone, lydite, touchstone, siliceous rock
Lysimeter *n*, Versickerungsmesser *m* = lysimeter
~ mit gewachsenem Boden, Versickerungsmesser *m* ~ ~ ~, Bodenmonolith-Lysimeter, Bodenmonolith-Versickerungsmesser = monolith lysimeter

M

Mäander *m* → Flußschlinge *f*
~ = meander [*An ornamental pattern of winding or crisscrossing lines*]
Macco-Seitendüse *f* = Macco side door choke
Machart, Drahtseil~ = (wire rope) construction
Mächtigkeit *f* (Geol.) = thickness, depth, richness, size, power, substance, width
Madenschraube *f* = set screw
Magazin *n*, Lager *n* = store
~bau *m* (⚒) = shrinkage stoping
~beschickung *f*, Magazinaufgabe *f* = hopper feed
~halter *m*, Lagerhalter = storekeeper
Magerbeton *m* → → Beton
~tragschicht *f* = lean concrete base
~tragschicht *f* **mit hohem Wasser-Zement-Faktor** = wet lean concrete base
magere Kohle *f* → Magerkohle
magerer Mörtel *m* → Magermörtel
Mager|gas *n*, mageres Gas = lean gas
~gemisch *n*, Magermischung *f* = lean mix(ture)
~kalk *m* → Graukalk
~kohle *f*, magere Kohle = lean coal
~kohle *f*, magere Kohle = semi-anthracite
~mischung *f*, Magergemisch *n* = lean mix(ture)
~mörtel *m*, Sparmörtel, Füllmörtel, magerer Mörtel = lean(-mixed) mortar

Magern *n*, Magerung *f* [*von Mörteln und Tonen durch Beigabe von Sand*] = making lean by adding sand
Mager|ton *m*, magerer Ton = lean clay
~walzbeton *m*, Walzmagerbeton = rolled lean concrete
Magistrat *m* = city council
Magma *n*, Schmelzfluß *m*, (magmatische) Schmelze *f*, Schmelzlösung *f* = magma
~rest *m* = magmatic residuum
magmatisches Gestein *n* → Eruptivgestein
Magnel-Verankerung *f* [*Spannbeton*] = Magnel anchoring system, ~ anchorage ~
Magnesia *f* = magnesia
~bindemittel *n* → Sorelzement *m*
~estrich *m*, Magnesitestrich = magnesite (composition) floor(ing), ~ (~) floor finish, ~ screed
~härte *f* = magnesia hardness
~kalk *m* [*dieser Ausdruck ist veraltet*]; Dolomitkalk [*DIN 1060*] = magnesian lime [*B. S. 890*]
~mörtel *m*, Magnesitmörtel = magnesite composition [*Composition made of (magnesium) oxychloride cement and filler*]
~stuck *m*, Magnesitstuck = magnesite stucco
~tongranat *m* = magnesium-alumin(i)um garnet
~treiben *n* = expansion tendency due to magnesia, action of magnesia
~zement *m*, Magnesitbinder *m* = (magnesium) oxychloride cement, ~ oxychloride [*When magnesium chloride is added to magnesium oxid(e) a cementing material, magnesium oxychloride, is formed. The best known is Sorel('s) cement*]
Magnesit *m*, $MgCO_3$ = magnesite
~binder *m* → Magnesiazement *m*
~-Chromstein *m* = magnesite chrome brick
~estrich *m* → Magnesiaestrich
~gestein *n* = magnesite rock
~mörtel *m* → Magnesiamörtel

Magnesitstein — Mahlhilfe

~stein m = magnesite brick, magnesia ~, ~ refractory
~stein m mit 96% MgO = periclase brick
~stuck m, Magnesiastuck = magnesite stucco
Magnesium n, ~legierung f = magnesium alloy
~, Mg = magnesium [*chemical element*]
Magnesiumchlorid n = magnesium chloride, $MgCl_2 \cdot 6H_2O$
~lösung f, Chlormagnesiumlauge f = liquid magnesium chloride
Magnesium|gußlegierung f = magnesium casting alloy
~hydrosilikat n, wasserhaltiges Mg-Silikat = hydrous silicate of magnesia, ~ magnesium silicate
~karbonat n = magnesium carbonate
~(legierung) n, (f) = magnesium alloy
~olivin m (Min.) → Forsterit m
~oxyd n = magnesium oxide, MgO
~silikat n, $3 SiO_3Mg \cdot 5H_2O$ = silicate of magnesium, magnesium silicate
~sulfat n = magnesium sulphate (Brit.); ~ sulfate (US)
Magnet|abscheider m → Magnetscheider
~band n = magnetic tape
~bremse f = magnetic brake
~eisenerz n = magnetic iron ore
~eisensand m = magnetic ironsand, iserine
~felderregung f = field excitation
~feldspule f = field coil
~feldstromkreis m = field circuit
~feldwindung f = field winding
~flußdichte f = magnetic flux density
~greifer m = (electric) lifting magnet(s) with tines (US) / tynes (Brit.); crane ~ ~ ~
magnetische Permeabilität f, ~ Durchlässigkeit f = magnetic permeability
magnetischer Bremslüfter m = brake lifting magnet
magnetisches (Aufschluß)Verfahren n → Geomagnetik f

magnetisierende Röstung f = magnetizing roast
Magnetit|beton m = magnetite concrete
~filter m, n, Laughlinfilter = Laughlin filter
~zuschlag(stoff) m = magnetite aggregate
Magnet|kies m, FeS (Min.) = magnetic pyrite(s), pyrrhotine, pyrrhotite
~kupplung f = electric clutch
Magneto|hydrodynamik f = magnetohydrodynamics
~meter n = magnetometer
magnetomotorische Kraft f = magnetomotive force
Magnet|paar n = pair of magnets
~rüttler m → Elektrorüttler
~scheider m, Magnetabscheider, Magnetausscheider = magnetic separator
~sieb n [*Motor*] = magnetic strainer
~ventil n = solenoid valve
~verstärkung f des Anlassers = starter solenoid
~vibrator m → Elektrorüttler m
Magnuseffekt-Durchflußmesser m = magnus effect flowmeter
Mahl|bahn f [*Hammermühle*] = hammer path
~barkeit f, Mahlwiderstand m = grindability [*The response of a material to grinding effort. Can be measured in a number of ways. A current grindability test measures it in grams of product material produced per revolution of a test mill. The more grams produced per revolution, the easier a material is to grind and the higher is the grindability*]
~barkeitprüfer m = grindability tester, ~ machine
~busen m = noria basin
Mahlen n, Mahlung f [*(Hart)Zerkleinerung in Mühlen*] = grinding
Mahl|feinheit f = fineness of grinding
~hilfe f, Mahlhilfsmittel n, Mahlzusatz m = grinding aid, ~ additive [*Certain chemical additives which aid in tube mill grinding by reducing ball*

Mahlkalk — manometrische Druckhöhe

coating or by dispersing the finely ground product]
~kalk *m* → ungelöschter gemahlener Kalk
~körper *mpl* = grinding media [*Hard, free-moving charge in a ball or tube mill between which particles of raw material, coal, or clinker are reduced in size by attrition or impact. Usually of steel, and spherical in shape with graded sizes, the maximum in a ball mill being about 3 to 4 times the maximum feed size*]
~körper *m* = grinding element
~kugel *f* = grinding ball
~ring *m*, kreisförmige Mahlbahn *f* = grinding ring
~stufe *f* = grinding stage
~trocknung *f* [*Feuchtes Aufgabegut wird während des Mahlvorganges getrocknet*] = hot-air drying
Mahlung *f* → Mahlen *n*
Mahl|widerstand *m* → Mahlbarkeit *f*
~zusatz *m* → Mahlhilfe *f*
Mähmaschine *f* = mowing machine
Maifeld *n*, Groden *m* [*Fläche an der Nordsee mit annähernd fester Grasnarbe wo Wasser steht*] = false channel
~ [*Bezeichnung für Bodenstreifen auf dem ein Deich geschüttet ist*] = foundation area of a dyke (or dike)
Makadam *m* = macadam
~-Ausleger *m* → Schotterverteiler
~bauweise *f* = macadam road construction method
~belag *m* → Makadamdecke *f*
~decke *f*, Makadambelag *m*, Kompressionsdecke, Kompressionsbelag = macadam surfacing, ~ pavement; ~ paving (US)
~(misch)anlage *f* = coated macadam plant
~-Unterbau *m* = macadam substructure
Mäkler *m*, Läuferrute *f*, Laufrute *m* = leader
~abstecker *m* = catch-bolt
~ansatzstück *n*, Mäklerverlängerung *f* = leader extension

~trommel *f*, Läuferrutentrommel, Laufrutentrommel = leader drum
Makroniederschlag *m* = macroprecipitation
malen, anstreichen = to paint
Maler *m*, Anstreicher *m* = painter, house ~
~-Abbrennlampe *f* = painter's torch
~arbeiten *fpl*, Anstricharbeiten = painting work
~eimer *m* = paint kettle [*A cylindrical pail used by painters for holding paint and dipping the brush into*]
~gerüst *n*, Malerrüstung *f*, Anstreichergerüst, Anstreicherrüstung = scaffold(ing) for painters
~pinsel *m* = paint brush
~rüstung *f* → Malergerüst *n*
~werkstatt *f* = paint shop
~werkzeug *n*, Anstreicherwerkzeug = (house) painter's tool
Malteserkreuz *n* = Maltese cross
Mammutpumpe *f* = mammoth pump [*The air-lift pump within the hollow boring rods of a trepan*]
Mangan|blende *f* → Glanzblende
~spat *m*, Rhodochrosit *m* (Min.) = rhodochrosite, manganese spar
~stahlplatte *f* [*Gleiskettenlaufwerk*] = manganese shoe
Maniak *n* → Glanzpech *n*
Manila(-Hanf)seil *n* = manila hawser; grass rope (US)
MAN-Laschenkette *f* = MAN type chain
Mannloch *n* → Kontrollschacht *m*
~ [*verschließbare Einstiegöffnung in Kessel, Behälter oder Schotte*] = manhole
Mannschaft *f* → Kolonne *f*
Mannschafts|aufzug *m* = personnel-carrying hoist
~wagen *m* = personnel carrier
Manometer *m, s*, Druckmesser *m* = manometer
~flüssigkeit *f* = manometer liquid
manometrische Druckhöhe *f* [*DIN 4047*] = manometric pressure head

manometrische Förderhöhe — (Markt)Halle

~ **Förderhöhe** *f* [*DIN 4044*] = manometric delivery head
~ **Saughöhe** *f* = manometric suction head
Manövrierfähigkeit *f* → Freibeweglichkeit
manövrierunfähig = immobilized
Mansarde *f* = mansard
Mansard(en)|dach *n* = mansard roof, curb ~ (Brit.); gambrel ~ (US) [*It slopes in two directions, but there is a break in the slope on each side*]
~**-Flachdach** *n* = deck roof (US); mansard flat ~ (Brit.) [*It slopes in four directions, but has a deck at the top*]
~**-Zeltdach** *n*, Zelt-Mansardendach = mansard roof
~**zimmer** *n* = attic room
Mantel *m* = jacket [*outer covering*]
~, Strecken~ (⚒) = surrounding rock
~ [*KELLER-Tiefenrüttler*] = cylinder
~, Bandage *f*, Stahl~ = (steel) facing
~**beton** *m* = haunching
~**druck** *m* (⚒) = surface pressure
~**elektrode** *f* = covered electrode
~**fläche** *f*, Flechtwerkmantel *m* [*Kuppel*] = trellis casing, surface shell
~**fläche** *f* [*Pfahl*] = skin surface
~**pfahl** *m*, Hülsenpfahl = shell pile, cased ~
~**reibung** *f*, Seitenreibung [*Pfahl*] = = skin friction, side ~, lateral ~
~**rohr** *n* → Pfahlrohr
~**rohr** *n* [*Einfachkernrohr*] = hollow cylinder
~**rohr** *n* [*beim Filterbrunnen*] = (perforated) (well) casing
~**rohr** *n* kleiner als $2^1/_2$ **Zoll Durchmesser** [*Grundwasser(ab)senkung*] = well point
~**rohrbrunnen** *m* = cased well
~**rohrpfahl** *m*, Blechhülsenpfahl = cased pile, shell ~
~**rohr-Preßpfahl** *m*, Blechhülsen-Preßpfahl = shell pressure pile, cased ~ ~

Marcus-Wurffförderrinne *f* → Propellerrinne
Mareogramm *n* = mareogram
marine Abrasion *f* → Abrasion
~ **Erosion** *f* → Brandungserosion
~ **Sedimente** *npl* = marine deposits
Marine|leim *m* = marine glue
~**luftstützpunkt** *m* = naval air station, ~ ~ base
~**pionier** *m* = Seabee (US)
~**probe** *f* = navy tear test
~**stützpunkt** *m* = naval base
markanter Bohrfortschrittwechsel *m* [*Tiefbohrtechnik*] = clear drilling (time) "break"
Markasit *m* (Min.) = marcasite, white iron pyrite, fool's gold
Marken(bau)stoff *m* = proprietary (construction) material, ~ building ~
Markierung *f* = marking
Markierungs|beton *m* = concrete for carriageway markings
~**farbe** *f* = marking paint
~**kopf** *m*, Spurnagel *m*, Verkehrsnagel, Straßennagel = traffic stud, street marker, roadstud
~**-Leucht-Nagel** *m* → Straßen-Leucht-Nagel
~**linie** *f* = marker line
~**linie** *f* → Verkehrslinie
~**maschine** *f* = marking machine
~**masse** *f* = marking compound
~**streifen** *m* → Verkehrslinie
~**streifen** *m* aus heiß aufgebrachter plastischer Masse = hot-melt plastic stripe
~**strich** *m* → Verkehrslinie *f*
Markscheidekunde *f* = mine surveying
Markscheider *m* = mining surveyor
~**gehilfe** *m*, Vermessungssteiger *m* = deputy mining surveyor
markscheiderische Messungen *fpl* = survey measurements
Markscheide|sicherheitspfeiler *m* = boundary pillar
~**wesen** *n*, Markscheidekunde *f* = mine surveying
(Markt)|Halle *f* = covered market

~platz *m* = market place

Marmor *m* = marble [*The term strictly applies to a granular crystalline limestone, but in a loose sense it includes any calcareous or other rock of similar hardness that can be polished for decorative purposes*]

~arbeiten *fpl* = marble work

~boden *m*, Marmorfußboden = marble floor finish, ~ floor(ing), ~ floor covering

~bruch *m* = marble quarry

~fliese *f*, Marmorplatte *f* = marble tile

~(fuß)boden *m* = marble floor(ing), ~ floor finish, ~ floor covering

~gips *m* → Alaungips

Marmorierung *f* = marbled effect

Marmor|kalkstein *m*, technischer Marmor *m*, polierbarer dichter Kalkstein = compact polishable limestone

~körnung *f* = marble grain mix(ture)

~mosaik *n* = marble mosaic

~platte *f* = marble slab

~platte *f*, Marmorfliese *f* = marble tile

~poliermaschine *f* = marble-polishing machine

~poliermittel *n* = marble polishing material

~sägewerk *n*, Marmorsägerei *f* = marble sawing works

~splitt *m* = marble chip(ping)s

~sperre *f* = marble dam [*Little known and inaccessible are two architectural wonders of India, the dams at Jai Samand and Raj Samand, in the Aravalli hills of India's State of Rajasthan. The most unusual feature of these 17th century structures is that they are of pure, polished white marble. — They are probably the most beautiful dams ever built*]

~tritt(stufe) *m*, (*f*) = marble slab (stair) tread

~verkleidung *f*, Marmorauskleidung, Marmorbekleidung = marble facing, ~ lining

~zement *m* → Alaungips

~zuschlag(stoff) *m* = marble aggregate

Maronage *f* → feine Rißbildung

Marsch|boden *m* = bog soil

~fahrt *f* = travel from one site to another

marschige Niederung *f*, Marsch *f* = marshland

Marsch|kompaß *m*, Prismakompaß = prismatic compass

~land *n* = marsh-land

~ton *m* = sea-clay

Marshall|-Fließwert *m*, Fließwert nach Marshall = Marhall flow value

~-Probekörper *m* = Marshall test specimen

~-Prüfpresse *f* = Marshall (cylindrical) test-head

~-Prüfung *f*, Marshall-Probe *f*, Marshall-Versuch *m* = Marshall test

~-Stabilität *f* = Marshall stability

Mascaret → Bore *f*

Masche *f* = mesh

Maschendraht *m* = screen wire, metallic fabric

~verzug *m* für Gesteinsanker = metallic fabric lagging for support between roof bolts

Maschen|gleichheit *f* = equal-sized meshes

~rechen *m* [*Am Freispiegelwasserschloß, Stolleneinlauf, Werkkanal oder Turbineneinlauf um Treibzeug abzufangen*] = screen

~rechenbecken *n* = screen basin

~rechengut *n*, Maschenrechenrückstand *m* = screenings

~sieb *n* = mesh screen

~teilung *f* = mesh pitch

~weite *f* = aperture width, mesh ~

~weitenfolge *f* = mesh series

~werk *n* = network of meshes

maschinell = mechanically

~ abgebauter Stoß *m* (⚒) = machine face

~ abgezogener Beton *m* = machine-finish concrete

~ eingebaut = machine-installed

~ ~ [*Straßenbelag*] = machine-laid, mechanically laid

maschinelle Straßendeckenherstellung *f* = machine-laid work

maschineller Aushub — Maschinengründung 656

maschineller Aushub m = mechanical excavation
~ Einbau m = machine laying
~~ von Gußasphalt = machine casting

maschinelles Aufbringen n [z. B. von Putz] = machine application, mechanical ~
Maschinenaggregat n → Aggregat

Maschinenberichtswesen n — equipment recordkeeping, ~ paperwork

  Arbeitszeitkarte f [für ein in der Werkstatt in Reparatur befindliches Gerät] — time card
  Aufzeichnung f der Werte auf der Baustelle — recording data
  Auswertung f der Berichte — analyzing data
  Baustellen-Unterhaltung f — field maintenance
  Bericht m mit Stundenkosten — cost-per-hour record
  Filterwechselzeitabschnitt m — filter-replacement period
  Geräteabteilung f — equipment division, ~ department
  Geräteeinsatztafel f — disposition board
  Grundüberholung f — major overhaul
  Karteikarte f — file card
  Lochen n eines Werkstattauftrages — punching in repair job
  Maschinenstammbuch n — home-office folder
  Monatsbericht m — recap sheet
  Ölwechselzeitabschnitt m — oil-change period
  Reparatur-Auftragnummer f — repair-order number
  Reparaturwagen m mit Gerätehandbüchern — equipment-manual library on wheels
  Schnellhefter m, Heftmappe f — file folder
  Schnellhefter m auf der Baustelle — travel(l)ing folder
  Standortkartei f — box of location cards
  Tagesberichtsblatt n, Tagesberichtsformular n — daily checklist form
  tägliche Maschinenberichte mpl — day-to-day equipment figures
  Überwachungsblatt n, Überwachungsformular n — checklist form
  Zusammenfassung f der Werte in der Zentralverwaltung — consolidating data
  Zweitausfertigung f — duplicate file

Maschinen|element n = machinery component
~fahrsteiger m (✹) = master mechanic
~fundament n, Maschinengründung f, Maschinenfundation f, Maschinengründungskörper m, Maschinenfundamentkörper = machine(ry) foundation (structure)
~fundamentkörper m → Maschinenfundament n
~fundation f → Maschinenfundament n
maschinengemischt = plant-mixed
(Maschinen)Grundplatte f = machine bed plate, ~ base ~
Maschinen|gründung f → Maschinenfundament n

Maschinengründungskörper — Massenrüttelbeton

~gründungskörper m → Maschinenfundament n
~halle f [Kraftwerk] = machine hall
~haus n, Krafthaus, Turbinenhaus = powerhouse
~hauskaverne f, Krafthauskaverne = powerhouse cavern
~ingenieur m = mechanical engineer
~-Isolierungen fpl = waterproofing and dampproofing of machines, ~ ~ ~ ~ machinery
~lärm m = machinery noise
~meister m = master mechanic
~mischen n = mechanical mixing, plant ~
~mischung f, Maschinengemisch n = plant-mix(ture)
~montage f = erection of machinery
~nietung f = machine riveting
~öl n = machine oil
~pumpe f = mechanically driven pump
~raum m = engine-room
~reibahle f = machine reamer
(~-)Satz m, (Maschinen-)Aggregat n = rig, outfit, set
~schlosser m, Reparaturschlosser = mechanic
~schotter m = machine-broken metal (Brit.)
~spachtelung f = mechanical screeding
~spritzpumpe f = power spraying pump
~stall m (⚒) = stable (hole)
~stunde f = machine-hour
~telegraph m = engine telegrapn
~torf m = machine-cut peat
~turm m → Kabelkran
~werkstatt f = engineering (work)shop
~wesen n = mechanical engineering
~ziegel m, Schnittziegel = wire-cut brick, machine-made ~
~zusammenbau m = fitting
Maschinist m = machinist, engine driver
Massageraum m = massage room
(Masse)Blatt n [zum Überdrehen. Steinzeugindustrie] = bat [for jiggering]

Massel|eisen n, Roheisen = pig (iron)
~gießmaschine f = pig-casting machine
Massen|asphalt m [Wasserbau] = mass-asphalt
~ausgleich m, Massenausgleichung f = balancing quantities, balanced excavation, earthwork(s) balance
~aushub m = bulk excavation
~auszug m = table of quantities
~bauzement m = cement for mass concrete
~berechnung f [Erdbau] = computing quantities of earthwork(s)
~berechnung f, Massenermitt(e)lung f = computation of quantities
~beschleunigung f = acceleration
~beton m → Massivbeton
Massenbeton m → Massivbeton
~bau m = massive concrete structures
~einbau m = mass pours
Massen|durchflußmesser m = mass (-rate) flowmeter
~dusche f → Brauseraum m
~entnahme f = borrow
~ermitt(e)lung f → Boden~
~ermitt(e)lung f → Massenberechnung
~förderer m → Stetigförderer
~gestein n → Erstarrungsgestein
~gut n, Schüttgut = bulk material, loose ~, material in bulk
~gutgreifer m, Schüttgutgreifer = bulk material grab, loose ~ ~
~gutlagerung f = bulk storage
~gutschiff n = bulk carrier, ~ vessel
~gutumlader m, Schüttgutumlader = bulk material transfer equipment
~gut-Verladeanlage f = bulk material loading plant
~gutwaage f, Schüttgutwaage = bulk material scale
~kalkstein m, körniger Kalkstein = granular limestone
~lader m → Dauerlader
~mittelpunkt m = centre of mass (Brit.); center ~ ~ (US)
~rüttelbeton m, Rüttelmassenbeton = vibrated mass concrete

~-Sieb- und Verladeanlage *f* = bulk screening and loading station, screened bulk loading ~

~summenverteilung *f* [*in der Kinetik des Zerkleinerns*] = size distribution

~verkehrsfahrzeug *n* = public-service vehicle

~verkehrstreppe *f* = public staircase (Brit.); ~ stair (US)

~verschiebung *f*, Massentransport *m* [*Die Übergangswellen sind aperiodisch. Die Bewegungen der Wasserteilchen erstrecken sich über die gesamte Wassertiefe, wobei die Wasserteilchen in Wellenfortpflanzungsrichtung verschoben werden*] = mass transport

~verteilung *f* → Erd~

~verzeichnis *n* → → Angebot *n*

~ware *f*, Massengut *n*, Massengüter *npl* = bulk goods

Masse|schlagmühle *f* = roller-kneading machine

~schraube *f* = earth screw

Massivbauwerkteil *m*, *n* = massive structure [*e. g. piers, abutments, dam sections, etc.*]

Massivbeton *m*, Massenbeton = mass(ive) concrete, bulk ~, concrete-in-mass

~decke *f*, Vollbetondecke = solid concrete floor

~pfeiler *m* = mass(ive) concrete pier

~platte *f*, Voll(beton)platte = solid (concrete) slab

~(stau)mauer *f*, Betonsperre, Staumauer, Talsperren(stau)mauer, (Beton)Sperrmauer = (massive) concrete dam

~widerlager *n* = mass concrete abutment

Massiv|blockmaschine *f*, Vollblockmaschine = solid-block machine

~bogen *m* = massive arch

~brücke *f* [*DIN 1075*] = solid bridge

~dach *n* = solid roof

~decke *f* = solid floor

massive Kernmauer *f*, Beton(dichtungs)kern *m*, Betonkernmauer [*Talsperre*] = concrete core wall

massives (Stau)Wehr *n* = solid weir

Massiv|gipsplatte *f* = solid gypsum board

~mauerblock *m*, Massivwandblock = solid walling block

~mauerwerk *n*, Vollmauerwerk = solid masonry

~-Ortbetonplatte *f* = solid in-situ slab

~pfahl *m* = solid pile

~platte *f* → Massivbetonplatte

~stufe *f*, Blockvollstufe = solid block step

~tragwerk *n* = solid structure

~trennwand *f* = solid partition

~treppe *f* = solid staircase (Brit.); ~ stair (US)

~wand *f* = solid wall

~wandblock *m*, Massivmauerblock = solid walling block

~-Zwischenwand *f* = solid partition

Massonet-Quasi-Verfahren *n* = Massonet quasi-method

Mast *m* = mast

~ → Beleuchtungsmast

~ausleger *m* [*Lichtmast*] = bracket; mast arm (US) [*An attachment to a lighting column from which a lantern (Brit.) / luminaire (US) is suspended*]

~-Betonrammpfahl *m*, Betonrammpfahl System Mast = Mast (impact-)driven concrete pile

Mast(en)form *f* = column mould (Brit.); mast mold (US)

Mastenkran *m* → Derrickkran

Mast|fuß *m* = lower end of the mast

~höhe *f* = height of mast

Mastix *m*, Asphalt~ = mastic asphalt (Brit.); asphalt mastic (US) [*B. S. 1446*]

~asphalt *m* → Gußasphalt

~brot *n*, Asphalt~ = mastic block

~eingußdecke *f* → Mastix-Vergußdecke

~fuge *f* = mastic joint

~-Kocher *m* → Gußasphaltkocher

~prisma *n*, Asphalt~ = asphalt mastic beam (US); mastic asphalt ~ (Brit.)

Mastixstrick — Materialpreisgleitklausel

~strick m = mastic cord
~-Vergußdecke f, Mastixeingußdecke, Asphalteingußdecke, Walzschotter-Gußasphalt m = mastic grouted surfacing (Brit.); ~ penetration pavement
~vergußmasse f = mastic filler
~zement m = mastic cement
Mast|schleuderanlage f, Schleudermastanlage = pole plant
~verlängerung f = mast extension
Masut n, abgetopptes russisches (Roh)Erdöl n = masut, mazut, heavy fuel
Maß n in fester Masse [*Der Rauminhalt von Erdreich oder Fels in ungestörter, natürlicher Lagerung*] = bank measure
~abweichung f → Abmaß n
maßeinheitlich = modular
(Maßein)Teilung f [*z. B. bei einem Skalenpegel*] = graduation(s)
maßgebender Niederschlag m, Berechnungs-Niederschlag = project precipitation, design ~
~ Regen m → Berechnungsregen
maßgebendes Hochwasser n → Berechnungs-Hochwasser
Maßgenauigkeit f, Maßhaltigkeit = dimensional accuracy
maßgleiche Darstellung f, isometrische ~ = isometric representation
Maßhaltigkeit f, Maßgenauigkeit = dimensional accuracy
mäßiger Gipsschlackenzement m, ~ Sulfathüttenzement = moderate sulphate resisting cement, ~ heat of hydration ~; Portland cement type II (US)
Maß|norm f → Abmessungsnorm
~ordnung f [*DIN 4172*] = modular co-ordination, ~ measure system
~ordnungsbreite f = module width
~ordnungssystem n = modular system
~skizze f = dimensional sketch
Maßstab m = scale
~-Beziehung f = scale relation
maßstabsgerechtes Modell n = scale model

Maß|stabwirkung f = scale effect
~toleranz f = dimensional tolerance
Maßwerk n = tracery
Material n für die Schweißnahtüberhöhung = reinforcing material
~, Werkstoff m = (engineering) material
~anfuhr f, Materialanlieferung f = material(s) delivery
~aufbereitung f = (materials) processing (or dressing, or preparation)
~aufgabe f → Aufgabe
~aufgabeapparat m → Aufgeber m
(~)Aufgabevorrichtung f → Aufgeber
~aufgeber m → Aufgeber
~aufzug m → (Lasten)Aufzug
~aufzugmast m, (Lasten)Aufzugmast = hoisting mast
~(be)förderung f, Materialumschlag m = material(s) handling
Beförderung von Massengütern = handling of bulk materials
Beförderung von Stückgütern = package handling
~behälter m, Fülltrichter m = (receiving) hopper, feed ~
~beschicker m → Aufgeber m
~beschickung f → Aufgabe f
~beschickungsapparat m → Aufgeber
(~)Beschickungsvorrichtung f → Aufgeber m
~bestellung f = ordering of material(s)
(~)Bunker m, Trichterbunker = hopper, bin
~flußbild n = nowsheet, flow diagram
~flußingenieur m = materials handling engineer
~förderer m → Förderer
~fördergerät n → Förderer m
~konstante f, Stoffkonstante = material constant
~kübel m [*Schachtbau*] = materials sinking bucket, ~ skip; ~ kibble, ~ bowk, ~ hoppit (Brit.)
~lager n = material store
(~)Lagerung f = (Auf)Speicherung
~preisgleitklausel f = materials variation clause, ~ fluctuation ~

42*

~**prüfreaktor** m = materials testing reactor, ~ ~ pile, MTR
~**prüf(ungs)anstalt** f, Materialprüf(ungs)amt n, MPA = material(s) testing lab(oratory), lab(oratory) for material(s) testing
~**schleuse** f = muck-lock, materials lock
(~)**Umschlaggeräte** npl, Geräte für die Material(be)förderung = material(s) handling equipment (or machinery)
~**verknappung** f, Materialknappheit f = material(s) shortage
mathematisches Modell n = mathematical model
Matratze f → Baggerrost m
~ → Sinkstück
Matrixrechnung f, Matrixmethode f [*Baustatik*] = matrix method
Matsch m, Schnee~, Schneebrei m = slush
Mattanstrichfarbe f = matt paint
Matte f [z. B. *aus Steinwolle*] = blanket ~, Asphalt~ = (asphalt) mattress
Matten|belag m = matting
~**bewehrung** f mit Längsmasche = oblong-mesh reinforcement
~**verleger** m, Bewehrungsmatten-Ausleger, Verlegegerät n für Bewehrungsmatten = fabric laying jumbo, mesh ~ ~
Matt|kohle f = dull coal, dulls
~**lack** m = flat varnish
Mauer f = wall
~ **in aufgelöster Bauweise** → Pfeilersperre f
~**abdeckstein** m, Mauerdeckelstein = coping stone, capping ~
~**abdeckung** f, Mauerdeckel m = coping, cope, capping
~**absatz** m = offset
~**anker** m → Deckenanker
~**baustelle** f → Talsperre f
~**binder** m, Mauermörtelzement m, Mauerwerkzement, PM-Binder, (Putz- und) Mauer(werk)binder = masonry cement, cementing material for masonry and brickwork
~**blende** f = flat niche of a wall

~**block** m, Wandblock = walling block
~**bogen** m = relieving arch, rough ~, discharging ~
~**deckel** m, Mauerabdeckung f = cope, coping, capping
~**deckelstein** m, Mauerabdeckstein = capping stone, coping ~
~**falz** m → Dammfalz
~**flucht** f = wall line
~**fräse** f = wall channeler
~**fraß** m, Salpeterfraß = efflorescence
~**fuge** f = wall joint
~**fuggerät** n = (wall-)pointing machine
~**fundament** n, Mauergründungskörper m, Mauergründung f, Mauerfundation f, Mauerfundament-Körper = wall foundation; ~ footing (US)
~**fundamentkörper** m → Mauerfundament n
~**fundation** f → Mauerfundament n
~**gründung** f → Mauerfundament n
~**gründungskörper** m → Mauerfundament n
~**haken** m → Deckenanker m
~**kanal** m = canal with masonry walls (on both banks)
~**klammer** f = wall clamp
~**klotz** m [*Widerlager*] = body of wall
~**kreuzung** f = crossing of walls
~**krone** f = wall crest, ~ top, ~ crown
~**latte** f = wall plate
~**lattenkranz** m = circle of wall plates
~**mörtel** m, Fugenmörtel, Fugungsmörtel, Mauerspeise f = masonry mortar, joint ~, pointing ~
~**mörtelzement** m → Mauerbinder m
mauern, auf~ = to brick up
Mauern n **mit (Beton)Blöcken**, ~ ~ (Beton)Steinen, ~ ~ (Beton)Formsteinen = blocklaying
~ ~ **Ziegeln** = bricklaying, bricking, laying bricks
Mauer|nagel m = masonry nail
~**nut** f = (wall) chase
~**nutfräse** f = (wall)chaser
~**öffnung** f = wall opening

~pfeiler *m* = wall pier
~pfeiler *m* [*im Luftraum unter Erdgeschoß/ußboden*] = sleeper wall, honeycomb ~
~quader *m* = ashlar
~säule *f* = wall column
~schale *f*, Hohl~, Schale = leaf
~schicht *f* = course
~schuttverwertung *f* → Trümmerverwertung
~sockel *m* = plinth [*A plinth is formed from ground level, rising a few courses. It is distinguished from the upper wall by a slight projection or by some difference in colour or material*]
~speise *f* → Mauermörtel *m*
Mauerstein *m*, Wandbaustein = masonry unit
~ **aus Beton**, Betonmauerstein = concrete wall block
~verband *m* → Mauerverband
Mauerung *f* → Aus~
Mauer|verband *m*, Mauersteinverband = (masonry) bond
~versatzung *f* = skew notch on wall
~-**Wellenbrecher** *m* = wall-breakwater
Mauerwerk *n* = masonry
~ **über Geländeoberkante** = above-grade masonry
~anstrichfarbe *f* = masonry paint
~auspressung *f* → Mauerwerkinjektion *f*
~bewehrung *f* = masonry (wall) reinforcement
~binder *m* → Mauerbinder
~bogen *m* = masonry arch
~-**Bogenbrücke** *f* = masonry arch(ed) bridge
~brücke *f* = masonry bridge
~-**Brückenpfeiler** *m* = masonry bridge pier, ~ ~ support
~-**Brunnengründung** *f* = masonry well foundation
~bunker *m* = masonry bin
~einpressung *f* → Mauerwerkinjektion *f*
~festigkeit *f* = masonry strength

~injektion *f*, Mauerwerkeinpressung *f*, Mauerwerkauspressung, Mauerwerkverpressung = masonry sealing
~öffnung *f* = masonry opening
~schleuse *f* [*Schiffahrtschleuse*] = masonry lock
~-**Schneidscheibe** *f* = masonry cutting blade
~sperre *f* → Talsperre
~stollen *m* = masonry gallery
~trümmer *f* = masonry ruins
~verband *m* = masonry bond
~verpressung *f* → Mauerwerkinjektion *f*
~-**Wasserturm** *m* = masonry water tower
~wehr *n* = masonry weir
~zement *n* → Mauerbinder *m*
Mauerziegel *m* = wall brick
Mauk|haus *n*, Sumpfhaus = souring house
~mischer *m* = tempering mixer
~turm *m* = tempering tower
Maul *n* → → Backenbrecher
~wurfschweißen *n* → Unterpulverschweißen
Maurer *m* = bricklayer (Brit.); brick mason (US)
~arbeiten *fpl* = bricklayer's work
~gerüst *n*, Maurerrüstung *f* = bricklayer's scaffold(ing)
~hammer *m* = brick(layer's) hammer
~handwerk *n* = handicraft of the bricklayer, bricklaying craft
~kelle *f* = brick trowel
~polier *m* = foreman bricklayer
~rüstung *f*, Maurergerüst *n* = bricklayer's scaffold(ing)
~schnur *f*, Fluchtschnur = bricklayer's line
~werkzeug *n* = bricklayer's tool
maurische Architektur *f* = Saracenic architecture, Moorish ~
maurischer Bogen *m* → Hufeisenbogen
„**Mausefalle**" *f* [*Fangwerkzeug für Tiefpump(en)gestänge*] = "mouse trap"
Mauseloch *n* [*Ein verrohrtes und etwa 10 m tiefes Bohrloch auf der Arbeitsbühne von Rotary-Bohrgerüsten zum*

Abstellen eines Bohrgestänges] = mouse hole
mäusesicher = mouseproof
Maximalbelastung *f*, Höchstbelastung = maximum loading
maximale Ringzugkraft *f* [*aus der Flüssigkeitsfüllung*] = maximum horizontal tensile strees, hoop ~ [*in the tank wall due to the liquid pressure*]
~ Zugkraft *f* **an Zughaken bei belastetem Schlepper** = maximum drawbar pull with adequate weight and traction
Maximalleistung *f* [*Motor*] = maximum horsepower
Maximumprinzip *n* = maximum principle
Mechanik *f* **des hydraulischen Grundbruches** = mechanics of (failure by) piping
Mechaniker *m* = mechanic
mechanisch angetriebener Schneidkopf *m*, Fräserkopf, Wühlkopf(-Fräser) *m* = revolving type of cutter head, suction-cutter apparatus, (revolving) cutter, mechanically operated cutter
~ verfestigte Straße *f* **mit Abdichtung des Planums** = membrane stabilized soil road
(~) versetzen, (~) verpacken (⚒) = to stow
mechanische Ähnlichkeit *f* = mechanical analogy
~ Beton(förder)pumpe *f*, **~ Förderpumpe** = piston type concrete pump
~ Bodenverfestigung *f*, Tonbeton *m* = mechanical (soil) stabilization, granular (~) ~, soil-aggregate surface
~ Eigenschaft *f* = mechanical property
~ Festigkeit *f* = mechanical strength
~ Flüssigkeits(ab)trennung *f* = liquid filtration
~ Förderpumpe *f*, **~ Beton(förder)pumpe** = piston type concrete pump

~ Haftung *f*, mechanischer Verbund *m* [*Haftung ohne Bindemittel oder Klebstoff*] = mechanical bond
~ Hand(trommel)winde *f* → Affe *m*
(~) Klärung *f* [*Abwasserwesen*] = settlement, sedimentation, settling, clarification
~ Korntrennung *f* = mechanical classification
(~) Kraftwinde *f* → Krafttrommelwinde
~ Kupp(e)lung *f* = mechanical clutch
~ Prallmühle *f*, Rotor-Prall-Mühle, Schnelläufermühle = (impeller) impact mill
~ Verwitterung *f* → physikalische ~
mechanischer Abbau *m* [*z. B. von Polybutylen in Lösungen von Mineralöl*] = mechanical breakdown
~ Aufgeber *m*, **~ Beschicker** *m* = mechanical feeder
~ Abweichungsmesser *m*, **~** Neigungsmesser [*Tiefbohrtechnik*] = mechanical drift indicator
~ Entstauber *m* = mechanical-type dust collector
~ Neigungsmesser *m*, **~** Abweichungsmesser [*Tiefbohrtechnik*] = mechanical drift indicator
~ Schwimmerschreibpegel *m*, selbst(auf)zeichnender Schwimmerpegel, [*Zeichen: Ss*] = float ga(u)ge, recording ~ ~, graphic water-stage register
~ Siebsatz-Schwingungserreger *m* → Siebsatzrüttler *m*
(~) Versatz *m* (⚒) = stowing
~ Vorschub *m* [*Bohren*] = screw feed
mechanisches Aufrauhen *n* **für den Verputz** = mechanical keying for renderings
~ Rührwerk *n* = mechanical agitator
(~) Schachtlot *n* = shaft plumb bob
~ Schwingsieb *n*, mechanischer Schwinger *n* = mechanically vibrated screen
~ Sedimentgestein *n* = mechanically deposited sedimentary rock
Mechanisierung *f* = mechanization
Mechanismus *m* = mechanism

Medizinalöl — Mehrfeldrahmen

~, Gelenkgetriebe n = mechanism, linkage
Medizinalöl n → Paraffinöl
Meerbusen m, Golf m = gulf
Meerdamm m → Deich m
~**bau** m → Deichbau
Meerdeich m → Deich
~**bau** m → Deichbau
Meer(es)|arm m = arm of the sea, sea inlet
~**bauwerk** n, Seebauwerk = offshore structure, marine ~
Meer(es)damm m → Deich m
~**bau** m → Deichbau
Meer(es)deich m → Deich
~**bau** m → Deichbau
Meer(es)|denudation f = marine denudation
~**enge** f = straits
~**grund** m, Seegrund = sea bottom, ocean floor, sea bed
~**höhe** f, Seehöhe, Höhe über Normalnull, H. ü. N. N. = altitude
~**klima** n = oceanic climate (US); marine ~ (Brit.)
~**kunde** f = oceanography
~**lebewesen** n = marine organism
~**luft** f, Seeluft = sea air
~**schlamm** m = ocean ooze
~**schlick** m, Seeschlick = sea mud
meeres|seitig, seewärts = seaward
~**seitige Böschung** f = seaward slope
Meeres|spiegel m, Seewasserspiegel = sea level
~**strömung** f [der ozeanische Strom] = marine current (US); ocean ~ (Brit.)
~**torf** m, Seetorf = sea peat
~**überschiebung** f = submergence by sea of land depression (Brit.); marine transgression (US)
~**wasser** n, Seewasser = seawater
~**wasseraufbereitungsanlage** f = sea water conversion plant
~**wasserbeton** m, Seewasserbeton = sea water concrete
Meer|kies m = marine gravel
~**schaum** m (Min.) = sepiolite, meerschaum
Mehl n → → Mo m

~**mühle** f = flour mill
~**sand** m → → Mo m
Mehrachs|-Anhänger m, Vollanhänger [Erdbaufahrzeug] = multi-axle trailer, full ~, multiwheeler ~, wagon
~**lenkung** f = multi-axle steering
Mehr|ausbruch m = overbreak(age) [*The amount excavated beyond the neat lines of a cutting or tunnel. This additional excavation is not paid for and its cost must therefore be included by the contractor in the price he tenders for the excavation within the neat lines*]
~**aushub** m = extra excavation
~**bogenbrücke** f = multi(ple)-arch bridge
~**decker(sieb)** m, (n) = (aggregate) grader, multideck screen
mehrere eingebaute Packer mpl = straddle packer
mehrfach abgespannter Mast m = multi-level guyed tower
Mehrfach|antrieb m = multiple drive
~**drahtgewebe** n [Siebtechnik] = multiplex weave, basket ~
~**dünger** m = multiple-nutrient fertilizer
mehrfacher Walzenkessel m = multiple cylindrical (or combined) boiler
Mehrfaches n = multiple
Mehrfach|kastenbauweise f = multiple box section construction
~**keilriemenantrieb** m = multiple V-belt drive
~**teerung** f = multiple tarring
~**verdampfer** m = multiple evaporator
~**verflechtung** f [Straßenverkehr] = multiple weaving
~**wand** f = multiple wall
mehrfeldrige kreuzweise bewehrte Platte f = continuous two-way slab
mehrfeldriger Fachwerk-Durchlaufträger m = multiple-span continuous truss
~ **Giebelrahmen** m = multiple span gabled frame
Mehr|feldrahmen m = multi-bay frame

~feldträger *m* = multiple span girder
~gang(boden)mischer *m* = multipass soil stabilizer
~geschoßbau *m*, Mehrgeschoßgebäude *n* = multi-stor(e)y building
~kammerrohrmühle *f* = multi-compartment mill
~kammerstromsichter *m* = hydrosizer
~kanalanzeige *f* = multi-channel indication
mehrlagig, mehrschichtig = multi-course, multi-layer
mehrlagige Dach(pappen)eindeckung *f* = built-up roof(ing)
~ Spannbewehrung *f* = multi-layer prestressed reinforcement
mehrlagiger Anstrich *m* → Anstrichsystem *n*
mehrläufige Treppe *f* = multiflight stair(case)
Mehrmotoren-Elektrobagger *m* = multiple-motor all-electric shovel
mehrmotoriger Elektroantrieb *m*, elektrischer Mehrmotorenantrieb = multiple (electric) motor drive
Mehr|muldenschütter *m* = dumper with lift-on lift-off (bulk) tipping containers
~pfeilerbrücke *f* = multi-piered bridge
~phasenmischer *m* = multiflow mixer
~plattenverdichter *m* = multi(ple)-pan compactor
mehrräd(e)riges Fahrgestell *n* [*Flugzeug*] = multiple-wheel assembly, ~ landing gear
Mehr|schalengreifer *m* = grapple, multi-bladed (circular) grab; grabs (US) [*A clamshell-type bucket having three or more jaws*]
~scheibenglas *n*, Verbund(-Sicherheits)glas = laminated (safety) glass
~scheibenkupp(e)lung *f*, Lamellenkupp(e)lung = multiple disc clutch; ~ disk ~ (US)
~schichtarbeit *f* = multishift work
~schichtbelag *m* = multi-layer cover(ing), multi-course ~

mehr|schichtig, mehrlagig = multi-layer, multi-course
~schichtige Einbauweise *f*, Straßenaufbauweise [*Straßenbau*] = multiple lift construction, stage ~ (US); multiple-course ~ (Brit.)
~schichtiger Anstrich *m* → Anstrichsystem *n*
~schiffig [*Kirche*] = multi-nave
Mehrseil|förderung *f* (⚒) = multi-rope winding
~greifer(krankorb) *m*, Mehrseilgreif(er)korb = multiple suspension grab(bing) (crane) bucket, ~ ~ grab
Mehrspannglied-Träger *m* = multitendon girder
mehrspurige Straße *f*, vielspurige ~ = multi-lane highway, ~ road
~ ~ mit Mittelstreifen = multi-lane divided road, ~ highway
Mehr|stoffbronze *f* = multi-compound bronze
~strangförderung *f* = multiple string completions
~stufenrißbildung *f* [*Verfahren der hydraulischen Rißbildung, um in einem Bohrloch Trägerabschnitte unterschiedlicher Permeabilität zu behandeln*] = multifrac
mehrstufige Druckturbine *f*, Druckturbine mit Druckabstufung = multi-stage action turbine, ~ impulse ~, action (or impulse) turbine with pressure stages
~ Durchlaufmahlung *f*, mehrstufiges Durchlaufmahlen *n* = multi-stage open-circuit grinding
~ (Hart)Zerkleinerung *f* = multi(ple-)stage comminution, ~ reduction
mehrteiliger Druckstab *m* = built-up compression member
Mehrwellen-Boden-Zwangsmischer *m* → → Bodenvermörtelungsmaschine *f*
mehr|zahniger Aufreißer *m* = multiple shank ripper
~zelliger Kastenquerschnitt *m* = multicellular hollow section

Mehrzonenförderung *f* [*Es wird gleichzeitig aus mehreren Erdölhorizonten gefördert*] = multiple (well) completion

Mehrzweck|ausbau *m* **eines Stromsystems** = multiple-purpose river basin development

~**-Auto(mobil)bagger** *m* → Auto(mobil)bagger

~**bagger** *m*, Universalbagger, Umbaubagger, Vielzweckbagger [*Zollbezeichnung: herkömmlicher Bagger*] = convertible excavator (Brit.); ~ shovel-crane (US); all-purpose ~, universal ~

Umbaueinrichtung = (convertible) front end attachment, interchangeable ~ ~ ~, front end conversion unit, work attachment

~**bauwerk** *n*, Verbundprojekt *n*, Verbundbauwerk, Vielzweckbauwerk, Mehrzweckprojekt [*Talsperre*] = multi(ple)-purpose structure

~**fahrgestell** *n* = multi(ple)-purpose carrier, ~ chassis

~**fahrzeug** *n* = multi-purpose vehicle

~**fett** *n* = multi-purpose lubricating grease

~**gerät** *n* = multi-purpose equipment item

~**-Industrieschlepper** *m* → Kompressorschlepper

~**-Inhibitor** *m* = multi-purpose inhibitor

~**löffel** *m* = multi-purpose bucket

~**öl** *n* = multi-purpose oil

~**projekt** *n* → Mehrzweckbauwerk *n*

~**schmierfett** *n*, Vielzweckschmierfett = multi-purpose lubricating grease

~**speicher** *m* = multi-purpose reservoir [*A multi-purpose reservoir is designed for two or more uses. For example, a reservoir located on the tributary of a major river might be designed to protect the downstream river towns and cities against disastrous floods, increase the dependable water supply, and generate hydroelectric energy*]

~**stein** *m*, Mehrzweckblock(stein) *m*, Mehrzweckkörper *m* = multi-purpose block, ~ tile

Mehrzylinder-Dieselmotor *m* = multi(ple)-cyl. diesel engine

Meilenstein *m* = milestone

Meiler *m* = charcoal kiln

~ → Feldofen *m*

Meißel *m* → Bohr~

~ = chisel

~**blatt** *n*, Bohr~ = bit blade

~**fanghaken** *m*, Bohr~ = bit hook

~**hammer** *m* = chipping hammer

~**lehre** *f*, Meißelschablone *f*, Bohr~ = bit ga(u)ge

meißeln [*mit Handmeißel bearbeiten*] = to chisel

~ [*mit dem Meißel bohren*] = to drill with a bit

Meißel|hammer *m* = chipping hammer

~**probe** *f*, Bohr~ [*am (Bohr)Meißel haftende Gebirgsprobe*] = bit sample

~**schablone** *f*, Meißellehre *f*, Bohr~ = bit ga(u)ge

~**schaft** *m*, Bohr~ = shank of bit, bit shank

~**schneide** *f* → (Bohr)Schneide

~**ung** *f*, Schlagbohren *n*, Schlagbohrung *f* = percussive drilling

~**wechsel** *m*, Bohr~ = changing of the bit

~**zapfen** *m*, Bohr~ = bit pin

Melaphyr *m*, schwarzer Porphyr [*DIN 52100*] = melaphyre, black porphyry

~**porphyr** *m* = porphyric melaphyre

Meldedruck *m* = signal pressure

Melilith *m*, Honigstein *m* = mellite, mellilite

Meliorationen *fpl* = land drainage, ~ reclamation

Meliorationsarbeiten *fpl* = land drainage work, ~ reclamation ~

meliorationsbedürftiger Standort *m* = area requiring soil improvements

Membran(e) *f* = diaphragm, membrane

~**balg** *m*, Faltbalg = membrane bellows

~-**Baupumpe** *f*, Dia(phragma)-Baupumpe = diaphragm contractor's pump, membrane ~ ~
~-**Bunkerstandanzeiger** *m* = diaphragm-operated (bin) material level indicator
~**filter** *n, m* = membrane filter
~**gleichnis** *n*, Seifenhautgleichnis [*Baustatik*] = membrane analogy
~**isolierung** *f* gegen Wasser = membrane waterproofing
~**kraft** *f* = membrane force
~**pumpe** *f*, Dia(phragma)-Pumpe = diaphragm pump, membrane ~
~**spannung** *f* = membrane stress
~**spannungszustand** *m* = membrane state of stress
~**theorie** *f* = membrane theory
~**ultrafilter** *m, n* = membrane ultrafilter
~**wirkung** *f* = membrane action
~**zugbruch** *m* = tensile membrane failure
Mengen|änderung *f* = quantity variation, ~ change
(~)**Ausbringen** *n* [*Siebtechnik*] = output, recovery, yield
~**ergebnis** *n* = quantitative result
~**regelventil** *n* [*Ölhydraulik*] = oil volume restrictor
~**verzeichnis** *n* → → Angebot *n*
Mennige *f*, Blei~, Bleisuperoxyd *n* = red lead
menschliche Siedlung *f* = agglomeration, community
Mergel *m* [*40–75% $CaCO_3$*] = marl
~**boden** *m* = marly soil
merg(e)liger Kalkstein *m*, Mergelkalk(stein) *m* [*90–96% $CaCO_3$*] = marly limestone
~ **Ton** *m* [*4–10% $CaCO_3$*] = marly clay
Mergelschiefer *m*, Schiefermergel *m* = marl slate, slaty marl, margode
Merg(e)lung *f* = marling
Meridian|durchgang *m* = meridian passage [*The culmination of a star*]
~**ebene** *f* = plane of the meridian
~**kordinate** *f* = meridian coordinate
~**normale** *f* = normal of meridian
Merkblatt *n* = Note

~ **für den Straßenbau** = Road Note
Merkuricyanat *n* → Knallquecksilber *n*
messen = to measure
Messen *n* = measuring
Messer *m* [*Für Flüssigkeiten oder Dampf oder Gas*] = flow meter
~ *n* → Hobel~
~**ablesung** *f* = meter reading
~**abschneider** *m*, Abschneider mit Messern [*Ziegelindustrie*] = knifecutter
~**mischen** *n*, Schar-Mischung *f* [*Durchmischen des Bodens mit Straßenhobel*] = blade mixing
~**pumpe** *f* = vane(-type) pump
~**pumpe** *f* **mit Druckplättchen** = insert vane pump
Messing|dorn *m* = brass drift
(~)**Form** *f* [*Streckbarkeitsmessung von Bitumen*] = ductility test mo(u)ld
~**rohr** *n* = brass tube
Messung *f* **der Oberflächengeschwindigkeit** *f* [*Abflußmengenmessung in der Gewässerkunde*] = velocity measurement at the surface
~ **von Gezeitenströmungen in Strommündungen während des Mondwechsels** = lunar-cycle measurement of estuarine flows
Messungen *fpl* **mit Gammastrahlen** [*Bohrlochüberwachung*] = gamma-ray logging, ~ survey(ing)
~ ~ **Neutronenstrahlen** [*Bohrlochüberwachung*] = neutron logging, ~ survey(ing)
Meß|anker *m* (⚒) = measurement bolt
~**band** *n*, Bandmaß *n* = measuring tape
~**brücken-Kreis** *m* = Wheatstone bridge circuit
~**dose** *f* = measuring cell
~**elektrode** *f* → Potentialelektrode
~**ergebnis** *n* = measuring result
~**fahrzeug** *n* **das zwangsweise bei niedriger Geschwindigkeit mit einem eingeschlagenen Rad in einer festgelegten Spur fährt** = Dunlop cornering force machine
~**fehler** *m* = measuring error
~**flügel** *m* → Wasser~

Meßgerät — metallisierte Faser

~gerät *n* für die Laufzeit eines Schallimpulses = soniscope
~instrument *n*, Meßgerät *n* = measuring instrument
~kanal *m* = measuring channel
~kiste *f* = batch box, gauge ~
~länge *f* [*Probestab*] = ga(u)ge length
~latte *f* = measuring rod
~profil *n* = measured profile
~rahmen *m* [*Aus Holz ohne Boden. Dient zum Abmessen von Sand, Kies, Splitt usw.*] = measuring frame
~schlauch *m*, Druckschlauch [*Zählgerät mit einem über die Fahrbahn gelegten Luftschlauch*] = pneumatic tube laid across the road surface
~stab *m* → Anzeigestab
~stelle *f* = point of measurement
~strecke *f* = (longitudinal) section used for gauging current velocity
~tisch *m* = plane-table, planchette
~tischaufnahme *f* = plane tabling, ~ table mapping
~uhr *f* = dial (gauge) (Brit.); ~ (gage) (US)
~verfahren *n* = measuring method
~vorrichtung *f* = measuring device
~vorrichtung *f* für Seitenführungskraft, nachlaufendes Meßrad *n* = fifth-wheel device, fifth wheel
~wagen *m*, Registrierwagen [*geophysikalischer Aufschluß*] = recording truck
Meßwehr *n* = measuring weir [*A measuring weir is a device or structure over which water falls in such a manner that the rate of flow can be computed from the depth or head on the crest of the weir. Measuring weirs may be standard, broad-crested, sharp, curved or flat; suppressed or contracted; rectangular, V-notch, trapezoidal, semicircular, or hyperbolic; free-falling or submerged; or other shapes or conditions*]
~ mit dreieckigem Ausschnitt = triangular weir, V-notch ~
~ ~ rechtwink(e)ligem Einschnitt in U-Form = notch ga(u)ge

~formel *f* = (measuring) weir formula
~zylinder *m* = measuring cylinder
Metall|-(auf-)Metall-Dichtung *f* = metal-to-metal seal
~barometer *n* → Aneroidbarometer
~bau *m* = metal construction
~binder *m*, Metalldachbinder, Metallfachwerkbinder = metal roof truss
~brennkammer *f* = metal combustion chamber
~dach *n* = metal roof deck
~(dach)binder *m*, Metallfachwerkbinder = metal roof truss
~(dach)eindeckung *f* = flexible-metal roofing
~deckentafel *f* = metal ceiling panel
~deckleiste *f* = metal cover strip
~eindeckung *f*, Metalldacheindeckung = flexible-metal roofing
~einlage *f* = metallic insert
~elektrode *f* = metal electrode
~ermüdung *f* = metal fatigue
~(fachwerk)binder *m*, Metalldachbinder = metal roof truss
~fenster *n* = metal window
~-Fenstersprosse *f* = metal glazing bar, ~ sash ~, ~ astragal
~filter *m, n* = metallic filter
~folie *f*, Blechfolie = metal foil
~folieneinlage *f* = metal foil insert
~folienschicht *f*, Metallfolienlage *f* = metal foil layer
~form *f* = metal mould (Brit.); ~ mold (US)
~-Fugendichtung *f* mit abgerundeter Kante = bull nose metal mo(u)ld
metallführend = metal-bearing, metalliferous
Metall|gewebe *n* als Putzträger = metallic lath(ing), metal ~ [*B. S. 1369*]
~gewinnung *f* = extractive metallurgy
~gummi *m* = metal-faced rubber
~(hütten)schlacke *f*, (Eisen)Hüttenschlacke = steel slag, steel-mill ~
~-Inertgas-Schweißen *n*, MIG-Schweißen = MIG welding, metal inert gas ~
metallisierte Faser *f* = metallized fibre (Brit.); ~ fiber (US)

metallkeramischer Reibbelag — Mikrobiologie

metallkeramischer Reibbelag m = metal-ceramic friction material
Metall|kitt m = iron-rust cement
~-Lichtbogenschweißung f, Lichtbogenschweißen n mit Metallelektrode = metal-arc welding
~-Metall-Dichtung f, Metall-auf-Metall-Dichtung = metal-to-metal seal
~platte f = metal plate
~profil n = metal section
~pulver n = metal powder
~putzträger m = metal lathing (for plastering)
~rahmenfenster n = metal framed window
~-Rolladen m = metal rolling shutter, ~ roller ~
~schalung f = metal formwork
~schalungsarbeiter m = metal formwork erector
~schiene f = metal rail
~schindel f = metal shingle
~schlauch m = metal hose, ~ flexible conduit
~schmelzofen m = metal-smelting furnace
~schmelzwerk n = metal melting works
~schrott m = scrap metal
~stab m = metal bar, ~ rod
~stäbchen n [beim Rolladen] = metal slat
~streifen m = metal strip
(~)Treibmittel n, Blähmittel, gaserzeugendes Mittel [Gasbeton] = gas-forming agent
~tür f = metal door
~-Unterkonstruktion f [Putztechnik] = metal furring, ~ firring
~zarge f = metal trim, ~ casing
~zylinder m [CBR-Gerät] = cylindrical mo(u)ld
metamikter Zirkon m = metamict zircon
metamorphe Fazies f (Geol.) = metamorphic facies
metamorphes Gestein n, Umwand(e)lungsgestein, Umprägungsgestein = metamorphic rock, transformed ~

Metamorphose f, Metamorphismus m, Umprägung f, Umwandlung f, Gesteins~ (Geol.) = metamorphism
metazentrischer Halbmesser m = metacentric radius
Metazentrum n = metacenter (US); metacentre (Brit.)
Methan n = methane
~absaugung f, Methanausgasung (⚒) = methane drainage, firedamp ~
Methyl|alkohol m, Methanol n = methyl alcohol, wood ~, methanol
~benzol n, Toluol n, $C_6H_5CH_3$ = toluene, methyl benzene, phenyl-methane
~chlorid n, Chlormethyl n, CH_3Cl = methyl(ic) chloride, chloride of methyl
metrisches Maß n = metric dimension
Mezzanin n, Zwischengeschoß n, Halbgeschoß, Beigeschoß = mezzanine (floor), ~ stor(e)y
Michaelis-Mast-Preßbeton(bohr)pfahl m = Michaelis Mast pressure pile
Michiganeinbruch m → Brennereinbruch
~-Bohrloch n, Brennereinbruch-Bohrloch = easer (Brit.); cut hole (US)
Miete f **aus Mutterboden** = stockpile of topsoil
~silo m [Landwirtschaft] = stack silo
Miet|gerätebestand m = hire fleet
~kauf m = rental purchase, hire ~
~kaufabkommen n, Mietkaufvertrag m = agreement for hire purchase
~satz m, Geräte~ = equipment rental rate, plant-hire ~
~vertrag m, Mietabkommen n = rental agreement
(~)Wohnhaus n = apartment house (US); block of flats (Brit.)
(~)Wohnung f = apartment (US); flat (Brit.)
~zeit f = rental period
MIG-Schweißen n, Metall-Inertgas-Schweißen = MIG welding, metal inert gas ~
Mikanit n = reconstructed mica, micanite
Mikro|biologie f = microbiology

Mikrogranit — Mineralbeständigkeitszahl

~granit m = microgranite
~härte f = microhardness
~härteprüfer m = microhardness tester
~meter n, Feinmeßschraube f = micrometer screw
Mikronfilter m, n = micron filter
Mikro|niederschlag m = microprecipitation
~pegmatit m (Geol.) = micropegmatite
~pore f = micropore
~riß n, Feinstriß = microcrack
~rißbildung f = microcracking
~schalter m = microswitch
~schnitt m = microsection
~sickerung f = microseepage
~sieb n = microsieve
~siebung f = microsieving
Milch|bucht f [im Kuhstall] = milking bay
~küche f = milking parlour
~säure f = lactic acid
mildes Gebirge n [Tunnelbau]; Hackboden m [Erdbau] = hacking (Brit.)
Militär|bauwesen n = military civil engineering work, infrastructure
~flugplatz m, Fliegerhorst m, Luftstützpunkt m = station, base
~straße f = military road, ~ highway
Millerit m (Min.) → Haarkies m
Millimeter|arbeit f [beim Absetzen einer Last] = inching
~papier n = millimeter paper
Milorigrün n, Chromgrün = chrome green
Minderung f, Abnahme f = diminution, decrease
Minderungsbeiwert m = reduction factor
Minderventil n, Reduzierventil = reducing valve
Mindest|abfluß m, Mindestdurchfluß = minimum discharge, ~ flow
~abmessung f = minimum dimension
~abstand m = minimum spacing
~abstreiffestigkeit f [Tiefbohrtechnik] = minimum joint strength
~armierung f → Mindestbewehrung

~bewehrung f, Mindest(stahl)einlagen fpl, Mindestarmierung = minimum reinforcement
~bruchdehnung f = minimum failure strain
~bruchfestigkeit f = minimum failure strength
~dicke f = minimum thickness
~druckfestigkeit f = minimum compression strength
~drucklinie f = line of least pressure
~durchfluß m, Mindestabfluß = minimum flow, ~ discharge
~einlagen fpl → Mindestbewehrung f
~festigkeit f = minimum strength
~fordernder m → → Angebot n
~gehalt m = minimum content
~größe f = minimum size
~leistungspreis m [Energieversorgung] = ratchet rate (US)
~luftbedarf m = minimum air requirement
~maß n = minimum dimension
~querschnitt m = minimum cross-section
~sichtweite f = minimum sight distance
~spannung f, Minimalspannung = minimum stress
~(stahl)einlagen fpl → Mindestbewehrung f
~stau(höhe) m, (f) → Absenk(ungs)ziel n
~wärme f = minimum heat
~wert m, Minimalwert = minimum value
~zementgehalt m, Mindestzementmenge f = minimum cement content
Minensuchgerät n = mine detector
Mineral n = mineral
~abstreuung f → Abstreuung
~ader f = mineral vein
~aufbau m [Straßenbaugemisch] = granulometric composition, granulometry
~bestand m = mineral constituents
~beständigkeitszahl f = mineral stability number [bitumen emulsion]

~bestandteil *m* = mineral constituent
~bestreuung *f* → Abstreuung
~beton *m* = wet mix aggregate
~bildner *m*, Mineralisator *m* = mineralizer, mineralizing agent
~boden *m* = mineral soil
~brunnen *m* = mineral well
~farbe *f* = (metallic) pigment

~faser *f* = mineral fibre (Brit.); ~ fiber (US)
~faserplatte *f* = mineral fibre board (Brit.); ~ fiber ~ (US)
~-Fettöl-Gemisch *n* = compounded oil
(~)Füller *m*, Füllstoff *m* = (mineral) filler, mineral dust, granular filler, granular dust (filler)

Mineralgemisch *n*, **Mineralmasse** *f*, **Gesteinsmasse** *f*, **Mineralgerüst** *n*, **Gesteinsgemisch**, **Gesteinsgemenge** *n*, **Gesteinsgerüst**, **Mineralgemenge**, **Mineralmischung** *f*, **(Gesteins)Kornmasse**, **(Gesteins)Kornmischung**, **(Gesteins)Korngemisch**
  geschlossen abgestuft

  offen abgestuft

(mineral) aggregate, mineral skeleton structure, granular framework

dense-graded [*abbrev. DGA*], close-graded
open-graded [*abbrev. OGA*]

Mineralisator *m*, **Mineralbildner** *m* = mineralizer, mineralizing agent
mineralische Bestreuung *f* [*Dachpappe*] = mineral granules
~ **Holzkohle** *f* = mineral charcoal, mother of coal, fusain, motherham
~ **und organische Stoffe** *mpl* [*Sie machen 0,1% des Abwassers aus; die restlichen 99,9% sind Wasser*] = pollutional matter [*partly suspended and partly dissolved*]
mineralisches Bindemittel *n*, mineralischer Binder *m* = mineral binder
Mineral|korn *n* = mineral grain
~**lager(stätte)** *n, (f)* = mineral deposit
~**öl** *n* = mineral oil
~**pulver** *n* = mineral powder, powdered mineral
~**quelle** *f* = mineral spring
~**umwand(e)lung** *f* = mineral inversion
~**vorräte** *mpl* = mineral reserves
~**wolle** *f* = mineral wool
~**wolledämmatte** *f*, Mineralwolleplatte *f* = mineral wool insulation mat
Mineur *m*, Gesteinshauer *m* = stoneman, hard-ground man, hard-heading man, rock header, borer, driller, stone drifter, tunneller, cutter, brancher; mine driver, stone miner [*Scotland*]
Miniaturtraktor *m*, Miniaturschlepper *m*, Miniaturtrecker *m* = pint-size economy tractor
Minimal|gewichtsbemessung *f* = minimum-weight design
~**spannung** *f*, Mindestspannung = minimum stress
~**wert** *m*, Mindestwert = minimum value
Minimum *n* **der Stahleinlagen** *fpl* **in rechteckigen Stahlbetonquerschnitten bei beliebigem Lastangriff** = minimum reinforcement for rectangular reinforced concrete sections in eccentric compression for any eccentricity of the compressive force
Ministerium *n* **für öffentliche Arbeiten** = Ministry of Public Works
Minium *n* → Bleimennige *f*
Minorannahme *f* = minor hypothesis
Minus-Temperatur *f* = below-zero temperature
minutliche Umlaufzahl *f* → Drehzahl

Miozänton *m* = miocene clay
Misch|anilinpunkt *m* [*Kohlenwasserstoff-Lösungsmittel*] = mixed anilin point
~anlage *f* = mixing plant
~arm *m* = mixing arm
~barkeit *f* = miscibility
~barkeitsprüfung *f* **mit Wasser** = miscibility with water test [*bitumen emulsion*]
~batterie *f* = bath-mixer
~behälter *m* = mixing channel [*water filtration plant*]
~behälter *m* = mixing tank [*for tar and bituminous compounds*]
~belag *m* → Mischdecke *f*
~bepflanzung *f* = mixed planting
~binder *m* = hydraulic binder manufactured according to DIN 4207
~blech *n*, Beton~ = concrete mixing plate
~bottich *m* = mixing tank
~bühne *f* = mixing platform
~bunker *m* = blending bunker
~dauer *f* → Mischzeit *f*
~decke *f*, Mischbelag *m*, Mischanlagendecke, Mischanlagenbelag [*Straße*] = pre-mix(ed) surfacing, plant-mix(ed) ~, ~ pavement
~druckförderung *f* [*Ölbohren*] = mixed drive
~dünger *m* = mixed fertilizer
mischen, anmachen = to mix
Mischen *n*, Anmachen = mixing
~ in Straßenbetonmischern = paver mixing
~ mit Straßenhobel = blade mixing
Mischentwässerung *f* → Mischkanalisation *f*
Mischer *m* **für Teppichbeläge** = retread mixer
~ mit Schurrenaustrag(ung), (Freifall-)Durchlaufmischer, Austragmischer = closed drum (concrete) mixer with tipping chute discharge
~fahrzeug *n*, Beton~ [*Ein Fahrzeug, das mit einer Einrichtung für intensives Mischen von Zement, Zuschlägen und Wasser ausgerüstet ist und den Beton bis zum Augenblick der Ablieferung in einwandfreiem Zustand hält*] = truckmixer
~führer *m*, (Mischer)Maschinist *m* = mixer driver, ~ operator
~mörtel *m* = machine-made mortar
~rahmen *m* = mixer frame
~schaufel *f* = mixer (or mixing) paddle (or blade)
(~)Segmentteil *m*, *n* = (mixer) liner segment, sectional liner
~-Stabilitäts-Prüfung *f* = plant mix test, ~ stability ~
~waage *f* [*Der Beschickerkübel dient als Wiegegefäß. Der untere Teil der Aufzugbahn, auf dem sich der Kübel während des Verwiegens abstützt, ist ein Bestandteil der Waage*] = mixer scale
~welle *f* = mixer shaft
Misch|gas *n* → Dowsongas
~gestein *n*, hybrides Gestein = hybrid rock
~gut *n* = mix(ture)
~gutbehälter *m* → → Schwarzbelageinbaumaschine *f*
~gutsorte *f* = type of mix(ture)
~gut(-Verlade)silo *m* = mixed material storage hopper
~gutzusammensetzung *f* = mix composition
~kammer *f* = mixing chamber
~kammer *f* [*Ein Betongefäß, in dem Erdreich und Wasser gemischt werden, bevor man sie nach einem Auftrag pumpt*] = hog box
~kanalisation *f*, Mischsystem *n*, Mischentwässerung *f* = combined system [*A system of sewers, in which sewage and storm water are carried*]
~kessel *m* [*Herstellung präparierten Teeres*] = mixer, mixing vessel
~kies *m* = graded gravel (mixture)
~kies *m* = coated gravel
~kiesanlage *f* = coated gravel plant
~kollergang *m* = mixing pan grinder
~kübel *m* → → Betonmischer
~leitung *f* = combined sewer [*A sewer designed to receive both storm water and sewage*]

Mischmakadam — Mischungskomponente

~makadam *m* = (pre-)mixed macadam; coated ~ (Brit.)
Teermakadam *m* = tarmacadam
Asphaltmischmakadam *m*, Steinschlagasphalt *m* = bitumen (or bituminous) macadam (Brit.); asphalt macadam (US) [*While every engineer will be familiar with the term "coated macadam," there may be a diversity of opinion regarding the range of "black top" materials which are included under this general description. The glossary of Highway Engineering Terms, B. S. 892, describes it as "a road material consisting of coarsely graded mineral aggregate that has been coated with a specified binder, such as road tar, bitumen or the like by a controlled process, having a preponderance of coarse aggregate and a substantial proportion of voids." According to this definition it obviously includes tarmacadam to B. S. 802 and 1242, bitumen macadam to B. S. 1621, cold asphalt to B.S.S. 1960, and similar mixtures.*
On the other hand, manufacturers and suppliers of bituminous surfacing materials, such as members of the Federation of Coated Macadam Industries, regard coated macadam as any material which can be made in their coating plants at mixing temperatures below about 220 deg. F. irrespective of grading, so fine cold asphalt, dense tar surfacing and dense bitumen macadam come within this definition.]
~maschine *f*, Mischer *m* = mixer, mixing machine
~maschinist *m* → Mischerführer *m*
~material *n* = premix [*road engineering*]
~mauerwerk *n*, Verblendmauerwerk = brick-lined masonry
~organ *n* = mixing element
~periode *f* → Mischzeit *f*
~pfanne *f* = mixing ladle

~phase *f* [*Erdölförderung*] = miscible phase
~phasen-Verdrängung *f* [*Sekundärförderung von Erdöl*] = miscible displacement, ~ drive
~polymerisat-Emulsionsfarbe *f* = copolymer emulsion paint
~reihenfolge *f* = mixing sequence
~rohr *n* = emulsion tube
~schaufel *f* → Mischerschaufel
~schleppe *f* = mixing drag
~schnecke *f* = broken-bladed conveyor, ~ conveyer, paddle worm ~
~schnecke *f* [*bei der Brikettierung*] = enclosed paddle worm
~spiel *n* = mixing cycle
Mischsplitt *m* = coated chip(ping)s
~-Aufbereitungsanlage *f* = bituminous mixing plant
~-Teppich *m* = chip(ping)s carpet (or mat)
Misch|stelle *f* = mixing site
~stellung *f* = mix position
~stern *m* = mixing star
~system *n* → Mischkanalisation *f*
~systemleitung *f* [*Abwasserwesen*] = combined sewer
~teer *m* = mixed tar
~teil *m, n*, Mischwerk [*eines Mischers*] = mixing unit
~teller *m* = cylindrical mixing pan
~temperatur *f* = mixing temperature
~teppich(belag) *m* → Straßenteppich
~trog *m* = mixing pan
~trommel *f* = mixer drum
~turm *m* = mixing tower
~turm *m* für bituminöse Belagmassen = tower-type bituminous mixing plant
~- und Einbauzug *m* für Tunnelbetonauskleidung = mixing-placing train for tunnel concreting
Mischung *f*, Gemisch *n*, Mischgut *n* = mix(ture)
Mischungs|aufbau *m*, Mischungszusammensetzung *f* = mix(ture) composition
~komponente *f*, Mischungsbestandteil *m* = mix ingredient

Mischungskonsistenz — Mitnehmerstange

~konsistenz *f* = mix consistency
~nebel *m*, Advektionsnebel = advection fog
~tabelle *f* = mixing table
~temperatur *f* = mixing temperature
~turbine *f* = mixture turbine
~- und Verwiegeanlage *f* für Baustellen = mixing and weigh-batching plant for construction sites
~zusammensetzung *f*, Mischungsaufbau *m* = mix(ture) composition
Misch|ventil *n* = (shower) mixing valve
~verfahren *n*, Mischweise *f* = mixing method
~verfahren *n* [*Abwasserwesen*] → Mischkanalisation *f*
~verhältnis *n*, Mischungsverhältnis = mix proportions [*ratio of cement (or binder) to aggregate*]
~verkehr *m* = mixed traffic
~vorgang *m* = mixing operation, ~ process, ~ action
~waage *f* → Gattierungswaage
Mischwasser *n*, Anmach(e)wasser = mix(ing) water, ga(u)ging ~, batch ~
~ = combined water [*Sewage and storm water combined*]
~erwärmung *f*, Anmach(e)wassererwärmung = mix(ing) water heating
~kanal *m* [*Abwasserwesen*] = combined system sewer
~waage *f*, Anmach(e)wasserwaage = water weighing device [*Central concrete mixing plant*]
Misch|weise *f*, Mischverfahren *n* = mixing method
~welle *f* = mixing shaft
~werk *n*, Mischteil *m*, *n* [*eines Mischers*] = mixing unit
~werk *n*, bituminöses ~ = central bituminous mixing plant
~werkzeug *n* = mixing element
~winkel *m* = mixing angle
~wirkungsgrad *m* = mixing efficiency
~zähler *m* = batchmeter
~zeit *f*, Mischperiode *f*, Mischdauer *f* = mixing time, ~ period, ~ time period

~zement *m*, Naturzement [*In den USA werden Naturzemente durch Brennen unterhalb der Sintergrenze gewonnen, sie sind also nach der deutschen Auffassung hydraulische Kalke*] = blended cement, natural ~
~zentrale *f* = central mixing plant
~zug *m* [*Bodenvermörtelung*] = mixing train [*soil stabilization*]
mißgriffsicher → narrensicher
Miß|pickel *m* (Min.) → Arsenkies *m*
~weisung *f*, D, Deklination *f* [*Die Abweichung gegenüber dem geodätischen Meridian*] = declination
mit Atomantrieb, atomgetrieben = with nuclear propulsion
~ Druckluftunterstützung = air boosted
~ gekrümmter Sichtfläche = curved-face
~ Gelenkbolzen versehen = pin ended
~ Gleitschalung betonieren = to slide
~ Luftfederung = air suspended
~ Luftpolsterung = air cushioned
~ Mittelaufhängung = center mounted (US); centre ~ (Brit.)
~ Sand bestreuen, (ab)sanden = to (cover with) sand
~ Sandsäcken verstopfen, Sandsäcke einbauen = to sandbag
~ Stahleinlagen, armiert, bewehrt = reinforced
~ umgelegtem Blechfalz [*Drahtgewebe*] = with folded edges and sheet metal
~ zwei Motoren = tandem powered
Mit|arbeiter *m* = co-worker
~berater *m* = co-consultant
mit|laufendes Zahnrad *n* → Zwischenrad
~nehmen [*Bergedamm im Bergbau*] = to carry [*dirt pack*]
Mitnehmer *m*, Mitnehmerschaufel *f* = (scraper) flight
~einsatz *m* = drive bushing
~förderer *m* = flight conveyor, ~ conveyer
Mitnehmerstange *f* [*Rotarybohren*] = kelly

Mitnehmerstange — Mittelmahlen 674

~ mit Muffengewinde und Zapfengewinde = box x pin kelly
~ beiderseits mit Muffengewinde versehen = box x box kelly
Mitnehmerstangen|büchse *f*, Rotarytischeinsatz *m* = kelly bushing, drill stem ~
~-Einsatzhalter *m* [*Rotarybohren*] = drill stem bushing holder
Mitnehmerteil *m*, *n* [*Mitnehmerstange einer Rotary-Bohranlage*] = drive section
Mittagspause *f* = midday break, lunch~
Mitte *f* = center (US); centre (Brit.)
~, Stützkörper *m*, Dammkörper, Kern *m* [*nicht zu verwechseln mit Dichtungskern beim Staudamm = core*] = body
Mittel *n*, arithmetisches ~, Mittelwert *m* = arithmetical mean
~**(ab)siebung** *f* = medium screening
~**abstand** *m* = average distance
~**binder** *m* = medium curing liquid asphaltic material, MC
~**blech** *n* = medium plate
~**blechwalzwerk** *n* = medium plate mill
~**bogen** *m* [*Brücke*] = center arch (span) (US); centre ~ (~) (Brit.)
~**brandziegel** *m* = medium baked brick
~**brechen** *n* = intermediate crushing
~**brecher** *m* = intermediate crusher, ~ breaker, ~ crushing machine, ~ breaking machine
~**damm** *m* (⚒) = intermediate pack
~**druckluftverdichter** *m* → zweistufiger Kompressor *m*
~**druckpumpe** *f* = medium pressure pump
~**druckregner** *m* = medium pressure sprinkler
~**fahrstreifen** *m* → Mittelstreifen
~**feld** *n*, Mittelöffnung *f*, Hauptfeld, Hauptöffnung [*Brücke*] = centre span (Brit.); center ~ (US); central ~, main ~, ~ opening
mittelflache Bodenplatte *f* [*Gleiskette*] = flat center shoe (US); ~ centre ~ (Brit.)

Mittel|flansch *m* der Laufrollenwelle [*Gleiskettenlaufwerk*] = shaft center flange (US); ~ centre ~ (Brit.)
~**fraktion** *f* → Mittelkörnung *f*
~**frequenz(beton)fertiger** *m* = medium frequency (concrete) finisher
~**gebäude** *n*, Mittelbau *m* = centre building (Brit.); center ~ (US)
~**gebirge** *n* = secondary mountain, average ~
mittelgekohlter Stahl *m* = medium carbon steel
Mittel|gelenk *n* = centre hinge (Brit.); center ~ (US)
~**geschwindigkeit** *f* [*Hydraulik*] = mean velocity
mittelhartes Steinkohlenteerpech *n*, Mittelhartpech *m* = medium-hard coal-tar pitch
Mittel|insel *f* [*Verkehrsinsel*] = central island
~**kies** *m* [*nach DIN 1179 30–7 mm; nach DIN 4022 30–5 mm; nach DIN 4220 20–7 mm; nach DEGEBO 10–5 mm; nach Fischer und Udluft 20–10 mm (heißt hier "großer Graup") und 10–5 mm (heißt hier "mittlerer Graup"); nach Dienemann 10–5 mm*] = medium gravel [*A.S.E.E. fraction 24.5 to 9.52 mm*]
~**kiessand** *m* = middle-sized gravelly sand
~**korn** *n*, Mittelkörner *npl* = middle-sized particles, ~ grains, ~ material
~**körnung** *f*, Mittelfraktion *f* = middle-sized grain mix(ture)
~**kraft** *f*, Ersatzkraft, Resultante *f*, Resultierende *f* = resultant (force)
~**kragbinder** *m* = central cantilever truss
~**lenkbügel** *m* = pivot arm
~**linie** *f*, Achse *f* = centre line (Brit.); center ~ (US); ~ axis
~**lochstegplatte** *f* [*Gleiskette*] = center punched grouser shoe (US); centre ~ ~ ~ (Brit.)
~**mahlen** *n*, Mittelmahlung *f* = intermediate grinding

~moment n, resultierendes Moment = resulting moment
~moräne f, Gufferlinie f = medial moraine, median ~
~öffnung f, Mittelfeld n [Brücke] = central span; centre ~ (Brit.); center ~ (US)
~pfeiler m = centre pier (Brit.); center ~ (US)
~pfette f = centre purlin(e) (Brit.); center ~ (US)
~produkt n [Magnetscheidung] = middling
~punktlenkung f = center point steering (US); centre ~ ~ (Brit.)
~rippe f = centre rib (Brit.); center ~ (US)
~sand m [nach DIN 1179 und DIN 4022 1,0–0,2 mm; nach Dücker (Jahr 1948) „feines Siebkorn IVa" 0,6 bis 0,2 mm] = medium sand [A.S.E.E. fraction 0.59–0.25 mm; British Standard 0.6–0.2 mm]
~schale f = central shell
~schenkel m (Geol.) = middle limb
~schiff n [Kirche] = nave
~schifter m = intermediate jack rafter
mittel|schlächtiges Wasserrad n = breast (water) wheel
~schnellabbindend [Verschnittbitumen] = medium-curing
~schnellbrechend, halbstabil [Bitumenemulsion] = medium-setting, normal-setting, semistable, medium-breaking
~schwer [Maschine] = medium-duty
~schwere (Straßen)Decke f, mittelschwerer (Straßen)Belag m = intermediate-type pavement, ~ surfacing
Mittel|spannungs-Überlandleitung f = middle voltage transmission line
~splitt m = medium-sized chip(ping)s
~steg m [Gleiskettenglied] = center strut (US); centre ~ (Brit.)
~stellung f [Schalthebel] = neutral position
~stiel m [Zweifeldrahmen] = centre support (Brit.); center ~ (US)

~streifen m, Mittelfahrstreifen [Nicht verwechseln mit dem nichtbefahrbaren Mittelstreifen, der zwei Fahrbahnen trennt. Der Ausdruck „Mittelspur" sollte nicht mehr verwendet werden] = median lane
~streifen m = central reserve, ~ reservation, ~ strip, median (strip), medial strip
~streifenbepflanzung f = median planting
~stromventilanordnung f = open center valve system (US); ~ centre ~ ~ (Brit.)
~stütze f = centre column (Brit.); center ~ (US)
~teil m, n [eines Bauwerkes] = central section
~temperaturzelle f, Mitteltemperaturelement n [Brennstoffzelle] = medium temperature fuel cell
~walzwerk n = rolling mill for medium-sized products, intermediate (rolling) mill
~wand f = centre wall (Brit.); center ~ (US)
Mittelwasser|anzeiger m = medimarimeter
~-Durchflußquotient m = mean discharge ratio
Mittel|welle f [Raupenbagger] = vertical propel shaft
~wert m, (arithmetisches) Mittel n = arithmetical mean, average value
~zahn m [Aufreißer] = centre tine, ~ tooth (Brit.); center ~ (US)
~zapfenlager n = center pin bearing (US); centre ~ (Brit.)
~zapfenlenkung f = center pin steering (US); centre ~ ~ (Brit.)
~zerkleinerung f [Korngröße 100 bis 10 mm] = intermediate reduction, ~ comminution [minimum grain size 3/8 in.]
Mitten|abstand m = centre-to-centre distance (Brit.); center-to-center ~ (US)
~durchbiegung f = deflection at midlength; deflexion ~ ~ (Brit.)
mittig = centrally, axially

mittige Normalkraft N — Modellherstellungsgeräte 676

mittige Normalkraft N f = centrally applied direct thrust N
mittiger Druck m = axial pressure
~ **Zug** m = axial tension
mittlere Förderung f, mittlerer Transport m [*Erdbau*] = medium-length haul
~ **Gezeitenhöhe** f = mean tide-level
~ **Nipptide** f = mean neap tide
~ **Ortszeit** f = local mean time, L. T.
~ **Springtide** f = mean spring tide
mittlerer Fallhöhenbereich m [*Wasserkraftanlage*] = medium-head range
~ **Fehler** m [*Siebtechnik*] = mean deviation, ~ error
~ **Geschiebetrieb** m = bed load movement above competence, general bed load movement
~ **Kolbenring** m, zweiter Kompressionsring [*Motor*] = intermediate ring
~ **Korndurchmesser** m, durchschnittlicher ~ = average grain diameter
~ **Tidehub** m = mean range
~ **Transport** m, mittlere Förderung f [*Erdbau*] = medium-length haul
~ **Wasserstand** m, durchschnittlicher ~, gewöhnlicher ~ = average stage
mittleres Drittel n = middle third
~ **Hochwasser** n = mean of high stages
~ **Kornvolumen** n = average volume of grain
~ **Mittelwasser** n [*Meer*] = mean level
~ **Niedrigwasser** n = mean of low stages
~ **Nipphochwasser** n, Höhe f des mittleren Hochwassers bei Nipptide = height of mean high water of neap tide
~ **Nippniedrigwasser** n, Höhe f des mittleren Niedrigwassers bei Nipptide = height of mean low water of neap tide
~ **Siebkorn IIIa** n [*nach Dücker 1948*] = granule [6–2 mm]
~ **Springhochwasser** n, Höhe f des mittleren Hochwassers bei Springtide = height of mean high water of spring tide

~ **Springniedrigwasser** n, Höhe f des mittleren Niedrigwassers bei Springtide = height of mean low water of spring tide
mitwirkende Plattenbreite f [*Brücke*] = co-operating slab width
~ **Stahlfahrbahn** f → Leichtfahrbahn
ml, Laufmeter n, laufendes Meter = Lin. M., linear metre (Brit.); linear meter (US)
Mo m [*nach Terzaghi 0,1–0,02 mm; nach DIN 4022 0,1–0,02 mm (heißt hier „Mehlsand"); nach Grengg (Jahr 1942) 0,1–0,02 mm (heißt hier „Mehl")*] = mo [0.1–0.02 mm]
Mobil|anlage f, fahrbare Anlage für Beton(auf)bereitung = low-profile batching and mixing plant, ~ (central) ~
~**bagger** m = mobile excavator
~**-Dosieranlage** f → fahrbare Dosieranlage
~**drehkran** m, Mobilschwenkkran = mobile rotary crane, ~ slewing ~
~**-Hydraulikbagger** m → Hydraulik-Mobilbagger
~**-Hydraulikkran** m → Hydraulik-Mobilkran
~**-Hydrobagger** m → Hydraulik-Mobilbagger
~**-Hydrokran** m → Hydraulik-Mobilkran
~**kran** m = mobile crane
~**schwenkkran** m, Mobildrehkran = mobile slewing crane, ~ rotary ~
Modell n = pattern; model
~ **mit beweglicher Sohle** = movable-bed model
~ ~ **fester Sohle** = fixed-bed model
~**bezeichnung** f, Katalogbezeichnung, Drahtwort n = code (word)
~**damm** m → Dammodell n
~**flüssigkeit** f = model liquid
~**(ge)rinne** f, (n) = flume model, model flume
~**gesetz** n = model law
~**herstellungsgeräte** npl für den Nickelcarbonyl-Prozeß = nickel carbonyl pattern equipment

Modellierraum m = model preparation room
Modellmaßstab m = model scale
Modellösung f [*Korrosionsprüfungsflüssigkeit*] = standardized test solution
Modell|pfahl m = model pile
~rinne f, **Modellgerinne** n = model flume, flume model
modellstatische Untersuchung f = static model investigation
Modell|untersuchung f = model study
~versuch m, **Modellprüfung** f, **Modellprobe** f = model experiment, ~ test, ~ analysis
Modul m [*Eine am Bau abgeleitete Maßeinheit, die als Grundlage für die Abmessungen aller Bauteile dient*] = module
~ von 10 cm → Grundmodul M
Modularordnung f = modular system
Modulatorventil n = pressure modulating valve
Modulbreite f = modular width
mögliche (ab)bauwürdige Vorräte mpl = possible reserves
~ Gleitfläche f = potential surface of sliding
möglicherweise vorhandenes Erz n = possible ore
Mohr'scher Satz m = Mohr's law, ~ analogy
~ (Spannungs)Kreis m = Mohr's circle (of stress)
  Bruchlinie f = strength envelope, envelope of rupture, envelope of failure, line of rupture, Mohr's envelope
Mohr'sches Diagramm n = Mohr's (rupture) diagram
Moiréseide f = watered silk
Moirette f [*Linoleumart*] = moire
Molasse f = molasses
~fels m (Geol.) = Molasse
Mole f = jetty [*A deck carried usually on piles at the water's edge and used as a landing stage*]
~ → Hafenaußenwerk n
Molekularanziehung f, innere Molekularspannung = molecular attraction

Möllerwagen m = scale car
Molybdän|disulfid-Schmiermittel n = molybdenum disulphide lubricant (Brit.); ~ disulfide ~ (US)
~glanz m (Min.) = molybdenite
molybdänhaltiger Austenitstahl m = molybdenum bearing austenitic steel
Moment n **der inneren Kräfte** = moment of resistance, resisting moment
Momenten|ausgleich m = moment distribution
~ausgleichverfahren n = moment distribution method
~einfluß m = moment influence
~einflußlinie f = moment influence line
~fläche f = moment diagram
~freiheit f = absence of moments
~gleichsetzen n = equating the moments
~gleichgewicht n = moment equilibrium
~knickung f = moment buckling
~kurve f, **Momentenlinie** f = moment curve
~linie f → Momentenkurve
~maßstab m = scale for moments
~mittellinie f = moment axis
~nullpunkt m = centre of moments (Brit.); center ~ ~ (US)
~pol m → Drehpunkt m
~punkt m → Drehpunkt
~-Schub-Verhältnis n = moment shear ratio
~überlagerung f = superposition of moments
~umlagerung f = redistribution of moments
~verlauf m = shape of the moment diagram, moment curvature
~verteilung f = moment distribution
Moment|(schraub)zwinge f, **Schnellschraubknecht** m = quick (action) clamp
~vermögen n = moment capacity
~zünder m → ~ 1. Schießen n; 2. Sprengen n

Monats|niederschlag m = monthly precipitation
~**regenmenge** f = (amount of) monthly rainfall
Mönch m = overtile
Mondzeit f **der Tide** = lunar time of the tide
Monelmetall n = Monel metal
Moniereisen n, **Monierstahl** m [*veraltete Ausdrücke für „Bewehrung"*] = reinforcement
Monoblocbohrer m = monobloc type drill rod
Monoblockpresse f = single mo(u)ld press
monodisperses Gut n, gleichkörniges ~ = uniformly sized grains, grains of equal size
Monolith|betonbauwerk n, fugenloses Betonbauwerk = monolithic concrete structure
~**decke** f = monolithic floor
monolithisch betoniert = monolithically poured, ~ concreted
monolithische Platte f = monolithic slab
monolithischer Stahlbeton m = monolithic reinforced concrete
monomolekulare Schicht f = monomolecular layer
Montage f, **Aufbau** m, **Aufstellen** n, **Aufstellung** f = erection, assembly
~ **des Siebbodens**, Siebmontage = fitting of screen bottom
~**arbeiten** fpl = erection work, assembling ~
~**armierung** f, Montagebewehrung, Montage(stahl)einlagen fpl = reinforcement for stresses in erection, erection reinforcement
~**balkendecke** f, Fertigbalkendecke = precast beam floor
~**band** n, Montage-Fließband = assembly line
~**bau** m, Fertigteilbau, Bauen n mit Fertigteilen = precast construction, prefab(ricated) ~, system building
~**bauweise** f, Fertigbauweise, Montagebauverfahren n, Fertigbauverfahren = prefab(ricated) construction method
~**bauweise** f **Camus** = Camus system
~**bauwerk** n, Fertigteilbauwerk = assembly-type structure, precast ~
~**betonteil** m, n → (Beton)Fertigteil
~**betonwand** f → Betonmontagewand
~**bewehrung** f, Montagearmierung, Montage(stahl)einlagen fpl = reinforcement for stresses in erection, erection reinforcement
~**bügel** m **für Reißzähne** = tooth mounting bracket
~**bühne** f, Arbeitsbühne, Montageplattform f, Arbeitsplattform = erecting platform, ~ deck
~**decke** f, Fertigdecke = precast floor
~**dreibaum** m → Dreibaum
~**fabrik** f, Montagewerk n = assembly plant
~**flansch** m = adapter, adaptor
~**(-Fließ)band** n = assembly line
~**folge** f = sequence of erection, erection sequence
~**gerät** n = erection equipment
~**gerüst** n, Hilfsgerüst, Abspanngerüst, Aufstell(ungs)gerüst [*Brückenbau*] = erecting tower, erection ~
~**gesamtzeit** f = overall erection time
montagegeschweißt, auf der Baustelle geschweißt = field-welded
Montage|gewicht n = weight for erection
~**halle** f → Montagewerkstatt f
~**haus** n = prefab(ricated) house
~**hilfe** f = erection aid
~**kolonne** f, Montagetrupp m, Montagemannschaft f = erection crew, ~ gang, ~ party, ~ team
~**kran** m = erection crane
~**mannschaft** f → Montagekolonne f
~**mast** m = erecting mast
~**meister** m → Chefmonteur m
~**phase** f, Montagestadium n = erection stage
~**plan** m = erection schedule
~**platte** f [*Spannverankerung beim Spannbeton*] = temporary erection plate

~platte f = assembly plate, mounting ~
~plattform f, Arbeitsplattform, Montagebühne f, Arbeitsbühne = erecting platform, ~ deck
~schweißnaht f = field weld
~schweißung f = field welding
~spannbetonwand f → Spannbetonmontagewand
~spannung f = erection stress
~stadium n, Montagephase f = erection stage
~stahlbetonwand f → Stahlbetonmontagewand
~stoß m = erection joint
~stützweite f [z. B. Omniaträger] = span during placing until hardening of the cast-in situ concrete
~teil m, n → (Beton)Fertigteil
~teil m, n → Fertigteil
~träger m, Fertigträger = prefab(ricated) girder
~träger I m, Doppel-T-Montageträger, Fertigträger I, Doppel-T-Fertigträger = double T prefab(ricated) girder
~tragfähigkeit f = load-bearing capacity of beam(s) without cast-in situ concrete
~trupp m → Montagekolonne f
~verfahren n = erection method
~vertrag m = erection contract
~wand f, Fertigwand = prefab(ricated) wall
~werk n, Montagefabrik f = assembly plant
~wohnungsbau m, Fertigteilwohnungsbau = prefab(ricated) housing
~zapfen m = mounting trunnion
Montan|geologe m = mining geologist
~geologie f [Ingenieurgeologie des Bergbaues] = mining geology
Monteur m = fitter, mechanic
~gerüst n = fitter's scaffold(ing), mechanic's ~
Montopore-Hartschaumplatte f = Montopore expanded polystyrene board
Monumentalbau m = monumental structure

Monzonit m = monzonite, syenodiorite
Moor|boden m = bog soil
~dränage f = bog drainage
~kanal m = bog canal
~kohle f = moor coal
~raupe f, Planier-Moorraupe [mit breiten (Gleis)Ketten zum Einsatz auf Moorböden] = wide-tracked caterpillar bulldozer for operation on boggy soil
~sprengung f, Schüttsprengverfahren n = bog blasting, peat ~, muck-blasting operation, blasting of peat, swamp shooting method, toe-shooting
~wasser n = bog water
Moos|heide f = moss heath
~mauer f = dry-stone wall with moss-filled joints
Mopedverkehr m = moped traffic
Moräne f = moraine
~kies m [durch Gletscher- und Wassertransport gerundeter Gesteinsschutt] = morainal gravel
Moränen|schutt m = glacial fill, morainal accumulations
~schuttboden m = glacial soil
~terrasse f = moraine terrace
Morgenspitzenverkehr m = morning peak traffic, ~ rush ~
Mörser m, Reibschale f [DIN 12906] = porcelain mortar
Mörtel m = mortar
~ mit Epoxyharzbindemittel = epoxy mortar
~(auf)bereitung f, Mörtelherstellung, Mörtelerzeugung = mortar fabrication
~auskleidung f → Mörtelverkleidung
~auspressung f → Mörteleinpressung
~behälter m [Mauerfuggerät] = mortar tank
~bekleidung f → Mörtelverkleidung
~belag m → Putz m
~belagdicke f, Putzdicke = plaster thickness
~bereitung f, Mörtelaufbereitung, Mörtelherstellung f, Mörtelerzeugung = mortar fabrication

Mörtelbetrieb — Mosaik(fußboden)platte

~betrieb m, Mörtelwerk n = mortar works, ~ factory
~bett n, Mörtelstreifen m = mortar bed
~deckung f = mortar covering
~eingußdecke f [*turbulente Mörtelfertigung*] = cement-bound surfacing, COLCRETE, colloidal concrete, cement penetration method, cement (-bound) macadam, mortar-bound macadam
   Zementschotterdecke zweilagig = sandwich process macadam, COLCRETE constructed in the sandwich process
~einpressen n am tiefliegenden Ende oder an den Tiefpunkten des Spannkanals = "uphill" grouting
~einpressung f, Mörtelinjektion f, Mörtelverpressung, Mörtelauspressung = mortar intrusion
~erzeugung f → Mörtel(auf)bereitung
~festigkeit f = mortar strength
~förderung f, Mörteltransport m = mortar handling, ~ transport
~fuge f = mortar joint
~fugenbewehrung f, Mörtelfugenarmierung = mortar joint reinforcement
~gruppe f = mortar class
~haue f = rake, batter
~herstellung f → Mörtelbereitung
~injektion f → Mörteleinpressung f
~kriechen n = creep of mortar, mortar creep
mörtellos, trocken = dry
Mörtel|mauerwerk n = mortar masonry
~mischer m, Mörtelmischmaschine f = mortar mixer, ~ mixing machine
~mischungsverhältnis n = mortar mix(ing) ratio
~nest n = mortar pocket
~prisma n = mortar prism
~probe f, Mörtelprobekörper m = mortar specimen
~pumpe f = mortar pump
~pumpen n = pumping of mortar, mortar pumping

~(pump)rohr n = mortar pumping tube
~-Pump- und -Spritzanlage f = mortar pumping and air placing machine
mörtelreich = rich in mortar
Mörtel|sand m = mortar sand
~schicht f = mortar layer, ~ bedding
~schlitten m = mortar sledge
~spritzer m = mortar dropping
~spritzmaschine f = mortar gun
~stoff m = mortar material
~streifen m, Mörtelbett n = mortar bed
~transport m, Mörtelförderung f = mortar transport, ~ handling
~trog m, Mörtelwanne f, Speiskübel m = hod
~unterlage f = mortar base
~verkleidung f, Mörtelauskleidung, Mörtelbekleidung = mortar lining
~verkrustung(en) f(pl) = built-up mortar
~verpressung f → Mörteleinpressung
(~)Verputz m = (mortar) rendering
~verteiler m = mortar spreader
~wanne f → Mörteltrog m
~weichmacher m = mortar plasticizer
~werk n, Mörtelbetrieb m = mortar factory, ~ works
~würfel m = mortar cube
~zusammensetzung f = mortar composition
~zusatz(mittel) m, (n) = mortar additive
~zwangsmischer m = mortar mill
~zylinder m = mortar cylinder
Mosaik n, musivische Kunst f = mosaic
~außenhaut f = mosaic external finish
~boden m → Mosaikfußboden
~fenster n = mosaic window
~fliese f, Mosaikplatte f = mosaic tile
~(fuß)boden m = mosaic floor(ing), ~ floor finish, ~ floor covering
~(fußboden)platte f, Mosaikfliese f [*Surrogat echten Mosaiks*] = mosaic tile

~pflaster(decke) n, (f) = mosaic (sett) paving
~pflasterstein m = mosaic (paving) sett
~platte f, Mosaikfliese f = mosaic tile
~platte f = mosaic panel
~splitt m → Terrazzokörnung f
~verband m (Geol.) → Bienenwabenverband
Motor|armaturenbrett n = engine panel
~aufhängung f = engine support
~auflager n, Motorträger m = engine mounting
~ausfall m = engine breakdown
~-Betonkarre(n) f, (m), Motor-Beton(rund)kipper m, Motor-Japaner-(karren) m, Motor-Japaner-(Kipp)-Karre(n), Motor-Kipp-Betonkarre(n), Motor-Betonvorderkipper, Motor-Kipp(er)karre(n) = power concrete buggy, ~ ~ cart, motorized buggy, motorized cart, motobug
~-Beton(rund)kipper m → Motor-Betonkarre(n) f, (m)
~-Betonvorderkipper m → Motor-Betonkarre(n) f, (m)
~block m = engine block
~-Bodenfräse f → Bodenfräse
~bohrhammer m = motor drill
~bohr- und Aufbruchhammer m = motor drill and breaker
~bootrennstrecke f = speed boat course
~bremse f, Auspuff-Klappenbremse = exhaust brake
~drehkraft f = engine power
~drehzahl f = engine speed
~drehzahl f bei max. Drehmoment = RPM at max. torque
~drehzahl f bei Vollast = RPM governed at full load
~drehzahlmesser m = engine tachometer
~einlauf m = engine break-in
Motoren|benzin n → Benzin
~benzol n = motor benzole
~bestückung f = motor complement
~gußteil m, n = cast-iron engine part

~öl n = engine oil
~petroleum n, Kraftstoffkerosin n = power kerosene, ~ kerosine
~schlosser m = fitter
~stahlteil m, n = steel engine part
Motor|-Entlüftungsanlage f = engine breathing system
~-Erdhobel m → Motor-Straßenhobel
~fähre f = self-propel(l)ing ferry
~fahrrad n = motor-assisted pedal cycle
~fahrzeug n → Kraftfahrzeug
~fahrzeugtunnel m → Autotunnel
~fahrzeugverkehr m → Autoverkehr
~-Fegemaschine f, Motor-Kehrmaschine = power-driven rotary sweeper, engine-driven ~ ~
~frachtkahn m → Motorlastkahn
~-Fußwegkehrmaschine f → Fußweg-Motorkehrmaschine
~generator m = motor generator (set)
~greifer(korb) m, Hakengreifer m = hook-on bucket
~grundplatte f = engine baseplate
~-Handsäge f = power hand saw
~haube f = engine hood, ~ cowl
motorisierter Schürfkübel m mit Einachs-Traktor → Autoschrapper m mit 2-Rad-Schlepper
Motorisierung f = motorization
Motor|-Japaner(karren) m → Motor-Betonkarre(n) f, (m)
~-Kehrmaschine f → Motor-Fegemaschine
~-Kipp-Betonkarre(n) f, (m) → Motor-Betonkarre(n)
~kipper m → Muldenkipper
~-Kipp(er)karre(n) f, (m) → Motor-Betonkarre(n)
~klopfen n = engine knocking
~kocher m, Gußasphalt-~ = engine-driven mastic asphalt mixer
~kompressor-Spritzmaschine f = spraying machine (or sprayer) with power air compressor and hand lance(s)
~kraftstoff m = motor fuel
~kran m = power crane
~kübelwagen m → Autoschütter m

~kühlwasser *n* = engine (cooling) water

~kupp(e)lung *f* = engine clutch

~lastkahn *m*, Selbstfahrer *m*, Motorfrachtkahn = self-propelling barge, powered ~, self-propelled ~

(~)Leistung *f* = rated horsepower

~leistung *f* an der Schwungscheibe = flywheel horsepower

~ölkühler *m* = engine oil cooler

~ölwanne *f* = crankcase

~platte *f* [*Elektromotor*] = motor bedplate

~platte *f* = engine bedplate

~ponton *m* = propulsion pontoon

~pumpe *f* = power(-driven) pump, motor-driven ~, power-operated ~

~reg(e)lung *f* = engine regulation system

~rüttler *m*, Motorvibrator *m* = power vibrator

~schürf(kübel)wagen *m* mit Einachsschlepper → Autoschrapper *m* mit 2-Rad-Schlepper

~schürf(kübel)wagen *m* mit zusätzlichem Heckantrieb → doppelmotoriger Radschrapper *m*

~-Schürfzug *m* → Traktor-Schrapper

~-Schürfzug *m* → Autoschrapper *m* mit 2-Rad-Schlepper

~-Schürfzug *m*, Autoschrapper *m* mit Radschlepper = tractor-scraper, motor(ized) scraper (with four-wheel(ed) traction)

~schutzschalter *m* = motor-protecting switch

~seitenverkleidung *f* = engine side plates

~sockel *m* = engine base

~-Splittstreumaschine *f*, Motor-Splittstreuer *m* = power gritting machine, ~ gritter, ~ chip(ping)s spreader

~-Spritzmaschine *f* [*Verspritzen durch Pumpe*] = sprayer (or spraying machine) for hand power spraying with hand lance(s)

~-Spritzmaschine *f* kombiniert mit Kocher = heater and sprayer (or spraying machine) for hand and power spraying with hand spray unit

~-Spritzwagen *m* → Bitumen-Sprengwagen

~stillstandzeit *f* = engine down time

~-Straßenhobel *m*, Motor-Wegehobel, Motor-Erdhobel, Motor-Straßenplanierer *m* = motor(ized) grader, power ~, tractor ~, self-propelled (blade) ~

~-Straßenplanierer *m* → Motor-Straßenhobel *m*

~tragbolzen *m* = pivot pin

~träger *m*, Motorauflager *n* = engine mounting

~traverse *f* = engine saddle

~vibrator *m*, Motorrüttler *m* = power vibrator

~walze *f* → → Walze

~-Wegehobel *m* → Motor-Straßenhobel

~welle *f* = motor shaft

~welle *f* = engine shaft

~wick(e)lung *f* = motor winding

~winde *f* = engine-driven winch, power ~

~zange *f*, automatische Gestängezange [*Rotarybohren*] = power tongs

~-Zweiganggetriebe *n* = two speed gearbox

~zylinder *m* [*Kompressor*] = power cylinder

Mud-Jack-Verfahren *n*, (Beton-)Deckenhebeverfahren durch Einpressen von einem aus Feinsand, Mo, Schluff und Kolloidmaterial als Zuschlag, Zement als Bindemittel und Wasser als Verflüssiger bestehendem Gemisch bei Setzung von Betonfahrbahnplatten = mud-jack(ing) (of pavement slabs) [*i.e. the pressing (by compressed air) of a fluid mixture of cement, clay, fine sand and water under the slabs*]

Muffe *f* [*becherartig erweitertes Ende eines Rohres*] = socket (Brit.); bell (US)

Muffen|anschluß *m* [*Mitnehmerstange einer Rotary-Bohranlage*] = box connection

Muffen(bohr)gestänge — Muldenkipper

~(bohr)gestänge n = collared (drill) pipes
~druckrohr n = socket pressure pipe
~einschweißfitting m = socket weld fitting
~-Elevator m [*Tiefbohrtechnik*] = collar type elevator
~gestänge n, Muffenbohrgestänge = collared (drill) pipes
muffenlose Verrohrung f, eingezogene Bohrrohre npl = inserted joint casing
Muffen|metall n = bell metal (US); socket ~ (Brit.)
~mindesttiefe f [*Rohr*] = minimum depth of socket
~pump(en)stange f [*Tiefbohrtechnik*] = collared sucker rod, ~ pump ~
~rohr n = socket pipe
~rohrbogen m = socket bend
~stirnfläche f [*Tiefbohrtechnik*] = face of coupling [*casing*]
~tiefe f [*Rohr*] = depth of socket
Mugelkohle f, Kugelkohle = pebble coal, coal pebbles
Mühlbach m, Mühlgerinne n, Mühlgraben m, Stoßgerinne = millrace
Mühle f = (grinding) mill, grinder
Mühlen|-Aufgabegut n = unground material
~beschicker m, Mühlenaufgeber, Mühlenspeiser = (grinding) mill feeder
~-Fertiggut n = ground product
~mantel m = (mill) shell, grinding ~
Mühl|gerinne n → Mühlbach m
~graben m → Mühlbach m
Mulchen n = mulching
Mulde f [*Einschienen-Transportbahn*] = carrying skip
~, Förderwagen-Behälter m, Lastmulde, Tragwanne f, Nutzlast-Wanne = (dump) body
~, Trog m = trough
~, Wanne f [*LKW*] = body
~ für hohe Entladung = high discharge skip
~ mit Bodenentleerung = bottom discharge skip

Mulden|abstützung f [*Muldenkipper*] = (dump) body prop
~auflager n [*Muldenkipper*] = (dump) body support pad
~aufzug m → Muldenbauaufzug
~auslösung f [*Autoschütter*] = skip release mechanism
~band n, Muldengurt m, Trogband, Troggurt = trough belt, ~ band
~bandförderer m → Muldengurtförderer
~(bau)aufzug m, Kippkübelaufzug, Kübel(bau)aufzug, Betonkübelaufzug, Betonmuldenaufzug [*(Bau-) Aufzug mit Kippmulde im Fahrkorb die oben selbsttätig kippt*] = skip-hoist
~beheizung f [*Erdbauwagen*] = (dump) body heating
(~)Beschickungskran m, Chargierkran = charging crane
~bildung f (Geol.) = synclinal formation
~bodenneigung f [*Muldenkipper*] = (dump) body bottom plate slope
~-Erdbaufahrzeug n → Muldenkipper m
~falzziegel m = trough gutter tile
muldenförmig = troughed
Mulden|gewölbe n = trough vault
~gurtförderer m, Muldenbandförderer, Trogbandförderer, Troggurtförderer = trough belt conveyor, ~ band ~, ~~ conveyer
~kipper m → Feldbahnlore f
~kipper m, geländegängiger luftbereifter Erdtransportwagen m, gummibereiftes Erdbau-Lastfahrzeug n, geländegängiger Förderwagen in Kraftwagen-Bauart, Groß-Förderwagen, gleisloser Förderwagen, Motorkipper m, Erdtransportfahrzeug für Fremdbeladung, Erdbaulast(kraft)wagen, Großmuldenkipper, Förderwagen mit Hinterkippwanne, Mulden-Erdbaufahrzeug = rubber-tired (US) / rubber-tyred (Brit.) off-(the-)highway hauling unit (or earth-moving vehicle), off-road hauler, off-highway hauler

(Mulden)Kipplore — Murgang

(~)Kipplore f → Feldbahnlore
~positionslicht n [*Muldenkipper*] = (dump) body position light
~presse f [*Muldenkipper*] = (dump) body hoist cylinder
~rippe f [*Muldenkipper*] = (dump) body rib
~rost m = trough grate
~rostfeuerung f = trough grate furnace
~stein m = trough(ed) block
~tiefstes n (Geol.) = trough of a syncline
~tragbandstation f, Muldentragrollensatz m = carrier idler set for trough belt conveyor (or conveyer)
~tragrolle f = troughing roller
~wagen m → Feldbahnlore f
~warnlicht n [*Muldenkipper*] = (dump) body warning light
Müll m, Abfall m = refuse
~, Stadt~ = (town) refuse, (~) rubbish
(~)Abfuhr f = refuse cartage, ~ collection
~abfuhrwagen m → Müllwagen
~abladeplatz m = elevation of well head
~abwurfschacht m → Müllschlucker m
~auffüllung f = sanitary landfill
~behälter m, Abfallbehälter = refuse hopper
~beseitigung f, Abfallbeseitigung = disposal of refuse, refuse disposal
~einwurfschacht m, Füllschacht [*zum Füllen einer Müllverbrennungsanlage*] = charging well, charge ~
~grube f = refuse pit
~kübel m, Mülltonne f = dustbin, refuse storage container
~schacht m → Müllschlucker m
~schlucker m, Müll(abwurf)schacht m = garbage disposer (Brit.); dispose-all (US)
~tonne f → Müllkübel m
~tonnenaufzug m = dustbin lift (Brit.); ~ elevator (US)
~tonnenraum m, Müllkübelraum = dustbin room

~verbrennung f, Abfallverbrennung = refuse destruction, ~ incineration
~verbrennungsanlage f → Abfallverbrennungsanlage
~vernichtungsanlage f = destructor
~vernichtungsofen m → Abfallverbrennungsanlage f
~verwertung f = refuse utilization
~wagen m, Müllabfuhrwagen = collecting vehicle, refuse collector
~zerkleinerer m, Abfallzerkleinerer = rubbish grinder
Multiklon-Entstauber m, Multizyklon-Staubabscheider = multiclone (dust) collector, multiclone
Mund m, Eingang m, Portal n [*Stollen; Tunnel*] = portal, entrance
münden, ein~ [*Fluß*] = to empty into
Mund|loch n, Stollen~ = gallery opening
~loch n, Mündung f, Tagkranz m [*einer Bohrung*] = mouth of the (bore) hole, opening ~ ~ (~) ~, well head
~lochhöhe f einer Bohrung, Seehöhe des Ansatzpunktes einer Bohrung = elevation of well head
~stück n [*Strangpresse*] → Form~
Mündung f, Mundloch n, Tagkranz m [*einer Bohrung*] = opening of the (bore) hole, mouth ~ ~ (~) ~, well head
~ [*Strom; Fluß*] = estuary
Mündungs|baggerung f = esturial dredging
~barre f = mouth bar
~fahrrinne f = estuary channel
~kegel m, Delta n = delta
Munilager n, Munitionsdepot n = explosives area, bomb dump area, ammunition depot
Munitionsmagazin n, Munitionsschuppen m = ammunition warehouse
Münster m = minster
Muntzmetall n = muntz metal
Murgang m, Mure f, Rüfe f [*Vom Wildbach zum Tal mitgenommenes Geröll, das gesamten Wasserlaufquerschnitt*

ausfüllt, auch Gelände überschüttet] = wet landslide, mud-lava

Muschel|gestein *n* = shell rock
~mergel *m* = shell marl
~sand *m* = shell sand
~sandstein *m* = shell(y) sandstone, beach rock
~schieber *m* [*Dampfmaschine*] = three port slide valve
Musik|bücherei *f* = music library
~springbrunnen *m* = musical waters
~zimmer *n* = music room
musivische Kunst *f*, Mosaik *n* = mosaic
Muskovit *m*, K-Al-Glimmer *m*, tonerdereicher Kaliglimmer (Min.) = muscovite, Muscovy glass
~-Bergbau *m* = muscovite mining
~-Biotit-Granit *m* = muscovite-biotite granite, two mica ~
~granit *m* = muscovite granite
Musselinglas *n*, gemustertes Glas = muslin glass
Muster *n*, Dekor *n* [*auf einer Platte*] = pattern
mustern [*Linoleum*] = to pattern
Muster|probe(stück) *f*, *(n)* = type sample
~sieblinie *f* = type grading curve
Mutter *f*, Schrauben~ = nut
~bett *n*, Hauptbett = main bed
~boden *m*, Bodenkrume *f* [*die oberste, Humus und Kleinlebewesen enthaltende Bodenschicht*] = top soil, surface ~
(Mutterboden)Abtrag *m*, (Mutterboden)Abhub *m* = stripping (of top soil), top soil stripping
Mutterboden|andeckung *f*, Mutterbodenauftrag *m* = replacing top soil
~miete *f* = stockpile of topsoil
Mutter|form *f* [*Ziegelindustrie*] = upper mo(u)ld, master ~
~gestein *n* (Geol.) = host rock
~gestein *n* [*Erdöl*] = source rock, ~ bed, mother rock
~pause *f* = negative, transparency, transparent positive original

~scheibe *f* = plain washer
~schiene *f* → Backenschiene
~werkstoff *m*, unedles Metall *n*, Grundmetall = base metal
MZ *m* → Vollziegel

N

Nabe *f*, Rad~ = (wheel) hub
~, Ausguß *m*, Auge *n* = boss
Nabenuntersetzung *f* = hub reduction
nach oben aufbrechen [*Schacht*] = to excavate in upward direction
Nach|arbeit *f* = rework
~ausfugen *n*, Nachverfugen, Nachausstreichen = repointing
~ausstreichen *n*, Nachausfugen, Nachverfugen = repointing
~backenbrecher *m* → Backenfeinbrecher
Nachbar|bahn *f* → Anschlußbahn
~pfahl *m* = neighbouring pile, adjacent ~
~schaftseinheit *f* = neighbourhood (unit), residential neighbourhood [*Is an area consisting of dwellings and land for ancillary uses, such as schools, shops and open spaces of limited size, it may have physical boundaries such as arterial roads, or regional open spaces. Each unit, while essentially but a single part of a greater whole, becomes a comprehensible entity in itself]*
~streifen *m* → Anschlußbahn *f*
Nachbehandlung *f* [*Abwasser*] = secondary process
~ [*Beton*] = curing, after-treatment
~ ohne Feuchthaltung = dry curing, ~ after-treatment
Nachbehandlungs|film *m* = curing membrane
~mittel *n*, Beton~ = (concrete) curing agent
~platz *m* = curing yard
~wasser *n* = curing water
~zeitraum *m*, Beton~ = (concrete) curing period

Nach|bestellung f = repeat order
~bohren n → Nachnehmen
Nachbrechen n, Zweitbrechen = secondary crushing [*The second crushing stage*]
~, Drittbrechen = tertiary crushing [*The tertiary crushing stage*]
~ → Abräumen (⚒)
Nachbrecher m, Zweitbrecher = secondary crusher, **~** breaker
~, Drittbrecher = tertiary crusher, **~** breaker
Nach|bruch m → Abräumen n (⚒)
~dichtung f [*von Staumauern, Staudämmen und Spundwänden*] = re-sealing
~eichung f, Spur f, Ausrichtung = alignment (Brit.); alinement (US)
~eilen n **der Tide** → Alter n **~** **~**
~(ein)rütt(e)lung f = re-vibration
~fall m, Einsturz m = cave(-in), caving
~fall m (⚒) = fall(s) of stone
nachfallen = to cave (in)
Nach|fall(packen) m (⚒) = draw roof, clod
~faulraum m [*Abwasser*] = secondary digestion chamber
~faulung f = secondary digestion
~filterelement n = after-filter element
~filterung f = secondary filtration
(~)Folgebohrung f = follow-up well
nachfüllen = to re-fill, to add
Nachgeben n [*Boden*] = yield
nachgemahlen = reground
Nachgewitter n = subsequent thunderstorm
nach|gezogene Gleitschalung f, Schleppschalung = trailing forms
~giebig, drucknachgiebig = yielding (under pressure)
~giebige Eckverbindung f [*Stahlrahmen*] = semi-rigid connection
nachgiebiger Gelenkbogen m, drucknachgiebiger **~** (⚒) = articulated yielding (roadway) arch
~ Streckenbogen m, drucknachgiebiger **~** (⚒) = yielding (roadway) arch

Nachhall m = reverberation
~raum m = reverberant room
~zeit f [*Raumakustik*] = reverberation time
Nachhärtung f = re-hardening
Nachholbedarf m = backlog of needs, inadequacy gap
~ [*Straßenunterhaltung*] = arrears of road maintenance
Nachklärbecken n [*Abwasserwesen*] = final settling basin [*A basin through which the effluent of a trickling filter, or other oxidizing device, passes for the purpose of removing the settleable solids before its discharge*]
~ [*einer Tropfkörperanlage*] = humus tank
Nach|klärbrunnen m [*Abwasserwesen*] = final settling tank [*Same as final settling basin, but deeper and of less area*]
~klärung f, Nachklären n = final clarification, secondary **~**
~knäppern n → Knäpperschießen
nachkonsolidieren = to re-consolidate
Nach|kühler m = aftercooler
~kühlung f = aftercooling
nachlassen [*Rotarybohren*] = to feed off the line from the drawworks, to pay out the line
Nachlaßschraube f, Nachlaßspindel f [*pennsylvanisches Seilbohren*] = temper screw
~, Nachlaßspindel f [*Rotarybohren*] = feed screw
Nachlaßspindel f, Nachlaßschraube f [*pennsylvanisches Seilbohren*] = temper screw
~, Nachlaßschraube f [*Rotarybohren*] = feed screw
Nachlaßvorrichtung f [*am Hebewerk einer Rotarybohranlage angebaut*] = feeding device
nachlaufendes Meßrad n, Meßvorrichtung f für Seitenführungskraft = fifth wheel, fifth-wheel device
Nach|läufer m, Anhänger = trailer
~löschbunker m [*Kalk*] = caving bin
Nachmahlen n, Nachmahlung f, Zweitmahlen, Zweitmahlung = second-

ary grinding [*The secondary grinding operation*]
~, Nachmahlung *f*, Drittmahlen, Drittmahlung = tertiary grinding [*The tertiary grinding operation*]
nachmessen = to re-measure
Nachmessen *n* = re-measuring
Nachmischen *n*, Rühren [*Transportbeton*] = agitating
~ = remixing
Nach|mischer *m* → (Beton)Rührfahrzeug *n*
~**mittags-Spitzenstunde** *f* = afternoon peak hour, ~ rush ~
~**mühle** *f* = regrind mill
Nachnehme|bohrer *m* → Nachnehme(bohr)meißel *m*
~**(bohr)meißel** *m*, Erweiterungs(bohr)meißel, (Auf)Räumer *m*, Nachnehmer *m*, Nachnehmebohrer *m* [*Rotarybohren*] = reaming bit, enlarging ~, reamer
nachnehmen [*Förderseil im Rotarybohrbetrieb*] = to cut [*drilling line*]
~ **der Sohle** (⚒) = to dint
Nach|nehmen *n*, Räumen *n*, Aufräumen *n*, Nachbohren [*Rotarybohren*] = reaming
~**nehmer** *m*, Nachnehmebohrmeißel *m*, (Auf)Räumer = reamer, reaming bit
nachölen = to reoil
Nach|prüfung *f* = confirmatory check
~**pumpen** *n* = re-pumping
~**rammen** *n* = redriving
~**rechnung** *f* = checking
~**reinigungszelle** *f* [*In die Nachreinigungszelle wird das Vorkonzentrat der Hauptflotation, zuweilen auch das Konzentrat der Bergereinigerzellen zu einer erneuten Flotation aufgegeben*] = cleaner
~**reißstelle** *f* (⚒) = ripping lip, canch
~**rütt(e)lung** *f*, Nacheinrütt(e)lung = re-vibration, re-vibrating
~**schaltgetriebe** *n*, Stufengetriebe = range transmission
~**schaltheizfläche** *f* [*Kessel*] = convection (heating) surface
nachschießen [*Streckenvortrieb im Bergbau*] = to trim

Nach|schlagebuch *n* = reference book
~**schubbasis** *f* = supply base
~**schweißen** *n* = rewelding
~**schwindung** *f*, Nachschwinden *n* = secondary shrinkage, post-shrinkage, re-shrinkage
~**schwindungs-Prüfung** *f* [*feuerfester Stein*] = reheat test [*fire-brick*]
~**senkung** *f* (⚒) = delayed subsidence
nachsetzen [*Tiefbohrtechnik*] = to add a new joint to the drill stem
Nach|setzung *f*, Nachsetzen *n* = secondary settlement, subsequent ~
~**sichtung** *f* = secondary air classification
nachspannen [*Spannbeton*] = to post-stress
Nach|spannen *n*, Nachspannung *f* [*Spannbeton*] = secondary tensioning, re-stressing
~**spannung** *f* → Vorspannung mit nachträglichem Verbund
~**spannvorrichtung** *f* [(*Gleis*)*Kette*] = slack adjuster
~**spannweg** *m* [(*Gleis*)*Kette*] = allowance
~**steinbrecher** *m*, Feinsteinbrecher = fine stone crusher, reduction ~ ~, secondary ~ ~, breaker
nachstellen = to re-adjust
Nachstell|platte *f* = adjusting plate
~**schnecke** *f* = adjusting worm
~**schraube** *f* = adjusting screw
Nach|streichen *n* = repainting
~**tanken** *n* = re-fuelling
~**teerung** *f* = re-tarring
~**trieb** *m* [*Tunnelbau*] = secondary heading
Nacht|anflug *m* = nighttime approach
~**arbeit** *f* = night work
Nachteerung *f* = re-tarring
nachtleuchtender Stoff *m* → Leuchtfarbe *f*
Nacht|nebel *m* = night-time fog
~**schicht** *f* = night shift
~**sichtbarkeit** *f*, Sichtverhältnisse *npl* bei Nacht = night visibility
~**strom** *m* = night power
~**tarif** *m* [*Strom*] = "off-peak" tariff

Nachtverkehr — Nagelschleppe 688

~verkehr m = night traffic
~verkehrsunfall m = night traffic accident
~wächter m = night watchman; dark horse (US)
nachträglicher Verbund m → Vorspannung mit nachträglichem Verbund
Nachunternehmer m, Subunternehmer = subcontractor
~vertrag m = subcontract
nachverdichten = to recompact
Nachverdichtung f = re-compaction
~ [*Pfahlgründung*] = reconsolidation
~ unter dem Verkehr = additional consolidation under traffic, kneading
Nach|verfugen n → Nachausfugen
~vermessung f = resurvey
nachverstemmen = to reca(u)lk
Nach|verteilung f → Umlagerung
~wachsversuch m, Nachwachsprobe f, Nachwachsprüfung f = reheating test [*on firebricks*]
nachweisen [*Baustatik*] = to check (up)
Nach|wirkung f, Hysteresis f = hysteresis
~zahlung f = retroactive pay, back-pay
~zerkleinerung f, Zweitzerkleinerung = secondary comminution, ~ reduction
~zerkleinerung f, Drittzerkleinerung = tertiary comminution, ~ reduction
~ziehen n = retightening [*bolt*]
nackt, blank, nicht umhüllt = uncoated, bare
nackter Boden m = bare ground
~ Schweißdraht m = bare filler rod
Nadel f [*Nadelwehr*] = needle
~, Pilz m [*Freistahlturbine*] = spear
~ausleger m = needle type jib (Brit.); ~ ~ boom (US)
~blende f, prismatische ~ (Min.) = red antimony ore
~bock m [*Nadelwehr*] = frame
~eindringgerät n, Nadeleindringungsmesser m, Nadelpenetrometer m, n = needle penetrometer
~düse f [*Freistrahlturbine*] = nozzle with spear

~eis n = spicular ice
~eisenstein m (Min.) = needle ironstone
nadelförmig → → Kornform
Nadel(gerät) f, (n) [*Zementprüfung*] = needle (apparatus)
~ nach Gillmore = Gillmore apparatus, ~ needles [*A pair of weighted, flat-ended needles of different cross section for determing initial and final setting times of hydraulic cements*]
~ ~ Vicat = Vicat needle (apparatus), ~ apparatus
Nadel|holzteer m → Teer
~kohle f = needle coal
~lager n = needle bearing
~penetrometer m, n → Nadeleindringgerät n
~rüttler m → Innenrüttler
~(stau)wehr n = needle weir
~stein m (Min.) = needle-stone, rutilated quartz
~vibrator m → Innenrüttler m
~wehr n, Nadelstauwehr = needle weir
~zinnerz n (Min.) = needle-tin
Nadiraufnahme f → Senkrechtaufnahme
Nagel(aus)zieher m → Nageleisen n
nagel|bar = nailable
~barer Beton m → Nagelbeton
Nagel|bauweise f = nailed construction method
~befestigung f = fixing by nails
~beton m, nagelbarer Beton m = nailing concrete [*Aggregates which will produce nailing concrete into which nails can be driven and maintain their grip include asbestos fibre, sawdust, and cinders*]
~binder m = nailed truss
~eisen n, Geißfuß m, Hampelmann m, Nagel-Klaue f, Nagel(aus)zieher m, Nagelheber m = nail puller, pry bar
~heber m → Nageleisen n
~klaue f → Nageleisen n
~kopf m = nail head
~ort m, Plattbohrer m = bradawl
~reihe f = row of nails
~schleppe f = nail drag, spike ~

Nagelverbindung — Naß(aufgabe)gut

~verbindung *f* = nailed connection
~ziehen *n* = nail pulling
~zieher *m* → Nageleisen *n*
Nah|(ablese)kompaß *m* = direct-reading compass
~ansicht *f* [*z. B. einer Straßendecke*] = close-up view
~beben *n* = neighbouring earthquake
Näherung *f*, An~ = approximation
Näherungs|annahme *f*, An~ = approximation hypothesis
~bruch *m* [*Mathem.*] = convergent
~formel *f* → An~
~gleichung *f* → An~
~lösung *f*, An~ = approximate solution
~probe(stück) *f*, *(n)*, An~ = approximate sample
~rechnung *f* → An~
~rechnung *f* nach Cross = Cross moment distribution method
~verfahren *n*, An~ = approximate method
~wert *m*, An~ = approximate value
Nahkompaß *m*, Nahablesekompaß = direct-reading compass
Nährstoff *m* = nutrient
Naht *f*, Schweiß~ = (welding) seam
~ [*Schwarzdecke*] = joint
~erwärmer *m* [*Schwarzdecke*] = joint heater
~fläche *f* [*Schwarzdecke*] = joint area
~instandsetzungserwärmer *m* [*Schwarzdecke*] = joint repair heater
nahtlose Verrohrung *f* [*Bohrloch*] = seamless casing, ~ lining
nahtloses Bohrgestänge *n* = seamless drill pipe
~ (Stahl)Rohr *n* [*Aus einem vollen Block hergestelltes Stahlrohr ohne Schweißnaht*] = seamless tube, ~ pipe
Naht|schluß *m* durch Walzverdichtung [*Schwarzdecke*] = rolling a joint
~schweißen *n*, Nahtschweißung *f* = seam welding
~überhöhung *f*, Gewölbeüberhöhung [*Schweißtechnik*] = arcing
Nahverkehr *m* = local transportation

Nahverkehrsbereich *m*, Abfertigungsanlage *f*, Abfertigungsvorfeld *n* [*Flugplatz. Fläche für Bewegungsvorgänge 2. Ordnung, wo die verkehrs- und betriebstechnische Abfertigung der Flugzeuge vorgenommen wird*] = terminal area
~überwachung *f* [*Flughafen*] = terminal air traffic control
NaOH → Ätznatron *n*
Naphthalin *n* = naphthalene, naphthalin(e)
~öl *n* = naphthalenic oil
~salz *n* = naphthalate
~säure *f* = naphthalic acid
naphthenbasisches Erdöl *n* = naphthene-base crude petroleum
Narbe *f*, Korrosions~ = saucer-shaped pit, scar
narrensicher, mißgriffsicher = foolproof
Nase *f*, Aufhänge~ [*Dachziegel*] = nib
~, Anschlag *m* [*(Gleis)Kettenlaufwerk*] = lug
~ zum Festhalten [*Jalousieschütz(e)*] = peg, top (Brit.); stop pin (US)
nasse Destillation *f*, Abdampfung *f*, Überdampfung = boiling down
~ (Zylinder)Laufbüchse *f* = wet cylinder liner
nasser Gaszähler *m* = drum(-type) gas meter
nasses Erdgas *n*, ~ Naturgas, feuchtes ~, Naßgas = wet natural gas, combination ~, casing-head ~, natural gas rich in oil vapo(u)rs
~ Naturgas *n* → ~ Erdgas
Naß|-Abbauhammer *m* = wet pick (hammer)
~abscheider *m*, Naß(staubab)scheider, Naßentstauber = water type dust collector, wet ~
~(ab)siebung *f*, (Ab)Sieben *n* von feuchtem Gut = wet screening
~aufbereitung *f* [*Zement*] = wet manufacture
~aufbereitungsanlage *f* [*Zementherstellung*] = wet process installation
~(aufgabe)gut *n* = wet feed

Naßbagger — Natriumsilikat 690

~bagger *m*, Schwimmbagger = dredge (US); dredger (Brit.); dredging craft
~bagger *m* (oder Schwimmbagger) mit zwei oder mehreren Typen von Baggeraggregaten = compound dredger (Brit.); ~ dredge (US)
~baggerfirma *f* = dredging firm
~baggergut *n* = dredged material, ~ spoil, dredging ~
~baggermaschine *f* = dredging machine
~baggerung *f*, Naßbaggerei *f*, Naßbaggern *n* = dredging, dredge(r) work
~betonierung *f* = underwater concreting, ~ pour(ing)
~drehofen *m*, Naßrotierofen = wet process-type rotary kiln
~engobierung *f* = wet slip coating
~entstauber *m* → Naßabscheider
~festigkeit *f* = wet strength
~gas *n* → nasses Erdgas
~gründung *f* = wet foundation
~gut *n*, Naßaufgabegut = wet feed
~haltung *f* → Feuchthaltung
~kollergang *m* = wet pan grinder
~kugelmühle *f* = wet ball (grinding) mill
~kupp(e)lung *f*, Ölkupp(e)lung = oil clutch
~löffelbagger *m*, Schwimmlöffelbagger, Löffel-Naßbagger, Löffel-Schwimmbagger = dipper (bucket) dredger, spoon ~ [*US = dredge*]
~löschverfahren *n*, Einsumpfen *n* [*Branntkalk*] = wet slaking process
~-Magnetscheider *m* = wet magnetic separator
~-Magnetscheidung *f* = wet magnetic separation
~mahlanlage *f* = wet grinding plant, ~ ~ installation
~mahlen *n*, Naßmahlung *f* = wet grinding
Naßmischen *n*, Naß(nach)mischung *f* [*Zementverfestigung*] = moist mixing
~ während der Fahrt *f* [*Transportbeton*] = shrink-mixing, partial mixing

Naß|pochwerk *n*, Naßstampfmühle *f* = wet stamp
~putz *m* [*im Gegensatz zur Trockenbauweise*] = wet plaster
~rohrmühle *f* = wet tube (grinding) mill
~rotierofen *m*, Naßdrehofen = wet process-type rotary kiln
~(-Sand-)Verfahren *n*, Kalkverfahren = wet-sand process, wet sand-binder construction, (hydrated) lime process, wet aggregate process, wet sand mix
~saugbagger *m* → Saugbagger
~schlamm *m* [*Abwasser*] = wet sludge
~schleifmaschine *f* = wet grinding machine
~siebung *f* = wet sieving [*determination of grain size*]
~spritzverfahren *n* mit kolloidalem Mörtel und Beton, hergestellt nach dem Colcrete-Verfahren, Colgunite-Verfahren = Colgunite
~staubabscheider *m* → Naßabscheider
~verfahren *n* [*Zementherstellung*] = wet process
~zerkleinerung *f* = wet comminution, ~ reduction
~zerkleinerungsmaschine *f* = wet comminuter
~zyklon *m*, Aufschwimmklassierer *m*, Zyklonnaßklassierer = cyclone classifier
Natrium|alginat *n* = sodium alginate
~-Aluminium-Silikat *n* = sodium alumin(i)um silicate
~-Benzoesalz *n* = sodium benzoate
~chlorid *n*, Kochsalz *n*, NaCl = sodium chloride
~chloridsole *f* = sodium chloride brine
~dampflampe *f*, Natriumdampfleuchte *f* = sodium vapo(u)r lamp, ~ (discharge) ~
~karbonat *n* = mineral alkali [*old name*], sodium carbonate
~karbonat wasserfrei *n* → kalzinierte Soda *f*
~oxyd *n*, Na_2O = sodium oxide
~silikat *n*, Wasserglas *n*, Na_2SiO_3 = sodium silicate

Natrolith *m* = mesotype, natrolite
Natron|feldspat *m* (Min.) = soda fel(d)spar
~glimmer *m* (Min.) = sodium mica
~kalkfeldspat *m* → Plagioklas *m*
~kalkglas *n* = soda lime glass
~kalkkieselsäureglas *n* = soda-lime-silica glass
~(seifen)fett *n*, natronverseiftes Fett = sodium grease
~stein *m* (Min.) = soda zeolite
~wasserglas *n* = soda water glass, ~ soluble ~
Natur *f* [*als Gegensatz zum Modell bei wasserbaulichen Ähnlichkeitsversuchen*] = prototype
(Natur)Asphalt *m*, natürliches Bitumen-Mineral-Gemisch *n*, natürliche Bitumen-Mineral-Mischung *f* [*das natürlich vorkommende Bitumen mit seinem Begleitgestein*] = (natural) asphalt (Brit.)
~gestein *n* = (natural) rock asphalt [*It usually consists of a rather porous limestone, less often sandstone, naturally impregnated with (asphaltic) bitumen (Brit.)/asphalt (US)*]
~mastix *m* = natural rock asphalt mastic
~-Mastixbrot *n*, Asphaltbrot = mastic block
~rohmehl *n* = powdered natural rock asphalt
Naturbaryt *m*, Bariumsulfaterz *n* = barytes ore
~zuschlag(stoff) *m* = barytes ore aggregate
Natur|baustein *m* = natural building stone
~benzin-Anlage *f* = natural gasoline plant, ~ gasolene ~ (US)
„Naturbeton" *m* → Brekzie ~ → Konglomerat *n*
Natur|bims *m* = (natural) pumice
(~)Bimsbeton *m* = (natural) pumice concrete
~bitumen *n*, natürlich vorkommendes Bitumen = natural(ly occurring) (asphaltic) bitumen (Brit.); native asphalt (US)

~brücke *f* (Geol.) = natural bridge
~feinkorn *n* = natural fine grain
~feinsandzuschlag(stoff) *m* = natural-sand fine aggregate
Naturgas *n*, Erdgas = natural gas, rock ~
~benzin *n*, Erdgasbenzin, Naturgasolin *n* = casing-head gasoline, natural ~, ~ gasolene (US)
~brenner *m* → Erdgasbrenner
~-Durchschlag *m* **aus dem Ringraum in die Steigerohre**, Erdgas-Durchschlag ~ ~ ~ ~ ~ ~ = surging
~feld *n*, Erdgasfeld = gas reservoir
~kraftwerk *n*, Erdgaskraftwerk = natural gas power station (or plant)
~leitung *f* = natural gas pipeline
~-Lok(omotive) *f* = natural gas locomotive
Naturgasolin *n*, Erdgasbenzin *n*, Naturgasbenzin = natural gasoline, casing-head ~, ~ gasolene (US)
Naturgas|schwefel *m*, Erdgasschwefel = sulfur from natural gas (US); sulphur ~ ~ ~ (Brit.)
~-Tanker *m*, Erdgas-Tanker, Natur-gas-Tankschiff *n*, Erdgas-Tankschiff = methane ship
Natur|gestein *n*, Naturstein *m* = natural rock
~gummimehl *n* = natural rubber powder
~hafen *m* = natural harbo(u)r
~harz *n* → Harz
~katastrophe *f* = act of God
natürlich belüfteter Kühlturm *m*, selbstventilierender ~ = natural draught (Brit.) draft (US) cooling tower
~ feuerfester Baustoff *m* = naturally refractory construction material, ~ ~ building ~
natürliche Austrocknung *f*, natürliches Austrocknen *n* = natural drying out, ~ desiccation
(~) Baugrundverdichtung *f* = ground consolidation
~ Bitumen-Mineral-Mischung *f* → (Natur)Asphalt *m*
~ Bodenverdichtung *f* → Konsolidierung

natürliche Entwässerung — Nebenfeld

~ **Entwässerung** f = natural drainage
~ **Geländehöhe** f = natural ground level
~ **Größe** f = natural size
(~) **Kolmation** f → Auflandung f
~ **Wasserstraße** f = natural waterway
natürlicher Anhydrit m = natural anhydrite
~ **Böschungswinkel** m, Ruhewinkel = angle of repose (of the natural slope), ~ ~ rest
~ **Feinzuschlag(stoff)** m, Betonsand m [*Durchgang durch das Rundlochsieb 7 nach DIN 1170*] = natural (concreting) sand, ~ concrete ~, ~ fine aggregate
~ **Grobzuschlag(stoff)** m, Betonkies m [*Rückstand auf dem Rundlochsieb 7 nach DIN 1170*] = natural coarse aggregate, ~ gravel [*This term refers to an aggregate mainly retained on a 3/16 in. B. S. test sieve*]
~ **Schlackensand** m → vulkanischer Sand
~ **Wasserlauf** m = natural water course, ~ waterway
natürliches Austrocknen n, natürliche Austrocknung f = natural drying out, ~ desiccation
~ **Bitumen-Mineral-Gemisch** n → (Natur)Asphalt m
~ **Schleifmittel** n [*in Steinbrüchen gewonnen, zerkleinert und gesiebt*] = natural abrasive
~ **Uran** n = natural uranium
Natur|material n → → (Beton-)Zuschlagstoffe mpl
~**mauerwerkstein** m = walling stone
~**pflasterstein** m = natural (paving) sett
~**salzsole** f = natural brine
~**sand** m = natural sand
~**schutz** m = preservation of natural beauty
~**schutzgebiet** n = wild life sanctuary, nature reserve
(Natur) Stein|bearbeitungsmaschine f = natural stone dressing machine
~-**Bekleidung** f = natural stone facing
~**bogen** m = stone arch
~**bordstein** m = (natural) stone curb; ~ ~ kerb (or kirb) (Brit.); ~ curbstone
~**decke** f = paving of natural setts
~**geländer** n = stone railing
~**gewinnung** f = winning of natural stone
~**gewölbe** n = stone vault
~-**Industrie** f = natural stone industry
~-**Innenbekleidung** f = natural stone internal facing
~**mauerwerk** n = masonry
~**mauerwerkbau** m = masonry construction
~-**Pflasterplatte** f = natural (paving) flag, ~ flagstone
~**platte** f = natural stone slab
~**säule** f = stone column, ~ pillar [*used for ornamental purposes*]
~**splitt** m = natural stone chip(ping)s
~**straße** f = road with paving of natural setts, highway ~ ~ ~ ~
~**verkleidung** f = natural stone lining
~**wandpfeiler** m = stone pier, ~ pillar
Natur|versuch m [*Korrosionsprüfung*] = field-test
~**zement** m → Mischzement
nautische Karte f, Seekarte = (marine) chart, sea ~
ND m, Nenndruck = rated pressure
Nebelhorn n, Nebelsirene f = fog siren
Nebelscheinwerfer m = fog lamp
~**strahl** m = fog beam
Nebel|signal n = fog signal
~**sirene** f, Nebelhorn n = fog siren
~**zerstreuer** m = fog disperser, ~ dispeller
Neben|anlage f = secondary installation, ~ plant
~**bewehrung** f, Nebenarmierung, Neben(stahl)einlagen fpl = secondary reinforcement
~**deich** m → Sommerdeich
~**erzeugnis** n → Nebenprodukt n
~**feld** n, Nebenöffnung f, Nachbarfeld, Nachbaröffnung [*Brücke*] = adjacent span, ~ opening

Nebenflugplatz — Nenn-Mischungsverhältnis

~flugplatz m → Hilfslandeplatz
~fluß m = affluent (Brit.); tributary
~gestein n (⚒) = country rock
~gleis n = side track, siding
~kanal m, Seitenkanal = lateral canal
~küche f = secondary kitchen
~landeplatz m → Hilfslandeplatz
~leistungen fpl = necessary work auxiliary to the accomplishment of the contract
~lenkarm m = follow-up arm
~produkt n, Nebenerzeugnis n = by-product
~pumpwerk n = secondary pumping station
~spannung f = secondary stress
~(stahl)einlagen fpl → Nebenbewehrung f
~straße f, Seitenstraße, Kleinverkehrsstraße = minor (or subsidiary, or side) road
~strom m [*Maschinen-Hydraulik*] = by-pass
~stromleitung f [*Maschinen-Hydraulik*] = by-pass line
~stromventil n = by-pass valve
~tür f = secondary door
~verwerfung f → Begleitverwerfer m
~zeit f, Griffzeit [*bei der Zeitnahme*] = downtime
neblig-dunstig, nebelig-dunstig = smoggy
Neck m → Schlot m
negative Auflagerkraft f = uplift
~ Ladung f = negative charge
~ Mantelreibung f = negative (skin) friction
negativer artesischer Brunnen m = negative artesian well
negatives Delta n, trichterförmige Mündung f = funnel shaped estuary
~ Zusatzmoment n, Umlagerungsmoment = secondary (bending) moment
Nehrung f, Landzunge f = spit (of land)
neigbares Einrammen n = driving on the rake

neigende Scheinlagerung f (Geol.) = false bedding
Neigung f, Anzug m, Dossierung f [*gemessen gegen die Vertikale (oder Lotrechte)*] = rake, batter [*deprecated: slope, incline*]
~ → Steigung f
~ [*des Längsprofils Steigungsverhältnis*] = gradient, incline [*deprecated: grade*]
~ [*einer Störungsfläche im Bergbau*] = hade [*of a fault plane*]
Neigungs|gleiche f, Isokline f = isoclinic line
~grenze f = inclination limit
~messer m, Abweichungsmesser [*Tiefbohrtechnik*] = drift recorder, driftmeter, inclinometer
~messung f, Abweichungsmessung [*Tiefbohrtechnik*] = deviation measurement, verticality survey
~verstellung f [*Motor-Straßenhobel*] = pitch adjustment
~winkel m = angle of inclination
NE-Metall n, Nichteisenmetall = non-ferrous metal
Nenn|abmessung f = nominal dimension
~belastung f = rated loading
~bohrung f = nominal bore
~drehzahl f, Solldrehzahl = rated speed, ~ RPM
~druck m, ND = rated pressure
~durchmesser m = nominal diameter
~fassungsvermögen n = rated capacity
~größe f = nominal size
~inhalt m, Nutzinhalt, Füllinhalt [*Betonmischer*] = dry capacity per batch, unmixed capacity
~innendurchmesser m = nominal internal diameter
~last f = nominal load
~leistung f [*Maschine*] = rated capacity
~leistung f [*Motor*] = nominal horsepower, ~ H. P., ~ h. p.
~maß n, Sollmaß = nominal measure
~-Mischungsverhältnis n = nominal mix

~querschnitt m = nominal cross-section

~-Reifenkraftschluß m = rated rimpull

~schubspannung f = nominal shear stress

~stromstärke f = rated current

~-Verdichtungsverhältnis n [Dieselmotor] = nominal compression ratio

~versuch m, Nennprobe f, Nennprüfung f = rating test

~weite f [kreisförmiges Rohr] = nominal internal diameter, ~ bore

~wert m = nominal value

~würfelfestigkeit f = nominal cube strength

Neonröhre f → Leuchtstoffröhre

Neoprenlager n = neoprene bearing

~platte f = neoprene bearing pad

Neopren|platte f = neoprene sheet

~werk n = neoprene production plant

Nephelin m (Min.) = nepheline, nephelite

~basalt m = nepheline-basalt

Nephelinit m (Min.) = nephelinite

Nephelinsyenit m (Geol.) = nephelinesyenite

neptunisches Sedimentgestein n, aquatisches ~ = water-borne sediment

neréitische Sedimente npl = neritic sediments

Nest n [im Beton] = pocket

Nesterbildung f [Beton] = formation of pockets

Netto|-Bohrfortschritt m = net drilling rate

~querschnitt m = net section

Netz n, Leitungs~ = line network
~, Ordinaten~ = grid

~angabe f [Karte] = map reference, grid ~, MR

~(anschluß)gerät n = power supply unit (or set)

~armierung f → Armierungsnetz n

~ausfall m [Elektrotechnik] = mains failure

~betriebsdienst m [Energieversorgung] = system operator

netzbewehrt, netzarmiert = mesh-reinforced

Netz|bewehrung f → Armierungsnetz n

~druck m [Gas; Wasser] = mains pressure

~druckunterschied m = mains pressure differential

~gewölbe n, Rautengewölbe = reticulated vault

~haftmittel n, Haftanreger m, adhäsionsfördernder Zusatzstoff m, Adhäsionsverbesserer m, Haftmittel, Haftfestigkeitsverbesserer = adhesion (promoting) agent, non-stripping ~, anti-stripping admixture (or additive), dope, bonding additive, activator, anti-stripping agent

~kabel n = main cable

~linie f, Systemlinie [Träger] = system line

~plantechnik f = critical path method, CPM

~rißbildung f → feine Rißbildung

~strom m = mains electricity, grid ~, ~ current

Netzwerk n = latticework

~träger m = latticework girder

Neubaugebiet n = developing area

neubohren [ein neues Loch bei starker Abweichung] = to redrill, to drill a new hole

Neuentwick(e)lung f = novelty, new development, newcomer

Neuland n = reclaimed land

~gewinnung f = land reclamation, reclamation of land

neutrale Achse f, Nullinie f = neutral axis, zero line

~ Druckspannung f, ~ Flächenlast f [Bodenmechanik] = neutral pressure

~ Flächenlast f, ~ Druckspannung f [Bodenmechanik] = neutral pressure

~ Ofenatmosphäre f = neutral conditions, ~ atmosphere

~ Spannung f [Bodenmechanik] = neutral stress

~ Temperaturtiefe f = depth of constant temperature

neutrales Abbeizmittel n, lösendes ~, Abbeizfluid n = solvent-based paint remover
Neutralisier|gefäß n = acid knock-out drum
~ventil n = neutralizer valve
Neutronen|aktivierungsanalyse f [*Bohrloch-Überwachung*] = neutron activation logging
~sonde f **zur Feuchtigkeitsbestimmung von Böden** = neutron soil moisture gauge
n-fach geschakt [*Eimerkettenbagger*] = number of links per bucket
n-freie Berechnungsweise f, **n-freies Verfahren** n = design method not involving the use of the modular ratio
Nichols-(Etagen)Ofen m = Nichols sludge incinerator
nichtabgewalzt = unrolled
nicht absetzbarer Stoff m = nonsettleable solid, ~ matter
nichtabspritzendes Öl n = nonfluid oil
nicht angeschlossener (Polygon)Zug m, freier ~ = open traverse
~ aufgehende Zahl f, irrationale ~ = irrational number, surd ~, ~ quantity
nicht|ausgepreßtes Spannglied n = unbonded tendon, ungrouted ~
~ballend, nichtverstopfend [*Siebtechnik*] = nonclogging
~beständige Strömung f, unstationäre ~ = nonsteady flow, unsteady ~
~bewehrt, unbewehrt = nonreinforced, plain
~bindig, nichtkohäsiv, inkohärent, kohäsionslos, rollig [*Boden*] = noncohesive, cohesionless, friable, nonplastic, frictional
~blasend [*Bohrhammer*] = non-blowing
~detonatives Sprengen n, Druckluftschießen n = bursting by compressed air
Nichteisenmetell n = non-ferrous metal
nicht|gängig [*Maschinenteil*] = not free
~gebrochen [*Gestein; Erz*] = uncrushed, unbroken

~genormtes Profil n [*Stahl oder Aluminium*] = non-standard shape, ~ section
~gespannte Bewehrung f → schlaffe ~
Nicht|geradlinigkeit f = non-linearity
~-Hochdruckschmiermittel n = non-extreme pressure lubricant
nichtkanalisierte Strecke f = section without canalization
Nichtkarbonathärte f = permanent hardness, non-carbonate ~
nicht|klastisches Sedimentgestein n = nonclastic sedimentary rock
~lasttragend = nonloadbearing
Nichtleiter m = non-conductor
nicht|linear = nonlinear
~lineares Regelsystem n = non-linear system
~massiver Beton m, Bauwerkbeton, Konstruktionsbeton = structural concrete
~ pendelnd, starr = nonoscillating
~prismatischer Stab m = nonprismatic member
Nicht-Raumbeständigkeit f [*Zement*] = unsoundness
nicht|registrierender Pegel m, nicht(auf)zeichnender ~, nichtschreibender ~ = nonrecording gage (US); ~ gauge (Brit.)
~rostender Stahl m, chemisch beständiger ~ = stainless steel
Nicht|schwimmerbecken n = shallow pool for non-swimmers
~seifenfett n = non-soap grease
nicht|starre Decke f, schmiegsame ~, nichtstarrer Belag m, schmiegsamer Belag [*Straße*] = flexible pavement, ~ surfacing, non rigid ~
~staubend = non-dusting
~ strukturempfindlicher Ton m = nonsensitive clay
Nichtüberhol(ungs)strecke f = no-passing zone
nichtumhüllt, blank, nackt = uncoated, bare
nicht umkehrbares Kriechen n = irreversible creep

nichtunterkellert — Niederschraub(zapf)hahn 696

~**unterkellert** = basementless
~**verstopfend, nichtballend** [*Siebtechnik*] = nonclogging
~ **vollständiger Brunnen** *m*, partieller ~ = shallow(-dug) well, incomplete ~
~**vulkanisiert** = unvulcanized
Nickel|bad *n* = nickel plating bath
~**chrom** *n* = nickel-chromium (alloy)
Nickelin *n* (Min.) → Rotnickelkies *m*
Nickelkupferlegierung *f* = nickel-copper alloy
niederbringen → (ab)teufen
Niederbringen *n* → (Ab)Teufen
Niederdruck|anlage *f* → Lauf(kraft)werk
~**bagger** *m* = low-pressure hydraulic excavator
~**-Dampfheizung** *f* = low-pressure steam heating
~**einpressung** *f*, Niederdruckauspressung, Niederdruckinjektion *f*, Niederdruckverpressung = low-pressure grouting (or injection)
~**heißwasserheizung** *f* = low-pressure hot water heating system
~**heißwasserkessel** *m* = low-pressure hot water boiler
~**heizung** *f* = low-pressure heating
~**-Hydraulik** *f* = low-pressure hydraulics
~**kanal** *m* [*Ölhydraulik*] = low-pressure gallery
~**kompressor** *m* = low-pressure (air) compressor
~**-Kreiselpumpe** *f* = low-pressure centrifugal pump
~**lagerbehälter** *m* = near atmospheric storage tank
~**-Naßdampf** *m* = low-pressure wet steam
~**pumpe** *f* = low-pressure pump
~**querschnitt-Reifen** *m* **mit einem Höhen-Breiten-Verhältnis von 0,82** = extra low profile tire (US); ~ ~ ~ tyre (Brit.)
~**regner** *m* = low-pressure sprinkler
~**reifen** *m* = low-pressure tire (US); ~ tyre (Brit.)

~**-Spritzgerät** *n* = low-pressure spraying unit
~**stollen** *m* = low-pressure tunnel
~**turbine** *f* = low-head turbine
~**-Warmwasserheizung** *f* = low-pressure hot water (heating) system
~**-Wasserkraftwerk** *n* → Lauf(kraft)werk
Niedermoor *n* = low moor
~**torf** *m* = low moor peat
Nieder|querschnitt-Reifen *m* [*Höhen-Breiten-Verhältnis 0,88*] = low section-height tire (US); ~ ~ tyre (Brit.)
~**schachtofen** *m* = low shaft furnace
Niederschlag *m*, atmosphärischer ~ = precipitation [*The general term used to include rainfall, snow, hail, etc.*]
~ **und Abfluß** *m* = precipitation and run-off
~**aufzeichnung** *f* = precipitation record
~**elektrode** *f* [*Elektroabscheider*] = collecting electrode
~**feuchtigkeit** *f* = meteorological moisture, ~ humidity
~**gebiet** *n* = precipitation area
~**gleiboden** *m* = gley soil due to atmospheric precipitation
~**heftigkeit** *f* → Niederschlagstärke
~**höhe** *f* = precipitation depth
~**intensität** *f* → Niederschlagstärke *f*
~**menge** *f* = amount of precipitation
~**meßstelle** *f* = precipitation-gaging station (US); precipitation-gauging ~ (Brit.)
~**meßwerte** *mpl* = precipitation data
niederschlagspendende Wolke *f*, Niederschlagwolke = precipitating cloud
Niederschlag|stärke *f*, Niederschlagheftigkeit *f*; Niederschlagintensität *f* [*Schweiz*] = intensity of precipitation, precipitation intensity
~**statistik** *f* = precipitation record
~**wasser** *n*, Tagwasser = meteorological water
Niederschraub(zapf)hahn *m* = screwdown (pattern) draining tap [*B.S. 2879*]

Niederspannungs|mast *m* = low-voltage distribution pole, ~ ~ mast
~-Überlandleitung *f* = low voltage transmission line
Niederspülabort *m* [*Zur Vermeidung der Geräuschentwicklung beim Entleeren sowie zur Herabsetzung der Spülwassermenge wird der Spülkasten niedrig angeordnet*] = close-coupled suite, ~ closet
Niederung *f* = flat, bottom land
Niederungsgebiet *n* [*Fluß*] = back-swamp area
Nieder(ungs)moor *n* → Flachmoor
Niederwasser *n* → Niedrigwasser
niedrig [*Träger von geringer Bauhöhe*] = shallow
niedrige Abbindewärme *f* = low heat of hydration (US); ~ ~ ~ setting (Brit.); ~ ~ generated while taking initial set
niedriger Gang *m* = low gear
niedrigeres Hochwasser *n*, NThw = lower high water
niedriges Tidenniedrigwasser *n*, NTnw = lower low water
niedrigster bekannter Wasserstand *m* = lowest recorded level
~ Leerlauf *m* = low idle
Niedrigstwasser *n*, niedrigstes Niederwasser, niedrigstes Niedrigwasser, NNW = lowest recorded stage, ~ ~ water
niedrigviskos = of low viscosity
Niedrigwasser *n*, Niederwasser [*unterer Grenzwert der Abflüsse als Höhenmarkierung innerhalb eines bestimmten Zeitraumes. DIN 4049*] = low water, ~ stage
~ der Springtide, Spring(tide)niedrigwasser = low water spring tide, ~ ~ level of ~ ~, L. W. S. T.
~abfluß *m*, Niedrigwasserdurchfluß = low water flow, ~ ~ discharge
~bett *n* = low-water bed, minor ~
~marke *f* = low-water mark
~pegel *m* = low-water ga(u)ge
~zeit *f* = low-water season
~zeitraum *m* = low-water period
Nieseln *n* → Rieselregen *m*

Niet *m* = rivet
~behälter *m* = riveted (steel) (storage) tank
~bohrrohr *n*, Nietfutterrohr = riveted casing tube
~brücke *f* = riveted bridge
~durchmesser *m* = rivet diameter
~feuer *n* = rivet hearth, ~ fire
~futterrohr *n*, Nietbohrrohr = riveted casing tube
~griff *m* = riveting handle
~hammer *m* = riveting hammer
~kolonne *f*, Niettrupp *m* = riveting gang, ~ crew, ~ team, ~ party
Nietkopf *m* = rivet head
~schneider *m* = rivet cutter
~verstemmer *m* = rivet ca(u)lking tool
Niet|loch *n* = rivet hole
~maschine *f* = riveting machine, riveter
~naht *f* = riveted seam
~presse *f* = compression riveter
~querschnitt *m* = rivet section
~reihe *f* = row of rivets
~schaft *m* = rivet shank
~schraube *f* = screwed rivet
~spaltmeißel *m* = splitting chisel
~stahl *m* = rivet steel
~stock *m* → Anpreßstempel *m* der Nietmaschine
(~)Teilung *f* = rivet pitch, pitch (of the rivets)
~träger *m* = riveted (plate) girder
~verbindung *f* = riveted joint
~verrohrung *f* → genietete Verrohrung
~winde *f* = screw-dolly
~wippe *f* = dolly bar
~zange *f* = riveting tongs
Ni-Hard-Mahlkugel *f* = Ni-Hard grinding ball
Nilos|-Verbinder *m* = Nilos belt hook
~-Zange *f* = Nilos belt fastening nip
Nippel *m* = connector, nipple
~schweißmuffe *f* = butt weld joint with backing ring
Nippflut *f*, Nipptide *f*, taube Flut, taube Tide = neap tide, dead ~, neaps

Nipptide *f* → Nippflut *f*
~ zur Zeit der Äquinoktien = equinoctial neap tide, ~ dead ~
~hub *m* = neap range
Nische *f* = niche [*A small recess in a wall, not extending to the floor*]
~ [*in einem Wehrpfeiler*] = pocket
Nischenbogen *m* → Blendbogen
Nitro|benzol *n*, Mirbanöl *n* = nitrobenzene
~glyzerin *n* → → 1. Schießen *n*; 2. Sprengen *n*
nitroses Gas *n* = nitrous fume [*reddish fume of NO_2 and N_2O_3 which is produced when nitroglycerine explosives burn instead of detonating*]
Nitrozelluloselack *m* = nitrocellulose lacquer, cellulose nitrate lacquer
Niveau *n*, Höhe *f*, Kote *f* = level
~ der Grabensohle, Graben-Niveau = level of trench bottom; ~ ~ ditch ~
~differenz *f*, Gefälle *n*, Druckhöhe *f*, Druckgefälle, Gefällhöhe, Fallhöhe, Niveauunterschied *m* = head, fall
niveaueben → bündig mit
Niveaufläche *f*, Potentialfläche = equipotential surface
niveaugleiche Kreuzung *f* → plangleiche ~
~gleicher Bahnübergang *m*, schienengleiche Niveaukreuzung *f*, schienengleicher Niveauübergang = level (railway) crossing (Brit.); ~ (railroad) ~ (US)
Niveau|regler *m* = level controller
~unterschied *m* → Fallhöhe *f*
Nivellieren *n* → geometrische Höhenmessung *f*
Nivellier(instrument) *n* = level
Nivellier|masse *f*, Ausgleichsmasse, Füllmasse, Planiermasse, Spachtel-~ [*Fehlnamen sind: Nivellierzement, Ausgleichzement, Planierzement, Füllzement, Spachtel-~*] = level(l)ing compound
~tachymeter *n* = level with compass
~zement *m* [*Fehlname*] → Nivelliermasse *f*

NN, Normalnull *n* = mean sea level
Nocke *f* = cam
nockenbetätigt = cam-operated
Nocken|bügel *m* [*Rotary-Zange*] = latch lug jaw
~lagerspiel *n* = cam bearing clearance
~schalter *m* = cam throttle [*(compressed-)air throttle control*]
~stahl *m* = deformed bars
~steuerrad *n* = camshaft phasing gear
~steuerung *f* = cam control
~welle *f*, Daumenwelle = camshaft
~welle *f* **der Einspritzelemente** [*Motor*] = injection camshaft
~wellenantrieb *m* = camshaft drive
~wellenlager *n* = camshaft bearing
~wellen-Stampfmaschine *f* = cam-operated (tamping) (concrete) block machine
Nomogramm *n* = nomogram
Nonius *m* = vernier
Nonne *f* [*Dachziegel*] = Italian tile
Nord-Ostsee-Kanal *m*, Kaiser-Wilhelm-Kanal = Kiel Canal, North Sea and Baltic ~
Noria *f* [*hölzernes Schöpfrad*] = noria
Norm *f* = standard (specification), ~ spec.
Normalbauweise *f*, Regelbauweise = standard construction method
Normalbeton *m*, Schwerbeton = normal concrete
~block *m* → Schwerbetonblock
~decke *f*, Schwerbetondecke = normal concrete floor
~element *n* → Schwerbetonelement
~pfahl *m*, Schwerbetonpfahl = normal concrete pile
~zuschlag(stoff) *m*, Schwerbetonzuschlag(stoff) = normal aggregate
Normal|betrieb *m* = normal operation
~binder *m* → Normalzement
~breite *f*, Regelbreite = normal width
normale geothermische Tiefenstufe *f* = normal reciprocal geothermal gradient

~ **Proctordichte** *f* = standard AASHO density
~ **Stadthauptstraße** *f*, Oberflächen(stadt)hauptstraße = surface artery, ~ arterial
~ **Stadtstraße** *f*, Oberflächen(stadt)straße = surface street, ordinary city ~
normaler Maßstab *m* = arithmetic scale
~ **Ringschieber** *m* = conventional-type needle valve, orthodox ~, ordinary ~
normales Stauziel *n* → → Talsperre *f*
~ **Streckmetall** *n* = plane expanded metal, ~ lath
~ **Wehr** *n*, gerades ~ = rectangular weir
Normalien *fpl* → Abnahmevorschriften *fpl*
Normal|keller *m* = upper basement
~**komponente** *f* = normal component
~**korn** *n* [*in die richtige Kornklasse gelangter Kornanteil*] = correctly sized product, correct ~
~**null** *n*, NN, Normalnullfläche *f* = mean sea level
~**profil** *n* = standard section
~**sand** *m* = normal sand
~**spannbeton** *m* = prestressed normal concrete
~**spannung** *f* = normal stress
~**spur** *f* → Normspur
~**stahlbeton** *m* = normal reinforced concrete
~**tiefe** *f* = normal depth
~**verschiebung** *f* = normal displacement
~**versuch** *m* **mit der Raupe im Zug**, Faltversuch = face bend test
~**verteilung** *f* [*Siebtechnik*] = normal distribution
~**würfel** *m* → Regelwürfel
~**zahlreihe** *f* **Renard**, Renardreihe = Renard series
~**zement** *m*, Normalbinder *m* = normal-hardening cement
normannisches Gewölbe *n*, angelsächsisches ~, Trichtergewölbe, Fächergewölbe, Palmengewölbe, Strahlengewölbe = fan vaulting
normen = to standardize
Normen|ausschuß *m* = committee on standardization, standards committee
~**brandversuch** *m*, Normenbrandprüfung *f*, Normenbrandprobe *f* = standard fire test
~**eigenschaft** *f* = standard property
~**emulsion** *f* = standard emulsion
~**felge** *f* = standard rim
~**fenster** *n* = standard window
~**festigkeit** *f* = standard strength
~**konsistenz** *f*, Normensteife *f* = standard consistency
~**mörtel** *m* = standard mortar
~**prisma** *n* = standard prism
~**prüfsieb** *n* = standard test sieve
~**prüfung** *f* = standard test
~**sand** *m* = standard sand, cement testing ~
~**sieb** *n*, genormtes Sieb = standard screen
~**steife** *f*, Normenkonsistenz *f* = standard consistency
~**straßenteer** *m* = standard road tar
~**tabelle** *f* = table of standards
~**treppe** *f* = standard staircase (Brit.); ~ stair (US)
~**verschnittbitumen** *n* = standard cut-back ((asphaltic) bitumen) (Brit.); ~ ~ (asphalt) (US)
~**zement** *m* = standard cement
~**ziegel** *m* = standard brick
Normschallpegel *m* = standard sound level
~**unterschied** *m* = standard sound levels difference
Normspur *f*, Regelspur, Vollspur, Normalspur *f* = standard gage (US); ~ gauge (Brit.)
Normung *f* = standardization
Normzahl *f* = preferred number
Norton-Brunnen *m*, Abessinier(brunnen), amerikanischer Brunnen, Rammbrunnen [*Dieser Brunnen wird gerammt und Rammrohr und Filterrohr sind eins. Der Name „Abessinier(brunnen)" stammt aus*

dem Feldzug der Engländer 1868 gegen Abessinien. Im Jahre 1815 wurde diese Brunnenart von Nigge in Deutschland eingeführt. Der Amerikaner Norton benutzte diese Brunnenart in den 60er Jahren des 19. Jahrhunderts] = Abyssinian well (pump), ~ pump, hollow ram pump, driven well

Not|abschließung *f* = emergency closure

~**ausgang** *m* = emergency exit

~**auslaß** *m* = emergency outlet

~**auslaß** *m*, Regen(wasser)überfall *m* = storm(-water) overflow [*A weir, orifice, or other device for permitting the discharge from a combined sewer of that part of the flow in excess of that which the sewer is designed to carry*]

~**ausschaltvorrichtung** *f* = emergency cut-out

~**beleuchtung** *f* = emergency lighting

~**bremse** *f* = emergency brake

notbremsen = to emergency-brake

Not|bremsung *f*, **Notbremsen** *n* = emergency(-type) braking

~**brücke** *f*, **Behelfsbrücke** = emergency bridge

~**halt** *m* = emergency(-type) stop

~**leiter** *f* = fire escape ladder

~**schieber** *m* = emergency valve

~**schieberstollen** *m* = emergency valve tunnel, ~ ~ gallery

~**standgebiet** *n* = depressed area

~**stromanlage** *f* = standby generating set

~**stromgebäude** *n* = standby power building

~**stromversorgung** *f* = standby power

~**treppe** *f* = emergency stair (US); ~ staircase (Brit.)

~**unterkunft** *f* = emergency accomodation

~**verband** *m* = first aid dressing

~**verschluß** *m* = emergency (guard) gate, guard ~

n-stieliger Rahmen *m* = (portal) frame with n-column

NThwh, niedrigeres Tidehochwasser *n* = lower high water

NTnw *n*, niedrigeres Tideniedrigwasser *n* = lower low water

nukleare Bohrloch-Überwachung *f*, nukleares Bohrloch-Logging *n* = nuclear (oil) well logging, ~ (~) ~ survey(ing)

Null|ablesung *f* = zero reading

~**abweichung** *f* = zero drift

~**druckfläche** *f* = zone of zero stress

~**druckhöhe** *f* = zero head

~**härte** *f* = zero hardness

~**isochrone** *f* = zero isochrone

~-**Last** *f* = zero load

~-**Linie** *f* → neutrale Achse *f*

~-**Linieneinfluß** *m* = deadline influence

~**meridian** *m* = Prime Meridian

~**punkt** *m* = zero point

~-**Zähigkeits-Umwand(e)lungstemperatur** *f* = nil ductility transition temperature

Nummern/Zoll-Bezeichnung *f* [*Siebtechnik*] = designation by number/inch

Nummulitensandstein *m* = nummulitic sandstone

Nuß *f* [*beim Nußband*] = centre (Brit.); center (US)

~**band** *n* = counter-flap hinge

~**kleinsumpf** *m*, **Körnersumpf** = collecting sump for undersize of the nut sizing screens

~**kohle** *f* = nut coal, nuts

~**schalenreifen** *m* = nutshell tyre, walnut ~ (Brit.); ~ tire (US)

Nut *f* = groove

Nutenmeißel *m* = grooving chisel

Nut|ringdichtung *f* = chevron ring packing

~**rolle** *f* = sheave

nutzbare Arbeitsfläche *f* [*Siebtechnik*] = effective working area, useful ~ ~, ~ ~ surface

~ **Fahrtiefe** *f* → Wassertiefe des Fahrwassers

~ **Fallhöhe** *f* [*Wasserkraft*] = effective head; ~ fall

~ **Pferdestärke** *f* → effektive ~

nutzbare Siebfläche — (obere) Asphalttragschicht

~ **Siebfläche** *f*, wirksame ~ = effective screen area
~ **Wassertiefe** *f* = depth of water (in berth); draught (Brit.)
nutzbarer Ausbruchquerschnitt *m* (⚒) [*Strecke*] = finished section
Nutz|barkeitsfaktor *m* [*Verhältnis der mittleren Leistungsfähigkeit zur Höchstleistungsfähigkeit eines Wasserkraftwerkes*] = capacity factor, plant(-use) ~ [*Is the ratio of the average output of a hydropower plant to the plant capacity*]
~**belag** *m* → Nutzschicht *f*
~**breite** *f* = useful width
~**druck** *m* (⚒) = effective pressure
nutzen, ausbauen [*einen Fluß durch Wasserkraftanlage(n)*] = to develop
Nutzfahrzeug *n* = commercial vehicle
~**achse** *f* = commercial vehicle axle
Nutz|fallhöhe *f* → Nutzgefälle *n*
~**fläche** *f* [*Gebäude*] = floors
~**gefälle** *n*, Nutzfallhöhe *f*, Werkgefälle [*Höhenunterschied der Energielinien vor und hinter Turbinen oder Rohrleitungen*] = useful head
~**höhe** *f* = useful height
~**inhalt** *m*, Füllinhalt, Nenninhalt [*Betonmischer*] = unmixed capacity, dry capacity per batch
~**ladungsbeiwert** *m* [*Erdbewegung*] = load factor
Nutzlast *f* → Hakentragkraft *f*
~, Gebrauchslast, Verkehrslast [*z. B. Brücke; Hochbaudecke*] = live load(ing), (incidental) superimposed load, additional ~, working ~
~ **bei abgestütztem Kran**, Tragkraft *f* ~ ~ ~ [*Autokran*] = lifting capacity on outriggers, blocked capacity
~ ~ **unabgestütztem Kran**, Tragkraft *f* ~ ~ ~ [*Autokran*] = free-on-wheels capacity, lifting capacity free-on-wheels
~**faktor** *m* = payload-to-vehicle-weight ratio
~**-Wanne** *f* → Mulde *f*
Nutz|querschnitt *m* [*Tunnel*] = pay-section

~**schicht** *f*, Nutzbelag *m*, (Fuß)Boden-~ = (flooring) wearing surface
~**- und Trinkwasserspeicherbecken** *n* = service reservoir
Nutzungs|dauer *f* → Lebensdauer
~**wert** *m* **von Gebäuden** = efficiency of buildings
Nutzwasser *n* = available water
Nylon|cord = nylon cord
~**garn** *n* = nylon yarn
~**-Schußgarn** *n* = nylon-weft yarn

oben offene Fachwerkbrücke *f* = pony truss bridge
~**liegende Nockenwelle** *f* = overhead camshaft
~**liegender Drücker** *m* → Ballendrücker
Oben|öl *n* = upper cylinder lubricant
~**schalung** *f* = top formwork
Oberbach-Schlämme *f* = Schlämme type slurry
Oberbau *m* = superstructure
~, Eisenbahn~, Gleis~ = permanent way, p. w.
~ [*Theodolit*] = upper plate, vernier ~
Ober|becken *n*, Hochbecken [*Pumpspeicherwerk*] = upper reservoir, ~ pool, ~ (storage) basin
~**begriff** *m* = all-embracing term
~**beton(schicht)** *m*, (*f*) = top (course) concrete, ~ concrete layer
~**block** *m* → Kronenblock
~**boden** *m* → A-Horizont *m*
~**bohrmeister** *m* [*Tiefbohrtechnik*] = tool pusher
oberdevonisch (Geol.) = of Upper Devonian age
obere Abbaustrecke *f*, Kopfstrecke (⚒) = upper development road
(~) Asphalttragschicht *f*, (~) Bitumentragschicht = asphalt base (course) [*The asphalt layer used in a pavement system to reinforce and protect the subgrade or subbase, and to support a wearing course*]

(~) **Betontragschicht** *f* = concrete base (course) [*The layer used in a pavement system to reinforce and protect the subgrade or subbase, and to support a wearing course of other material*]
(~) **Bitumentragschicht** *f* → (~) Asphalttragschicht
~ **Faser** *f* = top fibre (Brit.); ~ fiber (US)
~ **Scheinfuge** *f* = top surface dummy joint
~ **(Schütz)Tafel** *f*, Oberschütz(e) *n*, (*f*), Obertafel = top (gate) leaf
~ **Strömung** *f* [*Meer*] = upper current
~ **Tafel** *f* → ~ Schütztafel
(~) **Tragschicht** *f* = base (course) [*The layer used in a pavement system to reinforce and protect the subgrade or subbase, and to support a wearing course*]
oberer Grenzzustand *m* [*Boden*] = passive state [*soil*]
~ **Heizwert** *m*, Verbrennungswärme V *f*, H_0 = higher calorific value, H.C.V.
~ **(Seil)Strang** *m* [*Seiltransporteur*] = upper rope
~ **Totpunkt** *m*, O.T. = top center (US); ~ centre (Brit.)
~ **Windverband** *m* = top lateral bracing
obererdig, oberirdisch = above-ground
Oberflächen|abdichtungsschicht *f* = surface membrane
~**abschluß** *m* [*Straßenbau*] = surface seal(ing)
~**abstreifer** *m* = surface scraper
oberflächenaktiver Stoff *m* = surface-active agent
Oberflächen|bahn *f* = ground level railway (Brit.); ~ ~ railroad (US); surface ~
~**behälter** *m* = above-ground tank
~**behandlung** *f*, OB [*Schwarzdeckenbau*] = surface treatment, seal coating
~**beries(e)lung** *f* **mit Abwasser** → Bewässerung mit Abwasser
~**beschaffenheit** *f* = surface state

~**bestimmung** *f* [*Zerkleinerungstechnik*] = measurement of specific surface
~**bewässerung** *f* = surface irrigation
~**dichtung** *f*, (Ab)Dichtungsvorlage *f*, (Ab)Dichtungsteppich *m*, (Ab)Dichtungsschürze *f*, Schürze, Außen(haut)dichtung = impervious blanket, (waterproof) ~
~**(ein)rüttlung** *f* = surface vibration
~**eis** *n* = surface ice
~**entwässerung** *f* = surface drainage, storm ~
~**entwässerung** *f* **durch Dränage** = sub-drain type surface drainage, ~ ~ storm ~
~**entwässerung** *f* **durch offene Gräben** = open storm drainage, ~ surface ~
~**erneuerungsverfahren** *n* → Retread-Verfahren
~**feinstbehandlung** *f* = super-finish
~**frost** *m* = surface frost
~**geber** *m* (⚒) = surface strain indicator
~**gemisch** *n*, Oberflächenmischung *f* [*Kiesstraße*] = surface mix(ture)
~**geschwindigkeit** *f* [*Abflußmengenmessung in der Gewässerkunde*] = surface velocity
~**geschwindigkeit** *f* = surface velocity
~**gestaltung** *f* [*Beton*] = architectural finish
~**gestein** *n* → Ergußgestein
~**grundwasser** *n* = shallow ground water
~**härteprüfung** *f* **mit einer Diamant-Pyramide** = diamond pyramid hardness test
~**härter** *m* = surface hardener
~**inhalt** *m* = surface area
~**kennzahl** *f*, spezifische Oberfläche *f*, OK [*Siebtechnik*] = specific surface (factor)
~**kontur** *f* = surface contour
~**-Krafthaus** *n*, Freiluft-Krafthaus = surface power house
~**kraftwerk** *n*, freistehendes Kraftwerk, Übertagekraftwerk = overground (power) station, ~ (~) plant, surface (~) ~

oberflächen|nah [*Grundbau*] = within a few feet of the surface
~nahe Schicht *f*, **~ Lage** *f* = near-surface layer
~nahes Aufreißen *n*, **~ Aufbrechen, Flachaufreißen, Flachaufbrechen** = shallow loosening, ~ breaking up, ~ disintegrating, ~ disintegration, scarifying, scarification
Oberflächenrauhigkeit *f*, Absolutrauhigkeit [*Hydraulik*] = surface roughness
(Ober)Flächenrüttlung *f* = surface vibration
Oberflächen|rüttler *m*, Oberflächenvibrator *m* = surface vibrator
~schaden *m* = surface damage
~schluß *m* [*Beton*] = (concrete) finish
~schutz *m* = surface protection
~schutzschicht *f* = surface protection layer
~spannung *f* des Wassers = surface tension of the water
~(stadt)straße *f*, normale Stadtstraße = surface street
~(stadt)straßennetz *n* = surface street network
~strömung *f* [*Meer*] = superficial current, surface ~
~struktur *f* = texture
~teerung *f* = surface tarring, ~ dressing with tar
~ton *m* = surface clay
oberflächentrocken = surface-dry
Oberflächen|verdichter *m* = surface compactor
~verdichtung *f* = superficial compaction
~verformung *f* = surface deformation
~verkehr *m* = ground traffic
~vibrator *m*, Oberflächenrüttler *m* = surface vibrator
~vorbereitung *f* = preparation of a surface
~wasser *n* = surface water
~wasser *n* [*Grundbau*] = open water
~welle *f* [*Erdbebenwelle*] = surface wave
~welle *f* [*Meereswelle*] = surface wave
Ober|flansch *m* → Obergurt *m*

~flurhydrant *m* = above-ground hydrant
~graben *m* → Oberwasserkanal *m*
Obergurt *m*, Oberflansch *m* = top boom, ~ chord, ~ flange, upper ~
~ [*Bandförderer*] = carrier side
~gelenkknotenverbindung *f* = pin connection in top boom, ~ ~ ~ ~ flange, ~ ~ ~ ~ chord
~knotenblech *n* = top boom junction plate, ~ flange ~ ~, ~ chord ~ ~
~kraft *f* = top chord force, upper ~ ~, ~ boom ~, ~ flange ~
~stab *m* = top boom member, ~ chord ~, ~ flange ~, ~ ~ rod, ~ ~ bar
~winkel *m* = top flange angle, ~ boom ~, ~ chord ~, upper ~ ~
Oberhaupt *n* [*Schleuse*] = upper gates, inner ~
oberirdisch, obererdig = above-ground
~ gewinnen [*Bodenschätze; Steine*] = to quarry
~ verlegte Feld(rohr)leitung *f* [*Beregnung*] = overhead sprinkling irrigation system, ~ sprinkler installation, ~ sprinkler system
(oberirdischer) Abfluß *m* [*jener Anteil des Niederschlages, der oberirdisch abfließt*] = run-off, surface ~
~ Grabensilo *m* → Fahrsilo
~ Wasservorrat *m* = surface water resource
oberirdische Wasserscheide *f* = surface watershed line, ~ divide
oberirdisches Bauwerk *n* = surface structure
Ober|kante *f*, OK. = top
~kreide *f* (Geol.) = top chalk
oberlastig = top-heavy
Oberlauf [*Strom; Fluß*] = upper course
~charakter *m* [*Strom; Fluß*] = upper-course type
~-Kettentransporteur *m* = overhead chain conveyor, ~ ~ conveyer
Ober|leiter *f*, Ausleger *m* [*Eimerkettenbagger*] = top ladder, jib
~leitung *f* = overhead line

Oberleitungsdraht — offen abgestuft

~leitungsdraht *m*, Fahrdraht, Fahrleitung *f* = contact wire, troll(e)y ~
~leitungsmast *m* = overhead line mast
~leitungs(omni)bus *m*, Obus = troll(e)y bus, electric troll(e)y; troll(e)y coach (US)
Oberlicht *n* = skylight
~pfette *f* = skylight purlin
Ober|pegel *m*, OP = upstream ga(u)ge
~putz *m*, Feinputz [*Oberschicht des zweilagigen Putzes, die im besonderen Maße die ästhetische Wirkung des Putzes bestimmt. Bei Außenputzen muß sie witterungsbeständig sein*] = finishing coat, setting ~, skimming ~, fining ~
~putzgips *m* = gauging plaster for finish coat
oberschlächtig = overshot [*water wheel*]
Ober|schütz(e) *n*, *(f)*, Obertafel *f*, obere (Schütz)Tafel = top (gate) leaf
~schwelle *f*, Sturz *m* = lintel, lintol
~stempel *m*, Innenstempel (⚒) = upper prop, ~ peg
~stromseite *f* [*Spundwand*] = upstream surface
~tafel *f* → Oberschütz(e) *n*, *(f)*
obertägige Geräte *npl* = surface plant
Ober|teil *m*, *n* → Aufsatzteil
~tor *n* [*Schleuse*] = head gate
~trum *m*, *n* = upper strand
~turas *m* = head sprocket (wheel)
~turas *m* [*Naßbagger*] = upper tumbler, top ~
~wagen *m* → Aufsatzteil *m*, *n*
Oberwasser *n* = top water
~haltung *f*, Oberwasser *n* = upstream reach
~horizont *m*, Oberwasserseite *f* = upstream side, ~ water
~kanal *m*, Obergraben *m*, Oberkanal = head race
~seite *f*, Oberwasserhorizont *m* = upstream side, ~ water, ~ face
oberwasserseitig, bergseitig = upstream
Oberwasserspiegelhöhe *f*, Oberwasserstand *m* = upstream (water) level

obsequentes Tal *n* [*verläuft in der Richtung f der Abdachung f*] = obsequent valley
Obst|anbau *m* = fruit growing
~anlage *f* = orchard
~baum *m* = fruit tree
~bewässerung *f* = orchard irrigation
~bewässerung *f* mit Becken = basin irrigation [*A method of irrigating orchards by which the tree is surrounded by a border, to form a pool when water is applied*]
~(garten)bewässerung *f* = orchard irrigation, irrigation of orchards
Obus *m* → Oberleitungs(omni)bus
Ochsenauge *n* = bull's eye
Oedometer *n*, Druckgerät *n*, Verdichtungsapparat *m* = consolidation (test) apparatus, oedometer
~versuch *m*, Oedometerprobe *f*, Oedometerprüfung *f* = oedometric test
Ofen *m* = kiln
~ für ununterbrochenen Betrieb = continuous kiln
(~)Anflug *m*, Verschmauchung *f* = kiln scum
~beschicker *m* = kiln feeder
~betrieb *m*, Ofengang *m* = kiln operation, ~ run
~brennen *n* = kilning
~gang *m*, Ofenbetrieb *m* = kiln run, ~ operation
~heizung *f* = stove heating
~kopf *m* [*Zementofen*] = firing hood
~mantel *m* = kiln shell
(~)Manteltemperatur *f* = kiln-shell temperature
~material *n* [*Kalkindustrie*] = kiln stone
~rohr *n* = stove pipe, ~ tube
~sau *f*, Bodensau [*Hochofen*] = salamander, furnace sow
~überzug *m* [*Emaille*] = stoving finish
offen [*Strecke im Bergbau*] = open
~ [*Verschleißschicht*] = open-textured [*wearing course*]
~ abgestuft = open-graded

offen abgestuftes Mineralgemisch — Öffnung

~ **abgestuftes Mineralgemisch** n = open-graded (mineral) aggregate, OGA
~ **halten (✖)** = to maintain, to keep open [*road*]
offene Brücke f **mit versenkter Fahrbahn** → Trogbrücke
~ **Kabine** f = open cabin
~ **Kornabstufung** f = open gradation
~ **Mischung** f, **offenes Gemisch** n = open mix(ture), coarse ~
~ **Reede** f, **Außenreede** = open roadstead
(~) **Schappe** f, **Löffelbohrer** m, **Probelöffel** m = sampling spoon, spoonauger
~ **Schicht** f [*Bodenmechanik*] = open layer [*If the consolidating layer is free to drain through both its upper and lower surfaces, the layer is called an open layer*]
~ **Siebfläche** f [*Gesamtquerschnitt der Sieböffnungen bezogen auf die gesamte Siebfläche*] = open area, discharge area of screen, open screening area, percentage open area of screen
~ **Stütze** f = open column
~ **Tafel** f [*Eisenbahnbrücke*] = open deck
~ **Wasserhaltung** f = sump drainage, pumping from a sump pit
offener Baumischbelag m, **Baumischbelag nach Art des Makadam** = macadam aggregate type road mix surfacing (or pavement)
~ **Bogenwickel** m = open spandrel
~ **Bohrlöffel** m = clay auger
~ **Gang** m → Laubengang
~ **Graben** m, **Abflußgraben** = (open) ditch, field ~, drainage ~
~ **Güterwagen** m, ~ **Waggon** m = open top railway car (Brit.); ~ ~ railroad ~ (US); gondola car
(~) **Kanal** m → Gerinne n
~ **Pfahl** m, **Pfahl mit offenem Ende** = open-end(ed) pile, open-foot ~
~ **Polygonzug** m, ~ **Standlinienzug**, **offenes Polygon** n = unclosed traverse

~ **Querschnitt** m = open section
~ **Senkkasten** m → Bagger-Senkkasten
~ **Tiegel** m [*Flammpunktprüfung*] = open cup
~ **Viersitzer** m [*PKW*] = touring car
~ **Zweisitzer** m [*PKW*] = roadster
offenes dreistufiges Brechen n = three-stage open-circuit crushing
~ **Gemisch** n, **offene Mischung** f = open mix(ture), coarse ~
~ **Gerinne** n = open channel
~ **Parkhaus** n = open deck car park, open-sided multi-stor(e)y ~ ~
öffentliche Ausschreibung f → → Angebot n
~ **Bauarbeiten** fpl, **Bauarbeiten der öffentlichen Hand** = public works
~ **Beleuchtung** f = public lighting
~ **Gesundheit** f, ~ **Hygiene** = public health
(~) **Versorgungsleitung** f = utility line
öffentlicher Fernsprecher m = public telephone
~ **Massenverkehr** m, **Beförderung** f **von Menschenmassen** = public transit, mass ~, ~ transport(ation)
~ ~ **mit Rollbürgersteigen** = moving pavement public transport
~ **(Omni)Busbetrieb** m = mass transportation bus service
~ **Parkplatz** m = public parking place
~ **Schnellverkehr** m = rapid transit
(~) **Verkehrsbetrieb** m = transit authority; common carrier (US)
~ **Versorgungsbetrieb** m = public utility (undertaking)
öffentliches Ausrufungsnetz n [*Flughafen*] = public address system
~ **Bauvorhaben** n = public works project
~ **Gebäude** n = public building [*A building used principally for public assembly or public service*]
Öffnen n **des Durchflusses** [*in einer Rohrleitung*] = opening a flow
Öffnung f, **Feld** n, **Brücken~** = (bridge) span, (~) opening
~, **Auge** n = eye

Öffnung — ölführendes Gebiet

~ [*Wehr*] = bay
~ = opening
Öffnungswange *f* → Freiwange
ohne Einlaßbetätigung *f* **am Griff** = without throttle [*(compressed-)air throttle control*]
~ **Verbund** *m* = unbonded
OK., Oberkante *f* = top
Okrat-Beton *m* = concrete treated with SiF$_4$
Okratierung *f* = SiF$_4$ treatment
Oktanzahl *f*, Oktanziffer *f*, Klopffestigkeitszahl = octane number
Okular *n* = eye-piece, eye-glass
~**auszug** *m* = eye-piece draw tube
Öl *n* → Erd~
Öl *n* = oil
Ölablaßpfropfen *m*, Ölablaßstopfen *m*, Ölablaßschraube *f* = drain plug for oil sump, oil (pan) drain plug, waste oil screw
Ölabscheider *m* [*Grundstückentwässerung*] = oil trap (Brit.); ~ interceptor (US)
Ölabschöpfer *m* = oil scoop
Ölabstreifring *m*, Ölabstreifer *m* = scraper ring, oil ~ ~, oil (control) ~
Ölabziehstein *m* = oil stone
Ölbad *n* = oil bath
~**luftfilter** *m, n* = oil-bath air filter, oil-washed air cleaner
~**schmierung** *f* = oil-bath lubrication
ölbasische Spülung *f* → Spülung mit hohem Widerstand
Ölbehälter *m*, Öltank *m* = oil (storage) tank
ölbeständig = oil-resistant
(Öl)Bohrinsel *f*, Bohrinsel = offshore well drilling platform, (oil) drilling island, oil drilling platform
~ **mit verschiebbarem Bohrturm** = multi-well (offshore) (oil) drilling platform
Ölbohrung *f* → Erd~
Ölbohrzement *m* → Erdölzement
Ölbrenner *m* = oil(-fired) burner
Ölbrunnen *m* → Erdölbohrung *f*
Ölbüchse *f*, Öler *m* = oil cup
Öldampfentlüfter *m* = oil breather

öldicht = oiltight
Öldocht *m* = oil wick
Öldruck *m* = oil pressure
öldruckbetätigt = oil-hydraulic operated
Öldruck|-Turbinenregler *m* = oilpressur (turbine) governor, opengovernor oil pressure system
~**zylinder** *m*, Drucköl-Zylinder = ram
öldurchtränkt → ölgetränkt
Öleinfüll|schraube *f* = oiling hole plug
~**sieb** *n* [*Hydraulik*] = oil filler screen
~**stutzen** *m*, Ölstutzen = oil filler (cap)
Ölemulsion *f* = oil emulsion
Ölentnahme *f* = withdrawel of oil
Ölersatzmethode *f* [*Nachprüfung der Verdichtung f*] = heavy oil method
Oleum *n*, rauchende Schwefelsäure *f* = fuming sulfuric acid
Ölexploration *f* **unter dem Meeresspiegel** = submarine exploration for oil
Ölfalle *f* = oil trap
Ölfang *m* = oil collector
Ölfarbe *f* = oil paint
Ölfeld *n*, Erd~ = oil field
~ **im Küstenvorland**, Erd~ ~ ~ = offshore oil field
~**drahtseil** *n*, Bohrseil = rotary drilling line
ölfest = oil-resistant
Ölfeuerung *f* = oil furnace
Ölfilmwirbel *m* [*vom Schmierfilm angefachte Wellenschwingungen*] = resonant whip
Ölfilter *m, n* = oil filter
Ölförderung *f* **mittels unterirdischer Verbrennung** = oil production by underground combustion
Ölfraktion *f* [*Herstellung präparierten Teeres*] = (oil) distillate fraction, tar distillation ~, tar-oil ~
ölfrei, nicht ölführend = barren of oil
ölführende Schicht *f* → Ölträger *m*
ölführendes Gebiet *n*, erdölführendes ~ [*In diesem Gebiet ist die Ölführung nachgewiesen*] = proved oil land, ~ acreage, ~ area

ölführendes Gebiet — Ölsikkativ

~ ~ **unter dem Meeresgrund** = submarine oil formation
ölfündige Bohrung f → Erdölbohrung
Ölgas n, Fettgas, transportables Gas = oil gas
ölgefeuert = oil-fired
ölgekühlt = oil-cooled
ölgeschmierter Kolbenkompressor m = lubricated reciprocating compressor, ~ piston ~
ölgetränkt, ölimprägniert, öldurchtränkt = oil-impregnated
Ölgewinnung f **durch thermische Verfahren** = thermal methods oil recovery, oil recovery thermal methods, recovery of crude oil by heat
Ölhafen m, Erd~ = oil port
Ölheizer m = oil heater, ~ heating unit
Ölheizung f = oil-fired central heating
ölhöffiges Gebiet n, erdölhöffiges ~ [*Ein Gebiet in dem Ölführung möglich ist oder ein Gebiet, das eine möglicherweise ölführende Struktur überlagert*] = prospective oil land, ~ acreage, area overlying a possibly oil-bearing structure
Ölhöhe f, Ölstand m = oil level
Ölhorizont m, Erdölhorizont, erdölführender geologischer Horizont = oil horizon
Ölhydraulik f = pressurized oil system, oil hydraulic (power) ~
ölhydraulisch = oil hydraulic
(ölhydraulischer) Umformer m = service
Oligoklas m (Min.) = oligoclase
~**granit** m (Geol.) = miarolyte
ölimprägniert → ölgetränkt
Ölingenieur m = petroleum engineer
Öl-in-Wasser-Emulsion f = oil-in-water emulsion
Olivin m, Olivenstein m (Min.) = olivine
~**bombe** f (Geol.) = olivine-nodule
~**fels** m, Dunit m = dunite, olivine-rock
Ölkabel n = oil-filled cable
Ölkalk(stein) m = oil-impregnated limestone

Ölkitt m → Glaserkitt
Ölkreis m [*Hydraulikanlage*] = oil circuit
Ölkuchen m, Leinkuchen = oil cake, linseed ~
Ölkühler m = oil cooler
Ölkupp(e)lung f, Naßkupp(e)lung = oil clutch
Öllager n = oil depot
~**raum** m, Heiz~ = (fuel) oil store, heating ~ ~
~**tank** m, Heiz~ = (heating) oil storage tank, fuel ~ ~ ~
Öllamellenbremse f = oil disc brake (Brit.); ~ disk ~ (US)
Ölleitung f = oil line
Öllenkkupp(e)lung f = oil steering clutch
ölloses Lager n = oilless bearing
Öllösungsmittel n = oil solvent
Ölluftheizer m = oil-fired unit heater
Ölmole f = oil tanker jetty
Ölmolekül n = petroleum molecule
Ölpumpe f = oil pump
Ölpumpenhaus n = oil pump house
Ölraffination f = oil refining
Ölsammelsystem n = oil-gathering system
Ölsand m, Erdölsand = oil sand, "pay" (~)
~**stein** m = oil sandstone
Ölschalter m **für Warnlicht** = oil switch
Ölschiefer m → bituminöser ~
Ölschlamm m = oil mud
Ölschleuderring m = oil slinger
(öl)schlußvergüteter Kohlenstoffstahl m = high-carbon steel
Ölschmierpumpe f, Schmierölpumpe = oil lubricating pump
Ölschmierung f = oil lubrication
Ölschraube f = level plug
Ölschutzring m = oil catcher, thrower, oil shield
Ölsichtkontrolle f = oil sight-feed ga(u)ge
Ölsieb n = oil strainer
Ölsikkativ n = drier soluble in dryin oil

Ölsonde f → Erdölbohrung f
Ölspritz|blech n = oil splasher
~düse f, Einspritzdüse [*Dieselbär*] = fuel jet in piston head
Ölspülung f → Spülung mit hohem Widerstand
Ölspur f = oil showing, ~ trace
Ölstand m, Ölhöhe f = oil level
~anzeiger m = oil level indicator, ~ ~ gauge
~kontrollschraube f = plug, oil level ~
Ölstation f = oil station
Ölstein m (Min.) → Elaeolith m
Ölstoßdämpfer m = air-oil suspension
Ölstutzen m, Öleinfüllstutzen = oil filler (cap)
Ölsuche f → Erd~
Ölsucher m = oil prospector
Ölsumpf m, Ölwanne f = (oil) sump, (~) pan, lube ~
Öltank m → (Heiz)Öllagertank
~, Ölbehälter m = oil (storage) tank, ~ reservoir
Öltanker m = oil tanker
Ölträger m, Erd~, (erd)ölführende Schicht f = oil-bearing stratum, petroliferous bed
Öltropfschale f, Öltropfenfänger m = oil drip pan
Ölumlaufschmierung f = oil circulation lubricating system
Ölumschlaganlage f = oil (storage and unloading) terminal, ~ tanker terminal, bulk oil terminal
Ölverschmutzungen fpl [*auf einer Fahrbahn*] = oil drippings
Ölversorgungsanlage f, Heiz~ = (heating) oil supply system, fuel ~ ~ ~
Ölverteiler m [*Hydrauliksystem*] = flow divider, oil manifold
~ventil n = flow divider valve, oil manifold ~
Ölwanderung f = migration of oil
Ölwanne f → Ölsumpf m
Ölwirbel m = oil whirl
Öl-Zentralheizung f, Heiz~ = (heating) oil-fired central heating, fuel ~ ~ ~

Ölzerstäuber m = oil atomizer
Ölzusatz m, Wirkstoff m, Additiv n = additive
Omnia|dach n = Omnia roof
~-Decke f, Verbundgitterträgerdecke = Omnia (concrete) floor
~-Rippendecke f = Omnia rib floor
Omnibus m, (Auto)Bus m = bus
~ mit Oberdeck = double-deck bus
~ ohne Oberdeck = single deck bus
~bahnhof m → Busbahnhof
~haltestelle f, (Auto)Bushaltestelle = bus stop
~station f = oil station
~reise f, (Auto)Busreise = bus ride
~-Unterstell- und -wartungshalle f, Autobus-Unterstell- und -wartungshalle = bus depot
~werkstatt f, (Auto)Buswerkstatt = bus overhaul works
Onyx(marmor) m, Kalkonyx m = onyx marble
oolithischer Kalk(stein) m, Eisteinkalk m, Oolithkalk, Kalkoolith m = oölite, oölitic limestone
opake Glasur f, Schmelzglasur, Emaille f = opaque (glazed) finish
Opakglas n = opaque glass
~wandverkleidung f, Opakglaswandauskleidung, Opakglaswandbekleidung = opaque glass wall facing, ~ ~ ~ lining
Opalglas n = opal glass
Operationsgebäude n [*Krankenhaus*] = operating theatre
Ophikalzit m (Geol.) = forsteritemarble, ophicalcite
ophitische Textur f → divergentstrahlig-körnige ~
Oppermann-Klappen(stau)wehr n = Oppermann wicket (dam) (US); ~ shutter (~) (Brit.)
optimaler Wassergehalt m, ~ Feuchtigkeitsgehalt = optimum moisture content
optimieren = to optimize
optisch-akustisches Hilfsmittel n = audio visual aid
optische Achsenebene f = optical orientation [*mineralogy*]
~ Anzeige f = visual observation

optische Führung — örtliche Befunde

~ **Führung** f **des Verkehrs** = visual guidance of traffic
~ **Lotung** f = instrument plumbing
~ **Zeichenerkennung** f = optical character recognition
optisches Signal n = visible signal
~ **Warnsignal** n, **Warnlicht** n = alarm light, warning ~
Orbitalbewegung f = orbital motion, ~ movement
Orchesterstand m = orchestra floor
Ordinate f, **Höhe** f, **Kote** f = level
Ordinatenachse f = vertical scale
Ordnungs|gleis n = classification track
~**zahl** f **der Spannungsverteilung** f → Konzentrationsfaktor m nach Fröhlich
organische Materie f (Geol.) = organic matter
~ **Verbindung** f = organic compound
~ **Verunreinigungen** fpl, ausglühbare Bestandteile mpl = organic matter (present), ~ impurities (~)
organisches Bindemittel n, **organischer Binder** m = organic binder
Organozinnverbindung f = organotin compound
orientalischer Alabaster m = oriental alabaster
Originalmaß n **der (Gleis)Kettenteilung** = initial pitch
O-Ring-Dichtung f = O-ring seal
Ornamentglas n, **Zierglas** = pattern(ed) glass, ornamental ~
orogener Druck m, **Gebirgsdruck** (Geol.) = pressure of mountain mass
Orogenese f → Gebirgsbildung f
Orographie f, **Gebirgsbeschreibung** f = orography
Ortbeton m [*Ist der auf der Baustelle hergestellte Beton. (Der ,,Deutsche Beton-Verein" versteht unter ,,Ortbeton" solchen ,,Beton, der am Einbau-Ort in die Schalung eingebracht wird".)*] = cast-in-place concrete, (cast-)in-situ ~, poured-in-place ~
~**bau** m = in-situ construction, poured ~

~**bauwerk** n = poured-concrete structure
~**decke** f = cast-in-place floor, (cast-)in-situ ~, poured-in-place ~
~**kern** m [*eines Gebäudes*] = cast-in-place core, (cast-)in-situ ~, poured-in-place ~
~**-Massivplatte** f = solid (cast-)in-situ slab, ~ cast-in-place ~, ~ poured-in-place ~
~**pfahl** m → Ortpfahl
~**-Pfahlkopfplatte** f = poured-in-place (pile) cap
~**pfeiler** m = cast-in-situ pier
~**platte** f = poured-in-place (concrete) slab, (cast-)in-situ (~) ~
~**rammpfahl** m = in-situ pile formed by driving
~**rippe** f = poured-in-place (concrete) rib, (cast-)in-situ (~) ~
~**rohr** n = cast-in-place concrete pipe
~**-Rohrleitung** f = poured-in-place pipeline
~**schale** f = in-situ concrete shell
~**-Verkleidung** f = cast-in-place concrete lining
Orterde f = friable iron pan
Orthogneis m, **Orthogestein** n = orthogneiss
orthogonales Kurvennetz n, **rechtwinkliges** ~ = orthogonal curve system
Orthokieselsäure f = orthosilicic acid
Orthoklas m, **Kalium-Feldspat** m, $K_2Al_2Si_6O_{16}$ = orthoclase
orthotrop = orthotropic, orthogonally anisotropic
orthotrope Brückenkonstruktion f [*als Bauweise, nicht als deren Ergebnis*] = orthotropic deck bridge design
~ **(Brücken)Tafel** f → Leichtfahrbahn f
orthotroper (Brücken)Überbau m → Leichtfahrbahn f
Ortleichtbeton m = in-situ lightweight-aggregate concrete
örtlich = local
örtliche Befunde mpl = field observations

örtliche Verhältnisse — Oxydschicht

~ Verhältnisse *npl* = site conditions
örtlicher Boden *m* = locally available soil
~ **Verlust** *m* [*Hydraulik*] = local loss
~ **Widerstand** *m* [*Hydraulik*] = local resistance
örtliches Abscheren *n* = local shear failure
~ **Bindemittel** *n*, örtlicher Binder *m* = local binder
~ **Knicken** *n* = local buckling
~ **Material** *n* = local material, nearby ~
Ortmischpfahl *m* = mixed-in-place pile
Ortpfahl *m*, Ortbetonpfahl, Betonortpfahl = cast-in-place pile, cast-in-situ ~, cast-in-the-ground ~, situ-cast ~, in-situ concrete pile
~, Ortpflock *m* (⚒) = cornerstake
~ **mit Mantelrohr** = shell pile, cased ~, cast-in-place ~ ~
~ **ohne Mantelrohr** = shell-less pile, uncased ~, cast-in-place ~ ~
Ortrammpfahl *m* → gerammter Ortpfahl
Ortsbau|plan *m* → Bebauungsplan
~**satzung** *f* → Bauordnung
Orts|beben *n* = local earthquake
~**beweglichkeit** *f* = portability
(~)**Brust** *f*, Vortriebstelle *f*, Ausbruchstelle [*Tunnel- und Stollenbau*] = (working) face
~**entwässerung** *f*, Kanalisation *f* = system of sewerage, ~ ~ sewers, sewerage system [*A collecting system of sewers and appurtenances*]
~**fahrbahn** *f* → Anliegerfahrbahn
ortsfest, stationär = permanent, non-removable, static, stationary, fixed
Orts|höhenunterschied *m* [*Bodenmechanik*] = position head
~**lage** *f*, Standort *m* = site, location
~**schild** *n*, Ortstafel *f* = place-name sign
Ort-Stahlbeton *m* = reinforced in-situ concrete, in-situ reinforced ~
Ortstein *m* = hardpan
ortsveränderliche Einzellast *f* = variable concentrated load
Orts|verkehr *m* = local traffic

~**wechsel** *m* → Baustellenwechsel
Ortzugpfahl *m* = cast-in-place tension pile
Os *m*, Esker *m* (Geol.) = esker
Öse *f* = ear, eye, lug
Ösenschraube *f*, Schrauböse *f* = eye bolt
Osmose *f* = osmosis
~**verfahren** *n* = osmotic method
osmotischer Druck *m* = osmotic pressure
österreichische Bauweise *f* → Aufbruchbauweise
~ **Methode** *f* → Aufbruchbauweise *f*
österreichisches Verfahren *n* → Aufbruchbauweise *f*
Ostwald-Viskosimeter *n* = U tube viscosimeter
oszillatorische Welle *f*, oszillierende Schwingungswelle = wave of oscillation
Otto|-Kraftstoff *m* → Benzin *n*
~**(-Vergaser)motor** *m* → Benzinmotor
Ottrelithschiefer *m* = ottrelite slate
ovaler Kolben *m* = cam-shaped piston
oval-gerippter Draht *m* = oval ribbed wire
Ovalstütze *f* = oval-shaped column
Oxalsäure *f*, Kleesäure, $(C_2O_2H)_2 \times 2H_2O$ = oxalic acid
Oxychloridzement *m* = oxychloride cement
Oxydation *f* = oxidation
Oxydationszone *f* (Geol.) = oxidizing zone
Oxydhaut *f*, Eloxierhaut = anodic film
oxydierende Ofenatmosphäre *f* = oxidizing conditions, ~ atmosphere
Oxydiernessel *f* [*Linoleumherstellung*] = muslin scrim
oxydiertes Bitumen *n* → geblasenes ~
~ **Leinöl** *n*, Linoxyn *n* = oxidized linseed oil, linoxyn
oxydische Kochsalzlösung *f* = sodium chloride solution containing hydrogen peroxide
Oxydschicht *f*, Oxydfilm *m* = oxid(e) film

Ozeandampfer — Papierlage

Ozeandampfer *m* = ocean-going steamship
ozeanisches Flutkraftwerk *n* → Flutkraftwerk
Ozonanlage *f* = electrical ozone installation

P

paarweise angebrachte Bolzen *mpl* = paired bolts
Packeis *n* = pack-ice
Packerei *f* [*Zementfabrik*] = bag packing plant, bagging ~
Paketstapeln *n*, Steinpaketierung *f* [*z.B. Ziegel*] = packaging
(Pack)Faschinat *n*, Buschpackwerk *n*, (Faschinen)Packwerk, Faschinenlage *f* = fascine work
Pack|lage *f* allgemeiner Begriff für entweder „Setzpacklage *f*" oder „Schüttpacke *f*"
~**maschine** *f* [*z.B. für Zement*] = packer, packing machine
~**silo** *m* [*Einsackmaschine*] = supply bin, ~ tank
~**tisch** *m* = packing table
~**werk** *n* → (Pack)Faschinat *n*
Paddelradbelüftung *f*, Paddelraddurchlüftung = paddle-wheel aeration
Paketierung *f* = parcelling, packaging
Palette *f* = pallet
Paletten|fertiger *m* = pallet-type block (making) machine
~**gut** *n* = palletised load
~**reiniger** *m* **und -öler** = pallet cleaner and oiler
~-**Reinigungsmaschine** *f* = pallet cleaner
~**rückführung** *f* = pallet return
~**walze** *f*, Verteilerwalze [*Beton-(bahn)Automat*] = rotating grading screed, ~ leveller
Palingenese *f* → Anatexis *f*
Palisade *f* = palisade
Palmengewölbe *n* → Fächergewölbe
Palynologie *f* = palynology
palynologisches Vielkornpräparat *n* = multi-grain palynological slide

panallotriomorph → panidiomorphkörnig
Pancakeeis *n*, Pfannkuchen *m* = pancake ice
Paneel *n* → Täfelung
panidiomorphkörnig, panallotriomorph, autallotriomorph (Geol.) = panidiomorphic
Panne *f*, (Geräte)Ausfall *m* = (equipment) breakdown
Pantograph *m*, Storch(en)schnabel *m* = pantograph, pentograph
Panzer|förderer *m* → Doppelketten(kratzer)förderer
~**glas** *n* = armo(u)r-plate glass
~**(gleis)kette** *f* = tank track
~**(gleis)kettenschuh** *m* = tank shoe
~**graben** *m* = anti-tank ditch
~-**Hartbetonplatte** *f*, Panzer-Hartbetonfliese *f* [*mit metallischer Deckschicht für Industrie(fuß)böden*] = armo(u)red paving tile
~**kabel** *n* → bewehrtes Kabel
~**platte** *f* = armo(u)r plate
~**plattenwalzwerk** *n* = armo(u)r-plate rolling mill
~**pumpe** *f* = armo(u)red pump
~**rohr** *n* [*für Triebwasserleitungen*] = penstock pipe
~**schießplatz** *m* = tank gunnery range
~**straße** *f* = road for tank traffic
~**übungsgelände** *n* = tank-training area (or ground)
~**verkehr** *m* = tank traffic
~**verschleiß-Versuchsmaschine** *f* [*Ermittlung der Verschleißwirkung von Panzer(gleis)ketten auf Straßendecken im Laboratorium*] = tank wear machine
~**wanne** *f* [*Motor*] = crankcase guard
Papier|abrollwagen *m* [*Betonstraßenbau*] = paper cart
~**bahn** *f* [*Betonnachbehandlung*] = paper blanket
~**fabrikabwasser** *n* = paper mill waste
~**holzflachgabel** *f* = low profile pulp wood fork
~**kabel** *n* = paper(-insulated) cable
~**kohle** *f* = paper coal
~**lage** *f*, Papierschicht *f* = paper layer

Papierschicht — Parallelwerk

~schicht *f*, Papierlage *f* = paper layer
~torf *m* = paper peat
~trennschicht *f*, Papiertrennlage *f* = paper separator
~unterlage *f*, (Autobahn-)Unterlagspapier *n*, Straßenbaupapier = concrete subgrade paper, concreting ~, underlay ~, subsoil ~, road lining
~verbrenner *m* = paper destructor
~verkleidung *f* [*Strohplatte*] = paper liner
Pappe *f* = cardboard
Papprohr *n* = paper tube [*used as core to form holes in concrete*]
Parabel|belastung *f*, parabolische Belastung *f* = parabolic loading
~bogen *m* = parabolic arc
parabelförmige Krone *f* [*Straßenquerschnitt*] = parabolic crown, continuous parabolic curve
Parabel|-Querschnitt *m* = parabolic cross section
~schnitt *m* = segment of a parabola
~träger *m* = parabolic truss
~verteilung *f* = parabolic distribution
parabolisch exzentrisch verlaufende Bewehrung *f* = parabolic steel eccentricity [*concrete beam*]
parabolische Hyperboloidschale *f* = parabolic hyperboloid shell
~ **Rippe** *f* = parabolic rib
~ **Schale** *f* = parabolic shell
~ **Sohldruckverteilung** *f* = parabolic distribution of bearing pressure, ~ bearing pressure distribution
parabolischer Obergurt *m* = parabolic top chord, ~ ~ boom, ~ ~ flange, ~ upper ~
Paraboloid|dach *n* = paraboloid roof
~schale *f* = paraboloid shell
Paraffin *n* = paraffin wax; coal-oil (US)
paraffinbasisches Erdöl *n* = paraffinicbase (crude) petroleum, ~ crude
Paraffin|gatsch *m* = (paraffin) slack wax
~gehalt *m* = paraffinicity
paraffinhaltiges Destillat *n* = paraffindistillate, wax oil

Paraffin|-Kohlenwasserstoff *m* = paraffin
~öl *n*, Medizinöl = liquid paraffin, medicinal oil
~öl *n* **von niedrigem Schmelzpunkt zur Imprägnierung** *f* **von Streichhölzern** = match wax
~schaber *m*, Paraffinkratzer *m* = paraffin scraper
~schuppe *f* = paraffin scale
Paragneis *m*, Sedimentgneis, Renchgneis = paragneiss
Paraklase *f*, Absenkung *f*, Sprung *m*, Bruch *m*, Verwerfung, Abschiebung [*Grabenrand*] (Geol.) = fault (plane), geological fault, faulting
parallel laufende Kanäle *mpl* [*Einlaß- und Auslaßventile*] = parallel porting
Parallel(bahn)system *n* [*Flugplatz*] = parallel runways, ~ (runway) system
~ **mit versetzten Pisten** [*Flugplatz*] = staggered parallel runways, ~ (runway) system
Parallel|binder *m* [*siehe Anmerkung unter ,,Binder"*] = parallel roof truss
~fachwerkträger *m* = parallel truss
parallelfahrbarer Kabelkran *m* = tautline cableway with both towers travel(l)ing on parallel tracks
Parallelflügel *m*, Stirnflügel = parallel wing, spandrel wall
Parallelogrammgestänge *n* = parallel linkage
Parallel|schaltung *f* [*Ölhydraulik*] = parallel circuit design
~schlag *m* → Albertschlag
~schnitt *m* [*Baggerung*] = parallel cut
parallelseitiger Fertigbetonpfahl *m*, zylindrischer ~ = parallel-sided precast foundation pile, ~ ~ (load-)bearing
Parallel|system *n* → Parallelbahnsystem
~träger *m* = parallel flanges girder, girder of constant depth
~verschiebung *f* [*Kristallographie*] = parallel gliding
~werk *n*, Längswerk, Leitwerk, Richtwerk, Streichwerk = training

embankment; longitudinal dyke (Brit.); longitudinal dike (US)

Pardun(e) n, (f) = guy-rope

Parisergelb n, Chromgelb, chromsaures Blei n, $PbCrO_4$ = chrome yellow, lead chromate

Pariser Leiste f, Putzleiste aus Stuckgips = screed of stucco

Park|bau m → Parkhaus

~bremse f = parking brake

~dach n = car parking roof

Parken n, Abstellen [*Fahrzeug*] = parking

parkender Verkehr, ruhender ~ = standing traffic, stationary ~

Parkerisieren n = Parkerizing

Park|etage f → Parkgeschoß n

~fläche f, Parkplatz m = parking lot, ~ place

~geschoß n, Parketage f, Parkstock(werk) m, (n) = parking floor, ~ level

~haus n, Parkbau m = multi-stor(e)y car park, car park building, car storage park, park(ing) garage, park(ing) block

~hochhaus n = high-rise (parking) garage, multi-storied car park

~möglichkeit f = possibility of parking

~platz m, Parkfläche f = parking place, ~ lot

~platzbefestigung f [*siehe Anmerkung unter ,,Befestigung"*] = car park pavement, ~ ~ surfacing, parking place ~ ~, parking lot ~

~raumnot f = street sclerosis (US)

~stock(werk) m, (n) → Parkgeschoß n

~straße f = park highway, parkway

~turm m, Autosilo m = parking tower

Parkuhr f = parking meter

~ die im Wagen mitgeführt wird = CARPAM, mobile parking meter

Parkverbotsschild n = no-parking marker

Parshallkanalmesser m = Parshall(measuring) flume

partielle mechanische Ähnlichkeit f = partial mechanical similarity, ~ ~ similitude

partieller Brunnen m, nicht vollständiger ~ = incomplete well, shallow (-dug) ~

~ Geschiebetrieb m, feinkörniger ~ = partial bed load movement

Partikelchen n → Einzelkorn n

Partikularintegral n, partikuläres Integral = particular integral

Parzelle f, Grundstück n = plot; lot (US)

Passagier m, Reisender m, Fahrgast m = passenger

~brücke f [*beim Zweiebenenfinger. Flughafen*] = passenger bridge

~durchlauf m = passenger flow

~hafen m → Personenhafen

~schalter m = passenger-handling counter

~schiff n → Fahrgastschiff

~verkehr m → Fahrgastverkehr

~weg m, Passagierdurchlaufroute f = passenger flow route

Passat(wind) m = geostrophic wind

passiver Bruch m = passive rupture

~ Erddruck m → Erdwiderstand m

~ Rankine'scher Zustand m = passive Rankine state

Passivieren n, Passivierung f = passivation

Passivität f = passivity

Passung f = fit

Pastell|farbe f = pastel colo(u)r

~ton m = pastel shade

pastös = pasty

pastöse Paraffin-Öl-Emulsion f = (sucker-)rod wax

Paß m, Gebirgs~ = (mountain) pass

~blech n, Beilagscheibe f = shim

~gletscher m = through-glacier

~kontrolle f = passport control

~punktbestimmung f **aus der Luft** [*Luftbildmessung*] = airborne control system

~schraube f = turned bolt, tight-fitting ~, bright ~

~stift m = dowel pin

Paßstraße — Pelton-Rad

~straße f = (mountain) pass road, (~) ~ highway
~stück n = adapter, fitting piece, adaptor
~tal n = through-valley
Patent|anmeldung f = patent application
~inhaber m = patentee
patentverschlossenes Drahtseil n = locked-wire strand cable
Paternoster(-Aufzug) m, Umlaufaufzug, Elevator m = paternoster
Patina f, Edelrost m = patina
Patrone f = cartridge
Patsche f → Schlagbrett
Pauschal|gebühr f = flat-rate tariff
~summe f = lump sum
Pause f, Licht~ = print
~ = break
Paus|leinen n = tracing cloth
~papier n = tracing paper, blue print ~
Pavillon m = pavillon
Pech n = base-tar, pitch
~blende f → Uran~
~grieß m = pitch grit, ~ cake
~harz n = pitch resin
~kohle f = pitch coal
~koks m = pitch coke
~mastix m = pitch mastic
~schotter m = pitch macadam
~see m, Asphaltsee = pitch lake, asphalt ~
~stein m = pitch-stone
~torf m = pitch peat
Pedal n → Fuß~
~lenkung f = pedal steer(ing)
Pedologie f, Bodenkunde f = pedology
Pegel m, Wasserstandanzeiger m = gage (US); gauge (Brit.) [*A river or stream water-level measuring device which may be an elaborate recording instrument or merely a stick with feet or metres of depth painted on it*]
~beziehungslinie f = line of corresponding stages
~bogen m, Diagrammbogen [*Registrierpapier für Schreibpegel*] = record sheet

~höhe f = gage height (US); gauge ~ (Brit.) [*The elevation of a water surface above or below a datum corresponding to the zero of the staff or other type of gage by which the height is indicated*]
~latte f → Skalenpegel m
~(meß)stelle f, Pegel(meß)station f = flood measuring post, (river) ga(u)ging station [*A gaging station is the place were gage heights of a stream or river are observed, and at, or near, which discharge measurements are made*]
~null n, PN [*Bezugshorizont eines Pegels*] = ga(u)ge datum
~standdauer f = stage duration
~stand(gang)linie f = hydrograph of river stages, ~ ~ stream ~
~standhäufigkeit f = frequency of stages
~stelle f [*Meßstelle des Wasserstandes an einem Wasserlauf durch einen Pegel*] = gaging station, ~ point (US); gauging ~ (Brit.) [*A place on a water course where data are gathered by which continuous discharge records may be developed*]
Pegmatit m, grobkörniger Granit m = pegmatite
Peilen n, Peilung f [*lotrechte Wassertiefenmessung*] = sounding
Peil|latte f, Sonde f = sounding pole
~leine f, Peilseil n = sounding line, lead ~
~rohr n = tube for measurement of water level in a well
~schiff n = sounding vessel
~seil n, Peilleine f = lead line, sounding ~
~stab m → Anzeigestab
Peine|-Bohle f = Peine steel sheet pile
~-Profil n = Peine section
Peinerträger m = Peine girder
Peitschenmast m = whip-shaped lamp post
Pelletisieranlage f = pelletizing plant
Pelletisieren n = pelletizing
Pelton-Rad n, Freistrahlturbine f, Freistrahlrad, Peltonturbine =

Pelton (water) wheel, impulse (~) ~, tangential wheel (or turbine)
Pendel n, m = pendulum
~achse f = oscillating axle
pendelartig wirken [*Stütze*] = to function as a rocker [*support*]
Pendel|becherwerk n = gravity tipping conveyor, ~ ~ conveyor, tilting bucket ~, gravity bucket ~, swinging bucket elevator
~betrieb m = shuttle (service) [*A back and forth motion of a machine which continues to face in one direction*]
~blech n, Gelenkblatt n [*bei einem Pendelgelenk*] = pendulum plate, ~ leaf
~dynamometer n = cradle dynamometer
~gelenk n = pendulum joint
~glätter m = reciprocating screed
~joch n = rocker bent [*A bent generally of steel, though sometimes of timber, hinged at either one or both ends so as to provide for the expansion and contraction of the span supported*]
~kugellager n = self-aligning ball bearing
~lager n = self-aligning bearing
~lager n [*für Hoch- und Tiefbauten*] = pendulum bearing, tumbler ~
~lenkachse f = oscillating steering axle
~mäkler m, Pendelläuferrute f, Pendellaufrute = pendulum leader [*Is used for driving side-raking piles*]
~mühle f, Fliehkraftmühle mit Walzen, Fliehkraft-Rollenmühle = suspended roller mill
~pfeiler m = rocking pier, hinged ~
~portal n = rocking portal
~rad n = floating wheel
~rahmen m [*Brückenbau*] = rocking frame
~rollenlager n, Rollenpendellager = self-aligning roller bearing, spherical ~ ~
~scheibe f [*Brückenbau*] = flat rocker column

~schieber m [*zur Entleerung des Löffels beim Hochlöffelbagger. Im Gegensatz zur Klappe, die den Löffelboden sofort freigibt, gibt der Pendelschieber die Bodenöffnung erst allmählich frei*] = (shovel) dipper slide
~schlagwerk n = pendulum impact tester
~schlitten m [*Klappbrücke*] = pendulum guide
~schrapper m = cable excavator with dragline type bucket and running rail type carrier
~(stahl)stütze f = socketed stanchion, ~ steel column
~transport m [*z. B. zwischen Bagger und Kippe*] = shuttle haul(age)
~tür f, Durchschlagtür = double acting door
~ventil n = shuttle valve
~verkehr m = shuttle traffic
~verzögerungsmesser m = pendulum decelerometer
~wagen m = shuttle-car
~wand f [*Rahmensystem*] = rocker wall [*portal system*]
~wickelmaschine f = shuttle winding machine
~wurfrinne f → Torpedorinne
Pendler m, ständiger Benutzer (eines Verkehrsmittels) = commutor
~fahrt f = commuting trip
Penetration f, Eindringung f = penetration
Penetrometer n → Eindring(ungs)messer
pennsylvanisches Bohren n, ~ Bohrverfahren n, Seilbohren mit Schwengelantrieb, Schwengelantrieb-Seilbohren = Pennsylvanian drilling system, ~ ~ method
Perchlorat-Sprengstoff m = perchlorate explosive
perfekte Flüssigkeit f → ideale ~
perfekter Rahmen m = perfect frame
Perforieren n [*Tiefbohrtechnik*] = formation fracturing
Pergola f = pergola
Perimeterbeitrag m → Anliegerbeitrag
Perimetralfuge f → → Talsperre f

Periodendruck — „Perücke"

Perioden|druck *m*, Setzdruck, Hauptdruck (✗) = periodic weight
~-Mischanlage *f* → absatzweise arbeitende Mischanlage
~mischer *m* → → Betonmischer
periodische Strömung *f* [*Meer*] = periodic current
periodischer Ofen *m* = intermittent kiln, periodic ~
~ Wasserlauf *m*, zeitweiliger ~ = temporary water course
Periskopuntersuchung *f* von Bohrlöchern = periscopic inspection of drill holes
Perlenleim *m* = Scotch glue in pearl form
Perlit *m* [*ein vulkanisches Glas*] = perlite [*a volcanic glass*]
Perlit *m* [*Gefügebestandteil des langsam abgekühlten Kohlenstoffstahls*] = perlite
~beton *m* = perlite concrete
~kies *m* = perlite gravel
~putz *m* = perlite plaster
Perl|kies *m*, Erbskies = pea gravel
~kohle *f* → Grießkohle
~koks *m* = pea coke
~moos *n*, Knorpeltang *m*, irländisches Moos = Irish moss, carragheen
~pech *n*, Kugelpech = pellet(ed) pitch
Permafrostzone *f* [*Schweiz*] → dauernd gefrorener Boden *m*
permanente Härte *f*, bleibende ~, Dauerhärte = permanent hardness
~ Strömung *f* → stationäre ~
Persenning *f*, geteertes Segeltuch *n*, Presenning = tarp(aulin)
persisches Wasserrad *n* = Persian wheel
Personal *n*, Belegschaft *f* = personnel, staff
~abbau *m*, Personalverminderung *f* = cut-down of staff
~prüfstelle *f* = psychological testing centre (Brit.); ~ ~ center (US)
~wechsel *m*, Arbeitnehmerwechsel = employee turnover
(Personen)Aufzug *m*, Fahrstuhl *m* = passenger elevator (US); ~ lift (Brit.)

Personen|auto *n*, Personen(kraft)wagen *m*, PKW *m* = (passenger) automobile, (~) car
~bahnhof *m*, Fahrgastbahnhof = passenger railway station (Brit.); ~ railroad ~ (US)
~bandförderer *m* = passenger conveyor belt, ~ conveyer ~
~(be)förderung *f* = passenger transport
~-(Druck)Luftschleuse *f* → Personenschleuse
~fähre *f* = passenger ferry
~fahrzeug *n* = passenger vehicle
~förderung *f* → Personenbeförderung
~hafen *m*, Fahrgasthafen, Passagierhafen, Verkehrshafen = passenger harbo(u)r
~kammer *f* → Personenschleuse *f*
~kilometer *n* = passenger-kilometer, passenger-kilometre
~(kraft)wagen *m* → Personenauto *n*
~-Landungssteg *m* = passenger berth
~-Luftschleuse *f* → Personenschleuse
~-Meile *f* = passenger-mile
~-Meilen *fpl* = passenger-mileage
~schadenunfall *m* = (personal-)injury accident
~schiff *n* → Fahrgastschiff
~schleuse *f*, Personenkammer *f*, Personen-(Druck)Luftschleuse = manlock
~seilbahn *f* = passenger ropeway
~transport *m*, Fahrgasttransport = transportation of passengers
~- und Gepäckabfertigung *f* = processing passengers and baggage
~unfall *m* = accident to person(s)
~verkehr *m* → Fahrgastverkehr
~wagen *m* → Personenauto *n*
perspektivisch darstellen = to sketch in perspective, ~ draw ~ ~
Perspektiv|zeichnen *n*, perspektivisches Zeichnen = perspective drawing
~zeichnung *f*, perspektivische Zeichnung = perspective drawing
„Perücke" *f*, „Perückenkopf" *m*, Aufspalten *n* = brooming [*The crushing and spreading of the top of*

a wooden pile when driven into hard ground]
Petrischale *f* = Petri dish
Petrochemikalie *f* = petroleum chemical
petro|chemisch = petrochemical
~chemische Anlage *f* = petroleum chemical plant
Petrographie *f*, **Lehre** *f* **der Festgesteine** = petrography
petro|graphische Aufbereitung *f* = petrographic preparation
~graphische Deutung *f* = petrographic interpretation
~graphisches Merkmal *n* [*Zuschlag-(stoff)*] = petrographic characteristic
Petroleum *n* → Erdöl *n*
~beständigkeit *f* = kerosine resistance, Kerosene ~
~-Beton *m* [*Durch Zusatz von Petroleum oder Ölrückständen zum angemachten Zement soll die Dichte gesteigert werden. Erfolg ist zweifelhaft*] = concrete with oil addition
~feld *n* → Erdölfeld
Petrolkoks *m* = petroleum coke, oil ~
Petrologie *f*, **Gesteinsentstehungslehre** *f* = petrology
Petrolpech *n* = petroleum pitch, oil ~
Pfahl *m*, **Halte~** [*Naßbagger*] = spud
~, **Gründungs~** = pile, foundation ~
~ mit gerader Ziffer = pile with even number
~ ~ offenem Ende, **offener Pfahl** = open-foot pile
~ ~ ungerader Ziffer = pile with odd number
~abmessung *f* = pile dimension
~abschnitt *m*, **Pfahlschuß** *m* = pile section
~abstand *m* = pile spacing
~achse *f* = pile centre line (Brit.); ~ center ~ (US)
~anordnung *f* = arrangement of piles
~anzahl *f* = number of piles
~armierung *f*, **Pfahlbewehrung**, **Pfahl(stahl)einlagen** *fpl* = pile reinforcement
~art *f* = type of pile

~ausbildung *f*, **Pfahlform** *f* = pile shape
~baustoff *m* = pile material
~bauten *f*, **Pfahldorf** *n*, **Pfahlsiedlung** *f* = lake dwelling
~beanspruchung *f*, **Pfahlkraft** *f* = pile stress
~belastungsversuch *m*, **Pfahlbelastungsprüfung** *f*, **Pfahlbelastungsprobe** *f* = pile-loading test
~bemessung *f* = pile sizing
~beton *m* = pile concrete
~betonieren *n*, **Pfahlbetonierung** *f* = pile casting, ~ pour(ing), ~ concreting
~betonschleuder *f* = concrete pipe spinning unit
~bewehrung *f* → Pfahlarmierung
~biegemoment *n* = pile bending moment
~bohrgerät *n*, **Pfahlbohranlage** *f* = pile boring rig, ~ ~ plant
~bolzen *m* = pile bolt
~brücke *f* = pile bridge [*A bridge carried on piles or pile bents*]
~bündel *n* → Dalbe *m*
~dichtungswand *f* → Pfahlwand
~dicke *f* = pile thickness
~dorf *n*, **Pfahlbauten** *f*, **Pfahlsiedlung** *f* = lake dwelling
~druckkraft *f*, **Pfahldruckbeanspruchung** *f* = pile compressive stress
~durchmesser *m* = pile diameter
~einbringverfahren *n* = pile-placing method
~eindringung *f* = pile penetration
~einlagen *fpl* → Pfahlarmierung *f*
pfählen, **rammen von Pfählen** = to drive piles
Pfahl|ende *n* = pile foot [*The lower extremity of a pile*]
~entlastung *f* = pile unloading
~fabrik *f*, **Pfahlwerk** *n*, **Pfahlfertigungsplatz** *m* = pile yard
~fertigungsplatz *m* → Pfahlfabrik *f*
~form *f*, **Pfahlausbildung** *f* = pile shape
~form *f* = pile mo(u)ld
~formel *f* = static pile-driving formula

Pfahlfüllbeton — Pfahlschuh

~füllbeton *m* = concrete for pile filling
~fundament *n* → Pfahlgründung *f*
~fundamentkörper *m* → Pfahlgründung *f*
~foundation *f* → Pfahlgründung *f*
~fuß *m* = pile foot
~gerüst *n* = pile trestle
~gerüst *n* [*Naßbagger*] = spud frame
~gewicht *n* = pile weight
~größe *f* = pile size
~gründung *f*, Pfahlfundation *f*, Pfahlfundament *n*, Pfahlgründungskörper *m*, Pfahlfundamentkörper [*Österreich: Pilotengründung*] = pile(d) foundation (structure)
~gründungskörper *m* → Pfahlgründung
~gruppe *f* → Dalbe *m*
~gruppe *f* = pile group
~herstellung *f* = pile manufacture, ~ making
~joch *n* = pile bent [*A row of piles arranged transverse to the longitudinal axis of a structure and fastened together by a capping beam or transom and sometimes diagonal bracing*]
~jochbrücke *f* = pile bent bridge
~konstruktion *f* = piling
~kopf *m* = pile head, ~ butt [*The upper end of a pile*]
~kopfhöhe *f* = pile head elevation, ~ butt ~
(~)Kopfplatte *f*, Pfahlrostplatte = pile cap [*A slab, usually of reinforced concrete, covering the tops of a group of piles for the purpose of tying them together and transmitting to them as a group the load of the structure which they are to carry; also a metal plate often placed on top of steel piles to distribute the load to the concrete from the pile*]
~kopfplattenschalung *f* = pile cap formwork
~kopfschutz *m* = pile head protection
~kraft *f*, Pfahlbeanspruchung *f* = pile stress
~lagerplatz *m*, Pfahlliegeplatz = pile storage yard

~länge *f* = pile length
~last *f* = pile load
~loch *n* = pile hole
~mantel *m* = pile skin
~moment *n* = pile moment
~neigung *f* = pile rake
~oberfläche *f* = pile surface
~paar *n* = pair of piles
~prüflast *f* = pile test load
~prüfung *f*, Pfahlprobe *f*, Pfahlversuch *m* = pile test
~querschnitt *m* = pile section
~querschnittfläche *f* = cross-sectional area of a pile
~rammanlage *f* → Ramme *f*
~ramme *f* → Ramme
~rammeinrichtung *f* → Ramme *f*
~rammen *n*, Pfahlrammung *f* = pile driving
~rammhammer *m* = pile driving hammer
~rammschlauch *m* = piledriver hose
~reihe *f* = row of piles
~ring *m* = pile ring, ~ hoop, driving band [*A steel band fitted round the head of a timber pile to prevent brooming*]
~rohr *n*, Mantelrohr, Stahl(blech)rohr, Blechhülse *f* = (pile) casing (tube), (~) lining (~), (sheet-steel) shell, sheet-steel casing
Pfahlrost *m*, Pfahlwerk *n* = pile foundation grillage
~bauwerk *n* = pile foundation grillage structure
~berechnung *f* = pile foundation grillage design
~gründung *f* = pile grillage foundation
~pfeiler *m* = piled pier [*A piled pier is a pier of masonry or reinforced concrete carried on a group of piles which are capped at the lowest level to which it is convenient and economical to carry the pier*]
~platte *f* → Pfahlkopfplatte
Pfahl|saugbagger *m* = spud-type suction dredger (Brit.); ~~ dredge (US)
~schaft *m* = pile shaft
~schuh *m*, Rammspitze *f* = pile shoe

~schürze f → Pfahlwand f
~schuß m, Pfahlabschnitt m = pile section
~schute f [*Schute zum Transport von Pfählen*] = pile barge
~seil n, Halte~ [*Naßbagger*] = spud rope, ~ cable, ~ line
~setzung f = pile settlement
~siedlung f, Pfahlbauten f, Pfahldorf n = lake dwelling
~sperrwand f → Pfahlwand
~spitze f = pile point, ~ tip [*The lower end of a pile*]
~(stahl)einlagen fpl → Pfahlarmierung f
~stellung f = pile location
~tiefe f, Eindring(ungs)tiefe, Einbindelänge f = depth of penetration, embedded length
~tragfähigkeit f = supporting power of a pile, bearing ~ ~ ~, carrying ~ ~ ~ ~, ~ capacity ~ ~ ~
~trommel f [*Ramme*] = pile drum
~unternehmer m = piling contractor
~verankerung f = pile anchoring
Pfahlwand f → Bohr~
~, Pfahldichtungswand, Pfahlsperrwand, Dicht(ungs)wand aus Pfählen, Sperrwand aus Pfählen, Pfahlschürze f, Schürze aus Pfählen = pile diaphragm, ~ (~) wall, ~ cutoff
Pfahlwerk n, Pfahlfabrik f, Pfahlfertigungsplatz m = pile yard
~, Pfahlrost m = pile foundation grillage
Pfahl|ziehen n = pile extracting, withdrawing piles
~zieher m = pile extractor, pilepuller, piledrawer
~zugkraft f, Pfahlzugbeanspruchung f = pile tensile stress
~zwiebel f = pile bulb
Pfanne f, Gieß~ = (casting) ladle, pouring ~

~ [*Dacheindeckung*] = pantile
Pfannen|blech n [*Dacheindeckung*] = pantile plate
~dach n = pantiled roof
~gips m = pan-calcined gypsum
~kippstuhl m = ladle tilter
~latte f = pantile lath
~meer n = transgression sea
~stein m, Gießpfannenziegel m = ladle brick
~transportwagen m = ladle car
Pfannkuchen m → Pancakeeis n
Pfeifenton m = pipe clay, ball ~, potters' ~
Pfeil m → Bogen~
Pfeiler m = pier
~ (⚒) [*1.) Ein in oder zwischen Abbauräumen stehenbleibender, zu späterem Abbau bestimmter Lagerstättenteil von meist regelmäßiger Form mit der Aufgabe, das Dach für eine bestimmte Zeit zu stützen. 2.) Ein durch verhältnismäßig nahe beieinanderliegende Strecken, Kammern oder Örter zum Abbau vorgerichteter Lagerstättenteil*] = pillar
~ **mit Anlauf**, ~ ~ **Anzug**, ~ ~ **Dossierung** = tapered pier
~ mpl **mit Zwischenwänden** → Pfeilerfundament
pfeilerartiges Abbauverfahren n (⚒) = pillar (method of) mining
Pfeiler|basis f [*Grundplatte, auf die ein Pfeiler steht*] = pier base (plate)
~bau m (⚒) = pilar method (of mining), ~ mining
~beton m = pier concrete
~betonieren n = placing the concrete to form the pier
Pfeilerbruchbau m (⚒) = room and pillar caving
~ **mit diagonalen Pfeilerstrecken auf steilen Steinkohlenflözen** = slant method

---

Pfeiler-Bruch-Verfahren n [*Asbesterzabbau*]
    Fördergang m
    Förderrutsche f

block-caving method

haulage drift
chute raise

Grubenrostgang *m*
Randgang *m*
Sammelpunkt *m*
senkrecht aufwärts führender Gang *m*
Zusammenrutschen *n* der ungestützten Last

grizzly drift
advance fringe drift
grizzly
corner raise

caving

Pfeilerbrücke *f* = pier bridge
Pfeilerfundament *n*, Pfeilergründung *f*, Pfeilerfundation *f*, Pfeilergründungskörper *m*, Pfeilerfundamentkörper, Pfeiler *mpl* mit Zwischenwänden = pier foundation (structure)
~**körper** *m* → Pfeilerfundament *n*
Pfeiler|fundation *f* → Pfeilerfundament *n*
~**fuß** *m* = pier base
~**gewölbe(stau)mauer** *f*, Vielfachbogensperre *f*, Gewölbereihen(stau)mauer, Bogenpfeiler(stau)mauer = multiple arch(ed) dam
~**gründung** *f* → Pfeilerfundament *n*
~**gründungskörper** *m* → Pfeilerfundament *n*
~**höhe** *f* = pier height
~**hohlraum** *m* = pier cavity
Pfeilerkopf *m*, Pfeilervorhaupt *n*, Strom~ = cutwater of pier (Brit.); nose ~ ~ (US)
~, Rundkopf [*Talsperre*] = segmental-headed counterfort (Brit.); round-head(ed) buttress
~**sperre** *f* → → Talsperre
~**(stau)mauer** *f* → → Talsperre *f*
Pfeiler|kuppel(stau)mauer *f*, Vielfachkuppelsperre *f* = multiple-dome dam
~**last** *f* = pier load
~**loch** *n* = pier hole
pfeilerlos [*Architektur*] = astylar
Pfeilermauer *f* = buttressed wall
~ → Pfeilersperre *f*
Pfeiler|mauerwerk *n* = pier masonry
~**moment** *m* = pier moment
~**nische** *f* [*Wehr*] = pier pocket
~**paar** *n* = pair of piers
Pfeilerplattensperre *f* → Plattensperre

~ **Bauweise Ambursen** → Ambursen(stau)mauer *f*
~ ~ **Ransom**, Ransomsperre, Ransom(stau)mauer *f* = Ransom (type) dam [*This flat-slab type of dam was introduced by W. M. Ransom about 1908*]
Pfeiler|platten(stau)mauer *f* → Plattensperre *f*
~**querschnitt** *m* = pier section
~**riegel** *m* [*Brücke*] = pier strut
~**schaft** *m* = pier shaft
~**schalung** *f* = pier formwork
~**schwindfuge** *f* [*Pfeilersperre*] = buttress contraction joint
Pfeiler|sperre *f*, Pfeiler(stau)mauer *f*, aufgelöste (Stau)Mauer, (Stau)Mauer in aufgelöster Bauweise, aufgelöste Gewichtssperre = buttress (type) dam, buttressed ~ (US); counterfort (type) ~ (Brit.) [*The principal structural elements of a buttress dam are the water-supporting upstream face, or deck, and the buttresses*]
~**stau** *m*, Brückenstau = backwater effect of bridge pier
~**(stau)mauer** *f* → Pfeilersperre *f*
~**unterkolkung** *f* = scour under a pier
~**verband** *m* [*Träger*] = double triangulated bracing system, arrow point bracing
~**verlust** *m*, Pfeilerwiderstand *m* = pier loss
~**vorhaupt** *n* → Pfeilerkopf *m*
~**widerstand** *m*, Pfeilerverlust *m* = pier loss
Pfeil|höhe *f*, Stichhöhe = rise
~**verhältnis** *n* = rise-span ratio
Pfennig *m* (Geol.) = nummulite
Pferde|fuhrwerk *n*, Pferdewagen *m* = horse-drawn cart

~stärke *f*, P.S. = metric horsepower
~walze *f* = horse-drawn roller
Pfette *f*, **Dach**~ = binding rafter, purlin(e)
Pfetten|abstand *m* = spacing of purlin(e)s
~anordnung *f* = purlin arrangement
~anschluß *m* = purlin(e) connection
Pfettendach *n* = purlin(e) roof
~ **mit doppelt stehendem Stuhl** = purlin(e) roof with queen post
~ ~ ~ ~ ~ **und Versenkung** = purlin(e) roof with two posts and wind filling
~ ~ **dreifach stehendem Stuhl** = purlin(e) roof with three posts
~ ~ ~ ~ **und Versenkung** = purlin(e) roof with three posts and wind filling
~ ~ **liegendem Stuhl** = purlin(e) roof with sloped studs
~ ~ ~ **und stehenden Stuhl und Versenkung** = purlin(e) roof with post and struts and wind filling
~ ~ ~ **Säulen und Versenkung** = purlin(e) roof with inclined struts and wind filling
~ ~ **nach den Außenwänden gerichteten liegenden Säulen** = purlin(e) roof with struts inclined away from the centre
~ ~ **innen gerichteten liegenden Säulen** = purlin(e) roof with struts inclined towards the centre
~ ~ **stehendem Stuhl** = purlin(e) roof with king post
~ ~ ~ ~ **und Versenkung** = purlin(e) roof with king post and wind filling
~ ~ ~ **und liegenden Säulen** = purlin(e) roof with king post and slanting studs
Pfetten|last *f* = purlin(e) load
~plan *m* = purlin(e) layout
~stoß *m* = purlin(e) joint
~überstand *m* = purlin(e) projection
~verankerung *f* = purlin(e) anchoring
Pflanze *f* = plant
Pflanzen|bewuchs *m* = Bodenbewachsung *f*, Pflanzendecke *f* = vegetable cover, vegetation cover, vegetative cover, plant cover floor formwork
~decke *f*, Pflanzenbewuchs *m*, Bodenbewachsung *f* = vegetative cover, vegetation ~, vegetable ~, plant ~, mantle of vegetation
~faser *f* = plant fibre (Brit.); ~ fiber (US)
~krankheit *f* = plant disease
~leim *m*, pflanzlicher Leim = vegetable glue
~nährstoff *m* = plant nutrient
~öl *n* = vegetable oil
~physiologie *f* = plant physiology
~schale *f* = plant bowl
~verdunstung *f*, Transpiration *f* = plant transpiration
Pflanzung *f* = plantation
Pflaster *n* = sett paving
~arbeiten *fpl* = sett paving work
~aufrauhgerät *n* = sett roughening machine
~ausbesserung *f* = sett paving repair
~bett *n*, Pflastersandschicht *f*, Pflastersandlage *f*, Pflastersandbett = sand layer under a sett paving
~(decke) *n*, (*f*), Steinpflaster *n*, Pflasterung *f*, Pflasterstraßendecke [*Schweiz: Pflästerung*] = (stone-) sett paving (Brit.); [*in Scotland: causeway*]; block pavement (US)
Pflasterer *m*, Steinsetzer *m*, Pflastersetzer *m* = pavior (Brit.); paver (US)
Pflaster(er)hammer *m*, Steinsetzerhammer *m*, Pflasterschlägel *m* = paver's hammer (US); pavior's ~ (Brit.)
Pflastererhelfer *m* = pavior's labourer, rammer man
Pflaster|fuge *f* = paving joint
~(fugen)kitt *m*, Pflaster(fugen)vergußmasse *f* = sett joint filler, ~ ~ sealing compound
~fugenverfüllung *f* → Ausfüllen *n*
~(fugen)verguß *m* = sett joint sealing
~(fugen)vergußmasse *f* → Pflasterkitt *m*
~hammer *m* → Pflastererhammer
~holz *n* = paving wood

Pflasterkies — Pfostenzieher 722

~kies m = gravel for sett pavings
~kitt m = Pflasterfugenkitt
~klinker m → Pflasterziegel m
~kriechen n = creep of sett paving
pflastern = to sett pave
Pflastern n, Pflasterung f = 1. causeway(ing) [*in Scotland*]; 2. paving with setts
Pflaster|platte f = paving flag, flagstone
~ramme f = (sett) paving rammer
~rinne f, Rinnstein m = paved gutter
~rüttler m, Pflastervibrator m = sett paving vibrator
~sand m = sett joint sand
~sandbett n → Pflasterbett
~sandschicht f → Pflasterbett n
~schaufel f [*GRADALL-Anbauvorrichtung*] = sett-handling bucket
~schlägel m → Pflaster(er)hammer m
~setzer m → Pflasterer
~stampfer m = (sett) paving tamper
Pflasterstein m = (paving) sett
~herstellung f = sett-making
~maschine f = paving sett machine
~-Prüfmaschine f = (paving) sett tester, (~) ~ testing machine
Pflaster|straße f, Steinstraße = sett-paved road
~straßendecke f → Pflaster(decke) n, (f)
~streifen m = sett-paved strip
~ufer n = paved bank
Pflasterung f → Pflaster(decke)
Pflaster|verband m (Geol.) → Bienenwabenverband
~verguß m, Pflasterfugenverguß = sett joint sealing
~vergußmasse f → Pflaster(fugen)kitt m
~vibrator m, Pflasterrüttler m = sett paving vibrator
~würfel m = cube sett
~ziegel m, Pflasterklinker m, Straßenbauklinker, Straßenbauziegel = paving (clinker) brick
Pflege f = preventative maintenance
~ des Hangenden (⚒) = roof control
Pflichtwasser n, Anliegerrechtswasser, garantiertes Brauchwasser = compensation water

Pflock m → Absteck~
Pflugbagger m, Flachbagger-Lader m, Schürfbagger, Förderband(-Anhänge)-Schürfwagen m = elevator loader-excavator [*designates both types*]
~ (oder Flachbagger-Lader m, oder Schürfbagger, oder Förderband(Anhänge)Schürfwagen m) mit kreisförmiger Diskuspflugschar (oder Pflugteller) = elevating grader, belt loader
~ (oder Flachbagger-Lader m, oder Schürfbagger) mit keilförmiger Schar = (elevating) loader
Pförtner m = gatekeeper [*at a factory*]
~ = porter [*in a luxury block of flats*]
~haus n = gatekeeper's house
~loge f = gatekeeper's lodge [*The gatekeeper lives in a "lodge" if he keeps the gates at the entrance of a large county estate (Schloß)*]
~wohnung f = porter's flat
Pfosten m, Lotrechte f, Vertikale f, Vertikalstab m [*Fachwerkbinder*] = vertical web member, ~ strut ~, ~ (rod)
~, Stiel m, Ständer m, Säule f [*Fach(werk)wand*] = post
~, Stange f, Ständer m = post
~, Gerüst~ = standard, (scaffold) pole [*An upright of a scaffold whether of wood or metal*]
~fachwerk n, Kreuzverband m [*mit gekreuzten Streben*] = cross bracing
~(fachwerk)träger m = girder with cross bracing
~(fachwerk)trägerbrücke f = bridge with cross-bracing girders
~form f, Beton~ = post mo(u)ld
~haus n, Ständerbauhaus = post and beam house
~lochbohrer m, Pfostenlochbohrmaschine f = post hole borer
~ramme f = post driver
~schwenkkran m, Schwenkkran-Schnellbauaufzug m = crane hoist
~setzen n = post setting
~träger m, Pfostenfachwerkträger m = girder with cross bracing
~zieher m = post puller

Pfropfen — Pilaster

Pfropfen *m*, Plombe *f*, Stöpsel *m* = plug
P-Grenze *f* → Proportionalitätsgrenze
Phantasieholz *n* = fancy wood
Phasen|diagramm *n* = phase diagram
~gleichgewicht *n* = phase equilibrium
~kontrastmikroskopie *f* = phase microscopy
~trennung *f* = phase separation
~verschiebung *f* = phase displacement
Phenol *n*, Karbolsäure *f*, Phenylalkohol *m*, Steinkohlenteerkreosot *n*, C_6H_5OH = phenol, carbolic acid
~harz *n* = phenolic resin
~pech *n* → Karbolpech
Philharmonie *f* = philharmonic hall
Phlogopit *m*, hellbrauner Glimmer *m* (Min.) = phlogopite, amber mica
Phon *n* = phon
Phosphatieren *n*, Phosphatierung *f* = phosphating
Phosphatschmelze *f* = phosphate melt
Phosphorbronze *f* = phosphor bronze
Phosphorit *m* (Min.) = phosphorite, rock phosphate
Phosphor|kupfer *n* = phosphor copper
~-Leuchtfarbe *f* = phosphorescent paint
~pentoxyd *n*, Phosphorsäureanhydrid *n*, wasserfreie Phosphorsäure *f*, P_2O_5 = anhydrous phosphoric acid
~salz *n* = salt of phosphorus
Photochemie *f* = photochemistry
photoelektrischer Schwebestoffmesser *m* = photoelectric siltmeter
Photo|geologie *f* = photogeology
~grammetrie *f*, Bildmessung = photogrammetry, photogrammetric survey(ing)
photogrammetrische Kartierung *f* = photogrammetric mapping
Photo|montage *f* = photograph montage
~tachymetrie *f* = photo-tacheometry
~theodolit *m* = photo-theodolite
ph-Wert *m*, ph-Zahl *f*, Wasserstoffzahl H = pH-value
physikalische Höhenmessung *f*, barometrische ~ = barometric levelling

~ Konstante *f* = physical property
~ Verwitterung *f*, Gesteinszerfall *m*, mechanische Verwitterung = mechanical weathering, physical ~, rock disintegration
physiographische Geologie *f* = physiographic geology
physiologische Schmerzschwelle *f* [*Lautstärke*] = painfully loud sound
P. I. *m* → Plastizitätsindex
Piassavabesen *m* = bass broom
Picke *f*, Bicke, (Erd)Hacke *f* = pick
Pickhammer *m* → Abbauhammer
~gewinnung *f* → Abbauhammergewinnung
Pickprobe *f*, Schlitzprobe, Streifenprobe (✗) = channel sample
Pier *f*, *m*, Ladezunge *f*, Kaizunge; Höft *n* [*Nordsee*] [*manchmal fälschlicherweise „Einfahrtmole" genannt*] = pier, (cargo) jetty
~finger *m* [*Flughafen*] = pier finger
piezoelektrischer Seismograph *m*, Hydrophon *n* = hydrophone
Piezometer(rohr) *n*, Druckmesser *m* = piezometer [*Instrument for measuring pressure head, usually consisting of a small pipe tapped into the side of a conduit and flush with the inside, connected with a pressure ga(u)ge, mercury water column, or other device for indicating pressure head*]
piezometrische Linie *f* → Drucklinie
piezometrisches Gefälle *n* = Druckgefälle
Pigment *n* [*Farbkörper ohne Füller, wird mit Binder zu Anstrichfarbe. Beschwerende Zusätze: Kreide, Schwerspat, Ton, Gips*] = pigment
pigmentiert = pigmented
Pikrinsäure *f*, Trinitrophenol *n* = picric acid
Pikrolith *m*, Bastardasbest *m* = picrolite
Pilaster *m* [*Aus Mauerflucht hervortretender Wandstreifen oder hervortretende Mauerverstärkung. Im Gegensatz zur Lisene hat Pilaster, Kapitell und Basis*] = pilaster, attached pier

Pilasterfassade — Plandarre

~fassade *f* = pilaster façade
Pilgern *n* [*von Rohren*] = reciprocating rolling
Pilgerwalzwerk *n*, Pilgerschritt-Walzwerk = step-by-step-type seamless tube rolling mill
Pilotengründung *f* [*Österreich*] → Pfahlgründung
Pilz *m*, Nadel *f* [*Freistrahlturbine*] = spear
Pilzdecke *f* [*kreuzweise bewehrte Betonplatte, auf Stahlbetonstützen ohne Balken*] = flat slab floor, beamless ~, mushroom ~, mushroom slab, mushroom construction (Brit.); two-way-system flat slab (US)
Pilzdecken|schalung *f* = flat slab floor formwork
~stütze *f* = mushroom-head column
Pilzfelsen *m* = mushroom rock
Pilzkopf *m* [*bei einer Pilzdecke*] = (column) capital, ~ head
~platte *f*, Verstärkungsplatte, Pilzkopfverstärkung *f*, Deckplatte [*liegt zwischen Pilzkopf und Pilzdecke*] = drop (panel)
Pilzventil *n*, Tellerventil = poppet valve
Pinge *f* (⚒) = local depression
Pinnenträger *m*, Trägerstift *m* = pivot post, bearing ~, centre pin support
Pinselauftrag *m* = brush application
Pionier|bohrung *f* → Aufschließungsbohrung
~park *m* = engineer stores depot
~schacht *m* → Aufschließungsschacht
~stollen *m* → Aufschließungsstollen
~zwischenpark *m* = intermediate field depot for engineer tools and materials
Pisé|bau *m*, Pisémauerwerk *n* = pisé de terre [*Walling made of cob*]
~stein *m* [*Kalkmörtelstein aus Branntkalk und Sand (1:8) gestampft*] = lime-sand brick 1:8
Pisolith *m*, Erbsenstein *m* = pisolitic limestone, pisolite
Piste *f*, Start- und Landebahn *f* = runway
Pisten|anordnung *f*, Start- und Landebahnanordnung = runway configuration
~befeuerung *f*, Start- und Landebahnbefeuerung = runway lighting
~grundlänge *f* = basic runway length
~schwelle *f*, Start- und Landebahnschwelle = runway threshold
Pistill *n* = pestle
Pistole *f*, Spritz~ = (spray) gun
Pistolen(hand)griff *m* = pistol grip
Pistonieren *n*, Kolben = swabbing [*A method of getting oil from a well which has ceased to flow before permanent pumps or other lifting devices are installed. A rubber swab, which fits the casing closely, is run up and down the well on a wire rope*]
Pißbecken *n* = urinal
Pitotrohr *n*, Staurohr, Staudruckmesser *m* = Pitot tube [*Device for observing the velocity head of flowing water. It has an orifice held to point upstream in flowing water and connected with a tube by which the rise of water in the tube above the water's surface may be observed. It may be constructed with an upstream and downstream orifice and two water columns, the difference of water levels being an index of the velocity head*]
PKW *m* → Personenauto *n*
PKW-Streifen *m* [*Autobahn*] = fast lane, light vehicle ~
PKW-Verkehr *m* = light vehicle traffic
Plagge *f*, Rasen~, (Rasen)Sode *f* = (turf) sod
Plaggenschneider *m* → (Gras)Sodenschneider
Plagioklas *m*, Natron-Kalkfeldspat *m*, Schieferspalter *m* (Min.) = plagioclase, soda-lime fel(d)spar, limesoda fel(d)spar, plagioclase fel(d)spar, soda-lime plagioclase
plan [*Platte. Im Gegensatz zu gewellt*] = flat
Plan *m*, Zeichnung *f* = drawing
~darre *f*, Darrboden *m* = hot floor dryer; ~ ~ drier (US)

~decke *f* von Koenen = Koenen floor
Plane *f* = tarp
planebene Oberfläche *f* = level plane surface
Planebenheit *f*, Ebenflächigkeit *f*, Ebenheit = accuracy of levels, evenness; surface smoothness (US)
planender Ingenieurbau *m* → Tiefbau
Planer *m* = planner
Pläner Kalk *m* = Plauen limestone
Planeten|achse *f* = planetary axle
~antrieb *m* → Planetenradantrieb
~antriebachse *f* → Achse mit Planetenantrieb
~endantrieb *m* = planetary final drive
~enduntersetzung *f* = planetary final reduction
~getriebe *n*, Umlaufgetriebe = planetary gear system, ~ transmission, range transmission
~getriebeantrieb *m*, Planeten(rad)antrieb = planetary power drive
~getriebe-Steuerventil *n* = range transmission hydraulic control valve
~innenrad *n*, inneres Planetenrad = inner planetary gear
~käfig *m* = planet cage
~-Lastschaltgetriebe *n* = power shift transmission
~-Lenkachse *f* = planetary steer(ing) axle
~lenkung *f* = planetary steer(ing)
~mischer *m*, Planeten-Zwangsmischer = planetary mixer
~-Misch- und Knetmaschine *f* = planetary mixing and kneading machine
~rad *n*, Planet *m* = planetary gear
~(rad)antrieb *m*, Planetengetriebeantrieb = planetary power drive
~satz *m* = planetary (gear) set
~träger *m* = planetary carrier
~untersetzung *f* = planetary reduction
~walzwerk *n* = planetary mill
~(zwangs)mischer *m* = planetary mixer

plangleiche Kreuzung *f*, niveaugleiche ~, ebenerdige ~ = grade crossing, intersection at grade, at-grade intersection
Planie *f* → Planum *n*
Planier|anbauvorrichtung *f* = grading attachment, level(l)ing ~
~arbeiten *fpl* → Planieren
~bagger *m*, Flachlöffelbagger, Planierlöffelbagger [*früher:* Schichtbagger] = skimmer shovel
planieren → einebnen
Planieren *n* [*mit Straßenhobel*] = planing
~, Ein~, Planier(ungs)arbeiten *fpl*, Planierung *f* = grading (work), level(l)ing ~, ~ operations
Planier|fräse *f* → Vielzweck-Schnecke
~gerät *n*, Flachbagger *m* = planer (Brit.); level(l)er
~-(Gleis)Kettengerät *n* → → Bulldozer
~-(Gleis)Kettengerät *n* mit Schwenkschild → Trassenschäler
~kübel *m* → Flachlöffel *m*
~kübelausrüstung *f* → Flachlöffelausrüstung
~löffel *m* → Flachlöffel
~löffelausrüstung *f* → Flachlöffelausrüstung
~löffelbagger *m* → Planierbagger
~masse *f* → Nivelliermasse
~-Moorraupe *f* → Moorraupe
~pflug *m*, Kippenräumer *m*, Gleis-Planierpflug, Einebnungspflug, (Abraum)Kippenpflug = spreaderditcher
~raupe *f* → → Bulldozer *m*
~reifenschlepper *m* → → Bulldozer *m*
~schar *f* → Hobelmesser *n*
~schaufel *f* → Hobelmesser *n*
~schaufel *f* → Flachlöffel *m*
~schaufelausrüstung *f* → Flachlöffelausrüstung
~schild *n* → → Bulldozer
~schildzylinder *m* = bulldozer cylinder
(~)Schleppe *f* → Abziehschleppe
~schlepper *m* → Bulldozer *m*

(Planier-)Schneckenfräse — plastischer Grenzzustand 726

(~-)Schneckenfräse f → Vielzweck-Schnecke
~traktor m → Bulldozer
~trecker m → Bulldozer
~- und -Ladegerät n → (Front-)Ladeschaufel f
Planierung f → Planieren n
Planier(ungs)arbeiten fpl → Planieren
Planier|werkzeug n = grading tool, level(l)ing ~
~winkel m = grading angle
~zement m [Fehlname] → Nivelliermasse f
Planimeter n, Flächenmesser m = planimeter
Planimetrie f, Geometrie der Ebene, ebene Geometrie = planimetry, plane geometry
plankreuzungsfreie Kreuzung f, überschneidungsfreie ~, Kreuzungsbauwerk n, niveaufreie Kreuzung = fly over (junction) (Brit.); grade separation structure, grade-separated junction
planmäßig festgelegter Zeitraum m = planned period
planmäßige Absenkung f (⚒) = planned subsidence
planmäßiges Abbrennen n von Bodenüberzügen = controlled burning
Planquadrat n = map square
(Plan)Rätter m = ratter, gyratory screen, reciprocating (wedge wire) screen, griddle
Planrost m = flat grate
~feuerung f = flat grate furnace
Planschbecken n = paddling pool, children's ~
Planschleifmaschine f = face grinder, surface ~, plane ~
Plansieb n, Flachsieb [Siebmaschine mit Gutbewegung in der Siebebene] = flat screen, level ~, horizontal ~
Planum n, Erd~; Planie f [Schweiz] = formation (Brit.); (sub)grade (US)
~breite f, Erd~ = formation width (Brit.); (sub)grade ~ (US)
~fertiger m → Erd~
~herstellung f → Erd~

~höhe f = formation level (Brit.); (sub)grade ~, grade line ~ (US)
~material n = (sub)grade material
~modul m → Bettungszahl f
~schutzschicht f = formation protection layer (Brit.); (sub)grade ~ ~ (US)
Planung f = planning
Planungs|arbeiten fpl = planning work
~behörde f = development board
~gebiet n = planning region
~grundlage f = planning basis
~methode f = planning method
~norm f = planning standard spec(ification)
~rechnung f = mathematical programming
~richtlinie f = planning directive
~- und Bauauftrag m = "design and construct" basic contract
~unterlagen fpl = planning data
~wesen n = planning
Plasmaphysik f = plasma physics
plastifizieren = to plasticize
Plastifizierung f = plasticizing
Plastifizierungsmittel n → Betonverflüssiger m
plastische Biegung f = plastic bending
~ Drehung f = plastic rotation
~ Formänderung f, ~ Verformung, bleibende ~, bildsame ~ = plastic deformation
~ Knickung f = plastic buckling
~ Schwindung f = plastic shrinkage
~ Straßenmarkierung f = plastic white line composition
~ Verformung f → ~ Formänderung
~ Verformung f, ~ Formgebung [Schamotteindustrie] = stiff-mud method
~ Zone f → bildsame ~
plastischer Bereich m → bildsame Zone f
~ feuerfester Formling m = plastic refractory
~ Gleichgewichtszustand m = state of plastic equilibrium
~ Grenzzustand m = plastic limit state

plastischer Markierungsstoff — Plattenbewehrung

~ **Markierungsstoff** m = plastic marking material
~ **Markierungsstreifen** m, Kunststoff-Markierungsstreifen = plastic roadline, ~ strip
~ **Ton** m = plastic clay
~ **Zustand** m → bildsamer ~
plastisches Bindemittel n, plastischer Binder m = plastic binder
~ **Fließen** n [*Bodenmechanik*] = plastic flow
~ **Gebirge** n (⚒) = plastic strata, ~ ground
~ **Gelenk** n = plastic hinge
~ **Gleichgewicht** n = plastic equilibrium
~ **Moment** n = plastic moment
~ **Verfahren** n = plastic method
Plastizität f, Bildsamkeit f = plasticity
Plastizitäts|bereich m → bildsame Zone f
~**grenze** f → → Konsistenzgrenzen fpl nach Atterberg
~**index** m, Bildsamkeitsindex, Plastizitätszahl f, Bildsamkeitszahl, Bildsamkeit f, P. I., w_{fa} = plasticity index, index of plasticity
~**messer** m = plasticimeter
~**steigerung** f = plasticizing, plasticization
~**theorie** f, Bildsamkeitstheorie = theory of plasticity
~**theorie in der Baustatik** = plastic methods of structural analysis
~**zahl** f → Bildsamkeitszahl
~**zustand** m → bildsamer Zustand
Plastosphäre f, plastische Zone f, Fließzone (Geol.) = zone of flowage
Platinenwalzwerk n = sheet bar mill
Plattbohrer m, Nagelort m = bradawl
Platte f = plate
~, Bau~ = building board
~, Belag~ [*früher: Fliese f*] = tile
~ [*Ebenes Flächentragwerk. Es ist durch normal zur Mittelebene wirkende Belastung belastet und dadurch auf Biegung und Drillung beansprucht*] = (flat) slab
~ → angelenkte ~ [*Gliederband*]

~ **für schiefe Brücke** = skew bridge slab
Platten|abziehband n, Plattenabzugband = drawing (plate) apron conveyor, ~ (~) ~ conveyer
~**-Abzugkanal** m, Platten-Durchlaß m, gedeckter Abzugkanal, gedeckter Durchlaß = slab culvert
~**armierung** f → Plattenbewehrung
~**auskleidung** f → Plattenverkleidung
Plattenbalken m → ~träger m
~**brücke** f = slab-and-beam bridge, T-beam ~
~**decke** f = slab-and-beam floor, T-beam ~
~**querschnitt** m = T-beam cross-section
~**(träger)** m = T-beam
Plattenband n → ~förderer m
~**aufgeber** m → Aufgabeplattenband
~**beschicker** m → Aufgabeplattenband n
~**(förderer)** n, (m), Blechgliederband, Gliederband(förderer) = (plate) apron conveyor, (~) ~ conveyer
~**speiser** m → Aufgabeplattenband n
Plattenbasalt m = laminated basalt
Plattenbau m, Beton~ = slab construction, concrete ~
~**art** f, Großtafelbauweise f, Plattenbauweise f = slab method
~**gebäude** n [*Außen- und Innenwände des Gebäudes werden aus geschoßhohen, mit Wärme- und Schalldämmschichten vorgefertigten Platten aus Gips, Beton, Anhydrit usw. zusammengefügt*] = slab block
~**weise** f → Plattenbauart f
~**werk** n = slab structure
Platten|bekleidung f → Plattenverkleidung
~**belag** m, Plattendecke f [*Straßenbau*] = slab pavement, ~ surfacing; ~ paving (US)
~**belastungsversuch** m → Lastplattenversuch
~**bewehrung** f, Plattenarmierung, Platten(stahl)einlagen fpl = slab reinforcement

Plattenbiegung — Plattenpresse

~biegung *f*, Plattendurchbiegung = slab deflection (Brit.); ~ deflexion (US)
~boden *m*, Plattenfußboden = tile flooring
~brücke *f* = slab bridge
~decke *f*, Plattenbelag *m* [*Straßenbau*] = slab pavement, ~ surfacing
~decke *f* [*Hochbau*] = slab floor
~dicke *f* = slab thickness
~druckfestigkeit *f* = slab compressive strength
~druckprüfung *f* → Lastplattenversuch *m*
~druckversuch *m* → Lastplattenversuch
~(durch)biegung *f* = slab deflection (Brit.); ~ deflexion (US)
~-Durchlaß *m*, Platten-Abzugkanal *m*, gedeckter Durchlaß, gedeckter Abzugkanal = slab culvert
~ecke *f* [*Betonplatte*] = corner of slab
~einlagen *fpl* → Plattenbewehrung *f*
~element *n*, Fertigbetonplatte *f* = precast slab
~ende *n* = slab end
~fabrik *f* → Plattenwerk *f*
~fertiger *m* → Plattenpresse
~festigkeit *f* = slab strength
~form *f* = slab mould (Brit.); ~ mold (US)
plattenförmige (Gesteins)Absonderung *f* = platy structure
Platten|frischling *m* = green slab, ~ panel
~fuge *f* = slab joint
Plattenfundament *n*, Plattengründung *f*, Plattenfundation *f*, Plattengründungskörper *m*, Plattenfundamentkörper, Gründungsplatte *f*, Grundplatte, Fundamentplatte, Fundationsplatte, (Stahl)Beton~, großflächiges (Stahl)Betonfundament, bewehrte Fundamentplatte = (reinforced-concrete) mat, (~) raft, mat foundation, raft foundation [*A single, heavy, reinforced concrete slab covering the entire area of the foundation and supporting all the structure loads. A foundation of this type is often called a "floating foundation", although this term is commonly applied to all types of spread foundations*]
~ auf Erdreich = mat foundation supported directly on the soil, raft ~ ~ ~ ~ ~ ~
~ ~ Pfählen = mat foundation supported on piles, raft ~ ~ ~ ~, reinforced concrete mat foundation ~ ~ ~, reinforced concrete raft foundation ~ ~ ~
~körper *m* → Plattenfundament *n*
Platten|fundation *f* → Plattenfundament *n*
~(fuß)boden *m* = tile flooring
~glimmer *m* = book mica, sheet mica
~greifer *m* = slab grab(bing) equipment
~gründung *f* → Plattenfundament *n*
~gründungskörper *m* → Plattenfundament *n*
~heben *n* → Beton~
~heizkörper *m* = metal radiant panel
~kalk(stein) *m*, Kalkschiefer *m* = laminated limestone, platy ~, slabby ~
~kohle *f* = laminated sapropelic coal containing numeros animal remains
~lager *n* [*für Hoch- und Tiefbauten*] = plate bearing
~leger *m* [*Straßenbau*] = street mason [*sometimes wrongly referred to as "mason flagger"*]
~-Luftverteiler *m* = plaque
~moment *n* = slab moment
~oberseite *f* [*Betonplatte*] = top of slab
~papier *n*, Beton~ = slab paper
Plattenpfeiler(stau)mauer *f* → Plattensperre *f*
~ mit auf Pfeilerverbreiterungen liegender Platte = simple slab deck dam
~ ~ Pfeilervorkragungen = cantilever deck dam
~ ~ ~ und Einhängeplatte = cantilever deck dam with suspended slab
Plattenpresse *f*, Plattenfertiger *m* = slab press

Plattenpresse — Plattenwaschmaschine

~, Fliesenpresse, Plattenfertiger *m*, Fliesenfertiger = tile press
~ nach Freyssint → (Preß)Kissen
Platten|pumpen *n*, Beton~ = slab pumping, mud-pumping, subgrade erosion, pavement (slab) pumping, pumping of pavements [*The ejection of mixtures of water and subgrade or subbase material along transverse or longitudinal joints and cracks, and along pavement edges caused by downward slab movement activated by the passage of loads over the pavement*]
~querschnitt *m* = slab cross section
~reibung *f*, Beton~, Bodenreibung = (frictional) restraint of (the) subgrade, subgrade restraint
~rüttler *m* → Plattenvibrator *m*
~schalung *f* = slab formwork
~schleifautomat *m*, Kunststein~ = auto(matic) slab and tile grinder, ~ ~ ~ ~ grinding machine
~schleifautomat *m*, Fliesenschleifautomat = auto(matic) tile grinder, ~ ~ grinding machine
~schleifmaschine *f* = slab and tile grinder, ~ ~ ~ grinding machine
~schneidemaschine *f*, Fliesentrennmaschine, Plattentrennmaschine, Fliesenschneidemaschine, Plattenschneider *m*, Fliesenschneider = tile cutting machine, ~ cutter
~schuh *m* → Wechselschuh
~siebboden *m* [*aus Feinblech*] = (pierced metal) sheet-type screening medium
~siebboden *m* [*aus Grobblech*] = (pierced metal) plate-type screening medium
~spannung *f* = slab stress
~sperre *f*, Plattenpfeiler(stau)mauer *f*, Pfeilerplattensperre, Pfeilerplatten(stau)mauer, Plattenstaumauer = slab and buttress dam, flat slab ~
~(stahl)einlagen *fpl* → Plattenbewehrung *f*
~stampfmaschine *f* [*Betonsteinindustrie*] = plate-tamping machine
~staumauer *f* → Plattensperre *f*
~steg *m* [*Gleiskette*] = grouser
~steifigkeit *f* = slab rigidity
~stein *m* → Tonhohlkörper *m*
~stoß *m* = slab joint
~temperatur *f* = temperature of slab, slab temperature
~theorie *f*, Scheibentheorie = plate theory
~theorie *f* (⚒) = theory of fixed beams
~trennmaschine *f* → Plattenschneidemaschine
~-Tunnelofen *m*, Kleintunnelofen = pusher-type kiln
~überstand *m* = slab projection
~- und Bordsteinpresse *f*, Bordstein- und Plattenpresse = slab and kerb press (Brit.); ~ ~ curb ~ (US)
~- und Fliesenpresse *f*, Fliesen- und Plattenpresse = slab and tile press, tile and slab ~
~unterseite *f* [*Betonplatte*] = bottom of slab
~verblendmauer *f* = panel wall [*A wall faced with (stone) facing slabs*]
~verblendung *f* = (stone) slab facing
~verdampfer *m* = plate evaporator
~verdichter *m* → Plattenvibrator *m*
~verkleidung *f*, Plattenauskleidung, Plattenbekleidung = tile lining
~verkleidung *f* = slab facing
~verschiebung *f* = slab displacement
~versuch *m* → Last~
~vibrator *m*, Plattenrüttler *m*, Plattenverdichter *m* = pan vibrator, plate ~, slab ~, vibrating plate compactor, vibrating pan compactor, vibrating slab compactor, vibratory base plate compactor, vibratory base pan compactor, vibratory base slab compactor
~vorschub *m*, Fliesenvorschub [*Spachtelmaschine*] = tile feed
~wand *f* **mit Nut und Feder** = slab wall tongued and grooved
~waschmaschine *f*, Auswaschmaschine für Waschbetonplatten = slab washer, panel ~, ~ washing machine

~werk *n*, Fliesenwerk, Plattenfabrik *f*, Fliesenfabrik = tile making factory, ~ ~ works
~ziegel *m* = flat tile
plattes Korn *n*, flaches ~, scherbiges ~ [*Zuschlag(stoff)*] = flaky grain
Plattform *f*, Grundplatte *f*, Deck *n* [*Bagger; Kran*] = deck
~ = platform
~-Anhänger *m* = platform trailer
~-Aufl(i)eger *m*, Plattform-Halbanhänger = platform semitrailer
~(bau)aufzug *m* = platform hoist
~-Halbanhänger *m*, Plattform-Aufl(i)eger = platform semitrailer
~-Hochhubkarre(n) *f*, *(m)*, Plattform-Hochhubwagen *m* = high-lift platform truck
~-Hubkarre(n) *f*, *(m)*, Plattform-Hubwagen *m* = lift platform truck
~karre(n) *f*, *(m)* = platform truck
~kipper *m* [*Waggonkipper*] = gravity tip
~-Last(kraft)wagen *m*, Plattform-Lkw *m* = platform truck (US); ~ lorry (Brit.)
~-Lkw *m*, Plattform-Last(kraft)wagen *m* = platform lorry (Brit.); ~ truck (US)
~-Niederhubkarre(n) *f*, *(m)*, Plattform-Niederhubwagen *m* = low-lift platform truck
~waage *f* = platform scale
Platthacke *f*, Breithaue *f*, Breithacke = mattock
plattige Absonderung *f* (Geol.) = laminated jointing
plattiger Habitus *m* [*Kristall*] = bladed habit, lath-like ~
plattiges Gestein *n* = platy rock
Plattigkeit *f* [*gebrochenes Gestein*] = slabbiness
Plattnagel *m* = clout nail
Platz *m* [*in einer Stadt*] = square
~belader *m* = stockyard transporter
~regen *m* → Regenguß
Pleochroismus *m* = pleochroism
pleochroitischer Hof *m* (Min.) = pleochroic halos

Pleuel *n*, Pleuelstange *f* = connecting rod
~lager *n* = connecting rod bearing, big end ~
~lagerzapfen *m* = crank journal
~stange *f*, Pleuel *n* = connecting rod
~stange *f*, Aufhängestange [*Klappenwehr mit unterer Achse*] = suspension link
Plexiglas *n* = Plexiglass
Pliester|geflecht *n* = Pliester lath(ing)
~nagel *m* = Pliester nail for stucco
~-(Schilf)Rohrgewebe *n* = Pliester reed lath(ing)
Pliozänton *m* = pliocene clay
Plombe *f* → Pfropfen *m*
plötzliche Belastung *f* = sudden loading, instantaneous ~
~ Entlösung *f* [*von Gas aus Erdöl*] = flush liberation
Plunger(kolben) *m*, Tauchkolben *m* = pump plunger
Plus-Minus-Waage *f* = over-and-under scale
plutonisch, abyssisch = plutonic, abyssal
plutonischer Pfropfen *m* = plutonic plug
plutonisches Gestein *n*, Tiefengestein, Plutonit *m*, subkrustales Gestein, älteres Eruptivgestein, abyssisches Gestein, Intrusivgestein = intrusive rock, (igneous) plutonic ~, abyssal ~, abysmal ~
Plutonit *m* → plutonisches Gestein *n*
PM-Binder *m* → Mauerbinder
PMz *m*, Porenziegel = porous brick
Pneu|-Anhängewalze *f*, Pneu-Schleppwalze = towed pneumatic-tired roller (US); ~ pneumatic-tyred ~ (Brit.)
~bagger *m* = pneumatic-tyred excavator (Brit.); pneumatic-tired ~ (US)
~-Drehkran *m*, Pneu-Schwenkkran, Luftreifen-Drehkran, Luftreifen-Schwenkkran = pneumatic-tyred slewing crane (Brit.); pneumatic-tired ~ ~ (US)
~fahrladegerät *n* → Pneuladeschaufel

~**fahrlademaschine** *f* → Pneuladeschaufel
~**fahrlader** *m* → Pneuladeschaufel
~**fahrzeug** *n*, Luftreifenfahrzeug = pneumatic-tyred vehicle (Brit.); pneumatic-tired ~ (US)
~**gerät** *n*, Luftreifengerät = pneumatic-tyred machine (Brit.); pneumatic-tired ~ (US)
~**hochlader** *m* → Pneuladeschaufel
~**hochladeschaufel** *f* → Pneuladeschaufel
~**hublader** *m* → Pneuladeschaufel
~**-Jumbo** *m*, Luftreifen-Jumbo = pneumatic-tyred (drilling) jumbo (Brit.); pneumatic-tired (~) ~ (US)
~**kran** *m*, Luftreifenkran = pneumatic-tyred crane (Brit.); pneumatic-tired ~ (US)
~**kübelauflader** *m* → Pneuladeschaufel
~**-Lader** *m*, Pneu-Lademaschine *f*, Luftreifen-Lader, Luftreifen-Lademaschine = pneumatic-tyred loader, ~ loading machine (Brit.); pneumatic-tired loader, pneumatic-tired loading machine (US)
~**ladeschaufel** *f*, Pneufahrladegerät *n*, Pneuschaufellader *m*, Pneuhublader, Pneukübelauflader, Pneuhochlader, Pneuhochladeschaufel, Pneufahrlader, Pneufahrlademaschine *f* [*als Oberbegriff für Nur-Hublader, Schürflader und Wurfschaufellader auf Luftreifen*] = pneumatic-tyred shovel loader, ~ loading shovel (Brit.); pneumatic-tired ~ ~ (US)
~**-Löffelbagger** *m*, Luftreifen-Löffelbagger = pneumatic-tyred shovel (Brit.); pneumatic-tired ~ (US)
Pneumatik *f*, ~anlage *f* = pneumatic system
~ = pneumatic engineering
pneumatisch aufgebrachter Bitumen-Emulsions-Mörtel *m*, Bitumen-Spritz-Verfahren *n* = asphalt gunite process, pneumatically applied asphalt emulsion morter (US); asphaltic bitumen gunite process (Brit.)

~ **dosieren** = to batch pneumatically
~ **gesteuerte Siloschnauzen** *fpl* **für das Gestein** = pneumatically controlled aggregate bin gates
pneumatische Abmeßanlage *f* → ~ Dosieranlage
~ **Bedienung** *f* → Druckluftschaltung
~ **Besatzmaschine** *f*, ~ Verdämmmaschine, Druckluft-Verdämmmaschine, Druckluft-Besatzmaschine = pneumatic tamping machine
~ **Betonförderung** *f* → Druckluftbetonförderung
~ **Betonförderanlage** *f*, Druckluft-Betonförderer *m* = pneumatic concrete placer, air ~
~ **(Bohr)Hammerstütze** *f*, Druckluft-Bohrknecht *m* = air leg, pneumatic feed leg for rock drill, compressed-air leg
~ **Bremse** *f*, Druckluftbremse = air brake
~ **Dosieranlage** *f*, ~ Zumeßanlage, ~ Zuteilanlage, ~ Abmeßanlage = pneumatically operated batching plant, ~ ~ proportioning ~, ~ ~ measuring ~
~ **Druckförderanlage** *f*, Druckluftförderanlage = compressed-air conveying system
~ **Entstaubung** *f* = pneumatic dedusting
~ **Fernsteuerung** *f* → Druckluft-Fernsteuerung
~ **Förderrinne** *f* = pneumatic conveyor, ~ conveyer, air slide
~ **Förderschnecke** *f* → pneumatischer Schneckenförderer
~ **Förderung** *f* = pneumatic conveying
~ **Glättkelle** *f*, Druckluft-Glättkelle = air trowel, ~ float, pneumatic ~, compressed-air ~
~ **Gründung** *f* → (Druckluft-)Senkkastengründung
~ **Hammerstütze** *f* → ~ Bohrhammerstütze
~ **Nietmaschine** *f*, Druckluft-Nietmaschine = pneumatic riveter, ~ riveting machine

pneumatische Prallmühle — Pneutraktor 732

~ **Prallmühle** *f*, Strahlmühle, Luftstrahl-Prallzerkleinerer *m* = jet pulverizer, micronizer jet mill, reductionizer
~ **Reinigung** *f* = air cleaning
~ **Saugförderanlage** *f* = pneumatic suction conveying system
~ **Schaltung** *f* → Druckluftschaltung
~ **Schleuse** *f* → → pneumatische Gründung
~ **Schraubmaschine** *f* = air-operated impact wrench
~ **Verdämmaschine** *f* → ~ Besatzmaschine
~ **Wasserdosiervorrichtung** *f* = pneumatic water measuring equipment [*truck mixer*]
~ **Zementförderung** *f*, Preßluft-Zementförderung, Druckluft-Zementförderung = pneumatic cement handling
~ **Zumeßanlage** *f* → ~ Dosieranlage
~ **Zuteilanlage** *f* → ~ Dosieranlage
pneumatischer Abbauhammer *m* → Abbauhammer
~ **Abbruchhammer** *m*, Druckluft-Abbruchhammer = pneumatic demolition pick (hammer), air ~ ~ (~)
~ **Bär** *m*, Druckluftbär = pneumatic (pile) hammer
~ **Betonförderer** *m* → Betondruckluftförderer
~ **Betontransport** *m* → Druckluftbetonförderung
~ **Fernanzeiger** *m* = pneumatic remote indicator
~ **Förderer** *m*, pneumatische Förderanlage *f*, pneumatisches Fördergerät *n* = pneumatic conveyor, ~ transport system, air activator
~ **Förderhammer** *m* → Abbauhammer
~ **Getreideheber** *m* = pneumatic grain handling unit
~ **Höhenförderer** *m* = pneumatic elevating conveyor, ~ ~ conveyer
~ **Pickhammer** *m* → Abbauhammer
~ **Rüttler** *m* → Druckluftrüttler
~ **Schildvortrieb** *m*, Druckluftschildvortrieb = compressed-air shield driving, pneumatic ~ ~

~ **Schneckenförderer** *m*, pneumatische Förderschnecke *f* = pneumatic screw conveyor, ~ ~ conveyer, ~ conveying screw
~ **Senkkasten** *m* → Caisson *m*
~ **Spatenhammer** *m*, Druckluft-Spatenhammer = pneumatic spade, air ~
~ **Vibrator** *m* → Druckluftrüttler *m*
pneumatisches Baustellenwerkzeug *n* → Baustellen-Druckluftwerkzeug
~ **Betonfördergerät** *n*, pneumatischer Betonförderer *m*, Druckluft-Betonfördergerät, Druckluft-Betonförderer = pneumatic concrete placer, ~ ~ placing machine, ~ ~ handling ~, air placer
~ **Fördergerät** *n* → pneumatischer Förderer
(~) **Förderrohr** *n* → Gebläseluft-Förderrohr
~ **Hebezeug** *n*, Druckluft-Hebezeug = pneumatic elevating gear
~ **Kommandosystem** *n* = pneumatic control system, ~ controls
~ **Ladegerät** *n*, Druckluftlader *m* = pneumatic loader, ~ loading machine
pneumatisch-hydraulisch = pneumohydraulic
pneumatisch-hydraulischer Hebebock *m*, pneumatisch-hydraulische Winde *f* = air hydraulic jack
Pneumatolyse *f* = pneumatolysis
Pneu|-Mobilkran *m*, Luftreifen-Mobilkran = pneumatic-tyred mobile crane (Brit.); pneumatic-tired ~ ~ (US)
~**rad** *n* = pneumatic-tired wheel (US); pneumatic-tyred ~ (Brit.)
~**schaufellader** *m* → Pneuladeschaufel
~**schlepper** *m*, Pneutraktor *m*, Pneutrecker, Luftreifenschlepper, Luftreifentraktor, Luftreifentrecker = pneumatic-tired tractor (US); pneumatic-tyred ~ (Brit.)
~**-Schleppwalze** *f* → Pneu-Anhängewalze
~**-Schwenkkran** *m* → Pneu-Drehkran
~**traktor** *m* → Pneuschlepper

~trecker m → Pneuschlepper
~walze f → → Walze
Poch|mühle f → Pochwerk
~sand m, Splittsand, Quetschsand = stamp sand
~stempel m = stamp
~werk n, Pochmühle f = stamp battery, gravity stamp
~werkschuh m = stamp battery shoe, gravity stamp ~
Podest n, m, Absatz m, Treppen~ = landing [*Resting space usually arranged at the top of any flight of stairs*]
~balken(träger) m = landing joist
~breite f, Treppen~ = landing width
~länge f, Treppen~ = landing length
~platte f, Treppen~ = landing slab
Podsol|boden m, Bleicherde f = podzol
~horizont m = podzolic horizon
Podsolierung f, Auswaschung des Bodens = podzolization
podsoliger Moorboden m = podzolic peat
Poetsch-Verfahren n = Poetsch freezing process
Poirée-Nadelwehr n = Poirée dam, ~ weir
Poise f = poise
polare Festlandsluft f = polar continental air mass
~ Strömung f = polar current
polares Trägheitsmoment n = polar moment of inertia
Polarisationsmikroskop n = petrographic microscope, polarizing ~
polarisiertes Auflicht n = polarized reflected light
Polar|koordinate f = polar coordinate
~-Koordinatensystem n = polar coordinate system
~planimeter n = polar planimeter
~schichtung f, indirekte Wärmeschichtung = polar temperature lamination, indirect ~ ~
~winkel m [*Mathematik*] = vectorial angle, polar ~
Polder m, Ko(o)g m = polder
~mühle f = wind mill for drainage
Polhorn n [*Kreuzbandscheider*] = horn-shaped pole

Polier m = foreman
polierbar, polierfähig = polishable
polierbarer dichter Kalkstein m → technischer Marmor m
Polieren n = polishing
~ der Mineralkörner an der Straßenoberfläche durch den Verkehr = polishing of the road stones by traffic
polierfähig, polierbar = polishable
Polier|fähigkeit f [*Steinbaustoff einer Straßenoberfläche*] = resistance to polishing
~maschine f = polishing machine
~stange f und Stopfbüchse f [*einer Erdölpumpsonde*] = polish rod and stuffing box
~stein m, Abziehstein [*zum Feinnacharbeiten (= Polieren) von Werkzeugschneiden*] = oilstone
polizeiliches Kennzeichen n = license plate
Polizei|posten m [*z. B. an einer Autobahn*] = police post
~streife f = police patrol
Poller m, Hafen~, Haltepfahl m = bollard
~fundament n = bollard foundation
~zug m = bollard pull
polnischer Verband m, gotischer ~ = double Flemish bond
Poloplatz m = polo ground
Polschuh m [*Tiefbohrtechnik*] = (field) pole piece, pole shoe
Polster|gründung f [*Sie ermöglicht eine gleichmäßige Setzung des Bauwerkes und wird bei unterschiedlichem Baugrund vorgesehen. Ein Polster aus nachgiebigem Boden wird über dem tragfähigerem Teil des Untergrundes eingebaut, so daß die Bauwerksetzung gleichmäßig ist*] = wedge-type foundation of yielding soil
~tür f = padded door
Polstrahl m, Pollinie f, Seilstrahl, Seillinie = funicular line, polar ~
polumschaltbarer Motor m = pole-changing motor
Polumschalter m = polarity reversing switch

Polyester|beton m [*Wird der Zementleim im Beton durch ungesättigtes Polyesterharz ersetzt, so erhält man den Gießharzbeton oder, genauer gesagt, den Polyesterbeton*] = polyester (resin) concrete
~**cord** n = polyester cord
~**faser** f = polyester fibre (Brit.); ~ fiber (US)
~**harz** n = polyester resin
~**harzbeschichtung** f = polyester resin coating
~**tor** n = polyester gate
Polygon n im Gegen-Uhrzeigersinn = counter-clockwise polygon
~ Uhrzeigersinn = clockwise polygon
polygonal, vieleckig = polygonal
polygonaler Profilstahlbogen m [*Streckenausbau*] = polygonal steel arch
Polygonal|-Ring m = polygonal ring
~**verband** m, Vieleckverband = polygonal bond
Polygonhohlstein m = polygonal hollow block
Polygonierung f = traversing
Polygon|mauerwerk n = polygonal (random rubble) masonry
~**winkel** m = traverse angle
~**(zug)** n, (m), Standlinienzug = traverse
polymerer Viskositätsindex-Verbesserer m = polymeric viscosity modifier
Polymerisationsturm m = cat poly reactor
polimerisieren = to harden by polymerization
Polymeröl n = polymer modified oil
Polypgreifer m, Polypgreif(er)korb m = grabs (US); multi-blade grab (Brit.)
Polystyrol n = polystyrene
~**-Hartschaumplatte** f, Hartschaumplatte aus Polystyrol = expanded polystyrene board
~**-Hartschaumstoffplatte** f, Hartschaumstoffplatte aus Polystyrol = expanded polystyrene felt covered panel

Polysulfid|-Gummidichtungsmittel n = polysulphide rubber sealant (Brit.); polysulfide ~ ~ (US)
~**-Kautschuk** m, Thiokol n = polysuphide liquid polymer (Brit.); polysulfide ~ ~ (US)
Polyurethan|-Hartschaumschichtstoffbauplatte f = polyurethane rigid foam laminated building board
~**schaum** m = expanded polyurethane, polyurethane foam
Poncelet-Meßwehr n, Poncelet-Überfall m = Poncelet measuring weir
Ponton m = pontoon
~ mit Stahlbrücke = floating steel crane
~**-Betonieranlage** f, schwimmende Betonieranlage auf Ponton = pontoon-mounted concreting plant
~**brücke** f = pontoon bridge
~**drehbrücke** f = pontoon swing bridge
~**kran** m = pontoon crane
~**ramme** f = pontoon (pile) driving plant, ~ piling ~
Pore f, Hohlraum m = void, pore, interstice
Poren|art f = type of pore
~**auspressung** f → Poreninjektion f
Porenbeton m, Zellenbeton, aufgeblähter Beton, Blähbeton = cellular(-expanded) concrete, aerated ~
~**bauplatte** f = cellular(-expanded) concrete slab, aerated ~ ~
~**stein** m = cellular(-expanded) concrete block, aerated ~ ~
~**werk** n, Zellenbetonwerk = aerated concrete works, cellular(-expanded) ~
Poren|dichtung f = pore sealing
~**einpressung** f → Poreninjektion f
~**flüssigkeit** f = pore liquid
~**füller** m, porenschließender Sperrstoff m, Porenfüllstoff = void filler
~**füllung** f = filling of voids
~**gefüge** n = pore-structure
~**gehalt** m [*Gesamtinhalt der Hohlräume je Raumeinheit des Bodens*] = voids content, volume of voids, volume of pores, pores content

Porengips — Portal(verband)

~gips *m* = cellular(-expanded) gypsum, aerated ~
~gipsplatte *f* = aerated (gypsum) plasterboard
~injektion *f*, Porenauspressung *f*, Porenverpressung, Poreneinpressung [*Injektion in Lockergesteinen, wobei je nach den örtlichen Verhältnissen Suspensionen, Lösungen oder Bitumenemulsionen eingepreßt werden*] = soil injection
~kalkstein *m* = porous limestone
~mörtel *m*, Zellenmörtel, aufgeblähter Mörtel, Blähmörtel = cellular (-expanded) mortar, aerated ~
~prozentsatz *m* = percentage of voids, ~ ~ pores
~querschnitt *m* = pore cross-section
~-Saugwasser *n* → Kapillarwasser
~saugwirkung *f*, Rohrsaugkraft *f*, Kapillarität *f*, Kapillarattraktion *f* = capillarity, capillary attraction
porenschließender Sperrstoff *m*, Porenfüller *m*, Porenfüllstoff = void filler
Porenschluß *m*, Oberflächenabsieg(e)lung *f*, Deckenversieg(e)lung [*Straße*] = surface-dressing (treatment), seal(ing) (coat) [*deprecated: flush coat, squegee coat*]
~emulsion *f* = seal coat emulsion
~mittel *n*, Absieg(e)lungsmittel, Versieg(e)lungsmittel [*Straßenbau*] = surface sealant
~schicht *f*, Absieg(e)lungsschicht, Versieg(e)lungsschicht [*Straßenbau*] = seal(ing) coat
Poren|-Spannbeton *m* = prestressed cellular concrete
~system *n* = void system
~verpressung *f* → Poreninjektion *f*
~verteilung *f* = pore distribution
Porenwasser *n* = pore water
~druck *m*, Hohlraumdruck = pore (water) pressure
~druckhöhe *f* = pore (water) head
~druckmeßdose *f* = pore (water) pressure cell
~spannung *f* = neutral stress
~überdruck *m* = hydrostatic excess pressure, excess pore (water) ~

Poren|zement *m* (Geol.) = pore matrix, ~ cement(ing material)
~ziegel *m*, PMz = porous brick
~ziffer *f* [*Verhältnis zwischen dem Rauminhalt der Poren und demjenigen der Körner des Bodens oder Gesteins*] = pore ratio, void ~
~zwickelwasser *n* = contact moisture, internal water
porig, porös = porous
Porigkeit *f*, Porosität *f* = porosity
poröse Lava *f* → Lavaschlacke *f*
~ ~, Lavakies *m* = foamed lava gravel
~ Masse *f* [*Luftrutsche*] = porous medium
Porosität *f*, Porigkeit *f* = porosity
Porphyr *m* = porphyry
Porphyrit *m* = porphyrite
porphyroblastischer Verband *m* (Geol.) = porphyroblastic texture
Porphyrtuff *m* = porphyritic tuff
Portal *n*, Eingang *m*, Mund *m* [*Stollen; Tunnel*] = entrance, portal
~ → Portalverband *m*
~ = portal
~, Kran~ = portal, gantry
~bagger *m*, Torbagger = portal bucket ladder excavator
~-Bohrwagen *m* = portal-type wagon drill
~gebäude *n* [*Tunnel*] = portal building
~kran *m*, Torkran, Voll~ = portal (jib) crane, gantry ~ [*A jib crane carried on a fourlegged portal. The portal is built to run on rails*]
~pfeiler *m* [*Brücke*] = portal pier
~rahmen *m* = portal (frame)
~rahmenbauwerk *n* = portal (frame) structure
~rahmenbrücke *f* = portal (frame) bridge
~rahmengebäude *n* = portal (frame) building
~strecke *f* [*Tunnel- und Stollenbau*] = end section
~turmdrehkran *m* = portal tower crane
~(verband) *n*, (*m*), Endrahmen *m* [*Brückenbau*] = portal bracing

Portalwagen — Prallmühle

~wagen *m* → Torladewagen
Portikus *m* = portico
Portlandstein *m* = Portland stone [*An oolitic freestone*]
Portlandzement *m*, PZ = Portland cement
~ **mit sehr geringer Abbindewärme** = low heat (of hydration) cement, slow hardening cement; Portland cement Type IV (US)
~**klinker** *m* = Portland cement clinker
~**mörtel** *m* = Portland cement mortar
~**putz** *m*, Portlandzementmörtelbelag *m* [*Siehe Anmerkung unter „Putz"*] = 1. Portland cement plaster, ~ ~ plaster(ing) finish; 2. ~ ~ rendering (finish)
~**-Puzzolangemisch** *n* = Portland-pozzolan cement
Porzellanton *m*, Porzellanerde *f*, (Roh-)Kaolin *m*, Weißerde *f* = kaolin, china clay, porcelain clay, porcelain earth
Position *f* → → Angebot *n*
positiver artesischer Brunnen *m*, Steigbrunnen = positive artesian well, flowing ~
positives Momentvermögen *n* = positive moment capacity
Postkabel *n* = post-office telephone cable
Postkarten|methode *f* → Befragungsmethode durch Postkarten
~**zählung** *f* → Befragungsmethode *f* durch Postkarten
Potamologie *f* = potamology
Potential *n* = potential
~ [*Bodenmechanik*] = velocity potential
~**elektrode** *f*, Meßelektrode, Spannungsmeßsonde *f* [*geophysikalisches (Aufschluß)Verfahren*] = potential electrode
~**fläche** *f*, Niveaufläche = equipotential surface
~**funktion** *f* = potential function
~**gefälle** *n*, Potentialunterschied = potential drop, ~ difference

~**geschwindigkeit** *f* = potential velocity
~**kraft** *f* = potential force
~**linie** *f* = potential line
~**strömung** *f* = irrotational flow
~**unterschied** *m*, Potentialgefälle *n* = potential drop, ~ difference
~**verfahren** *n* = potential-drop (-ratio) method, ~ exploration, ~ prospecting
potentielle Energie *f* → Energie der Lage
Pottasche *f* → Kaliumkorbonat *n*
Prachtstraße *f* = boulevard
Prägeplatte *f*, Hartplatte mit eingepreßtem Muster = embossed (hard-)board
Prahm *m* = johnboat (US)
Praker *m* → Schlagbrett *n*
praktische (Förder)Leistung *f* [*Eimerkettenbagger*] = actual output
~ **Strömungslehre** *f* = engineering hydraulics
prall, abschüssig, schroff, steil abstürzend, abgerissen [*Küste*] = steep, bold
Prallblech *n*, Abweisungsblech [*im Fangkessel beim Druckluftbetonförderer*] = (central) baffle
~ → Ablenkblech
Prallbrechen *n* = impact breaking
Prallbrecher *m* **mit einer Schlagwalze**, Einwalzen-Prallbrecher, Einrotorenprallbrecher, Prallbrecher mit einem Rotor = single impeller impact breaker, ~ ~ ~ crusher
~ **zwei Rotoren** → Doppel-Rotoren-Prallbrecher
Prallmahlung *f* = impact pulverizing
Prallmühle *f* [*Das Mahlgut wird mit großer Geschwindigkeit gegen Prallplatten geschleudert und durch den Anprall feinzerkleinert. Die erforderliche Geschwindigkeit wird dem Mahlgut 1. mechanisch durch schnell laufende Rotoren (oder einen Rotor), oder 2. pneumatisch durch einen Luft- oder Gasstrom aufgezwungen*] = pulverator, impactor, (swinghammer) pulverizer, impact mill

Prallmühle mit senkrechter Achse — Preßbeton(bohr)pfahl

~ mit senkrechter Achse = vertical impact pulverizer, VIP
Prall|platte *f* → Ablenkplatte
~**platte** *f* [*zur Entstaubung und Aufbereitung allerfeinsten Sandkorns*] = impact separator
~**tellermühle** *f* = turbo miller
~**zerkleinerung** *f* = comminution by impact, reduction ~ ~
Prämienlohnsystem *n*, Prämiengedingesystem [*Tunnel- und Stollenbau*] = premium system
präparierter Straßenteer *m* = road tar produced by blending with distillate fractions
Pratt-Träger *m* = Pratt truss, Linville ~, Whipple-Murphy ~, N-truss
Pratzenkran *m* = claw crane
Präzipitatgestein *n* → chemisches Sediment(gestein)
Präzisions|-Anflugradar *n* = precision approach radar, PAR, ground control approach, GCA
~**aushublöffel** *m* = square hole bucket
~**drehbank** *f* = precision lathe
~**instrument** *n* = precision instrument
~**kette** *f* = precision chain
~**nivellement** *n*, Feinnivellement = precise levelling, precision ~
~**-Schmiedestück** *n* = precision forging
PREFLEX-Träger *m* = preflex girder
Preis|absprache *f* **unter Bietern** = collusion among bidders (or tenderers), price fixing, combining to eliminate competition
~**gefüge** *n* = price structure
~**gericht** *n* = jury
~**kalkulation** *f* = pricing
~**niveau** *n* = price level
~**verzeichnis** *n* = schedule of prices
Prell|bock *m* = bunter (US)
~**pfahl** *m* = mooring pile
~**scheibe** *f* [*Pfahlramme*] = bottom head
~**schlag** *m* = rebound blow
~**stein** *m*, Leitstein, Abweisstein, Radabweiser *m* = spur post
~**wand** *f*, Leitwand = baffle [*Deflector of wood, metal, or masonry placed in flowing liquid, to divert, guide, or agitate the flow of such liquid*]
Preload-Maschine *f* = Preload (prestressing) machine, wire-winding ~
Prepaktbeton *m*, vorgepackter Beton, Schlämmbeton, Skelettbeton = intruded (aggregate) concrete, grout-intruded ~, prepacked ~, Prepakt ~
Presenning *f* → Persenning
Presse *f*, Spann~ = jack, puller
~ → Hebebock *m*
~ [*Maschine zur Verformung von Werkstoffen*] = press
~**kabine** *f* [*Stadion*] = press box
pressen eines Kabels in Kreisform = to compact a cable into a circular cross section
Pressen|turm *m* = jack-handling tower
~**zylinder** *m*, Spann~ = jack cylinder, puller ~
~**zylinderhub** *m*, Spann~ = jack cylinder range, puller ~ ~
Pressolit *m* → Strangpreßbeton *m*
~**-Fertigteil** *m, n* = extruded concrete (precast) unit
~**-Profil** *n* = extruded concrete section
~**-Strangpresse** *f* = concrete extrusion press
~**träger** *m* = extruded concrete girder
Pressung *f*, Zusammendrückung = compression
Pressungsumwandlung *f* → Dynamometamorphose *f*
Preßabhebeformmaschine *f* = press mo(u)lding machine with lift-off
Preßbeton *m* [*mit Druckluft, 6–10 atü, wird Beton in Schalungen und Spalten, Risse u. ä. gepreßt. Hauptsächlich bei beschädigten Bauwerken verwendet*] = concrete placed by compressed air
~**(bohr)pfahl** *m* = pressure pile
~**(bohr)pfahl** *m* **Grün & Bilfinger**, Grün & Bilfinger-Preßbeton(bohr)pfahl = Grün & Bilfinger pressure pile

Preßbeton(bohr)pfahl — Preventergarnitur 738

~(bohr)pfahl *m* System Brechtel, Brechtel-Preßbeton(bohr)pfahl = Brechtel pressure pile
~rohr *n* = pressed concrete pipe
Preß|blech *n* = pressed (sheet) steel
~braunkohle *f*, Braunkohlenbrikett *n* = brown coal briquette
~dachziegel *m* = pressed (roof) tile
~falzziegel *m*, Preßfalzstein *m* = pressed gutter tile
~flüssigkeit *f* → Druckflüssigkeit
~fuge *f* = joint which does not allow expansion
~gasförderverfahren *n*, Gastrieb-Verfahren, Druckgasförderverfahren = gas-lift
~glas *n* = pressed glass
~holz *n* = pressed wood
~kammer *f*, Strang~ [*einer Strangpresse*] = moulding chamber (Brit.); molding ~ (US)
(~)Kissen *n*, Plattenpresse *f* nach Freyssinet, (Freyssinet-)Kapselpresse, Druckkissen = (Freyssinettype) flat jack
~kohle *f* → Brikett *n*
~kohlenfabrik *f* → Brikettfabrik
~kohlenwerk *n* → Brikettfabrik *f*
~kolben-Hebewerk *n* [*Schiffhebewerk*] = vertical lift with hydraulic rams
~kopf *m* [*Betonrohrpresse*] = concrete pipe press head
~kork *n* = compressed cork
~korkplatte *f* = compressed cork slab, ~ corkboard
Preßluft → Druckluft *f*
~gründung *f* → pneumatische Gründung
~nietung *f*, Druckluftnietung, pneumatische Nietung = (compressed-) air riveting, pneumatic ~
~senkkasten *m* → Caisson *m*
Preßmetell *n* = pressed metal
Preßöl *n*, Hydrauliköl, Drucköl = pressure oil, hydraulic ~
~motor *m* → Druckölmotor
~presse *f* → Druckölpresse
~pumpe *f* → Druckölpumpe
~schmierung *f*, Druckschmierung = pressure lubrication, forced ~
~zufluß *m* → Druckölzufluß
~zylinder *m* → Druckölzylinder
Preßplatte *f*, Druckplatte [*Gleitschalungsfertiger*] = pressure plate, (con)forming ~
(**Preß**)**Schnecke** *f* [*Strangpresse*] = auger
Preß|schraube *f*, Treibschraube = forcing screw
~schweißen *n* = pressure welding
~sitz *m* = press fit
~sperrholz *n*, PSP = high density plywood, superpressed ~
~stahl *m* = pressed steel
~stahltreppe *f* = pressed steel stair(case)
~stahl(treppen)wange *f* = pressed steel string
~stein *m*, Preßziegel *m* = pressed brick
~stempel *m* [*einer Stufenpresse*] = ram
~stroh *n* = compressed straw
~teil *m, n*, Strang~ [*einer Strangpresse*] = molding unit (US); moulding ~ (Brit.)
~teil *m, n* [*Kunststofferzeugnis*] = mo(u)lded article, ~ part, ~ item
~tisch *m* = press table
~torf *m* = pressed peat
~vollholz *n*, PVH = densified wood
~wagen *m* [*einer Stufenpresse*] = cross-head
~wasser *n*, Druckwasser, Hydraulikwasser = hydraulic water, pressure ~, pressurized ~
~ziegel *m*, Preßstein *m* = pressed brick
Preußischblau = Prussian blue, ferrocyanide ~, Chinese ~
preußische Kappe *f*, preußisches Kappengewölbe *n* [*Ein Tonnengewölbe, dessen Kämpferlinien außerhalb des zu überwölbenden Raumes liegen*] = Prussian vault
Preventer *m*, Absperrorgan *n* am Bohrloch, Schieber *m* = (blowout) preventer, cellar control
~garnitur *f* [*Tiefbohrtechnik*] = blowout preventer safety drilling hookup

Primärdampf m → Erstdampf
primärer Niederschlag m = initial precipitation
Primär|(förder)verfahren n [*Erdöl*] = primary recovery method
~löß m → Urlöß
~sprengen n = primary blasting
~sprengstoff m = primary blasting explosive
Primzahl f = prime number
Prinzipskizze f → Schemazeichnung f
Prisma n = prism
~ [*Zementprüfung*] = standard briquette
~ zur Ermitt(e)lung der Biegefestigkeit, Biegekörper m = bending test specimen
~grundfläche f, Prismabasisfläche = base of a prism
~kompaß m, Marschkompaß = prismatic compass
~schlitten m = vee slide, V ~
prismatische Nadelblende f (Min.) = red antimony ore
~ Schale f → Faltwerk n
~ ~ aus Fertigteilen, Fertigteil-Faltwerk n = precast folded plate
~ Stahlbetonschale f → Stahlbetonfaltwerk n
prismatischer Balken m, Balken mit gleichbleibendem Querschnitt = prismatic beam
~ Stahlbetonkörper m = reinforced concrete prism
prismatisches Bett n = prismatic bed
~ Dach n = prismatic roof
~ Faltwerk n, prismatische Schale f, Faltwerk n = prismatic structure, prismatic shell, folded plate (structure), folded slab
~ Mittelstück n [*dreiteiliger Steinwolf*] = central parallel piece
~ Stahlbetonfaltwerk n → Stahlbetonfaltwerk
Prismen-Biegeversuch m, Prismen-Biegeprobe f, Prismen-Biegeprüfung f = prism bending test
~druckfestigkeit f = prismatic beam compressive strength, ~ ~ crushing ~

~festigkeit f = prismatic beam strength
~glas n = prism glass
~kuppel f = prismatic dome
Pritschbläuel m → Schlagbrett n
Pritsche f, Grundplatte f [*z.B. zur Aufnahme eines Drucklufterzeugers und Motors. Diese Pritsche ist dann meistens auf ein Einachsfahrgestell montiert*] = saddle
~ = stake body [*A platform body with removable stakes all-round*]
Pritschenanhänger m = stake body trailer
privater Bahnanschluß m → Anschlußgleis n
Privat|anschlußgleis n → Anschlußgleis
~bahn f = private railway (Brit.); ~ railroad (US)
~garage f = private garage
~gleisanschluß m → Anschlußgleis n
~haus n = privat home
~parkplatz m = private car park, ~ parking place, ~ parking lot
Probe f → ~körper
~, Versuch m, Prüfung f = test
~ auf Reinheit, Absetzprobe = silt content test for fine aggregate, test for silt
~absenkung f [*Grundwasserabsenkung*] = test lowering
~balken m → Biegebalken
~belastung f, Prüfbelastung, Versuchsbelastung = test loading
~betrieb m = test operation
~bohrung f → Schürfbohrung
~brunnen m, Versuchsbrunnen = pilot well, test ~, experimental ~, trial ~
~charge f, Versuchscharge = trial batch, test ~
~druck m, Prüfdruck, Versuchsdruck = test pressure
~(ent)nahme f = sampling
~(ent)nahmeapparat m, Probe(ent)nahmegerät n, Probenehmer m = sample taker, sampler
~fahrt f = trial run
~fuge f, Versuchsfuge, Prüffuge = experimental joint

Probegemisch — Profil 740

~gemisch n, Probemischung f = trial mix(ture)

(~)Kern m, Bohrkern, Kernprobe f [*aus Straßendecke entnommen*] = test core, (drill) ~

~körper m, Prüfkörper, Probe(stück) f, (n), Prüfling m = (test) specimen

~körperform f, Prüfkörperform = test specimen mo(u)ld

~körperherstellung f, Prüfkörperherstellung = test specimen preparation

~last f, Prüflast, Versuchslast = test load

~lauf m [*Maschine*] = trial run

~löffel m, Löffelbohrer m, (offene) Schappe f = sampling spoon, spoonauger

~material n = trial material

~mischung f, Probegemisch n = trial mix(ture)

~montage f [*von Brückenbauteilen im Werk*] = test assembly, trial ~

~nahme f, Probeentnahme = sampling

~nahmegerät n → Probe(ent)nahmeapparat

~nahmestation f = sampling installation

~nehmer m, Prüfgerät n [*bringt eine Ölprobe aus der Tiefe unter dem natürlichen dort unten herrschenden Druck*] = bottom hole sampler

Proben|lagerung f = storage of test specimen(s)

~zylinder m [*eines Bodenprobenehmers*] = liner

Probe|pfahl m, Prüfpfahl, Versuchspfahl = test pile

~prisma n, Prüfprisma = test prism

~rammung f = test (pile) driving

~schüttung f = test fill(ing)

~schweißung f = test welding

~stab m, Prüfstab, Versuchsstab = test bar

~strecke f → Versuchsstrecke

~(stück) f, (n) → Probekörper

~verdichtung f = test compaction

probeweise = on trial

Probewürfel m → (Beton)Würfel

~presse f → Druckpresse

Probe|zeit f = trial period

~zylinder m → Beton~

~zylinder-Prüfmaschine f = cylinder tester, ~ testing machine

Probier|glas n, Reagenzglas = testtube

~hahn m, Prüfhahn, Wasserstand~ [*DIN 33038 und 33039*] = test cock

Proctor|form f, Proctor-Metallzylinder m = Proctor mo(u)ld

~-Gerät n [*zur Messung der Verdichtbarkeit von Böden*] = Proctor-type compaction tester

~kurve f, Proctorlinie f = dry density/moisture content graph, Proctor (compaction) curve

~nadel f, (Proctor'sche) Prüfnadel, Proctor'sche Sonde f, Proctor'sche Plastizitätsnadel = Proctor (penetration) needle, plasticity ~

~-Verfahren n = Proctor method [*A method for determining the optimum moisture content of soil*]

~-Versuch m → AASHO-Versuch

Produktions|anfall m = output

~bohrung f, Gewinnungsbohrung, Produktionssonde f, Gewinnungssonde = recovery well, production ~

~kreuz n, Produktionskopf m, Eruptionskreuz = christmas tree

~maschine f = production machine [*A machine that is normally set up for long runs, for economic production. It can be semi- or fully-automatic, but not necessarily so*]

~packer m = tubing packer

~sonde f → Produktionsbohrung f

(produzierende) Erdgassonde f, fördernde Gasbohrung f = producing gas well

Profanbau m = profane building

Profil n [*Die Umrißlinie beim Querschnitt durch einen Körper*] = profile

~, Bau~ = (structural) section [*Building material formed to a definite cross section but of unspecified*

length. Sections are usually manufactured by a continuous process such as rolling, drawing, extruding or machining. Examples are angles, bars, tubes, battens, wire, cable]
~blech n = formed plate
~draht m, Formdraht = section wire
Profile npl = section material
Profil|eisen n [*Fehlname*] → Trägerstahl
~fertiger m = contour paver [*It forms and compacts bituminous materials in special contours such as: gutters, drainage ditches, median strips, sidewalks, etc., without the use of forms*]
profilgebender Rüttelstampfer m = profiling vibro-stamper
Profilgebung f, Profilieren n, Profilherstellung [*Erdbau*] = trimming, shaping, tru(e)ing
profilgemäß [*Straßenquerschnitt*] = true to cross section, line and level (or grade)
Profil|halbmesser m → hydraulischer Radius
~herstellung f → Profilieren
~höhe f, Bau~ = depth of (structural) section
Profilieren n, Profilgebung f, Profilherstellung [*Erdbau*] = trimming, shaping, tru(e)ing
profilierter Draht m = indented wire
Profil|latte f → Profillehre f
~lehre f, Formlineal n, Profillatte f = template, templet, camber board
~löffel m = purpose-made bucket
~löffel m für Gräben, Trapezlöffel = trapezoidal bucket
profilmäßige Aufnahme f (Geol.) = sectional survey
Profilmundstück n, (Form)Mundstück [*Strangpresse*] = (preformed) die
Profilograph m → Profilzeichner m
Profilradius m → hydraulischer Radius
Profilstahl m, Fassonstahl = section steel
~ → Trägerstahl
~bewehrung f = section reinforcement

~-Rahmen m [*Baumaschine*] = section steel frame
~richtpresse f = straightening press for steel sections
~walzwerk n → Fassonstahlwalzwerk
Profil|stein m, Profilziegel m, Formstein, Formziegel = purpose-made brick
~stufe f = purpose-made step
~taster m, Kontrollrechen m = (sub-)grade tester, ((sub)grade) scratch template
~träger m = section girder
~werk n = section (rolling) mill, structural ~
~zeichner m, Profilograph m = profilometer, profilograph [*an instrument for recording the shape of irregularities in a road surface*]
~ziegel m, Profilstein m, Formziegel, Formstein = purpose-made brick
Programmsteuer(ungs)gerät n = sequence programmer, program(me) control (system)
progressive Welle f, fortschreitende ~ = propagating wave, translation ~
Progressivsystem n [*Straßenverkehrstechnik*] = progressive system
Projektbearbeitung f, Projektausarbeitung = planning and design work
projektieren = to plan and design
Projektierung f = planning and design
Projektierungs|arbeiten fpl = planning and design work
~büro n = planning and design office
Projektionslampe f = projector lamp
Pro-Kopf-Verbrauch m = per capita consumption, consumption per head
Promenade f [*breiter Spazierweg*] = promenade
Propan n, C_3H_8 = propane
~flasche f = propane bottle
~gas(-Leucht)boje f = propane gas-lighted buoy
Propeller|-Hochdruck-Ventilator m → Axial-Hochdruck-Ventilator
~pumpe f = propeller pump, axial-flow ~
~rad n mit festen Laufschaufeln = fixed propeller runner

~rinne *f*, Marcus-Wurfförderrinne = Marcus through conveyor, ~ ~ conveyer

~rührer *m* = propeller stirrer, ~ agitator

~-Saugzuganlage *f* = propeller-fan induced draught system (Brit.); ~ ~ draft ~ (US)

~schlitten *m*, Luftschrauben-Schlitten = propeller sled

~turbine *f* = propeller turbine

~ventilator *m* → Axialventilator

Proportionalität *f* = proportionality

Proportionalitätsgrenze *f*, P-Grenze = limit of proportionality, proportional limit

Proportional-Überfall *m*, linearer Überfall = proportional-flow weir, sutro ~, weir with linear discharge relation

prospektieren, suchen = to prospect, to search

Prospektion *f*, Suche *f* = (mineral) prospecting

~ im Küstenvorland → Aufsuchen *n* ~ ~

Prospektions|arbeiten *fpl*, Sucharbeiten = prospecting work

~bohrung *f* → Aufschließungsbohrung

~schacht *m* → Aufschließungsschacht

~stollen *m* → Aufschließungsstollen

~vorhaben *n* → Aufschließungsvorhaben

Proteinleim *m* = protein glue, ~ adhesive

Protektor *m*, Laufstreifen *m* = sidewall protection [*pneumatic tyre*]

Prototyp *m* = prototype, development-type

Proustit *m*, lichtes Rotgültigerz *n*, Ag_3AsS_3 (Min.) = proustite

provisorische Maueröffnung *f* = wall run [*An opening temporarily left in the wall, especially for the passage of wheel barrows*]

Prozentsatz *m*, Anteil *m*, Prozentgehalt *m* = percentage

prozentweise = percentage-wise

Prüf|alter *n* = age at test

~anstalt *f*, Prüfamt *n* = testing institute

~balken *m* → Biegebalken

~befund *m* → Prüfbericht *m*

~belastung *f*, Probebelastung, Versuchsbelastung = test loading

~bericht *m*, Prüf(ungs)befund *m*, Prüfungsbericht = test report

~bescheid *m* = test certificate

~dauer *f* → Prüfzeit

~druck *m*, Probedruck, Versuchsdruck = test pressure

Prüfen *n* = testing

~ der Kornzusammensetzung durch Siebversuch, Siebprobe *f*, Siebversuch *m*, Siebanalyse *f* = sieve analysis, test for ~ ~, sieve (analysis) test; screen analysis [*deprecated*]

~ mit eindringendem Farbstoff = dye penetrant inspection

Prüf|fehler *m* = test error

~feld *n* = test(ing) ground

~feldversuch *m* = test(ing) ground test

~fuge *f*, Probefuge, Versuchsfuge = experimental joint

~gang *m* = test operation

~gemisch *n*, Prüfmischung *f* = test mix(ture)

~gerät *n* = tester

~gerät *n* für die Siedeanalyse von Verschnittbitumen = cut-back (asphaltic) bitumen distillation apparatus (Brit.); ~ asphalt ~ ~ (US)

~geschwindigkeit *f* = test rate

~hahn *m* → Probierhahn

~ingenieur *m* = test engineer

~kaliber *n* = master ga(u)ge

~korngröße *f* [*Siebtechnik*] = basic size used for testing

~körper *m* → Probekörper

~körperform *f*, Probekörperform = test specimen mo(u)ld

~körperherstellung *f*, Probekörperherstellung = test specimen preparation

~labor(atorium) *n* = testing lab(oratory)

~last *f*, Probelast, Versuchslast = test load

Prüflehre — Puffer

~lehre *f* = checking gauge
Prüfling *m* → Probekörper
Prüf|maschine *f* = test(ing) machine
~methode *f* → Prüfverfahren *n*
~mischung *f*, Prüfgemisch *n* = test mix(ture)
~nadel *f* → Proctornadel
~pfahl *m*, Probepfahl, Versuchspfahl = test pile
~presse *f* → Druckpresse
~prisma *n*, Probeprisma = test prism
~rad *n* = test wheel
~schacht *m* → Kontrollschacht
Prüfsieb *n* = test(ing) sieve
~belag *m* [*als Gewebe*] = test(ing) sieve cloth
~belag *m* [*als Lochblech*] = test(ing) sieve plate
~ergebnis *n* = test sieving result
~gewebe *n* = test(ing) sieve cloth
~gewebenorm *f* = test(ing) sieve cloth standards
~gewebereihe *f* = test(ing) sieve cloth series
~maschine *f* = test(ing) sieve, test sieving machine
~normung *f* = test(ing) sieve standardization
~reihe *f* = test(ing) sieve series
~satz *m* = nest of (test(ing)) sieves, stack ~ (~) ~
Prüf|siebung *f* = test-sieving
~stab *m*, Sondierstab = sounding rod
~stab *m*, Probestab, Versuchsstab = test bar
~stab *m* Künzel, Sondierstab ~ = Künzel sounding rod
~stand *m* = test stand, ~ bench, ~ bed
~stelle *f* = testing institute
~strecke *f* = test section
~teich *m* [*zur Prüfung von Pionierbooten*] = testing tank
Prüfung *f*, Probe *f*, Versuch *m* = test
~ **auf Flüchtigkeit** = volatilization test
~ **der Wassersperre** [*Tiefbohrtechnik*] = water shut off test
~ ~ **Zerstörung von Hochofensteinen durch Kohlenstoffablagerung** = carbon disintegration test of blast furnace brick
~ **des Nachschwindens** [*feuerfester Stein*] = reheat test
~ **von Bauteilen auf Widerstandsfähigkeit gegen Feuer** = fire tests on structural elements
Prüfungsalter *n* = age at test
Prüf(ungs)befund *m* → Prüfbericht *m*
Prüfungs|bericht *m* → Prüfbericht
~pflicht *f* = testing obligation
Prüf(ungs)vorschriften *fpl* = test(ing) spec(ification)s
Prüf|verfahren *n*, Prüfmethode *f*, Versuchsverfahren, Versuchsmethode = test method, testing procedure
~wassermesser *m* = test water meter
~wert *m* = test figure
~würfel *m* → (Beton)Würfel
~zeit *f*, Prüfdauer *f* = test(ing) time, ~ period
~zeugnis *n* = test certificate
Prügelweg *m* → Knüppelweg
PR-Zahl *f* = ply-rating
P.S. *f*, Pferdestärke = metric horsepower, ~ H. P., ~ h. p.
PS an der Welle = shaft horsepower, s.h.p.
Pseudokieselgur *f*, Gaize *f* = gaize
~zement *m*, Gaizezement = gaize cement
pseudoplastisch = pseudo-plastic
Pseudo-Subkornstruktur *f* = pseudo-subgrain structure
Psilomelan *m*, schwarzer Glaskopf *m* (Min.) = psilomelane, black iron ore
PSP, Preßsperrholz *n* = superpressed plywood, high density ~
Psychodafliege *f*, Psychoda alternata, Tropfkörperfliege = filter fly
Psychrometer *n*, *m* = psychrometer
Puddelluppe *f*, Rohrschiene *f* = muck bar [*The bar made by the first rolling of the bloom*]
Puddingstein *m* = puddingstone
Puddle *m* → Lehmschlag
Pudersiebhammer *m* = powder sieve vibrator
Puffer *m* = buffer

Pufferbestand — Pumpenbagger

~bestand m, Reservebestand = buffer stock
~bunker m, Reservebunker = surge bin
~silo m, Reservesilo = surge silo
~zone f = buffer zone
Pulsator m [*Schwingungsprüfung von Asphaltbelägen auf Stahlplatten*] = pulsating machine
Pulserbagger m, Pulsometerbagger = pulsometer dredger (Brit.); ~ dredge (US)
pulsierende Last f = pulsating load
~ **Springquelle** f → Geysir m
pulsierender Geschiebetrieb m, sprungweiser ~ = saltation
Pulsierung f = pulsation
Pulsometerpumpe f, Dampfdruckpumpe = (steam) pulsometer pump
Pulszahlempfänger m = pulse code receiver
Pultdach n, Flugdach, Halbdach, Schleppdach, einhängiges Dach = shed roof, lean-to ~
~binder m, Flugdachbinder, Halbdachbinder, Schleppdachbinder, Pultbinder = shed roof truss, lean-to ~ ~
~ziegel m, Pultdachstein m, Flugdachziegel, Flugdachstein, Schleppdachziegel, Schleppdachstein = shed roof tile, lean-to ~ ~
pulver(art)ig → pulverförmig
Pulver|(brannt)kalk m = air-slaked lime, powdered ~, powdered calcium carbonate
~brennschneidverfahren n mit Kieselsäurepulver = flame-cutting process using silica powder
~erde f = dust earth
~form f = powdery form
pulver|förmig, gepulvert, pulver(art)ig = powdered, in powder form
~förmiges Schüttgut n = bulk material in powder form
Pulverisieren n = pulverizing
pulverisierter Gummi m = powdered rubber
pulverisiertes Lötzinn n, ~ Lot n = powdered solder

Pulver|kalkstein m = pulverized limestone
~lanze f = powder lance
~lanzenschneiden n, Schneiden mit Pulverlanze = powder lancing
~leim m = powdered glue, ~ adhesive
~metallurgie f = powder metallurgy
~schnee m = powder snow
~sprengstoff m = powder (blasting) explosive
~spritzpistole f = powder spray gun
~trichter im [*Schweißen*] = welding composition hopper, hopper for the Unionmelt
~zündschnur f → Bickford-Zündschnur
Pump|anlage f = pumping system
~ausrüstung f = pumping equipment
~beton m = pumping concrete, pumped ~, pumpcrete
~bewässerung f = irrigation by pumps
Pumpe f = pump
~ **für die Bauwirtschaft** → Baupumpe
~ ~ **Wasserhaltung** = dewatering pump, unwatering ~
~ **mit Flüssigkeitsgetriebe** = fluid-driven pump
~ ~ **geteiltem Gehäuse** = split-casing pump
~ ~ **hin- und hergehendem Arbeitsorgan** = reciprocating pump [*Reciprocating pumps may be of either the piston type or the plunger type*]
~ ~ ~ ~ ~ **Kolben** = reciprocating piston pump
~ ~ **Kammern** = section pump
~ **ohne Wellendichtungen** = sealless pump
Pumpeinrichtung f [*Pumpspeicherwerk*] = pumping plant
pumpen = to pump
Pumpen n = pumping
Pump(en)anlage f = pumping installation, ~ plant
Pumpenauslaß m = pump outlet
Pumpenbagger m → Saugbagger
~ **mit hydraulischer Bodenlösung durch Abspritzen unter hohem**

Pumpen|bock *m* = pumping jack
~**bühne** *f* = pump platform
~**deckel** *m* = pump cover
~**druck** *m* = pump pressure
~**einlaß** *m* = pump inlet
~**einsatz** *m* = pump cartridge
~**flügelrad** *n* = pump impeller
~**(förder)leistung** *f* = pump output, ~ capacity
~**gehäuse** *n* = pump housing
~**gestänge** *n* = pump rods, sucker ~
~**gestänge-(Dreh)Schlüssel** *m* = sucker rod wrench
~**gestängefahrstuhl** *m*, Hebebügel *m* für Pump(en)gestänge [*Tiefbohrtechnik*] = sucker rod elevator
~**gestängehalter** *m* = sucker rod hook
~**getriebe** *n* = pumping gear
~**haus** *n* = pump house, ~ building
~**heizung** *f* = accelerated heating system
~**-Hydraulik** *f* = power hydraulics
~**kolben** *m* = pump piston
~**körper** *m* = pump body
~**-Kreuzgelenkantrieb** *m* = pump universal joint drive
~**leistung** *f*, Pumpenförderleistung = pump capacity, ~ output
~**(naß)bagger** *m* → Saugbagger
~**rad** *n* = impeller
~**radnabe** *f* = impeller hub
~**satz** *m* = pumping set
~**saughöhe** *f* = pumping lift
~**schuh** *m* = sucker
~**schwengel** *m* = pumping beam
~**stange** *f*, Pumpstange [*Tiefbohrtechnik*] = sucker rod, pump ~
~**stiefel** *m* = barrel of a pump
(~)**Sumpf** *m* [*ist einer Pumpe vorgeschaltet*] = (pump) sump, ~ well, sump hole, sump pit; sump well [*deprecated*]
~**umwälzung** *f* = pumped circulation
~**ventil** *n* = pump valve
~**-Warmwasserheizung** *f* = warm water pump heating
~**wellendichtung** *f* = pump shaft seal
~**zuführung** *f* = pump lead
~**zylinder** *m* = pump barrel
pumpfähig = pumpable
Pump|fähigkeit *f* [*Beton*] = pumpability
~**förderung** *f* von Beton = concrete pumping
~**geschwindigkeit** *f* = pumping speed
~**gestänge** *n*, Pumpstangen *fpl*, Pumpstäbe *mpl* [*Tiefpumpenförderung*] = sucker rods
~**kosten** *f* = pumping cost
~**laden** *n* [*Radschrapper*] = pump loading, pumping
~**probe** *f*, Pumpversuch *m*, Pumpprüfung *f* = pumping test
~**prüfung** *f* → Pumpprobe *f*
~**schacht** *m* = pumping shaft
~**speicher** *m*, Becken *n* = reservoir, pool, (storage) basin
~**speicher(kraft)werk** *n*, Pumpspeicher(kraft)anlage *f* = pumped storage hydropower plant, ~ ~ (hydroelectric) ~
~**speichersatz** *m* = pumping/generating set
~**speicherung** *f*, hydraulische Energiespeicherung = pumped-storage
~**speicherwerk** *n*, Pumpspeicheranlage *f* = pumped-storage scheme
~**stäbe** *mpl* → Pumpgestänge *n*
~**stangen** *fpl* → Pumpgestänge *n*
~**turbine** *f* = turbine pump, (reversible) pump-turbine, reversible unit
~**turbine** *f* mit beweglichen Leitschaufeln = adjustable blade pump turbine, movable blade machine
~**versuch** *m* → Pumpprobe *f*
~**wasser** *n* [*Pumpspeicherkraftwerk*] = pumping water
~**wassermenge** *f* = quantity of pumping water
~**werk** *n* = pumping station
Punkt|belastung *f* = point loading
~**eruption** *f*, Zentraleruption = explosion-pipe eruption, central ~
~**feuer** *n* [*Flugplatz*] = point light
punkt|förmig gestützt = supported in place

~**geschweißte Mattenbewehrung** *f* = welded fabric, ~ reinforcing mesh
Punkt|haus *n* = point block, ~ building
~**hochhaus** *n*, **Punktturmhaus** = point tower block, ~ ~ building
Punktiereisen *n*, **Handkörner** *m* = prick punch
Punkt|lager *n* = point bearing
~**last** *f* = point load
~**paar** *n* (✹) = pair of convergence points
~**schweißanlage** *f* = spot welding installation
~**schweißen** *n* = spot welding
~**schweißkopf** *m* = spot welding head
~**schweißmaschine** *f* = spot welding machine
~**turmhaus** *n*, **Punkthochhaus** = point tower building, ~ ~ block
~**vermarkung** *f* (✹) = marking of points
~**wanderung** *f* (✹) = displacement of observation points
Pupinspulenkasten *m* = loading-coil case
Putz *m*, **Mörtelbelag** *m* = 1.) plaster, plaster(ing) finish; 2.) rendering (finish) [*There is apt to be some confusion over the terms "plaster" and "rendering" for the former is sometimes used when referring to an external finish, while the latter is quite often used to describe an undercoat of internal "plaster". The more generally accepted meanings are "plastering" for internal finishes and "rendering" for external finishes*]
~**anwendung** *f* = application of plaster, use ~ ~
~**arbeiten** *fpl* = plastering (work)
~**art** *f* = type of plaster
~**aufbau** *m* = composition of plaster
~**ausbesserung** *f* = plastering repair
~**bau** *m* [*Bauwerk mit Außenputz*] = rendered structure
~**bewehrung** *f*, **Putzarmierung** *f* = plaster reinforcement
~**binder** *m* = cementing material for plasterings and renderings

~**bohlenfertiger** *m* → **Glättbohlenfertiger**
~**decke** *f* = plastered ceiling
~**deckel** *m* = inspection cover, rodding ~
~**dicke** *f*, **Mörtelbelagdicke** = plaster thickness
Putzen *n*, **Verputzen** *n* = rendering, plastering
Putzer *m* = plasterer
~**gerüst** *n*, **Putzerrüstung** *f* = scaffold(ing) for plasterers
~**rüstung** *f*, **Putzergerüst** *n* = scaffold(ing) for plasterers
~**werkzeug** *n* = plasterer's tool, plasterers' ~
Putz|fläche *f* → **Putz(unter)grund** *m*
~**gewebe** *n* = wire lath
~**gips** *m* = gypsum plaster
~**grund** *m* → **Putzuntergrund**
~**haftung** *f* = adhesion of plaster
~**hammer** *m* = fettling hammer
~**kalk** *m* = lime for plastering
~**kelle** *f*, **Putzerkelle** = plastering trowel
~**lage** *f*, **Putzschicht** *f* = plaster coat, coat of plaster
~**leder** *n*, **Waschleder** = washleather
~**lederverglasung** *f*, **Waschlederverglasung** = washleather glazing, glazing bedded in washleather
~**leiste** *f* **aus Mörtel** [*schmales Mörtelband an Decken und Wänden, nach der Sollage der Putzoberfläche vor dem Verputzen als Richtfläche hergestellt*] = screed of coarse stuff
~**leiste** *f* **aus (Stuck)Gips**, **Pariser Leiste** = screed of stucco
~**maschine** *f* → **Putzwerfer** *m*
~**meißel** *m*, **Abgratmeißel** = fettling chisel
~**mörtel** *m*, **Putzmasse** *f* = stuff
~**mörtelträger** *m* → **Putzträger**
~**öffnung** *f* = access eye, inspection fitting [*An opening in a drain pipe closed by a plate held on to the opening by a bolt or wedge. It is often provided at pipe bends to enable the pipe to be rodded. It is therefore sometimes called a rodding eye*]

~riß m = plaster crack
~rohr n [hat ein Loch welches mit einem Deckel verschlossen ist] = pipe with access eye, ~ ~ inspection fitting
~sand m = plaster sand [B.S. 1198 and B.S. 1199]
~schicht f = coat of plastering
~träger m, Putzmörtelträger = plaster base, base for plaster(ing), (back) ground for plaster [It may be wood, metal lath(ing), brickwork, masonry, insulating board, or gypsum lath(ing)]
~- und Mauer(werk)binder m → Mauerbinder
~(unter)grund m, (Unter)Grund für Putzarbeiten, Putzfläche f [Fläche, die geputzt werden soll] = surface to be plastered
~verfahren n, Putzweise f = plastering method
~wand f = plaster wall [the opposite of a dry wall]
~weise f, Putzverfahren n = plastering method
~werfer m, Putz-Spritz-Apparat m, (Ver)Putzmaschine, Verputzanlage f = plaster-throwing machine, plastering ~
~wolle f = cotton waste
Puzzolan n = pozz(u)olana, puzzolane
~analyse f = pozzolana investigation
~erde f → Bröckeltuff m
~mörtel m = pozzolan(ic) mortar
~zement m = pozzolan(ic) cement
PVA-Beton m = polyvinyl acetate concrete
PVA-Emulsion f = polyvinyl acetate emulsion
PVA-Haftmittel n, Polyvinylalkohol-Haftmittel = PVA bonding agent, polyvinyl alcohol
PVA-Mörtel m = polyvinyl acetate mortar
PVC-Platte f, PVC-Fliese f = p.v.c. tile, PVC ~
PVH, Preßvollholz n = densified wood

Pyknometer n, m = (fruit jar) pycnometer, density bottle, pyknometer
Pylone f, (Kabel)Turm m, Pylon m = tower, pylone, suspension ~
Pylonen|kopf m = pylon head, tower ~
~pfeiler m = tower pier, pylon ~
Pyramidalgeröll(e) n, Kantenkiesel m, Kantengeröll(e) n, Kantengeschiebe n, Facettengeschiebe n, Pyramidalgeschiebe = (wind-)faceted pebble(s)
Pyramidendach n, Turmdach = pyramid(al) roof, polygonal broach ~, spire ~, steeple
Pyrargyrit m (Min.) → Antimonsilberblende f
Pyrit m → Schwefelkies m
pyroklastische Gesteine npl (Geol.) = pyroclastic rocks
Pyrometer n, Strahlungsthermometer = (heat) radiation pyrometer
~schutzrohr n für Thermoelement = pyrometer tube
Pyromorphit m, Grünbleierz n (Min.) = pyromorphite
PZ m, Portlandzement = Portland cement

Q

Q-Decke f [Deckenbauart mit I-Trägern] = Q-floor
Quader m, Block m, Quaderstein m = ashlar, cut-stone, ashler
~mauerwerk n, Blockmauerwerk = ashlar masonry, cut-stone ~, ashler ~
~pflaster(decke) n, (f), Blockpflaster(decke) = ashlar paving, cut-stone ~, ashler ~
~sandstein m = upper cretaceous sandstone
~stein m → Quader m
~verblendung f, Quaderverkleidung = ashlar facing, cut-stone ~, ashler ~
Quadrant m → Viertelkreis
Quadrantenverfahren n → Viertelung f
Quadratgründungssockel m, Quadratsockelgründung f = square (single) base, ~ individual ~

quadratische Mitnehmerstange *f* [*Rotarybohren*] = square kelly
~ **Umschnürung** *f* [*Säule*] = square hooping, ~ hoop(s)
quadratischer Gründungsbrunnen *m* → ~ Senkbrunnen
~ **Kübel** *m* [*Betontransport*] = square skip
~ **Pilzkopf** *m* = square (column) capital
~ **Senkbrunnen** *m*, ~ Gründungsbrunnen = square open caisson, ~ drop shaft
Quadrat|lochöffnung *f* [*Siebtechnik*] = square aperture, ~ punching
~**masche** *f* = square mesh
~**maschengewebe** *n* = square mesh screening cloth, ~ opening wire ~
~**öffnung** *f* = square hole, ~ opening
~**pfahl** *m* = square pile
~**pfeiler** *m* = square pier
~**platte** *f* = square slab
~**querschnitt** *m* = square cross section
~**schornstein** *m* = square chimney
~**sockelgründung** *f*, Quadratgründungssockel *m* = square (single) base, ~ individual ~
~**spundung** *f* = square grooving and tonguing
~**stahl** *m*, Vierkantstahl [*Flußstahl, gewalzt nach DIN 1014*] = square bar steel, ~ bar(s)
~**zelle** *f* [*Bagger-Senkkasten*] = square cell
Qualitäts|abkommen *n* = quality agreement
~**erzeugnis** *n* = quality product
~**überwachung** *f*, Güteüberwachung = quality control
Qualmwasser *n* → Körwasser
quantitative spektrochemische Bestimmung *f* = quantitative spectrochemical analysis
Quartier *m* = quarter brick
Quartowalzwerk *n* = four-high mill
Quarz *m*, SiO_2 (Min.) = quartz
~**diorit** *m* = quartz-diorite
~**fadenwaage** *f* = quartz spiral balance

~**feinsand** *m* = silica sand fines, ~ fine sand
quarzfreier Porphyr *m* = quartz-free porphyry
Quarz|gang *m* = quartz vein
~**glaskörper** *m* im Schlickergießverfahren hergestellt = slip-cast fused silica
~**glaswolle** *f* = quartz cloth
Quarzit *m* = quartzite
~**stein** *m*, Silikastein, Quarzkalkziegel *m* [*früher: Dinasstein*] = silica brick, acid firebrick; gannister brick (Brit.); ganister brick (US)
Quarz|kalkziegel *m* → Quarzitstein *m*
~**keil** *m* = quartz wedge
~**keratophyr** *m* = quartz keratophyre
~**kies** *m* = quartz gravel
~**mehl** *n*, Quarzstaub *m*, Quarzpulver *n* = siliceous dust, silica ~
~**porphyr** *m* = quartz porphyry
~**pulver** *n* → Quarzmehl
~**sandstein** *m* = silica sandstone
~**schieferton** *m* = arenaceous shale
~**schmelze** *f* = siliceous melt
~**staub** *m* → Quarzmehl
~**trachyt** *m* = quartz-trachyte
~**wachstum** *n* = quartz crystallization, growth of quartz
~**wolle** *f* = quartz wool
~**zuschlag(stoff)** *m* = quartz aggregate
quasi-gleichförmige Strömung *f* = quasi uniform flow
Quecksilber|barometer *n*, *m* = mercury barometer
~**chlorid** *n*, Sublimat *n* = corrosive sublimate, mercuric chloride
~**cyanat** *n* → Knallquecksilber *n*
~**dampf-Hochdrucklampe** *f*, Hochdruck-Quecksilberdampflampe = high-pressure mercury (discharge) lamp
~**dampflampe** *f* = mercury discharge lamp
~**dampf-Niederdruck-Leuchtstofflampe** *f* = colour-corrected mercury lamp
~**-Diffusionspumpe** *f*, Diffusionsluftpumpe = mercury-vapo(u)r pump

~hornerz *n* (Min.) = horn quicksilver
~-Jod-Lampe *f* = mercury iodine lamp
~lebererz *n* (Min.) = hepatic cinnabar
~oxydulsalz *n* = mercurous salt
Quell|beton *m*, Expansivbeton, Schwellbeton = self-stressed concrete
~boden *m*, Schwellboden = expansive soil, swell ~
~druck *m* [*Ton*] = swelling pressure
Quelle *f*, Austrittpunkt *m* von Grundwasser = spring
quellen, auf~ = to swell
Quellen *n* → Aufquellung *f*
~ der Sohle, Hebung *f* ~ ~, ~ des Liegenden (⚒) = floor-heave, floor-creep, floor-lift
~ des Liegenden → ~ der Sohle
Quell|ergiebigkeit *f*, Schüttung *f* = yield of a spring
~ergiebigkeit *f* [*Straßenverkehrstechnik*] = effective range of traffic originating in a given area
~erscheinung *f* = swelling phenomenon
~fassung *f* = catchment of spring, beheading ~ ~, spring protection
~fassung *f* [*Bauwerk zur Sammlung des Quellwassers an einer Austrittstelle*] = spring intercepting structure
~gebiet *n* = headwaters
~gut *n*, Quellwasser *n* = piping water
Quellinie *f*, Quell-Linie = spring front
Quell|kammer *f* = chamber of a spring
~punkt *m* [*Straßenverkehr*] = point of origin
~stoff *m* = swelling material
~teich *m* = spring-fed pond
~ton *m* = expansive clay
~trichter *m* = valley head
Quellung *f* → Auf~
Quell|vorgang *m* = swelling process
~wasser *n*, Quellgut *n* = piping water
~zement *m*, Expansivzement, Schwellzement, Dehnzement = high-expansion cement, expanding ~, expansive ~, self-stressing ~

quer belasteter Druckstab *m* = beam column
~ geneigte Ebene *f*, Quer-Schrägaufzug *m* [*Schiffhebewerk*] = inclined plane with transversely travel(l)ing caissons
Quer|arbeitsfuge *f* = transverse construction joint
~armierung *f* → Querbewehrung
~aufschleppe *f*, Quer(-Schiffauf)-Schleppe, Querschlipp *m*, Querslip = traversing slipway
(quer)aussteifen → (ab)bölzen
(Quer)Aussteifung *f* → (Ab)Bölzung
Quer|bahnsteig *m* = cross station platform
~balken *m* des Aufreißers = multiple shank beam
~balken(träger) *m* = cross beam
~bau(werk) *m*, (*n*), Querwerk = projecting structure [*groyne; jetty*]
~belastung *f* = transverse loading
~(be)lüftung *f* = transverse ventilation
Querbewehrung *f*, Querarmierung, Quer(stahl)einlagen *fpl* = transverse reinforcement, cross ~
~, Querarmierung, Quer(stahl)einlagen *fpl* [*Stütze*] = circumferential reinforcement, lateral ~
~ in Ringform → Umschnürung
~ nach der Schraubenlinie → Spiralbewehrung
Quer|biegemoment *n* = transverse bending moment
~biegen *n*, Querbiegung *f* = transverse bending
~biegeversuch *m*, Querbiegeprobe *f*, Querbiegeprüfung *f* = transverse bending test
~bohle *f* [*Betondeckenfertiger*] = transverse screed
~dehnung *f* = lateral strain, transverse ~
~(dehnungs)zahl *f* = Poisson's ratio [*Is the ratio of transverse strain to longitudinal strain under load, e. g. the ratio of transverse contraction to longitudinal extension under tension*]
~deich *m* = cross dyke; dike ~ (US)

Querdüne — Querschnittabnahme

~düne f, Walldüne = transverse dune
~einlagen fpl → Querbewehrung
~(erd)beben n = transverse earthquake
~festigkeit f = transverse strength, lateral ~
~förderband n = transverse belt conveyor, ~ ~ conveyer
~förderer m = transverse conveyor, ~ conveyer
~fuge f = transverse joint
~fugenrand m = transverse joint edge
~fugenverdübelung f [*Fugenbrett, Rundeisendübel und Baustahlgewebebügel*] = dowel load-transfer unit
~gebäude n = cross building
~gefälle n, Querneigung f = cross fall, transverse gradient, transversal gradient
~gefällebezug m = cross-slope reference
~gelenk n, Gestängekulisse f = cross link assy
quergerippter Betonformstahl m → (Beton)Rippenstahl
Quer|griff m = cross bar handle
~haupt n [*Prüfmaschine*] = cross head
~kanal m, Querleitungskanal = cross duct
~kette f [*Teil einer Reifenkette*] = cross chain
~kontraktion f, Querkürzung f = contraction of area, transverse contraction, lateral contraction
~last f = transverse load
~leiste f [*Brettertür*] = ledge
~(leitungs)kanal m = cross duct
~lüftung f → Querbelüftung
~maß n [*z.B. einer Stütze*] = lateral dimension
~mauer f = cross wall
~moment n = transverse moment
~naht f [*Schwarzdecke*] = transverse joint, lateral ~
~neigung f, Quergefälle n = cross fall, transverse gradient, transversal gradient
~neigungssteuerung f = control of cross fall [*asphalt paver*]

~neigungssteuervorrichtung f = cross fall controller [*asphalt paver*]
Querprofil n = cross section [*A vertical section of the ground along the centre line of a route*]
~aufnahme f = cross-sectioning
~zeichner m, Querprofilograph m = transverse profilometer, ~ profilograph
Quer|rahmen m = bent
~raumfuge f, Raumquerfuge = transverse expansion joint
~riegel m = cross arm
~riegel m [*Gerüst*] = putlog
~riegelloch n = putlog hole [*Hole left in brickwork for a putlog*]
~rinne f [*Straße*] = water splash
~rippe f = cross rib
~rippenstahl m [*Bewehrung*] = deformed bars with transverse projections
~riß m = transverse crack
~rohr n [*Motor*] = cross tube
~ruck m [*Straßenfahrzeug*] = sway
~(rund)stab m [*Betonstahlmatte*] = transverse (round) bar
~scheibe f = transverse stiffener
~scheinfuge f = transverse contraction joint
~(-Schiffauf)schleppe f → Queraufschleppe
~schild m, n, festes Schild n, fester Schild m [*Schürfschlepper*] = straight blade
~schild-Abstützung f → → Bulldozer
~schlag m (⚒) [*querschlägig zum Streichen der Lagerstätte verlaufende Gesteinsstrecke*] = crosscut
~-Schleppe f → Queraufschleppe
~schlipp m → Queraufschleppe f
~schlitz m = cross slot
~schnecke f = transverse screw
Querschnitt m = cross section, transverse ~ [*The shape of a body cut transversely to its length*]
Querschnitt|abmessung f **in Feldmitte** = size of the beam at the middle
~abnahme f, Querschnittverminderung f, Querschnittverringerung, Querschnittschwächung = reduc-

Querschnittänderung — Querverteiler

tion of the cross section, weakening ~ ~ ~ ~, contraction
~änderung *f* = change of cross section
~-Faktor-Verfahren *n* = intersection-factor method
~fläche *f* = cross-sectional area, CSA
~geschwindigkeit *f* [*Straßenverkehrstechnik*] = spot speed
~gestaltung *f*, Querschnittform *f* = shape of section
~gewicht des Rahmens = section weight of frame
querschnittgleich = of equal cross sections
Querschnitt|kern *m* = centre of section (Brit.); center ~ ~ (US)
~mitte *f* = cross section centre (Brit.); ~ ~ center (US)
~schwächung *f* → Querschnittabnahme
~verlust *m* = loss of cross section
~verminderung *f* → Querschnittabnahme *f*
~verringerung *f* → Querschnittabnahme *f*
~verringerung *f* [*als Stelle eines Rohres*] = throat
~wechsel *m* = change of the cross section
~wert *m* [*Profilstahl*] = section property, property of section
~zählung *f* = classification count
Quer|schott *n* = transverse bulkhead
~-Schrägaufzug *m*, quer geneigte Ebene *f* [*Schiffhebewerk*] = inclined plane with transversely travel(l)ing caissons
~schwelle *f* [*Eisenbahn*] = transverse sleeper
~slip *m* → Queraufschleppe *f*
~spaltsieb *n* = lateral-slot screen
~spannglied *n* = transverse (strand prestressing) tendon
~spannung *f* = transverse stress
~stab *m* = cross bar
~stab *m*, Querrundstab [*Betonstahlmatte*] = transverse (round) bar
~stabilisator *m* = stabilizer bar
~stabilität *f* = crosswise rigidity
~stabrost *m* [*Siebtechnik*] = lateral-bar screen, lateral-rod deck
~(stahl)einlagen *fpl* → Querbewehrung
~stange *f* = putlog [*Putlogs are short horizontal bearers, which carry scaffold boards in a bricklayer's scaffold. The putlog rests in a small hole left in the brickwork at one end, and on the ledger at the other end*]
~steg *m* [*Hohlblockstein*] = transverse web
~steife *f*, Stempel *m*, (Graben) Steife, Sprieße *f* [*in Österreich: Sprießel n*] = strut, brace [*The horizontal cross member in bracing of shallow cuts*]
~steifigkeit *f*, Quersteife *f*, Quersteifheit *f* = transverse rigidity
~stollen *m* = cross tunnel
~stoß *m*, Stegblech ~, Werkstattstoß = web plate transverse joint
~straße *f* = cross road
~strebe *f* = ranger, w(h)aler [*A horizontal bracing member used in form construction*]
~strebe *f* im Rahmen [*Baumaschine*] = frame cross member
~stromofen *m* für Ölschiefer-Schwelung = gas-flow oil-shale retort
~stromschachtofen *m* = cross flow (shaft) kiln [*for burning limestone*]
~strömung *f* [*Meer*] = cross current
~strömung *f*, Sekundärströmung [*Hydraulik*] = secondary flow
~stromwindsichter *m* = material collector
~träger *m* = cross girder
~traverse *f* [*(Gleis)Kettenlaufwerk*] = equalizer bar
~traversensattel *m* [*(Gleis)Kettenlaufwerk*] = equalizer saddle
~tunnel-Einmündung *f* = intersection of cross drift
~verband *m* = sway bracing
querverschiebbar = slideable in transverse direction
Quer|versteifung *f* = transverse stiffening
~verteiler *m* → Betonverteilungswagen *m*

~vorspannung *f* = transverse prestressing, ~ stretching
~-Walzträger *m* = rolled cross girder
~wand *f* = cross wall
~werk *n*, Querbau(werk) *m*, (*n*) = projecting structure [*groyne; jetty*]
~zahl *f* → Querdehnungszahl
~zahl *f* = bulk modulus [*The change in stress per unit change in volume*]
~zug *m* = transverse pull
~zugfestigkeit *f* = transverse strength
Quetsch|grenze *f* [*Baustahl*] = compressive yield point
~holz *n*, Einlage *f* (⚒) = crusher block, lid [*A short horizontal wood or steel piece placed over a single post to support the roof*]
~kies *m*, Brechkies = broken gravel, crushed ~
~sand *m* → Pochsand
Quick|sand *m*, Schwimmsand, Fließsand = quick sand
~ton *m*, Fließton = quick clay

R

RAB-Brückenmischer *m* → Brückenmischer
Rabitz|draht *m*, Rabitznetz *n*, Rabitzgewebe *n* = Rabitz type steel-wire plaster fabric
~rohrgewebe *n*, Rabitzrohrmatte *f* = Rabitz type reed lath(ing)
Rad *n* **mit beweglichen Leitschaufeln** [*Turbine*] = feathering blade runner
~abweiser *m* → Prellstein
(~)**Achse** *f* = axle
~-Aufladegerät *n* → Radlader
~-Auflademaschine *f* → Radlader
~auflader *m* → Radlader
~auswuchtgestell *n* = static wheel balancer
~belastung *f* = wheel loading
~bremse *f* = wheel brake
~dampfer *m*, Schaufel~ = paddle (wheel) steamer
~druck *m* = wheel pressure
~durchmesser *m* = wheel diameter

Rad(e)ber *f* → Schiebkarre(n) *f*, (*m*)
Radeindruck *m* → Spur
Rädelerz *n*, Spießglanzbleierz, Schwarzspießglanzerz (Min.) = wheel-ore, bournonite, cog-wheel ore
Räder|tierchenschlamm *m* → Radiolarienschlamm
~trecker *m* → Luftreifenschlepper
~vorgelege *n* = back gears, reduction ~, intermediate gear, back gear(ing) arrangement
~vorgelege *n* mit Fest- und Losscheibe = spur gearing fitted with fast and loose pulley
Radfahrer *m* = (pedal) cyclist
Rad|-Fahrgestell *n*, Rad-Untergestell = wheeled chassis
~-(Fahrlade-)Gerät *n* → Radlader
~-(Fahrlade-)Maschine *f* → Radlader
~-Fahrlader *m* → Radlader
~fahrstreifen *m* = cycle strip (on carriageway) [*A strip of the carriageway adjacent to the curb for the use of pedal cyclists*]
~fahrverkehr *m* = pedal cyclist traffic
~(fahr)weg *m*, Radfahrbahn *f* = (bi)cycle track, ~ way, ~ path
~fahrwerk *n*, Radfahrgestell *n*, Raduntergestell = wheel(ed) carrier, ~ chassis, ~ carriage
~fahrzeug *n* = wheeled vehicle
~flansch *m* = wheel flange
~-Gerät *n* → Radlader
radial verfahrbare Verladebrücke *f* = radial transporter
~ **verfahrbarer Kabelkran** *m* → Schwenkkabelkran
radiale Fugenverschiebung *f* = radial displacement of the joints
radialer Formstein *m* → Brunnenstein
Radial|balken(träger) *m* = radial beam
~fluß *m* = radial flow [*Direction of flow across a circular tank, from centre to periphery, or vice versa*]
~lüfter *m* → Fliehkraftlüfter
~sieb *n* = conical wedge-wire dewatering screen, radial ~

Radialsiebanlage — Radsatz

~siebanlage *f* = radial screens
~stadt *f* = radial city
~stapelförderer *m* = radial stacker-conveyer, ~ stacker-conveyer
~stein *m* → Brunnenstein
~stein *m* → Radialziegel
~straße *f* [*Schweiz*] → Ausfallstraße
~turbine *f* = radial flow turbine
~ventilator *m*, Zentrifugalventilator, Schleuderventilator = radial-flow fan, centrifugal ~
~ziegel *m*, Radialstein *m* = compass brick, radial ~, radiating ~ [*It tapers in at least one direction*]
Radiant *m* = radian
Radiator *m*, (Glieder)Heizkörper *m* = radiator [*The heating industry has accepted the definition of the term "radiator" as a heating unit exposed to view within a room or space heated, although such units transfer heat by convection as well as radiation*]
~-**Absperrventil** *n*, Heizkörper-Absperrventil = radiator shut-off valve
~**gitter** *n* = radiator grille
~-**Regulierventil** *n*, Radiator-Regelventil, Heizkörper-Regulierventil, Heizkörper-Regelventil = radiator control valve
~**ventil** *n*, Heizkörperventil = radiator valve
radiaktive Abfallflüssigkeit *f* = radioactive liquid waste
radioaktiver Abfall *m* = radioactive waste
Radioaktivierungsanalyse *f* [*Bohrloch-Überwachung*] = radioactivity logging, ~ survey
Radioaktivität *f* = radioactivity
Radio|entstörer *m* = radio suppressor
~**entstörung** *f* = radio interference suppression
~**isotop** *n* = radio isotope
Radiolarienschlamm *m*, Rädertierchenschlamm, Radiolarienschlick *m* = radiolaria(n) ooze
Radiometrie *f* [*geophysikalisches Aufschlußverfahren*] = radioactivity

(**Radius**)**Schürfbecherwerk** *n* [*Es eignet sich für den Einsatz in verschiedenen Steine und Erden-Betrieben wie auch auf der Baustelle. Der Mischmaschine oder in Betonwerken dem Mischerkübel werden kontinuierlich Kies und weitere Zuschlagstoffe ohne manuelle Bedienung zugeführt. Es ist dem jeweiligen Einsatzwerk anpaßbar. Kombiniert mit Silo und Waage hat man eine automatische Dosierung bzw. Schüttung in den Kipper zur Mischmaschine. Der Bedienungsmann der Mischmaschine hat nur eine Dreiknopfbedienung zu betätigen, um die Zuschlagstoffe mit Zement direkt in den Kübel zu fördern. Während das Mischgut in der Betonmischmaschine verarbeitet wird, wiederholt sich der Vorgang*] = scraper-type bucket elevator
Rad|kappe *f* = wheel cap
~**ladegerät** *n* → Radlader
~-**Lademaschine** *f* → Radlader
~**lader** *m*, Reifenlader, Rad(-Auf)ladegerät *n*, Reifen-Aufladegerät, Rad(-Auf)lademaschine *f*, Reifen-Aufladmaschine, Radauflader, Reifenauflader, Rad-Fahrlader, Reifen-Fahrlader, Rad(Fahrlade-)Gerät, Reifen-(Fahrlade-)Gerät, Rad-(Fahrlade-)Maschine, Reifen-(Fahrlade-)Maschine, Rad-Lademaschine, Reifen-Lademaschine = wheeled loading shovel, wheel-mounted ~, wheeled loader, wheel-mounted loader
~**lager** *n* = wheel bearing
~**last** *f* = wheel load
~**lastgröße** *f* = wheel load intensity
~-**Maschine** *f* → Radlader
~**mutternschlüssel** *m* = lug wrench
(~)**Nabe** *f* = (wheel) hub
~**nabenantrieb** *m* = wheel hub drive
~**nabenkappe** *f* = wheel hub cap
~**nabenlager** *n* = hub bearing
~**rennbahn** *f* = (bi)cycle racing track
~**rücksprungtheorie** *f* = wheel bound theory
~**satz** *m* = wheel set, set of wheels

Radschlepper — Rahmensteifigkeit 754

Radschlepper *m* → Luftreifenschlepper
~-Bulldozer *m* → → Bulldozer
~-Frontlader *m*, Radschlepper-Frontladeschaufel *f*, Radschlepper-Frontschaufellader, Radschlepper-Front-Hochlader, Radschlepper-Vorderlader, Radschlepper-Frontladegerät *n* = wheel tractor front-end loader, ~ ~ head-end ~
~-Schaufellader *m*, Radschlepper-Ladeschaufel *f*, Radschlepper-Kübelauflader, Radschlepper-Hochlader, Radschlepper-Hochladeschaufel = wheel tractor loader (or shovel)
~-Vorderlader *m* → Radschlepper-Frontlader
Radschlupf *m*, Durchdrehen *n* = (wheel) spin, (~) slip
Radschrapper *m* = scoop pan selfloading scraper, (wheel) scraper; trailing scoop grader, pan scraper, drag road-scraper (US); buck scraper (Brit.)
~deichsel *f* = scraper hitch
~hydraulikpumpe *f* = scraper pump
Rad|spur *f* → Spur
~stand *m* [*Fehlname: Achsstand. DIN 70 020*] = wheel base
~sturz *m* = wheel lean
~sturzantriebwelle *f* = front wheel lean shaft
~sturzsegment *n* = front wheel lean rack
~sturzzylinder *m* = front wheel lean cylinder
~traktor *m* → Luftreifenschlepper
~traktorkran *m*, Radschlepperkran, Radtreckerkran = wheeled tractor crane
~trecker *m* → Luftreifenschlepper
~treckerkran *m* → Radtraktorkran
~überweg *m* = cycle-track crossing
~-Untergestell *n*, Rad-Fahrgestell = wheeled chassis
~verkehr *m* = bicycle traffic
~wagen *m* → Selbstfahrwerk *n*
~waschvorrichtung *f* = wheel washer
Radweg *m* → Radfahrweg
~kante *f* [*Erzeugnis der Betonsteinindustrie*] = cycle path edging

Raffinerie|abwasser *n* = refinery waste
~bau *m* = refinery construction
~schieber *m*, Heißölschieber = process valve
~schlamm *m* = refinery sludge
raffiniertes Erdöl *n* = refined petroleum
~ Leinöl *n* = refined linseed oil
Rähm *f*, Rahmholz *n* [*Fach(werk)wand*] = head (rail)
Rahmen *m* = framework [*The load-carrying frame of a structure, generally of reinforced concrete, steel, timber, or occasionally light alloy*]
~, Grund~, Fahrgestell *n*, Untergestell, Chassis *n* = chassis, carrier
~ausbau *m* (⚒) = frame support
~bauteil *m*, *n*, Skelettbauteil = frame component
~bau(werk) *m*, (*n*) = frame(d) structure
~binder *m* = frame truss
~brücke *f* = frame bridge
~ecke *f* = frame corner
~form *f* = frame mould (Brit.); ~ mold (US)
~formel *f* = frame formula
~gelenk *n* = frame hinge
~gleis *n* [*Feldbahn*] = carryable track
~glied *n* = frame member
~-Holztür *f* = framed wood(en) door
~joch *n* = framed bent
~konstruktion *f* = frame(d) construction
~längsstück *n* [*Eisenbahnwagen*] = solebar
rahmenloser Ziegelpaket-Transport *m* = palletless package brick transporting
Rahmen|pfosten *m* = framed stanchion
~querträger *m* [*Fahrzeugrahmen*] = chassis cross member
~riegel *m* = horizontal member
~schergerät *n* → Scherkastengerät nach Casagrande
~schwenklager *n* [*Gleiskettenlaufwerk*] = frame pivot bearing
~steifigkeit *f* = frame rigidity

~stiel m = frame leg
~stützweite f = frame span
~träger m [Baumaschinenrahmen] = frame beam
träger m, Vierendeelträger = Vierendeel girder
~trägerbrücke f, Vierendeelträgerbrücke = Vierendeel girder bridge
~tragwerk n = frame(d) loadbearing system
~tür f = framed dor
~verbiegung f = frame crippling
~wirkung f = frame action
Rahmholz n, Rähm f [Fach(werk)wand] = head (rail)
Rajolpflug m → Grabenpflug
(Raketen)Abschußstelle f = launching site
Ramm|achse f = centre line of driving (Brit.); center ~ ~ ~ (US)
~anlage f → Ramme f
~arbeiten fpl = driving work
~ausrüstung f, Rammvorrichtung = driving attachment
(Ramm)Bär m, (Ramm)Hammer m = (pile) hammer
~ mit halbautomatischer Steuerung, (Ramm)Hammer m ~ ~ ~ = single-acting (pile) hammer
~trommel f, (Ramm)Hammertrommel = (pile) hammer drum
~-Ziehgerät n, Rammhammer-Pfahlzieher m, schlagender Pfahlzieher = double-duty double-acting hammer
Ramm|beanspruchung f = driving stress
~brücke f → Rammwagen m
~brunnen m → Norton-Brunnen
~bühne f [allgemein] = pile-driving platform
~bühne f → Rammwagen m
~-Dampfwinde f, Dampf-Rammwinde = steam piling winch
Ramme f, Rammanlage f, Rammeinrichtung f, Pfahl~ = (pile) driving plant, ~ driver, piling plant
~ [zur Verdichtung] = rammer
Ramm|einrichtung f → Ramme f
~-Elektrowinde f, Elektro-Rammwinde = electric piling winch

rammen, ein~, eintreiben, einschlagen = to drive
Rammen n [zur Verdichtung] = ramming
~, Ein~, Eintreiben, Einschlagen = [Pfahl; Spundbohle] = driving
~ mit Vorbohrloch = soft driving [Piles are driven more easily in predrilled holes]
~ ohne Vorbohrloch = hard driving
Ramm|energie f = energy used in driving [Determined from the product of the weight of the hammer and the vertical distance through which it moves before striking the pile]
~erschütterungen fpl = driving vibrations
~fallbär m, Fallhammer m, (Frei-)Fallbär, indirekt wirkender Bär = (tripped) drop hammer, drop pile ~, winch type ~, (pile-driving) monkey, tup
~formel f = dynamic pile-driving formula
~fortschritt m, Rammgeschwindigkeit f = rate of driving
~gerüst n = pile frame, piling ~, (pile) driving ~
~geschwindigkeit f, Rammfortschritt m = rate of driving
(Ramm)Hammer m, (Ramm)Bär m = (pile) hammer
~ mit halbautomatischer Steuerung, (Ramm)Bär m ~ ~ ~ = single-acting (pile) hammer
~trommel f, (Ramm)Bärtrommel = (pile) hammer drum
~-Pfahlzieher m, schlagender Pfahlzieher, Rammbär-Ziehgerät n = double-duty double-acting hammer
Ramm|haube f = pile helmet (Brit.); cushion head (US)
~hülsenpfahl f, Hülsenrammpfahl = (impact-)driven cased pile, ~ shell ~

(Ramm)Jungfer f, Schlagjungfer [Wenn Pfähle unter die Schlagtiefe des Rammbären oder unter Wasserspiegel gerammt werden, wird auf den Pfahl eine Verlängerung, die Jung-

fer, aufgesetzt, wodurch aber die Schlagwirkung stark verringert wird] = follower, long dolly, puncheon, sett

~ [*Dieselbär*] = anvil block
Ramm|kraft *f* = driving force
~kurve *f* = pile driving curve
~leichtgerüst *n*, Leichtrammgerüst = light type (pile) frame, ~ ~ piling ~
~-Meister *m* = pile-foreman
~-Ortpfahl *m* → gerammter Ortpfahl
Rammpfahl *m* = (impact-)driven pile [*A pile of steel, wood, reinforced concrete or prestressed concrete which is forced into the ground by blows from a pile hammer*]
~ aus Spannbeton → Spannbetonrammpfahl
~ ~ Stahlbeton mit schlaffer Bewehrung → Stahlbetonrammpfahl
~ ~ ~ ~ vorgespannter Bewehrung → Spannbetonrammpfahl
Ramm|platte *f* [*Pfahlramme*] = anvil (-block)
~plattenbagger *m* → Freifall-Kranstampfer *m*
~plattenverdichtung *f* = dropping weight method of compaction
~ponton *m* = pile-driving pontoon
~richtung *f* = driving direction
~ring *m* [*Der Kopf eines Rammpfahles aus Holz muß gegen Aufspalten beim Rammen durch einen Rammring gesichert werden*] = driving band, pile hoop, pile ring [*A steel band fitted round the head of a timber pile to prevent brooming*]
~rohr *n* = drive-pipe
~schiff *n* = driving vessel
~schlag *m* = blow, stroke
(~)Schlitten *m* → Rammwagen *m*
~sonde *f*, Schlagsonde = driving rod, drop-penetration sounding apparatus, percussion probe, dynamic sounding rod
~sondierung *f*, Rammsondieren *n* = drop-penetration testing, percussion penetration method, driving test, dynamic (subsurface) sounding

~sondierung *f* **mit Versuchspfählen** = pre-piling
~spitze *f*, Pfahlschuh *m* = pile shoe
~tiefe *f* = depth of penetration
~träger *m* = driven vertical pilot beam
~versuch *m*, Rammprüfung *f*, Rammprobe *f* = driving test
~vorgang *m* = driving process, ~ operation
~vorrichtung *f*, Rammausrüstung = driving attachment
~wagen *m*, Brücke *f*, Bühne *f*, Rammbrücke, Rammbühne, (Ramm-)Schlitten *m* = additional transverse carriage on rails
~widerstand *m*, dynamischer Pfahl~ = dynamic pile-driving resistance
~winde *f* = piling winch
Rampe *f*, Auffahrt *f* = accommodation ramp, approach ~, raised approach
~ = ramp
~ → schiefe Ebene *f*
~ vor Abfertigungsgebäude, ~ ~ Empfangsgebäude [*Flughafen*] = passenger loading apron, ramp
Rampen|garage *f* → Rampen-Parkhaus *n*
~-Parkhaus *n*, Rampen(-Park)garage *f*, Rampen-Parkbau *m* = ramp-type garage, ~ parking structure
~-Tankspritzmaschine *f* → Bitumen-Sprengwagen *m*
Rand *m* [*allgemein*] = edge
~, Schulter *f* [*Damm*] = edge
~abstand *m* = edge-to-edge distance
~armierung *f*, Randbewehrung, Rand(stahl)einlagen *fpl* = edge reinforcement
~balken(träger) *m* = edge beam, marginal ~
~bedingung *f* = boundary condition, marginal ~, fringe ~, edge ~
randbelastet = edge loaded
Rand|belastung *f* = edge loading
~bewehrung *f* → Randarmierung
~bohrung *f*, Randsonde *f*; Randschacht *m* [*in Galizien*] [*Bohrung am Rande eines Ölfeldes*] = marginal well, ~ bore, ~ boring, outstep ~, edge ~

Randeinfassung — Rasensodenandeckung

~einfassung *f*, Abschluß *m* [*Fahrbahn*] (carriageway) edging
~einfassung *f* aus Beton → Randarmierung *f*
~einlagen *fpl* → Randarmierung *f*
~feuer *n*, Umrandungsfeuer = boundary light
~form *f*, Einfassungsform = edging mo(u)ld
~glied *n* [*Baustatik*] = boundary member, ~ element, edge ~
~integral *n* = boundary integral
~last *f* = edge load
~linie *f* = pavement edge line, surfacing ~ ~
~markierung *f*, Seitenmarkierung [*Straße*] = lateral marking
~moräne *f* → Seitenmoräne
~naht *f* = edge seam
~pfette *f*, Saumpfette = verge purlin(e)
~platte *f* = edge slab
~schacht *m* [*in Galizien*] → Randbohrung *f*
~sonde *f* → Randbohrung *f*
~spannung *f* = extreme stress, (edge stress of) extreme fibre stress
~(stahl)einlagen *fpl* → Randarmierung *f*
~stein *m*, Beton~ = edging, concrete ~
~störung *f* = edge disturbance
~streifen *m* = Leitstreifen
~streifenfertiger *m* = marginal concrete strip finisher
~träger *m* = edge girder
randverstärkte Betondecke *f* = thickened edge type concrete pavement
Rand|verstärkung *f* = edge thickening
~wassertrieb *m* [*Ölfeld*] = edge water drive
~wert *m*, Grenzwert = boundary value
~wertdiagramm *n* = boundary value diagram
~wertproblem *n*, Randwertaufgabe *f* = boundary value problem
~zone *f* = boundary zone
Rangier|bahnhof *m* → Verschiebebahnhof
~fahrzeug *n*, Verschiebefahrzeug = shunting vehicle, marshalling ~, shunter

~gleis *n*, Verschiebegleis = shunt track, yard ~, classification ~
~lok(omotive) *f* → Verschiebelok(omotive)
~schlepper *m*, Verschiebeschlepper = shunting tractor, marshalling ~
~verkehr *m*, Rangierbetrieb *m*, Verschiebeverkehr, Verschiebebetrieb = shunting traffic
Rankine'sche Erddrucktheorie *f* = Rankine's earth-pressure theory
~ Zone *f* = Rankine zone
Rankine'scher Kreis *m* = Rankine cycle
~ Zustand *m* = Rankine state
Ransomsperre *f* → Pfeiler-Plattensperre Bauweise Ransom
Rapputz *m* = (cement) mortar rendering
Rasen *m* = turf
~abtrag *m*, Abschälen *n* des Rasens = turf stripping
~ansaat *f* = turfing by seeding
~band *n* [*mit Rasenziegelschneider geschnitten*] = grass strip, turf ~
~beil *n*, Wiesenbeil = hand turf stripper, ~ ~ cutter
~böschung *f* = turf slope
~decke *f*, Rasennarbe *f* = turf cover
~drän *m* = turf drain
~fläche *f* = turf area; greenery (US)
~fräse *f* für Bankettunterhaltung, Bankettfräse = verge cutter
~hängebank *f* (⚒) = pit-bank level
~kantenschneiden *n* = verge cutting
~mäher *m* = lawn mower
~narbe *f*, Rasendecke *f* = turf cover
~plagge *f* → Plagge
(~)Plaggenschneider *m* → (Gras)Sodenschneider
~regner *m* = lawn sprinkler
~rollfeld *n*, Grasnarbenrollfeld = grass landing ground, ~ ~ area
~schneider *m* → (Gras)Sodenschneider
~schneidpflug *m* → (Gras)Sodenschneider *m*
(~)Sode *f* → Plagge
~sodenandeckung *f* = turfing by sodding, planting sod(s)

~sodenarbeiten *fpl* = sodding (work)
~sodenplanum *n* = sod bed
~sodenschneider *m* → (Gras)Sodenschneider
~stapel *m* = turf sod stockpile, sod stack
~streifen *m* = turf strip
~stück *n* → (Rasen)Sode *f*
~torf *m* = turf peat
~trennstreifen *m* → Grünstreifen
~weg *m* = turf path
~ziegel *m* → (Rasen)Sode *f*
~ziegelschneider *m* → (Gras)Sodenschneider
rasiermesserdünn = micro-thin
Raspe *f* = rasp
Raste *f*, Federhaltestift *m* = detent
Rasten|segment *n* = vernier
~stellung *f* = detent position
Raster *m* [*Beim architektonischen Entwerfen legt ein Raster ein in Grundriß oder Aufriß stets wiederkehrendes Grundmaß, z.B. die Knotenpunkte einer Skelettkonstruktion, fest*] = planning grid
~netz *n* = grid plan, reference grid [*A plan in which setting-out lines called grid lines coincide with the most important walls and other building components. Prefabricated buildings are usually designed to fit a grid plan. A grid plan is not a planning grid = Raster*]
~stufe *f* [*zur Verstellung*] = notched stage
Rast|haus *n* = large motorway restaurant
~hof *m* = motel for drivers and trucks (US); ~ ~ ~ ~ lorries (Brit.)
~platz *m* = lay-by(e), passing place, waiting-bay, halting place, picnic site (Brit.); roadside rest (US); rest area
~stätte *f* = small motorway restaurant
Rathaus *n* = city hall (US); town ~
rationale Zahl *f*, aufgehende ~ = rational quantity, ~ number
Rationalisierung *f* = rationalization

~ im bituminösen Straßenbau = streamlined construction of asphalt paving, rationalized making of bituminous paving
Ratsche *f*, Bohr~, Knarre *f* = ratchet
Ratschzug *m* = snatch block
Rattenschwanz *m* = rat-tail file
Rätter *m* → Plan~
rauben (⚒) = to draw off props, ~ ~ ~ posts, to withdraw
Raub|schicht *f* (⚒) = prop-rdawing shift, post-drawing ~
~ventil *n*, Löseventil (⚒) = release valve
~vorrichtung *f* (⚒) = Sylvester, donkey, prop drawer, post drawer
Rauch *m* = smoke
rauchdicht = smoke-proof
Rauch|dichte *f* = smoke density
Rauchgas *n*, Verbrennungsgas, Feuergas = flue gas [*The smoke from a boiler fire*]
~entstaubung *f* = flue gas dust removal
~entstaubungs- und Saugzuganlage *f* = integrated flue gas dust removal and induced draft fan system
~gewicht *n* = flue gas weight
~kühlung *f* = flue gas cooling
~prüfer *m* = flue gas tester
~schieber *m* = flue gas valve
~thermometer *n* = flue gas thermometer
~trommeltrockner *m* = direct-heat cylindrical dryer; ~ ~ drier (US)
~wäscher *m* = flue gas washer
Rauch|grenze *f* = smoke stop
~kammer *f* = smoke box
Rauchkanal *m* = flue [*A passage for smoke in a chimney*]
~ eines Gasfeuerungsgerätes = gas flue, flue for gas appliance
~auslaß *m* = flue outlet
~stein *m* = flue block [*A precast hollow block which, with others, forms a flue*]
~auskleidung *f* = flue lining
~auskleidungsstein *m* = chimney block [*Precast, circular concrete pipe used as flue lining*]

~raum m = flueway [*The clear space for flue gases within a flue or flue lining*]
Rauch|quarz m (Min.) = cairngorm, smoky quartz
~rohr n = flue pipe [*A metal or asbestos-cement pipe which leads smoke from a slow-combustion stove to the flue*]
~schieber m = smoke valve
~schutztafel f [*Brücke*] = blast plate
~topas m = smokestone
Rau(c)hwacke f, Zellendolomitgestein n = crystallized dolomite
Rauchzimmer n = smoking room; smoke ~ (Brit.)
Räude f = scabbing [*The loss of aggregate from a surface dressing in patches, leading to exposure of the original road surface*]
Rauh|bank f → Langhobel m
~belag m, Rauhdecke f = friction course, non-skid surfacing
~decke f → Rauhbelag m
rauhe Turbulenz f = rough turbulence
Rauheit f, Rauhigkeit = roughness
Rauhigkeit f, Rauheit = roughness
Rauhigkeits|beiwert m, Rauhigkeitszahl f = roughness coefficient, ~ factor
~grad m = degree of roughness
~verzerrung f = distortion of roughness
~wert m = roughness value
~zahl f → Rauhigkeitsbeiwert m
rauhsandig = gritty
Rauh|schalung f = rough formwork
~schlämmebelag m für steile Bergstraßen = friction seal coat for steep-gradient roads
~spund m = rough T and G boarding
~stampfasphalt m = compressed (natural) rock asphalt with coarse surface
~wacke f → Rauhwacke
~werk n → Unterputz
~wurf m → Steinschüttung f
Raum m für Mutter und Kind [*Flughafen*] = nursery for passengers with small children

~abmessung f → Raum(teil)zumessung
~abmessung f = spatial dimension
~abnahme f, Raumverminderung f = volume decrease
~akustik f, Bauakustik = architectural acoustics, room ~
~änderung f = volume change
~aufteilungstechnik f = spacemanship [*The art of making fullest use of available space; overcoming waste of space while enhancing its appearance in commercial, industrial, institutional and residential buildings*]
~ausnutzung f = utilization of space
~bedarf m = space requirement
~bedarf m bei Versand, Versandabmessungen fpl [*Maschine*]=shipping dimensions
~bemessung f = three-dimensional design
~beständigkeit f [*Zement; Mörtel usw.*] = soundness
~beständigkeit f = volume stability, constancy of volume
~beständigkeitsprüfung f, Raumbeständigkeitsprobe f, Raumbeständigkeitsversuch m [*Zement*] = soundness test
raumbildende Tafel f → Ausfachungstafel
Räumboot n [*Ein mit Winde und Greifzange ausgerüstetes Boot, das vor einem Naßbagger herfährt und Hindernisse aus dem Weg räumt*] = snag boat
Raum|dosierung f → Raum(teil)zumessung
~dreieck n = spherical triangle
~einsparung f = saving in space
Räumen n, Auf-~, Nachnehmen n = reaming
~ und Roden n → Räumung(sarbeiten)
Räumer m → Nachnehmer m
~ → (Schlamm)Kratzer m
Raumfachwerk n = three-dimensional framework

Raumfuge *f*, (Aus)Dehnungsfuge, eigentliche Raumfuge [*Betonstraßenbau*] = expansion joint (during the day's work), running ∼
∼ → Trennfuge
Raumfugenkonstruktion *f* = expansion joint assembly
raumfugenlos = without expansion joint(s)
Raum|gefühl *n* = feeling of space
∼gerade *f* = line of space
∼getriebe *n* = space linkage, linkage in three dimensions
Raumgewicht *n*, Schüttgewicht, Rohwichte *f* = (dry) loose unit weight, unit weight
∼ der Volumeneinheit, Rohdichte *f* = volume weight, (bulk) density, bulk specific gravity [*deprecated: box weight*]
∼ unter Auftrieb = submerged unit weight
Raumgewinn *m* = gain in space
raumgroß = room-sized
Raum|heizer *m* = space heater
∼heizung *f* = space-heating, room-heating
Rauminhalt *m*, Kubatur *f*, Kubikinhalt = cubage, cubic content, volumetric content
∼ der Körner des Bodens (oder Gesteins) = volume of the solid substance
∼ ∼ Poren = volume of intergranular space, ∼ ∼ voids, ∼ ∼ pores
∼element *n* = volume element
Raum|klima *n* = indoor climate
∼koordinate *f* = space coordinate
Räumkraft *f*, wirksame Schleppkraft [*Strom; Fluß*] = effective tractive force
Raum|krümmung *f* = space curvature
∼kühlgerät *n* = room-cooling unit
∼kurve *f* = three-dimensional curve
räumlich verlegte Rohrleitung *f*, dreidimensionale ∼ = three-dimensional pipe line
räumliche Biegung *f* = three-dimensional bending
∼ Pfette *f* = space purlin(e)

räumlicher Ausdehnungsbeiwert *m*, kubischer ∼ = coefficient of cubic expansion
∼ Spannungszustand *m* = three-dimensional state of stress
∼ Stab *m* = space rod
∼ Vier-Säulen-Rahmen *m* **als Achteckträger** = octagonal girder four columns space frame
∼ Wechselsprung *m* = spatial hydraulic jump
räumliches Element *n*, dreidimensionales ∼ = three-dimensional element
∼ Fachwerksystem *n* → Raumfachwerk *n*
∼ Tragwerk *n*, Raumtragwerk = space structure
Raumluft *f* = room air, inside ∼
∼kühler *m* = room air cooler, inside ∼ ∼
∼temperatur *f* → Raumtemperatur
Raumlüftung *f* = room ventilation
Räummaschine *f*, Schnee∼ = (snow) remover
Raum|meter *n* [*in Süddeutschland* „Ster" *m*] = stacked cubic metre of timber, stere [*35.3 stacked cu.ft.*]
∼perspektive *f* = space-perspective
∼planung *f* = regional planning
∼prozent *n* → Raumteil *m*, *n*
∼querfuge *f*, Querraumfuge = transverse expansion joint
Räumschaufel *f*, Schrapperschaufel, Handschaufel = manually guided drag skip
∼ → Handschrapper
Räumschild *n* = blade; moldboard (US)
raumsparend = economical of space, space-saving
raumsparende Tür *f* = space-saver door
Raumteil *m*, *n*, Rtl., Raumprozent *n* = part per volume, ∼ by ∼, volumetric part
∼dosierer *m* = volume(tric) batcher
Raum(teil)dosierung *f* → Raum(teil)zumessung
Raumteile *mpl* = percentage by volume

Raum(teil)zumessung *f*, Raum(teil)abmessung, Raum(teil)dosierung = batching by volume, proportioning ~ ~, measuring ~ ~

Raumtemperatur *f*, Innentemperatur, Raumlufttemperatur = indoor temperature, room ~ [*Ambient temperature in enclosed buildings*]

~steuerung *f* = room temperature control

Raumthermostat *m* = room thermostat

raumtiefes Bild *n* = background image

Raum|tragwerk *n*, räumliches Tragwerk = space structure

~unbeständigkeit *f* = inconstancy of volume

Räumung(sarbeiten) *f*, Freilegung von Baugelände, Räumen *n* und Roden [*Entfernen aller Hindernisse pflanzlicher Natur*] = clearing (work) (or operations), land clearing stripping, site-clearing

Raum|verminderung *f*, Raumabnahme *f* = volume decrease

~-Zeit-Diagramm *n* = space-time diagram

~zumessung *f* → Raumteilzumessung

~zunahme *f*, Raumvergrößerung *f* = volume increase

Raupe *f*, (Gleis)Kettenfahrzeug *n*, Raupenfahrzeug = crawler-type vehicle, tracked ~

~, Schweiß~ = (weld(ing)) pass

~ → Kettenschlepper

~ → Raupenkette *f*

Raupenanhängerwagen *m* → Raupenwagen

Raupenantriebs|kette *f*, (Gleis)Kettenantriebskette = crawler drive chain

~kupp(e)lung *f*, (Gleis)Kettenantriebskupp(e)lung = crawler drive clutch

~turas *m*, (Gleis)Kettenantriebsturas = crawler drive sprocket

~turaswelle *f*, (Gleis)Kettenantriebsturaswelle = crawler pivot shaft

Raupen|-Aufladegerät *n* → Raupenlader

~-Auflademaschine *f* → Raupenlader

~-Auflader *m* → Raupenlader

~-Auflader *m* für Schüttgut → Becherwerk(auf)lader auf (Gleis)Ketten

~-Aufnehmer *m* → Becherwerk(auf)lader auf (Gleis)Ketten

~bagger *m*, (Gleis)Kettenbagger = crawler (type) excavator

~band *n* → Raupenkette *f*

Raupen-Becher|auflader *m* → Becherwerk(auf)lader auf (Gleis)Ketten

~(be)lademaschine *f* → Becherwerk(auf)lader auf (Gleis)Ketten

~werk(auf)lader *m* → Becherwerk(auf)lader auf (Gleis)Ketten

~werkfahrlader *m* → Becherwerk(auf)lader auf (Gleis)Ketten

Raupen|-(Beton)Mischer *m* = crawler-mounted concrete mixer

~-Bohrwagen *m*, (Gleis)Kettenbohrwagen, Raupen-Bohrmaschine *f*, (Gleis)Ketten-Bohrmaschine = crawler wagon drill

~bolzen *m*, (Gleis)Kettenbolzen = track roller shaft

~-Bulldozer *m* → Raupen-Planierschlepper

~-Drehkran *m* → Raupen-Schwenkkran

~-Eimerkettenbagger *m*, (Gleis)Ketten-Eimerkettenbagger = crawler-mounted bucket ladder excavator

~-(Eimerketten-)Fahrlader *m* → Becherwerk(auf)lader auf (Gleis)Ketten

~-Eimerketten-Lademaschine *f* → Becherwerk(auf)lader auf (Gleis-)Ketten

~-Eimerketten-Lader *m* → Becherwerk(auf)lader auf (Gleis)Ketten

raupenfahrbar, (gleis)kettenfahrbar = crawler-mounted

Raupen|fahrer *m*, Gleiskettengerätfahrer = cat-jockey

~-Fahrgestell *n* → Raupenfahrwerk

~-Fahrlade-Gerät *n* → Becherwerk(auf)lader auf (Gleis)Ketten

~-Fahrladegerät *n* → Raupenlader

~-Fahrlade-Maschine *f* → Becherwerk(auf)lader auf (Gleis)Ketten

Raupen-Fahrlade-Maschine — Raupenschuhplatte

~-**Fahrlade-Maschine** *f* → Raupenlader
~-**Fahrlader** *m* → Becher(werk)auflader auf (Gleis)Ketten
~-**Fahrlader** *m* → Raupenlader
~**fahrwerk** *n*, Zweiraupenfahrwerk, Raupen(unter)wagen *m*, (Gleis)Kettenfahrwerk, (Raupen-)Traktorenlaufwerk, Kettenlaufwerk, Raupenlaufwerk, (Gleis)Kettenlaufwerk, Raupenunterteil *m*, *n*, Raupen-Fahrgestell *n* = creeper undercarriage, crawler unit, crawlers, track-type undercarriage
~**fahrzeug** *n* → Raupe *f*
~**fertiger** *m* → (Gleis)Kettenfertiger
~**förderband** *n* = caterpillar belt conveyor, crawler ~ ~, ~ ~ conveyer
~-**Fronträumer** *m* → Raupen-Planierschlepper
~**geräte** *npl* = tracked plant, caterpillar equipment, caterpillars, tracklayers
~(**glieder**)**band** *n* → Raupenkette *f*
~-**Hoch**(**löffel**)**bagger** *m* → Hochlöffel-Raupenbagger
~-**Hublader** *m* → Raupen-Schaufellader
~**kette** *f*, Raupen(glieder)band *n*, Traktorenkette, (Gleis)Kette, Raupe *f*, Schlepperkette, Treckerkette = crawler, caterpillar track, (tracklaying) track, crawler track, creeper track, chain track, track chain
~**kraftschluß** *m*, (Gleis)Kettenkraftschluß = traction of tracks [*Gripping action between tracks and the surface*]
~**kran** *m*, (Gleis)Kettenkran = crawler-mounted crane
~**kupp(e)lung** *f*, (Gleis)Kettenkupp(e)lung = crawler clutch
~**ladegerät** *n* → Raupenlader
~**lademaschine** *f* → Raupenlader *m*
~**lader** *m*, Raupenlademaschine *f*, (Gleis)Kettenlader, (Gleis)Kettenlademaschine, Raupen-Ladegerät *n*, Raupen-Auflademaschine, Raupen-Auflader, Raupen(-Fahr)ladegerät, Raupen-Fahrlade-Maschine, Raupen-Fahrlader = crawler loader, ~ loading machine
~-**Ladeschaufel** *f* → (Gleis)Ketten-Ladeschaufel
~(**löffel**)**bagger** *m*, (Gleis)Ketten(löffel)bagger = crawler-mounted (mechanical) shovel
~-**Mehrzweckbagger** *m*, Raupen-Umbaubagger, (Gleis)Ketten-Umbaubagger, Raupen-Universalbagger, (Gleis)Ketten-Mehrzweckbagger, (Gleis)Ketten-Universalbagger, Raupen-Vielzweckbagger, (Gleis-)Ketten-Vielzweckbagger = crawler-mounted universal excavator
~-**Mischer** *m*, Raupen-Betonmischer = crawler-mounted concrete mixer
~-**Planierschlepper** *m*, Raupen-Bulldozer *m*, Raupen-Planiertraktor *m*, Raupen-Planiertrecker *m*, Raupen-Schürftrecker, Raupen-Schürftraktor, Raupen-Schürfschlepper, Raupen-Fronträumer = crawler-mounted bulldozer
~**platte** *f* → Bodenplatte
~**rahmen** *m*, (Gleis)Kettenrahmen = crawler frame, track ~
~-**Schaufellader** *m*, (Gleis)Ketten-Schaufellader, (Gleis)Ketten-Hublader, Raupen-Hublader = caterpillar loading shovel, crawler-mounted ~ ~
~-**Schaufelradbagger** *m*, (Gleis)Ketten-Schaufelradbagger = crawler-wheel excavator
~-**Schaufelradgrabenbagger** *m*, (Gleis-)Ketten-Schaufelradgrabenbagger = crawler-wheel trencher
Raupenschlepper *m* → Kettenschlepper
~ **mit Greifschild** = bullclam shovel
~ ~ **Sägenase**, Baumsäger *m* = saw nose dozer, tree saw
~ ~ **Sägeschild** = tree cutter
~ ~ **Seitenausleger** = side-boom crawler tractor
Raupen|-Schneckenfräse *f* → → Bodenvermörtelungsmaschine *f*
~**schuhplatte** *f*, Kettenschuh *m*, (Gleis)Kettenglied *n* = (track) shoe, crawler ~, tread plate

~-Schürfschlepper m → Raupen-Planierschlepper
~-Schürftraktor m → Raupen-Planierschlepper
~-Schürftrecker m → Raupen-Planierschlepper
~schütz(e) n, (f) = caterpillar gate
~-Schwenkkran m, Raupen-Drehkran, (Gleis)Ketten-Schwenkkran, (Gleis)Ketten-Drehkran = crawler-mounted revolving crane, ~ revolver ~, ~ slewing ~
~-Selbstauflader m → Becherwerk-(auf)lader auf (Gleis)Ketten
~spannen n, (Gleis)Kettenspannen = track tensioning, crawler ~
~spanner m, (Gleis)Kettenspanner = crawler take-up, track ~
~steuerung f, (Gleis)Kettensteuerung = crawler steering, track ~
~straßen(beton)mischer m → → Straßen(beton)mischer
~traktor m → Kettenschlepper
(~-)Traktorenlaufwerk n → Raupenfahrwerk
~trecker m → Kettenschlepper
~turas m, (Gleis)Kettenturas = crawler sprocket
~-Turmdrehkran m, (Gleis)Ketten-Turmdrehkran = crawler-mounted (mobile) tower crane
~-Umbaubagger m → Raupen-Mehrzweckbagger
~-Universalbagger m → Raupen-Mehrzweckbagger
~unterteil m, n → Raupenfahrwerk
~(unter)wagen m → Raupenfahrwerk n
~-Vielzweckbagger m → Raupen-Mehrzweckbagger
~wagen m → Raupenfahrwerk
~wagen m, (Gleis)Kettenwagen, Raupenanhängerwagen, (Gleis)Kettenanhängerwagen, Großraumlore f auf (Gleis)Ketten, Großraumlore auf Raupen, Großraum-Raupenwagen, Großraum-(Gleis)Kettenwagen = caterpillar tread wagon, track-type ~
Raute f = lozenge

Rauten|blech n = diamond plate
~dach n = lozenge roof
~form f = lozenge shape
rautenförmig = lozenge-shaped
rautenförmiger Schnellstraßenanschluß m = diamond interchange
Rauten|fries m = lozenge frieze
~gewölbe n → Netzgewölbe
Raymond|-Betonpfahl m aus teleskopartigen Futterrohren und glatten Rohren von größerer Wanddicke = Raymond pipe step taper pile
~-Standard-Betonpfahl m = Raymond standard concrete pile
Re, Reynolds'sche Zahl f = (critical) Reynolds number
Reagenzglas n, Probierglas = test-tube
~versuch m, Probierglasversuch, Reagenzglasprobe f, Reagenzglasprüfung f, Probierglasprüfung = tube test
Reaktions|geschwindigkeit f = reaction velocity
~gleichgewicht n = reaction equilibrium
~kinetik f = reaction kinetics
~kraft f = reaction force
~rohr n = tubular reactor
~turbine f → Überdruckturbine
Reaktor|-Druckschale f, Reaktor-Sicherheitsbehälter m, Reaktor-Druckgefäß n, Reaktor-Druckbehälter = containment, reactor pressure vessel
~-Gebäude n, Reaktorhalle f Meilerhaus n = reactor building
~katastrophe f = reactor disaster
~spaltstoff m = reactor fuel
(~-)Strahlenschutzanlage f, Schutzanlage, (Strahlen)Abschirmungsanlage = (reactor) shield, structure for shielding atomic plants
reale Flüssigkeit f, tatsächliche ~ = actual liquid
Realgar m, Rotrauschgelb n (Min.) = realgar
Rechen m, Harke f = rake
~ [Abwasserwesen] = screen [*A device with openings, generally of*

uniform size, used to retain coarse sewage solids. The screening element may consist of parallel bars, rods, or wires, grating, wire mesh, or perforated plate, and the openings may be of any shape, generally circular or rectangular slots]

~, Stab~ = rack [*A screen composed of parallel bars, either vertical or inclined, from which the screenings may be raked*]

~beispiel n, Berechnungsbeispiel = example of calculation

~fehler m = error of calculation

~gang m, Berechnungsgang [*Baustatik*] = design operation

~gut n, Rechenrückstand m = screenings [*Material removed from sewage by screens and racks*]

~klassierer m = trough classifier, rake ~

~kontrolle f = check calculations

~maschine f = calculating machine

~reiniger m, Rechenreinigungsmaschine f [*Maschine zum Freimachen eines Rechens von Rechengut und Eis*] = (trash) rack cleaner

~reinigung f = (trash) rack cleaning

~rückstand m → Rechengut n

~schacht m = screen well

~schema n = design constants

~schieber m, Rechenstab m = slide rule

~stab m, Rechenschieber m = slide rule

~tafel f, Berechnungstafel = chart

~verlust m = rack loss

~widerstand m = rack resistance

rechnerisch (ermittelt) = calculated, computed

rechnerisch vorgesehene Last f, Lastannahme f, angenommene Last = assumed load, design ~

~ Bemessung f = rational design

~ Festigkeit f → geforderte ~

Rechnungslegung f = billing

Rechteck|balken(träger) m = rectangular beam

~becken n = rectangular basin

~bunker m = rectangular bunker

~deckel m = rectangular cover

~-Dränung f = gridiron drainage

~dübel m = rectangular tie

~fliese f = rectangular tile

~gerinne n = rectangular channel

~hohlstütze f = hollow rectangular stanchion, ~ ~ column

rechteckige Masche f = rectangular mesh

~ Muffe f [*Am Ende eines vorgefertigten Röhrenelementes beim Tunnelbau*] = rectangular collar

~ Umschnürung f [*Säule*] = rectangular hooping, ~ hoop(s)

rechteckiger Ausschnitt m [*In einem Meßwehr zur Messung der überlaufenden Wassermengen*] = rectangular notch

~ Gründungsbrunnen m, ~ Senkbrunnen = rectangular open caisson, ~ drop shaft

~ Pilzkopf m = rectangular (column) capital

~ Senkbrunnen m → ~ Gründungsbrunnen

Rechteck|meßwehr n = rectangular weir

~pfeiler m = prismatic (foundation) pier, rectangular (~) ~

~platte f = rectangular slab

~querschnitt m = rectangular cross section

~schale f = rectangular shell

~sockelgründung f, Rechteckgründungssockel m = rectangular (single) base, ~ individual ~

~stahl m = rectangular steel

~welle f [*Vibration*] = square wave

rechter Pfahl m, Arbeitspfahl [*Naßbagger*] = working spud

rechtes Seitenwindenseil n, ~ Schwenkseil [*Naßbagger*] = right swing line, starboard ~

Rechts|abbiegen n = right-hand turn

~abbiegeverbot n = right-turn ban

~abbiegung f [*Straßenverkehr*] = right turn

~fahren n = driving on the right side

rechtsgängiges Gewinde n, Rechtsgewinde = right-hand thread

Rechts|geländer n = right-hand handrail
~händer m = right-handed person
~kurve f = right-hand curve
rechtsläufig, im Uhrzeigersinn = clockwise
Rechts|schloß n = right-hand lock
~treppe f = right-hand stairway
~tür f = right-hand door
rechtwink(e)liges Kurvennetz n, **orthogonales ~** = orthogonal curve system
Reckmaschine f = plate stretcher
Redestillat n = re-run oil
Redeverstärkerverfahren n = speech reinforcement system
Reduktion f (Geol.) = reduction, deoxidation
Reduktions|getriebe n = reduction gear
~tachymeter m, n = direct reading tacheometer, (self-)reducing ~
reduzierende Ofenatmosphäre f = reducing atmosphere, ~ conditions
Reduzier|flansch m = reducing flange
~nippel m = reducing nipple
~rohr n, verkehrtes Übergangsrohr = reducer [*A taper pipe reducing in diameter in the direction of flow*]
~stück n, Staubogen m [*Druckluftbetonförderer*] = adapter bend
~transformator m → Abspanner m
Reduzierung f **von Eisenerzen** = reduction of iron ores
Reduzier|ventil n, Druckminderventil = pressure reducing valve
~walzwerk n = sinking mill, reducing ~
Redwood-Viskosimeter I n = Redwood viscosimeter
~ II n = Admiralty viscosimeter
Reede f = roadstead
reemulgieren = to become reemulsified
Referat n, Beitrag m = paper
reflektierend = reflectorizing, reflecting
reflektierende Verkehrsfarbe f, **~ Markierungsfarbe, Reflexfarbe** = reflect(oriz)ing traffic paint, reflective ~ ~

Reflektions|verfahren n = seismic reflection method
~vermögen n = reflecting power
Refraktions|seismik f → Brechungsverfahren
~seismograph m = refraction seismograph
~verfahren n → Brechungsverfahren
Regal n = shelf
Regale npl = shelving, shelves
Regalkonsole f = shelf bracket
Regel|automat m, automatischer Regler m = auto(matic) governor
~bauweise f, Normalbauweise = standard construction method
~breite f, Normalbreite = normal width
~größe f, Normalgröße = standard size
~kammer f = control chamber
~kompaß m, Einstellkompaß = adjustment compass
~länge f, Normallänge = normal length, typical ~
~last f, Normallast = normal load, typical ~
regel|mäßige Schichtung f (Geol.) = regular bedding
~mäßiges Schichtmauerwerk n = regular coursed ashlar masonry, ~ ~ ashler ~
Regel|organ n **für den Kraftstoffdruck, Regulierorgan ~ ~ ~** = fuel pressure control
~profil n, Normalprofil = normal profile, typical ~
~pumpe f = variable displacement pump
~querschnitt m, Normalquerschnitt = typical cross section, normal ~ ~
~ring m [*Francis-Turbine*] = regulating ring
~spur f, Vollspur, Normspur = standard gauge (Brit.); ~ gage (US)
~stauvermögen n, Normalstauvermögen = normal storage capacity
Reg(e)lung f, Regulierung [*Bei der Regulierung bleibt das vorhandene Gefälle mit Ausnahme der Durchstiche bestehen. Bei der Kanalisierung wird das Gefälle treppenartig aufge-*

Regelventil — Regenwasserkanal

löst und damit bei jeder Wasserführung die erforderliche Mindesttiefe gewährleistet] = regulation

Regel|ventil *n*, Reguliertventil *[DIN 3141]* = regulating valve

~würfel *m*, Normalwürfel = typical cube, normal ~ *[concrete test cube]*

Regen *m* **mit Schnee** = sleet

~ablaufrohr *n* → Regen(fall)rohr

~abwasser *n* = storm water, ~ sewage

regenabweisend = rain-repellent

Regen|auslaß *m*, Regenauslauf *m* = storm-overflow

~auflauf *m*, Regenauslaß *n* = storm-overflow

~beiwert *m* = coefficient of rainfall, rainfall coefficient

~dauer *f* = duration of storm, ~ ~ rainfall, time ~ ~

regendicht = raintight

Regenerativofen *m*, Speicherofen = regenerative furnace

Regenerierbarkeit *f*, Regenerierfähigkeit *[Aufschweißen]* = rebuildability

regenerierter Gummi *m*, aufgearbeiteter ~ = reclaimed

regeneriertes Öl *n*, aufgearbeitetes ~ = reclaimed oil

Regen|erzeugung *f* = rainmaking, cloud-seeding

~fall *m* = rainfall

~(fall)rohr *n*, Fallrohr, Abfallrohr (für Regenwasser), Regenablaufrohr = downcomer, downpipe, downspout, fall pipe

~fang *m* *[Motor]* = rain trap

~front *f* = rain front

~gleiche *f*, Isohyete *f* = isohyetal line (Brit.); line of equal rainfall; isohyetal (US)

~guß *m*, Regenschauer *m*, Platzregen *m*, Sturzregen = (torrential) downpour; fence-lifter (US)

~haube *f* *[auf einem Schornstein]* = rainproof hood

~heftigkeit *f* → Regenstärke *f*

~hütchen *n*, Regenkappe *f* = rain cap

~index *m* = rainfall index

~intensität *f* *[(Feld)Beregnung]* = application rate

~intensität *f* → Regenstärke *f*

~kappe *f*, Regenhütchen *n*, Regenhaube *f* = rain cap

~karte *f* *[Karte, in der die Niederschlaghöhen durch Regengleichen (Isohyeten) dargestellt sind]* = isohyetal map

~kläranlage *f* = storm water clarification plant

~menge *f* = amount of rainfall

~messer *m* = rain gauge (Brit.); ~ gage (US)

~messung *f* = rain gauging (Brit.); ~ gaging (US)

~meßstelle *f* = rain gauging station (Brit.); ~ gaging ~ (US)

~rinne *f* *[am Dach]* = rainwater gutter along the eaves, eaves gutter

~rinne *f* (Geol.) = rain furrow

~rohr *n*, Regenfallrohr, Fallrohr, Abfallrohr (für Regenwasser), Regenablaufrohr = downcomer, downpipe, downspout, fall pipe

~sammler *m*, Regenwassersammler = storm (water) intercepting sewer

~schatten *m* = rain shadow

~schauer *m* = rain shower

~schirmschale *f* = umbrella shell

~schreiber *m* → Schreibregenmesser

~schutz *m* *[Motor]* = rain cap

regensicher = rain-proof

Regen|stärke *f*, Regenheftigkeit *f*; Regenintensität *f* *[Schweiz]* = intensity of rainfall, rain intensity

~statistik *f* = rainfall record

~tag *m* = rainfall day

Regenwasser *n* = storm water, rain ~ *[Excess water during rainfall or continuously following and resulting therefrom]*

~becken *n* = storm-water basin

~eindringung *f* = rain penetration

~kanal *m*, Flutdrän *m*, Regenwasserleitung *f* = storm sewer, ~ drain, storm-water ~, surface water ~, storm water conduit-type ~ *[A sewer which carries storm and surface water, street wash and other wash*

waters, or drainage, but excludes sewage]
~kanalisation *f* = storm drainage
~leitung *f* → Regenwasserkanal *m*
~(leitungs)netz *n* = strom sewer network
~netz *n*, Regenwasserleitungsnetz = storm water network
Regen(wasser)|sammler *m* = storm (water) intercepting sewer
~überfall *m* → Notauslaß *m*
Regenwasser|verdunstung *f* = f off [*The removal of rainwater by evaporation, as opposed to run off or cut-off (Brit.)/interception (US)*]
~vorfluter *m* = receiving water for storm water
Regen|wetter *n* = rainy weather
~zeit *f* = rainfall period
~zentrum *n* = area of maximum rainfall
Regie|arbeit(en) *f(pl)* = force(-)account (construction), ~ work
~arbeitskräfte *fpl* = direct (employed) labour, DEL (Brit.); state forces, public forces (US); day labour [*Australia*]
Regional|metamorphose *f*, Regionalumprägung *f*, Versenkungsumprägung = regional metamorphism
~planung *f* = regional planning
Registrier|apparat *m* → Registriergerät *n*
~-Barometer *n, m* → Barograph *m*
~boot *n*, Registrierschiff *n* [*Seeseismik*] = recording ship, ~ vessel
Registrieren *n*, Aufzeichnen = recording
registrieren, aufzeichnen = to record
registrierender Geschwindigkeitsmesser *m* = recording speedmeter
Registrier|gerät *n*, Registrierapparat *m* = recorder, recording apparatus (or meter)
~kasse *f* = cash totalizer
~manometer *n, m*, Druckschreiber *m*, Schreibmanometer = registering manometer, pressure recorder
~papier *n* = recording paper

~pegel *m*, selbstzeichnender Schwimmerpegel, Schreibpegel = recording gage, water-stage recorder
~rechenmaschine *f* = printing calculator
~schiff *n*, Registrierboot *n* [*Seeseismik*] = recording vessel, ~ ship
Registrierung *f*, Aufzeichnung = recording
Registrier|waage *f* = recording scale, ~ weighing machine
~wagen *m*, Meßwagen [*geophysikalischer Aufschluß*] = recording truck
Regler *m* = governor
~ [*Kaplanturbine*] = regulating gear
~feder *f* [*Motor*] = governor spring
~gestänge *n* [*Motor*] = governor linkage
~tauchkolben *m* = regulating plunger
Reglung *f* → Regelung
Regner *m*, Beregnungsgerät *n* = sprinkler
~düse *f* = sprinkler jet
~leitung *f* = lateral (pipe) line containing sprinklers
~mundstück *n* = sprinkler nozzle
reguläres System *n* [*Kristall*] = cubic system, regular ~
Regulieren *n* **von abgesunkenen Betonfahrbahnplatten** → (Beton)Plattenheben
Regulier|organ *n* **für den Kraftstoffdruck**, Regelorgan ~ ~ ~ = fuel pressure control
~schieber *m* → Durchflußschieber
~- **& Druckreduzierventil** *n* = regulating and pressure reducing valve
Regulierung *f* → Reg(e)lung
Regulierungs|bauwerk *n* = regulator, regulating works
~stau *m* = regulating storage
Regulierventil *n*, Regelventil [*DIN 3141*] = regulating valve
Rehbock|-Formel *f* = Rehbock (weir) formula
~-Meßwehr *n* = Rehbock measuring weir
Rehbock'sche Zahnschwelle *f* = Rehbock dental

Reib|ahle *f*, Räumer *m*, Aufreiber *m* = reamer
~antrieb *m* = friction drive
Reib(e)|brett *n* [*Brett mit Handgriff zum Verreiben des Putzes*] = plasterer's float
~holz *n*, (Holz)Fender *m* = timber fender
reibender Brechvorgang *m* → → Bakkenbrecher *m*
Reibepfahl *m* = fender pile
Reib|fläche *f* = friction area
~korrosion *f* = fretting corrosion
~kupp(e)lung *f* → Friktionskupp(e)lung
~oxydation *f* [*Korrosion unter mechanischer Beanspruchung. DIN 50900*] = fretting corrosion [*Corrosion under mechanical stress*]
Reibrad *n* = friction wheel
~betonmischer *m* = friction wheel-drive concrete mixer
~winde *f*, Reib(ungs)winde, Friktionswinde = friction hoist
Reib|rolle *f* = friction roller
~schale *f*, Mörser *m* [*DIN 12 906*] = porcelain mortar
~scheibe *f* = friction disc; ~ disk (US)
~scheibenkupp(e)lung *f* = disk friction clutch (US); disc ~ ~
~trieb *m*, Reibradtrieb = friction wheel drive
Reibung *f* = friction
~ zwischen Reifen und Straßenoberfläche = road friction
reibungs|bedingter Spannkraftverlust *m* = frictional prestress loss
~behinderter Behälterfuß *m*, ~ Fußring *m* = sliding (tank) (base) joint
Reibungs|beiwert *m*, Reibungsziffer *f*, Reibungszahl *f* = friction(al) coefficient, ~ factor, coefficient of friction
~belastung *f* = friction(al) loading
~brekzie *f* = crush-breccia
~bremse *f*, mechanische ~ = friction brake
~druck *m* = friction pressure
~druckverlust *m* → Reibungshöhe
~eigenschaft *f* = frictional property

~fläche *f* = friction surface
~gefälle *n* = friction slope
~höhe *f*, Reibungsdruckverlust *m* [*Die zum Überwinden des Reibungswiderstandes in einer Rohrleitung notwendige Druckhöhe*] = friction head, ~ loss
~kegel *m* = friction cone
~koeffizient *m* → Reibungsbeiwert *m*
~konglomerat *n* = crush conglomerate
~kopf *m* (⚒) = friction cap
~kraft *f* = friction(al) force
~kreis *m* [*Bodenmechanik*] = friction circle
~kupp(e)lung *f* → Friktionskupp(e)lung
~kurve *f* = friction curve
reibungs|los = frictionless
~lose Flüssigkeit *f* → ideale ~
Reibungs|messer *m* = friction meter
~messung *f* [*Straße*] = skidding test, measurement of slipperiness
~pfahl *m* = friction pile, floating ~
~richtung *f* = direction of friction
~schluß *m* = friction grip
~spannung *f* = stress due to friction, friction stress
~stempel *m* (⚒) = friction prop, ~ post
~strömung *f* = viscous flow
~verankerung *f* = friction-type anchorage, anchoring by friction
~verlust *m* = friction(al) loss
~widerstand *m* = friction(al) resistance, ~ drag
~winkel *m*, innerer ~, Winkel der inneren Reibung *f* = angle of internal friction
~wirkung *f* = frictional effect
~zahl *f* → Reibungsbeiwert *m*
~ziffer *f* → Reibungsbeiwert *m*
reiches Erz *n* = rich mineral
reichlich bemessen = of ample dimensions, ~ generous ~
Reichnaturgas *n* → feuchtes Erdgas
Reichweite *f* → Bereich *m*
~, Ausladung *f* [*Lader*] = dumping reach (at 45° discharge angle)
Reifen *m* **mit Stahldraht-Zwischenlage** = shredded wire undertread tyre (Brit.); ~ ~ ~ tire (US); SWU ~

Reifenabmessung — Reihenpflaster

~abmessung *f* = tire dimension (US); tyre ~ (Brit.)
~-Aufladegerät *n* → Radlader
~-Auflademaschine *f* → Radlader
~auflader *m* → Radlader
(~)Aufstandfläche *f* = contact area
~bauer *m* = tyre designer (Brit.); tire ~ (US)
~befeuchtungsanlage *f* [*Gummiradwalze*] = tire sprinkling system (US); tyre ~ ~ (Brit.)
~bewehrung *f*, Reifenarmierung *f* = hoop reinforcement
~bezeichnung *f* = size marking [*tyres*]
~cord *m*, Reifenkord = tyre cord (Brit.); tire ~ (US)
~druck *m* = tire pressure (US); tyre ~ (Brit.)
~druckmesser *m* = tire pressure gage (US); tyre ~ gauge (Brit.)
~eindruck *m* = tyre impression (Brit.); tire ~ (US)
~-(Fahrlade-)Maschine *f* → Radlader
~-Fahrlader *m* → Radlader
~fahrwerk *n* = wheel-type chassis
~fahrzeug *n* = tyred vehicle (Brit.); tired ~ (US)
~fender *m*, Gummifender = (rubber) tyre fender (Brit.); (~) tire ~ (US)
~gewebe *n* = tyre fabric (Brit.); tire ~ (US)
~griffigkeit *f* = tyre grip (Brit.); tire ~ (US)
~hersteller *m* = tyre manufacturer (Brit.); tire ~ (US)
~innendruck *m* = inflation pressure
~kette *f* → Gleitschutzkette
~-Kompressor *m* = (air) compressor for inflating tires (US)/tyres (Brit.)
~kord *m* → Reifencord
~kraft *f* = tyre force (Brit.); tire ~ (US)
~kraftschluß *m* = rimpull, grip between tyre and road [*The force available between the tire and the ground to propel the vehicle forward*]
~kran *m* = rubber-tired crane (US); rubber-tyred ~ (Brit.)
~-Lademaschine *f* → Radlader *m*
~lader *m* → Radlader

~lagen *fpl* = ply rating
~lauffläche *f* = tyre tread (Brit.); tire ~ (US)
(~)Laufflächenprofil *n* = tyre tread pattern (Brit.); tire ~ ~ (US)
~-Maschine *f* → Radlader
~planiergerät *n* → → Bulldozer
~platzen *n* = tyre burst (Brit.); tire ~ (US)
~profil *n* = tyre section, ~ tread pattern (Brit.); tire section (US)
~pumpe *f* = tire pump (US); tyre ~ (Brit.)
~pumpgerät *n* = tire inflation kit (US); tyre ~ ~ (Brit.)
~schlauch *m* → Luft~
~schlepper *m* → Luft~
~schlupf *m* = tyre slip (Brit.); tire ~ (US)
~(schutz)kette *f* = tyre chain, driving ~ (Brit.); tire ~ (US)
~trecker *m* mit Planierschild → → Bulldozer *m*
~walze *f* = tyre roller (Brit.); tire ~ (US)
~wandung *f* = tyre sidewall (Brit.); tire ~ (US)
Reihe *f*, arithmetische ~ = arithmetical progression
~, geometrische ~ = geometrical progression
~, Bau~, Typen~ [*Maschine*] = series
Reihen|becherwerk *n*, Vollbecherwerk, Vollelevator *m*, Becherwerk in Becher-an-Becher-Ausführung = continuous (bucket) (type) elevator
~bepflanzung *f*, Baumreihe *f* = avenue planting
~folge *f* des Abbaues, Abbaureihenfolge (⚒) = sequence of getting
~form *f* = gang mould (Brit.); ~ mold (US)
~garagen *fpl* = battery garages
~haus *n* = terrace house
~pflanzen *fpl* = row crops
~pflaster *n* = coursed sett paving, peg top ~, straight course ~, stone sett paving laid in rectilinear pattern, paving in rows

~ramme *f* = frame type pile driving plant for driving in row arrangement, ~ ~ piling ~ ~ ~ ~ ~ ~
~schlußmotor *m* = series(-wound) motor
~silo *m* = rectangular multiple-compartment bin
(~)Siloanlage *f* → Abmeßanlage für Betonzuschlagstoffe
~spritzvorrichtung *f* → Sprengrampe *f*
~stempel *m* (⚒) = breaker prop, ~ post
reine Biegung *f* = simple flexure
~ Bohrzeit *f*, „Bohrhitze" *f* = actual drilling time, ~ boring ~
~ Fahrgeschwindigkeit *f* [*Straßenverkehrstechnik*] = running speed
~ Fahrzeit *f* = running time
~ Zugfestigkeit *f* = direct tension (or tensile) strength
reiner Druck *m* → Alleindruck
~ Hagel(fall) *m* = isolated hail
Rein|aluminium *n*, Hüttenaluminium = pure alumin(i)um
~erzausbringen *n* = recovery
~gas *n* = clean gas
~haltung *f* [*Wasser; Luft*] = pollution prevention
~haltungsverordnung *f* = pollution code
Reinheit *f* = purity
Reinheits|grad *m* = degree of purification [*A measure of the removal and oxidation of the objectionable and putrescible contents of sewage*]
~schicht *f* → Sauberkeitsschicht
reinigen [*Luft*] = to purify, to clean
Reinigung *f* **fluorhaltiger Abgase** = control of fluoride emissions
Reinigungs|anstalt *f* [*für Bekleidung*] = dry cleaning shop
~bürste *f* = cleaning brush
~gerät *n* [*für Kanalisationen*] = cleaning epuipment
~klappe *f* [*Lüftung*] = access door
~lösung *f* = cleaning solution
~luke *f* **für Einstieg** = manhole [*An access hole to a tank or boiler drum just large enough for a man to enter. It is normally covered with a cast-iron or steel plate called the manhole cover*]
~lukendeckel *m* = manhole cover
~mittel *n*, Reinigungsstoff *m* = cleaning agent
~schacht *m*, Einsteigschacht [*zur Kanalkontrolle und Kanalreinigung*] = manhole
~spritze *f* = spray-down equipment, spray cleaner [*for cleaning construction machinery*]
~tür *f*, Kamintür = soot door (Brit.); ashpit ~, cleanout ~ (US)
~vorgang *m* = process of purification, purification process
Reinluft *f* = purified air, clean ~
Reinwasser *n* = purified water, clean ~
~behälter *m* = purified water tank, clean ~ ~
~stollen *m* = purified water gallery, clean ~ ~
Reisender *m*, Fahrgast *m*, Passagier *m* = passenger
Reisezug|bau *m* = railway coachbuilding (Brit.); railroad ~ (US)
~wagen *mpl* = passenger stock
Reiß|barkeit *f* → Reißfähigkeit
~brett *n*, Zeichenbrett = drawing board
~brettstift *m* = drawing pin
~brettzeichnen *n* = mechanical drawing
reißen = to crack
Reißen *n* = cracking
~ von Gestein = ripping of soft rock
Reiß|fähigleit *f*, Reißbarkeit [*Straßendecke; Boden*] = rippability
~(farb)lack *m* = brittle lacquer
~feder *f*, Zeichenfeder = ruling pen, drawing ~
~kraft *f* [*Bagger*] = tooth effect, tearout force, biting force, breakout force
~lack *m*, Reißfarblack = brittle lacquer
~maß *n* → Streichmaß
~nagel *m*, Reißzwecke *f* = thumb tack
Reißschenkel *m* = ripper shank, scarifier ~

Reißschenkelbolzen — Reserve-Reifenkraftschluß

~bolzen m = shank pin
~halter m [Aufreißer] = shank socket
Reiß|schiene f = tee-square
~streifen m [Aufreißer] = ripping path
~tiefe f → Arbeitstiefe
~winkel m [Aufreißer] = shank angle, ripping ~
Reißzahn m [Bagger] = breaking tooth
~ [Aufreißer] = ripper tooth, scarifier ~
~stellung f = shank position
Reiß|zeug n = drawing instruments, ~ set
~zwecke f = drawing pin (Brit.); thumb tack (US)
Reiter|sparren m = dormer rafter
~standbild n = equestrian statue
Reithdach n → Rethdach
Reitweg m = bridle path
Reklameschild n = advertising sign
Rekristallisation f, Alterung f [Schraube; Niet] = crystallization
Relativbewegung f = relative motion
~ [bei der Bewegungsreibung] = relative sliding movement
relative Feuchtigkeit f, ~ Feuchte f = ralative humidity, R. H.
(Relativ)Gefälle n [Hydraulik] = gradient [Change of elevation, velocity, pressure, or other characteristic per unit length; slope]
Relativrauhigkeit f = relative roughness
Relaxation f [Spannungsabfall im Stahl bei gleichbleibender Dehnung] = relaxation
Relaxationsverfahren n, Relaxationsmethode f = relaxation method
Relief|modell n = relief model
~muster n = relief pattern
Reliktmineral n = relic mineral
Remorquer m [Österreich] → Schlepper
Renaissance-Kreuzung f → Kleeblatt-Kreuzung
Renardreihe f, Normalzahlreihe Renard = Renard series
Renchgneis m, Sedimentgneis, Paragneis = paragneiss

Rendzinaboden m, Humuskarbonatboden = chalk humus soil, humus carbonate ~, rendzinic ~, humic carbonated ~, raw humus rendzina
Renn|bahn f = race track
~platztribüne f = race-course stand
Rentabilität f, Wirtschaftlichkeit f = profitability, rentability, economy
Reparatur f → Instandsetzung f
~dienst m = repair service
~glaser m = repairing glazier
~grube f, Arbeitsgrube, Untersuchungsgrube = repair pit
~hafen m, Ausbesserungshafen = repair port
~haken m → Ausbesserungshaken
~kai m, Ausbesserungskai = repair quay
~masse f [für Betonreparaturen] = patching compound
~park m = repair depot
~satz m = changeover kit
~schlosser m, Maschinenschlosser = mechanic
~werft f, Ausbesserungswerft = repair shipyard
~werkstatt f = repair (work)shop
Reproduktionsriß m = reflection crack [Crack in old concrete pavements reappearing as crack in the black top surface]
reproduzierbarer Wert m = reproducible value
Reproduzierbarkeit f = reproducibility
Repulsionsmotor m mit Dämpferwick(e)lung = repulsion motor with damper winding
Reserve|analage f = stand-by plant
~behälter m = make-up tank
~bestand m, Pufferbestand = buffer stock
~bunker m, Pufferbunker = surge bin
~halde f, Reservematerialhalde = reserve stockpile
~kapazität f = stand-by capacity
~mischer m = stand-by mixer
~pumpe f = stand-by pump, spare ~
~-Reifenkraftschluß m = reserve rimpull

Reserveseil — Reynolds'sche Zahl

~seil n, Hilfsseil [Seilschwebebahn] = emergency cable, ~ rope
~silo m, Puffersilo = surge silo
Resonanz f = resonance
~bedingung f = resonance condition
~bereich m = resonance range
~förderer m = resonance conveyor, ~ conveyer
~frequenz f = resonant frequency, resonance ~
~siebmaschine f = resonance (vibrating) screen, ~ vibro-screen
~spannung f = resonance potential
~tisch m = resonance table vibrator
~verfahren n [Schwingungsberechnung] = resonance method, resonating ~
Restaurantgeschoß n, Restaurantstockwerk n = restaurant floor
Rest|chlor n = residual chlorine
~druck m = residual pressure
~feuchtigkeit f = residual moisture
~härte f [Die nach der Enthärtung im Wasser noch vorhandene Härte] = residual hardness
~-Entleerung f = full discharge, complete ~
~pfeiler m (⚒) = coal pillar
~sauerstoff m = residual oxygen
~schwindung f = residual shrinkage
~spannung f = residual stress
~wassermenge f [strömt im Flutraum hin und her und behindert den Eintritt der Flutwelle] = residual water mass, oscillating ~ ~, opposing ~ ~
Resultante f → Mittelkraft f
Resultierende f → Mittelkraft f
resultierende Kohäsion f = resultant cohesion
resultierendes Moment n, Mittelmoment = resulting moment
(~) Wasserdruckdiagramm n = diagram of normal unit pressures
Rethdach n, Reetdach, Reithdach, Rieddach = reed roof
Retorte f = retort
Retreadverfahren n, Oberflächenerneuerungsverfahren = retread

Retrometamorphose f → rückläufige Metamorphose
Rettungs|arbeiten fpl = rescue work
~bohrloch n, Rettungsbohrung f = rescue borehole
~boje f = life buoy
~boot n = lifeboat
~leiter f = escape ladder
~trupp m (⚒) = rescue party
~wache f = rescue station
Reverberierofen m, Flammofen = reverberatory furnace
Reversierband n, Umkehrband = reversing belt
reversierbarer (Schwing)Rohrförderer m, reversierbares (Schwing) Förderrohr n = reversing vibrating circular pipe-line
Reversier|getriebe n, Umkehrgetriebe, Wendegetriebe = reversing gears, reverse ~, forward-reverse transmission
~mischer m, Umkehr(trommel)(beton)mischer, Freifallmischer mit Umkehraustragung, Mischer mit Reversier-Mischtrommel, Chargen-Betonmischer mit Wendetrommel, Umkehrtrommel-Freifallmischer, Freifall-Betonmischer mit Rücklaufentleerung, Reversier-Betonmischer = closed drum (concrete) mixer with reverse discharge, (free fall type) non-tilt(ing) (drum) mixer, N. T. mixer
~turbine f → Umkehrturbine
Revisions|schacht m = inspection shaft
~-Zeichnung f → Abnahme-Zeichnung
Reynolds'sche Grenzgeschwindigkeit f [Hydromechanik] = Reynolds critical velocity [Is that at which the flow changes from laminar to turbulent, and where friction ceases to be proportional to the first power of the velocity and becomes proportional to a higher power – practically the square]
~ **Zahl** f, Re = (critical) Reynolds number

Reyon n, m = rayon
~-**Hersteller** m = rayon manufacturer, ~ producer
~-**Kord** m = rayon cord
~-**Reifengarn** n = rayon tyre yarn (Brit.); ~ tire ~ (US)
Rezept n, Misch~, Mischungsformel f = mix formula
~**wähler** m [*Betonaufbereitung*] = mix selection mechanism, mix-selector
reziproker Kräfteplan m, Bow'scher ~ = reciprocal force polygon, Bow's ~ ~
~ **Wert** m, Kehrwert = inverse value, reversed ~, reciprocal ~
Rhede f → Reede f
rheologische Eigenschaft f = rheological property
Rhodochrosit m (Min.) → Manganspat m
Rhombenporphyr m, rautenförmiger Purpurstein m = rhomb porphyry
Richt|aufbauweise f, Aufkipp-Bauweise = tilt-up construction (or method)
~**auftafel** f = tilt-up panel
~**einbaustreifen** m [*Straßenbau*] = pilot (paving) lane (US); ~ pavement ~
richten, ab~ [*auf genaue Form bringen*] = to dress
Richt|fest n = builder's treat, topping-out ceremony
~**feuer** n = heading light, range ~
~**funkbake** f, Richtfunkfeuer n = directional radio beacon
~**funkempfänger** m = radio direction finding receiver
~**keil** m, Ablenkkeil, Abweichungskeil [*zum Ablenken von Bohrlöchern*] = whipstock
~**linien** fpl = code of recommended practice, directives
~**maschine** f = straightening machine
~**pfahl** m = guiding pile, guide ~
~**platte** f, Beilage f [*Gleiskettenlaufwerk*] = adjustment plate
~**schacht** m = pilot shaft
~**scheit** n = straightedge
~**stellung** f = aiming position

~**stollen** m, Richtvortrieb m = pilot heading, ~ tunnel, monkey drift
~**strecke** f (⚒) = lateral
~- **und Spannbett** n = straightening and prestressing bed
Richtungs|gleis n = forwarding track
~**kupp(e)lung** f = direction clutch
~**planetenträger** m [*Kraftübertragung*] = directional carrier
~**schalthebel** m = forward-reverse lever
~**schild** n = advance sign, distant ~
~**steuerventil** n = direction selector valve
~- **und Gangsteuergruppe** f = direction and range valve group
~**vorschriften** fpl = regulations concerning the direction in which traffic is allowed to travel
~**wechsel** m = change of direction
Richt|vortrieb m, Richtstollen m = pilot tunnel, ~ heading, monkey drift
~**werk** n → Parallelwerk
~**wert** m = guide value
Ridley-Scholes-Verfahren n = Ridley-Scholes process, sand flotation
Rieddach n → Rethdach
Riefe f → Kannelur f
Riefelung f, Kannelierung = fluting
Riegel m [*Rahmen*] = spanning member
~ [*Hakenschütz(e)*] = horizontal lifting beam
~ [*Schloß*] = latch
~- **und Ständertor** n **mit Gewichtsausgleich** = balanced gate with multiple transoms and multiple vertical frames
~**wand** f, Fachwerkwand = framed wall
Riemchen n → Viertelstein
Riemen m = belt
~**(fuß)boden** m → Schiffboden
~**pumpe** f = belt driven pump
~**scheibe** f = belt pulley
~**scheibe** f [*pennsylvanisches Bohrverfahren*] = bandwheel
~**scheibenantrieb** m = belt pulley drive

~scheiben-Übersetzungsverhältnis *n* = pulley ratio
~tang *m*, Laminarie *f* = laminaria
~trieb *m*, Bandtrieb = belt drive
~verbinderdraht *m* = belt fastener wire
(~)Vorgelege *n* = countershaft
Riesel-Entgaser *m* [*Wasseraufbereitung*] = degassing column
rieselfähiges Gut *n* = free-flowing material
Riesel|feld *n* = irrigated sewage field
~furche *f*, Bewässerungsfurche = irrigation furrow
~gut *n* = sewage farm
~kohle *f* → Formkohle
~regen *m*, Sprühregen, Nieseln *n* = drizzle, drizzling rain
~rohr *n* [*gelochtes Tonrohr zur Untergrundberies(e)lung*] = subirrigation pipe
~schutt *m* (Geol.) = fine rolling rock débris
~verfahren *n*, (Be)Ries(e)lung *f* = flooding (method) [*irrigation*]
~werk *n* → Gradierwerk
Riesen|luftreifen *m*, großvolumiger Luftreifen = giant (or oversize) pneumatic tire (US)/tyre (Brit.)
~reifen *m*, großvolumiger Reifen = giant (or oversize) tire (US)/tyre (Brit.)
~stock *m* (Geol.) → Batholith *m*
~topf *m* (Geol.) = giants' kettle
Riff *n* = reef
Riffel|bildung *f* (Geol.) = ripple formation, sand wave ~
~blech *n*, geripptes Blech = channeled plate, checker(ed) ~
Riffelung *f*, Riffeln *n* [*Betondeckenbau*] = texturing, grooving, scoring, scarifying [*It is a process of cutting grooves in hardened concrete surfaces*]
Riffelwalze *f* = indenting roller, branding iron, crimper
Riffkalkstein *m* = reef limestone
Rigole *f*, Sauggraben *m*, Sickergraben [*mit Kies ausgepackt*] = gravel-filled drain trench

~, Stein~, Sauggraben *m*, Sickergraben [*mit Steinen ausgepackt*] = stone-filled drain trench
~ mit Dränrohr, Sickergraben *m* ~ ~, Sauggraben ~ ~ = French drain
Rigolpflug *m* → Grabenpflug
Rille *f* = groove
Rillen|bohrer *m* = grooved soil sampler
~kugellager *n* = grooved ball bearing
~profil *n* = rib tread
~putzgeflecht *n* = corrugated lath(ing)
Ring *m* = annulus, ring
~ [*einer Flachfurchenwalze*] = collar
~ [*Ein auf einer Welle angebrachter, verschiebbarer Ring, der nicht mit der Welle umläuft. Hauptsächlich in Kupplungen und Getrieben gebraucht*] = collar
~analyse *f* = structural group analysis
~armierung *f* → Umschnürung
~aussteifung *f* = ring stiffener
~aussteifung *f*, Spinne *f* [*Gleitschalungsbau*] = (deck) spider
~balken(träger) *m* = ring(-shaped) beam, circular ~
~bewehrung *f* → Umschnürung
~bildung *f* = ring formation
~dehner *m* = ring expander
~dränage(system) *f*, (*n*) = circumferential drainage system
~dübel *m* „Alligator", Zahn~ ~, Alligator-(Zahn)Ringdübel = toothed-ring
~einlagen *fpl* → Umschnürung
~fläche *f* = annular surface
ringförmiger Frostkörper *m* [*Gefrierverfahren*] = ice ring, ~ annulus
~ Querschnitt *m*, Ringquerschnitt = annular section
ringförmiges Konsolgerüst *n* [*Schornsteinbau im Gleitschalungsverfahren*] = circumferential scaffold(ing)
Ring|kugellager *n* = annular ball bearing
~-Kugel-Verfahren *n*, Ring und Kugelmethode *f* = Ring-and-Ball test (or method), R. and B. ~ (~ ~)

~**leitung** *f* = ring main
~**mühle** *f* → Ringwalzenmühle
~**ofen** *m* = circular kiln
~**polygon** *n* = closed traverse
~**querschnitt** *m*, ringförmiger Querschnitt = annular section
~**rad** *n*, Glockenrad = ring gear
~**raum** *m* = annular space
~**riff** *n*, Atoll *n*, Lagunenriff = atoll
~**rillenlager** *n* = ring groove bearing
~**rost** *m* = circular grate
~**schale** *f* = toroid shell
~**schergerät** *n*, Kreisringscherapparat *m* = ring shear apparatus
~**schieber** *m* = needle valve
~**schmierlager** *n* = bearing for ring lubrication
~**schmierung** *f* = ring lubrication
~**spannglied** *n* = circumferential tendon
~**spannung** *f* = hoop stress
~**(stahl)einlagen** *fpl* → Umschnürung
~**straße** *f* = ring road, ~ highway
~**straße** *f* [*Flugplatz*] = perimeter
~**stück** *n* = banjo fitting
~**träger** *m* = ring(-shaped) girder, circular ~
~**träger** *m* [*Motor*] = ring band
(**ring**)**umschnürte Säule** *f*, ~ Stütze [*Stahlbetonstütze mit kreisförmigem Kernquerschnitt und Ringbewehrung*] = hooped column
Ring|umschnürung *f* → Umschnürung
~**- und Kugel-Verfahren** *n*, Ring-Kugel-Methode *f*, R. u. K. = ring and ball method, ~ ~ ~ test, R. and B. ~
~**ventil** *n* = annular valve
~**vorspannung** *f*, ringförmige Vorspannung = circumferential prestressing
~**(walzen)mühle** *f*, Walzenringmühle = ring (roll(er)) mill
~**wasserleitung** *f* = ring water main
~**zwischenraum** *m* = ring gap, annulus
Rinne *f* → Gerinne *n*
~ [*zur Abführung des Oberflächenwassers einer Straße*] = gutter
~ [*Gußbetonanlage*] = chute, spout

~, halbiertes Rohr *n* [*Kanalisationssteinzeug*] = channel
~ [*Kläranlage*] = trough, gutter, drain
~ **in der Sohle eines Abwasser-Einsteigschachtes** = flowing-through channel
Rinnen|beton *m* → → Beton
~**eisen** *n*, Rinneisen, Rinnenbügel *m* = gutter bearer
~**förderer** *m* → Förderrinne *f*
~**gefälle** *n* [*zur Oberflächenwasserabführung bei einer Straße*] = gutter gradient
~**ofen** *m* = channel-type induction furnace
~**pflaster** *n* = gutter paving
~**schelle** *f* = gutter bracket
~**stahl** *m*, tragförmiger Formstahl = trough section, troughing
~**stein** *m* [*Straßenbau*] = gutter (paving) sett
~**träger** *m* = gutter bearer
Rinn|kessel *m* = eaves trough, ~ trow
~**sal** *n* = small creek, ~ brook, rill
Rinnstein *m*, Bordrinne *f* = gutter, channel
~, Pflasterrinne *f* = paved gutter
~**form** *f*, Bordrinnenform = channel mould, gutter ~ (Brit.); ~ mold (US)
~**reinigungsmaschine** *f* **mit Aufladebecherwerk** = channel scraper and elevator
Rippe *f* = rib
~ = cleat [*Arranged transversely on a conveyor belt to make possible high angles of inclination*]
Rippel(marke) *f*, Wellenfurche *f* = ripple marking
Rippen|armierung *f* → Rippenbewehrung
~**balken(träger)** *m* = ribbed girder
~**beton** *m* = rib concrete
~**bewehrung** *f*, Rippenarmierung, Rippen(stahl)einlagen *fpl* = reinforcement in ribs, steel ~ ~
~**blech** *n* = ribbed plate
~**bogen** *m* = ribbed arch
~**decke** *f* = ribbed(-slab) floor

Rippendecken(-Form)stein — Ritzel

~decken(-Form)stein m → Füllkörper
~deckenschalung f = rib-span formwork
~dübel m = ribbed dowel
~einlagen fpl → Rippenbewehrung f
~feld n = ribbed panel
~gewölbe n = ribbed vault
~gurt m = grip-face belt [belt conveyor]
~kuppel f = ribbed dome
~platte f = ribbed slab, ~ panel [A panel composed of a thin slab reinforced by a system of ribs in one or two directions, usually orthogonal]
~putzträger m = ribbed lath(ing)
~rohr n, Lamellenrohr, beripptes Rohr = fin(ned) pipe, ~ tube, gilled ~, ribbed ~, grilled ~
~rohr-Wärme(aus)tauscher m = fin tube heat exchanger
~stab m = ribbed bar
~stahl m → Beton~
~(stahl)einlagen fpl → Rippenbewehrung f
Rippstreckmetall n = rib lath, rib-type expanded metal, rib mesh
~(-Fertigungs)straße f = rib lath line
~-Putzträger m = ribbed expanded metal lath(ing) for plastering
rippenversteift = rib-strengthened
Risse|ausbreitung f, Rissefortpflanzung = crack propagation
~bild n = crack(ing) pattern
~fortpflanzung f, Risseausbreitung = crack propagation
~freiheit f, Rißfreiheit = absence of cracks
~prüfer m = crack tester
risseverhütend = crack-preventing
Risseverteilung f = distribution of cracks, crack distribution
rissig, gerissen = cracked
Rissigwerden n, Rißbildung f = crack formation
Riß m = crack
~ [Bei technischen Zeichnungen werden die verschiedenen Projektionen als Risse bezeichnet] = elevation
~armierung f → Rißbewehrung

~ausbreitung f = crack extension
~beherrschung f = control of cracking
~belastung f = crack loading
rißbewehrt, rißarmiert [Beton] = crack-reinforced [concrete]
Riß|bewehrung f, Rißarmierung, Riß(stahl)einlagen fpl = anticrack reinforcement, crack control ~
~bildung f, Rissigwerden n = crack formation
~breite f = crack width
~dehnung f = cracking strain
~einlagen fpl → Rißbewehrung f
~festigkeit f [Beton] = resistance to cracking, extensibility
~fortpflanzung f = propagation of cracks, crack propagation
rißfrei = crack-free
Riß|freiheit f, Rissefreiheit = absence of cracks
~gefahr f = risk of cracking
~last f = cracking load
~moment n = cracking moment
~momentvermögen n = cracking moment capacity
~neigung f = tendency to crack
~probe f, Rißprüfung f, Rißversuch m = cracking test
~sicherheit f = crack resistance
~sicherung f = precaution against cracking
~spannung f = crack(ing) stress
~(stahl)einlagen fpl → Rißbewehrung
~unempfindlichkeit f = resistance to welding
~vergußmasse f = crack sealer
~versuch m, Rißprobe f, Rißprüfung f = cracking test
~wanderung f = crack wander
~weitung f = crack widening
~zeit f = cracking time
~zone f = fissured zone
Ritterdach n, Kronendach = high-pitched roof
Ritzel n, Antriebszahnrad n, Triebling m = drive gear, pinion, driving gear
~ und Tellerrad n spiralverzahnt = spiral bevel pinion and ring gear

Ritzelgehäuse — Rohbau

~gehäuse n = pinion housing
~welle f = pinion shaft
~wellenlager n = pinion shaft bearing
Ritz|härteprüfer m = sclerometer

~probe f → Ritzprüfung f
~prüfung f, Ritzprobe f, Ritzversuch m [*Härtebestimmung*] = scratch test
~versuch m → Ritzprüfung f

Road-Mix-Verfahren n [*ist gewissermaßen eine Abart der „Bodenvermörtelung" und kann ebenfalls nach dem System „mix-in-place" arbeiten*]
    Reinigen n
    Grundieren n und Oberflächenbehand(e)lung f
    Aufbringen n
    Fremd-Zuschlag(stoff) m
    Schwaden m, Längsreihe f
    Trocknen n der Zuschlagstoffe, ~
    ~ Zuschläge
    Mischen n
    Verteilen n
    Verdichtung f
    Mischen n mit Straßenhobel, Schar-Mischung f, Messermischen
    An-Ort-Mischverfahren n, Mischverfahren ohne Anheben n des Mischgutes [*auch mitunter als „Bodenvermörtelung" bezeichnet*]

road-mix (method)

    sweeping
    priming

    aggregate application
    commercial material
    windrow
    aggregate drying

    mixing
    spreading
    compaction
    blade mixing

    mix-in-place (method)

Robertson-Stahlzellendecke f = Q-floor
Rodeland n = cull-land
Rödeldraht m → Bindedraht
~anker m, Bindedrahtanker = twisted wire tie
~-Vorratstrommel f → Bindedraht-Vorratstrommel
Rodemaschine f, Roder m = grubbing machine
 ~ mit Teleskopausleger = tree stinger
 ~ mit Walzen = tree crusher
roden = to grub
Roden n → Rodung f
Roderechen m, Unterholz~, Gestrüppharke f = clearing rake, brush ~
Rodung f, Aus~, Ausstockung, Roden n [*Beseitigung des gesamten Baumbestandes einschließlich der Stöcke und Wurzeln*] = clear felling and stump-grubbing, ~ cutting ~ ~

~, Roden n, Stock~, Stockholzgewinnung, Rodungsarbeit f = stumpgrubbing
Rofen m → (Dach)Sparren m
Rogenstein m, Rogenkalkstein m, Bernburger Grauwacke f = roestone
Roggenboden m = rye earth
roh, unverputzt = unplastered
~ [*Straßenbaugestein ohne Bindemittelumhüllung*] = uncoated
rohe Schraube f = rough screw, unfinished ~
~ ~ mit Mutter, schwarze ~ ~ ~ = black bolt
rohes Abwasser n, Rohabwasser = crude sewage, raw ~ [*Sewage that has received no treatment*]
Rohasbest m = raw asbestos
Rohbau m = carcass; carcase [*deprecated*]; shell [*Canada*] [*A building*

or structure that is structurally complete but otherwise unfinished]
~arbeiten *fpl* = carcass work
Rohblock *m*, Rohgesteinsblock = rough block, quarried ~
Rohboden *m* → C-Horizont *m*
Rohdecke *f* = unplastered ceiling
Rohdichte *f* → Raumgewicht *n* der Volumeneinheit *f*
Roheisen *n* = pig iron
~gießmaschine *f* = pig machine
Roh(erd)öl *n* → Erdöl
Rohfallhöhe *f*, Bruttogefälle *n*, Rohgefälle = total head
Rohfilzpappe *f* = felt
Rohgefälle *n* → Rohfallhöhe *f*
Rohgestein *n* = raw stone [*lime industry*]
Roh(gesteins)block *m* = rough block, quarried ~
Rohgewicht *n*, Bruttogewicht = gross weight
Rohgips *m*, Dihydrat *n*, Doppelhydrat, Rohgipsstein *m* = (raw) gypsum, dihydrate, natural gypsum
Rohglas *n* = raw glass
Rohgut *n*, Rohhaufwerk *n*, Ausgangsgut [*Siebtechnik*] = raw material
Rohhaufwerk *n*, Ausgangsgut *n*, Rohgut [*Siebtechnik*] = raw material
Rohkies *m* = raw gravel
Rohkohle *f*, Förderkohle = run-of-the-mine coal, rough ~
~kord *m* = raw cord
Rohkreide *f* = natural chalk
Rohleinöl *n* = raw linseed oil
Rohmaterial *n* = raw material
~vorkommen *n* [*Steine- und Erden-Industrie*] = raw material deposit
Rohmehl *n* = raw meal
~mahlanlage *f* [*Zementherstellung*] = raw mill
~mischanlage *f* [*Zementwerk*] = raw meal mixing plant
Rohmischung *f*, Zement~ = raw mix(ture)
Rohöl *n* → Erdöl
~ mit geringem Schwefelgehalt = sweet crude (oil)
~destillation *f* = topping destillation
~empfangstation *f* = crude terminal
~-Emulsion *f* mit Wasser oder anderen Stoffen = oilfield emulsion
~leitung *f* = crude oil line
~ofen *m* = still
Rohperlit-Gestein *n* = perlite ore
(Roh)Petroleum *n* → Erdöl
Rohplanum *n* = rough (sub)grade (US); ~ formation (Brit.)
Rohr *n* → Schilf *n*
~ = pipe
~ mit Falz und Nut = rebated pipe
~ ~ Fuß = pipe with base
~ ~ Mantel = jacketed pipe
~ansatz *m*, (Rohr)Stutzen *m* = flanged socket; ~ nozzle (US)
~anschluß *m* → Rohrverbindung *f*
~art *f* = type of pipe
~aufwalzgerät *n* = tube expander
~auskleidung *f* = pipe lining
~auskleidungsmaschine *f* = pipelining machine
~auslegerkran *m* = pipe-boom crane (US); pipe-jib ~ (Brit.)
~außendurchmesser *m* = pipe O.D., ~ outside diameter
~bandagierung *f* = pipe wrapping
~befestigungsstück *n* = pipe support
~bemessung *f* = pipe sizing
~bemessungstabelle *f* = pipe sizing chart
~biegemaschine *f* = tube bending machine, pipe ~ ~
~biegewerkzeug *n* = pipe bending tool
~birne *f*, Treibbirne [*birnenförmiger Stahldorn zum Ausbeulen zusammengedrückter und eingebeulter Rohrfahrten*] = casing swedge
(~)Bogen *m*, (Rohr)Krümmer *m*, Bogenrohr *n* = (pipe) bend
~bruchsicherung *f* → Rohrbruchwächter *m*
~bruchwächter *m*, Rohrbruchsicherung *f*, Selbstschlußventil *n* = isolation valve
~brücke *f* → Rohrleitungsbrücke
~brückenbalken *m* = pipe bridge beam
~brunnen *m* [*veralteter Begriff*]; Bohrbrunnen = drilled well, bore-well, tube well

Rohrbündel-Wärmeübertrager — Röhrenerhitzer

~bündel-Wärmeübertrager *m* = tubular heat exchanger
~bürste *f* = pipe brush
~dach *n* = thatch roof
~damm *m* = pipe bank
~damm *m* im Freigelände [*Erdölwerk*] = yard bank
~dämmung *f* = pipe insulation; ~ lagging (Brit.)
~deckung *f* = pipe cover(ing)
~deichsel *f*, Rohrzugstange *f* = tubular towing pole, ~ ~ bar, ~ drawbar, ~ towbar
~diagonale *f* → Rohrschräge *f*
~diagonalstab *m* → Rohrschräge *f*
~dichtung *f* = jointing of pipes
~dichtungsband *n* = (pipe) jointing tape
~dichtungsring *m* = (performed) joint ring, pipe jointing ~
~drän *m* = pipe drain; tile ~ (US)
~dränage *f* = pipe drainage; tile ~ (US)
~drehkopf *m*, Rohrwirbel *m* [*Tiefbohrtechnik*] = casing swivel
~drückgerät *n* → Rohrdurchstoßgerät
(~)Düker *m* = sag pipe [*It is also called, very often but inaccurately, an "inverted siphon"*]
~durchlaß *m* = pipe culvert
~durchmesser *m* = pipe diameter

Rohrdurchstoßgerät *n*, Rohrdrückgerät, Horizontal-Preßanlage *f*, (Damm-) Durchstoßgerät; Röhrenstoßgerät [*Schweiz*] [*zur Unterfahrung von Straßen, Dämmen, Brücken usw.*]	pipe pusher, pipe-forcing system, pipe-jacking system
ausfahren [*Presse*]	to extend
Druckglied *n*	steel spreader plate
Durchdrückkraft *f*	thrusting force
Durchdrücklänge *f*	length of thrust(ing)
hydraulisches Preßverfahren *n*	pipe-jacking method
Preßgrube *f*	thrust(ing) pit, jacking ~
Preßschild *m*	jacking shield
Preßwagen *m*	jacking frame
Rohrdurchstoßen *n*, Rohrdurchdrücken *n*	pipe-jacking, pipe-pushing
Schild *m*	shield
Steuerpresse *f*	steering jack
Stoßpresse *f*, Preßkolben *m*	hydraulic jack, thrust(ing) ~
Transportwagen *m*, Sand~	troll(e)y
Widerlager *n*	thrust(ing) wall, jacking ~
Zwischenpresse *f*	auxiliary-station jack
Zwischen-Preßstation *f*	intermediate jacking station
Zwischenpreßverfahren *n*	auxiliary jacking system

Rohr|einbau-Empfehlungen *fpl* von **API** [*Tiefbohrtechnik*] = API casing landing recommendations
~einwalzapparat *m* → Siederohrdichtmaschine *f*
~elevator *m* [*Tiefbohrtechnik*] = casing elevator
~ende *n* = pipe end
Röhren|erhitzer *m*, Röhrenofen *m* = pipestill

Röhrenheizplatte — Rohrgerüstbau

~heizplatte *f*, Heizröhrenplatte = radiant panel
~kühler *m* = tubular cooler, ~ equipment
~ofen *m*, Röhrenerhitzer *m* = pipestill

~stoßgerät *n* [*Schweiz*] → Rohrdurchstoßgerät
~-Streifenwalzwerk *n* = skelp mill
~trockener *m* → Dampf~

Rohrfabrik *f* **auf (Gleis)Kettenfahrzeugen montiert, Rohrwerk** *n* ~ ~ ~, **fahrbare Rohrfabrik, fahrbares Rohrwerk**

  Blechrolle *f*
  Diesel-Wechselstromgenerator *m*
  Fertigungsstation *f*
  hochfrequente Widerstandsschweißung *f*
  Kraftstation *f* und Steuereinrichtung *f* auf einem gemeinsamen (Gleis)Kettenfahrwerk

  Krümmung *f*
  Kühlvorrichtung *f*
  Oszillatorröhre *f*
  Schweißanlage *f*
  Startmotor *m*
  Ultraschallprüfgerät *n*
  Walze *f*

crawler-mounted pipe factory, travel(l)ing ~ ~, two-unit power plant and pipe mill [*It turns coiled sheet steel into continuous welded pipe sections*]
  coil of sheet steel
  diesel-electric alternator
  rear unit, mill
  high-frequency electrical resistance welding
  lead unit incorporating power plant and steering section

  bend
  refrigeration unit
  oscillator tube
  welder
  starting motor
  ultrasonic flaw detector
  (forming) roll

Rohr|fachwerk *n*, **Rohrskelett** *n* = tubular framework, ~ skeleton, ~ framed structure
~**fahrstuhl** *m* [*Tiefbohrtechnik*] = casing elevator
~**fahrt** *f*, Rohrtour *f*, Rohrkolonne *f* [*Tiefbohrtechnik*] = string of casing, ~ ~ well tubing
~**fänger** *m* → (Auf)Fangkessel *m*
~**fänger** *m* [*Tiefbohrtechnik*] = casing spear, bull dog ~
~**fertiger** *m*, Rohrherstellungsmaschine *f* = pipe-making machine
~**(fertigungs)maschine** *f* → Betonrohrmaschine
(Rohr)Festpunkt *m* = pipe support, ~ pier, thrust block
~**beton** *m* = pipe support concrete, ~ pier ~, thrust block ~

~**mauerwerk** *n* = thrust block masonry, pipe support ~, pipe pier ~
Rohr|förderer *m* → Förderrohr *n*
~**förderung** *f* **/ von Beton** = pipe transport of concrete
~**form** = pipe mould (Brit.); ~ mold (US)
~**formmaschine** *f* → Betonrohrmaschine
(~)Formstück *n* = fitting
~**führung** *f* = pipe location
(~)Fuß *m* = base
Rohrgerüst *n*, **Rohrrüstung** *f* = tubular scaffold(ing)
~ **aus Alu(minium)legierung**, Rohrrüstung *f* ~ ~ = tubular scaffold(ing) in aluminium alloy (Brit.); ~ ~ ~ aluminum ~ (US)
~**bau** *m* = tubular scaffold(ing) construction

Rohrgerüstbauer — Rohr(leitungs)brücke

~bauer *m* = tubular scaffolder
~dampframme *f* = tubular frame type steam pile driving plant, ~ ~ ~ ~ piling plant
~-Elektrokran *m*, Elektro-Rohrgerüstkran = electric tubular crane
~ramme *f* = tubular frame (pile) driving plant, ~ ~ piling ~
Rohrgewebe *n* als Putz(mörtel)träger, Rohrmatte *f* ~ ~, Schilf~ ~ ~ = reed lath(ing)
Rohrgewinde *n* = pipe thread
~bohrer *m*, Gasgewindebohrer, Handgewindebohrer mit Rohrgewinde = pipe tap
~schneidkluppe *f* = pipe stock
Rohr/glätter *m* [*Betondeckenbau. Zum Beseitigen des überschüssigen Wassers auf einer Betondecke*] = pipe float
~graben *m* = pipe trench, piping ~
~grabenbau *m* = pipe trench construction
~griff *m* = tubular handle
~haken *m* = pipe hook
~-Handlauf *m*, Rohr-Handläufer *m* = tubular handrail, pipe ~
~hängebrücke *f* = pipe line suspension bridge
~herstellung *f* = pipe manufacture
~herstellungsmaschine *f* → Betonrohrmaschine
~hydraulik *f* = pipe hydraulics
~installation *f* = pipe installation
~kanal *m* → Rohrleitungskanal
~-Kanalauslaß *m* = pipe channel outlet
~kappe *f* = cap [*A cover, with internal threads, screwed over the end of a pipe. It thus closes, or caps, the pipe*]
~karre(n) *f*, (*m*), Hand~ = pipe (hand) truck [*for the transport of pipes in precasting factories*]
~karren *m* [*am Bohrturm*] = casing wagon
~keilklemme *f* → Abfangkeil *m*
~keilklemmen *fpl* mit auswechselbaren Schalen → Rohrklemmkeile *mpl* ~ ~ ~

~keiltopf *m* [*Tiefbohrtechnik*] = casing spider
~keller *m*, Rohrleitungskeller = pipework basement
~kennzeichnung *f* = pipe line identification
~kern *m* [*Betonrohrstampfmaschine*] = steel core
~kern *m*, Kernrohr *n* [*Teil eines Betonrohres*] = pipe core, core pipe
~kitt *m* = jointing compound for pipes
~klemmkeil *m* → Abfangkeil
~klemmkeile *mpl* mit auswechselbaren Schalen, Rohrkeilklemmen *fpl* ~ ~ ~, Abfangkeile ~ ~ ~ [*Rotarybohren*] = (rotary) slips with interchangeable inserts
~kolbenpumpe *f* = hollow piston pump
~kolonne *f* → Rohrfahrt *f*
~kompensator *m*, Rohrausgleicher *m* = pipe expansion bend
~konsole *f* = pipe bracket
~konstruktion *f* = tubular structure, ~ construction
Rohrkopf *m*, Verrohrungskopf [*Tiefbohrtechnik*] = casing-head
~benzin *n* = casing-head gasoline, ~ gasolene (US)
Rohr/-Kreisrippe *f* = annular fin
(~)Krümmer *m*, (Rohr)Bogen *m*, Bogenrohr *n*, Knierohr = (pipe) bend
~kupp(e)lung *f* = (pipe) union
~lagerung *f* = bedding of pipes, support for the bottom of the pipes
~last *f* = pipe load
Rohrlege/anlage *f* → Rohrverlegeanlage
~barke *f* = pipe-laying barge
~plan *m* → Rohrverlegeplan
~vorrichtung *f* = pipe-laying device
~winde *f*, Rohrverlegewinde = pipe-laying winch
Rohr/lehrgerüst *n*, Rohrrüstung *f* = scaffold tubing falsework
~leitung *f* = piping, pipe line
~leitungen *fpl* für Veredelungsanlagenteile [*z.B. in einem Erdölwerk*] = process piping, ~ pipe line
~(leitungs)brücke *f* = pipe (line) bridge

Rohrleitungsgrabenaushub — Rohrschwingmühle 782

~leitungsgrabenaushub *m* = pipe line trenching
~(leitungs)kanal *m* = pipeway, pipe duct, pipe subway
~(leitungs)keller *m* = pipework basement
~leitungspumpe *f* = pipe line pump
~leitungsschaden *m* = pipe line damage
~lichtweite *f* = internal pipe dia., ~ ~ diameter
~lieferant *m* = pipe supplier
~lieferungsvertrag *m* = pipe supply contract
~locher *m*, Verrohrungsperforator *m* = casing perforator
~maschine *f* → Beton~
~mast *m* = tubular mast
~material *n*, Rohrwerkstoff *m* = pipe material
~matte *f* als Putz(mörtel)träger → Rohrgewebe *n* ~ ~
~mühle *f*, Grießmühle = tube mill
~mutter *f* = pipe nut
~nagel *m* = pipe nail
~netz *n* = pipe network
~netzplan *m* = pipe network drawing
Rohrnippel *m* = pipe connector, ~ nipple
~ **mit beiden Enden flach** = pipe nipple BEP, ~ ~ both ends plain
Rohr|öffnung *f* = pipe opening
~pfahl *m*, Stahl~ = (steel-)pipe pile
~pfette *f*, Dach~ = tubular purlin(e)
~plan *m* = pipe drawing
~post *f* = pneumatic tube installation, carrier air tube system
~presse *f* = pipe press
~prüfgerät *n* [*Tiefbohrtechnik. Eine Art Schöpfbüchse*] = casing tester
~prüfmaschine *f* = pipe testing machine
~putz *m* [*Putz auf Rohrgewebe*] = plaster on reed lath(ing)
~querschnitt *m* = pipe cross-section
~querstrebe *f* = tubular cross member
~querträger *m* für Zylinder = tubular cylinder support
~rahmen *m* = tubular frame

~reibung *f* = pipe friction
~reinigungsgerät *n* = pipe cleaning device
~rippe *f* = fin
~rüstung *f*, Rohrlehrgerüst *n* = scaffold tubing falsework
~rüstung *f*, Rohrgerüst *n* = tubular scaffold(ing)
~rüstung *f* aus Alu(minium)legierung → Rohrgerüst *n* ~ ~
~säge *f* = pipe saw
~saugkraft *f*, Kapillarität *f*, Porensaugwirkung *f*, Kapillarattraktion *f* = capillarity, capillary attraction
~schacht *m* = pipe shaft
~schaden *m* = pipe damage
(~)Schaft *m* = (pipe) barrel [*Is that portion of a pipe throughout which the internal diameter and thickness of wall remain uniform*]
(~)Schale *f* = split pipe, half-section ~, channel [*A pipe cut lengthwise*]
~schelle *f* = pipe clamp
~schiene *f*, Puddelluppe *f* = muckbar [*The bar made by the first rolling of the bloom*]
~schlange *f* = pipe coil
~schleudergeräte *npl* = pipe spinning equipment
~schlitz *m* = pipe run, ~ chase
~schlosser *m* = pipe fitter
~schlüssel *m* = pipe wrench
~schneidemaschine *f* = pipe-cutting machine
~schneider *m* = pipe cutter
~schräge *f*, Rohrdiagonale *f*, Rohrdiagonalstab *m* = tubular diagonal (rod)
Rohrschuh *m*, Bohr~ = casing shoe
~ **mit Rückschlagventil**, Schwimmschuh [*Tiefbohrtechnik*] = float shoe
~rückschlagventil *n* [*Tiefbohrtechnik*] = casing float
Rohr|schuß *m* = pipe section, individual section of pipe
~schutzmuffe *f* [*Gewindeschutz für Ölfeldrohre*] = casing protector
~schwingmühle *f* = vibrating tube mill, vibratory ~ ~

Rohr(sicker)drän — Rollbetrieb

~(sicker)drän m = pipe drain, subdrain, subsoil drain, stone-filled trench with pipe, pipe subdrain

~skelett n, Rohrfachwerk n = tubular framework, ~ skeleton, ~ framed structure

~stampfmaschine f → Zement~

~stempel m (⚒) = tubular prop, ~ post

~stollen m = pipe drift (Brit.); ~ (small-diameter) tunnel, ~ gallery, ~ heading

~stopfen m = pipe plug

~stoß m = pipe joint

(~)Strang m = (pipe) run, piping ~

~stütze f = pipe column, tubular ~ [*Is a steel pipe filled with concrete*]

(~)Stutzen m, Rohransatz m = flanged socket; ~ nozzle (US)

~sumpf m = reed swamp

~tour f → Rohrfahrt f

~tragfähigkeit f = supporting strength of pipe

~trum m (⚒) = compartment for pipes

~turbinenkraftwerk n, Kraftwerk mit liegender (Turbinen)Welle, Kraftwerk mit waag(e)rechter (Turbinen-)Welle, Kraftwerk mit horizontaler (Turbinen)Welle = horizontal-shaft type power plant

~turm m = pipe tower [*Tower to accommodate pipework*]

~unterbrecher m = back-siphonage preventer

~unterkante f = bottom of pipe

~verbindung f, Rohranschluß m = pipe connection

~(ver)legeanlage f = pipe-laying plant

~(ver)legekahn m = laybarge

~(-Ver)legekran m [*Arbeitsgerät beim (Gleis)Kettenschlepper*] = pipe laying attachment

~(ver)legeplan m [*als Fortschrittsplan*] = pipe-laying progress chart, progress chart for pipe laying

~(ver)legeplan m [*als Zeichnung*] = pipe-laying drawing

~(ver)legewinde f = pipe-laying winch

~verlegung f = pipe laying

~verlust m = pipe loss

~walze f → Siederohrdichtmaschine f

~wand(ung) f = pipe wall

~wendelförderer m = vibratory spiral pipe elevator, vertical magnetic pipe transporter

~werkstoff m, Rohrmaterial n = pipe material

~werkzeug n = pipe tool

~widerstand m = pipe resistance

~wirbel m, Rohrdrehkopf m [*Tiefbohrtechnik*] = casing swivel

~-Zementmörtelauskleidung f = cement mortar pipe lining

~zubehör m, n = pipe fittings

~zugstange f → Rohrdeichsel f

~zugwinde f [*pennsylvanisches Bohrverfahren*] = calfwheel

Roh|sand m = raw sand

~schlacke f = raw slag

Rohschlamm m [*Abwasserwesen*] = raw sludge

~ [*Zementherstellung*] = slurry

~-Mischer m = slurry mixer

Roh|splitt m = uncoated chip(ping)s, dry ~

~teer m = crude tar

~ton m → Ton

~wand f = unplastered wall

Rohwasser n, Brauchwasser = nonpotable water, raw ~

~stollen m = raw-water tunnel, nonpotable water ~

Rohwichte f → Schüttgewicht

Rolladen m = rolling shutter, roller ~

Rollänge f, Roll-Länge, Rollweg m [*eines Flugzeuges*] = taxi distance

Rollbahn f → Rollweg m

~befeuerung f, Rollwegbefeuerung [*Flugplatz*] = taxiway lighting

~feuer n, Rollwegfeuer = taxiway light

Roll|bandpegel m = tape depth-ga(u)ge

~behälter m = castered floor container

~betrieb m, Rollverkehr m [*Flugplatz*] = taxiing traffic

Rollboden — Rollmaterial

Rollboden m [*Radschrapper*] = retractable floor, ~ bottom, bottom door
~**folgeventil** n [*Radschrapper*] = retractable floor sequence valve, ~ bottom ~ ~, bottom door ~ ~
~**zylinder** m [*Radschrapper*] = retractable floor jack, ~ bottom ~, bottom door ~
Roll|brücke f, Schiebebrücke = rolling drawbridge, pull-back ~
~**bürgersteig** m, fahrbarer Bürgersteig = passenger conveyor, travolator, passenger conveyer
Rolle f [*Seilzug*] = pulley
~ [*Ermittlung der Ausrollgrenze eines Bodens*] = (soil) thread
rollen [*Flugzeug*] = to taxi
Rollen n [*Flugzeug*] = taxiing
~**bahn** f → Rollenförderer m
~**bahnweiche** f = conveyor switch, conveyer ~
~**bock** m → Rollenstation f
rollende Eimerkette f [*Eimerkettenbagger*] = roller type bucket ladder
~ **Geschiebeführung** f = rolling transport
~ **Reibung** f = rolling friction
Rollen|flansch m [*Gleiskette*] = roller flange
~**förderer** m, Rollenbahn f, Rollentransporteur m = roller conveyor, ~ conveyer
rollengelagertes Rad n = roller-bearing wheel
Rollen|höhe f = height of boom sheave, ~ ~ jib ~
~**kettenantrieb** m = roller chain drive
~**kranz** m, Laufkranz = circular roller path
~**lager** n [*Maschine; Fahrzeug*] = roller bearing
~**lager** n, Walzenlager [*für Hoch- und Tiefbauten*] = roller bearing
~**länge** f [*Scharnier*] = knuckle length
~**lappen** m [*Scharnier*] = joint section
~**laufkranz** m = roller race
rollenloser Behälter m = skid container

Rollen|mast m [*Tiefen- und Langstreckenförderer*] = headmast, headpost
~**meißel** m → Kegel~
~**nahtschweißen** n = seam welding
~**pendellager** n, Pendelrollenlager = spherical roller bearing, self-aligning ~ ~
~**rost** m [*Siebtechnik*] = roller-bar grizzly, multi-roll sizer
~**schmierdüse** f [*Gleiskette*] = roller lube nozzle
~**schwinge** f = roller rocker
~**station** f, Rollenbock m, Abtragstation [*Bandförderer*] = idler set
~**tisch** m = roller table
~**transporteur** m → Rollenförderer m
~**turm** m der Schürze, Schürzen-Rollenturm [*Radschrapper*] = apron lift sheave tower
~**wagen** m [*Schützenwehr*] = rolling truck, series of fixed wheels
~**wagen** m, Raupenschlepperfahrwerk n = track frame
~**zug** m [*Stoney-Schütz(e)*] = roller train
Roll|erz n = float-ore
~**feld** n [*Fläche für Bewegungsvorgänge 1. Ordnung, wo sich Starten und Landen vollziehen*] = landing area
(~)**Feldhütte** f [*Militärflugplatz*] = crew hut, dispersal ~
~**feldringkabel** n = perimeter cable
~**geschwindigkeit** f [*Flugzeug*] = taxiing speed
~**gitter** n = rolling grille
~**grenze** f, Aus~, obere Plastizitätsgrenze, Wr [*Boden*] = plasticity limit
rollig → nichtbindig
Roll|keilschütz(e) n, (f) = vertical trapezoidal sluice
~**kies** m, Rundkies = round gravel
~**klappbrücke** f, Abrollbrücke, Schaukelbrücke = roller bascule bridge
~**krone** f → (Kegel)Rollenmeißel m
~**maschine** f → Siederohrdichtmaschine
~**material** n, Eisenbahn-~ = rolling stock

~membran(e) *f* [*Schmierölpumpe*] = rolling diaphragm
~ponton *m* [*Schleuse*] = caisson gate, rolling (lock) ~
~reibung *f* = rolling friction
~sand *m*, Rundsand = round sand
~schicht *f* = upright course
~schuhbahn *f* = roller-skating rink
~stein *m*, Rollstück *n* [*auf Schmalseite stehend vermauerter Ziegel*] = upright brick
~stück *n* → Rollstein *m*
~tor *n*, Schiebetor = sliding gate
~treppe *f*, Fahrtreppe, Treppenaufzug *m* = motorstair, escalator, moving stair(case)
~treppenschrägschacht *m* = inclined escalator shaft
~trommel *f*, Granuliertrommel, Granalientrommel = nodulizing drum
~verkehr *m*, Rollbetrieb *m* [*Flugplatz*] = taxiing traffic
~verlust *m* = rolling loss
~wagen *m*, Blockwagen [*für den Blocktransport*] = block bogie
Rollweg *m*, Roll-Länge *f*, Rollänge [*eines Flugzeuges*] = taxi distance
~, Zubahn *f*, (Zu)Rollbahn [*Flugplatz*] = taxiway
~ zum Nahverkehrsbereich, Rollbahn *f* ~ ~ [*Flugplatz*] = taxiway leading to the terminal area
~befeuerung *f*, Rollbahnbefeuerung [*Flugplatz*] = taxiway lighting
~feuer *n*, Rollbahnfeuer = taxiway light
Roll|welle *f* = roll wave
~widerstand *m* = rolling resistance between tyre (Brit.)/tire (US) and road surface
~widerstandsbeiwert *m* = rolling resistance factor, rr
Roman|kalk *n* [*veraltete Bezeichnung*]; hochhydraulischer Kalk = eminently hydraulic lime
~zement *m* = Roman cement, Parker's ~
Röntgen|abteilung *f* = radiography department, X-ray ~
~bild *n* = radiograph

~-Daten *f* der Zementminerale = x-ray powder diffraction data of cement minerals
~prüfung *f* = x-ray test
~untersuchung *f* = radiography
Rosenfenster *n* = rose window
Rosette *f* [*Beschlagteil*] = rose [*A decorative plate through which a door handle passes*]
Rosin-Rammler-Bennet'sche Kornverteilungskurve *f* = R.R.B. size distribution curve
Rost *m*, Gründungs~ = grillage, foundation ~
~ = rust
~, Kessel~, Feuer~ = (boiler) grate
~ auf den sich Schiffe bei fallender Tide setzen = careening grid, gridiron
~ aus (Faschinen)Würsten, (Faschinen)Wippenrost = grillage of fascine poles
~abdeckung *f* [*beim Schlammfang*] = iron grid
~balken(träger) *m* = grillage beam [*A steel beam of I section, used in broad foundations on yielding soil, where the central load must be distributed*]
rost|beständig = rust-resistant
~bildend = rust forming
Rost|bildung *f* = rust formation
~boden *m* [*Viehstall*] = slatted floor
~einlauf *m* = grated inlet
Rosten *n*, Rostung *f* = rusting
Rösten *n* [*Verhüttung*] → Kalzinieren
Rost|entfernen *n* = rust removal
~film *m* = rust film
~fläche *f* = grate area
~flecken *mpl* = rust staining
rostfrei = stainless
Rost|gründung *f*, Rostfundament *n* = grillage foundation
~kühler *m* = grate(-type) cooler
Röstofen *m*, Brennofen, Kalzinierofen [*Verhüttung*] = calcining kiln
Rostschicht *f* = rust layer
Rostschutz *m* = rust protection
~anstrich *m* = rust protection coat(ing), ~ preventing ~, ~ inhibitive ~, ~ proofing ~, anti-rust ~

Rostschutzbehand(e)lung — Rotationsverdichter

~behand(e)lung *f* = rust-preventative treatment

~farbe *f* = rust protection paint, ~ preventing ~, ~ proofing ~, ~ inhibitive ~, anti-rust ~

~fett *n* = rust preventing grease, ~ protection ~, ~ inhibitive ~, ~ proofing ~, anti-rust ~

~mittel *n* = rust preventing agent, ~ inhibitive ~, ~ protection ~, ~ proofing ~, anti-rust ~

~öl *n* = slushing oil

~verbindung *f* = rust preventing compound, ~ protection ~, ~ inhibitive ~, ~ proofing ~, anti-rust ~

rostsicher = rustproof

Rost|sieb *n* → Stabrost *m*

~stab *m* = grate bar

Rostung *f*, **Rosten** *n* = rusting

Röstzone *f* → Brennzone

Rotameter *n*, *m* = purgemeter

Rotary|bohren *n* = rotary drilling

~bohrgerät *n*, Rotarybohrmaschine *f* = rotary drilling outfit, ~ ~ rig

~-(Bohr)Meißel *m* = rotary (cutting) bit

~bohrseil *n* = rotary drilling line

~bohrtisch *m* = rotary table

~bohrung *f* = rotary(-drilled) hole, ~ well

~bohrverfahren *n* = rotary drilling system

~haken *m*, Zughaken = rotary hook

(~-)Hebewerk *n*, (Rotary-)Windwerk, (Rotary-)Bohrwerk, Zugwerk, Winde *f*, Antriebs- und Hebewerk [*Windwerk für Rotary-Bohranlagen mit mehreren mechanischen Geschwindigkeitsstufen oder stufenlosen Getrieben, das in Verbindung mit einem Flaschenzug die Hebevorrichtung für den Bohr- und Futterrohrstrang bildet*] = (rotary) draw works, ~ drilling unit, hoist draw works for rotary

(~-)Hebewerktrommel *f*, (Rotary-)Bohrwerktrommel, (Rotary-)Windwerktrommel = (rotary) draw works drum

~kette *f* = rotary chain

(~-)Kettenhebewerk *n*, (Rotary-)Kettenwindwerk, (Rotary-)Kettenbohrwerk = (rotary) chain draw works

~-Meißel *n*, Rotary-Bohrmeißel = rotary bit, ~ cutting ~

~schlauch *m* → Bohrschlauch

~tisch *m*, Drehtisch, Bohrtisch = rotary table, turn ~, rotary machine

~tischeinsatz *m*, Mitnehmerstangenbüchse *f* = drill stem bushing, kelly ~

~-Zange *f* = rotary tongs

Rotations|-Drucklufterzeuger *m* → Rotationskompressor

~einspritzpumpe *f* = rotary injection pump

~ellipsoid *n* = ellipsoid of revolution, spheroid

~hyperboloid *n* = hyperboloid of revolution

~-Kern-Gerät *n*, Doppelrohrentnahmegerät [*für ungestörte Bodenproben*] = double tube sampler

~kompressor *m*, Rotations(luft)verdichter, Rotations-Drucklufterzeuger = rotary (air) compressor

~-Kondensation *f* = rotating condensation

~kopf *m*, Drehkopf, Spülkopf, Bohrwirbel *m* [*Rotarybohren*] = rotary swivel

~pumpe *f* → Drehkolbenpumpe

~rüttler *m*, Rotationsvibrator *m* = rotary (type) vibrator

~schale *f* = shell of revolution

~sonde *f* nach Carlson = Carlson rotating auger

~spülbohren *n*, Rotationsspülbohrung *f* = hydraulic rotary drilling (or boring), rotary mud flush boring (or drilling)

rotationssymmetrische Schale *f* = axially symmetric shell

Rotations|vakuumpumpe *f* = rotary vacuum pump

~verdichter *m* → Rotationskompressor *m*

Rotationsvibrator — Rückkühlung

~vibrator *m*, Rotationsrüttler *m* = rotary (type) vibrator

~-(Zylinder-)Viskosimeter *n* = rotating cylinder viscosimeter, rotational ~

Rotbraunstein *m* → Kieselmangan *n*
rote Mineralfarbe *f* = red pigment
roter Verwitterungston *m* = residual red clay
rotierende Glättscheibe *f* = mechanically operated float
rotierender dampfbeheizter Röhrentrockner *m* → (Dampf)Röhrentrockner

~ **Exzenter** *m*, Kreisschwinger *m* [*Rüttelwalze*] = eccentrically loaded rotating shaft

~ **Schneidkörper** *m* = rotary cutter
rotierendes Trommelsieb *n* = rotary screen

Rotier|ofen *m* → Drehofen
~**scheibe** *f* = spinner

Rotnickelkies *m*, Kupfernickel *n*, Nikkelin *n*, NiAs (Min.) = nicolite, copper nickel [*old term: nickeline*]

Roto-Clone-Naßabscheider *m* = wet Roto-clone dust collector

Rotor *m*, Schlagwalze *f* [*Prallbrecher*] = impeller

~, bewegliche Schaufeln *fpl* [*Bohrturbine*] = rotor

~, Läufer *m*, Verteilerfinger *mpl* [*Motor*] = rotor

~, Schlagwerk *n* [*Hammerbrecher und Hammermühle*] = rotor

~-**Hammermühle** *f* → Hammermühle
~**(Prall-)mühle** *f*, Schnelläufermühle, mechanische Prallmühle = (impeller) impact mill

~**streuer** *m* → Rotor-Streugutverteiler
~-**Streugutverteiler** *m*, Rotorstreuer *m*, Tellerstreuer = spinner (spreader), rotary disc type gritter

Rot|rauschgelb *n* (Min.) → Realgar
~**spat** *m* → Kieselmangan *n*
Rottenarbeiter *m* = platelayer
Rotzinkerz *n*, Zinkit *n*, ZnO (Min.) = zincite, red oxid(e) of zinc, spartalite, sterlingite

R.R.B.-Netz *n*, Körnungsnetz = grain size distribution diagram, system of coordinates for representation of size distribution curves R.R.B.

Rtl. *m, n*, Raumteil = part per volume
Ruberoiddach(ein)deckung *f* = Ruberoid roofing

Rück|ansicht *f* = rear veiw
~**bau** *m* (✠) = retreating longwall system
~**blick** *m* [*Nivellieren*] = backsight
~**druckturbine** *f* → Überdruckturbine
~**druckverfahren** *n*, Gaseinpreßverfahren [*Erdöllagerstätte*] = repressuring (method)

Rücken *m* [*die obere, äußere Fläche eines Bogens oder Gewölbes*] = extrados, back

~**berg** *m* (Geol.)→ Drumlin *m*
~**lehne** *f* des Sitzes = seat back
~**schild** *n* [*Erzbergbau*] = beetle back
Rückfahr|licht *n* = back-up light
~**warnhorn** *n* = back-up alarm
Rückfenster *n* [*LKW*] = back window
Rückfluß *m* = reverse flow
~**verhinderer** *m* = closure to prevent reverse flow, backwater closure

Rück|gang *m*, Abfall *m* [*z. B. Temperatur; Leistung usw.*] = drop
~**gewinnung** *f*, Wiedergewinnung = recovery
~**gewinnungsanlage** *f*, Wiedergewinnungsanlage = reclaiming plant (or installation, or facility)
~**haltebecken** *n* = retention basin
~**haltewinde** *f* = holdback winch
~**holfeder** *f* = return spring
~**holgestänge** *n* **der Hydrauliklenkung** = (steering) follow-up linkage
~**(hol)seil** *n*, Leerseil [*Tiefen- und Langstreckenförderer*] = outhaul cable, ~ line, backhaul ~, pull back ~
~**kippen** *n* [*Schaufel*] = back tilting
~**kippstellung** *f* = tilt back position
~**kühlanlage** *f* = cooling back installation
~**kühlung** *f* = cooling back

~lader m, Überkopflader = overhead loader
~lauf m = return
~laufentleerung f [Betonmischer] = reverse discharge
rückläufige Metamorphose f, Retrometamorphose, Diaphthorese f (Geol.) = retrogressive metamorphism
Rücklauf|leitung f = return line
~öl n [Erdölwerk] = recycling stock
~öl n [Hydraulikanlage] = return oil
~rohr n = return (pipe)
~schlamm m, Rücknahmeschlamm = return sludge
~schlauch m = return hose
~sicherung f [Bauaufzug] = device to eliminate "running back"
~sperre f [Bauaufzug] = ratchet to eliminate "running back"
~verhältnis n [thermisches Krackverfahren] = recycle ratio
Rück|licht n, Schlußlicht = tail light, rear ~
~lieferung f = return shipment
Rückluft f = return air
~anlage f = return air system
~rost m = return air grille
Rücknahme|feder f, Rückzugfeder = return spring
~schlamm m, Rücklaufschlamm = return sludge
Rückprall|hammer m → Betonprüfhammer
~härte f = scleroscope hardness
~-Härteprüfer m → Betonprüfhammer
~härteprüfung f, Rückprallhärteversuch m, Rückprallhärteprobe f = scleroscope hardness test, rebound ~
Rück|presse f [zum Gleisrücken] = shifting ram
~pumpen n = return pumping
~pumpwerk n, Rückpumpstation f = return pumping station
~scheibenwischer m = rear window wiper
~schlag m, Rückstoß m = recoil
~schlagkappe f = flap valve [A check valve with a hinged disc which opens when the flow is normal and is closed by gravity or by the flow when the flow tends to go backwards]
Rückschlagventil n = check valve, clack ~, non-return ~, reflux ~, one-way ~ [A valve in a pipe which allows flow in one way only]
~ für Bohrgestänge = drill pipe float valve
~ über dem Rohrschuh [Tiefbohrtechnik] = casing float
rückschreitende Erosion f = backward erosion
rückschreitender Bruch m = retrogressive slide
Rück|seil n → Rückholseil
~seite n = back
~seite f = back [The inside surface of a wall]
~speisepumpe f des Wandlers = torque converter recirculation pump
~spiegel m = rear vision mirror, driving ~
rückspringende Ecke f = re-entrant corner
Rücksprunghärte f = hardness given by the rebound height
~ nach Shore, Skleroskophärte = Shore hardness
Rück|sprunghöhe f = rebound height
~spülpumpe f = scavenger pump
~spülung f, Rückspülen n = backwash(ing)
~standöl n = topped crude, reduced ~
~standsieb n = retaining screen
~stau m = backflow
rückstauen = to dam [To keep back water by means of a dam]
Rückstauen n = damming [Keeping back water by means of a dam]
Rückstau|klappe f, Rückstauverschluß m = back-water gate [A device for preventing the back flow of a liquid]
~wasser n, totes Wasser = backwater
Rück|stoß m, Rückschlag m = recoil
~strahler m = reflector
~strahlungskonus m = reradiating cone [tubular furnace]

Rückstromöl n = make-up oil
Rückströmung f **am Grunde** = deep return current
Rück|stromventil n = make-up valve
~**stufensieb** n = screen for two screening directions
~**verflüssigung** f = reliquefaction
~**verladegerät** n → Aufnahmegerät
~**verladegreifer** m = dumping grab (for rehandling)
~**verladekratzer** m, Aufnahmekratzer = reclaiming scraper, rehandling ~
~**verlademaschine** f → Aufnahmegerät n
~**verladen** n, Aufnehmen = rehandling, reclaiming
~**verlader** m → Aufnahmegerät n
~**verladetrichter** m = reclaim hopper
~**vermischung** f = back-mixing
~**wand** f [*Waschbecken*] = splash panel
rückwärtige Schutzsperre f → Talsperre
rückwärtiger Kämpferdruck m, hinterer ~ (✕) = rear abutment pressure, back ~ ~
rückwärts schreitende Moräne f = recessional moraine
Rückwärts|aufreißer m = reverse ripper
~**einschneiden** n = backsight reading
~**fahrt** f = reverse run, ~ travel
~**gang** m = reverse gear
~**ganghebel** m = reverse (gear) lever
~**kippen** n **der Schaufel** = backward tilting of the bucket
Rückwärtskipper m, Hinterkipper = rear-dump lorry (Brit.); ~ truck (US)
~**(-Anhänger)** m → Hinterkipp(er)anhänger
~**-Aufl(i)eger** m → Halbanhänger-Rückwärtskipper
~**-Halbanhänger** m → Halbanhänger-Rückwärtskipper
Rückwärts|kippmulde f → Hinterkippmulde
~**kippung** f → Hinterkippung
~**kippwanne** f → Hinterkippmulde f
~**kupp(e)lung** f = reverse clutch
~**optimierung** f = feedback optimalizing
~**planetensatz** m = reverse gear train
~**vorlegerad** n = reverse idler gear
ruckweises Nachgeben n (✕) = load shedding
Rück|zementation f [*Tiefbohrtechnik*] = plug-back cementing
~**zugtrommel** f [*Tiefen- und Langstreckenförderer*] = backhaul drum
~**zylinderbefestigung** f (✕) = ram attachment
Rüfe f, Mure f, Murgang m = mud lava
Ruhedruck m = earth pressure at rest
~**beiwert** m = coefficient of earth pressure at rest
Ruhelage f, Ruhestellung f = stationary position
~ **des Meeresspiegels** → Ruhewasserspiegel
ruhende Belastung f = static loading
~ **Düne** f, befestigte ~ = stabilized dune
~ **Last** f = quiescent load
~ ~ ~ ständige ~
ruhender Boden m = repose soil
~ **Draht** m [*Kabelspinnen beim Brückenbau*] = dead wire
~ **Tiefengranit** m = subjacent body
~ **Verkehr** m, parkender ~ = standing traffic, stationary ~
Ruhe|penetration f [*Schmierfett*] = unworked penetration
~**reibung** f = static friction
~**stellung** f, Ruhelage f = stationary position
~**wasserspiegel** m, Ruhelage f des Meeresspiegels = still water level, undisturbed ~ ~
~**wert** m, Ruhebeiwert = at-rest value
~**winkel** m, natürlicher Böschungswinkel = angle of repose (of the natural slope), ~ ~ rest
~**zeit** f = rest period
~**zustand** m = state of rest
ruhiges Fließen n, strömender Abfluß m, Strömen n = tranquil flow, streaming ~, flowing

Rühr|arm *m* = agitating arm, stirring ~
~**behälter** *m* = agitation vessel
rühren = to agitate, to stir
Rühren *n*, Nachmischen [*Transportbeton*] = agitating
Rührer *m* → Rührwerk *n*
Rühr|fahrzeug *n* → Beton~
~**mischer** *m* → Rührwerk *n*
~**stern** *m* = stirring star, star-shaped agitator
~**versuch** *m*, Rührprüfung *f*, Rührprobe *f* = stirring test
~**wagen** *m* → (Beton)Rührfahrzeug *n*
~**welle** *f* = stirring shaft
Rührwerk *n* [*Emulsionsherstellung*] = high-speed mixer
~, Rührer *m*, Rührmischer *m* = stirring gear, agitator, (mechanical) stirrer
~**kugelmühle** *f* = agitator ball mill, ~ ~ grinder
~**schaber** *m* = scraping stirrer
Rumpelkammer *f* (⚒) = gloryhole
~ = loft [*A storage space entirely within the roof-space of a building*]
Rumpffläche *f* (Geol.) → Denudationsebene *f*
runder Bewehrungskorb *m*, ~ Armierungskorb = circular reinforcing cage
~ **Gründungsbrunnen** *m* **kleiner Abmessung**, ~ Senkbrunnen ~ ~, kreisförmiger ~ ~ ~ = cylinder [*A cylinder has a single wall, and is sunk by the addition of kentledge, by hammer driving or by jacking down; it is of relatively small cross-section compared with a caisson*]
~ **Pilzkopf** *m* = circular (column) capital
Rund|-Abwasserleitung *f*, Kreis-Abwasserleitung = circular sewer
~**aufgeber** *m* → Tellerspeiser
~**bau** *m*, Rundgebäude *n* = circular building
~**becken** *n*, Rundkratzerbecken = circular tank
~**behälter** *m*, Kreisbehälter = circular tank

~**beschicker** *m* → Tellerspeiser
~**biegemaschine** *f* **für Bleche**, Blechwalze *f* = plate roll
~**blick** *m* = all-round vision, ~ view
~**blickkanzel** *f*, Vollsichtkanzel, Allsichtkanzel, Rundblickkabine *f*, Vollsichtkabine, Allsichtkabine = full vision cab(in), space-view ~, superview ~
~**bogen** *m*, voller Bogen, Halbkreisbogen, Zirkelbogen = semi-circular arch
~**brecher** *m* → Kegelbrecher
~**bundhaspel** *f*, *m*, Bandstahlhaspel für Rundbunde = strip steel coiler for round coils
~**deckel** *m*, Kreisdeckel = circular cover
~**draht** *m* = round wire
~**(draht)litze** *f* = round strand
~**drahtsiebgewebe** *n* = round wire screen cloth
~**dübel** *m* = round dowel
~**eindicker** *m* = circular thickener
Rundeisen *n* → Beton~
~**bewehrung** *f*, Rundeiseneinlagen *fpl*, Rundeisenarmierung = round bar reinforcement
~**bündel** *n* → Beton~
~**stab** *m* → Beton~
Runderneuerung *f* = recap job
~ **von Schulter zu Schulter** = retreading
~ **von Wulst zu Wulst** = remoulding, rebuilding
rundes Betonrohr *n* → Betonrundrohr
~ **Korn** *n*, abgerundetes ~ = round particle, ~ grain
~ **Rohr** *n* → Rundrohr
Rund|etage *f* → Rundgeschoß *n*
~**fahrt** *f*, Arbeitsrunde *f*, Umlauf *m* [*Radschrapper*] = round trip
~**fahrzeit** *f* [*Radschrapper*] = round-trip time
Rundfunk|studio *n* = sound broadcasting studio
~**turm** *m* = radio tower
Rundgang *m* = circular corridor
~ [*das Ein- und Ausbauen des Bohrstranges beim Rotary-Bohrverfahren*] = round trip

Rundgebäude — Rungenwagen

Rund|gebäude n, Rundbau m = circular building
~**geschoß** n, Kreisgeschoß, Rundetage f, Kreisetage, Rundstockwerk n, Kreisstockwerk = circular floor, round ~
~**hänger** m, Rundhängeeisen n [*Hängedecke*] = round hanger
~**herd** m [*Sinteranlage*] = rotary hearth
(~)**Holzgreifkorb** m, Holzklammer f = (pulp)wood grapple (or grab)
~**holzlehrgerüst** n, Rundholzrüstung f = round timber falsework
~**holzstütze** f = round timber support
~**hülse** f = round sheath
~**kanal** m für Abwasser, Kreiskanal ~ ~ = circular conduit-type sewer
~**-Kanaltunnel** m, Kreis-Kanaltunnel = circular canal tunnel
~**kant** m = half-round nosing
~**keil** m = round wedge
~**kies** m, Rollkies = round gravel
Rundkipper m, Kreiskipper = rotating tip, rotary tippler
~**waage** f = portable tip wagon type concrete batcher scale
Rundkopf m, Pfeilerkopf [*Talsperre*] = round-head(ed) buttress; segmental-headed counterfort (Brit.)
~**mauer** f → → Talsperre f
~**niet** m = round-head rivet
~**schraube** f = button-head(ed) screw, half-round ~
Rundkorn n = round grain
rund(körnig) = round
Rund|(kratzer)becken n = circular tank
~**lauf** m = circular track
~**laufkolbenpumpe** f = runner piston pump, rotating ~ ~
~**litze** f, Runddrahtlitze = round strand
~**litzenseil** n = round strand cable
Rundloch|blech n = round hole plate
~**prüfsieb** n = round hole test sieve
Rund|(loch)sieb n = round hole screen
~**lochung** f [*Siebtechnik*] = round holes

~**lochweite** f = round hole diameter
~**mast** m = round mast
~**pfahl** m = cylinder pile
~**pfeiler** m = cylindrical (foundation) pier, circular (~) ~
~**querschnitt** m, Kreisquerschnitt = circular cross section
~**rohr** n, Kreisrohr, kreisförmiges Rohr, rundes Rohr = cylindrical pipe, circular ~
~**sand** m, Rollsand = round sand
~**säule** f = round column
~**(säulen)schalung** f = round column formwork
~**schale** f, Kreisschale = circular shell
~**schalung** f → Rundsäulenschalung
~**schieber** m [*Zuschlagsilo*] = radial type fill valve
~**schornstein** m = circular chimney
~**sichtführerhaus** n → Vollsichtkanzel
~**sichtkuppel** f = full-vision dome
~**sichtradar** f = airport surveillance radar, ASR
~**sieb** n, Rundlochsieb = round hole screen
~**silo** m, Kreissilo = cylindrical silo
~**spannkopf** m [*Spannbeton*] = circular stressing head
~**spaten** m = half-round spade
~**speiser** m → Tellerspeiser
~**stab** m → (Beton)Rundeisenstab
Rundstahl m → (Beton)Rundeisen n
~**bündel** n → (Beton)Rundeisenbündel
~**dübel** m → Fugendübel
~**schneider** m = bar-cropping machine
Rund|(stahl)stab m → (Beton)Rundeisenstab
~**stockwerk** n → Rundgeschoß n
~**stütze** f = round column
~**stützenschalung** f = round column shuttering, ~ ~ formwork
Rundung f, Krümme f = rounding
Rundverkehr m = gyratory traffic, rotary ~, round-about ~
Rungen fpl, Lattengitter n [*am LKW*] = stakes
~**wagen** m, Rungenwaggon m = **flat** car, platform ~

Runse f → Felseinschnitt m
Ruschelzone f → Bruchzone
russischer Mörtelestrich m, Kalkestrich = lime mortar flooring
Rustika f, Bossage-Mauerwerkbildung f, Rustikamauerwerk n = rusticated ashlar, rustication, rusticated ashler
~**quader** m = rustic quoin
Rüstträger m → Schalungs(rüst)träger
Rüstung f → (Bau)Gerüst n
~ → Lehrgerüst n
Rüstungsbetrieb m = defense factory (US); armament ~ (Brit.)
Ruß m = soot
Rußbläser m = sootblower
Rußkohle f = sooty coal
Rußnebel m, Industriedunst m, Stadtdunst = smog
Rutilbergbau m = rutile mining
Rutsch m = skid
~**asphalt** m = slippery asphalt(ic) pavement, ~ ~ surfacing
Rutsche f, Rutschbahn f = shoot, chute
Rutschen n [*Die durch nicht begrenztes (unkontrolliertes) Gleiten der Reifen auf der Fahrbahndecke hervorgerufene Bewegung des Fahrzeuges in seiner Fahrtrichtung*] = forward skidding
rutschfest, gleitsicher, griffig = non-slip(pery), non-skid, skid-proof
Rutsch|festigkeit f = resistance to skidding
~**fläche** f [*Böschung*] = slide-failure surface
rutschgefährlich = skid-inducing
(Rutsch)Harnisch m, Rutschspiegel m, Gleitfläche f, glänzende Schubfläche, spiegelhafte Gleitfläche, spiegelnde Gleitfläche = slickenside
Rutsch|masse f, Gleitmasse [*Bodenmechanik*] = sliding mass
~**pflaster** n = slippery sett paving
~**reibung** f = sliding friction
~**schere** f, Bohrschere, Schlagschere [*pennsylvanisches Seilbohren*] = (drilling) jars
rutschsichere Farbe f = anti-slip paint
Rutsch|sicherheit f = skidding resistance
~**spiegel** m → (Rutsch)Harnisch m
~**unfall** m = accident due to skidding
Rutschung f, Böschungs~, Gleitflächenbruch m = slide
~ → Bergrutsch m
~, Erd~, Absetzung = (land)slide [*Earth slip upon a large scale*]
~, Böschungs~ = slide
~ → Bodenfließen n
~ **der Verankerung** = anchorage slip
Rutsch|verhütung f = skid prevention
~**weg** m [*Fahrzeug*] = skidding distance
~**winkel** m [*des Füllgutes in Silos und Bunkern*] = sliding angle
Rüttel|beton m, vibrierter Beton = vibrated concrete
~**(beton)rohr** n = vibrated concrete pipe
~**(beton)stütze** f = vibrated column
~**bohle** f, Schwing(ungs)bohle, Vibrationsbohle, Vibratorbohle, Vibrierbohle = vibrating beam, vibratory ~
~**(bohlen)fertiger** m, schienengeführter Oberflächenrüttler m [*mit Glättelement*] = power-propelled surface vibrating and finishing machine, vibratory finishing machine for concrete pavements, concrete finishing road vibrator, vibratory concrete compacting and finishing machine, road vibrating and finishing machine
~**(bohlen)fertiger** m ohne Glättelement = road vibrating machine [*the finishing screed is omitted*]
~**dauer** f, Rüttelzeit f = vibration period, ~ time
~**druck** m = vibratory pressure
~**druckverdichtung** f = compaction by vibration and compression, ~ combined with vibration
~**druckverfahren** n, KELLER-Verfahren [*Verdichtung nichtbindiger Böden in großen Massen und beliebiger Tiefe und durch Tiefenrüttler*] = vibrator-jetting deep compaction, vibroflotation (soil-compaction)
~**fertiger** m → Rüttelbohlenfertiger

Rüttelflasche — Sackbalken

~flasche *f*, Vibrierflasche, Rüttelflasche, Vibratorflasche = vibrating cylinder, ~ head, poker(-shaped case)
~förderer *m*, Rütteltransporteur *m* = vibratory conveyor, ~ conveyer
~form *f* = vibrating mould (Brit.); ~ mold (US)
~fußpfahl *m* = vibroflotation-type point-bearing concrete pile
~gerät *n*, Vibrationsgerät = vibratory device, vibrating ~
~grenze *f* = vibration limit
~grobbeton *m* = coarse vibrated concrete
~kraft *f* = output force, vibratory power
~lücke *f* = gap for insertion of vibrator
~massenbeton *m*, Massenrüttelbeton = vibrated mass concrete
~motor *m*, Benzin-~ = vibrating petrol engine (Brit.); ~ gas(olene) ~, ~ gasoline ~ (US)
~motor *m*, Elektro-~ = vibrating motor
rütteln = to vibrate
Rütteln *n* = vibrating
Rüttel|nadel *f* → Vibrationsnadel
~platte *f*, Vibratorplatte, Schwing(ungs)platte, Vibrationsplatte, Fußplatte [*Teil eines Plattenrüttlers*] = vibrating (base) plate, ~ (~) pan, ~ (~) slab, vibration (~) ~, vibratory (~) ~
~presse *f*, Vibrationspresse = vibrating press
~-Schaffußwalze *f* → Vibrationsschaffußwalze
~schotter *m* = dry-bound macadam
~schuh *m* → Wechselschuh
~sieb *n* = vibratory screen
~stampfen *n* = vibro-tamping
~stampfer *m*, Vibrostampfer, Vibrationsstampfer [*Straßen- und Erdbau*] = vibrating tamper, vibratory ~
~stampfer *m* → Rüttelstampfmaschine
~stampfmaschine *f*, Rüttelstampfer *m*, Vibrostampfmaschine, Vibrostampfer = blockmaking machine compacting by vibration (50 cycles per minute = 3.000 vibrations/min.) and impact tamping
~stampfwalze *f* → Vibrations-Schaffußwalze
~-Strangpreßverfahren *n* = vibroextrusion method
~stütze *f*, Rüttelbetonstütze = vibrated column
~tisch *m*, Schwingtisch, Tischrüttler *m* [*Betonverdichtung*] = vibrating table, vibratory ~, table vibrator
~transporteur *m*, Rüttelförderer *m* = vibratory conveyer, ~ conveyor
~verdichter *m* → Vibrationsverdichter
~verdichtung *f* → Vibrationsverdichtung
~verfahren *n* = vibratory method
~walze *f*, Vibrationswalze = vibratory roller, vibrating ~
~walzen *n*, Vibrationswalzen = vibratory rolling
~zeit *f*, Rütteldauer *f* = vibration time, ~ period
Rüttler *m*, Vibrator *m* = vibrator
~, Vibrator *m*, Beton~ = vibrator, concrete ~
~flasche *f* → Rüttelflasche

S

Sabine'scher Schluckgrad *m*, ~ Absorptionsgrad [*Schallschutz im Hochbau*] = P.E. Sabine absorption coefficient
Sach|schaden *m* = property damage
~schadenunfall *m* = damage-only accident
sach- und fachgemäß = in accordance with the established rules of good workmanship
Sachverständigen|gebühr *f* = expert fee, fee for expert opinion
~untersuchung *f*, Begutachtung = expertise, study, expert opinion
Sack|bahnhof *m* → Kopfbahnhof
~balken *m* → Ballastarm *m*

Sack-Beton — Salzkonzentration

∼-Beton m = prepackaged concrete
∼förderer m = bag conveyor, ∼ conveyer
∼füllmaschine f → Abfüllmaschine für Säcke
∼gasse f, Sackstraße f = blind alley, cul-de-sac (street), close [*A local street open at one end only and with special provision for turning around*]
∼gasse f ohne Wendemöglichkeit = dead-end street [*a local street open at one end only without special provision for turning around*]
∼kalk m = bagged lime
∼karre(n) f, (m) = bag truck
∼leinen-Deckenschluß m = burlap drag finish
∼leinenschleppe f = burlap drag
∼maß n, Ausbreitmaß, Setzmaß, Fließmaß [*Betonprüfung. DIN 1048*] = slump
∼salz n = bagged salt
∼schüttung f [*Betoneinbringung*] = depositing underwater concrete in bags
∼silo m = sack silo
∼stapel m = bag pile
∼stapler m = bag lift truck
∼straße f → Sackgasse f
Sackung f → Senkung
Sack|wall m = sack dam
∼wendelrutsche f, Sachwendelrutschbahn f = gravity sack shoot
∼zement m, eingesackter Zement, abgesackter Zement = sacked cement, bagged ∼
säen, an∼ = to sow, to seed
S. A. E. Nennleistung f = S. A. E. rated
Sägedach n, Sägezahndach, Sheddach = saw-tooth roof, north-light ∼
∼binder m → Shed(dach)binder
Säge|gewinde n [*Tiefbohrtechnik*] = buttress-thread
∼spänebeton m [*Fehlname: Sägemehlbeton*] = sawdust concrete
∼spanreifen m = sawdust tyre (Brit.); ∼ tire (US)
∼(zahn)dach n, Sheddach = saw-tooth roof, north-light ∼

∼(zahn)dachbinder m → Shed(dach)binder
∼zahnfinger m [*Flughafen*] = saw-tooth finger
∼(zahn)schalendach n, Shedschalendach = saw-tooth shell roof, north-light ∼
∼zahnwand f = saw-tooth wall
Saisonspeicherung f = seasonal storage
Saitenbeton m → Stahl ∼
Sakral|architektur f = religious architecture
∼bau(werk) m, (n) = religious building
Salinenwasser n = saline water
Salinität f, Salzhaltigkeit f, Salzgehalt m = salinity, salt content
Salpeter|fraß m, Mauerfraß = efflorescence
∼säure f, Stickstoffsäure, HNO_3 = nitric acid
salpetersaure Lösung f = nitric acid solution
salpetersaurer Kalk m → Kalksalpeter m
salpet(e)rige Säure f = nitrous acid
Salz|bergbau m = salt mining
∼bergwerk n, Salzgrube f = salt mine
∼brunnen m → Salzquelle f
∼dom m = salt dome
∼feste f = salt barrier pillar
∼gehalt m → Salzhaltigkeit f
∼geschwindigkeitsverfahren n [*oben Einspritzung einer Salzlösung; unten Messung des Zeitpunktes der Änderung der elektrischen Leitfähigkeit des Wassers*] = salt-velocity method
∼gitterbohrgerät n = Salzgitter reverse circulation drilling rig
∼gitterverfahren n = Salzgitter reverse circulation method
∼glasieren n = salt glazing
salzglasiert = salt-glazed
Salz|glasur f = salt glaze
∼grube f, Salzbergwerk n = salt mine
∼haltigkeit f, Salinität f, Salzgehalt m, Salzigkeit f = salinity, salt content
Salzigkeit f → Salzhaltigkeit
Salz|konzentration f = salt concentration

Salzlager(stätte) — Sand

~**lager(stätte)** *n, (f)*, Salzvorkommen *n*, Salzvorkommnis *n* = salt deposit
~**lösungsmethode** *f* **für Abflußmessungen** = salt-dilution method of flow measurement
~**marsch** *f*, Salzsumpf *m* = salt swamp
~**mischungsverfahren** *n*, Salzverdünnungsverfahren [*Messung von Wasserströmungen*] = salt-dilution method
~**molekül** *n* = salt molecule
~**mutter** *f* = mother of salt
~**pfeiler** *m* = salt pillar
~**quelle** *f*, Sole *f*, Salzbrunnen *m* = saline spring
salzsaure Kupferchlorürlösung *f*, $Cu_2Cl_2 + HCl + aq$ = hydrochloric solution of cuprous chloride
Salz|see *m* = salt lake
~**stock** *m* = salt dome
~**streuen** *n* = salting
~**streugerät** *n*, Salzstreuer *m* = salt spreader, ~ distributor
~**streuung** *f* = salt treatment, ~ application
~**sumpf** *m*, Salzmarsch *f* = salt swamp
~**verdünnungsverfahren** *n*, Salzmischungsverfahren [*Messung von Wasserströmungen*] = salt dilution method
~**verfestigung** *f* = salt stabilization
~**vorkommen** *n* → Salzlager(stätte)
Salzwasser|korrosion *f* = salt water corrosion
~**masse** *f* = body of salt water
~**ton** *m* = salt water clay
Sammel|bandförderer *m* = collecting belt conveyor, ~ conveyer
~**becken** *n* → → Talsperre *f*
~**becken** *n* **der Sedimentation** (Geol.) [*z. B. das Meer*] = sedimentary basin
~**drän** *m* → Sammler *m*
~**graben** *m* → Auffanggraben
~**heizung** *f*, Zentralheizung = central heating
~**kanal** *m* = intercepting conduit-type sewer
~**leitung** *f* → Hauptabwasserleitung
~**rinne** *f* = collecting channel
~**rohr** *n* = collecting pipe
~**schacht** *m* = collecting manhole
~**schacht** *m* [*Grundwasserfassung*] = central shaft
~**schiene** *f* = bus-bar [*Main conductor on a switchboard, forming a common terminal to which cables may be connected*]
~**schienenschrank** *m* = bus-bar chamber
~**schmierung** *f* = centralized oiling
~**spur** *f* [*Verkehr*] = storage lane
~**stollen** *m* = collecting gallery
~**trichter** *m* = collecting hopper
Sammler *m*, Abfang~, Sammeldrän *m* [*Entwässerung*] = collecting drain, outfall ~, intercepting ~, catch (-water) ~, interceptor, collector
~ [*Wassergewinnung*] = collector
~ = intercepting sewer [*A sewer which receives the dry-weather flow from a number of transverse sewers or outlets, with or without a determined quantity of storm water if from a combined system*]
~ → Hauptabwasserleitung *f*
~-**Zusammenflußanlage** *f* [*Abwasserwesen*] = sewer interchange
Samson-Stripper *m* (⚒) [*schälende Gewinnungsmaschine*] = Samson stripper
„**Sand**" *m* = grit [*The heavy mineral matter contained in sewage, such as sand, gravel, cinders, etc.*]
Sand *m* [*nach Atterberg, Fischer und Udluft, DIN 4220, International Society of Soil Science 2–0,02 mm; nach DIN 4022, DEGEBO, Terzaghi 2–0,1 mm; nach Casagrande (Jahr 1947), British Standards Institution, Massachusetts Institut of Technology 2–0.06 mm; nach Gallwitz (Jahr 1939) 2–0,2 mm; nach Grengg (Jahr 1942) 2–0,1 mm; nach U.S. Bureau of Soils 2–0.05 mm*] = sand
~ **mit hellen und dunklen Bestandteilen** = (salt and) pepper sand

Sandabdeckung — Sandgewinnung

~abdeckung *f*, Nachbehandlung mit feuchtem Sand [*Beton*] = sand curing

~abscheider *m* [*Grundstückentwässerung*] = sand trap (Brit.); ~ interceptor (US)

~anteil *m* = sand content

~anteil-Untersuchungsverfahren *n* = sand equivalent test

~asphalt *m* = sand carpet (Brit.); sheet asphalt (US)

~aufbereitungsanlage *f* = sand processing (or dressing, or producing) plant (or installation)

~aufbruch *m* [*Grundbuch bei kritischem Gefälle*] = boiling of sand, boil

~(auf)schüttung *f*, Sandauffüllung = sand fill

~aufspülung *f* = hydraulic placement of sand, ~ sand filling

~ausgleichschicht *f*, Sandausgleichlage *f* = regulating carpet of sand, ~ sand carpet

~auslauf *m* [*Waschmaschine*] = sand outlet

~ausscheidungsschnecke *f* → Sandschnecke

~(-Bagger)greifer *m*, Sand-Baggerkorb *m*, Sand-Greifer-Korb = sand grab(bing bucket)

~bahn *f* = sand race track

Sandballast *m* = sand ballast

~-Walz(en)rad *n* = sand ballast type roll

Sand|bank *f*, Sandbarre *f* = sand bar, ~ bank, ~ head

~barre *f*, Sandbank *f* = sand bank, ~ bar, ~ head

~beigabe *f*, Sandzusatz *m* = addition of sand, sand addition

~bett(ung) *n*, (*f*), Sand(unter)bett(ung) = sand underlay, ~ bed(ding course), ~ cushion, ~ blanket

~boden *m*, sandiger Boden = sandy soil

~böschung *f* = sand slope

~brunnen *n* → vertikale Sanddränage *f*

~-Brunnengründung *f* = sand well foundation

~damm *m* = sand embankment

~drän *m* → vertikale Sanddränage *f*

~düne *f* = sand-dune, down

~eindicker *m* = sand thickener

~einschluß *m* = inclusion of sand

sanden → ab~

Sand|entwässerung *f* = dewatering of sand

~entwässerungsschnecke *f*, Wasserausscheidungsschnecke *f* = sand dewatering screw, ~ dehydrator ~

~ersatzmethode *f* [*Nachprüfung der Verdichtung f*] = calibrated sand method

Sandfang *m* = detritus chamber, ~ tank [*A detention chamber larger than a grit chamber, usually with provision for removing the sediment without interrupting the flow of sewage; a settling tank of short detention period primarily, to remove heavy settleable solids*]

~ = grit chamber, ~ compartment [*A small detention chamber or an enlargement of a sewer designed to check the velocity of the sewage enough to permit the heaviest solid matter, such as grit, to be deposited with a view to its frequent and easy removal*]

~ = grit catcher [*A chamber usually placed at the upper end of a depressed sewer, or at other points of protection on combined or storm-water sewers, of such shape and dimensions as to reduce the velocity of flow and thus permit the settling out of grit*]

~ → Entsander *m*

~kanal *m* = grit-chamber channel

Sand|feld *n* → Kiesfeld

~feuchtigkeit *f*, Sandfeuchte *f* = moisture of sand

~filter *m*, *n* = sand filter

~filterung *f* = sand fitration

~förderung *f* = sand transport

~füllung *f* [*Kastenfang(e)damm; Zellenfang(e)damm*] = sand fill

~geröll(e) *n* (Geol.) = moving sand débris

sandgeschlämmt = blinded with sand

Sand|gewinnung *f* = sand winning

Sand-Greiferkorb — Sandschnecke

~-Greiferkorb *m* → Sand(-Bagger)-greifer

~-Grobzuschlag-Verhältnis *n* = sand-coarse aggregate ratio

~grube *f* = sand pit

Sandguß *m* = sand casting

~(eisen)rohr *n* = static-cast pipe, sand-cast ~

~teil *m, n* = sand casting

sandhaltiger Ton *m* = sandy clay

Sand|hinterfüllung *f* = sand backfill

~hose *f* = sand pillar, ~ spout

sandig = sandy, arenaceous

sandiger Kies *m*, Sandkies = hoggin (Brit.); sandy gravel, path gravel, gravel-sand-clay

~ Ton *m*, magerer ~, Magerton = sandy clay

Sand|inselmethode *f*, Sandinselschüttung *f* [*Gründung*] = sand island method

~kammer *f* [*Abwasserwesen*] = grit chamber

~kies *m* → sandiger Kies

~klassieranlage *f* = sand classifying plant, ~ ~ installation

~klassierer *m* = sand classifier

~kohle *f*, Magerkohle, magere Kohle, harte Kohle = non-caking coal, dry burning ~, hard ~; free-ash ~ (US)

~korn *n*, Sandteilchen *n* = sand grain

~-Kornabstufung *f* = sand grading

~körper *m* = mass of sand, sand mass

~kratzer(kette) *m*, (*f*) = sand drag, dewatering flight conveyor, washbox, dewatering flight conveyer

~(kreisel)pumpe *f* = sand sucker, ~ pump

~lage *f*, Sandschicht *f* = sand layer, ~ course

~lager *n* → Sandvorkommen

~lager *n* [*auf einer Baustelle*] = sand dump

~lagerstätte *f* → Sandvorkommen *n*

~linse *f* = lens of sand

~mergel *m* = sandy marl, lime-gravel, clay-grit

~mörtel *m* = sand mortar

~mudde *f* = sandy mud

~mühle *f* = sand mill, ~ grinder

~muschel *f* = clam

~naßbaggerei *f*, Sandnaßbaggerung *f* = sand dredging

~nest *n* = sand pocket

~papier *n*, Glaspapier = sand paper

~papieroberfläche *f*, Sandpapierrauheit *f* = sand paper surface, ~ ~ finish, ~ ~ non-skid texture [*road surface texture*]

~pfahl *m* = sand pile [*A sand pile is formed in a hole prepared by a driven pile or jumper and filled with sand*]

~prisma *n* = prism of sand

~pumpe *f*, Sandkreiselpumpe = sand sucker, ~ pump

~pumpe *f* → Stauchbohrer *m*

~reichtum *m* [*in Mörtel*] = sand concentration

~rückgewinnung *f*, Sandwiedergewinnung = sand recovery, ~ reclamation

~rückgewinnungsanlage *f*, Sandwiedergewinnungsanlage = sand reclaiming plant, ~ recovery ~, ~ ~ installation

~rückgewinnungsmaschine *f* = sand reclaiming machine, ~ recovery ~

~rückgewinnungsschnecke *f* = sand reclaiming screw, ~ recovery ~

~sack *m* = sand bag

~sackabdämmung *f* → Abdämmung durch Sandsäcke

~sackabsperrung *f* → Abdämmung durch Sandsäcke

~sackwall *m* = sand bag embankment, ~ ~ bank

~schicht *f*, Sandlage *f* = sand layer, ~ course

~schicht *f* (Geol.) = sand stratum

~schiefer *m* = schistous sandstone, foliated grit(-stone), arenaceous shale

~schlämme *f* = sand slurry

~schliff *m* (Geol.) → Korrosion *f*

~schmitze *f*, Sandstreifen *m* = sand streak, streak of sand

~schnecke *f*, Waschschnecke, Spiral-(naß)klassierer *m*, Schneckenwäsche

Sandschüttung — Sandstein

f, Sandausscheidungsschnecke = spiral classifier, (spiral) screw washer, screw-type sand classifier, washing screw
~schüttung *f* → Sandaufschüttung
~sieb *n* = sand screen

~siebmaschine *f* = sand screening machine
~siebtrommel *f* = rotary sand screen
~silo *m* = sand silo
~sortierung *f* = sand separation
~staub *m*, Staubsand *m* = dust sand

Sandstein	sandstone
Arkose~, feldspatreicher ~, Arkose *f*	arkose (~), arkosic grit
Asphaltsand(stein), bituminöser ~	sand asphalt (US); asphaltic sand (stone), bituminous sandstone
Bindemittel *n*, Kitt *m* [*Zwischenmasse, die dem Sandstein dauernd eine mehr oder minder große Festigkeit verleiht*]	cementing agent
Bunt~	mottled ~; variegated ~ (Brit.); bunter ~
Eisen~, eisenschüssiger ~	iron-~, ferrugin(e)ous ~; Hastings sand (Brit.)
feinkörniger ~	fine-grained ~
Füllmasse *f* [*Zwischenmasse, die lediglich die Zwischenräume zwischen den Felstrümmern ausfüllt, ohne sie fest und dauernd zu verkitten*]	filling agent having no cementing power
Glaukonit~	glauconitic ~
Glimmer~	mica(ceous) ~
grauer ~	grayband(s)
grobkörniger ~, Kristall~	coarse-grained ~, gritstone
Kalk~	calcareous ~, lime-cemented ~
kieselhaltiger ~, verkieselter ~	siliceous ~
konglomeratischer ~	conglomeratic ~
Kristall~ → grobkörniger ~	
Mergel~	marly ~
mittelkörniger ~	medium-grained ~
Muschel~	shell(y) ~, beach rock
Nummuliten~	nummulitic ~
ölgesättigter ~	oil-saturated ~
ölgetränkter ~	oil-impregnated ~
Quarz~	silica ~
Schiefer~	shaly ~
streifiger ~	linsey
~ vom Craig-yr-Hesg Quarry, Pontypridd, England	pennant grit
~schiefer *m* [*der Porenkitt enthält Glimmer*]	slaty ~
Ton~	argillaceous ~, clayey ~
verkieselter ~ → kieselhaltiger ~	

Sandsteinbruch — Sandzunge

Sandstein|bruch m = sandstone quarry
~mehl n = sandstone meal
~platte f = sandstone slab
~quader m = sandstone ashlar
~verblendung f = sandstone facing
Sand|stoß m = sand face, face of sand
~strahl m = sand blast
sandstrahlen, absanden = to sandblast
Sand|strahlen n, **Absanden** = sand blasting [*A system of cutting or abrading a surface such as concrete by a stream of sand ejected from a nozzle at high speed by compressed air; often used for cleanup of horizontal construction joints*]
~strahlentrostung f = rust removal by sand blasting
~strahlgebläse n = sand blaster, ~ blast(ing) machine
~strand m = sand beach
~streifen m, **Sandschmitze** f = sand streak, streak of sand
~streifenbildung f, **Sandschmitzenbildung** = sand streaking
(~)Streuautomat m = auto(matic) sand distributor, ~ ~ spreader
sandstreuen = to spread sand
Sand|streuen n = sand spreading
~streuer m, **Sandstreugerät** n, **Sandstreumaschine** f = sand spreader, ~ distributor
~streu-Nachläufer m = towed-type sand spreader, ~ ~ distributor
~strichziegel m, **Sandstrichstein** m = sand-struck brick
~teilchen n, **Sandkorn** n = sand grain, ~ particle
~-Ton-Boden m = sand-clay soil
~-Ton-Mischung f, **Sand-Ton-Gemisch** n = sand-clay mix(ture)
~-Ton-Straße f = sand-clay road
~topf m [*zum Absenken von Lehrgerüsten*] = sand holder
~trockner m = sand dryer; ~ drier (US)
~- und Kiesgreifer m, **Zweischalengreifer für Sand und Kies** = scraper grab

~- und Kieswaschmaschine f = sand and gravel washer, ~ ~ ~ washing machine
~- und Kieswerk n = sand and gravel plant
~(unter)bett(ung) n, (f) → Sandbett(ung)
~unterlage f = sand underlay
sandverdämmte Bohrung f, ~ **Sonde** f; **sandverdämmter Schacht** m [*in Galizien*] = sand-tamped well, ~ ~ bore, ~ boring
~ Sonde f → ~ **Bohrung** f
sandverdämmter Schacht m [*in Galizien*] → sandverdämmte Bohrung f
Sand|verwehung f, **Sandwehe** f = sand drift, drift of sand
~vorkommen n, **Sandvorkommnis** n, **Sandlagerstätte** f, **Sandlager** n, **Sandfundstätte** f = sand deposit
~waage f = sand scale
~walzwerk n → → Walzwerk
~wanderung f = sand migration
~wäsche f, **Sandwaschmaschine** f = sand washing machine, ~ washer
~wäsche f = sand washing [*filter operation*]; grit ~ [*sewage works operation*]
~weg m = sand path
~wehe f, **Sandverwehung** f = sand drift, drift of sand
~welle f = sand wave
~werk n, **Sandaufbereitungsanlage** f = sand plant
Sandwich|-Kabel n [*Spannbeton*] = sandwich cable
~-Platte f [*Spannbeton*] = locking plate, sandwich ~
~verfahren n = sandwich method
Sand|wiedergewinnung f, **Sandrückgewinnung** = sand recovery, ~ reclamation
~wiedergewinnungsanlage f, **Sandrückgewinnungsanlage** = sand recovery plant, ~ reclaiming ~
~wüste f = sandy desert
~-Zement-Schlämme f, **Sand-Zement-Schlempe**, **Sand-Zement-Schlämpe** = sand cement grout
~zunge f = sand-spit

Sandzusatz — Sattel-Tiefladeanhänger

~zusatz m, Sandbeigabe f = addition of sand, sand addition
Sanierung f, Erneuerung [*Städtebau*] = clearance, rehabilitation
Sanierungsgebiet n, Erneuerungsgebiet [*Städtebau*] = clearance zone, rehabilitation~
sanitäre Installation f, Sanitärinstallation = sanitary plumbing
Sanitär|arbeiten /pl = sanitary works
~**keramik** f, sanitäre Keramik = (china) sanitary ware [*Wash tubs, sinks, tanks, and ordinary bathroom equipment formed of clay, baked and glazed*]
~**rohr** n = plumbing pipe
~**steingut** n, Sanitärsteinzeug n = sanitary stoneware
~**technik** f = public health engineering in the wider sense
~**ware** f = sanitary ware
Sanitäter m = dresser; aid (US)
Sanitätsdienstzimmer n = first-aid room
Santorinerde f [*Schlackenartige Erde vulkanischen Ursprungs, als hydraulischer Zuschlag zu Luftmörtel verwendet. Nach der griechischen Insel Santorin benannt*] = Santorin earth
Sargdeckel m (⚒) = detached "coffin" in the roof
Sargossatang m, Beerentang = sargassum
Satellit|gebäude n [*Flughafen*] = satellite (building)
~**system** n, System zusätzlicher Abfertigungsgebäude [*Flughafen*] = satellite system
Satinweiß n = satin white
satt anliegend [*Kappe im Bergbau*] = flush (with the roof)
Sattdampf m, Naßdampf, gesättigter Dampf = saturated steam
~**zylinderöl** n = saturated steam cylinder oil
Sattel m, Ankupp(e)lung f = hitch
~ **der Faltung**, Sattelfalte f, Gewölbe n, Antikline f, Antiklinale f = anticlinal fold, anticline

~**anhänger** m, Auf~, Aufl(i)eger, aufgesatteltel Anhänger, Halbanhänger, Sattelfahrzeug n [*Fehlname: Sattelaufl(i)eger*] = semitrailer
~**aufl(i)eger** m [*Fehlname*] → Sattelanhänger
Satteldach n = gable(d) roof [*It slopes in two directions*]
~ **ohne Bundbalken** = couple roof
Sattel|fahrzeug n → Sattelanhänger m
~(**fahrzeug**)-**Bodenentleerer** m → Halbanhänger mit Bodenentleerung
~**falte** f → Sattel der Faltung
~**flügel** m (Geol.) = side of an anticline
~**haus** n, gleichseitiges Gewächshaus = greenhouse with two equal sides
~**höchstes** n = crown of an anticline
~**holz** n = bolster, corbel piece, head tree, crown plate, saddle [*A short timber cap over a post to increase the bearing area under a beam*]
~**kipper** m → Kipp-Halbanhänger m
~**linie** f = crest line of an anticline
~**oberlicht** n = double-inclined skylight
~**platte** f = cradle plate, swing ~, banjo, saddle strap [*in a launching cradle*]
~**rost** m = saddle grate
~**schlepper** m, Zugmaschine f für Halbanhänger = truck tractor for semitrailers, tractor-truck ~ ~
~**schlepperbauart** f = semitrailer type construction
~-**Schürfwagen** m → Schürfkübel m
~**schwelle** f, Saumschwelle, Brustschwelle = girt [*A rail or intermediate beam in wooden-framed buildings, often carrying floor joists*]
~**spalte** f (Geol.) = anticlinal fissure
~**tal** n, Antiklinaltal = saddle valley, anticlinal ~
~-**Tiefladeanhänger** m, Sattel-Tieflader m, Sattel-Tiefbett-Anhänger, Sattel-Tiefladewagen m = semi-low-loader, semi-low-bed trailer, semi-low-load trailer, semi-deck trailer; semi-low-boy (trailer) (US)

Sattelwagen — Saugbeton

~wagen *m*, Selbstentleerer *m* mit Sattelboden = saddle bottomed wagon

~wendung *f* = point at which the contour line of an anticline changes direction

Sättigung *f* = saturation

Sättigungs|beiwert *m* = saturation coefficient

~druck *m* = saturation pressure

~geschwindigkeit *f* = saturation velocity

~grad *m* [*Bodenmechanik*] = degree of saturation

~grenze *f* = saturation limit

~konzentration *f* = saturated concentration

~korrektur *f* = saturation correction

~linie *f*, Sättigungskurve *f* = line of saturation

~punkt *m* = saturation point

~zone *f*, gesättigte Zone [*eines Ölträgers*] = zone of saturation

Satz *m*, Maschinensatz, (Maschinen-) Aggregat *n* = rig, outfit, set

~ **Pläne** = set of plans

~ **Werkzeuge**, Werkzeugsatz = set of tools, tool set

satzweise (Hart)Zerkleinerung *f* = batch comminution, ~ reduction

~ **Mahlung** *f*, satzweises Mahlen *n*, Chargenmahlung, Chargenmahlen = batch grinding, intermittent ~

Sauberkeitsschicht *f*, Reinheitsschicht, verbesserter Untergrund *m* [*In Deutschland auch als „Unterbau" bezeichnet, wenn die darüberliegende Schicht „Tragschicht" genannt wird*]; Schüttung *f* [*in der Schweiz und Österreich; „Schüttung " und „verbesserter Untergrund" ergeben zusammen den „Unterbau"*] = (granular) sub-base (course), improved subgrade

Sauberspülen *n* [*Tiefbohrtechnik*] = pumping the hole clean

Säuberungsmaschine *f*, Reinigungsmaschine [*für Betonfugen*] = joint cleaner

sauer [*Gestein*] = acidic

Sauerquelle *f* = aerated spring

Sauerstoff|(atmungs)gerät *n* = oxygen breathing apparatus

~aufnahme *f* = oxygen uptake, uptake of oxygen

~-Azetylen-Schweißen *n* = oxy-acetylene welding

~bedarf *m* = oxygen requirement

~bilanz *f* = oxygen balance

~bohren *n*, Bohren mit Sauerstofflanze = oxygen lancing

~-Erzeugungsanlage *f* = oxygen plant

~-Flasche *f* = oxygen cylinder

~gebläse *n* = oxygen blower

~gehalt *m* = oxygen content

~haushalt *m* = oxygen economy

~hobelgerät *n* = (oxygen) flame-grooving equipment

~ionen-Beweglichkeit *f* = oxygen ion mobility

~lanze *f* = oxygen lance

~linie *f*, Sauerstoffkurve *f* = oxygen sag

sauerstofflos, luftlos = anaerobic

Sauerstoff|mangel *m*, Sauerstofffehlbetrag *m* = oxygen deficiency

~salz *n* = oxysalt, salt of oxyacid

~schneiden *n* = flame cutting, torch ~, torch burning

~ventil *n*, Schneid~ = oxygen valve

~verbrauch *m* = oxygen consumption

~zehrung *f* = oxygen depletion

~zufuhr *f* = supply of oxygen

Saug|anschluß *m* = suction connection

~apparat *m* = suction device

Saugbagger *m*, Pumpen(naß)bagger, Schwimmsaugbagger, Naßsaugbagger, Saugpumpenbagger [*früher „Grundsauger" genannt*] = suction dredger, (sand-)pump ~, hydraulic (suction) ~ [*US = dredge*]

~ **mit Schneidkopf** → Cutter(saug)bagger

Saug|becken *n* [*Pumpspeicherkraftwerk*] = suction pool

~behandlung *f*, Absaugen *n*, Vakuumbehandlung [*Beton*] = vacuum treatment

~beton *m* → → Beton

Saugbetonverfahren — Saugkrümmer

~betonverfahren n, Unterdruck-Oberflächenbehand(e)lung f = vacuum concrete process

~bewetterung f → Absaugebewetterung

~bohrgerät n = reverse circulation portadrill

~bohrverfahren n [*Die Saugbohrmethode fördert das Bohrklein im Gegensatz zur Rotary-Bohrmethode im Innern des Bohrgestänges. Damit wurde die Voraussetzung geschaffen, ohne Veränderung des Spülstromes große Bohrloch-Durchmesser herzustellen. Bohranlagen mit einem lichten Durchmesser von 150 mm im Gestänge sind für Bohrloch-Durchmesser von 400 bis 1500 mm wirtschaftlich einsetzbar. Ist die Anlage für 200 mm lichten Durchgang ausgelegt, so können die Bohrloch-Durchmesser zwischen 500 und 2500 mm gewählt werden. Die größte Saugbohranlage hat einen lichten Durchgang von 300 mm und ist für Bohrloch-Durchmesser zwischen 750 und 5000 mm geeignet.*
Bei Arbeiten im Lockergebirge ist außer einem Standrohr mit einer Länge von 2 bis 5 m in allgemeinen keine Hilfsverrohrung erforderlich, wodurch Bohrfortschritte von 4 bis 10 m/h, in Abhängigkeit vom Bohrloch-Durchmesser und vom Gebirge, erreicht werden. Bei Verwendung von Rollenmeißeln werden im leichten und mittleren Fels Bohrfortschritte von 0,5 bis 4 m/h möglich] = portadrill (reverse circulation) method

~brunnen m = suction well, well pumped by suction

~drän m → Sauger m

~-Druckförderung f = conveying by combined airsuction and airpressure

~düse f = suction jet

~düsenmundstück n = suction nozzle

Saug(e)kopf m [*Saugbagger*] = suction head

saugen, ab~ [*Späne und Staub*] = to exhaust, to extract, to suck

saugende Bewetterung f → Absaugbewetterung

saugender Injektor m → Dampfstrahlelevator m

Sauger m, Saugdrän m = branch drain, subsidiary ~, feeder ~
~ → Saugkorb m

Saug|fähigkeit f = suction property

~filter m, n, Saugzellenfilter, Vakuumfilter = vacuum filter

~förderer m = suction conveyor, ~ conveyer

~form f → Vakuumform

~gas n, Generatorgas = suction gas, generator ~, producer ~, power ~

~gebläse n = suction blower

~glocke f [*Motor*] = suction bell

~graben m, Rigole f, Sickergraben [*mit Kies ausgepackt*] = gravel-filled drain trench

~graben m, (Stein) Rigole f, Sickergraben [*mit Steinen ausgepackt*] = stone-filled drain trench

~graben mit Dränrohr, Sickergraben ~ ~, Rigole f ~ ~ = French drain

~haube f, Ab~ = suction hood

~heber m = siphon

~höhe f, An~ = suction lift, ~ head

~hub m, Ansaughub, Saugtakt m, Ansaugtakt = suction stroke

~kammer f = suction chamber

~kanal m = suction duct

~kegel m, Saugkonus m = suction cone

~kehrmaschine f, Kehrsaugmaschine = suction scavenging machine, ~ scavenger

~konus m, Saugkegel m = suction cone

Saugkopf m, Saugekopf [*Saugbagger*] = suction head

~ mit hydraulischer Bodenlösung, Frühling-Saugkopf, Schleppkopf Bauart Frühling = draghead with hydraulic jet

Saug|korb m, Seiher m, Sauger m, (Pumpen)Fußsieb n, Fußventil n für Saugleitung = (pump) strainer

~krümmer m, An~ [*Motor*] = induction manifold, intake ~, inlet ~

Saug-Kugelventil — Säulen-Drehbohrmaschine

~-**Kugelventil** *n*, Kugel-Saugventil [*Pumpe*] = ball suction valve
~**lader** *m* = suction loader
~**leistung** *f*, An~, angesaugte Luftmenge *f*, Ansaug(e)menge [*Kompressor*] = actual volume of the cylinder
~**leitung** *f* = suction piping, ~ line
~**leitungssieb** *n* = suction line screen
~**loch** *n* [*Abteufpumpe*] = snore-hole
~**luft** *f*, Ansaugluft = suction air
~**lüfter** *m* = suction fan
~**luftförderanlage** *f*, Saugluftförderer *m* = suction (or vacuum) pneumatic conveyor (or conveyer, or conveying system), vacuum pump transporter
~**lüftung** *f* = suction ventilation
~**matte** *f* = vacuum mat
~**motor** *m* = aspirating engine
~**napf** *m* = sucker
~**pumpe** *f* = suction pump, sucking ~
~**pumpenbagger** *m* → Saugbagger
~**raum** *m* [*Pumpe*] = inlet chamber
~**raum** *m*, Kapillarsaum, Saugsaum, Kapillarzone *f* = capillary fringe
Saugrohr *n*, An~ = suction tube, ~ pipe
~ [*Saugüberfall*] = siphon barrel
~ = draft tube (US); suction ~ [*An expanding tube connecting the passage of a turbine with the tail water*]
~**kappe** *f* = air cleaner cap
Saug|saum *m* → Saugraum
~**schlauch** *m* = suction hose
~**schlauchfilter** *m*, *n* = suction type cloth filter dust collector
~**schleppkopf** *m*, Schleppsaugkopf = drag suction head
~**seite** *f*, An~ [*Pumpe*] = suction side, ~ part
~**strahlpumpe** *f* = sucking jet pump
~-**Straßenkehrmaschine** *f* = suction scavenger, ~ scavenging machine
~**stutzen** *m* [*Motor*] = intake
~**stutzen** *m*, Ab~ = suction tube
~**takt** *m*, Saughub *m*, Ansaugtakt, Ansaughub = suction stroke
~**überfall** *m* → Heberüberfall

~- **und Druckleitung** *f* = suction and delivery line
~- **und Druckpumpe** *f* = (combined) suction and force pump, lift and force ~
~- **und Druckwindkessel** *m* [*Membran(e)pumpe*] = air balancing chamber
~**ventil** *n* [*Kompressor*] = inlet valve
~**ventil** *n* = suction valve
~**ventilator** *m* = suction fan
~**verfahren** *n* = surging
~**vermögen** *n* = suction property
~**vorrichtung** *f* zum Fertigteiltransport, Vakuum(an)heber *m* = vacuum lifter pad
~**wasserspiegel** *m* = level of water being sucked
~**wirkung** *f* = suction effect
~**(zellen)filter** *m*, *n*, Vakuumfilter = vacuum filter
Saugzug *m* = induced draft (US); ~ draught (Brit.) [*Flow of gases through the kiln created by a suction fan*]
~**ventilator** *m* = induced draft fa' (US); ~ draught ~ (Brit.)
Säule *f* = pillar, column [*Round pillar, including the base and capital. The proportions vary according to the style or order*]
~, Ständer *m*, Stiel *m*, Pfosten *m* [*Fach(werk)wand*] = post
Säulen|armierung *f* → Säulenbewehrung
~**basalt** *m* = columnar basalt
~**basis** *f* = column base (plate), pillar ~ (~)
~**beton** *m* = column concrete, pillar ~
~**bewehrung** *f*, Säulenarmierung, Säulen(stahl)einlagen *fpl* = pillar reinforcement, column ~
~**bohrmaschine** *f* = column drill
~**bruchlast** *f* = column failure load, pillar ~
~**deckplatte** *f*, Abakus *m*, Kapitellplatte = raised table, abacus [*A slab atop the capital of a column*]
~-**Drehbohrmaschine** *f* = column-mounted drifter (drill)

Säulendrehkran — Saussuritgabbro 804

~**drehkran** *m* = slewing pillar crane, ~ column ~
~**einlagen** *fpl* → Säulenbewehrung *f*
~**festigkeit** *f* = column strength, pillar ~
~**fundament** *n* = column foundation, pillar ~
~**fuß** *m* = base of column, ~ ~ pillar
~**gang** *m* = portico
~**halle** *f* → Hypostylos *m*
~**halle** *f*, Kolonnade *f* = colonnade [*In architecture, a series of columns*]
~**kopf** *m* = head of column, ~ ~ pillar
~**kran** *m* = pillar crane
~**laubwerk** *n*, Akanthus *m* = acanthus
säulenlos [*Architektur*] = astylar
Säulen|ordnung *f* = column arrangement, pillar ~
~**pfette** *f* = queen post purlin
~**rippe** *f* = rib of column, ~ ~ pillar
~**schaft** *m* = shaft of column, ~ ~ pillar
~**schalung** *f*, Schalungskasten *m* = pillar formwork
~**(stahl)einlagen** *fpl* → Säulenbewehrung *f*
~**struktur** *f*, Stengelgefüge *n* = columnar structure
~**trommel** *f* = drum of column, ~ ~ pillar
säulige Absonderung *f* (Geol.) = columnar jointing
säuliger Habitus *m* [*Kristall*] = short-columnar habit
Saum|graben *m* → Auffanggraben
~**lade** *f*, Saumlatte *f* [*Dach*] = chantlate, eaveslath
~**pfad** *m* = bridle-path, bridle-road, bridle-way, driftway, drove(-way) packway
~**pfette** *f*, Randpfette = verge purlin
~**schwelle** *f* → Sattelschwelle
~**stein** *m* → Bordziegel *m*
~**winkel** *m*, Eckwinkel = corner angle, angle bracket
~**ziegel** *m* → Bordziegel

saure Emulsion *f*, kationische ~ = acid emulsion, cationic ~
saures Bessemerverfahren *n* = acid Bessemer process
~ **Gestein** *n* = acidic rock [*Igneous rock containing over 65% of silica*]
säureabweisend = acid repelling
Säure|angriff *m* = acid attack
~**anstrich** *m* = acid-resisting coat(ing)
~**austausch** *m* = acid exchange
~**bad** *n* = acid bath
~**ballon** *m* = acid carboy
~**bau** *m* = acid-resisting construction
~**behälter** *m* = acid tank
~**behand(e)lung** *f*, Säuern *n* = acid treatment
~**behand(e)lung** *f* **unter Druck** = pressure acidizing
säurebeständig, säurefest = acid-proof, acid-resisting
Säure|bindungsfähigkeit *f*, Säurebindungsvermögen *n* = acid combining capacity
~**dampf** *m* = acid vapo(u)r
säurefest, säurebeständig = acid-resisting, acid-proof
säurefester Stein *m*, säurebeständiger ~ = acid-resisting brick, acid-proof ~ [*B.S. 3679*]
Säure|flasche *f* = acid bottle
~**gehalt** *m*, Azidität *f*, Säuregrad *m*, Neutralisationszahl *f* = acid content, acidity
~**grad** *m* → Säuregehalt *m*
säurehaltig = acid-laden
Säure|index *m* = index of acidity
~**leitung** *f* = line for acids
~**mauerwerk** *n* = acid-resisting masonry, acid-proof ~
~**messer** *m* = acidimeter
~**pumpe** *f* = acid-handling pump
~**rückgewinnungsanlage** *f* = acid recovery plant, ~ reclamation ~, ~ reclaiming ~
saures Kondensat *n* = sour condensate
Säure|schornstein *m* = acid-resisting chimney, acid-proof ~
~**schutz** *m* = protection against acids
Saussuritgabbro *m* = euphotide

SB-Laden *m*, Selbstbedienungsladen = self-service shop
Schaber *m*, Kratzer *m* = scraper
Schablone *f* → Bohrplan *m*
Schablonen-Schneidmaschine *f* = profile cutting machine
Schachbrettverband *m* = checker board bond
Schacht *m*; Kamin *m* [*Schweiz*] [*Druckluftgründung*] = access shaft of the caisson
~ = shaft
~, Gerüst *n* [*Bauaufzug*] = scaffold tower
~ → Kontrollschacht
~ [*in Galizien*]; Bohrung *f*, Sonde *f* = well
~ im Stadium der Anfangsspitzenförderung [*in Galizien*] → Bohrung *f* ~ ~ ~ ~
~ mit Hilfe von Druckgas fördernd [*in Galizien*] → Bohrung *f* ~ ~ ~ ~ ~
~ ~ ~ ~ Druckluft fördernd [*in Galizien*] → Bohrung *f* ~ ~ ~ ~ ~
~ ohne Filterrohre [*in Galizien*] → Bohrung *f* ~ ~
~ zum Pumpen vorrichten [*in Galizien*] → Bohrung *f* ~ ~ ~
(~)Abbohren *n* = shaft boring
(~)Abbohrverfahren *n* = shaft boring method
~abbohrverfahren *n* **Kind-Chaudron** = Kind-Chaudron method of shaft sinking
~abdeckung *f*, Schachtdeckel *m*, Kanalabdeckung, Kanaldeckel = manhole cover
~abseigerung *f* = shaft plumbing
(~)(ab)teufung *f*, Schacht(ab)teufen *n* = shaft sinking
~(ab)teufung *f* **im Gefrierverfahren**, Schacht(ab)teufen *n* ~ ~ = frozen shaft sinking
(~)Aufzugseil *n*, (Schacht)Förderseil = winding rope, hoist ~
~ausbau *m* = shaft lining, ~ support
~ausbau *m* **in Holz**, Holzausbau = shaft timbering
~auskleidung *f* **mit Mauerwerk** (⚒) = coffering
~bagger *m* → Schachtpumpenbagger
~bau *m* = shaft construction
~baufirma *f* = shaft-sinking company
~bohrer *m* = trepan
~brunnen *m* → Kesselbrunnen
~deckel *m* → Schachtabdeckung *f*
~deckel *m* **und -rahmen** *m* = manhole head [*The cast-iron fixture surmounting a manhole. It is made up of two parts: a frame, which rests on the masonry of the shaft, and a removable cover. Frames are either fixed or adjustable in height. Covers are "tight", "ventilated", or "anti-rattling"*]
~dichtung *f*, Schachtabdichtung = shaft sealing
~entwässerung *f* = shaft drainage
~feste *f* (⚒) = shaft pillar
~förderanlage *f* = winding installation
~fördergerüst *n* (⚒) = headframe
~fördergerüst *n* **einschließlich Ausrüstung** (⚒) = headgear [*The headframe and plant in it*]
~-Fördermaschine *f* (⚒) = mine winding engine
(~)Förderseil *n*, (Schacht)Aufzugseil = winding rope, hoist ~
~form *f* = vertical mould (Brit.); ~ mold (US)
(~)Führung *f*, Spurlatte *f*, Leitbaum *m* (⚒) = guide
~fundation *f* → Schachtgründung
~gerüst *n* → Turmgerüst
~(gerüst)-Bauaufzug *m*, Turmgerüst-Bauaufzug = scaffold tower hoist
~gründung *f*, Schachtfundation *f* [*Nach Bauweise unterschieden in 1. Brunnengründung; 2. Druckluftgründung und 3. Pfeilergründung*] = No equivalent English term. If „Schachtgründung" is used as a heading and is thus the collective term for 1., 2. and 3. above, it must be translated by "cylinders and caissons". If it is used for either 1., 2. or 3. above, the corresponding

Schachtheber(überfall) — Schalenelement 806

English terms for 1., 2. and 3. above must be used

~heber(überfall) *m*, Schacht-Saugüberfall = volute siphon
~kabelkasten *m* = manhole junction box
~kranz *m* = collar of a shaft
~lot *n*, mechanisches ~ = shaft plumb bob
~lotdraht *m* = shaft-plumbing wire
~lüftung *f* = manhole ventilation
~meister *m* = ganger
Schachtofen *m* = vertical kiln, shaft ~
~ **mit zentralem Rohr** = center shaft kiln, ~ vertical ~ (US); centre ~ ~ (Brit.)
~kalk *m* = shaft kiln lime, vertical ~ ~
~klinker *m* = shaft kiln clinker, vertical ~ ~
~material *m* = vertical kiln stone, shaft ~ ~ [*lime industry*]
Schacht|(pumpen)bagger *m*, Hopperbagger = hopper dredger (Brit.); ~ dredge (US)
~rahmen *m* = manhole frame
~ring *m* = shaft ring
~rohr *n*; Kaminrohr [*Schweiz*] [*Druckluftgründung*] = vertical air-lock tube
~schalung *f* = shaft formwork
~scheibe *f* = shaft cross section
~schleuse *f* = shaft (navigation) lock, ~ inland ~
~signalanlage *f* = shaft signalling system
~signalgebung *f* = shaft signalling
~sohle *f* = shaft bottom, ~ floor
~sohle *f* [*Einsteig(e)schacht*] = manhole floor, ~ bottom
~stein *m* → Brunnenstein
~teufe *f* = depth of a shaft
~teufung *f* → Schachtabteufung
~turm *m* (⚒) = shaft compartment
~überfall *m* = glory-hole spillway, shaft ~
~umtrieb *m* (⚒) = pass-by
~ventilator *m*, Grubenlüfter *m*, Abteuflüfter *m*, Abteufventilator *m* = mine fan

~verzug *m* = interval of a shaft, bay ~ ~ ~
~wand(ung) *f* = shaft wall
~winde *f* = shaft hoist
Schad(en)stelle *f* [*Straßendecke*] = point of failure, failed area
Schaden|umfang *m* = level of damge, degree ~
~ursache *f* = cause of damage
Schad|gas *n* = noxious gas, harmful ~
~stelle *f*, Schadenstelle [*Straßendecke*] = failed area, point of failure
~stoff *m*, angreifender Stoff = aggressive substance, ~ matter
~wasser *n*, Aggressivwasser, aggressives Wasser, angreifendes Wasser = aggressive water
Schaffernak-Schwingung *f* = Schaffernak's vibration
Schaffuß *m*, Stampffuß, Druckstempel *m* [*Walze*] = sheepsfoot
~walze *f* → ~ Walze
Schaft *m* → Rohr~
~ = shaft
~schale *f*, Pfeilerschaft *m* = hollow shaft, pier ~
Schäkel *m* = "U" bolt
Schalararbeit *f* = shuttering work
Schälblatt *n* → Hobelmesser *n*
Schalblech *n* = skin
Schalbrett *n*, Schalungsbrett = shutter(ing) board
Schale *f* = shell
~, (Hohl)Mauer~ = leaf
~ [*Fließgrenzengerät nach A. Casagrande*] = (brass) dish, cup
~, Rohr~ = half-sectoin pipe, split ~, channel
Schalemulsion *f* → Beton~
Schalen|aufgabe *f* = shell problem
~bau *m* = shell construction
~bauweise *f* = shell construction method
~bau(werk) *m*, (*n*) = shell structure
~bemessung *f* = shell design
~beton *m* = shell concrete
~dach *n* = shell roof
~dicke *f* = shell thickness
~element *n* = shell element

Schalenform — Schallschluckbauweise

~form *f*, Schalengestalt *f* = shell shape
schalenförmige Ablösungen *fpl* (⚒) = scalings
schalenförmiger Sitz *m* = bucket-type seat
Schalen|gestalt *f*, Schalenform *f* = shell shape
~gewölbe *n* = shell vault
~gleichung *f* = shell equation
~knickung *f* = shell buckling
~kraft *f* = shell force
~krümmung *f* = shell curvature
~kuppel *f* = shell dome
~kupp(e)lung *f* [*starre Kupp(e)lung, bei der zwei Schalenhälften durch 2 bis 5 Schrauben auf jeder Seite auf die Wellenenden so gepreßt werden, daß durch die Reibung allein das Drehmoment übertragen wird*] = clamp coupling, split ~
~last *f* = shell load
~mauer *f* → Hohlmauer
~mauer *f* → → Talsperre
~problem *n* **über die klassische Theorie hinaus** = non-classical shell problem
~punkt *m* = shell point
~scheitel *m* = shell apex, ~ key, ~ top, ~ crown, ~ vertex
~statik *f* = analysis of shells, statics ~ ~
~(stau)mauer → → Talsperre
~theorie *f* = theory of shells, shell theory
~träger *m* = shell support
~tragwerk *n* = shell system, ~ construction
schalenüberdacht = shell-roofed
Schalenwirkung *f* = shell support
schales Abwasser *n* = stale sewage [*Sewage containing little or no oxygen, but as yet free from putrefaction*]
Schal|fett *n*, Schalungsfett = formwork grease
~fläche *f* = contact area
~fuge *f*, Schalungsfuge = formwork joint
~gerüst *n* = falsework, lining form
~holz *n*, Schalungsholz = shutter boards
~holzreiniger *m*, Schalholzreinigungsmaschine *f* = machine for cleaning shutter boards
Schall *m* = sound
~absorption *f*, Schallschluckung *f* = sound absorption
~absorptionsgrad *m*, Schallschluckgrad = sound-reduction factor, acoustical reduction ~
~ausbreitung *f*, Schallfortpflanzung = sound propagation
~bohrwerkzeug *n*, Infra~ = sonic drill
schalldämmend = sound-absorbing
Schalldämm|platte *f* = sound baffle
~stoff *m* = sound proofing material, ~ insulation ~
Schall|dämmung *f* = sound (transmission) insulation, ~ quieting, acoustic ~
~dämpfer *m* → Auspufftopf *m*
schalldicht, schallundurchlässig = impervious to sound(s)
Schall|dichte *f*, Schallundurchlässigkeit *f* = imperviousness to sound(s)
~druck *m* = sound pressure
~energie *f*, akustische Energie = sound energy
~entstaubung *f* = sonic dust removal
~entstehung *f* = sound generation
~feld *n* = sound field
~fortpflanzung *f*, Schallausbreitung = sound propagation
~geräusch *n* = sound in the audible range
~geschwindigkeit *f* = sound velocity
~mauer *f* = sound barrier
~pegel *m* = sound level
~(pegel)meßgerät *n* = noise (level) measuring device
~-Peilgerät *n* → Echolot *n*
~reflektion *f* = sound reflection
Schallschluck|auskleidung *f*, Schallschluckbekleidung, Schallschluckverkleidung = acoustic lining
~baustoff *m*, akustischer Baustoff = acoustic material
~bauweise *f*, Akustikbauweise = acoustic construction (method)

~beiwert m = sound-absorbing coefficient
~decke f, Akustikdecke = acoustic ceiling
~deckenplatte f, Akustikdeckenplatte = acoustic ceiling baord
~eigenschaft f, Schallschluckvermögen n = sound absorption property
~farbe f = anti-noise paint
~faserplatte f = acoustic fibre building board (Brit.); ~ fiber ~ ~ (US)
~folie f, Akustikfolie = sound absorbing foil, acoustic ~
~grad a m, Schallabsorptionsgrad a [*ist das Verhältnis der nicht reflektierten („geschluckten") zur auffallenden Energie. Bei vollständiger Reflexion ist a = 0, bei vollständiger Schallschluckung ist a = 1. Die Schallschluckung wird über eine Aufzeichnung des Nachhalls eines Raumes bestimmt, in dem das zu prüfende Material an einer Raumfläche angebracht ist. Der so bestimmte Schallschluckgrad a kann mitunter auch Werte über 1 annehmen; die verwendete Formel basiert auf einer Näherungslösung; wird hier nicht behandelt)*] = acoustical reduction factor, sound-reduction ~
~kammer f = anechoic chamber
~konstruktion f = sound-absorbing system
~matte f, Akustikmatte = acoustic blanket
~mittel n [*eine Tafel oder eine Platte*] = sound absorber
~platte f, Akustikplatte = acoustic board, ~ tile
~putz m, Akustikputz = acoustic plaster
~stoff m, Schluckstoff = sound absorbing material
~tafel f, Akustiktafel = acoustic panel
Schall|schluckung f, Schallabsorption f = sound absorption
~schluckvermögen n eines Raumes [*Wird als äquivalente Schallschluckfläche (Schallabsorptionsfläche) A angegeben. Es ist dies die Fläche mit vollkommener Schluckung, die den gleichen Anteil an Schallenergie schlucken würde wie die gesamte Oberfläche des Raumes und die in ihm befindlichen Gegenstände und Personen tatsächlich schlucken*] = total absorbing power of the walls and materials in a room
~schluckwirkung f = sound absorption efficiency
~schluckwandplatte f, Akustikwandplatte = acoustic wallboard
~schutz m = protection against sound
~schwingung f = sound vibration
~siebmaschine f = sonic screening machine
~sperre f [*als Konstruktion*] = sound barrier
~übertragung f = sound transmission
schallundurchlässig, schalldicht = impervious to sound(s)
Schall|undurchlässigkeit f, Schalldichte f = imperviousness to sound(s)
~verfahren n = sonic method
~vibration f = sonic vibration
~welle f = sound wave
~wellenverlauf m in zwei Schichten [*geophysikalisches (Aufschluß)Verfahren*] = wave front advance in two layers
~zeichen n [*Seezeichen*] = audible signal
(Schäl)Messer n → Hobelmesser
Schal|mittel n → Beton~
~nagel m, Schalungsnagel = formwork nail
~öl n → Entschalungsöl
~öl-Sprühgerät n → Entschalungsöl-Sprühgerät
~paste f → Beton~
~platte f → Schal(ungs)platte
~-Reiniger m → (Beton)Schalungs-Reiniger
~(rüst)träger m → Schalungs(rüst)träger
~rüttler m → (Beton)Schalungsrüttler
~schiene f, Schalungsschiene [*Betondeckenbau*] = side rail

~steinsilo *m* → (Beton)Schalungssteinsilo
~stocherbeton *m* → → Beton
~tafel *f* = formwork panel
Schalt|anlage *f*, Schaltstation *f* = switching station
~arm *m* = shifter arm
~arretierhebel *m*, Sicherungshebel = safety lever
~bild *n*, Schaltplan *m*, Schaltschema *n*, Stromlaufplan = wiring diagram
~bremse *f* [*Vorrichtung zum Abbremsen einer Vorgelegewelle, um raschere Gangschaltung zu ermöglichen*] = clutch brake
~brett *n* → Armaturenbrett
~brücke *f* = shifter gate
~-Mudulatorventil *n* = clutch pressure modulating valve
Schalter *m* = switch
~ = switch throttle [*(compressed-)air throttle control*]
~ [*Bank; Post usw.*] = counter, teller
~halle *f* [*Bahnhof*] = booking hall
~halle *f*, Schalterraum *m* [*Bank*] = banking hall, ~ room
~öl *n* = switch oil
~sperrgruppe *f* = interlock assembly
~sperrwelle *f* = interlock shaft
Schalt|folge *f* = shift sequence
~gabel *f*, Bügel *m* = fork, shifting ~, shifter ~
~gebäude *n*, Schalthaus *n* = switch building
~gerät *n* [*Ein Gerät, das die Signalbilder schaltet und von einem Gruppensteuergerät oder einer Zentralsteueranlage abhängig ist*] = controller [*B.S. 505*]
~getriebe *n* = range transmission, shift ~
~getriebe-Eingangswelle *f* = input shaft forward reverse
~haus *n*, Schaltgebäude *n* = switch building
~hof *m*, Freiluftschaltanlage *f*, Umspannwerk *n*, Freiluftschaltstation *f* = outdoor switchgear, switchyard

~kasten *m* = switchbox
~knopf *m* [*wird gedreht*] = switching knob
~knopf *m*, Druckknopf = pushbutton
~kontakt *m*, Unterbrecherkontakt = contact point
~kreis *m*, Schaltung *f* [*Ölhydraulik*] = (valve) circuit
~leuchtanzeiger *m* = shift indicator light
~muffe *f* = shift collar
~mutter *f* = index nut, control ~
~plan *m*, Schaltbild *n*, Schaltschema *n*, Stromlaufplan = wiring diagram
~pult *n* = switching desk, ~ console
~rad *n* = rack wheel
Schaltträger *m*, Schal(ungs)rüstträger, Rüstträger
Schalt|raum *m*, Schaltwarte *f* = switch room
~regler *m* = shift governor
~schema *n*, Schaltplan *m*, Schaltbild *n*, Stromlaufplan = wiring diagram
~schrank *m* = switching cabinet
~schrankwagen *m* = mobile switching cabinet
~station *f*, Schaltanlage *f* = switching station
~steuergruppe *f* = selector control group
~stufe *f* [*beim Getriebe*] = speed, gear
~tafel *f* → Armaturenbrett *n*
Schaltung *f* = shift
~, Schaltkreis *m* [*Ölhydraulik*] = circuit, valve ~
Schalt|ventil *n* = "on-off" valve, shift ~
~warte *f*, Schaltraum *m* = switch room
~welle *f* = shifter shaft
Schalung *f*, Beton~ = formwork [*deprecated: shuttering*]
~ für Sichtbauelement = face (unit) formwork
~ mit Sperrholzauskleidung = plywood-lined formwork

Schalungsanker — Scharhubstange

Schalungs|anker m = formwork tie
~arbeiten fpl = formwork work
~arbeiter m, Schalungssetzer, Einschaler = formwork setter
~auskleidung f = formwork lining
~boden m = formwork bottom
~brett n → Schalbrett
~druck m, Frischbetondruck = concrete pressure, pressure developed by concrete on formwork
~(ein)rüttlung f → Außen(ein)rüttlung

Schal(ungs)|farbe f = formwork paint
~fett n = formwork grease
~fuge f = formwork joint
~geräte npl, Beton~ = concrete forming equipment
~holz n = formwork boards

Schalungskasten m, Säulenschalung f = column formwork, pillar ~
schalungslos = formwork-less
Schal(ungs)|nagel m = formwork nail
~öl n, Entschalungsöl = formwork oil
~paste f = formwork compound
~plan m = formwork plan
~platte f, Beton~ = formwork panel
schalungsrauh [Beton] = board-marked, natural
Schal(ungs)|-Reiniger m → Beton~
~reinigung f, Schalungssäuberung = formwork cleaning
~(rüst)träger m, Rüstträger, Schal(rüst)träger = service girder
Schal(ungs)rüttler m → Beton~
Schalungsrütt(e)lung f → Außen(ein)rüttlung
Schal(ungs)schiene f [Betondeckenbau] = side rail
Schalungssetzer m → Schalungsarbeiter
Schal(ungs)steinsilo m → Beton~
Schalungs|träger m → Schalungsrüstträger
~transportwagen m, Schalwagen = formwork transport wagon
~vibration f, Schalungsrüttlung f = formwork vibration
Schal(ungs)vibrator m → Beton~

Schalungszeichnung f = formwork (layout) drawing
Schal|vibrator m → (Beton)Schal(ungs)rüttler
~wagen m → Schalungstransportwagen

Schamotte f, feuerfester Ton m = refractory (clay), fireclay, structural clay
~, gemahlene ~ = grog
~auskleidung f = refractory (lining)
~beton m = fire-clay concrete, refractory ~
~erzeugnis n, Schamottekörper m = refractory product
~körper m, Schamotteerzeugnis n = refractory product
~mauerwerk n = refractory masonry
~mörtel m = fireclay-base refractory mortar, fire-proof cement
~stein m, Schamotteziegel m = fire clay brick, refractory ~
~ton m → Schamotte

Schappe f, Bohrschappe, Schappenbohrer m = mud auger, (shell) ~
~, offene ~, Löffelbohrer m, Probelöffel m = sampling spoon, spoonauger
Schappenbohrer m, Schappe f, Bohrschappe = auger, mud ~, shell ~

Schar f → Hobelmesser n
~ mit hydraulischer Seitenverstellung = power blade
~arretierbolzen m und Sperre f = blade locking plunger and camshaft
~arretierung f = blade lock
~einschnittende n, Scharspitze f = blade point, ~ toe
~ende n = blade heel
scharfkantig = angular, sharp(-edged)
~ zuschneiden = to cut true and square
scharfkantiges Meßwehr n = sharp-crested measuring weir
Schargestänge n = blade linkage
Scharhub m mit Planetenuntersetzung = planetary blade lift
~arm m = blade lift arm
~ritzel n = blade lift control pinion
~stange f = blade lift link

~vorrichtung f = blade lift mechanism
~welle f = blade lift shaft
Scharkipp|hydraulik f = blade tilt hydraulic system
~segment n = blade tilt strap
Schar|markierungswulst m, f = blade marker
~mittelstellung f = blade center position (US); ~ centre ~ (Brit.)
Scharnier n = hinge
~-Abfangkeil m für Bohrrohre mit Ein-Mann-Handhabung = hinged casing (rotary) slip for one-man handling
~band n = flap hinge
~rolle f = hinge knuckle
~rollenlappen m = hinge joint section
~stift m = hing pin, pin of a hinge, pintle
Scharriereisen n = boaster; drove [*Scotland*]
scharrieren = to boast
Scharrieren n = boasting; droving [*Scotland*]
Scharseiten|schub m = blade center shift (US); ~ centre ~ (Brit.)
~verstellung f = blade side shift
Schar|spitze f, Schareinschnittende n = blade toe, ~ point
~steuerantrieb m = power control
~steuerantriebgehäuse n = power control housing
~steuerung f [*Straßenhobel*] = blade control
~verlängerung f = blade extension
~versteifungsschiene f = blade shift beam
~zylinder m [*Motor-Straßenhobel*] = blade cylinder
Schattenseite f (Geol.) = ubac
Schau|bild n → Diagramm n
~deckel m = inspection door, ~ cover
Schaufel f, Schippe f = shovel
~ [*Mischer*] = blade, paddle
~, Kübel m, Ladeschaufel, Hubschaufel [*Schürflader*] = bucket, loading ~
~, Wende~, Verteiler~, Schild n [*Wendeschaufelverteiler*] = transverse spreading blade

~ [*Turbine*] = vane
~anschlag m = bucket control stop
~arm m = bucket arm
~armabsenkzeit f = bucket arm lowering time
~armventil n = bucket arm control valve
~bedienungshebel m = bucket control lever
~(beton)verteiler m → Wendeschaufelverteiler
~boden m = bucket floor
~einsteller m = bucket positioner assembly
~einstellschenkel m = bucket positioner bellcrank
~einstellsperre f = bucket positioner latch
~fassungsvermögen n, Schaufelinhalt m = bucket payload capacity
~gestänge n = bucket linkage
~höhe f = bucket-level
~hubhebel m = bucket lift control lever
~hydraulik f = bucket hydraulics
~inhalt m, Schaufelfassungsvermögen n = bucket payload capacity
~kippanschlag m = bucket dump stop
~kipphebel m = bucket tilt control lever
~kufe f = skid loader shoe
~lader m, Hochlader, Hochladeschaufel f [*wenn dieses Gerät auch schürft, dann heißt es ,,Schürflader''*] = front end loader, mobile mechanical shovel
~mischer m, Schaufelmischwerk n = paddle mixer
~(mischer)welle f = paddle (mixer) shaft
~mischvorgang m = (mixing) blade action
Schaufeln n, Schippen n = shovel(l)ing
Schaufelrad n, Becherrad, Eimerrad, Kreiseimerleiter f [*Bagger*] = bucket wheel
~ = impeller
~bagger m, Becherradbagger = bucket wheel excavator

Schaufelradgebläse — Schaumverhütungsmittel 812

~gebläse n = bucket wheel blower
~-Grabenbagger m → → Grabenbagger mit Eimern
~schlepper m = side paddle tug
Schaufel|schneidmesser n = bucket cutting edge
~seitenwand f = bucket side
~stellungsanzeiger m = bucket position indicator
~steuerhebel m = bucket control lever
~steuerorgan n = bucket control
~steuerung f = bucket control
~steuerventil n = bucket control valve
~verteiler m → Wende~
~welle f, Schaufelmischerwelle = paddle (mixer) shaft
~wellenmischer m = paddle-type mixer
~winkelanzeiger m = bucket-angle indicator
~zahn m = bucket tooth
Schaufenster n = display window
~front f, Ladenfront = store front
Schaufler m → (Front-)Ladeschaufel
Schau|glas n = glass sight gauge, jar
~kasten m = showcase
Schaukel|brücke f, Abrollbrücke, Rollklappbrücke = roller bascule bridge
~ofen m → Kippofen
Schaumbecken n = skimming tank [*A chamber so arranged that floating matter rises and remains on the surface of the sewage until removed, while the liquid flows out continuously under partitions, curtain walls, or scum boards*]
Schaumbeton m = foamed concrete
~estrich m = foamed concrete screed
~mischer m = foamed concrete mixer
~platte f = foamed concrete slab
~stein m = foamed concrete block
~wandbauplatte f = foamed concrete wall slab
~wandbaustein m = foamed concrete wall block
Schaum|bildner m, Schaummittel n, schaumbildendes Mittel [*Beton*] = foaming agent

~bildung f, Schäumen n = foaming
~block m = foam block
~dämpfer m → Schaumverhütungsmittel n
Schäumen n, Schaumbildung f = foaming
schäumende Wirkung f = foaming action
Schäumer m = frother [*A flotation agent for making froth in froth flotation*]
Schaum|feuerlöscher m → Schaumlöscher
~gummi m = foam rubber
~gummisitz m = foam rubber seat
schaumige Hochofenschlacke f → aufgeblähte ~
Schaum|kalk m, Wellenkalk(stein) m = aragonitic lime(stone)
~kunststoff m = foamed plastic
~lage f, Schaumschicht f = foam layer
~lava f → Basaltlava
~lava f → Lavaschlacke f
~löscher m, Schaumfeuerlöscher = foam type fire extinguisher
~mittel n, Schaumbildner m [*für Schaumbeton*] = foaming chemical, ~ agent
~mittelmischer m = apparatus for mixing foaming agent, ~ ~ ~ ~ chemical
~regulierungsmittel n = foam control agent
~schicht f, Schaumlage f = foam layer
~schlacke f = foamed slag
~schwimmverfahren n = froth flotation
~silikat n = foam-silicate
~silikatbeton m = foam-silicate concrete
~silikatbetonplatte f = foam-silicate concrete slab
~ton m = foamclay
Schäumungsbeginn m = onset of foaming
Schaum|verhütungsmittel n, Antischaummittel, Schaumdämpfer m = antifoam(ing agent), defoamant

~zelle *f* = foam cell
Schauöffnung *f* = inspection opening
Scheel|bleierz *n*, Scheelbleispat *m* (Min.) = stolzite, scheeletite
~erz *n* (Min.) = scheelite
Scheelit|flotation *f* = scheelite flotation
~schwüle *f* = scheelite spud
Scheibe *f*, Unterleg~ = washer
~ [*Faltwerk*] = fold
Scheiben|(bohr)meißel *m*, Disken(bohr)meißel = disc bit; disk ~ (US)
~brecher *m* = disc breaker, ~ crusher; disk ~ (US)
~bremse *f* = disc brake; disk ~ (US)
~egge *f*, Diskusegge = disc harrow; disk ~ (US)
~feder *f* = Woodroof key
~filter *n, m* = disc filter; disk ~ (US)
~kupp(e)lung *f* = flange coupling
~meißel *m*, Scheibenbohrmeißel, Disken(bohr)meißel = disc bit; disk ~ (US)
~mühle *f* = disc mill; disk ~ (US)
~nabe *f* = disc boss; disk ~ (US)
~pfahl *m* = disc pile; disk ~ (US)
~rechen *m* = disk screen (US); disc ~ [*A screen in the form of a circular disk rotating about an axis perpendicular to its centre*]
~rost *m* [*Siebtechnik*] = rotary-disc grizzly; rotary-disk ~ (US)
~scheider *m* = disc separator; disk ~ (US)
~schneider *m*, Spanbrecher *m* [*Spezial-Tunnelmaschine*] = disc cutter; disk ~ (US)
~theorie *f*, Plattentheorie = plate theory
~wäscher *m* = windshield washer, windscreen ~
~wischer *m* = windscreen wiper, windshield ~
Scheide|konus *m* = separatory cone
~moräne *f*, Einscharungsmoräne = subglacial moraine
Scheider *m*, Ab~ [*Grundstückentwässerung*] = interceptor (US); trap (Brit.)
Scheide|tal *n*, Isoklinaltal = isoclinal valley

~wand *f* → Trennwand
~wand *f* [*Bagger-Senkkasten*] = cross wall
~wandbauplatte *f* → Trennwandbauplatte
~wasser *n* = aqua fortis
scheinbare Haftfestigkeit *f* → Adhäsion *f*
~ **Haftung** *f* [*Umhüllung der Gesteinsoberfläche durch einen Bitumenfilm, ohne daß dieser fest haftet*] = coverage
~ **Kohäsion** *f* = apparent cohesion, ~ cohesiveness
~ **Porosität** *f* → Luftporenanteil *m*
Schein|flugplatz *m*, Scheinlandeplatz = dummy airfield, ~ landingground
~fuge *f* = dummy joint, concealed ~, contraction ~, shrinkage ~
~gewölbe *n* = blind vault
~landeplatz *m* → Scheinflugplatz
~leistung *f* [*Elektrotechnik*] = apparent power, ~ volt-amperes
~querfuge *f* = transverse dummy joint, ~ concealed ~, ~ contraction ~, ~ shrinkage ~
~werfer *m* = floodlight, searchlight
~werfer *m* [*Auto*] = head lamp
~werferbeleuchtung *f* = flood-lighting
~werferblendung *f* = headlight glare
Scheitel *m* → ~punkt *m*
~drucklast *f*, Scheitelbruchlast [*Rohr*] = crushing (proof) load
~druckpresse *f* [*Rohrprüfmaschine*] = load pipe tester
~druckprüfung *f*, Scheiteldruckprobe *f*, Scheiteldruckversuch *m* [*Rohr*] = crushing (proof) test
~form *f* [*Betonsteinindustrie*] = apex mould (Brit.); ~ mold (US)
~fuge *f* = crown joint
~gelenk *n* = crown hinge
~haltung *f*, Scheitelstrecke *f* = summit level reach
~höhe *f* = crown level
~kanal *m* = summit canal [*A canal crossing a summit which therefore needs to have water pumped to it*]
~pfette *f*, Firstpfette = ridge purlin(e)

~platte *f* [*Oberste Platte beim Stollenausbau mit vorgefertigten, gelenkig verbundenen Stahlbetonplatten*] = top slab
~(punkt) *m*, Bogenscheitel *m* = vertex, apex, key, top, crown
~querschnitt *m* = crown cross section
~senkung *f* = crown sag
~strecke *f*, Scheitelhaltung *f* = summit level reach
~stück *n* = apex piece
~überhöhung *f* = crown hog
~zulauf *m* [*Rohrteil*] = right-angled junction, top ~
scheitrechter (oder gerader) Bogen *m* = straight arch, jack ~, flat ~
Schelfeis *n* = shelf ice
Schellack *m* = shellac
Schelle *f* = clip
Schell|eisen *n* = lug
~hammer *m* = rivet set
Schemazeichnung *f*, Schemaskizze *f*, schematische Darstellung *f*, Prinzipskizze = schematic diagram, diagrammatic arrangement, diagrammatic illustration, diagrammatic drawing
Schenkel *m* [*Winkelstahl*] = leg
~abstand *m* [*Aufreißer*] = shank spacing
~schutz *m* [*Aufreißer*] = shank guard
Scherbe *f* → Scherben *m*
Scherben *m*, keramischer ~, Scherbe *f*, keramische Scherbe = ceramic body
~index *m* [*Kies*] = flakiness index
~kobalt *m* (Min.) = native arsenic
scherbiges Korn *n* → plattes ~
Scher|bruch *m* = shear failure
~büchse *f* → Scherkastengerät *n* nach Casagrande
~-Druck-Bruch *m* = shear compression failure
Scher(en)zapfen *m*, Gabelzapfen = forked mortise and tenon joint, ~ mortice ~ ~
Scherentreppe *f* = loft ladder, folding ~, disappearing stair(case)
Scher|festigkeit *f* = shear strength
~gerät *n* = soil shear test apparatus

~gerät *n* = shear machine
~geschwindigkeit *f* = shear rate
~kastengerät *n* nach Casagrande, Scherapparat *m* ~, Scherbüchse *f*, Rahmenschergerät = Casagrande shear test apparatus, box shear apparatus, shear box
~kraft *f* = shear force
~probe *f* → Scherversuch *m*
~prüfung *f* → Scherversuch *m*
~querschnitt *m* = shear section
~riß *m* = shear crack
~spannung *f* = shear stress
~stift *m* = shear pin
~verhalten *n* = shear behavio(u)r
~versuch *m*, Scherprobe *f*, Scherprüfung *f* = shear test
~widerstand *m* = shear resistance
~widerstandsspitze *f* = peak shear resistance
Scheuer|bürste *f* = scrubbing brush
~leiste *f* → Fußleiste
~leistenheizung *f* → Fußleistenheizung
Scheune *f* = barn
Scheunenbinder *m* = barn truss
Schicht *f* (Geol.) = bed, stratum
~, Arbeitsschicht = shift
~, Einbau~ = layer, course, lift
~, Auftrag *m* [*Putz*] = coat
~aufrichtung *f* (Geol.) = tilting of strata
~bagger *m* → Planierbagger
~beiwert *m* → Schicht(en)faktor *m*
~dicke *f* [*Bodenverfestigung*] = depth of processing
~dicke *f* = layer thickness
schichten [*Siebtechnik*] = to stratify the material bed
Schichtenbauweise *f* → Lagenbauweise
Schicht(en)|beiwert *m* → Schicht(en)faktor *m*
~faktor *m*, Schicht(en)koeffizient *m*, Schicht(en)beiwert *m* = coefficient of relative strength, equivalent factor [*road construction*]
Schichtenfolge *f* (Geol.) = sequence of strata, succession ~ ~, stratigraphic sequence of the beds, stratigraphic succession of the beds

Schicht(en)koeffizient m → Schicht(en)faktor m
Schichten|kunde f → Formations- und Schichtenkunde
~paket n = series of strata
~plan m → Höhenlinienkarte
~profil n, geologisches Profil = strata profile, geological ~
~reihe f (Geol.) = series of strata
~schnitt m = strata section
~verzeichnis n = section sheet, ~ diagram
~wasser n = artesian water
schichtenweise, lagenweise = in layers
Schichtenzusammenhang m = strata in undisturbed condition
Schicht|faktor m → Schichtenfaktor
~führer m = shift boss
~gestein n → Ablagerungsgestein
~höhe f = height of course
~koeffizient m → Schicht(en)faktor m
~konstruktion f [*Straßenbau*] = layered (paving) system (US); ~ (pavement) ~
~korrosion f = layer-corrosion
~kunststoffe mpl **als Wärmeschutz** = plastic laminate heat shields
~linie f, Höhen~, Höhenlinie = contour
~quelle f = strata spring
~silikat n, Tonmineral n = clay mineral, layer silicate
~-Silikat-Struktur f = layer silicate structure, clay mineral ~
~stoff m **mit Glasfasereinlage** = fibrous glass-reinforced plastics laminate
~stoffplatte f, Kunst~ = laminate, (all-)paper ~
(~)Streichen n (Geol.) = strike
Schichtung f **des Baugrundes**, ~ ~ **Untergrundes** = stratification of the ground, ~ ~ ~ supporting medium, ~ ~ ~ foundation (material)
Schichtwechsel m = change of shift
~ [*Boden*] = change of layer
schichtweises Betonieren n, lagenweises ~ = pouring in lifts, concreting ~ ~

Schiebe|brett n = squeegee [*A device, generally with a soft rubber edge, used for dislodging and removing deposited sewage solids from the walls and bottoms of settling tanks*]
~brücke f, Rollbrücke = pull-back drawbridge, rolling ~
~bühne f [*Ramme*] = sliding bridge
~bühne f [*Ziegelindustrie*] = transfer car
~deckel m = sliding cover
~fenster n, Schiebeflügelfenster = (sliding) sash window
~fenster n **horizontal** = horizontal (sliding) sash window
~fenster n **vertikal** = vertical (sliding) sash window
~(fenster)flügel m = sash
~gazefenster n = sliding sash screen
~kappe f (⚒) = slide (roof) bar
~lehre f → Schieblehre
~luke f = sliding hatch(way)
Schieben n [*Walze*] = shoving [*roller*]
~ [*Radschrapper mit Schubmaschine*] = push-loading
Schiebenabe f = splined hub
Schieber m, Preventer m, Absperrorgan n am Bohrloch = (blowout) preventer, cellar control
~ [*am Rechenschieber*] = cursor
~ = valve
Schiebe|rad n = sliding gear
~raupe f = tracked bulldozer for scraper push-loading
Schieber|betätigung f = valve manipulation
~bewegung f = valve movement
~haus n → Apparate(n)haus [*Schweiz*]
~kammer f; Apparate(n)kammer [*Schweiz*] = valve chamber
~schacht m [*Trinkwasserbecken*] = reservoir water tower, valve ~, draw-off ~, reservoir water shaft, valve shaft, draw-off shaft
~schließgeschwindigkeit f = rate of valve closure
~schütz(e) n, (f), Abzugschieber m = sluice valve
~verlust m = valve loss
~widerstand m = valve resistance

Schiebesitz — Schieferung

Schiebe|sitz *m* = sliding seat
~tachymeter *m* = direct reading tacheometer with mechanically-operated stadia lines
~tischpresse *f* = sliding table press
~tor *n*, Rolltor = sliding gate
~trennwand *f* = sliding partition
~treppe *f*; Aufzugtreppe [*Schweiz*] = disappearing stairway
Schiebetür *f* = sliding door
~ mit einem Flügel, Einfach-Schiebetür = single sliding door
~ ~ zwei Flügeln, Doppel-Schiebetür = double sliding door
~beschläge *mpl* = sliding door furniture
~mechanismus *m* = sliding door gear
~schloß *n* = sliding door lock
Schiebezahnrad *n* = sliding gearwheel
Schieb|karre(n) *f*, (*m*), Schubkarre(n) [*ostdeutsch und mitteldeutsch: Rad(e)ber f; in der Schweiz: Benne f*] = (wheel-)barrow, single-wheel barrow
~lehre *f*, Schublehre, Schiebelehre = caliper square
Schieds|gericht *n* = arbitration board
~gerichtsbarkeit *f* = arbitration
~gerichtsklausel *f* = arbitration clause
~richter *m* = arbiter, arbitrator
schiefe Anordnung *f* = skew
~ Biegung *f* = bending in two planes
~ Brücke *f* = skew(ed) bridge
~ Ebene *f*, schräge Bahn *f* = inclined plane
~ Ebene *f*, geneigte **~**, Schrägaufzug *m*, Rampe *f*, Schiffhebewerk *n* mit geneigter Bahn [*Schiffhebewerk, bei dem die Schiffe auf Wagen gesetzt und auf einer geneigten Ebene vom Unter- zum Oberwasser und umgekehrt befördert werden*] = inclined plane
~ Hauptzugspannung *f* = diagonal tension [*In reinforced or prestressed concrete the principle tensile stress due to horizontal tension and vertical shear*]

~ Platte *f* = skew slab, **~** plate
~ Plattenbrücke *f* = skew slab bridge, **~** plate **~**
~ Straßeneinmündung *f*, **~** Straßengab(e)lung *f* = Y junction
schiefer Turm *m* = leaning tower
Schiefer|bruch *m* = slate quarry
~dach *n* = slate roof
~decker *m* = slater(-and-tiler)
~deckung *f*, Schieferdachdeckung = slate roofing
~eindeckungsarbeiten *fpl* = slating (work)
~gips *m* = foliated gypsum
schieferige Grauwacke *f* = greywacke schist, graywacke **~**
Schiefer|kalkstein *m* = sparry limestone
~kehle *f* = slate valley
~kohle *f* = banded coal [*formerly called "slate coal"*]
~kreide *f* = graphitic clay
~letten *m* [*unreiner Schieferton*] = impure (clay-)shale
~mehl *n* = slate dust
~mergel *m* → Mergelschiefer *m*
~nagel *m* = slate nail
~niere *f* = reniform slate
~öl *n* = (crude) shale oil
~platte *f* = slate slab
~sandstein *m* = shaly sandstone
~schneider *m* = slate cutter
~spaltmaschine *f* = slate cleaning machine
~splitt *m* [*zur Bestreuung von Dachpappe*] = crushed slate
~talk *m* = indurated talc
~ton *m* [*Schieferton ist aus Ton entstanden, der durch geophysikalische Vorgänge entwässert und gepreßt worden ist. Durch stärkere Pressung (Druck und Schub) entstand der festere Tonschiefer (slate). Bei Wasserlagerung zerfließt der Ton, der Schieferton quillt stark auf, und Tonschiefer vergrößert sein Volumen nur noch wenig*] = (clay-)shale
Schieferung *f* (Geol.) = foliation

Schieferungsfläche — Schienenkran

Schieferungsfläche *f* (Geol.) = foliation plane, schistosity ~, plane of foliation, plane of schistosity
schiefes Bohrloch *n* **mit scharfem Knick** = "dog leg" well, "~~" hole
~ Gewölbe *n* = skew(ed) arch

Schief|lage *f* [*Bodensenkung*] = tilt, subside slope
~spalter *m* → Plagioklas *m*
~stellung *f* [*Schacht*] = tilt, deviation from plumb
schiefwink(e)lig = skew

Schiene *f*
Radlenker *m*, Leitschiene, Zwangsschiene
Zungen~, Weichen~
Mutter~, Stamm~
Zahn~
Breitfuß~

Doppelkopf~, Stuhl~
Schienenstoß *m*

rail
check ~, guard ~, safety ~, side ~, rail guard, safeguard
switch ~, switch blade
stock ~
rack ~
flanged ~, foot~ ~, flat-bottomed ~
bull-head(ed) ~, double headed ~
rail joint

Schiene *f*, Falzleiste *f* [*kittlose Verglasung*] = glazing bar
Schienen|-Auslegerkran *m*, Gleis-Auslegerkran [*fahrbarer Auslegerkran auf Schienen mit nicht drehbarem Ausleger*] = rail-mounted crane with fixed jib
~auto *n*, Schienen(omni)bus *m* = railbus
~bagger *m* → Gleisbagger
~band *n*, Schienenförderband = rail-mounted (belt) conveyor, ~ (~) conveyer
~befestigung *f* = rail fastening
~befestigungsmittel *n* = rail fastener
~biegemaschine *f*, Schienenbiegepresse *f* = rail bending machine
~bieger *m* = rail bender
~biege- und -richtmaschine *f* = rail bending and straightening machine
~bohrmaschine *f* = rail drilling machine
~-Bohrwagen *m*, Gleis-Bohrwagen = track-mounted wagon drill, rail-mounted ~ ~
~bus *m*, Schienenauto *n*, Schienenomnibus = railbus
~-Dampflöffelbagger *m*, Schienen-Dampfschaufel *f* = rail-mounted steam shovel

~-Drehkran *m*, Schienen-Schwenkkran, Gleis-Drehkran, Gleis-Schwenkkran = rail-mounted slewing crane
~druck *m* = rail pressure
~fahrwerk-(Band)Absetzer *m* = rail-mounted stacker
~fahrzeug *n* → Gleisfahrzeug
~(förder)band *n* = rail-mounted (belt) conveyor, ~ (~) conveyer
~förderung *f*, Schienentransport *m* = rail transport
~fugenvergußmasse *f* = rail joint filling compound
~führung *f* = rail guide
schienen|geführt → gleisgebunden
~gleiche Niveaukreuzung *f*, niveaugleicher Bahnübergang *m* = level crossing
(Schienen)Hängebahn *f*, Schwebebahn = suspension rail conveying system
Schienen|(hoch)löffelbagger *m*, Gleishochlöffelbagger = rail-mounted (power) shovel
~höhe *f* = rail-level
~kipper *m* = rail-mounted tip wagon
~kopf *m* = rail head
~kran *m*, Gleiskran = rail-mounted crane

Schienenkurve — 1. Schießen; 2. Sprengen

~kurve *f* = rail curve
~(lauf)rad *n* = rail wheel
~löffelbagger *m* → Schienenhochlöffelbagger
~nagel *m* [*Feldbahn*] = rail fixing nail, ~ spike, dog spike, track spike
~nagelhammer *m* [*Feldbahn*] = track layer's hammer, ~ ~ mallet
~oberkante *f*, SOK = rail top
~(omni)bus *m*, Schienenauto *n* = railbus
~rad *n*, Schienenlaufrad = rail wheel
~richtmaschine *f* = rail straightening machine
~säge *f* = rail saw
~-Schaufelradbagger *m* = rail-mounted bucket wheel excavator
~schleifwagen *m* = rail grinding car
~schotter *m* → Bettungsschotter
~-Schwarzbelageinbaumaschine *f*, Schienen-Schwarz(decken)verteiler *m*, Schienen-Verteilerfertiger *m*, Schienen-Schwarzdeckenfertiger, Schienen-Schwarzdeckeneinbaumaschine *f* = rail-mounted (or rail-guided) asphalt finisher (or asphalt paver, or bituminous paving machine, or bituminous spreading-and-finishing machine, or black-top spreader, or asphalt(ic) concrete paver, or bituminous road surfacing finisher, or bituminous paver-finisher, or asphalt and coated macadam finisher, or paver-finisher)
~schwelle *f*, Gleisschwelle, (Eisen-) Bahnschwelle = tie (US); sleeper, cross-sill (Brit.); cross-tie
~-Schwenkkran *m* → Schienen-Drehkran
~steg *m* = rail web
~-Stetigbagger *m*, Gleis-Stetigbagger = rail-mounted continuous excavator
~stoß *m* = rail joint
~stoß-Schleifmaschine *f* = rail joint grinding machine, ~ ~ grinder
~strang *m* → Gleis *n*
~transport *m*, Schienenförderung *f* = rail transport
~triebwagen *m* = railcar
~übergang *m* = rail crossing
~verkehr *m* = rail-mounted traffic
~walzwerk *n* = rail mill
~wandern *n* = rail creep
~weg *m* = Gleis *n*
~zange *f* = rail grip, ~ clamp
Schieß|baumwolle *f* → → 1. Schießen *n*; 2. Sprengen *n*
~berechtigter *m* = shot-firer
~boot *n*, Schießschiff *n* [*Seeseismik*] = ship (or vessel) producing the noise source by explosive charges

---

1. Schießen *n*; *2.* Sprengen *n*
(Ab)Brenngeschwindigkeit *f*
Abtun *n*, Abfeuern *n*
Alu(minium)sprengkapsel *f*

Ammonsalpetersprengstoff *m*

Anwürgen *n*
Anwürgzange *f* → Sprengkapselzange
Andzündlitze *f*
Auflegersprengung *f*

Ausbläser *m*
Besatz *m*, Verdämmung *f*
Bohrlochbesatzstoff *m*, Sprenglochbesatzstoff
Bohrpatrone *f* → Schlagpatrone

1. shoot(ing); *2.* blast(ing)
burning speed of fuse
firing (the blast)
lead azide alumin(i)um detonator tetryl-azide ~
ammonium nitrate (blasting) explosive
crimping

ignitor cord
mudcapping, bulldozing (US); plaster shooting (Brit.)
blown-out shot
stemming, tamping
stemming (or tamping) material

1. Schießen; 2. Sprengen

Bohr- und Schießdaten	drilling and blasting data
Brenngeschwindigkeit f → Abbrenngeschwindigkeit	
brisanter Sprengstoff m → detonierender ~	
Chloratit n, Chlorat-Sprengstoff m	chlorate blasting explosive
Detonation f [*Die chemische Zersetzung der Stoffe und der physikalische Vorgang*]	detonation
Detonationsgeschwindigkeit f [*Höchstmaß der Geschwindigkeit der explosiven Zersetzung*]	velocity of detonation
Detonationswelle f	wave of detonation
detonierende Zündschnur f → Knallzündschnur	
detonieren	to detonate
detonierender Sprengstoff m, (hoch-)brisanter ~	high (strength) blasting explosive
Drahtschutznetz n	mesh blasting net
Dynamit n, m	dynamite
Dynamitsprengung f	dynamiting
elektrischer Momentzünder m mit Sprengkapsel → Sprengzünder	
elektrischer Zünder m	electric detonator
elektrischer Zünder m mit einem Widerstand von 0,9–1,3 Ohm	low tension detonator [*The low-tension fuse-head has a resistance of 0.9 to 1.3 ohms*]
elektrischer Zünder m mit einem Widerstand von 1500–50000 Ohm	high tension detonator [*The high-tension fuse-head has a resistance of 1,500 to 50,000 ohms*]
elektrischer Zündschnurzünder m	electric powder fuse
Elektrosprengzünder m → Sprengzünder	
Erstsprengung f, Erstsprengen n	primary (drilling and) blasting
Explosionsgase npl → Sprenggase	
Explosivstoff m	explosive
Fehlschuß m, Fehlzündung f, Versager m	misfire(d shot), misfiring
Fehlschußloch n	blow(-out)
flüssige Luft f, Sprengluft f	liquid air
flüssiger Sauerstoff m	liquid oxygen
Fortpflanzung f	propagation
Gelatine-Sprengstoff m, Sprenggelatine f	blasting gelatin(e)
gelatinös-plastische Sprengstoffgemenge npl, gelatinöse Ammonsalpetersprengstoffe mpl	gelatinous (blasting) explosives, ammonium nitrate gelatin(e)
gewerblicher Sprengstoff m	commercial explosive
Guttaperchazündschnur f, wasserdichte Zündschnur	white counteres gutta-percha waterproof fuse

1. Schießen; 2. Sprengen

(hoch)brisanter Sprengstoff *m*, detonierender ~	high (strength) blasting explosive
Hohlladung *f*	hollow charge
Initialladung *f*	detonator
Intervallzünder *m*, Zeitzünder	delay detonator
Kammersprengverfahren *n*	coyote tunneling method (US)
Kernfels *m* → Sprengfels	
Kesselschießen *n*, Vorkesseln *n*, Kesseleinbruch *m*	sprung bore-hole method, springing, enlarging the hole with explosives, chambering, squibbing
Knallzündschnur *f*, detonierende Zündschnur, Sprengschnur	detonating fuse (or cord), Cordeau
Knäpper *m*	block of stone drilled for blasting
Knäpper(bohr)hammer *m*	block holer
Knäpperschießen *n* → Nachknäppern	
Knäppersprengung *f* → Nachknäppern	
Kollodiumwolle *f*, lösliche Schießbaumwolle	collodion cotton
Kondensatorzündmaschine *f*, Kondensationszündmaschine, Kondensatorzündapparat *m*, Kondensationszündapparat	condenser discharge (type) blasting machine, CD (~) ~ ~
Kupfersprengkapsel *f*	fulminate detonator
Laden *n*	charging
Leitungsprüfer *m*	circuit tester
losgesprengter Fels *m*	blasted rock
lösliche Schießbaumwolle *f*, Kollodiumwolle	collodion cotton
Luftdruck *m*	air blast
Millisekunden(-Verzögerungs)sprengen *n*, Millisekundenschießen	millisecond delay blasting, ~ ~ shooting
Millisekundenzünder *m*	millisecond delay electric blasting cap (or detonator)
Momentzünder *m*; Momentanzünder [*Schweiz*]	instantaneous electric detonator
Nachknäppern *n*, Knäpperschießen *n*, Sprengen *n* mit Knäpperschuß, Knäppersprengung *f*	boulder blasting,
Nitroglyzerin *n*, Sprengöl *n*	nitro-glycerin(e), nitroleum, blasting oil, fulminating oil, explosive oil, soup
Nitropulver *n*	nitro-powder
Pappenzünderhülse *f*	paper tube
Parallelschaltung *f*	connection in parallel, parallel series
Patrone *f* → Schlagpatrone	
Perchlorat-Sprengstoff *m*	perchlorate explosive
Pulversprengstoff *m* → Schießstoff	

1. Schießen; 2. Sprengen

Pulverzündschnur *f*	powder fuse
Schießbaumwolle *f*	gun cotton
Schießfels *m*	rock to be shot down
Schießkabel *n*, Zündkabel	(shot-)firing cable
Schießschwaden *mpl* → Sprenggase	
Schießstoff *m*, deflagrierender Sprengstoff, Pulversprengstoff, pulverförmiger Sprengstoff	(low strength) blasting explosive, powder ~
Schlagpatrone *f*, Bohrpatrone Sprengstoffpatrone, Patrone	prime(r) cartridge, blasting ~
schlagwettersichere Zündschnur *f*, Sicherheitszündschnur	safety fuse
Schnell-Zeit(spreng)zünder *m*	short(-period) delay (electric) (blasting) cap, split-second delay ~
Schußloch *n* → Sprengloch	
Schwaden *mpl* → Sprenggase	
Schwarzpulver *n*, Sprengpulver	black (blasting) powder, gunpowder
Schwarzpulver *n* in Pulverform, Sprengpulver ~ ~	blasting powder
Schwarzpulver *n* in Tablettenform, Sprengpulver ~ ~	blasting pellet, pellet powder
Schwarzpulverzündschnur *f*, Sprengpulverzündschnur	black powder fuse
Schwarzpulverzündschnurring *m* mit Initialladung, Sprengpulverzündschnurring ~ ~	coiled capped fuse
schwergefrierbares Dynamit *n*, ungefrierbares ~	uncongealable dynamite
seismischer Sprengstoff *m*	seismic explosive
Seitenwirkung *f*	side-shattering effect
Serienschaltung *f*	connection in series, straight series
Serienschüsse *mpl*	series firing
Sicherheitszündschnur *f* → schlagwettersichere Zündschnur	
sprengen	to dynamite, to blast
Sprengen *n* mit Knäpperschuß → Nachknäppern	
Sprengfels *m*, Kernfels, gewachsener Fels, anstehendes Gestein *n*, anstehender Fels	rock to be blasted, solid rock
Sprenggase *npl*, Schwaden *mpl*, Sprengschwaden, Schießschwaden, Explosionsgase	fumes
Sprenggelatine *f*, Gelatine-Sprengstoff *m*	blasting gelatin(e)
Sprengkammer *f*	coyote hole
Sprengkapsel *f*, Zündkapsel	blasting cap

Sprengkapselzange f, Würgezange, Anwürgzange	cap crimper
Sprengladung f	(explosive) charge, bursting ~
Sprengloch n, Schußloch, Sprengbohrloch	shot-hole, blast-hole
Sprenglochbohrgerät n, Sprenglochbohrmaschine f	blast-hole drill
Sprengluft f, flüssige Luft	liquid air
Sprengmeister m [früher: Schießmeister]	blasting technician; blaster (US)
Sprengöl n → Nitroglyzerin n	
Sprengpulver n → Schwarzpulver	
Sprengschnur f → Knallzündschnur	
(Spreng)Schuß m	shot
Sprengschwaden mpl → Sprenggase	
Sprengstoff m	blasting explosive, ~ agent
Sprengstoffmagazin n	explosives magazine
Sprengstoffpatrone f → Schlagpatrone	
Sprengtrupp m, Sprengkolonne f, Sprengmannschaft f	firing crew, ~ team, ~ gang, ~ party
Sprengung f	(explosive) blast
Sprengzünder m, elektrischer Momentzünder mit Sprengkapsel, Elektrosprengzünder	electric blasting cap, instantaneous cap
Stückung f, Stückigkeit f	fragmentation
Tetryl n [Tetranitromethylanilin ist von Anilin abgeleitet]	tetryl
Trinitrotoluol n	trinitrotoluene, TNT
Tunnelsprengstoff m	tunnel(l)ing blasting explosive
Übertagesprengen n	open face blasting
Umgang m mit Sprengstoffen	explosives handling
Untertagesprengen n	underground blasting
Unterwasser-Sprenggelatine f	submarine blasting gelatin(e)
Unterwassersprengung f	submarine (or underwater, or subaqueous) blasting
Unterwasserzünder m	submarine detonator
Verdämmung f, Besatz m	stemming, tamping
Versager m → Fehlschuß m	
vorgekesseltes Loch n	sprung borehole
vorkesseln	to spring a borehole
Vorkesseln n → Kesselschießen n	
wasserdichte Zündschnur f → Guttaperchazündschnur f	
Würgezange f → Sprengkapselzange	
Zeitzünder m, Intervallzünder	delay detonator
Zeitzündung f	delay action firing
Zünderkopf m	fuse-head
Zündkabel n, Schießkabel	(shot-)firing cable
Zündkreis m, Zündleitung f	(shot-)firing circuit

Zündmaschine *f*, Zündapparat *m* — (multi-shot) exploder, blasting machine

Zündsatz *m* — detonating charge, priming compound, primer, cap charge

Zündschnur *f* — fuse

Zündschnur *f* zum Knäpperschießen [*sie brennt 30 cm/sec mit äußerer Flamme*] — QUARRYCORD [*Trademark*]

Zündstrom *m* — firing current

Zündung *f* der Teilladungen *fpl* mit Verzögerungsintervallen — decking, deck initiation

Zweitsprengung *f* mit Bohrlochladung — block holing (US); pop shooting (Brit.)

Schießen *n*, schießender Abfluß *m* = shooting flow, rapid ~, rushing
schießender Abfluß *m*, Schießen *n* = rapid flow, shooting ~, rushing
Schieß|**fels** *m* = hard rock, bedrock
~**loch** *n* → Bohrloch
~**lochabstand** *m* → Bohrlochabstand
~**lochbesatzstoff** *m* → Bohrlochbesatzstoff
~**lochdurchmesser** *m* → Bohrlochdurchmesser
~**platz** *m* = gunnery base
~**schiff** *n* → Schießboot *n*
~**stand** *m* = shooting range
Schiff *n*, Hallen~ = bay
~ [*in einer Kirche*] = nave
Schiffahrt *f* = shipping
~**befeuerung** *f* = navigational lighting
~**kanal** *m* = navigational canal
~**öffnung** *f* = ship channel, navigation span
~**straße** *f*, schiffbarer Wasserlauf *m* = navigable water course
~**weg** *m* → Fahrrinne *f*
~**zeichen** *n* = navigation signal
Schiffanlege|**platz** *m* → Anlegestelle *f*
~**stelle** *f* → Anlegestelle
(**Schiffauf**)**Schleppe** *f* → Aufschleppe
schiffbar, befahrbar = navigable; boatable (US)
schiffbare Wassertiefe *f* = navigable depth
schiffbarer Wasserlauf *m*, Schiffahrtstraße *f* = navigable water course

Schiffbarkeit *f* = navigability
Schiff|**becherwerk** *n* → Getreideelevator *m*
~**beladeeinrichtung** *f* = shiploader
~**bewuchs** *m* [*Anwuchs von Balaniden usw.*] = marine fouling, fouling of ship bottoms
~**boden** *m*, Riemen(fuß)boden = (wood-)strip flooring, ~ floor finish
~**brücke** *f* [*auf Schwimmkörpern, z.B. Fässern, Kähnen, Pontons, gelagerte Brücke*] = floating bridge
~**einladebrücke** *f* → Einladebrücke
~**elevator** *m* → Getreideelevator
~**festmacheplatz** *m* → Anlegestelle *f*
~**haltevorrichtung** *f*, Anlegevorrichtung = mooring (device)
~**hebeanlage** *f* = barge lift
(**Schiff**)**Hebewerk** *n* = mechanical lift
~ **für Naßförderung**, Trogschleuse *f* = trough lift
~ ~ **Trockenförderung** = mechanical platform barge lift
~ **mit Gegengewichten**, lotrechter Aufzug *m* = vertical lift with suspension gear
Schiff|**körper** *m* = hull
~**motor** *m* = marine engine
~**schleuse** *f* = navigation lock
(~)**Schleusung** *f*, (Schiff)Schleusen *n*, Durchschleusen, Durchschleusung = lockage, locking
~**stoß** *m* = berthing impact
~**tunnel** = canal tunnel

Schiffverkehr — Schizolith

~verkehr *m*, Schiffahrt *f* [*auf einem Strom oder Fluß*] = river traffic

~wand *f* [*Kirche*] = nave wall

~zugschleuse *f*, Schleppzugschleuse = multiple (navigation) lock, ~ inland ~

Schiftsparren *m*, Schifter *m* = jack rafter

Schiftung *f*, Ab~, (Ab)Schiften *n*, Anschiftung, Anschiften, Lotschiften, Lotschiftung [*Zimmerei*] = shifting

Schikane *f*, Bremspfeiler *m* [*Kolkschutz*] = energy dissipator
~ = baffle

Schild *m* [*Tunnel- und Stollenbau*] = shield

~ → Schlüssel~

~, Wendeschaufel *f*, (Verteiler)Schaufel [*Wendeschaufelverteiler*] = transverse spreading blade

~armträger *m* = trunnion support

~bauweise *f* [*Tunnel- und Stollenbau*] = shield driving method

~(beton)verteiler *m* → Wendeschaufelverteiler

~bogen *m* → Blendbogen

~bogenrippe *f* = diagonal rib of wall arch

Schilderbeschriften *n* = signwriting

Schild|höhe *f* = blade height

~hubgeschwindigkeit *f* [*Straßenhobel*] = blade lift speed

~hubzylinder *m* [*Schürfschlepper*] = bulldozer lift cylinder

~mauer *f*, Stirnmauer = face wall

~pflug *m* → Schildschneepflug

~(schnee)pflug *m*, Schneepflug mit Frontschar, Einseitenräumer *m*, einseitiger Pflug, Seitenpflug, Einseitenpflug = blade-type snow plough, straight-blade ~ ~, snow plough with angling blade, (reversible) side plough [*US = plow*]

~schürfwinkel *m* = blade pitch

~schwenkwinkel *m* = blade angle

~verstärkung *f* = blade reinforcement

~verteiler *m* → Wendeschaufelverteiler

~vortrieb *m* = shield driving

~vortrieb *m* mit Druckluft, Druckluft-Schildvortrieb = shield driving with compressed air

~vortriebtunnel *m* = shield-driven tunnel

~vulkan *m* = shield volcano, lava dome

Schilf *n*, Rohr *n*, Teichrohr, Schilfrohr = (common) reed

~glaserz *n* (Min.) = freieslebenite

~kohle *f* = rush coal

(~)Rohrgewebe *n* als Putz(mörtel)träger = reed lath(ing)

~rohrplatte *f* = reed panel

Schill *m* = oyster shells

Schillerspat *m*, Bastit *m* (Min.) = schillerspar, bastite

Schindel *f* = shingle

~dach *n* = shingle roof

~(dach)(ein)deckung *f* = shingle roofing

~kehle *f* = shingle valley

~lage *f* → Schindelschicht

~reihe *f* = row of shingles

~schicht *f*, Schindellage *f* = shingle layer

~verkleidung *f*, Schindelbekleidung, Schindelauskleidung = shingle lining

~wandbekleidung *f*, Schindelwandverkleidung, Schindelwandauskleidung = shingle wall lining

~werk *n* = shingle works

Schippe *f*, Schaufel *f*, Handschaufel = hand shovel, digging ~

Schippen *n*, Schaufeln *n* = shovel(l)ing

~band *n* = hinge strap

Schipper *m*, Schüpper *m* = hand shovel(l)er, ~ shovel worker

Schirm|blech *n*, Lüfterhaube *f* = shroud

~gewölbe *n* = umbrella vault

Schirrhof *m*, Fuhrpark *m* = motor park

Schizolith *m*, Spaltungsgestein *n* = diaschistic rock, differentiated dike (or dyke) ~

Schlabber|höhe *f* = spill-over level [*The level at which water in a sanitary appliance will first spill over if the rate of inflow exceeds the rate of outflow through the outlet and any overflow*]
~**ventil** *n*, Überströmventil = overflow valve
Schlacht|halle *f* = slaughter hall
~**haus** *n*, Schlachthof *m* = slaughter house, abattoir
~**hof** *m* → Schlachthaus *n*
Schlacke *f*, Hochofen~ = (blast-furnace) slag
~, Verbrennungs~, Kessel~ [*Rückstand der Feuerungen*] = (furnace) clinker (Brit.); (~) cinder (US) [*It is sometimes mistakenly called "breeze"*]
Schlacken|aufbereitungsanlage *f* = slag processing plant
~**basalt** *m* = scoriaceous basalt
Schlackenbeton *m*, Hochofen~ = slag concrete
~ = clinker concrete (Brit.); cinder ~ (US)
~**-Hohl(block)stein** *m* = hollow clinker block (Brit.); ~ cinder ~ (US)
~**stein** *m* = clinker block (Brit.); cinder ~ (US) [*A cheap, strong building block of clinker concrete*]
~**-Vollstein** *m* = solid clinker block (Brit.); ~ cinder ~ (US)
Schlacken|bett(ung) *n*, (*f*) = slag bed
~**brechanlage** *f* = slag crushing plant
~**brecher** *m* = slag crusher, ~ breaker
~**einschluß** *m* = slag inclusion
~**faser** *f*, Hüttenfaser = slag fibre (Brit.); ~ fiber (US)
~**füllung** *f* = slag fill
~**granulierung** *f* = slag granulation
~**halde** *f* = slag pile, ~ stockpile, ~ dump, ~ tip
~**hinterfüllung** *f* = slag backfill
~**kammer** *f* = slag pocket
~**kegel** *m* (Geol.) = cinder cone
~**(kühl)grube** *f* = slag (cooling) pit
~**mehl** *n* → Thomasmehl
~**ofen** *m* = slag furnace
~**pfannenwagen** *m* = slag ladle car
~**(pflaster)stein** *m* = slag (paving) sett
~**rohsplitt** *m* = slag chip(ping)s, uncoated ~ ~
~**sand** *m* → granulierte Hochofenschlacke *f*
~**schaufel** *f* = slag bucket
~**schotter** *m* = crushed blast-furnace slag 30/70 mm
~**sinter** *m*, Sinterbims *m* = sintered slag
~**sinter-Wand(bau)platte** *f*, Sinterbims-Wand(bau)platte = sintered slag wallboard
~**splitt** *m* [*gebrochene Hochofenschlacke 7 bis 30 mm Körnung*] = slag chip(ping)s 7/30 mm
~**stand** *m* = slag line
Schlackenstein *m*, Schlackenpflasterstein = slag (paving) sett
~ [*Fehlname: Schlackenziegel m*] = slag brick [*made from crushed blast-furnace slag, sand and a hydraulic binder by pressing in mo(u)lds*]
~**presse** *f* = slag brick press
Schlacken|stelle *f* = slaggy patch
~**überlauf** *m* = skimmer
~**vorlauf** *m* [*Schweißen*] = slag flowing ahead of the molten pool
~**weg** *m* [*aus Kesselschlacke*] = clinker track (Brit.); cinder ~ (US)
~**weg** *m* [*aus Hochofenschlacke*] = slag track
~**wolle** *f* → Hüttenwolle
~**zahl** *f* → Basengrad *m*
~**zement** *m*, Hüttenzement = slag cement, artificial pozzolana ~
~**ziegel** *m* → Schlackenstein *m*
~**ziffer** *f* → Basengrad *m*
~**zinn** *n* = tin extracted from slag, prillion
schlaff bewehrt, ~ armiert = conventionally reinforced
schlaffe Bewehrung *f*, ~ Armierung, nichtgespannte ~, ~ (Stahl)Einlagen *fpl* = untensioned bar reinforcement, non-prestressed ~ ~, conventional ~ ~, mild steel

schlaffe Hauptschräge — Schlagkolben

~ **Hauptschräge** *f*, ~ Hauptdiagonale *f* = loose principal diagonal
schlaffer Bewehrungsstab *m*, ~ Armierungsstab = mild steel bar
Schlaffseil-Kabel|bagger *m*, Seilbahn-Schwebeschrapper = slackline (cableway) (excavator) with dragline-type bucket
~**schrapper** *m*, Seilbahn-Scbleppschrapper = track cable scraper, slackline (excavator) with bottomless bucket, slackline cableway with bottomless bucket
Schlafzimmer *n* = bedroom
~**fenster** *n* = bedroom window
~**tür** *f* = bedroom door
Schlag *m* [*Drahtseil*] = lay
~ [*bei Werksteinbearbeitung ein am Rand des Quaders mit Schlageisen in Breiten von 3 bis 4 cm geführter ebener Streifen*] = stroke
~ = impact load, shock [*A load applied suddenly, e. g. a blow from a falling weight or swinging pendulum*]
~**anarbeiten** *n* [*Werksteinbearbeitung*] = batting, broad tooling, angle dunting
schlagartige Zerkleinerung *f*, Schlagzerkleinerung = impact reduction
Schlagbiege|festigkeit *f* = impact bending strength
~**probe** *f*, Schlagbiegeversuch *m*, Schlagbiegeprüfung *f* = impact bending test
Schlag|bohle *f* → Bohle
~**bohlenfertiger** *m*, Stampfbohlenfertiger = tamping beam finisher, tamper ~
~**bohranlage** *f* = percussion drilling rig
~**bohren** *n*, Schlagbohrung, Meißelung *f* = percussive drilling
~**bohrgerät** *n*, Schlagbohrmaschine *f* = percussion drill
~**bohrhammer** *m* = percussion drill hammer
Schlagbohrkopf *m* = percussion drill point
~ **mit Kreuzschneide** = cross bit
~ ~ **X-Schneide** = X bit

Schlag|bohrmaschine *f*, Schlagbohrgerät *n* = percussion drill
~**(bohr)meißel** *m* [*Tiefbohrtechnik*] = cable drilling bit
~**brecher** *m* = jaw crusher (or breaker) with inclined crushing chamber
~**brett** *n*, Tatsche *f*, Praker *m*, Patsche *f*, Pritschbläuel *m* = muller and plate
~**drehbohren** *n*, Drehschlagbohren = rotary impact drilling, percussive rotary ~
~**drehbohrmaschine** *f* → Drehschlagbohrmaschine
Schlageisen *n* = hammer-headed chisel [*Any mason's chisel with a flat conical steel head, which is struck by a hammer and not by a mallet*]
~ **für Sandsteinbearbeitung** = batting tool, broad ~ [*A mason's chisel 3 to $4^{1}/_{2}$ in. wide for surfacing sandstones*]
Schläge/min. *mpl* = blows per min.
schlagen → (ab)teufen
~, (ein)rammen = to drive
Schlagen *n* → (Ab)Teufen
~, (Ein)Rammen = driving
schlagender (Niet)Gegenhalter *m* = hammer type holder-on
~ **Pfahlzieher** *m*, Rammbär-Ziehgerät *n*, Rammhammer-Pfahlzieher = double-duty double-acting hammer
schlagendes Wetter *n* → Schlagwetter
Schlag|energie *f* [*Ramme*] = striking energy
~**festigkeit** *f*, Stoßwiderstand *m*, (Schlag)Zähigkeit [*Widerstand eines Gesteins gegen Bruch unter Einwirkung von Stößen*] = resistance to impact
~**folge** *f* = succession of blows
~**gewicht** *n* [*(Hand-)Explosionsstampfer*] = tamping weight
~**gleisstopfer** *m*, Schlagschotterstopfer = percussion-type tie tamper (US); ~ track ~ (Brit.)
~**hammer** *m* = impact hammer
~**jungfer** *f* → (Ramm)Jungfer
~**kolben** *m* = percussion piston
~**kolben** *m* [*Aufbruchhammer*] = hammer piston

~**kolbenwerkzeug** *n* = percussive piston tool
~**laden** *m*, An~ [*Fenster*] = hinged (window) shutter
~**lawine** *f*, Grundlawine = ground avalanche
~**leiste** *f* [*Prallbrecher*] = impeller bar
~**leiste** *f* [*beim zweiflügeligen Fenster bzw. bei der zweiflügeligen Tür*] = rabbet ledge
~**loch** *n* = pothole; cahot, pitch-hole, (US); thank-you-madam (*US slang*)
~**lochbildung** *f* = potholing
~**lochflicken** *n* = pot-hole patching
~**lochflickgerät** *n*, Schlaglochflicker *m* = pot-hole patcher, ~ patching machine
~**meißel** *m* → Schlagbohrmeißel
~**-Nietmaschine** *f* = yoke riveter
~**nietung** *f*, Schlagnieten *n* = percussion riveting
~**patrone** *f* → → 1. Schießen *n*; 2. Sprengen *n*
~**probe** *f*, Schlagversuch *m*, Schlagprüfung *f* = impact test
~**prüfung** *f*, Schlagprobe *f*, Schlagversuch *m* = impact test
~**prüfungswert** *m* [*Straßenbaugestein*] = aggregate impact value
~**ramme** *f* = pile driver
~**regen** *m* = pelting rain, driving ~
~**regenindex** *m* = driving-rain index, pelting-rain ~
~**säule** *f* [*Schleusentor*] = mitre post; miter ~ (US)
~**schere** *f* → Rutschschere
~**schrauber** *m* = impact wrench
~**schwelle** *f* [*bei einem Holzdrempel*] = pointing sill (Brit.); miter ~ (US)
~**seil** *n* **zum Beginn der pennsylvanischen Bohrung** [*von der Kurbel aus betätigt*] = jerk line
~**sonde** *f* → Rammsonde
~**stück** *n* [*Dieselramme mit Schlagzerstäubung*] = impact piece
~**tür** *f* = swinging door
~**verdichter** *m* = impact soil-compacting device
~**verdichtung** *f* = soil compaction by impact

~**versuch** *m*, Schlagprobe *f*, Schlagprüfung *f* = impact test
~**walze** *f*, Rotor *m* [*Prallbrecher*] = impeller
~**wasser** *n* = bilge water
~**weite** *f*, Funken~, Entladeweite, Funkenstrecke *f* = sparking distance, striking ~, spark gap
~**werk** *n*, Rotor *m* [*Hammerbrecher und Hammermühle*] = rotor
~**werkzeug** *n* = percussive tool
~**werkzeug** *n* **mit Umsetzvorrichtung** = percussive pneumatic tool with rotation
~**wetter** *n*, schlagendes Wetter = explosive atmosphere, fire-damp, weatherdamp; filty (Brit.)
~**wetterentzündung** *f* = ignition of gas, ~ ~ firedamp, gas ignition, firedamp ignition
~**wetterexplosion** *f* (⚒) = colliery explosion
schlagwetter|gefährdete Grube *f* = fiery mine
~**geschützte Kapselung** *f* = flame-proof enclosure
Schlagwirkung *f* = impact action
(Schlag)Zähigkeit *f* → Schlagfestigkeit
Schlag|zahl *f* = number of blows per min.
~**zerkleinerung** *f*, schlagartige Zerkleinerung = impact reduction
~**zerstäubung** *f* [*(Pfahl)Ramme*] = kick-atomizing, impact-atomizing
~**zündung** *f* = percussion priming
Schlamm *m* [*Abwasserwesen*] = sludge [*The accumulated suspended solids of sewage deposited in tanks or basins, mixed with more or less water to form a semi-liquid mass*]
~ → → Feinschluff *m*
~ = slurry [*cement manufacture*]; pulp [*ore dressing*]
~**ablagerung** *f* = sullage [*Mud deposited by flowing water*]
~**ablaßventil** *n* = sludge valve
Schlämm|analyse *f* → Absetzversuch *m*
~**anlage** *f*, Ab~ [*Ziegelindustrie*] = settling basin

Schlämmapparat — Schlammindex 828

~apparat m → Entschlämmungsapparat
Schlämmaschine f, Schlämm-Maschine [Zementherstellung] = slurry tank
Schlamm|aufbereitung f = sludge conditioning
~ausfaulung f → Schlammfaulung
(~)Ausräumer m → (Schlamm)Kratzer m
~becken n → Schlammkule f
~behandlung f [Abwasserschlamm] = sludge conditioning
Schlämmbeton m → Prepaktbeton
Schlamm|boden m = mud soil
~brunnen m = sludge well
~büchse f → Stauchbohrer m
~büchse f mit Klappenventil = disc valve bailer; disk ~ ~ (US)
~büchse f mit Stoßventil = dart valve bailer
Schlammbüchsen|ventil n = bailer valve
~wirbel m = bailer swivel
Schlämmdichtung f [Kanal] = canal sealing by deposition of silt
Schlämme f nach Oberbach, Oberbach-Schlämme = Schlämme type slurry
~absieg(e)lung f, Schlämmeversieg(e)lung = slurry seal (coat)
~beton m [Ein nach dem Betonprinzip zusammengestelltes Mineralgemisch von hoher innerer Reibung, das durch Zusatz von Schlämme auf Kohlenwasserstoffbasis zusätzlich kohäsive Eigenschaften erhält] = slurry-type concrete
~emulsion f = slurry seal emulsion
Schlamm|eimer m, Schlammkasten m [beim Schlammfang] = grit bucket, mud ~
~eindicker m [Aufbereitungstechnik] = slurry concentrator
Schlämmemischung f = slurry mix
schlämmen [abscheiden von Sand aus Ton der dadurch fetter wird] = to decant
~, löffeln, schöpfen [Tiefbohrtechnik] = to bail
Schlämmen n, Schöpfen, Löffeln [Tiefbohrtechnik] = bailing

~ [Bodenprobe zur Bestimmung der Kornverteilungslinie] = wet mechanical analysis, elutriation
Schlamm|entwässerung f = dewatering of sludge
~erhitzer m, Zement~ = (cement) slurry heater
Schlämme|tankwagen m = self-contained slurry machine
~verteiler m = slurry spreader, grout ~
Schlamm|fang m [Straßen- oder Hofablauf mit Rostabdeckung und aushebbarem Schlammeimer] = grit trap, mud ~, ~ gull(e)y (trap)
(Schlamm)Faul|becken n = septic tank [A settling tank intended to retain the sludge in immediate contact with the sewage flowing through the tank for a sufficient period to secure a satisfactory decomposition of organic solids by anaerobic bacterial action]
~behälter m, (Schlamm)Faulraum m = hydrolyzing tank, septic ~
~behälter m, (Schlamm)Faulraum m [einer zweistöckigen mechanischen Kläranlage] = digestion tank, ~ chamber
~raum m, (Schlamm)Faulbehälter m = hydrolyzing tank, septic ~
~raum m, (Schlamm)Faulbehälter m [einer zweistöckigen mechanischen Kläranlage] = digestion chamber, ~ tank
~turm m = digestion tower
Schlamm|faulung f, (Aus)Faulung, (anaerober) Abbau m, Fäulnis f = (sewage) sludge digestion, (sludge) ~, putrefaction, anaerobic decomposition [The biochemical decomposition of organic matter resulting in the formation of mineral and simpler organic compounds]
~fladen m = flake of sludge
~generierung f = sludge activation
~heber m = sludge lifting machine
schlammiger Ton m = slushy clay
Schlamm|index m = sludge index

~inhibitor m = detergent, dispersant, antiflocculant

~kanal m = desilting channel

~kasten m, Schlammeiner m [beim Schlammfang] = grit bucket, mud ~

(~)Kratzer m, (Schlamm)Räumer m, (Schlamm)Ausräumer = sludge collector, ~ scraper

Schlämmkreide f = whiting

Schlamm|kreuz n = mud cross [rotary drilling]

~kuchen m = sludge cake [A mass resulting from sludge pressing or vacuum filtering]

~kule f, Schlammbecken n, Spülgrube f [Rotarybohren] = mud pit, slush ~, ~ pond, sump

~leitung f = discharge pipe [hydraulic fill]

~-Menge f, Abwasser~ = quantity of sludge

~presse f, Filterpresse [Abwasserwesen] = sludge press

~pressen n = sludge pressing [The process of dewatering sludge by subjecting it to pressure, the solids being retained, usually by a cloth fabric through which the water passes]

~pumpe f, Dickstoffpumpe = slush pump, mud ~, solids-handling ~

~pumpe f, Spülpumpe [Rotarybohren] = mud pump, slush ~, circulation ~

(~)Räumer m → (Schlamm)Kratzer m

~rinne f, Spülrinne, Spülungskanal m [Rotarybohren] = mud flue, ~ ditch, ~ trough, flume

~rinne f [Abwasserwesen] = sludge trough

~rückführung f [Abwasserschlamm] = sludge return

~sammelraum m [Abwasserwesen] = sludge collector

~sauger m = gull(e)y emptier

~saugtank m [Rotarybohren] = mud suction tank

Schlämmschicht f, Beton~, Zementmilch f, Feinschlämme f = laitance

Schlamm|schiff n = sludge boat

(~)Schleuder m = (sludge) centrifuge [A device in which sludge is dewatered by rapid rotation and automatically discharged]

Schlämmseil n, Schöpfseil, Löffelseil [pennsylvanisches Seilbohren] = bailing line, ~ rope

~haspel m, f, Schlämmhaspel, Löffel(seil)haspel = sand reel

~haspelbremse f, Schlämmhaspelbremse, Löffel(seil)haspelbremse [pennsylvanisches Seilbohren] = back brake, post ~

~haspelhebel m, Schlämmhaspelhebel, Löffel(seil)haspelhebel = sand reel lever

~rolle f [Tiefbohrtechnik] = bailing pulley, ~ sheave

Schlämm(seil)|rolle f, Schlämmseilscheibe f, Löffel(seil)rolle, [Tiefbohrtechnik], Schöpf(seil)scheibe = sand pump pulley, sand-line ~, sand sheave, bailing pulley, bailing sheave

~trommel f, Schöpf(seil)trommel, Löffel(seil)trommel [Tiefbohrtechnik] = sand reel drum, bailing ~, sand-line spool

Schlämmtrommel f → Schlämmseiltrommel

~trommel f [Tiefbohrtechnik] = bailing drum

Schlamm|stöpsel m = sludge plug

~strom m (Geol.) = mud-flow, mudstream, mud-spate

~sumpf m = sludge sump

~teich m [Tiefbohrtechnik] = tailings pond

~teich m, Klärteich, Absetzteich, Auflandungsteich = sludge lagoon

~trennung f [Aufbereitungstechnik] = slurry settling

(~)Trockenbeet n = (sludge) drying bed

~trockenplatz m = (sludge) drying area

~trocknung f = sludge drying [The process of drying sludge by drainage or evaporation, by exposure to the air, or by application of heat]

Schlämmung — Schleifen 830

Schlämmung f, Ab~ [Korngrößenbestimmung] = wet analysis [*The mechanical analysis of soil particles smaller than 0.06 mm (the smallest convenient sieve, BS 200 mesh)*]
Schlamm|ventil n = sludge valve
~verbrennung f = incineration of sludge
Schlämm|verfahren n zur Herstellung von Weichpreßziegeln = slurry process for making soft-mud brick
~verkleidung f des Bohrloches = mud cake
~verschiffung f = sludge disposal in the sea
~versuch m, Schlämmprobe f, Schlämmprüfung f [*Bodenmechanik*] = wet mechanical analysis, elutriation
Schlamm|verwertung f = sludge utilization
~wasser n [*Abwasserwesen*] = sludge liquor
~wasser n [*Eindickung*] = supernatant sludge
Schlange f = coil
schlängelnde Gleiskette f, ~ (Raupen)kette = snaky track
Schlangen|bohrstange f = turbine drill rod
~hohlbohrstange f = hollow turbine drill rod
~kühler m [*Erdölwerk*] = coil in box
~sumpf m = snake-infested swamp
Schlankheit f = slenderness
Schlankheitsgrad m, Schlankheitsverhältnis n [*Beziehung zwischen Stablänge I und Trägheitsmoment J*] = slenderness ratio, ratio of slenderness
~grenze f = slenderness limit
Schlauch m = flexible conduit, hose
~ → Luftreifen~
~ansatz m am Betonkübel zum Schütten in engen Schalungen = rubber elephant trunk
~anschluß m = hose connection
~brause f → Handdusche f
~bremse f = tube-type brake
~dusche f → Handdusche

~heber(überfall) m, Schlauch-Saugüberfall = hose siphon
~kupp(e)lung f = hose coupling
~leitung f = hose line, flexible tubing
schlauchloser Reifen m = tubeless tire (US); ~ tyre (Brit.)
Schlauch|öler m = hose oiler
~reifen m = tubed tire (US); ~ tyre (Brit.)
~schutz m = hose guard, ~ protector
~trommel f = hose reel
~verbindung f = hose connection
~waage f = rubber tube level
~welle f, Biegewelle, biegsame Welle = hose shaft, flexible ~
~wellenantrieb m, Biegewellenantrieb = flexible drive
Schlauder f, Balkenanker m, Kopfanker = beam tie
Schlaufe f = loop
Schlaufen|probe f, Schlaufenprüfung f, Schlaufenversuch m [*Korrosionsprüfung*] = loop-test bar
~wagen m, Schleifenwagen = tripper
Schlechte f = cleat, slip [*The plane along which the coal breaks most easily*]
Schlechtenrichtung f = cleat direction, cleavage, slip direction
Schlechtluftkanal m = foul-air flue, ~ duct
Schlechtwetter n = bad weather, adverse ~, inclement ~
Schlegel m = mallet
Schleier|auspressung f → Schleierverpressung
~injektion f → Schleierverpressung
~verpressung f, Schleierauspressung, Schleierinjektion f, Schleiereinpressung = curtain grouting
Schleifautomat m = auto(matic) grinder, ~ grinding machine
Schleife f = loop
Schleifemaillelack m = dull finish lacquer
Schleifen n [*Oberflächenbearbeitung mit Schleifmittel(n)*] = grinding
~, Holz~, Verschleifen = pulpwood grinding, grinding of pulpwood

Schleifen — schleppen

~, Holz~, (Holz)Schliff *m* [*Feinbearbeitung der gehobelten Oberfläche des Holzes. Man unterscheidet Handschliff und Maschinenschliff. Als Schleifmittel dienen Schleifpapier, Bimsstein, Bimsmehl, Sepaischale und Wachsstein*] = (wood) sand-(paper)ing
~ [*von Werkzeugen*] = grinding
~ **von Werksteinen** = rubbing (US)
schleifenförmiger Rundstahlanker *m* = anchor(ing) loop of reinforcing steel
Schleifen|oszillograph *m* = loop oscillograph
~**wagen** *m*, Schlaufenwagen = tripper
Schleiferei *f* = grinding department
Schleif|kontakt *m* = sliding contact
~**lack** *m* = sanding lacquer
~**lackfarbe** *f* = flat paint
~**leinen** *n* = abrasive cloth
~**leistung** *f* = grinding performance
~**leitung** *f*, Schleifkabel *n* = trailing cable
~**maschine** *f* [*Natur- und Betonsteinindustrie*] = grinder, grinding machine
~**maschine** *f* **mit Biegewelle** = flexible shaft grinding machine, ~ ~ grinder
~**papier** *n* = abrasive paper
~**ring** *m* = slip ring, collector ~
~**ringläufer** *m* = slip-ring rotor, collector-ring ~
~**ringläufer(motor)** *m* = slip-ring induction motor, collector-ring ~ ~
~**scheibe** *f* = abrasive disc; ~ disk (US)
~**scheibe** *f* [*Einschleifen von Betonfugen*] = abrasive blade
~**schlitten** *m* [*einer Stufenfräse*] = traversible carriage
~**spindel** *f* = grinding spindle
~**spur** *f* [*vom Betonschleifgerät auf einer Betondecke hinterlassen*] = cut(ting) mark
~**stein** *m* = grindstone
~**tisch** *m* = grinding table

~**trog** *m* [*einer Stufenfräse*] = trough-shaped base accommodating a grinding table
~**- und Poliermaschine** *f* = grinding and polishing machine, grinder and polisher
Schlenge *f*, Brunnenkranz *m* = drum curb, cutting ~, shoe [*A curb on which a drop shaft is built*]
Schlepp|anzeige(r)dämpfer *m* = drag dampening device
~**aufreißer** *m* → Aufreißer
~**ausführung** *f*, Anhängeausführung = towing model, mobile model (trailer type)
~**bagger** *m*, Anhängebagger = trailer excavator
~**barkasse** *f* = towing launch, launch tug
~**-Besenwalze** *f* → Anhänge-Bürstenwalze
~**-Betonmischer** *m*, Anhänge-Betonmischer = trailer-type concrete mixer
~**blech** *n* [*Brücke*] = cover plate
~**-Bodenentleerer** *m* → Anhänge-Bodenentleerer
~**-Bodenfräse** *f* → Anhänge(Motor)-Bodenfräse
~**boot** *n* → Schlepper
~**bremse** *f* [*Bei einem Bagger mit drehbarem Oberwagen die Bremse, die die Schlepptrommel anhält und festhält*] = drag brake
~**-Bürstenwalze** *f* → Anhänge-Bürstenwalze
~**dach** *n*, einhängiges Dach, Pultdach, Flugdach, Halbdach = shed roof
~**dampfer** *m*, Dampfschlepper = steam tug(boat), ~ towboat, ~ towing tug
~**-Doppelseitenkipper** *m* → Anhänge-Doppelseitenkipper
~**-Dreiseitenkipper** *m*, Anhänge-Dreiseitenkipper, Dreiseitenkippanhänger *m* [*Erdbaufahrzeug*] = three-way dump trailer wagon
Schleppe *f* → Abzieh~
~ → Auf~
schleppen, ziehen = to tow

Schleppen n, Schlepperei f [*Kähne*] = towage

Schlepper m, Schleppboot n [*Österreich: Remorqueur* m] = tug(boat), towboat, towing tug

~, Trecker m, Traktor m = tractor

~ **mit hydraulischer Greifervorrichtung** = hydro-clam

~ ~ **Tieflöffel(vorrichtung)** = tractor-operated trench hoe

~-**Anbaugerät** n, Schlepper-Austauschwerkzeug n, Schlepper-Zusatzwerkzeug, Schlepper-Anbauvorrichtung f, Schlepper-Arbeitsgerät, Schlepper-Austauschgerät, Schlepper-Zusatzgerät, Schlepper-Arbeitsvorrichtung, Schlepper-Zusatzvorrichtung, Schlepper-Zusatzteil m, n = tractor attachment

~-**Anbauvorrichtung** f → Schlepper-Anbaugerät n

~-**Arbeitsgerät** n → Schlepper-Anbaugerät

~-**Arbeitsvorrichtung** f → Schlepper-Anbaugerät n

(~-)**Aufbau-Bagger** m = tractor-mounted excavator

~-**Austauschgerät** n → Schlepper-Anbaugerät

~-**Austauschwerkzeug** n → Schlepper-Anbaugerät n

~-**Drehkran** m, Traktor-Drehkran, Trecker-Drehkran, Traktoren-Drehkran, Schlepperkran, Treckerkran, Traktor(en)kran = tractor (revolving) crane

~-**Eimerkettenaufzug** m, Trecker-Eimerkettenaufzug, Traktor-Eimerkettenaufzug, Schlepper-Becherwerk n, Traktor(en)-Becherwerk, Trecker-Becherwerk = tractor elevator

~**führer** m, Treckerführer, Traktorführer = tractor operator

~**gerät** n = tractor-based unit

~**geräte** npl = tractor-based equipment [*Tractor-based equipment is designed either as attachments to normal tracked or wheeled tractors, or as machines in which the attachments* and the tractor are designed as a single integrated unit]

~-**Grabenbagger** m, Trecker-Grabenbagger, Traktor(en)-Grabenbagger = tractor-mounted trench excavator

~**hydraulik** f = tractor hydraulics

~**kette** f → Raupenkette

~**motor** m = tractor engine

~(-**Schaufel**)**lader** m, Trecker(-Schaufel)lader, Traktore(n)(-Schaufel)lader = tractor-bucket machine

~**seilwinde** f, Schleppertrommelseilwinde, Traktoren(trommel)seilwinde = tractor (cable, or rope) winch

~-**Tieflöffel(bagger)** m = tractor backhoe

~-**Zusatzgerät** n → Schlepper-Anbaugerät

~-**Zusatzteil** m, n → Schlepper-Anbaugerät n

~-**Zusatzvorrichtung** f → Schlepper-Anbaugerät n

~-**Zusatzwerkzeug** n → Schlepper-Anbaugerät n

Schlepp|-**Erdförderwagen** m, Anhänge-Erdförderwagen = trailer wagon

~-**Erdhobel** m → Anhänge-Straßenhobel

~**fahrgestell** n → Anhängechassis n

~**fahrzeug** n → Anhängefahrzeug

~-**Fegemaschine** f → Anhänge-Kehrmaschine

~-**Fegewalze** f → Anhängebürstenwalze

~-**Förderwagen** m, Anhänge-Förderwagen [*Erdbau*] = wagon

~**gaupe** f = shed dormer

~**gerät** n, Anhängegerät = towed unit, trailer-type ~

~**geschwindigkeit** f, Anhängegeschwindigkeit = towing speed

~**grabenbagger** m, Anhängegrabenbagger = detachable ditcher

~-(**Grund**)**Rahmen** m → Anhängechassis n

~-**Gummiradwalze** f → Anhängegummiradwalze

~-**Gummi(reifenvielfach)walze** f → Anhängegummiradwalze

(Schlepp-)Hinterkipper — Schlepp-Streugutverteiler

(∼-)**Hinterkipper** m → Hinterkippp(er)anhänger
∼**kabel** n = trailing cable
∼-**Kehrmaschine** f → Anhänge-Kehrmaschine
∼-**Kehrwalze** f → Anhänge-Bürstenwalze
∼**keil** m (⚒) = drag wedge
∼**kopf** m **Bauart Frühling** → Saugkopf mit hydraulischer Bodenlösung
∼**kraft** f, Geschiebekraft, Schleppspannung f = tractive force
∼**kran** m, Anhängekran = trailer crane
∼**kurve** f = minimum turning radius
∼-**Lift** m = ski lift; ∼ tug (US)
∼**löffel** m → Schleppschaufel f
∼**löffel-Ausleger** m → Eimer-Ausleger
∼**löffel-Ausrüstung** f → Eimer-Ausrüstung
∼**löffelbagger** m → Schürfkübelbagger
∼**löffel-Schreitbagger** m → Schleppschaufel-Schreitbagger
∼**löffel-Vorrichtung** f → Eimer-Ausrüstung
∼-**Luftreifenwalze** f → Anhänge-Luftreifenwalze
∼**maschine** f, Anhängemaschine = trailer-type machine, towed ∼
∼**mischer** m → Anhängemischer
∼-**Mischmaschine** f → Anhängemischer m
∼-(**Motor**)**Bodenfräse** f → Anhänge-(Motor)Bodenfräse
∼-**Platte** f, Schlepplatte [*Brücke*] = sliding slab
∼-**Radschrapper** m → Anhänge-Radschrapper
∼-**Rahmen** m → Anhängechassis n
∼**reise** f, schwimmender Antransport m [*z.B. schwimmende Bohrinsel*] = tow
∼-**Rüttelwalze** f → Anhänge-Rüttelwalze
∼-**Sandstreuer** m, Anhänge-Sandstreuer = trailer (type) gritter
∼**saugkopf** m, Saugschleppkopf = drag suction head
∼**schalung** f, nachgezogene Gleitschalung = trailing forms

Schleppschaufel f, Schlepplöffel m, Eimerseilkübel m, Zugkübel, Eimer m, Schürfkübel, Schleppschaufelkübel = drag(line) bucket, ∼ scoop
∼-**Ausleger** m → Eimer-Ausleger
∼-**Ausrüstung** f → Eimer-Ausrüstung
∼**bagger** m → Schürfkübelbagger
∼-**Schreitbagger** m, Eimerseil-Schreitbagger, Schlepplöffel-Schreitbagger = walking dragline
∼-**Vorrichtung** f → Eimer-Ausrüstung
Schleppschiffahrt f [*Schiff wird an auf Flußsohle verlegtem Drahtseil oder Kette verholt, zum Beispiel auf dem Neckar bis 1930. Heute ist das Wort „Schleppschiffahrt" anwendbar auf Schlepper und Schleppkahn*] = towing traction
Schleppschrapper m, (gewöhnlicher) Schrapper, (gewöhnliche) Schrapperanlage f = (power) drag scraper (machine), ∼ scraper excavator
∼ **mit zwei fahrbaren Türmen** = tower machine, ∼ excavator
∼**anlage** f **mit 3-Seil-Anordnung**, Drei-Seil-Schrapper m = rapid shifting drag scraper machine, rapid-shifter
Schlepp|-Schürf(kübel)wagen m → Anhänge-Radschrapper m
∼-**Schwing(ungs)walze** f → Anhänge-Rüttelwalze
∼**seilwinde** f, Schlepptrommelseilwinde = towing (cable, or rope) winch
∼-**Seitenkipper** m → Anhänge-Seitenentleerer
∼**spannung** f → Schleppkraft
∼-**Straßenbesen** m → Anhänge-Kehrmaschine f
∼-**Straßenhobel** m → Anhänge-Straßenhobel
∼-**Straßenkehrmaschine** f → Anhänge-Kehrmaschine
∼-**Straßenplanierer** m → Anhänge-Straßenhobel m
∼-**Streugutverteiler** m, Anhänge-Streugutverteiler, Anhänge-Streuer m, Schleppstreuer = towed-type gritter

Schlepptau — Schleudergußrohr 834

~**tau** *n*, Abschleppseil *n* = towing rope
~**träger** *m* = bridge seating girder
~**-Untergestell** *n* → Anhängechassis *n*
~**verdichter** *m*, Anhängeverdichter = trailer compactor
~**versuch** *m*, Schlepp-Probe *f*, Schlepp-Prüfung *f* [*Straßengriffigkeit*] = trailer-type test
~**verteiler** *m* **für Schotter, Teer- und Asphaltbeton**, ~ für bituminöse Stoffe, Schotter und Splitt = towed paver (type) aggregate spreader, pull(-type) bituminous concrete and aggregate ~, finish ~
~**vibrationswalze** *f* → Anhänge-Rüttelwalze
~**-Vibrierwalze** *f* → Anhänge-Rüttelwalze
~**-Wasserwagen** *m*, Anhänge-Wasserwagen = water trailer
~**-Wegehobel** *m* → Anhänge-Straßenhobel
~**widerstand** *m* [*Widerstand gegen die Bewegung einer geschleppten bzw. gezogenen Last*] = draft
~**winde** *f* = towing winch
Schleppzug *m* = train of barges
~**schleuse** *f*, Schiffzugschleuse = multiple (navigation) lock, ~ inland ~
Schlepp-Zweiseitenkipper *m* → Anhänge-Doppelseitenkipper
Schleuder *f* → Schlamm~
~, Zentrifuge *f* = centrifuge
~**anlage** *f* = spinning plant
~**auskleidung** *f* = centrifugal lining
~**band** *n*, Wurf-Transporteur *m* = jet conveyor, ~ conveyer
Schleuderbeton *m* [*In durch besondere Maschinen in Umdrehung versetzte einwandige Schalungen wird der mit Hilfe eines langen Löffels gleichmäßig verteilt eingebrachte flüssige Beton infolge der Fliehkraft (Zentrifugalkraft) unter Aussonderung des Überschußwassers stark verdichtet. Der Schleuderbeton wird zur Herstellung von Schleuderbetonrohren und Schleuderbetonmasten verwendet*] = spun concrete
~**anlage** *f* → Beton~
~**form** *f* = centrifugally cast concrete mo(u)ld, spun ~ ~
~**(-Licht)mast** *m*, Schleuderbeton-Beleuchtungsmast, Schleuderbeton-(Straßen)Leuchtenmast = centrifugally-cast concrete (lighting) column, (centrifugally-)spun ~ (~) ~ (Brit.); centrifugally-cast concrete (lighting) mast, (centrifugally-)spun concrete (lighting) mast (US)
~**mast** *m* Betonschleudermast = centrifugally-cast concrete column, (centrifugally-)spun ~ ~ (Brit.); centrifugally-cast concrete mast, (centrifugally-)spun concrete mast (US)
~**(ramm)pfahl** *m*, Betonschleuder(ramm)pfahl = centrifugally-cast concrete driven pile, (centrifugally-)spun ~ ~ ~
~**rohr** *n*, Betonschleuderrohr = centrifugally-cast concrete pipe, (centrifugally-)spun ~ ~
~**rohrdurchlaß** *m* = spun concrete pipe culvert
~**rohr-Verlegeanlage** *f* = spun concrete pipe laying plant
~**verfahren** *n*, Schleuderverfahren = spun(-concrete) process
~**vorspannrohr** *n* → Betonschleudervorspannrohr
~**werk** *n* → Betonschleuderwerk
Schleuder|eindickung *f* = centrifugal thickening
~**entladung** *f*, Zentrifugentladung = centrifugal discharge
~**entwässerung** *f* = centrifugal dewatering
~**form** *f* = spinning mould (Brit.); ~ mold (US)
~**gebläse** *n*, Zentrifugalgebläse = centrifugal blower
~**guß** *m* = centrifugal casting
~**gußkokille** *f* = centrifugal casting mo(u)ld
~**gußrohr** *n* = centrifugal cast-iron pipe, spun ~ ~, spun-iron ~

Schleuderlüfter — Schleusenhafen

~lüfter m → Fliehkraftlüfter
~maschine f → Schleudervorrichtung f
~mastanlage f, Mastschleuderanlage = pole plant
~mischer m = centrifugal mixer
~mühle f, Fliehkraftmühle = centrifugal (force) mill
schleudern [*eine Form*] = to spin
Schleudern n [*Rohrherstellung*] = centrifugal action, spinning
~ [*Die durch nicht begrenztes (unkontrolliertes) Gleiten der Reifen auf der Fahrbahndecke hervorgerufene Bewegung des Fahrzeuges seitlich zu seiner Fahrtrichtung*] = sideway skidding
Schleuder|pfahlform f = spun pile mould (Brit.); ~ ~ mold (US)
~pumpe f, Zentrifugalpumpe, Kreiselpumpe = centrifugal pump
~ring m = slinger
~rohrausrüstung f = pipe spinning equipment
~rohrform f = spun pipe mo(u)ld
~schmierung f = centrifugal lubrication

Schleudersichter m | centrifugal air separator
Streuteller m | distributor plate, distributing ~, lower ~
Mahlgut n | material
Mehl n | finished product
Grieß m | tailings
Ventilator m, Lüfter m | fan
Strömungsgeschwindigkeit f | velocity of the air
Einlauftrichter m | feed opening
Kegelräderpaar n | bevel gearing
Ablaufschurre f | tailings spout
Gehäuse n | outer shell, ~ casing
Auslauf m | outlet

Schleuder|-Splittverteiler m, Splitt-Schleuderverteiler = spinner gritter, spinning ~, ~ gritting machine
~ventilator n, Radialventilator, Zentrifugalventilator = centrifugal fan, radial-flow ~
~verfahren n = centrifugal method
~verhalten n [*Material beim Schleuderverfahren*] = centrifugability
~versatz m (⚒) = mechanical stowing with slingers
~vorrichtung f, Schleudermaschine f [*Betonherstellung*] = spinning unit
~werk n [*für Schleuderbetonwaren*] = spinning plant
Schleuse f [*Sinkgutabzug am Trichterende eines Chancekegels über eine Schleuse*] = rubbish lock
~, Schiffahrt~ = (navigation) lock

Schleusen n, Schleusung f [*Druckluftgründung*] = locking, lockage
~, Schleusung f, Schiff~, Durchschleusen, Durchschleusung = locking, lockage
~boden m → (Schleusen)Kammersohle
~drempel m = lock sill, mitre~, clap ~ [*That part of the floor of a lock chamber against which the gates bear when shut*]
(~)Fallhöhe f, (Schleusen)Gefälle n, Schleusenfall m = fall (of a lock), lift (~ ~ ~)
~füllanlage f = lock-filling system
~füllzeit f = lock-filling time
(~)Gefälle n → (Schleusen)Fallhöhe f
~hafen m [*durch Schiffschleuse abgeschlossenes Hafenbecken*] = entrance lock dock

Schleusenhaupt — Schlittenwinde 836

~haupt *n* = lock gates
~hub *m* = lift of a lock
~kammer *f* = lock chamber
(~)Kammersohle *f*, (Schleusen)Kammerboden *m*, Schleusenboden, Schleusensohle = floor (of lock chamber), invert (~ ~ ~)
(~)Kammerwand *f* = (lock)side wall
~kanal *m*, Kanal mit Schleusen = canal with locks
~kopf *m* [*Bewässerung*] = check gate [*Is a gate placed across a water course from which it is desired to divert water*]
~mauer *f* = lock wall
~meister *m*, Schleusenwärter = lockkeeper
~oberhaupt *n* = upper lock gates
~sohle *f* → (Schleusen)Kammersohle
~tor *n* = lock gate
~treppe *f* = staircase locks, chain of locks, stairway of locks, series of locks, flight of locks
~unterhaupt *n* = lower lock gates
~wand *f* = lock side wall
~wärter *m*, Schleusenmeister *m* = lockkeeper
Schleuse *f* und Kraftwerk *n* = station and lock unit
Schleusung *f*, Schleusen *n* [*Druckluftgründung*] = lockage, locking
~, Schleusen *n*, Schiff~, Durchschleusen, Durchschleusung = locking, lockage
Schleusungs|verlust *m* = loss from lockage, ~ ~ locking
~wasser *n* = water used in lockage, ~ ~ ~ locking
~zeit *f* = locking time, lockage ~
Schlich *m* = slick
Schlick *m*, schlickiger Ton *m* = clay containing silt, mud
~bank *f*, Schlickbarre *f* = silt bar, ~ bank, mud ~
Schlicker *m*, feiner Emailbrei *m* = slip
schlickergegossen = slip-cast
Schlick|sandbank *f* = bank of muddy sand, bar ~ ~ ~
~strand *m* = mud beach

~wanderung *f* = movement of mud
~zaun *m*, Flechtwerk *n*, Flecht(werk)-zaun = wicker (work) fence
Schliere *f* = streak
Schließ|backen-Preventer *m* = ram-type blowout preventer
~bewegung *f*, Absperrbewegung [*Rohrleitung*] = shutting off movement, ~ ~ motion
~blech *n* [*Schloß*] = lock front, foreend
~dauer *f*, Schließzeit *f* = closing period, ~ time
~druck *m* [*Erdölsonde*] = build up pressure
Schließen *n*, Absperren [*Rohrleitung*] = shutting off
Schließ|kasten *m* = lock case, ~ casing
~kontakt *m* = "make" contact
~kopf *m* [*Niet*] = snap head
~kraft *f* [*Greifer*] = closing jaw pressure
~kraft *f* [*Automatisches Schließen von Türen*] = closing force
~seil *n* = closing rope, ~ cable, ~ line
~trommel *f* = grab closing drum
~ventil *n* [*Tiefbohrtechnik*] = closed-in pressure valve, CIP ~
~zeit *f*, Schließdauer *f* = closing time, ~ period
Schlingbandbremse *f* = wrap-around band brake
Schlingen|bildung = looping, loop formation
~probe *f*, Schlingenprüfung *f*, Schlingenversuch *m* [*Draht*] = sharl test
~spanner *m* = looper
Schlingern *n* [*Eisenbahnzug*] = train sway, lurching, side sway
Schlingerverband *m* = sway bracing
Schlipp *m* → Aufschleppe *f*
Schlitten *m* = slide, sled runner
~, Vorschub~, Bohrlafette *f*, Vorschublafette [*Bohrwagen*] = slide [*drill wagon*]
~ → Rammwagen *m*
~ [*für Putzprofile*] = template
~winde *f*, Schlittenhebebock *m* = sliding jack, traversing ~

Schlittweg m, Ziehweg [*Holzbringung*] = sledge-way
Schlitz m [*Probenahme im Bergbau*] = (sample) channel
~ [*Naßbagger*] = dredging well, ladder ~
~ → → Steinbruch
~bunker m = slit hopper
~drän m, Fräsrillendrän = mole drain
~dränherstellung f, Fräsrillendränherstellung = moling
~dränung f, Fräsrillendränung = mole drainage
~fassung f, Sickergalerie f = filter gallery
~hülse f, geschlitzte Hülse = split sleeve
~keilanker m (⚒) = slot-and-wedge bolt
~-Liner m = slotted liner
~masche f, Harfenmasche = harp mesh
~meißel m = gouging chisel
~naht f, Schlitzschweißnaht = slot weld
~probe f, Pickprobe, Streifenprobe, Hackprobe (⚒) = channel sample
~probenahme f, Pickprobenahme, Streifenprobenahme, Hackprobenahme (⚒) = channel sampling
~rohr n = slotted tube
~schweißung f = slot welding
~winkel m = slotted angle
Schloß n → → Spundwand
~, Stempel~ (⚒) = yoke
~, Tür~ = (door) lock
~blech n = lock-plate
~eisen n → → Spundwand f
~kasten m = lock case
~keil m (⚒) = locking-wedge [*In yokes of friction props*]
~reibung f [*Stahlspundwand*] = interlock friction
Schlot m, Esse f, Durchschlagröhre f, Neck m, Eruptionskanal m, Schußröhre, Stielgang m, Hals m = pipe (like conduit), (volcanic) neck
~ → Schornstein m
Schlotte f (Geol.) = sink

Schlucht f = (rock) gorge
Schluchten(stau)mauer f = gorge dam
Schluck|brunnen m = injection well, inverted ~, drainage ~
~fähigkeit f [*größtmöglicher Durchfluß einer Wasserturbine in* m^3/s] = maximum theoretical flow
~grad m, Absorptionsgrad [*Schallschutz im Hochbau*] = absorption coefficient
~stoff m → Schall~
~zone f in einer Bohrung = thief zone in a well
~zonensuche f in einer Bohrung = thief zone location
Schluff m, feinster Staubsand m [*nach Atterberg, Terzaghi, DIN 4022, Fischer und Udluft, Gallwitz, Grengg 0,02 bis 0,002 mm*] = silt, rock flour [*British Standards Institution 0.06 to 0.002; Massachusetts Institute of Technology 0.06–0.002 mm; United States Public Road Administration 0.05–0.005; International Society of Soil Science 0.02–0.002 mm*]
~ablagerung f = silt deposition
~anteil m → Schluffkörnung
~bereich m = silt range
~boden m = silty soil
~einpreßverfahren n = silt injection method
~gehalt m = silt content
schluffig = silty
schluffiger Feinsand m = silty fine sand
~ Ton m = silty clay
Schluff|korn n = silt grain
~linse f = lens of silt
~sand m = silty sand
Schlundbach m = subterranean brook
Schlupf m = slip(page)
~ [*Begrenztes Gleiten der Reifen auf der Fahrbahndecke in Fahrtrichtung*] = tyre slip (Brit.); tire ~ (US)
~freiheit f = freedom from slip(page)
~kausche f = reeving thimble
~kette f = disappearing chain
schlüpfrig, glatt = slippery
Schlüpfrigkeit f → Glätte f

Schlüssel *m*
  Bart *m*
  Einschnittiefe *f*
  Halm *m*
  Musterschlüssel *m*
  Nachtschlüssel *m*
  Profil *n*
  Ring *m*
  Rücken *m*
  scharfe Kante *f*
  Zahn *m*

key
  bitting
  depth of cut
  stem, shank
  sample key
  nightlatch key
  profile
  head, bow
  lower edge of (key) profile
  knife edge
  tooth of a pin

---

schlüsselfertig = turn-key
Schlüssel|komponente *f* = key component
~kurve *f* → Abfluß(mengen)kurve
~loch *n* = keyhole
~lochdeckel *m* → Schlüsselschild *n*
(~)Lochsäge *f* → Spitzsäge
~platte *f* [*Rotarybohren*] = break-out plate
~schild *n*, Schlüssellochdeckel *m*, Schild = (e)scutcheon, key plate
Schluß|abnahme *f* = final acceptance
~anstrich *m* = final coat
~licht *n* = tail light
~reparatur *f* = final repair
~ring *m*, Steinkranz *m* [*Gewölbe*] = soffit cusp, open ~
~stein *m* [*am Giebel*] = crow-stone
~stein *m* [*Bogen- und Gewölbebau*] = keystone
~steinschicht *f* [*Bogen- und Gewölbebau*] = keystone layer
~teufe *f*, Endteufe [*Tiefbohrtechnik*] total depth, T. D.
~übergang *m* [*Fertiger; Walze*] = finishing pass
schlußvergüteter Kohlenstoffstahl *m*, öl~ ~ = high-carbon steel
~ Stahldraht *m* = final-hardened and tempered steel wire
Schlußzahlung *f* = final payment
Schmalabbaufront *f* (⚒) = short wall
schmalblätt(e)rige Pflanze *f* = narrow-leaved plant
schmale Bodenplatte *f* [*Gleiskette*] = narrow shoe

~ Fundamentgrube *f*, Fundamentgraben *m* = foundation trench, footing ~, ~ ditch
~ Kuppelbogensperre *f* **mit waag- und senkrechter Krümmung** → → Talsperre
Schmalspur *f* = narrow ga(u)ge
~bahn *f* = narrow-gauge railway (Brit.); narrow-gage railroad (US)
Schmand|büchse *f* → Stauchbohrer *m*
~löffel *m* → Stauchbohrer *m*
Schmauchen *n*, Durchwärmung *f*, Vorwärmung *f* [*Ziegelbrennen*] = initial dehydration, water-smoking
Schmelz|barkeit *f* = fusibility
~bohren *n* → Feuerstrahlbohren
schmelzflüssige Hochofenschlacke *f* = molten iron blast furnace slag
Schmelz|fluß *m* (Geol.) → Magma *n*
~fluß *m* = complete fusion
~glasur *f*, opake Glasur, Emaille *f* = opaque (glazed) finish
~kegel *m* = fusion cone, pyrometric ~
~kessel *m* → Kocher *m*
~kessel *m* = melting pot
~kessel *m* **für Fugenvergußmassen**, Vergußmasseofen *m* = heating tank for melting joint sealing compounds, melting furnace (or melter) for joint sealers, kettle for heating joint filler, (joint) compound-melting furnace, compound heater
~kessel *m* **für Trinidad-(Roh)Asphalt** = refiner
~körper *m* = fusion pryometer
~magnesit *m* = sintered magnesite

Schmelzofen — Schmitz

~ofen *m* mit aushebbarem Tiegel = lift-out crucible furnace
~punkt *m* = melting point
~schweißung *f*, Schmelzschweißen *n* = fusion welding
~sicherung *f* = fusible plug
~sintern *n*, Sintern mit flüssiger Phase = liquid phase sintering
~stein (Min.) = mizzonite, dipyre, dipyrite
~- und Vergußgerät *n* für gummihaltige Vergußmassen = combined melter applicator for rubberized joint sealers
~wärme *f* = heat of fusion
~wasser *n* = melt water, melted snow and ice
~wasserbach *m* = superglacial stream
~zement *m* → Tonerdezement
Schmiede|hammerfundament *n* = forge-hammer foundation
~manipulator *m* = forging manipulator
Schmieden *n* = forging, smithing
Schmiege *f*, Stellwinkel *m* = sliding square, set ~, folding ~
schmiegsam, nichtstarr = flexible, non rigid
Schmier|dienst *m* = lubricating service
~druck *m* = lube pressure
~düse *f* = lube oil spray jet
schmieren, ab~, ein~ = to lubricate
~, ein~, ab~, (ein)fetten = to grease
Schmieren *n* → Ab~
Schmierer *m*, Ab~ = oiler, greaseboy, lube man
Schmier|ergiebigkeit *f* = lubricating efficiency
~fähigkeit *f* = lubricity, oiliness
~fähigkeitsverbesserer *m* [*Öl-Zusatz*] = oiliness agent
~fett *n* = lubricating grease; petroleum ~ (US)
~fettverdickungsmittel *n* = lubricating grease thickener
~film *m* [*Straße*] = slippery film
~gerät *n* = lubricating device
~kohle *f* = impure, earthy variety of brown coal
~mittel *n*, Schmierstoff *m* = lubricant

~mittelbehälter *m*, Schmierstoffbehälter = lubricant container
~mittelpumpe *f*, Schmierstoffpumpe = lubricant pump
~nippel *m* = lube fitting, lubricating ~
~nut *f* = oil groove
Schmieröl *n* = lubricating oil, lube (oil)
~ D = dark luboil
~ für bewegliche Maschinenteile = machine(ry) oil
~ mit Fettölzusatz, compoundiertes Schmieröl = compounded oil
~filter *m, n* = lube oil filter
~pumpe *f*, Ölschmierpumpe = oil lubricating pump, lubricating oil ~
~rückstand *m* = gum, carbon deposit, residue
~rückspülpumpe *f* = lube scavenge pump
~verdünnung *f* = dilution
~zusatz *m* = lubricant additive
Schmier|plan *m* = lubrication drawing
~pressenfett *n* = pressure gun grease
~pumpe *f* = lube pump
~seife *f* = soft soap
~stelle *f* = lube point, point of lubrication
~stoff *m*, Schmiermittel *n* = lubricant
~system *n* = lube system
~tabelle *f* = lubrication chart, ~ timetable, ~ plan, ~ form, lube ~
~überdruckventil *n* = lube relief valve
Schmierung *f* → (Ab)Schmieren *n*
~ auf Lebenszeit = lifetime lubrication, one-shot ~
~ mit inertem Gas = inert gas lubrication
Schmierungstheorie *f* = theory of lubrication
Schmier|ventil *n* = lube valve
~wagen *m* = lube truck
Schmirgel *m* = emery
~leinwand *f* [*in Österreich und Bayern*]; Schmergelleinwand [*in Preußen*] = emery cloth
Schmitz *m* (⚒) = coal band

Schmutzabscheider *m* [*Grundstückentwässerung*] = dirt trap (Brit.); ~ interceptor (US)

schmutzabweisend = dirt-repellent

Schmutz|fahne *f* [*An hellen Wand- und Deckenflächen infolge ungleichem Wärmeschutz in den betreffenden Zonen*] = ghost marking

~streifen *m* = dust streak

Schmutzwasser *n* = sanitary sewage

~abfluß *m*, Schmutzwasserablauf *m* = sanitary sewage flow, ~ ~ discharge

~(abfluß)menge *f* = sanitary sewage quantity

~kanal *m* = sanitary conduit-type sewer

~leitung *f* = sanitary sewer [*A sewer which carries sewage and excludes storm, surface and ground water*]

~netz *n* = sanitary sewer network

~pumpe *f*, Abwasserpumpe = sewage handling pump

Schnapp|ring *m*, Sprengring = snap ring, lock ~

~verschluß *m*, Schnäpper *m* = spring catch

Schnecke *f*, Preß~ [*Strangpresse*] = auger

~ [*Die Volute am ionischen Kapitell*] = scroll

~ [*Teil des Schneckengetriebes*] = worm

~ → Schneckenförderer *m*

Schnecken|antrieb *m* der Scharsteuerung = power control shaft worm gear

~aufgeber *m* → Beschickungsschnecke *f*

~austrag *m*, Schneckenaustragen *n*, Schneckenaustragung *f* = screw discharge

~beschicker *m* → Beschickungsschnecke *f*

~beschickung *f*, Schneckenaufgabe *f* = worm feed, screw ~

~(beton)verteiler *m* mit Glättelement *n* → → Beton(decken)verteiler

~dosierer *m* → Dosier(ungs)schnecke *f*

~-Erdbohrer *m*, Bohrschnecke *f* = soil auger, earth ~

~förderer *m*, (Förder)Schnecke, Transportschnecke *f* = worm conveyor, ~ conveyer, screw ~, conveyor screw, conveyer screw

~fräse *f* → Vielzweck-Schnecke

~gang *m* = worm thread

~getriebe *n* = worm gearing

~gewölbe *n* = helical barrel vault

~klassierer *m* = classifier using screw principle

~lenkung *f* = worm steer(ing)

~mischer *m* mit kontinuierlicher Wirkung = continuous (or constant-flow) screw-type mixer

(~-)Planierfräse *f* → Vielzweck-Schnecke

~presse *f* = auger-type extrusion unit, ~ machine, extrusion auger

~pumpe *f* = spindle drag pump

~rad *n* = worm gear

~speiser *n* → Beschickungsschnecke *f*

~trieb *m* → Schneckenantrieb

~trog *m* = screw trough

~untersetzung *f*, Schneckenuntersetzungsgetriebe *n* = worm reduction unit, ~ ~ gearbox, ~ and gear reducer

~verteiler *m*, Beton~ = screw spreader, ~ distributor, ~ spreading machine

~verteiler *m* mit Glättelement, Beton- ~ ~ ~ = combination screw-screed spreader, ~ ~ distributor, ~ ~ spreading machine

~wäsche *f* → Sandschnecke *f*

~welle *f* = auger shaft

~wendegetriebe *n* = worm reversing gear

~winde *f* = worm geared winch

~zubringer *m* → Beschickungsschnecke *f*

~zuteiler *m* → Dosier(ungs)schnecke *f*

Schnee|antriebsrad *n* [*Gleiskette*] = snow sprocket

~(auf)lader *m* = snow loader

~beseitigung *f* = snow clearing

~beseitigungsmaschinen *fpl* [*Dieser Begriff umfaßt "Schneeräummaschi-*

nen" und „Schneepflüge"] = snow clearing machinery, ~ ~ equipment, ~ handling ~
~**brei** *m*, (Schnee)Matsch *m* = slush
~**brett** *n* = windslab
~**dichte** *f* = snow density
~**flügel** *m* = snow wing
~**forschung** *f* = snow research
~**frässchleuder** *f*, Schneefräse *f* = rotary snow plough (Brit.); ~ ~ plow (US)
~**gestöber** *n*, Schneetreiben *n* = driving snow, blowing ~
~**glätte** *f* = compacted snow
~**glättebildung** *f* = compacted snow formation
~**grenze** *f*, Schneelinie *f* = snow limit, ~ line
~**haufen** *m*, Schneewall *m* = snow bank, ~ pile
~**hydrologie** *f* = snow hydrology
~**keilpflug** *m* → Keilpflug
~**kette** *f* → Gleitschutzkette
~**kübel** *n*, Schneeladekübel *m* = snow bucket
~(**lade**)**kübel** *n* = snow bucket
~**lader** *m*, Schneeauflader = snow loader
~**last** *f* = snow load
~**lawine** *f* = snow-avalanche, snowslide
~**lawinenschutz** *m* = snow-slide defense (US); ~ defence (Brit.)
~**linie** *f* Schneegrenze *f* = snow line, ~ limit
(~)**Matsch** *m*, Schneebrei *m* = slush
~**metamorphose** *f* = snow metamorphosis
Schneepflug *m* = snow plough (Brit.); ~ plow (US)

~ **mit Frontschar** → Schild(schnee)pflug
Schneeprobe *f* = snow sample
~**nehmer** *m* = snow sampler
(**Schnee**)**Räummaschine** *f* = (snow) remover
Schnee|**räumung** *f* = snow removal, ~ clearing
~**reifen** *m* = snow tyre (Brit.); ~ tire (US)
~**sack** *m* = snow trap [*roof*]
~**schleuder** *f*, Seiten~ = rotary snow plough, snow blowing machine, blower-type snow plough [*US = plow*]
~**schmelzbehälter** *m* = snow melting tank
~**schmelze** *f* = snow melt
~**schmelzer** *m* = snow melter
~**schmelzfahrzeug** *n* = snow melter
~**schmelzwärme** *f* = heat of fusion of snow
~**schmelzwasser** *n* = thaw water
~**schuhläufer-Aufzug** *m*, Ski(schlepp)lift *m* = ski-hoist
Schneeschutz *m* = protection against snow
~**anlage** *f* = snow protection system
~**bauteil** *m*, *n* = snow shed unit
~**kette** *f* → Gleitschutzkette
Schnee|**sturm** *m* = snow storm
~**sublimation** *f* = sublimation of snow
~**treiben** *n* → Schneegestöber *n*
(~)**Verwehung** *f* = (snow) drifting
~**wall** *m*, Schneehaufen *m* = snow pile, ~ bank
~**wasser** *n* = snow water
(~)**Wehe** *f* = (snow) drift

Schneezaun *m*	snow fence
Leit-~	leading ~ ~
Rückhalte-~	collecting ~ ~
Horizontallattenzaun	scissors' ~
Vertikallattenzaun	vertical slatted ~
Wirbelbereich *m*	eddy region
Schneezaun aus Papierbahnen	paper ~ ~

schneidbrennen — Schnelligkeitszahl 842

schneidbrennen, brennschneiden = to flame cut
Schneidbrenner *m*, Brennschneider *m* = autogenous cutting blowlamp, ~ cutter
Schneide *f*, Schneidekante *f* = cutting edge
~ → Bohr~
schneiden, gewinde~ [*Außengewinde*] = to thread
~, gewinde~ [*Innengewinde*] = to tap
Schneiden *n* **mit Pulverlanze,** Pulverlanzenschneiden = powder lancing
schneidenartiger Pol *m* **des oberen Magneten** [*Kreuzbandscheider*] = upper wedge-shaped pole
Scheiden|last *f*, Linienlast, Streckenlast = collinear load, line ~
~**meißel** *m* → Blattmeißel
~**system** *n* [*Waage*] = system of knife edges
Schneid|flüssigkeit *f* = cutting liquid
~**gut** *n* = 1. material to be cut; 2. cut material
~**kante** *f* = cutting edge
Schneidkopf *m* [*Naßbagger*] = cutter head, cutting~
~**(saug)bagger** *m* → Cutter(saug)-bagger
~**welle** *f* [*Naßbagger*] = cutter shaft
Schneid|öl *n* = cutting oil, ~ compound
(~)**Sauerstoffventil** *n* = oxygen valve
~**saugkopf** *m* = suction-cutter head
~**scheibe** *f*, Trennscheibe = blade
~**widerstand** *m* = cutting resistance
~**zahn** *m* = cutting tooth
~**zeug** *n* **für Rohrschneider** [*Tiefbohrtechnik*] = cutting head
Schnellabbindemittel *n* = quick set(ting) admix(ture)
schnellabbindend [*Verschnittbitumen*] = rapid-curing
Schnellanzapfstecker *m* **für Lampen** = prick-through type lampholder
schnellaufender Kabelkran *m* = fast cableway
Schnelläufer|-Drehtisch *m*, Schnellläufer-Rotarytisch = high speed rotary table

~**mühle** *f*, Rotor-Prall-Mühle, mechanische Prallmühle = impact mill, impeller ~ ~
Schnellaufzug *m* = express lift
Schnellbahn *f* = rapid transit system
~**bau** *m* = construction of rapid transit systems
~**netz** *n* = rapid transit network
Schnell|bauaufzug *m* = rapid building hoist
~**bemessung** *f*, Schnelldimensionierung = rapid design method
~**bestimmung** *f* = rapid determination
Schnellbinder *m*, Schnellbindemittel *n* = rapid curing liquid asphaltic material, RC
~, ~**zement** *m* = rapid-setting cement
schnellbrechend, unstabil, labil [*Bitumenemulsion*] = rapid-setting, labile, quick-breaking
Schnell|bremsbetätigung *f* = rapid brake application, quick ~ ~
~**-Brutreaktor** *m* = fast breeder (power reactor)
~**dimensionierung** *f*, Schnellbemessung = rapid design method
schnelle Zugfolge *f* **in der Spitzenzeit** = rush-hour services
schnellerhärten = to rapid harden
Schnellerhärter *m* → hochwertiger Portlandzement *m*
schnellfahrender Autokran *m*, Autokran mit Chassis auf Eigenantrieb = truck crane
Schnell|fahrwerk *n*, Schnellfahrgestell = speed-mobile carriage
~**fahrzeug** *n* = fast vehicle
~**faßzange** *f* = quick acting tongs
~**festspanner** *m* = quick-release clamp
~**filter** *m*, *n* = rapid filter
schnellfließendes Wasser *n* = high-velocity water
Schnell|gang *m* = speed gear; hi ~ (US)
~**hobel** *m* (⚒) = rapid plough (Brit.); ~ plow (US)
Schnelligkeitszahl *f* → Geschwindigkeitsbeiwert *m*

Schnell|korrosionsversuch *m*, Schnellkorossionsprüfung *f*, Schnellkorrosionsprobe *f* = accelerated corrosion test
~kugelmühle *f* = rapid ball grinder, ~ ~ mill
~kupp(e)lungsrohr *n* = quick-coupler pipe
~messung *f* = quick measurement
~methode *f* → abgekürztes Verfahren *n*
~mischer *m* = speedline mixer
~montagekran *m* = rapid erection crane
~montageverfahren *n* [*Bauen mit Fertigteilen*] = rapid-assembly method
~probe *f*, Schnellprüfung *f*, Schnellversuch *m*, Kurzprobe, Kurzprüfung, Kurzversuch, abgekürzte Probe, abgekürzte Prüfung, abgekürzter Versuch = accelerated test
~prüfung *f* → Schnellprobe *f*
~prüfung *f* **auf Haltbarkeit** = accelerated durability test
~reißschenkel *m* [*Aufreißer*] = speed shank
~(rohr)kupp(e)lung *f* = quick-acting coupling, quick coupler
~rücklauf *m* [*Umkehrgetriebe*] = fast reverser
~schlagbär *m* → Schnellschlagrammbär
~schlagbrecher *m* = rapid action jaw crusher (or breaker) with inclined crushing chamber
~schlag(ramm)bär *m*, Schnellschlag(ramm)hammer *m* = rapid-stroke (pile) hammer
~schleppen *n* = high-speed (road) towing, rapid (~) ~, fast (~) ~, ~ (~) trailing
~schlußventil *n* = rapid-closing valve, quick-closing ~, quick-close ~
~-Schneekeilpflug *m* **an LKW angebaut** = speed plough (Brit.); ~ plow (US) [*not equipped with wings*]
~senkventil *n* = quick drop valve
~siebmaschine *f* = rapid screening machine

~spannschraubenschlüssel *m* = crescent wrench
~spur *f* [*Autobahn*] = high-speed lane
~straßenring *m* = circumferential expressway
~stufe *f* = overdrive speed
~stufe *f*, Fahrstufe [*(Gleis)Kettenfahrzeug*] = high range
~stufenkupp(e)lung *f* = overdrive clutch
~stufenrad *n* = overdrive gear
~stufenschaltventil *n* = overdrive shift valve
~transport-Spritzmaschine *f* = speedline spraying outfit
~-Triebwagen *m* = rapid transit self-propel(l)ed railcar
schnelltrocknend = rapid-drying
Schnell|verfahren *n*, Kurzverfahren, abgekürztes Verfahren = accelerated method
~verkehr *m* = express traffic
~verkehr *m* [*öffentlicher Massenverkehr*] = rapid transit
Schnellverkehrs|(auto)bus *m*, Schnellverkehrsomnibus = express bus
~straße *f* [*Eine Kraftfahrzeugstraße mit mindestens zwei Fahrspuren je Richtung, vollständiger Zufahrtbeschränkung sowie in der Regel mit Knotenpunkten in mehreren Ebenen; höhengleiche Knotenpunkte sind so ausgebildet, daß ihre Leistungsfähigkeit derjenigen der freien Strecke entspricht*] = freeway
~strecke *f* [*Bahn*] = high-speed track
~untergrundbahn *f* = rapid transit subway
Schnell|verschluß *m* [*Rohrleitung*] = quick-acting coupling
~versuch *m* → Schnellprobe *f*
~-Versuch *m*, Schnell-Prüfung *f*, Schnell-Probe *f* [*zur Bestimmung der Reibungsfestigkeit bindiger Böden*] = undrained (shear) test, quick ~
~versuch *m* **mit konsolidiertem Boden** [*zur Bestimmung der Reibungsfestigkeit bindiger Böden*] = consolidated quick test

~-**Zeitsprengzünder** *m* → → 1. Schießen *n*; 2. Sprengen *n*
~-**Zubringer** *m* [*Bandauflader*] = full-floating feeder
~**zug-Dampflok(omotive)** *f* = express steam locomotive
Schnitt *m* = cut
~ = section(al elevation)
~ **in Feldmitte** = mid-span section
~**geschwindigkeit** *f* = cutting rate, ~ speed
~**holzlehrgerüst** *n*, Schnittholzrüstung *f* = sawn timber falsework, sawed ~ ~
~**ling** *m* [*halber Dachziegel*] = half tile
~**punkt** *m* = intersecting point, point of intersection
~**punkt** *m* **des geometrischen Ortes von Springniedrigwasser mit dem geometrischen Ort von Nippniedrigwasser** = point of intersection of the L.W. loci
~**unempfindlichkeit** *f* = resistance to cutting damage [*tyre*]
~**verletzung** *f* [*Gummireifen*] = cutting damage
~**winkel** *m* = intersecting angle
~**zeichnung** *f* = sectional drawing
~**ziegel** *m* → Maschinenziegel
Schnitzler *m* = shredder
Schnur|gerüst *n* → Visiergerüst
~**klemme** *f* = cord grip [*screwed on top of a lampholder*]
~**lot** *n*, Bleilot, Senkblei *n* = plummet, plumb-bob
~**schlag** *m* = lining out (by chalked line)
~**trieb** *m* **mit Kugeln** = flexible drive by ball and string
schocken = to jolt
Schocken *n*, schockende (Ein)Rütt(e)lung *f* = jolting
Schock|rüttler *m* = jolting vibrator
~**tisch** *m* = jolting table
~**verdichtung** *f* = compaction by jolting
Scholle *f* (𖣠) = block
Schollen|eis *n* = floes, floe ice
~**gebirge** *n* = mountain formed by plateau-forming movements

~**lava** *f*, Blocklava, Zackenlava, Aprolith *m*, Spratzlava = block lava, aa (~)
Schöpf|becher *m* [*Waschmaschine*] = dewatering bucket, dredging ~
~**becher** *m* = scoop
~**becherwerk** *n* → Teilbecherwerk
~**bütte** *f* [*Tiefbohrtechnik*] = bailing tub
schöpfen, schlämmen, löffeln [*Tiefbohrtechnik*] = to bail
Schöpfen *n*, Schlämmen *n*, Löffeln *n* [*Tiefbohrtechnik*] = bailing
Schöpfer *m*, Schlämmbüchse *f*, Klappsonde *f* [*Tiefbohrtechnik*] = bailer
Schöpfrad *n* = scoop wheel, water raising current ~
~, Becherrad [*Waschmaschine*] = dewatering wheel, dredging ~
Schöpf|seil *n*, Löffelseil, Schlämmseil = sand rope, ~ line, bailing ~ [*rope used for removing the sand and fluid from deep wells while being drilled with cable tools*]
~**(seil)scheibe** *f* → Schlämm(seil)rolle *f*
~**(seil)trommel** *f*, Schlämm(seil)trommel. Löffel(seil)trommel [*Tiefbohrtechnik*] = bailing drum, sand reel ~, sandline spool
~**sonde** *f* [*im Schöpfbetrieb fördernde Bohrung*] = bailing well
~**trommel** *f* → Schöpfseiltrommel
~**versuch** *m*, Schöpfprobe *f*, Schöpfprüfung *f* = bailing test
~**werk** *n* [*gemäß DIN 1184*] = pumping station for pumping large quantities of water at low heads
~**winde** *f* [*pennsylvanisches Bohrverfahren*] = sandreel
Schornstein *m*, Kamin *m* = chimney [*A construction containing one or more flues*]
~ **aus Lehm** = catted chimney, cat-and-clay ~ (US)
~**aufsatz** *m* → Kaminaufsatz
~**auskleidung** *f* = chimney lining
~**backstein** *m*, Kaminbackstein, Kaminziegel(stein) *m*, Schornsteinziegel(stein) = chimney brick

Schornsteinbauwinde — Schotterverteiler

~bauwinde *f* = chimney construction hoist
~bemessung *f* = chimney design
~formstein *m*, Kamin(form)stein = chimney block
~gas *n* = chimney gas
~haube *f* → Kaminaufsatz *m*
~höhe *f* = height of chimney
~kehren *n* = chimney-sweeping
~kopf *m* = chimney head
~(putz)tür *f*, Kamin(putz)tür = soot door of chimney, cleaning ~ ~ ~
~querschnitt *m* = cross-section of chimney
~rohr *n* = chimney tube, ~ pipe
~schaft *m* = chimney shaft [*A chimney stack that is of substantial unsupported height and usually contains a flue of large cross section*]
~steigeisen *n* = access hook for chimneys [B. S. 3678 : 1963]
~tür *f* → Schornsteinputztür
~verband *m*, Läuferverband = chimney bond, stretching ~
~wand *f* = chimney shell (wall)
~ziegel *m* = chimney brick
~zug *m* = chimney draught (Brit.); ~draft (US)
~zugregler *m* = chimney draught regulator (Brit.); ~ draft ~ (US)
Schott *n*, Schotte *f* = bulkhead
~blech *n* [*Kastenträger*] = transverse diaphragm
Schotter *m* (Geol.) → → Grobkies *m*
~, Gleis~, Schienen~ = ballast
~, Steinschlag *m*, Straßen(bau)schotter = broken rock, ~ stone, crushed ~
~aufreißer *m* [*Gleisbau*] = ballast scarifier
~ausgleichschicht *f*, Schotterausgleichlage *f*, Steinschlagausgleichschicht, Steinschlagausgleichlage = crushed-stone level(l)ing course, crushed-rock ~ ~, broken-stone ~ ~, broken-rock ~ ~
~belag *m* → Steinschlagdecke
~beton *m* → → Beton

~bett *n*, Schotterbettung *f* [*für Gleisanlagen*] = ballast bed [*for track systems*]
~bett(ung) *n*, (*f*), Steinschlagbett(ung) [*Straßenbau*] = crushed-stone bed, crushed-rock ~, brokenstone ~, broken-rock ~
~brecher *m* = stone breaker, ~ crusher
~decke *f* → Steinschlagdecke
~entlader *m* [*Gleisbaumaschine*] = ballast unloader
~falle *f*, Schotterfang *m*, Geschiebefalle, Geschiebefang = bed load trap
~fang *m* → Schotterfalle *f*
~gabel *f*, Stein(schlag)gabel = stone (picker) fork
~grube *f* = ballast pit
~kegel *m* (Geol.) = alluvial cone
~lage *f*, Schotterschicht *f*, Steinschlaglage, Steinschlagschicht = crushed-stone course, crushed-rock ~, broken-stone ~, broken-rock ~
Schottern *n* [*Schweiz*] → Auf~
Schotter|-Räummaschine *f* [*Gleisbau*] = cribling machine
~reiniger *m* [*Gleisbau*] = ballast cleaning machine
~reinigungs-Arbeiten *fpl* = ballast cleaning
~sand *m* → Grobsand
~schicht *f* → Schotterlage
~selbstentladewagen *m* [*Gleisbaumaschine*] = ballast self-unloader
~stampfmaschine *f* → Gleisstampfmaschine
~stopfer *m*, Gleisstopfer = ballast tamper
~straße *f*, Steinschlagstraße = metalled road (Brit.); crushed-rock ~, crushed-stone ~, broken-rock ~, broken-stone ~
~straßendecke *f* → Steinschlagdecke
~tragschicht *f*, Steinschlagtragschicht = (crushed) stone base, broken ~ ~, rock ~, macadam ~
~verteiler *m*, Makadam-Ausleger, Steinschlagverteiler = rock spreader, stone ~, aggregate ~, base paver

Schotterwerk — Schrägpfahl

~werk n = commercial stone-crushing plant

~werkanlagen fpl = equipment for commercial stone-crushing plants

schraffieren = to hatch

schraffiert = hatched

Schraffur f = hatching

(schräg)aussteifen → (ab)bölzen

schräg gerammt, geneigt ~ = driven on the rake

schräge Stahlbeton-Rahmenbrücke f = reinforced-concrete skewed rigid frame bridge

schräger Versatz m [Holzverbindung] = skew notch, ~ gain

schräges Blatt n → Schrägblatt

~ **Hakenblatt** n, französisches Blatt [Holzverbindung] = oblique scarf (joint), French ~ (~)

Schräg|armierung f → Schrägbewehrung

~aufnahme f, Schrägbild n = oblique photograph [aerial surveying]

~aufstellung f [Aufstellung der Fahrzeuge in einem spitzen Winkel zur Fahrtrichtung] = saw-tooth parking arrangement

~aufzug m → schiefe Ebene f [Schiffhebewerk]

~aufzug m → Schrägbauaufzug

~aufzug m [Drahtseil-Standbahn zur Überwindung starker Steigungen] = direct-rope haulage (Brit.); engine plane (US)

(~)**Aussteifung** f → (Ab)Bölzung

~bahn f, geneigte Laufbahn [Walzenwehr] = sloping rack

~bandförderer-Verlängerung f, Flieger m = extension flight conveyor, ~ ~ conveyer, swivel (~) ~, thrower belt unit

~bandförderung f → → Bandförderer m

~(bau)aufzug n = inclined (contractors') hoist

~becherwerk n = inclined bucket elevator

~bewehrung f, Schrägarmierung, Schräg(stahl)einlagen fpl, Schrägeisen npl = sloping reinforcement

~bild n, Schrägaufnahme f = oblique photograph [aerial surveying]

~blatt n, schräges Blatt [Holzverbindung] = oblique halved joint with butt ends

~bodenbunker m = sloping-bottom bin, slanting-bottom ~

~bohren n = drilling on the rake

~bügel m = inclined stirrup

Schräge f, Diagonale f, Diagonalstab m [Fachwerkbinder] = diagonal (rod), ~ web member

Schräg|einlagen fpl → Schrägbewehrung f

~eisen npl → Schrägbewehrung f

~elevator m = inclined bucket conveyor, ~ ~ conveyer

~fassung f [Wasserbau] = inclined intake

~fassung f [Grundwassergewinnung] = inclined collector system

~festigkeit f = oblique strength

~flügelwiderlager n = wing abutment

~kettenbahn f, Anhebekettenbahn = creeper [An endless chain with upward projecting arms which catch the axles of mine tubs and drag them up an incline]

~lauf m [Fahrzeug beim Bremsen] = angular deviation

~laufrichtung f [Fahrzeug beim Bremsen] = direction of (angular) deviation

~laufwinkel m [Fahrzeug beim Bremsen] = angular deviation - degrees

~mauer f = battered wall

~parken n = angle parking, diagonal ~

~parkstreifen m [Ein Parkstreifen für Schrägaufstellung] = saw-tooth parking reservation

~pegel m [schräger Lattenpegel mit neigungsgerecht verzerrter Teilung]; Treppenpegel [schräger Lattenpegel mit verzerrter getreppter Teilung] = inclined gage [A staff gage on a slope graduated to read vertical heights above the datum]

~pfahl m = batter pile, raking ~, raker

~pfahljoch n = batter pile bent, raking ~ ~, raker ~
schrägrammen = to drive with a rake
Schräg|(ramm)gerüst n = raking (pile) frame, ~ piling ~
~rammung f = driving on the rake, ~ raking piles
~rissebildung f, Auftreten n von Schrägrissen = oblique cracking
~riß m = sloping crack
~rollenlager n = tapered roller bearing
~rostkühler m = inclined grate cooler [*An enclosed, inclined system of moving grates. Clinker dropped on the high end moves progressively to the lower, discharge end while cool air is forced through the grates and load from below*]
~schacht m (✗) = slope
~schulterfelge f = tapered rim
~schwerkraftelevator m, Schrägschwerkraftbecherwerk n = inclined continuous bucket elevator
Schrägseil n = diagonal cable, ~ rope, inclined ~
~brücke f, Zügelgurtbrücke = stayed girder bridge, tied cantilever ~
Schräg|sieb n = inclined screen
~spundwand f = batter sheet pile wall, ~ ~ piling
~stab m = inclined bar
~stab m, Diagonalstab = diagonal member
~(stahl)einlagen fpl → Schrägbewehrung f
~stapelung f = diagonal stacking
schrägstehendes Kühlrohr n [*Motor*] = canted radiator tube
Schräg|steife f, Graben~, Strebe f = diagonal brace (US); ~ strut (Brit.)
~steife f, Spreize f (✗) = sprag [*Generally a short sloping timber placed with one end on the floor and the other end tight against a rock which may fall*]
~stempel m = raking prop
~stollen m = inclined gallery
~stoß m (✗) = under-tipped face

~-Stützplatte f [*beim Stollenausbau mit vorgefertigten, gelenkig verbundenen Stahlbetonplatten*] = inclined side slab
Schrägungswinkel m = angle of skew
Schrägverband m, Diagonalverband [*Träger*] = diagonal bracing
~ [*Wellenbrecherbau*] = sliced work, sloping bond
Schräg|verglasung f = sloping glazing
~verzahnung f, Schraubenverzahnung = helical gearing, ~ gearwheels
~walzen n [*von Rohren*] = cross rolling, skew ~, slant ~
~walzwerk n = skew-rolling mill, cross-rolling ~, slant-rolling ~
~wand f = battered wall, battering ~
~zahnrad n, Schraubenrad, Zahnrad mit Schrägverzahnung = helical gear
~zug m [*die waag(e)rechte Seitenkraft des Seilzuges eines Hebezeuges, z. B. Laufkran, Verladebrücke usw.*] = oblique pull
~zugangstollen m = sloping adit
Schrämbagger m, Kratzbagger, Fräserbagger [*zum spanartigen Abtrag im Hochschnitt von steilen Lehmwänden, Tonwänden, Braunkohlenflözen (bis 40 m) und dgl.*] = cutter chain excavator
Schrammbord n, (Hoch)Bordstein m, Bordschwelle f = curb; kerb (Brit.)
Schramme f (Geol.) = scar
Schranke f = gate, lifting barrier
Schränken n, Sägen~ = (saw) setting
Schrappen n [*einer Furche mit einem Tiefen- und Langstreckenförderer*] = channeling (US); chanelling (Brit.)
Schrapper m [*allgemeiner Ausdruck*] = scraper
~, gewöhnlicher ~, Schleppschrapper = power scraper excavator, (~) drag scraper (machine)
~ mit (Gleis)Kettentraktor → Traktor-Schrapper
~ ~ gummibereiftem Einachstraktor → Autoschrapper mit 2-Rad-Schlepper

Schrapper mit Raupe — Schraubenschaft

~ ~ **Raupe** → Traktor-Schrapper
~**anlage** *f* = (drag) scraper installation
Schrapp(er)gefäß *n* → Schrapp(er)kübel *m*
Schrapperhaspel *m, f* → Schrapperwinde *f*
Schrapp(er)kübel *m*, Schrapp(er)gefäß *n* [*beim Tiefen- und Langstreckenförderer*] = scraper bucket
Schrapper|schaufel *f*, Räumschaufel, Handschaufel = manually guided drag skip
~**versatz** *m* (⚒) = stowing with scraper buckets
~**wagen** *m*, (Anhänge-)Schürf(kübel-)wagen = four-wheel(ed) scraper, crawler tractor-drawn scraper, rubber mounted tractor-drawn scraper, hauling scraper, pull(-type) scraper, scraper trailer
~**winde** *f*, Schrapperhaspel *f, m* = (drag) scraper hoist
~**winde** *f* **mit Handschaufel**, Handschrapper *m* = handscraper, drag loader scraper; mixer-drive shovel
Schrapp|gefäß *n* → Schrapp(er)kübel *m*
~**gefäß** *n* [*bei einem Handschrapper*] = drag loader shovel
~**haspel** *m, f* (⚒) = double-drum haulage engine
~**kasten** *m* (⚒) = box-shaped scraper bucket, scraper box
~**kübel** *m* → Schrapperkübel
Schrapplader *m*, Fahrschrapper, Verlade(-Fahr)schrapper = drag scraper and loader, scraperloader
~ (⚒) = scraper (loader) (Brit.); slusher (US)
Schrapp|-Prüfung *f*, Schrapp-Probe *f*, Schrappversuch *m* = scraping test
~**weg** *n* [*Tiefen- und Langstreckenförderer*] = bucket path
Schraub|arbeit *f* [*mit einer Rotary-Zange*] = make-up job
~**beton** *m* = screwcrete
Schraube *f* [*Schraubenpfahl*] = screw, helix
~ **ohne Mutter** = screw
Schrauben|anker *m* = mooring screw
~**anzugsmoment** *n* = bolt torque
~**bock** *m* → Schraubenhebebock
~**bolzen** *m*, Stiftschraube *f* [*Schraube ohne Kopf, beide Enden haben Gewinde gleicher oder verschiedener Ausführung*] = stud (bolt), ~ screw
~**draht** *m*, Schraubenwalzdraht = wire rod for screws
~**druckfeder** *f* = compression spring
schraubenförmiges Fließen *n* = helical flow
Schrauben|gang *m* = pitch
~**gewinde** *n* = screw thread
~**(hebe)bock** *m*, (Schrauben)Winde *f*, Schraubenspindel *f*, Spindel, Bockwinde *f*, Hebebock *m*, Hubspindel = screw jack, jack(ing) screw, screwed spindle
~**-Hochdruck-Ventilator** *m* → Axial-Hochdruck-Ventilator
Schraub(en)|hohlpfahl *m*, Hohlschraub(en)pfahl = hollow screw pile
~**-Holzpfahl** *m*, Holz-Schraub(en)pfahl = wood(en) screw pile
Schrauben|kolbenpumpe *f* = pump with helicoidal piston
~**kopf** *m* = screw head
~**linie** *f* = helical line, helix (~), spiral ~, screw ~
~**linienarmierung** *f* → Spiralbewehrung
~**linienbewehrung** *f* → Spiralbewehrung
~**lüfter** *m* → Axialventilator *m*
~**material** *n* = screw steel, bolt stock
(~)**Mutter** *f* = nut
~**nagel** *m* = threaded nail, screw ~
Schraub(en)ortpfahl *m* **aus Stahlbeton** = reinforced concrete in-situ screw pile
Schraub(en)pfahl *m* = screw pile
~**gründung** *f*, Schraub(en)pfahlfundation *f* = screw pile foundation
~**schaft** *m* = screw pile shaft
Schrauben|pumpe *f* = propeller pump
~**rad** *n* = Schrägzahnrad
~**rampe** *f* = helicoidal ramp
~**schaft** *m* = screw body, ~ shank, ~ barrel

848

~schlepper m = screw (propelled) tug (boat), ~ (~) towboat
~schlüssel m = spanner
~schneidwerkzeug n = stock and die
~spindel f → Schrauben(hebe)bock m
Schraube(nspindel) f → Gerüstspindel
Schraub(en)spülpfahl m → Spülschraub(en)pfahl
Schrauben|tunnel m = helicoidal road tunnel, ~ highway ~
~tute f, Fangglocke f, Fangmuffe f = overshot
~ventilator m → Axialventilator
~verankerung f [*Spannbeton*] = screw anchoring system, ~ anchorage ~
~verbindung f = screw(ed) connection
~verdichter m = rotary screw compressor
~verzahnung f → Schrägverzahnung
~(walz)draht m = wire rod for screw
(~)Winde f → Schraubenspindel f
Schraub|gewinde n = screw thread
~hohlpfahl m = hollow screw pile
~kappe f = screw cap
~loch n = screw hole
~maschine f = impact wrench
~muffe f, Gewindemuffe = threaded sleeve, screwed socket
~öse f, Ösenschraube f = eye bolt
~pfahl m → Schraubenpfahl
~stempel m (⚒) = screw jack prop, ~ ~ post
Schraubung f [*Kristallographie*] = spiral axis
Schrebergarten m, Kleingarten = allotment
Schreib|barometer m, n → Barograph
~manometer n, m, Druckschreiber m, Registriermanometer = registering manometer, pressure recorder
~pegel m → Registrierpegel
~pegelnull n = datum of a recording ga(u)ge
~regenmesser m, Regenschreiber m = recording rain gauge (Brit.); ~ ~ gage (US)
~stift m [*Registriergerät*] = recording pen
~stube f = company office

~thermometer n, m, Temperaturschreiber m = temperature recorder
~tidepegel m = recording marigraph, registering tide ga(u)ge
~trommel f = drum with record sheet
~zimmer n = writing room [*in a club, hotel, on board ship*]
Schrein m = shrine
Schreinerarbeiten *fpl*, Tischlerarbeiten, Bau~ = joinery work
Schreit|bagger m = walker, walking excavator
~brechanlage f = walking crushing plant
schreitender Ausbau m (⚒) = walking support
Schreit|pfahl m, linker Pfahl [*Naßbagger*] = walking spud
~ramme f, Schreitpfahlramme = walking driver
~-Schlepplöffelbagger m, Schreit-Schleppschaufelbagger = walking dragline
~schuh m = walking shoe
~werk n, Schreitvorrichtung f = walking mechanism
Schrift|erz n (Min.) = sylvanite; graphic tellurium [*this term is obsolete*]
~granit m = graphic granite
schriftliche Befragungsmethode f → Befragungsmethode durch Postkarten
Schrittschalten n, Tippschalten = inching, jogging
schrittweise Annäherung f = successive approximation
schroff, abgerissen, steil abstürzend, abschüssig, prall [*Küste*] = bold, steep
schroffes Ufer n = bluff
Schroppenpflasterstein m, Wildpflasterstein = irregular (paving) sett
Schrot m, n = shot
~ → → Steinbruch
~beton m = shot concrete
~bohren n, Schrotbohrung f = shot drilling, calyx ~
~bohrer m = shot drill, calyx ~
~(bohr)krone f = (chilled) shot-bit

Schrotbohrloch — Schubkurbelpresse

~bohrloch *n* = shot drill hole
~hammer *m*, Stielschrot *m* = spalling hammer
~keil *m*, Steinspeidel *m* = spalling wedge
~korn *n* = shot grain
~krone *f* → Schrotbohrkrone
~meißel *m*, Kaltmeißel = cold chisel
~strahlen *n* = (steel) shot blasting, steel-grit ~
Schrott *m* = scrap
~greifer *m* = scrap grapple
~händler *m* = scrap dealer
~(lager)platz *m* = scrap yard
~muldenkran *m* = scrap charging box handling crane
~paketierpresse *f* = scrap baling press, ~ bundling ~
~schlagkran *m* = scrap drop crane
~zerkleinerungsanlage *f* = scrap crushing plant
Schrühbrandscherben *m*, Biskuit *m, n* = biscuit
Schrumpfen *n* [*unabgebundener Beton*] = contraction
Schrumpf|erzkammer *f* (⚒) = shrinkage stope
~grenze *f* → → Konsistenzgrenzen nach Atterberg
~ring *m* = shrunk-on ring
~riß *m* = shrinkage crack
~sitz-Gestängeverbinder *m* = (Reed) super shrink grip tool joint
~spannung *f* = shrinkage tension
~verbinder *m* [*Tiefenbohrtechnik*] = shrink grip tool joint
Schruppdrehbank *f* = roughing lathe
Schub *m* = wenn „*thrust*" und „*shear*" im gleichen Satz vorkommen, so ist *thrust* = Horizontalschubkomponente und „*shear*" or „*diagonal tension*" = Vertikalschubkomponente; sonst sind beide gleichartig
~anschlag *m* = thrust button, ~ pin
~arm *m* [*Schürflader*] = push arm
~armierung *f* → abgebogene (Stahl-) Einlagen *fpl*
~aufbiegung *f* → abgebogene (Stahl-) Einlagen

~aufgeber *m*, Stoßaufgeber, Schubspeiser, Stoßspeiser = reciprocating feeder
~ausmitte *f* = eccentricity of thrust
~bedingung *f* = shear condition
~bemessung *f* = calculation of shear, shear design
~bereich *m* = shear range
~bewehrung *f* → abgebogene (Stahl-) Einlagen *fpl*
~block *m* = push block
~bruch *m*, Schubversagen *n* = diagonal tension failure, shear ~
~-Bruchfestigkeit *f* = ultimate strength in shear
~bruchgefahr *f* = shear failure risk
~deckung *f* = allowance for shear
~druckfestigkeit *f* = shear-compression strength
~dübel *m* = stud shear connector
~eigenschaft *f* = shear property
~einlagen *fpl* → abgebogene (Stahl-) Einlagen
~eisen *npl* → abgebogene (Stahl)Einlagen
~fahrzeug *n* → An~
~festigkeit *f* → Schubwiderstand *m*
~fließgrenze *f* = yield stress in shear
~gerät *n* → (An)Schubfahrzeug *n*
~-(Gleis)Kettenschlepper *m* → (An-) Schubraupe *f*
~(hilfe)gerät *n* → (An)Schubfahrzeug *n*
~karre(n) *f*, (*m*) → Schiebkarre(n)
~karrenbahn *f*, Schubkarrensteg *m*, Brettersteg = (wheel)barrow run
~karrenfahrer *m*, Schubkarrenarbeiter *m* = wheeler
~karrenförderung *f*, Schubkarrentransport *m* = wheeling
(~)Karrenmischer *m*, Kleinmischer ohne Beschickungsaufzug = baby concrete mixer
~-Kettenschlepper *m* → (An)Schubraupe *f*
~kraft *f* = shear force
~kraftdiagramm *n* = shear force diagram
~kurbelpresse *f* = crank-operated press

Schubladen — Schürfbohrung

~laden n, Laden mit Schubhilfe = push loading
~laden-Hochdruck-Azetylen-Entwickler m = band case high pressure acetylene generator
~last f = shear load
~lehre f → Schieblehre
~linie f = line of thrust
~maschine f → (An)Schubfahrzeug n
~modul m = shear modulus
~moment n = shear moment
~moräne f, Stauchungsmoräne, Aufpressungsmoräne = push moraine, shove ~
~platte f = push plate
~problem n = shear problem
~-Radschlepper m → An~
~rahmen m = push frame
~raupe f → An~
~riß m = shear crack, diagonal (tension) ~, diagonal tensile ~
~rißbereich m = shear crack zone
~schiff n = pusher barge
~schiffahrt f = pusher barge navigation
~schild m, Stoßplatte f = pusher blader
~sicherung f = safety against shear failure
~spannung f = shear stress, diagonal tension ~, diagonal tensile ~
~spannweite f = shear span
~speiser m → Schubaufgeber
~(stahl)einlagen fpl → abgebogene (Stahl)Einlagen
~stange f = drag link, push rod, pitman
~stange f; Stoßbarren m [in Österreich] = pusher bar [tree-dozer]
~stangenlager n, Zugstangenlager [pennsylvanisches Seilbohren] = pitman bearing
~übertragung f = shear transfer
~verbinder m = shear connector
~verformung f = deformation due to shear
~versagen n, Schubbruch m = shear failure, diagonal tension ~
~versuch m, Schubprobe f, Schubprüfung f = shear test

~widerstand m, Schubfestigkeit f = shear strength, ~ resistance, diagonal tension ~, diagonal tensile ~
~winkel m = channel shear connector
~zug m [im Gegensatz zum Schleppzug in der Binnenschiffahrt] = train of pusher barges

Schuh m, Brems~, Hemm~ = brake shoe
~platte f [Gleiskette] = crawler shoe, track ~

Schul|bau m, Schulgebäude n = school building
~gebäude n, Schulbau m = school building
~hof m = school yard

Schulter f [Eisenbahnbau und Straßenbau] = shoulder (US)
~, Rand m [Damm] = edge

Schulungshalle f = (seat) lecture hall

Schulwarnzeichen n = school warning sign

Schuppen m = shed
~blech n = imbricated plate
~(ziegel)dach n = scale tile roof

Schüpper m → Schipper

Schurf m → Aufgraben n [Baugrundaufschluß]
~ (des Wassers) → Wassererosion f

Schürfarbeit(en) f(pl) → Aufgraben n [Baugrundaufschluß]
~ [zur Untersuchung von Lagerstätten] = prospecting work

Schürf|bagger m → Pflugbagger
~becherwerk n → Radius~
~bohren n [Bohren zum Zwecke der Untersuchung von Lagerstätten] = prospecting drilling
~bohrmaschine f [Bohrmaschine zum Abteufen von Schürfbohrungen] = prospecting drilling machine
~bohrung f, Probebohrung f [Bohrung zum Zwecke der Untersuchung von Lagerstätten. Sie soll Unterlagen über Inhalt und Lagerungsverhältnisse, Störungen usw. des erbohrten Vorkommens oder Gesteins liefern] = prospecting borehole

schürfen = to prospect [*To look for valuable mineral*]

~, aufgraben [*Baugrundaufschluß*] = to test by digging, to investigate ~

~ → ab~

Schürfen *n* → Aufgraben [*Baugrundaufschluß*]

~ = costeaning [*Searching for lodes by trenching down to 6 ft. deep at right angles to the supposed line of the outcrop*]

~ → Abziehen

~ [*mit Radschrapper*] = digging, scraping

~ **auf Rampe für LKW-Beladung** = (bull)dozing to ramp for truck filling (US); ~ ~ ~ ~ lorry ~ (Brit.)

~ **mit dem Brustschild** = (bull)dozing

~ ~ **zwei Schürfschleppern nebeneinander** = side by side (bull)dozing

Schürfer *m* = prospector

Schürf|erlaubnis *f* = prospecting license

~**gang** *m*, Ladegang [*Radschrapper*] = load gear

(Schürfgang-)|Schubmaschine *f* → (An)Schubfahrzeug *n*

~**Stoßmaschine** *f* → (An)Schubfahrzeug *n*

Schürf|gefäß *n*, Fördergefäß, Kübel *m* [*Radschrapper*] = (scraper) bowl, (skimmer) scoop

~**grube** *f* [*Baugrundaufschluß*] = test pit, investigation ~

~**kasten** *m* → kastenartiges Schrapp(er)gefäß *n*

~**kraft** *f* [*Schürflader*] = digging power, ~ force

Schürfkübel *m*, Sattel-Schürfwagen *m*, Schürfkasten *m* = two-wheel(ed) scraper, (tractor-pulled) carrying ~, wagon ~; pan (US)

~ [*als Teil eines Radschrappers*] = scraper bowl

~ **für Gabelstapler** = fork truck scoop

~**bagger** *m*, Eimerseilbagger, Schleppschaufelbagger, Schlepplöffelbagger, Zugkübelbagger = dragline (excavator); boom-dragline (US); dragline-type shovel

~**heckmotor** *m* = (wheel) scraper rear engine

Schürf(kübel)|raupe *f*, Schürfkübel-(gleis)kettenschlepper *m*, MENCK-(BENTELER-)Schürfraupe = scraper-dozer

~**wagen** *m* → Schrapperwagen

~**wagenzug** *m* → Traktor-Schrapper *m*

Schürf|kübel-Zugraupe *f*, Schürfkübel-(Gleis)Kettenschlepper *m* = crawler for towing wheel scrapers

~**laden** *n* = dozer shovel(l)ing

~**lader** *m* → Schaufellader

~**lader-Heckbagger-Kombination** *f* → Baggerlader *m*

~**leistung** *f* [*Schürfschlepper*] = bulldozing output

~**loch** *n* [*zur Untersuchung einer Lagerstätte*] = prospecting hole

~**raupe** *f* → Schürfkübelraupe

~**-Raupe** *f* → Bulldozer

~**schacht** *m* [*zur Untersuchung einer Lagerstätte*] = prospecting shaft

~**schacht** *m* [*Baugrundaufschluß*] = test shaft, investigation ~

~**schlepper** *m* → Bulldozer *m*

~**seil** *n* → Zugseil

~**seiljoch** *n*, Seilschloß *n* = drag line yoke

~**stollen** *m* [*zur Untersuchung einer Lagerstätte*] = prospecting gallery

~**tiefe** *f* **unter Planum** = blade drop below ground

Schürfung *f* → Abziehen *n*

~ → Aufgraben [*Baugrundaufschluß*]

Schürfwagen *m* → Schrapperwagen

~**zug** *m* → Traktor-Schrapper *m*

~**(zug)** *m* **mit besonderem Schürfkübelmotor** → doppelmotoriger Radschrapper *m*

Schürfzeit *f* [*Radschrapper*] = digging time, dig-and-turn ~

Schurre *f*, Rutsche *f* = chute, shoot ~

Schurscheibe *f* [*waag(e)recht kreisende Stahlscheibe zum Schleifen von Werksteinen*] = rotating disc, rubbing bed, revolving metal table

Schurzholzwand *f*, einfache Bohl(en)wand = single plank wall

Schürze *f,* (Ab)Dichtungsvorlage *f,* (Ab)Dichtungsteppich *m,* (Ab-)Dichtungsschürze, Außen(haut)-dichtung *f,* Oberflächendichtung *f* = impervious blanket, (waterproof) ~
~, Falle *f* [*Radschrapper*] = apron
~ aus Pfählen → Pfahlwand *f*
Schürzen|arm *m* [*Radschrapper*] = apron arm
~**folgeventil** *n* [*Radschrapper*] = apron sequence valve
~**hebel** *m* [*Radschrapper*] = apron lever
~**hub** *m* [*Radschrapper*] = apron lift
~**hubarm** *m* [*Radschrapper*] = apron lift arm
~**leitungen** *fpl* [*Radschrapper*] = apron lines
~**kante** *f* [*Radschrapper*] = apron lip
~**öffnung** *f,* Fallenöffnung [*Radschrapper*] = apron opening
~**-Rollenturm** *m,* Rollenturm der Schürze [*Radschrapper*] = apron lift sheave tower
~**steuerventil** *n* [*Radschrapper*] = apron control valve
~**zylinder** *m* [*Radschrapper*] = apron jack
Schüssel|-Klassierer *m* = bowl classifier
~**wagen** *m* → Beton~
Schuß *m* [*Drahtgewebe*] = weft
~ [*Pfahl; Rohr*] = section
~**draht** *m* = weft wire, tranverse wire of the cloth
~**gerät** *n* → Bolzenschießgerät
~**(ge)rinne** *f, (n)* = race, spill channel
Schußloch *n* → Bohrloch
~**abstand** *m* → Bohrlochabstand
~**besatzstoff** *m* → Bohrlochbesatzstoff
~**bohrer** *m,* Sprenglochbohrer = blasthole drill
~**durchmesser** *m* → Bohrlochdurchmesser
Schuß|richtung *f* [*Drahtgewebe*] = direction of weft
~**rinne** *f,* Schußgerinne *n* = spill channel, race

~**röhre** *f* → Schlot *m*
~**schweißen** *n* = shot welding
schußsicheres Glas *n,* kugelsicheres ~ = bullet-resisting glass
Schute *f,* Abfuhr~, Bagger~ = (hopper) barge (Brit.); (~) scow (US)
~ **mit Derrick(kran)** = derrick boat; ~ scow (US)
~ ~ **festem Boden für Elevatorbetrieb** = well barge
~ ~ ~ **Deck** = decked barge, flat pontoon (Brit.); cargo-box barge (US)
Schutt *m* (Geol.) → Gesteinstrümmer
~ = rubble [*Broken bricks, old plaster, and similar material*]
Schütt|arbeiten *fpl* → (Auf)Schüttungsarbeiten
~**beton** *m,* entfeinter Beton, Beton ohne Feinkorn = no-fines concrete (Brit.); popcorn ~ (US); rubble ~
~**boden** *m* → (Auf)Füllboden
~**dämmung** *f* = (loose) fill insulation
Schüttel|band *n,* ausziehbarer Stahlbandförderer *m* = shaker conveyor, ~ conveyer
~**rost** *m* = shaking grate
~**rostkühler** *m* = shaking grate-type cooler
~**rutsche** *f* = jigging conveyor, ~ conveyer, shaker ~, jigger ~, jigging chute, jigging shoot
~**sieb** *n* = shaking screen, shaker ~, jig(ging)~
~**sieb-Auffangkasten** *m* [*Rotarybohren*] = discharge box for the mud screen
~**-Stangen(transport)rost** *m,* Schüttelstangenaufgeber *m* = shaking (bar) grizzly, ~ ~ screen, ~ screen of bars [*mounted on eccentrics so that a forward-and-backward motion is given the entire bar assembly at a speed of 80–100 strokes per minute*]
schütten → auf~
Schütten *n,* An~, Auftragen, Auffüllen, Aufschütten = depositing, the fill(ing), placing ~ ~, ~ ~ fill material
~ [*Steine*] = random dumping

Schüttentleerung f = dumping
Schutter m, Ausbruch m, Haufwerk n = muck (pile), broken rock
~**bunkeranlage** f [*Tunnel- und Stollenbau*] = muck stage, spoil ~, ~ bins
~**gabel** f = muck fork
Schüttergebiet n (Geol.) = region of disturbance
Schutter|gut n, Schuttermassen fpl = tunnel spoil
~**massen** fpl, Schuttergut n = tunnel spoil
Schuttern n → (Auf)Schottern n [*Schweiz*]
Schutterschlitz m = mucking slot
Schutterung f → (Auf)Schottern n [*Schweiz*]
Schutterwagen m, Stollenwagen, Förderwagen [*früher: Hund; Hunt*] = skip
Schütt|fläche f → (Auf)Schütt(ungs-)gelände n
~**gelände** n → (Auf)Schüttungsgelände
~**gewicht** n, Raumgewicht, Rohwichte f = (dry) loose unit weight, unit weight
Schüttgut n, Massengut = bulk material, loose ~, material in bulk
~**entladevorrichtung** f, Schüttgutentlader m = bulk material unloader
~**greifer** m, Massengutgreifer = bulk material grab
~**lader** m, Schüttgutladevorrichtung f = bulk material loader
~**-Schleuder** m, fahrbare ~, Entladeschleuder für Schüttgüter = truck dump piler (US); lorry ~ = ~ (Brit.)
~**umlader** m, Massengutumlader = bulk material transfer equipment
~**waage** f, Massengutwaage = bulk material scale
Schütthalbmesser m, Aus~, Abwurfhalbmesser = dumping radius
Schutt|hang m, Gehängeschutt m = detrital slope
~**haufen** m = heap of rubble, pile ~ ~, rubble pile, rubble heap

Schütthöhe f → Auftraghöhe
~, Aus~, Ausleerhöhe, Kipphöhe, Abwurfhöhe [*Schaufellader*] = dumping height
Schüttkasten m, Schüttrumpf m, Schüttrichter m [*Eimerkettenbagger*] = discharge chute, hopper
Schuttkegel m (Geol.) = talus-fan
~, Unterlauf m [*unterster Teil eines Wildbaches, Zone der Auflandung*] = heap of débris
Schütt|körper m [*Erdbau*] = fill
~**lage** f → Auftragschicht f
~**lagenhöhe** f [*Höhe der einzelnen Schüttlage*] = depth of (fill) lift
Schuttlast f [*die von einem Wasserlauf fortbewegte Materialmenge*] = load
Schütt|masse f → Auftragmaterial n
~**material** n = granular material
~**material** n → Auftragmaterial
~**packe** f, Schüttpacklage f, verfestigte ~ = hardcore
Schutträumung(sarbeiten) f(pl) → Enttrümmerung(sarbeiten)
Schüttreichweite f [*Schaufellader*] → Aus~
Schüttrichter m → Schüttkasten m
Schütt|rinne f mit Gegengewicht [*Dosiersilo*] = counterbalanced subchute
~**rumpf** m → Schüttkasten
Schuttrutsche f = rubble chute, ~ shoot
Schüttschicht f [*Siebtechnik*] = material bed
~ → Auftragschicht
Schütt|schiff n mit Selbstentladung = self-unloading ship
~**sintern** n, Sintern von losem Pulver = pressureless sintering, loose powder ~
~**sprengverfahren** n → Moorsprengung f
~**steine** mpl, Steinmaterial n = stony material, rip-rap
~**steinverguß** m [*Er verbindet die Steine zu einer kompakten Masse und füllt die Hohlräume zwischen den Schüttsteinen ganz oder teilweise aus*] = rip-rap sealing compound

Schüttung *f,* Quellergiebigkeit *f* = yield of a spring

~, Auf~, Auftrag *m,* Anschüttung, Auffüllung, Dammaufschüttung bis zukünftiges Planum, Damm *m* [*Schüttung für einen Verkehrsweg, dessen Gradiente so hoch oberhalb der Geländeoberfläche liegt, daß die Böschungen beiderseits der Krone abwärts gerichtet sind*] = fill(ing)

~, Auf~, An~, aufgefülltes Gelände *n,* Auftrag *m,* Auffüllung, aufgeschüttetes Gelände, angeschüttetes Gelände, aufgetragenes Gelände = filled ground, filled-up ~, made-up ~, fill

~ [*in der Schweiz und Österreich*] → Sauberkeitsschicht *f*

~ **aus großen Bruchsteinen** = large rubble fill

~ ~ **kleinen Bruchsteinen** = small rubble fill

~ ~ **mittleren Bruchsteinen** = medium rubble fill, hand ~ ~

~ ~ **sortierten Bruchsteinen** = classified rubble fill, sorted ~ ~

~ **der Bohrung** = well flow

Schütt(ungs)|arbeiten *fpl* → Auf~

~**boden** *m* → (Auf)Füllboden

~**fläche** *f* → (Auf)Schütt(ungs)gelände *n*

~**gelände** *n* → Auf~

~**höhe** *f* → Auftraghöhe

~**lage** *f* → Auftragsschicht *f*

~**masse** *f* → Auftragmaterial *n*

~**material** *n* → Auftragmaterial

~**schicht** *f* → Auftragsschicht

~**zone** *f* → Auffüllzone

Schütt|verlust *m* = spillage

~**volumen** *n* = (dry) loose volume

~**winkel** *m* [*Schaufellader*] → Aus~

~**zeit** *f* [*Schaufellader*] → Aus~

Schütz *n* → (Schütz)Tafel *f*

Schutz *m* [*Hafenbau*] = shelter

~ **des Landschaftsbildes** = landscape protection

~**anlage** *f* → Abschirmungsanlage

~**anstrich** *m* = protective coating

~**anzug** *m* = coverall

~**bau** *m* = civil defense construction (US); ~ defence ~ (Brit.)

~**bau(werk)** *m, (n)* [*baulicher Luftschutz*] = civil defense structure (US); ~ defence ~ (Brit.)

~**(bau)werk** *n* = protective structure

~**blech** *n,* Kotflügel *m,* Fangblech = mudguard

~**brille** *f,* Schweißbrille = welding goggles

~**dach** *n* → Sonnendach

~**dampfheizung** *f* = protective steam heating

Schütz(e) *n, (f)* → (Schütz)Tafel *f*

Schützenwehr *n,* Schütz(tafel)wehr, Fallenwehr [*Bewegliches Wehr mit in senkrechter oder annähernd senkrechter Richtung beweglicher plattenartiger Stauvorrichtung, die in gesenktem Zustand in Wehrschwelle liegt*] = vertical-lift gate weir

(Schutz)Erdung *f* = earthing

Schutz|fläche *f* = protective area

~**flöz** *n* (⚒) = relieving working

~**folie** *f,* Bauten~ = protective foil

~**fußbekleidung** *f* = protective footwear

Schutzgas *n* [*Schweißen*] = shielding gas, ~ atmosphere, protective ~

~**-Handschweißanlage** *f* = inert gas hand welding installation

~**-Lichtbogenschweißen** *n* = shielded arc welding

~**lötung** *f* = hydrogen brazing

~**schneiden** *n* = shielded cutting

~**-Wurzellagenschweißung** *f* = inert gas root pass welding, ~ ~ ~ run ~

Schutz|gehäuse *n* [*Strom(erzeugungs-)aggregat*] = sheet-steel casing

~**helm** *m* = protective helmet, skullguard

~**kappe** *f* = grommet

~**kappe** *f* **für das Fahrerhaus,** Fahrerhaus-Überdachung *f* [*Erdbauwagen*] = full-width cab-protection plate

~**keller** *m* [*baulicher Luftschutz*] = basement shelter

~**kette** *f* → Gleit~

~**kleidung** *f* = protective clothing

Schutzmaßnahme — schwarzer Glaskopf

~maßnahme f = precaution
~mauer f [Hafenbau] = screen wall, shelter ~
~mittel n, Bauten~ = preservative for structures and buildings
~mittel n, Konservierungsmittel = preservative
~mittel n gegen alkaliempfindliche Zuschlagstoffe = alkali-aggregate expansion inhibitor
~netz n = protective net(ting)
~paste f = protective paste
~pfahl m [Hafenbau] = fender pile
~raumbau(werk) m, (n) = civil defense shelter (US); ~ defence ~ (Brit.)
~rohr n = protective pipe
~schicht f = protective layer
~schiene f, Streichschiene = check rail, safety ~, side ~, guard ~
~schild m → Schweiß(er)schild
~schlauch m = protective hose
~schlauch m [Einer biegsamen Welle] = (rubber-covered) outer casing
~spannung f = safety tension
~sperre f auf dem Mittelstreifen = center barrier (US); centre ~ (Brit.)
~streifen m = protective strip
(Schütz)Tafel f, Schütz(e) n, (f), Falle f [hölzerne oder stählerne Abschlußtafel bei einem Schützenwehr] = (gate) leaf
Schütz(tafel)wehr n → Schützenwehr
Schutz|überzug m = protective coat(ing)
~vorrichtung f = guard
~wall m = revetment
~wand f = protective wall
Schützwehr n → Schützenwehr
Schutzwerk n, Hafen~ = protective (harbo(u)r) structure
~, Schutzbauwerk = protective structure
Schutzwulst f, Ansatz m = protective lug
Schwabber m, Gummischieber m = squeegee
Schwachbrandziegel m, Weichbrandziegel = soft (burnt) brick, underburned ~

schwacher Geschiebetrieb m = light bed load movement, ~ ~ ~ transport, bed load movement below competence, bed load transport below competence
Schwachstrom|kabel n = cable for communication circuits
~leitung f = communication line
Schwächung f des Betons durch die Kabelaussparungen [Spannbeton] = reduction of the cross-sectional area of the concrete due to the holes for the cables
Schwaden m, Wrasen m, Brüden m = water vapo(u)r
~ → Längsreihe f
~beseitiger m = windrow eliminator
~glätter m, Langmahdplanierungsschablone f = windrow evener
~verteilung f [Straßenbau] = blading back [Pushing soil in a windrow back with a grader to the position from which it came]
Schwalbenschwanz|blatt n [Holzverbindung] = dovetail halving, ~ halved joint
~fenster n, Fledermausluke f = oval luthern
Schwall m = surge
~schacht m → → Talsperre
Schwamm|gummi m = sponge rubber
~verputzen n [Steinzeugindustrie] = sponging
Schwanenhals m = gooseneck
~ → Brückenhals
~ [Dränagewerkzeug] = swan-necked drainage tool
~-Auslegerkran m = gooseneck type crane
Schwankung f = variation
Schwanzende n, Spitzende [Muffenrohr] = spigot end
schwarze Mineralfarbe f = black pigment
~ Schraube f = black bolt
schwarzer (Bau)Stoff m → bituminöser ~
~ Diamant m → Carbonado m
~ Glaskopf m (Min.) → Psilomelan m

857 schwarzer Porphyr — Schwarzbelageinbaumaschine

~ **Porphyr** m, Augitporphyr = augite porphyry
~ **schwedischer Granit** m [*Fehlname*] → Basalt m

~ **Stoff** m → bituminöser (Bau)Stoff
„**schwarzes Brett**" n → Anschlagbrett
Schwarzbelag m → bituminöse Decke f

Schwarzbelageinbaumaschine f, Schwarz-(decken)verteiler m, Verteilerfertiger m, Schwarzdeckenfertiger, Schwarzdecken-Einbaumaschine

bituminous road surfacing finisher, asphalt finisher, asphalt paver, bituminous paving machine, bituminous spreading-and-finishing machine, black-top spreader, asphaltic concrete paver, (bituminous) paver-finisher, asphalt and coated macadam finisher, spreader finisher, black top paver

abkippen	to tilt down [*screed*]
Abziehbohle f, Glättbohle	(screed)plate
Anbauteil m, n	attachment
ankippen	to tilt up [*screed*]
Antriebskette f zum Fahrantriebsturas	crawler pivot shaft drive chain
Anzeigeskala f für Querneigung	crossfall indicator plate
Anzeigevorrichtung f für Deckendicke	mat depth indicator
Auflockerungsschnecke f → Verteilerschnecke	
Automatiksteuerung f für Längsneigung Einbauteil	auto(matic) grade control for screed
Bitumen-Mörtelschmiere f	"fat"
Bohlenwinkel m	screed planing angle
Brenner m mit Gebläse	heater and blower unit
Brückenbildung f [*von Mischgut*]	buildup of material
dauernder Schichtausgleich m	continuous course correction
Deckenschluß m	finish
Dickeneinstellspindel f	thickness control screw
Dickeneinstellung f	(mat) thickness control
Dickenmesser m	thickness ga(u)ge, depth ~
Dickensteuervorrichtung f	level controller
Dosierschild m → Durchlaßschieber	
Drehpunkt m	pivot point
Druckring m	operating ring
Drückrollen fpl für LKW-Räder, (LKW-)Schubrollen	bumper rollers, push ~
Durchlaßschieber m, Mengenregulierschieber, Dosierschild m	cutoff gate, hopper gate
Einbaustreifen m, Einbaubahn f	strip
Einbauteil m, n	(floating) screed
Einbauteil-Hubseil n	(floating) screed hoist rope
Einbauteil-Hubventil n	(floating) screed hoist valve
Einbauteil-Hubvorrichtung f	(floating) screed hoist

Einbauteil-Hubzylinder *m*	(floating) screed hoist ram
Einbauteil-Motor *m*	(floating) screed engine
Einbauvermögen *n*, Einbauleistung *f*	laying capacity, paving ~
Einstellhandrad *n* für Deckendicke	thickness control handwheel
Einstellorgan *n*	control
Einstellspindel *f*	adjusting screw
Einstellung *f* der Bohlenneigung	pivot point adjustment
elektrische Heizvorrichtung *f* für die Glättbohle	electric screed heater, ~ ~ heating unit
Fahrkupp(e)lung *f*	traction clutch
Fahrteil *m, n*	traction unit, tractor ~
Feineinregulierung *f*, Feineinstellung	fine adjustment
Flanschenrohrstück *n*	flanged tubular extension
Führungsstück *n* für Seitenbegrenzungsblech	guide bar
gebogene Ablenkungsplatte *f*	curved deflector plate
Gehäuseglocke *f*	transmission bell housing
Glättbohle *f*, Abziehbohle	(screed) plate
Größe *f* der Kronenüberhöhung *f*	amount of crown
Grundglättbohle *f*, Grundabriebbohle	standard (screed) plate, main (~) ~
Grundstampfbohle *f*	standard tamper (bar), ~ bar
Grundverteilerschnecke *f*	standard auger
Hälfte *f* [*Verteilerschnecke; Glättbohle; Stampfbohle*]	section
Handradwelle *f* zum Einstellen des Einbauteils	(floating) screed lift shaft
Haupteinstellorgan *n*	primary control
Hauptlagerdurchmesser *m*	diameter of main journal
Hinterkante *f* [*Glättbohle*]	rear edge, back ~, trailing ~
Höhensteuerung *f*	control of level(s)
Hydraulikölbehälterentlüfter *m*	hydraulic oil tank breather
Kettenrad *n*	sprocket
klebrige Mischung *f*	tacky mix
Kraftstoffanlage *f*	fuel system
Kraftstoffeinspritzpumpe *f*	fuel injection pump
Kraftstoffversorgungspumpe *f*	fuel lift pump
Kronenanzeiger *m*	crown control indicator
Kroneneinstellung *f*	crown adjustment
Kroneneinstellzahnrad *n*	crown adjustment sprocket
Kronenüberhöhung *f* im Verhältnis zur Einbaubreite	crown in relation to the width of mat being laid
Längsneigungswinkel *m*	forward angle
Längsnivellierung *f*	longitudinal level(l)ing
Latte *f*	(feeder) bar
Lattenrost *m* → Stangenzubringer	
Lattenrost(antriebs)kette *f*	(bar) feeder (drive) chain
Lattenrostantriebswelle *f*	(bar) feeder headshaft

Schwarzbelageinbaumaschine

Lattenrostkupp(e)lung f, Stangenzubringerkupp(e)lung	conveyor clutch, conveyer ~
Lattenrostspanner m	(bar) feeder take-up
Lattenrosttunnel m	(bar) feeder tunnel
Laufsteg m	rear platform
Leerlauf m [Hebelstellung]	neutral
lenkbares Vorderrad n	front steerable wheel
Lenkerhebel m, Schwenkarm m	screed side arm, level(l)ing ~, pivoting ~
Lenkerhebelanlenkpunkt m, Schwenkarmanlenkpunkt	level(l)ing arm pivot
Lenkkupplungshebel m	steering lever
letzte Getriebestufe f	final drive
LKW-Schubrollen fpl, Drückrollen für LKW-Räder	bumper rollers, push ~
Mengenregulierschieber m, Durchlaßschieber, Dosierschild m	hopper gate, cutoff ~
Mindestarbeitsbreite f	minimum operating width
Mindestöldruck m bei Leerlaufdrehzahl	minimum oil pressure with engine idling
Mischgutbehälter m, Aufnahmebehälter	(receiving) hopper
Mischgutfluß m	flow of mix, ~ ~ material
mittlere Dicke f	average thickness
Muldenanlenkpunkt m	hopper pivot shaft
Nachbarstreifen m, Anschlußbahn f	adjacent strip
Naht f	joint
Nahtanschlußgerät n	joint matcher
Nahtanschlußherstellung f	joint matching
Nivellierung f	levelling
normale Motordrehzahl f [Dieselmotor]	piston speed
Öldruck m bei voller Drehzahl und warmem Motor	oil pressure at full throttle with warm engine
Ölwanneninhalt m	wet sump capacity
Pleuellagerdurchmesser m	diameter of big end pin journal
Profilregulierung f	control of (floating) screed
Pumpenleistung f pro Minute	pump delivery per minute
Quernaht f	transverse joint
Querneigungssteuerung f	control of crossfall
Querneigungssteuervorrichtung f	crossfall controller
Querneigungswinkel m	side angle
Quernivellierung f	transverse level(l)ing
Reduzierschuh m, Reduzierstück n	cutoff shoe
Regelorgan n für den Kraftstoffdruck	fuel pressure control
rotierende Auflockerungswalze f	agitator (raker bar)
Schauloch n	inspection hatch
Schichtdicke f	thickness of mat, mat thickness

Schwarzbelageinbaumaschine 860

Schlagbohle *f*, Stampfer *m*, Stampfbohle *f*	tamper
Schräglage *f* auf der Hinterkante *f*	riding heel of screed
Schräglage *f* auf der Vorderkante *f*	riding toe of screed
Schwenkarm *m*, Lenkerhebel *m*	level(l)ing arm, (screed) side ∼, pivoting ∼
Seitenbegrenzungsblech *n*	end plate
Seitenblech *n* des Mischgutbehälters	hopper side plate
Spannrolle *f*	take-up idler
Spannschloß *n*	turnbuckle
Spannschuh *m*	take-up shoe
Spindel *f* für Mengenregulierschieber	central gate adjusting screw
Spitzenleistung *f* bei kurzzeitigem Betrieb [*Motor*]	B. H. P. intermittent
Spitzenleistung *f* bei 12stündigem Dauerbetrieb [*Motor*]	B. H. P. 12-hr. rating
Stampfbohlenantrieb *m*, Stampferantrieb	tamper transmission
Stampfer *m*, Stampfbohle *f*, Schlagbohle	tamper
Stampferabschnitt *m*	tamper bar, ∼ section
Stampfer(antrieb)motor *m*	tamper (drive) engine
Stampferantriebswelle *f*	tamper drive shaft
Stampfereinstellung *f*	tamper adjustment
Stampferexzenterwelle *f*	tamper eccentric shaft
Stampfergelenk *n*	tamper connection link
Stampferhebel *m*	tamper lever
Stampfer-Overdrive *m*	tamper overdrive
Stampfermikroschalter *m*	tamper micro-switch
Stampferrahmen *m*	tamper frame
Stampferschlagzahl *f*	tamper speed
Stampferüberdruckventil *n*	tamper relief valve
Stangenzubringer *m*, Lattenrost *m*	(flight) conveyor, (∼) conveyer, bar ∼, (bar) feeder
steigen und fallen [*Einbauteil*]	to ride up and down
Stellvorrichtung *f* für Bohlenneigung	side arm actuating unit
Strebe *f*	stay bar
Taster *m*	sensing device
Teilfuge *f*	mounting slot
Turaswelle *f*	sprocket shaft
Überschüttung *f*	surcharge
Umlenkturaswelle *f* für Lattenrost	(bar) feeder foot shaft
Unterbau *m*, Vorprofil *n*	base
unteres Trum *n* [*Lattenrost*]	return run bars
vergleichmäßigen	to stretch out [*the variations in level*]
Verlängerungsmuffe *f* der Verteilerschnecke	extension socket of the standard auger
Verlängerungsstück *n*	extension (section)
Versuchsbelag *m*, Versuchsdecke *f*	trial mat

Verteilerschnecke *f*, Auflockerungsschnecke — spreading screw, spreader ~, (cross) auger, screw conveyor, feed screw, agitator, screw conveyer

Verteilerschnecken-Antriebswelle *f* — auger drive shaft
Verteilerschneckenkette *f* — auger chain, screw conveyor (or conveyer) drive chain
Verteilerschneckenlager *n* — auger bearing
verzögernde Wirkung *f* des Einbauteils — delayed screed action
Visierstange *f* — guide rod
Vorabstreifer *m* — tamper shield
Vorderkante *f* [*Glättbohle*] — front edge, leading ~
Vorderradlenkung *f* — front steering
Vorgelegekupp(e)lung *f* — transmission clutch
Vorprofil *n*, Unterbau *m* — base
Zwischenvorgelege-Antriebsketten *fpl* für Lattenrost und Verteilerschnecke — (bar) feeder and screw (conveyor) (or conveyer) countershaft drive chains
Zwischenvorgelegewellen *fpl* für Lattenrost und Verteilerschnecke — (bar) feeder and screw (conveyor) (or conveyer) countershafts

---

Schwarz|(belag)einbaumasse *f* → bituminöses Mischgut *n*
~beton *m*, Kohlenwasserstoffbindemittelbeton, bituminöser Beton = bituminous concrete
~bindemittel *n* → Bindemittel auf Kohlenwasserstoffbasis
~binder *m* → bituminöses Bindemittel *n*
~blech *n* [*Fehlname: Eisenblech*] = black plate; ~ sheet
~decke *f* → bituminöse Decke
~decken-Einbaumaschine *f* → Schwarzbelageinbaumaschine
~(decken)einbaumasse *f* → bituminöses Mischgut
~deckenfertiger *m* → Schwarzbelageinbaumaschine *f*
~(decken)gemisch *n* → bituminöses Mischgut *n*
~(decken)gut *n* → bituminöses Mischgut
~decken-Instandsetzungsmaschine *f*, Flickgerät *n*, Flickmaschine = patching unit, patcher
~deckenmischanlage *f*, bituminöse Mischanlage = asphalt and coated macadam mixing plant, dual-purpose mixing plant (or machine) (Brit.); combination hot or cold mix plant (US); (asphalt and) bituminous mixing plant, high and low temperature drying and mixing plant, aggregate bituminizing plant, premixing plant, coating plant
~(decken)(misch)gut *n* → bituminöses Mischgut *n*
~(decken)mischung *f* → bituminöses Mischgut *n*
~deckenstraße *f* = flexible road, ~ highway
~(decken)verteiler *m* → Schwarzbelageinbaumaschine
~einbaumasse *f* → bituminöses Mischgut *n*
~erdboden *m* = black-earth soil
~gemisch *n* → bituminöses Mischgut *n*
~gut *n* → bituminöses Mischgut
~kalkzementmörtel *m* = black mortar
~kohle *f* = bituminous coal [*as distinct from brown coal*]
~kupfer *n* = black copper
~manganerz *n* → Glanzbraunstein *m* (Min.)

Schwarzmaterial — (schwedische) Leiterbohrmethode

~material *n* → bituminöses Mischgut *n*
~(misch)gut *n* → bituminöses Mischgut
~mischung *f* → bituminöses Mischgut *n*
~pulver *n*, Sprengpulver = gunpowder, black (blasting) powder
~sand *m* = black sand
~spießglanzerz *n* (Min.) → Rädelerz
~straßenbau *m*, bituminöser Straßenbau = flexible road construction, bituminous ~ ~, ~ highway ~
~streif *m* → Kohleneisenstein *m*
~torf *m* = black peat
~-Tragschicht *f* → bituminöse Tragschicht
~verteiler *m* → Schwarzbelageinbaumaschine
Schweb *m* → Schwebstoff(e)
Schwebe|bahn *f*, (Schienen)Hängebahn = suspension rail conveying system
~fähre *f*, Fahrbrücke *f* = aerial ferry
~gas-Wärme(aus)tauscher *m* = suspension type preheater
~körper *m*, Gehänge *n* [*Flußbau*] = tree retard
schweben [*Planierschild*] = to float
schwebende Pfahlgründung *f*, schwimmende ~ = pile(d) foundation distributing the load into the soil, piles in soil, floating (pile(d)) foundation
schwebender Pfahl *m*, schwimmender ~ = floating pile
~ Streb *m* (⚒) = rise face
Schwebestaub *m* = airborne dust
Schweb|stoff(e) *m (pl)*, Schweb *m* = suspended matter, ~ material
~stoffmenge *f* = sediment content
~stoffteilchen *n* = particle of suspended matter, ~ ~ material
Schwebungsfeuer *n* [*Flugplatz*] = oscillating beacon

(schwedische) Leiterbohrmethode *f*	(Swedish) ladder drilling method
Abschlag *m*	round
Ankerloch *n*	roofhole
Ansetzen *n* der Bohrungen *fpl*	collaring
Bohrbühne *f*, Arbeitsplattform *f*	drilling platform
Bohrfortschritt *m*	drilling rate
Bohrmeter *n*	drilled metre, ~ meter
Bohrstange *f*, Bohrstahl *m*	drill steel, drill rod
Bohrstangenunterlage *f*, Bohrstahlunterlage	drill steel support, ~ rod ~
(Bohr)Stütze *f*	(pusher) leg
einziehbare (Bohr)Stütze *f*	retractable (pusher) leg
Einziehen *n*	retraction
Felsanker *m*	rock bolt
hinterer Querbaum *m*	rear bar, ~ beam, aft bar, aft beam
Klauenfuß *m*	claw foot
Laufstange *f*	bar runner
Leiter-Bohrbühne, Leiter-Arbeitsplattform *f*	ladder drilling platform
Leiterbohrer *m*, leitermontierter Bohrhammer *m*, Bohrhammer mit Leitervorschub	ladder drill, ladder-mounted ~, ladder feed machine
Metallnase *f*	lug
Mineur *m*	drill operator, driller
Netto-Bohrfortschritt *m*	net drilling rate

Ortsbrust *f*	rock face
Querbaum *m*	bar, beam
Rückhalt *m*	grip
Schlitten *m*	(sliding) cradle
(schwedische) Bohrleiter *f*	(Swedish) drilling ladder
Sprosse *f*	rung
Stahlleiter *f*	steel ladder
übermäßiger Ausbruch *m*	overbreak
vorderer Querbaum *m*	front bar, ∼ beam, fore bar, fore beam
Winkeleisen *n*	angle iron

schwedisches Verfahren *n*, schwedische Gleitkreistheorie *f* [*Bestimmung der Rutschgefahr unter Annahme einer kreisförmigen Rutschfläche (oder Gleitfläche)*] = Swedish cylindrical-surface method, ∼ circular-arc ∼

Schwedler|kuppel *f* = Schwedler dome

∼träger *m* = Schwedler truss

Schwefel|blüte *f* = flowers of sulphur (Brit.); ∼ = sulfur (US)

∼dioxyd-Anzeige *f* **mit Bleiperoxyd-Zylindern** = monitoring sulfur dioxide with lead peroxide cylinders

∼gruppenanalyse *f* = group sulfur analysis (US); ∼ sulphur ∼ (Brit.)

∼kalzium *n* = calcium sulfide

∼karbolsäure *f* = sulphocarbolic acid

∼kies *m*, Eisenkies, Pyrit *m*, FeS$_2$ = (iron) pyrites, yellow ∼, fool's gold

∼kies in Kohle = coal brass

∼kohlenstoff *m*, CS$_2$ = carbon bisulphide, ∼ disulphide

∼kupfer *n*, Cu$_2$S = sulphide of copper

∼pocke *f* = sulphur pockmark

∼quelle *f* = sulphur spring

schwefelsaure Tonerde *f*, Aluminiumsulfat *n* = sulphate of alumina

Schwefelsäure *f* = sulphuric acid

∼anhydrid *n* = sulphuric anhydride

∼elektrolytverfahren *n* = sulphuric acid anodising process

schwefelsaures Kalzium *n*, CaSO$_4$ = sulphate of calcium

Schwefel|tellurwismut *n* (Min.) → Tetradymit *m*

∼trioxyd *n* = sulphur trioxide

∼vergußmasse *f* → Schwefelzement

∼wasserstoff *m*, H$_2$S = hydrogen sulfide

schwefelwasserstoffhaltiges Erdgas *n* = sour natural gas

schwefelwasserstofffreies Naturgas *n*, ∼ **Erdgas** = "sweet" natural gas, "∼" rock ∼

Schwefel|zement *m*, Schwefelvergußmasse *f* [*säurefester Kitt aus Asphalt, Schwefelblüte, Graphit und Eisenoxyd*] = sulphur cement

∼zink *n*, Zinksulfid *n*, Zinkblende *f*, ZnS = (zinc) blende, black jack, sphalerite

Schweifung *f*, Bogenkrümmung = bowing

Schweinstein *m* = swinestone [*Fetid bituminous limestone*]

Schweiß|anlage *f* = welding installation

∼apparat *m* **mit einer Hand tragbar** = portable electrical welding equipment

∼automat *m*, automatische Schweißmaschine *f* = auto(matic) welding machine, ∼ welder

∼barkeit *f* = weldability

∼behälter *m*, Schweißgefäß *n*, geschweißter Behälter, geschweißtes Gefäß [*Aufnahmegefäß für Dämpfe, Gase, Flüssigkeiten*] = welded tank, ∼ vessel

∼brenner *m* = welding burner, ∼ torch

∼brille *f*, Schutzbrille = welding goggles

∼brücke *f* = welded bridge

Schweißbühne — Schwellenbefeuerung

~bühne *f* = welding positioner
~draht *m* = welding rod
Schweißen *n*, Schweißung *f* = welding
~ in **Normallage**, Sechs-Uhr-Schweißen = six o'clock-welding
~ mit **Rollelektrode(n)** = roll welding
Schweißerei *f*, Schweißwerkstatt *f* = welding (work)shop
Schweißer|koje *f* [*Schweiz*]; Schweißkabine *f* = welding booth
~prüfung *f* = welder test
Schweiß(er)schild *m*, Handschild, Schutzschild = face screen, ~ shield, hand ~
Schweiß|faktor *m* = welding factor
~fehler *m* = welding defect
~flamme *f* = welding flame
~gefäß *n* → Schweißbehälter *m*
~generator *m* = welding generator
~gerät *n* = welding unit
~gitter *n* [*Siebtechnik*] = electrically welded weave
~gleichrichter *m* = welding rectifier
~grat *m* im Innern = penetration bead
~helfer *m* = welders' helper
~ingenieur *m* = welding engineer
~kabel *n* = welding cable
~kabine *f* [*Schweiz: Schweißerkoje f*] = welding booth
~kastenträger *m* = welded steel box girder
~konstruktion *f* = welded construction
~kopf *m* = welding head
~(-Licht)mast *m*, Schweiß-Beleuchtungsmast = (lantern) column of welded sheet construction
~naht *f* = welding seam
~nahtfläche *f*, F$_{Schw}$ = welding seam area
~ofenschlacke *f* = puddling cinder, tap ~
~plattenkonstruktion *f* = welded plate construction
~profil *n*, geschweißtes Profil = welded section
~pulver *n*, gekörnte Masse *f* [*Ellira-Schweißen*] = granulated welding composition, ~ material

(~)**Raupe** *f* = (weld(ing)) pass
~rost *m* = welded grating, ~ grate
~schild *m* → Schweißerschild
~spannung *f* = welding voltage
~spritzer *m* = weld spatter
~-Stahlkastenträger *m* = welded-steel box-girder
~stahlkonstruktion *f* = welded steelwork
~-Stahlskelett *n*, Schweiß-Stahlfachwerk *n* = welded steel framework, ~ skeleton
~stelle *f* = weld
~-Straßenbrücke *f* = welded road bridge, ~ highway ~
~strom *m* = welding current
~stromstärke *f* in Ampere = current strength in amps, welding current ~ ~
~technik *f* = welding engineering
~tisch *m* = welding bench
~träger *m* = welded (plate) girder
~transformator *n* = welding transformer
Schweißung *f*, Schweißen *n* = welding
Schweiß|verbindung *f* = welded connection
~verfahren *n* = welding method
~werkstatt *f*, Schweißerei *f* = welding (work)shop
~zone *f* = weld zone
~zustand *m* = as-welded condition
Schwelkohle *f* = tar coal
Schwell|beton *m*, Expansivbeton, Quellbeton = self-stressed concrete
~bewegung *f* → Schwellung
~boden *m*, Quellboden = expansive soil, swell ~
Schwelle *f* [*einer Dachkonstruktion*] = pole plate
~ [*Schleuse*] = mitre sill, clap ~, lock ~; miter ~ (US)
~, Anschwellung *f*, Bodenwelle *f* [*Geomorphologie*] = hillock
~ → Holz~
~ → Bahn~
Schwellen *n* → Schwellung
~befeuerung *f* = (runway) threshold lighting

Schwellenbohrmaschine — Schwenkbremse

~bohrmaschine *f* = sleeper drilling machine (Brit.); tie ~ ~ (US)
~feuer *n* [*Flugplatz*] = threshold light
~fuge *f* = sleeper joint [*It has no value as a load-transfer device and only reduces by a small amount the intensity of loading on the subgrade at the joint*]
~kranz *m* [*einer Dachkonstruktion*] = pole plate system
~nagel *m* = track spike
~schraube *f*, Schienenschraube = screw spike (US); railfastening coach screw (Brit.)
~stopfmaschine *f* = ballast tamper; tie ~ (US); track ~ (Brit.)
~verlegemaschine *f* = sleeper laying machine (Brit.); tie ~ ~(US)
~verlust *m* [*Hydraulik*] = sill loss
~verzug *m* (⚒) = sleeper lagging

Schwellfähigkeit *f*, Schwellvermögen *n* = expansibility
Schwellung *f*, Schwellen *n*, (Auf)Quellung, (Auf)Quellen, Schwellbewegung = bulking, bulkage
Schwell|vermögen *n*, Schwellfähigkeit *f* = expansibility
~zement *m* → Quellzement
Schwelretortenteer *m* = retort tar
Schwelwassertank *m*, Schwelwasserbehälter *m* = gas liquor tank

Schwemm|gut *n* → Schwemmzeug
~kanalisation *f* → Abwassernetz *n*
~(land)boden *m*, Wasserabsatzboden = alluvial soil, transported ~
~landdoline *f* = alluvial dolina
~landebene *f* = alluvial plain, alluvian ~
Schwemmsel *m* → Schwemmzeug
Schwemm|stein *m* [*nur aus Naturbims gefertigt*] = pumice block
~stoff *m* → Schwemmzeug
~theorie *f* = allochthonous theory [*coal formation*]
~torf *m* = floating peat
~verfahren *n* [*Hausmüllbeseitigung*] = water-carrier method [*of garbage disposal*]

~zeug *n*, Schwemmgut *n*, Schwemmstoff *m*, Schwemmsel *n* [*Fluß*] = floating débris
Schwengel *m*, Balanzier *m* = beam ~ → Bohr~
~antrieb-Seilbohren *n*, Seilbohren mit Schwengelantrieb, pennsylvanisches Bohren, pennsylvanisches Bohrverfahren *n* = Pennsylvanian drilling system, ~ ~ method
Schwengelbock *m* [*pennsylvanisches Seilbohren*] = sam(p)son post
~-Jochbolzen *m* = walking beam saddle pin
Schwengel|lager *n* = center irons
~schnellschlagbohrverfahren *n* = spring pole and cable tool method
~(schub)stange *f*, Zugstange = (beam) pitman
~(schub)stangenlager *n* = adjuster board

(Schwenk)|Absetzer *m* → Bandabsetzer
~antrieb *m* = slewing drive
~arm *m* = swivelling arm
~arm *m* [*Radschrapper*] = cross arm
~aufreißer *m* = hinge-type ripper, pivot-beam ~, swivel ~
~ausleger *m*, Schwenkarm *m* [*Kran*] = rotating boom, swing(ing) ~ (US); ~ jib (Brit.)
~bagger *m*, Drehbagger = slewing excavator
~bagger *m*, Drehbagger [*Eimerkettenbagger mit um 360° schwenkbarem Oberteil*] = slewing bucket-ladder excavator
schwenkbar gelagert = swivel mounted
schwenkbare Räder *npl*, Schwenk-Radsatz *m* = swivel wheels [*belt conveyor*]
schwenkbarer Oberteil *m* → Aufsatzteil *m, n*
~ Oberwagen *m* → Aufsatzteil *m, n*
Schwenk|Baukran *m* → Bau-Drehkran
~baum *m* = derrick boom
~bereich *m* = Arbeitsbereich
~block *m* [*Hydraulikleitung*] = swivel joint
~bremse *f* = slewing brake

Schwenkbügel — schwer siebbares Gut

~**bügel** *m* = swinging clevis
~**bühne** *f* = swinging platform
~**-Dampframme** *f* → Dampf-Drehramme
~**-Elektromotor** *m* → Elektro-Schwenkmotor
~**geschwindigkeit** *f* = slewing rate
~**getriebe** *n*, Schwenkmechanismus *m*, Schwenkwerk *n* = slewing mechanism, ~ gear, swing mechanism (group)
~**halbmesser** *m* → Arbeitsbereich *m*
~**hebel** *m* [*Bagger*] = swing lever
~**kabelkran** *m*, kreisfahrbarer Kabelkran mit einem ortsfesten und einem radial verfahrbaren Turm, radial verfahrbarer Kabelkran = tautline cableway with one stationary tower and one radially travel(l)ing tower, radial travel(l)ing cableway
~**keil** *m* [*beim Servo-Stempel*] = swinging wedge
~**kran** *m*, Drehkran, Ausleger~ = revolving crane, revolver ~, slewing ~; swing-jib ~ (Brit.); swing (-boom) ~ (US)
~**kran-Schnellbauaufzug** *m*, Pfostenschwenkkran *m* = crane hoist
~**kranz** *m*, Drehkranz = slewing ring
~**kübel** *m* → Drehkübel
~**kupp(e)lung** *f* = swing clutch
~**kupp(e)lung** *f* [*bei der Deichsel*] = swivel hitch [*on pole*]
~**lader** *m*, Schwenkschaufler *m*, Fahrlader-Schwenkschaufler, Schwenkschaufel-(Fahr)lader, Front-Schwenkschaufler, Schwenk-Frontschaufel-Fahrlader, Schwenk-Schürflader = swing loader, ~ shovel
~**lager** *n* = swivel bearing
~**prisma** *n* = swivelling V-block
~**punkt** *m* = pivot point
~**rad** *n* = tumbler gear
~**radius** *m* → Arbeitsbereich *m*
~**-Radsatz** *m*, schwenkbare Räder *npl* = swivel wheels [*belt conveyor*]
~**ramme** *f*, Drehramme = slewing and raking pile driving plant

~**regner** *m*, Drehregner = rotary sprinkler
~**rohr** *n* = swing pipe
~**rolle** *f*, drehbare Laufrolle = caster (wheel), castor (~)
~**rutsche** *f*, Schwenkschurre *f* = swivelling chute
~**säule** *f*, Drehsäule [*Anbaubagger und -lader*] = slew post
~**schaufel-(Fahr)lader** *m* → Schwenklader
~**schaufler** *m* → Schwenklader
~**schild** *m*, Seitenräumer-Planierschild, winkelbarer Schild, Schrägschild, Winkel-Planierschild, Schwenk-Vorbauschild = angledozer, angling blade
~**-Schnecken(an)trieb** *m* = swing worm drive
~**schrappen** *n* = swing-scraping
~**-Schürflader** *m* → Schwenklader
~**schurre** *f*, Schwenkrutsche *f* = swivelling chute
~**sitz** *m*, Drehsitz = swivel seat
~**teil** *m, n* → Aufsatzteil
~**tisch** *m* = swivel table
~**trieb** *m*, Drehtrieb [*Bagger; Kran*] = slewing drive, rotating ~
~**umformer** *m* [*Baggerhydraulik*] = slewing service
~**umformerleitung** *f* [*Baggerhydraulik*] = slewing service line
~**ventil** *n* [*Bagger; Kran*] = swing valve
~**-Vorbauschild** *m* → Schwenkschild
~**vorrichtung** *f*, Schwenkwerk *n* [*Drehkran; Drehbagger*] = slewing unit
~**werk** *n*, Schwenkvorrichtung *f* [*Drehbagger; Drehkran*] = slewing unit
~**winkel** *m* → Drehwinkel
~**winkel** *m* [*Motor-Straßenhobel*] = circle reverse
~**zylinder** *m*, Drehzylinder [*Bagger; Kran*] = slewing ram, rotating ~
schwer [*Maschine*] = heavy-duty
~ **siebbares Gut** *n*, siebschwieriges ~ = hard-to-screen material, difficult-to-screen ~

Schwere *f*, Schwerkraft *f*, Gravitation *f* = gravity, gravitational force, force of gravitation

schwere Fähre *f* **mit Turbinenantrieb** = heavy ferry with turbine propulsion

~ Rammung *f* = driving into hard ground

~ (Straßen)Decke *f*, Dauerdecke, Dauerbelag *m*, schwerer (Straßen-)Belag = high-type pavement, heavy-duty ~, ~ surfacing; ~ paving (US)

schwerer Schlägel *m*, Treibfäustel *m* = striking hammer

~ (Schnee)Keilpflug *m* **an LKW angebaut** = heavy-duty plough (Brit.); ~ plow (US)

~ (Straßen)Belag *m* → schwere (Straßen)Decke *f*

~ Ton *m* = fat clay

~ Verkehr *m* = heavy traffic

schweres Abwalzen *n* = heavy rolling

~ Heizöl *n* = heavy fuel oil

Schwerbeton *m* → → Beton

~abschirmung *f*, Schwerbetonschild *m* = heavy-aggregate shield

~block *m*, Normalbetonblock = dense concrete block [*Dense blocks are made with "dense" aggregates, that is ordinary gravel and sand or crushed stone and cement. But, except for some special facing blocks, dense blocks, because of the way they are made, are still not as heavy as ordinary concrete*]

~decke *f*, Normalbetondecke = normal concrete floor

~element *n*, Normalbetonelement = normal concrete unit

~pfahl *m*, Normalbetonpfahl = normal concrete pile

~zuschlag(stoff) *m*, Normalbetonzuschlag(stoff) = normal aggregate

Schwerbleierz *n* (Min.) = plattnerite

schwerbrennbar = slow-burning

Schwere|-Beschleunigung g *f* → Fallbeschleunigung g

~feld *n* = gravitational field

~-Gradient *m* = gradient of gravity

~losigkeit *f* = weightlessness

~messer *m*, Schweremeßgerät *n* = gravity meter

~messung *f*, gravimetrische Messung [*geophysikalisches (Aufschluß)Verfahren*] = gravimetric survey

~mittelpunkt *m*, Massenmittelpunkt, Schwerpunkt = center of gravity (US); centre ~ ~ (Brit.); centroid

schwerentflammbar = flame-resistant

Schwere|potential *n* = gravity potential

~störung *f* = gravity disturbance

~überschuß *m* = gravity surplus

~welle *f* [*Wasserwelle, die auf großen Wasserflächen durch Wind erzeugt wird und dem Einfluß der Schwerkraft unterliegt*] = gravity wave

Schwer|fahrzeug *n* = heavy vehicle

~flüssigkeit *f*, schwere Flüssigkeit = heavy liquid, liquid of high density

~flüssigkeitsaufbereitung *f*, Sink-Schwimm-Aufbereitung = sink float process, heavy media separation

schwergefrierbares Dynamit *n*, ungefrierbares ~ = uncongealable dynamite

Schwergeräte *npl* [*Baumaschinen*] = heavy construction equipment

Schwergewicht|aufgeber *m*, Schwergewichtspeiser, Schwergewichtbeschicker = gravity feeder

~bauwerk *n* = gravity type structure

~entleerung *f* = gravity discharge

~kaimauer *f*, Schwergewichtkajenmauer = gravity quay wall

~-Kippmulde *f* = gravity tipping skip

~mauer *f* → → Talsperre *f*

~-Pendelfender *m* = swinging gravity fender

~speiser *m*, Schwergewichtaufgeber, Schwergewichtbeschicker = gravity feeder

~sperre *f* → → Talsperre

~staumauer *f* → → Talsperre

~-Stützmauer *f* = gravity retaining wall

Schwergewicht-Talsperre — Schwerzuschlag(stoff)

~-Talsperre *f* → → Talsperre
~-Wasserrad *n* = gravity (water) wheel
~wehr *n* = gravity type weir
~widerlager *n* = gravity type abutment
Schwergüter-Anlegeplatz *m* = heavy lift berth
Schwerkraft *f* = (force of) gravity
~anreicherung *f* = gravity concentration
~bahn *f* → Schwerkraftrollenbahn
~bahn *f* [*Seilbahn*] = gravity cable(way)
~-Eindicker *m* = gravity type thickener
~entladung *f*, Eigengewichtentladung = gravity discharge
~förderer *m* = gravity conveyor, ~ conveyer, unpowered ~
~heizung *f* = gravity heating
~komponente *f* = component of gravity
~mühle *f* = tumbling mill
~(rollen)bahn *f*, Schwerkraftrollenförderer *m* = gravity roller runway, ~ ~ conveyor, ~ ~ conveyer
~verteilung *f* = gravity distribution
~-Warmwasserheizung *f* = gravity warm water heating
Schwer|last *f* = heavy load
~lastanhänger *m* = heavy-duty trailer
~lastausleger *m* = boom for heavy loads (US); jib ~ ~ ~ (Brit.)
~lastchassis *n* = heavy-duty chassis
~lastfahrzeug *n* = heavy-duty vehicle
~lastkipper *m* [*Lkw*] = heavy-duty tipper
~last(kraft)wagen *m*, Schwerlaster *m* = heavy lorry (Brit.); ~ truck (US)
~lastkran *m* = heavy-duty crane
~lastpfahl *m* = heavy-load pile
~lastschlepper *m* = heavy-duty tractor
~lasttransporter *m* = heavy equipment transporter, HET
~materialschaufel *f* = heavy material bucket

~punkt *m*, Massenmittelpunkt, Schweremittelpunkt = centroid; center of gravity (US); centre of gravity (Brit.)
~spat *m*, Baryt *m*, $BaSO_4$ (Min.) = heavy spar, barytes, native barium sulphate, barite
~spatbrekzie *f* = barite breccia, barytes ~
~spatmehl *n*, fein gemahlener Baryt *m* = finely ground barite, ~ ~ barytes
~spatzusatz *m* = addition of barytes, ~ ~ barite
~spülung *f* [*Tiefbohrtechnik*] = high weight mud
~stange *f*, Bohrkragen *m* = drill collar
~stange *f*, untere ~, Bohrstange [*pennsylv. Bohrverfahren*] = drill stem, auger ~, sinker bar
~stange *f* in glatter Ausführung = plain (type) drill collar
~stange *f* mit seichter Eindrehung am oberen Ende zum Abfangen durch Abfangkeile = drill collar with slight recess for application of retaining slip
~stangenführung *f*, Bohrkragenführung = (drill collar) stabilizer
Schwerst|beton *m* → → Beton
~bohrhammer *m*, Hammerbohrmaschine *f*, überschwerer Bohrhammer = rock drill drifter
~einsatzplatte *f* [*Gleiskettenlaufwerk*] = extreme service shoe
~fahrzeug *n* = super-heavy vehicle
~-Verdichter *m* → → Walze *f*
Schwert *n* [*Schwertwäsche*] = paddle, blade [*log washer*]
~ [*Ramme*] = control-bar
Schwertantalerz *n* (Min.) = tantalite
Schwertauflöser *m*, Schwert(er)wäsche *f* = log washer
Schwert(er)welle *f* = log
Schwer|wasserreaktor *m* = heavy-water reactor, ~ pile
~welle *f* = (ordinary) wave
~zuschlag(stoff) *m*, Beton-~ = heavy-weight (concrete) aggregate

Schwestern(wohn)heim *n*, Schwesternhaus *n* = nurses' hostel, ∼ home
Schwibbe *f* → Sturmlatte *f*
Schwibbogen *m*, Strebebogen = flying buttress, arched ∼
Schwimm|achse *f* [*Hydraulik*] = axis of flotation
∼aufbereitung *f*, Flotation *f* = flotation
Schwimmbad *n* = swimming bath
∼, Schwimmbecken *n* = swimming pool
∼filter *m*, *n* = swimming pool filter
∼-Forschungsreaktor *m*, MERLIN = medium energy research light water moderated industrial nuclear reactor
∼gebäude *n* = (swimming) bath building
Schwimm|bagger *m*, Naßbagger = dredger (Brit.); dredge (US); dredging craft
∼band *n* = floating belt conveyor, ∼ ∼ conveyer
∼bauwerk *n* [*z. B. beim Schwimmdock*] = floating structure
∼becken *n* = swimming pool
∼beckenhalle *f* = swimming-pool hall
∼-Betonanlage *f* = floating concrete plant, barge-mounted ∼ ∼
∼brücke *f* = floating bridge
∼decke *f*, Schwimmschicht *f*, Schwimmschlamm *m* = scum [*A mass of sewage solids, buoyed up by entrained gas, grease, or other substance, which floats at the surface of sewage*]
∼derrick(kran) *m* = floating derrick [*A movable derrick erected on a special boat, barge, or vessel*]
∼dock *n* = floating (dry) dock
∼ebene *f* = floating plane
Schwimmen *n* [*Pigment*] = flooding
schwimmend einbauen mit Rückschlagventil [*Rohrfahrt in der Tiefbohrtechnik*] = to float-in a string of casing
∼ gelagerter Kolbenbolzen *m* = full floating gudgeon pin

∼ verlegte (Rohr)Leitung *f* = floating (pipe)line
schwimmende Betonfabrik *f* = floating concrete factory
∼ Betonieranlage *f* **auf Ponton**, Ponton-Betonieranlage = pontoon-mounted concreting plant
∼ Deckscholle *f* (Geol.) = floating fault block
∼ Insel *f* = exotic block [*overthrust*]
∼ Pfahlgründung *f* → schwebende ∼
∼ Pflanzeninsel *f* = raft of vegetation, river raft
schwimmender Antransport *m*, Schleppreise *f* = tow
∼ Estrich *m* = floating floor(ing), ∼ floor finish, ∼ screed, ∼ inner floor(ing)
∼ Kippentlader *m* (für den Wasserbau), Steinschütter [*Ein Schiff, das auf Deck Kippmulden hat, die das Material über an den Kippmuldenöffnungen rotierende Verteilerwalzen abkippen, wodurch eine gleichmäßige Lagensschüttung bei in Bewegung befindlichem Schiff erreicht wird*] = water-borne dumper
∼ Pfahl *m*, schwebender ∼ = floating pile
schwimmendes Eisfeld *n* = (sea) floe
∼ Gebirge *n* → → Tunnelbau
∼ Gleitlager *n* = floating bearing
∼ Moor *n* = floating moor
∼ Seezeichen *n* = floating marine navigation aid
Schwimmer *m* = float
∼absaugung *f* = floating suction
∼armstück *n* = float arm
∼becken *n* [*Badeanstalt*] = swimming and diving pool
∼hebewerk *n*, Schwimmerschiffhebewerk = mechanical float barge lift
∼messung *f* = float gaging (US); ∼ gauging (Brit.) [*Measurement of the discharge of water by floats to determine velocities*]
∼nadel *f* = float needle
∼nadelventil *n* = float needle valve
∼pegel *m* = float gage (US); ∼ gauge (Brit.)

Schwimmerpyrometer — Schwind(ungs)messer

~pyrometer m, n = floating pyrometer
~schacht m [*Schwimmer(schiff)hebewerk*] = float shaft
~schalter m = float switch
~(schiff)hebewerk n = mechanical float barge lift
~schleuse f = floating lock
~stand m = float level
~ventil n = float valve
Schwimm|gerät n = floating rig, barge-mounted ~ [*e.g. a floating crane, a floating pile driver and so forth*]
~greifer m mit Zweischalen-Greif(er)-korb, Greif(er)-Naßbagger m, Greif(er)-Schwimmbagger = clamshell (bucket) dredger (Brit.); ~ (~) dredge (US)
~greiferanlage f [*zur Sand- und Kiesgewinnung*] = floating grab(bing) installation, ~ ~ plant
Schwimmittel n [*Flotation*] = reagent
Schwimm|-Kahnentladebecherwerk n = floating barge elevator, marine leg
~kasten m = box caisson [*open at the top and closed at the bottom*]
~kastengründung f, Schwimmkastenfundation f = box caisson foundation
~körper m = floating body, waterborne ~
~kraft f, Auftrieb m = buoyant effect, buoyancy
~kran m = floating crane
~kruste f [*Erdkruste*] = flotation crust
~lage f = floating position
~lager n → hydrostatisches Lager
~-Lagerhaus n = floating warehouse [*Is used to increase docking facilities in crowded harbo(u)rs*]
~löffelbagger m → Naßlöffelbagger
~pegel m = float-operated ga(u)ge
~(pfahl)ramme f = floating (pile) driving plant, ~ piling ~
~pfeilergründung f [*Pfeilergründung mittels Schwimmkasten*] = pier foundation by box caisson
~poller m = floating bollard

~probe f zur Ermittlung der Viskosität f nicht zu harter Bitumen nach A. S. T. M. D. 139-27 = float test
~-Pumpanlage f = floating pump station
~ramme f → Schwimmpfahlramme
~sand m, Quicksand, Fließsand = quick sand
~saugbagger m → Saugbagger
~schicht f → Schwimmdecke f
~schlamm m → Schwimmdecke f
~schuh m, Rohrschuh mit Rückschlagventil [*Tiefbohrtechnik*] = float shoe
~sinkscheider m = sink-float separator
~-Sink-Verfahren n = sink-float separation, ~ process
~stellung f [*Schaufel beim Schürflader*] = "float" position
~stoff m [*Abwasserwesen*] = scum
~stoffablenker m → Tauchbrett n
~stoffabstreicher m = scum collector
~stoffabweiser m → Tauchbrett n
~stoffe *mpl*, Treibzeug n [*Feststoffe, meist organischer Art, die leichter als Wasser sind und daher auf ihm schweben*] = floating debris, ~ trash
~tor n = ship caisson
~tunnel m = floating tunnel
Schwinden n, Schwindung f = shrinkage, shrinking
schwindfrei = nonshrink
Schwind|fuge f; Kontraktionsfuge [*Schweiz*] = shrinkage joint
~geschwindigkeit f = shrinkage rate
~grenze f = shrinkage limit
~kraft f = shrinkage force
~kriechen n = shrinkage creep
~maß n = shrinkage value
~neigung f = tendency to shrink
~riß m = shrinkage crack
~rißbildung f = contraction cracking
~schwankung f = shrinkage variation
~spannung f = shrinkage stress
Schwindung f [*Beton*] = curing shrinkage
Schwind(ungs)messer m = shrinkage meter, ~ measuring device

Schwind|unterschied m = differential shrinkage
~versuch m, Schwindprobe f, Schwindprüfung f = shrinkage test
~wirkung f = shrinkage action
~zugspannung f = tensile shrinkage stress
~zugspannungsriß m = tensile shrinkage stress crack
Schwing|achse f = oscillating axle
~aufgeber m, Schwingbeschicker, Schwingspeiser = vibratory feeder
~beschicker m, Schwingaufgeber, Schwingspeiser = vibratory feeder
~boden m → Schwingfußboden
~bohle f → Rüttelbohle
~bohrer m = vibration drill, vibratory ~
Schwingenlager n [*für Hoch- und Tiefbauten*] = rocker bar bearing
Schwingentwässerer m = vibrating dehydrator, vibratory ~
Schwingflügel m = centre-hung sash
~fenster n = centre-hung sash window
Schwing|förderer m → Vibrationsfördergerät
~fördergerät n → Vibrationsfördergerät
(~)Förderrinne f, (Schwing)Rinnenförderer m, Vibrationsrinne, Vibrierrinne, Schwingrinne = vibrating trough conveyor, ~ ~ conveyer
(~)Förderrohr n, (Schwing)Rohrförderer m, Vibrationsrohr = vibrating circular pipe-line, vibratory ~ ~
~(fuß)boden m = spring flooring, ~ floor finish
~gerät n, (Druck)Rüttler m [*Rütteldruckverfahren*] = vibroflot (machine)
~gleisstopfer m, Schwingschotterstopfer = vibrating type tie tamper (US); ~ ~ track ~ (Brit.)
~hebel m [*Ventil*] = rocker
~mahlen n = vibration milling, ~ grinding
~moor n, Staumoor = trembling moor
~mühle f, Vibromühle = vibrating grinding mill, vibratory ~ ~

~nadel f → Vibrationsnadel
~pfosten m [*für Erdölpumpenzuggestänge*] = swing post
~platte f → Rüttelplatte
~rahmen m [*Sieb*] = vibrating frame
~rohrförderer m → Schwingförderrohr
~saiten-Dehnungsmesser m, Schwingdrahtdehnungsmesser = vibrating wire-strain ga(u)ge, vibratory ~ ~
~sieb n = oscillating screen
~sitz m = vibration-free seat
~speiser m, Schwingaufgeber, Schwingbeschicker = vibratory feeder
~tisch m = shaking table [*A concentrating device*]
~tisch m, Rütteltisch, Tischrüttler m [*Betonverdichtung*] = vibrating table, vibratory ~, table vibrator
Schwingung f = vibration
~ infolge Wind = wind vibration
Schwingungen fpl [*von Hochspannungsleitungen*] = galloping
Schwingungs|becken n → Talsperre
~berechnung f = vibration calculation
~bohle f → Rüttelbohle
~dämpfer m, Vibrationsdämpfer = vibration absorber, ~ dampener
~dämpfung f = vibration isolation
~eigenschaft f = vibrational property
~erreger m = oscillator, exciter (or exciter) (of oscillations), vibrator unit
~festigkeit f = vibration strength
schwingungs|freie Aufstellung f [*einer Maschine*] = isolation mounting
~freier Griff m = anti-vibration handle
Schwingungs|isolierung f = vibration insulation
~messung f = vibration measurement
~nadel f → Vibrationsnadel
~platte f → Rüttelplatte
~stoß m = vibratory impact
~untersuchung f = vibration investigation
~verdichter m → Vibrationsverdichter

Schwingungsverdichtung — Sedimentierversuch 872

~verdichtung *f* = compaction by vibration
~walze *f* → → Walze
~weite *f*, Vibrationsamplitude *f*, Ausschlag(weite) *m*, *(f)*, Amplitude *f* = amplitude of vibration
~welle *f*, oszillierende Welle, oszillatorische Welle = wave of oscillation
~zahl *f*, Frequenz *f* = frequency
Schwing|verdichter *m* → Vibrationsverdichter
~versuch *m*, Schwingprobe *f*, Schwingprüfung *f* = vibrating test
~walz *f* → → Walze
~wäscher *m*, Schwingwaschmaschine *f* = shaking-screen washer, ~ washing machine
Schwitzen *n*, Bluten, Ausschwitzen [*Schwarzdecke*] = bleeding, fatting (-up), ponding
Schwitzwasser *n* → Kondenswasser
~falz *m* = condensation groove
~korrosion *f* = formation of water by condensation in corrosive services, corrosion by condensation
~rinne *f* = condensation gutter, ~ sinking, ~ channel
~verhütung *f* = anti-condensation
Schwundriß *m* → → Riß
Schwung *m*, Wucht *f* = momentum
~, Bagger-~ [*Naßbagger*] = digging swing
~gewicht *n* → Unbalance *f*
~rad-Speicherlok(omotive) *f* = electrogyro loco(motive)
~scheibe *f*, Schwungrad *n* = flywheel
~scheibengehäuse *n* = flywheel housing
sechseckiges Drahtgeflecht *n* = hexagon(al) wire netting, ~ ~ mesh
Sechseck|(pflaster)stein *m*, Sechskantblock *m* = hexagon(al) (paving) sett
~platte *f* = hexagon(al) tile
sechsfach geschakte Eimerkette *f*, 6-teilige ~ = bucket chain line with a bucket fitted every sixth link [*bucket ladder excavator*]
Sechs|-Füllungstür *f* = six-panel door
~ganggetriebe *n* = six-speed gear

Sechskant|block *m*, Sechseck(pflaster)stein *m* = hexagon(al) (paving) sett
~kopf *m* = hexagon(al) head
~messing *n* = hexagon(al) brass
~-Mitnehmerstange *f* = hexagon(al) kelly
~mutter *f*, Sechseckmutter = hexagon(al) nut
~profil *n* = hexagon(al) section
~querschnitt *m* = hexagon(al) section
~schornstein *m* = hexagon(al) chimney, ~ stack
~schraube *f* = hexagon(al) bolt
~stahl *m* = hexagon(al) steel
~stange *f*, Sechskantstab *m* = hexagon(al) bar
~-Vollstab *m*, Sechskant-Vollstange *f* = hexagon(al) solid bar
Sechs-Minuten-Zählverfahren *n* [*Straßenverkehrstechnik*] = six-minute sample count
Sechsrad|antrieb *m* = six-wheel drive
~-Fahrzeug *n* = 6-wheeler
~-Lenkung *f*, Allrad-Lenkung = six-wheel steer(ing)
Sechs|-Uhr-Schweißen *n*, Schweißen in Normallage = six o'clock-welding
~-Zylinder-Motor *m* = six-cylinder engine
Sediment *n*, Ablagerung *f* (Geol.) = sediment
Sedimentärgestein *n* → Sedimentgestein
Sedimentation *f* → Ablagerung *f*
Sedimentations|probe *f* → Absetzversuch *m*
~prüfung *f* → Absetzversuch *m*
~versuch *m* → Absetzversuch
Sediment|geochemie *f* = sedimentary geochemistry
~gestein *n*, Ablagerungsgestein, Absetzgestein, Absatzgestein, Bodensatzgestein, Sedimentärgestein = sedimentary rock, bedded ~
~gneis *m* → Paragneis
~hydraulik *f* = sediment hydraulics
Sedimentier|probe *f* → Absetzversuch *m*
~prüfung *f* → Absetzversuch *m*
~versuch *m* → Absetzversuch

sedimentierter Staub — Seewasser

sedimentierter Staub *m* = settled dust
Sedimenttransport *m* = sediment-transport
See *f*, **Meer** *n* = sea
See *m*, **Binnen~** = lake
~asphalt *m* = lake asphalt, ~ pitch [*In this type of naturally occurring asphalt the mineral matter is finely divided and dispersed trough the (asphaltic) bitumen (Brit.)/asphalt (US) which is the major component*]
~bagger *m* = marine dredge (US); ~ dredger (Brit.); sea-going ~
~baggerung *f* = marine dredging
~bahnhof *m*, Hafenbahnhof = marine terminal, sea ~
~bau *m* = marine construction, coastal engineering
~bauingenieur *m* = coastal engineer
~bauten *f* = marine (or off-shore-works (or structures, or installations)
~bauwerk *n*, Meer(es)bauwerk = offshore structure, marine ~
~beben *n* = submarine earthquake
~becken *n* = lake basin [*The basin filled by the water of a lake*]
~damm *m* → Deich *m*
~dammbau *m* → Deichbau
~deich *m* → Deich
~deichbau *m* → Deichbau
~-Erz *n* = lake ore
~funkfeuer *n* = marine radio beacon
seegehend = ocean-going
See|grund *m* → Meer(es)grund
~hafen *m* = sea port
~höhe *f* → Meer(es)höhe
~höhe *f* **des Ansatzpunktes einer Bohrung**, Mundlochhöhe einer Bohrung = elevation of well head
~kabel *n* = (sub)marine cable
~kanal *m* = ship-canal
~kanaltrasse *f* = ship-canal route
~karte *f*, nautische Karte = (marine) chart, sea ~
~kartenaufnahme *f* = hydrographic survey
~kliff *n*, Küstenkliff = coastal cliff
~kreide *f* = lake-marl, bog lime

Seele *f*, **Kern** *m*, **Einlage** *f* [*Drahtseil*] = core
Seeluft *f*, Meer(es)luft = sea air
Seenablagerung *f* = lake deposit, lacustrine ~
Seenavigation *f* = marine navigation
Seen|boden *m* = lacustrine soil
~gebiet *n* = lake area
~gürtel *m* = lake belt
~kette *f* = chain of lakes
~kunde *f* → Limnologie *f*
~wasser *n* = lake water
~wasserfassung *f* = lake intake
See|pumpwerk *n* = lake pumping station
~raum *m*, bestrichene Windbahn *f* = fetch
~retention *f*, Seerückhalt *m* [*Die ausgleichende Wirkung auf den Abfluß durch Seen oder seeartige Erweiterungen eines Gewässerbettes, auch Zurückhaltung von Wasser in einem Staubecken und dgl.*] = lake storage
~rückhalt *m* → Seeretention
~sand *m*, Meer(es)sand = sea sand
~schiff *n* = ocean-going vessel
~schiffahrt *f* = ocean-going shipping
~schlepper *m* = sea tug(boat), ~ towboat [*Sea tugs are large tugs, fitted out for sea journeys of several days*]
~schleuse *f* = sea lock
~schlick *m*, Meer(es)schlick = sea mud
~schlick *m* [*beim Binnensee*] = lake-bottom mud
~schutzbauten *f* = protection works; sea defence ~ (Brit.); sea defense ~ (US)
~seismik *f* = sea seismic method
seeseismische Prospektion *f* **vom Flugzeug aus** = overwater search
See|ton *m* = lacustrine clay
~torf *m*, Meer(es)torf = sea peat
~tüchtigkeit *f* = seaworthiness
~ufer *n* = lake bank
seewärts = meeresseitig
~, seeseitig = lakeward(s)
Seewasser *n*, Binnen~ = lake water
~, Meer(es)wasser = seawater

Seewasserbeton — Seilbohrung

~beton *m*, Meer(es)wasserbeton = seawater concrete
~-Einlaufstollen *m* = sea water inlet tunnel
Seewasserspiegel *m* → Meeresspiegel
~ = lake water level
See|wind *m*, auflandiger Wind = onshore wind
~zeichen *n* = marine navigation aid
~zollgebiet *n* = customs water
Segeltuch *n* = canvas
~schlauch *m* = canvas duct
Segerkegel *m*, SK = Seger cone
Segment|bogen *m*, Stichbogen = segmental arch
~form *f* = segment mo(u)ld
~krümmer *m* = segmental (pipe) bend
~strahl *m* = segmental jet
~teil *m, n*, Mischer~ = (mixer) liner segment, sectional liner
~träger *m*, Bogensehnenträger = segmental girder, polygonal bowstring ~
~wehr *n* = Tainter gate, ~ weir
~zellenfang(e)damm *m* = segmental type cellular cofferdam
Sehenswürdigkeit *f* = tourist attraction, sight
Sehlinie *f* = sight line
Sehne *f* [*Raumlehre*] = chord
Seiche *f* = seiche [*In an inlet or harbour where the topographical features are favourable, ground swell waves are often reflected standing waves or resonant oscillations*]
seicht, untief, geringe Tiefe = shallow
Seichtwasserzone *f* = shallow-water zone
Seiden|gaze *f* = silk gauze, fine mesh silk
~glimmer *m* (Min.) → Serizit *m*
Seife *f* (Geol.) = placer
seif(en)fest = soapproof
Seifen|gold *n*, Alluvialgold, Waschgold = placer gold, alluvial ~, gulch ~
~haut *f* = soap bubble, membrane
~hautgleichnis *n*, Membran(e)gleichnis [*Baustatik*] = membrane analogy, soap bubble ~

~-Kohlenwasserstoff-Gel *n* = soaphydrocarbon gel
~mineral *n* = placer mineral
~mulde *f* = soap dish
~pulver *n* = powdered soap
~sand *m* = placer sand
~stein *m* = soapstone
~zinn *n* (Min.) = stream tin
seiffest, seifenfest = soapproof
seiger (⚒) = vertical, upright
Seiher *m*, Saugkorb *m*, Filtersieb *n* = strainer, filtering screen
Seihtuch *n*, Filtertuch = filter(ing) cloth, ~ fabric
Seil *n*, Kabel *n* = rope, cable
seilabgespannter Derrick(kran) *m*, Seilderrick(kran) = guy(ed) derrick
Seil|anschlag *m* [*Schlaffseil-Kabelbagger*] = automatic dump
~(an)trieb *m* = cable drive, rope ~
~aufwickler *m*, Seilführung *f* = fairlead, cable guide, rope guide
~bagger *m* = cable-operated excavator
Seilbahn *f* → Seilschwebebahn
~ → Seil-Standbahn
~-Schleppschrapper *m* → Schlaffseil-Kabelschrapper
~-Schwebeschrapper *m*, Schlaffseil-Kabelbagger = slackline (cableway) (excavator) with dragline-type bucket
~seil *n* = ropeway cable
seilbetätigt = cable-operated, cable-powered, cable-controlled
Seil|bohranlage *f* = cable tool drilling equipment
~bohren *n*, Seilschlagbohren = well drilling, churn ~, cable ~
~bohren *n* mit Schwengelantrieb, Schwengelantrieb-Seilbohren, pennsylvanisches Bohren, pennsylvanisches Bohrverfahren *n*, pennsylvanisches Seilbohren = Pennsylvanian drilling method, ~ ~ system
~bohrgerät *n*, Seilschlagbohrgerät = well drill, cable ~, churn ~
~bohrung *f* = cable tool well, ~ ~ hole

~bohrverfahren *n* = cable tool (method of) drilling, American system of drilling, cable system
~bohrwinde *f* = bullwheel
~brücke *f* → Seilhängebrücke
~derrick(kran) *m*, seilabgespannter Derrick(kran) = guy(ed) derrick
~draht *m*, Litzendraht = strand(ing) wire
~durchhang *m*, Kabeldurchhang = cable sag, rope ~
~durchmesser *m* = rope diameter, cable ~
~eck *n*, Seilpolygon *n* = funicular polygon, string ~
~eckgleichung *f* → Seilpolygongleichung
~fähre *f* = cable ferry, rope ~
~fänger *m* [*Tiefbohrtechnik*] = rope grab, cable ~
~fanggabel *f* [*Tiefbohrtechnik*] = devil's pitch fork
~fassungsvermögen *n* der Trommel = drum capacity
~fett *n* = cable grease, rope ~
(~)Flasche *f*, (Seil)Kloben *m*, Block *m*, Unterflasche = (rope) block, fall
~förderanlage *f* = cable haulage machine, rope ~ ~
~förderung *f*, Seiltransport *m* = cable haulage, rope ~
~führung *f* → Seilaufwickler *m*
~führungsrolle *f* = fairlead sheave
~führungsrollenverkleidung *f* = fairlead sheave shroud
~fußstück *n*, Seilhülse *f* [*pennsylvanisches Bohrverfahren*] = rope-socket, cable-socket
~geschwindigkeit *f* = cable speed, rope ~, line ~
~hängebahn *f* → Seilschwebebahn
~(hänge)brücke *f*, Kabel(hänge)brücke = rope suspension bridge, cable ~
~hängedach *n*, Kabelhängedach = cable(-suspended) roof, rope(-suspended) ~
~hängeflachdach *n*, Kabelhängeflachdach = cable-suspended flat roof, rope-suspended ~ ~

~haspel *m, f*. Wickeltrommel *f* = rope reel, cable ~
~hebewerk *n* → lotrechter Aufzug *m*
~hülse *f*, Seilfußstück *n* [*pennsylvanisches Bohrverfahren*] = rope-socket, cable-socket
~kappvorrichtung *f* = cable cutter, rope ~
~kernen *n* = wire line core drilling
~kernrohr *n* = wire line core barrel
~klemme *f* = cable clamp, rope ~
(~)Kloben *m*, (Seil)Flasche *f*, Unterflasche, Block *m* = (rope) block, fall
~kraftausgleich *m* [*Bergbauförderseil*] = condition of balance
~lage *f*, Kabellage = layer of strands, strand layer
~laufkatze *f* = rope-operated crab, cable-operated ~
~linie *f* → Seilstrahl
~litze *f* = (rope) strand, cable ~
~machart *f* = rope construction, cable ~
~messer *n* = rope knife, cable ~
~nut *f*, Seilführungsnut = cable groove, rope ~
~polygon *n*, Seileck *n* = funicular polygon, string ~
~polygongleichung *f*, Seileckgleichung = funicular polygon equation, string ~ ~
~prüfungen *fpl* = examination of ropes, ~ ~ cables
~rille *f* = rope groove, cable ~
~ring *m*, Augenring = grommet, rope eyelet, cable eyelet
~rolle *f* [*eine kleine Seilscheibe*] = small sheave, ~ pulley
~rollenblock *m* = small sheave nest, ~ ~ block, ~ pulley ~
~rollenlager *n* = small sheave bearing, ~ pulley ~
~rollenträger *m* [*Radschrapper*] = small sheave carrier, ~ pulley ~
~rollenverkleidung *f* [*Radschrapper*] = small sheave nest guard, ~ pulley ~ ~
~rutsch *m*, Seilschlupf *m* = rope slip(page), cable ~

Seilscheibe — Seil(schwebe)bahn

~scheibe f [kreisrunde Scheibe, um deren Umfang ein Seil gelegt ist. Dient zur Seilführung, Umlenkung, oder Kraftübertragung durch Reibung] = sheave, pulley
~schlag m = rope lay
~(schlag)bohrgerät n, Seilschlaggerät = churn drill, cable ~, well ~
~schleppschiffahrt f = cable towing traction, rope ~ ~
~schloß n, Schürfseiljoch n = drag line yoke

~schlupf m, Seilrutsch m = rope slip(page), cable ~
~schmiermittel n = rope lubricant, cable ~
Seilschrapper m, Kabelschrapper = cable excavator, ~ scraper
~kasten m, Kabelschrapperkasten = cable-hauled bucket, scraper
Seilschutzverkleidung f = cable guard, rope ~

Seil(schwebe)bahn f, Luftseilbahn	aerial ropeway (Brit.); ~ tramway (US)
Abspannseil n	guy rope, guy cable
Augenspleiß m	eye splice
Doppel~, Zwei~ [sogenannte deutsche Bauart]	double ~
Einmann-Sessel m	single chair
Ein~ [sogenannte englische Bauart]	single ~
Einspurbahn f	single track
Fahrwerk n, Radsatz m	bogie
Fahrwerk n, Wagen m	carriage
Führerhaus n	operator's cab(in)
Gegenwagen m, Gegenfahrwerk n	counter-carriage
Knotenseil n	knotted rope, knotted cable
Maschinenhaus n	engine house
Material-~	material ~
Personen-~	passenger ~
Radsatz m, Fahrwerk n	bogie
Schmierwagen m	oiling car
Schutznetz n	protection net
Seilablenker m	rope saddle, cable saddle
Seilklemmapparat m	rope grip, cable grip
Seilklemme f	rope clip, cable clip
Sessellift m	chair-lift
Spannscheibe f	tension sheave, stretching ~
Spannstation f	tension station, stretching ~
Spannwagen m	stretching car, tension ~
Stütze f [bei großen Seilbahnen]	head mast
Stütze f [bei kleinen Seilbahnen]	line trestle
Tragseil n [Leerseite]	empty side carrying rope (or cable)
Tragseil n [Vollseite]	full side carrying rope (or cable)
Umlenkscheibe f	return sheave
Vier-Rollenfahrwerk n	under-type four-wheel carriage
Vorrichtung f zum Einziehen des Seiles mit Feder	spring block

Seilspeer — seismischer Sprengstoff

Wagen *m*, Fahrwerk *n* — carriage
Zugseil *n* — haulage rope, haulage cable
Zweimann-Sessel *m* — double chair
Zweiradlaufwerk *n* für Oberseil — overhead type two-wheel carriage
Zweispurstrecke *f* — double track

Seil|speer *m* [*Seil-Fangwerkzeug*] = rope spear [*fishing tool for rope*]
~**spinnen** *n* → Kabelspinnen
~-**Standbahn** *f*, (Stand)Seilbahn, Boden-Seilbahn = endless-rope haulage system, endless-cable ~ ~
~**steg** *m* → Kabelsteg
~-**Steigungswinkel** *m*, Flechtwinkel = angle of twist
~**strahl** *m*, Polstrahl, Seillinie *f*, Pollinie = funicular line, polar ~
~**strebe** *f*, Kabelstrebe [*Hängebrücke*] = cable stay, rope ~
~-**Streben-Derrick(kran)** *m* [*mit kombinierter Seil- und Strebenverspannung*] = combined guy and Scotch derrick (Brit.); ~ ~ ~ stiff leg ~ (US)
~**tragrolle** *f* [*Baudrehkran*] = top pulley
~**transport** *m*, Seilförderung *f* = cable haulage, rope ~
~**transportanlage** *f*, Seiltransporteur *m* [*Betonsteinindustrie*] = rope conveyor (take-off), ~ conveyer (~)
~**transportanlage** *f*, Seiltransporteur *m* = cableway transporter
~**trieb** *m*, Seilantrieb = rope drive, cable ~
~**trommel** *f*, Kabeltrommel = rope drum, cable ~
~**umlenkrolle** *f* = guide block
~**verankerung** *f*, Kabelverankerung [*Brückenbau*] = cable anchorage, rope ~
~**verankerung** *f* [*Radschrapper*] = cable dead end, rope ~ ~
~**verschleiß** *m* = rope wear, cable ~
~**verspannung** *f* → Abfangen *n* mit Trossen
~**vorschub** *m*, Seilvorstoß(en) *m*, (*n*) [*Hochlöffel*] = rope crowd(ing), cable ~

~**wandern** *n* = rope wander, cable ~
~**wickler** *m*, Seilaufwickler, Seilführung *f* = fairlead
~**winde** *f* → Kanalwinde
~**winde** *f* = cable winch, rope ~
~**wippen** *n* = rope luffing, cable ~
~**zug** *m* [*Bewegungsübertragung durch Seil*] = cable control, rope ~
~**zug** *m* → Flaschenzug
~**züge** *mpl* [*Eytelwein'sche Beziehung für Seilrutsch bei der Treibscheibenförderung*] = loaded side and unloaded side
~**zuggruppe** *f* [*z. B. bei einem Radschrapper, Bulldozer usw.*] = cable control unit, rope ~ ~

Seismik *f*, Seismologie *f*, Erdbebenkunde *f* = seismology
~, seismische Prospektion *f*, seismischer Aufschluß *m* = seismic exploration, ~ survey, ~ prospecting
~, seismisches Verfahren *n*, seismische Methode *f*, seismische Untersuchung *f* = seismic method
seismische Bodenforschung *f* nach dem Refraktionsverfahren → Brechungsverfahren
~ **Methode** *f* → seismisches Verfahren *n*
~ **Prospektion** *f*, seismischer Aufschluß *m*, Seismik *f* = seismic survey, ~ prospecting, ~ exploration
~ **Untersuchung** *f* → seismisches Verfahren *n*
~ **Wellengeschwindigkeit** *f* = seismic wave velocity
seismischer Aufschluß *m* → seismische Prospektion *f*
~ **Integrator** *m* = seismic magnetic integrator
~ **Sprengstoff** *m* = seismic (blasting) explosive

seismisches Verfahren *n*, seismische Methode *f*, Seismik *f*, seismische Untersuchung *f*, Geoseismik = seismic method
Seismograph *m* = seismograph
Seismometer *n*, Geophon *n* = pick-up
Seiten|ablage(rung) *f* [*wird vorgenommen, wenn die Einschnittmassen die Auftragmassen überwiegen*] = side-casting
~**ablauf** *m* [*die Öffnung liegt in der Bordschwelle*] = side-entrance gulley
(~)**Absetz(förder)band** *n* → Absetzband
~**abstreifer** *m* [*Spachtelmaschine*] = edge scraper
~**altar** *m* = side altar
~**ansicht** *f*, Seitenaufriß *m* = side elevation
~**antrieb** *m* [*Gleiskettenlaufwerk*] = final drive
~**antriebsrad** *n* [*Gleiskettenlaufwerk*] = final drive gear
~**antriebsschutz** *m* [*Gleiskettenlaufwerk*] = final drive guard
~**antriebs-Verschleißschutz** *m* [*Gleiskettenlaufwerk*] = final drive wear guard
~**arm** *m* [*zwischen Pfeiler und Verankerung am Ufer*] = anchor arm [*of a normal cantilever bridge*]
~**arm** *m* [*Radschrapper*] = draft tube side arm
~**aufriß** *m*, Seitenansicht *f* = side elevation
~**auslaß** *m* = side outlet
~**ausleger** *m*, seitlicher Ausleger, seitlicher Schwenkarm *m* [*am Schlepper angebaut*] = sideboom
~**ausleger-Zusatzvorrichtung** *f* **für Rohrverlegung** = sideboom pipelaying attachment
~**bau** *m*, Seitengebäude *n* = side building
~**beschleunigung** *f* = lateral acceleration
~**bogen** *m* [*Brücke*] = side arch (span)
~**böschung** *f* = side slope
~**dehnung** *f* = lateral strain

~**dichtung** *f* [*Wehr*] = side seal
~**drän** *m* = side drain, lateral ~
~**druck** *m* = lateral pressure
~**druck** *m* **auf Pfähle** = lateral earth pressure against piles
~**eingang** *m* = side entrance
~**einschnitt** *m* = side cut(ting) [*deprecated: side long cut*]
seitenentleerender Erdtransportwagen *m* → Anhänge-Seitenentleerer *m*
Seiten|entleerer *m* (⚒) = side dump mine car
~**entleerung** *f* = side dump discharge
~**entleerungsanhänger** *m* = side dump trailer
~**entleerungsschaufel** *f* [*Frontlader*] = side dump bucket [*front-end loader*]
~**entnahme** *f* [*Gewinnung von in Trasse einer Straße befindlichem einbaufähigen Boden seitlich derselben, wenn Massenausgleich beim Bau nicht möglich ist*] = side borrow
~**feld** *n*, Seitenöffnung *f* [*Brücke*] = side span, approach ~
~**führungskraft** *f* = sideway force
~**führungskraftbeiwert** *m* → Seitenkraftbeiwert
~**(gabel)stapler** *m* = side-loading fork-lift truck, sideloader
~**gebäude** *n*, Seitenbau *m* = side building
seitengesteuerter Motor *m* = valve-in-block engine
Seiten|graben *m* = side ditch
~**hubstapler** *m* = side-lift truck
~**kanal** *m*, Lateralkanal, Nebenkanal = lateral canal
~**kapelle** *f* = side chapel
~**kippanhänger** *m* → Anhänge-Seitenentleerer
~**kippen** *n*, Seitenkipp(entleer)ung *f* = side dumping
~**kipper** *m* [*Wagen wird um Längsachse um etwa 135–150 Grad gedreht*] = side tippler
~**kipper** *m* [*Erdbaufahrzeug*] = side dump wagon, side-tipping ~
~**kipper-Anhänger** *m* = side dump trailer, side-tipping ~

~**kipper-Halbanhänger** *m*, Halbanhänger-Seitenkipper *m*, Aufl(i)eger-Seitenkipper, Seitenkipper-Aufl(i)eger = side dump semi-trailer, side-tipping ~

~**kippgerät** *n* [*Erdbau*] = side dumping equipment, side-tipping ~

~**kipp(lade)schaufel** *f*, Seitenkippkübel *m* [*Schürflader*] = side-tipping (loading) bucket, dumping (~) ~

~**kippmulde** *f* [*Erdbaufahrzeug*] = side dumping body, ~ tipping ~

~**kippschaufellader** *m* = side dumping bucket loader, ~ tipping ~ ~

~**kippung** *f* → Seitenkippen

~**kippwaggon** *m* = side dump car, ~ tipping ~

~**kippwinkel** *m* = side dumping angle, ~ tipping ~

~**klappenpritsche** *f* [*LKW*] = drop-side body

~**kolonne** *f* [*Erdölwerk*] = stripping column

~**kraft** *f*, Zweigkraft, Komponente *f* [*Baustatik*] = component

~**kraftbeiwert** *m*, Seitenführungskraftbeiwert *m* [*Der bei Übertragung einer Seitenkraft vom Fahrzeug auf die Straße auftretende Kraftschlußbeiwert (Beiwert der Seitenführungskraft)*] = sideway force coefficient

~**kraftgleichung** *f*, Komponentengleichung ~ = equation connecting the components

~**kran** *m* = side boom

~**kranraupe** *f* = side boom crawler tractor

~**last** *f* = lateral load

~**markierung** *f*, Randmarkierung [*Straße*] = lateral marking

~**messer** *n* [*Radschrapper*] = routing bit, router ~

~**moräne** *f*, Wallmoräne, Randmoräne, Ganzdecke *f* = lateral moraine

~**öffnung** *f*, Seitenfeld *n* = side span, approach ~

~**öffnung** *f* [*zwischen Pfeiler und Verankerung am Ufer*] = anchor span, side ~ [*of a normal cantilever bridge*]

~**pflug** *m* → Schild(schnee)pflug

~**raum** *m* → (befestigter) Seitenstreifen *m*

~**raum** *m* → (unbefestigter) Seitenstreifen *m*

~**räumer** *m* → Trassenschäler

~**räumer-Planierschild** *m* → → Bulldozer

~**reibung** *f* = lateral friction, side ~

~**reibung** *f*, Mantelreibung [*Pfahl*] = side friction, skin ~, lateral ~

~**rippe** *f* = nervure [*The side rib of vaulted roofs, to distinguish from diagonal ribs*]

~**schalung** *f* → Betondeckenschalung

~**schalungsfertiger** *m* = side form finisher, ~ finishing machine

~**schalungsinnenrüttler** *m*, Seitenschalungsinnenvibrator *m* = side form vibrator

~**schiff** *n* = aisle

~**schneeschleuder** *f* → Schneeschleuder

~**schneide** *f* [*am Tieflöffel*] = side wing

~**schub** *m*, achsrechter Schub, Achsialschub = axial thrust

~**schubantriebwelle** *f* = center shift drive shaft (US); centre ~ ~ ~ (Brit.)

~**schubritzel** *n* = center shift pinion (US); centre ~ ~ (Brit.)

~**schubsegment** *n* = center shift rack (US); centre ~ ~ (Brit.)

~**schubstange** *f* = center shift link (US); centre ~ ~ (US)

~**schubvorrichtung** *f* = center shift (US); centre ~ (Brit.)

~**schurf** *m*, Lateralerosion *f* (Geol.) = lateral erosion by water action

~**schüttung** *f* [*Erdbau*] = side fill(ing)

~**stabilität** *f* = lateral stability

~**stapler** *m*, Seitengabelstapler ~ = side-loading fork-lift truck, sideloader

~**steifigkeit** *f* = lateral rigidity, ~ stiffness

~**stollen** *m* → Abstiegstollen

~**straße** *f* → Nebenstraße

~**streifen** *m*, befestigter ~ = shoulder

Seitenstreifen — selbstfahrender Schürfwagen 880

~streifen *m* → unbefestigter ~
(~)Stütze *f* → Hilfsstütze
~teil *m, n* = side section, ~ part
~turm *m* [*Erdölwerk*] = stripper
~überdeckung *f* [*Dachziegel*] = side (over)lap
~wagen *m*, Beiwagen = side-car
~wand *f* = side wall
~wind *m* = side wind
~winde *f* [*Naßbagger*] = swing winch
~zulauf *m* [*Rohrteil*] = oblique-angled junction, side ~
seitlich verstellbar = provided with lateral adjustment
seitliche Ausrichtung *f* [*Gleiskettenlaufwerk*] = lateral alignment
~ Belastung *f* = thrust loading, lateral ~
~ Bohrlochabweichung *f* = lateral drift of a (bore) hole, horizontal ~ ~ ~ (~) ~
~ Zufahrt *f* [*zu einer Hauptstraße*] = marginal access
seitliches Aufhören *n* → Auskeilen
Sektionsbohrturbine *f* = sectional turbodrill
Sektor *m*, Stauwand *f*, Sektorschütz(e) *n*, (*f*) [*Sektor(stau)wehr*] = sector (Brit.); gate (US)
~platte *f*, Sektorscheibe *f* = sectorial plate
~wehr *n* = drum gate (US); sector ~
Sekundär|förderung *f* [*Erdöl*]=secondary recovery
~(förder)verfahren *n* [*Erdöl*] = secondary recovery method
~strömung *f*, Querströmung [*Hydraulik*] = secondary flow
selbstadjungiertes Randwertproblem *n* = self-adjoint boundary value problem
selbständige Fahrbahndecke *f* aus Zement-Ton-Beton → Zement-Ton-Betonstraße
selbst|ansaugen = to self-prime
~ansaugend = self-priming
~ansaugender Motor *m* = naturally aspirated engine
Selbst|auflader *m* → Becher(werk)auflader

~aufnehme-Kehrmaschine *f*, Selbstaufnehme-Fegemaschine = (road) sweeper-collector
selbstaufnehmendes Becherwerk *n* = self-loading elevator
~ Förderband *n* → Becher(werk)auflader *m*
selbst(auf)zeichnender Luftdruckmesser *m* → Barograph *m*
~ Schwimmerpegel *m*, mechanischer Schwimmerschreibpegel [*Zeichen: Ss*] = float ga(u)ge, recording ~ ~, graphic water stage register
Selbstbedienungs|laden *m*, SB-Laden = self-service shop
~-Tankstelle *f* = gasateria (US)
Selbstbohrdübel *m* = rapid core drill anchor
selbsteinstellender Regelkreis *m* = adaptive control system
Selbst|entlader *m* = self-unloader
~entleerung *f*, automatische Entleerung = auto(matic) emptying
selbst|entwässernd = self-draining
~erregend = self-exciting
~fahrbar, selbstfahrend = self-propelled, self-propelling
~fahrbare (oder selbstfahrende) Motor-Spritzmaschine *f* = self-propelled sprayer (or spraying machine) for hand and power spraying with hand lance
~fahrbarer (Auf)Lader *m*, selbstfahrbare (Auf)Lademaschine *f* = self-propelled loader, self-propelling ~, ~ loading machine
~fahrbarer Bagger *m* = self-propelled excavator, self-propelling ~
~fahrbarer Beobachtungsturm *m* = self-propelled tower, self-propelling ~

~fahrender Abwurfwagen *m* [*Vom Förderband angetrieben fährt er zwischen zwei einstellbaren Umkehrstellen hin und her und erzielt so gleichmäßige Aufschüttung des Fördergutes*] = travel(l)ing tripper
~fahrender Schürfwagen *m* → Autoschrapper *m* mit 2-Rad-Schlepper

Selbst|fahrer m [*Fahrzeug mit Eigenantrieb*] = self-propelled vehicle, self-propelling ~

~fahrer m, Motorlastkahn m, Motorfrachtkahn = self-propelling barge, self-propelled ~

~fahrer(aufzug) m = lift with automatic push-button control (Brit.); elevator ~ ~ ~ ~ (US)

~fahrgerät n [*z. B. eine Walze*] = self-propelled rig, self-propelling ~

~fahrwerk n, Straßenräder-Unterwagen m, Radwagen [*Baggermotor gleichzeitig Antriebsmotor des Fahrwerks*] = rubber-tire (or rubber-tyre) carrier (or chassis) propelled by turntable engine

~hemmung f [*Stempel*] (⚒) = self-locking [prop]

~induktion f = self-induction

Selbstkosten-Erstattungsvertrag m = cost-plus-fee contract, value-cost ~, cost-value ~

~ mit begrenzter Höhe der zuschlagberechtigten Kosten = cost-plus-fixed fee contract, cost-plus-profit ~

selbstladend = self-loading

Selbst|laufrinne f [*Hydromechanisierung*] = flume [*hydraulicking*]

~montagekran m = self-erecting crane

~muldenkipper m, Stahlkastenselbstkipper = auto(matic) side tipping wagon, ~ ~ dumping ~

selbstregelnder Generator m = self-regulating generator

Selbst|reinigung f [*Abbau zugeführter Verunreinigungen, überwiegend biologischer Art, im Vorfluter durch Zusammenwirken von Bakterien, Pflanzen und Tieren*] = self-purification

~reinigungskraft f [*Abwasser*] = recuperative power

~schlußventil n → Rohrbruchwächter

~schmierlager n = self-lubricating bearing

selbst|schwimmende Taucherglocke f = floating diving bell

~sichernde (Schrauben)Mutter f = self-locking nut

Selbstspannung f [*Beton*] = self-stressing

selbsttätig, automatisch = auto(matic)(ally)

~ spannender Beton m = self-stressing concrete, self-stressed ~

~ ~ Zement m = self-stressing cement, self-stressed ~

selbsttätiges Klappen(stau)wehr n **mit Gegengewicht** = auto(matic) counterweighted gate, auto(matic)-leaf ~ ~

~ Wehr n = auto(matic) weir

selbst|ventilierender Kühlturm m, natürlich belüfteter Kühlturm = natural draught cooling tower (Brit.); ~ draft ~ ~ (US)

~verstärkende Bremse f = self energizing brake, ~ activating ~

~zeichnender Schwimmerpegel m, Registrierpegel, Schreibpegel = recording gage, water-stage recorder

selektive Flotation f, sortenweise ~ = differential flotation

~ Korrosion f [*Die bevorzugte Korrosion bestimmter Gefügebestandteile*] = selective corrosion

Seltene-Erden-Oxyd n = rare-earth oxid(e)

Semi-Dieselverfahren n = mixed cycle (method)

Semifusit m (Geol.) = semifusain

semilogarithmisches Netz n = semilogarithmic scale

Sendeschallkopf m = transmitting transducer

Senk|bewegung f = lowering motion, ~ movement

~blei n, Bleilot n, (Senk)Lot, Senkel m = lead line, sounding ~

~bohrung f, Ansenkung = counterbore

~bremse f = lowering brake

~bremsensteuerung f = lowering brake control

~bremsschaltung f = lowering brake connections for cranes in three-phase installations

Senkbrunnen *m*, Gründungsbrunnen = drop shaft, open caisson [*open both at the top and bottom*]

~, Sickerschacht *m* [*senkrechter Schacht zum Einleiten des Dränwassers in durchlässige Schichten des tieferen Untergrundes*] = absorbing well, well drain

~ **mit Scheidewänden**, Gründungsbrunnen ~ ~, Bagger-Senkkasten *m* = open caisson with cross walls, drop shaft ~ ~ ~

~**fundament** *n* → Brunnengründung *f*

~**fundamentkörper** *m* → Brunnengründung *f*

~**fundation** *f* → Brunnengründung *f*

~**gründung** *f* → Brunnengründung *f*

~**gründungskörper** *m* → Brunnengründung *f*

Senke *f* → Gelände~

Senkel *m* → Senkblei *n*

senken, ab~ [*Grundwasserspiegel*] = to lower [*ground-water table*]

~ = to counterbore, to rebore [*to enlarge a hole by drilling*]

Senken *n*, Ab~ [*Grundwasserspiegel*] = lowering [*ground-water table*]

Senkfaschine *f* [*Faschine mit Steinfüllung, 3—6 m lang, 0,6—1,0 m dick*] = sunk fascine

Senkfundament *n* **mit Brunnen** → Brunnengründung *f*

~**körper** *m* **mit Brunnen** → Brunnengründung *f*

Senk|fundation *f* **mit Brunnen** → Brunnengründung *f*

~**geschwindigkeit** *f* [*Kran*] = lowering speed

~**grube** *f* = cesspool [*A pit into which household sewage or other liquid waste is discharged and from which the liquid leaches into the surrounding soil or is otherwise removed*]

~**gründung** *f* **mit Brunnen** → Brunnengründung

~**gründungskörper** *m* **mit Brunnen** → Brunnengründung *f*

~**hammer-Bohrmaschine** *f* **mit Drehmotor** = down the hole type rock drill

~**kammer** *f* → Senkkasten *m*

Senkkasten *m*, Senkkammer *f*, Druckluft~ = (pneumatic) caisson, (compressed-)air ~

~ **mit Wasserdurchlässen**, Caisson *m* ~ ~ = water-passing caisson

~**arbeiter** *m*, Caissonarbeiter = sandhog

~**decke** *f*, Caissondecke = caisson ceiling

~**gründung** *f*, Druckluft~ = (pneumatic) caisson foundation, (compressed-)air ~ ~

~**krankheit** *f* = caisson disease, bends, diver's palsy, diver's paralysis, compressed-air disease, screws

~**schneide** *f*, Kastenschneide, Caissonschneide = caisson edge

Senk|kopf *m* [*Schraube; Nagel; Niet*] = countersunk head

~**lot** *n* → Senkblei *n*

~**niet** *m* = counter-sunk rivet

~**nietdöpper** *m* = rivet snap for countersunk head rivets

~**pumpe** *f* → Abteufpumpe

senkrechte Kraft *f* = vertical force

~ **Steife** *f* = dead shore

senkrechter Einschnitt *m* [*Erdbau*] = vertical cut

~ **Schweregradient** *m* = vertical gradient (of gravity)

Senkrecht|abweichungswinkel *m*, Senkrechtausschlagwinkel = angle of deviation from the vertical, ~ ~ deflection ~ ~ ~; ~ ~ deflexion ~ ~ ~ (Brit.)

~**aufnahme** *f*, Nadiraufnahme = vertical photograph [*aerial surveying*]

~**aufzug** *m* → Senkrechtbauaufzug

~**ausschlagwinkel** *m*, Senkrechtabweichungswinkel = angle of deflection from the vertical, ~ ~ deviation ~ ~ ~; ~ ~ deflexion ~ ~ ~ (Brit.)

~**(bau)aufzug** *m* = vertical (contractors') hoist

~**becherwerk** *n*, Senkrechtelevator *m* = vertical (bucket) elevator

Senkrechtbeleuchtung — Servobremsband

~**beleuchtung** *f*, Vertikalbeleuchtung = vertical lighting
~**beschickung** *f* = vertical feed(ing)
~**förderer** *m* = vertical transporter
~**(-Förder)schnecke** *f*, Senkrecht-Schneckenförderer *m* = (vertical) screw elevator, (~) ~ lift, worm elevator, worm lift, Archimedean screw elevator, vertical screw conveyor, vertical screw conveyer
~**pfahl** *m* = vertical pile
~**schweißung** *f* = vertical welding
~**transport** *m* = vertical transportation
~**verband** *m* = vertical bracing
~**verglasung** *f*, Stehverglasung, Vertikalverglasung = vertical glazing
Senk|röhre *f* [*röhrenförmiger Senkbrunnen aus Holz, Stahl oder Stahlbeton für Gründungszwecke*] = cylinder
~**rücken** *m* = saddle-back
~**sperre** *f* [*Heber*] = trip
~**stange** *f* [*pennsylvanisches Seilbohren*] = sinker bar
Senkung *f*, Setzung *f*, Sackung *f*, Setzen *n*, Absackung *f*, Absacken *n* = settlement [*downward movements of the soil or of the structure which it supports due to the consolidation of the subsoil*]
~, Absenkung, Grundwasser~ = ground-water lowering
~ = subsidence
~ **des Wasserspiegels durch Wind**, Ab~ ~ ~ ~ ~, ~ der Wasseroberfläche ~ ~ = lowering of the water level by wind
Senkungs|betrag *m*, Setzungsbetrag = rate of (ground) settlement
~**gebiet** *n* = subsidence site
~**geschwindigkeit** *f* = rate of subsidence
~**kurve** *f* (⚒) = subsidence curve
~**kurve** *f* → Senkungslinie *f*
~**linie** *f*, Senkungskurve *f* = dropdown curve [*A particular form of the surface curve of a stream of water, which is convex upward; for example, where a flume discharges freely into the air. The depth at all points is greater than Belanger's critical depth and less than the normal depth, and velocities increase downstream*]
~**schaden** *m*, Bergschaden = subsidence damage, mining ~
~**schutz** *m* **im Bergbau** = protection against mining subsidence
~**schwerpunkt** *m* = focal point of subsidence
~**trichter** *m* = cone of depression
~**trog** *m*, Senkungswanne *f*, Geosynklinale *f* (Geol.) = geosyncline
~**trog** *m* (⚒) = subsidence trough
~**wanne** *f* (Geol.) → Senkungstrog
~**-Zeit-Kurve** *f* = time-subsidence curve
Senk|waage *f*, Aräometer *n*, Spindel *f* = hydrometer
~**winde** *f* = lowering jack
Separator *m*, Abscheider *m* = separator
~ **zur Verlagerung der Trennung von Öl und Gas nach untertage** [*unterhalb der Tiefpumpe mit Packer abgesetzt*] = bottom hole separator
Septarienton *m* = septarian clay
Serien|herstellung *f*, Serienfertigung *f*, Reihenfertigung *f*, Reihenherstellung = full scale production, ~ ~ fabrication, serial ~
~**schaltung** *f* [*Elektrotechnik*] = connection in series, straight series
~**schaltung** *f*, Hintereinanderschaltung [*Ölhydraulik*] = series circuit design
~**schüsse** *mpl* → → 1. Schießen; 2. Sprengen
Serizit *m*, Seidenglimmer *m* (Min.) = sericite
~**schiefer** *m* = sericite schist
Serpa|-Draht *m* = Serpa wire
~**-Harfensieb** *n* = Serpa harp screen
Serpentin *m* [*grüner Porphyr, schlangenartig gefärbt*] = serpentine
~**asbest** *m* (Min.) = serpentine asbestos
Serpentine(nstraße) *f* = serpentine road, ~ highway
Servo|bremsband *n*, Bremsband mit Selbstverstärkungswirkung = self

Servobremse — Setzungsgeschwindigkeit 884

activating brake band, ~ energizing ~ ~

~**bremse** *f* = booster brake

~**lenkpumpe** *f*, Lenkhilfspumpe = steering booster pump

~**lenkung** *f* → Lenkhilfe *f*

Servomotor *m*, Stellmotor = servomotor

~ **zur Verstellung der Laufschaufeln** = runner vane servomotor

~ ~ ~ ~ **Leitschaufeln** = guide vane servomotor

~**steuerung** *f* = servomotor control

Servo|pumpe *f* = servo-pump

~**-Stempel** *m* [*Ein Reibungsstempel, bei dem während des Einsinkens des Oberstempels die Schloßreibung zunimmt und dadurch eine Servo-Wirkung (Hilfswirkung), d.h. eine selbsttätig herbeigeführte Steigerung der Lastaufnahme erreicht wird*] = servo-prop, servo-post

~**vorrichtung** *f* **für Gaspedal** = accelerator pedal booster

Sessellift *m* = chair lift

Setz|becher *m* [*Ausbreitversuch*] = slump cone, conical shell; mould (Brit.); mold (US)

~**druck** *m*, Hauptdruck, Periodendruck [*Der beim Absenken des Haupthangenden auf den in einen Hohlraum eingebrachten Versatz oder auf die beim Bruchbau hereingebrochenen Dachschichten auftretende Druck*] = squeeze, crush, weight, pressure, periodic weight

setzen [*Stempel usw. im Bergbau*] = to erect, to set

Setzen *n* **eines Bauwerkes** → Bauwerksetzung *f*

~ **von Pflanzen** = planting of plants

Setzer *m* → Stemmer

Setz|kolonne *f* (⚒) = prop-setters, arch-setters, post-setters

~**kopf** *m* [*Niet*] = die bead, set ~

~**latte** *f* = rule [*A straightedge for working plaster or dots to a plane surface or for other purposes*]

~**latte** *f* [*zum Nivellieren*] = staff

~**latte** *f*, Wiegelatte [*wird mit der Wasserwaage zusammen verwendet*] = plumb rule

~**maß** *n*, Ausbreitmaß, Sackmaß, Fließmaß [*Betonprüfung. DIN 1048*] = slump

~**packe** *f*, Handpacke = hand-packed rubble

~**packlage(schicht)** *f*, Gestück *n*, Setzpacklagenunterbau *m* = subbase of stone pitching, (hand-)pitched foundation, hand-set pitching, rough stone pitching, hand-packed bottoming, packed broken rock soling, hand-pitched stone subbase, blocking (Brit.); (hand-packed) Telford (type) subbase (US)

~**packlagestein** *m*, Vorlagestein, Stückstein = blockstone, hand-packed hardcore, pitching stone, pitcher, hand-packed stone, hand-placed stone, hand-pitched stone (Brit.); Telford stone, base stone (US)

~**pfosten** *m* → Losständer *m*

~**probe** *f* → Ausbreit(ungs)versuch *m*

~**prüfung** *f* → Ausbreit(ungs)versuch

~**riß** *m* → Setzungsriß

~**steinverguß** *m* = sealing compound for hand-set pitching

~**stufe** *f*, Stoßtritt *m*, Futterstufe [*wenn aus Holz = Futterbrett n*] = riser [*The member forming the vertical face of a step*]

Setzung *f* → Senkung

~ [*Bauwerk*] = sinkage

~ **durch Konsolidierung** = settlement due to consolidation

~ **in Bergwerkgebieten** → Bergsenkung

Setzungs|aufgabe *f* = settlement problem

~**beobachtung** *f* = settlement observation

~**berechnung** *f* [*Bodenmechanik*] = settlement computation

~**fuge** *f* = settlement joint

~**geschwindigkeit** *f* [*Bodenmechanik*] = rate of settlement, settlement rate

Setzungsklassierung *f* durch flüssige **Medien** = sorting, classifying
~ ~ **Luft**, (Wind)Sichtung *f*, Stromsichtung = air classification, (~) separation
~ ~ **Wasser oder Luft** = classifying, classification
Setzungs|kurve *f* = settlement curve
~**profil** *n* = settlement contour
~**riß** *m*, Setzriß = crack resulting from settlement, settlement crack
~**riß** *m*, Setzriß (⚒) = roof break
~**trichter** *m* = settlement crater
~**unterschied** *m*, ungleich(mäßig)e Setzungen *fpl* = differential settlement(s)
~**vorhersage** *f* = forecast of settlement, prediction ~ ~, settlement forecast, settlement prediction
Setz|versuch *m* → Ausbreit(ungs)versuch
~**vorrichtung** *f* (⚒) = setting device
~**winde** *f* (⚒) = setting winch
S-Farbe *f*, Silikatfarbe = silicate paint
Sgraffito *m*, Kratzgrund *m* = (s)graffito, scratch work
Shed|bau *m* = saw-tooth roof building, north-light ~ ~
~**binder** *m* → Sheddachbinder
~**dach** *n*, Säge(zahn)dach = sawtooth roof, north-light ~
~**(dach)binder** *m*, Shedfachwerkbinder, Säge(zahn)dachbinder = sawtooth truss, north-light ~
~**fachwerkbinder** *m* → Shed(dach)binder
~**-Firstziegel** *m* = saw tooth roof tile
~**schalendach** *n*, Säge(zahn)schalendach = north-light shell roof, sawtooth ~ ~
Shelby-Stutzen *m* = Shelby tube sampler, ~ (soil) ~
sich ablösen = to strip
~ **absetzen** → abschlämmen
(~) ~ (⚒) = to settle
~ **abstützen (gegen)** = to bear against
~ **auflegen** [*das Hangende*] = to begin to bear on
~ **aufrichten** = to rear up
(~) **aufwölben** = to warp upward

~ **brechende See** *f*, Brecher *m* = breaking sea
~ **hereinschieben** [*Gestein*] = to flow (into)
~ **setzen** [*Gebirge im Berg-, Tunnel- und Stollenbau*] = to settle
~ **verjüngen** = to taper
Sichausspitzen *n* → Auskeilen
Sichel|düne *f* → Bogendüne
~**pumpe** *f* = crescent pump, sickle ~
sichere (ab)bauwürdige Vorräte *mpl* = proved reserves
~ **Arbeitsweise** *f*, sicherer Betrieb *m* = safe working
~ **Mineralvorräte** *mpl* = assured mineral, ~ reserves, ore in sight
sicherer Betrieb *m*, sichere Arbeitsweise *f* = safe working
Sicherheit *f* gegen Umkippen = safety against overturning
Sicherheits|ausschaltung *f* = safety kick-out
~**beiwert** *m*, Sicherheitsgrad *m*, Sicherheitszahl *f*, Sicherheitsfaktor *m* [*Bruchbeanspruchung dividiert durch zulässige Beanspruchung*] = factor of safety, safety factor
~**belange** *fpl* = safety aspects
~**beleuchtung** *f* = safety lighting
~**bestimmung** *f*, Sicherheitsvorschrift *f* = safety regulation
~**faktor** *m* → Sicherheitsbeiwert
~**filterelement** *n* = safety filter element
~**flasche** *f* [*Bindemittelrückgewinnungsapparat*] = water trap
~**gitter** *n* = safety grille
~**glas** *n* = safety glass
~**grad** *m* → Sicherheitsbeiwert
~**haken** *m* = safety hook
~**kette** *f* = safety chain
~**lampe** *f* = safety lamp
~**last** *f* = safe load
(~)**Leitplanke** *f* [*Straße*] = guard rail
~**linie** *f* [*Straße*] = safety line
~**maßnahme** *f*, Sicherheitsvorkehrung *f* = safety measure
~**netz** *n* = safety net(ting)
~**norm** *f* = safety standard spec(ification)

Sicherheitspfeiler — Sichtlochkartei

~pfeiler *m* (⚒) = safety pillar
~schloß *n* = safety-cylinder lock
~spanne *f*, Sicherheitszuschlag *m* = margin of safety, safety margin
~sperre *f* = safety pawl
~strebe *f* [*Gelenkrahmen*] = safety brace [*articulated frame*]
~streifen *m* = safety strip
~theorie *f* [*in der Baustatik*] = safety theory
~tor *n* [*verhindert das Ausfließen großer Wassermengen bei Dammbrüchen oder Beschädigungen an Kunstbauten, namentlich an Kanalbrücken, Dükern usw.*] = safety gate
~ventil *n* = safety valve
~verbinder *m*, Gestänge-~ = safety joint
~-Verrohrungskopf *m*, Antieruptions-Verrohrungskopf = control casing head
~vorkehrung *f*, Sicherheitsmaßnahme *f* = safety measure
~vorrichtung *f* = safety device
~vorschrift *f*, Sicherheitsbestimmung *f* = safety regulation
~wand *f* = safety wall
~zahl *f* → Sicherheitsbeiwert
~zündschnur *f*, schlagwettersichere Schwarzpulverzündschnur = safety fuse
~zuschlag *m*, Sicherheitsspanne *f* = margin of safety, safety margin
Sicherung *f*, Verbau *m* [*Graben; Baugrube*] = well curbing, pit boards (US); sheeting [*used to keep earth out of a building pit*]
Sicherungs|blech *n* [*nach dem Festziehen der Verbindung formschlüssige Schraubensicherung. Sichert Schraube und Mutter gegen Lösen, insbesondere bei nicht erschütterungsfreier Belastung*] = locking plate
~draht *m* = locking wire
~hebel *m*, Schaltarretierhebel = safety lever
~mutter *f* [*DIN 7967*] = locking nut
~rückführventil *n* = safety reset valve
~scheibe *f* = retainer
~schraube *f* = lock screw
~sperre *f* = safety pawl
~stift *m* = captive pin
~ventil *n* = safety valve
Sicht *f* = vision, sight
sichtbare Decke *f*, Deckenuntersicht *f* = floor soffit
~ Strahlung *f* = visible radiation
~ Unterraupe *f* [*Schweißen*] = penetration bead
sichtbarer Horizont *m* → Kimme *f*
sichtbares Erz *n* = visible ore
Sicht|barkeit *f* = visibility
~bauelement *n* = face unit
~behinderung *f* = restriction of visibility
Sichtbeton *m*, Architekturbeton = exposed concrete, fair-faced ~
~fertigteil *m*, *n*, Sichtbetonmontageteil = precast (concrete) face member
~schalung *f* = exposed concrete formwork, fair-faced ~ ~
Sichten *n*, Sichtung *f* [*Klassieren im Luftstrom*] = air sifting, ~ separation
Sichter *m*, Wind~, Strom~ = separator, air ~, air classifier
Sichtermühle *f* = air-separating mill
~ mit Schleudersichtung im geschlossenen Kreislauf = air separating mill operating with a straight-air current
~ ~ Stromsichtung im geschlossenen Kreislauf = air separating mill with centrifugal separation in a closed circuit
Sicht|fläche *f* = (exposed) face, ~ surface
~flächenmaterial *n* = facing [*Is the material which forms the face*]
~freiheit *f* = unrestricted visibility
~höhe *f* [*Kranführer*] = crane operator's eye level
~höhe *f*, Augenhöhe = eye level
~länge *f* → Sichtweite *f*
~linie *f* = sight line, vision ~
~loch *n* = inspection hole
~lochkartei *f* = peek-a-boo system, ~ file

Sichtplatte — Sickerwasser

~platte *f* = face slab
~stein *m* [*Steinmauerwerk*] = face stone
Sichtung *f*, Windsichtung, Setzungsklassierung *f* durch Luft, Stromsichtung, Luftsichtung = separation, classification [*by air*]
Sicht|verhältnisse *f* = visibility conditions
~weite *f*, Sichtlänge *f*, Sichtweg *m* [*Die Fahrbahnstrecke, die der Fahrer eines Personenkraftwagens bei einer festgelegten Augenhöhe über Fahrbahn übersehen kann, wenn seine Sicht nicht durch den Verkehr und durch andere vorübergehenden Einflüsse (z. B. Witterung) behindert ist. Verkehrseinschränkende Sichtweite = restricted sight distance. Eine Sichtweite, die so kurz ist, daß sie eine Herabsetzung der Geschwindigkeit bewirkt und die freie und sichere Abwicklung des Verkehrs unter den vorherrschenden Bedingungen beeinträchtigt*] = sight distance, vision ~, visibility ~, seeing ~
~winkel *m* = visibility angle
Sicke *f* = knuckle
Sicker|becken *n* = detritus tank in the form of a shallow pit
~brunnen *m* [*Wasserversorgung*] = percolation well
~drän *m*, Filterdrän = blind drain, spall ~, filter ~, underdrain
~füllung *f*, Filterfüllung = porous backfill(ing)
~galerie *f*, Schlitzfassung *f*, Sickerstrang *m* = filter gallery
~geschwindigkeit *f* [*Bodenmechanik*] = seepage velocity, ~ rate
~graben *m*, Rigole *f*, Sauggraben [*mit Kies ausgepackt*] = gravel-filled drain trench
~graben *m*, (Stein)Rigole *f*, Sauggraben [*mit Steinen ausgepackt*] = stone-filled drain trench
~graben *m* mit Dränrohr, Sauggraben ~~, Rigole *f* ~~ = French drain

~grube *f* → Sickerschacht
~leitung *f* = drain pipeline allowing infiltration
~linie *f* [*Senkrechter Schnitt durch die Wasseroberfläche in einem durchflossenen festen porigen Körper*] = saturation line, ~ surface, seepage ~
Sickern *n* [*unterirdisches Wasser*] = seepage
Sicker|packung *f* → → Tunnelbau
~parabel *f* = seepage parabola
~rohr *n* [*Beton- oder Steinzeugrohr, das auf dem oberen halben Umfang geschlitzt oder gelocht ist*] = perforated pipe
~schacht *m*, Sickergrube *f* = soakage pit, soaking ~, soakaway, sump; dry well (US); rummel [*Scotland*] [*A pit which may be empty or filled with large stone and is lined (if at all) with stone or bricks laid without mortar. Surface water is drained into it to soak away into the ground*]
~schacht *m*, Senkbrunnen *m* [*senkrechter Schacht zum Einleiten des Dränwassers in durchlässige Schichten des tieferen Untergrundes*] = well drain, absorbing well
~schicht *f* → Filterschicht
~schlitz *m* → Entwässerungsschlitz
~stollen *m*, Sammelgalerie *f*, Sammelleitung *f*, Sammelstollen = infiltration gallery
~strang *m* → Sickergalerie *f*
~strömung *f*, Sickerwasserströmung = seepage flow
~strömungskraft *f* = seepage force
Sickerung *f* [*laminare Bewegung des Wassers in einem festen porigen Körper*] = percolation
Sicker|verlust *m* = seepage loss
~wasser *n* = meteoric water, cut-off ~ [*That part of rainfall which soaks into the ground, as opposed to run-off water which is disposed of by surface drainage and fly-off water which is disposed of by evaporation*]
~wasser *n* [*Wasser, das aus einer Stauhaltung, einem Kanal usw. in*

Sicker(wasser)strömung — Siebextraktionsapparat

Mauern, Dämme, Untergrund oder Talhänge eindringt und sie durchzieht] = seepage loss water
~(wasser)strömung *f* = seepage flow
~wasserverlust *m* [*Talsperre*] = seepage loss
~weg *m* = path of seepage
siderischer Tag *m* = sidereal day
Siderit *m* → Eisenspat *m*
Sieb *n* = screen
~analyse *f* → Prüfen der Kornzusammensetzung durch Siebversuch
~anlage *f* = screening plant
~ascheanalyse *f* = size analysis of ash
~austrag *m* = screen discharge, discharge of a screen
~band *n* → Abwasserbandrechen *m*
~batterie *f*, Sieberei *f* = battery of screens
~belag *m* [*als Gewebe*] = screen cloth
~belag *m* [*als Lochblech*] = screen plate
~bereich *m* = envelope of grading, grading envelope
~bewegung *f* = screen motion, ~ movement
~blech *n* = screen plate
Siebboden *m*, arbeitende Siebfläche *f*, Siebdeck *n* = screen(ing) deck, ~ bottom
~ **mit Gewölbespannung** = crown deck, ~ bottom
~beheizung *f* = screen bottom heating, ~ deck ~
Sieb|bruch *m* = screen failure

~bürste *f* = screen brush
~deck *n* → Siebboden *m*
~diagramm *n* = plot of screen test (results), screen test result diagram
~durchgang *m*, Siebdurchlauf *m*, Siebfeines *n* = undersize, sub-size, throughs; thrus (US)
~durchgangkurve *f*, Siebdurchlaufkurve = cumulative size distribution curve of the underflow
~durchlauf *m* → Siebdurchgang *m*
~durchsatz *m* = throughput of screen
~durchsatzleistung *f* = throughput of screen per hour
~ebene *f* = screening plane
~einheit *f* = screening arrangement
~einrichtung *f* = screen assembly
sieben, ab~ = to screen
Sieben *n*, Siebung *f* = screening
~ **bei erhöhter Materialeigenlast**, Siebung *f* ~ ~ ~ = screening by increased weight of the material
~ **unter Druck**, Siebung *f* ~ ~, Drucksieben, Drucksiebung = forced screening
siebendrähtiges Spannkabel *n* = seven-wire strand
Siebentschlämmung *f* = deslurrying by screens
Sieberei *f*, Siebbatterie *f* = battery of screens
Sieb|erfolg *m* → Siebgüte(grad) *f*, (*m*)
~erzeugnis *n* = screened product, ~ material, screen sized ~

Siebextraktionsapparat *m* [*Apparatur für die Schnellanalyse von Straßenproben mit bit. Bindemitteln durch gleichzeitiges Extrahieren und Sieben*]	sieving extractor
Motor *m* mit Getriebe	geared motor unit
Drehpunkt *m* der Schüttelbewegung *f*	pivot
Hebelklammer *f*	lever clamp
Ring *m* mit 4 Armen	clamping ring with 4 legs
Bewegungswinkel *m*	angle of oscillation
Metallhaube *f* mit Preßdeckel	metal head with press cap
Siebsatz *m*	nest of sieves
Ablaßhahn *m*	solution drain valve

Verbindungsstange *f*
Schwungrad *n*

connecting rod
flywheel

Sieb|feines *n* → Siebdurchgang *m*
~**feinleistung** *f* [*Austrag an Siebfeinem in t/h*] = discharge of undersize in t/h
~**feld** *n* [*Abschnitt des Siebbodens*] = section of screen(ing) deck, ~ ~ ~ bottom
~**filterrohr** *n* = screen pipe
~**fläche** *f* = screen(ing) surface, ~ area
~**flächenbelastung** *f* = screen surface load, ~ ~ charge, ~ area ~
~**folge** *f*, Siebreihe *f*, Siebskala *f* = mesh scale, screen ~, screen series; mesh gage (US)
~**fraktion** *f*, Siebkorngruppe *f*, Siebkornklasse *f* = screen fraction
~**gerüst** *n* = screen structure
(Sieb)Gewebe *n*, gewebter Siebboden *m* = (screen) cloth; ~ fabric (US)
~ **mit Rahmen** = framed (screen) cloth; ~ (~) fabric (US)
~ ~ **rechteckigen Maschen** = rectangular-mesh (woven-wire) screen (cloth)
~**ausführung** *f* = finish of cloth
~**güte** *f*, (Sieb)Gewebequalität *f* = quality of screen cloth
~**liste** *f* = catalog of screen cloth
~**nummer** *f* = mesh number, sieve ~
~**qualität** *f*, (Sieb)Gewebegüte *f* = quality of screen cloth
Sieb|grenze *f* = limiting curve
~**grobes** *n*, Sieböberlauf *m*, Siebrückstand *m*, Siebübergang *m* = screen reject, screening refuse, screen oversize, screen overflow, plus size of a screen, coarse of a screen
~**gut** *n*, Siebstoffe *mpl*, Siebschlamm *m* [*Abwasserwesen*] = screenings
~**gut** *n* → Aufgabegut
~**güte(grad)** *f*, (*m*), Sieberfolg *m*, Siebtrennungsgrad = screening efficiency, grading ~

~**gutteilchen** *n* = particle of material to be screened
~**hersteller** *m* = screen manufacturer
~**hilfe** *f* [*Kugeln gegen Verstopfen*] = balls
~**kasten** *m*, Siebrahmen *m* = screen frame
~**kennziffer** *f* **für Schwingsiebe** = maximum acceleration of screens in terms of g
~**kies** *m* = screened gravel, graded ~
~**klassierer** *m* = sizer, sizing screen, classifying screen
~**klassierung** *f* [*trockenes oder nasses Klassieren durch Siebe*] = screen sizing
~**kohle** *f* = screened coal, graded ~
~**korngruppe** *f* → Siebfraktion *f*
~**kornklasse** *f* → Siebfraktion *f*
~**kraft** *f* = screening force
~**kugelmühle** *f* = screen-discharge ball mill
~**kurve** *f*, Sieblinie *f* = (aggregate) grading curve, particle-size distribution ~, grain-size distribution ~
~**leistung** *f* = screening capacity
~**linie** *f* → Siebkurve *f*
~**linie** *f* **der gewünschten Endprodukte** = nominal grading curve
~**linie** *f* **des Rohmaterials**, Ist-Linie = actual grading curve
~**lochung** *f* = perforation of screen, punching ~ ~, hole pattern
~**lochweite** *f* = screen hole size
~**löffel** *m* [*Sitzventil*] = atomizer
~**masche** *f* = screen mesh
~**maschine** *f* = screening machine
~**maschine** *f* **mit zwangläufigem Antrieb** = positive screening machine
~**modul** *m* = screen modulus
~**montage** *f*, Montage des Siebbodens = fitting of screen bottom
~**neigung** *f* = screen inclination, ~ slope, slope of screen deck, inclination of screen deck

Siebplatte — Siegellackholz

~platte f [*Abwasserwesen*] = screen plate, perforated ~
~probe f, Siebversuch m, Siebprüfung f = sieve test
~prüfung f → Siebprobe f
~rahmen m, Siebkasten m = screen frame
~rechen m [*Abwasserwesen*] = mesh screen [*A screen composed of woven fabric, usually of wire*]
~reihe f → Siebfolge f
~rückstand m → Siebgrobes n
~rückstandkurve f, Siebrückstanddiagramm n = cumulative size distribution curve of sieve oversize
~sand m = screened sand
Siebsatz m = set of screens, nest ~ ~, screen set, screen nest
~ [*Labor(atorium)*] = set of sieves, nest ~ ~, sieve set, sieve nest
~rüttler m, mechanischer Siebsatz-Schwingungserreger m = (mechanical) sieve shaker, ~ vibrator for laboratory test sieves, test sieve shaker, test sieve vibrator
Sieb|schlamm m → Siebgut n
~schnitt m, Trennschnitt = screen cut
siebschwieriges Gut n, schwer siebbares ~ = hard-to-screen material, difficult-to-screen ~
Sieb|seite f, Arbeitsseite = screening side
~skala f → Siebfolge f
~skalenkoeffizient m, Siebskalenquotient m = screen (scale) ratio
~spannrahmen m = screen (surface) support frame
Siebspannun f in Längsrichtung = longitudinal screen tension
~ ~ Querrichtung = transverse screen tension
Sieb|station f, Sieberei f = screening plant
~stoffe mpl, Siebgut n [*Abwasserwesen*] = screenings
Siebstraße f mit fallender Masche [*am Anfang Grobkornabsiebung und am Ende Feinkornabsiebung*] = screen surface with decreasing dimension of the clear opening towards the end

~ ~ steigender Masche [*am Anfang Feinkornabsiebung f und am Ende Grobkornabsiebung*] = screen surface with increasing dimension of the clear opening towards the end
Sieb|technik f = screening practice
~trennschärfe f = precision of screen separation
~trennungsgrad m → Siebgüte(grad)
~trommel f [*Abwasserwesen*] = drum screen
~trommel f → Trommelsieb n
~trommel f → Trommelrechen m
~übergang m → Siebgrobes n
~übergangsleistung f = discharge of overflow in t/h
~überlauf m → Siebgrobes n
Siebung f, Sieben n = screening
~ durch spezifisches Gewicht, Sieben n ~ ~ ~ = separating
Sieb|unterbau m = screen sub-structure
~versuch m → Prüfen der Kornzusammensetzung durch Siebversuch
~vorgang m = screening process
~wäsche f, Waschsieb n = rinsing screen
~weite f = size of mesh, screen size
~welle f = screen shaft
siebwilliges Gut n = easy-to-screen material
Sieb|wirkungsgrad m = screening efficiency, grading ~, ~ quality
~zylinder m [*Waschtrommel*] = perforated cylinder
Siede|rohrdichtungsmaschine f, Rohreinwalzapparat m, Rohrwalze f, Rollmaschine = tube beader
~schwanz m = heavy tails, ~ ends
~stein m, Zeolith m (Min.) = zeolite
~wasserreaktor m = boiling water reactor, BWR
Siedlung f = (housing) estate; nesting (US)
Siedlungs|projekt n = (housing) estate development
~straße f = (housing) estate road
Siegel n = floor seal
~lackholz n = sealing-wax wood

Siegwart-Decke f = Siegwart (precast) floor
Siemens-Martin-Stahl m = Siemens-Martin steel, open-hearth ~
Sierozemboden m = Sierozem soil
Sietland n [*niederdeutscher Ausdruck*] = low-lying marsh land
Sigma-Schweißen n, **Sigma-Schweißung** f = self-adjusting arc welding, Argonaut ~, Sigma ~
Signal n = signal
~**bild** n [*Farbzeichen, Farbzeichenkombination oder Formzeichen, die in einer Signaleinheit jeweils angezeigt werden*] = aspect
signal|gesteuert = signalized
~**gesteuerter Knoten** m = signal-controlled intersection
signalisierter Zugverkehr m = signalled train movement
Signalisierung f → Lichtsignalsteuerung
~ **eines Fahrwassers** = marking of a channel
Signal|kopf m [*Der Teil des Verkehrssignals, in dem die Signalgeber zusammengefaßt sind; er besitzt eine oder mehrere Signalsichtflächen*] = signal head
~**lampe** f, optisches Signal n = signal light
~**reg(e)lung** f → Lichtsignalsteuerung
~**schutz** m der Fußgänger = safe accommodation of pedestrian movements at signal-controlled intersections
~**sichtfläche** f [*Die einer bestimmten Verkehrsrichtung zugewandte Fläche des Signalkopfes, die der Regelung des Verkehrs aus dieser Richtung dient*] = signal face
~**-Steuerung** f → Lichtsignalsteuerung
~**wechsel** m [*Verkehrssignal-Steuerung*] = changing of aspect of traffic control signals
Signierfarbe f, Kennfarbe = marking paint
~ [*Schweiz*] → Straßen-Markierungsfarbe
Signierkreide f = marking chalk
Sikkativ n = soluble drier, liquid ~
Silage|grube f = (en)silage pit
~**technik** f = (en)silage technique, ~ technic
Silben|reinheit f = articulation
~**verwischung** f [*Raumakustik*] = inarticulation
Silber|erz n = silver ore
~**glanz** m, Argentit m, Ag_2S (Min.) = argentite, silver glance
~**hornerz** n (Min.) → Hornsilber n
~**lamé** n = silver lame
~**lot** n = silver solder
~**nitrat** n → Höllenstein m
~**schmied-Industrie** f = silversmithing industry
~**vitriol** n = silver sulphate
Silhouette f = sky-line
Silika|erzeugnis n = silica refractory
~**gel** n = silica gel
~**mörtel** m = silica mortar
~**stein** m → Quarzitstein
~**stoff** m = silica material
Silikat n = silicate
~**belag** m → Silikatdecke
~**beton** m, Kalksandbeton = silicate concrete

Silikatbildner m, Hydraulefaktor m, Wasserbindner m — matter forming silicates
   Kieselsäure f, Siliziumdioxyd n, SiO_2 — silicic acid, silica
   Tonerde f, Aluminiumoxyd n, Al_2O_3 — alumina
   Eisenoxyd n, Fe_2O_3 — ferric oxide, iron oxide

Silikat|decke f, Silikatbelag m, Silikatmakadam m, Wasserglasdecke, Wasserglasbelag, Wasserglasmakadam, Wasserglas-Straße f, Silikat-Straße = silicated (road) surfacing, ~ (~) pavement

Silikatester — Silowand 892

~ester *m* = silicate ester
~farbe *f*, S-Farbe = silicate paint
~glas *n* = silicate glass
~schmelze *f* = silicate melt
~-Straße *f* → Silikatdecke
Silikon|harz *n* = silicone resin
~imprägniermittel *n* = silicone varnish
~kautschuk *m* = silicone rubber
~kühlung *f* = silicone cooling
~schmierfett *n* = silicone grease
Silikose *f*, Staublungenerkrankung *f* = silicosis
Silizium|dioxid *n*, SiO_2 = silica, dioxide of silicon, silicon dioxide
~gleichrichter *m* = silicon rectifier
Siliziumkarbid *n* → Karborund
~mörtel *m* = silicon carbide mortar
~platte *f* = silicon carbide slab
~schneidscheibe *f* = silicon carbide cutting blade
~stein *m* = silicon-carbide brick
Siliziumoxynitridbindung *f* = nitride bond containing silicon oxynitride
Sillanmatte *f* = rock-wool blanket SILLAN
Sillimanit *m* → Faserkiesel *m*
~stein *m*, Mullitstein = sillimanite brick
Silo *m* → Lager~
~abmeßanlage *f* → Abmeßanlage für Betonzuschlagstoffe
~abzug *m* = drawing from silo(s)
~abzugrinne *f* = silo drawing channel
~aktivität *f* = free-running silo capacity, live ~ ~
~anlage *f* = silo plant, ~ installation
~auslauf *m* = silo outlet
~batterie *f* = battery of silos
~bau *m* = silo construction
~beheizung *f* = silo heating
~beschickungsboden *m* = conveyor house, conveyer ~
~boden *m* = silo bottom
~bunker *m*, Trogbunker = trough-type bin
~entleerung *f* = silo discharge
~fahrzeug *n*, Silowagen *m*, Behälterfahrzeug, Behälterwagen = bulk transporter

~kraftwagen *n*, Behälterkraftwagen, Silo-Straßenfahrzeug *n*, Behälter-Straßenfahrzeug = truck-type bulk transporter (US); lorry-type ~ ~ (Brit.)
~lagerung *f* = silo storage
siloloser Schwarzdeckenfertiger *m*, silolose Schwarzbelageinbaumaschine *f*, siloloser Schwarz(decken)verteiler *m*, siloloser Verteilerfertiger, silolose Schwarzdeckeneinbaumaschine = bituminous road surfacing finisher without hopper, asphalt finisher ~ ~, asphalt paver ~ ~, bituminous spreading-and-finishing machine ~ ~, black-top spreader ~ ~, asphaltic concrete paver ~ ~, (bituminous) paver finisher ~ ~, asphalt and coated macadam finisher ~ ~, spreader finisher ~ ~, black top bituminous paver ~ ~, bituminous paving machine ~ ~
Silo|schalung *f* = silo formwork
~standanzeiger *m*, Silostandwächter *m*, Anzeigevorrichtung *f* für Silofüllungen = (silo) level detector, storage (~) ~ ~
~standwächter *m* → Silostandanzeiger *m*
~stein *m* = silo block
~-Straßenfahrzeug *n* → Silokraftwagen *m*
~tasche *f* → Silozelle *f*
~transport *m* [*Zement*] = cement transport in bulk transporters
~verschluß *m* = 1. bin door fill valve, bin gate [*controls the flow of materials from the storage bin into the batcher hopper*]; 2. discharge gate [*controls the flow of materials into transport vehicle*]
~waage *f* = bin weigh(ing) batcher scale
~wabe *f* → Bunkertasche *f*
~wagen *m* → Silofahrzeug *n*
~wagen *m* für Zement = cement bulk transporter
~wand *f* = silo wall

Silozelle — Sintern

~zelle *f*, Speicherzelle *f*, Silotrichter *m*, Silotasche *f* = silo compartment, ~ hopper, ~ bin
~zement *m* → Behälterzement

Silt *m* [*nach Fischer und Udluft (1936) und Gallwitz (1939) 0,2–0,02 mm*], Auelehm *m* = fine sand and coarse silt [*0.2–0.02 mm, Massachusetts Institute of Technology and British Standards Institution*]

silurisch = siluric, silurian

Simmerring *m* = grease retainer seal, seal cup

Simplex|-(Beton)Pfahl *m* = Simplex pile
~pumpe *f* = Simplex pump

Sims *m, n*, Gesims *n* = corbel
~brett *n* → Fensterbrett
~träger *m* → Atlant *m*

Sink *m* → → Feinschluff *m*

sinken, abschwellen, fallen [*Wasserspiegel*] = to fall

Sink|geschwindigkeit *f* **von Teilchen in einer Flüssigkeit**, Abschlämmgeschwindigkeit = settling rate, ~ velocity
~gut *n* [*beim Schwimm-Sink-Scheider*] = rubbish
~gutaustrag *m* = rubbish discharge
~kasten *m* = catch basin [*A chamber or well designed to prevent the admission of grit and detritus into a sewer*]
~matte *f* → Sinkstück
~-Schwimm-Aufbereitung *f*, Schwerflüssigkeitsaufbereitung = heavy media separation, sink float process
~stoff *m*, Ausfällung *f* = deposit, sediment
~stoffe *mpl* [*Feststoffe, die vor dem Absinken geschwebt haben*] = silt [*Water-borne sediment. The term is generally confined to fine earth, sand, or mud, but is sometimes broadened to all material carried, including both suspended and bed load*]
~stoffe *mpl*, absetzbare Stoffe = settling solids, settleable ~ [*Suspended solids which will subside in quiescent sewage in a reasonable period. Two hours is a common arbitrary period*]
~stück *n*, Matratze *f*, Faschinenfloß *n*, Sinkmatte *f* = (fascine) mattress, brushwood ~, framed ~, honeycombed fascine raft

Sinn *m*, Vorzeichen *n* [*Baustatik*] = sign
~ **der Tide** = set of tide
~bild *n* = symbol

Sinnestäuschung *f* = error of judgment [*road accident*]

Sinter|anlage *f* = sinter plant
~band *n* = continuously moving hearth, sinter ~
~belagkupp(e)lung *f* = sintered faced clutch

Sinterbims *m*, Schlackensinter *m* = sintered slag
~-Wand(bau)platte *f*, Schlackensinter-Wand(bau)platte = sintered slag wallboard

Sinter|blähhochofenschlacke *f* = sintered expanded slag
~-Blähton *m* = sintered expanded clay; ~ bloating ~ (Brit.)
~bronzebelag *m* = sintered bronze facing
~dolomit *m* = dead-burned refractory dolomite, sintered ~
~gemisch *n*, Sintermischung *f* = sintering mix(ture)
~haut *f* = sinter skin
~karbid *n* = cemented carbide, sintered ~
~kies *m* = sintered gravel
~kohle *f* = free burning coal, open ~ ~, cherry ~
~magnesit *m* = sintered magnesite
~maschine *f* = sinter machine
~metall *n* = sintered metal
~metall-Lager *n* = sintered bearing, powder metal ~
~mischung *f*, Sintergemisch *n* = sintering mix(ture)

Sintern *n* = sintering
~ **mit flüssiger Phase**, Schmelzsintern = liquid phase sintering
~ **von losem Pulver**, Schüttsintern = loose powder sintering, pressureless ~

Sinterofen — Sockelgründung

Sinter|ofen m = sintering furnace
~rost m = sintering grate
~stahl m = sintered steel
~ton m = sintered clay
~tonerde f = sintered alumina
Sinterung f, Sintern n = sintering
Sinter|verfahren n = sintering method
~vorgang m = sintering operation
~zone f = sintering zone
Sinusbogen m = sinusoidal arch
Sinusoiden|last f = sinusoidal load
~zugbrücke f = sinusoidal draw bridge
Siphon m → Geruchverschluß m
~ → Syphon
Sitz m = seat
~ausgleichventil n = seat level(l)ing valve
~badewanne f = hip bath
~bank f = seat (Brit.); bench (US)
sitzen [*auf den Schlechten*] = to work with the cleats lying back
~ [*unter den Schlechten*] = to work with the cleats lying forward
Sitz|kissen n = cushion seat
~polster n [*Österreich: m*] = seat cushion
~schalter m = seat switch
~-Spülabort m = commode-type closet
~unterbau m = seat frame
~verstellung f = seat adjustment
~winkel m = seat angle
SK m, Segerkegel m = Seger cone
Skagliolplatte f = scagliola panel
Skala f = scale
skalares Feld n = scalar field
~ Produkt n, inneres ~ = scalar product [*of two vectors*]
Skalenpegel m, Pegellatte f, Lattenpegel [*Zeichen: L*] = staff gage (US); ~ gauge (Brit.)
Skelett n, Fachwerk n = skeleton, (structural) framework, framed structure
~bau m = skeleton construction
~bau(werk) m, (n) = skeleton structure
~beton m → Prepaktbeton
~bodenplatte f [*Gleiskettenlaufwerk*] = skeleton shoe

~-Gesteinsschaufel f = skeleton rock bucket
~-Teil m = frame member, skeleton ~
Skikuli m, Ski-Kurz-Lift m = small type ski hoist
Ski(schlepp)lift m → Schneeschuhläuferaufzug m
Skisprung-Überlauf m → → Talsperre
Skleroskophärte f, Rücksprunghärte nach Shore = Shore hardness
Skolezit m, Wurmsiedestein m (Min.) = scoleccite
S-Kurve f → Gegenbogen m
Slip m → Aufschleppe f
Smithdecke f = Smith floor
Sockel m [*Im Deutschen wird unter „Sockel" der unterste sichtbare Teil eines Bauwerkes oder Bauteiles verstanden. Eine treffende englische Übersetzung hierfür gibt es nicht. Der Ausdruck "plinth" kommt dem deutschen Ausdruck „Sockel" am nächsten. Er ist in British Standard 3589:1963 wie folgt definiert: Plinth = A distinct feature forming the lowest part of a masonry column or wall. It usually projects from the surface above and may be of a more durable material. In den USA ist "plinth" (auch "base course" genannt) wie folgt definiert: The base course or plinth is usually a course of stone placed just above the ground level. A variety of stone is chosen which will resist the severe weathering action which occurs at this point*]
~ → Sockelgründung f
~ [*Zementdrehofen*] = kiln pier
~brett n = skirting (board); base plate [*Scotland*]; mopboard, washboard, scrub board, base (US)
~fundament n → Sockelgründung f
~fundamentkörper m → Sockelgründung f
~fundation f → Sockelgründung f
~gründung f, Sockelfundament n, Sockelfundation f, Sockel(fundamentkörper) m, Sockelgründungskörper, Gründungssockel m, Grund-

sockel, Fundamentsockel, Fundationssockel = (single) base, individual ~

~gründung *f* in Bruchsteinmauerwerk = rubble masonry (single) base, ~ ~ individual ~

~gründung *f* in Ziegelmauerwerk = brickwork (single) base, ~ individual ~

~gründungskörper *m* → Sockelgründung *f*

~mauerwerk *n* [*siehe Erläuterung unter ,,Sockel''*] = plinth masonry

Soda|stein *m*, Sodalith *m* (Min.) = sodalite

~verfahren *n* [*Wasserenthärtung*] = soda process

Sode *f* → Plagge

Soden|fläche *f*, Gras~ = sodded area

~schneider *m* → Gras~

Sofortdehnung *f* = instantaneous strain

soforttragend [*Grubenstempel*] = immediate-bearing [*prop*]

Sohlbank *f*, Fensterbank, Fenstersohlbank = window sill, ~ cill

~riegel *m* → Brustriegel

Sohldruck *m* = bearing pressure

~verteilung *f* = distribution of bearing pressure, bearing pressure distribution

Sohle *f* = invert [*Originally, the inverted arch of a masonry-lined sewer. By derivation, the floor, bottom, or lowest point in the internal cross section of a sewer*]

~ [*Stollen; Tunnel*] = invert, floor, bottom

~ [*Graben*] = bottom

~ [*Fluß; Strom; Kanal*] = bed, bottom

~ [*Schleuse*] = floor

~ [*Vertiefung vor einer Kaimauer*] = (dredged) berth

~, Bruch~, Steinbruch~ = floor, quarry ~

~ → → → Spundwand *f*

Sohlen|aufladung *f* [*Hydraulik*] = bed sedimentation

~auspflasterung *f* = bed stabilization with (paving) setts, bottom ~ ~ (~) ~

~befestigung *f* [*Kanal*] = bed stabilization, bottom ~

~dichtung *f* [*Wehr*] = bottom seal

(~)Dichtungsbalken *m*, hölzerne Dichtungsleiste *f* [*Walzen(stau)wehr*] = timber stop

~druck *m* → Aufquellen *n* des Liegenden (⚒)

~entwässerung *f* = floor drainage, invert ~, bottom ~

~falz *m* [*Schwimmtor*] = floor groove, ~ recess

~fracht *f* [*Fluß*] → Geschiebe(masse)

~gefälle *n* [*Wasserlauf*] = bed slope, bottom ~

~geschwindigkeit *f* [*Hydraulik*] = bottom velocity, bed ~

~gewölbe *n* = inverted arch

~glätter *m* [*Grabenbagger*] = crumbling shoe

~hebung *f* [*Tunnel; Stollen*] = floor-lift, bottom-lift, invert-lift

~höhe *f* [*Abwasserleitung*] = invert level

~höhe *f* [*Schleuse*] = floor level

~höhe *f* [*Fluß; Strom; Kanal*] = level of the bed, ~ ~ ~ bottom

~höhe *f* [*Stollen; Tunnel*] = level of the bottom, ~ ~ ~ floor, ~ ~ ~ invert

~material *n*, Bettmaterial = (stream) bed material, bottom ~

~ort *n* und Firste *f* → Aufbruchbauweise *f*

~quader *m* [*der Dammfalze einer einfachen Kammerschleuse*] = foot-block

~stein *m*, Sohlstein = invert block [*A voussoir-shaped hollow tile built into the invert of a masonry sewer*]

~strömung *f* = bed current, bottom ~

~tiefe *f*, Graben~ = bottom depth

~wasserdruck *m*, Auftrieb *m* = foundation water pressure, uplift

Sohl|fläche *f* [*untere Begrenzung einer geologischen Schicht*] = subface

~fuge *f* = foundation joint

~gewölbe *n*, Sohlengewölbe = inverted arch

Söhligbohrung *f*, Horizontalbohrung = horizontal drilling

söhlige Abbaustrecke *f* (⚒) = level advance line

~ Strebfront *f* (⚒) = level face

Sohl|platte *f*, Lagerplatte = sole plate, bed ~

~schwelle *f* → Drempel *m*

~schwelle *f* → Grundschwelle

~spannung *f* [*Bodenmechanik*] = contact pressure

~spannungsverteilung *f* = distribution of contact pressure

~stein *m* → Sohlenstein

~stollen *m*, Förderstollen = bottom drift

~stollen *m* → → Tunnelbau

~vortrieb *m* → → Tunnelbau

~wasserdruck *m* → Sohlenwasserdruck

Sojaöl *n* = soybean oil

SOK *f*, Schienenoberkante = rail top

Sol *n* = sol

Solar|konstante *f* = solar constant

~öl *n*, Fusel-Öl Nr. 2 = solar oil

Sole *f* = brine

~, Salzquelle *f* = saline spring

~kühlanlage *f*, Solekühler *m* [*Gefrierverfahren*] = brine cooling system

~pumpe *f* [*Gefrierverfahren*] = brine pump

Solfatara *f* (Geol.) = solfatara

Solfatarenstadium *n* = solfatara stage

Solifluktion *f* → Bodenfließen *n*

Soll|bruchstelle *f* = plane of weakness

~dicke *f* = nominal thickness

~drehzahl *f*, Nenndrehzahl = rated speed, ~ RPM

~höhe *f* = nominal height

~höhe *f* = theoretical elevation

~korngröße *f*, nominelle Korngröße = nominal grain size

~-Lage *f*, zeichnungsmäßige Lage = theoretical location

~-Leistung *f* = target output

~-Linie *f*, Sollinie = theoretical line

~maß *n*, Nennmaß = real measure, nominal size

~profil *n* = theoretical profile, nominal ~

~sieblinie *f*, Idealsieblinie = ideal grading curve

~tiefe *f* = theoretical depth

Solnhofener Schiefer *m*, ~ Plattenkalk *m*, ~ Kalkschiefer = Solnhofen (platy) (lime)stone

solodiziert [*Boden*] = solodized

Solquelle *f* [*fördert gelöstes Salz aus tiefen Schichten*] = brine spring

Solventnaphtha *n* = solvent naphtha

Sommer|beton *m* = summer-placed concrete

~bewässerung *f* = summer irrigation

~deich *m*, Nebendeich, Überlaufdeich [*Kehrt nur die HW*] = summer dyke; ~ dike (US)

~energie *f* = summer (electrical) energy

~flickarbeit *f* = summer patching

~hochwasser *n* = summer flood

~hydrant *m* = summer hydrant

~-Klimaanlage *f* = summer air conditioning system

~müll *m* = summer refuse

~öl *n* = summer oil

~polder *m* [*ist durch einen Sommerdeich geschützt*] = summer polder

~verkehr *m* = summer traffic

~weg *m* [*Landstraße*] = soft shoulder

~zeit *f* = daylight-saving-time

Sonde, Bohrung *f*; Schacht *m* [*in Galizien*] = bore, boring, well

~ = sounding apparatus, soil penetrometer

~, Peillatte *f* = sounding pole

~ im Stadium der Anfangsspitzenförderung → Bohrung *f* ~ ~ ~ ~

~ mit Hilfe von Druckgas fördernd → Bohrung *f* ~ ~ ~ ~ ~

~ ~ ~ ~ ~ Druckluft fördernd → Bohrung *f* ~ ~ ~ ~ ~

~ nach Barentsen = coneshaped sounding apparatus

~ ohne Filterrohre → Bohrung *f* ~ ~

~ zum Pumpen vorrichten → Bohrung *f* ~ ~ ~

Sonden|arbeit *f* [*Tiefbohrtechnik*] = formation testing operations

~bohrer *m*, Bodensonde *f* = probing staff, pricker ~, soil test probe

Sondenbohrung — Sonnenschutz

~bohrung *f*, Sondenpressung = probing, pricking

~rammung *f* = driver probing, ~ pricking

~stab *m* = sounding rod

Sonder|antrieb *m*, Spezialantrieb = special drive

~baustahl *m*, Spezialbaustahl = special structural steel

~bau(werk) *m*, *(n)*, Spezialbau(werk) = special structure

~bereifung *f* = optional tires (US); ~ tyres (Brit.)

~betonstahl *m* = special reinforcing steel

~bitumenemulsion *f*, Spezialbitumenemulsion = special asphalt emulsion (US); ~ asphaltic bitumen ~ (Brit.)

~block *m* = special block

~brücke *f*, Spezialbrücke = special bridge

~emulsion *f* = special emulsion

~fahrzeug *n*, Spezialfahrzeug = special vehicle

~fahrzeugachse *f*, Spezialfahrzeugachse = special vehicle axle

~fall *m*, Einzelfall = special problem

~(fertig)teil *m, n* = special (prefab(ricated)) element

~form *f*, Spezialform = special purpose mould (Brit.); ~ ~ mold (US) [*moulds for bridge beams, multistorey buildings, industrialized buildings etc.*]

~form *f* = special shape

~gewebe *n*, Spezialgewebe [*Siebtechnik*] = special screen cloth

~glas *n*, Spezialglas [*läßt die ultravioletten Strahlen mehr durch als gewöhnliches Glas*] = special glass

~gußeisen *n* **Ni-Hard mit martensitisch-weißem Gefüge** = Ni-Hard martensitic white cast iron

~güte *f* = high grade

~kran *m*, Spezialkran = special crane

~kübel *m*, Spezialkübel = bucket for special uses

~lack *m*, Speziallack = special varnish

~lager *n*, Speziallager = special bearing

~muffe *f*, Wahlmuffe [*Tiefbohrtechnik*] = special clearance coupling

~öl *n* = special lubeoil

~pfahl *m*, Spezialpfahl = special pile

~profil *n*, Spezialprofil = special section

~prüfmaschine *f*, Spezialprüfmaschine = special testing machine

~schalung *f*, Spezialschalung = special formwork

~schaufel *f*, Spezialschaufel = special application bucket

~sitz *m*, Spezialsitz = special seat

~stahl *m* = special steel

~teil *m, n* → Sonderfertigteil

~ziegel *m* = special brick

~zubehör *m, n* = options

Sondierbohrung *f* → Bohrsondierung

Sondieren *n* = sounding

Sondier|schacht *m* = exploratory shaft, exploring ~

~stab *m*, Prüfstab = sounding rod

~stab *m* Künzel, Prüfstab ~ = Künzel sounding rod

~stollen *m* = exploratory gallery, exploring drift

~-Tachygraph *m* = sounding tachygraph

Sondierung *f* = sounding, subsurface ~, penetration testing

~ [*zur vorherigen Festlegung der Rammtiefe von Pfählen für die Pfahlherstellung*] = static penetration test

~ **mit direktem Verfahren** → Bohrsondierung

Sondier(ungs)stollen *m* = exploration tunnel

Sonnen|blende *f* = sun screen, ~ breaker

~dach *n*, Schutzdach, Sonnenschutzdach = concrete protection tent, curing ~

~energie *f* = solar energy

~rad *n* = sun gear

~ritzel *n* = sun pinion

~schutz *m* [*im Hochbau*] = solar shading

Sonnenschutz-Baustoff — Spaltenquelle

~schutz-Baustoff *m* [*im Hochbau*] = sun control (device)

~(schutz)dach *n* [*Betonnachbehandlung*] = concrete protection tent, curing ~

~(schutz)dach *n* aus Zeltplane = low level canvas cover

~schutzglas *n* = anti-sun glass

~schutzvorrichtung *f*, ~ control = sun protective device

~seite *f* (Geol.) = adret

~terrasse *f* = sun-bathing terrace

~wärme *f* = solar heat

~welle *f* = solar wave

~wendtide *f* = solstitial tide

Sonometrie *f* [*Messung und Analyse von Geräuschen*] = sonometry

Sorel|mörtel *m* = Sorel('s) composition [*Composition made of Sorel('s) cement and filler*]

~zement *m*, Sorel'scher Zement, Magnesiabindemittel *n* = Sorel('s) cement, plastic magnesia

Sorte *f* [*(Bau)Stoff*] = grade

sortenweise Flotation *f*, selektive ~ = differential flotation

Sortier|anlage *f* = separating plant

~band *n* → Klaubeband

Sortieren *n* = separating

Sortier|gleis *n*, Zerlegungsgleis, Ausziehgleis = sorting track

~konus *m*, konischer Siebzylinder *m* = conical-screen (arrangement), perforated cone

~maschine *f* [*Aufbereitungstechnik*] = separating machine

~sieb *n* = separating screen

~trommel *f* = separating drum

sozialer Wohnungsbau *m* = social housing, low-cost ~

Sozial|wohnung *f* = subsidized dwelling

~wohnungsbauprojekt *n* = housing scheme

S.P., Eigenpotential *n* = spontaneous potential

Spa(ch)tel *n*, *f*, ~eisen *n* = spatula

Spachtel *m* → Spachtelmasse *f*

~-Ausgleichmasse *f* → Nivelliermasse

~-Ausgleichzement *m* [*Fehlname*] → Nivelliermasse *f*

~automat *m*, automatische Spachtelmaschine *f* [*zum Spachteln von Terrazzoplatten*] = auto(matic) grouting machine

~boden *m* → fugenloser (Fuß)Boden

~-Füllmasse *f* → Nivelliermasse

~-Füllzement *m* [*Fehlname*] → Nivelliermasse *f*

~(fuß)boden *m* → fugenloser (Fuß-)Boden

~-Getriebemotor *m* = geared grouting motor

~maschine *f* [*zum Spachteln von Terrazzoplatten*] = grouting machine

~masse *f* [*zum Maschinenspachteln von Terrazzoplatten*] = filling compound, grouting ~

~masse *f*, Spachtel *m* = screed(ing) compound) [*Is a means of level(l)ing uneven surfaces*]

spachteln [*Terrazzoplatten*] = to grout [*terrazzo tiles*]

Spachtel|-Nivelliermasse *f* → Nivelliermasse

~-Nivellierzement *m* [*Fehlname*] → Nivelliermasse *f*

~-Planiermasse *f* → Nivelliermasse

~-Planierzement *m* [*Fehlname*] → Nivelliermasse *f*

~trog *m* [*einer Spachtelmaschine*] = filler trough, grout ~

~werkzeug *n* [*einer Spachtelmaschine*] = filler tool, grouting ~

Spalier *n* = espalier

~latte *f* = espalier lath

~zaun *m* → Ästelzaun

Spalt *m* [*Siebtechnik*] = slot

~barkeit *f*, Spaltfähigkeit *f* = cleavage (property) [*of rocks and minerals*]

~destillation *f* = cracking

~druck *m* [*Steinspaltmaschine*] = blade thrust, ~ pressure

~ebene *f* [*Mineral; Gestein*] = cleavage plane, ~ crack [*mineral; rock*]

Spalten|bildung *f* = fissuring

~(erd)öl *n* = crevice oil

~quelle *f* → Kluftquelle

Spaltenwasser — Spannbetonelement

~**wasser** *n*, Kluftwasser = fissure water, crack ~
Spalt|federbolzen *m* = quick change pin
~**festigkeit** *f* = cleavage strength
~**hütte** *f* [*Gewinnung von Dachschieferplatten*] = splitting shanty
~**keil** *m* = splitting wedge
~**keilanker** *m* = split wedge anchor
~**korrosion** *f* [*DIN 50 900*] = crevice corrosion
~**probe** *f* → Spaltversuch
~**prüfung** *f* → Spaltversuch
~**sieb** *n*, Feinspaltrost *m* = slotted (-hole) screen
~**siebbelag** *m* = wedge-wire mesh
~**stein** *m* = split block
Spaltung *f* in den Spalthütten = hand method, shanty ~ [*manufacture of roof(ing) slate*]
Spaltungsgestein *n* → Schizolith *m*
Spalt|versuch *m*, Spaltprüfung *f*, Spaltprobe *f* = split(ting) test
~**wasserabdichtung** *f* [*Pumpe*] = scavenge water seal
~**weite** *f* [*Siebtechnik*] = width of opening, ~ ~ slot, ~ ~ gap
~**zugfestigkeit** *f* = tensile splitting strength, splitting tensile ~
~**zug(festigkeits)versuch** *m*, Spaltzug(festigkeits)prüfung *f*, Spaltzug(festigkeits)probe *f* = tensile splitting test
spanischer Reiter *m* = cheveaux de frise
Spann|anker *m* = (prestressing) anchorage fixture
~**armierung** *f* → Spannbewehrung
~**bahn** *f* = prestressing line, ~ bed
~**balken** *m* → Trambalken
Spannbeton *m*, vorgespannter Beton = prestressed (reinforced) concrete
~ **mit nachträglichem Verbund** = posttensioned prestressed (reinforced) concrete
~ ~ **Verbund** = pretensioned prestressed concrete
~**anlage** *f* = prestressing plant
~**arbeiten** *fpl* = prestressed work
~**aufgabe** *f* = prestressed concrete problem
~**balken(träger)** *m* = prestressed concrete beam
~**balken(träger)brücke** *f* = prestressed concrete beam bridge
~**bau** *m* = prestressed concrete construction
~**(bau)platte** *f* = prestressed concrete building panel, ~ ~ ~ slab
~**bau(werk)** *m*, (*n*) = prestressed concrete structure
~**becken** *n* = prestressed concrete reservoir
~**behälter** *m* = prestressed concrete reservoir, ~ ~ tank
~**belag** *m* → Spannbetondecke *f*
~**bemessung** *f* = prestress design
~**brett** *n* = prestressed concrete plank [*Used in Scotland for floor construction*]
~**biegebalken** *m*, Spannbetonprobebalken = prestressed flexure test beam
~**brücke** *f* = prestressed (concrete) bridge
~-**Brückenfertigteil** *m*, *n* = prestressed (concrete) bridge member
~**dachbalken(träger)** *m* = prestressed concrete roof-beam
~**(dach)binder** *m* = prestressed concrete (roof) truss
~**dachplatte** *f* = prestressed concrete roof(ing) slab
~**decke** *f*, Spannbetonbelag *m* = prestressed concrete pavement, ~ (rigid) ~, ~ (~) surfacing
~**decke** *f* = prestressed concrete floor
~**deckenplatte** *f* = prestressed floor slab
~-**Deckenplattenelement** *n* = prestressed floor-slab unit
~-**Doppel-T-Träger** *m*, Spannbeton-I-Träger = prestressed concrete I-beam
~-**Dreieckfachwerkträger** *m* = prestressed triangular truss
~**druckrohr** *n* = prestressed concrete pressure pipe
~**element** *n* = prestressed (concrete) unit

Spannbeton-Fachwerkbrücke — Spannbeton-Schleuderrohr

~-**Fachwerkbrücke** *f* = prestressed concrete truss bridge
~-**Fertigdecke** *f* → Spannbeton-Montagedecke
~-**Fertig-Deckenhohlplatte** *f* = hollow precast prestressed floor slab
~-**Fertigplatte** *f* = prestressed precast concrete slab
~-**Fertigteil** *m, n*, Spannbeton-Montageteil = precast prestressed concrete member, prestressed cast-concrete ~, prestressed (concrete) casting
~**fertigwand** *f* → Spannbetonmontagewand
~-**Flugplatzdecke** *f*, Spannbeton-Flugplatzbelag *m* = prestressed airport concrete pavement, ~ airfield ~ ~, ~ ~ (rigid) ~, ~ ~ ~ (~) surfacing
~**gebäude** *n* = prestressed (concrete) building
~**geräte** *npl* = prestressed concrete equipment
~(**gleis**)**schwelle** *f* → Spannbetonschienenschwelle
~-**Hängedach** *n* = prestressed suspended roof
~**herstellungsanlage** *f* = prestressed concrete plant
~**herstellungsplatz** *m* = prestress (casting) yard [*for making precast prestress(ed) members*]
~-**Hochdruckrohr** *n* = prestressed concrete high-pressure pipe
~-**Hochstraße** *f* = prestressed concrete elevated road
~-**Hohlblockträger** *m* = prestressed block-beam
~**hubplatte** *f* = prestressed (concrete) lift slab
~-**Hubplattenbauweise** *f*, Spannbeton-Hubplattenverfahren *n* = prestressed lift-slab construction
~-**I-Träger** *m*, Spannbeton-Doppel-T-Träger = prestressed concrete I-beam
~**kabel** *n* → Bündel aus Drähten
~**kabeldraht** *m* → Bündeldraht

~-**Kassettenhubplatte** *f* = prestressed concrete cored lift slab, ~ ~ waffle ~ ~
~**kastenträger** *m* = prestressed concrete box girder
~-**Kragtreppe** *f* = precast concrete cantilever stair(case)
~**lagerbehälter** *m*, Spannbetonlagertank *m* = prestressed concrete tank
~-**Leitungsmast** *m* = prestressed concrete power pole, ~ ~ ~ mast
~(-**Licht**)**mast** *m*, Spannbeton-Beleuchtungsmast, Spannbeton-(Straßen)Leuchtenmast = prestressed concrete (lighting) column [*B.S. 1308*]; ~ (~) mast (US)
~-**Montagedecke** *f*, Spannbeton-Fertigdecke = precast prestressed (concrete) floor
~-**Montageteil** *m, n* → Spannbeton-Fertigteil
~**montagewand** *f*, Spannbetonfertigwand, Montagespannbetonwand = prefab(ricated) prestressed concrete wall
~-**Parallelfachwerkträger** *m* = prestressed parallel truss
~**pfahl** *m* = prestressed concrete pile
~**platte** *f*, Spannbetonbauplatte = prestressed concrete building panel, ~ ~ ~ slab
~**plattenbrücke** *f* = prestressed concrete slab bridge
~**rammpfahl** *m*, Rammpfahl aus Stahlbeton mit vorgespannter Bewehrung, Rammpfahl aus Spannbeton = (impact-)driven prestressed concrete pile
~**rohr** *n* = prestressed concrete pipe
~**rohr-Triebwasserleitung** *f*, Spannbetonrohr-Druckleitung = prestressed concrete pipe penstock
~**schale** *f* = prestressed shell
~**schienenschwelle** *f*, Spannbeton-(gleis)schwelle = prestressed concrete sleeper (Brit.); ~ ~ tie (US)
~-**Schleuderrohr** *n*, Schleuderbeton-Vorspannrohr = prestressed concrete spun pipe

~schornstein *m* = prestressed concrete chimney, ~ ~ stack
~schwelle *f* → Spannbetonschienenschwelle
~-Schwimmkasten *m* = prestressed concrete box caisson
~(spund)bohle *f* = prestressed concrete sheet pile
~spundwand *f* = prestressed concrete sheet pile wall, ~ ~ sheeting, ~ ~ bulkhead, ~ ~ piling
~stahl *m* = prestressing steel
~startbahn *f* = prestressed runway
~straße *f* = prestressed concrete road
~-Straßenbrücke *f* = prestressed concrete road bridge
~straßendecke *f*, Spannbetonstraßenbelag *m* = prestressed (concrete) road pavement, ~ (~) highway ~, ~ rigid ~ ~, ~ (~) ~ surfacing
~stütze *f* = prestressed concrete column
~talsperre *f* = prestressed dam
~technik *f* = prestressed concrete design and construction
~träger *m* = prestressed concrete girder
~träger *m* mit Vorspannung im Verbund = pretensioned prestressed (concrete) beam
~trägerdecke *f* = prestressed concrete girder floor
~tragwerk *n* [*Teil eines Bauwerkes*] = prestressed loadbearing member, ~ ~ part
~tragwerk *n* [*Teil eines Gebäudes*] = prestressed structure [*The loadbearing part of a building*]
~treppe *f* = precast concrete stair(case)
~-Verbundbalken(träger) *m*, zusammengesetzter Spannbetonbalken(träger) = composite prestressed concrete beam
~verfahren *n* = prestressed clay method
~-Verfahren *n* mit Sandwich-Kabeln = sandwich-plate method
~versuchsstraße *f* = experimental prestressed concrete road, ~ ~ ~ highway
~-Vorfertigungsstelle *f*, bewegliches Montagespannbetonwerk *n* = precast prestressed concrete manufacturing yard
~werk *n* = pretensioning factory, ~ plant, prestressing ~
Spannbett *n* = (pre)stressing bed, ~ mo(u)ld
~verfahren *n* = (pre)stressing bed method, ~ mo(u)ld ~
spannbettvorgespannt = (pre)stressed against the mo(u)ld (or bed)
Spann|bett(vor)spannung *f* = (pre)stressing against the mo(u)ld, ~ ~ ~ bed
~bewehrung *f*, Spannarmierung, Spann(stahl)einlagen *fpl* = prestressing reinforcement, prestressed ~
~block *m* = (pre)stressing block, tensioning ~
~bohle *f* [*bei der Balkenverspreizung*] = strutting piece
~bolzen *m* (✂) = strut
~bolzen *m* = clamping bolt, tension ~
~bolzensicherung *f* [*(Gleis)Kette*] = securing wire of adjustment screw
~bündel *n* → Spannkabel
~(bündel)draht *m* → Bündeldraht
~draht *m* → Bündeldraht
~(draht)bündel *n* → Spannkabel
~einlagen *fpl* → Spannbewehrung *f*
~eisen *n*, Spannglied *n* = tendon, steel ~
Spannen *n*, An~ = stretching, tensioning
~ nach dem Erhärten des Betons = tensioning after hardening of concrete
~ vor dem Erhärten des Betons = tensioning prior to the pouring of concrete
Spann|ende *n* eines Spannbettes = jacking end of a bed
~feder *f*, Ausgleichfeder = equalizer spring, compensator ~
~geräte *npl* [*Spannbeton*] = (pre-)stressing equipment

Spannglied — Spannungsanhäufung

Spannglied *n*, Spanneisen *n* = tendon, steel ~
~ **für direkten Verbund** = pretensioning tendon
~**abstand** *m* = distance between strand prestressing tendons
~**kanal** *m* = tendon duct
~**verankerung** *f* → Spannkopf
~**verfahren** *n* = tendon method
Spann|greifvorrichtung *f* = grip-type locking device
~**größe** *f* = amount of prestress
~**hebel** *m* = tensioning lever
Spannkabel *n*, Spann(draht)bündel *n*, Bündel aus Drähten, Drahtbündel, Vorspannbündel, Vorspannkabel = prestressing cable, (strand) tendon
~**draht** *m* → Bündeldraht
~**schlupf** *m* = strand slip
Spann|kanal *m* = prestressing duct
~**kegel** *m*, Spannkonus *m* [*Spannbeton*] = interlocking cone
~**kolonne** *f*, Spanntrupp *m*, Spannmannschaft *f* = prestressing gang, ~ team, ~ party, ~ crew
~**konus** *m* → Spannkegel *m*
~**kopf** *m*, Spanngliedverankerung *f*, Verankerungs-Vorrichtung *f* = anchorage system, stressing head
~**kraft** *f*, Vor~ = stretching force, prestress(ed) ~, tensioning load, prestressing ~
~**kraftverlust** *m* = prestressing loss
~**mannschaft** *f* → Spannkolonne *f*
~**maschine** *f* = tensioning machine
~**mutter** *f* = tensioning nut
~**mutter** *f* [*Gleiskettenlaufwerk*] = spanner nut
~**platte** *f*, Vor~ = prestressing plate
~**presse** *f* = jack, puller
(~)**Pressenzylinder** *m* = jack cylinder, puller ~
(~)**Pressenzylinderhub** *m* = jack cylinder range, puller ~ ~
~**rad** *n*, Leerlaufturas *m* = idler
~**riegel** *m* [*Kantholz zur Aufnahme der Spannreaktion der Streben bei Hänge- und Sprengwerken*] = collar beam, top ~, span piece
~**rolle** *f* [*Kabelspinnen*] = spinning wheel
~**säule** *f* [*Hammerbohrmaschine*] = pillar support [*drifter*]
Spannschloß *n* = turnbuckle
~ [*Stahlbetonbau*] = coupling unit
Spannschraube *f* = take-up bolt
~ = clamp(ing) screw
Spann|schuh *m* [(*Gleis*)*Kette*] = dead idler
~**schütz(e)** *n*, (*f*) = pressure sluice (or shutter)
~**spindel** *f* = takeup [*belt conveyor*]
~**stab** *m*, Vorspannstab = prestressing bar
~**(stahl)einlagen** *fpl* → Spannbewehrung *f*
~**stahlspannung** *f* = prestressing steel stress
~**station** *f* [*Bandförderer*] = takeup set
spannsteife Form *f*, biegesteife ~ [*Elektrothermisches Spannen von Spannstahl*] = power form
Spann|stift *m* = dowel pin
~**trupp** *m* → Spannkolonne *f*
~**turm** *m* [*Kabelspinnen an einer Brückenverankerung*] = floating-sheave tower, compensating sheave ~
Spannungen *fpl* **durch Brems- und Beschleunigungskräfte** = stresses produced by braking and acceleration
Spannungs|abfall *m* **im Stahl bei gleichbleibender Dehnung** = relaxation (of steel stress), ~ of steel [*Decrease in stress in steel as a result of creep within the steel under prolonged strain; decrease in stress in steel as a result of decreased strain of the steel, such as results from shrinkage and creep of the concrete in a prestressed concrete unit*]
~**abweichung** *f* → Spannungsunterschied *m*
~**anhäufung** *f*, Druckanhäufung *f* [*Baugrund*] = bulb pressure
~**anhäufung** *f*, Spannungskonzentration *f* = stress concentration, concentration of stress

Spannungsaufbringung — Spannverschraubung

~aufbringung *f* = application of stress
~aufgabe *f* = stress problem
~ausgleich *m* [*Boden; Werkstoff*] = stress balancing
~bedingung *f* = stress condition
~berechnung *f* = stress analysis
~bereich *m* = range of stress
~bild *n* = diagram of stresses, stress diagram [*A skeleton drawing of a truss, upon which are written the stresses in the different members*]
~-Dehnungs-Linie *f* = stress-strain curve
~-Dehnungs-Messung *f* = stress-strain measurement
~diagramm *n*, Druckdiagramm [*Baugrunduntersuchung*] = pressure diagram
~ermitt(e)lung *f*, Spannungsberechnung = determination of stress(es)
~feld *n* = stress field
spannungs|frei = stress-relieved
~freie Pore *f* = stress-free pore, unstressed ~
~freies Überschußwasser *n*, Gravitationswasser = gravitational water
~freigeglüht = stress relieved, stress-free annealed
Spannungs|freiglühen *n* = stress relief
~freimachen *n* = stress relieving, relaxation
~gleiche *f*, Isostate *f* = isostatic line
~gleichgewicht *n* = stress equilibrium
~grenze *f* = limit of stress
~größe *f* = magnitude of stress
~intervall *n* = stress interval
~komponente *f*, Spannungsteilkraft *f* = stress component
~konzentration *f*, Spannungsanhäufung *f* = stress concentration, concentration of stress
~korrosion *f* = stress corrosion
~korrosionsbruch *m* = stress corrosion failure
~kreis *m* = circle of stress
~linie *f*, Spannungstrajektorie *f* = trajectory of stress
spannungs|los = stress-free
~lose Niveaufläche *f* (Geol.) = level of no strain

~loser Stab *m*, Blindstab, blinder Stab [*Gitterträger*] = unstrained member
Spannungs|messer *m* = stress meter
~meßsonde *f* → Potentialelektrode *f*
~moment *n* = stress moment
~optik *f* = photoelasticity
spannungs|optisch = photoelastic
~optisches Modell *n* = photoelastic model
Spannungs|resultierende *f* = stress resultant
~riß *m* = stress crack
~spiel *n* = stress repetition
~spitze *f* = stress riser
~system *n* = stress system
~teilkraft *f*, Spannungskomponente *f* = stress component
~tensor *m* = stress tensor
~trajektorie *f*, Spannungslinie *f* = trajectory of stress
~überlagerung *f* [*baustatische Theorie*] = superposition of stresses
~umkehr *f* = reversal of stress [*The changing of stress from tension to compression, or vice versa*]
~umlagerung *f* = redistribution of stress(es)
~unterschied *m*, Spannungsabweichung *f* = deviator stress
~verlust *m*, Vorspannverlust = loss of prestress
~verminderungsfaktor *m* = stress reduction factor
~verteilung *f* = stress distribution
~weg *m* = stress trajectory
~wert *m* = stress value
~zahl *f* = stress ratio
~zunahme *f* = stress increase
~zustand *m* [*Baustatik*] = state of stress
Spann|verankerung *f* = jacking anchorage
~verfahren *n*, Vorspannverfahren = prestressing method, ~ system
~verschluß *m*, Kistenverschluß = toggle bolt (Brit.); drawbolt (US); box closure [*generic term*]
~verschraubung *f* [*Druckluftwerkzeug*] = thread mounting

Spannvorrichtung — Speicher

~vorrichtung *f* = stressing device, tensioning ~, stretching ~

~vorrichtung *f* um bei Luftströmung oder Klemmung oder Durchhängen von Brückenteilen die Blechlagen fachgerecht zu ziehen = tensioner

~weg *m* [*Spannbeton*] = stretching distance, tensioning ~, prestressing ~

~weite *f*, lichte Weite = clear span

Spantenniethammer *m* = shell riveter

Spar|becken *n* [*Offenes Becken neben einer Kammerschleuse, in das zur Wasserersparnis beim Entleeren der Kammer ein Teil des Schleusungswassers eingelassen, und aus dem bei der nächsten Füllung der Schleuse das Wasser wieder in die Kammer zurückgeleitet wird*] = side pond

~beize *f* [*Korrosion unter Wasserstoffentwicklung*] = pickling bath containing an inhibitor [*hydrogen-evolution type of corrosion*]

~beizzusatz *m* [*Korrosion unter Wasserstoffentwicklung*] = pickling inhibitor [*hydrogen-evolution type of corrosion*]

~beton *m*, Magerbeton, Füllbeton, zementarmer Beton, magerer Beton = lean(-mixed) concrete

~düse *f* [*am Vergaser*] = (fuel and air) restrictor, economy jet

~flamme *f* = pilot light [*A small gas flame which on some gas water heaters is left always burning so as to ignite the main gas burners immediately the tap is opened*]

~gemisch *n*, Sparmischung *f* = lean mix(ture)

~kammer *f* [*überdecktes oder in die Wand einer Kammerschleuse eingebautes Becken, das denselben Zweck wie ein Sparbecken hat*] = storage chamber

~mischung *f*, Spargemisch *n* = lean mix(ture)

~mörtel *m* → Magermörtel

Sparren *m* → Dach~

~abstand *m* = distance between rafters

~dach *n* = rafter roof

~halter *m* = rafter clench, ~ cleat

~kopf *m* = rafter end

~nagel *m* = rafter nail

~neigung *f* = rafter slope

~schwelle *f*, Fußpfette *f* [*Pfettendach*] = inferior purlin

~verbindung *f* = rafter connection

Spar|schalung *f* = open formwork

~schleuse *f* [*Kammerschleuse mit Einrichtungen zur Wasserersparnis*] = lock with means for the economisation of water

~widerlager *n*, Hohlwiderlager, aufgelöstes Widerlager = hollow abutment

Spat *m* = spar

~eisenstein *m* = spathic iron (ore)

Spatel *m*, *f* = spatula

Spaten *m*, Grabscheit *n* = spade

~arbeit *f* = spade work

~boden *m* = soil diggable by spade

~(bohr)meißel *m* = drag bit

~(bohr)meißel *m* [*pennsylvanisches Seilbohren*] = spudding bit

~hammer *m* = clay digger, ~ spade, spader, spade hammer

~meißel *m* = drag bit, spudding ~

~spitze *f* [*Aufreiß(er)zahnspitze*] = spade tip

Spätherbstbewässerung *f* = late-fall irrigation

spätiger Gips *m* → Fasergips

Spatsand *m* → Arkose

spättragend [*Stempel*] = late-bearing [*prop*]

Speck|schmierung *f* = lard lubrication

~steinmehl *n* → Talkum *n*

~torf *m* = bacon peat, lard ~

Specularit *m* (Min.) = specular iron ore

Speichenrad *n*, Gitterrad = spoke(-type) wheel

Speicher *m*, Dachboden *m* = attic

~ [*hydraulische Anlage*] = accumulator [*An accumulator is used to store liquid at constant pressure*]

~, Lagerhaus *n* = warehouse

~ → → Talsperre *f*

~ablaß m → → Talsperre f
~becken n → → Talsperre f
~druck m [*Hydraulikanlage*] = accumulator pressure
~fähigkeit f → Fassungsvermögen n
~gas n = compressed gas
~gestein n, Trägergestein = reservoir rock, container ~
~heizkörper m = storage type heater
~heizung f = storage heating
~kessel m → Druckluftbehälter m
~kraftwerk n, Staukraftwerk, Speicherkraftanlage f, Staukraftanlage = storage scheme
speichern, auf~ [*Talsperre*] = to store (water)
~, auf~, einlagern, bevorraten = to store (up)
Speicher|ofen m, Regenerativofen = regenerative furnace
~puffer m, Arbeitspuffer [*Sieb*] = (working) rubber buffer
~raum m [*Wasserbau. Fassungsvermögen eines Speicherbeckens*] = storage capacity
~schleuse f [*Sparschleuse mit übereinanderliegenden in die Kammerwände eingebauten Sparkammern*] = lock with storage chambers
~sperre f → → Talsperre
Speicherung f → → Talsperre f
~ = storage
~ → Auf~
Speicher-Wasserheizer m = storage water heater
Speise|graben m, Zubringer m, Speisungsgraben [*zu einem Wasserlauf*] = feed(er) ditch, feeding ~, feeder
~kammer f = larder
~kanal m = feeder canal
speisen = aufgeben
Speisenaufzug m = dumb waiter (US); service lift (Brit.)
Speisepumpe f = feed pump
Speiser m, Beschickungsapparat m, (Material)Aufgeber m, Aufgabeapparat m, Beschickungsvorrichtung f, Beschicker m, Aufgabemaschine f = feeder, feeding device, charging feeder

Speise|rinne f, Beschickungsrinne, Aufgaberinne = feeder channel
~rohr n = Aufgaberohr
~rost m → Beschickungsrost
~salz n = table-salt
~schnecke f → Beschickungsschnecke
~transformatorenhaus n, Speisetrafohaus = (electricity) substation
~wagen m = diner (US); dining car (Brit.)
~walze f → Beschickungswalze
~wasser n = feed water
~wasserbehälter m = feed water tank
~zimmer n = dining room
Speiskübel m → Mörteltrog m
Speisungsgraben m → Speisegraben
Spektral|apparat m = spectroscope
~-Log n = spectral log
Spektrochemie f = spectrochemistry
Sperrad n, Sperr-Rad = ratchet (wheel)
Sperr|ballon m = barrage balloon
~beton m = waterproof concrete
~damm m → (Stau)Damm
~dammbau m, (Stau)Dammbau, Talsperrendammbau = embankment (type) dam construction, fill ~ ~
~dammbreite f → Staudammbreite
Sperre f → Talsperre
~ an der Seitenkippschaufel = side dump bucket latch
Sperr|griff m = locking handle
~holzabfall m = scrap plywood
~klinke f, Sperrzahn m, Auslöser m = (locking) pawl
~lage f → Sperrschicht f
~linie f [*Fahrbahnmarkierung*] = solid line
~mauer f, Beton~, Betonsperre f, (Massiv)Beton(stau)mauer, Stammauer, Talsperren(stau)mauer, Betonsperrmauer = (massive) concrete dam
~mörtel m = waterproof mortar
~pulver n = cement waterproofing powder
~putz m = waterproof plaster
~-Rad n, Sperrad = ratchet (wheel)
~schicht f, (Ab)Dichtungslage f, (Ab-)Dichtungsschicht, Sperrlage = impervious layer, ~ course, barrier

Sperrschieber — spezifischer Siebdurchsatz

~schieber m = abutment [*vane pump*]
~see f = dam lake
~segment n = locking quadrant
~stoff m, dichtender Zusatz m, Sperrmittel n [*Sperrbeton*] = cement waterproofing agent
~ventil n [*Motor*] = poppet valve
~wand f → Dichtwand
~wand f aus Pfählen → Pfahlwand
~wand f ICOS → ICOS-Wand
~werk n zum Schutz einer Flußmündung (oder Strommündung) vor Sturmfluten = hurricane barrier, ~ dam, ~ wall, storm ~
~zahn m → Sperrklinke f
~zeit f [*Straße*] = block
~zeitenbetrieb m = intermittent operation [*heating system*]
~zusatz m, (Beton)Dichtungsmittel n, wasserabweisendes Mittel, Sperrmittel = water-proofer, concrete waterproofing compound, water-repellent, water-repelling agent, densifier, densifying agent, damp-proofing and permeability reducing agent

Spesen f = entertainment expenses

Spezial|antrieb m, Sonderantrieb = special drive
~aufzug m, Sonderaufzug = purpose-made lift, special ~
~baustahl m, Sonderbaustahl = special structural steel
~bau(werk) m, (n), Sonderbau(werk) = special structure, purpose-made ~
~bitumenemulsion f, Sonderbitumenemulsion = special asphalt emulsion (US); ~ asphaltic bitumen ~ (Brit.)
~brücke f, Sonderbrücke = special bridge
~fahrzeug n, Sonderfahrzeug = special vehicle, purpose-made ~
~fahrzeugachse f, Sonderfahrzeugachse = special vehicle axle
~fenster n, Sonderfenster = purpose-made window, special ~
~form f → Sonderform
~getriebemotor m, Sondergetriebemotor = reduction motor
~gewebe n, Sondergewebe [*Siebtechnik*] = special screen cloth
~glas n → Sonderglas
~kleber m, Sonderkleber = purpose-made bonding agent, special ~ ~
~kran m, Sonderkran = special crane
~kübel m, Sonderkübel = bucket for special uses
~lack m, Sonderlack = special varnish
~lager n, Sonderlager = special bearing
~löffel m für Dränagearbeiten, Sonderlöffel ~ ~ = trench bucket, drainage ~
~pfahl m, Sonderpfahl = special pile
~profil n, Sonderprofil = special section, purpose-made ~
~prüfmaschine f, Sonderprüfmaschine = special testing machine
~schalung f, Sonderschalung = special, purpose-made ~ formwork
~schaufel f, Sonderschaufel = special application bucket
~schlauch m, Sonderschlauch = special hose
~sitz m, Sondersitz = special seat
~stahl m, Sonderstahl = special steel
~transportwagen m, Sondertransportwagen = special transporter

spezifische Austragsleistung f [*Siebtechnik*] = rated output per unit area of useful working surface
~ **Drehzahl** f [*Turbine*] = specific speed
~ **Energie** f → Energiehöhe f [*Hydraulik*]
~ **Oberfläche** f [*Oberfläche einer Körnung je Mengeneinheit des Gutes mit Angabe des Ermittlungsverfahrens* (cm^2/g)] = specific surface
~ ~ O_K, Oberflächenkennzahl f [*Siebtechnik*] = specific surface (factor)
~ **Siebübergangsleistung** f in t/h = specific discharge of overflow in t/h
~ **Wärme** f = specific heat
~ **Wasserführungsfähigkeit** f = discharge modulus

spezifischer Siebdurchsatz m, Leistung f in t/h m² nutzbare Siebfläche =

rated throughput in t/h m² effective screen area
S-Pfanne *f* = pantile
S-Pfannendeckung *f* = pantiling
sphärische Containerschale *f* = spherical containment shell [*reactor plant*]
Sphärolithgefüge *n* = spherulitic texture
Spiegel|(ab)senkung *f*, **Wasser~** = drawdown
~gewölbe *n* = cavetto vault
~glas *n* = plate glass [*This glass is different from the other types of flat glass in that, after rolling, it is mechanically ground and polished. This operation imparts a fine optically true surface which makes the plate glass useful for glazing large openings and display windows, for mirrors, and more recently, for glass clad buildings*]
~hebung *f* = (An)Stauung
~kurve *f*, **Wasserspiegellinie** *f* = surface profile, ~ curve
spiegelnde Gleitfläche *f* → (Rutsch-) Harnisch
~ Rückstrahlfläche *f* [*Reflektor*] = specular finish
Spiegel|probe *f*, **Abspiegeln** *n* [*Rohr*] = inspection with the aid of mirrors
~schwankung *f* = liquid level variation
~senkung *f* → Spiegelabsenkung
Spiegelungsglanz *m* = specular gloss
Spiel *n* [*(Gleis)Kettenfahrzeug*] = free travel
~ [*bei Passungen*] = play, clearance
~platz *m*, **Kinder~** = childrens' playground, games area
~platzgerät *n* = playground equipment
~straße *f* = play street
~zahl *f* = number of work cycles
Spierentonne *f* = spar buoy; dan ~ (US)
Spieß *m* [*Wenner'sche Vierpunktmethode*] = current stake
~glanzbleierz *n* → Rädererz
Spill *n* → Gang~
~ [*Rotary-Bohranlage*] = cat(head)

~ zum Anschrauben [*Rotary-Bohranlage*] = spinning cat(head)
~ zum Losschrauben [*Rotary-Bohranlage*] = break out cat(head)
~arbeit *f* [*Tiefbohranlage*] = cat(head) job
~bock *m* [*Tiefbohranlage*] = cat(head) installation
~hauptwelle *f* [*Tiefbohranlage*] = main cat(head) shaft
~kopf *m* [*Tiefbohrtechnik*] = cat(head)
~kupp(e)lung *f* = cat(head) clutch
~seil *n* [*Rotarybohren*] = cat(head) line
~seilrolle *f* [*Rotarybohren*] = catline sheave
~welle *f* [*Tiefbohranlage*] = cat(head) shaft
Spind *m* = locker
Spindel *f* → **Schrauben(hebe)bock** *m*
~ = spindle
~ → **Gerüst~**
~, **Aräometer** *n*, **Senkwaage** *f* = hydrometer
~boje *f*, **Spindeltonne** *f* = spindle buoy
~-Kniehebelverbundpresse *f* = screw and toggle press
~schieber *m* = spindle type valve
~treppe *f* [*Wendeltreppe auf Kreisgrundriß*] = spiral stair(case), helical ~ [*The term "helical stair" is correct since the shape is helical and not spiral, but this term is not usually employed*]
~vorschub *m* = spindle feed
Spinne *f*, **Kreuzstück** *n* **für Kardangelenk** = spider
~, **Ringaussteifung** *f* [*Gleitschalungsbau*] = spider, deck ~
spinnenförmige Ausdehnung *f* **einer Stadt** = urban sprawl
Spinn|kolonne *f*, **Spinnmannschaft** *f*, **Spinntrupp** *m* [*Kabelspinnen beim Brückenbau*] = spinning crew, ~ gang, ~ team, ~ party
~mannschaft *f* → **Spinnkolonne** *f*
~rad *n* [*Kabelspinnen beim Brückenbau*] = spinning wheel

Spinntrupp — Spitzendruck

~trupp *m* → Spinnkolonne *f*
Spion *m* = judas [*Spy-hole in a door panel*]
Spiralarmierung *f* → Spiralbewehrung
spiral|bewehrte Stütze *f*, spiralarmierte ~ = spirally reinforced column, hooped ~
~**bewehrter Schubverbinder** *m*, spiralarmierter ~ = spiral shear connector
Spiralbewehrung *f*, Spiralarmierung, Spiral(stahl)einlagen *fpl*, (Spiral-)Umschnürung, Spiralumwick(e)lung, Schraubenlinienbewehrung, Schraubenlinienarmierung, Querbewehrung nach der Schraubenlinie, Umschnürungseisen *npl* = spiral (reinforcement), ~ hoop(ing), ~ hoops, ~ bars, (steel) helix, helical reinforcement
Spiralbohrer *m* [*Bodenuntersuchung*] = earth auger, soil ~, ~ boring ~, ~ drill
~ [*für Gestein*] = spiral drill
~ [*für Metall*] = twist drill
Spiral|(bohr)meißel *m* = spiral (drill) bit
~**brenner** *m* = spiral burner
Spirale *f*, Wendelscheider *m* = spiral (concentrator)
~, Spiralgehäuse *n* [*Turbine*] = volute casing, spiral ~
Spiral|einlagen *fpl* → Spiralbewehrung *f*
~**feder** *f* = spiral spring
~**federschlauchschutz** *m* = spiral-type hose guard
~**förderer** *m* = spiral (feed) conveyor, ~ (~) conveyer, ~ feeder
spiral|förmig angeordnete Mitnehmer(schaufeln) *mpl*, (*fpl*) = spiral flights
~**geschweißtes Rohr** *n* = spiral welded pipe
Spiral|gehäuse *n*, Spirale *f* [*Turbine*] = volute casing, spiral ~
~**haken** *m* = spiral hook
~**kegelrad(an)trieb** *m* = spiral bevel gear drive
~**klassierer** *m* = spiral classifier
~**konzentrator** *m* = spiral concentrator
~**kratzer** *m* = spiral scraper
~**kurve** *f* = spiral (curve)
~**meißel** *m*, Spiralbohrmeißel = spiral bit
~**(naß)klassierer** *m* → Sandschnecke *f*
~**rampe** *f* = spiral ramp
~**seil** *n* = spiral rope
~**(stahl)einlagen** *fpl* → Spiralbewehrung *f*
(spiral)umschnürte Säule *f*, ~ Stütze [*Stahlbetonstütze mit kreisförmigem Kernquerschnitt und Schraubenlinienbewehrung*] = spirally reinforced column [*sometimes erroneously called "hooped column"*]
spiralumschnürter Beton *m* = spirally bound concrete
(Spiral)Umschnürung *f* → Spiralbewehrung
spiralumwickelt = spiral wrapped
Spiralumwick(e)lung *f* → Spiralbewehrung
Spiritus|beize *f* = spirit-based mordant
~**lack** *m* = spirit varnish
~**lack** *m* **auf Schellackbasis** = shellac spirit varnish
Spitalbau *m* [*Schweiz*]; Krankenhausbau = hospital construction
Spitz|balken *m* → Hahnenbalken
~**bogen** *m* → gotischer Bogen
~**bogenkuppel** *f* = pointed dome
~**(bohr)meißel** *m* [*Rotarybohren*] = diamond point(ed) bit, rotary drilling bit with pointed cutting edge
spitzen [*Werkstein-Bearbeitungsart*] = to point
Spitzen|abfluß *m*, ~**menge** *f* = peak flow, ~ discharge
~**ausleger** *m* = jib extension (Brit.); boom ~ (US)
~**bedarf** *m* = peak demand
~**bedarf** *m* **am Tage** = peak daytime demand
~**beiwert** *m* = peak coefficient
~**belastung** *f* = peak load
~**belastungszeit** *f* = peak load period
~**bügel** *m* [*Aufreißer*] = tip wrapper
Spitzende *n*, Schwanzende [*Muffenrohr*] = spigot end
Spitzendruck *m* = point bearing, end ~ [*pile*]

Spitzenpfahl — Splitt-Schleuderverteiler

~pfahl *m* = (point-)bearing pile, end-bearing ~
~sonde *f* = car-jack type sounding apparatus
Spitzen|energie *f* → Spitzenstrom
~geschwindigkeit *f*, Höchstgeschwindigkeit = maximum speed
~halter *m* [*Aufreißer*] = tip retainer, ~ holder
~-KWh-Kapazität *f* = peak generating capacity
~(kraft)werk *n* = peak-load power plant
~periode *f* → Spitzenzeit *f*
~pfahl *m*, Standpfahl [*durchdringt weiche Bodenschichten und steht auf festem Boden*] = (point-)bearing pile
~speicher *m* = peak reservoir
~spiel *n* [*von Zahnrädern*] = crest clearance
~strom *m*, Spitzenenergie *f* = peak (electrical) energy, ~ power
~stunde *f* = peak hour
~stundenpassagier *m* = peak-hour passenger
~temperatur *f* = peak temperature
~tragfähigkeit *f* [*Pfahl*] = point-bearing capacity, permissible point-load [*pile*]
~verbrauch *m* = peak consumption
~verkehr *m* = peak traffic
~werk *n*, Spitzenkraftwerk *f* = peak-load power plant
~wert *m* = peak value
~widerstand *m* [*Pfahl*] = point resistance
~zeit *f*, Spitzenperiode *f*, Stoßzeit, Stoßperiode = peak period
~zeitbahnhof *m* = rush-hour station
spitzer Trägeranschluß *m* = girder connection at an angle
Spitzfänger *m*, Fangdorn *m* [*Stahldorn mit gehärtetem Schneidgewinde zur Ausführung von Fangarbeiten*] = fishing tap(er)
~führungsrohr *n*, Fangdornführungsrohr = bowl for fishing tap(er)
Spitz|hacke *f*, Spitzhaue *f* = pick
~hammer *m* → Latthammer
~haue *f*, Spitzhacke *f*, = pick
~kasten *m* [*Ein nichtmechanischer Klassierer. Er bewirkt die Trennung von Sand und Schlamm ohne mechanische Austragvorrichtung*] = cell hopper, sorting ~, classifier pocket, sorting pocket
~kelle *f* = pointed trowel
~körper *m* [*Siebtechnik*] = twilled herringbone weave
~kuppel *f* = dome with spire
~masche *f* [*Siebtechnik*] = long diamond mesh
~meißel *m* → Spitzbohrmeißel
~säge *f*, Stichsäge, (Schlüssel)Lochsäge = keyhole saw, pad ~ [*used for cutting holes in the middle of the wood*]
spleißen → anscheren
Spleißen *n*, Anscheren = splicing
Spleißnadel *f* = splicing needle
Spliß = splice
~ [*Holzstück, Zinkblechstreifen oder Dachpappenstreifen zur Dichtung der Längsfugen bei einfacher Dachdeckung*] = slip, splinter
~dach *n* = split tiled roof
Splint *m* = cotter pin
~verbindung *f* = cotter joint
Splitt *m* = chipping, chip(ping)s, stone ~ [*deprecated: screening(s)*]
splittarmer Asphaltfeinbeton *m* → Topeka *m*
Splittbeton *m* → → Beton
splitten → verteilen von Splitt
Splitteppich *m*, Splitt-Teppich = chip(ping)s carpet, ~ mat
Splitter|graben *m* = slit trench (US)
~schutzboxe *f* = aircraft pen (Brit.): ~ revetment (US)
Splitt|lage *f*, Splittschicht *f* = layer of chip(ping)s
~-LKW *m* = chip(ping)s lorry
splittreich = stone-filled
Splitt|sand *m* → Pochsand
~schicht *f*, Splittlage *f* = layer of chip(ping)s
~-Schleuderverteiler *m*, Schleuder-Splittverteiler = spinner gritter, spinning ~, ~ gritting machine

Splittstreuen — Willkommen Sprenggelatine

~streuen *n* → Abdecken *n*

~streuer *m*, Splittstreugerät *n*, Splittstreumaschine *f* = (mechanical) gritter, gritting machine, chip(ping)s spreader

~streuer *m* mit Schleuderverteilung = spinner type distributor (or spreader) for chipping(s), spinning (or spinner) gritter

~streufahrzeug *n*, Absplittfahrzeug = grit spreading vehicle

~streugerät *n*, Splittstreuer *m* = gritting machine, (mechanical) gritter

~streukarre(n) *f. (m)*, Handsplittstreuer *m* = barrow-type chip(ping)s spreader, hand-operated ~ ~

~streu-LKW *m*, Absplitt-LKW = gritting truck (US); ~ lorry (Brit.)

~-Teppich *m*, Mischsplitt-Teppich = chip(ping)s carpet (or mat)

~ und Schotterverteiler *m* = aggregate spreader, stone ~

~verfüllschicht *f* [*Tränkmakadam*] = choker course (of aggregate)

~verteilung *f*, Absplitten *n*, Abdecken = chipping, grit blinding, gritting [*deprecated: blinding, dressing*]

~walzenbrecher *m*, Walzensplittbrecher = crushing rolls for chip(ping)s production

Spongiose *f*, Graphitierung *f* = graphitization, graphitic corrosion

Sporn *m*, Abschlußwand *f*, Herdmauer *f*, Fußmauer, (Ab)Dichtungsmauer, (Ab)Dichtungssporn, Trennmauer [*Talsperre*] = (toe) cutoff wall

~rad *n* = small pivoted wheel, swivel ~

Sport|feld *n*, Sportplatz *m* = sports ground

~halle *f* = sports hall

~palast *m* = sports palace

~platz *m*, Sportfeld *n* = sports ground

Spratzlava *f*, Blocklava, Schollenlava, Zackenlava, Aprolith *m* = block lava, aa (~)

Sprech|anlage *f* mit Hörmöglichkeiten, Wechselsprechanlage [*z.B. beim Bau von Hochhäusern eingesetzt*] = communication system to provide two-way conversation facilities

~funk *m*, Funksprechverkehr *m* = radio telephony

Spreizbolzen *m* = expansion bolt

Spreize *f*, Schrägsteife *f* (⚒) = sprag [*Generally a short sloping timber placed with one end on the floor and the other end tight against a rock which may fall*]

Spreiz|holmkurvenfahrwerk *n* = travel(l)ing gear with struts for running on curved rails

~hülsenanker *m* = expansion-shell bolt

~ringkupp(e)lung *f* = expanding-band clutch

~trompete *f* [*Spannbeton*] = widened end of cable duct

Spreizung *f* [*konzentriertes Spannglied*] = widened end of cable duct

Spreng|abschnitt *m*, Sprengstrecke *f*, Abschlag *m* [*Vortrieb je Angriff*] = round

~arbeit(en) *f(pl)* = blasting work

Spreng(bohr)loch *n* → Bohrloch

~besatzstoff *m* → Bohrlochbesatzstoff

~durchmesser *m* → Bohrlochdurchmesser

Sprengen *n* → → 1. Schießen; 2. Sprengen

~ mit Knäpperschuß, Knäppern *n*, Nachknäpperschießen = boulder blasting

Sprenger *m* = distributor [*A device used to apply sewage to the surface of a filter. There are two general types, fixed and movable. The fixed type may consist of perforated pipes or notched troughs, sloping boards, or sprinkler nozzles. The movable type may consist of rotating or reciprocating perforated pipes or troughs applying spray or a thin sheet of sewage*]

Spreng|fels *m* → anstehender Fels

~gelatine *f* = blasting gelatin(e)

Sprenggerät — Spritzasbest

~gerät *n* zum Trennen von Verschraubungen = back-off-tool [*deep drilling*]

~kammer *f* = demolition chamber, blast ~

~kammer *f* = coyote hole

~kapsel *f* = (fulminate) detonator, (blasting) cap

~leistung *f* → Sprengwirkung

Sprengloch *n* → Bohrloch

~abstand *m* → Bohrlochabstand

~besatzstoff *m* → Bohrlochbesatzstoff

~bohrer *m*, Schußlochbohrer = blasthole drill

~durchmesser *m* → Bohrlochdurchmesser

Spreng|öl *n*, Nitroglyzerin *n* = nitroglycerin(e), nitroleum, blasting oil, explosive oil, fulminating oil, soup

~pulver *n*, Schwarzpulver = black (blasting) powder, gunpowder

~rampe *f* [*Vibrationssieb*] = rinsing bar

~rampe *f*, Düsenstrang *m*, Spritzrampe, Reihenspritzvorrichtung *f*, Sprengstrang *m*; Balkenbrause *f* [*Schweiz*]; Spritzbarren *m* [*Österreich*] [*Motorspritzgerät*] = spray bar

~ring *m*, Schnappring = snap ring, lock ~

~ringnut *f* [*Radreifen*] = retaining ring groove

Sprengstoff *m* [*Ein Explosivstoff, der sich für sprengtechnische Zwecke eignet*] = blasting explosive

~ für Tränkungsschießen = pulsed infusion explosive

~dichte *f* = blasting explosive density [*g./cc.*]

Spreng|strang *m* → Sprengrampe

~strecke *f* → Sprengabschnitt *m*

~technik *f* = blasting practice

Sprengung *f* **von Gestein**, Sprengen *n* ~ ~ = rock blasting

Spreng|vorschriften *fpl* **für den Kohlenbergbau** = Coal Mines (Explosives) Regulations

~wagen *m*, Spritzwagen [*Betonnachbehandlung*] = water sprinkler

~wagen *m*, Spritzwagen [*zur Staubbekämpfung*] = sprinkler truck (US); ~ lorry (Brit.)

Sprengwerk *n* = truss

~balken *m* = strut-framed beam

~brücke *f* = trussed bridge

~dach *n* = strutted roof

Sprengwirkung *f*, Sprengleistung = intrinsic strength, strength per unit weight of the explosive, weight strength

Spreustein *m* (Min.) → Fasersiedestein

Sprieße *f* → Quersteife *f*

sprießen → (ab)bölzen

Sprießung *f* → (Ab)Bölzung

Springbrunnen *m* = fountain

(Spring)Brunnenbecken *n* = fountain basin

Springduowalzwerk *n*, Vorsturzwalzwerk = jump roughing mill

springende Geschiebeführung *f* = skipping transport, saltation ~

Springer *m* [*gewaltsamer Ausbruch einer unvorsichtig niedergebrachten Bohrung*] = gusher

Spring|flut *f*, Springtide *f* = spring tide, S.T.

~hochwasser *n* → Springtidehochwasser

~tide *f*, Springflut *f* = spring tide S.T.

~tide *f* zur Zeit der Äquinoktien = equinoctial spring tide

~(tide)hochwasser *n*, Hochwasser der Springtide = high water spring tide, ~ ~ (level) of ~ ~, H.W.S.T.

~tidehub *m* = spring range

~(tide)niedrigwasser *n*, Niedrigwasser der Springtide = low water spring tide, ~ ~ (level) of ~ ~, L.W.S.T.

~wasserstand *m* = water level at spring tide

~zeit *f* [*Zeitpunkt der größten Einwirkung von Mond und Sonne auf den Gezeitenhub*] = spring tide time

Sprinkleranlage *f* = stationary sprinkling system

Spritz|anstrich *m* = spray coat(ing)

~asbest *m* = limpet asbestos, sprayed ~

Spritzauftrag — Spritzwagen

~auftrag m = spray application
~auskleidung f, Spritzbekleidung, Spritzverkleidung = sprayed lining
~barren m → Sprengrampe f
~beizanlage f = spray pickling unit
~bekleidung f → Spritzauskleidung
Spritzbeton m, Torkretbeton [Man unterscheidet das Torkret-Verfahren und das Moser-Kraftbau-Verfahren. Beim Torkret-Verfahren wird mit der „Torkret-Kanone" unter einem Preßluftdruck von 2 bis 3 atü ein Gemisch von Zement und Sand bis 10 mm Korngröße, welches nicht vorgenäßt wird, durch eine beliebig lange Rohrleitung gepreßt, an deren Endmundstück die Benetzung des Feinbetons mit Hilfe einer Wasserdüse erfolgt. Die Feinbetonmasse wird durch den Luftkompressor gegen eine Schalung oder die zu torkretierende Wand geschleudert und bildet dort eine feste und dichte Kruste. Die Masse wird in mehreren Lagen übereinander aufgespritzt, die sich infolge des Preßdrucks gut miteinander verbinden und auch größere Wandstärken herzustellen gestatten. Mit dem Verfahren werden auch Stahleinlagen gut und rostsicher umhüllt. Das Verfahren findet bei der Auasbesserung beschädigter Bauteile, bei der Verstärkung von Stahlbetonkonstruktionsteilen, bei der feuersicheren Umkleidung von Stahlbauteilen und auch bei der Herstellung von Schalenkonstruktionen und Behältern sowie für sehr dichten Verputz Verwendung. Beim Moser-Kraftbau-Verfahren wird der Mörtel oder Feinbeton von vornherein mit Wasser versetzt, durch einen Trichter dem Endmundstück des Preßluftschlauches zugeführt, der im allgemeinen nur wesentlich kürzere Längen als beim Torkret-Verfahren haben darf, und von hier durch den Preßluftstrom hinausgeschleudert] = air-placed concrete, gunite, shotcrete, jetcrete, gun-applied concrete, pneumatically placed concrete, gunned concrete, cement-gun concrete, pneumatically-applied mortar
~auskleidung f, Torkretbetonauskeidung = shotcrete lining, gunite ~
~hilfe f = guniting aid
~maschine f, Torkretkanone f = compressed-air ejector
Spritzdruck m [Bindemittelverteilung] = pressure of application
(Spritz)Düse f = (spray) jet
Spritzen n mit Handpumpendruck und Handrohr = hand spraying [manual pumping and hand lance]
~ ~ Maschinenpumpendruck und Handrohr = power spraying with hand lance
~ ~ Sprengrampe = bar spraying
Spritzer m [Meer] = spray
~ = splash
~ [Arbeiter, der auf Spritzarbeiten spezialisiert ist] = spray painter
Spritz|flakon n, m → Zerstäuber m
~geräte npl [Straßenbau] = spraying machinery
~gießen n = injection mo(u)lding
~gußteil m, n → Druckgußteil
~isolierung f = sprayed insulation
~kanone f, Monitor m = monitor
~kitt m = gun putty
~kunststoff m = sprayed plastic
~maschine f = spraying machine, sprayer
~mastix m = gunned mastic
~mörtel m = shotcrete mortar, pneumatically applied ~
~pistole f = spray(ing) gun, ~ pistol
~rampe f → Sprengrampe
~rohr n → Sprengrohr
~schlauch m = spray hose
~schmierung f = splash lubrication
~verfahren n [Herstellung von Überzügen auf Metalloberflächen] = metallization
~verkleidung f → Spritzauskleidung
~versteller m, Einspritzzeitpunktversteller m = variable timing unit
~versuch m, Spritzprobe f, Spritzprüfung f = spraying test
~wagen m → Sprengwagen

Spritzwand — Spüldamm

~**wand** *f* [*Waschbecken*] = splashback
~**wasser** *n*, Überschlagwasser = oversplash [*The water that splashes over the top of a breakwater, seawall, etc.*]
Sprödbruch *m*, Bruchfestigkeit *f* in sprödem Zustand, Trennbruch = brittle failure, ~ fracture
sprödbruchunempfindlich = tough
Sprödbruchunempfindlichkeit *f* = toughness
Sprödigkeit *f* = brittleness
Sprödigkeitszone *f* → Bruchzone
Sprosse *f* [*Fenster*] = sash bar, glazing bar; muntin (US)
~ [*Leiter*] = rung, round, tread, rundle, stave
Sprossen|eisen *n* = iron sash bar, ~ glazing ~; ~ muntin (US)
~**fenster** *n* = sash bar window, glazing ~ ~; muntin ~ (US)
Sprudel|quelle *f* = bubbling spring
~**stein** *m* = flos ferri
Sprüh|brenner *m* = vaporizing burner
~**düse** *f* = atomizing jet
~**elektrode** *f* [*Elektroabscheider*] = discharge electrode, emitting ~, ionizing ~
Sprühen *n* → Versprühung
Sprüh|gerät *n* = atomizing device
~**regen** *m* → Rieselregen
~**rohr** *n* = atomizing pipe, ~ tube
~**versuch** *m* [*Korrosionsprüfung*] = (salt-)spray test
Sprung *m*, Paraklase *f*, Absenkung *f*, Abschiebung, Verwerfung, Bruch *m* [*Grabenrand*] (Geol.) = geological fault, faulting, fault (plane)
~ **ins Hangende**, Verwerfung *f* ~ ~ = upthrown fault
~ ~ **Liegende**, Verwerfung *f* ~ ~ = downthrown fault
~**brett** *n* [*Schwimmbad*] = diving board
~**rohr** *n* = offset
~**schanze** *f*, Skisprung-Überlauf *m* [*Talsperre*] = ski jump spillway
~**schanze** *f* [*für Skispringer*]= ski jump
~**wachs** *n* = crackwax
sprungweiser Geschiebetrieb *m*, pulsierender ~ = saltation

Sprung|weite *f* = distance of hydraulic jump, jump distance
~**welle** *f* → Bore *f*
Spül|abbau *m* = hydraulic mining, hydraulicking, sluicing [*Hydraulic excavation of a placer deposit followed by flow through flumes with riffles in the bottom (called sluices) which collect gold or other heavy minerals*]
~**abort** *m* → Spülklosett *n*
~**abtritt** *m* → Spülklosett *n*
~**anschüttung** *f* → Spülauftrag
~**(auf)füllung** *f* → Spülauftrag
~**aufschüttung** *f* → Spülauftrag
~**auftrag** *m*, Spül(auf)füllung *f*, Spülanschüttung, Spülaufschüttung = hydraulic filling
~**ausguß** *m* → Spüle *f*
~**aushub** *m* = hydraulic excavation [*Excavation by giants delivering a jet of water at high velocity against an earth or gravel bank and breaking it up*]
~**bagger** *m*, Spüler, Pumpenbagger mit hydraulischer Bodenlösung durch Abspritzen unter hohem Druck und gleichzeitigem Weiterspülen = dustpan (type of hydraulic) dredger, hydraulic erosion ~ [*US = dredge*]
~**becken** *n* → Spüle *f*
~**becken** *n* → Abortbecken
~**becken** *n*, Spülschleuse *f* = flushing basin, scour(ing) ~
~**behälter** *m*, Spülkammer *f* = flush tank, flushing chamber [*A chamber in which water or sewage is accumulated and discharged at intervals for flushing a sewer*]
~**bohren** *n*, Spülbohrung *f* = wash boring [*Sinking a casing or drive pipe to bedrock by a jet of water within it, sometimes helped by driving*]
~**bohrverfahren** *n* **mit indirekter Wasserspülung**, ~ ~ Verkehrtspülung = counter-flush method, reverse-rotary ~
~**brett** *n* = ablution board of sink
~**damm** *m* → gespülter (Erd)Damm

Spüldammaterial — Spülschieber

~**dammaterial** n, Spüldamm-Material = hydraulic fill (material) [*Embankment material carried by water in flumes or pipelines*]

~**druck** m = scavenge pressure [*I.C. engine*]

Spüle f, Spülstein m, Handstein, Spülausguß m, Küchenausguß, Spülbecken n = (kitchen) sink

Spüleinbaudamm m [*nur Spüleinbau*] = semi-hydraulic fill embankment; ~ ~ dam

spülen, drücken [*Naßbaggergut*] = to deliver mixtures of soil and water to a disposal point through a pipe line

~ = to flush [*To send a quantity of water down a pipe or channel to clean it or to flush a WC*]

Spülen n [*Pfahl*] → Ein~

~, Spülung f, Wasser~ [*für Reinigungszwecke*] = flushing

Spüler m → Spülbagger

Spuler m = spooler

Spül|feld n → Spülfläche f

~**fläche** f, Spülfeld n [*Bei Naßbaggerarbeiten eine Fläche, auf die die Flüssigkeit lange genug festgehalten wird, damit sich die Feinteile setzen können*] = pond, depositing site

~**flüssigkeit** f → Spülschlamm

~**füllung** f → Spülauftrag

~**grube** f → Schlammkule f

Spülicht n, Spülwasser n = (domestic) waste, sullage [*The drainage from sinks, baths, lavatory basins, etc.*]

Spül|kammer f → Spülbehälter m

~**kanal** m = flushing duct

~**kasten** m, Klosett~, Abort~ = (flushing) cistern, water-waste preventer, W. C. flushing cistern [*B.S. 1125*]

~**kernbohrung** f = core wash boring

~**kette** f = jetting chain

~**kippverfahren** n → Einspülverfahren

~**klosett** n, Spülabort m, Spülabtritt m, Wasser-Klosett = water closet, wc, WC

~**klosettraum** m = lavatory [*The term lavatory also means a wash basin (Waschbecken) and by extension the room containing the wash basin (Waschraum)*]

Spülkopf m, Drehkopf, Rotationskopf, Bohrwirbel m [*Rotarybohren*] = rotary swivel

~**knie** n → Spülkopfkrümmer m

~**krümmer** m, Spülkopfknie n, „Gänsehals" m [*Rotarybohren*] = "gooseneck" (bend)

Spül|küche f = scullery

~**lanze** f = jetting lance

~**löcher** npl **im Bohrmeißel** = holes (or perforations) in the (drilling) bit, openings in the sides of the cutting bit

~**luftkopf** m → Blaskopf

~**pfahl** m, eingespülter Pfahl = jetted pile

~**probe** f [*Rotarybohren; aus der Spülung entnommene Gebirgsprobe*] = ditch sample

~**probendiagramm** n = sampler log

~**pumpe** f, Schlammpumpe [*Rotarybohren*] = slush pump, circulation ~, mud ~

~**pumpe** f [*zum Einbringen von Pfählen*] = jetting pump

~**rinne** f, Schlammrinne, Spülungskanal m [*Rotarybohren*] = mud ditch, ~ flue, ~ trough, flume

Spülrohr n, Druckrohr [*Naßbaggerei*] = dredging pipe

~ [*Spülabort*] = flush pipe

~ [*Einbringen von Pfählen*] = jetting pipe

~ [*zum Reinigen*] = washout pipe, scour(ing) ~

~ = pipe for hydraulicking

~ [*Talsperre*] = scour(ing) pipe

Spül|säule f **im Gestänge** [*Lufthebe-Bohranlage*] = column of drilling fluid in the drill pipe

~**schacht** m = flushing manhole, ~ chamber [*A manhole provided with a gate so that sewage or water may be accumulated and then discharged rapidly for flushing a sewer*]

~**schieber** m = wahout valve [*A valve in a pipe line or dam, which is occasionally opened to release sediment or water*]

Spülschlamm — Spültisch

~schlamm *m*, Spülflüssigkeit *f*, Bohrschlamm, Bohrflüssigkeit = drilling mud, rotary ~
~schlauch *m* → Bohrschlauch
~schleuse *f*, Spülbecken *n* = scour(ing) basin, flushing ~
~-Schnellschlag-Bohren *n* = rapid percussive drilling, ~ percussion ~
~schraub(en)pfahl *m*, eingespülter Schraub(en)pfahl, Schraub(en)spülpfahl = jetted screw pile
~schütz(e) *n*, (*f*) = scour(ing) sluice, ~ gate, wash-out ~, flushing ~
~sonde *f* von Terzaghi = (Terzaghi) wash-point sounding apparatus, (~) ~ soil penetrometer

~spitze *f*, Stahl ~ [*Grundwasser(ab)senkung in Feinstsanden*] = jetting tip
~(spitzen)öffnung *f*, Einspülöffnung = jetting orifice
~spritzverfahren *n* → Einspülverfahren
~stange *f* = jetting rod
~stangenabstelloch *n* [*Rotarybohren*] = rat hole
~stangengerät *n* = jetting cutter rod
~(stau)damm *m* → gespülter (Erd-) Damm
~stein *m* → Spüle *f*
~strahl *m* [*Tiefbohrtechnik*] = drilling mud jet
~sumpf *m* [*Motor*] = scavenge sump

Spültisch *m*	sink unit
Abkanten *n*	folding the edges
Ablauffläche *f*, Abstellfläche	drainer, draining board, drainboard
Ablaufverbindung *f*	waste connection
Anschlußmaße *npl*	installation dimensions
Anschlußverschraubung *f*	screwed connection
Antidröhnbelag *m*	sound-deadening coat(ing)
Antidröhnmasse *f*	sound-deadening compound
Arbeitsfläche *f*	work top
Auslaufventil *n*	outlet
Beckenanordnung *f*	bowl arrangement
Beschichten *n*	coating
Doppelbecken *n*	twin bowls
Einloch-Standbatterie *f*	single-hole mixer
entdröhnt	sound-deadened
Entlüftungskiemen *m*	vent louvre
fugenloses Anformen *n* des Ausgußbeckens am Spültisch	seamless forming of slop bowl on sink unit
gebürstet	satin finish
gerillter Ablauf *m*	fluted drainer
Geschirrbrause *f*	dish spray
Geschirrspülkorb *m*	dishwashing basket
Gitterrost *m*	grid
Groß-Spültisch *m*, Gewerbe-Spültisch	sink unit for the trade
Herd-Spültisch-Kombination *f*	combination sink cabinet and cooker unit
hochglanzpoliert	mirror finished
Holzrahmen *m*	wooden underframe
Innenbearbeitung *f*	finishing
Kelchplatte *f*	waste (fitting), recessed waste ~

Spülung — Spülverfahren 916

Klapprost *m*	hinged grid
Lötstutzen *m*	brazing socket
Müllwolf *m*	(electric food) waste disposer
Müllwolfauslaufprägung *f*	piercing for (electric food) waste disposer
Prägen *n*	pressing
Reinigung *f* und Pflege *f*	cleaning and care
Rückleiste *f*, Stehbord *n*	back ledge
Schutzsieb *n*	perforated protector
schwenkbarer Auslauf *m*	swivel outlet head
Serien-Spültisch *m*	quantity-produced sink unit, series-produced ∼ ∼
Siebeinsatz *m*	basket strainer waste (unit)
Spülbecken *n*	sink bowl
Standbatterie *f*	sink mixer
Standrohr *n*	standpipe
Standrohrventil *n*	standpipe waste
Stegmitte *f*	centre of bowl (Brit.); center ∼ ∼ (US)
Stehbord *n*, Rückleiste *f*	back ledge
Stopfenventil *n*	waste plug
Tiefziehen *n*	deep drawing
Überlaufmutter *f*	union
Überlaufventil *n*	overflow
Umstellbrausekopf *m*	change-over spray head
Unterbau *m*	base unit
Volksspültisch *m*, Siedlungsspültisch	popular sink unit, economy ∼ ∼
Wandanschlußprofil *n*	fillet trim
Wasserfleck *m*	water stain

Spülung *f*, Spülen *n* [*Luftspülung beim Gesteinsbohren*] = puff blowing

∼, Spülen *n*, Wasser∼ [*für Reinigungszwecke*] = flushing

∼, Spülen *n* [*bei der Abwasserbeseitigung*] = flushing

∼, Spülen *n* [*Motor*] = scavenging

∼ → Bohr∼

∼, Auf∼, Spülverfahren *n* [*Herstellung von Dämmen*] = hydraulicking

∼ **auf Ölgrundlage** → ∼ mit hohem Widerstand

∼ **mit hohem Widerstand,** ∼ auf Ölgrundlage, ölbasische Spülung, Ölspülung [*Tiefbohrtechnik*] = oil-base mud

∼ **von Abraum** = hydraulic stripping [*Removal of overburden by hydraulicking*]

Spülungs|druck *m* [*Rotarybohren*] = mud pressure

∼**gewicht** *n* [*Rotarybohren*] = mud weight

∼**kanal** *m* → Spülrinne *f*

∼**säule** *f* [*Rotarybohren*] = column of mud, mud column

∼**umlauf** *m* [*Rotarybohren*] = circulation of mud, mud circulation

∼**verlust** [*Rotarybohren*] = loss of circulation, ∼ ∼ returns

∼**waage** *f* [*Tiefbohrtechnik*] = mud balance

Spülverfahren *n*, (Auf)Spülung *f* [*Herstellung von Dämmen*] = hydraulicking

∼ [*Rotarybohren*] = hydraulic circulating system

Spül|verlust *m*, Spülungsverlust [*Rotarybohren*] = loss of returns, ∼ ∼ circulation

∼**versatz** *m* (⚒) = hydraulic stowing, ∼ fill(ing); ∼ flushing, ∼ slushing, ∼ silting (US)

∼**vorrichtung** *f* [*Einspülen von Pfählen*] = jetting device

Spülwasser *n* [*Bohrhammer*] = injected water

∼ → Spülicht *n*

∼ [*zur Filterreinigung*] = filter cleaning water

∼ [*Klosett*] = flushing water

∼ [*Reinigung von Abwasseranlagen*] = rinse water, (back)wash ∼

∼, Bohrwasser = drilling water

∼ [*zum Einspülen*] = jetting water

∼**rest** *m* = after-flush [*The small quantity of water remaining in the cistern after a w. c. pan is flushed. It trickles slowly down and remakes the seal*]

(Spund)Bohle *f* → → Spundwand

Spundbohlenrammhaube *f* [*hydraulische Mehrzylinderspundwandramme*] = sheet pile connector mechanism

Spundbrett *n*, Nut- und ∼ = match board

Spundwand *f*

abgebaggerte ∼
Ankerhöhenverhältnis *n*
Ankerlinie *f*
Anker(spund)wand, verankerte ∼
Ankerzug *m*
Ankerzugstange *f*
Ankerzugverhältnis *n*
Auflastverhältnis *n*
ausgesteifte ∼
Ausrammung *f*, Einspundung *f*, Spundwandumschließung *f*
Biegelinienverfahren *n*
Biegemomentverhältnis *n*
Biegesteifigkeit *f*
Biegsamkeitszahl *f*
Bohle *f*, Spundbohle
Buckelblech∼
Drehpunkt *m* [*Steifenberechnung f*]
eingerostetes Schloß *n*
eingespannte ∼
Einspannung *f*
Einspundung *f*, Spundwandumschließung *f*, Ausrammung *f*
Ersatzbalkenverfahren *n*
Flachprofil *n*
freie untere Auflagerung *f*, untere freie Auflagerung
hinterfüllte ∼
Holz(spund)bohle *f*

Holz∼

sheet (pile) wall, (sheet) piling, (sheet pile) bulkhead, sheeting wall

dredge ∼
anchor level ratio
anchor line
anchored ∼, tied ∼
anchor pull
anchor tie
anchor pull-ratio
surcharge ratio
braced ∼
closed sheeting

elastic line method
bending-moment ratio
flexural rigidity
flexibility number
sheet pile
buckled plate ∼
point of inflection
frozen (inter)lock
fixed ∼
fixation, fixity
closed sheeting

equivalent beam method
straight web
free earth support

fill ∼
wood(en) sheet pile, timber sheet pile
wood ∼, timber ∼

Spundwand

Kastenspundbohle *f*	box section (sheet) pile, encased (sheet) pile
Kreiszelle *f*	circular cell
LARSSEN-(Spund)Bohle *f*	LARSSEN (sheet) pile
Momentenabminderungskurve *f*	moment-reduction curve
nachgebendes Gelenk *n*	yield hinge
Nachgiebigkeitsverhältnis *n*	yield ratio
paarig gerammte ∼	two sheets driven simultaneously
Profil LARSSEN V *n*	structural section LARSSEN V
Rammtiefe *f*	sheet-pile penetration, depth of penetration
Raumgewicht *n* unter Wasser	submerged unit weight
Schloß(eisen) *n*, Spundwandschloß *n*	(inter)lock
Schloßmodul *m*	interlocked modulus
Schloßspannung *f*	(inter)lock tension
senkrechter Abstand *m* der Sohle *f* von der Oberfläche *f* der Hinterfüllung *f*	vertical distance between the dredge line and the surface of the backfill
senkrechter Abstand *m* des Ankers *m* von der Sohle *f*	vertical distance from the anchor to the dredge line
senkrechter Abstand *m* des freien Wasserspiegels *m* von dem Wasserspiegel *m* in der Hinterfüllung *f*	vertical distance between the free water level and the water table in the backfill
Sohle *f*	dredge line
(Spund)Bohle *f*	sheet pile
∼ mit fester Einspannung *f*	fixed-support ∼
∼ mit unterer fester Einspannung *f*	∼ with fixed earth support
∼ mit unterer freier Auflagerung	∼ with free earth support
Spundwände *fpl* im Bruchzustand	sheet-pile walls at failure
∼bauwerk *n*, Spundwandfang(e)damm *m*	sheet-pile retaining wall, sheet piling cofferdam; pile dike (US)
∼berechnung *f*	∼ design
∼eisen *n*, Stahlspundbohle *f*, Spundwandstahlprofil *n*	steel sheet pile (section)
∼-Fang(e)damm-Ausbeulung *f*	cofferdam boil
∼gründung *f*	sheet pile foundation
∼ramme *f*	sheeting driver
∼rammung *f*	sheet pile driving
∼schloß *n*, Schloßeisen *n*	(inter)lock
∼-Stahlprofil *n*, Spundwandeisen *n*, Stahl(spund)bohle *f*	steel sheet pile (section)
∼-Stahlprofil *n* Bauart KRUPP	KRUPP (steel) sheet pile
∼umschließung *f*, Einspundung *f*, Ausrammung *f*	closed sheeting
∼zelle *f*	sheet pile cell
Stahl(spund)bohle *f* → Spundwand-Stahlprofil *n*	
Steg *m*	web
Steifenbeanspruchung *f*	strut load
Teileinspundung *f*	partial sheeting

Trägheitsmoment n des Spundbohlenquerschnittes m	moment of inertia of the cross section of a sheet pile
unverankerte ~	free ~, cantilever ~
verankerte ~, Anker(spund)wand	anchored ~, tied ~
Verhältnis n der Durchbiegung f zur Höhe f der Spundwand f	ratio of deflection to the height of a sheet-pile wall
Verhältnis n der freien Höhe f	free-height ratio
Verhältnis n zwischen Tiefe f der Ankerlinie f unter der Oberfläche f der Hinterfüllung f und der Länge f der Spundbohlen fpl	ratio between the depth of the anchor line below the surface of the backfill and the length of the sheet piles
Verhältnis n zwischen Tiefe f der Sohle f unter der Oberfläche f der Hinterfüllung f und der Länge f der Spundbohlen fpl	ratio between the depth of the dredge line below the surface of the backfill and the length of the sheet piles
Wandreibungswinkel m	angle of wall friction
Winkel m der teilweise mobilisierten inneren Reibung f	angle of partly mobilized internal friction
wirksamer Teil m [*Winkel der inneren Reibung*]	mobilized part
wirksames Nachgeben n	effective yield
wirksames Raumgewicht n	effective unit weight
zulässige Biegebeanspruchung f	allowable stress in bending

Spur f → Fahr~

~, ~streifen m, Fahrrinne f, Radspur, Radeindruck m = wheeler, rut, groove, furrow [*track made in the ground, earth road, or carriageway by the passage of wheeled vehicle(s)*]

~, Spurweite f = ga(u)ge

~, ~weite f, Gleisspur [*lichtes Maß zwischen den Schienenköpfen*] = (track) ga(u)ge

~, Nacheichung f, Ausrichtung f = alignment (Brit.); alinement (US)

~bildung f [*in einer Straßendecke*] = (development of) rutting

Spuren|analyse f = trace analysis

~element n = trace element

Spur|fahren n = tracking

~hebel m = guide lever

~kette f, Schneekette = non-skid chain, snow ~

~kranz m, Radkranz = wheel-flange

spurkranzloses Rad n = flangeless wheel

Spur|kranzrad n = flanged (rail) wheel

~latte f, (Schacht)Führung f, Leitbaum m (⚒) = guide

~lehre f **zur (Gleisketten)Laufwerkausrichtung** = alignment gauge

~leistungsfähigkeit f [*Straßenverkehr*] = lane capacity

~-Leucht-Nagel m → Straßen-Leucht-Nagel

~linie f, Fahr~, Verkehrs~ = (traffic) lane line

~nagel m → Markierungsknopf m

~streifen m → Spur

~streifenstraße f, befestigter Radspurstreifen m = strip(e) road, creteway, trackways

spur|treues Bremsen n = straight-line breaking

~versetzter Rahmen m [*Motor-Straßenhobel*] = offset frame

Spur(weite) f, Gleisspur f [*lichtes Maß zwischen den Schienenköpfen*] = (track) ga(u)ge

Stab m, Träger~, Gitter~, Füll(ungs-)~ = member, bar, rod

Stabablenkung — Stadtautobahn

~ablenkung *f* [*Rahmenträger*] = member slope, bar ~, rod ~
~abstand *m* = distance between bars, ~ ~ rods
~alge *f* → Diatomee *f*
~anordnung *f* = spacing of bars, ~ ~ rods
~anschluß *m* = bar connection, rod ~, member ~
~armierung *f* → Stabbewehrung
~bewehrung *f*, Stabarmierung, Stab(stahl)einlagen *fpl* = bar reinforcement, rod ~
Stäbchen *n* [*beim Rolladen*] = slat
Stab|ebene *f*, Träger~ = plane of members, ~ ~ bars, ~ ~ rods
~einlagen *fpl* → Stabbewehrung *f*
~endmoment *n* = moment at extremity of bar, ~ ~ ~ ~ rod, ~ ~ ~ ~ member
stabil, langsambrechend [*Bitumenemulsion*] = stable, slow-breaking, slow-setting
stabile Schwimm-Gleichgewichtslage *f* = stable buoyancy equilibrium
Stabilisator *m* = stabilizer [*added to an emulsion*]
~pulver-Verteiler *m* = powder spreader, bulk ~ [*A vehicle for distributing stabilizers in powder form*]
~stange *f* [*Kraftfahrzeugbau*] = stabilizer bar, stabilizing ~
Stabilisierkolonne *f* [*Erdölwerk*] = stabilizer
Stabilisierung *f* → (Boden)Verfestigung
Stabilisierungsstab *m* [*Stahlbau*] = stabilizing bar, stabilizer ~
Stabilität *f* = stability
Stabilitäts|grad *m* [*Emulsion*] = degree of stability
~grenze *f* = limit of stability
~theorie *f* = theory of stability
Stab|knickung *f* = buckling of bar(s), ~ rod(s), ~ ~ member(s)
~kugelmühle *f* = ball and rod grinder
~moment *n* = bar moment, rod ~, member ~

~mühle *f* → Stabrohrmühle
~querschnitt *m* = bar cross section, rod ~ ~
(~)Rechen *m* = rack [*A screen composed of parallel bars, either vertical or inclined, from which the screenings may be raked*]
~(rohr)mühle *f* = rod mill
~(rohr)mühle *f* mit offenem Kreislauf = open circuit rod mill
~(rohr)mühlenpanzerung *f* = rod mill lining, ~ ~ liners
~rost *m*, Stangen(sieb)rost, Rostsieb *n*, Stabsiebrost = bar grizzly, ~ screen, ~ grate, grizzly
~rührwerk *n*, Stabrührer *m* = finger blade agitator
~rüttler *m*, Stabvibrator *m* = internal vibrator with handle, needle ~ ~ ~, immersion ~ ~ ~, poker ~ ~ ~, pervibrator ~ ~
~siebrost *m* → Stabrost
~spannung *f* = bar stress, rod ~, member ~
~stahl *m* = bar-steel
~(stahl)einlagen *fpl* → Stabbewehrung *f*
~stahlwalzwerk *n* = bar mill
~vertauschung *f* [*Baustatik*] = exchange of members, ~ ~ bars, ~ ~ rods
~vibrator *m* → Stabrüttler *m*
~zahl *f*, Zahl der Stäbe = number of bars, ~ ~ members, ~ ~ rods
~zugverfahren *n* = circuit analysis
Stacheldraht *m* = barbed wire; bobwire (US)
~elektrode *f* = barbed wire electrode
~zaun *m* = barbed wire fence
Stachelwalze *f* → → Walze
Stadion *n* = stadium
~sitz *m* = stadium seat
Stadt *f* in mehreren Ebenen = multistor(e)y city
~abfluß *m* [*Jener Teil des Niederschlages einer Stadt, der oberirdisch abfließt*] = urban run-off
~autobahn *f* → innerstädtische Autobahn

~**autobahnnetz** n = urban expressway system, ~ ~ net(work)
~**autobus** m, Stadt(omni)bus = municipal bus
~**bad** n = town swimming bath, municipal pool
~**bahn** f = metropolitan railway (Brit.); ~ railroad (US); urban rapid transit system
~**bahngleis** n = metropolitan railway track (Brit.); ~ railroad ~ (US)
~**bahnhof** m = in-town station
~**bauarbeiten** fpl = urban construction
~**baukunst** f = civic design
~**bevölkerung** f = urban population
~**brücke** f = urban bridge
~**bus** m, Stadtautobus, Stadtomnibus = municipal bus
~**dunst** m, Industriedunst, Rußnebel m = smog
~**durchgangstunnel** m = crosstown tunnel [*A tunnel with no surface connection in the town*]
Städtebau m **und Raumordnung** f = town and country planning
~ ~ **Städteplanung** f = town construction and planning
Städtebauer m → Stadtplaner
städtebauliche Planung f → Stadtplanung
Stadt|entwässerung f = town drainage
~**entwick(e)lung** f = urban development
Städteplaner m → Stadtplaner
Städter m, Stadtbewohner m = town dweller, city ~
Stadt|erneuerung f = urban re-development
~**erweiterung** f = urban expansion
Stadtgas n, Dowsongas [*früher: Leuchtgas*] = town gas
~**heizung** f = heating by town gas
Stadt|gebiet n = urban area, city ~
~**gesundung** f = town improvement
~**grenze** f = town boundary
~**größe** f = size of a town (Brit.); ~ ~ ~ city (US)
~**heizung** f = urban heating
städtische Abwässer npl, **städtisches Abwasser** n = town sewage

~ **Müllverbrennungsanlage** f, ~ Abfallverbrennungsanlage = municipal destructor
~ **Wasserversorgung** f = municipal water supply
städtischer Tiefbau m = municipal engineering
Stadt|kern m = heart of the town, central area, nucleus; centre of the town (Brit.); center of the city (US)
~**landschaft** f = townscape
~**mitte** f = town centre (Brit.); city center (US)
(~)**Müll** m = (town) refuse, (~) rubbish
~**(omni)bus** m, Stadtautobus = municipal bus
~**park** m = city park
~**planer** m, Städteplaner, Städtebauer = city planner, town ~, municipal ~, urban ~
~**planung** f, Städteplanung, städtebauliche Planung = city planning, town ~, municipal ~, urban ~
~**planungsbehörde** f = town planning authority
~**rand** m = border of a town, periphery ~ ~ ~
~**randsiedlung** f = suburban housing estate
~**sammler** m = municipal intercepting sewer
~**schnellbahn** f = urban rapid transit system
~**schwimmbad** n = municipal swimming pool
~**straße** f = (city) street, municipal road, town ~, urban ~, urban thoroughfare, city road
~**straßenausrüstung** f = street furniture, ~ equipment
~**straßenbelag** m, Stadtstraßendecke f = street surfacing, municipal ~ ~ pavement; ~ paving (US)
(~)**Straßenbeleuchtung** f = street lighting
~**straßenbeton** m = municipal paving concrete (US); ~ pavement ~, ~ surfacing ~
~**straßenkreuzung** f = street crossing

(Stadt)Straßenlampe — Stahlbetonauskleidung

(~)Straßenlampe *f* = street lamp
(~)Straßenleuchte *f* = street lantern (Brit.); ~ luminaire (US); ~ lighting fixture
~straßennetz *n*, städtisches Straßennetz = (urban) street network, (~) ~ system
~straßenwesen *n* = design and layout of streets
~struktur *f* = structure and arrangement of a town (or city)
~tunnel *m* = urban tunnel
~umbau *m* = urban renewal, ~ redevelopment
~verkehr *m* = urban traffic
~zentrum *n* = town centre (Brit.); ~ center (US)

Staffel|befehlsstelle *f* [*Militärflugplatz*] = flight-line-bunker
~pegel *m* [*Zeichen: Lst*] = staff ga(u)ges [*several staff ga(u)ges for medium and high stages and for low-stage readings*]
staffelweise arbeiten [*z. B. Erdbaugeräte*] = to work in echelon

stagnierendes Gewässer *n* = stagnant water

Stahl|anker *m* = steel tie(-rod)
~auszug *m* = cutting list, summary of reinforcement [*A list of steel bars showing diameters and lengths only, from which the reinforcement is ordered. This list is prepared by the contractor from the bending schedules issued by the reinforced concrete designer with his detail drawings*]
~balken(träger) *m* = steel beam
~balken(träger)rost *m* = steel beam grillage
~ballast *m* = steel ballast
~baluster *m*, Stahldocke *f* = steel baluster
~band *n* [*beim Vollgummireifen*] = steel band
(~)Bandage *f*, (Stahl)Mantel *m* [*Walze*] = (steel) facing
~bandagen-Straßentandemwalze *f*, (Glattmantel-)Tandemwalze = smooth(-wheeled) tandem roller
~bandförderer *m* = steel band conveyor, ~ ~ conveyer
~bandmaß *n*, Stahlmeßband *n* = steel measuring tape

Stahlbau *m* = steel construction
~, Stahlbauwerk *n* = steel structure
~arbeiten *fpl* = steel construction work
~firma *f* = steel construction firm
~hilfsarbeiter *m* = steel erector's mate, ~ ~ labourer
~montagemeister *m* = foreman steel erector
~monteur *m* = steel erector, constructional fitter and erector, iron fighter
~profil *n* = structural-steel section

Stahl-Baustellen|lore *f* → Stahl-Muldenkipper *m*
~-Schienenwagen *m* → Stahl-Muldenkipper *m*

Stahl|bauteil *m*, *n* = structural steel member
~bauten *f* = structural steelwork
~bau(werk) *m*, (*n*) = steel structure
~bauwerkstatt *f* = steel workshop
~bedachung *f*, Stahldachhaut *f*, Stahldach(ein)deckung, Stahleindeckung = steel roofing, ~ roof decking
~bekleidung *f* → Stahlblechverkleidung
~bemessung *f*, Bemessung von Stahlbauten *f* = steel design

Stahlbeton *m* [*früher: Eisenbeton*] = (steel) reinforced concrete, R. C. [*deprecated: ferroconcrete*]
~(ab)deckung *f*, Stahlbetonbelag *m* [*Wasserseite eines Erdstaudammes*] = reinforced concrete facing (US); ~ ~ membrane (Brit.); R. C. ~
~anlegebrücke *f* = single berth reinforced-concrete jetty, R. C. jetty with single berth
~-Armierungsmatte *f* → Baustahlmatte
~aufgabe *f* = reinforced-concrete problem, R. C. ~
~auskleidung *f*, Stahlbetonverkleidung, Stahlbetonbekleidung = reinforced-concrete lining, R. C. ~

Stahlbetonbalkendecke — Stahlbetonfaltwerk

~balkendecke *f* = reinforced-concrete beam floor, R. C. ~ ~
~balken(träger) *m* = reinforced-concrete beam, R. C. ~
~bau *m* = reinforced-concrete construction, R. C. ~
~bau *m*, ~werk *n* = reinforced-concrete structure, R. C. ~
~bauteil *m, n*, Stahlbetonbauelement *n* = reinforced-concrete element, R. C. ~
~bauten *f* = reinforced-concrete structures, R. C. ~
~bau(werk) *m, (n)* = reinforced-concrete structure, R. C. ~
~-Beckenwand *f* = reservoir wall of reinforced concrete, R. C. reservoir wall
~belag *m* → Stahlbeton(ab)deckung *f*
~belag *m* → Stahlbetondecke *f*
~bemessung *f*, Bemessung im Stahlbetonbau = design in reinforced concrete, reinforced-concrete design, R. C. design
~bestimmungen *fpl* = Building Code Requirements for Reinforced Concrete
~-Bewehrungsmatte *f* → Baustahlmatte
~binder *m*, Stahlbetonfachwerkbinder, Stahlbetondachbinder = reinforced-concrete roof truss, R. C. ~ ~
~bogen *m* = reinforced-concrete arch, R. C. ~
~bogenbrücke *f* = reinforced-concrete arch(ed) bridge, R. C. ~ ~
~bohle *f*, Stahlbetonspundbohle = reinforced-concrete sheet pile, R. C. ~ ~
~bohrpfahl *m* = bored reinforced-concrete pile, ~ R. C. ~
~brücke *f* = reinforced-concrete bridge, R. C. ~
~brückenbau *m* = reinforced-concrete bridge construction, R. C. ~ ~
~-Brunnenkranz *m*, Stahlbeton-Schlenge *f* = reinforced-concrete cutting curb, ~ drum ~, ~ shoe, R. C. cutting curb, R. C. drum curb, R. C. shoe

~dachbinder *m* → Stahlbetonbinder
~dammbalken *m* = reinforced-concrete log, R. C. ~
~decke *f*, Stahlbetonbelag *m* = reinforced-concrete pavement, ~ surfacing, R. C. ~
~decke *f* = reinforced-concrete floor, R. C. ~
~deckung *f* → Stahlbetonabdeckung
~druckrohr *n* → Stahlbetonrohr für Druckleitungen
~-Durchlaufbalken(träger) *m* = continuous reinforced-concrete beam, ~ R. C. ~
~-Durchlaufrahmen *m* = continuous frame of reinforced concrete, R. C. continuous frame
~-Durchlaufträger *m* = continuous reinforced-concrete girder, ~ R. C. ~
~-Einzelfundament *n* = reinforced-concrete single base, ~ individual ~, R. C. ~ ~
~element *n* = reinforced-concrete unit, R. C. ~
~fabrik *f*, Stahlbetonwerk *n* = reinforced-concrete factory, ~ works R. C. ~
~fachwerkbinder *m* → Stahlbetonbinder
~-Fachwerkbrücke *f* = reinforced-concrete truss bridge, R. C. ~ ~
~fachwerkgebäude *n*, Stahlbetonskelettgebäude = R. C. framed building, ~ ~ skeleton ~; ~ ~ framework ~, reinforced-concrete ~ ~
~fachwerkrahmen *m* = reinforced-concrete truss, R. C. ~
~fachwerkträger *m* = reinforced-concrete truss, R. C. ~
~fahrbahntafel *f* = reinforced-concrete deck (slab), R. C. ~ (~)
~faltwerk *n*, prismatische Stahlbetonschale *f*, prismatisches Stahlbetonfaltwerk = reinforced-concrete folded slab, ~ ~ plate (structure), ~ prismatic shell, ~ hipped plate structure, ~ prismatic structure, ~ folded plate roof, R. C. folded **slab**

Stahlbetonfeld — Stahlbetonkuppel

~feld *n* [*Brückenbau*] = reinforced-concrete span, R. C. ~

~fenster *n* = reinforced-concrete window, R. C. ~

~-Fertigdecke *f*, Stahlbeton-Montagedecke = precast reinforced-concrete floor, ~ R. C. ~

~fertiggarage *f* = precast reinforced-concrete garage, ~ R. C. ~

~-Fertigpfahl *m* = precast reinforced-concrete pile, prefab(ricated) ~ ~, ~ R. C. ~

~-Fertigrahmen *m* = precast R. C. frame, ~ reinforced-concrete ~

~fertigsturz *m* = precast reinforced-concrete lintel, ~ ~ lintol, ~ R. C. ~

~fertigstütze *f* = precast reinforced-concrete column, ~ R. C. ~

~-Fertigteil *m, n,* Stahlbeton-Montageteil = precast reinforced-concrete member, reinforced cast-concrete ~, reinforced (concrete) casting, precast R. C. member

~fertigträger *m* = reinforced precast concrete beam, precast R. C. ~

~fertigträgerbrücke *f* = precast reinforced-concrete beam bridge, ~ R. C. ~ ~

~fertigwand *f* → Stahlbetonmontagewand

~-Flugplatzdecke *f*, Stahlbeton-Flugplatzbelag *m* = reinforced-concrete airport pavement, ~ airfield ~, ~ surfacing, R. C. ~ ~

~-Fundamentplatte *f* → Plattenfundament *n*

~fundamentsockel *m* → Stahlbetonsockelgründung *f*

~-Fundationsplatte *f* → Plattenfundament *n*

~fundationssockel *m* → Stahlbetonsockelgründung *f*

~-Gasbehälter *m* = reinforced-concrete gasholder, R. C. ~

~gebäude *n*, Stahlbetonhaus *n* = reinforced-concrete building, R. C. ~

~gelenk *n* = reinforced-concrete hinge, R. C. ~

~-Grundplatte *f* → Plattenfundament *n*

~grundsockel *m* → Stahlbetonsockelgründung *f*

~-Gründungsbrunnen *m*, Stahlbeton-Senkbrunnen = reinforced-concrete open caisson, ~ drop shaft, R. C. ~

~-Gründungsplatte *f* → Plattenfundament *n*

~gründungssockel *m* → Stahlbetonsockelgründung *f*

~haus *n*, Stahlbetongebäude *n* = reinforced-concrete building, R. C. ~

~-Hohlbalken(träger) *m* = reinforced-concrete hollow beam, R. C. ~ ~

~-Hohlkörperdecke *f*, Stahlbeton-Hohlblockdecke = reinforced-concrete hollow-tile floor, ~ hollowblock ~, ~ pot ~, R. C. ~ ~

~-Hohlpfahl *m* = reinforced-concrete hollow pile, R. C. ~ ~

(~)Hohlplatte *f*, Stahlbetonhohldiele *f* = (reinforced-concrete) hollow slab, R. C. ~ ~

(~)Hohlplattenrahmen *m* = (reinforced-concrete) hollow slab frame, R. C. ~ ~

~kasten *m* [*für Gründungen unter Wasser*] = reinforced-concrete caisson, R. C. ~

~kastenträger *m* = reinforced-concrete box girder, ~ ~ beam, R. C. ~

~-Kellerfenster *n* = reinforced-concrete cellar window, R. C. ~ ~

~kern *m*, Betonkern(bauwerk) *m*, (*n*), (Gebäude)Kern = reinforced-concrete core, R. C. ~

~-Kernmauer *f* → Talsperre *f*

~klumpfußpfahl *m* = R. C. solid bulb pile, reinforced-concrete ~ ~ ~

~konsole *f* = reinforced-concrete bracket, R. C. ~

~kuppel *f* = reinforced-concrete dome, R. C. ~

~**(-Licht)mast** *m*, Stahlbeton-Beleuchtungsmast, Stahlbeton-(Straßen)Leuchtenmast = reinforced-concrete (lighting) column, R. C. (~) ~, reinforced-concrete (lighting) mast, R. C. (lighting) mast

~**massivpfahl** *m* = reinforced-concrete solid pile, R. C. ~ ~

~**mast** *m* = reinforced-concrete pole, ~ mast, R. C. ~

~**mauer** *f* = reinforced-concrete wall, R. C. ~

~**mole** *f* = reinforced-concrete jetty, R. C. ~

~**mole** *f* **mit einem Anlegeplatz**, ~ ~ ~ Festmacheplatz = single berth reinforced-concrete jetty, R. C. jetty with single berth

~**-Montagedecke** *f*, Stahlbeton-Fertigdecke = precast reinforced-concrete floor, ~ R. C. ~

~**-Montageteil** *m, n* → Stahlbeton-Fertigteil

~**montagewand** *f*, Stahlbetonfertigwand, Montagestahlbetonwand = prefab(ricated) reinforced-concrete wall, ~ R. C. ~

~**pfahl** *m* = reinforced-concrete pile, R. C. ~

~**pfahlrost** *m* [*Fehlname: Stahlbetonrost*] = reinforced-concrete piled pier, R. C. ~ ~

~**pfeiler** *m* = reinforced-concrete (foundation) pier, R. C. (~) ~

~**pfette** *f* = reinforced-concrete purlin(e), R. C. ~

~**platte** *f* = reinforced-concrete slab, R. C. ~

~**platte** *f* [*Dock- und Kaimauerbau*] = reinforced-concrete deck, R. C. ~

~**-Plattenbalken** *m* = reinforced-concrete beam and slab, R. C. ~ ~

~**plattendecke** *f* = reinforced-concrete slab floor, R. C. ~ ~

~**-Plattenfundament** *n* → Plattenfundament

~**-Plattenfundamentkörper** *m* → Plattenfundament *n*

~**-Plattenfundation** *f* → Plattenfundament *n*

~**-Plattengründung** *f* → Plattenfundament *n*

~**-Plattengründungskörper** *m* → Plattenfundament *n*

~**plattenschalung** *f* [*als verlorene Schalung bei Massivbeton*] = permanent shuttering of reinforced-concrete slabs, ~ ~ ~ R. C. ~

~**pylone** *f*, Stahlbetonturm *m* [*Brücke*] = reinforced-concrete tower, ~ pylon, R. C. ~

~**querschnitt** *m* = reinforced-concrete section, R. C. ~

~**rammpfahl** *m*, Rammpfahl aus Stahlbeton mit schlaffer Bewehrung = (impact-)driven reinforced-concrete pile, ~ R. C. ~

~**raumtragwerk** *n* = reinforced-concrete space structure, R. C. ~ ~ ~

~**rippendecke** *f* **mit Füllkörpern** = tile-and-joist construction

~**rippendecke** *f* **ohne Füllkörper** = slab-and-joist construction, ribbed ~

~**rippenplatte** *f* = reinforced-concrete ribbed slab, R. C. ~ ~

~**rippenplatte** *f* **ohne Füllkörper** = joist slab

~**rohr** *n* = reinforced-concrete pipe, R. C. ~

~**rohr-Druckleitung** *f*, Stahlbetonrohr-Triebwasserleitung = reinforced-concrete pipe penstock, R. C. ~ ~

~**-Rollenlager** *n* [*für Hoch- und Tiefbauten*] = reinforced-concrete roller bearing, R. C. ~ ~

~**rost** *m* → Stahlbetonsockelgründung *f*

~**rost** *m* [*Fehlname*] → Stahlbetonpfahlrost

~**schacht** *m* = reinforced-concrete shaft, R. C. ~

~**schale** *f* = reinforced-concrete shell, R. C. ~

~**-Schlenge** *f* → Stahlbeton-Brunnenkranz *m*

Stahlbeton-Schleuderrohr — Stahlbetonwerk

~-Schleuderrohr *n* = reinforced-concrete spun pipe, R. C. ~ ~

~schornstein *m* = reinforced-concrete chimney, R. C. ~

~schott *n* = reinforced-concrete bulkhead, R. C. ~

~-Schraubenpfahl *m* = reinforced-concrete screw pile, R. C. ~ ~

~schutzanlage *f*, Stahlbeton(Reaktor)Strahlenschutzanlage, Stahlbeton-(Strahlen)Abschirmungsanlage = reinforced-concrete (reactor) shield, ~ structure for shielding atomic plants, R. C. (reactor) shield

~-Schwimmkasten *m* = reinforced-concrete box caisson, R. C. ~ ~

~-Senkbrunnen *m*, Stahlbeton-Gründungsbrunnen = reinforced-concrete open caisson, ~ drop shaft, R. C. ~ ~

~silo *m* = reinforced-concrete silo, R. C. ~

~skelettbau *m*, Stahlbetonfachwerkbau = reinforced-concrete framed construction, ~ skeleton ~, ~ framework ~, R. C. ~ ~

~skelettgebäude *n*, Stahlbetonfachwerkgebäude = R. C. framed building, ~ skeleton ~, ~ framework ~, reinforced-concrete ~

~sockelfundament *n* → Stahlbetonsockelgründung *f*

~sockel(fundamentkörper) *m* → Stahlbetonsockelgründung *f*

~sockelfundation *f* → Stahlbetonsockelgründung *f*

~sockelgründung *f*, Stahlbetonsockelfundament *n*, Stahlbetonsockelfundation *f*, Stahlbetonsockel(fundamentkörper) *m*, Stahlbetonsockelgründungskörper, Stahlbetongründungssockel *m*, Stahlbetongrundsockel, Stahlbetonfundamentsockel, Stahlbetonfundationssockel, Stahlbetonrost *m* = reinforced-concrete (single) base, ~ individual ~, R. C. (~) ~, ~ block

~sockelgründungskörper *m* → Stahlbetonsockelgründung *f*

~spirale *f* [*Überdruckturbine*] = reinforced-concrete spiral flume, R. C. ~ ~

~(spund)bohle *f* = reinforced-concrete sheet pile, R. C. ~ ~

~spundwand *f* = reinforced-concrete sheet (pile) wall, ~ (sheet)piling, ~ (sheet pile) bulkhead, ~ sheeting, ~ wall, R. C. sheet (pile) wall

~straßendecke *f*, Stahlbetonstraßenbelag *m* = reinforced-concrete road pavement, ~ highway ~, ~ ~ surfacing, R. C. ~ ~

~sturz *m* = reinforced-concrete lintel, ~ lintol, R. C. ~

~stütze *f* = reinforced-concrete column, R. C. ~

~stützmauer *f* = reinforced-concrete retaining wall, R. C. ~ ~

~technik *f* = reinforced-concrete design and construction, R. C. ~ ~, reinforced-concrete engineering, R. C. engineering

~-Tonnenschale *f* = R. C. barrel-vault shell, reinforced-concrete ~ ~

~träger *m* = reinforced-concrete girder, R. C. ~

~treppe *f* = reinforced-concrete stair(case), R. C. ~

~turm *m*, Stahlbetonpylone *f* [*Brücke*] = reinforced-concrete pylon, ~ tower, R. C. ~

~unterzug *m* = reinforced-concrete joist, R. C. ~

~-Vorfertigungsstelle *f*, bewegliches Montagestahlbetonwerk *n* = precast reinforced-concrete manufacturing yard, ~ R. C. ~ ~

~wand *f* = reinforced-concrete wall, R. C. ~

~ware *f* [*Erzeugnis aus Stahlbeton in Massenfertigung hergestellt*] = mass-produced reinforced-concrete product, ~ R. C. ~

~-Wasserbehälter *m* = reinforced-concrete water tank, R. C. ~ ~

~werk *n*, Stahlbetonfabrik *f* = reinforced-concrete works, ~ factory, R. C. ~

Stahl|biegeliste *f* → Biegeliste
~-Biegemaschine *f* = steel bending machine, ~ bender
~biegeplan *m* → Biegeplan
~biegeplatz *m* → Biegeplatz
~bieger *m* → Bieger
~binder *m*, Stahlfachwerkbinder, Stahldachbinder = steel roof truss
Stahlblech *n* = steel plate
~(brücken)tafel *f*, Stahlbrückentafel = steel deck plate
~fahrbahn *f* → Leichtfahrbahn
~fenster *n*, Stahlfenster = steel (plate) window
~form *f*, stählerne Form, Stahlform [*Betonsteinindustrie*] = steel mould (Brit.); ~ mold (US)
~mitnehmer *m*, Stahlmitnehmer = steel (plate) catch
~rohr *n* → Pfahlrohr
~schalung *f*, Stahl(tafel)schalung = steel formwork
~silo *m*, Stahlsilo = steel (plate) silo
~tafel *f*, Stahl(blech)brückentafel = steel deck plate
~verkleidung *f* Stahl(blech)auskleidung, (Stahl(blech)bekleidung = steel (plate) lining
~verzug *m*, Stahlverzug (⚒) = steel (plate) lagging
~bodenplatte *f* = steel floor tile
Stahlbogen(träger) *m* = steel arch(ed) girder)
~brücke *f*, = steel arch(ed girder) bridge
Stahl|bohle *f* → → Spundwand *f*
~bohrkrone *f* = steel (drill) bit
~bohrturm *m* = steel derrick
~bolzenkette *f* = steel pin chain
~brücke *f* = steel bridge
~brückenbau *m* = steel bridge building
~-Brückenfachwerkträger *m* = steel truss-span, ~ bridge truss
~-Brückenpfeiler *m* = steel bridge pier, ~ ~ support
~brückentafel *f*, Stahlblech(brücken)tafel = steel deck plate
~brunnen *m*, Stahlquelle *f*, Eisenquelle = chalybeate spring
~-Brunnenkranz *m*, Stahl-Schlenge *f* = steel cutting curb, ~ drum ~, ~ shoe
~bunker *m* = steel bin
~caisson *m*, Stahl-(Druckluft)Senkkasten *m* = pneumatic steel caisson
Stahldach *n* = steel deck (roof)
~binder *m* → Stahlbinder
~(ein)deckung *f* → Stahlbedachung
~haut *f* → Stahlbedachung
~konstruktion *f* = steel roof system
Stahl|(dach)pfanne *f* = steel (roof) tile
~dachplatte *f* = steel deck unit
~dalbe *m* = steel dolphin
Stahldecken|balken *m* [*für Balkendecken mit dicht verlegten Balken*] = (rolled-)steel joist, RSJ, r.s.j.
~balken-Bieger *m*, Stahldeckenbalken-Biegemaschine *f* = beam bender, bending machine [*A machine for straightening or bending rolled-steel joists*]
~stütze *f* = steel slab prop
Stahldocke *f*, Stahlbaluster *m* = steel baluster
Stahldraht *m* = steel wire
~besen *m* = steel wire brush
~bündel *n* → Bündel aus Drähten
~gewebe *n* = steel wire mesh
~gurt *m* = steel-wire belt
~-Harfensieb *n* = steel wire harp(-type) screen
~kabel *n* → Drahtseil
~korb *m*, (Fugen)Stützkorb = metal chair, wire ~, ~ cradle
~seil *n* → Drahtseil
Stahl-(Druckluft)Senkkasten *m*, Stahlcaisson *m* = pneumatic steel caisson
~druckrohrleitung *f*, Stahltriebwasserleitung = steel penstock
~eindeckung *f* → Stahlbedachung
~einlage *f*, Stahlseele *f*, Stahlkern *m* [*Drahtseil*] = steel core
~einlagen *fpl* → Bewehrung *f*
stählerne Form *f* → Stahl(blech)form
stählerner Wehrbock *m* = steel trestle
~ Zylinder *m* [*Walzenwehr*] = cylindrical plate-steel roller

stählernes Rollenlager *n* [*für Hoch- und Tiefbauten*] = steel roller bearing

Stahlfachwerk *n*, **Stahlskelett** *n* = steel framework, ~ skeleton

~**bau** *m* → Stahlskelettbau

~**binder** *m* → Stahlbinder

~**element** *n* [*Fehlname: Stahlfachwerkeinheit*] = steel framework unit, ~ skeleton ~

~**konstruktion** *f* → Stahlskelettkonstruktion

~**träger** *m* = steel truss

~**trägerbrücke** *f* = steel truss bridge

Stahl|-Fahrradstand *m* = steel (cycle) stand [*B.S. 1716*]

~**falle** *f*, Stahlgleitschütz(e) *n*, (*f*) = steel sliding gate, ~ slide ~

~**feilspäne** *mpl* = steel filings, ~ turnings

~**-Feldbahnlore** *f* → Stahl-Muldenkipper *m*

~**-Feldbahnwagen** *m* → Stahl-Muldenkipper *m*

~**fenster** *n*, Stahlblechfenster = steel (plate) window

~**fensterkitt** *m* = putty for steel (plate) windows

~**fertigteil** *m*, *n* [*Tunnel- und Stollenbau*] = liner plate

~**filterrohr** *n* = steel filter pipe

~**fitting** *m* → Stahlformstück

~**flachstraße** *f*, Stahlplattenstraße = auxiliary steel road, ~ ~ highway

~**flasche** *f*, Druckgasflasche [*für Propan oder Butan*] = steel cylinder

~**-Förderwagen** *m* → Stahl-Muldenkipper *m*

~**form** *f* → Stahlblechform

~**formstück** *n*, Stahlfitting *m* = steel fitting

~**(fuß)boden** *m* = steel flooring

~**garage** *f* = steel garage

~**gelenk** *n* = steel hinge

~**gerüst** *n* → Stahlrüstung

~**gerüst** *n* [*beim lotrechten Aufzug*] = overhead gantry

~**geviert** *n* (✵) = (rectangular) steel shaft set

~**gewebeeinlage** *f*, Bewehrungsnetz *n*, Armierungsnetz, Netzarmierung *f*, Netzbewehrung, Netzgewebeeinlage = reinforcing (road) mesh, mesh reinforcement, fabric reinforcement, reinforcing screen

~**-Gitterbogen(träger)brücke** *f* = steel lattice arch(ed girder) bridge

~**gittermast** *m* = lattice-steel mast

~**gitterstütze** *f* = steel lattice column, lattice-form steel ~

~**gitterträger** *m* = lattice steel girder, steel lattice ~

~**gleitschütz(e)** *n*, (*f*), Stahlfalle *f* = steel sliding gate, ~ slide ~

~**glied** *n* → (angelenkte) Blechplatte *f* [*Gliederband*]

~**gliederband** *n*, Trogbandförderer *m* = steel apron conveyor, ~ ~ conveyer, ~ plate ~

~**-Gründungsbrunnen** *m*, Stahl-Senkbrunnen = steel open caisson, ~ drop shaft

~**(gründungs)rost** *m* = steel (foundation) grillage

~**gurt** *m*, Stahl-Längstraverse *f*, Stahlleitbohle *f* [*Baugrubenverkleidung*] = steel wale (US); ~ waling (Brit.)

~**guß** *m* [*als Erzeugnis*] = steel casting

~**guß** *m* [*als Werkstoff*] = cast steel

~**güte** *f*, Stahlsorte *f* = steel grade

~**haken** *m* = steel hook

~**hauptträger** *m* = main steel girder

~**heizkörper** *m* = steel radiator

Stahlhochbau *m* [*als Bauweise*] steel building construction, ~ structural engineering

~ [*als Baukonstruktion*] = steel frame superstructure

~**kran** *m* = steel erecting crane

Stahl|hochbehälter *m* = elevated steel tank

~**hochstraße** *f*, Stahl-Brückenstraße, Stahl-Hochbrücke *f* = steel overhead roadway, ~ ~ highway, ~ elevated ~, ~ expressway; ~ skyway (US)

~**hohlpfahl** *m* = hollow steel pile

Stahlhohlrahmen — Stahlpfahl

~hohlrahmen m = hollow steel frame
~kabel n → Drahtseil
~kammer f = steel chamber
~kappe f (⚒) = steel bar
Stahlkasten m [für Gründungen unter Wasser] = steel caisson
~pfahl m = steel box pile
~selbstkipper m → Selbstmuldenkipper
~träger m = steel box girder
~trägerbrücke f = steel box girder bridge
Stahl|keil m = steel wedge
~kelle f = steel trowel
~-Kellerfenster n = steel cellar window
~kern m, Stahlseele f, Stahleinlage f [Drahtseil] = steel core
~kipplore f → Stahl-Muldenkipper m
~kletterschalung f = climbing steel formwork
~konsole f = steel bracket [A horizontal steel projection from a vertical or near vertical surface to support a load]
~konstruktion f = (structural) steelwork
~konvektor m = steel convector radiator
~korn n = steel grain
~kragbinder m = cantilever steel truss
~kragbrücke f = steel cantilever bridge
~kriechen n = steel creep
~kübel m, Stahlschaufel f [Hublader] = steel loading bucket
(~)Kugelrohrmühle f = (steel) ball tube mill
~lager n [für Hoch- und Tiefbauten] = steel bearing
~lagerplatte f → Stahlsohlplatte
~längsträger m = steel stringer
~-Längstraverse f → Stahlgurt
~-Lehrgerüst n = steel-supported formwork
~leichtbau m [Brücke] → Leichtfahrbahn f
~leichtbau m = light steel construction

~leichtträger m = light steel girder
~leichtträgerdecke f mit Hohlkörpern = lightweight composite floor of prefab(ricated) steel lattice girders and filler tiles
~-Leitbohle f → Stahlgurt
~leitplanke f → Stahl-(Sicherheits-)Leitplanke
~(-Licht)mast m, Stahl-Beleuchtungsmast, Stahl-(Straßen)Leuchtenmast = steel (lighting) column, ~ (~) mast
~litze f = steel strand
(~)Mantel m, (Stahl)Bandage f [Walze] = (steel) facing
~mantel-Dreiradwalze f, statische Dreiradwalze, Glattmantel-Dreiradwalze, Eigengewichts-Dreiradwalze = steel-faced threewheel roller
~mast m = steel mast
~mast m → Stahl-Lichtmast
~mattenbelag m = Square Mesh Tracking, SMT [Trademark]
~meßband n, Stahlbandmaß n = steel measuring tape
~meißel m = steel bit
~montage f = steel erection
~mörtel m = reinforced (cement) mortar [developed by Prof. P. L. Nervi in Italy]
~-Muldenkipper m, Stahl-Baustellenlore f, Stahl-Feldbahnlore, Stahl-Muldenwagen m, Stahl(-Mulden)kipplore, Stahl-Baustellen-Schienenwagen, Stahl-Förderwagen, Stahl-Feldbahnwagen = steel tip(ping) wagon, ~ skip, ~ (rocker) dump car, ~ side tip(ping) wagon; ~ jubilee wagon (or skip) (Brit.); ~ industrial rail car (US)
~(-Mulden)kipplore f → Stahl-Muldenkipper m
~-Muldenwagen m → Stahl-Muldenkipper m
~nagel m = steel nail
~nippel m, Band n [Schweißtechnik] = backing strip, ~ ring
~panzerung f = steel armo(u)r
~pfahl m = steel pile

Stahlpfahl — Stahlrohrstütze 930

~pfahl *m* in Kastenform = steel box-pile [*Steel box-piles are generally made up by welding together two steel sheetpiles. Other forms of steel box-piles comprise rolled steel channels with plates welded across the flanges*]
~pfette *f* = steel purlin(e)
~pfosten *m* = steel post
Stahlplatte *f* = steel plate
~ [*Steinformmaschine*] = steel pallet
~ → (angelenkte) Blechplatte [*Gliederband*]
~ → Stahlüberbau *m*
Stahlplatten|(ab)deckung *f*, Stahlplattenbelag *m* [*Wasserseite eines Erdstaudammes*] = steel facing (US); ~ ~ membrane (Brit.)
~band(förderer) *n*, (*m*) = steel plate conveyor, ~ ~ conveyer
~belag *m* → Stahlplatten(ab)deckung *f*
~deckung *f* → Stahlplattenabdeckung
~fahrbahn *f* [*Brücke*] = steel plate roadway
~straße *f*, Stahlflachstraße = auxiliary steel road, ~ ~ highway
~unterboden *m* = steel plates subfloor
~verkleidung *f* = armor plating (US); armour ~ (Brit.)
Stahl|ponton *m* = steel pontoon
~profil *n* → Trägerstahl
~profilwalzwerk *n* → Fassonstahlwalzwerk
~pylon *m* → Stahlturm *m* [*Hängebrücke*]
~quelle *f*, Stahlbrunnen *m*, Eisenquelle = chalybeate spring
~radiator *m* = steel radiator
~rahmen *m* = steel frame
~rahmenbau *m* = steel frame construction
~rammpfahl *m* = (impact-)driven steel pile, steel driving ~
~regal *n* = steel shelf
~riemenscheibe *f* = steel belt pulley
~ritzel *n* = steel pinion
~(roh)eisenschlacke *f* → Stahlschlakke

Stahlrohr *n* = steel pipe
~ → Pfahlrohr
~ **für Durchlässe**, ~ ~ Abzugkanäle = steel culvert
~**bau** *m* → Stahlrohrgerüstbau
~**-Beleuchtungsmast** *m* → Stahlrohr-(Licht)mast
~**binder** *m* = tubular steel truss
~**-Deckenstütze** *f*, Deckenstütze aus Stahl = tubular steel (slab) prop
~**-Druckleitung** *f*, Stahlrohr-Triebwasserleitung; Stahlrohrhangleitung [*Schweiz*] = steel pipe penstock
~**fenster** *n* = tubular steel window
~**(förder)band** *n* = tubular steel type belt conveyor, ~ ~ ~ conveyer
~**gerüst** *n*, Stahlrohrrüstung *f* = tubular steel scaffold(ing)
~**(gerüst)bau** *m* = tubular steel scaffold(ing) construction
(~)**Gerüstturm** *m* = tubular steel tower
~**gitterausleger** *m* [*Fehlname: Stahlrohrfachwerkausleger*] = steel tubing lattice jib (Brit.); ~ ~ ~ boom (US)
~**gitterturm** *m* [*Fehlname: Stahlrohrfachwerkturm*] = steel tubing lattice tower
~**griff** *m* = tubular steel handle
~**hangleitung** *f* → Stahlrohr-Druckleitung
~**karre(n)** *f*, (*m*) = tubular steel barrow
~**lehrgerüst** *n* = steel tube centering
~**-Leichtbaukran** *m* = steel tubing light building crane, tubular steel type ~ ~ ~
~**leitung** *f* = steel piping
~**(-Licht)mast** *m*, Stahlrohr-Beleuchtungsmast, Stahlrohr-(Straßen)-Leuchtenmast = tubular steel (lighting) column, ~ ~ (~) mast
~**nadel** *f* [*Wehr*] = steel pipe needle
~**pfette** *f* = tubular steel purlin(e)
~**rüstung** *f* → Stahlrohrgerüst
~**stütze** *f* [*Stahlgerüst*] = tubular steel leg
~**stütze** *f* = steel tube column

~-Triebwasserleitung *f* → Stahlrohr-Druckleitung
~(ver)legeanlage *f* = steel pipe laying plant
Stahlrollenlager *n* = steel roller bearing
Stahlrost *m*, Stahlgründungsrost = steel (foundation) grillage
~gründung *f* = steel grillage foundation
Stahl|rüstung *f*, Stahlgerüst *n* = steel scaffold(ing)
~saite *f* = piano wire
(~)**Saitenbeton** *m*, Bauweise *f* Hoyer = prestressed concrete with thin wires, Hoyer method
~saitenbetonträger *m* = Hoyer girder
~schalung *f* → Stahlblechschalung
~schalungs-Reinigungsmaschine *f* = steel formwork cleaning machine
~schaufel *f*, Stahlkübel *m* [*Hublader*] = steel loading bucket
~scheibenrad *n* = steel disc wheel; ~ disk ~ (US)
~schiene *f* = steel rail
~schlacke *f*, Stahl(roh)eisenschlacke = stahl-eisen slag, slag from pig iron for steel making purposes
~-Schlenge *f*, Stahl-Brunnenkranz *m* = steel drum curb, ~ cutting ~, ~ shoe
~schlitten *m* = steel skid
~schneidegerät *n* = steel cutting equipment
~schott *n* = steel bulkhead
~schraub(en)pfahl *m* = steel screw pile
~schrot *m*, *n* = steel grit, ~ shot
~schrott *m* = scrap steel
~schuh *m* = steel shoe
~schwelle *f* [*Eisenbahn*] = steel sleeper
~-Schwimmkasten *m* = steel box caisson
~seele *f*, Stahlkern *m*, Stahleinlage *f* [*Drahtseil*] = steel core
Stahlseil *n* → Drahtseil
~-Förderband *n* = steel wire rope belt conveyor, ~ ~ ~ ~ conveyor

Stahl|-Senkbrunnen *m*, Stahl-Gründungsbrunnen = steel open caisson, ~ drop shaft
~-(Sicherheits-)Leitplanke *f* = steel safety fence, ~ guard ~, ~ protection ~, ~ ~ rail
~sieb *n*, Federstahldrahtgewebesieb = spring steel wire screen
~silo *m*, Stahlblechsilo = steel (plate) silo
~skelett *n*, Stahlfachwerk *n* = steel skeleton, ~ framework
Stahlskelettbau *m*, Stahlfachwerkbau = steel skeleton construction, ~ framework ~
~, Stahlfachwerkbau [*mehrgeschossiger Bau, dessen Stützen, Unterzüge und Deckenträger zusammen ein Tragwerk aus Stahl bilden*] = steel framed building, ~ framework ~, ~ skeleton ~
Stahl|skelettkonstruktion *f*, Stahlfachwerkkonstruktion = steel framed construction, ~ framework ~, ~ skeleton ~
~sohlblech *n* → Stahlsohlplatte *f*
~sohlplatte *f*, Stahllagerplatte, Stahlsohlblech *n* [*Brücke*] = steel sole plate, ~ bed ~
~sonderprofil *n*, Stahlspezialprofil = special steel section, purpose-made ~ ~
~sorte *f*, Stahlgüte *f* = steel grade
~späne *mpl* = steel shavings
~spannung *f* = steel stress
~spirale *f* [*Überdruckturbine*] = steel spiral flume
~sprosse *f* [*Fenster*] = steel sash bar, ~ glazing ~; ~ muntin (US)
~sprosse *f* [*Leiter*] = steel rung, ~ round, ~ tread, ~ rundle
(~)**Spülspitze** *f* [*Grundwasser(ab)senkung in Feinstsanden*] = jetting tip
~(spund)bohle *f* → → Spundwand *f*
~spundwand *f* = steel sheet (pile) wall, ~ sheeting, ~ bulkhead, ~ (sheet) piling
~stab *m* = steel bar, ~ rod
~(stau)wehr *n* = steel (framework) weir

Stahlstemmtor — Stahlwellblech

~stemmtor *n* = steel mitre gate, ~ miter ~
~stempel *m*, stählerner (Abbau-)Stempel (⚒) = steel prop, ~ post
~stift *m* = steel brad
~straßenbrücke *f* = steel road bridge, ~ highway ~
~stütze *f* = stanchion (Brit.); (structural-)steel column
~tafel *f* → Stahlüberbau *m*
~tafelschalung *f*, Stahl(blech)schalung = steel formwork

Stahlton|brett *n*, vorgespanntes Ziegelbrett = (Stahlton) prestressed clay plank, ~ plank
~dach *n*, vorgespanntes Ziegeldach = Stahlton roof, roof in prestressed clay
~-Dachplatte *f* = Stahlton roof slab, roof slab in prestressed clay
~decke *f*, vorgespannte Ziegeldecke = Stahlton floor, floor in prestressed clay
~-Fenstersturz *m*, (vorgespannter) Ziegel-Fenstersturz = Stahlton plank window lintel, ~ ~ ~ lintol
~sturz *m* → Ziegel-Fertigsturz
~-Wandtafel *f* = Stahlton wall unit, ~ ~ slab, ~ ~ panel, wall unit in prestressed clay, wall panel in prestressed clay, wall slab in prestressed clay

Stahltor *n* = steel gate
Stahlträger *m* → Trägerstahl
~feld *n* = steel (girder) span
~-Hourdidecke *f*, Stahlträger-Tonhohlplattendecke, Stahlträger-Tonhohlkörperdecke = burnt-clay hollow-tile floor with steel girders, ~ hollow-block ~ ~ ~, ~ pot ~ ~ ~ ~
~pfahl *m*, H-Pfahl = H-pile, structural-steel pile
~-Verbundkonstruktion *f* = compound steel beam structure

Stahl|tragwerk *n* = steel load-bearing system
~traverse *f* = steel spreader bar [*used for picking up precast girders with a crane*]

~treppe *f* = steel stair(case)
~trümmer *f* = steel wreckage
~tür *f* = steel door

Stahlturm *m* = steel tower
~, Stahlpylon *m*, Stahlpylone *f* [*Hängebrücke*] = steel pylon, ~ tower
~ mit Hindernisleuchten [*Flugplatz*] = illumination tower

Stahltürzarge *f* = steel door trim, ~ ~ casing

Stahlüberbau *m* = steel (work) superstructure
~, Stahltafel *f*, Stahlplatte *f* [*Brücke*] = steel (bridge) deck(ing), ~ deck plate(s)

Stahl|unterlegscheibe *f* = steel washer
~verbrauch *m* = steel consumption
~verbund-Brückenträger *m* = composite bridge stringer
~verkleidung *f* → Stahlblechverkleidung
~verschluß *m* [*Wehr*] = steel gate

stahlverstärktes Präzisionslager *n* **aus Alu(minium)legierung** = special alloy alumin(i)um bearing with steel backing

Stahlverzug *m*, Stahlblechverzug (⚒) = steel (plate) lagging

Stahlvollwand|-Balkenbrücke *f* = steel plain web beam bridge, ~ solid ~ ~ ~
~-Verbundträgerbrücke *f* = steel plain web composite girder bridge, ~ solid ~ ~ ~ ~

Stahl|walze *f* → → Walze
~wangentreppe *f* = steel string stair(case)

Stahlwasserbau *m* = steel civil engineering hydraulics
~ = steel hydraulics structures
~werk *n* = steel hydraulics structure

Stahl|wasserrohr *n* = steel water pipe
~wehr *n*, Stahlstauwehr = steel (framework) weir
~wellblech *n* dessen beide Oberflächen durch Schichten von Asphalt, Asbestgeweben und Farbanstrichen geschützt sind = protected metal sheeting

~wendel *m* [*Spannbeton*] = steel helix
~werk *n* = steel plant
~winde *f* → Zahnstangenwinde
~windscheibe *f* = steel shear wall
~-Wolkenkratzer *m* = steel skyscraper
~wolle *f* = steel wool [*It consists of fine shreds of steel matted together for use in the same manner as sandpaper. It is available in several grades of fineness*]
~zahn *m* [*Aufreißer*] = steel tyne (Brit.); ~ tine (US)
~zarge *f* = steel trim, ~ casing
(~)Zellendecke *f* = cellular steel floor
Staket *n*, Lattenzaun *m*, Stakete *f*, Staketenzaun = pale fence, ~ fencing
Stakung *f* [*die Ausfüllung von Balkenfachen bei Holzbalkendecken oder Fachwerkwänden durch Windelpuppen*] = straw and loam pugging
Stalagmit *m*, Auftropfstein *m* = stalagmite
Stalaktit *m*, Abtropfstein *m* = stalactite
Stall *m*, Tierunterkunft *f* = animal shelter
~bodenplatte *f* = flooring slab for animal shelter
~decke *f* = floor for animal shelter
~fenster *n* = window for animal shelter
~fußboden *m* = floor(ing) for animal shelter
~wand *f* = animal shelter wall
Stamm|baum *m* = family tree
~kunde *m* = repeat customer
~leitung *f* → Stammrohrleitung
~personal *n*, Stammbelegschaft *f* = permanent staff
~(rohr)leitung *f* [*Beregnung*] = main line
~schiene *f* → Backenschiene

Stampfasphalt *m* = compressed (natural) rock asphalt [*B. S. 348*]
~aufbruch *m* = broken up compressed (natural) rock asphalt
~decke *f*, Stampfasphaltbelag *m* = compressed (natural) rock asphalt surfacing
~platte *f* [*besteht aus Naturasphaltrohmehl oder Kalksteinmehl und Bitumen*] = compressed (natural) rock asphalt tile
Stampf|bagger *m* → Freifall-Kranstampfer *m*
~bär *m*, Fallbär, Fallgewicht *n* [*Rüttelstampfmaschine*] = tamper
Stampfbeton *m* [*erdfeuchter oder weicher Beton in Schalung oder Graben, von Hand oder mit mechanischen Stampfern hineingestampft*] = tamped concrete
~rohr *n* = tamped concrete pipe
Stampf|bohle *f* → Bohle
~bohlenfertiger *m* → Schlagbohlenfertiger
Stampfe *f* → Holzstößel *m*
Stampf-Einrichtung *f* für Bagger → Freifall-Kranstampfer
Stampfer *m* [*zum Einstampfen von Formsand*] = sand rammer
~, Stampfgerät *n* = tamper
~ für Straßenbauschalung = form tamper
~abschnitt *m* [*Schwarzbelageinbaumaschine*] = (tamper) bar
Stampf|fallplatte *f* → Stampfplatte
~fertiger *m* = bridge tamper, road tamping machine
~fuge *f* = tamped joint
~fuß *m*, Schaffuß, Druckstempel *m* [*Walze*] = sheepsfoot
~gerät *n* → Stampfer
~gerät *n* zur Herstellung von Probewürfeln [*DIN 1996*] = compactor
~gewicht *n* = tamping weight

(Stampf)Hammerfertiger *m*, kombinierter Stampfbohlen- und -hammerfertiger	hammer tamping finisher
Abgleichbohle *f*, Abziehbohle *f*	oscillating levelling beam (Brit.); strike-off (US)

Stampfmaschine — ständiger Wasserlauf

Hammerreihe *f* — row of hammers, series of lozenge shaped hammers [*rhomboidal in shape, each weighing about 66 lbs. which fall freely on to the concrete surface*]

Nachstampfbohle *f* — heavy tamping beam, tamper [*delivering 160 blows per minute*]

Schwingungsschleifbalken *m*, Vibrationsschleifbalken — vibrating beam, ~ smoother

---

Stampf|maschine *f* → (Beton)Steinstampfmaschine
~platte *f*, Fallplatte, Fallstampfplatte, Stampffallplatte [*Freifall-Kranstampfer*] = tamping plate, fallingplate tamping unit
~schuh *m* = tamping shoe
~vorrichtung *f* für Bagger → Freifall-Kranstampfer
~werkzeug *n* = tamping tool
Standanzeiger *m* → Füll~
Standard|abweichung *f* [*Siebtechnik*] = standard deviation
~ausleger *m* = basic jib (Brit.); ~ boom (US)
~ausrüstung *f* = standard equipment
~bauweise *f* = standard construction method
~brückensystem *n* = unit-construction bridge system
~form *f* = standard mould (Brit.); ~ mold (US)
~länge *f*, Grundlänge = basic length
~muffe *f* [*Tiefbohrtechnik*] = regular coupling
~schaufel *f* = general purpose bucket
~stegplatte *f* [*Gleiskette*] = standard grouser shoe
~zeichnung *f* = typical drawing
Stand|bagger *m* = static excavator
~bahn *f* [*Auf Gleisen laufende Förderwagen werden durch ein endloses Zugorgan, Seil oder Kette, mitgenommen*] = haulage system with endless pulling device
~baum *m*, Standmast *m* = derrick mast
~behälter *m* → stehender Behälter
~belastung *f* = static loading
~bremse *f* = parking brake
~dauer *f*, Standzeit *f* [*von Schächten*] = age [*of shafts*]
~dauer *f*, Standzeit *f* [*von Stempeln usw.*] = standing time [*of props etc.*]
Ständer *m*, Stange *f*, Pfosten *m* = post
~ → Binderstiel *m*
~, Stiel *m*, Pfosten *m*, Säule *f* [*Fach(werk)wand*] = post
~bau *m* = post and beam construction
~bauhaus *n*, Pfostenhaus = post and beam house
~fachwerk(träger) *n*, (*m*) = vertical truss
~hydrant *m* = post hydrant
~konstruktion *f* [*Holzhaus*] = post and beam system
~schleifmaschine *f* = column grinder, upright ~, housing ~
Stand|festigkeit *f* = stability
~gefäß *n* → stehender Behälter *m*
~hahn *m* = pillar tap
ständige Last *f*, ruhende ~ [*Ist die Summe der unveränderlichen Lasten, also das Gewicht der tragenden oder stützenden Bauteile und der unveränderlichen, von den tragenden Bauteilen dauernd aufzunehmenden Lasten (z.B. Auffüllungen, Fußbodenbeläge, Putz u. dgl.)*] = dead load
~ Strömung *f*, stationäre ~, permanente ~, dauernde ~ = steady flow
ständiger Benutzer *m* (eines Verkehrsmittels), Pendler *m* = commutor
~ Wasserlauf *m* = perennial water course

Stand|kahn *m* = stationary storage barge
~**kessel** *m* = vertical steam boiler
~**kessel** *m* **mit waagerechten Quersiedern**, Lachapellekessel = vertical cross tube boiler
~**länge** *f* [*Bohrer*] = bit footage
~**licht** *n* = parking light
~**linienzug** *m*, Polygon(zug) *n*, (*m*) = traverse
~**mast** *m*, Standbaum *m* = derrick mast
~**meßgerät** *n* → (Füll)Standanzeiger *m*
~**öl** *n* = stand oil
Standort *m*, Ortslage *f*
~**faktor** *m* [*Alle Einflüsse, welche die Zusammensetzung der Pflanzendecke bestimmen*] = site factor
~**veränderung** *f* → Baustellenwechsel
~**wahl** *f* = location selection, site ~
~**wechsel** *m* → Baustellenwechsel
Stand|personal *n* [*Messestand*] = management staff
~**pfahl** *m* → Spitzenpfahl
~**pfeiler** *m* (⚒) = chock
~**platz** *m*, Abstellfläche *f*, Abstellplatz = hardstand, (hard-)standing, parking apron [*aerodrome*]
~**punkt** *m* [*Perspektivzeichnen*] = position of spectator
Standrohr *n* [*Am Ende oder im Zuge einer Wasser-Druckrohrleitung oder an beliebige Stellen des Ortsrohrnetzes angebrachtes, oben offenes Rohr mit Überlauf zum Ausgleich von Druckschwankungen oder zur Entlüftung*] = riser
~ → Beobachtungsbrunnen *m*
~ [*Aufsatzrohr auf Unterflurhydrant*] = stand pipe
~ [*Zum Schutz des Bohrlochanfanges gegen mechanische Beschädigungen*] = conductor pipe [*to protect the start of the hole against mechanical damage*]
~ → Leitrohrtour *f*
~**spiegelhöhe** *f*, Beobachtungsrohrspiegelhöhe = piezometric level, ~ head

Stand|schalung *f* = ordinary formwork, fixed ~
(~)**Seilbahn** *f* → Seil-Standbahn
~**sicherheit** *f* [*Bauwerk; Böschung*] = stability [*structure; slope*]
~**sicherheitsbeiwert** *m* = stability coefficient
~**sicherheitsberechnung** *f* = stability computation, ~ analysis
~**sicherheitsfaktor** *m* = stability factor
~**silo** *m* = vertical silo
~**spur** *f*, Standstreifen *m* [*Eine in der Fahrbahn liegende Spur, die dem ruhenden und fließendem Verkehr dient*] = shoulder
~(**stau**)**wehr** *n* → Losständerwehr
~**urinal** *n* = slab urinal
~**zeit** *f*, Standdauer *f* [*von Schächten*] = age [*of shafts*]
~**zeit** *f*, Standdauer *f* [*von Stempeln usw.*] = standing time [*of props etc.*]
~**zeit** *f* [*Bohrer*] = bit life
~**zeit** *f* → Topfzeit
Stange *f*, Ständer *m*, Pfosten *m* = post
Stangen|gerüst *n*, Stangenrüstung *f* = pole scaffold(ing)
~**kohle** *f* = columnar coal
~**mühle** *f* = rod grinder, ~ (grinding) mill
~**rost** *m* → Stabrost
~**rüstung** *f*, Stangengerüst *n* = pole scaffold(ing)
~**scharnier** *n* = continuous hinge
~**schere** *f* = bar cutter, bar-cutting [*A shearing machine that cuts metallic bars into lengths*]
~**schwimmer** *m*, Stockschwimmer = pole float, staff ~
~(**sieb**)**rost** *m* → Stabrost
~**zinn** *n* = bar tin
Stanzöl *n* = punching oil
Stapel|fläche *f*, Stapelplatz *m* = storage area
~**förderer** *m* → Stapler *m*
~**kran** *m* = stacker crane
~**lauf** *m* = launching ceremony
stapeln, auf~ = to stack, to pile (up)

Stapel|platz *m*, Stapelfläche *f* = storage area
~prüfung *f* **für Asbestspinnfaser** = array test for asbestos spinning fibre (Brit.)/fiber (US)
~tisch *m* = stacking table
~- und Abtragegerät *n* [*Betonsteinindustrie*] = rack loader and unloader [*loads green blocks, unloads cured blocks*]
Stapler *m* = stacker, tiering machine
~, Stapelförderer *m* = piler
stark ausgelaugter Boden *m* → degradierte Braunerde *f*
starke Flußwindung *f* → Flußschlinge *f*
starker Gipsschlackenzement *m*, **~ Sulfathüttenzement** = high sulphate resisting Portland cement, supersulphate ~; Portland cement type V (US)
Stärke *f* [*Fehlname*]; Dicke *f* = thickness
~ eines Schalles, Lautstärke = loudness (of sound), ~ ~ tone
~leim *m* = starch glue
Starkregen *m* = heavy rainfall
Starkstrom *m* = power current
~freileitung *f* = power current overhead transmission line
~kabel *n* = power current cable
~leitung *f* = power current transmission line
~mast *m* = power pole
starr → biegesteif
~, nicht pendelnd = non-oscillating
starre Decke *f* → ~ Straßendecke
~ Kreisschwingung *f* = positive circle-throw gyratory movement, ~ ~ ~ motion
~ Planiereinrichtung *f* = straight bulldozer
~ (Straßen)Decke *f*, starrer (Straßen-)Belag *m*, Beton(straßen)belag, Beton(straßen)decke = rigid pavement, concrete ~, ~ surfacing; ~ paving (US)
~ Verbindung *f* = fixity at the connection
starrer Bogen *m* = rigid arch
~ Kreisschwinger *m*, starres Kreisschwingsieb *n* = positive circle-throw type screen
~ Rahmen *m* = rigid frame
~ (Straßen)Belag *m* → starre (Straßen)Decke
starres Progressivsystem *n* [*Straßenverkehrstechnik*] = fixed progressive system
Starr|achse *f* = rigid-mounted axle
~heit *f* = rigidity
~heitsbedingung *f* = condition of rigidity
~punkt *m* [*Bitumen*] = shatter point, brittle ~
~verbindung *f* **mit dem Rahmen** = rigid mounting
Start *m*, Abflug *m* = take-off
Startbahn *f* = take-off runway
~grundlänge *f* = basic take-off runway length
~neigung *f* = take-off runway gradient
starten, anlassen = to start
Starter|batterie *f* = electric starter motor battery
~klappe *f* = choke valve
~motorbedienungsorgan *n* = starter engine control
~ritzel *n*, Anlasserritzel = starter pinion
Start|flüssigkeit *f*, Anlaßflüssigkeit = starting fluid
~freigabe *f*, Starterlaubnis *f* [*Flugzeug*] = take-off clearance
~hilfe *f*, Anlaßhilfe = starting aid
~hilfeknopf *m* = choke control knob
~kurbel *f* → Anwurfkurbel
~luftbehälter *m*, Anlaß(luft)behälter = starting air vessel
~motor *m* → Anlaßmotor
~patrone *f*, Anlaßpatrone [*Verbrennungsmotor*] = starting cartridge
~richtung *f* [*Flugzeug*] = direction of take-off
~schalter *m*, Anlaßschalter = starter switch
~stellung *f* = start position
~temperatur *f*, Anlaßtemperatur = starting temperature

Start- und Landebahn *f*, Piste *f* = runway
~ **die aus Gründen der Lärmminderung vorzugsweise benutzt werden soll** = preferential runway
~ ~ **in der vorherrschenden Windrichtung liegt** = prevailing wind runway
~**anordnung** *f*, Pistenanordnung = runway configuration
~**befeuerung** *f*, Pistenbefeuerung = runway lighting
~**feuer** *n* = runway light
~**schwelle** *f*, Pistenschwelle = runway threshold
Start- und Zielort(verkehrs)zählung *f* = origin and destination survey, O-D ~
Statik *f* = statics
Statiker *m* = designer
Station *f* [*Bezeichnung eines Festpunktes einer abgesteckten Linie*] = station, construction ~
~ [*Bahn*] = station
stationär, ortsfest = permanent, nonremovable, static, stationary
stationäre Mischanlage *f*, ortsfeste ~ = stationary mixing plant
~ **Strömung** *f*, ständige ~, permanente ~, dauernde ~ = steady flow
stationärer Abwurfwagen *m* [*Bandförderer*] = fixed tripper
(~) **Saugbagger** *m* **mit Schneidkopf** → Cutter(saug)bagger
(~) **Schneidkopf(saug)bagger** *m* → Cutter(saug)bagger
Stationärwelle *f*, stehende Welle = standing wave
Stationierung *f* = stationing
Stations|dichte *f* [*Geophysik*] = station density
~**kubikmeter** *n* [*Erdbau*] = station cubic metre (Brit.); ~ ~ meter (US)
statisch bestimmt = statically determinate, ~ determined, isostatic
~ **bewehrter Beton** *m*, ~ armierter ~ = statically reinforced concrete
~ **unbestimmt** = statically indeterminate, hyperstatic

~ **unbestimmter Rahmen** *m* = redundant frame, statically-indeterminate ~
~ **wirksame Geschoßdecke** *f* = structural floor
~ ~ **Walze** *f* → → Walze
~ **wirksamer Querschnitt** *m* = structural cross-section
statische Bedingungen *fpl* = conditions of statics
~ **Berechnung** *f* = structural analysis
~ ~ **auf Erdbebensicherheit** = seismic analysis
~ **Dreiradwalze** *f*, Stahlmantel-Dreiradwalze = steel-faced three-wheel roller
~ **Druckhöhe** *f* = (static) head, pressure ~
~ **(elektrische) Auflagung** *f* = static charge
~ **Festigkeit** *f* = static strength
~ **Kipplast** *f* = static tipping load rating
~ **Rammformel** *f* = static pile driving formula
~ **Unbestimmtheit** *f* = statical indeterminacy
~ **Walze** *f* → → Walze
statischer Druck *m* = static pressure
~ **Teil** *m*, statisches ~ *n* [*Rüttelverdichter*] = non-vibration part
statisches elastisches Verhalten *n* = static elastic behavio(u)r
~ **Frontgewicht** *n* = front-end static weight
~ **Moment** *n* = static moment
~ **System** *n* = static system
Stativ-Schnellrührer *m* = pedestal-type impeller agitator, ~ ~ mixer
Statuenmarmor *m*, Bildhauermarmor = statuary marble
Stau *m* → Anstauung *f*
~ [*Straßenverkehrstechnik*] = congestion
(~)**Abteilung** *f* [*Bewässerung*] = compartment, check [*A long rectangular area between borders, having a definite slope, is a "strip" and not a "check"*]

Stauanlage — Staubschwelung

~anlage *f*, Stau(bau)werk *n* = control work, ~ structure, water-retaining ~
Staub *m* = dust
~ → Blume *f*
~**ablagerung** *f* = dust deposit
~**absaugung** *f* = dust extraction
~**abscheideanlage** *f* = dust-collecting installation, ~ system
~**abscheider** *m* = dust collector
Staubalken *m*, Dammbalken = stop log
~**wehr** *n* → Dammbalkenwehr
Staub|anteil *m*, Staubgehalt *m* = dust content
~**ausblasung** *f* [*Bohrloch*] = puff blowing
~**auswerfer** *m* = dust ejector
Stau(bau)werk *n*, Stauanlage *f* = control work, ~ structure
Staubbekämpfung *f* = dust suppression
~ [*in der Kohlenaufbereitung*] = tipple dust control
staubbeladen, staubgeladen = dustladen
Staub|beutel *m* = dust bag
~**bildung** *f* = dust formation
~**bindemittel** *n* = dustproofer, dust palliative, ~ preventer
~**bindung** *f*, Staubfreimachung = dust alleviation, dust (al)laying, alleviation of dust, dust suppression, dust control, abatement of dusting
~**(brannt)kalk** *m* = pulverized lime
~**decke** *f* = dust cover
~**deckel** *m* = dust cap
staubdicht = dustproof, dust-tight
Staub|dichtigkeit *f* = tightness against dust
~**dichtung** *f* = dust seal
Staubecken *n* → → Talsperre
Stauben *n* = dusting
staubend = dusting
Stau|bereich *m* = impounding zone
~**berieselung** *f* [*Bewässerungsverfahren*] = border-strip flooding
staubfeines Gut *n* = pulverized material

Staubfilter *m*, *n* = dust filter
staub|förmiges Schüttgut *n* = pulverulent material
~**frei**, staublos = dustless
Staub|freimachung *f* → Staubbindung
~**gehalt** *m*, Staubanteil *m* = dust content
staubgeladen, staubbeladen = dustladen
Staub|granulierungsverfahren *n* = dust nodulizing process
~**haftung** *f* = dust bond
Staub|kalk *m*, Staubbranntkalk = lime powder, pulverized lime
~**kammer** *f* = velocity-reducing dust collector
~**kappe** *f*, Staubverschluß *m* = dust cap
~**korn** *n*, Staubkörnchen *n*, Staubteilchen *n* = dust grain
~**lawine** *f* [*In den Alpen „kalte Lawine" genannt*] = loose snow avalanche
staublos, staubfrei = dustless
Staub|luft *f* = dust-laden air
~**lungenerkrankung** *f*, Silikose *f* = silicosis
~**manschette** *f* = dust protection grummet, ~ boot
~**maske** *f* = dust respirator
~**messung** *f* = dust measurement
~**niederschlagung** *f* = dust precipitation
Stau|bogen *m*, Reduzierstück *n* [*Druckluftbetonförderer*] = adapter bend
~**bohle** *f* = flashboard
Staub|rückgewinnung *f* = dust recovery
~**-Sammelschnecke** *f* = dust collecting screw
~**sammler** *m* = dust collector
~**sand** *m* → Feinsand
~**schale** *f* [*Motor*] = dust cup
~**schutzbrille** *f* = dust goggles
~**schutzmittel** *n* **für Betonestrich** = agent to prevent dusting
~**schutzwand** *f* = dust protection wall
~**schwelung** *f* **in der Schwebe** *f* [*Schieferöl*] = fluidization

~technik *f* = dust engineering
~teilchen *n* → Staubkorn *n*
staubtrocken [*Anstrich*] = dry to permit handling
Staub|vermeidung *f* = dust prevention
~verschluß *m*, Staubkappe *f* = dust cap
~wolke *f* = dust cloud
Stauch|bohrer *m*, Ventilbüchse *f*, Schlammbüchse, Schmandbüchse, Kiespumpe *f*, Sandpumpe, Schlämmbüchse, Schmandlöffel, Ventilbohrer, Ventilschappe *f* [*In Bohrungen benutztes Rohr mit darin beweglichem Kolben und Fußventil. Beim Aufstoßen der Kisepumpe auf Bohrlochsohle tritt das kiesige Material durch das Ventil und wird durch den hochgehenden Kolben in das Innere der Kiespumpe gesaugt*] = bailer, shell with valve, sand pump, shell pump, sludger
~nagel *m* = clinch nail
~temperatur *f* [*Niet*] = upsetting temperature
Stauchung *f* [*Verkürzung der äußersten Randfaser des Betons eines Biegebalkens infolge der Biegedruckspannung im Bruchzustand. Die Stauchung beträgt etwa* $2^0/_{00}$] = compressive strain
Stauchungsmoräne *f*, Schubmoräne, Aufpressungsmoräne = push moraine, shove ~
(Stau)Damm *m*, Talsperrendamm, Sperrdamm = embankment (type) dam, fill ~
Staudamm|bau *m* → Sperrdammbau
~breite *f*, (Sperr)Dammbreite, Talsperrendammbreite = fill dam width, embankment (type) ~ ~
~untergrund *m*, (Sperr)Dammuntergrund, Talsperrendammuntergrund = fill dam subsoil, embankment (type) ~ ~
Staudruck *m* = pressure of the impounded water
~, Geschwindigkeitsdruck, dynamischer Druck = dynamic pressure
~ [*Windkanal*] = stagnation pressure

~messer *m* → Pitotrohr *n*
stauen, auf~ = to impound, to retain
Stauffer|büchse *f* = Stauffer cup, ~ lubricator
~fett *n* = Stauffer grease
Stau|haltung *f* → Haltung
~höhe *f* [*Höhe des Stauspiegels über dem ungestauten Wasserstand*] = surface level
~holz *n* → Garnierholz
~inhalt *m* → Fassungsvermögen
(~)Klappe *f*, (Stau)Tafel *f* [*Klappen-(stau)wehr*] = wicket (US); shutter (Brit.)
~körper *m*, (Wehr)Verschluß *m* = gate
~kraftwerk *n*, Speicherkraftwerk, Staukraftanlage *f*, Speicherkraftanlage = storage scheme
~kurve *f*, Staulinie *f* = back-water curve [*A particular form of the surface curve of a stream of water, which is concave upward. It is caused by an obstruction in the channel, such as an overflow dam; the depth is greater at all points than Belanger's critical, and the normal depth and the velocities diminish downstream. The term is also used in a generic sense to denote all water-surface curves*]
~linie *f* → Staukurve *f*
Staumauer *f*, Betonsperre *f*, (Massiv-)Beton(stau)mauer, Talsperren(stau)mauer, (Beton)Sperrmauer = (massive) concrete dam
~ in aufgelöster Bauweise → Pfeilersperre *f*
~beton *m* = dam concrete
Staumoor *n*, Schwingmoor = trembling moor
staunaß = waterlogged [*A condition of lands where the ground water stands at a level that is detrimental to plants. It may result from over-irrigation or seepage with inadequate drainage*]
Stau|quelle *f* = contact spring
~raum *m* → Fassungsvermögen *n*

Stauraumverlandung — Stegarmierung

~raumverlandung *f*, Stauraumanlandung = reservoir sedimentation, sedimentation of reservoirs
~reg(e)lung *f* → Kanalisierung
~rohr *n* → Pitotrohr
Staurolith *m*, Kreuzstein *m* (Min.) = staurolite, staurotide
Stau|scheibe *f*, Blende *f* = blind
~schild *m* [*Walzenwehr*] = shield
~schleuse *f* = check [*A structure designed to raise or control the water surface in a canal or ditch*]
~schwall *m* = back surge
~schwelle *f*, Grundwehr *n* = submerged weir, drowned ~
~schwelle *f* → Grundschwelle
~see *m* → → Talsperre *f*
~see *m*, Abdämmungssee, Abriegelungssee (Geol.) = ponded lake, obstruction ~
~spiegel *m* = level of storage water (Brit.); storage level (US)
~strecke *f* → Haltung *f*
~stufe *f* [*Besteht aus einem Wehr und dazugehörigen Zweckbauten*] = weir with lock, barrage ~ ~, ~ and ~
~stufenkraftwerk *n* = barrage power station
Stauß-Ziegelgewebe *n*, Drahtziegelgewebe = clay lath(ing) [*B.S. 2705*]
(Stau)Tafel *f*, (Stau)Klappe *f* [*Klappen(stau)wehr*] = shutter (Brit.); wicket (US)
Stauung *f* → An~
Stauwand *f* [*vom Oberwasser berührte Fläche eines Staubauwerkes*] = upstream face
~ [*Schwerttrommelwäsche*] = annular dam
~ [*(Stau)Wehr*] = gate [*general term*]
~ → Sektor
Stauwasser *n* = impounded water
~druck *m* = impounded water pressure
(Stau)Wehr *n*; Wuhr *n* [*Schweiz*] = weir, barrage
Stau|werk *n* → Stauanlage
~ziel *n* [*höchst zulässige Wasserspiegelhöhe*] = height of the maximum storage; capacity level (US)

Stech|beitel *m*, Stecheisen *n* = firmer chisel
~eisen *n*, Stechbeitel *m* = firmer chisel
~heber *m* = plunging siphon
~karre(n) *f*, (*m*) **zum Transport über Treppen** = stairwalking hand truck
~karte *f* = time card, clock ~
~schütze *f* [*Bewässerung*] = take-out gate
~uhr *f* = time clock, check ~
~zirkel *m* = dividers
Steck|achse *f* = half shaft, independent axle
~armierung *f* → Anschlußbewehrung
~bewehrung *f* → Anschlußbewehrung
~blende *f* [*DIN 19 206*] = line blind
~bolzen *m* = pin
~bolzenverbindung *f* = pin connection
~einlagen *fpl* → Anschlußbewehrung
~eisen *npl* → Anschlußbewehrung *f*
~mast *m* **aus Stahlblech** [*ein Lichtmast, bei dem die konischen Schüsse von 1,50 m Länge ineinander gesteckt werden*] = (lantern) column of tapered sheet construction
~schlüssel *m* = socket wrench; ~ spanner (Brit.)
~(stahl)einlagen *fpl* → Anschlußbewehrung
~stift *m* = socket pin
Steg *m* [*Verglasung*] = came [*An H-section strip of lead or soft copper, shaped to fix each piece of glass to the next one, in leaded lights or stained-glass windows*]
~, Schubkarren~ = runway
~, Träger~ = web
~, Stollen *m* [*Bodenplatte einer (Gleis-)Kette*] = grouser
~, Lauf~ = walkway, duckboard
~, Stegbreite *f* [*Siebtechnik*] = space between the holes
~ [*Hohlblockstein*] = web
~abstand *m* [*Schweißen*] = root opening
~anschluß *m*, Träger~ = web connection
~armierung *f* → Stegbewehrung

Stegbeton — Steifigkeitsbedingung

~beton m = web concrete
~bewehrung f, Stegarmierung, Steg(stahl)einlagen fpl = web bars, ~ steel, ~ reinforcement
Stegblech n, Stehblech = web plate
~aussteifung f, Stegblechversteifung = web plate stiffening
~dicke f = web plate thickness
~höhe f = web plate depth
~länge f = web plate length
~längsstoß m = web plate longitudinal joint
~querstoß m, Werkstattstoß, Querstoß = web plate transverse joint
~steife f = web plate stiffener
~stoß m = web plate joint
~versteifung f, Stegblechaussteifung = web plate stiffening
Stegbreite f = web width
~, Steg m [Siebtechnik] = space between the holes
Steg|brücke f, Brückensteg m = catwalk
~dicke f, Träger~ = web thickness
~diele f, Bimsbeton-Hohldiele = pumice concrete hollow slab
~einlagen fpl → Stegbewehrung f
~glied n = web member
~höhe f, Träger~ = web depth
~moment n, Träger~ = web moment
~platte f [Gleiskette] = grouser plate
~platten(gleis)kette f = grouser track
~schalung f, Träger~ = web formwork
~spannung f, Träger~ = web stress
~(stahl)einlagen fpl → Stegbewehrung f
~stoß m, Träger~ = web joint
~streifen m [Gleiskette] = grouser bar
~verbindungsstück n, Verkupp(e)lung f = separator
~zone f, Stegbereich m, Träger~ = web zone
Steh|blech n → Stegblech
~bolzen m = stay-bolt
~bolzendöpper m = rivet snap for stay-bolt riveting
~bolzenniethammer m = stud riveting hammer

stehen [von Bögen, Strecken usw. im Bergbau] = to stand
stehende Erdbaugeräte npl, ~ Erdbaumaschinen fpl = stationary earth-moving machinery, ~ ~ equipment
~ Pfahlgründung f [Die Bauwerklast wird durch Pfähle auf tieferliegende, tragfähige Bodenschichten übertragen] = pile foundation transmitting the load directly to hardpan or rock
~ Welle f, Stationärwelle = standing wave
stehender Behälter m, stehendes Gefäß n, Standgefäß, Standbehälter [Aufnahmegefäß für Dämpfe, Gase, Flüssigkeiten] = vertical tank, ~ vessel
~ Boden m = soil without cave-ins
~ Pfahl m = point-bearing pile
stehendes Gefäß n → stehender Behälter m
~ Wasser n = still water
~ ~ [auf einer Fläche] = standing water
Steh|lager n [für Hoch- und Tiefbauten] = pedestal bearing
~leiter f = standing ladder
~verglasung f, Vertikalverglasung, Senkrechtverglasung = vertical glazing
~zeit f, Wartezeit = stand-by time
steifer Anschluß m = rigid connection
~ Beton m = stiff concrete
Steife f → Quer~
~ = stiffener
~, Konsistenz f [Beton; Mörtel] = consistency
~beiwert m, Konsistenzbeiwert f = coefficient of consistency
~messer m = consistency meter
~mittelpunkt m [Bauwerk] = centre of rigidity (Brit.); center ~ ~ (US)
~prüfung f, Steifeprobe f, Steifeversuch m = consistency test
~zahl f → Steifeziffer f
~ziffer f, Steifezahl f [Boden] = modulus of volume change
Steifheit f, Steifigkeit f = stiffness, rigidity
Steifigkeits|bedingung f = stiffness condition, rigidity ~

Steifigkeitsverlust — Steilförderband

~verlust m = loss in stiffness, ~ ~ rigidity
steif|knotiges Faltwerk n = rigid-jointed prismatic structure
~**plastische Formgebung** f [*Feuerfestindustrie*] = stiff-plastic mo(u)lding
Steifrahmen m = rigid frame
~**brücke** f = rigid frame bridge [*The abutment and superstructure are designed as a frame*]
Steig|brunnen m, positiver artesischer Brunnen = flowing well, positive artesian ~
~**eisen** n, Kletereisen = step iron, hand ~, foot ~, access hook
~**-Elektroleitung** f = rising main [*An electrical power supply cable which passes up through one or more storeys of a building*]
steigen, anschwellen [*Wasserspiegel*] = to rise
~, an~ [*der HW-Zeit gegenüber dem mittleren HW-Intervall*] = to rise
steigender Bogen m, einhüftiger ~ = inclined arch, rising ~ rampant ~
steigendes Wasser n, an~ ~ = rising water
Steiger m (⚒) = assistant-foreman, unit foreman
~ [*Gießereitechnik*] = riser
Steigeschacht m = riser shaft, rising ~
Steig|fähigkeit f = ability to climb gradients, grade ability
~**geschwindigkeit** f, Kriechgang-Geschwindigkeit, Kriechgeschwindigkeit = creep speed
~**geschwindigkeit** f [*beim Betonieren aufgehender Bauwerke*] = rate of placing, vertical rise of concrete per hour
Steigleitung f → Steigrohr
~ **für das Kolben von Erdölbohrungen** = swabbing line
Steig|leitungsfußdüse f [*Tiefbohrtechnik*] = bottom hole choke, ~ ~ flow bean
~**naht** f, aufwärtsgeschweißte Naht = upward weld

Steigrohr n, Vereisungsrohr, Gefrierrohr [*Gefrierverfahren*] = freezing pipe, freeze ~
~, Steigleitung f, Steigstrang m = rising main [*A main gas pipe or water supply pipe which passes up through one or more storeys of a building*]
~ [*Wasserschloß*] = riser (pipe), standpipe
~**elevator** m, Förderstuhl m für Steigrohre = tubing elevator
~**fänger** m [*Tiefbohrtechnik*] = tubing catcher
~**kopf** m [*Tiefbohrtechnik*] = tubing head
~**kopfdruck** m [*Tiefbohrtechnik*] = tubing(-head) pressure
~**muffenzange** f [*Tiefbohrtechnik*] = tubing coupling tongs
~**-Programm** n [*Tiefbohrtechnik*] = tubing program(me)
~**startventil** n [*Erdöl*] = kickoff valve
~**wasserschloß** n, Differentialwasserschloß = differential surge tank
Steig|schacht m, Kühlturm~ = chimney, cooling tower ~
~**schachtkühlturm** m, Kaminkühlturm = chimney type cooling tower
~**spur** f [*auf Autobahnen für langsamfahrende LKWs bei Steigungen*] = creeper lane, climbing ~
~**strang** m → Steigrohr
Steigung f = uphill grade, adverse ~, Treppen~ = rise
~, Gewinde~, Ganghöhe f = pitch
Steigungs|strecke f = up-grade section, uphill ~
~**widerstand** m = negotiating resistance, grade ~
steil abstürzend, abgerissen, schroff, abschüssig, prall [*Küste*] = bold, steep
~ **geneigt** = steeply sloped
Steil|böschung f = steep slope
~**dach** n = pitched roof, double-pitch ~
~**flanke** f [*Schweißen*] = square edge
~**förderband** n = steep-incline belt conveyor, ~ ~ conveyer

Steilförderer — (Stein)Bruch

~**förderer** *m* = steep-incline conveyor, ~ conveyer, steep-angle ~, steeply inclined ~

~**förderer(zement)schnecke** *f* = cement screw for steep conveying

~**hang** *m* = steep rock slope

~**küste** *f*, Steilufer *n*, Kliff *n*, Kliffküste = high coast, bold shore, cliff, cliffed coast(line)

~**-Schleppschrapper** *m* = drag scraper with steeply inclined ramp

Stein *m* = stone

~, Block(stein) *m*, Körper *m* = tile, block [*The terms tile and block are synonymous, but the hollow clay and gypsum products are commonly called tile, and the concrete and glass products are usually called blocks. The terms tile and block are used to represent a single unit or collectively for a number of such units, as in the case of brick although the term blocks is usually used for the plural*]

~, (Ton)Ziegel *m*, Backstein *m* = (clay) brick

~ [*Metallurgie*] = matte, coarse metal [*An impure metal obtained in the smelting of various ores*]

~ **beim Prepaktbeton** = plum, displacer

~**abweiser** *m* = rock ejector

~**art** *f*, Blockart = type of tile, ~ ~ block

~**aufzug** *m*, Steinträgeraufzug = brick hoist

~**auskleidung** *f* → Steinbekleidung

~**aussonderungswalzwerk** *n*, Steinentferner *m* = stone separating mill

~**automat** *m* → (Beton)Stein(form)automat

~**balken** *m* = stone balcony

~**ballast** *m* = stone ballast

~**baluster** *m*, Steindocke *f* = stone baluster [*in Scotland: stone banister*]

~**baustoffe** *mpl* = stone building materials

~**bearbeitung** *f* = stone-working

~**bearbeitungsmaschine** *f* = stone-working machine

~**befestigung** *f* = metalling (Brit.); rubble work

~**bekleidung** *f*, Steinauskleidung, Steinverkleidung = stone lining

~**block** *m*, Felsblock = block of stone, ~ ~ rock

~**(block)-Greif(er)korb** *m*, Stein(block)-Greifer *m*, Fels(block)-Greif(er)korb, Fels(block)-Greifer = stone grapple

~**blume** *f* → Krabbe *f*

~**boden** *m*, Steinfußboden = stone flooring

~**boden** *m*, steiniger Boden = stony soil

~**bogen** *m*, Natur~ = stone arch

~**(bohr)krone** *f* = (drill) bit with stones embedded in the surface of the metal

~**-Bohrmaschine** *f* = natural stone drilling machine

~**brechanlage** *f* = stone-breaking plant, stone-crushing ~, rock-breaking ~, rock-crushing ~, ~ installation

~**brecher** *m* = rock crusher, ~ breaker, ~ crushing machine, ~ breaking machine

~**brecherfett** *n* = rock crusher lubricating grease, rock breaker ~ ~

(Stein)Bruch *m* — (stone) quarry, rock ~
Abbau *m*, Abbauen *n* — quarrying
Abbau *m* durch Sprengen, Abbauen *n* durch Sprengen — quarrying by the use of explosives
abbauen — to strip, to quarry
Abfall *m*, Steinbruchabfall — grout (US); quarry waste
Abkeilen *n* — bed lifting by wedging
Baustein *m* — structural stone, (building) stone

Steinbruch-Bohrgerät — Steinbuhne

Bearbeiten n von Kalkstein	milling of limestone
(Bruch)Sohle f, Steinbruchsohle	(quarry) floor
(Bruch)Wand f, Steinbruchwand, Abbauwand	(quarry) face
Dampf-Schrämmaschine f	steam channel(l)ing machine
Drehen n	turning
Elektro-Schrämmaschine f	electrical channel(l)ing machine
Gatter n, Gattersäge f	gang saw
gelernter Steinbrucharbeiter m	quarryman, quarry-worker
Gruben~	pit ~
Handabbau m, Abbauen n von Hand	hand quarrying, quarrying by hand
Hang~	hillside ~
Hobeln n	planing
lösen durch Abkeilen	to break free a block at the bottom (that is, at the floor level) by drilling holes and inserting wedges in the drilled holes
mechanisierter ~	motorized ~ (US)
Polieren n, feineres Schleifen n	polishing, buffing, final polishing
Roh(gesteins)block m	rough block, quarried block
Sägen n	sawing
Schleifen n von Werksteinen	rubbing (US)
Schram m, Schrot m, Schlitz m	channel
Schrämarbeit f	channel(l)ing
Schrämmaschine f	channel(l)ing machine
Schurscheibe f	rotating disc, rubbing bed, revolving metal table
(Steinbruch)Abfall m	quarry waste; grout (US)
Steinbruchbetrieb m	quarry practice
Steinbruch-Facharbeiter m	quarryman, quarry-worker
Steinbruchmaschinen fpl	stone quarrying machinery
Stollen~	mine ~
stillgelegt	inactive, disused
Wand f, (Stein)Bruchwand, Abbauwand	(quarry) face
Werkstein m	dimension stone
Zerteilen n	cutting

Steinbruch|-Bohrgerät n = quarry drill
~-**Groß-Förderwagen** m → Felsmuldenkipper m
~(**hochlöffel)bagger** m, Felsbagger = quarry shovel, rock ~, heavy duty ~
~-**LKW** m = quarry truck (US); ~ lorry (Brit.)
~**mulde** f, Felsmulde [*Erdbaulastfahrzeug*] = quarry body, rock ~
~-**Muldenkipper** m → Felsmuldenkipper
Stein|brücke f = stone bridge
~**brückenbau** m = stone bridge construction
~**buhne** f = stone built groyne (Brit.); ~ ~ groin (US)

Steindamm *m*, Steinschüttdamm = rock fill embankment
~, Steinschüttdamm, Steinsperre *f* = rock fill dam
Stein|decke *f*, Natur~ = paving of natural setts
~**deckung** *f* → Steindeckwerk
~**deckwerk** *n*, Steindeckung *f* = stone riprap, rock ~
~**docke** *f* → Steinbaluster *m*
~**drän** *m* → (Stein)Rigole
~**druckfestigkeit** *f*, Block(stein)druckfestigkeit = block compression strength, tile ~ ~
~**entferner** *m*, Steinaussonderungswalzwerk *n* = stone separating mill
Steine|spalten *n* = coping [*Splitting stones by drilling them and driving in steel wedges along a line*]
~- **und Erdenindustrie** *f* = non-metallic minerals industry, pit and quarry ~
Stein|fang(e)damm *m* = rock fill cofferdam
~**farbe** *f* = stone paint
~**faser** *f* = rock fibre (Brit.); ~ fiber (US)
~**fertiger** *m* → Blockfertiger
~**fertigung** *f* → Beton~
~**fertigungsautomat** *m* → (Beton-)Stein(form)automat
~**fertigungsmaschine** *f* → Blockfertiger *m*
~**festigkeit** *f*, Block(stein)festigkeit = block strength, tile ~
~**form** *f* → Beton~
~**(form)automat** *m* → (Beton)Stein(form)automat
~**formmaschine** *f*, Bausteinmaschine = block (making) machine
~**fräsmaschine** *f* [*für Naturstein und Betonstein*] = stone miller, ~ milling machine
~**(fugen)schnitt** *m* = cutting of stones
~**fülldamm** *m* → → Talsperre *f*
~**füllung** *f* [*einer Steinkiste*] = rock fill
~**(fuß)boden** *m* = stone flooring
~**gabel** *f* → Steinschlaggabel
~**garten** *m*, Felsgarten = rock garden
~**gehwegplatte** *f* = stone flag
~**geländer** *n*, Natur~ = stone railing
~**gelenk** *n* = stone hinge
~**gerüst** *n* [*Fahrbahndecke*] = mineral aggregate
~**gerüstdamm** *m*, Staudamm, Steinschüttdamm, Steinfülldamm, Felsschüttungsdamm, Damm aus Felsschüttung = rock fill dam
~**geschiebe** *n* = boulder shingle
~**gesims** *n*, Steinsims *m*, *n* = stone corbel
~**gewölbe** *n*, Natur~ = stone vault
~**greifer** *m*, Lastengreifer, Steinklammer *f* [*Betonsteinindustrie*] = loading clamp
~-**Greif(er)korb** *m* → Steinblock-Greif(er)korb
~**gut** *n*, Tongut [*Tonerzeugnis mit porösen Scherben, meist glasiert*] = earthenware
~**harke** *f* → Felsrechen *m*
steinhart = rock-hard
Stein|hersteller *m* → Beton~
~**herstellung** *f* → (Beton)Steinfertigung
~**herstellungmaschine** *f* → Blockfertiger *m*
Steinholz *n* [*Markenname: Xylolith*] = magnesite composition
~**(fuß)boden** *m* = magnesite floor finish, ~ floor(ing)
~**unterboden** *m* = magnesite subfloor(ing)
steiniger Boden *m*, Steinboden = stony soil
Steinindustrie *f* = rock industry
Steinkasten *m* → Steinkiste *f*
~**drän** *m* = stone(work) box drain
~**durchlaß** *m* = stone(work) box culvert
~**wehr** *n* → Steinkistenwehr
Stein|kette *f* → Kettensteinwurf *m*
~**kiste** *f*, Steinkasten *m* = (rock-fill) (timber) crib, timber rock fill ~, cribwork
Steinkisten|bauteil *m*, *n* = crib member, ~ unit
~**gründung** *f* = cribwork
~**sperre** *f* → Steinkistenwehr

Steinkistenstützmauer — Steinschlagbelag 946

~stützmauer *f* = crib (retaining) wall
~stützmauer *f* aus Stahl = bin-type retaining wall of steel
~stützmauer *f* aus vorgefertigten Stahlbetonelementen = crip retaining wall built with precast reinforced concrete members
~wabe *f* = crib cell
~wehr *n*, Steinkastenwehr, Steinkistensperre *f*, Steinkastensperre *f* = crib dam [*A barrier made of timber, forming bays or cells which are filled with stone or other suitable material*]
Stein|kitt *m* = stone putty
~kittung *f* = stone putty cementation
~klammer *f*, Lastengreifer *m*, Steingreifer [*Betonsteinindustrie*] = loading clamp
~knack *m*, Steinschutt *m* = rubble (Brit.)
Steinkohlenteer|kreosot *n* → Phenol *n*
~pech *n* = straight run coal tar, coal tar pitch
~sonderpech *n* = special straight run coal tar, ~ coal tar pitch
Stein|körnung *f* = nominal size of mineral aggregate, (size) fraction ~ ~ ~, (size) bracket ~ ~ ~
~kranz *m*, Schlußring *m* [*Gewölbe*] = open cusp, soffit ~
~kreissäge *f* = circular stonecutting saw
~krone *f*, Steinbohrkrone = (drill) bit with stones embedded in the surface of the metal
~lage *f*, Steinschicht *f* = stone course, ~ layer
~-Lehm-Kerndamm *m* → → Talsperre *f*
~maschine *f* → Blockfertiger *m*
~material *n*, Schüttsteine *mpl* = stony material
~matte *f* = stone matting
Steinmauerwerk *n* → Beton~
~ → (Natur-)Steinmauerwerk
Stein|mehl *n* → Blume *f*
~meißel *m* = stone chisel
~metz *m* = mason, stone ~

~metzwerkzeug *n* = masonry tool [*Implement for dressing and shaping stone*]
~mühle *f* = pebble mill
~nest *n* → Kiesnest
~packung *f* → Steinsatz *m*
~paketieren *n* = cubing
~paketiergerät *n* = cuber
~pfad *m* = stony track
~pflasterdecke *f* → Pflasterdecke
~platte *f* = stone slab
~plattenpflaster *n* = stone-block paving
~-Poliermaschine *f* = natural stone polishing machine
~presse *f*, Blockpresse [*Betonsteinindustrie*] = block press
~pritsche *f*, Steintransportkasten *m* [*Fördergefäß für Kran*] = brick cage
~prüfpresse *f* → Beton~
~putz *m* = stuc [*Plasterwork made to look like stone*]
~rechen *m*, Felsrechen, Steinharke *f*, Felsharke = rock rake
(~)Rigole *f*, Sickergraben *m*, Steindrän *m*, Sauggraben [*mit Steinen ausgepackt*] = stone-filled (drain) trench, rubble drain, rock drain
~säge *f* = stonecutting saw, masonry ~
~salz *n*, Chlornatrium *n*, NaCl = halite, common salt, rock salt
~sand *m* → Brechsand
~satz *m*, Steinpackung *f*, vorlageartiges Pflaster *n* = hand placed (stone) riprap, pitched slope
~satzdamm *m* → → Talsperre
Stein'scher Zement *m* → Eisenportlandzement
Steinschicht *f*, Steinlage *f* = stone layer, ~ course
Steinschlag *m*, Bergschlag, Felsabbruch *m*, Felssturz *m* = rock fall, popping rock
~, (Brech)Schotter *m*, Straßen(bau)schotter = broken rock, ~ stone, crushed ~
~ausgleichschicht *f* → Schotterausgleichschicht
~belag *m* → Steinschlagdecke

Steinschlagbett(ung) — (Stein)Wolfloch

~**bett(ung)** *n*, (*f*) → Schotterbett(ung)

~**decke** *f*, Schotterdecke, Steinschlagbelag *m*, Schotterbelag, Chaussierung *f* = broken-stone pavement, crushed-stone ~, broken-rock ~, crushed-rock ~, ~ surfacing

~**gabel** *f*, Steingabel, Schottergabel = stone (picker) fork

~**lage** *f* → Schotterlage

~**rinne** *f* (Geol.) = bergfall furrow, rock fall ~

~**schicht** *f* → Schotterlage

~**straße** *f* → Schotterstraße

~**tragschicht** *f* → Schottertragschicht

~**verteiler** *m* → Schotterverteiler

Stein|-Schleifmaschine *f* = natural stone rubbing machine (US)

~**schneider** *m* → (Beton)Steintrennmaschine *f*

~**schnitt** *m*, Steinfugenschnitt = cutting of stones

~**schotter** *m* → Kiesschotter

~**schraube** *f* = rag bolt

~**schutt** *m* → Steinknack *m*

Stein(schütt)damm *m* = rock fill embankment

~, Steinsperre *f* = rock fill dam

Steinschütter *m* → schwimmender Kippentlader (für den Wasserbau)

Steinschüttung *f* [*als Schutz gegen Unterspülung von gemauerten oder betonierten Fundamenten*] = rock fill to prevent scour

~ [*Talsperrenbau*] = rock fill

~, Steinwurf *m*, Rauhwurf [*Als Böschungsschutz von am Wasser entlang führenden Dämmen*] = rip-rap

~ [*als Fundament für Kaimauern und Molen (offene Gründung)*] = rock fill foundation

~ [*Zum Ausfüllen von Pfahlrosten gegen Knicken*] = rock fill to prevent buckling

Stein|schutz *m* [*am Muldenkipper*] = rock guard

~**setzer** *m* → Pflasterer *m*

~**setzerhammer** *m*, Pflaster(er)hammer = pavior's hammer (Brit.); paver's ~ (US)

~**sims** *m*, *n*, Steingesims *n* = stone corbel

~**spachtel(masse)** *m*, (*f*) = screed-(ing compound) for stone [*Is a means of level(l)ing uneven surfaces in stones*]

~**spalthammer** *m* = stone sledge, masons' hammer

~**spaltmaschine** *f* = block splitter, ~ splitting machine

~**speidel** *m*, Schrotkeil *m* = spalling wedge

~**sperre** *f*, Stein(schütt)damm *m* = rock fill dam

~**splitt** *m*, Natur~ = natural stone chip(ping)s

~**splitter** *m* = flake, spall, galet, stone splinter

~**stampfmaschine** *f* → Beton~

~**staub** *m* → Blume

~**straße** *f*, Pflasterstraße = sett paved road, ~ ~ highway

~**straße** *f*, Natur~ = road with paving of natural setts

~**strom** *m*, Blockstrom (Geol.) = rock stream, ~ flow

~(**träger**)**aufzug** *m* = brick hoist

~**transportkasten** *m*, Steinpritsche *f* [*Fördergefäß für Kran*] = brick cage

~**trennmaschine** *f* [*für Naturstein und Betonstein*] = stone cutter, ~ cutting machine

~**treppe** *f* = stone stair(case)

~**verklammerung** *f* = joggle jointing

~**verkleidung** *f* → Steinbekleidung

~**vibrationsmaschine** *f*, Beton~ = concrete block vibrating machine

~**vorlage** *f* = stone apron [*Used to protect the footing of a sea wall from the wash of the sea*]

~**wand** *f* = masonry wall

~**wehr** *n* = stone weir

Steinwerk *n*, Blockwerk, Beton~ = (concrete) block factory

~**platz** *m* → Beton~

(**Stein**)**Wolf** *m*, Keilklaue *f* = lewis, lifting ~, lifting pins

~**loch** *n* = (dovetail-shaped) hole, (~) aperture

Steinwolle f = rock wool
~**matte** f = mineral-wool quilt, rock-wool ~, ~ blanket
Stein|wurf m → Steinschüttung f
~**wüste** f = rocky desert
~**zange** f = stone (lifting) tongs, lifting ~
Steinzeug n = stoneware
~**filter** m, n = stoneware filter
~**formstück** n, Steinzeugfitting m = stoneware fitting
~**rohr** n = stoneware pipe
Stellitlegierung f = stellite
Stell|klappe f, Gegenklappe, Gegenschütz(e) n, (f) [*Trommelwehr*] = tailgate
~**macher** m, Wagner m, Wagenbauer m = cart-wright, wheel-wright
~**motor** m, Servomotor = servo-motor
~**mutter** f, Ver~ = adjusting nut, set ~
~**ring** m, Ver~ = adjusting ring, set ~
~**schraube** f, Ver~ = adjusting screw, set ~
~**schraube** f [*Motor*] = timing bolt
Stellung f = position
Stellungslinie f = earth-pressure line [*Culmann's graphical constuction*]
stellvertretendes Faltwerk n = equivalent prismatic roof
Stell|werk n = signal box
~**winkel** m, Schmiege f = set square, sliding ~, folding ~
~**zylinder** m = adjusting cylinder
Stelzenfundament n, hochliegender Pfahlrost m = elevated pile foundation grillage
Stemmarbeit(en) $f(pl)$ = cutting work
Stemmasse f, Stemm-Masse, Stemmstoff m [*Zur Dichtung von Fugen, Muffen und nassen Rissen an Bauwerken aller Art*] = ca(u)lking compound, ~ material
Stemmeisen n = mortise chisel
Stemmer m, Setzer m = pitching tool, pitcher [*A hammer-headed chisel bout 9 in. long with a thick broad edge (about $1/4$ in. thick)*]

Stemm|hammer m = ca(u)lking hammer
~**knüppel** m → Klopfholz n
~**stoff** m → Stemmasse f
~**tor** n [*Zweiflügeliges Schleusentor, dessen Flügel sich mit den Schlagsäulen zur Übertragung des Wasserdruckes gegeneinander stemmen*] = mitre gate; miter ~ (US)
Stemmuffe f = ca(u)lked joint
Stemmwerkzeug n = ca(u)lking tool
Stempel m (⚒) [*Ausbauelement, das in der Hauptsache in Längsrichtung auf Druck beansprucht wird*] = prop, post
~ → Quersteife f
~ [*CBR-Versuch*] = penetration piston
~**abstand** m (⚒) = prop spacing, post ~
~**dichte** f [*Wenn man von Stempeldichte schlechthin spricht, hat man stets die Stempelstützdichte St im Auge, worunter man die Anzahl der tragenden Stempel je m² offener Hangendfläche zu verstehen hat*] (⚒) = prop density, post ~
~**druckfestigkeit** f [*Bodenbeton*] = resistance to punching shear
~**einschubpresse** f, Stempelprüfpresse (⚒) = testing press for props, ~ ~ ~ posts
stempelfreier Streb m (⚒) = prop-free front, post-free ~
Stempel|kennlinie f, Lastwegkurve f (⚒) = load-yield curve, resistance-yield ~
~**prüfpresse** f, Stempeleinschubpresse (⚒) = testing press for props, ~ ~ ~ posts
~**rauben** n (⚒) = prop-drawing, post-drawing
~**raubkolonne** f (⚒) = prop-drawing gang, ~ team, ~ crew, ~ party, post-drawing ~
(~)**Schloß** n (⚒) = yoke, prop ~, post ~
~**-Setzkolonne** f (⚒) = prop-setters, post-setters
~**stützdichte** f → Stempeldichte (⚒)

Stengel *m* (Geol.) = stalk
stengeliger Habitus *m* [*Kristall*] = long-columnar habit [*This is the usual meaning of "columnar" in English-speaking countries*]
Stephanit *m*, Melanglanz *m* (Min.) = stephanite, brittle silver ore
Steppen|boden *m* = steppe soil
~**-Hochebene** *f* = barren plateau
Ster *m* → Raummeter *n*
Stereo|komparator *m* = stereo-comparator
~**photogrammetrie** *f* [*Erlaubt die gleichzeitige Darstellung von Situation und Topographie eines zu kartierenden Geländes*] = stereophotogrammetry
sterile Bohrung *f* → Fehlbohrung
~ **Sonde** *f* → Fehlbohrung *f*
steriler Schacht *m* → [*in Galizien*] Fehlbohrung *f*
Stern|anlage *f*, Sternzuteiler *m* = star batcher
~**balkenlage** *f* = system of beams and joists in star form
~**(bohr)meißel** *m* = star bit
~**gewölbe** *n* = lierne vault
~**meißel** *m*, Sternbohrmeißel = star bit
~**zuteiler** *m*, Sternanlage *f* = star batcher
Stethoskop *n* [*Zum Auffinden von Fehlern in Metallteilen, Maschinenelementen, Flüssigkeiten, Gasen, usw. Druck, Risse, schadhafte Verbindungsstellen, übermäßige Abnutzung u.ä. bringen unterschiedliche Töne hervor. Frequenzbereich 20 bis 7 000 Hz. Prüfen unter Spannung bis 6 000 V mit Isolationssonde*] = stethoscope
stetig arbeitender Zwangsmischer *m* → Dauer-Zwangsmischer
~ **arbeitendes Gleisrückgerät** *n* = continuous track shifter, ~ ~ shifting machine
stetige Belastung *f* → Dauerbelastung
~ **Betonförderung** *f*, stetiger Betontransport *m* = continuous concrete transport
~ **(Hart)Zerkleinerung** *f* = continuous comminution, ~ reduction
~ **Kornabstufung** *f*, kontinuierliche ~ = continuous grading
~ **Strömung** *f*, kontinuierliche ~ [*Kanal*] = permanent flow
~ **Welle** *f*, kontinuierliche ~ = continuous wave
stetiger Betontransport *m*, stetige Betonförderung *f* = continuous concrete transport
~ **Kräftezug** *m* = continuous line of forces
~ **Mischer** *m* → Fließmischer
~ **Umfahrungssinn** *m* [*Baustatik*] = continuous sense of direction
stetiges Kernen *n* = continuous core sampling, ~ sampling of cores
Stetig|-Abmeßvorrichtung *f*, Stetig-Dosiervorrichtung, Stetig-Zuteilvorrichtung, Stetig-Zumeßvorrichtung = continuous batcher
~**aufgeber** *m* → Dauerspeiser
(~)**Bagger** *m*, Dauerbagger = bucket excavator
(~)**Baggergleis** *n*, Dauerbaggergleis = bucket excavator track
(~)**Baggerkette** *f*, Dauerbaggerkette = bucket excavator chain
~**bahnsteuerung** *f* [*Werkzeugmaschinensteuerung*] = continuous path system
~**beschicker** *m* → Dauerspeiser
~**-Betoneinbringer** *m*, Stetig-Betoniergerät *n* = continuous concreting machine
~**(beton)verteiler** *m* = continuous (type)concrete (pavement) spreader, ~ (~) ~ (~) distributor, ~ (~) ~ (~) spreading machine
~**(beton)zwangsmischer** *m* → → Betonmischer
~**-Dieselbagger** *m*, Dauer-Dieselbagger = diesel bucket excavator
~**filter** *n*, *m* = continuous filter
~**förderer** *m*, Fließförderer, Massenförderer, Dauerförderer = continuous conveyor, ~ conveyer
~**förderung** *f* → Fließförderung
Stetigkeit *f* = continuity
Stetigkeitsbedingung *f* = continuity condition

Stetiglader — Stichkanal 950

Stetig|lader m → Dauerlader
~last f → Dauerlast
~mahlung f, Stetigmahlen n = constant-flow grinding, continuous ~
~-Mischanlage f, kontinuierliche Mischanlage, Durchlaufmischanlage = continuous volumetric type plant, ~ flow ~, ~ process ~, ~ mixing ~
~mischanlage f für Schwarzdeckenmaterial, Durchlaufmischanlage ~ ~ = continuous bituminous mixing plant
~mischer m → Fließmischer
~-Mischverfahren n, kontinuierliches Mischverfahren = continuous mixing, constant-flow ~
~speiser m → Dauerspeiser
~-Trockenbagger m = continuous excavator
~verteiler m → Stetigbetonverteiler
~-Zwangsmischer m → Dauer-Zwangsmischer
Steuer|anlage f = control equipment
~antrieb m = accessory drive
~druck m = control pressure
~element n → Steuerorgan n
~gehäuse n = control housing
~gerät n [Ein elektrisches Gerät, das die Signalbilder einer oder mehrerer Signaleinheiten steuert] = controller
~gruppe f = control group
~hebel m = control lever
~kabel n = control cable
~kasten m = control box
~kreis m = control circuit
steuern [Maschine] = to control
Steuer|organ n, Steuerelement n = control (element)
~organe npl in doppelter Ausführung, Steuerelemente npl ~ ~ ~ = duplicate controls, ~ control elements
~pedal n [Getriebeschaltdruckregler] = modulating pedal
~pult n, Steuerungspult = control desk, desk control unit

~ringkabel n [Flugplatz] = perimeter control cable
~saal m = control room
~schalter m = control switch
~schieber m → Durchflußschieber
~schieber m [Hydraulikanlage] = control valve piston
~schiene f [Ramme] = control-bar
~schrank m = control cabinet
~stand m = operator's station, control ~
~tafel f → Armaturenbrett
~tauchkolben m [Ölhydraulik] = control plunger
Steuerung f, Steuern n = control
Steuerungssystem n = control system
Steuer(ungs)tafel f → Armaturenbrett n
Steuerventil n = control valve
~ [Luftfederung] = level(l)ing valve
~block m = control valve block
Steuer|walze f → → Walze
~welle f = control shaft
~zeit f = control time
Stich m, Überhöhung f = camber, hog [Beams and trusses may be built with hog to counteract their sag]
~ → (Bogen)Pfeil
~ [Straßenquerschnitt] = rise
~axt f, Stoßaxt = mortise axe, mortice ~
~balken m = dragon beam, ~ piece
~boden m = soil cut by spade
~bogen m, Segmentbogen = segmental arch
stichfester Schlamm m = spadable sludge [Sludge dry enough to be shoveled from the drying bed. Ordinarily under 70% moisture]
Stich|flamme f = thin flame, narrow ~, blast ~
~gleis n = (railway) siding (Brit.); (railroad) ~ (US)
~graben m = off-set ditch
~höhe f, Pfeilhöhe = rise
~kabel n = connecting cable
~kanal m = branch canal

Stichkanal — Stirnradantrieb

~kanal m, Umleit(ungs)kanal [Dachwehr; Schleuse] = culvert
~kuppel f → Kugelgewölbe n
~loch-Bohrmaschine f = tap hole drilling machine
~lochhammer m = tap hole hammer
~probe f = random sample
~probe f, Stichprüfung f, Stichversuch m = random test
~probe(n)nahme f = random sampling
~säge f → Spitzsäge
~straße f [Verbindung zwischen Einbahnstraße und Stadtautobahn] = ramp
~torf m = dug peat
~wortverzeichnis n = subject index
Stickstoff m = nitrogen [former name "azote"]
~düngemittelfabrik f, Stickstoffdüngemittelwerk n = nitrogen fertilizer factory, ~ ~ works
Stickstoffflasche f = nitrogen bottle
Stickstofflade|gerät n = nitrogen charging apparatus
~ventil n = nitrogen valve
Stickstoff|oxyd n = oxide of nitrogen
~werk n, Stickstoffabrik f = nitrogen factory, ~ works
~zylinder m = nitrogen cylinder
Stiel m [Rahmen] = vertical member, supporting ~
~, Ständer m, Pfosten m, Säule f [Fach(werk)wand] = post
~gang m → Schlot m
~hammer m [Durch Kombination des Gradall-Auslegers mit einem Hydraulikhammer bzw. -meißel] = power hammer
~moment n = vertical member moment, supporting ~ ~
~schrot m, Schrothammer m = spalling hammer
~zylinder m, Löffel~ = (dipper) stick ram
Stift m = brad [A small headless nail]
~bolzen m = stud bolt
Stiftschleudermaschine f, Desintegrator m, Stiftmühle f = disintegrating mill, disintegrator

Stift|mühle f → Stiftenschleudermaschine
~nietung f = pin riveting
~schraube f → Schraubenbolzen
stillegen = to abandon permanently
Still|setzen n, Abstellen [Motor] = stopping
~setzung f, Abschaltung = shutdown [power plant]
Stillson-Rohrzange f = Stillson pipe wrench
Still|stand m = standstill
~standtag m, Brachtag [Maschine] = downtime day, stoppage ~
~standzeit f → Brachzeit
~wasser n [Beim Kentern der Strömungen] = still water [State of the tide at slack water]
Stillwasserpunkt m in der Nähe von Thw = slack water on the flood
~ ~ ~ ~ ~ Tnw = slack water on the ebb
Stink|kalk(stein) m → Asphaltkalkstein
~kohle f = stink coal
~mergel m = bituminous marl
Stipputz m, Besenputz = regrating skin
Stirn f, Außenfläche f, Haupt n [Quader] = face
~ [Bei Zapfen oder Versatzungen die auf Druck beanspruchte Hirnholzfläche. Auch die sichtbare Hirnholzfläche am Ende von Balken, Sparren usw., im Traufgesims mit dem Stirnbrett verkleidet] = end
~bogen m = face arch
~brett n, Traufbrett [Zur Verkleidung der Stirnseite vorspringender Balken- oder Sparrenköpfe] = fascia board
~fläche f → Kopffläche
~fläche f eines Rohres = end of a pipe
~flügel m, Parallelflügel = spandrel wall, parallel wing
~kipper m [Waggonkipper] = end tippler
~mauer f, Schildmauer = face wall
~naht f = edge weld
Stirnrad n = spur gear
~antrieb m = spur gear drive

Stirnradgetriebe — Störung

~getriebe n = spur gearing
~-Planetengetriebe n = spur planetary gearing
Stirn|seite f eines Bogens = face of an arch
~wind m, Gegenwind = headwind
Stocherbarkeit f = roddability [*The susceptibility of fresh concrete or mortar to compaction by means of a tamping rod*]
Stochern n [*Betonverdichtung*] = rodding, puddling, swording
Stock m → Anhäufung f (Geol.)
~ → Stumpf m
~ → Stockwerk
Stöckelpflaster n, Holzpflaster = wood(-block) paving, wooden block ~
Stocken n, Aufspitzen = bush hammering [*on concrete, brick, stone, etc.*]
Stock|fußwalze f → Walze
~hammer m = bush hammer
~holzgewinnung f → Rodung
Stockpunkt m [*DIN 51583; ASTM D 97–57*] = pour point
~erniedriger m = pour point depressant, ~ ~ depressor
Stock|roden n → Rodung f
~schiene f → Backenschiene
~schwimmer m, Stangenschwimmer = staff float, pole ~
Stockwerk n, Etage f, Geschoß n = storey (Brit.); story (US) [*The space between two floors or between a floor and a roof*]
~ (Geol.) = stockwork
~bau m → Etagenhaus n
~decke f, (Geschoß)Decke, Etagendecke = floor
~deckenteil m, n → Hochbaudeckenteil
~eigentum n, Eigentumswohnung f = horizontal property, freehold flat
~garage f = multi-stor(e)y car park
~gebäude n → Etagenhaus
~haus n → Etagenhaus
~heizung f → Etagenheizung
stockwerkhoch, geschoßhoch = of stor(e)y height
Stockwerkhöhe f → Geschoßhöhe

stockwerkhohe Gipsplatte f, geschoßhohe ~ = big plaster board
Stockwerk|kran m, Kletter(-Turmdreh)kran = climbing tower crane
~rahmen m, Geschoßrahmen = multi-stor(e)y portal structure
~-Spannbetondecke f, Gebäude-Spannbetondecke = prestressed concrete floor
~treppe f, Geschoßtreppe, Etagentreppe = main stair(case), principal ~
~wohnung f → Etagenwohnung
Stoffangtrichter m → Klärturm m
Stoff|einheitspreis m = material unit price, ~ ~ rate
~gleitklausel f = materials fluctuation clause
~konstante f, Materialkonstante = material constant
Stollen m, Steg m [*Bodenplatte einer (Gleis)Kette*] = grouser
~ → → → Tunnel- und Stollenbau
~gurt m = ribbed conveyor belt, ~ conveyer ~
Stoney-Schütz(e) n, (f) = Stoney gate
Stopfen m = plug
Stopfen n [*(Eisen)Bahnschwellen*] = boxing-up
~walzen n [*von Rohren*] = plug rolling
Stopfhacke f = tamping pick
Stöpsel m → Pfropfen m
Storch(en)schnabel m, Pantograph m = pantograph, pentagraph
Stör|anfälligkeit f = susceptibility to failures
~faktor m = disturbance factor
~körper m → → Talsperre
~pegel m = level of background noises
~pfeiler m, Energievernichtungspfeiler, Energieverzehrungspfeiler = baffle-pier [*A pier on the apron of an overflow dam, to dissipate energy and prevent scour*]
Störung f [*Im geologischen Sinne jede Unterbrechung des ursprünglichen Zusammenhanges der Gesteinsverbände in der Erdrinde als Folge von*

tektonischen oder atektonischen Vorgängen] = fault
~ [*Elektrofilter*] = disruptive effects
~ [*Bodenmechanik*] = remo(u)lding
Störungs|gewinn *m* [*Bodenmechanik*] = remo(u)lding gain
~**verlust** *m* [*Bodenmechanik*] = remo(u)lding loss
störungsfrei = trouble-free
Stoß *m* [*Holzstoßverbindung*] = joint
~, Kohlen~ = face, coal ~
~ → Abbau~
~ = impact
~**aufgeber** *m* → Schubaufgeber
~**axt** *f*, Stichaxt = mortise axe, mortice ~
~**barren** *m* [*in Österreich*] → Schubstange *f*
~**beiwert** *m* [*Pfahl*] = coefficient of restitution, restitution coefficient
~**belastung** *f* = shock loading, impulsive ~, dynamic ~
~**beton** *m*, Schockbeton [*250 Schläge/min.*] = concrete compacted by jolting
~**blech** *n* [*Stahlbau*] = splice plate
~**(boden)verdichtung** *f* = soil compaction by impact
~**(boden)verdichtungsgerät** *n* = impact soil-compaction device
~**dämpfer** *m* = shock absorber
~**deckung** *f* → Stoßüberdeckung
~**druck** *m* (⚒) = side pressure
Stößel *m*, Ausrückhebel = lifter
~**(schraube)** *m*, (*f*) = tappet
~**spiel** *n* = tappet clearance
~**teller** *m* = tappet head
stoßen, stumpf ~ = to butt
Stoß|festigkeit *f* = impact resistance
~**fläche** *f* [*einer Setzstufe*] = riser [*The vertical face of a step*]
~**fläche** *f* [*einer Trittstufe*] = nosing
~**fläche** *f* [*Mauerstein*] = vertical joint face
~**frei-Führungshandgriff** *m* = shock-absorbing guiding handle
~**fuge** *f* [*Mauerwerk*] = vertical joint
stoßgedämpfte Aufhängung *f* = shock mounting
Stoß|gerinne *n* → Mühlbach *m*

~**(gleis)kettenschlepper** *m* → Schubraupentraktor *m*
~**heber** *m* = Widder *m*
~**kernrohr** *n* = biscuit cutter
~**lasche** *f* = butt strap
~**last** *f* = dynamic load, impact ~
~**leiste** *f*, Trittleiste = kick(ing) strip [*A metal strip placed along the lower edge of a door to prevent the marring of the finish by shoe marks*]
~**maschine** *f* → (An)Schubfahrzeug *n*
~**-Mischanlage** *f* → absatzweise arbeitende Mischanlage
~**mischer** *m* = Chargenmischer
~**naht** *f* [*Schweiß*]; Stumpfnaht = butt weld
~**periode** *f* → Stoßzeit *f*
~**platte** *f*, Trittplatte = kick(ing) plate [*A metal plate placed along the lower edge of a door to prevent the marring of the finish by shoe marks*]
~**platte** *f*, Schubschild *m* = pusher blade
~**-Radschlepper** *m* → (An)Schub-Radschlepper
~**raupe** *f* → Schubraupentraktor *m*
~**-Schaufellader** *m* (⚒) = duckbill loader
~**segment** *n* (⚒) = side section [*of three-piece roadway arch*]
~**sieb** *n* = bumper type screen, impact ~ ~, cam ~ ~
~**speiser** *m* → Schubaufgeber
~**stange** *f* = crash bar, (front) bumper
~**tränkgerät** *n* (⚒) = face infusion pressure pipe
~**tränkmethode** *f*, Stoßtränkverfahren *n* [*Staubbekämpfung im Kohlenbergwerk*] = water infusion method
~**tränkung** *f* (⚒) = pulsed infusion
~**tränk(ungs)schießen** *n*, Tränk(ungs)schießen = pulsed-infusion shotfiring
~**tritt** *m* → Setzstufe *f*
~**(über)deckung** *f* [*Das Nebeneinanderordnen der Rundstähle beim Stahlbetonbau*] = overlapping of rebars

Stoßuntersuchungen — (Strahlungs)Kessel 954

~untersuchungen *fpl* an Betonbalken = impulsive testing of concrete beams
stoßweise [*Erdölförderung*] = by heads
~ frei ausfließen [*Erdöl*] = to flow by heads
Stoß|widerstand *m* [*Gestein*] → Schlagfestigkeit *f*
~zahl *f* = impact factor
~zeit *f*, Stoßperiode *f*, Spitzenzeit, Spitzenperiode = peak period
Strafanstaltbau *m* = prison building, ~ construction
Straffseil-Kabelbagger *m*, TEKA-Schrapper = cable excavator with bottom-dump bucket and tightened track cable
Strahlablenker *m* = jet deflector
(Strahl)Ablenker *m* [*Freistrahlturbine*] = diffuser
Strahl|ablösung *f*, Ablösung des Strahles = separation of the water layer (Brit.); freeing of the nappe (US)
~bohranlage *f* = jet drilling rig, jetting drill
~bohrung *f* → Feuerstrahlbohren *n*
~brenner *m* = jet burner
~dicke *f* [*Wasserturbine*] = size of the jet
Strahlen|abschirmungsanlage *f* → Abschirmungsanlage
~chemie *f* [*Ist nicht identisch mit dem vorgeschlagenen Oberbegriff „Radiationschemie", welcher für die beiden Begriffe Photo-Chemie und Strahlenchemie vorgeschlagen wurde. Im angelsächsischen Sprachgebrauch gibt es auch keinen Oberbegriff*] = radiation chemistry
~gewölbe *n* → Fächergewölbe
~gleis *n* = radiating track
~glimmer *m* (Min.) = striated mica
~kupfer *n* (Min.) = clinoclasite
Strahlenschutz *m*, Abschirmung *f* = (radiation) shielding
~anlage *f* → Abschirmungsanlage
~-Baustoff *m* = shielding (construction) material
~bemessung *f* = (radiation) shielding design

~beton *m*, Abschirmbeton = (radiation) shield concrete, (biological) shield(ing) ~
~block *m* → Abschirmblock
~fachmann *m* = (radiation) shielding expert
~konstrukteur *m* = (radiation) shielding designer
~schirm *m* = biological shield, radiation ~
~stein *m* → Abschirmblock *m*
~stoff *m* = (radiation) shielding material
~tür *f*, Abschirmtür = (radiation) shield door
Strahl|kies *m*, Markasit *m* (Min.) = marcasite
~mahlung *f*, Strahlmahlen *n*, Strahlzerkleinerung = jet pulverizing
~mühle *f* → pneumatische Prallmühle
~platte *f* = radiant panel
~pumpe *f* [*Das Wasser wird durch einen Druckwasser-, Dampf- oder Druckluftstrahl angesaugt und mitgerissen*] = jet pump
~regner *m* = sprinkler
~regneranlage *f* = sprinkler system [*It includes the sprinkler, the riser pipe, the lateral distribution pipe, the main line pipe, and, often, the pumping plant*]
~rohr *n* = jet pipe
~rohr *n* [*Atomreaktor*] = beam tube
~stein *m*, Aktinolith *m* = actinolite
~stoß *m* [*Hydraulik*] = jet pressure
~turbine *f* = impulse wheel, tangential ~, ~ turbine
Strahlungsaustauschgesetz *n* = law of radiation exchange
strahlungsbeständiges Schmierfett *n* = grease stable to radiation
(Strahlungs)|Deckenheizung *f* = ceiling panel heating
~energie *f* = radiation energy
~fühler *m* = radiation detector
~heizplatte *f* = radiant heating panel
~heizung *f* = radiant heating, panel ~
~Kessel *m* = radiant heat boiler

Strahlungsschutz *m*, Abschirmung *f* eines Reaktorkernes, Abschirmung eines Kernreaktors, Strahlenabschirmung = (reactor) shielding, radiation ~
~**bauten** *f* = civil defense structures for radiation protection (US); ~ defence ~ ~ ~ ~ (Brit.)
~**pfropfen** *m* = reactor shield plug
Strahlungs|thermometer *n*, Pyrometer = (heat) radiation pyrometer
~**verlust** *m* = radiant loss
~**wärme** *f* [*Raumheizung*] = radiant warmth, ~ heat
Strahl|verkehr *m* = jet (aricraft) traffic
~**vernichter** *m* → Strahlverzehrer
~**verzehrer** *m*, Strahlvernichter = jet disperser [*The release of water under high pressure through sluices or valves at the foot of a high dam may cause scour unless precautions are taken. Sometimes the toe of the dam is curved upwards to act as a deflector and a cushioning pool may be provided. A jet disperser, containing internal radial vanes, is often fixed to the end of the outlet pipe, which breaks up the jet into a conical shower of small drops, so that their energy is absorbed by the air*]
~**welle** *f* = jet wave
~**zerkleinerung** *f*, Strahlmahlen *n*, Strahlmahlung = jet pulverizing
Stramit|platte *f* = Stramit slab
~-**Zwischenwand** *f*, Stramit-Trennwand [*Zwischenwand aus Halmplankplatten aus gepreßtem Stroh, die allseitig mit Papier verkleidet sind*] = Stramit partition (wall) in timber framing
Strand *m*, Gestade *n* = shore, beach
~**bildung** *f* = beach building, shore ~
~**brandung** *f*, Uferbrandung = surgging
~**düne** *f*, Küstendüne = coastal dune, shore ~
~**hafen** *m* = dry harbour (Brit.); ~ harbor (US); stranding ~
~**hafer** *m* → Helm *m*

~**halde** *f* (Geol.) = (wave-built) shoreface terrace
~**kies** *m* = beach gravel, shingle
~**leiste** *f* → Strandterrasse *f*
~**linie** *f* = shore line, beach ~
~**mauer** *f* = sea wall
~**mulde** *f*, Strandrinne *f* = swale, furrow, slash
~**plattform** *f* → Strandterrasse *f*
~**promenade** *f* = boardwalk (US)
~**rinne** *f* → Strandmulde *f*
~**sand** *m* = beach sand
~**terrasse** *f*, Strandplattform *f*, Strandleiste *f*, Küstenterrasse, Brandungsplatte *f* = wave-cut platform, raised beach
Strang *m*, Rohr~ = (pipe) run, piping ~, (~) section
~**dachziegel** *m* = extruded tile
~**fertigungsverfahren** *n*, Strangpreßverfahren = extrusion method
strang|gefertigte Spannbetonplatte *f*, stranggepreßte ~ = extruded precast concrete slab
~**gepreßte (Fuß)Bodenplatte** *f* = extruded floor tile
~**gepreßtes Aluminiumprofil** *n* = extruded alumin(i)um alloy (structural) section, ~ ~ ~ (~) shape
Strang|moor *n*, Streifensumpf *m* = ropy peat-bog
~**presse** *f* = extrusion press, extruding machine, extruder, extrusion machine, extruding press
~**pressen** *n* = extruding, extrusion
~**pressenformung** *f* mit Nachpressung in halbtrockenem Zustand = plastic process [*manufacture of flooring tile and wall tile*]
~**preßbeton** *m* [*international geschützte Markenbezeichnung: Pressolit m*] = extruded concrete
(~)**Preßkammer** *f* [*einer Strangpresse*] = moulding chamber (Brit.); molding ~ (US)
(~)**Preßteil** *m, n* [*einer Strangpresse*] = moulding unit (Brit.); molding ~ (US)
~**preßteil** *m, n* für Bauzwecke = building extrusion (product)

Strangpreßverfahren — Straßenbaugestein

~preßverfahren *n* [*Ziegelherstellung*] = stiff mud process, wirecut brickmaking

Strapazreifen *m* = heavy-duty tyre (Brit.); ~ tire (US)

Straße *f* = road, highway

~ mit drei Spuren je Richtung = three-lane dual carriageway road, ~ ~ highway

~ ~ **Gegenverkehr** = two-way road, ~ highway

~ ~ **Mittelstreifen**, geteilte Straße = divided two-way road, ~ ~ highway

~ **ohne Mittelstreifen**, ungeteilte Straße = undivided two-way road, ~ ~ highway

~ ~ **ruhenden Verkehr** = clearway

~ ~ ~ ~ **in der Spitzenzeit** = peak-hour clearway

Straßen|ablauf *m*, Straßeneinlauf = street inlet, ~ gull(e)y, gull(e)y (trap) for draining a street

~achse *f* = centre line of road (Brit.); center ~ ~ ~ (US); ~ ~ ~ highway

~anhänger *m* = road trailer, highway ~

~anlage *f* = road layout, highway ~

~anlieferung *f*, LKW-Transport *m* = road delivery, highway ~

~arbeiten *fpl* = road work, highway~

~aufbauweise *f* → mehrschichtige Einbauweise

~aufbruchhammer *m*, Aufbruchhammer, (Straßen-)Aufreißhammer, (Straßen-)Aufbrechhammer = (road) breaker, pavement ~, paving ~; road ripper (Brit.)

~aufrauhmaschine *f*, Straßenfräse *f* = highway grooving machine, road ~ ~

~(auf)schüttung *f* = highway fill, road ~

~ausbau *m* = road improvement, highway ~

~ausbildung *f* = road layout, highway ~

~ausrüstung *f* = road furniture, ~ equipment, highway ~

Straßenbahn *f* = tramway

~brücke *f* = tramway bridge

~gleis *n* = tramway track

~körper *m*, eigener ~ = tramway right-of-way

~mast *m* = tramway mast

~netz *n* = tramway system, ~ network

~schiene *f* = tramway rail

~wagen *m* = tramcar

~-Wagenhalle *f* = car barn, ~ house (US); tramcar hangar

Straßen|-Balkenbrücke *f*, Balken-Straßenbrücke = highway beam bridge, road ~ ~

(~)Bankett *n* = (side) verge

Straßenbau *m* = road construction, roadmaking, road building, highway construction

~arbeiten *fpl* = road construction work, highway ~ ~

~ausstellung *f* = road show, highway ~

~behörde *f* = highway authority, road ~

~beton *m* = road concrete, highway ~

~betonwaren *fpl* = precast concrete products for roads and streets

(~)Binder *m*, (Straßen)Bindemittel *n*, Straßenbauhilfsstoff *m* = road binder, highway ~

~bitumen *n*, Straßenbitumen = road asphalt (US); ~ (asphaltic) bitumen (Brit.)

~bitumenemulsion *f* = road asphalt emulsion (US); ~ (asphaltic) bitumen ~ (Brit.)

~boden *m* = road construction soil, highway ~ ~

~emulsion *f* = road emulsion, emulsion for highway construction

Straßenbauer *m* = road builder, roadman

Straßenbau|finanzierung *f* = highway construction financing, road ~ ~

~firma *f*, Straßenbauunternehmen *n*, Straßenbauunternehmung *f* = road-building firm, road construction ~; paving company (US)

~gestein *n* = roadstone

Straßenbauhaushalt — Straßen(beton)mischer

~haushalt m = road construction budget, highway ~ ~
~hilfsstoff m → (Straßenbau)Binder m
~industrie f = highway industry, road ~
~ingenieur m = road engineer, highway ~
~klinker m → Pflasterziegel m
~kosten f = road cost, highway ~
~los n = road building contract section, highway ~ ~ ~
~maschinen fpl = road(-building) machinery, ~ equipment, road-making plant, mechanical highway equipment
~masse f = road-making mix(ture)
~mischer m = mixer for road construction, ~ ~ highway ~
~papier n, (Autobahn)Unterlagpapier = concreting paper, waterproof ~
~projekt n, Straßenbauvorhaben n = road construction project, highway ~ ~; paving ~ (US)
~-Prüfgerät n = road construction tester, highway ~ ~
~schalung f → Betondeckenschalung
~schotter m, Schotter, Steinschlag m, Straßenschotter = broken stone, ~ rock, crushed ~; road-metal (Brit.)
~splitt m = road chip(ping)s, ~ chipping, ~ stone chips
~stelle f = highway building site, road ~ ~, ~ construction ~
~stoff m = highway material, road ~
~technik f [als betriebstechnische Anwendung] = road building technique, highway ~ ~, ~ construction ~, ~ ~ technic
~technik f [als Ingenieurwissenschaft] = highway engineering, road ~ ~
straßenbautechnische Entwick(e)lung f = highway engineering progress, road ~ ~
Straßen(bau)teer m = road tar

Straßenbau|trockentrommel f → Trokken-Trommel
~trommeltrockner m → Trockentrommel f
~unternehmen n → Straßenbaufirma
~unternehmer m = road contractor, highway ~; paving ~ (US)
~verdichter m = road compaction machine, highway ~ ~
~vorhaben n → Straßenbauprojekt n
~zement m = cement for road and street construction
~ziegel m → Pflasterziegel
Straßen|befestigung f, Straßenkörper m = pavement (structure) [*One or more layers of specially processed materials overlying the embankment soil*]
~beförderungsmittel n = highway transportation means, road ~ ~ ~
~(be)heizung f = road heating, highway ~
~(be)heizungsanlage f = road heating installation, highway ~ ~, ~ system
(~)Belag m, (Straßen)Decke f = (road) surfacing, (~) pavement, highway ~
~belagbau m → Belagbau
~belageinbaumasse f → Belagmaterial n
~beleuchtung f, Land~ = highway lighting, road ~
~beleuchtung f, Stadt~ = street lighting
~bemessungsingenieur m = highway-design engineer, road-design ~
~benutzer m = highway user, road ~
~bepflanzung f = roadside planting
~beschilderung f = signing of roads, ~ ~ highways
~besen m = road broom, highway ~, street ~
~beton m = pavement concrete, road ~, highway ~

Straßen(beton)mischer m, Straßenbetoniermaschine f

rotating drum paver mixer, travel(l)ing (concrete) mixer (plant), combined mixing and paving machine, (com-

Straßen(beton)mischer

	bined) paver, (on-site) paving mixer, concrete paver, concrete mixer-paver, travel(l)ing mixing machine, main highway mixer
(RAB-)Brückenmischer, Autobahnbetonmischer, Autobahnbrückenmischer	bridge(-type travel(l)ing concrete)-mixer, bridge type travel(l)ing mixer (plant), super-highway-bridge mixer
selbstfahrbarer gummibereifter Doppeltrommel-Straßenbetonmischer, Straßenbetonmischer mit Reifenfahrwerk	self-propel(l)ed rubber-mounted twin batch paver
Raupenstraßen(beton)mischer, Freifallmischer auf Gleisketten für Betondeckenbau	track-laying type (combined) paver (Brit.); track (concrete) paver (US)
Doppel-Konus-Trommel *f*	double-cone-drum
Doppeltrommel(beton)straßenmischer, Doppeltrommel-Betoniermaschine	twin-batch paver, twin-drum type (concrete) paver, dual-drum type (concrete) paver, double-drum paver
Austrag(ungs)schurre *f*, Entladeschurre	discharge chute
Ausleger *m*, Schwenkausleger	(delivery or distributing) boom
Ausleger-Kübel *m*, Betonkübel	(self-spreading) bucket, boom bucket, operating bucket
Beschaufelung *f*	throw-over blades and pick-ups
(Chargen)Mischspielsteuerung *f*	batchmeter controlled autocycle operation
(Chargen)Mischzeitmesser *m*	batchmeter
Doppeltrommel *f*	two-compartment (mixing) drum
Fertig-Mischkammer *f*	second (mixing) compartment
Lader *m*	(charging) skip, paver skip
Laderaufzugwinde *f* mit gewichtsbelasteter Bremse	combination skip hoist clutch and brake
Transportschurre *f*, Überlaufschurre, drehbare Schurre	transfer chute
Vormischkammer *f*	first (mixing) compartment
Doppeltrommel(beton)straßenmischer, Doppeltrommel-Betoniermaschine, Fabrikat BLAW-KNOX CO, FOOTE DIVISION, NUNDA, N. Y., USA	DUOMIX 34E, DUOMIX 16E [*Trademarks*]
Doppeltrommel(beton)straßenmischer, Doppeltrommel-Betoniermaschine, Fabrikat CHAIN BELT Company, MILWAUKEE, WISCONSIN, USA	REX twinbatch paver
Eintrommelstraßen(beton)mischer	single-drum (type) (concrete) paver
Eintrommel(beton)straßenmischer Fabrikat BLAW-KNOX CO,	SINGLE MIX MULTIFOOTE PAVER 34E; SINGLE MIX

FOOTE DIVISION, NUNDA, N. Y., USA

MULTIFOOTE PAVER 27E [*Trademarks*]

Straßenbeton|platte *f* = (concrete) highway slab, (~) road ~, (~) ~ panel
~sand *m* = pavement concrete sand, highway ~ ~, road ~ ~
Straßen|bewehrungsmatte *f*, Straßenarmierungsmatte = road mesh, ~ fabric
~bitumen *n* → Straßenbaubitumen
~blendung *f* = road glare, highway ~, street ~
~-Bordkantenschneidgerät *n*, Abbordgerät = verge trimmer, ~ cutter
~breite *f* = road width, highway ~
~brücke *f* = road bridge, highway ~
~damm *m* = road embankment, highway ~
(~)Decke *f*, (Straßen)Belag *m* = (road) pavement (~) surfacing, highway ~
(Straßen)Decken|bau *m* → Belagbau
~baugeräte *npl* → Deckenbaugeräte
~baumaschinen *fpl* → Deckenbaugeräte *npl*
~beton *m* → → Beton
~einbaumasse *f* → Belagmaterial *n*
~einbruch *m* = pavement sinkage, surfacing ~
~erhitzer *m* = road heater, ~ heating machine [*A machine which heats a road surface by blowing flame or hot air on to it*]
~erneuerung *f* = resurfacing (operations)
~fertiger *m* → (Straßen-)Fertiger
~markierungsfarbe *f* = striping paint
~prüfgerät *n* = (road) pavement tester, (~) surfacing ~, highway ~

Straßeneinlauf *m* → Straßenablauf
~einfassung *f* = gull(e)y surround
Straßen|einmündung *f*, Straßengab(e)lung, Gabelkreuzung = fork junction
~entwässerung *f* = road drainage, highway ~

~entwässerungsrohr *n* = road drainage pipe, highway ~ ~
straßenfahrbare Pumpe *f* = road pump
Straßen|fahrt-Geschwindigkeit *f* [*Baumaschine*] = road speed
~fahrwerk *n* [*Turmdrehkran*] = road transport unit
~fahrzeug *n* = on-highway vehicle, road ~
~fahrzeugmotor *m* = road vehicle engine
~fahrzeugwaage *f* = road vehicle scale
(~)Fegemaschine *f* → (Straßen)Kehrmaschine
(~)Fertiger *m*, Deckenfertiger, Straßendeckenfertiger = (road) finisher, (laying and) finishing machine
~flickarbeiten *fpl* = patching (work), mending (~)
~fräse *f*, Straßenaufrauhmaschine *f* = road grooving machine, highway ~
~front *f* [*Gebäude*] = street façade
~fuge *f* = road joint, highway ~
~gab(e)lung *f* → Straßeneinmündung
straßengängig = on-the-highway
Straßen|geschwindigkeit *f*, Fahrgeschwindigkeit auf der Straße = road speed, highway ~
~glätte *f*, winterliche ~, Winterglätte = icing (condition)
~glättebekämpfung *f* → Winterglättebekämpfung
~gleitsicherheit *f* → Griffigkeit
~graben *m* = roadside ditch
~griffigkeit *f* → Griffigkeit
~-Hängebrücke *f* = suspension road bridge, ~ highway ~
~hecke *f* = roadside hedge, live fence
~hobel *m*, Wegehobel, Erdhobel, Straßen-Planierer *m* = (road) grader
~inanspruchnahme *f* = highway load, road ~

~karte *f* = road map, highway ~
~kehricht *m* = street sweepings
(~)Kehrmaschine *f*, (Straßen)Fegemaschine = rotary sweeper, scavenging machine, road sweeper, mechanical sweeper, street sweeper
~-Kernbohrgerät *n*, Straßen-Kernbohrmaschine *f* = road pavement coring machine, highway ~ ~ ~
~klinker *m* → Pflasterziegel *m*
~knoten *m* = intersection
~kocher *m* = road kettle, road maker's heater (US); road cooker (Brit.)
~koffer *m*, Straßenkasten *m*, Straßenbett *n* = roadbed
~körper *m*, Straßenbefestigung *f* = pavement (structure), road ~ [*one or more layers of specially processed materials overlying the embankment soil*]
~kreuzung *f* = road junction, ~ intersection, highway ~
~krone *f* = crown (of a carriageway)
~kurve *f* = road curve, highway ~
~lage *f* [*Kraftfahrzeug*] = roadability, road-holding
~lampe *f*, Stadt~ = street lamp
~landschaft *f* = highway landscape, road ~
~lärm *m* = external noise
~leistungsfähigkeit *f* bei Nacht = night capacity
~leistungsfähigkeit *f* bei Tage = day capacity
~leitpfosten *m* = indicator sign post
~leuchte *f* → Leuchte
~leuchte *f*, Land~ = highway lantern, road(way) ~ (Brit.); ~ luminaire (US); ~ lighting fixture
~leuchte *f*, Stadt~ = street lantern (Brit.); ~ luminaire (US); ~ lighting fixture
(~)Leuchtenmast *m* → Beleuchtungsmast
~-Leucht-Nagel *m*, Spur-Leucht-Nagel, Verkehrs-Leucht-Nagel, Markierungs-Leuchtnagel = reflectorizing traffic stud, ~ street marker, ~ road stud

~magnetreiniger *m* = magnetic road sweeper
~markierung *f*; Bodenmarkierung [*Schweiz*] = traffic-zoning, traffic-line marking
~-Markierungsfarbe *f*, Verkehrs(-Markierungs)farbe; Signierfarbe [*Schweiz*] = traffic(-zoning) paint, zone marking ~, road(-marking) ~, line-marking ~, road-line ~, traffic line ~

   Bestimmung *f* des Absetzgrades = evaluating degree of settling

   Lichtempfindlichkeit *f* = light sensitivity

   Straßendienstprüfung *f* = road service test

   Trockenzeit *f* bis zur Klebfreiheit = dry to no-pick up time

~markierungsmaschine *f* → Gerät *n* zum Anzeichnen der Verkehrslinien
~markierungsmaschine *f* für plastische Farbmassen = plastic line marking machine
~meister *m*, Wegemeister = road overseer
~meisterei *f* = maintenance compound
~mischer *m* → Straßen(beton)mischer
~nagel *m* → Markierungsknopf
~namenschild *n* = street nameplate
~netz *n* = road net(work), highway ~, ~ system
~nutzungsdauer *f* = road life, highway ~
~öl *n* = road oil, dust-laying ~
~ölung *f* = (surface) oiling
~planierer *m* → Straßenhobel *m*
~planung *f* = highway planning, road ~
~planungsreferat *n* = highway planning section, road ~ ~
~-(Prüf)Kernbohrgerät *n* = pavement core drilling machine, surfacing ~ ~ ~
~querschnitt *m* = cross section of road, ~ ~ highway, cross-sectional profile, profile of road, profile of highway

Straßenrad — Straßenverkehrsunfall

~rad n = road wheel
~räder-Unterwagen m → Selbstfahrwerk n
~randbepflanzung f = roadside planting
~reinigung f = street cleansing
~reklame f = outdoor advertising along highways, ~ ~ ~ roads
~scheitel m = vertex of a road, ~ ~ ~ highway; ~ ~ ~ street
~schleifmaschine f = road grinder, highway ~, ~ grinding machine
~schlepper m → Radschlepper
~schotter m, Straßenbauschotter, Steinschlag m, Schotter = crushed stone, ~ rock, broken ~; road-metal (Brit.)
~schüttung f, Straßenaufschüttung = road fill, highway ~
~seitenbepflanzung f = roadside vegetation, ~ planting
~seitenschutz m = roadside protection
~sperrtor n [*zum Verschließen einer Deichscharte*] = street gate
~sprengwagen m = street flusher, ~ sprinkler, ~ washer
~tanker m mit Absetzbehälter, ~ ~ Wechselbehälter = demountable road-rail tanker
~tankwagen m = road tanker, bowser (Brit.); tank truck (US)
Straßenteer m, Straßenbauteer = road tar
~emulsion f = emulsion of road tar, road tar emulsion
~viskosimeter n [*abgek. STV*] = Standard Tar Viscosimeter [*abbrev. STV*]
Straßen|teppich m, Teppich(belag) m, Mischteppich(belag) = road carpet, (~) mat, (pre-mix(ed)) carpet, carpet-coat, thin surfacing; veneer (Brit.)
~transport m = highway transportation, road ~
~transportfähigkeit f [*d. h., daß eine Baumaschine auf der Straße von Baustelle zu Baustelle gezogen werden kann oder mit eigener Kraft fährt*] = road portability
~tunnel m = highway tunnel, ~ tube, road ~
~überführung f = overpass
~übergangsbogen m = highway transition curve, road ~ ~
~umlegung f → Straßenverlegung
~- und Eisenbahnbrücke f = road-rail bridge, highway-rail ~
~- und (Eisen)Bahntunnel m → (Eisen)Bahn- und Straßentunnel
~unebenheitsmesser m = roughometer, roughness meter [*an instrument for providing a numerical estimate of the irregularity in a road surface*]
~-Unterbohrung f = horizontal drilling under a road surface
~unterführung f = underpass
~unterhaltung f = road maintenance, highway ~
~verbreiterung f = road widening, highway ~
~verbreiterung f mit Mittelstreifen = dualling
(Straßen)Verkehr m = (highway) traffic, road ~
Straßenverkehrs|erhebung → Verkehrsmessung
~form f = (street) travel pattern
~ordnung f = Highway Code
~recht n = Highway Traffic Law
~sicherheit f = road (traffic) safety, highway (~) ~
~sicherheitsforschung f = road safety research, highway ~ ~
~technik f = (highway) traffic engineering
    bauliche Maßnahmen fpl = engineering
    Verkehrserziehung f = education
    polizeiliche Maßnahmen fpl = enforcement

Straßenverkehrsunfall m
    unmittelbares Auffahren n eines Fahrzeuges auf ein stehendes oder

road traffic accident
    direct collision of a vehicle with another vehicle which is either

Straßenverkehrsunfall — (Straßen)Verkehrszeichen 962

vorausfahrendes Fahrzeug oder auf feste Objekte	stationary or is ahead of the first vehicle, or with a fixed object
seitliches Auffahren *n*	side-swipe
Auffahren *n*	rear-end collision
zu kurzes Überholen	overtaking short
Schneiden	cutting in
Rutschen *n* durch Quergefälle	side-slip due to crossfall
doppeltes Überholen *n*	double overtaking
Schleudern *n* eines Fahrzeuges mit nachfolgendem Sturz, Aufprall oder Abkommen *n*	skidding followed by overturning, collision or leaving the road
Schleudern *n* auf gerader Bahn	skidding on straight section
Schleudern *n* in der Kurve	skidding on a curve
Schleudern *n* auf glatter Bahn am Waldrand	skidding on a slippery surface at the edge of a forest
Schleudern *n* auf glatter Brückenfahrbahn	skidding on a slippery surface on a bridge
zu weites Überholen	overtaking too wide
doppeltes Überholen	double overtaking
allmähliches Abkommen *n* von der Fahrbahn	leaving the road gradually
Einschlafen *n* am Steuer	falling asleep
Abdrängen *n* durch Schneiden	cutting in of another vehicle
Abgleiten *n* in Gefällstrecken	skidding on downgrade
Seitenwindeinwirkung *f*	effect of side wind
Zusammenstoß zweier aus Gegenrichtungen oder Querrichtungen kommender Fahrzeuge	collision between vehicles proceeding in opposite directions or at right angles to each other
Wenden *n* über den Grünstreifen	turning across median
zu weites Überholen *n*	overtaking too wide
Überholen *n* bei Gegenverkehr	overtaking on section where roadway carries two-way traffic
Übergang *m* auf Gegenverkehr	collision at point where section with roadway carrying two-way traffic begins
Abkommen *n* auf die Gegenfahrbahn	crossing median strip
Überqueren der Fahrbahn an den Anschlußstellen	crossing roadway at an interchange

---

Straßenverkehrsunfall *m* **mit leichtem Personenschaden** = slight-injury road accident
~ ~ **schwerem Personenschaden** = serious-injury road accident
~ ~ **tödlichem Ausgang** = fatal road accident, ~ highway ~
Straßenverkehrszählgerät *n* = traffic counter, ~ counting apparatus

---

(Straßen)Verkehrszeichen *n*   **road sign, (roadside) traffic (control)** ~

Straßenverlegung — Strebeband

Gefahrenzeichen n, Warnzeichen
"Straßenarbeiten"
"Rutschgefahr"
"Straßenverengung"
"Kreuzung mit einer Straße ohne Vorrang"
"Doppelkurve, die erste nach links"
"Achtung – Kinder"
"Rinne"
"unbewachter Bahnübergang"
Verbotszeichen n
"Vorrangstraßenkreuzung"
"Wenden nach links verboten"
"Überholen verboten"
Befehlszeichen, Verpflichtungszeichen, Ordnungszeichen
"Richtung ist einzuhalten"
Hinweiszeichen, Informationszeichen

"Parken"
"Krankenhaus"
"Station für erste Hilfe"
"Tankstelle"
"Ende der Straße mit Vorfahrtsrecht"

danger sign, warning ~
"Road Works"
"Slippery Carriageway"
"Carriageway Narrows"
"Intersection with a Non-priority Road"
"Double Bend to the Left"

"Children"
"Gutter"
"Level Crossing without Gates"
prohibition sign, prohibitory ~
"Priority Road Ahead"
"Turning to the Left Prohibited"
"Overtaking Prohibited"
mandatory sign, regulatory ~

"Direction to be Followed"
indication sign, informative ~, guide ~
"Parking"
"Hospital"
"First-Aid Station"
"Filling Station"
"End of Priority"

Straßen|verlegung f, Straßenumlegung = diversion of road, ~ ~ highway
~vermessung f = highway survey, road ~
~versuch m = road test, highway ~
~verwaltung f = highway administration, road ~
~walze f = road roller
~walz(n)rad n = road roller wheel
~wärter m = road mender, (~) lengthsman, maintenance man, road surfaceman
~wesenforschung f = road research, highway ~
~zubehör m = road furniture, highway ~
(~)Zugmaschine f = tractor-truck, truck-tractor, motor tractor
stratifizieren (Geol.) = to stratify
Stratigraphie f → Formations- und Schichtenkunde f

Straußpfahl m = Strauß bore(d) pile, ~ drilled ~
Streb m (⚒) = longwall face
~ausbau m (⚒) = face supports
Strebbau m (⚒) = longwall (face) mining, ~ working
~ als Rückbau, ~ zum Schachte hin, Strebrückbau; Strebbau als Heimwärtsbau [Österreich] = longwall retreating
~ ~ Vorbau, ~ ~ Feldwärtsbau, ~ feldwärts, Strebvorbau (⚒) = longwall advancing
Strebe f, (Graben)Schrägsteife f = (diagonal) brace (US); (~) strut (Brit.)
~apparat m → Strebewerk n
~balken m **eines Bohrturmes** = derrick brace
~band n → Kopfband

Strebebogen — Streckgurtwinkeleisen 964

~bogen m, Schwibbogen = arched buttress, flying ~
Strebeeingang m (⚒) = end of longwall face
Strebenderrick(kran) m = Scotch derrick (Brit.); stiff leg ~ (US)
~ üblicher Bauart, einfacher Strebenderrick(kran) = three-legged derrick
Strebenfachwerk n [*Fachwerk mit nur geneigten Füllungsstäben, also ohne zum System gehörige Lotrechten außer gegebenenfalls den beiden Endpfosten. Als nicht zum System gehörig zählen Lotrechte zur Halbierung der Knicklänge von Gurtstäben oder zur Ermöglichung des Anschlusses von Querträgern*] = bracing of sloping members and without verticals forming par of the system
Strebe|pfeiler m → Bogenpfeiler
~schwarte f → Sturmlatte f
~werk n, Strebeapparat m [*Die Gesamtheit von Strebebogen und Strebepfeilern, wie sie vor allem die Gotik ausgebildet hat*] = system of buttresses and flying buttresses, ~ ~ ~ ~ arched ~
Streb|förderer m, Strebfördermittel n = (longwall) face conveyor, (~) ~ conveyer
~rückbau m → Strebbau als Rückbau
~vorbau m → Strebbau als Vorbau
Streck|barkeit f, Duktilität f, Dehnbarkeit [*Bitumen*] = ductility
~barkeitsmesser m, Duktilometer m, n [*Bitumenmessung*] = ductilometer
Strecke f, Teil~, Abschnitt m, Teilstück n = stretch, section
~ (⚒) = road(way)
~ [*Flugzeugführung im Hafen*] = route
~ in Gefälle, Gefällestrecke = downhill section, downgrade ~
~ zur Fahrung (⚒) = travelling road(way)
~ ~ Förderung (⚒) = haulage road(way)
~ ~ Wetterführung (⚒) = airway

strecken [*Streckmetall*] = to stretch, to expand, to draw out
Strecken|abschnitt m → Haltung f
~arbeiter m [*Eisenbahn*] = platelayer
~ast m, Abzweig m, Abzweigung f [*Bahn*] = branch line
~ausbau m (⚒) = roadway supports
~ausbauteil m, n (⚒) = roadway support
~band n → → Bandförderer
Streckenbogen m (⚒) = roadway arch
~ mit abnehmbarem Fußteil (⚒) = loose-footed roadway arch
Strecken|förderung f (⚒) = underground haulage
~(hohl)raum m (⚒) = roadway excavation
~isolierflansch m = mainline insulating flange
~last f, Linienlast = knife-edge load, line(ar) ~
~leuchtfeuer n [*Flugverkehr*] = airway beacon
(~)**Mantel** m (⚒) = surrounding rock
streckenmessende Triangulation f = trilateration
Strecken|netz n (⚒) = roadway system
~ort n (⚒) = road head
~schieber m = line valve
~stoß m (⚒) = roadway side
~unterhaltung f (⚒) = roadway maintenance
(~)**Vortrieb** m, Ausbruch m [*Tunnel-, Berg- und Stollenbau*] = progress, advance, heading
~vortriebmaschine f → Stollenvortriebmaschine
Strecker m, Binder(stein) m [*Steinbau*] = bonder, bondstone, header
~kopf m, Binderkopf = head
~schicht f → Binderschicht
~verband m → Binderverband
Streck|grenze f [*Baustahl*] = tensile yield point
~grenzverhältnis n [*Baustahl*] = ratio of yield point to tensile strength
~gurtwinkeleisen n = horizontal boom angle iron, ~ flange ~ ~, ~ chord ~ ~

Streckmetall *n* = expanded metal [*abbrev. XPM*]
~ **als Putzträger** = expanded metal lath(ing)
~**bewehrung** *f*, Streckenmetallarmierung, Streckmetalleinlagen *fpl* = expanded metal reinforcement
Streck|mittel *n* → Streckzusatz
~**spannung** *f* [*Stahl*] = yield stress
Streckung *f* [*Streckmetall*] = expansion
Streckzusatz *m*, Streckmittel *n* [*z.B. Farbe*] = extender
Streich|asphalt *m* → Gußasphalt
~**balken** *m*, Streifbalken [*Zur Balkenlage gehörig. Streichbalken liegen auf einer oder beiden Seiten parallel an einer Wand*] = wall plate
~**barkeit** *f* = brushability
~**brett** *n* → Holzspachtel
~**eisen** *n* → Fugenkelle *f*
streichen (Geol.) = to strike, to bear, to trend
Streichen *n*, Schicht~ (Geol.) = strike
~ → An~
~ **einer Verwerfung** (Geol.) = fault strike, trend of a fault
streichende Baulänge *f* (⚒) = life of face [*in terms of distance*]
~ **Verwerfung** *f*, Längsverwerfung (Geol.) = strike fault
streichender Ausbau *m* (⚒) = support parallel to the line of strike
~ **Streb** *m* **mit zusätzlichem Einfallen zum Alten Mann** (⚒) = strike face dipping towards the goaf
~ **Streb(bau)** *m* (⚒) = strike face
~ ~ **mit zusätzlichem Einfallen zum Kohlenstoß** (⚒) = strike face dipping towards the coal-face
Streichkonsistenz *f* [*Farbe*] = brushing consistency
Streichmaß *n*, Wurzelmaß, Reißmaß [*der Niete*] = distance from centre of rivet to outside of angle
~ = marking gauge, butt ~ [*It is used for marking lines parallel to that face of the wood which the block (Anschlag) travels along*]

Streich|pfahl *m* [*Pfahl zum Leiten von Schiffen vor und an Bauwerken*] = guiding pile
~**richtung** *f* (Geol.) = direction of the strike, bearing of the trend, strike direction
~**schiene** *f*, Schutzschiene = check rail, safety ~, side ~, guard ~
~**stange** *f* [*Waagerecht befestigte längslaufende Gerüststange, auf der die Querriegel liegen*] = ledger
~**torf** *m* = mo(u)lded peat
~**wehr** *n*, Streichstauwehr = side weir
~**werk** *n* → Parallelwerk
~**winkel** *m* (Geol.) = angle of strike
Streif|balken *m* → Streichbalken
~**boden** *m* → Einschub *m*
Streife *f* [*Polizei*] = patrol
Streifen *m* = strip
~, Bahn *f*, Einbau~ [*Schwarz- und Betondeckenbau*] = strip
~ → (Fahr)Spur *f*
~**art** *f* = banded-coal type
~**bildung** *f* [*Pigment*] = floating, flooding [*The re-arrangement or separation at the surface of a paint film of pigment grains*]
~**fundament** *n*, Gründungsstreifen *m*, Bankett *n* = strip footing, continuous ~
~**fundament** *n* **geringer Gründungstiefe**, durchlaufende Flachgründung *f* = shallow continuous footing
~**last** *f* = strip load
~**probe** *f*, Pickprobe, Schlitzprobe (⚒) = channel sample
~**probe** *f* [*Tiefbohrtechnik*] = strip specimen (of a casing)
~**sumpf** *m*, Strangmoor *n* = ropy peat-bog
~**wagen** *m* [*Polizei*] = patrol car
~**walzwerk** *n* = strip (rolling) mill
streifenweiser Geschiebetrieb *m* = striped bed load movement
Streif|haufen *m* → Längsreihe *f*
~**kohle** *f* = banded coal
strenge Biegetheorie *f* = strict bending theory
Streu|anhänger *m* = trailer gritter, ~ spreader, ~ distributor

Streuarbeit — Strohplatte

~arbeit *f*, Streubetrieb *m* [*Straßenwinterdienst*] = gritting work

~automat *m*, Sand~ = auto(matic) sand spreader, ~ ~ distributor

~bereich *m* [*von Versuchsergebnissen*] = (range of) scatter

~betrieb *m* → Streuarbeit

~dienst *m* [*Straßenwinterdienst*] = gritting service

~düse *f* [*Zur Abwasserverteilung über die Tropfkörperfläche*] = spray nozzle, sprinkler ~

Streuen *n*, Streuung *f* [*Straßenwinterdienst*] = gritting

Streuer *m* → Streugutverteiler *m*

~ mit Spezialaufbau → Streugutverteiler *m* ~ ~

~ zur Glatteisbekämpfung = frost gritter

Streu|fahrzeug *n* [*Straßenwinterdienst*] = gritting vehicle

~gerät *n* → Streugutverteiler *m*

~-Gleichstrom *m* = stray direct current

~gut *n* → abstumpfender Stoff *m*

~guthalle *f* [*einer Straßenmeisterei*] = abrasive (material) storage hangar, gritting ~ ~, abrading ~ ~ ~, grit(s) ~ ~

~gutlager *n* = abrasive (material) store, gritting ~ ~, abrading ~ ~, grit(s) ~ ~

~gutpritsche *f* = bulk gritter body

~gutsilo *m* = abrasive material bin, gritting ~ ~, abrading ~ ~

~gutverteiler *m*, Streugerät *n*, Streuer *m*, Streumaschine *f* = winter gritter, ~ grit spreader, abrasive (material) spreader

~gutverteiler *m* mit Spezialaufbau, Streuer ~ ~ = chassis-mounted gritter, ~ abrasive (material) spreader, ~ abrading material ~

~karre(n) *f*, (*m*) = barrow-type spreader, hand-operated ~

~(last)kraftwagen *m* = spreader truck (US); ~ lorry (Brit.)

~makadam *m* → Einstreudecke *f*

~maschine *f* → Streugutverteiler *m*

~material *n* → abstumpfender Stoff *m*

~pflicht *f* [*Straßenwartung*] = gritting duty

~salz *n* → (Auf)Tausalz

~sand *m* = gritting sand

~schicht *f* [*Dachpappe*] = mineral surfacing

~splitt *m* → (Ab)Decksplitt

~stoff *m* → abstumpfender Stoff

~strom *m* → Erdstrom

~strom-Elektrolyse *f* = stray-current electrolysis

~teller *m*, Tellerstreuer = disc spreader; disk ~ (US)

Streuung *f*, Streuen *n* [*Straßenwinterdienst*] = gritting

~ [*Versuchsergebnisse*] = scatter

Streuwalze *f*, Verteilerwalze = spreading roll

~ → Walzen(-Anbau)streuer

Strich|düne *f*, Längsdüne = linear dune

~linie *f* = dash line

~punktlinie *f* = dash-dotted line

~regen *m* = convectional rain

strichreines Mischen *n* = streakless mixing

Strich|zeichnung *f* = line drawing

~ziehmaschine *f* → Gerät zur Anzeichnung von Verkehrslinien

Stricklava *f*, Fladenlava, Wulstlava, Taulava = ropy lava, pahoehoe ~

Stroboskop *n* = stroboscope

Stroh|(ab)deckung *f* = straw cover(ing)

~ballen *m* = straw bale

~dach *n* = straw roof

~dämmplatte *f* = insulating straw board

~deckung *f* → Strohabdeckung

~faserplatte *f* = compressed straw fibre building slab (Brit.); ~ ~ fiber ~ ~ (US)

~lehm *m* = straw and loam

~-Lehm-Strick *m*, Windel(puppe) *f* = twisted straw

~matte *f* = straw mat

~platte *f* [*Stroh gepreßt, gebunden mit Stahldraht*] = compressed straw (building) slab

~**platten-Zwischenwand** *f*, Strohplatten-Trennwand = compressed straw slab partition (wall) in timber framing
~**seil** *n* = straw rope
~**stein** *m* = carpholite
Strom *m* = river
~, elektrischer ~, Energie *f* = electrical energy, (~) current
~ **mit Gezeiteneinwirkung** = tidal river
~ ~ **Sohlenwanderung** = river with shifting bed
~**ablagerung** *f* → Alluvium *n*
~**abnahme** *f* = current take-off
~**abnehmer** *m* [*Straßenbahn*] = grip (US)
stromab(wärts) = down river, downstream
~ **gerichtete Buhne** *f*, deklinante ~ = groyne pointing downstream (Brit.); groin ~ ~ (US)
Stromaggregat *n* → Aggregat
~ **mit Benzinmotorantrieb** = generating set with petrol engine (Brit.); ~ ~ gas(olene) ~, ~ ~ ~ gasoline ~ (US)
(Strom)Anschluß *m* = connection to electrical supply, grid connection
Strom|anzapfung *f*, Stromenthauptung = river beheading, ~ betrunking
~**arm** *m* = river arm, ~ branch
stromauf(wärts) = up river, upstream
~ **gerichtete Buhne** *f*, inklinante ~ = groyne pointing upstream (Brit.); groin ~ ~ (US)
Strom|ausbau *m* = river development
~**ausfall** *m* = current failure
~**ausfuhr** *f*, Elektrizitätsausfuhr, Energieausfuhr = power export, electricity ~
~**auslauf** *m* [*ohne Bauwerk*] = river outlet
~**bagger** *m* = river dredger (Brit.); ~ dredge (US)
~**bank** *f* = river bar
~**bau** *m* = river engineering
~**bauingenieur** *m* = river conservancy engineer
strombauliches Modell *n*, Strommodell = river model

Strom|bauwerk *n* = river structure
~**bedarf** *m*, Energiebedarf, Elektrizitätsbedarf = power requirement, electricity ~
~**bett** *n*, Stromschlauch *m* = river bed
~**brücke** *f* = river bridge
~**dichte** *f* [*Schweißen*] = current density
~**ecke** *f* = river corner
~**einfuhr** *f*, Elektrizitätseinfuhr, Energieeinfuhr = power import, electricity ~
~**einzuggebiet** *n* = river basin [*The total area drained by a river and its tributaries*]
~**elektrode** *f*, Feldelektrode [*geophysikalisches (Aufschluß)Verfahren*] = current electrode
Strömen *n*, ruhiges Fließen *n*, strömender Abfluß *m* = streaming flow, tranquil ~, flowing
strömender Abfluß *m*, ruhiges Fließen *n*, Strömen *n* = tranquil flow, streaming ~, flowing
Strom|enthauptung *f*, Stromanzapfung = river betrunking, ~ beheading
~**erosion** *f* → Erosion im engeren Sinne
~**erzeuger** *m* = generator, generating set
~**erzeuger** *m* **für Baustellen** = contractors' generator (or generating set)
~**erzeugung** *f*, Energieerzeugung, Elektrizitätserzeugung = power production, electricity ~
~**(erzeugungs)aggregat** *n* → Aggregat
~-, **Fluß- und Kanalbauingenieur** *m* = waterway engineer
~**geschwindigkeit** *f*, Strömungsgeschwindigkeit = flow velocity
~**hafen** *m* = river port
~**haltung** *f* = river reach
~**hydraulik** *f* = river hydraulics
~**kies** *m* = river gravel
~**kosten** *f* = power cost
~**kraftwerk** *n* → Lauf(kraft)werk
~**kreis** *m* [*Starkstrominstallation*] = power circuit

Stromkrümmung — Strömungsgeometrie

~krümmung *f* = river bend
~lauf *m* = river course
~laufplan *m*, Schaltschema *n*, Schaltbild *n* = wiring diagram
~linie *f* [*Bodenmechanik*] = flow line
~linienbild *n* → Stromliniennetz *n*
~linienform *f* = streamlined contours, ~ shape
~liniennetz *n*, Strömungsnetz, Stromlinienbild *n* = flow net
stromloses Verfahren *n* [*zur Herstellung von Überzügen auf Metalloberflächen*] = electroless plating
~ **Wasser** *n* = slack water
Strom|mitte *f* = mid-river
~modell *n* = river model
~modell *n* **aus Beton** = river model mo(u)lded in concrete
~morphologie *f* = river morphology
~mündung *f* = river mouth, mouth of a river
~pfeiler *m* = water pier, river ~
(~)**Pfeilerkopf** *m* = nose of pier (US); cutwater ~ ~ (Brit.)
~polizei *f*, Stromwache *f*, Wasserpolizei = river police
~regulierung *f* → Reg(e)lung
~rinne *f* → Fahrrinne
~röhre *f* [*Bodenmechanik. Der Raum zwischen den Stromlinien*] = flow channel [*The strip located between two adjacent flow lines*]
~rüttelverfahren *n* **nach Bernatzik** = Bernatzik method of sand compaction by simultaneously applied water injection and vibration
~schiene *f*, stromführende Schiene = live rail
~schlauch *m*, Strombett *n* = river bed
~schleuse *f* = river (navigation) lock
~schnelle *f* = bold water, dalle, dells (US); rapid
~seite *f* [*Brückenbau*] = channel side
~sichter *m*, (Wind)Sichter = air classifier, (~) separator
~sichtung *f* → Setzungsklassierung durch Luft
~sohle *f* = river floor, ~ bottom

~spaltung *f* = divarication of a river, branching ~ ~ ~, diffluent [*converse of confluent*]
~sperre *f* [*Flößerei*] = boom (US)
~stadium *n* = riverhood
~stärke *f* = amperage
~strecke *f* = river stretch
~strich *m* [*Der geometrische Ort der Schwerpunkte der durch die verschiedenen Querschnitte strömenden Wassermassen*] = main current line, drift ~
~system *n* = river system
~tal *n* = river valley
~terrasse *f* = terrace of the river, alluvial (river) bench, alluvial terrace
~trasse *f* = lie of a river
~überbau *m* = river superstructure
~übergabe *f* = power intake
~übergabestation *f* = power intake station
~übergang *m* = place where the current passes from one bank to the other
~ufer *n* = river bank
~uferschutz *m* = river bank protection
~umleitung *f* = river diversion
Strömung *f* = current
~ **des Wassers im Boden** = water circulation in the ground
~ **gleichmäßiger Geschwindigkeit**, gleichmäßige Strömung = uniform flow
Strömungs|beiwert *m* [*Meeresströmung*] = current coefficient
~bild *n*, hydrodynamisches Netz *n* = flow pattern, ~ net [*of water*]
~diagramm *n*, Gezeitenstromfigur *f*, Tide(n)stromfigur = tidal current diagram
~druck *m* [*Bodenmechanik*] = seepage pressure
~energie *f* [*Wasserturbine*] = kinetic energy
~energieverlust *m*, Druckverlust [*technische Hydromechanik*] = head loss
~geometrie *f* = flow geometry

~geschwindigkeit *f* [*Bohrspülung*] = velocity of mud circulation
~geschwindigkeit *f*, Stromgeschwindigkeit = flow velocity, current ~, ~ rate
~kanal *m* für Wasser = water tunnel
~kraft *f* = force of the current
~kreislauf *m*, Gezeitenstromfigur *f*, Tide(n)stromfigur = course of currents
~kupp(e)lung *f* = fluid clutch
~küsel *m* = current clapotis
~menge *f* = volumetric flow rate
~messer *m*, Meßflügel *m* = current meter, rotary ~ [*An instrument with a vane like a windmill which rotates when the instrument is kept head on to the current*]
~netz *n*, Stromliniennetz, Stromlinienbild *n* = flow net
~querschnitt *m*, wirksamer Querschnitt, effektiver Querschnitt = effective cross section
~regime *n* = current regime
~rose *f* = current rose
~verhältnisse *npl* am Meißel = bit hydraulics
~wirbel *m* = rips
Strom|unterquerung *f* = crossing below a river
~verband *m*, Festungsverband = raking bond, herringbone ~, diagonal ~, oblique ~
~verbrauch *m* = energy consumption
~verkehr *m* = river traffic
~versetzung *f* an der Küste, Küstenstromversetzung, Küstendrift *f* = littoral drift
~versetzung *f* bewirkt durch Brandungsbänke → Brandungsstromversetzung
~versorgung *f*, Kraftversorgung, Energieversorgung = power supply
~wache *f* → Strompolizei *f*
~wasser *n* = river water
~wasserfassung *f* = river intake
~zähler *m*, Energiezähler = electric current meter
~zählung *f* [*Verkehrszählung*] = cordon count

~zufuhr *f* = power transmission
Strontiumzement *m*, Edelzement = strontium cement
Strosse *f* → → Tunnel- und Stollenbau
strukturempfindlich, strukturanfällig [*Ton*] = sensitive
Strukturstörung *f* [*Ton*] = remoulding (Brit.); remolding (US)
strukturunempfindlich [*Ton*] = nonsensitive
Stubben *m* → Stumpf *m*
~roder *m* = grubber
Stuck *m* = stucco
~arbeit(en) *f(pl)* = stucco work
Stückeis *n* = crushed ice
Stuck|fläche *f* = stucco surface
~gips *m* = molding plaster (US); plaster of Paris (Brit.)
Stückgut *n* = part-load, piece good, parcelled good
~scheider *m*, Grobsieb *n* = scalping screen, scalper
~scheidung *f*, Grobsiebung = scalping
~gutwagen *m* = parcel van
Stück|kalk *m*, ungelöschter stückiger Branntkalk, Stückenkalk = lump lime
~kohle *f* = best coal, large ~
~kohle *f* → Knabbenkohle
~lohn *m*, Akkordlohn = piece wage
~lohnsatz *m*, Akkordlohnsatz = piece rate
~schlacke *f*, Eisenhochofen-~ = lump slag
~stein *m* = Setzpacklagestein
~zählung *f* [*Lagerbestand*] = physical check
Studentenwohnheim *n* = students' residential hostel
Studienreise *f* = study tour
Stufe *f*, Treppen~ = step
Stufen|ausbau *m* = provision of barrages and locks in a river
~belag *m* → Tritt~
~bildung *f* [*Betonplatten*] = stepping-off, faulting of transverse joints
stufenförmige Betonmauer *f* → abgestufte ~
Stufen|fräse *f*, Treppen~ = step grinder, ~ grinding machine

Stufengetriebe — Sturmflut

~getriebe n, Nachschaltgetriebe = range transmission

~kolbenpumpe f, Differentialpumpe = differential pump

~leiter f = stepladder

stufenlos regelbar = infinitely variable

stufenlose Drehzahlreg(e)lung f = infinite speed variation

stufenloser Antrieb m = infinitely variable speed drive

Stufen|meißel m = step bit, pilot ~

~platte f = step tile

~presse f, Treppen~ = step press

~rüttelpresse f, Treppen~ = step vibrating press, ~ vibratory ~

~-Schaltgetriebe n mit Druckluftschnellschaltung und Dauereingriff aller seiner Zahnräder = air-actuated multiple-disc clutches constant-mesh transmission

~schalthebel m = high-low lever, range selector ~

~schaubild n = histogram

~schleifmaschine f, Treppen~ = step grinder, ~ grinding machine

~sieb n = stepped screen

~stück n, Stufenzementiermuffe f [Tiefbohrtechnik] = stage-cementing collar

~tauchkolbenpumpe f = stage (or differential) plunger pump

~turbine f = stage turbine

~waschmaschine f, Stufenwäsche f = multi-compartment washer (or washing machine)

stufenweise Entlösung f [von Gas aus Erdöl] = differential eliberation

Stufenzementiermuffe f, Stufenstück n [Tiefbohrtechnik] = stage-cementing collar

Stuhl|rahmen m [Dach] = trussed purlin(e)

~säule f → Bindersteil m

~schiene f → Doppelkopfschiene

Stülp|decke f, gestülpte Holzdecke = clincher-built wood(en) ceiling, ~ timber ~

~schalung f [Schweiz: überluckte Schalung] = weatherboarding

~schalung f mit Wechselfalz = rebated weatherboarding; rabbetted ~ (US); ship-lap timber

~schalungsbrett n [zur Außenverschalung geeignetes Brett] = weatherboard; siding (US)

~wand f [Holzwand zur Baugrubenumschließung aus zwei Reihen von 4 bis 5 cm dicken, die Fugen überdeckenden Bohlen] = overlapping plank sheeting

Stumpf m, Baum~, Stubben m, Stock m = stump, stub

~-Ausreißer m → Wurzelzieher

stumpfer Spitzbogen m, gedrückter ~ = blunt arch, drop ~

stumpfmachen = to blunt

Stumpf|naht f [Schweiz: Stoßnaht] = butt weld

~schweißmaschine f = butt welding machine

~schweißung f = butt welding

~stoß m = butt joint

stunden (⚒) = to abandon

Stunden|geschwindigkeit f = miles per hour

~leistung f = hourly output

~lohn m = hourly wage

~lohnarbeit f, Tagelohnarbeit = daywork, open cost work, work outside scope of agreement

~(lohn)satz m = rate per working hour, hourly wage rate, daywork rate

~satz m, Stundenlohnsatz = rate per working hour, hourly wage rate, daywork rate

~-Spiel(an)zahl f, Anzahl der Spiele pro Stunde [z.B. Straßenhobel] = cycles per hour

~spitze f [Straßenverkehr] = peak hour value

~verdienst m = hourly earning

~zähler m = hourmeter

~zeichen n, Zeitzeichen = time signal

~zettel m [eines Maschinenführers] = time record

stündlicher Verbrauch m = hourly consumption

Sturm|flut f = hurricane tide

Sturmhaken — Stützenschalung

~haken m [*Für das Feststellen eines geöffneten Fensterflügels*] = casement stay, window ~

~laterne f = hurricane lamp

~latte f, Windrispe f, Strebeschwarte f, Schwibbe f = sprocket (piece), cocking ~

Sturz m, Oberschwelle f = lintel, lintol

~ der Vorderräder [*Motor-Straßenhobel*] = (front) wheel lean(ing)

~bett n, Absturzbecken n = (downstream) apron

~maschine f = lintel machine, lintol ~

~regen m, Regenguß m = (torrential) downpour

~riegel m [*Fachwerkwand*] = cross timber, intertie [*framed partition*]

Stürzrolle f, Bergerolle (⚒) = ore pass

Sturz|wehr n [*Festes Wehr, bei dem sich der Überfallstrahl an der Wehrkrone im Gegensatz zum Schußwehr vom Staukörper ablöst und frei abstürzt*] = free nappe weir

~welle f → Bore f

Stütz|arm m, Ab~, (Seiten)Stütze f [*zur Erhöhung der Standsicherheit von Baggern, Kränen und Schaufelladern*] = stabilizer, outrigger

~balken m [*Nadelwehr*] = supporting bar

~bauwerk n = retaining structure

~bolzen m für Hubarm = lift arm support pin

~boot n → Boot zur Stützung des Wehres

Stütze f, Decken~ = column [*An upright (vertical or near-vertical) loadbearing member whose length on plan is not more than four times its width*]

~ → Bohr~

~ [*Chanoine-Schütztafel*] = hinged tiebar (Brit.); horse (US)

~ → Stützarm m

~ → Hilfs~

Stutzen m, Rohrstutzen, Rohransatz m = flanged socket; ~ nozzle (US)

Stützen|anordnung f = column arrangement

~armierung f → Stützenbewehrung

~basis f = column base (plate)

~bemessung f = column design

~bereich m = column zone

~beton m = column concrete

~betonierautomat m, automatische Stützenbetoniermaschine f = auto(matic) concrete column pourer

~betoniergerät n = column pourer

~bewehrung f, Stützenarmierung, Stützen(stahl)einlagen fpl = column reinforcement

~druck m → Auflagerkraft f

~einlagen fpl → Stützenbewehrung f

~flansch m = column flange

~formel f = column formula

stützenfrei, stützenlos [*ohne Stahlstützen*] = stanchion-free, stanchionless

~, stützenlos [*ohne Stützen*] = column-free, column-less

Stützen|fundament n, Stützengründung f = pad foundation [*An isolated foundation for a separate column*]

~fuß m = column base

~gründung f → Stützenfundament n

~joch n = column bent [*A bent composed of columns and bracing in contradistinction to "pile bent"*]

~kern m [*Betonstütze*] = column core

~kopf m = column head

~länge f = column length

~last f = column load

stützen|los, stützenfrei = stanchion-free, stanchion-less

~loser Freiraum m = area free of supporting columns, clear space, clear area

Stützen|moment n [*Das Biegemoment über der Unterstützung eines Durchlaufträgers*] = moment at support

~querschnitt m = column cross section

~ramme f = oblique (pile) driver

~raster m [*Anordnung der Stützen im Rastersystem*] = column grid pattern

~reihe f = row of columns

~schaft m = column shaft

~schalung f = column box

Stützensenkung — Suchbohrung

~senkung f = column settlement
~(stahl)einlagen fpl → Stützenbewehrung f
~steg m = column web
~tangentiallagerung f = self-centreing column seating
~widerstand m → Auflagerkraft f
~zwinge f = column clamp
Stütz|fuß m, Abstützplatte f [*eines (Ab)Stützarmes zur Erhöhung der Standsicherheit von Baggern, Kränen und Schaufelladern*] = stabilizer base, outrigger ~, ~ plate
~insel f für Fußgänger, Fußgänger-Stützinsel = refuge (Brit.); ~ island, pedestrian island (US)
~knagge f [*Trommelwehr*] = guide
~korb m → Fugen~
~körper m, Dammkörper, Mitte f, Kern m [*nicht zu verwechseln mit Dichtungskern beim Staudamm = core*] = body, supporting shell
~kraft f → Auflagerkraft
~kuppel f → Kugelgewölbe n
~linie f [*Verbindungslinie aller Angriffpunkte der Druckkräfte längs der Stabachse. Meist für die Berechnung der Bogen und Gewölbe benützt*] = line of thrust, thrust line
~linienbogen m = thrust line arch
~liniengewölbe n = thrust line vault
Stützmauer f [*Oberbegriff für 1.) Böschungsmauer und 2.) Futtermauer*] = retaining wall [*The English term "retaining wall" covers "Böschungsmauer" only. Futtermauer = revetment wall*]
~ aus Bruchsteinmauerwerk = stone masonry retaining wall
~ mit Konsole = bracket-type retaining wall
Stütz|moment n → Stützenmoment
~pfosten m [*Seilschwebebahn*] = intermediate mast, supporting ~
~platte f [*Seitenplatte beim Stollenausbau mit vorgefertigten, gelenkig verbundenen Stahlbetonplatten*] = side slab
~presse f [*Bagger; Lader; Kran*] = outrigger jack

~punkt m, militärischer ~ = (military) base
~pyramide f [*bewegliche Brücke*] = tower pivot, supporting tower
~rahmen m = supporting frame
~reaktion f = reaction at support
~ring m = supporting ring
~rolle f [*Gleiskette*] = track carrier roller
~rollenträger m [*Gleiskette*] = carrier roller support, ~ ~ bracket
~säule f [*Nadelwehr*] = locking pin
~wand f = retaining wall
~weite f = effective span
~weitenverhältnis n, Spannweitenverhältnis = effective span length ratio
~weitenzwangspunkt m = span limit
Stylolithstruktur f = cone-in-cone structure
Styropor n, blähfähig hergestelltes Polystyrol n = Styropor expanded polystyrene
~-Hartschaumplatte f = Styropor expanded polystyrene board
~-Hartschaumstoffplatte f, Hartschaumstoffplatte aus Styropor = Styropor expanded polystyrene felt covered panel
subaquatische Rutschung f, Unterwasserrutschung, Subsolifluktion f = underwater solifluction (or solifluxion, or soil-creep)
Subkornstruktur f = subgrain structure
subkrustales Gestein n → plutonisches ~
Sublimat n → Quecksilberchlorid n
Submission f → → Angebot n
Submissionstermin m → → Angebot n
Subunternehmer m, Nachunternehmer = subcontractor
Such|arbeiten fpl, Prospektionsarbeiten = prospecting work, search ~
~bohrprogramm n = wildcatting campaign
~bohrung f → Aufschließungsbohrung
~bohrung f in einiger Entfernung von einer bekannten Erdöllagerstätte = semi-wildcat well

Suchbohrung — symmetrischer Bogen

~bohrung *f* nach tieferen Horizonten in einem Erdölfeld = deep test
Suche *f* → Prospektion *f*
~ im Küstenvorland → Aufsuchen *n*
~ ~
suchen, prospektieren = to search, to prospect [*To look for valuable mineral*]
Such|schacht *m* → Aufschließungsschacht
~scheinwerfer *m* = searchlight
~spule *f* = pick-up coil
~stollen *m* → Aufschließungsstollen
Südfenster *n* = south-facing window
Sulfat *n* = sulfate (US); sulphate (Brit.)
~angriff *m* = sulfate attack (US); sulphate ~ (Brit.)
sulfatbeständig = sulphate-resisting (Brit.); sulfate-resisting (US)
Sulfat|beständigkeit *f* = sulfate resistance (US); sulphate ~ (Brit.)
~bestimmung *f* nach der Trübungsmessmethode = turbidimetric sulfate determination (US); ~ sulphate ~ (Brit.)
~reaktionszahl *f* = sulphate reaction value (Brit.); sulfate ~ = (US)
~rückstand *m* = sulfated residue (US); sulphated ~ (Brit.)
~schwefel *m*, SO₃ = sulfur trioxide
Sulfid *n* = sulphide
sulfidischer Abgang *m* = sulphide tailings
Sulfidschwefel *m*, S = sulfide sulfur
Süll *m*, *n* → Drempel *m*
Sulphatangriff *m* = sulfate attack (US); sulphate ~ (Brit.)
sulphathaltiger Boden *m* = sulfate-laden soil (US); sulphate-laden ~ (Brit.)
Summe *f* aller Lasten → Lastensumme
Summen|kurve *f* [*Siebtechnik*] = cumulative curve
~linie *f* [*Hydraulik*] = summation curve, ~ hydrograph
~teilung *f* [*Zahnrad*] = cumulative circular pitch
Summer *m* = buzzer

Summier|gerät *n* = accumulating (traffic) counter
~wiegen *n* = cumulative weighing
Sumpf *m* [*Behälter zum Absetzen von Feststoffen*] = sump
~, eingespülter wasserdichter Kern *m* [*Erddamm*] = core pool
~ → Pumpen~
sümpfen [*bergmännischer Ausdruck für "entwässern"*] = to drain
Sumpf|gas *n* = marsh gas
~(gebiet) *m*, (*n*) = swamp (area), ~ land
~haus *n*, Maukhaus = souring house
~kalk *m*, eingesumpfter Kalk, Grubenkalk, Kalkbrei *m* = pit lime
~kohle *f* = bog coal
~moos *n*, Torfmoos = peat moss, bog ~, Spagnum ~
~pumpe *f*, Wasserhaltungspumpe = sump pump
~torf *m* = marsh peat
Super|gummiradwalze *f* → → Walze
~-Kraftstoff *m* = first grade fuel
~markt *m* = supermarket
~oberflächenschluß *m* = ultra-smooth finish
~position *f* von Wellenflächen = superposition of wave surfaces
~schmiermittel *n* = superior lubricant
supersulfatisierter Sulfathüttenzement *m* = gypsum slag cement (US); supersulphated ~ (Brit.)
suprakrustales Erstarrungsgestein *n* → Ergußgestein
Suspension *f* → Aufschlämmung *f*
~ von Zement in Wasser → Zementmilch *f*
Süßwasser *n* = fresh water
~becken *n* = fresh water reservoir
~fisch *m* = fresh water fish
~kalkstein *m* = fresh water limestone
~masse *f* = body of fresh water
~see *m* = fresh water lake
~träger *m* = fresh water resource
Symmetrieachse *f*, Bau(werk)achse = centre line (Brit.); center ~ (US)
symmetrischer Bogen *m* = symmetrical arch

Symons|(kegel)brecher m = (standard type) Symons (cone) crusher
  ununterbrochener Arbeitsgang m = continuous crushing action
  Brechmantel m = mantle
  schlagartige Zerkleinerung f, Schlagzerkleinerung = impact reduction, ~ comminution
  geschlossene Hubstellung f = close setting
  Umlaufdruckschmierung f = circulating pressure lubrication
~-**Kegelgranulator** m = short head type (Symons) cone crusher
Sympathie-Riß m = reflection crack [*Crack in old concrete pavements reappearing as crack in the asphalt surface*]
Synchron|getriebe n = synchromesh (transmission) [*A silent-shift transmission construction, in which hub speeds are synchronized before engagement by contact of leather cones*]
~-**Regler** m = synchronous governor
Synklinale f, Mulde f der Faltung f = syncline
Synklinaltal n = synclinal valley
synthetischer Diamant m, künstlicher ~ = man-made diamond, synthetic ~
~ **Graphit** m = synthetic graphite
~ **Quarz** m, künstlicher ~ = synthetic quartz, man-made ~
Syphon m, Flüssigkeitsheber m, Heberrohr n, Siphon = siphon
~ → Geruchsverschluß m
System n **der Betriebsplätze unmittelbar vor dem Abfertigungsgebäude** [*Flughafen*] = frontal system
~ **des offenen Vorfeldes** [*Flughafen*] = open apron system
~ **sich schneidender Pisten** [*Flughafen Zürich-Kloten*] = intersecting runway configuration, ~ pattern
~ **zusätzlicher Abfertigungsgebäude, Satellitsystem** [*Flughafen*] = satellite system
~**achse** f = system centre line (Brit.); ~ center ~ (US)

~**linie** f, Netzlinie [*Träger*] = system line
~**punkt** m, Knotenpunkt = system point, knot; system centre (Brit.); system center (US)
SZ-Diesel(pfahl)ramme f, Dieselramme mit Schlagzerstäubung = kick-atomizing pile driver, impact-atomizing ~ ~

T

Tabakspeicher m = tobacco warehouse
tabellarisch = in tabular form
Tabelle f, Zahlentafel f = table (of figures), numerical table
Tabellen|werk n = tabular compilation
~**wert** m = tabulated value
Tachograph m = tachograph
Tachometer n, m, Drehzahlmesser m, Tourenzähler m = tachometer, revolution counter
Tachymeter m, n = tacheometer
~**aufnahme** f, tachymetrische Aufnahme = tacheometric survey, ~ levelling, ~ contouring
~**theodolit** m = tacheometer theodolite
Tachymetrie f, kombinierte Lage- und Höhenmessung f = tacheometry
Tafel f → Schütz~
~, Tabelle f = table
~, Klappe f, Stau~ [*Klappen(stau)wehr*] = wicket (US); shutter (Brit.)
~ [*Brücke*] = deck(ing)
~ **für gerade Brücke** = right bridge deck(ing)
~**blei** n, Walzblei, Bleiblech n = sheet lead
~**glas** n = sheet glass
tafeliger Habitus m [*Kristall*] = tabular habit
Tafel|leim m = animal glue in cake form, ~ adhesive ~ ~ ~
~**platte** f [*Brücke*] = deck(ing) slab
~**schere** f, Blechschere = plate shears
~**schiefer** m = table-slate, school-slate
~**teil** m, n, Fahrbahn~, Brückentafelteil = roadway section
~**träger** m, Fahrbahn~, Brückentafelträger = deck girder

Täfelung *f*, Paneel *n*, Täfer *n*, Vertäfelung *f*, Getäfer *n*, Getäfel *n* = panelling; paneling (US)
Täfelungsplatte *f*, Ver~ = panel
Tafelwand *f* = precast concrete slab wall
Täfer *n* → Täfelung *f*
Taganflug *m* = daytime approach
Tagebau *m* = open-cut mining, surface ~, strip ~, opencast work(ing)
~**-Förderung** *f* = open pit haulage
~**grube** *f* = strip mine
~**-Lok(omotive)** *f* = open-cast mining loco(motive)
Tagelohnarbeit *f*, Stundenlohnarbeit = daywork, open cost work, work outside scope of agreement
Tages|(arbeits)fuge *f* = night header joint, day's work ~
~**ausgleichbecken** *n* → Tagesspeicher
~**ausgleichspeicher** *m* → Tagesspeicher
~**becken** *n* → Tagesspeicher
~**belichtung** *f* → Tages(licht)beleuchtung
~**bericht** *m* des Bohrmeisters → Bohrbericht
~**durchschnitt** *m*, Tagesmittel *n* = daily average
~**energie** *f*, Tagesstrom *m* = day power, ~ current
~**leistung** *f* = daily output
~**licht** *n* = day light, natural ~
~**(licht)beleuchtung** *f*, Tagesbeleuchtung = daylighting, natural lighting
~**mittel** *n*, Tagesdurchschnitt *m* = daily average
~**mitteltemperatur** *f* = daily mean temperature
~**nebel** *m* = daytime fog
~**oberfläche** *f* (⚒) = surface
~**rapport** *m* des Bohrmeisters → Bohrbericht *m*
~**raum** *m* = common room
~**riegel** *m* = day latch
~**schutzbezirk** *m* (⚒) = pillar-protected surface
~**schwankung** *f* = daily variation
~**sichtbarkeit** *f* = day visibility
~**speicher** *m*, Tages(ausgleich)becken *n*, Tagesausgleichspeicher [*Kleinspeicher zum Ausgleich von Wasserdargebot und Wasserbedarf über einen Tag*] = day-supply reservoir
~**speicherung** *f* = daily storage
~**spitze** *f* = daily peak
~**stollen** *m* → Abstiegstollen
~**strom** *m*, Tagesenergie *f* = day power, ~ current
~**tank** *m*, Tagesbehälter *m* = day-supply tank
~**tonnen** *fpl*, Tato *f* = tons per day
~**verbrauch** *m* = daily consumption
~**verkehr** *m* = daily traffic
Tagkranz *m*, Mündung *f*, Mundloch *n* [*einer Bohrung*] = well head, opening of the (bore) hole, mouth of the (bore) hole
tägliche Gezeitenungleichheit *f* = daily inequality of the tides, diurnal ~ ~ ~ ~
täglicher Bericht *m* des Bohrmeisters → Bohrbericht
~ **Tide(n)hub** *m*, ~ Gezeitenhub = diurnal range
Tagschicht *f* = daytime shift
Tagstollen *m* → Zugangsstollen
Tagverkehr *m* = daytime traffic
Taktarbeit *f* = repetitive operations (in building), repetition of site processes
Tagwasser *n*, Niederschlagswasser = meteorological water
Talbrücke *f*, Viadukt *m* = valley bridge, viaduct
Taleinschnitt *m* = section of a valley
Talfahrer *m* [*Binnenschiffahrt*] = descending barge
Talflanke *f*, Talhang *m* = valley slope, ~ side, ~ flank
Talgletscher *m* = valley glacier, mountain ~
Talhang *m*, Talflanke *f* = valley slope, ~ side, ~ flank
Talje *f* [*seemännischer Ausdruck für „Flaschenzug"* *m*] = (block and) tackle, pulley block, lifting block
Talk *m* (Min.) = talc
talkreiche keramische Masse *f* = high-talc body

Talkschiefer m = talc schist
Talkum n, Speaksteinmehl n = talcum powder, powdered talc, French chalk
Tallöl n, Kiefernöl = tall oil
Tallöß m = valley loess
Talmoor n = valley fen

Talniederschlag m = valley precipitation
Talquelle f = valley spring
Talquerschnitt m = valley cross section
~**schlucht** f = glen
~**schotter** m = valley gravel
~**sohle** f = valley bottom, ~ floor

Talsperre f, Sperre [*Eine Talsperre in ihrer Gesamtanlage umfaßt die folgenden Bauwerke: Die eigentliche Sperre, die als Mauer oder Damm ausgeführt sein kann, das Staubecken, dem häufig noch ein Vorbecken zugeordnet ist, einschließlich der Hochwasserentlastung. Ferner gehören dazu die Entnahmeeinrichtung, die als besonderes Bauwerk, als Grundablaß oder Überlauf mit Tosbecken ausgebildet ist, und das Maschinenhaus für die Erzeugung elektrischer Energie oder die Weiterleitung des Wassers*]   dam, hydro ~

   (Ab)Dichtungsgraben m, Verherdung f cutoff trench

   (Ab)Dichtungsvorlage f, (Ab)Dichtungsteppich m, (Ab)Dichtungsschürze f, Schürze, Außen(haut-)dichtung f, Oberflächendichtung   impervious blanket, (waterproof) ~

   Abflußleitung f discharge conduit
   abgesenkter Wasserspiegel m drawn-down water level
   Ablassen n, Entleeren n, Entleerung f, Absenkung, Absenken   emptying, drawdown

   Ablaß m auf mittlerer Höhe   middle-height discharge tunnel
   Ablaßregulierschütz(e) n, (f)   outlet control (or regulating) gate
   Ablaßrohr n discharge pipe
   Ablaßschieber m, Entleerungsschieber   outlet valve

   Ablaßstollen m, Entleerungsstollen, Ablaufstollen, Abflußstollen, Auslaufstollen   discharge tunnel, outlet ~

   Ablaßverschluß m emptying gate, outlet ~
   Ablösung f des Strahles, Strahlablösung   freeing of the nappe (US); separation of the water layer (Brit.)
   Abschlußschütz(e) n, (f), Grundablaßverschluß m lower gate, ~ sluice, sluice (gate), sluiceway gate

   Abschlußwand f → Herdmauer f
   Absenk(ungs)ziel n minimum water storage elevation, drawdown level

   Ambursen-(Stau)mauer f, Ambursen-Pfeiler-Plattensperre f, Am-   Ambursen-type dam

Talsperre

bursen-Pfeilerplattensperrmauer, Pfeiler-Plattensperre Bauweise Ambursen

Anzug *m* luftseits, Anlauf *m* ∼ — downstream batter

Anzug *m* wasserseits, Anlauf *m* ∼ — upstream batter

aufgelöste Gewichtssperre *f* System Noetzli, Pfeilerkopf(stau)mauer *f*, Rundkopf(stau)mauer von Noetzli, Pfeilerkopfsperre — dam with segmental-headed counterforts (Brit.); round-headed buttress dam

aufgelöste Staumauer *f* → Pfeiler-(stau)mauer *f*

Aufhöhung *f* — heightening

(Auf)Speicherung *f*, Stau *m*, (Wasser-)Einstau, Stauanhebung, Aufstau, Aufstauung — storage, pondage, filling, impounding

Ausgleichbecken *n*, Ausgleichweiher *m* — balancing reservoir, compensating ∼, equalizing ∼, regulating ∼

Auslaufbauwerk *n* → Betriebsauslaß

Auslaufstollen *m* [*von Turbine zum Ableitungstunnel*] — draft tube tunnel (US); draught ∼ ∼ (Brit.); discharge branch

Auslaufstollen *m* → Ablaßstollen

auspressen, einpressen, verpressen, injizieren — to inject, to grout under pressure

Außen(haut)dichtung *f* → Dichtungsvorlage *f*

Auszugrohr *n* — draft tube (US); draught ∼ (Brit.)

Außen(haut)dichtung *f* → Dichtungsvorlage *f*

Baubrücke *f* → Dienstbrücke

Baugrube *f* — excavation

Bausteg *m* → Dienstbrücke *f*

Bauzeithochwasser *n*, bauzeitliches Hochwasser — construction flood

Becken *n* → Sammelbecken

Bedienungskammer *f* — operating chamber

Bedienungsstation *f* — control cabin

Begrenzungsmauer *f* [*Tosbecken*] — training wall [*stilling basin*]

belüfteter Überfallstrahl *m* — aerated nappe (US); ∼ sheet of water (Brit.)

Beobachtungsschacht *m*, Kontrollschacht — inspection shaft

Beobachtungsstollen *m*, Kontrollgang *m* — inspection gallery, ∼ tunnel

bergseitig, oberwasserseitig, oberstrom — upstream

Beruhigungsbecken *n*, Tosbecken — stilling basin, absorption ∼ (Brit.); stilling pool (US)

Beruhigungskammer *f*, Toskammer — stilling chamber; absorption ∼ (Brit.)

Beton-Bogengewichts(stau)mauer *f* — concrete arch(ed) gravity dam

Talsperre

Betonkern *m*, massive Kernmauer *f*	concrete core (wall)
Beton-Kerndamm *m* → Kernmauerdamm	
Betonmauer *f* → (Massiv)Beton(stau)mauer	
Betonpfropfen *m*, Betonplombe *f*	concrete plug
Betonsperre *f* → Massivbetonmauer *f*	
Beton(stau)mauer *f* → (Massiv-)Beton(stau)mauer	
Betriebsauslaß *m*, Auslaufbauwerk *n*	river outlet, reservoir outlet (works)
Betriebswasserstollen *m*, Druck-(wasser)stollen, Triebwasserstollen, Zulaufstollen, (Wasser)Kraftstollen, Kraftwasserstollen [*Stollen, der im planmäßigen Betrieb voll und unter Scheiteldruck läuft*]	pressure tunnel, tunnel-type penstock, conversion tunnel, pressure gallery, hydro tunnel, hydro tube, power tunnel
bituminöse (Damm)Dichtung *f*	bituminous impervious element
Bogenpfeiler(stau)mauer *f* → Gewölbereihen(stau)mauer *f*	
Bogen(schwer)gewichts(stau)mauer *f*, Bogengewichtssperre *f*, Gewölbe-Gewichtssperre *f*	arch-gravity dam
Bogensehne *f*	chord
Bogensperrmauer *f*, Bogen(stau-)mauer *f*, Bogen(tal)sperre *f*, Gewölbe(stau)mauer, Einfach-Gewölbesperre, Gewölbesperrmauer, Einzelgewölbe(stau)mauer	arch(ed) dam
Bogenmauer *f* mit gleichem Bogenzentriwinkel, Jörgensen-Mauer, Gleichwinkel(stau)mauer, Bogen(stau)mauer mit gleichbleibendem Öffnungswinkel, Bogen(stau)mauer mit Festwinkel	constant angle arch dam, variable-radius ∼ ∼
Kuppel(stau)mauer *f*	dome-shaped dam, cupola-shaped ∼
Doppelbogen(stau)mauer *f*	double arch(ed) dam
Bogen(stau)mauer *f* mit gleichem Radius, kreiszylindrische Sperre	constant radius (arched) dam
Öffnungswinkel *m* der Krone *f*	angular width of arch at crest (Brit.); central angle at crest (US)
Querschnitt *m* im Scheitel	section at crown of arch (Brit.); ∼ ∼ key (or crown) (US)
Lastaufteilungsverfahren *n* des Bureau of Reclamation [*Der Wasserdruck wird auf die beiden statischen Hilfssysteme der Staumauer – die Bogen und die Kragträger – aufgeteilt*]	trial load method of analyzing arch dams

Gewölbering *m*, Bogen *m*	arch element
Kragträger *m*	cantilever element
Radius *m* der luftseitigen Kronenbegrenzung *f*	downstream radius of crest (circle)
Scheitelform *f* wasserseitig	upstream profile at crown of arch (Brit.); ∼ ∼ of section at key (or crown) (US)
Krümmungsradius *m*, Krümmungshalbmesser *m*	locus of centres (Brit.); line of centers (US)
Bogenüberlauf *m*	arch spillway
Bruchstein(stau)mauer *f* in Bogenform *f*	masonry arch gravity dam
Damm *m* → Staudamm	
Damm *m* aus Erdschüttung → Erd(schüttungs)(stau)damm	
Damm *m* aus Felsschüttung → Steinfülldamm	
Damm *m* aus Kiesschüttung → Kies(füll)damm	
Dammbalkennut *f*	stop-log groove
Dammbaustelle *f*, Staudamm-Baustelle, Talsperrendamm-Baustelle	fill dam site, embankment type ∼ ∼
(Damm)Dichtung *f*	impervious element
Dammkern *m* → → Erd(schüttungsstau)damm	
Dammkernmauer *f* → Kernmauer	
Dammkörper *m*, Stau∼	fill dam body
Dichtung *f*, Dammdichtung	impervious element
Dichtungsgraben *m* → Ab∼	
Dichtungshaut *f* → Dichtungsvorlage	
Dichtungskern *m* → → Erd(schüttungsstau)damm	
Dichtungsmaterial *n*, Dichtungsstoff *m*	impervious material
Dichtungsvorlage *f*, Dichtungsteppich *m*, (Dichtungs)Schürze *f*, Ab∼, Außen(haut)dichtung *f*, Oberflächendichtung, (Ab)Dichtungshaut *f*, Dichtungsschicht *f*	(waterproof) blanket, impervious ∼
Dichtungszone *f*	impervious section
Dienstbrücke *f*, Bausteg *m*, Baubrücke	service gangway (Brit.); distributing bridge, overhead track way, construction trestle (US)
Differentialwasserschloß *n*, Steigrohrwasserschloß	differential surge tank
direkter Auslaufstollen *m* [*von Schieberkammer zum Ableitungstunnel unter Umgehung des Krafthauses*]	penstock drain, drain(age) tunnel

Talsperre 980

Doppelbogen(stau)mauer *f* → → Bogensperrmauer

Doppel-Skisprung-Überlauf *m* über die Krafthausdecke, Doppel-Sprungschanze *f* — double (or two-jet) ski-jump crossing over the roof of the power house

Dränschicht *f*, Filterschicht, Entwässerungsschicht — filter layer, drainage ~

Druckgefälle *n* → Niveaudifferenz

Druckhöhe *f* → Niveaudifferenz

Druckleitung *f*, Triebwasserleitung — penstock

Druck(leitungs)schacht *m*, Fallschacht — penstock, pressure shaft

Druck(wasser)stollen *m* → Betriebswasserstollen

Druckwasserzuleitung *f* → Druckleitung

einfache Gewichts(stau)mauer *f* → Gewichts(stau)mauer

Einfach-Gewölbesperre *f* → Bogensperrmauer *f*

Einlaßbauwerk *n* zum Kraftwerk, Einlaufbauwerk ~ ~ [*als Sperre ausgebildet*] — power plant intake structure, penstock dam

Einlaßturm *m*, Turmeinlaß *m*, Einlaufturm, Turmeinlauf — intake tower

Einlaßverschluß *m*, Einlaufverschluß — head gate, intake ~

Einlaufbauwerkstollen *m* — penstock dam gallery

Einlaufkanal *m*, Einlauf(ge)rinne *f*, (*n*), Zulauf(ge)rinne, Zulaufkanal — inlet channel, intake ~

Einlauftrompete *f*, Einlaßtrompete — trumpet inlet

Einlaufturm *m* → Einlaßturm

einpressen, auspressen, verpressen, injizieren — to inject, to grout under pressure

(Ein)Stau *m* → (Auf)Speicherung *f*

Energievernichtung *f*, Energieverzehrung *f* — energy dissipation

Energievernichtungspfeiler *m*, Energieverzehrungspfeiler, Störpfeiler — baffle-pier

Entenschnäbel-Überlauf *m* — duck bills type spillway

Entlastungsstollen *m* — discharge tunnel, outlet ~

Entlastungsüberlauf *m* → Überlauf

Entleerungsstollen *m* → Ablaßstollen

Entnahmebauwerk *n*, Einlaufbauwerk, Einlauf *m*, Einlaßbauwerk — intake (or inlet) works (or structure), head works

Entnahme-Rohrstollen *m*, Einlauf-Rohrstollen — intake pipe gallery, inlet ~ ~

Entnahmestollen *m*, Einlaufstollen — intake gallery, inlet ~

Entwässerungsschicht *f*, Filterschicht, Dränschicht	filter layer, drainage ~
Entwässerungsschürze *f*	drainage curtain
Entwässerungsstollen *m*	drainage gallery
Erdkern *m*	earth core
erhöhen	to raise
Erhöhung *f*, Erhöhen *n*, Aufstockung	heightening
Erd(schüttungs)(stau)damm *m*, Staudamm, Damm aus Erdschüttung	earth fill dam, earth(work) ~, earthen ~
Aufbaudamm *m*	earth(en) dam of composite cross section
Aufbauquerschnitt *m*	composite (cross-)section
Beton-Fußmauer *f*	concrete cutoff wall
Dränschicht *f*, Filterschicht	pervious shell
(Dichtungs)Kern *m*, Dammkern, Innendichtung *f*, Kerndichtung	(impervious) core, ~ diaphragm
gespülter (Erd)Damm, (Voll-)Spül(stau)damm, gespülter Staudamm	hydraulic-fill dam
Randzone *f*	shell
Erddamm *m* mit Querschnittaufbau in Filterform *f*, ~ ~ symmetrischem Filteraufbau	earth (fill) dam with materials of various permeabilities, zoned earth (fill) dam
eingespülter wasserdichter Kern *m*, Sumpf *m*	core pool
Spüleinbaudamm *m*	semi-hydraulic fill dam
(ab)gewalzter Damm *m*, Walz-(stau)damm	rolled earth (fill) dam
Kiesfilter *m, n*	filter blanket of gravel
Einschlämmen *n*, Einspülen, Einspülverfahren *n*, Bodenverdichtung *f* durch Einschlämmen, Spülspritzverfahren, Spülkippverfahren, Wasserstrahlverfahren	sluicing
Schüttboden *m*	fill soil, fill earth
Steindeckwerk *n*, Stein(ab)deckung *f*, Steinbestürzung *f*	riprap, rock rip-rap, stone rip-rap
Steinsatz *m*, Steinpackung *f*	hand-placed (stone) riprap, pitched slope
Steinwurf *m*, Rauhwurf, geschüttetes Steindeckwerk *n*, geschüttete Stein(ab)deckung *f*, geschüttete Steinbestürzung	truck-dumped (US)/lorry-dumped (Brit.) riprap, dumped (stone) riprap
Fallhöhe *f* → Niveaudifferenz *f*	
Fallhöhenverlust *m*, Druckhöhenverlust, Druckgefälleverlust	head loss, fall ~

Fallschacht m, Druck(leitungs)schacht	penstock shaft
Fassungsvermögen n, Stauinhalt m	capacity
Felsschüttmaterial n	rock fill material, ~ material composing the embankment
Felsschüttungsstaudamm m → Steinfülldamm	
Filterschicht f → Entwässerungsschicht	
Flachschieber m	gate valve, flat slide ~
Floßgasse f, Floßrinne f	log chute, ~ way
freier Überfallstrahl m, ~ Strahl	free nappe (US); ~ sheet of water (Brit.)
Freistrahlüberlauf m, freier Überlauf	open spillway (US); free waste weir, free spillway
Freibord n	freeboard [*The difference in elevation between the top of the dam and the maximum reservoir level that would be attained during the spillway design flood*]
Freiluftschaltanlage f, Umspannwerk n, Schalthof m	switchyard, outdoor switchgear
Führungswand f [*Hochwasserüberfall*]	training wall
Gefälle n → Niveaudifferenz f	
Gefällhöhe f → Niveaudifferenz f	
Gegenmauer f, Überfallmauer [*Tosbecken*]	end wall [*stilling basin*]
gekrümmte Staumauer f	arch(ed) dam
gekrümmtes Saugrohr n	elbow (draft) tube (US); ~ (draught) ~ (Brit.)
gepackter Steindamm m → → Steinfülldamm	
Gesamtgefälle n	total head
getrenntes Grundablaßbauwerk n	separate sluiceway structure
Gewichts(stau)mauer f, (massive) Gewichtssperre f, einfache ~, Schwergewichts(stau)mauer	gravity dam
Beton-Gewichts(stau)mauer	concrete gravity dam
Bogen(schwer)gewichts(stau)mauer	arch gravity dam
Bruchstein-Gewichts(stau)mauer	masonry gravity dam
einfaches (oder einschnittiges) Lastaufteilungsverfahren n	simplified (or abridged) trial-load method
gerade Gewichts(stau)mauer f	straight-gravity dam
Gleitwiderstand m	sliding resistance
Gleitsicherheitsfaktor m	sliding factor
Scherreibungs-Sicherheitsfaktor m	shear-frication factor
Gewichtsmauer f (oder Schwergewichtssperre f) mit Überfall	gravity spillway dam

Ausrundung *f*	ogee curve (Brit.); bucket of ogee (US)
Leitwand *f*	guide wall (Brit.); training ∼ (US)
Pfeilerkopf *m*	upstream nose, cut-water (Brit.); nose (US)
Pfeilerrücken *m*	downstream nose (Brit.); ∼ pier nosing (US)
Schwelle *f*	sill (Brit.); gate seat (US)
Überfallrücken *m*	shaped face, profiled ∼ (Brit.); spillway face, profiled face (US)
Gewölbe-Gewichtssperre *f* → Bogen(schwer)gewichts(stau)mauer *f*	
Gewölbemauer *f* → Bogensperrmauer *f*	
Gewölbereihen(stau)mauer *f*, Vielfachbogensperre *f*, Bogenpfeiler(stau)mauer *f*, Pfeilergewölbe(stau)mauer	multiple arch(ed) dam, multi-arch ∼
Gewölbe(stau)mauer *f* → Bogensperrmauer	
Gleichgewichtsklappe *f*	balanced weir (or gate) (Brit.); automatic flap gate
Gleichwinkel(stau)mauer *f* → → Bogensperrmauer	
Grobfilterschicht *f*, Grobdränschicht, Grobentwässerungsschicht	coarse filter layer, coarse drainage ∼
Großspeicher *m*	large reservoir
Grundablaß *m*, Grundauslaß *m* [*Auslaß etwa in Höhe der Talsohle oder darunter*]	lower (or bottom) discharge tunnel, sluiceway, bottom emptying gallery, bottom outlet, scour outlet, undersluice
Grundablaßkanal *m*	sluicing channel
Grundablaßrohr *n*	bottom outlet pipe, ∼ discharge ∼
Grundablaßverschluß *m*, Abschlußschütz(e) *n*, (*f*)	sluice (gate), sluiceway gate, lower gate, lower sluice
Grundstrahl *m*	sluiceway flow
Gründungsfels *m*	rock foundation
halbdurchlässig	semipervious
Halbfreiluft-Kraftwerk *n*	semioutdoor-type power plant
halbphärisches Gewölbe *n*, Kuppel *f*	dome, cupola
Heberüberlauf *m*, Saugheberüberlauf	siphon(ic) spillway
Herdmauer *f*, Fußmauer, (Ab)Dichtungsmauer, (Ab)Dichtungssporn *m*, Trennmauer, Sporn, Abschlußwand *f*	(toe) cutoff wall
Hochwasserentlastung *f*, Hochwasserabführung	flood relief, ∼ discharge

Hochwasserentlastungsanlage f → Überlauf m	
Hochwasserentlastungsanlage f vom Stauwerk getrennt	(side-)channel (type) spillway
Hochwasserentlastungskanal m, Überlaufkanal	spillway channel, ~ chute
Hochwasserschutztalsperre f, Hochwasser(auffang)sperre	flood (control) dam
Hochwasserspeicherung f	flood control storage, storage of flood
Hochwasserstollen m → Überlaufstollen	
Hochwasserüberlauf → Überlauf	
(Hochwasser)Überlauf-Auslaufbauwerk n	spillway outlet structure
(Hochwasser)Überlaufverschluß m	spillway gate
Hohl(stau)mauer f	cavity dam (Brit.); hollow ~ (US)
Hubschütz(e) n, (f), Hubwehr n	vertical lift gate
hydraulischer Speicher m → Sammelbecken n	
injizieren, einpressen, auspressen, verpressen	to inject, to grout under pressure
Innendichtung f, Dammkern m, (Dichtungs)Kern	(impervious) core, impervions diaphragm
installierte Leistung f, Leistungssoll n	installed (nameplate) capacity
Kavernenkrafthaus n, Kraftwerkkaverne f, Maschinenkaverne, Untergrundkrafthaus, Kaverneneturbinenhaus	underground power station, ~ hydroelectric (power) plant
Kavernen-Wasserschloß n	underground surge tank
Kavitationsfestigkeit f [Tosbecken]	cavitation erosion resistance
Kern m	core
Kernbaugrube f	core pit
Kerndamm m	core-type dam
Kerndichtung f, Innendichtung, Dammkern m, (Dichtungs)Kern	impervious diaphragm, (~) core
Kernmauer f, Dammkernmauer	core wall
Kernmauerdamm m, Beton-Kerndamm	concrete core (wall) (type) dam
Kernneigung f	core slope
Kies(füll)damm m, Damm aus Kiesschüttung, Kiesschüttungsstaudamm, Kiesschüttdamm	gravel-fill dam
Kippen n	overturning (Brit.); tilting (US)
Kolkschutz m	protection against scour, erosion control
Kontaktzone f	area of contact
Kontrollgang m, Beobachtungsstollen m	inspection gallery, ~ tunnel

Kote *f*, Höhe *f*, Ordinate *f*	level (Brit.); el. (US)
Krafthaus *n* → Maschinenhaus	
Kraftstollen *m* → Betriebswasserstollen	
Kraftwerkgebäude *n* → Maschinenhaus *n*	
Kraftwerkkaverne *f* → Kavernenkrafthaus *n*	
kreiszylindrische Sperre *f*, Bogen(stau)mauer *f* mit gleichem Radius	constant radius (arched) dam
kritische Druckhöhe *f*, kritisches Gefälle *n*	critical head
Krümmer *m*	elbow bend, penstock elbow
Kugelschieber *m*	globe valve
Kühlwasserstollen *m*	condensation water tunnel, densing ~ ~
künstlicher See *m* → Sammelbecken *n*	
Kuppel *f*, halbsphärisches Gewölbe *n*	dome, cupola
Kuppel(stau)mauer *f* → → Bogensperrmauer	
Leistungssoll *n* → installierte Leistung *f*	
luftseitige Böschung *f*	downstream slope
Maschinenflur *m*	generator floor
Maschinenhaus *n*, Krafthaus, Kraftwerkgebäude *n* [*Gebäude, in dem die Turbinen und die Stromerzeuger oder Arbeitsmaschinen untergebracht sind*]	power house
Maschinenkaverne *f*, Kraftwerkkaverne, Kavernenkrafthaus *n* [*Krafthaus im Innern des Gebirges, bergmännisch oder im Einschnitt mit nachträglicher Überschüttung hergestellt*]	underground power station, ~ hydroelectric (power) plant
(Massiv)Beton(stau)mauer *f*, Betonsperre *f*, Staumauer *f*, Talsperren(stau)mauer, Sperrmauer, Betonsperrmauer	(massive) concrete dam
massive Kernmauer *f* → Betonkern	
Mauerbaustelle *f*	massive concrete dam site
Mauer *f* in aufgelöster Bauweise → Pfeiler(stau)mauer	
Mauerwerksperre *f* → → Steinfülldamm	
Mehrzweckbauwerk *n*, Verbundbauwerk, Vielzweckbauwerk, Verbundprojekt *n*, Mehrzweckprojekt	multi(ple)-purpose structure

Talsperre

Nadelschieber *m*	needle (control) (or regulating) valve
Nebenanlagen *fpl*	appurtenant works, ~ structures
Niedrigwasserverbesserung *f*, Niedrigwasserreg(e)lung *f*	regulation of low water flows
Niveaudifferenz *f*, Gefälle *n*, Druckhöhe *f*, Druckgefälle, Gefällhöhe, Fallhöhe, Niveauunterschied *m*	head, fall
Normalspiegel *m*, normales Stauziel *n*	normal op(erating) level, ~ water storage elevation, ~ full-pool elevation
Notverschluß *m*	emergency (guard) gate, guard ~
Nutzgefälle *n*, nutzbarer Höhenunterschied *m*	useful head, ~ fall
Oberflächendichtung *f* → Dichtungsvorlage *f*	
oberwasserseitig, bergseitig, oberstrom	upstream
oberwasserseitige Vorlage *f*	upstream apron
Perimetralfuge *f*, perimetrische Fuge, Umfangsfuge	perimetral joint
Pfeiler(stau)mauer *f*, Pfeilersperre *f* aufgelöste (Stau)Mauer, Mauer in aufgelöster Bauweise *f*, aufgelöste Gewichtssperre	buttress (type) dam, buttressed dam (US); counterfort (type) dam (Brit.)
Bogenpfeiler(stau)mauer *f*, Gewölbereihen(stau)mauer, Vielfachbogensperre *f*, Pfeilergewölbe(stau)mauer	multiple arch(ed) dam, multi-arch ~
Abstand *m* der Strebepfeilerachen *fpl*	buttress spacing (US); spacing of counterforts c. to c.
Bogenanfang *m* des Gewölbeinnern	intrados springing line (US); springing of intrados (or soffit) (Brit.)
Bogenanfang *m* des Gewölberückens	extrados springing line (US)
Erzeugende *f*, erzeugende Linie *f*	generator (US); generating line, generatrix (Brit.)
lichter Strebepfeilerabstand *m*, Spannweite *f* des Bogens	clear buttress spacing (US); ~ spacing of counterforts, arch span (Brit.)
schiefliegendes Gewölbe *n*	arch barrel (US); inclined barrel arch (Brit.)
Zellen(stau)mauer *f*	cellular dam
Rundkopf(stau)mauer *f* von Noetzli, Pfeilerkopfsperre *f*, Pfeilerkopf(stau)mauer, aufgelöste Gewichtssperre System Noetzli	dam with segmental-headed counterforts (Brit.); round-head(ed) buttress dam

Plattensperre f, Plattenpfeiler(stau)mauer f, Pfeiler-Plattensperre, Pfeilerplatten(stau)mauer	slab and buttress dam, flat slab ∼
Pfeilerkuppel(stau)mauer f, Vielfach-Kuppelsperre f	multiple-dome dam
Pfeiler(stau)mauer f mit gelenkiger Aussteifung	articulated buttress dam
Pfeiler(stau)mauer f mit starrer Aussteifung	rigid buttress dam
Pfeiler(stau)mauer f mit diamantkopfähnlichen Strebepfeilern	diamond-head buttress dam
Überfallpfeiler m	overflow buttress
Pumpspeicher(kraft)werk n	pumped-storage hydropower plant
Pumpenspeicherung f	pumped storage
Rechen m	(trash) rack (US); screen (rack), screen grillage (Brit.)
Rechenbauwerk n	(trash) rack structure
Rechenquerträger m	(trash) rack beam
Rechenreiniger m	screen cleaner (Brit.); (trash) rack rake (US)
Rechenstab m	trash bar
Regulierungsbauwerk n	control works, ∼ structure, headworks (structure)
Ringüberlauf m	shaft-and-tunnel spillway, morning-glory ∼, drop-inlet ∼
Ringüberlaufkrone f	circular spillway crest
Ringverschluß m	ring gate
Rohfallhöhe f	gross (available) head (or fall)
Rollschütz(e) n, (f)	wheel-mounted gate, fixed-wheel ∼, fixed axle ∼
Rollschütz(e) n, (f) mit endloser Rollenkette f	roller-mounted leaf gate (or sluice), coaster gate
Rückhaltebecken n	detention basin, ∼ reservoir [*Flood-control reservoir provided with outlet control*]
rückwertige Schutzsperre f	downstream cofferdam, ∼ temporary dam
Rundkopf(stau)mauer f von Noetzli → aufgelöste Gewichtssperre f System Nötzli	
Sammelbecken n, Staubecken, (Speicher)Becken, Speicher(see) m, Stausee, Talsperrenbecken, hydraulischer Speicher m, künstlicher See	storage reservoir, (impounding) ∼, artificial lake, storage pool, accumulation lake
Saugheberüberfall m → Heberüberfall	
Schalen(stau)mauer f	shell dam, thin arch ∼
Schalthof m, Umspannwerk n, Freiluftschaltanlage f	switchyard, outdoor switchgear

Talsperre

Schieber m	valve
Schieberkammer f; Apparatenkammer [*Schweiz*]	valve chamber
Schieberschacht m	valve tower
Schluchten(stau)mauer f	gorge dam
Schluckfähigkeit f [*Turbine*]	demand
schmale Kuppelbogensperre f mit waag- und senkrechter Krümmung	dome-shaped thin shell dam with the upstream face having pronounced curvature in both horizontal and vertical planes
Schrägschacht m	inclined shaft
Schürze f → Dichtungsvorlage f	
Schuß(ge)rinne f, (n)	race
Schußstrahl m → Überfallstrahl	
Schüttmaterial n, Schüttgut n, Schüttstoff m	fill material, material composing the embankment
Schützenbedienungskammer f	gate operating chamber
Schützensteuerorgan n	gate operating control
Schwallschacht m, Wasserschloßschacht, Schwallraum m	surge chamber, ∼ shaft
(Schwer)Gewichts(stau)mauer f mit dreieckigem Querschnitt	gravity dam of triangular section
Krone f	coping, crown (Brit.); crest (US)
Luftseite f	downstream face, air-side ∼
mittleres Drittel n	middle third
resultierende Kraft f	resultant of all forces
Spitze f des theoretischen Dreiecks	theoretical apex of the triangle (Brit.); intersection of upstream and downstream faces
Stauhöhe f, Stauziel n	top water level (Brit.); designing ∼ ∼
Wasserseite f	upstream face, water-side ∼
Wellenschlag m	wave action
Skisprung-Überlauf m, Sprungschanze f	ski jump spillway
Spannbeton-Staumauer f, vorgespannte Staumauer	prestressed concrete dam
Speicherablaß m	reservoir outflow
(Speicher)Becken n → Sammelbecken	
Speichersperre f	storage dam
Speicherung f → Stau m	
Speicherung f zur Energieerzeugung	power storage
Sperrenkörper m, Tal∼, Stauwerk n	(hydro) dam body
Sperrmauer f → (Massiv)Beton(stau)mauer	
Sperrstelle f, Talsperrenbaustelle	dam-site
Spiegelschwankung f	variation in storage level
Sporn m → Herdmauer f	

Talsperre

Sprungschanze *f*, Skisprung-Überlauf *m*	ski jump spillway
Spüldamm *m* → → Erd(schüttungsstau)damm	
Spülschütz(e) *n*, (*f*)	flushing gate (US); wash-out ∼ (Brit.); scour valve
Stahlbeton-Kernmauer *f*	reinforced concrete core wall
Stau *m* → (Auf)Speicherung	
Stauanlage *f*, Staustufe *f*	barrage
Staubecken *n* → Sammelbecken	
Staudamm *m*, Talsperrendamm, Damm	embankment type dam, fill ∼
Staudamm *m* → Erd(schüttungs)-(stau)damm	
Staudamm *m* → Steinfülldamm	
(Stau)Dammbaustelle *f* → Dammbaustelle	
(Stau)Dammkörper *m*	fill-dam body
Staugrenze *f*	storage limit, pondage ∼, filling ∼
Stauinhalt *m*, Fassungsvermögen *n*	capacity
Staumauer *f* → Massivbetonmauer	
Staudamm-Staumauer-Kombination *f*	combination type dam
Stausee *m* → Sammelbecken *n*	
Staustufe *f*, Stauanlage *f*	barrage
Staustufe *f*	fall
Stauwerk *n* → Sperrenkörper *m*	
Stauwerkkrone *f*, (Tal)Sperrenkrone	dam crest, hydro ∼ ∼
Stauwerkteil *m*, *n* neben der Überfallstrecke *f*	abutment section [*adjoining the (river) overflow (spillway) section*]
Stauziel *n*, Stauspiegel *m*	height of the maximum storage; level of storage water (Brit.); capacity level, storage level (US)
Steigrohrwasserschloß *n*, Differentialwasserschloß	differential surge tank
Steinfülldamm *m*, Damm aus Felsschüttung, Felsschüttungsstaudamm, Steinschüttdamm, Staudamm, Steingerüstdamm, Steindamm	rock fill dam
geschütteter Steindamm	dumped rock fill dam
Steinschüttdamm mit wasserseitiger Betondichtungsdecke	rock fill dam with concrete diaphragm on upstream face
Steinschüttdamm mit senkrechtem Erddichtungskern	rock fill dam with vertical earth core
gerüttelter Steindamm	vibrated rock fill dam
Mauerwerksperre *f*, Trockenmauerwerkdamm *m*, Steinsatzsperre, gepackter Steindamm	dry masonry dam, rubble ∼ ∼, dry rubble ∼, derrick-and handplaced stone rock fill ∼

Stein-Ton-Kern m	stone-clay core
Störkörper m, Energievernichter m	baffle, energy dissipator
Störpfeiler m → Energievernichtungspfeiler	
Sturzbett n	apron
Stützkörper m [*Staudamm*]	supporting shell
talseitig, unterwasserseitig, unterstrom	downstream
Talsperrenbaustelle f, Sperrstelle	dam-site
Talsperrenbauvertrag m	dam-building contract
Talsperrenbecken n → Sammelbecken	
Talsperrenbeton-Aufbereitungsanlage f	large dam (concrete) plant
Talsperrendamm m → Staudamm	
Talsperrenmauer f → Massivbetonmauer	
Talsperrenwärter m	dam warden
Ton(dichtungs)kern m	clay (puddle) core
Ton(dichtungs)schürze f	clay blanket
Tonkerndamm m	clay core type embankment (or dam)
Tonschlag m, Lehmschlag, Puddle m	puddle clay
Tosbecken n → Beruhigungsbecken	
Trennkeil m	energy dissipating wedge [*e. g. at the end of a diversion tunnel*]
Trennmauer → Herdmauer	
Triebwasserleitung f, Druckleitung	penstock
Triebwasserstollen m → Betriebswasserstollen	
Trinkwasserspeicher m, Trinkwasserbecken n	potable water reservoir, drinking ~ ~
Trockenmauerwerkdamm → → Steinfülldamm	
Turmeinlaß m → Einlaßturm	
Überfall m [*Abflußvorgang beim Überfließen des Wassers über ein Wehr*]	overflow
Überfallaufschlagbecken n → Überfall-Sturzbecken	
überfallfreie Strecke f	nonoverflow section
Überfallkrone f, Überlaufkrone	overflow crest, spillway ~
Überfallstrahl m, überfließender Strahl, Schutzstrahl, abfallender Strahl	(overflowing) sheet of water, nappe
Überfallstrecke f, Überlaufstrecke	(river) overflow (spill)way section, spillway section
Überfall-Sturzbecken n, Überfallaufschlagbecken	spillway bucket, roller-bucket type energy dissipator

Überfallwasser *n*	overflow water
Überfallwehr *n*, Überlaufwehr	overflow weir, overfall ~, spillway ~, waste ~, crest-control ~
Übergangszone *f*	transition zone
Überjahresspeicherbecken *n*	conservation (storage) reservoir
Überjahresspeicherung *f*	conservation storage
Überlauf *m*, Hochwasserüberlauf, Hochwasserentlastungsanlage *f*, Überlaufbauwerk *n*, Entlastungsüberlauf	spillway
Überlaufabführungsanlage *f*	discharge carrier [*is either an open channel or tunnel*]
Überlauf-Auslaufbauwerk *n*, Hochwasser-~	spillway outlet structure
Überlaufbauwerk *n* → Überlauf *m*	
Überlaufen *n*	overtopping
Überlaufkanal *m*, Hochwasserentlastungskanal	spillway channel, ~ chute, overflow ~
Überlaufoberfläche *f*	spillway surface
Überlaufstollen *m*, Hochwasserstollen	tunnel-type discharge carrier
Überlaufstrecke *f* → Überfallstrecke	
Überlauf *m* über die Stauwerkkrone	overflow(-type) spillway
Überlaufverschluß *m*, Hochwasser-~	spillway gate
Überlaufwehr *n* → Überfallwehr	
Überfallkraftwerk *n*	surface power plant
Umfangsfuge *f* → Perimetralfuge	
Umlaufstollen *m*, Umleit(ungs)stollen	diversion tunnel, by-pass ~
Umleit(ungs)kanal *m*	bye-wash, diversion cut, bye-channel, spillway; diversion flume (US)
Umleit(ungs)sperre *f* zur Bewässerung	diversion dam
Umleit(ungs)stollen *m*, Umlaufstollen	by-pass tunnel, diversion ~
Umspannwerk *n*, Freiluftschaltanlage *f*, Schalthof *m*	switchyard, outdoor switchgear
Unterwasseranstieg *m*	tailwater rise
unterwasserseitig, unterstrom, talseitig	downstream
Unterwassernotverschluß *m*	tailgate
unterwasserseitige Platte *f* mit (verkehrt)steigendem Karnies	ogee downstream slab, cyma ~ ~
unterwasserseitige Vorlage *f*	downstream apron
Unterwasserstollen *m*	tail (race) tunnel
unverschlossener Überlauf *m*	uncontrolled crest spillway

Verbindungsstollen *m* der Schützenbedienungskammern *fpl*	operating gallery
Verbundbauwerk *n* → Mehrzweckbauwerk	
Verherdung *f*, (Ab)Dichtungsgraben *m*	cutoff trench
Verlandebecken *n*	silt basin
verpressen, auspressen, einpressen, injizieren	to inject, to grout under pressure
Verteilerstollen *m*	penstock manifold, manifold tunnel
verschlossener Überlauf *m*	gated spillway, gate-controlled ∼, controlled crest ∼
Verzögerungsbecken *n*	(floodwater) retarding reservoir, flood detention ∼, flood-prevention ∼ [*Flood-control reservoir provided with uncontrolled outlets*]
Vielfachbogensperre *f* → → Pfeiler(stau)mauer	
Vielfachkuppelsperre *f* → → Pfeiler(stau)mauer	
Vielzweckbauwerk *n* → Mehrzweckbauwerk	
vorgespannte Staumauer *f*, Spannbeton-Staumauer	prestressed concrete dam
Vorsperre *f*	upstream cofferdam, ∼ temporary dam
Walzdamm *m* → → Erd(schüttungsstau)damm	
Walzdamm *m*, Walzstaudamm, (ab-)gewalzter Damm	rolled fill dam
Walzkern *m*	rolled core
(Wasser)Einstau *m* → (Auf)Speicherung *f*	
Wasserfassung *f*	water inlet (US); ∼ intake (Brit.)
Wasserfläche *f*	water-spread
(Wasser)Kraftstollen *m* → Betriebswasserstollen	
Wasserkraft-Talsperre *f*	hydroelectric dam
Wasserschloß *n*	surge tank
Wasserschloßschacht *m* → Schwallschacht	
Wasserseite *f*	upstream face
wasserseitige Betondichtungsdecke *f*	impervious concrete diaphragm on the upstream face
wasserseitige Böschung *f*	upstream slope, water ∼
Wasserumlenkung *f*	river diversion
Wehr *n* (oder Überlauf *m*) mit (verkehrt)steigendem Karnies	ogee weir (or spillway), cyma ∼ (∼ ∼), ogee(-shaped) dam, cyma (-shaped) dam

Talsperre — Tankerbauauftrag

Widerlagerkissen *n*
Zulaufkanal *m* → Einlaufkanal
Zuleitungsstollen *m*

saddle

feed gallery

Talsperre *f* **zur Energieerzeugung** = power dam
Talsperren|bruch *m* = dam failure
~damm *m* → Staudamm
~form *f* = shape of dam
~geologie *f* = dam geology
~gewicht *n* = weight of dam
~kraftwerk *n* = dam power plant
~mauer *f* = dam wall
Talstation *f* [*Bergbahn*] = valley station
Talterrasse *f* = valley terrace
Talweg *m* (Geol.) [*Verbindungslinie der tiefsten Punkte des Flußlaufes*] = thalweg, valley way, valley line [*The name frequently used for the longitudinal profile of the river, i. e. from source to mouth*]
Tambour *m* = tambour [*A circular wall carrying a dome*]
Tandem|achse *f* = tandem axle
~antrieb *m* = tandem drive

~antriebgehäuse *n* = tandem drive housing
~antriebkette *f* = tandem drive chain
tandemgetrieben = tandem-driven
Tandem|-Hinterachse *f* = tandem rear axle
~-Kettenantrieb *m* = tandem chain drive
~motor *m* = tandem engine
~reiß- und Schubblock *m* = combination tandem ripping and tandem pushing block
~-Rüttelwalze *f* → → Walze
~schlepper *m*, **Tandemtraktor** *m*, **Tandemtrecker** *m* = tandem tractor
~schub *m* = tandem push
~schubblock *m* **für Aufreißer** = tandem ripping push block
~schubladen *n* [*Radschrapper*] = tandem push loading
~-Vibrationswalze *f* → → Walze
~walze *f* = tandem roller

Tang *m*
  Meeresalge *f*
  Laminarie *f*, Riementang
  Sargossatang, Beerentang
  Blasentang, Fukus *m*

seaweed
  marine alga
  laminaria
  sargassum
  bladder wrack, fucus vesiculosus

Tangente *f* = tangent
Tangenteneinrückung *f* = tangent distance
tangentialer Zustrom *m* [*im Fangkessel beim Druckluftbetonförderer*] = division of the flow of concrete into two streams by central baffle inside the discharge box which are reunited at the far end of the box
Tangential|komponente *f* = tangential component
~kraft *f* = tangential force
~lager *n* [*für Hoch- und Tiefbauten*] = tangential bearing

Tank|anhänger *m*, **Behälter-Anhänger** = tank trailer
~anlage *f* → Tankstelle *f*
~boden *m* = tank bottom
~bodenwachs *n* = tank wax
~decke *f*, **(Behälter)Decke** = (tank) roof
~einfüllstutzen *m* = tank filler
Tanker *m*, **Tankschiff** *n* = tanker (ship)
~bau *m* = tanker building, ~ construction
~bauauftrag *m* = tanker order

Tank|schiff *n*, Tanker *m* = tanker (ship)
~spritzmaschine *f* → Bitumen-Sprengwagen *m*
~stand *m* = storage tank level
~standmessung *f* = tank gauging (Brit.); ~ gaging (US)
~stelle *f*, Tankanlage *f* = filling point, ~ station, service station; gas(oline) (or gasolene) station (US)
~wagen *m* [*zum Betanken von Flugzeugen usw.*] = fuel dispenser
~wart *m* = refuel(l)er
Tanz|fläche *f*, Tanzboden *m* = dance floor
~halle *f* = dance hall
Tapete *f* = wallpaper
Tapeten|schneider *m* = wallpaper trimmer
~tür *f* = secret door,
tapezieren = to hang (wall)paper
Tapezieren *n* = (wall)paperhanging, wall-papering
Tapezierer *m* = (wall)paperhanger
tapeziert = papered
Taragewicht *n*, Leergewicht *m* = tare
Tarier|balken *m* = tare beam
~waage *f* = tare balance
Tarn|anstrich *m* = camouflage coat(ing)
~farbe *f* = camouflage paint
~lack *m* = camouflage lacquer
Tasche *f*, Abteilung *f* [*Dosierapparat*] = compartment
Taschen|bunker *m* → Taschensilo *m*
~kolben *m* = pocket piston
~lampe *f* = flashlight (US); torch (Brit.)
~silo *m*, Wabensilo, Taschenzuteiler *m*, Wabenzuteiler, Taschenbunker *m*, Wabenbunker [*Beton-Dosieranlage*] = compartment bin, ~ storage hopper
~zuteiler *m* → Taschensilo *m*
Tast-Dehnungsmesser *m* = feeling elongation meter
Taster *m*, Fühler *m*, Höhen~ [*Gleitschalungsfertiger*] = grade sensor, sensing device
Tastrad *n* → Lenkrad

Tato *f*, Tagestonnen *fpl* = tons per day
tatsächliche Aufgabe *f* [*Siebtechnik*] = actual feed
~ Flüssigkeit *f*, reale ~ = actual liquid
~ Kohäsion *f*, wahre ~, wirkliche ~ = true cohesion
~ Kornscheide *f* [*Siebtechnik*] = effective screen cut-point
Tatsche *f* → Schlagbrett
Tau *m* → Frostaufgang *m*
~ [*Bildet sich durch Abkühlung des Wasserdampfes der Luft an erkalteten Gegenständen*] = dew
taube Flut *f*, ~ Tide *f*, Nippflut, Nipptide = dead tide, neap ~
tauber Kontaktgang *m* = barren contact
taubes Gestein *n* → Berg *m*
Tauch|-Auftrag-Schweißverfahren *n* [*Schutzgasschweißen ohne Spritzverluste*] = dip-transfer metal-arc welding method
~badschmierung *f* = spray lubrication
~brenner *m* = immersion gas burner
~brett *n*, Schwimmstoffabweiser *m* = scum board [*A vertical baffle dipping below the surface of sewage in a tank to prevent the passage of floating matter*]
~brücke *f*, überflutbare Brücke = submersible bridge, low level ~, causeway
~buhne *f*, Grundbuhne [*Vom Ufer ausgehende Grundschwelle*] = ground sill rooted in the bank
~druckversuch *m* [*Verformbarkeit bituminöser Straßenbaugemische*] = immersion-compression test
~elektrode *f* = dipped electrode
Tauchen *n* [*Werkstoffbehandlung*] = immersion (treatment)
Taucher *m* = diver
~anzug *m* = diving dress, ~ suit
~arbeit(en) *f(pl)* = diver's work
~glocke *f*, bewegliche Arbeitskammer *f*, wiedergewonnene Arbeitskammer = diving bell

~**glockengründung** *f* = diving bell foundation
Taucherhitzer *m* = immersion heater
~**automat** *m* = auto(matic) immersion heater
Taucherkrankheit *f* = diver's palsy, ~ paralysis, air embolism
Tauchgefäß *n* = immersion vessel
Tauchkolben *m*, Plunger(kolben) *m* = pump plunger
~**membran(e)pumpe** *f* = plunger diaphragm pump, ~ membrane ~
~**öse** *f* = plunger eye
~**pumpe** *f*, Verdrängerpumpe, Plungerpumpe = plunger pump, ram ~, displacement ~
~**weg** *m* = plunger travel
Tauch|konsistenz *f* [*Farbe*] = dipping consistency
~**körper** *m* → belüfteter ~
~**löten** *n* = dip brazing
~**patentieren** *n* = batch patenting
~**pumpe** *f*, Unterwassermotorpumpe, Tauchmotorpumpe = submersible pump [*Motor sealed against the entry of water*]
~**rohr** *n* = immersion pipe
~**rüttler** *m* → Innenrüttler
~**rüttler-Betonstraßen-Fertiger** *m*, Betondeckeninnenrüttler(-Fertiger), Betondeckeninnenvibrator(-Fertiger) = full depth internal concrete pavement (or paving/US) vibrator, ~ ~ ~ slab vibratior; paving vibrator (US); internal vibrating machine (Brit.)
~**rüttlung** *f*, Innenrüttlung = immersion vibration
~**schleuse** *f* **Bauweise Böhmler** = float barge lift type Böhmler
~**schleuse** *f* **Bauweise Rowley** = trough lift type Rowley
~**schmierung** *f*, Eintauchschmierung = flood lubrication
~**-Turbinenpumpe** *f* = submersible turbine pump
~**verdampfer** *m* = submerged evaporator
~**vibrator** *m* → Innenrüttler *m*

~**wägung** *f* = immersion weighing
~**wanne** *f* **für Fugeneisen** = immersion kettle for joint-forming metal strips
Tau-Frost-Prüfung *f*, Frost-Tau-Prüfung, Gefrier-Auftau-Prüfung = freezing and thawing test
Tau-Frostwechsel *m* → Gefrier- und Auftauzyklus *m*
Taukreuz *n* [*nach dem griechischen Buchstaben Tau genannt*], Antoniuskreuz, ägyptisches Kreuz = prechristian cross
Taulava *f* → Stricklava
Taumelmischer *m* = eccentric tumbling mixer, offset ~ ~
Taumelscheiben|getriebe *n* = swashplate mechanism
~**pumpe** *f* = axial cylinder swashplate pump
Tauperiode *f* = frost-melt period
Taupunkt *m* = dew point, condensation ~
~**temperatur** *f* = dew-point temperature
Tausalz *n* → (Auf)Tausalz
~**wasser** *n* = water molten by deicing agent(s)
Tauschaden *m* = damage due to thawing
Tausenkung *f* [*Straßendecke*] = settlement due to thawing out of frost
Tauwasser *n* → Kondenswasser
Tauwerkschlinge *f* = cordage sling
Taylor'scher Satz *m* = Taylor's series
T-Bau *m* [*fälschlich auch manchmal zweiflügeliger Streb genannt*] (✵) = T-support
Technik *f* [*als betriebstechnische Anwendung*] = technique, practice
~, Ingenieurwesen *n*, Ingenieurwissenschaft *f* = engineering (science)
technisch, betriebstechnisch = technical, practical
technische Daten *f* = specification(s), spec(s)
~ **Lebensdauer** *f* [*Maschine*] = physical life
~ **Mechanik** *f* = technical mechanics

Technische Nothilfe *f* = emergency management (US)

technische Reaktionsführung *f* **beim Ausarbeiten neuer Prozesse und ihrer Entwicklung zum produktionsreifen Verfahren** = chemical reaction engineering in process research and process development

~ **Richtlinien** *fpl*, ~ **Vorschriften** *fpl* = technical specifications, ~ specs

~ **Strömungslehre** *f* = theory of fluid flow

~ **Verbesserungen** *fpl* = engineering improvements

(~) **Versorgungsaggregate** *npl* [*in einem Wohnblock*] = utility equipment

~ **Vorschriften** *fpl* → ~ Richtlinien

Technischer Ausschuß *m* **des Deutschen Verbandes für Schweißtechnik** = Technical Sub-Committee of the German Welding Society

technischer Beratungsdienst *m* = engineering consultation (or counsel) service

~ **Betriebsversuch** *m* = plant test

~ **Kaufmann** *m* = technical sales executive

~ **Marmor** *m*, Marmorkalkstein *m*, polierbarer dichter Kalkstein = compact polishable limestone

~ **Staub** *m*, Industriestaub = industrial dust

~ **(Straßen)Überwachungstrupp** *m* = technical (road) patrol

technisches Gas *n* = heating gas

~ **Hilfspersonal** *n* = preprofessional grades (US)

Technologe *m* = technologist

Technologie *f* **der Feuerfeststoffe** = refractory technology

technologische Eigenschaft *f* [*Werkstoff*] = technological property

Teefabrik *f* = tea factory

Teeküche *f* = tea kitchen

Teer *m*
Absatz~, Absatz~
alterungsbeständiger Straßen~, Wetter~
Anthrazenöl~
Archangel~
aromatisch
Asphalt~
Bagasse~
Baumrinden~
Bernstein~
Bienenwachs~
Birkenholz~
Bitumenschiefer~, (Öl)Schiefer~
Blaugas~
Braunkohlenheizgenerator~
Braunkohlenkokerei~
Braunkohlenschwel~
Braunkohlen~
Braunkohlenur~
Buchen(holz)~
Coalite~
Del Monte~
destillierter ~
Fettgas~ → Ölgas~

tar
settled ~
weather-resistant ~

anthracene-oil ~
Archangel ~
aromatic
asphalt ~
bagasse ~
wood bark ~
amber ~
beeswax ~
birch wood ~
shale ~
blue gas ~
reducer gas brown coal ~
coke-oven brown coal ~
retort browncoal ~
brown-coal ~
low-temperature brown coal ~
beech (wood) ~
coalite ~
Delmonte ~
distilled ~

Fuselölgas~	fusel-oil ~
Gas(werks)~, Gasanstalts~	gas works coal ~, gas-house coal ~
Gummi~	rubberized ~
Hobelspäne~	wood shavings ~
Hochofen~	blast-furnace coal ~
hochviskoser ~	high-viscosity ~
Holzabfall~	wood waste ~
Holzsplitter~	wood splittings ~
Horizontal-Retorten~	horizontal-retort ~
Karbokohlen~	carbocoal ~
Kerzen~	candle ~
Kien~, Nadelholz~	pine ~
Knochen~	bone ~
Kohlenwassergas~	dehydrated water-gas ~
Kokerei~, Zechen~, Koksofen~, Steinkohlenkokerei~	coke oven coal ~
Kork~	cark ~
Laubholz~	hardwood ~
Leder~	leather ~
Lignin~	lignin ~
Lignit~	lignite ~
Linoleum~	linoleum ~
Mond~ [nach dem Erfinder Ludwig Mond]	Mond ~
Montanwachs~	montan-wax ~
Nadelholz~, Kien~	pine ~
niedrigviskoser ~	low-viscosity ~
Ölgas~, Fettgas~	oil gas ~
Ölschiefer~ → Bitumenschiefer~	
Ölwassergas~	carburetted water gas ~, oil-water-gas ~, fuel-oil gas ~, reformed-gas ~
präparierter ~	road ~ produced by blending with distillate fractions
Roh~	crude ~ [deprecated: green ~]
Rolle-Ofen~	Rolle-retort browncoal ~
Sägemehl~	sawdust ~
Schiefer~ → Bitumenschiefer~	
schottischer (Öl)Schiefer~, schottischer Bitumenschiefer~	Scottish shale ~
schwedischer (Schiffs)~ → Stockholmer ~	
Schwelretorten~	retort ~
Steinkohlen~	(bituminous) coal ~
Steinkohlen(heiz)generator~	producer-gas coal ~, gas-producer coal ~
Steinkohlenkokerei~ → Kokerei~	
Steinkohlen(teer)pech~	coal-tar pitch

Stockholmer ~, schwedischer (Schiffs)~ — Stockholm ~
Strohstoff~ — straw ~
Sulfitstoffablauge~ — sulfite cellulose ~, tall oil ~
Tabak~ — tobacco ~
Teerausbeute f — yield of ~
Torf-Retortenschwel~ — retort peat ~
Torf-Schwelgenerator~ — producer peat ~
Torf~ — peat ~
Torfur~ — low-temperature peat ~
Vertikal-Retorten~ — vertical-retort ~
Vinasse~ — vinasse ~
Wassergas~ — water gas~
Wetter~, alterungsbeständiger Straßen~ — weather-resistant ~
Zechen~ → Kokerei~

Teerbelag m, Teerdecke f = tar paving (US); ~ pavement, ~ surfacing
Teerbeton m = tar concrete
~**belag** m → Teerbetondecke f
~**decke** f, Teerbetonbelag m = dense tar surfacing, tar concrete pavement, tar concrete surfacing; tar concrete paving (US)
~ **-und Walzasphaltmischanlage** f → Walzasphaltmischanlage
Teerbitumen|emulsion f = emulsion of tar/(asphaltic) bitumen mix(ture) (Brit.); ~ ~ tar/asphalt ~ (US)
~**-Gemisch** n, Teer-Bitumen-Mischung f, VT-(Straßen)Teer m, Bitumenteer = pitch/(asphaltic) bitumen mix(ture) (Brit.); tar/asphalt ~ (US)
Teer|(dach)pappe f = tar rag felt
~**decke** f, Teerbelag m = tar pavement, ~ surfacing; ~ paving (US)
~**destillation** f = distillation of tar, refining ~ ~
~**destillationsanlage** f = tar distillation plant
~**eimer** m = tar pail
~**emulsion** f = tar emulsion, emulsion of tar
Teeren n, Teerung f = tarring, tarspraying
Teer|erzeugnis n = tar product

~**farbe** f = coal tar dye
~**faß** n = tar barrel
~**feinbeton** m = fine tar concrete
teergebunden = tarviated, tar-bound
Teer|grobbeton m = coarse tar concrete
~**-Kiespreßpappe** f = tar-gravel roofing
~**kocher** m = tar kettle
~**lösung** f = tar solution
~**magnesit** m = tar-bonded magnesite
Teermakadam m = tar(red) macadam
~**decke** f, Teermakadambelag m = tar(ed) macadam surfacing, ~ ~ pavement; ~ ~ paving (US)
~**(misch)anlage** f = tarmacadam plant
~**mischer** m = tarmacadam mixer
~**straße** f = tarred macadam road, ~ highway
~**teppich** m = tarmacadam carpet, coated-macadam carpet made with tar
Teer|mischanlage f = tar mixing plant
~**mischmakadam** m = plant-mixed tarmacadam
~**papier** n = tar paper
~**pappe** f, Teerdachpappe = tar rag felt
~**pech** n = pitch
~**pechemulsion** f = pitch emulsion
~**sand** m = tar sand

Teerschlacke — Teilstrahlungspyrometer

~schlacke *f* [*aus Hochofenschlacke*] = tar-coated slag
~schlämme *f* = tar slurry
~schlauch *m* = tar hose, ~ flexible conduit
~schöpfer *m* = tar dipping ladle
~schotter *m* = tarred stone
~schutzanstrich *m* = tar protective coat(ing)
~splitt *m* = chip(ping)s precoated with tar, tar-coated chip(ping)s, tarred chip(ping)s
~splitt-Teppich *m* = tar-coated chip(ping)s carpet
~spritzarbeiter *m* = tar sprayer [*One of a team of skilled men who spray tar on a road*]
~spritzgerät *n*, Teerspritzmaschine *f* = tarsprayer
~stabilisierung *f* = tar stabilization
~strick *m* = tarred cord
~teppich *m* = tar carpet
~tränkmakadam *m* [*früher: Teeraufguß-Beschotterung f*] = tar-grouted stone (Brit.); tar penetration macadam, road tar type penetration macadam
~überzug *m* = tar surface treatment
~- und Bitumenkocher *m* → Kocher
~- und Bitumenschmelzkessel *m* → Kocher
~- und Bitumen-Spritzmaschine *f* = tar and asphalt sprayer (or spraying machine) (US); ~ ~ asphaltic bitumen ~ (~ ~ ~) (Brit.)
Teerung *f*, **Teeren** *n* = tarspraying, tarring
Teerwerg *n* = tarred oakum
Teestube *f* = tea bar
Tegoleim *m*, **Tegofilm** *m* [*genormtes Kurzzeichen: T*] = Tego film
Teich|rohr *n*, Schilfrohr, Schilf *n*, Rohr = (common) reed
~verfahren *n*, Teichmethode *f*, Einsumpfen *n* [*Betonnachbehandlung*] = ponding
Teil *m, n* **einer belasteten Fläche** = element of a loaded area
teilbeaufschlagte Turbine *f* = partial admission turbine

Teil|becherwerk *n*, Schöpfbecherwerk = spaced (or intermittent, or dredger) bucket (type) elevator
~belastung *f* = partial loading
teilbewegliche (Feld)Beregnungsanlage *f* = semi-portable (agricultural) sprinkling system, ~ (~) ~ installation
Teilbrandung *f* = partial breaking of waves
Teilchen|ablagerung *f* = deposition of particles
~behinderung *f* **bei dichtester Packung im Betonzuschlag(stoff)** = particle interference in concrete mixes
~-Einbaugeschwindigkeit *f* [*Kristall*] = particle integration rate
~geschwindigkeit *f* = particle velocity
~wichte *f* = density of particle
Teil|dränung *f*, Teildränage *f*, Bedarfsdränung, Bedarfsdränage = partial drainage
~einspannung *f* = partial sheeting restraint
~einspundung *f* = partial sheeting
~(einwirkungs)fläche *f* (⚒) = sub-critical area of extraction
Teilfläche *f*, Teileinwirkungsfläche (⚒) = sub-critical area of extraction
~ **des Querschnitts** [*Abflußmengenmessung mit Meßflügel*] = partial area of the cross section
Teil|kosten *f* = component cost
~kreis *m* = pitch circle
~kreislinie *f* = pitch line
~last *f* [*Turbine*] = part load
~linie *f* = parting line
~montage *f* [*Maschine*] = sub assembly
~-Nr. *f* [*in einer Ersatzteilliste*] = part number
~rechteck *n* = subsidiary rectangle
~sättigung *f* = partial saturation
~senkung *f* (⚒) = partial subsidence
~skelettplatte *f* [*Gleiskettenlaufwerk*] = semi-skeleton shoe
~strahlungspyrometer *n, m* = optical pyrometer

(∼)Strecke *f*, Teilstück *n*, Abschnitt *m* = stretch, section
∼strich *m* [*Waage*] = graduation mark
∼strichwaage *f* = graduated scale
∼stück *n*, Abschnitt *m*, (Teil)Strecke *f* = stretch, section
Teilung *f* → Niet∼
∼ [*Scharnier*] = joint section
∼, Maßeinteilung [*z. B. bei einem Skalenpegel*] = graduation(s)
∼ **von Drahtmitte zu Drahtmitte** [*Siebtechnik*] = pitch, centre to centre of wires
Teilungs|korngröße *f*, Trennkorngröße T = separation size [*Effective separating size of a screen using the Tromp method*]
∼kurve *f*, Verteilungszahlenkurve [*Siebtechnik*] = Tromp curve, partition ∼
∼linie *f* → Lauflinie [*Treppe*]
∼zahl *f* → Verteilungszahl
Teil|vakuum *n* = partial vacuum
∼verglasung *f*, Klinkerung = partial vitrification
∼vorfertigung *f* = partial prefabrication
teilweise Betonummantelung *f* [*Rohr*] = (continuous monolithic concrete) cradle
∼ **Einspannung** *f* = partial restraint
∼ **vorgespannt**, beschränkt ∼ = partially prestressed
∼ **Vorspannung** *f* = partial prestressing
TEKA-Schrapper *m* → Straffseil-Kabelbagger
Tektogenese *f* → Gebirgsbildung *f*
tektonisch = tectonic
tektonischer und topographischer Sattel *m* (Geol.) = geoanticline
tektonisches Beben *n*, Dislokationsbeben = tectonic earthquake
Tektor *m* → Betonspritzmaschine *f*
Telamon *m* → Atlant *m*
Telefonzelle *f*, Telefonhäuschen *n* = telephone kiosk
Teleskop|arm *m*, Auszieharm = telescoping arm

∼-Ausleger *m*, Teleskop-Baggerausleger = telescoping (excavator) boom (US); ∼ (∼) jib (Brit.)
∼bagger *m* = telescoping excavator
∼-(Decken)rüstung *f* = telescopic (floor shuttering) center(s) (US); ∼ (∼ ∼) centre(s) (Brit.); ∼ (∼ ∼) cent(e)ring
∼deichsel *f* → Ausziehdeichsel
∼drehleiter *f* = telescopic revolving ladder
∼-Führungswagen *m* [*Brückenbau*] = telescopic guiding carriage
∼gelenkwelle *f* = telescopic multiple-part shaft, ∼ universal joint ∼
∼-(Gleis)Kettenbagger *m*, Teleskop-Raupenbagger = telescoping crawler excavator
∼-Hubwerk *n* = telescopic lifting mechanism
teleskopierbar, ausziehbar = telescopic
teleskopieren, ausziehen = to telescope
Teleskopierzylinder *m* = boom crowd ram (US) [*hydraulic crane*]
Teleskop|-Kettenbagger *m*, Teleskop-Raupenbagger, Teleskop-Gleiskettenbagger = telescoping crawler excavator
∼kran *m* = telescopic crane
∼-Laderampe *f* [*für Personen-Verladung auf Flughäfen*] = telescoping loading ramp
∼-Läuferrute *f*, Teleskop-Mäkler *m*, Teleskop-Laufrute = telescopic leader
∼-Leuchtturm *m* [*Besteht aus einem Beton-Senkkasten und einem senkrecht verschiebbaren Turm-Aufbau*] = telescopic lighthouse
∼-Löffelstiel *m*, Teleskopstiel = telescopic (dipper) stick
∼-Mäkler *m*, Teleskop-Läuferrute *f*, Teleskop-Laufrute = telescopic leader
∼presse *f* = telescoping hoist
∼-Raupenbagger *m*, Teleskop-(Gleis-)Kettenbagger = telescoping crawler excavator

~rohr n = telescoping tube
~rohr n [*Mobilkran*] = shipper
~schalung f, Ausziehschalung = telescoping formwork
~stiel m, Teleskop-Löffelstiel = telescopic (dipper) stick
~stütze f [*Zur Erhöhung der Standsicherheit von Baggern, Kränen und Schaufelladern*] = telescoping stabilizer, ~ outrigger
~stütze f = telescopic prop
~verrohrung f, Ausziehverrohrung = telescoping casing
~zugstange f → Ausziehdeichsel f
Teller|aufgeber m → Tellerspeiser
~beschicker m → Tellerspeiser
~bohrer m = disc auger; disk ~ (US)
~feder f = disc spring; disk ~ (US)
~rad n = bevel gear
~radgehäuse n = bevel gear case
~speiser m, Telleraufgeber, Tellerbeschicker, Rundspeiser, Rundaufgeber, Rundbeschicker = disc feeder, table ~; disk ~ (US)
~streuer m → Rotor-Streugutverteiler m
~ventil n, Pilzventil = poppet valve
~vibrator m, Tellerrüttler m = disc vibrator (Brit.); disk ~ (US)
~-Zwangsmischer m = turbomixer
Tellurometer m, n = tellurometer
Tempel m mit Säulengang → Hypostylos m
Temperatur|änderung f = temperature change
~anstieg-Meßverfahren n = temperature gradient method
~anzeige f = temperature indication
~dehnung f = temperature strain
~einfluß m = temperature influence
~fühler m = temperature senser
~gefälle n, Temperaturunterschied m = temperature gradient, ~ differential
temperaturgeregelt = temperature-controlled
Temperatur|kurve f = temperature curve
~leitfähigkeit f = thermal diffusivity

~messungen fpl [*Geothermik*] = surveying of the (ground) temperatures, temperature survey
~meßkurven-Auswertung f = interpretation of temperature logs
~riß m = crack due to temperature variations
~rückgang m, Temperaturabfall m = temperature drop
~schreiber m, Schreibthermometer n, m = temperature recorder
~schwankung f = temperature variation
~spannung f = stress due to temperature
~spannung f [*Betondecke*] = warping stress
~statistik f = temperature record
~unterschied m, Temperaturgefälle n = temperature differential, ~ gradient
~verlauf m = temperature development
~verteilung f = temperature distribution
~wechselbeanspruchung f = sensitivity to variations in temperature
Tempern n, Warmbehand(e)lung f = tempering, drawing
Tennisplatz m = tennis court
Tentor-Bewehrungsstahl m, Tentor-Armierungsstahl = Tentor (steel) (re-)bars, ~ (~) reinforcing bars
Teppich(belag) m, Straßenteppich m, Mischteppich(belag) = carpet-coat, (pre-mix(ed)) carpet, mat, thin surfacing, road carpet, road mat
Teppichfabrik f = carpet factory
termitenfest = termite treated
Termitenfestigkeit f = resistance to termite attack
Terrakotta f = terra-cotta
~platte f = terra-cotta slab
~stein m = terra-cotta block
Terrasse f = terrace
~ [*Steinbruch*] = bench
Terrassen|dach n, Dachterrasse f = terrace roof
~kies m (Geol.) = terrace gravel
Terrassieren n → Abtreppung f

Terrassierung *f* → Abtreppung
Terrazzo|-Asphaltplatte *f* [*Diese Platte hat eine Terrazzooberschicht und eine Asphalt-Unterschicht*] = terrazzo asphalt tile
~**estrich** *m* = terrazzo (jointless) floor(ing), ~ composition ~
~**fliese** *f*, Terrazzoplatte *f* = terrazzo tile
~**(fuß)boden** *m* = terrazzo floor(ing), ~ floor finish
~**handwerk** *n* = terrazzo trade
~**körnung** *f*, Terrazzomaterial *n* [*Das Wort „Terrazzosplitt" ist wenig gebräuchlich. In der Regel spricht man von „Terrazzokörnungen", die seit alters her meist in einem Kornbereich zwischen 0 und ca. 12 mm liegen. (Unterstufen in etwa 0/1 mm, 1/2, 2/3, 3/4, 4/6, 6/9, 9/12 mm). Es sind dies die Korngrößen, die der Terrazzoleger bei der Herstellung von Böden an Ort und Stelle bevorzugt.
Mit dem seit längeren Jahren zu beobachtenden Vordringen gröberer Körnungen (meist abgestuft in 0/7, 7/15, 15/30 mm und noch größer) hat sich die maschinelle Fertigung von Platten immer mehr durchgesetzt, die eben die Verwendung groben Kornes leichter zuläßt; diese Grobkorn-Arbeiten bezeichnet man üblicherweise als „Mosaik" und die hierzu verwendeten Splitte von weißem und buntem Marmor als „Mosaiksplitte".
Der Unterschied zwischen „Terrazzosplitt" und „Mosaiksplitt" ist streng genommen also eine Frage der Korngröße. Doch hat sich im Sprachgebrauch der Unterschied nicht scharf eingespielt, denn Mosaikplatten auch sehr groben Korns werden vielfach als „Terrazzoplatten" gehandelt und das hierzu verwendete Grobkorn einfach als „Terrazzokorn" oder „Terrazzosplitt" bezeichnet, obwohl „Mosaiksplitt" richtiger wäre*] = terrazzo aggregate
~**mischung** *f* = terrazzo mix(ture)
~**platte** *f*, Terrazzofliese *f* = terrazzo tile
~**plattenmaschine** *f*, Terrazzofliesenmaschine = terrazzo tile machine
~**plattenpresse** *f*, Terrazzofliesenpresse = terrazzo tile press
~**schleifmaschine** *f* = terrazzo grinding machine
~**splitt** *m* → Terrazzokörnung *f*
~**werk** *n*, Terrazzofabrik *f* = terrazzo tile plant, ~ ~ works, ~ ~ factory
terrestrische Photogrammetrie *f*, Erdbildmessung *f* = terrestrial photogrammetry
Territorialgewässer *n* = territorial waters
Testbenzin *n* = white spirit
Tetrachlorkohlenstoff *m*, CCl_4 = carbon tetrachloride
Tetradymit *m*, Schwefeltellurwismut *n*, Bi_2Te_3S (Min.) = tetradymite
Tetrakalziumaluminatferrit *n*, Brownmillerit *m*, $4\,CaO \times Al_2O_3 \times Fe_2O_3$ [*abgek.* C_4FA *oder* C_4AF] = tetracalcium aluminoferrite
Teufe *f* (⚒); Tiefe *f* = depth
teufen, ab~, herunterbringen, niederbringen [*Schacht; Bohrung*] = to drill, to sink, to put down, to bore [*shaft; well*]
Teufenanzeiger *m*, Tiefenanzeiger = depth indicator

Textileinlage *f* = textile reinforcement
textiles (Sieb)Gewebe *n* = textile screen cloth; ~ ~ fabric (US)
Textilfabrik *f* = textile mill
Textur *f* [*räumliche Anordnung der Mineralien sowie die Raumerfüllung*] = texture
texturierte Oberfläche *f* = textured surface
T-förmiger Windrichtungsanzeiger *m* = wind Tee

Theaterbau *m*   theatre construction (Brit.); theater ~ (US)

Theaterbau

Arbeitsgalerie *f*, Arbeitssteg *m*	fly floor, ~ gallery
Asbestvorhang *m*	asbestos curtain
Arenabühne *f*	arena stage
Ausgang *m*	exit
Beleuchterkabine *f*	lighting booth
Beleuchtungsbrücke *f*, Beleuchterbrücke	lighting gallery
Beleuchtungsschaltung *f*	lighting control
Beleuchtungsumstellung *f*	lighting change
Beries(e)lungsanlage *f*	sprinkler system
Bühnenbeleuchtung *f*	limelight (US)
Bühnenbild *n*	scenery, setting
Bühnenhaus *n*	stage and backstage
Bühnenhimmel *m*, Rundhorizont *m*	stage horizon, cyclorama, cyke
Bühnenhöhe *f*	stage level
Bühnenmechanismus *m*	stage mechanism
Bühnenturm *m*	stage shaft, ~ well
Bühnenwagen *m*	stage wagon
Bühnenwerkstatt *f*	theatre workshop
drehbarer Zuschauerraum *m*	revolving auditorium
Drehbühne *f*	revolving stage, turntable ~
Drehstuhl *m*	swivel chair
dritter Rang *m*, Galerie *f* [*volkstümlich: Heuboden m*]	gallery
Eingang *m*	entrance
eiserner Vorhang *m*	safety curtain, fire ~; shade (US)
Erfrischungsraum *m*	snack bar
erster Rang *m*	dress circle; balcony, (diamond) horsehoe (US)
Flutlicht *n*	floodlight
Foyer *n*	foyer
Galerie *f*, dritter Rang *m* [*volkstümlich: Heuboden m*]	gallery
Gang *m* [*im Zuschauerraum*]	aisle
Garderobe *f*, Kleiderablage *f*	coatroom, cloakroom; checkroom (US)
Guckkastenbühne *f*	peephole stage
Halle *f*	entrance foyer
Hinterbühne *f*, Rückbühne	backstage
Kleiderablage *f*, Garderobe *f*, Kleiderabgabe	cloakroom, coatroom; checkroom (US)
Magazin *n*	store
Malerwerkstatt *f*	painter's room
Notausgang *m*	emergency exit
Obermaschinerie *f*	flies
Orchesterversenkung *f*	orchestra pit
Parkett *n*	pit; parquet, parterre (US)
Probesaal *m*	rehearsal room
Prospekt *m*	back-cloth

Theaterbau — thermische Abschirmung

Proszenium n	proscenium
Punktlicht n	spotlight
Rampenlicht n	footlight
Raumbühne f	space stage
Rückbühne f, Hinterbühne	backstage
Rundhorizont m, Bühnenhimmel m	cyclorama, cyke, stage horizon
Schauspielergarderobe f	dressing room
Schiebewand f	sliding wall
Schnürboden m	gridiron
Sehverhältnisse f, Sichtverhältnisse	sight conditions
Sehwinkel m, Sichtwinkel	sight angle
Seitenbühne f	side stage
Sichtlinie f, Sehstrahl m, Sehlinie, Sichtstrahl	sight line
Signallampe f	call lamp
Sitz(platz) m	seat
Sitzreihe f	row of seats
Souffleurkasten m	prompt box
Spielfläche f	acting area
Szenenwechsel m	scene-shifting
Theatertechnik f	theatre engineering (Brit.); theater $\sim$ (US)
Totaltheater n	total theatre (Brit.); $\sim$ theater (US)
Unterbühne f	substage
Untermaschinerie f	substage machinery
Versenk- und Hebebühne f	elevator stage
Versenkungsöffnung f	lift opening
Versenkungspodium n	stage bridge
Versenkungstisch m	hydraulic lift
Vorbühne f	forestage
Wagenbühne f	wagon stage
Wandelgang m	promenade (US)
Werstatt f	workshop
Zugleine f	counterweighted rope
Zuschauerraum m	auditorium
Zuschauerraumbeleuchtung f	house lights
zweiter Rang m	upper circle; amphitheater (US)

Theodolit m = theodolite (Brit.); transit (US)
theoretisch ermittelte Verkehrsbedürfnisse npl [*Straßenverkehrstechnik*] = synthesized travel desires
theoretische Bodenmechanik f = theoretical soil mechanics
$\sim$ **(Förder)Leistung** f [*Eimerkettenbagger*] = theoretical output
$\sim$ **Mechanik** f = theoretical mechanics
$\sim$ **Physik** f = mathematical physics
Thermal|kernbolzen m [*Motor*] = heat plug
$\sim$**wasser** n = thermal water
Therme f, **thermale Quelle** f, **Thermalquelle** = hot spring, thermal $\sim$
thermische Abschirmung f [*Reaktoranlage*] = thermal shielding

thermische Aufrauhung — Tiden

~ **Aufrauhung** *f*, Abbrennen *n* von Basaltpflaster = heat treatment of basalt paving sets, sett burning
~ **Kolonne** *f* [*Reaktoranlage*] = thermal column
~ **Schutzanlage** *f*, ~ Abschirmungsanlage = thermal (reactor) shield, structure for thermal shielding of atomic plants
thermischer Ausdehnungskoeffizient *m* = thermal coefficient of expansion
~ **Stoß** *m* = thermal shock
~ **Wirkungsgrad** *m*, Wärmeleistung *f* = heat output, thermal ~
thermisches Bohren *n* → Feuerstrahlbohren
~ **Bohrgerät** *n*, Flammstrahlbohrgerät = jet piercer
~ **Gefälle** *n* = thermal gradient
~ **Kracken** *n* = liquid phase cracking
~ **Reformierungsverfahren** *n* [*Benzin*] = thermal reforming process
Thermit *n* = thermit(e), ignition powder
~-**Reaktion** *f* = thermite reaction
~**schweißen** *n*, aluminothermisches Schweißen = alumino-thermic welding, thermit(e) ~
~**schweißmasse** *f* = thermit(e) mix(ture)
Thermodynamik *f* = thermodynamics
thermo|dynamische Untersuchung *f* = thermodynamic investigation
~**elektrische Zentrale** *f* → Wärmekraftwerk
Thermo|element *n* = thermel
~**luminiszenz** *f* = thermoluminiscence
Thermometer *n*, *m* = heat indicator, temperature ~, ~ ga(u)ge
Thermo-Osmose *f* = thermo-osmosis
Thermopaar *n* = thermocouple
thermoplastische Kunststoff-Schalungseinlage *f* = thermoplastic sheeting
Thermo|säule *f* = thermopile
~**siphon** *m* **mit Hilfsflügelradpumpe** = impeller assisted thermo syphon
~**stat** *m* = temperature regulator

thermostatisches Heizkörperventil *n* = thermostatic radiator valve
thermostatisches Mischventil *n* = thermostatic (shower) mixing valve
Thermostat|steuerung *f* = thermostatic control
~**ventil** *n* = thermostatic valve
Thiokol *n*, Polysulfid-Kautschuk *m* = polysulfide liquid polymer (US); polysulphide ~ ~ (Brit.)
~-**Versieg(e)lungsmasse** *f*, Thiokol-Absieg(e)lungsmasse = polysulphide (liquid polymer) based sealant (Brit.); polysulfide (~ ~) ~ ~ (US)
thixotrope Flüssigkeit *f* = thixotropic liquid
Thixotropie *f* = thixotropy
Thomas|mehl *n*, Schlackenmehl, Thomasphosphat *n* = Thomas meal, slag flour
~**roheisen** *n* = basic Bessemer pig-iron, Thomas ~
~**roheisenschlacke** *f* = Thomas pig-iron slag, basic Bessemer pig (iron) slag
~**schlacke** *f*, basische Schlacke = basic slag, Thomas ~
~**stahl** *m* = basic converter steel, Thomas ~
Thomson-Meßwehr *n* = Thomson measuring weir
Tide *f* → Gezeit *f*
~**alter** *n* → Alter der Tide
~**becken** *n*, Gezeitenbecken = tidal basin
~(**erscheinung**) *f* → Gezeit *f*
~**gebiet** *n* = tidal zone
~**hafen** *m* → Fluthafen
~**hub** *m* → Flutgröße
~**kraftwerk** *n*, Gezeitenkraftwerk, Ebbe und Flut-Kraftwerk = tidal power plant
~**kurve** *f* = local curve (of a tide), ~ tide curve
~**kurvengestalt** *f* = shape of a tide curve
tidelos, gezeitenlos = tide-free
Tiden *fpl* → Gezeiten *fpl*

Tide(n)|energie *f*, Gezeitenenergie = tidal power
~stromfigur *f* → Strömungskreislauf *m*
~stromfigur *f* → Störungsdiagramm *n*
Tide|pegel *m* = tide gauge, tidal ~
~scheitel *m* = tidal wave crest
~scheitellinie *f* = crest line
~stieg *m* → Flut
~strom *m* [*Fehlname: Tidefluß*] = tidal river
~strömung *f* → Gezeitenströmung
~wechsel *m*, Gezeitenwechsel = tidal range
~zeit *f* = cotidal hour
Tief *n* → Tiefdruckgebiet
tiefaufbrechen, tiefaufreißen = to rip, to deep-cut
Tiefaufbrechen *n* → Tiefaufreißen
tiefaufreißen, tiefaufbrechen = to rip, to deep-cut
Tief|aufreißen *n*, Tiefaufbrechen = deep loosening, ~ breaking up, ~ disintegrating, ~ disintegration, ~ cutting, ripping
~aufreißer *m* → Aufreißer
~aufreiß(er)zahn *m* = ripping tooth, ~ tine
~aushub *m* = deep excavation
Tiefbagger *m* [*Bagger, der unter seiner Standebene baggert*] = excavator digging below ground level
~ → Löffeltiefbagger
Tief|baggerung *f*, Tiefschnitt *m* [*Eimerkettentrockenbagger*] = digging up, ~ in depth, excavating below rail or ground level
~bau *m*, planender Ingenieurbau = construction engineering
~baugrube *f* = deept cut, ~ excavation, ~ building pit
~bauingenieur *m* = civil engineer
~bauunternehmer *m* = civil engineering contractor
~becken *n*, Unterbecken [*Pumpspeicherwerk*] = lower pool, ~ reservoir, ~ (storage) basin, tail pond
~bett-Anhänger *m* → Tieflader *m*
~bohrbrunnen *m* = drilled well
~bohren *n* = deep drilling

Tiefbohr|geräte *npl* = deep drilling equipment
~kamera *f* = deep well camera
~technik *f* = deep drilling engineering
~zement *m* → Erdölzement
Tiefbohrung *f*, Tiefsonde *f*; Tiefschacht *m* [*in Galizien*] = deep well, ~ hole, ~ bore, ~ boring
(Tief)Bordstein *m*, Tiefbord *n* = flush kerb, kerb below ground (Brit.); flush curb, inverted curb, curb below ground
Tiefbrücke *f* = low-level bridge
Tiefbrunnen *m* [*Grundwasserabsenkung*] = filter well [*The pump can be submerged in the bottom of the well*]
~pumpe *f* [*Grundwasserabsenkung*] = deep-well pump
Tief|bunker *m* → Bunkergrube *f*
~(druckgebiet) *n* = low-pressure area
Tiefe *f* = depth
Tief|ebene *f* → Tiefland *n*
~-Eimerketten(trocken)bagger *m*, Eimerketten-Tiefbagger = bucket ladder excavator working below ground level
~einbau *m* = construction with removal of existing sub-grade
Tiefen|anzeiger *m*, Teufenanzeiger = depth indicator
~(ein)rüttlung *f* = deep vibration
~entwässerung *f*, Untergrundentwässerung, Bodenentwässerung = (sub-) soil (or subsurface) drainage, underdrainage, subdrainage, ground water drainage
~faktor *m* [*Standsicherheit von Böschungen*] = depth factor
~gestein *n* → plutonisches Gestein
~injektion *f* = deep injection
~karte *f* = bathymetric chart
~linienkarte *f* der Leitschicht = subsurface contour map of the key bed
~linienplan *m* = depth-contour map
~marke *f* am Schöpfseil [*Tiefbohrtechnik*] = target
~messer *m*, Tiefensondiergerät *n* = depth sounder
~rüttler *m* → Innenrüttler

Tiefenschwimmer — Tiefscholle

~schwimmer m = loaded float
~sondiergerät n, Tiefenmesser m = depth sounder
~sondierung f = deep sounding
~sondierung f mit stufenweiser Belastung = static penetration testing
~sprengung f [*Sprengmethode, durch welche weiche Baugrundmassen mit großer Wucht auseinandergetrieben und gestört werden so daß die aufgeschütteten tragfähigen Massen auf die tragenden Lockergesteine absacken können*] = underfill method
~strom m [*Ozean*] = bathy current
~stufe f, geothermische ~ = geothermal gradient, geothermic degree
~typ(us) m [*Gestein*] = deep-seated type
~- und Langstreckenförderer m = long range excavator-conveyor, ~ ~ excavator-conveyer, ~ ~ machine
~verdichtung f = deep compaction
~verwitterung f = downward weathering
~vibration f = deep vibration
~vibrator m → Innenrüttler m
~vulkan m = deep-seated volcano
~wasser n = water at greater depths
~wasser n [*Ozean*] = bottom water
~wirkung f = depth effect
~-Zeitkurve f [*Tiefbohrtechnik*] = depth-time curve
Tief|fundation f → Tiefgründung f
~furche f = deep-furrow
~gang m [*Schiff*] = draught (Brit.); draft (US)
~garage f, unterirdische Garage = underground garage
tief|gehendes Schiff n = deep-going vessel
~gelegene Sohlschwelle f = low-sill structure
Tief|gründung f, Tieffundation f = deep foundation
~kellergeschoß n = subbasement, lower basement
~kühlanlage f = intense cooling plant
~kühltruhe f = freezer

~ladeanhänger m → Tieflader
(~)Ladebrücke f [*Tiefladewagen*] = deck, platform
~ladelinie f [*Schiff*] = (deep) load-line
~lader m, Tiefladeanhänger, Tiefbett-Anhänger, Tiefladewagen m, Tiefbettlader = low-loader, low-bed trailer, low-load trailer, deck trailer; low boy (trailer) (US)
~lader m mit Kipp(tief)ladebrücke, Kipptieflader = tilt-deck trailer, tilting platform ~
~ladewagen m [*Eisenbahn*] = well wagon
~ladewagen m [*Straßenfahrzeug*] → Tieflader m
~land n, Tiefebene f = low-level flat, low-land, bottom land, low-lying land
~landsee m = lowland lake
tief|liegender (Gleit)Kreis m [*Bodenmechanik*] = mid-point circle
~liegendes Gelände n ~ Land n = low-lying tract of ground, low-level ground, low-lying land
Tiefloch|bohren n = deep-hole drilling
~wagen m = deep-hole wagon drill
Tieflöffel m, Bagger~, Tiefräumer m = (back)hoe dipper (US); backacter (Brit.)
~-Ausleger m = (back)hoe boom (US); backacter jib (Brit.)
~(bagger) m = Löffeltiefbagger
~(bagger)-Einscherung f = backhoe reeving
~führer m = backhoe operator
~vorrichtung f = backacter section
Tief|pfeiler m = deep (foundation) pier
~-Probe(ent)nahmeapparat m = deep sampler, ~ sample taker
~punkt m [*einer Rohrleitung*] = low point
~räumer m → Tieflöffel
~-Reinigungsschacht m, Tief-Einsteigschacht = deep manhole
~reißer m → Tiefaufreißer
~schacht m [*in Galizien*] → Tiefsonde
~schnitt m → Tiefbaggerung
~scholle f (Geol.) → Graben m

~schürfbohrung *f* = deep sampling

~schütter-Bandabsetzer *m* = stacker for filling below track level

Tiefsee *f* = deep sea, oceanic abyss

~**ablagerung** *f*, Tiefseesediment *n* = abysmal deposit, deposit of the deep sea

~**graben** *m*, Tiefseerinne *f* = trench, depressed trough

~**kabel** *n* = deep-sea cable

~**kern** *m* = deep-sea core

~**lot** *n* = deep-sea lead, ~ sounding apparatus

~**lotung** *f*, Tiefseemessung = bathymetry, deep-sea sounding

~**panzer** *m* = deep-sea diving outfit

~**rinne** *f*, Tiefseegraben *m* = depressed trough, trench

~**sand** *m* = deep-sea sand

~**schlamm** *m* → eupelagische Meeresablagerung *f*

~**sediment** *n*, Tiefseeablagerung *f* = deposit of the deep sea, absymal deposit

~**taucher** *m* = deep-sea diver

~**thermometer** *n*, *m* = sounding thermometer, deep-sea ~, submarine ~

~**ton** *m*, roter ~ = red clay

Tief|sonde *f*, Tiefbohrung *f*; Tiefschacht *m* [*in Galizien*] = deep hole, ~ well, ~ bore, ~ boring

~**straße** *f* = depressed highway, ~ road, sunken ~, sub-surface ~

~**tal** *n* = pronounced valley

~**terrasse** *f* (Geol.) = low-lying terrace

Tiefstbohren *n* = very deep drilling

Tief-U-Bahn *f*, Tief-Untergrundbahn = deep-tube railway

Tiefwasser|anlegeplatz *m* = deep-water berth

~**einfahrt** *f* = deep-water entrance

~**gründung** *f* = deep-water foundation

~**hafen** *m* = deep-water port

~**pfeiler** *m* = deep-water pier

~**rinne** *f* = deep-water channel

Tiegel|gußstahl *m* = crucible (cast) steel

~**schmelzofen** *m* = crucible furnace

Tier|kohle *f*, Knochenkohle, animalische Kohle = animal charcoal, animalic coal

~**unterkunft** *f*, Stall *m* = animal shelter

Tiger|auge *n* (Min.) = tiger('s) eye

~**sandstein** *m* = mottled sandstone

Tioäther *m* = thio-ether

Tipper *m* **am Vergaser** = carburet(t)or primer, ~ tickler (US)

Tippschalten *n*, Schrittschalten = jogging, inching

Tirolergrün *n*, Berggrün = mountain green

Tisch|drehwerk *n* = table-turning mechanism

~**(ein)rütt(e)lung** *f* [*Beton*] = table vibration

~**felsen** *m* = erosion pillar, rock ~, mushroom rock

Tischler|arbeiten *fpl*, Schreinerarbeiten, Bau~ = joinery work

~**leim** *m* = joiner's glue, ~ adhesive

Tisch|rütt(e)lung *f* = Tischeinrütt(e)lung [*Beton*] = table vibration

~**rüttler** *m* → Rütteltisch

titanberuhigter Stahl *m* = steel stabilized with titanium

Titan|dioxyd *n*, TiO_2 = titanium dioxide

~**karbid-Nickel-Cermet** *n* = titanium carbide-nickel cermet

Titanweiß *n* = titanium pigment

~ **mit Barium** = titanium-barium pigment

~ ~ **Kalzium** = titanium-calcium pigment

~ ~ **Magnesium** = titanium-magnesium pigment

tödlicher Unfall *m*, Unfall mit tödlichem Ausgang = fatal accident

Toiletten|raum *m* = toilet facility, ~ room

~**sitz** *m* = toilet seat

~**trennwand** *f*, Toilettenzwischenwand = toilet partition (wall), WC ~ (~)

~**wagen** *m* = toilet trailer, mobile toilet unit

Toleranz — Tonfräse

Toleranz *f* = tolerance [*It means the difference between the permitted oversize (upper) limit and the permitted undersize (lower) limit. This difference is always positive*]
Toluol *n* → Methylbenzol *n*
Ton *m*, Roh~ [*<0,002 mm*] = clay
~**abbau** *m*, Tongewinnung *f* = getting of clay, clay working
~**anfeuchtung** *f*, Tonbefeuchtung [*Ziegelindustrie*] = clay tempering
~**anhäufung** *f* = aggregation of clay
~**anteil** *m* → Tongehalt
~**anteil** *m* → Tonkomponente
~**aufbereitung** *f* = clay preparation, ~ processing
~**auspressung** *f* → Tonverpressung
~**auswaschebene** *f* = clay outwash plain
~**bagger** *m* = clay excavator
~**band** *n* (Geol.) = clay band
tonbasische Spülung *f* → Tonspülung
Ton|befeuchtung *f* → Tonanfeuchtung
~**beton** *m*, mechanische Bodenverfestigung *f* = granular soil stabilization, mechanical ~ ~, soil-aggregate surface
tonbildend = clay-forming
Tonboden *m* = clay soil
Tonbohr|flüssigkeit *f* → Tonspülung *f*
~**schlamm** *m* → Tonspülung *f*
~**spülung** *f* → Tonspülung
Ton|brecher *m* = clay crusher, ~ breaker
~**brei** *m* [*Keramik*] = clay slip
~**brennen** *n* = clay burning
~**brennkurve** *f* = clay burning curve
~**bruch** *m* = clay quarry
~**bunker** *m* = clay bin
~**(dach)ziegel** *m*, Tondachstein *m* = clay (roofing) tile
~**dichtung** *f* [*Wasserbau*] = clay puddle seal, ~ packing
~**(dichtungs)kern** *m* = clay core
~**(dichtungs)schürze** *f* = clay blanket
~**dränrohr** *n*, Dräntonrohr = clay drain(age) pipe; ~ (farm) ~ tile (US)
~**einpressung** *f* → Tonverpressung
~**einschluß** *m* = clay pocket
~**einschluß** *m* im Fels = stone gall
~**eisenstein** *m*, Sphärosiderit *m*, toniger Spateisenstein, Eisenton *m*, Toneisenerz *n* = argillaceous ironstone, clay ~, clay iron ore, iron clay, argillaceous iron ore
~**emulsion** *f* = clay emulsion
Tonerde *f*, Aluminiumoxyd *n*, Al_2O_3 = alumina, alumin(i)um oxide
~**fabrik** *f*, Tonerdewerk *n* = alumina plant
~**keramik** *f* = alumina ceramics, ~ keramics
~**-Kieselsäure-Erzeugnis** *n* = alumina-silica refractory
~**modul** *m* [*früher: Eisenmodul*] = iron-alumina ratio $\dfrac{[Fe_2O_3 \cdot Varies}{Al_2O_3}$
from plant to plant, depending on raw materials and type of cement being produced. At some plants the reciprocal is used for control]
~**mörtel** *m* = alumina mortar
~**natron** *n* = sodium aluminate
~**schamottestein** *m* = alumina firebrick
(~)**Schmelzzement** *m* → Tonerdezement
~**silikat** *n*, kieselsaure Tonerde *f* = silicate of alumina
~**sulfat** *n*, schwefelsaure Tonerde *f* = alumina sulphate, sulphate of alumina, alum
~**verbindung** *f*, Aluminat *n* = aluminate
~**werk** *n*, Tonerdefabrik *f* = alumina plant
~**zement** *m*, (Tonerde)Schmelzzement, Aluminoszement = (high-)alumina cement, aluminous ~, calcium aluminate ~
Ton|falzziegel *m* = clay interlocking tile
~**filmstudio** *n* = sound film studio
~**firstziegel** *m*, Tonfirststein *m* = burnt-clay ridge tile
~**fraktion** *f* → Tonanteil
~**fräse** *f*, Tonreißer *m*, Tonschnitzler = clay shredder

Tonfundstätte — Tonplatte

~fundstätte *f* → Tonlagerstätte
~galle *f* = clay gall, argillaceous ~, marl pellets
~gehalt *m*, Tonanteil *m* = clay content
~gestein *n* = argillaceous rock, clay ~
~gewinnung *f*, Tonabbau *m* = getting of clay, clay working
~gießwulst *f* [*Rohrverlegung*] = clay roll
~gips *m* = argillaceous gypsum
~glimmerschiefer *m* = phyllite
~grube *f* = clay pit, ~ quarry
~gruppe *f* (Geol.) = allophane group
~gut *n* → Steingut
tonhaltiger Sand *m* = clayey sand, argillaceous ~
Ton|häutchen *n*, Tonüberzug *m* = clay film, ~ coat(ing)
~hobel *m* [*Ziegelindustrie*] = shaleplaner
Tonhohl|körper *m*, Tonhohl(stein)-platte *f*, Hourdi *m*, Tonhohlstein *m*, Plattenstein [*DIN 278*] = (burnt-)clay hollow tile, ~ ~ block, hollow clay block, hollow clay tile
~kügelchen *n* = clay bubble
~platte *f* → Tonhohlkörper *m*
~plattendecke *f*, Tonhohlkörperdecke, Hourdidecke = (burnt-)clay hollowblock floor, ~ hollow-tile ~, ~ pot ~
~plattenwand *f*, Tonhohlkörperwand, Hourdiwand = (burnt-)clay hollowtile floor, ~ hollow-block ~, ~ pot ~
~stein *m* → Tonhohlkörper *m*
~(stein)platte *f* → Tonhohlkörper *m*
tonig, tonhaltig = clayey, argillaceous
Ton|injektion *f* → Tonverpressung
~kalk *m*, toniger Kalkstein *m*, tonhaltiger Kalkstein = argillaceous limestone, argillocalcite, clayey limestone
~kern *m*, Tondichtungskern = clay core
~kerndamm *m* → → Talsperre *f*
~klumpen *m* = clay lump
~komponente *f*, Tonanteil *m*, Tonfraktion *f* = clay fraction

~konglomerat *n* = argillaceous conglomerate, clayey ~
~korn *n*, Tonteilchen *n* = clay particle, ~ grain
~körperchen *n*, Tonstückchen [*Drahtziegelgewebe*] = clay pellet
~kugel *f* = clay pallet
~lage *f* [*an den Salbändern*] = claycourse
~lager *n*, Tonvorratslager = clay stor(ag)e
~lagerstätte *f*, Tonvorkommen *n*, Tonvorkommnis *n*, Tonfundstätte *f*, Tonlager *n* = clay deposit
~lamellen *fpl* = clay laminae
~lehm *m* = clay loam
~linse *f* = lenticle of clay, clay lens
~masse *f* = clay mass
~mehl *n* = clay powder, finely ground fire clay
~mergel *m* [*10—40% $CaCO_3$*] = clay marl, clayey ~, shaley ~
~mineral *n*, Schichtsilikat *n* = clay mineral, layer silicate
~mosaik *n* = pottery mosaic
Tonne *f* = barrel
Tonnen|blech *n* = arched plate
~dach *n* = barrel roof
~(flechtwerk)dach *n*, Tonnenflechtwerk *n* = barrel vaulted roof with trellis work
~gehalt *m* = tonnage
~gewinn *m* = ton-profit
~gewölbe *n* = barrel vault, wagon ~, tunnel ~, cylindrical ~
~lager *n* = barrel-shaped roller bearing
~meile *f* = ton-mile
~schale *f* = barrel-vault shell
~schalendach *n* = barrel-vault shell roof
~schalen-Sheddach *n* = north-light barrel-vault shell roof
tonnlägiger Schacht → donlägiger ~ [*von 45°–75° Neigung*]
Ton|oberfläche *f* = clay surface
~pfeiler *m* (Geol.) = clay pillar
~platte *f* [*kann Fußbodenplatte (Fliese) oder Wandplatte (Wandfliese) sein*] = clay tile

~pulver n = clay powder
~quirl m = blunger
tonreich = clay-rich
Ton|reiniger m = clay cleaner, wash mill, clay purifier
~reißer m, Tonfräse f, Tonschnitzler = clay shredder
~rohr n = clay pipe; ~ tile (US)
~sand m = argillaceous sand, clayey ~
~-Sand-Kies-Mischung f, Ton-Sand-Kies-Gemisch n = clay sand gravel mix(ture)
~sandstein m = clayey sandstone, argillaceous ~
~scherbe f, Tonscherben m = clay body
~schicht f, Tonlage f (Geol.) = clay stratum
~schiefer m = clay slate, argillaceous ~, argillite
~schiefer m mit viel Chlorit = chlorite slate
~schiefergebirge n [Tunnel- und Stollenbau] = (clay) slate ground
~schiefernädelchen n = slate needle [rutile]
~schlag m → Lehmschlag
~(schlag)ummantelung f → Lehm(schlag)ummantelung
~schlamm m (Geol.) = clayey mud, argillaceous ~
~schlamm m [Zementindustrie] = straight-clay slurry
~schlämme f, Tonschlempe f = clay grout
~schlammgestein n = mud rock, clay mudrock
~schlick m = clayey mud, argillaceous ~
~schnitzler m → Tonreißer
~schürze f, Tondichtungsschürze = clay blanket
~silo m = clay silo
~sinter m, gesinterter Ton m = sintered clay
~spaten m = clay spade
~splitt m → Ziegelsplitt
~spülflüssigkeit f → Tonspülung f
~spülung f, Tonbohrspülung, Tonbohrschlamm m, Tonbohrflüssigkeit f, Tonspülflüssigkeit, Tonspülschlamm, Spülung auf Tongrundlage, Dickspülung, tonbasische Spülung = clay-laden (rotary) liquid, clay base mud
~stechen n = clay digging
~stein m, schichtungsloser ~ [Durch Druck und Wasserverlust erhärteter Ton] = mudstone, clay-stone
~strang m [Ziegelherstellung] = column of clay
~stückchen n, Tonkörperchen [Drahtziegelgewebe] = clay pellet
~substanz f [Gestein] = argillaceous cement, clayey ~
~suspension f = clay suspension
~süßwasserspülung f = clay-water drilling liquid
~-Tagebau m = open-cut quarrying of clay
~teilchen n, Tonkorn n = clay particle, ~ grain
~trockner m = clay dryer; ~ drier (US)
~überzug m, Tonhäutchen n = clay film, ~ coat(ing)
~ummantelung f → Lehm(schlag)ummantelung
Tönung f, Farb~ = shade
Ton|unterlage f = clay base
~verpressung f, Tonauspressung, Toninjektion f, Toneinpressung = clay grouting
~verteilungsgerät n = clay slope revetment spreader
~verwischung f [Raumakustik] = blurring
~vorkommen n → Tonlagerstätte f
~(vorrats)lager n = clay stor(ag)e
~-Wasser-Beziehung f = clay-water relationship
~-Wasser-Gemisch n, Ton-Wasser-Mischung f = clay-water mix(ture)
~werk n = clay plant, ~ works
~zement m = clay-cement
~-Zement-Injektion f = clay-cement injection
(Ton)Ziegel m = (clay) brick
~, Tondachziegel = clay (roofing) tile
~dach n = clay tile roof

Topdestillation *f* = top distillation
Topeka *m*, splittarmer Asphaltfeinbeton *m* [*Verbunddecke von Sandasphalt und Splitt*] = Topeka (type asphaltic concrete)
Töpferton *m* → Backsteinton
Topf|ladesystem *n* [*Radschrapper*] = boiling bowl loading, ~ action ~
~pumpe, Pitcherpumpe = pitcher pump
~zeit *f*, Standzeit, Gebrauchsdauer *f*, Arbeitsdauer, Gelierzeit [*Leim*] = usable life, working ~, spreadable ~, potlife
Topographie *f*, Geländebeschaffenheit *f* = topography
topographische Aufnahme *f* = topo(graphic) survey
~ Karte *f* = topographic map
Topotaxie *f* [*Die Erscheinung, daß bei einer chemischen Reaktion zwischen festen Stoffen die Kristallorientierung des Reaktionsproduktes eine Korrelation mit der eines der Ausgangsstoffe aufweist*] = topotaxy
Toppzeichen *n* [*Seezeichen*] = topmark
Tor *n* = gate
~bagger *m* → Portalbagger
Torf *n* = peat
~bagger *m* = peat excavator
~boden *m*, Torferde *f* = peaty soil, ~ earth
~einschluß *m*, Torfnest *n* = concealed bed of peat, peat pocket
~erde *f*, Torfboden *m* = peaty earth, ~ soil
~faser *f* = peat fibre (Brit.); ~ fiber (US)
~faserkohle *f* = peaty fibrous coal
~grund *m* = peaty ground
~kohle *f* = peat coal
~koks *m* = peat coke
~kraftwerk *n* = peat burning power station, ~ ~ ~ plant, ~ ~ generating ~
~lage *f*, Torfschicht *f* = peat layer
~moor *n* = peatbog
~moos *m* → Sumpfmoos
~mull *m*, Torfmehl *n* = peat dust, ~ meal, ~ powder

~nest *n* → Torfeinschluß *m*
~pechkohle *f* = peaty pitch coal
~platte *f* = peat building slab
~schicht *f*, Torflage *f* = peat layer
~schlacke *f* = peat cinder
~schlamm *m*, Dy *m* = dy
~streu *f* = peat litter
~streuabort *m*, Latrine *f* = peat-litter privy
Torkran *m* → Portalkran
Torkret|auskleidung *f*, Spritzbetonauskleidung, Torkretbetonauskleidung = gunite lining, shotcrete ~
~beton *m* → Beton
~hilfe *f*, Torkretierhilfe = guniting aid
torkretieren [*Eine Wand mit Mörtel oder Beton durch Aufspritzen versehen*] = to gun
Torkret|kanone *f* = cement gun
~masse *f* = gunite
~schicht *f* mit Stahl(ein)lagen → bewehrte Torkretschicht
~verfahren *n*, (Torkret)Beton-Spritzverfahren, Torkretieren *n*, Torkretierung *f*, pneumatische Auftragung von (Putz)Mörtel, Auftorkretieren, Beton-Spritzarbeit *f* = guniting, cement gun work
Torladewagen *m*, Klemmgriff-Lastenträger *m*, Hublaster *m*, Portalwagen = straddle carrier (Brit.); ~ truck (US)
Tornister-Spritzgerät *n* = knapsack sprayer
Torpedierung *f*, Torpedieren *n* = torpedoeing, oil well shooting
Torpedorinne *f*, Wurfförderrinne mit Pendelantrieb, Pendelwurfrinne = torpedo-conveyor, torpedo-conveyer
Torsion *f*, (Ver)Drillung *f*, Verdrehung, Verwindung, Drehung = torsion
Torsions|armierung *f* → Drillbewehrung
~aufgabe *f* = torsion(al) problem
~bewehrung *f* → Drillbewehrung
~einfluß *m* = effect of torsion
~einlagen *fpl* → Drillbewehrung *f*
~elastizität *f* = torsion(al) elasticity

~federung *f* = torsion flex suspension
~festigkeit *f,* Verdrehfestigkeit = torsion(al) strength
~gleichgewicht *n* = torsion(al) equilibrium
torsionslos, torsionsfrei = torsionless
Torsions|moment *n,* Drill(ungs)moment = torsion(al) moment, twisting ~, moment due to torsion
~prüfmaschine *f* = torsion testing machine, ~ tester
~rohrverstrebung *f* = torsion-tube cross member
~schwingung *f* = torsion(al) oscillation, ~ vibration
~schwingungsdämpfer *m* = torsion(al) vibration damper
~spannung *f,* (Ver)Drillungsspannung, Verdrehungsspannung, Verwindungsspannung, Drehspannung = torsion(al) stress
~stab *m* = twisted bar
~(stahl)einlagen *fpl* → Drillbewehrung *f*
torsionssteif = torsionally stiff, ~ rigid
Torsions|steifigkeit *f,* Drillsteife *f,* Torsionssteife = torsional stiffness, ~ rigidity, stiffness in torsion, rigidity in torsion
~verband *m* = torsion bracing
torsionsversteift = torsionally braced
Torsions|viskosimeter *n, m* = torsion viscosimeter
~winkel *m,* Drill(ungs)winkel = angle of twist
Torstahl *m* [*Verwundener Rundstahl mit Längsrippen*] = Torsteel
Tosbecken *n,* Beruhigungsbecken = absorption basin, stilling ~ (Brit.); stilling pool (US)
Totalvolumen *n* [*Porenvolumen + Feststoffvolumen eines Bodens*] = total volume of the (soil) aggregate
Totarm *m* → Altwasser *n*
Totbrennen *n* = dead burning
toter Arm *m* → Altwasser *n*
~ Gang *m* = lost motion, idle ~
~ Griff *m* = dead handle

totes Wasser *n* [*Schachtabteufung*] = water standing at its normal level
totgebrannt = overburnt
totgebrannter Gips *m* → Anhydrit II *m*
totgelaufener Schuß *m,* abgerissener ~ = hangfire, cutoff shot
Totgewicht *n,* Totlast *f,* Eigengewicht, Eigenlast = dead weight
Totlast *f* → Totgewicht
Totpumpleitung *f* [*Tiefbohrtechnik*] = killing line
Totpunkt *m,* nach oberem ~ = after top dead center (US)/centre (Brit.), ATDC
~, nach unterem ~ = after bottom dead centre (Brit.)/center (US), ABDC
~, vor oberem ~ = before top dead center (US)/centre (Brit.), BTDC
~, vor unterem ~ = before bottom dead centre (Brit.)/center (US), BBDC
Totraum *m,* Wirbelraum [*Hydraulik*] = vortex zone, eddy ~
~ [*Bunker; Silo*] = dead space
totsöhlig (⚒) = absolutely horizontal
Totwasser *n* = adherent water
Totzeit *f* = dead time
Tourenzähler *m,* Tachometer *n, m,* Drehzahlmesser *m* = revolution counter, tachometer
Tovotefett *n* [*seltener gebrauchte Bezeichnung für Maschinenfett*] = lubricating grease, lube ~
Towne-Gitterträger *m* = Towne lattice truss
TP *m* → trigonometrischer Punkt
T-Querschnitt *m* = T section
Trachyt *m* = trachyte
~lava *f* = trachytic lava
Trafo|-Haus *n* → Trafo-Station
~-Station *f,* Transformatorenstation, Trafo-Haus *n,* Transformatorenhaus = transformer station
Trag|achse *f* = bearing axle
~arm *m,* Stützarm [*Segment(stau)wehr*] = supporting arm
Tragband *n* = carryable belt conveyor, ~ ~ conveyer
tragbar = carryable

tragbares (Feuer)Löschgerät *n*, tragbarer (Feuer)Löscher *m* = carryable (fire) extinguisher
Trage *f* = hand barrow [*A small platform with handles, for conveying goods by two men*]
~eisen *n* [*Hängedecke*] = 1.) bearer [*the main bearer*]; 2.) runner [*the secondary bearer*]
Trag(e)griff *m* = carrying handle
tragend, last~ = loadbearing
tragende Querwand *f*, last~ ~ = loadbearing cross wall
tragender Boden *m*, last~ ~ = supporting material, ~ soil, ~ earth
Träger *m* [*Linoleumherstellung*] = backing
~ [*Jeder Balken ist ein Träger, aber jeder Träger ist nicht notwendigerweise ein Balken; denn es gibt Balken(träger) und Bogen(träger)*] = girder
~anschluß *m* = girder connection
~auflager *n* = girder support
~bemessungsformel *f* = girder design formula
~biegepresse *f* = girder bending press
~brücke *f* = girder bridge
~brückenlager *n* = girder bridge bearing
~destillation *f* = carrier distillation
~drehung *f* = girder rotation
~einspannungen *fpl* **durch Wände** = girder restraints provided by walls with openings
~eisen *n* [*Fehlname*] → Trägerstahl
~element *n* = girder element
(~)Fachwerk *n* = truss, framework, latticework
~feldstück *n* = individual girder part
(~)Flansch *m*, (Träger)Fuß *m*, (Träger)Gurt *m* = chord, flange, boom
~frequenz *f* = carrier frequency
(~)Fuß *m*, (Träger)Flansch *m*, (Träger)Gurt *m* = flange, chord, boom
~gestein *m* → Speichergestein
~gewebe *n* = base fabric
(~)Gurt *m* → (Träger)Fuß *m*
~höhe *f* = depth of girder

~pfahl *m*, Stahl~, H-Pfahl = H-pile, structural-steel pile
~rost *m* = girder grille
~rostbrücke *f* = girder grille bridge
~schalung *f* = girder formwork
(~)Stab *m* = member, bar, rod
(~)Stabebene *f* = plane of members, ~ ~ bars, ~ ~ rods
~stahl *m*, Profilstahl, Stahlträger *m*, Stahlprofil *n* [*Fehlnamen: Trägereisen n, Profileisen. Von den Walzwerken hergestellter Formstahl mit I-, IP- und U-Querschnitt, der als Träger für Bauzwecke verwendet wird*] = steel girder, ~ section
(~)Steg *m* = web (of girder)
~stift *m*, Pinnenträger *m* = pivot post, bearing ~; centre pin support (Brit.); center pin support (US)
~stoß *m* = girder joint
~vorbiegung *f* = preflexion
~walzwerk *n* = girder rolling mill
Trag-Erdbohrgerät *n* = carryable soil auger, ~ earth ~
tragfähig [*Baugrund*] = bearing, firm
tragfähiger Boden *m* = natural foundation [*British standard name. B.S. 3589. Soil requiring no support or other foundation to support a building or structure*]
Tragfähigkeit *f*, Tragvermögen *n*, Tragleistung *f*, Tragkraft *f*, Tragwiderstand *m* [*Baugrund*] = bearing power, ~ capacity, ~ resistance, ~ property, supporting ~
~ [*Kran*] = lifting capacity
~ beim Bruch → Bruchlast *f*
~ der Spülung [*Tiefbohrtechnik*] = weight suspending properties of the mud
Tragfähigkeits|index *m* = bearing ratio
~messer *m* = bearing (capacity) apparatus
~tonne *f* = dead weight ton
~versuch *m*, Tragfähigkeitsprüfung *f*, Tragfähigkeitsprobe *f* = bearing test
~wert *m* [*Baugrund*] = bearing value, supporting ~

Trag|gabel *f* [*Reifenfahrwerk*] = A-frame
~glied *n*, **Tragteil** *m*, *n* [*Bauwerk*] = loadbearing member
~griff *m* = carrying handle
Trägheit *f* → Beharrungsvermögen *n*
Trägheits|ellipse *f* = ellipse of inertia
~halbmesser *m* = radius of inertia
~kraft *f* = inertia force
~moment *n* = moment of inertia
~vermögen *n* → Beharrungsvermögen
Tragkabel *n*, **Tragseil** *n* [*Tiefen- und Langstreckenförderer*] = track cable
~, Tragseil *n* [*Hängebrücke*] = carrying cable, ~ rope, ~ strand
Trag|konsole *f* = mounting bracket
~konstruktion *f* = load-bearing system
Tragkraft *f* [*Autoschütter*] = payload
~ [*Kran*] = lifting capacity
~ [*Baugrund*] → Tragfähigkeit *f*
~, Traglast *f* [*Pfahl*] = working load
~ → Biegefestigkeit *f*
~ bei abgestütztem Autokran, Nutzlast *f* ~ ~ ~ = blocked capacity, lifting capacity on outriggers
~ ~ unabgestütztem Kran, Nutzlast *f* ~ ~ ~ [*Autokran*] = lifting capacity free-on-wheels, free-on-wheel's capacity
~ des Reifens = tire capacity (US); tyre ~ (Brit.)
Trag|kreiselpumpe *f* = hand-carry centrifugal pump
~kreuz *n* [*Radschrapper*] = spider
~lager *n* **des Hinterachsgehäuse** = axle housing bracket cap
Traglast *f*, **Tragkraft** *f* [*Pfahl*] = working load
~verfahren *n* = load-factor method of design
Trag|leistung *f* → Tragfähigkeit *f*
~pfahl *m*, Bauwerkpfahl [*Im Gegensatz zum Verdichtungs- und Dalbenpfahl*] = structural pile, bearing ~
~platte *f* [*Außenvibrator*] = baseplate
~ring *m* = loadbearing ring
~rolle *f*, belastete Rolle = support(ing) roller, carrier (idler)
~rolle *f* [*Gleiskette*] = carrier roller

~rolle *f* **des Ausstoßers** = ejector carrier roller
~rollenstation *f* = carrier idler set
Tragschicht *f* = base (course) [*Base courses under rigid pavements are often called subbase courses*]
~ → obere ~
~ aus verfestigtem Material = (stabilized) soil (road) base (course)
~-Lage *f* = base layer
~material *n* = base material
Tragseil *n*, **Tragkabel** *n* [*Hängebrücke*] = carrying rope, ~ strand, ~ cable
~, Tragkabel *n* [*Tiefen- und Langstreckenförderer*] = track cable
~ [*Seilschwebebahn*] = rail rope
Tragstück *n*, Eingabelung *f*, Arm *m*, Halter *m* = bracket
~ = carrier assy
Trag|system *n* = loadbearing system
~teil *m*, *n*, Tragglied *n* [*Bauwerk*] = loadbearing member
~vermögen *n* → Tragfähigkeit *f*
~wanne *f* → Mulde *f*
Tragwerk *n* [*Teil eines Bauwerkes*] = loadbearing member, ~ part
~ [*Teil eines Gebäudes*] = structure [*The loadbearing part of a building*]
~ mit veränderlichen Trägheitsmomenten = non-uniform structural member
Trag|widerstand *m* → Tragfähigkeit *f*
~zylinder *m* [*Walzenwehr*] = shield-carrying cylinder (Brit.); ~ roller gate (US)
Traktor *m* → Schlepper *m*
traktorengezogene Geräte *npl* = tractor-allied equipment, tractor-drawn ~
Traktoren|kette *f* → Raupenkette *f*
~kraftstoff *m* = power kerosine, ~ kerosene
~kran *m*, Schlepperkran, Traktorkran = tractor crane
~schmieröl *n* = tractor luboil
~(trommel)seilwinde *f* → Schlepperseilwinde
Traktor-Schrapper *m*, Schrapper mit (Gleis)Kettentraktor, Schrapper mit

Trambalken — Transportbeton

Raupe, Schürf(kübel)wagenzug *m*, Motor-Schürfzug = crawler-scraper rig

Trambalken *m*, Zugbalken, Hauptbalken, Spannbalken [*eines einsäuligen Hängewerkes*] = main beam, tie ~

Tram|-Bauweise *f*, vorgebaute Lenkung *f* [*Fahrersitz befindet sich vor der Vorderachse*] = cab-over-engine design, full-forward-control cab ~

~fahrzeug *n* [*Dieses Fahrzeug hat eine glatte Vorderfront, es liegt keine Motorhaube (wie bei Haubenfahrzeugen) vor dem Fahrer*] = cab-over-engine vehicle, full-forward-control cab ~

Trampschiff *n* = "tramp" ship

Tränk|anlage *f* = impregnation plant

~bad *n* = impregnation bath

~belag *m* → Tränkdecke *f*

~decke *f*, Tränkbelag *m* [*Straßenbau*] = penetration pavement, ~ surfacing; ~ paving (US)

Tränken *n*, Einguß *m*, Tränkung *f* = penetration pavement, ~ surfacing

~, Imprägnieren = impregnation

Tränk|makadam *m*, getränkte Schotterdecke *f* = grouted macadam (Brit.); penetration ~

~trog *m* = watering trough

Tränkung *f*, Imprägnierung = impregnation

~, Tränken *n*, Einguß *m* [*Straße*] = grouting (Brit.); penetration

Tränkungsschießen *n*, Stoßtränkschießen = pulsed infusion shotfiring

Transduktor *m* = transducer

Transformator *m*, Umspanner *m*, Trafo *m* = transformer

Transformatorenstation *f* → Trafostation

Transit|hafen *m*, Durchganghafen = transit harbo(u)r

~schuppen *m* = transit shed

transkristalline Korrosion *f*, intrakristalline ~ = transcrystalline corrosion, intracrystalline ~

transkristalliner Riß *m*, intrakristalliner ~ [*Korrosion*] = transcrystalline crack, intercrystalline ~

translatorische Welle *f*, Übertragungswelle = wave of translation

Transmissions|öl *n* = transmission oil

~pumpe *f* = pump driven by power

~riemen *m* = transmission belt(ing)

transparente Kunststoff-Bauplatte *f* = translucent plastic panel, ~ ~ sheet

Transparentpapier *n* = transparent paper

Transparenz *f* = transparency

Transpiration *f*, Pflanzenverdunstung *f* = plant transpiration

Transport *m*, Beförderung *f* = transport, handling

~, Förderung *f* [*Erdbau*] = haulage, hauling

~, Verfrachtung *f* (Geol.) = transport

~ von Baustelle zu Baustelle → Ortswechsel *m*

transportables Gas *n* → Ölgas

Transport|anlage *f*, Förderanlage = conveying installation, ~ plant

~armierung *f* → Transportbewehrung

~bandanlage *f* → → Bandförderer *m*

~bandstraße *f* → → Bandförderer *m*

~beanspruchung *f* = handling stress [*precast concrete unit*]

~behälteranhänger *m* = container trailer

~behälterumschlag *m* = container handling

Transportbeton *m* [*Ist ein Beton, dessen Bestandteile in einem Transportbetonwerk nach Gewicht zugemessen werden und der entweder in Mischerfahrzeugen oder im Werk selbst gemischt und in geeigneten Fahrzeugen zur Verwendungsstelle, in der Regel zur Baustelle, transportiert und in einbaufertigem Zustand übergeben wird.*

Bemerkung: Das Kennzeichnende des Transportbetons ist die Betonherstellung in einer "Fabrik" abseits der Verwendungsstelle und daß, im Gegensatz zum Ortbeton, Herstellungs- und Verbrauchsort durch eine Transportleistung miteinander verbunden sind. Als Transportbeton gilt jeder

im frischen Zustand transportierte Beton, der einbaufertig übergeben wird und der von einer Produktionsanlage stammt, die sich nicht an der Verwendungsstelle des Betons befindet. In den meisten Fällen wird der Transportbeton als Handelsware verkauft. Die Bezeichnung Lieferbeton und Fertigbeton gibt es nur als Firmenbezeichnungen; sie gehen im Begriff Transportbeton auf und bedeuten weder eine besondere Herstellungsmethode noch eine besondere Lieferart. Diese o. a. Bezeichnungen müssen daher aus dem allgemeinen Sprachgebrauch verschwinden. Damit ist auch die Definition ,,Fertigbeton" in dem Werk ,,Begriffe und Begriffsbestimmungen im Bauwesen" nicht haltbar] = ready-mix(ed) concrete

Transportbeton(aufbereitungs)anlage f, Transportbetonwerk n, Betonlieferwerk = ready-mix(ed) (concrete) plant, ~ ~ depot

~ **zur Herstellung von Beton zum Verfahren durch Nachmischer** = ready-mix(ed) concrete plant of the central mix type

~ ~ ~ ~ ~ **Trockenbeton** = transit-mix(ing concrete) plant, ready-mix(ed) concrete plant of the transit-mix type; truck mixer plant (US)

Transportbeton|hersteller m = ready-mix(ed) concrete operator

~**mischer** m = ready-mix(ed) concrete mixer

~**-Nachmischer** m → (Beton)Rührfahrzeug n

Transport|-Betonpumpe f, Auto-Betonpumpe = self-propel(l)ed concrete pump

~**betonwerk** n → Transportbetonaufbereitungsanlage

~**bewehrung** f, Transportarmierung, Transport(stahl)einlagen fpl, Beförderungsbewehrung, Beförderungsarmierung, Beförderungs(stahl)einlagen = reinforcement for handling

~**birne** f, Beton~ = agitator

~**brücke** f, Verladebrücke = transporter bridge

~**einlagen** fpl → Transportbewehrung f

Transporter m **mit Absetzbehältern** = multi-skip dumper

Transport|fahrzeug n = transporting vehicle, haul(ing) unit, hauler

~**festigkeit** f = handling strength

~**gefäß** n, Behälter m = container

~**gerät** n = materials handling gear

~**geschwindigkeit** f [*Die Geschwindigkeit einer selbstfahrenden Baumaschine im Gegensatz zur Arbeitsgeschwindigkeit*] = working speed

~**karre(n)** f, (m) → Karre(n)

~**mischer** m, Liefermischer, Fahrmischer, Auto-Transportbeton-Mischer, Beton-Transportmischer [*zentrale Abmessung, Mischen während der Fahrt*] = truck mixer, transit-mixer (truck) (US); mixer lorry (Brit.)

~**mittel** n, Beförderungsmittel = means of transport

~**schnecke** f → Schneckenförderer m

~**(stahl)einlagen** fpl → Transportbewehrung f

~**stütze** f [*Radschrapper*] = shipping bracket

~**technik** f = transportation engineering

~**- und Verteileranhänger** m **mit Bodenentleerung** = bottom-dump hauling and spreading trailer

~**-unternehmer** m, Fuhrunternehmer = hauling contractor

~**verlust** m = conveyance loss [*Loss of water from a conduit, due to seepage, evaporation, or evapo-transpiration*]

~**wagen** m = transporter

~**weite** f → Förderweite [*Erdbau*]

~**zeit** f **hin und zurück minus Schürfzeit** [*Radschrapper*] = in-and-out haul

Trapez|binder m = trapezoidal truss

~**formel** f = trapezoidal formula

trapez|förmiger Kanal m = trapezoidal canal

~**förmiges Gerinne** *n* = trapezoidal channel
Trapez|löffel *m*, Profillöffel für Gräben = trapezoidal bucket
~**profil** *n* [*Hydromechanik*] = trapezoidal channel
~**rahmen** *m* = trapezoidal frame
~**träger** *m* = trapezoidal girder
~**wehr** *n* = trapezoidal weir
Trapp *m* = trap (rock)
~**tuff** *m* → Basalttuff
Traps *m* → Geruchverschluß *m*
Trasse *f* [*zu vermeiden: Trace f*] = route, location line
Trassen|schäler *m*, Planier-(Gleis)Kettengerät *n* mit schneepflugartiger Anordnung des Planierschildes, Seitenräumer *m*, Fronträumer mit Schwenkschild, Planier-(Gleis)Kettengerät mit Schwenkschild = roadbuilder, trailbuilder, gradebuilder
~**wahl** *f* = selection of route, ~ ~ location, route selection, location selection
trassieren = to locate
Trassierung *f*, Trassieren *n* = location
Trassierungs|merkmal *n* = geometric standard
~**ingenieur** *m* = location engineer
Traß *m*, feingemahlener vulkanischer Tuffstein *m*, gemahlener massiger Bimstuff *m* = (Rhenish) trass
~**hochofenzement** *m* = blast-furnace trass cement
~**kalk** *m* [*Bindemittel aus Traß und Kalkpulver oder Kalkteig*] = 1.) trass-lime powder mix(ture); 2.) ~ putty ~
~**mehl** *n* = trass powder
~**mörtel** *m* = trass mortar
~**zement** *m* = trass cement
Trauf|blech *n* = eaves flashing
~**brett** *n* → Stirnbrett
Traufe *f*, Trauflinie *f*, Traufkante *f*, Dachtraufe *f* = eaves
Trauf(en)pfette *f* = eaves purlin(e)
Traufenträger *m* = eaves girder
Trauf|kante *f* → Traufe
~**linie** *f* → Traufe
~**platte** *f* = eaves plate
~**ziegel** *m* = eaves-tile
Traverse *f* = spreader bar, yoke [*A stiff beam hanging from a crane hook having several ropes or chains hanging from different points along it. It is used for lifting long objects to prevent them breaking during lifting*]
Travisbecken *n* = Travis tank [*A two-stor(e)y hydrolytic or septic tank invented by Dr. Travis, consisting of an upper sedimentation chamber with steeply sloping bottom terminating in slots through which the deposited solids pass into a lower or sludge-digestion chamber through which a predetermined part of the sewage is allowed to pass for the purpose of seeding and maintaining bacterial life in the sludge and carrying away decomposition products. This is for the purpose of inducing digestion of the sludge attended by its reduction in volume*]
Trecker *m* → Schlepper *m*
~**kette** *f* → Raupenkette
Treib|bake *f* = floating beacon
~**birne** *f* → Rohrbirne
~**dorn** *m* = punch
~**eis** *n* = drift(ing) ice, floating ~, flow ~
Treiben *n* = expansion
Treiber *m* = discharge pusher [*for furnaces with side discharge*]
Treib|erscheinung *f* = expansion phenomenon
~**fäustel** *m*, schwerer Schlägel *m* = striking hammer
~**haus** *n*, Gewächshaus = greenhouse
~**kessel** *m*, Druckkessel, Förderkessel, einkammeriges Druckgefäß *n*, Druckluft-Treibkessel = pressure cylinder [*pneumatic concrete placer*]
~**kette** *f*, Antriebskette = driving chain, drive ~
~**mittel** *n* = gaserzeugendes Mittel
~**rad** *n* → Antriebsrad
~**riegel** *m* = espagnolette, cremorne bolt
~**riemen** *m*, Antriebsriemen = driving belt, drive ~

Treibriemenadhäsionsfett — Trennung

~riemenadhäsionsfett n = belt grease
~riß m = expansion crack
~sand m, Triebsand = drift sand
Treibscheibe f → Antriebsscheibe
~, Koepescheibe = Koepe pulley
Treibscheiben|achse f [*pennsylvanisches Seilbohren*] = band well shaft
~förderung f, Koepe-Förderung = Koepe winding, ~ system
~futter n = Koepe pulley groove lining
Treibschraube f, Preßschraube, Dichtschraube = forcing screw
Treibstoff m, Kraftstoff = fuel
treibstoffbeständige Fugenvergußmasse f = fuel-resistant joint sealing compound
Treibstoff|druck m, Kraftstoffdruck = fuel pressure
~einspritzpumpe f, Kraftstoffeinspritzpumpe = fuel injection pump
~-Filter m, n, Kraftstoff-Filter = fuel filter
~verschüttung f = fuel spillage
~zusatz m, Kraftstoffzusatz = fuel additive
Treib|stück n [*Handwerkzeug*] = driver
~zeug n, Schwimmstoffe mpl [*Feststoffe, meist organischer Art, die leichter als Wasser sind und daher auf ihm schwimmen*] = floating trash, ~ debris
Treidel|lok(omotive) f = shore-bound loco(motive)
~pfad m → Leinpfad
~weg m → Leinpfad m
Tremolit m (Min.) = tremolite (asbestos), Italian ~
Trennbruch m, Sprödbruch = brittle fracture
Trenneinlage f [*eines zweiteiligen Sakkes*] = separator
Trennen n mit Sauerstofflanze = oxygen lancing
~ von Beton = severing concrete
Trenn|entwässerung f → Trennsystem
~fertigwand f, Zwischenfertigwand = prefab(ricated) partition (wall)
~fuge f, Raumfuge, (Aus)Dehn(ungs)fuge [*absichtlich angelegte Fuge, um Ausdehnung der Baustoffe zu ermöglichen*] = expansion joint
~hohlwand f → Hohltrennwand
~insel f = divisional island, separator
~kanalisation f → Trennsystem
~konstruktion f [*Wände und Decken*] = partition [*walls and floors*]
Trennkorngröße f = effective separating size
~ HP, Ausgleichskorngröße nach Heidenreich-Paul = effective separating size of a screen using the Heidenreich-Paul method
~ T f → Teilungskorngröße
Trenn|lage f, Trennschicht f = separation layer, ~ course
~mauer f, Abschlußwand f = diaphragm
~mauer f, Sporn m, Abschlußwand f, Herdmauer, Fußmauer, (Ab)Dichtungsmauer, (Ab)Dichtungssporn [*Talsperre*] = (toe) cutoff wall
~mauer f (oder Trennungsmauer) der Schornsteinlängszüge = withe, midfeather (wall)
~mittel n = parting agent
~riß m = separation crack
~schärfe f [*Siebtechnik*] = precision of separation, accuracy ~ ~
~scheibe f, Schneidscheibe = blade
~schicht f, Trennlage f = separation course, ~ layer
~schnitt m, Siebschnitt = screen cut
~streifen m = dividing strip
~system n, Trennentwässerung f, Trennverfahren n, Trennkanalisation f = separate system [*A system of sewers in which sewage and storm water are carried in separate conduits*]
~systemleitung f = separate sewer
~transformator m = isolating transformer [*A transformer used to avoid direct connection with the high tension circuit*]
Trennung f von Fahrzeug- und Fußgängerverkehr = segregation of vehicles and pedestrians

Trennung mit schweren Medien — Tribüne

~ **mit schweren Medien** = heavy-media separation

(Trennungs)Damm *m* [*Überstauung zwecks Bewässerung*] = levee

Trennungs|grad *m*, Gütegrad [*Siebtechnik*] = separation sharpness, screening efficiency, efficiency of separation

~**siebung** *f* [*Entfernen des Überkorns oder Unterkorns*] = scalping

~**zulage** *f* = detachment allowance

Trennverfahren *n* → Trennsystem *n*

Trennwand *f*, Zwischenwand, Leichtwand, Scheidewand [*nichttragend*] = (interior) partition (wall)

~ **aus Kesselschlackenbetonplatten**, Zwischenwand ~~ = clinker concrete slab partition (wall)

~ **in Trockenbauweise**, Zwischenwand ~~ = dry (wall) partition

~(**bau)platte** *f*, Zwischenwandbauplatte, Leichtwandbauplatte, Scheidewandbauplatte = partition slab

~**block** *m* → Trennwandstein *m*

~**stein** *m*, Trennwandblock *m*, Zwischenwandstein, Zwischenwandblock = partition block

~**-Tonhohlkörper** *m* → Zwischenwand-Tonhohlkörper

~**verglasung** *f* = partition(ing) glazing

~**ziegel** *m*, Zwischenwandziegel = partition brick

Treppe *f* = stair(case) [*Is a number of steps leading from one floor to another*]

Treppen|absatz *m* → Podest *n*

~**-Abzugkanal** *m* → Treppen-Durchlaß *m*

~**anlage** *f* = stairs, stairway [*From the bottom floor to the top floor*]

(~)**Arm** *m* → (Treppen)Lauf *m*

~**art** *f* = type of stair(case)

~**auge** *n*, Treppenloch *n* = (stair) well [*A space around which a staircase is disposed*]

~**bau** *m* = stair(case) construction, construction of stair(case)s

~**belag** *m* → (Tritt)Stufenbelag

Treppen-Durchlaß *m*, Treppen-Abzugkanal *m*, Kaskaden-Durchlaß, Kaskaden-Abzugkanal = cascade culvert

treppenförmige Betonmauer *f* → abgestufte ~

Treppen|geländer *n* = stair railing

(~-)**Handlauf** *m* = hand-rail, stair-rail

(~)**Lauf** *m*, (Treppen)Arm *m* = (stair) flight, flight of stairs

~**loch** *n* → Treppenauge *n*

~**lochwange** *f* → Freiwange

~**pegel** *m* → → Schrägpegel

~**podest** *n* → Podest

(~)**Podestplatte** *f* = (stair) landing slab

~**schacht** *m* = stair(case) shaft

~**schutzleiste** *f* = stair nosing

~**spindel** *f* = newel

(~)**Steigung** *f* = rise

(~)**Stufe** *f*, Tritt *m* = step

~**stufenbau** *m* = construction of steps

(~)**Stufenfräse** *f* = step grinder, ~ grinding machine

~**stufenmaschine** *f* = step (making) machine

(~)**Stufenpresse** *f* = step press

(~)**Stufenrüttelpresse** *f* = step vibrating press, ~ vibratory ~

(~)**Stufenschleifmaschine** *f* = step grinder, ~ grinding machine

~**turm** *m* = stair tower

(~)**Wange** *f* = string

Tressengewebe *n* = corduroy cloth

Tria-Harfensieb *n* = Tria harp screen

Triangulierung *f*; Triangulation *f* [*Schweiz*] = triangulation

Trias *m* → geologische Formation *f*

Triaxial|gerät *n*, dreiaxiale Druckzelle *f*, triaxialer Druckapparat *m*, triaxiale Scherprüfzelle = triaxial cell, ~ test chamber, ~ load frame, ~ compression test machine, ~ test(ing) apparatus

~ **mit fest angeordneter Gummihülle** = fixed-sleeve cell

~ **mit frei angeordneter Gummihülle** = free-sleeve cell

~**versuch** *m*, Dreiachsenversuch = triaxial test

Tribüne *f* = grandstand

Trichter m = hopper [*A receiver or receptacle with bottom discharge in which substances are placed to be passed or fed to any equipment or part of any equipment*]

~, Einlauf~, Beton(einfüll)~ [*Betonpumpe*] = (pump) hopper, (concrete) receiving ~

~ (⚒) = glory hole, mill ~

Trichterbau m (⚒) = glory-hole mining, ~ milling

~ **in Tagebauen** = open-cut glory-hole mining, ~ ~ milling

~ **unter Tage**, untertägiger Trichterbau = underground glory-hole mining, ~ ~ milling

Trichter|bildung f [*bei der Entölung oder Entwässerung eines Sandes*] = coning effect

~**boden** m, Kegelboden [*Behälter*] = hoppered floor, ~ bottom, conical ~

~**bunker** m, (Material)Bunker = hopper, bin

~**einlauf** m = flaring inlet [*A funnel-shaped entrance to facilitate flow into a pipe or conduit*]

trichterförmige Mündung f, negatives Delta n = funnel shaped estuary

Trichter|gewölbe n, Fächergewölbe, Palmengewölbe, Strahlengewölbe, angelsächsisches Gewölbe, normannisches Gewölbe = fan vaulting

~**-Hochwasserentlastungseinlauf** m = glory-hole spillway, shaft ~, funnel-shaped ~

~**mühle** f = hopper mill, ~ grinder

~**(naß)klassierer** m = funnel classifier

~**öler** m = funnel type straight oil cup

~**rohr** n [*Kontraktorverfahren*] = tremie-pipe

Triebachse f → Antriebsachse

Triebling m, Antriebszahnrad n, Ritzel n = drive gear, pinion

Triebrad n → Antriebsrad

~ **mit Naben-Elektromotor**, ~ ~ Elektro-Nabenmotor = electric wheel

~**nabe** f = driving wheel hub

Trieb|sand m, Treibsand = drift sand

~**satz** m → (Zweirad-)Vorspänner m

~**satz** m, Antriebsgruppe f, Antrieb m = power unit, drive (~), driving gear

~**wagen** m = rail-car

Triebwasser n [*der Durchfluß, der von einer Turbine in einer Sekunde verarbeitet wird*] = turbine flow per second

~**kanal** m [*ein offenes Gerinne, das das Triebwasser von einer Abzweigstelle zum Werk oder von diesem zurück zum Hauptgewässer führt*] = (channel-type) race

~**leitung** f, Druckleitung = penstock

~**rohrleitung** f, Druckrohrleitung = pipe penstock

~**stollen** m, Betriebswasserstollen, Druck(wasser)stollen, Zulaufstollen, (Wasser-)Kraftstollen [*Stollen, der im planmäßigen Betrieb voll und unter Scheiteldruck läuft*] = tunnel-type penstock, conversion tunnel, pressure tunnel, pressure gallery, hydro tunnel, hydro tube, power tunnel

Triebwelle f, Antriebswelle [*die Eingangswelle bei Getrieben*] = input shaft

~, Antriebswelle [*eine Welle, die ein Fahrzeug in Bewegung setzt*] = driving shaft, drive ~

~ **der Nockenwelle**, Antriebswelle ~ ~ = drive shaft of the camshaft

Trigonometrie f = trigonometry

trigonometrische Höhenmessung f, Höhenbestimmung durch Messung von Vertikalwinkeln, trigonometrisches Nivellement n = trigonometrical levelling

trigonometrischer Punkt m, TP, Dreieckpunkt = trig(onometrical) station

Trikalziumaluminat n, aluminatische Schmelze f, 3 CaO x Al_2O_3 [*abgek.* C_3A] = tricalcium aluminate

Trikalziumsilikat n, 3 CaO x SiO_2 [*abgek.* C_3S] = tricalcium silicate

Trimetrogon-Bild n = trimetrogon photograph

Trimmballast m = trim ballast
Trinidad|-Asphaltsee m = (Trinidad) Pitch Lake, Trinidad ~
~bitumen n = Trinidad lake asphalt cement
~-Epuré n, gereinigter Trinidad-Asphalt m, Epurée- Asphalt = Trinidad épuré, ~ refined asphalt, parianite
~-(Roh)Asphalt m = (crude) Trinidad Lake Asphalt
Trinitrotoluol n = trinitrotoluene, TNT
trinkbar, genießbar = potable
Trinkbrunnen m = drinking fountain
Trinkwasser n = drinking water, potable ~
~aufbereitung f = processing of drinking water, ~ ~ potable ~
~becken n = drinking water reservoir, potable ~ ~
~behälter m = drinking water tank, potable ~ ~
~-Entsalzungsanlage f = demineralization plant
~versorgung f = potable water supply, drinking ~ ~
~zapfstelle f = bubbler (US)
Trio|straße f [*Walzstraße*] = three-high train
~walzwerk n = three-high mill
Tripel m, Polierschiefer m = tripoli
~wellblech n, Dreifachwellblech = triple corrugated (sheet) iron
Triplex|-Bohrhaken m = triplex drilling hook, ~ rotary ~
~-Schaffußwalze f [*wird erhalten durch Zusammenbau von 3 angetriebenen Einzel-Schaffußwalzen nach dem Baukasten-System*] = triplex sheepsfoot roller
Trippelbrett n [*Rotarybohrturm*] = thribble board, monkey ~
Tritt m, (Treppen)Stufe f = step
~belag m, (Tritt)Stufenbelag = (stair) tread cover(ing)
~blech n = tread plate
~fläche f → Auf~
~kante f [*Spaten*] = tread
~leiste f → Stoßleiste

~platte f → Stoßplatte
Trittschall m = footstep sound
~dämmplatte f = insulating slab against structure-borne sounds
~dämmung f = footstep (sound) insulation
~minderung f = footstep (sound) reduction
~pegel m = footstep (sound) level
~stärke f = footstep (sound) intensity
trittsicher [*Fußboden*] = non-slip
Tritt|spaten m = treaded spade [*Is used for digging compact soil which is not too hard or stony*]
~stufe f, Auftritt m, Auflagestufe = (stair) tread [*The member forming the horizontal top furface of a step*]
(~)Stufenbelag m, Trittbelag, Treppenbelag = (stair) tread cover(ing)
~verkehr m = foot traffic
Trochitenkalkstein m = entrochal limestone
trochoidale Dünung f, Gerstner-Welle f = trochoidal wave
trocken → dürr
~, mörtellos = dry
~ gelöschter Branntkalk m → Löschkalk
trockene Bohrung f → Fehlbohrung
~ (Zylinder)Laufbüchse f = dry cylinder liner
trockenes Erdgas n → Trockengas
~ Naturgas n → Trockengas
~ Vormischen n [*Betonbereitung*] = dry pre-mixing
Trocken|-Abbauhammer m = dry pick hammer
~abmeßanlage f → Trockendosieranlage
~abort m, Latrine f = privy
~(ab)siebung f = dry screening
~apparat m → Exsikkator m
~aufbereitung f [*Zement*] = dry manufacture
~aufbereitung f **von Kohle** = dry coal preparation
~aushub m = dry excavation
~bagger m [*auf dem Land fahrender Bagger*] = excavator

trockenbaggern, (ab)baggern, ausbaggern = to dig, to excavate
Trocken|ballast *m* = dry ballast
~bauweise *f* [*Verlegung von Platten anstelle von Verputzen*] = dry construction
~beet *n*, Schlamm~ = sludge(-drying) bed [*Natural or artificial layers of porous material upon which sludge is dried by drainage and evaporation*]
Trockenbeton *m* [*Gemisch ohne Wasserzugabe*] = (dry-)batched aggregate, ~ materials, matched ~, recombined ~
~-Mischungsverhältnis *n* = cement aggregate ratio
Trocken|bohren *n*, Trockenbohrung *f* = dry drilling
~bohrverfahren *n* = dry drilling system
~charge *f* [*Beton*] = dry batch
~dichte *f* = dry density
~dock *n* [*Ein Bauwerk, in das ein zu dockendes Schiff hineingefahren wird und auf dessen Boden es sich nach dem Lenzen des Dockes auf Kielstapeln absetzt und mit Kimmschlitten abgestützt wird*] = dry dock, graving ~
~dosieranlage *f*, Trockenzuteilanlage, Trockenabmeßanlage, Trockenzumeßanlage [*Betonaufbereitung*] = dry-batch(ing) plant
~drehofen *m*, Trockenrotierofen = dry process-type rotary kiln
~eis *n*, Kohlensäureschnee *m* = dry ice, solid carbon dioxide
~-Elektroabscheider *m* = dry type electric precipitator
~fähigkeit *f* [*Fehlname*] → Trocknungsfähigkeit
~farbe *f* = paint powder
~feldbau *m* = dry farming
~festigkeit *f* = dry strength
~filter *m*, *n* = dry filter
~fuge *f*, Knirschfuge = non-bonded joint, dry ~
~gas *n*, trockenes Erdgas, trockenes Naturgas [*Erdgas ohne höhere Kohlenwasserstoffe*] = dry (natural) gas

~gebiet *n* = arid region, ~ area, ~ zone
~gemisch *n*, Trockenmischung *f* = dry mix(ture)
~gewicht *n* = dry weight
~jahr *n*, wasserarmes Jahr = dry year
~kollergang *m*, Trockenläufermühle *f* = dry pan mill, edge-runner ~, chaser ~
~kupp(e)lung *f* = dry clutch
trockenlaufender Kolbenkompressor *m* = dry-piston compressor
Trockenläufermühle *f* → Trockenkollergang *m*
trockenlegen, entwässern = to dewater, to unwater, to drain
Trocken|legen *n*, Entwässern, Entwässerung *f*, Trockenlegung = dewatering, draining, unwatering
~legung *f* **der Baugrube** → Baugrubenentwässerung
~legungsarbeiten *fpl*, Entwässerungsarbeiten = drainage work
~löffelbagger *m*, Löffeltrockenbagger = shovel [*The term "mechanical shovel" is sometimes used to designate a "Trockenlöffelbagger" as distinct from a "shovel" = Schaufel. But "mechanical shovel" can also mean a "seilbetätigter Trockenlöffelbagger" as distinct from a "hydraulic shovel" = Hydraulik-Trockenlöffelbagger*]
~löschen *n* **von Kalk** = dry slaking
~lufreiniger *m*, Trockenluftfilter *m*, *n* = dry-type air cleaner, ~ ~ filter
~-Magnetscheider *m* = dry magnetic separator
~-Magnetscheidung *f* = dry magnetic separation
~mahlanlage *f* = dry grinding plant, ~ ~ installation
~mahlung *f*, Trockenmahlen *n* = dry grinding
Trockenmauer *f*, Bergemauer, Bergedamm *m* [*Eine aus Gesteinsstücken ohne Mörtel aufgesetzte Mauer*] (⚒)
~, Trockensteinmauer = dry-stone wall

~mauerwerk *n*, Trockensteinmauerwerk = dry masonry, ~ stone walling [*walling laid without mortar*]
~mauerwerkdamm *m* → → Talsperre
~mischung *f*, Trockengemisch *n* = dry mix(ture)
~mörtel *m* [*Fabrikmäßig hergestellter trocken gelieferter Mörtel*] = ready-made mortar
~mühle *f* = dry grinder
~platz *m* [*Betonwerk*] = curing area
~pochwerk *n*, Trockenpochmühle *f* = dry stamp battery, ~ gravity stamp
~polieren *n* = dry polishing
~pressen *n* = dry pressing
~preßverfahren *n* = dry press method
~raumgewicht *n* = bulk density, dry ~
~raumgewicht *n* eines Bodens = dry weight of the (soil) aggregate
~reibung *f*, Festkörperreibung = solid friction
~rotierofen *m*, Trockendrehofen = dry process-type rotary kiln
~rückstand *m* = dry residue
~scheibenkupp(e)lung *f* = dry plate clutch
trockenschleifen = to dry-ground
Trocken|schleifmaschine *f* = dry grinding machine, ~ grinder
~schleuse *f* → Dockschleuse
~schmiermittel *n* = dry-film lubricant
~schrank *m* = drying oven
~siebung *f* = dry mechanical grading
~(stein)mauer *f* = dry-stone wall
~steinmauerwerk *n* → Trockenmauerwerk
~stoff *m* = drier [*A material containing metallic compounds added to paints and painting materials for the purpose of accelerating drying*]
~tal *n*, Wadi *m* = wadi [*dried-up bed of stream*]
~-Trommel *f*, Trommel-Trockner *m*, Gesteinstrockner, Straßenbautrokkentrommel, Straßenbautrommeltrockner = aggregate drier (or dryer), revolving (or rotary) drier (for road-making aggregates)
~tundra *f* = dry tundra
~- und Mischanlage *f* = drying and mixing plant
~- und Naßmischen *n* während der Fahrt *f* [*Transportbeton*] = truck mixing
~verfahren *n* [*Zementherstellung*] = dry process
~vormischzeit *f* = dry mixing time
~wetterabfluß *m*, Trockenwettermenge *f* = dry-weather flow [*Flow of sewage in a sewer during dry weather*]
~zumeßanlage *f* → Trockendosieranlage
~zuteilanlage *f* → Trockendosieranlage
Trocknen *n* der Neubauten = drying out of new buildings
trocknendes Öl *n* = drying oil
trockner Abfallstoff *m* [*Kehricht; Küchenabfall; Asche usw.*] = dry waste
Trocknung *f*, Trocknen *n* = drying
Trocknungs|beginn *m* = commencement of drying
~brett *n* = drying pallet
~fähigkeit *f* [*Fehlname: Trockenfähigkeit*] = drying property [*The property of a coat of paint or varnish to dry by evaporation of the vehicle or chemical change (usually oxidation) or the two together*]
~geschwindigkeit *f* = rate of drying, drying rate
~gestell *n* = drying rack
~mechanismus *m* = mechanism of drying
~-Schrumpfung *f*, Erhärtungsschwindung [*Beton*] = drying shrinkage
Trog *m*, Wasch~ [*Schwertwäsche*] = tub, washer box, trough [*log washer*]
~, Mulde *f* = trough
~aufgeber *m* → Aufgabetrog *m*
~band *n* → Muldenband
~bandförderer *m* → Stahlgliederband
~brücke *f*, Brücke mit unten liegender Fahrbahn, offene Brücke mit versenkter Fahrbahn = trough bridge, open ~

~bunker m, Silobunker = trough-type bin
trogförmiger Formstahl m, Rinnenstahl = troughing, trough section
Trog|gurt m → Muldenband n
~kettenförderer m **System Redler** [*ist einstellbar und fördert eine bestimmte Menge pulverförmiger Stoffe in der Zeiteinheit*] = Redler conveyer, ~ conveyer
~mischer m = pan mixer
~platte f, Belegstahl m, Zoreseisen n = trough plate
~schleuse f, (Schiff)Hebewerk n für Naßförderung = trough lift
~tal n → U-Tal
Trommel f = drum
~ → → Bandförderer
~ mit Freilauf, Freilauftrommel [*Tiefbohrtechnik*] = free-wheeling drum
~ zur Tonkügelchenherstellung = drum pelletizer
~abscheider m = drum separator
~dosierer m → Dosier(ungs)trommel f
~drehzahl f = drum speed, ~ rotations per minute
~drehzahlzähler m = drum revolution counter
~durchmesser m = drum diameter
~(elektro)motor m → → Bandförderer
~filter m, n = rotating filtering drum
~flansch m = drum flange
~fördermaschine f (⚒) = drum winder
~förderung f (⚒) = drum winding
~kühler m, Kühltrommel f = cooling drum, rotary cooler
~-Löscher m → (Kalk)Löschtrommel f
~magnetabscheidung f [*Ein- oder mehrmaliges Abscheiden von magnetischen Bergen mit Hilfe von Trommelmagnetscheidern nach und zwischen den aufeinanderfolgenden groben Brechstufen*] = cobbing
~magnetscheider m = cobbing separator
~mantel m = drum shell
~mischer m = tumbling mixer
~mischer m → → Betonmischer
~motor m = → → Bandförderer
~mühle f = cylindrical batch mill
~rechen m = drum screen
~schmelzofen m = drum melting furnace
~(seil)winde f → Kanalwinde
~seilwindwerk n → Seilwinde
~sieb n, Wälzsieb, Siebtrommel f = trommel (screen), cylindrical ~, drum ~, rotary ~ [*A screen in the form of a cylinder or truncated cone rotating on its axis*]
~sinterofen m = rotary sintering kiln
~stauwehr n → Trommelwehr
~-Trockner m → Trocken-Trommel f
~trocknung f = rotary drying
~trocknungs- und -erwärmungsvorgang m = rotary drying and heating process
~ventil n = spool valve
~vibrationssieb n, Vibrations-Trommelsiebmaschine f = revolving vibrating screen, ~ vibratory ~
~vorwärmer m = drum preheater
~wäsche f → Waschtrommel
~waschsieb n = cylindrical washing screen
~wehr n, Trommelstauwehr = drum weir (Brit.) [*No US term. Drum gate in the USA means a sector gate*]
~welle f = drum shaft
~winde f → Kanalwinde
~zuteiler m → Dosier(ungs)trommel f
Trompeten-Abzweig m, Trompete f, Trompetenlösung f = trompet intersection, ~ type junction
Trompeter'sche Zone f (⚒) = Trompeter's zone [*Ogival stress-free zone around the roadways*]
Tropen|ausrüstung f = tropical outfit
~einsatz m = tropical use
~gewitter n = tropical storm
~krankenhaus n = tropical hospital
~kühler m = tropical radiator
~schichtung f, direkte Wärmeschichtung, lineare Wärmeschichtung = tropic temperature lamination, direct ~ ~
tropenverwendbar = tropic designed

Tropfdüsenmundstück *n* = drip nozzle
Tropfen *n* = dripping
~abfall *m* [*Rohr*] = leakage
tropfendicht = drop tight
Tropfkörper *m* = trickling filter, coarse-grained ~, percolating ~, sprinkling ~ [*An artificial bed of coarse material, such as broken stone, clinkers, slate, slats, or brush, over which sewage is distributed and applied in drops, films, or spray, from troughs or drippers, moving distributors, or fixed nozzles, and through which it trickles to the underdrains, giving opportunity for organic matter to be oxidized by biochemical agencies*]
~anlage *f* = purification plant with trickling filters
~fliege *f*, Psychodafliege, Psychoda alternata = filter fly
~material *n*, Brockenmaterial = contact material, filtering ~
~sprenger *m* = filter bed sprinkler
Tropf|öler *m* = drip oiler
~punkt *m* nach Ubbelohde [*Ist die Temperatur, bei der ein Tropfen durch sein Eigengewicht von einer gleichmäßig erwärmten Masse des Stoffes abfällt*] = Ubbelohde drop(ping) point
~wasser *n*, Ab~ = dripping water
tropfwassergeschützt [*Motor*] = dripproof
tropische Tide *f* [*Wenn der Mond die Wendekreise passiert*] = tropic tide
Trosse *f*, Abspannseil *n*, Ankerseil *n* = (back-)stay cable, guy ~, guy-rope, standing rope
Trossen|abspannung *f* → Abfangen *n* mit Trossen
~anker *m* = guy anchor
~-Derrick(kran) *m*, Seil-Derrick-(kran) *m* = guy derrick
~kranz *m* = guy table [*For a derrick crane*]
~verspannung *f* → Abfangen *n* mit Trossen
~zug *m* [*Schiff*] = rope pull, bollard ~

Trübekreislauf *m* = flotation circuit
Trübungs|messer *m* = turbidimeter
~punkt *m* [*Mineralöl*] = cloud point
Truhwasser *n* → Körwasser
Trümmer *f* = ruins
~beseitigung *f* = removal of ruins
~grundstück *n* = bomb-damaged site
~gut *n*, Trümmerschutt *m* = rubble from ruins
trümmerhaltiger Sand *m*, Trümmersand (Geol.) = detrital sand (Brit.); detritic ~ (US)
Trümmer|material *n* → Gesteinstrümmer *mpl*
~sand *m* → trümmerhaltiger Sand
~schutt *m*, Trümmergut *n* = rubble from ruins
~splitt *m* [*Splitt aus Mauerwerktrümmern*] = chip(ping)s from masonry ruins
~stück *n* (Geol.) → Fragment *n*
Trupp *m* → Kolonne *f*
T-Scharnier *n* = tee hinge, T-hinge
tschernosemartiger grauer Waldboden *m* → degradierte Schwarzerde *f*
TSR-Bauweise *f* [*Die Diagonalstäbe aus Holz werden mit Blechen versehen, die sich in den Anschlüssen unmittelbar überdecken und durchgenagelt werden*] = timber-steel-rivet method
T-Stahl *m* = tees
T-Stück *n*, Doppelabzweiger *m* 90° = tee, Tee, tee-piece
T-Träger *m* = T-section
Tsunanis *f*, Erdbebenflutwellen *fpl* = tsunamis, seismic sea waves
Tübbing *m* = segment
~ausbau *m* = tubbing
~-Einbaumaschine *f* = segment placer
Tuchlutte *f* (⚒) = canvas tube
Tudorbogen *m* → geschleifter Spitzbogen
Tulpennaht *f* = single U groove with vertical sides
Tünchbürste *f* = whitewash brush
Tünche *f*, Weiße *f*, Kalkmilch *f* = whitewash, whitening, limewash
tünchen → kalken

Tunnel- und Stollenbau

Tunnel- und Stollenbau *m*	construction of tunnels and galleries
abfangen [*den Druck*]	to resist
abfangen [*eine Schicht*]	to hold
abkippen [*Scholle*]	to tip
abklingen, ausklingen [*Geräusch; Erschütterung*]	to fade, to die away
Ablösung *f* der Schichten, Aufblätterung ~ ~, Ablösen *n* ~ ~, Aufblättern ~ ~	bed separation
Abluft *f*, Schlechtluft, verbrauchte Luft	exit air
Abluftkanal *m*, Schlechtluftkanal	exit air duct
Abluftventilator *m*, Ablüfter *m*	exit air fan
abplatzen	to spall
abreißen [*Firste*]	to break
Absaugebewetterung *f*, Saugbewetterung, saugende Bewetterung, Bewetterung durch Unterdruck	exhaust ventilation
Abschlag *m*, Sprengabschnitt *m*, Sprengstrecke *f*	round
Abschlaglänge *f*	length of a round, advance per ~
Abstiegstollen *m* → Zugangstollen	
Abstützung *f*	support
Abtreiben *n*, Nachbrechen, Nachbruch *m*, Nachreißen	barring (down), scaling [*Removal of loose rock(s) from the roof, walls, or invert*]
Abwasserstollen *m*	sewer tunnel
abwechselnde Saug- und Drucklüftung *f*, Wechsellüftung	plenum method of ventilation, push-pull ~
Akkord *m*, Gedinge *n*	piecework
Akkordarbeiter *m*, Gedingearbeiter	pieceworker, piece-rate man
Akkordzeit *f*, Vorgabezeit, Gedingezeit	allowed time, incentive ~
allseitiger Druck *m*	pressure acting in all directions, ~ from all sides
Anfangsschüttung *f* [*Wassereinbruch*]	initial yield
(an)geschärft [*Stempel*]	sharpened
angreifende Kräfte *fpl*	acting forces
Angriffstelle *f*	point of attack
Anker *m* nach dem Einsetzen zementiert	grouted bolt
Anker *m* nach der Zementierung eingesetzt	cemented bolt placed in previously filled hole
Ankerausbau *m*, Einziehen *n* von Deckenankern, Verbolzung *f* der Gewölbe-Kappen [*Schweiz: Nageln n*]	roof-bolting, roof-bolt installation, strata bolting
Ankerbalken *m*	anchorage beam

Tunnel- und Stollenbau

Ankerbolzen *m*, Verankerungsbolzen, (Gebirgs)Anker *m*	(rock) bolt, roof ~
Ankerloch-Bohrgerät *n*	roof-pinning rock drill
Ankerplatte *f*	roof bolt plate
Ankerstollen *m*	anchorage gallery
Anordnung *f* von Stempeln, Stempelanordnung	arrangement of posts, ~ ~ props
Anrisse zeigen	to show slight breaks
anschießen [*das Hangende*]	to blow down
Ansetzen *n*	collaring a hole
anstehend, gesund	solid
ansteigendes Ort *n*	incline
Arbeitsgeschwindigkeit *f*	rate of (face) advance, ~ ~ (~) progress
(Arbeits)Schacht *m*	(working) shaft, construction ~
Arbeitsspiel *n*	repetitive cycle
Arbeitsstufe *f*, Arbeitsteilvorgang *m*	sub-operation
Arbeitsunterbrechung *f*, Betriebsstörung	delay, interruption
Aufblätterung *f* der Schichten → Ablösung ~ ~	
Aufbruch *m*, hochgebrochener Schacht *m*, Hochbruch, Überbruch	raise(d shaft)
Aufbruchbauweise *f*, Dreizonenbauweise, österreichische Bauweise, österreichisches Verfahren *n*, österreichische Methode *f*, Sohlenort *n* und Firste *f*, Sohlenort-Firstenstrossenbau *m*	(English-)Austrian method, bottom heading-overhead bench
auffahren	to drive (headings), ~ ~ a heading
Auffahrung *f*, Vortrieb *m*, Auffahren *n*	(heading) driving, (face) advance, driving of headings
aufgenommene Last *f*	accepted load, sustained ~, carried ~
aufladen [*Das Haufwerk eines Abschlages durch die Lademaschine*]	to clean a round, to load out
Auflockerung *f*	breaking up
Auflockerungsdruck *m*	breaking up pressure
aufnehmen [*Last*]	to accept, to sustain, to carry
Aufschlitzung *f* in den First [*österreichische Bauweise*]	raise [*It is driven upwards from the bottom heading*]
aufwältigen [*Strecke*]	to re-open
Ausbau *m*, Ausmauerung *f* [*Ausbauformen sind: Spritzbetonabdeckung, Ausmauerung mit Mauerwerk, Ausmauerung mit Beton und Ausmauerung mit Stahlbeton*]	(permanent) lining
ausbauen, ausmauern	to line (permanently)
Ausbrechen *n* des Hangenden	flaking of the roof

Tunnel- und Stollenbau

Ausbruch m, Haufwerk n, Schutter m, Ausbruchmaterial n [*Schweiz: Schottergut n*]	muck (pile), debris, spoil
Ausbruch m, Ausbrechen n untertägiger Räume	underground excavation
Ausbrucharbeiten *fpl*	underground excavation work
Ausbruchdurchmesser m	excavation diameter
Ausbruchkubatur *f*	volume of excavation
Ausbruchprofil n	minimum excavation line
Ausbruchquerschnitt m, gesamtes Querprofil n, Vollprofil	full section
Ausbruchstelle *f* → Vortriebstelle	
ausgezimmert	timbered
Auskleidung *f*, Verkleidung	facing, lining
ausklingen, abklingen [*Geräusch; Erschütterung*]	to die away, to fade
Ausleger-Bohrwagen m, Bohrjumbo m, Großbohrwagen, Ausleger-Bohrjumbo	(tunnel) jumbo, drilling ∼
Auslösebalken m	chock release
ausmauern, ausbauen	to line (permanently)
Ausmauerung *f* → Ausbau	
Auspressung *f* → Einpressung	
Auspreßgut n → Injektionsgut	
Ausrichtungsarbeiten *fpl* im Gestein	development work
auswandern [*Gebirge*]	to creep
ausweiten [*österreichische Bauweise*]	to open (the section) outwards
Ausweiten n	opening-out
Außenring m	external ring
Ausziehgleis n	extension track
automatisch arbeitender Tunnel m	automatic tunnel
bankig, gebankt	bedded, stratified in thick beds
bankrecht	normal to the stratification
bankschräg	at an oblique angle to the stratification
Basisstollen m	base gallery
Basistunnel m	base tunnel
Baudränage *f*	construction drainage system
Baustollen m → Zugangsstollen	
Bedienungssteg m [*In einem Verkehrstunnel*]	service walkway
Beileitungsstollen m, Überleitungsstollen	trans-mountain water diversion gallery
belgische Bauweise *f*, Unterfang(ungs)bauweise, Firstenvortrieb m, belgisches Verfahren n, Firstenort-Sohlenstrossenbau m	Belgian method, top heading-underhead bench
Belüftung *f*, Bewetterung	ventilation
Belüftungsanlage *f*	ventilation system
Belüftungsschacht m	ventilation shaft

Tunnel- und Stollenbau

Belüftungsturm *m* mit Betriebsgebäude	ventilation station
bergmännisch vorgetrieben	driven by underground means (without disturbing the surface)
Bergmannspfahl *m*	forepoling board
Bergtunnel *m*, Felstunnel, Gebirgstunnel	rock tunnel
Bergwasser *n*	underground water
Betonauskleidung *f*	concrete facing
Beton(ier)zug *m*, fahrbare Stollenbetonieranlage *f*, ortsbewegliche Stollenbetonieranlage	tunnel concreting train
Betonkalotte *f*	concrete dome-shaped roof
Betonmantel *m*, Betonauskleidung *f*, Betonverkleidung, nicht tragende Betonmauerung geringer Dicke	concrete lining, ~ facing
Betonmauerung *f*	permanent concrete supporting
Betonquerschnitt *m*	concrete cross section
Betonsohle *f*	concrete invert, ~ floor
Betontransport- und -einbauwagen *m*	placer
Betriebsstörung *f*, Arbeitsunterbrechung	interruption, delay
bewehrte Spritzbetonkappe *f*, armierte ~	reinforced shotcrete (roof) bar
bewehrter Ortbeton *m*, armierter ~	cast-in-place R.C., cast-in-situ ~
Bewetterung *f*, Belüftung *f*, Bauventilation *f*	ventilation during construction
Bewetterung *f* durch Unterdruck → Absaugebewetterung	
Bewetterungskanal *m*, Lüftungskanal, Wetterlutte *f*	ventilation duct
Blasversatz *m*, Verblasen *n*	pneumatic stowing
Bodenverlust *m*	loss of ground
Bohr-Jumbo *m* → Ausleger-Bohrwagen	
Bohrleistung *f* pro Mineur und Schicht	drilling performance per driller and shift
Bohrleistung *f* pro Schicht	drilling performance per shift
Bohrplan *m*	drilling pattern
Bohrstahlwagen *m*	drill-steel carrier
Bohr- und Ladegeschwindigkeit *f*	rate of drilling and loading
Bohrunterwagen *m*	drilling carriage
Bohrloch *n*, Einpreßloch, Auspreßloch, Verpreßloch, Injektionsloch	drilled grout hole
Brennereinbruch *m*	burn cut, shutter ~, Michigan ~
Brust *f* → Vortriebstelle	
Brustschild *m*, Tunnelschild, Vortriebschild	(tunnel) shield

Tunnel- und Stollenbau

Brustverzug *m*	breasting of the face
Decke *f*, Firste *f*, Hangendes *n*, First *m*	roof
Deponie *f*	dump
deutsche Bauweise *f*	German method
Doppelgleistunnel *m*, zweispuriger Eisenbahntunnel, zweigleisiger Tunnel	double track tunnel
Doppelkeilanker *m*	sliding-wedge bolt
Doppel(rohr)tunnel *m* → Zwillingstunnel	
Dreizonenbauweise *f* → Aufbruchbauweise	
Druckanhäufung *f*	accumulation of pressure
druckhaft	squeezing
druckhaft werden	to develop (dangerous) pressure, ∼ ∼ squeeze
Drucklufttunnelbau *m*, Druckluftvortrieb *m*	compressed air tunnel(l)ing, plenum process of driving tunnels, compressed-air method of driving tunnels [*Usually air is used in conjunction with a shield, but numerous small tunnels have been driven using only liner plates or wood cants*]
Druckluftversorgung *f*	compressed air supply
dünnbankig	thinly stratified
Durchhieb *m*, Durchschlag *m*	cut-through
durchörtern, durchfahren	to drive through, to cross, to work through
Durchschlag *m*, Durchhieb *m*	cut-through
Einbau *m*, Verbau [*Bei nicht standfestem Gebirge sichert der Einbau den Hohlraum gegen das Herabfallen von Steinen, Abstürze und Drücke*]	support
Einbruch *m*	cut
Einfahrschacht *m*	access shaft
Einpressung *f*, Injektion *f*, Auspressung, Verpressung	pressure grouting, grout injection
Einziehen *n* von Deckenankern → Ankerausbau	
Einzonenbauweise *f* → englische Bauweise	
Eisenbahntunnel *m*	railway tunnel (Brit.); railroad ∼ (US)
Eisenrüstung *f*, Stahl(aus)zimmerung	steel tunnel supports, ∼ timbering
englische Bauweise *f*, Einzonenbauweise, Longarinenbauweise	English method, ∼ system of timbering

Tunnel- und Stollenbau 1032

Entaschungsstollen *m*	ash removal gallery
Entwässerungsstollen *m*	drainage tunnel
fahrbare Stollenbetonieranlage *f* → Beton(ier)zug	
Fahrung *f*, Gehzeit *f*	travel(l)ing time
Fallschacht *m*	downhole
Fehlabschlag *m*	missed round
Felstemperatur *f*	rock temperature
Felstunnel *m*, Bergtunnel, Gebirgstunnel	rock tunnel
Felsüberdeckung *f*, Felsüberlagerung	rock cover
Felsvortrieb *m*	rock tunnel(l)ing
Fensterstollen *m* → Zugangstollen	
Firste *f*, Decke *f*, Hangendes *n*, First *m*	roof
Firstendruck *m*	roof pressure
Firstenort-Sohlenstrossenbau *m* → belgische Bauweise	
Firstenstrossenbau *m*	heading-and-bench
Firstenvortrieb *m* → belgische Bauweise	
Firststollen *m*, Firstvortrieb *m*	top (pilot) heading
Firstverzug *m*	roof lagging
Fluchtschacht *m*, Notausstiegschacht	emergency shaft, escape ∼
Flucht(weg)stollen *m*, Notstollen	escape gallery, emergency ∼
Förderschacht *m*	extraction shaft
Förderstrecke *f*	haulage way
Förder- und Versorgungseinrichtungen *fpl*	haulage and supply equipment
Förderung *f* im Gleisbetrieb	haulage by rail-mounted vehicles
Freispiegelstollen *m*	free-flow gallery
Frischluftkanal *m*	fresh air duct
frühtragend [*Stempel*]	early-bearing
Fugenverfüllung *f*	mudding the joints [*To prevent loss of air between liner plates*]
Fußgängertunnel *m*	pedestrian subway
Fußteil *m*, *n* [*Streckenbogen*]	stilt
gebankt, bankig	stratified in thick beds, bedded
Gebirge *n*	(tunnel) ground
(Gebirgs)Anker *m* → Ankerbolzen	
Gebirgsdruck *m*	ground pressure
Gebirgstunnel *m* → Felstunnel	
gebräch	unstable, bad, friable
Gedinge *n*, Akkord *m*	piecework
Gedingeabzug *m*	contract account [*For explosives, drill steels, etc.*]

Tunnel- und Stollenbau

Gedingearbeiter *m*, Akkordarbeiter	piece-rate man, pieceworker
Gedingezeit *f* → Akkordzeit	
gegenläufige Schneidköpfe *mpl* [*Tunnelbohrmaschine*]	counter-rotating cutterheads
Gegenortbetrieb *m*	heading in reciprocal direction to another one, driving ∼ ∼ ∼ ∼ ∼
Gehzeit *f*, Fahrung *f*	travel(l)ing time
Gelenkbogen *m*	articulated (roadway) arch
Gelenkkappe *f*	link bar, hinged ∼, articulated ∼
gesamtes Querprofil *n*, Ausbruchquerschnitt *m*, Vollprofil	full section
geschärft, an∼ [*Stempel*]	sharpened
geschossenes Gestein *n*	shattered rock, broken ∼
gesund, anstehend	solid
Getriebezimmerung *f*	forepoling
Gewölbe *n*	arch
Gewölbemauerwerk *n*	arch masonry
Gewölbeschalung *f*	arch formwork
Gewölbescheitel *m*	arch apex
Gleitkappe *f*, Schiebekappe	slide bar
Gleitkappenausbau *m*	slide-bar support
Groß-Bohrwagen *m* → Ausleger-Bohrwagen	
Großgrubenwagen *m*	large capacity mine car
Hallinger-Schild *m*	Hallinger (tunnel) shield
Hangendes *n*, Decke *f*, Firste *f*, First *m*	roof
Haufwerk *n* → Ausbruch	
Herstellung *f* eines Tunnels in offener Baugrube	open-cut tunnel(l)ing
Hilfsstollen *m* für den Bau → Zugangstollen	
Hinterfüllungsbeton *m*	backfill concrete
Hochbruch *m* → Aufbruch	
hochgebrochener Schacht *m* → Aufbruch	
Hochpuffen *n* [*Sohle*]	boiling up [*floor*]
Holzeinbau *m*, Zimmerung *f*	timber support, wood(en) ∼
Holzspreize *f*	timber spreader
Holzverzug *m*, hölzerner Verzug	wood(en) lagging, timber ∼
Hufeisenform *f*	horse-shoe shape
Hund *m*, Schutterwagen *m*, Stollenwagen, Stollenkipper *m*	skip, car
Injektion *f* → Einpressung	
Injektionsgut *n*, Verpreßgut, Auspreßgut, Einpreßgut	grout
Innenring *m*	internal ring

italienische Bauweise f	Italian method
Kabelstollen m	cable tunnel
Kalotte f	crown
Kalottenkappe f	crown bar
Kammer f, Kaverne f	underground chamber
Kämpferhöhe f	springing level
Kämpferlinie f	springline
Kanaltunnel m, Schiff(ahrt)tunnel	canal tunnel
Kappe f	bar
Kappeneinbau m	bar timbering [*A method of supporting the roof and sides by a horizontal top timber (called "bar") supported at each side usually by posts*]
Kappenkette f, Kappenreihe f	row of bars
Kaverne f, Kammer f	underground chamber
Keileinbruch m	wedge cut, centre $\sim$, V-cut
Keilhülsenanker m	wedge-and-sleeve bolt
knistern	to crackle
kontinuierliches Stollenbohrgerät n	continuous miner, $\sim$ gallery machine
Kraftfahrzeugtunnel m	motor tunnel, automobile traffic $\sim$
Kreisquerschnitt m	circular section
Kreistunnel m	circular tunnel
Kugelgelenkkappe f	ball-joint bar
Kunz'sche Rüstung f für kleine Ausbruchquerschnitte in rolligem Gebirge	Kunz steel frame shoring for tunnels of small cross-section in unstable ground
Ladeband n	loading belt
Lademaschine f, Lader m	loading machine, loader
Laden n und Schießen n	loading and shooting
Ladespiel n	loading cycle
Landtunnelbau m	land tunnel(l)ing
Längsentlüftung f, Längsbelüftung	longitudinal ventilation system
lichter Eiquerschnitt m, Eiquerschnitt im Lichten	clear egg-shaped section
lichter Querschnitt m, Querschnitt im Lichten	clear section
Liegendes n, Sohle f	bottom, floor, invert
Longarinenbauweise f → englische Bauweise	
Luftausbruch m	blow
Luftaustrittstelle f	air leak
Lüfter m	fan, blower
Lüftung f → Bewetterung	
Lüftungskanal m, Bewetterungskanal, Wetterlutte f	ventilating duct
Luttenleitung f	air line, $\sim$ piping
Mannschaftsleiter f	manway ladder
Materialschleuse f	materials lock, muck-lock

Tunnel- und Stollenbau

Mehranfall m, Überprofil n	overbreak
Messerhohlraum m	(steel) poling plates cavity
Messerkammer f	(steel) poling plates chamber
Messervortrieb m	(steel) poling plates heading, (∼) ∼ ∼ driving
Messervortriebverfahren n	(steel) poling plates method
mildes Gebirge n, weiches ∼	soft (tunnel) ground
Misch- und Einbauzug m für Tunnelbetonauskleidung	mixing-placing train for tunnel concreting
Mundloch n	opening
Nachbrechen n → Abtreiben	
Nachbruch m → Abtreiben	
nachbrüchig	loose
Nachreißen n → Abtreiben	
nachschießen	to trim
nachträglicher Vollausbruch m	subsequent enlargement to the full section of the tunnel
Nachtrieb m, Nachbrechen n	secondary heading
Nageln n [Schweiz] → Ankerausbau	
Nettoladezeit f, Nettoschutterzeit	actual mucking time, ∼ loading ∼
Nische f	niche
Notausstiegschacht m, Fluchtschacht	escape shaft, emergency ∼
Notstollen m, Flucht(weg)stollen	emergency gallery, escape ∼
Nutzquerschnitt m	pay-section
obertägige Anlagen fpl	surface plant
offene Baugrube f	cut-and-cover
Ort n → Vortriebstelle	
ortsbewegliche Stollenbetonieranlage f → Beton(ier)zug	
(Orts)Brust f → Vortriebstelle	
österreichische Bauweise f → Aufbruchbauweise	
Pfändung f, Vor∼	spiling, forepoling, horse-heading
Pfeilerschacht m	pier shaft
Portal n, Tunnel∼	portal, tunnel ∼
Portalstrecke f, Portalzone f, Portalbereich m	end section, portal ∼
Probestollen m, Sondierstollen	investigation gallery
Profil n	section
Quarzstaublunge f, Silikose f	silicosis
Querprofil n	cross section
Querschlag m	cross-drift
Querschnitt m im Lichten, lichter Querschnitt	clear section
radiale Fugenverschiebung f	radial displacement of the joints
Rahmen m, Türstock m	frame
Regelquerschnitt m	standard section
Revisionsschacht m	inspection shaft
Richtschacht m	pilot shaft

Tunnel- und Stollenbau

Richtvortrieb *m*, Richtstollen *m*, Vortriebsstollen	pilot heading, ~ tunnel, monkey drift
Röhre *f*	tube
rollig	noncohesive
Sandfangkaverne *f*	sand catching underground chamber
satt anliegend [*Kappe*]	flush with the roof
Saugbewetterung *f* → Absaugebewetterung	
Schacht *m*, Arbeitsschacht	(working) shaft, conctruction ~
Schacht *m* für Förderkorb-Gegengewicht	cage counterweight well
Schachtbaugrube *f*	shaft-type construction pit
Scheibenschneider *m*, Spanbrecher *m* [*Tunnelbohrmaschine*]	disk cutter (US); disc ~
Scheitelzone *f*	apex zone
Schiebebühne *f*	pass-over
Schiebekappe *f*, Gleitkappe	slide bar
Schiff(ahrt)tunnel *m*, Kanaltunnel	canal tunnel
Schildbauweise *f*, Schildvortrieb *m*	shield tunnel(l)ing
Schildmantel *m*	cylindrical steel skin
Schlechte *f*	cleat
Schlechtenrichtung *f*	cleat direction
Schlechtluft *f*, Abluft, verbrauchte Luft	exit air, vitiated ~
Schlechtluftkanal *m*, Abluftkanal	exit air duct, vitiated ~ ~
Schleusen *n*	lockage
Schleusenbediener *m*	lock tender
Schlitzkeilanker *m*	slot-and-wedge bolt
Schneidrand *m*	cutting edge
Schottergut *n* [*Schweiz*] → Ausbruch	
Schrägstollen *m*	inclined gallery
Schußloch *n*, Spreng(bohr)loch	shot hole
Schutter *m* → Ausbruch	
Schuttergleis *n*	muck track
Schuttermaschine *f*, Schuttergerät *n*	mucking machnie, tunnel loading ~
Schutterschlitz *m*	muck(ing) slot
Schutterung *f*, Schuttern *n* [*Aufladen und Abtransport des Haufwerks*]	mucking (out)
Schutterwagen *m* → Hund	
Schutterzug *m*	mucking (out) train
Schüttungsbeiwert *m* [*für Haufwerk nach dem Sprengen*]	swelling (coefficient)
Schwaden *m* [*dichte Wolke aus Gas, Staub und Rauch*]	fumes
schwimmendes Gebirge *n*	waterlogged ground
Seitenstollen *m* → Zugangstollen	
Setzen *n* der Auszimmerung	setting ribs
sich auflegen [*Hangendes*]	to begin to bear on

Tunnel- und Stollenbau

Sicherungsarbeiten *fpl*	supporting work
Sicherungsmaßnahme *f*	supporting measure
Sickerpackung *f*	dry packing
Silikose *f*, Quarzstaublunge *f*	silicosis
Sohle *f*, Liegendes *n*	floor, bottom, invert
Sohlengewölbe *n*	invert arch, bottom ∼, floor ∼
Sohlenort *n* und Firste *f* → Aufbruchbauweise	
Sohlschwelle *f*	bottom sill
Sohlvertrieb *m*, Sohlstollen *m*	bottom (pilot) heading, floor (∼) ∼
Sondierstollen *m*, Probestollen	investigation gallery
Spanbrecher *m* → Scheibenschneider	
Sparverzug *m*	partial lagging
spiralförmiger Tunnel *m*	spiral tunnel
Spreizhülsenanker *m*	expansion-shell bolt
Sprengabschnitt *m*, Abschlag *m*, Sprengstrecke *f*	round
Sprengschwaden *mpl*	blasting fumes
Sprengstelle *f*	blasting point
Sprengstrecke *f*, Sprengabschnitt *m*, Abschlag *m*	round
Spritzbetonabdeckung *f*	shotcrete lining
Spritzbetonkappe *f*	shotcrete bar
Spülstollen *m*	scouring gallery, flushing ∼
Stahl(aus)zimmerung *f*, Eisenrüstung	steel timbering, ∼ tunnel supports
Stahlbetonankerbalken *m*	R.C. anchorage beam
Stahlbetonkappe *f*	R.C. bar
Stahlbogen *m*	steel roadway arch
Stahlbogeneinbau *m*	steel arches support
Stahldielenverzug *m*	steel girder lagging
stählerne Getriebezimmerung *f*	steel forepoling
Stahlschachtrahmen *m*	steel shaft frame
Stahlstreckenbogen *m*	steel roadway arch
Stahlverzug *m*	steel lagging
standfestes Gebirge *n*	unsupported (tunnel) ground
Stempelanordnung *f*, Anordnung von Stempeln	arrangement of posts, ∼ ∼ props
Stollen *m* [*Unterirdischer Gang auf dem Gebiete des Wasserbaues, der Entwässerung, der Wasserversorgung sowie der Be- und Entlüftung*]	drift (Brit.); (small-diameter) tunnel, gallery
Stollenaufzug *m*	cherry picker [*An overhead travel(l)ing crane fixed in the roof over a track at a point where it can lift an empty car off the track and set it down on the neighbouring parallel track*]

Tunnel- und Stollenbau

Stollenauskleidungsmaschine *f*, Tunnelauskleidungsmaschine	tunnel liner, ~ lining machine
Stollenbagger *m*	mucking shovel, ~ machine, mechanical muck-loader, (tunnel) mucker, (tunnel) mucking machine, tunnel shovel
Stollenbau *m*	gallery construction
Stollenbauarbeiten *fpl*	gallery work
(Stollen)Beton(ier)zug *m*	gallery concreting train
Stollenbohrgerät *n*	gallery machine, miner
Stollenbrand *m*	gallery fire
Stollenkammer *f*	gallery chamber
Stollenkanal *m*	gallery canal
Stollenkipper *m* → Hund	
Stollenportal *n*	gallery portal
Stollenpumpe *f*	gallery pump
Stollenregelquerschnitt *m*	standard gallery cross section
Stollenschalung *f*	gallery formwork,
Stollen-Schaufelradbagger *m*	bucket wheel excavator for tunnel(l)ing
Stollenvortriebmaschine *f*, Streckenvortriebmaschine	tunnel(l)ing machine
Stollenwagen *m* → Hund	
Stollenzimmerung *f*	gallery timbering
Straßentunnel *m*	road tunnel, vehicular ~, vehicular subway
Strecke *f*	roadway
Streckenbogen *m*	roadway arch
Stromsohlensicherung *f*	blanket [*Dumped on the river bed ahead of the tunnel to fight blows on subaqueous tunnel work*]
Strosse *f*	bench
Strossen *n* [*Schießen des Kranzes*]	stoping [*in mines*]; enlargement [*in civil engineering*]
Stückigkeit *f*	fragmentation
Stütze *f*, Stempel *m*	prop, post
Tagesstollen *m* → Zugangsstollen	
Temperaturzunahme *f*	temperature increase
Tunnel *m* [*Gebirgsdurchstich im Zuge von Verkehrswegen*]	tunnel; cut (US)
Tunnel *m* in offener Baugrube hergestellt	immersed tunnel
Tunnelauskleidungsmaschine *f*, Stollenauskleidungsmaschine	tunnel liner, ~ lining machine
Tunnelbahn *f*	underground railway
Tunnelbau *m*	tunnel(l)ing, tunnel work, tunnel construction, tunnel driving

Tunnel- und Stollenbau

Tunnelbauer m, Tunnelbauingenieur m	tunnel(l)er, tunnel(l)ing engineer, tunnel man
Tunnelbau-Geologie f	tunnel construction geology
Tunnelbaustelle f	tunnel site
Tunnelbautechnik f	tunnel(l)ing practice
Tunnelbeleuchtung f	tunnel lighting (system)
Tunnelbetonieren n	concreting of tunnel(s)
Tunnelblech n [Kölner Tunnelbauweise]	metal plate forepole, $\sim \sim$ spile
Tunnelfertigteil m, n aus Beton	precast tunnel section
Tunnelluft f	tunnel air
(Tunnel)Portal n	(tunnel) portal
Tunnelschild m → Brustschild	
Tunnelsegment n	tube segment, tunnel $\sim$
Tunneltrasse f	tunnel route
Tunnelverbindung f	tunnel link
Tunnelzufahrt f	tunnel approach
Türstock m, Rahmen m	frame
Überbruch m → Aufbruch	
Überlagerung f, Überdeckung	cover, overlying ground
Überleitungsstollen m, Beileitungsstollen	trans-mountain water diversion gallery
Überprofil n, Mehranfall m	overbreak
Ulme f	side wall
Umleitungsstollen m	diversion gallery
Umstellungszeit f [Für Abrüsten und Aufrüsten]	changeover time
Unterfang(ungs)bauweise f → belgische Bauweise	
Untergrundbahntunnel m	subway (tube)
Untertagebauwerk n	underground structure
Unterwasserstollen m [Liegt unterhalb einer Turbinenanlage]	tailwater gallery
Unterwassertunnel m	underwater tunnel, subaqueous $\sim$
Unterwassertunnelbau m	subaqueous tunnel(l)ing, underwater $\sim$
Verankerungsbolzen m → Ankerbolzen	
Verbau m → Einbau	
Verbau m der Ortsbrust	breasting
Verblasen n, Blasversatz m	pneumatic stowing
Verbolzung f der Gewölbekappen → Ankerausbau	
verbrauchte Luft f, Abluft, Schlechtluft	exit air
Verlegen n der Schienen des Bohrwagens	setting jumbo track and wall plate
Verpressung f → Einpressung	
Verpreßgut n → Injektionsgut	

Tunnel- und Stollenbau

Versenken *n* fertiger Tunnelstücke oder des ganzen Tunnels in eine vorher ausgebaggerte Rinne	trench method of subaqueous tunnel(l)ing, sunken-tube method (or construction)
Versorgungsstollen *m*	utility gallery
Versuchsstollen *m*	trial gallery
verziehen [*Ortsbrust; Firste; Ulmen*]	to lag
Verzug *m*	lagging
vierteilige Stahlrippe *f*	four-piece steel rib
Vollausbruch *m*	full-circle mining of tunnel heading, full-face work, full-face attack, full-face tunnel(l)ing
Vollprofil *n*, gesamtes Querprofil, Ausbruchquerschnitt *m*	full section
Voreinschnitt *m*	pre-cut
Vorgabezeit *f* → Akkordzeit	
vorgespannter Tunnel *m*	tunnel with anchorage by prestressing the ground with steel anchors
vorgetrieben, der Tunnel wird ∼	the tunnel advances
Vorort *n* [*Brust des Vortriebstollens*]	(working) face of the pilot heading, (∼) ∼ ∼ ∼ ∼ tunnel, (∼) ∼ ∼ ∼ monkey drift
(Vor)Pfändung *f*	forepoling, horse-heading, spiling
Vorspannanker *m*	steel anchor
Vorstollen *m*	foregallery
Vortrieb *m*, Auffahrung *f*, Auffahren *n*	(heading) driving, (face) advance, driving of headings
Vortrieb *m* → Vortriebstelle	
Vortriebarbeiten *fpl*	driving work, heading ∼
Vortriebmesser *n*	(steel) poling plate
Vortriebschild *m* → Brustschild	
Vortriebstelle *f*, (Orts)Brust *f*, Ausbruchstelle, Ort *n*, Vortrieb *m* [*Arbeitsplatz am Vortrieb; oft sind damit auch die Querflächen am Streckenende verstanden*]	(working) face, heading, face of the heading
Vortriebstollen *m* → Richtvortrieb	
Vortriebweise *f*	driving method, heading ∼
vorübergehender Ausbau *m*, ∼ Sicherungsverbau *m*	temporary ground support
vorziehen [*Vortriebmesser*]	to shove [*(steel) poling plate*]
Wagenwechsel *m*	car changing
wasserdichte Stahlauskleidung *f*	steel lagging
wasserführend	water-bearing
Wasserstollen *m*	water gallery
Wasserumleitungsstollen *m*	water diversion gallery
Wasserzutritt *m*, Wassereinbruch *m*, Wasserandrang *m*	ingress of water, inflow ∼ ∼, inrush ∼ ∼
wechselhaft	changeable [*geological structure*]
Wechsellüftung *f*, abwechselnde Saug- und Drucklüftung	plenum method of ventilation, push-pull ventilation

weiches Gebirge n, mildes ~	soft ground
Wetterlutte f, Bewetterungskanal m, Lüftungskanal	ventilating duct
Wetterstrom m	ventilating air flow
Wiederauffahren n	re-driving
Ziegelausbau m, Ziegelausmauerung f	brick lining
Zimmerung f, Holzeinbau m	wood(en) support, timber ~
Zugangstollen m, Fensterstollen, Seitenstollen, Tagesstollen, Abstiegstollen, Hilfsstollen für den Bau, Baustollen, Fenster n	access tunnel, (side) drift, (construction) adit, approach adit, access adit
zweispuriger Eisenbahntunnel m, Doppelgleistunnel, zweigleisiger Tunnel	double track tunnel
Zwillingstunnel m, Doppel(rohr)tunnel	twin(-bore) tunnel

Tunnelfinger m [*Flughafen*] = tunnel finger
Tunnelofen m mit Plattengleitbahn = slab kiln, sliding panel tunnel ~
~wagen m = tunnel kiln car
Tüpfelprobe f, Tüpfelprüfung f, Tüpfelversuch m = spot test
Tür f = door
(Tür)Anschlag m = (door) rabbet, (~) rebate; (~) check [*Scotland*]
Turas m [*bei Gleisketten: (Gleis)Kettenstern* m, *(Gleis)Kettenrad* n] = sprocket (wheel)
Türbank f → Türschwelle
Türbeschläge mpl = door hardware, ~ furniture
Turbinen|auslaß m = turbine outlet, outlet of turbines
~auto n = turbine-powered automobile
~bautechnik f = turbine building practice
~bohren n = turbodrilling
~bohrer m, Bohrturbine f, Turbo-Bohrmaschine f = turbodrill
~grube f, Turbinenkammer f = turbine pit
~halle f = turbine hall
~haus n, Krafthaus, Maschinenhaus = powerhouse
~kammer f, Turbinengrube f = turbine pit

~leistungsdiagramm n = turbine performance chart
~leistungskurve f = turbine performance curve
~pumpe f = pump turbine
~regler m = turbine governor
~tisch m = turbine platform
(~)Wirkungsgrad m = generating efficiency
Türblatt n, Türflügel m = wing of door, leaf ~ ~
Turbo|(auf)lader m → Turbolader
~-Bohrmaschine f → Turbinenbohrer m
~brenner m = turboburner
~gebläse n = turboblower
~generator m = turbogenerator
Turbolader m, Turboauflader = turbocharger
~-Drucklager n = turbocharger thrust bearing
~-Gehäusestrebe f = turbocharger housing strut
~kompressor m = turbocharger compressor
~-Ladedruckverhältnis n = turbocharger compression ratio
~lager n = turbocharger bearing
~-Leitkranzquerschnitt m = turbocharger nozzle size
~schmierventil n = turbocharger lube valve

Turboladerträger — Turmkran

~träger m = turbocharger mounting
~-Wärmeschirm m = turbocharger heat shield
Turbo|(luft)verdichter m, Turbokompressor m, Kreisel(luft)verdichter = turbocompressor
~mischer m = turbine mixer
~satz m = turbo(generator) set
turbulente Fließgeschwindigkeit f = turbulent velocity
~ Grenzschicht f = turbulent boundary layer
~ Strömung f → Flechtströmung
turbulentes Fließen n → Flechtströmung f
Turbulenz f = turbulence [*A state of flow in which the liquid is disturbed by eddies*]
turbulenzfreie Strömung f, wirbelfreie ~ = eddy-free flow, turbulence-free ~, vortex-free ~
Türdichtung f = door seal
Türdurchgang m = doorway [*An opening provided with a door for access through a wall*]
Türfeder f = door spring
Türflügel m, Türblatt n = wing of door, leaf ~ ~
(Tür)Füllung f = (door) panel
Türfutter n = lining of door frame, ~ ~ ~ casing
Türgewände n = door jamb
Türhalter m = door holder
Türklingel f = door bell
Türklinke f = door latch
Türklopfer m = door knocker
Türleibung f = door reveal
Turm m = tower
~ → Kranmast m
~ → Bohr~
~, Pylone f, Pylon m, Kabelturm = pylon, tower, suspension ~
~, Kolonne f [*Erdölwerk*] = column
~abmeßanlage f → Turmdosieranlage
~betonzentrale f → Betonturm m
~biber(schwanz) m = tapered plain tile
~dach n, Pyramidendach = pyramidal roof, polygonal broach ~, spire ~
~destillation f, Kolonnen-Destillation = flash distillation
~dosieranlage f, Turmabmeßanlage, Turmzumeßanlage, Turmzuteilanlage, Dosierturm m, Abmeßturm, Zumeßturm, Zuteilturm = upright batching plant, ~ measuring ~, ~ apportioning ~, ~ gauging ~
~drehkran m → Turmkran
Türmechanismus m = door gear
Turm|fundament n = tower base, ~ foundation
~garage f mit Lift, Autosilo m = autosilo [*Patented system of Messrs. SICOMATIC, ZÜRICH, SWITZERLAND*]
~gaststätte f = tower restaurant
~gebäude n [*als Teil eines Hotels*] = tower block
~gefängnis n = dungeon [*A prison in or under a tower*]
~gerüst n, Schachtgerüst [*Bauaufzug*] = cage, (elevator) tower, elevation tower
~gerüst(bau)aufzug m, Schachtgerüst(bau)aufzug, Fahrstuhlaufzug = cage hoist, (elevator) tower ~, scaffold tower ~
~gerüstkippkübel m, Schachtgerüstkippkübel = tower (hoist) bucket
~hochhaus n = tower block
~kopf m = tower head

Turmkran m, Turmdrehkran	tower crane
Abbau m	dismantling, stripping
(Ab)Senken n	lowering
Abtreiben n durch Wind	drifting of the crane during storms
Alarmanlage f	warning system
Aufbau m, Montage f	erection
Aufbauausleger m, Montageausleger	erecting jib
Aufbaufolge f, Montagefolge	sequence of erection
Aufbauseil n, Montageseil	erection cable, ~ rope

Turmkran

Ausladung *f*	rad., reach, radius
Ausleger *m*	jib (Brit.); boom (US)
Auslegerrolle *f*	jib pulley
Auslegerverlängerung *f*	jib extension
Auslegerverstellbewegung *f*	luffing motion, ~ movement
Auslegerverstellen *n*	luffing
Auslegerverstellmotor *m*	luffing motor
Auslegerverstellseil *n*	luffing rope, ~ cable
Auslegerverstellvorrichtung *f*	luffing gear, ~ unit
Auslegerverstellzeit *f*	luffing time
Außenschiene *f*	outer rail, outside ~
Ballastbehälter *m*	ballast tank, ~ box
ballastieren	to ballast
Begrenzung *f* für die Auslegerverstellung	limit switch for luffing motion
Betonkübel *m*	concrete silo
Betonschutt *m*	concrete scrap [*used as ballast*]
Bremsen *n*	braking
Diagonalfahrantrieb *m*	two diagonally opposite truck wheels individually driven through motors
Doppelspurkranz-Schienenrad *n*, Doppelspurkranz-Gleisrad	double-flanged rail wheel
Drehbewegung *f*	slewing motion, swivelling ~
Drehen *n*	slewing, swivelling
Drehgeschwindigkeit *f*	slewing speed, swivelling ~
Drehkranz *m*	slewing crown, swivelling ~
Drehkreis *m*	slewing circle, swivelling ~
Drehverbindung *f*	slewing point, ~ connection, swivelling ~
Drehwerk *n*	slewing gear, ~ device, ~ unit, swivelling ~
Drehwerkbremse *f*	slewing (gear) brake, swivelling (~) ~
Einscherung *f*, Seil~	rope reeving, reeving (of ropes)
Eisenballast *m*	iron ballast
Elektroanlage *f*	electric(al) equipment
Fahrbegrenzung *f*	limit switch for travel motion
Fahren *n*	travel(l)ing
Fahrgeschwindigkeit *f*	travel(ling) speed
Fahrmotor *m*	travelling (unit) motor
Fahrwerk *n*	travelling machinery, ~ unit, ~ device, ~ gear
Federkabeltrommel *f*	spring cable (winding) drum, spring-loaded cable winder, spring-loaded cable drum
Fernsteuerung *f*	remote control, distance ~
Fernsteuerungsautomatik *f*	auto(matic) remote control, ~ distance ~
Gegenausleger *m*	counter-jib (Brit.); counter-boom (US)

Turmkran

Gegengewicht n	counterweight
Gesamtanschlußwert m	total connected electric load, ∼ connection value, ∼ electrical power, connected load total
Getriebekasten n	gear case, gearbox
Gleisrad n, Schienenrad	rail wheel
Hakenhöhe f	hook height, height under hook
Hakenrolle f	hook pulley
Heben n	hoisting, lifting
Hubbegrenzung f	limit switch for hoist motion, ∼ ∼ ∼ lifting ∼
Hubbewegung f	hoisting motion, lifting ∼
Hubgeschwindigkeit f	hoisting speed, lifting ∼
Hubmotor m	hoist motor, lifting ∼
Hubseil n	hoist rope, lifting ∼, ∼ cable
Hubtrommel f	hoist drum
Hubwerk n	hoisting gear, ∼ unit, ∼ device
Hupe f	klaxon device
Ingenieurbau-Turmkran m	civil engineering tower crane [*suited for construction of dams, bridges, piers, etc.*]
Innenschiene f	inner rail, inside ∼
Kabeltrommel f	cable (winding) drum, ∼ winder
Katzausleger m	troll(e)y jib (Brit.); ∼ boom (US)
Katzfahrgeschwindigkeit f	traversing speed
Kiesballast m	gravel ballast
Kippmoment n	overturning moment
Kontroller m	controller
Kontrollersteuerung f	controller control
Kranbetrieb m	crane operation, operation of a crane
Kranseil n	crane rope
Kugeldrehkranz m	ball-bearing slewing crown, ∼ ∼ device, ∼ turntable [*connecting the upper and lower carriage*]
Kugeldrehverbindung f	ball-bearing slewing joint, ∼ connection
Kurvenfahrwerk n	curve going gear, ∼ travel(l)ing equipment, ∼ travel(l)ing mechanism
(Last)Katze f, Laufkatze	troll(e)y
Laststellungsanzeiger m	load position indicator
Leiter f	access ladder
Montage f, Aufbau m	erection
Montageausleger m, Aufbauausleger	erecting jib
Montagefolge f, Aufbaufolge	sequence of erection
Montageseil n, Aufbauseil	erection rope, ∼ cable
Rillentrommel f	grooved(d) drum
Rollenhöhe f	height of jib pulley, pulley height
Sandballast m	sand ballast

Säule *f*, Turm *m*	tower
Schallalarmanlage *f*	sound warning system
Schienenrad *n*, Gleisrad	rail wheel
Schienenzange *f*	rail tongs
Schleifringmotor *m*	slip-ring motor
(Seil)Einscherung *f*	reeving (of ropes), rope reeving
Seilgeschwindigkeit *f*	speed of rope, rope speed
Seilrolle *f*	rope pulley
Seiltrommel *f*	cable (winding) drum, ∼ winder
Seilverankerung *f*	rope anchorage, cable ∼
Selbstmontagekran *m*	self-erecting crane
Senkbremsschaltung *f*	lowering brake switching
Senken *n*, Ab∼	lowering
Sicherung *f*	safety device
Sichtalarmanlage *f*	visible warning system
Straßenfahrt *f*	road towing
Straßenrad *n*	road wheel
Straßentransportachse *f*	axle for road transport
Straßentransportvorrichtung *f*	road travel(l)ing gear, ∼ ∼ device, ∼ ∼ unit
Teleskopdrehsäule *f*, Teleskopdrehturm *m*	telescopic rotary tower, ∼ slewing ∼, ∼ swivel(l)ing ∼
Teleskopturm *m*, Teleskopsäule *f*	telescopic tower
Teleskopturmkran *m*	telescopic tower crane
Tragkraft *f*	safe working load, S. W. L.
Turm *m*, Säule *f*	tower
Turmdrehkran *m*	rotary tower crane, slewing ∼ ∼, swivelling ∼ ∼
Turmdrehkran *m* mit nichtschienengebundenem Fahrwerk, nichtgleisgebundener Turmdrehkran [*also mit Radfahrwerk oder (Gleis-) Kettenfahrwerk*]	mobile tower crane
Turmstück *n*	(tower) member, (∼) section
(Turm-)Zwischenstück *n*	(tower) intermediate member, (∼) ∼ section, ∼ extension
Überlast *f*	excessive load, overload
Überlastsicherung *f*	overload cut-out, ∼ protection
Überlastung *f*	overloading
Unterwagen *m*	lower carriage, undercarriage
unzerlegt	assembled
unzerlegt transportabel	crane telescopes together for short transport lengths, towed behind a vehicle completely assembled
Wanderrolle *f*	wandering lead
Zwischenbühne *f*	intermediate platform
Zwischenrolle *f*	intermediate pulley

Turm|mauer *f* = tower wall
∼**ofen** *m* = tower furnace
∼**pfeiler** *m* [*Hängebrücke*] = (suspension) tower pier

~rolle *f*, Bohr~ = crown block
~schaft *m* = tower shaft
~spitze *f*, Turmhelm *m* = (top of) spire
~uhr *f* = tower clock
~zumeßanlage *f* → Turmdosieranlage
~zuteilanlage *f* → Turmdosieranlage
Turnhalle *f* = gym(nasium)
Türnischenbogen *m* = door niche arch
turnusmäßig = in a rotary pattern
Türöffnung *f* = door opening
Türpfosten *m*, Türständer *m*, Türsäule *f* = door post, ~ jamb, ~ sidepiece, ~ cheek
Türrahmen *m* = door frame
Türriegel *m* = door head
Türsäule *f* → Türpfosten
Türscharnier *n* = door hinge
Türschild *n* = door plate [*A metal plate on the door of a house or apartment carrying the name of the occupant*]
Türschließer *m*, Türzuwerfer *m* = door closer, ~ check (and spring)
(Tür)Schloß *n* = (door) lock
Türschwelle *f*, Türbank *f* = door sill, ~ threshold, ~ cill; ~ abutment piece (US)
Türschwellenstein *m*, Türbankstein = door stone
Türständer *m* → Türpfosten
Türstufe *f* = door step
Türsturz *m* = door lintel, ~ lintol
Türzarge *f* = door trim, ~ casing
Türzuwerfer *m* → Türschließer
Tuschfeder *f*, Ziehfeder = drawing pen
Tyler-Siebsatz *m* = Tyler series
Type *f*, Bauart *f* [*Maschine*] = type
Typen|beschränkung *f* = simplification (US); variety reduction, limiting variety
~gebäude *n*, Typenhaus *n* = standardized building
~halle *f* = standardized hangar
~haus *n*, Typengebäude *n*, Typenbau *m* = standardized building
~rahmen *m* [*Eisenbahnwagenbau*] = standardized framing
(~)Reihe *f*, Baureihe [*Maschine*] = series

~schild *n*, Bezeichnungsschild = name plate
typischer Löß *m* → Urlöß
typisierter Fertigteil *m*, typisiertes ~ *n* = standardized structural element
Typisierung *f* = standardization of types

U

U-Bahn *f*, Untergrundbahn = underground railway, tube ~ (Brit.); underground railroad (US)
U-Bahnhof *m*, Untergrundbahnhof = tube station
U-Bahnsteig *m*, Untergrundbahnsteig = subway station platform
U-Bahn-Tunnel *m*, Untergrundbahntunnel = tube tunnel
Ubbelohde-Viskosimeter *n* = suspended level viscosimeter
Überspannen *n* [*Spannbeton*] = overstressing
Überbau *m* [*Brücke*] = superstructure
~platte *f* [*Brücke*] = deck slab
überbaut (⚒) = influenced by overlying workings
Überbau|unterkante *f*, Konstruktionsunterkante [*Brücke*] = deck soffit
~voranstrich *m* [*Brücke*] = deck prime coat(ing)
überbeanspruchen = to overstress
Über|beanspruchung *f* = excessive stress
~belegung *f* [*Wohnung*] = overcrowding
überbemessen = to overdesign
Überbesetzung *f* mit Arbeitskräften = featherbedding
überbestückt [*Baustelle mit Maschinen und Geräten*] = overplanted
Über|bewässerung *f* = over-irrigation
~bleibselverband *m* (Geol.) = palimpsest structure
überbohren, Gestänge ~ = to wash over (drill) pipe
~, Ölsand ~ [*Rotarybohren*] = to miss an oil sand, to overlook ~ ~ ~
Überbohrrohr *n* [*Tiefbohrtechnik*] = wash pipe

überbrochener Schacht m → Hochbruch m
Überbruch m → Hochbruch
überbrücken = to bridge
überbrückter Verkehrsweg m = bridged traffic route
Über|chlorung f = superchlorination
~dach n → Dachkappe f
überdachen = to roof (over)
Überdachen n, Überdachung f = roofing over
über|decken [*Dachziegel*] = to (over-)lap
~decktes Becken n = covered reservoir
Überdeckung f [*Dachziegel*] = (over-)lap
~ [*Abstand zwischen Rohrscheitel und Geländeoberfläche*] = cover
~ [*Tunnelbau*] = overlying ground
~ der Bewehrung, Umhüllung ~ ~, ~ ~ Armierung, ~ ~ (Stahl)Einlagen = concrete cover
Überdeckungshöhe f [*Rohrgrabenverfüllung*] = depth of cover
überdimensionieren = to overdesign
Über|dosis f [*Wasserreinigung*] = overdose
(über)drehen [*Steinzeugindustrie*] = to jigger
(Über)Drehen n [*Steinzeugindustrie*] = jiggering
Überdrehen n [*Motor*] = overspeeding
Überdruck m = relief pressure
Überdruckbewetterung f → Druckbewetterung
~entlastungsventil n = over-pressure release valve
~schalter m [*Kältemaschine*] = high-pressure cutout
~schutz m = overpressure protection
Überdruckturbine f, Reaktionsturbine, Rückdruckturbine = reaction turbine
~ mit Radialdampfströmung = radial flow reaction turbine
Überdruckventil n = (pressure) relief valve, unloader ~
~ des hydraulischen Kettenspanners = track adjuster relief valve

Über(einwirkungs)fläche f (✹) = super-critical area of extraction
überempfindlich = over-sensitive
Über|fahrung f → Übergang m
~fall m = overflow
überfallender Strahl m → abfallender ~
überfallendes Wasser n = overfalling water
Überfall|glocke f [*Tiefbohrtechnik*] = overshot
~höhe f [*Höhenunterschied zwischen Überfallkante und Oberwasser*] = overflow height
~messung f [*Wassermengenmessung vermittels Überfallwehr*] = measuring by a clear overfall weir
~sperre f = overflow dam
~strahl m → abfallender Strahl
~strecke f, Überlaufstrecke = overflow section, river (spillway) ~, spillway ~
~fallsturzbecken n → → Talsperre f
~wehr n = clear overfall weir
überfalzen = to rebate
überfein = superfine
Über|fläche, Übereinwirkungsfläche (✹) = supercritical area of extraction
~flurbewässerungsverfahren n = open field (method of) irrigation
~flurhydrant m = pillar hydrant, surface ~
über|flüssig, überzählig [*Baustatik*] = redundant
~flutbare Brücke f, Tauchbrücke = submersible bridge, low level ~, causeway
Über|flutung f → Überschwemmung
~führung f = overbridge, overpass, overspan bridge, overcrossing
~füllpumpe f = filling pump
~füllung f **in der Spitzenzeit** [*öffentliche Verkehrsmittel*] = rush-hour overcrowding
~gabebühne f = transfer platform
~gabestelle f = transfer station
~gabevorrichtung f = transfer system
~gabewalzwerk n = pass-over mill, pull-over hot ~

Übergang *m*, Fahrt *f*, Überfahrt *f* [*z.B. mit einer Walze*] = pass, traverse

~, ~zone *f* [*Erdstaudamm*] = transition (zone)

~ **vom Turbinenbetrieb (Stromerzeugung) zum Pumpbetrieb (Strombezug)** = reversal of operation from generating to pumping, change-over ~ ~ ~ ~

~ **zwischen Meißel und Bohrbär** [*pennsylvanisches Seilbohren*] = sub for bit and auger-stem

Übergangs|boden *m* = transition soil

~**bogen** *m* → Übergangskurve *f*

~**bogenlänge** *f* = transition length

~**bogenspirale** *f* = highway transition spiral, road ~ ~

~**düse** *f* [*Vergaser*] = transition jet

~**gebirge** *n* = transition rocks

~**kalk** *m* → Grauwackenkalk

~**kurve** *f*, Übergangsbogen *m* = transition curve, easement ~

~**lage** *f*, Übergangsschicht *f* = transition course, ~ layer

~**lasche** *f*, Reduzierlasche = cranked fishplate (Brit.); ~ joint-bar (US)

~**moor** *n* = transition wood moor

~**moorboden** *m* = transition wood moor soil, carr, medium moor soil

~**periode** *f* = changeover period

~**rohr** *n* = taper pipe

~**schicht** *f*, Übergangslage *f* = transition layer, ~ course

~**stelle** *f* [*Diagramm*] = changeover point

~**strecke** *f*, Furtstrecke = inflexion stretch

~**temperatur** *f* = transition temperature

~**zone** *f* (Geol.) = transition zone

~**zone** *f*, Bindezone = weld junction

Übergemengteil *m*, *n*, akzessorisches Mineral *n*, zufälliges Mineral = accessory mineral

Übergewicht *n* = out-of-balance weight

~ [*Absenken eines Tunnelröhrenelementes*] = negative buoyancy

übergroß = outsize

Überhang *m* = overhang

überhart = superhard

Über|hauen *n* → Aufhauen

~**hitze** *f* = superheat

~**hitzer** *m*, Dampf~ = superheater

~**hitzekühler** *m* = de-superheater

~**hitzerrohr** *n* = superheater tube

überhitzte Stelle *f* = hot spot [*An exterior area of the kiln shell, usually in the burning zone, which becomes heated to a temperature sufficient to cause the shell to be red hot or to glow. Usually caused by the loss of coating or lining*]

Über|hitzung *f* = overheating

~**hitzungstemperatur** *f* = superheat temperature

überhöhte Kurve *f* = superelevated turn

überhöhter Spitzbogen *m* → lanzettförmiger

Über|höhung *f*, einseitiger Querschnitt *m* [*Straße*] = superelevation, cant; banking [*deprecated*]

~**höhung** *f*, Stich *m* = hog, camber [*Beams and trusses may be built with hog to counteract their sag*]

~**holen** *n* [*Verkehr*] = overtaking; passing [*deprecated*]

~**holen** *n* [*Maschine*] = overhaul

~**holspur** *f* = overtaking lane; passing ~ [*deprecated*]

~**holungssichtweite** *f* = overtaking visibility distance

Überjahres-Speicher(becken) *m*, (*n*) → → Talsperre

(über)kippen = to overturn

über|kippte Antiklinale *f*, liegende ~ = recumbent anticline

~**kippter Stoß** *m* (⚒) = over-tipped face

Überkopf|gesteinsbohrer *m* = stoper [*A rock drill for upward drilling*]

~**laden** *n* = overhead loading

~**ladeschaufel** *f* → Über-Kopf-(Schaufel)Lader *m*

~**schaufel** *f* = overhead bucket, flipover ~

~**-(Schaufel)Lader** *m*, Wurfschaufellader, Überkopf-Fahrlader, Rück(wärts)lader, Überkopfladegerät *n*,

Überkopfladeschaufel *f* = rocker (type) shovel (loader), overshot loader, overhead loader, flip-over bucket loader, overloader, overhead shovel
~-Schweißen *n*, Zwölf-Uhr-Schweißen = welding in overhead position, twelve o'clock-welding, overhead welding
Überkorn *n* [*Siebtechnik*] = oversize
~anteil *m* = oversize percentage
~ausbringen *n*, Grobausbringen = amount of coarse material reporting in the screen overflow
Über|kreuznagelung *f* [*Waagerecht und senkrecht eingeschlagene Nägel kreuzen sich im Holz aneinander vorbei*] = cross nailing
~lagerung *f* (Geol.) → Abraum *m*
~lagerung *f* [*Baustatik*] = superposition
Überlagerungs|druck *m* [*Bodenmechanik*] = overburden pressure
~prinzip *n* = principle of superposition
~schicht *f* → Abraumschicht
~strömung *f* [*Meer*] = superimposed current
Überland|leitung *f* (für Strom) = transmission line
~landmast *m*, Freileitungsmast = electric power pylon, (powerline) tower, transmission line tower
~rohrleitung *f* = cross-country pipeline
~straße *f* = road, highway, rural route
~-Straßenbahn *f* = interurban tramway
~verkehr *m* = interurban traffic
Über|lappnaht *f* [*Schweiz*]; Kehl(schweiß)naht am Überlappungsstoß = lap-joint fillet weld, lapwelded seam
~lappung *f* = overlap
Überlappungs|nietung *f* = lap riveting
~schweißen *n* = lap welding
~stoß *m* = overlap joint
Überlast *f* = excess load, overload
über|lasten = to overload
~lastig = top-heavy

Über|lastkupp(e)lung *f* = overload clutch
~last(ungs)schutz *m* = overload protection, ~ kickout
Überlauf *m* = spillway, wasteway [*A passage for spilling surplus water*]
~ [*Siebtechnik*] = screen overflow
~ **mit Betriebsverschluß** = gated spillway, ~ wasteway
~abschnitt *m* = spillway section, wasteway ~
~bauwerk *n*, Überlaufanlage *f* = overflow
~deich *m* → Sommerdeich
Überlaufen *n* = overflowing
überlaufendes Wasser *n* = overflowing water
Überlauf|kanal *m*, Hochwasserentlastungskanal = overflow channel, ~ chute, ~
~krone *f* = spillway crest
~platte *f*, Materialauffangplatte = spill plate extension
~quelle *f* = depression spring
~rohr *n* = overflow pipe, ~ stand [*A standpipe in which water rises and overflows at the hydraulic grade line*]
~stollen *m*, Hochwasserstollen = tunnel-type discharge carrier
~strecke *f* → Überfallstrecke
~wehr *n* → Überfallwehr
Überleitung *f*, Beileitung [*Wasser von einem Flußgebiet in ein anderes*] = trans-mountain water diversion
überluckte Schalung *f* [*Schweiz*]; ·Stülpschalung = weatherboarding
Über|nahmebedingungen *fpl* [*Schweiz*] → Bauleistungsbuch *n*
~produktion *f* = overproduction
~querungsverkehr *m* [*über einen Fluß*] = cross-river traffic
~rollbrücke *f* = roller bridge sliding over the fixed part
übersättigt = supersaturated
Überschall|flugzeug *n* = supersonic aircraft
~methode *f* → Ultraschallmethode
~windkanal *m* = supersonic wind tunnel

Über|schiebung *f* (*Geol.*) → Wechsel *m*
~schlag *m* [*Sprengtechnik*] = self-excitation
überschläglicher Kostenanschlag *m* = spot estimate
Über|schlagwasser *n* → Spritzwasser
~schneidung *f* [*Holzbau*] = notching
~schneidung *f* [*Bewehrung*] = lap
über|schneidungsfreie Kreuzung *f* → plankreuzungsfreie ~
~schüssiger Aushub(boden) *m* = spoil, surplus earth (or soil), waste
Überschuß|energie *f*, Überschußstrom = secondary power, surplus ~, dump ~, excess ~, ~ energy
~gebiet *n* = overspill area [*Refers to population in a given area*]
~schlamm *m* [*Abwasserwesen*] = excess sludge, surplus ~
~strom *m*, Überschußenergie *f* = surplus power, secondary ~, dump ~, excess ~, ~ energy
~wasser *n* = excess water, surplus ~
Überschüttungshöhe *f* [*Betonverteilung*] = surcharge, excess of concrete
~ an der Tagesarbeitsfuge [*Betondeckenbau*] = end-of day surcharge
überschwemmen = to flood, to inundate
Über|schwemmung *f*, Überflutung = flood(ing), inundation
~schwemmungsschutz *m* = protection against floods, flood protection
~schwemmungszone *f*, Überschwemmungsgebiet *n* = flood zone
über|schwer = superheavy, extraheavy
~schwere Winde *f* = extraheavy-duty winch, superheavy-duty ~
~schwerer Bohrhammer *m*, Hammerbohrmaschine *f*, Schwerstbohrhammer = (rock drill) drifter, drifter drill
Über|seebecken *n* = transatlantic basin
Übersetzung *f* = (torque) multiplication
Übersetzungs|getriebe *n* = transmission gearing
~verhältnis *n* = gear ratio

Über|sichbrechen *n*, Aufstemmen = overhead stoping
~sichtsplan *m* = lay-out (plan)
~sichtstafel *f* = synoptic table
~sieb *n* [*Um eine Siebanlage platzsparend zu gestalten, werden Trommeln mit Übersieben versehen, d.h. mehrere Siebe mit verschiedenen Lochgrößen oder ein Sieb sind übereinander angeordnet. Bei Flachsieben ist der Etagenbau üblich*] = outer screen
übersintert = oversintered
Über|spannung *f* [*zum Anbringen von Straßenleuchten*] = span wire
~stand *m* = projecting length, excess ~, ~ end
~stand *m* [*Sonnenschutzvorrichtung eines Fensters*] = canopy
~stand *m* [*Rotarybohren*] = length of pipe above rotary table
~staubewässerung *f* **von Obstgärten** = basin flooding, ~ method
überstauter Wassersprung *m* = submerged hydraulic jump
Über|stauung *f* [*Bewässerungsverfahren*] = check flooding, ~ method (of irrigation)
~steckflansch *m* = slip-on flange
~steuern *n* [*Motor*] = overriding
~steuerungsventil *n* = override (selector) valve
über|streichen = to overpaint
~strichen = over-painted
Über|stömung *f* = submersion, submergence [*The ratio of the tailwater elevation to the head-water elevation, when both are higher than the crest, the overflow crest of the structure being the datum of reference*]
~strömventil *n* → Schlabberventil
~stunde *f* = overtime hour
~stunden *fpl*; Überzeit *f* [*Schweiz*] = overtime hours
(Übertage-) Arbeitsgerüst *n* → Abteufgerüst
Übertage|kraftwerk *n*, freistehendes Kraftwerk, Oberflächenkraftwerk = overground (power) station, ~ (~) plant, surface (~) ~
~sprengen *n* = open face blasting

übertägige (Herein)Gewinnung *f*, (Herein)Gewinnung über Tage = open cast getting
Übertiefe *f* = depth clearance, extra depth (Brit.); overdepth (US)
übertragen [*Last auf den Untergrund*] = to transmit
Übertragungs|länge *f* [*Spannbeton. Die für die Eintragung der Sollvorspannung notwendige Drahtlänge*] = transmission length, transfer ~
~**welle** *f*, translatorische Welle = wave of translation
Überverdichtung *f* = over-compaction
überwachen, lenken = to monitor
Überwachung *f* **betrieblicher Vorgänge mit Fernsehen** = industrial television, ITV
Über|waschkrone *f* [*Tiefbohrtechnik*] = washover shoe
~**wasserböschung** *f* = slope above the water
~**weg** *m* **für Fußgänger** = crosswalk
über|wölben = to vault
~**wölbte Kehlnaht** *f* = convex fillet weld
~**wölbter Abzugkanal** *m* → Bogen-Durchlaß *m*
~**wölbter Durchlaß** *m* → Bogen-Durchlaß
~**berwölbter Durchlaß** *m* → Bogen-Durchlaß
~**zählig**, überflüssig [*Baustatik*] = redundant
Über|zeit *f* [*Schweiz*]; Überstunden *fpl* = overtime hours
~**ziehen** *n* [*Brückenseil*] = pulling to place
~**ziehen** *n* [*für (Ab)Dichtungszwecke*] = coating
Überzug *m* = coat(ing)
~ = suspender beam
~**(decke)** *m*, (*f*) = overlay (pavement); topping, overlay paving (US)
U-Boot-Bunker *m* = U-boat pen
U-Eisen *n* = (iron) channel
~-**Ring** *m* = (iron) channel ring
Ufer *n* = bank [*The sloping ground along the edge of a river, stream, canal, or lake*]

(~)**Abbruch** *m* = washing of a bank, eroding ~ ~ ~, erosion ~ ~ ~
~**anlagen** *fpl* [*Brücke*] = shore structures
~**ausbesserung** *f* = bank reinstatement
~**bau** *m* = bank construction
~**befestigung** *f* = bank stabilization
~**bekleidung** *f* = bank revetment
~**böschung** *f* = bank slope
~**brandung** *f*, Strandbrandung = surging
~**deckwerk** *n* = riprap [*Broken stone placed on earth surfaces for their protection against the action of water; also applied to brush or pole mattresses or brush and stone, or similar materials used for protection*]
~**erosion** *f* = bank erosion
~**feld** *n*, Uferböschung *f*, Flutfeld, Flutöffnung, Vorlandfeld, Vorlandöffnung = shore span [*bridge*]
~**kippe** *f* = bank tip [*Where a dredger discharges the dredged material*]
~**linie** *f* [*Strom; Fluß*] = bank line, waterfront
~**öffnung** *f* → Uferfeld *n*
~**pfeiler** *m* = shore pier, land ~
~**pflaster** *n* = bank paving
~**portalkran** *m*, Hafenportalkran = harbo(u)r portal crane
Uferschutz *m* = bank protection
~ **unter Wasser** = underwater bank protection
~**(bau)werk** *n* = bank protection structure
Ufer|seite *f* = shore side
~**straße** *f* = shore boulevard
U-Flugplatz *m*, unterirdischer Flugplatz = underground airfield
U-Haken *m* = U-hook
Uhrenanlage *f* = clock system
Uhr(en)steuerung *f* [*Steuerung durch Uhr(en)*] = clock control
Uhrenturm *m* = clock tower
Uhrmacheröl *n* = watch-oil
Uhrwerk *n* = clockwork
Uintait *m* → Gilsonitasphalt *m*
UK *f*, Unterkante *f* = bottom
UKW-Funksprechgerät *n* = FM radio

Ultrahochfrequenzinnenrüttler *m*, Ultrahochfrequenzinnenvibrator *m* = ultra high-frequency poker, ~ ~ internal vibrator
Ultramarin *n* = ultramarine
~blau *n* = ultramarine blue
~farbe *f* = ultramarine pigment
Ultraschall *m*, Überschall, unhörbarer Schall = ultrasonic sound, supersonic ~, ultrasound
~betonprüfgerät *n* → → zerstörungsfreie Betonprüfung
~chemie *f* = ultrasonic chemistry, supersonic ~
~dickenmeßgerät *n* = ultrasonic thickness meter, supersonic ~ ~
~-Durchflußmesser *m* = ultrasonic flowmeter, supersonic ~
~-Echolot *n* = acoustic log
~entstaubung *f* = ultrasonic dedusting, supersonic ~, ~ dust removal
~gerät *n* → Ultraschall-Schwelle *f*
~impuls-Echo-Verfahren *n* → → zerstörungsfreie Betonprüfung
~-Impulsgerät *n* = ultrasonic impulse transmitter, supersonic ~ ~
~kopf *m*, Schallkopf = (ultrasonic) transducer, supersonic ~
~methode *f*, Überschallmethode = ultrasonic method, supersonic ~
~reinigung *f* = ultrasonic cleaning, supersonic ~
~-Reinigungsbehälter *m* = ultrasonic cleaning tank, supersonic ~ ~
~rütt(e)lung *f* = ultrasonic vibration, supersonic ~
~rüttler *m*, Ultraschallvibrator *m* = supersonic vibrator, ultrasonic ~
~schweißen *n* = ultrasonic welding, supersonic ~
~-Schwelle *f*, Ultraschall(zähl)gerät *n* = ultrasonic (type) detector, supersonic (~), ~
~steuerung *f* = ultrasonic control, supersonic ~
~technik *f* = ultrasonic engineering, supersonic ~
Ultraviolett|bestrahlung *f* = irradiation with ultraviolet light, ultraviolet irradiation
~-Spektroskopie *f* = ultraviolet spectrophotometry
~strahlung *f* = ultraviolet radiation
Umbau *m* = structural alteration, conversion
~arbeiten *fpl* = conversion work
~-Auto(mobil)bagger *m* → Auto(mobil)bagger
~bagger *m* → Mehrzweckbagger
umbauen [*Gleitschalung*] = to reset, to re-erect
~ [*Bauwerk*; *Maschine*] = to convert
Umbau|gerät *n* → Austauschgerät
~satz *m* = conversion unit
umbauter Raum *m* = (cubical) content
Umbau|vorrichtung *f* → Austauschgerät *n*
~werkzeug *n* → Austauschgerät *n*
Umbemessen *n* = redesigning
Umbemessung *f* = redesign
umbördeln → bördeln
Umbra *f* = umber
umdrehen [*oberes Ende nach unten bzw. umgekehrt*] = to up-end
Umdrehungen *fpl* **pro Minute**, Drehzahl *f*, Umdrehungszahl = r.p.m., speed, rotations per minute, RPM
Umfahrungsstraße *f* → Umgehungsstraße
(Umfang)Bügel *m* [*Säule*] = tie
Umfangs|fuge *f* → → Talsperre *f*
~geschwindigkeit *f* = peripheral speed
~länge *f* = peripheral length
~schweißnaht *f* = girth weld
~vorspannung *f* → Ringvorspannung
Umfassungsmauer *f* → Außenmauer
Umformer *m*, ölhydraulischer ~ = service
~kanal *m* [*Ölhydraulik*] = service gallery
~öffnung *f* [*Ölhydraulik*] = service port
umführbar = by-passable
Umfüll|-Halbanhänger *m* = semi-trailer (type) supply tank
~tank *m*, Umfüllbehälter *m* = supply tank
Umgebungstemperatur *f* = ambient temperature

umgehen [*Abbau im Bergbau*] = to take place
Umgehungs|autobahn *f*, Umfahrungsautobahn = by-pass motorway
~kanal *m* = by-pass canal
~straße *f*, Umfahrungsstraße = by-pass road, ~ highway, alternate route
~verkehr *m* = by-passable traffic
umgekehrte Emulsion *f*, Wasser-in-Öl-Emulsion = water-in-oil emulsion
umgekehrter doppelter Hängebock *m*, umgekehrtes doppeltes Hängewerk *n* = inverted queen (post) truss, trussed beam [*Is used for short spans in connection with wood construction*]
~ Filter *m*, umgekehrtes ~ *n*, Belastungsfilter [*Bodenmechanik*] = inverted filter, loaded ~, loaded inverted ~
~ Gurtbogen *m* → Konterbogen
~ T-Querschnitt *m*, T-Querschnitt = inverted T section, ~ tee ~
umgekehrtes einsäuliges Hängewerk *n*, ~ einfaches ~, umgekehrter einfacher Hängebock *m*, umgekehrter einsäuliger Hängebock *m* = inverted king (post) truss, trussed beam [*Is used for short spans in connection with wood construction*]
~ Gewölbe *n* → verkehrtes ~
ungelagerter Boden *m*, Absatzboden, Derivatboden, Kolluvialboden = transported soil, colluvial ~
Umgestaltung *f* = remodelling
Umhüllbarkeit *f*, Benetzbarkeit [*eines Gesteinskorns mit Bindemittel*] = wettability
umhüllen, benetzen [*mit Bindemittel*] = to wet, to coat
Umhüllen *n*, Benetzen, Bindemittelumhüllung *f* = wetting, coating; wet mixing (US)
umhüllter Splitt *m* = coated chip(ping)s
Umhüllung *f* **der Bewehrung**, Überdeckung ~ ~, ~ ~ Armierung, ~ ~ (Stahl)Einlagen = concrete cover

Umhüllungs|rohr *n* → Blechröhre
~wärme *f*, Benetzungswärme = wetting heat
Umkehr|band *n*, Reversierband = reversing belt
~getriebe *n*, Wendegetriebe, Reversiergetriebe = forward-reverse transmission
~mischer *m* → → Betonmischer
Umkehrpunkt *m* **des Wasserstandes bei Thw** = slack at high water, H. W.
~ ~ ~ ~ Tnw = slack at low water, L. W.
Umkehrschalt|getriebe *n*, Wendeschaltgetriebe = forward-reverse primary transmission
~kupp(e)lung *f* → Wendeschaltkupp(e)lung
Umkehr|(trommel)(beton)mischer *m* → Reversiermischer
~turbine *f*, Umsteuerungsturbine, Reversierturbine = reversible turbine, reversing ~
~walzwerk *n* = reversing mill, reversible ~
Umkippen *n* [*fahrbarer Drehkran*] = tipping
Umkleide|kabine *f*, Aus- und Ankleidekabine, Umkleidezelle *f*, Aus- und Ankleidezelle = dressing cab(in), changing ~
~raum *m*, Aus- und Ankleideraum = dressing room, changing ~
(um)krempen → bördeln
Umladen *n* = re-handling
Umlagerung *f*, Umlagern *n*, Nachverteilung, Nachverteilen [*Material im Erdbau*] = redistribution
Umlagerungsmoment *n*, negatives Zusatzmoment = secondary (bending) moment
Umlauf *m*, Arbeitsrunde *f*, Rundfahrt *f* [*Radschrapper*] = round trip
~ [*Trinkwasserbecken*] = culvert
~, Umlaufkanal *m* [*Schleuse*] = (longitudinal) culvert
~ [*von Anlieferungsfahrzeugen*] = hauling cycle
~aufzug *m*, Paternoster(-Aufzug), Elevator *m* = paternoster

umlaufender Staubabscheider *m* = rotary dust collector

Umlauf|gehäuse *n* = rotating housing

~**getriebe** *n* → Planetengetriebe

~**heizung** *f*, Umwälzheizung = circulation heating

Umläufigkeit *f* [*Durchtreten von Wasser durch die Talanschlüsse einer Talsperre*] = seepage through valley flanks

Umlauf|kanal *m* [*Ölhydraulik*] = by-pass gallery

~**(kanal)** *m* [*Schleuse*] = (longitudinal) culvert

~**kühlung** *f* = circulation cooling

~**last** *f* [*Mühle*] = circulating load

~**mahlung** *f*, Kreislaufmahlung, Umlaufmahlen *n*, Kreislaufmahlen = closed-circuit grinding

~**schmierung** *f* = circulation oiling

~**stollen** *m* → Umleitstollen

~**wasser** *n* = circulating water

umlegen [*einen Strebförderer*] = to turn over, to advance one turning

~ **nach oben**, abkanten ~ ~ = to tip up(wards), to fold

~ ~ **unten**, abkanten ~ ~ = to tip down(wards), to fold ~

Umlegung *f* = reorientation, relocation

umleiten [*Verkehr*] = to divert

~, ableiten [*Flußwasser*] = to divert

Umleit|kammer *f* → Umleitungskammer

~**kanal** *m*, Umleitungskanal = bywash, diversion cut, by-channel, spillway; diversion flume (US)

~**stollen** *m*, Umleitungsstollen, Umlaufstollen = by-pass tunnel, diversion ~

Umleitung *f*, Straßen~ = loop road, detour, by-pass; shoofly (US)

~ [*Verkehr*] = diversion [*traffic*]

~, Ableitung [*Talsperrenbau*] = diversion

Umleitungsbauwerk *n*, Ableitungsbauwerk = diversion structure

Umleit(ungs)|kammer *f* = diversion chamber, ~ manhole [*A chamber that contains a device for diverting all or a part of the flow*]

~**kanal** *m*, Stichkanal [*Dachwehr*] = culvert

~**kanal** *m* → → Talsperre *f*

~**kraftwerk** *n* = diversion power plant, ~ ~ station

~**sperre** *f* = diversion dam [*A barrier built for the purpose of diverting part or all of the water from a stream or river into a different course*]

~**sperre** *f* **zur Bewässerung** → → Talsperre

~**stollen** *m*, Umlaufstollen = diversion gallery, by-pass ~

~**ventil** *n* = transflow valve

~**wehr** *n* = diversion weir [*Diversion weirs designed with gates to pass flood flows are sometimes called "barrages"*]

Umlenk|platte *f* = baffle (plate), baffler

~**rolle** *f* → → Bandförderer

~**rolle** *f* = return sheave, ~ pulley

~**rolle** *f*, Turas *m* [*Eimerkettenbagger*] = sprocket (wheel)

~**station** *f* [*Zwei feste Türme beim Drei-Seil-Schrapper*] = bridle towers

~**stelle** *f* [*Spannglied*] = deflexion point (US); deflection ~ (Brit.)

~**turas** *m* = (crawler) foot shaft sprocket

~**turaswelle** *f* = (crawler) foot shaft

~**turas(wellen)lagerjoch** *n* = (crawler) foot shaft sprocket yoke

Umluft *f* = recirculated air

~**heizung** *f* = heating with recirculated air

~**kanal** *m* = recirculation duct

ummantelter Schweißstab *m* = covered metallic weldrod

Ummantelung *f* = jacketing

Ummantelungswand *f* [*Atomkraftwerk*] = envelope wall

Umplanung *f* = re-planning

Umprägung *f*, Umwand(e)lung, Metamorphose *f*, (Gesteins)Metamorphismus *m* (Geol.) = metamorphism

Umrandungsfeuer *n*, Randfeuer = boundary light

Umrechnungstabelle *f* = conversion table (or chart)

Umriß *m* = contour
~linie *f* = contour line
~verfahren *n*, Lageplan-Massenermitt(e)lung *f* [*Erdmassen werden aus dem Lageplan ermittelt*] = contour method of grade design and earthwork(s) calculation
~zeichnung *f* = outline drawing
Umrollen *n*, Abrollen [*Gleiskette*] = roll-over
Umrüstung *f* [*Zusatzvorrichtungen*] = changing attachments
Umsatzsteuer *f* = turnover tax
umschaltbarer Austauscher *m* = reversing exchanger
Umschaltventil *n* = crossover valve
Umschlag *m*, ~verkehr *m*, Warenverkehr, Güterumschlag [*Hafen*] = cargo traffic, goods ~
~, Umschlagen *n* [*von Gütern*] = handling
umschlagen, herumlegen [*Drahtseil*] = to twist
umschlagen, fördern = to handle
Umschlagen *n*, Umschlag *m* [*von Gütern*] = handling
Umschlag|gerät *n* **am Kai** = quayside appliance
~hafen *m* = port
~kapazität *f* [*Hafen*] = handling capacity
~verkehr *m* → Umschlag *m*
Umschließung *f* [*Baugrube*] = (en-)closure
Umschließungsdamm *m*, Umwallung *f* = closure embankment
Umschlingungs|festigkeit *f* = state of all-round tension, ~ ~ two-dimensional stress equality
~winkel *m* [*Treibscheibenförderung*] = angle of rope in contact with Koepe pulley in radians
~winkel *m* = angle of sling
Umschmelzaluminium *n* = secondary alumin(i)um
Umschmelzen *n* = remelting
Umschmelz|legierung *f* = remelting alloy
~ofen *m* = remelting furnace
umschnürte Säule *f* → spiral~ ~

~ Stütze *f* → (ring)umschnürte Säule
~ ~ → (spiral)umschnürte Säule
Umschnürung *f* → Spiralbewehrung
~, Ringbewehrung, Ringarmierung, Ring(stahl)einlagen *fpl*, Umwick(e)lung, Ringumschnürung, Querbewehrung in Ringform, Umschnürungseisen *npl*, Umschnürungsring(e) *m*, (*f*), Umschnürungsbewehrung, Umschnürungsarmierung, Umschnürungs(stahl)einlagen = hooping, hoop(s) [*Curved reinforcement such as the steel in a circular concrete tank which resists ring tension*]
Umschnürungseisen *npl* → Spiralbewehrung
~ → Umschnürung
Umschnürungsring(e) *m*, (*f*) → Umschnürung
umsetzbares Lehrgerüst *n* = on-the-ground non-travel(l)ing centering
Umsetzen *n* [*Fertiger im Straßenbau*] = lane move, ~ change, moving between lanes
Umsetzung *f*, chemische ~ [*Schwarzpulver*] = reaction
Umspanner *m*, Transformator *m*, Trafo *m* = transformer
Umspann(ungs)anlage *f*, Umspann(ungs)station *f* = grid substation
Umspannwerk *n*, Freiluftschaltanlage *f*, Schalthof *m* = outdoor switchgear, switchyard
umspunden, ausrammen [*Baugrube mit Spundbohlen*] = to close a building pit by means of sheet piling
umspundete Baugrube *f*, ausgerammte ~ = building pit closed by sheet piling
Umspundung *f* → Baugruben~
Umstampfung *f* **mit Lehm und Ton** → Lehm(schlag)ummantelung
Umstecklöffel *m* = reversible bucket
Umsteigebahnsteig *m* = interchange cross-platform
Umsteigen *n* = interchange (of passengers)
~ auf dem gleichen Bahnsteig = cross-platform interchange

Umsteigen über Fahrgasttunnel — Undichtigkeitsgrad

~ **über Fahrgasttunnel** = (pedestrian) subway interchange
umstellen = to reposition
Umstellung *f* = conversion
~ **auf eing(e)leisigen Betrieb** = switching over to single-track operation
~ **von Guß- in Schweißkonstruktionen** = redesigning castings to weldments
Umsteuerung *f*, Gangumkehr *f* = reversal
Umsteuer(ungs)schieber *m*, Wechselschieber = control(ling) valve
Umsteuerungsturbine *f* → Umkehrturbine
Umtrassierung *f* = relocation
Umwallung *f*, Umschließungsdamm *m* = closure embankment
umwälzen = to circulate
Umwälz|pumpe *f*, Umlaufpumpe, Zirkulationspumpe, Kreislaufpumpe = circulating pump
~-Sprengrampe *f*, Umlauf-Sprengrampe, Zirkulations-Sprengrampe = circulating spraybar
Umwälz(ungs)becken *n* = spiral-flow tank [*A tank used in carrying out the activated-sludge process in which a spiral motion is given to the sewage in its flow through the tank by the introduction of air through a line of diffusers placed on one side of the bottom*]
Umwand(e)lung *f* → Umprägung
Umwand(e)lungsgestein *n* → metamorphes Gestein
Umweltverhältnisse *npl* = ambient conditions
Umwickeln *n* = wrapping
Umwick(e)lung *f* → Umschnürung
Umzäunung *f*, Einfriedung, Einzäunung = fencing, boundary fence
unabgeschirmte Straßenleuchte *f* = non-cut-off lantern
unabhängige Stahlseele *f* = IWRC, independent wire rope core
unangenehmer nutzloser Dampf *m* = fume
unaufgeladen [*Motor*] = unsupercharged, normally aspirated, unblown
unaufgeschlossener Rückstand *m* = insoluble residue
unausgelastet [*Maschine*] = having not enough work to do
Unbalance *f*, Unwucht *f*, Schwunggewicht *n*, Unwuchtmasse *f* = out-of-balance weight, unbalance
unbebaut = not built upon
~ [*Acker*] = uncultivated
unbedampfter Beton *m* = unsteamed concrete
unbedeichte Anlandung *f* = un-embanked alluvial land
unbeeinflußter Wassersprung *m*, freier ~ = free hydraulic jump
unbefestigte Straße *f* = unmade road, ~ highway
(unbefestigter) Seitenstreifen *m*, Seitenraum *m*, Außenstreifen, Bankett *n* = verge, margin
Unbekannte *f* = unknown
unbelastet = unloaded
unbelüftet, belüftungslos = unventilated
unbeschottert, ungeschottert = unmetalled (Brit.)
unbeschrankter (Eisen)Bahnübergang *m*, unbewachter ~ = unprotected level-crossing; ~ grade-crossing (US)
unbewachter privater (Eisen)Bahnübergang *m*, unbeschrankter ~ ~ = occupation crossing [*A level-crossing for the private use of a landowner whose land is served by the railway line*]
unbewehrt, nicht bewehrt, nicht armiert = nonreinforced, plain
unbrechbare Fremdkörper *mpl* = tramp iron, uncrushable material, unbreakable material
unbrennbar = uncombustible
undetonierter Ladungsrest *m* = remains of a cartridge which had not detonated
undicht = leaky
undichte Stelle *f*, Leckstelle = leak(age)
Undichtigkeitsgrad *m*, Gesamtporosität *f*, wahre Porosität = total porosity

Undichtwerden n = leakage
undurchlässig, dicht = impervious
~ **machen, dicht** ~, **(ab)dichten** = to seal
undurchlässige Sohle f [*Talsperre*] = impervious foundation, impermeable ~
~ **Unterlage** f = impermeable base, impervious ~
Undurchlässigkeit f, **Dichtigkeit** ~ = imperviousness, impermeability
Undurchlässigkeitsbeiwert m = coefficient of imperviousness, ~ ~ impermeability [*The ratio, expressed decimally, of effectively impervious surface to the total catchment area*]
uneben, bewegt, unruhig [*Gelände*] = undulating, undulated
~ [*Straßendecke*] = rough, uneven
Unebenheit f = unevenness, irregularity
    ausgleichen = to true up, to average out
Unebenheiten fpl **in der Straßenoberfläche** = pavement roughness
Unebenheitsmeßanhänger m = bump integrator
unedles Metall n, **Grundmetall, Mutterwerkstoff** m = base metal
unendlich steif = infinitely rigid
Unfall m **durch Schleudern** [*Fahrzeug*] = skidding accident
~ **mit Personenschaden** = personal-injury accident
~ **ohne Personenschaden** = non-injury accident
unfallanfällige Stelle f **im Straßenverkehr** [*z.B. Kreuzung; Kurve; usw.*] = "black" spot
Unfall|anfälligkeit f = vulnerability to accidents, liability ~ ~, accident proneness
~**art** f = type of accident
~**auswertung** f = evaluation of accidents
~**entschädigung** f **der Arbeiter** = workmen's compensation
~**gefährdung** f = accident risk
~**kran** m, **Bergungskran** = wrecking crane, breakdown ~, accident ~

~**rate** f = accident toll, ~ rate
~**schutz** m = accident protection
unfallsicher = accident-proof
Unfall|situationsplan m = plan of the accident site
~**station** f = casualty station
~**statistik** f = accident record
~**stelle** f = accident site
~**tauglichkeit** f = crash-worthiness
~**tod** m = accidental death
~**ursache** f = cause of accident
~**verhütung** f = accident prevention
~**verhütungsvorschriften** fpl = accident prevention regulations
ungebrannter keramischer Scherben m = unfired ceramic body
ungedämpfte Seiche f [*in Häfen und Buchten*] = sustained seiche
ungelöschter gemahlener Kalk m, **Mahlkalk, gemahlener Branntkalk** [*fälschlich als „treibender Kalk" bezeichnet*] = unslaked and ground quicklime
ungelöster Abwasserstoff m = sewage solid
ungenauer Lauf m **der Schleppschaufel** = bucket wander
ungerichtetes Funkfeuer n = non-directional beacon light
ungesackter Zement m → **Behälterzement**
ungeschichtetes Bruchsteinmauerwerk n = random rubble masonry, uncoursed ~ ~ [*Masonry composed of roughly shaped stone laid without regularity of coursing, but fitted together to form well-defined joints*]
ungeschottert → **unbeschottert**
ungesicherte Leistung f [*Energieleistung, die nur zeitweise zur Verfügung steht*] = interruptable power, secondary ~
ungesiebter Grubenkies m = pit-run gravel
ungesiebtes gebrochenes Material n, **Brechgut** n = broken material, crushed ~
ungespannter Dampf m = atmospheric steam, steam vapo(u)r

ungestörte Bodenprobe f = undisturbed (soil) sample
$\sim$ **natürliche Lagerung** f, gewachsener Zustand m [*Boden*] = natural state
ungeteilte Straße f, Straße ohne Richtungstrennstreifen = undivided two-way road, $\sim\sim$ highway
ungewollte Arbeitsfuge f = cold joint
ungezieferfest, ungezieferbeständig = vermin-proof, vermin-resistant
unglasiert = unglazed
ungleichförmige Bewegung f = non-uniform movement
ungleichförmiger Sand m = non-uniform sand
Ungleichförmigkeitsgrad $U m = \dfrac{d_{60}}{d_{10}} =$
coefficient of uniformity, uniformity coefficient $C_u = \dfrac{D_{60}}{D_{10}}$
Ungleichgewicht n = unbalance
ungleich(mäßig)e Setzungen fpl, Setzungsunterschied m = differential settlement
Ungleichmäßigkeit f = non-uniformity
ungleichschenk(e)liger Auflagerwinkel m [*Stahlbau*] = unqual-leg angle
unhaltiges Gestein n → Berg m
unhältiges Gut n, Abgang m, Abgänge mpl [*Aufbereitungstechnik*] = tailing(s)
unhörbarer Schall m, Ultraschall, Überschall = ultrasonic sound, supersonic $\sim$, ultrasound
Universal|-Auto(mobil)bagger m → Auto(mobil)bagger
$\sim$**bagger** m → Mehrzweckbagger
$\sim$**-Bohrhammer** m → Universal-Gesteinsbohrhammer
$\sim$**dampframme** f = universal steam (pile) driving plant, $\sim\sim$ piling $\sim$
$\sim$**-Diesel-Elektro-Raupenbandbagger** m, Universal-Diesel-Elektro-Gleiskettenbagger = convertible diesel-electric caterpillar excavator
$\sim$**-Diesel-Raupen(band)bagger** m = crawler-mounted diesel excavator-crane
$\sim$**-Elektrowerkzeug** n = universal electric tool
$\sim$**feriger** m = universal finisher, $\sim$ finishing machine
$\sim$**fett** n = universal grease
$\sim$**-Gabelstapler** m = universal forklift, $\sim$ fork truck, $\sim$ high-lift fork stacking truck, $\sim$ fork lift truck type conveyancer
$\sim$**gerüstramme** f → Universalramme
$\sim$**-(Gesteins)Bohrhammer** m, Universal-Felsbohrhammer = a.c./d.c (hand) (hammer) (rock) drill, $\sim$ (rock) drill (hammer)
$\sim$**-(Gleis)Kettenbagger** m, Universal-Raupenbagger = universal crawler (type) excavator
$\sim$**-Grundbagger** m = universal base excavator
$\sim$**hebel** m = universal lever
$\sim$**-Kettenbagger** m → Universal-Raupenbagger
$\sim$**-Mobilbagger** m = universal mobile excavator
$\sim$**motor** m, Allstrommotor = universal AC/DC motor, $\sim$ (electric) a.c./d.c. $\sim$
$\sim$**pfahlramme** f → Universalramme
$\sim$**prüfmaschine** f = universal testing machine
$\sim$**ramme** f, Universalpfahlramme, Universalgerüstramme = universal (pile) driving plant, $\sim$ piling $\sim$ [*A universal frame can rake, rotate, adjust the leaders and travel under its own power. The description universal should not be taken to imply universal application but regarded as descriptive of the several forms of motion that can be obtained*]
$\sim$**-Raupenbagger** m, Universal-(Gleis-)Kettenbagger = universal crawler (type) excavator
$\sim$**-Schraubenschlüssel** m = adjustable spanner
$\sim$**stahl** m = universal mill plate
$\sim$**-Werkzeugmaschine** f = universal machine tool
unkartiert = unmapped

unkontrollierte Bewässerung — Unterbettungssand

unkontrollierte Bewässerung *f* = uncontrolled flooding, "wild" ~
Unkraut|abbrennmaschine *f* = weed burner
~bekämpfung *f*, Unkrautvertilgung = weed control
~bekämpfungsmittel *n* = weed killer, herbicide, weed destroyer
unlöslicher Anhydrit *m* → Anhydrit II
~ Emulgator *m*, fester ~ = colloidal emulsifier
~ Rückstand *m* = insoluble residue [*The material remaining after cement is treated successively with hydrochloric acid and sodium hydroxide solutions of specific concentrations for designated periods of time*]
unmagnetische Schwerstange *f* [*Tiefbohrtechnik*] = nonmagnetic drill collar
unmittelbarer Lichtbogenofen *m* = direct-arc melting furnace
~ Nachweis *m* **von Erdöl** = direct oil detection
~ Regen *m* = rain from warm clouds
unmittelbares Hangendes *n*, Dachschicht *f* (⚒) = immediate roof
unnachgiebig = unyielding
Unrat *m* = refuse
unregelmäßige Schichtung *f* (Geol.) = irregular bedding
unregelmäßiges Schichtmauerwerk *n* = alternating ~ coursed ashlar masonry, ~ ~ ashler ~
unrein = impure
unreines (Werk)Blei *n* = base bullion
unruhig, bewegt, uneben [*Gelände*] = undulating, undulated
Unrundheit *f* = out-of roundness
unsichtbar machen [*Fuge*] = to mask
unsichtbare Strahlung *f* = invisible radiation
unsortierter Brannkalk *m* = run-of-kiln lime
unstabil, labil, schnellbrechend [*Bitumenemulsion*] = labile, quick-breaking, rapid-setting
unstationäre Strömung *f*, nichtbeständige ~ = unsteady flow, nonsteady ~

unstetig abgestufte Mineralmasse *f*, diskontinuierlich ~ ~ = gap-graded aggregate
unstetige Kornabstufung *f*, diskontinuierliche ~ = discontinuous grading, gap ~, discontinuous granulometry
unsymmetrische Spannung *f* [*Elektrizität*] = out-of-balance potential
unten eingehender Bogen *m* → arabischer Bogen
Untenentleerung *f*, Bodenentleerung = bottom-dumping, bottom dump discharge
untenliegender Drücker *m* → Innendrücker
unter Federspannung stehend = spring loaded
Unter|abschnitt *m* **für den Abbau** → Bau-Unterabschnitt (⚒)
~(an)sicht *f* = underside view
~ausschuß *m* = subcommittee
Unterbau *m* [*Theodolit*] = tribrach
~ [*Straßenbau*] = substructure
~ [*Brücke*] = substructure
~ [*Der Teil eines Gebäudes oder Bauwerkes unter Geländeoberfläche, ausgenommen das Fundament. Gegenteil: Überbau = superstructure*] = substructure
unterbaut (⚒) = influenced by underlying workings
Unterbecken *n*, Tiefbecken [*Pumpspeicherwerk*] = lower reservoir, ~ pool, ~ (storage) basin, tail pond
unter|bemessen = to underdesign
~bestückt [*Baustelle mit Maschinen und Geräten*] = underplanted
Unterbeton *m* = seal-coat concrete, rough ~, sub-~, blinding (layer of) concrete [*Placed on the earth under foundations*]
~ [*Straßenbau*] = base course concrete
~ → (Hinter)Füll(ungs)beton
~schicht *f*, Unterbeton *m* = bottom concrete layer, (course) concrete
Unter|bettung *f* [*Pflasterdecke*] = bedding
~bettungssand *m* = bedding sand

(~)Bettungsschicht *f*, Bettschicht = bed(ding) course, underlying course, underlay, cushion course; sub-crust (Brit.)
unterbieten = to undercut
Unterboden *m* → B-Horizont *m*
~ = sub-floor
~ auf Unterzügen = joisted sub-floor
~spachtel(masse) *m*, (*f*) = surfacing compound
Unterbrecher|kontakt *m*, Schaltkontakt = contact point
~schalter *m* = disconnecting switch
Unter|bringung *f* [*von Bergen*] (⚒) = dirt disposal
~deck *n* [*Sieb*] = bottom deck
unterdimensionieren = to underdesign
Unterdruck *m*, Vakuum *n* = vacuum
~bremse *f*, Vakuumbremse = vacuum brake
~-Oberflächenbehand(e)lung *f*, Saugbetonverfahren *n*, Vakuumbetonverfahren = vacuum concrete method
~schalter *m* [*Kältemaschine*] = low-pressure cutout
unterdrücktes Widerlager *n*, verlorenes ~ = dead abutment
Unterdükerung *f* = sag crossing
untere Abbaustrecke *f*, Grundstrecke (⚒) = bottom development road
~ Asphalttragschicht *f*, ~ Bitumentragschicht = asphalt subbase [*The asphalt layer used in a pavement system between the subgrade and the base course*]
~ Bitumentragschicht *f* → ~ Asphalttragschicht
~ Faser *f* = bottom fibre (Brit.); ~ fiber (US)
~ feste Einspannung *f* [*Spundwand*] = fixed earth support
~ freie Auflagerung *f* [*Spundwand*] = free earth support
~ Fugenschalung *f* [*Absenken eines Tunnelelementes*] = trémie tub formwork
~ Geschosse *npl*, ~ Etagen *fpl*, ~ Stockwerke *npl* = bottom floors

~ Scheinfuge *f* = bottom surface dummy joint
~ (Schütz)Tafel *f*, Unterschütz(e) *n*, (*f*), Untertafel = bottom (gate) leaf
~ Tafel *f* → ~ Schütztafel
~ ~ ~ Gegenklappe *f*
~ Tragschicht *f* = subbase [*The layer used in a pavement system between the subgrade and the base course*]
unterer Grenzzustand *m* [*Boden*] = active state [*soil*]
(~) Heizwert *m*, H_u in kcal/Nm^3 = lower calorific value, L.C.V.
~ (Seil)Strang *m* [*Seiltransporteur*] = lower rope, bottom ~
unterfahren [*ein Flöz, eine Strecke*] (⚒) = to drive a roadway under [*a seam, a gallery, etc.*]
Unter|fahrungsstrecke *f* [*Trichterbau*] (⚒) = haulage road
~fangen *n*, Unterfangung *f* = dead shoring, vertical ~, needle ~, underpinning
~fang(ungs)bauweise *f* mit First- und Sohlstollen, belgische Bauweise, Firstenvortrieb *m*, Firstenort-Sohlenstrossenbau *m*, belgisches Verfahren *n* = Belgian method, top-heading ~
~fangungsmauer *f* = underpinning wall
~flasche *f*, (Seil)Flasche, (Seil)Kloben *m*, Block *m* = block [*The frame holding the pulley or pulleys of lifting tackle*]
~flur-Entwässerungssystem *n* = underfloor drainage system
~flurfeuer *n* [*Flugplatz*] = pancake light, button ~, inset ~
~flurhydrant *m* = ground hydrant
~flurhydrant *m* mit selbsttätiger Entleerung [*DIN 3221*] = self draining hydrant
~flurmotor *m* = underfloor engine
~flurzapfstelle *f* [*zum Betanken von Flugzeugen usw.*] = fuel pit
~führung *f* = underbridge, underpass, undercrossing
~füllen *n* von hohlliegenden Betonfahrbahnplatten mit Bitumen =

(pavement) undersealing, subsealing [*Pumping of a low penetration asphalt (US)/asphaltic bitumen (Brit.) under concrete pavement slabs*]
untergehängte Deckentafel *f* = ceiling panel
Unter|gestell *n* → Fahrgestell
~gleis-Förderschnecke *f* = undertrack screw conveyor, ~ ~ conveyer
~gleis-Silo *m* = undertrack hopper
Untergrund *m* → Baugrund
~ → Erdreich *n*
~ [*z.B. zur Aufnahme von Anstrichstoff*] = base (surface)
(Unter)Grund *m* **für Putzarbeiten** → Putz(unter)grund
~(ab)dichtung *f*, (Ab)Dichtung des Untergrundes = subsoil waterproofing
Untergrundauspressung *f* → Untergrundverpressung
Untergrundbahn *f*, U-Bahn = underground railway, tube ~ (Brit.); ~ railroad (US)
~baugrube *f*, U-Bahnbaugrube = underground railway building pit (Brit.); ~ railroad ~ ~ (US)
~linie *f* = underground line
~tunnel *m*, U-Bahntunnel = underground railway tunnel (Brit.); ~ railroad ~ (US)
Untergrund|behand(e)lung *f* = subsoil treatment
~beries(e)lung *f* = sub-irrigation
Untergrundbewässerung *f*, Grundanfeuchtung = sub-surface irrigation, (natural) sub-irrigation
~ **nach Jahnert** = artificial sub-irrigation
Untergrund|boden *m* = subgrade soil
~bohrung *f*, Baugrundbohrung = foundation drilling, ground ~
~(drän)pflug *m* → Maulwurf(drän)pflug
~einpressung *f* → Untergrundverpressung
~entwässerung *f*, Bodenentwässerung, Tiefenentwässerung = (sub)soil drainage, subsurface ~, underdrainage, sub-drainage, ground water drainage, subgrade drainage
~erosion *f* = subsurface erosion
~injektion *f* → Untergrundverpressung
~krafthaus *n* → Kavernenkrafthaus
~kraftwerk *n* → Kavernenkraftwerk
~parkplatz *m* = underground car park
~probenahme *f* = subsurface sampling
~rohrbewässerung *f* = artificial sub-irrigation
~setzung *f*, Untergrundsenkung *f* [*Senkrechte Verlagerung eines Punktes des Untergrundes; Senkungserscheinung am Baugrund*] = ground settlement
~spannung *f* → Baugrundspannung
~strömung *f* = underground flow
~verhältnisse *npl*, Baugrundverhältnisse = underground conditions
~verpressung *f*, Untergrundeinpressung, Untergrundauspressung, Untergrundinjektion *f* = subsoil grouting
Untergurt *m* = bottom chord, ~ boom, ~ flange
~gelenkknotenverbindung *f* = pin connection in bottom boom, ~ ~ ~ ~ flange, ~ ~ ~ chord
Untergurtknoten|blech *n* = bottom boom junction plate, ~ flange ~ ~, ~ chord ~ ~
~gelenk *n* = tie-bar joint in lower boom, multiple ~ ~ ~ ~
Untergurtstab *m* = bottom boom member, ~ chord ~, ~ flange ~, ~ ~ rod, ~ ~ bar
unterhalb der Siebgrenze = in the sub-screen range
Unterhals *m* → Hypotrachelion *n*
Unterhaltung *f*, Instandhaltung = upkeep, maintenance
laufende ~, Wartung = routine ~
Unterhaltungs|arbeiten *fpl* **im Vergabeverfahren** = contract(-performed) maintenance
~baggerung *f* = maintenance dredging, routine ~
~gehsteig *m* [*Brücke*] = maintenance sidewalk (US); ~ footway (Brit.)

Unterhaltungskosten — Unterlagpapier

~kosten f; Unterhalt(s)kosten [*Schweiz*] = maintenance outlay, ~ cost
~maschinen fpl und -geräte npl = maintenance machinery and equipment
~-Naßbaggerung f = maintenance dredging
~werkstatt f = maintenance (work-)shop, ~ depot
Unter|hängezimmerung f (✹) = hanger method, hanging bolt ~
~haupt n [*Schleuse*] = lower gates, outer ~
Unterholz n = brush(wood)
~pflug m, Unterholzräumer m = brush(wood) cutter, ~ chipper
~räumung f = brush(wood) clearing
~rechen m = clearing rake, brush-(wood) ~
unterirdisch = underground
unterirdische Anlagen fpl = underground installations
~ Bauarbeiten fpl = underground construction work
~ Be- und Entlüftungsstation f = underground ventilation plant
~ Bewässerung f = sub-irrigation [*(1) Watering plants by applying the water below the ground surface. (2) Irrigation by the water table rising within or near the root zone, often under control. Colloquially, "subbing"*]
~ Garage f = underground garage
~ Gasspeicherung f = underground gas storage
~ Leitungen fpl in einem Stadtgebiet = services below street level
~ Schienenschnellbahn f = high-speed underground railway (Brit.); ~ ~ railroad (US)
~ Städteplanung f = underground town planning
~ Tongewinnung f = underground clay mining, ~ mining of clay
~ Verkehrsader f = underground traffic artery
~ Wasserspeicherung f = underground water storage

unterirdischer Abfluß m = subsurface flow
~ Aushub m = underground excavation
~ Bunker m [*als Schutzbunker*] = underground shelter
~ Friedhof m, Katakombe f = underground cemetery, catacomb
~ Ladengang m = underground shopping arcade
~ Wasserlauf m = underground water course
unterirdisches Bauwerk n = underground structure
~ Kabel n = underground cable
~ Wasser n = subsurface water
Unterkante f, UK = bottom
unterkellert = with basement, ~ cellar
Unter|klappe f → Gegenklappe
~konstruktion f [*Putztechnik*] = furring, firring
~konstruktion f, Dachdecke f = roof deck
Unterkorn n = undersize
~anteil m, Unterkornprozentsatz m = undersize percentage
~ausbringen n → Fein(korn)ausbringen
unterkühlte Fahrbahn f = carriageway at a temperature below freezing
Unterkunft f = accomodation
Unterkunfts|gelände n = accomodation area
~lager n → Arbeiterlager
(Unterlag)Brett n [*Betonsteinindustrie*] = (wooden) pallet
~heber m [*Betonsteinindustrie*] = pallet lifter
~silo m [*Betonsteinindustrie*] = pallet feeder
Unterlage f = base, support
~ [*Linoleum*] = backing
~, Unterschicht f [*(Fuß)Boden*] = lower layer
~ [*Straßenbau*] = supporting medium
~ziffer f → Bettungszahl f
Unterlag|papier n, Autobahn~, Straßenbaupapier = concreting paper, waterproof ~

~platte *f* = bearing plate, bed ~, bedplate
~scheibe *f* → Unterlegscheibe
Unterlauf *m* [*Fluß; Strom*] = lower course
~, Schuttkegel *m* [*Unterster Teil eines Wildbaches, Zone der Auflandung*] = heap of débris
Unter|läufigkeit *f* [*Durchsickern von Wasser unter der Sohle eines Staubauwerkes*] = underflow
~legplatte *f* [*Schienenbefestigungsmittel*] = baseplate
~legscheibe *f*, Einlegescheibe, Unterlagscheibe = washer [*A steel ring placed under a bolt head or nut to distribute the pressure, particularly when the bolt passes through a timber joint. In this case the washer prevents the timber crushing*]
~leiter *f*, Eimerleiter [*Eimerkettenbagger*] = ladder, main ~
~liegerwerk *n*, Unterwasserkraftwerk = downstream power plant, ~ ~ station, ~ generating ~
~meeres-Stahlrohrleitung *f* = submarine steel pipeline
unter|meerische Strömung *f* = undercurrent, undertow
~meerisches Relief *n* = submarine relief
Unter|mieter *m* = sub-tenant
~moräne *f*, Grundmoräne = ground moraine
~nehmensforschung *f* = operations research
Unternehmer *m* = contractor
~verband *m* = contractor association
Unter|pallung *f* = timber packing, launching punt [*on which a caisson is launched*]
~pegel *m*, UP = downstream ga(u)ge
~pflaster(straßen)bahn *f*, U-Strab *f* = underground tram(way), subway ~, tunnel ~
~pflaster(straßen)bahntunnel *m*, U-Strab-Tunnel = underground tram(way) tunnel
~planumschnitt *m* = cutting below grade

~pressen *n* [*Straßendecke*] = underseal(ing), subseal(ing)
~pulverschweißen *n*, Maulwurfschweißen, U-P-Verfahren *n*, ELLIRA-Verfahren, Elektro-Linde-Rapidverfahren = UNIONMELT welding method
Unterputz *m*, Rauhwerk *n* = rendering coat, first ~, backing ~, undercoat, render-coat [*A first coat of plaster on a wall or ceiling. A first coat on lath(ing) is called a pricking-up coat*]
~ auf Putzträgergewebe = pricking-up coat [*The first coat on lath(ing)*]
~leitung *f* = concealed conduit
Unter|querung *f* = undercrossing
~räumen *n* [*Tiefbohrtechnik*] = underreaming
~rollbrücke *f* = roller bridge sliding under the fixed part
~schalung *f* = bottom form(work)
~schicht *f*, Unterlage *f* [(*Fuß*)*Boden*] = lower layer
~schiebung *m* [*Dach*] = footing piece
unterschiedliche Belüftung *f* [*Korrosion unter Sauerstoffverbrauch*] = differential aeration [*oxygen-consumption type of corrosion*]
~ Setzung *f* [*Untergrund*] = differential settlement, non-uniform movement
unter|schlächtig [*Wasserrad*] = undershot [*water wheel*]
~schlägige Seilanordnung *f* [*Ein Seil, das so auf eine Trommel aufgelegt und an ihr befestigt ist, daß es unter der Trommel herum zur Last geht*] = underwinding
Unterschneidung *f*, Wassernase *f* = throat, (water) drip, weather groove
unterschrämen = to undercut
Unter|schubfeuerung *f*, Rostbeschicker *m* = underfeed stoker
~schütz(e) *n*, (*f*) → untere (Schütz-)Tafel *f*
~schwelle *f* → Drempel *m*
~seil *n* [*Koepeförderung*] = balance rope
~seite *f* = underside
~setzung *f* = (torque) reduction

Untersetzungs|getriebe n = reduction gear
~ritzel n = secondary reduction pin
~verhältnis n = reduction ratio
Unter|sicht f, **Unteransicht** = underside view
~spannung f = undervoltage
unterspülen, unterwaschen, wegspülen, (aus)kolken, auswaschen = to underwash, to scour
Unter|spülung f → Auskolkung
~stopfen n **der Gleise** = boxing up [*Packing ballast under sleepers to raise sagging track*]
~stromseite f [*Spundwand*] = downstream surface, ~ side
~stützungsausbau m = standing supports [*i.e. conventional supports, as opposed to roof bolting*]
~suchung f = investigation, analysis
~suchung f **mittels Versuchen** = experimental analysis, ~ investigation
Untersuchungsbohrung f → Aufschließungsbohrung
~ in Küstengewässern [*um Erdöl oder Erdgas zu finden*] = marine-discovery well
Untersuchungs|ergebnis n = finding
~grube f → Reparaturgrube
~schacht m → Aufschließungsschacht
~stollen m → Aufschließungsstollen
~vorhaben n → Aufschließungsvorhaben
Unter|tafel f → untere (Schütz)Tafel
~tagebetrieb m = underground practice
~tagegrube f = mine
~tagesprengen n = underground blasting
~tageverbrennung f **von Öl nach dem Feuerflutverfahren** = fire flood recovery method
untertätige Triebwasserleitung f, **~ Druckleitung** = underground penstock
~ Geräte npl = underground plant
~ (Herein)Gewinnung f, (Herein)Gewinnung unter Tage = underground getting

untertägiger Abbau m → Abbau unter Tage
~ Trichterbau m, Trichterbau unter Tage = underground glory-hole milling, ~ ~ mining
untertägiges Abbauen n → Abbau m unter Tage
~ Bohren n = underground drilling
Unter|teil m, n, Unterwagen m, Bagger~ = travel unit, mounting ~
~tor n [*Schleuse*] = tail gate, lower ~
~-Triebwasserkanal m = (channel-type) tail race, tail (race) channel [*The tail race below the turbines is usually an open channel but may occasionally be a tunnel*]
~-Triebwasserstollen m = tunnel-type tail race
~turas m = bottom sprocket (wheel)
~wagen m, Fahrwerk n = (travelling) carriage, undercarriage, underbody
~wagen m, Unterteil m, n, Bagger~ = travel unit, mounting ~
unterwaschen, wegspülen, (aus)kolken, ausspülen, unterspülen = to underwash, to scour
Unterwaschung f → Auskolkung
Unterwasser n [*Unterhalb einer Stauanlage befindliches Wasser*] = tailwater
~abbruch m = underwater demolition [*The destruction or fragmentation of underwater obstacles by the use of explosive charges hand-placed by diver personnel*]
~anstrich m = underwater coat(ing)
~aushub m = underwater excavation
~beton m = underwater concrete
~betonieren n, Betonieren unter Wasser = underwater concreting
~bohren n = underwater drilling
~böschung f = underwater slope
~-Düsengrabenzieher m = jet-action trencher, jet-type ~
~-Erzfrachter m = submarine ore carrier
~fenster n [*Hallen(schwimm)bad*] = underwater window [*to assist swimming instruction and television facilities*]

~-Fernsehen n = underwater television
~greifer m = underwater bucket, ~ grab
~gußbeton m = tremie concrete
~haltung f = downstream reach
~horizont m, Unterwasserseite f = tailwater, downstream water, downstream side
~kraftwerk n = tail water power plant (or installation)
~mauer f = underwater wall
~motorpumpe f, Tauchpumpe = submersible pump [*Motor sealed against the entry of water*]
~pumpe f = borehole pump [*It is driven through rods from a motor on the surface*]
~quelle f = drowned spring
~notverschluß m = tailgate
~rohrleitung f = subaqueous pipeline
~schall m = underwater sound
~schallempfänger m = submarine sound receiver
~schneidgerät n, Unterwasserschneidbrenner m = underwater cutting equipment
~schüttung(sdamm) f, (m) = underwater rockfill (dam)
~-Schutzbelag n aus schweren Betonplatten = concrete block mattress
~seite f = downstream face
unterwasserseitig = downstream
Unterwasser|setzen n von Land = flowage (US)
~sieb n = underwater screen
~spiegelhöhe f = downsteam (water) level
~sprengung f → → 1. Schießen; 2. Sprengen
~stand m = downstream level
~stollen m, Abflußstollen [*Talsperre*] = tail (race) tunnel, discharge ~
~strand m = approach
~-Straßen(verkehrs)tunnel m = subaqueous vehicular (or road) tunnel
~tunnel m = subaqueous tunnel
~tunnelbau m = subaqueous tunnelling
~warngerät n = telltale [*Apparatus used in hydrographic surveying*]
~zünder m → → 1. Schießen; 2. Sprengen
Unterwerk n [*Stromversorgung*] = substation
Unterwerksbau m (⚒) = working under the main haulage level
Unterwindgebläse n = undergrate blower
Unterzug m → Decken~
untief, seicht, geringe Tiefe = shallow
Untiefe f = high bed, shoal
unüberdachter Tribünensitz m = bleacher
ununterbrochene Fahrrinne f = continuous channel
unverfestigt [*Tragschichtmaterial im Straßenbau*] = untreated
unverfestigte Ablagerung f, lockere ~ (Geol.) = unconsolidated deposit
unverformt = undeformed
unverkleidete Deckenunterschicht f = exposed ceiling
unvermeidliche Verlustzeit f = unavoidable lost-time
unverpackter Zement m → Behälterzement
unverputzt, roh = unplastered
unverritzt (⚒) = unworked, virgin
unverritztes Feld n [*Tiefbohrtechnik*] = maiden field
~ Gebiet n [*Tiefbohrtechnik*] = unexplored area, area without exposures and untested by drilling
unverrohrt = bare footed, open, uncased [*bore hole*]
unverschieblicher Knoten m = undisplaced joint
unverschiebliches Gelenk n = fixed hinge
unverschlossener Überlauf m → → Talsperre f
unverseifbar = unsaponifiable
unversteift [*Baugrube*] = unbraced
unversteifte Hängebrücke f = unstiffened suspension bridge
Unverträglichkeit f = incompatibility
unverwitterter Fels m = sound rock

unverzerrtes Modell n = undistorted model
unvollkommener Überfall m = drowning, submerging
Unvorhergesehenes n = contingencies
Unwucht f → Unbalance f
~achse f = eccentric weight shaft
U-Profilrahmen m = channel section frame
U-P-Verfahren n → Unterpulverschweißen
Uranerz n = uranium ore
~-Konglomerat n = uranium conglomerate
Uranglimmer m, Kalkuranit m = lime uranite
~, Kupferuranit m = copper uranite, cupro-uranite, tobernite
(Uran)Pechblende f, (Uran)Pecherz n (Min.) = pitchblende
Urantetrafluorid n = uranium tetrafluoride
Urboden m, jungfräulicher Boden = virgin soil, original ~
Urfestigkeit f = fundamental strength (of concrete) [*Is understood as a stage of concrete loading representing the limit the structure can withstand under given conditions and still perform its given function*]
Urgestein n, kompakter Fels m, dichtes Gestein, kompaktes Gestein = primitive rock
Urgneis m = fundamental gneiss
Urinal|becken n = urinal
~rinne f = urinal trough
Urkalk(stein) m, echter kristallinischer Marmor m = recrystallized limestone, true marble
Urlöß m, Primärlöß, typischer Löß, echter Löß = true loess
U-Rohrmanometer n = U-tube manometer
U-Rohr-Viskosimeter n = U-tube viscosimeter
Ursprungs|festigkeit f = original strength
~gewicht n, Anfangsgewicht = original weight
U-Stahl m = steel channel(s)

U-Strab f, Unterpflaster(-Straßen)-bahn = tunnel tram(way), subway ~, underground ~
~-Tunnel m = tram(way) subway, ~ tunnel
U-Tal n, Trogtal = U-shaped glaciated valley
Uwarowit m, Kalkchromgranat m, $Ca_3Cr_2Si_3O_{12}$ (Min.) = uwarowite, ouvarovite

V

V8-Motor m = V-8 engine
vadoses Wasser n = vadose water
vagabundierender Bahnstrom m = stray traction current
~ Strom m → Erdstrom
Vakuum|anheber m, Saugvorrichtung f zum Fertigteiltransport = vacuum lifter pad
~anlage f = vacuum plant, ~ installation
~anlage f **zur Grundwasser(ab)senkung** = wellpoint drainage installation
~behand(e)lung f, Absaugen n, Saugbehand(e)lung f [*Beton*] = vacuum treatment
~behand(e)lung f **nach Billner**, Billner-Verfahren n [*Vakuumbeton*] = Billner method
~beton m = vacuum concrete
~betonrohr n = vacuum concrete pipe
~bremse f → Unterdruckbremse
~brunnen m = vacuum method wellpoint
~-Drehfilter m, n = rotary vacuum filter
~filter m, n, Saug(zellen)filter = vacuum filter
~filterung f, Vakuumfiltern n = vacuum filtration
~form f, Saugform [*Mit Saugräumen bzw. Saugfiltern ausgerüstete Form*] = vacuum mo(u)ld
~kammer f = vacuum chamber
~kern m [*Als Saugkörper ausgebildeter Hohlraumbildner*] = vacuum core

~kran m = vacuum crane
~lampe f = vacuum lamp
~messer m, Vakuummeter m, n = vacuum gauge, vacuometer
~pumpe f = vacuum pump, air ~, evacuator
~schmelzen n = vacuum melting
~schweißen n = vacuum welding
~-Strangpresse f = vacuum extrusion press, ~ extruding machine, deairing extruder
~-Tonschneider m = vacuum pugmill
~trommelfilter m, n = continuous vacuum filter
~verfahren n [*Zur Entwässerung und Verfestigung von Feinsanden*] = vacuum method
Van-Dyk-Braun n, Kasselerbraun = Cassel brown, Vandyk ~
Variante f = alternative design
Vektor|betrag m, Vektorsumme f = vector sum
~feld n = vector(ial) field
vektorielles Produkt n, äußeres ~ = vector product [*of two vectors*]
Veloutieren n = velour finishing
Ventil n = valve
Ventilator m, Lüfter m = fan
~antrieb m = fan drive
~haube f = fan shroud
~-Kühlturm m, Lüfter-Kühlturm m = forced draught cooling tower, mechanical ~ ~ ~
~windabweiser m = fan blast deflector
Ventil|block m = valve block
~bohrer m → Stauchbohrer
~bohrung f = valve bore
~büchse f → Stauchbohrer m
~deckel m = valve cover
~deckelunterbau m, Ventildeckelunterteil m, n = valve cover base
~dreher m = valve rotator
~durchgang m = valve passage
~einsatz m = valve seat insert, ~ core
~einstellschraube f = valve adjustment screw
~fahne f = valve mask

~feder f = valve spring
~(feder)heber m = valve spring lifter
~federspannung f = valve spring force
~führung f = valve bushing, ~ guide
~führungsabstand m = valve guide clearance
~gehäuse n = valve core housing
~kanal m = valve port
~kappe f = valve cap
~kipphebel m = valve rocker arm
~kipphebelei f, Ventilkipphebel mpl = valve rocker gear
~kopf m = valve head
~körper m = valve body
ventillos = valveless
Ventil|nut f, Dämpfungsrille f = throttle slot
~öffnung f = valve orifice
~pumpe f = valve pump
~sackfüllmaschine f, Ventil(sack)packmaschine = valve bag filling machine, ~ ~ packer
~schaft m = valve stem
~schappe f → Stauchbohrer m
~schleifmaschine f = valve grinder, ~ grinding machine
~sitzring m = valve seat
~spiel n = valve clearance
~stößel m = valve plunger, ~ follower, tappet, cam follower
~tellerwinkel m = valve face angle
~verpichung f = valve gumming
~vorlage f = valve block
~weg m = valve travel
Venturi|kanalmesser m = Venturi flume [*A type of open flume with a contracted throat that causes a drop in the hydraulic grade line; used for measuring flow*]
~rohr n; Venturimeter n, m [*Schweiz*] ~ Venturi tube [*generalized term*]; ~meter [*proprietary name*]
Veranda f = veranda(h); porch (US)
Veränderliche f = variable
Verankern n, Verankerung f = anchoring
verankerte einfache Spundwand f = single-walled anchored sheet pile bulkhead

verankerte Spundwand — Verbindung komplett

~ **Spundwand** *f* = anchored sheet (pile) wall, tied ~ (~) ~, ~ bulkhead

Verankerung *f*, **Verankern** *n* = anchoring

~ [*Gesamtheit der Ankereisen einer Betondeckenfuge*] = tie bars, tieing

~ [*Spanndrähte beim Spannbeton*] = securing of the wires

~, End~ [*Spannbeton*] = (end) anchoring system, (~) anchorage ~

~ **der Schichten im Hangenden** = roof bolting, suspension support installation

Verankerungs|block *m*, Ankerblock = anchor block, stay ~

~**bock** *m* = anchoring jack

~**bruch** *m* = anchorage failure, anchoring ~

~**-Einpreßmörtel** *m* = anchor grout

~**ende** *n* **eines Spannbettes** = anchor end of a bed

~**fundament** *n*, Verankerungsgründungskörper *m* = anchorage foundation

~**kammer** *f*, Ankerkammer [*Hängebrücke*] = anchorage chamber

~**kegel** *m*, Ankerkegel, Ankerkonus *m* = anchoring cone

~**länge** *f* [*Spannbeton*] = anchorage length, anchoring ~

~**loch** *n*, Ankerloch = anchor hole

~**mast** *m* = anchor(age) mast

~**mutter** *f*, Ankermutter = tie nut

~**pfahl** *m* → Ankerpfahl

~**pfeiler** *m*, Ankerpfeiler [*Brücke*] = anchor(age) pier

~**ring** *m*, Ankerring = anchor loop (Brit.); U-bolt (US)

~**rolle** *f* = anchor link roller

~**scheibe** *f*, Ankerscheibe = form anchor (US); anchor ring (Brit.)

~**schiene** *f*, Ankerschiene = anchoring rail

~**schlitz** *m*, Ankerschlitz = anchor slot

~**seil** *n* = anchor(age) rope

~**stab** *m* → Ankerstab

~**stahl** *m* = anchorage steel

~**stange** *f* → Ankerstab *m*

~**teil** *m, n* [*Spannbeton*] = anchorage element

~**turm** *m* = anchorage tower

~**-Vorrichtung** *f*, Spannkopf *m*, Spanngliedverankerung *f* [*Spannbeton*] = anchorage system

~**wand** *f*, Ankerwand = anchored sheet (pile) wall, ~ ~ piling, anchor sheeting

~**widerlager** *n* = anchorage abutment, anchoring ~

veranschlagen = to estimate

Veranschlagung *f*, Veranschlagen *n* = estimating

Verarbeitbarkeit *f*, Beton~ = workability (of concrete)

verarbeiten, einbauen, einbringen [*Baustoff*] = to place

Verarbeitungs|konsistenz *f*, Verarbeitungssteife *f* = application consistency

~**stelle** *f*, Einbaustelle [*Baustoff*] = point of placement

~**zeit** *f*, Topfzeit [*Betonkleber*] = potlife

Verband *m*, Backstein~, Ziegel~ = (brick) bond

~ = braces

Verbands|kasten *m* = first aid box, ~ ~ kit

~**raum** *m* = first aid room

Verbau *m*, Sicherung *f* [*Baugrube; Graben*] = sheeting; well curbing, pit boards (US) [*used to keep earth out of a building pit*]

verbesserte Proctordichte *f* = modified Proctor (dry) density, ~ AASHO ~

verbesserter AASHO-Versuch *m*, ~ Proctor-Versuch = modified Proctor test, ~ AASHO ~

Verbesserung *f* = improvement

verbinden [*Dieselmotor durch elastische Kupplung mit Generator*] = to couple [*A generator through a flexible coupling to diesel engine*]

Verbinden *n* **durch Vorspannen**, Zusammenspannen von Fertigteilen = stressing together of precast units

Verbindung *f* = connection

~ [*Chemie*] = compound

~ **komplett** → Kreuzgelenk *n*

Verbindung — Verbundbauweise

~ von Bewehrungsstählen mit Gießharz zur Kraftübertragung = glued joint of steel reinforcement
Verbindungs|festigkeit f [*Tiefbohrtechnik*] = joint strength (of a casing)
~**gelenk** n = connecting link
~**glied** n = connecting link
~**kanal** m, Anschlußkanal = connecting canal, connection ~
~**pfahl** m, Anschlußpfahl = connection pile
~**rampe** f [*Eine getrennt geführte Fahrbahn, die die in verschiedenen Ebenen liegenden Fahrbahnen eines Knotenpunktes miteinander verbindet*] = slip road
~**rollbahn** f, Verbindungsrollweg m [*Flughafen*] = interconnecting taxiway
~**schlauch** m = connecting hose
~**straße** f = connecting road, ~ highway, connection ~
~**streben** n, Affinität f = affinity, liking
~**stück** n = jointing element
~**teil** m, n, Anschlußteil, Zwischenstück n [*Gleiskettenlaufwerk*] = adapter, mounting, adaptor
~**weg** m = connection path, connecting ~
verblasen (⚒) = to stow pneumatically
Verblasen n, Blasversatz m (⚒) = pneumatic stowing
Verblassen n = fading [*Bleaching of a colour by ageing or weathering. Chalking looks like fading, but the colour can be restored by a coat of varnish*]
verbleit = leaded
Verblender m, Verblendklinker m, Verblendziegel m, Verblendstein m, Blendziegel, Blendstein = hardburnt facing brick, ~ face ~
Verblend|mauerwerk n = (stone-)faced masonry
~**platte** f = (stone) facing slab [*Stone facing slabs may be of natural, precast or artificial stone*]
Verblendung f = facing
verblichen, verblaßt [*Farbe*] = faded

verbolzen, befestigen mit Bolzen = to bolt
verbolzter (Stahl)Behälter m = bolted (steel) (storage) tank
Verbrauch m pro Einwohner = consumption a head of the population
Verbraucherleitung f = lateral service piping
Verbrauchsbunker m = live bin
verbrauchte Luft f, schlechte ~ = vitiated air
Verbreiterung f = widening
verbrennbares Gift n [*Atomtechnik*] = burnable poison
Verbrennung f = combustion
~ [*Müll*] = incineration
Verbrennungs|gas n, Rauchgas, Feuergas = flue gas
~**geschwindigkeit** f = rate of combustion
~**kammer** f = combustion chamber
~**leistung** f [*Müllverbrennung*] = burning capacity
~**luft** f = air of combustion
~**motor** m, Brennkraftmotor = internal combustion engine, I. C. ~
~**rückstand** m, Kohlenstoffrückstand [*Motor*] = carbon deposit
~**schlacke** f → Schlacke
~**wärme** V f → oberer Heizwert m
Verbügelung f → Bügelbewehrung
Verbund m, Haftung f = bond
~**ablösung** f, Haftablösung = bond failure
~**antrieb** m → dieselelektrischer Antrieb
~**balken(träger)** m = composite beam
~**balken(träger)** m **Stahl/Beton** = composite beam (in steel and concrete)
~**bauweise** f, Verbundsystem n = composite construction [*Different materials in conjunction, for example facing and backing bricks in walls, reinforced concrete in-situ topping over a precast prestressed concrete floor beam, or brickwork carried on a concrete or steel beam and considered to form its compression flange. It can result in big economies*]

Verbundbauweise — Verdachung

~bauweise *f* Fertigteile/Ortbeton = composite construction precast with cast-in-place concrete

~bauweise *f* Stahl/Beton = composite construction (in steel and concrete)

~bauweise *f* zwischen dem Stahltragwerk und der Stahlbetonfahrbahnplatte = composite concrete-steel girders

~bauwerk *n*, Mehrzweckbauwerk = multi-purpose structure

~bauwerk *n* Stahl/Beton = composite structure (in steel and concrete)

~bruch *m*, Haftbruch = bond failure [*between steel and concrete*]

~brücke *f* Stahl/Beton = composite bridge (in steel and concrete)

~-Brückenüberbau *m*, Verbund-Brückentafel *f*, Verbund-Brückenplatte *f* = composite (bridge) deck(ing)

~(dampf)maschine *f* [*hat einen Hochdruck- und einen Niederdruckdampfzylinder, letzterer mit größeren Abmessungen*] = compound steam engine

~dampfpumpe *f* = compound steam pump

~decke *f* [*im Hochbau*] = composite floor

~deckenplatte *f* = composite floor panel, ~ ~ slab

~ermüdung *f*, Haftermüdung [*Beton*] = fatigue of bond

~fachwerkträgerbrücke *f* = composite truss bridge

~faltwerk *n* Fertigteil/Ortbeton = composite folded slab

~-Fang(e)damm *m* = composite cofferdam [*earth and sheet piling*]

~festigkeit *f*, Haftfestigkeit = bond strength

~gebäude *n* Stahl/Beton = composite building (in steel and concrete)

~gitterträgerdecke *f*, Omniadecke = Omnia floor

~glas *n*, Verbund-Sicherheitsglas = laminated (safety) glass

~kamin *m*, Verbundschornstein *m* = compound chimney

~konstruktion *f* = composite construction in steel and concrete

~kreiselpumpe *f* = two stage centrifugal pump

~länge *f* → Haftlänge

~maschine *f* → Verbunddampfmaschine

~-Massivtragwerk *n* = composite solid structure

~mühle *f* = combination mill, ~ grinder

~netz *n* = grid network, ~ system, integrated power grid [*for linking hydroelectric and thermal power stations*]

~pfahl *m* = composite pile

~platte *f* = composite slab

~projekt *n*, Mehrzweckprojekt = multi-purpose scheme

~querschnitt *m* = composite section

~-Rahmentragwerk *n* = composite framed structure

~rohrmühle *f* = combination tube mill, ~ ~ grinder

~schlupf *m*, Haftschlupf = bond slip

~schornstein *m*, Verbundkamin *m* = compound chimney

~(-Sicherheits)glas *n*, Mehrscheibenglas = laminated (safety) glass

~spannung *f*, Haftspannung = bond stress

~stoßstange *f* = integral bumper

~system *n* → Verbundbauweise *f*

~träger *m* Stahl/Beton = composite girder (in steel and concrete)

~tragwerk *n* = composite structure

~untersuchung *f*, Haftuntersuchung = bond study

~verhalten *n*, Haftverhalten = bond performance

~wirkung *f*, Haftwirkung = bond action

~wirkung *f* = composite action

~zerstörung *f*, Haftzerstörung = destruction of bond

~zustand *m* [*Verbundträger*] = composite state

verchromt = chrome plated

Verdachung *f* [*bei einer Flügeltür*] = pediment

Verdämmaschine *f* → Besatzmaschine
Verdämmstoff *m*, Besatzstoff = tamping material, stemming ~
verdämmte Bohrung *f*, ~ Sonde *f*; verdämmter Schacht *m* [*in Galizien*] = tamped well, ~ bore, ~ boring
Verdämmung *f*, Besatz *m* [*Schußloch*] = stemming, tamping
Verdichtbarkeit *f* → Verdichtungswilligkeit
Verdichten *n*, künstliches ~, (künstliche) Verdichtung *f* = compacting, compaction
Verdichter *m*, Verdichtungsmaschine *f*, Verdichtungsgerät *n* = compacting machine, compaction ~, ~ equipment, compactor
~, Kompressor *m* = compressor
~**station** *f*, Kompressor(en)station = compressor station
verdichtete Dicke *f* = compacted thickness
verdichtetes Gas *n*, Druckgas = compressed gas
Verdichtung *f*, künstliche ~ = compaction
~, natürliche ~ → Konsolidierung
~ **oberflächennaher Schichten** [*maschinell*] = surface compaction [*e. g. such as 6" to 18" depths*]
Verdichtungs|apparat *m* → Oedometer
~**beiwert** *m* [*Bodenmechanik*] = coefficient of compressibility, ~ ~ compactability
~**bohle** *f* → Vibrierbohle
~**bohle** *f* → Bohle
~**druck** *m* [*Dieselmotor*] = compression pressure
~**energie** *f* = compactive energy
~**fähigkeit** *f*, Zusammendrückbarkeit, Verdichtungswilligkeit, Verdichtbarkeit = compressibility, compactability
~**faktor** *m* [*Beton*] = compacting factor
~**feuchtigkeit** *f*, Verdichtungsfeuchte *f* = compaction moisture
~**feuer** *n* [*Flugplatz*] = range light
~**gang** *m*, Verdichtungsübergang = compacting pass
~**gerät** *n* → Verdichter *m*
~**grad** *m* = degree of compaction
~**höchstmaß** *n* = optimum compaction
~**hub** *m* = compression stroke
~**kurve** *f* = compaction curve
~**maschine** *f* → Verdichter *m*
~**nachprüfung** *f* = compaction control
~**pfahl** *m*, Boden~ = compaction pile, compacting ~
~**pumpe** *f* [*pneumatische Förderanlage*] = pneumatic transport pump
~**pumpenrad** *n* [*Motor*] = compressor impeller
~**setzung** *f* = consolidation settlement
~**theorie** *f* = compaction theory
~**tiefe** *f* = depth of compaction
~**verfahren** *n* [*Straßen- und Erdbau; Betonsteinindustrie*] = compaction method
~**verhältnis** *n* = compression ratio [*diesel engine*]
~**walze** *f* → Knetverdichter *m*
~**willigkeit** *f*, Verdichtungsfähigkeit, Zusammendrückbarkeit, Verdichtbarkeit = compactability, compressibility
~**wirkung** *f* = compacting effect
~**zylinder** *m*, Kompressorzylinder = compressor cylinder
Verdingungsunterlagen *fpl* → → Angebot *n*
verdornen einzelner Löcher von zu verbindenden Teilen = to line up holes with drift pins
Verdrahtung *f* = wiring
Verdränger|kolben *m* = recuperator piston
~**pumpe** *f* = displacement pump
Verdrängungs|bohrer *m* [*Bodenuntersuchung*] = displacement auger
~**lager(stätte)** *n*, *(f)* = replacement deposit
~**widerstand** *m*, Eindringungswiderstand [*Rammen*] = driving resistance, penetration ~
~**zähler** *m* = postive displacement meter

Verdreh|festigkeit *f*, Torsionsfestigkeit = torsional strength
~spannung *f* = twisting stress
verdrehte Schichtungen *fpl* (Geol.) = twisted strata
Verdrehung *f*, (Ver)Drillung = torsion
Verdrehungs|bewehrung *f*, Verdrehungsarmierung, Verdrehungs(stahl)einlagen *fpl* = torsion reinforcement
~spannung *f*, (Ver)Drillungsspannung = torsion(al) stress
Verdrehwinkel *m* = torsion angle
verdrillt = twisted
verdrilltes Kabel *n* [*für Spannbeton*] = stranded cable
(Ver)Drillung *f*, Verdrehung, Torsion *f* = torsion
Verdrückung *f* [*Straßenkörper*] = lateral displacement
~ (⚒) = pinch-out
verdübelte Fuge *f* = dowel (bar) joint, dowel(l)ed ~
Verdübelung *f* = dowel(l)ing
verdünnbar = thinnable
verdünnter Schmierstoff *m* = diluted lubricant
Verdünnung *f* = dilution [*1.) A method of disposing of sewage or effluent by discharging into a stream or other body of water 2.) The ratio of the volume of flow of a stream to the volume of sewage or effluent discharged into it*]
Verdünnungsmittel *n* [*organische Flüssigkeit zur Streichbarmachung von Anstrichfarben*] = solvent, (volatile) thinner
Verdunstung *f* = evaporation
Verdunstungs|geschwindigkeit *f* = evaporation rate
~höhe *f* [*Verdunstung von einer Gebietsfläche während einer anzugebenden Zeitspanne, unter Annahme gleichmäßiger Verteilung als Wasserhöhe in mm ausgedrückt*] = evaporation in mm; ~ ~ inches
~wasser *n* = fly-off water, evaporation ~ [*That part of rainfall which is disposed of by evaporation*]

Verdurstungserscheinung *f* [*Zementvermörtelung*] = deficiency of moisture at the surface
Verdüsungspulver *n*, verdüstes Pulver = atomized powder
Veredelung *f* **des Potentials** [*elektrochemische Korrosion*] = shift towards a more noble potential
Veredelungsanlagenteil *m*, *n* [*Erdölwerk*] = process unit
Vereinfachung *f* = simplification
vereinigte Achsial- und Radialpumpe *f* = mixed-flow pump
~ Achsial- und Radialturbine *f* = mixed-flow turbine
vereinigtes Hänge- und Sprengwerk *n*, zusammengesetztes Hängewerk, Hängesprengwerk = composite truss frame
vereist, eingeeist [*Hafen*] = ice-bound
~ [*Straße*] = icy
vereiste Wasserwolke *f* = mixed-state cloud
Vereisungsrohr *n*, Steigrohr, Gefrierrohr [*Gefrierverfahren*] = freezing pipe
(Ver)Fahren *n* [*Baumaschine; Laufkatze*] = travel(l)ing
Verfahren *n* = method
~ mit galvanischer Elektrodenkopplung → Widerstandsmessungsverfahren
~ ~ Innenkabeln [*Spannbetonbehälterbau*] = internal cable method
~ ~ logarithmischer Spirale als Gleitfläche [*Bodenmechanik*] = logarithmic spiral method for determining passive earth pressure
Verfahrens|ablauf *m* → Arbeitsablauf
~folge *f* → Arbeitsablauf *m*
~gänge *mpl* → Arbeitsablauf *m*
~gang(folge) *m*, (*f*) → Arbeitsablauf *m*
~ingenieur *m* = process engineer
~stammbaum *m* = flow-sheet
~technik *f* = process engineering
~technologie *f* = process technology
Verfärbung *f* = discolo(u)ration
verfestigte Schicht *f* (Geol.) = concretionary horizon
verfestigter Schluffmergel *m* = siltstone

verfestigtes Wasser n = solidified water
Verfestigung f → Baugrund~
~ → Boden~
Verfestigungsgrad m, Konsolidierungsgrad [*Bodenmechanik*] = degree of consolidation
Verfilzbarkeit f = felting property
verfilzen = to felt
Verflechtungs|länge f, Einfädelungslänge [*Straßenverkehrstechnik*] = merging distance, ~ length, weaving ~
~**raum** m, Einfädelungsraum [*Straßenverkehrstechnik*] = merging space, ~ area, ~ section, weaving ~
verflüchtigend, flüchtig = volatile
verflüssigen [*Beton und Mörtel durch Zusatzmittel*] = to plasticize
~, aufschließen [*Steinzeugindustrie*] = to defloculate, to dissolve, to disintegrate
Verflüssiger m [*Betonwirkstoff*] = workability agent, plasticizer; wetting agent (Brit.)
verflüssigtes Gas n, Flüssiggas = liquefied petroleum gas, L-P ~
Verflüssigung f, Aufschluß m [*Steinzeugindustrie*] = deflocculation
Verflüssigungsmittel n, Aufschlußmittel [*Steinzeugindustrie*] = defloculent
Verformbarkeit f = deformability
verformen = to deform
Verformen n → Verformung f
Verformung f, Deformation f, Formänderung, Gestaltänderung, Verformen n [*Änderung einer Dimension eines Körpers unter dem Einfluß physikalischer Beanspruchung(en)*] = deformation
~ **der Bereifung**, Walkwiderstand m der Reifenseitenwände = tire flexing (US); tyre ~ (Brit.)
Verformungs|arbeit f, Formänderungsarbeit = deformation work
~**bedingung** f, Formänderungsbedingung = deformation condition
~**eigenschaft** f, Formänderungseigenschaft = deformation property
~**geschwindigkeit** f, Formänderungsgeschwindigkeit, Deformationsgeschwindigkeit = rate of deformation, deformation rate
~**mechanismus** m, Formänderungsmechanismus = deformation mechanism
~**messer** m → Deformationsmesser
~**spanne** f, Formänderungsspanne = margin of deformability
~**theorie** f = theory of deformation
~**unterschied** m, Formänderungsunterschied = differential deformation
~**vorgang** m = deformation process
~**widerstand** m, Formänderungswiderstand = resistance to deformation
Verfrachtung f, Transport m (Geol.) = transport
verfugen, ausstreichen, ausfugen = to point
(Ver)Fugen n → Ausfugen
verfugt, verstrichen, ausgefugt = pointed
Verfugung f → Ausfugen n
~, Ausfugung, Ausstrich m [*als Ergebnis des Ausfugens*] = pointing
verfüllen, einfüllen [*Graben*] = to refill, to backfill, to fill in
Verfüll|schicht f [*Setzpacke*] = racking course
~**schild** m, Erdschieber m [*Anbaugerät*] = planing blade, backfilling ~, backfiller ~
~**vorrichtung** f, Planierschild n zum Zuziehen (oder Zuschieben) von Gräben = backfiller attachment
Vergabe f, Auftragerteilung f = award of contract, letting ~ ~, contract award, contract letting
~**beamter** m = contracting officer
~**verfahren** n = contract-awarding procedure
Vergaser m = carburetor (US); carburettor, carburetter
~**motor** m → Benzinmotor
Vergasung f [*flüssiger Brennstoff im Vergaser*] = vaporization, gasification
Vergasungs|gas n = manufactured gas
~**mittel** n = gasifying agent

vergelt = gel-like
vergießen, ausgießen [*Fuge*] = to pour
Vergießen *n*, Ausgießen [*Fuge*] = pouring
~ [*Drahtseil*] → Verguß
vergifteter Anstrich *m*, anwuchsverhindernder Schiffbodenanstrich [*Schiffbodenfarben sind Schutzfarben. Sie sollen das Unterwasserschiff vor zerstörenden Einflüssen bewahren. Wir unterscheiden zwischen dem zerstörenden Einfluß des salzhaltigen Seewassers und der Bewuchsgefahr des Schiffbodens durch tierische und pflanzliche Organismen. Durch den Einfluß des Salzwassers tritt Korrosion ein, die zu schweren Schädigungen des Schiffbodens führt. Ein mehr oder weniger starker Bewuchs des Schiffbodens aber bedeutet für das Schiff eine Verringerung der Geschwindigkeit, erhöhten Energieverbrauch und Zeitverlust. Aus diesem Grunde benötigt das Schiff zwei Anstriche, einen Rostschutzanstrich und einen vergifteten Anstrich, der Anti-Fouling genannt wird*] = anti-fouling
verglasen = to glaze
Verglasen *n* = glazing [*The process of placing glass in the sash*]
verglast = glazed
verglaste Aussichtsplattform *f* = glassed-in observation deck
verglastes Gestein *n* = vitreous rock
Verglasung *f* [*Das Ergebnis des Verglasens*] = glazing
Verglasungs|arbeiten *fpl*, Glaserarbeiten = glazier's work
~kitt *m*, Glaserkitt = glazier's putty
vergleichende Untersuchung *f*, Vergleichsstudie *f* = comparative study
Vergleichs|beton *m* [*normal erhärtende Probekörper bei warmbehandeltem Beton*] = comparison concrete
~festigkeit *f* = comparison strength
~last *f* = equivalent load
~probekörper *m* = control specimen
~spannung *f* = comparison stress
~tafel *f* = comparison table

~untersuchung *f* = comparative study, comparison ~
~wert *m* = comparative value
~zement *m* = control cement
vergletschert = covered by glaciers
(Ver)Glühen *n* [*Verhüttung*] → Kalzinieren
(Ver)Glühzone *f* → Brennzone
vergossen [*Fuge*] = filled, sealed
Vergrasung *f* = invasion by grass
Verguß *m*, Vergießen *n* [*Das Ausgießen eines zu einem Drahtseilbesen aufgeflochtenen und entsprechend vorgerichteten Drahtseilendes in einer kegeligen Vergußmuffe durch geeignete Metalle, Legierungen oder Vergußmassen*] = capping
~fuge *f* = poured joint, grouted ~
~gerät *n* **mit druckhafter Füllung,** Fugen~ ~ ~ ~ = pressure applicator
Vergußmasse *f*, Fugen~, Fugenkitt *m*, Ausgußmasse = (joint-)sealing compound, joint filling composition, jointing compound, (joint) filler; paving joint sealer, top-sealer (US)
~, Füllmasse [*Zur Vermuffung von Rohren*] = sewer joint(ing) compound, pipe ~ ~, joint cement
~ aus Kautschukmilch, Zement und Zuschlägen = (rubber) latex-cement-aggregate mix(ture)
~ofen *m* → Schmelzkessel *m* für Fugenvergußmassen
Verguß|mörtel *m* = seal(ing) mortar
~spalt *m* = sealing groove, ~ slot
vergüten [*einen Stoff weiterverarbeiten*] = to modify
Vergüteöl *n*, Härteöl = annealing oil, hardening ~
Verhalten *n* = behavio(u)r
Verhältnis *n* = ratio
~ Anthrazenöl II : Anthrazenöl I = Anthracene II/Anthracene I ratio
~ Kettfestigkeit zur Schußfestigkeit = ratio of warp to weft strength
verhältnisgleich = proportional
Verhältniszahl *f* = ratio, proportion(ality) factor
verharzen = to resinify

Verharzung *f*, Verpichung = gumming, gum forming
Verharzungs|neigung *f* [*Krackbenzin*] = gum forming
~probe *f*, Verharzungsprüfung *f*, Verharzungsversuch *m* = gum test
Verherdung *f* → Herdmauer *f*
Verhieb *m* (⚒) = face advance
Verhiebsbreite *f* (⚒) = increment of face advance
Verhol|poller *m*, Landpoller = checking bollard
~seil *n* [*Kabelspinnen beim Brückenbau*] = hauling rope, ~ cable
Verhütung *f* **von Staubeintritt** = dust exclusion
Verhütungsmaßnahme *f* = preventive measure
verjüngen = to rejuvenate [*asphalt surface*]
~, sich ~, konisch zulaufen = to taper
Verjüngung *f* = taper(ing)
~ [*Fluß; Strom*] = rejuvenation
Verjüngungsverhältnis *n* **beim Kegel** → Kegelneigung *f*
Verkämmung *f*, Verkämmen *n* [*Holzverband im Fachwerkbau*] = cogging, cocking, corking
Verkaufs|automat *m* = automatic vending machine
~ingenieur *m* = sales engineer
~preis *m* **ab Zeche** = pithead price of coal
Verkehr *m* = traffic
~, Straßen~ = traffic, road ~, highway ~
verkehrsabhängige Steuerung *f* **(oder Signalisierung) durch Kontaktschwellen** = vehicle-actuated traffic control
Verkehrs|ablauf *m* = traffic flow
~ader *f*, Groß~, Hauptfernverkehrsstraße *f* = traffic artery
~ampel *f*, Verkehrslichtsignal *n* = traffic (control) (light) signal, ~ light
~analyse *f* = traffic analysis
~anlage *f* = traffic facility
~anlage *f* **mit Benutzungsgebühr** = toll facility

~anlagen *fpl* **neben der Straße** = roadside facilities
verkehrsarmer Zeitraum *m* = slack traffic period
Verkehrs|art *f* = type of traffic
~aufkommen *n* = traffic volume
~bau *m* = construction of traffic facilities
~bauten *mpl* = traffic structures
~bau(werk) *m*, (*n*) = traffic structure
~belastung *f* [*Straßendecke; Flugplatzdecke*] = externally applied load(s)
~betrieb *m*, öffentlicher ~ = transit authority
~bewährung *f*, Verkehrsverhalten *n* [*eines eingebauten Straßenbaustoffes*] = road performance, service behavio(u)r
~chaos *n* = traffic maze
(Verkehrs)Damm *m* = bank, embankment [*A ridge of rock or earth thrown up to carry a road, railway, or canal*]
~bau *m* = embankment construction, bank ~, embanking
~breite *f* = embankment width, bank ~
Verkehrs|dichte *f* = traffic concentration
~ebene *f* = traffic level
~einheit *f* = unit of traffic
~entstehung *f* = traffic generation
~entstehungsraum *m* = traffic generator
~erhebung *f* → Verkehrsmessung
~erziehung *f* = training in road sense, traffic safety education
~farbe *f* → Straßen-Markierungsfarbe
~fläche *f* = traffic area
~freigabe *f* = opening to traffic
~führung *f* = traffic handling
verkehrs|gebundene Schotterdecke *f* = traffic-bound macadam
~gesteuerte Signalanlage *f* = vehicle-actuated traffic signal installation
Verkehrs|hafen *m* → Personenhafen
~ingenieur *m* = traffic engineer
~ingenieurwesen *n* = traffic engineering
~insel *f* = traffic (island)

Verkehrskategorie — Verkehrsumlegung

~kategorie *f* = traffic category

~komprimierung *f*, Knetwirkung = kneading action of traffic, traffic compaction [*pushes the particles into a closer and more permanent fit*]

~kreisel *m*, Kreisverkehrsplatz *m* = traffic roundabout

~lärm *m* = traffic noise

~last *f* [*Die veränderliche oder bewegliche Last eines Bauteiles; z.B. Personen, Einrichtungsstücke, Lagerstoffe, Riemenantriebe, Kranlasten, Wind, Schnee*] = live load [*This English term does not include wind loads and snow loads, as the German term does. It is, therefore, not a true synonym for the German term*]

Verkehrsleistungsfähigkeit *f* = highway capacity, working ~
theoretische ~, grundlegende ~ = basic ~ ~
mögliche ~ = possible ~ ~
praktische ~ = practical ~ ~

Verkehrs|leitkegel *m*, Wegebake *f* aus Gummi = (rubber) traffic cone

~lenkung *f*, Verkehrssteuerung = traffic monitoring

~-Leuchtnagel *m* → Straßen-Leuchtnagel

~linie *f*, Verkehrsstrich *m*, Markierungslinie, Markierungsstrich, Farbstreifen *m*, Farbstrich, Farblinie, Markierungsstreifen, Verkehrsstreifen = traffic line, white ~, road-(way) stripe

~(-Markierungs)farbe *f* → Straßenmarkierungsfarbe

~messung *f*, Verkehrszählung, Verkehrserhebung = traffic survey, ~ count, ~ census

~minister *m* = Minister of Transport

~ministerium *n* = Ministry of Transport, M.O.T. (Brit.); Transport Ministry

~nagel *m* → Markierungsknopf *m*

~netz *n* = traffic network

~netzplanung *f* = traffic network planning

~opfer *n* = traffic victim

~planung *f* = traffic planning

~platz *m* = traffic square

~politik *f* = transportation policy

~polizei *f* = traffic police

~polizist *m* = (traffic) patrolman

~prognose *f* = traffic prognosis

~reg(e)lung *f* [*Maßnahme oder Maßnahmen, ausgenommen baulicher Art, die der Sicherheit und Leichtigkeit des Verkehrs dient bzw. dienen*] = traffic management

verkehrsreich = heavily trafficked

Verkehrs|schätzung *f* = estimate of traffic, traffic estimate

~schild *n* = traffic sign

~schwankung *f* = traffic fluctuation

~sicherheit *f* = traffic safety

~signal *n* = traffic (control) signal

(~)**Signal-Steuerung** *f*, Signalreg(e)lung, Signalisierung, Lichtsignalsteuerung = signal control, traffic ~ ~

~spitzenstunde *f* = rush hour

~spur *f* → (Fahr)Spur

(~)**Spurlinie** *f*, Fahrspurlinie = (traffic) lane line

~steuerung *f*, Verkehrslenkung = traffic monitoring

~stockung *f*, Verkehrsstau *m* = traffic congestion, ~ jam

~stoß *m* = traffic impact

~streife *f* = traffic patrol

~streifen *m* → (Fahr)Spur *f*

~streifen *m* → Verkehrslinie

~strich *m* → Verkehrslinie

~strom *m* = traffic stream

~stromlinie *f* = desire line

~sünder *m* = traffic offender

~toter *m* = fatality

~tunnel *m* = traffic tunnel

~übergabe *f*, Freigabe für den Verkehr = opening to traffic

~übertretung *f* = traffic offence

~überweg *m*, Zebrastreifen *m* = zebra crossing

~umlegung *f*, Verkehrsverteilung [*Die Verteilung des Verkehrs auf ein bestehendes oder geplantes Netz*] = traffic assignment

~umlegung *f* [*wegen Bauarbeiten*] = traffic dislocation
~unfall *m* = traffic accident
~unfallstatistik *f* = traffic accident record
~verbot *n* = traffic ban
~verhalten *n* → Verkehrsbewährung *f*
~verteilung *f* → Verkehrsumlegung
~weg *m* = (transportation) route, traffic ~
~wert *m*, Zeitwert [*Maschine*] = present value, trade-in ~
~zählgerät *n* = vehicle counter, traffic measurement device
~zählstelle *f* = observation point, traffic census ~
~zählung *f* → Verkehrsmessung
~zählung *f* von **Hand** = manual (traffic) counting
~zeichen *n* → Straßen~
~zentrum *n* für den Stückgutverkehr = railhead (Brit.)
~zuwachsrate *f* = traffic growth-rate
verkehrtes Gewölbe *n*, umgekehrtes ~, Grundgewölbe = invert arch [*Used to spread the loads of piers between openings evenly on the foundation*]
~ Übergangsrohr *n* → Reduzierrohr
verkeilen = to wedge, to key, to fix by wedges
Verkeilen *n*, Verklemmen mit Keilen = fixing by wedges, wedging
Verkeilung *f*, Wellenkeil *m*, Keilprofil *n* = (shaft) spline
Verkieseln *n*, Verkieselung *f*, Einkieselung = silicifying (US); silicating, treatment with silicate (Brit.)
Verkiesung *f* = gravel filling
Verkippung *f* der Abraummassen, Verstürzen *n* des Abraums = dumping of overburden
verkitten = to cement (together)
verkittete Zwischensubstanz *f* → Bindemittel *n*
verkitteter Boden *m* = cemented soil
Verkittung *f* [*Gestein*] = cementation
Verklappen *n* [*Naßbaggergut*] = discharge of dredged material by hopper barge (Brit.); ~ ~ ~ ~ ~ bottom-dump scow (US)

verkleben (mit) = to cement down, to glue ~, to bond ~, ~ ~ to
verkleidete Baugrube *f*, ausgekleidete ~, umspundete ~ = sheeted excavation, lined ~
verkleideter Bohrturm *m* = enclosed (drilling) derrick
Verkleidung *f*, Auskleidung, Bekleidung = lining
~, Bekleidung, Auskleidung [*Baugrube*] = lining, sheeting
Verkleidungs|arbeit(en) *f(pl)* → Auskleidungsarbeit(en)
~beton *m*, Auskleidungsbeton, Bekleidungsbeton = lining concrete
~-Betonfertigteil *m, n* → Auskleidungs-Betonfertigteil
~blech *n* → Auskleidungsblech
~bohle *f* → Auskleidungsbohle
~fliese *f* → Verkleidungsplatte *f*
~mauer *f* → Futtermauer
~platte *f*, Verkleidungsfliese *f* [*Für Wände und Decken, aber nicht für Fußböden*] = facing tile
~platte *f* → Auskleidungsplatte
~ziegel *m*, Auskleidungsziegel, Bekleidungsziegel = lining brick
Verkleinerungsverhältnis *n* = reduction ratio
Verklemmen *n* mit Keilen, Verkeilen *n* = fixing by wedges, wedging
Verknappung *f*, Knappheit *f* = shortage
Verkohlung *f* = carbonization
Verkokungsneigung *f* = coking tendency
verkrautet = weedy
Verkrautung *f* = water-weed growth
Verkröpfen *n*, Verkröpfung *f* = cranking, offsetting
verkröpft = cranked, offset
Verkrustung *f* = incrustation
verkupfern = to (coat with) copper, to copper plate
Verkupferung *f*, Kupferplattierung = copper plating, coppering
Verkupp(e)lung *f* → Stegverbindungsstück *n*
Verkürzung *f* = shortening

Verlade|anlage *f* = loading installation, ~ plant
~band *n* = loading belt
~brücke *f*, Transportbrücke, Brückenkran *m* = transporter bridge
~bühne *f* = loading platform
~drehkran *m* = revolving loading and unloading crane
~(-Fahr)schrapper *m* → Schrapplader
~gerät *n* → Auflader
~greifer *m*, Ladegreifer = loading grab(bing crane), ~ (grab) bucket crane
~kran *m* = loading and unloading crane
~rampe *f*, (Ver)Ladebühne *f*, Laderampe = loading and unloading ramp
~schrapper *m* → Schrapplader
~sieb *n* = loading screen
~siebung *f*, Verladesieben *n* = rescreening of graded sizes before loading
~silo *m*, Fertiggut-Silo, Mischgut-Verladesilo [*für Beton und bituminöse Gemische*] = loading bin, mixed material storage hopper
Verlandebecken *n* → Talsperre
verlandendes Ufer *n*, anlandendes ~, Anwachsufer = accreting bank
Verlandung *f* → Auflandung
~ → Verschlammung
Verlandungs|geschwindigkeit *f* = rate of deposit, ~ ~ sedimentation
~höhe *f* = height of deposit, ~ ~ sedimentation
~material *n* = sediment deposit(s), sediment
~sperre *f* = sediment barrier [*Is constructed above the head of a reservoir*]
verlängerter Portlandzementmörtel *m* = Portland-cement-lime mortar
~ **Zementmörtel** *m*, Zementkalkmörtel, Kalkzementmörtel, Magermörtel = lime cement mortar, cement lime ~, weak ~; gauged ~ (Brit.)
Verlängerung *f* für das (An)Saugrohr = air inlet extension
~ ~ ~ **Auspuffrohr** = exhaust pipe extension
Verlängerungen *fpl* für das Schaufelgestänge [*Schaufellader*] = bucket linkage extensions
Verlängerungs|bohrstahl *m*, Bohrstange *f* = (drill) steel extension
~bolzen *m* = extension bolt
~rohr *n*, Ansatzrohr = extension pipe
verlaschen = to bolt up fishplates
Verlaschen *n*, Verlaschung *f* = fishing [*Bolting up fishplates to rails or other members*]
Verlauf *m* einer Bohrung = course of a bore
Verlege|anleitung *f* = directions for laying, spec(ification)s ~ ~
~fläche *f* = laying area
~gerät *n* für Bewehrungsmatten, Mattenausleger *m* = mesh laying jumbo
~kitt *m* = bedding putty
~mörtel *m* = bedding mortar
Verlegen *n* → Versatz *m*
~, Verlegung *f* [*Im Sinne von „Einbauen"*] = laying
Verlegeschiff *n* [*Rohre; Kabel*] = lay barge
Verlegung *f*, Verlegen *n* [*Im Sinne von „Einbauen"*] = laying
(Ver)Leimen *n*, (Ver)Leimung *f* = glu(e)ing, bonding
verlorene Arbeitskammer *f* [*Druckluftgründung*] = permanent caisson
~ **Schalung** *f*, Dauerschalung = permanent formwork
~ **Schleusenfüllung** *f* = lock-full lost
verlorenes Bohrgestänge *n* [*Rotarybohren*] = lost drill pipe
~ **Rohr** *n* [*Ortpfahl*] = shell remaining in the ground, casing ~ ~ ~ ~
~ **Widerlager** *n*, unterdrücktes ~ = dead abutment
Verlust|multiplikator *m* = loss multiplier
~schleife *f*, Hysteresisschleife, Hystereseschleife = hysteresis curve, ~ loop
~zeit *f*, Verzögerung *f* [*Straßenverkehr*] = stopped time

Vermahlen — Verpreßgürtel

Vermahlen n, **Vermahlung** f = intergrinding, combined grinding
vermahlen mit = to intergrind with
vermarken [*Strecke im Bergbau*] = to set out observation points [*in a roadway*]
Vermauerungsautomat m [*Maschine die Ziegel automatisch vermauert*] = auto-mason
vermeidbare Verlustzeit f = avoidable downtime
Vermessen n = surveying
vermessingen = to (coat with) brass
Vermessung f = survey
Vermessungs|arbeiten fpl im Gelände = groundwork, field work, ground survey
(~)**Bildflug** m = mapping flight
(~)**Bildflugzeug** n = mapping (air-) plane
~**ingenieur** m = survey(ing) engineer
~**kompaß** m, Feldkompaß = surveying compass
~**mast** m [*Brückenbau*] = sighting tower
~**schiff** n = survey(ing) vessel, ~ ship
~**steiger** m, Markscheidergehilfe m = deputy mining surveyor
~**trupp** m, Vermessungskolonne f, Vermessungsmannschaft f = survey(ing) party, ~ gang, ~ team, ~ crew
~**wesen** n = survey(ing) practice
Vermikulit m [*Verwitterungsprodukt der Eisenmagnesiaglimmer*] = vermiculite
~**beton** m = vermiculite concrete
~**putz** m = vermiculite plaster
~**stein** m = vermiculite block
verminderte Tragfähigkeit f **des anstehenden Bodens** = reduced subgrade strength
verminderter Querschnitt m [*Blechträger*] = net sectional area
Verminderung f = reduction
Vermischen n **körniger Stoffe** = granular solids mixing
Vermischung f [*Strom; Fluß*] = bifurcation

Vermoderung f = moder formation, mouldering
Vermörteln n, **Vermörtelung** f [*Boden*] = processing, stabilization
Vermörtelungs|(bau)zug m = processing train, stabilization ~
~**maschine** f, Boden~ = soil stabilizer
vermutet = inferred [*resources*]
vernichten [*Energie in der Hydraulik*] = to dissipate
vernieten, befestigen mit Nieten = to rivet
Veröffentlichung f = publication
veröltes Wasser n, öliges ~ = oily water
verpacken, versetzen, mechanisch ~ (⚒) = to stow
~**, versetzen, von Hand** ~ (⚒) = to pack (by hand)
verpfählen → abpfählen
Verpfählen n → Abpfählen
verpflocken, abpflocken, abpfählen, abstecken = to peg out, to set ~, to stake ~
Verpflocken n → Abpfählen
Verpichung f, **Verharzung** = gum forming, gumming
verpressen, einpressen, auspressen, injizieren, abpressen = to grout under pressure, to inject
Verpressen n **in einem einzigen Arbeitsgang** [*Baugrundverfestigung*] = single-shot process
Verpressung f → Injektion f
Verpreß|anlage f, Einpreßanlage, Auspreßanlage, Injektionsanlage = grout(ing) plant, ~ installation
~**-Diaphragma** n → Einpreß-Dichtwand f
~**-Dicht(ungs)wand** f → Einpreß-Dichtwand
~**druck** m → Injektionsdruck
~**-Entlüftungsbohrung** f → Injektions-Entlüftungsbohrung
~**gang** m → Auspreßgang
~**gerät** n → Verpreßkessel m
~**gründung** f → Auspreßgründung
~**gürtel** m → Einpreß-Dichtwand f

Verpreßgut — Versatzdruck

~gut *n*, Einpreßgut, Auspreßgut, Injektionsgut = grout(ing) material
~kammer *f* → Injektionskammer
~kessel *m*, (Zementmörtel)Injektor *m*, (Zementmörtel-)Einpreßgerät *n*, Drucklufteinpreßgerät, (Druckluft-)Auspreßgerät, (Druckluft)Verpreßgerät, Druckluftinjektor = boojee pump, boogie box, grout(ing) machine, grout(ing) pan
~lanze *f* → Injektionslanze
~loch *n* → Injektionsloch
~mörtel *m* → Injektionsmörtel
~öffnung *f* → Injektionsloch *n*
~pumpe *f* → Injektionspumpe
~ring *m* → Injektionsring
~rohr *n* → Injektionsrohr
~schirm *m* → Einpreß-Dichtwand *f*
~schlauch *m* → Einpreßschlauch
~-Schleier *m* → Einpreß-Dichtwand *f*
~schürze *f* → Einpreß-Dichtwand *f*
~-Sperrwand *f* → Einpreß-Dichtwand
~stutzen *m*, Auspreßstutzen, Einpreßstutzen, Injektionsstutzen, Injizierstutzen = injection socket, grouting ~
verpreßte Schüttung *f* → ausgepreßte ~
verpreßter Alluvionsschleier *m* → (injizierter) ~
verpreßtes Spannglied *n*, ausgepreßtes ~ = bonded tendon, grouted ~
Verpreß|verfahren *n* → Injektionsverfahren
~zement *m* → Injektionszement
Verproviantierungshafen *m* = victualling port
Verputz *m* = plastering
~gerät *n* → Putzwerfer
~kelle *f* = gauging trowel
~maschine *f* → Putzwerfer *m*
verputzt = plastered
Verriegelung *f* = (inter)locking
~ = (inter)locking device
Verriegelungsventil *n* = lock valve
verringern [*Querschnitt*] = to contract
verrohren = to pipe
~, ausfuttern, ausfüttern [*Bohrloch*] = to case
Verrohrung *f* → (Bohrloch)Ausfütterung

Verrohrungs|bühne *f* = casing platform
~kopf *m*, Rohrkopf [*Tiefbohrtechnik*] = casing-head
~maschine *f*, Verrohrungsgerät *n* = casing machine
~perforator *m*, Rohrlocher *m* = casing perforator
~schuß *m* → Ausfütterungsschuß
~seil *n* [*pennsylvanisches Seilbohren*] = calf line, casing ~
~(seil)rolle *f* = casing line sheave, ~ pulley
~(seil)trommel *f*, Flaschenzughaspel *m*, *f* = calf wheel
~teufe *f* [*Bohrloch*] = casing point
verrußen = to soot
versagen [*Sprengtechnik*] = to misfire
Versagen *n* = failure
Versager *m*, Fehlschuß *m*, Fehlzündung *f* = misfiring, misfire(d shot), blown-out shot, blow, misfired cartridge
Versandabmessungen *fpl*, Raumbedarf *m* bei Versand [*Maschine*] = shipping dimensions
Versanden *n* → Versandung
Versand|gewicht *n* = shipping weight
~hafen *m*, Verschiffungshafen, Ladehafen = port of shipping, ~ ~ shipment
~kiste *f* = shipping box
~ort *m*, Absendeort = point of shipment, ~ ~ shipping
Versandung *f*, Versanden *n*, Ansanden, Ansandung = sand silting (Brit.); ~ filling (US); sanding up
Versatz *m*, Versetzen *n* [*von Betonfertigteilen. Bei Platten auch „Verlegen" genannt*] = placing
~, Hand~ (✕) = (hand) packing
~, mechanischer ~ (✕) = stowing
~böschung *f* (✕) = slope of the stowed material [*stowing*]; ~ ~ ~ packed ~ [*hand packing*]
~druck *m* [*auf dem Versatz lastend*] = pressure acting upon the stowed waste (area); ~ ~ ~ ~ ~ goaf, ~ ~ ~ ~ ~ gob [*Wales*]; ~ ~ ~ ~ ~ condie, ~ ~ ~ ~ ~ cundy

Versatzdruck — Verschleißprüfung

[*Scotland*] [*stowing*]; ~ ~ ~ ~ packed waste (area); ~ ~ ~ ~ ~ goaf, ~ ~ ~ ~ ~ gob [*Wales*]; ~ ~ ~ ~ ~ condie, ~ ~ ~ ~ ~ cundy [*Scotland*] [*hand packing*]

~**druck** *m* [*vom Versatz ausgeübt*] = pressure exerted by the stowing material [*power stowing*]; ~ ~ ~ ~ packing material [*hand packing*]

~**druckdose** *f* (✵) = pack-pressure dynamometer

~**feld** *n* [*mechanischer Versatz*] = stowed goaf, ~ gob [*Wales*]; ~ condie, ~ cundy [*Scotland*]; ~ waste (area)

~**gut** *n* [*mechanischer Versatz*] = stowing material

~**kante** *f* (✵) = pack line [*hand packing*]

~**schicht** *f* = stowing shift [*power stowing*]

~**schicht** *f* (✵) = packing shift [*hand packing*]

verschalen, einschalen = to form

verschiebbar, gleitend = sliding

Verschiebe|bahnhof *m*, Rangierbahnhof = (freight-)classification yard, marshalling ~, shunting ~, switchyard

~**block** *m*, Verschieberolle *f* [*Bei Hochlöffelbaggern der drehbare Block, durch den der Löffelstiel beim Vorschieben oder Einziehen gleitet*] = saddle block

~**dienst** *m* **auf einer Kohlengrube** = coalfield transfer traffic

~**einrichtungen** *fpl* **für Waggons** = marshalling equipment, shunting ~

~**fahrzeug** *n*, Rangierfahrzeug = marshalling vehicle, shunting ~, shunter

~**gleis** *n* → Rangiergleis

~**lok(omotive)** *f*, Rangierlok(omotive) = shunting loco(motive), rail shunter; dinkey (US)

~**schlepper** *m*, Rangierschlepper = marshalling tractor, shunting ~

~**verkehr** *m* → Rangierverkehr

~**welle** *f* [*Bei Hochlöffelbaggern die Welle, um die sich der Löffelstiel beim Anheben des Löffels dreht*] = shipper shaft

Verschieblichkeit *f* **des Knotenpunktes** [*Baustatik*] = movability of the point of intersection

Verschiebung *f* = displacement

Verschiebungs|ebene *f* (Geol.) = thrust plane (Brit.); displaced surface (US)

~**geber** *m* (✵) = displacement indicator

~**kurve** *f* (✵) = displacement curve

~**plan** *m* [*Zeichnerische Bestimmung der Verschiebung von Knotenpunkten kinematischer Ketten mit Hilfe der um 90° gedrehten Geschwindigkeiten*] = plane of transposition

Verschiffungshafen *m*, Versandhafen, Ladehafen = port of shipment, ~ ~ shipping

Verschlackungsbeständigkeit *f* = slag resistance

Verschlag *m*, Latten~, Lattenkiste *f* = crate

Verschlammung *f*, Verlandung = mudsilting, siltation, silting up (Brit.); mud filling (US)

Verschleiß *m*, Abnutzung *f* durch mechanische Einwirkung = wear (and tear)

~**beiwert** *m* = coefficient of wear

~**bekämpfung** *f* = wear(ing) control

~**betrag** *m* = amount of wear

~**blech** *n* = wear(ing) plate, renewable steel, lining plate, liner plate

~**eigenschaft** *f* = wear(ing) property

verschleißfest = hardwearing

Verschleiß|festigkeit *f*, Abnützungswiderstand *m* = wear(ing) resistance

~**freiheit** *f* = absence of wear

~**grenze** *f* = wear(ing) limit

~**hemmstoff** *m* = wear inhibitor

~**lehre** *f* = wear(ing) ga(u)ge

~**messung** *f* = wear(ing) measurement

~**platte** *f* = wear(ing) plate

~**prüfung** *f*, Verschleißprobe *f*, Verschleißversuch *m* = wear(ing) test

Verschleißschicht — Versenktiefe

~schicht *f*, Decklage *f* [*Straße*] = surface course, wearing carpet, wearing course, wearing surface, road surface; top course (US); coat (Brit.) [*deprecated: topping, crust, sheeting, carped, veneer, top*]
  offen = open-textured
  halb-offen = medium-textured
  geschlossen = close-textured
~schicht *f*, Decke *f* [*Tonbetonstraße, mechanisch verfestigt*] = surface course
~streifen *m* = wear(ing) strip
~teil *m, n* = wear(ing) part
~tiefe *f* = depth of wear
~versuch *m*, Verschleißprobe *f*, Verschleißprüfung *f* = wear(ing) test
~wirkung *f* = wear(ing) action
Verschlickung *f* → Auflandung
verschließbar = lockfast, lockable
verschlissen, abgenutzt = worn(-out)
verschlossener Überlauf *m* → → Talsperre
Verschluß *m*, Wehr~, Staukörper *m* = gate
~ → Absperrorgan *n* [*Talsperre*]
~führung *f* [*Wehr*] = gate guide
~kappe *f* = cap, end ~
~rahmen *m* [*Wehr*] = gate frame
~schicht *f* [*von Schotterdecken*] = seal(ing) coat
~stellung *f* [*Wehr*] = gate position
~windwerk *n* [*Wehr*] = gate hoist
Verschmauchung *f*, (Ofen)Anflug *m* = kiln scum
verschmutztes Wasser *n* mit großen Feststoffen = dirty water containing large solids
~ ~ mit kleinen Feststoffen = dirty water containing small solids
Verschmutzung *f* → Verunreinigung
Verschneiden *n* = thinning, cutting back
verschneit, eingeschneit = snowed-up, snow-bound
Verschnittbitumen *n* = cut-back ((asphaltic) bitumen) (Brit.); ~ (asphalt) (US)
~ mit Haftanreger, gedoptes Verschnittbitumen = doped cutback

~ *npl* und Straßenöle *npl* = liquid asphalts, ~ asphaltic (road) materials (US)
~emulsion *f* = cut-back (asphaltic) bitumen-emulsion (Brit.); ~ asphalt emulsion (US)
Verschnittmittel *n* [*Verschnittbitumen*] = volatile diluent
~ = thinning agent
Verschotterung *f* [*Fluß*] = choking
Verschränkung *f* [*Holzbau*] = tabled joint
Verschraubmoment *n* = make-up torque
verschraubte Bauart *f* [*Maschine*] = screwed type, ~ version
Verschraubung *f* [*Tiefbohrtechnik*] = joint
verschrotten = to reduce to scrap
Verschubwagen *m* [*Trägertransport im Brückenbau*] = lorry
Verschwenk-Gestänge *n* [*Motor-Straßenhobel*] = crank side shift
verseifbar = saponifiable
verseiftes Kiefernwurzelharz *n* [*Luftporenbildner*] = Vinsol Resin [*Trademark*]
Verseifung *f* = saponification
Verseifungs|probe *f*, Verseifungsprüfung *f*, Verseifungsversuch *m* = saponification test
~wert *m* = saponification value
versenken, absenken, ablassen [*Schwimmkasten*] = to sink
Versenken *n*, Ablassen, Absenken [*Schwimmkasten*] = sinking
Versenker *m* für Zylinderlaufbüchse = counterbore for cylinder liner
Versenk|kasten *m*, Beton~ = underwater concreting box
~rohr *n*, Beton~ = trémie pipe
versenkte Kehl(schweiß)naht *f* → Kehl(schweiß)naht mit Fugenvorbereitung
~ Naht *f* [*Schweißtechnik*] = grooved weld
versenkter Bunker *m* → Bunkergrube *f*
Versenktiefe *f* [*Schweißtechnik*] = depth of groove

Versenkungs|umprägung f → Regionalmetamorphose f
~**wand** f → Drempelwand
Versenkwalze f → Absenkwalze
versetzen, verpacken, mechanisch ~ (⚒) = to stow
~, verpacken, von Hand ~ (⚒) = to pack (by hand)
Versetzen n, Versatz m [*von Betonfertigteilen. Bei Platten auch „Verlegen" genannt*] = placing
versetzen der Fugen = to break joint [*Constructively, not to allow two joints to occur over each other*]
versetzt [*Sieblochung*] = staggered
versetzte Fuge f = breaking joint, staggered ~, break of ~
~ **Kreuzung** f = staggered junction
versetztes Gewölbe n → Gurtgewölbe
Versickerung f, Infiltration f = infiltration [*Rainfall reaching the ground moves through the soil surface, a process which is called infiltration*]
Versickerungs|gebiet n → Infiltrationsgebiet
~**messer** m, Lysimeter n, m = lysimeter
versiegeln, absiegeln = to seal
Versiegeln n, Absiegeln = seal coating
Versieg(e)lung f, Absieg(e)lung, Porenschluß m = seal(ing), finish
Versieg(e)lungs|mittel n, Absieg(e)lungsmittel, Porenschlußmittel, Versiegeler m = surface sealant, sealing agent, seal(er)
~**schicht** f, Absieg(e)lungsshicht, Porenschlußschicht = seal(ing) coat, finish ~
Versinken n [*Pfahl*] = downward plunging, sinking
Versorgungs|aggregate npl, technische ~ [*in einem Wohnblock*] = utility equipment
~**betrieb** m = public utility, utility undertaking
~**druck** m = distribution pressure
~**element** n = utility element [*e. g. water line, gas line, electrical conduit, etc. In prefab(ricated) construction, these elements are laid in the forms and encased when the slabs are poured*]
~**gebiet** n = supply zone
~**leitung** f = supply line
~**leitung** f, öffentliche ~ = utility line
~**netz** n = distribution system, supply ~
~**rohr** n **von einem Wasserbehälter aus** = distributing pipe [*A pipe conveying water from a cistern, and under pressure from that cistern*]
~**rohrleitung** f, öffentliche ~ = utility piping, ~ pipework
~**schiff** n = supply ship
~**zone** f, Druckzone [*Wasserversorgung*] = service district
Verspannung f, Ineinandergreifen n [*von Schüttsteinen*] = interlocking
versplinten = to cotter
Verspreizung f → (Ab)Bölzung
verspröden = to become brittle
Versprödung f = embrittlement
~ **von Metallen durch Wasserstoffgas** = hydrogen embrittlement of metals
Versprühung f, Zerstäubung, Zerstäuben n, (Ver)Sprühen = atomization, atomizing, mist-spraying, fog spray(ing)
verstärkt, versteift, ausgesteift = stiffened, reinforced
Verstärkungs|balken(träger) m, Versteifungsbalken(träger), Aussteifungsbalken(träger) = reinforcing beam, stiffening ~
~**element** n, Versteifungselement, Aussteifungselement = stiffener
~**platte** f → Pilzkopfplatte
~**platte** f → Gurtlamelle f
~**platte** f **in der Schildmitte** = blade center reinforcement plate (US); ~ centre ~ ~ (Brit.)
~**pumpe** f, Druck~ = booster pump
~**rahmen** m, Versteifungsrahmen = stiffening frame
~**rippe** f, Versteifungsrippe, Aussteifungsrippe = reinforcing rib, stiffening ~
~**scheibe** f, Versteifungsscheibe, Aussteifungsscheibe = reinforcing panel, stiffening ~

~wand *f*, Versteifungswand, Aussteifungswand = stiffening wall, reinforcing ~
~werkstoff *m* [*z. B. Glasfasern*] = reinforcing material
verstaubt, staubig = dusty
Versteifen *n* → Aussteifung *f*
versteift, verstärkt, ausgesteift = reinforced, stiffened
versteifte Hängebrücke *f*, ausgesteifte ~ = stiffened suspension bridge
versteifter (Stab)Bogen(träger) *m*, ausgesteifter ~ = stiffened arch(ed girder)
Versteifung *f* → Aussteifung
Versteifungs|balken(träger) *m*, Verstärkungsbalken(träger), Aussteifungsbalken(träger) = stiffening beam, reinforcing ~
~element *n*, Verstärkungselement, Aussteifungselement = stiffener
~rahmen *m*, Verstärkungsrahmen, Aussteifungsrahmen = stiffening frame, reinforcing ~
~ring *m*, Verstärkungsring, Aussteifungsring = stiffening ring, reinforcing ~
~rippe *f*, Verstärkungsrippe, Aussteifungsrippe = stiffening rib, reinforcing ~
~rippe *f* einer orthotropen (Brücken-)Tafel = stringer, plate stiffener, rib
~scheibe *f*, Verstärkungsscheibe, Aussteifungsscheibe = stiffening panel, reinforcing ~
~stab *m* [*Kastenträger*] = truss bar
~wand *f*, Aussteifungswand, Verstärkungswand = stiffening wall, reinforcing ~
~winkel *m*, Aussteifungswinkel, Verstärkungswinkel = stiffening angle, reinforcing ~
Versteigerungshalle *f*, Auktionshalle = auction hall
versteinerter Wald *m* = petrified forest, fossil ~
Versteinerungs|gründung *f* = artificial cementing
~verfahren *n* = artificial cementing method

verstellbar starrer Stempel *m* (⚒) = rigid-extensible prop, ~ post
verstellbarer Ausleger *m*, Wippausleger = raisable jib, luffing ~ (Brit.); ~ boom (US)
~ Lüfter *m* = reversible fan
verstellbares Lager *n* = adjustable bearing
Verstell|barkeit *f*, Einstellbarkeit = adjustability
~mutter *f* = adjusting nut
~platte *f*, Einstellplatte = adjusting plate
~-Schrägscheibe *f* [*hydrostatischer Antrieb*] = swash plate
Verstellung *f* = adjustment
verstemmen = to ca(u)lk
Verstemmen *n* → Verstemmung
Verstemm-Masse *f* = caulking compound (Brit.); calking ~ (US)
Verstemmung *f*, Verstemmen *n*, Kalfatern *n* = caulking, fullering (Brit.); calking (US)
verstopfen [*Sieb mit feuchtem Material*] = to blind, to lock, to plug, to clog (up)
Verstopfen *n*, Verstopfung *f* [*Sieb*] = blinding, locking, plugging, clogging
~ [*Erdöltechnik*] = plugging
Verstopfungs|anfälligkeit *f* [*Sieb*] = proneness to clogging, ~ ~ blinding, ~ ~ plugging, ~ ~ locking
~chemikal *n* [*Erdöltechnik*] = plugging agent
verstopf(ungs)frei = non-clogging
verstreichen, aufspachteln = to float
verstrichen, verfugt, ausgefugt = pointed
verstümmelter Boden *m* = truncated soil
Verstürzen *n* des Abraums, Verkippung *f* der Abraummassen = dumping of overburden
Versuch *m*, Prüfung *f*, Probe *f* = test
Versuchs|abschnitt *m* = test section
~anlage *f* = pilot plant
~balken *m* → Biegebalken
~baustelle *f* = experimental site

~belag m, Versuchsdecke f [*Straßenbau*] = trial pavement, test ~, ~ surfacing
~belastung f, Probebelastung, Prüfbelastung = test loading
~betonfläche f = experimental concrete surface
~boden m = test soil
~bohrung f, erste Bohrung = pilot boring
~(brenn)ofen m = pilot kiln
~brücke f = test bridge
~brunnen m, Probebrunnen = experimental well, test ~, pilot ~, trial ~
~charge f, Probecharge = test batch, trial ~
~dauer f, Versuchszeitraum m = test period
~decke f → Versuchsbelag
~druck m, Probedruck, Prüfdruck = test pressure
~durchführung f = test procedure
~einzelheit f = experimental detail, test ~
~ergebnis n = test result
~feld n, Versuchsgelände n = test ground, proving ~
~fahrzeug n [*z.B. Bestimmung der Bordsteinführung*] = design vehicle
~fertigung f = test production
~fläche f = experimental surface
~fuge f, Probefuge, Prüffuge = experimental joint
~gelände n, Versuchsfeld n = test ground, proving ~
~hof m = experimental farm
~-Isolieranstrich m = isolating test coating [*Is a paint coating to which certain outdoor conditions by a nearby chemical plant, or blast furnace(s), or cement factory etc. are admitted under observation, so that the effect of destroying the pigments (also by atomic fall-out) may be studied*]
~last f, Probelast, Prüflast = test load
~methode f, Versuchsverfahren n, Prüfverfahren, Prüfmethode = test(ing) method
~modell n [*Maschinenmodell zu Versuchszwecken vor Aufnahme der Serienfertigung*] = pre-production machine, prototype
~montage f = trial assembly
~mühle f = test mill, ~ grinder
~ofen m, Versuchsbrennofen = pilot kiln
~pfahl m, Prüfpfahl, Probepfahl = test pile
~platte f = test slab
~rad n = test wheel
~reihe f = test series, trial ~
~schüttung f = test fill(ing), trial ~
~stab m, Probestab, Prüfstab = test bar
~stadium n = test stage
~standuntersuchung f = trial-test investigation, proving-ground testing
~straße f = experimental road, ~ highway, test ~
~strecke f, Probestrecke, Beobachtungsstrecke, Versuchsabschnitt m = experimental (or test) section (or track)
~strecke f (⚒) = (dust-)explosion (testing) gallery
~technik f [*Die Art und Weise, wie ein Versuch durchgeführt wird*] = testing procedure
~verfahren n → Versuchsmethode
~verkehr m, künstlicher Verkehr = test traffic
~vorrichtung f = test device
~wand f = experimental wall, test ~
~wert m = test value
~würfel m → (Beton)Würfel
~zeitraum m, Versuchsdauer f = test period
~zylinder m → (Beton)Probezylinder
Vertäfelung f → Täfelung
(Ver)Täfelungs|platte f = panel
~-Sperrholz n = panelling plywood; paneling ~ (US)
Vertau-Boje f = can buoy
vertäuen [*Schiff*] = to tie up
Verteerungszahl f, VTZ = tar value
verteilen, ausbreiten [*eine Schicht*] = to spread
~ von Splitt, (ab)splitten, abdecken mit Splitt = to spread grits, to grit

Verteilen — Vertiefung

Verteilen *n*, Ausbreiten [*Schicht*] = spreading
Verteiler *m* [*Motor*] = distributor
~ [*Hydrauliksystem*] = manifold
~ [*Straßenbaumaschine zum Verteilen von Schotter, Kies, Sand oder Splitt. Die Maschine zum Verteilen von Mischgut heißt „finisher"*] = spreader
~ [*Straßenbaumaschine zum Aufbringen von Bindemitteln*] = distributor
~ → Decken~ [*Straßenbau*]
~ **für Tragschichtmaterial** = base spreader
~**armierung** *f* → Verteilereisen
~**bewehrung** *f* → Verteilereisen
~**block** *m*, Abzweigstück *n* = junction block [*of a power train*]
~**bohle** *f* = spreading beam
~**einlagen** *fpl* → Verteilereisen
~**eisen** *npl*, Verteilerbewehrung *f*, Verteilerarmierung, Verteiler(stahl)-einlagen *fpl*, VE = distribution steel, distributed ~ [*In a reinforced concrete slab, the subsidiary reinforcement placed at right angles to the main steel to hold it in place during concreting and to distribute concentrated loads over a large area of slab*]
~**fertiger** *m* → Schwarzbelageinbaumaschine *f*
~**finger** *mpl*, Läufer *m*, Rotor *m* [*Motor*] = rotor
~**getriebe** *n* = transfer case
~**kasten** *m*, Kastenverteiler *m* = spreader box, spreading box, box spreader, spreading hopper, drag (spreader) box, drag spreader
~**mundstücke** *npl* [*Dosiersilo*] = breeches chute
~**rohr** *n* **im Tank** = diffuser tube in tank
(~)**Schaufel** *f*, Wendeschaufel, Schild [*Wendeschaufelverteiler*] = transverse spreading blade
~**schnecke** *f*, Auflockerungsschnecke [*Schwarzbelageinbaumaschine*] = spreader screw, spreading ~, feed ~, (cross) auger, agitator
~**schurre** *f* = distributing chute, distribution ~, ~ shoot
~**(stahl)einlagen** *fpl* → Verteilereisen
~**stollen** *m* → → Talsperre
~**walze** *f*, Palettenwalze [*Beton(bahn)-Automat*] = rotating grading screed, ~ leveller
~**walze** *f*, Streuwalze = spreading roll
~**walze** *f* → Walzen(-Anbau)streuer
Verteil|gerät *n* → (Decken)Verteiler *m* [*Straßenbau*]
~**graben** *m* [*Furchenbewässerung*] = (earth-)supply ditch
~**planetensatz** *m* **des Differentialwandlers** = torque divider planetary set
Verteilung *f*, Aufbringen *n* [*Bindemittel im Straßenbau*] = application, distribution
~ [*Schotter, Splitt, Sand, Kies und Mischgut im Straßenbau*] = spreading
~, Dispersion *f* = dispersion [*A method of disposal of the suspended solids in sewage or effluent by scattering them widely in a stream or other body of water*]
~ **der Bodenpressung** [*Grundbau*] = distribution of subgrade reaction
~ **des Wasserdrucks** = hydrostatic load distribution
Verteilungs|platte *f*, Druckverteilungsplatte [*Spannbetonverfahren mit Sandwich-Kabeln*] = distributing plate, bearing ~
~**stab** *m* [*Senkrecht zur Tragbewehrung einachsiger Platten verlaufende Querbewehrung*] = distribution rod
~**vermögen** *n* [*Frostschutzmittel*] = dispersing characteristic, ~ property
~**zahl** *f*, Teilungszahl = distribution factor, partition ratio; distribution number (US)
~**zahlenkurve** *f*, Teilungskurve = partition curve, Tromp ~
vertiefen, austiefen = to deepen
Vertiefen *n*, Austiefen = deepening
Vertiefung *f* → (Gelände)Senke *f*

Vertikalbeleuchtung *f*, Senkrechtbeleuchtung = vertical lighting
Vertikale *f* → Lotrechte
vertikale Komponente *f* **der Fliehkraft**, lotrechte Fliehkraftkomponente = vertical component of the centrifugal force
~ **Krümmung** *f*, ~ **Kurve** *f* = vertical curve
~ **Sanddränage** *f*, (lotrechter) Sanddrän *m*, Sandbrunnen *m*, vertikale Sandeinschlämme *f* = vertical sand drain (or pile)
Vertikalität *f* = plumbness
Vertikal|klassierer *m* = vertical classifier
~**komponente** *f*, lotrechte Komponente = vertical component
~**kreis** *m* = vertical circle
~**projektion** *f*, Aufriß *m*, Ansicht *f* = elevation
~**pumpe** *f* = vertical pump
~**rührwerk** *n* → Knetmischer *m*
~**schub** *m* = vertical shear
~**schubkomponente** *f* = shear force, ~ component
~**-Seismograph** *m*, Geophon *n* = pickup
~**stab** *m* → Lotrechte *f*
~**strömungs-Trockenelektroabscheider** *m* = vertical gas flow dry precipitator
~**transport** *m*, Senkrechttransport = vertical transportation
~**verglasung** *f*, Stehverglasung, Senkrechtverglasung = vertical glazing
Vertorfung *f* = peat formation
verträglich = compatible
Verträglichkeit *f* = compatibility
Vertrags|blankett *n* = form of contract
~**bohrgerät** *n* = contract rig
~**erfüllung** *f* = contract performance
~**bedingung** *f* = contract condition, ~ provision
~**strafe** *f* = contract penalty
~**summe** *f*, (vertragliche) Bausumme = contract sum, ~ price
~**unterlagen** *fpl* = contract documents
~**zeichnung** *f* = contract drawing
~**zeit** *f* = contract time period

Verunedelung *f* **des Potentials** [*elektrochemische Korrosion*] = shift towards a less noble potential
Verunreinigung *f*, Verschmutzung = contamination [*The introduction into water, otherwise satisfactory, of bacteria, sewage, or other substances, which makes it unfit for any given use*]
~ **des Brunnenwassers von der Erdoberfläche her** = surface contamination
Verwaltungsgebäude *n*, Verwaltungsbau *m* = administration building
verwässerte Bohrung *f*, ~ **Sonde** *f*; verwässerter Schacht *m* [*in Galizien*] = drowned well, well gone to water
Verwässerung *f* **einer erdölführenden Schicht** = inundation of a petroliferous well
Verwehung *f*, Schnee~ = drifting, snow ~
Verweilzeit *f* = residence time
~**spektrum** *n* = residence time spectrum
~**verteilung** *f* = residence time distribution
Verwendungs|stelle *f* = point of use
~**zweck** *m* = purpose of use
Verwerfung *f* [*Betonplatten*] = warping
~, Bruch *m*, Abschiebung, Sprung *m*, Paraklase *f*, Absenkung [*Grabenrand*] (Geol.) = geological fault, faulting, fault (plane)
~ **ins Hangende**, Sprung *m* ~ ~ = upthrown fault
~ ~ **Liegende**, Sprung *m* ~ ~ = downthrown fault
Verwerfungs|böschung *f* = fault scarp
~**brekzie** *f* = fault breccia
~**fläche** *f*, Bruchfläche (Geol.) = fault plane
~**lette** *f* = gouge, flucan
~**linie** *f*, Bruchlinie (Geol.) = fault line
~**spaltenquelle** *f* → Kluftquelle
~**tal** *n* = fault valley
Verwertung *f*, Ausnutzung = utilization

Verwesung f = decay
(Ver)Wiege|anlage f = weigh(ing) plant, $\sim$ installation
~behälter m → Wiegebehälter
(Ver)Wiegen n → Abwiegen
(Ver)Wiegeturm m = weigh(ing) tower
Verwindung f → Torsion f
verwindungsfest, verwindungssteif = torsion-proof
Verwindungsspannung f = torsion(al) stress
verwittern = to weather
~, altern, „wettern" *[Steinzeugindustrie]* = to store, to age
Verwitterung f, Abwitterung *[Gestein]* = process of rock wastage, weathering
Verwitterungs|boden m, Eluvialboden = residual soil
~kruste f, Abwitterungsprodukte npl (Geol.) *[In mächtiger Form als „Rückstandsediment" bezeichnet]* = residual deposits *[Accumulations of rock waste resulting from disintegration in situ]*
~meßgerät n *[Dachpappe]* = weather-O-meter
~ton m = residual clay
verworfene Fuge f = stepped-off joint
Verwurf m (Geol.) = throw
verzahnt = indented
Verzahnung f = indentation
~ *[Holzbau]* = indented joint
(Ver)Zapfung f → Zapfenverbindung
verzerrter Maßstab m = non-uniform scale division
verzerrtes Modell n = distorted model
Verzerrungsmaß n = degree of distortion
verziehen *[Ortsbrust; Firste; Ulmen]* = to lag
~ *[Gleitschalung]* = to pull the slipforms out of plumb *[if done so, the structure will go out of plumb]*
Verziehen n **im Feuer, Brandverzug** m *[Keramik]* = deformation during burning
Verzierung f = ornament
Verzinken n = galvanizing

Verzinnen n = tinning
Verzögerer m = retarder, retarding admix(ture), $\sim$ additive
verzögerter Halbhydratgips m → Halbhydrat n
Verzögerung f = retardation
~, Drehzahlverminderung = deceleration
~ *[Straßenverkehr]* → Verlustzeit f
Verzögerungs|becken n → → Talsperre
~bogen m = deceleration curve
~bremse f, hydraulische $\sim$ = retarder, hydraulic $\sim$
~kraft f = retarding force
verzögerungslose Steuerung f = instantaneous control
Verzögerungs|streifen m *[Der Ausdruck „Verzögerungsspur" ist nicht mehr zu verwenden]* = deceleration lane
~zeit f = delay period, $\sim$ time
Verzug m (⚒) = lagging
~strecke f *[U-Bahnbau]* = transition section between levels
Verzunderung f = high temperature scaling
Verzweigungsgetriebe n = power splitting transmission
verzwicken, abdecken, auszwicken *[Packe]* = to key, to blind, to choke, to chink
Vestibül n = vestibule
V-Fuge f = V section joint
Viadukt m, Talbrücke f = viaduct, valley bridge
Vibrations|amplitude f, Schwing(ungs)weite f, Ausschlag(weite) m, (f), Amplitude = amplitude of vibration
~aufgeber m, Vibrationsspeiser m, Vibrationsbeschicker m = vibrating feeder, vibratory $\sim$
~bär m → Vibrationsrammbär
~beschicker m → Vibrationsaufgeber
~-Betonsteinmaschine f → Vibro-Bausteinmaschine
~bohle f → Rüttelbohle
~dämpfer m, Schwingungsdämpfer = vibration dampener, $\sim$ absorber
~flasche f, (Innen)Rüttelflasche = vibrating cylinder, $\sim$ head, vibratory $\sim$

Vibrationsfördergerät — Vibrationsverdichtung

~**fördergerät** *n*, Schwingfördergerät, Vibrationsförderer *m*, Schwingförderer, Förderschwinge *f* [*Oberbegriffe für Schwingförderrinne und Schwingförderrohr*] = vibrating conveying machine, vibratory ~ ~, ~ conveyer, ~ conveyor

vibrationsfreier Griff *m*, erschütterungsfreier ~ = anti-vibration handle

Vibrationsgerät *n*, Rüttelgerät = vibrating device, vibratory ~

vibrationsgeschliffene Fuge *f* = vibrated joint

Vibrations|glättbohle *f* = vibrating (or vibratory) smoother, ~ smoothing (or finishing) beam (or screed)

~**glätter** *m* = vibrating float, vibratory ~

~**-Gleisstopfer** *m*, Vibrations-Schotterstopfer, Vibrations-Schwellenstopfer = vibrating tie tamper, vibratory ~ ~

~**hammer** *m* → Vibrations(ramm)bär

~**kern** *m*, Vibrokern = vibratory core, vibrating ~

~**maschensieb** *n* = square mesh vibrating screen, ~ ~ vibratory ~

~**mischer** *m* = vibratory mixer, vibrating ~

~**mischer** *m* **für Beton**, Beton-Vibrationsmischer = vibratory concrete mixer, vibrating ~ ~

~**mörtelmischer** *m* = vibratory mortar mixer, vibrating ~ ~

~**nadel** *f*, Vibriernadel, Rüttelnadel, Schwing(ungs)nadel = immersion needle, (~) poker

~**platte** *f* → Rüttelplatte

~**presse** *f*, Rüttelpresse = vibrating press, vibratory ~

~**(ramm)bär** *m*, Vibrations(ramm)hammer *m* = vibrating (pile) hammer, vibrating (~) ~

~**rammen** *n* = pile driving by vibration

~**(ramm)hammer** *m* → Vibrations(ramm)bär *m*

~**rinne** *f* → (Schwing)Förderrinne

~**rohr** *n* → Förderrohr

~**rohrmaschine** *f* → Betonrohr-Vibriermaschine

~**-Schaffußwalze** *f*, Vibrationsstampfwalze, Vibro-Schaffußwalze, Vibrostampfwalze, Rüttel-Schaffußwalze, Rüttelstampfwalze = vibrating sheepsfoot roller, vibratory ~ ~

~**-Schotterstopfer** *m* → Vibrations-Gleisstopfer

~**schurre** *f* = vibrating chute, vibratory ~, ~ shoot

~**-Schwellenstopfer** *m* → Vibrations-Gleisstopfer

~**sieb** *n* = screen with vibratory (or vibrating) action applied direct to the screen cloth

~**-Schiebetischpresse** *f* = vibrating sliding table press, vibratory ~ ~

~**-Spannbett** *n* = vibrating stressing bed, vibratory ~ ~ [*for the production of prestressed concrete products*]

~**speiser** *m* → Vibrationsaufgeber

~**sprengung** *f* = vibratory explosion, vibrating ~

~**stampfer** *m*, Rüttelstampfer, Vibrostampfer [*Straßen- und Erdbau*] = vibratory tamper, vibrating ~

~**stampfmaschine** *f* → Rüttelstampfmaschine

~**stampfwalze** *f* → Vibrations-Schaffußwalze

~**-Trommelsiebmaschine** *f*, Trommelvibrationssieb *n* = revolving vibrating screen, ~ vibratory ~

~**verdichter** *m*, Schwingverdichter, Rüttelverdichter, Vibroverdichter = vibrating tamper, vibratory ~, ~ compactor [*sometimes erroneously called "consolidation vibrator"*]

vibrationsverdichtet = vibro-compacted

Vibrations|verdichtung *f*, Rüttelverdichtung, dynamische Verdichtung, Schwing(ungs)verdichtung, Einrüttelungsverdichtung, Bewegungsverdichtung = vibrator compaction, vibrating ~, forced vibratory ~, compaction by vibration [*sometimes*

Vibrationsverfahren — Viehdurchlaß

erroneously called "consolidation by vibration"]
~**verfahren** n = vibrational method
~**verhalten** n = vibrational behaviour (Brit.); ~ behavior (US)
~**viskosimeter** n = vibrating plate viscosimeter, vibratory ~ ~
~**walze** f, Rüttelwalze = vibrating roller, vibratory ~
~**walzen** n, Rüttelwalzen = vibratory rolling
Vibrator m, Rüttler m = vibrator
~, Rüttler m, Beton~ = vibrator, concrete ~
~**bohle** f → Rüttelbohle
~**flasche** f → Rüttelflasche
~**platte** f → Rüttelplatte
~**siebmaschine** f, Vibrationssieb n [*Eine präzise Abgrenzung für Vibrationssieb gibt es erstaunlicherweise nicht. Dieser kurze, einprägsame Begriff hat sich im Laufe der Jahre auf dem gesamten Weltmarkt durchgesetzt, und es werden ganz allgemein hierunter alle nicht zwangsgesteuerten Siebsysteme im Bereich von etwa 800 bis 3000 Schwingungen/min verstanden*] = vibrating screen, vibratory ~
Vibrier-Elektroschurre f, Elektrovibrierschurre = electrically vibrated chute, ~ ~ shoot
vibrierende Spirale f [*Sie ermöglicht, aus den Abwässern der Zinnindustrie noch einen Teil der sonst mit abfließenden Feinstanteile zurückzugewinnen*] = shaken helicoid
Vibrier|bohle f, Rüttelbohle, Vibratorbohle, Verdichtungsbohle, Schwingungsbohle = vibrating beam, ~ smoother, vibratory ~
~**flasche** f → Rüttelflasche
~**geräte** npl **zur Materialförderung** = vibrating materials handling equipment, vibratory ~ ~ ~
~**nadel** f → Vibrationsnadel
~**rinne** f → (Schwing)Förderrinne
~**rohr** n [*Gleitschalungsfertiger*] = (horizontal) tube (type) vibrator
~**tisch** m = vibrating table, vibratory ~

vibrierter Beton m, Rüttelbeton = vibrated concrete
Vibro|-Bausteinmaschine f, Vibro-Steinformmaschine, Vibrations-Betonsteinmaschine = vibrating (concrete) block (making) machine, vibratory (~) ~ (~) ~
~**bohren** n = soil drilling by vibration, vibrodrilling
~**förderer** m → Vibrotransporteur m
~**-Formationsbrechen** n [*Erdölgewinnung*] = vibration fracturing, vibratory ~, fracturing by vibration
~**kern** m, Vibrationskern = vibratory core, vibrating ~
~**mühle** f, Schwingmühle = vibrating grinding mill, vibratory ~ ~
~**pfahl** m = vibro pile
~**-Schaffußwalze** f → Vibrations-Schaffußwalze
~**schaufel-Betonmischer** m = concrete mixer with vibratory blades, ~ ~ ~ vibrating ~
~**stampfer** m, Vibrationsstampfer, Rüttelstampfer [*Straßen- und Erdbau*] = vibratory tamper, vibrating ~
~**stampfer** m → Rüttelstampfmaschine [*Betonsteinindustrie*]
~**stampfmaschine** f → Rüttelstampfmaschine [*Betonsteinindustrie*]
~**stampfwalze** f → Vibrations-Schaffußwalze
~**-Steinformmaschine** f → Vibro-Bausteinmaschine
~**transporteur** m, Vibroförderer m = vibratory conveyor, ~ conveyer, vibrating ~
~**verdichter** m → Vibrationsverdichter
Vicat-Nadel(apparat) f, (m), Vicat-Nadelgerät n = Vicat needle apparatus, ~ setting time ~
Vickers-Härteversuch m, Vickers-Härteprüfung f, Vickers-Härteprobe f = Vickers hardness test
Viehdurchlaß m, Viehtunnel m [*unter einer Straße oder Eisenbahn*] = cattle creep, ~ (under)pass, ~ subway, stock ~

viehsicher — Vierkantrohr

viehsicher [*Zaun*] = stock-proof
Viehtunnel *m* → Viehdurchlaß *m*
Viel|bechergerät *n*, Mehrbechergerät = multi-bucket appliance
~eckgewölbe *n* = polygonal vault
vieleckig, polygonal = polygonal
Vieleck|kuppel *f* = polygonal dome, dome of polygonal plan
~rahmen *m* = polygonal frame
~sprengwerk *n* = polygonal truss
~stab *m* = polygonal bar, ~ rod
~verband *m*, Polygonalverband = polygonal bond
Vielfach|-Abzugkanal *m* → Vielfach-Durchlaß *m*
~betonfugenschleifgerät *n* = multiple blade joint saw
~bogensperre *f*, Pfeilergewölbe(stau)mauer *f*, Bogenpfeiler(stau)mauer, Gewölbereihen(stau)mauer = multiple arch(ed) dam
~-Durchlaß *m*, Vielfach-Abzugkanal *m* = multi(ple)-opening culvert
~kuppelsperre *f*, Pfeilerkuppel(stau)mauer *f* = multiple-dome dam
~-Rüttelplatten-Verdichter *m* = multiple vibratory compactor, ~ vibrating ~
viel|geschossiger Stockwerkrahmen *m*, mehrgeschossiger ~ = multistor(e)y bent
~geschossiges Gebäude *n* → vielstöckiges ~
Viel|lochstein *m*, Viellochziegel *m* = perforated brick
~nutsteckachse *f* = splined half shaft, ~ independent axle
~punkt-Schweißmaschine *f* = multipoint welder, press ~
~rippendecke *f* = multi-ribbed floor
(~)Röhrenkessel *m* = multitubular boiler
~scheiben-Luftdruckbremse *f* = multiple-disc air-operated brake; multiple-disk ~ ~ (US)
~seitigkeit *f* = versatility
viel|spurige Straße *f* → mehrspurige ~
~stöckiges Gebäude *n*, vielgeschossiges ~, vieletagiges ~ = multistor(e)y building

Viel|strahl-Kondensator *m* = multi-jet (ejector) condenser, ejector ~
~trägerbrücke *f* = multi-girder bridge
~zellenrotationsverdichter *m* = compressor with multi-stage rotor
Vielzweck|-Auto(mobil)bagger *m* → Auto(mobil)bagger
~bagger *m* → Mehrzweckbagger
~bauwerk *n* → Talsperre
~schmierfett *n*, Mehrzweckschmierfett = multi-purpose (lubricating) grease
~-Schnecke *f*, (Planier-)Schneckenfräse *f*, (Schnecken-)Planierfräse [*für Schneebeseitigung, Grabenverfüllung, Planieren und zur Herstellung von Schwarzgemischen auf dem Straßenplanum*] = auger-backfiller
Vier|achsfahrzeug *n* = four-axle vehicle
~decker(sieb) *m*, (*n*) = four-deck grader, ~ screen
~drahtspannpresse *f* = four-wire prestressing jack
Vierendeelstab *m* = Vierendeel member, ~ rod
Vierendeelträger *m*, Rahmenträger = Vierendeel truss, ~ girder [*The Vierendeel truss does not satisfy the definition of a truss but is given that designation*]
~brücke *f*, Rahmenträgerbrücke = Vierendeel girder bridge
vierfach geschakte Eimerkette *f*, 4-teilige ~ [*Eimerkettenbagger*] = bucket chain line with a bucket fitted every fourth link
~ geschertes Seil *n* [*Ein einfaches Seil, das in der Weise über Seilscheiben eingescheert ist, daß vier Seillängen die Verbindung zwischen feststehender Oberflasche und beweglicher Unterflasche bilden*] = four-part line
Vier|-Füllungstür *f* = four-panel door
~fußkörper *m*, Vierfüßler *m*, Beton~ = (concrete) tetrapod
Vierkant|kupfer *n* = rectangular copper
~pfahl *m* = rectangular pile
~rohr *n* = rectangular pipe

Vierkant-Rohrausleger — Visieren

~-**Rohrausleger** *m* = rectangular pipe jib (Brit.); ~ ~ boom (US)
~**stahl** *m* = square bar (steel)
~**welle** *f* [*Schwertwäsche*] = angle steel log [*log washer*]
Vier|-Momente-Lehrsatz *m* = four moment theorem
~-**Motoren-Laufkran** *m* [*für elektrischen Betrieb*] = four-motor travel(l)ing crane
~**punktlagerung** *f* = four-point mounting
Vierrad|-Allradantrıeb-Schlepper *m*, Allradantrieb-Vierradschlepper = four-wheel(ed) drive tractor
~**antrieb-Lader** *m*, Vierradantrieb-Lademaschine *f* = four-wheel(ed) drive loader, ~ ~ loading machine
~**bremse** *f* = four-wheel(ed) brake
~**fahrgestell** *n*, Vierraduntergestell, Vierradchassis *n* = four-wheel(ed) chassis
~**gabelstapler** *m* = four-wheel(ed) fork lift truck
~(**lauf**)**katze** *f*, Vierradlastkatze, Vierrad-Fahrkatze = four-wheel(ed) crab
~-**Lenkung** *f*, Allrad-Lenkung = four-wheel(ed) steer(ing)
~**trecker** *m*, Zweiachsschlepper *m*, Vierradschlepper, Vierradtraktor *m*, Zwei-Achs-Reifen-Sattelschlepper, Zweiachs-Zugwagen *m* = four-wheel(ed) tractor
~-**Vorspänner** *m* = four-wheel(ed) prime mover
Vierrollen|lager *n* [*für Hoch- und Tiefbauten*] = four-roller bearing
~**meißel** *m* = four-roller bit
Vier|schichtendiode *f* = four layer diode
~**schneidenwaage** *f* [*(Beton-)Dosieranlage*] = four-knife edge weigh(ing) gear
~**seilgreifer(krankorb)** *m*, Vierseilgreif(er)korb = four-rope suspension grab(bing) (crane) bucket, ~ ~ grab
vier|seitig gelagert = supported on 4 sides

~**spuriger Einbau** *m* [*Betondeckenbau*] = 4-lane at a time paving
~**stöckiges Kreuzungsbauwerk** *n* = four-level interchange, ~ grade separating structure
Vier|stundenschicht *f* = four-hour shift
~**takt-Diesel(motor)** *m* = four-stroke diesel engine
~**takter** *m*, Viertaktmotor *m* = four-stroke cycle engine
~**taktverfahren** *n* = four-stroke cycle
vierteilige Schnittkante *f* = 4-section cutting edge
Vierteilung *f* = quartering
Viertel|dach *n* = roof with pitch 1:4
~**kreis** *m*, Quadrat *m* = quadrant [*A quarter circle; an arc of 90°*]
~**podest** *n*, *m* = quarter-landing; quarter-space landing [*deprecated*]
~**punkt** *m* = quarter-point
~**stab** *m* = quarter round
~**stein** *m*, Riemchen *n*, Viertelziegel *m* = one-quarter brick
vierteltägige Gezeit *f* = quarter-diurnal tide
Viertelung *f*, Quadrantenverfahren *n*, Kegelverfahren [*DIN 51701*] = coning and quartering
Viertelziegel *m* → Viertelstein
Vier|wabensilo *m* = four-compartment bin
~**weg(e)ventil** *n* = four-way valve
vierzelliger Dosierapparat *m* = four-compartment aggregate feeder
Vier|zimmerwohnung *f* = four-room(ed) flat (Brit.); ~ apartment (US)
~**zonenförderung** *f* [*Es wird gleichzeitig aus vier Erdölhorizonten gefördert*] = quadruple (well) completion
~**zugbauweise** *f* [*Dampferzeuger*] = four-pass design
~-**Zylinder-Motor** *m* = four cylinder engine
virtuelles Sohlengefälle *n* = virtual slope
Visieren *n*, Ausrichten = boning [*Operation of levelling or regulating the straightness of trenches by sighting along the tops of a series of tee-pieces, known as boning rods*]

Visieren — Vollgummireifen

~, Ein~ = sighting, boning(-in) [*Setting out intermediate levels on the straight line joining two given level pegs*]
Visier|gerüst *n*, Schnurgerüst = sight rail, ~ board, batter board
~linie *f* = line of sight
~mast *m* = sighting mast
~tafel *f*, Ausrichtetafel = boning rod
visko-elastische Eigenschaft *f* = visco-elastic property
Viskoelastizität *f* = visco-elasticity
Viskose|-Reyon = viscose rayon
~zellfaser *f* = viscose fibre (Brit.); ~ fiber (US)
~zellgarn *n*, ZGV *n* = viscose yarn
Viskosimeter *n* **mit rotierendem Zylinder** = rotating cylinder viscosimeter
Viskosimetrie *f* = viscometry
Viskosität *f*, Zähflüssigkeit *f* = viscosity
Viskositäts|-Brechung *f* = viscosity breaking
~-Dichte-Konstante *f* = viscosity gravity content (or constant)
~index *m* = viscosity index
~index-Verbesserer *m* = viscosity modifier
~spanne *f* = viscosity range
~stufe *f* = viscosity grade
Vitriol|bleierz *m* → Anglesit *m*
~torf *m* = vitriol peat [*Peat containing much iron sulphate*]
Vitrit *m* = vitrain
Vivianit *m* (Min.) → Blaueisenerde *f*
VOB, *f*, Verdingungsordnung *f* für Bauleistungen = German Contract Procedure in the Building Industry
Vogelperspektive *f* = bird's-eye view [*An oblique aerial photograph*]
Volkserholung *f* = (public) recreation
volle Betonummantelung *f* [*Rohr*] = (integral and monolithic concrete masonry) encasement
voller Bogen *m* → Rundbogen
vollangesaugter Strahl *m*, Adhäsionsstrahl [*Hydraulik*] = weir nappe, adherent ~
Vollanhänger *m* = full trailer, rigid vehicle

Vollast *f*, Voll-Last = full load
~betrieb *m* = full-load running
Voll|bahngleis *n* → Normalspurgleis
~bahn-Lok(omotive) *f* = main line loco(motive)
~becherwerk *n* → Reihenbecherwerk
~belastung *f* = full loading
~betondecke *f*, Massivbetondecke = solid concrete floor
~(beton)platte *f*, Massiv(beton)platte = solid (concrete) slab
voll|bewegliche (Feld)Beregnungsanlage *f* = fully portable (agricultural) sprinkling installation, ~ ~ (~) ~ system
~bezahlter Urlaub *m* = leave on full salary, ~ ~ ~ pay
Vollblock|form *f*, Massivblockform = solid-block mould (Brit.); ~ mold (US)
~maschine *f*, Massivblockmaschine = solid-block machine
~stein *m*, Beton-~ = solid concrete block
Vollbohr|kern *m* = solid drill core
~stahl *m* = solid drill steel
Voll|drehkran *m*, Vollschwenkkran = full-circle slewing crane
~einspannung *f* = full restraint, ~ fixity
~elektroantrieb *m* = all-electric drive
~elevator *m* → Reihenbecherwerk *n*
~fläche *f* (⚒) = critical area of extraction
vollflächige Schalung *f* = solid formwork
Vollförderschnecke *f* = closed spiral (worm) conveyor (or conveyer)
Vollgas *n* **bei Belastung** = full load
~ ~ Leerlauf = high idle
Vollgestänge *n* [*Tiefbohrtechnik*] = percussion rods, percussive ~
~bohrung *f*, kanadisches Bohren *n* = percussion rod drilling, percussive ~ ~
Voll|-Glas(bau)stein *m* = solid glass block
~gummikantenschutz *m* = edge protection with rubber
~gummireifen *m* → Elastikreifen

Vollgummireifenwalze — Vollwegeventil

~gummireifenwalze f → → Walze
Vollhydraulik|bagger m = all-hydraulic excavator
~-**Betonmischer** m = all-hydraulic concrete mixer
~**kran** m = all-hydraulic crane
~-**Lader** m = all-hydraulic loader
~-**Mobilbagger** m = all-hydraulic mobile excavator
~-**Raupenbagger** m, Vollhydraulik-(Gleis)Kettenbagger = all-hydraulic crawler-mounted excavator
voll kalibriertes Bohrloch n = full ga(u)ge (drill) hole, ~ ~ bore ~
voll|isoliert = all insulated
~**kantig** = full edged
Voll|kegel m, Kegelschaft m = taper shank
~**kernbohrung** f = solid core boring
~**kippanhänger** m = wagon [*A full trailer with a dump body*]
~**kolben(bohr)hammer** m = solid piston rock drill
vollkommene Flüssigkeit f → ideale ~
~ **Vorspannung** f = perfect prestress
vollkommener Überfall m = clear overfall
vollkörniger Geschiebetrieb m, allgemeiner ~ = general bed load transport
vollkristallines Erstarrungsgestein n = holocrystalline rock
Voll|-Last f, Vollast = full load
Vollmauer f, Vollwand f = solid wall
~**werk** n, Massivmauerwerk = solid masonry
Voll|pfahl m, Vollquerschnittpfahl = solid pile
~**platte** f → Vollbetonplatte
~**portalkran** m → Portalkran
~**profil** n = solid section
Vollquerschnitt m = full cross section
~**element** n = full cross section element
~**pfahl** m, Vollpfahl = solid pile
Voll|scheibe f = solid disc; ~ disk (US)
~**scheibenbremse** f = solid pulley brake, ~ disc ~; ~ disk ~ (US)
~**schwenkkran** m, Volldrehkran = full-circle slewing crane

~**seil** n → Zugseil
~**senkung** f = full subsidence
~**sicht** f, Rundblick m = all-round view, ~ vision, full ~
~**sichtkanzel** f, Rundblick-Führerhaus n, Vollsichtkabine f, Vollsicht-Führerstandkanzel, Rundblickkanzel, Rundsichtkanzel, Rundsichtführerhaus = full-vision cab(in), all-round view ~, all-round vision ~
(~)**Spül(stau)damm** m → gespülter (Erd)Damm
~**spur** f, Regelspur, Normspur = standard gauge (Brit.); ~ gage (US)
~**stab** m, Massivstab = solid rod
~**stahlbauweise** f, Ganzstahl-Bauweise = all-steel construction
~**stahlkette** f, Ganzstahlkette = all-steel chain
~**standanzeiger** m = high level indicator
~**steg** m = solid web
Vollstein m, Vollziegel m [*Kurzzeichen: MZ*] = solid brick
~, Vollblock(stein), Vollkörper m = solid block
~**gewölbe** n, Vollziegelgewölbe = solid brick vault
~-**Prüfmaschine** f = solid block tester, ~ ~ testing machine
Vollström|filter m, n = full flow filter
~**ventil** n = full-way valve
Voll|stufe f = solid step
~**tidefang(e)damm** m = full-tide cofferdam
~**torkran** m → Portalkran
volltransistorierter Spannungsregler m = transistor(ized) voltage regulator
Voll|turbine f = full admission turbine
~**verglasung** f = floor-to-ceiling glazing
~**versatz** m, Hand~ (⚒) = solid (hand) packing
Vollwand f, Vollmauer f = solid wall
~**balkenbrücke** f = plain beam bridge
~**balken(träger)** m = plain beam
~**träger** m, vollwandiger Träger = plain girder
Vollwegeventil n = full-way valve

Vollziegel *m*, Vollstein *m* [*Kurzzeichen:* MZ] = solid brick
~**gewölbe** *n*, Vollsteingewölbe = solid brick vault
~**wand** *f*, Vollziegelmauer *f* = solid brick wall
Volumen|dosierung *f*, Raumzumessung, Raumdosierung, Volumenzumessung, Raumteilmischung, Volumenzugabe *f* = (loose-)volume batching, bulk ~, volumetric ~, batching by volume
~**minderungsfaktor** *m* = shrinkage factor [*from dry materials to mixed*]
volumetrischer Wirkungsgrad *m* = volumetric efficiency
Volute *f* = scroll
Volutenkapitell *n* = scroll capital, voluted ~
Vorabbrausesieb *n* = preliminary rinsing screen
Vorabscheider *m* = primary separator
~ [*Motor*] = prescreener
Vorabscheidung *f* = pre-separation
Vorabsenkung *f* (⚒) = initial convergence
Vorabsiebung *f* = preliminary screening, scalping
Vorankergehen *n*, Ankern *n* = mooring, anchoring
Voranstrich *m* = prime coat(ing) ~, Voranstreichen *n* = priming
~**mittel** *n* = primer
Vorarbeiten *fpl* = preparatory work, preliminary ~
Vorarbeiter *m* = leading hand, ganger
Vorbau *m* (⚒) = advancing longwall system
~**abschnitt** *m*, Lamelle *f* [*Brückenbau*] = cantilever(ed) segment
~-**Derrick-Kran** *m* = erecting derrick (crane); erecting derricking jib crane (Brit.); erecting derricking boom crane (US)
~-**Harke** *f*, Vorbau-Rechen *m* = front rake
~**kran** *m* [*Brückenbau*] = creeper crane
~(**pfahl)ramme** *f* = cantilever (pile) driving plant, ~ piling ~

~**rüstträger** *m* = stepping service girder
~**rüstung** *f* = stepping formwork equipment
~**spitze** *f* [*Brückenbau*] = projecting end
~**wagen** *m* = cantilever construction carriage
Vorbecken *n* [*Am Einlauf einer Rohrleitung, besonders über Wasserkraftwerken*] = forebay
Vorbehandlung *f* [*Abwasser*] = primary process
~ = pre-treatment
Vorbelastung *f* = preloading
Vorbelüftung *f* **des Abwassers** = pre-aeration of sewage
Vorbenetzungsmittel *n* = pre-coating agent
Vorbereitung *f* **zum Rammen** = driving preliminaries
Vorberge *mpl*, Vorträge *mpl* [*eines Gebirgszuges*] = foothills
vorbeugende Instandhaltung *f* = preven(ta)tive maintenance
Vorbeugungsmaßnahme *f* = preven(ta)tive remedy
Vorblick *m* [*Nivellieren*] = fore-sight, minus sight
Vorblock *m* = bloom [*It is used as a stanchion base when the top surface is machined*]
Vorboden *m* = upstream floor, ~ apron
Vorbohren *n*, Anbohren [*Der Anfang einer Bohrung*] = collaring
Vorbohrer *m* = starter drill, starting ~
Vorböschung *f* = fore-slope
Vorbrechen *n*, Vorbruch *m* = primary crushing, ~ breaking [*The first crushing stage*]
Vorbrecher *m* = primary crusher, ~ breaker, ~ crushing machine, ~ breaking machine
Vorchlorung *f* = pre-chlorination
Vordach *n*, Wetter(schutz)dach, Abdach = canopy [*A roof-like covering usually projecting over an entrance or window or along the side of a wall*]

~kante *f* = canopy lip
Vorder|achsantrieb *m* = front axle drive
~achsaufhängung *f* = front axle suspension
~achse *f*, Frontachse = front axle
~achstragbolzen *m* = front axle pivot pin
~ansicht *f* = front elevation, ~ view
~balkon *m* = front balcony
vordere Kurbeldichtung *f* = crankshaft front seal
vorderer Kämpferdruck *m*, voreilender ~ (⚒) = front abutment pressure
~ Kraftanschluß-Stutzen *m* **für Anbaugeräte,** Frontzapfwellen-Anschluß *m* [*Schlepper*] = front power take-off
~ Planetenträger *m* [*Kraftübertragung*] = front carrier
vorderes Kettengehänge *n*, vorderer Kettenzaum *m* [*Schrapp(er)gefäß*] = front bridle chains
Vorder|gebäude *n*, Vorderhaus *n* = front building
~grund *m* = fore-ground
~haus *n*, Vordergebäude *n* = front building
~kipper *m* → Autoschütter *m*
~kipp-(Front)Ladeschaufel *f* = forward-tip(ping) bucket
~kübel *m*, Vorderschaufel *f* [*Schürflader*] = front end shovel, ~ ~ bucket
~lader *m* → Front-Ladeschaufel
~radantrieb *m*, Frontantrieb = front wheel drive
~radbremsschalter *m* = front wheel brake control, ~ brake limiter switch
~radnabe *f* = front wheel hub
~radsturz *m* = front wheel lean
~schaufel *f*, Vorderkübel *m* [*Schürflader*] = front end bucket, ~ ~ shovel
~walze *f* → Frontwalze
voreilender Kämpferdruck *m*, vorderer ~ (⚒) = front abutment pressure
Voreilwinkel *m* = lead-angle
Vorentwurf *m* = preliminary design
Vorfahrt *f*, ~recht *n* = right-of-way
Vorfeld *n*, Abbau~ = zone in front of the face
~, Hallen~ [*Luftfahrtgelände*] = (hangar) apron
~ebene *f*, Bodenebene [*Abfertigungsgebäude*] = ground level
~flutlichtbeleuchtung *f* = apron floodlighting
Vorfenster *n* → Kastenfenster
Vorfertigung *f* = prefabrication
~ [*aus Beton*] = prefabrication, precasting
~ in der Fabrik = factory prefabrication
~ von Fahrbahntafelteilen einer Brükke = roadway casting
Vorfilter *m*, *n*, Grobfilter = pre-filter
~element *n* = primary filter element
Vorflotation *f* **und mehrfache Nachreinigung des Schaumkonzentrates in Einsatzversuchen** = multiple cleaning test
Vorflotationszelle *f* = rougher
Vorfluter *m* [*Gewässer, das die Abflußmenge eines anderen aufnimmt*] = receiving water, recipient
Vorführer *m*, Anleitungsfachmann *m* = demonstrator
Vorfüllkasten *m* →→ Betonmischer *m*
Vorgabezeit *f* [*auf Zeitstudien aufgebaut*] = standard time
Vorgarten *m* [*Gelände zwischen der Bauflucht- und der Straßenfluchtlinie, das nicht zur Straße sondern zum Grundstück des Anliegers gehört*] = forecourt, front garden; front yard (US) [*An open space between a public way and a building*]
vorgebaute Lenkung *f*, Tram-Bauweise *f* [*Fahrersitz befindet sich vor der Vorderachse*] = cab-over-engine design, full-forward-control cab ~
Vorgebirge *n* = front range
vorgeblendete Fertigwand *f*, Vorhangwand = curtain wall
vorgedrückte Zugzone *f* = pre-compressed tension zone, ~ tensile ~
vorgefertigte Armierung *f* → Bewehrungskorb *m*

vorgefertigte Bewehrung — Vorklärbecken

~ **Bewehrung** *f* → Bewehrungskorb *m*
~ **Eiseneinlagen** *fpl* → Bewehrungskorb *m*
~ **Schalldämpf-Packung** *f* **für Klimaanlagen** = packaged sound attenuator for air-conditioning systems
~ **Stahleinlagen** *fpl* → Bewehrungskorb *m*
vorgefertigter Kabeltunnel *m*, ~ **Kabelstollen** *m* [*aus Beton*] = precast cable tunnel, ~ ~ gallery
vorgefertigtes Mannloch *n* = precast manhole
~ **Röhrenelement** *n*, ~ **Tunnelelement** = precast (tube) segment, ~ (~) section
vorgeformt [*Drahtseil*] = pre-formed
vorgeformtes drallarmes Drahtseil *n* = = TRU-LAY-ROPE [*Trademark*]
vorgekesseltes Loch *n* → → 1. Schießen *n*; 2. Sprengen *n*
vorgekragt, ausgekragt = cantilevered
Vorgelege *n*, **Räder**~ = back gear(ing) arrangement, intermediate gear, back gears, reduction gears ~, **Riemen**~ = countershaft
~**welle** *f*, **Vorgelege** *n* = countershaft
vorgepackter Beton *m* → Prepaktbeton
vorgepaßt = pre-adjusted
vorgepfändet = forepoled
vorgesetzte Abbaustrecke *f* (⚒) = advance heading
vorgespannte Schraubenverbindung *f* = friction grip bolt joint
~ **(Straßen)Decke** *f* = prestressed concrete road pavement, ~ ~ ~ surfacing, ~ ~ highway ~
~ **Verbund-Stahlbetonkonstruktion** *f* = compound prestressed reinforced concrete structure
~ **Ziegeldecke** *f*, **Stahltondecke** = floor in prestressed clay, Stahlton floor
vorgespannter Fertigsturz *m* → Ziegel-Fertigsturz
~ **Stollen** *m* = gallery with anchorage by prestressing the ground with steel anchors
(~) **Ziegelfenstersturz** *m* → Stahlton-Fenstersturz

~ ~ → Ziegel-Fertigsturz
vorgespanntes Spiegelglas *n* = tempered plate glass
~ **Ziegelbrett** *n*, **Stahltonbrett** = (Stahlton) prestressed clay plank, ~ plank
~ **Ziegeldach** *n*, **Stahltondach** = roof in prestressed clay, Stahlton roof
~ **Zugband** *n* = prestressed tieback
vorgewelltes (Sieb)Gitter *n*, Wellengitter [*Wellengitter werden meist aus relativ dünnen Drähten im Verhältnis zur Maschenweite hergestellt. Es befinden sich Zwischenwellen zwischen den Drahtkreuzungen, die dem Siebboden eine gewisse Steifigkeit verleihen. Meist handelt es sich um Gewebe mit Lang-, Breit- oder Schlitzmaschen (Harfensiebe), die nach diesem Verfahren hergestellt werden. Kett- und Schußdrähte werden vorgewellt*] = intermediate weave, corrugated ~, crimp ~
Vorhaben *n* → (Bau)Projekt
Vorhalte|holz *n* = builders' rough planks
~**kosten** *f* = commissioning cost
vorhalten [*Baumaschinen und -geräte*] = to commission [*To keep construction machinery and equipment in fit condition for use on a construction contract*]
Vorhalten *n* = commissioning
Vorhaltezeit *f* = commissioning time
vorhandene Festigkeit *f*, **errreichte** ~ = actual strength
Vorhängeschloß *n* = padlock
Vorhangwand *f*, **vorgeblendete Fertigwand** = curtain wall
vorherrschende Verwerfungslinie *f*, (Geol.) = dominant fault line
Vorhof *m* = forecourt
Vorkammer *f* = pre-chamber, pre-combustion chamber
Vorkesseln *n*, **Kesselschießen** *n* = sprungbore hole method, springing, enlarging the hole with explosives, chambering, squibbing
Vorklärbecken *n* = preliminary clarification tank, ~ ~ basin

Vorklärung — Vorratssilo

Vorklärung *f* = preliminary clarification, pre-sedimentation
Vorklassiersieb *n* = primary screen
Vorknicken *n* [*Scharnierherstellung*] = nipping
Vorkocher *m* = primary heater, ~ heating kettle, ~ melting tank; ~ boiler [*deprecated*]
Vorkommen *n*, Lager *n*, Lagerstätte *f*, Vorkommnis *n*, Fundstätte (Geol.) = deposit
Vorkommnis *n* → Vorkommen *n*
Vorkopf *m* [*z.B. Schleuse*] = roundhead
vorkragen → auskragen
Vorkragen *n*, Auskragen, Ausladen = cantilevering
vorkragender Träger *m* → Kragträger
Vorkragung *f*, Ausladung, Auskragung = cantilever
Vorkröpfgesenk *n* = snaker
Vorkühlung *f*, Vorkühlen *n* = pre-cooling
vorlageartiges Pflaster *n* → Steinsatz *m*
Vorlage *f* = apron
~stein *m* = Setzpacklagestein
Vorlagerungszeit *f* [*Dampfbehandlung von Beton*]= delay period, holding ~
Vorland *n* = foreland
~ = foreshore
~feld *n* → Uferfeld
~öffnung *f* → Uferfeld *n*
vorlasieren = to pre-glaze
Vorlaufbehälter *m* = header (tank) [*A surge tank in a process plant*]
vorläufige Abnahme *f* = initial acceptance
~ Richtlinien *fpl* = tentative spec(ification)s
vorläufiger Ausbau *m* (⚒) [*Dieser Ausbau sichert die Grubenbaue bis zum Einbringen des endgültigen Ausbaues*] = temporary support
Vorlauf|leitung *f*, Zulauf *m* = feed line
~rohr *n* = flow pipe
Vorlege|rad *n* → Zwischenrad
~welle *f*, Zwischenwelle = idler shaft
Vorloch *n* = pilot hole [*carpentry*]
Vormagnet *m* = primary magnet

Vormahlen *n*, Vormahlung *f* = primary grinding [*The first grinding operation*]
Vormauerziegel *m*, frostbeständiger Ziegel, Vormauerstein *m*, frostbeständiger Stein = frost-resistant brick
Vormischen *n* **der Betonbestandteile vor Einspeisen in den Liefermischer** [*Transportbeton*] = pre-shrink mixing
Vormischer *m* = pre-mixer
Vormischsilo *m* [*Betonherstellung*] = prebatching bin
Vormischsiloband *n* = cumulative bin conveyor (belt), ~ ~ conveyer (~)
Vormischung *f*, Vormischen *n* = pre-mixing
Vormittagsschicht *f* = morning shift
Vormittags-Spitzenstunde *f* = morning peak hour
Vormontage *f* = pre-assembly
Vormühle *f* = primary (grinding) mill
Vorort *m* = suburb
~verkehr-Triebzug *m* = multiple-unit suburban train
Vorpfändbohle *f*, Fangbohle = forepole, spile, forepoling board
vorpfänden (⚒) = to forepole
Vorpfändschiene *f*, Fangschiene = forepoling girder
Vorpfändung *f*, Vorpfänden *n* = forepoling, spiling, horseheading [*In tunnel timbering, driving forepoles ahead over the caps of the last four-piece set erected, to give temporary protection over miners who are putting up the next permanent timber set or are digging*]
Vorplastifizierung *f* = pre-plasticizing
Vorprofildecke *f* → Ausgleichschicht *f*
Vorratsbehälter *m* = storage tank, ~ reservoir
~ [*Wasserversorgung*] = storage reservoir, ~ tank
Vorrats|bunker *m* = storage bunker
(~)Halde *f*, Materialhalde = stockpile, (storage) pile, stock-heap
~silo *m* [*auf einer Baustelle*] = storage silo

~tunnel *m* → Haldentunnel
~wasserheizer *m*, Badespeicher *m* = storage-type geyser
Vorreiniger *m* = precleaner
~schutz *m* = precleaner guard
Vorrichtung *f* **zum Einsammeln Lübecker Hütchen** = pickerupper
~ **zur stufenlosen Regelung des Reifendruckes während der Fahrt vom Führerstand aus** [*Gummiradwalze*] = air-on-the-run system
~ ~ **Verlagerung höherer Achslasten auf die Antriebsräder** = traction control
Vorsatzbeton *m*, Vorsatzschicht *f* = face concrete
~-**Schalung** *f* = face concrete formwork
Vorsatz|material *n* = facing material
~mischung *f*, Vorsatzgemisch *n* = facing mix(ture), face ~
~schicht *f*, Vorsatzbeton *m* = face concrete
~zuschlag(stoff) *m* = facing aggregate
Vorschaltturbine *f* = primary high-pressure turbine
Vorschräge *f* = tapered haunch
Vorschreibung *f* → Angebot *n*
Vorschub *m*, Vorstoß(en) *m*, (*n*) [*Hochlöffel(bagger)*] = crowd(ing)
~gerät *n* [*Bohrhammer*] = feed unit
~hebel *m*, Vorstoßhebel = crowd lever [*shovel*]
~kette *f*, Vorstoßkette = crowd chain [*shovel*]
~lafette *f*, (Vorschub)Schlitten *m*, Bohrlafette [*Bohrwagen*] = slide [*drill wagon*]
~motor *m* [*Bohrwagen*] = feed motor
(~)**Schlitten** *m*, Bohrlafette *f*, Vorschublafette [*Bohrwagen*] = slide [*drill wagon*]
(~)**Stütze** *f* → Bohrstütze
~turas *m*, Vorstoßturas = crowd sprocket
~weg *m* [*Bohrhammer*] = feed travel
~werk *n* = crowd mechanism
~zylinder *m* = crowd ram

Vorschweiß|flansch *m* = welded on flange, flange with welded neck
~-**Gestängeverbinder** *m* [*Rotarybohren*] = flash-weld tool joint
Vorsetzstein *m* → Flügelanfänger *m*
Vorsichtung *f* = preliminary air classification, ~ (~) separation
Vorspachtelung *f* = preliminary screeding
Vorspannbewehrung *f*, Vorspannarmierung, Spannbewehrung, Spannarmierung = prestressing reinforcement
(**Vor**)**Spann|**(**bündel**)**draht** *m* → Bündeldraht
~draht *m* → Bündeldraht
~druck *m* [*Hydraulikanlage*] = precharge
vorspannen vor dem Erhärten des Betons = to pretension
Vorspannen *n* = prestressing
~ **ohne Verbund** = no-bond tensioning
Vorspänner *m* → (Zweirad)~
Vorspannkabel *n* → Bündel aus Drähten
(**Vor**)**Spannkraft** *f* = tensioning force, stretching ~, prestress(ed) ~
Vorspann|moment *n* = prestressing moment
~platte *f* = prestressing plate
~stab *m* = prestressing bar
~stab *m* **des Systems Lee McCall** = Lee McCall bar
~stahl *m* = prestressing steel
~technik *f* = prestressing engineering
Vorspannung *f* = prestress
~ **auf elektrischem Wege**, elektrothermisches Spannen *n* der Bewehrung = electrical prestressing
~ **im Verbund**, ~ **mit** ~ = pretensioning
~ ~ ~ **auf Langspannbetten** = longline pre-tensioning
~ **mit nachträglichem Verbund**, nachträglicher Verbund *m*, Nachspannung = posttensioning
Vorspannungsverteilung *f* = distribution of prestress

(Vor)Spann|verfahren n = prestressing method

~verlust m = loss of prestress [*The reduction of the prestressing force which results from the combined effects of creep in the steel and creep and shrinkage of the concrete. This term does not normally include friction losses but may include the effect of elastic deformation of the concrete*]

~wagen m → (Zweirad-)Vorspänner m

~werk n, Fertigungswerk für Spannbetonbauteile = prestressing plant

~wert m = prestressing value

Vorsperre f → → Talsperre

vorspringende Ecke f = salient corner

Vorsprung m [*eines Körpers*] = projection

vorstädtisch = suburban

Vorstand m, flache Meeresküste f = shoreface, inshore

Vorsteuerpumpe f = pilot pump

Vorstoß|hebel m, Vorschubhebel = crowd lever [*shovel*]

~kette f, Vorschubkette = crowd chain [*shovel*]

~seil n, Vorschubseil [*Hochlöffel(bagger)*] = crowd rope

~turas m, Vorschubturas = crowd sprocket

Vorstrand m, flache Meeresküste f = shoreface, inshore

Vorsturzwalzwerk n, Springduowalzwerk = jump roughing mill

Vorträge mpl, Vorberge mpl [*eines Gebirgszuges*] = foothills

Vortränkung f = pre-impregnation

Vortreiben n, Vortrieb m [*Material in einer Strangpresse*] = forcing

Vortreibrohr n, Bohrrohr [*gebohrter Ortpfahl*] = drilled casing (tube), **~** lining (**~**), bore(d) **~** (**~**)

Vortrieb m, Strecken**~**, Ausbruch m, Auffahren n [*Tunnel-, Berg- und Stollenbau*] = advance, progress, heading, driving

~, Vortreiben n [*Material in einer Strangpresse*] = forcing

~geschwindigkeit f [*Betondeckenfertiger*] = operating speed

~presse f = driving jack

~stelle f → Einbaustelle [*Straßenbau*]

~stelle f, Ausbruchstelle, (Orts)Brust f [*Tunnel- und Stollenbau*] = (working) face

~weise → → Tunnelbau

Vortrocknung f = pre-drying

vorübergehende Härte f = temporary hardness

vorumhüllt = precoated

Vorumhüllung f, Bindemittel**~** = precoating

Voruntersuchung f = preliminary investigation

Vorverdichter m, (Auf)Ladegebläse n, Auflader m, Aufladekompressor m = supercharger

Vorverdichtung f = pre-compaction

~ [*Pfahlgründung*] = pre-consolidation

~, Aufladung [*Dieselmotor*] = supercharging

Vorverdichtungsdruck m [*Pfahlgründung*] = pre-consolidation pressure

Vorvertrag m = pilot contract

Vorwahl f = pre-selection

Vorwähler m, Vorwahlgerät n = preselector

Vorwählgang m [*Eine Vorrichtung, mittels der ein Gangschalthebel bewegt werden kann, ohne daß der betreffende Gang eingerückt wird, bevor Kupplung oder Drosselklappe betätigt worden sind*] = pre-selective

Vorwärmbunker m = preheating bin

Vorwärmen n, Vorwärmung f = preheating

Vorwärmer m = preheater

Vorwärmung f, Durchwärmung, Schmauchen n [*Ziegelbrennen*] = initial dehydration, water-smoking

Vorwärmzone f = preheating zone, preheater [*Upper extension of a shaft kiln*]

Vorwärts|einschnitt m [*Eine der wichtigsten Methoden der trigonometrischen Punkteinschaltung*] = foresight reading

~fahrt f = forward travel, **~** run

~gang m = forward gear

~geschwindigkeit f = forward speed

~**kippen** *n* **der Schaufel** = forward tilting of the bucket
~**kupplung** *f* = forward clutch
~**optimierung** *f* = feedforward optimalizing
~**planetensatz** *m* = forward gear train
Vorwaschmaschine *f*, **Vorwäsche** *f* = pre-washing machine
Vorwegweiser *m* = advance direction sign
Vorwohlitdecke *f* → Asphalteingußdecke
Vorzeichen *n*, **Sinn** *m* [*Baustatik*] = sign
vorzeitig eingestellte Bohrung *f*, ~ ~ **Sonde** *f*; ~ **eingestellter Schacht** *m* [*in Galizien*] = abandoned drilling well, ~ ~ bore, ~ ~ boring
vorzeitiges Abbinden *n* → falsches ~
Vorzerkleinerung *f* [*manchmal ist damit auch das „Vorbrechen" gemeint*] = primary comminution, ~ reduction
Vorzerkleinerungs|anlage *f* → Brechanlage
~**maschine** *f* → Brecher
Vorzugstarif *m* = preferential tariff
Voutenbalken(träger) *m* = haunched beam
V-Pflug *m* → Keilpflug
V-Stütze *f* = V-shaped column
VT-(Straßen)Teer *m* → Teer-Bitumen-Gemisch *n*
Vulkan *m*, Feuerberg *m*, feuerspeiender Berg = volcano
~**fiberhammer** *m* = fibre hammer (Brit.); fiber ~ (US)
vulkanische Asche *f*, Feuerbergasche = volcanic ash, ~ cinder, ~ dust
~ **Schlacke** *f* → Lavaschlacke
vulkanischer Sand *m*, Feuerbergsand, Kratersand, natürlicher Schlackensand = volcanic sand
~ **Tuff** *m*, Feuerbergtuff, Durchbruchgesteintuff = (volcanic) tuff, tuffaceous rock
vulkanisches (Erd)Beben *n* = volcanic earthquake
~ **Glas** *n* = volcanic glass

Vulkanisieranstalt *f* = tire shop (US); tyre ~ (Brit.)
vulkanisieren = to vulcanize
Vulkanismus *m* = vulcanicity
Vulkankraftwerk *n* → Erddampfkraftwerk
VZ *m* → Abbindeverzögerer *m*

W

Waage *f* **für alle Gemengeteile** → Gattierungswaage
Waagebalken-Querschwinge *f* [(*Gleis*)-*Ketten-Fahrgestell*] = pivoted equalizer bar
waag(e)recht laufende Zahnräder *npl* = horizontal gears
waag(e)rechte Holzbohle *f* [*Baugrubenverbau*] = horizontal wood(en) plank, ~ timber ~
~ **Komponente** *f* **des Seilzugs** [*Hängebrücke*] = horizontal component of the cable tension
~ **Rollenbahn** *f* = level roller runway
~ **Steife** *f* = flying shore
waag(e)rechter Abstandhalter *m* = horizontal spacer
~ **(Graben)Verbau** *m* = horizontal trench sheeting and bracing
~ **Niederschlag** *m*, horizontaler ~ = horizontal precipitation
waag(e)rechtes Geschwindigkeitsverteilungsdiagramm *n* = horizontal velocity distribution curve
~ **Sturzbett** *n* = horizontal apron
Waagerecht|förderer *m* = horizontal conveyor, ~ conveyer
~**-Schiebefenster** *n*, Horizontal-Schiebefenster = horizontal sliding window
Waagscheit *n* = level(l)ing plank
Waagtrichter *m* → Wiegebehälter
Waben|bunker *m* → Taschensilo *m*
~**(dach)deckung** *f* = honeycomb slating
~**fenster** *n* = honeycomb window
~**gefüge** *n* = honeycombed structure
~**silo** *m*, Taschensilo, Taschenzuteiler *m*, Wabenzuteiler [*Beton-Dosier-*

Wabentextur 2. Ordnung — Wahrscheinlichkeitsrechnung

anlage] = compartment bin, ~ storage hopper
~**textur** *f* **2. Ordnung**, Krümelgefüge *n*, Flockengefüge = flocculent structure
~**zuteiler** *m* → Wabensilo *m*
Wache *f* = guardhouse
Wachposten *m* = watchman
Wachs|kohle *f* = wax coal, pyropissite
~**krone** *f*, Abdruckstempel *m*, Abdruckbüchse *f* = impression block [*rotary drilling*]
~**schiefer** *m* = wax shale, oil ~
Wadi *m*, Trockental *n* = wadi [*dried-up bed of stream*]
Waffelblech *n* = goffered plate
Waffenkammer *f* = armo(u)ry
Wägeband *n* → Wiegeband
Wägen *n*, Wiegen, Wiegung *f*, Wägung = weighing
Wagen *m* = wagon
~**bauer** *m* → Wagner
~-**Fähranlage** *f* = car-ferry terminal
~**fett** *n*, Wagenschmiere *f* = sett grease
~**halle** *f* = motor vehicle hangar
~**hebebock** *m* → Autowinde *f*
~**heber** *m* → Zahnstangenwinde
~**heber** *m* → Autowinde
~**rahmen** *m*, Fahrgestell *n* [*Bohrwagen*] = wheel mounted main frame, ~ ~ chassis [*wagon drill*]
~**straße** *f* = wagon road
~**umlauf** *m* **an der Hängebank** = surface mine car circulation
~**waschplatz** *m* = washdown yard
~**waschung** *f* = washing down of cars and lorries, car wash
~**winde** *f* → Zahnstangenwinde
~**winde** *f* → Autowinde
~**zug** *m* [*Erdbau; Tunnel- und Stollenbau*] = train of cars
Waggon *m* → Bahn~
~**(band)entlader** *m* = (conveyor-type) wagon unloader, conveyer-type ~ ~
~**bauanstalt** *f* = wagon plant
~**bauträger** *m* = wagon building channel
~**beladeanlage** *f* = wagon loading plant, ~ ~ installation
~**beladeeinrichtung** *f* = wagon loading equipment
~**beladeturm** *m* = wagon loading tower
~**blech** *n* = wagon plate
~**entladepumpe** *f* = wagon unloader pump
~**entlader** *m* = wagon unloader
~**entlader** *m* → Waggonbandentlader
~**hebeanlage** *f* = wagon hoist
~**kippaufzug** *m* = wagon tippler hoist, ~ dumper ~
~**kipper** *m* = wagon tippler, ~ dumper
~**niethammer** *m* = close quarter riveting hammer
~**schaufel** *f* **für Schüttgüter** → Entladeschaufel
~**schrapper** *m* → Entladeschaufel
~-**Verschiebevorrichtung** *f* = wagon shifter
~**waage** *f* → Gleiswaage
Wagner *m*, Wagenbauer *m*, Stellmacher = wheel-wright, cart-wright
Wählen *n* **einer Zahl auf einem Zifferblatt** [*wie beim Telefon*] = dialling a number
Wahl|muffe *f*, Sondermuffe [*Tiefbohrtechnik*] = special clearance coupling
~**position** *f* = alternative item
~**preis** *m* = alternative price
~**projekt** *n*, Wahlvorschlag *m* [*Ideenwettbewerb*] = alternative scheme
Wähltaste *f* = selection key
wahlweise, nach Wahl, freigestellt = optional
wahre Kohäsion *f*, tatsächliche ~, wirkliche ~ = true cohesion
~ **Porosität** *f*, Gesamtporosität, Undichtigkeitsgrad *m* = total porosity
wahrscheinliche (Mineral)Vorräte *mpl* = probable mineral, ~ reserves, ~ ore
~ **Strömung** *f* [*Meer*] = probable current
Wahrscheinlichkeits|rechnung *f* [*Hochwasser*] = probability method

~theorie *f* = probability theory
waldbestanden, bewaldet = forest-clad, forested
Waldgleiboden *m* = forest gley soil
Walk|penetration *f* [*Schmierfett*] = worked penetration
~**widerstand** *m* **der Reifenseitenwände,** Verformung *f* der Bereifung = tire flexing (US); tyre ~ (Brit.)
Wall|düne *f*, Querdüne = transverse dune
~**fahrtskapelle** *f* = pilgrimage chapel
~**moräne** *f* → Seitenmoräne
~**riff** *n*, Barriereriff, Dammriff = barrier reef
Walm *m*, Walmfläche *f* [*Dreieckige Dachfläche an der Schmalseite eines Walmdaches*] = hip(ped end) [*Scotland: piend*]
~**anfänger** *m* = hip starting tile
~**dach** *n* = mansard roof (US); hip(ped) ~ (Brit.)
~**gaupe** *f* = hip(ped) dormer
~**kappe** *f* → Endfirstziegel *m*
~**ziegel** *m* = hip tile

Walzasphalt *m* = rolled asphalt [*formerly called "hot-rolled asphalt"*]
~**decke** *f*, Walzasphaltbelag *m* = rolled asphalt pavement, ~ ~ surfacing
~**(misch)anlage** *f*, Heißmischanlage, Teerbeton- und Walzasphalt-Mischanlage, Asphaltaufbereitungsanlage, Asphalt(misch)anlage = hot-mix (asphalt) plant, central mixing plant for the preparation of high-type hot mix(tur)es; asphalt (mix) plant
Walz|belag *m*, Walzdecke *f* = rolled pavement, ~ surfacing
~**beton** *m* = rolled concrete
~**blei** *n*, Tafelblei, Bleiblech *n* = sheet lead
Walzdamm *m*, (ab)gewalzte (Auf-)Schüttung *f* = rolled fill
~, (ab)gewalzte Erd(auf)schüttung *f* = rolled earth fill
~ → Walzstaudamm
Walz|decke *f*, Walzbelag *m* = rolled pavement, ~ surfacing
~**draht** *m* = wire rod

Walze *f*	roller
Abstreicher *m*	scraper
(ab)walzen	to roll
(Ab)Walzen *n*, Druckverdichtung *f*, Walzverdichtung, Walzung, Walzkompression *f*	(compaction by) rolling
(Ab)Walzen *n* mit Stahlmantelwalzen	steel rolling
(Ab)Walzen *n* mit Gummiwalzen	pneumatic rolling
Advance-Dieselmotor-(Straßen)~ [*Dreiradwalze mit Differentialgetriebe und geteilter Hinterachse*]	advance diesel (road) ~
Anhänge~ → Schlepp~	
Arbeitsgang *m*, Walzgang, Übergang, Walzenübergang, (Walz-) Fahrt *f*, Überrollung *f*	pass, coverage
Arbeitsgewicht *n* → Dienstgewicht	
Aufreiß(er)einrichtung *f* mit 3 Zähnen	three-tine scarifying attachment
Aufstandfläche *f*, Standfläche	contact area
Ballast *m*	ballast
Ballastbehälter *m*, Ballastkasten *m*	weight box, ballast container, ballast box

Walze

Ballastierung *f*	ballasting
Ballastwegnahme *f*	ballast removal
Bandage *f*	rim
Benzin~, Ottomotor~	petrol(-driven) ~ (Brit.); gasoline-driven (or gasolene-driven) ~ (US)
Beries(e)lungsanlage *f*	sprinkling device, wetting system
Betriebsgewicht *n* → Dienstgewicht	
Bodenverfestigungs~ → Verdichtungs~	
Böschungs~	slope ~
Dampf(straßen)~	steam (road) ~
Dampf-Tandem-~	tandem steam ~
Deckenschluß~	black-top finishing ~
Dienstgewicht *n*, Betriebsgewicht, Arbeitsgewicht	rolling weight, service weight, working weight, operating weight
Diesel-Motor-Dreirad-~	diesel engine three-wheel(ed) ~
Dieselmotor-Tandem~	diesel tandem ~
Diesel(-Straßen)~	diesel (road) ~
Doppelvibrations~, Doppelschwing(ungs)~, Doppelvibrier~, Doppelrüttel~	double vibrating ~, double vibratory ~
Dreiachs(-Tandem)~ [*Dreiradwalze mit drei hintereinander liegenden Radsätzen*]	three-axle (tandem) ~
Dreiraddampf~	three-wheel(ed) steam ~
Dreiradglatt~	three-wheel(ed) static ~
Dreirad(straßen)~ [*Zweiachswalze mit vorne einem meist geteilten Rad und hinten zwei Rädern, wobei die hinteren Raddurchmesser größer als die vorderen sind*]	macadam ~ (US); three-wheel(ed) (all-purpose) ~, three-roll type machine, three-legged ~, three-roll ~
Druckausgleichvorrichtung *f*	pressure balancing device
Druckverdichtung *f* → (Ab)Walzen	
Eigengewichts~ → Glatt~	
einachsige Gummirad~	single-axle compactor
Einrad-Motor-~ → handgeführte ~ mit Motor	
Einrad-Rüttel~, Einrad-Vibrations-~, Einrad-Schwing(ungs)~, Einrad-Vibrier~	single-wheel(ed) vibrating ~, single-wheel(ed) vibratory ~
Einrad~, Einachs~ [*mit und ohne Stützrad*]	single-wheel(ed) ~
Einrad~ mit Gummireifenfahrwerk	wheel(ed) ~
Einzel~ → Walzrad	
Einzweck~	single-purpose ~
Eisenballast *m*	iron ballast
Erdbau~ → Verdichtungs~	
Fahrt *f* → Arbeitsgang	
Festaufreißer *m*	fixed scarifying attachment

Walze

Fußboden~	floor ~
Fußweg~, Gehsteig~, Gehweg~	sidewalk ~ (US); footpath ~ (Brit.)
Gewicht n der Hinterräder von 60 kg je 1 cm linearer Berührungsfläche	330 pounds per linear inch of tread of rear wheels
gezogene ~ → Schlepp~	
Gitter~	grid ~
Glattmantel-Schlepp~	plain towed ~
Glattmantel~, Eigengewichts~, statische ~, statisch wirksame ~, Stahlbandagen-Straßen~, Stahlmantel~	smooth(-wheeled) ~, static ~, flat-wheel(ed) ~; smooth-tired ~, flat-steel ~ (US); steel-faced ~, steel-wheel ~
Glattvibrations~, Vibrationsglatt~	vibratory smooth(-wheeled) ~
Gleisbettungs~	permanent way ~
Graben~	trench ~
Grundgewicht n	basic weight
Gummi(reifenvielfach)~, Gummi(viel)rad(-Verdichtungs)~	multi-rubber-tire ~ (US); multi-tyred ~ (Brit.); rubber-tired (US)/tyred (Brit.) ~, multiwheel(ed) ~
entweichende Luft f und Feuchtigkeit f	escaping air and moisture
Gummirad~ mit Vibration, Fabrikat IOWA MFG. CO., CEDAR RAPIDS, IOWA, USA	CEDARAPIDS COMPACTOR [*Trademark*]
~ mit nicht oszillierenden Rädern	straight (or fixed, or rigid, or nonoscillating) wheel (pneumatic-tyred) ~
~ mit oszillierenden Rädern	wobbly (or wobbled) wheel(ed) (pneumatic tire) ~, wobbly-wheel(ed) compactor
Pneu~, Luftreifen~	pneumatic-tire ~ (US); pneumatic-tyre ~ (Brit.)
schwingende Gummi(reifenvielfach)~, gummibereifte ~ mit Vibration, Gummirad~ mit Vibration	vibratory pneumatic-tired (US)/tyred (Brit.) ~; pneumatic-tired vibrating compacting ~ (US)
senkrechte Verdichtungswirkung f in der Arbeitsrichtung	directional depth penetration
Unwucht f	pair of unbalanced weighted shafts
Vollgummireifen~	solid rubber-tire ~ (US); solid rubber-tyre ~ (Brit.)
Gürtelrad~, Gürtelradbodenverdichter m	~ with movable shoes fixed to the wheels
handgeführte ~ mit Motor, Einrad-Motor~, handgeführte kraftgetriebene Einrad~, Motorhand~	hand-guided motor ~, power driven single-wheel(ed) ~
Hand~ mit variablem Gewicht	ballast type hand ~

Walze

Hand~ mit nicht variablem Gewicht	fixed weight type hand ~
Hand~ mit Wasserballast	water ballast hand ~
Hinter~, Hinterwalzrad *n*, Hinterwalzenzylinder *m*, Hinter-Walzentrommel *f*	rear roll, rear roller wheel
Kleinmotor~	light duty power ~
Klein~	midget ~
Knet~ → Verdichtungs~	
Kraft~, selbstfahrende ~, Motor~	power-driven ~, self-propelled ~
Lenkgabel *f*	steering fork
Lenk~, Lenkwalzrad *n*, Lenkwalzenzylinder *m*, Lenkwalzentrommel *f*, Steuer~	steerable roll(er wheel)
Luftreifen~, Pneu~	pneumatic-tire ~ (US); pneumatic-tyred ~ (Brit.)
Motor-Straßen~	power-driven road ~, self-propelled road ~
Motor~ → Kraft~	
Öffnung *f* zum Einfüllen von Sand- oder Wasserballast	ballast cleanout door
Ottomotor~ → Benzin~	
Pferde~	horse-drawn ~
Pneu~, Luftreifen~	pneumatic-tire ~ (US); pneumatic-tyred ~ (Brit.)
Rasen~	grass ~, lawn ~
Raupenschlepper~, (Gleis)Kettenschlepper~	tracked ~ [*A tractor with smooth closely-fitting track plates*]
Riesengummi(reifenvielfach)~, Super-Gummirad~, Schwerstverdichter *m*	supercompactor
Riffel~	indented ~, indenting ~, branding iron, crimper
Rillen~	disc ~; disk ~ (US)
Rüttel~ → Vibrations~	
Sandballast *m*	sand ballast
Schaffuß~	sheepsfoot ~, tamper (~), (sheepsfoot) tamping ~
ALBARET-~, Fabrikat ALBARET, RANTIGNY (OISE), FRANKREICH	TOURNEPIEDS [*Trademark*]
Doppelzylinder-~ mit gelenkiger Verbindung	two-unit roll with the units connected together with a pivoting link
in einer Linie gezogen	towed in line
in Einfachanordnung *f* gezogen	towed by tractor singly
in Mehrfachanordnung *f* gezogen	towed by tractor in multiples
Klein-~	midget sheepsfoot ~
Riesen~, Riesenstampf~	giant weight tamper

Walze

Schaffuß *m*, Walzen-Stempel *m*, Stampffuß	projecting foot, tamping foot
Schaffußtrommel *f*	(sheepsfoot tamping) roller drum
~ mit zapfenförmigen Stacheln	peg-foot ~
(Schaf)Klumpfuß~	club-foot (type sheepsfoot) ~
(Schaf)Stockfuß~, Igel~, Stachel~, ~ mit konischen Stacheln (oder Druckstempeln) versetzt gezogen	spiked ~, taper(ed)-foot (type sheepsfoot) ~, pyramid feet ~ towed offset
zylindrischer Walzkörper *m*	cylindrical drum
Schieben *n*	shoving
Schlepp~, Anhänge~, gezogene ~	tractor-drawn ~, towed-type ~
schwere einachsige Gummirad~	heavy compactor
schwere ~	heavy-duty ~
Schwing(ungs)~ → Vibrations~	
selbstfahrende Graben-Vibrations~	self-propel(l)ed vibratory trench ~
selbstfahrende ~ → Kraft~	
Sportplatztandem~	sports ground type tandem ~
Sportplatz~	sportsfield ~
Stahlbandagen-Straßen~ → Glatt~	
Stahlmantel *m* mit Gummibelag	steel rim with rubber cover(ing)
Stahlmantel~ → Glattmantel~	
Stampf~	tamping ~, tamper
Standfläche *f*, Aufstandfläche	contact area
statische Tandem~	static tandem ~
statische ~ → Glatt~	
Steuer~ → Lenk~	
Straßen~	road ~, highway ~
Tandem-Vibrations~, Vibrationstandem~, Tandem-Schwing(ungs)~	vibratory tandem ~, vibrating ~ ~
Tandem~	tandem ~
Trieb~, Trieb-Walzrad *n*, Trieb-Walzenzylinder *m*, Trieb-Walzentrommel *f*, Treib~	drive roller wheel
Übergang *m*, Walzgang, Arbeitsgang, Überrollung *f*, Fahrt *f*, Walzenübergang, Walzfahrt	pass, coverage
Überrollung *f* → Übergang	
Umbau~	conversion type ~
Verdichtungs~, Bodenverfestigungs~, Erdbau~, Knet~	compaction ~, compactor
Vibrations~, Schwing(ungs)~, Vibrier~, Rüttel~	vibrating ~, vibratory ~
handgeführte kraftbewegte (oder selbstfahrende) ~	hand-guided power-propelled ~ ~
handgezogene ~, von Hand verfahrbare ~	hand-propelled ~ ~

Walze

Schlepp~, Anhänge~	trailer (type) ~ ~
Vibrationsglatt~, Glattvibrations~	vibratory (or vibrating) smooth(-wheeled) ~
Vibrationstandem~, Tandem-Vibrations~, Tandem-Schwing(ungs)~	vibratory tandem ~, vibrating tandem ~
Vibrationstandem~ selbstfahrend, selbstfahrende Tandem-Schwing(ungs)~ (oder Tandem-Vibrations~) (oder Tandem-Rüttel~)	self-propelled vibratory (or vibrating) tandem ~
Vollgummireifen~	solid rubber-tire ~ (US); solid rubber-tyre ~ (Brit.)
Vorder~, Vorderwalzrad *n*, Vorder-Walzenzylinder *m*, Vorderwalzentrommel *f*	front roll(er wheel)
Waffel~	segmented ~, island ~
Walzarbeiten *fpl*	rolling work
Walzbreite *f*	rolling width
Walzdruck *m*	rolling pressure
Walze *f* → Walzrad	
Walze mit gummibereiften Hinterrädern	portable ~
Walzen *n* → Abwalzen	
Walzenführer *m*, Walzenfahrer *m*, Walzenbediener *m*	rollerman, roller operator
Walzenachse *f*	roller axle
Walzenrahmen *m*	roller frame
Walzenübergang *m* → Übergang	
(Walz)Fahrt *f* → Übergang	
Walzgang *m* → Übergang	
Walzgrund *m*	surface to be rolled
Walzgut *n*	material to be rolled
Walzkompression *f* → (Ab)Walzen	
Walzprofil *n*	rolled shape, rolled section
Walzrad *n*, Walzenzylinder *m*, Walzentrommel *f*, Walze *f*, Einzelwalze	roller wheel, roll
Walzung *f* → (Ab)Walzen	
Walzverdichtung *f* → (Ab)Walzen	
Wasserballast *m*	water ballast
Wege~	path ~
Wendekreis *m*, Drehkreis	turning circle
Zweiachs-(Tandem)~	two axle (tandem) ~
Zweirad-Graben~	dual-compression trench ~
Zweizweck~	dual-purpose ~

Walze *f*, Koller *m*, Läufer *m* [*Kollergang*] = runner

~ [*Walzenwehr*] = cylinder (Brit.); roller gate (US)

~ mit vorgesetztem Stauschild, ~ ~ Schnabelansatz [*Walzenwehr*] = small roller with shields

Walzen|(-Anbau)streuer *m*, Verteilerwalze *f*, Streuwalze = gritting roll (attachment)

~**aufgeber** *m*, Walzenspeise, Walzenbeschicker = roll feeder

~**beschicker** *m*, Walzenaufgeber, Walzenspeiser = roll feeder

~**brecher** *m* → → Walzwerk

~**fettbrikett** *n* [*DIN 51824*] = neck grease briquet

walzenförmig = cylindrical

Walzen|lager *n*, Rollenlager [*für Hoch- und Tiefbauten*] = roller bearing

~**mantel** *m* → Brech~

~**mühle** *f* = roll(er) mill, ~ grinder

~**pumpe** *f* = drum pump

~**rad** *n* = roller wheel

~**ringmühle** *f*, Ring(walzen)mühle = ring (roll(er)) mill

~**schrämlader** *m* = coalcutter with rotating cutting drum and chute

~-**Schrämmaschine** *f* = coalcutter with rotating cutting drum

~**speiser** *m*, Walzenaufgeber, Walzenbeschicker = roll feeder

~**splittbrecher** *m*, Splittwalzenbrecher = crushing rolls for chip(ping)s production

~**streuer** *m* → Walzen-Anbaustreuer

~**streuer** *m* → Walzen-Streugutverteiler *m*

~-**Streugutverteiler** *m*, Walzenstreuer *m* = hopper type gritter with feed roll

~**wehr** *n* = cylindrical barrage (Brit.); roller dam (US)

Wälzgelenk *n* = rolling contact joint

Walzkern *m*, (ab)gewalzter Erdkern [*Felsschüttungs(stau)damm*] = rolled earth(en) core

Wälz|klappbrücke *f* = rolling lift bridge

~**lager** *n* [*für Hoch- und Tiefbauten*] = rolling contact bearing

~**lager** *n* = anti-friction bearing

Walz|magerbeton *m*, Magerwalzbeton = rolled lean concrete

~**profil** *n*, Walzträger *m* = rolled section, ~ steel girder, ~ beam

~**rad** *n*, Walzenrad = roller wheel

~**sand** *m* = blinding sand

~**schotter-Gußasphalt** *m* → Mastix-Vergußdecke *f*

~**schüttung** *f* = rolled fill

Wälz|sieb *n* → Trommelsieb

~**siebung** *f* = drum screening, trommel ~

Walz|splitt *m* = rolled-in chip(ping)s

~**stahl** *m* = rolled steel

~**stahlprofil** *n* = rolled steel section

~**stahl-Schweißkonstruktion** *f* = all-welded rolled steel construction

~(**stau**)**damm** *m*, (ab)gewalzter Damm = rolled earth (fill) dam, ~ earth(work) ~, ~ earthen ~

~**temperatur** *f*, Ab~ [*Straßenbau*] = rolling temperature

~**träger** *m*, Walzprofil *n* = rolled steel girder, ~ section, ~ beam

~**werk** *n* = rolling mill

Walzwerk *n*, Brech~ [*In der Hartzerkleinerungsindustrie Sammelbegriff für Walzenbrecher und Walzen(mahl)mühlen*]

tertiary crusher with rolls [*Strictly speaking, the tertiary crusher is the machine used in the third stage of crushing. Often, in large scale operation, the cone crusher is used in this manner. Before the advent of the cone crusher, however, the crushing rolls enjoyed a near monopoly as a tertiary crusher*]

Anpaßfeder *f*, Pufferfeder

compression spring

Beilageeisen *n*

shim

Wand — Wandelgang 1110

bewegliche Walze *f*	moveable roll
Brechring *m*	(crushing) roll segment
Brechwalze *f*	crushing roll
(Brech)Walzenmantel *m*	(crushing) roll shell
Dreiwalzenbrecher *m*	triple roll crusher, ~ ~ breaker
Eisenstücke *npl*	tramp iron
Fremdmaterial *n*	uncrushable material
geriffelte (Brech)Walze *f*	corrugated (crushing) roll
Glattwalzenbrecher *m*, Glattwalzwerk *n*	smooth-shell crushing rolls
Körnung *f* des Fertiggutes	product size, size of product
Manteloberfläche *f*	shell face
Nabe *f*	core
Pufferfeder *f* → Anpaßfeder	
Pyramidenwalzenbrecher *m*	pyramidal-toothed crushing rolls
Sandwalzenbrecher *m*, Sandwalzwerk *n*	crushing rolls for sand production
Spaltweite *f* zwischen den beiden Walzen, Walzenabstand *m*	setting between rolls, feed opening, roll feed
Splittwalzenbrecher *m*, Walzensplittbrecher	crushing rolls for chip(ping)s production
Stahltrommel *f*	steel drum
Walzenabstreifer *m*	roll scraper
Walzenbrecher *m* mit Brechplatte und Brechwalze	single roll crusher, ~ ~ breaker
Walzenbrecher *m* mit zwei Brechwalzen, Zweiwalzenbrecher	crushing rolls, (double) roll crusher (or breaker), twin roll crusher (or breaker)
Walzen(mahl)mühle *f*	roll(er) (grinding) mill, roll grinder
Zweiwalzenbrecher *m* → Walzenbrecher mit zwei Brechwalzen	

Wand *f*, Steinbruch~, Abbau~, Bruch~ = face, quarry ~
~ → (Abbau)Stoß *m*
~**(anstrich)farbe** *f* = wall paint
~**auskleidung** *f* → Wandverkleidung
~**balken** *m*, Mauerbalken = wall beam
~**(bau)platte** *f*, Wandelement *n* = wallboard
~**(bau)platte** *f* **aus Gips** → Gipswand(bau)platte
~**baustein** *m*, Mauerstein = masonry unit
~**baustoff** *m*, Wandbaumaterial *n* = wall material
~**(bau)tafel** *f* = wall unit, ~ slab, ~ panel
~**bekleidung** *f* → Wandverkleidung
~**belag** *m* = wall covering
~**betonierung** *f* = wall pour
~**block** *m*, Mauerblock = walling block
~**bogen** *m*, Gurtbogen = wall arch
~**brause** *f* = wall-mounted shower
~**deckenleiste** *f* = cover strip on the wall
~**dicke** *f* = wall thickness
~**dicke** *f* [*Schornstein*] = shell thickness
~**dickenveränderung** *f* [*Gleitschalungsbau*] = thickness taper
~**dichtung** *f* [*Unterkellerung*] = wall skin
~**element** *n*, Wand(bau)platte *f* = wallboard
Wandelgang *m* = promenade gallery

Wander|düne *f* → lebendige Düne
~fisch *m* [*Ein Fisch, der seinen Standort unter Zurücklegung großer Strecken in regelmäßiger Wiederkehr wechselt*] = migratory fish
~form *f* = travelling mould (Brit.); traveling mold (US)
~kasten *m* (⚒) = cog, chock (Brit.); crib, pigsty (US) [*A method of ground support consisting of a rectangular crisscross or square or round timbers laid in pairs parallel to each other and at right angles to the row above or below. The cog may be filled with stone to strengthen it*]
~last *f*, bewegliche Last = moving load
~mischer *m* = travel(ling) mixer
wandern [*von Punkten*] (⚒) = to move
Wander|rost *m* = travel(l)ing grate
~rost *m* → Wander-Stangen(transport)rost
~rostsintern *n* = travel(l)ing-grate sintering
~schalung *f* = moving formwork, travel(l)ing ~ [*for tunnel linings*]
~sprenger *m*, Fahrsprenger [*Abwasserwesen*] = movable distributor, travel(l)ing ~, reciprocating ~
~-Stangen(transport)rost *m*, Wander-Stangenaufgeber *m*, Wander(sieb)rost = travel(l)ing grizzly feeder, ~-bar grizzly [*lateral instead of longitudinal bars are used*]
~stempel *m* (⚒) = self-advancing chock, hydraulic ~, ~ cog, ~ roof support, hydraulic walking chock
~welle *f* = travel(l)ing wave
Wand|farbe *f*, Wandanstrichfarbe = wall paint
~faserplatte *f*, Faserwandplatte = fibre wallboard; fiber ~ (US)
Wand|fertigteil *n*, *m* aus Beton = precast concrete wall unit
~festigkeit *f* = wall strength
~flächenheizung *f* = wall panel heating
~fliese *f*, Wandplatte *f* = wall tile
~fuge *f* = wall joint
~hahn *m* = bib(cock), bib tap

~-Handlauf *m* [*Treppe*] = grab rail [*A hand rail along a wall*]
~heizkörper *m*, Wandradiator *m* = wall radiator
~hohlstein *m* = hollow wall block
~hohltafel *f* = hollow (core) wall slab, ~ (~) ~ panel
~hydrant *m*, Zapfhahn *m*, Feuerhahn = wall hydrant
~innenseite *f* = interior wall surface
~isolierschicht *f* = wall damp-proof course
~isolierung *f* = wall insulation
~kies *m*, Grubenkies = pit-run gravel
~-Kleiderschrank *m*, Einbau-Kleiderschrank = built-in wardrobe
~konstruktion *f* = wall system
~kran *m*, Konsolkran = wall crane
~(kratz)schaufel *f* [*Betonmischer*] = wall scraper blade
~kreuzung *f* = wall junction
Wandler *m*, Drehmoment~ = (torque) converter
~antrieb *m*, Wandlerstufe *f* = (torque) converter drive
~flüssigkeit *f* = (torque) converter liquid
~gehäuse *n* = (torque) converter housing
~haltegang *m* = (torque) converter hold gear
~ladedruck *m* = (torque) converter charging pressure
~ladedruckventil *n* [*Auslaßseite*] = (torque) converter outlet relief valve
~ladedruckventil *n* [*Einlaßseite*] = (torque) converter inlet relief valve
~ladeflüssigkeit *f* = (torque) converter charging liquid
~ladeventil *n* = (torque) converte charging valve
~öldruckmesser *m* = (torque) converter oil pressure ga(u)ge
~ölverlust *m* = (torque) converter leakage
~-Planetengetriebe *n* = (torque) converter planetary drive
~stufe *f*, Wandlerantrieb *m* = (torque) converter drive

Wandlerstufenhalteventil — Wärmeabschirmung 1112

~stufenhalteventil n = (torque) converter hold valve
~stufenkupp(e)lung f = (torque) converter drive clutch
~stufenschaltventil n = (torque) converter speed valve
~thermometer n, m = (torque) converter heat ga(u)ge
wandloses Gebäude n, wandloser Bau m = open-sided building
Wand|mörtelbelag m, Wandputz m = wall plaster
~pfeiler m = wall pier
~platte f, Wandbauplatte, Wandelement n = wallboard
~platte f, Wandfliese f = wall tile
~platte f mit kunstharzgetränktem Dekorpapier beschichtet = decorative plastic-faced wall panel
~plattenverkleidung f = wall-panel(l)ing
~putz m, Wandmörtelbelag m = wall plaster
~putzmörtel m = wall stuff
~radiator m, Wandheizkörper m = wall radiator
~rauhigkeit f = wall roughness
~reibung f = wall friction
~reibungswinkel m [*Bodenmechanik*] = angle of wall friction
~rückseite f = back of a wall
~säule f, Mauersäule = wall column
~schale f = wall shell
~schalung f = wall formwork
~scheibe f, dünne ~ = diaphragm
~schleifmaschine f = wall-mounted grinder, ~ grinding machine
~schrank m, Einbauschrank = built-in cupboard
~spachtel(masse) m, (f) zum Vorbehandeln von Wand- und Deckenflächen zur Erzielung planebener, glatter und spannungsfreier Untergründe = wall surfacing compound
~stütze f = wall column
~tafel f, Wandbautafel = wall unit, ~ slab, ~ panel
~trockner m = wall dryer; ~ drier (US)
~urinal n = wall urinal

~verkleidung f, Wandbekleidung, Wandauskleidung = wall lining, ~ facing
~wange f = wall string
~winde f, Mauerwinde = wall-mounted winch
Wange f, Treppen~ = string
Wangen|schmiege f → Backenschmiege
~treppe f = string stair(case)
Wankelmotor m = Wankel engine
Wanne f [*Maschine*] = sump, pan
~ = sag [*A small valley between ranges of low hills*]
~ = sag [*The hollow or depression formed by the junction of two falling gradients*]
~, Mulde f [*LKW*] = body
~ [*Bauwerk(ab)dichtung*] = tanking
Wannen|ausrundung f [*Straße*] = sag curve, concave transition between gradients
~isolierung f → Grundwasserdichtungsschicht f
wannen- und kuppenreiche Straße f, kuppen- und wannenreiche ~ = undulating road, ~ highway
Ward-Leonard-Schaltungssatz m, Ward-Leonard-Schaltung f = Ward-Leonard set, ~ control
Waren|hafen m, Güterhafen = cargo harbour (Brit.); ~ harbor (US)
~haus n, Kaufhaus = departmental store
~haus-Personenaufzug m = departmental store passenger lift (Brit.); ~ ~ elevator (US)
~speicher m, Güterschuppen m = warehouse
~verkehr m, Güterumschlag m, Umschlag(verkehr), Warenumschlag [*Hafen*] = goods traffic, cargo ~
Warm|behand(e)lung f, Tempern n = tempering, drawing
~behand(e)lungszeit f, Warmbehand(e)lungsdauer f [*Beton*] = curing time at maximum temperature
Wärme|abgabe f = heat release
~abschirmung f, Wärmeschild m = thermal shield

~aufbereitung f = pyro-processing
~aufnahme f = heat absorption
~(auf)speicherung f = heat accumulation, ~ storage
~(aus)dehnung f = thermal expansion
~(aus)dehn(ungs)zahl f, Wärme(aus)dehn(ungs)beiwert m, Wärme(aus)dehn(ungs)koeffizient m = coefficient of thermal expansion
~ausnutzung f = heat utilization
~(aus)tausch m = heat exchange
~(aus)tauscher m = heat exchanger
~bedarf m = heat requirement
~behaglichkeit f = thermal comfort
wärmebehandelt, dampfbehandelt [Beton] = cured by atmospheric steam, low-pressure steam-cured
~ = heat treated
~ nach dem Schweißen = post weld heat treated
Wärme|behand(e)lung f = heat treatment
~behand(e)lung f, Dampfbehandlung [es wird Dampf mit Temperaturen unter 100° C als Wärmeträger benutzt] = steam curing at atmospheric pressure, low-pressure steam curing [concrete]
wärmebeständig = heat-proof
Wärme|beständigkeit f = heat resistance
~bilanz f, Wärmehaushalt m = heat balance, thermal ~ [A method of accounting for all the heat units supplied, transferred, utilized in, and lost from a kiln]
~brücke f = heat build-up
~dämmbeton m = heat-insulating concrete
~dämmplatte f = heat-insulating slab
~dämmschicht f, Wärmedämmlage f = heat-insulating layer (or course)
~dämmstein m = heat-insulating block
~dämmstoff m = heat-insulating material, thermal insulator
~dämmung f = thermal insulation, heat ~

~dämmwert m = thermal insulation value, heat ~ ~
~dämmwirkung f = thermal insulating efficiency, heat-insulating ~
~dehnung f = heat expansion, thermal ~
~dehn(ungs)zahl f → Wärmeausdehn(ungs)zahl
~durchgang m → Wärmeübertragung
~durchgangszahl f = overall heat transfer
~durchlässigkeit f = thermal diffusivity
~durchlaßwiderstand m [Der Wärmeschutz eines Raumes wird bestimmt vom Wärmedurchlaßwiderstand von Wänden und Decke] = thermal property, heat insulating ~
~durchlaßzahl f, Wärmedurchleitungszahl = U-value, thermal transmittance, air-to-air heat-transmission coefficient
~eigenschaft f = thermal property
~entwick(e)lung f = heat evolution, ~ build-up
~fluß m = heat flux, ~ flow
~freigabe f = liberation of heat
~gefälle n, Wärmeunterschied m = heat gradient
~gewinn m = heat gain
~haushalt m → Wärmebilanz
~impulsschweißen n = thermal impulse welding
Warmeinbaubelag m, Warmeinbaudecke f = warm-laid surfacing, ~ pavement
Wärme|isolierung f = heat insulation
~kammer f [Für die Betonbehandlung in ungespanntem Dampf bzw. warmer feuchter Luft unter 100° C, meist 40° bis 70°] = steam vapo(u)r curing chamber
~kraft f = thermal power
~kraftlehre f = thermodynamics
~kraftwerk n, thermisches Kraftwerk, thermoelektrische Zentrale f = thermal(-electric) station, ~ plant
~leistung f, thermischer Wirkungsgrad m = heat output, thermal ~

Wärmeleitfähigkeit — Warmluftleistung

~**leitfähigkeit** *f*, Wärmeleitvermögen *n* = thermal conductivity
~**mengenmesser** *m* = heat balancer
~**pumpe** *f* = heat pump
~**riß** *m* = thermal crack
~**rißbildung** *f* = thermal cracking
~**rückgewinnung** *f* = heat recovery
~**rückstrahlung** *f* = heat reflectivity
~**schalter** *m* = thermal switch
~**schild** *m*, Wärmeabschirmung *f* = thermal shield
~**schlange** *f* → Heizschlange
~**schutz** *m* = thermal protection
~**schutzglas** *n* = heat absorbing (plate) glass
~**schwindung** *f* = thermal shrinkage
~**spannungsriß** *m* = thermo-fracture
wärmesparend = heat-economizing, heat-saving
Wärme|speicherfähigkeit *f* = thermal (storage) capacity of a building material
~**speicherung** *f*, Wärmeaufspeicherung = heat accumulation, ~ storage
~**speichervermögen** *n* = thermal storage capacity
~**sperre** *f* = thermal barrier
~**stoß** *m* = thermal shock
~**stoßfestigkeit** *f* = thermal shock resistance, ~ ~ behavio(u)r
~**strahlungsfühler** *m* = heat radiation sensing device
~**strömung** *f*, Konvektion *f* = convection
~**tausch** *m*, Wärmeaustausch = heat exchange
~**tauscher** *m*, Wärmeaustauscher = heat exchanger
~**tauscherturm** *m*, Wärmeaustauscherturm = heat exchanger tower
wärmetechnische Untersuchung *f* = thermal examination
~ **Verfestigung** *f* **bindiger Böden** = thermal treatment of cohesive soils
Wärme|träger *m*, Wärmeübertragungsmittel *n* = heat transfer medium, heating ~
~**träger-Umwälzpumpe** *f* = accelerator [*A pump for circulating the heating medium through a heating system*]

~**übergang** *m* = heat transfer
~**übergang** *m* **bei Kondensation** = condensing heat transfer
~**überträger** *m* **mit innerer Wärmequelle** = heat exchanger with internal heat source
~**übertragung** *f*, Wärmedurchgang *m* = thermal transmittance, heat transmission
~**übertragungsmittel** *n*, Wärmeträger *m* = heat-transfer medium
~**umlauf** *m* = heat circulation
~**umwand(e)lung** *f*, Wärmeumprägung, Wärmemetamorphose *f* (Geol.) = thermal metamorphism
~**unterschied** *m*, Wärmegefälle *n* = heat gradient
wärmeverfestigter Boden *m* = thermally reinforced weak soil
Wärmeverlust *m* = heat loss
~**berechnung** *f* = estimate of heat loss
~**geschwindigkeit** *f* = rate of heat loss
Wärme|vorreg(e)lung *f* = heat pre-control
~**widerstand** *m* = thermal resistance
~**wirkungsgrad** *m* = thermal efficiency
~**wirtschaft** *f* = heat economy
wärmewirtschaftliches Bauen *n* = economically sound thermal design
Wärmezufuhr *f* = heat input
warm|feste Legierung *f* = heat-resistant alloy
~**gebogener Rohrbogen** *m* = hot bent bend
Warmhalteofen *m* = holding furnace
Warmluft *f* = warm air
~**(bau)austrocknung** *f* = dehumidifying by warm air, drying(-out) ~ ~ ~
~**gerät** *n*, Lufterhitzer *m*, Warmlufterzeuger *m* = (air) heater, space ~
~**heizung** *f* = warm air (central) heating
~**heiz(ungs)anlage** *f* = warm air heating system
~**kanal** *m* = warm air duct
~**leistung** *f* = heated air output, warm ~ ~

Warm|pressen n [*von Rohren*] = hot pressing
~richtmaschine f **für Bleche** = hot levelling machine
~versprödung f = hot embrittlement
~walzwerk n = hot rolling mill
Warmwasser|boiler m, **Warmwasserbereiter** m = calorifier
~hauptleitung f = warm water main
~heizung f = warm water heating
~umwälzpumpe f = accelerator
Warnblinker m, **Blinklaterne** f = flash lamp
warnen [*Gebirge im Tunnel-, Stollen-, Schacht- und Bergbau*] = to crackle
Warn|hupe f = waring horn, alarm ~
~licht n, **optisches Warnsignal** n = warning light, alarm ~
~posten m, **Winkerposten** [*Straße*] = signal man, flagman
~signal n = warning signal, alarm ~
~summer m = warning buzzer
~vorrichtung f = warning device
~zentrale f = warning station
Warren-Träger m = Warren truss
Warte|gleis n → Ausweichgleis
~halle f = waiting hall
~platz m = lay-by [*A part of a road out of the traffic lanes, where vehicles may wait*]
~platz m [*Flugplatz*] = warm-up pad, run-up ~, holding apron
~raum m, **Wartesaal** m = waiting room
~tag m, **ausgefallener Arbeitstag** = shutdown day
~zeit f, **Stehzeit** = stand-by time
Wartung f, **laufende Unterhaltung** = routine maintenance
wartungsarm = making low maintenance demands
Wartungseinrichtungen fpl [*Flughafen*] = service facilities
wartungsfrei = service-free
Wartungs|messer m = service meter
~öffnung f = service opening
~punkt m = service point
~zeitraum m = (maintenance) task period
Warzenblech n = button plate

Wasch|anlage f, **Wäsche** f = washing plant, ~ installation
~becken n = wash basin, lavatory, washbowl [*The term lavatory has now come to mean the room containing the wash basin (Waschraum), and by extension the room containing a WC (Spülklosettraum)*]
~benzin n **mit Siedegrenze** f **zwischen 100 und 200° C** = cleaners' naphtha, ~ solvent
Waschbeton m [*Mit ausgewaschenen Sichtflächen*] = concrete with exposed aggregates by washing
~maschine f, **Auswaschmaschine** für Waschbetonplatten [*Mit Einzel- oder Ringbürsten*] = scrubbing machine for concrete tiles
~oberfläche f = washed (concrete) surface, ~ (~) finish
~platte f = concrete tile with exposed aggregates by washing
Wäsche f, **Waschmaschine** f, **Wäscher** m = washing machine, washer
~, **Waschanlage** f = washing plant, ~ installation
~leinestange f = clothesline pole
Waschen n = washing [*Removing dust, mud, clay or other similar matter from ore, coal, or stone by rinsing it with clean water*]
Wäscher m, **Waschmaschine** f, **Wäsche** f = washing machine, washer
Wäscherutsche f = laundry chute, clothes ~ [*A duct from a bathroom to a lower floor, into which dirty clothes are dropped*]
waschfest = washproof
Wasch|filtern n = filtration washing
~gold n → Seifengold
~gut n = material to be washed
~kessel m = laundry tray, ~ tub (US) [*A deep, wide sink fixed to a wall for washing clothes*]
~kies m, **gewaschener Kies** = washed gravel
~kopf m [*Waschbetonmaschine mit Bürsten*] = scrubbing head
~kühler m = conventional gas washer

~leder *n*, Putzleder = washleather
~lederverglasung *f*, Putzlederverglasung *f* = glazing bedded in washleather, washleather glazing
~maschine *f*, Wäsche *f*, Wäscher *m* = washing machine, washer
~nische *f* = washing recess
~öl *n* = wash oil
~putz *m* = scrubbed plaster
~raum *m* = lavatory [*The term lavatory also means a wash basin (Waschbecken) and by extension the room containing a WC (Spülklosettraum)*]
~sand *m* = washed sand
~schnecke *f* = washing screw
~sieb *n* = washing screen, screening washer
~splitt *m* = washed chip(ping)s
~trog *n* → Trog
~trommel *f*, Trommelwäsche *f*, Trommelwaschmaschine *f* = drum scrubber
~- und Siebmaschine *f* = washing and screening machine
~wasser *n* [*Für das Waschen von Erzen, Kohle und Gestein*] = rinsing water
~zyklon *m* = cyclone washer

Wasser|(ab)dichtung *f* = water sealing
~ablassen *n* [*Ablassen des mitgeförderten Wassers aus dem Erdölmeßbehälter*] = "bleeding", draining off the water
~ableitungsbauwerk *n* = water-diverting structure
~abmessung *f* → Wasserdosierung
~abmeßvorrichtung *f* [*Betonmischer*] = water ga(u)ge
~absatzboden *m* → Schwemm(land)boden
~abscheider *m* = water separator
~abschirmung *f*, Wasserschild *m* [*Atomreaktorbau*] = water shield
~absenkungsbrunnen *m*, Wasserentzugsbrunnen = drainage well
~absonderung *f* → Wasserabstoßen *n*
~absperren *n* = water shutoff
~abstoßen *n*, Bluten, Wasserabsonderung *f*, Betonschwitzen = bleeding, sweating, water gain

wasser|abweisend, hydrophob, wasserabstoßend = water-repellent, water repelling
~abweisendes Mittel *n*, Sperrzusatz *m*, (Beton)Dichtungsmittel = waterproofer, concrete water-proofing compound, water-repellent, water-repelling agent, densifier, densifying agent, dampproofing and permeability reducing agent
Wasser|abweisung *f*, hydrophobierende Wirkung = water repellency
~andrang *m* → Wasserzutritt *m*
wasseranfällig = susceptible to water
Wasser|anlagerungsvermögen *n* → Adsorption *f*
~ansammlung *f* [*z. B. auf dem Planum*] = lodg(e)ment of water [*e.g. on the subgrade*]
~ansammlung *f* [*Bleibt bei Ebbe auf dem Strand zurück*] = tidal pool
~anstieg *m* = rise of water
wasserarmes Jahr *n*, Trockenjahr = dry year
Wasser|aufbereitung *f*, Aufbereitung von Wasser = water purification, ~ treatment
~aufbereitungsanlage *f*, Aufbereitungsanlage für Wasser = water purification plant, ~ treatment ~
~aufnahme *f* = absorption [*e.g. of soils or building bricks*]
~aufnahmevermögen *n*, Wasseraufnahmefähigkeit *f* = absorptive capacity for water
~auftrieb *m* = water buoyancy
~aufwand *m* = expenditure of water
~ausscheidungsschnecke *f* → Sandentwässerungsschnecke
~automat *m* [*Betonmischer*] = auto(-matic) water ga(u)ge
~bad *n* = water bath
~bad-Druckprüfverfahren *n*, Wasserlagerungs-Druckfestigkeitsprüfung *f* = immersion-compression test
~badkühlung *f* = water bath cooling
~ballastpumpe *f* = ballast pump
~ballast-Walz(en)rad *n* = water ballast type roll

Wasserbau m = civil engineering hydraulics
~**geräte** npl = civil engineering hydraulics equipment
~**ingenieur** m = hydraulics engineer
~**klinker** m = clinker brick for hydraulics engineering
~**labor(atorium)** n = hydraulics lab(oratory)
wasserbauliche Forschung f, Wasserbauforschung = hydraulics research
~ **Versuchsanstalt** f = hydraulics research station
wasserbaulicher Versuch m = hydraulics test
wasserbauliches Modell n = hydraulics model
~ ~ **mit beweglicher Sohle** = movable-bed scale model, loose-boundary ~
~ **Versuchswesen** n = hydraulics research
Wasserbau|projekt n = hydraulic engineering project
~**werk** n = hydraulic engineerings structure
~**wesen** n = water engineering practice, hydraulics ~ ~
Wasser|becken n = water reservoir, ~ basin
~**bedarf** m = water demand [*A schedule of the water requirements for a particular purpose, as for irrigation, power, municipal supply, plant transpiration, storage, etc.*]
wasserbedientes Kalorimeter n, Wasserkalorimeter = calorimeter with water connections
Wasserbedürfnis n = water requirement [*The total quantity of water, regardless of its source, required by crops for their normal growth under field conditions*]
~ **einer Pflanze** = plant requirement
Wasser|behälter m = (storage) cistern [*A container for water having a free water surface at atmospheric pressure*]
~**behälter** m [*auf einem Betonmischer*] = water tank

~**beigabe** f, Wasserzusatz m, Wasserzugabe = addition of water
(~)**Beries(e)lungsanlage** f [*Walze*] = sprinkler system
wasserbeständig, wasserfest = water-resistant, waterproof
Wasser|beständigkeit f = water-resistance
~**bevorratung** f = storage of water, water storage
~**bezieher** m = water user
~**brunnen** m = water well
~**damm** m (🗙) = water dam
(~)**Dampf** m = steam
(**wasser)dampfdurchlässig** = permeable to steam
(**Wasser)Dampf|durchlässigkeit** f = permeability to steam
~**gehalt** m = steam content
(**wasser)dampfgesättigt** = steam saturated
Wasser|dargebot n = minimum reliable yield, safe ~
~**desinfektion** f = water sterilization, ~ disinfection
wasserdicht, wasserundurchlässig = watertight
Wasser|dichtigkeit f, Wasserdichtheit, Wasserdichte f, Wasserundurchlässigkeit = watertightness, imperviousness to water, impermeability to water
~**dichtung** f, Wasserabdichtung = water sealing
~**dosiergerät** n, Wasserdosator m = batch water meter
~**dosierung** f, Wasserzuteilung, Wasserzumessung, Wasserabmessung, [*Betonmischer*] = water batching, ~ ga(u)ging, ~ measuring, ~ proportioning
Wasserdruck m = water pressure, ~ load
~**diagramm** n, resultierendes ~ = diagram of normal unit pressures
~**förderung** f [*Erdölförderung*] = hydrostatic drive
~**probe** f, kalte Druckprobe = water test, hydraulic ~
Wasserdurchgang m = passage of water

wasserdurchlässig — wassergebundene Schotterdecke 1118

wasserdurchlässig = permeable to water
Wasserdurch|lässigkeit *f* = permeability to water
~lässigkeitsprüfung *f*, **Wasserdurchlässigkeitsprobe** *f*, **Wasserdurchlässigkeitsversuch** *m* = water permeability test
~leitungsrecht *n* = water-piping right
Wasser|egel *m* → Heberüberfall *m*
~eigenschaft *f* = water property
~einbruch *m* → Wasserzutritt *m*
~einbruch *m* **mit Gesteinsschutt** = debacle [*An inrush of debris-laden waters*]
~einlaufstollen *m* = water inlet gallery
(~)Einstau *m* → → Talsperre
~einwirkung *f* = action of water
~enthärtung *f* = water softening
~enthärtungsmittel *n* = water softener
~entziehung *f*, De-Hydrierung = dehydration
~entzug *m*, Dränage *f*, Entwässerung *f* [*Der Vorgang der im Boden zum Endzustand kapillaren Gleichgewichtes führt*] = drainage
~entzug *m* **durch Austrocknen** = drainage by desiccation
~entzugsbrunnen *m*, Wasserabsenkungsbrunnen *m* = drainage well
~erosion *f*, Schurf *m* (des Wassers) = erosion by water action, geological erosion by stream ~
~fahrzeug *n* = floating craft
~fall *m* = waterfall
~farbe *f* = water paint, (washable) distemper (~) [*A paint having water as a vehicle*]
~farbe *f*, Aquarellfarbe = water colo(u)r
~fassung *f* = water intake
~fassung *f* **eines Stollens** = capacity of a gallery
~faß *n* = water barrel
~feinkalk *m* = hydrated lime [*Is sold in powder form*]

wasserfest, wasserbeständig = water-resistant, waterproof
Wasser|festigkeit *f* = resistance to water
~film *m* → Wasserhäutchen
~filtrationsmethode *f* [*Eine Klassifizierungsmethode für Asbest mit dem Apparat nach Bauer-McNett*] = water elutriation
~findepapier *n* = water finder
~fitting *m* = water fitting [*Water fitting includes pipes (other than mains), taps, cocks, valves, ferrules, meters, cisterns, baths, water-closets, soil pans and other similar apparatus used in connection with the supply and use of water. Definition from British Standard Code of Practice 310:1965*]
~fläche *f* [*Staubecken*] = waterspread
~fläche *f* = water area
~flutverfahren *n*, Wasserflutung *f* [*Ölausbeute*] = water flood method [*oil recovery*]
~förderung *f* = transport of water
wasser|frei → anhydrisch
~freier Gips *m* = anhydrous gypsum plaster [*Entspricht nach Verwendung dem Putzgips, nach Herstellung dem Estrichgips*]
~freier Gipsstein *m*, Anhydrit *m*, $CaSO_4$ = anhydrite
~führend = water-bearing
~führende (Boden)Schicht *f*, ~ Formation *f*, Grundwasserleiter *m*, Wasserträger *m* = water-bearing (soil) stratum, aquifer, water-bearing formation, aquafer
~führendes Gebirge *n* = waterlogged ground
Wasser|führung *f* = regime
~gas *n* = water gas
wassergebundene Schotterdecke *f*, wassergebundener Makadam *m* = water-bound macadam [*A road material consisting of crushed stone or gravel in which an appropriate quantity of clay and water forms the binder*]

Wassergehalt — Wasserkraftbau

Wassergehalt *m*, Feuchtigkeitsgehalt [*Boden*] = moisture content, water ~

wasser|gekühlt = water-cooled
~gekühlter Kondensator *m* = water-cooled condenser
Wassergeld *n* = water charge
wasser|gesättigt, wasserdurchsetzt = waterlogged
~gesättigter Schluff *m* = saturated silt
~geschmiertes Lager *n* = water-lubricated fluid-film bearing
~getränkt = water-impregnated
Wasser|gewinnung *f* = water winning
~gewinnungsbohren *n* = drilling for water supply
Wasserglas *n* = water glass, soluble ~ [*A glassy or stony substance consisting of silicates of sodium or potassium, or both, soluble in water, forming a viscous liquid*]
~, Kaliumsilikat *n*, K_2SiO_3 = potassium silicate
~, Natriumsilikat *n*, Na_2SiO_3 = sodium silicate
~anstrich *m* = water glass coat, soluble ~ ~
~belag *m* → Silikatdecke
~decke *f* → Silikatdecke
wasserglasgebundener Sand *m*, natriumsilikatgebundener ~ = sodium silicate bonded sand
Wasserglas|kitt *m* = water glass putty, soluble ~ ~
~makadam *m* → Silikatdecke *f*
~-Straße *f* → Silikatdecke *f*
wassergranulierter Schlackensand *m* = water-spray granulated slag sand
wasserhaltiges borsaures Natrium *n*, $Na_2B_7O_4 \times 10H_2O$ = hydrated sodium borate
~ Magnesia-Tonerde-Silikat *n* = hydrous iron-magnesium alumin(i)um silicate
~ Mg-Silikat *n* → Magnesiumhydrosilikat
Wasser|haltung *f* [*Baugrube*] = predraining, predrainage, dewatering, unwatering
~haltungspumpe *f*, Sumpfpumpe = sump pump
~hauptrohr *n* → Hauptrohr
~haushalt *m* = water regime
~haushaltbilanz *f* = water balance
~haushaltpflege *f* = water conservation
~häutchen *n*, Wasserfilm *m* = lubricating film of water
~hebemaschine *f* = water-raising machine, water-lifting ~
~hebeschnecke *f* = Archimedes-screw water lift
~hebung *f* = raising of water, lifting ~ ~
~heizer *m* = water heater
wasserhelles Hydrier-Leuchtöl *n* = water white high-grade burning oil
Wasser|hochbehälter *m* → Hochbehälter
~höhe *f*, Wasserspiegellage *f* = water level, elevation of water surface
wasserhydraulische Beton(förder)-pumpe *f*, ~ Förderpumpe = water-hydraulic concrete pump
wässerig, wäßrig = aqueous
Wasser|-ın-Öl-Emulsion *f*, umgekehrte Emulsion = water-in-oil emulsion
~isolierschicht *f*, Wassersperre *f* = water seal, ~ barrier
~kalk *m* = hydraulic lime
~kanal *m* **zur Erzwäsche** = boom (US)
~kasten *m* [*beim Kühler*] = radiator tank
~kegelbildung *f* **in Öllagerstätten** = water coning in oil reservoirs
~klemme *f*, Wassermangel *m* = deficiency of water, water shortage
~klosett *n* → Spülklosett
~körper *m* = water body
Wasserkraft *f*, „weiße Kohle" *f*, Binnenwasserkraft = hydro power, water ~
~anlage *f* = hydro plant
~anlage *f* **mit großer Fallhöhe** = high-head hydro plant
~-Ausbauplan *m*, Wasserkraftbauprogramm *n* = hydroelectric scheme
~bau *m* = hydroelectric construction

wasserkrafterzeugter Strom *m* = hydroelectric energy
Wasserkraft|erzeugung *f* = hydroelectric power generation
~ingenieur *m* = hydroelectric engineer
~kanal *m* = power canal
~maschine *f* = hydraulic engine
~nutzung *f* = water-power development
~projekt *n* = water power scheme, ~~development, hydroelectric ~
~reserve *f* = hydropower resource
~stollen *m* → Triebwasserstollen
~system *n* = hydroelectric scheme
~-Talsperre *f* = hydroelectric dam
~technik *f* = water power engineering
~werk *n* = hydro(electric) power station, ~ ~ plant, ~ generating ~
~werk *n* **mit niedriger Fallhöhe** = low-head installation, low-fall ~
Wasser|kraut *n* = water weed
~kreislauf *m* = continual cycle of water in its round from atmosphere to earth and return by evaporation from land and water surface
~kultur *f*, **Hydroponik** *f* = hydroponics [*The science of growing plants without soil in water and chemicals*]
~lagerung *f* [*Probekörper*] = water immersion
~lagerungsprüfung *f*, Wasserlagerungsprobe *f*, Wasserlagerungsversuch *m* [*Zur Bestimmung der Haftung des Bindemittels am Gestein*] = stripping test in the presence of water
~last *f* = water load
~lauf *m* = water course
~leitung *f* = water line
~leitungsbaugeräte *npl* = water line construction equipment
~leitungsbrücke *f* = aqueduct, water-carrying bridge; water aqueduct [*Australia*]
~leitungsnetz *n* = water line network
~(leitungs)stollen *m*, Wasserversorgungsstollen = tunneled aqueduct, aqueduct gallery

wasserliebend = hydrophibic
Wasser|linie *f* [*Schiff*] = water-line, water-level
~linienfläche *f*, Wasserriß *m* = plane surface circumscribed by the water line
~loch *n* = water hole
wasserlösliches Alkali *n* = water-soluble alkali
~ Öl *n* = emulsifiable oil
~ Salz *n* = water-soluble salt
Wasser|löslichkeit *f* = water solubility
~-Luft-Gemisch *n* = water-air mix(ture)
~mangel *m*, Wasserklemme *f* = deficiency of water, water shortage
~mantel *m* [*Motor mit Wasserkühlung*] = water jacket
~mantelverfahren *n* = water-jacketing
~masse *f* = mass of water, body ~ ~
wassermeidend = hydrophobic
Wassermenge *f* → Abfluß *m*
~ = quantity of water
~ die den Flutraum eines Hafens füllt = tidal prism
Wassermengen|ganglinie *f*, Abflußganglinie = flow hydrograph
~kurve *f* → Abfluß(mengen)kurve
~linie *f* → Abfluß(mengen)kurve *f*
~messung *f* = measurement of water quantities
Wasser|messer *m* [*mißt in der Zeiteinheit die durchfließende Wassermenge (z. B. m3/h oder l/s)*] = water flow meter
~messung *f*, Abfluß(mengen)messung, Durchfluß(mengen)messung [*Fluß; Strom; Kanal*] = flow measurement, discharge ~
(~)Meßflügel *m*, hydrometrischer Flügel [*Anzeigegerät für Anströmgeschwindigkeit*] = screw current meter
~meßwesen *n*, Hydrometrie *f* = hydrometry [*The measurement and analysis of the flow of water*]
~mörtel *m*, hydraulischer Mörtel = hydraulic mortar
~mörtel *m* **aus hydraulischem Kalk** = hydraulic lime mortar

~nase *f*, Unterschneidung *f* = (water) drip, throat, weather groove

~nest *n* = water pocket [*On the surface of concrete*]

~nutzung *f* = use of water

~nutzungsrecht *n* = water right [*A legal right to the use of water*]

~oberfläche *f*, Wasserspiegel *m* = water level [*A water surface; also its elevation above any datum; gage height; stage*]

~pfeiler *m* = water pier

~polizei *f* → Strompolizei

~polizeistation *f*, Strompolizeistation = river police station

~polster *n* [*In Österreich: m*] = water cushion, cushioning pool [*A pool of water maintained to take the impact of water overflowing a dam, chute, drop, or other spillway structure*]

~pore *f* [*Beton*] = water void

~probe *f* = water sample

~probenehmer *m* = water sampler

~pumpe *f* = water pump

~pumpendichtung *f* = water pump seal

~rad *n* = water-wheel [*According to English usage, a water-wheel is a small slow-speed machine, but in America a water-wheel implies any kind of rotary machine, up to the largest size of turbine*]

~reinhaltung *f* = water pollution prevention

Wasserreinigung *f*, Wasseraufbereitung	water treatment, ~ conditioning, ~ purification
Absetzerklärung *f* nach der Behandlung mit Flockungsmitteln	sedimentation [*the process of clarifying water by settling after coagulation by chemical treatment*]
Absetzklärung ohne Flockungsmittel	plain sedimentation, ~ subsidence
Schwebestoffe *mpl*	suspended matter, ~ solids
Absetzbecken *n* für kontinuierlichen Betrieb	continuous settling basin, ~ subsidence tank
Absetzbecken für aussetzenden Betrieb	intermittent settling basin, ~ subsidence tank
Aufenthaltsdauer *f*, Durchflußdauer	detention period, retention ~
Trübe *f*	turbidity
Schlamm *m*	sludge
Algenbekämpfung *f*	algae control, removal of algae
Algenbekämpfungsmittel *n*	algaecide
Belüftung *f*	aeration
Beseitigung *f* übler Geruchs- und Geschmackstoffe	taste and odour removal, ~ ~ ~ control
Brackwasseraufbereitung *f*	brackish water conversion
Enteisenung *f*	iron removal
Entgasung *f*	deaeration, degassing, degasification
thermischer Entgaser *m*	deaerator, deaerating heater
Rieseler *m*	tray-type degasifier, spray-type ~

Wasserreinigung

Enthärtung *f* — softening
  Säureimpfung *f* — acid treatment, ~ injection
  Kalk-Soda-Verfahren *n* — lime soda (softening) process
  Kalkenthärtung *f* — lime softening process
  Zeolithe *npl* — zeolites
  Basenaustausch *m* — base exchange, zeolite process
  Kesselspeisewasser *n* — boiler feedwater
Entkeimung *f* — desinfection, sterilization
  Abkochen *n* — heating
  Chlorung *f* — chlorination
  direkte Chlorung *f* — gas chlorination
  indirekte Chlorung *f* — solution feed chlorination
  Hochchlorung *f* — superchlorination
  Knickpunktchlorung *f* — break-point chlorination
  starker Chlorüberschuß *m* — excessive chlorine residual
  Entchlorung *f* — dechlorination
  Aktivkohle *f* — active carbon
  Oxydationsmittel *n* — oxidizing agent
  Chloramin *n* — chloramine
  Ultraviolett-Bestrahlung *f* — UV-ray treatment
Entmanganung *f* — mangenese removal
  Manganbakterien *fpl* — manganese bacteria
Entsalzen *n* — neutralization, pH-control
Filterung *f*, Filtration *f* — filtration
  Langsamfilterung, biologische Filterung — slow sand filtration
  Langsamfilter *n, m* — slow sand filter
  Schnellfilterung, mechanische Filterung — rapid sand filtration
  Schnellfilter *n, m* — rapid sand filter
  geschlossenes Schnellfilter *n*, Druckfilter — pressure filter
  Filterspülung *f* — filter backwash
Flockung *f*, Fällung *f* — flocculation, precipitation, coagulation
  pH-Wert *m* — pH-value
  Flockungsmittel *n*, Fällmittel — coagulant, flocculant
    Aluminiumsulfat *n* — (filter) alum, aluminium sulphate
    Eisen-II-Sulfat *n* — ferrous sulphate
    Eisen-III-Sulfat *n* — ferric sulphate
    Eisenchlorid *n* — ferric chloride
    Natriumaluminat *n* — sodium aluminate
  Trockendosierung *f* — dry chemical feeding
  Flüssigdosierung *f* — solution feeding, liquid chemical
  Vermischung *f* — mixing
    Wassersprung *m* — hydraulic jump
  Flockenbildung *f* — floc formation, flocculation
Kaliumpermanganat *n* — potassium permanganate

Kleinlebewesen *n*	microorganism
Meereswasseraufbereitung *f*	sea water conversion
Reinwasserbehälter *m*	clear water basin, ~ ~ tank, clear well
Speicherung *f*	storage
Verbrauchschwankung *f*	fluctuation in demand
Vollentsalzung *f*	demineralization, deionization
Ionenaustausch *m*	ion exchange
Austauschharz *n*	exchange resin
Kationenaustausch *m*	cation exchange
Anionenaustausch *m*	anion exchange
Mischbettaustauschfilter *n, m*	mixed bed exchange unit
Entkieselung *f*	silica removal

Wasser|ring *m* = water ring [*gunite equipment*]

~**riß** *m*, Wasserlinienfläche *f* = plane surface circumscribed by the water line

~**rohr** *n*, Wasserleitungsrohr = water pipe

~**rohrkessel** *m* = water-tube boiler

~**rösche** *f* → Wasserseige

~**rückhaltevermögen** *n* = water retention

~**sättigungsbeiwert** *m* = water saturation coefficient

~**säule** *f* = water column

~**scheide** *f* = water parting, watershed [*The divide between drainage basins*]

~**schenkel** *m*, Wetterschenkel = water bar

~**schieber** *m* = water valve

~**schild** *m* → Wasserabschirmung *f*

~**schlag** *m*, Wasserschräge *f* = slope of a cornice

~**schlag** *m* → Wasserstoß

~**schloß** *n* → → Talsperre *f*

~**schmierung** *f* = water lubrication

~**schöpfwindmühle** *f* = water pump wind mill

~**schräge** *f*, Wasserschlag *m* = slope of a cornice

~**seige** *f*, Wasserrösche *f* (⚒) [*Eine in der Streckensohle ausgesparte Rinne zur Zuleitung der Grubenwasser zur Wasserhaltung*] = drainage channel

~**seite** *f* [*einer Stauanlage*] = upstream side

wasserseitige Böschung *f* = water-side slope

~ **Sohle** *f* = upstream bottom, ~ floor

Wasser|speicher *m* = water reservoir

~**speicherung** *f* = water storage

~**speier** *m* = gargoyle [*A projecting stone spout, usually carved with a grotesque figure*]

~**spende** *f* → Abflußspende

~**spender** *m* = source of water supply

~**sperre** *f*, Wasserisolierschicht *f* = water seal, ~ barrier

wassersperrende Rohrfahrt *f*, ~ Rohrtour *f* [*Tiefbohrtechnik*] = water string

Wasserspiegel *m*, Wasseroberfläche *f* = water level [*A water surface; also its elevation above any datum; gage height; stage*]

~ → Grund~

~**absenkung** *f* = drawdown

~**breite** *f* = water level width

~**gefälle** *n* [*Lotrechter Höhenunterschied zwischen zwei Punkten eines Wasserspiegels, bezogen auf deren horizontale Projektion oder deren Abstand*] = difference in water levels

~**hebung** *f* → (An)Stauung

~**höhe** *f* = water level (elevation)

~**lage** *f*, Wasserhöhe *f* = elevation of water surface, water level

Wasserspiegellinie — Wasserumlauf

~linie f = water line
~linie f, Spiegelkurve f = surface curve, ~ profile
~-Meßanlage f für Tiefbrunnen = deep well water level measuring equipment
Wasser|spiel(raum) n, (m) = stage fluctuation range
~sprengwagen m, Wasserspritzwagen = water sprinkler car, ~ ~ wagon
~spritzkühlung f = water spray cooling
~sprühverfahren n = water fog spray method
~sprung m, Wechselsprung = hydraulic jump [*The sudden and usually turbulent passage of water from low stage below critical depth to high stage above critical depth during which the velocity passes from supercritical to subcritical. It represents the limiting condition of the surface curve wherein it tends to become perpendicular to the stream bed*]
~sprung m mit Deckwalze = submerged hydraulic jump
(~)Spülen n, (Wasser)Spülung f [*für Reinigungszwecke*] = flushing
~spülkasten m = flushing cistern
~spülung f [*Toilette*] = water flushing
~spülung f [*Gesteinsbohren*] = water injection, injection of water
Wasserstand m = water level
~achse f [*einer Tidekurve*] = axis of heights, ~ ~ water levels
~anzeiger m → Pegel m
~glas n = water ga(u)ge glass, graduated glass ga(u)ge-tube
~messung f = water level measurement
~-Prüfhahn m → Probierhahn
Wasserstoff|gehalt m [*g/100 g. DIN 51721*] = hydrogen content
~-Ionen-Konzentration f [*g/l. DIN 4049*] = hydrogen-ion concentration
~krankheit f des Kupfers = hydrogen-embrittlement of copper
~superoxyd n plus Chlorkalk m [*Gasbildner beim Gasbeton*] = hydrogen peroxide and calcium chloride
~zahl H f → pH-Wert m
Wasserstollen m = water gallery
~ → Wasserversorgungsstollen
Wasserstoß m, Wasserschlag m, Druckstoß = water hammer [*Any sudden very high pressure in a pipe caused by stopping the flow too rapidly*]
Wasserstrahl m = water jet
~ [*(Stau)Wehr*] = nappe
~-Kondensator m = jet condenser
~pumpe f = water-operated vacuum pump
~regler m [*Hahn*] = built-in antisplash [*tap*]
~verfahren n → Einspülverfahren
Wasser|straße f = waterway, water road; barge line (US)
~straßenbauamt n = river catchment board
~straßennetz n = waterway system, ~ network
~strichziegel(stein) m = water-struck brick
~strom m [*Wasserturbine*] = (water) flow
~tarif m = water rate
wassertemperaturabhängige Abstellvorrichtung f = water temperature shut-off
Wasser|temperaturregler m, Kühlthermostat m = water temperature regulator
~tiefe f des Fahrwassers, nutzbare Fahrtiefe = depth of the navigable channel
~träger m → wasserführende (Boden-) Schicht
~transport m = water transportation
~treibverfahren n, Fluten n [*Erdölförderung*] = water flooding, ~ drive
~turbine f = water turbine, hydroturbine, hydraulic turbine
~turm m → Hochbehälter m
~überschuß m = surplus of water, excess ~ ~
~umlauf m = circulation of water

~umlenkung *f* → → Talsperre
~- und Energie-Entwicklungsbehörde *f* = water and power development authority
wasserundurchlässig, wasserdicht = watertight
Wasserundurchlässigkeit *f* → Wasserdichtigkeit
wasserunlöslich = insoluble in water
Wasser|untersuchung *f* = water examination
~ventil *n* [*Betonmischer*] = water valve
~verbrauch *m* = water consumption
~verbrauch *m* einer **Pflanze** = plant consumption [*The water used by a plant in the process of growth. It includes that stored in the body of the plant and that dissipated from its leaf and body surfaces by transpiration*]
~verbraucher *m* = water consumer
wasserverdünnbar = water-thinnable
Wasser|verlust *m* = water loss
~vernebelung *f* = atomized water spray
~verschluß *m* = air trap [*A water-sealed trap which prevents foul air or foul odours from rising from sinks, wash basins, drain pipes, and sewers*]
~verschmutzung *f* = water pollution
~versorgung *f* = water supply, provision of water
Wasserversorgungs|anlage *f*, Wasserversorgungsbauwerk *n* = water supply plant, ~ ~ installation
~gebiet *n* = water supply area
~ingenieur *m* = water supply engineer
~leitung *f* = water supply line
~prahm *m*, Wasser(versorgungs)boot *n* = water boat [*A small, self-propelled tank boat used in harbours for supplying fresh water to seagoing vessels*]
~recht *n* = water supply law
~stollen *m*, Wasser(leitungs)stollen = tunneled aqueduct, aqueduct gallery
~system *n*, Wasserversorgungsanlage *f* = water supply scheme

~technik *f* = water supply engineering
~unternehmen *n* = water undertaker, ~ undertaking
Wasser|versprühung *f*, Wasservernebelung = atomization of water, fog spraying ~ ~
~verteilung *f* = water distribution
~vorkommen *n* = body of water
~vorräte *mpl* = water resources
~waage *f* = (mechanic's) level, spirit ~, water ~
~wagen *m* = water car, ~ wagon
Wässerwasser *n* → Bewässerungswasser
~bedarf *m* → Bewässerungswasserbedarf
~kanal *m*, Bewässerungskanal = irrigation canal
Wässerwehr *n* → Bewässerungswehr
Wasserwelle *f* = water wave
Wasserwerk *n* = water (treatment) works
~bau *m* = waterworks construction
~entwurf *m* = waterworks design
~ingenieur *m* = waterworks engineer
~planung *f* = waterworks planning
Wasserwirtschaft *f* = watershed management
wasserwirtschaftliche **Großplanung** *f* = integrated river basin development
Wasserwirtschafts|behörde *f* = water authority
~politik *f* = water resources policy
~verband *m* = water board
Wasser|zähler *m* [*Zählt die in beliebiger Zeit durchgeflossenen Mengeneinheiten (z. B. m³ oder l)*] = integrating water flow meter
~zapfbrunnen *m*, Hydrantbrunnen = hydrant well (Brit.); combination hydrant and fountain (US)
~-Zement-Faktor *m*, Wasser-Zementwert *m*, Wasser-Zement-Verhältnis *n*, WZV, W/Z-Faktor [*Wassergewicht geteilt durch Zementgewicht einer Betonmischung*] = water-cement ratio, water/cement ~
~zugabe *f* → Wasserbeigabe

~zumessung f → Wasserdosierung
~zusatz m, Wasserbeigabe f = addition of water
~zuteilung f → Wasserdosierung
~zutritt m, Wasserandrang m, Wassereinbruch m = inflow of water, ingress ~ ~, inrush ~ ~
wäßrig, wässerig = aqueous
Watbagger m = amphibious (mechanical) shovel
Watson'sches Vierpunkt-Verfahren n = Watson four electrode system method
Watt n = tidal mud flat
~verbrauch m = wattage consumption
W.C.-Becken n = w.c. pan
Webebreite f [Siebtechnik] = width of screen cloth
Weber'sche Hohlräume mpl = bed separation cavities, Weber's ~
Webkante f [Sieb] = selvedge, selvage
Wechsel m, Überschiebung f (Geol.) = overthrust fault
~ → Wechselholz n
~(balken) m → Wechselholz n
~beanspruchung f = alternating stress
~behälter m, Absetzbehälter = demountable tank, detachable ~
~behälter m → Absetz(kipp)behälter
~beziehung f = correlation
~feldeinbau m, Herstellung f benachbarter Plattenfelder in zeitlichem Abstand [Deckenbeton] = alternate (concrete) bay construction, ~ bay method
~getriebe n = sliding gear transmission
~holz n, Wechsel(balken)m = header (joist) (US); trimmer (~) (Brit.) [A beam which carries the ends of beams which are cut off in framing around an opening]
~kippbehälter m → Absetzkippbehälter
~(kipp)mulde f → Absetzkippbehälter m
~kübel m → Absetz(kipp)behälter
wechsellagern, abwechseln (Geol.) = to alternate

Wechsel|lagerung f (Geol.) = alternating sequence
~mulde f → Absetzkippbehälter m
Wechseln n der Gänge, Gangwechsel m = gear changing
wechselnde Belastung f = alternate loading
Wechsel|schieber m, Umsteuer(ungs)schieber = control(ling) valve
~schuh m, Plattenschuh, Rüttelschuh [Zur Vergrößerung der Arbeitsfläche einer Rüttelplatte] = vibrating (base) plate extension, ~ (~) pan ~, ~ (~) slab ~, vibration (~) ~ ~, vibratory (~) ~ ~
~spannung f = alternating stress
~sparren m = valley jack rafter
~sprechanlage f, Sprechanlage mit Hörmöglichkeiten [z.B. beim Bau von Hochhäusern eingesetzt] = communications system to provide two-way conversation facilities, intercom(munication system)
~sprung m → Wassersprung
~spurverkehr m, Flut(straßen)verkehr = tidal traffic
~stab m, Gegenschräge f, Gegendiagonale f = counter diagonal
~streifen m, Wechselverkehrsstreifen = reversible (traffic) lane
Wechselstrom|anlage f [einer Maschine] = A.C. equipment, alternating current ~
~(-Aufschluß)verfahren n = alternating current method, ~ ~ prospecting, ~ ~ exploration
~erzeuger m = alternating current generator A.C. ~
~-Rohrschieber m [Druckluftwerkzeug] = alternating flow hollow spool valve
~verfahren n, Wechselstrom-Aufschlußverfahren = alternating current method, ~ ~ exploration, ~ prospecting
Wechsel|tauchversuch m = alternate immersion test
~(verkehrs)streifen m = reversible (traffic) lane
~wanne f → Absetz(kipp)behälter

~wirkungsformel *f* = interaction formula
Wegbenutzungsrecht *n* → Wegerecht
Wegebake *f* **aus Gummi**, Verkehrsleitkegel *m* = (rubber) traffic cone
Wegebau *m* = construction of unclassified roads
~zug *m* = train of equipment for farm road construction
Wegegeld *n*, **Wegezoll** *m*, **Wegegebühr** *f* [*Gebühr für Benutzung einer Straße, einer Brücke oder eines Tunnels*] = toll
~autobahn *f*, Gebührenautobahn, gebührenpflichtige Autobahn, Zollstraße *f* für Schnellverkehr = toll road, ~ (super)highway, (~) (turn)pike
~brücke = toll bridge
Wege|hobel *m* → Straßenhobel
~meister *m* → Straßenmeister
~recht *n*, Wegbenutzungsrecht = right-of-way
~recht *n* (oder **Wegbenutzungsrecht**) **mit Erlaubnis zum Legen von Rohr- und Kabelleitungen** = wayleave
~zoll *m* → Wegegeld
wegnehmen, entfernen, abnehmen = to remove, to detach, to withdraw
Wegräumen *n* = clearing away
wegspülen, unterwaschen, (aus)kolken, auswaschen, unterspülen = to scour, to underwash, to undermine
Wegspülung *f* → Auskolkung
Wegweiser *m* = direction post, supplementary direction sign, guide ~ [*deprecated:* direction sign, sign-post, finger post]
Wegwerf-Filterelement *n* = throwaway filter element
Wegzeit|diagramm *n* (✗) = yield-time diagram
~kurve *f*, Laufzeitkurve = travel time curve
Wehe *f*, Schnee~ = drift, snow ~
Wehr *n*, Stau~; **Wuhr** *n* [*Schweiz*] = weir
~bock *m* [*Nadel(stau)wehr*] = frame
~formel *f*, Meß~ = weir formula, measuring ~ ~

~kammer *f* [*Dachwehr*] = chamber
~krone *f* = weir crest
~pfeiler *m* = weir pier
~rücken *m*, Abfallmauer *f*, Abfallwand *f* [*Die vom überfallenden Wasser mit oder ohne Berührung verdeckte Luftseite einer Stauanlage*] = downstream face of a weir
~schwelle *f* = weir sill, ~ cill
~sohle *f* = weir floor, ~ bottom
(~)Verschluß *m*, Staukörper *m* [*Beweglicher Teil eines Wehres zur Regelung des Oberwasserstandes*] = gate
~wange *f* [*Landpfeiler eines Wehres, der den wasserdichten Anschluß an das Ufer herstellt*] = abutment, sidewall
weich [*Gestein*] = soft
~ [*Stempel*] (✗) = late-bearing
~ **machen** [*Beton*] = to butter
weiche Kohle *f* → backende ~
weicher Kalkstein *m* = soft limestone
Weich|beton *m* → → Beton
~bitumen *n* = soft bitumen
~brandziegel *m*, Schwachbrandziegel = soft (burnt) brick
Weiche *f* = point, switch
Weicheisendraht *m*, Reineisendraht = soft iron wire
weichen, durch~, auf~ = to soften
Weichenheizer *m* = switch heater
Weich|gestein *n* = soft rock
~lot *n* = soft solder
~löten *n*, Weichlötung *f* = (soft) soldering
~macher *m* [*Betonwirkstoff*] = wetting agent, plasticizer
~pech *n* = soft pitch
weichplastisch = soft-plastic
Weich|preßziegel *m*, Weichschlammziegel = soft-mud brick
~putz *m* = soft plaster
~putzmörtel *m* = soft stuff
~ton *m* = soft clay
Weiden|faschine *f* = willow fascine, ~ faggot
~geflecht *n* [*Flußuferbefestigung*] = hurdle work, wattle ~
Weinhold-Gefäß *n* → Dewargefäß

Weißbetonmarkierungen *fpl* = white concrete carriageway markings

Weißblech *n* [*Stahlblech mit Zinnüberzug*] = tinplate

~band *n* = tinplate strip

~stift *m* = tinplate brad

~walzwerk *n* = tin plate mill

Weiße *f*, Kalkmilch *f*, Tünche *f* = whitening, limewash, whitewash

„**weiße Kohle**" *f* → Wasserkraft *f*

weiße Mineralfarbe *f* = white pigment

weißen → kalken

weißer Bol(us) *m*, weiße Boluserde *f* = potters' clay, ball ~

~ Horizont *m* = white-out

~ Mauerstein *m* → Kalksandstein

~ Portlandzement *m* = white Portland cement

~ Rost *m* = white rust

weißes Petroleum *n* → Petroleum

Weiß|gußeisen *n* = white cast iron

~kalk *m* [*früher: Fettkalk*] = high-calcium lime, white ~, rich ~, fat ~, pure, ~, plastic high-calcium hydrated ~, non-hydraulic ~, white chalk

~moostorf *m* = white moss peat

~ofen *m* = refining furnace

~pigment *n* = white pigment

weißrandiger Reifen *m*, Weißwand-Reifen = white-wall tire (US)/tyre (Brit.)

Weißzement *m* = white (Portland) cement

weite Klassierung *f* [*Siebtechnik*] = wide sizing, screening into wide size ranges

weiten, auf~, aufbördeln [*Rohr*] = to expand

Weiterentwick(e)lung *f* [*Maschine*] = improved design, advanced ~

weitervergeben [*Bauauftrag*] = to sub(let), to subcontract, to sub out to

weit|gespannt = wide span

~maschig = coarse-mesh

Weitstrahlregner *m*, Drehstrahlregner = rotating-head sprinkler, circular-spray ~

Weitungen *fpl* im Alten Mann (⚒) = gaps in the waste (area); ~ ~ ~ goaf, ~ ~ ~ gob [*Wales*]; ~ ~ ~ condie, ~ ~ ~ cundy [*Scotland*]

Weitwinkel|beleuchtung *f* = high-angle (beam) lighting

~(-Straßen)leuchte *f* = high-angle (beam) lantern (Brit.); ~ (~) luminaire (US); ~ (~) fixture

Well|alu(minium) *n* = corrugated alumin(i)um

~asbest *m* = corrugated asbestos

~asbestzementplatte *f* = corrugated asbestos cement sheet

~bahn *f* [*Für Terrassen-Überdachungen, Wind-, Wetter-, Licht- und Sonnenschutz*] = corrugated panel, ~ sheet

~baum *m* → Kanalwinde *f*

Wellblech *n* = corrugated iron sheet

~(dach)(ein)deckung *f* = corrugated sheet roof covering

~rohr *n* = corrugated sheet pipe

~walzwerk *n* = corrugated sheet rolling mill

Welle *f* [*Zur Übertragung von Drehmomenten zwischen sich drehenden Maschinenteilen*] = shaft

~ [*Schwertwäsche*] = log [*log washer*]

~ [*Pumpe*] = spindle [*vertical or horizontal*]

~ [*beim Rolladen*] = roller

Wellen|auflauf *m*, Wellenauslauf = wave run-up

~ausbreitung *f*, Wellenfortpflanzung = wave propagation

~becken *n* = wave basin

~bewegung *f* = wave agitation

~bildung *f* [*Erdstraße*] = formation of washboard waves, corrugations, washboarding, corduroy effect [*deprecated: road waves, deformations, creep*]

wellenbrechendes Becken *n*, Beruhigungsbecken [*Hafen*] = stilling basin

Wellenbrecher *m* [*Einer Mole ähnliches und dem gleichen Zweck dienendes Bauwerk ohne Verbindung mit dem Lande*] = breakwater

~ **mit flachen Böschungen** = mound-breakwater

Wellen|breite *f* [*Welle in einer Erdstraße*] = (corrugation) amplitude

(~)Bund *m* = collar, shaft ~

~**dämpfer** *m* = wave suppressor

~**druck** *m* = wave pressure

~**düne** *f* = shore dune [*deposition by sea waves*]

~**endspiel** *n* = shaft end clearance

~**energie** *f* = wave energy

~**erscheinung** *f* = wave phenomenon

~**erzeuger** *m* = wave generator

~**fortpflanzung** *f*, Wellenausbreitung = wave propagation

~**fortpflanzungsrichtung** *f*, Wellenausbreitungsrichtung = direction of wave propagation

~**front** *f* = wave front

~**furche** *f*, Rippel(marke) *f* = ripple marking

~**fuß** *m* = wave foot

~**gitter** *n* → vorgewelltes (Sieb)Gitter

~**höhe** *f* [*Unterschied zwischen Wellenberg und Wellental*] = wave height

~**kalk(stein)** *m*, Schaumkalk(stein) = aragonitic lime(stone)

~**kamm** *m* = wave crest

~**kammlinie** *f* = wave crest line

~**keil** *m*, Keilprofil *n*, Verkeilung *f* = (shaft) spline

~**kennzeichnung** *f* = shafting identification

~**kraft** *f* = wave force

~**krafttheorie** *f* **nach Sainflou** = Sainflou theory (for forces on breakwaters)

~**länge** *f* [*Strecke von einem Wellenberg bis zum nächsten*] = wave length

~**maschine** *f* = wave-machine

~**mischer** *m* = shaft-type mixer

~**muster** *n*, Wellenschema *n* = wave pattern

~**nagel** *m* = corrugated fastener

~**scheitel** *m* = wave crest

~**schema** *n*, Wellenmuster *n* = wave pattern

~**schlag** *m* = wave action

~**schutzhülse** *f* = flinger

~**spektrum** *n* = wave spectrum

~**strömung** *f*, Brandungsstrom *m* = wave current

~**tal** *n* = wave trough

~**tätigkeit** *f* = wave action

~**theorie** *f* = wave theory

~**tiefe** *f* [*Welle in einer Erdstraße*] = (corrugation) pitch

~**träger** *m* = shaft carrier

~**verfärbung** *f* = shaft discolo(u)ring

~**zapfen** *m* = knuckle trunnion

~**zug** *m* = wave train

Well(hüll)rohr *n* [*Dient zur Aufnahme der Spannbetonstähle*] = corrugated (sheet-)metal sheath

Welligkeit *f* = waviness

Wellplatte *f* = corrugated sheet

Wellrohr|bogen *m* = corrugated pipe bend

~**dehnungsausgleicher** *m* = bellows type expansion joint

Wellstahl|blech *n* = corrugated steel plate

~**kappe** *f* (⚒) = corrugated (steel) bar

Wellung *f* [*Siebboden*] = crimp

Wende|becken *n* [*Teil eines Hafenbeckens zum Wenden der ein- oder ausfahrenden Schiffe*] = turning basin

~**fläche** *f* = turning area

~**flügel** *m* = vertical pivoted sash

~**flügelfenster** *n* = vertical pivot hung window

~**getriebe** *n*, Umkehrgetriebe = forward-reverse transmission

~**getriebe-Schalthebel** *m* = (transmission) forward-reverse lever, forward-reverse transmission shift ~

Wendekreis *m*, Drehkreis = turning circle

~**durchmesser** *m* = turning diameter

~**halbmesser** *m* = turning radius

Wendekupp(e)lung *f* → Wendeschaltkupp(e)lung

Wendel *m* = helix

wendelartige Flachstahlrippe *f* [*Beseitigt einwandfrei eine regelmäßige Ablösung von Windwirbeln als Schwingungserreger*] = helical strake

Wendel|förderer m = vibratory spiral elevator, vertical magnetic transporter
~-Parkhaus n = spiral ramp car park
~rampe f = spiral ramp
~rollenbahn f = roller-fitted spiral (gravity) shoot (or chute)
~rutsche f = helical shoot, spiral (gravity) ~
~scheider m, Spirale f = spiral (concentrator)
~schwingrinne f = vibratory spiral elevator, vibrating ~ ~
~stufe f = winder, wheel step
~treppe f **auf Kreisgrundriß** → Spindeltreppe
Wende|platz m [*Schiffahrtstraße*] = turning basin
~säule f [*In Tornische von Kammerschleusen und Docks befindliche Säule, um die sich Stemm- oder Drehtor dreht*]
~schaltgetriebe n, Umkehrschaltgetriebe = forward-reverse primary transmission
~schalthebel m, Gangumkehrhebel = forward-reverse lever
~schaltkupp(e)lung f, Reversierschaltkupp(e)lung, Umkehrschaltkupp(e)lung = reversing clutch (coupling)
Wendeschaufel f, (Verteiler)Schaufel, Schild m [*Wendeschaufelverteiler*] = transverse spreading blade
~verteiler m, Betonverteiler mit hin- und hergehender (Verteiler)Schaufel, Betonverteiler mit Wendeschaufel, Schaufel(beton)verteiler, Schild(beton)verteiler = transverse spreading blade concrete spreader
Wende|sitz m, Drehsitz, Schwenksitz = swivel(ling) seat
~spur f = turning path
Wendigkeit f = manoeuvrability (Brit.); maneuverability (US)
Wendung f = turn
Wenner'sche Vierpunkt-Methode f, Wenner'sches Verfahren n = four electrode system, Wenner method, four-point method

Werft f = (ship)yard, naval yard
~ [*Auf einem Luftfahrtgelände*] = maintenance hangar(s), service ~, tech(nical) ~
~kran m = (ship)yard crane
Werg n = oakum
Werk n, Fabrik f, Betrieb m = factory, mill, works, plant, establishment
~ausrüstung f, Fabrikausrüstung = plant equipment, works ~, factory ~, mill ~
~bahn f = factory railway (Brit.); ~ railroad (US)
~bank f → Werktisch m
~besichtigung f = works visit
~boden m → Betriebsboden
~(fuß)boden m → Betriebsboden
~gebäude n, Fabrikgebäude = factory building
~gefälle n → Nutzgefälle
~gelände n, Fabrikgelände = factory ground
werk|gemischt [*z. B. Transportbeton*] ready-mixed
~geschmiert = lubricated at the factory
Werk|graben m, Werkkanal m [*Laufkraftwerk*] = race
~hauptstraße f = factory main road
~kanal m, Werkgraben m [*Laufkraftwerk*] = race
~kantine f = works canteen
~leimung f = factory bonding
~naht f = mill weld
~platz m, Herstellungsplatz, Betonier(ungs)platz [*für Fertigteile*] = casting yard
~prüfung f = manufactures' test
~sirene f = brummer (US); factory hooter, factory blower
Werkstatt f (work)shop
~anstrich m = (work)shop coat
~ausrüstung f = (work)shop equipment
~fertigung f = (work)shop manufacture
werkstattgenietet = (work)shop-welded
Werkstatt|halle f, Werkhalle, Werkshalle = (work)shop hangar

Werkstattkran — Wetterlampe

~kran *m* = (work)shop crane
~montage *f* = (work)shop assembly
~monteur *m* = (work)shop fitter
~niet *m* = (work)shop rivet
~raum *m* = (work)shop room
~schweißung *f* = (work)shop welding
~stoß *m*, (Stegblech)Querstoß = web plate transverse joint
~stütze *f* = (work)shop column
~wagen *m* = on-the-job service unit, field service ~
~zeichnung *f* = (work)shop drawing [*A drawing of a machine or structure showing all parts and dimensions so that the shop can actually build what is indicated on the drawing without other information*]
Werkstein *m*, Haustein [*Ein vom Steinmetz bearbeiteter natürlicher Stein*] = cut stone, ashlar, ashler
Werksteinmauerwerk *n* → Hausteinmauerwerk
~ mit Bruchsteinhintermauerung, Hausteinmauerwerk ~ ~ = rubble ashler (masonry), ~ ashlar (~) [*An ashlar-faced wall, backed with rubble*]
Werkstein|setzen *n* → Hausteinsetzen
~vorsatz *m* → Hausteinvorsatz
Werkstoff *m*, Material *n* = (engineering) material
~ höchster Warmfestigkeit = high temperature (engineering) material
~bestandteil *m, n* = material component
Werkstoffestigkeit *f*, Materialfestigkeit = material strength
Werkstoff|konstante *f* = material constant
~kosten *f* = materials cost
~mindergüte *f* = material failure
~prüfmaschine *f* = materials testing machine
~technik *f* = materials engineering
Werk|stoß *m* = milled joint [*steel structure*]
~tisch *m*, Arbeitstisch, Werkbank *f*, Arbeitsbank [*z. B. für Schlosser, Schreiner, Dreher usw.*] = bench
Werkzeug|anhänger *m* = tool trailer
~auflast *f* → Bohrdruck *m*

~halter *m* = tool holder
~kasten *m*, Werkzeugkiste *f* = tool box, ~ chest, ~ kit
~macher *m* [*Rotarybohren*] = tool dresser
~satz *m*, Satz Werkzeuge = tool set, set of tools
~schrank *m* = tool cabinet
~schublade *f* = tool drawer
~spindel *f* = mounting spindle
~stahl *m* = tool steel
Wert|analyse *f* [*Eine Methode, die das Ziel hat, die weder zur Güte, zum Nutzwert, zur Lebensdauer, zur äußeren Erscheinung und zu anderen vom Kunden erwünschten Eigenschaften beitragenden Kosten zu senken*] = value engineering
~gegenstand *m* = article of value
wesentlicher Gemengeteil *m* = essential mineral
Westfenster *n* = west-facing window
Wetherill-Scheider *m* = Wetherill separator
Wettbewerbs|entwurf *m* = competitive design
~fähigkeit *f* = competitive position
Wetter *f*, Grubenluft *f* = mine air
~ = weather
~ausziehstrecke *f* → Wetterstrecke (⚒)
~beständigkeit *f*, Wetterfestigkeit = resistance to weather, weathering quality, weather resistance
~dach *n* → Vordach
~dichtigkeit *f* = weather-tightness
wetterecht [*Anstrichfarbe*] = weatherfast
(Wetter)|Einziehschacht *m*, einziehender (Wetter)Schacht (⚒) = downcast shaft
~fahne *f* = weather vane
~festigkeit *f* → Wetterbeständigkeit
~führung *f* → Bewetterung (der Grubenbaue)
~gardine *f* (⚒) = brattice (cloth)
~glas *n*, Barometer *n* = barometer
~hahn *m* = weathercock
~karte *f* = weather map
~lampe *f* (⚒) = safety lamp

Wetterlampenbenzin — Wiederanwachsen der Tide

~**lampenbenzin** n = safety lamp gasoline
~**mann** m (⚒) = gas-testing safety man, ~ deputy, ~ examiner, fireman, gas watchman
„**wettern**", altern, verwittern [*Steinzeugindustrie*] = to store, to age
Wetter|schacht m (⚒) = ventilation shaft
~**schenkel** m, Wasserschenkel = water bar
~**(schutz)dach** n → Vordach
~**seite** f = weather face, ~ side
~**sprengstoff** m (⚒) = permitted explosive (Brit.); permissible ~ (US)
~**station** f, Wetterwarte f = weather station, meteorological ~
~**strecke** f, Wetterweg m, Wetterausziehstrecke (⚒) = ventilation road
~**teer** m, alterungsbeständiger Straßenteer = weather-resistant tar
~**teilstrom** m (⚒) = (air-)split
~**teilung** f (⚒) = (air-)split system
wetterunempfindlich = weatherproof
Wetterung f → Bewetterung (der Grubenbaue)
Wetter|versorgung f → Bewetterung (der Grubenbaue)
~**warte** f = meteorological station
~**weg** m → Wetterstrecke f (⚒)
Wetzschiefer m = hone-stone
WHK-Leim m, Klemmleim [*Nach dem Erfinder Klemm benannter Kaltleim*] = Klemm glue, ~ adhesive, ~ cement
Wickel|feder f, Schraubenfeder = helical spring, coil
~**maschine** f, Behälter~ = (tank) winding machine
~**trommel** f, Seilhaspel m, f = rope reel
~**verfahren** n [*Spannbetonbehälterbau*] = wire-wrapping method
Widder m, hydraulischer ~, Stoßheber m, automatische Wasserhebemaschine f = hydraulic ram
Widerlager n = abutment
~ [*Probebelastung von Pfählen*] = kentledge

~ **aus einem Holzkasten** [*Probebelastung von Pfählen*] = kentledge placed on a platform bearing directly on the test pile
~ ~ **zwei Holzkästen** [*Probebelastung von Pfählen*] = two boxes of kentledge placed on opposite sides of the test pile
~**bogen** m, Endbogen = abutment arch
~**hinterfüllung** f = abutment fill
~**verschiebung** f = displacement of abutment
Widerstand m **in der Dampfleitung**, Dampfleitungswiderstand = resistance to stream flow
Widerstands|beiwert m = coefficient of resistance
~**-Buckelschweißen** n, Widerstands-Dellenschweißen = resistance projection welding
~**elektrode** f = soil (electrical) resistivity rod
~**fähigkeit** f [*Werkstoff*] = strength
~**messungsverfahren** n, Verfahren mit galvanischer Elektrodenkopplung, geoelektrisches Verfahren, elektrische Widerstandsmessung f, geoelektrische Methode f [*Wenner; Hummel; Cish e Rooney; Megger usw.*] = resistivity reconnaissance, ~ prospecting, ~ exploration, ~ survey, ~ (measurement) method, electrical resistivity method, electrical resistivity survey
~**moment** n = resisting moment
~**-Nahtschweißen** n = resistance seam welding
~**-Preßschweißen** n = resistance (percussive) welding
~**-Punktschweißen** n = resistance spot welding
~**-Schweißmaschine** f = resistance welding machine
~**-Stumpfschweißen** n = resistance butt-welding
wieder aufbereiten [*Bohrschlamm*] = to recondition [*drilling mud*]
Wieder|anwachsen n **der Tide**, ~ ~ Gezeit = turn of the tide

~aufbau *m* = reconstruction
~auffahren *n* [*Tunnel; Stollen*] = re-driving
wiederausbaubarer Packer = retrievable packer
Wieder|belastung *f*, Wiederbelasten *n* = recharge
~einleitungsbauwerk *n* [*Zur Rückgabe von Kühlwasser oder Turbinenwasser an den Fluß*] = river outlet, stream ~
~entdeckung *f* = rediscovery
~flottmachen *n* [*steckengebliebenes Fahrzeug*] = debogging
wiedergefrieren = to re-freeze
Wiedergewinnung *f*, Rückgewinnung = recovery
wiedergewonnene Arbeitskammer *f*, bewegliche ~, Taucherglocke *f* = diving bell
~ Verrohrung *f* [*nach Einstellung der Bohrung*] = recovered casing
Wieder|herstellung *f* der Oberfläche [*nach einem Aufbruch auf Grund von Kanalisationsarbeiten u. dgl.*] = surface restoration, pavement ~
~herstellung(sarbeiten) *f(pl)* = remedial works, reconstruction (~), reconditioning
~holbarkeit *f* = repeatability
wiederholte Belastung *f* = repeated loading
Wieder|kehrdach *n* → Dach mit Wiederkehr
~verdampfer *m*, Zwischenverdampfer = re-boiler
~verwendung *f* = re-use
~verwendungsgerät *n* = multi-use equipment item
~verwendungsmöglichkeit *f* = reusability
~zulassung *f* = requalification
~zusammenbau *m* = reassembly
Wiege|-Abmessung *f*, Gewichtsabmessung = weigh-batching
~-Abmeß-Anlage *f*, Gewichtsabmeßanlage = weigh-batching plant
~anlage *f*, Ver~ = weigh(ing) plant, ~ installation

~aufgeber *m*, Wiegespeiser, Wiegebeschicker = weigh(ing) feeder
~automatik *f* = auto(matic) weigh(ing) system
~balken *m* = weigh(ing) beam
~band *n*, Bandwaage *f*, Wägeband = conveyor type scale, conveyer ~ ~
~behälter *m*, Wiegegefäß *n*, Waagtrichter *m*, Verwiegebehälter *m*, Wiegekübel *m*, Wiegedosiergefäß = weigh(ing) bucket, weigh(ing) box, batcher hopper, weigh(ing) hopper, scale hopper
~beschicker *m*, Wiegeaufgeber, Wiegespeiser = weigh(ing) feeder
~bunker *m* = weighing bin
~dosieranlage *f* = weigh-batching plant, ~ installation
~dosierapparat *m* = weigh-batching service
~dosiergefäß *n* → Wiegebehälter
~einrichtung *f* = weigh(ing) plant
~fähigkeit *f* = weighing capacity
~gerät *n* = weigh(ing) appliance
~gleis *n* = scale track
~kübel *m* → Wiegebehälter
~latte *f*, Setzlatte [*wird mit der Wasserwaage zusammen verwendet*] = plumb rule
~meister *m* = scale master
Wiegen *n* → Ab~
Wiege|schein *m* = weigh bill
~skala *f* = weigh(ing) dial
~speiser *m*, Wiegeaufgeber, Wiegebeschicker = weigh(ing) feeder
~trichter *m* → Behälterwaage *f*
~turm *m*, Ver~ = weigh(ing) tower
~vorrichtung *f* = weigh(ing) gear
Wiegung *f* → Abwiegen *n*
Wieland|fuge *f* = Wieland joint
~fugeneisen *n* = Wieland (jointing) strip, ~ joint bar, ~ metal bar
Wiener Kalk *m* = dolomitic lime, Pennine ~, Vienna ~
Wiesen|beil *n*, Rasenbeil = hand turf cutter, ~ ~ stripper
~gleiboden *m* = meadow gley soil
~torf *m* = meadow peat

WIG-Verfahren n, Wolframinertgasverfahren, Argonarc-Verfahren, Wolfram-Inert-Schweißen, Argonarcschweißen = tungsten inert gas welding, TIG ~, argon-arc ~

wild ausgebrochene Bohrung f, ~ ~ Sonde f; ~ ausgebrochener Schacht m [*in Galizien*] = wild well, ~ bore, ~ boring

Wildbach m = (mountain) torrent

Wildbett n = natural bed

wilde (Be)Rieselung f, Hang(be)rieselung = "wild" flooding, uncontrolled ~

~ Rißbildung f, ~ Rissebildung = random cracking

Wild|pflasterstein m, Schroppenpflasterstein = irregular (paving) sett

~reservat n = wildlife refuge

„**Wildwechsel**" m = "Beware of Deer", deer crossing

Williot|-Diagramm n = Williot (deflection) diagram

~-Plan m = Williot graphical method [*for the analysis of deflections of frameworks*]

willkürliche Annahme f = arbitrary assumption

Wimpelstange f = pennant rod

Wind|absatzboden m → windtransportierter Boden

~abtragung f → Deflation f

~angriffläche f, Windangriff-Fläche = wind load area

~brett n, Windfeder f = barge board, verge ~, gable ~

Winddruck m = wind pressure, pressure due to wind

~kessel m → Druckluftbehälter m

Winddüne f, Festlanddüne, Inlanddüne, Binnendüne = inland dune [*deposition by wind*]

Winde f = winch

~ → (Rotary-)Hebewerk n

~, Schrauben~, Schraubenspindel f, Hebebock m = screw jack

~, hydraulische ~, Hebebock m, Hydraulikwinde, Hydrowinde = hydraulic jack

~ **mit Einfachtrommel**, Einfach-Trommel(seil)winde = single-drum winch

Windel(puppe) f, Stroh-Lehm-Strick m = twisted straw

Winden|antrieb m = winch drive

~arbeit f = winching

~aufzug m = winch hoist

~bediener m, Windenführer = winch operator

~führer m, Windenbediener = winch operator

~getriebe n = winch transmission

~haus n = hoist house

~trommel f = winch drum

Wind|erhitzer m, Cowper m = Cowper

~erosion f (Geol.) = (geological) erosion by wind action, aeolation

~feder f → Windbrett n

windgepeitscht = wind-swept

Wind|geschwindigkeit f = wind velocity

~hose f = wind spout

~kanal m = wind tunnel

~kanalgebläse n = wind tunnel fan

~kessel m → Druckluftbehälter m

(~)Korrasion f, Sandschliff m (Geol.) = corrasion

Windkraft f = wind power

~, Windlast f [*DIN 1055*] = wind force, ~ load [*The force on a structure or building due to wind pressure multiplied by the area of the structure or building acting at right angles to the pressure*]

~maschine f = wind power machine

Wind|last f → Windkraft

~laterne f = barn lantern

~messer m, Anemometer n, m, Windstärkemesser m, Schalenkreuz n = anemometer, wind ga(u)ge

~richtung f = wind direction

~richtungsanzeiger m = wind direction indicator

~rispe f → Sturmlatte f

~rose f = windrose

~sack m = wind cone, ~ sleeve, ~ sock

~schatten m = wind shield

~scheibe f = shear wall

Windschliff — Winkelstufe

~schliff *m* = wind-polish rock
~schreiber *m* = anemograph [*A recording anemometer*]
~**schutzhaube** *f* **am Dunstrohr** = cage
~schutzscheibe *f* = windshield
~schwingung *f* = wind excited oscillation
~sediment *n* → äolisches Sediment
~seite *f* = wind side
windseitig = windward
Wind|sichter *m*, (Strom)Sichter = air classifier, (~) separator, sifter
(~)Sichtung *f*, Setzungsklassierung *f* durch Luft, Stromsichtung = air classification, (~) separation, particle selection by air flotation, particle selection by air blast
~sog *m* = wind suction
~spannung *f* = wind stress
~stau *m* = heaping up of waters by wind, (water) level raised by wind, wind-raised (water) level
~strebe *f* = wind brace
~strebengurtung *f* = wind braced boom
~strömung *f* [*Meer*] = wind current
~ „T" *n* = wind tee
windtransportierter Boden *m*, Windabsatzboden = (wind-)blow(n) soil, (a)eolian ~, wind-borne ~
Windung *f* = loop
Windverband *m*, Windversteifung *f*, Windverspannung = wind bracing
~anschluß *m* = wind bracing connection
Wind|verhältnisse *f* = wind conditions
~verspannung *f* → Windverband
~versteifung *f* → Windverband
~weg *m* = fetch
~welle *f* = wind wave
~werk *n* → (Rotary-)Hebewerk
~werk *n* [*Bagger*] = hoist machinery
~werktrommel *f*, Hebewerktrommel, Bohrwerktrommel, Rotary-~ = (rotary) draw works drum
Winkel *m* [*Zeichengerät*] = set-square
~ **der inneren Reibung**, (innerer) Reibungswinkel = angle of internal friction

~ ~ **Scherfestigkeit** = angle of shearing resistance
~ **mit 90°, 60° und 30°** [*Zeichengerät*] = 60-degree set-square
~ ~ ~, **45° und 45°** [*Zeichengerät*] = 45-degree set-square
~abweichung *f* = angular deviation
~aussteifung *f*, Winkelversteifung = angle stiffening
~band *n* → Bug *m*
winkelbarer Schild *m* → → Bulldozer
Winkel|bewegung *f* = angular movement, ~ motion
~dach *n* = roof with 45° pitch
~eckleiste *f* = nosing
~eisen *n* = angle iron
~eisenschmied *m* = angle iron smith
~führung *f* = angle guide
~geschwindigkeit *f* = angular velocity [*rate of motion in a fixed rotational direction*]
~**getriebe Vorgelege** *n* [*Tiefbohranlage*] = miter gear compound, mitre ~ ~ [*deep drilling rig*]
~halbierende *f*, Halbierungslinie *f* eines Winkels = bisecting line of an angle
~hebel *m* [*Ein Hebel, dessen Arme am Drehpunkt einen Winkel bilden, oder eine um eine Spitze drehbare, dreieckige Platte*] = bell crank
(~)Greifer *m* [*Gleiskette*] = grouser
Winkeligkeit *f*, Kantigkeit, Eckförmigkeit = angularity
Winkel|messer *m* = protractor
~planierschild *m* → → Bulldozer
~profil *n* = angle section, ~ (bar)
~rad *n* → Kegelrad
~schalung *f* = L-shaped forms
~schleifmaschine *f*, Winkelschleifer *m* = angle drive grinder
~spaten *m* = corner spade
~spiegel *m* = periscopic sight
~stahl *m* = angle steel (section), ~ ~ bar
~stahlstütze *f* = steel angle stanchion
~steg *m* = angle web
~stift *m* = corner pin
Winkelstufe *f* = angle(-type) step

~ **mit Bartprofil** = angle-type step with nosing
~ ~ **unterschnittenem Profil** = overhanging angle-type step
~ **ohne Profil** = regular angle-type step
Winkel|stützmauer *f* = angular retaining wall
~**(tür)schwelle** *f* = angle threshold
~**verformung** *f* = angular deformation
~**verlaschung** *f* = angle butt strap
~**verschiebung** *f* = angular displacement
~**versteifung** *f*, Winkelaussteifung, Winkelverstärkung = angle stiffening
Winker *m* [*Fahrzeug*] = direction indicator
~**posten** *m* → Warnposten
~**schalter** *m* = direction signal switch
Winter|arbeit *f* = winter work
~**bau** *m* = winter building
~**baugeräte** *npl* = winter construction equipment
~**bemessungstemperatur** *f* = winter design temperature
~**beton** *m* = winter-placed concrete
~**betonieren** *n* = winter concreting
~**bewässerung** *f* = winter irrigation
~**deich** *m*, Hauptdeich, Banndeich = main dyke, ~ dike
~**energie** *f* = winter energy
~**emulsion** *f* = winter emulsion
~**fenster** *n* → Kastenfenster
~**flickarbeit** *f* = winter patching
~**glätte** *f*, (winterliche) Straßenglätte = icing (condition)
~**glättebekämpfung** *f*, Straßenglättebekämpfung = treatement of icy pavements, highway ice-control, road ice-control, control of ice and frost
~**hafen** *m* = winter harbo(u)r
~**hydrant** *m* = winter hydrant
~**-Klimaanlage** *f* = winter air conditioning system
~**kraftwerk** *n* = winter power plant, ~ ~ station
~**müll** *m* = winter refuse

~**niederschlag** *m* = winter precipitation
~**sandstreudienst** *m* = sanding service, sand spreading [*for highway ice control*]
~**schmieröl** *n* = winter oil
~**schutzkleidung** *f* = protective winter clothing
Wipp|ausleger *m* = (level-)luffing jib (Brit.); ~ boom (US)
~**brücke** *f*, Klappbrücke = balance bridge, bascule ~
~**drehkran** *m*, Wippschwenkkran = (level-)luffing slewing crane
Wippe *f* = rocker
~ = swinging bracket [*On which a motor is mounted*]
~ → Faschinen~
wippen = to luff
~ [*Auslegerkran*] = to luff out (the jib)
Wippen *n*, Ausleger~ = derrick(ing) motion, (level) luffing
~**rost** *m* → Faschinen~
Wipp|kran *m* = luffing crane
~**seil** *n* = luffing rope
~**werk** *n* [*Auslegerkran*] = luffing mechanism
~**werk-Sicherheitsventil** *n* = luffing mechanism safety valve
~**zylinder** *m* [*Hydraulikkran*] = luffing ram
Wirbel *m*, Bogenströmung *f* = eddy, whirl, vortex
~**bett** *n* = fluidized bed
~**brenner** *m* = turbulent burner
wirbelfreie Strömung *f*, turbulenzfreie ~ = vortex-free flow, eddy-free ~, turbulence-free ~
Wirbel|haken *m* = swivel hook
~**kammer** *f* = rotochamber, swirl chamber [*I. C. engine*]
~**kammermahlen** *n* **von Zementklinker** = vortex-chamber grinding of cement clinker
~**mischer** *m* = impeller, whirling mixer
~**raum** *m*, Totraum [*Hydraulik*] = eddy zone, vortex ~
~**rohr** *n* [*Motor*] = cyclon tube

Wirbelschicht — Wismutocker

~schicht f [*Wirbelschicht-Röstverfahren zum Abrösten schwefelhaltiger Kiese*] = fluidized bed, turbulent layer

~senken-Sichter m = vertical separator [*U. S. Patent 2 967 618 dated 10th January 1961. Inventor: Z. Vane*]

~verlust m = eddy loss [*The energy lost (converted into heat) by swirls, eddies, and impact, as distinguished from friction loss*]

~wind m = eddy wind, whirlwind

~windschwingung f = vortex-excited oscillation

wirken [*Kraft*] = to act [*force*]
~ [*Last*] = to act [*load*]

wirksame Bremsenfläche f → Bremsennutzfläche

~ Druckspannung f, ~ Flächenlast f [*Bodenmechanik*] = effective pressure

~ Flächenlast f, ~ Druckspannung f [*Bodenmechanik*] = effective pressure

~ Korngröße f = effective size (Hazen) (D_{10}) [*The grain size on a mechanical analysis curve corresponding to $W\% = 10$*]

~ Schleppkraft f, Räumkraft [*Strom; Fluß*] = effective tractive force

~ Siebfläche f, nutzbare ~ = effective screen area

~ Spannung f [*Bodenmechanik*] = effective stress

wirksamer Auftrieb m = positive buoyancy

~ Korndurchmesser m = effective grain diameter

~ Querschnitt, effektiver ~, Strömungsquerschnitt = effective cross section

Wirkstoff m, Zusatzmittel n = additive, agent, admix(ture)

~ → Beton~

~ zur Beeinflussung des Viskositäts-Temperatur-Verhaltens [*Öl-Zusatz*] = VI improver

Wirkung f = action

Wirkungsgrad m = efficiency

~, Turbinen~ = generating efficiency

~ des Wandlers im Anfahrpunkt = torque converter stall ratio

~kurve f = efficiency-load curve [*turbine*]

Wirkungshalbmesser m der Molekularanziehung = radius of molecular action

wirtschaftlich produktive Bohrung f, ~ ~ Sonde f; ~ produktiver Schacht m [*in Galizien*] = commercially productive well, ~ ~ bore, ~ ~ boring

wirtschaftliche Querschnittgestaltung f = economical shape of cross section

~ und soziale Rechtfertigung f des Ausbaues der Straßen = economic and social justification for road development

Wirtschaftlichkeit f, Rentabilität f = rentability, profitability, economy

~ (⚒) = payout

Wirtschaftlichkeits|berechnung f = economy calculation

~faktor m, Einflußfaktor auf die Wirtschaftlichkeit = economic factor

~untersuchung f = profitability investigation, ~ study

Wirtschafts|gebäude n = farm building

~geographie f = economic geography

~plan m → Flächennutzungsplan

~weg m, ländlicher Weg = farm track

~wegebau m, ländlicher Wegebau = construction of farm tracks

Wisch|blatt n [*Scheibenwischer*] = wiper blade

~hebel m [*Scheibenwischer*] = wiper arm

Wismut|blende f (Min.) = eulytite, bismuth blende

~glanz m, Bismuthin m, Bi_2S_3 (Min.) = bismuthinite, bismuth glance

~nickel(kobalt)kies m (Min.) = grunauite

~ocker m, Bismit n, Bi_2O_3 (Min.) = bismuth ochre, bismite

Wismut-Zonenschmelzen — Wölbung

~-Zonenschmelzen n, Zonenschmelzen von Wismut = zone refining of bismuth

Witterungs|beständigkeit f, Wetterfestigkeit = weather-resistance, weather-proofness, resistance to the action of weather

~einfluß m = climatic effect, effect of climatic conditions, atmospheric action

Wochen|öltank m, Wochenölbehälter m = week-supply oil tank

~tank m, Wochenbehälter m = week-supply tank

~zeitplan m [*Durchführung von Bauarbeiten*] = schedule of work to be performed during a week

Wohn|baracke f = living hut

~bauten f = residential buildings

~dichte f = residential density

~etage f → Wohngeschoß n

~fläche f = flat floor space

~gebiet n, Wohnviertel n = residential area, residences, populated area

~geschoß n, Wohnstockwerk n, Wohnetage f = apartment floor, residential ~, ~ stor(e)y

Wohnhaus n → Miet~

~bau m = dwelling construction, building of dwellings

~fenster n = residential window

~treppe f = domestic stair(case)

Wohn|heim n = hostel

~hochhaus n [*manchmal „Wohnmaschine" genannt*] = apartment tower, (tall) block of flats, high-rise apartment building

~küche f = dwelling kitchen

~maschine f → Wohnhochhaus

~(schlaf)wagen(anhänger) m → Wohnwagen(anhänger)

~schlafzimmer n = bed-sitting room

~siedlung f = housing estate

~stockwerk n → Wohngeschoß n

~straße f = residential (local) street

~terrasse f = living terrace

~- und Bürohaus n, Büro- und Wohnhaus = office-and-flat block, flats-and-offices building

Wohnung f [*Die Summe der Räume, welche die Führung eines Haushaltes ermöglichen, darunter stets eine Küche oder ein Raum mit Kochgelegenheit*] = self-contained dwelling

~, Miet~ = flat (Brit.); apartment (US)

~ ohne Komfort = cold water flat (Brit.); ~ ~ apartment (US)

Wohnungsbau m = housing

~genossenschaft f = home building co-operative society

~gesellschaft f = home building society

Wohnungs|hilfswerk n = emergency housing scheme

~mangel m = housing shortage

~politik f = housing policy

~projekt n = housing scheme

~tür f [*Etagenwohnung*] = landing door

Wohn|viertel n, Wohngebiet n = residences, residential area, populated area

~wagen(anhänger) m, Wohnschlafwagen(anhänger) = sleeping caravan, accommodation trailer, house trailer; mobile home (US)

~wagenplatz m = sleeping caravan standing (Brit.); accommodation trailer ~ (US); house trailer ~

~zimmer n = keeping room (US); sitting ~

~zone f = residential area

Wölber m → Wölbstein

Wölb|fuge f = voussoir joint

~kehlnaht f [*Deutschland*]; Vollkehlnaht [*Österreich*]; konvexe Kehlnaht [*Schweiz*] = convex fillet weld, reinforcing fillet

~linie f = intrados soffit

~-Naturstein m, Keil-Naturstein [*(Natur)Steinbogen*] = arch-stone

~stein m, Keilstein, Wölber m, Gewölbstein = voussoir [*An arch-stone in a stone arch or an arch-brick in a brick arch or a well lining*]

Wölbung f, kreisförmige Krone f [*Straße*] = (barrel) camber

~ → Bogenlinie f

Wölbziegel *m*, Keilziegel, Gewölbeziegel = arch-brick [*A wedge-shaped brick for building an arch or lining a well*]
Wolf *m* → Stein~
~**loch** *n* → Stein~
Wolfram *n* = wolfram, tungsten
~**inertgasverfahren** *n*, WIG-Verfahren = tungsten inert gas welding
~**-Inert-Schweißen** *n* → Argonarc-Verfahren *n*
Wolfsholzpfahl *m* = Wolfsholz bore(d) pile, ~ drilled ~
Wolken|decke *f* = cloud deck
~**kratzer** *m* = skyscraper
~**scheinwerfer** *m* = ceiling light projector
Wollfett-Olein *n* = distilled-grease olein, dégras oil
~**säure** *f* = dégras acid
~**-Stearin** *n* = dégras stearin
Wollfilzpappe *f* = rag felt
~ **mit Bleifolie** = rag felt with lead foil
Woll|sackabsonderung *f*, brotlaibige Absonderung (Geol.) = loaf-like jointing, pillow-jointing
~**teppich** *m* = wool carpet
~**wäscherei** *f* = wool-washing plant, ~ works, wool-cleaning ~
Woltman-Flügel *m*, hydrometrischer Flügel = Woltman current meter
Woodward-Regler *m* = Woodward governor
Wrack|bezeichnung *f* = marking of wrecks
~**gut** *n* = derelict, wreck
Wrasen *m* → Schwaden *m*
Wucht *f*, Schwung *m* = momentum
~**förderer** *m* = spring-supported vibrating conveyor, ~ ~ conveyer
Wühl|kopf(-Fräser) *m* → mechanisch angetriebener Schneidkopf
~**tier** *n* = burrowing animal
Wuhr *n* [*Schweiz*]; (Stau)Wehr *n* = weir
Wulfenit *m*, Gelbbleierz *n*, $PbMoO_4$ (Min.) = wulfenite
Wulst mit = bulb
~|**band** *n* = bead seat band
~**eisen** *n* = bulb-iron
~**felge** *f* = clincher rim
~**lava** *f* → Stricklava
~**reifen** *m* = beaded tyre (Brit.); ~ tire (US)
~**winkel** *m* = bulb-angle
~**winkelstahl** *m* = bulb-angle steel
Wünschel|rute *f* = pointer [*e.g. a twig, a metal rod, or a piece of wire*]
~**rutengänger** *m* = dowsing practitioner
~**rutenverfahren** *n* = dowsing
~**ruten-Wassersuche** *f* = water divining
Wurf *m* [*Sieb*] = throw, amount of eccentricity
~ [*Handschacht*] = throw
~**beschicker** *m* [*Kesselfeuerung*] = mechanical stoker
Würfel *m* → Beton~
~(**druck**)**festigkeit** *f*, Druckfestigkeit = cube (crushing) strength, (~) compressive ~, crushing ~
~(**druck**)**probe** *f*, Würfel(druck)versuch *m*, Würfel(druck)prüfung *f* = cube (crushing) test
~**erz** *n*, Pharmakosiderit *m* (Min.) = pharmacosiderite
~**festigkeit** *f* = cube strength
~**festigkeit** *f*, W_{28} = (compressive) cube strength at 28 days
~**form** *f* = cube mo(u)ld
~**pflaster(decke)** *n*, (*f*) = cube sett paving
~(**pflaster**)**stein** *m* = cube (paving) sett
~**presse** *f* → Druckpresse
~**probe** *f* → Würfeldruckprobe
~**probe** *f*, Probewürfel *m* = cube test specimen, test cube
~**prüfmaschine** *f* → Druckpresse *f*
~(**prüf**)**presse** *f* → Druckpresse
~**-Schlagfestigkeit** *f*, Schlagfestigkeit an Würfeln ermittelt = cube impact strength
~**stein** *m*, Würfelpflasterstein = cube (paving) sett
Wurf|förderrinne *f* **mit Pendelantrieb** → Torpedorinne

~radschaufel *f* [*Schneeschleuder*] = rotary blade
~schaufellader *m* → Über-Kopf-Lader
~siebung *f* = throw screening
~-Transporteur *m*, Schleuderband *n* = jet conveyor, ~ conveyer
Würgezange *f*, Sprengkapselzange, Anwürzzange = cap crimper
„**Wurstmaschine**" *f* [*zur Herstellung von Einfassungen bei bituminösen Straßendecken*] = curbing paver
Wurzel *f*, Landanschluß *m* [*z.B. einer Buhne*] = root
wurzelabweisend = root-resisting
Wurzel|-Beseitigung *f* = uprooting
~einbrand *m* = root penetration
~harke *f*, Wurzelrechen *m* = root rake, land-clearing ~
~hinterlegung *f* [*Schweißtechnik*] = root backing, ~ reinforcement
~lage *f* = stringer bead, root pass, root run, initial run
~lagenschweißer *m* = stringer bead welder, root pass ~, ~ ~ operator
~lagenschweißung *f* = root pass welding, ~ run ~, initial ~ ~, stringer bead ~
~loch *n*, Wurzelhöhlung *f* = root hole
~maß *n* → Streichmaß
~pflug *m* = root cutter; ~ plow (US)
~rechen *m*, Wurzelharke *f* = root rake, land-clearing ~
~torf *m*, filziger Torf, Fasertorf = fibrous peat
~- und Felsrechen *m*, Wurzel- und Felsharke *f* = land-clearing and rock rake, root ~ ~ ~
~zone *f* = rooting zone
Wüstensteppe *f* = semidesert
W/Z-Faktor *m* → Wasser-Zement-Faktor
WZV *n* → Wasser-Zement-Faktor

X

xenomorph, fremdgestaltig = xenomorphic
Xenon-Entladungslampe *f* = Xenon filled quartz discharge tube
Xylol *n* = xylene
Xylolith *n* [*Markenname*]; Steinholz *n* = magnesite composition

Y

Y-Grundriß *m* = Y-plan

Z

Zacken|lava *f*, Schollenlava, Blocklava, Aprolith *m*, Spratzlava = block lava, aa (~)
~schere *f* = pinking shears
zähe Widerstandsfähigkeit *f* **gegen die Sprödbruchbildung** = fracture toughness
Zähflüssigkeit *f* → Viskosität *f*
Zähigkeit *f* → Schlagfestigkeit [*Gestein*]
Zähigkeits|-Erschöpfung *f* = ductility exhaustion
~verteilung *f* = density-spread [*After the confluence of streams differing in density*]
~wechsel *m* = densimetric exchange [*between parallel flows*]
Zahlen|beispiel *n* = numerical example
~tafel *f*, Tabelle = table (of figures), numerical table
~vergleich *m* = numerical comparison
~verhältnis *n* = numerical ratio
~verhältnis *n* [*Proportion*] = numerical proportion
Zähler *m* [*für Flüssigkeiten oder Dampf oder Gas*] = integrating flow meter
~ [*Mathematik*] = numerator
Zähl|gerät *n* = counting device
~posten *m* = observer, checker
Zählung *f* = count
~ **für Fahrzeugeinteilungszwecke** = vehicle classification count
Zählwerk *n* = counter
Zahn|abstand *m* = tooth spacing
~eisen *n* [*Steinhauerwerkzeug*] = notched chisel, indented ~

~halter *m* = tooth adapter, ~ adaptor
~kloben *m* = toothed lock
~kontakt *m* = tooth contact
~kupp(e)lung *f* = toothed gear type coupling
~profil *n* = (gear) tooth profile
Zahnrad *n* = gear
~ mit Schrägverzahnung = helical gear
~antrieb *m* = gear drive
~eingriffeinstellung *f* = mesh adjustment
~fett *n* = gear wheel lubricating grease, crater compound
~nabe *f* = gear hub
~pumpe *f* = gear(-type) pump
~-Schleifgerät *n* = gear grinder
~spiel *n* = gear backlash
~-Tenderlok(omotive) *f* = rack tank loco(motive)
~übersetzung *f* = gear train
~winde *f* → Zahnstangenwinde
~zahn *m* = gear tooth
Zahn|riemen *m* = toothed belt
(~)Ringdübel *m* „Alligator", Alligator-(Zahn)Ringdübel = toothed-ring
~scheibenmühle *f* = toothed disc mill; ~ disk ~ (US)
~schiene *f* = rack rail
~schwelle *f* = dental, dentated sill [*A toothlike projection on an apron, or other surface, to deflect or break the force of flowing water; a form of baffle*]
~spitze *f* = tooth tip
~stange *f* = rack [*A toothed bar*]
Zahnstangen|(an)trieb *m* = rack gearing, ~ gear mechanism, ~ gear drive
~aufzug *m* [*für Baustellen*] = rack building hoist
~bahn *f* = rack railway (Brit.); ~ railroad (US)
~begrenzer *m* = rack limiter
~hebebock *m* → Zahnstangenwinde
~hebel *m* = rack lever
~schlitten *m* [*Klappbrücke*] = rack guide, ~ carriage
~winde *f*, Zahnstangenhebebock *m*, Stahlwinde, Zahnradwinde, Wagenwinde, Bockwinde, Wagenheber *m* = rack-and-pinion jack
Zahn|walzenbrecher *m* = toothed roll crusher, ~ ~ breaker
~winkel *m* [*Aufreißer*] = tip angle
Zange *f* [*Rechteckiges oder halbrundes Holz für Verstrebungen und zum Zusammenhalten von Holzkonstruktionen*] = wale (piece), horizontal timber, waling (Brit.); whaler, transverse plank, binding piece
~ [*Dachverbandholz*] = tie
~ [*dient zur unteren Führung von Spundbohlen*] = bottom wale piece (of the piling)
Zapf|behälter *m* = dispenser
~einrichtung *f* [*für Kraftstoff*] = draw-off unit
Zapfen *m* [*Holzverbindung*] = tenon [*The end of a member shaped to fit a mortise*]
~anschluß *m* [*Rollenmeißel*] = pin shank
~fuge *f* → Zapfenverbindung
~verbindung *f*, (Ver)Zapfung, Zapfenfuge *f* [*von Hölzern*] = mortise and tenon joint, mortice ~ ~ ~, tenon jointing
Zapfhahn *m* [*DIN 3271 bis 3279*] = draining tap
~, Feuerhahn, Wandhydrant *m* = wall hydrant
Zapfstelle *f* [*für Kraftstoff*] = dispense point, refuel(l)ing hydrant
~ zur Druckprüfung = pressure tap
Zapfung *f* → Zapfenverbindung
Zapfwelle *f* = power take-off
~ mit Direktantrieb = live power take-off
Zapfwellen|-Anschluß *m* für Anbaugeräte, hinterer Kraftanschluß-Stutzen *m* ~ ~ [*Schlepper*] = rear power take-off
~antrieb *m* = tractor power take-off
Zarge *f* = trim (Brit.); casing (US)
Zaun *m* = fence

Zaunpfosten — Zeitspanne

~pfosten *m*, Zaunpfahl *m* = fence post, ~ stake, fencing ~
~zubehör *m*, *n* = fencing accessories
Zebrastreifen *m*, Verkehrsüberweg *m* = zebra crossing
Zeche *f*, Kohlenbergwerk *n*, Kohlengrube *f* = colliery, coal mine
Zechstein *m* (Geol.) = Zechstein, marine Permian, Upper Permian [*The higher of the two series into which the Permian System of Germany is divided. The whole of the English Permian of Durham and Yorkshire is probably of Zechstein age*]
~kalk *m* = Permian limestone
zehnjähriges Hochwasser *n* = 10 years' flood
Zeichen *n* [*DIN 1350*] = symbol
~brett *n*, Reißbrett = drawing board
~büro *n* = drawing office (Brit.); drafting ~ (US)
~erklärung *f* → Zeichenschlüssel *m*
~erklärungsschild *n* = legend plate
~gerät *n* [*Bildmessung*] = plotter
~geräte *npl* und -material *n* = drawing equipment and material
~papier *n* = drawing paper
~schlüssel *m*, Zeichenerklärung *f*, Legende *f* = legend, key
Zeichner *m* = draughtsman (Brit.); draftsman (US)
zeichnerische Bestimmung *f*, ~ Lösung, graphische ~ = graphical solution
zeichnerisches Verfahren *n*, graphisches ~ = graphical construction, ~ method, ~ procedure
Zeichnerlehrling *m* = apprentice draughtsman, drawing office apprentice
Zeichnung *f*, Plan *m* = drawing
zeichnungsmäßige Lage *f*, Soll-Lage = theoretical location
Zeiger *m* = pointer
~ablesung *f* = pointer reading
~zählwerk *n* = counting dial
Zeis-Bosshardt-Tachymeter *m*, Doppelbildtachymeter = double image tacheometer, Bosshardt-Zeiss reducing ~

zeitabhängig = time-dependent
Zeit|abstand *m* = time interval, interval of time
~achse *f* [*einer Tidekurve*] = axis of time, time axis
zeitanteilig = time-sharing
Zeit|ball *m* = time ball
~dauer *f*, Zeitraum *m*, Zeitspanne *f* = period of time, time period, length of time
~diagramm *n* = time diagram
~einheit *f* = time unit
~faktor *m* = time factor
~faktorkurve *f* = time factor curve
~folge *f* = time sequence
zeit|genössische Architektur *f* = contemporary architecture
~gesteuert = timed
Zeitlohn *m* = time wage
~arbeit *f* = time work
~stundenanteil *m* = time on daywork, "hours per hour", hours paid at hourly rate
Zeit|lücke *f* [*Straßenverkehrstechnik*] = headway
~lückenmethode *f* nach Greenshields = Greenshields method [*It is based on the observed "green time requirement" for a single line of passenger cars entering an intersection at minimum headways after starting from standstill*]
~lückenverteilung *f* = headway distribution
~nehmer *m* = time keeper
~plan *m* = time schedule
~raum *m* → Zeitdauer *f*
~schaltautomat *m* = auto(matic) timing device
~schreiber *m* = time recorder
~schriftenraum *m* = periodical room
~-Setzungs-Linie *f*, Zeit-Setzungs-Kurve *f* = time/settlement graph, ~ curve, ~ diagram
~-Schwind-Kurve *f*, Zeit-Schwind-Linie *f* = time/shrinkage curve, ~ graph, ~ diagram
~spanne *f* → Zeitdauer

Zeitungsstand — Zement(dach)ziegel

Zeitungsstand m = newspaper kiosk
Zeit-Versatzdruckkurve f (⚒) = time-pressure curve for stowed waste (area); ~ ~ ~ ~ goaf, ~ ~ ~ ~ gob [*Wales*]; ~ ~ ~ ~ condie, ~ ~ ~ ~ cundy [*Scotland*]
zeit|weilige Umleitung f → provisorische ~
~weiliger Wasserlauf m, periodischer ~ = temporary water course
Zeit|wert m, Verkehrswert [*Maschine*] = trade-in value, present ~
~zeichen n, Stundenzeichen = time signal
~zünder m → → 1. Schießen n; 2. Sprengen n
~zündersprengen n, Zeitzündersprengung f = delay-firing, delay-blasting
Zellen|bauweise f = cellular design
~beton m → Blähbeton
~betonwerk n → Porenbetonwerk
~decke f, Stahl~ = cellular steel floor
~dolomitgestein n → Rau(c)hwacke f
~fang(e)damm m = cellular cofferdam
~kalkstein m = cellular limestone
~mörtel m → Porenmörtel
~pfeiler m = cellular pier
~radaufgeber m, Zellenradbeschicker, Zellenradspeiser = rotary vane feeder, rotating ~ ~
~radzuteiler m, Zellenraddosierer = rotary vane batcher, rotating ~ ~
~reihe f = series of cells
~(stau)mauer f → → Talsperre
~tiefofen m = cell type soaking pit
~turm m mit Abwiegung = combined aggregate and cement weigh batching unit
zellig-löch(e)riger Dolomit m → Rau(c)hwacke
Zelt|dach n [*Walmdach auf quadratischem Grundriß mit Firstpunkt*] = broach roof
~-Mansardendach n, Mansarden-Zeltdach = mansard roof
Zement m [*hydraulisches Bindemittel*] = (hydraulic) cement, water ~, cement matrix

~abfüllwaage f → Bindemittelwaage
~-Abmeßanlage f → Dosier(ungs)anlage für Zement
~abmeßschnecke f → Abmeßschnecke für Zement
~(abzug)schnecke f = screw conveyor for bulk cement, ~ conveyer ~ ~ ~, cement screw (conveyer), cement screw conveyor
~-Anhänger m = cement trailer unit
~anstrich m = cement coat
~aufgeber m, Zementbeschicker, Zementspeiser = cement feeder
zement|ausgekleidet [*gegen Korrosion*] = cement-lined
~auspressen → zementeinpressen
Zement|auspressung f → Zementeinpressung
~auspreßgut n → Zementeinpreßgut
~bazillus m, Kalktonerdesulfat n = Michaelis' salt, Candlot's ~
~becherwerk n, Zementelevator m = cement (bucket) elevator
~behälter m = cement container
~behälterwagen m = cement container car, ~ ~ wagon
~beschicker m → Zementaufgeber
~bestandteil m, Zementkomponente f = cement constituent
(~)Beton m → Beton
~betonfugenvergußmasse f = cement concrete joint sealing compound
(~)Betonstraße f = concrete highway, ~ road
(~)Betonstraßenbau m = concrete highway construction, ~ road ~
~bett n [*Blaine-Gerät*] = cement bed, test bed of cement
~brand m, Zementbrennen n = cement burning
~brei m, frischer ~ [*Zement plus Wasser*] = wet paste, (~) cement ~
~chemie f = cement chemistry
~(dach)stein m → Betondachstein
~dachsteinmaschine f → Dachsteinmaschine
(~)Dachsteinfarbe f → Betondachsteinfarbe
~(dach)ziegel m → Betondachstein **m**

Zement(dach)ziegelfarbe — Zementierung

~**(dach)ziegelfarbe** *f* → Betondachsteinfarbe
~**dosator** *m*, Zementdosierer *m* = cement batcher
~**dosierschnecke** *f* → Abmeßschnecke für Zement
~**-Dosier(ungs)anlage** *f* → Dosier(ungs)anlage für Zement
~**drehofen** *m*, Zementrotierofen = rotary cement kiln
~**druckfestigkeit** *f* = cement compressive strength
~**einsackmaschine** *f* = cement bagging machine
zementeinpressen, zementverpressen, zementauspressen, zementinjizieren = to cement-grout
Zement|einpressung *f*, Zementverpressung, Zementauspressung, Zementinjektion *f* = cement grouting, ~ injection
~**einpreßgut** *n*, Zementverpreßgut, Zementauspreßgut, Zementinjektionsgut = cement grout [*A fluid mixture of cement and water or of cement, sand, and water*]
~**einpreßmaschine** *f* → Zement-Injektionspumpe *f*
~**einspritzapparat** *m* → Zement-Injektionspumpe *f*
~**-Endmühle** *f* = finish mill
~**ersatz** *m* = cement replacement [*e. g. fly ash*]
~**estrich** *m* = cement (floor) screed, ~ ~ topping, ~ topping finish [*A screed of cement mortar laid on a floor, particularly on a concrete slab*]
~**estrich** *m* **mit Granitsplitt** = granolithic (screed), ~ paving
~**estrichmörtel** *m* = cement screed mortar
~**estrichstreicher** *m* = trowel hand
Zementfabrik *f*, Zementwerk *n* = cement mill, ~ plant
~**nach dem Naßverfahren** = all-wet cement plant, ~ ~ mill
~ ~ ~ **Trockenverfahren** = all-dry cement plant, ~ ~ mill
Zement|faktor *m* → Zementgehalt

~**farbe** *f* = cement paint; ~ color (US)
~**feinheit** *f* = cement fineness
~**förderer** *m* = cement conveyor, ~ conveyer
~**förderrohr** *n* = cement conveying pipe, ~ ~ tube
~**(förder)schnecke** *f* = cement screw, screw conveyor for bulk cement, screw conveyer for bulk cement
~**förderung** *f*, Zementumschlag *m* = cement handling
~**frostschutzmittel** *n* = cement antifreeze compound
~**fuge** *f* = cement joint
~**-Füller-Schlämme** *f* = cement-filler grout, ~ slurry
zement|gebunden = cement-bound
~**gebundener feuerfester Schamottestein** *m* = fireclay grog refractory
Zement|gehalt *m*, Zementfaktor *m* [*Beton*] = cement content, ~ factor
~**gel** *n* = cement gel
~**hydratation** *f* = cement hydration
~**hydratationserzeugnis** *n* = cement hydration product
Zementier|ausrüstung *f* [*Tiefbohrtechnik*] = cementing outfit
~**büchse** *f*, Betonbüchse, Betonlöffel *m*, Zementierlöffel [*Tiefbohrtechnik*] = dump bailer, cement dump
zementieren, ein~ [*Tiefbohrtechnik*] = to cement
Zementier|kopf *m* [*Tiefbohrtechnik*] = cementing head
~**loch** *n* = cementing hole
~**löffel** *m* → Zementierbüchse *f*
~**pumpe** *f* [*Tiefbohrtechnik*] = cementing pump
~**schirm** *m* [*Tiefbohrtechnik*] = cementing basket
~**schlauch** *m* = cementing hose
~**schuh** *m* [*Tiefbohrtechnik*] = cement casing shoe, cementing ~
~**stopfen** *m* [*Tiefbohrtechnik*] = cementing plug
Zementierung *f*, Auszementieren *n* des Bohrloches = cementation, cementing job

Zementierungs|index *m* = cementation index
~wert *m* = cementing value
Zementierverfahren *n* = cementation method
Zement|injektion *f* → Zementeinpressung *f*
~injektionsgut *n* → Zementeinpreßgut
~-Injektionspumpe *f*, Zementeinpreßmaschine *f*, Zementeinspritzapparat *m*, Zementmilch(injektions)pumpe *f*, Zementinjektor *m* = cement injection pump, grout ~
zementinjizieren → zementeinpressen
Zementit *m* → Eisenkarbid *n*
Zement|kahn *m*, Zementschiff *n*, Spezialschiff für Zementtransport = cement barge
~kalkmörtel *m* → verlängerter Zementmörtel
~kalksandmörtel *m* = cement-lime-sand mortar
~kanone *f* → Betonspritzmaschine *f*
~kelle *f* = cement trowel
~klinker *m* = cement clinker
(~)Klinkerlager *n* = clinker storage area, ~ store
~klinkermahlanlage *f* → Klinkermühle
~klinkermühle *f* → Klinkermühle
~komponente *f*, Zementbestandteil *m* = cement constituent
~kuchen *m* = circular-domed pat of cement, cement pat, pat of cement-water paste, soundness test pat
~kugelmühle *f* = preliminator
~kühler *m* = cement cooler
~labor(atorium) *n* = cement lab(oratory)
~lagerhalle *f* = cement hangar
~lagerung *f* = cement storage
~-Lastkraftwagen *m* = bulk cement truck (US); ~ ~ lorry (Brit.)
~-Latex-Estrich *m* = cement-rubber latex (jointless) floor(ing), ~ ~ composition ~, fleximer
~leim *m*, flüssiger Zement *m* = cement paste
Zementmilch *f*, Feinschlämme *f*, (Beton)Schlämmschicht *f* = laitance

~, Suspension *f* von Zement in Wasser, Zementschlämme *f*, Zementsuspension = (neat) cement-water grout, (~) ~ mix(ture), (~) cement slurry, cement grout
~pumpe *f* → Zement-Injektionspumpe
~verpressung *f*, Zementmilcheinpressung, Zementmilchinjektion *f*, Zementmilchauspressung = grouting with cement slurry
Zement|mineral *n* = cement mineral
~mischprüfung *f*, Zementmischversuch *m*, Zementmischprobe *f* [*Bitumenemulsion*] = cement mixing test
Zementmörtel *m* = (hydraulic-)cement mortar
~auskleidung *f* = cement mortar lining
~belag *m*, Zement(mörtel)putz *m* = cement plaster
~einpressung *f* = cement mortar injection
(~-)Einpreßgerät *n* → Verpreßkessel
(~)Injektor *m* → Verpreßkessel *m*
~kanone *f* → Beton-Spritzmaschine *f*
~probe *f* = cement mortar specimen
~putz *m*, Zementmörtelbelag *m*, Zementputz = cement plaster
~-Spritzapparat nach dem Torkretverfahren → Beton-Spritzmaschine *f*
~zusatz *m* = cement mortar additive
Zement|mühle *f* = cement (grinding) mill
~oberfläche *f* = cement surface
~ofen *m* = cement kiln
~phase *f* = cement phase
~plattenpflaster *n* = cement block paving
~probe *f* = cement sample
~prüfung *f*, Zementprobe *f*, Zementversuch *m* = cement test
~pumpe *f*, Fuller(-Kinyon)-Pumpe, Fuller-Zementpumpe, Zementstaubpumpe, Fuller-Entladepumpe = Fuller-Kinyon unloader (pump), F-K (dry) pump
~putz *m*, Zementmörtelbelag *m*, Zementmörtelputz = cement plaster

Zementputzmörtel — Zement- und Kalk-Verteiler

~putzmörtel *m* = cement stuff
~-Quarz-Mischung *f* = cement-silica mix(ture)
zementreicher Beton *m* = rich concrete
Zement|rohmehl *n* = cement raw meal
(~)Rohmischung *f* = raw mixture
~rohr *n*, Betonrohr = concrete pipe
~rohrform *f* → Betonrohrform
~rohrpresse *f*, Betonrohrpresse = concrete pipe press
~rohrstampfmaschine *f*, (Beton)Rohrstampfmaschine, Stampfmaschine für Betonrohrherstellung = concrete pipe tamping machine
~-Rohschlamm *m* = cement slurry
~rotierofen *m* → Zementdrehofen
~-Sand-Schlämme *f* = cement-sand grout, sand-cement ~
~-Sand-Schlämme-Mischer *m*, Colcrete-Mischer = Colcrete mixer, cement slurry ~, sand-cement grout ~, cement-sand grout ~
~-Saugförderung *f* = vacuum-handling of cement
~schachtofen *m* = vertical cement kiln
~schieber *m* = cement valve
~schiff *n*, Zementkahn *m*, Spezialschiff für Zementtransport = cement barge
~schlämme *f*, Zementsuspension *f*, Zement-Wasser-Gemisch *n*, Zement-Wasser-Mischung *f*, Zementmilch = (neat) cement-water grout, (~) ~ mix(ture), (~) cement slurry, cement grout
~schnecke *f*, Zementförderschnecke = screw conveyor for bulk cement, screw conveyer ~ ~ ~, cement screw
~schotterdecke *f*, Zementschotterbelag *m*, Zementschotterstraße *f* = sandwich process macadam, COLCRETE constructed in the sandwich process
~schuppen *m* = cement shed
~silo *m*, Bindemittelsilo = cement silo
~-Silofahrzeug *n* → Zementtanker

~(silo)waage *f*, Bindemittelwaage, Zementabfüllwaage, Abfüllwaage für Zement = cement batcher scale, ~ weigh(ing) batcher
~speiser *m* → Zementaufgeber
~spritzschlauch *m* = cement grout discharge hose
~staub *m* = cement dust
~stein *m*, erhärteter Zementleim *m* = hardened cement paste
~stein *m* → Betondachstein
~steinfarbe *f* → Betondachsteinfarbe
~suspension *f* → Zementschlämme *f*
~tanker *m*, Zementtransportfahrzeug *n*, Zement-Silofahrzeug = cement tanker, ~ haulage unit, ~ hauler, bulk cement transporter
~tanker *m* mit Druckluftentladung = pressurized cement tanker
~tankerlastzug *m* = bulk cement truck-and-trailer unit
~tanker-LKW *m* mit Druckluftentladung = pressurized (cement) lorry (Brit.); ~ (~) truck (US)
~teilchen *n* = cement grain, ~ particle
~-Ton-Betonstraße *f*, selbständige Fahrbahndecke *f* aus Zement-Ton-Beton = soil-cement surface course, ~ road
~-Ton-Injektion *f*, Zement-Ton-Auspressung *f*, Zement-Ton-Einpressung, Zement-Ton-Verpressung = cement-clay injection
~transportanhänger *m* mit Doppelschnecke = twin-screw bulk cement trailer
~transportanhänger *m* mit Gebläseluft-Förderrinne = air-slide bulk cement trailer
~transportfahrzeug *n* → Zementtanker
~-Traßmörtel *m* = cement-trass mortar
zementumhüllt = cement-coated
Zement|umschlaganlage *f* = cement handling installation, ~ ~ plant
~- und Kalk-Verteiler *m* → Kalk- und Zement-Verteiler

zement|verfestigte Tragschicht f = cement-treated base, CTB
~verfestigter Untergrund m = cement-treated subgrade
Zement|verfestigung f, Bodenbeton m, Erdbeton, Bodenzement m, Erdzement = cement stabilization, soil-cement
~vermahlung f = cement grinding
zementverpressen → zementeinpressen
Zement|verpressung f → Zementeinpressung
~verpreßgut n → Zementeinpreßgut
~verteiler m = cement spreader, ~ distributor
~verteilung f **in Säcken** = spreading and spotting cement
~(ver)wiegebehälter m = cement weigh(ing) hopper
~(ver)wiegung f = cement weighing
~waage f, Bindemittelwaage, Zementsilowaage, Abfüllwaage für Zement, Zementabfüllwaage = cement weigh(ing) batcher, ~ batcher scale
~warenform f, Betonwarenform = concrete products mo(u)ld
~warenwerk n [*Schweiz*]; Betonwarenwerk = concrete product plant
~-Wasser-Gemisch n → Zementschlämme f
~-Wasser-Verhältnis n [*Beton*] = cement water ratio
~werk n, Zementfabrik f = cement mill, ~ plant
~ziegel m → Betondachstein m
(~)Ziegeldach n, Betonsteindach = concrete tile roof
~ziegelfarbe f → Betondachsteinfarbe
~-Zumeßanlage f → Dosier(ungs)anlage für Zement
~zumeßschnecke f → Abmeßschnecke für Zement
~zusatz m = cement additive
~-Zuteilanlage f → Dosier(ungs)anlage für Zement
~zuteilschnecke f → Abmeßschnecke für Zement
zentrale Betonmischanlage f → Betonzentrale f
~ Betriebsanlage f = "packaged" plant, central engine ~
zentraler Austrag m, zentrale Austragung f [*Stabrohrmühle*] = centre peripheral discharge (Brit.); center ~ ~ (US)
~ Austritt m **durch Überlauf** [*Stabrohrmühle*] = overflow
~ Bedienungsstand m = central control platform
~ Betonstahlverarbeitungsplatz m, Betonstahlverarbeitungsanlage f = reinforcement yard
~ Doppeltrommelmischer m = dual-drum central-mixer
~ Ein-Trommelmischer m = single-drum central-mixer
zentrales Abfertigungsgebäude n [*Flughafen*] = main terminal building
~ Hydraulik- und Schmiersystem n = closed center hydraulic and lube system (US); ~ centre ~ ~ ~ (Brit.)
Zentral|bad n = central baths
~dach n = central roof
~eruption f, Punkteruption = central eruption, explosion-pipe ~
~-(Heiz)Öllagertank m = central(ized) (fuel oil) storage tank, ~ heating oil ~ ~
~heizung f, Sammelheizung = central heating
~heizungsschornstein m = central heating chimney
zentralisierte Steuerung f, ~ Bedienung = centralized controls
Zentral|lenkung f, Knicklenkung = center pivot steer(ing) (US); centre ~ ~ (Brit.)
~mischanlage f, stationäre Großmischanlage = central mixing plant
~prüfamt n = inspection directorate
~-Reparaturwerkstatt f = central repair (work)shop
~(schalt)warte f = central switch room
~schmierung f = central lubrication
~-Steuerstand m = central operating platform

zentralsymmetrisch = axisymmetric
Zentralwerkstatt *f* = central(work)shop
Zentrier|feder *f* = centering spring
~lager *n* = pilot bearing
~mutter *f* = centering nut
~stift *m* = centering pin
Zentrierung *f* = centering
zentrifugaler Elevator *m*, zentrifugales Becherwerk *n* = centrifugal discharge elevator, spaced bucket ~, boot loading ~
~ Schwerkraftelevator *m*, zentrifugales Schwerkraftbecherwerk *n* = chain type spaced bucket elevator with centrifugal gravity discharge
Zentrifugal|entladung *f*, Schleuderentladung = centrifugal discharge
~-Feuchtigkeits-Äquivalent *n* = centrifuge moisture equivalent
~moment *n* = centrifugal moment
~pumpe *f*, Kreiselpumpe, Schleuderpumpe = centrifugal pump
~-Staubabscheider *m*, Fliehkraft-Staubabscheider = centrifugal (dust) collector, cyclone (~) ~, cyclonic (~) ~ [*It accomplishes dust removal through combined velocity reduction and the action of centrifugal forces and is effective principally on coarser particles*]
~ventilator *m*, Radialventilator, Schleuderventilator = centrifugal fan, radial-flow ~
Zentrifuge *f*, Schleuder *f* = centrifuge
zentrisch vorgespannter Betonstab *m* = gravity centre (Brit.)/center (US) prestressed concrete bar
Zeolith *m*, Siedestein *m* = zeolite
Zeresin(wachs) *n*, Kunstwachs = ceresine wax
Zerfall *m*, Brechen *n* [*Emulsion*] = breaking, breakdown
~, Zerrüttung *f*, Auflösung = disintegration
zerfallen, brechen [*Emulsion*] = to break (down)
Zerfallschlacke *f* = disintegrating slag, slaking ~
Zerfallswert-Bestimmung *f* nach **Weber und Bechler** = breaking time test of Weber and Bechler [*bitumen emulsion (Brit.)*]
zerfließen = to deliquesce [*To dissolve gradually and become liquid by absorption of moisture from the air, as certain salts*]
zerkleinerter Feinzuschlag(stoff) *m*, Betonbrechsand *m* = crushed (concreting) sand, ~ concrete ~, ~ fine aggregate
Zerkleinerung *f*, Hart~ = comminution, reduction
~ im Durchlauf → Hart~ ~ ~
~ mit Umlauf → Hart~ ~ ~
Zerkleinerungs|gesetz *n* nach v. **Rittinger** [*Die aufgewendete Zerkleinerungsarbeit steht in proportionalem Verhältnis zur erzeugten Oberfläche*] = Rittinger's law [*Basic law of crushing and pulverizing which states that the work done is in proportion to the area of the new surfaces developed*]
~kinetik *f* = kinetics of comminution
~maschine *f*, Hart~ = comminuter
~stufe *f*, Hart~ = comminution step, reduction ~
~theorie *f* = theory of comminution, ~ ~ reduction
zerklüftetes Gelände *n* = jagged terrain
Zerknallgefahr *f* = risk of igniting explosive mix(tur)es of pit gases and air [*sewer*]
zerlegen, abmontieren, demontieren, auseinandernehmen, abbauen = to disassemble, to strip down, to dismantle
Zerlegen *n*, Zerlegung *f*, Abbauen, Abbau *m*, Demontage *f* = dismantling
Zerlegung *f* → Zerlegen *n*
Zerlegungsgleis *n*, Sortiergleis, Ausziehgleis = sorting track
zermahlen = to break down by grinding
zermürbter Ton *m*, geklüfteter ~ = shattered clay
Zerreiß|festigkeit *f* → Zugfestigkeit
~prüfmaschine *f* = tension testing machine, tensile ~ ~
Zerrüttung *f*, Zerfall *m*, Auflösung *f* = disintegration

Zerrüttungszone — zerstörungsfreie Betonprüfung

Zerrüttungszone f → Bruchzone
zersetzen = to decompose
zersetzter Granit m = decomposed granite, gowan
Zersetzung f = decomposition
~, chemische ~ [*Sprengstoff*] = decomposition
Zersetzungsdestillation f → Entgasung
Zerspratzen n [*Bohrtechnik*] = decrepitation
zerspringen, abplatzen [*Keramikindustrie*] = to chip, to spall
Zerspringen n → Abplatzen
Zerstäuben n, Versprühen = fog spray, atomizing, mist-spraying
Zerstäuber m, Spritzflakon n, m = atomizer

~, Bindemittel-Eindüsapparatur f = binder spray device
Zerstäubung f, Versprühung = atomization
Zerstäubungs|brenner m = atomizing burner
~trockner m = spray dryer; ~ drier (US)
~trocknung f = spray drying
zerstörende Betonprüfung f = destructive testing of concrete
Zerstörung f des Betons, Betonkorrosion f = concrete corrosion
~ von Förderbändern durch Bakterien = microbial destruction of conveyors, ~ ~ ~ conveyers

---

zerstörungsfreie Betonprüfung f	non-destructive testing of concrete
Ankopp(e)lungsmittel n	acoustic coupling agent, couplant
dynamischer Elastizitätsmodul m	dynamic modulus of elasticity
dynamisches Prüfverfahren n	dynamic testing technique (or technic)
ankoppeln	to couple
Durchlaufzeit f	time of propagation, propagation time, transit time, pulse-transmission time, time of transmittal
Empfangsverstärker m	receiver amplifier
Fortpflanzungsgeschwindigkeit f	velocity of propagation
Geschwindigkeitsmessung f	(longitudinal) wave-velocity method (or measurement), pulse-velocity technique (or technic)
Gummimembran(e) f	rubber diaphragm
Impulsverfahren n	pulse-echo technique (or technic)
Kathodenstrahlröhre f	cathode-ray tube
Kugelschlagprüfung f	dynamic ball-impact test method
Laufweglänge f	path length
Knallfunkengeber m	firing-type pulse-velocity measuring device
Longitudinalwelle f	longitudinal wave
mechanisches Hammergerät n	hammer-blow type pulse-velocity measuring device
piezoelektrischer Empfangsschallkopf m	piezoelectric crystal receiving transducer (or head)
piezoelektrischer Sendeschallkopf m	piezoelectric crystal transmitting transducer (or head)
Resonanzfrequenzmessung f	resonant-frequency technique (or technic), lowfrequency vibration method
Schallkopf m	transducer, pickup

zerstörungsfreie Probe — Ziegelgebäude 1150

Schallbereich *m*	sonic range
Schallverfahren *n*	sonic method
Schallvibration *f*	sonic vibration
Soniskop *n*	soniscope
Ultraschall *m*, Überschall, unhörbarer Schall	ultrasonic sound, ultrasound
Ultraschallimpuls *m*	ultrasonic pulse
Ultraschallimpuls-Echo-Verfahren *n*	ultrasonic pulse-echo technique (or technic)
Ultraschall(impuls)methode *f*	ultrasonic (pulse) method
Ultraschallbetonprüfgerät *n*	ultrasonic concrete tester

zerstörungsfreie Probe *f*, ∼ Prüfung *f*, zerstörungsfreier Versuch *m* = nondestructive test
Zerstörungs|kraft *f* = destructiveness
∼produkt *n* (Geol.) = product of destruction
zerstückelter Draht *m* = shredded wire
Zerteilen *n* [*Gesteinsrohblöcke*] = cutting
Zertrümmerungs|kugel *f* = ball breaker, breaking ball
∼ramme *f* = drop hammer for shattering old pavements
ZGV *n*, Viskosezellgarn *n* = viscose yarn
Zick-Zack-Duostraße *f*, Einstichstraße = staggered mill
zick-zack-förmiges Schubsicherungseisen *n* [*Hochkant auf die Brückentafel geschweißt*] = zig-zag anchor bar
Ziegel *m*, Ton∼ (Back)Stein *m* = (clay) brick
∼ für nichttragendes Mauerwerk, ∼ ∼ nichttragende Wände = nonloadbearing brick
∼ ∼ tragendes Mauerwerk, ∼ ∼ tragende Wände = loadbearing brick
∼aufzug *m* = brick lift
∼auskleidung *f* → Backsteinauskleidung
∼balken(träger) *m* = brick beam
∼bekleidung *f* → Backsteinauskleidung
∼betonplatte *f* [*Mit Ziegeln verblendete Betonplatte*] = brick-lined concrete slab

∼boden *m*, Ziegelfußboden = brick flooring
∼bogen *m*, Backsteinbogen = brick arch
∼brecher *m*, Backsteinbrecher = brick crusher, ∼ breaker
∼brennen *n*, Ziegelbrand *m* = brick burning
∼brennerei *f* → Ziegelei
∼(brenn)ofen *m* = brick kiln
∼bruch *m*, Ziegelschotter *m* = broken bricks
Ziegeldach *n* = tile roof
∼, Zement∼, Betonsteindach = concrete tile roof
∼, Ton∼ = clay tile roof
∼decker *m* = tiler
Ziegeldecke *f* = brick floor
Ziegelei *f*, Ziegelwerk *n*, Ziegelbrennerei *f*, Ziegelfabrik *f* = brickworks
Ziegel|-Eikanal *m* = egg oval brick sewer, ∼ ∼ ∼ drain
∼(-Einsteig)schacht *m*, Ziegelmannloch *n* = brick manhole
∼fabrik *f* → Ziegelei
∼fabrikation *f*, Ziegelherstellung *f* = brickmaking
∼-Fenstersturz *m* → Stahlton-Fenstersturz
∼-Fertigsturz *m*, vorgespannter ∼, vorgespannter Fertigsturz, Stahltonsturz = Stahlton plank lintel, ∼ lintol
∼fertigteil *m*, *n* = precast brick unit
∼(fuß)boden *m* = brick flooring
∼gebäude *n*, Ziegelhaus *n* = brick building

~gesims n, Ziegelsims m, n = brick corbel
~gewölbe n = brick vault
~gewölbedecke f = brick vault floor
~greifer m = brick grab, ~ gripping device
~gut n, Ziegelton m = brickearth(s), brick clay
~haus n, Ziegelgebäude n = brick building
~herstellung f, Ziegelfabrikation f = brickmaking
~hohlmauer f = brick cavity wall
~-Hohlmauerwerk n = cavity brickwork
~kammerofen m = stationary continuous brick kiln
~kehle f = tile valley
~mannloch n → Ziegel(-Einsteig-)schacht m
~mauer f, Ziegelwand f = brick wall
Ziegelmauerwerk n = brickwork; brick masonry (US)
~-Bewehrung f, Ziegelmauerwerk-Armierung = brickwork reinforcement; brick masonry ~ (US)
~-Bewehrungsmatte f = brick reinforcement fabric
~muster n = brickwork pattern; brick masonry ~ (US)
Ziegel|mehl n, Ziegelstaub m = brickdust, ground brick, brick flour
~/Mörtel-Grenzfläche f = brick/mortar interface
~ofen m, Ziegelbrennofen = brick kiln
~paket n = brick package
~pfeiler m = brick (foundation) pier
~pflaster(decke) n, (f) = brick paving
~presse f = brick-pressing machine
~-Prüfmaschine f = brick tester, ~ testing machine
~-Rauchkanal m = brick flue
~rollenbahn f = roller conveyor for bricks, ~ conveyer ~ ~
~rollschicht f = brick-on-edge
~säge f = brick saw
~schacht m → Ziegeleinsteigschacht
~schicht f = course of brickwork

~schleuder f, Ziegelwurfmaschine f; SEEGERS-Steinschleuder [Trademark] = brick-throwing machine
~schneider m = brick cutter
~schornstein m = brick(-built) chimney
~schotter m, Ziegelbruch m = broken bricks
~schutt m = brick rubble
~sortieren n = culling [Sorting brick for size, colour and quality]
~splitt m, Tonsplitt [Splitt aus Ziegeltrümmern] = chip(ping)s from brick ruins
~stapel m = stacked bricks
~stapler m = brick stacker
~staub m → Ziegelmehl
~stein m, (Ton)Ziegel m = clay brick
~(stein)schornstein m, Ziegel(stein)-kamin m, Backsteinschornstein, Backsteinkamin = brick chimney
~sturz m = brick lintel, ~ lintol
~ton m → Ziegelgut
~(transport)karre(n) f, (m) = brick truck
~(transport)wagen m = brick carrier
~trümmer f = brick ruins
~umfassungswand f = external brick wall
~verband m, (Backstein)Verband = (brick) bond
~verkleidung f → Backsteinauskleidung
~wand f, Ziegelmauer f = brick wall
~werk n → Ziegelei
~wurfmaschine f → Ziegelschleuder
Zieh|brunnen m = draw well, open bucket ~ (Brit.); sweep ~ (US)
~düse f [Preload-Maschine] = die
ziehen, (her)aus~ [Pfahl; Spundbohle] = to pull out, to extract
~, schleppen = to tow
Ziehen n [Gleitschalung] = slide (operation), sliding
ziehen des Bohrzeuges, auf~ ~ ~, herausziehen ~ ~, ~ der Bohrgarnitur f [pennsylvanisches Seilbohren] = to pull the drilling tools out, to hoist ~ ~ ~ ~

Zieh|feder *f*, Tuschfeder = drawing pen
~**fett** *n* = drawing grease
~**gehänge** *n* [*(Pfahl)Ramme*] = grip
~**gerät** *n* = towing unit
~**gerät** *n* [*Ziehen von Pfählen und Spundbohlen*] = puller, pulling tool
~**geschwindigkeit** *f* [*Gleitschalung*] = speed of slide (operation), rate ~ ~ (~), ~ ~ sliding
~**harmonikatür** *f* = accordion door
~**vorrichtung** *f* = puller device
~**walzwerk** *n* = coiler tension rolling mill
~**weg** *m*, Schlittweg [*Holzbringung*] = sledge-way
Ziel|achse *f*, Kollimationsachse [*Theodolit*] = line of collimation
~**verkehrszählung** *f* = destination survey
~**verkehr** *m* **zur Innenstadt** = downtown terminal traffic
Zier|beton *m*, Dekorativbeton = decorative concrete, ornamental ~
~**bogen** *m* = decorative arch, ornamental ~
~**eisen** *n* = decorative iron, ornamental ~
~**folie** *f* = decorative foil, ornamental ~
~**fuge** *f* = ornamental joint, decorative ~
~**glas** *n*, Ornamentglas = ornamental glass, decorative ~
~**konsole** *f* = ornamental bracket, decorative ~
~**oberfläche** *f*, Dekor(ations)oberfläche = decorative surface, ornamental ~
~**rille** *f* = glyph [*In architecture, a short, vertical ornamental channel or groove*]
~**stahl** *m* = decorative steel, ornamental ~
~**verband** *m* [*Mauerwerk*] = decorative bond, ornamental ~
~**wand** *f* = decorative wall, ornamental ~
~**ziegelmauerwerk** *n* = decorative brickwork, (ornamental) pattern ~; decorative brick masonry, (ornamental) pattern brick masonry (US)

Zifferblatt|-Abmeßwaage *f* **mit Waagenbehälter** = dial batcher
~**waage** *f*, Kreiszeigerwaage = circular dial-type scale
Zigarettenglut *f* = cigarette burn
Zimmer|arbeiten *fpl* = carpenter's work
~**bad** *n* [*Hotel*] = private bathroom
~**mann** *m*, Zimmerer *m* = carpenter
Zimmermanns|axt *f* = carpenter's axe; ~ ax (US)
~**(blei)stift** *m* = lumber crayon (US)
Zimmer|platz *m*, Zulage *f*, Zimmerwerkplatz = carpenter's yard
~**polier** *m* = foreman carpenter
~**tür** *f*, Stubentür = room door
Zimmerung *f* = timbering
Zink|abweis(e)blech *n* = zinc flashing
(~)**Blende** *f*, Zinksulfid *n*, Schwefelzink *n*, ZnS = black jack, sphalerite, (zinc) blende; zinc sulphide (Brit.); zinc sulfide (US)
~**druckgußstück** *n* = zinc die casting
~**(ein)deckung** *f* = zinc roofing
~**elektrolyse** *f* = zinc electrolysis
~**erz** *n* = zinc ore
~**farbe** *f* = zinc pigment
~**feinblech** *n* = zinc sheet
~**gelb** *n* = zinc chromate primer
~**glas(erz)** *n*, Kieselzink(erz) *n*, Kieselgalmei *m* (Min.) = siliceous calamine, willemite
~**grobblech** *n* = zinc plate
~**grube** *f* = zinc mine
Zinkit *n*, Rotzinkerz *n*, ZnO (Min.) = zincite, red oxid(e) of zinc, spartalite, sterlingite
Zinkoxyd *n* = zinc oxid(e)
~**farbe** *f* = zinc oxid(e) pigmented paint
Zinkschindel *f* = zinc shingle
Zinkstaub *m* = zinc dust [*Powdered zinc which is used in priming paints for use on galvanized iron*]
~**fällung** *f* [*Cyanidlaugung*] = precipitation with zinc dust
~**(grundier)farbe** *f* = zinc dust primer
Zink|vitriol *n, m*, Goslarit *m* (Min.) = goslarite, white vitriol, white copperas

Zinkweiß — Zufall-Schwingung

~weiß n, ZnO = zinc white, Chinese ~, zinc oxid(e)
~wellblech n = corrugated zinc sheet
Zinn n = tin
~-Blei-Lot n, Blei-Zinn-Lot = lead-tin solder, tin-lead ~
~folie f = tin foil
~lot n, Lötzinn n = fine solder [An alloy of $^2/_3$ tin, $^1/_3$ lead]
~stein m, SnO_2 (Min.) = cassiterite, tinstone
Zinsöl n = royalty petroleum
Zirkel m mit **Stellschraube und auswechselbarer Bleimine und Tuschfeder** = spring bows, compasses with extension bar for use with pen or pencil
~bogen m → Rundbogen
Zirkulations|-Sprengrampe f, Umlauf-Sprengrampe, Umwälz-Sprengrampe = circulating spraybar
~ventil n [*Tiefbohrtechnik*] = circulating sub
Zirkustal n, Kar n = corrie, cirque
Zisterne f = rain water storage tank
ziviler Bevölkerungsschutz m = civil defense (US); ~ defence (Brit.)
Zoll|abfertigung f = customs control, ~ clearance
~abfertigungsbereich m [*Flughafen*] = customs control area, ~ clearance ~
~büro n = customs control
(zoll)freier Hafen m → Freihafen
Zoll|hafen m = customs harbo(u)r
~stock m, Glieder-Maßstab m = folding (pocket) rule
Zone f der Öffnungen → Bruchzone
Zonen|bauordnung f = use-zoning
~bauweise f [*Erdstaudamm*] = "zoned" type of construction
~gewölbe n → Gurtgewölbe
~klimatisierung f = zoned air-conditioning
~schmelzen n = zone-melting
~schmelzen n von Wismut, Wismut-Zonenschmelzen = zone refining of bismuth
Zoreseisen n, Belagstahl m, Trogplatte f [*DIN 1023*] = trough plate

Z-Schneiden-Bohrkrone f = z (drill) bit
Z-Stahl m = Z-bar, zee-bar [*A Z-shaped bar used as a wall tie*]
Zubahn f → Rollweg m
Zubehör m, n = accessories
~teil m, n = accessory part
Zubringer m, Speisegraben m [*zu einem Wasserlauf*] = feed(er) ditch, feeding ~, feeder
~band n [*zum Abwurfband*] = feed belt
~förderer m = feeder conveyor, ~ conveyer, supply ~
~kanal m [*Wasserkraftwerk*] = head race, headwater channel
~schnecke f → Beschickungsschnecke
~straße f, Zufahrtstraße, Zubringer m [*Verbindungsstraße zwischen Hauptdurchgangstraßennetz, Städten oder Stadtteilen und besonderen Verkehrszielen*] = feeder street
~walze f → Beschickungswalze
(Zu)Bruchwerfen n → Hereinbrechen
Zucker|rübengabel f = beet fork
~rübenlöffel m = beet bucket
~silo m = sugar silo
zudecken, abdecken, bedecken = to cover (over)
zuerst abbauen [*Flöz*] = to work first
Zufahrt f = access way
~, Brücken~, Zufahrtrampe f = approach (viaduct)
~damm m zu einer Brücke = approach embankment
~gradiente f = approach gradient
~rampe f, (Brücken)Zufahrt f = approach (viaduct)
~reg(e)lung f = access control
~rinne f [*z.B. zu einem Dock*] = shipping lane
~stelle f → Anschlußpunkt m
~straße f → Zubringerstraße
zufällige Strömung f = accidental current
zufälliges Mineral n, akzessorisches ~, Übergemengteil m, n = accessory mineral
Zufall-Schwingung f = random vibration

Zufluß *m* [*Ölhydraulik*] = supply
~ = inflow
~ = influent [*Sewage, raw or partly treated, flowing into any sewage-treatment device*]
~**begrenzer** *m* = flow limiter
~**regler** *m* = flow control
~**regulierventil** *n* [*Hydrauliksystem*] = flow control valve
~**steigschacht** *m*, (Kühlturm)Steigschacht = (cooling tower) chimney
Zug *m* = tension
~ = draught (Brit.); draft (US)
~ = train
~**(ab)dichtung** *f* = weather strip, wind stop; air lock, weatherstripping (US) [*Used to reduce the leakage of air through the joints of a closed door or window*]
Zugabe *f* [*Beim Bewehrungsstumpfschweißen erforderliche Überlänge des Stahls, die beim Schweißen verloren geht*] = overlength
Zugang *m* = access
~ **zu einem U-Bahnhof von der Straße aus** = pavement entrance
Zugänglichkeit *f* = accessibility
Zugang|schacht *m*, Einsteigschacht = access shaft
~**stollen** *m* → Abstiegstollen
(Zug)Anker *m* → Ankerstab *m*
Zugarmierung *f* → Zugbewehrung
Zugbalken *m* → Trambalken
Zugband *n* = tieback
Zugbewehrung *f*, Zugarmierung, Zug(stahl)einlagen *fpl* = tensile steel, ~ reinforcement, reinforcement for tension, steel for tension
Zugbruch *m* = tensile failure, tension ~
Zugbrücke *f* = draw-bridge
Zugdehnung *f* = tensile strain
Zugdeichsel *f*, (Fahr)Deichsel = towbar
Zugdiagonale *f*, Zugschräge *f*, gezogene Diagonale, gezogene Schräge [*Fachwerkbinder*] = diagonal tie, ~ in tension, tension diagonal
Zugdichtung *f* → Zugabdichtung
Zug-Druck-Wechselbeanspruchung *f* = reversed direct stress
Zugeinlagen *fpl* → Zugbewehrung

zugelegtes Gerüst *n* → abgebundenes ~
Zügelgurt|bauweise *f*, Schrägseilbauweise [*Brücke*] = stayed girder construction, tied cantilever ~
~**brücke** *f*, Schrägseilbrücke = tied cantilever bridge, stayed girder ~
zugeordnete Gerade *f*, konjugierte ~ = conjugate line
Zugermüdung *f* = fatigue in tension
Zugermüdungsgrenze *f* = limit of fatigue in tension
zugeschnittenes Bauholz *n* → Bindeholz
Zugfaser *f* = fibre in tension (Brit.); fiber ~ ~ (US)
Zugfeder *f* = tension spring
Zugfestigkeit *f*, Zerreißfestigkeit = tensile strength, tension ~
Zugfestigkeitsmesser *m* = tensile strength tester, tension ~
Zugfestigkeitsprüfung *f*, Zugfestigkeitsprobe *f*, Zugfestigkeitsversuch *m* = tensile strength test, tension ~ ~
Zugfreiheit *f* [*Lüftung*] = absence of draught (Brit.); ~ ~ draft (US)
Zuggestänge *n* [*Tiefbohrtechnik*] = pull rods
~**aufhänger** *m* [*Tiefbohrtechnik*] = pull rods hanger
~**strang** *m* [*Zwischen Erdölpumpenbock und zentralem Antrieb*] = pull line, rod ~, shackle ~
~**träger** *m* = pull rods carrier
~**verbindung** *f* = pull rods connection
Zugglied *n*, Zugstab *m* [*Zur Übertragung von Zugkräften bei Tragwerken*] = tension member, ~ bar, bar in tension, member in tension
Zuggurt(ung) *m*, (*f*) = tension flange, ~ boom, ~ chord
Zughaftung *f* = tensile bond
Zughaken *m*, Rotaryhaken = rotary hook
~**leistung** *f*, Arbeitsleistung = drawbar horsepower, ~ performance
~**prüfung** *f*, Zughakenprobe *f*, Zughakenversuch *m* = drawbar test
~**-PS** *f* = drawbar horsepower, ~ H.P., ~ HP, ~ h.p., ~ hp

zügige Linienführung f = flowing alignment (Brit.)/alinement (US)
Zugkomponente f = tensile component, tension ~
Zugkraft f [*beansprucht*] = tensile force, pull
~ = tractive power, ~ effort, pulling power, pulling force
~ **am Zughaken** = drawbar pull
~**übertragung** f = flotation
Zugkübelbagger m → Schürfkübelbagger
Zugkübel-Schreitbagger m, Eimerseil-Schreitbagger, Schürfkübel-Schreitbagger = walking dragline
Zuglasche f = tension butt strap
Zuglast f = tensile load
Zugleistung f = pull
Zugloch n, Luftloch = air vent, ~ hole, port
Zugluft f = draught air (Brit.); draft ~ (US)
Zugmaschine f, Straßen~ = tractor-truck, truck-tractor, motor tractor
Zugpfahl m = uplift pile, tension ~
~**versuch** m, Zugpfahlprobe f, Zugpfahlprüfung f = tension pile test
Zugprobe f, Zugprüfung f, Zugverversuch m = tension test
~(**körper**) f, (m) = tension test sample
Zugquerschnitt m = tensile section
Zugramme f = hand pile driver
Zugraupe f = towing caterpillar tractor, ~ crawler ~
Zugring m = tension ring
Zugriß m = tension crack
Zugschräge f → Zugdiagonale
Zugschwellfestigkeit f = fatigue strength under pulsating tensile stresses
Zugseil n = pull rope, ~ cable, hauling ~
~, Vollseil, Schürfseil [*Tiefen- und Langstreckenförderer*] = inhaul cable, ~ line, load ~, pull ~
~**bagger** m, Schürfkübelbagger, Eimerseilbagger = dragline
~**umlenkrolle** f [*Tiefen- und Langstreckenförderer*] = load-line guide block
~**winde** f = pull-rope winch
Zugspannung f = tensile stress
~ [*Raupenkette*] = traction stress
Zugspannungsriß m = tension crack
Zugstab m → Zugglied
Zug(stahl)einlagen fpl → Zugbewehrung
Zugstange f → Ankerstab m
~ = tow(ing) pole, ~ bar, drawbar
~ [*Dachkonstruktion*] = suspension rod
~, Schwengel(schub)stange [*pennsylvanisches Seilbohren*] = pitman, beam ~
Zugstangen|lager n, Schubstangenlager [*pennsylvanisches Seilbohren*] = (beam) pitman bearing
~**öse** f = eye for towing
Zugtierverkehr m, Fuhrwerkverkehr, Gespannverkehr = animal-drawn traffic
Zugtrommel f [*Tiefen- und Langstreckenförderer*] = load drum
Zugverlust m = draught loss (Brit.); draft ~ (US)
Zugversuch m, Zugprüfung f, Zugprobe f = tension test
Zugvorspannung f = tensile prestress
Zugwagen m, Radschrapper-~ = (wheel) scraper tractor
Zugwerk n → (Rotary-)Hebewerk
Zugwiderstand m = resistance to tension
Zugwinde f = tow(ing) winch
Zugzone f = tension(ed) zone, tensile ~
Zuhaltung f [*Schloß*] = tumbler, lever ~, lever
Zukaufschrott m = bought scrap, external ~
Zulage f, Zimmer(werk)platz m = carpenter's yard
~**bewehrung** f, Zulagearmierung, Zulage(stahl)einlagen fpl = secondary reinforcement
~**stab** m = secondary re-bar, reinforcing bar
zulässig [*Druck; Spannung usw.*] = safe, allowable, permissible

zulässige Nutzlast — Zündkapsel

zulässige Nutzlast *f* [*LKW*] = legal payload
Zulassungs|kennzeichen *n* [*für Sprenggerät*] = approved marking
~prüfung *f* [*z.B. Schweißer*] = performance qualification test
Zulauf *m* [*Rohrteil*] = junction
~, Vorlaufleitung *f* = feed line
~druck *m* [*Ölhydraulik*] = supply pressure
~(ge)rinne *f*, *(n)* → Einlaufkanal *m*
~stollen *m* → Triebwasserstollen
zulegen, abbinden [*Holzkonstruktion*] = to trim, to join
Zuleitung *f* vom Pumpwerk [*Beregnung*] = main (pipe) line
Zuleitungs|graben *m*, Zuleiter *m* [*Furchenbewässerung*] = (earth) supply ditch, (~) feed ~
~kabel *n* = supply cable, feed ~
~stollen *m* [*Talsperre*] = feed gallery
Zuluft *f* = supply air
~anlage *f* = air supply system
~gitter *n* = inlet grille, supply air ~
~kanal *m* = air supply duct
Zumahlung *f* = intergrinding
zumessen, zuteilen, abmessen, dosieren = to measure, to proportion, to batch
Zumessung *f* → Dosierung
Zumeßanlage *f* → Zuteilanlage
~ für Zement → Dosier(ungs)anlage ~ ~
~ ~ Zuschlagstoffe → Dosier(ungs-)anlage ~ ~
~ in Turmanordnung → Abmeßanlage ~ ~
~ mit dreimaligem Halten der Fahrzeuge → Abmeßanlage ~ ~ ~ ~
~ ~ einmaligem Halten der Fahrzeuge → Abmeßanlage ~ ~ ~ ~
~ ~ zweimaligem Halten der Fahrzeuge → Abmeßanlage ~ ~ ~ ~
Zumeß|apparat *m* → Dosier(ungs)gerät
~automat *m* → Dosier(ungs)automat
~automatik *f* → Abmeßautomatik
~band *n* → Abmeßband

~bandwaage *f* → Dosierbandwaage
~bunker *m* → Dosier(ungs)bunker
~förderschnecke *f* → Abmeßförderschnecke
~gefäß *n* → Abmeßgefäß
~gerät *n* → Dosier(ungs)gerät
~-Hygrometer *m, n* → Abmeß-Hygrometer
~karre(n) *f*, *(m)* für (Beton)Zuschläge → Abmeßkarre(n) ~ ~
~kiste *f* → Abmeßkiste
~programmwähler *m* → Dosier(ungs-)programmwähler
~pumpe *f* → Abmeßpumpe
~rinne *f* → Dosier(ungs)rinne
~schleuse *f* → Abmeßschleuse
~schnecke *f*, Zuteilschnecke, Dosier(ungs)schnecke, Abmeßschnecke, Schneckenzuteiler *m* = proportioning worm conveyor, ~ screw ~, ~ ~ conveyer
~schnecke *f* für Zement → Abmeßschnecke ~ ~
~silo *m* → Dosier(ungs)silo
~spiel *n*, Zuteilspiel, Abmeßspiel, Dosier(ungs)spiel = batching cycle
~stern *m* → Dosier(ungs)stern
~teil *m, n* → Abmeßteil
~trommel *f* → Dosier(ungs)trommel
~turm *m* → Turmdosieranlage *f*
~vorrichtung *f* → Dosier(ungs)gerät
~waage *f* → Dosier(ungs)waage
Zumischung *f* → Beimischung
Zünd|apparat *m* → Zündmaschine
~beschleuniger *m* [*Dieselkraftstoff*] = dope, ignition accelerator
~draht *m* = ignition wire
Zünden *n* → Abtun *n* [*Sprengschüsse*]
Zunder *m*, Walzhaut *f*, Hammerschlag *m*, Abbrand *m* = scale
~brecher *m* = scale breaker
~brechgerüst *n* = scale breaker (stand)
zunderfrei = scale-free
Zunderschicht *f* = oxid(e) layer, scale ~
Zünd|folge *f* = firing order
~hebel *m* [*Motor*] = magneto wires
~kabel *n*, Schießkabel = shot-firing cable
~kapsel *f*, Sprengkapsel = blasting cap

~kerze *f* = spark(ing) plug
~kerzenprüfgerät *n* = (spark) plug tester, sparking ~ ~
~knopf *m* = ignition button
~kontakt *m* [*Motor*] = point
~kreis *m*, Zündleitung *f* [*Sprengen*] = firing circuit
~leitung *f*, Zündkreis *m* [*Sprengen*] = firing circuit
~magnet *m* = (ignition) magneto
~maschine *f*, Zündapparat *m* = (multi-shot) exploder, blasting machine
~papier *n* für Diesel = touch paper, ignition ~
~satz *m* [*Cardoxrohr*] = exploder
~satz *m* = detonating charge, cap ~
~schalter *m* = igniter switch, ignition ~
~schnur *f* = fuse
Zündung *f* = ignition
Zünd|verzug *m* = ignition delay
~willigkeit *f* = ignition performance
~zeitpunkt *m* = ignition time
~(zeitpunkt)verstellung *f*, Zündeinstellung = sparking advance, ignition timing
Zunge *f* [*Trenndamm zwischen zwei Hafenbecken*] = pier [*sometimes: jetty*]
Zungenstein *m* → Biberschwanz *m*
~dach *n* → Flachziegeldach
(Zu)Rollbahn *f* → Rollweg *m*
Zurücklegung *f* des Ufers = setting back of the bank
zurückpumpen = to pump back
Zurückschaltbereich *m* = downshift range
zurück|schalten, herunterschalten = to shift down
~schnellen [*Pfahl*] = to bounce
Zusammenbau *m* = assembly
~werkstatt *f*, Montagewerkstatt, Montagehalle *f* = assembly (work)shop
Zusammenbrechen *n* → Hereinbrechen
zusammendrückbar = compressible
Zusammen|drückbarkeit *f*, Verdichtungsfähigkeit, Verdichtungswilligkeit = compressibility, compactability

~drückung *f*, Pressung = compression
~drückversuch *m* → Kompressionsversuch
~drückversuch *m* mit beschränkter Seitenausdehnung = compression test with confined lateral expansion
~fassen *n* [*Bündel zu einem Tragkabel für eine Hängebrücke*] = compacting
~fluß *m* [*Flüsse*] = confluence (US); confluent (Brit.); junction
zusammengefaßtes Spannkabel *n* = multi-strand steel prestressing cable
zusammengehörige Tiefen *fpl* = conjugate depths, corresponding ~
~ Wasserstände *mpl* = corresponding stages, conjugate ~
zusammengesetzte Biegung *f* = compound flexure
~ Gleitflächen *fpl* = composite surfaces of sliding
~ Rohrfahrt *f*, kombinierte ~ [*Tiefbohrtechnik*] = combined string of casing
~ Spannung *f* = combined stress
zusammengesetzter Bauteil *m*, zusammengesetztes Bauteil *n* [*Aus verschiedenen Bauelementen zusammengesetzt*] = compound unit [*Building material which is formed as a composite article complete in itself but which is intended to be part of a complete building or structure. Examples are door with frame, window, sink unit*]
~ (Erz)Gang *m* = (ore) lode
~ Pfahl *m* [*Zusammengesetzte Pfähle bestehen aus zwei oder mehr vorgefertigten Pfahlabschnitten (Teillängen) mit*
a) gleicher Querschnittform und Größe bei gleichem Baustoff,
b) unterschiedlicher Querschnittform und Größe bei gleichem Baustoff,
c) wechselnder Querschnittform und Größe bei verschiedenen Baustoffen.
Die Teillängen können sowohl vor als auch während der Rammung zusam-

zusammengesetzter Spannbetonbalken — Zuschlagsiloanlage 1158

mengefügt werden] = composite pile in separate lengths
~ **Spannbetonbalken(träger)** *m*, Spannbeton-Verbundbalken(träger) = composite prestressed concrete beam
zusammen|gesetztes Hängewerk *n*, vereinigtes Hänge- und Sprengwerk, Hängesprengwerk = composite truss frame
~**klappbare Stahlschalung** *f* = collapsible formwork
~**legbar** = collapsible
~**legbarer Behälter** *m* = collapsible tank
~**nieten** = to assemble by welding
Zusammenschub *m*, Einsinkweg *m* (⚒) = yield [*of props, arches, etc.*]
~**widerstand** *m*, Einsinkwiderstand (⚒) = resistant to yield
Zusammensetzung *f*, Aufbau *m* [*Mörtel; Beton*] = composition
~ = formulation [*e. g. adhesives are formulated of epoxy resin, a plasticizer, and a curing agent*]
Zusammensetzverfahren *n* [*Berechnung von Faltwerken*] = method of combination
zusammenspannen [*Fertigteile durch Vorspannung miteinander verbinden*] = to stress together, to joint by prestressing, to tension together
Zusammen|spannen *n* von Fertigteilen, Verbinden durch Vorspannen = stressing together of precast units
~**wachsung** *f* = concretion
~**zählregel** *f*, Additionsregel = rule of addition
Zusatz *m* → (Beton)Wirkstoff *m*
~ **zur Hemmung der Schmierölalterung** = antioxidant
~**abschalthebel** *m* = emergency stop lever
~**armierung** *f* → Zusatzbewehrung
~**ausleger** *m* = fly jib (Brit.); ~ boom (US)
~**bewehrung** *f*, Zusatzarmierung, Zusatz(stahl)einlagen *fpl* = additional reinforcement, ~ steel

~**boden** *m*, Kornverbesserungsboden = complementary soil
~**einlagen** *fpl* → Zusatzbewehrung *f*
~**einrichtung** *f*, Anbaugerät *n*, Arbeitsgerät, Zusatzvorrichtung [*vorn angebaut*] = attachment, rig, front
~**gerät** *n* → Austauschgerät
~**heizung** *f* = additional heating
~**-Ladegerät** *n* = loading attachment
~**last** *f* = additional load
zusätzlich [*in einem Angebot*] = "extra over"
zusätzliche Arbeit *f* = addition, extra work
Zusatzluft *f* = secondary air
Zusatzmittel *n*, Wirkstoff *m*, Zusatzstoff = agent, additive, admix(ture)
~ → (Beton)Wirkstoff *m*
Zusatz|moment *n*, Moment aus Zusatzkräften = transient moment
~**spannung** *f* = additional stress
~**(stahl)einlagen** *fpl* → Zusatzbewehrung *f*
~**stoff** *m* → Zusatzmittel
~**teil** *m*, *n*, Anbau-Gerät *n*, Zusatzvorrichtung *f*, Zusatzwerkzeug *n* = attachment, rig
~**vorrichtung** *f* → Austauschgerät *n*
~**werkstoff** *m* [*Schweißen*] = filler metal
~**werkzeug** *n* → Austauschgerät *n*
~**werkzeug** *n* → Zusatzteil
Zuschalten *n* [*von einer Maschine*] = engagement
~ **der Pumpe** = pump engagement
Zuschauerraum *m* = auditorium (space)
Zuschieben *n* **eines Grabens**, Zuziehen *n* ~ ~ = backfilling of a trench by means of a backfiller attachment, re-filling ~ ~ ~ ~ ~ ~ ~
Zuschlag *m* [*Baustatik*] = addition
~ [*einem Zngebot den Zuschlag erteilen*] = award
~, ~**stoff** *m* = aggregate
~**angebot** *n* → → Angebot
~**silo** *m* **mit Dosiereinrichtung** = bin and batcher plant
Zuschlagsiloanlage *f* → Abmeßanlage für Betonzuschlagstoffe

Zuschlagsiloanlage — Zuteilschnecke

~ mit Wiegebalken = (bin and) troll(e)y batcher plant
Zuschlag(stoff) *m* = aggregate
~**-Abmeßanlage** *f* → Dosier(ungs)anlage für Zuschlagstoffe
~**abteil** *n* [*Chargen-LKW*] = batchtruck compartment (US); batchlorry ~ (Brit.)
~**-Aufbereitungsanlage** *f*, Aufbereitungsanlage für Zuschlagstoffe = aggregate(s) preparation plant, ~ ~ installation
~**-Aufbereitungsmaschine** *f*, Aufbereitungsmaschine für Zuschlagstoffe = aggregate(s) preparation machine
~**aufgeber** *m*, Zuschlagstoffspeiser, Zuschlagstoffbeschicker = aggregate feeder, ~ feeding unit
~**-Becherwerk** *n*, Mineralmasse-Becherwerk = aggregate (bucket) elevator
~**dosierapparat** *m* = aggregate batcher
~**-Dosier(ungs)anlage** *f* → Dosier(ungs)anlage für Zuschlagstoffe
Zuschlagstoffe *mpl* → (Beton)Zuschlagstoffe
Zuschlag(stoff)|eigenschaft *f* = aggregate property
~**-Erwärmung** *f* = aggregate heating
~**-Festigkeit** *f* = aggregate strength
~**gemenge** *n*, Gesamtzuschlag(stoff) *m*, Zuschlag(stoff)gemisch *n* = combined aggregate, total ~
~**korn** *n*, Zuschlagstoffteilchen *n* = aggregate particle, ~ grain
~**-Tunnel** *m* = aggregate tunnel
~**waage** *f* → Gesteinswaage *f*
~**-Zumeßanlage** *f* → Dosier(ungs)anlage für Zuschlagstoffe
~**-Zuteilanlage** *f* → Dosier(ungs)anlage für Zuschlagstoffe
Zuschneider *m* [*Gewinnung von Dachschieferplatten*] = trimmer
Zuschußwasser *n* = supplemental water
Zusickern *n*, zusitzen = infiltration
zusitzendes Grundwasser *n*, zusickerndes ~ = infiltration water
Zustand *m*, Beschaffenheit *f* = condition

Zustandsänderung *f* **nach Zeit** *f* [*z.B. Geschwindigkeit; Druck; Temperatur*] = rate
Zustellgleis *n* = shunting track
zutage treten → ausgehen (Geol.)
Zuteilanlage *f*, Abmeßanlage, Zumeßanlage, Dosieranlage = batch(ing) plant, proportioning ~, measuring ~, ~ installation
~ **für Zement** → Dosier(ungs)anlage ~ ~
~ **für Zuschlagstoffe** → Dosier(ungs)anlage ~ ~
~ **in Turmanordnung** → Abmeßanlage ~ ~
~ **mit dreimaligem Halten der Fahrzeuge** → Abmeßanlage ~ ~ ~ ~
~ ~ **zweimaligem Halten der Fahrzeuge** → Abmeßanlage ~ ~ ~ ~
~ ~**einmaligem Halten der Fahrzeuge** → Abmeßanlage ~ ~ ~ ~
Zuteil|apparat *m* → Dosier(ungs)gerät *n*
~**automat** *m* → Dosier(ungs)automat
~**automatik** *f* → Abmeßautomatik
~**band** *n* → Abmeßband
~**bandwaage** *f* → Dosierbandwaage
~**bunker** *m* → Dosier(ungs)bunker
zuteilen, zumessen, dosieren, abmessen = to measure, to batch, to proportion
Zuteil|förderschnecke *f* → Abmeßförderschnecke
~**gefäß** *n* → Abmeßgefäß
~**gerät** *n* → Dosier(ungs)gerät
~**-Hygrometer** *m* → Abmeß-Hygrometer
~**karre(n)** *f, (m)* **für (Beton)Zuschläge** → Abmeßkarre(n) ~ ~
~**kiste** *f* → Abmeßkiste
~**programmwähler** *m* → Dosier(ungs)programmwähler
~**pumpe** *f* → Abmeßpumpe
~**rinne** *f* → Dosier(ungs)rinne
~**schleuse** *f* → Abmeßschleuse
~**schnecke** *f* → Dosier(ungs)schnecke
~**schnecke** *f* **für Zement** → Abmeßschnecke ~ ~

Zuteilsilo — zweietagiges Kellergeschoß

~silo m → Dosier(ungs)silo
~spiel n → Zumeßspiel
~stern m → Dosier(ungs)stern
~trommel f → Dosier(ungs)trommel
~turm m → Turmdosieranlage f
Zuteilung f → Dosierung
Zuteil|vorrichtung f → Dosier(ungs)gerät n
~waage f → Dosier(ungs)waage
Zuverlässigkeitsindex m einer Quelle = reliability index of a spring
Zuwachsrate f = growth-rate
Zuziehen n eines Grabens → Zuschieben ~ ~
Zwanglauf m = positive movement
zwangläufige Kupp(e)lung f = positive clutch (coupling)
zwangläufiger Exzenterschwinger m = positive throw eccentric type screen
Zwanglauf|lehre f → Bewegungslehre
~trieb m = positive drive
kraftschlüssiger Antrieb m = nonpositive drive
Zwangs|auflader m, Lademaschine f mit gegen das Haufwerk fahrender Schaufelkette und Abwurfband = force-feed loader
~auslösung f = forced release
~betonmischer m → → Betonmischer
~bewirtschaftung f → Bewirtschaftung
~entleerung f, Zwangsentladung = positive discharge
zwangsgemischt, geknetet = pugmill-mixed, pressure-mixed, kneaded
Zwangs|halt m = compulsory stop
~mischen n, Kneten = pugmill mixing, pressure ~, kneading, rolling and mixing action, nonlift mixing action
Zwangsmischer m = compulsory type mixer
~ mit kurzen und langen Mischwerkzeugen = double zone (pug mill) mixer
~ ~ vertikalem Rührwerk → Knetmischer
~schaufel f = pugmill paddle (or blade)
Zwangs|schiene f → Leitschiene

~umwälzung f = forced circulation
Zwanzigflach n → Ikosaeder n
Zweck|bindung f = non-diversion
~dienlichkeit f = usefulness
~entfremdung f = diversion
zwei durch Mittelstreifen getrennte dreispurige Richtungsfahrbahnen f pl = three-lane dual carriageways
Zweiachsgestell n, Zweiachsunterwagen m [Bagger] = two-axle carrier
zweiachsige Biegung f = bi-axial bending
zweiachsiger Motor-Straßenhobel m, Abziehverteilgerät n = (auto)patrol (grader), motor patrol, blade maintainer, maintainer (scraper), road patrol, road maintainer
Zweiachs|-Motorstraßenhobel m mit Allradantrieb und All(rad)lenkung = four wheel drive and steer grader
~schlepper m → Vierradtrecker
~unterwagen m → Zweiachsgestell
~-Zugwagen m → Vierradtrecker
zweiarmiger Auslauf m → Hosenrohr n
~ Hebel m, an zwei Punkten drehbar gelagerter Verstellhebel = fulcrum lever
Zweibahnstraße f = two-way road
Zweibettzimmer n = two-bed room
~ [Krankenhaus] = two-bed room, ~ ward
Zweibohlenfertiger m = two-screed (or two-beam) finisher (or finishing machine)
Zwei(brenn)stoffmotor m = dual fuel engine
Zweidecker m → Doppeldecker
Zweiebenen|betrieb m, Doppelebenenbetrieb [Flughafen] = two-level operation
~finger m [Flughafen] = two-stor(e)y finger system
~system n, Doppelebenensystem [Passagier- und Gepäckabfertigung auf Flughäfen] = two-level (building operational) system
zweietagige Kreuzung f → zweigeschossige ~
zweietagiges Kellergeschoß n → zweistöckiges ~

zweifach geschertes Seil n [*Ein Einzelseil, das so in einer Seilscheibe eingeschert ist, daß zwei Längen dieses Seils die Last tragen*] = two part line

Zweifach-Rollenkette f = two strand roller chain

Zweifamilienhaus n = duplex (US)

Zweifeldträger m = two-span girder

Zweiflächengleitlager n = elliptical bearing

zweiflügelige Schlagtür f = double swinging door

~ **Tür** f, Flügeltür = double wing door

zweiflügeliger Abbau m (⚒) = working two faces away from a central roadway

~ **Streb** m → T-Bau m

Zwei-Füllungstür f = two-panel door

zweigängiger Schraub(en)pfahl m = screw pile with two turns

Zweigang-Schleifringmotor m = two-speed slipring motor

Zweigelenkbogen m = two-centred arch (Brit.); two-centered ~ (US)

~**scheibenbrücke** f = two-pinned arch bridge

~(**träger**) m, Bogen(träger) mit Kämpfergelenken = two-hinged arch(ed girder), double-hinged ~, arch(ed girder) hinged at the abutments

Zweigelenk|fachwerkrahmen m = trussed frame with two hinges

~**rahmen** m = frame with two hinges, two-hinged frame

zweigeschossige Brücke f, Etagenbrücke = double-deck bridge

~ **Kreuzung** f, zweietagige ~, zweistöckige ~ = two-level junction

Zweiggleis n = branch track

Zweigkraft f, Seitenkraft, Komponente f [*Baustatik*] = component

zweigleisige Bahnlinie f, doppelgleisige ~ = double-track line

Zweigleitung f = branch line

zweiglied(e)rig → binär

Zweigut-Versuch m [*Gewinnung von je einem Schaumkonzentrat und Flotationsrückstand pro Einsatzversuch*] = two-product test

Zweihandstein m = two-hand block

Zweihebelsteuerung f = two-lever control

2-in-1-Behälter-Halbanhänger m = two-in-one tank semi-trailer, two-tanks-in-one bulk liquids ~

Zweikammer|-Hohlkörper m, Zweikammer-Hohlblock m, Zweikammer-Hohl(block)stein m = two-cell hollow block, ~ ~ tile, ~ pot

~**-Rohrmühle** f = tow-compartment tube mill

Zwei-Karrenaufzug m = two-barrow hoist (concrete elevator)

Zweiketten-Greifer(krankorb) m, Zweiketten-Greif(er)korb = twin-chain suspension grab(bing) (crane) bucket, ~ ~ grab

Zweikomponenten|-Dichtungsmittel n = two-part sealant

~**kitt** m = two-part putty

~**-Masse** f = two-part formulation

Zweikopf-Ellira-Schweißmaschine f, Doppelkopf-Ellira-Schweißmaschine = twin-head submerged-arc welding machine

zweilagig, doppellagig = two-layer, double-layer

zweilagige Dach(pappen)eindeckung f, doppellagige ~ = 2-ply roofing

~ **Teerpappe** f = two-ply tarpaper

zweilagiger Einbau m, doppellagiger ~ = two-layer construction, double-layer ~

~ **Putz** m, doppellagiger ~ = two-coat plaster

~ ~ **auf Putzträgergewebe**, doppellagiger ~ ~ ~ = two-coat work [*The first, or "pricking up" coat of coarse stuff is scored with a birch broom or undercut with a lath, and a thin finishing coat of fine stuff or gauged stuff added. Also called "lath, plaster, and set"*]

Zwei-Lösemittel-Verfahren [*Schmieröl*] = duo-sol process

Zweimann-U-Boot n [*für die Kontrolle von Unterwasserausrüstungen und Schweißarbeiten*] = diving saucer

Zweimassen|-Schrägwurfsieb n = horizontal counter-balanced action shaking screen

~-Vibrator m = two-mass vibrator

zweiminutige Zugfolge f **in der Spitzenzeit** = rush-hour services at two-minutes intervals

Zweimotoren|antrieb m = twin-powered drive

~bagger m, Doppelmotorenbagger = twin-powered excavator

zweimotorig, doppelmotorig = twin-powered

zweimotoriger Hubschrauber m = twin-engine helicopter

Zweipendel|backenbrecher m → → Backenbrecher

~lager n [*für Hoch- und Tiefbauten*] = double pendulum bearing, ~ tumbler

Zweiquartier m, Halbstein m, Halbziegel m = two quarters, half bat

Zweirad|achse f = two-wheel axle

~-Bohrwagen m = two-wheel wagondrill

~-Trecker m, Einachs-Zugwagen m, Einachs(rad)schlepper m, Zweirad-Schlepper, Zweirad-Traktor m = two-wheel(ed) tractor

(~)Vorspänner m, (Zweirad)Vorspannwagen m, (Zweirad)Triebsatz m, Zweiradtrecker m, Ein-Achs-Reifen-Sattelschlepper m, Zweirad-Traktor m = prime mover

'(Zwei)Raupenfahrwerk n → Raupenfahrwerk

Zweirohr-Durchlaß m, Zweirohr-Abzugkanal m, Doppelrohr-Durchlaß, Doppelrohr-Abzugkanal = two pipe culvert, twin ~ ~, double ~ ~

Zweirollenlager n, Doppelrollenlager [*für Hoch- und Tiefbauten*] = two roller bearing

Zweischalengreifer m = clamshell bucket, ~ basket (US); ~ grab, pair of halfscoops (Brit.)

~ für Sand und Kies, Sand- und Kiesgreifer = scraper grab

~ ~ Stichboden und Gesteinstrümmer = whole-tine grab

~ mit Rammwirkung = clamshell bucket with pile driver action, impact clam

Zweischalen|-Kippmulde f [*schwimmender Kippentlader*] = clamshell hopper [*water-borne dumper*]

~mauer f, Zweischalenwand f, Hohlmauer, Hohlwand = cavity wall, hollow ~

zweischalige Tafelwand f = two-leaf precast concrete slab wall

Zweischeiben|-Glättmaschine f [*Beton*] = double mechanical trowel, ~ ~ float

~-Trockenkupplung f = double-plate clutch, twin-plate ~, twin dry plate ~, double dry plate ~

Zweischichten|arbeit f, Zweischichtenbetrieb m = two-shift work

~bauweise f [*Straßenbau*] = two-layer construction

~problem n [*geophysikalisches (Aufschluß)Verfahren*] = two(-)layer(-ed) problem

zweischneidige aufklappbare Bohrschappe f = split spoon sampler

~ Schappe f = soil sampler with two cutting edges

Zweischwingenbrecher m → → Bakkenbrecher

Zweiseil|greifer(krankorb) m, Zweiseilgreif(er)korb, Doppelseilgreifer(krankorb), Doppelseilgreif(er)korb = two-rope (or double-rope, or two-line, or double-line) suspension grab(bing) (crane) bucket (or suspension grab)

~schwebebahn f, Doppelseilschwebebahn [*sogenannte deutsche Bauart*] = double ropeway, ~ cableway

Zweiseiten|entleerung f = two-way dumping, ~ discharge

~-Kippanhänger m = two-way dump trailer

~-Kippmulde f = two-way dump body

zweiseitig eingespannter Balken(träger) m = fixed beam (US); beam fixed at both ends
~ ~ Träger m = fixed girder (US); girder fixed at both ends
zweiseitiger Korbbogen m = two-centred compound curve
Zweispitz m → Doppelspitzhacke f
zweispurig [*Fahrbahn*] = two-lane
zweispurige Straße f = undivided two-way road with two lanes altogether
Zweiständerblechschere f = guillotine plate shear
zweistegig = double-webbed, twin-webbed, two-webbed
zweistegiger Plattenbalken m = twin-webbed T-beam, double-webbed ~, two-webbed ~
Zweistegplatte f [*Gleiskette*] = double grouser track shoe
zweistieliger symmetrischer Stockwerkrahmen m = symmetrical two-legged multi-storied frame
zweistöckige Kreuzung f → zweigeschossige ~
zweistöckiger Grundwasserträger m = two-stor(e)y aquifer
zweistöckiges Kellergeschoß n, zweietagiges ~, zweigeschossiges ~ = two-stor(e)y basement
Zweistoff ... → binär
~motor m, Zweibrennstoffmotor = dual fuel engine
Zweistufen|abscheider m = two-stage separator
~-Brechen n = two-stage crushing, ~ breaking
~-Destillation f [*Erdöl*] = double-flash distillation
zweistufige Entsalzungsanlage f [*Wasseraufbereitung*] = two-stage demineralizer
~ (Luft)Verdichtung f = two-stage compression
zweistufiger Backenbrecher m = two-stage jaw crusher, ~ ~ breaker
~ Kompressor m, ~ Luftverdichter m, ~ Druckluzterzeuger m, Mitteldruckkompressor = two-stage (air) compressor

~ Luftreiniger m = two-stage air cleaner
zweistufiges Sieb n → Doppeldecker m
zweisymmetrischer Träger m = bisymmetrical girder
Zweitbrechen n, Nachbrechen = secondary crushing [*The second crushing stage*]
Zweitbrecher m, Nachbrecher = secondary crusher, ~ breaker
zweite Putzlage f, ~ Putzschicht f = second coat
~Verkehrsebene f = second level
zweiteilig = two-piece
zweiteiliger Behälter m = twin compartment tank
~ Bogen m (✕) = two-element arch
~ Dieseltriebwagen m = two-car diesel set
~ Steinwolf m = two-leg(ged) lewis
zweiteiliges Netzwerk n [*Träger*] = double bracing, ~ triangulated system
zweiter Kompressionsring m, mittlerer Kolbenring [*Motor*] = intermediate ring
Zweitmahlen n, Zweitmahlung f, Nachmahlen, Nachmahlung = secondary grinding [*The secondary grinding operation*]
Zweitrog-Zwangsbetonmischer m → → Betonmischer
zweitrommeliger Haspel m, Zweitrommel-Schrapperwinde f = two-drum (drag) scraper hoist
Zweitrommelwindwerk n = double-drum hoisting gear
Zweitsprengung f mit **Bohrlochladung** = block holing (US); pop shooting (Brit.)
Zweitumschlag m = rehandling
Zweitzerkleinerung f, Nachzerkleinerung = secondary reduction, ~ comminution
Zweiwalzen|-Prallbrecher m → Doppel-Rotoren-Prallbrecher
~-Prallmühle f, Doppel-Rotoren-Prallmühle = double impeller impact mill

zweiwandige Gurtung *f* = boom of the open box girder type, chord ~ ~ ~ ~ ~ ~, flange ~ ~ ~ ~ ~ ~ ~
zweiwandiger Blechbogen(träger) *m* = double-webbed plate arch(ed girder)
Zweiwege-Güterfahrzeug *n* für Schiene und Straße = combined rail-and-road vehicle
Zweiweg|-Hydraulikzylinder *m* = two-way ram
~rückschlagventil *n* = double check valve
Zweiwellen|-Knetmischer *m* → Zweiwellen-Zwangsmischer
~-Zwangsmischer *m*, Zweiwellen-Knetmischer, Doppelwellen-Knetmischer, Doppelwellen-Zwangsmischer = twin shaft pug mill, twin-pug-mill mixer, twin-pug mixer, twin pug mill
Zweiwellrohrkessel *m* = Lancashire boiler with corrugated flues
zweizelliger Hohlkastenträger *m* = box girder with two compartments
Zweizonenförderung *f* [*Es wird gleichzeitig aus zwei Erdölhorizonten gefördert*] = dual (well) completion
2:1-Untersetzungsgetriebe *n* = half speed gear
Zweizylinder|-Dieselmotor *m* = 2-cyl. diesel engine
~-Motor *m* = two-cylinder engine
Zwerg|schlepper *m*, Zwergtraktor *m*, Zwergtrecker *m* = mini-tractor
~zündkerze *f* = dwarf plug
Zwicke *f* [*für Packe*] = (rock) spalls
Zwickel *m* = spandrel
~silo *m* = interstice bin [*A bin formed by the exterior walls of three or more silos or bins, so placed as to enclose the adjacent space on all sides*]
Zwickstein *m*, Aus~ [*für Packe*] = (rock) spall
Zwiebel|dach *n*, Zwiebelhaube *f*, Kaiserdach [*geschweiftes Turmdach*] = imperial roof
~haube *f* → Zwiebeldach *n*
~kuppel *f* = imperial dome, Moorish ~
Zwillings|bereifung *f* = dual tyres (Brit.); ~ tires (US)

~-Endantrieb *m* = dual final drives
~fahrgestell *n*, Zwillingsradpaar *n* [*Flugzeug*] = dual-wheel assembly
~fördersonde *f* [*Bohrung, die gleichzeitig aus zwei getrennten Horizonten produziert*] = dual producer
~form *f* = dual mould (Brit.); ~ mold (US)
~gashebel *m* = twin throttle
~(gleis)kettenschlepper *m* = caterpillar twin D 8 tractor
~kolbenpumpe *f* = twin cylinder piston pump
~lamellierung *f* [*Natronkalkfeldspat*] = twin lamellae [*plagioclase*]
~packer *m* [*Tiefbohrtechnik*] = dual packer
~pumpe *f* = twin pump
~rad *n* → Doppelrad
~reifenrad *n*, Doppelreifenrad = dual-tyre wheel (Brit.); dual-tire ~ (US)
~schleuse *f*, Doppelschleuse = twin locks
~träger *m* = twin girder
~ventil *n* = double valve
Zwischen|anschlag *m* (✹) = (shaft) inset
~behälter *m* = auxiliary air receiver
~durchmesser *m* = fractional diameter
~fertigwand *f*, Trennfertigwand = prefab(ricated) partition (wall)
~festigkeit *f* [*Beton*] = intermediate strength
~geschoß *n*, Mezzanin *n*, Halbgeschoß, Beigeschoß = mezzanine
~getriebe *n* = secondary transmission
~glied *n* = transition link
~glied *n* = (cylindrical steel) adaptor, (~ ~) adapter [*LEE-McCALL method*]
~haupt *n* [*Schleuse*] = intermediate head
~kühler *m* = intercooler
~kühlung *f* = intercooling
~lage *f* [*Förderband*] = ply
~lage *f*, Zwischenstellung *f* = intermediate position

Zwischenlage — Zyklopenmauerwerk

~lage *f* [*hält den Asphalt von der vollflächigen Verklebung auf waag(e)rechten Flächen frei*] = black sheeting felt [*British Standards 747*]
~lagerung *f* = intermediate storage
~mittel *n* → Bergemittel
~pfahl *m* = intermediate pile
~pfeiler *m* = intermediate pier
~pfette *f* = middle purlin(e)
~podest *n* = intermediate landing
~querträger *m* [*Brücke*] = intermediate floor beam, ~ cross girder
~querverband *m* [*Brücke*] = intermediate transverse frame
~rad *n*, Vorlegerad, mitlaufendes Zahnrad = idler (gear), transfer ~
~raum *m* [*Theater*] = proscenium
~schicht *f* = intermediate course, ~ layer
~silo *m* = surge bin
~spannweg *m* [*Spannbeton*] = stretching distance increment, prestressing ~ ~, tensioning ~ ~
~sparren *mpl* → Leergespärre *n*
~steife *f* [*Blechträger*] = intermediate stiffener
~stellung *f*, Zwischenlage *f* = intermediate position
~stück *n*, Ausleger-~ = (jib) section, (~) extension
~stück *n*, Anschlußteil *m, n*, Verbindungsteil [*Gleiskettenlaufwerk*] = adapter, mounting, adaptor
~stufe *f* = intermediate stage
~stütze *f* = intermediate column
~tauperiode *f* = intermediate thaw(ing) period
~tor *n* [*Schleuse*] = intermediate gate
~transport(last)kraftwagen *m* = transfer (motor) truck (US); ~ lorry (Brit.)
~verdampfer *m*, Wiederverdampfer = re-boiler
~verrohrung *f* [*Tiefbohrtechnik*] = intermediate string of casing
Zwischenwand *f*, Trennwand [*nichttragend*] = partition (wall)
~ aus Kesselschlackenbetonplatten, Trennwand ~ ~ = clinker concrete slab partition (wall)
~ in Trockenbauweise, Trennwand ~ ~ = dry (wall) partition
~bauplatte *f* → Trennwandbauplatte
~block *m* → Zwischenwandstein *m*
~platte *f* aus Gips, Gipszwischenwandplatte = gypsum partition board
~stein *m*, Zwischenwandblock, Trennwandstein, Trennwandblock = partition block
~-Tonhohlkörper *m*, Zwischenwand-Tonhohl(stein)platte *f*, Zwischenwand-Hourdi *m*, Zwischenwand-Tonhohlstein *m*, Trennwand-Tonhohlkörper *m*, Trennwand-Tonhohl(stein)platte, Trennwand-Hourdi, Trennwand-Tonhohlstein = hollow (burnt-)clay partition block, ~ ~ ~ tile
~ziegel *m*, Trennwandziegel = partition brick
Zwischen|welle *f*, Vorlegewelle = idler shaft
~wellung *f*, Zwischenkrippung, Zwischenwellen *n* [*vorgewelltes (Sieb-) Gitter*] = intermediate crimp
~wirkungsdiagramm *n* = interaction diagram
~zahnrad *n* = idler gear ~ sprocket (wheel)
Zwölf-Uhr-Schweißen *n*, Überkopfschweißen = welding in overhead position, twelve o'clock-welding, overhead welding
Zyanlaugung *f* = cyaniding
zyklische Kurve *f* = cyclic(al) curve
.~ Symmetrie *f* = cycle symmetry
~ Veränderung *f* = cyclic(al) variation
~ Vertauschung *f* = cyclic(al) permutation
Zyklonnaßklassierer *m*, Aufschwimmklassierer, Naßzyklon *m* = cyclone classifier
Zyklopenbeton *m* = cyclopean concrete
~-Stein *m* = plum
Zyklopen|mauerwerk *n* = cyclopean masonry, polygonal ~

Zyklopenpflaster — zylindrisch-konische Fördertrommel

~pflaster n = crazy paving
Zylinder|aufhängung f = cylinder support
~befestigung f = cylinder bracket
~block m = cylinder block
~bohrung f = cylindrical boring, cylinder bore
~büchse f, (Zylinder)Laufbüchse = cylinder liner
~deckel m = cylinder cover
~dichtung f = cylinder seal
~druck m = cylinder pressure
~druckfestigkeit f = cylinder compressive strength
~-Druck(prüf)presse f, Beton~ = (concrete) cylinder tester, (~) ~ testing machine
~einsteckschloß n = cylinder mortice lock, ~ mortise ~
~festigkeit f = cylinder strength
~gelenk n = cylindrical hinge
~innendruck m = cylinder pressure
~-Koordinatensystem n = cylindrical coordinate system
Zylinderkopf m = cylinder head
~ mit abnehmbarem Ansaugkrümmer = open inlet manifold cylinder head
~ ~ angegossenem Ansaugkrümmer = cylinder head with integral inlet manifold
~ ~ Zwischenboden = shelf-type cylinder head
~deckel m = cylinder head cover
~dichtung f = cylinder head gasket
~ende n = cylinder head end
Zylinder|körper m [*Schloß*] = cylinder barrel, ~ body
(~)Laufbüchse f, Zylinderbüchse = cylinder liner
~mantel m = cylinder barrel
~rohr n = cylindrical pipe
~rollenmeißel m = Reed roller bit
~schale f → Zylindertonne
~schalendach n = cylindrical concrete shell roof
~schleifkopf m [*Betondeckenschleifer*] = cylinder cutting head
~schloß n = cylinder lock, cylindrical ~
~schütz(e) n, (f) = cylindrical sluice-gate
~stangenende n = cylinder rod end
~tonne f, Zylinderschale f = cylindrical shell
~tragjoch n = cylinder support yoke
~tragkonsole f = cylinder bracket
~walm m = cylindrical hip(ped end)
~wand(ung) f = cylinder wall
~zapfenkipplager n [*für Hoch- und Tiefbauten*] = pin rocker bearing
zylindrische Trockentrommel f, zylindrischer Trommeltrockner m = cylindrical drier (or dryer)
zylindrischer Fertigbetonpfahl m, parallelseitiger ~ = parallel-sided precast foundation pile, ~ ~ (load-) bearing ~
~ Kübel m [*Betontransport*] = circular skip
~ Pfeiler m = cylindrical pier
~ Probekörper m → Probezylinder m
~ Trommeltrockner m, zylindrische Trockentrommel f = cylindrical dryer (or drier)
zylindrisches Gewinde n = cylindrical thread, straight ~
zylindrisch-konische Fördertrommel f (⚒) = cylindro-conical drum

APPENDIX
ANHANG

Conversion Table Umrechnungstabelle

Multiply by Multipliziere mit	To convert Zur Umwandlung von	To In	
2.54	Inches; Zoll	Centimetres; cm	.39371
30.48	Feet; Fuß	Centimetres; cm	.03281
.03937	Millimetres; mm	Inches; Zoll	25.40
.9144	Yards	Metres; mtr	1.094
1,609.31	Miles; Meilen	Metres; mtr	.000621
3.281	Metres; mtr.	Feet; Fuß	.3048
39.37	Metres; mtr.	Inches; Zoll	$2.540 \cdot 10^{-2}$
1,853.27	Nautical Miles; Seemeilen	Metres; mtr.	.00054
6.45137	Square inches; Zoll²	Sq. cms; cm²	.15501
.093	Square feet; Fuß²	Sq. metres; m²	10.7643
.83610	Square yards; Yard²	Sq. metres; m²	1.19603
2.58989	Sqare miles; Meile²	Sq. kilometres; km²	.38612
16.38618	Cubic inches; Zoll³	Cub. cms.; cm³	.06103
28.33	Cubic feet; Fuß³	Litres; Liter	.0353
.02832	Cubic feet; Fuß³	Cub. metres; m³	35.31658
6.24	Cubic feet; Fuß³	Imperial Gallons	.1602
.76451	Cubic yards; Yards³	Cub. metres; m³	1.30802
.3732	Pounds (Troy)	Kilogrammes; kg	2.68
31.10	Ounces (Troy)	Grammes; g	.03216
.4536	Pounds (Avoir.)	Kilogrammes; kg	2.2045
7,000.00	Pounds (Avoir.)	Grains (Troy)	.00014
28.35	Ounces (Avoir.)	Grammes; g	.0352
.065	Grains	Grammes; g	15.38
50.80238	Cwt.	Kilogrammes; kg	.01968
1,016.04754	Tons; Tonnen	Kilogrammes; kg	.00098
907.18	Short tons	Kilogrammes; kg	$1.102 \cdot 10^{-3}$
4.54346	Imperial Gallons	Litres; Liter	.22010
3.785	US Gallons	Litres; Liter	.264
.56793	Pints	Litres; Liter	1.76077
264.2	Cubic metres; m³	US Gallons	$3.785 \cdot 10^{-3}$
36.34766	Bushels	Litres; Liter	.02751
10.00	Imp. Gall of water	Pounds	.1
.454	Pounds of water	Litres; Liter	.2202
70.31	Lb. per sq. in. (psi.)	Gm./sq. cms	.01422
.07031	Lb. per sq. in. (psi.)	kg/cm²	14.22272

	To obtain Zur Errechnung obiger Einheiten	From von obigen	Multiply by above ist mit obengenanntem Wert zu multiplizieren

Conversion Table Umrechnungstabelle
Continued Fortsetzung

Multiply by Multipliziere mit	To convert Zur Umwandlung von	To In	
4.88241	Lb. per sq. foot; Pfund/Fuß²	Kilogrammes per sq metres; kg/m²	.20482
.00049	Lb. per sq. foot; Pfund/Fuß²	Kilogrammes per sq. cm; kg/cm²	2,049.1807
.24803	Lb. per fathom	kg/mtr.	4.0318
1.48816	Lb. per foot; Pfund/Fuß	kg/mtr.	.6719
.49606	Lb. per Yard; Pfund/Yard	kg/mtr.	2.0159
3,333.4784	Tons per foot; Tonne/Fuß	kg/mtr.	.0003
2.3	Lb. per sq. in. (psi.)	Head of water (ft.)	.434
.7	Lb. per sq. in. (psi.)	Head of water (M.)	1.4285
.068	Lb. per sq. in. (psi.)	Atmospheres; Atmosphären	14.7
157.4944	Tons per sq. in.; Tonnen/Zoll²	kg/cm²	.0063
1.574944	Tons per sq. in.; Tonnen/Zoll²	kg/mm²	.635
10,936.59840	Tons per sq. foot; Tonnen/Fuß²	kg/mtr²	.0001
.03613	Gramm/cm³	lb. per cub. inch; Pfund/Zoll³	27.68
62.43	Gramm/cm³	lb. per cub. foot; Pfund/Fuß³	.01602
.593	Lb. per cub. yard; Pfund/Yard³	kg/mtr³	1.686
16.02	Lb. per cub. foot; Pfund/Fuß³	kg/mtr³ (m³/kg)	.0624
.4047	Acres	Hektar	2.471
.00155	Sq. mm.; mm²	Square inch; Zoll²	645.2
.0998	Lb. per Imp. Gallon	Kgm./litre; kg/Liter	10.02
.13825	Foot-lb.; Fuß-Pfund	K'grammetres; mkg	7.2332
.33	Foot-tons	Tonnen-Meter	3.00
309.680	Foot-tons	kilogramm-metres	.0032
25.80667	Inch-tons	kilogramm-metres	.0388

To obtain Zur Errechnung obiger Einheiten	From von obigen	Multiply by above ist mit obengenanntem Wert zu multiplizieren

Conversion Table Umrechnungstabelle
Continued Fortsetzung

Multiply by Multipliziere mit	To convert Zur Umwandlung von	To In	
41.62314	Inches^4	cm^4	.0240
.62138	Kilometre	Miles	1.6093
3,280.8693	Kilometre	Foot; Fuß	219.98
1.60931	Miles per hour; Meilen pro Stunde	kilometres per hour; km/Stde	.6214
1.467	Miles/h; Meilen/Stde.	Fuß/s	.6818
.869	Miles/h; Meilen/Stde.	Knoten	1.151
1.014	Horse-Power	Force de cheval; Chevaux-vapeur; PS	.9863
746.00	Horse-Power	Watts	.00134
3.415	Wattstunde	B.T.U.	.293
76.00	H.P.	kg-m./sec.	.01316
.1	Watts	Kg.-m./sec.	10.00
860.5	Kilowattstunde (kWh)	kcal	$1.162 \cdot 10^{-3}$
.447	Pounds per H.P.	Kilogrammes per Cheval-vapeur; kg/PS	2.235
3.968	kcal; Kg. Calories	B.T.U.	.252
4,184.00	kcal; Kg. Calories	Joules	.00024
.738	Joules	Foot-pound	1.356
.1124	kcal/m^3	B.T.U./foot^3	8.9
1.8	kcal/Kilogramm (kg)	B.T.U./pound	.5556
14.7	Atmospheres	psig, pounds per square inch gauge	.068
2.713	B.T.U./sq. foot	Kg. Calories/m^2	.369
.293	B.T.U.	Wattstunde	3.415
.03342	Zoll-Quecksilbersäule; Inch-Mercury-Column	atm	29.92
13.6	Zoll-Quecksilbersäule; Inch-Mercury-Column	Zoll-Wassersäule; Inch-Water-Col.	.0735
.4912	Zoll-Quecksilbersäule; Inch-Mercury-Column	pound/inch^2	2.036
.9	German candles	English candles	1.1111
9.55	Carcels	Candles	.1047
88.00	Miles/hour	Ft./min.	.01134

	To obtain	From	Multiply by above
	Zur Errechnung obiger Einheiten	von obigen	ist mit obengenanntem Wert zu multiplizieren

Conversion Table Umrechnungstabelle
Continued Fortsetzung

Multiply by Multipliziere mit	To convert Zur Umwandlung von	To In	
197.00	Metres/sec.	Ft./min.	.00508
.208	Centipoise	Lb. force sec./sq. ft.	4.8
5.43	Imp. Gall./sq. yard	Liter/m²	.1845
	To obtain Zur Errechnung obiger Einheiten	From von obigen	Multiply by above ist mit obengenanntem Wert zu multiplizieren

British Weights and Measures

Lineal Measure

4	Inches	make	1 Hand	1 Inch	=	.08	Ft.
9	,,	,,	1 Span		=	.207	Yard
12	,,	,,	1 Foot	1 Link	=	7.92	Inches
3	Feet	,,	1 Yard	1 Foot	=	.333	Yard
5	,,	,,	1 Pace	1 Yard	=	36.00	Inches
6	,,	,,	1 Fathom	1 Chain	=	100.00	Links
5.5	Yards	,,	1 Rod, Pole or Perch		=	22.00	Yards
					=	.0125	Miles
4	Poles	,,	1 Chain	1 Furlong	=	220.00	Yards
10	Chains	,,	1 Furlong		=	.125	Miles
8	Furlongs	,,	1 Mile	1 Mile	=	80.00	Chains
3	Miles	,,	1 League		=	1,760.00	Yards
1.151	,,	,,	1 Nautical Mile				

A Knot is a speed of 1 Nautical Mile per hour

Square or Land Measure

144	Sq. Inches	=	1 Sq. Foot
9	Sq. Feet	=	1 Sq. Yard
30.25	Sq. Yards	=	1 Sq. Pole
40	Poles	=	1 Rood
4	Roods	=	1 Acre
640	Acres	=	1 Sq. Mile

An Acre equals 4.840 Square Yards
1 Sq. Yard = .4047 Hectares
1 Sq. Link = 62.75 Sq. Inches (approx.)
1 Sq. Chain = 10,000.00 Sq. Links = 484 Sq. Yards
10 Sq. Chains = 1 Acre = 100,000.0 Sq. Links
33 Sq. Yards = 1 Rod of Building = 27.6 Sq. Metre
100 Sq. Feet = Square of Flooring of Roofing = 9.3 Sq. Metre
272.25 Square Feet = Rod of Bricklayer's Work = 25.4 Sq. Metre

Cubic or Solid Measure

Cubic Foot = 0.037 cubic yard = 1,728 Cub. Inches = 6.23 Imp. Gall. = 7.48 US-Gallons
 = 28,317 Cub. centimetre = .0283 Cub. metre = 28.3 Litres
Cubic Yard = 27 Cub. Feet = 168 Imp. Gall. = 202 US-Gallons = 21.033 Bushels = .7645 Cubic metre = 764.5 Litres
1 Cubic Inch = 0.00058 Cubic Foot = 16,38618 cm³
Stack of Wood = 108 Cubic Feet = 3.06 Cubic metre
Shipping ton = 40 Cubic Feet of merchandise = 1.13 Cubic metre
Shipping ton = 42 Cubic Feet of timber = 1.18 Cubic metre
One ton or load = 50 Cubic Feet of hewn timber = 1.42 Cubic metre
Ton of displacement of a ship = 35 Cubic Feet = 1.02 Cubic metre

Fluid Memoranda

1 Imp. Gallon of water = 10 lb.
1 Cubic Foot of water = 6.23 Imp. Gall. = 62.3 lb = 7.48 US-Gallons
 = .0283 Cub. metre = 1,728.00 Cub. inches
 = 28,317.00 Cub. centimetre = .037 Cubic yard
 = 28.317 Litres

1 lb. water at 62° F. = .016 Cub. Foot
1 Imp. Gallon = 1.2 US-Gallon = 8 Pints = 277.418 Cub. inches = 4.546 Litres
1 Quart = 2 Pints = .25 Imp. Gall. = 69.35 Cub. inches = 1.136 Litres
1 Pint = 4 Gills = .125 Imp. Gall. = 34.682 Cub. inches = .02 Cub. Ft. = 568.3 Cub. centimetre = .5683 Litres
1 Firkin = 9 Imp. Gall. = 41 Litres
1 Kilderkin = 2 Firkins = 82 Litres
1 British Barrel = 4 Firkins = 36 Imp. Gall. = 1.028 US-Barrel = 288 Pints = .215 Cub. Yard = 5.77 Cub. Ft. = 163.566 Litres
1 Gill = .25 Pint = 8.67 Cub. inches = .142 Litres
1 Inch of Rainfall = 22,622.00 Imp. Gallons per Acre = 100.00 tons (approximately)

Avoirdupois Weight

1 Oz. = 28.35 Grammes = .063 lb. = 16 Drams = 437.5 Grains
1 lb. = 16 Oz. = 453.6 Grammes = .4536 kg = 7,000.00 Grains
1 cwt. (Hundredweight) = 112 lb. = 50.80238 kg
1 Stone = 14 lb. = 6.35 kg
1 Quarter = 28 lb.
1 Butcher Stone = 8 lb.
1 Grain = .0648 Grammes
1 Long ton = 2,240.00 lb. = 1.120 Short ton = 20 cwt. = 1,016.05 kg = 1.01605 Metric tons
1 Short ton = 2,000.00 lb. = .893 Long ton = 907.19 kg = .90719 Metric ton

Miscellaneous

1 US-Gallon = 231 Cub. inches = .1337 Cub. feet = 3.785 Litres = .833 Imp. Gallon
1 PSh = 2512 Btu = 0.986 HPh = 633 kcal = 0.736 kWh
1 kWh = 3415 Btu = 861 kcal = 1.36 PSh
1 Btu (Brit. thermal unit) = 0.252 kcal = 108 mkg = 0.00029 kWh = 0.0004 PSh
1 Imperial Bushel = 36.37 Litres
1 Winchester Bushel (USA) = 35.257 Litres
1 Bale = 205 kg
1 Sack of Cement (USA) = 42.676 kg
1 Acre-foot = 43,560 Cubic feet = 1,232.6 Cubic metre (this unit relates to irrigation computations)
1 US-Barrel = .972 British Barrel = 42 US-Gallons = .208 Cubic Yards = 5.62 Cub. Ft. = 158.98 Litres
1 preußischer Morgen = 2,500.00 Sq. metre = 25 Ar = .25 Hektar = .0025 Sq. kilometre = .61775 Acres
1 Zentner = 50 kg = 110.225 lb.
1 metrisches Pfund = 500 Grammes = 1.1 lb.
1 Cental = .05 Short ton = 45.36 kg
1 Dram = 3 Scruples = 1.77 Grammes
1 morgen-foot = 2,608.847204 Cubic metre (this unit relates to irrigation computations in South Africa)
1 morgen (South-Afrika) = 2.11654 Acres
1 geographische Meile = 4 Seemeilen = 7.240 Kilometer

1 Seemeile (Knoten) = 1,000.00 Faden = 1.852 km
1 preußische Landmeile = 7.532 km
1 Cubic foot/second per 1,000 acres = 0,00708 m³/Sekunde per km²
1 Day-second-foot (dsf.) = 86.400 cu. ft.
1 russische Werst = 1,500.00 Arschinen = 1.06678 km
1 deutsche Quadratmeile = 56.25 Square kilometre (km²)
1 geographische Quadratmeile = 55.06 Square kilometre (km²)
1 russische Quadratwerst = 1.138 km²
1 bbl = 376 pds. = 170.704 kg
1 bbl. = 4 Cubic feet = 0.112 m³
1 mbm (mille feet board measure) = 2.36 m³

Cubic Measures and Weights per Unit of Area
Raummaße und Gewichte pro Flächeneinheit

1 kg/cm² = 14.226 lb./sq. inch (psi) = 2.050 lb./sq. ft. = 18,450.00 lb./sq. yard = 10 tons sq. metre
1 kg/m² = .205 lb./sq. ft. = 1.844 lb./sq. yard
1 lb./sq. inch = .0703 kg/cm² = 144 lb./sq. ft.
1 lb./sq. ft. = 9 lb./sq. yard = 4.882 kg/mtr.² = .0069 lb./sq. inch
1 lb./sq. yard = .11 lb./sq. ft. = .542 kg/mtr.²
1 Litre/mtr.² = .020 Imp. Gall./ft² = .184 Imp. Gall./yard² = .025 US-Gall./ft.² = .221 US-Gallon/yard²
1 Imp. Gall./ft.² = 9 Imp. Gall./yard² = 48.94 Litres/mtr.²
1 Imp. Gall./yard² = .111 Imp. Gall./ft.² = 5.44 Litres/mtr.²
1 US-Gallon/ft.² = 9 US-Gallons/yard² = 40.77 Litres/mtr.²
1 US-Gallon/yard² = .111 US/Gallon-ft.² = 4.53 Litres/mtr.²
1 Litre/mtr.² = .093 Litre/ft.² = .836 Litre/yard².
1 Litre/ft.² = 9 Liter/yard² = 9.18 Litre/mtr.²
1 Litre/yard² = .111 Litre/ft.² = 1.02 Litre/mtr.²

Velocities Geschwindigkeiten

1 mtr./sec. = 3.6 km/h = 1.093 Yards/sec. = 2.237 Miles/h = 3.281 Ft./sec.
1 km/h = .278 mtr./sec. = .304 Yards/sec. = .622 Miles/h = 0.91178 Ft./sec.
1 Yard/sec. = .915 mtr./sec. = 3.292 km/h = 2.040 Miles/h
1 Mile/h = .447 mtr./sec. = 1.609 km/h = .489 Yards/sec.

Temperatures Temperaturen

Celsius (°C) = 5/9 (°F—32) = 5/4 °R
Réaumur (°R) = 4/5 °C = 4/9 (°F—32)
Fahrenheit (°F) = 9/5 °C + 32 = 9/4 °R + 32

°C	°R	°F	°C	°R	°F
— 40	— 32	— 40	+ 60	+ 48	+ 140
— 35	— 28	— 31	+ 65	+ 52	+ 149
— 30	— 24	— 22	+ 70	+ 56	+ 158
— 25	— 20	— 13	+ 75	+ 60	+ 167
— 20	— 16	— 4	+ 80	+ 64	+ 176
— 17.8	— 14.2	0	+ 85	+ 68	+ 185
— 15	— 12	+ 5	+ 90	+ 72	+ 194
— 10	— 8	+ 14	+ 95	+ 76	+ 203
— 5	— 4	+ 23	+ 100	+ 80	+ 212
0	0	+ 32	+ 110	+ 88	+ 230
+ 5	+ 4	+ 41	+ 120	+ 96	+ 248
+ 10	+ 8	+ 50	+ 130	+ 104	+ 266
+ 15	+ 12	+ 59	+ 140	+ 112	+ 284
+ 20	+ 16	+ 68	+ 150	+ 120	+ 302
+ 25	+ 20	+ 77	+ 175	+ 140	+ 347
+ 30	+ 24	+ 86	+ 200	+ 160	+ 392
+ 35	+ 28	+ 95	+ 225	+ 180	+ 437
+ 40	+ 32	+ 104	+ 250	+ 200	+ 482
+ 45	+ 36	+ 113	+ 275	+ 220	+ 527
+ 50	+ 40	+ 122	+ 300	+ 240	+ 572
+ 55	+ 44	+ 131	+ 350	+ 280	+ 662

U. S. Standard Sieve Data (A. S. T. M.)

Bureau of Standard Sieve Number		Specified sieve opening lichte Maschenweite		Specified wire diameter	
		Inches	Millimeters	Inches	Millimeters
(3/16 in.)	4	—.187	4.76	—.050	1.27
	5	—.157	4.00	—.044	1.12
	6	—.132	3.36	—.040	1.02
	7	—.111	2.83	—.036	—.92
	8[1]	—.0937	2.38	—.0331	—.84
	10	—.0787	2.00	—.0299	—.76
	12	—.0661	1.68	—.0272	—.69
	14	—.0555	1.41	—.0240	—.61
	16[2]	—.0469	1.19	—.0213	—.54
	18	—.0394	1.00	—.0189	—.48
	20	—.0331	—.84	—.0165	—.42
	25	—.0280	—.71	—.0146	—.37
	30[3]	—.0232	—.59	—.0130	—.33
	35	—.0197	—.50	—.0114	—.29
	40	—.0165	—.42	—.0098	—.25
	45	—.0138	—.35	—.0087	—.22
	50[4]	—.0117	—.297	—.0074	—.188
	60	—.0098	—.250	—.0064	—.162
	70	—.0083	—.210	—.0055	—.140
	80	—.0070	—.177	—.0047	—.119
	100[5]	—.0059	—.149	—.0040	—.102
	120	—.0049	—.125	—.0034	—.086
	140	—.0041	—.105	—.0029	—.074
	170	—.0035	—.088	—.0025	—.063
	200	—.0029	—.074	—.0021	—.053
	230	—.0024	—.062	—.0018	—.046
	270	—.0021	—.053	—.0016	—.041
	325	—.0017	—.044	—.0014	—.036

[1] B.S. Sieve No. 7   0.0949 Aperture size (in.)
[2] B.S. Sieve No. 14 0.0474 Aperture size (in.)
[3] B.S. Sieve No. 25 0.0236 Aperture size (in.)
[4] B.S. Sieve No. 52 0.0116 Aperture size (in.)
[5] B.S. Sieve No. 100 0.0060 Aperture size (in.)

Die gebräuchlichsten Siebsysteme mit ihren wichtigsten Daten

Amerikanischer ASTM Maschensiebsatz		Britischer STANDARD Maschensiebsatz		Maschensiebe DIN 1171 Ausgabe 1935			Rundlochsiebe DIN 1170 Ausgabe 1933	
Maschen je Zoll	lichte Maschenweite mm	Maschen je Zoll	lichte Maschenweite mm	Gewebe Nr.	Maschen je cm²	lichte Maschenweite mm	Lochdurchmesser mm	Umrechnung in lichte Maschenweite mm
—	—	—	—	100	10000	0,060		
200	0,074	200	0,076	80	6400	0,075		
—	—	—	—	70	4900	0,090²)		
100	0,149	100	0,152	40	1600	0,150	Umrechnung nach Rotfuchs	
80	0,177	85	0,178	35¹)	1225	0,177		
70	0,210	72	0,211	30	900	0,200		
60	0,250	60	0,251	24	576	0,250		
50	0,297	52	0,294	20	400	0,300		
—	—	44	0,353	—	—	—		
40	0,420	36	0,422	14	196	0,430		
30	0,590	25	0,599	10	100	0,600		
—	—	—	—	—	—	—	1	0,7
—	—	—	—	8	64	0,750		
20	0,840	18	0,853	—	—	—		
18	1,000	—	—	—	—	1,000		
16	1,190	14	1,200	5	—	1,200		
12	1,680	10	1,676	—	—	—		
10	2,000	8	2,057	—	—	2,000	3	2,3
5	4,000			3			5	3,8
³/₁₆″	(4) 4,760	²/₁₆″	4,760	5			7	5,4
¼″	6,35	¼″	6,35	6			8	6,2
				8			10	7,8
³/₈″	9,52	³/₈″	9,52	—			12	9,5
½″	12,7	½″	12,7	12			15	12
				15			20	16,4
¾″	19,050	¾″	19,050	18			25	20,8
1″	25,4	1″	25,4	25			30	25,2
1¼″	31,75	1¼″	31,75	30			40	34,0
1½″	38,100	1½″	38,100					
2″	50,800							
3″	76,200	3″	76,200					
6″	152,400							

Anmerkung: Unterstrichene Zahlen = Siebe für abgekürzte Siebungen.
¹) Ergänzungssieb.   ²) früher 0,088.

Soil Classification Diagram

Classification System	Particle Size Equivalent Diameter MM								
International Society of Soil Science	CLAY		SILT		FINE SAND		COARSE SAND		GRAVEL
United States Public Roads Administration	COLLOIDS	CLAY	SILT →			FINE SAND	COARSE SAND		GRAVEL
Massachusetts Institute of Technology	CLAY →		FINE SILT	MEDIUM SILT	COARSE SILT	FINE SAND	MEDIUM SAND	COARSE SAND	GRAVEL
British Standards Institution	CLAY →		FINE SILT	MEDIUM SILT	COARSE SILT	FINE SAND	MEDIUM SAND	COARSE SAND	GRAVEL
Particle Size	0.001	0.002	0.005 / 0.006		0.02	0.05 / 0.06	0.2	0.6	2.0 — 60 MM

British Standard Sieve: 200, 72, 25, 7, 3/16 in, 3/8 in, 3/4 in, 1 1/2 in, 2 1/2 in

Übersicht über verschiedene Einteilungsversuche der Lockergesteine (nach Dücker)

Bezeichnung und Korndurchmesser in mm (logar. Einteilung)

Jahr	Vorschläge angewandt durch	200	60	20	6	2,0	0,5	0,2	0,06 0,05 0,02 0,006	0,005	0,002	0,0006 0,0002
ungefähr 1890 bis 95	**U. S. Bureau of Soils** (A. Casagrande 1947)	Gravel					Sand			Silt		Clay	
1905 1913	**A. Atterberg** Internat. Bodenk. Gesellsch.	Steine und Geröll			Kies		Grobsand / Feinsand			Schluff		Ton	
1914 1931	**I. Kopecki**, Mass-Institute of Technology (M.I.T.) (n. A. Casagrande 1947)	Kies / Gravel					Sand			Schluff / Silt		Ton / Clay	
1925	**K. Terzaghi**					Sand			Mo	Schluff		Kolloid-schlamm	Ultra Ton
1936	**DIN 4022**			Steine	Kies		Sand		Mehl Sand	Schluff		Rohton	
1936	**S. Fischer u. H. Udluft**			Grand									
		Block	Brock	Graup			Sand / Gritt		Silt	Schmand Schlamm Sink / Schweb		
1938	**P. Niggli**			Gries								
		Block	Grobkies	Feinkies		Grobsand / Feinsand			Silt / Grobschluff / Feinschluff		Schlamm / Schweb	
1938	**U.S. Depart. of Agricul-ture** (n. A. Casagrande 1941)	Block	Schotter	Gravel		Sand			Silt		Clay	
1939	**H. Gallwitz**	Block	Grobst.-Boden	Grobschotter-Boden	Feinschotter-Boden	Kies		Sand		Schluff		Ton	
1942	**B. Grengg**	Gröbstes I	Grobschotter-Boden				Sand			Mehl	Schluff	Schlamm	Feinstes
												Siebkorn
1948	**A. Dücker**		IIa grobes IIb	IIIa mittleres IIIb	IVa feines IVb	Va grobes Vb	VIa mittleres VIb	VII feines	Schlammkorn			
		200	60	20	6 / 2,0	2,0 / 0,6	0,6 / 0,2	0,2 / 0,06	0,06 / 0,02 / 0,006 0,002	0,002	0,0006	0,0002

Herbert Bucksch

Wörterbuch für Bautechnik und Baumaschinen

Dictionary of Civil Engineering and Construction Machinery and Equipment

Band II: Englisch-Deutsch / English-German.
7. Auflage. 71 000 Stichwörter. 1219 Seiten, Format 12,5 × 17 cm.
Plastik DM 160,–

Französische Ausgabe:

Dictionnaire pour les Travaux Publics et l'Équipement des Chantiers de Construction

Band I: Deutsch-Französisch.
4. Auflage. Rund 54 000 Stichwörter. 875 Seiten DIN B 6. Plastik
DM 140,–

Band II: Französisch-Deutsch.
4. Auflage. Rund 56 000 Stichwörter. 911 Seiten DIN B 6. Plastik
DM 140,–

Englisch-Französische Ausgabe:

Band I: Englisch-Französisch.
5. Auflage. 420 Seiten, Format 12,5 × 16 cm. Ganzgewebe DM 38,–

Band II: Französisch-Englisch.
5. Auflage. 548 Seiten, Format 12,5 × 16 cm. Ganzgewebe DM 48,–

Spanische Ausgabe:

Diccionario para Obras Publicas, Edificacion y Maquinaria en Obra

Deutsch-Spanisch / Spanisch-Deutsch.
1114 Seiten DIN B 6. 68 000 Stichwörter. Ganzgewebe DM 108,–

BAUVERLAG GMBH · WIESBADEN, BERLIN

Herbert Bucksch

Wörterbuch für Architektur, Hochbau und Baustoffe
Dictionary of Architecture, Building Construction and Materials

Band I: Deutsch-Englisch.
2. Auflage. Rund 65 000 Stichwörter. 942 Seiten, Format 13,5 × 20,5 cm. Plastik DM 220,—

Band II: Englisch-Deutsch.
Rund 75 000 Stichwörter. 1137 Seiten, Format 13,5 × 20,5 cm. Plastik DM 220,—

Dictionnaire pour l'Architecture, le Bâtiment et les Matériaux de Construction

Band I: Deutsch-Französisch.
Rund 52 000 Stichwörter. 820 Seiten, Format 13,5 × 20,5 cm. Plastik DM 290,—

Band II: Französisch-Deutsch.
Rund 40 000 Stichwörter. 688 Seiten, Format 13,5 × 20,5 cm. Plastik DM 290,—

Getriebe-Wörterbuch
Dictionary of Mechanisms

Deutsch-Englisch und Englisch-Deutsch in einem Band.
Rund 16 000 Stichwörter. 286 Seiten mit über 250 Kleinzeichnungen. Format 13,5 × 20,5 cm. Plastik DM 165,—

Holz-Wörterbuch
Dictionary of Wood and Woodworking Practice

Zusammen über 40 000 Stichwörter.

Band I: Deutsch-Englisch. 461 Seiten DIN B 6. Plastik DM 75,—

Band II: Englisch-Deutsch. 536 Seiten DIN B 6. Plastik DM 84,—

BAUVERLAG GMBH · WIESBADEN, BERLIN

Gips-Wörterbuch
Gypsum and Plaster Dictionary
Dictionnaire du Gypse et du Plâtre

Deutsch-Englisch-Französisch.

Von Dipl.-Ing. K.-H. Volkart. Rund 3000 Stichwörter. 176 Seiten, Format 17 × 24 cm. Ganzgewebe DM 85,–

Zement-Wörterbuch
Dictionary of Cement

Herstellung und Technologie / Manufacture and Technology

Von Dipl.-Ing. C. van Amerongen. **Deutsch-Englisch / Englisch-Deutsch.** Neuauflage in Vorbereitung.

Pipeline Dictionary
Rohrfernleitungs-Wörterbuch
Dictionnaire des Canalisations à grande Distance

Englisch-Deutsch-Französisch.

Von H. Bucksch und A. P. Altmeyer. 4430 Stichwörter. 288 Seiten, Format 18 × 24 cm. Ganzgewebe DM 125,–

Internationales Wörterbuch für Metallurgie, Mineralogie, Geologie, Bergbau und die Ölindustrie
International Dictionary of Metallurgy, Mineralogy, Geology and the Mining and Oil Industries

Englisch-Französisch-Deutsch-Italienisch / English-French-German-Italian.

Zusammengestellt von A. Cagnacci-Schwicker. Rund 27 000 Stichwörter. 1530 Seiten, Format 15,5 × 23,5 cm. Ganzgewebe DM 110,–

BAUVERLAG GMBH · WIESBADEN, BERLIN

Englische und französische Fachsprache im Auslandsbau

von der Voranfrage bis zur Bauausführung

International Construction Contracts Terminology in French and English with German Vocabularies / L'anglais dans la terminologie de la construction et du bâtiment dans le monde avec glossaires allemand.

Von Prof. Dipl.-Ing. K. Lange, Dipl.-Ing. L. Ferval und Dipl.-Ing. Arch. K. Kellmann. 131 Seiten DIN A 5. Kartoniert DM 24,–

Bautechnisches Englisch im Bild

Illustrated Technical German for Builders.

Von W. K. Killer. 5. Auflage. 183 Seiten DIN B 5 mit zahlreichen Abbildungen. Texte Deutsch und Englisch. Kartoniert DM 24,–

Englisch für Baufachleute

Einführung in die bautechnische Fachsprache.
(Schulkenntnisse werden vorausgesetzt.)

Von Prof. Dipl.-Ing. G. Wallnig und H. Evered F.C.S.I. Format 17 × 24 cm. Kartoniert.

Band 1: 6. Auflage. 101 Seiten mit 35 Abbildungen. DM 16,–

Band 2: 2. Auflage. VIII, 192 Seiten mit Abbildungen. DM 38,–

Beide Bände sind ebenso für französischsprachige Leser geeignet.

Deutsch für Baufachleute
German for Building Specialists
L'Allemand dans le Bâtiment

Von Prof. Dipl.-Ing. G. Wallnig und H. Evered F.C.S.I. 102 Seiten, Format 17 × 24 cm, mit Abbildungen. Kartoniert DM 16,–

BAUVERLAG GMBH · WIESBADEN, BERLIN